Before you begin, review your basic study skills.

Study Skills Workshop ▶ Making Homework a Priority

Attending class and taking notes are important, but they are not enough. The only way that you are really going to learn algebra is by doing your homework.

WHEN TO DO YOUR HOMEWORK: Homework should be started on the day it is assigned, when the material is fresh in your mind. It's best to break your homework sessions into 30-minute periods, allowing for short breaks in between.

HOW TO BEGIN YOUR HOMEWORK: Review your notes and the examples in your text before starting your homework assignment.

GETTING HELP WITH YOUR HOMEWORK: It's normal to have some questions when doing homework. Talk to a tutor, a classmate, or your instructor to get those questions answered.

Now Try This ▶

1. Write a one-page paper that describes *when, where,* and *how* you go about completing your algebra homework assignments.
2. For each problem on your next homework assignment, find an example in this book that is similar. Write the example number next to the problem.
3. Make a list of questions that you have while doing your next assignment. Then decide whom you are going to ask to get those questions answered.

To get the most out of the ***Study Skills Workshops*** that begin each chapter, you may choose to review them in the early weeks of your course. Each one includes action items, in addition to simple suggestions that can put you on a clear path to success. Below, we have included a table of contents to aid you in locating these:

Options That SAVE Your Students Money

Online Learning with Cengage YouBook

UP TO 60% OFF

ENHANCED WebAssign

Save your students money and give them an interactive learning experience!

Choose a variety of exercises to assign as homework, and Enhanced WebAssign® will automatically grade the assignments for you and record the grades.

Learning resources include:

- **Master It** algorithmic tutorials provide unlimited practice of a similar problem.
- **Watch It** narrated solution videos of the problem found within the text.
- **Read It** icons direct students to the concept in the Cengage YouBook to learn more.

In addition, students get access to **Cengage YouBook**, an interactive eBook that contains embedded simulations and videos with note-taking, highlighting, bookmarking, and searching capabilities.

Stand-Alone Single Term Access
ISBN: 0-538-73810-3

Enhanced WebAssign with the eBook can be bundled with the text.

Contact your Cengage Learning representative to find out more about bundling options.

Book Rentals

UP TO 60% OFF

CENGAGE brain .com

Students have the CHOICE to purchase the eBook or rent the text at CengageBrain.com

OR

an eTextbook in PDF format is also available for instant access for your students at **www.coursesmart.com.**

CourseSmart

Students can rent Tussy and Gustafson's textbooks for 60% off list price.

Custom Solutions

FIT EVERY NEED

Contact your Cengage Learning representative to learn more about what custom solutions are available to meet your course needs.

- Adapt existing Cengage Learning content by adding or removing chapters.
- Incorporate your own materials.
- Add mathematics review for your less-prepared students.

Customize Tussy and Gustafson's textbooks with:

Student Solutions Manual

- *Elementary Algebra* 5E, ISBN: 1-111-98902-8
- *Intermediate Algebra* 5E, ISBN: 1-111-98758-0
- *Elementary and Intermediate Algebra* 5E, ISBN: 1-111-57847-8

Nolting's Math Study Skills Workbook, 4E, ISBN: 0-840-05309-6

5e

Elementary and Intermediate Algebra

Alan S. Tussy
Citrus College

R. David Gustafson
Rock Valley College

BROOKS/COLE
CENGAGE Learning™

Australia • Brazil • Japan • Korea • Mexico • Singapore • Spain • United Kingdom • United States

BROOKS/COLE
CENGAGE Learning™

Elementary and Intermediate Algebra,
Fifth Edition
Alan S. Tussy, R. David Gustafson

Publisher: Charlie Van Wagner

Senior Developmental Editor: Danielle Derbenti

Assistant Editor: Carrie Jones

Senior Editorial Assistant: Jennifer Cordoba

Media Editors: Heleny Wong, Guanglei Zhang

Marketing Manager: Laura McGinn

Marketing Assistant: Shannon Maier

Marketing Communications Manager: Darlene Macanan

Content Project Manager: Jennifer Risden

Design Director: Rob Hugel

Art Director: Vernon Boes

Print Buyer: Judy Inouye

Rights Acquisitions Specialist: Tom McDonough

Production Service: Chapter Two, Ellen Brownstein

Text Designer: Terri Wright

Photo Researcher: Bill Smith Group

Copy Editor: Chapter Two, Ellen Brownstein

Illustrator: Lori Heckelman

Cover Designer: Terri Wright

Cover Image: Artie Ng/www.flickr.com/photos/artiephotography

Compositor: Graphic World, Inc.

Library of Congress Control Number: 2010939897

Student Edition:
ISBN-13: 978-1-111-56768-2
ISBN-10: 1-111-56768-9

Loose-leaf Edition:
ISBN-13: 978-1-111-57827-5
ISBN-10: 1-111-57827-3

Brooks/Cole
20 Davis Drive
Belmont, CA 94002-3098
USA

Printed in the United States of America
1 2 3 4 5 6 7 15 14 13 12 11

In memory of my mother, Jeanene, and my dad, Bill.

—AST

To my wife, Carol,

with love and appreciation.

—RDG

About the Authors

Alan S. Tussy

Alan Tussy teaches all levels of developmental mathematics at Citrus College in Glendora, California. He has written nine math books—a paperback series and a hard-cover series. A meticulous, creative, and visionary teacher who maintains a keen focus on his students' greatest challenges, Alan Tussy is an extraordinary author, dedicated to his students' success. Alan received his Bachelor of Science degree in Mathematics from the University of Redlands and his Master of Science degree in Applied Mathematics from California State University, Los Angeles. He has taught up and down the curriculum from prealgebra to differential equations. He is currently focusing on the developmental math courses. Professor Tussy is a member of the American Mathematical Association of Two-Year Colleges.

R. David Gustafson

R. David Gustafson is Professor Emeritus of Mathematics at Rock Valley College in Illinois and coauthor of several best-selling math texts, including Gustafson/Frisk's *Beginning Algebra, Intermediate Algebra, Beginning and Intermediate Algebra: A Combined Approach, College Algebra,* and the Tussy/Gustafson developmental mathematics series. His numerous professional honors include Rock Valley Teacher of the Year and Rockford's Outstanding Educator of the Year. He earned a Master of Arts from Rockford College in Illinois, as well as a Master of Science from Northern Illinois University.

CONTENTS

CHAPTER 1 An Introduction to Algebra 1

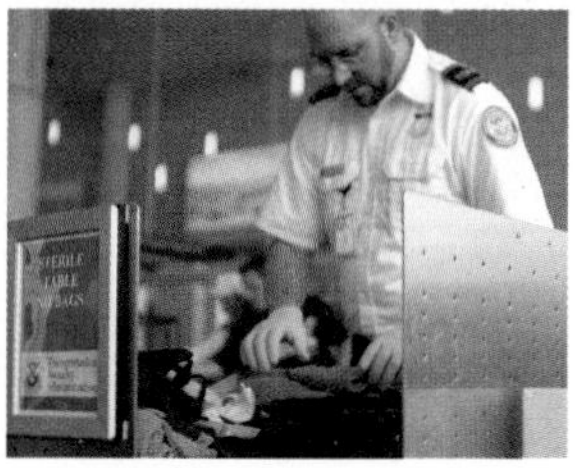

©Caroline K. Smith, M.D./Shutterstock.com

CHAPTER 2 Equations, Inequalities, and Problem Solving 101

©barang/Shutterstock.com

CHAPTER 3 Graphing Linear Equations and Inequalities in Two Variables; Functions 185

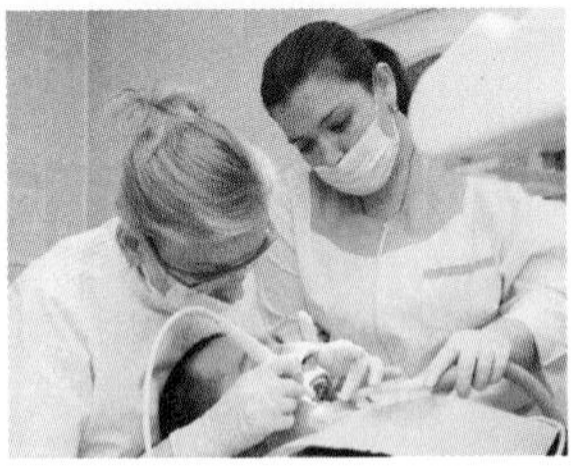

CHAPTER 4 Systems of Linear Equations and Inequalities 283

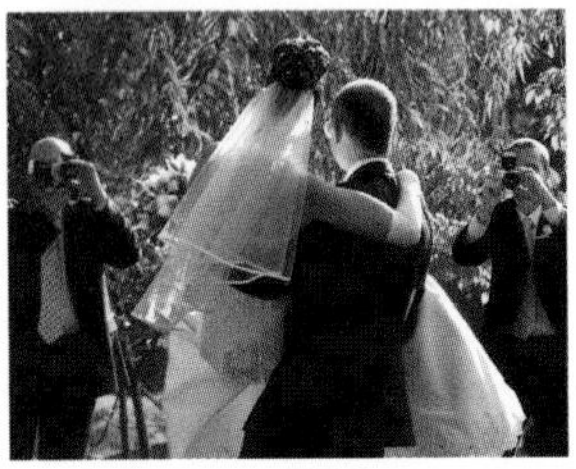

CHAPTER 5 Exponents and Polynomials 347

CHAPTER 6 Factoring and Quadratic Equations 431

©iStockPhoto.com/Neustockimages

CHAPTER 7 Rational Expressions and Equations 511

©BananaStock/Superstock

CHAPTER 8 Transition to Intermediate Algebra 599

©Ronen/Shutterstock.com

CHAPTER 9 Radical Expressions and Equations 727

©Goodluz/Shutterstock.com

CHAPTER 10 Quadratic Equations, Functions, and Inequalities 825

© Robert E Daemmrich/Getty Images

CHAPTER 11 Exponential and Logarithmic Functions 899

©iStockphoto/Chris Schmidt

CHAPTER 12 More on Systems of Equations 1001

©iStockphoto.com/CandyBox Photography

CHAPTER 13 Conic Sections; More Graphing 1075

©Andresr/Shutterstock.com

CHAPTER 14 Miscellaneous Topics 1121

©Michael Pettigrew/Shutterstock.com

APPENDIXES

PREFACE

We are excited to present the Fifth Edition of *Elementary and Intermediate Algebra.* We believe the revision process has produced an even stronger instructional experience for both students and teachers. First, we have fine-tuned several of the popular features of our series, including the *Strategy* and *Why* example structure, the problem-solving strategy, and the online homework.

Second, we have introduced some new features to further promote student understanding and success. The new *Look Alikes* problems in the *Study Sets* will help students improve their problem recognition. The *Are You Ready?* exercises that begin each section give students the opportunity to review necessary prerequisite skills before they are asked to apply them in the study of new topics. The additional *Campus to Careers* problems inserted in the *Study Sets* expand on the real-life applications of the chapter material. Finally, we trust that instructors will appreciate the new *Teaching Tips* placed in the margins and the new *Teaching Examples* that now accompany each worked example in the text.

We want to thank all of you throughout the country who have provided suggestions and input about the previous edition. Your insight has proven invaluable. Throughout this process, our fundamental belief has remained the same: Algebra is a language in its own right. And, as always, the prime objective of this textbook is to teach students how to read, write, speak, and think using the language of algebra.

A Focus on Transition

When designing the table of contents for a combination textbook, the task is two-fold. Care must be taken so as not to overwhelm beginning algebra students with intermediate algebra topics for which they are not adequately prepared. On the other hand, none of the traditional intermediate algebra topics should be deleted or downplayed. The question that arises is this: How do we maintain the pedagogy and integrity of an Elementary Algebra course and an Intermediate Algebra course when they are merged?

We believe the answer to this question is our Chapter 8: *Transition to Intermediate Algebra.* The objective of this chapter is to:

- Review some of the basic algebraic concepts studied in the first half of the book, such as equations, inequalities, factoring, rational expressions, graphing, and problem solving.
- Expand on those concepts, such as factoring, functions, and graphing, with more in-depth study at the intermediate algebra level.
- Offer students an opportunity to apply previously learned skills in new settings with topics such as compound inequalities, absolute value equations and inequalities, and variation.

New to the Fifth Edition

Sections That Begin with Review: *Are You Ready?*

Each section begins with a set of *Are You Ready?* problems. These problems review crucial prerequisite skills that students need to have mastered in order to be successful with the new topics of that section.

ARE YOU READY? *Are You Ready? exercises available online at www.webassign.net/brookscole*

The following problems review some basic skills that are needed to find the slope of a line.

1. Evaluate: $\frac{4-1}{8-3}$ $\frac{3}{5}$

2. Evaluate: $\frac{-10-1}{-4-(-4)}$ undefined

3. Multiply: $-\frac{7}{9}\cdot\frac{9}{7}$ -1

4. Simplify: $\frac{15}{18}$ $\frac{5}{6}$

Study Sets with More Problem-Recognition Practice: *Look Alikes*

After a poorer than expected performance on a test, students often tell their instructors, "I could do the homework each night, but when it comes to the test, I get confused." The new *Look Alikes* feature builds students' problem-recognition skills. It requires students to distinguish between similar looking problem types and then to select the correct strategy to solve the problem. Encountering such situations in the homework assignments will better prepare students for quizzes and tests.

Look Alikes . . .

Perform the indicated operations to simplify each expression, if possible.

105. a. $(x - 2) + (x^2 + 2x + 4)$ **b.** $(x - 2)(x^2 + 2x + 4)$
$x^2 + 3x + 2$ $x^3 - 8$

106. a. $(a + 3) + (a^2 - 3a + 9)$ **b.** $(a + 3)(a^2 - 3a + 9)$
$a^2 - 2a + 12$ $a^3 + 27$

107. a. $(6x^2z^5) - (-3xz^3)$ **b.** $(6x^2z^5)(-3xz^3)$
Does not simplify $-18x^3z^8$

108. a. $(-5r^4t^2) - (2r^2t)$ **b.** $(-5r^4t^2)(2r^2t)$
Does not simplify $-10r^6t^3$

109. a. $(2x^2 - x) - (3x^2 - 3x)$ **b.** $(2x^2 - x)(3x^2 - 3x)$

And Even More Problem-Recognition Practice: *Try It Yourself*

Designed to promote problem recognition, instructors and reviewers requested more *Try It Yourself* problems for the Fifth Edition. These problem types are thoroughly mixed, giving students an opportunity to practice decision making and strategy selection as they would when taking a test or quiz. With more than 80% more problems added, students will have even more opportunities to practice this essential skill.

TRY IT YOURSELF

Graph each equation. Solve for y first, when necessary.
See AIE Appendix 3.

53. $y = x$	**54.** $y = 4x$
55. $y = -x - 1$	**56.** $y = -x + 2$
57. $3y = 12x + 15$	**58.** $5y = 20x - 30$
59. $y = \frac{3}{8}x - 6$	**60.** $y = -\frac{3}{2}x + 2$
61. $y = 1.5x - 4$	**62.** $y = 0.5x + 3$
63. $8x + 4y = 16$	**64.** $14x + 7y = 28$
65. $y = -\frac{1}{2}x$	**66.** $y = \frac{3}{4}x$
67. $y = \frac{5}{6}x - 5$	**68.** $y = \frac{2}{3}x - 2$
69. $-6y = 30x + 12$	**70.** $-3y = 9x - 15$

Comprehensive Test Preparation: *Chapter Tests*

Instructors often assign an end-of-chapter test for students to use as a means to study for a classroom exam. However, after taking the exam, students often remark that the classroom exam included problem types that weren't in the *Chapter Test.* To address this issue, we have made sure that each *Chapter Test* is a comprehensive collection of problems that covers *all* of the topics discussed in the chapter. As a result, the *Chapter Tests* are lengthy. If your students have time to complete a *Chapter Test,* that would be optimal. If, because of time constraints, they are unable to do so, assign an appropriate subset of problems that reflects the types of problems that the students will see on your exam. This should alleviate the discrepancy between what your students practice and what they will see on the test.

2 CHAPTER TEST

1. Fill in the blanks.
   a. To solve an equation means to find all of the values of the variable that make the equation true.
   b. Percent means parts per one hundred.
   c. The distance around a circle is called its circumference.
   d. An inequality is a statement that contains one of the symbols $>, \geq, <$, or $\leq$.
   e. The multiplication property of equality says that multiplying both sides of an equation by the same nonzero number does not change its solution.
2. Use a check to determine if 3 is a solution of $5y + 2 = 12$. No

Solve each equation.

3. $3h + 2 = 8$ 2
4. $-22 = -x$ 22
5. $\frac{4}{5}t = -4$ −5
6. $\frac{11b - 11}{5} = \frac{3b - 2}{2}$ $\frac{12}{7}$
7. $0.8(x - 1{,}000) + 1.3 = 2.9 + 0.2x$ 1,336
8. $2(y - 7) - 3y = -(y - 3) - 17$ All real numbers (an identity)
9. $\frac{m}{2} - \frac{1}{3} = \frac{1}{4} + \frac{m}{6}$ $\frac{7}{4}$
10. $\frac{3}{4}(6n - 2) = 246$ 55

Solve for the specified variable.

23. $V = \pi r^2 h$ for h $h = \frac{V}{\pi r^2}$
24. $A = P + Prt$ for r $r = \frac{A - P}{Pt}$
25. $A = \frac{a + b + c + d}{4}$ for c $c = 4A - a - b - d$
26. $2x - 3y = 9$ for y $y = \frac{2}{3}x - 3$
27. **Irons.** Estimate the area of the soleplate of the iron. 20 in.^{2}

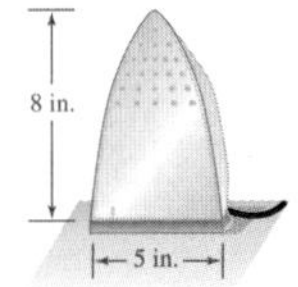

28. **Television.** In a typical 30-minute block of time on TV, the number of programming minutes are 2 less than three times the number of minutes of commercials. How many minutes of programming and how many minutes of commercials are there? Programming: 22 min; commercials: 8 min

Additional Relevant, Motivating Applications in the *Examples* and *Study Sets*

We have included many new applied examples and problems that involve relevant topics such as our environment and sustainability issues, energy savings, technology and social media, and recycling. For a complete list of topics, see the *Index of Applications* following the *Preface.* To see a sampling of new topics added to each chapter, see *Content Changes by Chapter* in this *Preface.*

EXAMPLE 6 **Water Management.** Lake Mead, on the Nevada-Arizona border, is the largest reservoir in the United States. It is formed by Hoover Dam across the Colorado River. In 2000, the water level in Lake Mead was 89 feet above drought level. By 2010, the water level was 27 feet below drought level. Find the change in the water level over that time span. (Source: Bureau of Reclamation)

Strategy We can represent a water level above drought level using a positive number and a water level below drought level using a negative number. To find the change in the water level, we will subtract.

Why In general, *to find the c… later value.*

Solution $-27 - 89 = -27 + (-$ …

$= -116$

The negative result indicates th…

Teaching Example 6 Find the change in water level for a week that started at 6 feet above normal and went to 8 feet below normal level.
Answer: −14 ft

Self Check 6 Find the cha… normal and …

Now Try Problem 97

APPLICATIONS

73. **Social Networks.** Refer to the graph below. In what month did Facebook and MySpace have the same number of unique visitors? Estimate the number. May '09; about 70 million

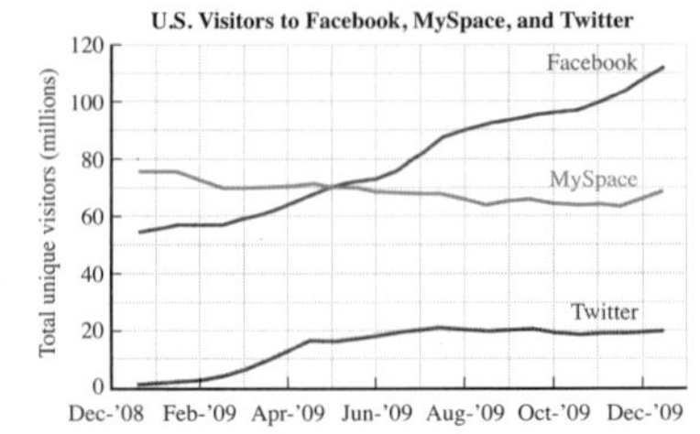

Source: Comscore Media Metrix

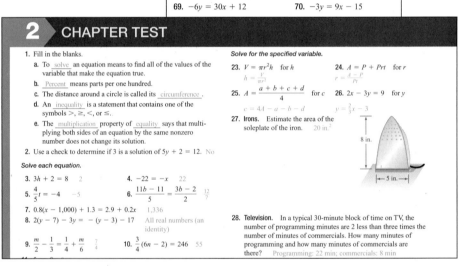

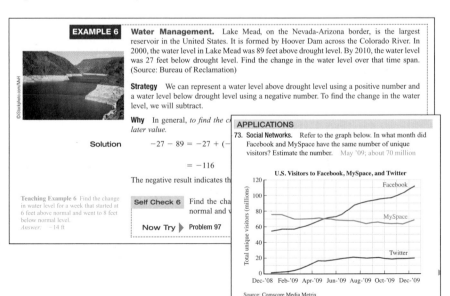

A More Precise Problem-Solving Strategy

In an effort to better describe the problem-solving strategy used in this book, we have inserted a new second step in what was formerly a five-step process. This additional step (*Assign a variable to represent an unknown value in the problem.*) better delineates the thought process students should use as they solve application problems. The six steps of the problem-solving strategy are now: *Analyze the problem*, *Assign a variable*, *Form an equation*, *Solve the equation*, *State the conclusion*, and *Check the result.*

EXAMPLE 11 **Grades.** A student has scores of 72%, 74%, and 78% on three exams. What percent score does he need on the last exam to earn a grade of no less than B (80%)?

The Language of Algebra

Some phrases that suggest an inequality are:

surpass: > | at least: ≥
not exceed: ≤ | at most: ≤
is no less than: ≥
is no greater than: ≤
between: ☐ < ☐ < ☐

Analyze We know three scores. We are to find what the student must score on the last exam to earn a grade of B or higher.

Assign Let x = the score on the fourth (and last) exam.

Form To find the average grade, we add the four scores and divide by 4. To earn a grade of *no less than* B, the student's average must be *greater than or equal to* 80%.

The average of the four grades	must be no less than	80.
$\frac{72 + 74 + 78 + x}{4}$	$\geq$	80

Solve

$$\frac{224 + x}{4} \geq 80 \quad \text{Combine like terms in the numerator: } 72 + 74 + 78 = 224.$$

$$4\left(\frac{224 + x}{4}\right) \geq 4(80) \quad \text{To clear the inequality of the fraction, multiply both sides by 4.}$$

$$224 + x \geq 320 \quad \text{Simplify each side.}$$

$$x \geq 96 \quad \text{To isolate } x \text{, undo the addition of 224 by subtracting 224 from both sides.}$$

State To earn a B, the student must score 96% or better on the last exam.

Check Pick several exam scores that are 96% or better and verify that the student's average will be 80% or greater. For example, a score of 96% gives the student an average that is exactly 80%.

$$\frac{72 + 74 + 78 + 96}{4} = \frac{320}{4} = 80$$

Teaching Example 11 Grades. A student has scores of 68%, 67%, and 72% on three exams. What percent score does he need on the last test to earn a grade of no less than a C (70%)? *Answer:* 73% or better

Self Check 11 **Grades.** A student has scores of 78%, 82%, and 76% on three exams. What percent score does he need on the last test to earn a grade of no less than a B (80%)? 84% or better

Now Try Problem 103

More Emphasis on "When Will I Use This?"

Each chapter now has three *Campus to Careers* problems that explore the mathematical connections to careers that are presented at the beginning of each chapter.

79. from Campus to Careers

Photographer

Photographers often use the *rule of thirds* to add more interest to a photo rather than simply centering the subject. They imagine two horizontal and two vertical lines dividing the

Systems of Linea[r] Equations and Inequalities

55. from Campus to Careers

Photographer

In some cameras, the image that you see in the viewfinder does not exactly match the image that will be recorded through the lens. To understand this difference, graph the solutions of the system

$$\begin{cases} y \leq \frac{1}{4}x + 2 \\ y \geq -\frac{1}{4}x + 2 \end{cases}$$

29. from Campus to Careers

Photographer

Suppose you are a wedding photographer and you sell:

Package 1: one 10 × 14 and ten 8 × 10 color photos for $239.50

Package 2: one 10 × 14 and five 8 × 10 color photos for $134.50

A newlywed couple buys Package 1 and decides that they want one more 10 × 14 and one more 8 × 10 photograph. At the same prices, what should you charge them for each additional photograph? $29.50 for a 10 × 14; $21 for an 8 × 10

from Campus to Careers

Photographer

Photographers record our surroundings, the special events in our lives, and people, so that all can be remembered in pictures. Some specialize in weddings and portraits, some photograph landscapes and fashion, while others work on location as photojournalists. Their job responsibilities require a variety of mathematical skills such as: scheduling appointments, keeping financial records, pricing photographs, purchasing supplies, billing customers, and operating digital equipment.

Problem 79 in **Study Set 4.1**, **problem 29** in **Study Set 4.4**, and **problem 55** in **Study Set 4.5** involve situations that a photographer might encounter on the job. The mathematical concepts discussed in this chapter can be used to solve those problems.

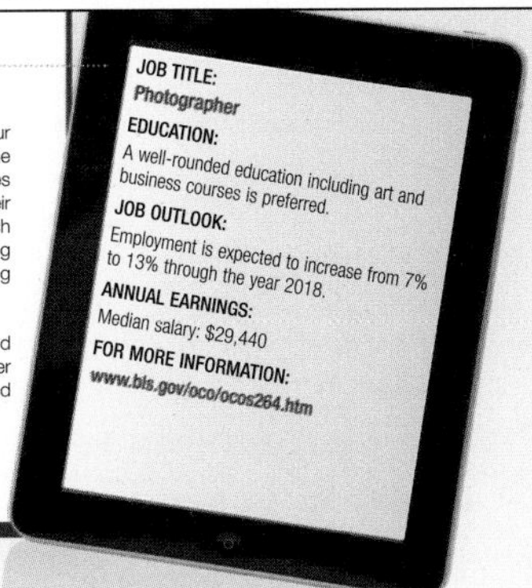
JOB TITLE:
Photographer
EDUCATION:
A well-rounded education including art and business courses is preferred.
JOB OUTLOOK:
Employment is expected to increase from 7% to 13% through the year 2018.
ANNUAL EARNINGS:
Median salary: $29,440
FOR MORE INFORMATION:
www.bls.gov/oco/ocos264.htm

283

Reading the Language of Algebra

> **Product Rule for Exponents** To multiply exponential expressions that have the same base, keep the common base and add the exponents.
> For any number x and any natural numbers m and n,
> $$x^m \cdot x^n = x^{m+n}$$ *Read as "x to the mth power times x to the nth power equals x to the m plus nth power."*

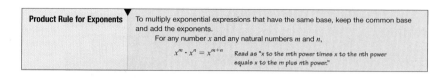

Students often have difficulty reading the mathematical notation of algebra and, as a consequence, their understanding suffers. To provide assistance in this area, we have inserted notes that explain how to read newly introduced notation. Your students will appreciate the ever-present "Read as ..." statements that follow algebraic symbolism that they encounter for the first time.

Added Teaching Tools

> **EXAMPLE 6** Solve: $\frac{1}{6}x + \frac{5}{2} = \frac{1}{3}$
>
> **Teaching Tip:** It is wise to have students identify the number of terms on each side of the equation before multiplying both sides by the LCD.
>
> **Strategy** To clear the equation of fractions, we will multiply both sides by the LCD of all the fractions in the equation.
>
> Teaching Example 6
> Solve: $\frac{1}{10}x + \frac{4}{5} = \frac{9}{2}$
> *Answer:* 37
>
> **Self Check 6** Solve: $\frac{1}{4}x + \frac{1}{2} = -\frac{1}{8}$ $-\frac{5}{2}$
>
> **Now Try** Problem 33

Teaching Tips have been placed at strategic places in the margins to offer instructional suggestions that can be employed during classroom presentations. In addition, each worked example has an accompanying *Teaching Example* that instructors can use as part of a classroom lecture.

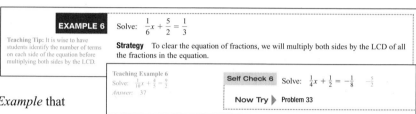

New to *Enhanced WebAssign*

Tools to Gauge Student Readiness: *Are You Ready?*

With *all* the *Are You Ready?* problems available in *Enhanced WebAssign,* now teachers can easily gauge student readiness before beginning instruction on a new section. Using *Enhanced WebAssign,* the *Are You Ready?* problems can be assigned as a short quiz to be submitted before class. Student results are reported electronically and give instructors an indication of the classes' command of the prerequisite skills necessary for success with the new topics.

Better Understanding of Assigned Reading: *Vocabulary*

Learning algebra is like learning a foreign language. Therefore, familiarity with some of the key vocabulary *before* hearing it spoken in class can help to reinforce recognition and understanding. Now that all of the section *Vocabulary* problems are available in *Enhanced WebAssign,* instructors can assign these problems before class to encourage student completion of reading assignments and introduce section vocabulary.

Assign More *Guided Practice*

With many more *Guided Practice* problems available to assign through *Enhanced WebAssign,* instructors will have a complete representation of the various problem types that are explained in the worked examples. This also will make it easier to build short quizzes, covering each problem type for each section.

Tools to Practice Problem Recognition: *Look Alikes* and *Try It Yourself*

With a broad selection of *Look Alikes* and *Try It Yourself* problems available in *Enhanced WebAssign,* short quizzes and homework assignments can be assigned easily, offering strategy-selection practice.

More *Challenges*

At least one *Challenge Problem* per section is now available in *Enhanced WebAssign,* providing a better representation of all parts of the *Study Sets* in the online homework.

More Problems to Assign

Across all three books in the series, we now offer approximately 70% of end-of-section problems in *Enhanced WebAssign.* With so many problems to choose from, instructors will have more choice than ever for the Fifth Edition. Throughout the life of an edition, we are continually adding problems to *Enhanced WebAssign.* While existing problems at the time of publication are marked with a blue triangle in the section exercises of the *Annotated Instructor's Edition,* visit www.webassign.net/brookscole and search by title for the most up-to-date list of *Enhanced WebAssign* problems.

Trusted Features

- **Examples That Tell Students Not Just How, But WHY:** Why? That question is often asked by students as they watch their instructors solve problems in class and as they are working on problems at home. It's not enough to know how a problem is solved. Students gain a deeper understanding of the algebraic concepts if they know why a particular approach was taken. This instructional truth was the motivation for adding a Strategy and Why explanation to each worked example.
- **Examples That Offer Immediate Feedback:** Each worked example includes a Self Check. These can be completed by students on their own or as classroom lecture examples, which is how Alan Tussy uses them. Alan asks selected students to read the Self Check problems aloud as he writes what the student says on the board. The other students, with their books open to that page, can quickly copy the Self Check problem to their notes. This speeds up the note-taking process, teaches students how to read mathematical symbols, and encourages student participation in his lectures. Each Self Check answer is printed adjacent to the corresponding problem in the *Annotated Instructor's Edition* for easy reference. Students can find Self Check solutions in the *Answers to Selected Exercises* in the back of the Student Edition text.
- **Examples That Ask Students to Work Independently:** Each worked example ends with a Now Try problem. These are the final step in the learning process. Each one is linked to a similar problem found within the *Guided Practice* sections of the *Study Sets.*
- **Study Sets** found in each section offer a multifaceted approach to practicing and reinforcing the concepts taught in each section. They are designed for students to build their knowledge of the section concepts methodically, from basic recall to increasingly complex problem solving, through reading, writing, and thinking mathematically.

  Vocabulary—Each *Study Set* begins with the important *Vocabulary* discussed in that section. The fill-in-the-blank vocabulary problems emphasize the main concepts taught in the chapter and provide the foundation for learning and communicating the language of algebra.

  Concepts—In *Concepts,* students are asked about the specific subskills and procedures necessary to successfully complete the *Guided Practice* and *Try It Yourself* problems that follow.

  Notation—*Notation* problems review the new symbols introduced in a section. Often, students are asked to fill in steps of a sample solution. This strengthens their ability to read and write mathematics and prepares them for the *Guided Practice* problems by modeling solution formats.

  Guided Practice—The *Guided Practice* section of each *Study Set* consistently provides 1-to-1 linking for each problem type to a single worked example or objective (*i.e.*, See Example 1, See Example 2, See Example 3, and so on). Students will appreciate this 1-to-1 linking as opposed to the all-encompassing linking statements such as See Examples 2–8 or See Example 7–11 that are found in some textbooks.

  Try It Yourself—To promote problem recognition, the *Try It Yourself* problems are thoroughly mixed and are *not* linked to worked examples, giving students an opportunity to practice decision making and strategy selection as they would when taking a test or quiz.

  Applications—The *Applications* provide students the opportunity to apply their newly acquired algebraic skills to relevant and interesting real-life situations.

  Writing—The *Writing* problems help students build mathematical communication skills.

Review—The *Review* problems consist of randomly selected problems from previous chapters. These problems are designed to keep students' successfully mastered skills up-to-date before they move on to the next section.

Challenge Problems—The *Challenge Problems* provide students with an opportunity to stretch themselves and develop their skills beyond the basics. Instructors often find these to be useful as extra-credit problems.

- **Detailed Author Notes** that guide students along in a step-by-step process appear in the solutions to every worked example.
- **The Language of Algebra** boxes draw connections between mathematical terms and everyday references to reinforce the language of algebra thread that runs throughout the text.
- **The Notation, Success Tips, Caution,** and **Calculators** boxes offer helpful tips to reinforce correct mathematical notation, improve students' problem-solving abilities, warn students of potential pitfalls and increase clarity, and offer tips on using scientific calculators.
- **Chapter Tests,** at the end of every chapter, can be used as preparation for the class exam.
- **Cumulative Reviews** follow the end-of-chapter material and keep students' skills current before moving on to the next chapter. Each problem is linked to the associated section from which the problem came for ease of reference. The final *Cumulative Review* often is used by instructors as a final exam review.
- **Using Your Calculator** is an optional feature that is designed for instructors who want to use calculators as part of the instruction in this course. This feature introduces keystrokes and shows how scientific and graphing calculators can be used to solve problems. In the *Study Sets,* icons are used to denote problems that may be solved using a calculator.

Content Changes by Chapter

Based on feedback from colleagues and users of the Fourth Edition, the following changes have been made in an effort to further streamline and update the text.

Chapter 1

- New example and exercise applications include topics such as snowboarding, Lake Mead water levels, melting glaciers, calories burned doing housework, lost luggage, iPhone signal strength, U.S. Federal Budget Deficit/Surplus, and average wait time in airport security lines.
- New instructional features include a screened color 1 that is used when explaining how to build and simplify fractions, additional notes explaining how the notation is read, a worked example demonstrating uses of the commutative and associative properties, a more comprehensive *Chapter Test,* additional cautions, and an upgraded *Group Project.*

Chapter 2

- New example and exercise applications include topics such as Twitter, how tire pressure affects gas mileage, rainforest deforestation, Craigslist ads, *Harry Potter* box office revenue, calculating horsepower, iPhone apps, water usage, consignment shops, signing bonuses, and target heart rates.
- New instructional features include a comparison of linear and nonlinear equations, a new worked example of solving formulas for a specified variable, a new step added to the problem-solving strategy (*Assign a variable.*), additional cautions, additional explanation of consecutive integers, a worked example showing how to clear an inequality of fractions to solve it, and a more comprehensive *Chapter Test.*

Chapter 3

- New example and exercise applications include topics such as the Hollywood sign, René Descartes, endangered species, the U.S. Space Program, Honda Insight gas mileage, renewable energy, dental-assistant programs, firefighting, printing presses, calculating the cost to use an iPad, online games, U.S. credit card debt, the cost of raising a family, managing dental appointment times, and the amount of carbon dioxide in the Earth's atmosphere.
- New instructional features include a comparison of linear and nonlinear equations, tips for constructing a table of solutions for equations of the form $Ax + By = 0$, a comparison of one- and two-dimensional graphs, lines with slopes 1 and -1, additional cautions, a summary table of the forms of linear equations in two variables, and a more comprehensive *Chapter Test.*

Chapter 4

- New example and exercise applications include topics such as social networking, using the *rule of thirds* when taking photographs, the cost of changing CFL light bulbs, the number of women awarded Bachelor's degrees, sources of electricity, newspaper readership, greenhouse gas emissions, and lung cancer statistics.
- New instructional features include additional emphasis on what to write when both variables drop out when using the substitution method, a new step added to the problem-solving strategy (*Assign variables to the unknowns.*), additional cautions, and a comparison of one-variable and two-variable approaches to solving application problems.

Chapter 5

- New example and exercise applications include topics such as threshold hearing, supercomputers, replacing a fan belt, and diabetes diagnoses.
- New instructional features include additional notes explaining how exponential notation is read, additional cautions, a verification that the definitions of zero and negative exponents are consistent with students' previous experience with exponents, a more visible explanation showing how to multiply and divide numbers written in scientific notation, a comparison of polynomials and expressions that are not polynomials, an additional example of polynomial subtraction (vertical form), and more detailed author notes explaining polynomial long division.

Chapter 6

- New example and exercise applications include topics such as making crayons, staining a front door, Grammy nominations, antique shows, and rate of change of Wikipedia entries.
- New instructional features include a worked example in which the terms of a polynomial are rearranged to facilitate factoring by grouping, additional factoring tips, additional insight into factoring using the key number method, more information about how to recognize perfect-square trinomials, alternate factoring approaches, additional *Language of Algebra* boxes, and a more extensive list of types of quadratic equations.

Chapter 7

- New example and exercise applications include topics such as preparing an operating room, number of Tweets per ten seconds, building design, computer hard drives, and exercise equipment depreciation.
- New instructional features include a screened color 1 that is used when explaining how to build and simplify fractions, a worked example in which the common factor $x + 1$ in the numerator and $1 + x$ in the denominator are removed, and increased use of the term *rational expression* in place of the word *fraction.*

Chapter 8

- New example and exercise applications include topics such as websites, packaging, breathing capacity, airline seating, health care, earthquakes, and currency exchange.
- New instructional features include several new *Success Tips,* a more detailed discussion of identities and contradictions, increased emphasis on polynomial functions and their graphs, a more detailed explanation of finding the domain of a function, a more in-depth discussion of rational functions and their graphs, new examples and *Study Set* problems about addition, subtraction, multiplication, and division of functions, a screened color 1 is used when explaining how to build and simplify fractions, and a more comprehensive *Chapter Test.*

Chapter 9

- New example and exercise applications include topics such as matting art and beach pollution.
- New instructional features include some new *Language of Algebra* and *Caution* features, a more in-depth discussion of radical functions and their graphs, additional notes explaining how the notation is read, a screened color 1 is used when explaining how to rationalize expressions, additional worked examples and *Study Set* problems about addition, subtraction, and multiplication of radical expressions, a new worked example and *Study Set* problems about solving equations containing fractional exponents, a more detailed discussion of the complex number system, the midpoint formula has been moved from Section 8.7 to Section 9.6, and a more comprehensive *Chapter Test.*

Chapter 10

- New example and exercise applications include topics such as physics, shopping centers, women in law enforcement, crowd control, hospital emergency departments, comparing job offers, and oceanography.
- New instructional features include a new worked example and *Study Set* problems about quadratic functions, some new *Language of Algebra, Caution,* and *Success Tip* features, a new worked example and *Study Set* problems about quadratic equations with complex-number solutions, a revamped strategy for solving quadratic equations, and a more comprehensive *Chapter Test.*

Chapter 11

- New example and exercise applications include topics such as medication absorption, atmospheric carbon dioxide concentrations, book publishing, U.S. poverty rates, stocking lakes, the Richter scale, social services case loads, buying advertising time, bed bugs, cross-country skiing, weight training progress, gross domestic product of U.S. states, and human growth hormone.
- New instructional features include additional notes explaining how the notation is read, a new worked example and *Study Set* problems about evaluating sum, difference, product, quotient, and composition functions, comparison and contrast of other types of functions to exponential functions, a more detailed discussion of exponential growth and decay models and their graphs, new visuals relating exponential and logarithmic forms, a new visual explaining logarithmic function notation, a new worked example and *Study Set* problems about logarithmic growth, the former Fourth Edition Sections 11.4 and 11.6 were combined to create a new *Section 11.5: Base-e Exponential and Logarithmic Functions,* updated calculator keystrokes instructions, a revamped strategy for solving quadratic equations, more author notes and steps shown in the solutions for worked examples solving logarithmic equations, and a more comprehensive *Chapter Test.*

Chapter 12

- New example and exercise applications include topics such as footwear imports, area codes, investing, interior design, fashion design, physical therapy, hair color treatments, production planning, manufacturing, nutrition, NBA centers, television vs. internet video viewing, student loans, gourmet fruit, weight training, and railroad safety.

- New instructional features include moving the application problems to the end of the chapter in Sections 12.4 and 12.5 so that any of the methods (substitution, elimination, matrices, Cramer's rule) could be used to solve the system of equations, updated graphing calculator keystrokes instructions, a more precise definition of a set of dependent equations, reduced row-echelon form and Gauss-Jordan elimination, more *Notation* and *Language of Algebra* features, additional use of tables for problem solving, and a more comprehensive *Chapter Test.*

Chapter 13

- New example and exercise applications include topics such as civil engineering, landscaping, nuclear power, and advertising.
- New instructional features include several new *Success Tips* about determining h and k, a more in-depth discussion of the role of the center when graphing conics, a discussion of the curvature of the branches of a hyperbola, and a more comprehensive *Chapter Test.*

Chapter 14

- New example and exercise applications include topics such as union membership, manufacturing costs, and flight times.
- New instructional features include several new *Notation* and *Success Tip* features explaining notation, several new *Language of Algebra* features, a discussion of the curvature of the branches of a hyperbola, and a more comprehensive *Chapter Test.*

Instructor Resources

Print Ancillaries

Instructor's Resource Binder (0-538-73675-5)
Maria H. Andersen, *Muskegon Community College*
Each section of the main text is discussed in uniquely designed *Teaching Guides* containing instruction tips, examples, activities, worksheets, overheads, assessments, and solutions to all worksheets and activities.

Complete Solutions Manual (1-111-57848-6)
Alexander Lee, *Hinds Community College* and Kristy Hill, *Hinds Community College*
The *Complete Solutions Manual* provides worked-out solutions to all of the problems in the text.

Annotated Instructor's Edition (1-111-57836-2)
The *Annotated Instructor's Edition* provides the complete student text with answers next to each respective exercise. New to this edition, *Teaching Tips* have been placed at strategic teaching moments and each worked example has an accompanying *Teaching Example* for use during the classroom lecture.

Complete Course Notebook (1-133-36448-9)
Ann Ostberg
NEW! Get your students up to speed with study skills while they learn course-specific material. This notebook introduces students to the steps they need to take to ensure college success, and then guides them step-by-step through each section of their textbook. Each chapter contains a chapter readiness assessment, fill-in-the-blank course notes, structured note-taking guides, questions for reflection, activities, and end-of-chapter test prep. The notebook also contains general tip sheets on note-taking, studying, assessing their test scores, and much more.

Enhanced WebAssign: Start Smart Guide for Students (0-495-38479-8)
The *Enhanced WebAssign* Student Start Smart Guide helps students get up and running quickly with *Enhanced WebAssign* so they can study smarter and improve their performance in class.

Electronic Ancillaries

Enhanced WebAssign (0-538-73810-3)
Instant feedback and ease of use are just two reasons why *WebAssign* is the most widely used homework system in higher education. *WebAssign's* homework delivery system allows you to assign, collect, grade, and record homework assignments via the web. And now this proven system has been enhanced to include a multimedia eBook, video examples, and problem-specific tutorials. *Enhanced WebAssign* is more than a homework system—it is a complete learning system for math students.

Solution Builder
This online instructor database offers complete worked solutions to all exercises in the text, allowing you to create customized, secure solutions printouts (in PDF format) matched exactly to the problems you assign in class. For more information, visit www.cengage.com/solutionbuilder.

PowerLecture with ExamView® (1-111-98824-2)
This CD-ROM provides the instructor with dynamic media tools for teaching. Create, deliver, and customize tests (both print and online) in minutes with *ExamView® Computerized Testing Featuring Algorithmic Equations.* Easily build solution sets for homework or exams using *Solution Builder's* online solutions manual. Microsoft® PowerPoint® lecture slides including *all* examples from the text, figures from the book, and easy-to-use PDF test banks, in electronic format, are also included on this CD-ROM.

Text Specific Videos
Rena Petrello, *Moorpark College*
These 10- to 20-minute problem-solving lessons cover nearly every learning objective from each chapter in the Tussy/Gustafson text. Recipient of the Mark Dever Award for Excellence in Teaching, Rena Petrello presents each lesson using her experience teaching online mathematics courses. It was through this online teaching experience that Rena discovered the lack of suitable content for online instructors, which caused her to develop her own video lessons—and ultimately to create this video project. These videos have won four awards: two Telly Awards, one Communicator Award, and one Aurora Award (an international honor). Students will love the additional guidance and support when they have missed a class or when they are preparing for an upcoming quiz or exam. These videos are available in *Enhanced WebAssign* and *CourseMate.*

Printed Access Card for CourseMate with eBook for *Elementary and Intermediate Algebra,* Fifth Edition (1-4282-7475-8)

Instant Access Card for CourseMate with eBook for *Elementary and Intermediate Algebra,* Fifth Edition (1-4282-7474-X)
Complement your text and course content with study and practice materials.
Cengage Learning's *Developmental Mathematics CourseMate* brings course concepts to life with interactive learning, study, and exam preparation tools that support the printed textbook. Watch student comprehension soar as your class works with the printed textbook and the textbook-specific website. *Developmental Mathematics CourseMate* goes beyond the book to deliver what you need!

Student Resources

Print Ancillaries

Student Workbook (1-111-98783-1)
Maria H. Andersen, *Muskegon Community College*
The *Student Workbook* contains all of the assessments, activities, and worksheets from the *Instructor's Resource Binder* for classroom discussions, in-class activities, and group work.

Student Solutions Manual (1-111-57847-8)
Alexander Lee, *Hinds Community College* and Kristy Hill, *Hinds Community College*
The *Student Solutions Manual* provides worked-out solutions to the odd-numbered problems in the text.

Electronic Ancillaries

Enhanced WebAssign (0-538-73810-3)
Enhanced WebAssign (assigned by the instructor) provides instant feedback on homework assignments to students. This online homework system is easy to use and includes a multimedia eBook, video examples, and problem-specific tutorials.

Website www.cengagebrain.com or www.cengage.com/math/tussy
Students can visit us on the web and search by title to access a wealth of learning resources, including *Study Skills Workshop* materials, tutorials, final exams, chapter outlines, chapter reviews, web links, videos, flashcards, study skills handouts, and more.

Printed Access Card for CourseMate with eBook for *Elementary and Intermediate Algebra,* Fifth Edition (1-4282-7475-8)

Instant Access Card for CourseMate with eBook for *Elementary and Intermediate Algebra,* Fifth Edition (1-4282-7474-X)
The more students study, the better the results. Students can make the most of their study time by accessing everything they need to succeed in one place: read the textbook, take notes, review flashcards, watch videos, and take practice quizzes—online with *CourseMate.*

Acknowledgments

We want to express our gratitude to our accuracy checkers, Diane Koenig and Steve Odrich, as well as many others for their help with this project: Steve Odrich, Maria H. Andersen, Sheila Pisa, Alexander Lee, Paul McCombs, Ed Kavanaugh, Karl Hunsicker, Cathy Gong, Dave Ryba, Terry Damron, Marion Hammond, Lin Humphrey, Doug Keebaugh, Robin Carter, Tanja Rinkel, Bob Billups, Jeff Cleveland, Jo Morrison, Sheila White, Jim McClain, Paul Swatzel, Brandon Tussy, Liz Tussy, Dan Davison, Marshall Dean, Dennis Korn, Matt Greenbeck, Joyce Low, Ralph Tippins, Mohamad Trad, and the Citrus College library staff (including Barbara Rugeley) for their help with this project. Your encouragement, suggestions, and insight have been invaluable to us.

We also would like to express our thanks to the Cengage Learning editorial, marketing, production, and design staff for helping us craft this new edition: Charlie Van Wagner, Danielle Derbenti, Gordon Lee, Carrie Jones, Jennifer Cordoba, Heleny Wong, Guanglei Zhang, Maureen Ross, Sam Subity, Jennifer Risden, Vernon Boes, Terri Wright, Ellen Brownstein, Helen Walden, Lori Heckelman, and Graphic World.

Additionally, we would like to say that authoring a textbook is a tremendous undertaking. A revision of this scale would not have been possible without the thoughtful feedback and support from the colleagues listed below. Their contributions to this edition have shaped this revision in countless ways.

Alan S. Tussy
R. David Gustafson

Advisory Reviewers

Ashish Gupta, *William Paterson University*
Katrina Keating, *Diablo Valley College*
Roger Larson, *Anoka Ramsey Community College*
Lorraine Lopez, *San Antonio College*
Paul J. Vroman, *St. Louis Community College at Florissant Valley*

Reviewers of the Fourth and Fifth Editions

Andrea Adlman, *Ventura College*
Khadija Ahmed, *Monroe Community College*
Rodney Alford, *Calhoun Community College*
Maria Andersen, *Muskegon Community College*
Hamid Attarzadeh, *Jefferson Community and Technical College*
Victoria Baker, *University of Houston–Downtown*
Betty Barks, *Lansing Community College*
Scott Barnett, *Henry Ford Community College*
Susan Beane, *University of Houston–Downtown*
David Behrman, *Somerset Community College*
Chad Bemis, *Riverside Community College*

John F. Beyers, *University of Maryland University College*
Barbara Blass, *Oakland Community College*
Candace Blazek, *Anoka Ramsey Community College*
Jennifer Bluth, *Anoka Ramsey Community College*
A. Elena Bogardus, *Camden Community College*
Carilynn Bouie, *Cuyahoga Community College*
Charles A. Bower, *St. Philip's College*
Jeanne Bowman, *University of Cincinnati*
Kim Brown, *Tarrant Community College*
Kirby Bunas, *Santa Rosa Junior College*
Shawna M. Bynum, *Napa Valley College*
Kim Caldwell, *Volunteer State Community College*
Carole Carney, *Brookdale Community College*
Edythe L. Carter, *Amarillo College*
Joe Castillo, *Broward Community College*
Sandra Chandler, *Tidewater Community College*
Carol Cheshire, *Macon State College*
John Close, *Salt Lake Community College*
Chris Copple, *Northwest State Community College*
Tony Craig, *Paradise Valley Community College*
Patrick Cross, *University of Oklahoma*
Mary Deas, *Johnson County Community College*
Suzanne Doviak, *Old Dominion University*
Archie Earl, *Norfolk State University*
Melody Eldred, *State University of New York at Cobleskill*
Peter Embalabala, *Lincoln Land Community College*
Joan Evans, *Texas Southern University*
Mike Everett, *Santa Ana College*
Betsy Farber, *Bucks County Community College*
Rita Fielder, *University of Central Arkansas*
Maggie Flint, *Northeast State*
Anissa Florence, *Jefferson Community and Technical College*
Pat Foard, *South Plains College*
Nancy Forrest, *Grand Rapids Community College*
Tom Fox, *Cleveland State Community College*
Heng Fu, *Thomas Nelson Community College*
Douglas Furman, *SUNY Ulster Community College*
Abel Gage, *Skagit Valley College*
John Garlow, *Tarrant Community College–Southeast Campus*
Vicki Gearhart, *San Antonio College*
Radu Georgescu, *Prince George's Community College*
Rebecca Giles, *Jefferson State Community College*
Alketa Gjikuria, *Cecil College*
Megan Goodwin, *Anoka Ramsey Community College*
Kim Gregor, *Delaware Technical Community College–Wilmington*
Thomas Grogan, *Cincinnati State*
Sally Haas, *Angelina College*
Paula Jean Haigis, *Calhoun Community College*
Haile Kebede Haile, *Minneapolis Community and Technical College*
Kelli Jade Hammer, *Broward Community College*
Mehdi Hakim Hashemi, *Normandale Community College*
Julia Hassett, *Oakton Community College*
Jennifer Hastings, *Northeast Mississippi Community College*
Alan Hayashi, *Oxnard College*
Kristy Hill, *Hinds Community College*
Jim Hodge, *Mountain State University*
Amy Hoherz, *Johnson County Community College*
Laura Hoye, *Trident Technical College*
Becki Huffman, *Tyler Junior College*
Jeffrey Hughes, *Hinds Community College*
Vera Hu-Hyneman, *SUNY–Suffolk Community College*
Angela Jahns, *North Idaho College*
Cassandra Johnson, *Robeson Community College*
Cynthia Johnson, *Heartland Community College*
Leslie Johnson, *John C. Calhoun State Community College*
Pete Johnson, *Eastern Connecticut State University*
Shelbra Jones, *Wake Technical Community College*
Ed Kavanaugh, *Schoolcraft College*
Leonid Khazanov, *Borough of Manhattan Community College*
MC Kim, *Suffolk County Community College*
Lynette King, *Gadsden State Community College*
Mike Kirby, *Tidewater Community College*
Alex Kolesnik, *Ventura College*
Patricia Kopf, *Kellogg Community College*
Elena Kravchuk, *University of Alabama–Birmingham*
Marlene Kutesky, *Virginia Commonwealth University*
Fred Lang, *Art Institute of Washington*
Hoat Le, *San Diego Community College*
Alexander Lee, *Hinds Community College, Rankin Campus*
Wayne (Paul) Lee, *Saint Philip's College*
Richard Leedy, *Polk Community College*
Mary Legner, *Riverside Community College*
Lamar Lider-Manuel, *Seminole Community College*
Daniel Lopez, *Brookdale Community College*
Ann Loving, *J. Sargeant Reynolds Community College*
Yixia Lu, *South Suburban College*
Keith Luoma, *Augusta State University*
Julie L. Mays, *Angelina College*
Mikal McDowell, *Cedar Valley College*
Marcus McGuff, *Austin Community College*
Owen Mertens, *Missouri State University*
Susan Meshulam, *Indiana University/Purdue University Indianapolis*
James Metz, *Kapi'olani Community College*
Trudy Meyer, *El Camino College*
Pam Miller, *Phoenix College*
Molly Misko, *Gadsden State Community College*
Catherine Moushon, *Elgin Community College*
Tania Munding, *Ohlone College*
Charlie Naffziger, *Central Oregon Community College*
Oscar Neal, *Grand Rapids Community College*
Doug Nelson, *Central Oregon Community College*
Elsie Newman, *Owens Community College*
Charlotte Newsom, *Tidewater Community College*
Katrina Nichols, *Delta College*
Randy Nichols, *Delta College*
Stephen Nicoloff, *Paradise Valley Community College*
Megan Nielsen, *St. Cloud State University*
Charles Odion, *Houston Community College*
Jason Pallett, *Longview Community College*
Mary Beth Pattengale, *Sierra College*
Naeemah Payne, *Los Angeles Community College*
Fred Peskoff, *Borough of Manhattan Community College*
Sheila Pisa, *Riverside Community College–Moreno Valley*

Carol Ann Poore, *Hinds Community College*
Jill Rafael, *Sierra College*
Pamela Reed, *North Harris Montgomery Community College*
Pamelyn Reed, *Cy-Fair College*
Nancy Ressler, *Oakton Community College*
Elaine Richards, *Eastern Michigan University*
Harriette Roadman, *New River Community College*
Lilia Ruvalcaba, *Oxnard College*
Jeffrey Saikali, *San Diego Miramar College*
Fary Sami, *Harford Community College*
Emma Sargent, *Tennessee State University*
Ned Schillow, *Lehigh Carbon Community College*
Joe Sedlacek, *Kirkwood Community College*
Wendiann Sethi, *Seton Hall University*
Debra Shafer, *University of North Carolina*
Hazel Shedd, *Hinds Community College*
Patty Sheeran, *McHenry Community College*
Karen Smith, *Nicholls State University*
Christa Solheid, *Santa Ana College*
Donald Solomon, *University of Wisconsin–Milwaukee*
Frankie Solomon, *University of Houston–Downtown*
Jim Spencer, *Santa Rosa Junior College*
John Squires, *Cleveland State Community College*
Michael Stack, *South Suburban College*
Kristen Starkey, *Rose State College*
Robin Steinberg, *Pima Community College*
Kristin Stoley, *Blinn College*
Eleanor Storey, *Front Range Community College–Westminster Campus*
Teresa Sutcliffe, *Los Angeles Valley College*
Eden Thompson, *Utah Valley State College*
Cindy Thore, *Central Piedmont Community College*
Rose Toering, *Kilian Community College*
Fariheh Towfiq, *Palomar College*
James Vallade, *Monroe County Community College*
Gowribalan "Ana" Vamadeva, *University of Cincinnati*
Maggie Pasqua Viz, *Brookdale Community College*
Beverly Vredevelt, *Spokane Falls Community College*
Andreana Walker, *Calhoun Community College*
Carol Walker, *Hinds Community College*
Cynthia Wallin, *Central Virginia Community College*
John Ward, *Kentucky Community and Technical College–Jefferson Community College*
Richard Watkins, *Tidewater Community College*
Diane Williams, *Northern Kentucky University*
Antoinette Willis, *St. Philip's College*
Jackie Wing, *Angelina College*
Judith Wood, *Central Florida Community College*
Nazar Wright, *Guilford Technical Community College*
Valerie Wright, *Central Piedmont Community College*
Shishen Xie, *University of Houston–Downtown*
Catalina Yang, *Oxnard College*
Heidi Young, *Bryant and Stratton College*
Mary Young, *Brookdale Community College*
Ghidei Zedingle, *Normandale Community College*
Loris Zucca, *Kingwood College*

Index of Applications

Examples that are applications are shown with **boldface** page numbers.
Exercises that are applications are shown with lightface page numbers.

Politics, Government, and Military

An Introduction to Algebra

1

©Carolina K. Smith, M.D./Shutterstock.com

from Campus to Careers

Lead Transportation Security Officer

Since 9/11, Homeland Security is one of the fastest-growing career choices in the United States. A lead transportation security officer works in an airport where he or she searches passengers, screens baggage, reviews tickets, and determines staffing requirements. The job description calls for the ability to perform arithmetic computations correctly and solve practical problems by choosing from a variety of mathematical techniques such as formulas and percentages.

Problem 113 in **Study Set 1.2, Problem 99** in **Study Set 1.5,** and **Problem 117** in **Study Set 1.7** involve situations that a lead transportation security officer might encounter on the job. The mathematical concepts discussed in this chapter can be used to solve those problems.

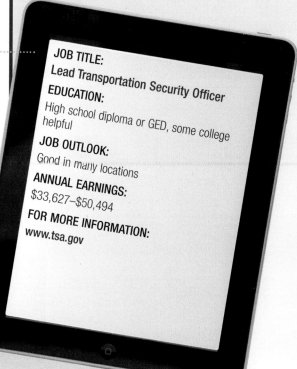

JOB TITLE:
Lead Transportation Security Officer
EDUCATION:
High school diploma or GED, some college helpful
JOB OUTLOOK:
Good in many locations
ANNUAL EARNINGS:
$33,627–$50,494
FOR MORE INFORMATION:
www.tsa.gov

Study Skills Workshop ▸ Committing to the Course

Teaching Tip: Study Skills Workshop topics may be taught in any order, at any time during the course. See the Index for a complete list to determine if and when other Study Skills Workshop topics should be discussed with your students.

Starting a new course is exciting, but it might also make you a bit nervous. In order to be successful in your algebra class, you need a plan.

MAKE TIME FOR THE COURSE: As a general guideline, 2 hours of independent study time is recommended for every hour in the classroom.

KNOW WHAT IS EXPECTED: Read your instructor's syllabus thoroughly. It lists class policies about attendance, homework, tests, calculators, grading, and so on.

BUILD A SUPPORT SYSTEM: Know where to go for help. Take advantage of your instructor's office hours, your school's tutorial services, the resources that accompany this textbook, and the assistance that you can get from classmates.

Now Try This ▸

Each of the forms referred to below can be found online at: www.cengage.com/math/tussy.

1. To help organize your schedule, fill out the *Weekly Planner Form.*
2. Review the class policies by completing the *Course Information Sheet.*
3. Use the *Support System Worksheet* to build your course support system.

Teaching Tip: The *Are You Ready?* feature reviews crucial prerequisite skills that students should have mastered already if they are to be successful with the new topics in this section. All of these problems can be found in Enhanced WebAssign, allowing you to require them to be submitted before class in order to gauge student readiness.

SECTION 1.1

OBJECTIVES

1. Read tables and graphs.
2. Use the basic vocabulary and notation of algebra.
3. Identify expressions and equations.
4. Use equations to construct tables of data.

Introducing the Language of Algebra

ARE YOU READY? *Are You Ready? exercises available online at www.webassign.net/brookscole*

The following problems review some basic arithmetic skills that are needed in this section. Answers to the Are You Ready? *problems are located in Appendix 3 at the back of the book.*

1. Add: $125 + 85$ 210

2. Subtract: $2{,}400 - 650$ 1,750

3. Multiply: $78 \cdot 14$ 1,092

4. Divide: $243 \div 27$ 9

Algebra is the result of contributions from many cultures over thousands of years. The word *algebra* comes from the title of the book *Ihm Al-jabr wa'l muqābalah,* written by an Arabian mathematician around A.D. 800. We can think of algebra as a language with its own vocabulary and notation. In this section, we begin to explore the language of algebra by introducing some of its basic components.

1 Read Tables and Graphs.

In algebra, we often use **tables** to show relationships between quantities. For example, the table below lists the number of calories a 160-pound adult burns during 10, 20, 30, and 40 minutes of snowboarding. For a workout of, say, 30 minutes, we locate 30 in the left column and then scan across the table to see that 300 calories are burned.

©Ipatov/Shutterstock.com

Minutes snowboarding	Calories burned
10	100
20	200
30	300
40	400

The information in the table also can be presented in a **bar graph,** as shown on the next page, on the left. The **horizontal axis** of the graph is labeled "Minutes snowboarding," and it

The Language of Algebra

Horizontal is a form of the word *horizon.* Think of the sun setting over the *horizon.* **Vertical** means in an upright position. Professional basketball player LeBron James' *vertical* leap measures more than 40 inches.

is scaled in units of 10 minutes. The **vertical axis,** labeled "Calories burned," is scaled in units of 50 calories. The height of a bar indicates the number of calories burned. For example, the bar over 40 minutes extends upward to 400. This means 400 calories are burned during a 40-minute snowboarding workout.

Another way to present the snowboarding information is with a **line graph.** Instead of using a bar to represent the number of calories burned, we use a dot drawn at the correct height. After drawing the data points for workouts of 10, 20, 30, and 40 minutes, we connect them with line segments to create the graph shown below, on the right.

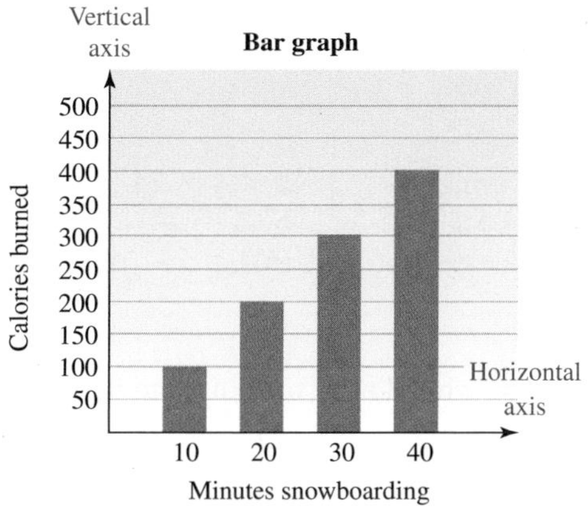

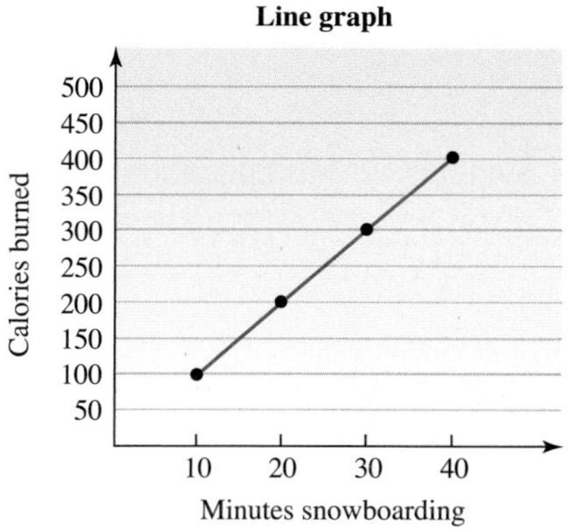

EXAMPLE 1

Teaching Tip: Highlight the structure of the worked examples by first explaining the word *strategy* to your students. Then point out that they will gain a deeper understanding of the concepts studied in this course if they know *why* a particular approach was taken.

Fitness. Use the line graph above to find the number of calories burned during a 25-minute snowboarding workout.

Strategy We will start at 25 on the horizontal axis of the graph. Then we will scan up to the red line, and over, to read the number of calories burned on the vertical axis.

Why We start on the horizontal axis because that scale gives the number of minutes of snowboarding. We scan up and over to the vertical axis because that scale gives the number of calories burned.

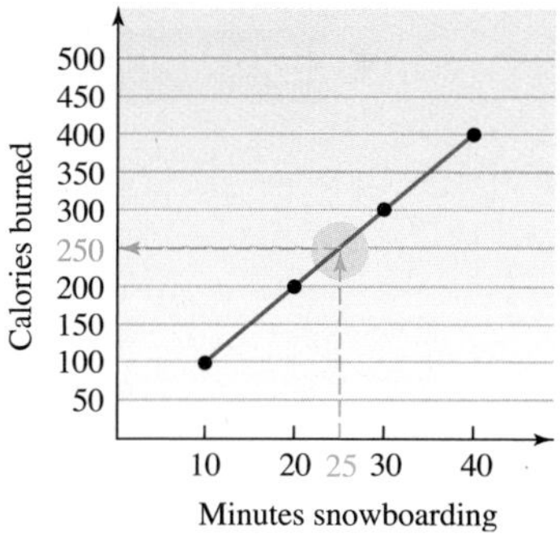

Solution

We locate 25 minutes (between 20 and 30 minutes) on the horizontal axis and draw a dashed line upward to intersect the red line. From the point of intersection, we then draw a dashed line to the left that points to the vertical axis at 250. This means that a 25-minute snowboarding workout burns 250 calories.

Success Tip

Answers to the Self Check problems are given in Appendix 3, at the back of the book.

Self Check 1 **Fitness.** Use the graph to find the number of calories burned during a 35-minute snowboarding workout. 350 calories

Now Try Problem 29

Teaching Example 1 Fitness. Use the graph to find the number of calories burned during a 15-minute snowboarding workout.
Answer: 150 calories

Teaching Tip: *Self Check* problems can be completed by students on their own, or you may use them as lecture examples. Since they appear in the book, you can have selected students read them aloud so that they become comfortable reading the language of algebra.

Teaching Tip: Tell your students that each *Now Try* problem is forward-linked to a similar problem in the *Guided Practice* of the *Study Set.*

2 Use the Basic Vocabulary and Notation of Algebra.

From the table and graphs, we see that there is a relationship between the number of calories burned and the number of minutes snowboarding. Using words, we can express this relationship as a **verbal model:**

"The number of calories burned is ten times the number of minutes snowboarding."

Since the word **product** indicates the result of a multiplication, we can also write:

"The number of calories burned is the *product* of ten and the number of minutes snowboarding."

The Language of Algebra

The collection of symbols and write-up forms used in this course is called the **notation** of algebra.

To indicate other arithmetic operations, we will use the following words.

- A **sum** is the result of an addition: The sum of 5 and 6 is 11.
- A **difference** is the result of a subtraction: The difference of 3 and 2 is 1.
- A **quotient** is the result of a division: The quotient of 6 and 3 is 2.

Many symbols used in arithmetic are also used in algebra. For example, a + symbol is used to indicate addition, a − symbol is used to indicate subtraction, and an = symbol means *is equal to.*

Since the letter x is often used in algebra and could be confused with the multiplication symbol ×, we usually write multiplication using a **raised dot** or **parentheses.**

Symbols for Multiplication

×	Times symbol	$6 \times 4 = 24$
·	Raised dot	$6 \cdot 4 = 24$
( )	Parentheses	$(6)4 = 24$ or $6(4) = 24$ or $(6)(4) = 24$

In algebra, the symbol most often used to indicate division is the **fraction bar.**

Symbols for Division

÷	Division symbol	$24 \div 4 = 6$
$\overline{)}$	Long division	$4\overline{)24}$ with quotient 6
—	Fraction bar	$\frac{24}{4} = 6$

EXAMPLE 2 Write each statement in words, using one of the words *sum, product, difference,* or *quotient:*

a. $\frac{22}{11} = 2$ **b.** $22 + 11 = 33$

Strategy We will examine each statement to determine whether addition, subtraction, multiplication, or division is being performed.

Why The word that we should use (*sum, product, difference,* or *quotient*) depends on the arithmetic operation that we have to describe.

Solution

a. Since the fraction bar indicates division, we have: The quotient of 22 and 11 equals 2.

b. The + symbol indicates addition: The sum of 22 and 11 equals 33.

Teaching Example 2 Write the following statement in words: $67 - 41 = 26$
Answer: The difference of 67 and 41 equals 26.

Self Check 2 Write the following statement in words: $22 \cdot 11 = 242$
The product of 22 and 11 equals 242.

Now Try Problems 33 and 35

3 Identify Expressions and Equations.

The Language of Algebra

Since the number of minutes snowboarding can *vary,* or change, it is represented using a **variable.**

Another way to describe the relationship between calories burned and snowboarding time uses *variables.* **Variables** are letters that stand for numbers. If we let the letter m represent the number of minutes snowboarding, then the number of calories burned is ten times m, written $10m$. In this notation, the number 10 is an example of a **constant** because it does not change value.

When multiplying a variable by a number, or a variable by another variable, we can omit the symbol for multiplication. For example,

$10m$ means $10 \cdot m$ $\qquad$ xy means $x \cdot y$ $\qquad$ $8abc$ means $8 \cdot a \cdot b \cdot c$

We call $10m$, xy, and $8abc$ *algebraic expressions.*

Algebraic Expressions Variables and/or numbers can be combined with the operations of addition, subtraction, multiplication, and division to create **algebraic expressions.**

The Language of Algebra

Algebraic expressions are often simply called **expressions.**

Here are some other examples of algebraic expressions.

$4a + 7$	This expression is a combination of the numbers 4 and 7, the variable *a*, and the operations of multiplication and addition.
$\frac{10 - y}{3}$	This expression is a combination of the numbers 10 and 3, the variable *y*, and the operations of subtraction and division.
$15mn(2m)$	This expression is a combination of the numbers 15 and 2, the variables *m* and *n*, and the operation of multiplication.

The Language of Algebra

The equal symbol = can be represented by verbs such as:

is *are* *gives* *yields*

The symbol ≠ is read as *"is not equal to."*

In the snowboarding example, if we let the letter *c* stand for the number of calories burned, we can translate the verbal model to mathematical symbols.

The number of calories burned	is	ten	times	the number of minutes snowboarding.
c	$=$	10	$\cdot$	m

The statement $c = 10 \cdot m$, or more simply, $c = 10m$, is called an *equation.* An **equation** is a mathematical sentence that contains an = symbol. The = symbol indicates that the expressions on either side of it have the same value. Other examples of equations are

$$3 + 5 = 8 \qquad x + 5 = 20 \qquad 17 - 2r = 14 + 3r \qquad p = 100 - d$$

EXAMPLE 3

©iStockPhoto.com/MvH

Stormy Weather. One way to estimate your distance (in miles) from a lightning strike is to count the number of seconds between the flash of lightning and the sound of thunder and divide by five. Translate this verbal model into an equation.

Strategy We will represent the two unknown quantities using variables and we will use symbols to represent the words *is* and *divided by.*

Why To translate a verbal (word) model into an equation means to write it using mathematical symbols.

Solution Let d = your distance (in miles) from the lightning strike and s = the number of seconds between the lightning and the thunder. Then we have:

Your distance (in miles) from the lightning strike	is	the number of seconds between the lightning and thunder	divided by	five.
d	$=$	s	$\div$	5

If we write the division using a fraction bar, then the verbal model translates to the equation $d = \frac{s}{5}$.

Success Tip

Throughout this course you will be working with *expressions* and *equations.* It is important to know the difference between them. An equation contains an = symbol. An expression does not.

Self Check 3 Translate into an equation: The number of unsold tickets is the difference of 500 and the number of tickets that have been purchased.

$u = 500 - p$ (Answers may vary.)

Now Try Problems 41 and 45

Teaching Example 3 Translate into an equation: The number of decades is the number of years divided by ten.

Answer: $d = \frac{y}{10}$

In the snowboarding example, we have seen that a table, a graph, and an equation can be used to describe the relationship between calories burned and workout time. The equation $c = 10m$ has one major advantage over the other methods. It can be used to accurately determine the number of calories burned during a snowboarding workout of *any* length of time.

EXAMPLE 4 **Fitness.** Use the equation $c = 10m$ to find the number of calories burned during a 36-minute snowboarding workout.

Strategy In $c = 10m$, we will replace m with 36. Then we will multiply 36 by 10 to obtain the value of c.

Why The equation $c = 10m$ indicates that the number of calories burned is found by multiplying the number of minutes snowboarding by 10.

Solution

$c = 10m$ This is the describing equation.

$c = 10(36)$ Replace m, which stands for the number of minutes snowboarding, with 36. Use parentheses to show the multiplication. We also could write $10 \cdot 36$.

$c = 360$ Do the multiplication.

A snowboarding workout of 36 minutes will burn 360 calories.

Teaching Example 4 Fitness. Use the equation $c = 10m$ to find the number of calories burned during a 28-minute snowboarding workout.
Answer: 280 calories

Self Check 4 **Fitness.** Use the equation $c = 10m$ to find the number of calories burned during a 48-minute snowboarding workout. 480 calories

Now Try Problem 53

4 Use Equations to Construct Tables of Data.

Equations such as $c = 10m$, which express a relationship between two or more variables, are called **formulas.** Some applications require the repeated use of a formula.

EXAMPLE 5 **Fitness.** Find the number of calories burned during snowboarding workouts of 18 minutes and 65 minutes. Present the results in a table.

Strategy We need to use the formula $c = 10m$ twice.

Why There are two different workouts: one that is 18 minutes long and another that is 65 minutes long.

Solution

Step 1: We construct a two-column table and enter the workout times in the first column, as shown below in red.

$c = 10m$

m	c
18	180
65	650

Since m represents the number of minutes snowboarding, we use it as the heading of the first column.

Since c represents the number of calories burned, we use it as the heading of the second column.

Step 2: We substitute 18 and 65 for m in $c = 10m$ and find each corresponding value of c. The results are entered in the second column of the table, as shown above.

$c = 10m$
$c = 10(18)$
$c = 180$

$c = 10m$
$c = 10(65)$
$c = 650$

The Language of Algebra

To **substitute** means to put or use in place of another, as with a *substitute* teacher. Here, we *substitute* 18 and 65 for m.

Teaching Example 5 Fitness. Find the number of calories burned during snowboarding workouts of 22 minutes and 85 minutes. Present the results in a table.
Answer:

m	c
22	220
85	850

Success Tip

Answers to the odd-numbered problems in each Study Set can be found at the back of the book in Appendix 3, beginning on page A-9.

Self Check 5 **Fitness.** Find the number of calories burned during snowboarding workouts of 8 minutes and 75 minutes. Present the results in a table.

m	c
8	80
75	750

Now Try Problem 55

Teaching Tip: Each Study Set begins with Vocabulary exercises. Educational research has shown that vocabulary plays a crucial role in the process of concept formation in students. All Vocabulary problems can be found in Enhanced WebAssign so that you can require them to be submitted before class to promote student readiness.

SECTION 1.1 STUDY SET

VOCABULARY

Fill in the blanks.

1. A sum is the result of an addition. A difference is the result of a subtraction. A product is the result of a multiplication. A quotient is the result of a division.

2. Variables are letters (or symbols) that stand for numbers.

3. A number, such as 8, is called a constant because it does not change.

4. Variables and numbers can be combined with the operations of addition, subtraction, multiplication, and division to create algebraic expressions.

5. An equation is a mathematical sentence that contains an = symbol. An algebraic expression does not.

6. An equation such as $c = 10m$, which expresses a relationship between two or more variables, is called a formula.

7. The horizontal axis of a graph extends left and right and the vertical axis extends up and down.

8. The word algebra comes from the title of a book written by an Arabian mathematician around A.D. 800.

CONCEPTS

Classify each item as an algebraic expression or an equation.

9. a. $m + 18 = 23$ Eqn **b.** $m + 18$ Alg exp

10. a. $30x$ Alg exp **b.** $30x = 600$ Eqn

11. a. $\frac{c - 7}{5}$ Alg exp **b.** $\frac{c - 7}{5} = 7c$ Eqn

12. a. $r = \frac{2}{3}$ Eqn **b.** $\frac{2}{3}r$ Alg exp

13. What arithmetic operations does the expression $\frac{12 + 9t}{25}$ contain? What variable does it contain? Addition, multiplication, division; t

14. What arithmetic operations does the equation $4y - 14 = 5(6)$ contain? What variable does it contain? Subtraction, multiplication; y

15. Construct a line graph using the data in the following table.

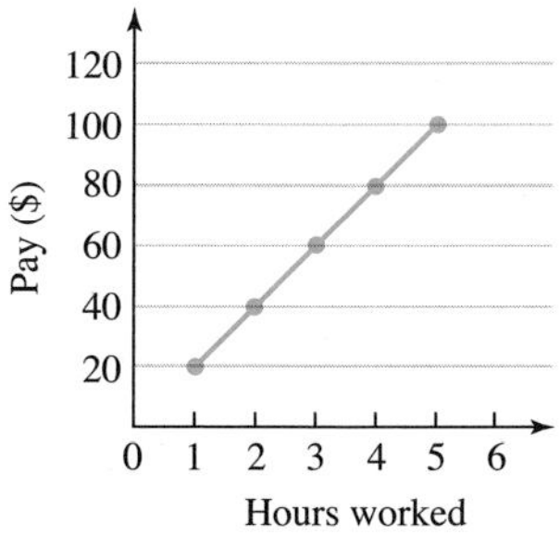

Hours worked	Pay (dollars)
1	20
2	40
3	60
4	80
5	100

Teaching Tip: Tell your students that if they encounter any difficulties with these *Guided Practice* problems, the problems are linked to worked examples that appeared earlier in the section.

16. Use the data in the graph to complete the table.

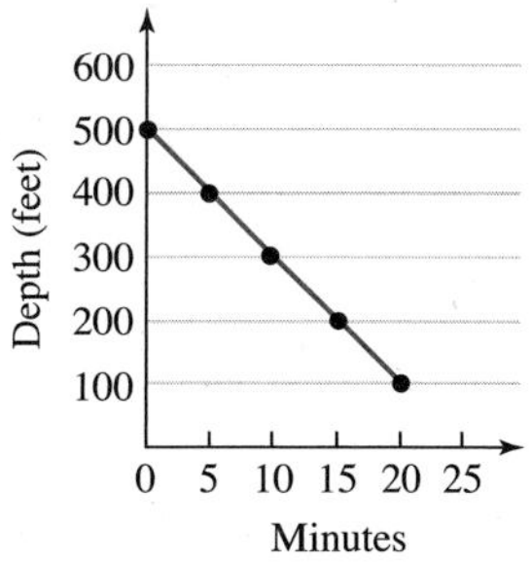

Minutes	Depth (feet)
0	500
5	400
10	300
15	200
20	100

NOTATION

Fill in the blanks.

17. The symbol ≠ means is not equal to.

18. The symbols ( ) are called parentheses.

19. Write the multiplication 5×6 using a raised dot and then using parentheses. $5 \cdot 6$, $5(6)$

20. Give four verbs that can be represented by an equal symbol =. Is, are, gives, yields

Write each expression without using a multiplication symbol or parentheses.

21. $4 \cdot x$ $4x$

22. $P \cdot r \cdot t$ Prt

23. $2(w)$ $2w$

24. $(x)(y)$ xy

Write each division using a fraction bar.

25. $32 \div x$ $\frac{32}{x}$

26. $30\overline{)90}$ $\frac{90}{30}$

27. $5\overline{)55}$ $\frac{55}{5}$

28. $h \div 15$ $\frac{h}{15}$

GUIDED PRACTICE

Use the given line graphs to answer the following questions. **See Example 1.**

29. Accounting. Explain what the dashed lines in the graph below help us find. 15-year-old machinery is worth $35,000.

30. Accounting. What is the value of 35-year-old machinery? $15,000

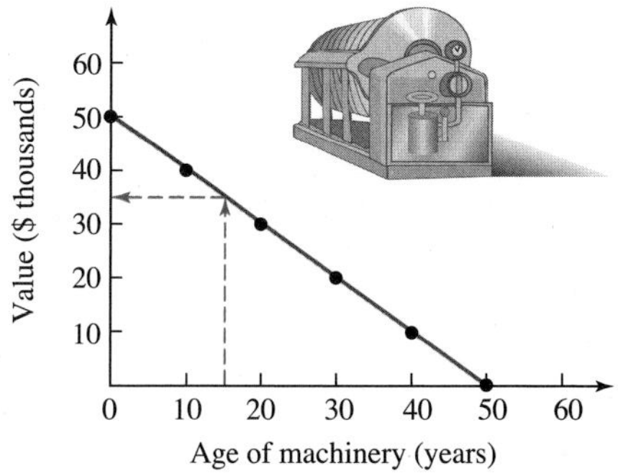

31. **Business.** Refer to the graph below. Find the income received from 30 customers. $250

32. **Business.** Refer to the graph below. Find the income received from 70 customers. $450

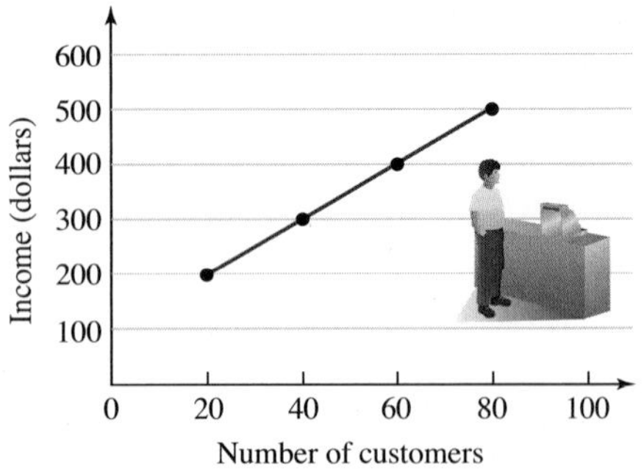

***Express each statement using one of the words* sum, product, difference, *or* quotient. See Example 2.**

33. $8(2) = 16$ The product of 8 and 2 equals 16.

34. $45 \cdot 12 = 540$ The product of 45 and 12 equals 540.

35. $11 - 9 = 2$ The difference of 11 and 9 equals 2.

36. $65 + 89 = 154$ The sum of 65 and 89 equals 154.

37. $x + 2 = 10$ The sum of x and 2 equals 10.

38. $16 - t = 4$ The difference of 16 and t equals 4.

39. $\frac{66}{11} = 6$ The quotient of 66 and 11 equals 6.

40. $12 \div 3 = 4$ The quotient of 12 and 3 equals 4.

***Translate each verbal model into an equation. (Answers may vary, depending on the variables chosen.)* See Example 3.**

41. [The sale price] is [$100] minus [the discount.] $p = 100 - d$

42. [The cost of dining out] equals [the cost of the meal] plus [$7 for parking.] $c = m + 7$

43. [7] times [the age of a dog in years] gives [the dog's equivalent human age.] $7d = h$

44. [The number of centuries] is [the number of years] divided by [100.] $c = \frac{y}{100}$

45. The amount of sand that should be used is the product of 3 and the amount of cement used. $s = 3c$

46. The number of waiters needed is the quotient of the number of customers and 10. $w = \frac{c}{10}$

47. The weight of the truck is the sum of the weight of the engine and 1,200. $w = e + 1{,}200$

48. The number of classes still open is the difference of 150 and the number of classes that are closed. $n = 150 - c$

49. The profit is the difference of the revenue and 600. $p = r - 600$

50. The distance is the product of the rate and 3. $d = 3r$

51. The quotient of the number of laps run and 4 gives the number of miles run. $\frac{l}{4} = m$

52. The sum of the tax and 35 gives the total cost. $t + 35 = c$

***Use the formula to complete each table.* See Examples 4 and 5.**

53. $d = 360 + L$

Lunch time (minutes) L	School day (minutes) d
30	390
40	400
45	405

54. $b = 1{,}024k$

Kilobytes k	Bytes b
1	1,024
5	5,120
10	10,240

55. $t = 1{,}500 - d$

Deductions d	Take-home pay t
200	1,300
300	1,200
400	1,100

56. $w = \frac{s}{12}$

Inches of snow s	Inches of water w
12	1
24	2
72	6

Use the data in the table to complete the formula.

57. $d = \frac{e}{12}$

Eggs e	Dozens d
24	2
36	3
48	4

58. $p = 2c$

Canoes c	Paddles p
6	12
7	14
8	16

59. $I = 2c$

Couples c	Individuals I
20	40
100	200
200	400

60. $t = \frac{p}{5}$

Players p	Teams t
5	1
10	2
15	3

APPLICATIONS

61. **Exercise.** The number of calories that a 125-pound adult burns doing general house cleaning chores is three times the number of minutes spent cleaning.

   a. Write a verbal model using the word *product* that describes the relationship between calories burned and minutes cleaning. The number of calories burned is the product of 3 and the number of minutes cleaning.

   b. Write a formula using the variables c and m that describes the relationship between calories burned and minutes cleaning. $c = 3m$

c. Use your answer to part b to complete the following table.

m	10	20	30	40	50	60
c	30	60	90	120	150	180

d. Use the data from the table to construct a line graph. Scale the horizontal axis in units of 10 minutes. Scale the vertical axis in units of 30 calories. See AIE Appendix 3.

62. Traffic Safety. As the railroad crossing guard drops, the measure of angle 1 (written $\angle 1$) increases while the measure of $\angle 2$ decreases. At any instant the *sum* of the measures of the two angles is 90°. Complete the table. Then use the data to construct a line graph. Scale each axis in units of 15°. See AIE Appendix 3.

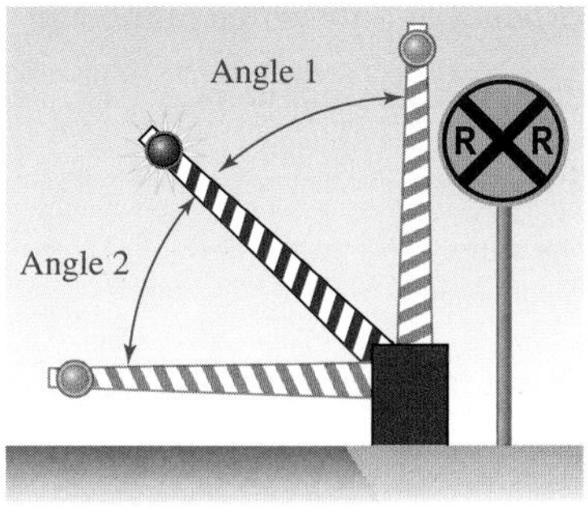

Angle 1 (degrees)	Angle 2 (degrees)
0	90
15	75
30	60
45	45
60	30
75	15
90	0

WRITING

▶ **63.** Many students misuse the word *equation* when discussing mathematics. What is an equation? Give an example.

▶ **64.** Explain the difference between an algebraic expression and an equation. Give an example of each.

65. In this section, four methods for describing numerical relationships were discussed: tables, verbal models (words), graphs, and equations. Which method do you think is the most useful? Explain why.

66. In your own words, define *horizontal* and *vertical.*

CHALLENGE PROBLEMS

67. Complete the formula. $t = 3s + 1$

s	*t*
18	55
33	100
47	142

▶ **68.** Suppose $h = 4n$ and $n = 2g$. Complete the following formula: $h = 8g$

Teaching Tip: The *Are You Ready?* feature reviews crucial prerequisite skills that students should have mastered already if they are to be successful with the new topics in this section. All of these problems can be found in Enhanced WebAssign, allowing you to require them to be submitted before class in order to gauge student readiness.

SECTION 1.2 Fractions

OBJECTIVES

1. Factor and prime factor natural numbers.
2. Recognize special fraction forms.
3. Multiply and divide fractions.
4. Build equivalent fractions.
5. Simplify fractions.
6. Add and subtract fractions.
7. Simplify answers.
8. Compute with mixed numbers.

ARE YOU READY? *Are You Ready? exercises available online at www.webassign.net/brookscole*

The following problems review some basic skills that are needed when working with fractions.

1. What is the value of $\frac{8}{8}$? 1

2. Multiply: $2 \cdot 3 \cdot 5 \cdot 5$ 150

3. Is 42 divisible by 3? Yes

4. Write the fraction $\frac{4}{5}$ in words. four-fifths

In arithmetic, we add, subtract, multiply, and divide **natural numbers:** 1, 2, 3, 4, 5, and so on. Assuming that you have mastered those skills, we will now review the arithmetic of fractions.

1 Factor and Prime Factor Natural Numbers.

To compute with fractions, we need to know how to *factor* natural numbers. To **factor** a number means to express it as a product of two or more numbers. For example, some ways to factor 8 are

$$1 \cdot 8, \qquad 4 \cdot 2, \qquad \text{and} \qquad 2 \cdot 2 \cdot 2$$

The numbers 1, 2, 4, and 8 that were used to write the products are called *factors* of 8. In general, a **factor** is a number being multiplied.

Sometimes a number has only two factors, itself and 1. We call such numbers *prime numbers.*

The Language of Algebra

When we say “factor 8,” we are using the word **factor** as a verb. When we say “2 is a *factor* of 8,” we are using the word **factor** as a noun.

Prime Numbers and Composite Numbers

A **prime number** is a natural number greater than 1 that has only itself and 1 as factors. The first ten prime numbers are 2, 3, 5, 7, 11, 13, 17, 19, 23, and 29.

A **composite number** is a natural number, greater than 1, that is not prime. The first ten composite numbers are 4, 6, 8, 9, 10, 12, 14, 15, 16, and 18.

Teaching Tip: Remind students that 0 and 1 are neither prime nor composite. Also note that 2 is the smallest prime number, and the only prime number that is even.

Every composite number can be factored into the product of two or more prime numbers. This product of these prime numbers is called its **prime factorization.**

EXAMPLE 1 Find the prime factorization of 210.

Strategy We will use a series of steps to express 210 as a product of only prime numbers.

Why To *prime factor* a number means to write it as a product of prime numbers.

Solution First, write 210 as the product of two natural numbers other than 1.

$210 = \mathbf{10 \cdot 21}$ The resulting prime factorization will be the same no matter which two factors of 210 you begin with.

The Language of Algebra

Prime factors often are written in **ascending** order. To **ascend** means to move upward.

Neither 10 nor 21 are prime numbers, so we factor each of them.

$210 = \mathbf{2 \cdot 5 \cdot 3 \cdot 7}$ Factor 10 as $2 \cdot 5$ and factor 21 as $3 \cdot 7$.

Writing the factors in order, from least to greatest, the **prime-factored form** of 210 is $2 \cdot 3 \cdot 5 \cdot 7$. Two other methods for prime factoring 210 are shown below.

Teaching Tip: Remind your students that the resulting prime factorizations will be the same no matter how they begin the factor tree:

$2 \cdot 105 = 210$ $3 \cdot 70 = 210$
$5 \cdot 42 = 210$ $6 \cdot 35 = 210$
$7 \cdot 30 = 210$ $10 \cdot 21 = 210$
$14 \cdot 15 = 210$

Work downward. Factor each number as a product of two numbers (other than 1 and itself) until all factors are prime. Circle prime numbers as they appear at the end of a branch.

Factor tree *Division ladder*

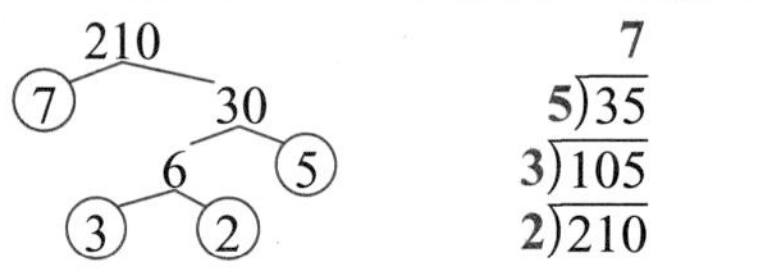

Work upward. Perform repeated division until the final quotient is a prime number. It is helpful to start with the smallest prime, 2, as a trial divisor. Then, in order, try larger primes as divisors: 3, 5, 7, 11, and so on.

Success Tip

The following divisibility rules are helpful when prime factoring.
A whole number is divisible by
- 2 if it ends in 0, 2, 4, 6, or 8
- 3 if the sum of the digits is divisible by 3
- 5 if it ends in 0 or 5
- 10 if it ends in 0

Either way, the factorization is $2 \cdot 3 \cdot 5 \cdot 7$. To check it, multiply the prime factors. The product should be 210.

Self Check 1 Find the prime factorization of 189. $189 = 3 \cdot 3 \cdot 3 \cdot 7$

Now Try Problems 15 and 23

Teaching Example 1 Find the prime factorization of 104.
Answer: $2 \cdot 2 \cdot 2 \cdot 13$

2 Recognize Special Fraction Forms.

A **fraction** describes the number of equal parts of a whole. For example, consider the figure below with 5 of the 6 equal parts colored red. We say that $\frac{5}{6}$ (five-sixths) of the figure is shaded.

In a fraction, the number above the **fraction bar** is called the **numerator,** and the number below is called the **denominator.**

The Language of Algebra

The word **fraction** comes from the Latin word *fractio* meaning "breaking in pieces."

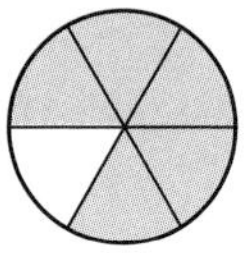

Fraction bar ⟶ $\frac{5}{6}$ ⟵ numerator / ⟵ denominator

Fractions are also used to indicate division. For example, the fraction bar in $\frac{8}{2}$ indicates that the numerator, 8, is to be divided by the denominator, 2:

$$\frac{8}{2} = 8 \div 2 = 4$$ We know that $\frac{8}{2} = 4$ because of its related multiplication statement: $2 \cdot 4 = 8$.

If the numerator and denominator of a fraction are the same nonzero number, the fraction indicates division of a number by itself, and the result is 1. Each of the following fractions is, therefore, a **form of 1.**

$$1 = \frac{1}{1} = \frac{2}{2} = \frac{3}{3} = \frac{4}{4} = \frac{5}{5} = \frac{6}{6} = \frac{7}{7} = \frac{8}{8} = \frac{9}{9} = \ldots$$

If a denominator is 1, the fraction indicates division by 1, and the result is simply the numerator. For example, $\frac{5}{1} = 5$ and $\frac{24}{1} = 24$.

Special Fraction Forms

For any nonzero number a,

$$\frac{a}{a} = 1 \quad \text{and} \quad \frac{a}{1} = a$$

3 Multiply and Divide Fractions.

The rule for multiplying fractions can be expressed in words and in symbols as follows.

Multiplying Fractions

To multiply two fractions, multiply the numerators and multiply the denominators. For any two fractions $\frac{a}{b}$ and $\frac{c}{d}$,

$$\frac{a}{b} \cdot \frac{c}{d} = \frac{a \cdot c}{b \cdot d}$$

EXAMPLE 2 Multiply: $\frac{7}{8} \cdot \frac{3}{5}$

Teaching Tip: Multiplication of fractions is explained first because it is used later to justify the methods for building fractions and simplifying fractions.

Strategy To find the product, we will multiply the numerators, 7 and 3, and multiply the denominators, 8 and 5.

Why This is the rule for multiplying two fractions.

Solution

$$\frac{7}{8} \cdot \frac{3}{5} = \frac{7 \cdot 3}{8 \cdot 5}$$ Multiply the numerators. Multiply the denominators.

$$= \frac{21}{40}$$

Self Check 2 Multiply: $\frac{5}{9} \cdot \frac{2}{3}$ $\frac{10}{27}$

Now Try Problem 27

Teaching Example 2 Multiply: $\frac{5}{9} \cdot \frac{2}{3}$

Answer: $\frac{10}{27}$

Success Tip

Every number, except 0, has a reciprocal. Zero has no reciprocal, because the product of 0 and a number cannot be 1.

One number is called the **reciprocal** of another if their product is 1. To find the reciprocal of a fraction, we invert its numerator and denominator.

$\frac{3}{4}$ is the reciprocal of $\frac{4}{3}$, because $\frac{3}{4} \cdot \frac{4}{3} = \frac{12}{12} = 1$.

$\frac{1}{10}$ is the reciprocal of 10, because $\frac{1}{10} \cdot 10 = \frac{10}{10} = 1$.

We use reciprocals to divide fractions.

Dividing Fractions

To divide two fractions, multiply the first fraction by the reciprocal of the second.

For any two fractions $\frac{a}{b}$ and $\frac{c}{d}$, where $c \neq 0$,

$$\frac{a}{b} \div \frac{c}{d} = \frac{a}{b} \cdot \frac{d}{c}$$

EXAMPLE 3 Divide: $\frac{1}{3} \div \frac{4}{5}$

Strategy We will multiply the first fraction, $\frac{1}{3}$, by the reciprocal of the second fraction, $\frac{4}{5}$.

Why This is the rule for dividing two fractions.

Solution

$\frac{1}{3} \div \frac{4}{5} = \frac{1}{3} \cdot \frac{5}{4}$ Multiply $\frac{1}{3}$ by the reciprocal of $\frac{4}{5}$, which is $\frac{5}{4}$.

$= \frac{1 \cdot 5}{3 \cdot 4}$ Use the rule for multiplying fractions. Multiply the numerators. Multiply the denominators.

$= \frac{5}{12}$

Teaching Example 3

Divide: $\frac{1}{7} \div \frac{2}{3}$

Answer: $\frac{3}{14}$

Self Check 3 Divide: $\frac{6}{25} \div \frac{1}{2}$ $\frac{12}{25}$

Now Try Problem 31

4 Build Equivalent Fractions.

The two rectangular regions on the right are identical. The first one is divided into 10 equal parts. Since 6 of those parts are red, $\frac{6}{10}$ of the figure is shaded.

The second figure is divided into 5 equal parts. Since 3 of those parts are red, $\frac{3}{5}$ of the figure is shaded. We can conclude that $\frac{6}{10} = \frac{3}{5}$ because $\frac{6}{10}$ and $\frac{3}{5}$ represent the same shaded portion of the figure. We say that $\frac{6}{10}$ and $\frac{3}{5}$ are *equivalent fractions.*

$\frac{6}{10}$

$\frac{3}{5}$

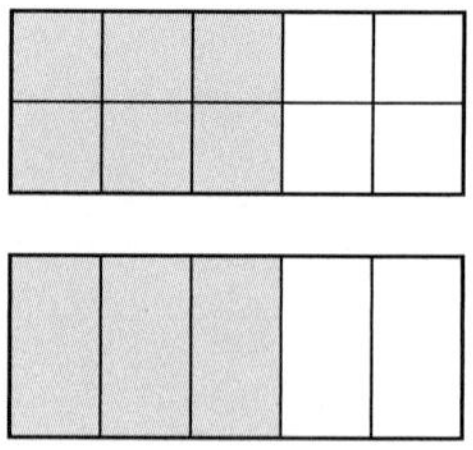

Equivalent Fractions

Two fractions are **equivalent** if they represent the same number. **Equivalent fractions** represent the same portion of a whole.

Writing a fraction as an equivalent fraction with a larger denominator is called **building** the fraction. To build a fraction, we multiply it by a form of 1. Since any number multiplied by 1 remains the same (identical), 1 is called the **multiplicative identity element.**

Multiplication Property of 1

The product of 1 and any number is that number.
For any number a,

$$1 \cdot a = a \quad \text{and} \quad a \cdot 1 = a$$

EXAMPLE 4 Write $\frac{3}{5}$ as an equivalent fraction with a denominator of 35.

Strategy We will compare the given denominator to the required denominator and ask, "By what must we multiply 5 to get 35?"

Why The answer to that question helps us determine the form of 1 to be used to build an equivalent fraction.

Solution We need to multiply the denominator of $\frac{3}{5}$ by 7 to obtain a denominator of 35. It follows that $\frac{7}{7}$ should be the form of 1 that is used to build $\frac{3}{5}$. Multiplying $\frac{3}{5}$ by $\frac{7}{7}$ changes its appearance but does not change its value, because we are multiplying it by 1.

Success Tip

Multiplying $\frac{3}{5}$ by $\frac{7}{7}$ changes its appearance, but does not change its value, because we are multiplying it by a form of 1.

$$\frac{3}{5} = \frac{3}{5} \cdot \frac{7}{7} \quad \text{Multiply } \tfrac{3}{5} \text{ by a form of 1: } \tfrac{7}{7} = 1$$

$$= \frac{3 \cdot 7}{5 \cdot 7} \quad \text{Multiply the numerators. Multiply the denominators.}$$

$$= \frac{21}{35}$$

Teaching Example 4 Write $\frac{8}{11}$ as an equivalent fraction with a denominator of 66.
Answer: $\frac{48}{66}$

Self Check 4 Write $\frac{5}{8}$ as an equivalent fraction with a denominator of 24. $\frac{15}{24}$

Now Try Problem 35

Building Fractions

To build a fraction, multiply it by 1 in the form of $\frac{c}{c}$, where c is any nonzero number.

To build an equivalent fraction in Example 4, we multiplied $\frac{3}{5}$ by 1 in the form of $\frac{7}{7}$. As a result of that step, the numerator and the denominator of $\frac{3}{5}$ were multiplied by 7:

$$\frac{3 \cdot 7}{5 \cdot 7} \begin{array}{l} \leftarrow \text{The numerator is multiplied by 7.} \\ \leftarrow \text{The denominator is multiplied by 7.} \end{array}$$

This process illustrates the following property of fractions.

The Fundamental Property of Fractions

If the numerator and denominator of a fraction are multiplied by the same nonzero number, the resulting fraction is equivalent to the original fraction.

Since multiplying the numerator and denominator of a fraction by the same nonzero number produces an equivalent fraction, your instructor may allow you to begin your solution to problems like Example 4 as shown above.

5 Simplify Fractions.

The Language of Algebra

The word **infinitely** is a form of the word **infinite,** which means endless.

Every fraction can be written in infinitely many equivalent forms. For example, some equivalent forms of $\frac{10}{15}$ are:

$$\frac{2}{3} = \frac{4}{6} = \frac{6}{9} = \frac{8}{12} = \frac{10}{15} = \frac{12}{18} = \frac{14}{21} = \frac{16}{24} = \frac{18}{27} = \frac{20}{30} = \ldots$$

Of all of the equivalent forms in which we can write a fraction, we often need to determine the one that is in *simplest form.*

Simplest Form of a Fraction

A fraction is in **simplest form,** or **lowest terms,** when the numerator and denominator have no common factors other than 1.

To **simplify a fraction,** we write it in simplest form by removing a factor equal to 1. For example, to simplify $\frac{10}{15}$, we note that the greatest factor common to the numerator and denominator is 5 and proceed as follows:

$$\frac{10}{15} = \frac{2 \cdot 5}{3 \cdot 5}$$ To prepare to simplify the fraction, factor 10 and 15. Note the form of 1 highlighted in red.

$$= \frac{2}{3} \cdot \frac{5}{5}$$ Use the rule for multiplying fractions in reverse: write $\frac{2 \cdot 5}{3 \cdot 5}$ as the product of two fractions, $\frac{2}{3}$ and $\frac{5}{5}$.

$$= \frac{2}{3} \cdot 1$$ Any nonzero number divided by itself is equal to 1: $\frac{5}{5} = 1$.

$$= \frac{2}{3}$$ Any number multiplied by 1 remains the same.

To simplify $\frac{10}{15}$, we removed a factor equal to 1 in the form of $\frac{5}{5}$. The result, $\frac{2}{3}$, is equivalent to $\frac{10}{15}$.

We can easily identify the greatest common factor of the numerator and the denominator of a fraction if we write them in prime-factored form.

EXAMPLE 5 Simplify each fraction, if possible: **a.** $\frac{63}{42}$ **b.** $\frac{33}{40}$

Strategy We will begin by prime factoring the numerator and denominator of the fraction. Then, to simplify it, we will remove a factor equal to 1.

Why We need to make sure that the numerator and denominator have no common factors other than 1. If that is the case, then the fraction is in *simplest form.*

The Language of Algebra

It is best to use the phrase "**simplify** a fraction" as opposed to "reduce a fraction." The word *reduce* suggests getting smaller, and might be confusing. When we simplify a fraction, the result is an equivalent fraction with the same value as the original fraction, not a smaller value.

Solution **a.** After prime factoring 63 and 42, we see that the greatest common factor of the numerator and the denominator is $3 \cdot 7 = 21$.

$$\frac{63}{42} = \frac{3 \cdot 3 \cdot 7}{2 \cdot 3 \cdot 7}$$ To prepare to simplify the fraction, write 63 and 42 in prime-factored form.

$$= \frac{3}{2} \cdot \frac{3 \cdot 7}{3 \cdot 7}$$ Write $\frac{3 \cdot 3 \cdot 7}{2 \cdot 3 \cdot 7}$ as the product of two fractions, $\frac{3}{2}$ and $\frac{3 \cdot 7}{3 \cdot 7}$.

$$= \frac{3}{2} \cdot 1$$ Any nonzero number divided by itself is equal to 1: $\frac{3 \cdot 7}{3 \cdot 7} = 1$.

$$= \frac{3}{2}$$ Any number multiplied by 1 remains the same.

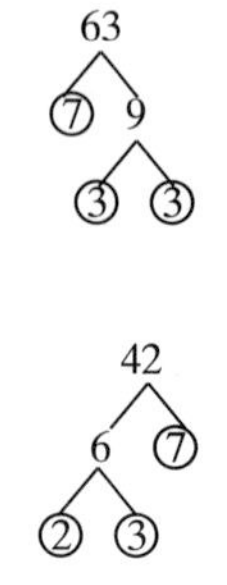

b. To attempt to simplify the fraction, prime factor 33 and 40.

$$\frac{33}{40} = \frac{3 \cdot 11}{2 \cdot 2 \cdot 2 \cdot 5}$$

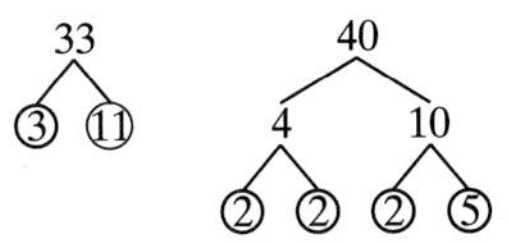

Since the numerator and the denominator have no common factors other than 1, the fraction $\frac{33}{40}$ is in simplest form (lowest terms).

Teaching Example 5 Simplify each fraction, if possible:
a. $\frac{39}{54}$ b. $\frac{8}{27}$
Answers: $\frac{13}{18}$; In simplest form

Self Check 5 Simplify each fraction, if possible: **a.** $\frac{24}{56}$ $\frac{3}{7}$ **b.** $\frac{16}{125}$ in simplest form

Now Try Problems 45 and 49

To streamline the simplifying process, we can replace pairs of factors common to the numerator and denominator with the equivalent fraction $\frac{1}{1}$.

EXAMPLE 6 Simplify: $\frac{90}{105}$

Strategy We will begin by prime factoring the numerator, 90, and denominator, 105. Then we will look for any factors common to the numerator and denominator and remove them.

Why When the numerator and/or denominator of a fraction are large numbers, such as 90 and 105, writing their prime factorizations is helpful in identifying any common factors.

Solution

$$\frac{90}{105} = \frac{2 \cdot 3 \cdot 3 \cdot 5}{3 \cdot 5 \cdot 7}$$ *To prepare to simplify the fraction, write 90 and 105 in prime-factored form.*

$$= \frac{2 \cdot \overset{1}{\cancel{3}} \cdot 3 \cdot \overset{1}{\cancel{5}}}{\underset{1}{\cancel{3}} \cdot \underset{1}{\cancel{5}} \cdot 7}$$ *Slashes and 1's are used to show that $\frac{3}{3}$ and $\frac{5}{5}$ are replaced by the equivalent fraction $\frac{1}{1}$. A factor equal to 1 in the form of $\frac{3 \cdot 5}{3 \cdot 5} = \frac{15}{15}$ was removed.*

$$= \frac{6}{7}$$ *Multiply the remaining factors in the numerator: $2 \cdot 1 \cdot 3 \cdot 1 = 6$. Multiply the remaining factors in the denominator: $1 \cdot 1 \cdot 7 = 7$.*

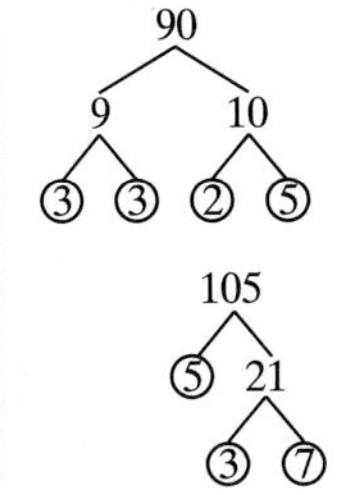

Teaching Example 6 Simplify: $\frac{180}{315}$
Answer: $\frac{4}{7}$

Self Check 6 Simplify: $\frac{126}{70}$ $\frac{9}{5}$

Now Try Problem 53

We can use the following steps to simplify a fraction.

Simplifying Fractions

1. Factor (or prime factor) the numerator and denominator to determine their common factors.
2. Remove factors equal to 1 by replacing each pair of factors common to the numerator and denominator with the equivalent fraction $\frac{1}{1}$.
3. Multiply the remaining factors in the numerator and in the denominator.

6 Add and Subtract Fractions.

In algebra as in everyday life, we can add or subtract only objects that are similar. For example, we can add dollars to dollars, but we cannot add dollars to oranges. This concept is important when adding or subtracting fractions.

Consider the problem $\frac{2}{5} + \frac{1}{5}$. When we write it in words, it is apparent we are adding similar objects.

two-**fifths** + one-**fifth**

Similar objects

Because the denominators of $\frac{2}{5}$ and $\frac{1}{5}$ are the same, we say that they have a **common denominator.**

Adding and Subtracting Fractions That Have the Same Denominator

To add (or subtract) fractions that have the same denominator, add (or subtract) their numerators and write the sum (or difference) over the common denominator. Simplify the result, if possible.

For any fractions $\frac{a}{d}$ and $\frac{b}{d}$,

$$\frac{a}{d} + \frac{b}{d} = \frac{a+b}{d} \quad \text{and} \quad \frac{a}{d} - \frac{b}{d} = \frac{a-b}{d}$$

Caution

We **do not** add fractions by adding the numerators and adding the denominators!

$$\frac{2}{5} + \frac{1}{5} = \frac{2+1}{5+5} = \frac{3}{10}$$

(crossed out)

The same caution applies when subtracting fractions.

For example,

$$\frac{2}{5} + \frac{1}{5} = \frac{2+1}{5} = \frac{3}{5} \quad \text{and} \quad \frac{18}{23} - \frac{9}{23} = \frac{18-9}{23} = \frac{9}{23}$$

Now we consider the problem $\frac{2}{5} + \frac{1}{3}$. Since the denominators are not the same, we cannot add these fractions in their present form.

two-**fifths** + one-**third**

Not similar objects

To add (or subtract) fractions with different denominators, we express them as equivalent fractions that have a common denominator. The smallest common denominator, called the **least** or **lowest common denominator,** is always the easiest common denominator to use.

Least Common Denominator (LCD)

The **least** or **lowest common denominator (LCD)** for a set of fractions is the smallest number each denominator will divide exactly (divide with no remainder).

Success Tip

To determine the LCD of two fractions, list the multiples of the larger denominator. The first number in the list that is exactly divisible by the other denominator is their LCD. For $\frac{2}{5}$ and $\frac{1}{3}$, the multiples of the larger denominator, 5, are

5, 10, (15), 20, 25, . . .

Since 15 is the first number in the list that is exactly divisible by the second denominator, 3, the LCD is 15.

The denominators of $\frac{2}{5}$ and $\frac{1}{3}$ are 5 and 3. The numbers 5 and 3 divide many numbers exactly (30, 45, and 60, to name a few), but the smallest number that they divide exactly is 15. Thus, 15 is the LCD for $\frac{2}{5}$ and $\frac{1}{3}$.

To find $\frac{2}{5} + \frac{1}{3}$, we build equivalent fractions that have denominators of 15 and we use the rule for adding fractions.

$$\frac{2}{5} + \frac{1}{3} = \frac{2}{5} \cdot \frac{3}{3} + \frac{1}{3} \cdot \frac{5}{5}$$

Multiply $\frac{2}{5}$ by 1 in the form of $\frac{3}{3}$. Multiply $\frac{1}{3}$ by 1 in the form of $\frac{5}{5}$.

We need to multiply this denominator by 5 to obtain 15. It follows that $\frac{5}{5}$ should be the form of 1 used to build $\frac{1}{3}$.

We need to multiply this denominator by 3 to obtain 15. It follows that $\frac{3}{3}$ should be the form of 1 used to build $\frac{2}{5}$.

$$= \frac{6}{15} + \frac{5}{15}$$

Multiply the numerators and multiply the denominators. Note that the denominators are now the same.

$$= \frac{6 + 5}{15}$$ Add the numerators. Write the sum over the common denominator, 15.

$$= \frac{11}{15}$$ Since 11 and 15 have no common factors other than 1, this fraction is in simplest form.

When adding (or subtracting) fractions with unlike denominators, the least common denominator is not always obvious. Prime factorization is helpful in determining the LCD.

Finding the LCD Using Prime Factorization

1. Prime factor each denominator.
2. The LCD is a product of prime factors, where each factor is used the greatest number of times it appears in any one factorization found in step 1.

EXAMPLE 7 Subtract: $\frac{3}{10} - \frac{5}{28}$

Strategy We will begin by expressing each fraction as an equivalent fraction that has the LCD for its denominator. Then we will use the rule for subtracting fractions with *like* denominators.

Why To add or subtract fractions, the fractions must have like denominators.

Solution To find the LCD, we find the prime factorization of both denominators and use each prime factor the *greatest* number of times it appears in any one factorization.

$$\left.\begin{aligned} 10 &= 2 \cdot 5 \\ 28 &= 2 \cdot 2 \cdot 7 \end{aligned}\right\} \text{LCD} = 2 \cdot 2 \cdot 5 \cdot 7 = 140$$

2 appears twice in the factorization of 28.
5 appears once in the factorization of 10.
7 appears once in the factorization of 28.

Since 140 is the smallest number that 10 and 28 divide exactly, we write $\frac{3}{10}$ and $\frac{5}{28}$ as fractions with the LCD 140.

$$\frac{3}{10} - \frac{5}{28} = \frac{3}{10} \cdot \frac{14}{14} - \frac{5}{28} \cdot \frac{5}{5}$$ We must multiply 10 by 14 to obtain 140. We must multiply 28 by 5 to obtain 140.

$$= \frac{42}{140} - \frac{25}{140}$$ Multiply the numerators and multiply the denominators. Note that the denominators are now the same.

$$= \frac{42 - 25}{140}$$ Subtract the numerators. Write the difference over the common denominator, 140.

$$= \frac{17}{140}$$ Since 17 and 140 have no common factors other than 1, this fraction is in simplest form.

Teaching Example 7 Subtract: $\frac{5}{12} - \frac{4}{21}$
Answer: $\frac{19}{84}$

Self Check 7 Subtract: $\frac{11}{48} - \frac{7}{40}$ $\frac{13}{240}$

Now Try Problem 65

We can use the following steps to add or subtract fractions with different denominators.

Adding and Subtracting Fractions That Have Different Denominators

1. Find the LCD.
2. Rewrite each fraction as an equivalent fraction with the LCD as the denominator. To do so, build each fraction using a form of 1 that involves any factors needed to obtain the LCD.
3. Add or subtract the numerators and write the sum or difference over the LCD.
4. Simplify the result, if possible.

7 Simplify Answers.

When adding, subtracting, multiplying, or dividing fractions, remember to express the answer in simplest form.

EXAMPLE 8 Perform the operations and simplify: **a.** $45\left(\frac{4}{9}\right)$ **b.** $\frac{5}{12}+\frac{3}{2}-\frac{1}{4}$

Strategy We will perform the indicated operations and then make sure that the answer is in simplest form.

Why Fractional answers should always be given in simplest form.

Solution

a. $45\left(\frac{4}{9}\right)=\frac{45}{1}\left(\frac{4}{9}\right)$ Write 45 as a fraction: $45=\frac{45}{1}$.

$=\frac{45\cdot 4}{1\cdot 9}$ Multiply the numerators. Multiply the denominators.

$=\frac{5\cdot \overset{1}{\cancel{9}}\cdot 4}{1\cdot \underset{1}{\cancel{9}}}$ To simplify the result, factor 45 as $5\cdot 9$. Then remove the common factor 9 in the numerator and denominator.

$=\frac{20}{1}$ Multiply the remaining factors in the numerator. Multiply the remaining factors in the denominator.

$=20$ Any number divided by 1 is the number itself.

b. Since the smallest number that 12, 2, and 4 divide exactly is 12, the LCD is 12.

$\frac{5}{12}+\frac{3}{2}-\frac{1}{4}=\frac{5}{12}+\frac{3}{2}\cdot\frac{6}{6}-\frac{1}{4}\cdot\frac{3}{3}$ $\frac{5}{12}$ already has a denominator of 12. Build $\frac{3}{2}$ and $\frac{1}{4}$ so that their denominators are 12.

$=\frac{5}{12}+\frac{18}{12}-\frac{3}{12}$ Multiply the numerators and multiply the denominators. The denominators are now the same.

$=\frac{20}{12}$ Add the numerators, 5 and 18, to get 23. From that sum, subtract 3. Write that result, 20, over the common denominator.

$=\frac{\overset{1}{\cancel{4}}\cdot 5}{3\cdot \underset{1}{\cancel{4}}}$ To simplify $\frac{20}{12}$, factor 20 and 12, using their greatest common factor, 4. Then remove $\frac{4}{4}=1$.

$=\frac{5}{3}$

Caution

Remember that an LCD is **not needed** when multiplying or dividing fractions.

The Language of Algebra

Fractions such as $\frac{5}{3}$, with a numerator greater than or equal to the denominator, are called **improper fractions**. In algebra, such fractions are often preferable to their equivalent mixed number form.

Teaching Example 8 Perform the operations and simplify: a. $54\left(\frac{11}{6}\right)$ b. $\frac{8}{9}+\frac{1}{3}-\frac{1}{18}$

Answers: a. 99 b. $\frac{7}{6}$

Self Check 8 Perform the operations and simplify: **a.** $24\left(\frac{7}{6}\right)$ 28 **b.** $\frac{1}{15}+\frac{31}{30}-\frac{3}{10}$ $\frac{4}{5}$

Now Try Problems 67 and 71

8 Compute with Mixed Numbers.

A **mixed number** represents the sum of a whole number and a fraction. For example, $5\frac{3}{4}$ means $5+\frac{3}{4}$ and $179\frac{15}{16}$ means $179+\frac{15}{16}$.

EXAMPLE 9 Divide: $5\frac{3}{4} \div 2$

Strategy We begin by writing the mixed number $5\frac{3}{4}$ and the whole number 2 as fractions. Then we use the rule for dividing two fractions.

Why To multiply (or divide) with mixed numbers, we first write them as improper fractions, and then multiply (or divide) as usual.

Solution

$$5\frac{3}{4} \div 2 = \frac{23}{4} \div \frac{2}{1}$$

Write $5\frac{3}{4}$ as an improper fraction by multiplying its whole-number part by the denominator: $5 \cdot 4 = 20$. Then add the numerator to that product: $3 + 20 = 23$. Finally, write the result, 23, over the denominator 4. Write 2 as a fraction: $2 = \frac{2}{1}$.

$$= \frac{23}{4} \cdot \frac{1}{2}$$

Multiply by the reciprocal of $\frac{2}{1}$, which is $\frac{1}{2}$.

$$= \frac{23}{8}$$

Multiply the numerators.
Multiply the denominators.

$$= 2\frac{7}{8}$$

Since the original problem involves a mixed number, we will express the answer in mixed-number form. Write $\frac{23}{8}$ as a mixed number by dividing the numerator, 23, by the denominator, 8.

$$\begin{array}{r} 2 \\ 8\overline{)23} \\ -\,16 \\ \hline 7 \end{array}$$

Teaching Example 9
Divide: $2\frac{4}{5} \div 3$
Answer: $\frac{14}{15}$

Self Check 9 Multiply: $1\frac{1}{8} \cdot 9$ $\quad \frac{81}{8} = 10\frac{1}{8}$

Now Try Problem 77

EXAMPLE 10 **Freeway Signs.** How far apart are the Downtown San Diego and Sea World Drive exits?

Strategy We can find the distance between exits by finding the difference in the mileages on the freeway sign: $6\frac{1}{2} - 1\frac{3}{4}$.

Why The word *difference* indicates subtraction.

Solution

$$\begin{array}{rcrcrcr} 6\frac{1}{2} & = & 6\frac{2}{4} & = & 5\frac{2}{4} + \frac{4}{4} & = & 5\frac{6}{4} \\ -1\frac{3}{4} & = & -1\frac{3}{4} & = & -1\frac{3}{4} & = & -1\frac{3}{4} \\ \hline & & & & & & 4\frac{3}{4} \end{array}$$

Using vertical form, express $\frac{1}{2}$ as an equivalent fraction with denominator 4. Then, borrow 1 in the form of $\frac{4}{4}$ from 6 to subtract the fractional parts of the mixed numbers.

The Downtown San Diego and Sea World Drive exits are $4\frac{3}{4}$ miles apart.

Success Tip

This problem could also be solved by writing the mixed numbers $6\frac{1}{2}$ and $1\frac{3}{4}$, as improper fractions and subtracting them. However, answers to real-world problems are most often given as mixed numbers instead of improper fractions, because mixed numbers are easier to understand.

Self Check 10 Subtract: $9\frac{1}{8} - 2\frac{2}{3}$ $\quad 6\frac{11}{24}$

Now Try Problem 114

Teaching Example 10 Subtract:
$6\frac{1}{4} - 1\frac{2}{3}$
Answer: $4\frac{7}{12}$

Teaching Tip: Each Study Set begins with Vocabulary exercises. Educational research has shown that Vocabulary plays a crucial role in the process of concept formation in students. All Vocabulary problems can be found in Enhanced WebAssign so that you can require them to be submitted before class to promote student readiness.

SECTION 1.2 STUDY SET

VOCABULARY

Fill in the blanks.

▶ **1.** A factor is a number being multiplied.

▶ **2.** Numbers that have only 1 and themselves as factors, such as 23, 37, and 41, are called prime numbers.

▶ **3.** When we write 60 as $2 \cdot 2 \cdot 3 \cdot 5$, we say that we have written 60 in prime-factored form.

▶ **4.** The numerator of the fraction $\frac{3}{4}$ is 3, and the denominator is 4.

▶ **5.** Two fractions that represent the same number, such as $\frac{1}{2}$ and $\frac{2}{4}$, are called equivalent fractions.

▶ **6.** $\frac{2}{3}$ is the reciprocal of $\frac{3}{2}$, because their product is 1.

▶ **7.** The least or lowest common denominator for a set of fractions is the smallest number each denominator will divide exactly.

▶ **8.** The mixed number $7\frac{1}{3}$ represents the sum of a whole number and a fraction: $7 + \frac{1}{3}$.

CONCEPTS

Complete each fact about fractions. Assume there are no divisions by 0.

▶ **9. a.** $\frac{a}{a} = 1$ **b.** $\frac{a}{1} = a$

c. $\frac{a}{b} \cdot \frac{c}{d} = \frac{a \cdot c}{b \cdot d}$ **d.** $\frac{a}{b} \div \frac{c}{d} = \frac{a \cdot d}{b \cdot c}$

e. $\frac{a}{d} + \frac{b}{d} = \frac{a+b}{d}$ **f.** $\frac{a}{d} - \frac{b}{d} = \frac{a-b}{d}$

10. What two equivalent fractions are shown? $\frac{4}{12} = \frac{1}{3}$

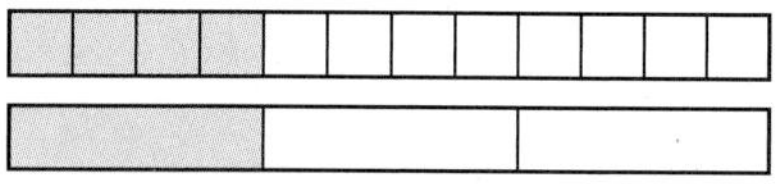

11. Complete each statement.

a. To simplify a fraction, we remove factors equal to 1 in the form of $\frac{2}{2}$, $\frac{3}{3}$, or $\frac{4}{4}$, and so on.

b. To build a fraction, we multiply it by 1 in the form of $\frac{2}{2}$, $\frac{3}{3}$, or $\frac{4}{4}$, and so on.

12. What is the LCD for fractions having denominators of:

a. 3 and 7? 21

b. 4 and 6? 12

NOTATION

Fill in the blanks.

13. a. Multiply $\frac{5}{6}$ by a form of 1 to build an equivalent fraction with denominator 30.

$$\frac{5}{6} \cdot \frac{5}{5} = \frac{25}{30}$$

b. Remove common factors to simplify $\frac{12}{42}$.

$$\frac{12}{42} = \frac{2 \cdot 2 \cdot 3}{2 \cdot 3 \cdot 7} = \frac{2}{7}$$

▶ **14. a.** Write $2\frac{15}{16}$ as an improper fraction. $\frac{47}{16}$

b. Write $\frac{49}{12}$ as a mixed number. $4\frac{1}{12}$

GUIDED PRACTICE

Find the prime factorization of each number. See Example 1.

▶ **15.** 75 $3 \cdot 5 \cdot 5$

16. 20 $2 \cdot 2 \cdot 5$

17. 28 $2 \cdot 2 \cdot 7$

18. 54 $2 \cdot 3 \cdot 3 \cdot 3$

▶ **19.** 81 $3 \cdot 3 \cdot 3 \cdot 3$

20. 125 $5 \cdot 5 \cdot 5$

▶ **21.** 117 $3 \cdot 3 \cdot 13$

22. 147 $3 \cdot 7 \cdot 7$

▶ **23.** 220 $2 \cdot 2 \cdot 5 \cdot 11$

24. 270 $2 \cdot 3 \cdot 3 \cdot 3 \cdot 5$

25. 1,254 $2 \cdot 3 \cdot 11 \cdot 19$

26. 1,144 $2 \cdot 2 \cdot 2 \cdot 11 \cdot 13$

Multiply. See Example 2.

27. $\frac{5}{6} \cdot \frac{1}{8}$ $\frac{5}{48}$

▶ **28.** $\frac{2}{3} \cdot \frac{1}{5}$ $\frac{2}{15}$

▶ **29.** $\frac{7}{11} \cdot \frac{3}{5}$ $\frac{21}{55}$

30. $\frac{13}{9} \cdot \frac{2}{3}$ $\frac{26}{27}$

Divide. See Example 3.

31. $\frac{3}{4} \div \frac{2}{5}$ $\frac{15}{8}$

▶ **32.** $\frac{7}{8} \div \frac{6}{13}$ $\frac{91}{48}$

▶ **33.** $\frac{6}{5} \div \frac{5}{7}$ $\frac{42}{25}$

34. $\frac{4}{3} \div \frac{3}{2}$ $\frac{8}{9}$

Build each fraction or whole number to an equivalent fraction with the indicated denominator. See Example 4.

▶ **35.** $\frac{1}{3}$, denominator 9 $\frac{3}{9}$

36. $\frac{3}{8}$, denominator 24 $\frac{9}{24}$

▶ **37.** $\frac{4}{9}$, denominator 54 $\frac{24}{54}$

38. $\frac{9}{16}$, denominator 64 $\frac{36}{64}$

▶ **39.** 7, denominator 5 $\frac{35}{5}$

40. 12, denominator 3 $\frac{36}{3}$

41. 5, denominator 7 $\frac{35}{7}$

42. 6, denominator 8 $\frac{48}{8}$

Simplify each fraction, if possible. See Examples 5 and 6.

▶ **43.** $\frac{6}{18}$ $\frac{1}{3}$

44. $\frac{6}{9}$ $\frac{2}{3}$

▶ **45.** $\frac{24}{28}$ $\frac{6}{7}$

▶ **46.** $\frac{35}{14}$ $\frac{5}{2}$

47. $\frac{15}{40}$ $\frac{3}{8}$

48. $\frac{22}{77}$ $\frac{2}{7}$

▶ **49.** $\frac{33}{56}$ Simplest form

50. $\frac{26}{21}$ Simplest form

51. $\frac{26}{39}$ $\frac{2}{3}$

52. $\frac{72}{64}$ $\frac{9}{8}$

▶ **53.** $\frac{36}{225}$ $\frac{4}{25}$

54. $\frac{175}{490}$ $\frac{5}{14}$

Perform the operations and, if possible, simplify. See Objective 6 and Example 7.

▶ **55.** $\frac{3}{5} + \frac{3}{5}$ $\frac{6}{5}$

56. $\frac{4}{9} - \frac{1}{9}$ $\frac{1}{3}$

▶ **57.** $\frac{6}{7} - \frac{2}{7}$ $\frac{4}{7}$

▶ **58.** $\frac{5}{13} + \frac{6}{13}$ $\frac{11}{13}$

▶ **59.** $\frac{1}{6} + \frac{1}{24}$ $\frac{5}{24}$

▶ **60.** $\frac{17}{25} - \frac{2}{5}$ $\frac{7}{25}$

61. $\frac{7}{10} - \frac{1}{14}$ $\frac{22}{35}$

62. $\frac{9}{8} - \frac{5}{6}$ $\frac{7}{24}$

▶ 63. $\frac{2}{15} + \frac{7}{9}$ $\frac{41}{45}$

64. $\frac{7}{25} + \frac{3}{10}$ $\frac{29}{50}$

65. $\frac{13}{28} - \frac{1}{21}$ $\frac{5}{12}$

66. $\frac{13}{24} - \frac{3}{40}$ $\frac{7}{15}$

Perform the operations and, if possible, simplify. See Example 8.

▶ 67. $16\left(\frac{3}{2}\right)$ 24

▶ 68. $30\left(\frac{5}{6}\right)$ 25

69. $18 \cdot \frac{2}{9}$ 4

70. $14 \cdot \frac{3}{7}$ 6

71. $\frac{2}{3} + \frac{5}{18} - \frac{1}{6}$ $\frac{7}{9}$

▶ 72. $\frac{3}{5} + \frac{7}{20} - \frac{7}{10}$ $\frac{1}{4}$

73. $\frac{5}{12} + \frac{1}{3} - \frac{2}{5}$ $\frac{7}{20}$

74. $\frac{7}{15} + \frac{1}{5} - \frac{4}{9}$ $\frac{2}{9}$

Perform the operations and, if possible, simplify. See Examples 9 and 10.

▶ 75. $4\frac{2}{3} \cdot 7$ $32\frac{2}{3}$

76. $7 \cdot 1\frac{3}{28}$ $7\frac{3}{4}$

▶ 77. $8 \div 3\frac{1}{5}$ $2\frac{1}{2}$

▶ 78. $15 \div 3\frac{1}{3}$ $4\frac{1}{2}$

▶ 79. $8\frac{2}{9} - 7\frac{2}{3}$ $\frac{5}{9}$

80. $3\frac{4}{5} - 3\frac{1}{10}$ $\frac{7}{10}$

81. $3\frac{3}{16} + 2\frac{5}{24}$ $5\frac{19}{48}$

▶ 82. $15\frac{5}{6} + 11\frac{5}{8}$ $27\frac{11}{24}$

TRY IT YOURSELF

Perform the operations and, if possible, simplify.

83. $\frac{3}{5} + \frac{2}{3}$ $\frac{19}{15}$

84. $\frac{4}{3} + \frac{7}{2}$ $\frac{29}{6}$

85. $21\left(\frac{10}{3}\right)$ 70

86. $28\left(\frac{4}{7}\right)$ 16

87. $6 \cdot 2\frac{7}{24}$ $13\frac{3}{4}$

88. $3\frac{1}{2} \cdot \frac{1}{5}$ $\frac{7}{10}$

▶ 89. $\frac{2}{3} - \frac{1}{4} + \frac{1}{12}$ $\frac{1}{2}$

90. $\frac{3}{7} - \frac{2}{5} + \frac{2}{35}$ $\frac{3}{35}$

▶ 91. $\frac{21}{35} \div \frac{3}{14}$ $\frac{14}{5}$

92. $\frac{23}{25} \div \frac{46}{5}$ $\frac{1}{10}$

93. $\frac{4}{3}\left(\frac{6}{5}\right)$ $\frac{8}{5}$

94. $\frac{21}{8}\left(\frac{2}{15}\right)$ $\frac{7}{20}$

95. $\frac{4}{63} + \frac{1}{45}$ $\frac{3}{35}$

▶ 96. $\frac{5}{18} + \frac{1}{99}$ $\frac{19}{66}$

▶ 97. $3 - \frac{3}{4}$ $\frac{9}{4}$

98. $4 - \frac{7}{3}$ $\frac{5}{3}$

99. $\frac{1}{5} \cdot \frac{3}{5}$ $\frac{3}{25}$

▶ 100. $\frac{3}{4} \cdot \frac{5}{7}$ $\frac{15}{28}$

▶ 101. $3\frac{1}{3} \div 1\frac{5}{6}$ $1\frac{9}{11}$

102. $2\frac{1}{2} \div 1\frac{5}{8}$ $1\frac{7}{13}$

103. $\frac{11}{21} - \frac{8}{21}$ $\frac{1}{7}$

104. $\frac{19}{35} - \frac{12}{35}$ $\frac{1}{5}$

105. $\frac{7}{30} + \frac{1}{50} - \frac{19}{75}$ 0

106. $\frac{11}{12} - \frac{7}{15} - \frac{9}{20}$ 0

107. $1\frac{31}{32} \cdot 7\frac{1}{9}$ 14

108. $3\frac{1}{16} \cdot 4\frac{4}{7}$ 14

Look Alikes . . .

109. a. $\frac{4}{9} + \frac{3}{7}$ $\frac{55}{63}$ b. $\frac{4}{9} - \frac{3}{7}$ $\frac{1}{63}$ c. $\frac{4}{9} \cdot \frac{3}{7}$ $\frac{4}{21}$ d. $\frac{4}{9} \div \frac{3}{7}$ $\frac{28}{27}$

▶ 110. a. $4\frac{1}{8} + 1\frac{5}{6}$ $5\frac{23}{24}$ b. $4\frac{1}{8} - 1\frac{5}{6}$ $2\frac{7}{24}$

c. $4\frac{1}{8} \cdot 1\frac{5}{6}$ $7\frac{9}{16}$ d. $4\frac{1}{8} \div 1\frac{5}{6}$ $2\frac{1}{4}$

APPLICATIONS

111. **Forestry.** A ranger cut down a tree and measured the widths of the outer two growth rings.

a. What was the growth over this 2-year period? $\frac{7}{32}$ in.

b. What is the difference in the widths of the rings? $\frac{3}{32}$ in.

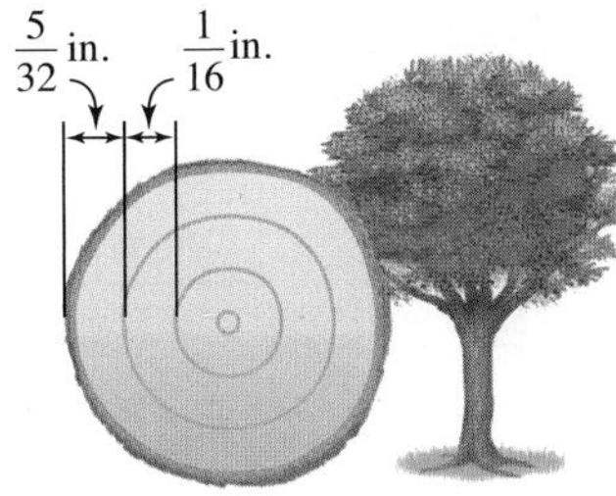

▶ 112. **Hardware.** To secure the bracket to the stock, a bolt and a nut are used. How long should the threaded part of the bolt be? $7\frac{9}{16}$ in.

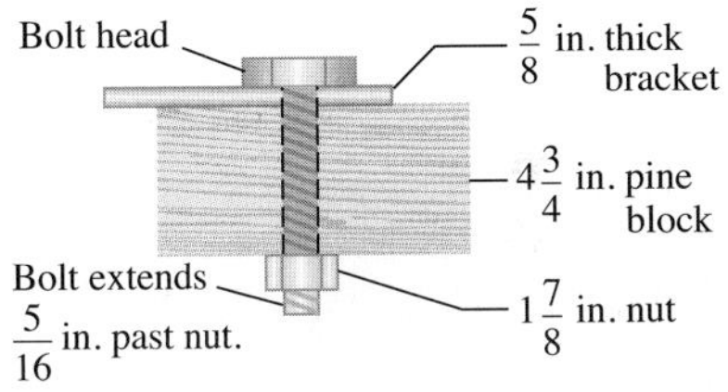

113. ▶ **from Campus to Careers**

Lead Transportation Security Officer

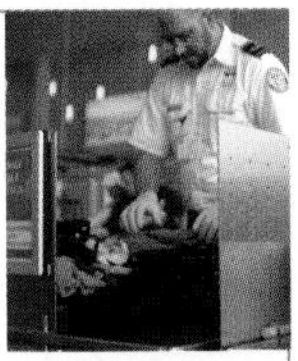

Each year, the Transportation Security Administration (TSA) screens more than 500 million pieces of luggage. On many flights, airlines do not accept luggage whose total dimension (length + width + height) exceeds 62 inches. What is the total dimension figure for the suitcase shown below? $63\frac{1}{8}$ in.

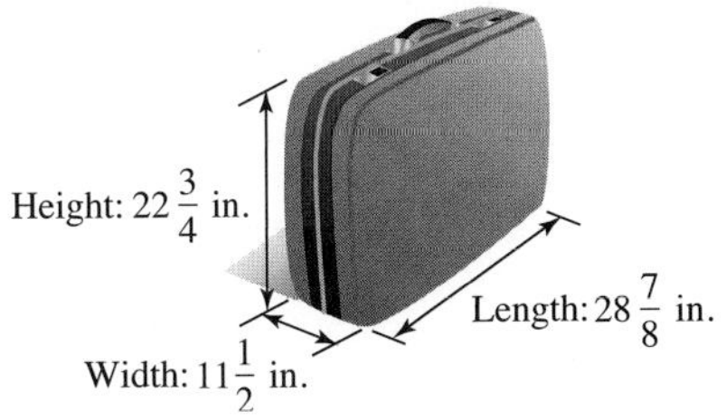

▶ 114. **Cooking.** How much butter is left in a $10\frac{1}{2}$-pound tub of butter if $4\frac{3}{4}$ pounds are used to make a wedding cake? $5\frac{3}{4}$ lb

▶ **115. Frames.** How many inches of molding are needed to make the square picture frame? $40\frac{1}{2}$ in.

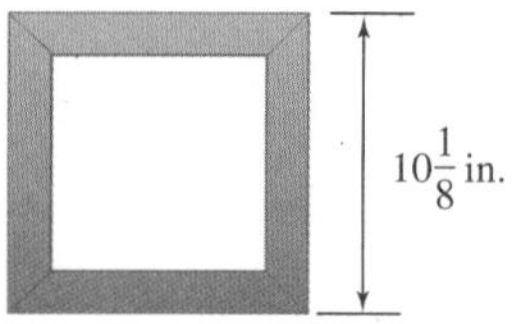

▶ **116. Decorating.** The materials used to make a pillow are shown below. Examine the inventory list to decide how many pillows can be manufactured in one production run with the materials in stock. 147

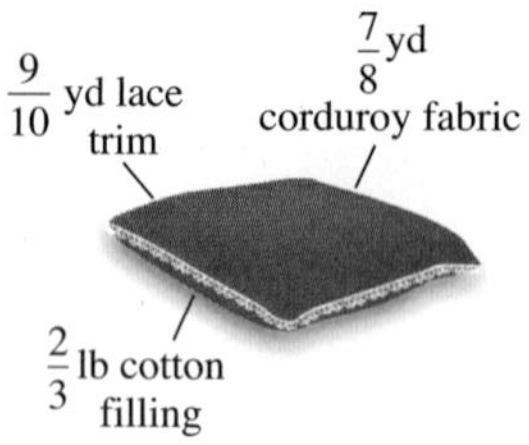

Materials	Amount in stock
Lace trim	135 yd
Corduroy fabric	154 yd
Cotton filling	98 lb

WRITING

117. Explain the error made below in simplifying $\frac{15}{45}$.

$$\frac{15}{45} = \frac{3 \cdot 5}{3 \cdot 3 \cdot 5} = \frac{0}{3} = 0$$

▶ **118.** To multiply two fractions, must they have like denominators? Explain.

119. What are equivalent fractions? Give an example.

120. Explain the error in the following addition.

$$\frac{4}{3} + \frac{3}{2} = \frac{4+3}{3+2} = \frac{7}{5}$$

REVIEW

Fill in the blanks.

121. Variables are letters (or symbols) that stand for numbers.

122. A number, such as 10, is called a constant because it does not change.

CHALLENGE PROBLEMS

123. Which is larger: $\frac{11}{12}$ or $\frac{8}{9}$? $\frac{11}{12}$

▶ **124.** If the circle represents a whole, find the missing value. $\frac{1}{20}$

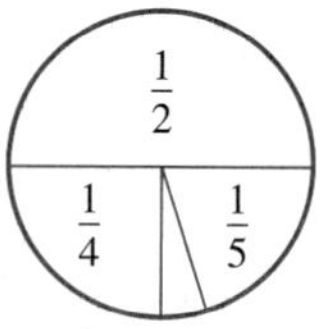

SECTION 1.3 The Real Numbers

OBJECTIVES

1. Define the set of integers.
2. Define the set of rational numbers.
3. Define the set of irrational numbers.
4. Classify real numbers.
5. Graph sets of real numbers on the number line.
6. Find the absolute value of a real number.

ARE YOU READY? *Are You Ready? exercises available online at www.webassign.net/brookscole*

The following problems review several types of numbers that we use in everyday life.

1. Count the number of letters in the word *antidisestablishmentarianism.* 28
2. What number represents a temperature that is 10 degrees below zero? −10
3. What type of number is used to express a grade point average (GPA)? a decimal (3.5 GPA)
4. Suppose a recipe calls for only part of a full cup of sugar. What type of number is normally used to describe such an amount? a fraction ($\frac{3}{4}$ cup of sugar)

A **set** is a collection of objects, such as a set of golf clubs or a set of dishes. In this section, we will define some important sets of numbers that are used in algebra.

1 Define the Set of Integers.

Natural numbers are the numbers that we use for counting. To write this set, we list its **members** (or **elements**) within **braces** { }.

Natural Numbers	The set of natural numbers is $\{1, 2, 3, 4, 5, \ldots\}$. *Read as "the set containing one, two, three, four, five, and so on."*

The natural numbers, together with 0, form the set of **whole numbers.**

Whole Numbers

The set of **whole numbers** is $\{0, 1, 2, 3, 4, 5, \ldots\}$.

Notation

The symbol . . . used in the previous definitions is called an **ellipsis** and it indicates that the established pattern continues forever.

Whole numbers are not adequate for describing many real-life situations. For example, if you write a check for more than what's in your account, the account balance will be less than zero.

We can use the **number line** below to visualize numbers less than zero. A number line is straight and has uniform markings. The arrowheads indicate that it extends forever in both directions. For each natural number on the number line, there is a corresponding number, called its *opposite,* to the left of 0. In the diagram, we see that 3 and -3 (negative three) are opposites, as are -5 (negative five) and 5. Note that 0 is its own opposite.

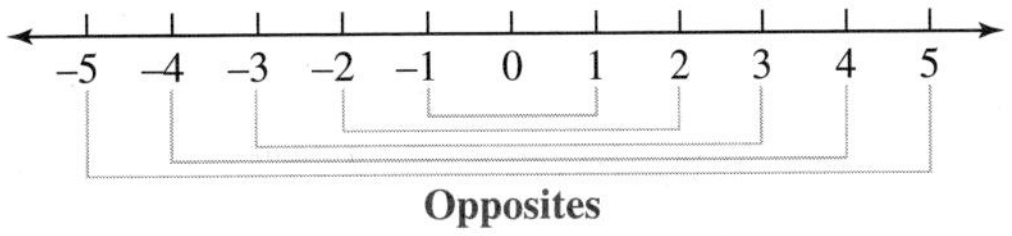

Opposites

Two numbers that are the same distance from 0 on the number line, but on opposite sides of it, are called **opposites.**

The whole numbers, together with their opposites, form the set of **integers.**

Integers

The set of **integers** is $\{\ldots, -4, -3, -2, -1, 0, 1, 2, 3, 4, \ldots\}$.

The Language of Algebra

The **positive integers** are: $1, 2, 3, 4, 5, \ldots$
The **negative integers** are: $-1, -2, -3, -4, -5, \ldots$
The **nonnegative integers** are: $0, 1, 2, 3, 4, 5, \ldots$

On the number line, numbers greater than 0 are to the right of 0. They are called **positive numbers.** Positive numbers can be written with or without a **positive sign** $+$. For example, $2 = +2$ (positive two). They are used to describe such quantities as an elevation above sea level (+3,000 ft) or a pay raise ($25).

Numbers less than 0 are to the left of 0 on the number line. They are called **negative numbers.** Negative numbers are always written with a **negative sign** $-$. They are used to describe such quantities as an overdrawn checking account ($-$\$75) or a below-zero temperature ($-12°$).

Positive and negative numbers are called **signed numbers.**

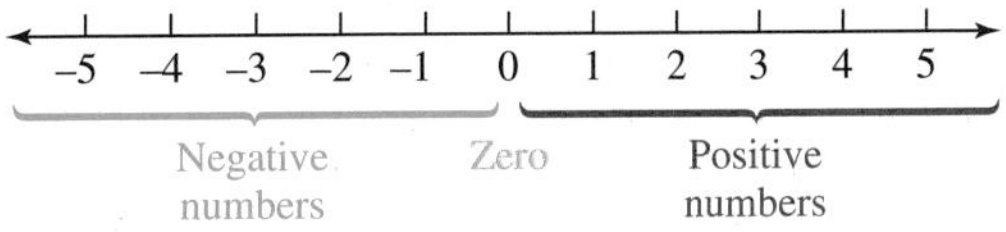

2 Define the Set of Rational Numbers.

We use fractions to describe many situations in daily life. For example, a morning commute might take $\frac{1}{4}$ hour or a recipe might call for $\frac{2}{3}$ cup of sugar. Fractions such as $\frac{1}{4}$ and $\frac{2}{3}$, which are quotients of two integers, are called *rational numbers.*

Rational Numbers

A **rational number** is any number that can be expressed as a fraction (ratio*) with an integer numerator and a nonzero integer denominator.

*Ratios are discussed in more detail in Section 7.8.

The Language of Algebra

Rational numbers are so named because they can be expressed as the *ratio* (quotient) of two integers.

Some other examples of rational numbers are

$$\frac{3}{8}, \quad \frac{41}{100}, \quad \frac{25}{25}, \quad \text{and} \quad \frac{19}{12}$$

To show that negative fractions are rational numbers, we use the following fact.

Negative Fractions

For any numbers a and b where b is not 0,

$$-\frac{a}{b} = \frac{-a}{b} = \frac{a}{-b}$$

To illustrate this rule, we consider $-\frac{11}{16}$ (read as negative eleven-sixteenths). It is a rational number because it can be written as $\frac{-11}{16}$ or as $\frac{11}{-16}$.

Positive and negative mixed numbers are also rational numbers because they can be expressed as fractions. For example,

$$7\frac{5}{8} = \frac{61}{8} \quad \text{and} \quad -6\tfrac{1}{2} = -\frac{13}{2} = \frac{-13}{2}$$

Any natural number, whole number, or integer can be expressed as a fraction with a denominator of 1. For example, $5 = \frac{5}{1}$, $0 = \frac{0}{1}$, and $-3 = \frac{-3}{1}$. Therefore, every natural number, whole number, and integer is also a rational number.

Many numerical quantities are written in decimal notation. For instance, a candy bar might cost \$0.89, a dragster might travel at 203.156 mph, or a business loss might be −\$4.7 million. These decimals are called **terminating decimals** because their representations terminate (stop). As shown below, terminating decimals can be expressed as fractions. Therefore, terminating decimals are rational numbers.

The Language of Algebra

To **terminate** means to bring to an end. In the movie *The Terminator,* actor Arnold Schwarzenegger plays a heartless machine sent to Earth to bring an end to his enemies.

$$0.89 = \frac{89}{100} \qquad 203.156 = 203\frac{156}{1{,}000} = \frac{203{,}156}{1{,}000} \qquad -4.7 = -4\frac{7}{10} = \frac{-47}{10}$$

Decimals such as 0.3333 . . . and 2.8167167167 . . . , which have a digit (or a block of digits) that repeats, are called **repeating decimals.** Since any repeating decimal can be expressed as a fraction, repeating decimals are rational numbers.

The set of rational numbers cannot be listed in the same way as the natural numbers, the whole numbers, and the integers. Instead, we use **set-builder** notation.

Rational Numbers

The set of rational numbers is

$$\left\{\frac{a}{b} \,\middle|\, a \text{ and } b \text{ are integers, with } b \neq 0.\right\}$$

Read as "the set of all numbers of the form $\frac{a}{b}$, such that a and b are integers, with $b \neq 0$."

A fraction and its **decimal equivalent** are different forms that represent the same value. To find the decimal equivalent for a fraction, we divide its numerator by its denominator. For example, to write $\frac{1}{4}$ and $\frac{5}{22}$ as decimals, we proceed as follows:

$$\begin{array}{r} 0.25 \\ 4\overline{)1.00} \\ -\ 8 \\ \hline 20 \\ -20 \\ \hline 0 \end{array}$$

Write a decimal point and additional zeros to the right of 1.

The remainder is 0.

$$\begin{array}{r} 0.22727\ldots \\ 22\overline{)5.00000} \\ -4\,4 \\ \hline 60 \\ -44 \\ \hline 160 \\ -154 \\ \hline 60 \\ -44 \\ \hline 160 \end{array}$$

Write a decimal point and additional zeros to the right of 5.

60 and 160 continually appear as remainders. Therefore, 2 and 7 will continually appear in the quotient.

The Language of Algebra

Since every natural number belongs to the set of whole numbers, we say the set of natural numbers is a **subset** of the set of whole numbers. Similarly, the set of whole numbers is a *subset* of the set of integers, and the set of integers is a *subset* of the set of rational numbers.

The decimal equivalent of $\frac{1}{4}$ is 0.25 and the decimal equivalent of $\frac{5}{22}$ is 0.2272727 . . . . We can use an **overbar** to write repeating decimals in more compact form: $0.2272727\ldots = 0.2\overline{27}$. Here are more fractions and their decimal equivalents.

Terminating decimals	*Repeating decimals*
$\frac{1}{2} = 0.5$	$\frac{1}{6} = 0.166666\ldots$ or $0.1\overline{6}$
$\frac{5}{8} = 0.625$	$\frac{1}{3} = 0.333333\ldots$ or $0.\overline{3}$
$\frac{3}{4} = 0.75$	$\frac{5}{11} = 0.454545\ldots$ or $0.\overline{45}$

3 Define the Set of Irrational Numbers.

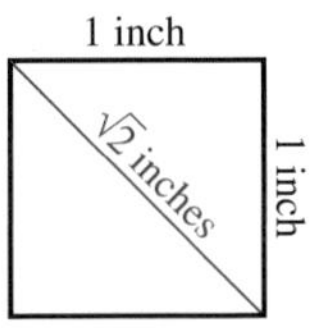

Not all numbers are rational numbers. One example is the square root of 2, written $\sqrt{2}$. It is the number that, when multiplied by itself, gives 2. That is, $\sqrt{2} \cdot \sqrt{2} = 2$. It can be shown that $\sqrt{2}$ *cannot* be written as a fraction with an integer numerator and an integer denominator. Therefore, it is not rational; it is an *irrational number*. It is interesting to note that a square with sides of length 1 inch has a diagonal that is $\sqrt{2}$ inches long.

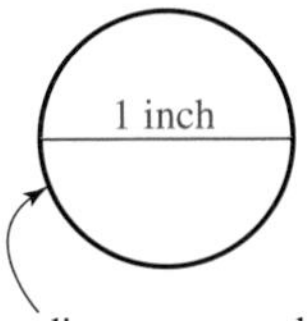

The distance around the circle is π inches.

The number represented by the Greek letter π (pi) is another example of an irrational number. A circle, with a 1-inch diameter, has a circumference of π inches.

Expressed in decimal form,

$$\sqrt{2} = 1.414213562\ldots \quad \text{and} \quad \pi = 3.141592654\ldots$$

These decimals neither terminate nor repeat.

Irrational Numbers

An **irrational number** is a nonterminating, nonrepeating decimal. An irrational number cannot be expressed as a fraction with an integer numerator and an integer denominator.

The Language of Algebra

The prefix **ir** means *not*. For example,
irresponsible ↔ not responsible
irregular ↔ not regular
irrational ↔ not rational

We have seen that $\sqrt{2}$ and π are irrational numbers. Other examples of irrational numbers are:

$$\sqrt{3} = 1.732050808\ldots \qquad -\sqrt{5} = -2.236067977\ldots$$

$$-\pi = -3.141592654\ldots \qquad 3\pi = 9.424777961\ldots$$

3π means $3 \cdot \pi$.

We can use a calculator to approximate the decimal value of an irrational number. To approximate $\sqrt{2}$ using a scientific calculator, we use the square root key $\sqrt{\ }$. To approximate π, we use the *pi* key π.

$$\sqrt{2} \approx 1.414213562 \quad \text{and} \quad \pi \approx 3.141592654$$

Read $\approx$ as "is approximately equal to."

Rounded to the nearest thousandth, $\sqrt{2} \approx 1.414$ and $\pi \approx 3.142$.

4 Classify Real Numbers.

The set of **real numbers** is formed by combining the set of rational numbers and the set of irrational numbers. Every real number has a decimal representation. If it is rational, its corresponding decimal terminates or repeats. If it is irrational, its decimal representation is nonterminating and nonrepeating.

The Real Numbers

A **real number** is any number that is a rational number or an irrational number.

The following diagram shows how various sets of numbers are related. Note that a number can belong to more than one set. For example, -6 is an integer, a rational number, and a real number.

The Language of Algebra

The symbol $\mathbb{R}$ is used to represent the set of real numbers.

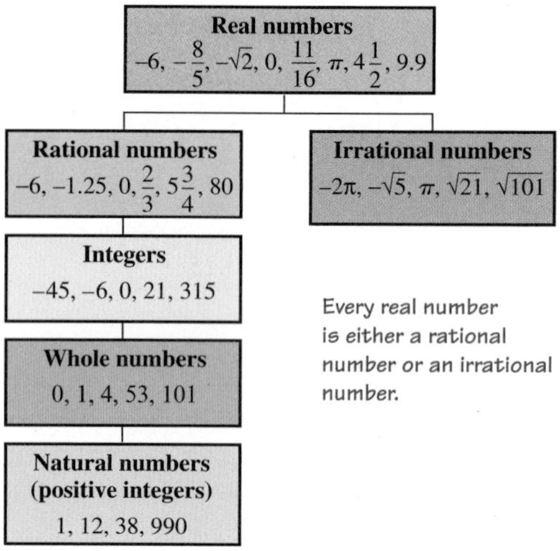

EXAMPLE 1 Which numbers in the following set are natural numbers, whole numbers, integers, rational numbers, irrational numbers, real numbers? $\left\{-3.4, \frac{2}{5}, 0, -6, 1\frac{3}{4}, \pi, 16\right\}$

Teaching Tip: Point out the possibility that a set might not include one (or more) of the types of numbers discussed in this section. For example, the set in Self Check 1 does not contain an irrational number.

Strategy We begin by scanning the given set, looking for any natural numbers. Then we scan it five more times, looking for whole numbers, for integers, for rational numbers, for irrational numbers, and finally, for real numbers.

Why We need to scan the given set of numbers six times, because numbers in the set can belong to more than one classification.

Solution

Natural numbers: 16 — *16 is a member of $\{1, 2, 3, 4, 5, \ldots\}$.*

Whole numbers: 0, 16 — *0 and 16 are members of $\{0, 1, 2, 3, 4, 5, \ldots\}$.*

Integers: $0, -6, 16$ — *$0, -6$, and 16 are members of $\{\ldots, -3, -2, -1, 0, 1, 2, 3, \ldots\}$.*

Rational numbers:

$-3.4, \frac{2}{5}, 0, -6, 1\frac{3}{4}, 16$ — *A rational number can be expressed as a ratio of two integers: $-3.4 = \frac{-34}{10}$, $0 = \frac{0}{1}$, $-6 = \frac{-6}{1}$, $1\frac{3}{4} = \frac{7}{4}$, and $16 = \frac{16}{1}$.*

Irrational numbers: π — *$\pi = 3.1415\ldots$ is a nonterminating, nonrepeating decimal.*

Real numbers:

$-3.4, \frac{2}{5}, 0, -6, 1\frac{3}{4}, \pi, 16$ — *Every natural number, whole number, integer, rational number, and irrational number is a real number.*

Teaching Example 1
Which numbers in the following set are natural numbers, whole numbers, integers, rational numbers, irrational numbers, real numbers? $\left\{-22, \pi, \frac{6}{11}, 0, 36, 5\frac{1}{8}, -1.9\right\}$
Answers: Natural: 36; whole: 0, 36; integers: -22, 0, 36; rational: $-22, \frac{6}{11}, 0, 36, 5\frac{1}{8}, -1.9$; irrational: π; real: all

Self Check 1 Use the instructions for Example 1 with: $\left\{0.1, -\frac{2}{7}, 45, -2, \frac{13}{4}, -6\frac{7}{8}\right\}$

Natural: 45; whole: 45; integers: 45, -2; rational: $0.1, -\frac{2}{7}, 45, -2, \frac{13}{4}, -6\frac{7}{8}$; irrational: none; real: all

Now Try Problem 27

5 Graph Sets of Real Numbers on the Number Line.

Every real number corresponds to a point on the number line, and every point on the number line corresponds to exactly one real number. As we move right on the number line, the values of the numbers increase. As we move left, the values decrease. On the following number line, we see that 5 is greater than -3, because 5 lies to the right of -3. Similarly, -3 is less than 5, because it lies to the left of 5.

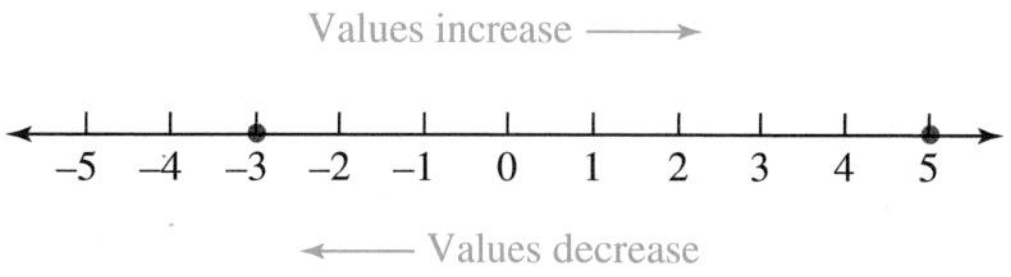

The Language of Algebra

The prefix **in** means *not.* For example:

inaccurate ↔ not accurate
inexpensive ↔ not expensive
inequality ↔ not equal

The **inequality symbol** $>$ means "is greater than." It is used to show that one number is greater than another. The inequality symbol $<$ means "is less than." It is used to show that one number is less than another. For example,

$5 > -3$ Read as "5 is greater than -3."

$-3 < 5$ Read as "-3 is less than 5."

To distinguish between these inequality symbols, remember that each one points to the smaller of the two numbers involved.

$5 > -3$ $\qquad$ $-3 < 5$

Points to the smaller number.

EXAMPLE 2 Use one of the symbols $>$ or $<$ to make each statement true:

a. $-4 \quad 4$ **b.** $-2 \quad -3$ **c.** $4.47 \quad 12.5$ **d.** $\frac{3}{4} \quad \frac{5}{8}$

Strategy To pick the correct inequality symbol to place between a given pair of numbers, we need to determine the position of each number on a number line.

Why For any two numbers on a number line, the number to the *left* is the smaller number and the number to the *right* is the larger number.

Solution

a. Since -4 is to the left of 4 on the number line, we have $-4 < 4$.

b. Since -2 is to the right of -3 on the number line, we have $-2 > -3$.

c. Since 4.47 is to the left of 12.5 on the number line, we have $4.47 < 12.5$.

d. To compare fractions, express them in terms of the same denominator, preferably the LCD. If we write $\frac{3}{4}$ as an equivalent fraction with denominator 8, we see that $\frac{3}{4} = \frac{3}{4} \cdot \frac{2}{2} = \frac{6}{8}$. Therefore, $\frac{3}{4} > \frac{5}{8}$.

To compare the fractions, we also could convert each to its decimal equivalent. Since $\frac{3}{4} = 0.75$ and $\frac{5}{8} = 0.625$, we know that $\frac{3}{4} > \frac{5}{8}$.

Teaching Example 2 Use one of the symbols $>$ or $<$ to make each statement true:
a. $-6 < 1$ b. $-9 > -12$
c. $3.8987 < 4.1$ d. $\frac{5}{6} > \frac{13}{18}$

Self Check 2 Use one of the symbols $<$ or $>$ to make each statement true:

a. $1 > -1$ **b.** $-5 < -4$ **c.** $6.7 > 4.999$ **d.** $\frac{3}{5} < \frac{2}{3}$

Now Try Problems 37 and 42

To **graph a number** means to mark its position on the number line.

EXAMPLE 3 Graph each number in the set: $\left\{-2.43, \sqrt{2}, 1, -0.\overline{3}, 2\frac{5}{6}, -\frac{3}{2}\right\}$

Strategy We locate the position of each number on the number line, draw a bold dot, and label it.

Why To *graph a number* means to make a drawing that represents the number.

Solution It is helpful to approximate the value of a number or to write the number in an equivalent form to determine its location on a number line.

- To locate -2.43, we round it to the nearest tenth: $-2.43 \approx -2.4$.
- To locate $\sqrt{2}$, we use a calculator: $\sqrt{2} \approx 1.4$.
- To locate $-0.\overline{3}$, we recall that $0.\overline{3} = 0.333\ldots = \frac{1}{3}$. Therefore, $-0.\overline{3} = -\frac{1}{3}$.
- In mixed-number form, $-\frac{3}{2} = -1\frac{1}{2}$. This is midway between -1 and -2.

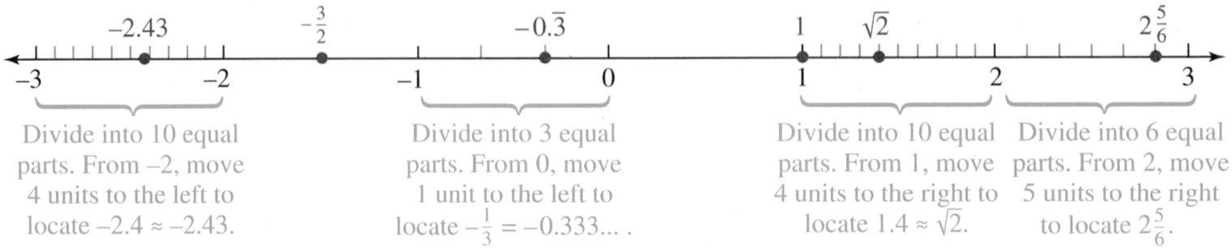
–2.43
$-\frac{3}{2}$
$-0.\overline{3}$
1
$\sqrt{2}$
$2\frac{5}{6}$
–3 –2 –1 0 1 2 3
Divide into 10 equal parts. From –2, move 4 units to the left to locate –2.4 ≈ –2.43.
Divide into 3 equal parts. From 0, move 1 unit to the left to locate $-\frac{1}{3} = -0.333\ldots$.
Divide into 10 equal parts. From 1, move 4 units to the right to locate 1.4 ≈ $\sqrt{2}$.
Divide into 6 equal parts. From 2, move 5 units to the right to locate $2\frac{5}{6}$.

Self Check 3 Graph each number in the set: $\left\{1.7, \pi, -1\frac{3}{4}, 0.\overline{6}, \frac{5}{2}, -3\right\}$

See AIE Appendix 3.

Now Try Problem 57

Teaching Example 3 Graph each number in the set: $\left\{3.2, \frac{\pi}{2}, -1\frac{1}{2}, -0.333\ldots, 2\frac{3}{4}\right\}$

Answer:

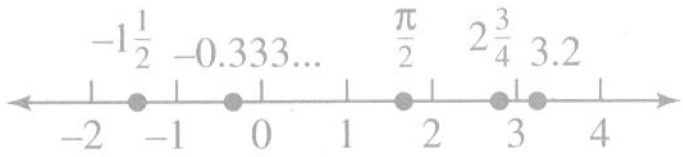
$-1\frac{1}{2}$ –0.333... $\frac{\pi}{2}$ $2\frac{3}{4}$ 3.2
–2 –1 0 1 2 3 4

6 Find the Absolute Value of a Real Number.

A number line can be used to measure the distance from one number to another. For example, on the number line below, we see that the distance from 0 to -4 is 4 units and the distance from 0 to 3 is 3 units.

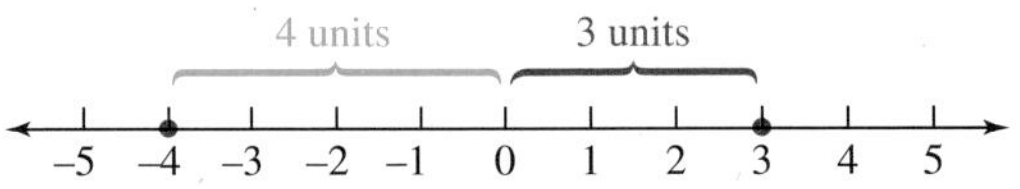
4 units
3 units
–5 –4 –3 –2 –1 0 1 2 3 4 5

To express the distance that a number is from 0 on a number line, we can use *absolute value*.

Absolute Value

The **absolute value** of a number is its distance from 0 on the number line.

To indicate the absolute value of a number, we write the number between two vertical bars. From the figure above, we see that $|-4| = 4$. This is read as "the absolute value of negative 4 is 4" and it tells us that the distance from 0 to -4 is 4 units. It also follows from the figure that $|3| = 3$.

EXAMPLE 4 Find each absolute value: **a.** $|18|$ **b.** $\left|-\frac{7}{8}\right|$ **c.** $|98.6|$ **d.** $|0|$

Strategy We need to determine the distance that the number within the vertical absolute value bars is from 0.

Why The absolute value of a number is the distance between 0 and the number on a number line.

Solution **a.** Since 18 is a distance of 18 from 0 on the number line, $|18| = 18$.

b. Since $-\frac{7}{8}$ is a distance of $\frac{7}{8}$ from 0 on the number line, $\left|-\frac{7}{8}\right| = \frac{7}{8}$.

Success Tip

Since absolute value expresses distance, the absolute value of a number is always positive or zero, but never negative.

Teaching Example 4
Find each absolute value:

a. $|49|$ b. $\left|-\frac{24}{25}\right|$ c. $|11.09|$ d. $|0|$

Answers:
a. 49 b. $\frac{24}{25}$ c. 11.09 d. 0

c. Since 98.6 is a distance of 98.6 from 0 on the number line, $|98.6| = 98.6$.

d. Since 0 is a distance of 0 from 0 on the number line, $|0| = 0$.

Self Check 4 Find each absolute value: **a.** $|100|$ 100 **b.** $|-4.7|$ 4.7 **c.** $|\sqrt{2}|$ $\sqrt{2}$

Now Try Problems 61 and 67

SECTION 1.3 STUDY SET

VOCABULARY

Fill in the blanks.

1. The set of whole numbers is {0, 1, 2, 3, 4, 5, . . .}.
2. The set of natural numbers is {1, 2, 3, 4, 5, . . .}.
3. The figure 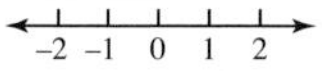 is called a number line.
4. The set of integers is {. . . , −2, −1, 0, 1, 2, . . .}.
5. Positive and negative numbers are called signed numbers.
6. 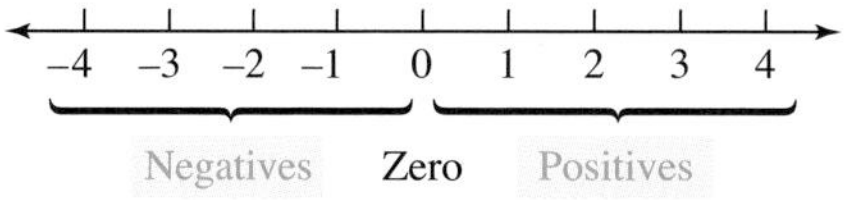

7. The symbols < and > are inequality symbols.
8. A rational number is any number that can be expressed as a fraction with an integer numerator and a nonzero integer denominator.
9. 0.25 is called a terminating decimal and 0.333 . . . is called a repeating decimal.
10. An irrational number cannot be expressed as a quotient of two integers.
11. An irrational number is a nonterminating, nonrepeating decimal.
12. The absolute value of a number is the distance on the number line between the number and 0.

CONCEPTS

13. Represent each situation using a signed number.
    a. A loss of \$15 million −\$15 million
    b. A building foundation $\frac{5}{16}$ inch above grade $\frac{5}{16}$ in. or $+\frac{5}{16}$ in.
14. Show that each of the following numbers is a rational number by expressing it as a fraction with an integer numerator and a nonzero integer denominator: 6, −9, $-\frac{7}{8}$, $3\frac{1}{2}$, −0.3, 2.83.
    $\frac{6}{1}, \frac{-9}{1}, \frac{-7}{8}, \frac{7}{2}, \frac{-3}{10}, \frac{283}{100}$
15. Give the opposite of each number.
    a. 20 −20
    b. $-\frac{2}{3}$ $\frac{2}{3}$
16. What two numbers are a distance of 8 away from 5 on the number line? 13 and −3
17. What two numbers are a distance of 5 away from −9 on the number line? −14 and −4
18. Refer to the number line below. Use an inequality symbol, < or >, to make each statement true.
    a. $a < b$
    b. $b > a$
    c. $b > 0$ and $a < 0$
    d. $|a| > |b|$

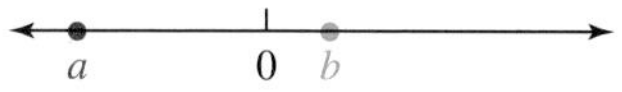

NOTATION

Fill in the blanks.

19. $\sqrt{2}$ is read "the square root of 2."
20. $|-15|$ is read "the absolute value of −15."
21. The symbol ≈ means is approximately equal to.
22. The symbols { } are called braces.
23. The symbol π is a letter from the Greek alphabet.
24. To find the decimal equivalent for $\frac{2}{3}$, we perform the following long division: $3\overline{)2}$
25. $-\frac{4}{5} = \frac{-4}{5} = \frac{4}{-5}$
26. Write each repeating decimal using an overbar.
    a. 0.666 . . . $0.\overline{6}$
    b. 0.2444 . . . $0.2\overline{4}$
    c. 0.717171 . . . $0.\overline{71}$
    d. 0.456456456 . . . $0.\overline{456}$

GUIDED PRACTICE

***Place check marks in the table to show the set or sets to which each number belongs. For example, the check shows that $\sqrt{2}$ is irrational.* See Example 1.**

27.

	5	0	−3	$\frac{7}{8}$	0.17	$-9\frac{1}{4}$	$\sqrt{2}$	π
Real	✓	✓	✓	✓	✓	✓	✓	✓
Irrational							✓	✓
Rational	✓	✓	✓	✓	✓	✓		
Integer	✓	✓	✓					
Whole	✓	✓						
Natural	✓							

28. Which numbers in the following set are natural numbers, whole numbers, integers, rational numbers, irrational numbers, real numbers? $\{67, \frac{4}{13}, -5.9, 11\frac{2}{3}, \sqrt{2}, 0, -3, \pi\}$
Natural: 67; whole: 67, 0; integers: 67, 0, −3; rational: 67, $\frac{4}{13}$, −5.9, $11\frac{2}{3}$, 0, −3; irrational: $\sqrt{2}$, π; real: all

***Determine whether each statement is true or false.* See Example 1.**

▶ **29.** Every whole number is an integer. True
▶ **30.** Every integer is a natural number. False
▶ **31.** Every integer is a whole number. False
▶ **32.** Every real number is either a rational number or an irrational number. True
▶ **33.** Irrational numbers are real numbers. True
▶ **34.** Every whole number is a rational number. True
▶ **35.** Every rational number can be written as a fraction (ratio) of two integers. True
▶ **36.** Every rational number is a whole number. False

***Use one of the symbols < or > to make each statement true.* See Example 2.**

▶ **37.** $0 > -4$
▶ **38.** $0 < 32$
▶ **39.** $917 < 971$
40. $898 > 889$
41. $-2 > -3$
▶ **42.** $-5 < -4$
▶ **43.** $-\frac{5}{8} < -\frac{3}{8}$
▶ **44.** $-19\frac{2}{3} < -19\frac{1}{3}$
▶ **45.** $\frac{2}{3} > \frac{3}{5}$
46. $\frac{3}{4} < \frac{5}{6}$
47. $-6.19 < -5.8$
▶ **48.** $-2.27 > -5.25$

***Write each fraction as a decimal. If the result is a repeating decimal, use an overbar.* See Objective 2.**

▶ **49.** $\frac{5}{8}$ 0.625
50. $\frac{3}{32}$ 0.09375
51. $\frac{1}{30}$ $0.0\overline{3}$
52. $\frac{7}{9}$ $0.\overline{7}$
53. $\frac{1}{60}$ $0.01\overline{6}$
▶ **54.** $\frac{5}{11}$ $0.\overline{45}$
▶ **55.** $\frac{21}{50}$ 0.42
▶ **56.** $\frac{2}{125}$ 0.016

***Graph each set of numbers on a number line.* See Example 3.** See AIE Appendix 3.

▶ **57.** $\{-\pi, 4.25, -1\frac{1}{2}, -0.333\ldots, \sqrt{2}, -\frac{35}{8}, 3\}$
58. $\{-2\frac{1}{8}, \pi, 2.75, -\sqrt{2}, \frac{17}{4}, 0.666\ldots, -3\}$
59. The integers between −5 and 2
▶ **60.** The whole numbers less than 4

***Find each absolute value.* See Example 4.**

61. $|83|$ 83
▶ **62.** $|29|$ 29
63. $|\frac{4}{3}|$ $\frac{4}{3}$
▶ **64.** $|\frac{9}{16}|$ $\frac{9}{16}$
▶ **65.** $|-11|$ 11
66. $|-14|$ 14
67. $|-6.1|$ 6.1
▶ **68.** $|-25.3|$ 25.3

***Insert one of the symbols >, <, or = in the blank to make each statement true.* See Examples 2 and 4.**

69. $|3.4| > -3$
▶ **70.** $0.08 > 0.079$
71. $|-1.1| < 1.2$
▶ **72.** $-5.5 = -5\frac{1}{2}$
73. $|-\frac{15}{2}| = 7.5$
▶ **74.** $|-2\frac{2}{3}| > \frac{7}{3}$
75. $\frac{99}{100} = 0.99$
▶ **76.** $|2| = |-2|$
77. $0.3 < 0.333\ldots$
▶ **78.** $-0.666\ldots < -0.6$
79. $1 > |-\frac{15}{16}|$
▶ **80.** $\sqrt{2} < \pi$

APPLICATIONS

▶ **81. Drafting.** Which dimensions of the aluminum bracket shown below are natural numbers, whole numbers, integers, rational numbers, irrational numbers, and real numbers?
Natural, whole, integers: 9; rational: 9, $\frac{15}{16}$, $3\frac{1}{8}$, 1.765; irrational: 2π, 3π, $\sqrt{89}$; real: all

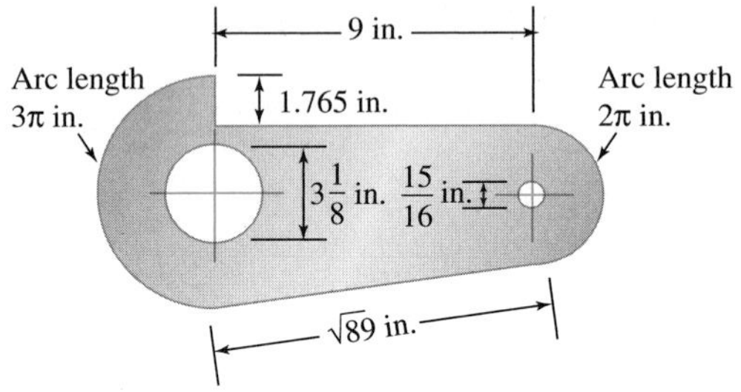

▶ **82. History.** Refer to the time line shown below.
a. What basic unit was used to scale the time line? 500 years
b. What symbolism is used to represent zero? B.C./A.D.
c. Which numbers could be thought of as positive and which as negative? Pos: A.D.; neg: B.C.
d. Express the dates for the Maya civilization using positive and negative numbers. −500 to 1697

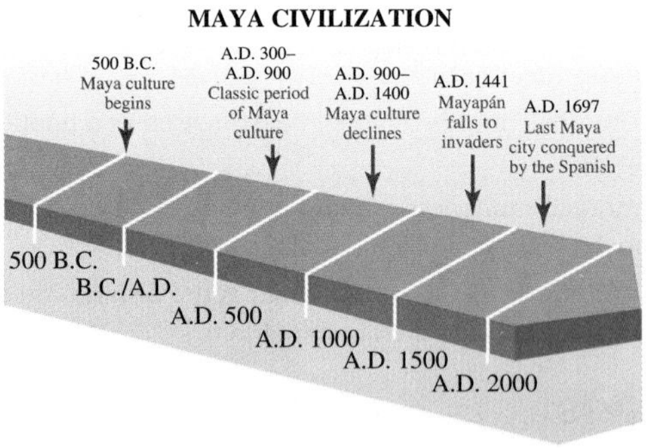

Based on data from *People in Time and Place, Western Hemisphere* (Silver Burdett & Ginn, 1991), p. 129.

83. iPhones. You can get a more accurate reading of an iPhone's signal strength by dialing *3001#12345#*. Field test mode is then activated and the standard signal strength bars (in the upper left corner of the display) are replaced by a negative number. The closer the negative number is to zero, the stronger the signal. Which iPhone shown below is receiving the strongest signal? iii

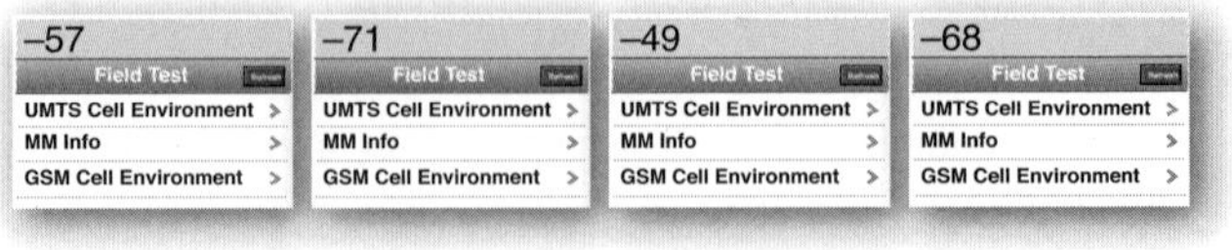

(i) (ii) (iii) (iv)

84. Drafting. On an architect's scale, the edge marked 16 divides each inch into 16 equal parts. Find the decimal form for each fractional part of one inch that is highlighted on the scale. 0.0625, 0.375, 0.5625, 0.9375

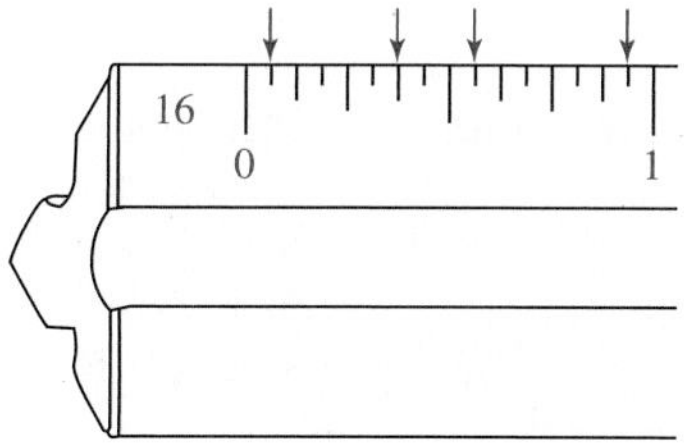

85. Trade. Each year from 1994 through 2009, the United States imported more goods and services from Japan than it exported to Japan. This caused trade *deficits*, which are represented by negative numbers on the following graph.

a. In which year was the deficit the worst? Express that deficit using a signed number. 2006: −\$90 billion

b. In which year was the deficit the smallest? Express that deficit using a signed number. 2009: −\$45 billion

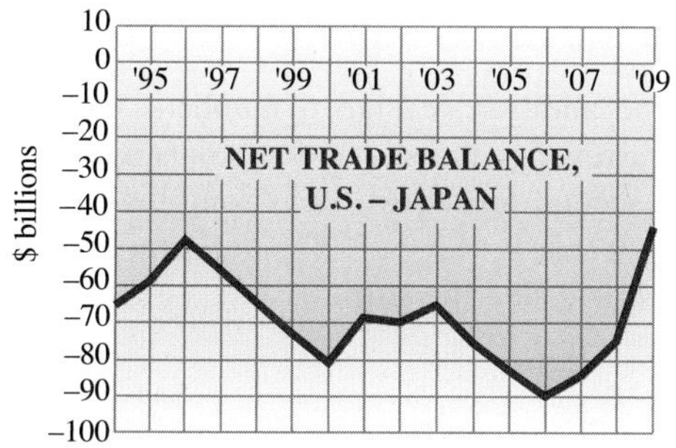

Source: U.S. Bureau of the Census

86. U.S. Budget. A budget *deficit* is a negative number that indicates the government spent more money than it took in that year. A budget *surplus* is a positive number that indicates the government took in more money than it spent that year.

a. Refer to the graph in the next column that shows the U.S. Federal Budget Deficit/Surplus for the years 1980 through 2009. For how many of those years was there a budget surplus? 4 years

b. Consider the years in which there was a budget deficit. For how many of those years was it smaller than \$300 billion? 21 years

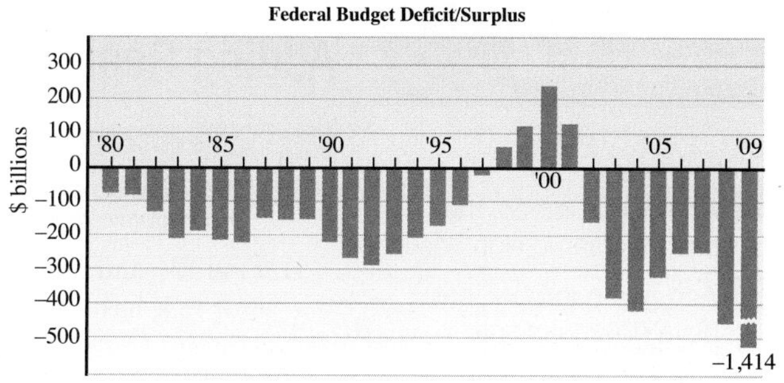

Source: U.S. Bureau of the Census

WRITING

87. Explain the difference between a rational and an irrational number.

88. Can two different numbers have the same absolute value? Explain.

89. Explain how to find the decimal equivalent of a fraction.

90. What is a real number?

91. *Pi Day* (or *Pi Approximation Day*) is an unofficial holiday held to celebrate π. Why do you think Pi Day is observed each year on March 14?

92. Explain why $0.1\overline{333}$ is not the simplest way to represent 0.1333. . . .

REVIEW

93. Simplify: $\frac{24}{54}$ $\frac{4}{9}$

94. Multiply: $\frac{3}{4}\left(\frac{8}{5}\right)$ $\frac{6}{5}$

95. Divide: $5\frac{2}{3} \div 2\frac{5}{9}$ $2\frac{5}{23}$

96. Add: $\frac{3}{10} + \frac{2}{15}$ $\frac{13}{30}$

CHALLENGE PROBLEMS

97. How many integers have an absolute value that is less than 1,000? 1,999

98. Is 0.10100100010000 . . . a repeating decimal? Explain. No. There is a pattern, but the given number does not consist of a block of digits that repeats.

Find a rational number between each pair of numbers.

99. $\frac{1}{8}$ and $\frac{1}{9}$ $\frac{17}{144}$

100. $1.7\overline{1}$ and $1.7\overline{2}$ 1.712

SECTION 1.4

OBJECTIVES

1. Add two numbers that have the same sign.
2. Add two numbers that have different signs.
3. Use properties of addition.
4. Identify opposites (additive inverses).

Adding Real Numbers; Properties of Addition

ARE YOU READY? *Are You Ready? exercises available online at www.webassign.net/brookscole*

The following problems review some basic concepts that are important when adding positive and negative real numbers.

1. Find $|3|$ and $|-5|$. Which number, 3 or -5, has the larger absolute value? 3, 5; -5 has the larger absolute value.

2. Add: $4.37 + 2.8$ 7.17

3. Subtract: $710 - 89$ 621

4. Subtract: $\frac{4}{5} - \frac{2}{3}$ $\frac{2}{15}$

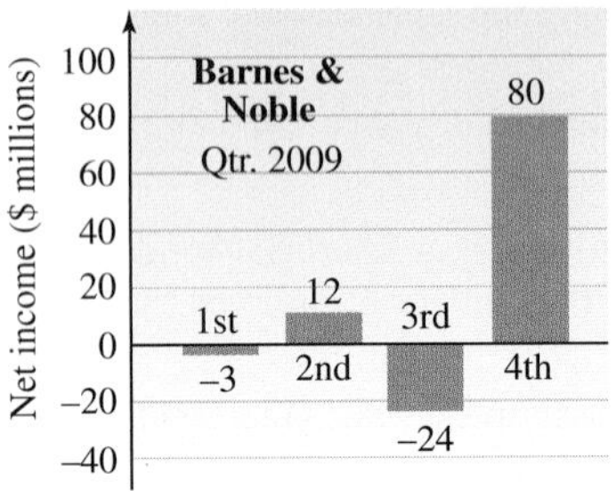

Source: msnbc.msn.com

In the graph to the left, signed numbers are used to show the financial performance of Barnes and Noble Corporation for the year 2009. Positive numbers indicate *profits* and negative numbers indicate *losses*. To find the company's 2009 net income (in millions of dollars), we need to calculate the following sum:

$$\text{Net income} = -3 + 12 + (-24) + 80$$

In this section, we discuss how to perform this addition and others involving signed numbers.

1 Add Two Numbers That Have the Same Sign.

A number line can be used to explain the addition of signed numbers. For example, to compute $5 + 2$, we begin at 0 and draw an arrow five units long that points right. It represents 5. From the tip of that arrow, we draw a second arrow two units long that points right. It represents 2. Since we end up at 7, it follows that $5 + 2 = 7$. The numbers that we added, 5 and 2, are called **addends,** and the result, 7, is called the **sum.**

Teaching Tip: Stress that this number line addition process always begins at 0 (the origin).

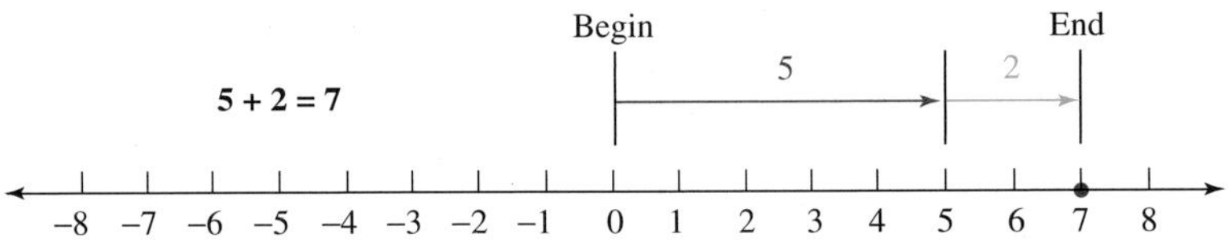

To compute $-5 + (-2)$, we begin at 0 and draw an arrow five units long that points left. It represents -5. From the tip of that arrow, we draw a second arrow two units long that points left. It represents -2. Since we end up at -7, it follows that $-5 + (-2) = -7$.

Notation

To avoid confusion, we write negative numbers within parentheses to separate the negative sign $-$ from the addition symbol $+$.

$$-5 + (-2)$$

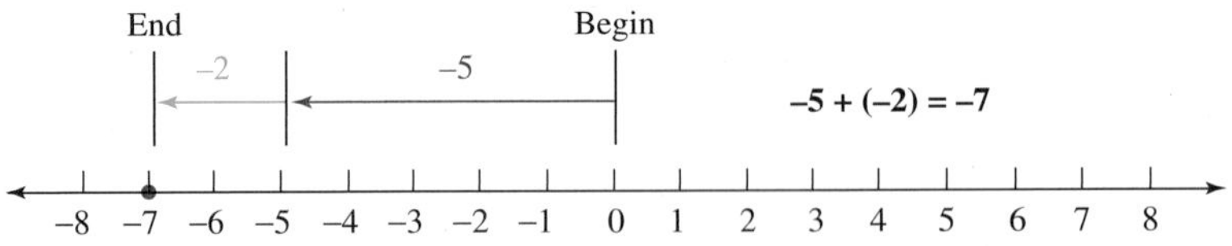

To check this result, think of the problem in terms of money. If you lost \$5 (-5) and then lost another \$2 (-2), you would have lost a total of \$7 (-7).

When we use a number line to add numbers with the same sign, the arrows point in the same direction and they build upon each other. Furthermore, the answer has the same sign as the numbers that we added. These observations suggest the following rules.

Adding Two Numbers That Have the Same (Like) Signs

1. To add two positive numbers, add them as usual. The final answer is positive.
2. To add two negative numbers, add their absolute values and make the final answer negative.

EXAMPLE 1 Add: **a.** $-20 + (-15)$ **b.** $-7.89 + (-0.6)$ **c.** $-\frac{1}{3} + \left(-\frac{1}{2}\right)$

Strategy We will use the rule for adding two numbers that have the same sign.

Why In each case, we are asked to add two negative numbers.

Solution **a.** $-20 + (-15) = -35$ Add their absolute values, 20 and 15, to get 35. Then make the final answer negative.

The Language of Algebra

Two negative numbers, as well as two positive numbers, are said to have **like** signs.

b. To find $-7.89 + (-0.6)$, add their absolute values, 7.89 and 0.6.

$$\begin{array}{r} 7.89 \\ +0.6 \\ \hline 8.49 \end{array}$$ Remember to align the decimal points when adding decimals.

Then make the final answer negative: $-7.89 + (-0.6) = -8.49$.

Success Tip

The sum of two positive numbers is *always* positive. The sum of two negative numbers is *always* negative.

c. To find $-\frac{1}{3} + \left(-\frac{1}{2}\right)$, add their absolute values, $\frac{1}{3}$ and $\frac{1}{2}$.

$$\frac{1}{3} + \frac{1}{2} = \frac{2}{6} + \frac{3}{6}$$ The LCD is 6. Build each fraction: $\frac{1}{3} \cdot \frac{2}{2} = \frac{2}{6}$ and $\frac{1}{2} \cdot \frac{3}{3} = \frac{3}{6}$.

$$= \frac{5}{6}$$ Add the numerators and write the sum over the LCD.

Then make the final answer negative: $-\frac{1}{3} + \left(-\frac{1}{2}\right) = -\frac{5}{6}$.

Teaching Example 1
Add: **a.** $-62 + (-2)$
b. $-26.4 + (-0.48)$ **c.** $-\frac{1}{4} + \left(-\frac{2}{5}\right)$
Answers:
a. -64 **b.** -26.88 **c.** $-\frac{13}{20}$

Self Check 1 Add: **a.** $-51 + (-9)$ -60 **b.** $-12.3 + (-0.88)$ -13.18
c. $-\frac{1}{4} + \left(-\frac{2}{3}\right)$ $-\frac{11}{12}$

Now Try Problems 15, 21, and 25

2 Add Two Numbers That Have Different Signs.

To compute $5 + (-2)$, we begin at 0 and draw an arrow five units long that points right. From the tip of that arrow, we draw a second arrow two units long that points left. Since we end up at 3, it follows that $5 + (-2) = 3$. In terms of money, if you won \$5 and then lost \$2, you would have \$3 left.

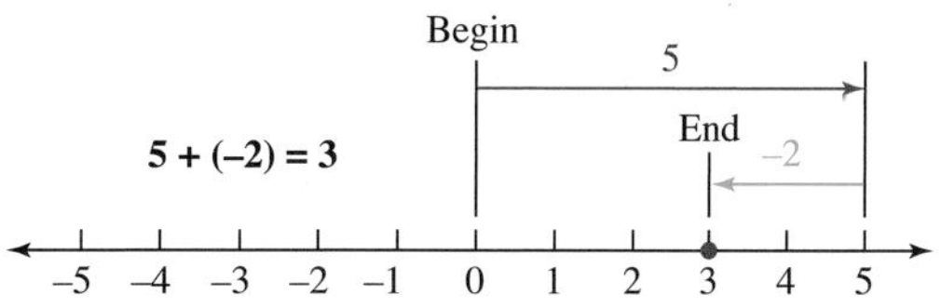

The Language of Algebra

A positive number and a negative number are said to have **unlike** signs.

To compute $-5 + 2$, we begin at 0 and draw an arrow five units long that points left. From the tip of that arrow, we draw a second arrow two units long that points right. Since we end up at -3, it follows that $-5 + 2 = -3$. In terms of money, if you lost \$5 and then won \$2, you have lost \$3.

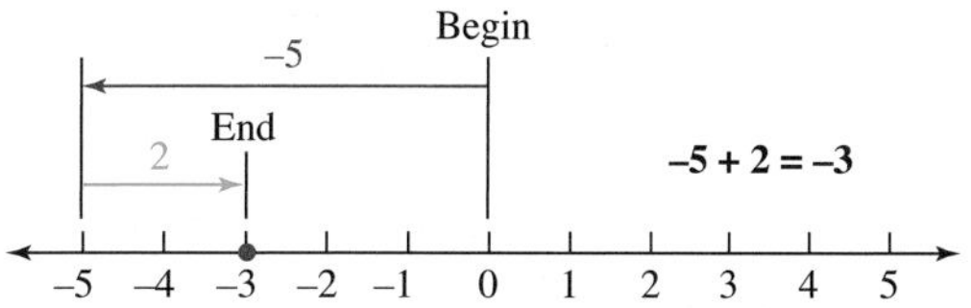

When we use a number line to add numbers with different signs, the arrows point in opposite directions and the longer arrow determines the sign of the answer. If the longer arrow represents a positive number, the sum is positive. If it represents a negative number, the sum is negative. These observations suggest the following rules.

Adding Two Numbers That Have Different (Unlike) Signs

To add a positive number and a negative number, subtract the smaller absolute value from the larger.

1. If the positive number has the larger absolute value, the final answer is positive.
2. If the negative number has the larger absolute value, make the final answer negative.

EXAMPLE 2 Add: **a.** $-20 + 32$ **b.** $5.7 + (-7.4)$ **c.** $-\frac{19}{25} + \frac{2}{5}$

Strategy We will use the rule for adding two numbers that have different (unlike) signs.

Why In each case, we are asked to add a positive number and a negative number.

Solution

a. $-20 + 32 = 12$ Subtract the smaller absolute value from the larger: $32 - 20 = 12$. The positive number, 32, has the larger absolute value, so the final answer is positive.

b. To find $5.7 + (-7.4)$, subtract the smaller absolute value, 5.7, from the larger, 7.4.

$$\begin{array}{r} 7.4 \\ -5.7 \\ \hline 1.7 \end{array}$$

Remember to align the decimal points when subtracting decimals.

Since the negative decimal, -7.4, has the larger absolute value, make the final answer negative: $5.7 + (-7.4) = -1.7$.

c. Since $\frac{2}{5} = \frac{10}{25}$, the fraction $-\frac{19}{25}$ has the larger absolute value. To find $-\frac{19}{25} + \frac{2}{5}$, we subtract the smaller absolute value from the larger:

$$\frac{19}{25} - \frac{2}{5} = \frac{19}{25} - \frac{10}{25}$$ The LCD is 25. Build $\frac{2}{5}$: $\frac{2}{5} \cdot \frac{5}{5} = \frac{10}{25}$.

$$= \frac{9}{25}$$ Subtract the numerators and write the difference over the LCD.

Since the negative fraction $-\frac{19}{25}$ has the larger absolute value, make the final answer negative: $-\frac{19}{25} + \frac{10}{25} = -\frac{9}{25}$.

Success Tip

The sum of two numbers with different signs may be positive or negative. The sign of the sum is the sign of the number with the greater absolute value.

Calculators

Entering negative numbers
We don't do anything special to enter positive numbers on a calculator. To enter a negative number, say -7.4, some calculators require the $-$ sign to be entered before entering 7.4 while others require the $-$ sign to be entered after entering 7.4. Consult your owner's manual to determine the proper keystrokes.

Teaching Example 2
Add: a. $-23 + 58$ b. $6.3 + (-8.1)$ c. $-\frac{11}{20} + \frac{9}{10}$
Answers:
a. 35 b. -1.8 c. $\frac{7}{20}$

Self Check 2 Add: **a.** $63 + (-87)$ -24 **b.** $-6.27 + 8$ 1.73 **c.** $-\frac{1}{10} + \frac{1}{2}$ $\frac{2}{5}$

Now Try Problems 29, 33, and 35

EXAMPLE 3 **Accounting.** Find the net income of Barnes and Noble Corporation for the year 2009 using the data in the graph on page 32.

Strategy To find the net income, we will add the quarterly profits and losses (in millions of dollars), performing the additions as they occur from left to right.

Why The phrase *net income* means that we should combine (add) the quarterly profits and losses to determine whether there was an overall profit or loss that year.

Solution

$$-3 + 12 + (-24) + 80 = 9 + (-24) + 80 \quad \text{Add: } -3 + 12 = 9.$$
$$= -15 + 80 \quad \text{Add: } 9 + (-24) = -15.$$
$$= 65$$

In 2009, Barnes and Noble's net income was $65 million.

The Language of Algebra

Net refers to what remains after all the deductions (losses) have been accounted for. *Net income* is a term used in business that often is referred to as the *bottom line.* Net income indicates what a company has earned (or lost) in a given period of time (usually one year).

Self Check 3 Add: $650 + (-13) + 87 + (-155)$ 569

Now Try Problem 43

Teaching Example 3
Add: $-15 + 36 + (-48) + 90$
Answer: 63

3 Use Properties of Addition.

The addition of two numbers can be done in any order and the result is the same. For example, $8 + (-1) = 7$ and $-1 + 8 = 7$. This example illustrates that addition is **commutative.**

The Commutative Property of Addition

Changing the order when adding does not affect the answer.
For any real numbers a and b,

$$a + b = b + a$$

The Language of Algebra

Commutative is a form of the word *commute,* meaning to go back and forth. *Commuter* trains take people to and from work.

In the following example, we add $-3 + 7 + 5$ in two ways. We will use grouping symbols ( ), called **parentheses,** to show this. Standard practice requires that the operation within the parentheses be performed first.

We read $(-3 + 7) + 5$ as "The quantity of –3 plus 7" pause slightly, and then say "plus 5." We read $-3 + (7 + 5)$ as "–3" pause slightly, and then say "plus the quantity of 7 plus 5." The word **quantity** alerts the reader to the parentheses that are used as grouping symbols.

Method 1: Group **−3** *and* **7**	*Method 2: Group* **7** *and* **5**
$(-3 + 7) + 5 = 4 + 5$	$-3 + (7 + 5) = -3 + 12$
$= 9$	$= 9$

It doesn't matter how we group the numbers in this addition; the result is 9. This example illustrates that addition is **associative.**

The Associative Property of Addition

Changing the grouping when adding does not affect the answer.
For any real numbers a, b, and c,

$$(a + b) + c = a + (b + c)$$

Teaching Tip: Point out that in the statement of the associative property of addition above, the *order* of the addends, a, b, and c, does not change.

Sometimes, an application of the associative property can simplify a computation.

EXAMPLE 4 Find the sum: $98 + (2 + 17)$

Strategy We will use the associative property to group 2 with 98. Then, we evaluate the expression by performing the addition within the parentheses first.

Why It is helpful to regroup because 98 and 2 are a pair of numbers that are easily added.

Solution

$98 + (2 + 17) = (\mathbf{98 + 2}) + 17$ *Use the associative property of addition to regroup. Note that the order of the addends, 98, 2, and 17, is not changed.*

$= \mathbf{100} + 17$ *Do the addition within the parentheses first.*

$= 117$

The Language of Algebra

Associative is a form of the word *associate*, meaning to join a group. The WNBA (Women's National Basketball Association) is a group of 14 professional basketball teams.

Self Check 4 Find the sum: $(39 + 25) + 75$ 139

Now Try Problem 49

Teaching Example 4
Find the sum: $27 + (3 + 89)$
Answer: 119

EXAMPLE 5 **Game Shows.** A contestant on *Jeopardy!* correctly answered the first question to win \$100, missed the second to lose \$200, correctly answered the third to win \$300, and missed the fourth to lose \$400. What is her score after answering four questions?

Strategy We can represent money won by a positive number and money lost by a negative number. Her score is the sum of 100, −200, 300, and −400. Instead of doing the additions from left to right, we will use another approach. Applying the commutative and associative properties, we will add the positives, add the negatives, and then add those results.

Why It is easier to add numbers that have the same sign than numbers that have different signs. This method minimizes the possibility of an error, because we have to add numbers that have different signs only once.

Solution

$100 + (-200) + 300 + (-400)$

$= (\mathbf{100 + 300}) + [(\mathbf{-200}) + (\mathbf{-400})]$ *Reorder the numbers. Group the positives together. Group the negatives together using brackets [ ].*

$= \mathbf{400} + (\mathbf{-600})$ *Add the positives. Add the negatives.*

$= -200$ *Add the results.*

After four questions, her score was −\$200, which represents a loss of \$200.

Teaching Example 5
Add: $-23 + 2 + (-9) + (-8) + 16$
Answer: −22

Self Check 5 Add: $-6 + 1 + (-4) + (-5) + 9$ −5

Now Try Problem 45

The Language of Algebra

Identity is a form of the word *identical,* meaning the same. You probably have seen *identical* twins.

Whenever we add 0 to a number, the result is the number. Therefore, $8 + 0 = 8$, $2.3 + 0 = 2.3$, and $0 + (-16) = -16$. These examples illustrate the **addition property of 0.** Since any number added to 0 remains the same, 0 is called the **identity element** for addition.

Addition Property of 0 (Identity Property of Addition)

When 0 is added to any real number, the result is the same real number.

For any real number a,

$$a + 0 = a \quad \text{and} \quad 0 + a = a$$

4 Identify Opposites (Additive Inverses).

Recall that two numbers that are the same distance from 0 on a number line, but on opposite sides of it, are called **opposites.** To develop a property for adding opposites, we will find $-4 + 4$ using a number line. We begin at 0 and draw an arrow four units long that points left, to represent -4. From the tip of that arrow, we draw a second arrow, four units long that points right, to represent 4. We end up at 0; therefore, $-4 + 4 = 0$.

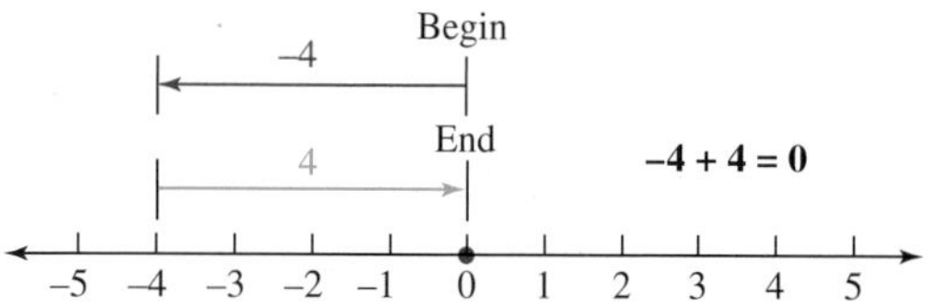

This example illustrates that when we add opposites, the result is 0. Therefore, $1.6 + (-1.6) = 0$ and $-\frac{3}{4} + \frac{3}{4} = 0$. Also, whenever the sum of two numbers is 0, those numbers are opposites. For these reasons, opposites also are called **additive inverses.**

Addition Property of Opposites (Inverse Property of Addition)

The sum of a number and its opposite (additive inverse) is 0.

For any real number a and its opposite or additive inverse $-a$,

$$a + (-a) = 0$$ *Read $-a$ as "the opposite of a."*

EXAMPLE 6 Add: $12 + (-5) + 6 + 5 + (-12)$

Strategy Instead of working from left to right, we will use the commutative and associative properties of addition to add pairs of opposites.

Why Since the sum of a number and its opposite is 0, it is helpful to identify such pairs in an addition.

Solution

opposites (12 and −12)

$$\mathbf{12 + (-5) + 6 + 5 + (-12)} = \mathbf{0 + 0} + 6$$
$$= 6$$

opposites (−5 and 5)

Teaching Example 6
Add: $6 + (-7) + 18 + 7 + (-6)$
Answer: 18

Self Check 6 Add: $8 + (-1) + 6 + 5 + (-8) + 1$ 11

Now Try ▶ Problem 73

SECTION 1.4 STUDY SET

VOCABULARY

Fill in the blanks.

▶ **1.** In the addition statement $-2 + 5 = 3$, the result, 3, is called the _sum_.

▶ **2.** Two numbers that are the same distance from 0 on a number line, but on opposite sides of it, are called _opposites_ or additive _inverses_.

▶ **3.** The _commutative_ property of addition states that changing the order when adding does not affect the answer. The _associative_ property of addition states that changing the grouping when adding does not affect the answer.

▶ **4.** Since any number added to 0 remains the same (is identical), the number 0 is called the _identity_ element for addition.

CONCEPTS

5. For each pair of numbers, which one has the larger absolute value?

a. 6 or 5 6 **b.** 8.9 or -9.2 -9.2

6. Determine whether each statement is true or false.

a. The sum of a number and its opposite is always 0. True

b. The sum of two negative numbers is always negative. True

c. The sum of two numbers with different signs is always negative. False

7. For each addition, just determine the sign of the answer.

a. $39.6 + (-64.9)$ Negative **b.** $-18.9 + 19.8$ Positive

8. Complete each property of addition. Then give its name.
   a. $a + (-a) =$ 0 Add. prop. of opp. (inv. prop. of add.)
   b. $a + 0 =$ a Ident. prop. of add.
   c. $a + b = b +$ a Comm. prop. of add.
   d. $(a + b) + c = a +$ $(b+c)$ Assoc. prop. of add.
9. Use the commutative property of addition to complete each statement.
   a. $-5 + 1 =$ $1 + (-5)$
   b. $15 + (-80.5) =$ $-80.5 + 15$
   c. $-20 + (4 + 20) = -20 + ($ $20 + 4$ $)$
   d. $(2.1 + 3) + 6 = ($ $3 + 2.1$ $) + 6$
10. Use the associative property of addition to complete each statement.
   a. $(-6 + 2) + 8 =$ $-6 + (2 + 8)$
   b. $-7 + (7 + 3) =$ $(-7 + 7) + 3$
11. What properties were used in Step 1 and Step 2 of the solution?

$$(99 + 4) + 1 = (4 + 99) + 1 \quad \text{Step 1}$$
$$= 4 + (99 + 1) \quad \text{Step 2}$$
$$= 4 + 100$$
$$= 104$$

1. Comm. prop. of add.
2. Assoc. prop. of add.

▶ 12. Consider: $-3 + 6 + (-9) + 8 + (-4)$
   a. Add all the positives in the expression. 14
   b. Add all of the negatives. −16
   c. Add the results from parts **a** and **b**. −2

NOTATION

13. a. Express the commutative property of addition using the variables x and y. $x + y = y + x$
   ▶ b. Express the associative property of addition using the variables x, y, and z. $(x + y) + z = x + (y + z)$
14. Fill in the blank: We read $-a$ as "the opposite of a."

GUIDED PRACTICE

Add. See Example 1.

15. $-8 + (-1)$ −9
16. $-3 + (-2)$ −5
▶ 17. $-5 + (-12)$ −17
18. $-4 + (-14)$ −18
▶ 19. $-29 + (-45)$ −74
20. $-23 + (-31)$ −54
▶ 21. $-4.2 + (-6.1)$ −10.3
22. $-5.1 + (-5.1)$ −10.2
23. $-\frac{3}{4} + \left(-\frac{2}{3}\right)$ $-\frac{17}{12}$
24. $-\frac{1}{5} + \left(-\frac{3}{4}\right)$ $-\frac{19}{20}$
25. $-\frac{1}{4} + \left(-\frac{1}{10}\right)$ $-\frac{7}{20}$
▶ 26. $-\frac{3}{8} + \left(-\frac{1}{3}\right)$ $-\frac{17}{24}$

Add. See Example 2.

▶ 27. $-7 + 4$ −3
28. $-9 + 7$ −2
▶ 29. $50 + (-11)$ 39
▶ 30. $27 + (-30)$ −3
31. $15.84 + (-15.84)$ 0
32. $9.19 + (-9.19)$ 0
33. $-6.25 + 8.5$ 2.25
▶ 34. $21.37 + (-12.1)$ 9.27
35. $-\frac{7}{15} + \frac{3}{15}$ $-\frac{4}{15}$
36. $-\frac{8}{11} + \frac{3}{11}$ $-\frac{5}{11}$
▶ 37. $\frac{1}{2} + \left(-\frac{1}{8}\right)$ $\frac{3}{8}$
▶ 38. $\frac{5}{6} + \left(-\frac{1}{4}\right)$ $\frac{7}{12}$

Add. See Examples 3 and 5.

▶ 39. $8 + (-5) + 13$ 16
▶ 40. $17 + (-12) + (-23)$ −18
▶ 41. $21 + (-27) + (-9)$ −15
▶ 42. $-32 + 12 + 17$ −3
43. $-27 + (-3) + (-13) + 22$ −21
▶ 44. $53 + (-27) + (-32) + (-7)$ −13
45. $-60 + 70 + (-10) + (-10) + 205$ 195
▶ 46. $-100 + 200 + (-300) + (-100) + 200$ −100

Apply the associative property of addition to find the sum. See Example 4.

47. $-99 + (99 + 215)$ 215
▶ 48. $67 + (-67 + 127)$ 127
▶ 49. $(-112 + 56) + (-56)$ −112
▶ 50. $(-67 + 5) + (-5)$ −67
51. $\frac{1}{8} + \left(\frac{7}{8} + \frac{2}{3}\right)$ $1\frac{2}{3}$
▶ 52. $\left(\frac{1}{2} + \frac{9}{16}\right) + \frac{7}{16}$ $1\frac{1}{2}$
53. $(12.4 + 1.9) + 1.1$ 15.4
▶ 54. $87.6 + (2.4 + 1.7)$ 91.7

Add. See Example 6.

55. $-1 + 9 + 1$ 9
▶ 56. $5 + 8 + (-5)$ 8
▶ 57. $-8 + 11 + (-11) + 8 + 1$ 1
▶ 58. $2 + 15 + (-15) + 8 + (-2)$ 8

TRY IT YOURSELF

Add.

59. $-9 + 81 + (-2)$ 70
60. $11 + (-21) + (-13)$ −23
61. $0 + (-6.6)$ −6.6
62. $0 + (-2.14)$ −2.14
▶ 63. $-\frac{9}{16} + \frac{7}{16}$ $-\frac{1}{8}$
64. $-\frac{3}{4} + \frac{1}{4}$ $-\frac{1}{2}$
65. $-6 + (-8)$ −14
66. $-4 + (-3)$ −7
67. $-167 + 167$ 0
▶ 68. $-25 + 25$ 0
69. $-20 + (-16) + 10$ −26
70. $-13 + (-16) + 4$ −25
71. $19.35 + (-20.21) + 1.53$ 0.67
72. $33.12 + (-35.7) + 2.98$ 0.4
▶ 73. $-7 + 5 + (-10) + 7$ −5
74. $-3 + 6 + (-9) + (-6)$ −12
75. $19.2 + (-41.3)$ −22.1
76. $57.93 + (-93.27)$ −35.34
77. $2{,}345 + (-178)$ 2,167
78. $-4{,}061 + 5{,}000$ 939
79. $-2.1 + 6.5 + (-8.2) + 2.1$ −1.7
▶ 80. $0.9 + 0.5 + (-0.2) + (-0.9)$ 0.3
81. $3 + (-6) + (-3) + 74$ 68
82. $4 + (-3) + (-4) + 5$ 2
83. $-\frac{1}{4} + \left(-\frac{2}{7}\right)$ $-\frac{15}{28}$
84. $-\frac{3}{32} + \left(-\frac{1}{2}\right)$ $-\frac{19}{32}$
85. $-0.2 + (-0.3) + (-0.4)$ −0.9
86. $-0.9 + (-1.9) + (-2.9)$ −5.7

Look Alikes . . .

87. a. $12 + 15$ 27
    b. $-12 + 15$ 3
    c. $-12 + (-15)$ −27
    d. $12 + (-15)$ −3
▶ 88. a. $432 + 67$ 499
    b. $-432 + 67$ −365
    c. $-432 + (-67)$ −499
    d. $432 + (-67)$ 365
89. a. $\frac{1}{2} + \frac{2}{9}$ $\frac{13}{18}$
    b. $-\frac{1}{2} + \frac{2}{9}$ $-\frac{5}{18}$
    c. $-\frac{1}{2} + \left(-\frac{2}{9}\right)$ $-\frac{13}{18}$
    d. $\frac{1}{2} + \left(-\frac{2}{9}\right)$ $\frac{5}{18}$
90. a. $0.87 + 0.29$ 1.16
    b. $-0.87 + 0.29$ −0.58
    c. $-0.87 + (-0.29)$ −1.16
    d. $0.87 + (-0.29)$ 0.58

APPLICATIONS

91. Military Science. During a battle, an army retreated 1,500 meters, regrouped, and advanced 2,400 meters. The next day, it advanced another 1,250 meters. Find the army's net gain. 2,150 m

▶ **92. Health.** Find the point total for the six risk factors (in blue) on the medical questionnaire. Then use the table at the bottom of the form to determine the patient's risk of contracting heart disease in the next 10 years. 4%

Age		Total Cholesterol	
Age	Points	Reading	Points
35	−4	280	3
Cholesterol		**Blood Pressure**	
HDL	Points	Systolic/Diastolic	Points
62	−3	124/100	3
Diabetic		**Smoker**	
	Points		Points
Yes	4	Yes	2

10-Year Heart Disease Risk			
Total Points	Risk	Total Points	Risk
−2 or less	1%	5	4%
−1 to 1	2%	6	6%
2 to 3	3%	7	6%
4	4%	8	7%

Source: National Heart, Lung, and Blood Institute

93. Golf. The leaderboard below shows the top four finishers from the 2009 PGA Championship Golf Tournament. Scores for each round are compared to *par,* the standard number of strokes necessary to complete the course. A score of −2, for example, indicates that the golfer used two strokes less than par to complete the course. A score of 5 indicates five strokes more than par. Determine the tournament total for each golfer.

Leaderboard

	Round 1	2	3	4	Total
Y.E. Yang	+1	−2	−5	−2	−8
Tiger Woods	−5	−2	−1	+3	−5
Lee Westwood	−2	0	+1	−2	−3
Rory McIlroy	−1	+1	−1	−2	−3

▶ **94. Submarines.** A submarine was cruising at a depth of 1,250 feet. The captain gave the order to climb 550 feet. Compared to sea level, find the new depth of the sub. −700 ft

▶ **95. Credit Cards.** Refer to the monthly statement. What is the new balance? 1,242.86

Previous Balance	New Purchases, Fees, Advances & Debts	Payments & Credits	New Balance
3,660.66	1,408.78	3,826.58	

Billing Date	Date Payment Due	Credit Line
04/21/11	05/16/11	9,100

▶ **96. Politics.** The following proposal to limit campaign contributions was on the ballot in a state election, and it passed. What will be the net fiscal impact on the state government? A gain of $2.2 million

212 Campaign Spending Limits YES ☐ NO ☐

Limits contributions to $200 in state campaigns. Fiscal impact: Costs of $4.5 million for implementation and enforcement. Increases state revenue by $6.7 million by eliminating tax deductions for lobbying.

97. Movie Losses. According to the Numbers Box Office Data website, the movie *Stealth,* released in 2005 by Sony Pictures, cost about $176,350,000 to produce, promote, and distribute. It reportedly earned back just $76,700,000 worldwide. Express the dollar loss suffered by Sony as a signed number. −$99,650,000

▶ **98. Stocks.** The last entry on the line for June 12 indicates that one share of Walt Disney Co. stock lost $0.81 in value that day. How much did the value of a share of Disney stock rise or fall over the 5-day period from June 12 through June 16? It fell $0.38.

June 12	43.88	23.38	Disney	.21	0.5	87	−43	40.75	−.81
June 13	43.88	23.38	Disney	.21	0.5	86	−15	40.19	−.56
June 14	43.88	23.38	Disney	.21	0.5	87	−50	41.00	+.81
June 15	43.88	23.38	Disney	.21	0.5	89	−28	41.81	+.81
June 16	43.88	23.38	Disney				−15	41.19	−.63

Based on data from the *Los Angeles Times*

99. Chemistry. An atom is composed of protons (with a charge of +1), neutrons (with no charge), and electrons (with a charge of −1). Two simple models of atoms are shown. What is the overall charge of each atom? −1, 3

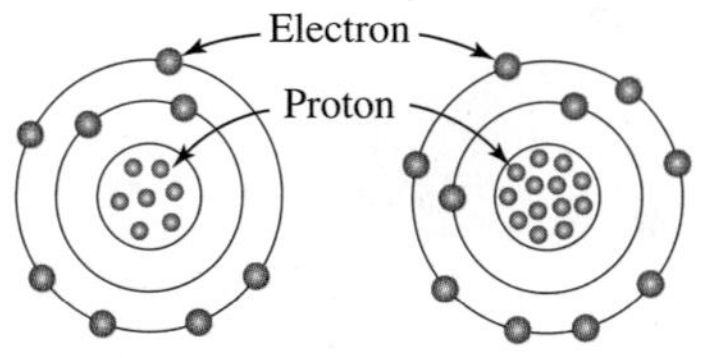
Electron
Proton

(a) (b)

100. Physics. In the illustration, arrows show the two forces acting on a lamp hanging from a ceiling. What is the sum of the forces? 0

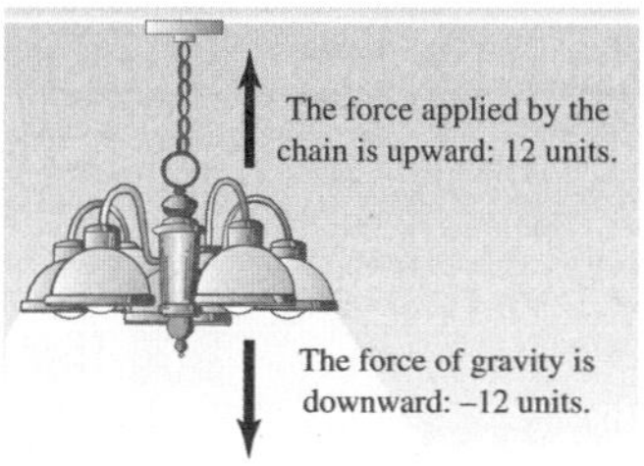
The force applied by the chain is upward: 12 units.
The force of gravity is downward: −12 units.

101. The Big Easy. The city of New Orleans lies, on average, 6 feet below sea level. What is the elevation of the top of an 85-foot tall building in New Orleans? 79 feet above sea level

102. Electronics. A closed circuit contains two batteries and three resistors. The sum of the voltages in the loop must be 0. Is it? Yes

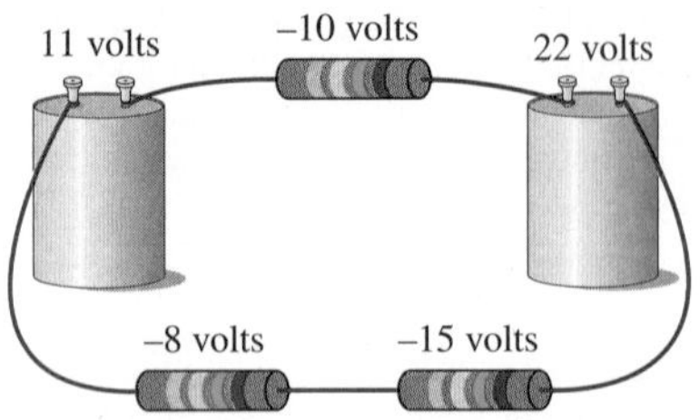

103. Accounting. The 2009 quarterly profits and losses of the Bank of America are shown in the table. Losses are denoted using parentheses. Calculate the company's total net income for 2009. $6,276 million

Quarter	Net income ($ million)
1st	4,247
2nd	3,224
3rd	(1,001)
4th	(194)

Source: www.scottrade.com

104. Politics. Six months before an election, the incumbent trailed the challenger by 18 points. To overtake her opponent, the incumbent decided to use a four-part strategy. Each part of the plan is shown below, with the expected point gain. With these gains, will the incumbent overtake the challenger on election day? No

- TV ads +10 pts
- Voter mailing +3 pts
- Union endorsement +2 pts
- Telephone calls +1 pts

WRITING

105. Explain why the sum of two positive numbers is always positive and the sum of two negative numbers is always negative.

106. Explain why the sum of a negative number and a positive number is sometimes positive, sometimes negative, and sometimes zero.

REVIEW

107. True or false: Every real number can be expressed as a decimal. True

108. Multiply: $\frac{1}{3} \cdot \frac{1}{3}$ $\frac{1}{9}$

109. What two numbers are a distance of 6 away from -3 on the number line? -9 and 3

110. Graph: $\left\{-2.5, \sqrt{2}, \frac{11}{3}, -0.333\ldots, 0.75\right\}$
See AIE Appendix 3.

CHALLENGE PROBLEMS

111. A set is said to be *closed under addition* if the sum of any two of its members is also a member of the set. Is the set $\{-1, 0, 1\}$ a closed set under addition? Explain.
No. $1 + 1 = 2$, and 2 is not a member of the set.

112. Think of two numbers. First, add the absolute value of the two numbers, and write your answer. Second, add the two numbers, take the absolute value of that sum, and write that answer. Do the two answers agree? Can you find two numbers that produce different answers? When do you get answers that agree, and when don't you?

SECTION 1.5 Subtracting Real Numbers

OBJECTIVES

1. Use the definition of subtraction.
2. Solve application problems using subtraction.

ARE YOU READY? *Are You Ready? exercises available online at www.webassign.net/brookscole*

The following problems review some basic concepts that are important when subtracting positive and negative real numbers.

1. What is the opposite of 6? What is the opposite of -15? -6, 15
2. Write *twenty-two minus six* in symbols. $22 - 6$
3. If 8 is subtracted from 20, what is the result? 12
4. Add: $-11 + 2$ -9

In this section, we discuss a rule to use when subtracting signed numbers.

1 Use the Definition of Subtraction.

A minus symbol $-$ is used to indicate subtraction. However, this symbol is also used in two other ways, depending on where it appears in an expression.

$5 - 18$	This is read as "five minus eighteen."
-5	This usually is read as "negative five." It also could be read as "the additive inverse of five" or "the opposite of five."
$-(-5)$	This usually is read as "the opposite of negative five." It also could be read as "the additive inverse of negative five."

In $-(-5)$, parentheses are used to write the opposite of a negative number. When such expressions are encountered in computations, we simplify them by finding the opposite of the number within the parentheses.

$-(-5) = 5$ *Read as "the opposite of negative five is five."*

This observation illustrates the following rule.

Opposite of an Opposite

The opposite of the opposite of a number is that number. For any real number a,

$$-(-a) = a$$ *Read as "the opposite of the opposite of a is a."*

EXAMPLE 1 Simplify each expression: **a.** $-(-45)$ **b.** $-(-h)$ **c.** $-|-10|$

Strategy To simplify each expression, we will use the concept of opposite.

Why In each case, the outermost $-$ symbol is read as "the opposite."

Solution

a. The number within the parentheses is -45. Its opposite is 45. Therefore, $-(-45) = 45$.

b. The opposite of the opposite of h is h. Therefore, $-(-h) = h$.

c. The notation $-|-10|$ means "the opposite of the absolute value of negative ten." Since $|-10| = 10$, we have:

$$-|-10| = -10$$ *The absolute value bars do not affect the $-$ symbol outside them. Therefore, the result is negative.*

Teaching Example 1
Simplify:
a. $-(-27)$ b. $-(-x)$ c. $-|-36|$
Answers:
a. 27 b. x c. -36

Self Check 1 Simplify each expression: **a.** $-(-1)$ 1 **b.** $-(-y)$ y **c.** $-|-500|$ -500

Now Try Problems 15, 17, and 19

To develop a rule for subtraction, we consider the following illustration. It represents the subtraction $5 - 2 = 3$.

The Language of Algebra

The names of the parts of a subtraction fact are:

Minuend, Subtrahend: $5 - 2 = 3$, Difference

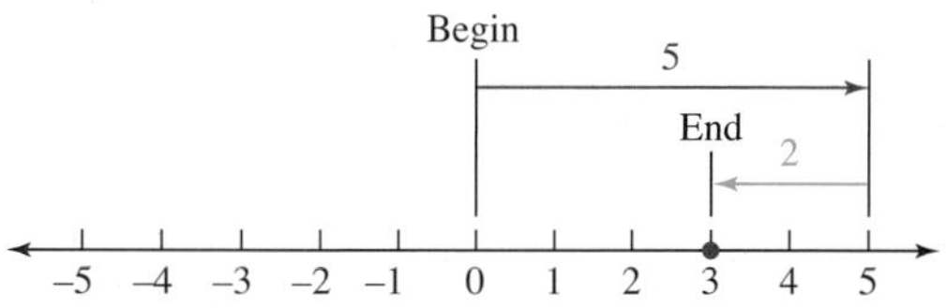

The illustration above also represents the addition $5 + (-2) = 3$. We see that

Subtracting 2 from 5 *is the same as* *adding the opposite of 2 to 5.*

$5 - 2 = 3$ $\quad\quad$ $5 + (-2) = 3$

The results are the same.

This observation suggests the following definition.

Subtraction of Real Numbers

To subtract two real numbers, add the first number to the opposite (additive inverse) of the number to be subtracted.

For any real numbers a and b,

$$a - b = a + (-b)$$ Read as "a minus b equals a plus the opposite of b."

EXAMPLE 2 Subtract and check the result:

a. $-13 - 8$ **b.** $-7.6 - (-4.5)$ **c.** $\frac{1}{4} - \left(-\frac{1}{8}\right)$

Teaching Tip: Have the students apply the subtraction rule to the following "look alikes." Remind them that the first number is not changed.

$4 - 8$
$4 - (-8)$
$-4 - 8$
$-4 - (-8)$

Strategy To find each difference, we will apply the rule for subtraction: Add the first number to the opposite of the number to be subtracted.

Why It is easy to make an error when subtracting signed numbers. We will probably be more accurate if we write each subtraction as addition of the opposite.

Solution

a. We read $-13 - 8$ as "negative thirteen *minus* eight." Subtracting 8 is the same as adding -8.

Change the subtraction to addition.

$$-13 - \mathbf{8} = -13 + (\mathbf{-8}) = -21$$ Use the rule for adding two numbers with like signs.

No change — Change the number being subtracted to its opposite.

To check, we add the *difference*, -21, and the *subtrahend*, 8, to obtain the *minuend*, -13.

Check: $-21 + 8 = -13$

b. We read $-7.6 - (-4.5)$ as "negative seven point six *minus* negative four point five." Subtracting -4.5 is the same as adding 4.5.

Add . . .

$$-7.6 - (\mathbf{-4.5}) = -7.6 + \mathbf{4.5} = -3.1$$ Use the rule for adding two numbers with unlike signs.

No change — . . . the opposite

Check: $-3.1 + (-4.5) = -7.6$

c. $\frac{1}{4} - \left(-\frac{1}{8}\right) = \frac{2}{8} - \left(-\frac{1}{8}\right)$ The LCD is 8. Build $\frac{1}{4}$ so that the denominator is 8: $\frac{1}{4} \cdot \frac{2}{2} = \frac{2}{8}$.

$= \frac{2}{8} + \frac{1}{8}$ To subtract, add the opposite. Do not change $\frac{2}{8}$.

$= \frac{3}{8}$

Check: $\frac{3}{8} + \left(-\frac{1}{8}\right) = \frac{2}{8} = \frac{1}{4}$

The Language of Algebra

When we change a number to its opposite, we say we have **changed** (or *reversed*) its sign.

The Language of Algebra

The rule for subtracting real numbers is often summarized as: **Subtracting a number is the same as adding its opposite.**

Calculators

The subtraction key
When using a calculator to subtract signed numbers, be careful to distinguish between the *subtraction* key [−] and the keys that are used to enter negative values: [+/−] on a scientific calculator and [(−)] on a graphing calculator.

Teaching Example 2 Subtract and check the result:
a. $-9 - 11$ **b.** $2.6 - (-5.3)$
c. $-\frac{1}{3} - \frac{1}{6}$

Answers:
a. -20 **b.** 7.9 **c.** $-\frac{1}{2}$

Self Check 2 Subtract and check the result: **a.** $-32 - 25$ -57
b. $1.7 - (-1.2)$ 2.9 **c.** $-\frac{1}{3} - \left(-\frac{3}{4}\right)$ $\frac{5}{12}$

Now Try Problems 25, 39, and 43

EXAMPLE 3 **a.** Subtract 0.5 from 4.6 **b.** Subtract 4.6 from 0.5

Strategy We will translate each phrase to mathematical symbols and then perform the subtraction. We must be careful when translating the instruction to subtract one number *from* another number.

Why The order of the numbers in each word phrase must be reversed when we translate it to mathematical symbols.

Solution **a.** The number to be subtracted is 0.5.

Subtract 0.5 from 4.6 — To translate, reverse the order in which 0.5 and 4.6 appear in the sentence.

$$4.6 - 0.5 = 4.1$$

b. The number to be subtracted is 4.6.

Subtract 4.6 from 0.5 — To translate, reverse the order in which 4.6 and 0.5 appear in the sentence. Add the opposite of 4.6.

$$0.5 - 4.6 = 0.5 + (-4.6)$$
$$= -4.1$$

Caution

Notice from parts **a** and **b** that $4.6 - 0.5 \neq 0.5 - 4.6$. This result illustrates an important fact: Subtraction is *not* commutative. When subtracting two numbers, it is important that we write them in the correct order, because, in general, $a - b \neq b - a$.

Teaching Example 3
a. Subtract 2.1 from 4.1
b. Subtract 4.1 from 2.1
Answers:
a. 2 **b.** −2

Self Check 3 **a.** Subtract 2.2 from 4.9 2.7 **b.** Subtract 4.9 from 2.2 −2.7

Now Try Problem 47

EXAMPLE 4 Perform the operations: $-9 - 15 + 20 - (-6)$

Strategy This expression contains addition and subtraction. We will write each subtraction as addition of the opposite and then evaluate the expression.

Why It is easy to make an error when subtracting signed numbers. We probably will be more accurate if we write each subtraction as addition of the opposite.

Solution

$$-9 - \mathbf{15} + 20 - (\mathbf{-6}) = -9 + (\mathbf{-15}) + 20 + \mathbf{6}$$
$$= -24 + 26 \quad \text{Add the negatives. Add the positives.}$$
$$= 2 \quad \text{Add the results.}$$

Teaching Example 4 Perform the operations:
a. $-5 - 32 + 22 - (-4)$
b. $3 - 15 + 17 - (-3)$
Answers:
a. −11 **b.** 8

Self Check 4 Perform the operations: $-40 - (-10) + 7 - (-15)$ −8

Now Try Problem 51

2 Solve Application Problems Using Subtraction.

Subtraction finds the *difference* between two numbers. When we find the difference between the maximum value and the minimum value of a collection of measurements, we are finding the **range** of the values.

EXAMPLE 5 **U.S. Temperatures.** The record high temperature in the United States of 134°F was set in Death Valley, California, on July 10, 1913. The record low of −80°F was set at Prospect Creek, Alaska, on January 23, 1971. Find the temperature range for these extremes.

Strategy We will subtract the lowest temperature from the highest temperature.

Why The *range* of a collection of data indicates the spread of the data. It is the difference between the largest and smallest values.

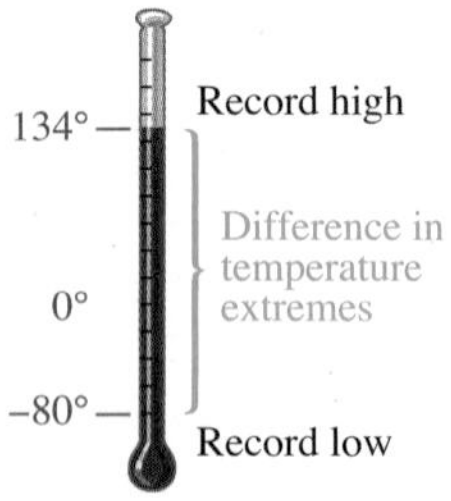

Solution

$$134 - (-80) = 134 + 80 \quad \text{134° is the higher temperature and −80° is the lower.}$$
$$= 214$$

The temperature range for these extremes is 214°F.

Self Check 5 Find the temperature range for a day that had a low of -5°F and a high of 36°F. 41°F

Now Try Problem 95

Teaching Example 5 Find the temperature range for a day that had a low of -15°F and a high of -2°F.
Answer: 13°F

Things are constantly changing in our daily lives. The amount of money we have in the bank, the price of gasoline, and our ages are examples. In mathematics, the operation of subtraction is used to measure change. To find the **change** in a quantity, we subtract the earlier value from the later value.

Change = later value − earlier value

EXAMPLE 6 **Water Management.** Lake Mead, on the Nevada-Arizona border, is the largest reservoir in the United States. It is formed by Hoover Dam across the Colorado River. In 2000, the water level in Lake Mead was 89 feet above drought level. By 2010, the water level was 27 feet below drought level. Find the change in the water level over that time span. (Source: Bureau of Reclamation)

©iStockphoto.com/MvH

Strategy We can represent a water level above drought level using a positive number and a water level below drought level using a negative number. To find the change in the water level, we will subtract.

Why In general, *to find the change in a quantity, we subtract the earlier value from the later value.*

Solution

$$-27 - 89 = -27 + (-89) \quad \text{The earlier water level in 2000 (89 ft) is subtracted from the later water level in 2010 (−27 ft).}$$
$$= -116 \quad \text{Do the addition.}$$

The negative result indicates that the water level *fell* 116 feet in that time span.

Teaching Example 6 Find the change in water level for a week that started at 6 feet above normal and went to 8 feet below normal level.
Answer: -14 ft

Self Check 6 Find the change in water level for a week that started at 4 feet above normal and went to 7 feet below normal level. -11 ft

Now Try Problem 97

SECTION 1.5 STUDY SET

VOCABULARY

Fill in the blanks.

▶ **1.** Subtraction finds the difference between two numbers.

▶ **2.** In the subtraction $-2 - 5 = -7$, the result of -7 is called the difference.

▶ **3.** The difference between the maximum and the minimum value of a collection of measurements is called the range of the values.

▶ **4.** To find the change in a quantity, subtract the earlier value from the later value.

CONCEPTS

5. Find the opposite (additive inverse) of each number.

   a. 12 -12 b. $-\frac{1}{5}$ $\frac{1}{5}$

   c. 2.71 -2.71 d. 0 0

6. Complete each statement.

   a. $a - b = a +$ $(-b)$

   To subtract two numbers, add the first number to the opposite of the number to be subtracted.

   b. $-(-a) =$ a

   The opposite of the opposite of a number is that number.

7. Apply the rule for subtraction and fill in the blanks.

   $1 - (-9) = 1 + 9 = 10$

8. Use addition to check this subtraction: $15 - (-8) = 7$. Is the result correct? No; $7 + (-8) \neq 15$

9. Write each subtraction in the following expression as addition of the opposite. $-10 + (-8) + (-23) + 5 + 34$

   $-10 - 8 + (-23) + 5 - (-34)$

10. For each subtraction, just determine the sign of the answer.

    a. $8.76 - 12.91$ Negative b. $8.76 - (-12.91)$ Positive

11. Circle any minus signs in each expression.

    a. $-6 - (-4)$ b. $7 + (-3) - 5 - (-2)$

▶ 12. In each case, determine what number is being subtracted.

    a. $5 - 8$ 8 b. $-5 - (-8)$ -8

NOTATION

▶ 13. Write each phrase using symbols. Then find its value.

    a. One minus negative seven $1 - (-7) = 8$

    b. The opposite of negative two $-(-2) = 2$

    c. The opposite of the absolute value of negative three $-|-3| = -3$

    d. Subtract 6 from 2 $2 - 6 = -4$

14. Write each expression in words.

    a. $-(-m)$ The opposite of the opposite of m

    b. $-2 - (-3)$ Negative two minus negative three

    c. $x - (-y)$ x minus the opposite of y

GUIDED PRACTICE

Simplify each expression. See Example 1.

15. $-(-55)$ 55
▶ 16. $-(-27.2)$ 27.2
17. $-(-x)$ x
▶ 18. $-(-t)$ t
▶ 19. $-|-25|$ -25
20. $-|-100|$ -100
21. $-\left|-\frac{3}{16}\right|$ $-\frac{3}{16}$
▶ 22. $-\left|-\frac{4}{3}\right|$ $-\frac{4}{3}$

Subtract. See Example 2.

▶ 23. $4 - 7$ -3
24. $1 - 6$ -5
▶ 25. $-6 - 4$ -10
26. $-3 - 4$ -7
▶ 27. $8 - (-3)$ 11
▶ 28. $17 - (-21)$ 38
▶ 29. $0 - 6$ -6
▶ 30. $0 - 9$ -9
▶ 31. $-1 - (-3)$ 2
32. $-1 - (-7)$ 6
▶ 33. $20 - (-20)$ 40
▶ 34. $30 - (-30)$ 60
▶ 35. $-2 - (-7)$ 5
36. $-9 - (-1)$ -8
▶ 37. $0 - (-12)$ 12
38. $0 - 12$ -12
39. $-1.4 - 5.5$ -6.9
▶ 40. $-1.3 - 4.7$ -6.0
41. $-1.5 - 0.81$ -2.31
42. $-1.57 - (-0.8)$ -0.77
▶ 43. $-\frac{1}{8} - \frac{3}{8}$ $-\frac{1}{2}$
44. $-\frac{3}{4} - \frac{1}{4}$ -1
▶ 45. $\frac{1}{3} - \frac{3}{4}$ $-\frac{5}{12}$
46. $\frac{1}{6} - \frac{5}{8}$ $-\frac{11}{24}$

Perform the indicated operation. See Example 3.

47. Subtract -5 from 17. 22
▶ 48. Subtract 45 from -50. -95
49. Subtract 12 from -13. -25
▶ 50. Subtract -11 from -20. -9

Perform the operations. See Example 4.

51. $-6 + 8 - (-1) - 10$ -7
▶ 52. $-4 + 5 - (-3) - 13$ -9
▶ 53. $61 - (-62) + (-64) - 60$ -1
▶ 54. $93 - (-92) + (-94) - 95$ -4

TRY IT YOURSELF

Perform the operations.

55. $244 - (-12)$ 256
56. $354 - (-29)$ 383
57. $-20 - (-30) - 50 + 40$ 0
▶ 58. $-24 - (-28) - 48 - 44$ -88
59. $-1.2 - 0.9$ -2.1
60. $-2.52 - 1.72$ -4.24
61. $\frac{1}{8} - \left(-\frac{5}{7}\right)$ $\frac{47}{56}$
62. $\frac{5}{8} - \left(-\frac{2}{9}\right)$ $\frac{61}{72}$
63. $-62 - 71 - (-37) + 99$ 3
64. $-17 - 32 - (-85) - 51$ -15
65. Subtract 47.5 from 0. -47.5
66. Subtract 30.3 from 0. -30.3
67. Subtract -137 from 12. 149
68. Subtract 512 from -47. -559
69. $-1{,}903 - (-1{,}732)$ -171
70. $-300 - (-11)$ -289
71. $2.83 - (-1.8)$ 4.63
72. $4.75 - (-1.9)$ 6.65
73. $-\frac{5}{6} - \frac{3}{4}$ $-\frac{19}{12}$
74. $-\frac{3}{7} - \frac{2}{5}$ $-\frac{29}{35}$
▶ 75. $8 - 9 - 10$ -11
76. $1 - 2 - 3$ -4
▶ 77. $-44 - 44$ -88
78. $-33 - 33$ -66
79. $-0.9 - 0.2$ -1.1
▶ 80. $-0.3 - 0.2$ -0.5
▶ 81. $-25 - (-50) - 75$ 50
▶ 82. 33 $(\ 22)$ 44 55
83. $6.3 - 9.8$ -3.5
84. $2.1 - 9.4$ -7.3
▶ 85. $-\frac{9}{16} - \left(-\frac{1}{4}\right)$ $-\frac{5}{16}$
86. $-\frac{1}{2} - \left(-\frac{1}{4}\right)$ $-\frac{1}{4}$
▶ 87. $0 - (-1)$ 1
▶ 88. $0 - (-8)$ 8
▶ 89. $2 - 15$ -13
90. $3 - 14$ -11

Look Alikes . . .

▶ 91. a. $-50 + (-3)$ -53 b. $-50 - (-3)$ -47

92. a. $-\frac{1}{16} + \frac{1}{4}$ $\frac{3}{16}$ b. $-\frac{1}{16} - \frac{1}{4}$ $-\frac{5}{16}$

93. a. $-\frac{5}{9} + \left(-\frac{1}{6}\right)$ $-\frac{13}{18}$ b. $-\frac{5}{9} - \left(-\frac{1}{6}\right)$ $-\frac{7}{18}$

▶ 94. a. $2.96 + (-1.78)$ 1.18 b. $2.96 - (-1.78)$ 4.74

APPLICATIONS

95. The Empire State. New York state's record high temperature of 108°F was set in 1926, and the record low of −52°F was set in 1979. What is the range of these temperature extremes? 160°F

96. Eyesight. Nearsightedness, the condition where near objects are clear and far objects are blurry, is measured using negative numbers. Farsightedness, the condition where far objects are clear and near objects are blurry, is measured using positive numbers. Find the range in the measurements shown. 6.85

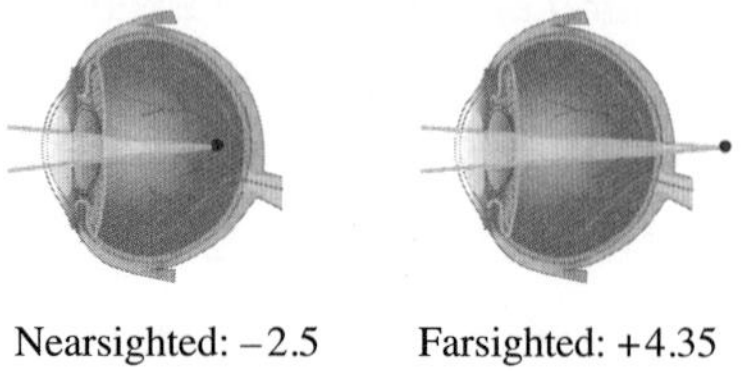

97. Law Enforcement. A burglar scored −18 on a lie detector test, a score that indicates deception. However, on a second test, he scored +3, a score that is inconclusive. Find the change in the scores. 21 points

98. Racing. To improve handling, drivers often adjust the angle of the wheels of their car. When the wheel leans out, the degree measure is considered positive. When the wheel leans in, the degree measure is considered negative. Find the change in the position of the wheel shown below. −5.75°

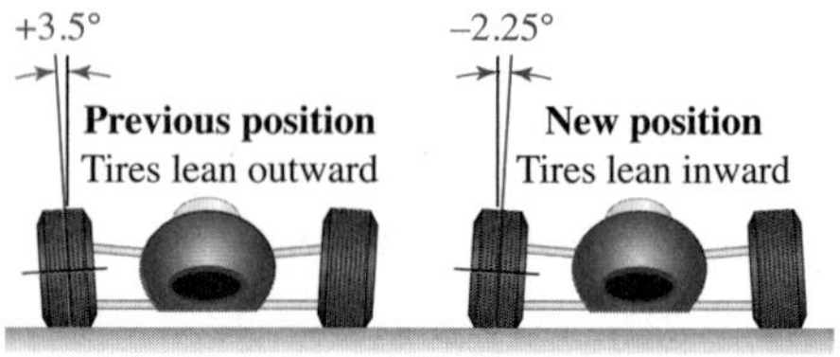

99. from **Campus to Careers**

Lead Transportation Security Officer

Determine the change in the number of passengers using each airport in 2009 compared with 2008. Orlando: −115,000 passengers; Ft Lauderdale: −50,000 passengers

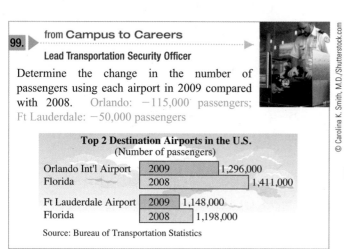

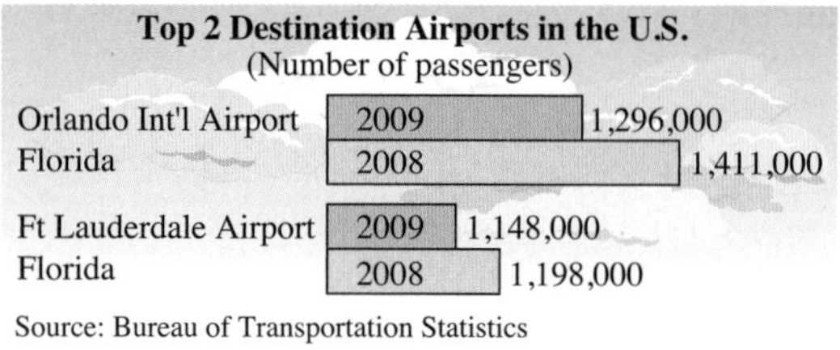

100. U.S. Jobs. The table lists the three occupations that are predicted to have the largest job declines from 2008 to 2018. Complete the column labeled "Change."

Occupation	Number of jobs 2008	2018	Change
Farmers/ranchers	985,900	906,700	−79,200
Sewing machine operators	212,400	140,900	−71,500
Order clerks	245,700	181,500	−64,200

Source: Bureau of Labor Statistics

101. Geography. The elevation of Death Valley, California, is 282 feet below sea level. The elevation of the Dead Sea in Israel is 1,312 feet below sea level. Find the difference in their elevations. 1,030 ft

102. Card Games. Gonzalo won the second round of a card game and earned 50 points. Matt and Hydecki had to deduct the value of each of the cards left in their hands from their score on the first round. Use this information to update the score sheet below. (Face cards are counted as 10 points, aces as 1 point, and all others have the value of the number printed on the card.)

Running point total	Round 1	Round 2
Matt	+50	+29
Gonzalo	−15	+35
Hydecki	−2	−23

103. World's Coldest Ice Cream. Dippin' Dots is an ice cream snack that was invented by Curt Jones in 1987. The tiny multicolored beads are created by flash freezing ice cream mix in liquid nitrogen at a temperature of −355°F. When they come out of the processor, they are stored at a temperature of −40°F. Find the change in temperature of Dippin' Dots from production to storage. (Source: fundinguniverse.com) an increase of 315°F

104. History. Plato, a famous Greek philosopher, died in 347 B.C. at the age of 81. When was he born? 428 B.C.

105. NASCAR. Complete the table below to determine how many points the third and fourth place finishers were behind the leader.

2009 Final Driver Standings			
Rank	Driver	Points	Points behind leader
1	Jimmie Johnson	6,652	. . .
2	Mark Martin	6,511	−141
3	Jeff Gordon	6,473	−179
4	Kurt Busch	6,446	−206

106. Gauges. With the engine off, the ammeter on a car reads 0. If the headlights, which draw a current of 7 amps, and the radio, which draws a current of 6 amps, are both turned on, what will be the new reading? −13

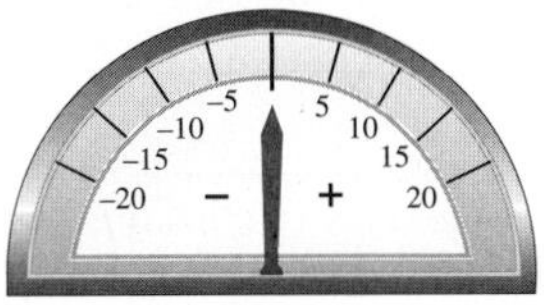

WRITING

107. Explain what it means when we say that subtraction is *not commutative.*

108. Why is addition of signed numbers taught before subtraction of signed numbers?

▶ **109.** Explain why we know that the answer to $4 - 10$ is negative without having to do any computation.

110. Is the following statement true or false? Explain.
Having a debt of $100 forgiven is equivalent to gaining $100.

REVIEW

111. Find the prime factorization of 30. $2 \cdot 3 \cdot 5$

112. Write the set of integers.
$\{\ldots, -3, -2, -1, 0, 1, 2, 3, \ldots\}$

113. True or false: $-4 > -5$? True

▶ **114.** Use the associative property of addition to simplify the calculation: $-18 + (18 + 89)$ $(-18 + 18) + 89 = 89$

CHALLENGE PROBLEMS

▶ **115.** Suppose x is positive and y is negative. Determine whether each statement is true or false.

a. $x - y > 0$ True **b.** $y - x < 0$ True
c. $|-x| < 0$ False **d.** $-|y| < 0$ True

116. Find:

$$1 - 2 + 3 - 4 + 5 - 6 + \ldots + 99 - 100$$ −50

SECTION 1.6

OBJECTIVES

1. Multiply signed numbers.
2. Use properties of multiplication.
3. Divide signed numbers.
4. Use properties of division.

Multiplying and Dividing Real Numbers; Multiplication and Division Properties

ARE YOU READY? *Are You Ready? exercises available online at www.webassign.net/brookscole*

The following problems review some basic concepts that are important when multiplying positive and negative real numbers.

1. Find $|-14|$ and $|6.75|$. 14, 6.75

2. Do the integers -3 and 24 have the same sign or different signs? Different signs

3. Multiply: $\frac{7}{8} \cdot \frac{5}{14}$ $\frac{5}{16}$

4. Divide: $2.22 \div 0.6$ 3.7

In this section, we will develop rules for multiplying and dividing positive and negative numbers.

1 Multiply Signed Numbers.

Multiplication represents repeated addition. For example, 4(3) is equal to the sum of four 3's.

$$4(3) = 3 + 3 + 3 + 3$$
$$= 12$$

This example illustrates that *the product of two positive numbers is positive.*

To develop a rule for multiplying a positive number and a negative number, we will find $4(-3)$, which is equal to the sum of four -3's.

$$4(-3) = -3 + (-3) + (-3) + (-3)$$
$$= -12$$

We see that the result is negative. As a check, think in terms of money. If you lose $3 four times, you have lost a total of $12, which is written −$12. This example illustrates that *the product of a positive number and a negative number is negative.*

The Language of Algebra

The names of the parts of a multiplication fact are:

Factor Factor Product
$4(3) = 12$

Multiplying Two Numbers That Have Different (Unlike) Signs

To multiply a positive real number and a negative real number, multiply their absolute values. Then make the final answer negative.

EXAMPLE 1 Multiply: **a.** $8(-12)$ **b.** $-151 \cdot 5$ **c.** $(-0.6)(1.2)$ **d.** $\frac{3}{4}\left(-\frac{4}{15}\right)$

Strategy We will use the rule for multiplying two numbers that have different signs.

Why In each case, we are asked to multiply a positive number and a negative number.

Solution

a. $8(-12) = -96$ Multiply the absolute values, 8 and 12, to get 96. Since the signs are unlike, make the final answer negative.

b. $-151 \cdot 5 = -755$ Multiply the absolute values, 151 and 5, to get 755. Since the signs are unlike, make the final answer negative.

c. To find the product of these two decimals with unlike signs, first multiply their absolute values, 0.6 and 1.2.

$$\begin{array}{r} 1.2 \\ \times\ 0.6 \\ \hline 0.72 \end{array}$$

Place the decimal point in the result so that the answer has the same number of decimal places as the sum of the number of decimal places in the factors.

Then make the final answer negative: $(-0.6)(1.2) = -0.72$.

Success Tip

The product of two numbers with unlike signs is *always* negative.

d. $\frac{3}{4}\left(-\frac{4}{15}\right) = -\frac{\overset{1}{\cancel{3}} \cdot \overset{1}{\cancel{4}}}{\underset{1}{\cancel{4}} \cdot \underset{1}{\cancel{3}} \cdot 5}$ Multiply the absolute values $\frac{3}{4}$ and $\frac{4}{15}$. Since the signs are unlike, make the final answer negative.

$= -\frac{1}{5}$ To simplify the fraction, factor 15 as $3 \cdot 5$. Remove the common factors 3 and 4 in the numerator and denominator.

Teaching Example 1
Multiply: **a.** $30(-3)$ **b.** $-4 \cdot 7$ **c.** $5.9(-1.8)$ **d.** $-\frac{9}{16} \cdot \frac{10}{27}$
Answers: **a.** -90 **b.** -28 **c.** -10.62 **d.** $-\frac{5}{24}$

Self Check 1 Multiply: **a.** $20(-3)$ -60 **b.** $-3 \cdot 5$ -15 **c.** $4.3(-2.6)$ -11.18 **d.** $-\frac{5}{8} \cdot \frac{16}{25}$ $-\frac{2}{5}$

Now Try Problems 21, 27, and 29

To develop a rule for multiplying two negative numbers, consider the following list, where we multiply -4 by factors that decrease by 1. We know how to find the first four products. Graphing those results on a number line is helpful in determining the last three products.

This factor decreases by 1 each time | Look for a pattern here

$$\begin{aligned} -4(\mathbf{3}) &= -12 \\ -4(\mathbf{2}) &= -8 \\ -4(\mathbf{1}) &= -4 \\ -4(\mathbf{0}) &= 0 \\ -4(\mathbf{-1}) &= ? \\ -4(\mathbf{-2}) &= ? \\ -4(\mathbf{-3}) &= ? \end{aligned}$$

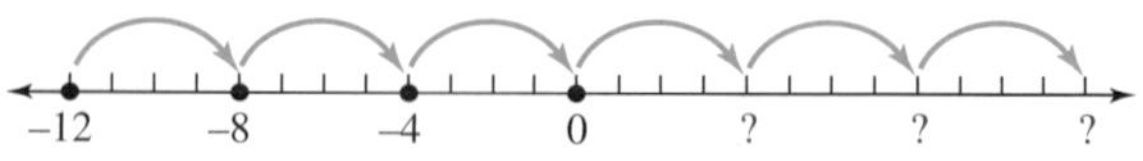

A graph of the products

From the pattern, we see that the product increases by 4 each time. Thus,

$$-4(-1) = 4, \qquad -4(-2) = 8, \qquad \text{and} \qquad -4(-3) = 12$$

These results illustrate that *the product of two negative numbers is positive.* As a check, think of losing four debts of \$3. This is equivalent to gaining \$12. Therefore, $-4(-\$3) = \$12$.

Since the product of two positive numbers is positive, and the product of two negative numbers is also positive, we can summarize the multiplication rule as follows.

Multiplying Two Numbers That Have the Same (Like) Signs

To multiply two real numbers that have the same sign, multiply their absolute values. The final answer is positive.

EXAMPLE 2 Multiply: **a.** $-5(-6)$ **b.** $\left(-\frac{1}{2}\right)\left(-\frac{5}{8}\right)$

Strategy We will use the rule for multiplying two numbers that have the same sign.

Why In each case, we are asked to multiply two negative numbers.

Solution

a. $-5(-6) = 30$ Multiply the absolute values, 5 and 6, to get 30. Since both factors are negative, the final answer is positive.

b. $\left(-\frac{1}{2}\right)\left(-\frac{5}{8}\right) = \frac{5}{16}$ Multiply the absolute values, $\frac{1}{2}$ and $\frac{5}{8}$, to get $\frac{5}{16}$. Since the two factors have the same sign, the final answer is positive.

Success Tip

The product of two numbers with like signs is *always* positive.

Teaching Example 2
Multiply: **a.** $-9(-4)$ **b.** $-\frac{1}{5}\left(-\frac{3}{10}\right)$
Answers: **a.** 36 **b.** $\frac{3}{50}$

Self Check 2 Multiply: **a.** $-15(-8)$ 120 **b.** $-\frac{1}{4}\left(-\frac{1}{3}\right)$ $\frac{1}{12}$

Now Try Problems 33 and 41

2 Use Properties of Multiplication.

The multiplication of two numbers can be done in any order; the result is the same. For example, $-9(4) = -36$ and $4(-9) = -36$. This illustrates that multiplication is **commutative.**

The Commutative Property of Multiplication

Changing the order when multiplying does not affect the answer.
For any real numbers a and b,

$$ab = ba$$

In the following example, we multiply $-3 \cdot 7 \cdot 5$ in two ways. Recall that the operation within the parentheses should be performed first. We read $(-3 \cdot 7)5$ as "the *quantity* of -3 times 7," pause slightly, and then say "times 5." We read $-3(7 \cdot 5)$ as "-3 times the *quantity* of 7 times 5." The word *quantity* alerts the reader to the parentheses that are used as grouping symbols.

Method 1: Group −3 and 7	***Method 2: Group 7 and 5***
$(-3 \cdot 7)5 = (-21)5$	$-3(7 \cdot 5) = -3(35)$
$= -105$	$= -105$

Teaching Tip: In the statement of the associative property of multiplication below, point out that the *order* of the factors, *a, b,* and *c,* does not change.

It doesn't matter how we group the numbers in this multiplication; the result is -105. This example illustrates that multiplication is **associative.**

The Associative Property of Multiplication

Changing the grouping when multiplying does not affect the answer.
For any real numbers a, b, and c,

$$(ab)c = a(bc)$$

EXAMPLE 3 Multiply: **a.** $-5(-37)(-2)$ **b.** $-4(-3)(-2)(-1)$

Strategy First, we will use the commutative and associative properties of multiplication to reorder and regroup the factors. Then we will perform the multiplications.

Why Applying one or both of these properties before multiplying can simplify the computations and lessen the chance of a sign error.

Solution Using the commutative and associative properties of multiplication, we can reorder and regroup the factors to simplify computations.

a. Since it is easy to multiply by 10, we will find $-5(-2)$ first.

$$\begin{aligned} -5(-37)(-2) &= -5(-2)(-37) && \text{Use the commutative property of multiplication.} \\ &= 10(-37) \\ &= -370 \end{aligned}$$

b. $$\begin{aligned} -4(-3)(-2)(-1) &= 12(2) && \text{Multiply the first two factors and multiply the last two factors.} \\ &= 24 \end{aligned}$$

Teaching Example 3
Multiply: **a.** $-2(-16)(-3)$
b. $-4(-6)(-8)(-1)$
Answers: **a.** -96 **b.** 192

Self Check 3 Multiply: **a.** $-25(-3)(-4)$ -300 **b.** $-1(-2)(-3)(-3)$ 18

Now Try ▶ Problems 43 and 47

In Example 3a, we multiplied three negative numbers. In Example 3b, we multiplied four negative numbers. The results illustrate the following fact.

Multiplying Negative Numbers

The product of an even number of negative numbers is positive. The product of an odd number of negative numbers is negative.

Recall that the product of 0 and any whole number is 0. The same is true for any real number. Therefore, $-6 \cdot 0 = 0$, $\frac{7}{16} \cdot 0 = 0$, and $0(4.51) = 0$.

Multiplication Property of 0

The product of 0 and any real number is 0.
For any real number a,

$$0 \cdot a = 0 \quad \text{and} \quad a \cdot 0 = 0$$

Whenever we multiply a number by 1, the number remains the same. Therefore, $1 \cdot 6 = 6$, $4.57 \cdot 1 = 4.57$, and $1(-9) = -9$. Since any number multiplied by 1 remains the same (is identical), the number 1 is called the **identity element** for multiplication.

Multiplication Property of 1 (Identity Property of Multiplication)

The product of 1 and any number is that number.
For any real number a,

$$1 \cdot a = a \quad \text{and} \quad a \cdot 1 = a$$

Whenever we multiply a number by -1, the result is the opposite of that number. For example, $-1 \cdot 12 = -12$ and $-\frac{5}{8}(-1) = \frac{5}{8}$.

Multiplication Property of −1

The product of −1 and any number is the opposite (or additive inverse) of that number.
For any real number a,

$$-1 \cdot a = -a \quad \text{and} \quad a(-1) = -a$$

The Language of Algebra

Don't confuse the words **opposite** and **reciprocal**. The opposite of 4 is −4. The reciprocal of 4 is $\frac{1}{4}$.

Two numbers whose product is 1 are **reciprocals** or **multiplicative inverses** of each other. For example, 8 is the multiplicative inverse of $\frac{1}{8}$, and $\frac{1}{8}$ is the multiplicative inverse of 8, because $8 \cdot \frac{1}{8} = 1$. Likewise, $-\frac{3}{4}$ and $-\frac{4}{3}$ are multiplicative inverses because $-\frac{3}{4}\left(-\frac{4}{3}\right) = 1$. All real numbers, except 0, have a multiplicative inverse.

Multiplicative Inverses (Inverse Property of Multiplication)

The product of any number and its multiplicative inverse (reciprocal) is 1.
For any nonzero real number a,

$$a\left(\frac{1}{a}\right) = 1$$

EXAMPLE 4 Find the reciprocal of each number: **a.** $\frac{2}{3}$ **b.** $-\frac{2}{3}$ **c.** -11

Strategy To find the reciprocal of a fraction, we invert the numerator and the denominator.

Why We want the product of the given number and its reciprocal to be 1.

Solution

a. The reciprocal of $\frac{2}{3}$ is $\frac{3}{2}$ because $\frac{2}{3}\left(\frac{3}{2}\right) = 1$. *To find the reciprocal of a fraction, invert the numerator and denominator.*

b. The reciprocal of $-\frac{2}{3}$ is $-\frac{3}{2}$ because $-\frac{2}{3}\left(-\frac{3}{2}\right) = 1$.

c. The reciprocal of -11 is $-\frac{1}{11}$ because $-11\left(-\frac{1}{11}\right) = 1$. *Think of −11 as $\frac{-11}{1}$ to find its reciprocal.*

Caution

Do not change the sign of a number when finding its reciprocal.

Teaching Example 4
Find the reciprocal of each number:
a. $\frac{23}{25}$ b. $-\frac{23}{25}$ c. -16
Answers: a. $\frac{25}{23}$ b. $-\frac{25}{23}$ c. $-\frac{1}{16}$

Self Check 4 Find the reciprocal of each number: **a.** $-\frac{15}{16}$ $-\frac{16}{15}$ **b.** $\frac{15}{16}$ $\frac{16}{15}$ **c.** -27 $-\frac{1}{27}$

Now Try Problems 51 and 53

3 Divide Signed Numbers.

Every division fact can be written as an equivalent multiplication fact.

Division

For any real numbers a, b, and c, where $b \neq 0$,

$$\frac{a}{b} = c \quad \text{provided that} \quad c \cdot b = a$$ *Quotient · divisor = dividend*

The Language of Algebra

The names of the parts of a division fact are:

Dividend $\frac{15}{5} = 3$ Quotient

Divisor

We can use this relationship between multiplication and division to develop rules for dividing signed numbers. For example,

$$\frac{15}{5} = 3 \qquad \text{because} \qquad 3(5) = 15$$

From this example, we see that *the quotient of two positive numbers is positive.*

To determine the quotient of two negative numbers, we consider $\frac{-15}{-5}$.

$$\frac{-15}{-5} = 3 \qquad \text{because} \qquad 3(-5) = -15$$

From this example, we see that the *quotient of two negative numbers is positive.*

To determine the quotient of a positive number and a negative number, we consider $\frac{15}{-5}$.

$$\frac{15}{-5} = -3 \qquad \text{because} \qquad -3(-5) = 15$$

From this example, we see that *the quotient of a positive number and a negative number is negative.*

To determine the quotient of a negative number and a positive number, we consider $\frac{-15}{5}$.

$$\frac{-15}{5} = -3 \qquad \text{because} \qquad -3(5) = -15$$

From this example, we see that *the quotient of a negative number and a positive number is negative.*

We summarize the rules from the previous examples and note that they are similar to the rules for multiplication.

Dividing Two Real Numbers

To divide two real numbers, divide their absolute values.

1. The quotient of two numbers that have the same (*like*) signs is positive.
2. The quotient of two numbers that have different (*unlike*) signs is negative.

EXAMPLE 5 Divide and check the result: **a.** $\frac{-81}{-9}$ **b.** $\frac{45}{-9}$ **c.** $-2.87 \div 0.7$

d. $-\frac{5}{16} \div \left(-\frac{1}{2}\right)$

Strategy We will use the rules for dividing signed numbers. In each case, we need to ask, "Is it a quotient of two numbers with the same sign or different signs?"

Why The signs of the numbers that we are dividing determine the sign of the result.

Solution **a.** $\frac{-81}{-9} = 9$ Divide the absolute values, 81 by 9, to get 9. Since the signs are like, the final answer is positive.

Multiply the quotient and the divisor to check the result: $9(-9) = -81$.

b. $\frac{45}{-9} = -5$ Divide the absolute values, 45 by 9, to get 5. Since the signs are unlike, make the final answer negative.

Check: $-5(-9) = 45$ Quotient · divisor = dividend

c. $-2.87 \div 0.7 = -4.1$ Since the signs are unlike, make the final answer negative.

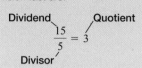

Check: $-4.1(0.7) = -2.87$ Quotient · divisor = dividend

Success Tip

To perform this decimal division, move each decimal point one place to the right.

$0.7\overline{)2.87}$

d. $-\frac{5}{16} \div \left(-\frac{1}{2}\right) = -\frac{5}{16}\left(-\frac{2}{1}\right)$ Multiply the first fraction by the reciprocal of the second fraction. The reciprocal of $-\frac{1}{2}$ is $-\frac{2}{1}$.

$= \frac{5 \cdot 2}{16 \cdot 1}$ Multiply the absolute values $\frac{5}{16}$ and $\frac{2}{1}$. Since the signs are like, the final answer is positive.

$= \frac{5 \cdot \overset{1}{\cancel{2}}}{\underset{1}{\cancel{2}} \cdot 8 \cdot 1}$ To simplify the fraction, factor 16 as $2 \cdot 8$. Then remove the common factor 2.

$= \frac{5}{8}$

Check: $\frac{5}{8}\left(-\frac{1}{2}\right) = -\frac{5}{16}$ Quotient · divisor = dividend

Teaching Example 5
Divide and check the result: **a.** $\frac{-66}{-11}$ **b.** $\frac{25}{-5}$ **c.** $-3.28 \div 0.4$ **d.** $-\frac{7}{6} \div \left(-\frac{1}{3}\right)$
Answers: **a.** 6 **b.** -5 **c.** -8.2 **d.** $\frac{7}{2}$

Self Check 5 Divide and check the result: **a.** $\frac{-28}{-4}$ 7 **b.** $\frac{75}{-25}$ -3 **c.** $0.32 \div (-1.6)$ -0.2 **d.** $\frac{3}{4} \div \left(-\frac{5}{8}\right)$ $-\frac{6}{5}$

Now Try Problems 55, 59, 63, and 67

EXAMPLE 6

Depreciation. Over an 8-year period, the value of a \$150,000 house fell at a uniform rate to \$110,000. Find the amount of depreciation per year.

The Language of Algebra

Depreciation is a form of the word *depreciate,* meaning to lose value. You've probably heard that the minute you drive a new car off the lot, it has *depreciated.*

Strategy The phrase *uniform rate* means that the value of the house fell the same amount each year, for 8 straight years. We can determine the amount it depreciated in one year (per year) by dividing the total change in value of the house by 8.

Why The process of separating a quantity into equal parts (in this case, the change in the value of the house) indicates division.

Solution First, we find the change in the value of the house.

$110{,}000 - 150{,}000 = -40{,}000$ Subtract the previous value from the current value.

The negative result represents a drop in value of \$40,000. Since the depreciation occurred over 8 years, we divide $-40{,}000$ by 8.

$\frac{-40{,}000}{8} = -5{,}000$ Divide the absolute values, 40,000 by 8, to get 5,000, and make the quotient negative.

The house depreciated \$5,000 per year.

Self Check 6 **Depreciation.** Over a 6-year period, the value of a \$300,000 house fell at a uniform rate to \$286,500. Find the amount of depreciation each year. The house depreciated \$2,250 per year.

Now Try Problem 109

Teaching Example 6 Depreciation. Over an 11-year period, the value of a \$250,000 house fell at a uniform rate to \$233,500. Find the amount of depreciation each year.
Answer: The house depreciated \$1,500 per year.

4 Use Properties of Division.

Whenever we divide a number by 1, the quotient is that number. Therefore, $\frac{12}{1} = 12$, $\frac{-80}{1} = -80$, and $7.75 \div 1 = 7.75$. Furthermore, whenever we divide a nonzero number by itself, the quotient is 1. Therefore, $\frac{35}{35} = 1$, $\frac{-4}{-4} = 1$, and $0.9 \div 0.9 = 1$. These observations suggest the following properties of division.

Division Properties

Any number divided by 1 is the number itself. Any number (except 0) divided by itself is 1.

For any real number a,

$$\frac{a}{1} = a \quad \text{and} \quad \frac{a}{a} = 1 \quad (\text{where } a \neq 0)$$

Caution

Division is *not commutative.* For example, $\frac{6}{3} \neq \frac{3}{6}$ and $\frac{-12}{4} \neq \frac{4}{-12}$. In general, $\frac{a}{b} \neq \frac{b}{a}$.

We will now consider division that involves zero. First, we examine division of zero. Let's look at two examples. We know that

$$\frac{0}{2} = 0 \quad \text{because} \quad 0 \cdot 2 = 0 \quad \text{and} \quad \frac{0}{-5} = 0 \quad \text{because} \quad 0(-5) = 0$$

These examples illustrate that *0 divided by a nonzero number is 0.*

To examine division by zero, let's look at $\frac{2}{0}$ and its related multiplication statement.

$$\frac{2}{0} = ? \quad \text{because} \quad ? \cdot 0 = 2$$

The Language of Algebra

When we say a division by 0, such as $\frac{2}{0}$, is **undefined,** we mean that $\frac{2}{0}$ does not represent a real number.

There is no number that can make $0 \cdot ? = 2$ true because any number multiplied by 0 is equal to 0, not 2. Therefore, $\frac{2}{0}$ does not have an answer. We say that such a division is **undefined.**

These results suggest the following division facts.

Division Involving 0

For any nonzero real number a,

$$\frac{0}{a} = 0 \quad \text{and} \quad \frac{a}{0} \text{ is undefined.}$$

EXAMPLE 7 Find each quotient, if possible: **a.** $\frac{0}{8}$ **b.** $\frac{-24}{0}$

Strategy In each case, we need to determine if we have division *of* 0 or division *by* 0.

Why *Division of 0* by a nonzero number is defined, and the result is 0. However, *division by 0* is undefined; there is no result.

Solution

a. $\frac{0}{8} = 0$ because $0 \cdot 8 = 0$. This is division of 0 by 8.

b. $\frac{-24}{0}$ is undefined. This is division of −24 by 0.

The Language of Algebra

Division of 0 by 0, written $\frac{0}{0}$, is called **indeterminate.** This form is studied in advanced mathematics classes.

Self Check 7 Find each quotient, if possible: **a.** $\frac{4}{0}$ Undefined **b.** $\frac{0}{17}$ 0

Now Try Problems 71 and 73

Teaching Example 7
Find each quotient, if possible: **a.** $\frac{0}{50}$ **b.** $\frac{-6}{0}$
Answers: **a.** 0 **b.** Undefined

SECTION 1.6 STUDY SET

VOCABULARY

Fill in the blanks.

1. The answer to a multiplication problem is called a product. The answer to a division problem is called a quotient.
2. The commutative property of multiplication states that changing the order when multiplying does not affect the answer.
3. The associative property of multiplication states that changing the grouping when multiplying does not affect the answer.
4. Division of a nonzero number by 0 is undefined.

CONCEPTS

Fill in the blanks.

5. a. The product or quotient of two numbers with like signs is positive.
   b. The product or quotient of two numbers with unlike signs is negative.
6. a. The product of an even number of negative numbers is positive.
   b. The product of an odd number of negative numbers is negative.
7. a. $\frac{-9}{3} = -3$ because $-3 \cdot 3 = -9$
   b. $\frac{0}{8} = 0$ because $0 \cdot 8 = 0$
8. Complete each property of multiplication.
   a. $a \cdot b = b \cdot a$
   b. $(ab)c = a(bc)$
   c. $0 \cdot a = 0$
   d. $1 \cdot a = a$
   e. $a\left(\frac{1}{a}\right) = 1$
   f. $-1 \cdot a = -a$
9. Complete each property of division.
   a. $\frac{a}{1} = a$
   b. $\frac{a}{a} = 1$
   c. $\frac{0}{a} = 0$
   d. $\frac{a}{0}$ is Undefined
10. Which property justifies each statement?
   a. $-5(2 \cdot 17) = (-5 \cdot 2)17$ Assoc. prop. of mult.
   b. $-5\left(-\frac{1}{5}\right) = 1$ Mult. inv.
   c. $-5 \cdot 2 = 2(-5)$ Comm. prop. of mult.
   d. $-5(1) = -5$ Mult. prop. of 1
   e. $-5 \cdot 0 = 0$ Mult. prop. of 0
11. For each multiplication or division, just determine the sign of the answer
   a. $-\frac{19}{37}\left(-\frac{51}{75}\right)$ Positive
   b. $\frac{45.568}{-2.56}$ Negative
   c. $-8.2(-4.1)(-6)(-9.3)(-1.5)$ Negative
12. Use multiplication to check this division: $\frac{-29.4}{7} = -4.1$. Is the answer correct? No; $-4.1 \cdot 7 \neq -29.4$
13. Complete each statement using the given property.
   a. $5 \cdot 8 = 8 \cdot 5$ Commutative property of multiplication
   b. $-2(6 \cdot 9) = (-2 \cdot 6)9$ Associative property of multiplication
   c. $5\left(\frac{1}{5}\right) = 1$ Inverse property of multiplication
   d. $1(-20) = -20$ Multiplication property of 1
14. Complete the table.

Number	Opposite (additive inverse)	Reciprocal (multiplicative inverse)
2	-2	$\frac{1}{2}$
$-\frac{4}{5}$	$\frac{4}{5}$	$-\frac{5}{4}$
1.75	-1.75	$\frac{4}{7}$

Let POS stand for a positive number and NEG stand for a negative number. Determine the sign of each result, if possible.

15. a. POS · NEG NEG
   b. POS + NEG Not poss. to tell
   c. POS − NEG POS
   d. $\frac{\text{POS}}{\text{NEG}}$ NEG
16. a. NEG · NEG POS
   b. NEG + NEG NEG
   c. NEG − NEG Not poss. to tell
   d. $\frac{\text{NEG}}{\text{NEG}}$ POS

NOTATION

Write each sentence using symbols.

17. The product of negative four and negative five is twenty. $-4(-5) = 20$
18. The quotient of sixteen and negative eight is negative two. $\frac{16}{-8} = -2$

GUIDED PRACTICE

Multiply. See Example 1.

19. $4(-1)$ -4
20. $6(-1)$ -6
21. $-2 \cdot 8$ -16
22. $-3 \cdot 4$ -12
23. $12(-5)$ -60
24. $(-9)(11)$ -99
25. $3(-22)$ -66
26. $-8 \cdot 9$ -72
27. $1.2(-0.4)$ -0.48
28. $(-3.6)(0.9)$ -3.24
29. $\frac{1}{3}\left(-\frac{3}{4}\right)$ $-\frac{1}{4}$
30. $\left(-\frac{3}{4}\right)\left(\frac{4}{5}\right)$ $-\frac{3}{5}$

Multiply. See Example 2.

31. $(-1)(-7)$ 7
32. $(-2)(-5)$ 10
33. $(-6)(-9)$ 54
34. $(-8)(-7)$ 56

35. $-3(-3)$ 9

36. $-1(-1)$ 1

▶ 37. $63(-7)$ -441

▶ 38. $43(-6)$ -258

39. $-0.6(-4)$ 2.4

▶ 40. $-0.7(-8)$ 5.6

41. $\left(-\frac{7}{8}\right)\left(-\frac{2}{21}\right)$ $\frac{1}{12}$

▶ 42. $\left(-\frac{5}{6}\right)\left(-\frac{2}{15}\right)$ $\frac{1}{9}$

Multiply. See Example 3.

43. $3.3(-4)(-5)$ 66

▶ 44. $(-2.2)(-4)(-5)$ -44

45. $-2(-3)(-4)(-5)(-6)$ -720

46. $-9(-7)(-5)(-3)(-1)$ -945

47. $(-41)(3)(-7)(-1)$ -861

▶ 48. $56(-3)(-4)(-1)$ -672

49. $(-6)(-6)(-6)$ -216

50. $(-5)(-5)(-5)$ -125

Find the reciprocal of each number. Then find the product of the given number and its reciprocal. See Example 4.

51. $\frac{7}{9}$ $\frac{9}{7}$; 1

▶ 52. $-\frac{8}{9}$ $-\frac{9}{8}$; 1

▶ 53. -13 $-\frac{1}{13}$; 1

54. $\frac{1}{8}$ 8; 1

Divide. See Example 5.

55. $-30 \div (-3)$ 10

56. $-12 \div (-2)$ 6

57. $-6 \div (-2)$ 3

▶ 58. $-36 \div (-9)$ 4

59. $\frac{85}{-5}$ -17

▶ 60. $\frac{-84}{7}$ -12

▶ 61. $\frac{-110}{-110}$ 1

▶ 62. $\frac{-200}{-200}$ 1

63. $\frac{-10.8}{1.2}$ -9

64. $\frac{-13.5}{-1.5}$ 9

65. $\frac{0.5}{-100}$ -0.005

▶ 66. $\frac{-1.7}{10}$ -0.17

▶ 67. $-\frac{1}{3} \div \frac{4}{5}$ $-\frac{5}{12}$

68. $-\frac{2}{3} \div \frac{7}{8}$ $-\frac{16}{21}$

▶ 69. $-\frac{9}{16} \div \left(-\frac{3}{20}\right)$ $\frac{15}{4}$

70. $-\frac{4}{5} \div \left(-\frac{8}{25}\right)$ $\frac{5}{2}$

TRY IT YOURSELF

Perform the operations.

71. $\frac{0}{150}$ 0

72. $\frac{0}{-12}$ 0

▶ 73. $\frac{-17}{0}$ Undefined

▶ 74. $\frac{225}{0}$ Undefined

▶ 75. $\frac{24}{-6}$ -4

▶ 76. $\frac{-78}{6}$ -13

77. $\frac{17}{-17}$ -1

78. $\frac{-24}{24}$ -1

▶ 79. $(-2)(-2)(-2)(-2)$ 16

80. $(-3)(-3)(-3)(-3)$ 81

▶ 81. $-3(-4)(0)$ 0

▶ 82. $15(0)(-22)$ 0

83. $\frac{-23.5}{5}$ -4.7

84. $\frac{-337.8}{6}$ -56.3

85. $-5.2 \cdot 100$ -520

▶ 86. $-1.17 \cdot 1{,}000$ $-1{,}170$

▶ 87. $\frac{1}{2}\left(-\frac{1}{3}\right)\left(-\frac{1}{4}\right)$ $\frac{1}{24}$

88. $\frac{1}{3}\left(-\frac{1}{5}\right)\left(-\frac{1}{7}\right)$ $\frac{1}{105}$

89. $\frac{550}{-50}$ -11

90. $\frac{440}{-20}$ -22

91. $-3\frac{3}{8} \div \left(-2\frac{1}{4}\right)$ $1\frac{1}{2}$

▶ 92. $-3\frac{4}{15} \div \left(-2\frac{1}{10}\right)$ $1\frac{5}{9}$

93. $7.2(-2.1)(-2)$ 30.24

94. $4.6(-5.4)(-2)$ 49.68

95. $\frac{1}{2}\left(-\frac{3}{4}\right)$ $-\frac{3}{8}$

96. $\frac{1}{3}\left(-\frac{5}{16}\right)$ $-\frac{5}{48}$

97. $-\frac{16}{25} \div \frac{64}{15}$ $-\frac{3}{20}$

98. $-\frac{15}{16} \div \frac{25}{8}$ $-\frac{3}{10}$

▶ 99. $\frac{-24.24}{-0.8}$ 30.3

100. $\frac{-55.02}{-0.7}$ 78.6

101. $-1\frac{1}{4}\left(-\frac{3}{4}\right)$ $\frac{15}{16}$

▶ 102. $-1\frac{1}{8}\left(-\frac{3}{8}\right)$ $\frac{27}{64}$

Look Alikes . . .

103. a. $2.7 + (-0.9)$ 1.8 b. $2.7 - (-0.9)$ 3.6
c. $2.7(-0.9)$ -2.43 d. $\frac{2.7}{-0.9}$ -3

▶ 104. a. $-\frac{5}{3} + \left(-\frac{9}{25}\right)$ $-\frac{152}{75}$ b. $-\frac{5}{3} - \left(-\frac{9}{25}\right)$ $-\frac{98}{75}$
c. $-\frac{5}{3}\left(-\frac{9}{25}\right)$ $\frac{3}{5}$ d. $-\frac{5}{3} \div \left(-\frac{9}{25}\right)$ $\frac{125}{27}$

Use the associative property of multiplication to find each product.

▶ 105. $-\frac{1}{2}(2 \cdot 67)$ -67

106. $\left(-\frac{5}{16} \cdot \frac{1}{7}\right)7$ $-\frac{5}{16}$

107. $-0.2(-10 \cdot 3)$ 6

108. $-1.5(-100 \cdot 4)$ 600

APPLICATIONS

▶ 109. **Real Estate.** Over a 5-year period, the value of a \$200,000 lot fell at a uniform rate to \$160,000. What signed number indicates the amount of depreciation per year? $-$\$8,000 per year

110. **Tourism.** The ocean liner Queen Mary cost \$22,500,000 to build in 1936. The ship was purchased by the city of Long Beach, California, in 1967 for \$3,450,000. It now serves as a convention center. What signed number indicates the annual average depreciation of the ship over the 31-year period from 1936 to 1967? Round to the nearest dollar. $-$\$614,516

▶ 111. **Fluid Flow.** In a lab, the temperature of a fluid was decreased 6° per hour for 12 hours. What signed number indicates the change in temperature? $-72°$

112. **Stress on the Job.** A health care provider for a company estimates that 75 hours per week are lost by employees suffering from stress-related illness. In one year, how many hours are lost? Use a signed number to answer. $-3{,}900$ hr

113. **Weight Loss.** As a result of a diet, Tom has been steadily losing $4\frac{1}{2}$ pounds per month.

a. Which expression below can be used to determine how much heavier Tom was 8 months ago? ii

i. $-4\frac{1}{2} \cdot 8$ ii. $-4\frac{1}{2}(-8)$
iii. $4\frac{1}{2}(-8)$ iv. $-4\frac{1}{2} - 8$

b. How much heavier was Tom 8 months ago? 36 lb

▶ 114. **Astronomy.** The temperature on Pluto gets as low as -386°F. This is twice as low as the lowest temperature reached on Jupiter. What is the lowest temperature on Jupiter? -193°F

▶ **115. Car Radiators.** The instructions on a container of antifreeze state, "A 50/50 mixture of antifreeze and water protects against freeze-ups down to −34°F, while a 60/40 mix protects against freeze-ups down to one and one-half times that temperature." To what temperature does the 60/40 mixture protect? −51°F

116. Accounting. For 2010, the net income of Rite Aid Corporation (the drugstore chain) was about −$508 million. The previous year, the company's net income was even worse, by a factor of about 5.75. What signed number represents Rite Aid's net income in 2009? (Source: moneycentral .msn.com) −$2,921 million or −$2,921,000,000

▶ **117. Airlines.** In the 2009 income statement for Delta Air Lines below, numbers within parentheses represent a loss. Complete the statement given these facts. The second quarter loss was about 6.4 times the first quarter loss. The fourth quarter loss was about 5 times the second quarter loss. The third quarter loss was about $\frac{5}{16}$ of the fourth quarter loss. (160), (250), (800)

DELTA INCOME STATEMENT				2009
All amounts in millions of dollars	1st Qtr (25)	2nd Qtr (?)	3rd Qtr (?)	4th Qtr (?)

Source: dailyfinance.com

▶ **118. Computers.** The formula = A1*B1*C1 in cell D1 of the spreadsheet instructs the computer to multiply the values in cells A1, B1, and C1 and to print the result *in place of the formula* in cell D1. (The symbol * represents multiplication.) What value will be printed in the cell D1? What values will be printed in cells D2 and D3? 340, −9,240, −40,800

File Edit View Insert Format Tools Data Window

	A	B	C	D
1	4	−5	−17	= A1*B1*C1
2	22	−30	14	= A2*B2*C2
3	−60	−20	−34	= A3*B3*C3
4				
5				

119. Physics. An oscilloscope displays electrical signals as wavy lines on a screen. By switching the magnification dial to ×2, for example, the height of the "peak" and the depth of the "valley" of a graph will be doubled. Use signed numbers to indicate the height and depth of the display for each setting of the dial.

a. normal 5, −10 **b.** ×0.5 2.5, −5
c. ×1.5 7.5, −15 **d.** ×2 10, −20

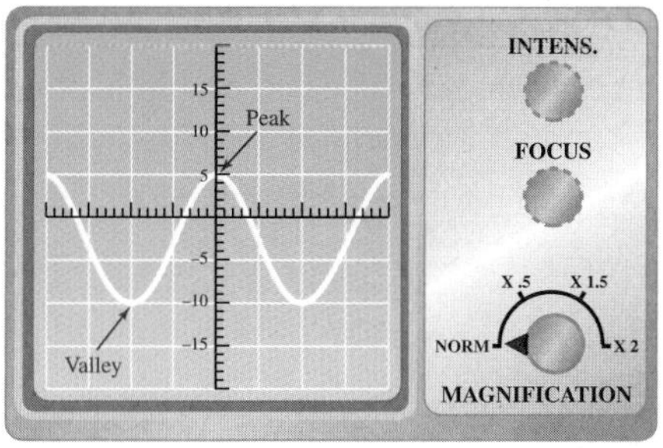

▶ **120. Light.** Water acts as a selective filter of light. In the illustration, we see that red light waves penetrate water only to a depth of about 5 meters. How many times deeper does

a. yellow light penetrate than red light? 6 times
b. green light penetrate than orange light? 4 times
c. blue light penetrate than yellow light? 2.5 times

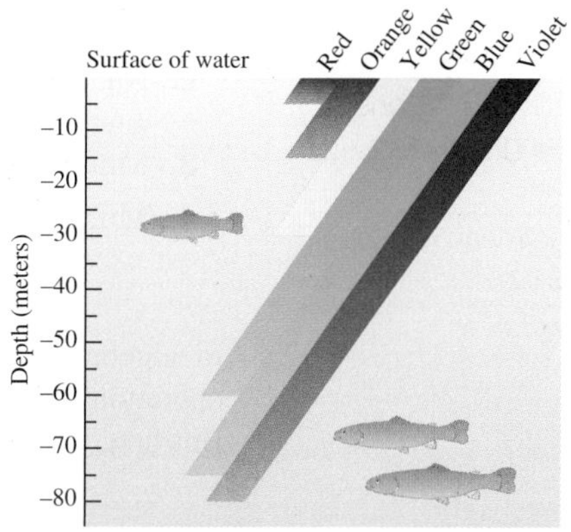

WRITING

121. Explain why $\frac{16}{0}$ is undefined.

122. The commutative property states that changing the order when multiplying does not change the answer. Are the following activities commutative? Explain.

a. Washing a load of clothes; drying a load of clothes
b. Putting on your left sock; putting on your right sock

123. What is wrong with the following statement?

A negative and a positive is a negative.

124. If we multiply two different numbers and the answer is 0, what must be true about one of the numbers? Explain your answer.

REVIEW

125. Add: $-3 + (-4) + (-5) + 4 + 3$ −5

126. Write $-3 - (-5)$ as addition of the opposite. $-3 + 5$

127. Find $\frac{1}{2} + \frac{1}{4} + \frac{1}{3}$. Answer in decimal form. $1.08\overline{3}$

128. Which integers have an absolute value equal to 15? 15, 15

CHALLENGE PROBLEMS

129. If the product of five numbers is negative, how many of them could be negative? Explain. An odd number of factors must be negative.

▶ **130.** Suppose a is a positive number and b is a negative number. Determine whether the given expression is positive or negative.

a. $-a(-b)$ Negative **b.** $\frac{-a}{b}$ Positive

c. $\frac{-a}{a}$ Negative **d.** $\frac{1}{b}$ Negative

SECTION 1.7

Exponents and Order of Operations

OBJECTIVES

1. Evaluate exponential expressions.
2. Use the order of operations rule.
3. Evaluate expressions containing grouping symbols.
4. Find the mean (average).

ARE YOU READY? *Are You Ready? exercises available online at www.webassign.net/brookscole*

The following problems review some basic concepts that are important when working with numerical expressions.

1. Name the operations that are involved in the expression: $50 - 2(3)$ Subtraction, multiplication
2. Name the operations that are involved in the expression: $\dfrac{2 + 6 \cdot 8}{4 + 6}$ Addition, multiplication, division
3. Multiply: $3 \cdot 3 \cdot 3 \cdot 3$ 81
4. Multiply: $(-5)(-5)(-5)$ -125

In algebra, we often have to find the value of expressions that involve more than one operation. In this section, we introduce an order of operations rule to follow in such cases. But first, we discuss a way to write repeated multiplication using *exponents.*

1 Evaluate Exponential Expressions.

In the expression $3 \cdot 3 \cdot 3 \cdot 3 \cdot 3$, the number 3 repeats as a factor five times. We can use **exponential notation** to write this product in a more compact form.

Exponent and Base

An **exponent** is used to indicate repeated multiplication. It is how many times the **base** is used as a factor.

The Language of Algebra

5^2 represents the area of a square with sides 5 units long. 4^3 represents the volume of a cube with sides 4 units long.

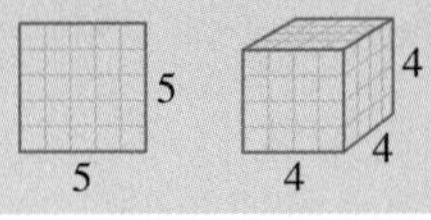

$$\underbrace{3 \cdot 3 \cdot 3 \cdot 3 \cdot 3}_{\text{Five repeated factors of 3.}} = 3^5$$

The exponent is 5.

The base is 3.

In the **exponential expression** 3^5, the base is 3, and 5 is the exponent. The expression is called a power of 3. Some other examples of exponential expressions are:

5^2 Read as "5 to the second power" or "5 squared."

4^3 Read as "4 to the third power" or "4 cubed."

$(-2)^5$ Read as "-2 to the fifth power."

EXAMPLE 1 Write each product using exponents: **a.** $7 \cdot 7 \cdot 7$ **b.** $(-5)(-5)(-5)(-5)(-5)$ **c.** $8 \cdot 8 \cdot 15 \cdot 15 \cdot 15 \cdot 15$ **d.** $a \cdot a \cdot a \cdot a \cdot a \cdot a$ **e.** $4 \cdot \pi \cdot r \cdot r$

Strategy We need to determine the number of repeated factors in the expression.

Why An exponent can be used to represent repeated multiplication.

Solution

a. The factor 7 is repeated 3 times. We can represent this repeated multiplication with an exponential expression having a base of 7 and an exponent of 3: $7 \cdot 7 \cdot 7 = 7^3$.

b. The factor -5 is repeated five times: $(-5)(-5)(-5)(-5)(-5) = (-5)^5$.

c. $8 \cdot 8 \cdot 15 \cdot 15 \cdot 15 \cdot 15 = 8^2 \cdot 15^4$

d. $a \cdot a \cdot a \cdot a \cdot a \cdot a = a^6$

e. $4 \cdot \pi \cdot r \cdot r = 4\pi r^2$

Teaching Example 1
Write each product using exponents:
a. $8 \cdot 8 \cdot 8$
b. $(-20)(-20)(-20)(-20)(-20)$
c. $5 \cdot 5 \cdot 2 \cdot 2 \cdot 2 \cdot 2$
d. $m \cdot m \cdot m \cdot m \cdot m \cdot m$
e. $4 \cdot s \cdot s \cdot s \cdot t$
Answers:
a. 8^3 b. $(-20)^5$ c. $5^2 \cdot 2^4$ d. m^6
e. $4s^3t$

Self Check 1 Write each product using exponents: **a.** $(12)(12)(12)$ 12^3 **b.** $2 \cdot 9 \cdot 9 \cdot 9 \cdot 9$ $2 \cdot 9^4$ **c.** $(-30)(-30)$ $(-30)^2$ **d.** $y \cdot y \cdot y \cdot y \cdot y \cdot y$ y^6 **e.** $8 \cdot b \cdot b \cdot b \cdot c$ $8b^3c$

Now Try Problems 15 and 21

To **evaluate** (find the value of) an exponential expression, we write the base as a factor the number of times indicated by the exponent. Then we multiply the factors.

EXAMPLE 2 Evaluate each expression: **a.** 5^3 **b.** $\left(-\frac{2}{3}\right)^3$ **c.** 10^1 **d.** $(0.6)^2$ **e.** $(-3)^4$ **f.** $(-3)^5$

Strategy We will rewrite each exponential expression as a product of repeated factors, and then perform the multiplication. This requires that we identify the base and the exponent.

Why The exponent tells the number of times the base is to be written as a factor.

Solution

a. $5^3 = 5 \cdot 5 \cdot 5$ Write the base, 5, as a factor 3 times.
$= 125$ Multiply, working left to right. We say 125 is the *cube* of 5.

Because $5^3 = 125$, we say that 125 is a **power** of 5.

b. $\left(-\frac{2}{3}\right)^3 = \left(-\frac{2}{3}\right)\left(-\frac{2}{3}\right)\left(-\frac{2}{3}\right)$ Since $-\frac{2}{3}$ is the base and 3 is the exponent, we write $-\frac{2}{3}$ as a factor 3 times.
$= \frac{4}{9}\left(-\frac{2}{3}\right)$ Work from left to right: $\left(-\frac{2}{3}\right)\left(-\frac{2}{3}\right) = \frac{4}{9}$.
$= -\frac{8}{27}$

c. $10^1 = 10$ The base is 10. Since the exponent is 1, we write the base once.

d. $(0.6)^2 = (0.6)(0.6)$ Write the base, 0.6, as a factor 2 times.
$= 0.36$ We say 0.36 is the *square* of 0.6.

e. $(-3)^4 = (-3)(-3)(-3)(-3)$ Write the base, -3, as a factor 4 times.
$= 9(-3)(-3)$ Work from left to right.
$= -27(-3)$
$= 81$

f. $(-3)^5 = (-3)(-3)(-3)(-3)(-3)$ Write the base, -3, as a factor 5 times.
$= 9(-3)(-3)(-3)$ Work from left to right.
$= -27(-3)(-3)$
$= 81(-3)$
$= -243$

Caution

Don't make the common mistake of multiplying the base and the exponent:

$$5^3 \neq 5 \cdot 3$$

Calculators

Finding a power
The squaring key x^2 can be used to find the square of a number. To raise a number to a power, we use the y^x key on a scientific calculator and the ^ key on a graphing calculator.

The Language of Algebra

A number or a variable has an **understood** exponent of 1. For example,

$$8 = 8^1 \quad \text{and} \quad x = x^1$$

Teaching Example 2
Evaluate: a. 6^3 b. $\left(-\frac{2}{5}\right)^3$ c. 3^1
d. $(0.7)^2$ e. $(-4)^4$ f. $(-4)^5$
Answers:
a. 216 b. $-\frac{8}{125}$ c. 3 d. 0.49
e. 256 f. $-1{,}024$

Self Check 2 Evaluate: **a.** 2^5 32 **b.** $\left(-\frac{3}{4}\right)^3$ $-\frac{27}{64}$ **c.** 9^1 9 **d.** $(-0.3)^2$ 0.09 **e.** $(-6)^2$ 36 **f.** $(-5)^3$ -125

Now Try Problems 23, 29, and 33

In Example 2e, we raised -3 to an even power; the result was positive. In part f, we raised -3 to an odd power; the result was negative. These results illustrate the following rule.

Even and Odd Powers of a Negative Number

When a negative number is raised to an even power, the result is positive.

When a negative number is raised to an odd power, the result is negative.

The Language of Algebra

Read $(-4)^2$ as "negative four squared" and -4^2 as "the opposite of the square of four."

Although the expressions $(-4)^2$ and -4^2 look alike, they are not. When we find the value of each expression, it becomes clear that they are not equivalent.

$$(-4)^2 = (-4)(-4) = 16$$
The base is -4, the exponent is 2.

$$-4^2 = -(4 \cdot 4) = -16$$
The base is 4, the exponent is 2.

Different results

Any real number can be used as a base. However, the base of an exponential expression *does not include* the negative sign unless parentheses are used.

-7^3 Positive base: 7

$(-7)^3$ Negative base: -7

EXAMPLE 3 Evaluate: -2^4

Strategy We will rewrite the expression as a product of repeated factors and then perform the multiplication. We must be careful when identifying the base. It is 2, not -2.

Why Since there are no parentheses around -2, the base is 2.

Solution

$$-2^4 = -(2 \cdot 2 \cdot 2 \cdot 2)$$
Read as "the opposite of the fourth power of two."

$$= -16$$
Do the multiplication within the parentheses to get 16. Then write the opposite of that result.

Teaching Example 3
Evaluate: -8^2
Answer: -64

Self Check 3 Evaluate: -5^4 -625

Now Try Problem 35

2 Use the Order of Operations Rule.

Suppose you have been asked to contact a friend if you see a Rolex watch for sale when you are traveling in Europe. While in Switzerland, you find the watch and send the text message shown on the left. The next day, you get the response shown on the right.

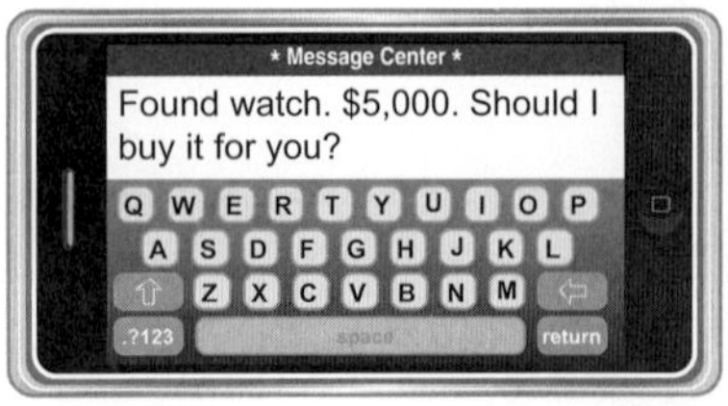

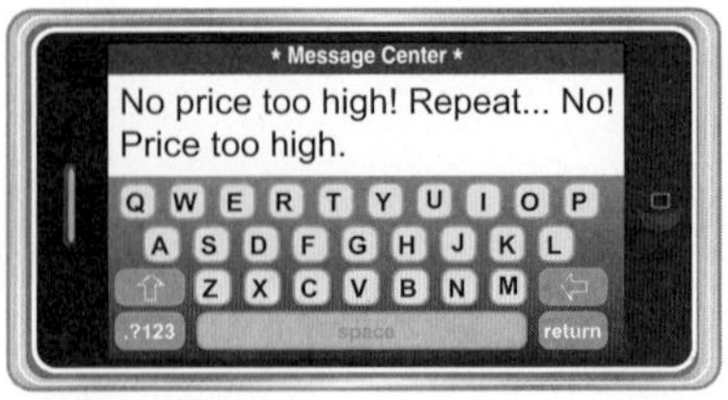

Something is wrong. The first part of the response (No price too high!) says to buy the watch at any price. The second part (No! Price too high.) says not to buy it, because it's too expensive. The placement of the exclamation point makes us read the two parts of the response differently, resulting in different meanings. When reading a mathematical statement, the same kind of confusion is possible. For example, consider the expression

$$2 + 3 \cdot 6$$

We can evaluate this expression in two ways. We can add first, and then multiply. Or we can multiply first, and then add. However, the results are different.

$2 + 3 \cdot 6 = \mathbf{5} \cdot 6$	Add 2 and 3 first.	$2 + \mathbf{3 \cdot 6} = 2 + \mathbf{18}$	Multiply 3 and 6 first.
$= 30$	Multiply 5 and 6.	$= 20$	Add 2 and 18.

Different answers

If we don't establish a uniform order of operations, the expression has two different values. To avoid this possibility, we will always use the following set of priority rules.

Order of Operations

1. Perform all calculations within parentheses and other grouping symbols following the order listed in Steps 2–4 below, working from the innermost pair of grouping symbols to the outermost pair.
2. Evaluate all exponential expressions.
3. Perform all multiplications and divisions as they occur from left to right.
4. Perform all additions and subtractions as they occur from left to right.

When grouping symbols have been removed, repeat Steps 2–4 to complete the calculation.

If a fraction is present, evaluate the expression above and the expression below the bar separately. Then simplify the fraction, if possible.

Teaching Tip: Tell your students to memorize the order of operations rule now because it is used throughout the course.

It isn't necessary to apply all of these steps in every problem. For example, the expression $2 + 3 \cdot 6$ does not contain any parentheses, and there are no exponential expressions. So we look for multiplications and divisions to perform and proceed as follows:

$$2 + \mathbf{3 \cdot 6} = 2 + \mathbf{18} \quad \text{Do the multiplication first.}$$
$$= 20 \quad \text{Do the addition.}$$

EXAMPLE 4 Evaluate: **a.** $3 \cdot 2^3 - 4$ **b.** $-30 - 4 \cdot 5 + 9$ **c.** $24 \div 6 \cdot 2$
d. $160 - 4 + 6(-2)(-3)$

The Language of Algebra

Sometimes, for problems like these, the instruction **simplify** is used instead of *evaluate.*

Strategy We will scan the expression to determine what operations need to be performed. Then we will perform those operations, one-at-a-time, following the order of operations rules.

Why If we don't follow the correct order of operations, the expression can have more than one value.

Solution

a. Three operations need to be performed to evaluate this expression: multiplication, raising to a power, and subtraction. By the order of operations rules, we evaluate 2^3 first.

$$3 \cdot \mathbf{2^3} - 4 = 3 \cdot \mathbf{8} - 4 \quad \text{Evaluate the exponential expression: } 2^3 = 8.$$
$$= 24 - 4 \quad \text{Do the multiplication: } 3 \cdot 8 = 24.$$
$$= 20 \quad \text{Do the subtraction.}$$

b. This expression involves subtraction, multiplication, and addition. The order of operations rule tells us to multiply first.

$$-30 - \mathbf{4 \cdot 5} + 9 = -30 - \mathbf{20} + 9$$ Do the multiplication: $4 \cdot 5 = 20$.

$$= -50 + 9$$ Working from left to right, do the subtraction: $-30 - 20 = -30 + (-20) = -50$.

$$= -41$$ Do the addition.

Caution

A common mistake is to forget to work from left to right and incorrectly perform the multiplication before the division. This produces the wrong answer, 2.

~~$24 \div 6 \cdot 2 = 24 \div 12 = 2$~~

c. Since there are no calculations within parentheses nor are there exponents, we perform the multiplications and divisions as they occur from left to right. The division occurs before the multiplication, so it must be performed first.

$$\mathbf{24 \div 6} \cdot 2 = \mathbf{4} \cdot 2$$ Working left to right, do the division: $24 \div 6 = 4$.

$$= 8$$ Do the multiplication.

d. Although this expression contains parentheses, there are no operations to perform within them. Since there are no exponents, we will perform the multiplications as they occur from left to right.

Teaching Tip: Remind students that they are to copy the entire expression on each line of their solution, not just the portion that they are currently evaluating.

$$160 - 4 + \mathbf{6(-2)}(-3) = 160 - 4 + \mathbf{(-12)}(-3)$$ Do the multiplication, working left to right: $6(-2) = -12$.

$$= 160 - 4 + 36$$ Complete the multiplication: $(-12)(-3) = 36$.

$$= 156 + 36$$ Working left to right, do the subtraction before the addition.

$$= 192$$ Do the addition.

Teaching Example 4
Evaluate: **a.** $3 \cdot 4^2 + 9$ **b.** $-20 - 7 \cdot 8 + 31$ **c.** $30 \div 5 \cdot 2$ **d.** $180 - 3 + 4(-3)(-2)$
Answers: **a.** 57 **b.** −45 **c.** 12 **d.** 201

Self Check 4 Evaluate: **a.** $2 \cdot 3^2 + 17$ 35 **b.** $-40 - 9 \cdot 4 + 10$ −66 **c.** $18 \div 2 \cdot 3$ 27 **d.** $240 - 8 + 3(-2)(-4)$ 256

Now Try Problems 39, 43, 45, and 49

3 Evaluate Expressions Containing Grouping Symbols.

Grouping symbols serve as mathematical punctuation marks. They help determine the order in which an expression is to be evaluated. Examples of grouping symbols are parentheses ( ), brackets [ ], braces { }, absolute value symbols | |, and the fraction bar —.

EXAMPLE 5 Evaluate each expression: **a.** $(6 - 3)^2$ **b.** $5^3 + 2(-8 - 3 \cdot 2)$

Strategy We will perform the operation(s) within the parentheses first. When there is more than one operation to perform within the parentheses, we follow the order of operations rule.

Why This is the first step of the order of operations rule.

Solution **a.** We read $(6 - 3)^2$ as "the square of the quantity of 6 minus 3."

$$\mathbf{(6 - 3)}^2 = \mathbf{3}^2$$ Do the subtraction within the parentheses: $6 - 3 = 3$.

$$= 9$$ Evaluate the exponential expression.

Notation

Multiplication is indicated when a number is next to a parenthesis or bracket.

$5^3 + 2(-8 - 3 \cdot 2)$

b. We begin by performing the operations within the parentheses in the proper order: multiplication first, and then subtraction.

$5^3 + 2(-8 - \mathbf{3 \cdot 2}) = 5^3 + 2(-8 - \mathbf{6})$ Do the multiplication: $3 \cdot 2 = 6$.

$= 5^3 + 2(-14)$ Do the subtraction: $-8 - 6 = -14$.

$= 125 + 2(-14)$ Evaluate 5^3.

$= 125 + (-28)$ Do the multiplication: $2(-14) = -28$.

$= 97$ Do the addition.

Teaching Example 5
Evaluate: **a.** $(14 - 4)^2$
b. $3^3 + 4(-1 - 9 \cdot 4)$
Answers: **a.** 100 **b.** −121

Self Check 5 Evaluate: **a.** $(12 - 6)^3$ 216 **b.** $1^3 + 6(-6 - 3 \cdot 0)$ −35

Now Try Problems 53 and 55

Expressions can contain two or more pairs of grouping symbols. To evaluate the following expression, we begin within the innermost pair of grouping symbols, the parentheses. Then we work within the outermost pair, the brackets.

Innermost pair

$-4[2 + 3(4 - 8^2)] - 2$

Outermost pair

EXAMPLE 6 Evaluate: $-4[2 + 3(4 - 8^2)] - 2$

Strategy We will work within the parentheses first and then within the brackets. At each stage, we follow the order of operations rules.

Why By the order of operations, we must work from the *innermost* pair of grouping symbols to the *outermost.*

Solution

$-4[2 + 3(4 - \mathbf{8^2})] - 2$

$= -4[2 + 3(4 - \mathbf{64})] - 2$ Evaluate the exponential expression within the parentheses: $8^2 = 64$.

$= -4[2 + 3(-60)] - 2$ Do the subtraction within the parentheses: $4 - 64 = 4 + (-64) = -60$.

$= -4[2 + (-180)] - 2$ Do the multiplication within the brackets: $3(-60) = -180$.

$= -4[-178] - 2$ Do the addition within the brackets: $2 + (-180) = -178$.

$= 712 - 2$ Do the multiplication: $-4[-178] = 712$.

$= 710$ Do the subtraction.

The Language of Algebra

When one pair of grouping symbols is inside another pair, we say that those grouping symbols are **nested,** or **embedded.**

Teaching Example 6
Evaluate: $-3[1 + 5(7^2 - 40)] - 26$
Answer: −164

Self Check 6 Evaluate: $-5[4 + 2(5^2 - 15)] - 10$ −130

Now Try Problem 61

EXAMPLE 7 Evaluate: $\dfrac{-3(3+2)+5}{17-3(-4)}$

Strategy We will evaluate the expression above and the expression below the fraction bar separately. Then we will simplify the fraction, if possible.

Why Fraction bars are grouping symbols. They group the numerator and denominator. The expression could be written $[-3(3+2)+5] \div [17-3(-4)]$.

Solution

$$\frac{-3(3+2)+5}{17-3(-4)} = \frac{-3(5)+5}{17-(-12)}$$ In the numerator, do the addition within the parentheses. In the denominator, do the multiplication.

$$= \frac{-15+5}{17+12}$$ In the numerator, do the multiplication. In the denominator, write the subtraction as the addition of the opposite of −12, which is 12.

$$= \frac{-10}{29}$$ Do the additions.

$$= -\frac{10}{29}$$ Write the − sign in front of the fraction: $\frac{-10}{29} = -\frac{10}{29}$. The fraction does not simplify.

Calculators

Order of operations
Calculators have the order of operations built in. A left parenthesis key (and a right parenthesis key) should be used when grouping symbols, including a fraction bar, are needed.

Self Check 7 Evaluate: $\dfrac{-4(-2+8)+6}{8-5(-2)}$ -1

Now Try Problem 73

Teaching Example 7
Evaluate: $\dfrac{-5(-6+9)+4}{1-8(-3)}$
Answer: $-\frac{11}{25}$

EXAMPLE 8 Evaluate: $10|9-15| - 2^5$

Strategy The absolute value bars are grouping symbols. We will perform the calculation within them first.

Why By the order of operations, we must perform all calculations within parentheses and other grouping symbols (such as absolute value bars) first.

Solution

$10|\mathbf{9-15}| - 2^5 = 10|\mathbf{-6}| - 2^5$ Subtract: $9 - 15 = 9 + (-15) = -6$.

$= 10(6) - 2^5$ Find the absolute value: $|-6| = 6$.

$= 10(6) - 32$ Evaluate the exponential expression: $2^5 = 32$.

$= 60 - 32$ Do the multiplication: $10(6) = 60$.

$= 28$ Do the subtraction.

Notation

Multiplication is indicated when a number is next to an absolute value symbol.

$10|9-15| - 2^5$

Self Check 8 Evaluate: $10^3 + 3|24-25|$ 1,003

Now Try Problem 77

Teaching Example 8
Evaluate: $2^5 + 4|16-17|$
Answer: 36

4 Find the Mean (Average).

The **arithmetic mean** (or simply **mean**) of a set of numbers is a value around which the values of the numbers are grouped. The mean is also commonly called the **average.**

Finding an Arithmetic Mean

To find the mean of a set of values, divide the sum of the values by the number of values.

When a value in a set appears more than once, that value has a greater "influence" on the mean than another value that only occurs a single time. To simplify the process of finding a mean, any value that appears more than once can be "weighted" by multiplying it by the number of times it occurs. A mean that is found in this way is called a **weighted mean.**

EXAMPLE 9 **Hotel Reservations.** In an effort to improve customer service, a hotel electronically recorded the number of times the reservation desk telephone rang before it was answered by a receptionist. The results of the week-long survey are shown in the table. Find the average (mean) number of times the phone rang before a receptionist answered.

Number of rings	Number of calls
1	11
2	46
3	45
4	28
5	20

Strategy First, we will determine the total number of times the reservation desk telephone rang during the week. Then we will divide that result by the total number of calls received.

Why To find the *average* value of a set of values, we divide the sum of the values by the number of values.

Solution To find the total number of rings, we multiply each *number of rings* (1, 2, 3, 4, and 5 rings) by the respective number of occurrences and add those subtotals.

$$\text{Total number of rings} = 11(1) + 46(2) + 45(3) + 28(4) + 20(5)$$

The total number of calls received was $11 + 46 + 45 + 28 + 20$. To find the average, we divide the total number of rings by the total number of calls.

$$\text{Average} = \frac{11(1) + 46(2) + 45(3) + 28(4) + 20(5)}{11 + 46 + 45 + 28 + 20}$$

$$= \frac{11 + 92 + 135 + 112 + 100}{150}$$ In the numerator, do the multiplications. In the denominator, do the additions.

$$= \frac{450}{150}$$ Do the addition.

$$= 3$$ Simplify the fraction.

The average number of times the phone rang before it was answered was 3.

Teaching Example 9 Surveys.
For a recent survey, the responses of 1 = satisfied, 2 = no opinion, and 3 = dissatisfied were recorded. The results are shown in the table. Find the average (mean) rating of satisfaction for these responses.

Survey options	Number of responses
1	15
2	1
3	4

Answer: 1.45

Self Check 9 **Evaluations.** On the first question of an instructor's evaluation, 14 students marked 1 for *strongly agree,* 10 students marked 2 for *agree,* 6 students marked 3 for *disagree,* and 4 students marked 4 for *strongly disagree.* What was the average (mean) response for the first question on the evaluation? 2

Now Try Problem 119

SECTION 1.7 STUDY SET

VOCABULARY

Fill in the blanks.

1. In the exponential expression 7^5, 7 is the base, and 5 is the exponent. 7^5 is the fifth power of seven.
2. 10^2 can be read as ten squared, and 10^3 can be read as ten cubed.
3. An exponent is used to represent repeated multiplication.
4. To evaluate the expression $2(-1 + 4^2)$ means to find its value.
5. The rule for the order of operations guarantees that an evaluation of a numerical expression will result in a single answer.
6. To find the arithmetic mean or average of a set of values, divide the sum of the values by the number of values.

CONCEPTS

7. To evaluate each expression, what operation should be performed first?
   a. $24 - 4 + 2$ Subtraction b. $32 \div 8 \cdot 4$ Division
   c. $8 - (3 + 5)^2$ Addition d. $65 \cdot 3^3$ Power

8. To evaluate $\frac{36 - 4(7)}{2(10 - 8)}$, what operation should be performed first in the numerator? In the denominator? Mult.; sub.

NOTATION

9. a. Give the name of each grouping symbol: ( ), [ ], { }, | |, and —. Parentheses, brackets, braces, absolute value symbols, fraction bar
   b. In the expression $-8 + 2[15 - (-6 + 1)]$, which grouping symbols are innermost, and which are outermost? Innermost: parentheses; outermost: brackets

10. What operation is indicated?

    ↓
    $2 + 9|5 - (2 + 4)|$ Multiplication

11. a. In the expression $(-5)^2$, what is the base? -5
    b. In the expression -5^2, what is the base? 5

▶ 12. Write each expression using symbols. Then evaluate it.
   a. Negative two squared $(-2)^2 = 4$
   b. The opposite of the square of two $-2^2 = -4$

Complete the evaluation of each expression.

13. $-19 - 2[(1 + 2)^2 \cdot 3] = -19 - 2[3^2 \cdot 3]$
    $= -19 - 2[9 \cdot 3]$
    $= -19 - 2[27]$
    $= -19 - 54$
    $= -73$

14. $\frac{46 - 2^3}{-3(5) - 4} = \frac{46 - 8}{-15 - 4}$
    $= \frac{38}{-19}$
    $= -2$

GUIDED PRACTICE

Write each product using exponents. See Example 1.

▶ 15. $8 \cdot 8 \cdot 8$ 8^3
16. $(-4)(-4)(-4)(-4)$ $(-4)^4$
▶ 17. $7 \cdot 7 \cdot 7 \cdot 12 \cdot 12$ $7^3 12^2$
18. $5 \cdot 5 \cdot 5 \cdot 5 \cdot 5 \cdot 5 \cdot 7 \cdot 7 \cdot 7$ $5^6 7^3$
19. $x \cdot x \cdot x$ x^3
20. $b \cdot b \cdot b \cdot b$ b^4
▶ 21. $r \cdot r \cdot r \cdot r \cdot s \cdot s$ $r^4 s^2$
22. $m \cdot m \cdot m \cdot n \cdot n \cdot n \cdot n$ $m^3 n^4$

Evaluate each expression. See Example 2.

23. 7^2 49
24. 9^2 81
25. 6^3 216
▶ 26. 6^4 1,296
▶ 27. $(-5)^4$ 625
▶ 28. $(-5)^3$ -125
▶ 29. $(-0.1)^2$ 0.01
30. $(-0.8)^2$ 0.64
31. $\left(-\frac{1}{4}\right)^3$ $-\frac{1}{64}$
32. $\left(-\frac{1}{3}\right)^4$ $\frac{1}{81}$
33. $\left(\frac{2}{3}\right)^3$ $\frac{8}{27}$
▶ 34. $\left(\frac{3}{4}\right)^3$ $\frac{27}{64}$

Evaluate each expression. See Example 3.

▶ 35. $(-6)^2$ and -6^2 36, -36
36. $(-4)^2$ and -4^2 16, -16
37. $(-8)^2$ and -8^2 64, -64
▶ 38. $(-9)^2$ and -9^2 81, -81

Evaluate each expression. See Example 4.

39. $3 - 5 \cdot 4$ -17
40. $-4 \cdot 6 + 5$ -19
41. $32 - 16 \div 4 + 2$ 30
▶ 42. $60 - 20 \div 10 + 5$ 63
▶ 43. $3^2 \cdot 5 - 6 \div 3$ 43
44. $2^3 \cdot 5 - 4 \div 2$ 38
45. $12 \div 3 \cdot 2$ 8
46. $18 \div 6 \cdot 3$ 9
47. $-22 - 15 + 3$ -34
▶ 48. $-33 - 8 + 10$ -31
49. $-2(9) - 2(5)(10)$ -118
▶ 50. $-6(7) - 3(-4)(-2)$ -66

Evaluate each expression. See Example 5.

51. $-4(6 + 5)$ -44
▶ 52. $-3(5 - 4)$ -3
53. $(9 - 3)(9 - 9)^2$ 0
54. $-(-8 - 6)(6 - 6)^2$ 0
55. $(-1 - 3^2 \cdot 4)2^2$ -148
▶ 56. $-1(28 - 5^2 \cdot 2)3^2$ 198
57. $1 + 5(10 + 2 \cdot 5) - 1$ 100
58. $14 + 3(7 - 5 \cdot 3)$ -10

Evaluate each expression. See Example 6.

59. $(-1)^9[-7^2 - (-2)^2]$ 53
▶ 60. $[-9^2 - (-8)^2](-1)^{10}$ -145
61. $64 - 6[15 + 2(-3 + 8)]$ -86
62. $4 - 2[26 + 2(5 - 3)]$ -56
63. $-2[2 + 4^2(8 - 9)]^2$ -392
64. $-3[5 + 3^2(4 - 5)]^2$ -48
▶ 65. $3 + 2[-1 - (4 - 5)]$ 3
66. $4 + 2[-7 - (3 - 9)]$ 2

Evaluate each expression. See Example 7.

67. $\frac{-2 - 5}{-7 + (-7)}$ $\frac{1}{2}$
68. $\frac{-3 - (-1)}{-2 + (-2)}$ $\frac{1}{2}$
▶ 69. $\frac{2 \cdot 2^5 - 60 + (-4)}{5^4 - (-4)(-5)}$ 0
▶ 70. $\frac{(6 - 5)^8 - 1}{(-9)(-3) - 4}$ 0
71. $\frac{2(-4 - 2 \cdot 2)}{3(-3)(-2)}$ $-\frac{8}{9}$
▶ 72. $\frac{3(-3^2 + 2 \cdot 2^2)}{(5 - 8)(7 - 9)}$ $-\frac{1}{2}$
▶ 73. $\frac{72 - (2 - 2 \cdot 4)}{10^2 - (9 \cdot 10 + 2^2)}$ 13
▶ 74. $\frac{13^2 - 5^2}{-3(5 - 3^2)}$ 12

Evaluate each expression. See Example 8.

▶ 75. $10 - 2|4 - 8|$ 2
76. $45 - 5|1 - 8|$ 10
77. $-|7 - 2^3(4 - 7)|$ -31
78. $-|9 - 5(1 - 2^3)|$ -44
▶ 79. $\frac{(3 + 5)^2 + |-2|}{-2(5 - 8)}$ 11
80. $\frac{|-25| - 8(-5)}{2^4 - 29}$ -5
81. $\frac{|6 - 4| + 2|-4|}{226 - 6^3}$ 1
82. $\frac{4|9 - 7| + |-7|}{6^3 - 211}$ 3
83. $-(2 \cdot 3 - 2^2)^5$ -32
▶ 84. $-(3 \cdot 5 - 2 \cdot 6)^4$ -81
85. $2 \cdot 5^2 + 4 \cdot 3^2$ 86
▶ 86. $5 \cdot 3^3 - 4 \cdot 2^3$ 103
87. $-2(-1)^2 + 3(-1) - 3$ -8
88. $-4(-3)^2 + 3(-3) - 1$ -46
▶ 89. $8 - 3[5^2 - (7 - 3)^2]$ -19
90. $3 - [3^3 + (3 - 1)^3]$ -32

TRY IT YOURSELF

Evaluate each expression.

91. $[6(5) - 5(5)]^3(-4)$ -500
92. $5 - 2 \cdot 3^4 - (-6 + 5)^3$ -156
93. $8 - 6[(130 - 4^3) - 2]$ -376
94. $91 - 5[(150 - 3^3) - 1]$ -519

95. $-2\left(\frac{15}{-5}\right) - \frac{6}{2} + 9$ 12

▶ 96. $-6\left(\frac{25}{-5}\right) - \frac{36}{9} + 1$ 27

97. $-5(-2)^3 - |-2 + 1|$ 39

98. $-6(-3)^3 - |-6 + 5|$ 161

99. $\frac{18 - [2 + (1 - 6)]}{16 - (-4)^2}$ Undef.

100. $\frac{6 - [6(-1) - 88]}{4 - 2^2}$ Undef.

101. $-|-5 \cdot 7^2| - 30$ −275

102. $2 + |-3 \cdot 2^2 \cdot 8^2 \cdot 1^2|$ 770

▶ 103. $(-3)^3\left(\frac{-4}{2}\right)(-1)$ −54

104. $(-2)^3\left(\frac{-6}{2}\right)(-1)$ −24

▶ 105. $\frac{1}{2}\left(\frac{1}{8}\right) + \left(-\frac{1}{4}\right)^2$ $\frac{1}{8}$

106. $-\frac{1}{9}\left(\frac{1}{4}\right) + \left(-\frac{1}{6}\right)^2$ 0

107. $\frac{-5^2 \cdot 10 + 5 \cdot 2^5}{-5 - 3 - 1}$ 10

108. $\frac{(-6^2 - 2^4 \cdot 2) + 5}{-4 - 3}$ 9

▶ 109. $-\left(\frac{40 - 1^3 - 2^4}{3(2 + 5) + 2}\right)$ −1

▶ 110. $-\left(\frac{8^2 - 10}{2(3)(4) - 5(3)}\right)$ −6

Look Alikes . . .

111. a. $(-7 - 4)(-2)$ 22　b. $(-7 - 4) - 2$ −13

▶ 112. a. $2 \cdot 3^3$ 54　b. $(2 \cdot 3)^3$ 216

113. a. $-100 \div 5 \cdot 2$ −40　b. $-100 \div (5 \cdot 2)$ −10

114. a. $8 + 3[-2 - (6 + 1)]$ −19　b. $(8 + 3)[-2 - (6 + 1)]$ −99

APPLICATIONS

▶ **115. Light.** As light energy passes through the first unit of area, 1 yard away from the bulb, it spreads out. How much area does that light energy cover 2 yards, 3 yards, and 4 yards from the bulb? Express each answer using exponents. 2^2 square units, 3^2 square units, 4^2 square units

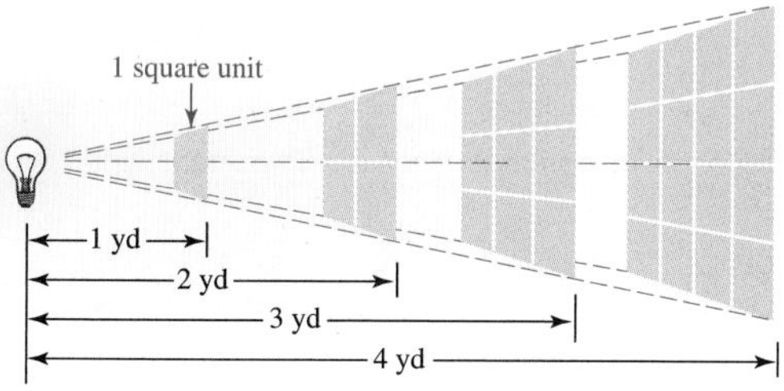

▶ **116. Chain Letters.** A woman sent two friends a letter with the following request: "Please send a copy of this letter to two of your friends." Assume that all those receiving letters responded and that everyone in the chain received just one letter. Complete the table and then determine how many letters will be circulated in the 10th level. $2^{10} = 1{,}024$

Level	Number of letters circulated
1st	$2 = 2^1$
2nd	$4 = 2^2$
3rd	$8 = 2^3$
4th	$16 = 2^4$

117 from **Campus to Careers**

Lead Transportation Security Officer

To determine the average afternoon wait time in security lines at an airport, officials monitored four passengers, each at a different gate. The time that each passenger entered a security line and the time the same passenger cleared the checkpoint was recorded, as shown below. Find the average (mean) wait time for these passengers. 12 min

	Time entered	Time cleared
Passenger at Gate A	3:05 pm	3:21 pm
Passenger at Gate B	3:03 pm	3:13 pm
Passenger at Gate C	3:01 pm	3:09 pm
Passenger at Gate D	3:02 pm	3:16 pm

▶ **118. Energy Usage.** Find the average number of therms of natural gas used per month. 31.5 therms

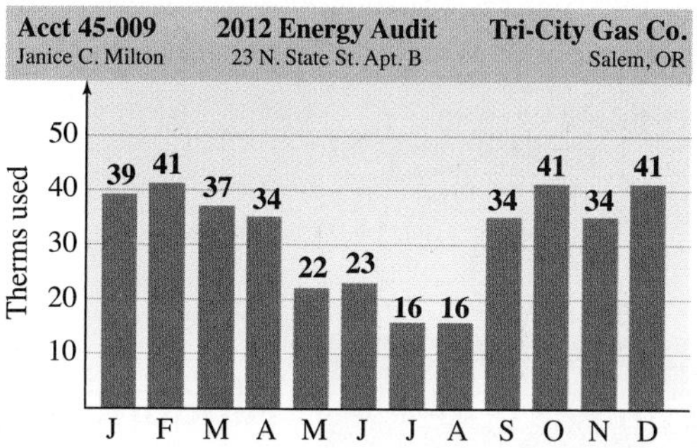

119. Cash Awards. A contest is to be part of a promotional kickoff for a new children's cereal. The prizes to be awarded are shown.

a. How much money will be awarded in the promotion? $11,875

b. What is the average cash prize? $95

YouTube Video Contest

Grand prize: Disney World vacation plus $2,500

Four 1st place prizes of $500
Thirty-five 2nd place prizes of $150
Eighty-five 3rd place prizes of $25

▶ **120. Surveys.** Some students were asked to rate their college cafeteria food on a scale from 1 to 5. The responses are shown on the tally sheet. Find the average rating. 4

Poor 1	2	Fair 3	4	Excellent 5
	|||	|||	卌	卌 ||||

▶ **121. Wrapping Gifts.** How much ribbon is needed to wrap the package if 15 inches of ribbon are needed to make the bow? 81 in.

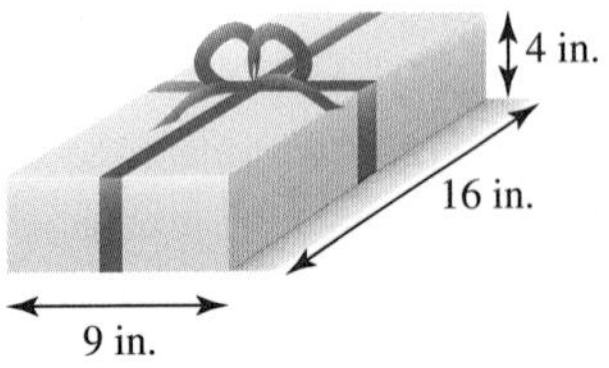

▶ **122. Scrabble.** Write an expression to determine the number of points received for playing the word QUARTZY and then evaluate it. (The number on each tile gives the point value of the letter.) $3[10 + 1 + 1 + 1 + 1 + 2(10) + 4] = 114$

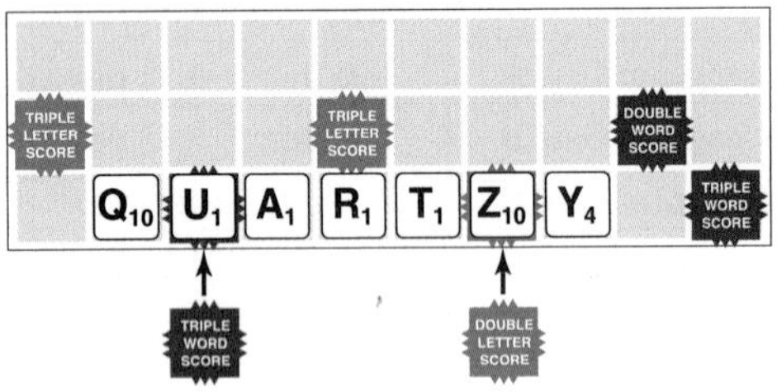

WRITING

123. Explain the difference between 2^3 and 3^2.

124. Why is the order of operations rule necessary?

125. Explain the error. What is the correct answer?

$$40 \div 4 \cdot 2 = 40 \div 8$$
$$= 5$$

126. Explain the error. What is the correct answer?

$$5 + 3(2 - 6) = 5 + 3(-4)$$
$$= 8(-4)$$
$$= -32$$

REVIEW

127. What numbers are a distance of 6 away from -11 on a number line? $-17, -5$

128. Fill in the blank with $>$ or $<$: $0.3 < \frac{1}{3}$

CHALLENGE PROBLEMS

129. Using each of the numbers 2, 3, and 4 only once, what is the greatest value that the following expression can have? 6,561

$$(\square^{\square})^{\square}$$

▶ **130.** Insert a pair of parentheses into $4 \cdot 3^2 - 4 \cdot 2$ so that it has a value of 40. $4 \cdot (3^2 - 4) \cdot 2$

Translate the set of instructions to an expression and then evaluate it.

131. Subtract the sum of -9 and 8 from the product of the cube of -3 and the opposite of 4. 109

132. Increase the square of the reciprocal of -2 by the difference of -0.25 and -1. 1

SECTION 1.8 Algebraic Expressions

OBJECTIVES

1. Identify terms and coefficients of terms.
2. Translate word phrases to algebraic expressions.
3. Analyze problems to determine hidden operations.
4. Evaluate algebraic expressions.

ARE YOU READY? *Are You Ready? exercises available online at www.webassign.net/brookscole*

The following problems review some basic concepts that are important when working with algebraic expressions.

Write each expression in simpler form.

1. $9 \cdot x$ $9x$ **2.** $1 \cdot m$ m **3.** $-1 \cdot t$ $-t$ **4.** $\frac{7}{8} \cdot y$ $\frac{7}{8}y$

Identify each of the following expressions as either a sum, difference, product, or quotient.

5. $\frac{x}{12}$ Quotient **6.** $45 - a$ Difference **7.** $2bc$ Product **8.** $d + 5$ Sum

Since problems in algebra are often presented in words, the ability to interpret what you read is important. In this section, we will introduce several strategies that will help you translate English words into algebraic expressions.

1 Identify Terms and Coefficients of Terms.

Recall that variables and/or numbers can be combined with the operations of arithmetic to create **algebraic expressions.** Addition symbols separate expressions into parts called *terms*. For example, the expression $x + 8$ has two terms.

$$x + 8$$

First term Second term

Since subtraction can be written as addition of the opposite, the expression $a^2 - 3a - 9$ has three terms.

$$a^2 - 3a - 9 = \underset{\text{First term}}{a^2} + \underset{\text{Second term}}{(-3a)} + \underset{\text{Third term}}{(-9)}$$

In general, a **term** is a product or quotient of numbers and/or variables. A single number or variable is also a term. Examples of terms are:

Notation

By the commutative property of multiplication, $r6 = 6r$ and $-15b^2a = -15ab^2$. However, we usually write the numerical factor first and the variable factors in alphabetical order.

$$4, \quad y, \quad 6r, \quad -w^3, \quad 3.7x^5, \quad \frac{3}{n}, \quad -15ab^2$$

The numerical factor of a term is called the **coefficient** of the term. For instance, the term $6r$ has a coefficient of 6 because $6r = 6 \cdot r$. The coefficient of $-15ab^2$ is -15 because $-15ab^2 = -15 \cdot ab^2$. More examples are shown below.

A term such as 4, that consists of a single number, is called a **constant term.**

The Language of Algebra

Terms such as x and y have **implied** coefficients of 1. *Implied* means suggested without being precisely expressed.

Term	Coefficient	
$8y^2$	8	
$-0.9pq$	-0.9	
$\frac{3}{4}b$	$\frac{3}{4}$	This term also could be written as $\frac{3b}{4}$.
$-\frac{x}{6}$	$-\frac{1}{6}$	Because $-\frac{x}{6} = -\frac{1x}{6} = -\frac{1}{6} \cdot x$
x	1	Because $x = 1x$
$-t$	-1	Because $-t = -1t$
27	27	

EXAMPLE 1 Identify the coefficient of each term in the expression: $7x^2 - x + 6$

Strategy We will begin by writing the subtraction as addition of the opposite. Then we will determine the numerical factor of each term.

Why Addition symbols separate expressions into terms.

Solution If we write $7x^2 - x + 6$ as $7x^2 + (-x) + 6$, we see that it has three terms: $7x^2$, $-x$, and 6. The numerical factor of each term is its coefficient.

- The coefficient of $7x^2$ is **7** because $7x^2$ means $\mathbf{7} \cdot x^2$.
- The coefficient of $-x$ is $\mathbf{-1}$ because $-x$ means $\mathbf{-1} \cdot x$.
- The coefficient of the constant 6 is 6.

Teaching Example 1 Identify the coefficient of each term in the expression: $-3x^2 + x - 5$
Answer:
$-3, 1, -5$

Self Check 1 Identify the coefficient of each term in the expression: $p^3 - 12p^2 + 3p - 4$ $\quad 1, -12, 3, -4$

Now Try Problem 19

It is important to be able to distinguish between the *terms* of an expression and the *factors* of a term.

EXAMPLE 2 Is m used as a *factor* or a *term* in each expression? **a.** $m + 6$ **b.** $8m$

Strategy We will begin by determining whether m is involved in an addition or a multiplication.

Why Addition symbols separate expressions into *terms.* A *factor* is a number being multiplied.

Solution

a. Since m is added to 6, m is a term of $m + 6$.

b. Since m is multiplied by 8, m is a factor of $8m$.

Teaching Example 2 Is a used as a factor or a term in each expression?
a. $32a$ **b.** $a + 9$
Answers:
a. Factor **b.** Term

Self Check 2 Is b used as a *factor* or a *term* in each expression?
a. $-27b$ Factor **b.** $5a + b$ Term

Now Try Problems 21 and 23

2 Translate Word Phrases to Algebraic Expressions.

The four tables below show how key phrases can be translated into algebraic expressions.

Caution

Be careful when translating subtraction. Order is important. For example, when a translation involves the phrase *less than,* note how the terms are reversed as we translate from English to mathematical symbols.

18 less than w

$w - 18$

Addition	
the sum of a and 8	$a + 8$
4 plus c	$4 + c$
16 added to m	$m + 16$
4 more than t	$t + 4$
20 greater than F	$F + 20$
T increased by r	$T + r$
exceeds y by 35	$y + 35$

Subtraction	
the difference of 23 and P	$23 - P$
550 minus h	$550 - h$
18 less than w	$w - 18$
7 decreased by j	$7 - j$
M reduced by x	$M - x$
12 subtracted from L	$L - 12$
5 less f	$5 - f$

Multiplication	
the product of 4 and x	$4x$
20 times B	$20B$
twice r	$2r$
double the amount a	$2a$
triple the profit P	$3P$
three-fourths of m	$\frac{3}{4}m$

Division	
the quotient of R and 19	$\frac{R}{19}$
s divided by d	$\frac{s}{d}$
the ratio of c to d	$\frac{c}{d}$
k split into 4 equal parts	$\frac{k}{4}$

Be careful when translating division. As with subtraction, order is important. For example, s divided by d is *not* written $\frac{d}{s}$.

EXAMPLE 3 Write each phrase as an algebraic expression:

a. one-half of the profit P **b.** 5 less than the capacity c

c. the product of the weight w and 2,000, increased by 300

Strategy We will begin by identifying any key phrases.

Why Key phrases can be translated to mathematical symbols.

Solution

a. Key phrase: *One-half of* **Translation:** multiplication by $\frac{1}{2}$

The algebraic expression is: $\frac{1}{2}P$.

Caution

$5 < c$ is the translation of the statement 5 ***is*** *less than the capacity c*. It is not the translation of 5 *less than the capacity c*.

b. Key phrase: *less than* **Translation:** subtraction

Sometimes thinking in terms of specific numbers makes translating easier. Suppose the capacity was 100. Then 5 *less than* 100 would be $100 - 5$. If the capacity is c, then we need to make it 5 less. The algebraic expression is: $c - 5$.

The Language of Algebra

The word **of** often indicates multiplication, as in $\frac{1}{2}$ of 8, but it also is used in phrases that indicate other operations, such as the sum *of* 8 and c, or the difference *of* x and 1.

c. **Key phrase:** *product of* **Translation:** multiplication
Key phrase: *increased by* **Translation:** addition

In the given wording, the comma after 2,000 means w is first multiplied by 2,000; then 300 is added to that product. The algebraic expression is: $2{,}000w + 300$.

If there is no comma, the phrase *the product of the weight w and 2,000 increased by 300* translates to: $w(2{,}000 + 300)$.

Teaching Example 3
Write each phrase as an algebraic expression:
a. 17 less than the weight w
b. $\frac{3}{4}$ of the population p
c. the difference of twice d and 9, cubed.
Answers: **a.** $w - 17$ **b.** $\frac{3}{4}p$ **c.** $(2d - 9)^3$

Self Check 3 Write each phrase as an algebraic expression:
a. 80 less than the total t $t - 80$ **b.** $\frac{2}{3}$ of the time T $\frac{2}{3}T$
c. the difference of twice a and 15, squared $(2a - 15)^2$

Now Try Problems 25, 31, and 35

To solve application problems, we often let a variable represent an unknown quantity.

EXAMPLE 4 **Swimming.** A pool is to be sectioned into 8 equally wide swimming lanes. Write an algebraic expression that represents the width of each lane.

Strategy We will begin by letting $x =$ the width of the swimming pool in feet. Then we will identify any key phrases.

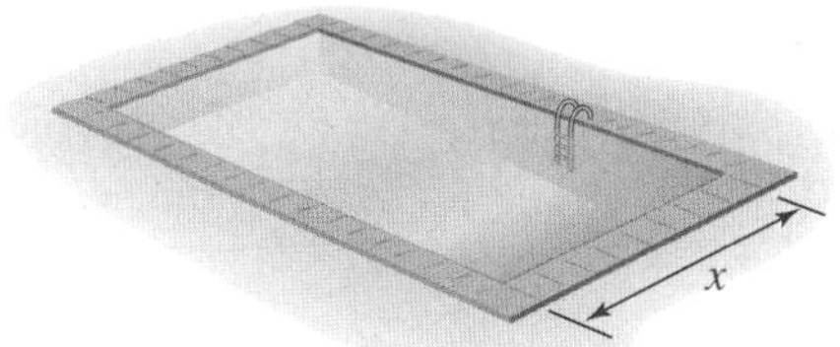

Why The width of the pool is unknown.

Solution The key phrase, *sectioned into 8 equally wide lanes,* indicates division.

Therefore, the width of each lane is $\frac{x}{8}$ feet.

Teaching Example 4 Educational costs. The semester tuition cost for a total of 15 credit hours is x dollars. Write an expression that represents the cost of each credit hour.
Answer: $\frac{x}{15}$

Self Check 4 **Commuting.** It takes Val m minutes to get to work by bus. If she drives her car, her travel time exceeds this by 15 minutes. How many minutes does it take her to get to work by car? $(m + 15)$ min

Now Try Problem 61

EXAMPLE 5 **Painting.** A 10-inch-long paintbrush has two parts: a handle and bristles. Choose a variable to represent the length of one of the parts. Then write an expression to represent the length of the other part.

Strategy There are two approaches. We can let $h =$ the length of the handle or we can let $b =$ the length of the bristles.

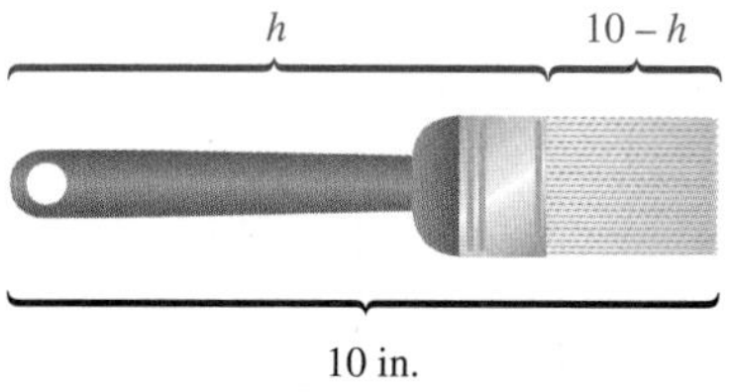

Why Both the length of the handle and the length of the bristles are unknown.

Solution Refer to the drawing on the top. If we let $h =$ the length of the handle (in inches), then the length of the bristles is $10 - h$.

Now refer to the drawing on the bottom. If we let $b =$ the length of the bristles (in inches), then the length of the handle is $10 - b$.

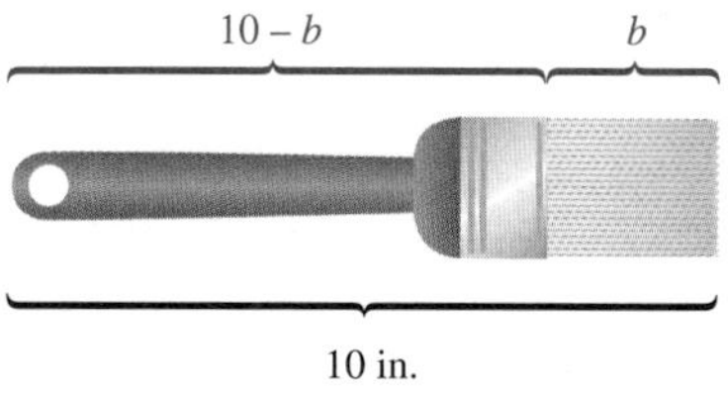

Teaching Example 5 Investing. Part of a \$2,000 investment is to be invested in an account paying $3\frac{1}{2}\%$, and the rest in an account paying 4%. Choose a variable to represent the amount invested at $3\frac{1}{2}\%$. Write an expression that represents the amount invested at 4%.
Answer: x = amount invested at $3\frac{1}{2}\%$, $2{,}000 - x$ = amount invested at 4%

Self Check 5 **Scholarships.** Part of a \$900 donation to a college went to the scholarship fund, the rest to the building fund. Choose a variable to represent the amount donated to one of the funds. Then write an expression that represents the amount donated to the other fund. s = amount donated to scholarship fund in dollars; $900 - s$ = amount donated to building fund in dollars

Now Try Problem 13

EXAMPLE 6 **Enrollments.** Second semester enrollment in a nursing program was 32 more than twice that of the first semester. Let x represent the enrollment for one of the semesters. Write an expression that represents the enrollment for the other semester.

Strategy We will begin by letting x = the enrollment for the first semester.

Why Because the second-semester enrollment is related to the first-semester enrollment.

Solution

Key phrase: *more than* **Translation:** addition

Key phrase: *twice that* **Translation:** multiplication by 2

The second semester enrollment was $2x + 32$.

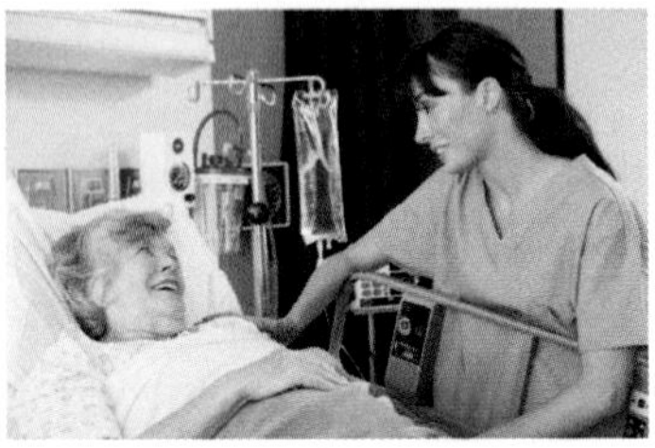

Teaching Example 6 Tuition. The cost of the tuition at a small private college is \$500 more than 3 times the tuition at a state university. Write an expression that represents the tuition cost at the small private college.
Answer:
x = the tuition at the state university, $3x + 500$ = the tuition at the small private college

Self Check 6 **Politics.** In an election, the incumbent received 55 fewer votes than three times the challenger's votes. Let x represent the number of votes received by one candidate. Write an expression that represents the number of votes received by the other. x = number of votes received by the challenger; $3x - 55$ = number of votes received by the incumbent

Now Try Problem 109

3 Analyze Problems to Determine Hidden Operations.

Many applied problems require insight and analysis to determine which mathematical operations to use.

EXAMPLE 7 **Vacations.** Disneyland, in California, was in operation 16 years before the opening of Disney World in Florida. Euro Disney, in France, was constructed 21 years after Disney World. Write algebraic expressions to represent the ages (in years) of each Disney attraction.

Strategy We will begin by letting x = the age of Disney World.

Why The ages of Disneyland and Euro Disney are both related to the age of Disney World.

Solution In carefully reading the problem, we see that Disneyland was built 16 years before Disney World. That makes its age 16 years more than that of Disney World. The key phrase *more than* indicates addition.

Attraction	Age
Disneyland	$x + 16$
Disney World	x
Euro Disney	$x - 21$

$x + 16$ = the age of Disneyland

Euro Disney was built 21 years *after* Disney World. That makes its age 21 years less than that of Disney World. The key phrase *less than* indicates subtraction.

$x - 21$ = the age of Euro Disney

Teaching Example 7 Ages. Tom is 3 years older than Joan. Carly is 5 years younger than Joan. Write algebraic expressions to represent the ages of Tom, Joan, and Carly.
Answer:
Tom: $x + 3$
Joan: x
Carly: $x - 5$

Self Check 7 **Tax Forms.** Kayla worked 5 more hours preparing her tax return than she did on her daughter's return. Kayla's son's return took her 2 more hours to prepare than her daughter's. Write expressions to represent the hours she spent on each return. Daughter's: x; Kayla's: $x + 5$; Son's: $x + 2$

Now Try Problem 111

EXAMPLE 8 How many months are in x years?

Strategy There are no key phrases so we must analyze the problem carefully. We will begin by considering some specific cases.

Why It's often easier to work with specifics first to get a better understanding of the relationship between the two quantities. Then we can generalize using a variable.

Solution Let's calculate the number of months in 1 year, 2 years, and 3 years. When we write the results in a table, a pattern is apparent.

The number of months in x years is $12 \cdot x$ or $12x$.

Number of years	Number of months
1	12
2	24
3	36
x	$12x$

We multiply the number of years by 12 to find the number of months.

Teaching Example 8 How many seconds are in x minutes?
Answer: $60x$ seconds

Self Check 8 How many days is h hours? $\frac{h}{24}$ days

Now Try Problems 7 and 67

In some problems, we must distinguish between *the number of* and *the value of* the unknown quantity. For example, to find the value of 3 quarters, we multiply the number of quarters by the value (in cents) of one quarter. Therefore, the value of 3 quarters is $3 \cdot 25$ cents $= 75$ cents.

The same distinction must be made if the number is unknown. For example, the value of n nickels is not n cents. The value of n nickels is $n \cdot 5$ cents $= 5n$ cents. For problems of this type, we will use the relationship

Number · value = total value

EXAMPLE 9 Find the total value of: **a.** five dimes **b.** q quarters **c.** $x + 1$ half-dollars

Teaching Tip: Give several other examples to illustrate the number-value relationship. For instance, hold up a pencil and tell your students that it costs c cents. Ask them what a dozen pencils would cost.

Strategy To find the total value (in cents) of each collection of coins, we multiply the number of coins by the value (in cents) of one coin, as shown in the table.

Why Number · value = total value

Solution

Type of coin	Number	Value	Total value	
Dime	5	10	50	Multiply: $5 \cdot 10 = 50$.
Quarter	q	25	$25q$	Multiply: $q \cdot 25$ can be written $25q$.
Half-dollar	$x + 1$	50	$50(x + 1)$	Multiply: $(x + 1) \cdot 50$ can be written $50(x + 1)$.

Teaching Example 9 Find the value of:
a. 7 nickels **b.** t \$50 bills
c. $x + 3$ \$20 bills
Answers:
a. 35¢ **b.** \$50t **c.** \$20($x + 3$)

Self Check 9 Find the value of: **a.** six \$50 savings bonds \$300
b. t \$100 savings bonds \$100t
c. $x - 4$ \$1,000 savings bonds \$1,000($x - 4$)

Now Try Problems 14 and 69

4 Evaluate Algebraic Expressions.

To evaluate an algebraic expression, we substitute given numbers for each variable and perform the necessary calculations in the proper order.

EXAMPLE 10 Evaluate each expression for $x = 3$ and $y = -4$: **a.** $y^3 + y^2$ **b.** $-y - x$ **c.** $|5xy - 7|$ **d.** $\frac{y - 0}{x - (-1)}$

Strategy We will replace each x and y in the expression with the given value of the variable, and evaluate the expression using the order of operation rule.

Why To *evaluate an expression* means to find its numerical value, once we know the value of its variable(s).

Solution

Caution

When replacing a variable with its numerical value, we must often write the replacement number within parentheses to convey the proper meaning.

a. $y^3 + y^2 = (-4)^3 + (-4)^2$ Substitute −4 for each y. We must write −4 within parentheses so that it is the base of each exponential expression.

$= -64 + 16$ Evaluate each exponential expression.

$= -48$ Do the addition.

b. $-y - x = -(-4) - 3$ Substitute −4 for y and 3 for x. Don't forget to write the − sign in front of (−4).

$= 4 - 3$ Simplify: $-(-4) = 4$.

$= 1$ Do the subtraction.

c. $|5xy - 7| = |5(3)(-4) - 7|$ Substitute 3 for x and −4 for y.

$= |-60 - 7|$ Do the multiplication: $5(3)(-4) = -60$.

$= |-67|$ Do the subtraction: $-60 - 7 = -60 + (-7) = -67$.

$= 67$ Find the absolute value of −67.

d. $\frac{y - 0}{x - (-1)} = \frac{-4 - 0}{3 - (-1)}$ Substitute 3 for x and −4 for y.

$= \frac{-4}{4}$ In the denominator, do the subtraction: $3 - (-1) = 3 + 1 = 4$.

$= -1$ Simplify the fraction.

Teaching Example 10
Evaluate each expression for $x = -5$ and $y = 3$:
a. $x^3 + x^2$ b. $-x - y$
c. $|4xy - 12|$ d. $\frac{y - 15}{x - (-3)}$
Answers: a. −100 b. 2 c. 72 d. 6

Self Check 10 Evaluate each expression for $a = -2$ and $b = 5$:
a. $|a^3 + b^2|$ 17 **b.** $-a + 2ab$ −18 **c.** $\frac{a + 2}{b - 3}$ 0

Now Try Problems 79 and 91

EXAMPLE 11

Rocketry. If a toy rocket is shot into the air with an initial velocity of 80 feet per second, its height (in feet) after t seconds in flight is approximated by $-16t^2 + 80t$. How many seconds after the launch will it hit the ground?

Strategy We can substitute positive values for t, the time in flight, until we find the one that gives a height of 0.

Why When the height of the rocket is 0, it is on the ground.

Solution We begin by finding the height after the rocket has been in flight for 1 second ($t = 1$).

$$-16t^2 + 80t = -16(\mathbf{1})^2 + 80(\mathbf{1}) \quad \text{Substitute 1 for } t.$$
$$= 64$$

Peter Barrett/Shutterstock.com

As we evaluate $-16t^2 + 80t$ for several more values of t, we record each result in a table. The columns of the table can also be headed with the terms **input** and **output.** The values of t are the inputs into the expression $-16t^2 + 80t$, and the resulting values are the outputs.

t	$-16t^2 + 80t$
1	64
2	96
3	96
4	64
5	0

Evaluate for $t = 2$: $-16t^2 + 80t = -16(2)^2 + 80(2) = 96$
Evaluate for $t = 3$: $-16t^2 + 80t = -16(3)^2 + 80(3) = 96$
Evaluate for $t = 4$: $-16t^2 + 80t = -16(4)^2 + 80(4) = 64$
Evaluate for $t = 5$: $-16t^2 + 80t = -16(5)^2 + 80(5) = 0$

Input	Output
1	64
2	96
3	96
4	64
5	0

The height of the rocket is 0 when $t = 5$. The rocket will hit the ground 5 seconds after being launched.

Teaching Example 11 The height of the rocket is given by the expression $160t - 16t^2$. Complete the table to find out how many seconds after launch it would hit the ground.

t	$160t - 16t^2$
1	144
5	400
10	0

Answer: 10 seconds

Self Check 11 In Example 11, suppose the height of the rocket is given by $-16t^2 + 112t$. What will be the height of the rocket 6 seconds after launch? 96 ft

Now Try Problem 97

SECTION 1.8 STUDY SET

VOCABULARY

Fill in the blanks.

1. Variables and/or numbers can be combined with the operations of arithmetic to create algebraic expressions.
2. A term is a product or quotient of numbers and/or variables. Examples are: $8x$, $\frac{t}{2}$, and $-cd^3$.
3. Addition symbols separate algebraic expressions into parts called terms.
4. A term, such as 27, that consists of a single number is called a constant term.
5. The coefficient of the term $10x$ is 10.
6. To evaluate $4x - 3$ for $x = 5$, we substitute 5 for x and perform the necessary calculations.

CONCEPTS

7. Complete the table below on the left to determine the number of days in w weeks.
8. Complete the table below on the right to determine the number of minutes in s seconds.

Number of weeks	Number of days
1	7
2	14
3	21
w	$7w$

Number of seconds	Number of minutes
60	1
120	2
180	3
s	$\frac{s}{60}$

9. The knife shown below is 12 inches long. Write an expression that represents the length (in inches) of the blade. $12 - h$

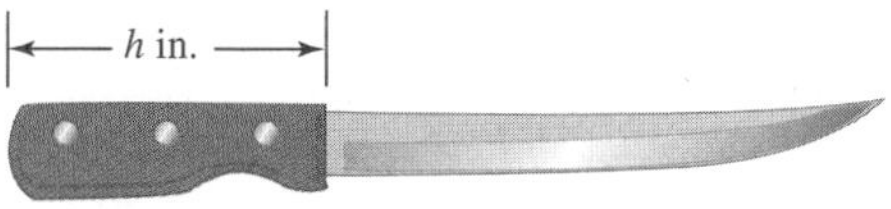

10. A student inherited $5,000 and deposits x dollars in American Savings. Write an expression that represents the number of dollars left to deposit in a City Mutual account. $5{,}000 - x$

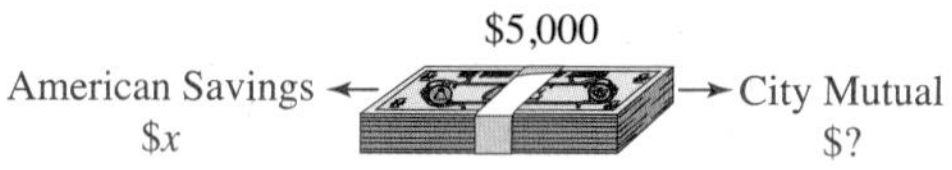

11. Solution 2 is poured into solution 1. Write an expression that represents the number of ounces in the mixture. $x + 20$

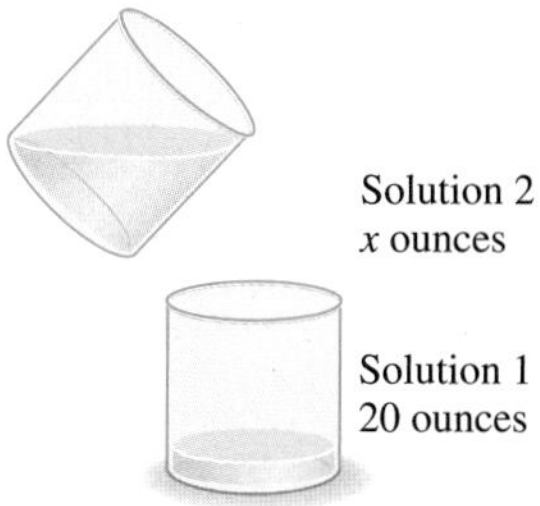

12. Peanuts were mixed with c pounds of cashews to make 100 pounds of a mixture. Write an expression that represents the number of pounds of peanuts that were used. $100 - c$

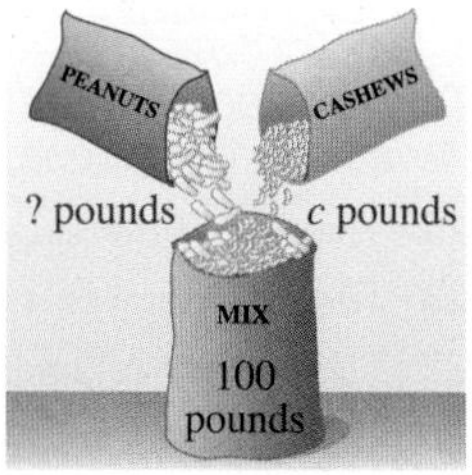

13. a. Let b = the length of the beam shown below (in feet). Write an expression that represents the length of the pipe. $b - 15$

    b. Let p = the length of the pipe (in feet). Write an expression that represents the length of the beam. $p + 15$

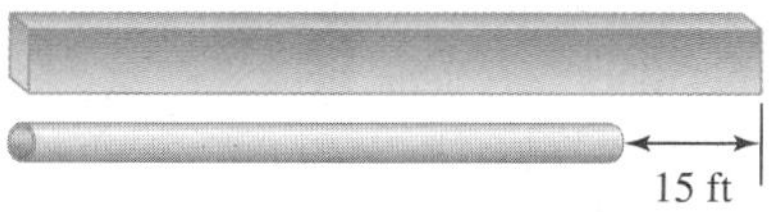

14. Complete the table. Give each value in cents.

Coin	Number ·	Value =	Total value
Nickel	6	5	30
Dime	d	10	$10d$
Half-dollar	$x + 5$	50	$50(x + 5)$

NOTATION

Complete each solution. Evaluate each expression for a = 5, x = −2, and y = 4.

15. $9a - a^2 = 9(5) - (5)^2$
$= 9(5) - 25$
$= 45 - 25$
$= 20$

16. $-x + 6y = -(-2) + 6(4)$
$= 2 + 24$
$= 26$

▶ 17. Write each term in standard form.
a. $y8$ $8y$
b. $d2c$ $2cd$
c. What property of multiplication did you use? Commutative

18. Fill in the blanks.
a. $\frac{w}{2} = \frac{1}{2}w$
b. $\frac{2}{3}m = \frac{2m}{3}$

GUIDED PRACTICE

See Example 1.

19. Consider the expression $3x^3 + 11x^2 - x + 9$.
a. How many terms does the expression have? 4
b. What is the coefficient of each term? 3, 11, −1, 9

▶ 20. Complete the following table.

Term	$6m$	$-75t$	w	$\frac{1}{2}bh$	$\frac{x}{5}$	t
Coefficient	6	−75	1	$\frac{1}{2}$	$\frac{1}{5}$	1

***Determine whether the variable c is used as a factor or as a term.* See Example 2.**

21. $c + 32$ Term
▶ 22. $-24c + 6$ Factor
23. $5c$ Factor
▶ 24. $a + b + c$ Term

***Translate each phrase to an algebraic expression. If no variable is given, use x as the variable.* See Example 3.**

▶ 25. The sum of the length l and 15 $l + 15$
▶ 26. The difference of a number and 10 $x - 10$
27. The product of a number and 50 $50x$
▶ 28. Three-fourths of the population p $\frac{3}{4}p$
29. The ratio of the amount won w and lost l $\frac{w}{l}$
30. The tax t added to c $c + t$
31. P increased by two-thirds of p $P + \frac{2}{3}p$
▶ 32. 21 less than the total height h $h - 21$
33. The square of k, minus 2,005 $k^2 - 2{,}005$
▶ 34. s subtracted from S $S - s$
▶ 35. 1 less than twice the attendance a $2a - 1$
36. J reduced by 500 $J - 500$
37. 1,000 split n equal ways $\frac{1{,}000}{n}$
▶ 38. Exceeds the cost c by 25,000 $c + 25{,}000$
39. 90 more than twice the current price p $2p + 90$
▶ 40. 64 divided by the cube of y $\frac{64}{y^3}$
41. 3 times the total of 35, h, and 300 $3(35 + h + 300)$
42. Decrease x by −17 $x - (-17)$
▶ 43. 680 fewer than the entire population p $p - 680$
44. Triple the number of expected participants $3x$
45. The product of d and 4, decreased by 15 $4d - 15$
▶ 46. The quotient of y and 6, cubed $\left(\frac{y}{6}\right)^3$
▶ 47. Twice the sum of 200 and t $2(200 + t)$
▶ 48. The square of the quantity 14 less than x $(x - 14)^2$
49. The absolute value of the difference of a and 2 $|a - 2|$
▶ 50. The absolute value of a, decreased by 2 $|a| - 2$
51. One-tenth of the distance d $0.1d$ or $\frac{1}{10}d$
52. Double the difference of x and 18 $2(x - 18)$

***Translate each algebraic expression into an English phrase. (Answers may vary.)* See Example 3.**

53. $\frac{3}{4}r$ Three-fourths of r
54. $\frac{2}{3}d$ Two-thirds of d
▶ 55. $t - 50$ 50 less than t
56. $c + 19$ 19 more than c
57. xyz The prod. of x, y, and z
58. $10ab$ The prod. of 10, a, and b
▶ 59. $2m + 5$ Twice m, incr. by 5
60. $2s - 8$ Twice s, decr. by 8

***Answer with an algebraic expression.* See Example 4.**

▶ 61. A model's skirt is x inches long. The designer then lets the hem down 2 inches. What is the length (in inches) of the altered skirt? $x + 2$
▶ 62. A soft drink manufacturer produced c cans of cola during the morning shift. Write an expression for how many six-packs of cola can be assembled from the morning shift's production. $\frac{c}{6}$
▶ 63. The tag on a new pair of 36-inch-long jeans warns that after washing, they will shrink x inches in length. What is the length (in inches) of the jeans after they are washed? $36 - x$
64. A caravan of b cars, each carrying 5 people, traveled to the state capital for a political rally. How many people were in the caravan? $5b$

***Answer with an algebraic expression.* See Example 8.**

65. How many minutes are there in h hours? $60h$
▶ 66. How many feet are in y yards? $3y$
67. How many feet are in i inches? $\frac{i}{12}$
▶ 68. How many centuries in y years? $\frac{y}{100}$

***Answer with an algebraic expression.* See Example 9.**

69. A sales clerk earns \$$x$ an hour; how much does he earn in an 8-hour day? \$$8x$
70. A cashier earns \$$d$ an hour; how much does she earn in a 40-hour week? \$$40d$
▶ 71. If a car rental agency charges 49¢ a mile, what is the rental fee if a car is driven x miles? $49x$¢
72. If one egg is worth e cents, find the value (in cents) of one dozen eggs. $12e$¢
73. A ticket to a concert costs \$$t$. What would a pair of concert tickets cost? \$$2t$
▶ 74. If one apple is worth a cents, find the value (in cents) of 20 apples. $20a$¢
▶ 75. Tickets to a circus cost \$25 each. What will tickets cost for a family of x people if they also pay for two of their neighbors? \$$25(x + 2)$
▶ 76. A certain type of office desk that used to sell for \$$x$ is now on sale for \$50 off. What will a company pay if it purchases 80 of the desks? \$$80(x - 50)$

***Evaluate each expression, for x = 3, y = −2, and z = −4.* See Example 10.**

77. $-y$ 2
▶ 78. $-z$ 4

79. $-z + 3x$ 13

80. $-y - 5x$ −13

81. $3y^2 - 6y - 4$ 20

82. $-z^2 - z - 12$ −24

83. $(3 + x)y$ −12

84. $(4 + z)y$ 0

85. $(x + y)^2 - |z + y|$ −5

86. $[(z - 1)(z + 1)]^2$ 225

87. $\dfrac{2x + y^3}{y + 2z}$ $-\frac{1}{5}$

88. $\dfrac{2z^2 - x}{2x - y^2}$ $-\frac{29}{2}$

Evaluate each expression. See Example 10.

89. $b^2 - 4ac$ for $a = -1, b = 5$, and $c = -2$ 17

90. $(x - a)^2 + (y - b)^2$ for $x = -2, y = 1, a = 5$, and $b = -3$ 65

91. $a^2 + 2ab + b^2$ for $a = -5$ and $b = -1$ 36

92. $\dfrac{a - x}{y - b}$ for $x = -2, y = 1, a = 5$, and $b = 2$ −7

93. $\dfrac{n}{2}[2a + (n - 1)d]$ for $n = 10, a = -4.2$, and $d = 6.6$ 255

94. $\dfrac{a(1 - r^n)}{1 - r}$ for $a = -5, r = 2$, and $n = 3$ −35

95. $(27c^2 - 4d^2)^3$ for $c = \frac{1}{3}$ and $d = \frac{1}{2}$ 8

96. $\dfrac{-b^2 + 16a^2 + 1}{2}$ for $a = \frac{1}{4}$ and $b = -10$ −49

Complete each table. See Example 11.

97.

x	$x^3 - 1$
0	−1
−1	−2
−3	−28

98.

g	$g^2 - 7g + 1$
0	1
7	1
−10	171

99.

s	$\frac{5s + 36}{s}$
1	41
6	11
−12	2

100.

a	$2{,}500a + a^3$
2	5,008
4	10,064
−5	−12,625

101.

Input x	Output $2x - \frac{x}{2}$
100	150
−300	−450

102.

Input x	Output $\frac{x}{3} + \frac{x}{4}$
12	7
−36	−21

103.

x	$(x + 1)(x + 5)$
−1	0
−5	0
−6	5

104.

x	$\frac{1}{x + 8}$
−7	1
−9	−1
−8	Undefined

TRY IT YOURSELF

Look Alikes . . .

Translate each phrase to mathematical symbols. Let x represent the unknown number.

105. a. The sum of a number and 7 squared $x + 7^2$

b. The sum of a number and 7, squared $(x + 7)^2$

106. a. 19 less than a number $x - 19$

b. 19 is less than a number $19 < x$

107. a. 4 times a number increased by 2 $4(x + 2)$

b. 4 times a number, increased by 2 $4x + 2$

108. a. Twice a number decreased by 3 $2(x - 3)$

b. Twice a number, decreased by 3 $2x - 3$

APPLICATIONS

109. **Vehicle Weights.** A Hummer H2 weighs 340 pounds less than twice a Honda Element.

a. Let x represent the weight of one of the vehicles. Write an expression for the weight of the other vehicle. Let x = weight of the Element, $2x - 340$ = weight of the Hummer

b. If the weight of the Element is 3,370 pounds, what is the weight of the Hummer? 6,400 lb

110. **Sod Farms.** The expression $20{,}000 - 3s$ gives the number of square feet of sod that are left in a field after s strips have been removed. Suppose a city orders 7,000 strips of sod. Evaluate the expression and explain the result. −1,000 means the sod farm is short 1,000 ft^2 to fill the city's order.

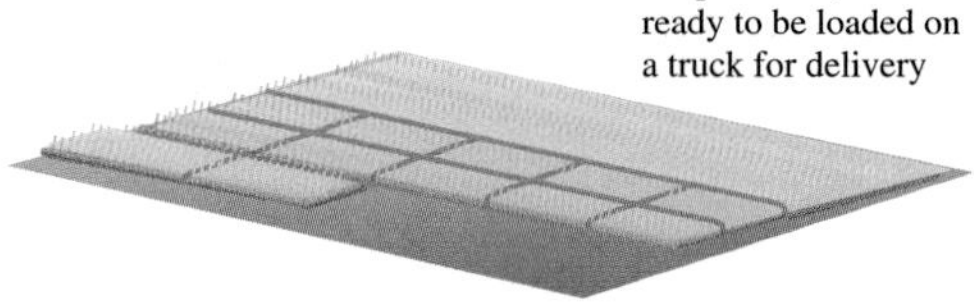

Strips of sod, cut and ready to be loaded on a truck for delivery

111. **Computer Companies.** IBM was founded 80 years before Apple Computer. Dell Computer Corporation was founded 9 years after Apple.

a. Let x represent the age (in years) of one of the companies. Write expressions to represent the ages (in years) of the other two companies. Let x = age of Apple; $x + 80$ = age of IBM; $x - 9$ = age of Dell

b. On April 1, 2008, Apple Computer Company was 32 years old. How old were the other two computer companies then? IBM: 112 yr; Dell: 23 yr

112. **Thrill Rides.** The distance in feet that an object will fall in t seconds is given by the expression $16t^2$. Find the distance that riders on "Drop Zone" will fall during the times listed in the table.

©sahua d/Shutterstock

Time (seconds)	Distance (feet)
1	16
2	64
3	144
4	256

WRITING

113. What is an algebraic expression? Give some examples.
114. Explain why 2 *less than* x does not translate to $2 < x$.
115. In this section, we substituted a number for a variable. List some other uses of the word *substitute* that you encounter in everyday life.
116. Explain why d dimes are not worth d¢.

REVIEW

117. Find the LCD for $\frac{5}{12}$ and $\frac{1}{15}$. 60
118. Simplify: $\frac{3 \cdot 3 \cdot 5}{3 \cdot 5 \cdot 5 \cdot 11}$ $\frac{3}{55}$
119. Evaluate: $\left(\frac{2}{3}\right)^3$ $\frac{8}{27}$
120. Find the result when $\frac{7}{8}$ is multiplied by its reciprocal. 1

CHALLENGE PROBLEMS

121. Evaluate: $(8 - 1)(8 - 2)(8 - 3) \ldots (8 - 49)(8 - 50)$
0: $(8 - 8) = 0$ is a factor of the expression.
122. Translate to an expression: The sum of a number decreased by six, and seven more than the quotient of triple the number and five. $(x - 6) + \left(\frac{3x}{5} + 7\right)$

SECTION 1.9

Simplifying Algebraic Expressions Using Properties of Real Numbers

OBJECTIVES

1 Use the commutative and associative properties.
2 Simplify products.
3 Use the distributive property.
4 Identify like terms.
5 Combine like terms.

ARE YOU READY? *Are You Ready? exercises available online at www.webassign.net/brookscole*

The following problems review some basic concepts that are important when simplifying expressions using properties of real numbers.

1. How do the expressions $3 + x$ and $x + 3$ differ? The terms are in a different order.
2. How do the expressions $1 + (7 + x)$ and $(1 + 7) + x$ differ? The position of the parentheses is different.
3. Evaluate $5(2 + 4)$ and $5 \cdot 2 + 5 \cdot 4$ and compare the results. 30, 30; same result
4. How do the terms $4x$ and $4y$ differ? What do they have in common? Different variable factors (x and y); the same coefficient (4)

In algebra, we frequently replace one algebraic expression with another that is equivalent and simpler in form. That process, called *simplifying an algebraic expression,* often involves the use of one or more properties of real numbers.

1 Use the Commutative and Associative Properties.

Recall the commutative and associative properties discussed in Sections 1.4 and 1.6.

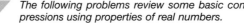

Commutative Properties

Changing the order when adding or multiplying does not affect the answer.

$a + b = b + a$ and $ab = ba$

Associative Properties

Changing the grouping when adding or multiplying does not affect the answer.

$(a + b) + c = a + (b + c)$ and $(ab)c = a(bc)$

These properties can be applied when working with algebraic expressions that involve addition and multiplication.

EXAMPLE 1 Use the given property to complete each statement.

a. $9 + x =$ ______ *Commutative property of addition*

b. $t \cdot 5 =$ ______ *Commutative property of multiplication*

c. $(a + 92) + 8 =$ ______ *Associative property of addition*

d. $6(10n) =$ ______ *Associative property of multiplication*

Strategy For problems like these, it is important to memorize the properties by name. To fill in each blank, we will determine the way in which the property enables us to rewrite the given expression.

Why We should memorize the properties by name because their names remind us how to use them.

Solution

a. To *commute* means to go back and forth. The commutative property of addition enables us to change the order of the terms, 9 and x.

$$9 + x = \underline{x + 9} \quad \text{Commutative property of addition}$$

b. We change the order of the factors, t and 5.

$$t \cdot 5 = \underline{5 \cdot t} \quad \text{Commutative property of multiplication}$$

c. To *associate* means to group together. The associative property of addition enables us to group the terms in a different way.

$$(a + 92) + 8 = \underline{a + (92 + 8)} \quad \text{Associative property of addition}$$

d. We change the grouping of the factors, 6, 10, and n.

$$6(10n) = \underline{(6 \cdot 10)n} \quad \text{Associative property of multiplication}$$

Success Tip

In part c, note that the order of the terms does not change. In part d, note that the order of the factors does not change.

Teaching Example 1
Use the given property to complete each statement.
a. $24 + r =$ ______ Comm. prop. of add.
b. $a \cdot 33 =$ ______ Comm. prop. of mult.
c. $(x + 37) + 3 =$ ______ Assoc. prop. of add.
d. $9(4d) =$ ______ Assoc. prop. of mult.
Answers: a. $r + 24$ b. $33a$ c. $x + (37 + 3)$ d. $(9 \cdot 4)d$

Self Check 1 Use the given property to complete each statement.

a. $5x + (3x + 1) = \underline{(5x + 3x) + 1}$ *Associative property of addition*

b. $a(-15) = \underline{-15a}$ *Commutative property of multiplication*

Now Try Problems 15 and 17

2 Simplify Products.

The commutative and associative properties of multiplication can be used to simplify certain products. For example, let's simplify $8(4x)$.

$$\begin{aligned} 8(4x) &= 8 \cdot (4 \cdot x) && \text{Rewrite } 4x \text{ as } 4 \cdot x. \\ &= (8 \cdot 4) \cdot x && \text{Use the associative property of multiplication to group 4 with 8.} \\ &= 32x && \text{Do the multiplication within the parentheses.} \end{aligned}$$

We have found that $8(4x) = 32x$. We say that $8(4x)$ and $32x$ are **equivalent expressions** because for each value of x, they represent the same number. For example, both expressions have the value 320 if $x = 10$, and both have the value -96 if $x = -3$.

If $x = 10$

$$\begin{aligned} 8(4x) &= 8[4(\mathbf{10})] & 32x &= 32(\mathbf{10}) \\ &= 8(40) & &= 320 \\ &= 320 \end{aligned}$$

If $x = -3$

$$\begin{aligned} 8(4x) &= 8[4(-3)] & 32x &= 32(-3) \\ &= 8(-12) & &= -96 \\ &= -96 \end{aligned}$$

EXAMPLE 2 Multiply: **a.** $-9 \cdot 3b$ **b.** $15a(6)$ **c.** $3(7p)(-5)$ **d.** $\frac{8}{3} \cdot \frac{3}{8}r$ **e.** $35\left(\frac{4}{5}x\right)$

Strategy We will use the commutative and associative properties of multiplication to reorder and regroup the factors in each expression.

Why We want to group all of the numerical factors of an expression together so that we can find their product.

Solution

a. $-9 \cdot 3b = (-9 \cdot 3)b$ Use the associative property of multiplication to group −9 and 3.

$= -27b$ Do the multiplication within the parentheses.

b. $15a(6) = 15(6)a$ Use the commutative property of multiplication to reorder the factors.

$= 90a$ Do the multiplication, working from left to right: 15(6) = 90.

c. $3(7p)(-5) = [3(7)(-5)]p$ Use the commutative and associative properties of multiplication to reorder and regroup the factors.

$= -105p$ Do the multiplication within the brackets.

d. $\frac{8}{3} \cdot \frac{3}{8}r = \left(\frac{8}{3} \cdot \frac{3}{8}\right)r$ Use the associative property of multiplication to regroup the factors.

$= 1r$ Multiply within the parentheses. The product of a number and its reciprocal is 1.

$= r$ The coefficient 1 need not be written.

e. $35\left(\frac{4}{5}x\right) = \left(35 \cdot \frac{4}{5}\right)x$ Use the associative property of multiplication to regroup the factors.

$= \left(\frac{\overset{1}{\cancel{5}} \cdot 7 \cdot 4}{\underset{1}{\cancel{5}}}\right)x$ Factor 35 as 5 · 7 and then remove the common factor 5.

$= 28x$

Success Tip

By the commutative property of multiplication, we can *change* the order of factors. By the associative property of multiplication, we can change the *grouping* of factors.

Teaching Tip: Verify several of these results. Have a student select a value for the variable and evaluate the given expression and the simplified expression. Show students that the values are the same.

Self Check 2 Multiply: **a.** $9 \cdot 6s$ $54s$ **b.** $-4(6u)(-2)$ $48u$ **c.** $\frac{2}{3} \cdot \frac{3}{2}m$ m **d.** $36\left(\frac{2}{9}y\right)$ $8y$

Now Try Problems 23, 29, 31, and 33

Teaching Example 2
Multiply: a. $6 \cdot 5x$ b. $20n(7)$ c. $-2(-3m)(-9)$ d. $\frac{15}{16} \cdot \frac{16}{15}t$ e. $72\left(\frac{7}{8}b\right)$
Answers: a. $30x$ b. $140n$ c. $-54m$ d. t e. $63b$

3 Use the Distributive Property.

Another property that is often used to simplify algebraic expressions is the **distributive property.** To introduce it, we will evaluate 4(5 + 3) in two ways.

Use the order of operations:	*Distribute the multiplication:*
$4(5 + 3) = 4(8)$	$4(5 + 3) = 4 \cdot 5 + 4 \cdot 3$
$= 32$	$= 20 + 12$
	$= 32$

Each method gives a result of 32. This observation suggests the following property.

The Language of Algebra

To **distribute** means to give from one to several. You have probably *distributed* candy to children coming to your door on Halloween.

The Distributive Property

For any real numbers a, b, and c,

$a(b + c) = ab + ac$ Read as "a times the quantity of b plus c."

The Language of Algebra

Formally, it is called the **distributive property of multiplication over addition.** When we use it to write a product, such as $5(x + 2)$, as a sum, $5x + 10$, we say that we have **removed** or **cleared** the parentheses.

To illustrate one use of the distributive property, let's consider the expression $5(x + 3)$. *Since we are not given the value of x, we cannot add x and 3 within the parentheses.* However, we can distribute the multiplication by the factor of 5 that is outside the parentheses to x and to 3 and add those products.

$$5(x + 3) = 5 \cdot x + 5 \cdot 3 \quad \text{Distribute the multiplication by 5.}$$
$$= 5x + 15 \quad \text{Do the multiplication.}$$

EXAMPLE 3 Multiply: **a.** $8(m + 9)$ **b.** $-12(4t + 1)$ **c.** $6\left(\frac{x}{3} + \frac{9}{2}\right)$

Strategy In each case, we will distribute the multiplication by the factor *outside* the parentheses over each term *within* the parentheses.

Why In each case, we cannot simplify the expression within the parentheses. To multiply, we must use the distributive property.

Solution

The Language of Algebra

We read $8(m + 9)$ as "eight times the **quantity** of m plus nine." The word *quantity* alerts us to the grouping symbols in the expression.

a. $8(m + 9) = 8 \cdot m + 8 \cdot 9$ Read as "8 times the quantity of m plus 9." Distribute the multiplication by 8.

$= 8m + 72$ Do the multiplication.

b. $-12(4t + 1) = -12 \cdot 4t + -12 \cdot 1$ Distribute the multiplication by −12.

$= -48t + (-12)$ Do the multiplication.

$= -48t - 12$ Write the result in simpler form. Recall that adding −12 is the same as subtracting 12.

c. $6\left(\frac{x}{3} + \frac{9}{2}\right) = 6 \cdot \frac{x}{3} + 6 \cdot \frac{9}{2}$ Distribute the multiplication by 6.

$= \frac{2 \cdot \overset{1}{\cancel{3}} \cdot x}{\underset{1}{\cancel{3}}} + \frac{\overset{1}{\cancel{2}} \cdot 3 \cdot 9}{\underset{1}{\cancel{2}}}$ Factor 6 as $2 \cdot 3$ and then remove the common factors 3 and 2.

$= 2x + 27$

Teaching Example 3
Multiply: **a.** $8(y + 4)$
b. $-15(2x + 6)$ **c.** $45\left(\frac{m}{9} + \frac{4}{5}\right)$
Answers: **a.** $8y + 32$ b. $-30x - 90$
c. $5m + 36$

Self Check 3 Multiply: **a.** $7(m + 2)$ $7m + 14$ **b.** $-80(8x + 3)$ $-640x - 240$
c. $24\left(\frac{y}{6} + \frac{3}{8}\right)$ $4y + 9$

Now Try Problems 35, 37, and 39

Since subtraction is the same as adding the opposite, the distributive property also holds for subtraction.

$$a(b - c) = ab - ac$$

EXAMPLE 4 Multiply: **a.** $3(3b - 4)$ **b.** $-6(-3y - 8)$ **c.** $-1(t - 9)$

Strategy In each case, we will distribute the multiplication by the factor *outside* the parentheses over each term *within* the parentheses.

Why In each case, we cannot simplify the expression within the parentheses. To multiply, we must use the distributive property.

Solution a. $3(3b - 4) = 3 \cdot 3b - 3 \cdot 4$ Distribute the multiplication by 3.

$= 9b - 12$ Do the multiplication.

Caution

A common mistake is to forget to distribute the multiplication over each of the terms within the parentheses.

$3(3b - 4) = 9b - 4$

b. $-6(-3y - 8) = -6(-3y) - (-6)(8)$ Distribute the multiplication by −6.

$= 18y - (-48)$ Do the multiplication.

$= 18y + 48$ Write the result in simpler form. Add the opposite of −48.

Another approach is to write the subtraction within the parentheses as addition of the opposite. Then we distribute the multiplication by −6 over the addition.

$-6(-3y - 8) = -6[-3y + (-8)]$ Add the opposite of 8.

$= -6(-3y) + (-6)(-8)$ Distribute the multiplication by −6.

$= 18y + 48$ Do the multiplication.

Success Tip

Notice that distributing the multiplication by −1 *changes the sign* of each term within the parentheses.

c. $-1(t - 9) = -1(t) - (-1)(9)$ Distribute the multiplication by −1.

$= -t - (-9)$ Do the multiplication.

$= -t + 9$ Write the result in simpler form. Add the opposite of −9.

Self Check 4 Multiply: **a.** $5(2x - 1)$ $10x - 5$ **b.** $-9(-y - 4)$ $9y + 36$ **c.** $-1(c - 22)$ $-c + 22$

Now Try Problems 43, 47, and 49

Teaching Example 4
Multiply: a. $7(7a - 6)$
b. $-30(-x - 8)$ c. $-1(n - 87)$
Answers: a. $49a - 42$
b. $30x + 240$ c. $-n + 87$

Caution

The distributive property applies only to expressions in which multiplication is distributed over addition (or subtraction). For example, to simplify $6(5x)$, do not use the distibutive property.

Correct

$6(5x) = (6 \cdot 5)x = 30x$

Incorrect

$6(5x) = 30 \cdot 6x = 180x$

The distributive property can be extended to several other useful forms. Since multiplication is commutative, we have:

$(b + c)a = ba + ca$ $(b - c)a = ba - ca$

For situations in which there are more than two terms within parentheses, we have:

$a(b + c + d) = ab + ac + ad$ $a(b - c - d) = ab - ac - ad$

EXAMPLE 5 Multiply: **a.** $(6x + 4)\frac{1}{2}$ **b.** $2(a - 3b)8$ **c.** $-0.3(3a - 4b + 7)$

Strategy We will multiply each term within the parentheses by the factor (or factors) outside the parentheses.

Why In each case, we cannot simplify the expression within the parentheses. To multiply, we must use the distributive property.

Solution a. $(6x + 4)\frac{1}{2} = (6x)\frac{1}{2} + (4)\frac{1}{2}$ Distribute the multiplication by $\frac{1}{2}$.

$= 3x + 2$ Do the multiplication.

b. $2(a - 3b)8 = 2 \cdot 8(a - 3b)$ Use the commutative property of multiplication to reorder the factors.

$= 16(a - 3b)$ Multiply 2 and 8 to get 16.

$= 16a - 48b$ Distribute the multiplication by 16.

c. $-0.3(3a - 4b + 7) = -0.3(3a) - (-0.3)(4b) + (-0.3)(7)$

$= -0.9a + 1.2b - 2.1$ Do each multiplication.

Teaching Example 5
Multiply: **a.** $(-16r - 40)\frac{1}{4}$
b. $3(x - 2y)12$
c. $-0.3(4a + b - 9)$
Answers: **a.** $-4r - 10$
b. $36x - 72y$ **c.** $-1.2a - 0.3b + 2.7$

Self Check 5 Multiply: **a.** $(-6x - 24)\frac{1}{3}$ $-2x - 8$ **b.** $6(c - 2d)9$ $54c - 108d$
c. $-0.7(2r + 5s - 8)$ $-1.4r - 3.5s + 5.6$

Now Try Problems 53, 55, and 57

We can use the distributive property to find the opposite of a sum. For example, to find $-(x + 10)$, we interpret the $-$ symbol as a factor of -1, and proceed as follows:

$-(x + 10) = -1(x + 10)$ Replace the $-$ symbol with -1.

$= -1(x) + (-1)(10)$ Distribute the multiplication by -1.

$= -x - 10$ Do the multiplication.

In general, we have the following property of real numbers.

The Opposite of a Sum

The opposite of a sum is the sum of the opposites.
For any real numbers a and b,

$$-(a + b) = -a + (-b)$$

EXAMPLE 6 Simplify: $-(-9s - 3)$

Strategy We will multiply each term within the parentheses by -1.

Why The $-$ outside the parentheses represents a factor of -1 that is to be distributed.

Solution

$-(-9s - 3) = -1(-9s - 3)$ Replace the $-$ symbol in front of the parentheses with -1.

$= -1(-9s) - (-1)(3)$ Distribute the multiplication by -1.

$= 9s + 3$ Do the multiplication.

Success Tip

Notice that the $-$ symbol in front of the parentheses changes the sign of each term within the parentheses.

Teaching Example 6 Simplify: $-(-4a - 5)$
Answer: $4a + 5$

Self Check 6 Simplify: $-(-5x + 18)$ $5x - 18$

Now Try Problem 59

4 Identify Like Terms.

Before we can discuss methods for simplifying algebraic expressions involving addition and subtraction, we need to introduce some new vocabulary.

Like Terms

Like terms are terms containing exactly the same variables raised to exactly the same powers. Any constant terms in an expression are considered to be like terms. Terms that are not like terms are called **unlike terms**.

Success Tip

When looking for like terms, don't look at the coefficients of the terms. Consider only the variable factors of each term. If two terms are like terms, only their coefficients may differ.

Here are several examples.

Like terms		*Unlike terms*	
$4x$ and $7x$	Same variable	$4x$ and $7y$	The variables are not the same.
$-10p^2$ and $25p^2$	Same variable to the same power	$-10p$ and $25p^2$	Same variable, but different powers
$\frac{1}{3}c^3d$ and c^3d	Same variables to the same powers	$\frac{1}{3}c^3d$ and c^3	The variables are not the same.

EXAMPLE 7 List the like terms in each expression: **a.** $7r + 5 + 3r$ **b.** $6x^4 - 6x^2 - 6x$ **c.** $-17m^3 + 3 - 2 + m^3$

Strategy First, we will identify the terms of the expression. Then we will look for terms that contain the same variables raised to exactly the same powers.

Why If two terms contain the same variables raised to the same powers, they are like terms.

Solution

a. $7r + 5 + 3r$ contains the like terms $7r$ and $3r$.

b. Since the exponents on x are different, $6x^4 - 6x^2 - 6x$ contains no like terms.

c. $-17m^3 + 3 - 2 + m^3$ contains two pairs of like terms: $-17m^3$ and m^3 are like terms, and the constant terms, 3 and -2, are like terms.

Teaching Example 7
List the like terms: **a.** $8a - 8b + 4b$
b. $x^2 - 1 + 23x^2 + 30$
Answers: **a.** $-8b$ and $4b$
b. x^2 and $23x^2$; -1 and 30

Self Check 7 List the like terms: **a.** $2x - 2y + 7y$ $-2y$ and $7y$
b. $5p^2 - 12 + 17p^2 + 2$ $5p^2$ and $17p^2$; -12 and 2

Now Try Problem 65

5 Combine Like Terms.

To add or subtract objects, they must be similar. For example, fractions that are to be added must have a common denominator. When adding decimals, we align columns to be sure to add tenths to tenths, hundredths to hundredths, and so on. The same is true when working with terms of an algebraic expression. They can be added or subtracted only if they are like terms.

This expression can be simplified because it contains like terms.	This expression cannot be simplified because its terms are not like terms.
$3x + 4x$	$3x + 4y$

Recall that the distributive property can be written in the following forms:

$$(b + c)a = ba + ca \qquad (b - c)a = ba - ca$$

We can use these forms of the distributive property in reverse to simplify a sum or difference of like terms. For example, we can simplify $3x + 4x$ as follows:

The Language of Algebra

Simplifying a sum or difference of like terms is called **combining like terms.**

$$3x + 4x = (3 + 4)x \quad \text{Use } ba + ca = (b + c)a.$$
$$= 7x$$

We can simplify $15m^2 - 9m^2$ in a similar way:

$$\mathbf{15m^2 - 9m^2 = (15 - 9)m^2} \quad \text{Use } ba - ca = (b - c)a.$$
$$\mathbf{= 6m^2}$$

In each case, we say that we **combined like terms.** These examples suggest the following general rule.

Combining Like Terms

Like terms can be combined by adding or subtracting the coefficients of the terms and keeping the same variables with the same exponents.

EXAMPLE 8

Simplify by combining like terms, if possible: **a.** $2x + 9x$ **b.** $-8p + (-2p) + 4p$ **c.** $0.5s^3 - 0.3s^3$ **d.** $4w + 6$ **e.** $\frac{4}{9}b + \frac{7}{9}b$

Strategy We will use the distributive property in reverse to add (or subtract) the coefficients of the like terms. We will keep the same variables raised to the same powers.

Why To *combine like terms* means to add or subtract the like terms in an expression.

Solution

a. Since $2x$ and $9x$ are like terms with the common variable x, we can combine them.

$2x + 9x = 11x$ Think: $(2 + 9)x = 11x$.

Success Tip

Just as 2 apples plus 9 apples is 11 apples, $2x + 9x = 11x$.

b. $-8p + (-2p) + 4p = -6p$ Think: $[-8 + (-2) + 4]p = -6p$.

c. $0.5s^3 - 0.3s^3 = 0.2s^3$ Think: $(0.5 - 0.3)s^3 = 0.2s^3$.

d. Since $4w$ and 6 are not like terms, they cannot be combined. $4w + 6$ doesn't simplify.

e. $\frac{4}{9}b + \frac{7}{9}b = \frac{11}{9}b$ Think: $\left(\frac{4}{9} + \frac{7}{9}\right)b = \frac{11}{9}b$.

Teaching Example 8
Simplify, if possible: **a.** $4d + 5d$ **b.** $-20b + (-9b) + 12b$ **c.** $1.8a^4 - 5.9a^4$ **d.** $19x - 4$ **e.** $\frac{4}{15}c + \frac{3}{15}c$
Answers: **a.** $9d$ **b.** $-17b$ **c.** $-4.1a^4$ **d.** Doesn't simplify **e.** $\frac{7}{15}c$

Self Check 8 Simplify, if possible: **a.** $3x + 5x$ $8x$ **b.** $-6y + (-6y) + 9y$ $-3y$ **c.** $4.4s^4 - 3.9s^4$ $0.5s^4$ **d.** $4a - 2$ Doesn't simplify **e.** $\frac{10}{7}c - \frac{4}{7}c$ $\frac{6}{7}c$

Now Try Problems 67, 69, 71, and 73

EXAMPLE 9

Simplify by combining like terms: **a.** $16t - 15t$ **b.** $16t - t$ **c.** $15t - 16t$ **d.** $16t + t$

Teaching Tip: Remind your students that the coefficient of t is 1 and the coefficient of $-t$ is -1.

Strategy As we combine like terms, we must be careful when working with the terms such as t and $-t$.

Why Coefficients of 1 and -1 are usually not written.

Solution

a. $16t - 15t = t$ Think: $(16 - 15)t = 1t = t$.

b. $16t - t = 15t$ Think: $16t - 1t = (16 - 1)t = 15t$.

c. $15t - 16t = -t$ Think: $(15 - 16)t = -1t = -t$.

d. $16t + t = 17t$ Think: $16t + 1t = (16 + 1)t = 17t$.

Teaching Example 9
Simplify: **a.** $10g - 9g$ **b.** $10g + g$ **c.** $10g - g$ **d.** $9g - 10g$
Answers: **a.** g **b.** $11g$ **c.** $9g$ **d.** $-g$

Self Check 9 Simplify: **a.** $9h - h$ $8h$ **b.** $9h + h$ $10h$ **c.** $9h - 8h$ h **d.** $8h - 9h$ $-h$

Now Try Problems 75 and 77

EXAMPLE 10

Simplify: $6a^2 + 54a - 4a - 36$

Strategy First, we will identify any like terms in the expression. Then we will use the distributive property in reverse to combine them.

Why To *simplify* an expression, we use properties of real numbers to write an equivalent expression in simpler form.

Solution We can combine the like terms that involve the variable a.

$$6a^2 + \mathbf{54a} - \mathbf{4a} - 36 = 6a^2 + \mathbf{50a} - 36 \quad \text{Think: } (54 - 4)a = 50a.$$

Teaching Example 10
Simplify: $16x^2 + 14x - 8x + 14$
Answer: $16x^2 + 6x + 14$

Self Check 10 Simplify: $7y^2 + 21y - 2y - 6$ $7y^2 + 19y - 6$

Now Try Problem 85

EXAMPLE 11 Simplify: $4(x + 5) - 5 - (2x - 4)$

Strategy First, we will use the distributive property to remove the parentheses. Then we will identify any like terms and combine them.

Why To *simplify* an expression, we use properties of real numbers, such as the distributive property, to write an equivalent expression in simpler form.

Solution

$4(x + 5) - 5 - (2x - 4) = 4(x + 5) - 5 - 1(2x - 4)$ Replace the − symbol in front of $(2x - 4)$ with −1.

$= 4x + 20 - 5 - 2x + 4$ Distribute the multiplication by 4 and −1.

$= 2x + 19$ Think: $(4 - 2)x = 2x$. Think: $(20 - 5 + 4) = 19$.

Success Tip

Here, the distributive property is used both *forward* (to remove parentheses) and in *reverse* (to combine like terms).

Teaching Example 11
Simplify: $5(2a - 7) + 11 - (-4a + 9)$
Answer: $14a - 33$

Self Check 11 Simplify: $6(3y - 1) + 2 - (-3y + 4)$ $21y - 8$

Now Try Problem 87

SECTION 1.9 STUDY SET

VOCABULARY

Fill in the blanks.

1. To simplify the expression $5(6x)$ means to write it in simpler form: $5(6x) = 30x$.
2. $5(6x)$ and $30x$ are equivalent expressions because for each value of x, they represent the same number.
3. To perform the multiplication $2(x + 8)$, we use the distributive property.
4. We call $-(c + 9)$ the opposite of a sum.
5. Terms such as $7x^2$ and $5x^2$, which have the same variables raised to exactly the same power, are called like terms.
6. When we write $9x + x$ as $10x$, we say we have combined like terms.

CONCEPTS

7. a. Fill in the blanks to simplify the expression.

$4(9t) = (4 \cdot 9)t = 36t$

b. What property did you use in part a? Associative property of multiplication

8. a. Fill in the blanks to simplify the expression.

$-6y \cdot 2 = -6 \cdot 2 \cdot y = -12y$

b. What property did you use in part a? Commutative property of multiplication

9. Fill in the blanks.

a. $2(x + 4) = 2x + 8$ b. $2(x - 4) = 2x - 8$

c. $-2(x + 4) = -2x - 8$ d. $-2(-x - 4) = 2x + 8$

10. Fill in the blanks to combine like terms.

a. $4m + 6m = (4 + 6)m = 10m$

b. $30n^2 - 50n^2 = (30 - 50)n^2 = -20n^2$

c. $12 + 32d + 15 = 32d + 27$

d. Like terms can be combined by adding or subtracting the coefficients of the terms and keeping the same variables with the same exponents.

11. Simplify each expression, if possible.

a. $5(2x)$ $10x$ b. $5 + 2x$ Can't be simplified

c. $6(-7x)$ $-42x$ d. $6 - 7x$ Can't be simplified

e. $2(3x)(3)$ $18x$ f. $2 + 3x + 3$ $3x + 5$

12. Fill in the blanks: Distributing multiplication by -1 changes the sign of each term within the parentheses.

$-(x + 10) = -1(x + 10) = -x - 10$

NOTATION

13. Translate to symbols.
   a. Six times the quantity of h minus four. $6(h - 4)$
   b. The opposite of the sum of z and sixteen. $-(z + 16)$

▶ 14. Write an equivalent expression for the given expression using fewer symbols.
   a. $1x$ x b. $-1d$ $-d$ c. $0m$ 0
   d. $5x - (-1)$ $5x + 1$ e. $16t + (-6)$ $16t - 6$

GUIDED PRACTICE

***Use the given property to complete each statement.* See Example 1.**

15. $8 + (7 + a) = (8 + 7) + a$ Associative property of addition
▶ 16. $-2(5b) = (-2 \cdot 5)b$ Associative property of multiplication
17. $y \cdot 11 = 11y$ Commutative property of multiplication
▶ 18. $x + x^2 = x^2 + x$ Commutative property of addition
19. $(8d \cdot 2)6 = 8d(2 \cdot 6)$ Associative property of multiplication
▶ 20. $(-1 + 3a) + 7a = -1 + (3a + 7a)$ Associative property of addition
21. $9t + (4 + t) = 9t + (t + 4)$ Commutative property of addition
▶ 22. $(x - 2)3 = 3(x - 2)$ Commutative property of multiplication

***Simplify each expression.* See Example 2.**

23. $3 \cdot 4t$ $12t$
24. $9 \cdot 3s$ $27s$
25. $5(-7q)$ $-35q$
▶ 26. $-7(5t)$ $-35t$
27. $(-5.6x)(-2)$ $11.2x$
▶ 28. $(-4.4x)(-3)$ $13.2x$
29. $5(4c)(3)$ $60c$
30. $9(2h)(2)$ $36h$
31. $\frac{5}{3} \cdot \frac{3}{5}g$ g
▶ 32. $\frac{9}{7} \cdot \frac{7}{9}k$ k
33. $12\left(\frac{5}{12}x\right)$ $5x$
34. $15\left(\frac{4}{15}w\right)$ $4w$

***Multiply.* See Example 3.**

▶ 35. $5(x + 3)$ $5x + 15$
▶ 36. $4(x + 2)$ $4x + 8$
37. $-3(4x + 9)$ $-12x - 27$
38. $-5(8x + 9)$ $-40x - 45$
39. $45\left(\frac{x}{5} + \frac{2}{9}\right)$ $9x + 10$
▶ 40. $35\left(\frac{y}{5} + \frac{8}{7}\right)$ $7y + 40$
▶ 41. $0.4(x + 4)$ $0.4x + 1.6$
42. $2.2(2q + 1)$ $4.4q + 2.2$

***Multiply.* See Example 4.**

43. $6(6c - 7)$ $36c - 42$
▶ 44. $9(9d - 3)$ $81d - 27$
45. $-6(13c - 3)$ $-78c + 18$
▶ 46. $-2(10s - 11)$ $-20s + 22$
▶ 47. $-15(-2t - 6)$ $30t + 90$
▶ 48. $-20(-4z - 5)$ $80z + 100$
49. $-1(-4a + 1)$ $4a - 1$
50. $-1(-2x + 3)$ $2x - 3$

***Multiply.* See Example 5.**

▶ 51. $(3t + 2)8$ $24t + 16$
52. $(2q + 1)9$ $18q + 9$
53. $(3w - 6)\frac{2}{3}$ $2w - 4$
▶ 54. $(2y - 8)\frac{1}{2}$ $y - 4$
▶ 55. $4(7y + 4)2$ $56y + 32$
56. $8(2a - 3)4$ $64a - 96$
57. $2.5(2a - 3b + 1)$ $5a - 7.5b + 2.5$
58. $5(9s - 12t - 3)$ $45s - 60t - 15$

***Multiply.* See Example 6.**

59. $-(x - 7)$ $-x + 7$
▶ 60. $-(y + 1)$ $-y - 1$
61. $-(-5.6y + 7)$ $5.6y - 7$
▶ 62. $-(-4.8a - 3)$ $4.8a + 3$

***List the like terms in each expression, if any.* See Example 7.**

▶ 63. $3x + 2 - 2x$ $3x$ and $-2x$
▶ 64. $3y + 4 - 11y + 6$ $3y$ and $-11y$, 4 and 6
▶ 65. $-12m^4 - 3m^3 + 2m^2 - m^3$ $-3m^3$ and $-m^3$
▶ 66. $6x^3 + 3x^2 + 6x$ No like terms

***Simplify by combining like terms.* See Example 8.**

67. $3x + 7x$ $10x$
▶ 68. $12y - 15y$ $-3y$
69. $-7b^2 + 27b^2$ $20b^2$
70. $-2c^3 + 12c^3$ $10c^3$

***Simplify by combining like terms.* See Example 9.**

71. $36y + y - 9y$ $28y$
▶ 72. $32a - a + 5a$ $36a$
73. $\frac{3}{5}t + \frac{1}{5}t$ $\frac{4}{5}t$
▶ 74. $\frac{3}{16}x - \frac{5}{16}x$ $-\frac{1}{8}x$
75. $13r - 12r$ r
▶ 76. $25s + s$ $26s$
77. $43s^3 - 44s^3$ $-s^3$
78. $8j^3 - 9j^3$ $-j^3$

***Simplify by combining like terms.* See Example 10.**

79. $15y - 10 - y - 20y$ $-6y - 10$
80. $9z - 7 - z - 19z$ $-11z - 7$
81. $3x + 4 - 5x + 1$ $-2x + 5$
▶ 82. $4b + 9 - 9b + 9$ $-5b + 18$
83. $9m^2 - 6m + 12m - 4$ $9m^2 + 6m - 4$
84. $6a^2 + 18a - 9a + 5$ $6a^2 + 9a + 5$
85. $4x^2 + 5x - 8x + 9$ $4x^2 - 3x + 9$
▶ 86. $10y^2 - 8y + y - 7$ $10y^2 - 7y - 7$

***Simplify.* See Example 11.**

87. $2z + 5(z - 3) - 10$ $7z - 25$
88. $12(m + 11) - 11 + m$ $13m + 121$
▶ 89. $2(s^2 - 7) - (s^2 - 2)$ $s^2 - 12$
90. $4(d^2 - 3) - (d^2 - 1)$ $3d^2 - 11$

TRY IT YOURSELF

Simplify each expression, if possible.

91. $-\frac{7}{16}x - \frac{3}{16}x$ $-\frac{5}{8}x$
92. $-\frac{5}{18}x - \frac{7}{18}x$ $-\frac{2}{3}x$
93. $-9.8c + 6.2c$ $-3.6c$
▶ 94. $-5.7m + 4.3m$ $-1.4m$
95. $-4(-6)(-4m)$ $-96m$
▶ 96. $-5(-9)(-4n)$ $-180n$
▶ 97. $-4x + 4x$ 0
▶ 98. $-16y + 16y$ 0
99. $-0.2r - (-0.6r)$ $0.4r$
100. $-1.1m - (-2.4m)$ $1.3m$
101. $8\left(\frac{3}{4}y\right)$ $6y$
102. $27\left(\frac{2}{3}x\right)$ $18x$

103. $-9(3r - 9) - 7(2r - 7)$ $-41r + 130$

104. $-6(3t - 6) - 3(11t - 3)$ $-51t + 45$

105. $9(7m)$ $63m$

106. $12n(8)$ $96n$

▶ 107. $6 - 4(-3c - 7)$ $12c + 34$

▶ 108. $10 - 5(-5g - 1)$ $25g + 15$

109. $5t \cdot 60$ $300t$

▶ 110. $70a \cdot 10$ $700a$

▶ 111. $36\left(\frac{2}{9}x - \frac{3}{4}\right) + 36\left(\frac{1}{2}\right)$ $8x - 9$

112. $40\left(\frac{3}{8}y - \frac{1}{4}\right) + 40\left(\frac{4}{5}\right)$ $15y + 22$

▶ 113. $-4r - 7r + 2r - r$ $-10r$

114. $-v - 3v + 6v + 2v$ $4v$

115. $24\left(-\frac{5}{6}r\right)$ $-20r$

116. $\frac{3}{4} \cdot \frac{1}{2}g$ $\frac{3}{8}g$

117. $a + a + a$ $3a$

118. $t - t - t - t$ $-2t$

▶ 119. $60\left(\frac{3}{20}r - \frac{4}{15}\right)$ $9r - 16$

120. $72\left(\frac{7}{8}f - \frac{8}{9}\right)$ $63f - 64$

121. $4a + 4b + 4c$ Doesn't simplify

▶ 122. $2x + 2y + 2z$ Doesn't simplify

123. $-(c + 7) + 2(c - 3)$ $c - 13$

▶ 124. $-(z + 2) + 5(3 - z)$ $-6z + 13$

125. $a^3 + 2a^2 + 4a - 2a^2 - 4a - 8$ $a^3 - 8$

126. $c^3 - 3c^2 + 9c + 3c^2 - 9c + 27$ $c^3 + 27$

Look Alikes . . .

127. a. $2(7x)5$ $70x$ b. $2(7x + 5)$ $14x + 10$

▶ 128. a. $-3(-4a)(-2)$ $-24a$ b. $-3(-4a) - 2$ $12a - 2$

APPLICATIONS

In Exercises 129 and 130, recall that the perimeter of a figure is equal to the sum of the lengths of its sides.

▶ 129. **First Aid.** Each side of the red cross has length x inches. Write an algebraic expression that represents the perimeter of the cross. $12x$ in.

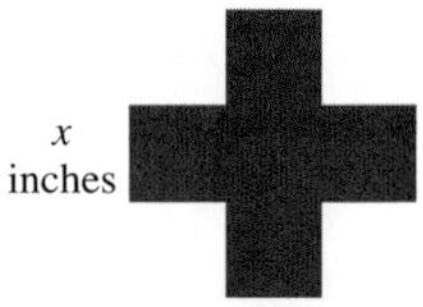

▶ 130. **Billiards.** Write an algebraic expression that represents the perimeter of the table. $6x$ ft

WRITING

131. Explain why the distributive property applies to $2(3 + x)$ but not to $2(3x)$.

132. Explain each error. Then give the correct answer.

a. $9(4b - 2) = 36b - 2$ b. $3(2x) = 6 \cdot 3x = 18x$

c. $-(23c + 2) = -23c + 2$ d. $(5n + 1)2 = 5n + 2$

REVIEW

Evaluate each expression for $x = -3$ and $y = -5$.

133. $\frac{x - y^2}{2y - 1 + x}$ 2

▶ 134. $\frac{2y + 1}{x} - x$ 6

CHALLENGE PROBLEMS

135. Fill in the blanks: $-17(11x - 7) = -187x + 119$

Simplify.

▶ 136. $2\{-2[x + 4(2x + 1)] - 5[x + 2(3x + 4)]\} + 106x$ -96

Teaching Tip: Remind your students that this Summary and Review is an excellent way to study for their test. Suggest that they arrive on campus early and, as a final preparation, read through the section summaries before taking the exam.

1 Summary & Review

SECTION 1.1 Introducing the Language of Algebra

DEFINITIONS AND CONCEPTS	EXAMPLES			
Tables, bar graphs, and **line graphs** are used to describe numerical relationships.	See pages 2 and 3 for examples of tables and graphs.			
A **sum** is the result of an addition. A **difference** is the result of a subtraction. A **product** is the result of a multiplication. A **quotient** is the result of a division.	$3 + 15 = \mathbf{18}$ sum	$16 - 1 = \mathbf{15}$ difference	$7 \cdot 8 = \mathbf{56}$ product	$\frac{63}{9} = \mathbf{7}$ quotient

A **variable** is a letter (or symbol) that stands for a number.	Variables: x, a, and y
Algebraic expressions contain variables and numbers combined with the operations of addition, subtraction, multiplication, and division.	Expressions: $5y + 7$, $\frac{12 - x}{5}$, and $8a^2(b - 3)$
An **equation** is a statement that two expressions are equal.	Equations: $3x - 4 = 12$ and $\frac{t}{9} = 12$ — Equations contain an = sign. Expressions do not.
Equations that express a relationship between two or more variables are called **formulas.**	$A = lw$ (The formula for the area of a rectangle)

REVIEW EXERCISES

The line graph shows the number of cars in a parking structure from 6 P.M. to 12 midnight on a Saturday.

1. What units are used to scale the horizontal and vertical axes? 1 hr; 100 cars
2. How many cars were in the parking structure at 11 P.M.? 100
3. At what time did the parking structure have 500 cars in it? 7 P.M.
4. When was the structure empty of cars? 12 A.M. (midnight)

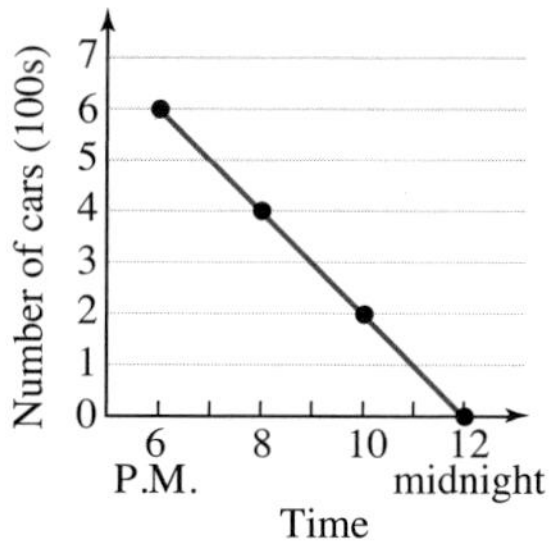

Express each statement in words, using one of these words: sum, difference, product, or quotient.

5. $15 - 3 = 12$ The difference of 15 and 3 equals 12.
6. $15 + 3 = 18$ The sum of 15 and 3 equals 18.
7. $15 \div 3 = 5$ The quotient of 15 and 3 equals 5.
8. $15 \cdot 3 = 45$ The product of 15 and 3 equals 45.
9. **a.** Write the multiplication 4×9 with a raised dot and then with parentheses. $4 \cdot 9$; $4(9)$

   b. Write the division $9 \div 3$ using a fraction bar. $\frac{9}{3}$
10. Write each multiplication without a multiplication symbol.

    a. $8 \cdot b$ $8b$ **b.** $P \cdot r \cdot t$ Prt
11. Classify each item as either an expression or an equation.

    a. $5 = 2x + 3$ Equation **b.** $2x + 3$ Expression
12. Use the formula $n = b + 5$ to complete the table.

Brackets (b)	Nails (n)
5	10
10	15
20	25

SECTION 1.2 Fractions

DEFINITIONS AND CONCEPTS	EXAMPLES
A **factor** is a number being multiplied.	$8 \cdot 9 = 72$ (8: Factor, 9: Factor)
A **prime number** is a natural number that is greater than 1 that has only itself and 1 as factors. A **composite number** is a natural number, greater than 1, that is not prime.	Primes: $\{2, 3, 5, 7, 11, 13, 17, 19, 23, \ldots\}$ Composites: $\{4, 6, 8, 9, 10, 12, 14, 15, \ldots\}$
Any composite number can be factored into the product of two or more prime factors. The product of these prime numbers is called its **prime factorization.**	Find the prime factorization of 98. $98 = 2 \cdot 49 = 2 \cdot 7 \cdot 7$ 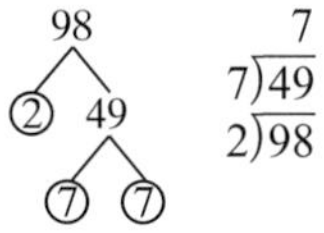
In a fraction, the number above the **fraction bar** is the **numerator** and the number below the fraction bar is called the **denominator.**	$\frac{11}{15}$ 11 ← Numerator, 15 ← Denominator
Two fractions are **equivalent** if they represent the same number.	Equivalent fractions: $\frac{1}{2} = \frac{2}{4} = \frac{3}{6} = \frac{4}{8} = \ldots$
To **multiply two fractions,** multiply their numerators and multiply their denominators.	Multiply: $\frac{5}{8} \cdot \frac{3}{4} = \frac{15}{32}$

One number is the **reciprocal** of another if their product is 1. To **divide two fractions,** multiply the first fraction by the reciprocal of the second fraction.	The reciprocal of $\frac{4}{5}$ is $\frac{5}{4}$ because $\frac{4}{5} \cdot \frac{5}{4} = 1$. Divide: $\frac{4}{7} \div \frac{5}{8} = \frac{4}{7} \cdot \frac{8}{5} = \frac{32}{35}$ *The reciprocal of $\frac{5}{8}$ is $\frac{8}{5}$.*
Multiplication property of 1: The product of 1 and any number is that number.	$1 \cdot 5 = 5$ and $\frac{7}{8} \cdot 1 = \frac{7}{8}$
To **build a fraction,** multiply it by a form of 1 such as $\frac{2}{2}, \frac{3}{3}, \frac{4}{4}, \ldots$.	Write $\frac{3}{4}$ as an equivalent fraction with a denominator of 20. $\frac{3}{4} = \frac{3}{4} \cdot \frac{5}{5} = \frac{15}{20}$
To **simplify a fraction,** remove pairs of factors common to the numerator and the denominator. A fraction is in **simplest form,** or **lowest terms,** when the numerator and denominator have no common factors other than 1.	Simplify: $\frac{12}{18} = \frac{2 \cdot \overset{1}{\not{6}}}{3 \cdot \underset{1}{\not{6}}} = \frac{2}{3}$ *Factor 12 and 18. Remove the common factor 6 from the numerator and denominator. Slashes and 1's are used to show that $\frac{6}{6}$ is replaced by the equivalent fraction $\frac{1}{1}$. A factor equal to 1 in the form of $\frac{6}{6}$ was removed.*
To find the **LCD** of two fractions, prime factor each denominator and find the product of the prime factors, using each factor the greatest number of times it appears in any one factorization.	Find the LCD of $\frac{5}{12}$ and $\frac{7}{8}$. $\left.\begin{array}{l} 12 = 2 \cdot 2 \cdot \mathbf{3} \\ 8 = \mathbf{2 \cdot 2 \cdot 2} \end{array}\right\}$ LCD $= \mathbf{2 \cdot 2 \cdot 2 \cdot 3} = 24$
To **add (or subtract) fractions that have the same denominator,** add (or subtract) the numerators and write the sum (or difference) over the common denominator. Simplify, if possible. To **add (or subtract) fractions that have different denominators,** rewrite each fraction as an equivalent fraction with the LCD as the denominator. Then add (or subtract) as usual. Simplify, if possible.	Add: $\frac{5}{12} + \frac{7}{8} = \frac{5}{12} \cdot \frac{2}{2} + \frac{7}{8} \cdot \frac{3}{3}$ *The LCD is 24. Build each fraction.* $= \frac{10}{24} + \frac{21}{24}$ *The denominators are now the same.* $= \frac{31}{24}$ *This result does not simplify.*
A **mixed number** represents the sum of a whole number and a fraction. In some computations, it is necessary to write mixed numbers as improper fractions.	Mixed numbers: $6\frac{1}{3} = 6 + \frac{1}{3}$ and $1\frac{3}{4} = \frac{7}{4}$

REVIEW EXERCISES

13. **a.** Write 24 as the product of two factors. $2 \cdot 12, 3 \cdot 8$ (Answers may vary)

b. Write 24 as the product of three factors. $2 \cdot 2 \cdot 6$ (Answers may vary)

c. List the factors of 24. 1, 2, 3, 4, 6, 8, 12, 24

14. What do we call fractions, such as $\frac{1}{8}$ and $\frac{2}{16}$, that represent the same number? Equivalent

Give the prime factorization of each number, if possible.

15. 54 $2 \cdot 3^3$

16. 147 $3 \cdot 7^2$

17. 385 $5 \cdot 7 \cdot 11$

18. 41 Prime

Simplify each fraction.

19. $\frac{20}{35}$ $\frac{4}{7}$

20. $\frac{24}{18}$ $\frac{4}{3}$

Build each number to an equivalent fraction with the indicated denominator.

21. $\frac{5}{8}$, denominator 64 $\frac{40}{64}$

22. 12, denominator 3 $\frac{36}{3}$

What is the LCD for fractions having the following denominators?

23. 10 and 18 90

24. 21 and 70 210

Perform each operation and simplify, if possible.

25. $\frac{1}{8} \cdot \frac{7}{8}$ $\frac{7}{64}$

26. $\frac{16}{35} \cdot \frac{25}{48}$ $\frac{5}{21}$

27. $\frac{1}{3} \div \frac{15}{16}$ $\frac{16}{45}$

28. $16\frac{1}{4} \div 5$ $3\frac{1}{4}$

29. $\frac{17}{25} - \frac{7}{25}$ $\frac{2}{5}$

30. $\frac{8}{11} - \frac{1}{2}$ $\frac{5}{22}$

31. $\frac{17}{24} + \frac{11}{40}$ $\frac{59}{60}$

32. $4\frac{1}{9} - 3\frac{5}{6}$ $\frac{5}{18}$

33. The Internet. A popular website averaged $1\frac{3}{4}$ million hits per day during a 30-day period. How many hits did it receive during that time? $52\frac{1}{2}$ million

34. Machine Shops. How much must be milled off the $\frac{17}{24}$-inch-thick steel rod so that the collar will slip over it? $\frac{17}{96}$ in.

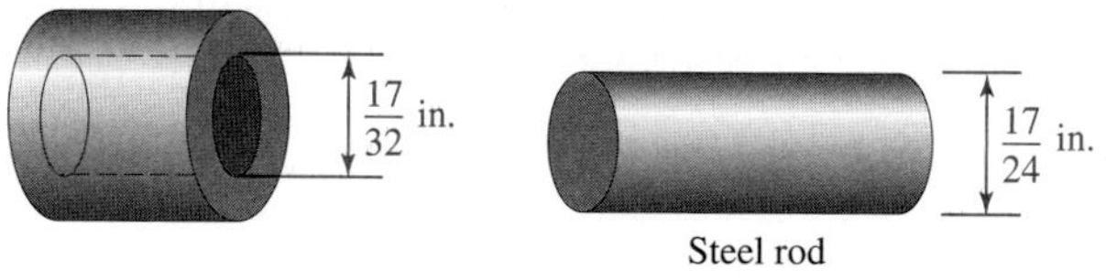

SECTION 1.3 The Real Numbers

DEFINITIONS AND CONCEPTS	EXAMPLES
To write a set, we list its **elements** within **braces** { }.	In the English alphabet, the set of vowels is {a, e, i, o, u}.
The **natural numbers** are the numbers we count with.	Natural numbers: {1, 2, 3, 4, 5, 6, . . .}
The **whole numbers** are the natural numbers together with 0.	Whole numbers: {0, 1, 2, 3, 4, 5, 6, . . .}
Two numbers are called **opposites** if they are the same distance from 0 on the number line but are on opposite sides of it.	The opposite of 3 is −3 and the opposite of −21 is 21.
The **integers** include the whole numbers and their opposites.	Integers: {. . . , −3, −2, −1, 0, 1, 2, 3, . . .}
The **rational numbers** are numbers that can be expressed as fractions with an integer numerator and a nonzero integer denominator.	Rational numbers: $-6, -3.1, -\frac{1}{2}, 0, \frac{11}{12}, 9\frac{4}{5}$, and 87 $-6 = \frac{-6}{1}$, $-3.1 = \frac{-31}{10}$, $0 = \frac{0}{1}$, $9\frac{4}{5} = \frac{49}{5}$, $87 = \frac{87}{1}$
Terminating and **repeating decimals** can be expressed as fractions and are, therefore, rational numbers.	Terminating decimal: $-0.25 = -\frac{1}{4}$ Repeating decimal: $0.\overline{6} = \frac{2}{3}$
Negative fractions: $-\frac{a}{b} = \frac{-a}{b} = \frac{a}{-b}$	$-\frac{3}{4} = \frac{-3}{4} = \frac{3}{-4}$ and $\frac{-1}{8} = \frac{1}{-8} = -\frac{1}{8}$
An **irrational number** is a nonterminating, nonrepeating decimal. An irrational number cannot be expressed as a fraction with an integer numerator and a nonzero integer denominator. A **real number** is any number that is either a rational or an irrational number. Every real number corresponds to a point on the **number line,** and every point on the number line corresponds to exactly one real number.	Irrational numbers: $\sqrt{2}$, π, and $-\sqrt{7}$ Graph the numbers in the set $\left\{-2, 4, -0.75, 1\frac{3}{4}, \pi, 0, \frac{7}{8}\right\}$ on a number line and classify them. 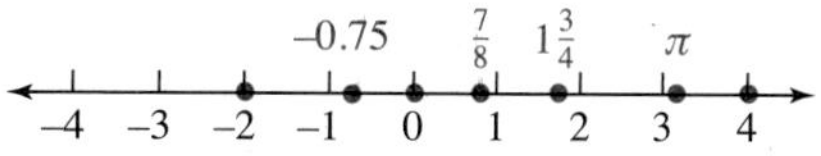 Natural numbers: 4 Whole numbers: 4, 0 Integers: −2, 4, 0 Rational numbers: −2, 4, −0.75, $1\frac{3}{4}$, 0, $\frac{7}{8}$ Irrational numbers: π Real numbers: all
Inequality symbols: > is greater than < is less than	25 > 15 and −2 > −7 3.3 < 9.7 and −10 < −9
The **absolute value** of a number is the distance on the number line between the number and 0.	$\|5\| = 5$, $\|-7\| = 7$, and $-\left\|-\frac{5}{9}\right\| = -\frac{5}{9}$

REVIEW EXERCISES

35. a. Which number is a whole number but not a natural number? 0
b. Write the set of integers. $\{\ldots, -2, -1, 0, 1, 2, \ldots\}$

36. Represent 206 feet below sea level with a signed number. -206 ft

37. Use one of the symbols > or < to make each statement true.
a. $0 < 5$ b. $-12 > -13$

38. Show that each of the following numbers is a rational number by expressing it as a ratio (quotient) of two integers.
a. 0.7 $\frac{7}{10}$ b. $4\frac{2}{3}$ $\frac{14}{3}$

Write each fraction as a decimal. Use an overbar if the result is a repeating decimal.

39. $\frac{1}{250}$ 0.004

40. $\frac{17}{22}$ $0.7\overline{72}$

41. Graph each number on a number line:
$\{\pi, 0.333\ldots, 3.75, \sqrt{2}, -\frac{17}{4}, \frac{7}{8}, -2\}$
See AIE Appendix 3.

42. Determine which numbers in the given set are natural numbers, whole numbers, integers, rational numbers, irrational numbers, and real numbers. $\{-\frac{4}{5}, 99.99, 0, \sqrt{2}, -12, 4\frac{1}{2}, 0.666\ldots\ 8\}$
Natural: 8; whole: 0, 8; integers: 0, -12, 8; rational: $-\frac{4}{5}$, 99.99, 0, -12, $4\frac{1}{2}$, 0.666. . . , 8; irrational: $\sqrt{2}$; real: all

Determine whether each statement is true or false.

43. All integers are whole numbers. False

44. π is a rational number. False

45. The set of real numbers corresponds to all points on the number line. True

46. A real number is either rational or irrational. True

Insert one of the symbols >, <, or = in the blank to make each statement true.

47. $|-6| > |5|$

48. $-9 < |-10|$

SECTION 1.4 Adding Real Numbers; Properties of Addition

DEFINITIONS AND CONCEPTS	EXAMPLES
To **add two real numbers with like signs:**	
1. To add two positive numbers, add them as usual. The final answer is positive.	Add: $3 + 5 = 8$
2. To add two negative numbers, add their absolute values and make the final answer negative.	Add: $-5 + (-11) = -16$
To **add two real numbers with unlike signs:**	
1. Subtract their absolute values (the smaller from the larger).	Add: $-8 + 6 = -2$ -8 has the larger absolute value.
2. To that result, attach the sign of the number with the larger absolute value.	Add: $12 + (-5) = 7$ 12 has the larger absolute value.
Properties of Addition	
Commutative property: $a + b = b + a$ *Changing the order when adding does not affect the answer.*	$5 + (-9) = -9 + 5$ Reorder.
Associative property: $(a + b) + c = a + (b + c)$ *Changing the grouping when adding does not affect the answer.*	$(3 + 7) + 5 = 3 + (7 + 5)$ Regroup. Note that the order of the addends, 3, 7, and 5, does not change.
Addition property of 0: $a + 0 = a$ and $0 + a = a$	$-6 + \mathbf{0} = -6$ 0 is the additive identity element.
Addition property of opposites: $a + (-a) = 0$ and $(-a) + a = 0$	$\mathbf{11 + (-11)} = 0$ 11 and -11 are additive inverses.

REVIEW EXERCISES

Add.

49. $-45 + (-37)$ -82

50. $25 + (-13)$ 12

51. $0 + (-7)$ -7

52. $-7 + 7$ 0

53. $12 + (-8) + (-15)$ -11

54. $-9.9 + (-2.4)$ -12.3

55. $\frac{5}{16} + \left(-\frac{1}{2}\right)$ $-\frac{3}{16}$

56. $35 + (-13) + (-17) + 6$ 11

57. Determine what property of addition is shown.
a. $-2 + 5 = 5 + (-2)$ Comm. prop. add
b. $(-2 + 5) + 1 = -2 + (5 + 1)$ Assoc. prop. add.
c. $80 + (-80) = 0$ Add. prop. opp. (inv. prop. add.)
d. $-5.75 + 0 = -5.75$ Add. prop. 0 (ident. prop. add.)

58. **Temperatures.** Determine Washington State's record high temperature if it is 166° greater than the state's record low temperature of -48°F. 118°F

SECTION 1.5 Subtracting Real Numbers

DEFINITIONS AND CONCEPTS	EXAMPLES
The **opposite of the opposite of a number** is that number. For any real number a, $-(-a) = a$.	$-(-13) = 13$ and $-(-x) = x$
To **subtract two real numbers,** add the first to the opposite (additive inverse) of the number to be subtracted. For any real numbers a and b, $a - b = a + (-b)$ To **check** a subtraction, the difference plus the subtrahend should equal the minuend.	Subtract: $4 - 7 = 4 + (-7) = -3$ The opposite of 7 is −7. $6 - (-8) = 6 + 8 = 14$ The opposite of −8 is 8. $-1 - (-2) = -1 + 2 = 1$ The opposite of −2 is 2. To check $-6 - 2 = -8$, verify that $-8 + 2 = -6$.

REVIEW EXERCISES

Write the expression in simpler form.

59. a. The opposite of 10 -10
    b. The additive inverse of -3 3

60. a. $-\left(-\frac{9}{16}\right)$ $\frac{9}{16}$
    b. $-|-4|$ -4

Perform the operations.

61. $45-64$ -19

62. Subtract $\frac{1}{3}$ from $-\frac{3}{5}$ $-\frac{14}{15}$

63. $-7 - (-12)$ 5

64. $3.6 - (-2.1)$ 5.7

65. $0 - 10$ -10

66. $-33 + 7 - 5 - (-2)$ -29

67. **Geography.** The tallest peak on Earth is Mount Everest, at 29,028 feet, and the greatest ocean depth is the Mariana Trench, at $-36{,}205$ feet. Find the difference in these elevations. Check the result. 65,233 ft; $65{,}233 + (-36{,}205) = 29{,}028$

68. **History.** Archimedes, a famous Greek mathematician, died in 212 B.C. (-212) at the age of 75. When was he born? Check the result. 287 B.C. (-287); $-287 + 75 = -212$

SECTION 1.6 Multiplying and Dividing Real Numbers; Multiplication and Division Properties

DEFINITIONS AND CONCEPTS	EXAMPLES
To **multiply two real numbers,** multiply their absolute values. 1. If the numbers have **like signs,** the final answer is positive. 2. If the numbers have **unlike signs,** the final answer is negative.	 Multiply: $-5(-7) = 35$ and $14(3) = 42$ Multiply: $6(-6) = -36$ and $-11(5) = -55$
Properties of multiplication **Commutative property:** $ab = ba$ *Changing the order when multiplying does not affect the answer.*	$-8(12) = 12(-8)$ Reorder.
Associative property: $(ab)c = a(bc)$ *Changing the grouping when multiplying does not affect the answer.*	$(-4 \cdot 9) \cdot 7 = -4(9 \cdot 7)$ Regroup. Note that the order of the factors, −4, 9, and 7, does not change.
Multiplication property of 0: $0 \cdot a = 0$ and $a \cdot 0 = 0$	$0 \cdot (-7) = 0$ and $6(5)0 = 0$
Multiplication property of 1: $1 \cdot a = a$ and $a \cdot 1 = a$	$1 \cdot 32 = 32$ 1 is the multiplicative identity.
Multiplication property of −1: $-1 \cdot a = -a$ and $a(-1) = -a$	$-1(8) = -8$, $-1(-32) = 32$, $-x = -1 \cdot x$
Multiplicative inverse property: $a\left(\frac{1}{a}\right) = 1$ and $\frac{1}{a}(a) = 1$	$4\left(\frac{1}{4}\right) = 1$ 4 and $\frac{1}{4}$ are multiplicative inverses.

To **divide two real numbers,** divide their absolute values. **1.** If the numbers have **like signs,** the final answer is positive. **2.** If the numbers have **unlike signs,** the final answer is negative.	Divide: $\frac{16}{8} = 2$ and $\frac{-25}{-5} = 5$ Divide: $\frac{-36}{9} = -4$ and $\frac{56}{-7} = -8$
For any real number, $\frac{a}{1} = a$ and $\frac{a}{a} = 1$, where $a \neq 0$.	$\frac{25}{1} = 25$ and $\frac{-32}{-32} = 1$
Division of zero by a nonzero number is 0. **Division by zero** is undefined.	$\frac{0}{17} = 0$ but $\frac{2}{0}$ is undefined because no number multiplied by 0 gives 2.
To **check** the division $\frac{a}{b} = c$, verify that $c \cdot b = a$.	To check $\frac{6}{-2} = -3$, verify that $-3(-2) = 6$. *Quotient · divisor = dividend.*

REVIEW EXERCISES

Multiply.

69. $-8 \cdot 7$ -56

70. $-9\left(-\frac{1}{9}\right)$ 1

71. $2(-3)(-2)$ 12

72. $(-4)(-1)(-3)$ -12

73. $-1.2(-5.3)$ 6.36

74. $0.002(-1{,}000)$ -2

75. $-\frac{2}{3}\left(\frac{1}{5}\right)$ $-\frac{2}{15}$

76. $-6(-3)(0)(-1)$ 0

77. Electronics. The picture on the screen can be magnified by switching a setting on the monitor. What would be the new high and low if every value changed by a factor of 1.5? High: 3, low: -4.5

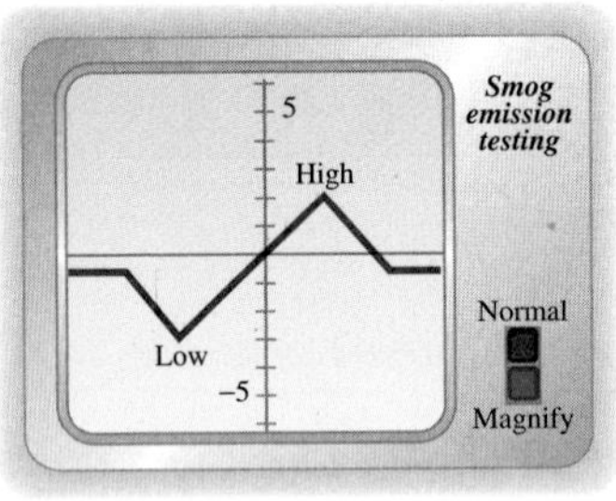

78. Determine what property of multiplication is shown.

a. $(2 \cdot 3)5 = 2(3 \cdot 5)$ Assoc. prop. mult.

b. $(-5)(-6) = (-6)(-5)$ Comm. prop. mult.

c. $-6 \cdot 1 = -6$ Mult. prop. 1 (Iden. prop. of mult.)

d. $\frac{1}{2}(2) = 1$ Inv. prop. mult.

Perform each division, if possible.

79. $\frac{44}{-44}$ -1

80. $\frac{-272}{16}$ -17

81. $\frac{-81}{-27}$ 3

82. $-\frac{3}{5} \div \frac{1}{2}$ $-\frac{6}{5}$

83. $\frac{-60}{0}$ Undefined

84. $\frac{-4.5}{1}$ -4.5

85. Fill in the blanks: $\frac{0}{18} = 0$ because $0 \cdot 18 = 0$.

86. Gemstones. A 3-carat yellow sapphire stone valued at $3,000 five years ago is now worth $1,200. What signed number indicates the average annual depreciation of the sapphire? $-\$360$

SECTION 1.7 Exponents and Order of Operations

DEFINITIONS AND CONCEPTS	EXAMPLES
An **exponent** represents repeated multiplication.	$8^5 = 8 \cdot 8 \cdot 8 \cdot 8 \cdot 8$ *The exponent 5 indicates that 8 is to be used as a factor 5 times.*
In a^n, a is the **base** and n is the **exponent.**	In 7^4, the base is 7 and 4 is the exponent.

Order of Operations 1. Perform all calculations within grouping symbols, working from the innermost to the outermost in the following order. 2. Evaluate all exponential expressions. 3. Perform all multiplications and divisions as they occur from left to right. 4. Perform all additions and subtractions as they occur from left to right. In fractions, evaluate the numerator and denominator separately. Then simplify the fraction. **Grouping Symbols:** Innermost parentheses $2[7 + 3(1 - 4)]$ Outermost brackets	Evaluate: $\frac{3(6 - 4^3) - 2^4 + 4}{8 \div 4 \cdot 3} = \frac{3(6 - 64) - 2^4 + 4}{2 \cdot 3}$ Evaluate: $4^3 = 64$. Divide: $8 \div 4 = 2$. $= \frac{3(-58) - 2^4 + 4}{6}$ Subtract: $6 - 64 = -58$. $= \frac{3(-58) - 16 + 4}{6}$ Evaluate: $2^4 = 16$. $= \frac{-174 - 16 + 4}{6}$ Multiply: $3(-58) = -174$. $= \frac{-190 + 4}{6}$ Subtract: $-174 - 16 = -190$. $= \frac{-186}{6}$ Add: $-190 + 4 = -186$. $= -31$ Divide: -186 by 6.
$\text{Mean} = \frac{\text{sum of values}}{\text{number of values}}$ The mean is also called the average.	Find the mean of the test scores of 74, 83, 79, 91, and 73. $\text{Mean} = \frac{74 + 83 + 79 + 91 + 73}{5} = 80$ The mean is a value around which 74, 83, 79, 91, and 73 are grouped.

REVIEW EXERCISES

87. Write each expression using exponents.

a. $8 \cdot 8 \cdot 8 \cdot 8 \cdot 8$ 8^5 b. $9 \cdot \pi \cdot r \cdot r$ $9\pi r^2$

88. Evaluate each expression.

a. 9^2 81 b. $\left(-\frac{2}{3}\right)^3$ $-\frac{8}{27}$

c. 2^5 32 d. 50^1 50

Evaluate each expression.

89. $2 + 5 \cdot 3$ 17

90. $-24 \div 2 \cdot 3$ -36

91. $-(16 - 3)^2$ -169

92. $43 + 2(-6 - 2 \cdot 2)$ 23

93. $10 - 5[-3 - 2(5 - 7^2)] - 5$ -420

94. $\frac{-4(4 + 2) - 4}{2|-18 - 4(5)|}$ $-\frac{7}{19}$

95. $(-3)^3\left(\frac{-8}{2}\right) + 5$ 113

96. $\frac{2^4 - (4 - 6)(3 - 6)}{12 + 4[(-1)^8 - 2^2]}$ Undefined

97. Write each expression in symbols and then evaluate it.

a. Negative nine squared $(-9)^2 = 81$

b. The opposite of the square of nine $-9^2 = -81$

98. Walk-A-Thons. Use the data in the table to find the average (mean) donation to a charity walk-a-thon. \$20

Donation	\$5	\$10	\$20	\$50	\$100
Number received	20	65	25	5	10

SECTION 1.8 Algebraic Expressions

DEFINITIONS AND CONCEPTS	EXAMPLES
Addition symbols separate algebraic expressions into **terms.** In a term, the numerical factor is called the **coefficient.** A term that consists of a single number is called a **constant** term.	Since $a^2 + 3a - 5$ can be written as $a^2 + 3a + (-5)$, it has three terms. The coefficient of a^2 is 1, the coefficient of $3a$ is 3, and the coefficient of the constant term -5 is -5.
Key phrases can be translated to algebraic expressions.	*5 more than x* can be expressed as $x + 5$. *25 less than twice y* can be expressed as $2y - 25$. One-half *of c* can be expressed as $\frac{1}{2}c$.
Number · value = total value	The total value (in cents) of n nickels is $n \cdot 5 = 5n$ cents.

To **evaluate algebraic expressions,** we substitute the values of its variables and use the order of operations rule.	Evaluate $\frac{x^2 - y^2}{x + y}$ for $x = 2$ and $y = -3$. $\frac{x^2 - y^2}{x + y} = \frac{2^2 - (-3)^2}{2 + (-3)}$ Substitute 2 for x and −3 for y. $= \frac{4 - 9}{-1}$ Evaluate: $2^2 = 4$ and $(-3)^2 = 9$. Add: $2 + (-3) = -1$. $= \frac{-5}{-1}$ Subtract: $4 - 9 = -5$. $= 5$ Divide −5 by −1.

REVIEW EXERCISES

99. How many terms does each expression have?

a. $3x^2 + 2x - 5$ 3 **b.** $-12xyz$ 1

100. Identify the coefficient of each term of the given expression.

a. $16x^2 - 5x + 25$ 16, −5, 25 **b.** $\frac{x}{2} + y$ $\frac{1}{2}$, 1

Write each phrase as an algebraic expression.

101. 25 more than the height h $h + 25$

102. 15 less than triple the cutoff score s $3s - 15$

103. 6 less than one-half of the time $\frac{1}{2}t - 6$

104. The absolute value of the difference of 2 and the square of a $|2 - a^2|$

105. Hardware. Let n represent the length of the nail in inches. Write an algebraic expression that represents the length of the bolt (in inches). $n + 4$

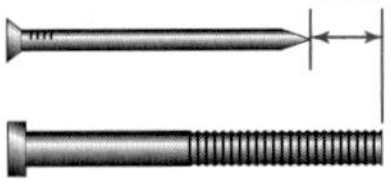

106. Hardware. Let b represent the length of the bolt in inches. Write an algebraic expression that represents the length of the nail (in inches). $b - 4$

107. How many years are in d decades? $10d$

108. Five years after a house was constructed, a patio was added. How old, in years, is the patio if the house is x years old? $x - 5$

109. Complete the table below. The units are cents.

Coin	Number	Value	Total value
Nickel	6	5	30
Dime	d	10	$10d$

110. Complete the table.

x	$20x - x^3$
0	0
1	19
−4	−16

Evaluate each algebraic expression for the given values of the variables.

111. $b^2 - 4ac$ for $b = -10$, $a = 3$, and $c = 5$ 40

112. $\frac{x + y}{-x - z}$ for $x = 19$, and $y = 17$, and $z = -18$ −36

SECTION 1.9 Simplifying Algebraic Expressions Using Properties of Real Numbers

DEFINITIONS AND CONCEPTS	EXAMPLES
We often use the *commutative property of multiplication* to reorder factors and the *associative property of multiplication* to regroup factors when **simplifying expressions.**	Simplify: $-5(3y) = (-5 \cdot 3)y = -15y$ −5(3y) and −15y are called equivalent expressions. $-45b\left(\frac{5}{9}\right) = \left(-45 \cdot \frac{5}{9}\right)b = -25b$
The **distributive property** can be used to *remove parentheses:* $a(b + c) = ab + ac$ $a(b - c) = ab - ac$ $a(b + c + d) = ab + ac + ad$	Multiply: $7(x + 3) = 7 \cdot x + 7 \cdot 3 = 7x + 21$ $-0.2(4m - 5n - 7) = -0.2(4m) - (-0.2)(5n) - (-0.2)(7)$ $= -0.8m + n + 1.4$
The **opposite of a sum:** $-(a + b) = -a + (-b)$	$-(3x + 4) = -3x - 4$ and $-(-2y - 12) = 2y + 12$

Like terms are terms with exactly the same variables raised to exactly the same powers.	$3x$ and $-5x$ are like terms. $-4t^3$ and $3t^2$ are unlike terms because the variable t has different exponents. $0.5xyz$ and $3.7xy$ are unlike terms because they have different variables.
Simplifying the sum or difference of like terms is called **combining like terms.** Like terms can be combined by adding or subtracting the coefficients of the terms and keeping the same variables with the same exponents.	Simplify: $4a + 2a = 6a$ Think: $(4 + 2)a = 6a$. $5p^2 + p - p^2 - 9p = 4p^2 - 8p$ Think: $(5 - 1)p^2 = 4p^2$ and $(1 - 9)p = -8p$. $2(k - 1) - 3(k + 2) = 2k - 2 - 3k - 6 = -k - 8$

REVIEW EXERCISES

Use the given property to complete each statement.

113. $a \cdot 150 =$ $150a$ Commutative property of multiplication

114. $9 + (1 + 7y) =$ $(9 + 1) + 7y$ Associative property of addition

115. $2.7(10b) =$ $(2.7 \cdot 10)b$ Associative property of multiplication

116. $x + 2x^2 =$ $2x^2 + x$ Commutative property of addition

Simplify each expression.

117. $-4(7w)$ $-28w$

118. $3(-2x)(-4)$ $24x$

119. $0.4(5.2f)$ $2.08f$

120. $\frac{7}{2} \cdot \frac{2}{7}r$ r

Use the distributive property to remove parentheses.

121. $5(x + 3)$ $5x + 15$

122. $-(2x + 3 - y)$ $-2x - 3 + y$

123. $\frac{3}{4}(4c - 8)$ $3c - 6$

124. $-2(-3c - 7)(2.1)$ $12.6c + 29.4$

Simplify each expression by combining like terms.

125. $8p + 5p - 4p$ $9p$

126. $-5m + 2 - 2m - 2$ $-7m$

127. $n + n + n + n$ $4n$

128. $5(p - 2) - 2(3p + 4)$ $-p - 18$

129. $55.7k^2 - 55.6k^2$ $0.1k^2$

130. $8a^3 + 4a^3 + 2a - 4a^3 - 2a - 1$ $8a^3 - 1$

131. $\frac{3}{5}w - \left(-\frac{2}{5}w\right)$ w

132. $36\left(\frac{1}{9}h - \frac{3}{4}\right) + 36\left(\frac{1}{3}\right)$ $4h - 15$

133. $-(7.6t - 1.9) + (1.4t - 1.2)8 + t$ $4.6t - 7.7$

134. Write an equivalent expression for the given expression using fewer symbols.

a. $1x$ x **b.** $-1x$ $-x$

c. $4x - (-1)$ $4x + 1$ **d.** $4x + (-1)$ $4x - 1$

Teaching Tip: Because this Chapter Test is a comprehensive collection of problems that covers all of the topics discussed in Chapter 1, it is lengthy. If your students have time to complete it, that would be optimal. If, because of time constraints, they are unable to do so, assign an appropriate subset of problems that reflects the types of problems that the students will see on your exam.

1 CHAPTER TEST

1. Fill in the blanks.

a. Two fractions, such as $\frac{1}{2}$ and $\frac{5}{10}$, that represent the same number are called equivalent fractions.

b. The result of a multiplication is called a product.

c. $\frac{8}{7}$ is the reciprocal of $\frac{7}{8}$ because $\frac{8}{7} \cdot \frac{7}{8} = 1$.

d. $9x^2$ and $7x^2$ are like terms because they have the same variable raised to exactly the same power.

e. For any nonzero real number a, $\frac{a}{0}$ is undefined.

2. Security Guards. The graph shows the cost to hire a security guard.

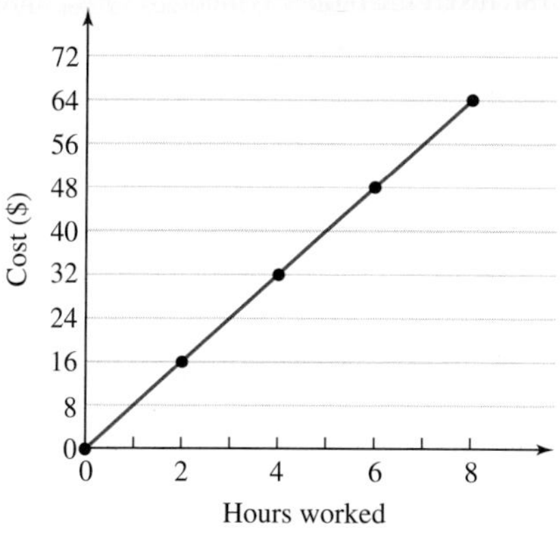

a. What will it cost to hire a security guard for 3 hours? $24

b. If a school was billed $40 for hiring a security guard for a dance, for how long did the guard work? 5 hr

3. Use the formula $f = \frac{a}{5}$ to complete the table.

Square miles (a)	Fire stations (f)
15	3
100	20
350	70

4. Give the prime factorization of 180.
$2 \cdot 2 \cdot 3 \cdot 3 \cdot 5 = 2^2 \cdot 3^2 \cdot 5$

5. Simplify: $\frac{42}{105}$ $\frac{2}{5}$

6. Divide: $\frac{15}{16} \div \frac{5}{8}$ $\frac{3}{2} = 1\frac{1}{2}$

7. Add: $\frac{7}{10} + \frac{1}{14}$ $\frac{27}{35}$

8. Subtract: $8\frac{2}{5} - 1\frac{2}{3}$ $6\frac{11}{15}$

9. **Shopping.**
   a. What is the weight of the oranges? Answer using a mixed number. $4\frac{1}{4}$ lb
   b. Find the cost of the oranges. Answer in dollars. \$3.57

10. Write $\frac{5}{6}$ as a decimal. $0.8\overline{3}$

11. a. Graph the set of numbers on a number line.

$$\left\{-1\tfrac{1}{4},\ 0,\ \sqrt{2},\ -3.75,\ 2,\ \tfrac{7}{2},\ 0.5,\ -3\right\}$$

See AIE Appendix 3.

   b. Determine which numbers in the set are natural numbers, whole numbers, integers, rational numbers, irrational numbers, and real numbers. Natural: 2; whole: 0, 2; integers: 0, 2, −3; rational: $-1\frac{1}{4}$, 0, −3.75, 2, $\frac{7}{2}$, 0.5, −3; irrational: $\sqrt{2}$; real: all

12. Determine whether each statement is true or false.
   a. Every integer is a rational number. True
   b. Every rational number is an integer. False
   c. π is an irrational number. True
   d. 500 is a whole number. True
   e. A real number is any number that is either a rational or an irrational number. True

13. Insert the proper symbol, > or <, in the blank.
   a. $-2 > -3$
   b. $-|-9| < 8$
   c. $|-4| < -(-5)$
   d. $\left|-\frac{7}{8}\right| > 0.5$

14. **Television.** During "sweeps week," networks try to gain viewers by showing flashy programs. Use the data to determine the average daily gain (or loss) of ratings points by a network for the 7-day "sweeps period." A gain of 0.6 of a rating point

Day	M	T	W	Th	F	Sa	Su
Point loss/gain	0.6	−0.3	1.7	1.5	−0.2	1.1	−0.2

Perform the operations.

15. $-5.6 + (-2)$ −7.6

16. $(-6) + 8 + (-4)$ −2

17. $-\frac{1}{2} + \frac{7}{8}$ $\frac{3}{8}$

18. a. $-10 - (-4)$ −6
   b. Show a check of the result. $-6 + (-4) = -10$

19. a. $\frac{-12.6}{-0.9}$ 14
   b. Show a check of the result. $14(-0.9) = -12.6$

20. $(-2)(-3)(-5)$ −30

21. $-6.1(0.4)$ −2.44

22. $\frac{0}{-3}$ 0

23. $\left(-\frac{3}{5}\right)^3$ $-\frac{27}{125}$

24. $3 + (-3)$ 0

25. $0 - 3$ −3

26. $-30 + 50 - 10 - (-40)$ 50

27. **Astronomy.** *Magnitude* is a term used in astronomy to describe the brightness of planets and stars. Negative magnitudes are associated with brighter objects. By how many magnitudes do a full moon and the sun differ? 14

Object	Magnitude
Sun	−26.5
Full moon	−12.5

28. **Glaciers.** In 2005, the South Cascade Glacier in Washington State gained approximately 87 inches in thickness because of snowfall. That same year it lost approximately 165 inches of thickness due to melting. Was there a net gain or loss in the glacier's thickness that year? How much? (Source: U.S. Geological Survey) a net loss of 78 in.

29. **Inventory.** It was discovered that fifteen MP3 players were missing from the stockroom of an electronics store. If the players cost \$85 each, what signed number represents the store's financial loss? −\$1,275

30. Use the given property to complete each statement.
   a. $(-12 + 97) + 3 =$ $-12 + (97 + 3)$
   Associative property of addition
   b. $2(x + 7) =$ $2x + 14$
   Distributive property
   c. $-2(m)5 =$ $-2(5)m$
   Commutative property of multiplication
   d. $\frac{1}{8}(8) =$ 1
   Inverse property of multiplication
   e. $0 + 15x =$ $15x$
   Identity property of addition

31. Write each product using exponents:
   a. $9(9)(9)(9)(9)$ 9^5
   b. $3 \cdot x \cdot x \cdot z \cdot z \cdot z$ $3x^2z^3$

32. Complete the table.

x	$2x - \frac{30}{x}$
5	4
10	17
−30	−59

Evaluate each expression.

33. $8 + 2 \cdot 3^4$ 170

34. $\frac{3(40 - 2^3)}{-2(6 - 4)^2}$ −12

35. $-10^2 - 5 + 6$ −99

36. $9 - 3[45 - 5^2(1^5 - 4)]$ −351

37. $|-50 \div 5 \cdot 2|$ 20

38. Evaluate $3(10x - y) - 5(x + y^2)$ for $x = 2$ and $y = -5$. -60

39. Translate to an algebraic expression: seven less than twice the width w. $2w - 7$

40. a. Music. A band recorded x songs for a CD. However, two of the songs were not included in the album because of poor sound quality. Write an algebraic expression that represents the number of songs on the CD. $x - 2$

b. Money. Find the value of q quarters in cents. $25q$¢

41. How many terms are in the expression $4x^2 + 5x - 7$? 3

42. What is the coefficient of each term of $a^3 - 6a^2 - a + 10$? 1, −6, −1, 10

Simplify each expression.

43. $5(-4x)$ $-20x$

44. $-8(-7t)(4)$ $224t$

45. $\frac{4}{5}(15a + 5) - 16a$ $-4a + 4$

46. $-1.1d^3 - 3.8d^3 - d^3$ $-5.9d^3$

47. $9x + 2(7x - 3) - 9(x - 1)$ $14x + 3$

48. $m^2 + 4m^2 + 5m - 2m^2 - 3m - 4$ $3m^2 + 2m - 4$

49. Tell whether each statement is true or false.

a. $-\frac{3}{5} = \frac{-3}{5} = \frac{3}{-5}$ True

b. $-|-8| = -(-8)$ False

c. $-5^2 = (-5)^2$ False

d. $-(4a + 7) = -4a + 7$ False

e. $(-9)^{14}$ is a positive number. True

f. The reciprocal of $-\frac{4}{7}$ is $\frac{7}{4}$. False

Group Project

WRITING FRACTIONS AS DECIMALS

Overview: This is a good activity to try at the beginning of the course. You can become acquainted with other students in your class while you review the process for finding decimal equivalents of fractions.

Instructions: Form groups of 6 students. Select one person from your group to record the group's responses on the questionnaire. Express the results in fraction form and in decimal form.

What fraction (decimal) of the students in your group . . .	Fraction	Decimal
■ have the letter *a* in their first names?		
■ have a birthday in January or February?		
■ work full-time or part-time?		
■ have ever been on television?		
■ have downloaded music in the last week?		
■ log into Facebook at least once a day?		
■ send 30 or more text messages a day?		
■ have downloaded at least 10 applications to their cell phones?		

Equations, Inequalities, and Problem Solving

2

from Campus to Careers

Automotive Service Technician

Anyone whose car has ever broken down appreciates the talents of automotive service technicians. To work on today's high-tech cars and trucks, a person needs strong diagnostic and problem-solving skills. Courses in automotive repair, electronics, physics, chemistry, English, computers, and mathematics provide a good educational background for a career as a service technician.

Problem 47 in **Study Set 2.3, problem 85** in **Study Set 2.4,** and **problem 54** in **Study Set 2.5** involve situations that an automotive service technician might encounter on the job. The mathematical concepts discussed in this chapter can be used to solve those problems.

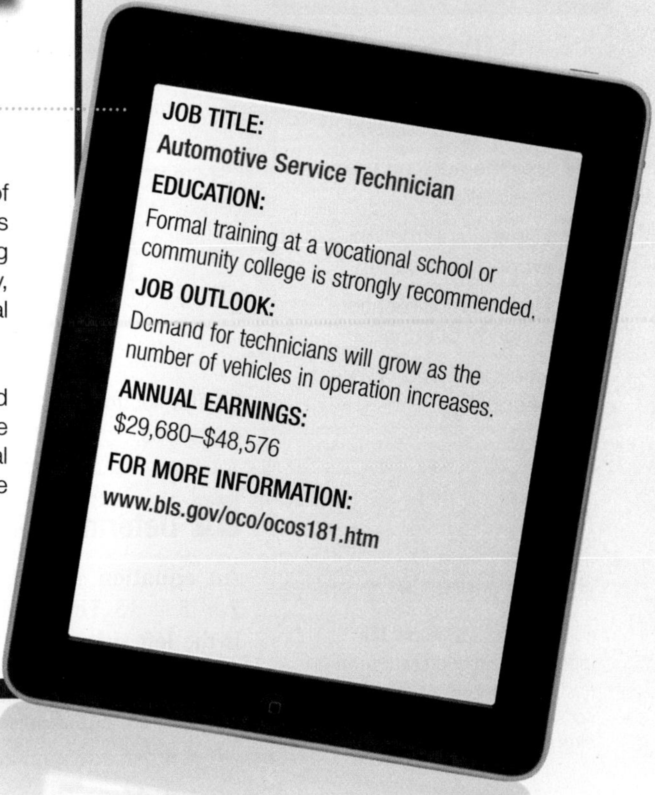
JOB TITLE:
Automotive Service Technician
EDUCATION:
Formal training at a vocational school or community college is strongly recommended.
JOB OUTLOOK:
Demand for technicians will grow as the number of vehicles in operation increases.
ANNUAL EARNINGS:
$29,680–$48,576
FOR MORE INFORMATION:
www.bls.gov/oco/ocos181.htm

Study Skills Workshop ▶ Preparing to Learn

Many students think that there are two types of people—those who are good at math and those who are not—and that this cannot be changed. This isn't true! Here are some suggestions that can increase your chances for success in algebra.

Teaching Tip: Study Skills Workshop topics may be taught in any order, at any time during the course. See the Index for a complete list to determine if and when other Study Skills Workshop topics should be discussed with your students.

DISCOVER YOUR LEARNING STYLE: Are you a visual, verbal, or audio learner? Knowing this will help you determine how best to study.

GET THE MOST OUT OF THE TEXTBOOK: This book and the software that comes with it contain many student-support features. Are you taking advantage of them?

TAKE GOOD NOTES: Are your class notes complete so that they are helpful when doing your homework and studying for tests?

Now Try This ▶

1. To determine what type of learner you are, take the *Learning Style Survey* found online at http://www.metamath.com/multiple/multiple_choice_questions.html. Then, write a one-page paper explaining what you learned from the survey results and how you will use the information to help you succeed in the class.
2. To learn more about the student-support features of this book, take the *Textbook Tour* found online at www.cengage.com/math/tussy.
3. Rewrite a set of your class notes to make them more readable and to clarify the concepts and examples covered. If they are not already, write them in outline form. Fill in any information you didn't have time to copy down in class and complete any phrases or sentence fragments.

Teaching Tip: The *Are You Ready?* feature reviews crucial prerequisite skills that students should have mastered already if they are to be successful with the new topics in this section. All of these problems can be found in Enhanced WebAssign, allowing you to require them to be submitted before class in order to gauge student readiness.

SECTION 2.1 Solving Equations Using Properties of Equality

OBJECTIVES

1. Determine whether a number is a solution.
2. Use the addition property of equality.
3. Use the subtraction property of equality.
4. Use the multiplication property of equality.
5. Use the division property of equality.

ARE YOU READY? *Are You Ready? exercises available online at www.webassign.net/brookscole*

The following problems review some basic skills that are needed when solving equations. Fill in the blanks.

1. $8 - 8 = 0$ **2.** $-1.6 + 1.6 = 0$ **3.** $\frac{7}{7} = 1$

4. $3 \cdot \frac{1}{3} = 1$ **5.** $\frac{2}{5} - \frac{2}{5} = 0$ **6.** $-\frac{9}{8}\left(-\frac{8}{9}\right) = 1$

In this section, we introduce four fundamental properties of equality that are used to solve equations.

1 Determine Whether a Number Is a Solution.

The Language of Algebra

It is important to know the difference between an **equation** and an **expression**. An equation contains an = symbol and an expression does not.

An **equation** is a statement indicating that two expressions are equal. An example is $x + 5 = 15$. The equal symbol = separates the equation into two parts: The expression $x + 5$ is the **left side** and 15 is the **right side.** The letter x is the **variable** (or the **unknown**). The sides of an equation can be reversed, so we can write $x + 5 = 15$ or $15 = x + 5$.

- An equation can be true: $6 + 3 = 9$
- An equation can be false: $2 + 4 = 7$
- An equation can be neither true nor false. For example, $x + 5 = 15$ is neither true nor false because we don't know what number x represents.

An equation that contains a variable is made true or false by substituting a number for the variable. If we substitute 10 for x in $x + 5 = 15$, the resulting equation is true: $10 + 5 = 15$. If we substitute 1 for x, the resulting equation is false: $1 + 5 = 15$. A number that makes an equation true when substituted for the variable is called a **solution** and it is said to **satisfy** the equation. Therefore, 10 is a solution of $x + 5 = 15$, and 1 is not. The **solution set** of an equation is the set of all numbers that make the equation true.

EXAMPLE 1 Check to determine whether 9 is a solution of $3y - 1 = 2y + 7$.

Strategy We will substitute 9 for each y in the equation and evaluate the expression on the left side and the expression on the right side separately.

Why If a true statement results, 9 is a solution of the equation. If we obtain a false statement, 9 is not a solution.

Solution

The Language of Algebra

Read $\stackrel{?}{=}$ as "is possibly equal to."

Evaluate the expression on the left side. | Evaluate the expression on the right side.

$$\begin{aligned} 3y - 1 &= 2y + 7 \\ 3(9) - 1 &\stackrel{?}{=} 2(9) + 7 \\ 27 - 1 &\stackrel{?}{=} 18 + 7 \\ 26 &= 25 \end{aligned}$$

Since $26 = 25$ is false, 9 is not a solution of $3y - 1 = 2y + 7$.

Teaching Example 1 Check to determine whether 15 is a solution of $20 - x = 50 - 3x$.
Answer: Yes

Self Check 1 Check to determine whether 25 is a solution of $10 - x = 35 - 2x$. Yes

Now Try ▶ **Problem 19**

2 Use the Addition Property of Equality.

To **solve an equation** means to find all values of the variable that make the equation true. We can develop an understanding of how to solve equations by referring to the scales shown on the right.

The first scale represents the equation $x - 2 = 3$. The scale is in balance because the weights on the left side and right side are equal. To find x, we must add 2 to the left side. To keep the scale in balance, we must also add 2 to the right side. After doing this, we see that x is balanced by 5. Therefore, x must be 5. We say that we have solved the equation $x - 2 = 3$ and that the solution is 5.

In this example, we solved $x - 2 = 3$ by transforming it to a simpler *equivalent equation*, $x = 5$.

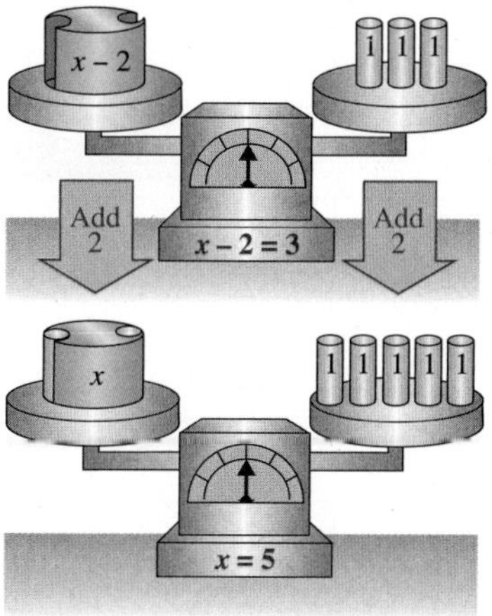

Equivalent Equations	Equations with the same solutions are called **equivalent equations.**

The procedure that we used above suggests the following property of equality.

Addition Property of Equality	Adding the same number to both sides of an equation does not change its solution. For any real numbers a, b, and c, if $a = b$, then $a + c = b + c$

When we use this property, the resulting equation is *equivalent to the original one.* We will now show how it is used to solve $x - 2 = 3$ algebraically.

EXAMPLE 2 Solve: $x - 2 = 3$

Strategy We will use a property of equality to isolate the variable on one side of the equation.

Why To solve the original equation, we want to find a simpler equivalent equation of the form $x =$ **a number**, whose solution is obvious.

Solution

$x - 2 = 3$	This is the equation to solve.
$x - 2 + 2 = 3 + 2$	Use the addition property of equality to isolate x on the left side of the equation. Undo the subtraction of 2 by adding 2 to both sides.
$x + 0 = 5$	The sum of a number and its opposite is zero: $-2 + 2 = 0$.
$x = 5$	When 0 is added to a number, the result is the same number.

The Language of Algebra

We solve equations by writing a series of steps that result in an equivalent equation of the form

x = a number

or

a number = x

We say the variable is **isolated** on one side of the equation. *Isolated* means alone or by itself.

Since 5 is obviously the solution of the equivalent equation $x = 5$, the solution of the original equation, $x - 2 = 3$, is also 5. To check this result, we substitute 5 for x in the original equation and simplify.

$x - 2 = 3$	
$5 - 2 \stackrel{?}{=} 3$	Substitute 5 for x.
$3 = 3$	True

Since the resulting statement is true, 5 is the solution of $x - 2 = 3$. A more formal way to present this result is to write the solution within braces as a solution set: $\{5\}$.

Teaching Example 2
Solve: $x - 12 = 9$
Answer: 21

Self Check 2 Solve: $n - 16 = 33$ 49

Now Try ▶ **Problem 37**

EXAMPLE 3 Solve: **a.** $-19 = y - 7$ **b.** $-27 + y = -3$

Strategy We will use a property of equality to isolate the variable on one side of the equation.

Why To solve the original equation, we want to find a simpler equivalent equation of the form $y =$ **a number** or **a number** $= y$, whose solution is obvious.

Solution

a.

$-19 = y - 7$	This is the equation to solve.
$-19 + 7 = y - 7 + 7$	To isolate y on the right side, use the addition property of equality. Undo the subtraction of 7 by adding 7 to both sides.
$-12 = y$	The sum of a number and its opposite is zero: $-7 + 7 = 0$.

Notation

We may solve an equation so that the variable is isolated on either side of the equation. Note that $-12 = y$ is equivalent to $y = -12$.

Check:

$-19 = y - 7$	This is the original equation.
$-19 \stackrel{?}{=} -12 - 7$	Substitute -12 for y.
$-19 = -19$	True

Since the resulting statement is true, the solution is -12. The solution set is $\{-12\}$.

b.

$-27 + y = -3$	The equation to solve.
$-27 + y + 27 = -3 + 27$	To isolate y, use the addition property of equality. Eliminate -27 on the left side by adding its opposite (additive inverse) to both sides.
$y = 24$	The sum of a number and its opposite is zero: $-27 + 27 = 0$.

Caution

After checking a result, be careful when stating your conclusion. Here, it would be incorrect to say:

~~The solution is −3.~~

The number we were checking was 24, not −3.

Teaching Example 3
Solve: **a.** $-6 = x - 4$
b. $-31 + x = -9$
Answers: **a.** −2 **b.** 22

Check:

$$-27 + y = -3 \quad \text{This is the original equation.}$$
$$-27 + \mathbf{24} \stackrel{?}{=} -3 \quad \text{Substitute 24 for } y.$$
$$-3 = -3 \quad \text{True}$$

The solution is 24. The solution set is {24}.

Self Check 3 Solve: **a.** $-5 = b - 38$ 33 **b.** $-20 + n = 29$ 49

Now Try Problems 41 and 43

3 Use the Subtraction Property of Equality.

Since any subtraction can be written as an addition by adding the opposite of the number to be subtracted, the following property is an extension of the addition property of equality.

Subtraction Property of Equality

Subtracting the same number from both sides of an equation does not change its solution.

For any real numbers a, b, and c,

if $a = b$, then $a - c = b - c$

When we use this property, the resulting equation is equivalent to the original one.

EXAMPLE 4 Solve: **a.** $x + \frac{1}{8} = \frac{7}{4}$ **b.** $54.9 + x = 45.2$

Strategy We will use a property of equality to isolate the variable on one side of the equation.

Why To solve the original equation, we want to find a simpler equivalent equation of the form x **= a number**, whose solution is obvious.

Solution **a.**

$$x + \frac{1}{8} = \frac{7}{4} \quad \text{This is the equation to solve.}$$
$$x + \frac{1}{8} - \frac{1}{8} = \frac{7}{4} - \frac{1}{8} \quad \text{To isolate } x \text{, use the subtraction property of equality. Undo the addition of } \tfrac{1}{8} \text{ by subtracting } \tfrac{1}{8} \text{ from both sides.}$$
$$x = \frac{7}{4} - \frac{1}{8} \quad \text{On the left side, } \tfrac{1}{8} - \tfrac{1}{8} = 0.$$
$$x = \frac{7}{4} \cdot \frac{2}{2} - \frac{1}{8} \quad \text{To prepare to subtract the fractions, build } \tfrac{7}{4} \text{ so that it has a denominator of 8.}$$
$$x = \frac{14}{8} - \frac{1}{8} \quad \text{Multiply the numerators and multiply the denominators.}$$
$$x = \frac{13}{8} \quad \text{Subtract the numerators. Write the result over the common denominator 8. The fraction is in simplest form.}$$

Verify that $\frac{13}{8}$ is the solution by substituting it for x in the original equation and simplifying.

Success Tip

We could also isolate x by adding the additive inverse of $\frac{1}{8}$, which is $-\frac{1}{8}$, to both sides:

$$x + \frac{1}{8} + \left(-\frac{1}{8}\right) = \frac{7}{4} + \left(-\frac{1}{8}\right)$$

Success Tip

It is not necessary to write the solution $\frac{13}{8}$ as a mixed number. In algebra, a solution may be expressed as an improper fraction, as long as it is in simplest form. (The numerator and denominator have no common factors other than 1.)

b.

$$54.9 + x = 45.2$$ This is the equation to solve.

$$54.9 + x - \mathbf{54.9} = 45.2 - \mathbf{54.9}$$ To isolate x, use the subtraction property of equality. Undo the addition of 54.9 by subtracting 54.9 from both sides.

$$x = -9.7$$ On the left side, $54.9 - 54.9 = 0$.

Check:

$$54.9 + x = 45.2$$ This is the original equation.

$$54.9 + (-9.7) \stackrel{?}{=} 45.2$$ Substitute −9.7 for x.

$$45.2 = 45.2$$ True

The solution is −9.7. The solution set is $\{-9.7\}$.

Teaching Example 4
Solve: **a.** $x + \frac{1}{12} = \frac{5}{3}$ **b.** $47.8 + x = 29.1$
Answers: **a.** $\frac{19}{12}$ **b.** -18.7

Self Check 4 Solve: **a.** $x + \frac{4}{15} = \frac{11}{5}$ $\frac{29}{15}$ **b.** $0.7 + a = 0.2$ -0.5

Now Try Problems 45 and 47

4 Use the Multiplication Property of Equality.

To develop another property of equality, consider the first scale shown on the right that represents the equation $\frac{x}{3} = 25$. The scale is in balance because the weights on the left side and right side are equal. To find x, we must triple (multiply by 3) the weight on the left side. To keep the scale in balance, we also must triple the weight on the right side. After doing this, we see in the second illustration that x is balanced by 75. Therefore, x must be 75.

The procedure that we used to keep the scale balanced suggests the following property of equality.

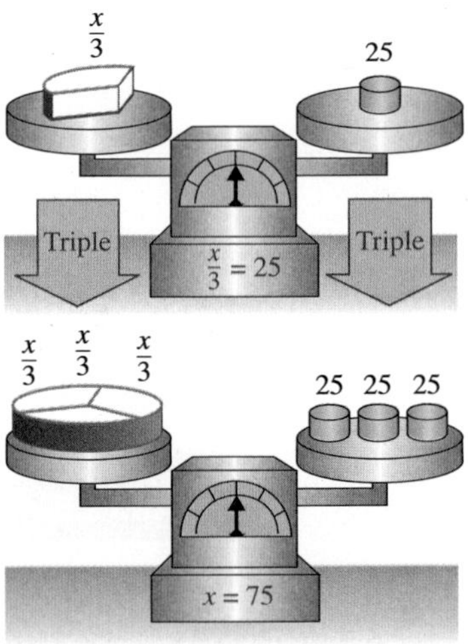

Multiplication Property of Equality

Multiplying both sides of an equation by the same nonzero number does not change its solution.

For any real numbers a, b, and c, where c is not 0,

if $a = b$, then $ca = cb$

When we use this property, the resulting equation is equivalent to the original one. We will now show how it is used to solve $\frac{x}{3} = 25$ algebraically.

EXAMPLE 5 Solve: $\frac{x}{3} = 25$

Strategy We will use a property of equality to isolate the variable on one side of the equation.

Why To solve the original equation, we want to find a simpler equivalent equation of the form $x =$ **a number**, whose solution is obvious.

Solution

$\frac{x}{3} = 25$ This is the equation to solve.

$3 \cdot \frac{x}{3} = 3 \cdot 25$ To isolate x, use the multiplication property of equality. Undo the division by 3 by multiplying both sides by 3.

$\frac{3x}{3} = 75$ Do the multiplication.

$1x = 75$ Simplify $\frac{3x}{3}$ by removing the common factor of 3 in the numerator and denominator: $\frac{3}{3} = 1$.

$x = 75$ The coefficient 1 need not be written since $1x = x$.

If we substitute 75 for x in $\frac{x}{3} = 25$, we obtain the true statement $25 = 25$. This verifies that 75 is the solution. The solution set is $\{75\}$.

Teaching Example 5
Solve: $\frac{a}{27} = 4$
Answer: 108

Self Check 5 Solve: $\frac{b}{24} = 3$ 72

Now Try Problem 49

Since the product of a number and its reciprocal (or multiplicative inverse) is 1, we can solve equations such as $\frac{2}{3}x = 6$, where the coefficient of the variable term is a fraction, as follows.

EXAMPLE 6 Solve: **a.** $\frac{2}{3}x = 6$ **b.** $-\frac{5x}{4} = \frac{3}{16}$

Strategy We will use a property of equality to isolate the variable on one side of the equation.

Why To solve the original equation, we want to find a simpler equivalent equation of the form $x =$ **a number**, whose solution is obvious.

Solution **a.** Since the coefficient of x is $\frac{2}{3}$, we can isolate x by multiplying both sides of the equation by the reciprocal of $\frac{2}{3}$, which is $\frac{3}{2}$.

$\frac{2}{3}x = 6$ This is the equation to solve.

$\frac{3}{2} \cdot \frac{2}{3}x = \frac{3}{2} \cdot 6$ To undo the multiplication by $\frac{2}{3}$, multiply both sides by the reciprocal of $\frac{2}{3}$.

$\left(\frac{3}{2} \cdot \frac{2}{3}\right)x = \frac{3}{2} \cdot 6$ Use the associative property of multiplication to group $\frac{3}{2}$ and $\frac{2}{3}$.

$1x = 9$ On the left side, $\frac{3}{2} \cdot \frac{2}{3} = 1$. On the right side, $\frac{3}{2} \cdot 6 = \frac{18}{2} = 9$.

$x = 9$ The coefficient 1 need not be written since $1x = x$.

Check:

$\frac{2}{3}x = 6$ This is the original equation.

$\frac{2}{3}(9) \stackrel{?}{=} 6$ Substitute 9 for x in the original equation.

$6 = 6$ On the left side, $\frac{2}{3}(9) = \frac{18}{3} = 6$.

Since the resulting statement is true, 9 is the solution. The solution set is $\{9\}$.

Notation

Variable terms with fractional coefficients can be written in two ways. For example:

$-\frac{5x}{4} = -\frac{5}{4}x$

b.

$$-\frac{5x}{4} = \frac{3}{16}$$ This is the equation to solve.

$$-\frac{5}{4}x = \frac{3}{16}$$ Write $-\frac{5x}{4}$ as $-\frac{5}{4}x$.

$$-\frac{4}{5}\left(-\frac{5}{4}x\right) = -\frac{4}{5}\left(\frac{3}{16}\right)$$ To isolate x, undo the multiplication by $-\frac{5}{4}$ by multiplying both sides by the reciprocal of $-\frac{5}{4}$.

$$1x = -\left(\frac{4 \cdot 3}{5 \cdot 16}\right)$$ On the left side, $-\frac{4}{5}\left(-\frac{5}{4}\right) = 1$. On the right side, multiply the numerators and multiply the denominators. Since the signs of the fractions are unlike, make the final answer negative.

$$x = -\frac{\overset{1}{\cancel{4}} \cdot 3}{5 \cdot \underset{1}{\cancel{4}} \cdot 4}$$ On the left side, the coefficient 1 need not be written since 1x = x. On the right side, simplify the fraction by factoring 16 as 4 · 4 and removing the common factor 4.

$$x = -\frac{3}{20}$$ Multiply the remaining factors in the numerator and in the denominator.

Verify that $-\frac{3}{20}$ is the solution by checking.

Caution

If the solution of an equation is a fraction, make sure it is written in simplest form. (The numerator and denominator have no common factors other than 1.)

Teaching Example 6
Solve: **a.** $\frac{4}{9}m = 16$ **b.** $-\frac{2m}{7} = \frac{9}{49}$
Answers: **a.** 36 **b.** $-\frac{9}{14}$

Self Check 6 Solve: **a.** $\frac{7}{2}x = 21$ 6 **b.** $-\frac{3b}{8} = \frac{11}{56}$ $-\frac{11}{21}$

Now Try Problems 53 and 55

5 Use the Division Property of Equality.

Since any division can be rewritten as a multiplication by multiplying by the reciprocal, the following property is a natural extension of the multiplication property of equality.

Division Property of Equality

Dividing both sides of an equation by the same nonzero number does not change its solution.

For any real numbers a, b, and c, where c is not 0,

$$\text{if } a = b, \text{ then } \quad \frac{a}{c} = \frac{b}{c}$$

When we use this property, the resulting equation is equivalent to the original one.

EXAMPLE 7 Solve: **a.** $2t = 80$ **b.** $-6.02 = -8.6t$

Strategy We will use a property of equality to isolate the variable on one side of the equation.

Why To solve the original equation, we want to find a simpler equivalent equation of the form t = **a number** or **a number** = t, whose solution is obvious.

Solution

Success Tip

Since division by 2 is the same as multiplication by $\frac{1}{2}$, we also can solve $2t = 80$ using the multiplication property of equality. We can isolate t by multiplying both sides by the *multiplicative inverse* of 2, which is $\frac{1}{2}$:

$$\frac{1}{2} \cdot 2t = \frac{1}{2} \cdot 80$$

a. $2t = 80$ This is the equation to solve.

$\frac{2t}{2} = \frac{80}{2}$ To isolate t on the left side, use the division property of equality. Undo the multiplication by 2 by dividing both sides of the equation by 2.

$1t = 40$ Simplify $\frac{2t}{2}$ by removing the common factor of 2 in the numerator and denominator: $\frac{2}{2} = 1$.

$t = 40$ The product of 1 and any number is that number: $1t = t$.

If we substitute 40 for t in $2t = 80$, we obtain the true statement $80 = 80$. This verifies that 40 is the solution. The solution set is $\{40\}$.

Success Tip

It is usually easier to multiply on each side if the coefficient of the variable term is a *fraction*, and divide on each side if the coefficient is an *integer* or *decimal*.

b. $-6.02 = -8.6t$ This is the equation to solve.

$\frac{-6.02}{-8.6} = \frac{-8.6t}{-8.6}$ To isolate t on the right side, use the division property of equality. Undo the multiplication by -8.6 by dividing both sides by -8.6.

$0.7 = t$ Do the division: $8.6\overline{)6.02}$. The quotient of two negative numbers is positive.

The solution is 0.7. Verify that this is correct by checking.

Self Check 7 Solve: **a.** $16x = 176$ 11 **b.** $10.04 = -0.4r$ −25.1

Now Try Problems 57 and 59

Teaching Example 7
Solve: **a.** $14n = 154$
b. $15.26 = -0.7h$
Answers: **a.** 11 **b.** −21.8

EXAMPLE 8 Solve: $-x = 3$

Strategy The variable x is not isolated, because there is a − sign in front of it. Since the term $-x$ has an understood coefficient of -1, the equation can be written as $-1x = 3$. We need to select a property of equality and use it to isolate the variable on one side of the equation.

Why To find the solution of the original equation, we want to find a simpler equivalent equation of the form **x = a number**, whose solution is obvious.

Solution To isolate x, we can either multiply or divide both sides by -1.

Multiply both sides by −1:

$-x = 3$ The equation to solve

$-1x = 3$ Write: $-x = -1x$

$(-1)(-1x) = (-1)3$

$1x = 3$

$x = -3$ $1x = x$

Divide both sides by −1:

$-x = 3$ The equation to solve

$-1x = 3$ Write: $-x = -1x$

$\frac{-1x}{1} = \frac{3}{1}$

$1x = -3$ On the left side, $\frac{-1}{-1} = 1$.

$x = -3$ $1x = x$

Either way, we get the same result, −3.

Check:

$-x = 3$ This is the original equation.

$-(-3) \stackrel{?}{=} 3$ Substitute −3 for x.

$3 = 3$ On the left side, the opposite of −3 is 3.

Since the resulting statement is true, −3 is the solution. The solution set is $\{-3\}$.

Self Check 8 Solve: $-h = -12$ 12

Now Try Problem 61

Teaching Example 8
Solve: $-t = 6$
Answer: −6

Teaching Tip: Each Study Set begins with Vocabulary exercises. Educational research has shown that vocabulary plays a crucial role in the process of concept formation in students. All Vocabulary problems can be found in Enhanced WebAssign so that you can require them to be submitted before class to promote student readiness.

SECTION 2.1 STUDY SET

VOCABULARY

Fill in the blanks.

▶ **1.** A statement indicating that two expressions are equal, such as $x + 1 = 7$, is called an equation.

▶ **2.** Any number that makes an equation true when substituted for the variable is said to satisfy the equation. Such numbers are called solutions.

▶ **3.** To solve an equation means to find all values of the variable that make the equation true.

▶ **4.** To solve an equation, we isolate the variable on one side of the equation.

▶ **5.** Equations with the same solutions are called equivalent equations.

▶ **6.** To check the solution of an equation, we substitute the value for the variable in the original equation and determine whether the result is a true statement.

CONCEPTS

7. Given $x + 6 = 12$,

a. What is the left side of the equation? $x + 6$

b. Is this equation true, false, or neither? Neither

c. Is 5 the solution? No

d. Does 6 satisfy the equation? Yes

▶ **8.** For each equation, determine what operation is performed on the variable. Then explain how to undo that operation to isolate the variable.

a. $x - 8 = 24$ Subtraction of 8; add 8

b. $x + 8 = 24$ Addition of 8; subtract 8 or add -8

c. $\frac{x}{8} = 24$ Division by 8; multiply by 8

d. $8x = 24$ Multiplication by 8; divide by 8 or multiply by $\frac{1}{8}$

9. Complete the following properties of equality. If $a = b$, then

a. $a + c = b + c$ and $a - c = b - c$

b. $ca = cb$ and $\frac{a}{c} = \frac{b}{c}$ (where $c \neq 0$)

▶ **10. a.** To solve $\frac{h}{10} = 20$, do we multiply both sides of the equation by 10 or 20? 10

b. To solve $4k = 16$, do we subtract 4 from both sides of the equation or divide both sides by 4? Divide both sides by 4.

11. Simplify each expression.

a. $x + 7 - 7$ x

b. $y - 2 + 2$ y

c. $\frac{5t}{5}$ t

d. $6 \cdot \frac{h}{6}$ h

12. a. To solve $-\frac{4}{5}x = 8$, we can multiply both sides by the reciprocal of $-\frac{4}{5}$. What is the reciprocal of $-\frac{4}{5}$? $-\frac{5}{4}$

b. What is $-\frac{5}{4}\left(-\frac{4}{5}\right)$? 1

NOTATION

Complete each solution to solve the equation.

13.
$$x - 5 = 45$$
$$x - 5 + 5 = 45 + 5$$
$$x = 50$$

Check:
$$x - 5 = 45$$
$$50 - 5 \stackrel{?}{=} 45$$
$$45 = 45 \quad \text{True}$$

50 is the solution.

14.
$$8x = 40$$
$$\frac{8x}{8} = \frac{40}{8}$$
$$x = 5$$

Check:
$$8x = 40$$
$$8(5) \stackrel{?}{=} 40$$
$$40 = 40 \quad \text{True}$$

5 is the solution.

▶ **15. a.** What does the symbol $\stackrel{?}{=}$ mean? Is possibly equal to

b. If you solve an equation and obtain $50 = x$, can you write $x = 50$? Yes

16. Fill in the blank: $-x = -1x$

GUIDED PRACTICE

Check to determine whether the number in red is a solution of the equation. **See Example 1.**

17. 6, $x + 12 = 28$ No

▶ **18.** 110, $x - 50 = 60$ Yes

19. -8, $2b + 3 = -15$ No

▶ **20.** -2, $5t - 4 = -16$ No

21. 5, $0.5x = 2.9$ No

▶ **22.** 3.5, $1.2 + x = 4.7$ Yes

23. -6, $33 - \frac{x}{2} = 30$ No

▶ **24.** -8, $\frac{x}{4} + 98 = 100$ No

▶ **25.** -2, $|c - 8| = 10$ Yes

▶ **26.** -45, $|30 - r| = 15$ No

27. 12, $3x - 2 = 4x - 5$ No

▶ **28.** 5, $5y + 8 = 3y - 2$ No

29. -3, $x^2 - x - 6 = 0$ No

30. -2, $y^2 + 5y - 3 = 0$ No

▶ **31.** 1, $\frac{2}{a + 1} + 5 = \frac{12}{a + 1}$ Yes

▶ **32.** 4, $\frac{2t}{t - 2} - \frac{4}{t - 2} = 1$ No

33. $\frac{3}{4}$, $x - \frac{1}{8} = \frac{5}{8}$ Yes

▶ **34.** $\frac{7}{3}$, $-4 = a + \frac{5}{3}$ No

35. -3, $(x - 4)(x + 3) = 0$ Yes

▶ **36.** 5, $(2x + 1)(x - 5) = 0$ Yes

Use a property of equality to solve each equation. Then check the result. **See Example 2.**

▶ **37.** $a - 5 = 66$ 71

▶ **38.** $x - 34 = 19$ 53

▶ **39.** $9 = p - 9$ 18

40. $3 = j - 88$ 91

Use a property of equality to solve each equation. Then check the result. **See Example 3.**

41. $-16 = y - 4$ -12

▶ **42.** $-23 = y - 19$ -4

43. $-3 + a = 0$ 3

▶ **44.** $-1 + m = 0$ 1

Use a property of equality to solve each equation. Then check the result. **See Example 4.**

45. $x + \frac{1}{10} = \frac{6}{5}$ $\frac{11}{10}$

▶ **46.** $x + \frac{1}{14} = \frac{9}{7}$ $\frac{17}{14}$

▶ **47.** $3.5 + f = 1.2$ -2.3

48. $9.4 + h = 8.1$ -1.3

Use a property of equality to solve each equation. Then check the result. **See Example 5.**

▶ **49.** $\frac{x}{15} = 3$ 45
▶ **50.** $\frac{y}{7} = 12$ 84
51. $\frac{d}{8} = -6$ −48
▶ **52.** $\frac{n}{9} = -2$ −18

Use a property of equality to solve each equation. Then check the result. **See Example 6.**

53. $\frac{4}{5}t = 16$ 20
▶ **54.** $\frac{11}{15}y = 22$ 30
55. $-\frac{7r}{2} = \frac{5}{12}$ $-\frac{5}{42}$
▶ **56.** $-\frac{4a}{5} = \frac{11}{15}$ $-\frac{11}{12}$

Use a property of equality to solve each equation. Then check the result. **See Example 7.**

▶ **57.** $4x = 16$ 4
58. $5y = 45$ 9
▶ **59.** $-1.7 = -3.4y$ 0.5
60. $-1.26 = -2.1x$ 0.6

Use a property of equality to solve each equation. Then check the result. **See Example 8.**

61. $-x = 18$ −18
▶ **62.** $-y = 50$ −50
63. $-n = \frac{4}{21}$ $-\frac{4}{21}$
▶ **64.** $-w = \frac{11}{16}$ $-\frac{11}{16}$

TRY IT YOURSELF

Solve each equation. Then check the result.

▶ **65.** $63 = 9c$ 7
▶ **66.** $40 = 5t$ 8
67. $d - \frac{1}{9} = \frac{7}{9}$ $\frac{8}{9}$
▶ **68.** $\frac{7}{15} = b - \frac{1}{15}$ $\frac{8}{15}$
69. $0 = \frac{v}{11}$ 0
▶ **70.** $\frac{d}{49} = 0$ 0
71. $x - 1.6 = -2.5$ −0.9
72. $y - 1.2 = -1.3$ −0.1
73. $\frac{2}{3}c = 10$ 15
▶ **74.** $\frac{9}{7}d = 81$ 63
▶ **75.** $-100 = -5g$ 20
76. $-80 = -5w$ 16
▶ **77.** $s + \frac{1}{5} = \frac{4}{25}$ $-\frac{1}{25}$
▶ **78.** $\frac{1}{6} = h + \frac{4}{3}$ $-\frac{7}{6}$
79. $\frac{d}{-7} = -3$ 21
80. $\frac{c}{-2} = -11$ 22
81. $8h = 0$ 0
82. $9a = 0$ 0
▶ **83.** $\frac{y}{0.6} = -4.4$ −2.64
▶ **84.** $\frac{y}{0.8} = -2.9$ −2.32
85. $23b = 23$ 1
▶ **86.** $16 = 16h$ 1
87. $-\frac{5}{4}h = -5$ 4
88. $-\frac{3}{8}t = -3$ 8
89. $8.9 = -4.1 + t$ 13
▶ **90.** $7.7 = -3.2 + s$ 10.9
91. $-2.5 = -m$ 2.5
92. $-1.8 = -b$ 1.8
93. $-\frac{9}{8}x = 3$ $-\frac{8}{3}$
94. $-\frac{14}{3}c = 7$ $-\frac{3}{2}$
95. $\frac{2}{3}n = -\frac{7}{8}$ $-\frac{21}{16}$
96. $\frac{4}{9}m = -\frac{3}{5}$ $-\frac{27}{20}$
97. $-10 = n - 5$ −5
▶ **98.** $-8 = t - 2$ −6
99. $\frac{h}{-40} = 5$ −200
▶ **100.** $\frac{x}{-7} = 12$ −84
101. $-\frac{15}{16}a = -\frac{5}{4}$ $\frac{4}{3}$
102. $-\frac{20}{27}b = -\frac{4}{9}$ $\frac{3}{5}$
▶ **103.** $-15x = -60$ 4
▶ **104.** $-14x = -84$ 6

Look Alikes . . .

105. a. $d + \frac{1}{10} = \frac{3}{4}$ $\frac{13}{20}$
b. $d - \frac{1}{10} = \frac{3}{4}$ $\frac{17}{20}$
c. $\frac{1}{10}d = \frac{3}{4}$ $\frac{15}{2}$
d. $10d = \frac{3}{4}$ $\frac{3}{40}$

▶ **106. a.** $x + 4.2 = -18.9$ −23.1
b. $x - 4.2 = -18.9$ −14.7
c. $4.2x = -18.9$ −4.5
d. $\frac{x}{4.2} = -18.9$ −79.38

APPLICATIONS

107. Synthesizers. To find the unknown angle measure, which is represented by x, solve the equation $x + 115 = 180$. 65°

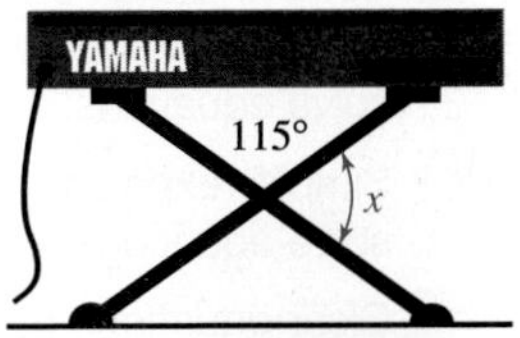

▶ **108. Stop Signs.** To find the degree measure of one angle of the stop sign, which is represented by x, solve the equation $8x = 1{,}080$. 135°

109. Sharing the Winning Ticket. When a Florida Lotto Jackpot was won by a group of 16 nurses employed at a Southwest Florida Medical Center, each received \$375,000. To find the amount of the jackpot, which is represented by x, solve the equation $\frac{x}{16} = 375{,}000$. \$6,000,000

▶ **110. Social Networks.** In 2009, the annual revenue generated per employee at Twitter was \$142,857. This is \$777,143 less than the annual revenue generated per employee at Facebook. To find the annual revenue generated per employee at Facebook, which is represented by x, solve the equation $x - 777{,}143 = 142{,}857$. (Source: businessinsider.com) \$920,000

WRITING

111. What does it mean to solve an equation?

112. Explain the error in the following work.

Solve:
$$-6.4x = 1.1$$
$$-6.4x + 6.4 = 1.1 + 6.4$$
$$x = 7.5$$

113. Explain the error in the following work.

Solve:
$$x + 2 = 40$$
$$x + 2 - 2 = 40$$
$$x = 40$$

▶ **114.** After solving an equation, how do we check the result?

REVIEW

115. Evaluate $-9 - 3x$ for $x = -3$. 0

116. Evaluate: $-5^2 + (-5)^2$ 0

117. Translate to symbols: Subtract x from 45 $45 - x$

118. Evaluate: $\dfrac{2^3 + 3(5 - 3)}{15 - 4 \cdot 2}$ 2

CHALLENGE PROBLEMS

119. If $a + 81 = 49$, what is $a - 81$? -113

120. Find two solutions of $\frac{1}{4}|x + 1| = 25$. 99, -101

Teaching Tip: The *Are You Ready?* feature reviews crucial prerequisite skills that students should have mastered already if they are to be successful with the new topics in this section. All of these problems can be found in Enhanced WebAssign, allowing you to require them to be submitted before class in order to gauge student readiness.

SECTION 2.2 More about Solving Equations

OBJECTIVES

1. Use more than one property of equality to solve equations.
2. Simplify expressions to solve equations.
3. Clear equations of fractions and decimals.
4. Identify identities and contradictions.

ARE YOU READY? *Are You Ready? exercises available online at www.webassign.net/brookscole*

The following problems review some basic skills that are needed when solving equations.

1. Simplify: $4x - 12 - 4x$ -12

2. Simplify: $2a + 2 - 2$ $2a$

3. Simplify: $5m - 3(4m - 6)$ $-7m + 18$

4. Multiply: $5\left(\frac{3}{5}x\right)$ $3x$

5. Multiply: $18\left(\frac{4}{3}n\right)$ $24n$

6. Multiply: $100 \cdot 0.08$ 8

We have solved simple equations by using properties of equality. We now will expand our equation-solving skills by considering more complicated equations. We want to develop a general strategy that can be used to solve any kind of *linear equation in one variable.*

Linear Equation in One Variable

A **linear equation in one variable** can be written in the form

$$ax + b = c$$

where a, b, and c are real numbers and $a \neq 0$. Read $\neq$ as "is not equal to 0."

Some examples of linear and nonlinear equations in one variable are shown below.

Linear equations in one variable (x):		***Not linear equations in one variable:***	
$3x + 1 = 4$	Think: $3x^1 + 1 = 4$	$x^2 - x - 6 = 0$	The exponent on x is not 1.
$-\frac{5}{3}x - 7 = 0$	Think: $-\frac{5}{3}x^1 - 7 = 0$	$\frac{1}{2} + \frac{3}{x} = \frac{5}{6}$	x is in the denominator.

1 Use More Than One Property of Equality to Solve Equations.

Sometimes we must use several properties of equality to solve an equation. For example, on the left side of $2x + 6 = 10$, the variable x is multiplied by 2, and then 6 is added to that product. To isolate x, we use the order of operations rule in reverse. First, we undo the addition of 6, and then we undo the multiplication by 2.

$2x + 6 = 10$	This is the equation to solve.
$2x + 6 \mathbf{- 6} = 10 \mathbf{- 6}$	To undo the addition of 6, subtract 6 from both sides.
$2x = 4$	Do the subtraction.
$\frac{2x}{\mathbf{2}} = \frac{4}{\mathbf{2}}$	To undo the multiplication by 2, divide both sides by 2.
$x = 2$	Do the division.

The solution is 2.

Success Tip

Recall that:
- Subtraction undoes addition.
- Addition undoes subtraction.
- Division undoes multiplication.
- Multiplication undoes division.

EXAMPLE 1 Solve: $-12x + 5 = 17$

Strategy First, we will use a property of equality to isolate the *variable term* on one side of the equation. Then we will use a second property of equality to isolate the *variable* itself.

Why To solve the original equation, we want to find a simpler equivalent equation of the form $x =$ **a number**, whose solution is obvious.

Solution On the left side of the equation, x is multiplied by -12, and then 5 is added to that product. To isolate x, we undo the operations in the opposite order.

The Language of Algebra

We subtract 5 from both sides to isolate the **variable term**, $-12x$. Then we divide both sides by -12 to isolate the **variable**, x.

- To isolate the variable term, $-12x$, we subtract 5 from both sides to undo the addition of 5.
- To isolate the variable, x, we divide both sides by -12 to undo the multiplication by -12.

$$-12x + 5 = 17$$ This is the equation to solve.

$$-12x + 5 - 5 = 17 - 5$$ Use the subtraction property of equality: Subtract 5 from both sides to undo the addition and isolate the variable term, $-12x$.

$$-12x = 12$$ Do the subtractions: $5 - 5 = 0$ and $17 - 5 = 12$.

$$\frac{-12x}{-12} = \frac{12}{-12}$$ Use the division property of equality: Divide both sides by -12 to undo the multiplication and isolate the variable x.

$$x = -1$$ Do the division.

Caution

When checking solutions, always use the original equation.

Check:

$$-12x + 5 = 17$$ This is the original equation.

$$-12(-1) + 5 \stackrel{?}{=} 17$$ Substitute -1 for x.

$$12 + 5 \stackrel{?}{=} 17$$ Do the multiplication on the left side.

$$17 = 17$$ True

Since the resulting statement is true, the solution is -1. The solution set is $\{-1\}$.

Teaching Example 1
Solve: $-9x + 7 = 34$
Answer: -3

Self Check 1 Solve: $8x - 13 = 43$ 7

Now Try ▶ **Problem 13**

EXAMPLE 2 Solve: $\frac{5}{8}m - 2 = -12$

Strategy We will use properties of equality to isolate the variable on one side of the equation.

Why To solve the original equation, we want to find a simpler equivalent equation of the form $m =$ **a number**, whose solution is obvious.

Solution We note that the coefficient of m is $\frac{5}{8}$ and proceed as follows.

- To isolate the variable term, $\frac{5}{8}m$, we add 2 to both sides to undo the subtraction of 2.
- To isolate the variable, m, we multiply both sides by $\frac{8}{5}$ to undo the multiplication by $\frac{5}{8}$.

Teaching Tip: To isolate x, some students may be inclined to divide both sides by $\frac{5}{8}$ to undo the multiplication by $\frac{5}{8}$. Show them that approach introduces complex fractions on each side of the equation and that multiplying both sides by the reciprocal is easier.

$$\frac{\frac{5}{8}m}{\frac{5}{8}} = \frac{-10}{\frac{5}{8}}$$

$\frac{5}{8}m - 2 = -12$	This is the equation to solve.
$\frac{5}{8}m - 2 + 2 = -12 + 2$	Use the addition property of equality: Add 2 to both sides to undo the subtraction and isolate the variable term, $\frac{5}{8}m$.
$\frac{5}{8}m = -10$	Do the additions: $-2 + 2 = 0$ and $-12 + 2 = -10$.
$\frac{8}{5}\left(\frac{5}{8}m\right) = \frac{8}{5}(-10)$	Use the multiplication property of equality: Multiply both sides by $\frac{8}{5}$ (which is the reciprocal of $\frac{5}{8}$) to isolate the variable, m.
$m = -16$	On the left side: $\frac{8}{5}\left(\frac{5}{8}\right) = 1$ and $1m = m$. On the right side, multiply: $\frac{8}{5}(-10) = -\frac{8 \cdot 2 \cdot \overset{1}{\cancel{5}}}{\underset{1}{\cancel{5}}} = -16$.

The solution is -16. Verify this by substituting -16 into the original equation. The solution set is $\{-16\}$.

Teaching Example 2
Solve: $\frac{4}{9}r - 3 = -23$
Answer: -45

Self Check 2 Solve: $\frac{7}{12}a - 6 = -27$ -36

Now Try Problem 17

EXAMPLE 3 Solve: $-0.2 = -0.8 - y$

Strategy First, we will use a property of equality to isolate the variable term on one side of the equation. Then we will use a second property of equality to isolate the variable itself.

Why To solve the original equation, we want to find a simpler equivalent equation of the form **a number** $= y$, whose solution is obvious.

Solution

$-0.2 = -0.8 - y$	This is the equation to solve.
$-0.2 + \mathbf{0.8} = -0.8 - y + \mathbf{0.8}$	To isolate the variable term $-y$ on the right side, we eliminate -0.8 by adding 0.8 to both sides.
$0.6 = -y$	Do the addition.

Since the term $-y$ has an understood coefficient of -1, the equation can be written as $0.6 = -1y$. To isolate y, we can either multiply both sides or divide both sides by -1.

$0.6 = -1y$	If it is helpful, write $-y$ as $-1y$.
$\frac{0.6}{-1} = \frac{-1y}{-1}$	To isolate y, undo the multiplication by -1 by dividing both sides by -1.
$-0.6 = y$	

The solution is -0.6. Verify this by substituting -0.6 into the original equation.

Success Tip

We also can multiply both sides of the equation by -1 to isolate y:

$$-1(0.6) = -1(-1y)$$
$$-0.6 = y$$

Teaching Example 3
Solve: $-0.1 = -0.3 - y$
Answer: -0.2

Self Check 3 Solve: $-6.6 - m = -2.7$ -3.9

Now Try Problem 23

2 Simplify Expressions to Solve Equations.

When solving equations, we should simplify the expressions that make up the left and right sides before applying any properties of equality. Often, that involves using the distributive property to remove parentheses and/or combining like terms.

EXAMPLE 4 Solve: **a.** $3(k + 1) - 5k = 0$ **b.** $10a - 2(2a - 7) = 68$

Strategy We will use the distributive property along with the process of combining like terms to simplify the left side of each equation.

Why It's best to simplify each side of an equation before using a property of equality.

Solution

a.

$3(k + 1) - 5k = 0$	This is the equation to solve.
$3k + 3 - 5k = 0$	Distribute the multiplication by 3.
$-2k + 3 = 0$	Combine like terms: $3k - 5k = -2k$.
$-2k + 3 - 3 = 0 - 3$	To isolate the variable term $-2k$, undo the addition of 3 by subtracting 3 from both sides.
$-2k = -3$	Do the subtraction: $3 - 3 = 0$ and $0 - 3 = -3$.
$\frac{-2k}{-2} = \frac{-3}{-2}$	To isolate the variable k, undo the multiplication by -2 by dividing both sides by -2.
$k = \frac{3}{2}$	Simplify the fraction: $\frac{-3}{-2} = \frac{3}{2}$.

Success Tip

To check a result, evaluate each side of the equation following the order of operations rule.

Check:

$3(k + 1) - 5k = 0$	This is the original equation.
$3\left(\frac{3}{2} + 1\right) - 5\left(\frac{3}{2}\right) \stackrel{?}{=} 0$	Substitute $\frac{3}{2}$ for k.
$3\left(\frac{5}{2}\right) - 5\left(\frac{3}{2}\right) \stackrel{?}{=} 0$	Do the addition within the parentheses. Think of 1 as $\frac{2}{2}$ and then add: $\frac{3}{2} + \frac{2}{2} = \frac{5}{2}$.
$\frac{15}{2} - \frac{15}{2} \stackrel{?}{=} 0$	Do the multiplication.
$0 = 0$	True

The solution is $\frac{3}{2}$ and the solution set is $\left\{\frac{3}{2}\right\}$.

b.

$10a - 2(2a - 7) = 68$	This is the equation to solve.
$10a - 4a + 14 = 68$	Distribute the multiplication by -2.
$6a + 14 = 68$	Combine like terms: $10a - 4a = 6a$.
$6a + 14 - 14 = 68 - 14$	To isolate the variable term $6a$, undo the addition of 14 by subtracting 14 from both sides.
$6a = 54$	Do the subtraction.
$\frac{6a}{6} = \frac{54}{6}$	To isolate the variable a, undo the multiplication by 6 by dividing both sides by 6.
$a = 9$	Do the division.

Use a check to verify that 9 is the solution.

Teaching Example 4
Solve:
a. $4(x + 2) - 7x = 0$
b. $25a - 9(2a - 4) = 8$
Answers: a. $\frac{8}{3}$ b. -4

Self Check 4 Solve: **a.** $4(a + 2) - a = 11$ 1 **b.** $19x - 5(3x - 9) = 1$ -11

Now Try ▶ Problems 25 and 27

When solving an equation, if variables appear on both sides, we can use the addition (or subtraction) property of equality to get all variable terms on one side and all constant terms on the other.

EXAMPLE 5 Solve: $3x - 15 = 4x + 36$

Strategy There are variable terms ($3x$ and $4x$) on both sides of the equation. We will eliminate $3x$ from the left side of the equation by subtracting $3x$ from both sides.

Why To solve for x, all the terms containing x must be on the same side of the equation.

Solution

$3x - 15 = 4x + 36$	This is the equation to solve.
$3x - 15 - 3x = 4x + 36 - 3x$	Subtract 3x from both sides to isolate the variable term on the right side.
$-15 = x + 36$	Combine like terms: $3x - 3x = 0$ and $4x - 3x = x$.
$-15 - 36 = x + 36 - 36$	To isolate x, undo the addition of 36 by subtracting 36 from both sides.
$-51 = x$	Do the subtraction.

Success Tip

We could have eliminated $4x$ from the right side by subtracting $4x$ from both sides. The same solution results.

$$3x - 15 - 4x = 4x + 36 - 4x$$
$$-x - 15 = 36$$

However, it is usually easier to isolate the variable term on the side that will result in a *positive* coefficient.

Check:

$3x - 15 = 4x + 36$	The original equation.
$3(-51) - 15 \stackrel{?}{=} 4(-51) + 36$	Substitute −51 for x.
$-153 - 15 \stackrel{?}{=} -204 + 36$	Do the multiplication.
$-168 = -168$	True

The solution is -51 and the solution set is $\{-51\}$.

Teaching Example 5
Solve: $5x - 4 = 6x + 10$
Answer: -14

Self Check 5 Solve: $30 + 6n = 4n - 2$ -16

Now Try Problem 29

3 Clear Equations of Fractions and Decimals.

Equations are usually easier to solve if they don't involve fractions. We can use the multiplication property of equality to clear an equation of fractions by multiplying both sides of the equation by the least common denominator of all the fractions that appear in the equation.

EXAMPLE 6 Solve: $\frac{1}{6}x + \frac{5}{2} = \frac{1}{3}$

Teaching Tip: It is wise to have students identify the number of terms on each side of the equation before multiplying both sides by the LCD.

Strategy To clear the equation of fractions, we will multiply both sides by the LCD of all the fractions in the equation.

Why It's easier to solve an equation that involves only integers.

Solution

$\frac{1}{6}x + \frac{5}{2} = \frac{1}{3}$	This is the equation to solve.
$6\left(\frac{1}{6}x + \frac{5}{2}\right) = 6\left(\frac{1}{3}\right)$	Multiply both sides by the LCD of $\frac{1}{6}$, $\frac{5}{2}$, and $\frac{1}{3}$, which is 6. Don't forget the parentheses on the left side.
$6\left(\frac{1}{6}x\right) + 6\left(\frac{5}{2}\right) = 6\left(\frac{1}{3}\right)$	On the left side, distribute the multiplication by 6.
$x + 15 = 2$	Do each multiplication: $6\left(\frac{1}{6}\right) = 1$, $6\left(\frac{5}{2}\right) = \frac{30}{2} = 15$, and $6\left(\frac{1}{3}\right) = \frac{6}{3} = 2$. The fractions have been cleared.
$x + 15 - 15 = 2 - 15$	To undo the addition of 15, subtract 15 from both sides.
$x = -13$	

Success Tip

Some students find it helpful to perform each fraction multiplication as shown below:

$$\overset{1}{\cancel{6}}\left(\frac{1}{\underset{1}{\cancel{6}}}x\right) + \overset{3}{\cancel{6}}\left(\frac{5}{\underset{1}{\cancel{2}}}\right) = \overset{2}{\cancel{6}}\left(\frac{1}{\underset{1}{\cancel{3}}}\right)$$

Check the solution by substituting -13 for x in $\frac{1}{6}x + \frac{5}{2} = \frac{1}{3}$.

Teaching Example 6
Solve: $\frac{1}{10}x + \frac{4}{5} = \frac{9}{2}$
Answer: 37

Self Check 6 Solve: $\frac{1}{4}x + \frac{1}{2} = -\frac{1}{8}$ $-\frac{5}{2}$

Now Try Problem 33

If an equation contains decimals, it is often convenient to multiply both sides by a power of 10 to change the decimals in the equation to integers.

EXAMPLE 7 Solve: $0.04(12) + 0.01x = 0.02(12 + x)$

Strategy To clear the equation of decimals, we will multiply both sides by a carefully chosen power of 10.

Why It's easier to solve an equation that involves only integers.

Solution The equation contains the decimals 0.04, 0.01, and 0.02. Since the greatest number of decimal places in any one of these numbers is two, we multiply both sides of the equation by 10^2 or 100. This changes 0.04 to 4, and 0.01 to 1, and 0.02 to 2.

Success Tip

Recall that multiplying a decimal by 10 moves the decimal point 1 place to the right, multiplying it by 100 moves it 2 places to the right, and so on.

$0.04(12) + 0.01x = 0.02(12 + x)$	This is the equation to solve.
$\mathbf{100}[0.04(12) + 0.01x] = \mathbf{100} \cdot 0.02(12 + x)$	Multiply both sides by 100. Don't forget the brackets.
$\mathbf{100} \cdot 0.04(12) + \mathbf{100} \cdot 0.01x = \mathbf{100} \cdot 0.02(12 + x)$	Distribute.
$4(12) + 1x = 2(12 + x)$	Multiply each decimal by 100 by moving its decimal point 2 places to the right.
$48 + x = 24 + 2x$	Distribute the multiplication by 2.
$48 + x - \mathbf{24} - \mathbf{x} = 24 + 2x - \mathbf{24} - \mathbf{x}$	Subtract 24 and x from both sides.
$24 = x$	Simplify each side.
$x = 24$	

Success Tip

When we write the decimals in the original equation as fractions, it becomes more apparent why it is helpful to multiply both sides by the LCD, 100.

$$\frac{4}{100}(12) + \frac{1}{100}x = \frac{2}{100}(12 + x)$$

The solution is 24. Check by substituting 24 for x in the original equation.

Teaching Example 7
Solve:
$0.03x + 0.03(15) = 0.04(5 + x)$
Answer: 25

Self Check 7 Solve: $0.08x + 0.07(15{,}000 - x) = 1{,}110$ 6,000

Now Try Problem 37

The previous examples suggest the following strategy for solving equations. It is important to note that not every step is needed to solve every equation.

Strategy for Solving Linear Equations in One Variable

1. **Clear the equation of fractions or decimals:** Multiply both sides by the LCD to clear fractions or multiply both sides by a power of 10 to clear decimals.
2. **Simplify each side of the equation:** Use the distributive property to remove parentheses, and then combine like terms on each side.
3. **Isolate the variable term on one side:** Add (or subtract) to get the variable term on one side of the equation and a number on the other using the addition (or subtraction) property of equality.
4. **Isolate the variable:** Multiply (or divide) to isolate the variable using the multiplication (or division) property of equality.
5. **Check the result:** Substitute the possible solution for the variable in the *original* equation to see if a true statement results.

EXAMPLE 8 Solve: $\frac{7m + 5}{5} = -4m + 1$

Strategy We will follow the steps of the equation-solving strategy to solve the equation.

Why This is the most efficient way to solve a linear equation in one variable.

$$\frac{7m + 5}{5} = -4m + 1$$ This is the equation to solve.

Step 1 $$5\left(\frac{7m + 5}{5}\right) = 5(-4m + 1)$$ Clear the equation of the fraction by multiplying both sides by 5.

Step 2 $$7m + 5 = -20m + 5$$ On the left side, remove the common factor 5 in the numerator and denominator. On the right side, distribute the multiplication by 5.

Step 3 $$7m + 5 + \mathbf{20m} = -20m + 5 + \mathbf{20m}$$ To eliminate the term $-20m$ on the right side, add $20m$ to both sides.

$$27m + 5 = 5$$ Combine like terms: $7m + 20m = 27m$ and $-20m + 20m = 0$.

$$27m + 5 - \mathbf{5} = 5 - \mathbf{5}$$ To isolate the term $27m$, undo the addition of 5 by subtracting 5 from both sides.

$$27m = 0$$ Do the subtraction.

Step 4 $$\frac{27m}{\mathbf{27}} = \frac{0}{\mathbf{27}}$$ To isolate m, undo the multiplication by 27 by dividing both sides by 27.

$$m = 0$$ 0 divided by any nonzero number is 0.

Step 5 Substitute 0 for m in $\frac{7m + 5}{5} = -4m + 1$ to check that the solution is 0.

Self Check 8 Solve: $6c + 2 = \frac{18 - c}{9}$ 0

Now Try Problem 41

Success Tip

We can remove the common factor 5 from the numerator and the denominator in the following way:

$$\frac{\overset{1}{\cancel{5}}}{1}\left(\frac{7m + 5}{\underset{1}{\cancel{5}}}\right)$$

Caution

Remember that when you multiply one side of an equation by a nonzero number, you must multiply the other side of the equation by the same number.

Teaching Example 8
Solve: $\frac{4x + 9}{2} = 3x + 1$
Answer: $\frac{7}{2}$

4 Identify Identities and Contradictions.

Each of the equations that we solved in Examples 1 through 8 had exactly one solution. However, not every linear equation in one variable has a single solution. Some equations are made true by *any* permissible replacement value for the variable. Such equations are called **identities.** An example of an **identity** is

$$x + x = 2x$$ If we substitute -10 for x, we get the true statement $-20 = -20$. If we substitute 7 for x, we get $14 = 14$, and so on.

Since we can replace x with any number and the equation will be true, all real numbers are solutions of $x + x = 2x$. This equation has infinitely many solutions.

Another type of equation, called a **contradiction,** is false for all replacement values for the variable. An example is

$$x = x + 1$$ No number is equal to 1 more than itself.

Since this equation is false for any value of x, it has no solution.

EXAMPLE 9 Solve: $3(x + 8) + 5x = 2(12 + 4x)$

Strategy We will follow the steps of the equation-solving strategy to solve the equation.

Why This is the most efficient way to solve a linear equation in one variable.

Solution

$3(x + 8) + 5x = 2(12 + 4x)$	This is the equation to solve.
$3x + 24 + 5x = 24 + 8x$	Distribute the multiplication by 3 and by 2.
$8x + 24 = 24 + 8x$	Combine like terms: $3x + 5x = 8x$. Note that the sides of the equation are identical.
$8x + 24 - \mathbf{8x} = 24 + 8x - \mathbf{8x}$	To eliminate the term $8x$ on the right side, subtract $8x$ from both sides.
$24 = 24$	Combine like terms on both sides: $8x - 8x = 0$.

Success Tip

At the step $8x + 24 = 24 + 8x$, we know that the equation is an identity because both sides are exactly the same.

In this case, the terms involving x drop out and the result is true. This means that any number substituted for x in the original equation will give a true statement. Therefore, *all real numbers* are solutions and this equation is an identity. Its solution set is written as {all real numbers} or using the symbol $\mathbb{R}$.

Self Check 9 Solve: $3(x + 5) - 4(x + 4) = -x - 1$ All real numbers; the equation is an identity.

Now Try Problem 45

Teaching Example 9
Solve:
$4(x - 3) + 3x = 12(x - 1) - 5x$
Answer: All real numbers; the equation is an identity.

EXAMPLE 10 Solve: $3(d + 7) - d = 2(d + 10)$

Strategy We will follow the steps of the equation-solving strategy to solve the equation.

Why This is the most efficient way to solve a linear equation in one variable.

Solution

$3(d + 7) - d = 2(d + 10)$	This is the equation to solve.
$3d + 21 - d = 2d + 20$	Distribute the multiplication by 3 and by 2.
$2d + 21 = 2d + 20$	Combine like terms: $3d - d = 2d$.
$2d + 21 - \mathbf{2d} = 2d + 20 - \mathbf{2d}$	To eliminate the term $2d$ on the right side, subtract $2d$ from both sides.
$21 = 20$	Combine like terms on both sides: $2d - 2d = 0$.

The Language of Algebra

Contradiction is a form of the word *contradict,* meaning conflicting ideas. During a trial, evidence might be introduced that *contradicts* the testimony of a witness.

In this case, the terms involving d drop out and the result is false. This means that any number that is substituted for d in the original equation will give a false statement. Therefore, this equation has *no solution* and it is a contradiction. Its solution set is the **empty set**, which is written as { } or using the symbol $\varnothing$.

Self Check 10 Solve: $-4(c - 3) + 2c = 2(10 - c)$ No solution; the equation is a contradiction.

Now Try Problem 47

Teaching Example 10
Solve:
$3(y + 2) - 2y = 4(y - 3) - 3y$
Answer: No solution; the equation is a contradiction.

Teaching Tip: Each Study Set begins with Vocabulary exercises. Educational research has shown that vocabulary plays a crucial role in the process of concept formation in students. All Vocabulary problems can be found in Enhanced WebAssign so that you can require them to be submitted before class to promote student readiness.

SECTION 2.2 STUDY SET

VOCABULARY

Fill in the blanks.

▶ **1.** $3x + 8 = 10$ is an example of a linear equation in one variable.

▶ **2.** To solve $\frac{s}{3} + \frac{1}{4} = -\frac{1}{2}$, we can clear the equation of the fractions by multiplying both sides by 12.

▶ **3.** A linear equation that is true for any permissible replacement value for the variable is called an identity.

▶ **4.** A linear equation that is false for all replacement values for the variable is called a contradiction.

CONCEPTS

Fill in the blanks.

5. a. To solve $3x - 5 = 1$, we first undo the subtraction of 5 by adding 5 to both sides. Then we undo the multiplication by 3 by dividing both sides by 3.

▶ **b.** To solve $\frac{x}{2} + 3 = 5$, we can undo the addition of 3 by subtracting 3 from both sides. Then we can undo the division by 2 by multiplying both sides by 2.

6. a. Combine like terms on the left side of $6x - 8 - 8x = -24$. $-2x - 8 = -24$

b. Distribute and then combine like terms on the right side of $-20 = 4(3x - 4) - 9x$. $-20 = 3x - 16$

7. Use a check to determine whether -2 is a solution of $6x + 5 = 7$. No

8. Multiply.

▶ **a.** $20\left(\frac{3}{5}x\right)$ $12x$ **b.** $100 \cdot 0.02x$ $2x$

9. a. By what must you multiply both sides of $\frac{2}{3} - \frac{1}{2}b = -\frac{4}{3}$ to clear it of fractions? 6

b. By what must you multiply both sides of $0.7x + 0.3(x - 1) = 0.5x$ to clear it of decimals? 10

Look Alikes . . .

10. a. Simplify: $3x + 5 - x$ $2x + 5$

b. Solve: $3x + 5 = 9$ $\frac{4}{3}$

c. Evaluate $3x + 5 - x$ for $x = 9$ 23

d. Check: Is -1 a solution of $3x + 5 - x = 9$? No

NOTATION

Complete the solution.

11. Solve:

$$2x - 7 = 21$$
$$2x - 7 + 7 = 21 + 7$$
$$2x = 28$$
$$\frac{2x}{2} = \frac{28}{2}$$
$$x = 14$$

Check:

$$2x - 7 = 21$$
$$2(14) - 7 \stackrel{?}{=} 21$$
$$28 - 7 \stackrel{?}{=} 21$$
$$21 = 21$$

14 is the solution.

12. A student multiplied both sides of $\frac{3}{4}t + \frac{5}{8} = \frac{1}{2}t$ by 8 to clear it of fractions, as shown below. Explain his error in showing this step.

$8\left(\frac{3}{4}t + \frac{5}{8}\right) = 8 \cdot \frac{1}{2}t$

~~$8 \cdot \frac{3}{4}t + \frac{5}{8} = 8 \cdot \frac{1}{2}t$~~

GUIDED PRACTICE

Solve each equation and check the result. **See Example 1.**

13. $-8x + 1 = 73$ -9

▶ **14.** $-7y + 4 = 60$ -8

▶ **15.** $-5q - 2 = 23$ -5

16. $-4p + 3 = 43$ -10

Solve each equation and check the result. **See Example 2.**

▶ **17.** $\frac{5}{6}k - 5 = 10$ 18

▶ **18.** $\frac{2}{5}c - 12 = 2$ 35

19. $-\frac{7}{16}h + 28 = 21$ 16

▶ **20.** $-\frac{5}{8}h + 25 = 15$ 16

Solve each equation and check the result. **See Example 3.**

21. $-6 - y = -2$ -4

▶ **22.** $-1 - h = -9$ 8

▶ **23.** $-1.7 = 1.2 - x$ 2.9

24. $0.6 = 4.1 - x$ 3.5

Solve each equation and check the result. **See Example 4.**

▶ **25.** $3(2y - 2) - y = 5$ $\frac{11}{5}$

26. $2(-3a + 2) + a = 2$ $\frac{2}{5}$

27. $6a - 3(3a - 4) = 30$ -6

▶ **28.** $16y - 8(3y - 2) = -24$ 5

Solve each equation and check the result. **See Example 5.**

29. $7a - 12 = 8a + 9$ -21

▶ **30.** $10m - 14 = 11m + 13$ -27

31. $60r - 50 = 15r - 5$ 1

▶ **32.** $100f - 75 = 50f + 75$ 3

Solve each equation and check the result. **See Example 6.**

33. $\frac{5}{6}x + \frac{2}{9} = \frac{1}{3}$ $\frac{2}{15}$

▶ **34.** $\frac{2}{3}x + \frac{2}{3} = \frac{3}{4}$ $\frac{1}{8}$

35. $\frac{1}{8}y - \frac{1}{2} = \frac{1}{4}$ 6

▶ **36.** $\frac{1}{15}x - \frac{4}{5} = \frac{2}{3}$ 22

Solve each equation and check the result. **See Example 7.**

37. $0.02(62) - 0.08s = 0.06(s + 9)$ 5

▶ **38.** $0.04(50) + 0.16x = 0.08(x + 50)$ 25

39. $0.09(t + 50) + 0.15t = 52.5$ 200

▶ **40.** $0.08(x - 100) = 44.5 - 0.07x$ 350

Solve each equation and check the result. **See Example 8.**

▶ **41.** $\frac{10 - 5s}{3} = -s + 6$ -4

42. $\frac{40 - 8s}{5} = -2s + 8$ 0

43. $t = \frac{7t - 9}{16}$ -1

▶ **44.** $-3 = \frac{11r + 68}{3}$ -7

Solve each equation, if possible. **See Examples 9 and 10.**

▶ **45.** $8x + 3(2 - x) = 5x + 6$ All real numbers

▶ **46.** $5(x + 2) = 5x - 2$ No solution

47. $-3(s + 2) = -2(s + 4) - s$ No solution

▶ **48.** $21(b - 1) + 3 = 3(7b - 6)$ All real numbers

TRY IT YOURSELF

Solve each equation, if possible. Check the result.

49. $3x - 8 - 4x - 7x = -2 - 8$ $\frac{1}{4}$

50. $-6t - 7t - 5t - 1 = 12 - 3$ $-\frac{5}{9}$

51. $\frac{t}{3} + 2 = 6$ 12

52. $\frac{x}{5} - 5 = -12$ -35

53. $4(5b) + 2(6b - 1) = -34$ -1

54. $9(x + 11) + 5(13 - x) = 0$ -41

▶ 55. $2x + 5 = 17$ 6

56. $3x - 5 = 13$ 6

▶ 57. $\frac{5}{6}(1 - x) = -x + 1$ 1

▶ 58. $\frac{3}{8}(14 - u) = -3u + 6$ $\frac{2}{7}$

▶ 59. $0.05a + 0.01(90) = 0.02(a + 90)$ 30

60. $0.03x + 0.05(2{,}000 - x) = 99.5$ 25

61. $\frac{7}{2} + \frac{3}{2}d = -9 + 1.5d$ No solution

62. $x + 7 = \frac{2x + 6}{2} + 4$ All real numbers

▶ 63. $-(19 - 3s) - (8s + 1) = 35$ -11

64. $2(3x) - 5(3x + 1) = 58$ -7

65. $5x = 4x + 7$ 7

66. $3x = 2x + 2$ 2

▶ 67. $\frac{3(b + 2)}{2} = \frac{4b - 10}{4}$ -11

68. $\frac{2(5a - 7)}{4} = \frac{9(a - 1)}{3}$ -1

▶ 69. $8y - 2 = 4y + 16$ $\frac{9}{2}$

▶ 70. $7 + 3w = 4 + 9w$ $\frac{1}{2}$

▶ 71. $\frac{1}{6}y + \frac{1}{4}y = -1$ $-\frac{12}{5}$

72. $\frac{1}{3}x + \frac{1}{4}x = -2$ $-\frac{24}{7}$

73. $0.7 - 4y = 1.7$ -0.25

74. $0.3 - 2x = -0.9$ 0.6

75. $-33 = 5t + 2$ -7

76. $-55 = 3w + 5$ -20

▶ 77. $-3p + 7 = -3$ $\frac{10}{3}$

▶ 78. $-2r + 8 = -1$ $\frac{9}{2}$

▶ 79. $2(-3) + 4y = 14$ 5

▶ 80. $4(-1) + 3y = 8$ 4

81. $0.06(a + 200) + 0.1a = 172$ 1,000

82. $0.03x + 0.05(6{,}000 - x) = 280$ 1,000

83. $8.6y + 3.4 = 4.2y - 9.8$ -3

84. $9.1y + 3.6 = 6.5y - 30.2$ -13

85. $\frac{2}{3}y + 2 = \frac{1}{5} + y$ $\frac{27}{5}$

▶ 86. $\frac{2}{5}x + 1 = \frac{1}{3} + x$ $\frac{10}{9}$

87. $0.4b - 0.1(b - 100) = 70$ 200

▶ 88. $0.105x + 0.06(20{,}000 - x) = 1{,}740$ 12,000

89. $\frac{1}{4}(10 - 2y) = 8$ -11

90. $\frac{1}{3}(7 - 7x) = -21$ 10

91. $2 - 3(x - 5) = 4(x - 1)$ 3

92. $2 - (4x + 7) = 3 + 2(x + 2)$ -2

93. $2n - \frac{3}{4}n = \frac{1}{2}n + \frac{13}{3}$ $\frac{52}{9}$

▶ 94. $\frac{5}{6}n + 3n = -\frac{1}{3}n - \frac{11}{9}$ $-\frac{22}{75}$

95. $10.08 = 4(0.5x + 2.5)$ 0.04

▶ 96. $-3.28 = 8(1.5y - 0.5)$ 0.06

97. $\frac{3}{4}(d - 8) = \frac{2}{3}(d + 1)$ 80

▶ 98. $\frac{3}{2}(c - 2) = \frac{2}{5}(2c + 3)$ 6

99. $2d + 5 = 0$ $-\frac{5}{2}$

100. $3c + 8 = 0$ $-\frac{8}{3}$

101. $3(A + 2) = 2(A - 7)$ -20

▶ 102. $9(T - 1) = 6(T + 2) - T$ $\frac{21}{4}$

103. $4(a - 3) = -2(a - 6) + 6a$ No solution

104. $9(t + 2) = -6(t - 3) + 15t$ All real numbers

▶ 105. $4(y - 3) - y = 3(y - 4)$ All real numbers

106. $5(x + 3) - 3x = 2(x + 8)$ No solution

107. $-(4 - m) = -10$ -6

▶ 108. $-(6 - t) = -12$ -6

109. $-\frac{2}{3}z + 4 = 8$ -6

110. $-\frac{7}{5}x + 9 = -5$ 10

Look Alikes . . .

Simplify each expression and solve each equation.

111. a. $-2(9 - 3x) - (5x + 2)$ $x - 20$
 b. $-2(9 - 3x) - (5x + 2) = -25$ -5

▶ 112. a. $4(x - 5) - 3(12 - x)$ $7x - 56$
 b. $4(x - 5) - 3(12 - x) = 7$ 9

113. a. $0.6 - 0.2(x + 1)$ $0.4 - 0.2x$
 b. $0.6 - 0.2(x + 1) = 0.4$ 0

114. a. $2(6n + 4) + 4(3n + 2) + 2$ $24n + 18$
 b. $2(6n + 5) = 4(3n + 2) + 2$ All real numbers

WRITING

115. To solve $3x - 4 = 5x + 1$, one student began by subtracting $3x$ from both sides. Another student solved the same equation by first subtracting $5x$ from both sides. Will the students get the same solution? Explain why or why not.

▶ 116. What does it mean to clear an equation such as $\frac{1}{4} + \frac{1}{2}x = \frac{3}{8}$ of the fractions?

117. Explain the error in the following solution.

Solve: $2x + 4 = 30$

$$\frac{2x}{2} + 4 = \frac{30}{2}$$
$$x + 4 = 15$$
$$x + 4 - 4 = 15 - 4$$
$$x = 11$$

118. a. Write an equation that is an identity. Explain why every real number is a solution.
 b. Write an equation that is a contradiction. Explain why no real number is a solution.

REVIEW

Name the property that is used.

119. $x \cdot 9 = 9x$ Com. prop. mult.

120. $4 \cdot \frac{1}{4} = 1$ Mult. inv. prop.

121. $(x + 1) + 2 = x + (1 + 2)$ Assoc. prop. add.

122. $2(30y) = (2 \cdot 30)y$ Assoc. prop. mult.

CHALLENGE PROBLEMS

123. Solve: $\frac{5}{6}\left(-\frac{3}{4}m + 1\right) = -\frac{2}{3}\left(\frac{1}{2}m - 1\right)$ $\frac{4}{7}$

▶ 124. In this section, we discussed equations that have no solution, one solution, and an infinite number of solutions. Do you think an equation could have exactly two solutions? If so, give an example. $|x| = 2$ or $x^2 = 4$; 2 and -2 are solutions.

SECTION 2.3

Applications of Percent

OBJECTIVES

1. Change percents to decimals and decimals to percents.
2. Solve percent problems by direct translation.
3. Solve applied percent problems.
4. Find percent of increase and decrease.
5. Solve discount and commission problems.

ARE YOU READY? *Are You Ready? exercises available online at www.webassign.net/brookscole.*

The following problems review some basic skills that are needed to solve percent problems.

1. Multiply: $100 \cdot 0.61$ 61

2. Multiply: $100 \cdot 0.02$ 2

3. Write $\frac{1}{2}$ and $\frac{3}{4}$ as decimals. 0.5, 0.75

4. Divide: $\frac{27}{100}$ 0.27

5. Multiply: $124 \cdot 0.03$ 3.72

6. Divide: $\frac{17.82}{0.36}$ 49.5

In this section, we will use translation skills from Chapter 1 and equation-solving skills from Chapter 2 to solve problems involving percents.

1 Change Percents to Decimals and Decimals to Percents.

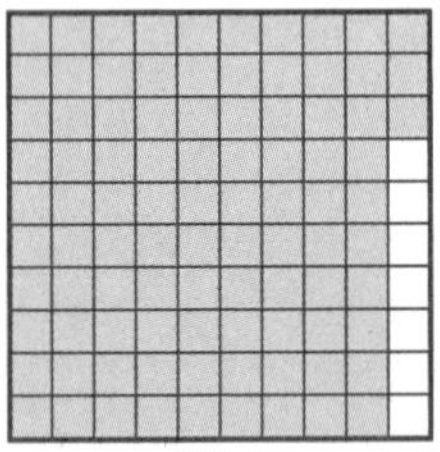

93% or $\frac{93}{100}$ or 0.93 of the figure is shaded.

The word **percent** means parts per one hundred. We can think of the percent symbol % as representing a denominator of 100. Thus, $93\% = \frac{93}{100}$. Since the fraction $\frac{93}{100}$ is equal to the decimal 0.93, it is also true that $93\% = 0.93$. In general, $n\% = \frac{n}{100}$.

When solving percent problems, we must often convert percents to decimals and decimals to percents. To change a percent to a decimal, we drop the % symbol and *divide the given number by 100 by moving the decimal point 2 places to the left.* For example,

$$31\% = 31.0\% = 0.31$$

To change a decimal to a percent, we *multiply the decimal by 100 by moving the decimal point 2 places to the right, and then inserting a % symbol.* For example,

$$0.678 = 67.8\%$$

2 Solve Percent Problems by Direct Translation.

There are three basic types of percent problems. Examples of these are:

Type 1 What number is 8% of 215?

Type 2 102 is 21.3% of what number?

Type 3 31 is what percent of 500?

Every percent problem has three parts: the *amount,* the *percent,* and the *base.* For example, in the question *What number is 8% of 215?,* the words "what number" represent the **amount,** 8% represents the **percent,** and 215 represents the **base.** In these problems, the word "is" means "is equal to," and the word "of" means "multiplication."

What number	is	8%	of	215?
↓	↓	↓	↓	↓
Amount	=	**Percent**	·	**Base**

EXAMPLE 1 What number is 8% of 215?

Strategy We will translate the words of this problem into an equation and then solve the equation.

Why The variable in the translation equation represents the unknown number that we are asked to find.

Solution

The Language of Algebra

Translate the word
- **is** to an equal symbol =
- **of** to multiplication
- **what** to a variable

In this problem, the phrase "what number" represents the amount, 8% is the percent, and 215 is the base.

What number	is	8%	of	215?	
↓	↓	↓	↓	↓	
x	=	0.08	·	215	Change the percent to a decimal: 8% = 0.08.
x	=	17.2			Do the multiplication.

Thus, 8% of 215 is 17.2.

To check, we note that 17.2 out of 215 is $\frac{17.2}{215} = 0.08 = 8\%$.

Teaching Example 1
What number is 6% of 450?
Answer: 27

Self Check 1 What number is 5.6% of 40? 2.24

Now Try ▶ **Problem 13**

We will illustrate the other two types of percent problems with application problems.

3 Solve Applied Percent Problems.

One method for solving applied percent problems is to use the given facts to write a **percent sentence** of the form

______ is ______ % of ______ ?

We enter the appropriate numbers in two of the blanks and the words "what number" or "what percent" in the remaining blank. As before, we translate the words into an equation and solve it.

EXAMPLE 2

Aging Populations. By the year 2050, a study by the Pew Research Center predicts that about 81 million residents of the U.S. will be age 65 or older. The **circle graph** (or **pie chart**) indicates that age group will make up 18.5% of the population. If this prediction is correct, what will the population of the United States be in 2050? (Round to the nearest million.)

Projection of the 2050 U.S. Population by Age

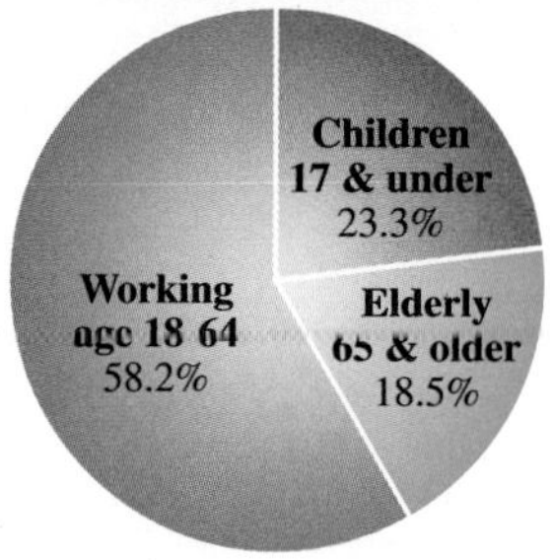

Source: Pew Research Center (2008)

Strategy To find the predicted U.S. population in 2050, we will translate the words of the problem into an equation and then solve the equation.

Why The variable in the translation equation represents the unknown population in 2050 that we are asked to find.

Teaching Tip: Show your students that they do not have to write 81 million as 81,000,000. They can perform the calculations to solve this problem using 81, as long as the units of million are attached to the answer.

Solution

In this problem, 81 is the amount, 18.5% is the percent, and the words "what number" represent the base. The units are millions of people.

Success Tip

Don't forget to change percents to decimal form (or fraction form) before performing any calculations.

81	is	18.5%	of	what number?
↓	↓	↓	↓	↓
81	=	0.185	·	x

$$\frac{81}{\mathbf{0.185}} = \frac{0.185x}{\mathbf{0.185}}$$ To isolate x, undo the multiplication by 0.185 by dividing both sides by 0.185.

$437.8 \approx x$ Do the division.

$438 \approx x$ Round 437.8 million to the nearest million.

The U.S. population is predicted to be about 438 million in the year 2050. We can check using estimation: 81 million out of a population of 438 million is about $\frac{80 \text{ million}}{400 \text{ million}}$, or $\frac{1}{5}$, which is 20%. Since this is close to 18.5%, the answer 438 million seems reasonable.

Teaching Example 2 U.S. Population. By the year 2050, it is predicted that 102 million, or 23.3%, of the U.S. residents will be age 17 or younger. If that prediction is correct, what will the population of the United States be in 2050? (Round to the nearest million.)
Answer: 438 million

Self Check 2 **Aging Populations.** By the year 2100, it is predicted that 131 million, or 23%, of the U.S. residents will be age 65 or older. If the prediction is correct, find the population in 2100. (Round to the nearest million.) 570 million

Now Try Problem 17

We pay many types of taxes in our daily lives, such as sales tax, gasoline tax, income tax, and Social Security tax. **Tax rates** usually are expressed as percents.

EXAMPLE 3 **Taxes.** A maid makes \$500 a week. One of the deductions from her weekly paycheck is a Social Security tax of \$31. Find her Social Security tax rate.

Strategy To find the tax rate, we will translate the words of the problem into an equation and then solve the equation.

Why The variable in the translation equation represents the unknown tax rate that we are asked to find.

Solution

31	is	what percent	of	500?	
↓	↓	↓	↓	↓	
31	=	x	·	500	31 is the amount, x is the percent, and 500 is the base.

Success Tip

One way to check an answer is using *estimation.* If your estimate and your answer are close, you can be reasonably sure that your answer is correct.

$\frac{31}{500} = \frac{500x}{500}$ To undo the multiplication by 500, divide both sides by 500.

$0.062 = x$ Do the division.

$6.2\% = x$ Change the decimal 0.062 to a percent.

The Social Security tax rate is 6.2%.

We can use estimation to check: \$31 out of \$500 is about $\frac{30}{500}$ or $\frac{6}{100}$, which is 6%. Since this is close to 6.2%, the answer seems reasonable.

Teaching Example 3 State Income Tax. A cabinetmaker earns \$980 a week. One of the deductions from his weekly paycheck is a state income tax of \$24.50. Find his state income tax rate.
Answer: 2.5%

Self Check 3 **Medicare Tax.** The maid mentioned in Example 3 also has \$7.25 of Medicare tax deducted from her weekly paycheck. Find her Medicare tax rate. 1.45%

Now Try Problem 23

4 Find Percent of Increase and Decrease.

Percents often are used to describe how a quantity has changed. For example, a health care provider might increase the cost of medical insurance by 3%, or a police department might decrease the number of officers assigned to street patrols by 10%. To describe such changes, we use **percent of increase** or **percent of decrease.**

EXAMPLE 4

IDENTITY THEFT
Federal Trade Commission (2010)

Year	2008	2009
Number of Complaints	314,000	278,000

Identity Theft. The Federal Trade Commission receives complaints involving the theft of someone's identity information, such as a credit card, Social Security number, or cell phone account. Refer to the data in the table. What was the percent of decrease in the number of complaints from 2008 to 2009? (Round to the nearest percent.)

Strategy First, we will subtract to find the *amount of decrease* in the number of complaints. Then we will translate the words of the problem into an equation and solve it.

Why A percent of decrease problem involves finding the *percent of change,* and the change in a quantity is found using subtraction.

Solution To find the *amount of decrease,* we find the difference of 314,000 and 278,000.

$$314{,}000 - 278{,}000 = 36{,}000$$

Caution

The percent of increase (or decrease) is a percent of the *original* number, that is, the number before the change occurred.

Now we translate the percent sentence to an equation, and solve it.

The decrease	is	what percent	of	the number of complaints in 2008?
↓	↓	↓	↓	↓
36,000	=	x	·	314,000

36,000 is the amount, x is the percent, and 314,000 is the base.

$$\frac{36{,}000}{314{,}000} = \frac{314{,}000x}{314{,}000}$$ To undo the multiplication by 314,000, divide both sides by 314,000.

$$0.114649681 \approx x$$ Do the division using a calculator.

$$11.4649681\% \approx x$$ Change the decimal to a percent.

Rounding to the nearest percent, we find that the number of identity theft complaints decreased by about 11% from 2008 to 2009.

A 10% decrease would be 0.10(314,000) or 31,400 fewer complaints. It seems reasonable that 36,000 fewer complaints is an 11% decrease.

Teaching Example 4 PC Games. U.S. retail sales in the PC game software industry were $538 million in 2009. In 2008, retail sales were $695 million. Find the percent decrease in retail sales from 2008 to 2009. Round to the nearest percent. (Source: NPD Group)
Answer: 23%

Self Check 4 **Identity Theft.** In 2007, there were 259,000 complaints of identity theft. Find the percent increase from 2007 to 2008. (Round to the nearest percent.) 21%

Now Try Problem 43

5 Solve Discount and Commission Problems.

When the price of an item is reduced, we call the amount of the reduction a **discount.** If a discount is expressed as a percent, it is called the **rate of discount.**

EXAMPLE 5

Health Club Discounts. A 30% discount on a 1-year membership for a fitness center amounted to a $90 savings. Find the cost of a 1-year membership before the discount.

Strategy We will translate the words of the problem into an equation and then solve the equation.

Why The variable in the translation equation represents the unknown cost of a 1-year membership before the discount.

Solution We are told that \$90 is 30% of some unknown membership cost.

©barang/Shutterstock.com

90	is	30%	of	what number?
↓	↓	↓	↓	↓
90	=	0.30	·	x

90 is the amount, 30% is the percent, and x is the base.

$$\frac{90}{\mathbf{0.30}} = \frac{0.30x}{\mathbf{0.30}}$$ To undo the multiplication by 0.30, divide both sides by 0.30.

$$300 = x$$ Do the division.

A one-year membership cost \$300 before the discount.

Teaching Example 5 Discounts. A 40% discount on a wedding dress amounted to a \$500 savings. Find the cost of the dress before the discount. *Answer*: \$1,250

Self Check 5 **Discounts.** A shopper saved \$6 on a pen that was discounted 5%. Find the original cost. \$120

Now Try Problem 51

Instead of working for a salary or at an hourly rate, many salespeople are paid on **commission.** An employee who is paid a commission is paid a percent of the price of goods or services that he or she sells. We call that percent the **rate of commission.**

EXAMPLE 6 **Commissions.** A real estate agent earned \$14,025 for selling a house. If she received a $5\frac{1}{2}$% commission, what was the selling price?

Strategy We will translate the words of the problem into an equation and then solve the equation.

Why The variable in the translation equation represents the unknown selling price of the house that we are asked to find.

Solution We are told that \$14,025 is $5\frac{1}{2}$% of some unknown selling price of a house.

Success Tip

It is helpful to write percents that involve fractions in an equivalent decimal form. For example,

$1\frac{1}{4}\% = 1.25\%$

$5\frac{1}{2}\% = 5.5\%$

$7\frac{3}{4}\% = 7.75\%$

\$14,025	is	5.5%	of	what number?
↓	↓	↓	↓	↓
14,025	=	0.055	·	x

Write $5\frac{1}{2}$% as 5.5%.

14,025 is the amount, 5.5% is the percent, and x is the base.

$$\frac{14{,}025}{\mathbf{0.055}} = \frac{0.055x}{\mathbf{0.055}}$$ To undo the multiplication by 0.055, divide both sides by 0.055.

$$255{,}000 = x$$ Do the division.

The selling price of the house was \$255,000.

Teaching Example 6 Commissions. An insurance salesperson earns \$2.75 on each monthly premium paid by one of her clients. If she receives a $2\frac{1}{2}$% commission, what is the amount of the client's monthly premium? *Answer*: \$110

Self Check 6 **Commissions.** A jewelry store clerk receives a $4\frac{1}{4}$% commission on all sales. What was the price of a gold necklace sold by the clerk if his commission was \$25.50? \$600

Now Try Problem 53

SECTION 2.3 STUDY SET

VOCABULARY

Fill in the blanks.

▶ **1.** Percent means parts per one hundred.

▶ **2.** In the statement "10 is 50% of 20," 10 is the amount, 50% is the percent, and 20 is the base.

▶ **3.** In percent questions, the word *of* means multiplication, and the word is means equals.

▶ **4.** An employee who is paid a commission is paid a percent of the cost of goods or services that he or she sells.

CONCEPTS

5. Represent the amount of the figure that is shaded using a fraction, a decimal, and a percent.
$\frac{51}{100}$, 0.51, 51%

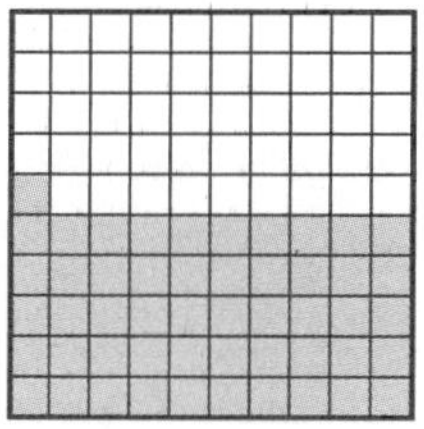

6. Fill in the blanks.

a. To change a percent to a decimal, drop the % symbol and move the decimal point 2 places to the left.

b. To change a decimal to a percent, move the decimal point 2 places to the right and insert a % symbol.

7. Fill in the blanks using the words *percent, amount,* and *base.*

Amount = percent · base

8. Translate each sentence into an equation. **Do not solve.**

▶ **a.** 12 is 40% of what number? $12 = 0.40 \cdot x$

b. 99 is what percent of 200? $99 = x \cdot 200$

c. What is 66% of 3? $x = 0.66 \cdot 3$

9.

Number of Earthquakes in the U.S.

2008	2009
3,618	4,257

Source: U.S. Geological Survey

a. Find the *amount* of increase in the number of earthquakes.
639

b. Fill in blanks to find the percent of increase in earthquakes:
639 is what % of 3,618 ?

10. Use estimation to determine if each statement is reasonable.

a. 18 is 48% of 93. No **b.** 47 is 6% of 206. No

NOTATION

11. Change each percent to a decimal.

a. 35% 0.35 **b.** 8.5% 0.085

c. 150% 1.5 **d.** $2\frac{3}{4}\%$ 0.0275

e. 9.25% 0.0925 **f.** $1\frac{1}{2}\%$ 0.015

▶ **12.** Change each decimal to a percent.

a. 0.9 90% **b.** 0.99 99%

c. 0.999 99.9% **d.** 9 900%

GUIDED PRACTICE

See Example 1.

▶ **13.** What number is 48% of 650? 312

▶ **14.** What number is 60% of 200? 120

▶ **15.** What number is 92.4% of 50? 46.2

▶ **16.** What number is 2.8% of 220? 6.16

See Example 2.

▶ **17.** 75 is 25% of what number? 300

▶ **18.** 78 is 6% of what number? 1,300

19. 128.1 is 8.75% of what number? 1,464

▶ **20.** 1.12 is 140% of what number? 0.8

See Example 3.

21. 78 is what percent of 300? 26%

▶ **22.** 143 is what percent of 325? 44%

23. 0.42 is what percent of 16.8? 2.5%

▶ **24.** 199.92 is what percent of 2,352? 8.5%

APPLICATIONS

25. Antiseptics. Refer to the label on the bottle in figure (a) below. Find the amount of pure hydrogen peroxide in the bottle.
0.48 oz

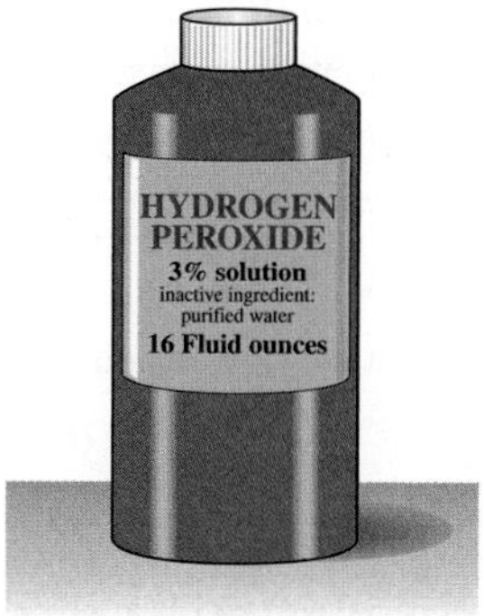

(a)

(b)

▶ **26. Dining Out.** Refer to the sales receipt in figure (b) above. Compute the 15% tip (*rounded up* to the nearest dollar). Then find the total cost of the meal. $12.00, $87.18

▶ **27. U.S. Federal Budget.** The circle graph shows how the government spent $2,980 billion in 2008. How much was spent on

a. Social Security/Medicare? $1,102.6 billion

b. Defense/Veterans? $715.2 billion

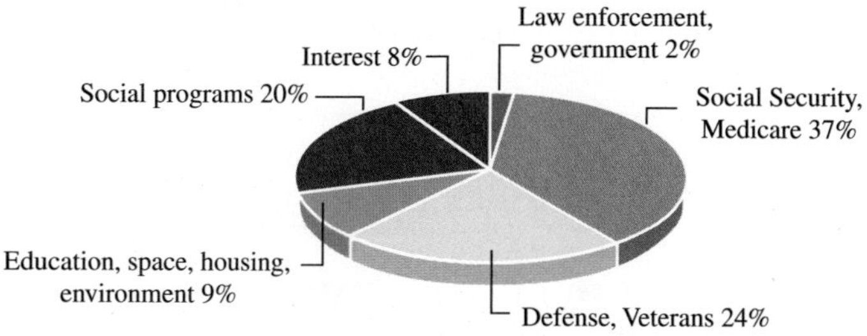

Based on 2009 Federal Income Tax Form 1040

28. Tax Tables. Use the Schedule X table below to compute the amount of federal income tax to be paid on an income of \$39,909. \$6,164.75

If your income is over—	But not over—	Your income tax is—	of the amount over—
\$0	\$8,350	.... 10%	\$0
8,350	33,950	\$835.00 + 15%	8,350
33,950	82,250	4,675.00 + 25%	33,950

29. PayPal. Many e-commerce businesses use PayPal to perform payment processing for them. For certain transactions, merchants are charged a fee of 2.9% of the selling price of the item plus \$0.30. What would PayPal charge an online art store to collect payment on a painting selling for \$350? \$10.45

30. eBay. When a student sold a Play Station 3 on eBay for \$195, she was charged a two-part final value fee: 8% of the first \$50 of the selling price plus 5% of the remainder of the selling price over \$50. Find the fee to sell the Play Station 3 on eBay. \$4.00 + \$7.25 = \$11.25

31. Price Guarantees. Home Club offers a "10% Plus" guarantee: If the customer finds the same item selling for less somewhere else, he or she receives the difference in price plus 10% of the difference. A woman bought miniblinds at the Home Club for \$120 but later saw the same blinds on sale for \$98 at another store. How much can she expect to be reimbursed? \$24.20

32. Room Taxes. A guest at the San Antonio Hilton Airport Hotel paid \$180 for a room plus a 9% city room tax, a $1\frac{3}{4}$% county room tax, and a 6% state room tax. Find the total amount of tax that the guest paid on the room. \$30.15

33. Computer Memory. The *My Computer* screen on a student's computer is shown on the right. What percent of the storage capacity on the hard drive Local Disk (C:) of his computer is used? What percent is free? (GB stands for gigabytes.) 60%, 40%

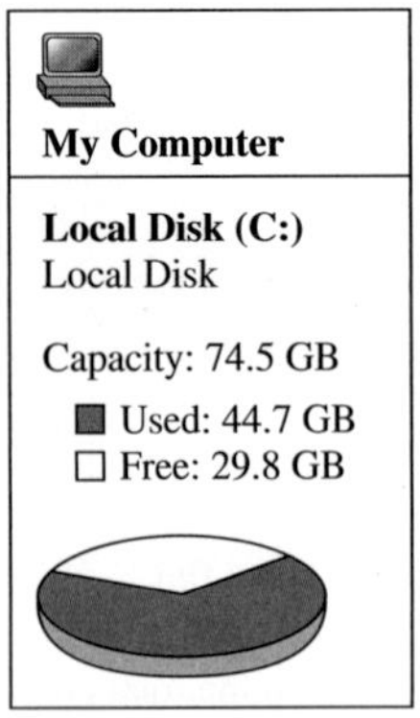

34. Genealogy. Through an extensive computer search, a genealogist determined that worldwide, 180 out of every 10 million people had his last name. What percent is this? 0.0018%

35. Dentistry. Refer to the dental record. What percent of the patient's teeth have fillings? Round to the nearest percent. 19%

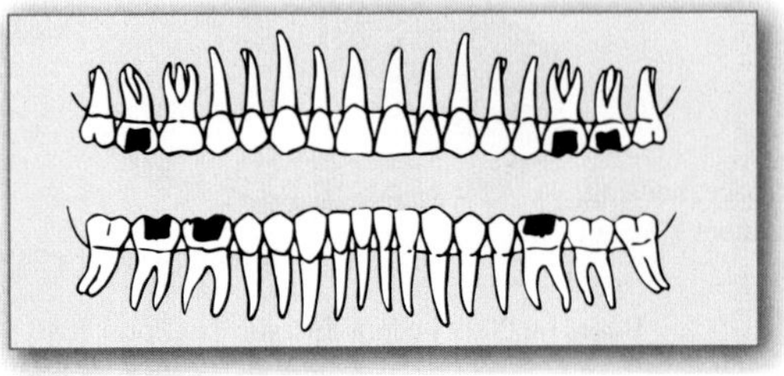

36. Test Scores. The score 175/200 was written by an algebra instructor at the top of a student's test paper. Write the test score as a percent. 87.5%

37. DMV Written Test. To obtain a learner's permit to drive in Nevada, a score of 80% (or better) on a 50-question multiple-choice test is required. If a teenager answered 33 questions correctly, did he pass the test? No (66%)

38. iPods. The settings menu screen of an Apple iPod is shown. What percent of the memory capacity is still available? Round to the nearest percent. (GB stands for gigabytes.) 56%

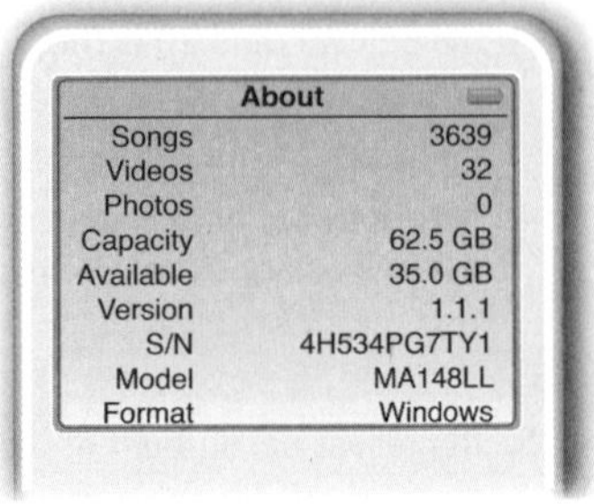

About
Songs 3639
Videos 32
Photos 0
Capacity 62.5 GB
Available 35.0 GB
Version 1.1.1
S/N 4H534PG7TY1
Model MA148LL
Format Windows

39. Child Care. After the first day of registration, 84 children had been enrolled in a day care center. That represented 70% of the available slots. Find the maximum number of children the center could enroll. 120 children

40. Racing Programs. One month before a stock car race, the sale of ads for the official race program was slow. Only 12 pages, or just 30% of the available pages, had been sold. Find the total number of pages devoted to advertising in the program. 40

41. Nutrition. The Nutrition Facts label from a can of clam chowder is shown.

a. Find the number of grams of saturated fat in one serving. What percent of a person's recommended daily intake is this? 5 g; 25%

b. Determine the recommended number of grams of saturated fat that a person should consume daily. 20 g

Nutrition Facts

Serving Size 1 cup (240mL)
Servings Per Container about 2

Amount per serving	
Calories 240	Calories from Fat 140
	% Daily Value*
Total Fat 15 g	**23%**
Saturated Fat 5 g	**25%**
Cholesterol 10 mg	**3%**
Sodium 980 mg	**41%**
Total Carbohydrate 21 g	**7%**
Dietary Fiber 2 g	**8%**
Sugars 1 g	
Protein 7 g	

42. Commercials. Jared Fogle credits his tremendous weight loss to exercise and a diet of low-fat Subway sandwiches. His current weight (about 187 pounds) is 44% of his maximum weight (reached in March of 1998). What did he weigh then? 425 lbs

43. Rainforests. Refer to the graph on the next page, which shows the number of square miles of Brazilian Amazon rainforest that has recently been deforested.

a. Find the percent of increase in the number of square miles of the Brazilian Amazon rainforest that was cleared from 2007 to 2008. Round to the nearest percent. 11%

b. Find the percent of decrease in the number of square miles of the Brazilian Amazon rainforest that was cleared from 2008 to 2009. Round to the nearest percent. 46%

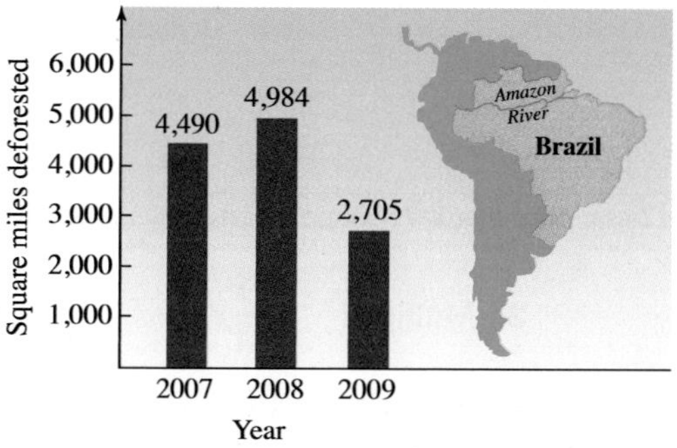

Source: mongabay.com

▶ **44. Auctions.** A pearl necklace of former First Lady Jacqueline Kennedy Onassis, originally valued at $700, was sold at auction in 1996 for $211,500. Find the percent of increase in the value of the necklace. (Round to the nearest percent.) 30,114%

▶ **45. Insurance Costs.** A college student's good grades earned her a student discount on her car insurance premium. Find the percent of decrease to the nearest percent if her annual premium was lowered from $1,050 to $925. 12%

▶ **46. U.S. Life Expectancy.** Use the following life expectancy data for 1900 and 2010 to find the percent of increase for males and for females. Round to the nearest percent. Males: 64%; females: 67%

Years of life expected at birth		
	Male	**Female**
1900	46.3	48.3
2010	75.8	80.8

Source: *The World Fact Book*, CIA

47 from **Campus To Careers**

Automotive Service Technician

© barang/Shutterstock.com

A *single* automobile tire that is underinflated by 5 pounds per square inch (psi) decreases the car's gas mileage by about 1.5%. (Source: auto-buying-tips.com and popularmechanics.com)

a. If all four tires of a car are underinflated by 5 psi, by what percent will its gas mileage be decreased? 6%

b. A 2010 Chevrolet Camaro with properly inflated tires gets 25 miles per gallon on the highway. If each of its tires is underinflated by 5 psi, what will be its highway mileage? 23.5 mpg

▶ **48. Food Labels.** To be labeled "Reduced Fat," foods must contain at least 25% less fat per serving than the regular product. One serving of the original Jif peanut butter has 16 grams of fat per serving. The new Jif Reduced Fat product contains 12 grams of fat per serving. Does it meet the labeling requirement? 25% decrease; yes

▶ **49. TV Shopping.** Jan bought a toy from the QVC home shopping network that was discounted 20%. If she saved $15, what was the original price of the toy? $75

▶ **50. Discounts.** A 12% discount on a watch saved a shopper $48. Find the price of the watch before the discount. $400

51. Sales. The price of a certain model patio set was reduced 35% because it was being discontinued. A shopper purchased two of them and saved a total of $210. Find the price of a patio set before the discount. $300

▶ **52. Craigslist.** A post on the classifieds website Craigslist advertised a mountain bike selling for $800 less than its retail price. If this is a 40% savings, what is the retail price of the bike? $2,000

53. Real Estate. The $3\frac{1}{2}$% commission paid to a real estate agent on the sale of a condominium earned her $3,325. Find the selling price of the condo. $95,000

▶ **54. Consignment.** An art gallery agreed to sell an artist's sculpture for a commission of 45%. What must be the selling price of the sculpture if the gallery would like to make $13,500? $30,000

▶ **55. Stockbrokers.** A stockbroker charges a 2.5% commission to sell shares of a stock for a client. Find the value of stock sold by a broker if the commission was $640. $25,600

▶ **56. Agents.** A sports agent made one million dollars by charging a 12.5% commission to negotiate a long-term contract for a professional athlete. Find the amount of the contract. $8,000,000

WRITING

57. Explain the error: What is 5% of 8?

~~$x = 5 \cdot 8$~~

~~$x = 40$~~

40 is 5% of 8.

▶ **58.** Write a real-life situation that could be described by "9 is what percent of 20?"

59. Explain why 150% of a number is more than the number.

60. Why is the problem "What is 9% of 100?" easy to solve?

REVIEW

61. Divide: $-\frac{16}{25} \div \left(-\frac{4}{15}\right)$ $\frac{12}{5} = 2\frac{2}{5}$

▶ **62.** What two numbers are a distance of 8 away from 4 on the number line? 12 and −4

63. Is −34 a solution of $x + 15 = -49$? No

64. Evaluate: $2 + 3[24 - 2(2 - 5)]$ 92

CHALLENGE PROBLEMS

65. Soaps. A soap advertises itself as $99\frac{44}{100}$% pure. First, determine what percent of the soap is impurities. Then express your answer as a decimal. $\frac{56}{100}\% = \frac{14}{25}\%$; 0.0056

▶ **66.** Express $\frac{1}{20}$ of 1% as a percent using decimal notation. 0.05%

SECTION 2.4

OBJECTIVES

1. Use formulas from business.
2. Use formulas from science.
3. Use formulas from geometry.
4. Solve for a specified variable.

Formulas

ARE YOU READY? *Are You Ready? exercises available online at www.webassign.net/brookscole*

The following problems review some basic skills that are needed when working with formulas.

1. How many variables does each equation contain?

a. $4x + 3 = 15$ one

b. $P = 2l + 2w$ three

2. Simplify: $a + d - a$ d

3. Multiply: $4 \cdot \frac{1}{4}x$ x

4. Multiply: $c(8 - x)$ $8c - cx$

5. If $2 = t$, is it also true that $t = 2$? Yes

6. Simplify: $\frac{b}{b}$ 1

A **formula** is an equation that states a mathematical relationship between two or more variables. The variables of a formula represent quantities such as cost, time, and perimeter. Formulas are used in fields such as business, science, and geometry.

1 Use Formulas from Business.

A formula for retail price: To make a profit, a merchant must sell an item for more than he or she paid for it. The price at which the merchant sells the product, called the **retail price,** is the *sum* of what the item cost the merchant plus the **markup.** Using r to represent the retail price, c the cost, and m the markup, we can write this formula as

$$r = c + m \qquad \text{Retail price} = \text{cost} + \text{markup}$$

A formula for profit: The **profit** a business makes is the *difference* between the **revenue** (the money it takes in) and the cost. Using p to represent the profit, r the revenue, and c the cost, we can write this formula as

$$p = r - c \qquad \text{Profit} = \text{revenue} - \text{cost}$$

If we are given the values of all but one of the variables in a formula, we can use our equation-solving skills to find the value of the remaining variable.

EXAMPLE 1

Teaching Tip: Point out that after the given values for two of the variables are substituted into the formula, the result is a linear equation in one variable. Then, the methods from Sections 2.1 and 2.2 are used to solve for the one remaining variable.

Films. Estimates are that Warner Brothers made a $219 million profit on the film *Harry Potter and the Half-Blood Prince.* If the studio received $469 million in worldwide box office revenue, find the cost to make and distribute the film. (Source: www.thenumbers.com, June 2010)

Strategy To find the cost to make and distribute the film, we will substitute the given values in the formula $p = r - c$ and solve for c.

Why The variable c in the formula represents the unknown cost.

Solution

The film made $219 million (the profit p) and the studio took in $469 million (the revenue r). To find the cost c, we proceed as follows.

$$p = r - c \qquad \text{This is the formula for profit.}$$

$$219 = 469 - c \qquad \text{Substitute 219 for } p \text{ and 469 for } r.$$

$$219 - 469 = 469 - c - 469 \qquad \text{To eliminate 469 on the right side, subtract 469 from both sides.}$$

$$-250 = -c \qquad \text{Do the subtraction.}$$

$$\frac{-250}{-1} = \frac{-c}{-1} \qquad \text{To solve for } c \text{, divide (or multiply) both sides by } -1.$$

$$250 = c \qquad \text{The units are millions of dollars.}$$

It cost $250 million to make and distribute the film.

Teaching Example 1 **Fundraisers.** The sale of fruit baskets by the students in a music department generated \$17,747 in revenue. If a profit of \$12,115 was made, how much did the fruit cost the music department?
Answer: \$5,632

Self Check 1 **Fundraisers.** A PTA spaghetti dinner made a profit of \$275.50. If the cost to host the dinner was \$1,235, how much revenue did it generate? \$1,510.50

Now Try Problem 13

The Language of Algebra

The word **annual** means occurring once a year. An *annual* interest rate is the interest rate paid per year.

A formula for simple interest: When money is borrowed, the lender expects to be paid back the amount of the loan plus an additional charge for the use of the money, called **interest.** When money is deposited in a bank, the depositor is paid for the use of the money. The money the deposit earns is also called interest.

Interest is calculated in two ways: either as **simple interest** or as **compound interest.** Simple interest is the *product* of the principal (the amount of money that is invested, deposited, or borrowed), the annual interest rate, and the length of time in years. Using I to represent the simple interest, P the principal, r the annual interest rate, and t the time in years, we can write this formula as

$$I = Prt \quad \text{Interest} = \text{principal} \cdot \text{rate} \cdot \text{time}$$

EXAMPLE 2 **Retirement Income.** One year after investing \$15,000, a retired couple received a check for \$1,125 in interest. Find the interest rate their money earned that year.

Strategy To find the interest rate, we will substitute the given values in the formula $I = Prt$ and solve for r.

Why The variable r in the formula represents the unknown interest rate.

Solution The couple invested \$15,000 (the principal P) for 1 year (the time t) and made \$1,125 (the interest I). To find the annual interest rate r, we proceed as follows.

$I = Prt$ This is the formula for simple interest.

$1{,}125 = 15{,}000r(1)$ Substitute 1,125 for I, 15,000 for P, and 1 for t.

$1{,}125 = 15{,}000r$ Simplify the right side.

$\frac{1{,}125}{15{,}000} = \frac{15{,}000r}{15{,}000}$ To solve for r, undo the multiplication by 15,000 by dividing both sides by 15,000.

$0.075 = r$ Do the division. This is the rate expressed as a decimal.

$7.5\% = r$ To write 0.075 as a percent, multiply 0.075 by 100 by moving the decimal point two places to the right and inserting a % symbol.

Caution

When using the formula $I = Prt$, always write the interest rate r (which is given as a percent) as a decimal (or fraction) before performing any calculations.

The couple received an annual rate of 7.5% that year on their investment. We can display the facts of the problem in a table as shown on the right.

	P	$\cdot\ r$	$\cdot\ t =$	I
Investment	15,000	0.075	1	1,125

Teaching Example 2 **Saving for College.** A student deposited \$5,000 in a savings account. After one year, she received a check for \$137.50 in interest. Find the annual interest rate her money earned that year.
Answer: 2.75%

Self Check 2 **Home Loans.** A father loaned his daughter \$12,200 at a 2% annual simple interest rate for a down payment on a house. If the interest on the loan amounted to \$610, for how long was the loan? 2.5 yr

Now Try Problem 17

2 Use Formulas from Science.

A formula for distance traveled: If we know the average rate (of speed) at which we will be traveling and the time we will be traveling at that rate, we can find the distance traveled. Using d to represent the distance, r the average rate, and t the time, we can write this formula as

$$d = rt \quad \text{Distance} = \text{rate} \cdot \text{time}$$

EXAMPLE 3 **Whales.** As they migrate from the Bering Sea to Baja California, gray whales swim for about 20 hours each day, covering a distance of approximately 70 miles. Estimate their average swimming rate in miles per hour. (Source: marinebio.net)

Strategy To find the swimming rate, we will substitute the given values in the formula $d = rt$ and solve for r.

Why The variable r in the formula represents the unknown average swimming rate.

Solution The whales swam 70 miles (the distance d) in 20 hours (the time t). To find their average swimming rate r, we proceed as follows.

Caution

When using the formula $d = rt$, make sure the units are consistent. For example, if the rate is given in miles per hour, the time must be expressed in hours.

$$d = rt \quad \text{This is the formula for distance traveled.}$$

$$\mathbf{70} = r(\mathbf{20}) \quad \text{Substitute 70 for } d \text{ and 20 for } t.$$

$$\frac{70}{\mathbf{20}} = \frac{20r}{\mathbf{20}} \quad \text{To solve for } r \text{, undo the multiplication by 20 by dividing both sides by 20.}$$

$$3.5 = r \quad \text{Do the division.}$$

The whales' average swimming rate is 3.5 miles per hour (mph). The facts of the problem can be displayed in a table, as shown on the right.

	r	$\cdot\ t$	$= d$
Gray whale	3.5	20	70

Teaching Example 3 Glaciers. A typical glacier moves about 304 feet in one year (365 days). Estimate the speed of a typical glacier in feet per day. Round to the nearest hundredth. (Source: "Fastest Moving Glacier" on Vimeo)
Answer: 0.83 feet per day

Self Check 3 **Elevators.** An elevator travels at an average rate of 288 feet per minute. How long will it take the elevator to climb 30 stories, a distance of 360 feet? 1.25 min

Now Try Problem 21

A formula for converting temperatures: In the American system, temperature is measured on the Fahrenheit scale. The Celsius scale is used to measure temperature in the metric system. The formula that relates a Fahrenheit temperature F to a Celsius temperature C is:

$$C = \frac{5}{9}(F - 32)$$

EXAMPLE 4 **Temperature Conversion.** Convert the temperature shown on the City Savings sign to degrees Fahrenheit.

Strategy To find the temperature in degrees Fahrenheit, we will substitute the given Celsius temperature in the formula $C = \frac{5}{9}(F - 32)$ and solve for F.

Why The variable F represents the unknown temperature in degrees Fahrenheit.

Solution The temperature in degrees Celsius is 30°. To find the temperature in degrees Fahrenheit F, we proceed as follows.

$$C = \frac{5}{9}(F - 32) \quad \text{This is the formula for temperature conversion.}$$

$$\mathbf{30} = \frac{5}{9}(F - 32) \quad \text{Substitute 30 for } C \text{, the Celsius temperature.}$$

$$\frac{\mathbf{9}}{\mathbf{5}} \cdot 30 = \frac{\mathbf{9}}{\mathbf{5}} \cdot \frac{5}{9}(F - 32) \quad \text{To undo the multiplication by } \tfrac{5}{9}\text{, multiply both sides by the reciprocal of } \tfrac{5}{9}.$$

$$54 = F - 32 \quad \text{Do the multiplication.}$$

$$54 + \mathbf{32} = F - 32 + \mathbf{32} \quad \text{To isolate } F \text{, undo the subtraction of 32 by adding 32 to both sides.}$$

$$86 = F \quad \text{Do the addition.}$$

30°C is equivalent to 86°F.

The Language of Algebra

In 1724, Daniel Gabriel **Fahrenheit,** a German scientist, introduced the temperature scale that bears his name. The Celsius scale was invented in 1742 by Swedish astronomer Anders **Celsius.**

Teaching Example 4 **Planets.** The minimum temperature on Mars is about $-140°C$. Convert the temperature to degrees Fahrenheit. (Source: universetoday.com)
Answer: $-220°F$

Self Check 4 **Planets.** Change $-175°C$, the temperature on Saturn, to degrees Fahrenheit. (Source: universetoday.com) $-283°F$

Now Try Problem 27

3 Use Formulas from Geometry.

The Language of Algebra

When you hear the word **perimeter**, think of the distance around the "rim" of a flat figure.

To find the **perimeter** of a plane (two-dimensional, flat) geometric figure, such as a rectangle or triangle, we find the distance around the figure by computing the sum of the lengths of its sides. Perimeter is measured in American units, such as inches, feet, yards, and in metric units, such as millimeters, meters, and kilometers.

EXAMPLE 5

Perimeter formulas

$P = 2l + 2w$ (rectangle)
$P = 4s$ (square)
$P = a + b + c$ (triangle)

Flags. The largest flag ever flown was an American flag that had a perimeter of 1,520 feet and a length of 505 feet. It was hoisted on cables across Hoover Dam to celebrate the 1996 Olympic Torch Relay. Find the width of the flag.

Strategy To find the width of the flag, we will substitute the given values in the formula $P = 2l + 2w$ and solve for w.

Why The variable w in the formula represents the unknown width of the flag.

Solution

©AP Photo/Las Vegas Review Journal/Carol Ferguson

The perimeter P of the rectangular-shaped flag is 1,520 ft and the length l is 505 ft. To find the width w, we proceed as follows.

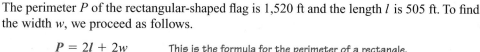

$P = 2l + 2w$ This is the formula for the perimeter of a rectangle.

$\mathbf{1{,}520} = 2(\mathbf{505}) + 2w$ Substitute 1,520 for P and 505 for l.

$1{,}520 = 1{,}010 + 2w$ Do the multiplication.

$510 = 2w$ To undo the addition of 1,010, subtract 1,010 from both sides. This step is done mentally and is not shown.

$255 = w$ To isolate w, undo the multiplication by 2 by dividing both sides by 2. This step is done mentally and is not shown.

The width of the flag is 255 feet. If its length is 505 feet and its width is 255 feet, its perimeter is $2(505) + 2(255) = 1{,}010 + 510 = 1{,}520$ feet, as given.

Teaching Example 5 **LED Diamond Vision.** One of the rectangular screens that hangs from the roof of the Dallas Cowboys Stadium has a perimeter of 464 feet and a length of 160 feet. Find the width of the screen.
Answer: 72 ft

Self Check 5 **Flags.** The largest flag that consistently flies is the flag of Brazil in Brasilia, the country's capital. It has a perimeter 1,116 feet and length 328 feet. Find its width. 230 ft

Now Try Problem 29

Area formulas

$A = lw$ (rectangle)
$A = s^2$ (square)
$A = \frac{1}{2}bh$ (triangle)
$A = \frac{1}{2}h(B + b)$ (trapezoid)

The **area** of a plane (two-dimensional, flat) geometric figure is the amount of surface that it encloses. Area is measured in square units, such as square inches, square feet, square yards, and square meters (written as in.^{2}, ft^2, yd^2, and m^2, respectively).

EXAMPLE 6

a. What is the circumference of a circle with diameter 14 feet? Round to the nearest tenth of a foot. **b.** What is the area of the circle? Round to the nearest tenth of a square foot.

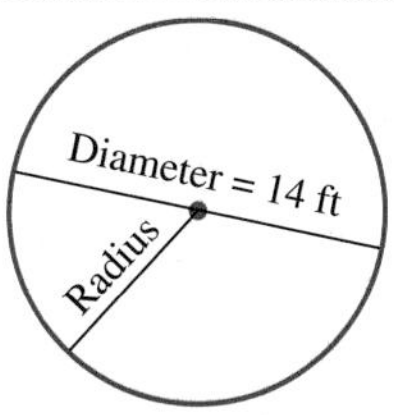

Strategy To find the circumference and area of the circle, we will substitute the proper values into the formulas $C = \pi D$ and $A = \pi r^2$ and find C and A.

Why The variable C in the formula represents the unknown circumference of the circle and A represents the unknown area.

Solution

Circle formulas

$D = 2r$ (diameter)
$r = \frac{1}{2}D$ (radius)
$C = 2\pi r = \pi D$ (circumference)
$A = \pi r^2$ (area)

a. Recall that the circumference of a circle is the distance around it. To find the circumference C of a circle with diameter D equal to 14 ft, we proceed as follows.

$C = \pi D$	This is the formula for the circumference of a circle. πD means $\pi \cdot D$.
$C = \pi(14)$	Substitute 14 for D, the diameter of the circle.
$C = 14\pi$	The exact circumference of the circle is 14π.
$C \approx 43.98229715$	To use a calculator to approximate the circumference, enter π × 14 =. If you do not have a calculator, use 3.14 as an approximation of π. (Answers may vary slightly depending on which approximation of π is used.)

The circumference is exactly 14π ft. Rounded to the nearest tenth, this is 44.0 ft.

b. The radius r of the circle is one-half the diameter, or 7 feet. To find the area A of the circle, we proceed as follows.

$A = \pi r^2$	This is the formula for the area of a circle. πr^2 means $\pi \cdot r^2$.
$A = \pi(7)^2$	Substitute 7 for r, the radius of the circle.
$A = 49\pi$	Evaluate the exponential expression: $7^2 = 49$. The exact area is 49π ft^2.
$A \approx 153.93804$	To use a calculator to approximate the area, enter 49 × π =.

The area is exactly 49π ft^2. To the nearest tenth, the area is 153.9 ft^2.

Notation

Since π is an irrational number, its decimal representation has an infinite number of decimal places. When an approximation of π is used in a calculation, it produces an approximate answer. Use an *is approximately equal to* symbol $\approx$ in your solution to show that.

Teaching Example 6 The diameter of a circle is 18 cm.
a. Find the circumference of the circle. Round to the nearest tenth.
b. What is the area of the circle? Round to the nearest tenth.
Answers: **a.** 18π cm ≈ 56.5 cm **b.** 81π cm$^2 \approx 254.5$ cm^2

Self Check 6 Find the circumference of a circle with radius 10 inches. Round to the nearest hundredth of an inch. 62.83 in.

Now Try Problem 30

The **volume** of a three-dimensional geometric solid is the amount of space it encloses. Volume is measured in cubic units, such as cubic inches, cubic feet, and cubic meters (written as in.^{3}, ft^3, and m^3, respectively).

EXAMPLE 7 Find the volume of the cylinder. Round to the nearest tenth of a cubic centimeter.

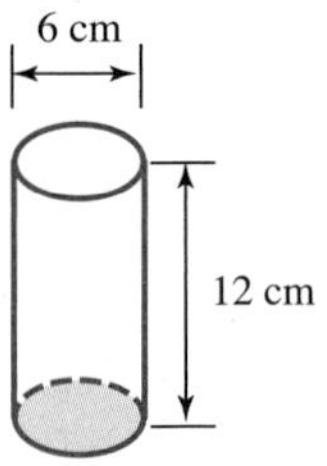

Success Tip

A complete list of geometric figures and formulas appears on the inside of the back cover of this textbook.

Strategy To find the volume of the cylinder, we will substitute the proper values into the formula $V = \pi r^2 h$ and find V.

Why The variable V in the formula represents the unknown volume.

Solution Since the radius of a circle is one-half its diameter, the radius r of the circular base of the cylinder is $\frac{1}{2}(6 \text{ cm}) = 3$ cm. The height h of the cylinder is 12 cm. To find the volume V of the cylinder, we proceed as follows.

Volume formulas

$V = lwh$ (rectangular solid)
$V = s^3$ (cube)
$V = \frac{4}{3}\pi r^3$ (sphere)
$V = \pi r^2 h$ (cylinder)
$V = \frac{1}{3}\pi r^2 h$ (cone)

$V = \pi r^2 h$	This is the formula for the volume of a cylinder. $\pi r^2 h$ means $\pi \cdot r^2 \cdot h$.
$V = \pi(3)^2(12)$	Substitute 3 for r and 12 for h.
$V = \pi(9)(12)$	Evaluate the exponential expression.
$V = 108\pi$	Multiply. The exact volume is 108π cm^3.
$V \approx 339.2920066$	Use a calculator to approximate the volume.

To the nearest tenth, the volume is 339.3 cubic centimeters. This can be written as 339.3 cm^3.

Teaching Example 7 Find the volume of a cylinder with a diameter of 10 in. and a height of 15 in. Round to the nearest tenth.
Answer: 375π in.$^3 \approx 1{,}178.1$ in.^{3}

Self Check 7 Find the volume of a cone whose base has radius 12 meters and whose height is 9 meters. Round to the nearest tenth of a cubic meter. Use the formula $V = \frac{1}{3}\pi r^2 h$. 1,357.2 m^3

Now Try Problem 31

4 Solve for a Specified Variable.

The Language of Algebra

The word **specified** is a form of the word *specify,* which means to select something for a purpose. Here, we select a variable for the purpose of solving for it.

Suppose a shopper wants to calculate the markup m on several items, knowing their retail price r and their cost c to the merchant. It would take a lot of time to substitute values for r and c into the formula for retail price $r = c + m$ and then repeatedly solve for m. A better way is to solve the formula for m first, substitute values for r and c, and then compute m directly.

To **solve a formula for a specified variable** means to isolate that variable on one side of the equation, with all other variables and constants on the opposite side.

EXAMPLE 8 Solve the retail-price formula $r = c + m$ for m.

Teaching Tip: Show your students that solving $r = c + m$ for m is just like solving $8 = 3 + m$ for m.

Strategy To solve for m, we will focus on it as if it is the only variable in the equation. We will use a strategy similar to that used to solve linear equations in one variable to isolate m on one side. (See page 117 if you need to review the strategy.)

Why We can solve the formula as if it were an equation in one variable because all the other variables are treated as if they were numbers (constants).

Solution

We will isolate m on this side of the equation.

$r = c + m$

$r - c = c + m - c$ To isolate m, undo the addition of c by subtracting c from both sides.

$r - c = m$ Simplify the right side: $c - c = 0$.

$m = r - c$ Reverse the sides of the equation so that m is on the left.

The Language of Algebra

We say that the formula is **solved for m** because m is alone on one side of the equation and the other side does not contain m.

The resulting formula, $m = r - c$, indicates that the markup on an item is the difference between its retail price and its cost to the merchant.

Teaching Example 8
Solve $T = a + b$ for a.
Answer: $a = T - b$

Self Check 8 Solve the profit formula $p = r - c$, for c. $c = r - p$

Now Try Problem 33

EXAMPLE 9 Solve the area of a triangle formula $A = \frac{1}{2}bh$ for b.

Strategy To solve for b, we will treat b as the only variable in the equation and use properties of equality to isolate it on one side. We will treat the other variables as if they were numbers (constants).

Why To solve for a specified variable means to isolate it on one side of the equation.

Solution We use the same steps to solve an equation for a specified variable that we use to solve equations with only one variable.

Success Tip

To solve for b, think of it as the only variable in the equation. Treat A and h as if they were numbers. It is often helpful to first circle the variable in the given formula that you are solving for:

$A = \frac{1}{2}\textcircled{b}h$

We will isolate b on this side of the equation.

$A = \frac{1}{2}bh$

$2 \cdot A = 2 \cdot \frac{1}{2}bh$ To clear the equation of the fraction, multiply both sides by 2.

$2A = bh$ Simplify the right side: $2 \cdot \frac{1}{2} = 1$.

$\frac{2A}{h} = \frac{bh}{h}$ bh means $b \cdot h$. To isolate b, undo the multiplication by h by dividing both sides by h.

$\frac{2A}{h} = b$ On the right side, remove the common factor of h: $\frac{b\not{h}^{1}}{\not{h}_{1}} = b$.

$b = \frac{2A}{h}$ Reverse the sides of the equation so that b is on the left.

Teaching Example 9
Solve $V = \frac{1}{3}Bh$ for B.
Answer: $B = \frac{3V}{h}$

Self Check 9 Solve $A = \frac{1}{2}r^2a$ for a. $a = \frac{2A}{r^2}$

Now Try Problem 37

EXAMPLE 10 Solve the perimeter of a rectangle formula $P = 2l + 2w$ for l.

Strategy To solve for l, we will treat l as the only variable in the equation and use properties of equality to isolate it on one side. We will treat the other variables as if they were numbers (constants).

Why To solve for a specified variable means to isolate it on one side of the equation.

Solution

We will isolate l on this side of the equation.

$$P = 2l + 2w$$

$$P - \mathbf{2w} = 2l + 2w - \mathbf{2w}$$ To undo the addition of $2w$, subtract $2w$ from both sides.

$$P - 2w = 2l$$ Combine like terms: $2w - 2w = 0$.

$$\frac{P - 2w}{\mathbf{2}} = \frac{2l}{\mathbf{2}}$$ To isolate l, undo the multiplication by 2 by dividing both sides by 2.

$$\frac{P - 2w}{2} = l$$ Simplify the right side: $\frac{\overset{1}{\cancel{2}}l}{\underset{1}{\cancel{2}}} = l$.

$$l = \frac{P - 2w}{2}$$ Reverse the sides of the equation so that l is on the left.

Caution

Do not try to simplify the result this way:

$$l = \frac{P - \overset{1}{\cancel{2}}w}{\underset{1}{\cancel{2}}}$$

This step is incorrect because 2 is not a factor of the entire numerator.

Teaching Example 10
Solve $h = vt - 16t^2$ for v.
Answer: $v = \frac{h + 16t^2}{t}$

Self Check 10 Solve $B = 3c + 4d$ for c. $c = \frac{B - 4d}{3}$

Now Try Problem 41

EXAMPLE 11 In Chapter 3, we will work with equations that involve the variables x and y, such as $3x + 5y = 10$. Solve this equation for y.

Teaching Tip: Show your students that solving $3 + 2y = 4$ for y is just like solving $3x + 2y = 4$ for y.

Strategy To solve for y, we will treat y as the only variable in the equation and use properties of equality to isolate it on one side.

Why To solve for a specified variable means to isolate it on one side of the equation.

Solution

We will isolate y on this side of the equation.

$$3x + 5y = 10$$

$$3x + 5y - \mathbf{3x} = 10 - \mathbf{3x}$$ To eliminate $3x$ on the left side, subtract $3x$ from both sides.

$$5y = 10 - 3x$$ Combine like terms: $3x - 3x = 0$.

$$\frac{5y}{\mathbf{5}} = \frac{10 - 3x}{\mathbf{5}}$$ To isolate y, undo the multiplication by 5 by dividing both sides by 5.

$$y = \frac{10}{5} - \frac{3x}{5}$$ Write $\frac{10 - 3x}{5}$ as the difference of two fractions with like denominators, $\frac{10}{5}$ and $\frac{3x}{5}$.

$$y = 2 - \frac{3}{5}x$$ Simplify: $\frac{10}{5} = 2$. Write $\frac{3x}{5}$ as $\frac{3}{5}x$.

This result can be written in the following equivalent form:

$$y = -\frac{3}{5}x + 2$$ On the right side, write the x-term first.

Success Tip

When solving for a specified variable, there is often more than one way to express the result. Keep this in mind when you are comparing your answers with those in the back of the text.

Teaching Example 11
Solve $5x + 3y = 9$ for y.
Answer: $y = 3 - \frac{5}{3}x$ or $y = -\frac{5}{3}x + 3$

Self Check 11 Solve $x + 3y = 12$ for y. $y = 4 - \frac{1}{3}x$ or $y = -\frac{1}{3}x + 4$

Now Try ▶ Problem 45

Sometimes the distributive property is used to solve a formula for a specified variable.

EXAMPLE 12 Solve $S = 2\pi(r^2 + rh)$ for h.

Strategy To solve for h, we will treat it as the only variable in the equation and isolate it on one side.

Why To solve for a specified variable means to isolate it on one side of the equation.

Solution We will isolate h on the right side of the equation. To begin, we use the distributive property to remove the parentheses.

$$S = 2\pi(r^2 + rh)$$ This is the given formula.

$$S = 2\pi r^2 + 2\pi rh$$ Distribute the multiplication by 2π.

$$S - 2\pi r^2 = 2\pi r^2 + 2\pi rh - 2\pi r^2$$ To eliminate $2\pi r^2$ on the right side, subtract $2\pi r^2$ from both sides.

$$S - 2\pi r^2 = 2\pi rh$$ On the right side, combine like terms: $2\pi r^2 - 2\pi r^2 = 0$.

$$\frac{S - 2\pi r^2}{2\pi r} = \frac{2\pi rh}{2\pi r}$$ $2\pi rh$ means $2 \cdot \pi \cdot r \cdot h$. To isolate h, undo the multiplication by $2\pi r$ by dividing both sides by $2\pi r$.

$$\frac{S - 2\pi r^2}{2\pi r} = h$$ On the right side, remove the common factors of 2, π, and r: $\frac{\overset{1}{\cancel{2}}\overset{1}{\cancel{\pi}}\overset{1}{\cancel{r}}h}{\underset{1}{\cancel{2}}\underset{1}{\cancel{\pi}}\underset{1}{\cancel{r}}} = h$.

$$h = \frac{S - 2\pi r^2}{2\pi r}$$ Reverse the sides of the equation so that h is on the left.

Teaching Example 12
Solve $r = 4b(d^2 + dz)$ for z.
Answer: $z = \frac{r - 4bd^2}{4bd}$

Self Check 12 Solve $A = -xy(3 - s)$ for s. $s = \frac{A + 3xy}{xy}$

Now Try ▶ Problem 49

SECTION 2.4 STUDY SET

VOCABULARY

Fill in the blanks.

▶ **1.** A <u>formula</u> is an equation that states a mathematical relationship between two or more variables.

▶ **2.** The distance around a plane geometric figure is called its <u>perimeter</u>, and the amount of surface that it encloses is called its <u>area</u>.

▶ **3.** The <u>volume</u> of a three-dimensional geometric solid is the amount of space it encloses.

▶ **4.** The formula $a = P - b - c$ is <u>solved</u> for a because a is isolated on one side of the equation and the other side does not contain a.

CONCEPTS

5. Use variables to write the formula relating:

a. Time, distance, rate $d = rt$

b. Markup, retail price, cost $r = c + m$

c. Costs, revenue, profit $p = r - c$

d. Interest rate, time, interest, principal $I = Prt$

▶ Selected exercises available online at www.webassign.net/brookscole

6. Complete the table.

	Principal ·	rate ·	time =	interest
Account 1	\$2,500	5%	2 yr	\$250
Account 2	\$15,000	4.8%	1 yr	\$720

7. Complete the table to find how far light and sound travel in 60 seconds. (*Hint:* mi/sec means miles per second.)

	Rate ·	time =	distance
Light	186,282 mi/sec	60 sec	11,176,920 mi
Sound	1,088 ft/sec	60 sec	65,280 ft

8. Determine which concept (perimeter, area, or volume) should be used to find each of the following. Then determine which unit of measurement, ft, ft^2, or ft^3, would be appropriate.
   a. The amount of storage in a freezer Volume; ft^3
   b. The amount of ground covered by a sleeping bag lying on the floor Area; ft^2
   c. The distance around a dance floor Perimeter; ft

NOTATION

Complete the solution.

9. Solve $Ax + By = C$ for y.

$$Ax + By = C$$
$$Ax + By - Ax = C - Ax$$
$$By = C - Ax$$
$$\frac{By}{B} = \frac{C - Ax}{B}$$
$$y = \frac{C - Ax}{B}$$

10. Approximate 98π to the nearest hundredth. 307.88
11. a. Write $\pi \cdot r^2 \cdot h$ in simpler form. $\pi r^2 h$
    b. In the formula $V = \pi r^2 h$, what does r represent? What does h represent? The radius of a cylinder; the height of a cylinder
12. a. What does 45°C mean? 45 degrees Celsius
    b. What does 15°F mean? 15 degrees Fahrenheit

GUIDED PRACTICE

***Use a formula to solve each problem.* See Example 1.**

13. **Hollywood.** As of 2010, the movie *Titanic* had brought in \$1,842 million worldwide and made a gross profit of \$1,602 million. What did it cost to make the movie? (Source: the numbers.com/movies) \$240 million
14. **Valentine's Day.** Find the markup on a dozen roses if a florist buys them wholesale for \$12.95 and sells them for \$47.50. \$34.55
15. **Service Clubs.** After expenses of \$55.15 were paid, a Rotary Club donated \$875.85 in proceeds from a pancake breakfast to a local health clinic. How much did the pancake breakfast gross? \$931
16. **New Cars.** The factory invoice for a minivan shows that the dealer paid \$16,264.55 for the vehicle. If the sticker price of the van is \$18,202, how much over factory invoice is the sticker price? \$1,937.45

See Example 2.

17. **Entrepreneurs.** To start a mobile dog-grooming service, a woman borrowed \$2,500. If the loan was for 2 years and the amount of interest was \$175, what simple interest rate was she charged? 3.5%
18. **Savings.** A man deposited \$5,000 in a credit union paying 6% simple interest. How long will the money have to be left on deposit to earn \$6,000 in interest? 20 yr
19. **Loans.** A student borrowed some money from his father at 2% simple interest to buy a car. If he paid his father \$360 in interest after 3 years, how much did he borrow? \$6,000
20. **Banking.** Three years after opening an account that paid simple interest of 6.45% annually, a depositor withdrew the \$3,483 in interest earned. How much money was left in the account? \$18,000

See Example 3.

21. **Swimming.** In 1930, a man swam down the Mississippi River from Minneapolis to New Orleans, a total of 1,826 miles. He was in the water for 742 hours. To the nearest tenth, what was his average swimming rate? 2.5 mph
22. **Parades.** Rose Parade floats travel down the 5.5-mile-long parade route at a rate of 2.5 mph. How long will it take a float to complete the route if there are no delays? 2.2 hr
23. **Hot-Air Balloons.** If a hot-air balloon travels at an average of 37 mph, how long will it take to fly 166.5 miles? 4.5 hours
24. **Air Travel.** An airplane flew from Chicago to San Francisco in 3.75 hours. If the cities are 1,950 miles apart, what was the average speed of the plane? 520 mph

See Example 4.

25. **Frying Foods.** One of the most popular cookbooks in U.S. history, *The Joy of Cooking,* recommends frying foods at 365°F for best results. Convert this to degrees Celsius. 185°C
26. **Freezing Points.** Saltwater has a much lower freezing point than freshwater does. For saltwater that is saturated as much as it can possibly get (23.3% salt by weight), the freezing point is −5.8°F. Convert this to degrees Celsius. −21°C
27. **Biology.** Cryobiologists freeze living matter to preserve it for future use. They can work with temperatures as low as −270°C. Change this to degrees Fahrenheit. −454°F
28. **Metallurgy.** Change 2,212°C, the temperature at which silver boils, to degrees Fahrenheit. Round to the nearest degree. 4,014°F

See Examples 5–7. *If you do not have a calculator, use 3.14 as an approximation of π. Answers may vary slightly depending on which approximation of π is used.*

29. **Energy Savings.** One hundred inches of foam weather stripping tape was placed around the perimeter of a rectangular-shaped window. If the length of the window is 30 inches, what is its width? 20 in.
30. **Rugs.** Find the amount of floor area covered by a circular throw rug that has a radius of 15 inches. Round to the nearest square inch. 707 $in.^2$
31. **Straws.** Find the volume of a 150 millimeter-long drinking straw that has an inside diameter of 4 millimeters. Round to the nearest cubic millimeter. 1,885 mm^3

32. **Rubber Bands.** The world's largest rubber band ball is $5\frac{1}{2}$ ft tall and was made in 2006 by Steve Milton of Eugene, Oregon. Find the volume of the ball. Round to the nearest cubic foot. (*Hint:* The formula for the volume of a sphere is $V = \frac{4}{3}\pi r^3$.) (Source: timesunion.com) 87 ft^3

***Solve for the specified variable.* See Example 8.**

33. $r = c + m$ for c — $c = r - m$
34. $p = r - c$ for r — $r = p + c$
35. $P = a + b + c$ for b — $b = P - a - c$
36. $a + b + c = 180$ for a — $a = 180 - b - c$

***Solve for the specified variable.* See Example 9.**

37. $V = \frac{1}{3}Bh$ for h — $h = \frac{3V}{B}$
38. $C = \frac{1}{7}Rt$ for R — $R = \frac{7C}{t}$
39. $E = IR$ for R — $R = \frac{E}{I}$
40. $d = rt$ for t — $t = \frac{d}{r}$

***Solve for the specified variable.* See Example 10.**

41. $T = 2r + 2t$ for r — $r = \frac{T - 2t}{2}$
42. $y = mx + b$ for x — $x = \frac{y - b}{m}$
43. $Ax + By = C$ for x — $x = \frac{C - By}{A}$
44. $A = P + Prt$ for t — $t = \frac{A - P}{Pr}$

***Solve for y.* See Example 11.**

45. $2x + 7y = 21$ — $y = -\frac{2}{7}x + 3$
46. $3x + 4y = 20$ — $y = -\frac{3}{4}x + 5$
47. $9x - 2y = -8$ — $y = \frac{9}{2}x + 4$
48. $5x - 6y = -12$ — $y = \frac{5}{6}x + 2$

***Solve for the specified variable.* See Example 12.**

49. $T = 4b(a + am)$ for m — $m = \frac{T - 4ab}{4ab}$
50. $f = 7n(d + dz)$ for z — $z = \frac{f - 7dn}{7dn}$
51. $G = g(4r - 1)$ for r — $r = \frac{G + g}{4g}$
52. $F = f(9n - 1)$ for n — $n = \frac{F + f}{9f}$

TRY IT YOURSELF

Solve for the specified variable or expression.

53. $A = \frac{a + b + c}{3}$ for c — $c = 3A - a - b$
54. $x = \frac{a + b}{2}$ for b — $b = 2x - a$
55. $3x + y = 9$ for y — $y = -3x + 9$
56. $-5x + y = 4$ for y — $y = 5x + 4$
57. $K = \frac{1}{2}mv^2$ for m — $m = \frac{2K}{v^2}$
58. $V = \frac{1}{3}\pi r^2 h$ for h — $h = \frac{3V}{\pi r^2}$
59. $C = 2\pi r$ for r — $r = \frac{C}{2\pi}$
60. $V = \pi r^2 h$ for h — $h = \frac{V}{\pi r^2}$
61. $\frac{M}{2} - 9.9 = 2.1B$ for M — $M = 4.2B + 19.8$
62. $\frac{G}{0.5} + 16r = -8t$ for G — $G = -4t - 8r$
63. $w = \frac{s}{f}$ for f — $f = \frac{s}{w}$
64. $P = \frac{ab}{c}$ for c — $c = \frac{ab}{P}$
65. $-x + 3y = 9$ for y — $y = \frac{1}{3}x + 3$
66. $5y - x = 25$ for y — $y = \frac{1}{5}x + 5$
67. $A = \frac{1}{2}h(b + d)$ for b — $b = \frac{2A}{h} - d$ or $b = \frac{2A - hd}{h}$
68. $C = \frac{1}{4}s(t - d)$ for t — $t = \frac{4C + sd}{s}$ or $t = \frac{4C}{s} + d$
69. $c^2 = a^2 + b^2$ for a^2 — $a^2 = c^2 - b^2$
70. $x^2 + y^2 + z^2 = d^2$ for y^2 — $y^2 = d^2 - x^2 - z^2$
71. $\frac{7}{8}c + w = 9$ for c — $c = \frac{72 - 8w}{7}$
72. $\frac{3}{4}m - t = 5b$ for m — $m = \frac{20b + 4t}{3}$
73. $m = 70 + t(a + b)$ for b — $b = \frac{m - 70 - at}{t}$
74. $B = 50 + r(x + y)$ for y — $y = \frac{B - 50 - rx}{r}$
75. $V = lwh$ for l — $l = \frac{V}{wh}$
76. $I = Prt$ for r — $r = \frac{I}{Pt}$
77. $2E = \frac{T - t}{9}$ for t — $t = T - 18E$
78. $D = \frac{C - s}{n}$ for s — $s = C - Dn$
79. $s = 4\pi r^2$ for r^2 — $r^2 = \frac{s}{4\pi}$
80. $E = mc^2$ for c^2 — $c^2 = \frac{E}{m}$

Look Alikes . . .

81. Solve $A = R + ab$
    a. for R — $R = A - ab$
    b. for a — $a = \frac{A - R}{b}$
82. Solve $m = (a + d)T$
    a. for T — $T = \frac{m}{a + d}$
    b. for d — $d = \frac{m - aT}{T}$ or $d = \frac{m}{T} - a$
83. Solve $S = 2(2lw + wh)$
    a. for h — $h = \frac{S - 4lw}{2w}$
    b. for l — $l = \frac{S - 2wh}{4w}$
84. Solve $t = -40 + 9(r + az)$
    a. for r — $r = \frac{t + 40 - 9az}{9}$
    b. for z — $z = \frac{t + 40 - 9r}{9a}$

APPLICATIONS

85. **from Campus to Careers**

Automotive Service Technician

© barang/Shutterstock.com

One of the formulas that is often used by automotive technicians who service engines is:

$$\text{Torque} = \frac{5{,}252 \cdot \text{Horsepower}}{\text{RPM}}$$

RPM stands for revolutions per minute. Solve the formula for horsepower. $\text{Horsepower} = \frac{\text{RPM} \cdot \text{Torque}}{5{,}252}$

86. Properties of Water. Refer to the illustration below. Use the temperature formula from this section to find the boiling point of water in degrees Fahrenheit and the freezing point of water in degrees Celsius. 212°F, 0°C

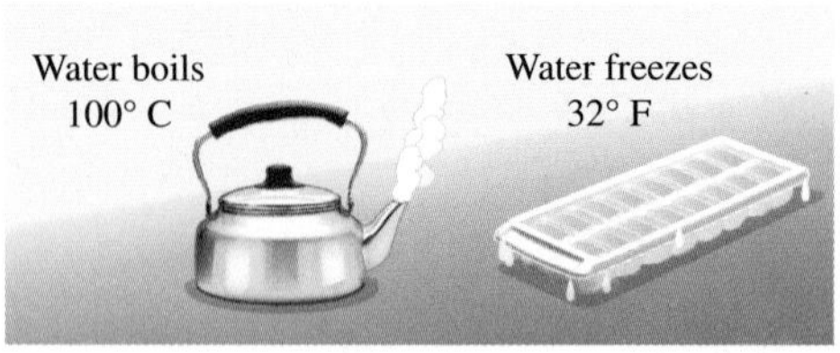

87. Avon Products. Complete the financial statement.

Income statement (dollar amounts in millions)	Quarter ending March '09	Quarter ending March '10
Revenue	2,186.9	2,490.4
Cost of goods sold	2,018.5	2,297.6
Operating profit	168.4	192.8

Source: Avon Products, Inc.

88. Credit Cards. The finance charge that a student pays on his credit card is 19.8% APR (annual percentage rate). Determine the finance charges (interest) the student would have to pay if the account's average balance for the year was $2,500. $495

89. Campers. The perimeter of the window of the camper shell is 140 in. Find the length of one of the shorter sides of the window. 14 in.

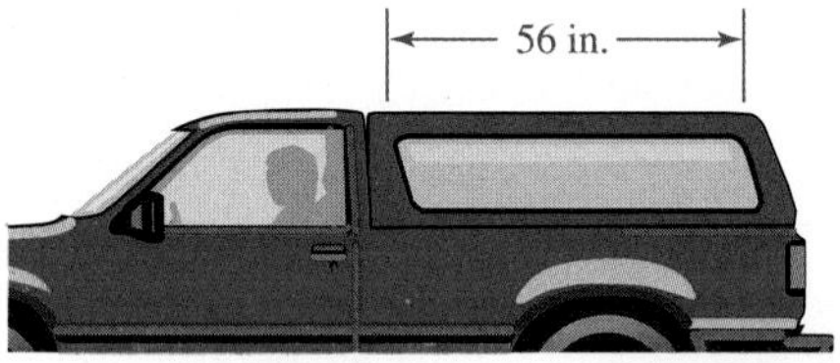

90. Flags. The flag of Eritrea, a country in east Africa, is shown. The perimeter of the flag is 160 inches.

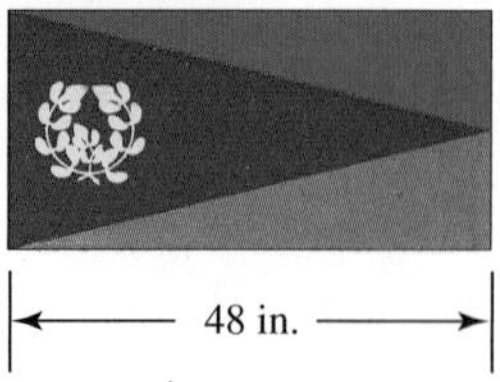

a. What is the width of the flag? 32 in.

b. What is the area of the red triangular region of the flag? 768 in.^{2}

91. Kites. 650 in.^{2} of nylon cloth were used to make the kite shown. If its height is 26 inches, what is the wingspan? 50 in.

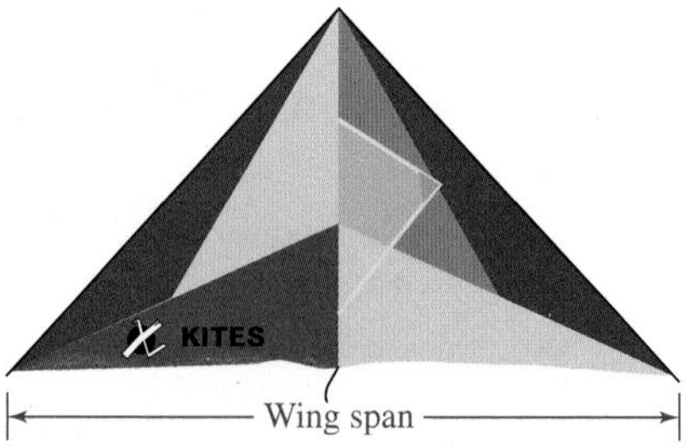

92. Memorials. The Vietnam Veterans Memorial is a black granite wall recognizing the more than 58,000 Americans who lost their lives or remain missing. Find the total area of the two triangular-shaped surfaces on which the names are inscribed. 2,450 ft^2

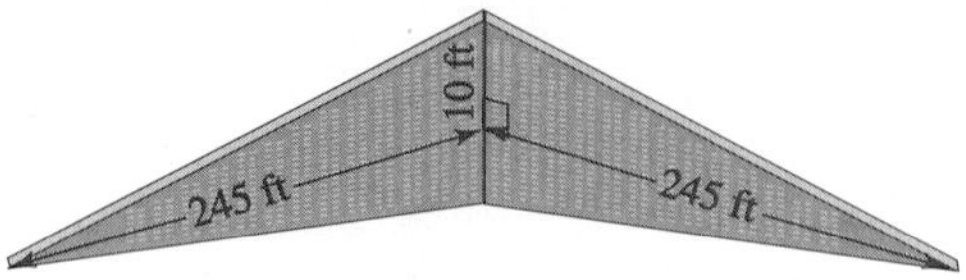

93. Wheelchairs. Find the diameter of the rear wheel and the radius of the front wheel. 25 in., 2.5 in.

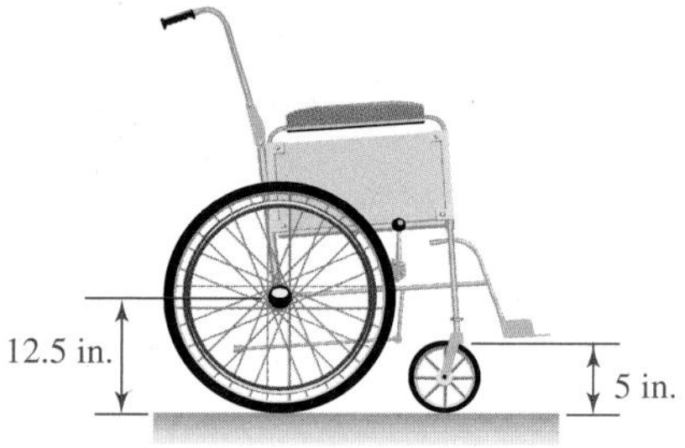

94. Archery. The diameter of a standard archery target used in the Olympics is 48.8 inches. Find the area of the target. Round to the nearest square inch. 1,870 in.^{2}

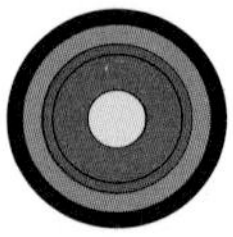

95. Bulls-Eye. See Exercise 94. The diameter of the center yellow ring of a standard archery target is 4.8 inches. What is the area of the bulls-eye? Round to the nearest tenth of a square inch. 18.1 in.^{2}

96. Geography. The circumference of the Earth is about 25,000 miles. Find its diameter to the nearest mile. 7,958 mi

97. Horses. A horse trots in a circle around its trainer at the end of a 28-foot-long rope. Find the area of the circle that is swept out. Round to the nearest square foot. 2,463 ft^2

98. Yo-Yos. How far does a yo-yo travel during one revolution of the "around the world" trick if the length of the string is 21 inches? About 132 in.

99. History. The Inca Empire (1438–1533) was centered in what is now called Peru. A special feature of Inca architecture was the trapezoid-shaped windows and doorways. A standard Inca window was 70 cm high, 50 cm at the base, and 40 cm at the top. Find the area of a window opening. 3,150 cm^2

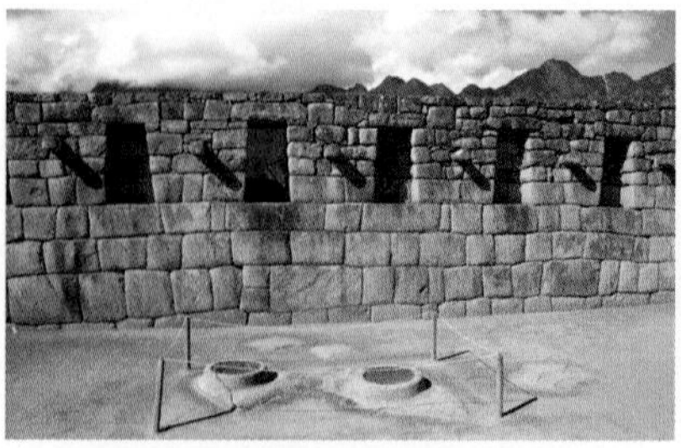

©DIOMEDIA/Alamy

100. Hamster Habitats. Find the amount of space in the tube. About 85 in.^{3}

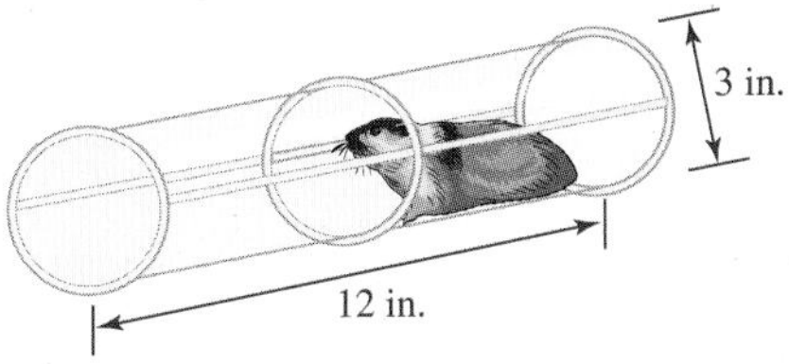

101. Tires. The road surface footprint of a sport truck tire is approximately rectangular. If the area of the footprint is 45 in.^{2}, about how wide is the tire? 6 in.

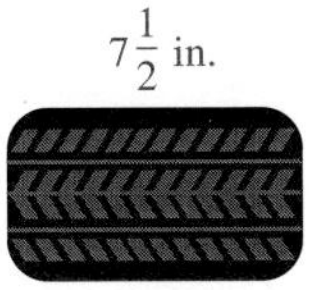

102. Softball. The strike zone in fast-pitch softball is between the batter's armpit and the top of her knees, as shown. If the area of the strike zone for this batter is 442 in.^{2}, what is the width of home plate? 17 in.

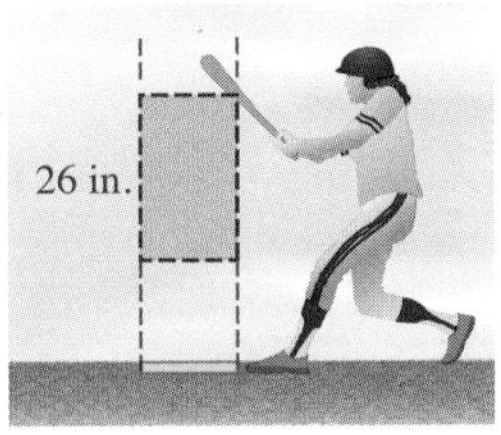

103. Firewood. The cord of wood shown occupies a volume of 128 ft^3. How long is the stack? 8 ft

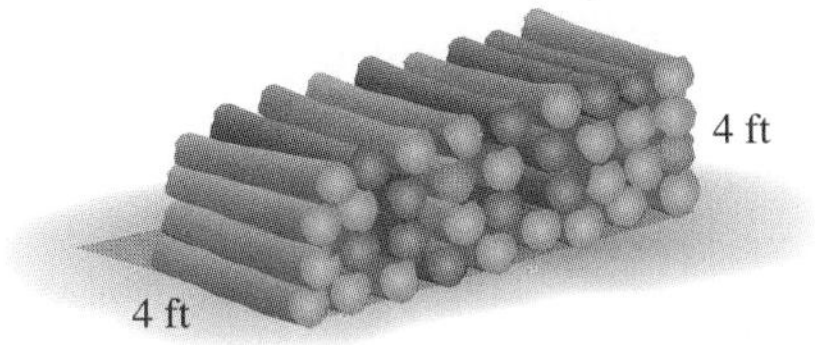

104. Teepees. The teepees constructed by the Blackfoot Indians were cone-shaped tents about 10 feet high and about 15 feet across at the ground. Estimate the volume of a teepee with these dimensions, to the nearest cubic foot. 589 ft^3

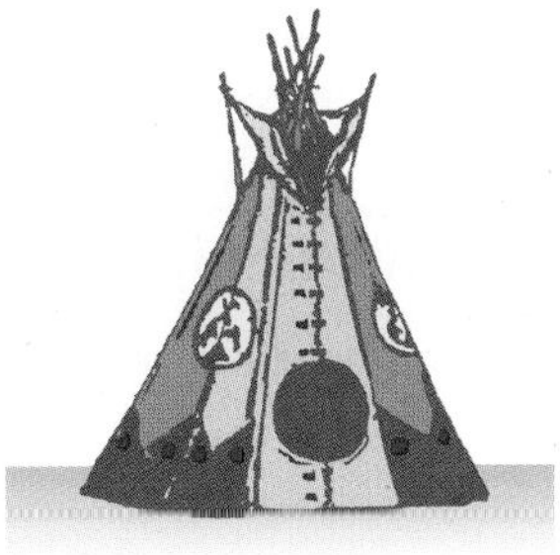

105. Igloos. During long journeys, some Canadian Eskimos built winter houses of snow blocks stacked in the dome shape shown. Estimate the volume of an igloo having an interior height of 5.5 feet to the nearest cubic foot. 348 ft^3

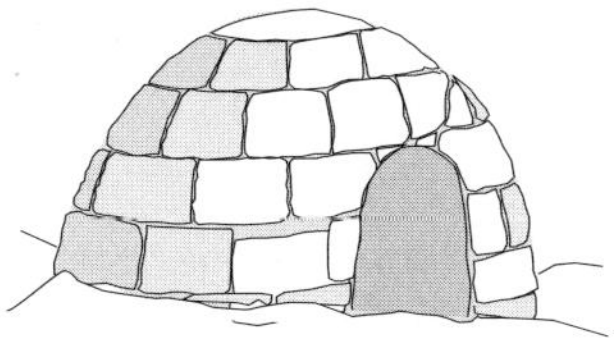

106. Pyramids. The Great Pyramid at Giza in northern Egypt is one of the most famous works of architecture in the world. Find its volume to the nearest cubic foot. (*Hint:* The formula to use is on the inside back cover.) 85,503,750 ft^3

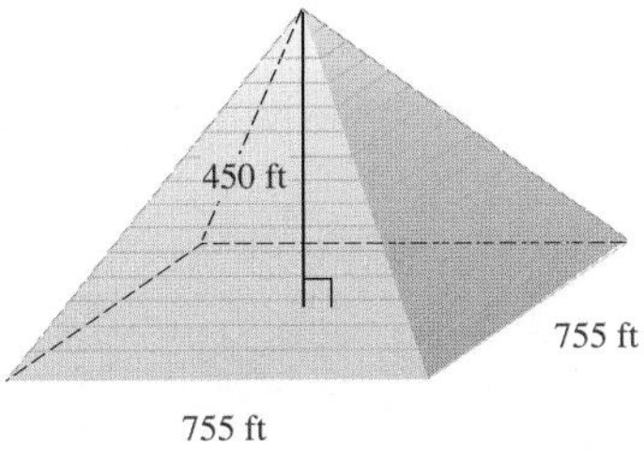

107. Cooking. If the fish shown in the illustration is 18 inches long, what is the area of the grill? Round to the nearest square inch. 254 in.^{2}

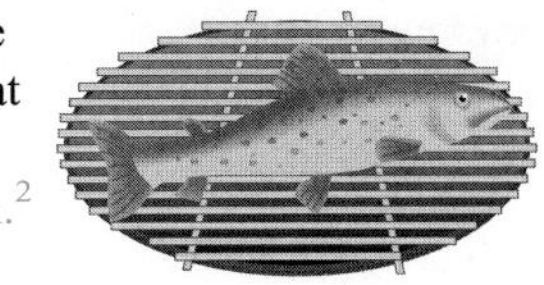

108. Skateboarding. A half-pipe ramp is in the shape of a semicircle with a radius of 8 feet. To the nearest tenth of a foot, what is the length of the arc that the rider travels on the ramp? 25.1 ft

109. Pulleys. The approximate length L of a belt joining two pulleys of radii r and R feet with centers D feet apart is given by the formula $L = 2D + 3.25(r + R)$. Solve the formula for R. $R = \frac{L - 2D - 3.25r}{3.25}$ or $R = \frac{L - 2D}{3.25} - r$

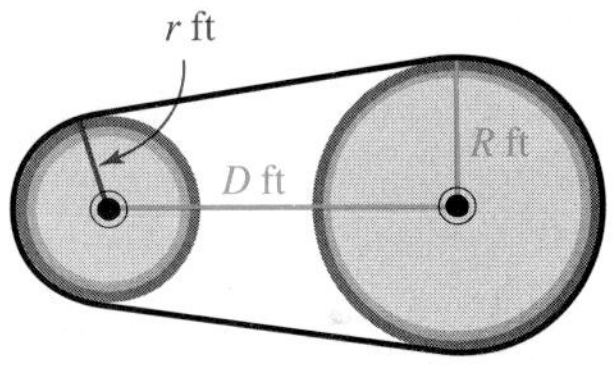

110. Thermodynamics. The Gibbs free-energy formula is given by $G = U - TS + pV$. Solve the formula for the pressure p. $p = \frac{G - U + TS}{V}$

WRITING

111. After solving $A = B + C + D$ for B, a student compared her answer with that at the back of the textbook. Could this problem have two different-looking answers? Explain why or why not.

Student's answer: $B = A - C - D$

Book's answer: $B = A - D - C$

112. A student solved $x + 5c = 3c + a$ for c. His answer was $c = \frac{3c + a - x}{5}$. Explain why the equation is not solved for c.

113. Explain the difference between what perimeter measures and what area measures.

114. Explain the error made below.

$$a = \frac{3b + \overset{1}{\cancel{e}}}{\underset{1}{\cancel{e}}}$$

REVIEW

115. Find 82% of 168. 137.76

116. 29.05 is what percent of 415? 7%

117. What percent of 200 is 30? 15%

118. Shopping. A woman bought a coat for \$98.95 and some gloves for \$7.95. If the sales tax was 6%, how much did the purchase cost her? \$113.31

CHALLENGE PROBLEMS

119. In mathematics, letters from the Greek alphabet are often used as variables. Solve the following equation for α (read as "alpha"), the first letter of the Greek alphabet.

$$-7(\alpha - \beta) - (4\alpha - \theta) = \frac{\alpha}{2} \qquad \alpha = \frac{14\beta + 2\theta}{23}$$

120. Solve $B = R - \frac{1}{16}(c - 3D)$ for D. $D = \frac{16B - 16R + c}{3}$

SECTION 2.5 Problem Solving

OBJECTIVES

1. Apply the steps of a problem-solving strategy.
2. Solve consecutive integer problems.
3. Solve geometry problems.

ARE YOU READY? *Are You Ready? exercises available online at www.webassign.net/brookscole*

The following problems review some basic skills that are needed to solve the application problems in this section.

1. Simplify: $x + x + 1 + 2x + 3$ $4x + 4$

2. If $x = 8$, find $6x + 1$. 49

3. If staplers cost \$4.35 each, what is the cost of 9 staplers? \$39.15

4. Simplify: $x - 0.72x$ $0.28x$

5. What is the formula for the perimeter of a rectangle? $P = 2l + 2w$

6. What is the sum of the measures of the angles of a triangle? 180°

7. Translate to symbols: *8 less than twice a number x* $2x - 8$

8. Write 6% as a decimal. 0.06

In this section, you will see that algebra is a powerful tool that can be used to solve a wide variety of real-world problems.

1 Apply the Steps of a Problem-Solving Strategy.

To become a good problem solver, you need a plan to follow, such as the following six-step strategy.

Strategy for Problem Solving

1. **Analyze the problem** by reading it carefully to understand the given facts. What information is given? What are you asked to find? What vocabulary is given? Often, a diagram or table will help you understand the facts of the problem.
2. **Assign a variable** to represent an unknown value in the problem. This means, in most cases, to let x = what you are asked to find. If there are other unknown values, represent each of them using an algebraic expression that involves the variable.
3. **Form an equation** by translating the words of the problem into mathematical symbols.
4. **Solve the equation** formed in step 3.
5. **State the conclusion clearly.** Be sure to include the units (such as feet, seconds, or pounds) in your answer.
6. **Check the result** using the original wording of the problem, not the equation that was formed in step 3.

EXAMPLE 1 **California Coastline.** The first part of California's magnificent 17-Mile Drive begins at the Pacific Grove entrance and continues to Seal Rock. It is 1 mile longer than the second part of the drive, which extends from Seal Rock to the Lone Cypress as shown in the map below. The third and final part of the drive winds through Pebble Beach, eventually returning to the entrance. This part of the drive is 1 mile longer than four times the length of the second part. How long is each part of 17-Mile Drive?

©michalis/Shutterstock.com

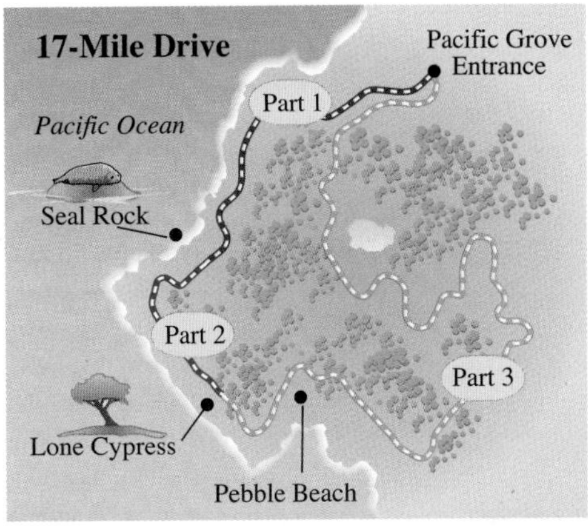

Success Tip

When there is more than one unknown value in a problem, let the variable represent the unknown value on which any other unknown values are based.

Analyze The drive is composed of three parts. We need to find the length of each part. We can straighten out the winding 17-Mile Drive and model it with a line segment.

Assign Since the lengths of the first part and of the third part of the drive are related to the length of the second part, we will let x represent the length of the second part. We then express the other lengths in terms of x.

$x =$ the length of the second part of the drive (in miles)

$x + 1 =$ the length of the first part of the drive (in miles)

$4x + 1 =$ the length of the third part of the drive (in miles)

Teaching Tip: Point out how the three parts of the drive are compared in pairs: The first is compared to the *second*, and the third compared to the *second*. We let x represent the part that the comparisons have in common, which is the second part.

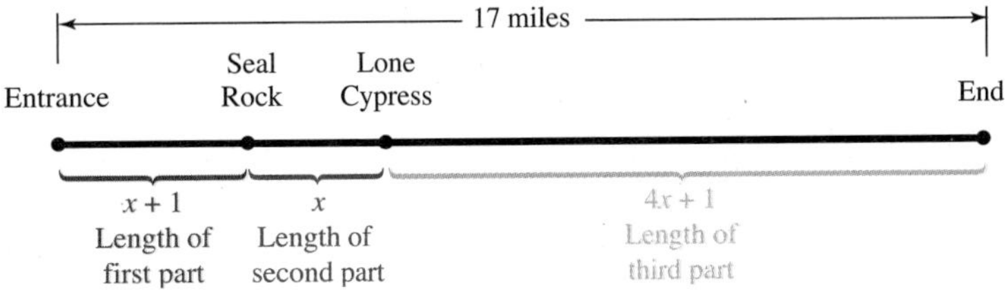

Form Now we translate the words of the problem to an equation.

The length of part 1	plus	the length of part 2	plus	the length of part 3	equals	the total length of the drive.
$x + 1$	$+$	x	$+$	$4x + 1$	$=$	17

Solve

$$x + 1 + x + 4x + 1 = 17$$

$$6x + 2 = 17 \quad \text{Combine like terms: } x + x + 4x = 6x \text{ and } 1 + 1 = 2.$$

$$6x = 15 \quad \text{To undo the addition of 2, subtract 2 from both sides.}$$

$$\frac{6x}{6} = \frac{15}{6} \quad \text{To isolate } x\text{, undo the multiplication by 6 by dividing both sides by 6.}$$

$$x = 2.5 \quad \text{Do the division.}$$

Recall that x represents the length of the second part of the drive. To find the lengths of the first and third parts, we evaluate $x + 1$ and $4x + 1$ for $x = 2.5$.

Caution

Check the result using the original wording of the problem, not by substituting it into the equation. Why? The equation may have been solved correctly, but the danger is that you may have formed it incorrectly.

First part of drive	*Third part of drive*	
$x + 1 = \mathbf{2.5} + 1$	$4x + 1 = 4(\mathbf{2.5}) + 1$	Substitute 2.5 for x.
$= 3.5$	$= 11$	The units are miles.

State The first part of the drive is 3.5 miles long, the second part is 2.5 miles long, and the third part is 11 miles long.

Check Since 3.5 mi + 2.5 mi + 11 mi = 17 mi, the results check.

Teaching Example 1 Mountaineering. A mountain climber wants to cut a 213-foot rope into three pieces. If each piece is to be 2 feet longer than the previous one, how long is each piece?
Answer: 69 ft, 71 ft, and 73 ft

Self Check 1 **Biking.** The Mountain-Bay State Park Bike Trail in Northeast Wisconsin is 76 miles long. A couple rode the trail in four days. Each day they rode 2 miles more than the previous day. How many miles did they ride each day? 16 mi, 18 mi, 20 mi, and 22 mi

Now Try Problems 13 and 15

EXAMPLE 2 **Computer Logos.** A trucking company had its logo embroidered on the front of baseball caps. It was charged \$8.90 per hat plus a one time set up fee of \$25. If the project cost \$559, how many hats were embroidered?

Analyze

- It cost \$8.90 to have a logo embroidered on a hat.
- The set up charge was \$25.
- The project cost \$559.
- We need to find the number of hats that were embroidered.

Assign Let x = the number of hats that were embroidered.

Success Tip

The *Form* step is often the hardest. To help, write a **verbal model** of the situation (shown here in the boxes) and then translate it into an equation.

Form If x hats are embroidered, at a cost of \$8.90 per hat, the cost to embroider all of the hats is $x \cdot \$8.90$ or $\$8.90x$. Now we translate the words of the problem into an equation.

The cost to embroider one hat	times	the number of hats	plus	the set up charge	equals	the total cost.
8.90	$\cdot$	x	+	25	=	559

Solve

$$8.90x + 25 = 559$$

$$8.90x = 534 \quad \text{To undo the addition of 25, subtract 25 from both sides.}$$

$$\frac{8.90x}{\mathbf{8.90}} = \frac{534}{\mathbf{8.90}} \quad \text{To isolate x, undo the multiplication by 8.90 by dividing both sides by 8.90.}$$

$$x = 60 \quad \text{Do the division.}$$

State The company had 60 hats embroidered.

Check The cost to embroider 60 hats is 60(\$8.90) = \$534. When the \$25 set up charge is added, we get \$534 + \$25 = \$559. The result checks.

Teaching Example 2 Uniforms. A youth soccer league had team names printed on colored shirts. It was charged \$18 per shirt plus a one time set up fee of \$20. If the total due was \$4,232, how many shirts were printed?
Answer: 234

Self Check 2 **T-shirts.** A school club had their motto screenprinted on the front of T-shirts. They were charged \$5 per shirt plus a one-time set up fee of \$20. If the project cost \$255, how many T-shirts were printed? 47 shirts

Now Try Problem 23

EXAMPLE 3 **Auctions.** A classic car owner is going to sell his 1959 Chevy Impala at an auction. He wants to make $46,000 after paying an 8% commission to the auctioneer. What should be the selling price (called the "hammer price") for the car owner to make this amount of money?

Analyze When the commission is subtracted from the selling price of the car, the owner wants to have $46,000 left. We need to find the selling price.

Assign Let x = the selling price of the car.

Form The amount of the commission is 8% of x, or $0.08x$. Now we translate the words of the problem to an equation.

The selling price of the car	minus	the auctioneer's commission	should be	$46,000.
x	$-$	$0.08x$	$=$	46,000

Solve

$$x - 0.08x = 46{,}000$$

$$0.92x = 46{,}000$$ Combine like terms: $1.00x - 0.08x = 0.92x$. We could begin with this equation because after the 8% commission is paid, 100%–8% or 92% of the selling price should be $46,000.

$$\frac{0.92x}{\mathbf{0.92}} = \frac{46{,}000}{\mathbf{0.92}}$$ To isolate x, undo the multiplication by 0.92 by dividing both sides by 0.92.

$$x = 50{,}000$$ Do the division. This is the selling price of the car.

State The owner will make $46,000 if the car sells for $50,000.

Check An 8% commission on $50,000 is 0.08($50,000) = $4,000. The owner will keep $50,000 − $4,000 = $46,000. The result checks.

The Language of Algebra

Here are some words and phrases that often translate to an equal symbol =.

is	are
should be	will be
yields	amounts to
represents	gives
is the same as	was

Teaching Example 3 **Commission.** A homeowner is planning on selling his home. He wants to get $240,975 after paying a $5\frac{1}{2}$% commission. What selling price is needed to meet his requirement?
Answer: $255,000

Self Check 3 **Cattle Auction.** A farmer is going to sell one of his Black Angus cattle at an auction and would like to make $2,597 after paying a 6% commission to the auctioneer. For what selling price will the farmer make this amount of money? $2,762.77

Now Try Problem 29

2 Solve Consecutive Integer Problems.

Integers that follow one another, such as 15 and 16, are called **consecutive integers.** They are 1 unit apart. **Consecutive even integers** are even integers that differ by 2 units, such as 12 and 14. Similarly, **consecutive odd integers** differ by 2 units, such as 9 and 11. When solving consecutive integer problems, if we let x = the first integer, then

- two consecutive integers are x and $x + 1$
- two consecutive even integers are x and $x + 2$
- two consecutive odd integers are x and $x + 2$
- three consecutive integers are x, $x + 1$, and $x + 2$
- three consecutive even integers are x, $x + 2$, and $x + 4$
- three consecutive odd integers are x, $x + 2$, and $x + 4$

EXAMPLE 4 **U.S. History.** The year George Washington was chosen president and the year the Bill of Rights went into effect are consecutive odd integers whose sum is 3,580. Find the years.

Analyze We need to find two consecutive odd integers whose sum is 3,580. From history, we know that Washington was elected president first and the Bill of Rights went into effect later.

Assign Let x = the first odd integer (the date when Washington was chosen president). The next odd integer is 2 *greater than* x, therefore $x + 2$ = the next larger odd integer (the date when the Bill of Rights went into effect).

The Language of Algebra

Consecutive means following one after the other in order. Elton John holds the record for the most *consecutive* years with a song on the Top 50 music chart: 31 years (1970 to 2000).

Form

The first odd integer	plus	the second odd integer	is	3,580.
x	$+$	$x + 2$	$=$	3,580

Solve

$$x + x + 2 = 3{,}580$$

$2x + 2 = 3{,}580$ Combine like terms: $x + x = 2x$.

$2x = 3{,}578$ To undo the addition of 2, subtract 2 from both sides.

$x = 1{,}789$ To isolate x, undo the multiplication by 2 by dividing both sides by 2.

State George Washington was chosen president in the year 1789. The Bill of Rights went into effect in $1789 + 2 = 1791$.

Check 1789 and 1791 are consecutive odd integers whose sum is $1789 + 1791 = 3{,}580$. The answers check.

Teaching Example 4 **Siblings.** The birth years of the three children in a family are consecutive even integers. The sum of the integers is 5,982. Find the year of each child's birth.
Answer: 1992, 1994, 1996

Self Check 4 **Dictionaries.** The definitions of the words *little* and *lobby* are on back-to-back pages in a dictionary. If the sum of the page numbers is 1,159, on what page can the definition of *little* be found? Page 579

Now Try Problem 37

3 Solve Geometry Problems.

EXAMPLE 5 **Crime Scenes.** Police used 400 feet of yellow tape to fence off a rectangular-shaped lot for an investigation. They used 50 fewer feet of tape for each width than for each length. Find the dimensions of the lot.

Analyze Since the yellow tape surrounded the lot, the concept of perimeter applies. Recall that the formula for the perimeter of a rectangle is $P = 2l + 2w$. We also know that the width of the lot is 50 feet less than the length.

The Language of Algebra

Dimensions are measurements of length and width. We might speak of the *dimensions* of a dance floor or a TV screen.

Assign Since the width of the lot is given in terms of the length, we let $l =$ the length of the lot. Then $l - 50 =$ the width.

Form Using the perimeter formula, we have:

2	times	the length	plus	2	times	the width	is	the perimeter.
2	$\cdot$	l	$+$	2	$\cdot$	$(l - 50)$	$=$	400

Success Tip

When solving geometry problems, a sketch is often helpful.

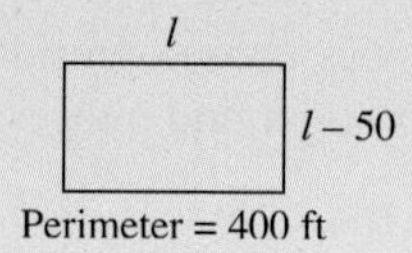

Solve

$2l + 2(l - 50) = 400$ Write the parentheses so that the entire expression $l - 50$ is multiplied by 2.

$2l + 2l - 100 = 400$ Distribute the multiplication by 2.

$4l - 100 = 400$ Combine like terms: $2l + 2l = 4l$.

$4l = 500$ To undo the subtraction of 100, add 100 to both sides.

$l = 125$ To isolate l, undo the multiplication by 4 by dividing both sides by 4.

State The length of the lot is 125 feet and width is $125 - 50 = 75$ feet.

Check The width (75 feet) is 50 less than the length (125 feet). The perimeter of the lot is $2(125) + 2(75) = 250 + 150 = 400$ feet. The results check.

Teaching Example 5 Dog Kennels. A woman has 36 meters of fencing to make a rectangular kennel. If the kennel is to be 10 meters longer than it is wide, find its dimensions.
Answer: 4 m by 14 m

Self Check 5 **Counters.** A rectangular counter for the customer service department of a store is 6 feet longer than it is wide. If the perimeter is 32 feet, find the outside dimensions of the counter. 5 ft by 11 ft

Now Try Problem 43

EXAMPLE 6 **Isosceles Triangles.** If the vertex angle of an isosceles triangle is 56°, find the measure of each base angle.

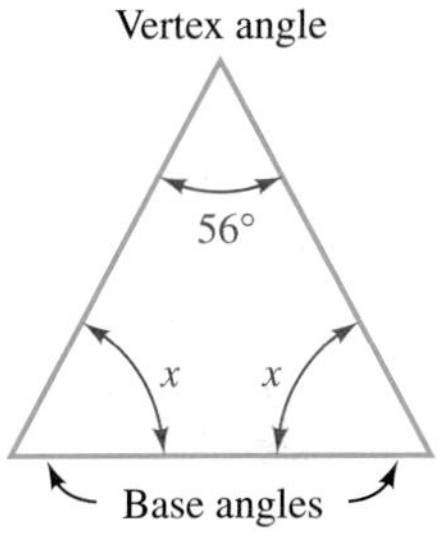

Analyze An **isosceles triangle** has two sides of equal length, which meet to form the **vertex angle.** In this case, the measurement of the vertex angle is 56°. We can sketch the triangle as shown. The **base angles** opposite the equal sides are also equal. We need to find their measure.

Assign If we let x = the measure (in degrees) of one base angle, the measure of the other base angle is also x.

Form Since the sum of the angles of any triangle is 180°, the sum of the base angles and the vertex angle is 180°. We can use this fact to form the equation.

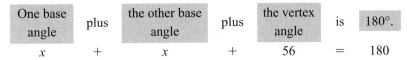

One base angle	plus	the other base angle	plus	the vertex angle	is	180°.
x	$+$	x	$+$	56	$=$	180

Solve

$$x + x + 56 = 180$$
$$2x + 56 = 180 \quad \text{Combine like terms: } x + x = 2x.$$
$$2x = 124 \quad \text{To undo the addition of 56, subtract 56 from both sides.}$$
$$x = 62 \quad \text{To isolate } x \text{, undo the multiplication by 2 by dividing both sides by 2.}$$

State The measure of each base angle is 62°.

Check Since $62° + 62° + 56° = 180°$, the answer checks.

Teaching Example 6 Geometry. If the vertex angle of an isosceles triangle is 106°, find the measure of each base angle.
Answer: 37°, 37°

Self Check 6 **Geometry.** The perimeter of an isosceles triangle is 32 cm. If the base is 8 cm, find the length of each remaining side. 12 cm, 12 cm

Now Try Problem 47

SECTION 2.5 STUDY SET

VOCABULARY

Fill in the blanks.

1. Integers that follow one another, such as 7 and 8, are called consecutive integers.
2. An isosceles triangle is a triangle with two sides of the same length.
3. The equal sides of an isosceles triangle meet to form the vertex angle. The angles opposite the equal sides are called base angles, and they have equal measures.
4. When asked to find the dimensions of a rectangle, we are to find its length and width.

CONCEPTS

5. A 17-foot pipe is cut into three sections. The longest section is three times as long as the shortest, and the middle-sized section is 2 feet longer than the shortest. Complete the diagram.

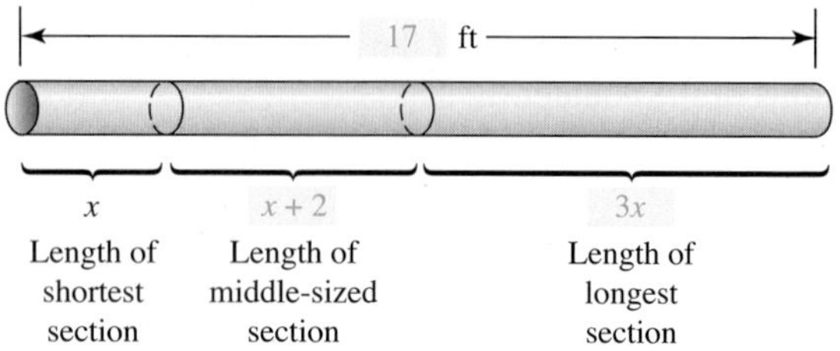

6. It costs \$28 per hour to rent a trailer. Write an expression that represents the cost to rent the trailer for x hours. \$28x

7. A realtor is paid a 3% commission on the sale of a house. Write an expression that represents the amount of the commission if a house sells for \$$x$. \$0.03x

8. The perimeter of the rectangle below is 15 feet. Fill in the blanks: $2(5x - 1) + 2x = 15$

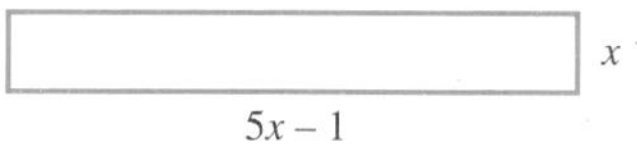

9. What is the sum of the measures of the angles of any triangle? 180°

10. Refer to the isosceles triangle on the right.
   a. Find the missing angle measure. 56°
   b. Find the missing side length. 7 ft

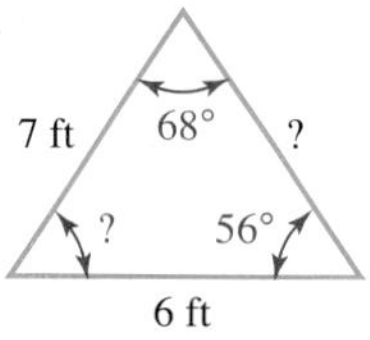

NOTATION

11. a. If x represents an integer, write an expression for the next largest integer. $x + 1$
   b. If x represents an odd integer, write an expression for the next largest odd integer. $x + 2$
   c. If x represents an even integer, write expressions for the next two largest even integers. $x + 2, x + 4$

12. What does 45° mean? 45 degrees

GUIDED PRACTICE

See Example 1.

13. A 12-foot board has been cut into two sections, one twice as long as the other. How long is each section? 4 ft, 8 ft

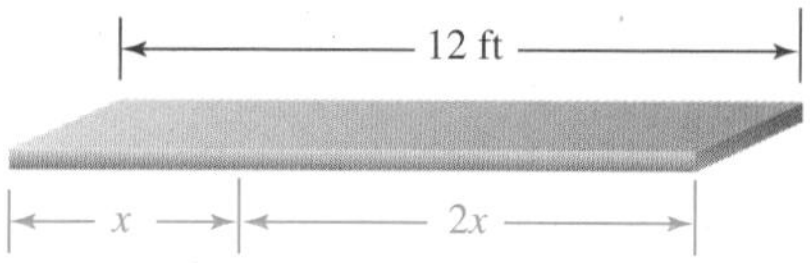

14. The robotic arm will extend a total distance of 18 feet. Find the length of each section. 5 ft, 9 ft, 4 ft

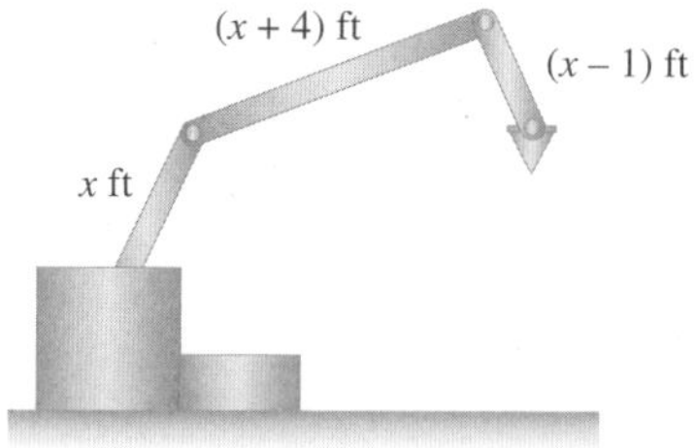

APPLICATIONS

15. **National Parks.** The Natchez Trace Parkway is a historical 444-mile route from Natchez, Mississippi, to Nashville, Tennessee. A couple drove the Trace in four days. Each day they drove 6 miles more than the previous day. How many miles did they drive each day? Day 1: 102 mi; day 2: 108 mi; day 3: 114 mi; day 4: 120 mi

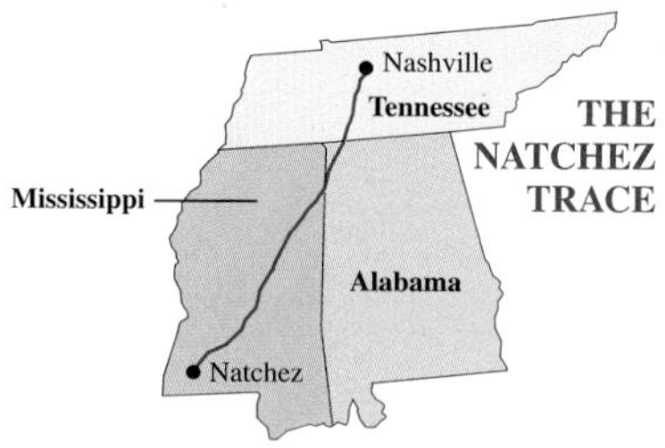

16. **Touring.** A rock group plans to travel for a total of 38 weeks, making three concert stops. They will be in Japan for 4 more weeks than they will be in Australia. Their stay in Sweden will be 2 weeks shorter than that in Australia. How many weeks will they be in each country? Australia: 12 wk; Japan: 16 wk; Sweden: 10 wk

17. **Solar Heating.** Two solar panels were installed side-by-side on a roof, as shown below. One panel is 3.4 feet wider than the other. Find the width of each panel. 7.3 ft, 10.7 ft

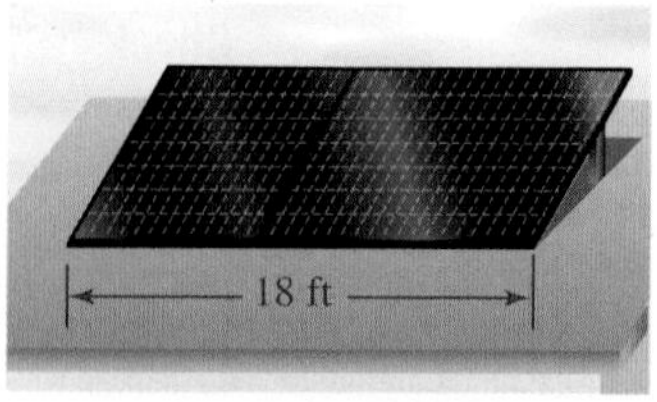

18. **Accounting.** Determine the 2010 income of Aeropostale Inc. for each quarter from the data in the graph below. In millions of dollars: Q1: \$32; Q2: \$39; Q3: \$63; Q4: \$97

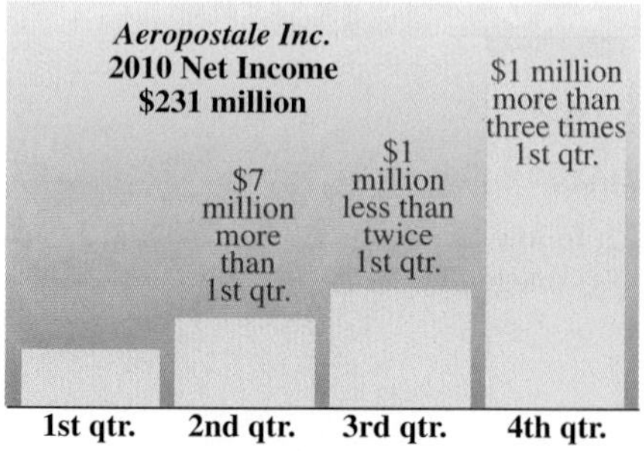

Source: moneycentral.msn.com

19. iPhone Apps. A student spent a total of \$52.97 on the purchase of three applications in the Apps Store on his iPhone. *Call of Duty: World at War: Zombies II* cost \$7 more than *Guitar Hero,* and *Tom Tom USA* cost \$4.11 more than twelve times *Guitar Hero.* Find the cost of each application. *Guitar Hero:* \$2.99; *Call of Duty: World at War: Zombies II:* \$9.99; *Tom Tom USA:* \$39.99

20. Water Usage. It takes about 3.8 times more gallons of water to produce one pound of grain-fed beef than it does to produce one pound of grain-fed chicken. If a combined total of 3,135 gallons of water are needed to produce one pound of each meat, how many gallons does it take to produce one pound of chicken? How many gallons does it take to produce one pound of beef? (Source: The Sierra Club: *The True Cost of Food*) 653.125 gal $= 653\frac{1}{8}$ gal; 2,481.875 gal $= 2{,}481\frac{7}{8}$ gal

▶ **21. Counting Calories.** A slice of pie with a scoop of ice cream has 850 calories. The calories in the pie alone are 100 more than twice the calories in the ice cream alone. How many calories are in each food? 250 calories in ice cream, 600 calories in pie

▶ **22. Waste Disposal.** Two tanks hold a total of 45 gallons of a toxic solvent. One tank holds 6 gallons more than twice the amount in the other. How many gallons does each tank hold? 13 gal, 32 gal

23. Concerts. The fee to rent a concert hall is \$2,250 plus \$150 per hour to pay for the support staff. For how many hours can an orchestra rent the hall and stay within a budget of \$3,300? 7 hr

▶ **24. Truck Mechanics.** An engine repair cost a truck owner \$1,185 in parts and labor. If the parts were \$690 and the mechanic charged \$45 per hour, how many hours did the repair take? 11 hr

25. Field Trips. It costs a school \$65 a day plus \$0.25 per mile to rent a 15-passenger van. If the van is rented for two days, how many miles can be driven on a \$275 budget? 580 mi

26. Decorations. A party supply store charges a set-up fee of \$80 plus 35¢ per balloon to make a balloon arch. A business has \$150 to spend on decorations for their grand opening. How many balloons can they have in the arch? (*Hint:* 35¢ = \$0.35.) 200 balloons

▶ **27. Tutoring.** High school students enrolling in a private tutoring program must first take a placement test (cost \$25) before receiving tutoring (cost \$18.75 per hour). If a family has set aside \$400 to get their child extra help, how many hours of tutoring can they afford? 20 hr

▶ **28. Data Conversion.** The *Books2Bytes* service converts old print books to Microsoft Word electronic files for \$20 per book plus \$2.25 per page. If it cost \$1,201.25 to convert a novel, how many pages did the novel have? 525 pages

▶ **29. Cattle Auctions.** A cattle rancher is going to sell one of his prize bulls at an auction and would like to make \$45,500 after paying a 9% commission to the auctioneer. For what selling price will the rancher make this amount of money? \$50,000

30. Listing Price. At what price should a home be listed if the owner wants to make \$567,000 on its sale after paying a 5.5% real estate commission? \$600,000

31. Selling Used Clothing. A *consignment shop* accepts an item of clothing that no longer fits (or one you have grown tired of) and sells it for you. The shop then charges you an agreed on percent of the selling price as their profit. Suppose the owner of a designer wool coat would like to make \$210 on its sale at a consignment shop. If there is a $12\frac{1}{2}\%$ consignment charge, for what price must the coat be sold? \$240

32. Finder's Fees. A *finder's fee* is an amount of money that is paid to someone who brings people together for business purposes. Suppose the owner of a software company needs to sell it and make \$9,950,000 to pay back creditors. If he expects to pay a finder's fee of $\frac{1}{2}\%$ of the selling price to find a qualified buyer, for what price must the company be sold? \$10,000,000

33. Savings Accounts. The balance in a savings account grew by 5% in one year, to \$5,512.50. What was the balance at the beginning of the year? \$5,250

▶ **34. Aluminum Cans.** Today's aluminum cans are much thinner and lighter than those of the past. From 1972 to 2010, the number of empty cans produced from one pound of aluminum has increased by about 45%. If 32 cans could be produced from one pound of aluminum in 2010, how many cans could be produced from one pound of aluminum in 1972? Round to the nearest can. (Source: cancentral.com) 22 cans

Consecutive integer problems

35 Soccer. Ronaldo of Brazil and Gerd Mueller of Germany rank 1 and 2, respectively, with the most goals scored in World Cup play. The number of goals Ronaldo and Mueller have scored are consecutive integers that total 29. Find the number of goals scored by each man. (Source: planetworldcup.com) Ronaldo: 15 goals; Mueller: 14 goals

36. Dictionaries. The definitions of the words *job* and *join* are on back-to-back pages in a dictionary. If the sum of those page numbers is 1,411, on what page can the definition of *job* be found? page 705

▶ **37. TV History.** *Friends* and *Leave It to Beaver* are two of the most popular television shows of all time. The number of episodes of each show are consecutive even integers whose sum is 470. If there are more episodes of *Friends,* how many episodes of each were there? (Source: angelfire.com) *Friends:* 236 episodes; *Leave It to Beaver:* 234 episodes

▶ **38. Time Off.** The table shows the average number of days off an employed adult receives for selected countries. Complete the table. The numbers of days are listed in descending order. Norway: 35 days; South Africa: 33 days

Average Number of Days Off per Year*	
Country	**Days**
Brazil	41
Norway	?
South Africa	?
U.S.	25

(Norway and South Africa: Consecutive odd integers whose sum is 68.)

* Employee has 10 years of service and works 5 days a week
Source: *The Wall Street Journal,* 2009.

39. Celebrity Birthdays. Selena Gomez, Jennifer Lopez, and Sandra Bullock have birthdays (in that order) on consecutive even-numbered days in July. The sum of the calendar dates of their birthdays is 72. Find each birthday. July 22, 24, 26

▶ **40. Locks.** The three numbers of the combination for a lock are consecutive integers, and their sum is 81. Find the combination. 26, 27, 28

Geometry problems

41. Tennis. The perimeter of a regulation singles tennis court is 210 feet and the length is 3 feet less than three times the width. What are the dimensions of the court? Width: 27 ft; length: 78 ft

42. Swimming Pools. The seawater Orthlieb Pool in Casablanca, Morocco, is the largest swimming pool in the world. With a perimeter of 1,110 meters, this rectangular-shaped pool is 30 meters longer than 6 times its width. Find its dimensions. 75 m by 480 m

43. Art. The *Mona Lisa* was completed by Leonardo da Vinci in 1506. The length of the picture is 11.75 inches shorter than twice the width. If the perimeter of the picture is 102.5 inches, find its dimensions. 21 in. by 30.25 in.

44. New York City. Central Park, which lies in the middle of Manhattan, is rectangular-shaped and has a 6-mile perimeter. The length is 5 times the width. What are the dimensions of the park? Width: 0.5 mi, length: 2.5 mi

45. Engineering. A truss is in the form of an isosceles triangle. Each of the two equal sides is 4 feet shorter than the third side. If the perimeter is 25 feet, find the lengths of the sides. 7 ft, 7 ft, 11 ft

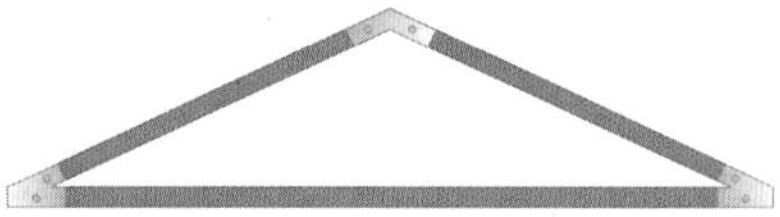

46. First Aid. A sling is in the shape of an isosceles triangle with a perimeter of 144 inches. The longest side of the sling is 18 inches longer than either of the other two sides. Find the lengths of each side. 60 in., 42 in., 42 in.

47. TV Towers. A TV tower is supported by several guy wires. Two of the guy wires are attached to the top of the tower to form an isosceles triangle with the ground, as shown on the right. The measure of each of the base angles of the triangle is 4 times the third angle (the vertex angle). Find the measure of the vertex angle. 20°

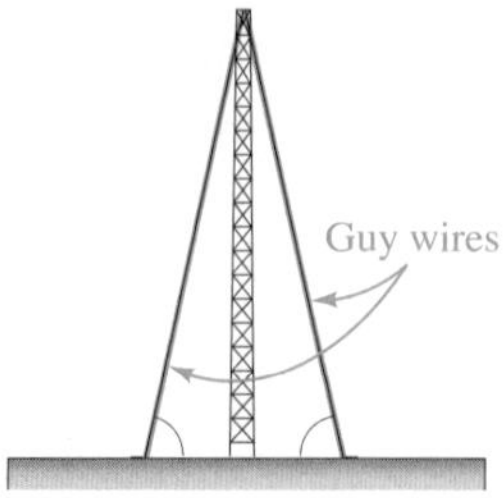

48. Clotheslines. A pair of damp jeans are hung in the middle of a clothesline to dry. Find x, the angle that the clothesline makes with the horizontal. 11°

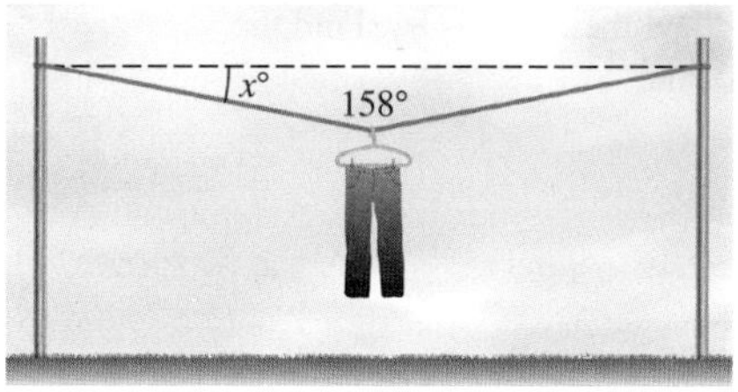

49. Mountain Bicycles. For the bicycle frame shown, the angle that the horizontal crossbar makes with the seat support is 15° less than twice the angle at the steering column. The angle at the pedal gear is 25° more than the angle at the steering column. Find these three angle measures. At steering column: 42.5°; at seat support: 70°; at pedal gear: 67.5°

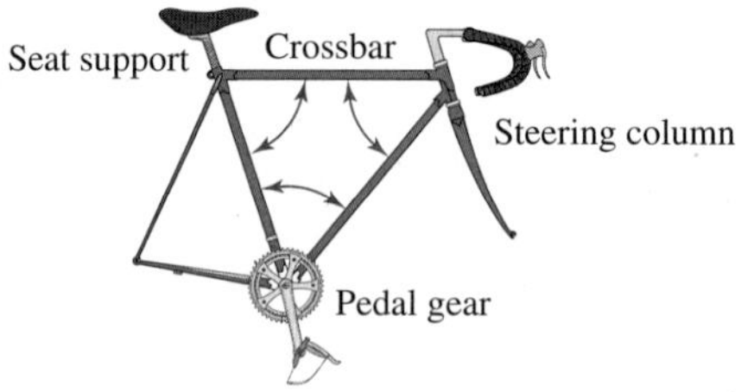

50. Triangles. The measure of $\angle 1$ (read as angle 1) of a triangle is one-half that of $\angle 2$. The measure of $\angle 3$ is equal to the sum of the measures of $\angle 1$ and $\angle 2$. Find each angle measure. $\angle 1$: 30°; $\angle 2$: 60°; $\angle 3$: 90°

51. Angles. Two angles are called ***complementary angles*** when the sum of their measures is 90°. Refer to the figure on the right. Find x. Then find the measures of the complementary angles. $x = 11°$; 22°, 68°

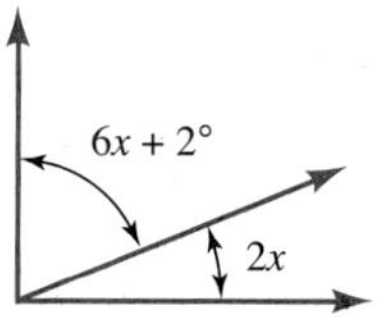

52. Angles. Two angles are called ***supplementary angles*** when the sum of their measures is 180°. Refer to the figure below. Find x. Then find the measures of the supplementary angles. $x = 25°$; 40°, 140°

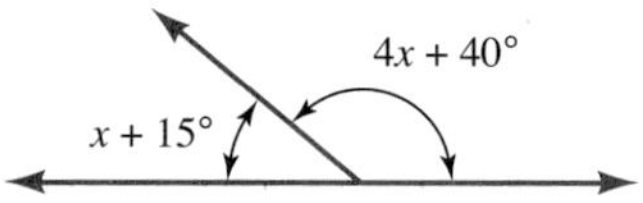

53. "Lightning Bolt." In 2010, Usain Bolt of Jamaica held the world record for the 100 meters and the 200 meters sprints. His *maximum stride angle* shown below is 5° less than 1.5 times its supplement. Find his maximum stride angle. You may need to refer to problem 52 to review the geometry involved. (Source: somaxsports.com) Maximum stride angle: 106°

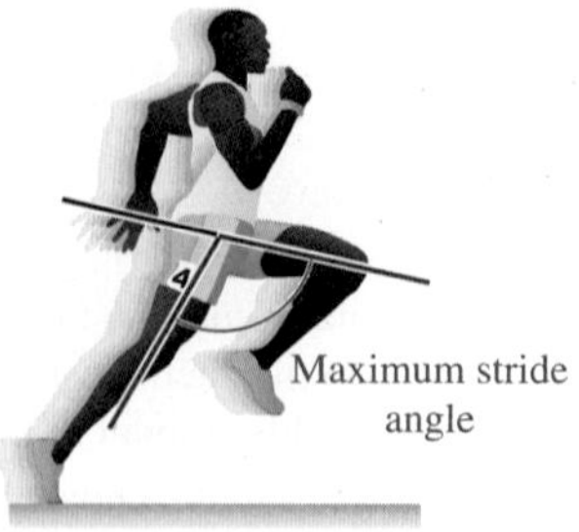

54. from **Campus To Careers**

Automotive Service Technician

© barang/Shutterstock.com

The *sweep angle* of a windshield wiper arm is 115° as shown below. Find x so that the area that is cleared by the wiper is centered on the car's windshield. 32.5°

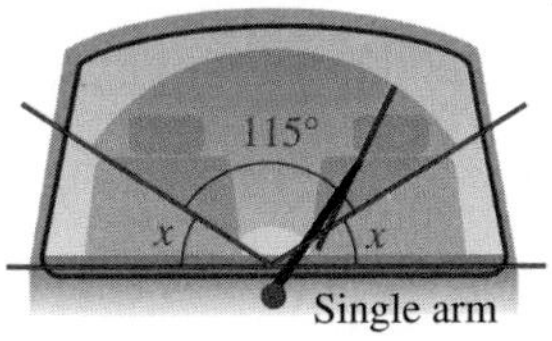

WRITING

55. Create a geometry problem that could be answered by solving the equation $2w + 2(w + 5) = 26$.

56. What information do you need to know to answer the following question?

A business rented a copy machine for \$85 per month plus 4¢ for every copy made. How many copies can be made each month?

57. Make a list of words and phrases that translate to an equal symbol =.

58. Define the word *strategy.*

REVIEW

Solve.

59. $\frac{5}{8}x = -15$ -24

60. $\frac{12x + 24}{13} = 36$ 37

61. $\frac{3}{4}y = \frac{2}{5}y - \frac{3}{2}y - 2$ $-\frac{40}{37}$

62. $4.2(y - 4) - 0.6y = -13.2$ 1

CHALLENGE PROBLEMS

63. What concept discussed in this section is illustrated by the following day and time?

Two minutes and three seconds past 1 A.M. on the 5th day of April, 2006

Consecutive integers; 01:02:03; 04/05/06

64. Manufacturing. A company has two machines that make widgets. The production costs are listed below.

Machine 1: Setup cost \$400 and \$1.70 per widget

Machine 2: Setup cost \$500 and \$1.20 per widget

Find the number of widgets for which the cost to manufacture them on either machine is the same. 200 widgets

SECTION 2.6 More about Problem Solving

OBJECTIVES

1. Solve investment problems.
2. Solve uniform motion problems.
3. Solve liquid mixture problems.
4. Solve dry mixture problems.
5. Solve number-value problems.

ARE YOU READY? *Are You Ready? exercises available online at www.webassign.net/brookscole*

The following problems review basic skills and some formulas that are needed to solve money, motion, and mixture applications.

1. Find the amount of interest earned by \$8,000 invested at a 5% annual simple interest rate for 1 year. \$400

2. At 45 miles per hour, how far will a car travel in 3 hours? 135 mi

3. A 12-gallon mixture of antifreeze and water is 30% antifreeze. How many gallons of the mixture is antifreeze? 3.6 gal of antifreeze

4. At \$2.45 per pound, what is the value of 8 pounds of ground beef? \$19.60

5. A couple invested \$6,000 of their \$20,000 lottery winnings in bonds. How much do they have left to invest in stocks? \$14,000

6. Multiply: $100(0.03x)$ $3x$

In this section, we will solve problems that involve money, motion, and mixtures. Tables are a helpful way to organize the information given in these problems.

1 Solve Investment Problems.

To find the amount of *simple interest* I an investment earns, we use the formula $I = Prt$, where P is the principal (the amount invested), r is the annual interest rate, and t is the time in years.

EXAMPLE 1 **Paying Tuition.** A college student wants to invest the \$12,000 inheritance he received and use the annual interest earned to pay his tuition cost of \$945. The highest rate offered by a bank is 6% annual simple interest. At this rate, he cannot earn the needed \$945, so he decides to invest some of the money in a riskier, but more profitable, investment offering a 9% return. How much should he invest at each rate?

Teaching Tip: You may want to use two class sessions to cover the material in this section: Investment, motion, and liquid mixture problems during one class meeting, and dry mixture and number-value problems at the next.

Similar interest, uniform motion, and mixture problems are solved using two variables in Section 4.4.

Analyze We know that \$12,000 will be invested for 1 year at two rates: 6% and 9%. We are asked to find the amount invested at each rate so that the total return would be \$945.

Assign Let x = the amount invested at 6%. Then $12,000 - x$ = the amount invested at 9%.

Form To organize the facts of the problem, we enter the principal, rate, time, and interest earned from each account in a table.

Step 1: List each investment in a row of the table.

Bank				
Riskier Investment				

Step 2: Label the columns using $I = Prt$ reversed and also write Total.

	P	$\cdot\ r$	$\cdot\ t =$	I
Bank				
Riskier Investment				
				Total:

Step 3: Enter the rates as decimals, the times, and the total interest.

	P	$\cdot\ r$	$\cdot\ t =$	I
Bank		**0.06**	**1**	
Riskier Investment		**0.09**	**1**	
				Total: **945**

Step 4: Enter each unknown principal.

	P	$\cdot\ r$	$\cdot\ t =$	I
Bank	$\mathbf{x}$	0.06	1	
Riskier Investment	**12,000 − x**	0.09	1	
				Total: 945

Teaching Tip: Show students that the sum of the amounts invested is:

$$x + 12,000 - x = 12,000$$

Also warn them that a common error is to write the amount invested at 9% incorrectly as $x - 12,000$.

Step 5: In the last column, multiply P, r, and t to obtain expressions for the interest earned.

	P	$\cdot\ r$	$\cdot\ t =$	I	
Bank	x	0.06	1	$\mathbf{0.06x}$	← *This is $x \cdot 0.06 \cdot 1$.*
Riskier Investment	$12,000 - x$	0.09	1	$\mathbf{0.09(12,000 - x)}$	← *This is $(12,000 - x) \cdot 0.09 \cdot 1$.*
				Total: 945	

Use the information in this column to form an equation.

The interest earned at 6%	plus	the interest earned at 9%	equals	the total interest.
$0.06x$	$+$	$0.09(12,000 - x)$	$=$	945

Solve

$$0.06x + 0.09(12,000 - x) = 945$$

$$100[0.06x + 0.09(12,000 - x)] = 100(945)$$ *Multiply both sides by 100 to clear the equation of decimals.*

$$100(0.06x) + 100(0.09)(12,000 - x) = 100(945)$$ *Distribute the multiplication by 100.*

$$6x + 9(12,000 - x) = 94,500$$ *Do the multiplications by 100.*

$$6x + 108,000 - 9x = 94,500$$ *Use the distributive property.*

$$-3x + 108,000 = 94,500$$ *Combine like terms.*

$$-3x = -13,500$$ *Subtract 108,000 from both sides.*

$$x = 4,500$$ *To isolate x, divide both sides by −3.*

Success Tip

We can clear an equation of decimals by multiplying both sides by a power of 10. Here, we multiply 0.06 and 0.09 by 100 to move each decimal point two places to the right:

$$100(0.06) = 6 \quad 100(0.09) = 9$$

State The student should invest \$4,500 at 6% and \$12,000 − \$4,500 = \$7,500 at 9%.

Check The first investment will earn 0.06(\$4,500), or \$270. The second will earn 0.09(\$7,500), or \$675. Since the total return will be \$270 + \$675 = \$945, the results check.

Teaching Example 1 **Investments.** A professor has \$15,000 to invest for 1 year, some at 8% and the rest at 7%. If she wants to earn \$1,110 from these investments, how much should she invest at each rate?
Answer: \$6,000 at 8%, \$9,000 at 7%

Self Check 1 **Investments.** A student invested a total of \$4,200 in certificates of deposit, one at 2% and the other at 3%. Find the amount invested at each rate if the first year combined interest income from the two investments was \$102. \$2,400 at 2%, \$1,800 at 3%

Now Try Problem 17

2 Solve Uniform Motion Problems.

If we know the rate r at which we will be traveling and the time t we will be traveling at that rate, we can find the distance d traveled by using the formula $d = rt$.

EXAMPLE 2 **Rescues at Sea.** A cargo ship, heading into port, radios the Coast Guard that it is experiencing engine trouble and that its speed has dropped to 3 knots. (This is 3 sea miles per hour.) Immediately, a Coast Guard cutter leaves port and speeds at a rate of 25 knots directly toward the disabled ship, which is 56 sea miles away. How long will it take the Coast Guard to reach the ship? (Sea miles are also called nautical miles.)

Success Tip

A sketch is helpful when solving uniform motion problems.

Analyze We know the *rate* of each ship (25 knots and 3 knots), and we know that they must close a *distance* of 56 sea miles between them. We don't know the *time* it will take to do this.

Assign Let $t =$ the time it takes the Coast Guard to reach the cargo ship. During the rescue, the ships don't travel at the same rate, but they do travel for the same amount of time. Therefore, t also represents the travel time for the cargo ship.

Form We enter the rates, the variable t for each time, and the total distance traveled by the ships (56 sea miles) in the table. To fill in the last column, we use the formula $r \cdot t = d$ twice to find an expression for each distance traveled: $25 \cdot t = 25t$ and $3 \cdot t = 3t$.

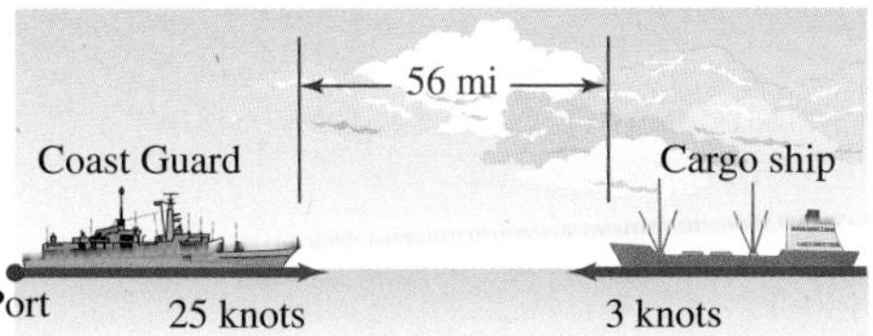

	r	$\cdot\ t$	$= d$
Coast Guard cutter	25	t	$25t$
Cargo ship	3	t	$3t$
			Total: 56

Multiply $r \cdot t$ to obtain an expression for each distance traveled.

Use the information in this column to form an equation.

Caution

A common error is to enter 56 miles as the distance traveled for each ship. However, neither ship traveled 56 miles. Together, they travel 56 miles.

	r	$\cdot\ t$	$= d$
Cutter			56
Cargo			56

The distance the cutter travels	plus	the distance the ship travels	equals	the original distance between the ships.
$25t$	$+$	$3t$	$=$	56

Solve

$$25t + 3t = 56$$

$$28t = 56 \quad \text{Combine like terms: } 25t + 3t = 28t.$$

$$t = \frac{56}{28} \quad \text{To isolate } t, \text{ divide both sides by 28.}$$

$$t = 2 \quad \text{Do the division.}$$

State The ships will meet in 2 hours.

Check In 2 hours, the Coast Guard cutter travels $25 \cdot 2 = 50$ sea miles, and the cargo ship travels $3 \cdot 2 = 6$ sea miles. Together, they travel $50 + 6 = 56$ sea miles. Since this is the original distance between the ships, the result checks.

Teaching Example 2 Road Trips. A car leaves Rockford traveling toward Wausau at the rate of 55 mph. At the same time, another car leaves Wausau traveling toward Rockford at the rate of 50 mph. How long will it take them to meet if the cities are 157.5 miles apart?

Answer: $1\frac{1}{2}$ hr

Self Check 2 **Rescues.** Two search-and-rescue teams leave base at the same time looking for a lost boy. The first team, on foot, heads north at 2 mph, and the other, on horseback, heads south at 4 mph. How long will it take them to search a distance of 21 miles between them? 3.5 hr

Now Try Problem 29

EXAMPLE 3 **Concert Tours.** While on tour, a country music star travels by bus. Her musical equipment is carried in a truck. How long will it take her bus, traveling 60 mph, to overtake the truck, traveling at 45 mph, if the truck had a $1\frac{1}{2}$-hour head start to her next concert location?

Analyze We know the rate of each vehicle (60 mph and 45 mph) and that the truck began the trip $1\frac{1}{2}$ or 1.5 hours earlier than the bus. We need to determine how long it will take the bus to catch up to the truck.

Assign Let $t =$ the time it takes the bus to overtake the truck. With a 1.5-hour head start, the truck is on the road longer than the bus. Therefore, $t + 1.5 =$ the truck's travel time.

Form We enter each rate and time in the table, and use the formula $r \cdot t = d$ twice to fill in the distance column.

	r ·	t =	d
Bus	60	t	$60t$
Truck	45	$t + 1.5$	$45(t + 1.5)$

Multiply $r \cdot t$ to obtain an expression for each distance traveled.

Enter this information first.

Use the information in this column to form an equation.

When the bus overtakes the truck, they will have traveled the same distance.

The distance traveled by the bus	is the same as	the distance traveled by the truck.
$60t$	$=$	$45(t + 1.5)$

Solve

$$60t = 45(t + 1.5)$$

$$60t = 45t + 67.5 \quad \text{Distribute the multiplication by 45: } 45(1.5) = 67.5.$$

$$15t = 67.5 \quad \text{Subtract } 45t \text{ from both sides: } 60t - 45t = 15t.$$

$$t = 4.5 \quad \text{To isolate } t\text{, divide both sides by 15: } \frac{67.5}{15} = 4.5.$$

Success Tip

We used 1.5 hrs for the head start because it is easier to solve $60t = 45(t + 1.5)$ than $60t = 45\left(t + 1\frac{1}{2}\right)$

State The bus will overtake the truck in 4.5 or $4\frac{1}{2}$ hours.

Check In 4.5 hours, the bus travels $60(4.5) = 270$ miles. The truck travels for $1.5 + 4.5 = 6$ hours at 45 mph, which is $45(6) = 270$ miles. Since the distances traveled are the same, the result checks.

Teaching Example 3 Rock Climbing. A group of rock climbers is traveling the Southwest by car. Their gear is carried in a minivan. How long will it take the car, traveling at 60 miles per hour, to overtake the minivan, traveling at 50 miles per hour, if the minivan had a half hour head start to the next climbing location?
Answer: 2.5 hr

Self Check 3 **Moving Day.** A moving van, packed with a family's belongings, left their old home for their new home, traveling at 40 miles per hour. Forty-five minutes $\left(\frac{3}{4}\text{ hour}\right)$ later, the family left for their new home, traveling by car at 60 miles per hour. How long did it take the family to overtake the moving van? 1.5 hr

Now Try Problem 33

3 Solve Liquid Mixture Problems.

We now discuss how to solve mixture problems. In the first type, a liquid mixture of a desired strength is made from two solutions with different concentrations (strengths).

EXAMPLE 4 **Mixing Solutions.** A chemistry experiment calls for a 30% sulfuric acid solution. If the lab supply room has only 50% and 20% sulfuric acid solutions, how much of each should be mixed to obtain 12 liters of a 30% acid solution?

Success Tip

The strength *(concentration)* of a mixture is always between the strengths of the two solutions used to make it.

Analyze The 50% solution is too strong and the 20% solution is too weak. We must find how much of each should be combined to obtain 12 liters of a 30% solution.

Assign If x = the number of liters of the 50% solution used in the mixture, the remaining $(12 - x)$ liters must be the 20% solution.

Form The amount of pure sulfuric acid in each solution is given by

Amount of solution · strength of the solution = amount of pure sulfuric acid

A table and sketch are helpful in organizing the facts of the problem.

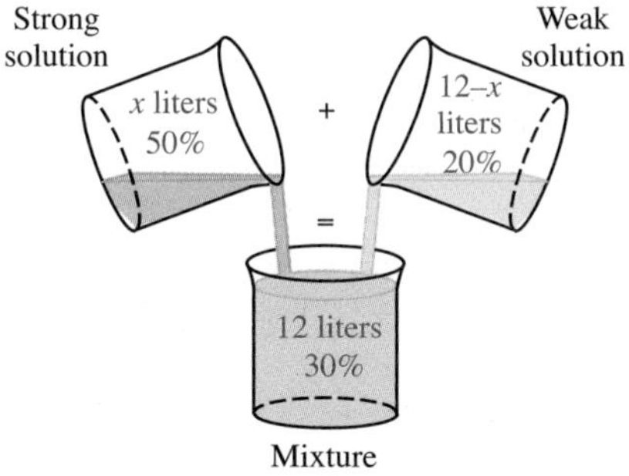

	Amount	· Strength =	Amount of pure sulfuric acid
Weak	$12 - x$	0.20	$0.20(12 - x)$
Strong	x	0.50	$0.50x$
Mixture	12	0.30	$12(0.30)$

Multiply amount · strength three times to fill in this column.

Enter this information first.

Use the information in this column to form an equation.

Teaching Tip: Use some specific examples to help your students understand that subtraction is used to write an expression for the amount of the weak 20% solution.
If there are 12 liters of the mixture:

Liters of strong	Liters of weak
1	12 − 1 = 11
4	12 − 4 = 8
10	12 − 10 = 2
x	$12 - x$

The sulfuric acid in the 20% solution	plus	the sulfuric acid in the 50% solution	equals	the sulfuric acid in the mixture.
$0.20(12 - x)$	+	$0.50x$	=	$12(0.30)$

Solve

$$0.20(12 - x) + 0.50x = 12(0.30)$$

$$2.4 - 0.2x + 0.5x = 3.6 \quad \text{Distribute the multiplication by 0.20.}$$

$$0.3x + 2.4 = 3.6 \quad \text{Combine like terms: } -0.2x + 0.5x = 0.3x.$$

$$0.3x = 1.2 \quad \text{Subtract 2.4 from both sides.}$$

$$x = 4 \quad \text{To isolate } x\text{, undo the multiplication by 0.3 by dividing both sides by 0.3: } \frac{1.2}{0.3} = 4.$$

Success Tip

We could begin by multiplying both sides of the equation by 10 to clear it of the decimals.

State 4 liters of 50% solution and $12 - 4 = 8$ liters of 20% solution should be used.

Check The amount of acid in 4 liters of the 50% solution is $0.50(4) = 2.0$ liters and the amount of acid in 8 liters of the 20% solution is $0.20(8) = 1.6$ liters. Thus, the amount of acid in these two solutions is $2.0 + 1.6 = 3.6$ liters. The amount of acid in 12 liters of the 30% mixture is also $0.30(12) = 3.6$ liters. Since the amounts of acid are equal, the results check.

Teaching Example 4 Dairy Products. Cream is approximately 22% butterfat. How many gallons of cream must be mixed with milk testing at 2% butterfat to get 20 gallons of milk containing 4% butterfat?
Answer: 2 gal

Self Check 4 **Mixing Solutions.** How many gallons of a 3% salt solution must be mixed with a 7% salt solution to obtain 25 gallons of a 5.4% salt solution? 10 gal

Now Try Problem 41

4 Solve Dry Mixture Problems.

In another type of mixture problem, a dry mixture of a specified value is created from two differently priced ingredients.

EXAMPLE 5 **Snack Foods.** Because cashews priced at \$9 per pound were not selling, a produce clerk decided to combine them with less expensive peanuts and sell the mixture for \$7 per pound. How many pounds of peanuts, selling at \$6 per pound, should be mixed with 50 pounds of cashews to obtain such a mixture?

Analyze We need to determine how many pounds of peanuts to mix with 50 pounds of cashews to obtain a mixture worth \$7 per pound.

Assign Let $x =$ the number of pounds of peanuts to use in the mixture. Since 50 pounds of cashews will be combined with the peanuts, the mixture will weigh $50 + x$ pounds.

Form The value of the mixture and of each of its ingredients is given by

Amount · the price = the total value

We can organize the facts of the problem in a table.

Caution

To find the number of pounds in the mixture, add the number of pounds of the ingredients:

$$50 + x$$

It would be incorrect to multiply:

~~$50 \cdot x$~~

	Amount	· Price	= Total value
Peanuts	x	6	$6x$
Cashews	50	9	450
Mixture	$50 + x$	7	$7(50 + x)$

Multiply amount · price three times to fill in this column.

Enter this information first.

Use the information in this column to form an equation.

The value of the peanuts	plus	the value of the cashews	equals	the value of the mixture.
$6x$	+	450	=	$7(50 + x)$

Solve

$$6x + 450 = 7(50 + x)$$

$6x + 450 = 350 + 7x$ Distribute the multiplication by 7.

$450 = 350 + x$ To eliminate the term $6x$ on the left side, subtract $6x$ from both sides: $7x - 6x = x$.

$100 = x$ To isolate x, subtract 350 from both sides.

State 100 pounds of peanuts should be used in the mixture.

Check The value of 100 pounds of peanuts, at \$6 per pound, is 100(\$6) = \$600 and the value of 50 pounds of cashews, at \$9 per pound, is 50(\$9) = \$450. Thus, the total value of these two ingredients is \$1,050. Since the value of 150 pounds of the mixture, at \$7 per pound, is also 150(7) = \$1,050, the result checks.

Teaching Example 5 Party Mix. A party mix is made by combining peanuts worth \$5 per pound and pretzels worth \$3 per pound. How many pounds of peanuts should be mixed with 50 pounds of pretzels to obtain a mix worth \$4.50 per pound? *Answer:* 150 lb peanuts

Self Check 5 **Mixing Candy.** Candy worth \$1.90 per pound is to be mixed with 60 lb of a second candy worth \$1.20 per pound. How many pounds of the \$1.90 per pound candy should be used to make a mixture worth \$1.48 per pound? 40 lb

Now Try Problem 47

5 Solve Number-Value Problems.

When problems deal with collections of different items having different values, we must distinguish between the *number of* and the *value of* the items. For these problems, we will use the fact that

Number · value = total value

EXAMPLE 6 **Dining Area Improvements.** A restaurant owner needs to purchase some tables, chairs, and dinner plates for the dining area of her establishment. She plans to buy four chairs and four plates for each new table. She also plans to buy 20 additional plates in case of breakage. If a table costs \$100, a chair \$50, and a plate \$5, how many of each can she buy if she takes out a loan for \$6,500 to pay for the new items?

Analyze We know the *value* of each item: Tables cost \$100, chairs cost \$50, and plates cost \$5 each. We need to find the *number* of tables, chairs, and plates she can purchase for \$6,500.

Teaching Tip: Review the concept of number and value using some specific examples.

Number of nickels	Total value (cents)
3	$3 \cdot 5 = 15$
10	$10 \cdot 5 = 50$
x	$x \cdot 5 = 5x$

Assign The number of chairs and plates she needs depends on the number of tables she buys. So we let t = the number of tables to be purchased. Since every table requires four chairs and four plates, she needs to order $4t$ chairs. Because 20 additional plates are needed, she should order $(4t + 20)$ plates.

Form We can organize the facts of the problem in a table.

	Number	· Value	= Total value
Tables	t	100	$100t$
Chairs	$4t$	50	$50(4t)$
Plates	$4t + 20$	5	$5(4t + 20)$
			Total: 6,500

Multiply number · value three times to fill in this column.

Enter this information first.

Use the information in this column to form an equation.

The value of the tables	plus	the value of the chairs	plus	the value of the plates	equals	the total value of the purchase.
$100t$	$+$	$50(4t)$	$+$	$5(4t + 20)$	$=$	6,500

Solve

$$100t + 50(4t) + 5(4t + 20) = 6{,}500$$

$$100t + 200t + 20t + 100 = 6{,}500$$ Do the multiplications and distribute.

$$320t + 100 = 6{,}500$$ Combine like terms: $100t + 200t + 20t = 320t$.

$$320t = 6{,}400$$ Subtract 100 from both sides.

$$t = 20$$ To isolate t, divide both sides by 320.

To find the number of chairs and plates to buy, we evaluate $4t$ and $4t + 20$ for $t = 20$.

Chairs: $4t = 4(20) = 80$ *Plates:* $4t + 20 = 4(20) + 20 = 100$ Substitute 20 for *t*.

State The owner needs to buy 20 tables, 80 chairs, and 100 plates.

Check The total value of 20 tables is 20($100) = $2,000, the total value of 80 chairs is 80($50) = $4,000, and the total value of 100 plates is 100($5) = $500. Because the total purchase is $2,000 + $4,000 + $500 = $6,500, the results check.

Teaching Example 6 Ticket Sales. Tickets to a concert cost $4 for students, $7 for adults, and $5 for seniors. Twice as many students as adults and half as many seniors as adults attended the concert. If the total receipts were $875, how many students attended the concert?
Answer: 100 students

Self Check 6 **Electronics.** A small electronics store buys iPods for $189, iPod skins for $32, and iTunes cards for $15. If they place an order for three times as many iPods as skins and 20 more iTunes cards than skins, how many of each did they order if the items totaled $2,756? 12 iPods, 4 skins, 24 cards

Now Try Problem 55

SECTION 2.6 STUDY SET

VOCABULARY

Fill in the blanks.

1. Problems that involve depositing money are called investment problems, and problems that involve moving vehicles are called uniform motion problems.
2. Problems that involve combining ingredients are called mixture problems, and problems that involve collections of different items having different values are called number-value problems.

CONCEPTS

3. Complete only the *principal column* given that part of $30,000 is invested in stocks and the rest in art. $30,000 - x$

	P	$\cdot\ r$	$\cdot\ t$	$= I$
Stocks	x			
Art	?			

4. A man made two investments that earned a combined annual simple interest of $280. Complete the table and then form an equation for this investment problem. $0.04x + 0.06(6,000 - x) = 280$

	P	$\cdot\ r$	$\cdot\ t$	$= I$
Bank	x	0.04	1	$0.04x$
Stocks	$6,000 - x$	0.06	1	$0.06(6,000 - x)$
				Total: 280

5. Complete only the *rate column* given that the east-bound plane flew 150 mph slower than the west-bound plane. $r - 150$

	r	$\cdot\ t$	$= d$
West	r		
East	?		

6. a. Complete only the *time column* given that a runner wants to overtake a walker and the walker had a $\frac{1}{2}$-hour head start. $t + 0.5$

	r	$\cdot\ t$	$= d$
Runner		t	
Walker		?	

b. Complete only the *time column* given that part of a 6-hour drive was in fog and the other part was in clear conditions. $6 - t$

	r	$\cdot\ t$	$= d$
Foggy		t	
Clear		?	

7. A husband and wife drive in opposite directions to work. Their drives last the same amount of time and their workplaces are 80 miles apart. Complete the table and then form an equation for this distance problem.
$35t + 45t = 80$

	r	$\cdot\ t$	$= d$
Husband	35	t	$35t$
Wife	45	t	$45t$
			Total: 80

8. a. How many gallons of acetic acid are there in barrel 2? 16.8 gal
b. Suppose the contents of the two barrels are poured into an empty third barrel. How many gallons of liquid will the third barrel contain? $x + 42$
c. Estimate the strength of the solution in the third barrel: 15%, 35%, or 60% acid? 35%

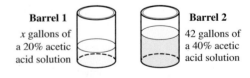

Barrel 1
x gallons of a 20% acetic acid solution

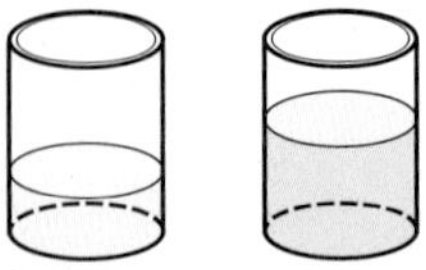

Barrel 2
42 gallons of a 40% acetic acid solution

9. a. Two antifreeze solutions are combined to form a mixture. Complete the table and then form an equation for this mixture problem. $0.25x + 0.50(6) = 0.30(6 + x)$

	Amount	· Strength	= Pure antifreeze
Weak	x	0.25	$0.25x$
Strong	6	0.50	$0.50(6)$
Mixture	$6 + x$	0.30	$0.30(6 + x)$

b. Two oil-and-vinegar salad dressings are combined to make a new mixture. Complete the table and then form an equation for this mixture problem. $0.03(10 - x) + 0.06x = 0.05(10)$

	Amount	· Strength	= Pure vinegar
Weak	$10 - x$	0.03	$0.03(10 - x)$
Strong	x	0.06	$0.06x$
Mixture	10	0.05	$0.05(10)$

10. The value of all the nylon brushes that a paint store carries is \$670. Complete the table and then form an equation for this number-value problem. $8x + 5x + 7(x + 10) = 670$

	Number	· Value	= Total value
1-inch	$2x$	4	$8x$
2-inch	x	5	$5x$
3-inch	$x + 10$	7	$7(x + 10)$
			Total: 670

NOTATION

11. Write 6% and 15.2% in decimal form. 0.06, 0.152

12. By what power of 10 should each decimal be multiplied to make it a whole number?

a. 0.08 100 **b.** 0.162 1,000

GUIDED PRACTICE

***Solve each equation.* See Example 1.**

13. $0.18x + 0.45(12 - x) = 0.36(12)$ 4

14. $0.12x + 0.20(4 - x) = 0.6$ 2.5

15. $0.08x + 0.07(15{,}000 - x) = 1{,}110$ 6,000

16. $0.108x + 0.07(16{,}000 - x) = 1{,}500$ 10,000

APPLICATIONS

***Investment problems.* See Example 1.**

17. **Corporate Investments.** The financial board of a corporation invested \$25,000 overseas, part at 4% and part at 7% annual simple interest. Find the amount invested at each rate if the first-year combined income from the two investments was \$1,300. \$15,000 at 4%; \$10,000 at 7%

18. **Loans.** A credit union loaned out \$50,000, part at an annual simple rate of 5% and the rest at an annual simple rate of 8%. They collected combined interest of \$3,400 from the loans that year. How much was loaned out at each rate? \$20,000 at 5%; \$30,000 at 8%

19. **Old Coins.** A salesperson used her \$3,500 year-end bonus to purchase some old gold and silver coins. She earned 15% annual simple interest on the gold coins and 12% annual simple interest on the silver coins. If she saw a return on her investment of \$480 the first year, how much did she invest in each type of coin? Silver: \$1,500; gold: \$2,000

20. **High-Risk Companies.** An investment club used funds totaling \$200,000 to invest in a bio-tech company and in an ethanol plant, with hopes of earning 11% and 14% annual simple interest, respectively. Their hunch paid off. The club made a total of \$24,250 interest the first year. How much was invested at each rate? \$125,000 at 11%; \$75,000 at 14%

21. **Retirement.** A professor wants to supplement her pension with investment interest. If she invests \$28,000 at 6% annual simple interest, how much would she have to invest at 7% annual simple interest to achieve a goal of \$3,500 per year in supplemental income? \$26,000

22. **Extra Income.** An investor wants to receive \$1,000 annually from two investments. He has put \$4,500 in a money market account paying 4% annual simple interest. How much should he invest in a stock fund that pays 10% annual simple interest to achieve his goal? \$8,200

23. **1099 Forms.** The form below shows the interest income Terrell Washington earned in 2011 from two savings accounts. He deposited a total of \$15,000 at the first of that year, and made no further deposits or withdrawals. How much money did he deposit in account 822 and in account 721? 822: \$9,000; 721: \$6,000

RECIPIENT'S name TERRELL WASHINGTON	USA HOME SAVINGS This is important tax information and is being furnished to the Internal Revenue Service.	2011
Account Number	**Annual Percent Yield**	**Interest earned**
822	5%	?
721	4.5%	?
FORM 1099		Total Interest Income \$720.00

24. **Investment Plans.** A financial planner recommends a plan for a client who has \$65,000 to invest. (See the chart at the right.) At the end of the presentation, the client asks, "How much will be invested at each rate?" Answer this question using the given information. \$42,200 at 12%, \$22,800 at 6.2%

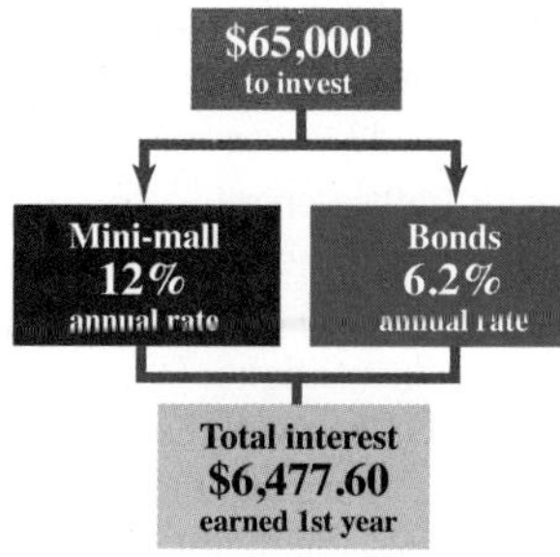

25. **Investments.** Equal amounts are invested in each of three accounts paying 7%, 8%, and 10.5% annual simple interest. If one year's combined interest income is \$1,249.50, how much is invested in each account? \$4,900

26. **Personal Loans.** Maggy lent her brother some money at 2% annual simple interest. She lent her sister twice as much money at half of the interest rate. In one year, Maggy collected combined interest of \$200 from her brother and sister. How much did she lend each of them? Brother: \$5,000, sister: \$10,000

27. **Bad Investments.** A newly hired MBA graduate received an \$18,000 signing bonus from her employer. She invested part of it in a credit union account earning 3% annual simple interest and the rest in utility stocks that suffered a 7% loss of value. The net income from both investments for the first year was only \$90. How much of her bonus was originally placed in each investment? Credit union: \$13,500; stocks: \$4,500

28. **Losses.** A financial planner invested a portion of his client's \$190,000 in a high-yield mutual fund that earned 11% annual simple interest. The remainder of the money was invested in a mini-mall development. Unfortunately, that investment lost 25% of its value the first year. Find the amount originally made in each investment if the first-year net income was \$6,500. Mutual fund: \$150,000; mini-mall: \$40,000

Uniform motion problems. See Example 2.

29. **Tornadoes.** During a storm, two teams of scientists leave a university at the same time in vans to search for tornadoes. The first team travels east at 20 mph and the second travels west at 25 mph. If their radios have a range of up to 90 miles, how long will it be before they lose radio contact? 2 hr

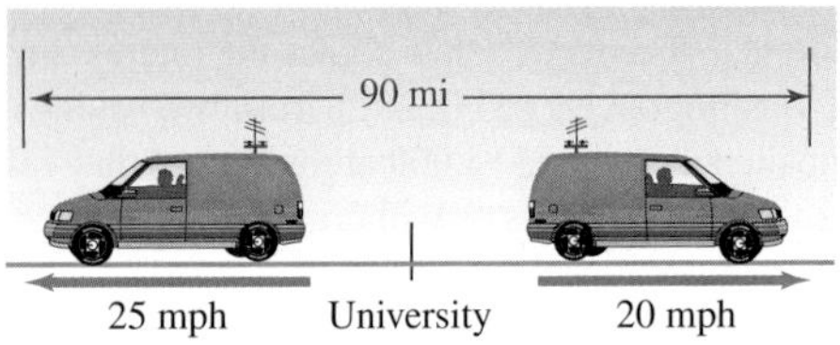

30. **Unmanned Aircraft.** Two remotely controlled unmanned aircraft are launched in opposite directions. One flies east at 78 mph and the other west at 82 mph. How long will it take the aircraft to fly a combined distance of 560 miles? 3.5 hr

31. **Hello/Goodbye.** A husband and wife work different shifts at the same plant. When the husband leaves from work to make the 20-mile trip home, the wife leaves their home and drives to work. They travel on the same road. The husband's driving rate is 45 mph and the wife's is 35 mph. How long into their drives can they wave at each other when passing on the road? $\frac{1}{4}$ hr = 15 min

32. **Air Traffic Control.** An airliner leaves Berlin, Germany, headed for Montreal, Canada, flying at an average speed of 450 mph. At the same time, an airliner leaves Montreal headed for Berlin, averaging 500 mph. If the airports are 3,800 miles apart, when will the air traffic controllers have to make the pilots aware that the planes are passing each other? 4 hr into the flights

33. **Cycling.** A cyclist leaves his training base for a morning workout, riding at the rate of 18 mph. One and one-half hours later, his support staff leaves the base in a car going 45 mph in the same direction. How long will it take the support staff to catch up with the cyclist? 1 hr

34. **Parenting.** How long will it take a mother, running at 4 feet per second, to catch up with her toddler, running down the sidewalk at 2 feet per second, if the child had a 5-second head start? 5 sec

35. **Road Trips.** A car averaged 40 mph for part of a trip and 50 mph for the remainder. If the 5-hour trip covered 210 miles, for how long did the car average 40 mph? 4 hr

36. **Cross-Training.** An athlete runs up a set of stadium stairs at a rate of 2 stairs per second, immediately turns around, and then descends the same stairs at a rate of 3 stairs per second. If the workout takes 90 seconds, how long does it take him to run up the stairs? 54 sec

37. **Winter Driving.** A trucker drove for 4 hours before he encountered icy road conditions. He reduced his speed by 20 mph and continued driving for 3 more hours. Find his average speed during the first part of the trip if the entire trip was 325 miles. 55 mph

38. **Speed of Trains.** Two trains are 330 miles apart, and their speeds differ by 20 mph. Find the speed of each train if they are traveling toward each other and will meet in 3 hours. 65 mph, 45 mph

Liquid mixture problems. See Example 3.

39. **Salt Solutions.** How many gallons of a 3% salt solution must be mixed with 50 gallons of a 7% solution to obtain a 5% solution? 50 gal

40. **Photography.** A photographer wishes to mix 2 liters of a 5% acetic acid solution with a 10% solution to get a 7% solution. How many liters of 10% solution must be added? $1\frac{1}{3}$ liters

41. **Making Cheese.** To make low-fat cottage cheese, milk containing 4% butterfat is mixed with milk containing 1% butterfat to obtain 15 gallons of a mixture containing 2% butterfat. How many gallons of each milk must be used? 4%: 5 gal; 1%: 10 gal

42. **Antifreeze.** How many quarts of a 10% antifreeze solution must be mixed with 16 quarts of a 40% antifreeze solution to make a 30% solution? 8 quarts

43. **Printing.** A printer has ink that is 8% cobalt blue color and ink that is 22% cobalt blue color. How many ounces of each ink are needed to make one-half gallon (64 ounces) of ink that is 15% cobalt blue color? 32 ounces of 8%; 32 ounces of 22%

44. **Flood Damage.** One website recommends a 6% chlorine bleach-water solution to remove mildew. A chemical lab has 3% and 15% chlorine bleach-water solutions in stock. How many gallons of each should be mixed to obtain 100 gallons of the mildew spray? 75 gallons of 3%; 25 gallons of 15%

45. **Interior Decorating.** The colors on the paint chip card below are created by adding different amounts of orange tint to a white latex base. How many gallons of Desert Sunrise should be mixed with 1 gallon of Bright Pumpkin to obtain Cool Cantaloupe? 6 gal

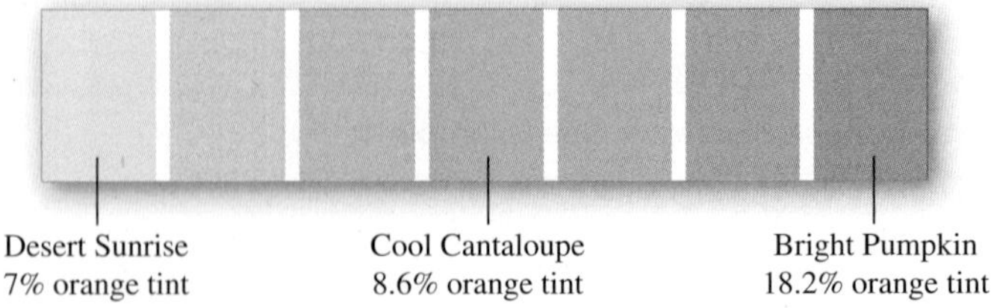

46. **Antiseptics.** A nurse wants to add water to 30 ounces of a 10% solution of benzalkonium chloride to dilute it to an 8% solution. How much water must she add? (*Hint:* Water is 0% benzalkonium chloride.) 7.5 oz

Dry mixture problems. See Example 4.

47. Lawn Seed. A store sells bluegrass seed for $6 per pound and ryegrass seed for $3 per pound. How much ryegrass must be mixed with 100 pounds of bluegrass to obtain a blend that will sell for $5 per pound? 50 lb

48. Coffee Blends. A store sells regular coffee for $8 a pound and gourmet coffee for $14 a pound. To get rid of 40 pounds of the gourmet coffee, a shopkeeper makes a blend to put on sale for $10 a pound. How many pounds of regular coffee should he use? 80 lb

49. Raisins. How many scoops of natural seedless raisins costing $3.45 per scoop must be mixed with 20 scoops of golden seedless raisins costing $2.55 per scoop to obtain a mixture costing $3 per scoop? 20 scoops

50. Fertilizer. Fertilizer with weed control costing $38 per 50-pound bag is to be mixed with a less expensive fertilizer costing $6 per 50-pound bag to make 16 bags of fertilizer that can be sold for $28 per bag. How many bags of cheaper fertilizer should be used? 5 bags

51. Packaged Salad. How many 10-ounce bags of Romaine lettuce must be mixed with fifty 10-ounce bags of Iceberg lettuce to obtain a blend that sells for $2.50 per ten-ounce bag? 15

52. Mixing Candy. Lemon drops worth $3.80 per pound are to be mixed with jelly beans that cost $2.40 per pound to make 300 pounds of a mixture worth $2.96 per pound. How many pounds of each candy should be used? 120 lb lemon drops, 180 lb jelly beans

53. Bronze. A pound of tin is worth $1 more than a pound of copper. Four pounds of tin are mixed with 6 pounds of copper to make bronze that sells for $3.65 per pound. How much is a pound of tin worth? $4.25

54. Snack Foods. A bag of peanuts is worth $0.30 less than a bag of cashews. Equal amounts of peanuts and cashews are used to make 40 bags of a mixture that sells for $1.05 per bag. How much is a bag of cashews worth? $1.20

Number-value problems. See Example 5.

55. Rentals. The owners of an apartment building rent equal numbers of 1-, 2-, and 3-bedroom units. The monthly rent for a 1-bedroom is $550, a 2-bedroom is $700, and a 3-bedroom is $900. If the total monthly income is $36,550, how many of each type of unit are there? 17

56. Warehousing. A store warehouses 40 more portables than big-screen TV sets, and 15 more consoles than big-screen sets. The monthly storage cost for a portable is $1.50, a console is $4.00, and a big-screen is $7.50. If storage for all the televisions costs $276 per month, how many big-screen sets are in stock? 12

57. Software. Three software applications are priced as shown. Spreadsheet and database programs sold in equal numbers, but 15 more word processing applications were sold than the other two combined. If the three applications generated sales of $72,000, how many spreadsheets were sold? 90

Software	Price
Spreadsheet	$150
Database	$195
Word processing	$210

58. Inventories. With summer approaching, the number of air conditioners sold is expected to be double that of stoves and refrigerators combined. Stoves sell for $350, refrigerators for $450, and air conditioners for $500, and sales of $56,000 are expected. If stoves and refrigerators sell in equal numbers, how many of each appliance should be stocked? 20 stoves, 20 refrigerators, 80 air conditioners

59. Piggy Banks. When a child emptied his coin bank, he had a collection of pennies, nickels, and dimes. There were 20 more pennies than dimes and the number of nickels was triple the number of dimes. If the coins had a value of $5.40, how many of each type coin were in the bank? 40 pennies, 20 dimes, 60 nickels

60. Wishing Wells. A scuba diver, hired by an amusement park, collected $121 in nickels, dimes, and quarters at the bottom of a wishing well. There were 500 nickels, and 90 more quarters than dimes. How many quarters and dimes were thrown into the wishing well? Dimes: 210, quarters: 300

61. Basketball. Epiphanny Prince, of New York, scored 113 points in a high school game on February 1, 2006, breaking a national prep record that was held by Cheryl Miller. Prince made 46 more 2-point baskets than 3-point baskets, and only 1 free throw. How many 2-point and 3-point baskets did she make? 2-pointers: 50, 3-pointers: 4

62. Museum Tours. The ticket prices for the Coca-Cola Museum in Atlanta are shown. A family purchased 3 more children's tickets than adult tickets, and 1 less senior ticket than adult tickets. The total cost of the tickets was $131. How many of each type did they purchase? Adult: 3, senior: 2, child: 6

WRITING

63. Create a mixture problem of your own, and solve it.

64. Write an investment problem to fit the following equation, and then solve it.

$$0.08x + 0.07(10{,}000 - x) = 770$$

65. Explain the error in each statement.

a. If 3 pounds of Mocha coffee are mixed with x pounds of Java coffee, there will be $3x$ pounds of the Mocha-Java blend.

b. A financial manager has a total of \$5,000 to invest in two accounts. If $\$x$ is invested in the first account, then $\$(x - 5{,}000)$ is left to be invested in the second account.

66. Is it possible to mix a 10% sugar solution with a 20% sugar solution to get a 30% sugar solution? Explain.

REVIEW

Multiply.

67. $-12(3a + 4b - 32)$ $-36a - 48b + 384$

68. $\frac{1}{2}(4b - 8)$ $2b - 4$

69. $3(5t + 1)2$ $30t + 6$

70. $2.9(4c - 12)$ $11.6c - 34.8$

CHALLENGE PROBLEMS

71. Evaporation. How much water must be boiled away to increase the concentration of 300 milliliters of a 2% salt solution to a 3% salt solution? 100 milliliters

72. Diluting Solutions. How much water should be added to 20 ounces of a 15% solution of alcohol to dilute it to a 10% alcohol solution? 10 oz

73. Financial Planning. A plumber has a choice of two investment plans:

- An insured fund that pays 11% interest
- A risky investment that pays a 13% return

If the same amount invested at the higher rate would generate an extra \$150 per year, how much does the plumber have to invest? \$7,500

74. Investments. The amount of annual interest earned by \$8,000 invested at a certain rate is \$200 less than \$12,000 would earn at a rate 1% lower. At what rate is the \$8,000 invested? 8%

SECTION 2.7 Solving Inequalities

OBJECTIVES

1. Determine whether a number is a solution of an inequality.
2. Graph solution sets and use interval notation.
3. Solve linear inequalities.
4. Solve compound inequalities.
5. Solve inequality applications.

ARE YOU READY? *Are You Ready? exercises available online at www.webassign.net/brookscole*

The following problems review some basic skills that are needed to solve inequalities.

1. Fill in the blanks: The symbol $<$ means "is less than."

2. Is $-5 > -6$ a true or false statement? True

3. Graph each number in the set $\left\{-4, -1.7, 2, \frac{13}{4}\right\}$ on a number line.

See AIE Appendix 3.

4. Express the fact that $10 > 0$ using an $<$ symbol. $0 < 10$

In our daily lives, we often speak of one value being *greater than* or *less than* another. For example, a sick child might have a temperature *greater than* 98.6°F or a granola bar might contain *less than* 2 grams of fat. In mathematics, we use *inequalities* to show that one expression is greater than or is less than another expression.

1 Determine Whether a Number Is a Solution of an Inequality.

An **inequality** is a statement that contains one or more of the following symbols.

Inequality Symbols

$<$ is less than	$>$ is greater than	$\neq$ is not equal to
$\leq$ is less than or equal to	$\geq$ is greater than or equal to	

An inequality can be true, false, or neither true nor false. For example,

- $9 \geq 9$ is true because $9 = 9$.
- $37 < 24$ is false.
- $x + 1 > 5$ is neither true nor false because we don't know what number x represents.

The Language of Algebra

Because < requires one number to be strictly less than another number and > requires one number to be strictly greater than another number, < and > are called **strict inequalities**.

An inequality that contains a variable can be made true or false depending on the number that is substituted for the variable. If we substitute 10 for x in $x + 1 > 5$, the resulting inequality is true: $\mathbf{10} + 1 > 5$. If we substitute 1 for x, the resulting inequality is false: $\mathbf{1} + 1 > 5$. A number that makes an inequality true is called a **solution** of the inequality, and we say that the number *satisfies* the inequality. Thus, 10 is a solution of $x + 1 > 5$ and 1 is not.

In this section, we will find the solutions of *linear inequalities in one variable.*

Linear Inequality in One Variable

A **linear inequality in one variable** can be written in one of the following forms where a, b, and c are real numbers and $a \neq 0$.

$$ax + b > c \qquad ax + b \geq c \qquad ax + b < c \qquad ax + b \leq c$$

EXAMPLE 1 Is 9 a solution of $2x + 4 \leq 21$?

Strategy We will substitute 9 for x and evaluate the expression on the left side.

Why If a true statement results, 9 is a solution of the inequality. If we obtain a false statement, 9 is not a solution.

Solution

$2x + 4 \leq 21$

$2(9) + 4 \overset{?}{\leq} 21$ Substitute 9 for x. Read $\overset{?}{\leq}$ as "is possibly less than or equal to."

$18 + 4 \overset{?}{\leq} 21$ Do the multiplication.

$22 \leq 21$ This inequality is false.

The statement $22 \leq 21$ is false because neither $22 < 21$ nor $22 = 21$ is true. Therefore, 9 is not a solution of $2x + 4 \leq 21$.

The Language of Algebra

A **linear inequality** in one variable is similar to a linear equation in one variable except that the equal symbol is replaced with an inequality symbol.

Equation	*Inequality*
$2x + 1 = 9$	$2x + 1 > 9$

Teaching Example 1 Is −3 a solution of $4x + 5 \leq -6$?
Answer: Yes

Self Check 1 Is 2 a solution of $3x - 1 \geq 0$? Yes

Now Try Problem 15

2 Graph Solution Sets and Use Interval Notation.

The **solution set** of an inequality is the set of all numbers that make the inequality true. Some solution sets are easy to find. For example, if we replace the variable in $x > -3$ with a number greater than -3, the resulting inequality will be true. Because there are infinitely many real numbers greater than -3, it follows that $x > -3$ has infinitely many solutions. Since there are too many solutions to list, we use **set-builder notation** to describe the solutions set.

$$\{x \mid x > -3\}$$

Read as "the set of all x such that x is greater than −3."

We can illustrate the solution set by **graphing the inequality** on a number line. To graph $x > -3$, a **parenthesis** or **open circle** is drawn on the endpoint -3 to indicate that -3 is not part of the graph. Then we shade all of the points on the number line to the right of -3. The right arrowhead is also shaded to show that the solutions continue forever to the right.

Notation

The parenthesis (opens in the direction of the shading and indicates that an endpoint is not included in the shaded interval.

Method 1: parenthesis

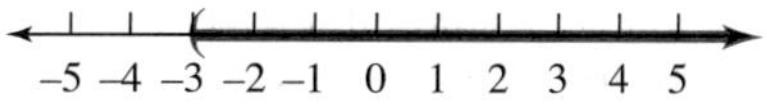

Method 2: open circle

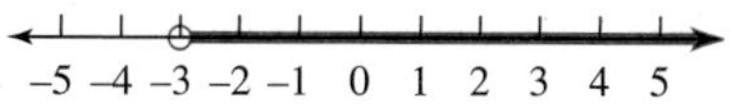

All real numbers greater than –3

Notation

The positive infinity symbol ∞ does not represent a real number. It indicates that an interval extends to the right without end.

The graph of $x > -3$ is an example of an **interval** on the number line. We can write intervals in a compact form called **interval notation.**

The interval notation that represents the graph of $x > -3$ is $(-3, \infty)$. As on the number line, a left parenthesis is written next to -3 to indicate that -3 is not included in the interval. The **positive infinity symbol** ∞ that follows indicates that the interval continues without end to the right. With this notation, *a parenthesis is always used next to an infinity symbol.*

The illustration below shows the relationship between the symbols used to graph an interval and the corresponding interval notation. If we begin at -3 and move to the right, the shaded arrowhead on the graph indicates that the interval approaches positive infinity ∞.

Teaching Tip: Remind your students that interval notation does not contain a variable.

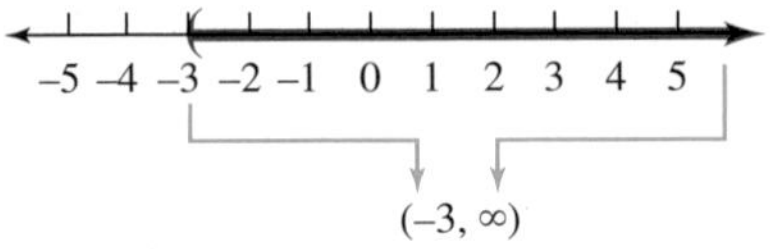

Success Tip

Think of interval notation as a way to tell someone how to draw the graph, from left to right, giving them only a "start" and a "stop" instruction.

We now have three ways to describe the solution set of an inequality.

Set-builder notation	*Number line graph*	*Interval notation*
$\{x \mid x > -3\}$	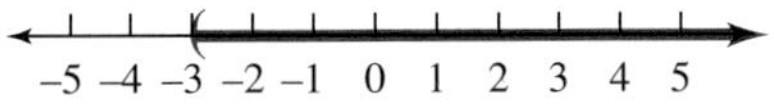	$(-3, \infty)$

EXAMPLE 2 Graph: $x \le 2$

Strategy We need to determine which real numbers, when substituted for x, would make $x \le 2$ a true statement.

Why To graph $x \le 2$ means to draw a "picture" of all of the values of x that make the inequality true.

Solution If we replace x with a number less than or equal to 2, the resulting inequality will be true. To graph the solution set, a **bracket** or a **closed circle** is drawn at the endpoint 2 to indicate that 2 is part of the graph. Then we shade all of the points on the number line to the left of 2 as well as the left arrowhead.

Notation

The *bracket*] opens in the direction of the shading and indicates that an endpoint is included in the shaded interval.

Method 1: bracket

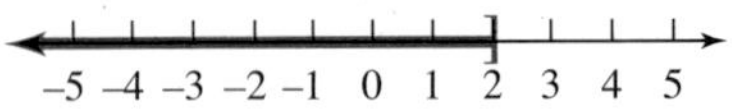

Method 2: closed circle

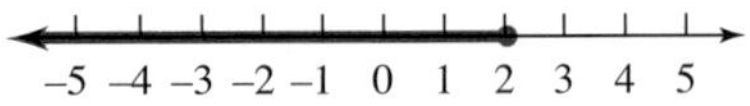

All real numbers less than or equal to 2

The interval is written as $(-\infty, 2]$. The right bracket indicates that 2 is included in the interval. The **negative infinity symbol** $-\infty$ shows that the interval continues forever to the left. The illustration below shows the relationship between the symbols used to graph the interval and the corresponding interval notation.

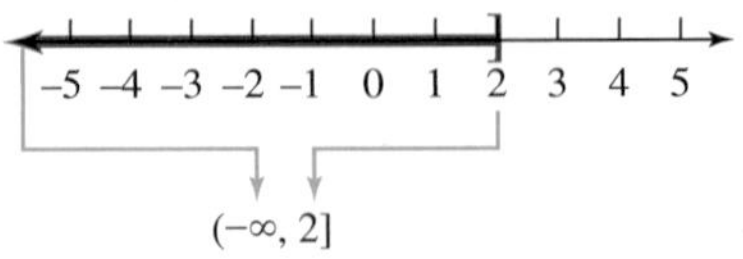

Teaching Example 2 Graph: $x \ge -1$
Answer:

Self Check 2 Graph: $x \ge 0$ [0, ∞) See AIE Appendix 3.

Now Try Problem 19

3 Solve Linear Inequalities.

To **solve an inequality** means to find all values of the variable that make the inequality true. As with equations, there are properties that we can use to solve inequalities.

Addition and Subtraction Properties of Inequality

Adding the same number to, or subtracting the same number from, both sides of an inequality does not change its solutions.

For any real numbers a, b, and c,

If $a < b$, then $a + c < b + c$. If $a < b$, then $a - c < b - c$.

Similar statements can be made for the symbols $\leq$, $>$, and $\geq$.

After applying one of these properties, the resulting inequality is equivalent to the original one. **Equivalent inequalities** have the same solution set.

Like equations, inequalities are solved by isolating the variable on one side.

EXAMPLE 3 Solve $x + 3 > 2$. Write the solution set in interval notation and graph it.

Strategy We will use a property of inequality to isolate the variable on one side.

Why To solve the original inequality, we want to find a simpler equivalent inequality of the form $x >$ **a number** or $x <$ **a number**, whose solution is obvious.

Success Tip

We solve linear inequalities by writing a series of steps that result in an equivalent inequality of the form

x > a number

or

x < a number

Similar statements apply to linear inequalities containing $\leq$ and $\geq$.

Solution We will use the subtraction property of inequality to isolate x on the left side of the inequality. We can undo the addition of 3 by subtracting 3 from both sides.

$$x + 3 > 2 \quad \text{This is the inequality to solve.}$$

$$x + 3 - \mathbf{3} > 2 - \mathbf{3} \quad \text{Subtract 3 from both sides.}$$

$$x > -1 \quad \text{Do the subtraction: } 3 - 3 = 0 \text{ and } 2 - 3 = -1.$$

All real numbers greater than -1 are solutions of $x + 3 > 2$. The solution set can be written in set-builder notation as $\{x \mid x > -1\}$ and in interval notation as $(-1, \infty)$. The graph of the solution set is shown below.

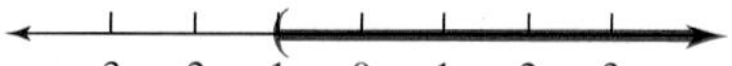

Teaching Tip: Ask your students: If we pick a number from the shaded portion of the number line and substitute for x in $x + 3 > 2$, will the result be true or false? Ask the same question for a number in the unshaded portion of the number line, and for the number -1.

Since there are infinitely many solutions, we cannot check all of them. As an informal check, we can pick some numbers in the graph, say 0 and 30, substitute each number for x in the original inequality, and see whether true statements result.

Check:

$x + 3 > 2$		$x + 3 > 2$	
$\mathbf{0} + 3 \overset{?}{>} 2$	Substitute 0 for x.	$\mathbf{30} + 3 \overset{?}{>} 2$	Substitute 30 for x.
$3 > 2$	True	$33 > 2$	True

The solution set appears to be correct.

Teaching Example 3 Solve $x - 5 \leq 3$. Write the solution set in interval notation and graph it.
Answer:
$x \leq 8$, $(-\infty, 8]$

Self Check 3 Solve $x - 3 < -2$. Write the solution set in interval notation and graph it. $(-\infty, 1)$ See AIE Apendix 3.

Now Try Problem 23

As with equations, there are properties for multiplying and dividing both sides of an inequality by the same number. To develop what is called *the multiplication property of inequality,* we consider the true statement $2 < 5$. If both sides are multiplied by a positive number, such as 3, another true inequality results.

$$2 < 5 \quad \text{This inequality is true.}$$

$$\mathbf{3} \cdot 2 < \mathbf{3} \cdot 5 \quad \text{Multiply both sides by 3.}$$

$$6 < 15 \quad \text{This inequality is true.}$$

Caution

If the inequality symbol is not reversed when both sides of a true inequality are multiplied by a negative number, the result is a false inequality. For example,

$$3 < 6 \qquad \text{True}$$
$$-3 \cdot 3 < -3 \cdot 6$$
$$-9 < -18 \qquad \text{False}$$

The same is true if we divide both sides of a true inequality by a negative number. Divide both sides of $3 < 6$ by -3 to see for yourself.

However, if we multiply both sides of $2 < 5$ by a negative number, such as -3, the direction of the inequality symbol must be reversed to produce another true inequality.

$2 < 5$ This inequality is true.

$-3 \cdot 2 > -3 \cdot 5$ Multiply both sides by -3 and reverse the direction of the inequality.

$-6 > -15$ This inequality is true.

The inequality $-6 > -15$ is true because -6 is to the right of -15 on the number line.

Dividing both sides of an inequality by the same negative number also requires that the direction of the inequality symbol be reversed.

$-4 < 6$ This inequality is true.

$\frac{-4}{-2} > \frac{6}{-2}$ Divide both sides by -2 and change $<$ to $>$.

$2 > -3$ This inequality is true.

These examples illustrate the **multiplication and division properties of inequality.**

Multiplication and Division Properties of Inequality

Multiplying or dividing both sides of an inequality by the same **positive number** does not change its solutions.

For any real numbers a, b, and c, where c is **positive**,

If $a < b$, then $ac < bc$. If $a < b$, then $\frac{a}{c} < \frac{b}{c}$.

If we multiply or divide both sides of an inequality by the same **negative number**, the direction of the inequality symbol must be reversed for the inequalities to have the same solutions.

For any real numbers a, b, and c, where c is **negative,**

If $a < b$, then $ac > bc$. If $a < b$, then $\frac{a}{c} > \frac{b}{c}$.

Similar statements can be made for the symbols $\le$, $>$, and $\ge$.

EXAMPLE 4 Solve each inequality. Write the solution set in interval notation and graph it.

a. $-\frac{3}{2}t \ge -12$ **b.** $-5t < 55$

Strategy We will use a property of inequality to isolate the variable on one side.

Why To solve the original inequality, we want to find a simpler equivalent inequality, whose solution is obvious.

Solution

a. To undo the multiplication by $-\frac{3}{2}$, we multiply both sides by the reciprocal, which is $-\frac{2}{3}$.

Teaching Tip: Many students reverse the inequality symbol, but they do it one line too late in their solutions. Remind them to reverse the inequality symbol on the line on which the multiplication or division by a negative number occurs.

$-\frac{3}{2}t \ge -12$ This is the inequality to solve.

$-\frac{2}{3}\left(-\frac{3}{2}t\right) \le -\frac{2}{3}(-12)$ Multiply both sides by $-\frac{2}{3}$. Since we are multiplying both sides by a negative number, reverse the direction of the $\ge$ symbol.

$t \le 8$ Do the multiplication: $-\frac{2}{3}\left(-\frac{3}{2}\right) = 1$ and $-\frac{2}{3}(-12) = 8$.

The solution set is $(-\infty, 8]$ and it is graphed as shown.

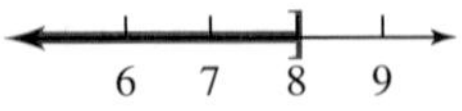

Success Tip

You may find it easier to graph the solution set first and then use the symbols that occur in that form of the result to write the interval notation.

b. To undo the multiplication by -5, we divide both sides by -5.

$-5t < 55$ This is the inequality to solve.

$\frac{-5t}{-5} > \frac{55}{-5}$ To isolate t, undo the multiplication by -5 by dividing both sides by -5. Since we are dividing both sides by a negative number, reverse the direction of the $<$ symbol.

$t > -11$ Do the division.

The solution set is $(-11, \infty)$ and it is graphed as shown.

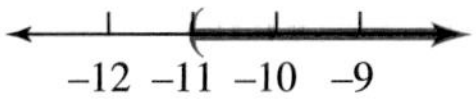

Teaching Example 4 Solve each inequality. Write the solution set in interval notation and graph it.
a. $-\frac{5}{3}x \ge 10$ **b.** $-6x < 12$
Answers:
a. $x \le -6, (-\infty, -6]$
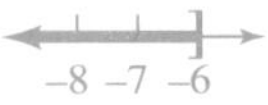

b. $x > -2, (-2, \infty)$
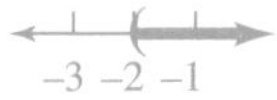

Self Check 4 Solve each inequality. Write the solution set in interval notation and graph it. **a.** $-\frac{h}{20} \le 10$ $[-200, \infty)$ **b.** $-12a > -144$ $(-\infty, 12)$ See AIE Appendix 3.

Now Try Problems 27 and 31

EXAMPLE 5 Solve $-5 > 3x + 7$. Write the solution set in interval notation and graph it.

Strategy First we will use a property of inequality to isolate the *variable term* on one side. Then we will use a second property of inequality to isolate the *variable* itself.

Why To solve the original inequality, we want to find a simpler equivalent inequality of the form $x >$ **a number** or $x <$ **a number**, whose solution is obvious.

Solution

$-5 > 3x + 7$ This is the inequality to solve.

$-5 - 7 > 3x + 7 - 7$ To isolate the variable term $3x$ on the right side, undo the addition of 7 by subtracting 7 from both sides.

$-12 > 3x$ Do the subtraction: $-5 - 7 = -12$ and $7 - 7 = 0$.

$\frac{-12}{3} > \frac{3x}{3}$ To isolate x, undo the multiplication by 3 by dividing both sides by 3.

$-4 > x$ Do the division.

Success Tip

Don't be confused by the negative number on the left side. We didn't reverse the $>$ symbol because we divided both sides by *positive* 3.

$$\frac{-12}{3} > \frac{3x}{3}$$

Teaching Tip: In $-4 > x$, the inequality symbol points to the smaller number, x. Show students that the same is true for the equivalent inequality, $x < -4$. The symbol points to the smaller number, x.

To determine the solution set, it is useful to rewrite the inequality $-4 > x$ in an equivalent form with the variable on the left side.

$x < -4$ If -4 is greater than x, it follows that x must be less than -4.

The solution set is $(-\infty, -4)$ and it is graphed as shown.

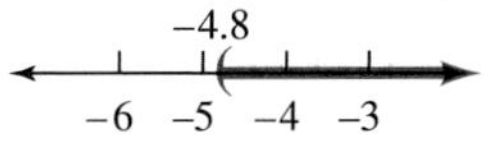

Teaching Example 5 Solve $3 > 4x + 11$. Write the solution set in interval notation and graph it.
Answer:
$x < -2, (-\infty, -2)$

Self Check 5 Solve $-13 < 2r - 7$. Write the solution set in interval notation and graph it. $(-3, \infty)$ See AIE Appendix 3.

Now Try Problem 35

EXAMPLE 6 Solve $5.1 - 3k < 19.5$. Write the solution set in interval notation and graph it.

Strategy We will use properties of inequality to isolate the variable on one side.

Why To solve the original inequality, we want to find a simpler equivalent inequality of the form $k >$ **a number** or $k <$ **a number**, whose solution is obvious.

Solution

$5.1 - 3k < 19.5$	This is the inequality to solve.
$5.1 - 3k - \mathbf{5.1} < 19.5 - \mathbf{5.1}$	To isolate $-3k$ on the left side, subtract 5.1 from both sides.
$-3k < 14.4$	Do the subtraction.
$\frac{-3k}{-3} > \frac{14.4}{-3}$	To isolate k, undo the multiplication by -3 by dividing both sides by -3 and reverse the direction of the $<$ symbol.
$k > -4.8$	Do the division.

The solution set is $(-4.8, \infty)$, and it is graphed as shown.

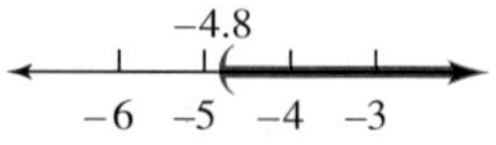

Teaching Example 6 Solve $-4.2 - 2x < 3.6$. Write the solution set in interval notation and graph it.
Answer: $x > -3.9, (-3.9, \infty)$

Self Check 6 Solve $-9n + 1.8 > -17.1$. Write the solution set in interval notation and graph it. $(-\infty, 2.1)$ See AIE Appendix 3.

Now Try Problem 39

Teaching Tip: This would be a good time to review the steps of the equation-solving strategy in Section 2.2 with the class.

The equation-solving strategy in Section 2.2 can be applied to inequalities. However, when solving inequalities, we must remember to *reverse the direction of the inequality symbol when multiplying or dividing both sides by a negative number.*

EXAMPLE 7 Solve $8(y + 1) \geq 2(y - 4) + y$. Write the solution set in interval notation and graph it.

Strategy We will follow the steps of the equation-solving strategy (adapted to inequalities) to solve the inequality.

Why This is the most efficient way to solve a linear inequality in one variable.

Solution

$8(y + 1) \geq 2(y - 4) + y$	This is the inequality to solve.
$8y + 8 \geq 2y - 8 + y$	Distribute the multiplication by 8 and by 2.
$8y + 8 \geq 3y - 8$	Combine like terms: $2y + y = 3y$.
$8y + 8 - \mathbf{3y} \geq 3y - 8 - \mathbf{3y}$	To eliminate $3y$ from the right side, subtract $3y$ from both sides.
$5y + 8 \geq -8$	Combine like terms: $8y - 3y = 5y$ and $3y - 3y = 0$.
$5y + 8 - \mathbf{8} \geq -8 - \mathbf{8}$	To isolate the variable term $5y$ on the left side, undo the addition of 8 by subtracting 8 from both sides.
$5y \geq -16$	Do the subtraction: $8 - 8 = 0$ and $-8 - 8 = -16$.
$\frac{5y}{5} \geq \frac{-16}{5}$	To isolate y, undo the multiplication by 5 by dividing both sides by 5. Do not reverse the direction of the $\geq$ symbol.
$y \geq -\frac{16}{5}$	

The solution set is $\left[-\frac{16}{5}, \infty\right)$ and it is graphed as shown.

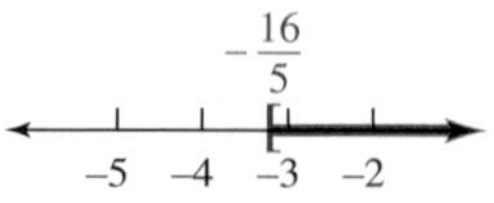

Success Tip

As an informal check, substitute a number on the graph that is shaded, such as -3, into $8(y + 1) \geq 2(y - 4) + y$. A true statement should result. Then substitute a number on the graph that is not shaded, such as -4, into the inequality. A false statement should result.

A solution involving an improper fraction is perfectly acceptable. Just make sure the fraction is in simplified form. To locate the endpoint of the interval on the graph, it is helpful to note that $-\frac{16}{5} = -3\frac{1}{5}$.

Teaching Example 7 Solve $4(2x - 2) \leq 2(x + 2) + x$. Write the solution set in interval notation and graph it.
Answer: $x \leq \frac{12}{5}, \left(-\infty, \frac{12}{5}\right]$

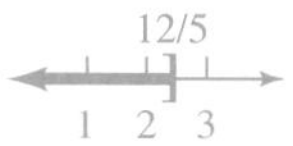

Self Check 7 Solve $5(b - 2) \geq -(b - 3) + 2b$. Write the solution set in interval notation and graph it. $\left[\frac{13}{4}, \infty\right)$ See AIE Appendix 3.

Now Try Problem 45

EXAMPLE 8 Solve $\frac{3}{4} + \frac{x}{2} > \frac{6}{7}$. Write the solution set in interval notation and graph it.

Strategy The first step of the equation-solving strategy (adapted to inequalities) is to clear the inequality of fractions by multiplying both sides by the LCD.

Why It's easier to solve an inequality that involves only integers.

Solution

$$\frac{3}{4} + \frac{x}{2} > \frac{6}{7}$$ This is the inequality to solve.

$$28\left(\frac{3}{4} + \frac{x}{2}\right) > 28\left(\frac{6}{7}\right)$$ Clear the inequality of fractions by multiplying both sides by the LCD of $\frac{3}{4}$, $\frac{x}{2}$, and $\frac{6}{7}$, which is 28.

$$28\left(\frac{3}{4}\right) + 28\left(\frac{x}{2}\right) > 28\left(\frac{6}{7}\right)$$ On the left side, distribute the multiplication by 28.

$$21 + 14x > 24$$ Multiply: $28\left(\frac{3}{4}\right) = 21$, $28\left(\frac{x}{2}\right) = 14x$, and $28\left(\frac{6}{7}\right) = 24$.

$$21 + 14x - 21 > 24 - 21$$ To isolate the variable term 14x on the left side, undo the addition of 21 by subtracting 21 from both sides.

$$14x > 3$$ Do the subtraction: $21 - 21 = 0$ and $24 - 21 = 3$.

$$\frac{14x}{14} > \frac{3}{14}$$ To isolate x, undo the multiplication by 14 by dividing both sides by 14.

$$x > \frac{3}{14}$$

The solution set is $\left(\frac{3}{14}, \infty\right)$ and it is graphed as shown.

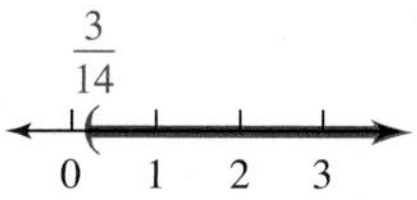

Teaching Example 8
Solve $\frac{5}{8} + \frac{x}{3} > \frac{9}{4}$. Write the solution set in interval notation and graph it.

Answer: $x > \frac{39}{8}$, $\left(\frac{39}{8}, \infty\right)$

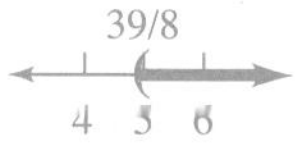

Self Check 8 Solve $\frac{1}{6} + \frac{2a}{9} < \frac{5}{2}$. Write the solution set in interval notation and graph it. $\left(-\infty, \frac{21}{2}\right)$ See AIE Appendix 3.

Now Try Problem 47

4 Solve Compound Inequalities.

The Language of Algebra

The word **compound** means made up of two or more parts. For example, a *compound* inequality has three parts. Other examples are: a *compound* sentence, a *compound* fracture, and a chemical *compound*.

Two inequalities can be combined into a **compound inequality** to show that an expression lies between two fixed values. For example, $-2 < x < 3$ is a combination of

$$-2 < x \quad \text{and} \quad x < 3$$

It indicates that x is greater than -2 and that x is also less than 3. The solution set of $-2 < x < 3$ consists of all numbers that lie between -2 and 3, and we write it as the interval $(-2, 3)$. The graph of the compound inequality is shown below.

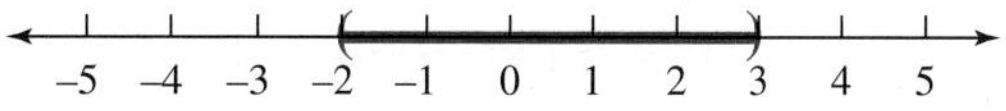

EXAMPLE 9 Graph $-4 \le x < 0$ and write the solution set in interval notation.

Teaching Tip: Show your students why compound inequalities such as $3 < x < -4$ and $-4 > x < 0$ do not make sense.

Strategy We need to determine which real numbers, when substituted for x, would make $-4 \le x < 0$ a true statement.

Why To graph $-4 \le x < 0$ means to draw a "picture" of all of the values of x that make the compound inequality true.

Solution If we replace the variable in $-4 \le x < 0$ with a number between -4 and 0, including -4, the resulting compound inequality will be true. Therefore, the solution set is the interval $[-4, 0)$.

Notation

Note that the two inequality symbols in $-4 \le x < 0$ point in the same direction and both point to the smaller number.

To graph the interval, we draw a bracket at -4, a parenthesis at 0, and shade in between, as shown.

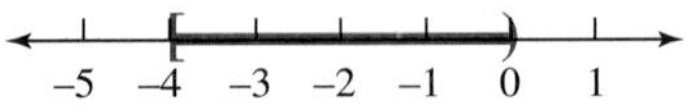

To check, we pick a number in the graph, such as -2, and see whether it satisfies the inequality. Since $-4 \le -2 < 0$ is true, the answer appears to be correct.

Self Check 9 Graph $-2 \le x < 1$ and write the solution set in interval notation. $[-2, 1)$ See AIE Appendix 3.

Now Try Problem 51

Teaching Example 9 Graph $-3 < x \le 4$ and write the solution set in interval notation.
Answer: $(-3, 4]$;

To solve these types of compound inequalities, we isolate the variable in the middle part of the inequality. To do this, we apply the properties of inequality to all *three* parts of the inequality: the left, the middle, and the right.

EXAMPLE 10 Solve $-4 < 2(x - 1) \le 4$. Write the solution set in interval notation and graph it.

Strategy We will use properties of inequality to isolate the variable by itself as the middle part of the inequality.

Why To solve the original inequality, we want to find a simpler equivalent inequality of the form **a number $< x \le$ a number**, whose solution is obvious.

Solution

$-4 < 2(x - 1) \le 4$	This is the compound inequality to solve.
$-4 < 2x - 2 \le 4$	In the middle, distribute the multiplication by 2.
$-4 + \mathbf{2} < 2x - 2 + \mathbf{2} \le 4 + \mathbf{2}$	To isolate the variable term 2x, undo the subtraction of 2 by adding 2 to all three parts.
$-2 < 2x \le 6$	Do the addition: $-4 + 2 = -2$, $-2 + 2 = 0$, and $4 + 2 = 6$.
$\frac{-2}{2} < \frac{2x}{2} \le \frac{6}{2}$	To isolate x, we undo the multiplication by 2 by dividing all three parts by 2.
$-1 < x \le 3$	Do the division.

The solution set is $(-1, 3]$ and it is graphed as shown.

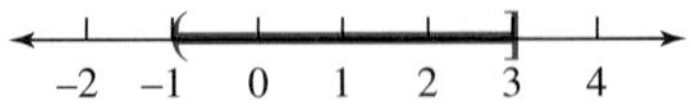

Teaching Example 10 Solve $-6 < 3(x + 1) \le 9$. Write the solution set in interval notation and graph it.
Answer:
$-3 < x \le 2$, $(-3, 2]$

Self Check 10 Solve $-6 \le 3(t + 2) \le 6$. Write the solution set in interval notation and graph it. $[-4, 0]$ See AIE Appendix 3.

Now Try Problem 59

5 Solve Inequality Applications.

When solving application problems, phrases such as "not more than," or "should exceed" suggest that the problem involves an inequality rather than an equation.

EXAMPLE 11 **Grades.** A student has scores of 72%, 74%, and 78% on three exams. What percent score does he need on the last exam to earn a grade of no less than B (80%)?

The Language of Algebra

Some phrases that suggest an inequality are:

surpass: $>$	**at least:** $\geq$
not exceed: $\leq$	**at most:** $\leq$
is no less than: $\geq$	
is no greater than: $\leq$	
between: $\square < \square < \square$	

Analyze We know three scores. We are to find what the student must score on the last exam to earn a grade of B or higher.

Assign Let $x =$ the score on the fourth (and last) exam.

Form To find the average grade, we add the four scores and divide by 4. To earn a grade of *no less than* B, the student's average must be *greater than or equal to* 80%.

The average of the four grades	must be no less than	80.
$\frac{72 + 74 + 78 + x}{4}$	$\geq$	80

Solve

$$\frac{224 + x}{4} \geq 80$$ Combine like terms in the numerator: $72 + 74 + 78 = 224$.

$$4\left(\frac{224 + x}{4}\right) \geq 4(80)$$ To clear the inequality of the fraction, multiply both sides by 4.

$$224 + x \geq 320$$ Simplify each side.

$$x \geq 96$$ To isolate x, undo the addition of 224 by subtracting 224 from both sides.

State To earn a B, the student must score 96% or better on the last exam.

Check Pick several exam scores that are 96% or better and verify that the student's average will be 80% or greater. For example, a score of 96% gives the student an average that is exactly 80%.

$$\frac{72 + 74 + 78 + 96}{4} = \frac{320}{4} = 80$$

Teaching Example 11 **Grades.** A student has scores of 68%, 67%, and 72% on three exams. What percent score does he need on the last test to earn a grade of no less than a C (70%)? *Answer:* 73% or better

Self Check 11 **Grades.** A student has scores of 78%, 82%, and 76% on three exams. What percent score does he need on the last test to earn a grade of no less than a B (80%)? 84% or better

Now Try Problem 103

SECTION 2.7 STUDY SET

VOCABULARY

Fill in the blanks.

1. An inequality is a statement that contains one of the symbols: $>$, $\geq$, $<$, or $\leq$. An equation is a statement that contains an $=$ symbol.
2. To solve an inequality means to find all the values of the variable that make the inequality true.
3. The solution set of $x > 2$ can be expressed in interval notation as $(2, \infty)$.
4. The inequality $-4 < x \leq 10$ is an example of a compound inequality.

CONCEPTS

Fill in the blanks.

5. a. Adding the same number to both sides of an inequality does not change the solutions.
   b. Multiplying or dividing both sides of an inequality by the same positive number does not change the solutions.
   c. If we multiply or divide both sides of an inequality by a negative number, the direction of the inequality symbol must be reversed for the inequalities to have the same solutions.
6. To solve $-4 \leq 2x + 1 < 3$, properties of inequality are applied to all three parts of the inequality: left, middle, and right.

Selected exercises available online at www.webassign.net/brookscole

7. Rewrite the inequality $32 < x$ in an equivalent form with the variable on the left side. $x > 32$

8. The solution set of an inequality is graphed below. Which of the four numbers, 3, −3, 2, and 4.5, when substituted for the variable in that inequality, would make it true? 3, 4.5

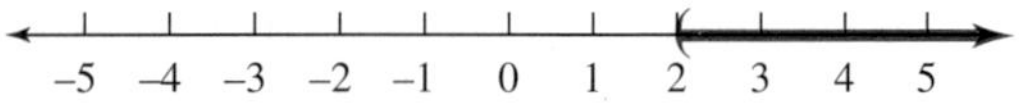

Write the inequality that is represented by each graph. Then describe the graph using interval notation.

9. a. 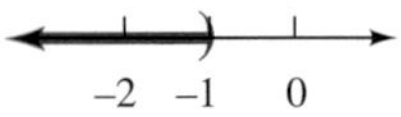

$x < -1$; $(-\infty, -1)$

b. 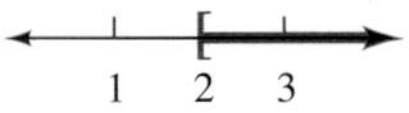

$x \geq 2$; $[2, \infty)$

10. a. 

$-7 < x \leq 2$; $(-7, 2]$

b. 

$4 < x \leq 6$; $(4, 6]$

NOTATION

11. Write each symbol.
 a. is less than or equal to $\leq$ b. infinity ∞
 c. bracket [or] d. is greater than $>$

12. Consider the graph of the interval [4, 8).
 a. Is 4 included in the graph? Yes b. Is 8 included in the graph? No

Complete the solution to solve each inequality.

13. $4x - 5 \geq 7$
$4x - 5 + 5 \geq 7 + 5$
$4x \geq 12$
$\frac{4x}{4} \geq \frac{12}{4}$
$x \geq 3$ Solution set: $[3, \infty)$

14. $-6x > 12$
$\frac{-6x}{-6} < \frac{12}{-6}$
$x < -2$ Solution set: $(-\infty, -2)$

GUIDED PRACTICE

See Example 1.

15. Determine whether each number is a solution of $3x - 2 > 5$.
 a. 5 Yes b. −4 No

16. Determine whether each number is a solution of $3x + 7 < 4x - 2$.
 a. 12 Yes b. 9 No

17. Determine whether each number is a solution of $-5(x - 1) \geq 2x + 12$.
 a. 1 No b. −1 Yes

18. Determine whether each number is a solution of $\frac{4}{5}a \geq -2$.
 a. $-\frac{5}{4}$ Yes b. −15 No

Graph each inequality and describe the graph using interval notation. **See Example 2.** See AIE Appendix 3.

19. $x < 5$ $(-\infty, 5)$
20. $x \geq -2$ $[-2, \infty)$
21. $-3 < x \leq 1$ $(-3, 1]$
22. $-4 \leq x \leq 2$ $[-4, 2]$

Solve each inequality. Write the solution set in interval notation and graph it. **See Example 3.** See AIE Appendix 3.

23. $x + 2 > 5$ $(3, \infty)$
24. $x + 5 \geq 2$ $[-3, \infty)$
25. $g - 30 \geq -20$ $[10, \infty)$
26. $h - 18 \leq -3$ $(-\infty, 15]$

Solve each inequality. Write the solution set in interval notation and graph it. **See Example 4.** See AIE Appendix 3.

27. $-\frac{3}{16}x \geq -9$ $(-\infty, 48]$
28. $-\frac{7}{8}x \leq 21$ $[-24, \infty)$
29. $\frac{2}{3}x \geq 2$ $[3, \infty)$
30. $\frac{3}{4}x < 3$ $(-\infty, 4)$
31. $-3y \leq -6$ $[2, \infty)$
32. $-6y \geq -6$ $(-\infty, 1]$
33. $8h < 48$ $(-\infty, 6)$
34. $2t > 22$ $(11, \infty)$

Solve each inequality. Write the solution set in interval notation and graph it. **See Example 5.** See AIE Appendix 3.

35. $64 < 9x + 1$ $(7, \infty)$
36. $32 > 4x + 8$ $(-\infty, 6)$
37. $-20 \geq 3m - 5$ $(-\infty, -5]$
38. $-29 \leq 7t - 1$ $[-4, \infty)$

Solve each inequality. Write the solution set in interval notation and graph it. **See Example 6.** See AIE Appendix 3.

39. $1.3 - 2x \geq 0.5$ $(-\infty, 0.4]$
40. $1.04 - 7x > 0.2$ $(-\infty, 0.12)$
41. $24.9 - 12a < -3.9$ $(2.4, \infty)$
42. $37.5 - 16t \leq 99.9$ $[-3.9, \infty)$

Solve each inequality. Write the solution set in interval notation and graph it. **See Example 7.** See AIE Appendix 3.

43. $9a + 4 > 5a - 16$ $(-5, \infty)$
44. $8t + 1 < 4t - 19$ $(-\infty, -5)$
45. $8(2n + 1) \leq 4(6n + 7) + 4n$ $\left[-\frac{5}{3}, \infty\right)$
46. $5(2 - d) \leq 3(d - 5) + 3d$ $\left[\frac{25}{11}, \infty\right)$

Solve each inequality. Write the solution set in interval notation and graph it. **See Example 8.** See AIE Appendix 3.

47. $\frac{1}{2} + \frac{n}{5} > \frac{3}{4}$ $\left(\frac{5}{4}, \infty\right)$
48. $\frac{1}{3} + \frac{c}{5} > -\frac{3}{2}$ $\left(-\frac{55}{6}, \infty\right)$
49. $\frac{1}{2} - \frac{x}{24} \geq -\frac{1}{8}$ $(-\infty, 15]$
50. $\frac{4}{3} - \frac{x}{5} \geq \frac{4}{15}$ $\left(-\infty, \frac{16}{3}\right]$

Graph each compound inequality and describe the graph using interval notation. See AIE Appendix 3.

51. $-2 \leq x < 3$ $[-2, 3)$
52. $-1 < x \leq 4$ $(-1, 4]$
53. $-\frac{7}{4} < x < 2$ $\left(-\frac{7}{4}, 2\right)$
54. $0 \leq x \leq \frac{11}{3}$ $\left[0, \frac{11}{3}\right]$

***Solve each compound inequality. Write the solution set in interval notation and graph it.* See Example 10.** See AIE Appendix 3.

55. $2 < x - 5 < 5$ $(7, 10)$

56. $-8 < t - 8 < 8$ $(0, 16)$

▶ **57.** $0 \le x + 10 \le 10$ $[-10, 0]$

58. $-9 \le x + 8 < 1$ $[-17, -7)$

59. $3 \le 2x - 1 < 5$ $[2, 3)$

▶ **60.** $4 < 3x - 5 \le 7$ $(3, 4]$

61. $-9 < 6x + 9 \le 45$ $(-3, 6]$

62. $-30 \le 10d + 20 < 90$ $[-5, 7)$

TRY IT YOURSELF

Solve each inequality or compound inequality. Write the solution set in interval notation and graph it. See AIE Appendix 3.

63. $\frac{6x + 1}{4} \le x + 1$ $\left(-\infty, \frac{3}{2}\right]$

▶ **64.** $\frac{3x - 10}{5} \le x + 4$ $[-15, \infty)$

▶ **65.** $17(3 - x) \ge 3 - 13x$ $(-\infty, 12]$

▶ **66.** $7x + 6 \ge -(x - 6)$ $[0, \infty)$

67. $0 < 5(x + 2) \le 15$ $(-2, 1]$

▶ **68.** $-18 \le 9(x - 5) < 27$ $[3, 8)$

69. $0.4x \le 0.1x + 0.45$ $(-\infty, 1.5]$

70. $0.9s \le 0.3s + 0.54$ $(-\infty, 0.9]$

▶ **71.** $-\frac{2}{3} \ge \frac{2y}{3} - \frac{3}{4}$ $\left(-\infty, \frac{1}{8}\right]$

72. $-\frac{2}{9} \ge \frac{5x}{6} - \frac{1}{3}$ $\left(-\infty, \frac{2}{15}\right]$

73. $\frac{m}{-42} - 1 > -1$ $(-\infty, 0)$

74. $\frac{a}{-25} + 3 < 3$ $(0, \infty)$

75. $6 - x \le 3(x - 1)$ $\left[\frac{9}{4}, \infty\right)$

76. $3(3 - x) \ge 6 + x$ $\left(-\infty, \frac{3}{4}\right]$

▶ **77.** $6 < -2(x - 1) < 12$ $(-5, -2)$

78. $4 \le -4(x - 2) < 20$ $(-3, 1]$

79. $-1 \le -\frac{1}{2}n$ $(-\infty, 2]$

80. $-3 \ge -\frac{1}{3}t$ $[9, \infty)$

81. $-m - 12 > 15$ $(-\infty, -27)$

▶ **82.** $-5x + 7 \le 12$ $[-1, \infty)$

83. $y - \frac{1}{7} \le \frac{2}{3}$ $\left(-\infty, \frac{17}{21}\right]$

84. $m - \frac{1}{9} \ge \frac{4}{5}$ $\left[\frac{41}{45}, \infty\right)$

85. $9x + 13 \ge 2x + 6x$ $[-13, \infty)$

▶ **86.** $7x - 16 < 2x + 4x$ $(-\infty, 16)$

87. $7 < \frac{5}{3}a + (-3)$ $(6, \infty)$

▶ **88.** $5 < \frac{7}{2}a + (-9)$ $(4, \infty)$

89. $-8 \le \frac{y}{8} - 4 \le 2$ $[-32, 48]$

▶ **90.** $-12 < \frac{b}{3} < 0$ $(-36, 0)$

91. $0.04x + 1.04 \le 0.01x + 1.085$ $(-\infty, 1.5]$

92. $0.005 + 2.08x \le 2.05x - 0.07$ $(-\infty, -2.5]$

▶ **93.** $\frac{5}{3}(x + 1) \ge -x + \frac{2}{3}$ $\left[-\frac{3}{8}, \infty\right)$

94. $\frac{5}{2}(7x - 15) \ge \frac{11}{2}x - \frac{3}{2}$ $[3, \infty)$

95. $\frac{4}{5}x < \frac{2}{5}$ $\left(-\infty, \frac{1}{2}\right)$

96. $\frac{11}{9}x > \frac{5}{9}$ $\left(\frac{5}{11}, \infty\right)$

97. $2x + 3(2x + 3) \le 7(x + 1) + 1$ $(-\infty, -1]$

98. $3(3x + 3) - 2 \le 2(2x - 1) + 6x$ $[9, \infty)$

Look Alikes . . .

Solve each equation and inequality. Write the solution set of each inequality in interval notation and graph it. See AIE Appendix 3.

99. **a.** $\frac{3}{8} + \frac{b}{3} > \frac{5}{12}$ $\left(\frac{1}{8}, \infty\right)$ **b.** $\frac{3}{8} + \frac{b}{3} = \frac{5}{12}$ $\frac{1}{8}$

▶ **100.** **a.** $7(a - 3) < 2(5a - 8)$ $\left(-\frac{5}{3}, \infty\right)$ **b.** $7(a - 3) = 2(5a - 8)$ $-\frac{5}{3}$

101. **a.** $4 \le 2x - 6$ $[5, \infty)$ **b.** $4 \le 2x - 6 < 18$ $[5, 12)$

102. **a.** $-16 < 4(x + 8) \le 8$ $(-12, -6]$ **b.** $-16 = 4(x + 8) + 8$ -14

APPLICATIONS

▶ **103. Grades.** A student has test scores of 68%, 75%, and 79% in a government class. What must she score on the last exam to earn a B (80% or better) in the course? 98% or better

▶ **104. Occupational Testing.** An employment agency requires applicants average at least 70% on a battery of four job skills tests. If an applicant scored 70%, 74%, and 84% on the first three exams, what must he score on the fourth test to maintain a 70% or better average? 52% or better

105. Gas Mileage. A car manufacturer produces three models in equal quantities. One model has an economy rating of 17 miles per gallon, and the second model is rated for 19 mpg. If government regulations require the manufacturer to have a fleet average that exceeds 21 mpg, what economy rating is required for the third model? More than 27 mpg

▶ **106. Service Charges.** When the average daily balance of a customer's checking account falls below $500 in any week, the bank assesses a $5 service charge. The table shows the daily balances of one customer. What must Friday's balance be to avoid the service charge? $869.20 or more

Day	Balance
Monday	$540.00
Tuesday	$435.50
Wednesday	$345.30
Thursday	$310.00

107. Geometry. The perimeter of an equilateral triangle is at most 57 feet. What could the length of a side be? (*Hint:* All three sides of an equilateral triangle are equal.) 19 ft or less

▶ **108. Geometry.** The perimeter of a square is no less than 68 centimeters. How long can a side be? 17 cm or more

109. Counter Space. A rectangular counter is being built for the customer service department of a store. Designers have determined that the outside perimeter of the counter (shown in red) needs to exceed 30 feet. Determine the acceptable values for x. More than 5 ft

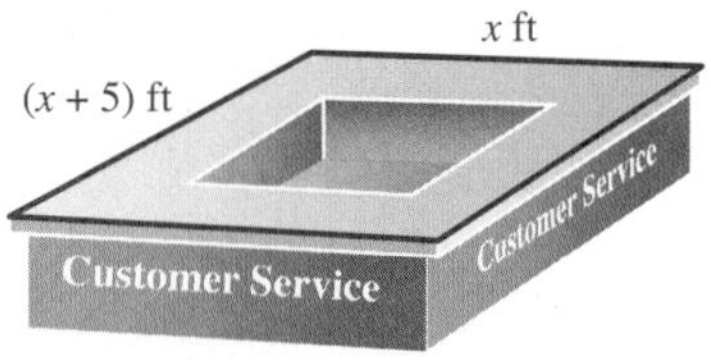

110. Number Puzzles. What numbers satisfy the condition: Four more than three times the number is at most 10? $x \le 2$

111. Graduations. It costs a student $18 to rent a cap and gown and 80 cents for each graduation announcement that she orders. If she doesn't want her spending on these graduation costs to exceed $50, how many announcements can she order? 40 or less

112. Telephones. A cellular telephone company has currently enrolled 36,000 customers in a new calling plan. If an average of 1,200 people are signing up for the plan each day, in how many days will the company surpass their goal of having 150,000 customers enrolled? 96 days

113. Windows. An architect needs to design a triangular-shaped bathroom window that has an area no greater than 100 in.^{2}. If the base of the window must be 16 inches long, what window heights will meet this condition? 12.5 in. or less

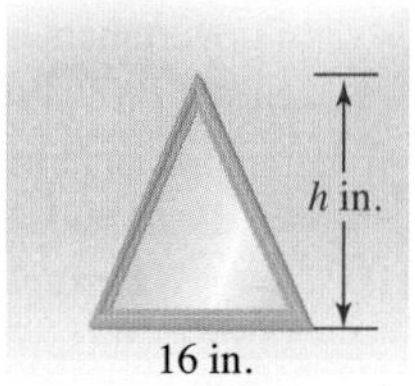

114. Room Temperatures. To hold the temperature of a room between 19° and 22° Celsius, what Fahrenheit temperatures must be maintained? *Hint:* Use the formula $C = \frac{5}{9}(F - 32)$. $66.2° < F < 71.6°$

115. Number Puzzles. What *whole* numbers satisfy the condition: Twice the number decreased by 1 is between 50 and 60? 26, 27, 28, 29, 30

116. Exercise. The graph in the next column shows the target heart beat range for different ages and exercise intensity levels. If we let b represent the number of beats per minute, then the compound inequality that estimates the heart beat rate range for a 30-year-old involved in a high-intensity workout is about $168 \le b \le 198$. Use a compound inequality to estimate the heart beat rate range for the following ages and zones. (Answers may vary slightly.)

a. 45-year-old, fat-burning zone $100 \le b \le 118$

b. 70-year-old, high-intensity zone $135 \le b \le 160$

c. 25-year-old, intermediate zone $135 \le b \le 172$

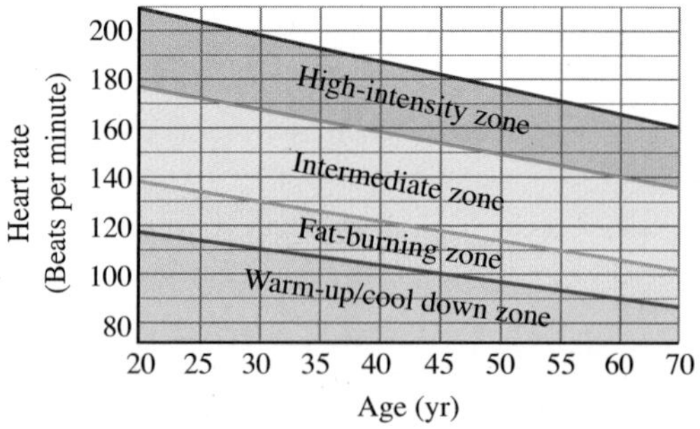

Source: elitefitness.co.nz

WRITING

117. Explain why multiplying both sides of an inequality by a negative number reverses the direction of the inequality.

118. a. What number is a solution of $3x - 26 \ge 4$ but is not a solution of $3x - 26 > 4$? Explain your reasoning.

b. What numbers are solutions of $3x - 26 \ge 4$ but not solutions of $3x - 26 = 4$? Explain your reasoning.

REVIEW

Complete each table.

119.

x	$x^2 - 3$
-2	1
0	-3
3	6

120.

x	$\frac{x}{3} + 2$
-6	0
0	2
12	6

CHALLENGE PROBLEMS

Solve each inequality. Write the solution set in interval notation and graph it. See AIE Appendix 3.

121. $3 - x < 5 < 7 - x$ $(-2, 2)$

122. $\frac{1}{x} > 1$. (*Hint:* Use a guess-and-check approach.) $(0, 1)$

123. $2(5x - 6) > 4x - 15 + 6x$ $(-\infty, \infty)$

124. $\frac{3a - 4}{-5} < \frac{3a + 15}{-5}$ No solution

Teaching Tip: Remind your students that this Summary and Review is an excellent way to study for their test. Suggest that they arrive on campus early and, as a final preparation, read through the section summaries before taking the exam.

2 Summary & Review

SECTION 2.1 Solving Equations Using Properties of Equality

DEFINITIONS AND CONCEPTS	EXAMPLES
An **equation** is a statement indicating that two expressions are equal. The equal symbol = separates an equation into two parts: the *left side* and the *right side*.	Equations: $2x + 4 = 10$ $\quad -5(a + 4) = -11a$ $\quad \frac{3}{2}t + 6 = t - \frac{1}{3}$
A number that makes an equation a true statement when substituted for the variable is called a **solution** of the equation.	Use a check to determine whether 2 is a solution of $x + 4 = 3x$. ***Check:*** $x + 4 = 3x$ $2 + 4 \stackrel{?}{=} 3(2)$ Substitute 2 for each x. $6 = 6$ True Since the resulting statement $6 = 6$ is true, 2 is a solution of $x + 4 = 3x$.
Equivalent equations have the same solutions.	$x - 2 = 6$ and $x = 8$ are equivalent equations because they have the same solution, 8.
To **solve an equation** isolate the variable on one side of the equation by undoing the operations performed on it using properties of equality. **Addition (Subtraction) property of equality:** If the same number is added to (or subtracted from) both sides of an equation, the result is an equivalent equation.	Solve: $x - 5 = 7$ $x - 5 + 5 = 7 + 5$ $x = 12$ The solution is 12. The solution set is {12}. Solve: $c + 9 = 16$ $c + 9 - 9 = 16 - 9$ $c = 7$ The solution is 7. The solution set is {7}.
Multiplication (Division) property of equality: If both sides of an equation are multiplied (or divided) by the same nonzero number, the result is an equivalent equation.	Solve: $\frac{1}{3}m = 2$ $3\left(\frac{1}{3}m\right) = 3(2)$ $m = 6$ The solution is 6. Solve: $10y = 50$ $\frac{10y}{10} = \frac{50}{10}$ $y = 5$ The solution is 5.

REVIEW EXERCISES

Use a check to determine whether the given number is a solution of the equation.

1. 84, $x - 34 = 50$ Yes
2. 3, $5y + 2 = 12$ No
3. −30, $\frac{x}{5} = 6$ No
4. 2, $|a^2 - a - 1| = 0$ No
5. −3, $5b - 2 = 3b - 8$ Yes
6. 1, $\frac{2}{y + 1} = \frac{12}{y + 1} - 5$ Yes

Fill in the blanks.

7. An equation is a statement indicating that two expressions are equal.
8. To solve $x - 8 = 10$ means to find all the values of the variable that make the equation a true statement.

Solve each equation and check the result.

9. $x - 9 = 12$ 21
10. $-y = -32$ 32
11. $a + 3.7 = -16.9$ −20.6
12. $100 = -7 + r$ 107
13. $120 = 5c$ 24
14. $t - \frac{1}{3} = \frac{3}{7}$ $\frac{16}{21}$
15. $\frac{4}{3}t = -12$ −9
16. $3 = \frac{q}{-2.6}$ −7.8
17. $6b = 0$ 0
18. $\frac{15}{16}s = -3$ $-\frac{16}{5}$

SECTION 2.2 More about Solving Equations

DEFINITIONS AND CONCEPTS	EXAMPLES
A five-step **strategy for solving linear equations:** 1. *Clear* the equation of fractions or decimals. 2. *Simplify* each side. Use the distributive property and combine like terms when necessary. 3. *Isolate the variable term.* Use the addition and subtraction properties of equality. 4. *Isolate the variable.* Use the multiplication and division properties of equality. 5. *Check* the result in the original equation.	Solve: $2(y+2)+4y=11-y$ $2y+4+4y=11-y$ Distribute the multiplication by 2. $6y+4=11-y$ Combine like terms: $2y+4y=6y$. $6y+4+y=11-y+y$ To eliminate $-y$ on the right, add y to both sides. $7y+4=11$ Combine like terms: $6y+y=7y$ and $-y+y=0$. $7y+4-4=11-4$ To isolate the variable term $7y$, undo the addition of 4 by subtracting 4 from both sides. $7y=7$ Simplify each side of the equation. $\frac{7y}{7}=\frac{7}{7}$ To isolate y, undo the multiplication by 7 by dividing both sides by 7. $y=1$ The solution is 1.
It is easier to solve an equation that involves only integers. **To clear an equation of fractions,** multiply both sides of an equation by the LCD. **To clear an equation of decimals,** multiply both sides by a power of 10 to change the decimals in the equation to integers.	To solve $\frac{1}{2}+\frac{x}{3}=\frac{3}{4}$, first clear the fractions by multiplying both sides by 12: $\mathbf{12}\left(\frac{1}{2}+\frac{x}{3}\right)=\mathbf{12}\left(\frac{3}{4}\right)$ The LCD of $\frac{1}{2}$, $\frac{x}{3}$, and $\frac{3}{4}$ is 12. To solve $0.5(x-4)=0.1x+0.2$, first clear the decimals by multiplying both sides by 10: $\mathbf{10}[0.5(x-4)]=\mathbf{10}(0.1x+0.2)$
Not every equation in one variable has a single solution. Some equations are made true by *any* permissible replacement value for the variable. Such equations are called **identities.** An equation that is not true for any value of its variable is called a **contradiction.**	When we solve $x+5+x=2x+5$, the variable drops out and we obtain a true statement $5=5$. All real numbers are solutions. When we solve $y+2=y$, the variable drops out and we obtain a false statement $2=0$. The equation has no solutions.

REVIEW EXERCISES

Solve each equation. Check the result.

19. $5x+4=14$ 2

20. $98.6-t=129.2$ -30.6

21. $\frac{n}{5}+(-2)=4$ 30

22. $\frac{b-5}{4}=-6$ -19

23. $5(2x-4)-5x=0$ 4

24. $-2(x-5)=5(-3x+4)+3$ 1

25. $\frac{3}{4}=\frac{1}{2}+\frac{d}{5}$ $\frac{5}{4}$

26. $\frac{5(7-x)}{4}=2x-3$ $\frac{47}{13}$

27. $\frac{3(2-c)}{2}=\frac{-2(2c+3)}{5}$ 6

28. $\frac{b}{3}+\frac{11}{9}+3b=-\frac{5}{6}b$ $-\frac{22}{75}$

29. $0.15(x+2)+0.3=0.35x-0.4$ 5

30. $0.5-0.02(y-2)=0.16+0.36y$ 1

31. $3(a+8)=6(a+4)-3a$ Identity; all real numbers

32. $2(y+10)+y=3(y+8)$ Contradiction; no solution

SECTION 2.3 Applications of Percent

DEFINITIONS AND CONCEPTS	EXAMPLES
To solve **percent problems,** use the facts of the problem to write a sentence of the form: ☐ is ☐% of ☐? Translate the sentence to mathematical symbols: *is* translates to an = symbol and *of* means multiply. Then solve the equation.	648 is 30% of what number? $648 = 30\% \cdot x$ Translate. $648 = 0.30x$ Change 30% to a decimal: 30% = 0.30. $\frac{648}{0.30} = \frac{0.30x}{0.30}$ To isolate x, undo the multiplication by 0.30 by dividing both sides by 0.30. $2{,}160 = x$ Do the division. Thus, 648 is 30% of 2,160.
To find the **percent of increase** or **the percent of decrease,** find what percent the increase or decrease is of the original amount.	**Sales Prices.** To find the percent of decrease when ground beef prices are reduced from \$4.89 to \$4.25 per pound, we first find the amount of decrease: $4.89 - 4.25 = 0.64$. Then we determine what percent 0.64 is of 4.89 (the original price). 0.64 is what % of 4.89? $0.64 = x \cdot 4.89$ Translate. $0.64 = 4.89x$ $\frac{0.64}{4.89} = \frac{4.89x}{4.89}$ To isolate x, undo the multiplication by 4.89 by dividing both sides by 4.89. $0.130879346 \approx x$ Do the division. $013.0879346\% \approx x$ Write the decimal as a percent. To the nearest tenth of a percent, the percent of decrease is 13.1%.

REVIEW EXERCISES

33. Fill in the blanks.

a. Percent means parts per one hundred.

b. When the price of an item is reduced, we call the amount of the reduction a discount.

c. An employee who is paid a commission is paid a percent of the goods or services that he or she sells.

34. 4.81 is 2.5% of what number? 192.4

35. What number is 15% of 950? 142.5

36. What percent of 410 is 49.2? 12%

37. Internet Users. The circle graph below shows the percent of the U.S. population that used the Internet in 2010.

a. What percent of the population did not use the Internet? 28.8%

b. If the U.S. population in 2010 was about 310 million, how many people used the Internet that year? Round to the nearest million. 221 million

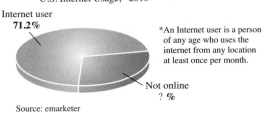

U.S. Internet Usage,* 2010

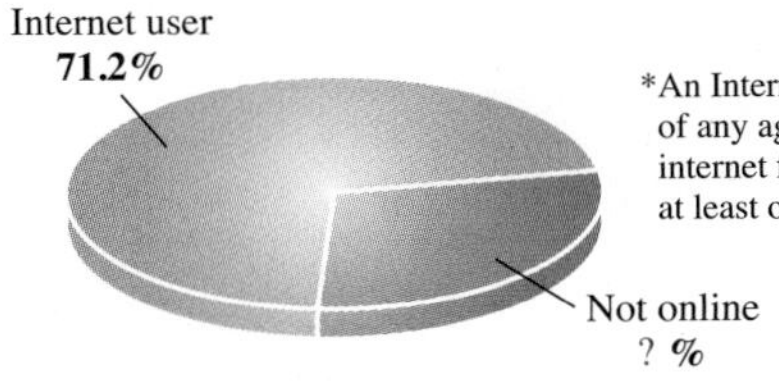

*An Internet user is a person of any age who uses the internet from any location at least once per month.

Source: emarketer

38. Cost of Living. A retired trucker receives a monthly Social Security check of \$764. If she is to receive a 3.5% cost-of-living increase soon, how much larger will her check be? \$26.74

39. Family Budgets. It is recommended that a family pay no more than 30% of its monthly income (after taxes) on housing. If a family has an after-tax income of \$1,890 per month and pays \$625 in housing costs each month, are they within the recommended range? No

40. Discounts. A shopper saved \$148.50 on a food processor that was discounted 33%. What did it originally cost? \$450

41. Tupperware. The hostess of a Tupperware party is paid a 25% commission on her in-home party's sales. What would the hostess earn if sales totaled \$600? \$150

42. Collectibles. A collector of football trading cards paid \$6 for a 1984 Dan Marino rookie card several years ago. If the card is now worth \$100, what is the percent of increase in the card's value? (Round to the nearest percent.) 1,567%

SECTION 2.4 Formulas

DEFINITIONS AND CONCEPTS | **EXAMPLES**

A **formula** is an equation that states a relationship between two or more variables.

Retail price: $r = c + m$ Profit: $p = r - c$

Simple Interest: $I = Prt$ Distance: $d = rt$

Temperature: $C = \frac{5}{9}(F - 32)$

The **perimeter** of a plane geometric figure is the distance around it. The **area** of a plane geometric figure is the amount of surface that it encloses. The **volume** of a three-dimensional geometric solid is the amount of space it encloses.

Rectangle: $P = 2l + 2w$, $A = lw$ Circle: $C = \pi D = 2\pi r$, $A = \pi r^2$

Rectangular solid: $V = lwh$ Cylinder: $V = \pi r^2 h$

*See inside the back cover of the text for more geometric formulas.

If we are given the values of all but one of the variables in a formula, we can use our equation-solving skills to find the value of the remaining variable.

Bedding. The area of a standard queen-size bed sheet is 9,180 in.^{2}. If the length is 102 inches, what is the width?

$A = lw$ This is the formula for the area of a rectangle.

$9{,}180 = 102w$ Substitute 9,180 for the area A and 102 for the length l.

$\frac{9{,}180}{102} = \frac{102w}{102}$ To isolate w, undo the multiplication by 102 by dividing both sides by 102.

$90 = w$ Do the division.

The width of a standard queen-size bed sheet is 90 inches.

To solve a formula for a specific variable means to isolate that variable on one side of the equation, with all other variables and constants on the opposite side. Treat the specified variable as if it is the only variable in the equation. Treat the other variables as if they were numbers (constants).

Solve the formula for the volume of a cone for h.

$V = \frac{1}{3}\pi r^2 h$ This is the formula for the volume of a cone.

$3(V) = 3\left(\frac{1}{3}\pi r^2 h\right)$ To clear the equation of the fraction, multiply both sides by 3.

$3V = \pi r^2 h$ Simplify.

$\frac{3V}{\pi r^2} = \frac{\pi r^2 h}{\pi r^2}$ To isolate h, undo the multiplication by πr^2 by dividing both sides by πr^2.

$\frac{3V}{\pi r^2} = h$ or $h = \frac{3V}{\pi r^2}$

REVIEW EXERCISES

43. **Shopping.** Find the markup on a CD player whose wholesale cost is $219 and whose retail price is $395. $176
44. **Restaurants.** One month, a restaurant had sales of $13,500 and made a profit of $1,700. Find the expenses for the month. $11,800
45. **Snails.** A typical garden snail travels at an average rate of 2.5 feet per minute. How long would it take a snail to cross a 20-foot-long flower bed? 8 min
46. **Certificates of Deposit.** A $26,000 investment in a CD earned $1,170 in interest the first year. What was the annual interest rate? 4.5%
47. **Jewelry.** Gold melts at about 1,065°C. Change this temperature to degrees Fahrenheit. 1,949°F
48. **Camping.**
    a. Find the perimeter of the air mattress. 168 in.
    b. Find the amount of sleeping area on the top surface of the air mattress. 1,440 in.^{2}
    c. Find the volume of the air mattress if it is 3 inches thick. 4,320 in.^{3}

49. Find the area of a triangle with a base 17 meters long and a height of 9 meters. 76.5 m^2

50. Find the area of a trapezoid with bases 11 inches and 13 inches long and a height of 12 inches. 144 in.^2

In Problems 51–53, the answers may vary slightly depending on which approximation of π is used.

51. **a.** Find the circumference of a circle with a radius of 8 centimeters. Round to the nearest hundredth. 50.27 cm

b. Find the area of the circle. Round to the nearest square centimeter. 201 cm^2

52. Find the volume of a 12-foot tall cylinder whose circular base has a radius of 0.5 feet. Give the result to the nearest tenth. 9.4 ft^3

53. Halloween. After being cleaned out, a spherical-shaped pumpkin has an inside diameter of 9 inches. To the nearest hundredth, what is its volume? 381.70 in.^3

54. Find the volume of a pyramid that has a square base, measuring 6 feet on a side, and a height of 10 feet. 120 ft^3

Solve each formula for the specified variable.

55. $A = 2\pi rh$ for h $\quad h = \frac{A}{2\pi r}$

56. $A - BC = \frac{G - K}{3}$ for G $\quad G = 3A - 3BC + K$

57. $C = \frac{1}{4}s(t - d)$ for t $\quad t = \frac{4C}{s} + d$

58. $4y - 3x = 16$ for y $\quad y = \frac{3}{4}x + 4$

SECTION 2.5 Problem Solving

DEFINITIONS AND CONCEPTS	EXAMPLES
To solve application problems, use the six-step problem-solving strategy. 1. **Analyze** the problem. 2. **Assign** a variable. 3. **Form** an equation. 4. **Solve** the equation. 5. **State** the conclusion. 6. **Check** the result. (See page 142 for the more detailed list of steps.) In this section, we solved application problems involving: ■ More than one unknown (Example 1) ■ Set-up fees (Example 2) ■ Commissions (Example 3) ■ Consecutive integers (Example 4) ■ Perimeter (Example 5) ■ Isosceles triangles (Example 6)	**Income Taxes.** After taxes, an author kept \$85,340 of her total annual earnings. If her earnings were taxed at a 15% rate, how much did she earn that year? **Analyze** The author earned some unknown amount of money. On that amount, she paid 15% in taxes. The difference between her total earnings and the taxes paid was \$85,340. **Assign** If we let x = the author's total earnings, the amount of taxes that she paid was 15% of x or $0.15x$. **Form** We can use the words of the problem to form an equation. Her total earnings — minus — the taxes that she paid — equals — the money that she kept. $x \quad - \quad 0.15x \quad = \quad 85{,}340$ **Solve** $x - 0.15x = 85{,}340$ $0.85x = 85{,}340$ *Combine like terms: $1x - 0.15x = 0.85x$.* $x = 100{,}400$ *To isolate x, divide both sides by 0.85.* **State** The author earned \$100,400 that year. **Check** The taxes were 15% of \$100,400 or \$15,060. If we subtract the taxes from her total earnings, we get \$100,400 − \$15,060 = \$85,340. The answer checks.

REVIEW EXERCISES

59. Sound Systems. A 45-foot-long speaker wire is to be cut into three pieces. One piece is to be 15 feet long. Of the remaining pieces, one must be 2 feet less than 3 times the length of the other. Find the length of the shorter piece. 8 ft

60. Signing Petitions. A professional signature collector is paid \$50 a day plus \$2.25 for each verified signature he gets from a registered voter. How many signatures are needed to earn \$500 a day? 200 signatures

61. Lottery Winnings. After taxes, a lottery winner was left with a lump sum of \$1,800,000. If 28% of the original prize was withheld to pay federal income taxes, what was the original cash prize? \$2,500,000

62. NASCAR. The car numbers of drivers Bobby Labonte and Kyle Petty are consecutive odd integers whose sum is 88. If Labonte's number is the smaller, find the numbers of each car. Labonte: 43; Petty: 45

63. Art History. *American Gothic* was painted in 1930 by Grant Wood. The length of the rectangular painting is 5 inches more than the width. Find the dimensions of the painting if it has a perimeter of 109.5 inches.
24.875 in. × 29.875 in. $\left(24\frac{7}{8}\text{ in.} \times 29\frac{7}{8}\text{ in.}\right)$

64. Geometry. Find the missing angle measures of the triangle.
76.5°, 76.5°

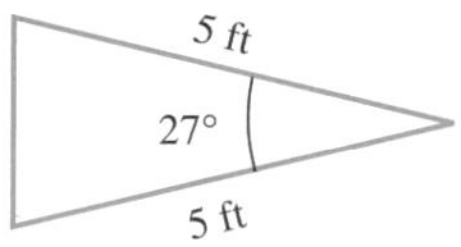

SECTION 2.6 More on Problem Solving

DEFINITIONS AND CONCEPTS

To solve application problems, use the six-step problem-solving strategy.

1. **Analyze** the problem.
2. **Assign** a variable.
3. **Form** an equation.
4. **Solve** the equation.
5. **State** the conclusion.
6. **Check** the result.

(See page 142 for the more detailed list of steps.)

Tables are a helpful way to organize the facts of a problem.

In this section, we solved application problems involving:

- Money/investments (Example 1)
- Motion (Example 2 and Example 3)
- Liquid mixture (Example 4)
- Dry mixture (Example 5)
- Number-value (Example 6)

EXAMPLES

Trucking. Two trucks leave from the same place at the same time traveling in opposite directions. One travels at a rate of 60 mph and the other at 50 mph. How long will it take them to be 165 miles apart?

Analyze We know that one truck travels at 60 mph and the other at 50 mph. Together, the trucks will travel a distance of 165 miles.

Assign Let t = the number of hours until the distance between the trucks is 165 miles.

Form We enter each rate in the table under the heading $\boldsymbol{r}$. Since the trucks travel for the same length of time, say t hours, we enter t for each truck under the heading $\boldsymbol{t}$. We enter the distances traveled under the heading $\boldsymbol{d}$ in the table.

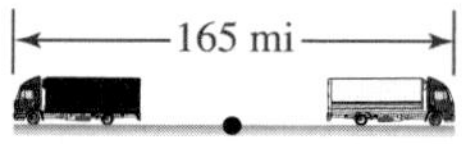

	r	$\cdot\ t$	$= d$
Truck 1	60	t	$60t$
Truck 2	50	t	$50t$
			Total: 165

Multiply $r \cdot t$ to obtain an expression for the distance traveled by each truck.

Use the information in this column to form an equation.

The distance the first truck travels	plus	the distance the second truck travels	is	165 miles.
$60t$	$+$	$50t$	$=$	165

Solve $60t + 50t = 165$

$110t = 165$ Combine like terms: $60t + 50t = 110t$.

$\frac{110t}{\mathbf{110}} = \frac{165}{\mathbf{110}}$ To isolate t, undo the multiplication by 110 by dividing both sides by 110.

$t = 1.5$ Do the division.

State The trucks will be 165 miles apart in 1.5 hours.

Check If the first truck travels 60 mph for 1.5 hours, it will go $60(1.5) = 90$ miles. If the second truck travels 50 mph for 1.5 hours, it will go $50(1.5) = 75$ miles. Since 90 miles + 75 miles = 165 miles, the result checks.

REVIEW EXERCISES

65. Investment Income. A woman has \$27,000. Part is invested for 1 year in a certificate of deposit paying 7% interest, and the remaining amount in a cash management fund paying 9%. After 1 year, the total interest on the two investments is \$2,110. How much is invested at each rate? \$16,000 at 7%, \$11,000 at 9%

66. Walking and Bicycling. A bicycle path is 5 miles long. A man walks from one end at the rate of 3 mph. At the same time, a friend bicycles from the other end, traveling at 12 mph. In how many minutes will they meet? 20 min

67. Airplanes. How long will it take a jet plane, flying at 450 mph, to overtake a propeller plane, flying at 180 mph, if the propeller plane had a $2\frac{1}{2}$-hour head start? $1\frac{2}{3}$ hr = 1 hr 40 min

68. Autographs. Kesha collected the autographs of 8 more television celebrities than she has of movie stars. Each TV celebrity autograph is worth \$75 and each movie star autograph is worth \$250. If her collection is valued at \$1,900, how many of each type of autograph does she have? TV celebrities: 12 autographs; movie stars: 4 autographs

69. Mixtures. A store manager mixes candy worth 90¢ per pound with gumdrops worth \$1.50 per pound to make 20 pounds of a mixture worth \$1.20 per pound. How many pounds of each kind of candy does he use? 10 lb of each

70. Eliminating Mildew. How many gallons of a 2% fungicide solution must be mixed with 4 gallons of a 5% fungicide solution to get a 4% fungicide solution? 2 gal

SECTION 2.7 Solving Inequalities

DEFINITIONS AND CONCEPTS | **EXAMPLES**

An **inequality** is a mathematical statement that contains an $>$, $<$, $\geq$, or $\leq$ symbol.

Inequalities: $3x < 8$ $\quad \frac{1}{2}y - 4 \geq 12$ $\quad 2z + 4 \leq z - 5$

A **solution of an inequality** is any number that makes the inequality true.

Determine whether 3 is a solution of $2x - 7 < 5$.

Check:

$2x - 7 < 5$

$2(3) - 7 \overset{?}{<} 5$ Substitute 3 for x.

$-1 < 5$ True

Since the resulting statement $-1 < 5$ is true, 3 is a solution of $2x - 7 < 5$.

We **solve inequalities** as we solve equations. However, if we **multiply or divide both sides by a negative number,** we must *reverse* the inequality symbol.

Solve: $-3(z - 1) \geq -6$

$-3z + 3 \geq -6$ Distribute the multiplication by -3.

$-3z \geq -9$ To isolate the variable term $-3z$, undo the addition of 3 by subtracting 3 from both sides.

$\frac{-3z}{-3} \leq \frac{-9}{-3}$ To isolate z, undo the multiplication by dividing both sides by -3. Reverse the $\geq$ symbol.

$z \leq 3$ Do the division.

Interval notation can be used to describe the solution set of an inequality. A **parenthesis** indicates that a number is not included in the solution set of an inequality. A **bracket** indicates that a number is included in the solution set.

The solution set is $(-\infty, 3]$ and it is graphed as shown.

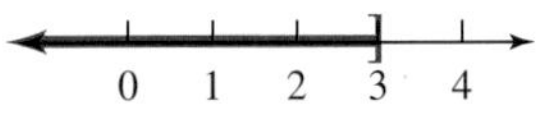

REVIEW EXERCISES

Solve each inequality. Write the solution set in interval notation and graph it. See AIE Appendix 3.

71. $3x + 2 < 5$ $(-\infty, 1)$

72. $-\frac{3}{4}x \geq -9$ $(-\infty, 12]$

73. $\frac{3}{4} < \frac{d}{5} + \frac{1}{2}$ $(\frac{5}{4}, \infty)$

74. $5(3 - x) \leq 3(x - 3)$ $[3, \infty)$

75. $\frac{t}{-5} - (-1.8) \geq -6.2$ $(-\infty, 40]$

76. $a + 5 - 2(10 - a) > 6$ $(7, \infty)$

77. $24 < 3(x + 2) < 39$ $(6, 11)$

78. $0 \leq 3 - 2x < 10$ $\left(-\frac{7}{2}, \frac{3}{2}\right]$

79. Sports Equipment. The acceptable weight w of Ping-Pong balls used in competition can range from 2.40 to 2.53 grams. Express this range using a compound inequality. $2.40 \text{ g} \leq w \leq 2.53 \text{ g}$

80. Signs. A large office complex has a strict policy about signs. Any sign to be posted in the building must be rectangular in shape, its width must be 18 inches, and its perimeter is not to exceed 132 inches. What possible sign lengths meet these specifications? $0 \text{ in.} < l \leq 48 \text{ in.}$; 48 in. or less

Teaching Tip: Because this Chapter Test is a comprehensive collection of problems that covers all of the topics discussed in Chapter 2, it is lengthy. If your students have time to complete it, that would be optimal. If, because of time constraints, they are unable to do so, assign an appropriate subset of problems that reflects the types of problems that the students will see on your exam.

2 CHAPTER TEST

1. Fill in the blanks.
   a. To solve an equation means to find all of the values of the variable that make the equation true.
   b. Percent means parts per one hundred.
   c. The distance around a circle is called its circumference.
   d. An inequality is a statement that contains one of the symbols $>$, $\geq$, $<$, or $\leq$.
   e. The multiplication property of equality says that multiplying both sides of an equation by the same nonzero number does not change its solution.
2. Use a check to determine if 3 is a solution of $5y + 2 = 12$. No

Solve each equation.

3. $3h + 2 = 8$ 2
4. $-22 = -x$ 22
5. $\frac{4}{5}t = -4$ -5
6. $\frac{11b - 11}{5} = \frac{3b - 2}{2}$ $\frac{12}{7}$
7. $0.8(x - 1{,}000) + 1.3 = 2.9 + 0.2x$ 1,336
8. $2(y - 7) - 3y = -(y - 3) - 17$ All real numbers (an identity)
9. $\frac{m}{2} - \frac{1}{3} = \frac{1}{4} + \frac{m}{6}$ $\frac{7}{4}$
10. $\frac{3}{4}(6n - 2) = 246$ 55
11. $5x = 0$ 0
12. $6a + (-7) = 3a - 7 + 2a$ 0
13. $9 - 5(2x + 10) = -1$ -4
14. $24t = -6(8 - 4t)$ No solution (a contradiction)
15. What is 15.2% of 80? 12.16
16. **Down Payments.** To buy a house, a woman was required to make a down payment of \$11,400. What did the house sell for if this was 15% of the purchase price? \$76,000
17. **Body Temperatures.** Suppose a person's body temperature rises from 98.6°F to a dangerous 105°F. What is the percent increase? Round to the nearest percent. 6%
18. **Commissions.** An appliance store salesperson receives a commission of 5% of the price of every item that she sells. What will she make if she sells a \$599.99 refrigerator? \$30
19. **Grand Openings.** On its first night of business, a pizza parlor brought in \$445. The owner estimated his profits that night to be \$150. What were the costs? \$295
20. Find the Celsius temperature reading if the Fahrenheit reading is 14°. -10°C
21. **Sound.** The speed of sound at sea level is about 1,108 feet per second. How far will sound travel in 1 minute? 66,480 feet
22. **Pets.** The spherical fishbowl is three-quarters full of water. To the nearest cubic inch, find the volume of water in the bowl. (*Hint:* The volume of a sphere is given by $V = \frac{4}{3}\pi r^3$.) 393 in.^{3}

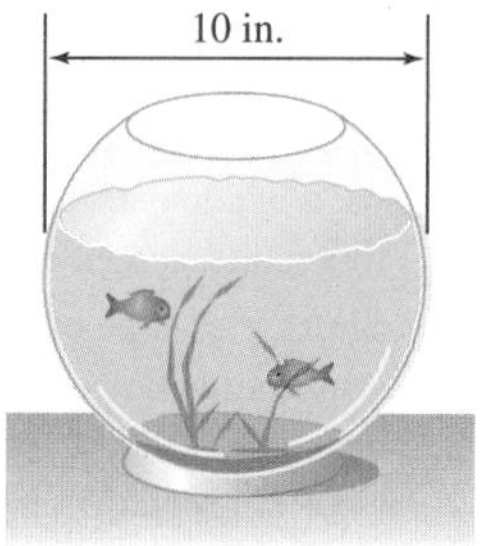

Solve for the specified variable.

23. $V = \pi r^2 h$ for h $h = \frac{V}{\pi r^2}$
24. $A = P + Prt$ for r $r = \frac{A - P}{Pt}$
25. $A = \frac{a + b + c + d}{4}$ for c $c = 4A - a - b - d$
26. $2x - 3y = 9$ for y $y = \frac{2}{3}x - 3$
27. **Irons.** Estimate the area of the soleplate of the iron. 20 in.^{2}

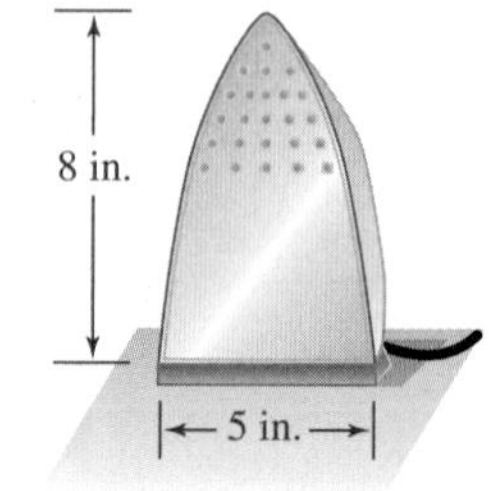

28. **Television.** In a typical 30-minute block of time on TV, the number of programming minutes are 2 less than three times the number of minutes of commercials. How many minutes of programming and how many minutes of commercials are there? Programming: 22 min; commercials: 8 min
29. **Plumbing Bills.** A part of the invoice for plumbing work shown below is torn away. What is the cost per hour for labor? \$40.55

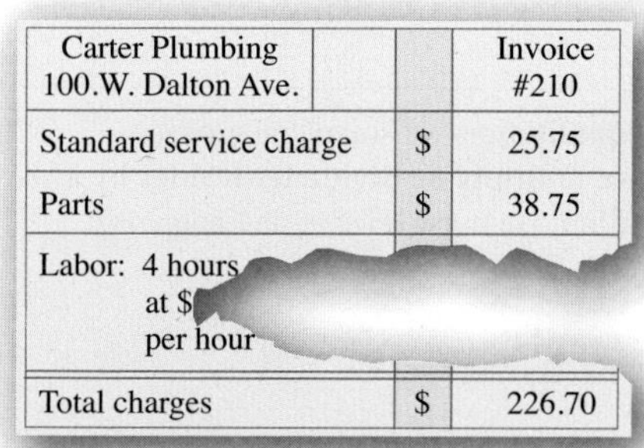

Carter Plumbing 100.W. Dalton Ave.		Invoice #210
Standard service charge	\$	25.75
Parts	\$	38.75
Labor: 4 hours at \$ per hour		
Total charges	\$	226.70

30. **Concert Seating.** Two types of seating were sold for a concert. Floor seats cost \$12.50 a ticket and balcony seats cost \$20.50. Ten times as many floor seats were sold as balcony seats. If the total receipts from the sale of both types of tickets were \$11,640, how many of each type of ticket were sold? 80 balcony seats, 800 floor seats
31. **Home Sales.** A condominium owner cleared \$114,600 on the sale of his condo, after paying a 4.5% real estate commission. What was the selling price? \$120,000
32. **Colorado.** The state of Colorado is approximately rectangular-shaped with perimeter 1,320 miles. Find the length (east to west) and width (north to south), if the length is 100 miles longer than the width. 380 mi, 280 mi
33. **Tea.** How many pounds of green tea, worth \$40 a pound, should be mixed with herbal tea, worth \$50 a pound, to produce 20 pounds of a blend worth \$42 a pound? Green: 16 lb; herbal: 4 lb

34. Reading. A bookmark is inserted between two page numbers whose sum is 825. What are the page numbers? 412, 413

35. Travel Times. A car leaves Rockford, Illinois, at the rate of 65 mph, bound for Madison, Wisconsin. At the same time, a truck leaves Madison at the rate of 55 mph, bound for Rockford. If the cities are 72 miles apart, how long will it take for the car and the truck to meet? $\frac{3}{5}$ hr

36. Pickles. To make pickles, fresh cucumbers are soaked in a salt water solution called *brine.* How many liters of a 2% brine solution must be added to 30 liters of a 10% brine solution to dilute it to an 8% solution? 10 liters

37. Exercise. How long will it take a bicyclist, traveling at 20 mph, to catch up with a jogger, traveling at 8 mph, if the jogger had a half-hour head start? $\frac{1}{3}$ hr = 20 min

38. Geometry. If the vertex angle of an isosceles triangle is 44°, find the measure of each base angle. 68°

39. Investments. Part of \$13,750 is invested at 9% annual interest, and the rest is invested at 8%. After one year, the accounts paid \$1,185 in interest. How much was invested at the lower rate? \$5,250

40. Use a check to determine whether -3 is a solution of $4 - 9w < -4w + 19$. No

Solve each inequality. Write the solution set in interval notation and graph it. See AIE Appendix 3.

41. $-8x - 20 \leq 4$ $[-3, \infty)$

42. $-8.1 > \frac{t}{2} + (-11.3)$ $(-\infty, 6.4)$

43. $-12 \leq 2(x + 1) < 10$ $[-7, 4)$

44. $\frac{1}{3}(a - 5) > \frac{1}{2}(a + 1)$ $(-\infty, -13)$

45. $-9(h - 3) + 2h \leq 8(4 - h)$ $(-\infty, 5]$

46. Awards. A city honors its citizen of the year with a framed certificate. An artist charges \$15 for the frame and 75 cents per word for writing out the proclamation. If a city regulation does not allow gifts in excess of \$150, what is the maximum number of words that can be written on the certificate? 180 words

Group Project

TRANSLATING KEY WORDS AND PHRASES

Overview: Students often say that the most challenging step of the six-step problem-solving strategy is forming an equation. This activity is designed to make that step easier by improving your translating skills.

Instructions: Form groups of 3 or 4 students. Select one person from your group to record the group's responses. Determine whether addition, subtraction, multiplication, or division is suggested by each of the following words or phrases. Then use the word or phrase in a sentence to illustrate its meaning. (If a word stumps everyone in your group, a dictionary can be helpful.)

deflate	recede	partition	evaporate	amplify
bisect	augment	hike	erode	boost
annexed	diminish	plummet	upsurge	wane
quadruple	corrode	taper off	trisect	broaden

COMPUTER SPREADSHEETS

Overview: In this activity, you will get some experience working with a spreadsheet.

Instructions: Form groups of 3 or 4 students. Examine the following spreadsheet, which consists of cells named by column and row. For example, 7 is entered in cell B3. In any cell you may enter data or a formula. For each formula in cells D1–D4 and E1–E4, the computer performs a calculation using values entered in other cells and prints the result in place of the formula. Find the value that will be printed in each formula cell. The symbol * means multiply, / means divide, and ^ means raise to a power.

	A	B	C	D	E
1	−8	20	−6	= 2*B1 − 3*C1 + 4	= B1 − 3*A1^2
2	39	2	−1	= A2/(B2 − C2)	= B3*B2*C2*2
3	50	7	3	= A3/5 + C3^3	= 65 − 2*(B3 − 5)^5
4	6.8	−2.8	−0.5	= 100*A4 + B4*C4	= A4/10 + A3/2*5

Graphing Linear Equations and Inequalities in Two Variables; Functions

3

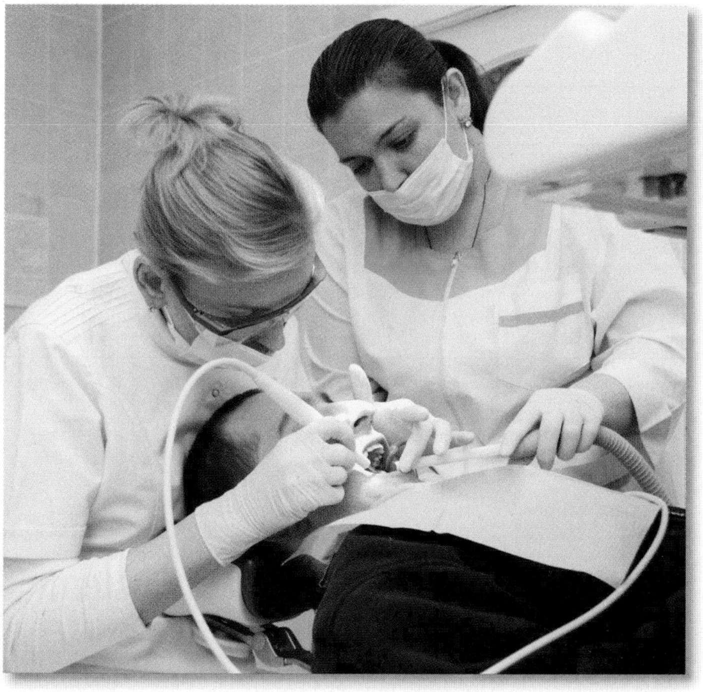

from Campus to Careers

Dental Assistant

A dental assistant is a valuable member of the dental health team who prepares patients for treatment, takes x-rays, sterilizes instruments, and keeps records. Part of the training of a dental assistant includes learning about a *coordinate system* that is used to identify the location of teeth in the mouth. This coordinate system is much like one used in algebra to graph points, lines, and curves.

Problem 33 in **Study Set 3.1, problem 105** in **Study Set 3.4,** and **problem 77** in **Study Set 3.7** involve situations that a dental assistant might encounter on the job. The mathematical concepts discussed in this chapter can be used to solve those problems.

JOB TITLE:
Dental Assistant
EDUCATION:
An associate's degree from an accredited program is required.
JOB OUTLOOK:
Excellent—One of the fastest-growing occupations in the health care industry.
ANNUAL EARNINGS:
Median salary $32,380
FOR MORE INFORMATION:
www.bls.gov/oco/ocos163.htm

Study Skills Workshop ▶ Successful Test Taking

Taking a math test doesn't have to be an unpleasant experience. Here are some suggestions that can make it more enjoyable and also improve your score.

Teaching Tip: Study Skills Workshop topics may be taught in any order, at any time during the course. See the complete list in the Index to determine if and when other Study Skills Workshop topics should be discussed with your students.

PREPARING FOR THE TEST: Begin studying several days before the test rather than cramming your studying into one marathon session the night before.

TAKING THE TEST: Follow a test-taking strategy so you can maximize your score by using the testing time wisely.

EVALUATING YOUR PERFORMANCE: After your graded test is returned, classify the types of errors that you made on the test so that you do not make them again.

Now Try This

1. Write a study session plan that explains how you will prepare on each of the 4 days before the test, as well as on test day. For some suggestions, see *Preparing for a Test.**
2. Develop your own test-taking strategy by answering the survey questions found in *How to Take a Math Test.**
3. Use the outline found in *Analyzing Your Test Results** to classify the errors that you made on your most recent test.

*Found online at: www.cengage.com/math/tussy

Teaching Tip: The *Are You Ready?* feature reviews crucial prerequisite skills that students should have mastered already if they are to be successful with the new topics in this section. All of these problems can be found in Enhanced WebAssign, allowing you to require them to be submitted before class to gauge student readiness.

SECTION 3.1 Graphing Using the Rectangular Coordinate System

OBJECTIVES

1. Construct a rectangular coordinate system.
2. Plot ordered pairs and determine the coordinates of a point.
3. Graph paired data.
4. Read line graphs.

ARE YOU READY? *Are You Ready? exercises available online at www.webassign.net/brookscole*

The following problems review some basic skills that are needed when graphing ordered pairs.

1. Graph each number in the set $\left\{\frac{7}{3}, -3, 0, 4, -1.5\right\}$ on a number line. See AIE Appendix 3.
2. **a.** What number is 8 units to the right of 0 on a number line? 8

   b. What number is 3.5 units to the left of 0 on a number line? -3.5
3. List the first four Roman numerals. I, II, III, IV
4. Write $\frac{9}{2}$ and $-\frac{11}{3}$ in mixed-number form. $4\frac{1}{2}, -3\frac{2}{3}$

It is often said, "A picture is worth a thousand words." This is certainly true in algebra, where we often use mathematical pictures called *rectangular coordinate graphs* to illustrate numerical relationships.

The Language of Algebra

A rectangular coordinate system is a **grid**—a network of uniformly spaced perpendicular lines. At times, some U.S. cities have such horrible traffic congestion that vehicles can barely move, if at all. The condition is called *gridlock.*

1 Construct a Rectangular Coordinate System.

When designing the Gateway Arch in St. Louis, architects created a mathematical model called a **rectangular coordinate graph.** This graph, shown on the next page, is drawn on a grid called a **rectangular coordinate system.** This coordinate system also is called a **Cartesian coordinate system,** after the 17th-century French mathematician René Descartes.

Teaching Tip: If the desks in your classroom are arranged in rows and columns, you can use that configuration to introduce graphing.
1. Pick a point of reference (an origin) and from it, number the rows and columns of desks.
2. Show the students that the location of any desk can be described by an ordered pair: (row, column).
3. Ask the student in row 3, column 2 to stand and the student in row 2, column 3 to stand to illustrate the difference between (3, 2) and (2, 3).
4. Have groups of students stand who meet specific conditions such as: all those in row 4, or all those whose column is an even number.

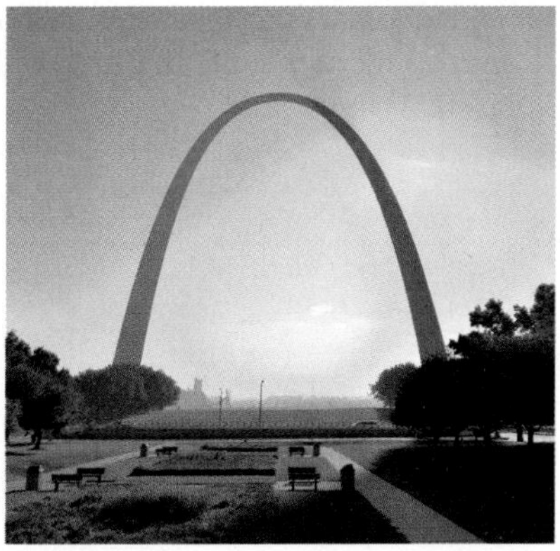

HHsu/Shutterstock.com

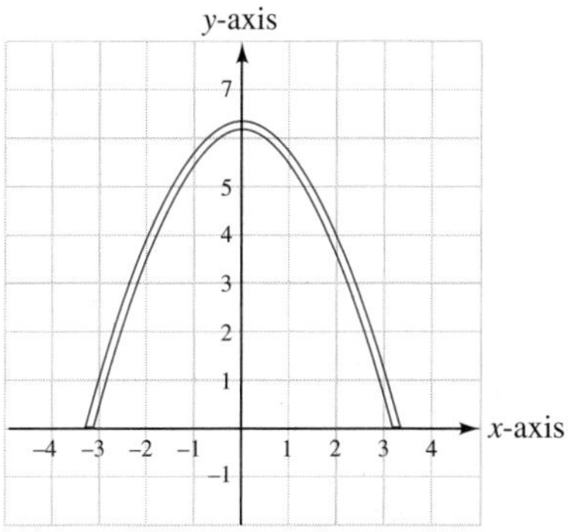

Scale: 1 unit = 100 ft

The Language of Algebra

The word **axis** is used in mathematics and science. For example, Earth rotates on its *axis* once every 24 hours. The plural of *axis* is **axes,** which is pronounced ak-seas.

A rectangular coordinate system is formed by two perpendicular number lines. The horizontal number line is usually called the ***x*-axis,** and the vertical number line is usually called the ***y*-axis.** On the x-axis, the positive direction is to the right. On the y-axis, the positive direction is upward. Each axis should be scaled to fit the data. For example, the axes of the graph of the arch are scaled in units of 100 feet.

The point where the axes intersect is called the **origin.** This is the zero point on each axis. The axes form a **coordinate plane,** and they divide it into four regions called **quadrants,** which are numbered counterclockwise using Roman numerals.

The Language of Algebra

A **coordinate plane** can be thought of as a perfectly flat surface extending infinitely far in every direction.

Points in quadrant II have a negative x- and positive y-coordinate.

Points in quadrant I have a positive x- and positive y-coordinate.

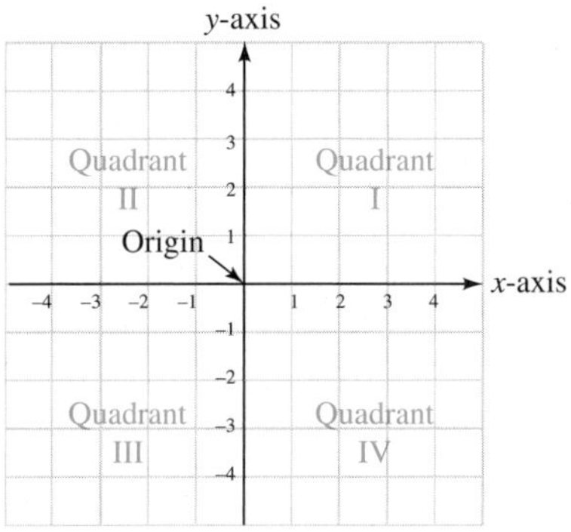

Points in quadrant III have a negative x- and negative y-coordinate.

Points in quadrant IV have a positive x- and negative y-coordinate.

Notation

Don't be confused by this new use of parentheses. $(3, -4)$ represents a point on the coordinate plane, whereas $3(-4)$ indicates multiplication. Also, don't confuse the ordered pair with interval notation.

Each point in a coordinate plane can be identified by an **ordered pair** of real numbers x and y written in the form (x, y). The first number, x, in the pair is called the ***x*-coordinate,** and the second number, y, is called the ***y*-coordinate.** Some examples of such pairs are $(3, -4)$, $\left(-1, -\frac{3}{2}\right)$, and $(0, 2.5)$.

$(3, -4)$ Read as "the point three, negative four" or as "the ordered pair three, negative four."

↑ The x-coordinate is listed first.

↑ The y-coordinate is listed second.

2 Plot Ordered Pairs and Determine the Coordinates of a Point.

The process of locating a point in the coordinate plane is called **graphing** or **plotting** the point. On the next page, we use blue arrows to show how to graph the point with coordinates $(3, -4)$. Since the x-coordinate, 3, is positive, we start at the origin and move 3 units to the *right* along the x-axis. Since the y-coordinate, -4, is negative, we then move *down* 4 units and draw a dot. This locates the point $(3, -4)$.

In the figure, red arrows are used to show how to plot the point $(-4, 3)$. We start at the origin, move 4 units to the *left* along the x-axis, then move *up* 3 units and draw a dot. This locates the point $(-4, 3)$.

The Language of Algebra

Note that the points $(3, -4)$ and $(-4, 3)$ have different locations. Since the order of the coordinates of a point is important, we call them **ordered pairs.**

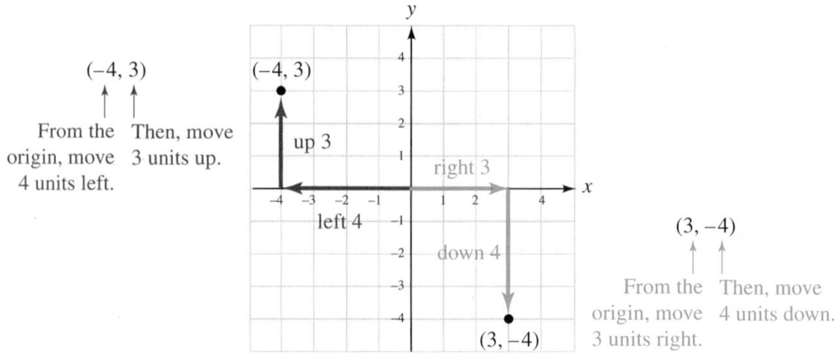

EXAMPLE 1 Plot each point. Then state the quadrant in which it lies or the axis on which it lies.

a. $(4, 4)$ **b.** $\left(-1, -\frac{7}{2}\right)$ **c.** $(0, 2.5)$ **d.** $(-3, 0)$ **e.** $(0, 0)$

Strategy After identifying the x- and y-coordinates of the ordered pair, we will move the corresponding number of units left, right, up, or down to locate the point.

Why The coordinates of a point determine its location on the coordinate plane.

Solution

a. Since the x-coordinate, 4, is positive, we start at the origin and move 4 units to the *right* along the x-axis. Since the y-coordinate, 4, is positive, we then move *up* 4 units and draw a dot. This locates the point $(4, 4)$. The point lies in quadrant I.

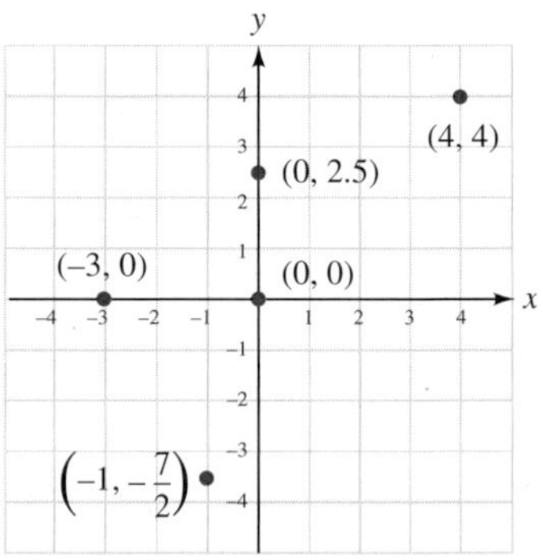

b. To plot $\left(-1, -\frac{7}{2}\right)$, we begin at the origin and move 1 unit to the *left,* because the x-coordinate is -1. Then, since the y-coordinate is negative, we move $\frac{7}{2}$ units, or $3\frac{1}{2}$ units, *down.* The point lies in quadrant III.

c. To plot $(0, 2.5)$, we begin at the origin and do not move right or left, because the x-coordinate is 0. Since the y-coordinate is positive, we move 2.5 units *up.* The point lies on the y-axis.

d. To plot $(-3, 0)$, we begin at the origin and move 3 units to the *left,* because the x-coordinate is -3. Since the y-coordinate is 0, we do not move up or down. The point lies on the x-axis.

e. To plot $(0, 0)$, we begin at the origin, and we remain there because both coordinates are 0. The point with coordinates $(0, 0)$ is the origin.

Caution

When drawing a rectangular coordinate system, always label each axis with the appropriate letter (or title) and **scale** each axis. Here, the scaling shows that each square in the grid is 1 unit long and 1 unit wide.

Success Tip

Points with an x-coordinate that is 0 lie on the y-axis. Points with a y-coordinate that is 0 lie on the x-axis. Points that lie on an axis are not considered to be in any quadrant.

Teaching Example 1 Plot each point. Then state the quadrant in which it lies or the axis on which it lies.

a. $(4, -5)$ **b.** $\left(-2, -\frac{5}{2}\right)$ **c.** $(4, 0)$ **d.** $(-1.5, 3)$

Answers:

a. QIV **b.** QIII **c.** x-axis **d.** QII

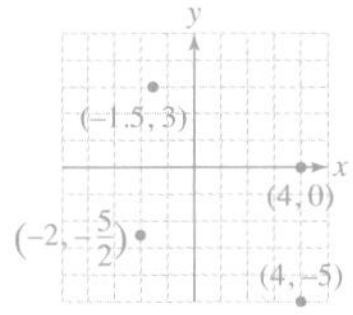

Self Check 1 Plot each point: $(2, -2)$, $(-4, 0)$, $\left(1.5, \frac{5}{2}\right)$, and $(0, 5)$

See AIE Appendix 3.

Now Try ▶ **Problem 17**

EXAMPLE 2 Find the coordinates of points A, B, C, D, E, and F plotted in figure (a) below.

Notation

Points are often labeled with capital letters. For example, the notation $A(2, 3)$ indicates that point A has coordinates $(2, 3)$.

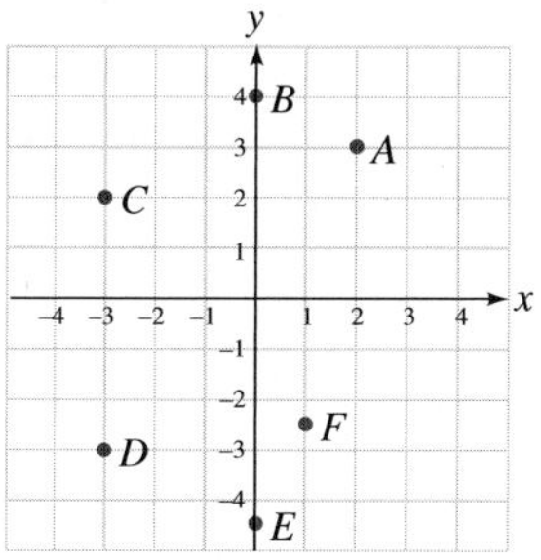

(a)

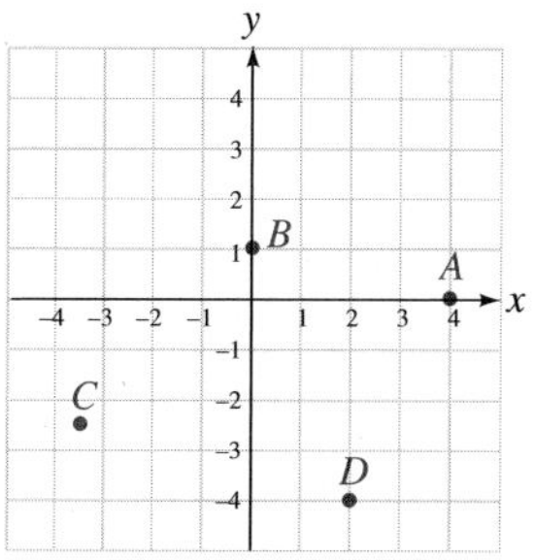

(b)

Strategy We will start at the origin and count to the left or right on the x-axis, and then up or down to reach each point.

Why The movement left or right gives the x-coordinate of the ordered pair and the movement up or down gives the y-coordinate.

Solution To locate point A, we start at the origin, move 2 units to the right on the x-axis, and then 3 units up. Its coordinates are $(2, 3)$. The coordinates of the other points are found in the same manner.

$$B(0, 4) \qquad C(-3, 2) \qquad D(-3, -3) \qquad E(0, -4.5) \qquad F(1, -2.5)$$

The coordinates of points E and F could also be given as $E\left(0, -4\frac{1}{2}\right)$ and $F\left(1, -2\frac{1}{2}\right)$.

Teaching Example 2 Find the coordinates of points A, B, C, D, and E.

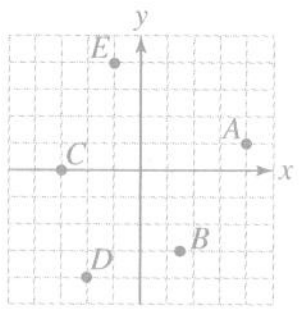

Answers:
$A(4, 1)$, $B(1.5, -3)$, $C(-3, 0)$, $D(-2, -4)$, $E(-1, 4)$

Self Check 2 Find the coordinates of each point in figure (b) above.
$A(4, 0)$; $B(0, 1)$; $C(-3.5, -2.5)$; $D(2, -4)$

Now Try Problem 20

3 Graph Paired Data.

Every day, we deal with quantities that are related:

- The time it takes to cook a roast depends on the weight of the roast.
- The money we earn depends on the number of hours we work.
- The sales tax that we pay depends on the price of the item purchased.

We can use graphs to visualize such relationships. For example, suppose a tub is filling with water, as shown below. Obviously, the amount of water in the tub depends on how long the water has been running. To graph this relationship, we can use the measurements that were taken as the tub began to fill.

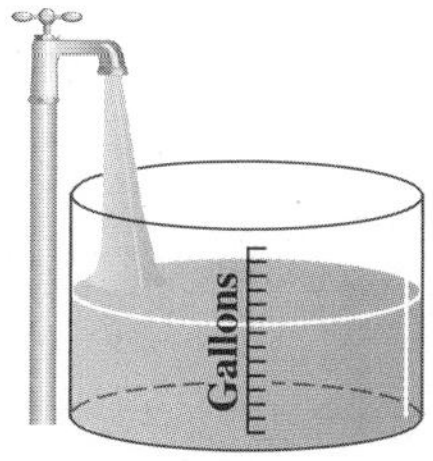

Time (min)	Water in tub (gal)	
0	0	← (0, 0)
1	8	← (1, 8)
3	24	← (3, 24)
4	32	← (4, 32)

↑ x-coordinate ↑ y-coordinate

The data in the table can be expressed as ordered pairs (x, y).

The data in each row of the table can be written as an ordered pair and plotted on a rectangular coordinate system. Since the first coordinate of each ordered pair is a time, we

The Language of Algebra

Data that can be represented as an ordered pair are called **paired data.**

label the x-axis *Time (min)*. The second coordinate is an amount of water, so we label the y-axis *Amount of water (gal)*. The y-axis is scaled in larger units (multiples of 4 gallons) because the size of the data ranges from 0 to 32 gallons.

After plotting the ordered pairs, we use a straightedge to draw a line through the points. As expected, the completed graph shows that the amount of water in the tub increases steadily as the water is allowed to run.

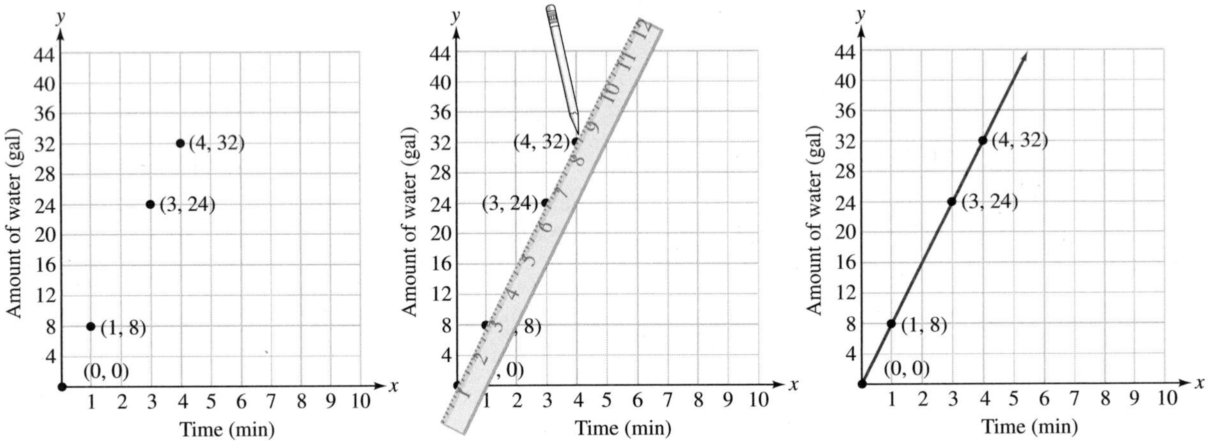

We can use the graph to determine the amount of water in the tub at various times. For example, the green dashed line on the graph at the right shows that in 2 minutes, the tub will contain 16 gallons of water. This process, called **interpolation,** uses known information to predict values that are not known but are *within* the range of the data. The blue dashed line on the graph shows that in 5 minutes, the tub will contain 40 gallons of water. This process, called **extrapolation,** uses known information to predict values that are not known and are *outside* the range of the data.

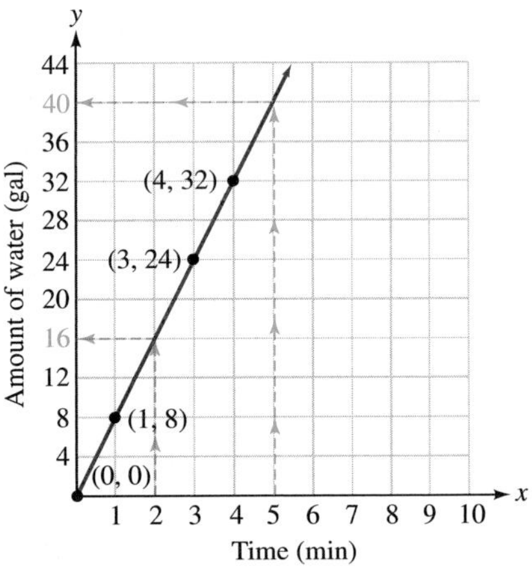

4 Read Line Graphs.

Since graphs are a popular way to present information, the ability to read and interpret them is very important.

EXAMPLE 3 **TV Shows.** The following graph shows the number of people in an audience before, during, and after the taping of a television show. Use the graph to answer the following questions.

©Scott Gries/Staff/Entertainment/Getty Images

a. How many people were in the audience when the taping began?

b. At what times were there exactly 100 people in the audience?

c. How long did it take the audience to leave after the taping ended?

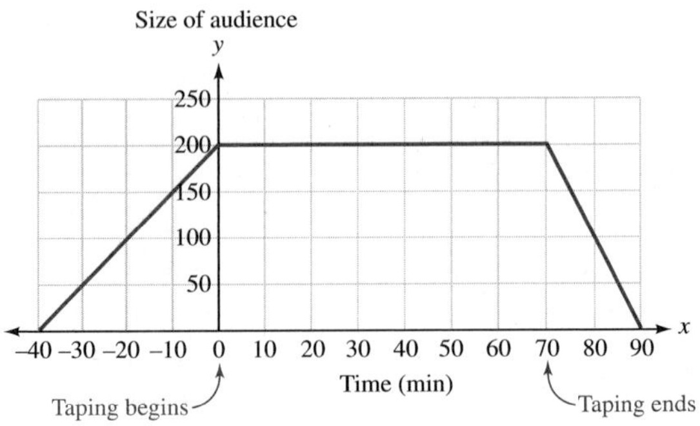

Strategy We will use an ordered pair of the form *(time, size of audience)* to describe each situation mentioned in parts (a), (b), and (c).

Why The coordinates of specific points on the graph can be used to answer each of these questions.

Solution

a. The time when the taping began is represented by 0 on the x-axis. The point on the graph directly above 0 is (0, **200**). The y-coordinate indicates that 200 people were in the audience when the taping began. This result is shown in the first row of the table in the margin.

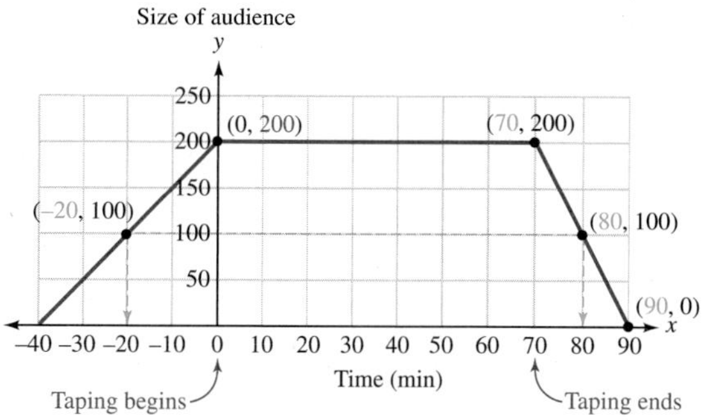

Time	Audience
0	200
−20	100
80	100
70	200
90	0

b. We can draw a horizontal line passing through 100 on the y-axis. Since the line intersects the graph twice, at (**−20**, 100) and at (**80**, 100), there are two times when 100 people were in the audience. These results are shown in the second and third rows of the table. The x-coordinates of the points tell us those times: 20 minutes before the taping began, and 80 minutes after.

c. The x-coordinate of the point (70, 200) tells us when the audience began to leave. The x-coordinate of (90, 0) tells when the exiting was completed. Subtracting the x-coordinates, we see that it took $90 - 70 = 20$ minutes for the audience to leave.

Teaching Example 3
Use the graph in Example 3 to answer the following questions.
a. How many people were in the audience 80 minutes after taping began?
b. At what time before taping were there exactly 150 people in the audience?
Answers:
a. 100 people
b. 10 min. before taping

Self Check 3 Use the graph in Example 3 to answer the following questions.

a. At what times were there exactly 50 people in the audience? 30 min before and 85 min after taping began

b. How many people were in the audience when the taping took place? 200

c. When were the first audience members allowed into the taping session? 40 min before taping began

Now Try ▶ **Problems 21 and 23**

Teaching Tip: Each Study Set begins with Vocabulary exercises. Educational research has shown that vocabulary plays a crucial role in the process of concept formation in students. All Vocabulary problems can be found in Enhanced WebAssign so that you can require them to be submitted before class to promote student readiness.

SECTION 3.1 STUDY SET

VOCABULARY

Fill in the blanks.

▶ **1.** (7, 1) is called an ordered pair.

▶ **2.** In the ordered pair (2, −5), the x-coordinate is 2 and the y-coordinate is −5.

▶ **3.** A rectangular coordinate system is formed by two perpendicular number lines called the x-axis and the y-axis. The point where the axes cross is called the origin.

▶ **4.** The x- and y-axes divide the coordinate plane into four regions called quadrants.

▶ **5.** The point with coordinates (4, 2) can be graphed on a rectangular coordinate system.

▶ **6.** The process of locating the position of a point on a coordinate plane is called graphing/plotting the point.

▶ Selected exercises available online at www.webassign.net/brookscole

CONCEPTS

Fill in the blanks.

▶ **7. a.** To plot (−5, 4), we start at the origin and move 5 units to the left and then move 4 units up.

b. To plot $\left(6, -\frac{3}{2}\right)$, we start at the origin and move 6 units to the right and then move $\frac{3}{2}$ units down.

▶ **8.** In which quadrant is each point located?

a. (−2, 7) II

b. $\left(\frac{1}{2}, \frac{15}{16}\right)$ I

c. (−1, −2.75) III

d. (50, −16) IV

▶ **9. a.** In which quadrants are the second coordinates of points positive? I and II

b. In which quadrants are the first coordinates of points negative? II and III

c. In which quadrant do points with a positive x-coordinate and a negative y-coordinate lie? IV

d. On what axis are the first coordinates of points zero? The y-axis

▶ **10. Farming.** The number of bushels of wheat produced per acre depends on the amount of water it receives. Plot the data in the table as ordered pairs and draw a straight line through the points. Use the graph to determine how many bushels per acre will be produced if

a. 6 inches of rain fall. 40 bushels per acre

b. 10 inches of rain fall. 60 bushels per acre

Inches of rain	Bushels per acre
2	20
4	30
8	50

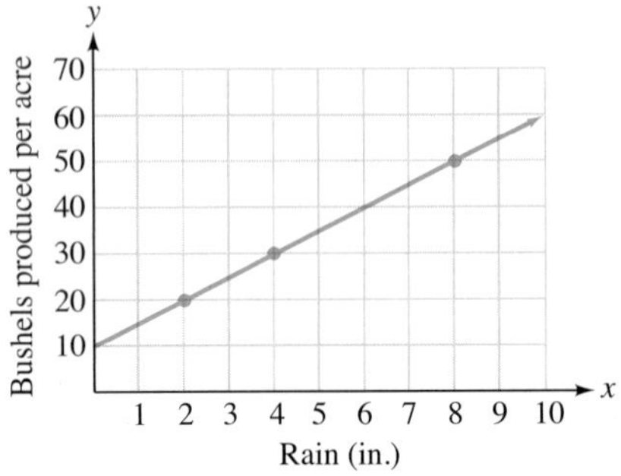

NOTATION

11. Explain the difference between (3, 5) and 3(5).
(3, 5) is an ordered pair, $3(5) = 3 \cdot 5$

▶ **12.** In a paired-data table, does the *first* or *second* column contain values associated with the vertical axis of a graph?
The second column

▶ **13.** Do these ordered pairs name the same point?

$$\left(2.5, -\frac{7}{2}\right), \left(2\frac{1}{2}, -3.5\right), \left(2.5, -3\frac{1}{2}\right) \quad \text{Yes}$$

▶ **14.** Do (3, 2) and (2, 3) represent the same point? No

15. In the ordered pair (4, 5), is the number 4 associated with the horizontal or the vertical axis? Horizontal

16. Fill in the blank: In the notation $P(4, 5)$, the capital letter P is used to name a <u>point</u>.

GUIDED PRACTICE

See Examples 1 and 2.

▶ **17.** Plot each point:
$(-3, 4), (4, 3.5), \left(-2, -\frac{5}{2}\right), (0, -4), \left(\frac{3}{2}, 0\right), (2.7, -4.1)$
See AIE Appendix 3.

18. Plot each point:
$(4, 4), (0.5, -3), (-3.9, -3.2), (0, -1), (0, 0), (0, 3), (-2, 0)$
See AIE Appendix 3.

▶ **19.** Complete the coordinates for each point in figure (a) below.
(4, 3), (0, 4), (−5, 0), (−4, −5), (3, −3)

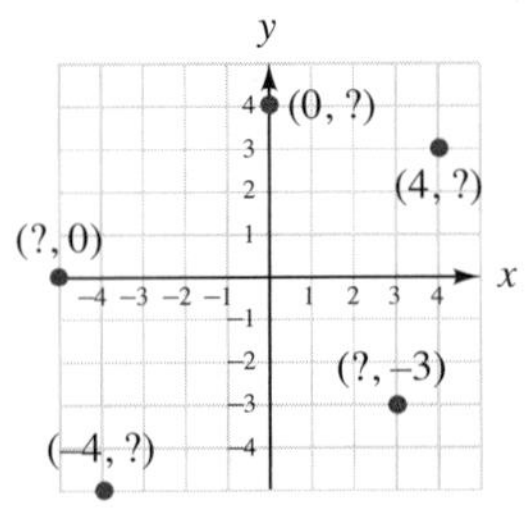

(a)

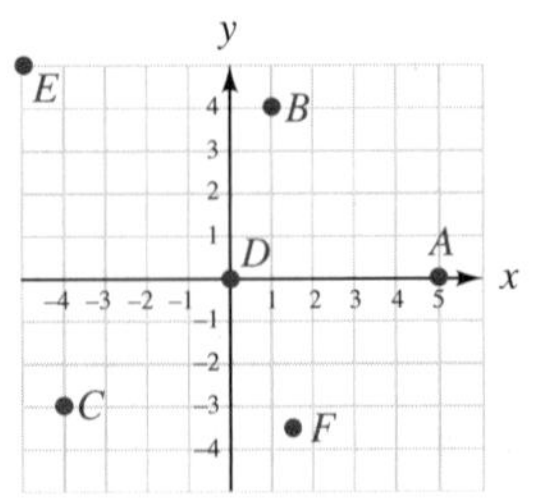

(b)

▶ **20.** Find the coordinates of points A, B, C, D, E, and F in figure (b) above.
$A(5, 0), B(1, 4), C(-4, -3), D(0, 0), E(-5, 5), F(1.5, -3.5)$

The following graph gives the heart rate of a woman before, during, and after an aerobic workout. Use it to answer Problems 21–24. **See Example 3.**

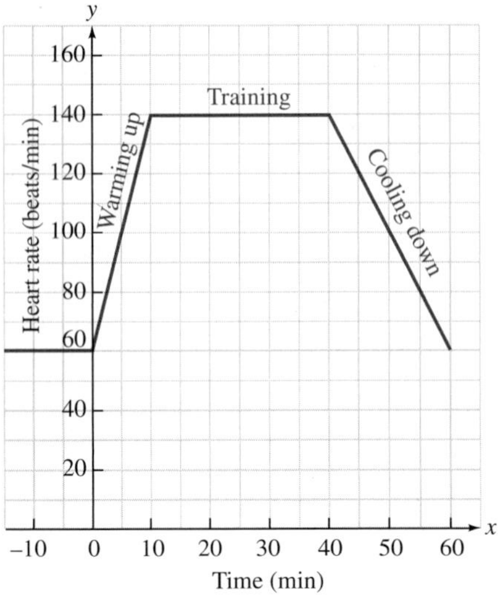

▶ **21. a.** What was her heart rate before beginning the workout? 60 beats/min

b. After beginning her workout, how long did it take the woman to reach her training-zone heart rate? 10 min

▶ **22. a.** What was the woman's heart rate half an hour after beginning the workout? 140 beats/min

b. For how long did the woman work out at her training zone? 30 min

▶ **23. a.** At what time was her heart rate 100 beats per minute? 5 min and 50 min after starting

b. How long was her cool-down period? 20 min

▶ **24. a.** What was the difference in the woman's heart rate before the workout and after the cool-down period? No difference

b. What was her approximate heart rate 8 minutes after beginning? About 130 beats/min

The following graph shows the depths of a submarine at certain times after it leaves port. Use the graph to answer Problems 25–28. **See Example 3.**

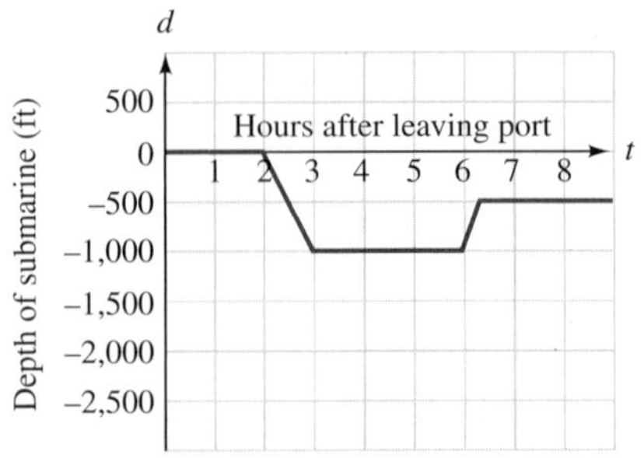

25. a. For how long does the sub travel at sea level? 2 hr

b. What is the depth of the sub 5 hours after leaving port? −1,000 ft

▶ **26. a.** Once the sub begins to dive, how long does it take to reach −1,000 feet in depth? 1 hr

b. For how long does the sub travel at a depth of 1,000 feet? 3 hr

27. a. Explain what happens 6 hours after the sub leaves port. It ascends (rises) 500 ft

b. What is the depth of the sub 8 hours after leaving port? -500 ft

▶ **28. a.** How long does it take the sub to first reach -500 feet in depth? 2.5 hr after leaving port

b. Approximate the time when the sub reaches -500 feet in depth for the second time. 6.2 hr after leaving port

APPLICATIONS

29. Bridge Construction. Find the coordinates of each rivet, weld, and anchor. Rivets: $(-60, 0), (-20, 0), (20, 0), (60, 0)$; welds: $(-40, 30), (0, 30), (40, 30)$; anchors: $(-60, -30), (60, -30)$

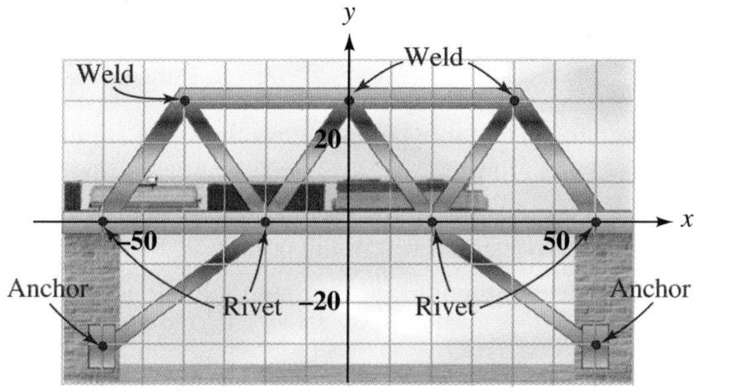

▶ **30. Golf.** A golfer is videotaped and then has her swing displayed on a computer monitor so that it can be analyzed. Give the coordinates of the three highlighted points in red. $(6, 10), (-7, 4.5), (-5, 11)$

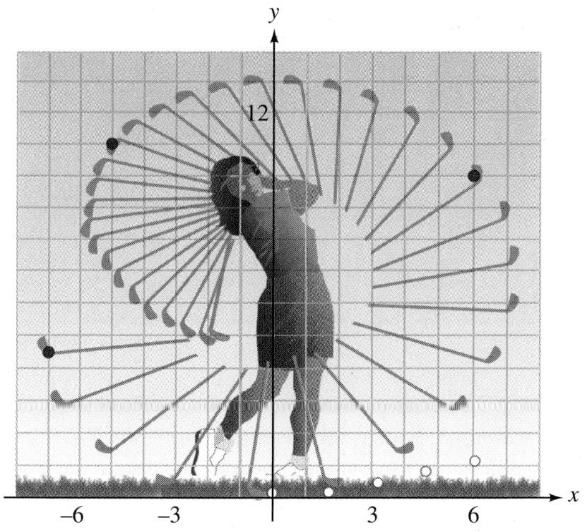

▶ **31. Games.** In the game *Battleship,* players use coordinates to drop depth charges from their ships to hit submarines. What coordinates should be used to make three hits on the submarine seen here? Express each answer in the form (letter, number). (G, 2), (G, 3), (G, 4)

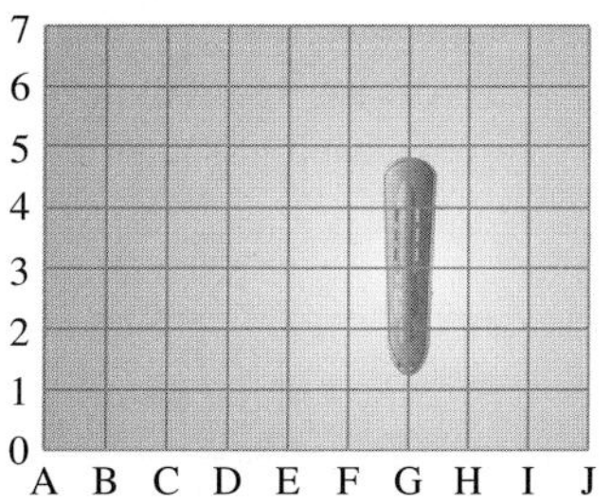

32. Maps. Use coordinates of the form (number, letter) to locate each of the following on the map: Tempe, Glendale, Paradise Valley, Sky Harbor Airport, and the intersection of Camelback Road and 7th Avenue. Tempe: (7, E), Glendale: (1, A), Paradise Valley: (7, A) , Sky Harbor Airport: (5, E), and the intersection of Camelback Road and 7th Avenue: (3, B)

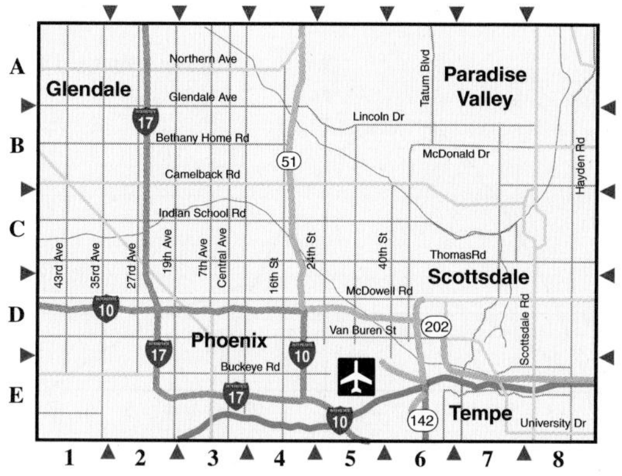

▶ **33.** ▶ from **Campus to Careers**

Dental Assistant

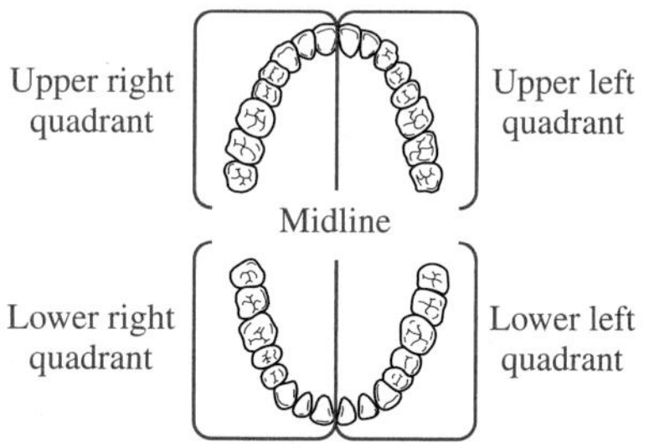

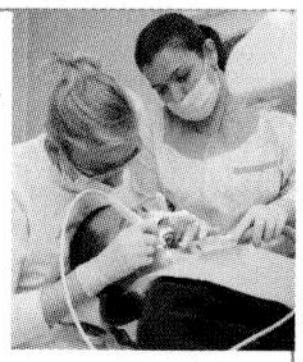

Dentists describe teeth as being located in one of four *quadrants* as shown above.

a. How many teeth are in each quadrant? 8 teeth

b. Why would the upper left quadrant appear on the right in the illustration? It represents the patient's left side.

34. Geography. A coordinate system that describes the location of any place on the surface of the Earth uses a series of *latitude* and *longitude* lines, as shown below. Estimate the location of the Deep Water Horizon (the oil drilling rig in the Gulf of Mexico that exploded in 2010) using an ordered pair of the form (latitude, longitude). (Source: sailwx.info) $(28°, -89°)$

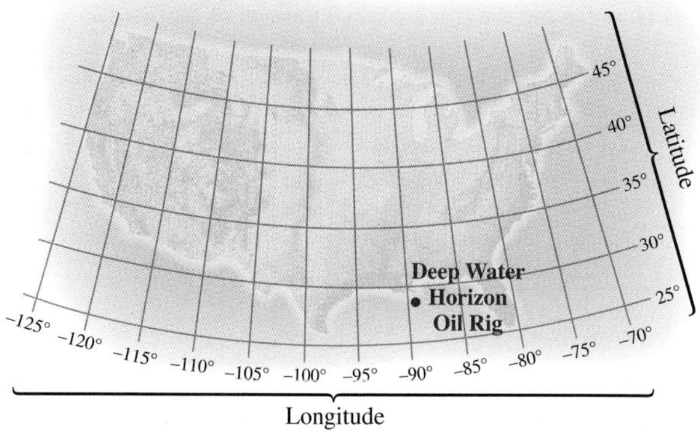

35. Water Pressure. The graphs show how the path of a stream of water changes when the hose is held at two different angles.

a. At which angle does the stream of water shoot up higher? How much higher? 60°; 4 ft

b. At which angle does the stream of water shoot out farther? How much farther? 30°; 4 ft

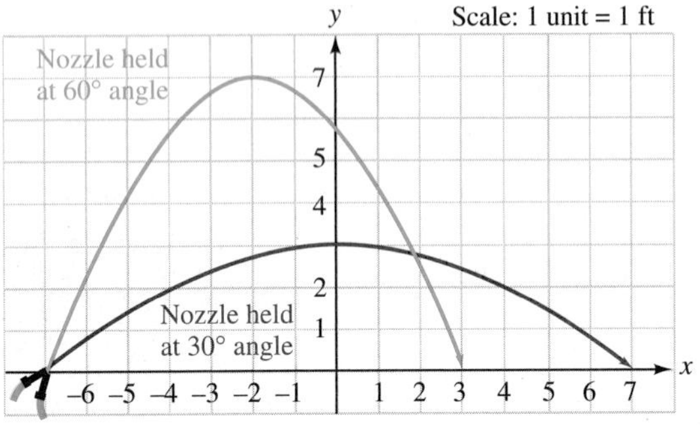

36. Area. Three vertices (corners) of a rectangle are (2, 1), (6, 1), and (6, 4). Find the coordinates of the fourth vertex. Then find the area of the rectangle. (2, 4); 12 sq. units

37. Area. Three vertices (corners) of a right triangle are (−1, −7), (−5, −7), and (−5, −2). Find the area of the triangle. 10 sq. units

38. Landmarks. A scale model of the block letter H in the Hollywood sign can be drawn by plotting the following points and connecting them: (0, 0), (13, 0), (13, 16), (26, 16), (26, 0), (39, 0), (39, 45), (26, 45), (26, 29), (13, 29), (13, 45), and (0, 45). The scale is 1 unit on the graph is equal to 1 foot on the actual sign. If a gallon of paint covers 350 square feet, how many gallons are needed to paint the front side of the letter H? Round to the nearest gallon. 4 gal (1,339 ft^2 to cover)

©Cary Kalscheuer/Shutterstock.com

39. Trucks. The table below shows the number of miles that an 18-wheel truck can be driven on a given number of gallons of diesel fuel. Plot the data in the table as ordered pairs. Then draw a straight line through the points.

a. How far can the truck go on 4 gallons of fuel? 20 mi

b. How many gallons of fuel are needed to travel a distance of 30 miles? 6 gal

c. How far can the truck go on 7 gallons of fuel? 35 mi

Fuel (gal)	Distance (mi)
2	10
3	15
5	25

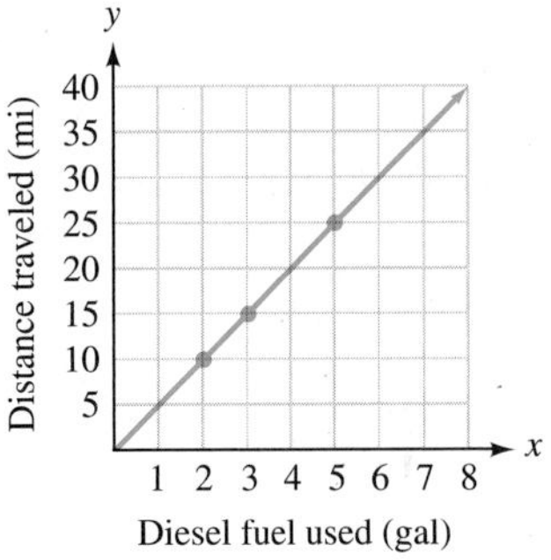

40. Boating. The table below shows the cost to rent a sailboat for a given number of hours. Plot the data in the table as ordered pairs. Then draw a straight line through the points.

a. What does it cost to rent the boat for 3 hours? $25

b. For how long can the boat be rented for $60? 10 hr

c. What does it cost to rent the boat for 9 hours? $55

Rental time (hr)	Cost ($)
2	20
4	30
6	40

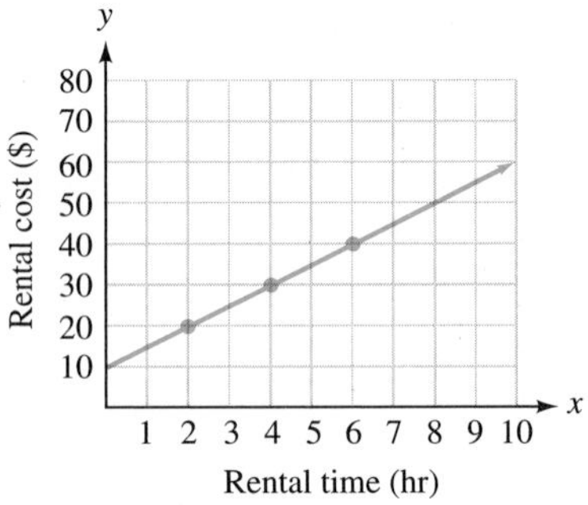

41. Depreciation. The table below shows the value (in thousands of dollars) of a color copier at various lengths of time after its purchase. Plot the data in the table as ordered pairs. Then draw a straight line passing through the points.

a. What does the point (3, 7) on the graph tell you? A 3-yr-old copier is worth $7,000.

b. Find the value of the copier when it is 7 years old. $1,000

c. After how many years will the copier be worth $2,500? 6 yr

Age (yr)	Value ($1,000)
3	7
4	5.5
5	4

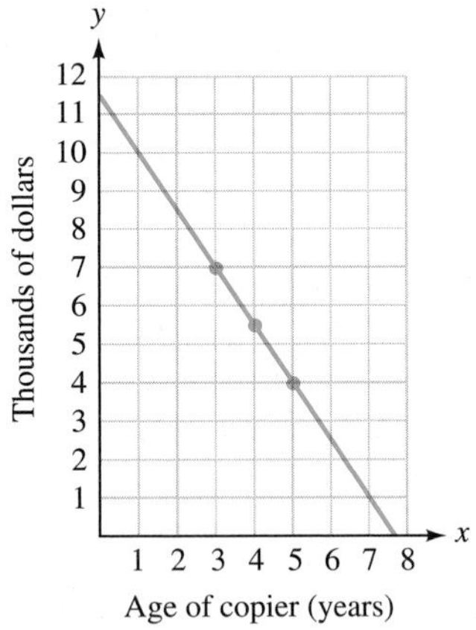

42. Swimming. The table below shows the number of people at a public swimming pool at various times during the day. (0 represents noon, 1 represents 1 PM, and so on.) Plot the data in the table as ordered pairs. Then draw a straight line passing through the points.

a. How many people will be at the pool at 6 P.M.? 50

b. At what time will there be 250 people at the pool? 2 P.M.

c. At what time will the number of people at the pool be half of what it was at noon? 3:30 P.M.

Time	Number of people
0	350
3	200
5	100

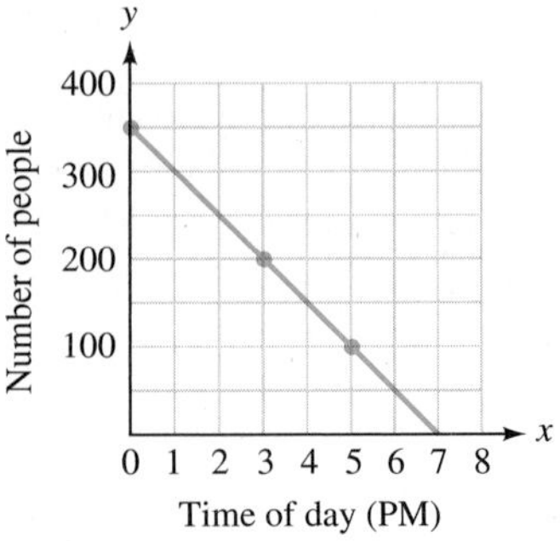

WRITING

43. Explain why the point $(-3, 3)$ is not the same as the point $(3, -3)$.

44. Explain how to plot the point $(-2, 5)$.

45. Use the Internet to perform a search of the name René Descartes. After reading about him, explain how a fly on his bedroom ceiling provided the inspiration for the concept of a rectangular coordinate system.

46. Explain this diagram.

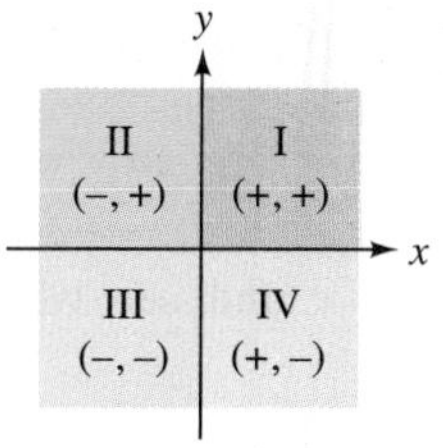

REVIEW

47. Solve $AC = \frac{2}{3}h - T$ for h. $h = \frac{3(AC + T)}{2}$ or $h = \frac{3AC + 3T}{2}$

48. Solve $5(x + 1) \le 2(x - 3)$. Write the solution set in interval notation and graph it. $\left(-\infty, -\frac{11}{3}\right]$ See AIE Appendix 3.

49. Evaluate: $\frac{-4(4 + 2) - 2^3}{|-12 - 4(5)|}$ -1

50. Simplify: $\frac{24}{54}$ $\frac{4}{9}$

CHALLENGE PROBLEMS

51. In what quadrant does a point lie if the *sum* of its coordinates is negative and the *product* of its coordinates is positive? III

52. Draw line segment $\overline{AB}$ with endpoints $A(6, 5)$ and $B(-4, 5)$. Suppose that the x-coordinate of a point C is the average of the x-coordinates of points A and B, and the y-coordinate of point C is the average of the y-coordinates of points A and B. Find the coordinates of point C. Why is C called the midpoint of $\overline{AB}$? $C(1, 5)$; it divides $\overline{AB}$ into two parts of equal length.

Teaching Tip: The *Are You Ready?* feature reviews crucial prerequisite skills that students should have mastered already if they are to be successful with the new topics in this section. All of these problems can be found in Enhanced WebAssign, allowing you to require them to be submitted before class in order to gauge student readiness.

SECTION 3.2 Graphing Linear Equations

OBJECTIVES

1. Determine whether an ordered pair is a solution of an equation.
2. Complete ordered-pair solutions of equations.
3. Construct a table of solutions.
4. Graph linear equations by plotting points.
5. Use graphs of linear equations to solve applied problems.

ARE YOU READY?

Are You Ready? exercises available online at www.webassign.net/brookscole

The following problems review some basic skills that are needed when graphing linear equations.

1. Is $2(4) + 3(-1) = 4$ a true or false statement? False
2. Evaluate $-6x - 7$ for $x = -2$. 5
3. Solve: $4(2) + 3y = 14$ 2
4. Solve for y: $4x + 5y = -15$ $y = -\frac{4}{5}x - 3$
5. Multiply: $\frac{3}{2}(2)$ 3
6. Multiply: $-\frac{5}{9}(-9)$ 5

In this section, we will discuss equations that contain two variables. Such equations are often used to describe algebraic relationships between two quantities. To see a mathematical picture of these relationships, we will construct graphs of their equations.

1 Determine Whether an Ordered Pair Is a Solution of an Equation.

We have previously solved **equations in one variable.** For example, $x + 3 = 9$ is an equation in x. If we subtract 3 from both sides, we see that 6 is the solution. To verify this, we replace x with 6 and note that the result is a true statement: $9 = 9$.

In this chapter, we extend our equation-solving skills to find solutions of **equations in two variables.** To begin, let's consider $y = x - 1$, an equation in x and y.

Notation

Equations in two variables often involve the variables x and y. However, other letters can be used. For example, $a - 3b = 5$ and $n = 4m + 6$ are equations in two variables.

A solution of $y = x - 1$ is a pair of values, one for x and one for y, that make the equation true. To illustrate, suppose x is 5 and y is 4. Then we have:

$y = x - 1$ This is the given equation.

$4 \stackrel{?}{=} 5 - 1$ Substitute 5 for x and 4 for y.

$4 = 4$ True

Since the result is a true statement, $x = 5$ and $y = 4$ is a solution of $y = x - 1$. We write the solution as the ordered pair (5, 4), with the value of x listed first. We say that (5, 4) **satisfies** the equation.

In general, a **solution of an equation in two variables** is an ordered pair of numbers that makes the equation a true statement.

EXAMPLE 1 Is $(-1, -3)$ a solution of $y = x - 1$?

Strategy We will substitute -1 for x and -3 for y and see whether the resulting equation is true.

Why An ordered pair is a solution of $y = x - 1$ if replacing the variables with the values of the ordered pair results in a true statement.

Solution

$y = x - 1$ This is the given equation.

$-3 \stackrel{?}{=} -1 - 1$ Substitute −1 for x and −3 for y.

$-3 = -2$ False

Since $-3 = -2$ is false, $(-1, -3)$ is not a solution of $y = x - 1$.

Teaching Example 1 Is $(-5, -4)$ a solution of $y = x - 1$?
Answer: No

Self Check 1 Is (9, 8) a solution of $y = x - 1$? Yes

Now Try Problem 17

2 Complete Ordered-Pair Solutions of Equations.

If only one of the values of an ordered-pair solution is known, we can substitute it into the equation to determine the other value.

EXAMPLE 2 Complete the solution $(-5, \quad)$ of the equation $y = -2x + 3$.

Strategy We will substitute the known x-coordinate of the solution into the given equation.

Why We can use the resulting equation in one variable to find the unknown y-coordinate of the solution.

Solution In the ordered pair $(-5, \quad)$, the x-value is -5; the y-value is not known. To find y, we substitute -5 for x in the equation and evaluate the right side.

$y = -2x + 3$ This is the given equation.

$y = -2(-5) + 3$ Substitute −5 for x.

$y = 10 + 3$ Do the multiplication.

$y = 13$ This is the missing y-coordinate of the solution.

The completed ordered pair is $(-5, 13)$.

Teaching Example 2 Complete the solution $(2, \quad)$ of the equation $y = 2x - 4$.
Answer: (2, 0)

Self Check 2 Complete the solution $(-2, \quad)$ of the equation $y = 4x - 2$. $(-2, -10)$

Now Try Problem 29

Solutions of equations in two variables are often listed in a **table of solutions** (or **table of values**).

EXAMPLE 3 Complete the table of solutions for $3x + 2y = 5$.

x	y	(x, y)
7		(7,)
	4	(, 4)

Strategy In each case we will substitute the known coordinate of the solution into the given equation.

Why We can solve the resulting equation in one variable to find the unknown coordinate of the solution.

Solution In the first row, we are given an x-value of 7. To find the corresponding y-value, we substitute 7 for x and solve for y.

Caution

When solving equations, we will no longer show all of the steps involving the use of a property of equality, as we did in Chapter 2. For example, on the right, we do not show the subtraction of 21 from both sides nor both sides being divided by 2. These calculations are done mentally.

$3x + 2y = 5$ This is the given equation.

$3(7) + 2y = 5$ Substitute 7 for x.

$21 + 2y = 5$ Do the multiplication.

$2y = -16$ To isolate the variable term 2y, subtract 21 from both sides.

$y = -8$ To isolate y, divide both sides by 2. This is the missing y-coordinate of the solution.

A solution of $3x + 2y = 5$ is $(7, -8)$. It is entered in the table on the left.

x	y	(x, y)
7	−8	(7, −8)

In the second row, we are given a y-value of 4. To find the corresponding x-value, we substitute 4 for y and solve for x.

$3x + 2y = 5$ This is the given equation.

$3x + 2(4) = 5$ Substitute 4 for y.

$3x + 8 = 5$ Do the multiplication.

$3x = -3$ To isolate the variable term 3x, subtract 8 from both sides.

$x = -1$ To isolate x, divide both sides by 3. This is the missing x-coordinate of the solution.

Another solution is $(-1, 4)$. It is entered in the table on the left.

x	y	(x, y)
7	−8	(7, −8)
−1	4	(−1, 4)

Teaching Example 3
Complete the table of solutions for $2x - 5y = 10$.

x	y	(x, y)
10	2	(10, 2)
5	0	(5, 0)

Self Check 3 Complete the table of solutions for $3x + 2y = 5$.

x	y	(x, y)
3	−2	(3, −2)
5	−5	(5, −5)

Now Try Problem 37

3 Construct a Table of Solutions.

To find a solution of an equation in two variables, we can select a number, substitute it for one of the variables, and find the corresponding value of the other variable. For example, to find a solution of $y = x - 1$, we can select a value for x, say, -4, substitute -4 for x in the equation, and find y.

$y = x - 1$

$y = \mathbf{-4} - 1$ Substitute −4 for x.

$y = -5$ Do the subtraction.

The ordered pair $(-4, -5)$ is a solution. We list it in the table on the left.

x	y	(x, y)
−4	−5	(−4, −5)

To find another solution of $y = x - 1$, we select another value for x, say, -2, and find the corresponding y-value.

$y = x - 1$

$y = -2 - 1$ Substitute -2 for x.

$y = -3$ Do the subtraction.

x	y	(x, y)
-4	-5	$(-4, -5)$
-2	-3	$(-2, -3)$

A second solution is $(-2, -3)$, and we list it in the table of solutions.

If we let $x = 0$, we can find a third ordered pair that satisfies $y = x - 1$.

$y = x - 1$

$y = 0 - 1$ Substitute 0 for x.

$y = -1$ Do the subtraction.

x	y	(x, y)
-4	-5	$(-4, -5)$
-2	-3	$(-2, -3)$
0	-1	$(0, -1)$

A third solution is $(0, -1)$, which we also add to the table of solutions.

We can find a fourth solution by letting $x = 2$, and a fifth solution by letting $x = 4$.

$y = x - 1$

$y = 2 - 1$ Substitute 2 for x.

$y = 1$ Do the subtraction.

$y = x - 1$

$y = 4 - 1$ Substitute 4 for x.

$y = 3$ Do the subtraction.

x	y	(x, y)
-4	-5	$(-4, -5)$
-2	-3	$(-2, -3)$
0	-1	$(0, -1)$
2	1	$(2, 1)$
4	3	$(4, 3)$

A fourth solution is $(2, 1)$ and a fifth solution is $(4, 3)$. We add them to the table.

Since we can choose any real number for x, and since any choice of x will give a corresponding value of y, it is apparent that the equation $y = x - 1$ has *infinitely many solutions.* We have found five of them: $(-4, -5)$, $(-2, -3)$, $(0, -1)$, $(2, 1)$, and $(4, 3)$.

4 Graph Linear Equations by Plotting Points.

It is impossible to list the infinitely many solutions of the equation $y = x - 1$. However, to show all of its solutions, we can draw a mathematical "picture" of them. We call this picture the *graph of the equation.*

Notation

The graph only shows a part of the line. The arrowheads indicate that it extends indefinitely in both directions.

To graph $y = x - 1$, we plot the ordered pairs shown in the table on a rectangular coordinate system. Then we draw a straight line through the points, because the graph of any solution of $y = x - 1$ will lie on this line. We also draw arrowheads on either end of the line to indicate that the solutions continue indefinitely in both directions, beyond what we can see on the coordinate grid. We call the line the **graph of the equation.** It represents all of the solutions of $y = x - 1$.

$y = x - 1$

x	y	(x, y)
-4	-5	$(-4, -5)$
-2	-3	$(-2, -3)$
0	-1	$(0, -1)$
2	1	$(2, 1)$
4	3	$(4, 3)$

↑ Select x ↑ Find y ↑ Plot (x, y)

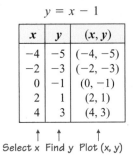

Construct a table of solutions.

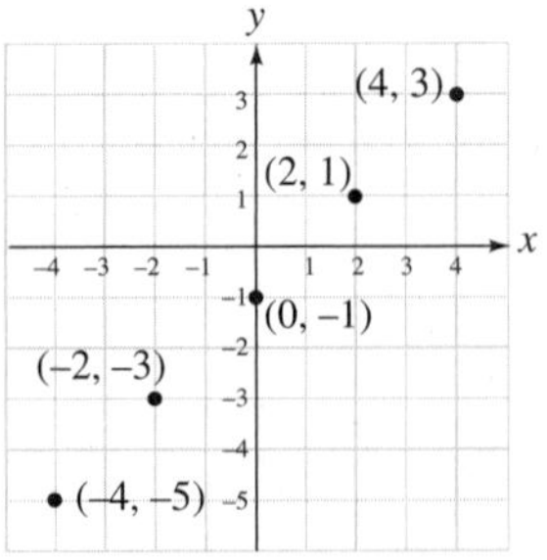

Plot the ordered pairs.

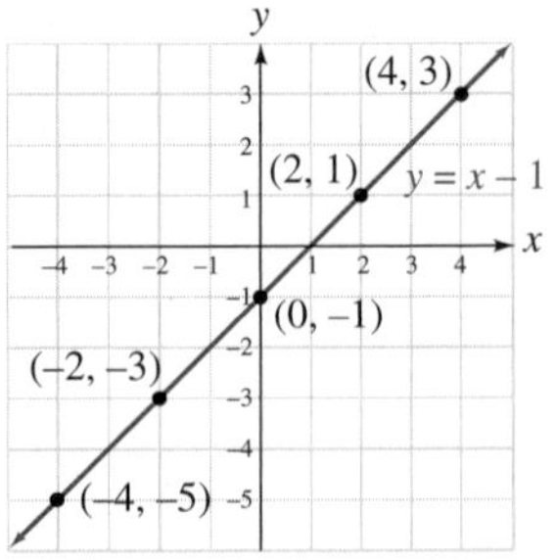

Draw a straight line through the points. This is the *graph of the equation.*

The equation $y = x - 1$ is said to be *linear* and its graph is a line. By definition, a linear equation in two variables is any equation that can be written in the following form, where the variable terms appear on one side of an equal symbol and a constant appears on the other.

Linear Equations

A **linear equation in two variables** is an equation that can be written in the form

$$Ax + By = C$$

where A, B, and C are real numbers and A and B are not both 0. This form is called **standard form.***

*In some textbooks, the definition of the standard form of a linear equation in two variables contains additional requirements, such as: A, B, and C are integers, $A > 0$, and the greatest common factor of A, B, and C is 1.

Every linear equation in two variables has an infinite number of ordered-pair solutions. The graph of a linear equation in two variables is a straight line. Every point on the line corresponds to a solution.

Some more examples of linear equations are

$$y = 2x + 4, \qquad 2x + 3y = 12, \qquad \text{and} \qquad 3x = 5y$$

The exponent on each variable of a linear equation in two variables is an understood 1. For example, $y = 2x + 4$ can be thought of as $y^1 = 2x^1 + 4$ and $2x + 3y = 12$ can be thought of as $2x^1 + 3y^1 = 12$.

Some examples of equations in two variables that are *not* linear are shown below. You will see later in this course (and in more advanced courses) that the graphs of these equations are not straight lines.

Teaching Tip: Have your students point out why each equation shown here is not a linear equation.

$$y = x^2 + 3, \qquad y = \sqrt{x}, \qquad y = 4^x, \qquad x^2 + y^2 = 25, \qquad \text{and} \qquad y = \frac{1}{x}$$

Linear equations can be graphed in several ways. Generally, the form in which an equation is written determines the method that we use to graph it. To graph linear equations solved for y, such as $y = 2x + 4$, we can use the following **point-plotting method.**

Graphing Linear Equations Solved for y by Plotting Points

1. Find three ordered pairs that are solutions of the equation by selecting three values for x and calculating the corresponding values of y.
2. Plot the solutions on a rectangular coordinate system.
3. Draw a straight line passing through the points. If the points do not lie on a line, check your calculations.

EXAMPLE 4 Graph: $y = 2x + 4$

Strategy We will find three solutions of the equation, plot them on a rectangular coordinate system, and then draw a straight line passing through the points.

Why To *graph* a linear equation in two variables means to make a drawing that represents all of its solutions.

Solution To find three solutions of this linear equation, we select three values for x that will make the calculations easy. Then we find each corresponding value of y.

Success Tip

When selecting x-values for a table of solutions, a rule of thumb is to choose a negative number, a positive number, and 0. When $x = 0$, the calculations to find y are usually quite simple.

If $x = -2$:	*If* $x = 0$:	*If* $x = 2$:
$y = 2x + 4$	$y = 2x + 4$	$y = 2x + 4$
$y = 2(-2) + 4$	$y = 2(0) + 4$	$y = 2(2) + 4$
$y = -4 + 4$	$y = 0 + 4$	$y = 4 + 4$
$y = 0$	$y = 4$	$y = 8$
$(-2, 0)$ is a solution.	$(0, 4)$ is a solution.	$(2, 8)$ is a solution.

Success Tip

Since two points determine a line, only two points are needed to graph a linear equation. However, we should plot a third point as a check. If the three points do not lie on a straight line, then at least one of them is in error.

We enter the results in a table of solutions and plot the points. Then we draw a straight line through the points and label it $y = 2x + 4$.

$y = 2x + 4$

x	y	(x, y)
-2	0	$(-2, 0)$
0	4	$(0, 4)$
2	8	$(2, 8)$

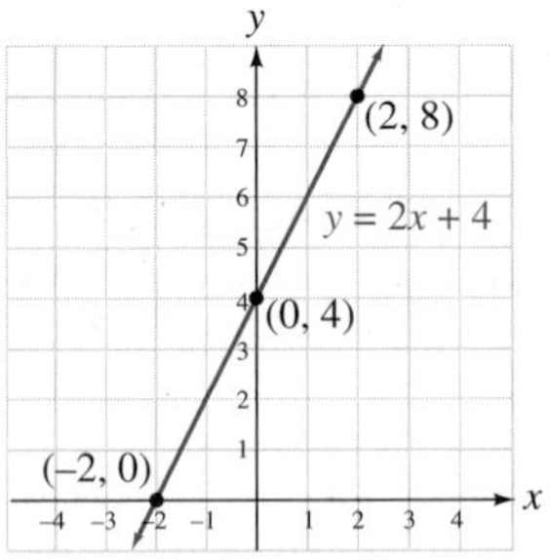

As a check, we can pick two points that the line appears to pass through, such as $(1, 6)$ and $(-1, 2)$. When we substitute their coordinates into the given equation, the two true statements that result indicate that $(1, 6)$ and $(-1, 2)$ are solutions and that the graph of the line is correctly drawn.

***Check* (1, 6):**
$$y = 2x + 4$$
$$6 \stackrel{?}{=} 2(\mathbf{1}) + 4$$
$$6 \stackrel{?}{=} 2 + 4$$
$$6 = 6 \quad \text{True}$$

***Check* (−1, 2):**
$$y = 2x + 4$$
$$\mathbf{2} \stackrel{?}{=} 2(\mathbf{-1}) + 4$$
$$2 \stackrel{?}{=} -2 + 4$$
$$2 = 2 \quad \text{True}$$

Teaching Example 4
Graph: $y = 2x + 3$
Answer:

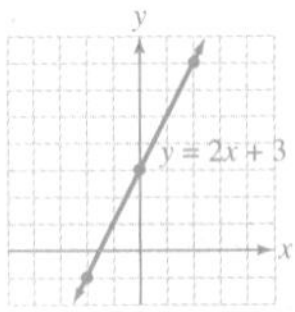

Self Check 4 Graph: $y = 2x - 2$ See AIE Appendix 3.

Now Try Problem 41

EXAMPLE 5 Graph: $y = -3x$

Strategy We will find three solutions of the equation, plot them on a rectangular coordinate system, and then draw a straight line passing through the points.

Why To *graph* a linear equation in two variables means to make a drawing that represents all of its solutions.

Solution To find three solutions, we begin by selecting three x-values: -1, 0, and 1. Then we find the corresponding values of y. If $x = -1$, we have

$y = -3x$ This is the equation to graph.

$y = -3(\mathbf{-1})$ Substitute −1 for x.

$y = 3$ Do the multiplication.

$(-1, 3)$ is a solution.

Notation

Tables of solutions can be written vertically, as shown in Examples 3–6, or they can be written horizontally, as shown below.

x	-1	0	1
y	3	0	-3

In a similar manner, we find the y-values for x-values of 0 and 1, and record the results in a table of solutions. After plotting the ordered pairs, we draw a straight line through the points and label it $y = -3x$.

$y = -3x$

x	y	(x, y)
-1	3	$(-1, 3)$
0	0	$(0, 0)$
1	-3	$(1, -3)$

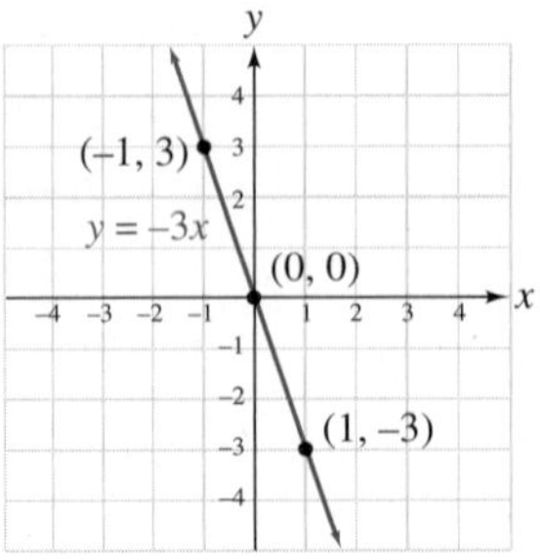

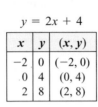

Teaching Example 5
Graph: $y = -5x$
Answer:

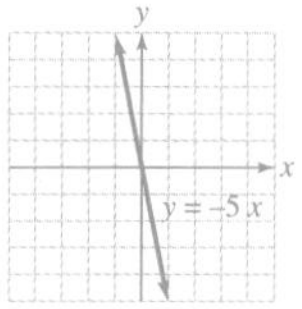

Self Check 5 Graph: $y = -4x$ See AIE Appendix 3.

Now Try ▶ **Problem 45**

To graph linear equations in x and y using the method discussed in this section, the variable y must be isolated on one side of the equation.

EXAMPLE 6 Graph $2x + 3y = -12$ by first solving for y.

Strategy We will use properties of equality to solve the given equation for y. Then we will use the point-plotting method of this section to graph the resulting equivalent equation.

Why The calculations to find several solutions of a linear equation in two variables are usually easier when the equation is solved for y.

Solution To solve for y, we proceed as follows.

$2x + 3y = -12$ This is the given equation.

$2x + 3y - \mathbf{2x} = \mathbf{-2x} - 12$ To isolate the variable term 3y on the left side, subtract 2x from both sides. When solving for y, it is common practice to write the subtraction (or addition) of a variable term *before* the constant term.

$3y = -2x - 12$ On the left side, combine like terms: $2x - 2x = 0$.

$\frac{3y}{\mathbf{3}} = \frac{-2x}{\mathbf{3}} - \frac{12}{\mathbf{3}}$ To isolate the variable y, undo the multiplication by 3 by dividing both sides, term-by-term, by 3.

$y = -\frac{2}{3}x - 4$ Write $\frac{-2x}{3}$ as $-\frac{2}{3}x$. Simplify: $\frac{12}{3} = 4$.

Teaching Tip: Explain this new approach with the subtraction property of equality. Point out that in future sections it is better if we write the x-term immediately after the equal symbol. Compare:

$2x + 3y - \mathbf{2x} = \mathbf{-2x} - 12$

to

$2x + 3y - \mathbf{2x} = -12 - \mathbf{2x}$

Success Tip

The division by 3 on the right side of the equation is done term-by-term instead of with a single fraction bar. We write:

$\frac{-2x}{3} - \frac{12}{3}$ not ~~$\frac{-2x - 12}{3}$~~

Since $y = -\frac{2}{3}x - 4$ is equivalent to $2x + 3y = -12$, we can use it to draw the graph of $2x + 3y = -12$.

To find solutions of $y = -\frac{2}{3}x - 4$, each value of x must be multiplied by $-\frac{2}{3}$. This calculation is made easier if we select x-values that are *multiples of the denominator 3,* such as -3, 0, and 6. For example, if $x = -3$, we have

$y = -\frac{2}{3}x - 4$ This is the equation to graph.

$y = -\frac{2}{3}(\mathbf{-3}) - 4$ Substitute −3 for x.

$y = 2 - 4$ Multiply: $-\frac{2}{3}(-3) = 2$.

$y = -2$ Do the subtraction.

Thus, $(-3, -2)$ is a solution.

Teaching Tip: Spend extra time here to explain this new application of the division property of equality. Show how we divide both sides, term-by-term, by the same number.

Success Tip

When we chose x-values that are multiples of the denominator 3, the corresponding y-values are integers, and not difficult-to-plot fractions.

Two more solutions, one for $x = 0$ and one for $x = 6$, can be found in a similar way, and entered in a table. We plot the ordered pairs, draw a straight line through the points, and label the line as $y = -\frac{2}{3}x - 4$ or as $2x + 3y = -12$.

$2x + 3y = -12$
or
$y = -\frac{2}{3}x - 4$

x	y	(x, y)
−3	−2	(−3, −2)
0	−4	(0, −4)
6	−8	(6, −8)

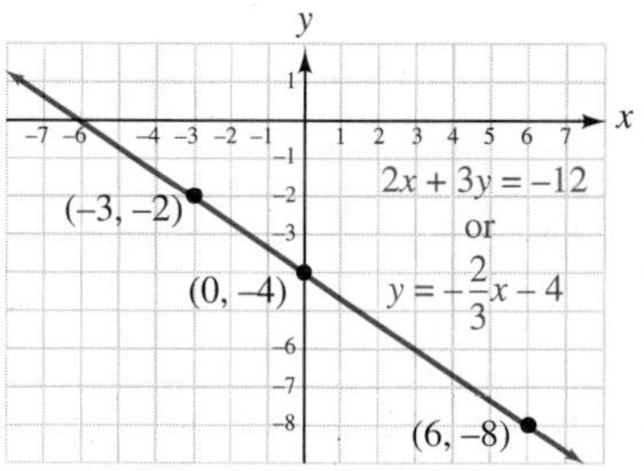

Teaching Example 6
Graph $5y = 2x - 15$ by first solving for y.
Answer: $y = \frac{2}{5}x - 3$

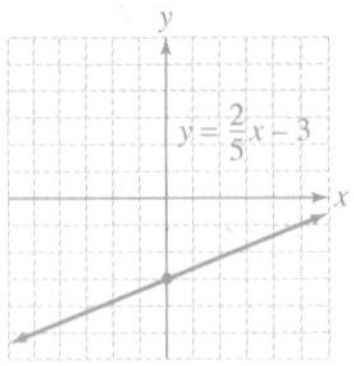

Self Check 6 Graph $5x - 2y = -2$ by first solving for y. $y = \frac{5}{2}x + 1$; See AIE Appendix 3.

Now Try Problem 49

5 Use Graphs of Linear Equations to Solve Applied Problems.

When linear equations are used to model real-life situations, they are often written in variables other than x and y. In such cases, we must make the appropriate changes when labeling the table of solutions and the graph of the equation.

EXAMPLE 7 **Cleaning Windows.** The linear equation $A = -0.03n + 32$ estimates the amount A of glass-cleaning solution (in ounces) that is left in the bottle after the sprayer trigger has been pulled a total of n times. Graph the equation and use the graph to estimate the amount of solution that is left after 500 sprays.

Strategy We will find three solutions of the equation, plot them on a rectangular coordinate system, and then draw a straight line passing through the points.

Why We can use the graph to estimate the amount of solution left after any number of sprays.

©Matthew Benoit/Shutterstock.com

Solution Since A depends on n in the equation $A = -0.03n + 32$, solutions will have the form (n, A). To find three solutions, we begin by selecting three values of n. Because the number of sprays cannot be negative, and the calculations to find A involve decimal multiplication, we select 0, 100, and 1,000. For example, if $n = 100$, we have

$A = -0.03n + 32$ This is the equation to graph.

$A = -0.03(\mathbf{100}) + 32$ Substitute 100 for n.

$A = -3 + 32$ Multiply by moving the decimal point in -0.03 two places to the right: $-0.03(100) = -3$.

$A = 29$ Do the addition.

Thus, (100, 29) is a solution. It indicates that after 100 sprays, 29 ounces of cleaner will be left in the bottle.

In the same way, solutions are found for $n = 0$ and $n = 1{,}000$ and listed in the table. Then the ordered pairs are plotted and a straight line is drawn through the points.

To graphically estimate the amount of solution that is left after 500 sprays, we draw the dashed blue lines, as shown. Reading on the vertical A-axis, we see that after 500 sprays, about 17 ounces of glass cleaning solution would be left.

Success Tip

It is often helpful, especially with applications, to scale the axes differently. Since we selected large n-values such as 100 and 1,000, the horizontal n-axis was scaled in units of 100. Since the corresponding A-values range from 2 to 32, the vertical A-axis was scaled in units of 4.

Teaching Tip: Ask your students to explain why only quadrant I is needed to graph this application problem.

$A = -0.03n + 32$

n	A	(n, A)
0	32	(0, 32)
100	29	(100, 29)
1,000	2	(1,000, 2)

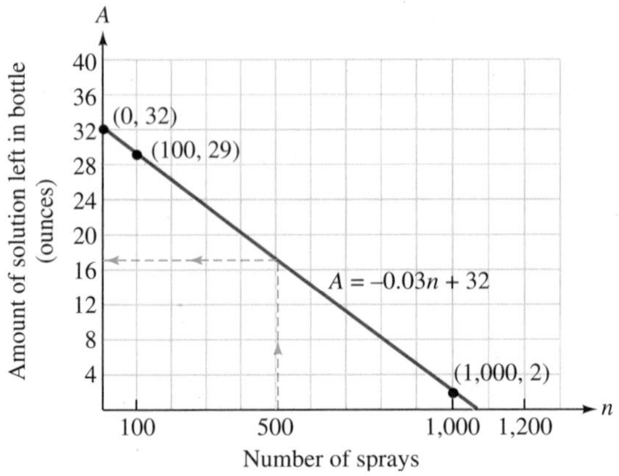

Teaching Example 7 Profit. The linear equation $p = 12.50c - 150$ estimates the weekly profit p (in dollars) that the owner of a nail salon makes if c customers are served. Graph the equation and use the graph to estimate the amount of profit if 80 customers are served in a week.
Answer: \$850 profit

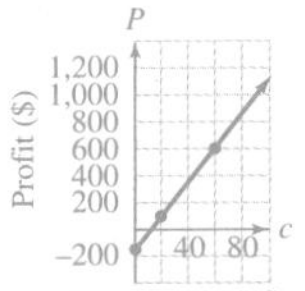

$P = 12.50c - 150$

Self Check 7 **Parties.** A laser tag business offers a party package that includes invitations, a party room, and 2 rounds of laser tag. The cost is \$15 plus \$10 per child. Write a linear equation that will give the cost for a party of any size, and then graph the equation. $c = 15 + 10n$; see AIE Appendix 3.

Now Try ▶ **Problems 87 and 89**

Teaching Tip: Each Study Set begins with Vocabulary exercises. Educational research has shown that vocabulary plays a crucial role in the process of concept formation in students. All Vocabulary problems can be found in Enhanced WebAssign so that you can require them to be submitted before class to promote student readiness.

SECTION 3.2 STUDY SET

VOCABULARY

Fill in the blanks.

▶ **1.** $y = 9x + 5$ is an equation in __two__ variables, x and y.

▶ **2.** A __solution__ of an equation in two variables is an ordered pair of numbers that makes the equation a true statement.

▶ **3.** Solutions of equations in two variables are often listed in a __table__ of solutions.

▶ **4.** The line that represents all of the solutions of a linear equation is called the __graph__ of the equation.

▶ **5.** $y = 3x + 8$ is a __linear__ equation and its graph is a line.

▶ **6.** The __standard__ form of a linear equation in two variables is $Ax + By = C$.

CONCEPTS

7. Consider: $y = -3x + 6$

a. How many variables does the equation contain? 2

b. Does $(4, -6)$ satisfy the equation? Yes

c. Is $(-2, 0)$ a solution of the equation? No

d. How many solutions does this equation have? Infinitely many

8. To graph a linear equation, three solutions were found, they were plotted (in black), and a straight line was drawn through them, as shown below.

▶ **a.** Looking at the graph, complete the table of solutions.

b. From the graph, determine three other solutions of the equation. $(-3, 3), (-2, 1), (0, -3)$

x	y	(x, y)
−4	5	(−4, 5)
−1	−1	(−1, −1)
1	−5	(1, −5)

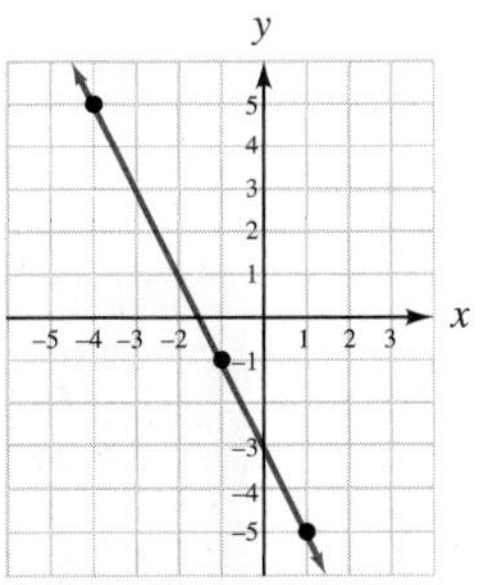

9. The graph of $y = -2x - 3$ is shown in Problem 8. Fill in the blanks: Every point on the graph represents an ordered-pair __solution__ of $y = -2x - 3$ and every ordered-pair solution is a __point__ on the graph.

▶ **10.** The graph of a linear equation is shown.

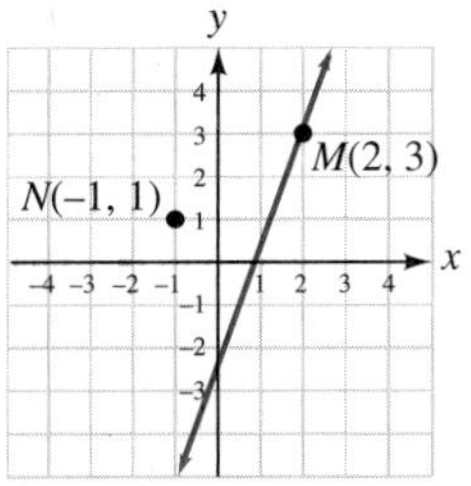

a. If the coordinates of point M are substituted into the equation, will the result be true or false? True

b. If the coordinates of point N are substituted into the equation, will the result be true or false? False

11. Suppose you are making a table of solutions for each given equation. What three x-values would you select to make the calculations for finding the corresponding y-values the easiest?

▶ **a.** $y = \frac{4}{5}x + 2$ −5, 0, 5 (Answers may vary)

b. $y = 0.6x + 500$ −10, 0, 10 (Answers may vary)

12. A table of solutions for a linear equation is shown below. When constructing the graph of the equation, how would you scale the x-axis and the y-axis? Multiples of 5; multiples of 100

x	y	(x, y)
−20	600	(−20, 600)
5	100	(5, 100)
35	−500	(35, −500)

NOTATION

Complete each solution.

13. Verify that $(-2, 6)$ is a solution of $y = -x + 4$.

$$y = -x + 4$$
$$6 \stackrel{?}{=} -(-2) + 4$$
$$6 \stackrel{?}{=} 2 + 4$$
$$6 = 6$$

14. Solve $5x + 3y = 15$ for y.

$$5x + 3y - 5x = -5x + 15$$
$$3y = -5x + 15$$
$$\frac{3y}{3} = \frac{-5x}{3} + \frac{15}{3}$$
$$y = -\frac{5}{3}x + 5$$

▶ **15. a.** In the linear equation $y = \frac{1}{2}x + 7$ what are the understood exponents on the variables? 1's

b. Explain why $y = x^2 + 2$ and $y = x^3 - 4$ are not linear equations. The exponent on x is not 1.

16. Complete the labeling of the table of solutions and the axes of the graph of $c = -a + 4$.

a	c	(a, c)
−1	5	(−1, 5)
0	4	(0, 4)
2	2	(2, 2)

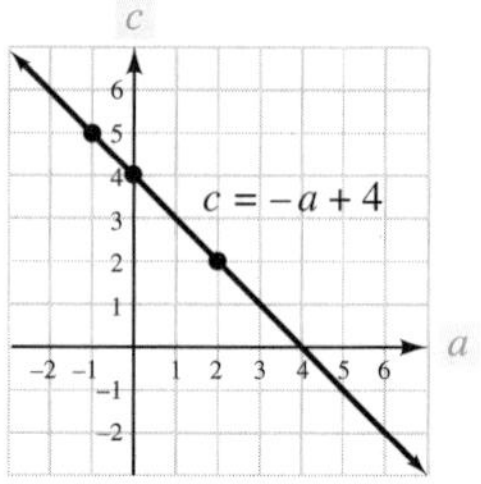

GUIDED PRACTICE

Determine whether each equation has the given ordered pair as a solution. **See Example 1.**

▶ **17.** $y = 5x - 4$; $(1, 1)$ Yes
▶ **18.** $y = -2x + 3$; $(2, -1)$ Yes
19. $7x - 2y = 3$; $(2, 6)$ No
▶ **20.** $10x - y = 10$; $(0, 0)$ No
21. $x + 12y = -12$; $(0, -1)$ Yes
▶ **22.** $-2x + 3y = 0$; $(-3, -2)$ Yes
23. $3x - 6y = 12$; $(-3.6, -3.8)$ Yes
▶ **24.** $8x + 4y = 10$; $(-0.5, 3.5)$ Yes
25. $y - 6x = 12$; $\left(\frac{5}{6}, 7\right)$ No
▶ **26.** $y + 8x = 4$; $\left(\frac{3}{4}, 2\right)$ No
▶ **27.** $y = -\frac{3}{4}x + 8$; $(-8, 12)$ No
28. $y = \frac{1}{6}x - 2$; $(-12, 4)$ No

For each equation, complete the solution. **See Example 2.**

▶ **29.** $y = -5x - 4$; $(-3, 11)$
▶ **30.** $y = 8x + 30$; $(-6, -18)$
▶ **31.** $4x - 5y = -4$; $(4, 4)$
▶ **32.** $7x + y = -12$; $(-2, 2)$
33. $y = \frac{x}{4} + 9$; $(16, 13)$
▶ **34.** $y = \frac{x}{6} - 8$; $(48, 0)$
35. $7x = 4y$; $\left(-\frac{8}{7}, -2\right)$
▶ **36.** $11x = 16y$; $\left(-\frac{48}{11}, -3\right)$

Complete each table of solutions. **See Example 3.**

▶ **37.** $y = 2x - 4$

x	y	(x, y)
8	12	(8, 12)
6	8	(6, 8)

▶ **38.** $y = 3x + 1$

x	y	(x, y)
−3	−8	(−3, −8)
−1	−2	(−1, −2)

▶ **39.** $3x - y = -2$

x	y	(x, y)
−5	−13	(−5, −13)
−1	−1	(−1, −1)

40. $5x - 2y = -15$

x	y	(x, y)
5	20	(5, 20)
−3	0	(−3, 0)

Construct a table of solutions and then graph each equation. **See Example 4.** See AIE Appendix 3.

▶ **41.** $y = 2x - 3$
▶ **42.** $y = 3x + 1$
43. $y = 5x - 4$
44. $y = 6x - 3$

Construct a table of solutions and then graph each equation. **See Example 5.** See AIE Appendix 3.

45. $y = -6x$
▶ **46.** $y = -2x$
47. $y = -7x$
48. $y = -8x$

Solve each equation for y and then graph it. **See Example 6.** See AIE Appendix 3.

49. $2x + 3y = -3$
▶ **50.** $2x + 3y = 9$
▶ **51.** $5y - x = 20$
▶ **52.** $4y - x = 8$

TRY IT YOURSELF

Graph each equation. Solve for y first, when necessary. See AIE Appendix 3.

▶ **53.** $y = x$
▶ **54.** $y = 4x$
▶ **55.** $y = -x - 1$
56. $y = -x + 2$
▶ **57.** $3y = 12x + 15$
58. $5y = 20x - 30$
59. $y = \frac{3}{8}x - 6$
▶ **60.** $y = -\frac{3}{2}x + 2$
61. $y = 1.5x - 4$
▶ **62.** $y = 0.5x + 3$
63. $8x + 4y = 16$
64. $14x + 7y = 28$
▶ **65.** $y = -\frac{1}{2}x$
66. $y = \frac{3}{4}x$
67. $y = \frac{5}{6}x - 5$
▶ **68.** $y = \frac{2}{3}x - 2$
69. $-6y = 30x + 12$
70. $-3y = 9x - 15$
71. $y = \frac{x}{3}$
▶ **72.** $y = -\frac{x}{3} - 1$
▶ **73.** $y = -2x + 1$
▶ **74.** $y = -3x + 2$
75. $7x - y = 1$
▶ **76.** $2x - y = -3$
77. $7y = -2x$
▶ **78.** $6y = -4x$
79. $y = -2.5x + 5$
80. $y = -3.5x + 4$

APPLICATIONS

▶ **81. Billiards.** The path traveled by the black 8-ball is described by the equations $y = 2x - 4$ and $y = -2x + 12$. Construct a table of solutions for $y = 2x - 4$ using the x-values 1, 2, and 4. Do the same for $y = -2x + 12$, using the x-values 4, 6, and 8. Then graph the path of the 8-ball. See AIE Appendix 3.

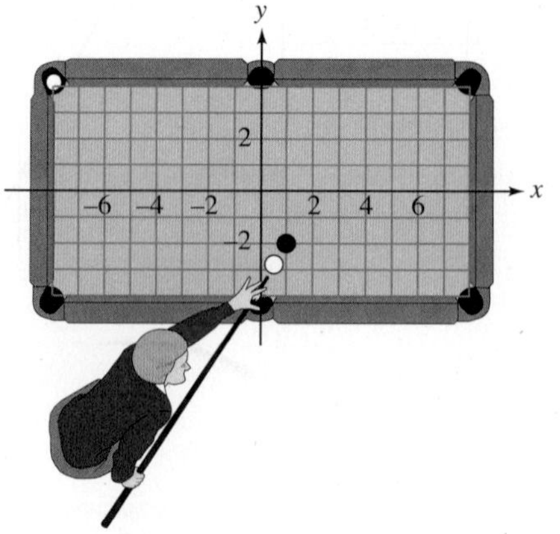

82. Ping-Pong. The path traveled by a Ping-Pong ball is described by the equations $y = \frac{1}{2}x + \frac{3}{2}$ and $y = -\frac{1}{2}x - \frac{3}{2}$. Construct a table of solutions for $y = \frac{1}{2}x + \frac{3}{2}$ using the x-values 7, 3, and -3. Do the same for $y = -\frac{1}{2}x - \frac{3}{2}$, using the x-values -3, -5, and -7. Then graph the path of the ball.

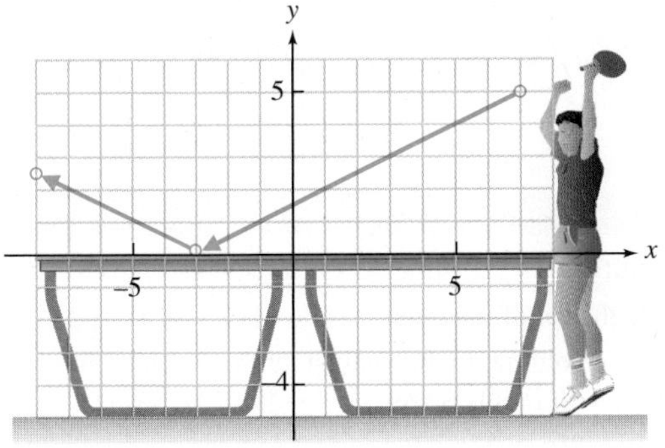

83. Defrosting Poultry. The number of hours h needed to defrost a turkey weighing p pounds in the refrigerator can be estimated by $h = 5p$. Graph the equation and use the graph to estimate the time needed to defrost a 25-pound turkey. (Source: helpwithcooking.com.) About 125 hr; see AIE Appendix 3.

84. Owning a Car. In 2010, the average cost c (in dollars) to own and operate a car was estimated by $c = 0.57m$, where m represents the number of miles driven. Graph the equation and use the graph to estimate the cost in 2010 of operating a car that is driven 25,000 miles. (Source: AAA Auto Club) About \$14,500; see AIE Appendix 3.

85. Housekeeping. The linear equation $A = -0.02n + 16$ estimates the amount A of furniture polish (in ounces) that is left in the bottle after the sprayer trigger has been pulled a total of n times. Graph the equation and use the graph to estimate the amount of polish that is left after 650 sprays. About 3 oz; see AIE Appendix 3.

86. Sharpening Pencils. The linear equation $L = -0.04t + 8$ estimates the length L (in inches) of a pencil after it has been inserted into a sharpener and the handle turned a total of t times. Graph the equation and use the graph to estimate the length of the pencil after 75 turns of the handle. About 5 in.; see AIE Appendix 3.

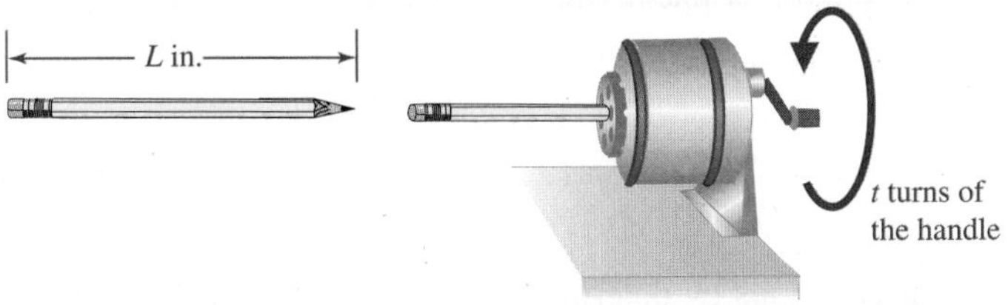

87. NFL Tickets. The average ticket price p to a National Football League game during the years 1990–2009 is approximated by $p = 2.7t + 20$, where t is the number of years after 1990. Graph this equation and use the graph to predict the average ticket price in 2020. (Source: Team Marketing Report, NFL.) About \$100; see AIE Appendix 3.

88. U.S. Automobile Accidents. The number n of lives saved by seat belts during the years 2000–2009 is estimated by $n = 170t + 13{,}800$, where t is the number of years after 2000. Graph this equation and use the graph to predict the number of lives that will be saved by seat belts in 2015. (Source: NHTSA National Center for Statistics.) About 16,500 lives; see AIE Appendix 3.

89. Raffles. A private school is going to sell raffle tickets as a fund raiser. Suppose the number n of raffle tickets that will be sold is predicted by the equation $n = -20p + 300$, where p is the price of a raffle ticket in dollars. Graph the equation and use the graph to predict the number of raffle tickets that will be sold at a price of \$6. About 180 tickets; see AIE Appendix 3.

90. Endangered Species. The number n of endangered plant and animal species in the U.S. during the years 2000–2010 is estimated by $n = 9t + 960$, where t is the number of years after 2000. Graph this equation and use the graph to predict the number of endangered species in the U.S. in 2022. (Source: *The World Almanac and Book of Facts, 2010*). About 1,150 species; see AIE Appendix 3.

91. U.S. Space Program. Since 1980, the Gallup Poll organization has surveyed Americans to see whether they think the space program has brought enough benefits to the country to justify its cost. The percent p responding "yes" is estimated by $p = \frac{3}{5}t + 40$, where t is the number of years after 1980. Graph the equation. If the polling trend continues, when will the percent that respond "yes" reach 70%? (Source: galluppoll.com) 2030; see AIE Appendix 3.

92. Gas Mileage. The mileage for a Honda Insight traveling between 55 mph and 75 mph is estimated by the equation $m = -\frac{3}{4}s + 95$, where s is the speed of the car (in mph) and m is the mileage (in miles per gallon). Graph the equation for s between 55 and 75. Estimate the speed at which the mileage of the car drops below 40 miles per gallon. (Source: *Consumer Reports* 9/10/2009) About 73 mph

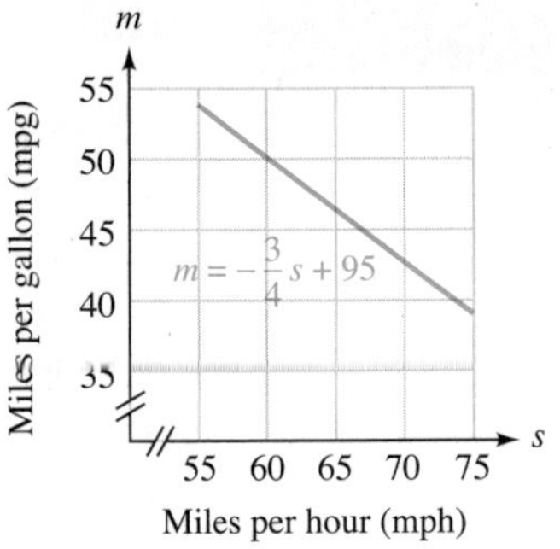

The symbol ⫲ is used to show the break in the scale on an axis. Such a break enables us to omit large portions of empty space on a graph.

WRITING

93. When we say that $(-2, -6)$ is a solution of $y = x - 4$, what do we mean?

94. What is a table of solutions?

95. What does it mean when we say that a linear equation in two variables has infinitely many solutions?

▶ **96.** A linear equation and a graph are two ways of describing a relationship between two quantities. Which do you think is more informative and why?

97. From geometry, we know that two points determine a line. Why is it a good practice when graphing linear equations to find and plot three solutions instead of just two?

98. A student found three solutions of a linear equation in two variables and plotted them as shown. What conclusion can be made about the location of the points?

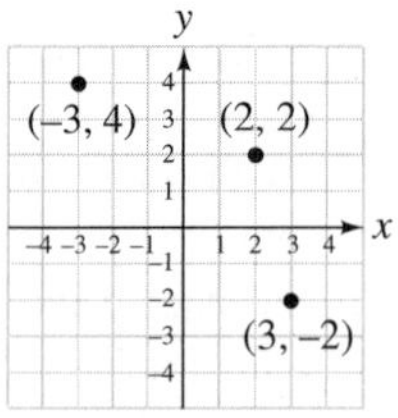

99. To graph $y = 3x - 1$, one student made the table of solutions on the left. Another student made the table on the right. The tables are different. Could they both be correct? Explain.

x	y	(x, y)
0	−1	(0, −1)
2	5	(2, 5)
3	8	(3, 8)

x	y	(x, y)
−2	−7	(−2, −7)
−1	−4	(−1, −4)
1	2	(1, 2)

100. Both graphs below are of the same linear equation $y = 10x$. Why do the graphs have a different appearance?

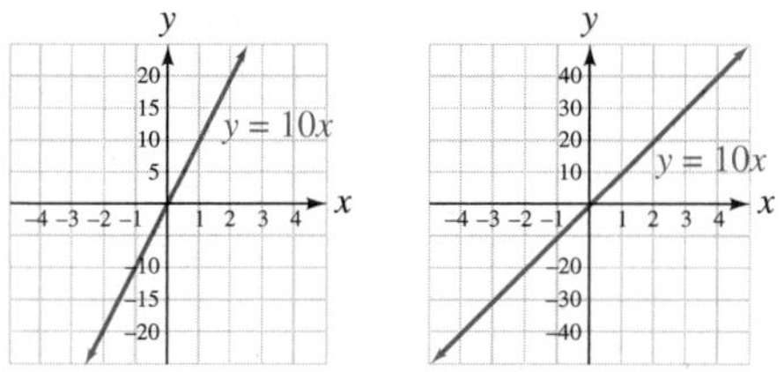

REVIEW

101. Simplify: $-(-5 - 4c)$ $5 + 4c$ or $4c + 5$

102. Write the set of integers. $\{. . . -2, -1, 0, 1, 2, . . .\}$

103. Find the volume, to the nearest tenth of a cubic foot, of a sphere with radius 6 feet. 904.8 ft^3

▶ **104.** Solve: $-2(a + 3) = 3(a - 5)$ $\frac{9}{5}$

CHALLENGE PROBLEMS

Graph each of the following* nonlinear *equations in two variables by constructing a table of solutions consisting of seven ordered pairs. These equations are called nonlinear, because their graphs are not straight lines. See AIE Appendix 3.

105. $y = x^2 + 1$

106. $y = x^3 - 2$

▶ **107.** $y = |x| - 2$

108. $y = (x + 2)^2$

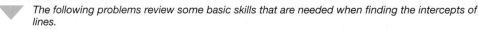

SECTION 3.3 Intercepts

OBJECTIVES

1. Identify intercepts of a graph.
2. Graph linear equations by finding intercepts.
3. Identify and graph horizontal and vertical lines.
4. Obtain information from intercepts.

ARE YOU READY? *Are You Ready? exercises available online at www.webassign.net/brookscole*

The following problems review some basic skills that are needed when finding the intercepts of lines.

1. Graph the points (2, 0), (−4, 0), (0, 1) and (0, −3). See AIE Appendix 3.

2. What is the x-coordinate of any point that lies on the y-axis? 0

3. What point lies on both the x-axis and the y-axis? The origin (0, 0)

4. Solve: $3(0) + 2y = 10$ 5

In this section, we will graph linear equations by determining the points where their graphs intersect the x-axis and the y-axis. These points are called the *intercepts* of the graph.

The Language of Algebra

Note the difference in spelling. The point where a line **intersects** the x- or y-axis is called an **intercept.**

1 Identify Intercepts of a Graph.

The graph of $y = 2x - 4$ is shown on the next page. We see that the graph intersects (crosses) the y-axis at the point $(0, -4)$; this point is called the **y-intercept** of the graph. The graph intersects (crosses) the x-axis at the point $(2, 0)$; this point is called the **x-intercept** of the graph.

Teaching Tip: Point out to your students that later in this course (and in more advanced courses) they will see graphs of equations in two variables that are not straight lines. Show them the intercepts of a parabola and a circle as examples.

Recall that the x-coordinate of any point on the y-axis is 0.

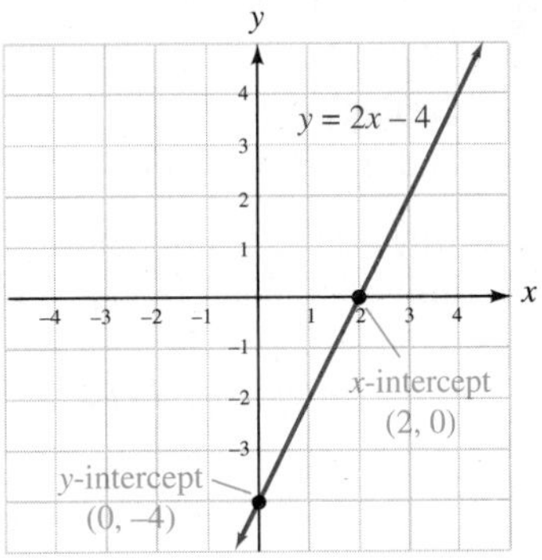

Recall that the y-coordinate of any point on the x-axis is 0.

EXAMPLE 1 For the graphs in figures (a) and (b), identify the x- and y-intercepts.

a.

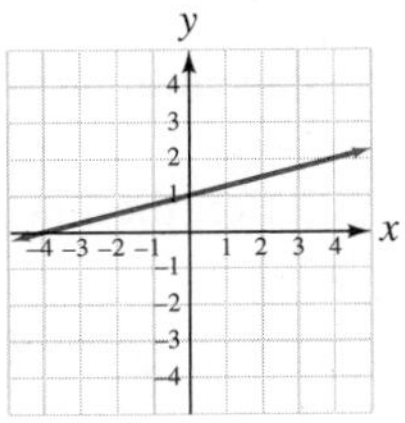

b.

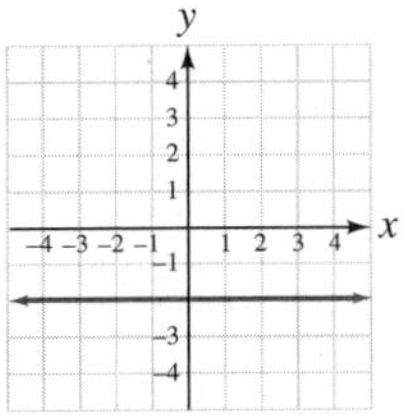

c.

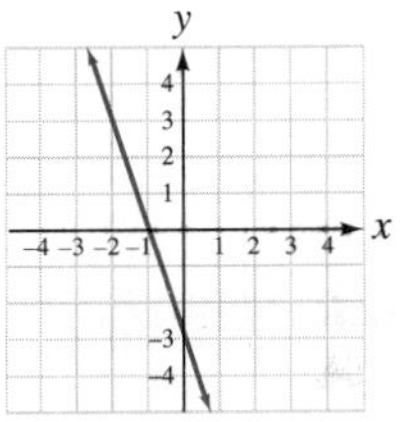

The Language of Algebra

Intercepts are ordered pairs. However, other language is sometimes used to describe them. When referring to an x-intercept of $(-4, 0)$, for example, some textbooks use a single number and say the x-intercept is at -4 on the x-axis.

Strategy We will determine where each graph (shown in red) crosses the x-axis and the y-axis.

Why The point at which a graph crosses the x-axis is the x-intercept and the point at which a graph crosses the y-axis is the y-intercept.

Solution

a. In figure (a), the graph crosses the x-axis at $(-4, 0)$. This is the x-intercept. The graph crosses the y-axis at $(0, 1)$. This is the y-intercept.

b. In figure (b), the horizontal line does not cross the x-axis; there is no x-intercept. The graph crosses the y-axis at $(0, -2)$. This is the y-intercept.

Teaching Example 1
Identify the x- and y-intercepts of the graphs of line 1 and line 2.

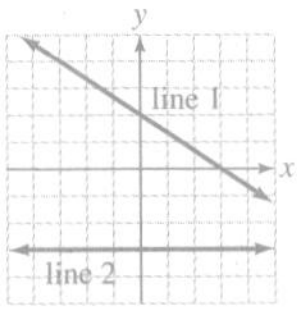

Answers: Line 1: $(3, 0)$, $(0, 2)$; line 2: no x-intercept; $(0, -3)$

Self Check 1 Identify the x- and y-intercepts of the graph in figure (c).
$(-1, 0)$; $(0, -3)$

Now Try Problem 11

From the previous examples, we see that a y-intercept has an x-coordinate of 0, and an x-intercept has a y-coordinate of 0. These observations suggest the following procedures for finding the intercepts of a graph from its equation.

Finding Intercepts	To find the y-intercept, substitute 0 for x in the given equation and solve for y. To find the x-intercept, substitute 0 for y in the given equation and solve for x.

2 Graph Linear Equations by Finding Intercepts.

Plotting the x- and y-intercepts of a graph and drawing a line through them is called the **intercept method of graphing a line.** This method is useful when graphing linear equations written in the standard form $Ax + By = C$.

EXAMPLE 2 Graph $x - 3y = 6$ by finding the y- and x-intercepts.

Teaching Tip: Point out to your students that the textbook no longer shows all of the steps involving the use of a property of equality when solving an equation. For example, the division on both sides by -3 is not shown here.

Strategy We will let $x = 0$ to find the y-intercept of the graph. We will then let $y = 0$ to find the x-intercept.

Why Since two points determine a line, the y-intercept and x-intercept are enough information to graph this linear equation.

Solution

***y*-intercept: Let $x = 0$**

$x - 3y = 6$

$0 - 3y = 6$ Substitute 0 for x.

$-3y = 6$ Subtract.

$y = -2$ Divide both sides by −3.

The y-intercept is $(0, -2)$.

***x*-intercept: Let $y = 0$**

$x - 3y = 6$

$x - 3(0) = 6$ Substitute 0 for y.

$x - 0 = 6$ Multiply.

$x = 6$ Subtract.

The x-intercept is $(6, 0)$.

Since each intercept of the graph is a solution of the equation, we enter the intercepts in the table of solutions below.

As a check, we find one more point on the line. We select a convenient value for x, say, 3, and find the corresponding value of y.

Success Tip

The check point should lie on the same line as the x- and y-intercepts. If it does not, check your work to find the incorrect coordinate or coordinates.

$x - 3y = 6$ This is the equation to graph.

$3 - 3y = 6$ Substitute 3 for x.

$-3y = 3$ To isolate the variable term −3y, subtract 3 from both sides.

$y = -1$ To isolate y, divide both sides by −3.

Therefore, $(3, -1)$ is a solution. It is also entered in the table.

We plot the intercepts and the check point, draw a straight line through them, and label the line as $x - 3y = 6$.

The Language of Algebra

Points that lie on the same line are said to be **collinear.**

$x - 3y = 6$

x	y	(x, y)	
0	−2	(0, −2)	← y-intercept
6	0	(6, 0)	← x-intercept
3	−1	(3, −1)	← Check point

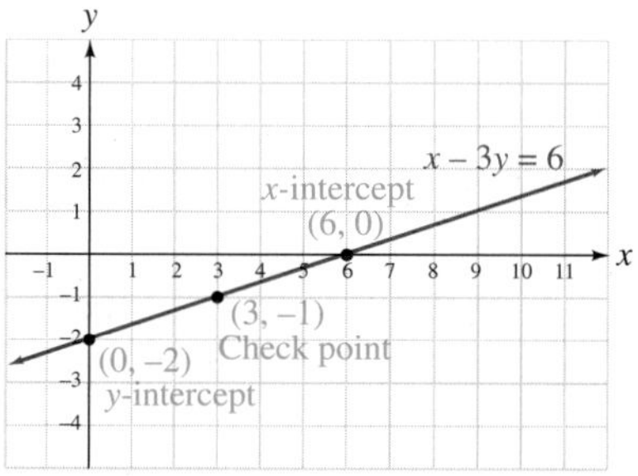

Self Check 2 Graph $x - 2y = 2$ by finding the intercepts. See AIE Appendix 3.

Now Try ▶ Problem 27

Teaching Example 2
Graph $2x - 3y = 6$ by finding the intercepts.
Answer:

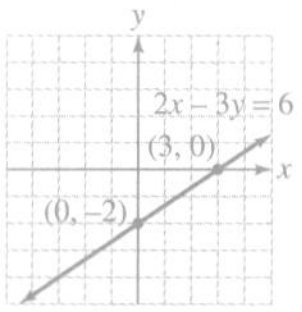

The calculations for finding intercepts can be simplified if we realize what occurs when we substitute 0 for y or 0 for x in an equation written in the form $Ax + By = C$.

EXAMPLE 3 Graph $40x + 3y = -120$ by finding the y- and x-intercepts.

Strategy We will let $x = 0$ to find the y-intercept of the graph. We will then let $y = 0$ to find the x-intercept.

Why Since two points determine a line, the y-intercept and x-intercept are enough information to graph this linear equation.

Solution When we substitute 0 for x, it follows that the term $40x$ will be equal to 0. Therefore, to find the y-intercept, we can cover over the $40x$ and solve the remaining equation for y.

$40x + 3y = -120$ If $x = 0$, then $40x = 40(0) = 0$. Cover over the $40x$ term.

$y = -40$ To solve $3y = -120$, divide both sides by 3.

The y-intercept is $(0, -40)$. This is entered in the table below.

When we substitute 0 for y, it follows that the term $3y$ will be equal to 0. Therefore, to find the x-intercept, we can cover over the $3y$ and solve the remaining equation for x.

$40x + 3y = -120$ If $y = 0$, then $3y = 3(0) = 0$. Cover over the $3y$ term.

$x = -3$ To solve $40x = -120$, divide both sides by 40.

The x-intercept is $(-3, 0)$. This is entered in the table below.

The Language of Algebra

This method to find the intercepts of the graph of a linear equation is commonly referred to as the **cover-over method.**

Caution

When using the cover-over method to find the y-intercept, be careful not to cover the sign in front of the y-term.

We can find a third solution by selecting a convenient value for x and finding the corresponding value for y. If we choose $x = -6$, we find that $y = 40$. The solution $(-6, 40)$ is entered in the table, and the equation is graphed as shown.

$40x + 3y = -120$

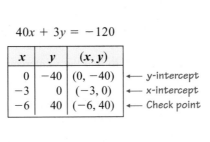

x	y	(x, y)	
0	-40	$(0, -40)$	← y-intercept
-3	0	$(-3, 0)$	← x-intercept
-6	40	$(-6, 40)$	← Check point

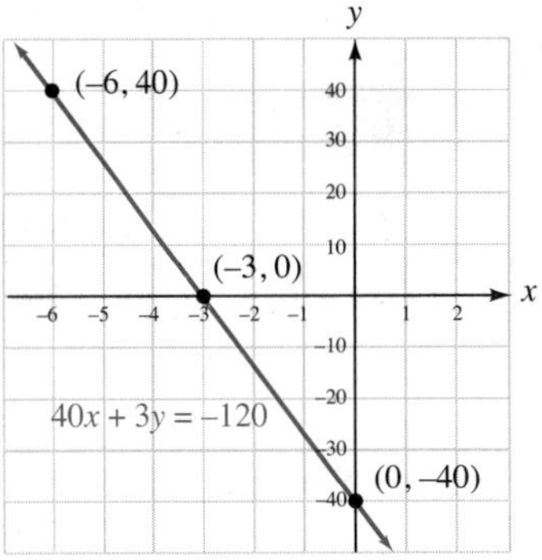

Success Tip

To fit y-values of 40 and -40 on the graph, the y-axis was scaled in units of 10.

If a different scale is used on the y-axis, the same equation $(40x + 30y = -120)$ will produce a graph with a somewhat different look.

Teaching Example 3
Graph $25x + 4y = -100$ by finding the intercepts.
Answer:

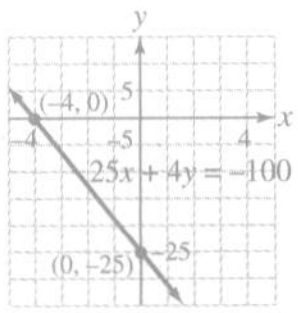

Self Check 3 Graph $32x + 5y = -160$ by finding the intercepts. See AIE Appendix 3.

Now Try Problem 35

EXAMPLE 4 Graph $3x = -5y + 8$ by finding the intercepts.

Strategy We will let $x = 0$ to find the y-intercept of the graph. We will then let $y = 0$ to find the x-intercept.

Why Since two points determine a line, the y-intercept and x-intercept are enough information to graph this linear equation.

Solution We find the intercepts and select $x = 1$ to find a check point.

***y*-intercept: Let $x = 0$**

$$3x = -5y + 8$$
$$3(0) = -5y + 8$$
$$0 = -5y + 8$$
$$-8 = -5y$$
$$\frac{8}{5} = y$$
$$1\tfrac{3}{5} = y$$

The y-intercept is $\left(0, 1\tfrac{3}{5}\right)$.

***x*-intercept: Let $y = 0$**

$$3x = -5y + 8$$
$$3x = -5(0) + 8$$
$$3x = 8$$
$$x = \frac{8}{3}$$
$$x = 2\tfrac{2}{3}$$

The x-intercept is $\left(2\tfrac{2}{3}, 0\right)$.

***Check point*: Let $x = 1$**

$$3x = -5y + 8$$
$$3(1) = -5y + 8$$
$$3 = -5y + 8$$
$$-5 = -5y$$
$$1 = y$$

A check point is $(1, 1)$.

Success Tip

When graphing, it is often helpful to write any coordinates that are improper fractions as terminating decimals or mixed numbers. For example:

$$\left(0, \frac{8}{5}\right) = \left(0, 1\frac{3}{5}\right) = (0, 1.6)$$

and

$$\left(\frac{8}{3}, 0\right) = \left(2\frac{2}{3}, 0\right)$$

The ordered pairs are plotted as shown, and a straight line is then drawn through them.

$3x = -5y + 8$

x	y	(x, y)	
0	$\frac{8}{5} = 1\frac{3}{5}$	$\left(0, 1\frac{3}{5}\right)$	← y-intercept
$\frac{8}{3} = 2\frac{2}{3}$	0	$\left(2\frac{2}{3}, 0\right)$	← x-intercept
1	1	(1, 1)	← Check point

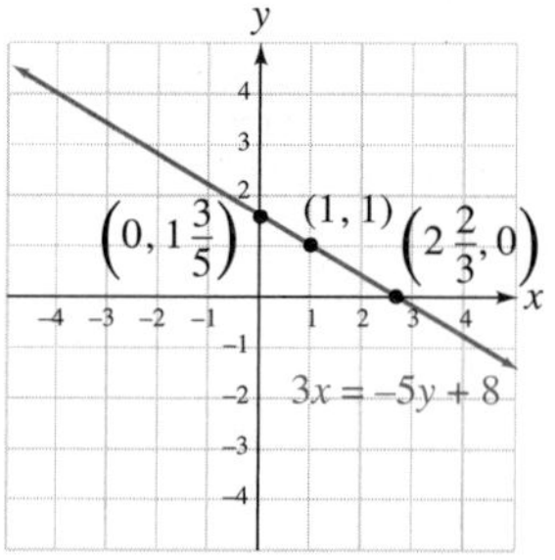

Teaching Example 4
Graph $6x = -8y + 13$ by finding the intercepts.
Answer:

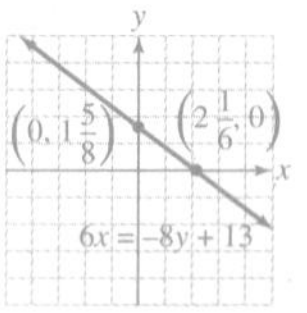

Self Check 4 Graph $8x = -4y + 15$ by finding the intercepts. See AIE Appendix 3.

Now Try ▶ Problem 39

EXAMPLE 5 Graph $2x + 3y = 0$ by finding the intercepts.

Strategy We will let $x = 0$ to find the y-intercept of the graph. We will then let $y = 0$ to find the x-intercept.

Why Since two points determine a line, the y-intercept and x-intercept are enough information to graph this linear equation.

Solution When we find the y- and x-intercepts (shown below), we see that they are both (0, 0). In this case, the line passes through the origin. Since we are using two points and a check point to graph lines, we need to find two more ordered-pair solutions.

If $x = 3$, we see that $(3, -2)$ is a solution. And if $x = -3$, we see that $(-3, 2)$ is also a solution. These two solutions and the origin are plotted and a straight line is drawn through them to give the graph of $2x + 3y = 0$.

***y*-intercept: Let $x = 0$**

$$2x + 3y = 0$$
$$2(\mathbf{0}) + 3y = 0$$
$$3y = 0$$
$$y = 0$$

The y-intercept is (0, 0).

***x*-intercept: Let $y = 0$**

$$2x + 3y = 0$$
$$2x + 3(\mathbf{0}) = 0$$
$$2x = 0$$
$$x = 0$$

The x-intercept is (0, 0).

The intercepts are the same.

***Let* $x = 3$**

$$2x + 3y = 0$$
$$2(\mathbf{3}) + 3y = 0$$
$$6 + 3y = 0$$
$$3y = -6$$
$$y = -2$$

$(3, -2)$ is a solution.

***Let* $x = -3$**

$$2x + 3y = 0$$
$$2(\mathbf{-3}) + 3y = 0$$
$$-6 + 3y = 0$$
$$3y = 6$$
$$y = 2$$

$(-3, 2)$ is a solution.

Teaching Tip: Point out that equations that can be written in the form $Ax + By = 0$, such as $2x + 3y = 0$, have graphs that pass through the origin. They have the same x- and y-intercept: namely, (0, 0).

$2x + 3y = 0$

x	y	(x, y)	
0	0	(0, 0)	← The x-intercept and y-intercept.
3	−2	(3, −2)	← A solution.
−3	2	(−3, 2)	← This solution serves as a check point.

Equations that can be written in the form $Ax + By = 0$, such as $2x + 3y = 0$, have graphs that pass through the origin. To find another point on such a line that has integer coordinates, select an x-value equal to the coefficient of y or the opposite of the coefficient of y. Then substitute that x-value into the equation and solve for y.

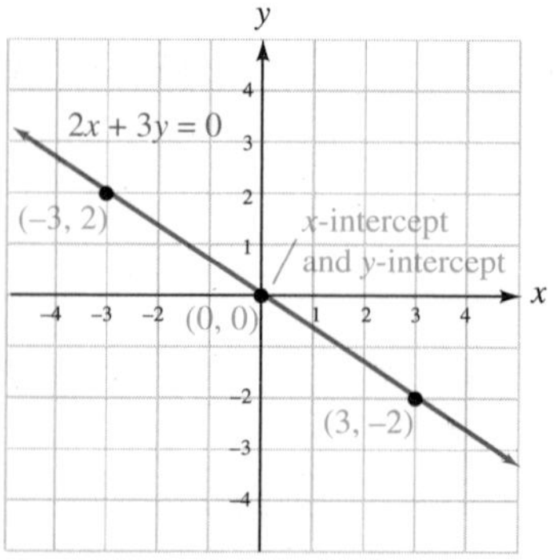

Teaching Example 5
Graph $3x - 2y = 0$ by finding the intercepts.
Answer:

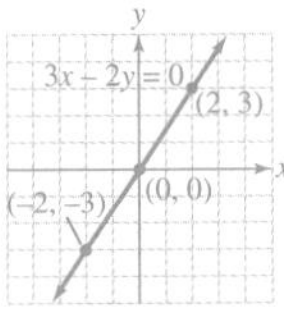

Self Check 5 Graph $5x - 2y = 0$ by finding the intercepts. See AIE Appendix 3.

Now Try ▶ Problem 47

3 Identify and Graph Horizontal and Vertical Lines.

Equations such as $y = 4$ and $x = -3$ are linear equations in two variables, because they can be written in the standard form $Ax + By = C$. For example, $y = 4$ is equivalent to $0x + 1y = 4$ and $x = -3$ is equivalent to $1x + 0y = -3$. We can graph these types of equations using point-plotting.

EXAMPLE 6 Graph: $y = 4$

Strategy To find three ordered-pair solutions of this equation to plot, we will select three values for x and use 4 for y each time.

Why The equation requires that $y = 4$.

Solution We can write the equation in standard form as $0x + y = 4$. Since the coefficient of x is 0, the numbers selected for x have no effect on y. The value of y is always 4. For example, if $x = 2$, we have

$$0x + y = 4 \quad \text{This is the given equation, } y = 4\text{, written in standard form: } Ax + By = C.$$

$$0(2) + y = 4 \quad \text{Substitute 2 for x.}$$

$$y = 4 \quad \text{Simplify the left side.}$$

One solution is (2, 4). To find two more solutions, we select $x = 0$ and $x = -3$. For any x-value, the y-value is always 4, so we enter (0, 4) and (−3, 4) in the table. If we plot the ordered pairs and draw a straight line through the points, the result is a horizontal line. The y-intercept is (0, 4) and there is no x-intercept.

Success Tip

An equation with only the variable y is a horizontal line that intersects the y-axis.

$y = 4$

x	y	(x, y)
2	4	(2, 4)
0	4	(0, 4)
−3	4	(−3, 4)

↑ Select any number for x. ↑ Each value of y must be 4.

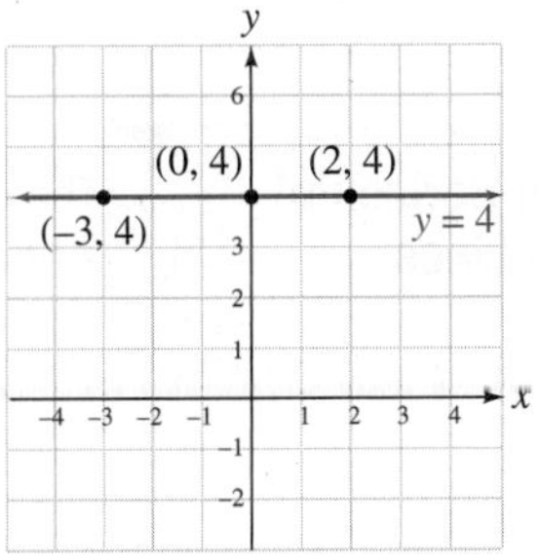

Teaching Example 6
Graph: $y = -3$
Answer:

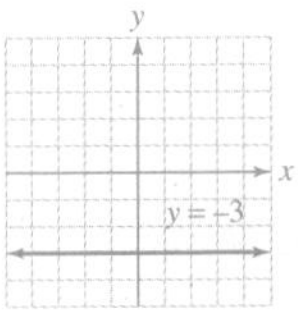

Self Check 6 Graph: $y = -2$ See AIE Appendix 3.

Now Try ▶ Problem 51

EXAMPLE 7 Graph: $x = -3$

Strategy To find three ordered-pair solutions of this equation to plot, we must select −3 for x each time.

Why The equation requires that $x = -3$.

Success Tip

If we graph $x = -3$ in one dimension (as in Chapter 2), the graph is a single point on a number line.

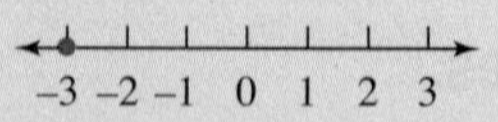

If we graph $x = -3$ in two dimensions (as in Chapter 3), the result is the line shown in Example 7 at the right.

Success Tip

An equation with only the variable x is a vertical line that intersects the x-axis.

Solution We can write the equation in standard form as $x + 0y = -3$. Since the coefficient of y is 0, the value of y has no effect on x. The value of x is always -3. For example, if $y = -2$, we have

$$x + 0y = -3 \quad \text{This is the given equation, } x = -3\text{, written in standard form: } Ax + By = C.$$
$$x + 0(-2) = -3 \quad \text{Substitute } -2 \text{ for } y.$$
$$x = -3 \quad \text{Simplify the left side.}$$

One solution is $(-3, -2)$. To find more solutions, we must again select -3 for x. Any number can be used for y. If $y = 0$, then a second solution is $(-3, 0)$. If $y = 3$, a third solution is $(-3, 3)$. The three solutions are entered in the table below. When we plot the ordered pairs and draw a straight line through the points, the result is a vertical line. The x-intercept is $(-3, 0)$ and there is no y-intercept.

$x = -3$

x	y	(x, y)
-3	-2	$(-3, -2)$
-3	0	$(-3, 0)$
-3	3	$(-3, 3)$

↑ Each value of x we select must be -3.
↑ Any number can be used for y.

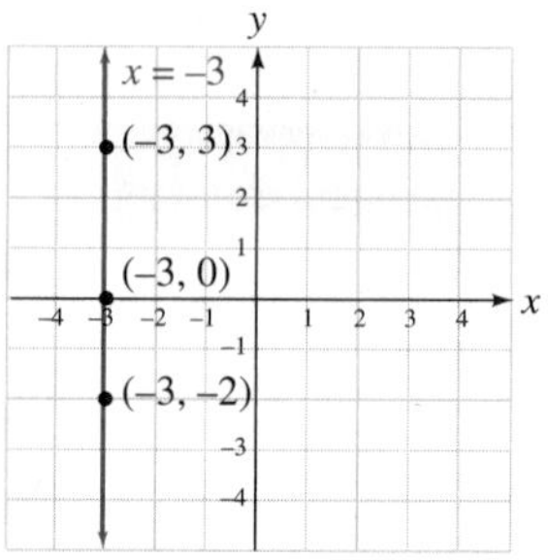

Teaching Example 7
Graph: $x = 2$
Answer:

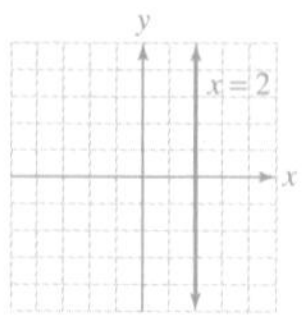

Self Check 7 Graph: $x = 4$ See AIE Appendix 3.

Now Try Problem 55

From the results of Examples 6 and 7, we have the following facts.

Equations of Horizontal and Vertical Lines

The graph of $y = b$ is a horizontal line with y-intercept $(0, b)$.

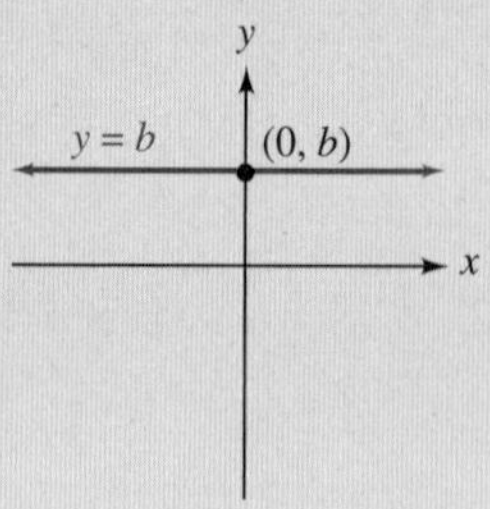

The graph of $x = a$ is a vertical line with x-intercept $(a, 0)$.

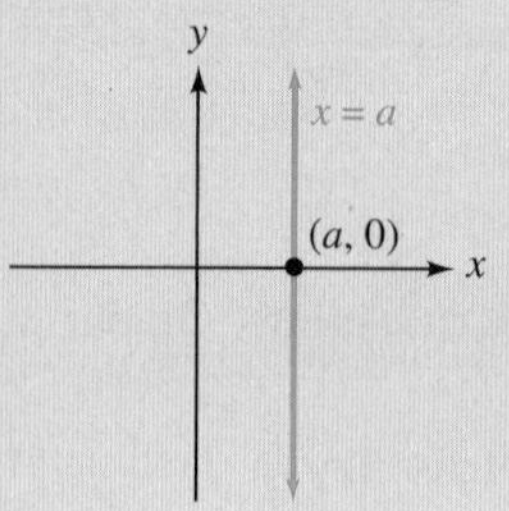

The graph of $y = 0$ is the x-axis. The graph of $x = 0$ is the y-axis.

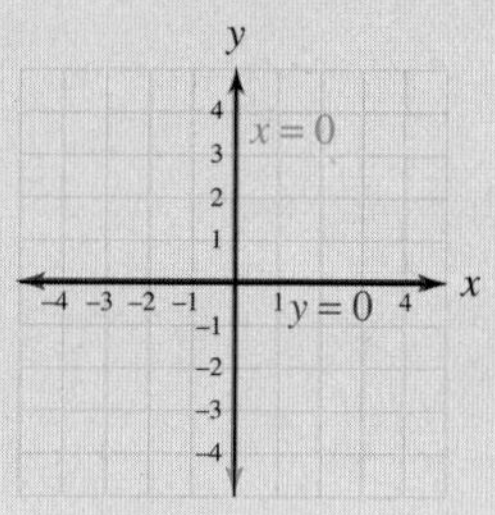

4 Obtain Information from Intercepts.

The ability to read and interpret graphs is a valuable skill. When analyzing a graph, we should locate and examine the intercepts. As the following example illustrates, the coordinates of the intercepts can give useful information.

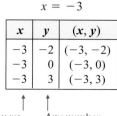

EXAMPLE 8 **Hybrid Mileage.** Figure (a) shows mileage data for a 2010 Toyota Prius Hybrid. What information do the intercepts give about the car?

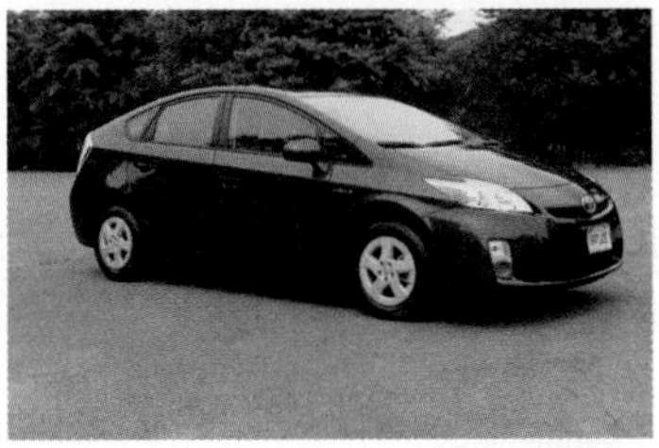

©Michael Doolittle/Alamy

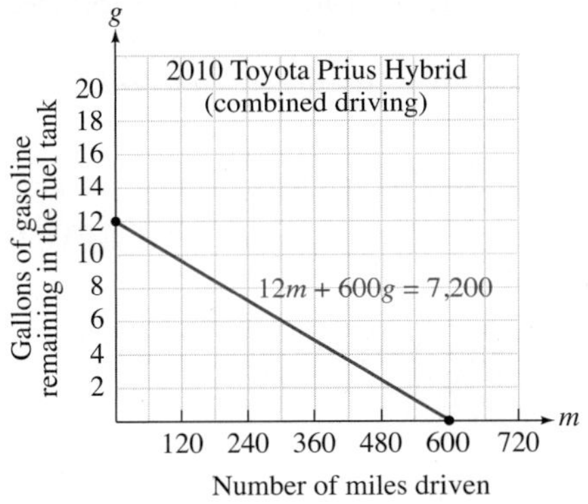

(a)

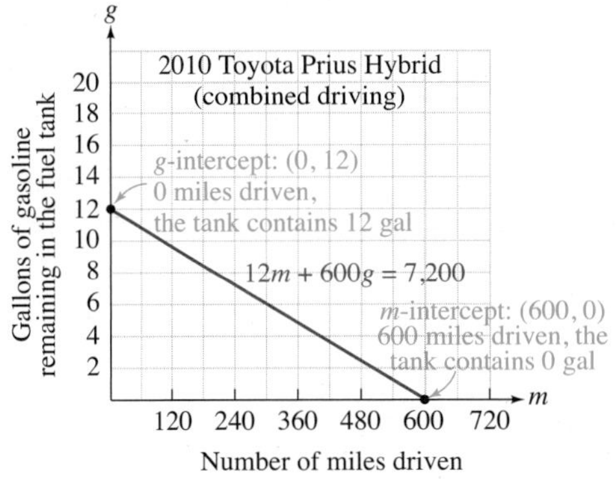

(b)

Teaching Example 8 Depreciation. The graph shows how the value of a piece of machinery decreased over its lifetime. What information do the intercepts give about the machinery?

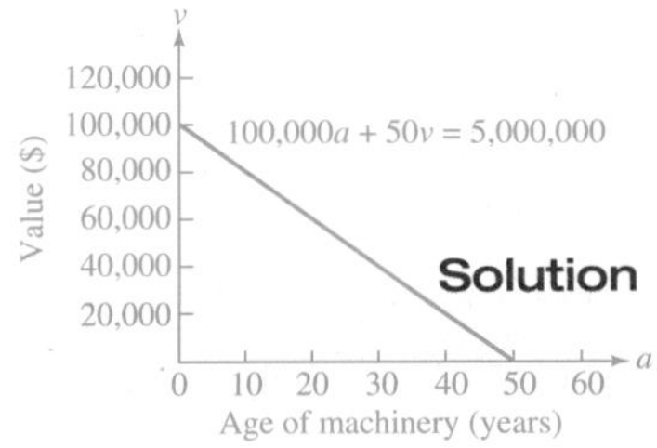

Answer: The v-intercept is (0, 100,000): When new, the machinery had a value of \$100,000. The a-intercept is (50, 0): After 50 years, the machinery had no value.

Strategy We will determine where the graph (the line in red) intersects the g-axis and where it intersects the m-axis.

Why Once we know the intercepts, we can interpret their meaning.

Solution

See figure (b). The g-intercept (0, 12) indicates that when the car has been driven 0 miles, the fuel tank contains 12 gallons of gasoline. That is, the Prius has a 12-gallon fuel tank.

The m-intercept (600, 0) indicates that after 600 miles of combined driving, the fuel tank contains 0 gallons of gasoline. Thus, 600 miles of combined driving can be done on 1 tank of gas in a Prius.

Now Try ▶ Problems 79 and 81

Using Your Calculator ▶ Use a Calculator to Graph Linear Equations. (Optional)

Courtesy of Texas Instruments Incorporated

So far, we have graphed linear equations by making tables of solutions and plotting points. A graphing calculator can make the task of graphing much easier. However, a graphing calculator does not take the place of a working knowledge of the topics discussed in this chapter. It should serve as an aid to enhance your study of algebra.

The Viewing Window The screen on which a graph is displayed is called the **viewing window.** The **standard window** has settings of

$$\text{Xmin} = -10, \quad \text{Xmax} = 10, \quad \text{Ymin} = -10, \quad \text{and} \quad \text{Ymax} = 10$$

which indicate that the minimum x- and y-coordinates used in the graph will be -10, and that the maximum x- and y-coordinates will be 10.

Graphing an Equation To graph $y = x - 1$ using a graphing calculator, we press the **Y =** key and enter $x - 1$ after the symbol Y_1. Then we press the **GRAPH** key to see the graph.

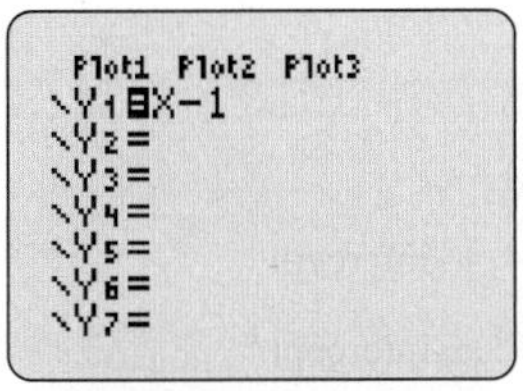

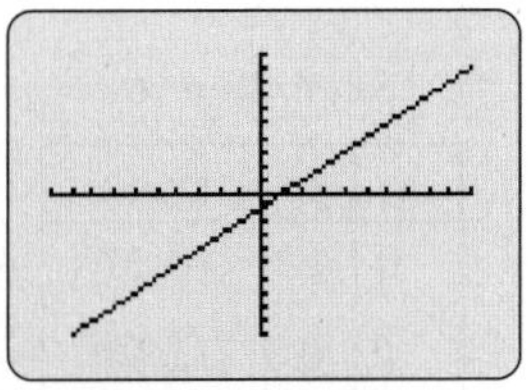

Change the Viewing Window We can change the viewing window by pressing the **WINDOW** key and entering -4 for the minimum x- and y-coordinates and 4 for the maximum x- and y-coordinates. Then we press the **GRAPH** key to see the graph of $y = x - 1$ in more detail.

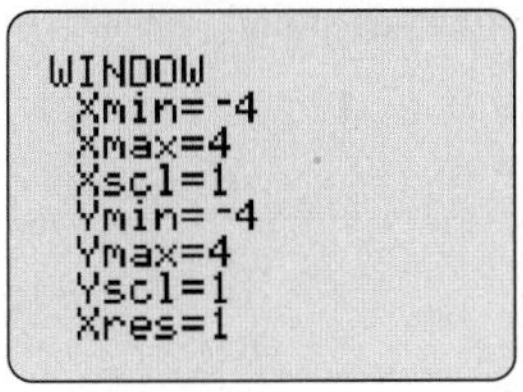

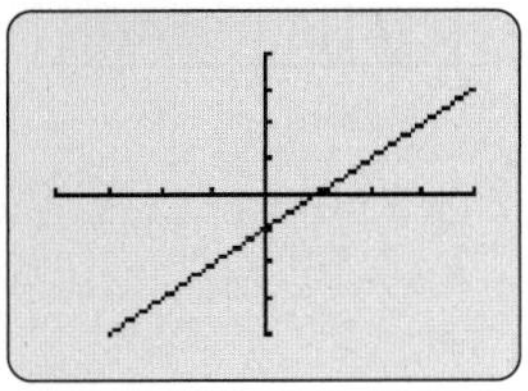

Solving an Equation for y To graph $3x + 2y = 12$, we must first solve the equation for y.

$$3x + 2y = 12$$
$$2y = -3x + 12 \quad \text{Subtract } 3x \text{ from both sides.}$$
$$y = -\frac{3}{2}x + 6 \quad \text{Divide both sides by 2.}$$

Next, we press the **WINDOW** key to reenter the standard window settings, press **Y =** and enter $-\frac{3}{2}x + 6$, as shown below. Then press **GRAPH** to see the graph.

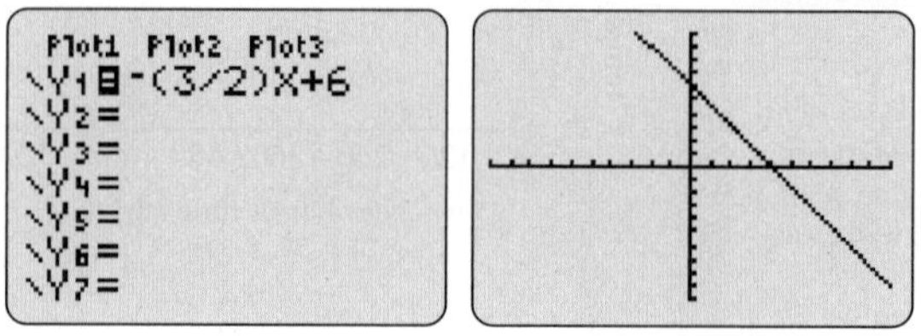

SECTION 3.3 STUDY SET

VOCABULARY

Fill in the blanks.

1. The x-intercept of a line is the point where the line intersects/crosses the x-axis.
2. The y-intercept of a line is the point where the line intersects/crosses the y-axis.
3. The graph of $y = 4$ is a horizontal line and the graph of $x = 6$ is a vertical line.
4. The intercept method is useful when graphing linear equations written in the standard form $Ax + By = C$.

CONCEPTS

5. Fill in the blanks.
   a. To find the y-intercept of the graph of a line, substitute 0 for x in the equation and solve for y.
   b. To find the x-intercept of the graph of a line, substitute 0 for y in the equation and solve for x.
6. Complete the table of solutions and fill in the blanks.

$3x + 2y = 6$

x	y	(x, y)	
0	3	(0, 3)	← y-intercept
2	0	(2, 0)	← x-intercept
−2	6	(−2, 6)	← Check point

7. a. Refer to the graph. Which intercept tells the purchase price of the machinery? What was that price?
   y-intercept: (0, 80,000); \$80,000
   b. Which intercept indicates when the machinery will have lost all of its value? When is that?
   x-intercept: (30, 0); 30 years after purchase

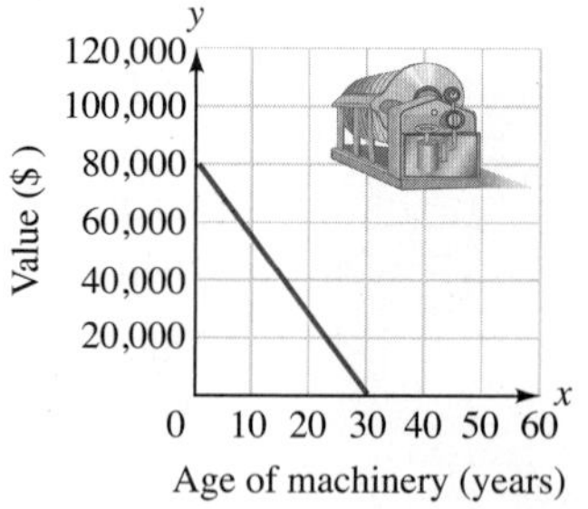

8. Match each graph with its equation.
   a. $x = 2$ ii
   b. $y = 2$ iv
   c. $y = 2x$ vi
   d. $2x - y = 2$ i
   e. $y = 2x + 2$ iii
   f. $y = -2x$ v

i.

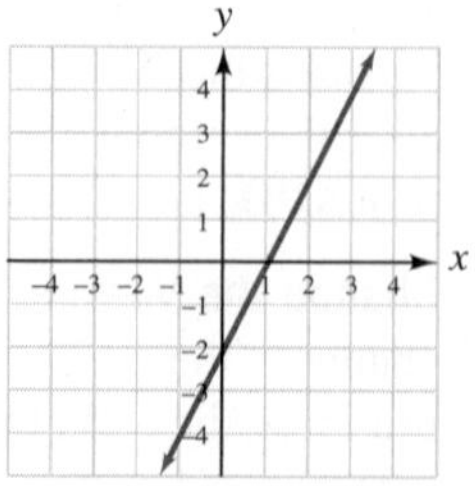

ii.

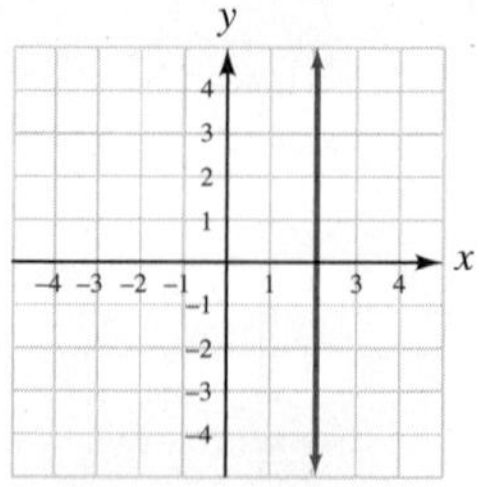

iii.

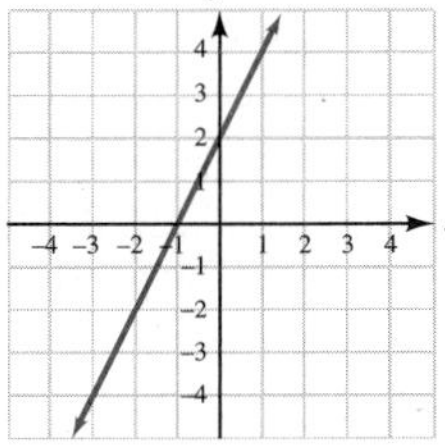

iv.

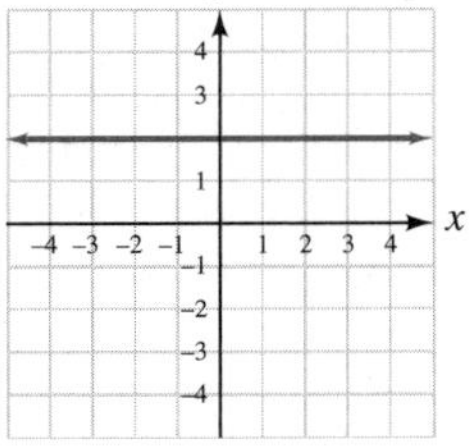

v.

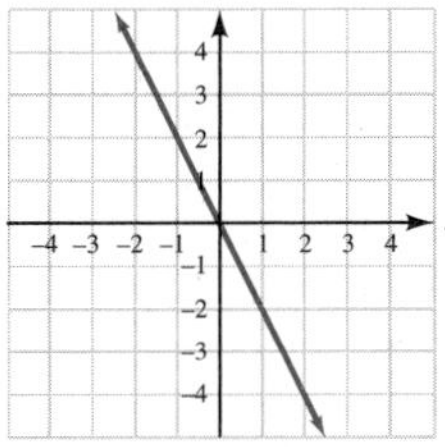

vi.

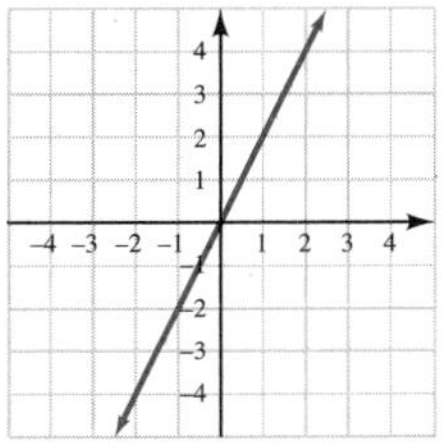

NOTATION

9. What is the equation of the x-axis? What is the equation of the y-axis? $y = 0; x = 0$

10. Write the coordinates that are improper fractions as mixed numbers.

a. $\left(\frac{7}{2}, 0\right)$ $\left(3\frac{1}{2}, 0\right)$

b. $\left(0, -\frac{17}{3}\right)$ $\left(0, -5\frac{2}{3}\right)$

GUIDED PRACTICE

Give the coordinates of the intercepts of each graph. **See Example 1.**

▶ 11.

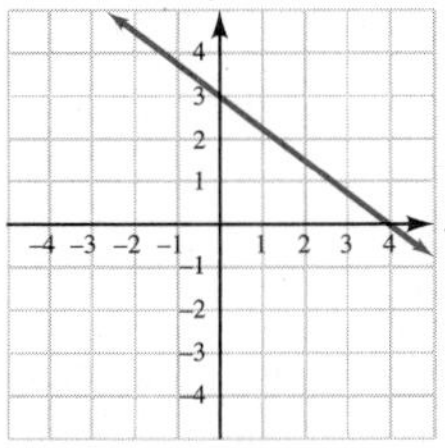

x-intercept: (4, 0)
y-intercept: (0, 3)

12.

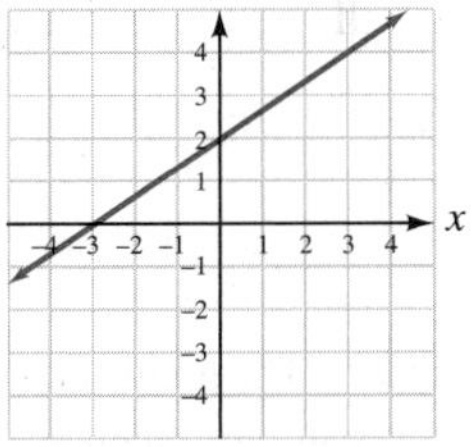

x-intercept: (−3, 0)
y-intercept: (0, 2)

▶ 13.

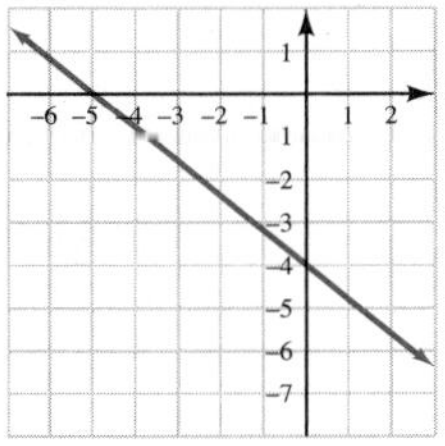

x-intercept: (−5, 0)
y-intercept: (0, −4)

14.

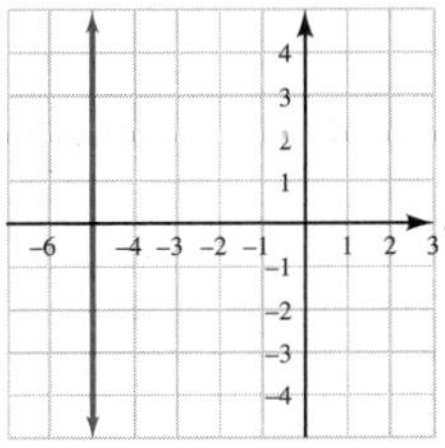

x-intercept: (−5, 0)
No y-intercept

15.

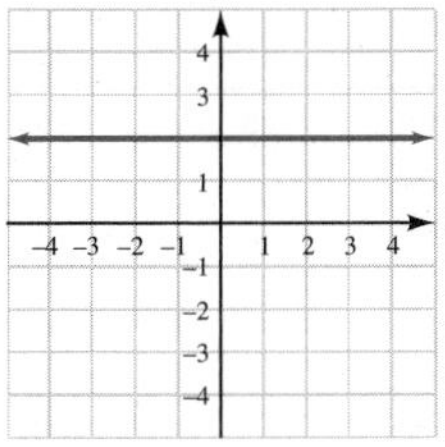

No x-intercept
y-intercept: (0, 2)

▶ 16.

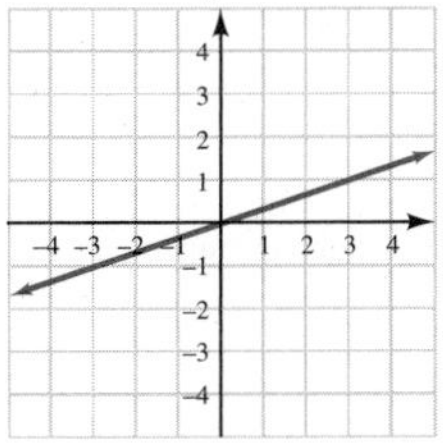

x-intercept: (0, 0)
y-intercept: (0, 0)

Estimate the coordinates of the intercepts of each graph. (Some are not integers.) **See Example 1. *(Answers may vary.)***

17.

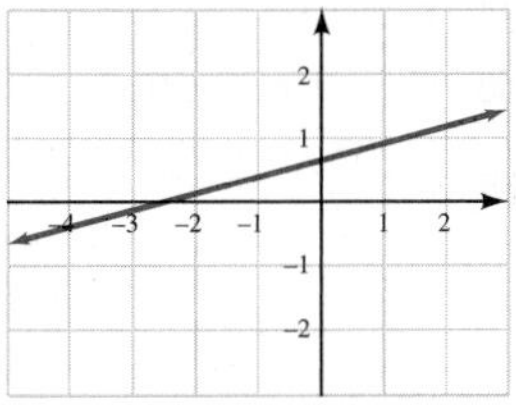

x-intercept: $\left(-2\frac{1}{2}, 0\right)$
y-intercept: $\left(0, \frac{2}{3}\right)$

▶ 18.

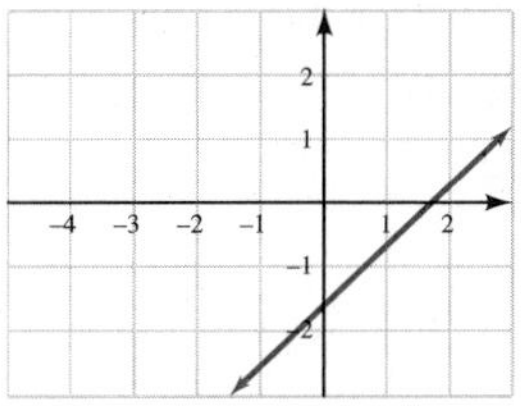

x-intercept: $\left(1\frac{2}{3}, 0\right)$
y-intercept: $\left(0, -1\frac{1}{2}\right)$

Find the x- and y-intercepts of the graph of each equation. Do not graph the line. **See Example 2.**

19. $8x + 3y = 24$ (3, 0); (0, 8)

▶ 20. $5x + 6y = 30$ (6, 0); (0, 5)

21. $7x - 2y = 28$ (4, 0); (0, −14)

22. $2x - 9y = 36$ (18, 0); (0, −4)

23. $-5x - 3y = 10$ $(-2, 0); \left(0, -\frac{10}{3}\right)$

▶ 24. $-9x - 5y = 25$ $\left(-\frac{25}{9}, 0\right); (0, -5)$

25. $6x + y = 9$ $\left(\frac{3}{2}, 0\right); (0, 9)$

26. $x + 8y = 14$ $(14, 0); \left(0, \frac{7}{4}\right)$

Use the intercept method to graph each equation. **See Example 2.** See AIE Appendix 3.

▶ 27. $4x + 5y = 20$

▶ 28. $3x + 4y = 12$

29. $5x + 15y = -15$

30. $8x + 4y = -24$

31. $x - y = -3$

▶ 32. $x - y = 3$

▶ 33. $x + 2y = -2$

▶ 34. $x + 2y = -4$

Use the intercept method to graph each equation. **See Example 3.** See AIE Appendix 3.

35. $30x + y = -30$

▶ 36. $20x - y = -20$

▶ 37. $4x - 20y = 60$

38. $6x - 30y = 30$

Use the intercept method to graph each equation. **See Example 4.** See AIE Appendix 3.

39. $3x + 4y = 8$

▶ 40. $2x + 3y = 9$

41. $-9x + 4y = 9$

▶ 42. $-5x + 4y = 15$

▶ 43. $3x - 4y = 11$

▶ 44. $5x - 4y = 13$

▶ 45. $9x + 3y = 10$

▶ 46. $4x + 4y = 5$

Use the intercept method to graph each equation. **See Example 5.** See AIE Appendix 3.

47. $3x + 5y = 0$

▶ 48. $4x + 3y = 0$

49. $2x - 7y = 0$

▶ 50. $6x - 5y = 0$

Graph each equation. **See Examples 6 and 7.** See AIE Appendix 3.

▶ 51. $y = 5$

▶ 52. $y = -3$

53. $y = 0$

▶ 54. $x = 0$

55. $x = -2$

▶ 56. $x = 5$

▶ 57. $x = \frac{4}{3}$

58. $y = -\frac{1}{2}$

▶ 59. $y - 2 = 0$ (*Hint:* Solve for y first.) $y = 2$

▶ 60. $x + 1 = 0$ (*Hint:* Solve for x first.) $x = -1$

61. $5x = 7.5$ (*Hint:* Solve for x first.) $x = 1.5$

▶ 62. $3y = 4.5$ (*Hint:* Solve for y first.) $y = 1.5$

TRY IT YOURSELF

Graph each equation. See AIE Appendix 3.

63. $7x = 4y - 12$
64. $7x = 5y - 15$
65. $4x - 3y = 12$
▶ 66. $5x - 10y = 20$
67. $x = -\frac{5}{3}$
68. $y = \frac{5}{2}$
69. $y - 3x = -\frac{4}{3}$
▶ 70. $y - 2x = -\frac{9}{8}$
71. $7x + 3y = 0$
72. $4x - 5y = 0$
▶ 73. $-4x = 8 - 2y$
74. $-5x = 10 + 5y$
▶ 75. $3x = -150 - 5y$
▶ 76. $x = 50 - 5y$
▶ 77. $-3y = 3$
▶ 78. $-2x = 8$

APPLICATIONS

▶ 79. **Chemistry.** The relationship between the temperature T and volume V of a gas kept in a sealed container at a constant pressure is graphed below. The T-intercept of this graph is a very important scientific fact. It represents the lowest possible temperature, called **absolute zero.**

a. Estimate absolute zero. About $-270°$C

b. What is the volume of the gas when the temperature is absolute zero? 0 milliliters

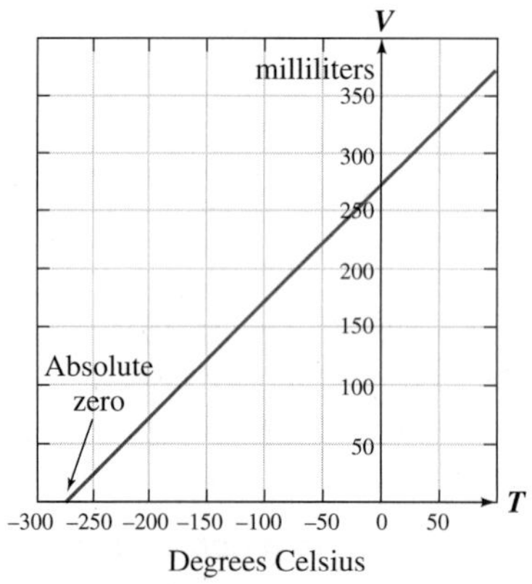

▶ 80. **Physics.** The graph shows the length L of a stretched spring (in inches) as different weights w (in pounds) are attached to it. What information about the spring does the L-intercept give? (0, 1.5): the spring is 1.5 inches long if no weight is attached.

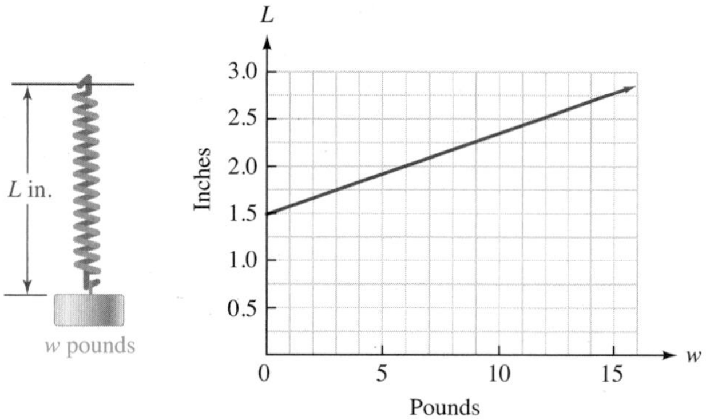

81. **Bottled Water Dispenser.** The graph below shows the number of gallons g of water remaining in a bottle after c six-ounce cups have been served from it. Find the intercepts of the graph. What information do they give? The g-intercept is (0, 5): Before any cups of water have been served from the bottle, it contains 5 gallons of water. The c-intercept is $\left(106\frac{2}{3}, 0\right)$: The bottle will be empty after $106\frac{2}{3}$ six-ounce cups of water have been served from it.

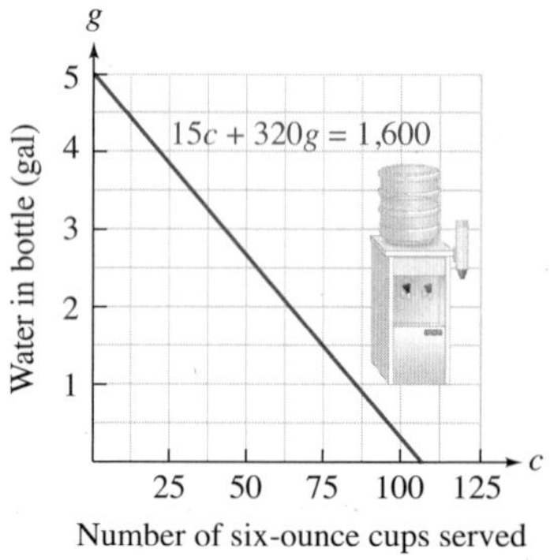

82. **Renewable Energy.** The equation $p = 50s - 300$ estimates the power output in watts from a propeller blade turbine driven by a wind of speed s miles per hour. What information does the s-intercept of the graph of the equation give? (Source: otherpower.com) The s-intercept is (6, 0). A wind speed of 6 mph does not produce any power.

▶ 83. **Landscaping.** A developer is going to purchase x trees and y shrubs to landscape a new office complex. The trees cost \$50 each and the shrubs cost \$25 each. His budget is \$5,000. This situation is modeled by the equation $50x + 25y = 5{,}000$. Use the intercept method to graph it. See AIE Appendix 3.

a. What information is given by the y-intercept? If only shrubs are purchased, he can buy 200.

b. What information is given by the x-intercept? If only trees are purchased, he can buy 100.

84. **Eggs.** The number of eggs eaten by an average American in one year has remained almost constant since the year 2000. See the graph below. Draw a horizontal line that passes through, or near, the data points. What is the equation of the line? $n = 253$ (Answers may vary slightly)

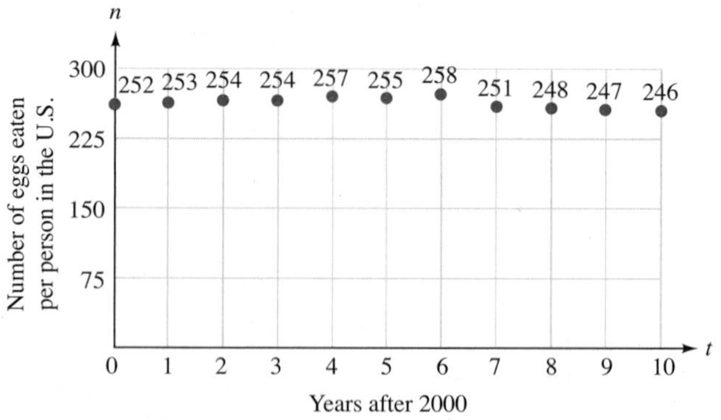

Source: United Egg

WRITING

85. To graph $3x + 2y = 12$, a student found the intercepts and a check point, and graphed them, as shown in figure (a). Instead of drawing a crooked line through the points, what should he have done?

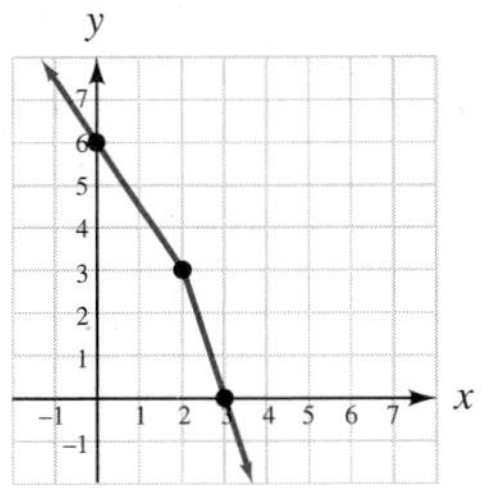

(a)

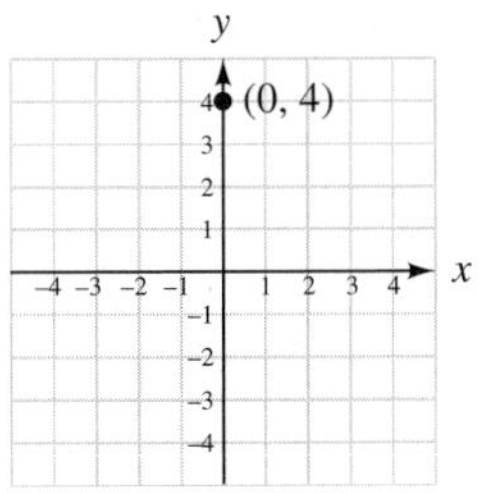

(b)

86. A student graphed the linear equation $y = 4$, as shown above in figure (b). Explain her error.

87. How do we find the intercepts of the graph of an equation without having to graph the equation?

88. In Section 3.2, we discussed a method to graph $y = 2x - 3$. In Section 3.3, we discussed a method to graph $2x + 3y = 6$. Briefly explain the steps involved in each method.

REVIEW

89. Simplify: $\frac{3 \cdot 5 \cdot 5}{3 \cdot 5 \cdot 5 \cdot 5}$ $\frac{1}{5}$

90. Simplify: $4\left(\frac{d}{2} - 3\right) - 5\left(\frac{2}{5}d - 1\right)$ -7

91. Translate: Six less than twice x $2x - 6$

▶ **92.** Is -5 a solution of $2(3x + 10) = 5x + 6$? No

CHALLENGE PROBLEMS

93. Where will the line $y = b$ intersect the line $x = a$? (a, b)

▶ **94.** Write an equation of the line that has an x-intercept of $(4, 0)$ and a y-intercept of $(0, 3)$. $3x + 4y = 12$ (Answers may vary)

95. What is the least number of intercepts a line can have? What is the greatest number a line can have? 1; infinitely many

96. On a rectangular coordinate system, draw a circle that has exactly two intercepts. See AIE Appendix 3.

SECTION 3.4 Slope and Rate of Change

OBJECTIVES

1. Find the slope of a line from its graph.
2. Find the slope of a line given two points.
3. Find slopes of horizontal and vertical lines.
4. Solve applications of slope.
5. Calculate rates of change.
6. Determine whether lines are parallel or perpendicular using slope.

ARE YOU READY? *Are You Ready? exercises available online at www.webassign.net/brookscole*

The following problems review some basic skills that are needed to find the slope of a line.

1. Evaluate: $\frac{4 - 1}{8 - 3}$ $\frac{3}{5}$

2. Evaluate: $\frac{-10 - 1}{-4 - (-4)}$ undefined

3. Multiply: $-\frac{7}{9} \cdot \frac{9}{7}$ -1

4. Simplify: $\frac{15}{18}$ $\frac{5}{6}$

In this section, we introduce a method to measure the steepness (or slant) of a line. We call this measure the *slope of the line,* and it can be found in several ways.

1 Find the Slope of a Line from Its Graph.

The **slope of a line** is a ratio that compares the vertical change with the corresponding horizontal change as we move along the line from one point to another.

As an example, let's find the slope of the line graphed on the right. To begin, we select two points on the line and call them P and Q. One way to move from P to Q is to start at point P and count upward 5 grid squares. Then, moving to the right, we count 6 grid squares to reach point Q. The vertical change in this movement is called the **rise.** The horizontal change is called the **run.** Notice that a right triangle, called a **slope triangle,** is created by this process.

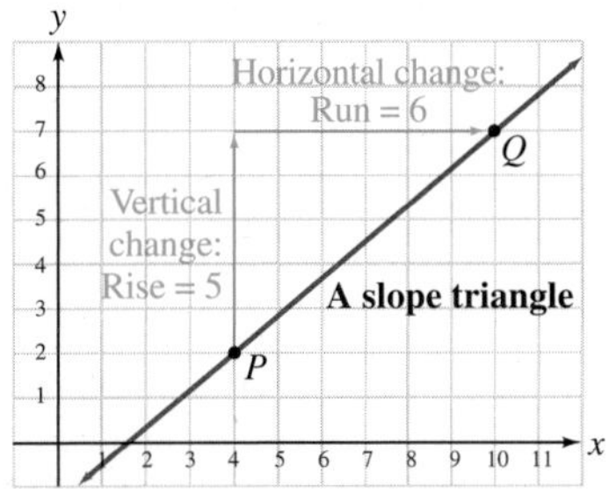

The Language of Algebra

A **ratio** is a comparison of two numbers using a quotient. Ratios are used in many settings. Mechanics speak of gear ratios. Colleges like to advertise their low student-to-teacher ratios.

The Language of Algebra

Many historians believe that *m* was chosen to represent the slope of a line because it is the first letter of the French word **monter,** meaning to ascend or climb.

Teaching Tip: To introduce the concept of slope, draw three lines on the board, as shown below, and ask your students to compare the steepness of the lines.

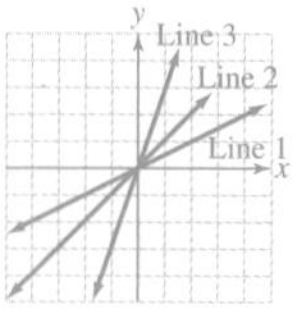

The slope of a line is defined to be *the ratio of the vertical change to the horizontal change.* By tradition, the letter *m* is used to represent slope. For the line graphed on the previous page, we have

$$m = \text{slope} = \frac{\text{vertical change}}{\text{horizontal change}} = \frac{\text{rise}}{\text{run}} = \frac{5}{6}$$ This ratio is a comparison of the rise and the run using a quotient.

The slope of the line is $\frac{5}{6}$. This indicates that there is a rise (vertical change) of 5 units for each run (horizontal change) of 6 units.

EXAMPLE 1

Find the slope of the line graphed in figure (a) below.

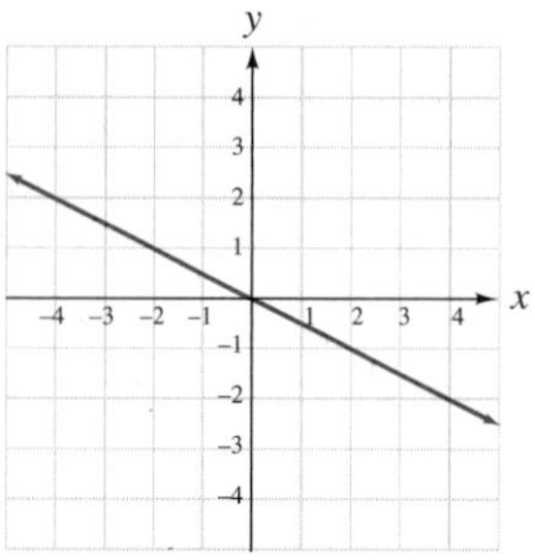

(a)

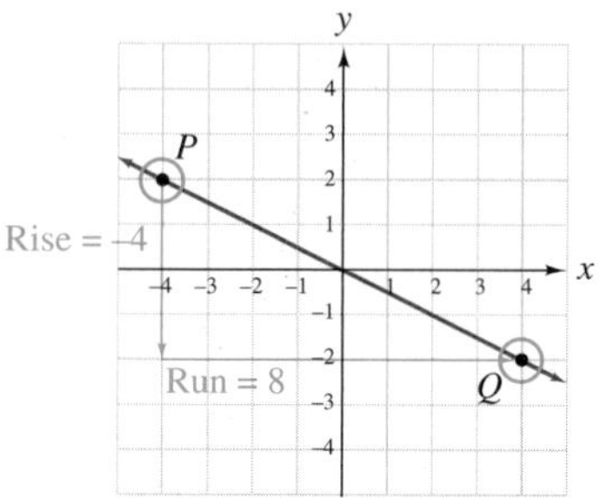

(b)

Pick two points on the line that also lie on the intersection of two grid lines.

Teaching Tip: Stress the words *downward* and *to the right* as you show your students one way to move from *P* to *Q*. Show them there is another way to move from *P* to *Q* (*to the right, downward*). Then have them describe in words the two ways to move from *Q* to *P* (use *to the left* and *upward*).

Strategy We will pick two points on the line, construct a slope triangle, and find the rise and run. Then we will write the ratio of the rise to the run.

Why The slope of a line is the ratio of the rise to the run.

Solution We begin by choosing two points on the line, *P* and *Q*, as shown in figure (b). One way to move from *P* to *Q* is to start at point *P* and count *downward* 4 grid squares. Because this movement is downward, the rise is −4. Then, moving right, we count 8 grid squares to reach *Q*. This indicates that the run is 8.

To find the slope of the line, we write a ratio of the rise to the run in simplified form.

$$m = \frac{\text{rise}}{\text{run}} = \frac{-4}{8} = -\frac{1}{2}$$ Always simplify slope fractions.

The slope of the line is $-\frac{1}{2}$.

Success Tip

When drawing a slope triangle, movement upward or to the right is positive. Movement downward or to the left is negative.
- Negative rise: downward
- Negative run: to the left

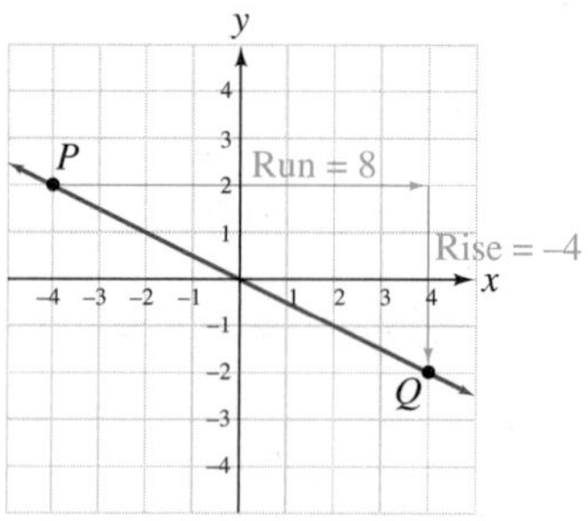

The movement from *P* to *Q* can be reversed. Starting at *P*, we can move to the right, a run of 8; and then downward, a rise of −4, to reach *Q*. With this approach, the slope triangle is above the line. When we form the ratio to find the slope, we get the *same result* as before:

$$m = \frac{\text{rise}}{\text{run}} = \frac{-4}{8} = -\frac{1}{2}$$

Arrowheads are used to show horizontal movement (left or right) and vertical movement (up or down).

Teaching Example 1
Find the slope of the line shown in the figure using (−2, 1) and (4, −2).
Answer: $-\frac{1}{2}$

Self Check 1 Find the slope of the line shown above using two points different from those used in the solution of Example 1. $-\frac{1}{2}$

Now Try Problem 21

The identical answers from Example 1 and its Self Check illustrate an important fact: *For any line, the same value will be obtained no matter which two points on the line are used to find the slope.*

2 Find the Slope of a Line Given Two Points.

The Language of Algebra

The prefix **sub** means below or beneath, as in submarine or subway. In x_1, x_2, y_1, and y_2, the **subscripts** 1 and 2 are written lower than the variable. They are not exponents, and they do not represent a calculation.

Success Tip

Recall that subtraction is used to measure change.

We can generalize the graphic method for finding slope to develop a slope formula. To begin, we select points P and Q on the line shown in the figure on the right. To distinguish between the coordinates of these two points, we use **subscript notation.** Point P has coordinates (x_1, y_1), which are read as "x sub 1 and y sub 1." Point Q has coordinates (x_2, y_2), which are read as "x sub 2 and y sub 2."

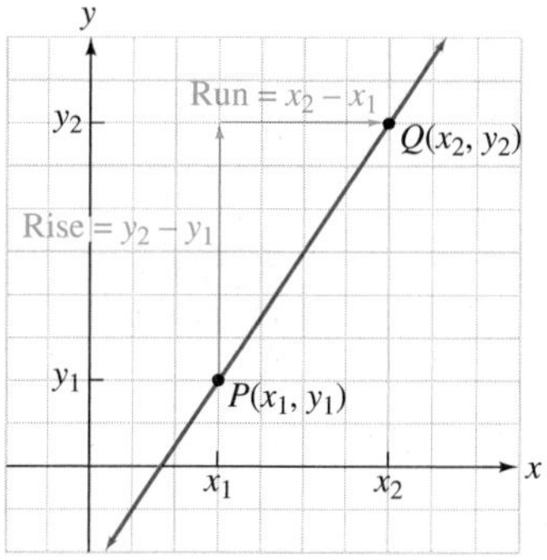

As we move from point P to point Q, the rise is the difference of the y-coordinates: $y_2 - y_1$. We call this difference the **change in y.** The run is the difference of the x-coordinates: $x_2 - x_1$. This difference is called the **change in x.** Since the slope is the ratio $\frac{\text{rise}}{\text{run}}$, we have the following formula for calculating slope.

Slope of a Line

The slope m of a line that passes through points (x_1, y_1) and (x_2, y_2) is

$$m = \frac{\text{vertical change}}{\text{horizontal change}} = \frac{\text{rise}}{\text{run}} = \frac{\text{change in } y}{\text{change in } x} = \frac{y_2 - y_1}{x_2 - x_1} \quad \text{if } x_2 \neq x_1$$

EXAMPLE 2 Find the slope of the line that passes through (1, 2) and (3, 8).

Strategy We will use the slope formula to find the slope of the line.

Why We know the coordinates of two points on the line.

Solution When using the slope formula, it makes no difference which point you call (x_1, y_1) and which point you call (x_2, y_2). If we let (x_1, y_1) be (1, 2) and (x_2, y_2) be (3, 8), then

Success Tip

When using the slope formula, you may find it helpful to begin your solution by labeling the coordinates of the points in this way:

$$\begin{array}{cc} (x_1, y_1) & (x_2, y_2) \\ \downarrow\ \downarrow & \downarrow\ \downarrow \\ (1, 2) & (3, 8) \end{array}$$

$m = \frac{y_2 - y_1}{x_2 - x_1}$ This is the slope formula.

$m = \frac{8 - 2}{3 - 1}$ Substitute 8 for y_2, 2 for y_1, 3 for x_2, and 1 for x_1.

$m = \frac{6}{2}$ Do the subtraction.

$m = 3$ Simplify. Think of this as a $\frac{3}{1}$ rise-to-run ratio.

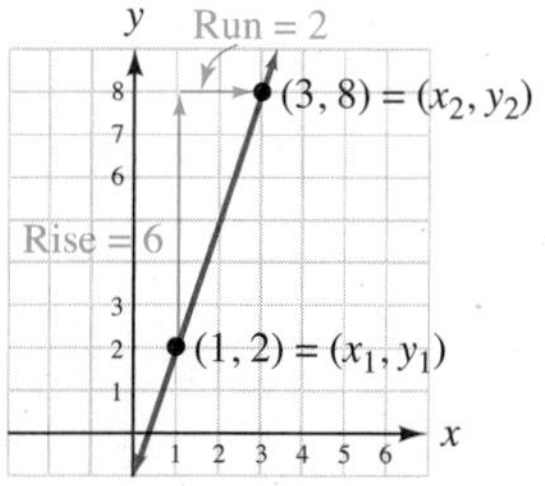

The slope of the line is 3. Note that we obtain the same value for the slope if we let $(x_1, y_1) = (3, 8)$ and $(x_2, y_2) = (1, 2)$.

$$m = \frac{y_2 - y_1}{x_2 - x_1} = \frac{2 - 8}{1 - 3} = \frac{-6}{-2} = 3$$

Success Tip

The slope formula is a valuable tool because it allows us to calculate the slope of a line without having to view its graph.

Although it is not necessary, the line passing through (1, 2) and (3, 8) has been graphed on the right. The graph of the line, including the slope triangle, verifies that the rise is 6 and the run is 2, and that $m = \frac{6}{2} = 3$.

CAUTION When using the slope formula, always subtract the y-coordinates and their corresponding x-coordinates in the same order. Otherwise, your answer will have the wrong sign. If we let (x_1, y_1) be (1, 2) and (x_2, y_2) be (3, 8):

$$\cancel{m = \frac{y_2 - y_1}{x_1 - x_2}} = \frac{8 - 2}{1 - 3} = \frac{6}{-2} = -3 \quad \text{and} \quad \cancel{m = \frac{y_1 - y_2}{x_2 - x_1}} = \frac{2 - 8}{3 - 1} = \frac{-6}{2} = -3$$

Teaching Example 2 Find the slope of the line that passes through (5, 1) and (7, 9).
Answer: 4

Self Check 2 Find the slope of the line that passes through (2, 1) and (4, 11). 5

Now Try Problem 33

EXAMPLE 3 Find the slope of the line that passes through $(-2, 4)$ and $(5, -6)$.

Strategy We will use the slope formula to find the slope of the line.

Why We know the coordinates of two points on the line.

Solution If we let (x_1, y_1) be $(-2, 4)$ and (x_2, y_2) be $(5, -6)$, then

Caution

Slopes normally are written as fractions, sometimes as decimals, but never as mixed numbers.

As with any fractional answer, always express slope in simplified form.

$$m = \frac{y_2 - y_1}{x_2 - x_1}$$ This is the slope formula.

$$m = \frac{-6 - 4}{5 - (-2)}$$ Substitute -6 for y_2, 4 for y_1, 5 for x_2, and -2 for x_1.

$$m = -\frac{10}{7}$$ Do the subtraction. We can write the result as $\frac{-10}{7}$ or $-\frac{10}{7}$.

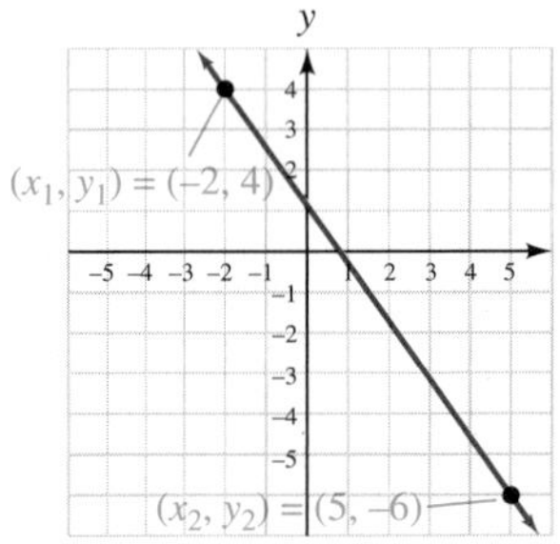

The slope of the line is $-\frac{10}{7}$.

If we graph the line by plotting the two points, we see that the line falls from left to right—a fact indicated by its negative slope.

Teaching Example 3 Find the slope of the line that passes through $(-3, -4)$ and $(0, -8)$.
Answer: $-\frac{4}{3}$

Self Check 3 Find the slope of the line that passes through $(-1, -2)$ and $(1, -7)$. $-\frac{5}{2}$

Now Try Problem 39

In Example 2, the slope of the line was positive. In Examples 1 and 3, the slopes of the lines were negative. In general, lines that rise from left to right have a positive slope. Lines that fall from left to right have a negative slope.

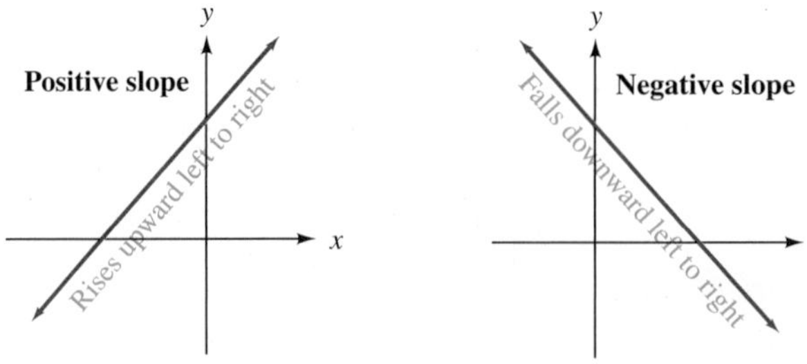

Success Tip

To classify the slope of a line as positive or negative, follow it from left to right, as you would read a sentence in a book.

In figure (a) on the next page, we see that a line with slope 3 is steeper than a line with slope $\frac{5}{6}$, and a line with slope $\frac{5}{6}$ is steeper than a line with slope $\frac{1}{4}$. In general, *the larger the absolute value of the slope, the steeper the line.*

Lines with slopes of 1 and -1 are graphed in figure (b) on the next page. In each case, there is a special relationship between the rise and the run. When $m = 1$, the rise and run are, of course, the same number. When $m = -1$ the rise and run are opposites. Note that both lines create a 45° angle with the horizontal x-axis.

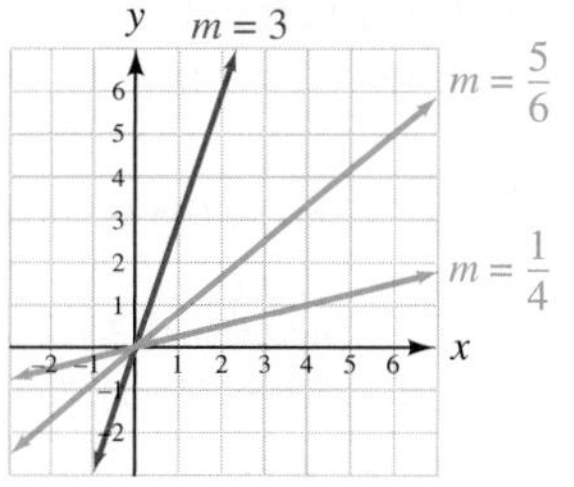

(a)

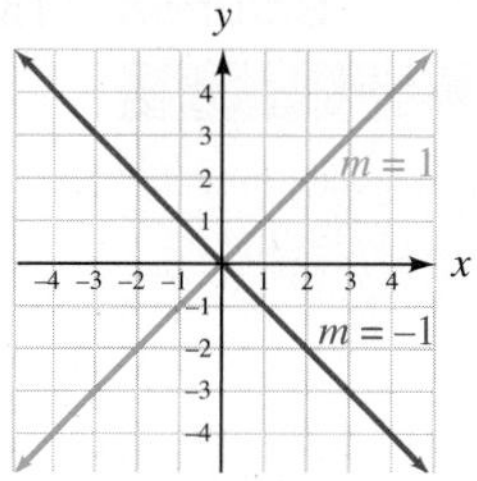

(b)

3 Find Slopes of Horizontal and Vertical Lines.

In the next two examples, we calculate the slope of a horizontal line and we show that a vertical line has no defined slope.

EXAMPLE 4 Find the slope of the line $y = 3$.

Strategy We will find the coordinates of two points on the line.

Why We can then use the slope formula to find the slope of the line.

Solution The graph of $y = 3$ is a horizontal line. To find its slope, we select two points on the line: $(-2, 3)$ and $(3, 3)$. If (x_1, y_1) is $(-2, 3)$ and (x_2, y_2) is $(3, 3)$, we have

$$m = \frac{y_2 - y_1}{x_2 - x_1}$$ This is the slope formula.

$$m = \frac{3 - 3}{3 - (-2)}$$ Substitute 3 for y_2, 3 for y_1, 3 for x_2, and −2 for x_1.

$$m = \frac{0}{5}$$ Simplify the numerator and the denominator.

$$m = 0$$ 0 divided by any nonzero number is equal to 0.

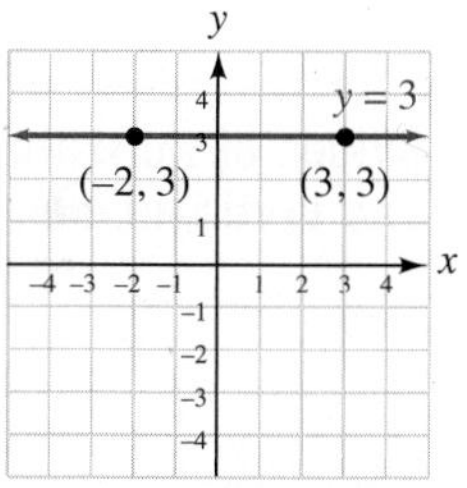

The rise = 0 for these two points.

The slope of the line $y = 3$ is 0.

Teaching Example 4 Find the slope of the line $y = 2$.
Answer: 0

Self Check 4 Find the slope of the line $y = -10$. 0

Now Try Problem 61

The y-coordinates of any two points on a horizontal line will be the same, and the x-coordinates will be different. Thus, the numerator of $\frac{y_2 - y_1}{x_2 - x_1}$ will always be zero, and the denominator will always be nonzero. Therefore, the slope of a horizontal line is 0.

EXAMPLE 5 Find the slope of the line $x = -2$.

Strategy We will find the coordinates of two points on the line.

Why We can then use the slope formula to find the slope of the line.

Notation

This example explains why the definition of slope includes the restriction that $x_1 \neq x_2$.

Teaching Tip: Discourage students from saying that a vertical line has "no slope." Encourage them to use more precise language: "A vertical line has no defined slope" or "the slope of a vertical line is undefined."

Solution The graph of $x = -2$ is a vertical line. To find its slope, we select two points on the line: $(-2, 3)$ and $(-2, -1)$. If (x_1, y_1) is $(-2, -1)$ and (x_2, y_2) is $(-2, 3)$, we have

$$m = \frac{y_2 - y_1}{x_2 - x_1}$$ This is the slope formula.

$$m = \frac{3 - (-1)}{-2 - (-2)}$$ Substitute 3 for y_2, -1 for y_1, -2 for x_2, and -2 for x_1. Note that $x_1 = x_2$.

$$m = \frac{4}{0}$$ Simplify the numerator and the denominator.

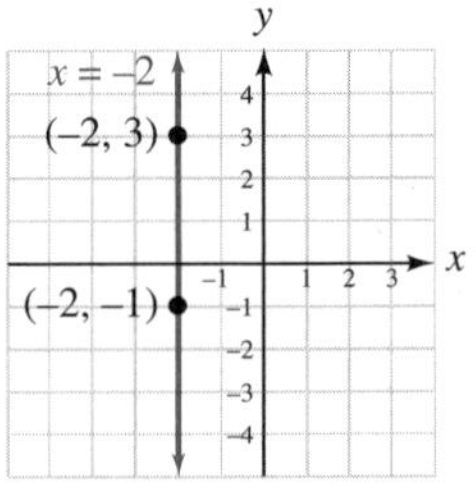

The run = 0 for these two points.

Since division by zero is undefined, $\frac{4}{0}$ has no meaning. The slope of the line $x = -2$ is undefined.

Teaching Example 5 Find the slope of the line $x = -4$.
Answer: Undefined slope

Self Check 5 Find the slope of the line $x = 12$. Undefined slope

Now Try Problem 67

The Language of Algebra

Undefined and **0** do not mean the same thing. A horizontal line has a defined slope; it is 0. A vertical line does not have a defined slope; we say its slope is *undefined.*

The y-coordinates of any two points on a vertical line will be different, and the x-coordinates will be the same. Thus, the numerator of $\frac{y_2 - y_1}{x_2 - x_1}$ will always be nonzero, and the denominator will always be 0. Therefore, the slope of a vertical line is undefined.

We now summarize the results from Examples 4 and 5.

Slopes of Horizontal and Vertical Lines

Horizontal lines (lines with equations of the form $y = b$) have **slope 0.**

Vertical lines (lines with equations of the form $x = a$) have **undefined slope.**

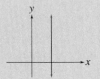

4 Solve Applications of Slope.

The concept of slope has many applications. For example, architects use slope when designing ramps and roofs. Truckers must be aware of the slope, or **grade,** of the roads they travel. Mountain bikers ride up rocky trails and snow skiers speed down steep slopes.

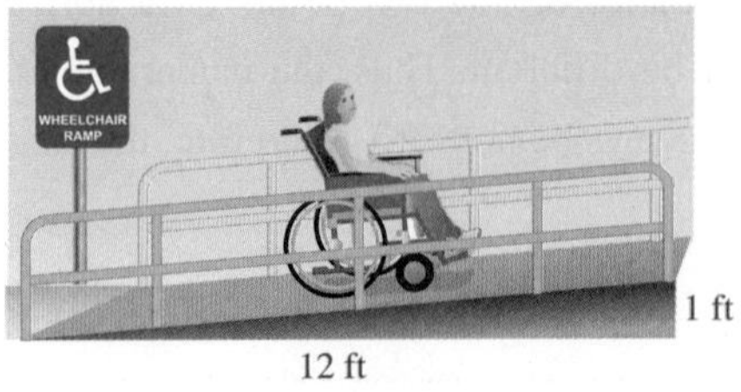

The *Americans with Disabilities Act* provides a guideline for the steepness of a ramp. The maximum slope for a wheelchair ramp is 1 foot of rise for every 12 feet of run: $m = \frac{1}{12}$.

The grade of an incline is its slope expressed as a percent: A 15% grade means a rise of 15 feet for every run of 100 feet: $m = \frac{15}{100}$, which simplifies to $\frac{3}{20}$. A grade is always expressed as a positive percent.

EXAMPLE 6 **Architecture.** **Pitch** is the incline of a roof written as a ratio of the vertical rise to the horizontal run. It is always expressed as a positive number. Find the pitch of the roof shown in the illustration.

Strategy We will determine the rise and the run of the roof from the illustration. Then we will write the ratio of the rise to the run.

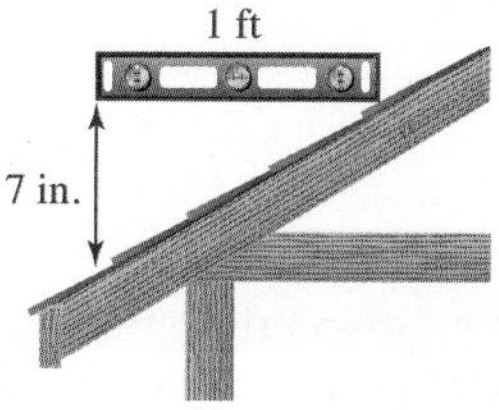

Why The pitch of a roof is its slope, and the slope of a line is the ratio of the rise to the run.

Solution A level is used to create a slope triangle. The rise of the slope triangle is given as 7 inches. Since a ratio is a quotient of two quantities with the *same* units, we will express the length of the one-foot-long level as 12 inches. Therefore, the run of the slope triangle is 12 inches.

$$m = \frac{\text{rise}}{\text{run}} = \frac{7}{12}$$ The roof has a $\frac{7}{12}$ pitch. This means that the roof rises 7 units for every 12 units in the horizontal direction.

Teaching Example 6 **Roofing.** Find the pitch of the roof.

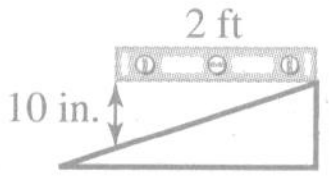

Answer: $\frac{5}{12}$

Self Check 6 **Roofing.** Find the pitch of the roof. $\frac{5}{12}$

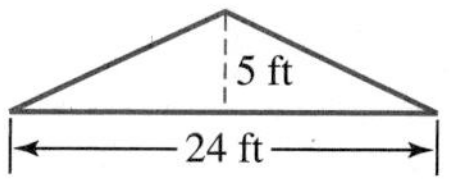

Now Try Problem 99

5 Calculate Rates of Change.

We have seen that the slope of a line compares the change in y to the change in x. This is called the **rate of change** of y with respect to x. In our daily lives, we often make many such comparisons of the change in one quantity with respect to another. For example, we might speak of snow melting at the rate of 6 inches per day or a tourist exchanging money at the rate of 12 pesos per dollar.

When finding rates of change in application problems, we attach units to the numerator and denominator in the slope calculation.

EXAMPLE 7 **Banking.** A bank offers a business account with a fixed monthly fee, plus a service charge for each check written. The relationship between the monthly cost y and the number x of checks written is graphed below. At what rate does the monthly cost change?

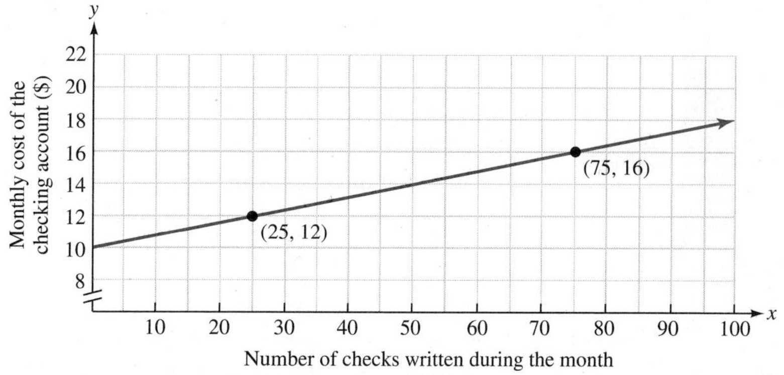

Notation

In the graph, the symbol ≉ indicates a break in the labeling of the vertical axis. The break enables us to omit a large portion of the grid that would not be used.

Strategy We will use the slope formula to calculate the slope of the line and attach the proper units to the numerator and denominator.

Why We know the coordinates of two points on the line.

Solution Two points on the line are (25, 12) and (75, 16). We will let $(x_1, y_1) = (25, 12)$ and $(x_2, y_2) = (75, 16)$, so we have

Teaching Tip: Remind your students to include the units when calculating rates of change. They often forget to do this.

$$\text{Rate of change} = \frac{(y_2 - y_1)\text{ dollars}}{(x_2 - x_1)\text{ checks}} = \frac{(16 - 12)\text{ dollars}}{(75 - 25)\text{ checks}} = \frac{4\text{ dollars}}{50\text{ checks}} = \frac{2\text{ dollars}}{25\text{ checks}}$$

Since the y-axis is scaled in dollars, we attach the units of "dollars" to the calculation of $y_2 - y_1$ in the numerator. Since x-axis indicates the number of checks written, we attach the units of "checks" to the calculation of $x_2 - x_1$ in the denominator. The monthly cost of the checking account increases $2 for every 25 checks written.

The Language of Algebra

The preposition **per** means for each, or for every. When we say the rate of change is 8¢ *per* check, we mean 8¢ for each check.

We can express $\frac{2}{25}$ in decimal form by dividing the numerator by the denominator. Then we can write the rate of change in two other ways, using the word *per,* which indicates division.

Rate of change = $0.08 per check or Rate of change = 8¢ per check

Teaching Example 7 Depreciation. Refer to the graph. Find the rate of change of the value of the sound equipment.

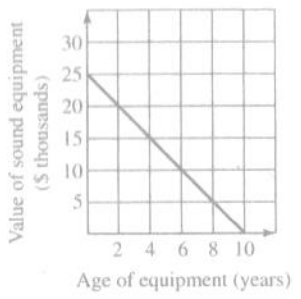

Answer: −$2,500 per year

Self Check 7 **Ethanol.** In 2005, the U.S. produced about 1,600 million bushels of corn for ethanol. By 2009, production had risen to 4,200 million bushels. Find the rate of change in the number of bushels of corn produced for ethanol over that time span. (Source: USDA, ERS Feed Outlook) An increase of 650 million bushels per year

Now Try ▶ Problem 103

6 Determine Whether Lines Are Parallel or Perpendicular Using Slope.

Two lines that lie in the same plane but do not intersect are called **parallel lines.** Parallel lines have the same slope and different y-intercepts. For example, the lines graphed in figure (a) are parallel because they both have slope $-\frac{2}{3}$.

The Language of Algebra

The words **parallel** and **perpendicular** are used in many settings. For example, the gymnast on the *parallel* bars is in a position that is *perpendicular* to the floor.

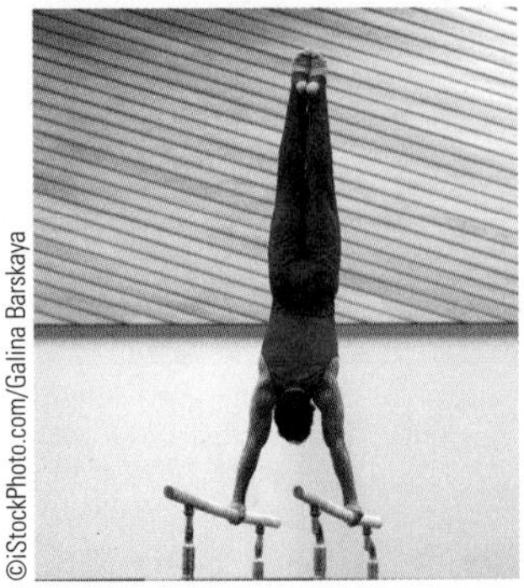

©iStockPhoto.com/Galina Barskaya

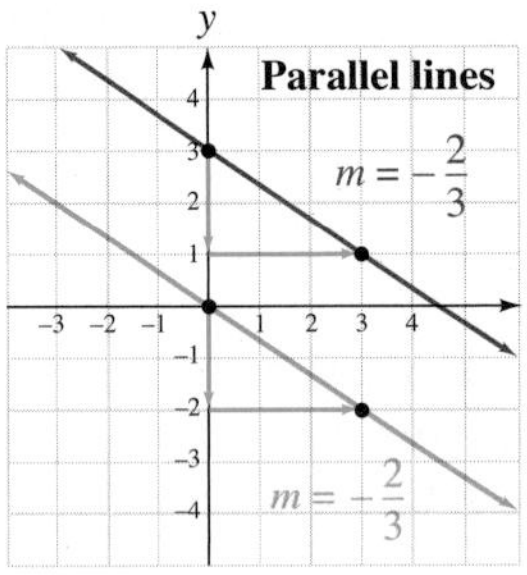

(a)

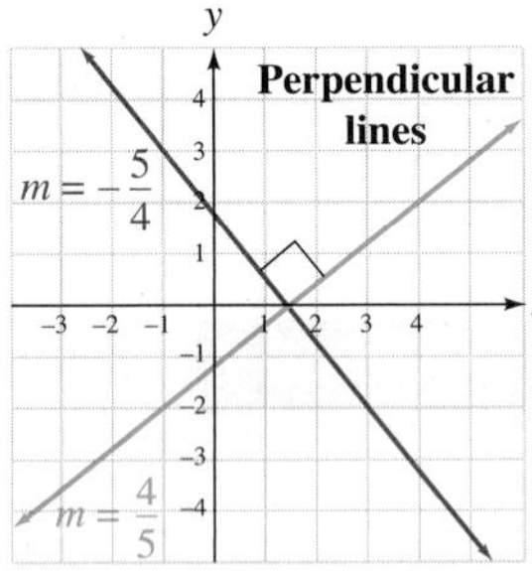

(b)

Lines that intersect to form four right angles (angles with measure 90°) are called **perpendicular lines.** If the product of the slopes of two lines is −1, the lines are perpendicular. This means that the slopes are **negative** (or **opposite**) **reciprocals.** In figure (b), we know that the lines with slopes $\frac{4}{5}$ and $-\frac{5}{4}$ are perpendicular because

$$\frac{4}{5}\left(-\frac{5}{4}\right) = -\frac{20}{20} = -1 \qquad \frac{4}{5} \text{ and } -\frac{5}{4} \text{ are negative reciprocals.}$$

Slopes of Parallel and Perpendicular Lines

1. Two lines with the same slope are parallel.
2. Two lines are perpendicular if the product of their slopes is −1; that is, if their slopes are negative reciprocals.
3. A horizontal line is perpendicular to any vertical line, and vice versa.

EXAMPLE 8 Determine whether the line that passes through (7, −9) and (10, 2) and the line that passes through (0, 1) and (3, 12) are parallel, perpendicular, or neither.

Strategy We will use the slope formula to find the slope of each line.

Why If the slopes are equal, the lines are parallel. If the slopes are negative reciprocals, the lines are perpendicular. Otherwise, the lines are neither parallel nor perpendicular.

Solution To calculate the slope of each line, we use the slope formula.

Success Tip

An example of two lines that are neither parallel nor perpendicular would be lines with slopes of $\frac{5}{2}$ and −2. When graphed, the lines intersect, but they do not form right angles.

Teaching Example 8 Determine whether the line that passes through (1, 4) and (5, −2) and the line that passes through (6, 2) and (9, 7) are parallel, perpendicular, or neither.
Answer: Neither

***The line through* (7, −9) *and* (10, 2):**

$$m = \frac{y_2 - y_1}{x_2 - x_1} = \frac{2 - (-9)}{10 - 7} = \frac{11}{3}$$

***The line through* (0, 1) *and* (3, 12):**

$$m = \frac{y_2 - y_1}{x_2 - x_1} = \frac{12 - 1}{3 - 0} = \frac{11}{3}$$

Since the slopes are the same, the lines are parallel.

Self Check 8 Determine whether the line that passes through (2, 1) and (6, 8) and the line that passes through (−1, 0) and (−5, 7) are parallel, perpendicular, or neither. Neither

Now Try Problems 73 and 75

EXAMPLE 9 Find the slope of a line perpendicular to the line that passes through (1, −4) and (8, 4).

Strategy We will use the slope formula to find the slope of the line passing through (1, −4) and (8, 4).

Why We can then form the negative reciprocal of the result to produce the slope of a line perpendicular to the given line.

Solution The slope of the line that passes through (1, −4) and (8, 4) is

Teaching Tip: Write several real numbers on the board and ask students to tell you the negative reciprocal of each.

$$m = \frac{y_2 - y_1}{x_2 - x_1} = \frac{4 - (-4)}{8 - 1} = \frac{8}{7}$$ We let $(x_1, y_1) = (1, -4)$ and $(x_2, y_2) = (8, 4)$.

The slope of a line perpendicular to the given line has slope that is the negative (or opposite) reciprocal of $\frac{8}{7}$, which is $-\frac{7}{8}$.

Teaching Example 9 Find the slope of a line perpendicular to the line that passes through (−3, 4) and (5, 7).
Answer: $-\frac{8}{3}$

Self Check 9 Find the slope of a line perpendicular to the line that passes through (−4, 1) and (9, 5). $-\frac{13}{4}$

Now Try Problem 85

SECTION 3.4 STUDY SET

VOCABULARY

Fill in the blanks.

▶ **1.** The ___slope___ of a line is a measure of the line's steepness. It is the ___ratio___ of the vertical change to the horizontal change.

▶ **2.** $m = \frac{\text{vertical change}}{\text{horizontal change}} = \frac{\text{rise}}{\text{run}} = \frac{\text{change in } y}{\text{change in } x}$

▶ **3.** The rate of ___change___ of a linear relationship can be found by finding the slope of the graph of the line and attaching the proper units.

▶ **4.** ___Parallel___ lines do not intersect. ___Perpendicular___ lines intersect to form four right angles.

CONCEPTS

5. Which line graphed has

a. a positive slope? Line 2

b. a negative slope? Line 1

c. zero slope? Line 4

d. undefined slope? Line 3

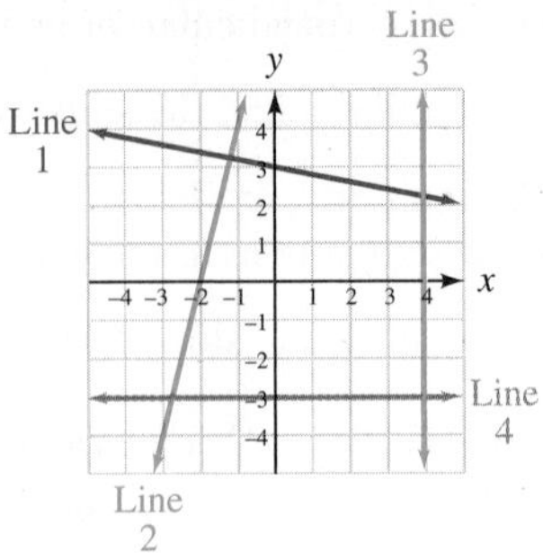

6. Consider each graph of a line and the slope triangle. What is the rise? What is the run? What is the slope of the line?

a.

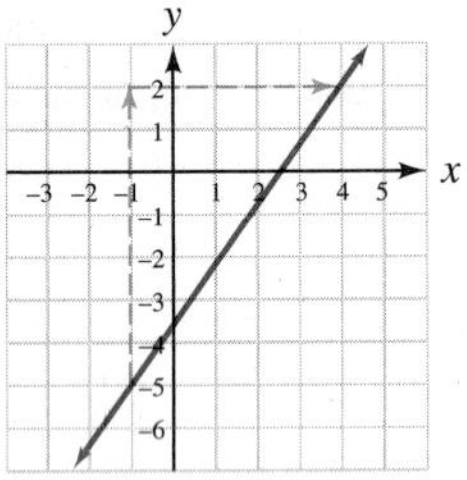

$7, 5, \frac{7}{5}$

b.

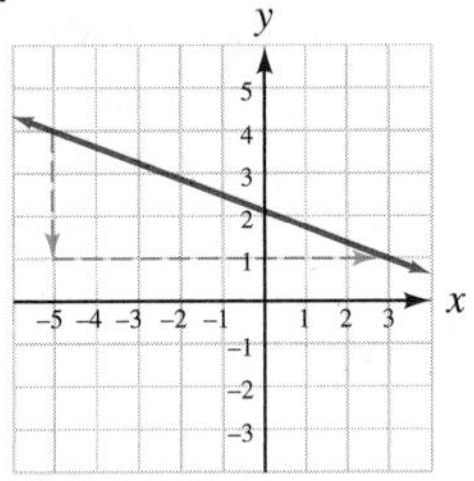

$-3, 8, -\frac{3}{8}$

7. For each graph, determine which line has the greater slope.

a.

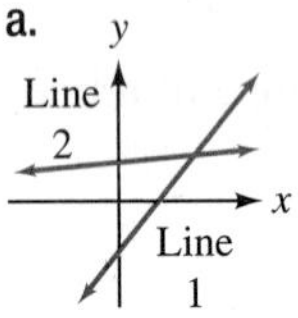

Line 1

b.

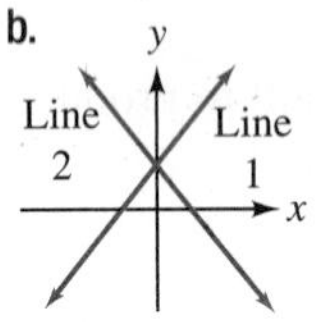

Line 1

c.

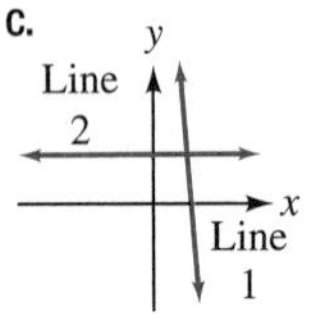

Line 2

8. Which two labeled points should be used to find the slope of the line? *B* and *F*

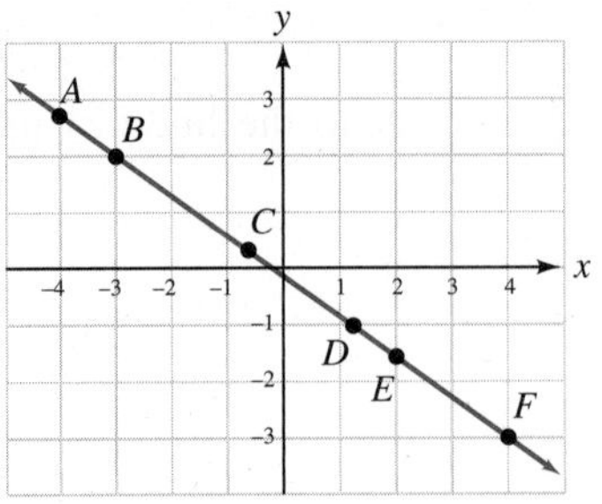

9. a. Find the slope of the line below using the points in black. $\frac{1}{3}$

b. Find the slope of the line using the points in green. $\frac{4}{12} = \frac{1}{3}$

c. Fill in the blank: When finding the slope of a line, the _same_ value will be obtained no matter which two points on the line are used.

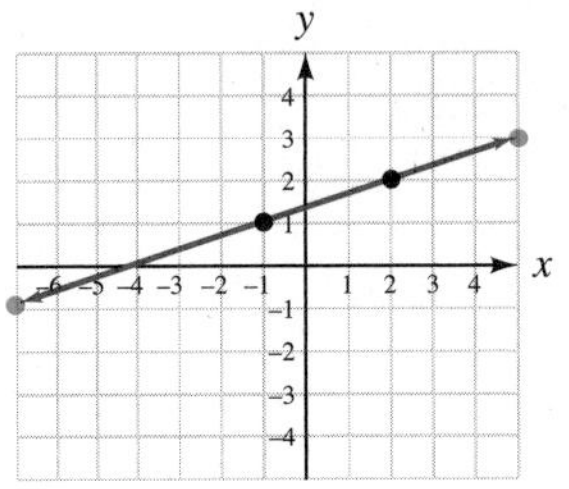

10. Evaluate each expression.

a. $\frac{10 - 4}{6 - 5}$ 6

b. $\frac{-1 - 1}{-2 - (-7)}$ $-\frac{2}{5}$

11. Write each slope in a better way.

a. $m = \frac{0}{6}$ 0

b. $m = \frac{8}{0}$ Undefined

c. $m = \frac{3}{12}$ $\frac{1}{4}$

d. $m = \frac{-10}{-5}$ 2

12. Fill in the blanks: _Horizontal_ lines have a slope of 0. Vertical lines have _undefined_ slope.

13. The *grade* of an incline is its slope expressed as a percent. Express the slope $\frac{2}{5}$ as a grade. 40%

14. Growth Rates. The graph shows how a child's height increased from ages 2 through 5. Fill in the correct units to find the rate of change in the child's height.

$$\text{Rate of change} = \frac{(40 - 31) \text{ in.}}{(5 - 2) \text{ yr}}$$

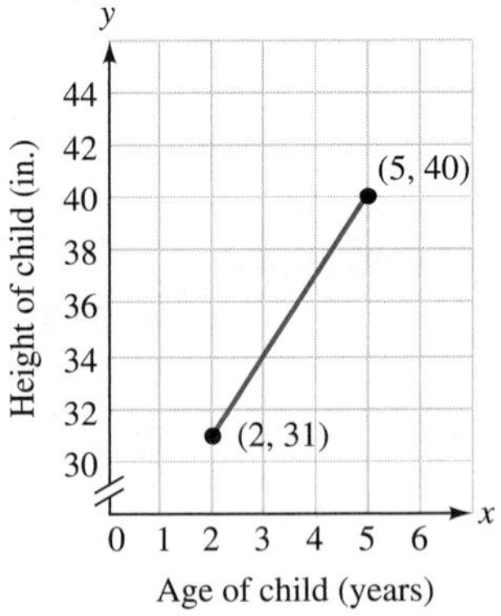

15. Find the negative reciprocal of each number.

a. 6 $-\frac{1}{6}$

b. $-\frac{7}{8}$ $\frac{8}{7}$

c. -1 1

16. Fill in the blanks.

a. Two different lines with the same slope are _parallel_.

b. If the slopes of two lines are negative reciprocals, the lines are _perpendicular_.

c. The product of the slopes of perpendicular lines is -1.

NOTATION

17. a. What is the formula used to find the slope of a line passing through (x_1, y_1) and (x_2, y_2)? $m = \frac{y_2 - y_1}{x_2 - x_1}$

b. Fill in the blanks to state the slope formula in words: *m* equals *y* _sub_ two minus *y* _sub_ one _over (divided by)_ *x* sub _two_ minus *x* sub _one_.

18. Explain the difference between y^2 and y_2. $y^2 = y \cdot y$ and y_2 is *y* sub 2.

19. Consider the points (7, 2) and (−4, 1). If we let $x_1 = 7$, then what is y_2? 1

20. The symbol ≑ is used when graphing to indicate a _break_ in the labeling of an axis.

GUIDED PRACTICE

Find the slope of each line. **See Example 1.**

21. $\frac{2}{3}$

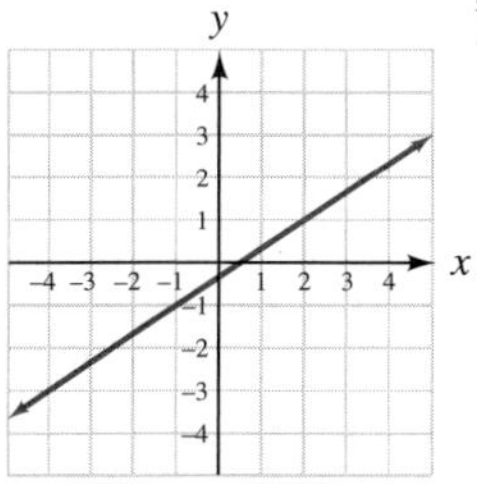

22. -1

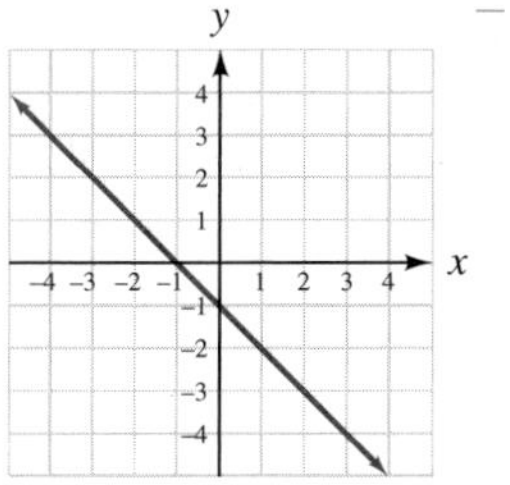

23. $\frac{4}{3}$

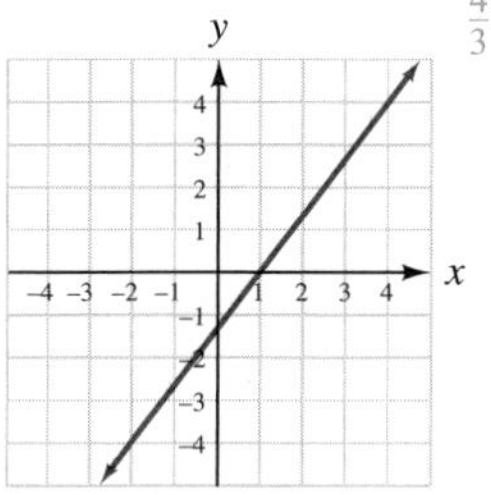

24. 4

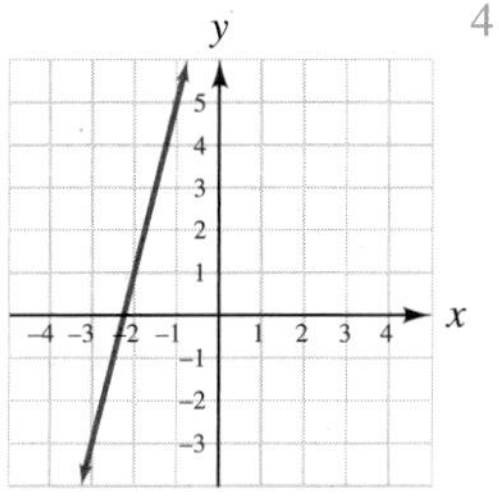

25. -2

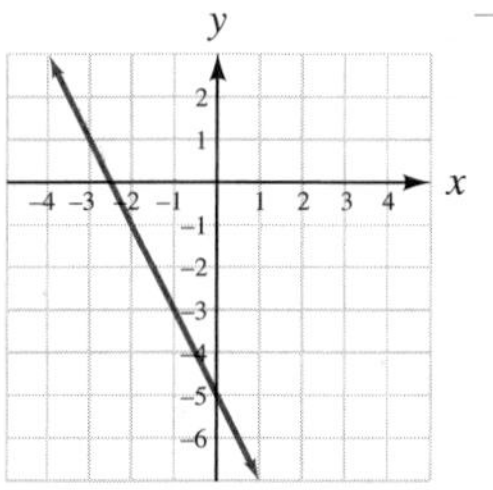

26. $-\frac{7}{8}$

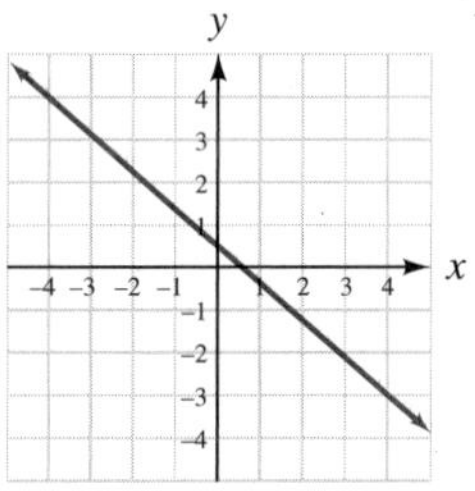

27. 0

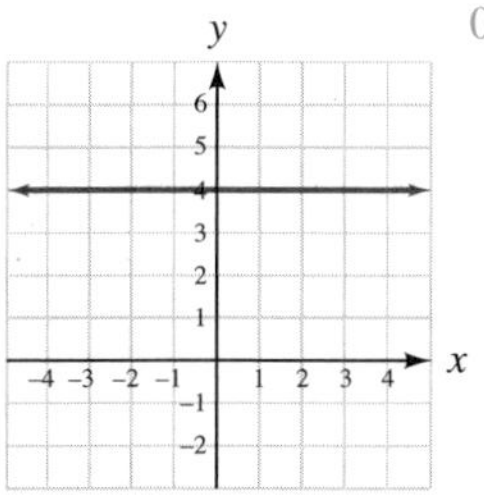

28. Undefined

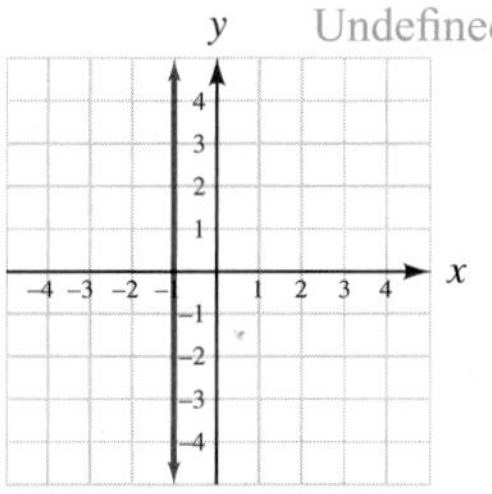

29. $-\frac{1}{5}$

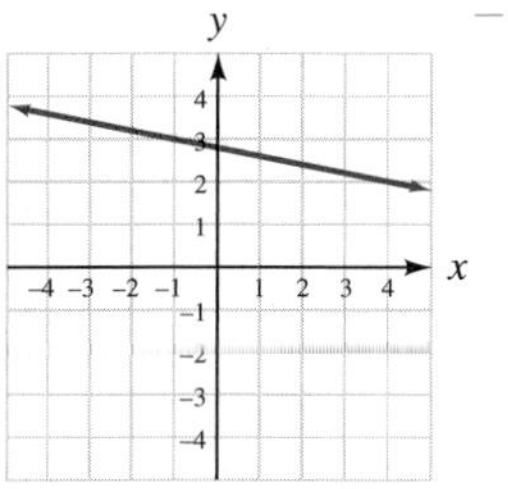

30. $\frac{4}{3}$

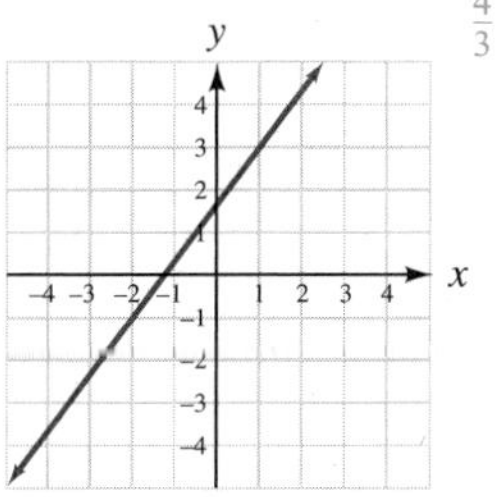

31. $\frac{1}{2}$

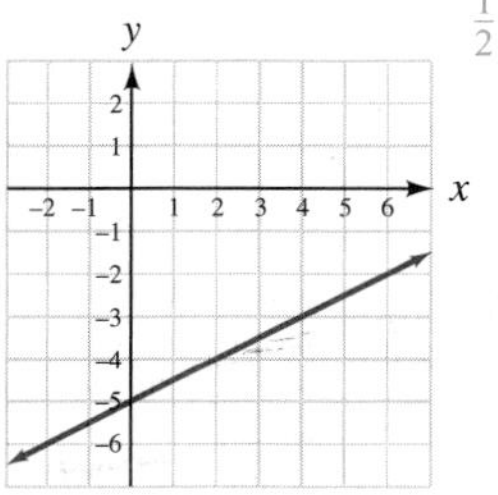

32. $-\frac{1}{7}$

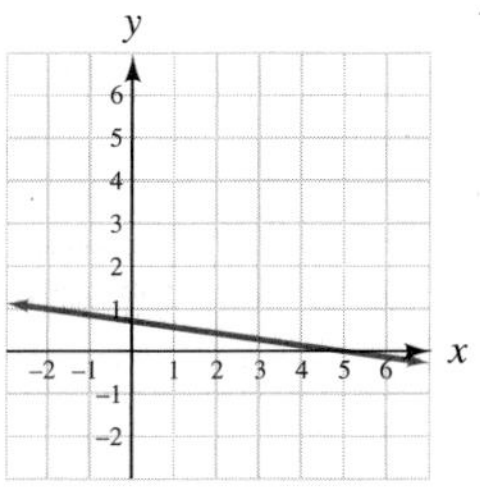

Find the slope of the line passing through the given points. **See Examples 2 and 3.**

33. $(1, 3)$ and $(2, 4)$ 1
34. $(1, 3)$ and $(2, 5)$ 2
35. $(3, 4)$ and $(2, 7)$ -3
36. $(3, 6)$ and $(5, 2)$ -2
37. $(0, 0)$ and $(4, 5)$ $\frac{5}{4}$
38. $(4, 3)$ and $(7, 8)$ $\frac{5}{3}$
39. $(-3, 5)$ and $(-5, 6)$ $-\frac{1}{2}$
40. $(6, -2)$ and $(-3, 2)$ $-\frac{4}{9}$
41. $(-2, -2)$ and $(-12, -8)$ $\frac{3}{5}$
42. $(-1, -2)$ and $(-10, -5)$ $\frac{1}{3}$
43. $(5, 7)$ and $(-4, 7)$ 0
44. $(-1, -12)$ and $(6, -12)$ 0
45. $(8, -4)$ and $(8, -3)$ Undef.
46. $(-2, 8)$ and $(-2, 15)$ Undef.
47. $(-6, 0)$ and $(0, -4)$ $-\frac{2}{3}$
48. $(0, -9)$ and $(-6, 0)$ $-\frac{3}{2}$
49. $(-2.5, 1.75)$ and $(-0.5, -7.75)$ -4.75
50. $(6.4, -7.2)$ and $(-8.8, 4.2)$ -0.75
51. $(-2.2, 18.6)$ and $(-1.7, 18.6)$ 0
52. $(4.6, 3.2)$ and $(4.6, -4.8)$ Undef.
53. $\left(-\frac{4}{7}, -\frac{1}{5}\right)$ and $\left(\frac{3}{7}, \frac{6}{5}\right)$ $\frac{7}{5}$
54. $\left(-\frac{4}{9}, -\frac{1}{8}\right)$ and $\left(\frac{5}{9}, \frac{3}{8}\right)$ $\frac{1}{2}$
55. $\left(-\frac{3}{4}, \frac{2}{3}\right)$ and $\left(\frac{4}{3}, -\frac{1}{6}\right)$ $-\frac{2}{5}$
56. $\left(\frac{1}{2}, \frac{3}{4}\right)$ and $\left(-\frac{11}{16}, -\frac{1}{2}\right)$ $\frac{20}{19}$

Determine the slope of the graph of the line that has the given table of solutions. **See Examples 2 and 3.**

57. $\frac{3}{4}$

x	y	(x, y)
-3	-1	$(-3, -1)$
1	2	$(1, 2)$

58. $-\frac{4}{3}$

x	y	(x, y)
-3	6	$(-3, 6)$
0	2	$(0, 2)$

59. 0

x	y	(x, y)
-3	6	$(-3, 6)$
0	6	$(0, 6)$

60.

x	y	(x, y)
4	-5	$(4, -5)$
4	0	$(4, 0)$

Undefined

Find the slope of each line. **See Examples 4 and 5.**

61. $y = -11$ 0
62. $y = -2$ 0
63. $y = 0$ 0
64. $x = 0$ Undefined
65. $x = 6$ Undefined
66. $x = 4$ Undefined
67. $x = -10$ Undefined
68. $y = 8$ 0
69. $y - 9 = 0$ 0
70. $x + 14 = 0$ Undefined
71. $3x = -12$ Undefined
72. $2y + 2 = -6$ 0

Determine whether the lines through each pair of points are parallel, perpendicular, or neither. **See Example 8.**

73. $(5, 3)$ and $(1, 4)$
$(-3, -4)$ and $(1, -5)$
Parallel

74. $(2, 4)$ and $(-1, -1)$
$(8, 0)$ and $(11, 5)$
Parallel

75. $(-4, -2)$ and $(2, -3)$
$(7, 1)$ and $(8, 7)$
Perpendicular

76. $(-2, 4)$ and $(6, -7)$
$(-6, 4)$ and $(5, 12)$
Perpendicular

77. $(2, 2)$ and $(4, -3)$
$(-3, 4)$ and $(-1, 9)$
Neither

78. $(-1, -3)$ and $(2, 4)$
$(5, 2)$ and $(8, -5)$
Neither

79. $(-1, 8)$ and $(-6, 8)$
$(3, 3)$ and $(3, 7)$
Perpendicular

80. $(11, 0)$ and $(11, -5)$
$(14, 6)$ and $(25, 6)$
Perpendicular

81. $(6, 4)$ and $(2, 5)$
$(-2, -3)$ and $(2, -4)$
Parallel

82. $(-3, -1)$ and $(3, -2)$
$(8, 2)$ and $(9, 8)$
Perpendicular

83. $(4, 2)$ and $(5, -3)$
$(-5, 3)$ and $(-2, 9)$
Neither

84. $(8, -3)$ and $(8, -8)$
$(11, 3)$ and $(22, 3)$
Perpendicular

Find the slope of a line perpendicular to the line passing through the given two points. **See Example 9.**

85. (0, 0) and (5, −9) $\frac{5}{9}$

86. (0, 0) and (5, 12) $-\frac{5}{12}$

87. (−1, 7) and (1, 10) $-\frac{2}{3}$

88. (−7, 6) and (0, 4) $\frac{7}{2}$

89. $\left(-2, \frac{1}{2}\right)$ and $\left(-1, \frac{3}{2}\right)$ -1

90. $\left(\frac{1}{3}, -1\right)$ and $\left(\frac{4}{3}, -2\right)$ 1

91. (−1, 2) and (−3, 6) $\frac{1}{2}$

92. (5, −4) and (−1, −7) -2

APPLICATIONS

93. Pools. Find the slope of the bottom of the swimming pool as it drops off from the shallow end to the deep end. $-\frac{2}{5}$

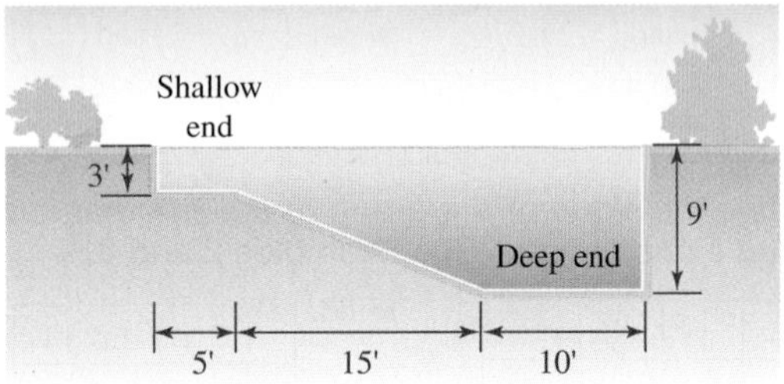

94. Drainage. Find the slope of the concrete patio slab using the 1-foot ruler, level, and 10-foot-long board shown in the illustration. (*Hint:* 10 feet = 120 in.) $\frac{1}{40}$

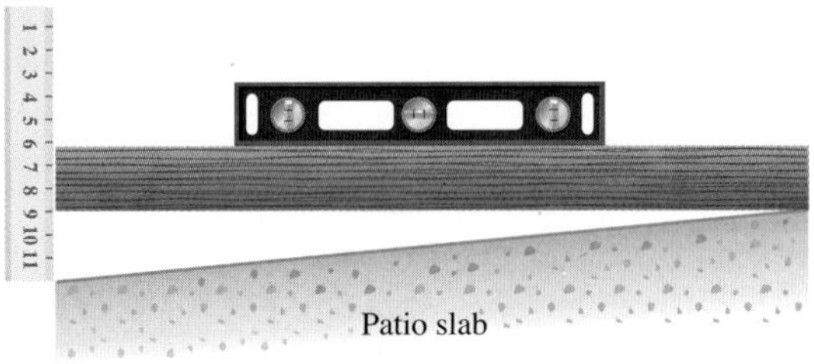

95. Grade of a Road. Refer to the illustration below. Find the slope of the decline and use that information to find the grade of the road. $\frac{1}{20}$; 5%

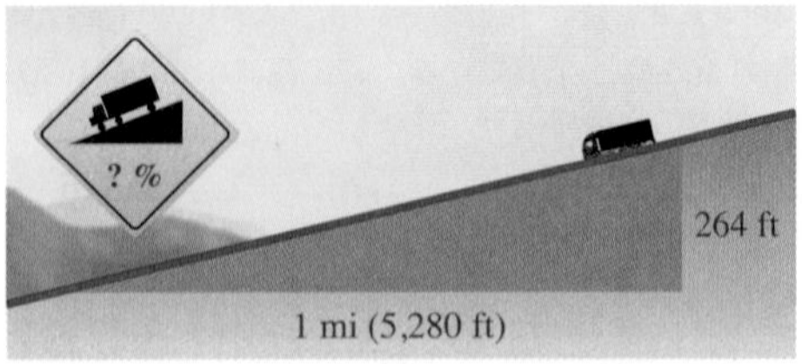

96. Streets. One of the steepest streets in the United States is Eldred Street in Highland Park, California (near Los Angeles). It rises approximately 220 feet over a horizontal distance of 665 feet. What is the grade of the street? About 33%

Vince Compagnone/©LAT

97. Treadmills. Find the slope of the jogging surface of the treadmill for a height setting of 6 inches. Then express the incline as a percent. $\frac{3}{25}$; 12%

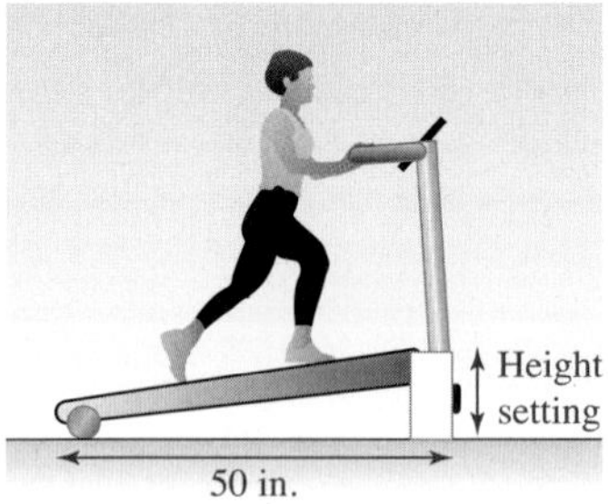

98. Architecture. Locate the coordinates of the peak of the roof if it is to have a pitch of $\frac{2}{5}$ and the roof line is to pass through the two given points in black. (10, 10)

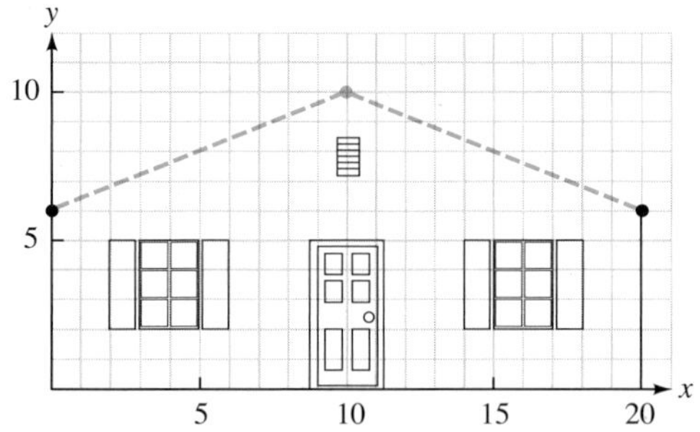

99. Carpentry. Find the pitch of each roof. Front: $\frac{3}{2}$; side: $\frac{3}{5}$

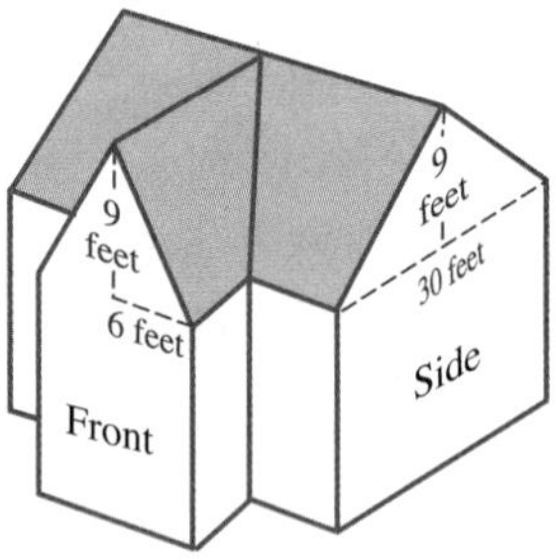

100. Doll Houses. Find x so that the pitch of the roof of the doll house is $\frac{4}{3}$. 1 ft 8 in.

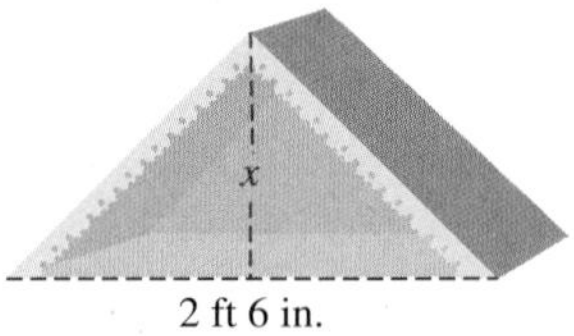

101. Irrigation. The graph on the next page shows the number of gallons of water remaining in a reservoir as water is used from it to irrigate a field. Find the rate of change in the number of gallons of water in the reservoir. A decrease of 875 gal per hour (−875 gal per hr)

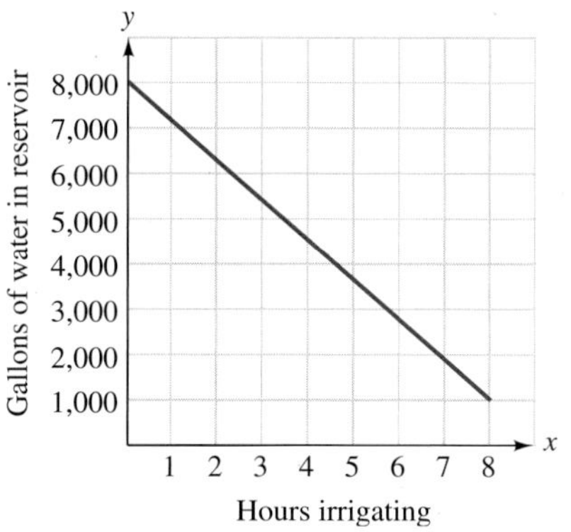

102. Commercial Jets. Examine the graph and consider trips of more than 7,000 miles by a Boeing 777. Use a rate of change to estimate how the maximum payload decreases as the distance traveled increases. a decrease of 15 lb per mi (−15 lb per mi)

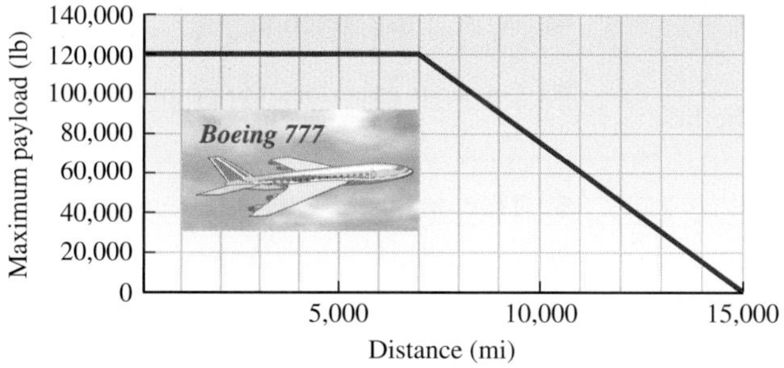

Based on data from Lawrence Livermore National Laboratory and *Los Angeles Times* (October 22, 1998).

103. Milk Production. The following graph approximates the amount of milk produced per cow in the United States for the years 1996–2009. Find the rate of change. 319 lb per yr

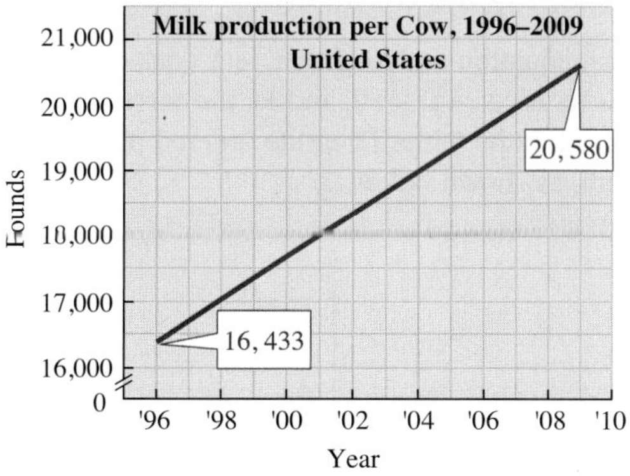

Source: USDA; Agricultural and Applied Economics, UW Madison

104. Wal-Mart. The graph in the next column approximates the sales revenue of Wal-Mart for the years 1991–2009. Find the rate of change in revenue for the years

a. 1991–1998 $12 billion per yr

b. 1998–2009 $26 billion per yr

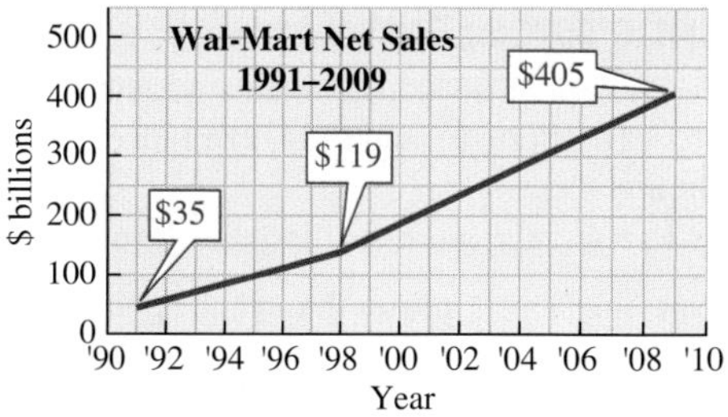

Source: wikinvest.com

105. from Campus to Careers

Dental Assistant

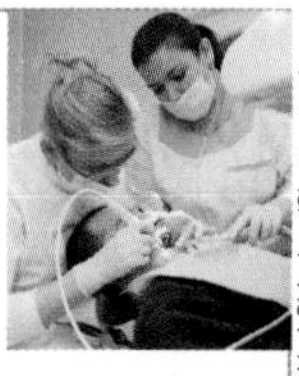

In 2000, there were approximately 6,600 students enrolled in dental assisting programs in the U.S. By 2008, that number had steadily increased to about 9,200 students. Find the rate of change in the number of students studying to be dental assistants from 2000 to 2008. (Source: American Dental Education Association) An increase of 325 students per yr

106. Firefighting. When flames are tilted due to effects of wind, firefighters measure what is called the **slope percent** of the flames. Calculate the slope percent of the flame shown below by expressing its "slope" as a percent. (Source math.fire.org) 75%

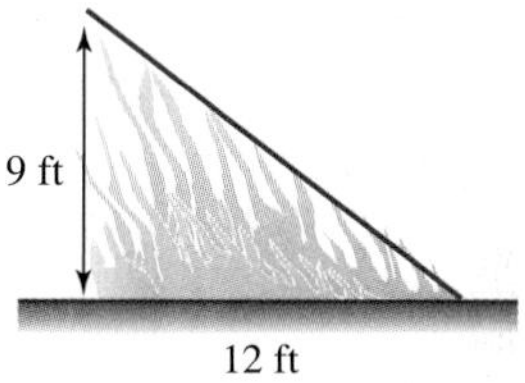

WRITING

107. Explain why the slope of a vertical line is undefined.

108. How do we distinguish between a line with positive slope and a line with negative slope?

109. Explain the error in the following solution: *Find the slope of the line that passes through* (6, 4) *and* (3, 1).

$$m = \frac{1-4}{6-3} = \frac{-3}{3} = -1$$

(crossed out)

110. Explain the difference between a rate of change that is positive and one that is negative. Give an example of each.

REVIEW

111. Halloween Candy. A candy maker wants to make a 60-pound mixture of two candies to sell for $2 per pound. If black licorice bits sell for $1.90 per pound and orange gumdrops sell for $2.20 per pound, how many pounds of each should be used? 40 lb licorice; 20 lb gumdrops

112. Medications. A doctor prescribes an ointment that is 2% hydrocortisone. A pharmacist has 1% and 5% concentrations in stock. How many ounces of each should the pharmacist use to make a 1-ounce tube? 0.75 oz of the 1%, 0.25 oz of the 5%

CHALLENGE PROBLEMS

113. Use the concept of slope to determine whether $A(-50, -10)$, $B(20, 0)$, and $C(34, 2)$ all lie on the same straight line.
$m_{\overline{AB}} = \frac{1}{7}$; $m_{\overline{BC}} = \frac{1}{7}$; $m_{\overline{AC}} = \frac{1}{7}$. Yes, they lie on the same line.

▶ **114.** A line having slope $\frac{2}{3}$ passes through the point $(10, -12)$. What is the y-coordinate of another point on the line whose x-coordinate is 16? -8

115. Subscripts are used in other disciplines besides mathematics. In what disciplines are the following symbols used?
a. H_2O and CO_2 Chemistry; water and carbon dioxide
b. C_7 and G_7 Music; chords
c. B_6 and B_{12} Medicine; vitamins

116. Evaluate $2a_2^2 + 3a_3^3 + 4a_4^4$ for $a_2 = 2$, $a_3 = 3$, and $a_4 = 4$. 1,113

SECTION 3.5 Slope–Intercept Form

OBJECTIVES

1. Use slope–intercept form to identify the slope and y-intercept of a line.
2. Write a linear equation in slope–intercept form.
3. Write an equation of a line given its slope and y-intercept.
4. Use the slope and y-intercept to graph a linear equation.
5. Recognize parallel and perpendicular lines.
6. Use slope–intercept form to write an equation to model data.

ARE YOU READY? *Are You Ready? exercises available online at www.webassign.net/brookscole*

The following problems review some basic skills that are needed when working with equations of lines in slope–intercept form.

1. a. Identify each term in the expression $3x - 6$. $3x, -6$
b. What is the coefficient of the first term? 3

2. Solve for y: $2x + 5y = 15$ $y = -\frac{2}{5}x + 3$

3. True or false: $\frac{x}{4} = \frac{1}{4}x$ True

4. Write 3 as a fraction. $\frac{3}{1}$

5. On what axis does the point (0, 6) lie? The y-axis

6. True or false: $-\frac{7}{8} = \frac{-7}{8} = \frac{7}{-8}$ True

Of all of the ways in which a linear equation can be written, one form, called *slope–intercept form,* is probably the most useful. When an equation is written in this form, two important features of its graph are evident.

1 Use Slope–Intercept Form to Identify the Slope and y-Intercept of a Line.

To explore the relationship between a linear equation and its graph, let's consider $y = 2x + 1$. To graph this equation, three values of x were selected (-1, 0, and 1), the corresponding values of y were found, and the results were entered in the table. Then the ordered pairs were plotted and a straight line was drawn through them, as shown below.

$y = 2x + 1$

x	y	(x, y)
-1	-1	$(-1, -1)$
0	1	(0, 1)
1	3	(1, 3)

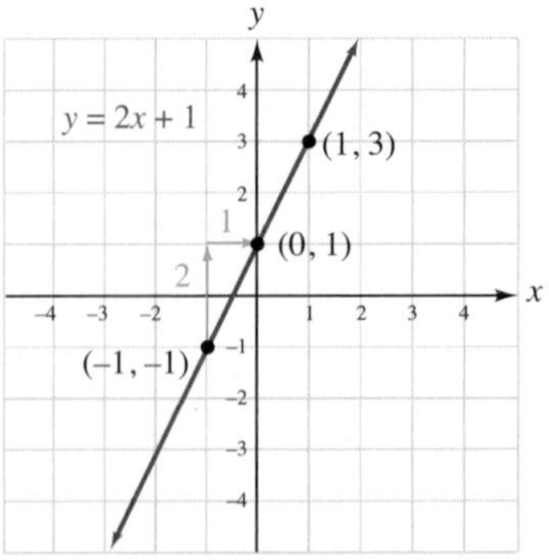

To find the slope of the line, we pick two points on the line, $(-1, -1)$ and $(0, 1)$, and draw a slope triangle and count grid squares:

$$\text{Slope} = \frac{\text{rise}}{\text{run}} = \frac{2}{1} = 2$$

From the equation and the graph, we can make two observations:

- The graph crosses the y-axis at 1. This is the same as the constant term in $y = 2x + 1$.
- The slope of the line is 2. This is the same as the coefficient of x in $y = 2x + 1$.

The Language of Algebra

Recall that a term that contains only a number is called a **constant term.**

This illustrates that the slope and y-intercept of the graph of $y = 2x + 1$ can be determined from the equation.

$$y = 2x + 1$$

The slope of the line is 2. (↑ under 2) The y-intercept is (0, 1). (↑ under 1)

These observations suggest the following form of an equation of a line.

Slope–Intercept Form of the Equation of a Line

If a linear equation is written in the form

$$y = mx + b$$

the graph of the equation is a line with slope m and y-intercept $(0, b)$.

When an equation of a line is written in slope–intercept form, the coefficient of the x-term is the line's slope and the constant term gives the y-coordinate of the y-intercept.

$$y = mx + b$$

Slope (↑ under m) y-intercept: (0, b) (↑ under b)

Caution

For equations in $y = mx + b$ form, the slope of the line is the *coefficient* of x, not the x-term. For the graph of $y = 6x - 2$:

$m = 6$
not
$m = 6x$

Linear equation	*Equation written in slope–intercept form*	*Slope*	*y-intercept*
$y = 6x - 2$	$y = 6x + (-2)$	6	$(0, -2)$
$y = -\frac{5}{4}x$	$y = -\frac{5}{4}x + 0$	$-\frac{5}{4}$	$(0, 0)$
$y = \frac{x}{2} + 3$	$y = \frac{1}{2}x + 3$	$\frac{1}{2}$	$(0, 3)$
$y = -\frac{7}{8} - x$	$y = -x + \left(-\frac{7}{8}\right)$	-1	$\left(0, -\frac{7}{8}\right)$

2 Write a Linear Equation in Slope–Intercept Form.

The equation of any nonvertical line can be written in slope–intercept form.

Slope-Intercept form

To write a linear equation in two variables in slope–intercept form, solve the equation for y.

EXAMPLE 1

Find the slope and y-intercept of the line with the given equation.

a. $8x + y = 9$ **b.** $x + 4y = 16$ **c.** $-9x - 3y = 11$

Strategy We will write each equation in slope–intercept form by solving for y.

Why When the equations are written in slope–intercept form, the slope and y-intercept of their graphs become apparent.

The Language of Algebra

When working with the slope–intercept form, use precise language: b is not the y-intercept of the graph of the equation $8x + y = 9$. b is the y-*coordinate of the y-intercept.*

Success Tip

Since we want the right side of the equation to have the form $mx + b$, we show the division by 4 as

$-\frac{x}{4} + \frac{16}{4}$ not $\frac{-x + 16}{4}$

Solution **a.** The slope and y-intercept of the graph of $8x + y = 9$ are not obvious because the equation is not in slope–intercept form. To write it in $y = mx + b$ form, we isolate y.

$8x + y = 9$ This is the given equation.

$8x + y - 8x = -8x + 9$ To isolate y on the left side, subtract 8x from both sides. Since we want the right side of the equation to have the form $mx + b$, we show the subtraction from that side as $-8x + 9$ rather than $9 - 8x$.

$y = -8x + 9$ On the left side, combine like terms: $8x - 8x = 0$.

Since $m = -8$ and $b = 9$, the slope is -8 and the y-intercept is $(0, 9)$.

b. To write the equation in slope–intercept form, we solve for y.

$x + 4y = 16$ This is the given equation.

$x + 4y - x = -x + 16$ To isolate the term 4y on the left side, subtract x from both sides. Write the subtraction before the constant term 16.

$4y = -x + 16$ On the left side. combine like terms: $x - x = 0$.

$y = \frac{-x}{4} + \frac{16}{4}$ To isolate y, undo the multiplication by 4 by dividing both sides by 4, term-by-term.

$y = -\frac{1}{4}x + 4$ Write $\frac{-x}{4}$ as $-\frac{1}{4}x$. Simplify: $\frac{16}{4} = 4$.

Since $m = -\frac{1}{4}$ and $b = 4$, the slope is $-\frac{1}{4}$ and the y-intercept is $(0, 4)$.

c. To write the equation in $y = mx + b$ form, we isolate y on the left side.

$-9x - 3y = 11$ This is the given equation.

$-3y = 9x + 11$ To isolate the term $-3y$ on the left side, add 9x to both sides. Write the addition before the constant term 11.

$\frac{-3y}{-3} = \frac{9x}{-3} + \frac{11}{-3}$ To isolate y, undo the multiplication by -3 by dividing both sides by -3, term-by-term.

$y = -3x - \frac{11}{3}$ Simplify.

Since $m = -3$ and $b = -\frac{11}{3}$, the slope is -3 and the y-intercept is $\left(0, -\frac{11}{3}\right)$.

Teaching Example 1 Find the slope and y-intercept of the line with the given equation.
a. $6x + y = 7$
b. $-x + 15y = -30$
c. $-9x - 3y = 4$
Answers:
a. $m = -6$; $(0, 7)$
b. $m = \frac{1}{15}$; $(0, -2)$
c. $m = -3$; $(0, -\frac{4}{3})$

Self Check 1 Find the slope and y-intercept of the line with the given equation.
a. $9x + y = 4$ **b.** $-x + 11y = -22$ **c.** $-10x - 2y = 7$
$m = -9$; $(0, 4)$ $m = \frac{1}{11}$; $(0, -2)$ $m = -5$; $\left(0, -\frac{7}{2}\right)$

Now Try Problems 11, 31, and 39

3 Write an Equation of a Line Given Its Slope and *y*-Intercept.

If we are given the slope and y-intercept of a line, we can write an equation of the line by substituting for m and b in the slope–intercept form.

EXAMPLE 2 Write an equation of the line with slope -1 and y-intercept $(0, 9)$.

Strategy We will use the slope–intercept form, $y = mx + b$, to write an equation of the line.

Why We know the slope of the line and its y-intercept.

Solution If the slope is -1 and the y-intercept is $(0, 9)$, then $m = -1$ and $b = 9$.

Teaching Tip: Compare the instructions in Example 1 to those in Example 2. Explain to your students that Example 2 asks them to work in reverse.

$y = mx + b$ This is the slope–intercept form.

$y = -1x + 9$ Substitute -1 for m and 9 for b.

$y = -x + 9$ Simplify: $-1x = -x$.

The equation of the line with slope -1 and y-intercept $(0, 9)$ is $y = -x + 9$.

Teaching Example 2 Write an equation of the line with slope 9 and y-intercept $(0, -5)$.
Answer: $y = 9x - 5$

Self Check 2 Write an equation of the line with slope 1 and y-intercept $(0, -12)$.
$y = x - 12$

Now Try Problem 43

EXAMPLE 3 Write an equation of the line graphed in figure (a).

Teaching Tip: Students often have difficulty picking an acceptable second point on the line. Show them what it means to *pick a point that lies on the intersection of two grid lines.*

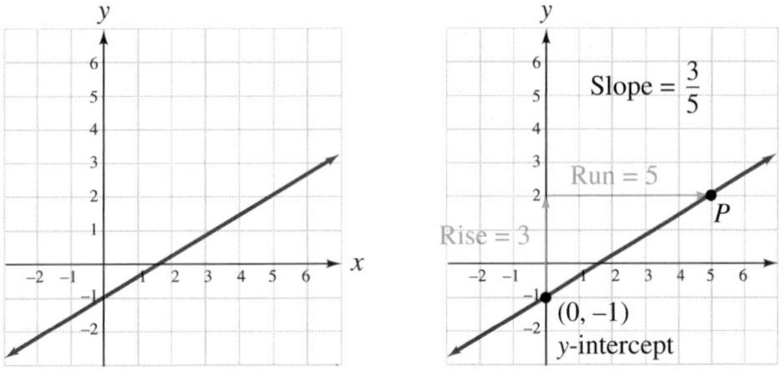

(a) (b)

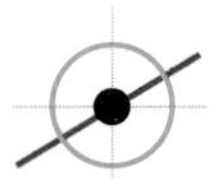

Pick a second point on the line that lies on the intersection of two grid lines.

Strategy We will use the slope–intercept form, $y = mx + b$, to write an equation of the line.

Why We can determine the slope and y-intercept of the line from the given graph.

Solution In figure (b), we highlight and label the y-intercept of the line, $(0, -1)$. Then we pick a convenient second point on the line and label it point P. Moving from the y-intercept to point P, we draw a slope triangle to find that the slope of the line is $\frac{3}{5}$. When we substitute $\frac{3}{5}$ for m and -1 for b into the slope–intercept form $y = mx + b$, we obtain an equation of the line: $y = \frac{3}{5}x - 1$.

Teaching Example 3 Write an equation of the line graphed below.

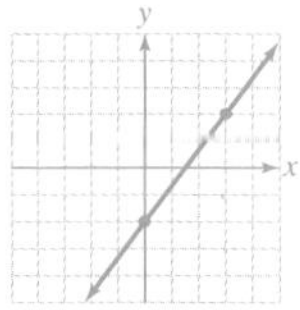

Answer: $y = \frac{4}{3}x - 2$

Self Check 3 Write an equation of the line graphed here.
$y = -\frac{3}{2}x + 2$

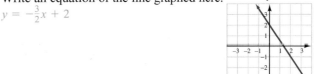

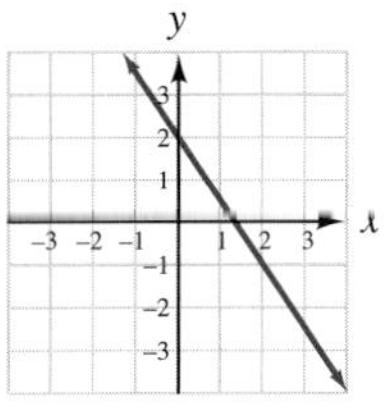

Now Try Problem 55

4 Use the Slope and y-Intercept to Graph a Linear Equation.

If we know the slope and y-intercept of a line, we can graph the line.

EXAMPLE 4 Use the slope and y-intercept to graph $y = 5x - 4$.

Strategy We will examine the equation to identify the slope and the y-intercept of the line to be graphed. Then we will plot the y-intercept and use the slope to determine a second point on the line.

Why Once we locate two points on the line, we can draw the graph of the line.

Solution Since $y = 5x - 4$ is written in $y = mx + b$ form, we know that its graph is a line with a slope of 5 and a y-intercept of $(0, -4)$. To draw the graph, we begin by plotting the y-intercept. The slope can be used to find another point on the line.

Notation

Slopes that are integers can be written in $\frac{\text{rise}}{\text{run}}$ form with the run equal to 1. For example, $m = 5 = \frac{5}{1}$.

If we write the slope as the fraction $\frac{5}{1}$, the rise is 5 and the run is 1. From $(0, -4)$, we move 5 units *upward* (because the numerator, 5, is positive) and 1 unit to the right (because the denominator, 1, is positive). This locates a second point on the line, $(1, 1)$. The line through $(0, -4)$ and $(1, 1)$ is the graph of $y = 5x - 4$.

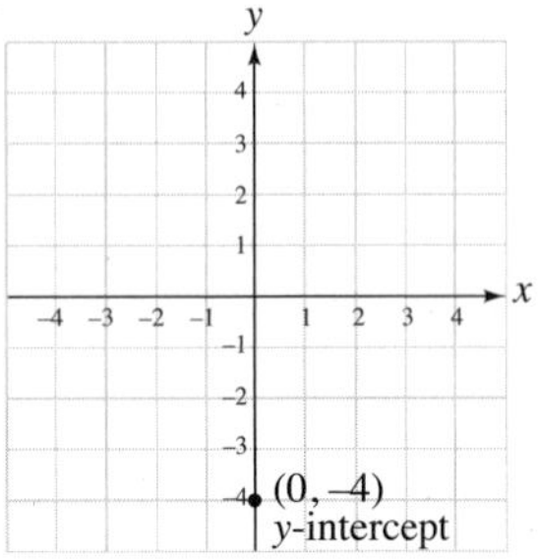

Plot the y-intercept.

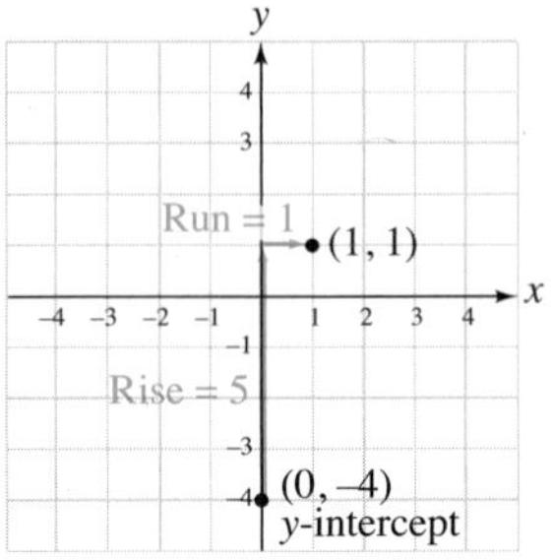

From (0, –4), draw the rise and run parts of the slope triangle for $m = \frac{5}{1}$ to find another point on the line.

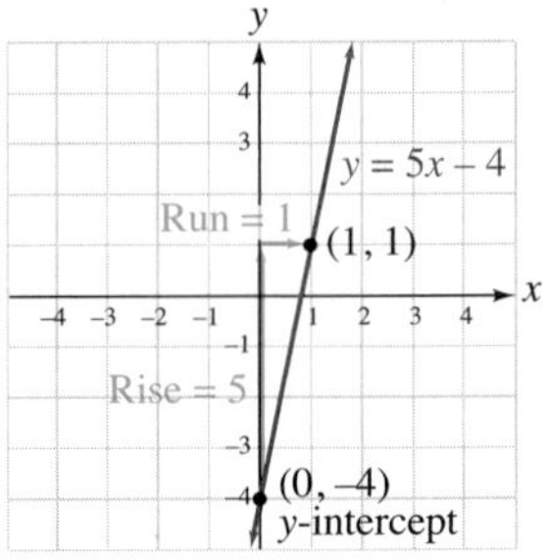

Use a straightedge to draw a line through the two points.

Caution

When using the y-intercept and the slope to graph a line, remember to draw the slope triangle from the y-intercept, **not the origin.**

Since $\frac{5}{1} = \frac{-5}{-1}$, an alternate way to find another point on the line is to write the slope as $\frac{-5}{-1}$. Like before, we begin at the y-intercept $(0, -4)$. Since the rise is negative, we move 5 units *downward,* and since the run is negative, we then move 1 unit to the *left.* We arrive at $(-1, -9)$, another point on the graph of $y = 5x - 4$.

Teaching Example 4 Use the slope and y-intercept to graph $y = 4x - 5$.
Answer:

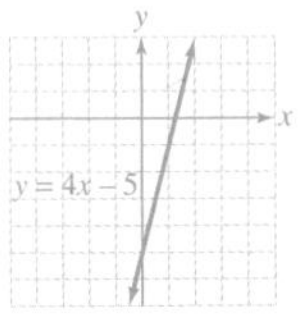

Self Check 4 Use the slope and y-intercept to graph $y = 2x - 3$. See AIE Appendix 3.

Now Try ▶ **Problem 63**

EXAMPLE 5 Use the slope and y-intercept to graph $4x + 3y = 6$.

Strategy We will write the equation of the line in slope–intercept form, $y = mx + b$. Then we will identify the slope and y-intercept of its graph.

Why We can use that information to plot two points that the line passes through.

Solution To write $4x + 3y = 6$ in slope–intercept form, we isolate y on the left side.

$4x + 3y = 6$ This is the given equation.

$3y = -4x + 6$ To isolate the term $3y$ on the left side, subtract $4x$ from both sides. Write the subtraction before the constant term 6.

$\frac{3y}{3} = \frac{-4x}{3} + \frac{6}{3}$ To isolate y, undo the multiplication by 3 by dividing both sides by 3, term-by-term.

$y = -\frac{4}{3}x + 2$ $m = -\frac{4}{3}$ and $b = 2$.

Notation

The negative sign in $-\frac{4}{3}$ can be attached to either the numerator or the denominator, since $-\frac{4}{3} = \frac{-4}{3} = \frac{4}{-3}$.

The slope of the line is $-\frac{4}{3}$ and the y-intercept is $(0, 2)$. To draw the graph, we begin by plotting the y-intercept. If we write the slope as $\frac{-4}{3}$, the rise is -4 and the run is 3. From $(0, 2)$, we then move 4 units *downward* (because the numerator is negative) and 3 units to the *right* (because the denominator is positive). This locates a second point on the line, $(3, -2)$.

Since $\frac{-4}{3} = \frac{4}{-3}$, we can find another point on the graph by writing the slope as $\frac{4}{-3}$. In this case, the rise is 4 and the run is -3. Again, we begin at the y-intercept (0, 2), but this time, we move 4 units *upward* because the rise is positive. Then we move 3 units to the *left,* because the run is negative, and arrive at the point $(-3, 6)$. The line that passes through (0, 2), $(3, -2)$, and $(-3, 6)$ is the graph of $4x + 3y = 6$.

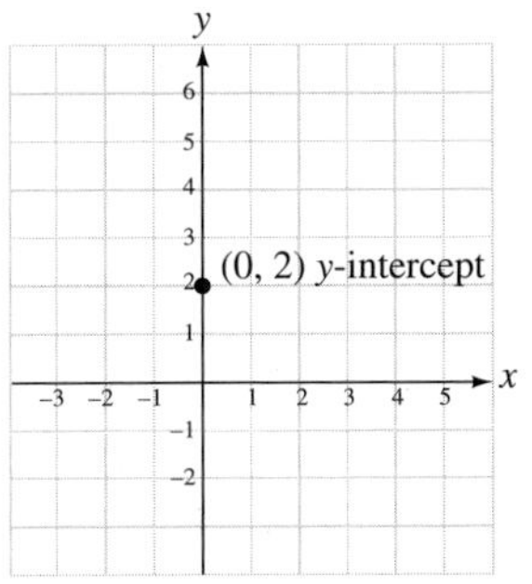

Plot the y-intercept.

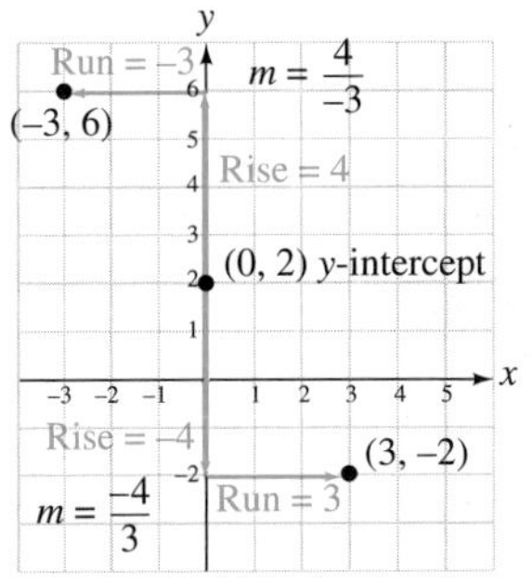

From (0, 2), draw the rise and run parts of the slope triangle for $m = \frac{-4}{3}$ $\left(\text{or } m = \frac{4}{-3}\right)$ to find another point on the line.

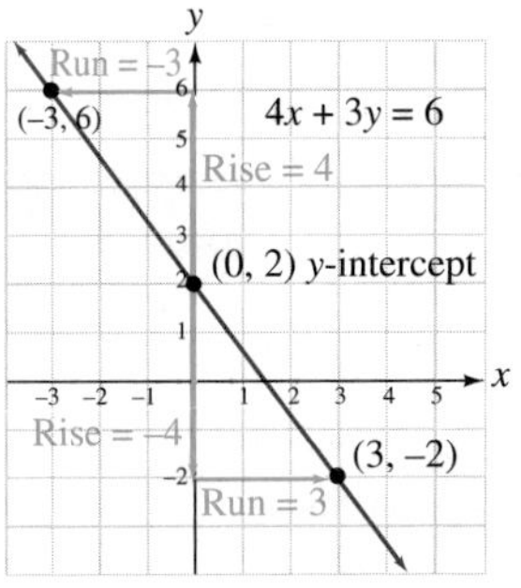

Use a straightedge to draw a line through the points.

To check the graph, verify that $(3, -2)$ and $(-3, 6)$ satisfy $4x + 3y = 6$.

Teaching Example 5 Use the slope and y-intercept to graph $5x + 2y = 6$.
Answer:

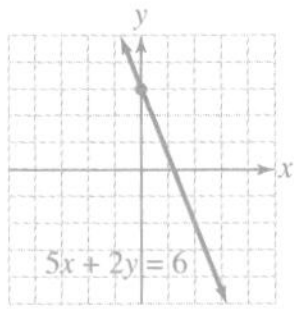

Self Check 5 Use the slope and y-intercept to graph $5x + 6y = 12$. See AIE Appendix 3.

Now Try Problem 71

5 Recognize Parallel and Perpendicular Lines.

The slope–intercept form enables us to quickly identify parallel and perpendicular lines.

EXAMPLE 6 Are the graphs of $y = -5x + 6$ and $x - 5y = -10$ parallel, perpendicular, or neither?

Strategy We will find the slope of each line and then compare the slopes.

Why If the slopes are equal, the lines are parallel. If the slopes are negative reciprocals, the lines are perpendicular. Otherwise, the lines are neither parallel nor perpendicular.

Solution The graph of $y = -5x + 6$ is a line with slope -5. To find the slope of the graph of $x - 5y = -10$, we will write the equation in slope–intercept form.

Success Tip

Graphs are not necessary to determine if two lines are parallel, perpendicular, or neither. We simply examine the slopes of the lines.

$$x - 5y = -10 \quad \text{This is the second given equation.}$$

$$-5y = -x - 10 \quad \text{To isolate the term } -5y \text{ on the left side, subtract } x \text{ from both sides.}$$

$$\frac{-5y}{-5} = \frac{-x}{-5} - \frac{10}{-5} \quad \text{To isolate } y\text{, undo the multiplication by } -5 \text{ by dividing both sides by } -5 \text{ term-by-term.}$$

$$y = \frac{x}{5} + 2 \quad m = \tfrac{1}{5} \text{ because } \tfrac{x}{5} = \tfrac{1}{5}x.$$

The graph of $y = \frac{x}{5} + 2$ is a line with slope $\frac{1}{5}$. Since the slopes -5 and $\frac{1}{5}$ are negative reciprocals, the lines are perpendicular.

Teaching Example 6 Determine whether the graphs of $y = -\frac{2}{3}x + 3$ and $3x - 2y = 14$ are parallel, perpendicular, or neither.
Answer: Perpendicular

Self Check 6 Determine whether the graphs of $y = 4x + 6$ and $x - 4y = -8$ are parallel, perpendicular, or neither. Neither

Now Try Problem 81

6 Use Slope–Intercept Form to Write an Equation to Model Data.

The concepts that we have studied in the first four sections of this chapter can be used to write equations that mathematically describe, or **model,** many real-world situations. To make the equation more descriptive of the given situation, we can replace the variables x and y in $y = mx + b$ with other letters.

EXAMPLE 7 **Group Discounts.** To promote group sales for an Alaskan cruise, a travel agency reduces the regular ticket price of \$4,500 by \$5 for each person traveling in the group.

a. Write a linear equation that gives the per-person cost c of the cruise, if p people travel together.

b. Use the equation to determine the per-person cost if 55 teachers travel together.

Strategy We will determine the slope and the y-intercept of the graph of the equation from the given facts about the cruise.

Why If we know the slope and y-intercept, we can use the slope–intercept form, $y = mx + b$, to write an equation to model the situation.

Solution

a. We will let p represent the number of people traveling in the group and c represent the per-person cost of the cruise. Since the cost depends on the number of people in the group, the linear equation that models this situation is

$c = mp + b$ — This is the slope-intercept form $y = mx + b$ with the variable c in place of y and the variable p in place of x.

Since the per-person cost of the cruise steadily decreases as the number of people in the group increases, the rate of change of $-\$5$ per person is the slope of the graph of the equation. Thus, m is -5.

If 0 people take the cruise, there will be no discount and the per-person cost of the cruise will be \$4,500. Written as an ordered pair of the form (p, c), we have $(0, 4{,}500)$. When graphed, this point would be the c-intercept. Thus, b is 4,500.

Substituting for m and b in the slope–intercept form $c = mp + b$, we obtain the linear equation that models the pricing arrangement.

A graph of the equation for groups of up to 100 $(c \leq 100)$ is shown on the right.

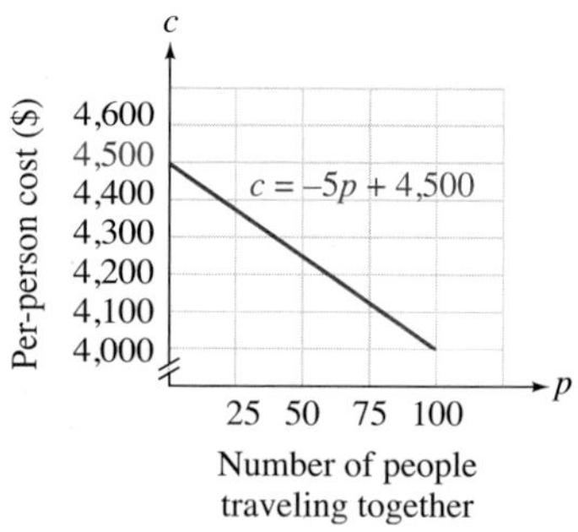

$c = -5p + 4{,}500$ — $m = -5$ and $b = 4{,}500$.

The Language of Algebra

To determine the slope m for modeling problems like this, look for a phrase that describes a **rate of change,** such as:

- \$5 for each person
- \$20 per unit
- 5° every minute
- 12 feet a year

b. To find the per-person cost of the cruise for a group of 55 people, we substitute 55 for p and evaluate the right side of the equation.

$c = -5p + 4{,}500$

$c = -5(\mathbf{55}) + 4{,}500$ — Substitute 55 for p.

$c = -275 + 4{,}500$ — Do the multiplication.

$c = 4{,}225$ — Do the addition.

If a group of 55 people travel together, the cruise will cost each person \$4,225.

Self Check 7 **Group Discounts.** Write a linear equation in slope–intercept form that gives the cost of the cruise of Example 7 if a \$10.50-per-person discount is offered for groups. $c = -10.50p + 4{,}500$

Now Try Problem 91

Teaching Example 7 Group Discounts. To promote group sales for an African safari, the tour company reduces the regular ticket price of \$15,850 by \$75 for each person traveling in the group.

a. Write a linear equation that gives the per-person cost of the safari if p people travel together.

b. Use the equation to determine the per-person cost if 8 co-workers travel together.

Answers:

a. $c = -75p + 15{,}850$ b. \$15,250

SECTION 3.5 STUDY SET

VOCABULARY

Fill in the blanks.

1. The equation $y = mx + b$ is called the slope–intercept form of the equation of a line.
2. The graph of the linear equation $y = mx + b$ has y-intercept $(0, b)$ and slope m.

CONCEPTS

3. Determine whether each equation is in slope–intercept form.
   a. $7x + 4y = 2$ No
   b. $5y = 2x - 3$ No
   c. $y = 6x + 1$ Yes
   d. $x = 4y - 8$ No
4. a. Fill in the blank: To write a linear equation in two variables in slope–intercept form, solve the equation for y.
   b. Solve $4x + y = 9$ for y. $y = -4x + 9$
5. Simplify the right side of each equation.
   a. $y = \frac{4x}{2} + \frac{16}{2}$ $y = 2x + 8$
   b. $y = \frac{15x}{-3} + \frac{9}{-3}$ $y = -5x - 3$
   c. $y = \frac{2x}{6} - \frac{6}{6}$ $y = \frac{1}{3}x - 1$
   d. $y = \frac{-9x}{-5} - \frac{20}{-5}$ $y = \frac{9}{5}x + 4$
6. Find the slope and y-intercept of each line graphed below. Then use that information to write an equation for that line.

a.

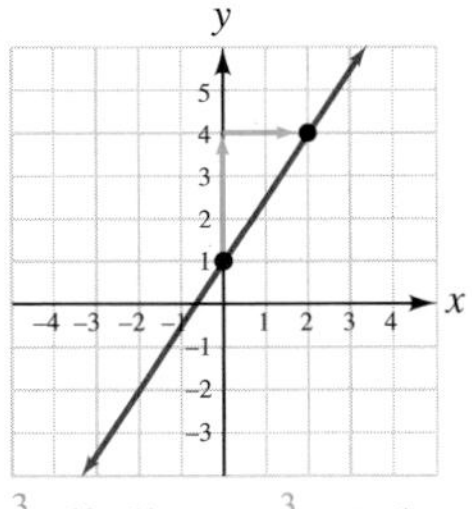

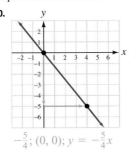

$\frac{3}{2}$; $(0, 1)$; $y = \frac{3}{2}x + 1$

b.

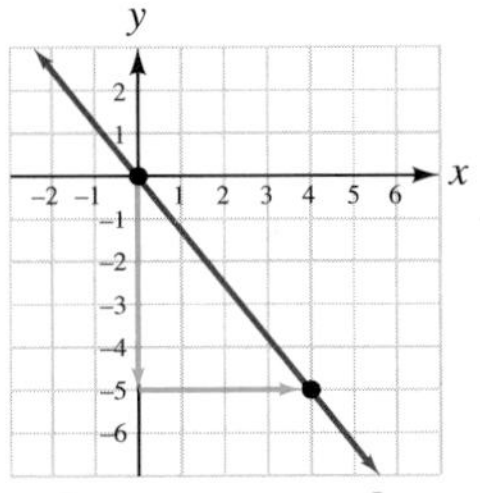

$-\frac{5}{4}$; $(0, 0)$; $y = -\frac{5}{4}x$

NOTATION

Complete the solution by solving the equation for y. Then find the slope and the y-intercept of its graph.

7.
$$2x + 5y = 15$$
$$2x + 5y - 2x = -2x + 15$$
$$5y = -2x + 15$$
$$\frac{5y}{5} = \frac{-2x}{5} + \frac{15}{5}$$
$$y = -\frac{2}{5}x + 3$$
The slope is $-\frac{2}{5}$ and the y-intercept is $(0, 3)$.
8. What is the slope–intercept form of the equation of a line? $y = mx + b$
9. Fill in the blanks: $-\frac{3}{2} = \frac{3}{-2} = \frac{-3}{2}$
10. Determine whether each statement is true or false.
   a. $\frac{x}{6} = \frac{1}{6}x$ True
   b. $\frac{5}{3}x = \frac{5x}{3}$ True

GUIDED PRACTICE

Find the slope and the y-intercept of the line with the given equation. **See Example 1.**

11. $y = 4x + 2$ $4, (0, 2)$
12. $y = 7x + 3$ $7, (0, 3)$
13. $y = -5x - 8$ $-5, (0, -8)$
14. $y = -4x - 2$ $-4, (0, -2)$
15. $y = 25x - 9$ $25, (0, -9)$
16. $y = 6x - 1$ $6, (0, -1)$
17. $y = 11 - x$ $-1, (0, 11)$
18. $y = 12 - 4x$ $-4, (0, 12)$
19. $y = \frac{1}{2}x + 6$ $\frac{1}{2}, (0, 6)$
20. $y = \frac{4}{5}x - 9$ $\frac{4}{5}, (0, -9)$
21. $y = \frac{x}{4} - \frac{1}{2}$ $\frac{1}{4}, \left(0, -\frac{1}{2}\right)$
22. $y = \frac{x}{15} - \frac{3}{4}$ $\frac{1}{15}, \left(0, -\frac{3}{4}\right)$
23. $y = -5x$ $-5, (0, 0)$
24. $y = 14x$ $14, (0, 0)$
25. $y = x$ $1, (0, 0)$
26. $y = -x$ $-1, (0, 0)$
27. $y = -2$ $0, (0, -2)$
28. $y = 30$ $0, (0, 30)$
29. $-5y - 2 = 0$ $0, \left(0, -\frac{2}{5}\right)$
30. $3y - 13 = 0$ $0, \left(0, \frac{13}{3}\right)$
31. $x + y = 8$ $-1, (0, 8)$
32. $x - y = -30$ $1, (0, 30)$
33. $6y = x - 6$ $\frac{1}{6}, (0, -1)$
34. $2y = x + 20$ $\frac{1}{2}, (0, 10)$
35. $-4y = 6x - 4$ $-\frac{3}{2}, (0, 1)$
36. $-6y = 8x + 6$ $-\frac{4}{3}, (0, -1)$
37. $2x + 3y = 6$ $-\frac{2}{3}, (0, 2)$
38. $4x + 5y = 25$ $-\frac{4}{5}, (0, 5)$
39. $3x - 5y = 15$ $\frac{3}{5}, (0, -3)$
40. $x - 6y = 6$ $\frac{1}{6}, (0, -1)$
41. $-6x + 6y = -11$ $1, \left(0, -\frac{11}{6}\right)$
42. $-4x + 4y = -9$ $1, \left(0, -\frac{9}{4}\right)$

Write an equation of the line with the given slope and y-intercept and graph it. **See Example 2.** See AIE Appendix 3.

43. Slope 5, y-intercept $(0, -3)$ $y = 5x - 3$
44. Slope -2, y-intercept $(0, 1)$ $y = -2x + 1$
45. Slope -3, y-intercept $(0, 6)$ $y = -3x + 6$
46. Slope 4, y-intercept $(0, -1)$ $y = 4x - 1$
47. Slope $\frac{1}{4}$, y-intercept $(0, -2)$ $y = \frac{1}{4}x - 2$
48. Slope $\frac{1}{3}$, y-intercept $(0, -5)$ $y = \frac{1}{3}x - 5$
49. Slope $-\frac{8}{3}$, y-intercept $(0, 5)$ $y = -\frac{8}{3}x + 5$
50. Slope $-\frac{7}{6}$, y-intercept $(0, 2)$ $y = -\frac{7}{6}x + 2$
51. Slope $\frac{6}{5}$, y-intercept $(0, 0)$ $y = \frac{6}{5}x$
52. Slope $\frac{5}{4}$, y-intercept $(0, 0)$ $y = \frac{5}{4}x$
53. Slope -2, y-intercept $\left(0, \frac{1}{2}\right)$ $y = -2x + \frac{1}{2}$
54. Slope -3, y-intercept $\left(0, -\frac{1}{2}\right)$ $y = -3x - \frac{1}{2}$

***Write an equation for each line shown below.* See Example 3.**

55. $y = 5x - 1$

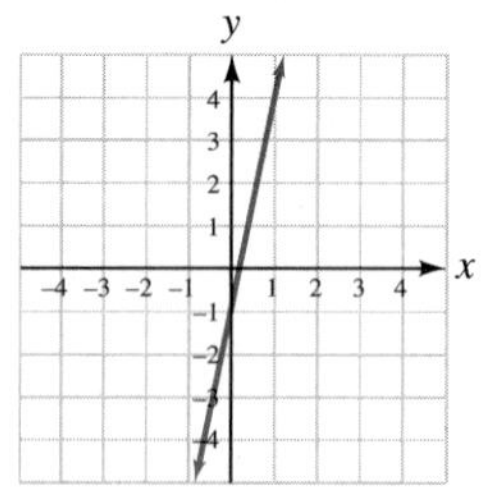

▶ 56. $y = 3x - 1$

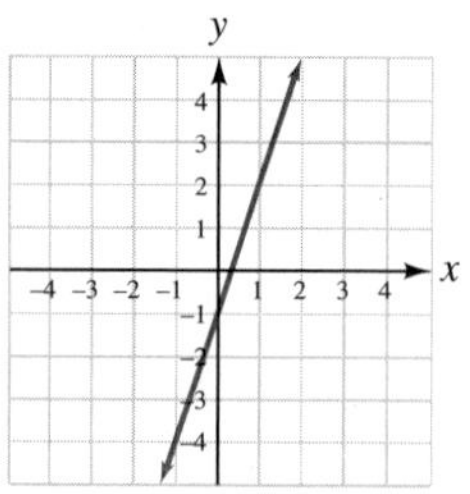

57. $y = -2x + 3$

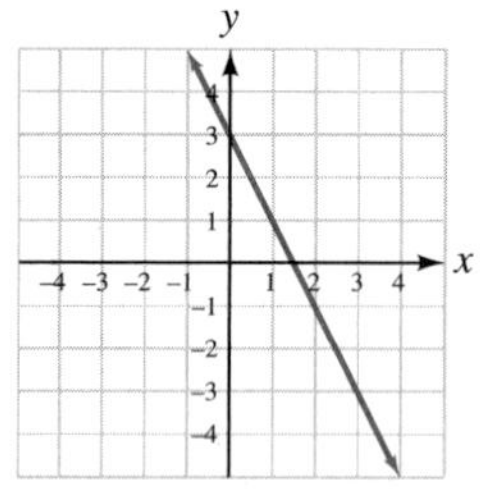

▶ 58. $y = -4x + 2$

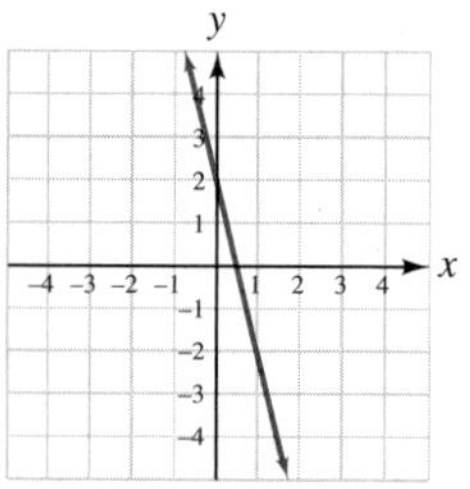

59. $y = \frac{4}{5}x - 2$

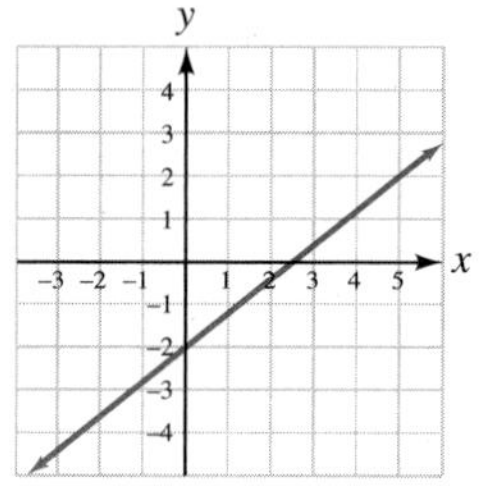

▶ 60. $y = \frac{3}{2}x - 4$

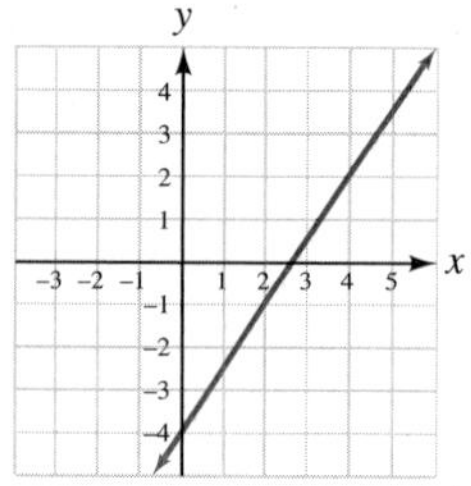

▶ 61. $y = -\frac{5}{3}x + 2$

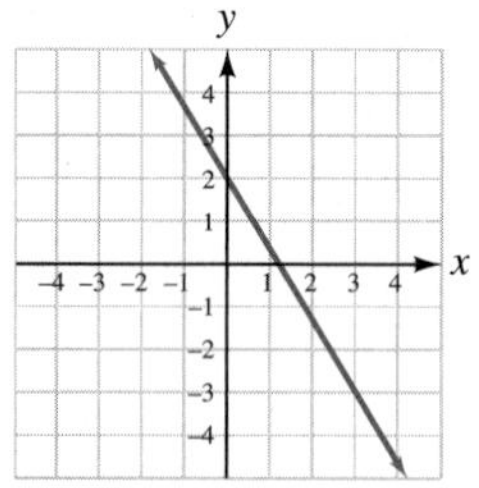

62. $y = -\frac{5}{6}x + 3$

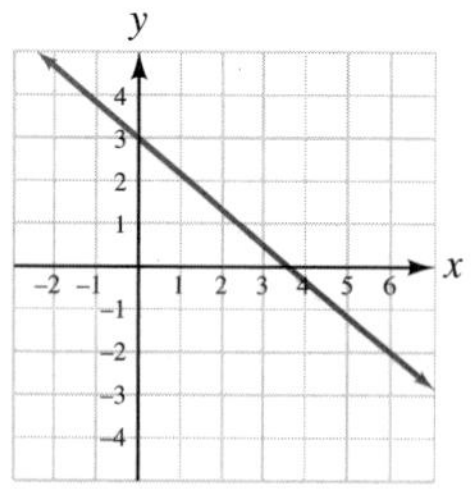

***Find the slope and the y-intercept of the graph of each equation and graph it.* See Examples 4 and 5.** See AIE Appendix 3.

▶ 63. $y = 3x + 3$ $3, (0, 3)$

64. $y = -3x + 5$ $-3, (0, 5)$

65. $y = \frac{1}{2}x + 2$ $\frac{1}{2}, (0, 2)$

▶ 66. $y = \frac{x}{3}$ $\frac{1}{3}, (0, 0)$

▶ 67. $y = -3x$ $-3, (0, 0)$

68. $y = -4x$ $-4, (0, 0)$

▶ 69. $4x + y = -4$ $-4, (0, -4)$

70. $2x + y = -6$ $-2, (0, -6)$

▶ 71. $3x + 4y = 16$ $-\frac{3}{4}, (0, 4)$

▶ 72. $2x + 3y = 9$ $-\frac{2}{3}, (0, 3)$

▶ 73. $10x - 5y = 5$ $2, (0, -1)$

▶ 74. $4x - 2y = 6$ $2, (0, -3)$

***For each pair of equations, determine whether their graphs are parallel, perpendicular, or neither.* See Example 6.**

▶ 75. $y = 6x + 8$, $y = 6x$ Parallel

76. $y = -9x - 3$, $y = -9x$ Parallel

▶ 77. $y = x$, $y = -x$ Perpendicular

▶ 78. $y = 3x$, $y = 4x$ Neither

79. $y = \frac{1}{2}x - \frac{4}{5}$, $y = 0.5x + 3$ Parallel

▶ 80. $y = 3x - 15$, $y = -\frac{1}{3}x + 4$ Perpendicular

▶ 81. $y = -2x - 9$, $2x - y = 9$ Neither

82. $y = \frac{3}{4}x + 1$, $4x - 3y = 15$ Neither

83. $3x = 5y - 10$, $5x = 1 - 3y$ Perpendicular

▶ 84. $-2y = 2 - x$, $2x - 3 = 4y$ Parallel

85. $x - y = 12$, $-2x + 2y = -23$ Parallel

86. $y = -3x + 1$, $3y = x - 5$ Perpendicular

87. $x = 9$, $y = 8$ Perpendicular

▶ 88. $-x + 4y = 10$, $2y + 16 = -8x$ Perpendicular

89. $-4x + 3y = -12$, $8x + 6y = 54$ Neither

▶ 90. $-5x + 2y = -8$, $15x + 6y = -48$ Neither

APPLICATIONS

▶ 91. Production Costs. A television production company charges a basic fee of \$5,000 and then \$2,000 an hour when filming a commercial.

a. Write a linear equation that describes the relationship between the total production costs c and the hours h of filming. $c = 2{,}000h + 5{,}000$

b. Use your answer to part a to find the production costs if a commercial required 8 hours of filming. \$21,000

▶ 92. College Fees. Each semester, students enrolling at a community college must pay tuition costs of \$20 per unit as well as a \$40 student services fee.

a. Write a linear equation that gives the total fees t to be paid by a student enrolling at the college and taking x units. $t = 20x + 40$

b. Use your answer to part a to find the enrollment cost for a student taking 12 units. \$280

93. Chemistry. A portion of a student's chemistry lab manual is shown on the right. Use the information to write a linear equation relating the temperature F in degrees Fahrenheit of the compound to the time t (in minutes) elapsed during the lab procedure. $F = 5t - 10$

> Chem. Lab #1 Aug. 13
> **Step 1:** Removed compound from freezer @ −10°F.
>
> **Step 2:** Used heating unit to raise temperature of compound 5° F every minute.

94. Rentals. Use the information in the newspaper advertisement to write a linear equation that gives the amount of income A (in dollars) the apartment owner will receive when the unit is rented for m months. $A = 500m + 250$

> **APARTMENT FOR RENT**
> 1 bedroom/1 bath, with garage
> \$500 per month +
> \$250 nonrefundable one-time security fee.

95. **Employment Services.** A policy statement of LIZCO, Inc., is shown below. Suppose a secretary had to pay an employment service $500 to get placed in a new job at LIZCO. Write a linear equation that tells the secretary the actual cost c of the employment service to her m months after being hired. $c = -20m + 500$

> **Policy no. 23452**—A new hire will be reimbursed by LIZCO for any employment service fees paid by the employee at the rate of $20 per month.

96. **Printing Presses.** Every three minutes, 100 feet of paper is used off of an 8,000 foot-roll to print the pages of a magazine. Write a linear equation that relates the number of feet of paper that remain on the roll and the number of minutes the printing press has been operating. $p = -\frac{100}{3}m + 8{,}000$

97. **Sewing Costs.** A tailor charges a basic fee of $20 plus $5 per letter to sew an athlete's name on the back of a jacket. Write a linear equation that will find the cost to have a name containing x letters sewn on the back of a jacket. $c = 5x + 20$

98. **Salad Bars.** For lunch, a delicatessen offers a "Salad and Soda" special where customers serve themselves at a well-stocked salad bar. The cost is $2.00 for the drink and 42¢ an ounce for the salad. Write a linear equation that will find the cost of a "Salad and Soda" lunch when a salad weighing x ounces is purchased. $c = 0.42x + 2.00$

99. **iPads.** When a student purchased an Apple iPad with Wi-Fi + 3G for $629.99, he also enrolled in a 250 MB data plan that cost $14.95 per month.
    a. Write a linear equation that gives the cost for him to purchase and use the iPad for m months. $c = 14.95m + 629.99$
    b. Use your answer to part a to find the cost to purchase and use the iPad for 2 years. $988.79

100. **Online Games.** A new Playstation 3 costs $310.50 and membership in an online videogame multiplayer network cost $18.49 per month.
    a. Write a linear equation that gives the cost for someone to buy the machine and belong to the online network for m months. $c = 18.49m + 310.50$
    b. Use your answer to part a to find the cost to buy the machine and belong to the network for 3 years. $976.14

101. **Navigation.** The graph below shows the recommended speed at which a ship should proceed into head waves of various heights.
    a. What information does the y-intercept of the line give? When there are no head waves, the ship can travel at 18 knots.
    b. What is the rate of change in the recommended speed of the ship as the wave height increases? $-\frac{1}{2}$ knot/ft
    c. Write an equation of the line. $y = -\frac{1}{2}x + 18$

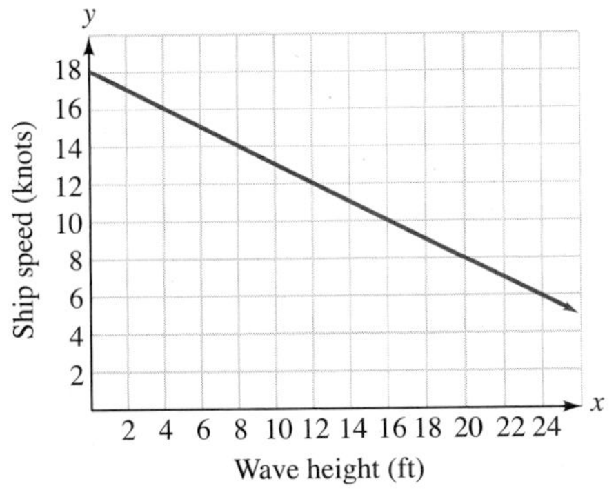

102. **Debt.** The graph below estimates the amount of credit card debt per U.S. household for the years 1990–2010.
    a. What information does the d-intercept of the line give? The credit card debt per household in 1990 was $2,500.
    b. What was the rate of change in the amount of debt per household? An increase of $275 per yr
    c. Write an equation of the line, where d is the approximate credit card debt and t is the number of years since 1990. $d = 275t + 2{,}500$

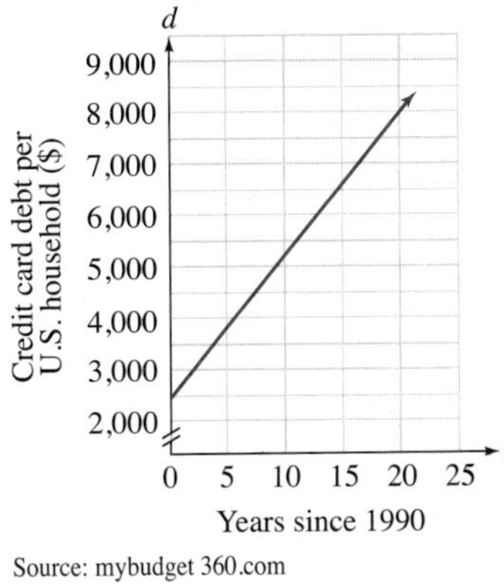

Source: mybudget 360.com

WRITING

103. Why is $y = mx + b$ called the slope–intercept form of the equation of a line?

104. On a quiz, a student was asked to find the slope of the graph of $y = 2x + 3$. She answered: $m = 2x$. Her instructor marked it wrong. Explain why the answer is incorrect.

REVIEW

105. **Cable TV.** A 186-foot television cable is to be cut into four pieces. Find the length of each piece if each successive piece is 3 feet longer than the previous one. 42 ft, 45 ft, 48 ft, 51 ft

106. **Investments.** Joni received $25,000 as part of a settlement in a class action lawsuit. She invested some money at 10% and the rest at a 9% simple interest rate. If her total annual income from these two investments was $2,430, how much did she invest at each rate? $18,000 at 10%, $7,000 at 9%

CHALLENGE PROBLEMS

107. If the graph of $y = mx + b$ passes through quadrants I, II, and IV, what can be known about the constants m and b? $m < 0, b > 0$

108. Which of the following equations has the steeper graph, $103x - 200y = -400$ or $17x - 33y = -66$? $17x - 33y = -66$

SECTION 3.6

Point–Slope Form

OBJECTIVES

1. Use point–slope form to write an equation of a line.
2. Write an equation of a line given two points on the line.
3. Write equations of horizontal and vertical lines.
4. Use a point and the slope to graph a line.
5. Write linear equations that model data.

ARE YOU READY? *Are You Ready? exercises available online at www.webassign.net/brookscole*

The following problems review some basic skills that are needed when working with equations of lines in point–slope form.

1. Find the slope of the line that passes through $(-2, 0)$ and $(-12, -8)$. $\frac{4}{5}$
2. Simplify: $x - (-5)$ $x + 5$
3. Solve $y + 2 = 6(x - 7)$ for y. $y = 6x - 44$
4. Add: $-\frac{3}{4} + 8$ $\frac{29}{4}$

If we know the slope of a line and its y-intercept, we can use the slope–intercept form to write the equation of the line. The question that now arises is, can *any* point on the line be used in combination with its slope to write its equation? In this section, we answer this question.

1 Use Point–Slope Form to Write an Equation of a Line.

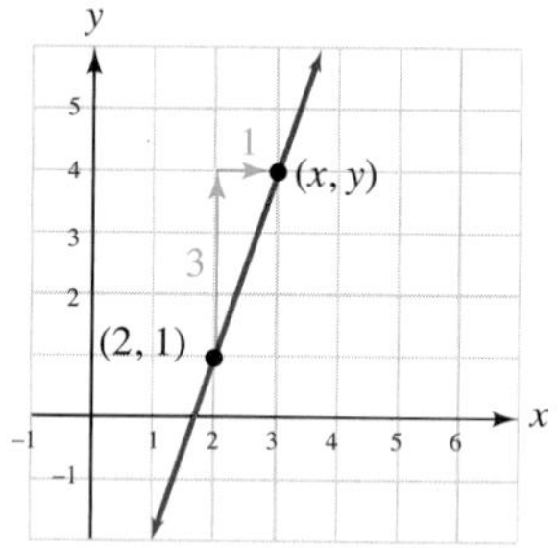

Refer to the line graphed on the left, with slope 3 and passing through the point (2,1). To develop a new form for the equation of a line, we will find the slope of this line in another way.

If we pick another point on the line with coordinates (x, y), we can find the slope of the line by substituting the coordinates of the points (x, y) and $(2, 1)$ into the slope formula.

$$\frac{y_2 - y_1}{x_2 - x_1} = m$$

$$\frac{y - 1}{x - 2} = m$$ Let (x_1, y_1) be $(2, 1)$ and (x_2, y_2) be (x, y). Substitute y for y_2, 1 for y_1, x for x_2, and 2 for x_1.

Since the slope of the line is 3, we can substitute 3 for m in the previous equation.

$$\frac{y - 1}{x - 2} = 3$$

We then multiply both sides by $x - 2$ to clear the equation of the fraction.

$$\frac{y - 1}{x - 2}(x - 2) = 3(x - 2)$$

$$y - 1 = 3(x - 2)$$ Simplify the left side. Remove the common factor $x - 2$ in the numerator and denominator: $\frac{y-1}{\cancel{x-2}_1} \cdot \frac{\cancel{x-2}^1}{1}$.

The resulting equation displays the slope of the line and the coordinates of one point on the line:

Slope of the line → 3

$$y - 1 = 3(x - 2)$$

1 ← y-coordinate of the point; 2 ← x-coordinate of the point

In general, suppose we know that the slope of a line is m and that the line passes through the point (x_1, y_1). Then if (x, y) is any other point on the line, we can use the definition of slope to write

$$\frac{y - y_1}{x - x_1} = m$$

If we multiply both sides by $x - x_1$ to clear the equation of the fraction, we have

$$y - y_1 = m(x - x_1)$$

This form of a linear equation is called **point–slope form.** It can be used to write the equation of a line when the slope and one point on the line are known.

Point–Slope Form of the Equation of a Line

If a line with slope m passes through the point (x_1, y_1), the equation of the line is

$$y - y_1 = m(x - x_1)$$ Read as "y minus y sub 1 equals m times the quantity x minus x sub 1."

EXAMPLE 1 Find an equation of a line that has slope -8 and passes through $(-1, 5)$. Write the answer in slope–intercept form.

Strategy Although the problem asks for an answer in slope–intercept form, we will begin with the point–slope form to write an equation of the line.

Why We know the slope of the line and the coordinates of a point that it passes through.

Solution The given point is $(-1, 5)$, so $x_1 = -1$ and $y_1 = 5$. The given slope is -8, so $m = -8$. We will subsitute these values into the point–slope form and simplify the right side of the equation.

Caution

When using the point–slope form, never substitute values for x or y.

$$y - y_1 = m(x - x_1)$$

Only substitute values for x_1, y_1, and m.

$y - y_1 = m(x - x_1)$ This is the point–slope form.

$y - 5 = -8[x - (-1)]$ Substitute -8 for m, -1 for x_1, and 5 for y_1. Brackets are used to enclose $x - (-1)$.

$y - 5 = -8(x + 1)$ Simplify the expression within the brackets.

To write an equivalent equation in slope–intercept form, we solve for y.

$y - 5 = -8(x + 1)$ This is the simplified point–slope form.

$y - 5 = -8x - 8$ Distribute the multiplication by -8.

$y - 5 + 5 = -8x - 8 + 5$ To isolate y, undo the subtraction of 5 by adding 5 to both sides.

$y = -8x - 3$ This is the requested slope–intercept form.

In slope–intercept form, the equation is $y = -8x - 3$.

To verify this result, we note that $m = -8$. Therefore, the slope of the line is -8, as required. To see whether the line passes through $(-1, 5)$, we substitute -1 for x and 5 for y in the equation. If this point is on the line, a true statement should result.

$$y = -8x - 3$$
$$5 \stackrel{?}{=} -8(-1) - 3$$
$$5 \stackrel{?}{=} 8 - 3$$
$$5 = 5 \quad \text{True}$$

Teaching Tip: Explain to your students that the line's equation can be written in several equivalent forms:

$y - 5 = -8(x + 1)$ point–slope form

$y = -8x - 3$ slope–intercept form

$8x + y = -3$ standard form

Teaching Example 1 Find an equation of the line that has slope -5 and passes through $(1, -7)$. Write the answer in slope–intercept form.
Answer: $y = -5x - 2$

Self Check 1 Find an equation of the line that has slope -2 and passes through $(4, -3)$. Write the answer in slope–intercept form. $y = -2x + 5$

Now Try Problems 13 and 19

2 Write an Equation of a Line Given Two Points on the Line.

In the next example, we show that it is possible to write the equation of a line when we know the coordinates of two points on the line.

EXAMPLE 2 Find an equation of the line that passes through $(-2, 6)$ and $(4, 7)$. Write the equation in slope–intercept form.

Strategy We will use the point–slope form, $y - y_1 = m(x - x_1)$, to write an equation of the line.

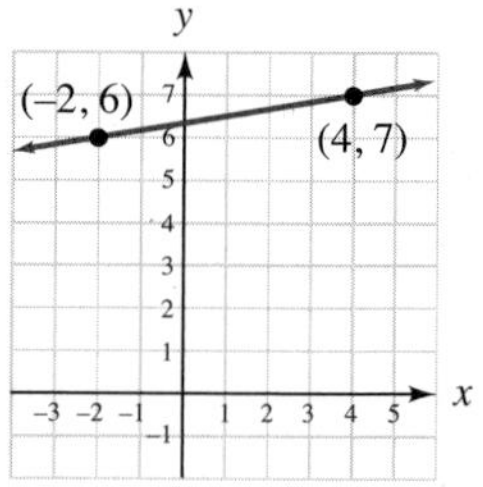

Why We know the coordinates of a point that the line passes through and we can calculate the slope of the line using the slope formula.

Solution To find the slope of the line, we use the slope formula.

$$m = \frac{y_2 - y_1}{x_2 - x_1} = \frac{7 - 6}{4 - (-2)} = \frac{1}{6}$$ Substitute 7 for y_2, 6 for y_1, 4 for x_2, and -2 for x_1.

Success Tip

In Example 2, either of the given points can be used as (x_1, y_1) when writing the point–slope equation. The results will be the same.

Looking ahead, we usually choose the point whose coordinates will make the calculations the easiest.

Teaching Tip: If time allows, show your students that the result is the same if the point $(-2, 6)$ is used as (x_1, y_1) when substituting into the point–slope form.

Success Tip

To check this result, verify that $(-2, 6)$ and $(4, 7)$ satisfy $y = \frac{1}{6}x + \frac{19}{3}$ using substitution.

Either point on the line can serve as (x_1, y_1). If we use $(4, 7)$, we have

$y - y_1 = m(x - x_1)$ This is the point–slope form.

$y - 7 = \frac{1}{6}(x - 4)$ Substitute $\frac{1}{6}$ for m, 7 for y_1, and 4 for x_1.

To write an equivalent equation in slope–intercept form, we solve for y.

$y - 7 = \frac{1}{6}x - \frac{2}{3}$ Distribute the multiplication by $\frac{1}{6}$: $\frac{1}{6}(-4) = -\frac{4}{6} = -\frac{2}{3}$.

$y - 7 + 7 = \frac{1}{6}x - \frac{2}{3} + 7$ To isolate y, add 7 to both sides.

$y = \frac{1}{6}x - \frac{2}{3} + \frac{21}{3}$ Simplify the left side. On the right side, express 7 as $\frac{21}{3}$ to prepare to add the fractions with the common denominator 3.

$y = \frac{1}{6}x + \frac{19}{3}$ Add the fractions: $-\frac{2}{3} + \frac{21}{3} = \frac{19}{3}$. This is slope–intercept form.

An equation of the line that passes through $(-2, 6)$ and $(4, 7)$ is $y = \frac{1}{6}x + \frac{19}{3}$.

Teaching Example 2 Find an equation of the line that passes through $(3, 3)$ and $(-5, 2)$. Write the equation in slope–intercept form.
Answer: $y = \frac{1}{8}x + \frac{21}{8}$

Self Check 2 Find an equation of the line that passes through $(-5, 4)$ and $(8, -6)$. Write the equation in slope–intercept form. $y = -\frac{10}{13}x + \frac{2}{13}$

Now Try Problem 29

3 Write Equations of Horizontal and Vertical Lines.

We have previously graphed horizontal and vertical lines. We will now discuss how to write their equations.

EXAMPLE 3 Write an equation of each line and graph it. **a.** A horizontal line that passes through $(-2, -4)$ **b.** A vertical line that passes through $(1, 3)$

Strategy We will use the appropriate form, either $y = b$ or $x = a$, to write an equation of each line.

Why These are the standard forms for the equations of a horizontal and a vertical line.

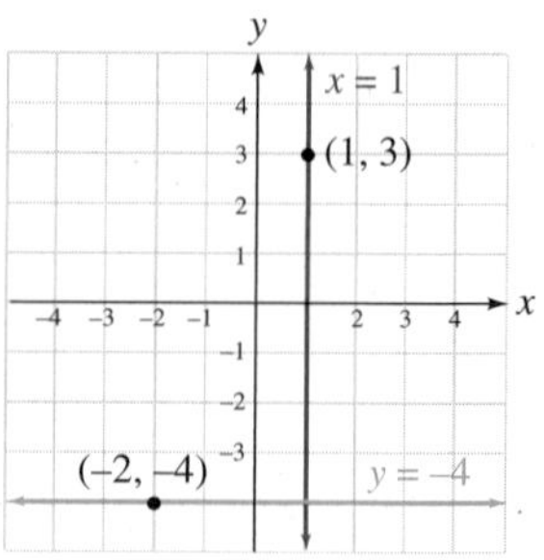

Solution

a. The equation of a horizontal line can be written in the form $y = b$. Since the y-coordinate of $(-2, -4)$ is -4, the equation of the line is $y = -4$. The graph is shown in the figure.

b. The equation of a vertical line can be written in the form $x = a$. Since the x-coordinate of $(1, 3)$ is 1, the equation of the line is $x = 1$. The graph is shown in the figure.

Teaching Example 3 Write an equation of each line:
a. a horizontal line that passes through $(2, -7)$
b. a vertical line that passes through $(5, 3)$
Answers: a. $y = -7$ b. $x = 5$

Self Check 3 Write an equation of each line. **a.** A horizontal line that passes through $(3, 2)$ $y = 2$ **b.** A vertical line passing through $(-1, -3)$ $x = -1$

Now Try Problems 41 and 43

4 Use a Point and the Slope to Graph a Line.

If we know the coordinates of a point on a line, and if we know the slope of the line, we can use the slope to determine a second point on the line.

EXAMPLE 4 Graph the line with slope $\frac{2}{5}$ that passes through $(-1, -3)$.

Strategy First, we will plot the given point $(-1, -3)$. Then we will use the slope to find a second point that the line passes through.

Why Once we determine two points that the line passes through, we can draw the graph of the line.

Solution To draw the graph, we begin by plotting the point $(-1, -3)$. From there, we move 2 units up (rise) and then 5 units to the right (run), since the slope is $\frac{2}{5}$. This locates a second point on the line, $(4, -1)$. We then draw a straight line through the two points.

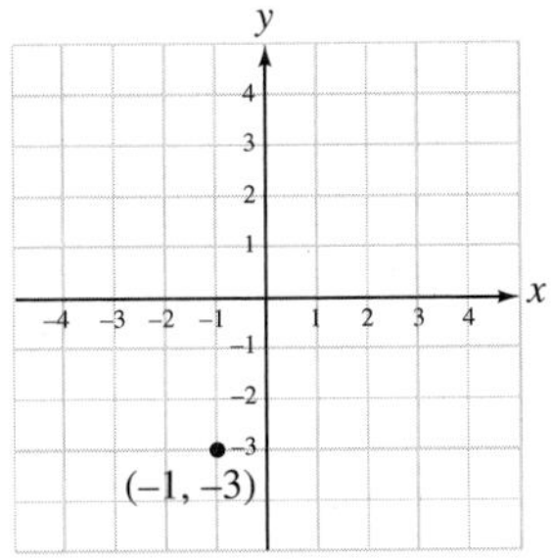

Plot the given point (–1, –3).

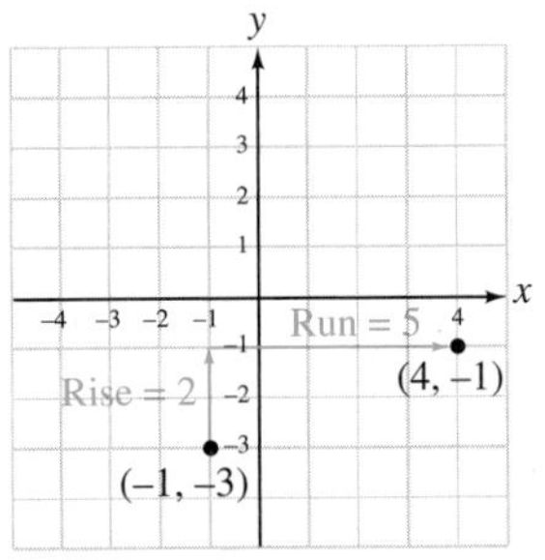

From (–1, –3) draw the rise and run parts of the slope triangle for $m = \frac{2}{5}$ to find another point on the line.

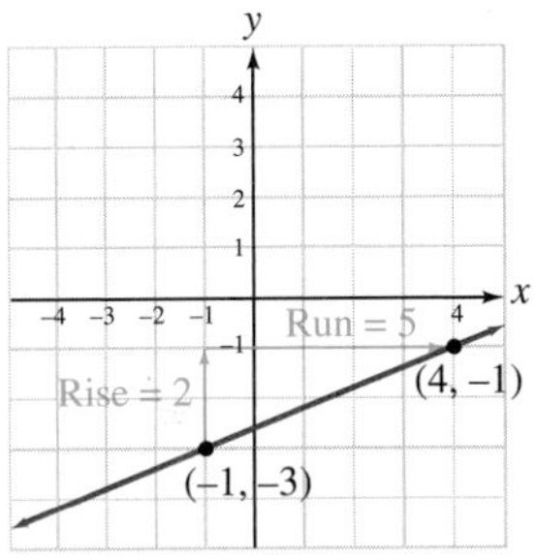

Use a straightedge to draw a line through the points.

Teaching Example 4 Graph the line with slope $\frac{4}{7}$ that passes through $(-4, -3)$.
Answer:

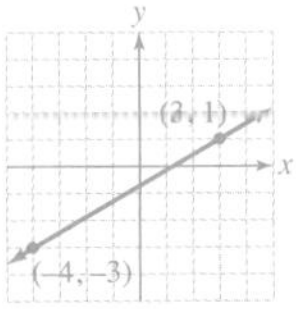

Self Check 4 Graph the line with slope -4 that passes through $(-4, 2)$. See AIE Appendix 3.

Now Try Problem 45

The following table summarizes what you should know about each form of an equation of a line.

Form	Example	Comments
Standard form $Ax + By = C$	$2x + 5y = 9$	■ To graph, find the x- and y-intercepts by letting $y = 0$ and finding x, and letting $x = 0$ and finding y. Also find a checkpoint.
Slope–intercept form $y = mx + b$	$y = \frac{5}{3}x + 4$	■ To graph, plot the y-intercept $(0, b)$. From there, draw a slope triangle using the rise and run to locate another point. ■ Use this form to write a line's equation if you know its slope and y-intercept.
Point–slope form $y - y_1 = m(x - x_1)$	$y - 1 = 6(x - 8)$	■ To graph, plot the point (x_1, y_1). From there, draw a slope triangle using the rise and run of the slope m to locate another point. ■ Use this form to write a line's equation if you know a point on the line and the slope. If two points on the line are known, find the slope and use it with one of the points to write the equation.
Horizontal line $y = b$	$y = 7$	■ To graph, draw a horizontal line ($m = 0$) with y-intercept $(0, b)$.
Vertical line $x = a$	$x = -1$	■ To graph, draw a vertical line (undefined slope) with x-intercept $(a, 0)$.

5 Write Linear Equations That Model Data.

In the next two examples, we will see how the point–slope form can be used to write linear equations that model certain real–world situations.

EXAMPLE 5 **Men's Shoe Sizes.** The length (in inches) of a man's foot is not his shoe size. For example, the smallest adult men's shoe size is 5, and it fits a 9-inch-long foot. There is, however, a linear relationship between the two. It can be stated this way: Shoe size increases by 3 sizes for each 1-inch increase in foot length.

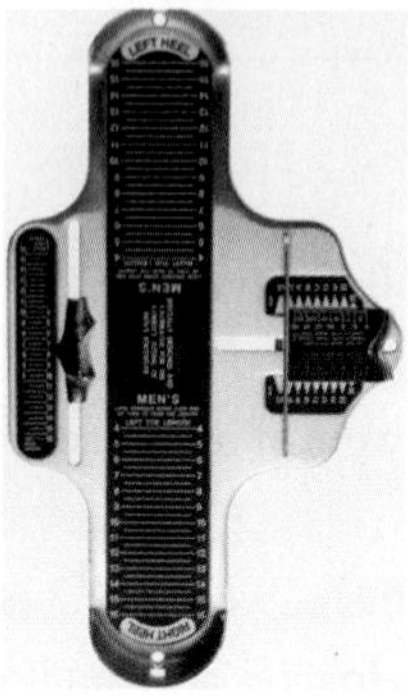

a. Write a linear equation that relates shoe size s to foot length L.

b. Shaquille O'Neal, a famous basketball player, has a foot that is about 14.6 inches long. Find his shoe size.

Strategy We will first find the slope of the line that describes the linear relationship between shoe size and the length of a foot. Then we will determine the coordinates of a point on that line.

Why Once we know the slope and the coordinates of one point on the line, we can use the point–slope form to write the equation of the line.

Solution **a.** Since shoe size s depends on the length L of the foot, ordered pairs have the form (L, s). Because the relationship is linear, the graph of the desired equation is a line.

- The line's slope is the rate of change: $\frac{3 \text{ sizes}}{1 \text{ inch}}$. Therefore, $m = 3$.
- A 9-inch-long foot wears size 5, so the line passes through $(9, 5)$.

We substitute 3 for m and the coordinates of the point into the point–slope form and solve for s.

$s - s_1 = m(L - L_1)$ This is the point–slope form using the variables L and s.

$s - 5 = 3(L - 9)$ Substitute 3 for m, 9 for L_1, and 5 for s_1.

$s - 5 = 3L - 27$ Distribute the multiplication by 3.

$s = 3L - 22$ To isolate s, add 5 to both sides: $-27 + 5 = -22$.

The equation relating men's shoe size and foot length is $s = 3L - 22$.

b. To find Shaquille's shoe size, we substitute 14.6 inches for L in the equation.

$s = 3L - 22$

$s = 3(14.6) - 22$

$s = 43.8 - 22$ Do the multiplication.

$s = 21.8$ Do the subtraction.

Since men's shoes come in only full- and half-sizes, we round 21.8 up to 22. Shaquille O'Neal wears size 22 shoes.

Self Check 5 **Comparing Temperature Scales.** Celsius and Fahrenheit measures of temperature are not the same. There is, however, a linear relationship between the two. Degrees Fahrenheit increase by 9° for each 5° increase in Celsius. If a 212° Fahrenheit temperature measure is the same as 100° Celsius, write a linear equation that relates Fahrenheit measure to Celsius measure. $F = \frac{9}{5}C + 32$

Now Try Problem 77

Teaching Example 5 Temperature Drop. A refrigeration unit lowers the temperature in a railroad car 6°F every 5 minutes. One day, the temperature in a car was 76°F after the cooler had run for 10 minutes. Find a linear equation that describes the relationship between the number of minutes the cooler has been running and the temperature in the car.
Answer: $t = -\frac{6}{5}m + 88$

EXAMPLE 6 **Studying Learning.** In a series of 40 trials, a rat was released in a maze to search for food. Researchers recorded the trial number and the time that it took the rat to complete the maze as ordered pairs on a **scatter diagram** shown below. All of the points fell on or near the line drawn in red. Write an equation of the line in slope–intercept form.

The Language of Algebra

The term *scatter diagram* is somewhat misleading. Often, the data points are not scattered loosely about. In this case, they fall, more or less, along an imaginary straight line, indicating a linear relationship.

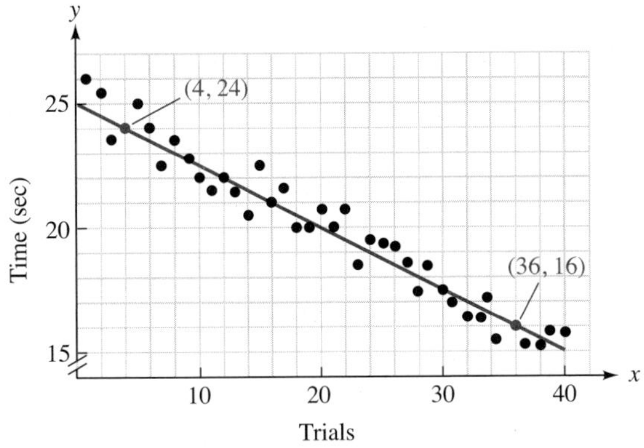

Strategy From the graph, we will determine the coordinates of two points on the line.

Why We can write an equation of a line when we know the coordinates of two points on the line. (See Example 2.)

Solution To write a point–slope equation, we need to know the slope of the line. The line passes through several points; we will use (4, 24) and (36, 16) to find its slope.

$$m = \frac{y_2 - y_1}{x_2 - x_1} = \frac{16 - 24}{36 - 4} = \frac{-8}{32} = -\frac{1}{4}$$

Any point on the line can serve as (x_1, y_1). We will use (4, 24).

$y - y_1 = m(x - x_1)$ This is the point–slope form.

$y - 24 = -\frac{1}{4}(x - 4)$ Substitute $-\frac{1}{4}$ for m, 4 for x_1, and 24 for y_1.

To write this equation in slope–intercept form, solve for y.

$y - 24 = -\frac{1}{4}x + 1$ Distribute the multiplication by $-\frac{1}{4}$: $-\frac{1}{4}(-4) = 1$.

$y = -\frac{1}{4}x + 25$ To isolate y, add 24 to both sides: $1 + 24 = 25$.

A linear equation that models the rat's performance on the maze is $y = -\frac{1}{4}x + 25$, where x is the number of the trial and y is the time it took, in seconds.

Self Check 6 **Awards.** Orders for awards to be given to math team members were placed on two separate occasions. The first order of 32 awards cost $172 and the second order of 5 awards cost $37. Write a linear equation that gives the cost for an order of any number of awards. $c = 5n + 12$

Now Try Problem 85

Teaching Example 6 **Trophies.** A youth soccer league placed an order for 273 trophies and paid $1,663. Near the end of the season, they placed a second order for 13 trophies and paid $103. Write a linear equation that gives the cost for an order of any number of trophies.
Answer: $c = 6n + 25$

SECTION 3.6 STUDY SET

VOCABULARY

Fill in the blanks.

1. $y - y_1 = m(x - x_1)$ is called the point–slope form of the equation of a line. In words, we read this as y minus y sub one equals m times the quantity of x minus x sub one.
2. $y = mx + b$ is called the slope–intercept form of the equation of a line.

CONCEPTS

3. Determine in what form each equation is written.
   a. $y - 4 = 2(x - 5)$ Point–slope
   b. $y = 2x + 15$ Slope–intercept
4. What point does the graph of each equation pass through, and what is the line's slope?
   a. $y - 2 = 6(x - 7)$ $(7, 2)$; 6
   b. $y + 3 = -8(x + 1)$ $(-1, -3)$; -8
5. Refer to the following graph of a line.
   a. What highlighted point does the line pass through? $(-2, -3)$
   b. What is the slope of the line? $\frac{5}{6}$
   c. Write an equation of the line in point–slope form.
   $y + 3 = \frac{5}{6}(x + 2)$

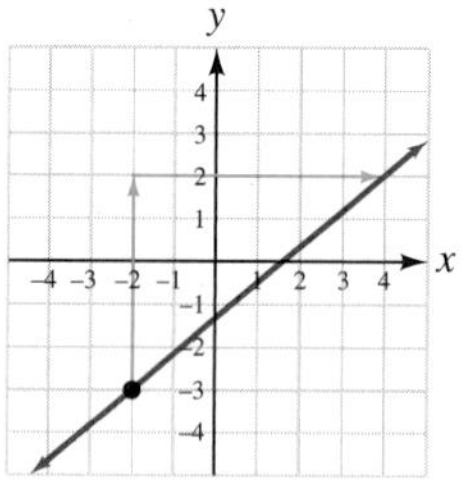

6. On a quiz, a student was asked to write the equation of a line with slope 4 that passes through $(-1, 3)$. Explain how the student can check her answer, $y = 4x + 7$. $m = 4$ and $3 = 4(-1) + 7$
7. Suppose you are asked to write an equation of the line in the scatter diagram below. What two points would you use to write the point–slope equation? $(67, 170)$, $(79, 220)$

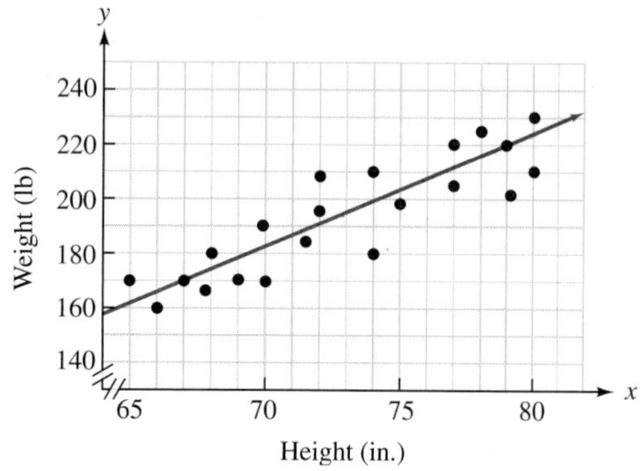

8. In each case, a linear relationship between two quantities is described. If the relationship were graphed, what would be the slope of the line?
   a. The sales of new cars increased by 15 every 2 months. $\frac{15}{2}$
   b. There were 35 fewer robberies for each dozen police officers added to the force. $-\frac{35}{12}$
   c. One acre of forest is being destroyed every 30 seconds. $-\frac{1}{30}$

NOTATION

Complete the solution.

9. Find an equation of the line with slope -2 that passes through the point $(-1, 5)$. Write the answer in slope–intercept form.

$$y - y_1 = m(x - x_1)$$
$$y - 5 = -2[x - (-1)]$$
$$y - 5 = -2[x + 1]$$
$$y - 5 = -2x - 2$$
$$y = -2x + 3$$

10. What is the point–slope form of the equation of a line?
    $y - y_1 = m(x - x_1)$
11. Consider the steps below and then fill in the blanks:

$$y - 3 = 2(x + 1)$$
$$y - 3 = 2x + 2$$
$$y = 2x + 5$$

The original equation was in point–slope form. After solving for y, we obtain an equation in slope–intercept form.

12. Fill in the blanks: The equation of a horizontal line has the form $y = b$ and the equation of a vertical line has the form $x = a$.

GUIDED PRACTICE

***Use the point–slope form to write an equation of the line with the given slope and point. Leave the equation in that form.* See Example 1.**

13. Slope 3, passes through $(2, 1)$ $y - 1 = 3(x - 2)$
14. Slope 2, passes through $(4, 3)$ $y - 3 = 2(x - 4)$
15. Slope $\frac{4}{5}$, passes through $(-5, -1)$ $y + 1 = \frac{4}{5}(x + 5)$
16. Slope $\frac{7}{8}$, passes through $(-2, -9)$ $y + 9 = \frac{7}{8}(x + 2)$

***Use the point–slope form to find an equation of the line with the given slope and point. Then write the equation in slope–intercept form.* See Example 1.**

17. Slope 2, passes through $(3, 5)$ $y = 2x - 1$
18. Slope 8, passes through $(2, 6)$ $y = 8x - 10$
19. Slope -5, passes through $(-9, 8)$ $y = -5x - 37$
20. Slope -4, passes through $(-2, 10)$ $y = -4x + 2$
21. Slope -3, passes through the origin $y = -3x$
22. Slope -1, passes through the origin $y = -x$
23. Slope $\frac{1}{5}$, passes through $(10, 1)$ $y = \frac{1}{5}x - 1$
24. Slope $\frac{1}{4}$, passes through $(8, 1)$ $y = \frac{1}{4}x - 1$
25. Slope $-\frac{4}{3}$,

x	y
6	−4

$y = -\frac{4}{3}x + 4$

26. Slope $-\frac{3}{2}$,

x	y
−2	1

$y = -\frac{3}{2}x - 2$

27. Slope $-\frac{11}{6}$, passes through $(2, -6)$ $y = -\frac{11}{6}x - \frac{7}{3}$

28. Slope $-\frac{5}{4}$, passes through $(2, 0)$ $y = -\frac{5}{4}x + \frac{5}{2}$

Find an equation of the line that passes through the two given points. Write the equation in slope–intercept form, if possible. **See Example 2.**

29. Passes through $(1, 7)$ and $(-2, 1)$ $y = 2x + 5$

30. Passes through $(-2, 2)$ and $(2, -8)$ $y = -\frac{5}{2}x - 3$

31. $y = -\frac{1}{2}x + 1$

x	y
-4	3
2	0

32. $y = x - 3$

x	y
-1	-4
1	-2

33. Passes through $(5, 5)$ and $(7, 5)$ $y = 5$

34. Passes through $(-2, 1)$ and $(-2, 15)$ $x = -2$

35. Passes through $(5, 1)$ and $(-5, 0)$ $y = \frac{1}{10}x + \frac{1}{2}$

36. Passes through $(-3, 0)$ and $(3, 1)$ $y = \frac{1}{6}x + \frac{1}{2}$

37. Passes through $(-8, 2)$ and $(-8, 17)$ $x = -8$

38. Passes through $\left(\frac{2}{3}, 2\right)$ and $(0, 2)$ $y = 2$

39. Passes through $\left(\frac{2}{3}, \frac{1}{3}\right)$ and $(0, 0)$ $y = \frac{1}{2}x$

40. Passes through $\left(\frac{1}{2}, \frac{3}{4}\right)$ and $(0, 0)$ $y = \frac{3}{2}x$

Write an equation of each line. **See Example 3.**

41. Vertical, passes through $(4, 5)$ $x = 4$

42. Vertical, passes through $(-2, -5)$ $x = -2$

43. Horizontal, passes through $(4, 5)$ $y = 5$

44. Horizontal, passes through $(-2, -5)$ $y = -5$

Graph the line that passes through the given point and has the given slope. **See Example 4.** See AIE Appendix 3.

45. $(1, -2)$, $m = -1$

46. $(-4, 1)$, $m = -3$

47. $(5, -3)$, $m = \frac{3}{4}$

48. $(2, -4)$, $m = \frac{2}{3}$

49. $(-2, -3)$, slope 2

50. $(-3, -3)$, slope 4

51. $(4, -3)$, slope $-\frac{7}{8}$

52. $(4, 2)$, slope $-\frac{1}{5}$

TRY IT YOURSELF

Use either the slope–intercept form (from Section 3.5) or the point–slope form (from Section 3.6) to find an equation of each line. Write each result in slope–intercept form, if possible.

53. Passes through $(5, 0)$ and $(-11, -4)$ $y = \frac{1}{4}x - \frac{5}{4}$

54. Passes through $(7, -3)$ and $(-5, 1)$ $y = -\frac{1}{3}x - \frac{2}{3}$

55. Horizontal, passes through $(-8, 12)$ $y = 12$

56. Horizontal, passes through $(9, -32)$ $y = -32$

57. Slope $-\frac{1}{4}$, y-intercept $\left(0, \frac{7}{8}\right)$ $y = -\frac{1}{4}x + \frac{7}{8}$

58. Slope $-\frac{9}{5}$, y-intercept $\left(0, \frac{11}{3}\right)$ $y = -\frac{9}{5}x + \frac{11}{3}$

59. Slope $-\frac{2}{3}$, passes through $(3, 0)$ $y = -\frac{2}{3}x + 2$

60. Slope $-\frac{2}{5}$, passes through $(15, 0)$ $y = -\frac{2}{5}x + 6$

61. Slope 8, passes through $(2, 20)$ $y = 8x + 4$

62. Slope 6, passes through $(1, -2)$ $y = 6x - 8$

63. Vertical, passes through $(-3, 7)$ $x = -3$

64. Vertical, passes through $(12, -23)$ $x = 12$

65. Slope 7 and y-intercept $(0, -11)$ $y = 7x - 11$

66. Slope 3 and y-intercept $(0, 4)$ $y = 3x + 4$

67. Passes through $(-2, -1)$ and $(-1, -5)$ $y = -4x - 9$

68. Passes through $(-3, 6)$ and $(-1, -4)$ $y = -5x - 9$

69. x-intercept $(7, 0)$ and y-intercept $(0, -2)$ $y = \frac{2}{7}x - 2$

70. x-intercept $(-3, 0)$ and y-intercept $(0, 7)$ $y = \frac{7}{3}x + 7$

71. Slope $\frac{1}{10}$, passes through the origin $y = \frac{1}{10}x$

72. Slope $\frac{9}{8}$, passes through the origin $y = \frac{9}{8}x$

73. Undefined slope, passes through $\left(-\frac{1}{8}, 12\right)$ $x = -\frac{1}{8}$

74. Undefined slope, passes through $\left(\frac{2}{5}, -\frac{5}{6}\right)$ $x = \frac{2}{5}$

75. Slope 1.7, y-intercept $(0, -2.8)$ $y = 1.7x - 2.8$

76. Slope 9.5, y-intercept $(0, -14.3)$ $y = 9.5x - 14.3$

APPLICATIONS

77. **Anatomy.** There is a linear relationship between a woman's height and the length of her radius bone. It can be stated this way: Height increases by 3.9 inches for each 1-inch increase in the length of the radius. Suppose a 64-inch-tall woman has a 9-inch-long radius bone. Use this information to find a linear equation that relates height h to the length r of the radius. Write the equation in slope–intercept form. $h = 3.9r + 28.9$

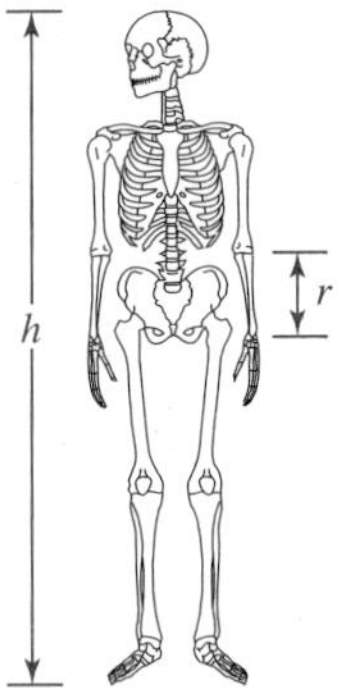

78. **Automation.** An automated production line uses distilled water at a rate of 300 gallons every 2 hours to make shampoo. After the line had run for 7 hours, planners noted that 2,500 gallons of distilled water remained in the storage tank. Find a linear equation relating the time t in hours since the production line began and the number g of gallons of distilled water in the storage tank. Write the equation in slope–intercept form. $g = -150t + 3{,}550$

79. **Pole Vaulting.** Find the equations of the lines that describe the positions of the pole for parts 1, 3, and 4 of the jump. Write the equations in slope–intercept form, if possible.
$y = -\frac{2}{5}x + 4$, $y = -7x + 70$, $x = 10$

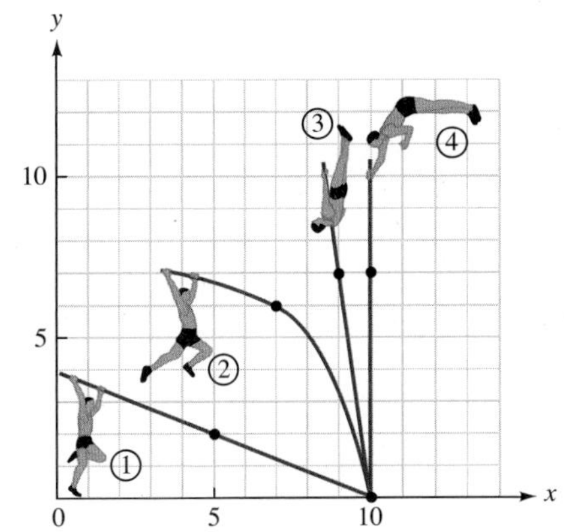

80. Freeway Design. The graph below shows the route of a proposed freeway.

a. Give the coordinates of the points where the proposed Freeway 133 will join Interstate 25 and Highway 40. $(-3, -4), (6, 2)$

b. Find an equation of the line that describes the route of the proposed freeway. Write the equation in slope–intercept form. $y = \frac{2}{3}x - 2$

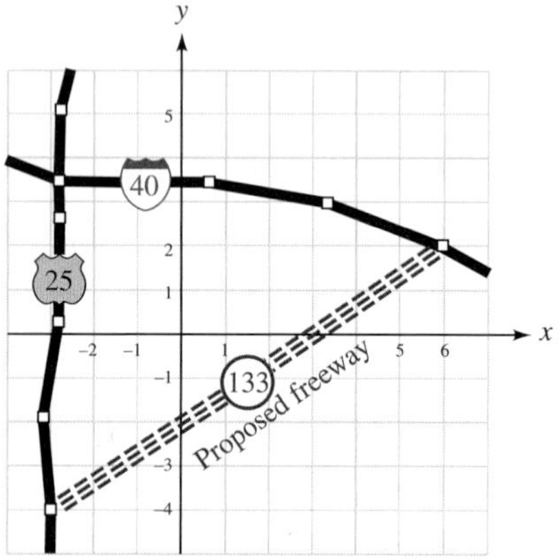

81. Toxic Cleanup. Three months after cleanup began at a dump site, 800 cubic yards of toxic waste had yet to be removed. Two months later, that number had been lowered to 720 cubic yards.

a. Find an equation that describes the linear relationship between the length of time m (in months) the cleanup crew has been working and the number of cubic yards y of toxic waste remaining. Write the equation in slope–intercept form. $y = -40m + 920$

b. Use your answer to part (a) to predict the number of cubic yards of waste that will still be on the site one year after the cleanup project began. 440 yd^3

82. Depreciation. To lower its corporate income tax, accountants of a company depreciated a word processing system over several years using a linear model, as shown in the worksheet.

a. Find a linear equation relating the years since the system was purchased, x, and its value, y, in dollars. Write the equation in slope–intercept form. $y = -15{,}000x + 90{,}000$

b. Find the purchase price of the system. $90,000

Tax Worksheet Method of depreciation: *Linear*

Property	Years after purchase	Value
Word processing system	2	*$60,000*
	4	*$30,000*

83. Trampolines. There is a linear relationship between the length of the protective pad that wraps around a trampoline and the radius of the trampoline. Use the data in the table to find an equation that gives the length l of pad needed for any trampoline with radius r. Write the equation in slope–intercept form. Use units of feet for both l and r. $l = \frac{25}{4}r + \frac{1}{4}$

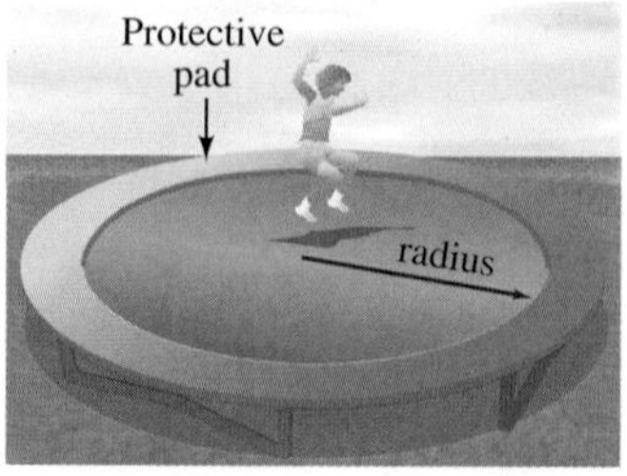

Radius	Pad length
3 ft	19 ft
7 ft	44 ft

84. Raising a Family. In the report *"Expenditures on Children by Families,"* the U.S. Department of Agriculture projected the yearly child-rearing expenditures on children from birth through age 17. For a child born in 2010 to a two-parent middle-income family, the report estimated annual expenditures of $10,808 when the child is 6 years old, and $14,570 when the child is 15 years old.

a. Write two ordered pairs of the form (child's age, annual expenditure). (6, 10,808), (15, 14,570)

b. Assume the relationship between the child's age a and the annual expenditures E is linear. Use your answers to part (a) to write an equation in slope–intercept form that models this relationship. $E = 418a + 8{,}300$

c. What are the projected child-rearing expenses when the child is 17 years old? $15,406

85. Got Milk? The scatter diagram shows the amount of milk that an average American drank in one year for the years 1980–2008. A straight line can be used to model the data.

a. Use the two highlighted points on the line to find its equation. Write the equation in slope–intercept form. $y = -\frac{3}{10}x + \frac{283}{10}$ or $y = -0.3x + 28.3$

b. Use your answer to part (a) to predict the amount of milk that an average American will drink in 2020. 16.3 gal

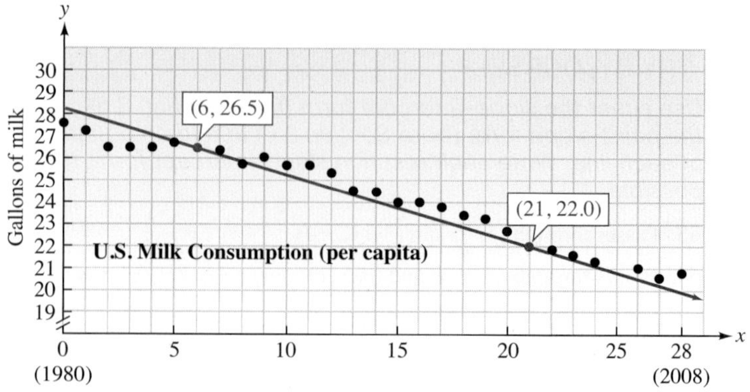

Source: United States Department of Agriculture.

▶ **86. Engine Output.** The horsepower produced by an automobile engine was recorded for various engine speeds in the range of 2,400 to 4,800 revolutions per minute (rpm). The data were recorded on the following scatter diagram. Find an equation of the line that models the relationship between engine speed s and horsepower h. Write the equation in slope–intercept form. $h = \frac{3}{40}s - 30$

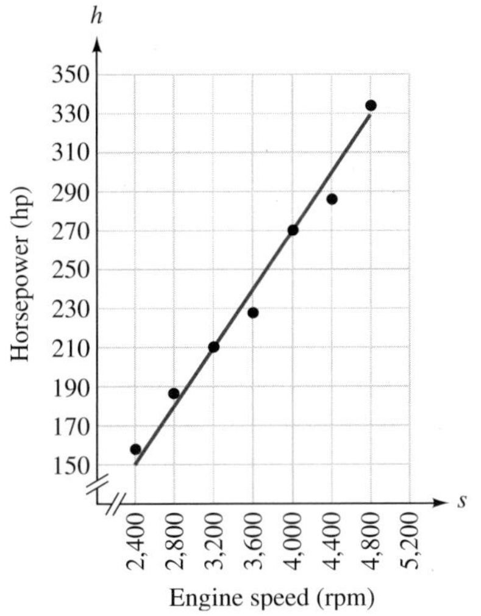

WRITING

87. Why is $y - y_1 = m(x - x_1)$ called the point–slope form of the equation of a line?

▶ **88.** If we know two points that a line passes through, we can write its equation. Explain how this is done.

89. Explain the steps involved in writing $y - 6 = 4(x - 1)$ in slope–intercept form.

▶ **90.** Think of the points on the graph of the horizontal line $y = 4$. What do the points have in common? How do they differ?

REVIEW

91. Frames. The length of a rectangular picture is 5 inches greater than twice the width. If the perimeter is 112 inches, find the dimensions of the frame. 17 in. by 39 in.

▶ **92. Speed of an Airplane.** Two planes are 6,000 miles apart, and their speeds differ by 200 mph. They travel toward each other and meet in 5 hours. Find the speed of the slower plane. 500 mph

CHALLENGE PROBLEMS

93. Find an equation of the line that passes through (2, 5) and is parallel to the line $y = 4x - 7$. Write the equation in slope–intercept form. $y = 4x - 3$

▶ **94.** Find an equation of the line that passes through $(-6, 3)$ and is perpendicular to the line $y = -3x - 12$. Write the equation in slope–intercept form. $y = \frac{1}{3}x + 5$

SECTION 3.7 Graphing Linear Inequalities

OBJECTIVES

1. Determine whether an ordered pair is a solution of an inequality.
2. Graph a linear inequality in two variables.
3. Graph inequalities with a boundary through the origin.
4. Solve applied problems involving linear inequalities in two variables.

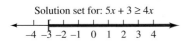

ARE YOU READY? *Are You Ready? exercises available online at www.webassign.net/brookscole*

The following problems review some basic skills that are needed when graphing linear inequalities.

1. True or false: $2(-3) + 1 > -4$ False

2. True or false: $8 \le 8$ True

3. Graph: $2x - 3y = 6$ See AIE Appendix 3.

4. Determine whether each of the following points lies *above, below,* or *on* the line graphed in problem 3.

a. $(2, -4)$ Below **b.** $(-3, -4)$ On **c.** $(0, 0)$ Above

Solution set for: $x + 6 < 8$

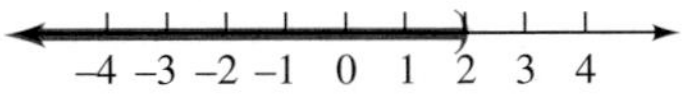

Solution set for: $5x + 3 \ge 4x$

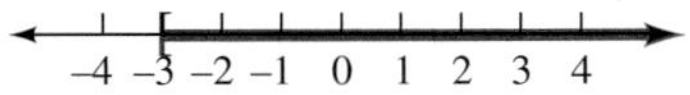

Recall that an **inequality** is a statement that contains one of the symbols $<$, $\le$, $>$, or $\ge$. Inequalities in one variable, such as $x + 6 < 8$ and $5x + 3 \ge 4x$, were solved in Section 2.7. Because they have an infinite number of solutions, we represented their solution sets graphically, by shading intervals on a number line. Two examples of this type of shading are shown in the left margin.

We now extend that concept to linear inequalities *in two variables,* as we introduce a procedure that is used to graph their solution sets.

1 Determine Whether an Ordered Pair Is a Solution of an Inequality.

If the = symbol in a linear equation in two variables is replaced with an inequality symbol, we have a **linear inequality in two variables.**

Linear Inequalities

A **linear inequality in two variables** is an inequality that can be written in one of the forms

$$Ax + By > C, \quad Ax + By < C, \quad Ax + By \geq C, \quad \text{or} \quad Ax + By \leq C$$

where A, B, and C are real numbers and A and B are not both 0.

Some examples of linear inequalities in two variables are

$$x - y \leq 5, \quad 4x + 3y < -6, \quad y > 2x \quad \text{and} \quad x < -3$$

As with linear equations, an ordered pair (x, y) is a **solution of an inequality** in x and y if a true statement results when the values of the variables are substituted into the inequality.

EXAMPLE 1 Determine whether each ordered pair is a solution of $x - y \leq 5$. Then graph each solution: **a.** $(4, 2)$ **b.** $(0, -6)$ **c.** $(1, -4)$

Strategy We will substitute each ordered pair of coordinates into the inequality.

Why If the resulting statement is true, the ordered pair is a solution.

Solution

Notation

The symbol $\overset{?}{\leq}$ is read as "is possibly less than or equal to."

a. For (4, 2):

$x - y \leq 5$ This is the given inequality.

$4 - 2 \overset{?}{\leq} 5$ Substitute 4 for x and 2 for y.

$2 \leq 5$ True

Because $2 \leq 5$ is true, (4, 2) is a solution of $x - y \leq 5$. We say that (4, 2) *satisfies* the inequality. This solution is graphed as shown on the right.

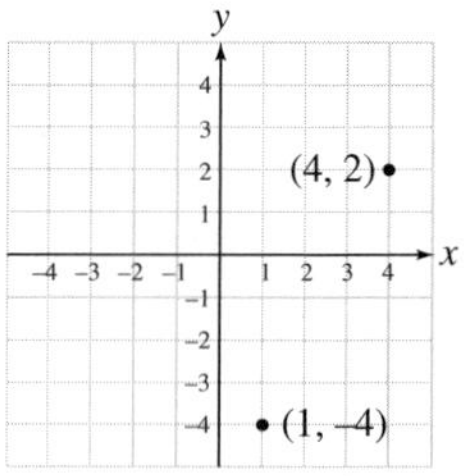

Two solutions of $x - y \leq 5$.

b. For (0, −6):

$x - y \leq 5$ This is the given inequality.

$0 - (-6) \overset{?}{\leq} 5$ Substitute 0 for x and −6 for y.

$6 \leq 5$ False

Because $6 \leq 5$ is false, (0, −6) is not a solution.

c. For (1, −4):

$x - y \leq 5$ This is the given inequality.

$1 - (-4) \overset{?}{\leq} 5$ Substitute 1 for x and −4 for y.

$5 \leq 5$ True

Because $5 \leq 5$ is true, (1, −4) is a solution, and we graph it as shown.

Self Check 1 Using the inequality in Example 1, determine whether each ordered pair is a solution: **a.** (8, 2) Not a solution **b.** (4, −1) Solution **c.** (−2, 4) Solution **d.** (−3, −5) Solution

Now Try Problem 19

Teaching Example 1 Determine whether each ordered pair is a solution of $x + y \leq 6$.
a. (0, 4) b. (−3, 1)
c. (9, −2) d. (−1, 7)
Answers:
a. Solution b. Solution
c. Not a solution d. Solution

In Example 1, we graphed two of the solutions of $x - y \leq 5$. Since there are infinitely more ordered pairs (x, y) that make the inequality true, it would not be reasonable to plot all of them. Fortunately, there is an easier way to show all of the solutions.

2 Graph a Linear Inequality in Two Variables.

The graph of a linear inequality is a picture that represents the set of all points whose coordinates satisfy the inequality. In general, such graphs are regions bounded by a line. We call those regions **half-planes,** and we use a two-step procedure to find them.

EXAMPLE 2 Graph: $x - y \leq 5$

Strategy We will graph the **related equation** $x - y = 5$ to establish a boundary line between two regions of the coordinate plane. Then we will determine which region contains points whose coordinates satisfy the given inequality.

Why The graph of a linear inequality in two variables is a region of the coordinate plane on one side of a boundary line.

Solution Since the inequality symbol $\leq$ includes an equal symbol, the graph of $x - y \leq 5$ includes the graph of $x - y = 5$.

Notation

The inequality $x - y \leq 5$ means

$x - y = 5$ or $x - y < 5$

Step 1: To graph $x - y = 5$, we use the intercept method, as shown in part (a) of the illustration below. The resulting line, called a **boundary line,** divides the coordinate plane into two half-planes. To show that the points on the boundary line are solutions of $x - y \leq 5$, we draw it as a solid line.

A table of solutions to graph the boundary

$x - y = 5$

x	y	(x, y)
0	-5	$(0, -5)$
5	0	$(5, 0)$
6	1	$(6, 1)$

Let $x = 0$ and find y.
Let $y = 0$ and find x.
As a check, let $x = 6$ and find y.

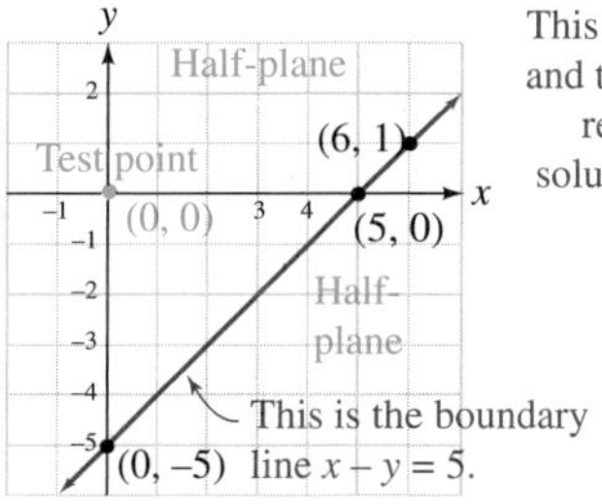

(a)

This shaded half-plane and the solid boundary represent all the solutions of $x - y \leq 5$.

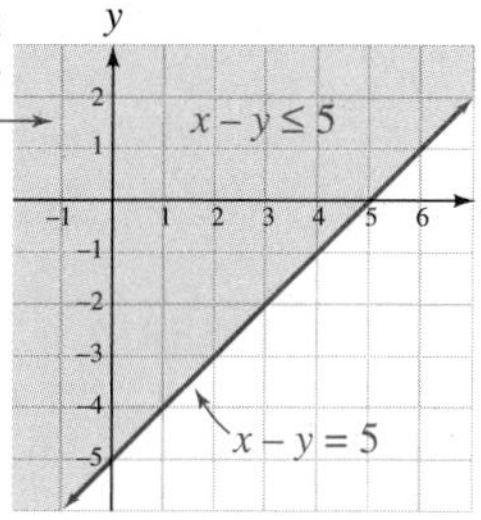

(b)

Step 2: Since the inequality $x - y \leq 5$ also allows $x - y$ to be less than 5, other ordered pairs, besides those on the boundary, satisfy the inequality. For example, consider the origin, with coordinates (0, 0). If we substitute 0 for x and 0 for y in the given inequality, we have

$x - y \leq 5$ This is the given inequality.

$0 - 0 \overset{?}{\leq} 5$ Substitute.

$0 \leq 5$ True

Because $0 \leq 5$, the coordinates of the origin satisfy $x - y \leq 5$. In fact, the coordinates of every point on the same side of the boundary as the origin satisfy the inequality. To indicate this, we shade in red the half-plane that contains the test point (0, 0), as shown in part (b). Every point in the shaded half-plane and every point on the boundary line satisfies $x - y \leq 5$. On the other hand, the points in the unshaded half-plane *do not* satisfy $x - y \leq 5$.

As an informal check, we can pick an ordered pair that lies in the shaded region and one that does not lie in the shaded region. When we substitute their coordinates into the inequality, we should obtain a true statement and then a false statement.

For* (3, 1), *in the shaded region:

$x - y \leq 5$

$3 - 1 \overset{?}{\leq} 5$ Substitute.

$2 \leq 5$ True

For* (5, −4), *not in the shaded region:

$x - y \leq 5$

$5 - (-4) \overset{?}{\leq} 5$ Substitute.

$9 \leq 5$ False

Success Tip

All the points in the unshaded region below the boundary line have coordinates that satisfy $x - y > 5$.

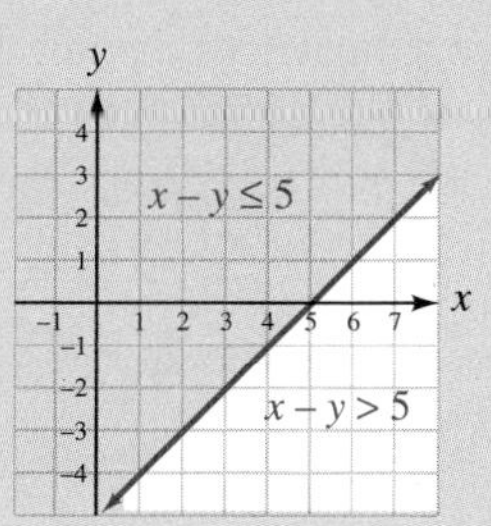

Teaching Example 2
Graph: $3x - y \leq 6$
Answer:

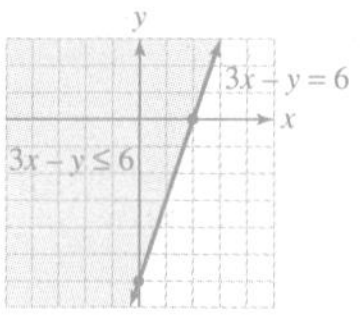

Self Check 2 Graph: $x - y \leq 2$ See AIE Appendix 3.

Now Try Problem 35

The previous example suggests the following **test-point method** to graph linear inequalities in two variables.

Graphing Linear Inequalities in Two Variables

1. Replace the inequality symbol with an equal symbol = and graph the boundary line of the region. If the original inequality allows the possibility of equality (the symbol is either ≤ or ≥), draw the boundary line as a solid line. If equality is not allowed (< or >), draw the boundary line as a dashed line.
2. Pick a test point that is on one side of the boundary line. (Use the origin if possible.) Replace x and y in the inequality with the coordinates of that point. If a true statement results, shade the side that contains that point. If a false statement results, shade the other side of the boundary.

EXAMPLE 3 Graph: $4x + 3y < -6$

Teaching Tip: After completing the explanation of this solution, ask your students what points could be used as a check. Also ask what inequality describes the *unshaded* region of the graph.

Strategy We will graph the related equation $4x + 3y = -6$ to establish the boundary line between two regions of the coordinate plane. Then we will determine which region contains points that satisfy the given inequality.

Why The graph of a linear inequality in two variables is a region of the coordinate plane on one side of a boundary line.

Solution To find the boundary line, we replace the inequality symbol with an equal symbol = and graph $4x + 3y = -6$ using the intercept method. Since the inequality symbol < does not include an equal symbol, the points on the graph of $4x + 3y = -6$ will not be part of the graph of $4x + 3y < -6$. To show this, we draw the boundary line as a dashed line. See part (a) of the illustration below.

A table of solutions to graph the boundary

$4x + 3y = -6$

x	y	(x, y)
0	−2	(0, −2)
$-\frac{3}{2}$	0	$\left(-\frac{3}{2}, 0\right)$
−3	2	(−3, 2)

Let $x = 0$ and find y.
Let $y = 0$ and find x.
As a check, let $x = -3$ and find y.

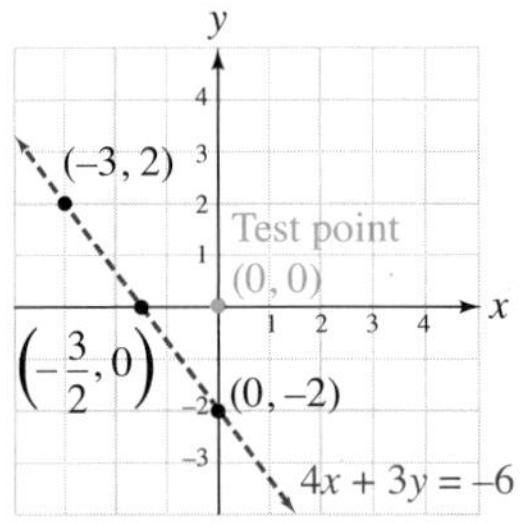

(a)

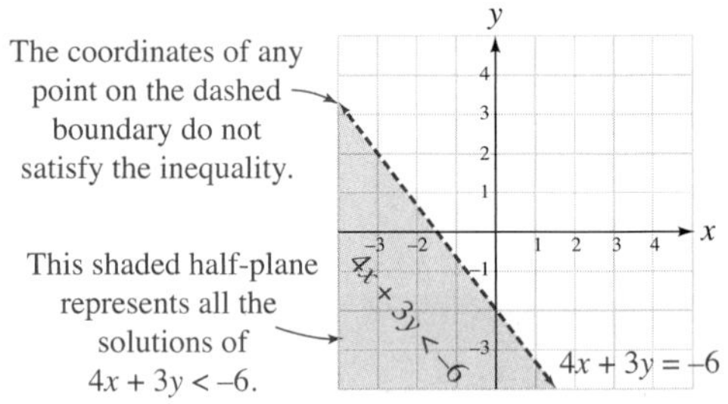

(b)

To determine which half-plane to shade, we substitute the coordinates of a point that lies on one side of the boundary line into $4x + 3y < -6$. We choose the origin (0, 0) as the test point because the calculations are easy when they involve 0. We substitute 0 for x and 0 for y in the inequality.

Caution

When using a test point to determine which half-plane to shade, remember to substitute the coordinates into the given inequality, not the equation for the boundary.

~~$4(0) + 3(0) = -6$~~

$4x + 3y < -6$ This is the given inequality.

$4(\mathbf{0}) + 3(\mathbf{0}) \overset{?}{<} -6$ The symbol $\overset{?}{<}$ is read as "is possibly less than."

$0 + 0 \overset{?}{<} -6$

$0 < -6$ False

Since $0 < -6$ is a false statement, the point (0, 0) does not satisfy the inequality. In fact, no point in the half-plane containing (0, 0) is a solution. Therefore, we shade the other side of the boundary line—the half-plane that does not contain (0, 0). The graph of the solution set of $4x + 3y < -6$ is the half-plane below the dashed line, as shown in part (b).

Teaching Example 3
Graph: $6x + 5y < -12$
Answer:

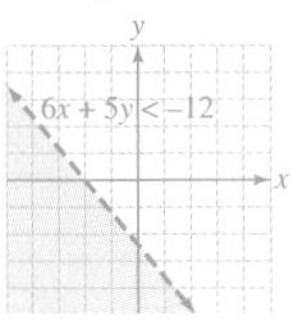

Self Check 3 Graph: $5x + 6y < -15$ See AIE Appendix 3.

Now Try Problem 37

3 Graph Inequalities with a Boundary through the Origin.

In the next example, the boundary line passes through the origin. In such cases, the ordered pair (0, 0) should not be used as a test point to determine which half-plane to shade.

EXAMPLE 4 Graph: $y > 2x$

Strategy We will graph the related equation $y = 2x$ to establish the boundary line between two regions of the coordinate plane. Then we will determine which region contains points that satisfy the given inequality.

Why The graph of a linear inequality in two variables is a region of the coordinate plane on one side of a boundary line.

Solution To find the boundary line, we graph $y = 2x$. Since the symbol $>$ does *not* include an equal symbol, the points on the graph of $y = 2x$ are not part of the graph of $y > 2x$. Therefore, the boundary line should be dashed, as shown in part (a) of the illustration below.

Success Tip

Draw a solid boundary line if the inequality has $\le$ or $\ge$. Draw a dashed line if the inequality has $<$ or $>$.

Teaching Tip: After completing the graph, pick several points from the shaded and unshaded regions, as well as on the boundary. Ask your students if the coordinates of those points, when substituted into the inequality, produce a true or false statement.

A table of solutions to graph the boundary

$y = 2x$

x	y	(x, y)
0	0	(0, 0)
−1	−2	(−1, −2)
1	2	(1, 2)

Select three values for x and find the corresponding values of y.

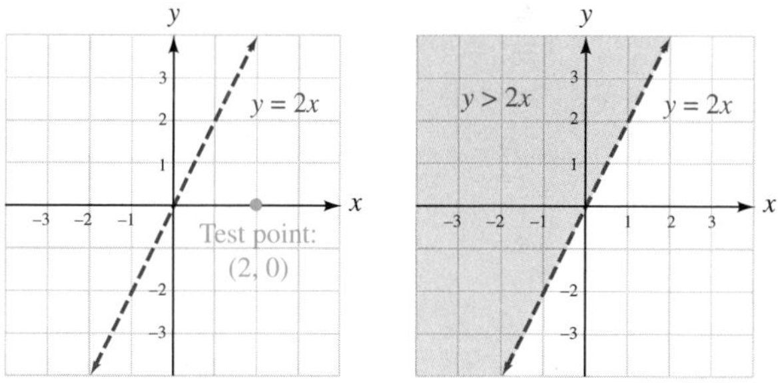

(a) (b)

Success Tip

The origin (0, 0) is a smart choice for a test point because calculations involving 0 are usually easy. If the origin is on the boundary, choose a test point not on the boundary that has one coordinate that is 0, such as (0, 1) or (2, 0).

To determine which half-plane to shade, we substitute the coordinates of a point that lies on one side of the boundary line into $y > 2x$. Since the origin is on the boundary, it cannot serve as a test point. One of the many possible choices for a test point is (2, 0), because it does not lie on the boundary line. To see whether it satisfies $y > 2x$, we substitute 2 for x and 0 for y in the inequality.

$y > 2x$ This is the given inequality.

$\mathbf{0} \overset{?}{>} 2(\mathbf{2})$ The symbol $\overset{?}{>}$ is read as "is possibly greater than."

$0 > 4$ False

Since $0 > 4$ is a false statement, the point (2, 0) does not satisfy the inequality. We shade the half-plane that does not contain (2, 0), as shown in part (b).

Teaching Example 4

Graph: $y > -\frac{1}{2}x$

Answer:

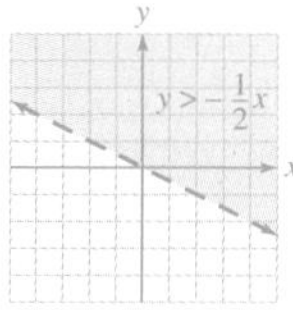

Self Check 4 Graph: $y < 3x$ See AIE Appendix 3.

Now Try Problem 55

The graphs of some linear inequalities in two variables have boundary lines that are horizontal or vertical.

EXAMPLE 5 Graph each linear inequality: **a.** $x < -3$ **b.** $y \ge 0$

Strategy We will use the procedure for graphing linear inequalities in two variables.

Why Since the inequalities can be written as $x + 0y < -3$ and $0x + y \ge 0$, they are linear inequalities in two variables.

Solution **a.** Because $x < -3$ contains an $<$ symbol, we draw the boundary, $x = -3$, as a dashed vertical line. See figure (a) below. We can use (0, 0) as the test point.

$x < -3$ This is the given inequality.

$0 < -3$ Substitute 0 for x. The y-coordinate of the test point (0, 0) is not used.

Since the result is false, we shade the half-plane that does not contain (0, 0), as shown in figure (b) below. Note that the solution consists of all points that have an x-coordinate that is less than -3.

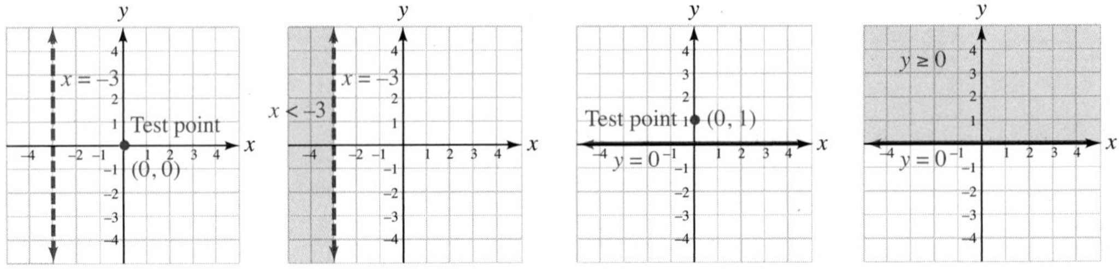

(a) (b) (c) (d)

b. Because $y \geq 0$ contains an $\geq$ symbol, we draw the boundary, $y = 0$, as a solid horizontal line. (Recall that the graph of $y = 0$ is the x-axis. See figure (c) above.) Next, we choose a test point not on the boundary. The point (0, 1) is a convenient choice. See figure (c) above.

$y \geq 0$ This is the given inequality.

$1 \geq 0$ Substitute 1 for y. The x-coordinate of the test point (0, 1) is not used.

Since the result is true, we shade the half-plane that contains (0, 1), as shown in part (d) above. Note that the solution consists of all points that have a y-coordinate that is greater than or equal to 0.

Teaching Example 5
Graph each linear inequality:
a. $x < -4$ **b.** $y \geq 1$
Answers:
a. b.

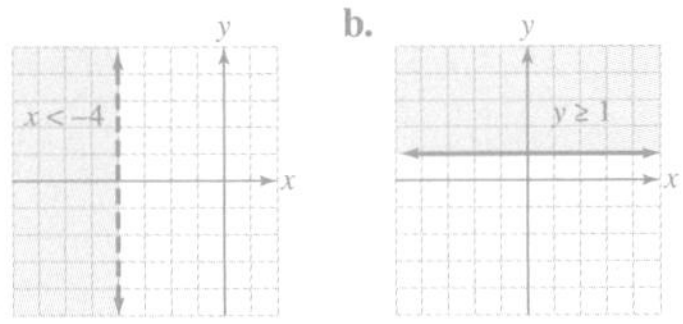

Self Check 5 Graph each linear inequality: **a.** $x \geq 2$ **b.** $y < 4$ See AIE Appendix 3.

Now Try Problems 63 and 65

4 Solve Applied Problems Involving Linear Inequalities in Two Variables.

When solving applied problems, phrases such as *at least, at most,* and *should not exceed* indicate that an inequality should be used.

EXAMPLE 6 **Working Two Jobs.** Carlos has two part-time jobs, one paying \$10 per hour and another paying \$12 per hour. If x represents the number of hours he works on the first job, and y represents the number of hours he works on the second, the graph of $10x + 12y \geq 240$ shows the possible ways he can schedule his time to earn at least \$240 per week to pay his college expenses. Find four possible combinations of hours he can work to achieve his financial goal.

Strategy We will graph the inequality and find four points whose coordinates satisfy the inequality.

Why The coordinates of these points will give four possible combinations.

Solution The graph of the inequality is shown below in part (a) of the illustration. Any point in the shaded region represents a possible way Carlos can schedule his time and earn \$240 or more per week. If each shift is a whole number of hours long, the red highlighted points in part (b) represent four of the many acceptable combinations.

Success Tip

Since Carlos cannot work a negative number of hours on job 1 or on job 2, we need to graph only the inequality in quadrant I.

(6, 24): 6 hours on the first job, 24 hours on the second job
(12, 12): 12 hours on the first job, 12 hours on the second job
(22, 4): 22 hours on the first job, 4 hours on the second job
(26, 20): 26 hours on the first job, 20 hours on the second job

To verify one combination, suppose Carlos works 22 hours on the first job and 4 hours on the second job. He will earn

$$\$10(\mathbf{22}) + \$12(\mathbf{4}) = \$220 + \$48$$
$$= \$268 \qquad \text{This is at least \$240 per week.}$$

A table of solutions to graph the boundary
$10x + 12y = 240$

x	y	(x, y)
0	20	(0, 20)
24	0	(24, 0)

Let $x = 0$ and find y.
Let $y = 0$ and find x.

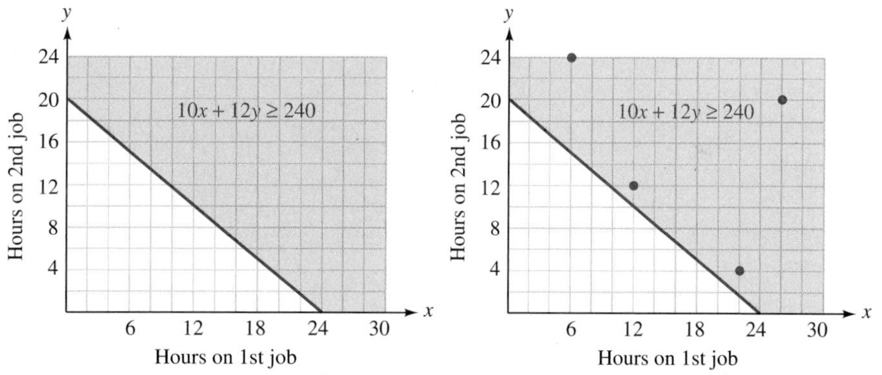

(a) (b)

Teaching Example 6
Entertainment. Carly is going to purchase some CDs and some DVDs. If each CD costs \$12 amd each DVD costs \$20, write an inequality to represent the number of CDs and DVDs she can buy if she wants to spend \$180 or less.
Answer: $12x + 20y \le 180$

Self Check 6 **iTunes.** Brianna and Ashley pool their money to purchase some songs and movies on iTunes. If songs cost \$1 and movies cost \$15, write an inequality to represent the number of songs and movies they can buy if they want to spend \$150 or less. $x + 15y \le 150$

Now Try Problem 79

SECTION 3.7 STUDY SET

VOCABULARY

Fill in the blanks.

1. $2x - y \le 4$ is a linear <u>inequality</u> in two variables.
2. An ordered pair (x, y) is a <u>solution</u> of a linear inequality in two variables if a true statement results when the values of the variables are substituted into the inequality.
3. (7, 2) is a solution of $x - y > 1$. We say that (7, 2) <u>satisfies</u> the inequality.
4. In the graph, the line $2x - y = 4$ is the <u>boundary</u> line.
5. In the graph, the line $2x - y = 4$ divides the coordinate plane into two <u>half-planes</u>.
6. When graphing a linear inequality, we determine which half-plane to shade by substituting the coordinates of a test <u>point</u> into the inequality.

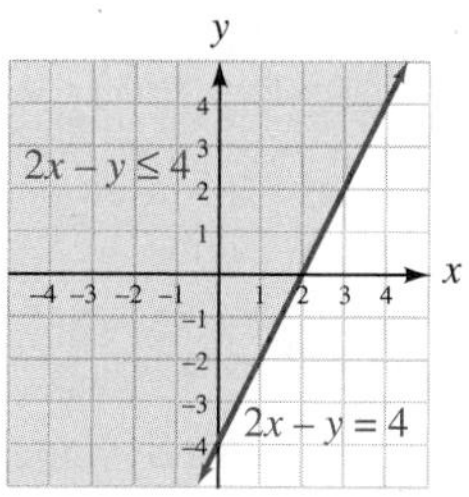

CONCEPTS

7. Determine whether $(-3, -5)$ is a solution of $5x - 3y \ge 0$. Yes
8. Determine whether $(3, -1)$ is a solution of $x + 4y < -1$. No

9. Fill in the blanks: A dashed line indicates that points on the boundary are not solutions and a solid line indicates that points on the boundary are solutions.

10. The boundary for the graph of a linear inequality is shown. Why can't the origin be used as a test point to decide which side to shade? The test point must be on one side of the boundary.

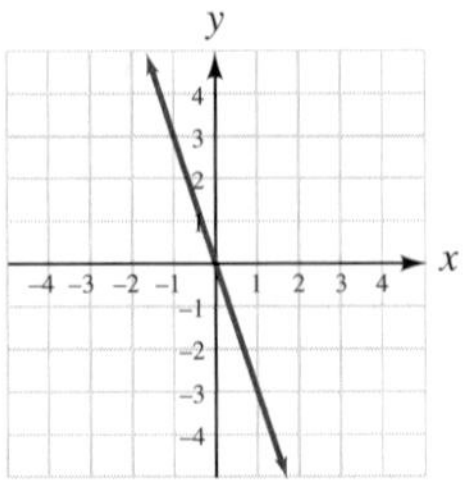

11. If a false statement results when the coordinates of a test point are substituted into a linear inequality, which half-plane should be shaded to represent the solution of the inequality? The half-plane opposite that in which the test point lies

12. A linear inequality has been graphed. Determine whether each point satisfies the inequality.

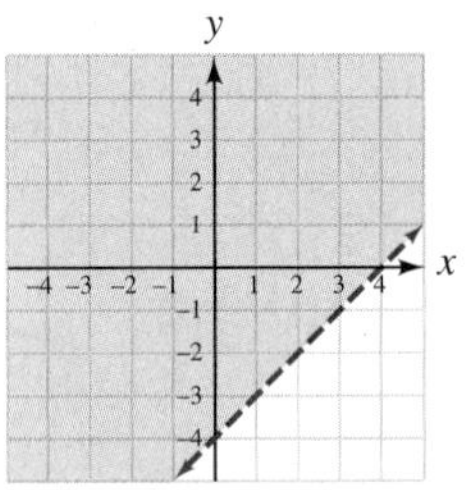

a. $(1, -3)$ No
b. $(-2, -1)$ Yes
c. $(2, 3)$ Yes
d. $(3, -4)$ No

13. A linear inequality has been graphed. Determine whether each point satisfies the inequality.

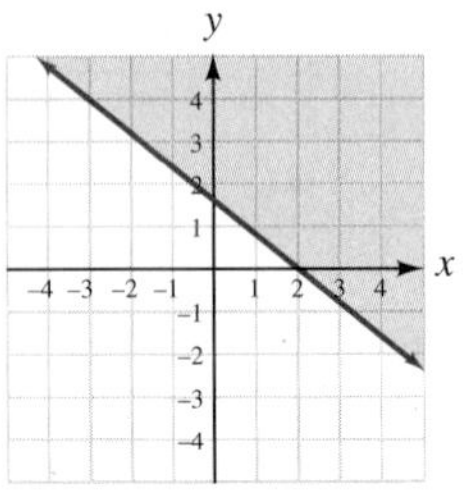

a. $(2, 1)$ Yes
b. $(-2, -4)$ No
c. $(4, -2)$ No
d. $(-3, 4)$ Yes

14. To graph linear inequalities, we must be able to graph boundary lines. Complete the table of solutions for each given boundary line.

a. $5x - 3y = 15$

x	y	(x, y)
0	-5	$(0, -5)$
3	0	$(3, 0)$
1	$-3\frac{1}{3}$	$\left(1, -3\frac{1}{3}\right)$

b. $y = 3x - 2$

x	y	(x, y)
-1	-5	$(-1, -5)$
0	-2	$(0, -2)$
2	4	$(2, 4)$

NOTATION

15. Write the meaning of each symbol in words.

a. $<$ Is less than
b. $\geq$ Is greater than or equal to
c. $\leq$ Is less than or equal to
d. $\overset{?}{>}$ Is possibly greater than

16. a. When graphing linear inequalities, which inequality symbols are associated with a dashed boundary line? $<, >$

b. When graphing linear inequalities, which inequality symbols are associated with a solid boundary line? $\leq, \geq$

17. Fill in the blanks: The inequality $4x + 2y \leq 9$ means $4x + 2y = 9$ or $4x + 2y < 9$.

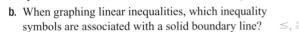

18. Fill in the blanks: The inequality $-x + 8y \geq 1$ means $-x + 8y = 1$ or $-x + 8y > 1$.

GUIDED PRACTICE

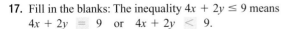

Determine whether each ordered pair is a solution of the given inequality. **See Example 1.**

19. $2x + y > 6$; $(3, 2)$ Yes
20. $4x - 2y \geq -6$; $(-2, 1)$ No
21. $-5x - 8y < 8$; $(-8, 4)$ No
22. $x + 3y > 14$; $(-3, 8)$ Yes
23. $4x - y \leq 0$; $\left(\frac{1}{2}, 1\right)$ No
24. $9x - y \leq 2$; $\left(\frac{1}{3}, 1\right)$ Yes
25. $-5x + 2y > -4$; $(0.8, 0.6)$ Yes
26. $6x - 2y < -7$; $(-0.2, 1.5)$ No

Complete the graph by shading the correct side of the boundary. **See Example 2.**

27. $x - y \geq -2$

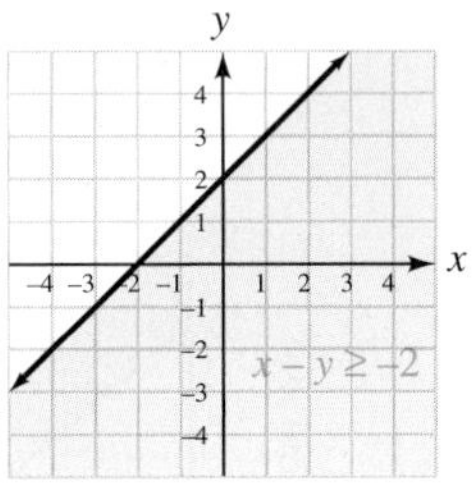

28. $x - y < 3$

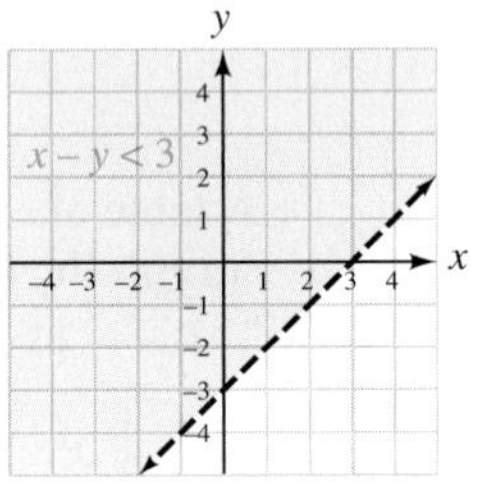

29. $y > 2x - 4$

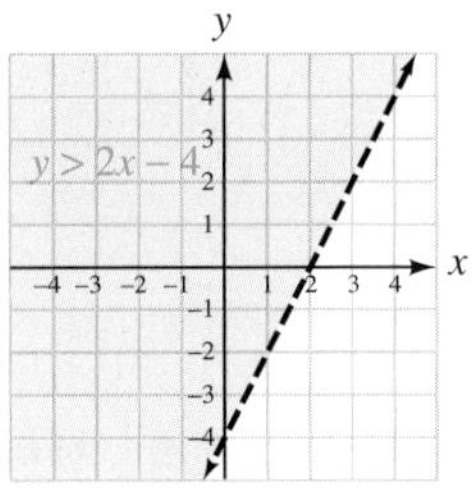

30. $y \leq -x + 1$

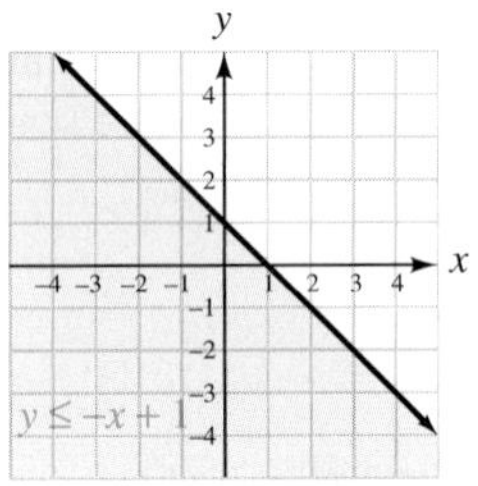

31. $x - 2y \geq 4$

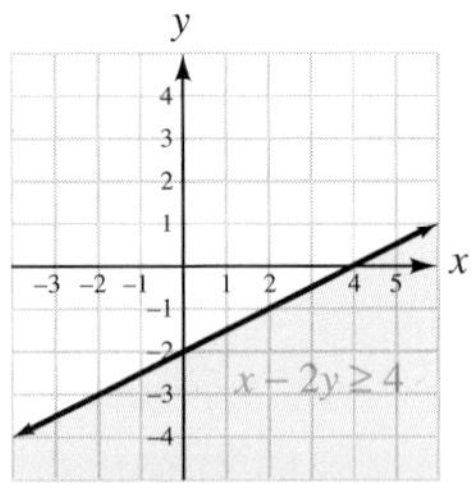

32. $3x + 2y > 12$

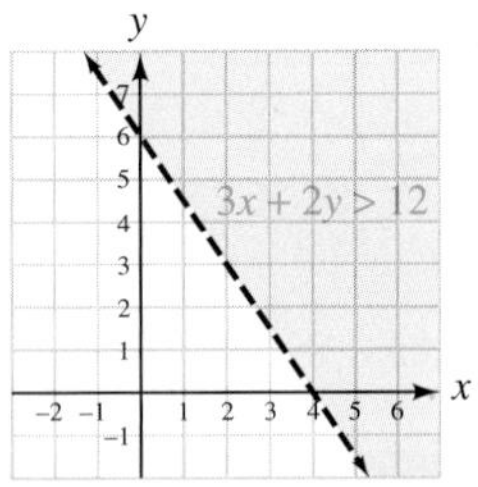

33. $y \leq 4x$

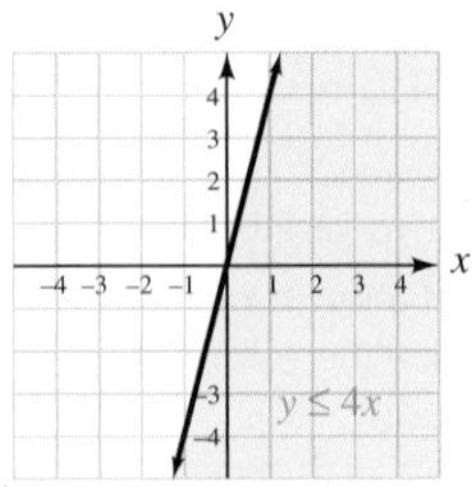

34. $y + 2x < 0$

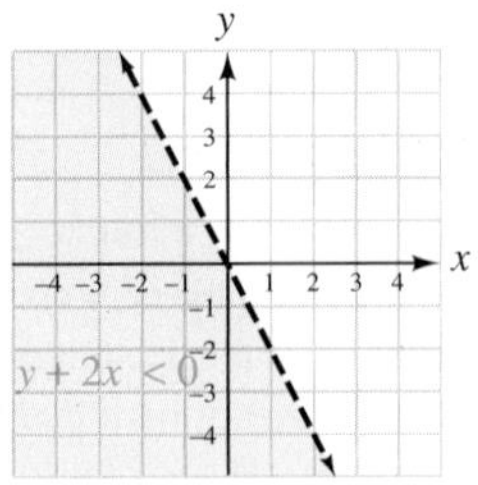

Graph each inequality. **See Examples 2 and 3.** See AIE Appendix 3.

35. $x + y \geq 3$
36. $x + y < 2$
37. $3x - 4y > 12$
38. $5x + 4y \geq 20$
39. $2x + 3y \leq -12$
40. $3x - 2y > 6$
41. $y < 2x - 1$
42. $y > x + 1$
43. $y < -3x + 2$
44. $y \geq -2x + 5$

45. $y \geq -\frac{3}{2}x + 1$

▶ 46. $y < \frac{x}{3} - 1$

47. $x - 2y \geq 4$

▶ 48. $4x + y \geq -4$

49. $2y - x < 8$

50. $y + 9x \geq 3$

51. $7x - 2y < 21$

▶ 52. $3x - 3y \geq -10$

53. $2x - 3y \geq 4$

54. $4x + 3y < 6$

Graph each inequality. See Example 4. See AIE Appendix 3.

55. $y \geq 2x$

▶ 56. $y < 3x$

57. $y < -\frac{x}{2}$

▶ 58. $y \geq x$

59. $y + x < 0$

▶ 60. $y - x < 0$

61. $5x + 3y < 0$

62. $2x + 5y > 0$

Graph each inequality. See Example 5. See AIE Appendix 3.

63. $x < 2$

▶ 64. $y > -3$

65. $y \leq 1$

▶ 66. $x \geq -4$

67. $y + 2.5 > 0$

▶ 68. $x - 1.5 \leq 0$

69. $x \leq 0$

▶ 70. $y < 0$

TRY IT YOURSELF

Look Alikes . . .

Graph the given inequality in part a. Then use your answer to part a to help you quickly *graph the associated inequality in part b. (*Hint: *If you spot the relationship between the inequalities, the graph in part b can be completed without having to use the test-point method.)* See AIE Appendix 3.

71. a. $5x - 3y \geq -15$ b. $5x - 3y < -15$

▶ 72. a. $y > -\frac{2}{3}x + 2$ b. $y \leq -\frac{2}{3}x + 2$

73. a. $y + 2x < 0$ b. $y + 2x \geq 0$

74. a. $y \leq \frac{1}{4}x$ b. $y > \frac{1}{4}x$

APPLICATIONS

▶ 75. **Deliveries.** To decide the number x of pallets and the number y of barrels that a truck can hold, a driver refers to the graph below. Can a truck make a delivery of 4 pallets and 10 barrels in one trip? No

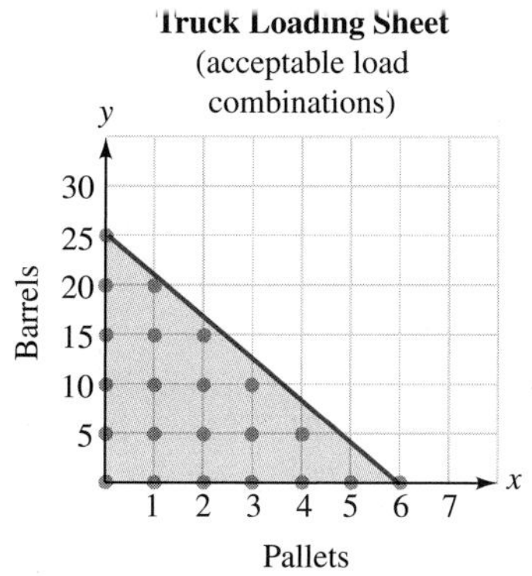

76. **Zoos.** To determine the allowable number of juvenile chimpanzees x and adult chimpanzees y that can live in an enclosure, a zookeeper refers to the graph. Can 6 juvenile and 4 adult chimps be kept in the enclosure? Yes

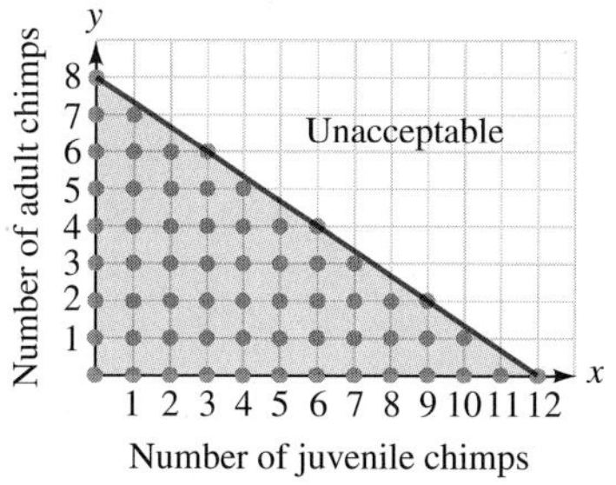

77. ▶ **from Campus to Careers**

Dental Assistant

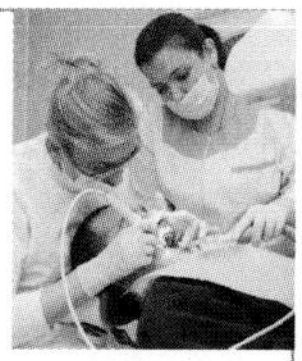

A dentist's office schedules 1-hour long appointments for adults and $\frac{3}{4}$-hour long appointments for children. The appointment times do not overlap. Let c represent the number of appointments scheduled for children and a represent the number of appointments scheduled for adults. The graph of $\frac{3}{4}c + a \leq 9$ shows the possible ways the time for seeing patients can be scheduled so that it does not exceed 9 hours per day. Graph the inequality. Label the horizontal axis c and the vertical axis a. Then find three possible combinations of children/adult appointments. (2, 7), (4, 6), (9, 2); answers may vary. See AIE Appendix 3.

78. **Rolling Dice.** The points on the graph represent all of the possible outcomes when two fair dice are rolled a single time. For example, (5, 2), shown in red, represents a 5 on the first die and a 2 on the second. Which of the following sentences best describes the outcomes that lie in the shaded area? ii

(i) Their sum is at most 6. **(ii)** Their sum exceeds 6.

(iii) Their sum does not exceed 6. **(iv)** Their sum is at least 6.

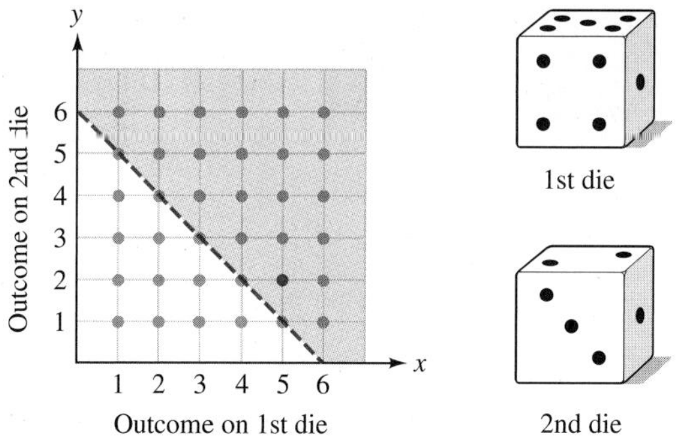

▶ 79. **Production Planning.** It costs a bakery \$3 to make a cake and \$4 to make a pie. If x represents the number of cakes made, and y represents the number of pies made, the graph of $3x + 4y \leq 120$ shows the possible combinations of cakes and pies that can be produced so that costs do not exceed \$120 per day. Graph the inequality. Then find three possible combinations of cakes and pies that can be made so that the daily costs are not exceeded. (10, 10), (20, 10), (10, 20); answers may vary. See AIE Appendix 3.

80. Hiring Babysitters. Mrs. Cansino has a choice of two babysitters. Sitter 1 charges \$6 per hour, and Sitter 2 charges \$7 per hour. If x represents the number of hours she uses Sitter 1 and y represents the number of hours she uses Sitter 2, the graph of $6x + 7y \le 42$ shows the possible ways she can hire the sitters and not spend more than \$42 per week. Graph the inequality. Then find three possible ways she can hire the babysitters so that her weekly budget for babysitting is not exceeded.
(2, 2), (4, 2), (3, 3); answers may vary. See AIE Appendix 3.

81. Inventories. A clothing store advertises that it maintains an inventory of at least \$4,400 worth of men's jackets at all times. At the store, leather jackets cost \$100 and nylon jackets cost \$88. If x represents the number of leather jackets in stock and y represents the number of nylon jackets in stock, the graph of $100x + 88y \ge 4{,}400$ shows the possible ways the jackets can be stocked. Graph the inequality. Then find three possible combinations of leather and nylon jackets so that the store lives up to its advertising claim.
(40, 30), (30, 40), (40, 20); answers may vary. See AIE Appendix 3.

82. Making Sporting Goods. A sporting goods manufacturer allocates at least 2,400 units of production time per day to make baseballs and footballs. It takes 20 units of time to make a baseball and 30 units of time to make a football. If x represents the number of baseballs made and y represents the number of footballs made, the graph of $20x + 30y \ge 2{,}400$ shows the possible ways to schedule the production time. Graph the inequality. Then find three possible combinations of production time for the company to make baseballs and footballs. (60, 60), (80, 40), (100, 40); answers may vary. See AIE Appendix 3.

WRITING

83. Explain how to decide which side of the boundary line to shade when graphing a linear inequality in two variables.

84. Why is the origin usually a good test point to choose when graphing a linear inequality?

85. Why is (0, 0) not an acceptable choice for a test point when graphing a linear inequality whose boundary passes through the origin?

86. Explain the difference between the graph of the solution set of $x + 1 > 8$, an inequality in one variable, and the graph of $x + y > 8$, an inequality in two variables.

REVIEW

87. Solve $A = P + Prt$ for t. $t = \frac{A - P}{Pr}$

88. What is the sum of the measures of the three angles of any triangle? 180°

89. Simplify: $40\left(\frac{3}{8}x - \frac{1}{4}\right) + 40\left(\frac{4}{5}\right)$ $15x + 22$

90. Evaluate: $-4 + 5 - (-3) - 13$ -9

CHALLENGE PROBLEMS

91. Find a linear inequality that has the graph shown. $3x - 2y \ge 6$

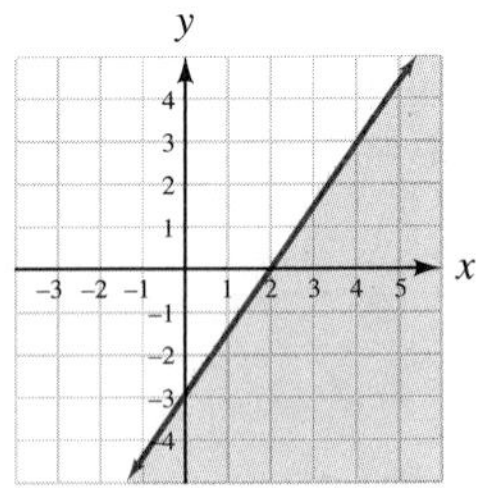

92. Graph the inequality: $4x - 3(x + 2y - 1) \ge -6\left(y - \frac{1}{2}\right)$
See AIE Appendix 3.

SECTION 3.8 An Introduction to Functions

OBJECTIVES

1. Find the domain and range of a relation.
2. Identify functions and their domains and ranges.
3. Use function notation.
4. Graph functions.
5. Use the vertical line test.
6. Solve applications involving functions.

ARE YOU READY? *Are You Ready? exercises available online at www.webassign.net/brookscole*

The following problems review some basic skills that are needed when working with functions.

1. If $y = 3x - 1$, find the value for y when $x = 6$. 17

2. Which of the following ordered pairs have the same x-coordinate?
(3, 5), (2, 9), (3, 0), (−1, 5) (3, 5), (3, 0)

3. Which of the following ordered pairs have the same y-coordinate?
(8, 4), (−7, 6), (0, 0), (−3, 4) (8, 4), (−3, 4)

4. Find each absolute value:
a. $|-5|$ 5 **b.** $|2|$ 2

In this section, we will discuss *relations* and *functions*. These two concepts are included in our study of graphing because they involve ordered pairs.

1 Find the Domain and Range of a Relation.

The following table shows the number of medals won by American athletes at several recent Winter Olympics.

USA Winter Olympic Medal Count

Year	1988	1992	1994*	1998	2002	2006	2010
Medals	6	11	13	13	34	25	37
	Calgary CAN	Albertville FRA	Lillehammer NOR	Nagano JPN	Salt Lake City USA	Turin ITA	Vancouver CAN

* The Winter Olympics were moved ahead two years so that the winter and summer games would alternate every two years.

Teaching Tip: Remind your students that the members of a set are written within braces { }.

We can display the data in the table as a set of ordered pairs, where the **first component** represents the year and the **second component** represents the number of medals won by American athletes:

$$\{(1988, 6), (1992, 11), (1994, 13), (1998, 13), (2002, 34), (2006, 25), (2010, 37)\}$$

A set of ordered pairs, such as this, is called a **relation.** The set of all first components is called the **domain** of the relation and the set of all second components is called the **range** of the relation.

EXAMPLE 1 Find the domain and range of the relation $\{(1, 7), (4, -6), (-3, 1), (2, 7)\}$.

Strategy We will examine the first and second components of the ordered pairs.

Why The set of first components is the domain and the set of second components is the range.

Solution The relation $\{(\mathbf{1}, 7), (\mathbf{4}, -6), (\mathbf{-3}, 1), (\mathbf{2}, 7)\}$ has the domain $\{\mathbf{-3, 1, 2, 4}\}$ and the range is $\{\mathbf{-6, 1, 7}\}$. The elements of the domain and range are usually listed in increasing order, and if a value is repeated, it is listed only once.

Teaching Example 1 Find the domain and range of the relation $\{(-3, 2), (5, 4), (-1, 6), (-4, 9), (10, 3)\}$.
Answers: Domain: $\{-4, -3, -1, 5, 10\}$, range: $\{2, 3, 4, 6, 9\}$

Self Check 1 Find the domain and range of the relation $\{(8, 2), (-1, 10), (6, 2), (-5, -5)\}$. Domain: $\{-5, -1, 6, 8\}$; range: $\{-5, 2, 10\}$

Now Try Problem 15

2 Identify Functions and Their Domains and Ranges.

An **arrow** or **mapping diagram** can be used to define a relation. The data from the Winter Olympics example are shown on the right in that form. Relations are also often defined using **two-column tables.**

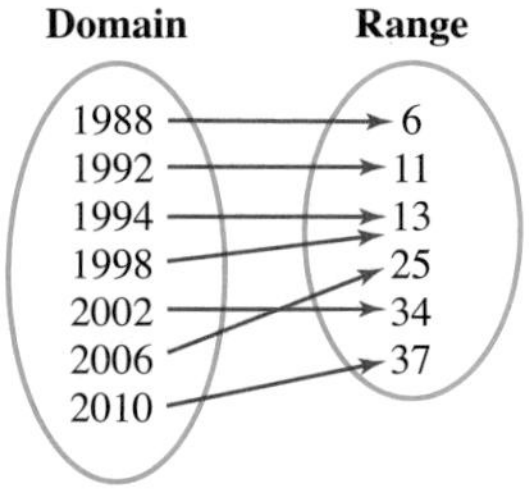

Notice that for each year, there corresponds exactly one medal count. That is, this relation assigns to each member of the domain exactly one member of the range. Relations that have this characteristic are called *functions.*

Function A **function** is a set of ordered pairs (a relation) in which to each first component there corresponds exactly one second component.

We may also think of a function as a rule that assigns to each value of one variable exactly one value of another variable. Since we often worked with sets of ordered pairs of the form (x, y), it is helpful to define a function in an alternate way using the variables x and y.

y Is a Function of x

Given a relation in x and y, if to each value of x in the domain there is assigned exactly one value of y in the range, then y is said to be a function of x.

EXAMPLE 2 Determine whether each relation defines y to be a function of x. If a function is defined, give its domain and range.

a.

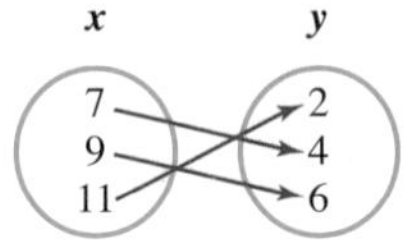

b.

x	y
2	3
5	7
2	1
6	5

c. $\{(0, 8), (3, 8), (4, 8), (9, 8)\}$

d.

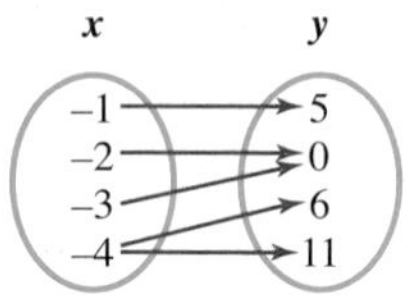

Strategy We will check to see whether each value of x is assigned exactly one value of y.

Why If this is true, then y is a function of x.

Solution

a. The arrow diagram defines a function because to each value of x there is assigned exactly one value of y: $7 \to 4$, $9 \to 6$, and $11 \to 2$.

The domain of the function is $\{7, 9, 11\}$ and the range is $\{2, 4, 6\}$.

b. The table does not define a function, because to the x value 2 there is assigned more than one value of y: $2 \to 3$ and $2 \to 1$.

c. Since to each number x exactly one value y is assigned, the set of ordered pairs defines y to be a function of x. It also illustrates an important fact about functions: *The same value of y can be assigned to different values of x.* In this case, each number x is assigned the y-value 8.

The domain of the function is $\{0, 3, 4, 9\}$ and the range is $\{8\}$.

d. The arrow diagram does not define a function, because to the x value -4 there is assigned more than one value of y: $-4 \to 6$ and $-4 \to 11$.

Success Tip

Every function is a relation, but not every relation is a function.

Self Check 2 Determine whether each relation defines y to be a function of x. If a function is defined, give its domain and range.

a.

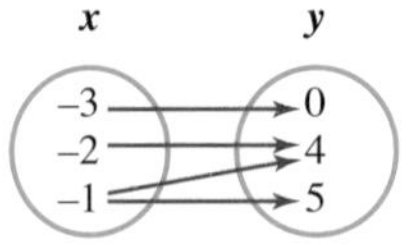

Not a function; $-1 \to 4$ and $-1 \to 5$

b.

x	y
−6	−6
5	5
4	8

Function; domain: $\{-6, 4, 5\}$; range: $\{-6, 5, 8\}$

c. $\{(1, 4), (4, 9), (9, 4), (3, 9)\}$

Function; domain: $\{1, 3, 4, 9\}$; range: $\{4, 9\}$

Now Try Problems 19, 25, and 31

Teaching Example 2
Determine whether each relation defines y as a function of x. If a function is defined, give its domain and range.

a.

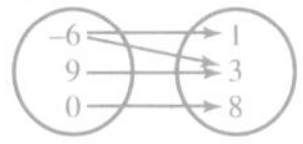

b.

x	y
12	4
13	6
14	12
15	9

c. $\{(3, -2), (5, -2), (-8, -2), (9, -2)\}$

d.

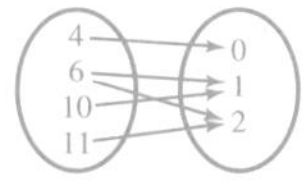

Answers:
a. Not a function
b. Function; domain: $\{12, 13, 14, 15\}$; range: $\{4, 6, 9, 12\}$
c. Function; domain: $\{-8, 3, 5, 9\}$; range: $\{-2\}$
d. Not a function

3 Use Function Notation.

A function can be defined by an equation. For example, $y = 2x - 3$ is a rule that assigns to each value of x exactly one value of y. To find the y-value that is assigned to the x-value 4, we substitute 4 for x and evaluate the right side of the equation.

$y = 2x - 3$

$y = 2(4) - 3$ Substitute 4 for x.

$= 8 - 3$ Evaluate the right side.

$= 5$

The function $y = 2x - 3$ assigns the y-value 5 to an x-value of 4. When making such calculations, the value of x is called an **input** and its corresponding value of y is called an **output.**

A special notation is used to name functions that are defined by equations.

Function Notation

The notation $y = f(x)$ indicates that the variable y is a function of x.

Since $y = f(x)$, the equations $y = 2x - 3$ and $f(x) = 2x - 3$ are equivalent. We read $f(x) = 2x - 3$ as "f of x is equal to $2x$ minus 3."

Notation

If y is a function of x, then y and $f(x)$ are interchangeable.

This is the variable used to represent the input value.

$$f(x) = 2x - 3$$

This is the name of the function.

This expression shows how to obtain an output from a given input.

Function notation provides a compact way of representing the value that is assigned to some number x. For example, if $f(x) = 2x - 3$, the value that is assigned to an x-value 5 is represented by $f(5)$.

$f(x) = 2x - 3$
$f(5) = 2(5) - 3$ Substitute the input 5 for each x.
$= 10 - 3$ Evaluate the right side.
$= 7$ The output is 7.

Caution

The symbol $f(x)$ **does not** mean $f \cdot x$. We read $f(x)$ as "f of x" or as "the value of f at x."

Thus, $f(5) = 7$. We read this as "f of 5 is 7." The output 7 is called a **function value.**

To see why function notation is helpful, consider these two sentences, which ask you to do the same thing:

1. If $y = 2x - 3$, find the value of y when x is 5.
2. If $f(x) = 2x - 3$, find $f(5)$.

Sentence 2, which uses $f(x)$ notation, is much more compact.

EXAMPLE 3 For $f(x) = 5x + 7$, find each of the following function values:
a. $f(2)$ **b.** $f(-4)$ **c.** $f(0)$

Strategy We will substitute 2, −4, and 0 for x in the expression $5x + 7$ and then evaluate it.

Why The notation $f(x) = 5x + 7$ indicates that we are to multiply each input (each number written within the parentheses) by 5 and then add 7 to that product.

Solution

a. To find $f(2)$, we substitute the number within the parentheses, 2, for each x in $f(x) = 5x + 7$, and evaluate the right side of the equation.

The Language of Algebra

Another way to read $f(2) = 17$ is to say "the value of f at 2 is 17."

$f(x) = 5x + 7$
$f(2) = 5(2) + 7$ Substitute the input 2 for each x.
$= 10 + 7$ Evaluate the right side.
$= 17$ The output is 17.

Thus, $f(2) = 17$.

b. $f(x) = 5x + 7$

$f(-4) = 5(-4) + 7$ Substitute the input −4 for each x.

$= -20 + 7$ Evaluate the right side.

$= -13$ The output is −13.

Thus, $f(-4) = -13$.

c. $f(x) = 5x + 7$

$f(0) = 5(0) + 7$ Substitute the input 0 for each x.

$= 0 + 7$ Evaluate the right side.

$= 7$ The output is 7.

Thus, $f(0) = 7$.

Teaching Example 3
For $f(x) = -5x + 7$, find each of the following function values: **a.** $f(3)$ **b.** $f(-9)$ **c.** $f(0)$
Answers: **a.** −8 **b.** 52 **c.** 7

Self Check 3 For $f(x) = -2x + 3$, find each of the following function values:
a. $f(4)$ −5 **b.** $f(-1)$ 5 **c.** $f(0)$ 3

Now Try Problem 35

We can think of a function as a machine that takes some input x and turns it into some output $f(x)$, as shown in part (a) of the figure below. In part (b), the function machine for $f(x) = x^2 + 2x$ turns the input 4 into the output $4^2 + 2(4) = 24$, and we have $f(4) = 24$.

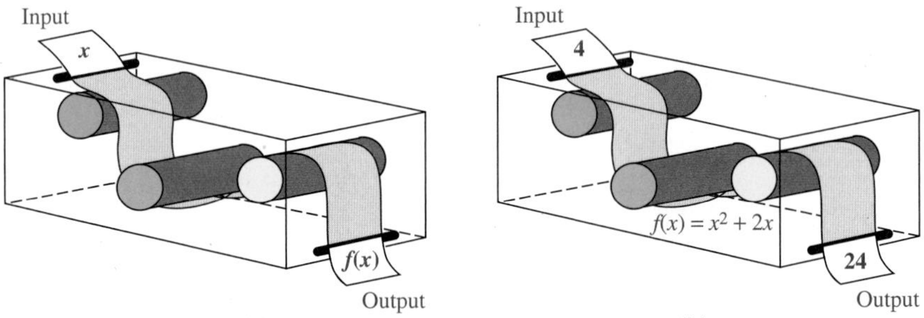

(a) (b)

The letter f used in the notation $y = f(x)$ represents the word *function.* However, other letters, such as g and h, can also be used to name functions.

EXAMPLE 4 For $g(x) = 3 - 2x$ and $h(x) = x^3 + x^2 - 1$, find: **a.** $g(3)$ **b.** $h(-2)$

Strategy We will substitute 3 for x in $3 - 2x$ and substitute −2 for x in $x^3 + x^2 - 1$, and then evaluate each expression.

Why The numbers 3 and −2, which are within the parentheses, are inputs that should be substituted for the variable x.

Solution **a.** To find $g(3)$, we use the function rule $g(x) = 3 - 2x$ and replace x with 3.

$g(x) = 3 - 2x$ Read g(x) as "g of x."

$g(3) = 3 - 2(3)$ Substitute the input 3 for each x.

$= 3 - 6$ Evaluate the right side.

$= -3$ The output is −3.

Thus, $g(3) = -3$.

b. To find $h(-2)$, we use the function rule $h(x) = x^3 + x^2 - 1$ and replace x with -2.

$$h(x) = x^3 + x^2 - 1 \quad \text{Read } h(x) \text{ as "}h \text{ of } x\text{."}$$
$$h(-2) = (-2)^3 + (-2)^2 - 1 \quad \text{Substitute the input } -2 \text{ for each } x.$$
$$= -8 + 4 - 1 \quad \text{Evaluate the right side.}$$
$$= -5 \quad \text{This is the output.}$$

Thus, $h(-2) = -5$.

Teaching Example 4
For $g(x) = 1 - 6x$ and $h(x) = x^3 - x^2 + 8$, find: **a.** $g(-8)$ **b.** $h(-3)$
Answers: **a.** 49 **b.** −28

Self Check 4 Find $g(0)$ and $h(4)$ for the functions in Example 4. 3, 79

Now Try Problem 41

4 Graph Functions.

We have seen that a function such as $f(x) = 4x + 1$ assigns to each value of x a single value $f(x)$. The input-output pairs generated by a function can be written in the form $(x, f(x))$. These ordered pairs can be plotted on a rectangular coordinate system to give the **graph of the function.**

EXAMPLE 5 Graph: $f(x) = 4x + 1$

Strategy We can graph the function by creating a table of function values and plotting the corresponding ordered pairs.

Why After drawing a line through the plotted points, we will have the graph of the function.

Solution To make a table, we choose several values for x and find the corresponding values of $f(x)$. If x is -1, we have

$$f(x) = 4x + 1 \quad \text{This is the function to graph.}$$
$$f(-1) = 4(-1) + 1 \quad \text{Substitute the input } -1 \text{ for each } x.$$
$$= -4 + 1 \quad \text{Evaluate the right side.}$$
$$= -3 \quad \text{This is the output.}$$

Thus, $f(-1) = -3$. This means when x is -1, $f(x)$ or y is -3, and that the ordered pair $(-1, -3)$ lies on the graph of $f(x)$.

Function notation	***Ordered-pair notation***
$f(-1) = -3$	$(-1, -3)$

Similarly, we find the corresponding values of $f(x)$ for x-values of 0 and 1. Then we plot the resulting ordered pairs and draw a straight line through them to get the graph of $f(x) = 4x + 1$. Since $y = f(x)$, the graph of $f(x) = 4x + 1$ is the same as the graph of the equation $y = 4x + 1$.

Notation

A table of function values is similar to a table of solutions, except that the second column is usually labeled $f(x)$ instead of y.

x	$f(x)$

x	y

$f(x) = 4x + 1$

x	$f(x)$	
−1	−3	→ (−1, −3)
0	1	→ (0, 1)
1	5	→ (1, 5)

The function generates these ordered pairs.

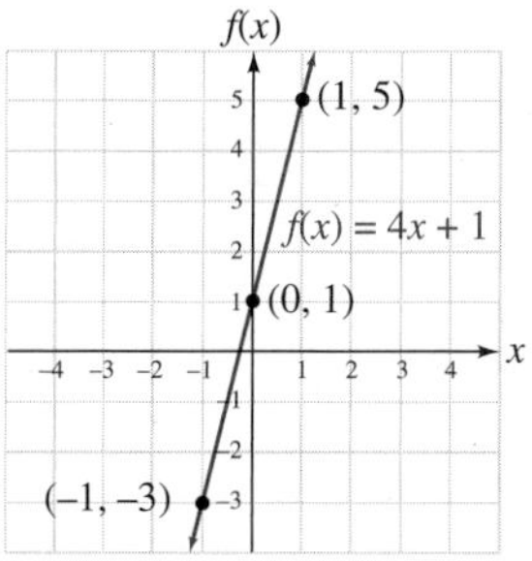

The vertical axis can be labeled y or $f(x)$.

Teaching Example 5
Graph: $f(x) = -3x + 2$
Answer:

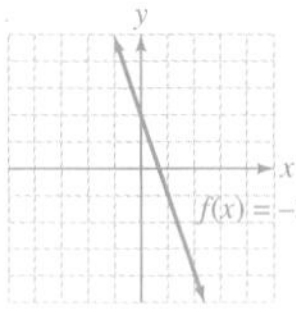

Self Check 5 Graph: $f(x) = -3x - 2$ See AIE Appendix 3.

Now Try Problem 47

We call $f(x) = 4x + 1$ from Example 5 a **linear function** because its graph is a nonvertical line. Any linear equation, except those of the form $x = a$, can be written using function notation by writing it in slope–intercept form ($y = mx + b$) and then replacing y with $f(x)$.

The graphs of some functions are not straight lines.

EXAMPLE 6 Graph: $f(x) = |x|$

Strategy We can graph the function by creating a table of function values and plotting the corresponding ordered pairs.

Why After drawing a "V" shape through the plotted points, we will have the graph of the function.

Solution To create a table of function values, we choose values for x and find the corresponding values of $f(x)$. For $x = -4$ and $x = 3$, we have

$$f(x) = |x| \qquad f(x) = |x|$$
$$f(-4) = |-4| \qquad f(3) = |3|$$
$$= 4 \qquad = 3$$

The results $f(-4) = 4$ and $f(3) = 3$ produce the ordered pairs $(-4, 4)$ and $(3, 3)$.

Similarly, we find the corresponding values of $f(x)$ for several other x-values. When we plot the resulting ordered pairs, we see that they lie in a "V" shape. We join the points to complete the graph as shown. We call $f(x) = |x|$ an **absolute value function.**

Teaching Tip: Show your students that the shape of the graph of the absolute value function can be seen in its name: absolute Value function.

$f(x) = |x|$

x	$f(x)$	
−4	4	→ (−4, 4)
−3	3	→ (−3, 3)
−2	2	→ (−2, 2)
−1	1	→ (−1, 1)
0	0	→ (0, 0)
1	1	→ (1, 1)
2	2	→ (2, 2)
3	3	→ (3, 3)
4	4	→ (4, 4)

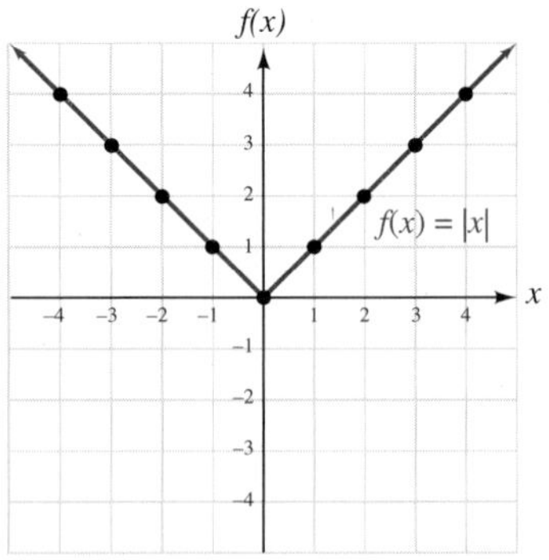

Teaching Example 6
Graph: $f(x) = |x| - 3$
Answer:

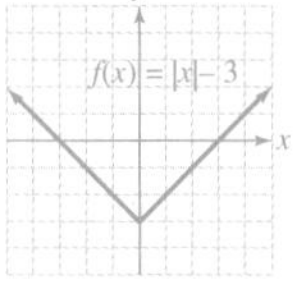

Self Check 6 Graph: $f(x) = |x| + 2$ See AIE Appendix 3.

Now Try Problem 49

5 Use the Vertical Line Test.

If any vertical line intersects a graph more than once, the graph cannot represent a function, because to one value of x there would correspond more than one value of y.

The Vertical Line Test	If a vertical line intersects a graph in more than one point, the graph is not the graph of a function.

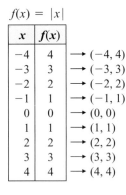

The graph shown on the right in red does not represent a function, because a vertical line intersects the graph at more than one point. The points of intersection indicate that the x-value -1 corresponds to two different y-values, 3 and -1.

When the coordinates of the two points of intersection are listed in a table, it is easy to see that the x-value of -1 is assigned two different y-values. Thus, this is not the graph of a function.

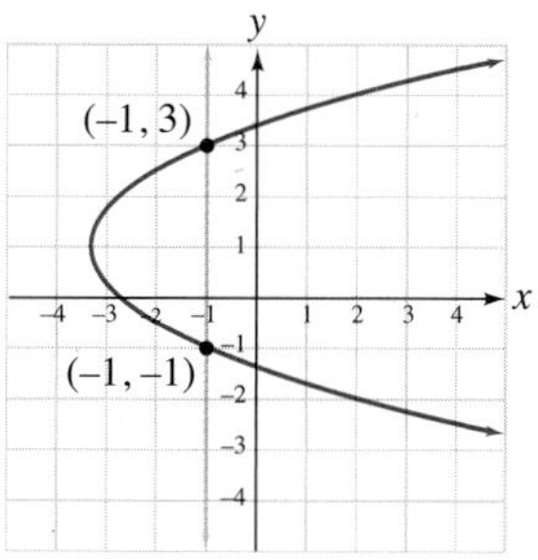

Not a function

x	y
-1	3
-1	-1

EXAMPLE 7 Determine whether each of the graphs shown in red is the graph of a function.

a.

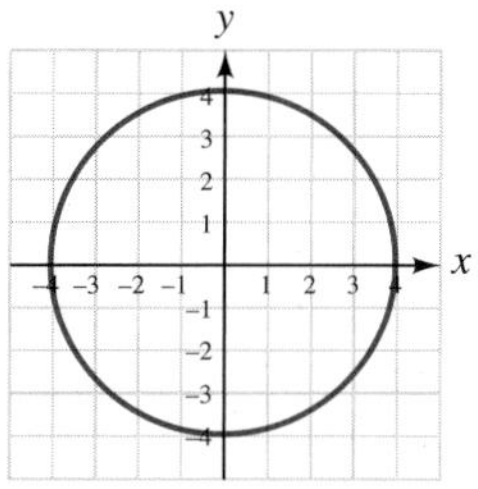

b.

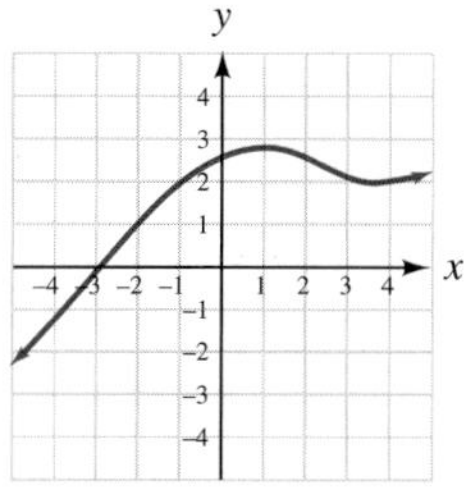

c.

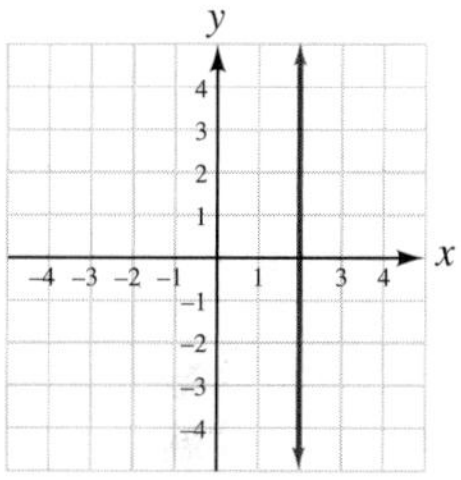

Teaching Tip: Show your students that they can use the side of a credit card or a pencil to apply the vertical line test. Simply hold the card (or pencil) vertically and run it across the graph from left to right.

Strategy We will check to see whether any vertical line intersects the graph more than once.

Why If any vertical line does intersect the graph more than once, the graph is not a function.

Solution

a. Refer to figure (a) below. The graph shown in red is not the graph of a function, because a vertical line can be drawn that intersects the graph at more than one point. The points of intersection of the graph and the line reveal that the x-value 3 is assigned two different y-values, 2.5 and -2.5.

b. Refer to figure (b) below. The graph shown in red is a graph of a function, because no vertical line can be drawn that intersects the graph at more than one point. Several vertical lines are drawn in blue to illustrate this.

c. Refer to figure (c) below. The graph shown in red is not the graph of a function, because a vertical line can be drawn that intersects the graph at more than one point. In fact, it intersects it at infinitely many points. From this example, we can conclude that any vertical line will fail the vertical line test. Thus, vertical lines are not functions.

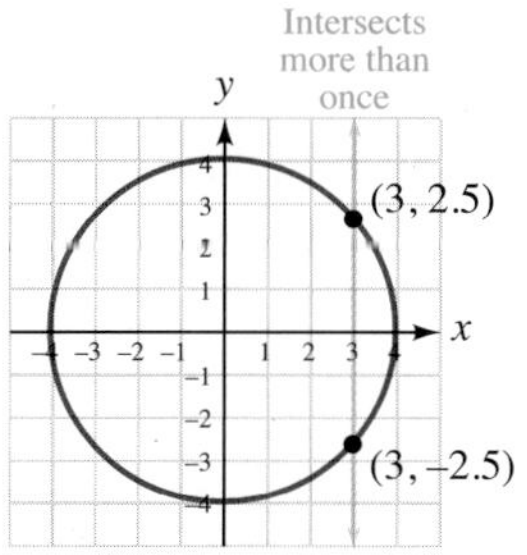

Not a function

(a)

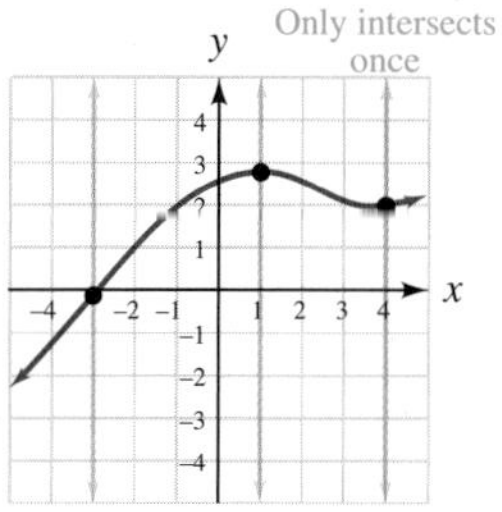

A function

(b)

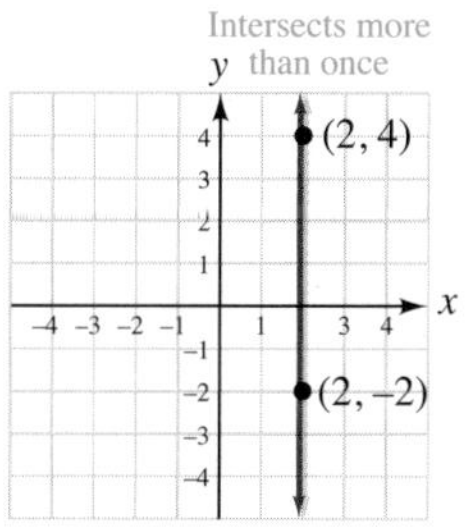

Not a function

(c)

Teaching Example 7
Determine whether each of the following is a graph of a function.

a.

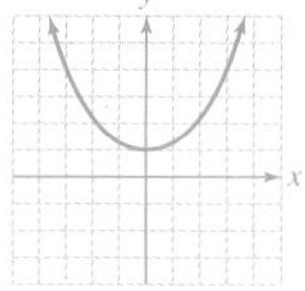

b.

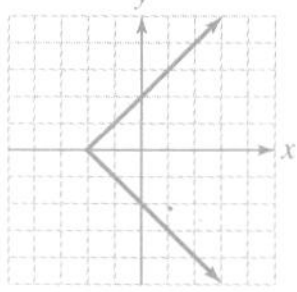

Answers: a. Function b. Not a function

Self Check 7 Determine whether each of the following is the graph of a function.

a.

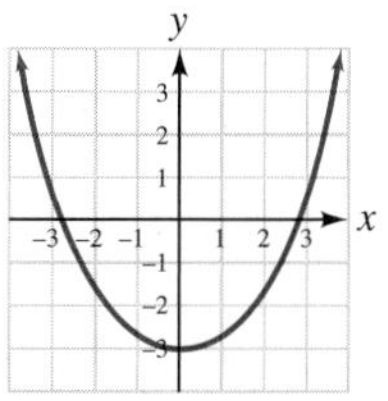

Function

b.

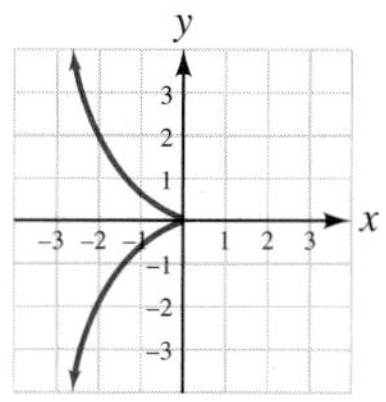

Not a function

Now Try Problems 55 and 57

6 Solve Applications Involving Functions.

Functions are used to describe certain relationships where one quantity depends on another. Letters other than f and x are often chosen to more clearly describe these situations.

EXAMPLE 8 **Bounce Houses.** The function $C(h) = 80 + 15(h - 4)$ gives the cost in dollars to rent an inflatable jumper for h hours. (The terms of the rental agreement require a 4-hour minimum.) Find the cost of renting the jumper for 10 hours.

Strategy To find the cost to rent the jumper for 10 hours, we will substitute 10 for each h in $C(h) = 80 + 15(h - 4)$ and evaluate the right side.

Why In $C(h) = 80 + 15(h - 4)$, the variable h represents the number of hours that the jumper is rented. We need to find $C(10)$.

Solution For this application involving hours and cost, the notation $C(h)$ is used. The input variable is h and the name of the function is C. If the jumper is rented for 10 hours, then h is 10 and we must find $C(10)$.

$C(h) = 80 + 15(h - 4)$ Read $C(h)$ as "C of h."

$C(\mathbf{10}) = 80 + 15(\mathbf{10} - 4)$ Substitute the input 10 for each h.

$= 80 + 15(6)$ Evaluate the right side.

$= 80 + 90$ Do the multiplication.

$= 170$ This is the output.

It costs \$170 to rent the jumper for 10 hours.

Teaching Example 8 Party Rentals. The function $C(h) = 40 + 5(h - 2)$ gives the cost in dollars to rent a cotton candy machine for h hours. (The terms of the rental agreement require a 2-hour minimum.) Find the cost of renting the machine for 24 hours.
Answer: \$150

Self Check 8 **Bounce Houses.** Find the cost of renting the jumper for 8 hours. \$140

Now Try Problem 65

SECTION 3.8 STUDY SET

VOCABULARY

Fill in the blanks.

1. A set of ordered pairs is called a relation.
2. A function is a rule that assigns to each x-value exactly one y-value.
3. The set of all input values for a function is called the domain, and the set of all output values is called the range.
4. We can think of a function as a machine that takes some input x and turns it into some output $f(x)$.
5. If $f(2) = -3$, we call -3 a function value.
6. The graph of a linear function is a straight line and the graph of an absolute value function is V-shaped.

CONCEPTS

7. **Federal Minimum Hourly Wage.** The following table is an example of a function. Use an arrow diagram to illustrate this. See AIE Appendix 3.

Year	1992	1994	1996	1998	2000	2002	2004	2006	2008	2010
Minimum wage (\$)	4.25	4.25	4.75	5.15	5.15	5.15	5.15	5.15	6.55	7.25

Source: infoplease.com

8. The arrow diagram describes a function. What is the domain and what is the range of the function? Domain: {5, 6, 7, 8}; range: {0, 1, 2}

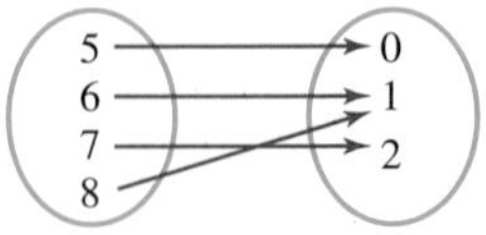

9. For the given input, what value will the function machine output? 33

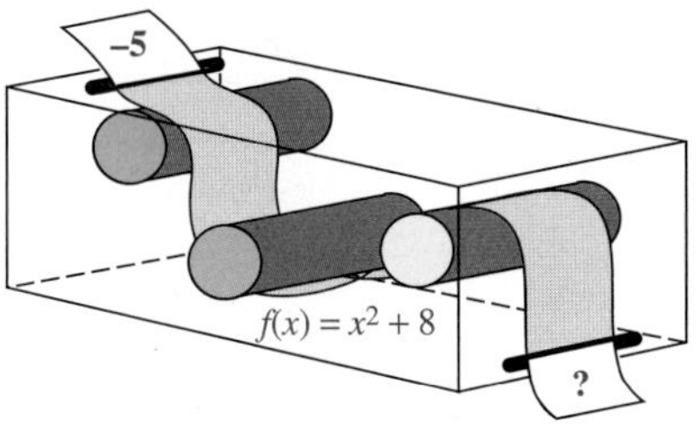

10. a. Fill in the blank: If a vertical line intersects a graph in more than one point, the graph is not the graph of a function.

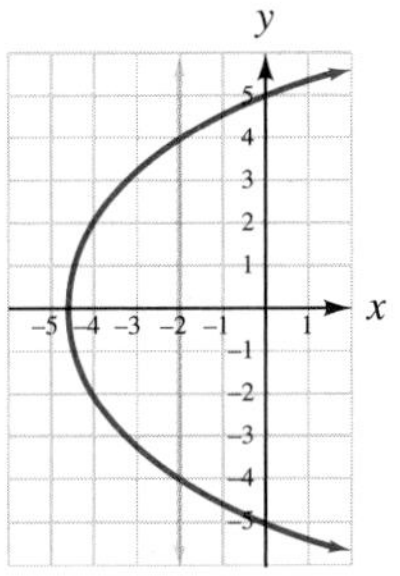

b. Give the coordinates of the points where the given vertical line intersects the graph.
$(-2, 4), (-2, -4)$

c. Is this the graph of a function?
No

NOTATION

Fill in the blanks.

11. We read $f(x) = 5x - 6$ as "f of x is $5x$ minus 6."

▶ 12. Since $y =$ $f(x)$, the following two equations are equivalent:

$$y = 3x + 2 \quad \text{and} \quad f(x) = 3x + 2$$

13. The notation $f(4) = 5$ indicates that when the x-value 4 is input into a function rule, the output is 5. This fact can be shown graphically by plotting the ordered pair (4, 5).

▶ 14. When graphing the function $f(x) = -x + 5$, the vertical axis of the coordinate system can be labeled $f(x)$ or y.

GUIDED PRACTICE

Find the domain and range of each relation. **See Example 1.**

15. $\{(6, -1), (-1, -10), (-6, 2), (8, -5)\}$
Domain: $\{-6, -1, 6, 8\}$; range: $\{-10, -5, -1, 2\}$

▶ 16. $\{(11, -3), (0, 0), (4, 5), (-3, -7)\}$
Domain: $\{-3, 0, 4, 11\}$; range: $\{-7, -3, 0, 5\}$

▶ 17. $\{(0, 9), (-8, 50), (6, 9)\}$
Domain: $\{-8, 0, 6\}$; range: $\{9, 50\}$

18. $\{(1, -12), (-6, 8), (5, 8)\}$
Domain: $\{-6, 1, 5\}$; range: $\{-12, 8\}$

Determine whether the relation defines y to be a function of x. If a function is defined, give its domain and range. If it does not define a function, find two ordered pairs that show a value of x that is assigned more than one value of y. **See Example 2.**

▶ 19.

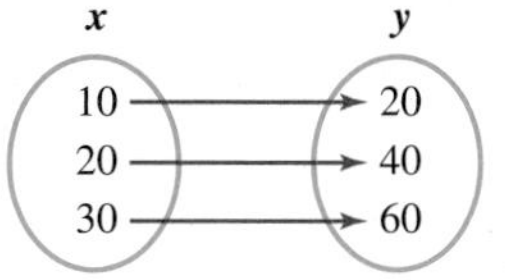

Yes; domain: $\{10, 20, 30\}$; range: $\{20, 40, 60\}$

▶ 20.

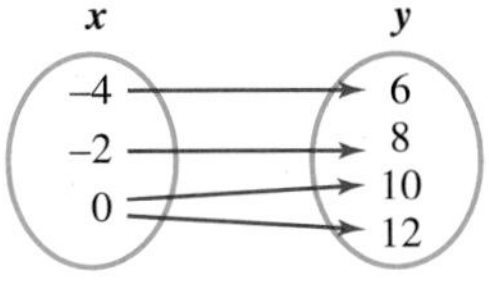

No; (0, 10), (0, 12)

21.

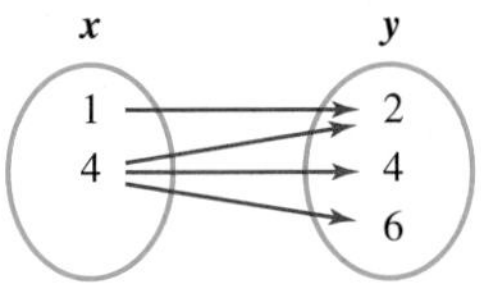

No; (4, 2), (4, 4), (4, 6) (Answers may vary)

▶ 22.

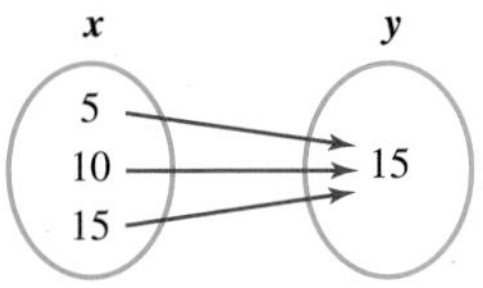

Yes; domain: $\{5, 10, 15\}$; range: $\{15\}$

▶ 23.

x	y
1	7
2	15
3	23
4	16
5	8

Yes; domain: $\{1, 2, 3, 4, 5\}$; range: $\{7, 8, 15, 16, 23\}$

▶ 24.

x	y
30	2
30	4
30	6
30	8
30	10

No; (30, 2), (30, 4) (Answers may vary)

25.

x	y
−4	6
−1	0
0	−3
2	4
−1	2

No; $(-1, 0), (-1, 2)$

26.

x	y
1	1
2	2
3	3
4	4

Yes; domain: $\{1, 2, 3, 4\}$; range: $\{1, 2, 3, 4\}$

27.

x	y
3	4
3	−4
4	3
4	−3

No; $(3, 4), (3, -4)$ or $(4, 3), (4, -3)$

▶ 28.

x	y
−1	1
−3	1
−5	1
−7	1
−9	1

Yes; domain: $\{-9, -7, -5, -3, -1\}$; range: $\{1\}$

29.

x	y
6	0
−3	−8
1	9
5	4

Yes; domain: $\{-3, 1, 5, 6\}$; range: $\{-8, 0, 4, 9\}$

▶ 30.

x	y
1.6	0
−3	−1
2.5	20
−7	0.1
1.6	19

No; (1.6, 0), (1.6, 19)

31. $\{(3, 4), (3, -4), (4, 3), (4, -3)\}$ No; $(3, 4), (3, -4)$ or $(4, 3), (4, -3)$

▶ 32. $\{(-1, 1), (-3, 1), (-5, 1), (-7, 1), (-9, 1)\}$
Yes; domain: $\{-9, -7, -5, -3, -1\}$; range: $\{1\}$

33. $\{(-2, 7), (-1, 10), (0, 13), (1, 16)\}$
Yes; domain: $\{-2, -1, 0, 1\}$; range: $\{7, 10, 13, 16\}$

▶ 34. $\{(-2, 4), (-3, 8), (-3, 12), (-4, 16)\}$ No; $(-3, 8), (-3, 12)$

Find each function value. **See Examples 3 and 4.**

▶ 35. $f(x) = 4x - 1$

a. $f(1)$ 3 b. $f(-2)$ −9 c. $f\left(\frac{1}{4}\right)$ 0 d. $f(50)$ 199

▶ 36. $f(x) = 1 - 5x$

a. $f(0)$ 1 b. $f(-75)$ 376 c. $f(0.2)$ 0 d. $f\left(-\frac{4}{5}\right)$ 5

▶ 37. $f(x) = 2x^2$

a. $f(0.4)$ 0.32 b. $f(-3)$ 18 c. $f(1{,}000)$ 2,000,000 d. $f\left(\frac{1}{8}\right)$ $\frac{1}{32}$

38. $g(x) = 6 - x^2$

a. $g(30)$ -894 b. $g(6)$ -30 c. $g(-1)$ 5 d. $g(0.5)$ 5.75

39. $h(x) = |x - 7|$

a. $h(0)$ 7 b. $h(-7)$ 14 c. $h(7)$ 0 d. $h(8)$ 1

40. $f(x) = |2 + x|$

a. $f(0)$ 2 b. $f(2)$ 4 c. $f(-2)$ 0 d. $f(-99)$ 97

41. $g(x) = x^3 - x$

a. $g(1)$ 0 b. $g(10)$ 990 c. $g(-3)$ -24 d. $g(6)$ 210

42. $g(x) = x^4 + x$

a. $g(1)$ 2 b. $g(-2)$ 14 c. $g(0)$ 0 d. $g(10)$ $10{,}010$

43. $s(x) = (x + 3)^2$

a. $s(3)$ 36 b. $s(-3)$ 0 c. $s(0)$ 9 d. $s(-5)$ 4

44. $s(x) = (x - 8)^2$

a. $s(8)$ 0 b. $s(-8)$ 256 c. $s(1)$ 49 d. $s(12)$ 16

45. If $f(x) = 3.4x^2 - 1.2x + 0.5$, find $f(-0.3)$. 1.166

46. If $g(x) = x^4 - x^3 + x^2 - x$, find $g(-12)$. $22{,}620$

Complete each table of function values and then graph each function. **See Examples 5 and 6.** See AIE Appendix 3.

47. $f(x) = -3x - 2$

x	$f(x)$
-2	4
-1	1
0	-2
1	-5

48. $f(x) = -2x + 8$

x	$f(x)$
-1	10
0	8
1	6
2	4

49. $h(x) = |1 - x|$

x	$h(x)$
-2	3
-1	2
0	1
1	0
2	1
3	2
4	3

50. $h(x) = |x + 2|$

x	$h(x)$
-5	3
-4	2
-3	1
-2	0
-1	1
0	2
1	3

Graph each function. **See Examples 5 and 6.** See AIE Appendix 3.

51. $f(x) = \frac{1}{2}x - 2$

52. $f(x) = -\frac{2}{3}x + 3$

53. $h(x) = -|x|$

54. $g(x) = |x| - 2$

Determine whether each graph is the graph of a function. If it is not, find ordered pairs that show a value of x that is assigned more than one value of y. **See Example 7.**

55.

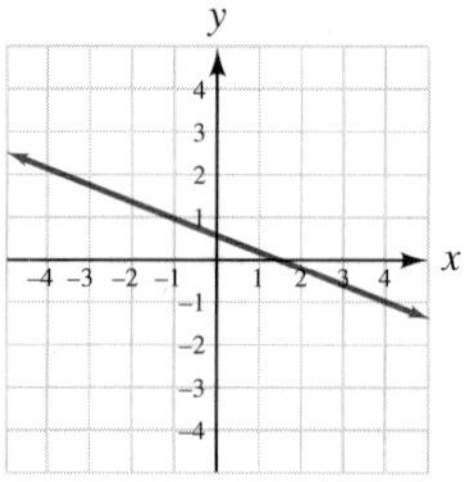

Yes

56.

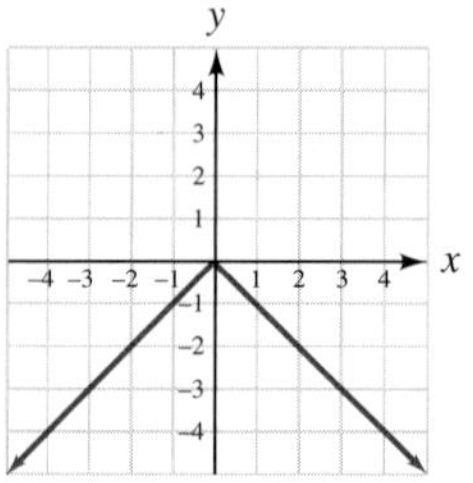

Yes

57.

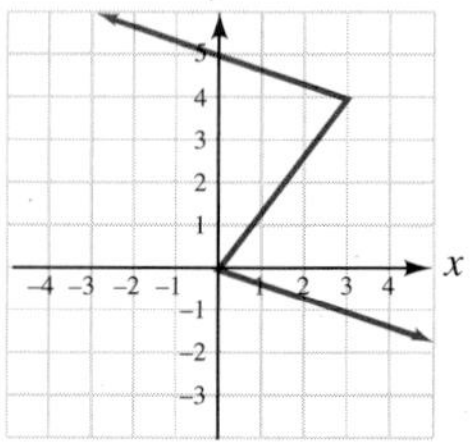

No; (3, 4), (3, −1) (Answers may vary)

58.

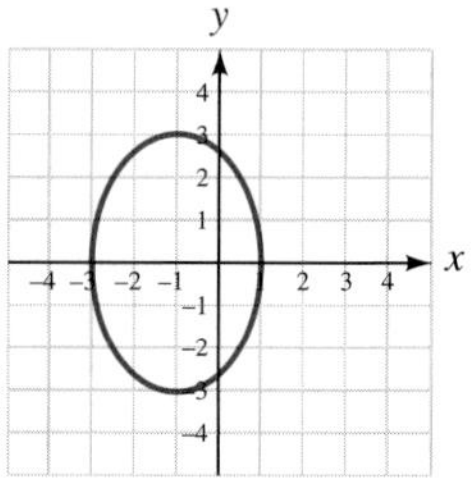

No; (−1, 3), (−1, −3) (Answers may vary)

59.

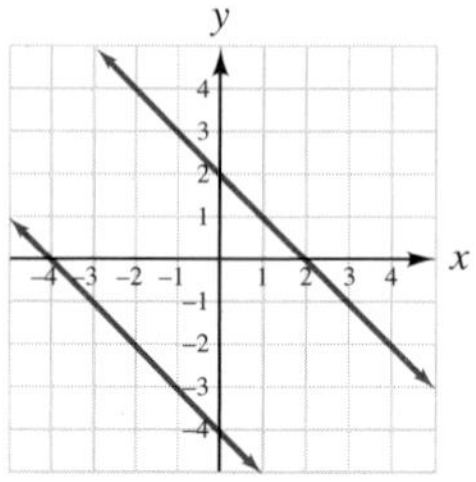

No; (0, 2), (0, −4) (Answers may vary)

60.

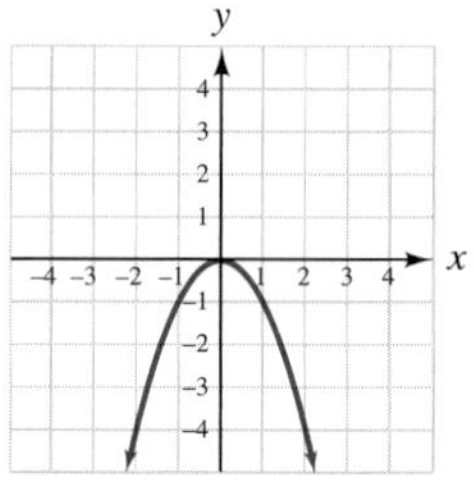

Yes

61.

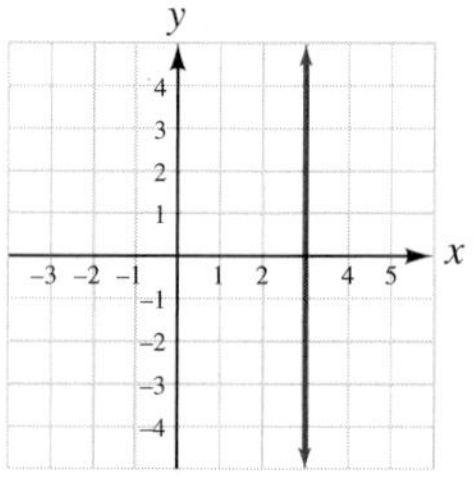

No; (3, 0), (3, 1) (Answers may vary)

62.

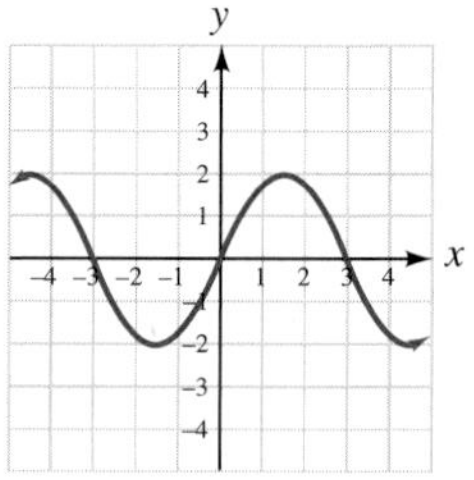

Yes

APPLICATIONS

63. **Reflections.** When a beam of light hits a mirror, it is reflected off the mirror at the same angle that the incoming beam struck the mirror. What type of function could serve as a mathematical model for the path of the light beam shown here? $f(x) = |x|$

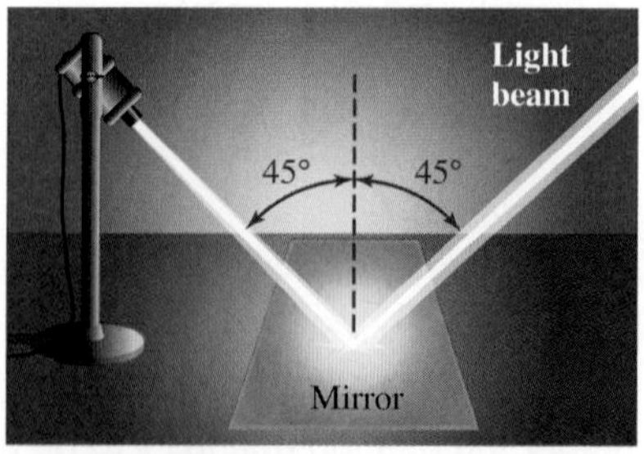

64. **Lightning.** The function $D(t) = \frac{t}{5}$ gives the approximate distance in miles that you are from a lightning strike, where t is the number of seconds between seeing the lightning and hearing the thunder. Find $D(5)$ and explain what it means. $D(5) = 1$; if it is 5 seconds between seeing the lightning and hearing the thunder, the lightning strike was 1 mile away.

65. Vacationing. The function $C(d) = 500 + 100(d - 3)$ gives the cost in dollars to rent an RV motor home for d days. (The terms of the rental agreement require a 3-day minimum.) Find the cost of renting the RV for a vacation that will last 7 days. \$900

66. Structural Engineering. The maximum safe load in pounds of the rectangular beam shown in the figure is given by the function $S(t) = \frac{1{,}875t^2}{8}$, where t is the thickness of the beam, in inches. Find the maximum safe load if the beam is 4 inches thick. 3,750 lb

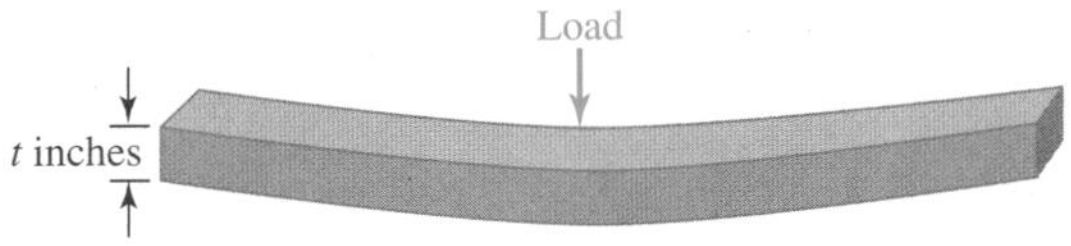

67. Lawn Sprinklers. The function $A(r) = \pi r^2$ can be used to determine the area in square feet that will be watered by a rotating sprinkler that sprays out a stream of water. Find $A(5)$ and $A(20)$. Round to the nearest tenth. 78.5 ft², 1,256.6 ft²

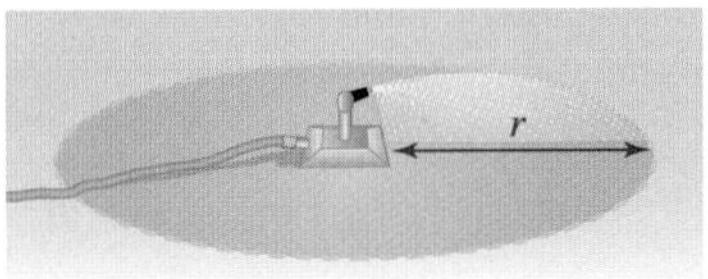

68. Parts Lists. The function $f(r) = 2.30 + 3.25(r + 0.40)$ approximates the length (in feet) of the belt that joins the two pulleys, where r is the radius (in feet) of the smaller pulley. Find the belt length needed for each pulley in the parts list.

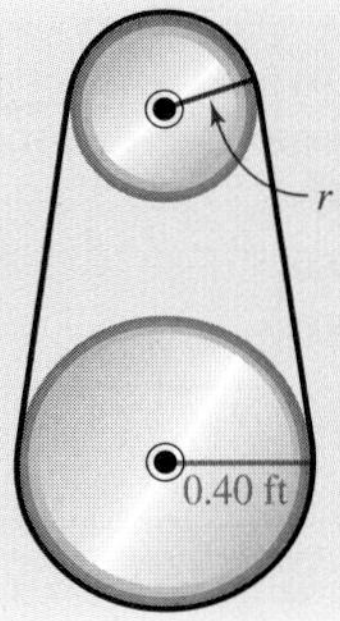

Parts list		
Pulley	r	**Belt length**
P-45M	0.32	4.64
P-08D	0.24	4.38

69. Postage. The **step graph** below shows how the cost of a first class U.S. postage stamp increased from 1990 through 2010. An open circle at the end of a line segment means the endpoint of the segment is not included. Is this the graph of a function? Yes

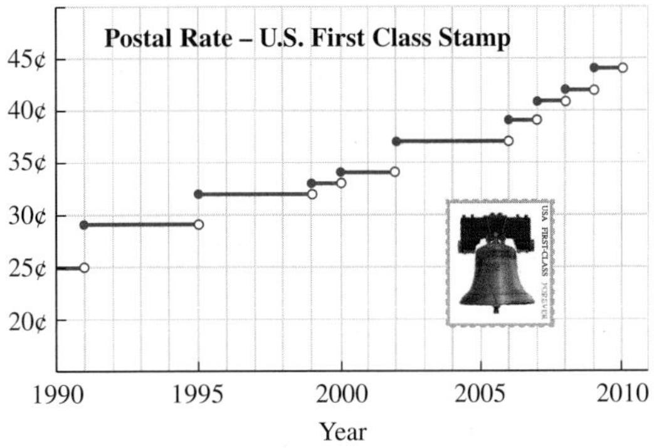

Source: U.S. Postal Service

70. Sound. We cannot see sound waves, but certain scientific instruments are used to draw mathematical models of them. Is the graph of a sound wave shown below a function? Yes

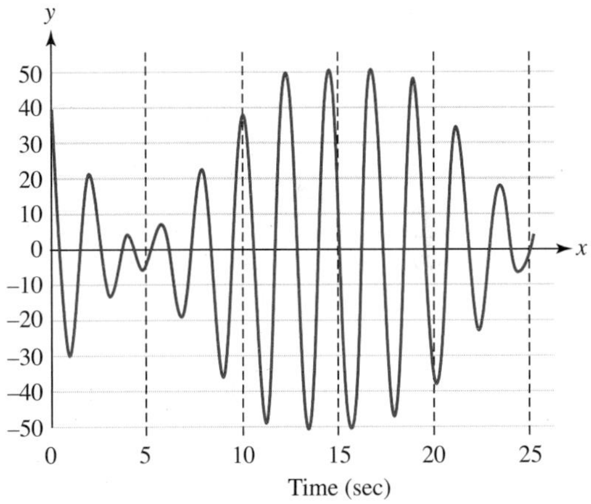

WRITING

71. In the function $y = -5x + 2$, why do you think the value of x is called the *input* and the corresponding value of y the *output*?

72. Explain what a politician meant when she said, "The speed at which the downtown area will be redeveloped is a function of the number of low-interest loans made available to the property owners."

73. Explain the following diagram:

$f(4) = 11$ $(4, 11)$

74. Explain the error in the following solution.
If $f(x) = x^2 + 7x + 1$, find $f(10)$.

$$f(10) = 10^2 + 7x + 1$$
$$= 100 + 7x + 1$$
$$= 101 + 7x$$

75. A student was asked to determine whether the graph on the right is the graph of a function. What is wrong with the following reasoning?

When I draw a vertical line through the graph, it intersects the graph only once. By the vertical line test, this is the graph of a function.

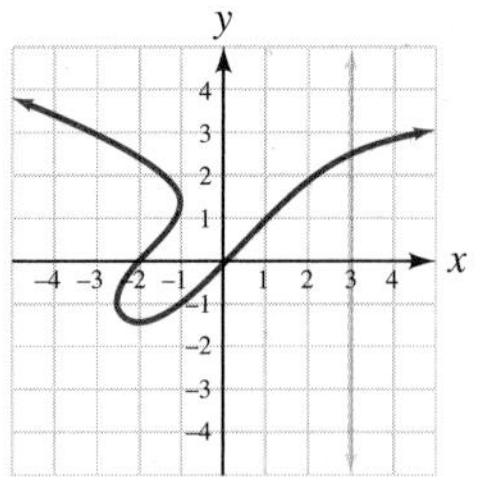

76. In your own words, what is a function?

REVIEW

77. **Coffee Blends.** A store sells regular coffee for \$4 a pound and gourmet coffee for \$7 a pound. To get rid of 40 pounds of the gourmet coffee, the shopkeeper plans to make a gourmet blend that he will put on sale for \$5 a pound. How many pounds of regular coffee should be used? 80 lb regular coffee

78. **Photographic Chemicals.** A photographer wishes to mix 2 liters of a 5% acetic acid solution with a 10% acetic solution to get a 7% acetic solution. How many liters of 10% acetic solution must be added? $1\frac{1}{3}$ liters 10% acetic solution

CHALLENGE PROBLEMS

79. Is the graph of $y \geq 3 - x$ a function? Explain. No

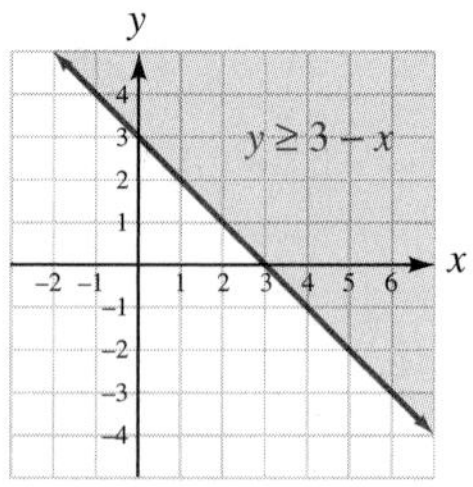

80. If $f(x) = x^2 + x$, find: $f\left(\frac{4}{5}r\right)$ $\frac{16}{25}r^2 + \frac{4}{5}r$

81. Let $f(x) = -2x + 5$. For what value of x is $f(x) = -7$? 6

82. Let $f(x) = x - 2$ and $g(x) = 3x$. Find $f(g(6))$. 16

Teaching Tip: Remind your students that this Summary and Review is an excellent way to study for their test. Suggest that they arrive on campus early and, as a final preparation, read through the section summaries before taking the exam.

3 Summary & Review

SECTION 3.1 Graphing Using the Rectangular Coordinate System

DEFINITIONS AND CONCEPTS	EXAMPLES
A **rectangular coordinate system** is composed of a horizontal number line called the ***x*-axis** and a vertical number line called the ***y*-axis.** The two axes intersect at the **origin.** To **plot** or **graph** ordered pairs means to locate their position on a rectangular coordinate system. The *x*- and *y*-axes divide the **coordinate plane** into four regions called **quadrants.**	Plot the points: $(2, 3), (-4, 2), (-3, -1), (0, -2.5), (4, -2)$ To graph each point, start at the origin and count the appropriate number of units in the *x*-direction and then the appropriate number of units in the *y*-direction. 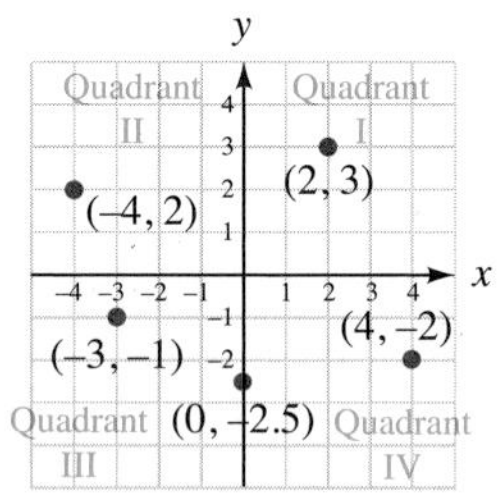

REVIEW EXERCISES

1. Graph the points with coordinates $(-1, 3)$, $(0, 1.5)$, $(-4, -4)$, $\left(2, \frac{7}{2}\right)$, and $(4, 0)$. See AIE Appendix 3.

2. **Hawaii.** Estimate the coordinates of Oahu using an ordered pair of the form (longitude, latitude). (158, 21.5)

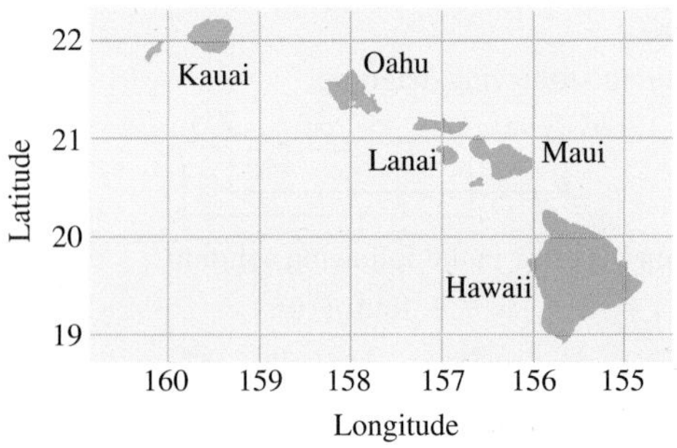

3. In what quadrant does the point $(-3, -4)$ lie? Quadrant III

4. What are the coordinates of the origin? (0, 0)

5. **Geometry.** Three vertices (corners) of a square are $(-5, 4)$, $(-5, -2)$, and $(1, -2)$. Find the coordinates of the fourth vertex and then find the area of the square. (1, 4); 36 square units

6. **College Enrollments.** The graph gives the number of students enrolled at a college for the period from 4 weeks before to 5 weeks after the semester began.

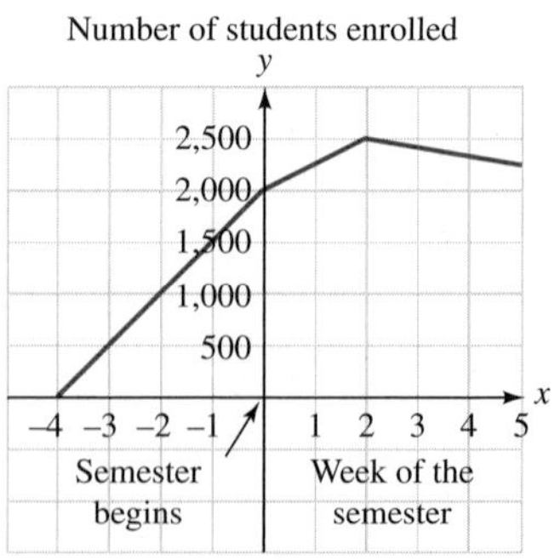

a. What was the maximum enrollment and when did it occur? 2,500; week 2

b. How many students were enrolled 2 weeks before the semester began? 1,000

c. When was the enrollment 2,250? 1st week and 5th week

SECTION 3.2 Graphing Linear Equations

DEFINITIONS AND CONCEPTS

A **solution of an equation in two variables** is an ordered pair of numbers that makes the equation a true statement when the numbers are substituted for the variables.

The **standard form** of a linear equation in two variables is $Ax + By = C$, where A, B, and C are real numbers and A and B are not both zero.

EXAMPLES

Determine whether $(2, -3)$ is a solution of $2x - y = 7$.

$2x - y = 7$	This is the given equation.
$2(2) - (-3) \stackrel{?}{=} 7$	Substitute 2 for x and −3 for y.
$4 + 3 \stackrel{?}{=} 7$	Evaluate the left side.
$7 = 7$	True

Since the result $7 = 7$ is true, $(2, -3)$ is a solution of $2x - y = 7$.

DEFINITIONS AND CONCEPTS

If only one coordinate of an ordered-pair solution is known:

1. Substitute it into the equation for the appropriate variable.
2. Solve the resulting equation to find the unknown coordinate.

EXAMPLES

To complete the solution (, 8) for $3x + y = -1$, we substitute 8 for y and solve the resulting equation for x.

$3x + y = -1$	This is the given equation.
$3x + 8 = -1$	Substitute 8 for y.
$3x = -9$	Subtract 8 from both sides.
$x = -3$	To isolate x, divide both sides by 3.

The solution is $(-3, 8)$.

DEFINITIONS AND CONCEPTS

To **graph a linear equation** solved for y:

1. Find three solutions by **selecting three values** of x and finding the corresponding values of y.
2. **Plot** each ordered-pair solution.
3. **Draw** a line through the points.

EXAMPLES

Graph: $y = -2x + 1$

$y = -2x + 1$

x	y	(x, y)
−1	3	(−1, 3)
0	1	(0, 1)
2	−3	(2, −3)

We construct a table of solutions, plot the points, and draw the line.

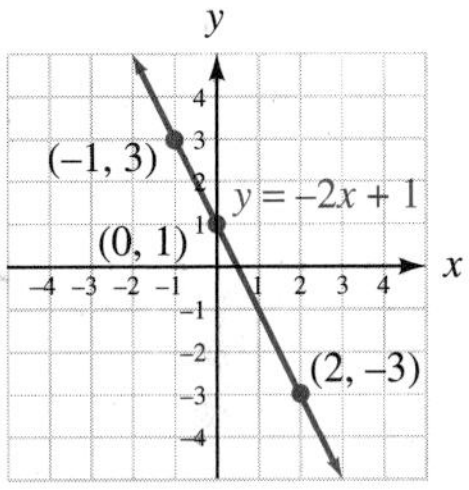

REVIEW EXERCISES

7. Is $(-3, -2)$ a solution of $y = 2x + 4$? Yes

8. Complete the table of solutions.

$3x + 2y = -18$

x	y	(x, y)
−2	−6	(−2, −6)
−8	3	(−8, 3)

9. Which of the following equations are not linear equations?

$8x - 2y = 6 \quad y = x^2 + 1 \quad y = x \quad y - x^3 = 0$

$y = x^2 + 1$ and $y - x^3 = 0$

10. The graph of a linear equation is shown.

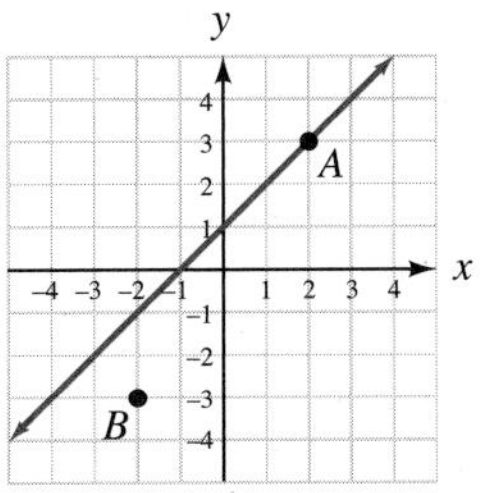

a. When the coordinates of point A are substituted into the equation, will a true or false statement result? True

b. When the coordinates of point B are substituted into the equation, will a true or false statement result? False

Graph each equation by constructing a table of solutions. See AIE Appendix 3.

11. $y = 4x - 2$

12. $y = \frac{3}{4}x$

13. $5y = -5x + 15$ (*Hint:* Solve for y first.)

14. $6y = -4x$ (*Hint:* Solve for y first.)

15. **Birthday Parties.** A restaurant offers a party package for children that includes everything: food, drinks, cake, and favors. The cost c, in dollars, is given by the equation $c = 8n + 50$, where n is the number of children attending the party. Graph the equation and use the graph to estimate the cost of a party if 18 children attend. About \$190; see AIE Appendix 3.

16. Determine whether each statement is true or false.

a. It takes three or more points to determine a line. False

b. A linear equation in two variables has infinitely many solutions. True

SECTION 3.3 Intercepts

DEFINITIONS AND CONCEPTS	EXAMPLES
The point where a line intersects the x-axis is called the **x-intercept.** The point where a line intersects the y-axis is called the **y-intercept.** To **find the y-intercept,** substitute 0 for x in the given equation and solve for y. To **find the x-intercept,** substitute 0 for y and solve for x. Plotting the x- and y-intercepts of a graph and drawing a line through them is called the **intercept method for graphing a line.**	Use the y- and x-intercepts to graph $3x + 4y = -6$. ***y*-intercept: $x = 0$** $3x + 4y = -6$ $3(0) + 4y = -6$ $4y = -6$ $y = -\frac{3}{2}$ ***x*-intercept: $y = 0$** $3x + 4y = -6$ $3x + 4(0) = -6$ $3x = -6$ $x = -2$ 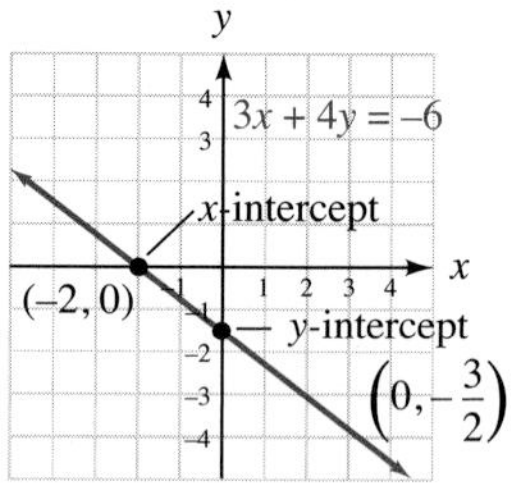 The y-intercept is $\left(0, -\frac{3}{2}\right)$ and the x-intercept is $(-2, 0)$.
The equation $y = b$ represents the **horizontal line** that intersects the y-axis at $(0, b)$. The equation $x = a$ represents the **vertical line** that intersects the x-axis at $(a, 0)$.	Graph: $y = 2$ and $x = -1$ on the same rectangular coordinate system. 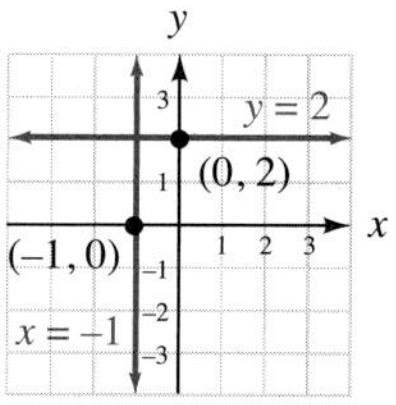

REVIEW EXERCISES

17. Identify the x- and y-intercepts of the graph shown in figure (a) below. $(-3, 0), (0, 2.5)$

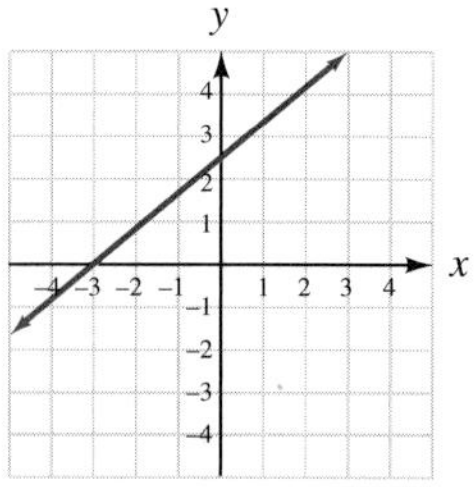

(a)

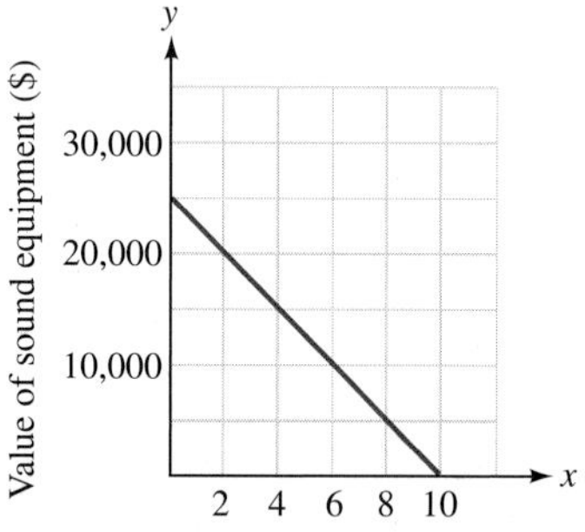

(b)

18. Depreciation. The graph in figure (b) above shows how the value of some sound equipment decreased over the years. Find the intercepts of the graph. What information do the intercepts give about the equipment? (0, 25,000); the equipment was originally valued at \$25,000. (10, 0); in 10 years, the sound equipment had no value.

Use the intercept method to graph each equation. See AIE Appendix 3.

19. $-4x + 2y = 8$ x-intercept: $(-2, 0)$; y-intercept: $(0, 4)$

20. $5x - 4y = 13$ x-intercept: $\left(\frac{13}{5}, 0\right)$; y-intercept: $\left(0, -\frac{13}{4}\right)$

21. Graph: $y = 4$ See AIE Appendix 3.

22. Graph: $x = -1$ See AIE Appendix 3.

SECTION 3.4 Slope and Rate of Change

DEFINITIONS AND CONCEPTS	EXAMPLES
The **slope** m of a line is a ratio that compares the vertical and horizontal change as we move along the line from one point to another. We can find the slope of a line graphically using the ratio $m = \frac{\text{rise}}{\text{run}}$. We also can find the slope of a line using the **slope formula:** $m = \frac{y_2 - y_1}{x_2 - x_1}$ if $x_1 \neq x_2$	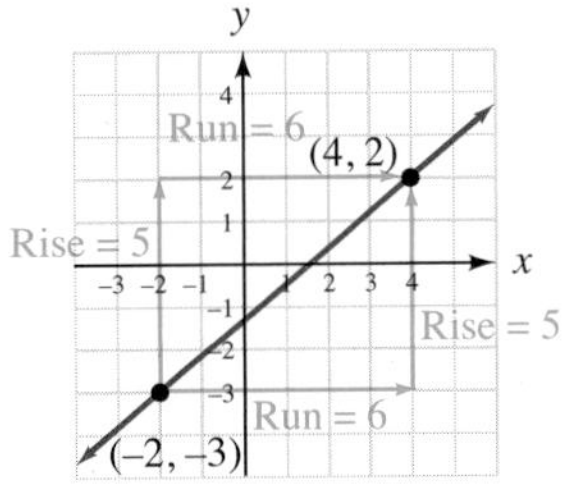 To find the slope of the line graphed on the left, we can use the slope triangle method: $m = \frac{\text{rise}}{\text{run}} = \frac{5}{6}$
Lines that rise from left to right have a **positive slope,** and lines that fall from left to right have a **negative slope.** Horizontal lines have **zero slope** and vertical lines have **undefined slope.**	To find the slope of the line that passes through the points $(-2, -3)$ and $(4, 2)$, we substitute into the **slope formula:** $m = \frac{y_2 - y_1}{x_2 - x_1} = \frac{2 - (-3)}{4 - (-2)} = \frac{5}{6}$
When units are attached to a slope, the slope is called a **rate of change.** The **grade** of an incline is its slope expressed as a percent.	An example of a rate of change is: $\frac{300 \text{ pounds}}{1 \text{ year}}$ *Read as "300 pounds per year."*
Parallel lines have the same slope. The slopes of **perpendicular lines** are negative reciprocals. The product of their slopes is -1.	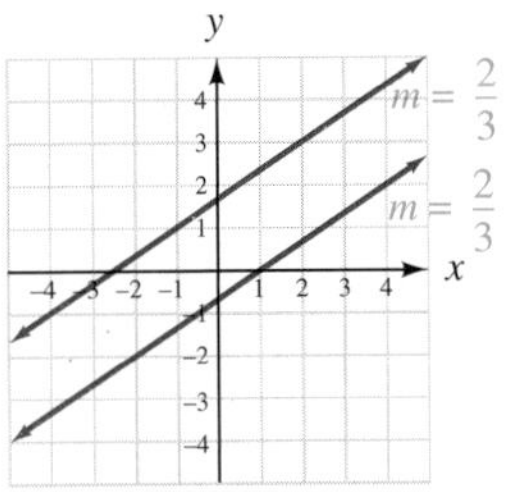 Parallel lines 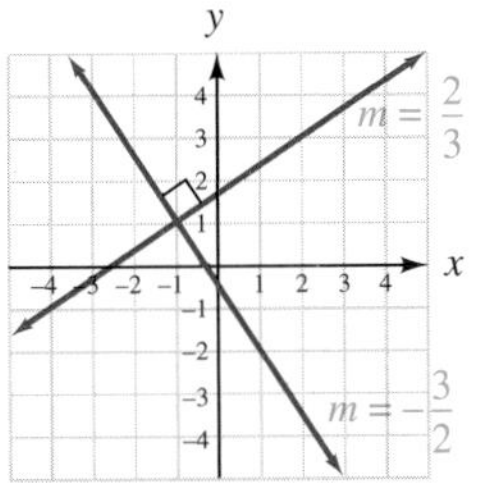 Perpendicular lines

REVIEW EXERCISES

In each case, find the slope of the line.

23. $\frac{1}{4}$

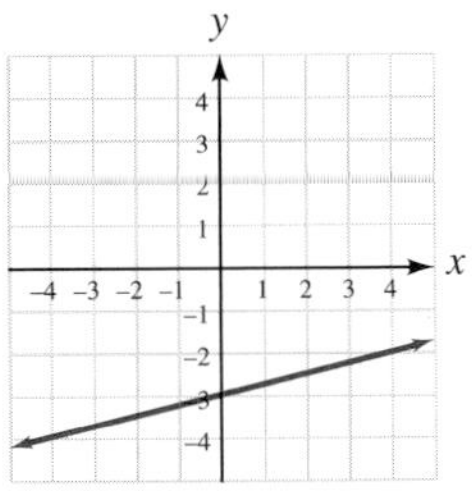

24. $-\frac{7}{8}$

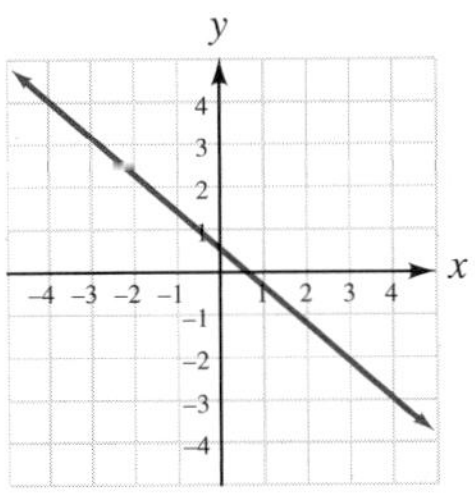

25. The line with this table of solutions: -7

x	y	(x, y)
2	-3	$(2, -3)$
4	-17	$(4, -17)$

26. The line passing through the points $(1, -4)$ and $(3, -7)$ $-\frac{3}{2}$

27. Draw a line having a slope that is

a. Positive **b.** Negative **c.** 0 **d.** Undefined

See AIE Appendix 3.

28. Carpentry. If a truss like the one shown below is used to build the roof of a shed, find the slope (pitch) of the roof. $\frac{3}{4}$

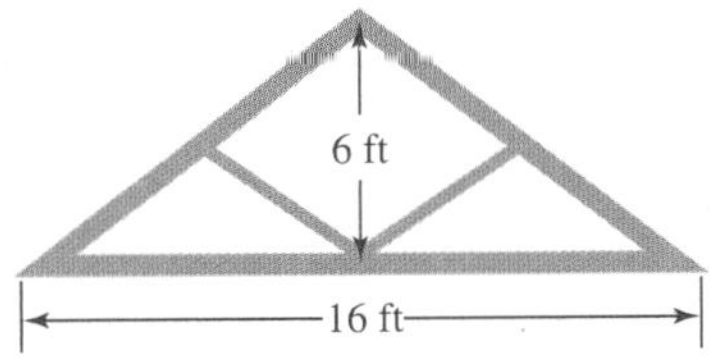

29. Ramps. Find the grade of the ramp shown below. Round to the nearest tenth of a percent. 8.3%

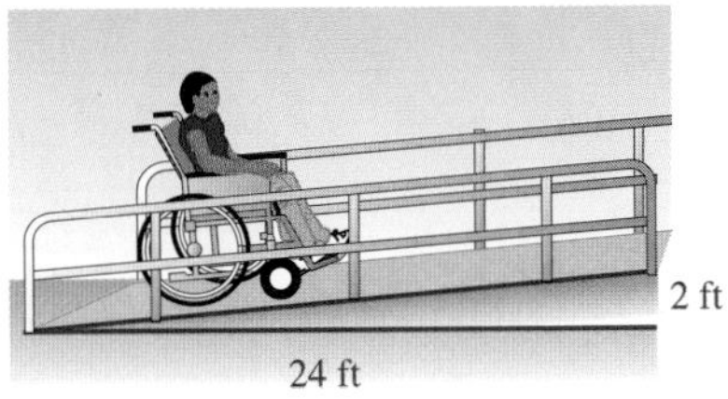

30. Bottled Water. Refer to the graph below. Find the rate of change in the amount of bottled water consumed per person in the U.S. from 2000 through 2008. 1.5 gal/yr

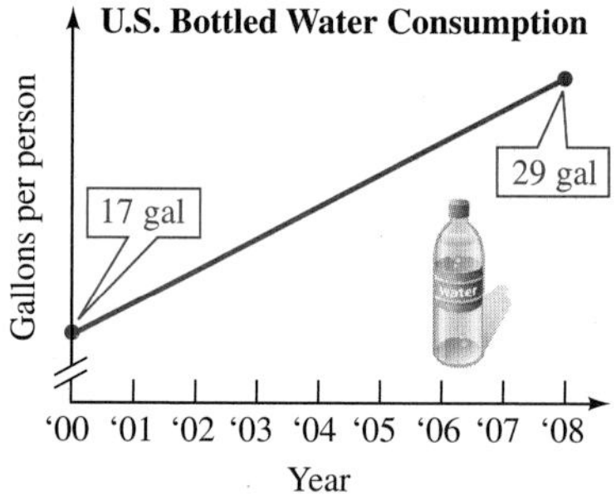

Source: Beverage Marketing Corporation

31. Without graphing, determine whether the line that passes through (6, 6) and (4, 2) and the line that passes through (2, −10) and (−2, −2) are parallel, perpendicular, or neither. They are neither.

32. Find the slope of a line perpendicular to the line passing through (−1, 9) and (−8, 4). $-\frac{7}{5}$

SECTION 3.5 Slope–Intercept Form

DEFINITIONS AND CONCEPTS	EXAMPLES
If a linear equation is written in **slope–intercept form** $y = mx + b$ the graph of the equation is a line with slope m and y-intercept $(0, b)$. To write a linear equation in two variables in slope–intercept form, solve the equation for y.	Find the slope and y-intercept of the line whose equation is $5x + 3y = 3$. To find the slope and y-intercept, we solve the equation for y. $5x + 3y = 3$ $3y = -5x + 3$ Subtract 5x from both sides. $y = -\frac{5}{3}x + 1$ To isolate y, divide both sides by 3. $m = -\frac{5}{3}$ and $b = 1$. The slope of the line is $-\frac{5}{3}$ and the y-intercept is $(0, 1)$.
To **graph a line in slope–intercept form,** plot the y-intercept and use the slope to determine a second point on the line.	Graph: $y = -\frac{5}{3}x + 1$ $y = -\frac{5}{3}x + 1$ $m = \frac{\text{rise}}{\text{run}} = \frac{-5}{3}$ $b = 1$ y-intercept: (0, 1) 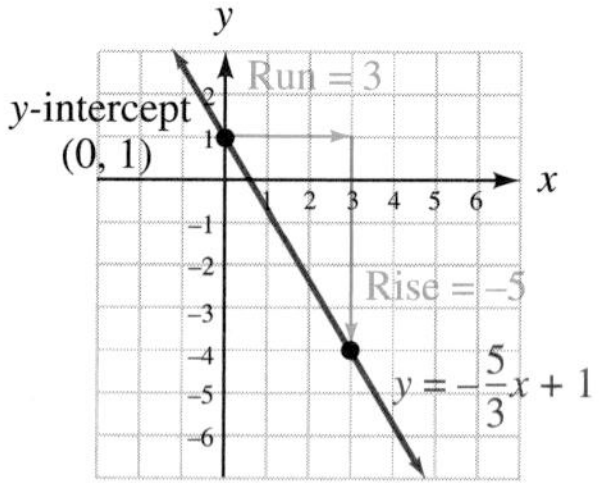
If we know the slope of a line and its y-intercept, we can write its equation.	The equation of a line with slope $\frac{1}{8}$ and y-intercept $(0, -5)$ is $y = \frac{1}{8}x - 5$.
Two different lines with the same slope are **parallel**. If the slopes of two lines are negative reciprocals, the product of their slopes is -1 and the lines are **perpendicular.**	Lines with equations $y = 3x + 4$ and $y = 3x - 12$ are parallel because each line has slope 3. Lines with equations $y = 3x + 4$ and $y = -\frac{1}{3}x - 12$ are perpendicular because their slopes, 3 and $-\frac{1}{3}$, are negative reciprocals.

REVIEW EXERCISES

Find the slope and the y-intercept of each line.

33. $y = \frac{3}{4}x - 2$ $m = \frac{3}{4}$; y-intercept: $(0, -2)$

34. $y = -4x$ $m = -4$; y-intercept: $(0, 0)$

35. $y = \frac{x}{8} + 10$ $m = \frac{1}{8}$; y-intercept: $(0, 10)$

36. $7x + 5y = -21$ $m = -\frac{7}{5}$; y-intercept: $\left(0, -\frac{21}{5}\right)$

37. Graph the line with slope -4 and y-intercept $(0, -1)$. Write an equation of the line. See AIE Appendix 3. $y = -4x - 1$

38. Write an equation for the line shown here. $y = \frac{3}{2}x - 3$

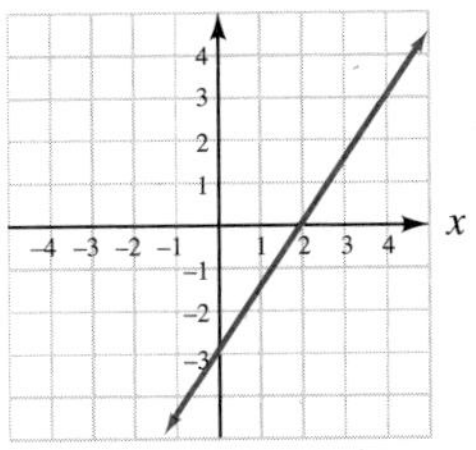

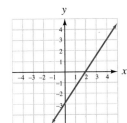

39. Find the slope and the y-intercept of the line whose equation is $9x - 3y = 15$. Then graph it.
$m = 3$; y-intercept: $(0, -5)$. See AIE Appendix 3.

40. Copiers. A business buys a used copy machine that has already produced 75,000 copies.

a. If the business plans to run 300 copies a week, write a linear equation that would find the number of copies c the machine has made in its lifetime after the business has used it for w weeks. $c = 300w + 75{,}000$

b. Use your result in part a to predict the total number of copies that will have been made on the machine 1 year, or 52 weeks, after being purchased by the business. 90,600 copies

Without graphing, determine whether graphs of the given pairs of lines are parallel, perpendicular, or neither.

41. $y = -\frac{2}{3}x + 6$

$y = -\frac{2}{3}x - 6$ Parallel

42. $x + 5y = -10$

$y - 5x = 0$ Perpendicular

SECTION 3.6 Point–Slope Form

DEFINITIONS AND CONCEPTS	EXAMPLES
If a line with slope m passes through the point with coordinates (x_1, y_1), the equation of the line in **point–slope form** is $y - y_1 = m(x - x_1)$	Find an equation of the line with slope -3 that passes through $(-2, 4)$. Write the equation in slope–intercept form. We substitute the slope and the coordinates of the point into the point–slope form. $y - y_1 = m(x - x_1)$ This is point–slope form. $y - 4 = -3[x - (-2)]$ Substitute. $y - 4 = -3(x + 2)$ Simplify within the brackets. $y - 4 = -3x - 6$ Distribute. $y = -3x - 2$ To isolate y, add 4 to both sides. This is slope–intercept form.
If we know **two points that a line passes through,** we can write its equation.	Find an equation of the line that passes through $(2, 5)$ and $(3, 7)$. Write the equation in slope–intercept form. The slope of the line is: $m = \frac{y_2 - y_1}{x_2 - x_1} = \frac{7 - 5}{3 - 2} = 2$ Either point on the line can serve as (x_1, y_1). If we use $(2, 5)$, we have: $y - y_1 = m(x - x_1)$ This is point–slope form. $y - 5 = 2(x - 2)$ Substitute: $x_1 = 2$, $y_1 = 5$, and $m = 2$. $y - 5 = 2x - 4$ Distribute. $y = 2x + 1$ To isolate y, add 5 to both sides. This is the slope–intercept form.

REVIEW EXERCISES

Find an equation of the line with the given slope that passes through the given point. Write the equation in slope–intercept form and graph the equation. See AIE Appendix 3.

43. $m = 3, (1, 5)$ $\quad y = 3x + 2$

44. $m = -\frac{1}{2}, (-4, -1)$ $\quad y = -\frac{1}{2}x - 3$

Find an equation of the line with the following characteristics. Write the equation in slope–intercept form.

45. Passing through $(3, 7)$ and $(-6, 1)$ $\quad y = \frac{2}{3}x + 5$

46. Horizontal, passing through $(6, -8)$ $\quad y = -8$

47. Car Registration. When it was 2 years old, the annual registration fee for a Dodge Caravan was \$380. When it was 4 years old, the registration fee dropped to \$310. If the relationship is linear, write an equation that gives the registration fee f in dollars for the van when it is x years old.
$f = -35x + 450$

48. The Atmosphere. The scatter diagram below shows the amount of carbon dioxide in the Earth's atmosphere as measured at Hawaii's Mauna Loa Observatory from 1960 through 2010. A straight line can be used to model the data.

a. Use the two highlighted points in red to write the equation of the line. Write the answer in slope–intercept form.
$P = \frac{3}{2}t + 310$

b. Use your answer to part a to predict the amount of carbon dioxide in the atmosphere in 2020. 400 parts per million

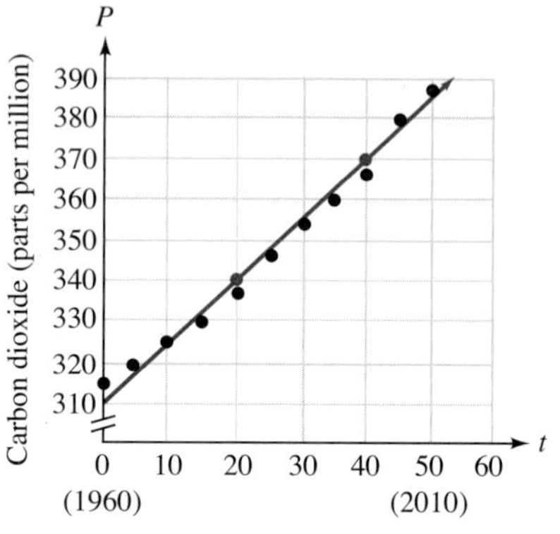

Source: National Oceanic and Atmospheric Administration

SECTION 3.7 Graphing Linear Inequalities

DEFINITIONS AND CONCEPTS	EXAMPLES
An ordered pair (x, y) is a **solution of an inequality** in x and y if a true statement results when the variables are replaced by the coordinates of the ordered pair.	Determine whether $(-2, 5)$ is a solution of $x + 3y > -6$. We substitute the coordinates into the inequality and see if a true statement results. $x + 3y > -6$ This is the given inequality. $-2 + 3(5) \overset{?}{>} -6$ Substitute. $13 > -6$ True Since the result $13 > -6$ is true, $(-2, 5)$ is a solution.
To graph a linear inequality: **1.** Replace the inequality symbol with an $=$ symbol and graph the **boundary line.** Draw a solid line if the inequality contains $\le$ or $\ge$ and a dashed line if it contains $<$ or $>$. **2.** Pick a **test point** not on the boundary. Substitute its coordinates into the inequality. If the inequality is satisfied, shade the half-plane that contains the test point. If the inequality is not satisfied, shade the other half-plane.	Graph: $2x - y \le 4$ **1.** Graph the boundary line $2x - y = 4$ and draw it as a solid line because the inequality symbol is $\le$. **2.** Test the point $(0, 0)$: $2x - y \le 4$ This is the given inequality. $2(0) - 0 \overset{?}{\le} 4$ Substitute. $0 \le 4$ True 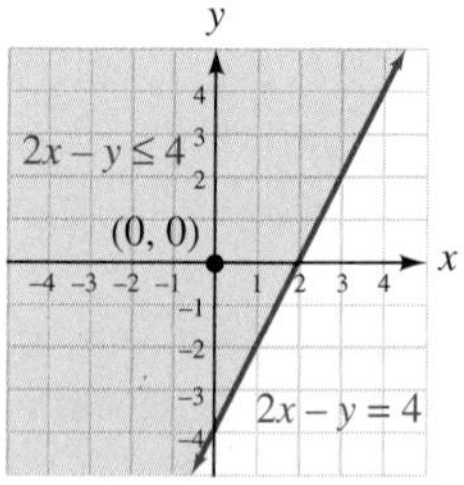 Since the coordinates of the test point satisfy the inequality, we shade the side of the boundary line that contains $(0, 0)$.

REVIEW EXERCISES

49. Determine whether each ordered pair is a solution of $2x - y \le -4$.

a. $(0, 5)$ Yes **b.** $(2, 8)$ Yes

c. $(-3, -2)$ Yes **d.** $\left(\frac{1}{2}, -5\right)$ No

50. Fill in the blanks: $2x - 3y \ge 6$ means $2x - 3y$ = 6 or $2x - 3y$ > 6.

Graph each inequality. See AIE Appendix 3.

51. $x - y < 5$ **52.** $2x - 3y \ge 6$

53. $y \le -2x$ **54.** $y < -4$

55. The graph of a linear inequality is shown in the next column. Would a true or a false statement result if the coordinates of

a. point A were substituted into the inequality? True

b. point B were substituted into the inequality? False

c. point C were substituted into the inequality? False

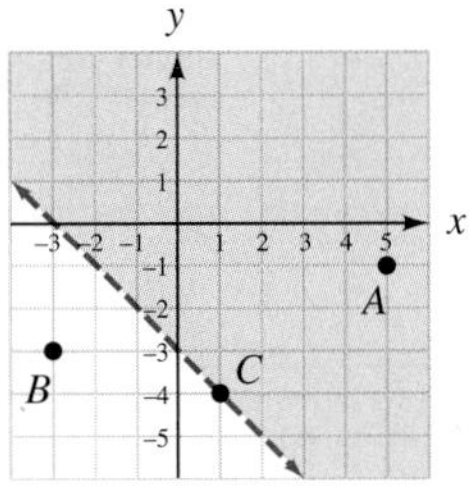

56. Work Schedules. A student told her employer that during the school year, she would be available for up to 30 hours a week, working either 3- or 5-hour shifts. If x represents the number of 3-hour shifts she works and y represents the number of 5-hour shifts she works, the inequality $3x + 5y \le 30$ shows the possible combinations of shifts she can work. Graph the inequality and find three possible combinations. See AIE Appendix 3. (2, 4), (5, 3), (6, 2); answers may vary

SECTION 3.8 An Introduction to Functions

DEFINITIONS AND CONCEPTS | **EXAMPLES**

A **relation** is a set of ordered pairs. The set of all **first components** is called the **domain** of the relation and the set of all **second components** is called the **range** of a relation.

The relation $\{(4, 7), (0, -3), (-3, 8), (1, 7)\}$ has the domain $\{-3, 0, 1, 4\}$ and the range is $\{-3, 7, 8\}$.

A **function** is a set of ordered pairs (a relation) in which to each first component there corresponds exactly one second component.

If to each value of x in the domain there is assigned exactly one value of y in the range, then **y is a function of x.**

x	y
1	2
5	3
6	8
5	9

y is not a function of x: $5 \rightarrow 3$ and $5 \rightarrow 9$

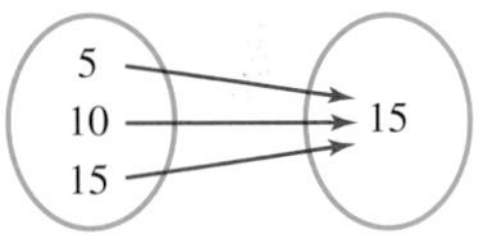

y is a function of x

$\{(4, 7), (7, 4), (4, 1)\}$

Not a function: $4 \rightarrow 7$ and $4 \rightarrow 1$

A function can be defined by an equation. The notation $y = f(x)$ indicates that the variable y is a function of x. It is read as "f of x."

We can think of a function as a machine that takes some **input** x and turns it into some **output** $f(x)$, called a **function value.**

For the function $f(x) = 8x + 5$, $f(-2)$ is the value of $f(x)$ when $x = -2$.

$$f(x) = 8x + 5$$
$$f(-2) = 8(-2) + 5 \quad \text{Substitute the input } -2 \text{ for each } x.$$
$$= -16 + 5 \quad \text{Evaluate the right side.}$$
$$= -11$$

Thus, $f(-2) = -11$.

The **vertical line test:** If a vertical line intersects a graph in more than one point, the graph is not the graph of a function.

Vertical lines intersect only once

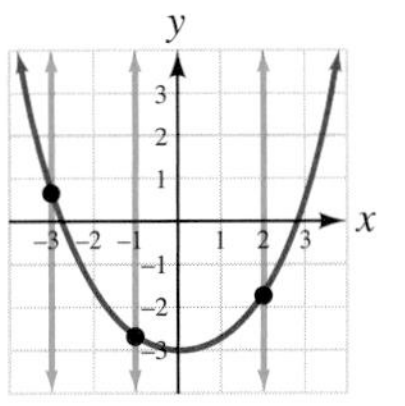

A function

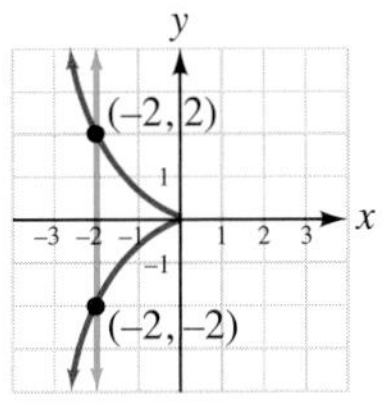

Not a function

A vertical line intersects more than once

x	y
−2	2
−2	2

The input-output pairs that a function generates can be written as ordered pairs and plotted on a rectangular coordinate system to give the **graph of a function.**

Function notation	***Ordered-pair notation***
$f(-3) = 5$	$(-3, 5)$

Graph the function: $f(x) = -\frac{2}{3}x + 3$

We make a table of function values, plot the points, and draw the graph.

$f(x) = -\frac{2}{3}x + 3$

x	$f(x)$	
-3	5	→ $(-3, 5)$
0	3	→ $(0, 3)$
3	1	→ $(3, 1)$

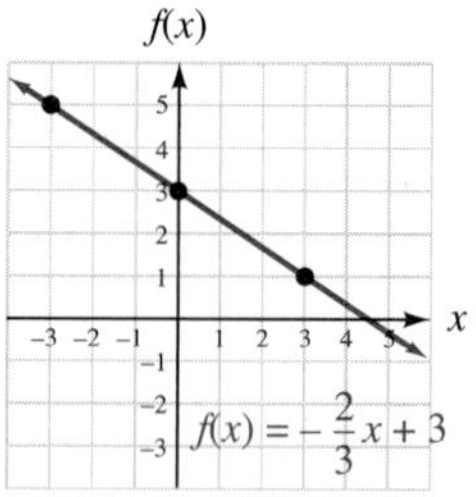

REVIEW EXERCISES

Find the domain and range of each relation.

57. $\{(7, -3), (-5, 9), (4, 4), (0, -11)\}$
Domain: $\{-5, 0, 4, 7\}$; range: $\{-11, -3, 4, 9\}$

58. $\{(2, -2), (15, -8), (-6, 9), (1, -8)\}$
Domain: $\{-6, 1, 2, 15\}$; range: $\{-8, -2, 9\}$

Determine whether each relation defines y to be function of x. If a function is defined, give its domain and range. If it does not define a function, find ordered pairs that show a value of x that corresponds to more than one value of y.

59. 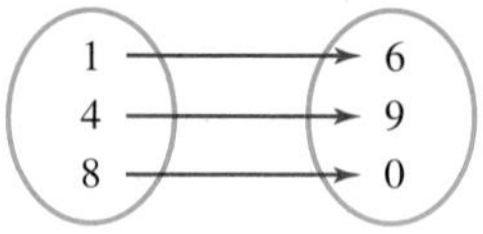

Yes; domain: $\{1, 4, 8\}$; range: $\{0, 6, 9\}$

60. 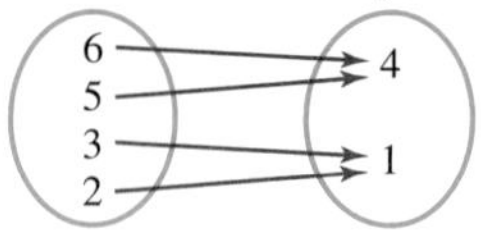

Yes; domain: $\{2, 3, 5, 6\}$; range: $\{1, 4\}$

61.

x	y
9	81
7	49
5	25
3	9

Yes; domain: $\{3, 5, 7, 9\}$; range: $\{9, 25, 49, 81\}$

62.

x	y
-1	2
0	3
-1	4
1	5

No; $(-1, 2), (-1, 4)$

63. $\{(-1, 6), (0, 6), (1, 6), (2, 6)\}$ Yes; domain: $\{-1, 0, 1, 2\}$; range: $\{6\}$

64. $\{(4, 4), (6, 4), (4, 6)\}$ No; $(4, 4), (4, 6)$

Fill in the blanks.

65. The set of all input values for a function is called the domain, and the set of all output values is called the range.

66. Fill in the blank: Since $y =$ $f(x)$, the equations $y = 2x - 8$ and $f(x) = 2x - 8$ are equivalent.

For $f(x) = x^2 - 4x$, find each of the following function values.

67. $f(1)$ -3 68. $f(0)$ 0 69. $f(-3)$ 21 70. $f\left(\frac{1}{2}\right)$ $-\frac{7}{4}$

For $g(x) = 1 - 6x$, find each of the following function values.

71. $g(1)$ -5 72. $g(-6)$ 37 73. $g(0.5)$ -2 74. $g\left(\frac{3}{2}\right)$ -8

Determine whether each graph is the graph of a function. If it is not, find two ordered pairs that show a value of x that corresponds to more than one value of y.

75. 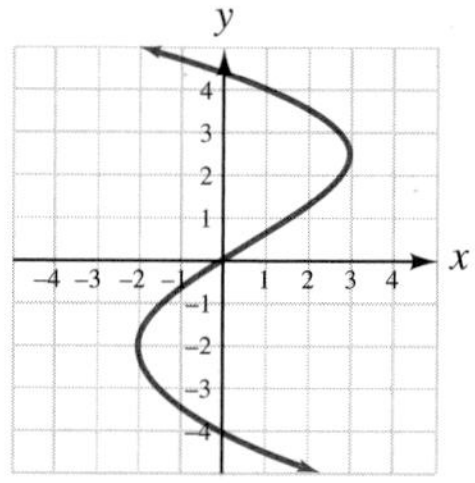

No; (1, 0.5), (1, 4), (answers may vary)

76. 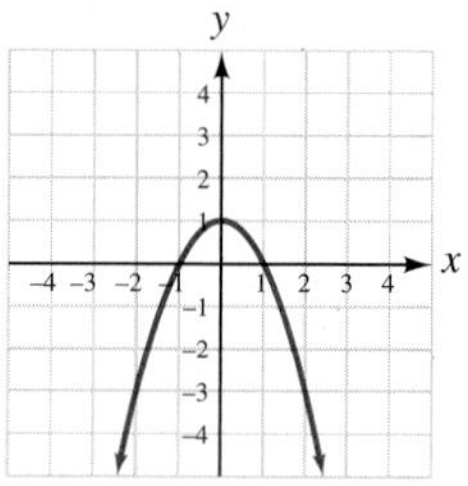

Yes

77. Complete the table of function values for $f(x) = 1 - |x|$. Then graph the function.
See AIE Appendix 3.

x	$f(x)$
0	1
1	0
3	-2
-1	0
-3	-2

78. **Aluminum Cans.** The function $V(r) = 15.7r^2$ estimates the volume in cubic inches of a can 5 inches tall with a radius of r inches. Find the volume of the can shown in the illustration.
1,004.8 in³

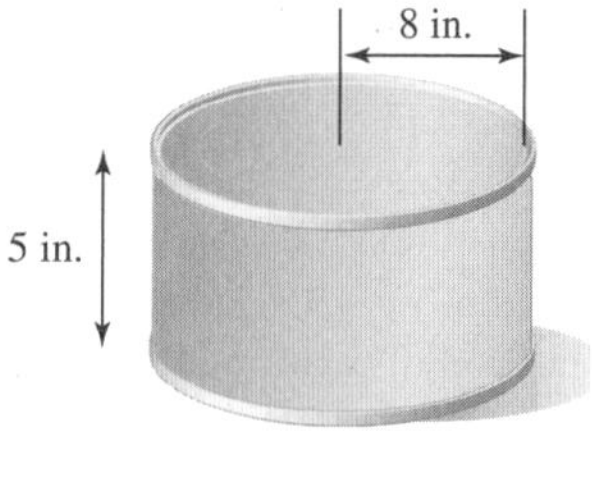

Teaching Tip: Because this Chapter Test is a comprehensive collection of problems that covers all of the topics discussed in Chapter 3, it is lengthy. If your students have time to complete it, that would be optimal. If, because of time constraints, they are unable to do so, assign an appropriate subset of problems that reflects the types of problems that the students will see on your exam.

3 CHAPTER TEST

1. Fill in the blanks.

a. A rectangular coordinate system is formed by two perpendicular number lines called the x- axis and the y- axis.

b. A solution of an equation in two variables is an ordered pair of numbers that makes the equation a true statement.

c. $3x + y = 10$ is a linear equation in two variables and its graph is a line.

d. The slope of a line is a measure of steepness.

e. A function is a set of ordered pairs in which to each first component there corresponds exactly one second component.

The graph shows the number of dogs being boarded in a kennel over a 3-day holiday weekend.

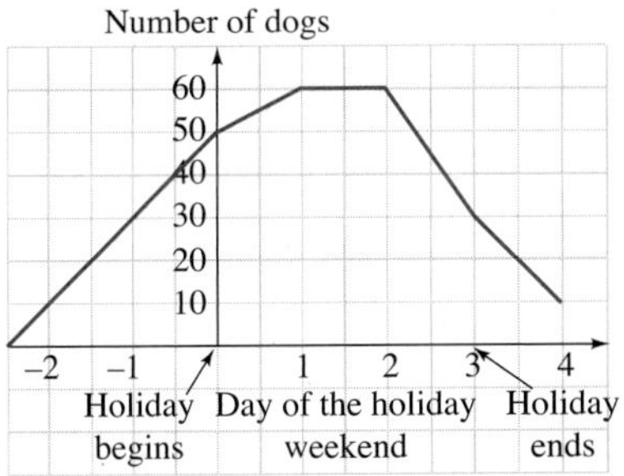

2. How many dogs were in the kennel 2 days before the holiday? 10 dogs

3. What is the maximum number of dogs that were boarded on the holiday weekend at any one time? 60 dogs

4. When were there 30 dogs in the kennel? 1 day before and the 3rd day of the holiday

5. What information does the y-intercept of the graph give? 50 dogs were in the kennel when the holiday began.

6. Plot each point on a rectangular coordinate system: $(1, 3), (-2, 4), (-3, -2), (3, -2), (-1, 0), (0, -1)$, and $\left(-\frac{1}{2}, \frac{7}{2}\right)$. See AIE Appendix 3.

7. Find the coordinates of each point shown in the graph. $A(2, 4)$, $B(-3, 3)$, $C(-2, -3)$, $D(4, -3)$, $E(-4, 0)$, $F(3.5, 1.5)$

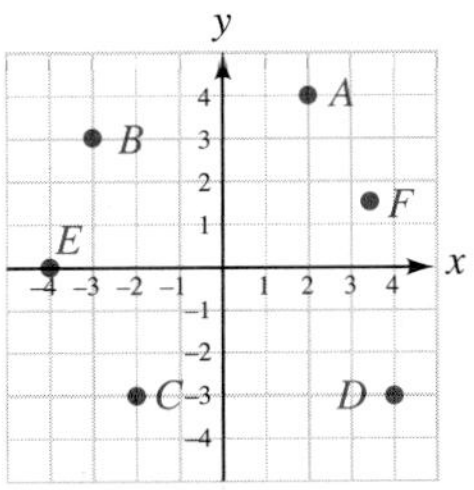

8. In which quadrant is each point located?

a. $(-1, -5)$ III **b.** $\left(6, -2\frac{3}{4}\right)$ IV

9. Is $(-3, -4)$ a solution of $3x - 4y = 7$? Yes

10. Complete the table of solutions for the linear equation $x + 4y = 6$.

x	y	(x, y)
2	1	(2, 1)
−6	3	(−6, 3)

11. The graph of a linear equation is shown.

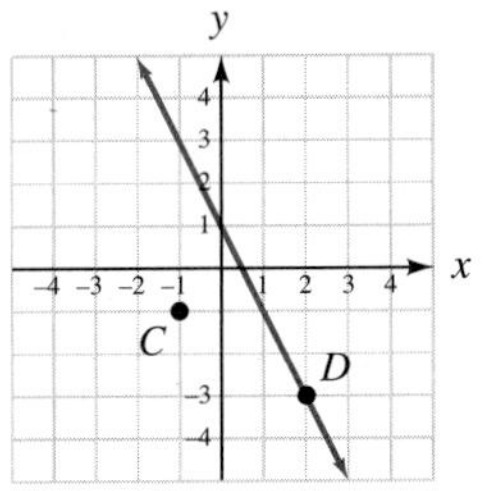

a. If the coordinates of point C are substituted into the equation, will the result be true or false? False

b. If the coordinates of point D are substituted into the equation, will the result be true or false? True

12. Graph: $y = \frac{x}{3}$ See AIE Appendix 3.

13. What are the x- and y-intercepts of the graph of $2x - 3y = 6$? x-intercept: $(3, 0)$; y-intercept: $(0, -2)$

14. Graph: $8x + 4y = -24$ See AIE Appendix 3.

15. Find the slope of the line. $\frac{8}{7}$

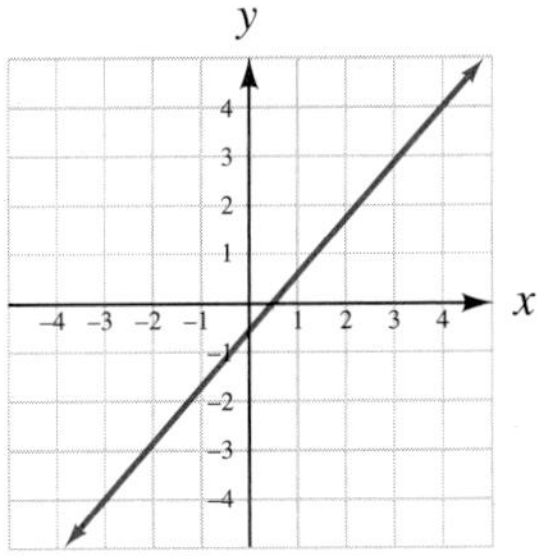

16. Find the slope of the line passing through $(-1, 3)$ and $(3, -1)$. -1

17. What is the slope of a horizontal line? 0

18. **Ramps.** Find the grade of a ramp that rises 2 feet over a horizontal distance of 20 feet. 10%

19. One line passes through $(9, 2)$ and $(6, 4)$. Another line passes through $(0, 7)$ and $(2, 10)$. Without graphing, determine whether the lines are parallel, perpendicular, or neither. Perpendicular

20. When graphed, are the lines $y = 2x + 6$ and $2x - y = 0$ parallel, perpendicular, or neither? Parallel

In Problems 21 and 22, refer to the illustration that shows the elevation changes for part of a 26-mile marathon course.

21. Find the rate of change of the decline on which the woman is running. −15 ft per mi

22. Find the rate of change of the incline on which the man is running. 25 ft per mi

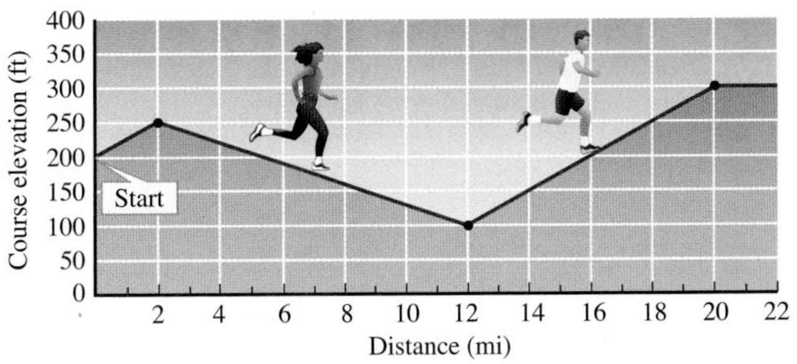

23. Graph: $x = -4$
See AIE Appendix 3.

24. Graph the line passing through $(-2, -4)$ having slope $\frac{2}{3}$.
See AIE Appendix 3.

25. Find the slope and the y-intercept of the graph of $x + 2y = 8$.
$m = -\frac{1}{2}$; $(0, 4)$

26. Find an equation of the line passing through $(-2, 5)$ with slope 7. Write the equation in slope–intercept form.
$y = 7x + 19$

27. Find an equation for the line shown. Write the equation in slope–intercept form.
$y = -2x - 5$

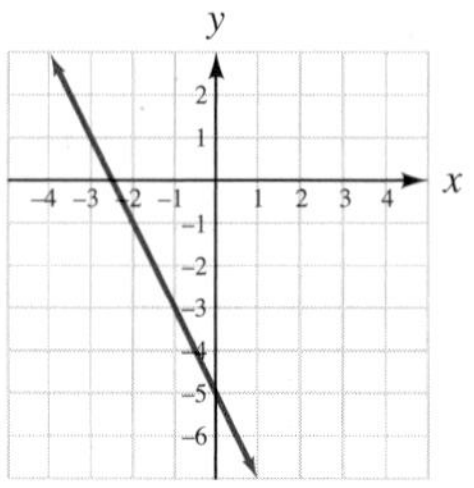

28. Depreciation. After it is purchased, a \$15,000 computer loses \$1,500 in resale value every year.

a. Write a linear equation that gives the resale value v of the computer x years after being purchased.
$v = -1{,}500x + 15{,}000$

b. Use your answer to part (a) to predict the value of the computer 8 years after it is purchased.
\$3,000

29. Check to determine whether (6, 1) is a solution of $2x - 4y \geq 8$. Yes

30. Water Heaters. The scatter diagram shows how excessively high temperatures affect the life of a water heater. Write an equation of the line that models the data for water temperatures between 140° and 180°. Let T represent the temperature of the water in degrees Fahrenheit and y represent the expected life of the heater in years. Give the answer in slope–intercept form.
$y = -\frac{1}{5}T + 41$

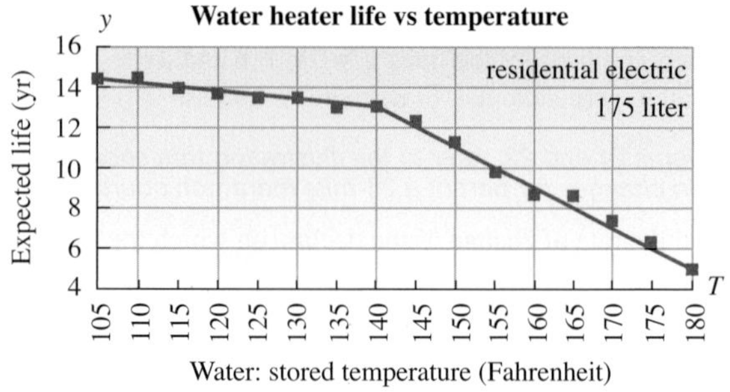

Source: www.uniongas.com/WaterHeating

31. A linear inequality has been graphed below. Determine whether each point satisfies the inequality.

a. $(-2, 3)$ Yes
b. $(3, -4)$ No
c. $(0, 0)$ No

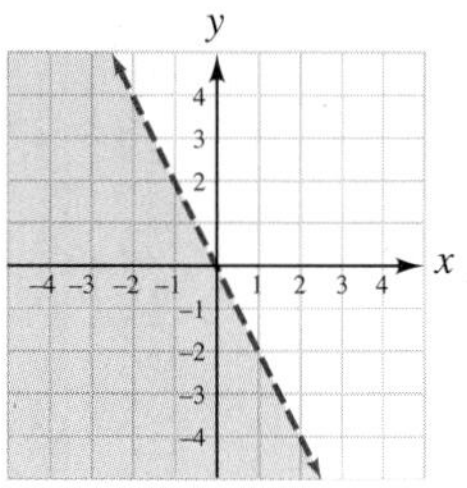

32. Is $(-20, -2)$ a solution of $\frac{1}{2}x - 3y \geq -4$? Yes

33. Graph the inequality: $2x - 5y \leq -10$
See AIE Appendix 3.

34. Find the domain and range of the relation: $\{(5, 3), (1, 12), (-4, 3), (0, -8)\}$
Domain: $\{-4, 0, 1, 5\}$; range: $\{-8, 3, 12\}$

Determine whether the relation defines y to be function of x. If a function is defined, give its domain and range. If it does not define a function, find ordered pairs that show a value of x that corresponds to more than one value of y.

35.

x	y
1	4
2	3
3	2
4	1

Yes; domain: $\{1, 2, 3, 4\}$; range: $\{1, 2, 3, 4\}$

36.

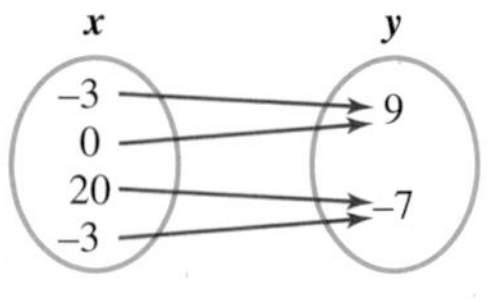

No; $(-3, 9), (-3, -7)$

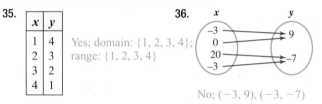

37. $\{(6, 5), (7, 5), (8, 5), (9, 5), (10, 5)\}$
Yes; domain: $\{6, 7, 8, 9, 10\}$; range: $\{5\}$

38. $\{(-9, 41), (2, 6), (4, -9), (2, 2)\}$ No; $(2, 6), (2, 2)$

39.

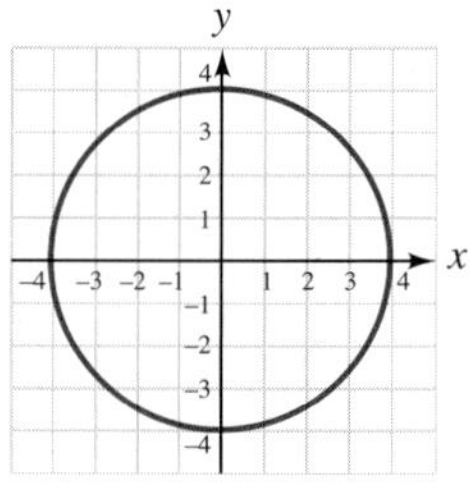

No; $(2, 3.5), (2, -3.5)$; (answers may vary)

40.

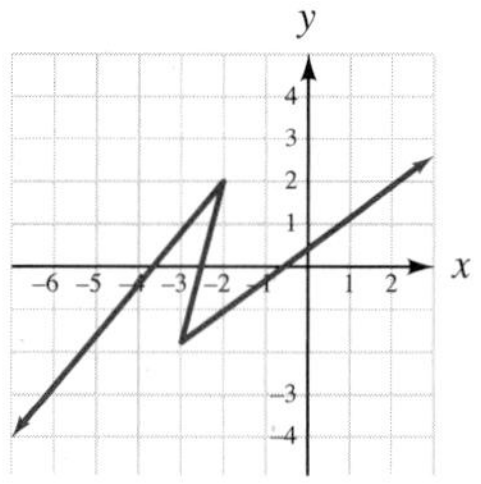

No; $(-2, 2), (-2, -1)$; (answers may vary)

41. If $f(x) = 2x - 7$, find: $f(-3)$ -13

42. If $g(x) = 3.5x^3$ find: $g(6)$ 756

43. Telephone Calls. The function $C(n) = 0.30n + 15$ gives the cost C per month in dollars for making n phone calls. Find $C(45)$ and explain what it means. $C(45) = 28.50$; it costs \$28.50 to make 45 calls.

44. Graph: $f(x) = |x| - 1$ See AIE Appendix 3.

Group Project

Overview: In this activity, you will explore the relationship between a person's height and arm span. Arm span is defined to be the distance between the tips of a person's fingers when his or her arms are held out to the side.

Instructions: Form groups of 5 or 6 students. Measure the height and arm span of each person in your group, and record the results in a table like the one shown below.

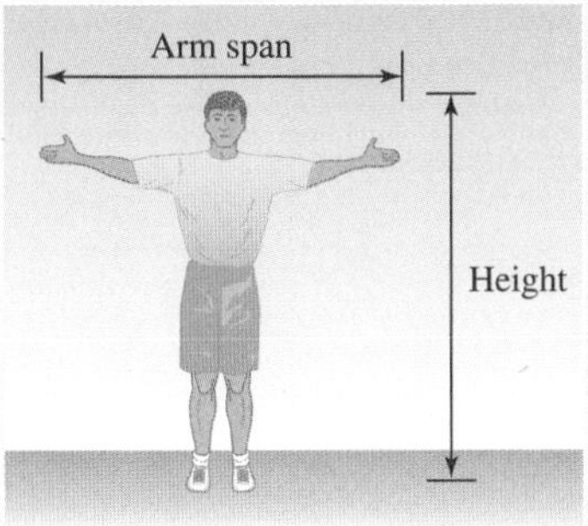

Name	Height (in.)	Arm span (in.)
1.		
2.		
3.		
4.		
5.		
6.		

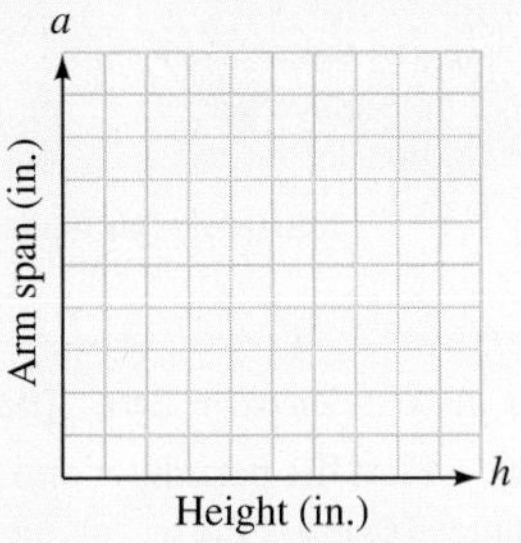

Plot the data in the table as ordered pairs of the form (height, arm span) on a graph like the one shown above. Then draw a straight-line model that best fits the data points.

Pick two convenient points on the line and find its slope. Use the point-slope form $a - a_1 = m(h - h_1)$ to find an equation of the line. Then, write the equation in slope–intercept form.

Ask a person from another group for his or her height measurement. Substitute that value into your linear model to predict that person's arm span. How close is your prediction to the person's actual arm span?

(From *Activities for Beginning and Intermediate Algebra* by Debbie Garrison, Judy Jones, and Jolene Rhodes)

CUMULATIVE REVIEW Chapters 1–3

1. Find the prime factorization of 108. [Section 1.2] $2^2 \cdot 3^3$
2. Write $\frac{1}{250}$ as a decimal. [Section 1.3] 0.004
3. Determine whether each statement is true or false. [Section 1.3]
   a. Every whole number is an integer. True
   b. Every integer is a real number. True
   c. 0 is a whole number, an integer, and a rational number. True

Perform the operations.

4. $-27 + 21 + (-9)$ [Section 1.4] -15
5. $-1.57 - (-0.8)$ [Section 1.5] -0.77
6. $-9(-7)(5)(-3)$ [Section 1.6] -945
7. $\dfrac{-180}{-6}$ [Section 1.6] 30
8. Evaluate: $\left|\dfrac{(6-5)^4 - (-21)}{-27 + 4^2}\right|$ [Section 1.7] 2
9. Evaluate $b^2 - 4ac$ for $a = 2$, $b = -8$, and $c = 4$. [Section 1.8] 32
10. Suppose x sheets from a 500-sheet ream of paper have been used. How many sheets are left? [Section 1.8] $500 - x$
11. How many terms does the algebraic expression $3x^2 - 2x + 1$ have? What is the coefficient of the second term? [Section 1.8] 3, -2
12. Multiply. [Section 1.9]
    a. $2(x + 4)$ $2x + 8$
    b. $-2(x - 4)$ $-2x + 8$

Simplify each expression. [Section 1.9]

13. $5a + 10 - a$ $4a + 10$
14. $-7(9t)$ $-63t$
15. $-2b^2 + 6b^2$ $4b^2$
16. $5(-17)(0)(2)$ 0
17. $(a + 2) - (a - 2)$ 4
18. $-4(-5)(-8a)$ $-160a$
19. $-y - y - y$ $-3y$
20. $\dfrac{3}{2}(4x - 8) + x$ $7x - 12$

Solve each equation. [Sections 2.1 and 2.2]

21. $3x - 5 = 13$ 6
22. $1.2 - x = -1.7$ 2.9
23. $\dfrac{2x}{3} - 2 = 4$ 9
24. $\dfrac{y - 2}{7} = -3$ -19
25. $-3(2y - 2) - y = 5$ $\frac{1}{7}$
26. $9y - 3 = 6y$ 1
27. $\dfrac{1}{3} + \dfrac{c}{5} = -\dfrac{3}{2}$ $-\frac{55}{6}$
28. $5(x + 2) = 5x - 2$ No solution, contradiction
29. $-x = 99$ -99
30. $3c - 2 = \dfrac{11(c - 1)}{5}$ $-\frac{1}{4}$

31. Pennies. A 2010 telephone survey of adults asked whether the penny should be discontinued from the national currency. The results are shown in the circle graph. If 869 people favored keeping the penny, how many took part in the survey? [Section 2.3] 1,100

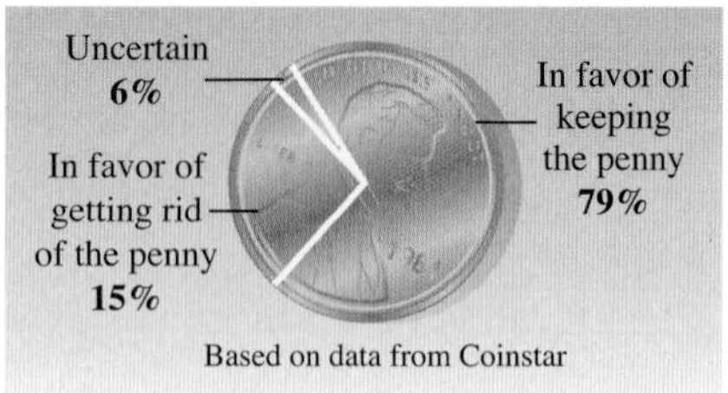

32. Solve for h: $S = 2\pi rh + 2\pi r^2$ [Section 2.4] $h = \frac{S - 2\pi r^2}{2\pi r}$

33. Band Aids. Find the perimeter and the area of the gauze pad of the bandage. [Section 2.4] $3\frac{1}{8}$ in., $\frac{39}{64}$ in.^{2}

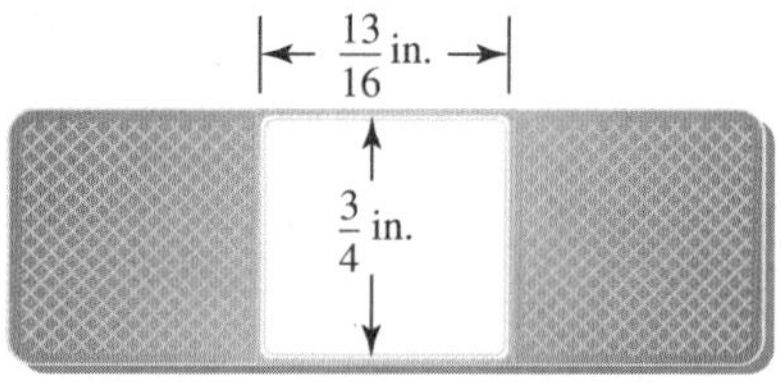

34. High Heels. Find the unknown angle measure represented by x. [Section 2.5] 45°

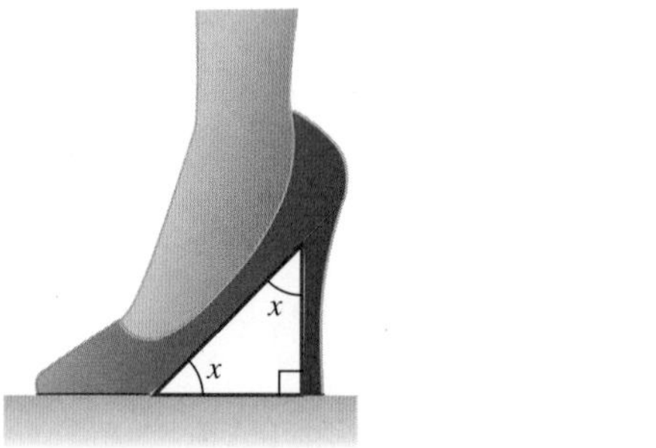

35. Complete the table. [Section 2.6]

	% acid	Liters	Amount of acid
50% solution	0.50	x	$0.50x$
25% solution	0.25	$13 - x$	$0.25(13 - x)$
30% mixture	0.30	13	0.30(13)

36. Road Trips. A bus, carrying the members of a marching band, and a truck, carrying their instruments, leave a high school at the same time. The bus travels at 60 mph and the truck at 50 mph. In how many hours will they be 75 miles apart? [Section 2.6] 7.5 hr

37. Mixing Candy. Candy corn worth \$2.85 per pound is to be mixed with black gumdrops that cost \$1.80 per pound to make 200 pounds of a mixture worth \$2.22 per pound. How many pounds of each candy should be used? [Section 2.6] 80 lb candy corn, 120 lb gumdrops

Solve each inequality. Write the solution set in interval notation and graph it. [Section 2.7]

38. $-\frac{3}{16}x \geq -9$ $(-\infty, 48]$ See AIE Appendix 3.

39. $8x + 4 > 3x + 4$ $(0, \infty)$ See AIE Appendix 3.

40. In which quadrants are the second coordinates of ordered pairs positive? [Section 3.1] I and II

41. Is $(-2, 4)$ a solution of $y = 2x - 8$? [Section 3.2] No

Graph each equation. See AIE Appendix 3.

42. $y = x$ [Section 3.2]

43. $2x + 4y = -8$ [Section 3.3]

44. What is the slope of the graph of the line $y = 5$? [Section 3.4] 0

45. What is the slope of the line passing through $(-2, 4)$ and $(5, -6)$? [Section 3.4] $-\frac{10}{7}$

46. Roofing. Find the pitch of the roof. [Section 3.4] $\frac{7}{12}$

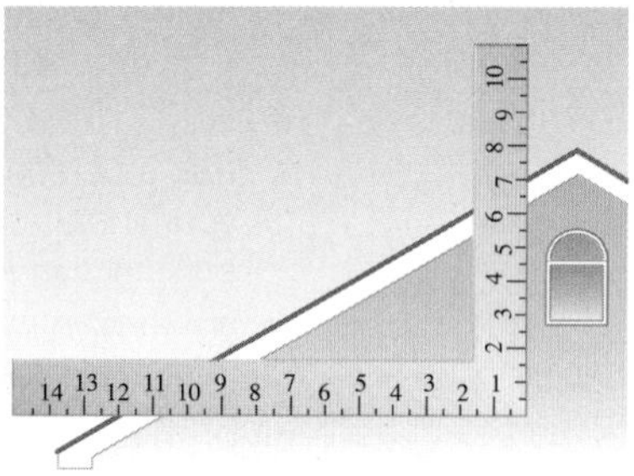

47. Find the slope and the y-intercept of the graph of the line described by $4x - 6y = -12$. [Section 3.5] $\frac{2}{3}$, $(0, 2)$

48. Write an equation of the line that has slope -2 and y-intercept $(0, 1)$. [Section 3.5] $y = -2x + 1$

49. Find an equation of the line that has slope $-\frac{7}{8}$ and passes through $(2, -9)$. Write the equation in point–slope form and in slope–intercept form. [Section 3.6] $y + 9 = -\frac{7}{8}(x - 2)$; $y = -\frac{7}{8}x - \frac{29}{4}$

50. Is $(-2, -4)$ a solution of $x + y \leq -6$? [Section 3.7] Yes

51. Graph: $y \geq x + 1$ [Section 3.7] See AIE Appendix 3.

52. Graph $x < 4$ on a rectangular coordinate system. [Section 3.7] See AIE Appendix 3.

53. If $f(x) = x^4 + x$, find: $f(-3)$ [Section 3.8] 78

54. Is this the graph of a function? [Section 3.8] No

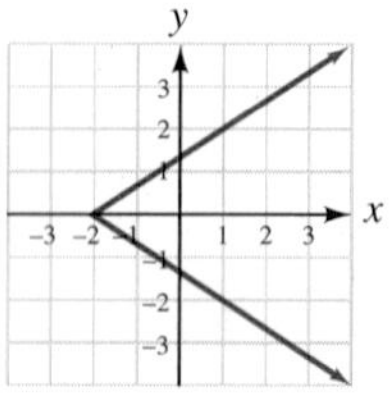

Systems of Linear Equations and Inequalities

4

from Campus to Careers

Photographer

Photographers record our surroundings, the special events in our lives, and people, so that all can be remembered in pictures. Some specialize in weddings and portraits, some photograph landscapes and fashion, while others work on location as photojournalists. Their job responsibilities require a variety of mathematical skills such as: scheduling appointments, keeping financial records, pricing photographs, purchasing supplies, billing customers, and operating digital equipment.

Problem 79 in **Study Set 4.1, problem 29** in **Study Set 4.4,** and **problem 55** in **Study Set 4.5** involve situations that a photographer might encounter on the job. The mathematical concepts discussed in this chapter can be used to solve those problems.

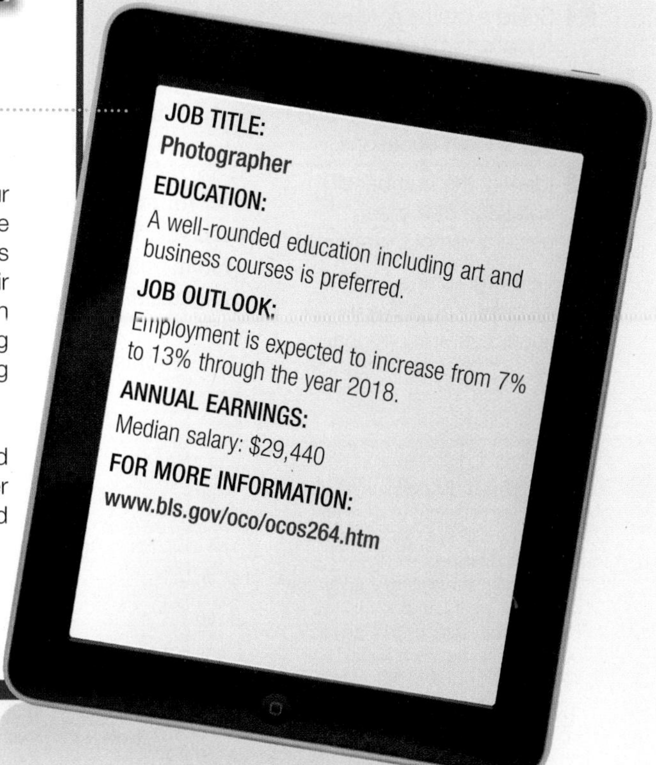

JOB TITLE:
Photographer
EDUCATION:
A well-rounded education including art and business courses is preferred.
JOB OUTLOOK:
Employment is expected to increase from 7% to 13% through the year 2018.
ANNUAL EARNINGS:
Median salary: $29,440
FOR MORE INFORMATION:
www.bls.gov/oco/ocos264.htm

Study Skills Workshop ▶ Making Homework a Priority

Teaching Tip: Study Skills Workshop topics may be taught in any order, at any time during the course. See the Index for a complete list to determine if and when other Study Skills Workshop topics should be discussed with your students.

Attending class and taking notes are important, but they are not enough. The only way that you are really going to learn algebra is by doing your homework.

WHEN TO DO YOUR HOMEWORK: Homework should be started on the day it is assigned, when the material is fresh in your mind. It's best to break your homework sessions into 30-minute periods, allowing for short breaks in between.

HOW TO BEGIN YOUR HOMEWORK: Review your notes and the examples in your text before starting your homework assignment.

GETTING HELP WITH YOUR HOMEWORK: It's normal to have some questions when doing homework. Talk to a tutor, a classmate, or your instructor to get those questions answered.

Now Try This ▶

1. Write a one-page paper that describes *when, where,* and *how* you go about completing your algebra homework assignments.
2. For each problem on your next homework assignment, find an example in this book that is similar. Write the example number next to the problem.
3. Make a list of questions that you have while doing your next assignment. Then decide whom you are going to ask to get those questions answered.

Teaching Tip: The *Are You Ready?* feature reviews crucial prerequisite skills that students should have mastered already if they are to be successful with the new topics in this section. All of these problems can be found in Enhanced WebAssign, allowing you to require them to be submitted before class to gauge student readiness.

SECTION 4.1 Solving Systems of Equations by Graphing

OBJECTIVES

1. Determine whether a given ordered pair is a solution of a system.
2. Solve systems of linear equations by graphing.
3. Use graphing to identify inconsistent systems and dependent equations.
4. Identify the number of solutions of a linear system without graphing.
5. Use a graphing calculator to solve a linear system (optional).

ARE YOU READY? *Are You Ready? exercises available online at www.webassign.net/brookscole*

The following problems review some basic skills that are needed when solving systems of equations by graphing.

1. Is $(-2, -6)$ a solution of $y = 3x - 1$? Not a solution
2. Use the slope and the y-intercept to graph $y = -4x + 2$. See AIE Appendix 3.
3. Graph $3x + 4y = 12$ by finding the x- and y-intercepts. See AIE Appendix 3.
4. Without graphing, determine whether the graphs of $y = \frac{1}{2}x - 3$ and $x - 2y = 2$ are parallel, perpendicular, or neither. Parallel

The following illustration shows the average amounts of chicken and beef eaten per person each year in the United States from 1985 to 2010. Plotting both graphs on the same coordinate system makes it easy to compare recent trends. The point of intersection of the graphs indicates that Americans ate equal amounts of chicken and beef in 1992—about 66 pounds of each, per person.

In this section, we will use a similar graphical approach to solve systems of equations.

Teaching Tip: If the desks in your classroom are arranged in rows and columns, you can use that configuration to introduce this section.
1. Have all the students in a given row raise their right hands.
2. Have all the students in a given column raise their left hands.
3. Ask: "Who raised both hands?"

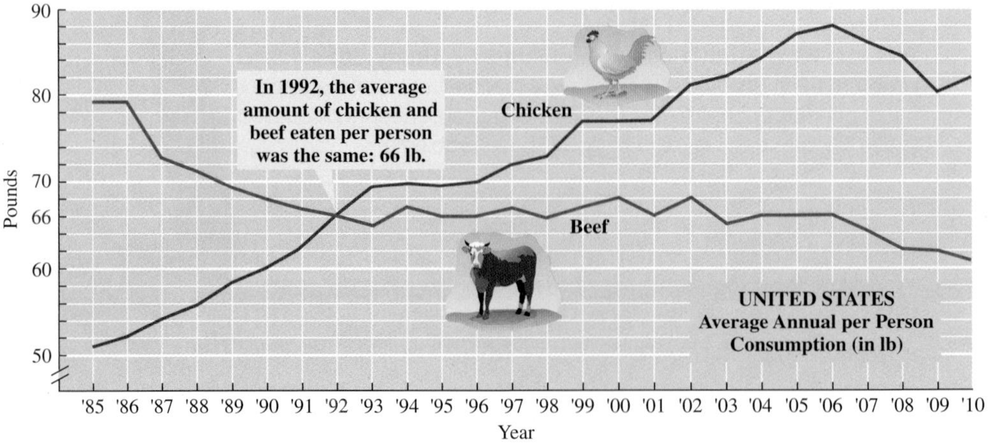

Source: U.S. Department of Agriculture

1 Determine Whether a Given Ordered Pair Is a Solution of a System.

We previously have discussed equations in two variables, such as $x + y = 3$. Because there are infinitely many pairs of numbers whose sum is 3, there are infinitely many pairs (x, y) that satisfy this equation. Some of these pairs are listed in table (a).

Now consider the equation $x - y = 1$. Because there are infinitely many pairs of numbers whose difference is 1, there are infinitely many pairs (x, y) that satisfy $x - y = 1$. Some of these pairs are listed in table (b).

The Language of Algebra

We say that (2, 1) **satisfies** $x + y = 3$, because the x-coordinate, 2, and the y-coordinate, 1, make the equation true when substituted for x and y: $2 + 1 = 3$. To *satisfy* means to make content, as in *satisfy* your thirst or a *satisfied* customer.

$x + y = 3$

x	y	(x, y)
0	3	(0, 3)
1	2	(1, 2)
2	**1**	**(2, 1)**
3	0	(3, 0)

(a)

$x - y = 1$

x	y	(x, y)
0	−1	(0, −1)
1	0	(1, 0)
2	**1**	**(2, 1)**
3	2	(3, 2)

(b)

From the two tables, we see that (2, 1) satisfies both equations.

When two equations with the same variables are considered simultaneously (at the same time), we say that they form a **system of equations.** Using a left brace { , we can write the equations from the previous example as a system:

$$\begin{cases} x + y = 3 \\ x - y = 1 \end{cases}$$ Read as "the system of equations $x + y = 3$ and $x - y = 1$."

Because the ordered pair (2, 1) satisfies both of these equations, it is called a **solution of the system.** In general, a system of linear equations can have exactly one solution, no solution, or infinitely many solutions.

EXAMPLE 1 Determine whether (−2, 5) is a solution of each system of equations.

a. $\begin{cases} 3x + 2y = 4 \\ x - y = -7 \end{cases}$ **b.** $\begin{cases} 4y = 18 - x \\ y = 2x \end{cases}$

Strategy We will substitute the x- and y-coordinates of (−2, 5) for the corresponding variables in both equations of the system.

Why If both equations are satisfied (made true) by the x- and y-coordinates, then the ordered pair is a solution of the system.

Solution

a. Recall that in an ordered pair, the first number is the x-coordinate and the second number is the y-coordinate. To determine whether (−2, 5) is a solution, we substitute −2 for x and 5 for y in each equation.

Check:

$$\begin{aligned} 3x + 2y &= 4 && \text{The first equation.} \\ 3(\mathbf{-2}) + 2(\mathbf{5}) &\stackrel{?}{=} 4 \\ -6 + 10 &\stackrel{?}{=} 4 \\ 4 &= 4 && \text{True} \end{aligned}$$

$$\begin{aligned} x - y &= -7 && \text{The second equation.} \\ \mathbf{-2} - \mathbf{5} &\stackrel{?}{=} -7 \\ -7 &= -7 && \text{True} \end{aligned}$$

Since (−2, 5) satisfies both equations, it is a solution of the system.

b. Again, we substitute −2 for x and 5 for y in each equation.

Check:

$$\begin{aligned} 4y &= 18 - x && \text{The first equation.} \\ 4(\mathbf{5}) &\stackrel{?}{=} 18 - (\mathbf{-2}) \\ 20 &\stackrel{?}{=} 18 + 2 \\ 20 &= 20 && \text{True} \end{aligned}$$

$$\begin{aligned} y &= 2x && \text{The second equation.} \\ \mathbf{5} &\stackrel{?}{=} 2(\mathbf{-2}) \\ 5 &= -4 && \text{False} \end{aligned}$$

Although (−2, 5) satisfies the first equation, it does not satisfy the second. Because it does not satisfy both equations, (−2, 5) is not a solution of the system.

The Language of Algebra

A system of equations is two (or more) equations that we consider **simultaneously**—at the same time. On June 3, 2007, in Kansas City, more than 1,680 guitarists set a world record for the most people playing the same song *simultaneously.* The song was Deep Purple's *Smoke on the Water.*

Teaching Tip: Use this analogy: Both the bride and groom must say "I do" during a wedding ceremony or there is no marriage. Similarly, both ordered pairs must satisfy the equations.

Teaching Example 1 Determine whether (6, −1) is a solution of the system of equations $\begin{cases} 2x + y = 11 \\ y = x - 7 \end{cases}$
Answer: A solution

Self Check 1 Determine whether (4, −1) is a solution of: $\begin{cases} x - 2y = 6 \\ y = 3x - 11 \end{cases}$ Not a solution

Now Try ▶ Problem 15

2 Solve Systems of Linear Equations by Graphing.

To **solve a system of equations** means to find all of the solutions of the system. One way to solve a system of linear equations is to graph the equations on the same rectangular coordinate system.

EXAMPLE 2 Solve the system of equations by graphing: $\begin{cases} 2x + 3y = 2 \\ 3x - 2y = 16 \end{cases}$

Strategy We will graph both equations on the same coordinate system.

Why Recall that the graph of a linear equation is a "picture" of its solutions. If both equations are graphed on the same coordinate system, we can see whether they have any common solutions.

Solution The intercept-method is a convenient way to graph equations such as $2x + 3y = 2$ and $3x - 2y = 16$, because they are in standard $Ax + By = C$ form.

Success Tip

Accuracy is crucial when using the graphing method to solve a system. Here are some suggestions for improving your accuracy:
- Use graph paper.
- Use a sharp pencil.
- Use a ruler or straightedge.

$2x + 3y = 2$

x	y	(x, y)
0	$\frac{2}{3}$	$\left(0, \frac{2}{3}\right)$
1	0	(1, 0)
−2	2	(−2, 2)

$3x - 2y = 16$

x	y	(x, y)
0	−8	(0, −8)
$\frac{16}{3}$	0	$\left(\frac{16}{3}, 0\right)$
2	−5	(2, −5)

To find the y-intercept, let x = 0 and solve for y.
To find the x-intercept, let y = 0 and solve for x.
As a check, pick another x-value, such as −2 or 2, and find y.

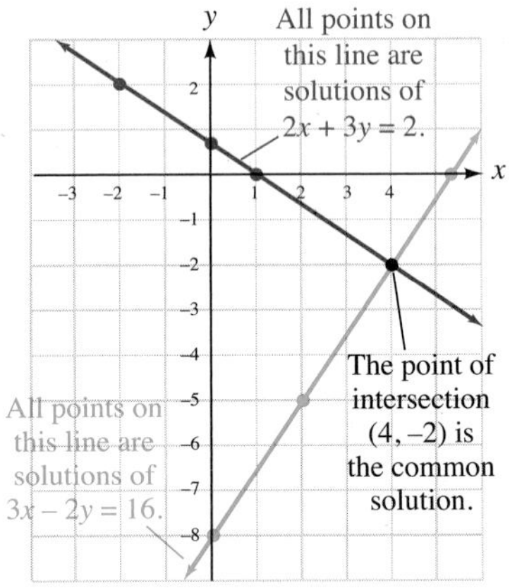

y
x
All points on this line are solutions of 2x + 3y = 2.
All points on this line are solutions of 3x − 2y = 16.
The point of intersection (4, −2) is the common solution.

Success Tip

When determining the coordinates of a point of intersection from a graph, realize that they are simply estimates. Only after algebraically checking a proposed solution can we be sure that it is an actual solution.

The coordinates of each point on the line graphed in red satisfy $2x + 3y = 2$ and the coordinates of each point on the line graphed in blue satisfy $3x - 2y = 16$. Because the point of intersection is on both graphs, its coordinates satisfy both equations.

It appears that the graphs intersect at the point (4, −2). To verify that it is the solution of the system, we substitute 4 for x and −2 for y in each equation.

Check:

$$2x + 3y = 2 \quad \text{The first equation.}$$
$$2(4) + 3(-2) \stackrel{?}{=} 2$$
$$8 + (-6) \stackrel{?}{=} 2$$
$$2 = 2 \quad \text{True}$$

$$3x - 2y = 16 \quad \text{The second equation.}$$
$$3(4) - 2(-2) \stackrel{?}{=} 16$$
$$12 - (-4) \stackrel{?}{=} 16$$
$$16 = 16 \quad \text{True}$$

Since (4, −2) makes both equations true, it is the solution of the system. The solution set is written as {(4, −2)}.

Teaching Example 2 Solve the system of equations by graphing: $\begin{cases} x + y = -4 \\ x - 2y = 2 \end{cases}$
Answer:
(−2, −2)

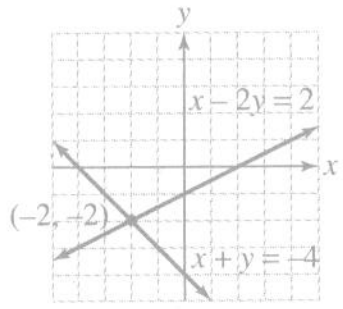

y
x
x − 2y = 2
(−2, −2)
x + y = −4

Self Check 2 Solve the system of equations by graphing: $\begin{cases} 2x - y = -5 \\ x + y = -1 \end{cases}$ (−2, 1) See AIE Appendix 3.

Now Try ▶ Problem 25

To solve a system of linear equations in two variables by graphing, follow these steps.

The Graphing Method

1. Carefully graph each equation on the same rectangular coordinate system.
2. If the lines intersect, determine the coordinates of the point of intersection of the graphs. That ordered pair is the solution of the system.
3. Check the proposed solution in each equation of the original system.

3 Use Graphing to Identify Inconsistent Systems and Dependent Equations.

A system of equations that has at least one solution, like that in Example 2, is called a **consistent system.** A system with no solution is called an **inconsistent system.**

EXAMPLE 3 Solve the system of equations by graphing: $\begin{cases} y = -2x - 6 \\ 4x + 2y = 8 \end{cases}$

Strategy We will graph both equations on the same coordinate system.

Why If both equations are graphed on the same coordinate system, we can see whether they have any common solutions.

Solution Since $y = -2x - 6$ is written in slope–intercept form, we can graph it by plotting the y-intercept $(0, -6)$ and then drawing a slope triangle whose rise is -2 and whose run is 1. We can graph $4x + 2y = 8$ using the intercept method.

Teaching Tip: Ask your students to describe the graph of the system:
1. if the lines have the same slope.
2. if the lines have different slopes.

$y = -2x - 6$

$m = -2 \quad b = -6$

Slope: $\frac{\text{Rise}}{\text{Run}} = \frac{-2}{1}$

y-intercept: $(0, -6)$

$4x + 2y = 8$

x	y	(x, y)
0	4	(0, 4)
2	0	(2, 0)
1	2	(1, 2)

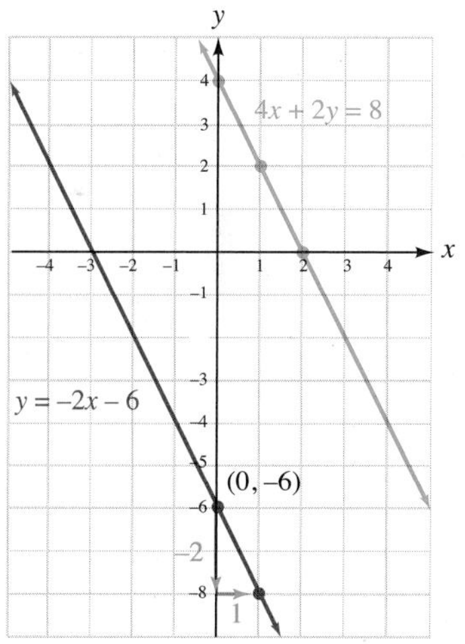

The lines in the graph appear to be parallel. We can verify this by writing the second equation in slope–intercept form and observing that the lines have the same slope, -2, and different y-intercepts, $(0, -6)$ and $(0, 4)$.

$y = -2x - 6 \qquad 4x + 2y = 8$

$2y = -4x + 8$ Subtract 4x from both sides.

$y = -2x + 4$ To isolate y, divide both sides by 2.

Different y-intercepts

Same slope

Caution

A common error is to graph the parallel lines, but forget to answer with the words *no solution.*

Because the lines are parallel, there is no point of intersection. Such a system has *no solution.* The solution set is the empty set, which is written as { } or as $\varnothing$.

Teaching Example 3 Solve the system of equations by graphing:

$\begin{cases} y = \frac{1}{2}x \\ x - 2y = 4 \end{cases}$

Answer:
No solution

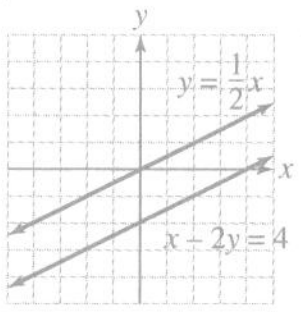

Self Check 3 Solve the system of equations by graphing: $\begin{cases} y = \frac{3}{2}x \\ 3x - 2y = 6 \end{cases}$

No solution See AIE Appendix 3.

Now Try Problem 29

Some systems of equations have infinitely many solutions.

EXAMPLE 4 Solve the system of equations by graphing: $\begin{cases} y = 2x + 4 \\ 4x + 8 = 2y \end{cases}$

Strategy We will graph both equations on the same coordinate system.

Why If both equations are graphed on the same coordinate system, we will be able to see if they have any solutions in common.

Solution To graph $y = 2x + 4$, we use the slope and y-intercept, and to graph $4x + 8 = 2y$, we use the intercept method.

$y = 2x + 4$: $m = 2$, $b = 4$

Slope: $\frac{\text{Rise}}{\text{Run}} = \frac{2}{1}$

y-intercept: (0, 4)

$4x + 8 = 2y$

x	y	(x, y)
0	4	(0, 4)
−2	0	(−2, 0)
−3	−2	(−3, −2)

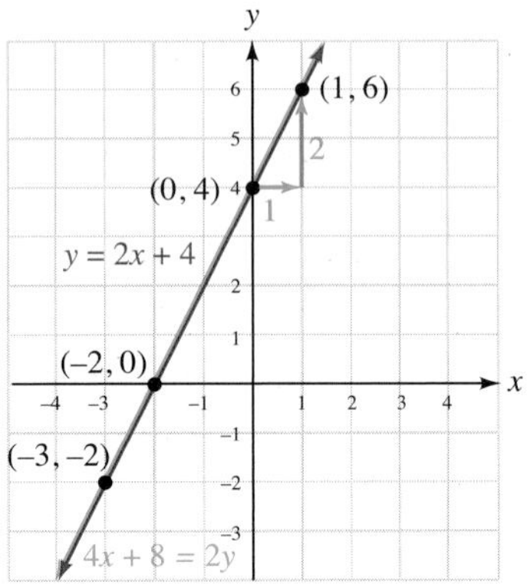

The Language of Algebra

The graphs of these lines **coincide.** That is, they occupy the same location. To illustrate this concept, think of a clock. At noon and at midnight, the hands of the clock coincide.

The graphs appear to be identical. We can verify this by writing the second equation in slope–intercept form and observing that it is the same as the first equation.

$y = 2x + 4$ The first equation.

$$4x + 8 = 2y \quad \text{The second equation.}$$
$$2y = 4x + 8 \quad \text{Reverse the sides.}$$
$$\frac{2y}{2} = \frac{4x}{2} + \frac{8}{2} \quad \text{To isolate } y \text{, divide both sides by 2.}$$
$$y = 2x + 4$$

This confirms that $y = 2x + 4$ and $4x + 8 = 2y$ are different forms of the same equation. Thus, the equations of this system are equivalent and their graphs are indeed identical.

Since the graphs are the same line, they have infinitely many points in common. All of the points that lie on the common line are solutions because the coordinates of each of those points satisfy both equations of the system. In cases like this, we say that there are *infinitely many solutions.*

From the graph, it appears that four of the infinitely many solutions are (−3, −2), (−2, 0), (0, 4), and (1, 6). Checks for two of these ordered pairs follow.

Check (−3, −2):

$$4x + 8 = 2y \qquad y = 2x + 4$$
$$4(-3) + 8 \stackrel{?}{=} 2(-2) \qquad -2 \stackrel{?}{=} 2(-3) + 4$$
$$-12 + 8 \stackrel{?}{=} -4 \qquad -2 \stackrel{?}{=} -6 + 4$$
$$-4 = -4 \qquad -2 = -2$$

Check (0, 4):

$$4x + 8 = 2y \qquad y = 2x + 4$$
$$4(0) + 8 \stackrel{?}{=} 2(4) \qquad 4 \stackrel{?}{=} 2(0) + 4$$
$$0 + 8 \stackrel{?}{=} 8 \qquad 4 \stackrel{?}{=} 0 + 4$$
$$8 = 8 \qquad 4 = 4$$

Caution

A common error is to graph the identical lines, but forget to answer with the words *infinitely many solutions.*

Self Check 4 Solve the system of equations by graphing: $\begin{cases} 6x - 4 = 2y \\ y = 3x - 2 \end{cases}$

Now Try Problem 33

Infinitely many solutions See AIE Appendix 3.

Teaching Example 4 Solve the system of equations by graphing: $\begin{cases} x + 3y = 6 \\ y = -\frac{1}{3}x + 2 \end{cases}$

Answer:
Infinitely many solutions

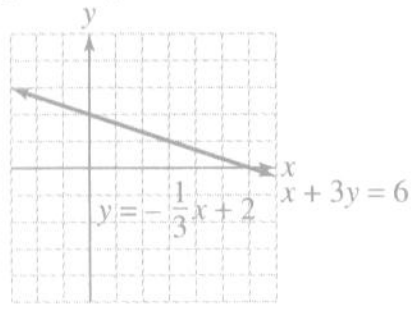

In Examples 2 and 3, the graphs of the equations of the system were different lines. We call equations with different graphs **independent equations.** The equations in Example 4 have the same graph and are equivalent. Because they are different forms of the same equation, they are called **dependent equations.**

There are three possible outcomes when we solve a system of two linear equations using the graphing method.

Solving Systems by Graphing

The two lines intersect at one point.	The two lines are parallel.	The two lines are identical.
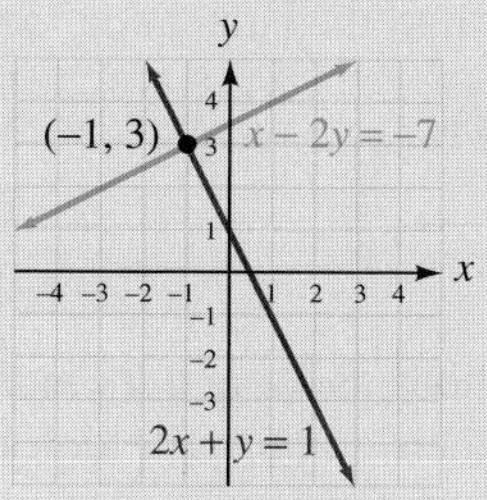	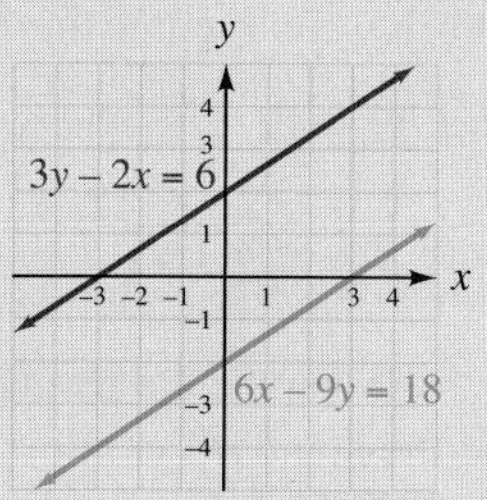	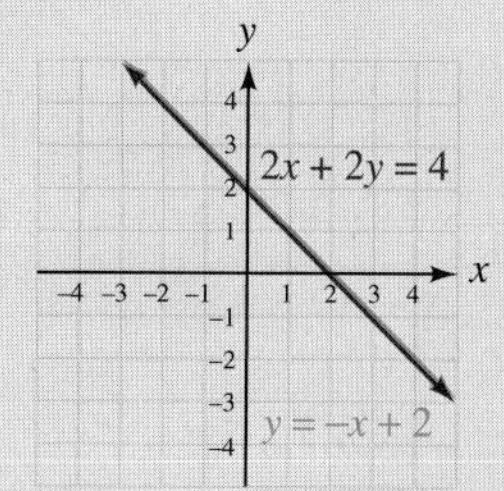
Exactly one solution (the point of intersection)	No solution	Infinitely many solutions (any point on the line is a solution)
Consistent system ***Independent equations***	***Inconsistent system*** ***Independent equations***	***Consistent system*** ***Dependent equations***

4 Identify the Number of Solutions of a Linear System Without Graphing.

We can determine the number of solutions that a system of two linear equations has by writing each equation in slope–intercept form.

- If the lines have different slopes, they intersect, and the system has one solution. (See Example 2.)
- If the lines have the same slope and different y-intercepts, they are parallel, and the system has no solution. (See Example 3.)
- If the lines have the same slope and same y-intercept, they are the same line, and the system has infinitely many solutions. (See Example 4.)

EXAMPLE 5 Without graphing, determine the number of solutions of: $\begin{cases} 5x + y = 5 \\ 3x + 2y = 8 \end{cases}$

Strategy We will write both equations in slope–intercept form.

Why We can determine the number of solutions of a linear system by comparing the slopes and y-intercepts of the graphs of the equations.

Solution To write each equation in slope–intercept form, we solve for y.

$$5x + y = 5 \quad \text{The first equation.}$$
$$y = -5x + 5$$

$$3x + 2y = 8 \quad \text{The second equation.}$$
$$2y = -3x + 8$$
$$y = -\frac{3}{2}x + 4$$

Different slopes

Since the slopes are different, the lines are neither parallel nor identical. Therefore, they will intersect at one point and the system has one solution.

Self Check 5 Without graphing, determine the number of solutions of:
$\begin{cases} 3x + 6y = 1 \\ 2x + 4y = 0 \end{cases}$ No solution

Now Try Problem 41

Teaching Example 5 Without graphing, determine the number of solutions of:

a. $\begin{cases} 2x + 3y = 6 \\ 7x + 5y = -10 \end{cases}$

b. $\begin{cases} 2x + y = 4 \\ 3y = -6x + 12 \end{cases}$

Answers:
a. One solution
b. Infinitely many solutions

5 Use a Graphing Calculator to Solve a Linear System (Optional).

A graphing calculator can be used to solve systems of equations, such as

$$\begin{cases} 2x + y = 12 \\ 2x - y = -2 \end{cases}$$

Before we can enter the equations into the calculator, we must solve them for y.

$2x + y = 12$ *The first equation.*

$y = -2x + 12$

$2x - y = -2$ *The second equation.*

$-y = -2x - 2$

$y = 2x + 2$

We enter the resulting equations as Y_1 and Y_2 and graph them on the same axes. If we use the standard window setting, their graphs will look like figure (a).

To find the solution of the system, we can use the INTERSECT feature found on most graphing calculators. With this feature, after pushing enter three times to identify each graph and a guess for the point, the cursor automatically moves to the point of intersection of the graphs and displays the coordinates of that point. In figure (b), we see that the solution is (2.5, 7).

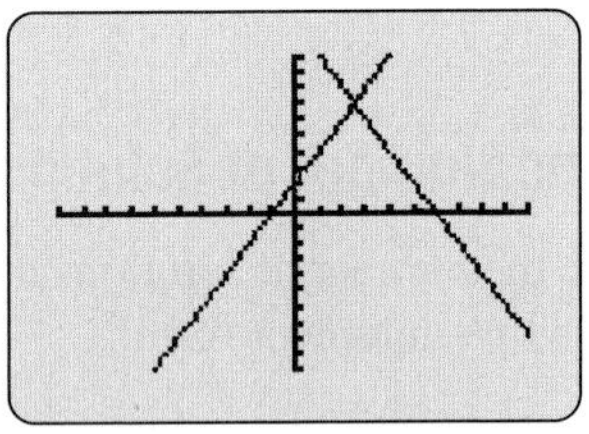

(a)

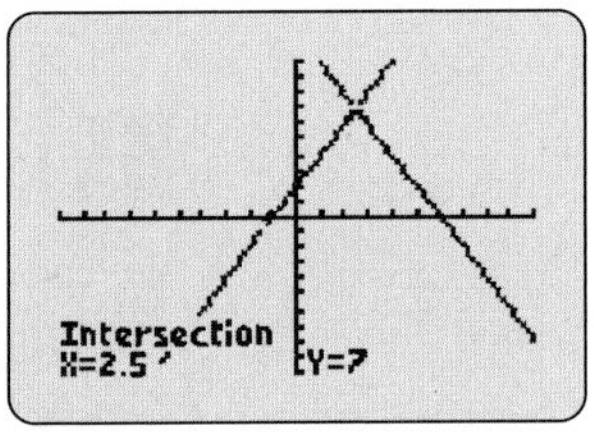

(b)

SECTION 4.1 STUDY SET

VOCABULARY

Fill in the blanks.

1. The pair of equations $\begin{cases} x - y = -1 \\ 2x - y = 1 \end{cases}$ is called a <u>system</u> of linear equations.
2. Because the ordered pair (2, 3) satisfies both equations in Problem 1, it is called a <u>solution</u> of the system of equations.
3. The point of <u>intersection</u> of the lines graphed in part (a) below is (1, 2).

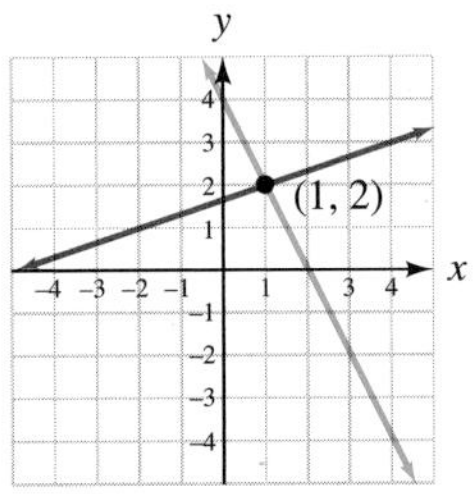

(a)

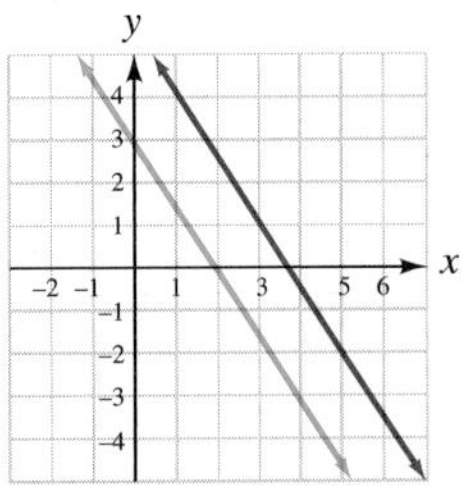

(b)

4. The lines graphed in part (b) above do not intersect. They are <u>parallel</u> lines.
5. A system of equations that has at least one solution is called a <u>consistent</u> system. A system with no solution is called an <u>inconsistent</u> system.
6. We call equations with different graphs <u>independent</u> equations. Because <u>dependent</u> equations are different forms of the same equation, they have the same graph.

CONCEPTS

7. Refer to the illustration below.
   a. If the coordinates of point A are substituted into the equation for Line 1, will the result be true or false? True
   b. If the coordinates of point C are substituted into the equation for Line 1, will the result be true or false? True
8. Refer to the illustration on the right.
   a. If the coordinates of point C are substituted into the equation for Line 2, will the result be true or false? True
   b. If the coordinates of point B are substituted into the equation for Line 1, will the result be true or false? False

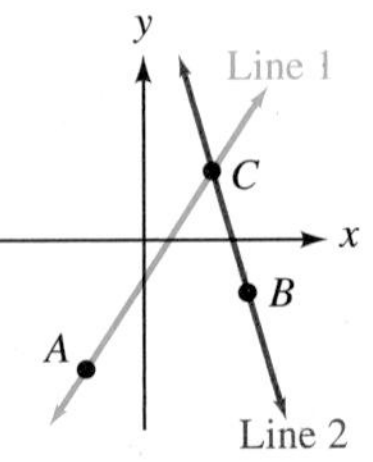

▶ 9. a. To graph $5x - 2y = 10$, we can use the intercept method. Complete the table.

x	y
0	−5
2	0

b. To graph $y = 3x - 2$, we can use the slope and y-intercept. Fill in the blanks.

Slope: $3 = \frac{3}{1}$ y-intercept: $(0, -2)$

10. What is the apparent solution of the system graphed in figure (a) below? Is the system consistent or inconsistent?
$(2, -3)$; consistent

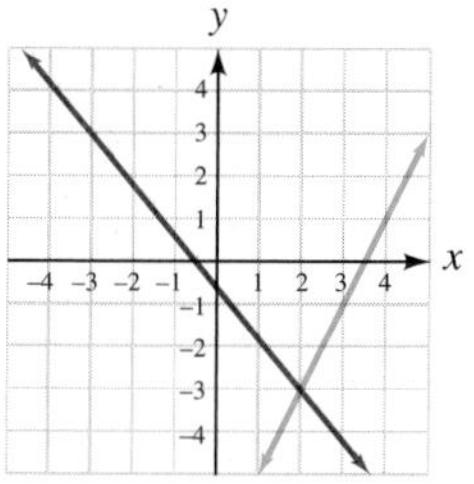

(a)

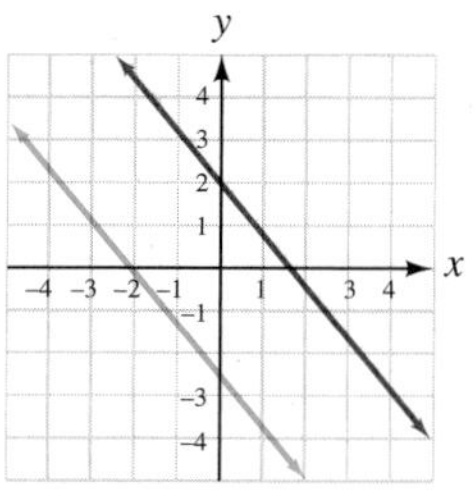

(b)

▶ 11. How many solutions does the system graphed in figure (b) above have? Are the equations dependent or independent?
No solution; independent

12. How many solutions does the system graphed on the right have? Give three of the solutions. Is the system consistent or inconsistent?
Infinitely many solutions; $(-3, 0)$, $(-2, -2)$, $(0, -6)$ (answers may vary); consistent

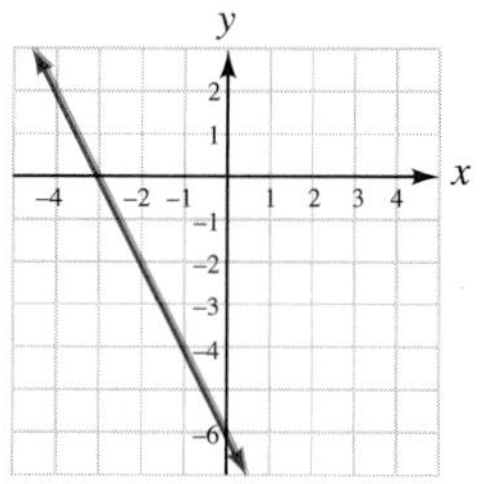

GUIDED PRACTICE

Determine whether the ordered pair is a solution of the given system of equations. **See Example 1.**

13. $(1, 1)$, $\begin{cases} x + y = 2 \\ 2x - y = 1 \end{cases}$ A solution

▶ 14. $(1, 3)$, $\begin{cases} 2x + y = 5 \\ 3x - y = 0 \end{cases}$ A solution

▶ 15. $(3, -2)$, $\begin{cases} 2x + y = 4 \\ y = 1 - x \end{cases}$ A solution

▶ 16. $(-2, 4)$, $\begin{cases} 2x + 2y = 4 \\ 3y = 10 - x \end{cases}$ A solution

17. $(12, 0)$, $\begin{cases} x - 9y = 12 \\ y = 10 - x \end{cases}$ Not a solution

▶ 18. $(15, 0)$, $\begin{cases} x - 2y = 15 \\ y = 16 - x \end{cases}$ Not a solution

19. $(-2, -4)$, $\begin{cases} 4x + 5y = -23 \\ -3x + 2y = 0 \end{cases}$ Not a solution

▶ 20. $(-5, 2)$, $\begin{cases} -2x + 7y = 17 \\ 3x - 4y = -19 \end{cases}$ Not a solution

▶ 21. $\left(\frac{1}{2}, 3\right)$, $\begin{cases} 2x + y = 4 \\ 4x - 11 = 3y \end{cases}$ Not a solution

22. $\left(2, \frac{1}{3}\right)$, $\begin{cases} x - 3y = 1 \\ -2x + 6 = -6y \end{cases}$ Not a solution

23. $(2.5, 3.5)$, $\begin{cases} 4x - 3 = 2y \\ 4y + 1 = 6x \end{cases}$ A solution

▶ 24. $(0.2, 0.3)$, $\begin{cases} 20x + 10y = 7 \\ 20y = 15x + 3 \end{cases}$ A solution

Solve each system of equations by graphing. **See Example 2.** See AIE Appendix 3.

25. $\begin{cases} 2x + 3y = 12 \\ 2x - y = 4 \end{cases}$ $(3, 2)$

▶ 26. $\begin{cases} 5x + y = 5 \\ 5x + 3y = 15 \end{cases}$ $(0, 5)$

▶ 27. $\begin{cases} x + y = 4 \\ x - y = -6 \end{cases}$ $(-1, 5)$

▶ 28. $\begin{cases} x + y = 4 \\ x - y = -2 \end{cases}$ $(1, 3)$

Solve each system by graphing. If a system has no solution or infinitely many solutions, so state. **See Example 3.** See AIE Appendix 3.

▶ 29. $\begin{cases} y = -\frac{1}{3}x - 4 \\ x + 3y = 6 \end{cases}$ No solution

30. $\begin{cases} y = -\frac{1}{2}x - 3 \\ x + 2y = 2 \end{cases}$ No solution

31. $\begin{cases} y = 3x \\ y - 3x = -3 \end{cases}$ No solution

▶ 32. $\begin{cases} y = -2x \\ 2x + y = -2 \end{cases}$ No solution

Solve each system by graphing. If a system has no solution or infinitely many solutions, so state. **See Example 4.** See AIE Appendix 3.

33. $\begin{cases} y = x - 1 \\ 3x - 3y = 3 \end{cases}$ Infinitely many solutions

▶ 34. $\begin{cases} y = -x + 1 \\ 4x + 4y = 4 \end{cases}$ Infinitely many solutions

35. $\begin{cases} 4x + 6y = 12 \\ 2x + 3y = 6 \end{cases}$ Infinitely many solutions

▶ 36. $\begin{cases} 2x - y = 0 \\ 2y - 4x = 0 \end{cases}$ Infinitely many solutions

Find the slope and the y-intercept of the graph of each line in the system of equations. Then, use that information to determine the number of solutions of the system. **See Example 5.**

37. $\begin{cases} y = 6x - 7 \\ y = -2x + 1 \end{cases}$ 1 solution

▶ 38. $\begin{cases} y = \frac{1}{2}x + 8 \\ y = 4x - 10 \end{cases}$ 1 solution

▶ 39. $\begin{cases} 3x - y = -3 \\ y - 3x = 3 \end{cases}$ Infinitely many solutions

40. $\begin{cases} x + 4y = 4 \\ 12y = 12 - 3x \end{cases}$ Infinitely many solutions

▶ 41. $\begin{cases} x + y = 6 \\ x + y = 8 \end{cases}$ No solution

▶ 42. $\begin{cases} 5x + y = 0 \\ 5x + y = 6 \end{cases}$ No solution

▶ 43. $\begin{cases} 6x + y = 0 \\ 2x + 2y = 0 \end{cases}$ 1 solution

▶ 44. $\begin{cases} x + y = 1 \\ 2x - 2y = 5 \end{cases}$ 1 solution

Use a graphing calculator to solve each system. **See Objective 5.**

45. $\begin{cases} y = 4 - x \\ y = 2 + x \end{cases}$ $(1, 3)$

▶ 46. $\begin{cases} 3x - 6y = 4 \\ 2x + y = 1 \end{cases}$ $\left(\frac{2}{3}, -\frac{1}{3}\right)$

47. $\begin{cases} 6x - 2y = 5 \\ 3x = y + 10 \end{cases}$ No solution

▶ 48. $\begin{cases} x - 3y = -2 \\ 5x + y = 10 \end{cases}$ $(1.75, 1.25)$

TRY IT YOURSELF

Solve each system of equations by graphing.

▶ 49. $\begin{cases} y = 3x + 6 \\ y = -2x - 4 \end{cases}$ $(-2, 0)$

50. $\begin{cases} y = x + 3 \\ y = -2x - 3 \end{cases}$ $(-2, 1)$

51. $\begin{cases} 2y = 3x + 2 \\ 3x - 2y = 6 \end{cases}$ No solution

▶ 52. $\begin{cases} 3x - 6y = 18 \\ x = 2y + 3 \end{cases}$ No solution

53. $\begin{cases} x + y = 2 \\ y = x - 4 \end{cases}$ $(3, -1)$

▶ 54. $\begin{cases} x + y = 1 \\ y = x + 5 \end{cases}$ $(-2, 3)$

▶ 55. $\begin{cases} x = 3 \\ 3y = 6 - 2x \end{cases}$ $(3, 0)$

56. $\begin{cases} x = 4 \\ 2y = 12 - 4x \end{cases}$ $(4, -2)$

57. $\begin{cases} y = \frac{3}{4}x + 3 \\ y = -\frac{x}{4} - 1 \end{cases}$ (−4, 0)

58. $\begin{cases} y = \frac{2}{3}x + 4 \\ y = -\frac{x}{3} + 7 \end{cases}$ (3, 6)

▶ **59.** $\begin{cases} y = -x - 2 \\ y = -3x + 6 \end{cases}$ (4, −6)

▶ **60.** $\begin{cases} y = 2x - 4 \\ y = -5x + 3 \end{cases}$ (1, −2)

61. $\begin{cases} -x + 3y = -11 \\ 3x - y = 17 \end{cases}$ (5, −2)

62. $\begin{cases} 2x - 3y = -18 \\ 3x + 2y = -1 \end{cases}$ (−3, 4)

▶ **63.** $\begin{cases} x + y = 2 \\ y = x \end{cases}$ (1, 1)

▶ **64.** $\begin{cases} x + y = 4 \\ y = x \end{cases}$ (2, 2)

▶ **65.** $\begin{cases} 4x - 2y = 8 \\ y = 2x - 4 \end{cases}$ Infinitely many solutions

66. $\begin{cases} 2y = -6x - 12 \\ 3x + y = -6 \end{cases}$ Infinitely many solutions

▶ **67.** $\begin{cases} x + 4y = -2 \\ y = -x - 5 \end{cases}$ (−6, 1)

▶ **68.** $\begin{cases} 3x + 2y = -8 \\ 2x - 3y = -1 \end{cases}$ (−2, −1)

▶ **69.** $\begin{cases} y = -3 \\ -x + 2y = -4 \end{cases}$ (−2, −3)

▶ **70.** $\begin{cases} y = -4 \\ -2x - y = 8 \end{cases}$ (−2, −4)

71. $\begin{cases} x + 2y = -4 \\ x - \frac{1}{2}y = 6 \end{cases}$ (4, −4)

▶ **72.** $\begin{cases} \frac{2}{3}x - y = -3 \\ 3x + y = 3 \end{cases}$ (0, 3)

APPLICATIONS

▶ **73. Social Networks.** Refer to the graph below. In what month did Facebook and MySpace have the same number of unique visitors? Estimate the number. May '09; about 70 million

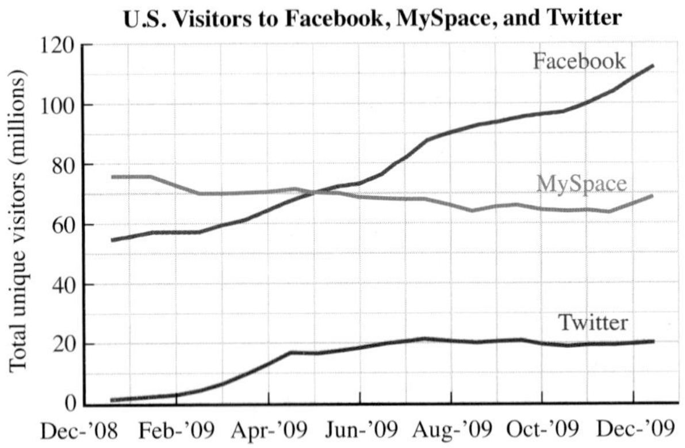

Source: Comscore Media Metrix

▶ **74. Beverages.** Refer to the graph below. In what year was average number of gallons of milk and carbonated soft drinks consumed per person the same? Estimate the number of gallons. 1975; about 29 gal

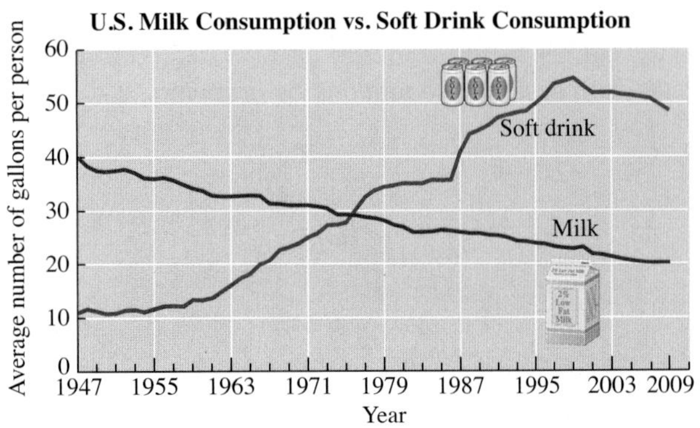

Source: USDA. Economic Research Service

▶ **75. Latitude and Longitude.** Refer to the following map.

a. Name three American cities that lie on a latitude line of 30°. Houston, New Orleans, St. Augustine

b. Name three American cities that lie on a longitude line of −90°. St. Louis, Memphis, New Orleans

c. What city lies on both lines? New Orleans

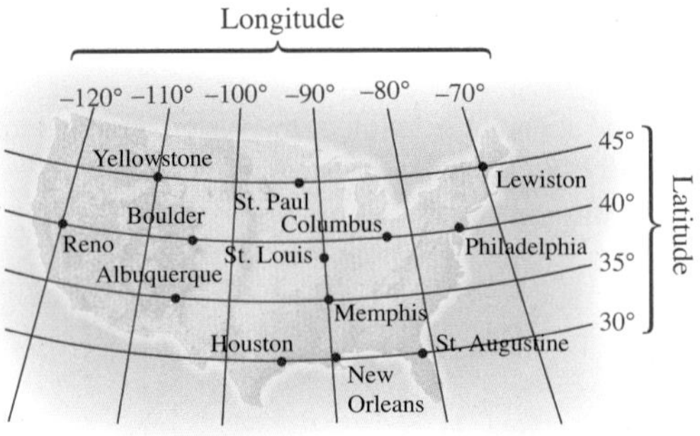

▶ **76. Economics.** The following graph illustrates the law of supply and demand.

a. Complete each sentence with the word *increases* or *decreases*. As the price of an item increases, the supply of the item increases. As the price of an item increases, the demand for the item decreases.

b. For what price will the supply equal the demand? How many items will be supplied for this price? $6; 30,000

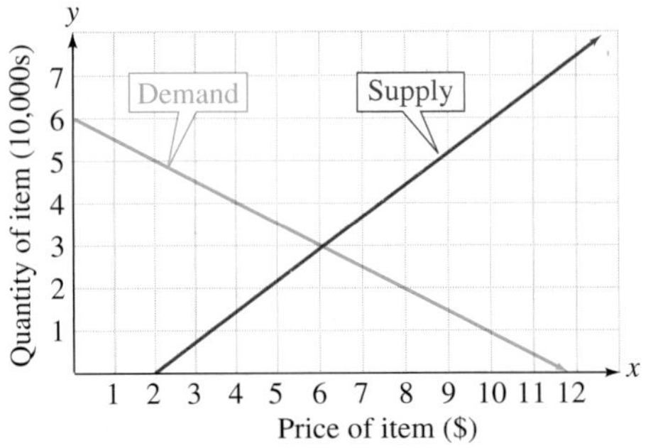

▶ **77. Daily Tracking Polls.** Refer to the graph below.

a. Which political candidate was ahead on October 28 and by how much? The incumbent; 7%

b. On what day did the challenger pull even with the incumbent? November 2

c. If the election was held November 4, who did the poll predict would win, and by how many percentage points? The challenger; 3

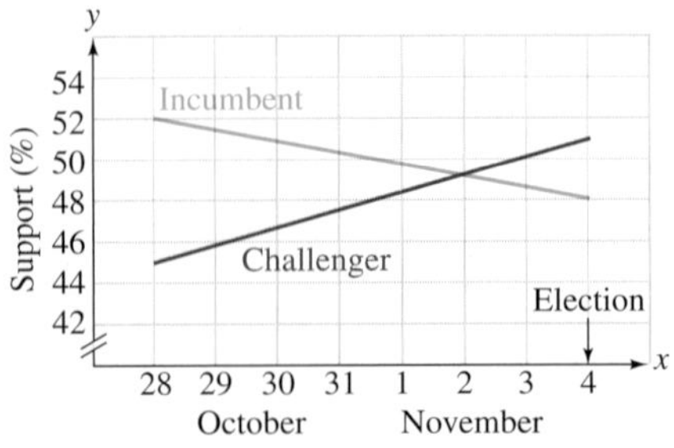

78. Air Traffic Control. The equations describing the paths of two airplanes are $y = -\frac{1}{2}x + 3$ and $3y = 2x + 2$. Graph each equation on the radar screen shown. Is there a possibility of a midair collision? If so, where? Yes; (2, 2)

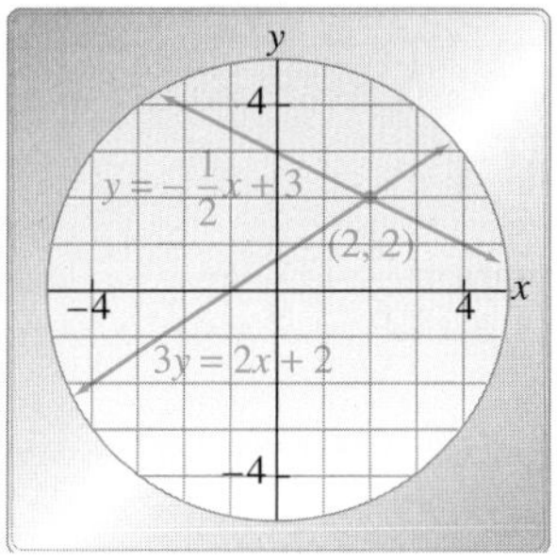

79. **from Campus to Careers**

Photographer

Photographers often use the *rule of thirds* to add more interest to a photo rather than simply centering the subject. They imagine two horizontal and two vertical lines dividing the image into a grid of 9 equal parts (like a tic-tac-toe game). The four points where the lines intersect are possible "sweet spots" for placing the subject of the picture. Give the coordinates of the four sweet spots for the image below. Is the sea gull properly placed? (6, 4), (6, 8), (12, 4), (12, 8); yes

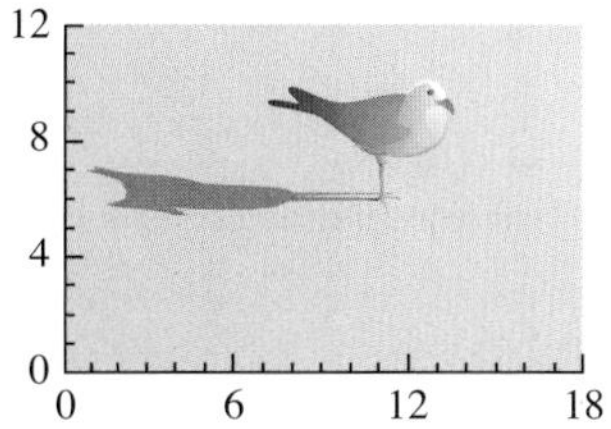

80. TV Coverage. A television camera is located at $(-2, 0)$ and will follow the launch of a space shuttle, as shown here. (Each unit in the illustration is 1 mile.) As the shuttle rises vertically on a path described by $x = 2$, the farthest the camera can tilt back is a line of sight given by $y = \frac{5}{2}x + 5$. For how many miles of the shuttle's flight will it be in view of the camera? 10 mi

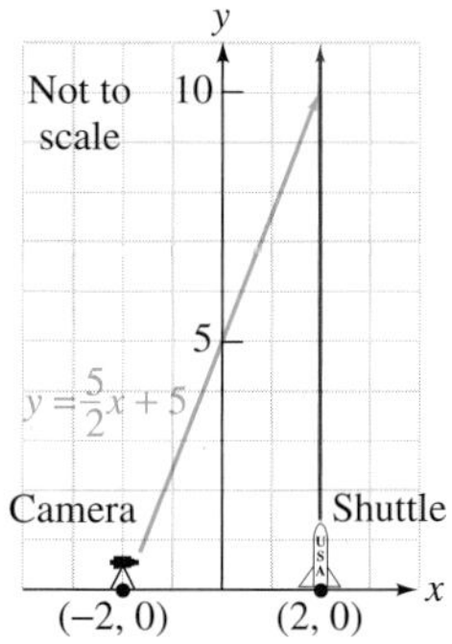

WRITING

81. Explain why it is difficult to determine the solution of the system in the graph.

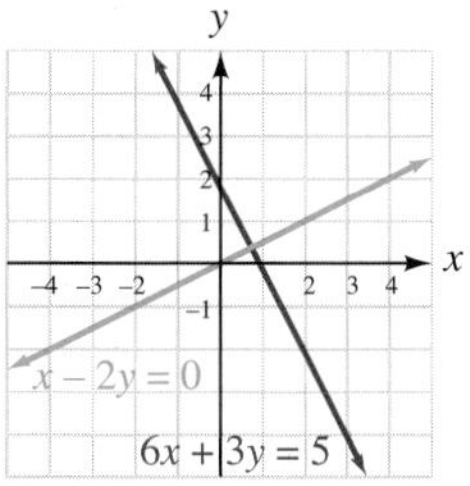

82. Without graphing, how can you tell that the graphs of $y = 2x + 1$ and $y = 3x + 2$ intersect?

83. Could a system of two linear equations have exactly two solutions? Explain why or why not.

84. What is an inconsistent system? What are dependent equations?

85. Suppose the graphs of the two linear equations of a system are the same line. What is wrong with the following statement? *The system has infinitely many solutions. Any ordered pair is a solution of the system.*

86. The Swine Flu. The graph below is for the month of January, 2010. It shows the total number of doses of the H1N1 Swine Flu vaccine that had been produced to date. It also shows the total number of doses that U.S. health officials had ordered to date. Did the total number of doses ordered ever equal or surpass the total number of doses produced? How can you tell?

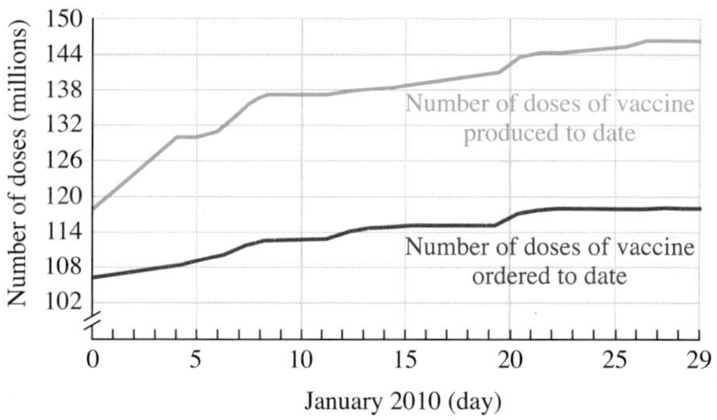

REVIEW

Solve each inequality. Write the solution set in interval notation and graph it. See AIE Appendix 3.

87. $-4(2y + 2) \le 4y + 28$ $[-3, \infty)$

88. $-5 < 3t + 4 \le 13$ $(-3, 3]$

89. $-1 \le -\frac{1}{2}n$ $(-\infty, 2]$

90. $\frac{1}{3} + \frac{c}{5} > -\frac{3}{2}$ $\left(-\frac{55}{6}, \infty\right)$

CHALLENGE PROBLEMS

91. Can a system of two linear equations in two variables be inconsistent but have dependent equations? Explain. No

92. Construct a system of two linear equations that has a solution of $(-2, 6)$. $\begin{cases} x + y = 4 \\ x - y = -8 \end{cases}$ (answers may vary)

93. Write a system of two linear equations such that (2, 3) is a solution of the first equation but is not a solution of the second equation. $\begin{cases} x + y = 5 \\ x - y = 10 \end{cases}$ (answers may vary)

94. Solve by graphing: $\begin{cases} \frac{1}{3}x - \frac{1}{2}y = \frac{1}{6} \\ \frac{2x}{5} + \frac{y}{2} = \frac{13}{10} \end{cases}$ (2, 1) See AIE Appendix 3.

SECTION 4.2

Solving Systems of Equations by Substitution

OBJECTIVES

1 Solve systems of linear equations by substitution.

2 Find a substitution equation.

3 Solve systems of linear equations that contain fractions.

4 Use substitution to identify inconsistent systems and dependent equations.

ARE YOU READY? *Are You Ready? exercises available online at www.webassign.net/brookscole*

The following problems review some basic skills that are needed when solving systems of equations by substitution.

1. In $8x + y = 2$, what is the coefficient of y? 1
2. Solve $2x - y = -3$ for y. $y = 2x + 3$
3. What is the LCD for the fractions in the equation $\frac{x}{6} + \frac{y}{9} = \frac{3}{2}$? 18
4. Substitute 4 for x in $y = 5x - 3$ and find y. 17
5. Multiply: $8\left(\frac{3x}{8}\right)$ $3x$

When solving a system of equations by graphing, it is often difficult to determine the coordinates of the intersection point. For example, a solution of $\left(\frac{7}{8}, \frac{3}{5}\right)$ would be almost impossible to identify accurately. In this section, we will discuss a second, more precise method for solving systems that does not involve graphing.

1 Solve Systems of Linear Equations by Substitution.

One algebraic method for solving a system of equations is the **substitution method.** It is introduced in the following example.

EXAMPLE 1 Solve the system: $\begin{cases} y = 3x - 2 \\ 2x + y = 8 \end{cases}$

Strategy Note that the first equation is solved for y. Because y and $3x - 2$ are equal (represent the same value), we will substitute $3x - 2$ for y in the second equation.

Why The objective is to obtain one equation containing only one unknown. When $3x - 2$ is substituted for y in the second equation, the result will be just that—an equation in one variable, x.

Solution Since the right side of $y = 3x - 2$ is used to make a substitution, $y = 3x - 2$ is called the **substitution equation.**

$$\begin{cases} y = 3x - 2 \\ 2x + y = 8 \end{cases}$$

To find the solution of the system, we proceed as follows:

$2x + y = 8$ This is the second equation of the system. It has two variables.

$2x + 3x - 2 = 8$ Substitute $3x - 2$ for y. This equation has only one variable.

The resulting equation can be solved for x.

$2x + 3x - 2 = 8$

$5x - 2 = 8$ Combine like terms: $2x + 3x = 5x$.

$5x = 10$ To isolate the variable term, $5x$, add 2 to both sides.

$x = 2$ Divide both sides by 5. This is the x-value of the solution.

Success Tip

Throughout the course, we have been substituting numbers for variables. With this method, we substitute a *variable expression for a variable.*

Caution

When using the substitution method, a common error is to find the value of one of the variables, say x, and forget to find the value of the other. Remember that a solution of a linear system of two equations is an ordered pair (x, y).

We can find the y-value of the solution by substituting 2 for x in either equation of the original system. We will use the substitution equation because it is already solved for y.

$y = 3x - 2$ This is the substitution equation.

$y = 3(\mathbf{2}) - 2$ Substitute 2 for x.

$y = 6 - 2$ Do the multiplication.

$y = 4$ This is the y-value of the solution. We would have obtained the same result if we had substituted 2 for x in $2x + y = 8$ and solved for y.

The ordered pair (2, 4) appears to be the solution of the system. To check, we substitute 2 for x and 4 for y in each equation.

Check:

$y = 3x - 2$	The first equation.	$2x + y = 8$	The second equation.
$4 \stackrel{?}{=} 3(\mathbf{2}) - 2$		$2(\mathbf{2}) + \mathbf{4} \stackrel{?}{=} 8$	
$4 \stackrel{?}{=} 6 - 2$		$4 + 4 \stackrel{?}{=} 8$	
$4 = 4$	True	$8 = 8$	True

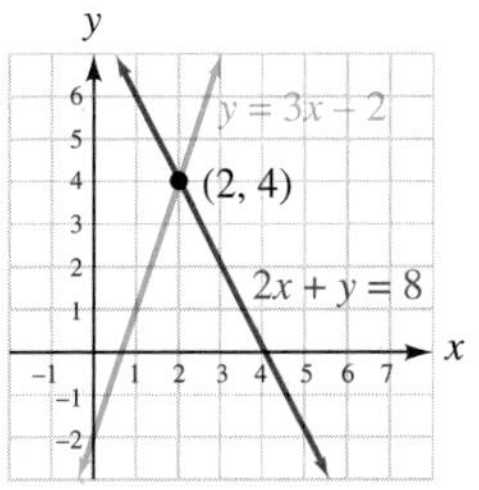

Since (2, 4) satisfies both equations, it is the solution. The solution set is written as $\{(2, 4)\}$. A graph of the equations of the system shows an intersection point of (2, 4). This illustrates an important fact: *The solution found using the substitution method will be the same as the solution found using the graphing method.*

Teaching Example 1 Solve the system:
$\begin{cases} y = 5x - 2 \\ 6x + y = 9 \end{cases}$
Answer: (1, 3)

Self Check 1 Solve the system: $\begin{cases} x + 4y = 7 \\ x = 6y - 3 \end{cases}$ (3, 1)

Now Try Problem 15

The substitution method works well for solving systems where one equation is solved, or can be easily solved, for one of the variables. To solve a system of equations in x and y by the substitution method, follow these steps.

The Substitution Method

1. Solve one of the equations for either x or y. If this is already done, go to step 2. (We call this equation the **substitution equation.**)
2. Substitute the expression for x or for y obtained in step 1 into the other equation and solve that equation.
3. Substitute the value of the variable found in step 2 into the substitution equation to find the value of the remaining variable.
4. Check the proposed solution in each equation of the original system. Write the solution as an ordered pair.

EXAMPLE 2 Solve the system: $\begin{cases} 4x + 27 = 7y \\ x = -5y \end{cases}$

Strategy We will use the substitution method to solve this system.

Why The substitution method works well when one of the equations of the system (in this case, $x = -5y$) is solved for a variable.

Solution

Step 1: Because x and $-5y$ represent the same value, we can substitute $-5y$ for x in the first equation.

Success Tip

The basic objective of this method is to use an appropriate substitution to obtain one equation in one variable.

$\begin{cases} 4x + 27 = 7y \\ x = \mathbf{-5y} \end{cases}$ This is the substitution equation.

Step 2: When we substitute $-5y$ for x in the first equation, the resulting equation contains only one variable, and it can be solved for y.

$4x + 27 = 7y$	This is the first equation of the system. It has two variables.
$4(\mathbf{-5y}) + 27 = 7y$	Substitute $-5y$ for x. Don't forget the parentheses. This equation has only one variable.
$-20y + 27 = 7y$	Do the multiplication.
$27 = 27y$	To eliminate $-20y$ on the left side, add $20y$ to both sides.
$1 = y$	Divide both sides by 27. This is the y-value of the solution.

Caution

We don't have to find the values of the variables in alphabetical order. Here, we found y first.

Step 3: To find x, substitute 1 for y in the equation $x = -5y$.

$x = -5y$	This is the substitution equation.
$x = -5(\mathbf{1})$	Substitute 1 for y.
$x = -5$	This is the x-value of the solution.

Step 4: The following check verifies that the solution is $(-5, 1)$.

Check:

$4x + 27 = 7y$	The first equation.	$x = -5y$	The second equation.
$4(\mathbf{-5}) + 27 \stackrel{?}{=} 7(\mathbf{1})$		$\mathbf{-5} \stackrel{?}{=} -5(\mathbf{1})$	
$-20 + 27 \stackrel{?}{=} 7$		$-5 = -5$	True
$7 = 7$	True		

Teaching Example 2 Solve the system: $\begin{cases} 3x - 12 = 6y \\ x = -2y \end{cases}$ *Answer:* $(2, -1)$

Self Check 2 Solve the system: $\begin{cases} 3x + 40 = 8y \\ x = -4y \end{cases}$ $(-8, 2)$

Now Try Problem 17

2 Find a Substitution Equation.

Sometimes neither equation of a system is solved for a variable. In such cases, we can find a substitution equation by solving one of the equations for one of its variables.

EXAMPLE 3 Solve the system: $\begin{cases} 4x + y = 3 \\ 3x + 5y = 15 \end{cases}$

Teaching Tip: Before beginning the solution, have the students tell you the coefficients of each of the variable terms in both equations.

Strategy Since the system does not contain an equation solved for x or y, we must choose an equation and solve it for x or y. We will solve for y in the first equation, because y has a coefficient of 1. Then we will use the substitution method to solve the system.

Why Solving $4x + y = 3$ for x, or solving $3x + 5y = 15$ for x or y, would involve working with cumbersome fractions.

Solution

Success Tip

To find a substitution equation, solve one of the equations of the system for one of its variables. If possible, solve for a variable whose coefficient is 1 or -1 to avoid working with fractions.

Step 1: To find a substitution equation, we proceed as follows:

$4x + y = 3$	This is the first equation of the system. Think: $4x + 1y = 3$.
$y = 3 - 4x$	To isolate y, subtract $4x$ from both sides. This is the substitution equation.

Because y and $3 - 4x$ are equal, we can substitute $3 - 4x$ for y in the second equation of the system.

$$\begin{cases} 4x + y = 3 \quad \rightarrow y = \mathbf{3 - 4x} \\ 3x + 5y = 15 \end{cases}$$

Caution

Here, use parentheses when substituting $3 - 4x$ for y so that the multiplication by 5 is distributed over both terms of $3 - 4x$.

$$3x + 5(3 - 4x) = 15$$

Caution

Always use the original equations when checking a solution. Do not use a substitution equation or an equivalent equation that you found algebraically. If an error was made, a proposed solution that would not satisfy the original system might appear to be correct.

Teaching Example 3 Solve the system:
$\begin{cases} 3x + 4y = -8 \\ 2x + y = 3 \end{cases}$
Answer: $(4, -5)$

Step 2: When we substitute for y in the second equation, the resulting equation contains only one variable and can be solved for x.

$3x + 5y = 15$	This is the second equation of the system. It has two variables.
$3x + 5(\mathbf{3 - 4x}) = 15$	Substitute $3 - 4x$ for y. Don't forget the parentheses.
$3x + 15 - 20x = 15$	Distribute the multiplication by 5.
$15 - 17x = 15$	Combine like terms.
$-17x = 0$	To isolate the variable term, $-17x$, subtract 15 from both sides.
$x = 0$	Divide both sides by -17. This is the x-value of the solution.

Step 3: To find y, substitute 0 for x in the equation $y = 3 - 4x$.

$y = 3 - 4x$	This is the substitution equation.
$y = 3 - 4(\mathbf{0})$	Substitute 0 for x.
$y = 3 - 0$	
$y = 3$	This is the y-value of the solution.

Step 4: The solution appears to be (0, 3). Check it in the original equations.

Check:

$4x + y = 3$	The first equation.	$3x + 5y = 15$	The second equation.
$4(\mathbf{0}) + \mathbf{3} \stackrel{?}{=} 3$		$3(\mathbf{0}) + 5(\mathbf{3}) \stackrel{?}{=} 15$	
$0 + 3 \stackrel{?}{=} 3$		$0 + 15 \stackrel{?}{=} 15$	
$3 = 3$	True	$15 = 15$	True

Self Check 3 Solve the system: $\begin{cases} 2x - 3y = 10 \\ 3x + y = 15 \end{cases}$ (5, 0)

Now Try Problem 21

EXAMPLE 4 Solve the system: $\begin{cases} 3a - 3b = 5 \\ 3 - a = -2b \end{cases}$

Strategy Since the coefficient of a in the second equation is -1, we will solve that equation for a. Then we will use the substitution method to solve the system.

Why If we solve for the variable with a numerical coefficient of -1, we can avoid having to work with fractions.

Solution ***Step 1:*** To find a substitution equation, we proceed as follows:

$3 - a = -2b$	This is the second equation of the system. Think: $3 - 1a = -2b$.
$-a = -2b - 3$	To isolate the variable term, $-a$, subtract 3 from both sides.

To obtain a on the left side, multiply both sides of the equation by -1.

$\mathbf{-1}(-a) = \mathbf{-1}(-2b - 3)$	Don't forget the parentheses.
$a = 2b + 3$	Do the multiplication. This is the substitution equation.

Caution

If you solve for a variable in the first equation, be sure to substitute into the second equation. If you solve for a variable in the second equation, be sure to substitute into the first equation.

Because a and $2b + 3$ represent the same value, we can substitute $2b + 3$ for a in the first equation.

$$\begin{cases} 3a - 3b = 5 \\ 3 - a = -2b \rightarrow a = 2b + 3 \end{cases}$$

Step 2: Substitute $2b + 3$ for a in the first equation and solve for b.

$3a - 3b = 5$	This is the first equation of the system. It has two variables.
$3(2b + 3) - 3b = 5$	Substitute $2b + 3$ for a. Don't forget the parentheses.
$6b + 9 - 3b = 5$	Distribute the multiplication by 3.
$3b + 9 = 5$	Combine like terms: $6b - 3b = 3b$.
$3b = -4$	To isolate the variable term, $3b$, subtract 9 from both sides.
$b = -\frac{4}{3}$	Divide both sides by 3. This is the b-value of the solution.

Step 3: To find a, substitute $-\frac{4}{3}$ for b in the equation $a = 2b + 3$.

$a = 2b + 3$	This is the substitution equation.
$a = 2\left(-\frac{4}{3}\right) + 3$	Substitute $-\frac{4}{3}$ for b.
$a = -\frac{8}{3} + \frac{9}{3}$	Do the multiplication. To add, we must write 3 as a fraction with a denominator of 3: $3 = \frac{9}{3}$.
$a = \frac{1}{3}$	Add the fractions. This is the a-value of the solution.

Step 4: The solution is $\left(\frac{1}{3}, -\frac{4}{3}\right)$. Check it in the original equations.

Notation

Unless told otherwise, list the values of the variables of a solution in *alphabetical* order. For example, if the equations of a system involve the variables a and b, write the solution in the form (a, b).

Teaching Example 4 Solve the system:
$$\begin{cases} 8a - b = -2 \\ 5a + 10b = 3 \end{cases}$$
Answer: $\left(-\frac{1}{5}, \frac{2}{5}\right)$

Self Check 4 Solve the system: $\begin{cases} 2s - t = 4 \\ 3s - 5t = 2 \end{cases}$ $\left(\frac{18}{7}, \frac{8}{7}\right)$

Now Try Problem 25

3 Solve Systems of Linear Equations that Contain Fractions.

It is usually helpful to clear any equations of fractions and combine any like terms before performing a substitution.

EXAMPLE 5 Solve the system: $\begin{cases} \frac{y}{4} = -\frac{x}{2} - \frac{3}{4} \\ 2x - y = -1 + y - x \end{cases}$

Strategy We will use properties of equality to write each equation of the system in simpler, equivalent form. Then we will use the substitution method to solve the resulting system.

Why The first equation will be easier to work with if we clear it of fractions. The second equation will be easier to work with if we eliminate the variable terms on the right side.

Solution We can clear the first equation of fractions by multiplying both sides by the LCD.

$\frac{y}{4} = -\frac{x}{2} - \frac{3}{4}$	This is the first equation of the system.
$4\left(\frac{y}{4}\right) = 4\left(-\frac{x}{2} - \frac{3}{4}\right)$	Multiply both sides by the LCD, 4. Don't forget the parentheses.
$4\left(\frac{y}{4}\right) = 4\left(-\frac{x}{2}\right) - 4\left(\frac{3}{4}\right)$	Distribute the multiplication by 4.
(1) $y = -2x - 3$	Simplify. Call this equation 1.

Notation

We number the equations (1) and (2) to help describe how the system is solved using the substitution method.

Notation

When a linear equation is written in standard $Ax + By = C$ form, the *variable terms* are on the left side of the equal symbol and the *constant* term is on the right side.

We can write the second equation of the system in standard $Ax + By = C$ form by adding x and subtracting y from both sides.

$$2x - y = -1 + y - x \quad \text{This is the second equation of the system.}$$
$$2x - y + x - y = -1 + y - x + x - y$$
$$(2) \quad 3x - 2y = -1 \quad \text{Combine like terms. Call this equation 2.}$$

Step 1: Equations 1 and 2 form an equivalent system, which has the same solution as the original one. To find x, we proceed as follows:

$$\begin{aligned}(1)\quad & \begin{cases} y = -2x - 3 \\ 3x - 2y = -1 \end{cases} \\ (2)\quad & \end{aligned} \quad \text{This is the substitution equation.}$$

Step 2: To find x, substitute $-2x - 3$ for y in equation 2 and proceed as follows:

$$3x - 2y = -1 \quad \text{This is equation 2. It has two variables.}$$
$$3x - 2(-2x - 3) = -1 \quad \text{Substitute } -2x - 3 \text{ for } y\text{. Don't forget the parentheses.}$$
$$3x + 4x + 6 = -1 \quad \text{Distribute the multiplication by } -2.$$
$$7x + 6 = -1 \quad \text{Combine like terms: } 3x + 4x = 7x.$$
$$7x = -7 \quad \text{To isolate the variable term, } 7x\text{, subtract 6 from both sides.}$$
$$x = -1 \quad \text{Divide both sides by 7. This is the } x\text{-value of the solution.}$$

Step 3: To find y, we substitute -1 for x in equation 1.

$$y = -2x - 3 \quad \text{This is equation 1.}$$
$$y = -2(-1) - 3 \quad \text{Substitute } -1 \text{ for } x.$$
$$y = 2 - 3 \quad \text{Do the multiplication.}$$
$$y = -1 \quad \text{Subtract. This is the } y\text{-value of the solution.}$$

Step 4: The solution is $(-1, -1)$. Check it in the original system.

Teaching Example 5 Solve the system:
$$\begin{cases} \frac{y}{18} = \frac{x}{3} - \frac{1}{9} \\ 3x + y = 14 - 2y + x \end{cases}$$
Answer: $(1, 4)$

Self Check 5 Solve the system: $\begin{cases} \frac{y}{6} = \frac{x}{3} + \frac{1}{2} \\ 2x - y = -3 + y - x \end{cases}$ $(-3, -3)$

Now Try ▶ Problem 29

4 Use Substitution to Identify Inconsistent Systems and Dependent Equations.

In the previous section, we solved inconsistent systems and systems of dependent equations graphically. We also can solve these systems using the substitution method.

EXAMPLE 6 Solve the system: $\begin{cases} 4y - 12 = x \\ y = \frac{1}{4}x \end{cases}$

Strategy We will use the substitution method to solve this system.

Why The substitution method works well when one of the equations of the system $\left(\text{in this case, } y = \frac{1}{4}x\right)$ is solved for a variable.

Solution To try to solve this system, substitute $\frac{1}{4}x$ for y in the first equation and solve for x.

$$4y - 12 = x \quad \text{This is the first equation of the system.}$$
$$4\left(\frac{1}{4}x\right) - 12 = x \quad \text{Substitute } \tfrac{1}{4}x \text{ for } y.$$
$$x - 12 = x \quad \text{Do the multiplication: } 4\left(\tfrac{1}{4}\right) = 1.$$
$$x - 12 - x = x - x \quad \text{To eliminate } x \text{ on the right side, subtract } x \text{ from both sides.}$$
$$-12 = 0 \quad \text{False}$$

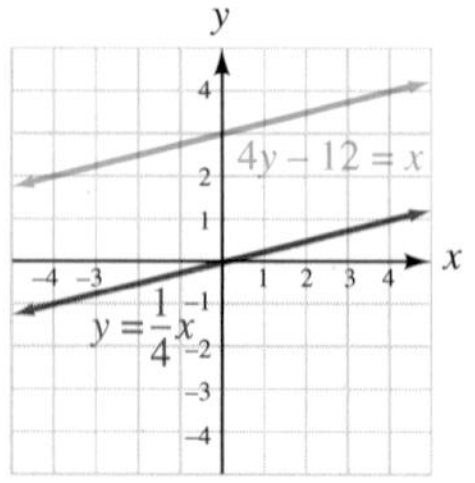

Here, the terms involving x drop out, and we get $-12 = 0$. This false statement indicates that the system has **no solution** and is inconsistent. The solution set is the empty set, $\varnothing$. The graphs of the equations of the system help to verify this; they are parallel lines.

Teaching Example 6 Solve the system:
$\begin{cases} 3y - 10 = x \\ y = \frac{1}{3}x \end{cases}$
Answer: No solution

Self Check 6 Solve the system: $\begin{cases} x - 4 = y \\ -2y = 4 - 2x \end{cases}$ No solution

Now Try Problem 33

EXAMPLE 7 Solve the system: $\begin{cases} x = -3y + 6 \\ 2x + 6y = 12 \end{cases}$

Strategy We will use the substitution method to solve this system.

Why The substitution method works well when one of the equations of the system (in this case, $x = -3y + 6$) is solved for a variable.

Solution To solve this system, substitute $-3y + 6$ for x in the second equation and solve for y.

$$2x + 6y = 12 \quad \text{This is the second equation of the system.}$$
$$2(-3y + 6) + 6y = 12 \quad \text{Substitute } -3y + 6 \text{ for } x. \text{ Don't forget the parentheses.}$$
$$-6y + 12 + 6y = 12 \quad \text{Distribute the multiplication by 2.}$$
$$12 = 12 \quad \text{True}$$

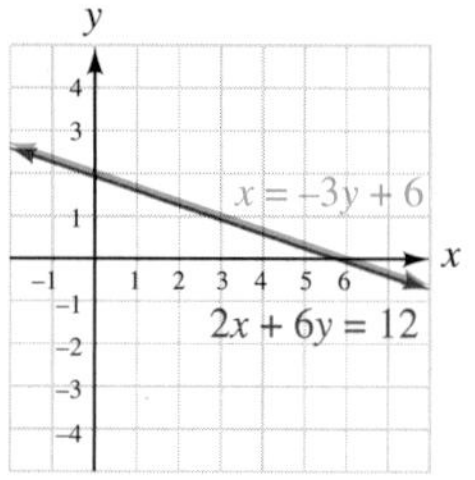

Here, the terms involving y drop out, and we get $12 = 12$. This true statement indicates that the two equations of the system are equivalent. Therefore, they are dependent equations and the system has **infinitely many solutions**. The graphs of the equations help to verify this; they are the same line.

Any ordered pair that satisfies one equation of this system also satisfies the other. To find several of the infinitely many solutions, we can substitute some values of x, say 0, 3, and 6, in either equation and solve for y. The results are: (0, 2), (3, 1), and (6, 0).

Teaching Example 7 Solve the system:
$\begin{cases} 4x + 3y = 5 \\ y = -\frac{4}{3}x + \frac{5}{3} \end{cases}$
Answer: Infinitely many solutions

Self Check 7 Solve the system: $\begin{cases} y = 2 - x \\ 3x + 3y = 6 \end{cases}$ Infinitely many solutions

Now Try Problem 37

Now we summarize the results from Examples 6 and 7. If both variables are eliminated in the substitution process, there are two possible outcomes.

- If the resulting equation is false, write "*no solution*" as the answer. The system is inconsistent. (*Caution:* Do not answer with "false.")
- If the resulting equation is true, write "*infinitely many solutions*" as the answer. The system's equations are dependent. (*Caution:* Do not answer with "true.")

SECTION 4.2 STUDY SET

VOCABULARY

Fill in the blanks.

1. To solve the system $\begin{cases} x = y + 1 \\ 3x + 2y = 8 \end{cases}$ using the method discussed in this section, we begin by substituting $y + 1$ for x in the second equation.
2. We say that the equation $y = 2x + 4$ is solved for y.

CONCEPTS

3. If the substitution method is used to solve $\begin{cases} 5x + y = 2 \\ y = -3x \end{cases}$, which equation should be used as the substitution equation? $y = -3x$
4. Suppose the substitution method will be used to solve $\begin{cases} x - 2y = 2 \\ 2x + 3y = 11 \end{cases}$. Find a substitution equation by solving one of the equations for one of the variables. $x = 2 + 2y$
5. Suppose $x - 4$ is substituted for y in the equation $x + 3y = 8$. Insert parentheses in $x + 3x - 4 = 8$ to show the substitution. $x + 3(x - 4) = 8$
6. Fill in the blank. With the substitution method, the objective is to use an appropriate substitution to obtain one equation in one variable.
7. A student uses the substitution method to solve the system $\begin{cases} 4a + 5b = 2 \\ b = 3a - 11 \end{cases}$ and finds that $a = 3$. What is the easiest way for her to determine the value of b? Substitute 3 for a in the second equation.
8. a. Clear $\frac{x}{5} + \frac{2y}{3} = 1$ of fractions. $3x + 10y = 15$
   b. Write $2x + y = x - 5y + 3$ in the form $Ax + By = C$. $x + 6y = 3$
9. Suppose $-2 = 1$ is obtained when a system is solved by the substitution method.
   a. Does the system have a solution? No
   b. Which of the following is a possible graph of the system? ii

i.
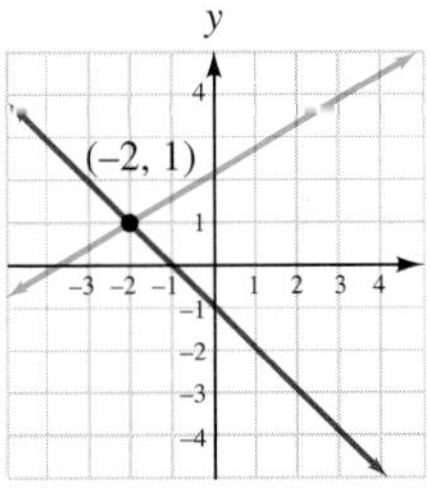

ii.
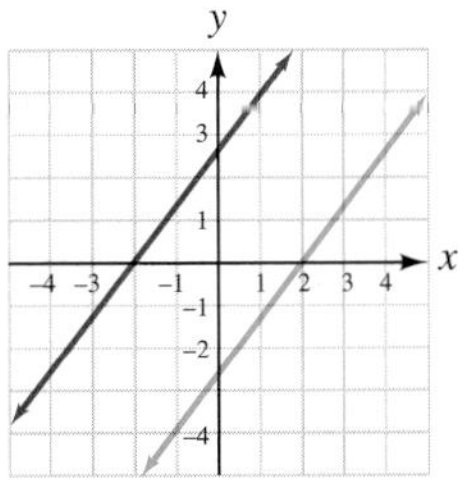

10. Suppose $2 = 2$ is obtained when a system is solved by the substitution method.
    a. Does the system have a solution? Yes
    b. Which graph below is a possible graph of the system? ii

i.
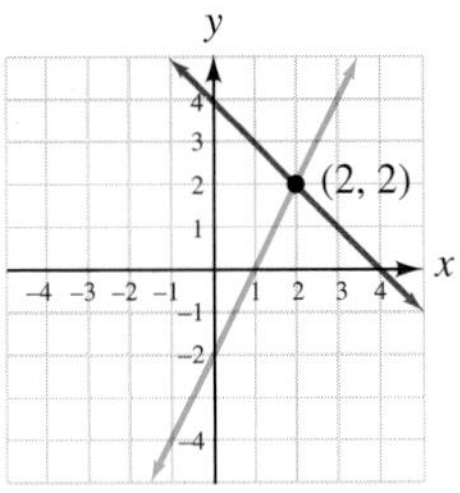

ii.
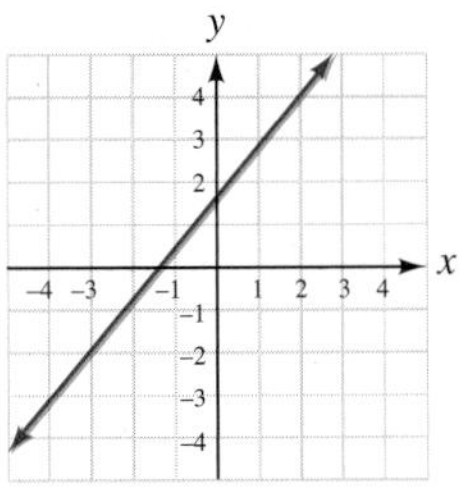

NOTATION

Complete the solution to solve the system.

11. Solve: $\begin{cases} y = 3x \\ x - y = 4 \end{cases}$

$x - y = 4$ *This is the second equation.*

$x - (3x) = 4$

$-2x = 4$

$x = -2$ *This is the x-value of the solution.*

$y = 3x$ *This is the first equation.*

$y = 3(-2)$

$y = -6$ *This is the y-value of the solution.*

The solution is $(-2, -6)$.

12. The system $\begin{cases} a = 3b + 2 \\ a + 3b = 8 \end{cases}$ was solved, and it was found that $b = 1$ and $a = 5$. Write the solution as an ordered pair. $(5, 1)$

GUIDED PRACTICE

Solve each system by substitution. **See Example 1.**

13. $\begin{cases} y = 2x \\ x + y = 6 \end{cases}$ $(2, 4)$
14. $\begin{cases} y = 3x \\ x + y = 4 \end{cases}$ $(1, 3)$
15. $\begin{cases} y = 2x - 6 \\ 2x + y = 6 \end{cases}$ $(3, 0)$
16. $\begin{cases} 2x + y = -7 \\ y = 3x + 8 \end{cases}$ $(-3, -1)$

Solve each system by substitution. **See Example 2.**

17. $\begin{cases} x + 3y = -4 \\ x = -5y \end{cases}$ $(-10, 2)$
18. $\begin{cases} x + 5y = -3 \\ x = -4y \end{cases}$ $(12, -3)$
19. $\begin{cases} y = -5x \\ x + 3y = -28 \end{cases}$ $(2, -10)$
20. $\begin{cases} y = -2x \\ 3x + 2y = -1 \end{cases}$ $(1, -2)$

Solve each system by substitution. **See Example 3.**

21. $\begin{cases} r + 3s = 9 \\ 3r + 2s = 13 \end{cases}$ $(3, 2)$
22. $\begin{cases} x - 2y = 2 \\ 2x + 3y = 11 \end{cases}$ $(4, 1)$
23. $\begin{cases} 4x + y = -15 \\ 2x + 3y = 5 \end{cases}$ $(-5, 5)$
24. $\begin{cases} 4x + y = -5 \\ 2x - 3y = -13 \end{cases}$ $(-2, 3)$

Solve each system by substitution. **See Example 4.**

25. $\begin{cases} 8x - 6y = 4 \\ 2x - y = -2 \end{cases}$ $(-4, -6)$
26. $\begin{cases} 5x + 4y = 0 \\ 2x - y = 0 \end{cases}$ $(0, 0)$
27. $\begin{cases} 4x + 5y = 2 \\ 3x - y = 11 \end{cases}$ $(3, -2)$
28. $\begin{cases} 5u + 3v = 5 \\ 4u - v = 4 \end{cases}$ $(1, 0)$

Solve each system by substitution. **See Example 5.**

29. $\begin{cases} \frac{x}{4} + \frac{y}{4} = -\frac{1}{2} \\ x - 2y = y - 24 - x \end{cases}$ $(-6, 4)$
30. $\begin{cases} 5a + 2b = 15 - 5a - b \\ \frac{a}{3} - \frac{2}{3}b = \frac{13}{3} \end{cases}$ $(3, -5)$

31. $\begin{cases} x - 5y = 20 - 4x - y \\ \dfrac{y}{3} = \dfrac{x}{2} - \dfrac{5}{2} \end{cases}$ $\left(10, \frac{15}{2}\right)$

▶ 32. $\begin{cases} x - 6y = 9 - 5x - 2y \\ \dfrac{x}{2} - \dfrac{3}{4} = 2y \end{cases}$ $\left(\frac{3}{2}, 0\right)$

***Solve each system by substitution.* See Example 6.**

33. $\begin{cases} 2a + 4b = -24 \\ a = 20 - 2b \end{cases}$ No solution

▶ 34. $\begin{cases} 8y = 15 - 4x \\ x + 2y = 4 \end{cases}$ No solution

35. $\begin{cases} 6 - y = 4x \\ 2y = -8x - 20 \end{cases}$ No solution

▶ 36. $\begin{cases} 2x - y = x + y \\ -2x + 4y = 6 \end{cases}$ No solution

***Solve each system by substitution.* See Example 7.**

▶ 37. $\begin{cases} y - 3x = -5 \\ 21x = 7y + 35 \end{cases}$ Infinitely many solutions

▶ 38. $\begin{cases} 3a + 6b = -15 \\ a = -2b - 5 \end{cases}$ Infinitely many solutions

39. $\begin{cases} x = -3y + 6 \\ 2x + 4y = 6 + x + y \end{cases}$ Infinitely many solutions

▶ 40. $\begin{cases} 9x = 3y + 12 \\ 4 = 3x - y \end{cases}$ Infinitely many solutions

TRY IT YOURSELF

Solve each system by substitution. If a system has no solution or infinitely many solutions, so state.

▶ 41. $\begin{cases} -y = 11 - 3x \\ 2x + 5y = -4 \end{cases}$ $(3, -2)$

▶ 42. $\begin{cases} -x = 10 - 3y \\ 2x + 8y = -6 \end{cases}$ $(-7, 1)$

43. $\begin{cases} \dfrac{x}{2} + \dfrac{y}{6} = \dfrac{2}{3} \\ \dfrac{x}{3} - \dfrac{y}{4} = \dfrac{1}{12} \end{cases}$ $(1, 1)$

44. $\begin{cases} \dfrac{c}{2} + \dfrac{d}{14} = 1 \\ \dfrac{c}{5} - \dfrac{d}{2} = -\dfrac{33}{10} \end{cases}$ $(1, 7)$

▶ 45. $\begin{cases} y + x = 2x + 2 \\ 6x - 4y = 21 - y \end{cases}$ $(9, 11)$

▶ 46. $\begin{cases} y - x = 3x \\ 2x + 2y = 14 - y \end{cases}$ $(1, 4)$

47. $\begin{cases} y - 4 = 2x \\ y = 2x + 2 \end{cases}$ No solution

48. $\begin{cases} x + 3y = 6 \\ x = -3y + 6 \end{cases}$ Infinitely many solutions

▶ 49. $\begin{cases} a + b = 1 \\ a - 2b = -1 \end{cases}$ $\left(\frac{1}{3}, \frac{2}{3}\right)$

50. $\begin{cases} 2b - a = -1 \\ 3a + 10b = -1 \end{cases}$ $\left(\frac{1}{2}, -\frac{1}{4}\right)$

51. $\begin{cases} 5x = \dfrac{1}{2}y - 1 \\ \dfrac{1}{4}y = 10x - 1 \end{cases}$ $\left(\frac{1}{5}, 4\right)$

▶ 52. $\begin{cases} \dfrac{x}{4} + y = \dfrac{1}{4} \\ \dfrac{y}{2} + \dfrac{11}{20} = \dfrac{x}{10} \end{cases}$ $\left(3, -\frac{1}{2}\right)$

▶ 53. $\begin{cases} x + 2y = -6 \\ x = y \end{cases}$ $(-2, -2)$

▶ 54. $\begin{cases} y = 2x - 9 \\ x + 3y = 8 \end{cases}$ $(5, 1)$

▶ 55. $\begin{cases} b = \dfrac{2}{3}a \\ 8a - 3b = 3 \end{cases}$ $\left(\frac{1}{2}, \frac{1}{3}\right)$

56. $\begin{cases} a = \dfrac{2}{3}b \\ 9a + 4b = 5 \end{cases}$ $\left(\frac{1}{3}, \frac{1}{2}\right)$

▶ 57. $\begin{cases} 6x - 3y = 5 \\ x + 2y = 0 \end{cases}$ $\left(\frac{2}{3}, -\frac{1}{3}\right)$

▶ 58. $\begin{cases} 5s + 10t = 3 \\ 2s + t = 0 \end{cases}$ $\left(-\frac{1}{5}, \frac{2}{5}\right)$

59. $\begin{cases} 2x + 3 = -4y \\ x - 6 = -8y \end{cases}$ $\left(-4, \frac{5}{4}\right)$

60. $\begin{cases} 5y + 2 = -4x \\ x + 2y = -2 \end{cases}$ $(2, -2)$

61. $\begin{cases} 2x + 5y = -2 \\ y = -\dfrac{x}{2} \end{cases}$ $(4, -2)$

▶ 62. $\begin{cases} y = -\dfrac{x}{2} \\ 2x - 3y = -7 \end{cases}$ $(-2, 1)$

63. $\begin{cases} 3x + 4y = -19 \\ 2y - x = 3 \end{cases}$ $(-5, -1)$

▶ 64. $\begin{cases} 5x - 2y = -7 \\ 5 - y = -3x \end{cases}$ $(-3, -4)$

65. $\begin{cases} 3(x - 1) + 3 = 8 + 2y \\ 2(x + 1) = 8 + y \end{cases}$ $(4, 2)$

▶ 66. $\begin{cases} 4(x - 2) = 19 - 5y \\ 3(x - 2) - 2y = -y \end{cases}$ $(3, 3)$

67. $\begin{cases} x = \dfrac{1}{3}y - 1 \\ x = y + 5 \end{cases}$ $(-4, -9)$

68. $\begin{cases} x = \dfrac{1}{2}y + 2 \\ x = y - 6 \end{cases}$ $(10, 16)$

69. $\begin{cases} 2a - 3b = -13 \\ -b = -2a - 7 \end{cases}$ $(-2, 3)$

▶ 70. $\begin{cases} a - 3b = -1 \\ -b = -2a - 2 \end{cases}$ $(-1, 0)$

71. $\begin{cases} x = 7y - 10 \\ 2x - 14y + 20 = 0 \end{cases}$ Infinitely many solutions

72. $\begin{cases} y - 1 = 5x \\ 10x - 2y = 2 \end{cases}$ No solution

73. $\begin{cases} 4x + 1 = 2x + 5 + y \\ 2x + 2y = 5x + y + 6 \end{cases}$ $(-10, -24)$

▶ 74. $\begin{cases} 6x = 2(y + 20) + 5x \\ 5(x - 1) = 3y + 4(x + 10) \end{cases}$ $(30, -5)$

75. $\begin{cases} 2a + 3b = 7 \\ 6a - b = 1 \end{cases}$ $\left(\frac{1}{2}, 2\right)$

76. $\begin{cases} 3a + 5b = -6 \\ 5b - a = -3 \end{cases}$ $\left(-\frac{3}{4}, -\frac{3}{4}\right)$

77. $\begin{cases} 2x - 3y = -4 \\ x = -\dfrac{3}{2}y \end{cases}$ $\left(-1, \frac{2}{3}\right)$

78. $\begin{cases} x = -\dfrac{3}{8}y \\ 8x - 3y = 4 \end{cases}$ $\left(\frac{1}{4}, -\frac{2}{3}\right)$

79. $\begin{cases} \dfrac{9x}{7} - \dfrac{3y}{7} = \dfrac{12}{7} \\ y - 3x = -4 \end{cases}$ Infinitely many solutions

80. $\begin{cases} a = 9 - 2b \\ 2(a + b) = 13 + a \end{cases}$ No solution

▶ 81. $\begin{cases} 4x + 5y + 1 = -8 + 3x \\ x - 3y + 2 = -3 - x \end{cases}$ $(-4, -1)$

▶ 82. $\begin{cases} 6x + y = -8 + 3x - y \\ 3x + y = 2y + x - 3 \end{cases}$ $(-2, -1)$

APPLICATIONS

83. **Offroading.** The *angle of approach* indicates how steep of an incline a vehicle can drive up without damaging the front bumper. The *angle of departure* indicates a vehicle's ability to exit an incline without damaging the rear bumper. The angle of approach a and the departure angle d for an H3 Hummer are described by the system $\begin{cases} a + d = 77 \\ a = d + 3 \end{cases}$. Use substitution to solve the system. (Each angle is measured in degrees.)
Angle of approach: 40°; angle of departure: 37°

84. High School Sports. The equations shown in the following graph model the number of boys and girls taking part in high school soccer programs. In both models, x is the number of years after 2000, and y is the number of participants. If the trends continue, the graphs will intersect. Use the substitution method to predict the year when the number of boys and girls participating in high school soccer will be the same. 2019

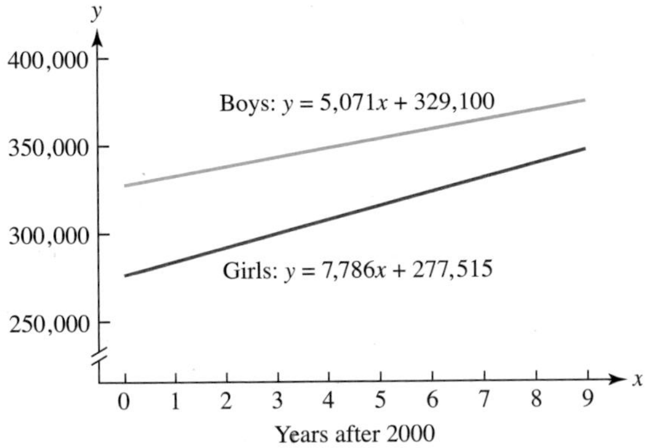

Source: National Federation of State High School Associations

85. Geometry. In the illustration, x and y represent the degree measures of angles. If $x + y = 90$ and $y = 3x$, find x and y. 22.5°, 67.5°

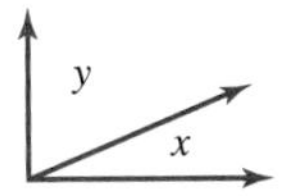

86. Geometry. In the illustration, x and y represent the degree measures of angles. If $x + y = 180$ and $x = 4y$, find x and y. 144°, 36°

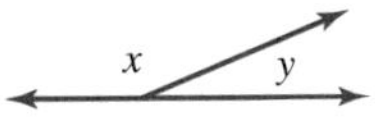

WRITING

87. What concept does this diagram illustrate?

$$\begin{cases} 6x + 5y = 11 \\ y = 3x - 2 \end{cases}$$

88. Explain the error.

Solve: $\begin{cases} 3a + 4b = 1 \\ a + 2b = 9 \end{cases}$

$a = 9 - 2b$ Solve for a in the second equation

$9 - 2b + 2b = 9$ Substitute $9 - 2b$ for a.

$9 = 9$ The b-terms drop out. True

The system has infinitely many solutions.

89. When using the substitution method, how can you tell whether

a. a system of linear equations has no solution?

b. a system of linear equations has infinitely many solutions?

90. When solving a system, what advantages are there with the substitution method compared with the graphing method?

91. Consider the equation $5x + y = 12$. Explain why it is easier to solve for y than it is for x.

92. Could the substitution method be used to solve the following system? Explain why or why not. If not, what method could be used? No; the graphing method

$$\begin{cases} y = -2 \\ x = 5 \end{cases}$$

REVIEW

93. Find the prime factorization of 189. $3^3 \cdot 7$

94. Complete each statement. For any nonzero number a,

a. $\frac{0}{a} =$ 0

b. $\frac{a}{0}$ is Undefined

95. Add: $\frac{5}{12} + \frac{1}{4}$ $\frac{2}{3}$

96. Divide: $\frac{1}{3} \div \frac{4}{5}$ $\frac{5}{12}$

CHALLENGE PROBLEMS

Use the substitution method to solve each system.

97. $\begin{cases} \frac{6x - 1}{3} - \frac{5}{3} = \frac{3y + 1}{2} \\ \frac{1 + 5y}{4} + \frac{x + 3}{4} = \frac{17}{2} \end{cases}$ (5, 5)

98. $\begin{cases} 0.5x + 0.5y = 6 \\ 0.001x - 0.001y = -0.004 \end{cases}$ (4, 8)

99. $\begin{cases} 2(2x + 3y) = 5 \\ 8x = 3(1 + 3y) \end{cases}$ $\left(\frac{3}{4}, \frac{1}{3}\right)$

100. The system $\begin{cases} \frac{1}{2}x = y + 3 \\ x - 2y = 6 \end{cases}$ has infinitely many solutions. Find three of them. (0, −3), (6, 0), (2, −2) (answers may vary)

SECTION 4.3 Solving Systems of Equations by Elimination (Addition)

OBJECTIVES

1. Solve systems of linear equations by the elimination method.
2. Use multiplication to eliminate a variable.
3. Use the elimination method twice to solve a system.
4. Use elimination to identify inconsistent systems and dependent equations.
5. Determine the most efficient method to use to solve a linear system.

ARE YOU READY? *Are You Ready? exercises available online at www.webassign.net/brookscole*

The following problems review some basic skills that are needed when solving systems of equations by elimination (addition).

1. In $4x - 6y = 7$, what is the *opposite* of the coefficient of y? 6
2. Add: $8x + (-8x)$ 0
3. Substitute 2 for y in $5x + 6y = 7$ and find x. -1
4. Multiply both sides of the equation $3x + 9y = 1$ by -4. $-12x - 36y = -4$

We have seen that graphing can be an inaccurate method for solving a system of equations because we must estimate the coordinates of the point of intersection. In Section 4.2, we learned that the substitution method is more precise. However, it too has its drawbacks. Recall that in the first step of the substitution method, we solve one of the equations for one of the variables. At times, this can be difficult, especially if none of the variables has a coefficient of 1 or -1. This is the case for the system

$$\begin{cases} 2x + 5y = 11 \\ 7x - 5y = 16 \end{cases}$$

Solving either equation for x or y involves working with cumbersome fractions. For example, if we solve the first equation for x, the resulting substitution equation is $x = \frac{11 - 5y}{2}$. Fortunately, we can solve systems like this one using an easier algebraic method called the **elimination** or the **addition method.**

1 Solve Systems of Linear Equations by the Elimination Method.

The elimination method for solving a system is based on the **addition property of equality:** *When equal quantities are added to both sides of an equation, the results are equal.* In symbols, if $A = B$ and $C = D$, then adding the left sides and the right sides of these equations, we have $A + C = B + D$. This procedure is called *adding the equations.*

Add the terms on the left sides.

$$\begin{array}{r} A = B \\ C = D \\ \hline A + C = B + D \end{array}$$

Add the terms on the right sides.

EXAMPLE 1 Solve the system: $\begin{cases} 2x + 5y = 11 \\ 6x - 5y = 13 \end{cases}$

Strategy Since the coefficients of the y-terms are opposites (5 and -5), we will add the left sides and the right sides of the given equations.

Why When we add the equations in this way, the result will be an equation that contains only one variable, x.

Solution Since $6x - 5y$ and 13 are equal quantities, we can add $6x - 5y$ to the left side and 13 to the right side of the first equation, $2x + 5y = 11$.

$$\begin{array}{rl} 2x + 5y &= 11 \\ 6x - 5y &= 13 \\ \hline 8x \quad\quad &= 24 \end{array}$$

To add the equations, add the like terms, column by column.

$11 + 13 = 24$

$5y + (-5y) = 0$

$2x + 6x = 8x$

The Language of Algebra

The **elimination** method, or **addition** method as it is also known, is so named because one of the variables is eliminated using addition.

Because the sum of the terms $5y$ and $-5y$ is 0, we say that the variable y has been eliminated. Since the resulting equation has only one variable, we can solve it for x.

$8x = 24$

$x = 3$ Divide both sides by 8. This is the x-value of the solution.

To find the y-value of the solution, substitute 3 for x in *either* equation of the original system.

Teaching Tip: Substitute 3 for x in the second equation, $6x - 5y = 13$, and show that the same value, 1, is obtained for y.

$2x + 5y = 11$ This is the first equation of the system.

$2(3) + 5y = 11$ Substitute 3 for x.

$6 + 5y = 11$ Multiply.

$5y = 5$ Subtract 6 from both sides.

$y = 1$ Divide both sides by 5. This is the y-value of the solution.

Now we check the proposed solution (3, 1) in the equations of the original system.

Check:

$2x + 5y = 11$	The first equation.	$6x - 5y = 13$	The second equation.
$2(3) + 5(1) \stackrel{?}{=} 11$		$6(3) - 5(1) \stackrel{?}{=} 13$	
$6 + 5 \stackrel{?}{=} 11$		$18 - 5 \stackrel{?}{=} 13$	
$11 = 11$	True	$13 = 13$	True

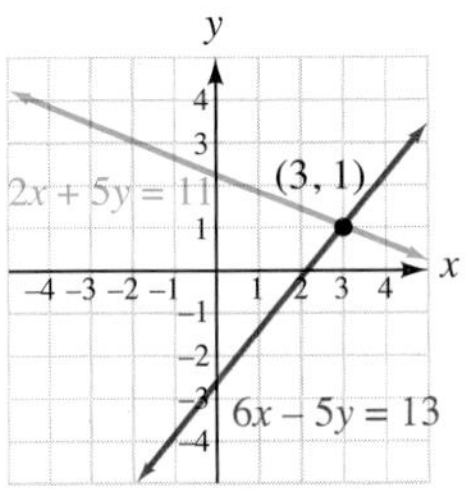

Since (3, 1) satisfies both equations, it is the solution. The graph on the left helps to verify this and it illustrates an important fact: *The solution found using the elimination (addition) method will be the same as the solution found using the graphing method.* The solution set is written $\{(3, 1)\}$.

Teaching Example 1 Solve the system:
$\begin{cases} 4x - 2y = 22 \\ 3x + 2y = 20 \end{cases}$
Answer: (6, 1)

Self Check 1 Solve the system: $\begin{cases} -4x + 3y = 4 \\ 4x + 5y = 28 \end{cases}$ (2, 4)

Now Try Problem 17

To solve a system of equations in x and y by the elimination (addition) method, follow these steps.

The Elimination (Addition) Method

1. Write both equations of the system in standard $Ax + By = C$ form.
2. If necessary, multiply one or both of the equations by a nonzero number chosen to make the coefficients of x (or the coefficients of y) opposites.
3. Add the equations to eliminate the terms involving x (or y).
4. Solve the equation resulting from step 3.
5. Find the value of the remaining variable by substituting the solution found in step 4 into any equation containing both variables. Or, repeat steps 2–4 to eliminate the other variable.
6. Check the proposed solution in each equation of the original system. Write the solution as an ordered pair.

2 Use Multiplication to Eliminate a Variable.

In Example 1, the coefficients of the terms $5y$ in the first equation and $-5y$ in the second equation were opposites. When we added the equations, the variable y was eliminated. For many systems, however, we are not able to eliminate a variable immediately by adding. In such cases, we use the multiplication property of equality to create coefficients of x or y that are opposites.

EXAMPLE 2 Solve the system: $\begin{cases} 2x + 7y = -18 \\ 2x + 3y = -10 \end{cases}$

Strategy We will use the elimination method to solve this system.

Why Since none of the variables has a coefficient of 1 or -1, it would be difficult to solve this system using substitution.

Solution ***Step 1:*** Both equations are in standard $Ax + By = C$ form. We see that neither the coefficients of x nor the coefficients of y are opposites. Adding these equations as written does not eliminate a variable.

Teaching Tip: Add the equations as given, to show that neither variable gets eliminated:

$$\begin{array}{r} 2x + 7y = -18 \\ 2x + 3y = -10 \\ \hline 4x + 10y = -28 \end{array}$$

Step 2: To eliminate x, we can multiply both sides of the second equation by -1. This creates the term $-2x$, whose coefficient is opposite that of the $2x$ term in the first equation.

$$\begin{cases} 2x + 7y = -18 \xrightarrow{\text{Unchanged}} \\ 2x + 3y = -10 \xrightarrow[\text{Multiply by } -1]{} \end{cases} \begin{array}{c} 2x + 7y = -18 \\ \mathbf{-1}(2x + 3y) = \mathbf{-1}(-10) \end{array} \begin{array}{c} \xrightarrow{\text{Unchanged}} \\ \xrightarrow[\text{Simplify}]{} \end{array} \begin{cases} 2x + 7y = -18 \\ -2x - 3y = 10 \end{cases}$$

Success Tip

It doesn't matter which variable is eliminated first. We don't have to find the values of the variables in alphabetical order. Choose the one that is the easier to eliminate. The objective of this method is to obtain two equations whose sum will be one equation in one variable.

Step 3: When the equations are added, x is eliminated.

$$\begin{array}{rl} 2x + 7y = -18 & \\ -2x - 3y = 10 & \text{In the left column: } 2x + (-2x) = 0. \\ \hline 4y = -8 & \text{This equation has only one variable.} \end{array}$$

Step 4: Solve the resulting equation to find y.

$$\begin{aligned} 4y &= -8 \\ y &= -2 \quad \text{Divide both sides by 4. This is the } y\text{-value of the solution.} \end{aligned}$$

Step 5: To find x, we can substitute -2 for y in either of the equations of the original system, or in $-2x - 3y = 10$. It appears the calculations will be simplest if we use $2x + 3y = -10$.

$$\begin{aligned} 2x + 3y &= -10 && \text{This is the second equation of the original system.} \\ 2x + 3(\mathbf{-2}) &= -10 && \text{Substitute } -2 \text{ for } y. \\ 2x - 6 &= -10 && \text{Multiply.} \\ 2x &= -4 && \text{Add 6 to both sides.} \\ x &= -2 && \text{Divide both sides by 2. This is the } x\text{-value of the solution.} \end{aligned}$$

The Language of Algebra

Recall that $2x$ and $-2x$ are called **additive inverses** because their sum is equal to 0:

$$2x + (-2x) = 0$$

Step 6: The solution is $(-2, -2)$. Check this result in the original equations.

Teaching Example 2 Solve the system: $\begin{cases} 6x + 5y = -49 \\ 6x + 7y = -59 \end{cases}$

Answer: $(-4, -5)$

Self Check 2 Solve the system: $\begin{cases} 2x + 7y = -27 \\ 3x + 7y = -30 \end{cases}$ $(-3, -3)$

Now Try Problem 21

EXAMPLE 3 Solve the system: $\begin{cases} 7x + 2y - 14 = 0 \\ 9x = 4y - 28 \end{cases}$

Strategy We will use the elimination method to solve this system.

Why Since none of the variables has coefficient 1 or -1, it would be difficult to solve this system using substitution.

Solution ***Step 1:*** To compare coefficients, write each equation in the standard $Ax + By = C$ form. Since each of the original equations will be written in an equivalent form, the resulting system will have the same solution as the original system.

$$\begin{cases} 7x + 2y = 14 & \text{Add 14 to both sides of } 7x + 2y - 14 = 0. \\ 9x - 4y = -28 & \text{Subtract } 4y \text{ from both sides of } 9x = 4y - 28. \end{cases}$$

Success Tip

We choose to eliminate y because the coefficient -4 is a *multiple* of the coefficient 2. The same cannot be said for 7 and 9, the coefficients of x.

Step 2: Neither the coefficients of x nor the coefficients of y are opposites. To eliminate y, we can multiply both sides of the first equation by 2. This creates the term $4y$, whose coefficient is opposite that of the $-4y$ term in the second equation.

$$\begin{cases} 7x + 2y = 14 \\ 9x - 4y = -28 \end{cases} \xrightarrow[\text{Unchanged}]{\text{Multiply by 2}} \begin{matrix} 2(7x + 2y) = 2(14) \\ 9x - 4y = -28 \end{matrix} \xrightarrow[\text{Unchanged}]{\text{Simplify}} \begin{cases} 14x + 4y = 28 \\ 9x - 4y = -28 \end{cases}$$

Caution

When using the elimination method, don't forget to multiply both sides of an equation by the appropriate number.

Multiply both sides by 2.
$2(7x + 2y) = 2(14)$

Step 3: When the equations are added, y is eliminated.

$$\begin{aligned} 14x + 4y &= 28 \\ 9x - 4y &= -28 \\ \hline 23x \quad &= 0 \end{aligned}$$

In the middle column: $4y + (-4y) = 0$.
This equation has only one variable.

Step 4: Solve the resulting equation to find x.

$23x = 0$

$x = 0$ Divide both sides by 23. This is the x-value of the solution.

Step 5: To find y, we can substitute 0 for x in any equation that contains both variables. It appears the computations will be simplest if we use $7x + 2y = 14$.

$7x + 2y = 14$ This is the first equation of the original system.

$7(0) + 2y = 14$ Substitute 0 for x.

$0 + 2y = 14$ Multiply.

$2y = 14$ Simplify the left side.

$y = 7$ Divide both sides by 2. This is the y-value of the solution.

Step 6: The solution is (0, 7). Check this result in the original equations.

Teaching Example 3 Solve the system:
$\begin{cases} 5x + 3y - 6 = 0 \\ 8x = 9y - 18 \end{cases}$
Answer: (0, 2)

Self Check 3 Solve the system: $\begin{cases} 3x = 10 - 2y \\ 5x - 6y + 30 = 0 \end{cases}$ (0, 5)

Now Try Problem 25

Sometimes we must apply the multiplication property of equality to both equations to create coefficients of one variable that are opposites.

EXAMPLE 4 Solve the system: $\begin{cases} 4a + 7b = -8 \\ 5a + 6b = 1 \end{cases}$

Strategy We will use the elimination method to solve this system.

Why Since none of the variables has coefficient 1 or -1, it would be difficult to solve this system using substitution.

Solution ***Step 1:*** Both equations are written in standard $Ax + By = C$ form.

Step 2: In this example, we must write *both* equations in equivalent forms to obtain like terms that are opposites. To eliminate a, we can multiply the first equation by 5 to create the term $20a$, and we can multiply the second equation by -4 to create the term $-20a$.

Teaching Tip: As you explain this example, purposefully make an error. Only multiply the left sides of the equations by the required numbers. See if any students catch the mistake.

$$\begin{cases} 4a + 7b = -8 \\ 5a + 6b = 1 \end{cases} \xrightarrow[\text{Multiply by } -4]{\text{Multiply by 5}} \begin{matrix} 5(4a + 7b) = 5(-8) \\ -4(5a + 6b) = -4(1) \end{matrix} \xrightarrow[\text{Simplify}]{\text{Simplify}} \begin{cases} 20a + 35b = -40 \\ -20a - 24b = -4 \end{cases}$$

Success Tip

We create the term $20a$ from $4a$ and the term $-20a$ from $5a$. Note that the *least common multiple* of 4 and 5 is 20:

4, 8, 12, 16, **20**, 24, 28, . . .
5, 10, 15, **20**, 25, 30, . . .

Step 3: When we add the resulting equations, a is eliminated.

$$\begin{aligned} 20a + 35b &= -40 \\ -20a - 24b &= -4 \\ \hline 11b &= -44 \end{aligned}$$

In the left column: $20a + (-20a) = 0$.
This equation has only one variable.

Step 4: Solve the resulting equation to find b.

$$\begin{aligned} 11b &= -44 \\ b &= -4 \end{aligned}$$

Divide both sides by 11. This is the b-value of the solution.

Success Tip

With this method, it doesn't matter which variable is eliminated. We could have created terms of $42b$ and $-42b$ to eliminate b. We will get the same solution, $(5, -4)$.

Step 5: To find a, we can substitute -4 for b in any equation that contains both variables. It appears the calculations will be simplest if we use $5a + 6b = 1$.

$5a + 6b = 1$	This is the second equation of the original system.
$5a + 6(-4) = 1$	Substitute -4 for b.
$5a - 24 = 1$	Multiply.
$5a = 25$	Add 24 to both sides.
$a = 5$	Divide both sides by 5. This is the a-value of the solution.

Step 6: Written in (a, b) form, the solution is $(5, -4)$. Check it in the original equations.

Teaching Example 4 Solve the system:
$\begin{cases} 3a + 5b = -1 \\ 4a + 3b = 6 \end{cases}$
Answer: $(3, -2)$

Self Check 4 Solve the system: $\begin{cases} 5a + 3b = -7 \\ 3a + 4b = 9 \end{cases}$ $(-5, 6)$

Now Try Problem 29

3 Use the Elimination Method Twice to Solve a System.

Sometimes it is easier to find the value of the second variable of a solution by using elimination a second time.

EXAMPLE 5 Solve the system: $\begin{cases} \frac{1}{6}x + \frac{1}{2}y = \frac{1}{3} \\ -\frac{x}{9} + y = \frac{5}{9} \end{cases}$

Strategy We will begin by clearing each equation of fractions. Then we will use the elimination method to solve the resulting equivalent system.

Why It is easier to create a pair of terms that are opposites if their coefficients are integers rather than fractions.

Solution ***Step 1:*** To clear the equations of the fractions, multiply both sides of the first equation by 6 and both sides of the second equation by 9.

$$\begin{cases} \frac{1}{6}x + \frac{1}{2}y = \frac{1}{3} \xrightarrow{\text{Multiply by 6}} 6\left(\frac{1}{6}x + \frac{1}{2}y\right) = 6\left(\frac{1}{3}\right) \xrightarrow{\text{Simplify}} \\ -\frac{x}{9} + y = \frac{5}{9} \xrightarrow{\text{Multiply by 9}} 9\left(-\frac{x}{9} + y\right) = 9\left(\frac{5}{9}\right) \xrightarrow{\text{Simplify}} \end{cases} \begin{cases} x + 3y = 2 \\ -x + 9y = 5 \end{cases}$$

Success Tip

Some students find it helpful to cross out the terms that are eliminated:

$$\begin{aligned} x + 3y &= 2 \\ -x + 9y &= 5 \\ \hline 12y &= 7 \end{aligned}$$

Step 2: The coefficients of x are opposites.

Step 3: The variable x is eliminated when we add the resulting equations.

$$\begin{aligned} x + 3y &= 2 \\ -x + 9y &= 5 \\ \hline 12y &= 7 \end{aligned}$$

In the left column: $x + (-x) = 0$.
This equation has only one variable.

Step 4: Solve the resulting equation to find y.

$$12y = 7$$

$$y = \frac{7}{12} \quad \text{Divide both sides by 12. This is the } y\text{-value of the solution.}$$

Step 5: We can find x by substituting $\frac{7}{12}$ for y in any equation containing both variables. However, that calculation could be complicated, because $\frac{7}{12}$ is a fraction. Instead, we can begin again with the system that is cleared of fractions, but this time, eliminate y. If we multiply both sides of the first equation by -3, this creates the term $-9y$, whose coefficient is opposite that of the $9y$ term in the second equation.

$$\begin{cases} x + 3y = 2 \\ -x + 9y = 5 \end{cases} \xrightarrow[\text{Unchanged}]{\text{Multiply by } -3} \begin{matrix} -3(x + 3y) = -3(2) \\ -x + 9y = 5 \end{matrix} \xrightarrow[\text{Unchanged}]{\text{Simplify}} \begin{cases} -3x - 9y = -6 \\ -x + 9y = 5 \end{cases}$$

When we add the resulting equations, y is eliminated.

$$\begin{aligned} -3x - 9y &= -6 && \text{In the middle column: } 9y + (-9y) = 0. \\ -x + 9y &= 5 \\ \hline -4x \quad &= -1 && \text{This equation has only one variable.} \end{aligned}$$

Now we solve the resulting equation to find x.

$$-4x = -1$$

$$x = \frac{1}{4} \quad \text{Divide both sides by } -4\text{. This is the } x\text{-value of the solution.}$$

Step 6: The solution is $\left(\frac{1}{4}, \frac{7}{12}\right)$. To verify this, check it in the original equations.

Teaching Example 5 Solve the system:

$$\begin{cases} \frac{5}{6}x + \frac{2}{3}y = \frac{7}{6} \\ \frac{10}{7}x - \frac{4}{9}y = \frac{17}{21} \end{cases}$$

Answer: $\left(\frac{4}{5}, \frac{3}{4}\right)$

Self Check 5 Solve the system: $\begin{cases} -\frac{1}{5}x + y = \frac{8}{5} \\ \frac{x}{8} + \frac{y}{2} = \frac{1}{4} \end{cases}$ $\left(-\frac{22}{9}, \frac{10}{9}\right)$

Now Try ▶ Problem 33

4 Use Elimination to Identify Inconsistent Systems and Dependent Equations.

We have solved inconsistent systems and systems of dependent equations by substitution and by graphing. We also can solve these systems using the elimination method.

EXAMPLE 6 Solve the system: $\begin{cases} 3x - 2y = 2 \\ -3x + 2y = -12 \end{cases}$

Strategy We will use the elimination method to solve this system.

Why The terms $3x$ and $-3x$ are immediately eliminated.

Solution

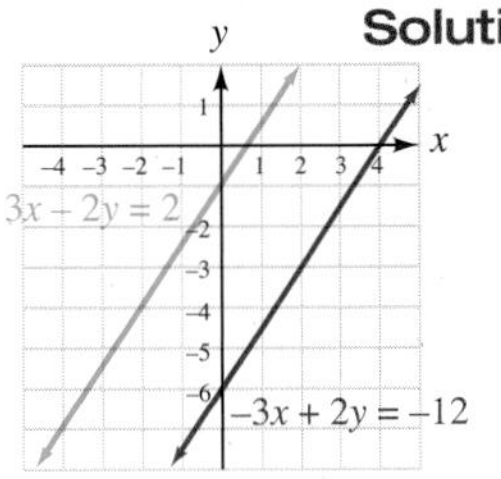

$$\begin{aligned} 3x - 2y &= 2 && \text{In the left column: } 3x + (-3x) = 0. \\ -3x + 2y &= -12 && \text{In the middle column: } -2y + 2y = 0. \\ \hline 0 &= -10 && \text{In the right column: } 2 + (-12) = -10. \end{aligned}$$

In eliminating x, the variable y is eliminated as well. The resulting false statement, $0 = -10$, indicates that the system has no solution and is inconsistent. The graphs of the equations help to verify this; they are parallel lines.

Teaching Example 6 Solve the system:
$\begin{cases} 6x - 5y = 4 \\ -6x + 5y = 1 \end{cases}$
Answer: No solution

Self Check 6 Solve the system: $\begin{cases} 2x - 7y = 5 \\ -2x + 7y = 3 \end{cases}$ No solution

Now Try Problem 37

EXAMPLE 7 Solve the system: $\begin{cases} \dfrac{2x - 5y}{15} = \dfrac{8}{15} \\ -0.2x + 0.5y = -0.8 \end{cases}$

Strategy We will begin by clearing the equations of fractions and decimals. Then we will use the elimination method to solve the resulting equivalent system.

Why In this form, the equations do not contain terms with coefficients that are opposites.

Solution We can multiply both sides of the first equation by **15** to clear it of fractions and both sides of the second equation by **10** to clear it of decimals.

$$\begin{cases} \dfrac{2x - 5y}{15} = \dfrac{8}{15} & \rightarrow & \mathbf{15}\left(\dfrac{2x - 5y}{15}\right) = \mathbf{15}\left(\dfrac{8}{15}\right) & \rightarrow & \begin{cases} 2x - 5y = 8 \\ -2x + 5y = -8 \end{cases} \\ -0.2x + 0.5y = -0.8 & \rightarrow & \mathbf{10}(-0.2x + 0.5y) = \mathbf{10}(-0.8) & \rightarrow & \end{cases}$$

We add the resulting equations to get

$$\begin{aligned} 2x - 5y &= 8 && \text{In the left column: } 2x + (-2x) = 0. \\ -2x + 5y &= -8 && \text{In the middle column: } -5y + 5y = 0. \\ \hline 0 &= 0 && \text{In the right column: } 8 + (-8) = 0. \end{aligned}$$

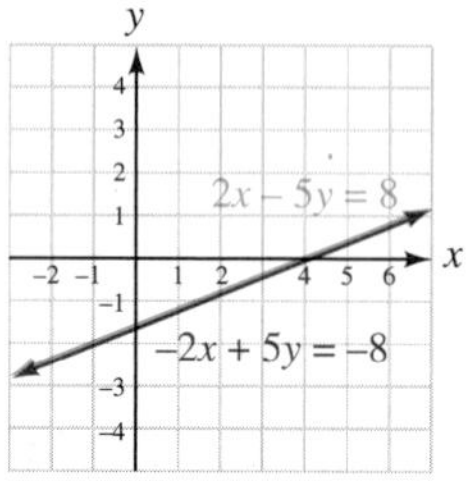

As in Example 6, both variables are eliminated. However, this time a true statement, $0 = 0$, is obtained. It indicates that the equations are dependent and that the system has infinitely many solutions. The graphs of the equations help to verify this; they are identical.

To find several of the infinitely many solutions, we can substitute some values of x, say -1, 4, and 9, in either equation and solve for y. The results are: $(-1, -2)$, $(4, 0)$, and $(9, 2)$.

Teaching Example 7 Solve the system:
$\begin{cases} \frac{4x - 3y}{2} = \frac{1}{5} \\ 0.20x - 0.15y = 0.02 \end{cases}$
Answer: Infinitely many solutions

Self Check 7 Solve the system: $\begin{cases} \dfrac{3x + y}{6} = \dfrac{1}{3} \\ -0.3x - 0.1y = -0.2 \end{cases}$ Infinitely many solutions

Now Try Problem 41

5 Determine the Most Efficient Method to Use to Solve a Linear System.

If no method is specified for solving a particular system of two linear equations, the following guidelines can be helpful in determining whether to use graphing, substitution, or elimination.

1. If you want to show trends and see the point that the two graphs have in common, then use the **graphing method.** However, this method is not exact and can be lengthy.
2. If one of the equations is solved for one of the variables, or easily solved for one of the variables, use the **substitution method.**
3. If both equations are in standard $Ax + By = C$ form, and no variable has a coefficient of 1 or -1, use the **elimination method.**
4. If the coefficient of one of the variables is 1 or -1, you have a choice. You can write each equation in standard $(Ax + By = C)$ form and use elimination, or you can solve for the variable with coefficient 1 or -1 and use substitution.

Here are some examples of suggested approaches:

$\begin{cases} 2x + 3y = 1 \\ y = 4x - 3 \end{cases}$ Substitution

$\begin{cases} 5x + 3y = 9 \\ 8x + 4y = 3 \end{cases}$ Elimination

$\begin{cases} 4x - y = -6 \\ 3x + 2y = 1 \end{cases}$ Elimination

$\begin{cases} x - 23 = 6y \\ 7x - 9y = -3 \end{cases}$ Substitution

Each method that we use to solve systems of equations has advantages and disadvantages.

Method	Advantages	Disadvantages
Graphing	■ You see the solution(s). ■ The graphs allow you to observe trends.	■ Inaccurate when the solutions are not integers or are large numbers off the graph
Substitution	■ Always gives the exact solutions ■ Works well if one of the equations is solved for one of the variables, or if it is easy to solve for one of the variables	■ You do not see the solution. ■ If no variable has a coefficient of 1 or −1, solving for one of the variables often involves fractions.
Elimination	■ Always gives the exact solutions ■ Works well if no variable has a coefficient of 1 or −1	■ You do not see the solution. ■ The equations must be written in the form $Ax + By = C$.

SECTION 4.3 STUDY SET

VOCABULARY

Fill in the blanks.

1. The coefficients of $3x$ and $-3x$ are opposites.
2. When the given equations are added, the variable y will be eliminated.

$$\begin{array}{r} 5x - 6y = 10 \\ -3x + 6y = 24 \end{array}$$

CONCEPTS

3. In the given system, which terms have coefficients that are opposites? $\begin{cases} 3x + 7y = -25 \\ 4x - 7y = 12 \end{cases}$
   $7y$ and $-7y$
4. Fill in the blank. The objective of the elimination method is to obtain two equations whose sum will be one equation in one variable.
5. Add each pair of equations.
   a. $\begin{array}{r} 2a + 2b = -6 \\ 3a - 2b = 2 \\ \hline 5a \quad = -4 \end{array}$
   b. $\begin{array}{r} x - 3y = 15 \\ -x - y = -14 \\ \hline -4y = 1 \end{array}$
6. a. Multiply both sides of $4x + y = 2$ by 3.
   $12x + 3y = 6$
   b. Multiply both sides of $x - 3y = 4$ by -2.
   $-2x + 6y = -8$
7. If the elimination method is used to solve

$$\begin{cases} 3x + 12y = 4 \\ 6x - 4y = 7 \end{cases}$$

   a. By what would we multiply the first equation to eliminate the variable x? -2
   b. By what would we multiply the second equation to eliminate the variable y? 3
8. Suppose the following system is solved using the elimination method and it is found that x is 2. Find the value of y. 1

$$\begin{cases} 4x + 3y = 11 \\ 3x - 2y = 4 \end{cases}$$

9. What algebraic step should be performed to
   a. Clear $\frac{2}{3}x + 4y = -\frac{4}{5}$ of fractions?
   Multiply both sides by 15.
   b. Clear $0.2x - 0.9y = 6.4$ of decimals?
   Multiply both sides by 10.
10. a. Suppose $0 = 0$ is obtained when a system is solved by the elimination method. Does the system have a solution? Which of the following is a possible graph of the system?
    Yes; ii
    b. Suppose $0 = 2$ is obtained when a system is solved by the elimination method. Does the system have a solution? Which of the following is a possible graph of the system?
    No; iii

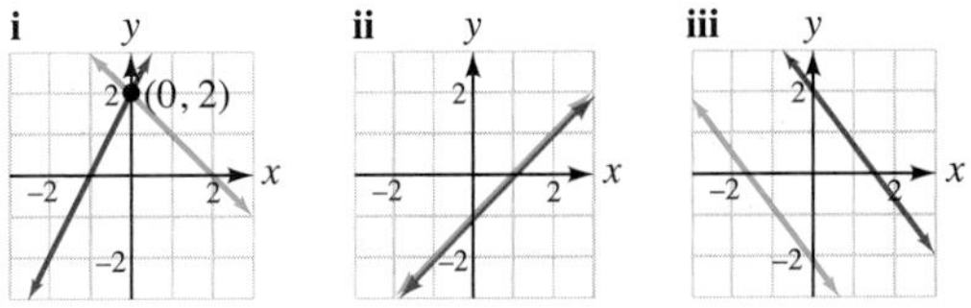

NOTATION

Complete the solution to solve the system.

11. Solve: $\begin{cases} x + y = 5 \\ x - y = -3 \end{cases}$

$$\begin{aligned} x + y &= 5 \quad \text{Add the equations.} \\ x - y &= -3 \\ \hline 2x &= 2 \\ x &= 1 \end{aligned}$$

$$\begin{aligned} x + y &= 5 \quad \text{This is the first equation.} \\ 1 + y &= 5 \\ y &= 4 \end{aligned}$$

The solution is $(1, 4)$.

12. Write each equation in $Ax + By = C$ form:

$\begin{cases} 7x + y + 3 = 0 \\ 8x + 4 = -y \end{cases} \rightarrow \begin{cases} 7x + y = -3 \\ 8x + y = -4 \end{cases}$

GUIDED PRACTICE

Use the elimination method to solve each system. See Example 1.

13. $\begin{cases} x + y = 5 \\ x - y = 1 \end{cases}$ $(3, 2)$

14. $\begin{cases} x - y = 4 \\ x + y = 8 \end{cases}$ $(6, 2)$

15. $\begin{cases} x + y = -5 \\ -x + y = -1 \end{cases}$ $(-2, -3)$

16. $\begin{cases} -x + y = -3 \\ x + y = 1 \end{cases}$ $(2, -1)$

17. $\begin{cases} 4x + 3y = 24 \\ 4x - 3y = -24 \end{cases}$ $(0, 8)$

18. $\begin{cases} -9x + 5y = -9 \\ -9x - 5y = -9 \end{cases}$ $(1, 0)$

19. $\begin{cases} 2s + t = -2 \\ -2s - 3t = -6 \end{cases}$ $(-3, 4)$

20. $\begin{cases} -2x + 4y = 12 \\ 2x + 4y = 28 \end{cases}$ $(4, 5)$

Use the elimination method to solve each system. See Example 2.

21. $\begin{cases} x + 3y = -9 \\ x + 8y = -4 \end{cases}$ $(-12, 1)$

22. $\begin{cases} x + 7y = -22 \\ x + 9y = -24 \end{cases}$ $(-15, -1)$

23. $\begin{cases} 7x - y = 10 \\ 8x - y = 13 \end{cases}$ $(3, 11)$

24. $\begin{cases} 6x - y = 4 \\ 9x - y = 10 \end{cases}$ $(2, 8)$

Use the elimination method to solve each system. See Example 3.

25. $\begin{cases} 7x + 4y - 14 = 0 \\ 3x = 2y - 20 \end{cases}$ $(-2, 7)$

26. $\begin{cases} 5x - 14y - 32 = 0 \\ -x = 6y + 20 \end{cases}$ $(-2, -3)$

27. $\begin{cases} 7x - 50y + 43 = 0 \\ x = 4 - 3y \end{cases}$ $(1, 1)$

28. $\begin{cases} x - 2y + 1 = 0 \\ 12x = 23 - 11y \end{cases}$ $(1, 1)$

Use the elimination method to solve each system. See Example 4.

29. $\begin{cases} 4x + 3y = 7 \\ 3x - 2y = -16 \end{cases}$ $(-2, 5)$

30. $\begin{cases} 3x - 2y = 20 \\ 2x + 7y = 5 \end{cases}$ $(6, -1)$

31. $\begin{cases} 5a + 8b = 2 \\ 11a - 3b = 25 \end{cases}$ $(2, -1)$

32. $\begin{cases} 7a - 5b = 24 \\ 12a + 8b = 8 \end{cases}$ $(2, -2)$

Use the elimination method to solve each system. See Example 5.

33. $\begin{cases} \frac{1}{8}x - \frac{1}{8}y = \frac{3}{8} \\ \frac{x}{4} + \frac{y}{4} = \frac{1}{2} \end{cases}$ $\left(\frac{5}{2}, -\frac{1}{2}\right)$

34. $\begin{cases} \frac{1}{8}x + \frac{1}{4}y = \frac{1}{4} \\ \frac{x}{2} + \frac{y}{4} = \frac{1}{2} \end{cases}$ $\left(\frac{2}{3}, \frac{2}{3}\right)$

35. $\begin{cases} \frac{3}{4}x - \frac{5}{8}y = \frac{1}{24} \\ \frac{5x}{6} - y = \frac{1}{4} \end{cases}$ $\left(-\frac{1}{2}, -\frac{2}{3}\right)$

36. $\begin{cases} \frac{1}{3}x + \frac{1}{2}y = \frac{5}{3} \\ \frac{x}{7} - \frac{y}{7} = -\frac{1}{7} \end{cases}$ $\left(\frac{7}{5}, \frac{12}{5}\right)$

Use the elimination method to solve each system. If there is no solution, or infinitely many solutions, so state. See Example 6.

37. $\begin{cases} 3x - 5y = -29 \\ 3x - 5y = 15 \end{cases}$ No solution

38. $\begin{cases} 2a - 3b = -6 \\ 2a - 3b = 8 \end{cases}$ No solution

39. $\begin{cases} 3x - 16 = 5y \\ -3x + 5y - 33 = 0 \end{cases}$ No solution

40. $\begin{cases} \frac{-18x + y}{2} = \frac{7}{2} \\ 18x = y \end{cases}$ No solution

Use elimination to solve each system. If there is no solution, or infinitely many solutions, so state. See Example 7.

41. $\begin{cases} 0.4x - 0.7y = -1.9 \\ -x + \frac{7y}{4} = \frac{19}{4} \end{cases}$ Infinitely many solutions

42. $\begin{cases} 0.1x + 2y + 0.2 = 0 \\ -\frac{x}{4} - 5y = \frac{1}{2} \end{cases}$ Infinitely many solutions

43. $\begin{cases} \frac{x - 6y}{2} = 7 \\ -x + 6y + 14 = 0 \end{cases}$ Infinitely many solutions

44. $\begin{cases} 2x + 5y - 13 = 0 \\ -2x + 13 = 5y \end{cases}$ Infinitely many solutions

TRY IT YOURSELF

Solve the system by either the substitution or the elimination method.

45. $\begin{cases} y = -3x + 9 \\ y = x + 1 \end{cases}$ $(2, 3)$

46. $\begin{cases} x = 5y - 4 \\ x = 9y - 8 \end{cases}$ $(1, 1)$

47. $\begin{cases} 4x + 6y = 5 \\ 8x - 9y = 3 \end{cases}$ $\left(\frac{3}{4}, \frac{1}{3}\right)$

48. $\begin{cases} 3a + 4b = 36 \\ 6a - 2b = -21 \end{cases}$ $\left(-\frac{2}{5}, \frac{93}{10}\right)$

49. $\begin{cases} 6x - 3y = -7 \\ y + 9x = 6 \end{cases}$ $\left(\frac{1}{3}, 3\right)$

50. $\begin{cases} 9x + 4y = 31 \\ y - 5 = 6x \end{cases}$ $\left(\frac{1}{3}, 7\right)$

51. $\begin{cases} x + y = 1 \\ x - y = 5 \end{cases}$ $(3, -2)$

52. $\begin{cases} x - y = -5 \\ x + y = 1 \end{cases}$ $(-2, 3)$

53. $\begin{cases} 4(x - 2y) = 36 \\ 3x - 6y = 27 \end{cases}$ Infinitely many solutions

54. $\begin{cases} 2(x + 2y) = 15 \\ 3x = 8 - 6y \end{cases}$ No solution

55. $\begin{cases} x = y \\ 0.1x + 0.2y = 1.0 \end{cases}$ $\left(\frac{10}{3}, \frac{10}{3}\right)$

56. $\begin{cases} x = y \\ 0.4x - 0.8y = -0.5 \end{cases}$ $\left(\frac{5}{4}, \frac{5}{4}\right)$

57. $\begin{cases} 2x + 11y = -10 \\ 5x + 4y = 22 \end{cases}$ $(6, -2)$

58. $\begin{cases} 3x + 4y = 12 \\ 4x + 5y = 17 \end{cases}$ $(8, -3)$

59. $\begin{cases} 7x = 21 - 6y \\ 4x + 5y = 12 \end{cases}$ $(3, 0)$

60. $\begin{cases} -4x = -3y - 13 \\ -6x + 8y = -16 \end{cases}$ $(4, 1)$

61. $\begin{cases} 9x - 10y = 0 \\ \frac{9x - 3y}{63} = 1 \end{cases}$ $(10, 9)$

62. $\begin{cases} 8x - 9y = 0 \\ \frac{2x - 3y}{6} = -1 \end{cases}$ $(9, 8)$

63. $\begin{cases} \frac{m}{4} + \frac{n}{3} = -\frac{1}{12} \\ \frac{m}{2} - \frac{5}{4}n = \frac{7}{4} \end{cases}$ $(1, -1)$

64. $\begin{cases} \frac{x}{2} - \frac{y}{3} = -2 \\ \frac{x}{3} + \frac{2}{3}y = \frac{4}{3} \end{cases}$ $(-2, 3)$

65. $\begin{cases} x - \frac{4}{3}y = \frac{1}{3} \\ 2x + \frac{3}{2}y = \frac{1}{2} \end{cases}$ $\left(\frac{7}{25}, -\frac{1}{25}\right)$

66. $\begin{cases} x + y = -\frac{1}{4} \\ x - \frac{y}{2} = -\frac{3}{2} \end{cases}$ $\left(-\frac{13}{12}, \frac{5}{6}\right)$

67. $\begin{cases} 4x - 7y + 32 = 0 \\ 5x = 4y - 2 \end{cases}$ (6, 8)

68. $\begin{cases} 6x = -3y \\ 5x + 15 = 5y \end{cases}$ (−1, 2)

69. $\begin{cases} 3(x + 4y) = -12 \\ x = 3y + 10 \end{cases}$ (4, −2)

70. $\begin{cases} 3x + 2y = 3 \\ y = 2(x - 8) \end{cases}$ (5, −6)

71. $\begin{cases} 4a + 7b = 2 \\ 9a - 3b = 1 \end{cases}$ $\left(\frac{13}{75}, \frac{14}{75}\right)$

72. $\begin{cases} 5a - 7b = 6 \\ 7a - 6b = 8 \end{cases}$ $\left(\frac{20}{19}, -\frac{2}{19}\right)$

73. $\begin{cases} 3a - b = 12.3 \\ 4a - b = 14.9 \end{cases}$ (2.6, −4.5)

74. $\begin{cases} -7x - y = 8.5 \\ 4x - y = -12.4 \end{cases}$ (−1.9, 4.8)

75. $\begin{cases} 5x - 4y = 8 \\ -5x - 4y = 8 \end{cases}$ (0, −2)

76. $\begin{cases} 2r + s = -8 \\ -2r + 4s = 28 \end{cases}$ (−6, 4)

77. $\begin{cases} 9a + 16b = -36 \\ 7a + 4b = 48 \end{cases}$ (12, −9)

78. $\begin{cases} 4a + 7b = -24 \\ 9a + b = 64 \end{cases}$ (8, −8)

79. $\begin{cases} 8x + 12y = -22 \\ 3x - 2y = 8 \end{cases}$ $\left(1, -\frac{5}{2}\right)$

80. $\begin{cases} 3x + 2y = 45 \\ 5x - 4y = 20 \end{cases}$ $\left(10, \frac{15}{2}\right)$

81. $\begin{cases} 6x + 5y + 29 = 0 \\ 0.02x = 0.03y - 0.05 \end{cases}$ (−4, −1)

82. $\begin{cases} 3x = 20y + 1 \\ 0.04x + 0.05y - 0.33 = 0 \end{cases}$ (7, 1)

83. $\begin{cases} c = d - 9 \\ 5c = 3d - 35 \end{cases}$ (−4, 5)

84. $\begin{cases} a = b + 7 \\ 3a - 15 = 5b \end{cases}$ (10, 3)

85. $\begin{cases} 0.9x + 2.1 = 0.3y \\ 0.4x = 0.7y + 1.9 \end{cases}$ (−4, −5)

86. $\begin{cases} 0.7x + 1.1 = 0.4y \\ 0.4x = 0.7y + 2.2 \end{cases}$ (−5, −6)

87. $\begin{cases} 5c + 2d = -5 \\ 6c + 2d = -10 \end{cases}$ (−5, 10)

88. $\begin{cases} 11c + 3d = -68 \\ 10c + 3d = -64 \end{cases}$ (−4, −8)

89. $\begin{cases} \frac{2}{15}x - \frac{1}{5}y = \frac{1}{3} \\ \frac{2}{15}x - \frac{1}{5}y = \frac{1}{10} \end{cases}$ No solution

90. $\begin{cases} \frac{1}{5}x + \frac{3}{5}y = \frac{4}{5} \\ \frac{1}{6}x + \frac{1}{2}y = \frac{2}{3} \end{cases}$ Infinitely many solutions

APPLICATIONS

91. **Education.** The graph shows educational trends during the years 1980–2009 for persons 25 years or older in the United States. The equation $9x + 11y = 352$ approximates the percent y that had less than high school completion. The equation $5x - 11y = -198$ approximates the percent y that had a Bachelor's or higher degree. In each case, x is the number of years since 1980. Use the elimination method to determine in what year the percents were equal. 1991

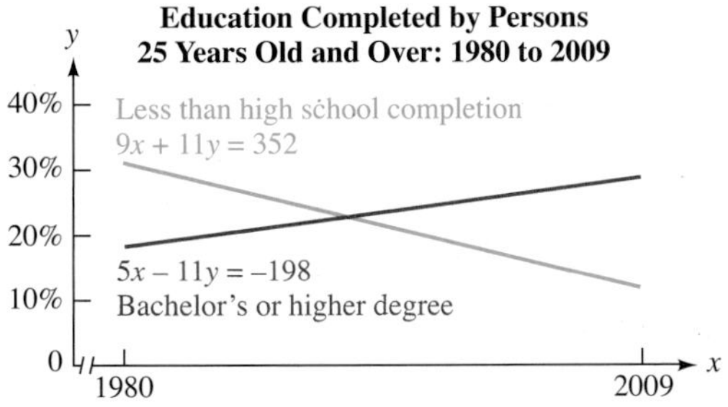

Source: U.S. Department of Commerce, Census Bureau

92. **Newspapers.** The graph shows the trends in the newspaper publishing industry during the years 1990–2008 in the United States. The equation $37x - 2y = -1{,}128$ models the number y of morning newspapers published and $31x + y = 1{,}059$ models the number y of evening newspapers published. In each case, x is the number of years since 1990. Use the elimination method to determine in what year there were an equal number of morning and evening newspapers being published. 2000

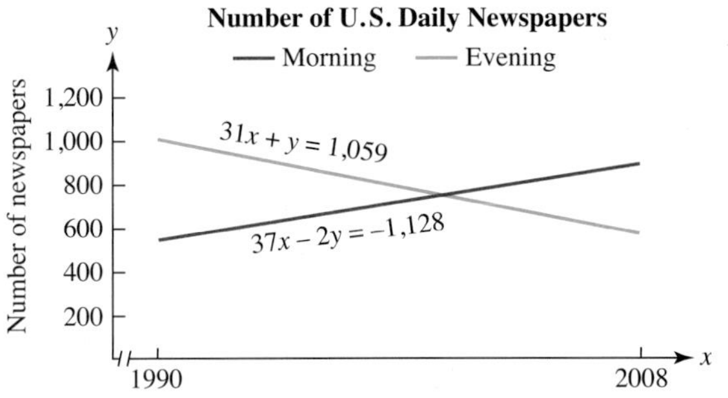

Source: Editor and Publisher Yearbook data

93. **CFL Bulbs.** The graph below shows how a more expensive, but more energy-efficient, compact fluorescent light bulb eventually costs less to use than an incandescent light bulb. The equation $60c - d = 96$ approximates the cost c (in dollars) to purchase and use a CFL bulb 8 hours a day for d days. The equation $15c - d = 6$ does the same for an incandescent bulb. Use the elimination method to determine after how many days the upgrade to a CFL bulb begins to save money. After 24 days

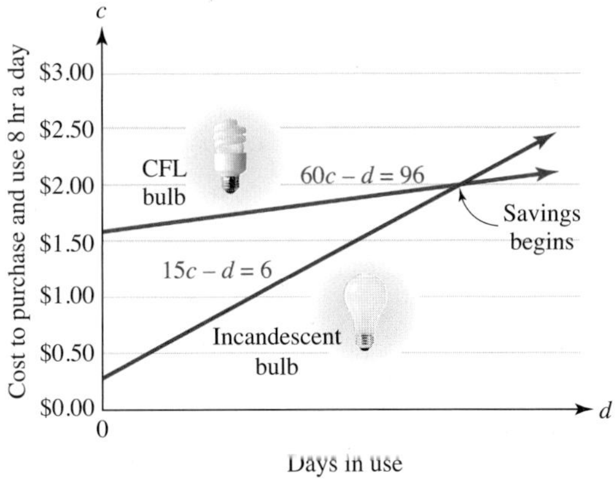

Source: Whitesites.com

94. **The Human Skeleton.** The equation $h + f = 53$ models the fact that the number of bones in the hand and foot totals 53. The equation $h - f = 1$ models the fact that the difference between the number of bones in the hand and foot is just 1. Use the elimination method to find h and f. h (bones in hand): 27; f (bones in foot): 26

WRITING

95. Why is the method for solving systems that is discussed in this section called the *elimination method?* Why is it also referred to as the *addition method?*

96. If the elimination method is to be used to solve this system, what is wrong with the form in which it is written?

$$\begin{cases} 2x - 5y = -3 \\ -2y - 10 = -5x \end{cases}$$

97. Can the system $\begin{cases} 2x + 3y = 13 \\ 7x - 3y = -5 \end{cases}$ be solved more easily using the elimination method or the substitution method? Explain.

98. Explain the error in the following work.

Solve: $\begin{cases} x + y = 1 \\ x - y = 5 \end{cases}$

$$\begin{aligned} x + y &= 1 \\ +x - y &= 5 \\ \hline 2x \quad &= 6 \\ \frac{2x}{2} &= \frac{6}{2} \\ x &= 3 \end{aligned}$$

The solution is 3.

REVIEW

99. Find an equation of the line with slope $-\frac{11}{6}$ that passes through $(2, -6)$. Write the equation in slope–intercept form. $y = -\frac{11}{6}x - \frac{7}{3}$

100. Solve $S = 2\pi rh + 2\pi r^2$ for h. $h = \frac{S - 2\pi r^2}{2\pi r}$

101. Evaluate: $-10(18 - 4^2)^3$ -80

102. Evaluate: -5^2 -25

CHALLENGE PROBLEMS

Use the elimination method to solve each system.

103. $\begin{cases} \frac{x - 3}{2} = \frac{11}{6} - \frac{y + 5}{3} \\ \frac{x + 3}{3} - \frac{y + 3}{4} = \frac{5}{12} \end{cases}$ (2, 2)

104. $\begin{cases} \frac{4(x + 1)}{34} = \frac{1}{2} - \frac{3(y - 1)}{34} \\ 0.2(x + 0.2) + 0.3(y - 0.3) = 0.75 \end{cases}$ (4, 0)

SECTION 4.4 Problem Solving Using Systems of Equations

OBJECTIVES

1. Assign variables to two unknowns.
2. Use systems to solve geometry problems.
3. Use systems to solve number-value problems.
4. Use systems to solve interest, uniform motion, and mixture problems.

ARE YOU READY? *Are You Ready? exercises available online at www.webassign.net/brookscole*

The following problems review some basic skills that are needed when solving application problems using systems of equations.

1. What is the formula for the perimeter of a rectangle? $P = 2l + 2w$
2. At \$12.95 per pound, what is the value of 3 pounds of crab legs? \$38.85
3. Find the amount of interest earned by \$55,000 invested at a 4% annual simple interest rate for 1 year. \$2,200
4. At 35 miles per hour, how far will a Canada goose travel in 12 hours? 420 mi
5. A 12-ounce mixture of window cleaner is 8% alcohol. How many ounces of the mixture is alcohol? 0.96 oz alcohol

Teaching Tip: For some classes, it might be beneficial to spend two class sessions on this section, with the money, motion, and mixture problems covered the second session.

In previous chapters, many applied problems were modeled and solved with an equation in one variable. In this section, the application problems involve two unknowns. It is often easier to solve such problems using a two-variable approach.

1 Assign Variables to Two Unknowns.

The following steps are helpful when solving problems involving two unknown quantities.

Problem-Solving Strategy

1. **Analyze the problem** by reading it carefully to understand the given facts. What information is given? What are you asked to find? What vocabulary is given? Often a diagram or table will help you understand the facts of the problem.
2. **Assign variables** to represent unknown values in the problem. This means, in most cases, to let x = one of the unknowns that you are asked to find, and y = the other unknown.
3. **Form a system of equations** by translating the words of the problem into mathematical symbols.
4. **Solve the system** of equations using graphing, substitution, or elimination.
5. **State the conclusion clearly.** Be sure to include the units.
6. **Check the results** using the words of the problem, not the equations that were formed in step 3.

EXAMPLE 1 **Motion Pictures.** Each year, Academy Award winners are presented with Oscars. The 13.5-inch statuette has a base on which a gold-plated figure stands. The figure itself is 7.5 inches taller than its base. Find the height of the figure and the height of the base.

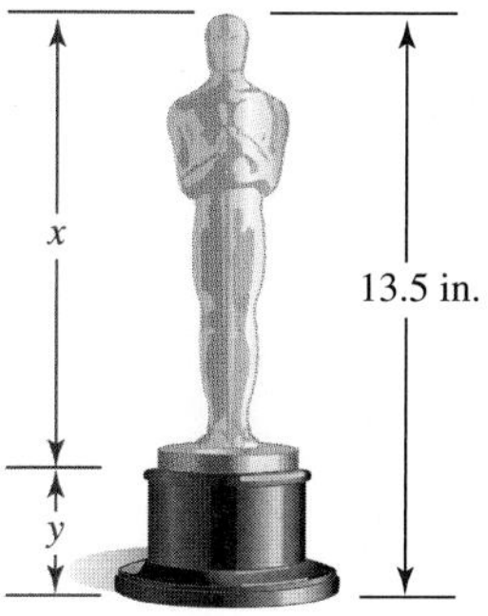

Analyze

- The statuette is a total of 13.5 inches tall.
- The figure is 7.5 inches taller than the base.
- Find the height of the figure and the height of the base.

Assign Let x = the height of the figure, in inches, and y = the height of the base, in inches.

Form We can translate the words of the problem into two equations, each involving x and y.

The height of the figure	plus	the height of the base	is	13.5 inches.
x	$+$	y	$=$	13.5

The height of the figure	is	the height of the base	plus	7.5 inches.
x	$=$	y	$+$	7.5

The resulting system is: $\begin{cases} x + y = 13.5 \\ x = y + 7.5 \end{cases}$

Solve Since the second equation is solved for x, we will use substitution to solve the system.

$$\begin{cases} x + y = 13.5 \\ x = y + 7.5 \end{cases}$$

$x + y = 13.5$ This is the first equation of the system.

$\mathbf{y + 7.5} + y = 13.5$ Substitute $y + 7.5$ for x.

$2y + 7.5 = 13.5$ Combine like terms: $y + y = 2y$.

$2y = 6$ Subtract 7.5 from both sides.

$y = 3$ Divide both sides by 2. This is the height of the base.

To find x, substitute 3 for y in the second equation of the system.

$x = y + 7.5$ This is the substitution equation.

$x = \mathbf{3} + 7.5$ Substitute 3 for y.

$x = 10.5$ This is the height of the figure.

State The height of the figure is 10.5 inches and the height of the base is 3 inches.

Check The sum of 10.5 inches and 3 inches is 13.5 inches, and the 10.5-inch figure is 7.5 inches taller than the 3-inch base. The results check.

Caution

If two variables are used to represent two unknown quantities, we must form a system of two equations to find the unknowns.

Teaching Tip: You may remind your students of the one-variable approach that was used in Chapter 2 for problems like this.

Let x = height of base
$x + 7.5$ = height of figure

$x + x + 7.5 = 13.5$
$2x = 6$
$x = 3$

$3 + 7.5 = 10.5$

Caution

In this problem we are to find two unknowns, the height of the figure and height of the base. Remember to give both in the *State* step of the solution.

Teaching Example 1 Plumbing. A plumber needs to cut a 16-foot pipe into two pieces. The longer piece is to be 1 foot longer than twice the shorter piece. Find the length of each piece.
Answers: 5 ft, 11 ft

Self Check 1 **Woodworking.** A carpenter wants to cut a 12-foot board into two pieces. The longer piece is to be twice as long as the shorter piece. Find the length of each piece. 4 ft, 8 ft

Now Try Problem 17

2 Use Systems to Solve Geometry Problems.

Two angles are said to be **complementary** if the sum of their measures is 90°. Two angles are said to be **supplementary** if the sum of their measures is 180°.

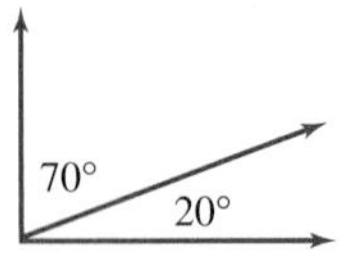

Complementary angles
$20° + 70° = 90°$

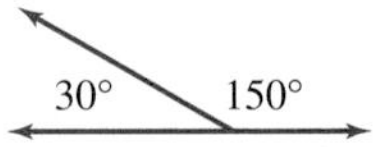

Supplementary angles
$150° + 30° = 180°$

EXAMPLE 2 **Angles.** The difference of the measures of two complementary angles is 6°. Find the measure of each angle.

Analyze

- Since the angles are complementary, the sum of their measures is 90°.
- The word *difference* indicates subtraction. If the measure of the smaller angle is subtracted from the measure of the larger angle, the result will be 6°.
- Find the measure of the larger angle and the measure of the smaller angle.

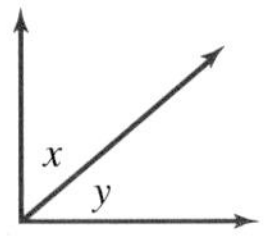

Assign Let x = the measure of the larger angle and y = the measure of the smaller angle.

Form We can translate the words of the problem into two equations, each involving x and y.

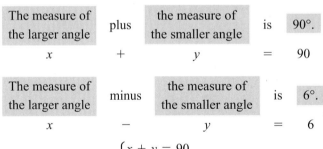

The measure of the larger angle	plus	the measure of the smaller angle	is	90°.
x	$+$	y	$=$	90

The measure of the larger angle	minus	the measure of the smaller angle	is	6°.
x	$-$	y	$=$	6

The resulting system is: $\begin{cases} x + y = 90 \\ x - y = 6 \end{cases}$

Solve Since the coefficients of y are opposites, we will use elimination to solve the system.

$$\begin{aligned} x + y &= 90 \\ x - y &= 6 \quad \text{Add the equations to eliminate } y. \\ \hline 2x \quad &= 96 \\ x &= 48 \quad \text{Divide both sides by 2. This is the measure of the larger angle.} \end{aligned}$$

To find y, substitute 48 for x in the first equation of the system.

$$\begin{aligned} x + y &= 90 \\ \mathbf{48} + y &= 90 \quad \text{Substitute 48 for } x. \\ y &= 42 \quad \text{Subtract 48 from both sides. This is the measure of the smaller angle.} \end{aligned}$$

State The measure of the larger angle is 48° and the measure of the smaller angle is 42°.

Check The sum of 48° and 42° is 90°, and the difference is 6°. The results check.

Self Check 2 **Angles.** The difference of the measures of two supplementary angles is 22°. Find the measure of each angle. 101°, 79°

Now Try Problem 13

Success Tip

The substitution method also could be used to solve this system. If we solve $x + y = 90$ for x, we obtain the substitution equation $x = 90 - y$.

Success Tip

When solving problems using the two-variable approach, be sure to check the results in two ways.

Teaching Example 2 **Angles.** The difference of the measures of two complementary angles is 64°. Find the measure of each angle.
Answers: 77°, 13°

EXAMPLE 3 **History.** In 1917, James Montgomery Flagg created the classic *I Want You* poster to help recruiting for World War I. The perimeter of the poster is 114 inches, and its length is 9 inches less than twice its width. Find the length and the width of the poster.

Analyze

- The perimeter of the rectangular poster is 114 inches.
- The length is 9 inches less than twice the width.
- Find the length and the width of the poster.

Assign Let l = the length of the poster, in inches, and w = the width of the poster, in inches.

Form The perimeter of a rectangle is the sum of two lengths and two widths, as given by the formula $\boldsymbol{P = 2l + 2w}$, so we have

2	times	the length of the poster	plus	2	times	the width of the poster	is	114 inches.
2	$\cdot$	l	$+$	2	$\cdot$	w	$=$	114

If the length of the poster is 9 inches less than twice the width, we have

The length of the poster	is	2	times	the width of the poster	minus	9 inches.
l	$=$	2	$\cdot$	w	$-$	9

The resulting system is: $\begin{cases} 2l + 2w = 114 \\ l = 2w - 9 \end{cases}$

Solve Since the second equation is solved for l, we will use substitution to solve the system.

$$2l + 2w = 114 \quad \text{This is the first equation of the system.}$$
$$2(\mathbf{2w - 9}) + 2w = 114 \quad \text{Substitute } 2w - 9 \text{ for } l\text{. Don't forget the parentheses.}$$
$$4w - 18 + 2w = 114 \quad \text{Distribute the multiplication by 2.}$$
$$6w - 18 = 114 \quad \text{Combine like terms: } 4w + 2w = 6w.$$
$$6w = 132 \quad \text{Add 18 to both sides.}$$
$$w = 22 \quad \text{Divide both sides by 6. This is the width of the poster.}$$

To find l, substitute 22 for w in the second equation of the system.

$$l = 2w - 9$$
$$l = 2(\mathbf{22}) - 9$$
$$l = 44 - 9$$
$$l = 35 \quad \text{This is the length of the poster.}$$

State The length of the poster is 35 inches and the width is 22 inches.

Check The perimeter is $2(35) + 2(22) = 70 + 44 = 114$ inches, and 35 inches is 9 inches less than twice 22 inches. The results check.

Self Check 3 **Gardening.** Tom has 150 feet of fencing to enclose a rectangular garden. If the garden's length is to be 5 feet less than 3 times its width, find the length and width of the garden. Length: 55 ft; width: 20 ft

Now Try Problem 23

Success Tip

In this section, to solve the application problems that involve two unknowns, we:

- Assign **two** variables.
- Form **two** equations.
- Solve a system of **two** equations.
- State **two** answers.
- Perform **two** checks.

Teaching Example 3 **Pets.** A pet owner wants to make a rectangular dog run from 28 feet of fencing. If the length of the dog run is to be 10 feet longer than the width, find the length and width of the area that is enclosed for the pet.
Answers: Length: 12 ft; width: 2 ft

3 Use Systems to Solve Number-Value Problems.

EXAMPLE 4 **Photography.** At a school, two picture packages are available, as shown in the illustration. Find the cost of a class picture and the cost of an individual wallet-size picture.

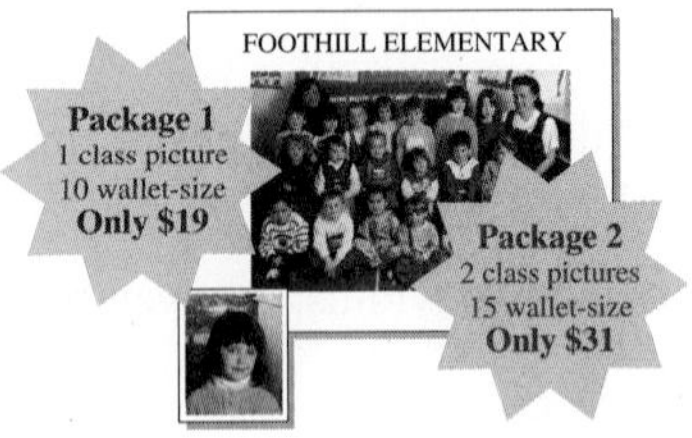

Analyze

- Package 1 contains 1 class picture and 10 wallet-size pictures.
- Package 2 contains 2 class pictures and 15 wallet-size pictures.
- Find the cost of a class picture and the cost of a wallet-size picture.

Assign Let c = the cost of one class picture and w = the cost of one wallet-size picture.

Form We can use the fact that **number · value = total value** to write an equation that models the first package. We note that (in dollars) the cost of 1 class picture is $1 \cdot c = c$ and the cost of 10 wallet-size pictures is $10 \cdot w = 10w$.

The cost of 1 class picture	plus	the cost of 10 wallet-size pictures	is	$19.
c	$+$	$10w$	$=$	19

To write an equation that models the second package, we note that (in dollars) the cost of 2 class pictures is $2 \cdot c = 2c$, and the cost of 15 wallet-size pictures is $15 \cdot w = 15w$.

The cost of 2 class pictures	plus	the cost of 15 wallet-size pictures	is	$31.
$2c$	$+$	$15w$	$=$	31

The resulting system is: $\begin{cases} c + 10w = 19 \\ 2c + 15w = 31 \end{cases}$

Solve We can use substitution or elimination to solve this system. If we use elimination, we can eliminate c as follows.

$$-2c - 20w = -38 \quad \text{Multiply both sides of } c + 10w = 19 \text{ by } -2.$$
$$2c + 15w = 31$$
$$-5w = -7 \quad \text{Add the equations to eliminate } c.$$
$$w = 1.4 \quad \text{Divide both sides by } -5. \text{ This is the cost of a wallet-size picture.}$$

To find c, substitute 1.4 for w in the first equation of the original system.

$$c + 10w = 19$$
$$c + 10(\mathbf{1.4}) = 19 \quad \text{Substitute 1.4 for } w.$$
$$c + 14 = 19 \quad \text{Multiply.}$$
$$c = 5 \quad \text{Subtract 14 from both sides. This is the cost of a class picture.}$$

State A class picture costs $5 and a wallet-size picture costs $1.40.

Check Package 1 has 1 class picture and 10 wallets: $5 +10($1.40) =$5 + $14 = $19. Package 2 has 2 class pictures and 15 wallets: 2($5) + 15($1.40) = $10 + $21 = $31. The results check.

Teaching Example 4 Team Pictures. Team Sports Photography offers two packages for their school sports pictures. The first package includes a team photo and 8 individual wallets for $29. The second package offers two team photos and one individual wallet for $35.50. Find the cost of a team photo and an individual wallet photo
Answers: Team photo: $17.00, wallet: $1.50

Self Check 4 **Sporting Goods.** Getting ready for soccer season, a coach purchased 6 practice cones and a portable goal for $23. On another trip to the sporting goods store, the coach purchased 4 more cones and 2 more goals, spending $42. Find the cost of a practice cone and a portable goal.

Practice cone: $0.50, portable goal: $20

Now Try Problem 27

4 Use Systems to Solve Interest, Uniform Motion, and Mixture Problems.

We can solve investment problems like those in Section 2.6 using a two-variable approach.

EXAMPLE 5

White-Collar Crime. Investigators discovered that a small business secretly moved \$150,000 out of the country to avoid paying income tax. Some of the money was invested in a Swiss bank account that paid 8% annual simple interest. The remainder was deposited in a Cayman Islands account, paying 7% annual simple interest. The investigation also revealed that the combined interest earned the first year was \$11,500. How much money was invested in each account?

Caution

It is incorrect to let

x = the amount invested in each account

This implies that *equal amounts* were invested in the Swiss and Cayman Island accounts. We do not know that.

Teaching Tip: You may want to remind your students of the one-variable approach that was used in Chapter 2 for investment problems like this.

Let x = amount at 8%
$150{,}000 - x$ = amount at 7%

$0.08x + 0.07(150{,}000 - x) = 11{,}500$

Analyze We are told that an unknown part of the \$150,000 was invested at an annual rate of 8% and the rest at 7%. Together, the accounts earned \$11,500 in interest.

Assign Let x = the amount invested in the Swiss account and y = the amount invested in the Cayman Islands account.

Form Because the total investment was \$150,000, we have

The amount invested in the Swiss account	plus	the amount invested in the Cayman Islands account	is	\$150,000.
x	$+$	y	$=$	150,000

We can use the formula $I = Prt$ to determine that x dollars invested for 1 year at 8% earns $x \cdot 0.08 \cdot 1 = 0.08x$ dollars. Similarly, y dollars invested for 1 year at 7% earns $y \cdot 0.07 \cdot 1 = 0.07y$ dollars. If the total combined interest earned was \$11,500, we have

The income on the 8% investment	plus	the income on the 7% investment	is	\$11,500.
$0.08x$	$+$	$0.07y$	$=$	11,500

The resulting system is: $\begin{cases} x + y = 150{,}000 \\ 0.08x + 0.07y = 11{,}500 \end{cases}$

Solve First, clear the second equation of decimals. Then we can use substitution or elimination to solve the system. If we use elimination, we can eliminate x as follows.

$$\begin{aligned} -8x - 8y &= -1{,}200{,}000 && \text{Multiply both sides of } x + y = 150{,}000 \text{ by } -8. \\ 8x + 7y &= 1{,}150{,}000 && \text{Multiply both sides of } 0.08x + 0.07y = 11{,}500 \text{ by } 100. \\ \hline -y &= -50{,}000 \\ y &= 50{,}000 && \text{Multiply both sides by } -1. \end{aligned}$$

To find x, substitute 50,000 for y in the first equation of the original system.

$$\begin{aligned} x + y &= 150{,}000 \\ x + \mathbf{50{,}000} &= 150{,}000 && \text{Substitute 50,000 for } y. \\ x &= 100{,}000 && \text{Subtract 50,000 from both sides.} \end{aligned}$$

State \$100,000 was invested in the Swiss bank account, and \$50,000 was invested in the Cayman Islands account.

Check

$$\begin{aligned} \$100{,}000 + \$50{,}000 &= \$150{,}000 && \text{The two investments total \$150,000.} \\ 0.08(\$100{,}000) &= \$8{,}000 && \text{The Swiss bank account earned \$8,000.} \\ 0.07(\$50{,}000) &= \$3{,}500 && \text{The Cayman Islands account earned \$3,500.} \end{aligned}$$

The combined interest is \$8,000 + \$3,500 = \$11,500. The results check.

Teaching Example 5 Investments. An executive invests \$22,000, some at 7% and the remainder at 6% annual simple interest. If he receives an annual income of \$1,420, how much is invested at each rate?
Answers: \$10,000 at 7%, \$12,000 at 6%

Self Check 5 **Investments.** A woman invested \$10,000, some at 9% and the remainder at 10% annual simple interest. The annual income from these two investments was \$975. How much was invested at each rate? \$2,500 at 9%, \$7,500 at 10%

Now Try ▶ Problem 35

EXAMPLE 6 **Boating.** A boat traveled 30 miles downstream in 3 hours and made the return trip in 5 hours. Find the speed of the boat in still water and the speed of the current.

Analyze Traveling downstream, the speed of the boat will be faster than it would be in still water. Traveling upstream, the speed of the boat will be slower than it would be in still water.

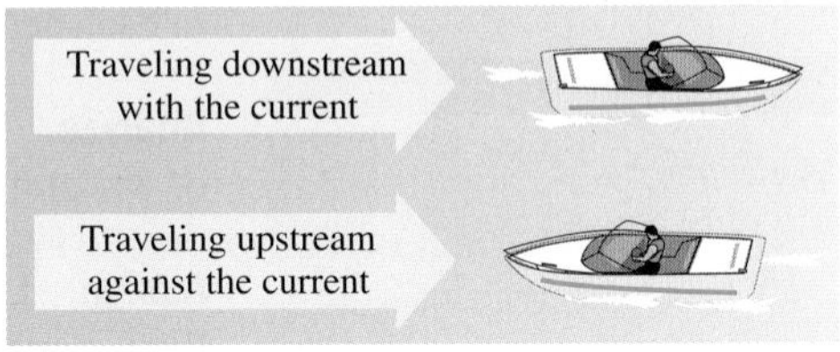

Assign Let $s =$ the speed of the boat in still water and $c =$ the speed of the current.

Form The speed of the boat going downstream is $s + c$ and the speed of the boat going upstream is $s - c$. Using the formula $\boldsymbol{d = rt}$, we find that $3(s + c)$ represents the distance traveled downstream and $5(s - c)$ represents the distance traveled upstream. We can organize the facts of the problem in a table.

	Rate	· Time	= Distance
Downstream	$s + c$	3	$3(s + c)$
Upstream	$s - c$	5	$5(s - c)$

Multiply $r \cdot t$ to obtain an expression for each distance traveled.

Enter this information first.

Set each of these expressions for distance traveled equal to 30 to form the system of equations.

Since each trip is 30 miles long, the Distance column of the table helps us to write two equations in two variables. To write each equation in standard form, use the distributive property.

$$\begin{cases} 3(s + c) = 30 \\ 5(s - c) = 30 \end{cases} \xrightarrow{\text{Distribute}} \begin{cases} 3s + 3c = 30 \\ 5s - 5c = 30 \end{cases}$$

Success Tip

At this stage, you also could divide both sides of $3(s + c) = 30$ by 3 and divide both sides of $5(s - c) = 30$ by 5.

Solve To eliminate c, we proceed as follows.

$15s + 15c = 150$ Multiply both sides of $3s + 3c = 30$ by 5.
$15s - 15c = 90$ Multiply both sides of $5s - 5c = 30$ by 3.
$30s \quad = 240$
$s = 8$ Divide both sides by 30. This is the speed of the boat in still water.

To find c, it appears that the calculations will be easiest if we use $3s + 3c = 30$.

$3s + 3c = 30$
$3(\mathbf{8}) + 3c = 30$ Substitute 8 for s.
$24 + 3c = 30$ Multiply.
$3c = 6$ Subtract 24 from both sides.
$c = 2$ Divide both sides by 3. This is the speed of the current.

State The speed of the boat in still water is 8 mph and the speed of the current is 2 mph.

Check With a 2-mph current, the boat's downstream speed will be $8 + 2 = 10$ mph. In 3 hours, it will travel $10 \cdot 3 = 30$ miles. With a 2-mph current, the boat's upstream speed will be $8 - 2 = 6$ mph. In 5 hours, it will cover $6 \cdot 5 = 30$ miles. The results check.

Teaching Example 6 **Boating.** A boat traveled 20 miles downstream in 2 hours and made the return trip in 5 hours. Find the speed of the boat in still water and the speed of the current.
Answers: Boat: 7 mph; current: 3 mph

Self Check 6 **Boating.** A boat traveled 24 miles downstream in 2 hours and made the return trip in 3 hours. Find the speed of the boat in still water and the speed of the current. Boat: 10 mph; current: 2 mph

Now Try Problem 41

The liquid and dry mixture problems that we studied in Section 2.6 can be solved with a two-variable approach.

EXAMPLE 7

Medical Technology. A laboratory technician has one batch of antiseptic that is 40% alcohol and a second batch that is 60% alcohol. She would like to make 8 fluid ounces of solution that is 55% alcohol. How many fluid ounces of each batch should she use?

Teaching Tip: You might want to remind students of the one-variable approach used to solve the liquid mixture problem in Example 4 on page 155 in Section 2.6.

Analyze Some 60% solution must be added to some 40% solution to make a 55% solution.

Assign Let $x =$ the number of ounces to be used from batch 1 and $y =$ the number of ounces to be used from batch 2.

Form The amount of alcohol in each solution is given by

Amount of solution · **strength of solution** = **amount of alcohol**

We can organize the facts of the problem in a table.

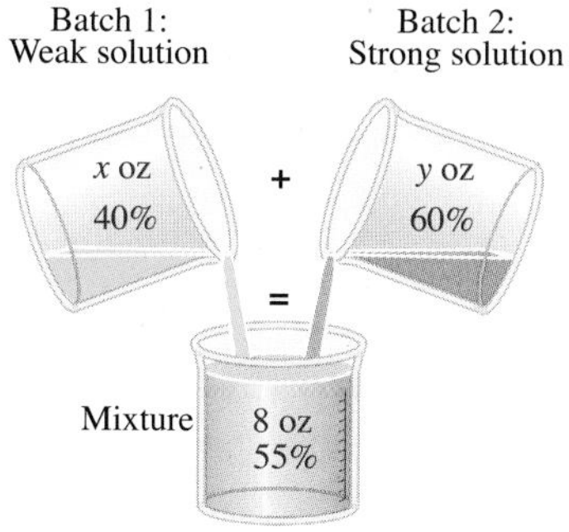

	Amount	· Strength	= Amount of alcohol
Batch 1 (too weak)	x	0.40	$0.40x$
Batch 2 (too strong)	y	0.60	$0.60y$
Mixture	8	0.55	$0.55(8)$

↑ One equation comes from information in this column. (Amount)
↑ 40%, 60%, and 55% have been expressed as decimals. (Strength)
↑ Another equation comes from information in this column. (Amount of alcohol)

The information in the table provides two equations.

$$\begin{cases} x + y = 8 \\ 0.40x + 0.60y = 0.55(8) \end{cases}$$

The number of ounces of batch 1 plus the number of ounces of batch 2 equals the total number of ounces in the mixture.

The amount of alcohol in batch 1 plus the amount of alcohol in batch 2 equals the total amount of alcohol in the mixture.

Solve We can solve this system by elimination. To eliminate x, we proceed as follows.

$$-40x - 40y = -320 \quad \text{Multiply both sides of the first equation by } -40.$$
$$40x + 60y = 440 \quad \text{Multiply both sides of the second equation by 100.}$$
$$20y = 120$$
$$y = 6 \quad \text{Divide both sides by 20. This is the number of ounces of batch 2 needed.}$$

To find x, we substitute 6 for y in the first equation of the original system.

$$x + y = 8$$
$$x + \mathbf{6} = 8 \quad \text{Substitute.}$$
$$x = 2 \quad \text{Subtract 6 from both sides. This is the number of ounces of batch 1 needed.}$$

State The technician should use 2 fluid ounces of the 40% solution and 6 fluid ounces of the 60% solution.

Check Note that 2 ounces + 6 ounces = 8 ounces, the required number. Also, the amount of alcohol in the two solutions is equal to the amount of alcohol in the mixture.

Alcohol in batch 1: $0.40x = 0.40(2) = 0.8$ ounces
Alcohol in batch 2: $0.60y = 0.60(6) = 3.6$ ounces
(Total: 4.4 ounces)
Alcohol in the mixture: $0.55(8) = 4.4$ ounces

The results check.

Teaching Example 7 Medications. How much 1% hydrocortisone cream and how much 5% hydrocortisone cream must be mixed to make 10 ounces of a 2% cream?
Answers: 1%: 7.5 oz; 5%: 2.5 oz

Self Check 7 **Dairy Products.** How much 1% milk and how much 4% milk must be combined to obtain 60 liters of 2% milk? 1%: 40 L; 4%: 20 L

Now Try ▶ Problem 45

EXAMPLE 8 **Breakfast Cereal.** One ounce of raisins (by weight) sells for 22¢ and one ounce of bran flakes (by weight) sells for 12¢. How many ounces of each should be used to create a 20-ounce box of raisin bran cereal that can be sold for 15¢ an ounce?

Analyze We will use a two-variable approach to solve this dry mixture problem.

Assign Let x = the number of ounces of raisins and y = the number of ounces of bran flakes that should be mixed.

Form The value of the mixture and the value of each of its components is given by

Amount · price = total value

Thus, the value of x ounces of raisins is $x \cdot 22$¢ or $22x$¢ and the value of y ounces of bran flakes is $y \cdot 12$¢ or $12y$¢. The sum of these values is also equal to the total value of the final mixture, that is $20 \cdot 15$¢ or 300¢. This information is shown in the table.

Teaching Tip: You might want to remind students of the one-variable approach used to solve the dry mixture problem in Example 5 on page 156 in Section 2.6.

	Amount	· Price	= Total value
Raisins	x	22	$22x$
Bran flakes	y	12	$12y$
Mixture	20	15	$20(15)$

Multiply amount · price three times to fill in this column.

One equation comes from this column. (Amount) One equation comes from this column. (Total value)

The facts of the problem give the following two equations:

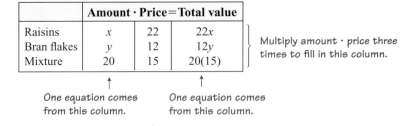

The number of ounces of raisins	plus	the number of ounces of bran flakes	is	20.
x	$+$	y	$=$	20

The value of the raisins	plus	the value of the bran flakes	is	the value of the mixture.
$22x$	$+$	$12y$	$=$	$20(15)$

Solve To find out how many ounces of raisins and bran flakes are needed we solve the following system:

$$\begin{cases} x + y = 20 \\ 22x + 12y = 300 \quad \text{Multiply: } 20(15) = 300. \end{cases}$$

To solve this system by substitution, we can solve the first equation for x:

$$x + y = 20$$
$$x = 20 - y \quad \text{This is the substitution equation.}$$

Then we substitute $20 - y$ for x in the second equation of the system and solve for y.

$$\begin{aligned} 22x + 12y &= 300 \\ 22(20 - y) + 12y &= 300 && \text{Substitute } 20 - y \text{ for } x. \\ 440 - 22y + 12y &= 300 && \text{Distribute the multiplication by 22.} \\ 440 - 10y &= 300 && \text{Combine like terms: } -22y + 12y = -10y. \\ -10y &= -140 && \text{Subtract 440 from both sides.} \\ y &= 14 && \text{Divide both sides by } -10. \text{ This is the number of ounces of bran flakes needed.} \end{aligned}$$

To find x, we substitute 14 for y in the substitution equation and simplify the right side.

$$\begin{aligned} x &= 20 - y \\ &= 20 - 14 && \text{Substitute 14 for } y. \\ &= 6 && \text{This is the number of ounces of raisins needed.} \end{aligned}$$

State To obtain 20 ounces of raisin bran cereal, 6 ounces of raisins and 14 ounces of bran flakes should be combined.

Check When 6 ounces of raisins and 14 ounces of bran flakes are combined, the result is 20 ounces of raisin bran cereal. The 6 ounces of raisins are valued at $6 \cdot 22¢ = 132¢$ and the 14 ounces of bran flakes are valued at $14 \cdot 12¢ = 168¢$. The sum of those values, $132¢ + 168¢ = 300¢$, is the same as the value of the mixture, $20 \cdot 15¢ = 300¢$. The results check.

Teaching Example 8 Celebrations. How much multi-colored confetti worth \$3 a pound should be mixed with metallic star confetti worth \$4 a pound to make 400 pounds of a mixture worth \$3.65 a pound?
Answers: Multi-colored confetti: 140 lb; metallic star confetti: 260 lb

Self Check 8 **Gardening.** How much planting mix, worth \$36 per cubic yard, and how much topsoil, worth \$24 per cubic yard, should be combined to make 200 cubic yards of a mixture worth \$31.20 per cubic yard?
Planting mix: 120 yd^3; topsoil: 80 yd^3

Now Try ▶ Problem 49

SECTION 4.4 STUDY SET

VOCABULARY

Fill in the blanks.

1. Two angles are said to be ___complementary___ if the sum of their measures is 90°. Two angles are said to be ___supplementary___ if the sum of their measures is 180°.

2. Problems that involve moving vehicles are called uniform ___motion___ problems. Problems that involve combining ingredients are called ___mixture___ problems. Problems that involve collections of different items having different values are called number-___value___ problems.

CONCEPTS

3. A length of pipe is to be cut into two pieces. The longer piece is to be 1 foot less than twice the shorter piece. Write two equations that model the situation. $x + y = 20$, $y = 2x - 1$

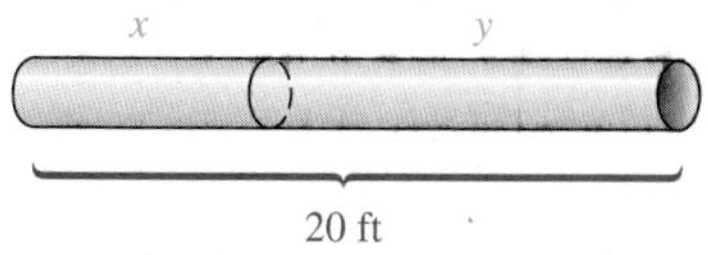

4. Two angles are complementary. The measure of the larger angle is four times the measure of the smaller angle. Write two equations that model the situation. $x + y = 90, x = 4y$

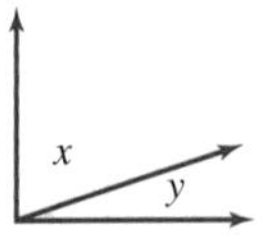

5. Two angles are supplementary. The measure of the smaller angle is 25° less than the measure of the larger angle. Write two equations that model the situation.
$x + y = 180, y = x - 25$

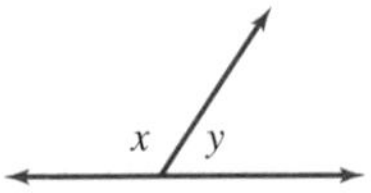

6. The perimeter of the following Ping-Pong table is 28 feet. The length is 4 feet more than the width. Write two equations that model the situation. $2l + 2w = 28, l = w + 4$

7. Let x = the cost of a chicken taco, in dollars, and y = the cost of a beef taco, in dollars. Write an equation that models the offer shown in the advertisement. $5x + 2y = 15$

8. a. Complete the following table.

	Principal	Rate	Time	Interest
City Bank	x	5%	1 yr	$0.05x$
USA Savings	y	11%	1 yr	$0.11y$

b. A total of $50,000 was deposited in the two accounts. Use that information to write an equation about the principal. $x + y = 50{,}000$

c. A total of $4,300 was earned by the two accounts. Use that information to write an equation about the interest. $0.05x + 0.11y = 4{,}300$

9. For each case on the right, write an algebraic expression that represents the speed of the canoe in miles per hour if its speed in still water is x mph. Downstream: $x + c$, upstream: $x - c$

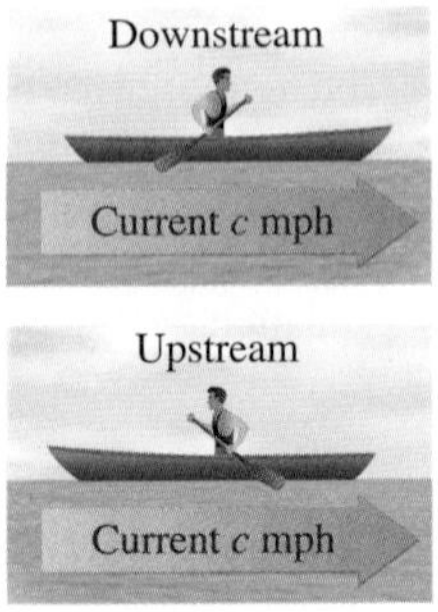

10. Complete the table, which contains information about an airplane flying in windy conditions.

	Rate	Time	Distance
With wind	$x + y$	3	$3(x + y)$
Against wind	$x - y$	5	$5(x - y)$

11. a. If the contents of the two test tubes are poured into a third tube, how much solution will the third tube contain? (mL stands for milliliter. A milliliter is about 15 drops from an eyedropper.) $(x + y)$ mL

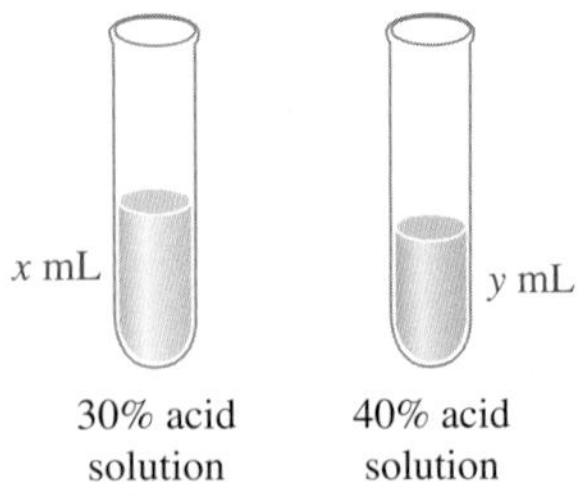

b. Which of the following strengths could the mixture possibly be: 27%, 33%, or 44% acid solution? 33%

12. a. Complete the table, which contains information about mixing two salt solutions to get 12 gallons of a 3% salt solution.

	Amount	Strength	Amount of salt
Weak	x	0.01	$0.01x$
Strong	y	0.06	$0.06y$
Mix	12	0.03	$12(0.03)$

b. Use the information from the Amount column to write an equation. $x + y = 12$

c. Use the information from the Amount of salt column to write an equation. $0.01x + 0.06y = 12(0.03)$

GUIDED PRACTICE

See Example 2.

13. **Complementary Angles.** Two angles are complementary. The measure of one angle is 10° more than three times the measure of the other. Find the measure of each angle. 20°, 70°

14. **Supplementary Angles.** Two angles are supplementary. The measure of one angle is 20° less than 19 times the measure of the other. Find the measure of each angle. 10°, 170°

15. **Supplementary Angles.** The difference of the measures of two supplementary angles is 80°. Find the measure of each angle. 50°, 130°

16. **Complementary Angles.** Two angles are complementary. The measure of one angle is 15° more than one-half of the measure of the other. Find the measure of each angle. 40°, 50°

APPLICATIONS

Write a system of two equations in two variables to solve each problem.

17. **Tree Trimming.** When fully extended, the arm on a tree service truck is 51 feet long. If the upper part of the arm is 7 feet shorter than the lower part, how long is each part of the arm? Upper arm: 22 ft, lower arm: 29 ft

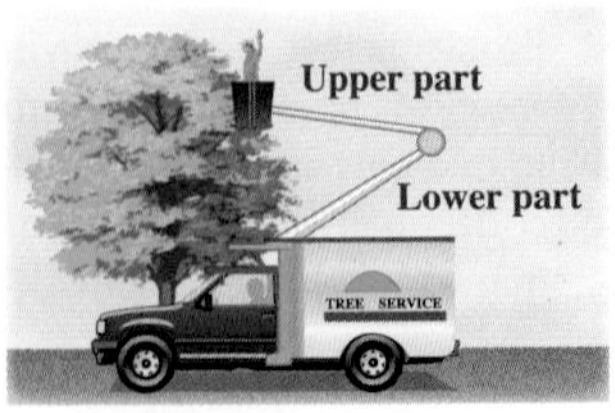

18. Alaska. Most of the 1,422-mile-long Alaskan Highway is actually in Canada. Find the length of the highway that is in Alaska and the length of the highway that is in Canada if it is known that the difference in the lengths is 1,020 miles. Alaska: 201 mi; Canada: 1,221 mi

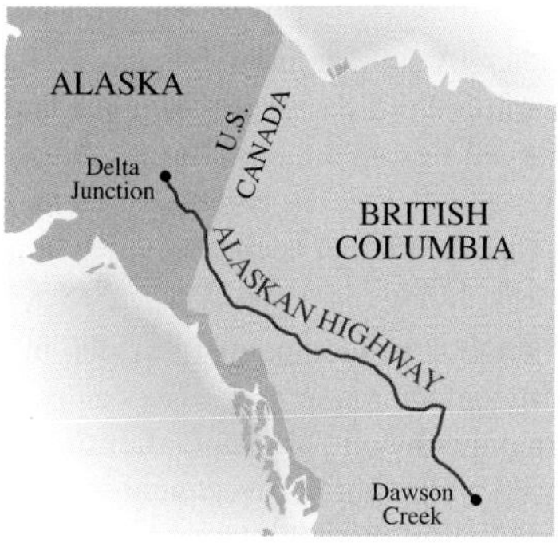

19. Government. The salaries of the president and vice president of the United States total $627,300 a year. If the president makes $172,700 more than the vice president, find each of their salaries. President: $400,000; vice president: $227,300

20. Causes of Death. According to the *National Vital Statistics Reports,* in 2007, the number of Americans who died from motor vehicle accidents was about twice the number who died from falls. If the total number of deaths from these two causes was approximately 66,000, how many Americans died from each cause in 2007? 22,000 from falls; 44,000 from motor vehicle accidents

21. Monuments. The Marine Corps War Memorial in Arlington, Virginia, portrays the raising of the U.S. flag on Iwo Jima during World War II. Find the measures of the two angles shown if the measure of $\angle 1$ is 15° less than twice the measure of $\angle 2$. $\angle 1$: 115°; $\angle 2$: 65°

22. Physical Therapy. To rehabilitate her knee, an athlete does leg extensions. Her goal is to regain a full 90° range of motion in this exercise. Use the information in the illustration to determine her current range of motion in degrees and the number of degrees of improvement she still needs to make. 72°, 18°

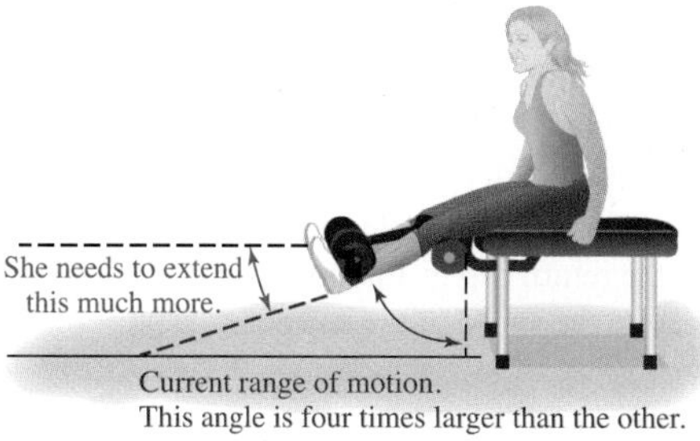

23. Theater Screens. At an IMAX theater, the giant rectangular movie screen has a width 26 feet less than its length. If its perimeter is 332 feet, find the length and the width of the screen. Length: 96 ft, width: 70 ft

24. Art. In 1770, Thomas Gainsborough painted *The Blue Boy.* The sum of the length and width of the painting is 118 inches. The difference of the length and width is 22 inches. Find the length and width. Length 70 in.; width 48 in.

© Huntington Library/Superstock

25. Geometry. A 50-meter path surrounds a rectangular garden. The width of the garden is two-thirds its length. Find the length and width. Length: 15 m, width: 10 m

26. Ballroom Dancing. A rectangular-shaped dance floor has a perimeter of 200 feet. If the floor were 20 feet wider, its width would equal its length. Find the length and width of the dance floor. Length: 60 ft, width: 40 ft

27. Empty Cartridges. A bank recycles its empty printer and copier cartridges. In January, the bank received $40 for recycling 5 printer and 2 copier cartridges. In February, the bank received $57 for recycling 6 printer and 3 copier cartridges. How much is the bank paid for an empty printer cartridge and for an empty copier cartridge? Printer: $2; copier: $15

28. Thanksgiving Dinner. There are a total of 510 calories in 6 ounces of turkey and one slice of pumpkin pie. There are a total of 580 calories in 4 ounces of turkey and two slices of pumpkin pie. How many calories are there in 1 ounce of turkey and in one slice of pumpkin pie? 55 calories in one ounce of turkey, 180 calories in one slice of pumpkin pie

29. from **Campus to Careers**

Photographer

© Andrei Contiu/Shutterstock.com

Suppose you are a wedding photographer and you sell:

Package 1: one 10 × 14 and ten 8 × 10 color photos for $239.50

Package 2: one 10 × 14 and five 8 × 10 color photos for $134.50

A newlywed couple buys Package 1 and decides that they want one more 10 × 14 and one more 8 × 10 photograph. At the same prices, what should you charge them for each additional photograph? $29.50 for a 10 × 14; $21 for an 8 × 10

30. Buying Painting Supplies. Two partial receipts for paint supplies are shown. (Assume no sales tax was charged.) Find the cost of one gallon of latex paint and the cost of one paint brush. Paint: $30 a gallon; brush: $10

31. Collecting Stamps. Determine the price of an Elvis Presley stamp and a Statue of Liberty stamp given the following information.

- One Elvis stamp and one Liberty stamp cost a total of 63¢.
- A sheet of 40 Elvis stamps and a sheet of 20 Liberty stamps cost a total of \$18.40. (*Hint:* \$18.40 = 1,840¢.) Elvis: 29¢; Liberty: 34¢

32. Recycling. A boy scout troop earned \$24 by recycling a total of 330 beverage containers. The recycling rates are shown below. How many of the small capacity containers and how many of the large capacity containers did they recycle? (*Hint:* 5¢ = \$.05 and 10¢ = \$0.10) Small: 180; large: 150

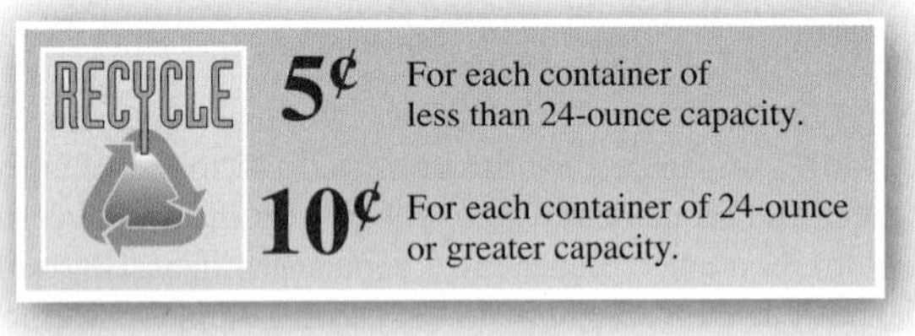

33. Selling Ice Cream. At a store, ice cream cones cost \$1.80 and sundaes cost \$3.30. One day the receipts for a total of 148 cones and sundaes were \$360.90. How many cones were sold? How many sundaes?
Cones: 85, sundaes: 63

34. Buying Tickets. The ticket prices for a movie are shown. Receipts for one showing were \$1,740 for an audience of 190 people. How many general admission tickets and how many senior citizen tickets were sold? General admission: 150; senior: 40

35. Student Loans. A college used a \$5,000 gift from an alumnus to make two student loans. The first was at 5% annual simple interest to a nursing student. The second was at 7% annual simple interest to a business major. If the college collected \$310 in interest the first year, how much was loaned to each student? Nursing: \$2,000, business: \$3,000

36. Financial Planning. In investing \$6,000 of a couple's money, a financial planner put some of it into a savings account paying 6% annual simple interest. The rest was invested in a riskier mini-mall development plan paying 12% annual simple interest. The combined interest earned for the first year was \$540. How much money was invested at each rate?
6%: \$3,000, 12%: \$3,000

37. Investing a Bonus. A businessman invested part of his \$40,000 end-of-the-year bonus in an international fund that paid an annual yield of 8%. The rest of the bonus was invested in an offshore bank that paid an annual yield of 9%. Find the amount of each investment if he made a total of \$3,415 in interest from them the first year. International fund: \$18,500; offshore bank: \$21,500

38. Pension Funds. A state employees' pension fund invested a total of one million dollars in two accounts that earned 3.5% and 4.5% annual simple interest. At the end of the year, the total interest earned from the two investments was \$39,000. How much was invested at each rate?
3.5%: 600,000; 4.5%: \$400,000

39. Losses. A CEO deposited part of \$22,000 in an account paying 4% annual simple interest. The rest of the money was invested in a biotech company that, after only one year, caused him to lose 3% of his initial investment in it. Find the amount of each investment if the net interest he earned the first year was only \$110.
4% account: \$11,000; biotech: \$11,000

40. Lottery Winnings. After winning \$60,000 in the lottery, a retired teacher gave \$10,000 of it to her grandchildren. She invested part of the remainder in a growth fund that earned 4.4% annually and the rest in certificates of deposit paying a 5.8% annual percentage yield. The interest that she received on these two investments totaled \$2,732 at the end of the first year. Find the amount of each investment. Growth fund: \$12,000; certificates of deposit: \$38,000

41. The Gulf Stream. The gulf stream is a warm ocean current of the North Atlantic Ocean that flows northward, as shown below. Heading north with the gulf stream, a cruise ship traveled 300 miles in 10 hours. Against the current, it took 15 hours to make the return trip. Find the speed of the ship in still water and the speed of the current. Still water: 25 mph, current: 5 mph

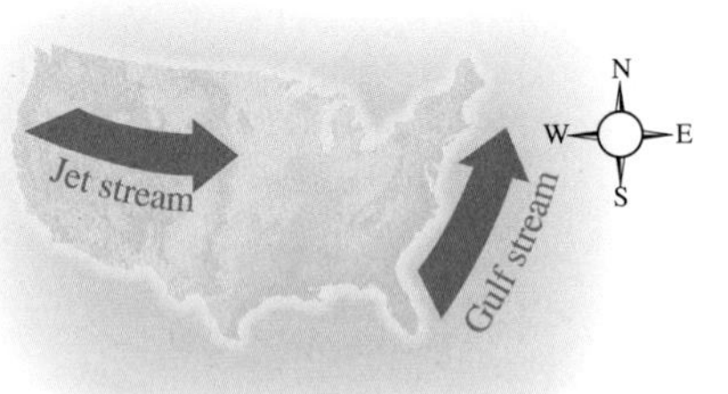

42. The Jet Stream. The jet stream is a strong wind current that flows across the United States, as shown above. Flying with the jet stream, a plane flew 3,000 miles in 5 hours. Against the same wind, the trip took 6 hours. Find the speed of the plane in still air and the speed of the wind current. Still air: 550 mph, wind: 50 mph

43. Aviation. An airplane can fly with the wind a distance of 800 miles in 4 hours. However, the return trip against the wind takes 5 hours. Find the speed of the plane in still air and the speed of the wind.
Still air: 180 mph, wind: 20 mph

44. Boating. A boat can travel 24 miles downstream in 2 hours and can make the return trip in 3 hours. Find the speed of the boat in still water and the speed of the current.
Still water: 10 mph, current: 2 mph

45. Marine Biology. A marine biologist wants to set up an aquarium containing 3% salt water. He has two tanks on hand that contain 6% and 2% salt water. How much water from each tank must he use to fill a 32-gallon aquarium with a 3% saltwater mixture? 8 gal 6% salt water, 24 gal 2% salt water

▶ **46. Commemorative Coins.** A foundry has been commissioned to make souvenir coins. The coins are to be made from an alloy that is 40% silver. The foundry has on hand two alloys, one with 50% silver content and one with a 25% silver content. How many kilograms of each alloy should be used to make 20 kilograms of the 40% silver alloy? 12 kg 50% alloy, 8 kg 25% alloy

▶ **47. Cleaning Floors.** A custodian is going to mix a 4% ammonia solution and a 12% ammonia solution to get 1 gallon (128 fluid ounces) of a 9% ammonia solution. How many fluid ounces of the 4% solution and the 12% solution should be used? 4% solution: 48 oz; 12% solution: 80 oz

48. Mouthwash. A pharmacist has a mouthwash solution that is 6% ethanol alcohol and another that is 18% ethanol alcohol. How many milliliters of each must be mixed to make 750 milliliters of a mouthwash that is 10% ethanol alcohol? 500 mL of 6%, 250 mL of 18%

▶ **49. Coffee Sales.** A coffee supply store waits until the orders for its special blend reach 100 pounds before making up a batch. Columbian coffee selling for \$8.75 a pound is blended with Brazilian coffee selling for \$3.75 a pound to make a product that sells for \$6.35 a pound. How much of each type of coffee should be used to make the blend that will fill the orders? 52 lb of the \$8.75, 48 lb of the \$3.75

▶ **50. Mixing Nuts.** A merchant wants to mix peanuts with cashews, as shown in the illustration, to get 48 pounds of mixed nuts that will be sold at \$6 per pound. How many pounds of each should the merchant use? 32 lb peanuts, 16 lb cashews

51. Gourmet Foods. A New York delicatessen sells marinated mushrooms for \$12 a pint and stuffed Kalamata olives for \$9 a pint. How many pints of each should be used to get 20 pints of a mixture that will sell for \$10 a pint?

$\frac{20}{3} = 6\frac{2}{3}$ pints of mushrooms; $\frac{40}{3} = 13\frac{1}{3}$ pints of olives

52. Herbs. Ginger root powder sells for \$6.50 a pound and ginkgo leaf powder sells for \$9.50 a pound. How many pounds of each should be used to make 15 pounds of a mixture that sells for \$7 a pound?

$\frac{25}{2} = 12\frac{1}{2}$ lb of ginger root powder; $\frac{5}{2} = 2\frac{1}{2}$ lb of ginkgo leaf powder

WRITING

53. Explain why a table is helpful in solving uniform motion and mixture problems.

▶ **54.** A man paid \$89 for two shirts and four pairs of socks. If we let x = the cost of a shirt, in dollars, and y = the cost of a pair of socks, in dollars, an equation modeling the purchase is $2x + 4y = 89$. Explain why there is not enough information to determine the cost of a shirt or the cost of a pair of socks.

REVIEW

Graph each inequality. Then describe the graph using interval notation. See AIE Appendix 3.

55. $x < 4$ $(-\infty, 4)$

56. $x \geq -3$ $[-3, \infty)$

57. $-1 < x \leq 2$ $(-1, 2]$

58. $-2 \leq x \leq 0$ $[-2, 0]$

CHALLENGE PROBLEMS

59. Three types of hardware items, nails, bolts, and nuts, are placed on a scale as shown below. On the last scale, how many nails will it take to balance 1 nut? 2 nails

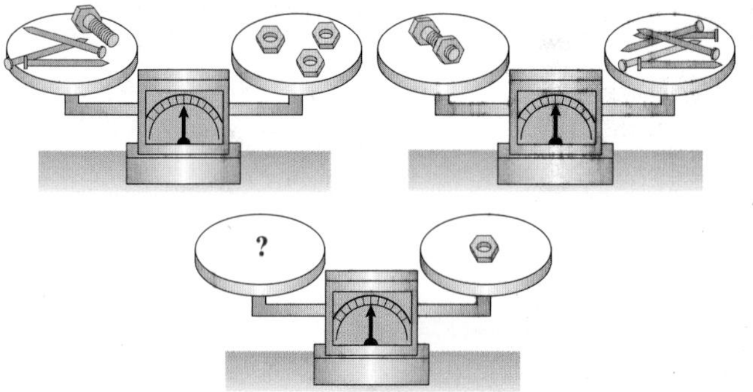

▶ **60. Farming.** In a pen of goats and chickens, there are 40 heads and 130 feet. How many goats and chickens are in the pen? 25 goats, 15 chickens

SECTION 4.5

Solving Systems of Linear Inequalities

OBJECTIVES

1 Solve a system of linear inequalities by graphing.

2 Solve application problems involving systems of linear inequalities.

ARE YOU READY? *Are You Ready? exercises available online at www.webassign.net/brookscole*

The following problems review some basic skills that are needed when solving systems of linear inequalities.

1. True or false: $3(-4) + 7 > -9$ True
2. Graph: $4x - 3y = -12$ See AIE Appendix 3.
3. Determine whether each of the following points lies *above, below,* or *on* the line graphed in problem 2.
   a. $(-2, 3)$ Above **b.** $(-4, -5)$ Below **c.** $(-3, 0)$ On
4. Is the boundary of the graph of $y < 3x - 1$ solid or dashed? Dashed

In Section 4.1, we solved systems of linear *equations* graphically by finding the point of intersection of two lines. Now we consider **systems of linear inequalities**, such as

$$\begin{cases} x + y \geq -1 \\ x - y \geq 1 \end{cases}$$

To solve systems of linear inequalities, we again find the points of intersection of graphs. In this case, however, we are not looking for an intersection of two lines, but an intersection of two regions.

1 Solve a System of Linear Inequalities by Graphing.

A **solution of a system of linear inequalities** is an ordered pair that makes each inequality true. *To solve a system of linear inequalities* means to find all of its solutions. This can be done by graphing each inequality on the same set of axes and finding the points that are common to every graph in the system.

EXAMPLE 1 Graph the solutions of the system: $\begin{cases} x + y \geq -1 \\ x - y \geq 1 \end{cases}$

Strategy We will graph the solutions of $x + y \geq -1$ in one color and the solutions of $x - y \geq 1$ in another color on the same coordinate system.

Why We need to see where the graphs of the two inequalities intersect (overlap).

Solution To graph $x + y \geq -1$, we begin by graphing the boundary line $x + y = -1$. Since the inequality contains an $\geq$ symbol, the boundary is a solid line. Because the coordinates of the test point (0, 0) satisfy $x + y \geq -1$, we shade (in red) the side of the boundary that contains (0, 0). See part (a) of the figure on the next page.

Graph the boundary: The intercept method

$x + y = -1$

x	y	(x, y)
0	-1	$(0, -1)$
-1	0	$(-1, 0)$

***Shading: Check the test point* (0, 0)**

$x + y \geq -1$

$0 + 0 \overset{?}{\geq} -1$ Substitute.

$0 \geq -1$ True

(0, 0) is a solution of $x + y \geq -1$.

The Language of Algebra

To solve a system of linear inequalities, we **superimpose** the graphs of the inequalities. That is, we place one graph over the other. Most video camcorders can *superimpose* the date and time over the picture being recorded.

In part (b) of the figure, we superimpose the graph of $x - y \geq 1$ on the graph of $x + y \geq -1$ so that we can determine the points that the graphs have in common. To graph $x - y \geq 1$, we graph the boundary $x - y = 1$ as a solid line. Since the test point (0, 0) does not satisfy $x - y \geq 1$, we shade (in blue) the half-plane that does not contain (0, 0).

Graph the boundary: The intercept method

$x - y = 1$

x	y	(x, y)
0	−1	(0, −1)
1	0	(1, 0)

***Shading: Check the test point* (0, 0)**

$x - y \geq 1$

$0 - 0 \overset{?}{\geq} 1$ Substitute.

$0 \geq 1$ False

(0, 0) is not a solution of $x - y \geq 1$.

Teaching Tip: Select a sampling of points from the red, blue, and purple regions, as well as on the boundaries. For each point, ask your students if its coordinates satisfy the first inequality or the second inequality, and if it is a solution of the system.

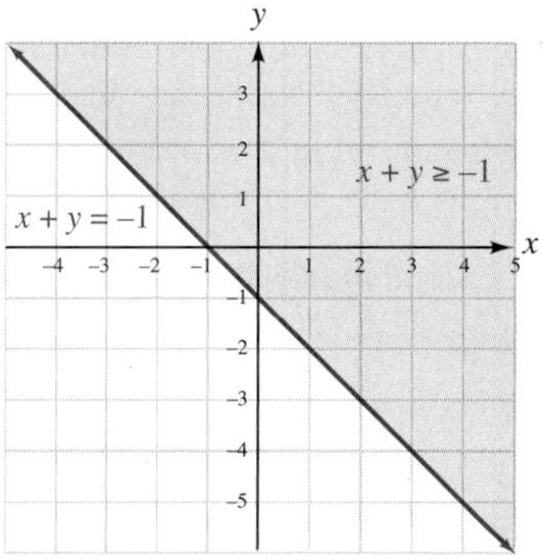

The graph of $x + y \geq -1$ is shaded in red.

(a)

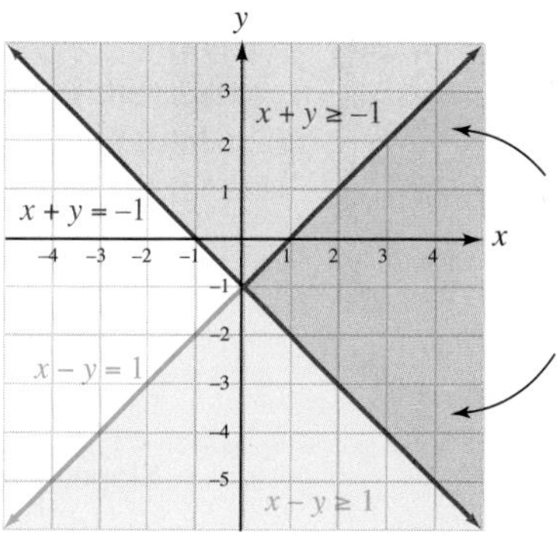

The solutions of the system are shaded in purple. The purple region is the intersection or overlap of the red and blue shaded regions. It includes portions of each boundary.

The graph of $x - y \geq 1$ is shaded in blue. It is drawn over the graph of $x + y \geq -1$.

(b)

Success Tip

Colored pencils often are used to graph systems of inequalities. A standard pencil also can be used. Just draw different patterns of lines instead of shading.

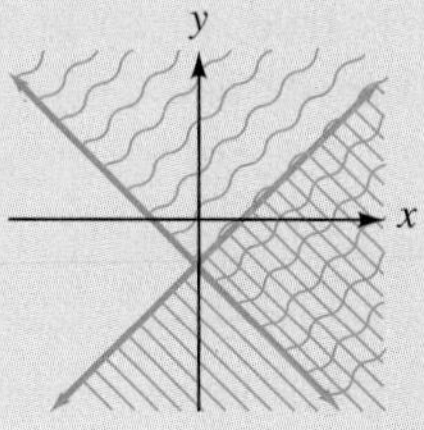

In part (b) of the figure, the area that is shaded twice represents the solutions of the given system. Any point in the **doubly shaded region** in purple (including the purple portions of each boundary) has coordinates that satisfy *both* inequalities.

Since there are infinitely many solutions, we cannot check each of them. However, as an informal check, we can select one ordered pair, say (4, 1), that lies in the doubly shaded region and show that its coordinates satisfy both inequalities of the system.

Check: $x + y \geq -1$ The first inequality. $\quad x - y \geq 1$ The second inequality.

$4 + 1 \overset{?}{\geq} -1 \qquad 4 - 1 \overset{?}{\geq} 1$

$5 \geq -1$ True $\qquad 3 \geq 1$ True

The resulting true statements verify that (4, 1) is a solution of the system. If we pick a point that is not in the doubly shaded region, such as (1, 3), (−2, −2), or (0, −4), the coordinates of that point will fail to satisfy one or both of the inequalities.

Teaching Example 1 Graph the solutions of the system:
$\begin{cases} x + y < 3 \\ x - y \leq -1 \end{cases}$
Answer:

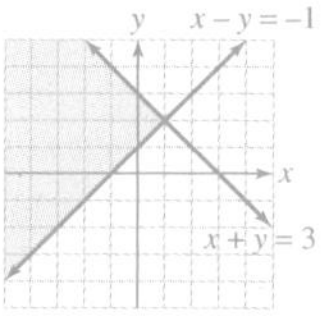

Self Check 1 Graph the solutions of the system: $\begin{cases} x - y \leq 2 \\ x + y \geq -1 \end{cases}$

Now Try Problem 15

See AIE Appendix 3.

In general, to solve systems of linear inequalities, we will follow these steps.

Solving Systems of Linear Inequalities

1. Graph each inequality on the same rectangular coordinate system.
2. Use shading to highlight the intersection of the graphs (the region where the graphs overlap). The points in this region are the solutions of the system.
3. As an informal check, pick a point from the region where the graphs intersect and verify that its coordinates satisfy each inequality of the original system.

EXAMPLE 2 Graph the solutions of the system: $\begin{cases} y > 3x \\ 2x + y < 4 \end{cases}$

Strategy We will graph the solutions of $y > 3x$ in one color and the solutions of $2x + y < 4$ in another color on the same coordinate system to see where the graphs intersect.

Why The solution set of the system is the set of all points in the intersection of the two graphs.

Solution To graph $y > 3x$, we begin by graphing the boundary line $y = 3x$. Since the inequality contains an $>$ symbol, the boundary is a dashed line. Because the boundary passes through (0, 0), we use (2, 0) as the test point instead. Since (2, 0) does not satisfy $y > 3x$, we shade (in red) the half-plane that does not contain (2, 0). See part (a) of the following figure.

Success Tip

You do not have to use the same test point when graphing the two inequalities.

Graph the boundary: Slope and y-intercept

$$y = 3x + 0$$

$m = 3$ $\quad b = 0$

$$\text{Slope: } \frac{\text{Rise}}{\text{Run}} = \frac{3}{1} \qquad y\text{-intercept: } (0, 0)$$

***Shading: Check the test point* (2, 0)**

$y > 3x$

$\mathbf{0} \stackrel{?}{>} 3(\mathbf{2})$ Substitute.

$0 > 6$ False

Since $0 > 6$ is false, (2, 0) is not a solution of $y > 3x$.

In part (b) of the figure, we superimpose the graph of $2x + y < 4$ on the graph of $y > 3x$ to determine the points that the graphs have in common. To graph $2x + y < 4$, we graph the boundary $2x + y = 4$ as a dashed line. Then we shade (in blue) the half-plane that contains (0, 0), because the coordinates of the test point satisfy $2x + y < 4$.

Success Tip

The ordered pairs that lie on a dashed boundary line are never part of the solution of a system of linear inequalities.

Graph the boundary: The intercept method

$$2x + y = 4$$

x	y	(x, y)
0	4	(0, 4)
2	0	(2, 0)

***Shading: Use the test point* (0, 0)**

$2x + y < 4$

$2(\mathbf{0}) + \mathbf{0} \stackrel{?}{<} 4$ Substitute.

$0 < 4$ True

Since $0 < 4$ is true, (0, 0) is a solution of $2x + y < 4$.

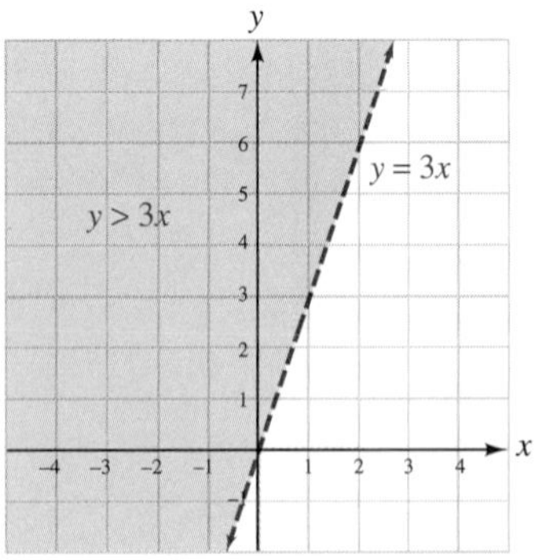
y
y = 3x
y > 3x
x

The graph of $y > 3x$ is shaded in red.

(a)

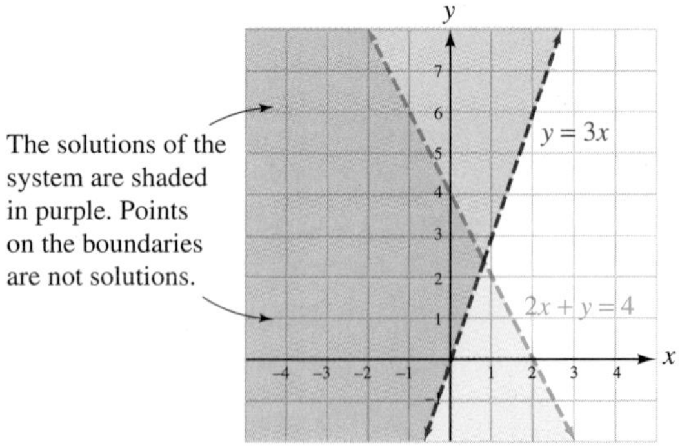
The solutions of the system are shaded in purple. Points on the boundaries are not solutions.
y
y = 3x
2x + y = 4
x

The graph of $2x + y < 4$ is shaded in blue. It is drawn over the graph of $y > 3x$.

(b)

In part (b) of the figure, the area that is shaded twice represents the solutions of the given system. Any point in the doubly shaded region in purple has coordinates that satisfy both inequalities. Pick a point in the region and show that this is true. Note that the region does not include either boundary; points on the boundaries are not solutions of the system.

Teaching Example 2 Graph the solutions of the system: $\begin{cases} 3x + 2y < 6 \\ y > 2x \end{cases}$
Answer:

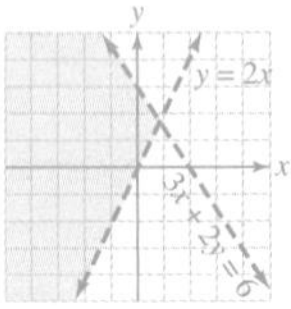

Self Check 2 Graph the solutions of the system: $\begin{cases} x + 3y < 3 \\ y > \frac{1}{3}x \end{cases}$

Now Try Problem 19

See AIE Appendix 3.

EXAMPLE 3 Graph the solutions of the system: $\begin{cases} x \le 2 \\ y > 3 \end{cases}$

Strategy We will graph the solutions of $x \le 2$ in one color and the solutions of $y > 3$ in another color on the same coordinate system to see where the graphs of the two inequalities intersect.

Why The solution set of the system is the set of all points in the intersection of the two graphs.

Solution The boundary of the graph of $x \le 2$ is the line $x = 2$. Since the inequality contains the symbol $\le$, we draw the boundary as a solid line. The test point (0, 0) makes $x \le 2$ true, so we shade the side of the boundary that contains (0, 0). See part (a) of the figure below.

Graph the boundary: A table of solutions

$x = 2$

x	y	(x, y)
2	0	(2, 0)
2	2	(2, 2)
2	4	(2, 4)

***Shading: Check the test point* (0, 0)**

$x \le 2$

$0 \le 2$ True

Since $0 \le 2$ is true, (0, 0) is a solution of $x \le 2$.

In part (b) of the figure, the graph of $y > 3$ is superimposed over the graph of $x \le 2$. The boundary of the graph of $y > 3$ is the line $y = 3$. Since the inequality contains the symbol $>$, we draw the boundary as a dashed line. The test point (0, 0) makes $y > 3$ false, so we shade the side of the boundary that does not contain (0, 0).

Graph the boundary: A table of solutions

$y = 3$

x	y	(x, y)
0	3	(0, 3)
1	3	(1, 3)
4	3	(4, 3)

***Shading: Check the test point* (0, 0)**

$y > 3$

$0 > 3$ False

Since $0 > 3$ is false, (0, 0) is not a solution of $y > 3$.

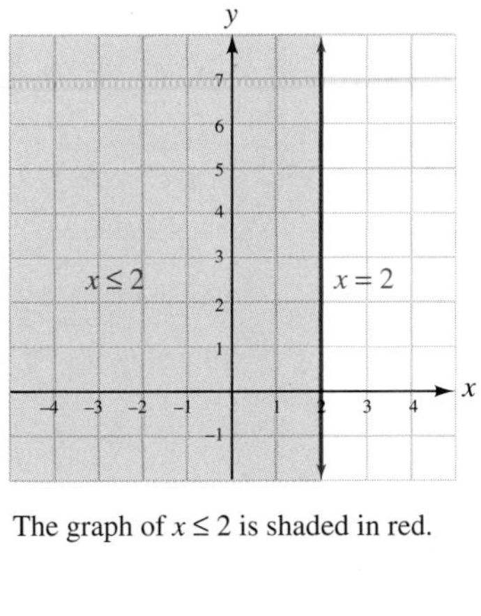

y
x
x ≤ 2
x = 2

The graph of $x \le 2$ is shaded in red.

(a)

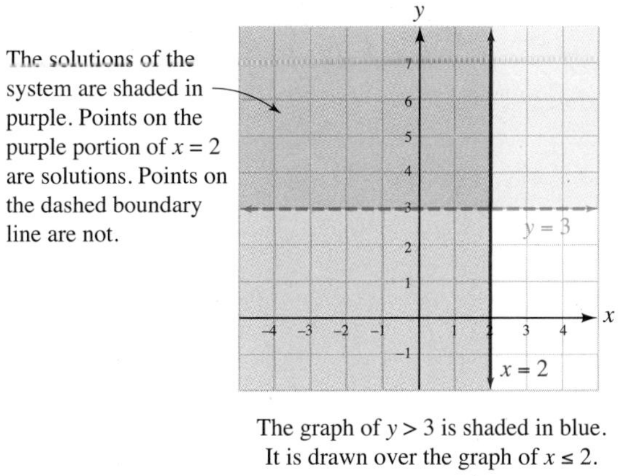

The solutions of the system are shaded in purple. Points on the purple portion of x = 2 are solutions. Points on the dashed boundary line are not.
y
x
y = 3
x = 2

The graph of $y > 3$ is shaded in blue. It is drawn over the graph of $x \le 2$.

(b)

The area that is shaded twice represents the solutions of the system of inequalities. Any point in the doubly shaded region in purple has coordinates that satisfy both inequalities, including the purple portion of the $x = 2$ boundary. Pick a point in the region and show that this is true.

Teaching Example 3 Graph the solutions of the system: $\begin{cases} y \geq -2 \\ x > 1 \end{cases}$

Answer:

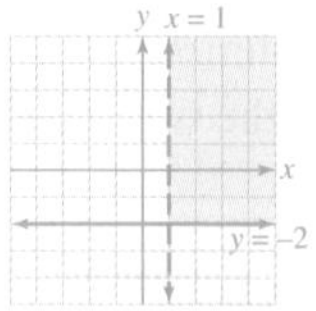

Self Check 3 Graph the solutions of the system: $\begin{cases} y \leq 1 \\ x > 2 \end{cases}$ See AIE Appendix 3.

Now Try Problem 23

EXAMPLE 4 Graph the solutions of the system: $\begin{cases} x \geq 0 \\ y \geq 0 \\ x + 2y \leq 6 \end{cases}$

Strategy We will graph the solutions of $x \geq 0$, $y \geq 0$, and $x + 2y \leq 6$ on the same coordinate system to see where all three graphs intersect (overlap).

Why The solution set of the system is the set of all points in the intersection of the three graphs.

Solution This is a system of three linear inequalities. If shading is used to graph them on the same set of axes, it can become difficult to interpret the results. Instead, we can draw directional arrows attached to each boundary line in place of the shading.

- The graph of $x \geq 0$ has the boundary $x = 0$ and includes all points on the y-axis and to the right.
- The graph of $y \geq 0$ has the boundary $y = 0$ and includes all points on the x-axis and above.
- The graph of $x + 2y \leq 6$ has the boundary $x + 2y = 6$. Because the coordinates of the origin satisfy $x + 2y \leq 6$, the graph includes all points on and below the boundary.

The solutions of the system are the points that lie on triangle OPQ and the shaded triangular region that it encloses.

$x + 2y = 6$

x	y	(x, y)
0	3	(0, 3)
6	0	(6, 0)

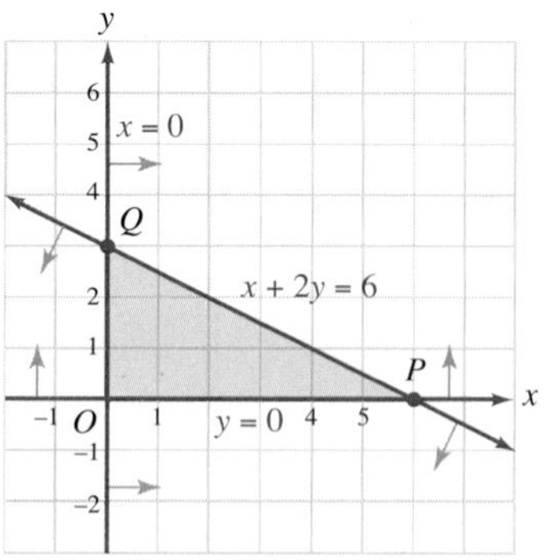

The solutions of the system are shaded purple.

Teaching Example 4 Graph the solutions of the system: $\begin{cases} x \leq 3 \\ y \leq 5 \\ 2x - 3y \leq 6 \end{cases}$

Answer:

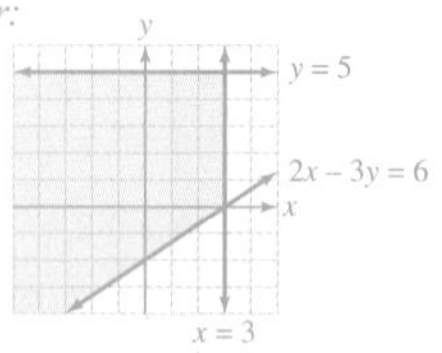

Self Check 4 Graph the solutions of the system: $\begin{cases} x \leq 1 \\ y \leq 2 \\ 2x - y \leq 4 \end{cases}$ See AIE Appendix 3.

Now Try Problem 27

2 Solve Application Problems Involving Systems of Linear Inequalities.

EXAMPLE 5

©iStockphoto/Chris Bernard

Landscaping. A homeowner budgets from \$300 to \$600 for trees and bushes to landscape his yard. After shopping around, he finds that good trees cost \$150 each and mature bushes cost \$75 each. What combinations of trees and bushes can he buy?

Analyze

- At least \$300 but not more than \$600 is to be spent for trees and bushes.
- Trees cost \$150 each and bushes cost \$75 each.
- What combination of trees and bushes can he buy?

Assign Let $x =$ the number of trees purchased and $y =$ the number of bushes purchased.

Form We can form the following system of inequalities:

The cost of a tree	times	the number of trees purchased	plus	the cost of a bush	times	the number of bushes purchased	should at least be	\$300.
\$150	$\cdot$	x	$+$	\$75	$\cdot$	y	$\geq$	\$300

The cost of a tree	times	the number of trees purchased	plus	the cost of a bush	times	the number of bushes purchased	should not be more than	\$600.
\$150	$\cdot$	x	$+$	\$75	$\cdot$	y	$\leq$	\$600

Solve To solve the following system of linear inequalities

$$\begin{cases} 150x + 75y \geq 300 \\ 150x + 75y \leq 600 \end{cases}$$

we use the graphing methods discussed in this section. Neither a negative number of trees nor a negative number of bushes can be purchased, so we restrict the graph to Quadrant I.

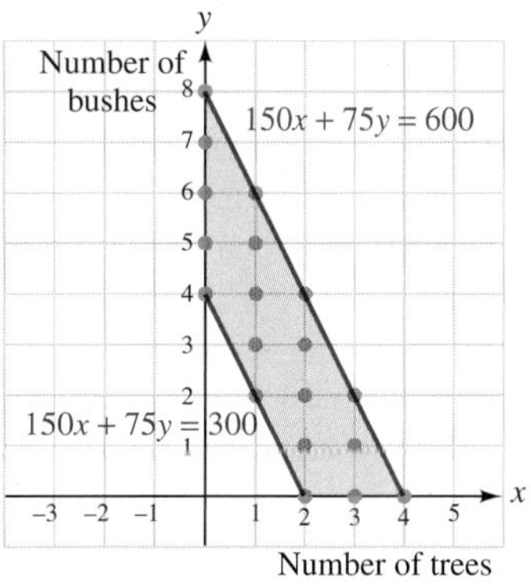

State The coordinates of each point highlighted in the graph give a possible combination of the number of trees, x, and the number of bushes, y, that can be purchased. Written as ordered pairs, these possibilities are

(0, 4), (0, 5), (0, 6), (0, 7), (0, 8),
(1, 2), (1, 3), (1, 4), (1, 5), (1, 6),
(2, 0), (2, 1), (2, 2), (2, 3), (2, 4),
(3, 0), (3, 1), (3, 2), (4, 0)

Check Suppose the homeowner picks the combination of 3 trees and 2 bushes, as represented by (3, 2). Show that this point satisfies both inequalities of the system.

Self Check 5 **Buying CDs and DVDs.** An electronics store sells CDs for \$10 and DVDs for \$20. Donna wants to spend at least \$100 but no more than \$200 on (x) CDs and (y) DVDs. What combinations of CDs and DVDs can she afford to buy? Any ordered pair in the shaded region with whole number coordinates is a possible combination. See AIE Appendix 3.

Now Try Problem 51

The Language of Algebra

Phrases such as **must be at least** and **cannot go below** can be represented by the inequality symbol $\geq$. Phrases such as **is not more than** and **should not surpass** can be represented by the inequality symbol $\leq$.

Teaching Example 5 Buying Stocks. Marcia has at least \$500 but no more than \$1,000 to invest in stocks from two companies. Each share of ABC stock costs \$5 per share, and each share of General Supply cost \$20 per share. What combination of shares of ABC and General Supply can she afford to buy?

Answer: Any ordered pair in the shaded region with whole number coordinates is a possible combination.

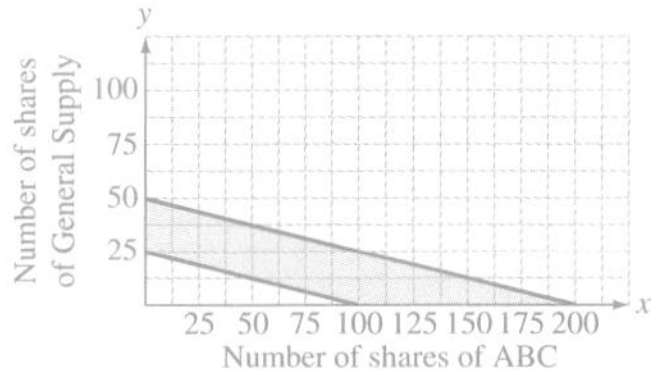

SECTION 4.5 STUDY SET

VOCABULARY

Fill in the blanks.

▶ **1.** $\begin{cases} x + y > 2 \\ x + y < 4 \end{cases}$ is a system of linear _inequalities_.

▶ **2.** To graph the linear inequality $x + y > 2$, first graph the _boundary_ $x + y = 2$. Then pick the test _point_ $(0, 0)$ to determine which half-plane to shade.

▶ **3.** To find the solutions of a system of two linear inequalities graphically, look for the _intersection_, or overlap, of the two shaded regions.

▶ **4.** The phrase *should not surpass* can be represented by the inequality symbol _$\le$_ and the phrase *must be at least* can be represented by the inequality symbol _$\ge$_.

CONCEPTS

5. **a.** What is the equation of the boundary line of the graph of $3x - y < 5$? $3x - y = 5$

b. Is the boundary a solid or dashed line? Dashed

▶ **6.** **a.** What is the equation of the boundary line of the graph of $y \ge 4x$? $y = 4x$

b. Is the boundary a solid or dashed line? Solid

c. Why can't $(0, 0)$ be used as a test point to determine what to shade? The boundary passes through $(0, 0)$.

▶ **7.** Find the slope and the y-intercept of the line whose equation is $y = 4x - 3$. Slope: $4 = \frac{4}{1}$, y-intercept: $(0, -3)$

8. Complete the table to find the x- and y-intercepts of the line whose equation is $8x - 3y = -24$.

x	y
0	8
-3	0

▶ **9.** The boundary of the graph of $2x + y > 4$ is shown.

a. Does the point $(0, 0)$ make the inequality true? No

b. Should the region above or below the boundary be shaded? Above

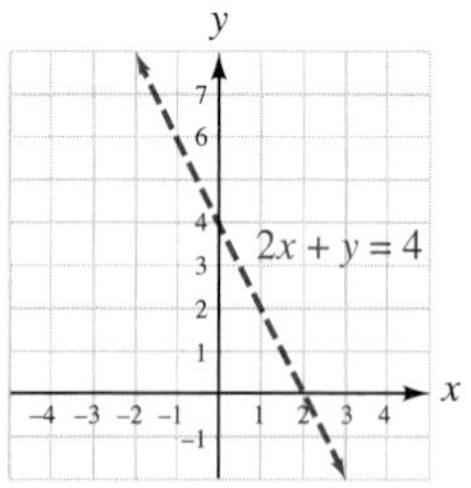

10. Linear inequality 1 is graphed in red below and linear inequality 2 is graphed in blue. Determine whether a true or false statement results when

▶ **a.** The coordinates of point A are substituted into inequality 1 True

▶ **b.** The coordinates of point A are substituted into inequality 2 False

c. The coordinates of point B are substituted into inequality 1 False

d. The coordinates of point B are substituted into inequality 2 True

e. The coordinates of point C are substituted into inequality 1 True

f. The coordinates of point C are substituted into inequality 2 True

Inequality 1 solutions are in **red**

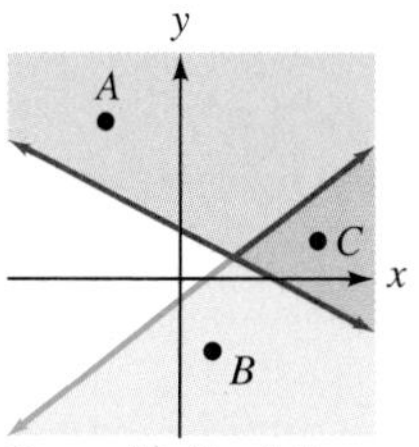

Inequality 2 solutions are in **blue**

▶ **11.** The graph of a system of two linear inequalities is shown. Determine whether each point is a solution of the system.

a. $(4, -2)$ Yes

b. $(1, 3)$ No

c. the origin No

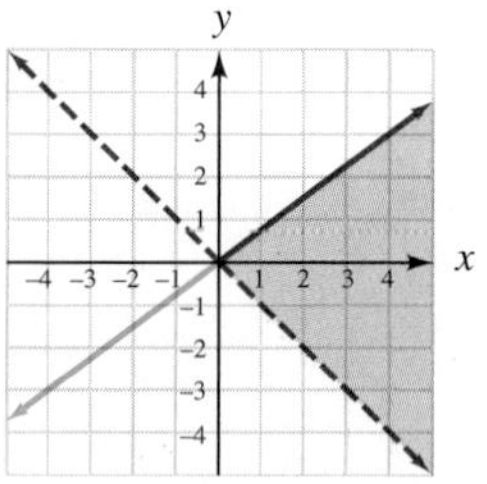

▶ **12.** Use a check to determine whether each ordered pair is a solution of the system.

$$\begin{cases} x + 2y \ge -1 \\ x - y < 2 \end{cases}$$

a. $(1, 4)$ Yes **b.** $(-2, 0)$ No

▶ **13.** Match each equation, inequality, or system with the graph of its solution.

a. $x + y = 2$ ii **b.** $x + y \ge 2$ iii

c. $\begin{cases} x + y = 2 \\ x - y = 2 \end{cases}$ iv **d.** $\begin{cases} x + y \ge 2 \\ x - y \le 2 \end{cases}$ i

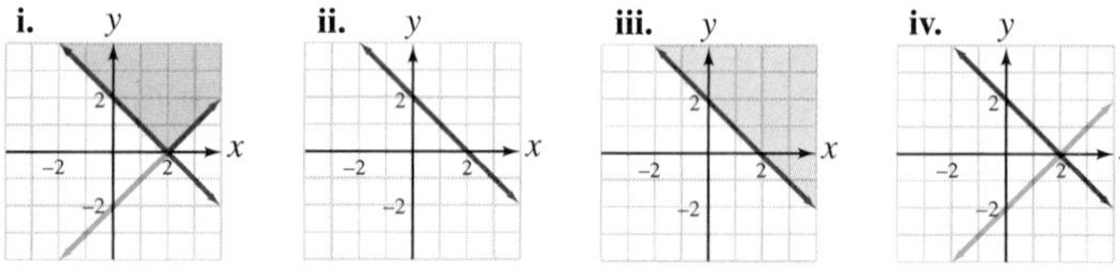

14. Match the system of inequalities with the correct graph.

a. $\begin{cases} x \ge 2 \\ y < 1 \end{cases}$ ii **b.** $\begin{cases} x > 2 \\ y \le 1 \end{cases}$ iii

c. $\begin{cases} x \ge 2 \\ y \ge 1 \end{cases}$ iv **d.** $\begin{cases} x > 2 \\ y > -1 \end{cases}$ i

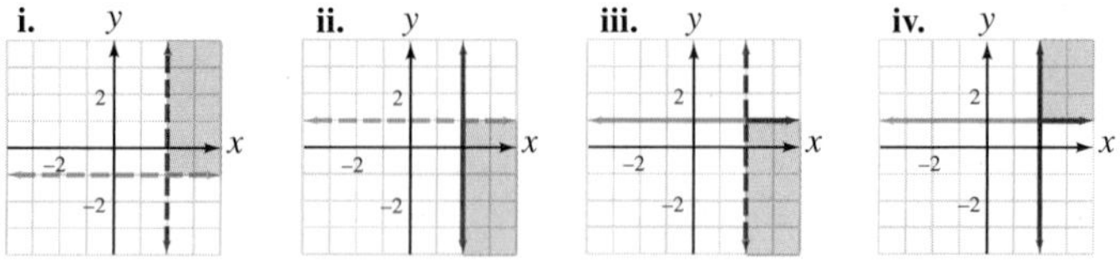

GUIDED PRACTICE

***Graph the solutions of each system.* See Example 1.**
See AIE Appendix 3.

▶ **15.** $\begin{cases} x + 2y \le 3 \\ 2x - y \ge 1 \end{cases}$

▶ **16.** $\begin{cases} 2x + y \ge 3 \\ x - 2y \le -1 \end{cases}$

▶ **17.** $\begin{cases} x + y < -1 \\ x - y > -1 \end{cases}$

▶ **18.** $\begin{cases} x + y > 2 \\ x - y < -2 \end{cases}$

***Graph the solutions of each system.* See Example 2.**
See AIE Appendix 3.

19. $\begin{cases} y > 2x \\ x + 2y < 6 \end{cases}$

▶ **20.** $\begin{cases} y \le 2x \\ x + y < 4 \end{cases}$

▶ **21.** $\begin{cases} y \ge x \\ y \le \frac{1}{3}x + 1 \end{cases}$

▶ **22.** $\begin{cases} y > 3x \\ y \le -x - 1 \end{cases}$

Graph the solutions of each system. **See Example 3.**
See AIE Appendix 3.

23. $\begin{cases} x \geq 2 \\ y \leq 3 \end{cases}$

24. $\begin{cases} x \geq -1 \\ y > -2 \end{cases}$

25. $\begin{cases} x > 0 \\ y > 0 \end{cases}$

26. $\begin{cases} x \leq 0 \\ y < 0 \end{cases}$

Graph the solutions of each system. **See Example 4.**
See AIE Appendix 3.

27. $\begin{cases} x \geq 0 \\ y \geq 0 \\ x + y \leq 3 \end{cases}$

28. $\begin{cases} x - y \leq 6 \\ x + 2y \leq 6 \\ x \geq 0 \end{cases}$

29. $\begin{cases} x - y < 4 \\ y \leq 0 \\ x \geq 0 \end{cases}$

30. $\begin{cases} 2x + y \leq 2 \\ y > x \\ x \geq 0 \end{cases}$

TRY IT YOURSELF

Graph the solutions of each system. See AIE Appendix 3.

31. $\begin{cases} 2x - 3y \leq 0 \\ y \geq x - 1 \end{cases}$

32. $\begin{cases} y > 2x - 4 \\ y \geq -x - 1 \end{cases}$

33. $\begin{cases} x + y < 2 \\ x + y \leq 1 \end{cases}$

34. $\begin{cases} y > -x + 2 \\ y < -x + 4 \end{cases}$

35. $\begin{cases} 3x + 4y \geq -7 \\ 2x - 3y \geq 1 \end{cases}$

36. $\begin{cases} 3x + y \leq 1 \\ 4x - y \geq -8 \end{cases}$

37. $\begin{cases} 2x + y < 7 \\ y > 2 - 2x \end{cases}$

38. $\begin{cases} 2x + y \geq 6 \\ y \leq 2(2x - 3) \end{cases}$

39. $\begin{cases} 2(x - 2y) > -6 \\ 3x + y \geq 5 \end{cases}$

40. $\begin{cases} 2x - 3y < 0 \\ 2x + 3y \geq 12 \end{cases}$

41. $\begin{cases} 3x - y + 4 \leq 0 \\ 3y > -2x - 10 \end{cases}$

42. $\begin{cases} 3x + 2y - 12 \geq 0 \\ x < -2 + y \end{cases}$

43. $\begin{cases} x \geq -1 \\ y \leq -x \\ x - y \leq 3 \end{cases}$

44. $\begin{cases} y > -2.5 \\ 2x - y \geq 2 \\ x \leq 2 \end{cases}$

45. $\begin{cases} x + y > 0 \\ y - x < -2 \end{cases}$

46. $\begin{cases} y + 2x \leq 0 \\ y \leq \frac{1}{2}x + 2 \end{cases}$

Look Alikes . . .

In part a, graph the solution of each system. Use your answer to part a to determine the solution of the system of equations in part b. (No new work is needed.)
See AIE Appendix 3.

47. a. $\begin{cases} x + y > -1 \\ y \geq x - 3 \end{cases}$ b. $\begin{cases} x + y = -1 \\ y = x - 3 \end{cases}$ (1, −2)

48. a. $\begin{cases} x - 2y \geq 6 \\ y < -\frac{1}{2}x + 1 \end{cases}$ b. $\begin{cases} x - 2y = 6 \\ y = -\frac{1}{2}x + 1 \end{cases}$ (4, −1)

APPLICATIONS

49. **Birds of Prey.** Parts (a) and (b) of the illustration show the fields of vision for each eye of an owl. In part (c), shade the area where the fields of vision overlap—that is, the area that is seen by both eyes.

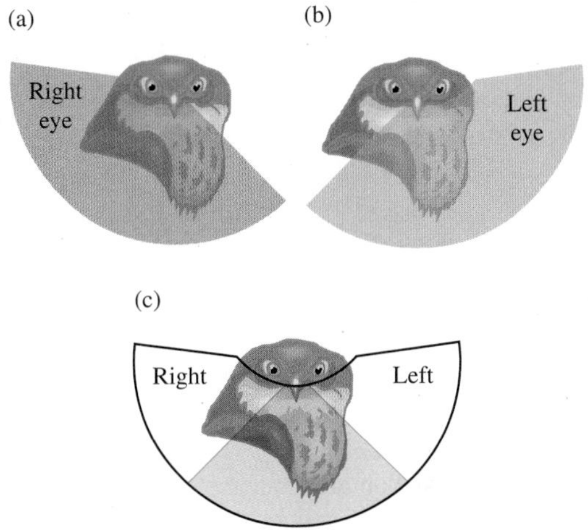

50. **Earth Science.** Shade the area of the earth's surface that is north of the Tropic of Capricorn and south of the Tropic of Cancer.

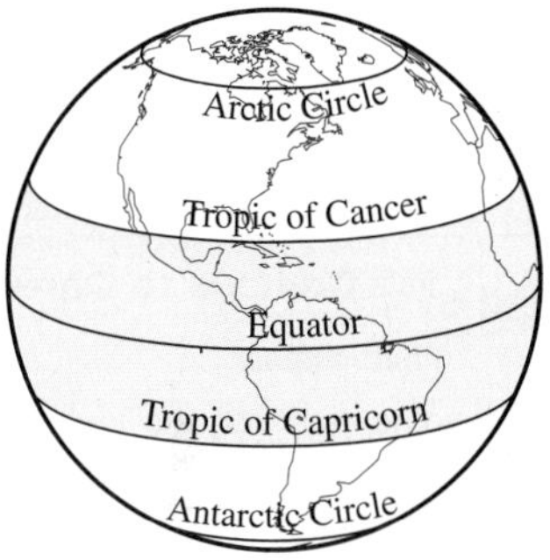

In Problems 51–54, graph each system of inequalities and give two possible solutions. **See Example 5.**

51. **Buying Compact Discs.** Melodic Music has compact discs on sale for either \$10 or \$15. If a customer wants to spend at least \$30 but no more than \$60 on CDs, graph a system of inequalities showing the possible combinations of \$10 CDs (x) and \$15 CDs (y) that the customer can buy. 1 \$10 CD and 2 \$15 CDs; 4 \$10 CDs and 1 \$15 CD (answers may vary)

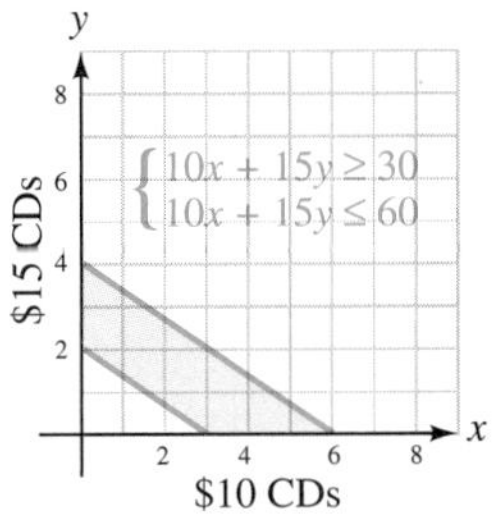

52. **Boats.** Boatworks wholesales aluminum boats for \$800 and fiberglass boats for \$600. Northland Marina wants to make a purchase totaling at least \$2,400 but no more than \$4,800. Graph a system of inequalities showing the possible combinations of aluminum boats (x) and fiberglass boats (y) that can be ordered.
4 alum. and 1 glass; 1 alum. and 4 glass (answers may vary)

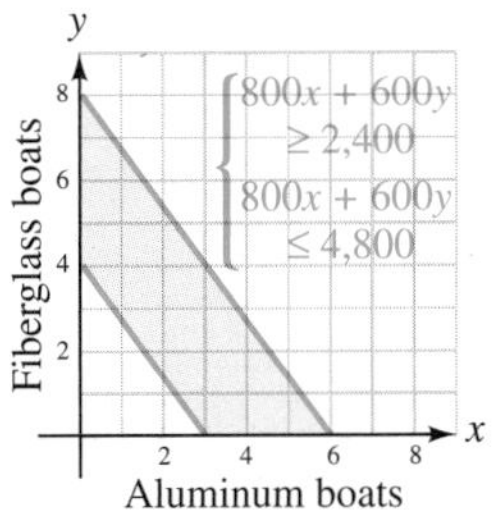

53. Furniture. A distributor wholesales desk chairs for \$150 and side chairs for \$100. Best Furniture wants its order to total no more than \$900; Best also wants to order more side chairs than desk chairs. Graph a system of inequalities showing the possible combinations of desk chairs (x) and side chairs (y) that can be ordered.

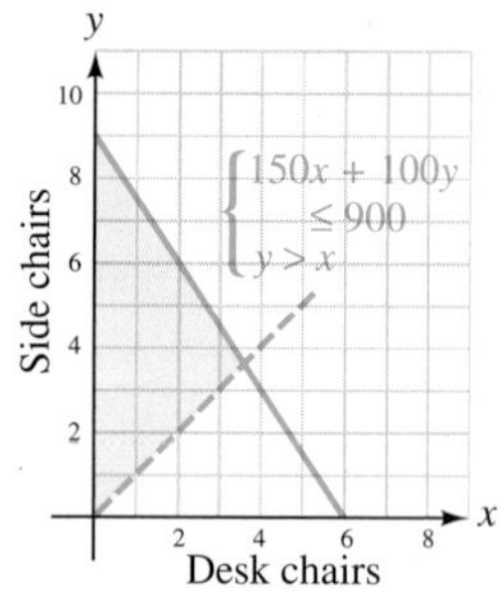

2 desk chairs and 4 side chairs; 1 desk chair and 5 side chairs (answers may vary)

54. Air Quality. J. Bolden Heating Company wants to order no more than \$2,000 worth of electronic air cleaners and humidifiers from a wholesaler that charges \$500 for air cleaners and \$200 for humidifiers. If Bolden wants more humidifiers than air cleaners, graph a system of inequalities showing the possible combinations of air cleaners (x) and humidifiers (y) that can be ordered. 1 air cleaner and 2 humidifiers; 2 air cleaners and 3 humidifiers (answers may vary)

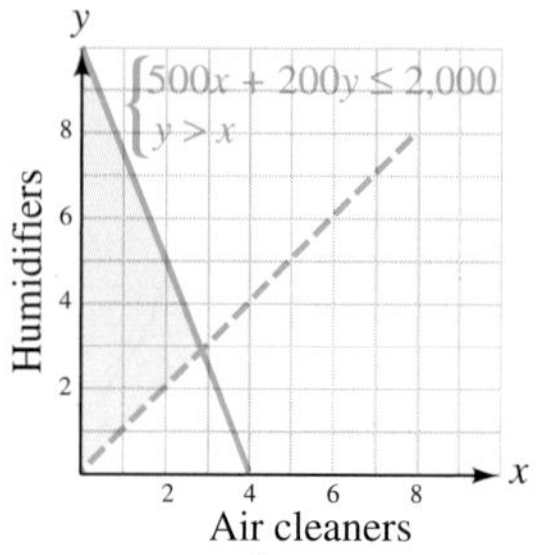

55. from **Campus to Careers**

Photographer

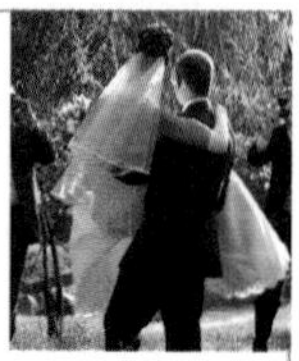

© Andrei Contiu/Shutterstock.com

In some cameras, the image that you see in the viewfinder does not exactly match the image that will be recorded through the lens. To understand this difference, graph the solutions of the system

$$\begin{cases} y \le \dfrac{1}{4}x + 2 \\ y \ge -\dfrac{1}{4}x + 2 \end{cases}$$

on the grid below. The shaded solution shows how the viewfinder image is slightly different from the lens image for extreme close-ups.

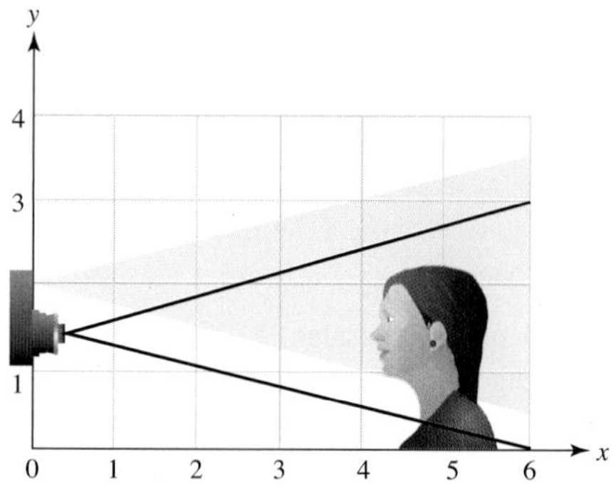

56. Pesticides. To get rid of fruit flies, helicopters sprayed an area of a city that can be described by $y \ge -2x + 1$ (within the city limits). Two weeks later, more spraying was ordered over the area described by $y \ge \frac{1}{4}x - 4$ (within the city limits). Show the part of the city that was sprayed twice.

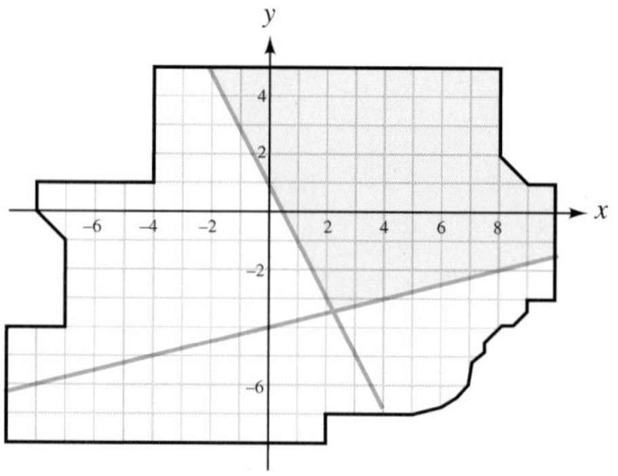

WRITING

57. Explain the error.

Graph: $\begin{cases} y > x \\ x + y < 1 \end{cases}$

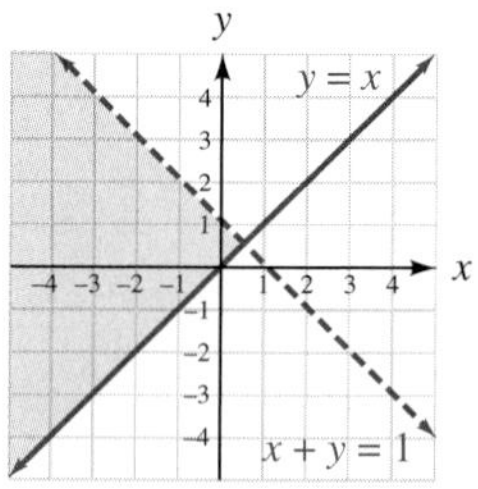

58. When a solution of a system of linear inequalities is graphed, what does the shading represent?

59. Describe how the graphs of the solutions of these systems are similar and how they differ.

$$\begin{cases} x + y = 4 \\ x - y = 4 \end{cases} \quad \text{and} \quad \begin{cases} x + y \ge 4 \\ x - y \ge 4 \end{cases}$$

60. Explain when a system of inequalities will have no solutions.

REVIEW

Simplify each expression.

61. $8\left(\frac{3}{4}t\right)$ $6t$

62. $-\dfrac{2}{3}(3w - 6)$ $-2w + 4$

63. $-\dfrac{7}{16}x - \dfrac{3}{16}x$ $-\frac{5}{8}x$

64. $60\left(\dfrac{3}{20}r - \dfrac{4}{15}\right)$ $9r - 16$

CHALLENGE PROBLEMS

Graph the solutions of each system. See AIE Appendix 3.

65. $\begin{cases} \dfrac{x}{3} - \dfrac{y}{2} < -3 \\ \dfrac{x}{3} + \dfrac{y}{2} > -1 \end{cases}$

66. $\begin{cases} 3x + y < -2 \\ y > 3(1 - x) \end{cases}$ No solutions

67. $\begin{cases} 2x + 3y \le 6 \\ 3x + y \le 1 \\ x \le 0 \end{cases}$

68. $\begin{cases} x \ge 0 \\ y \ge 0 \\ 9x + 3y \le 18 \\ 3x + 6y \le 18 \end{cases}$

4 Summary & Review

SECTION 4.1 Solving Systems of Equations by Graphing

DEFINITIONS AND CONCEPTS | **EXAMPLES**

When two equations are considered at the same time, we say that they form a **system of equations.**

A **solution of a system** of equations in two variables is an ordered pair that satisfies both equations of the system.

Is (4, 3) a solution of the system $\begin{cases} x + y = 7 \\ x - y = 5 \end{cases}$?

To answer this question, we substitute 4 for x and 3 for y in each equation.

$x + y = 7$	The first equation.	$x - y = 5$	The second equation.
$\mathbf{4} + \mathbf{3} \stackrel{?}{=} 7$	Substitute.	$\mathbf{4} - \mathbf{3} \stackrel{?}{=} 5$	Substitute.
$7 = 7$	True	$1 = 5$	False

Although (4, 3) satisfies the first equation, it does not satisfy the second. Because it does not satisfy both equations, (4, 3) is not a solution of the system.

To **solve a system graphically:**

1. Graph each equation on the same coordinate system.
2. Determine the coordinates of the **point of intersection** of the graphs. That ordered pair is the solution.
3. Check the solution in each equation of the original system.

Use graphing to solve the system: $\begin{cases} y = -2x + 3 \\ x - 2y = 4 \end{cases}$

Step 1: Graph each equation.

$\mathbf{y = -2x + 3}$

$m = \frac{\text{Rise}}{\text{Run}} = \frac{-2}{1}$

$b = 3$

y-intercept: (0, 3)

$x - 2y = 4$

x	y
0	−2
4	0

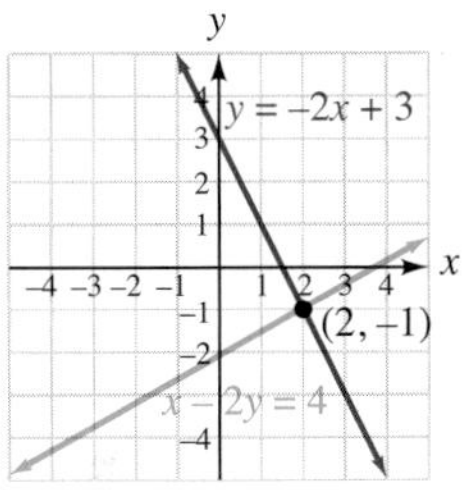

Step 2: It appears that the graphs intersect at the point (2, −1). To verify that this is the solution of the system, substitute 2 for x and −1 for y in each equation.

Step 3: Check

$y = -2x + 3$	The first equation.	$x - 2y = 4$	The second equation.
$\mathbf{-1} \stackrel{?}{=} -2(\mathbf{2}) + 3$	Substitute.	$\mathbf{2} - 2(\mathbf{-1}) \stackrel{?}{=} 4$	Substitute.
$-1 = -1$	True	$4 = 4$	True

Since (2, −1) makes both equations true, it is the solution of the system.

A system of equations that has at least one solution is called a **consistent system.** If the graphs of the equations of the system are parallel lines, the system has no solution and is called an **inconsistent system.**

Equations with different graphs are called **independent equations.** If the graphs of the equations in a system are the same line, the system has infinitely many solutions. The equations are called **dependent equations.**

We can determine the **number of solutions** that a system of two linear equations has by writing each equation in slope-intercept form, $y = mx + b$, and comparing the slopes and y-intercepts.

There are three possible outcomes when solving a system by graphing.

Consistent system Independent equations

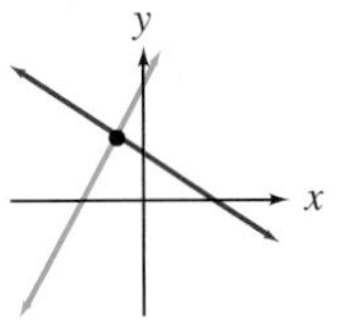

- Exactly one solution
- The lines have different slopes.

Inconsistent system Independent equations

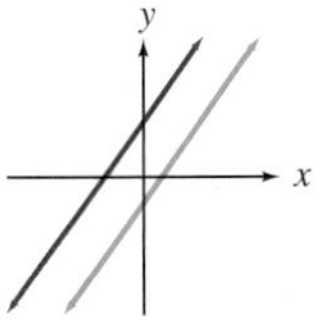

- No solution
- The lines have the same slope but different y-intercepts.

Consistent system Dependent equations

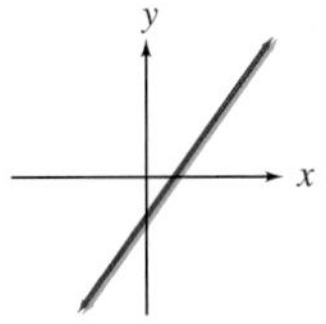

- Infinitely many solutions
- The lines have the same slope and same y-intercept.

REVIEW EXERCISES

Check to determine whether the ordered pair is a solution of the system.

1. $(2, -3)$, $\begin{cases} 3x - 2y = 12 \\ 2x + 3y = -5 \end{cases}$ Yes

2. $\left(\frac{7}{2}, -\frac{2}{3}\right)$, $\begin{cases} 3y = 2x - 9 \\ 2x + 3y = 6 \end{cases}$ No

Use the graphing method to solve each system. See AIE Appendix 3.

3. $\begin{cases} x + y = 7 \\ 2x - y = 5 \end{cases}$ (4, 3)

4. $\begin{cases} 2x + y = 5 \\ y = -\dfrac{x}{3} \end{cases}$ (3, −1)

5. $\begin{cases} 3x + 6y = 6 \\ x + 2y - 2 = 0 \end{cases}$ Infinitely many solutions

6. $\begin{cases} 6x + 3y = 12 \\ y = -2x + 2 \end{cases}$ No solution

7. Find the slope and the y-intercept of the graph of each line in the system $\begin{cases} y = -2x + 1 \\ 8x + 4y = 3 \end{cases}$. Then, use that information to determine the number of solutions of the system. No solution

8. **Bachelor's Degrees.** Estimate the point of intersection of the graphs below. Explain its significance. (1980, 480,000); In 1980, the number of men and women awarded Bachelor's degrees in the U.S. was the same, 480,000.

Bachelor's Degrees Awarded by U.S. Institutions

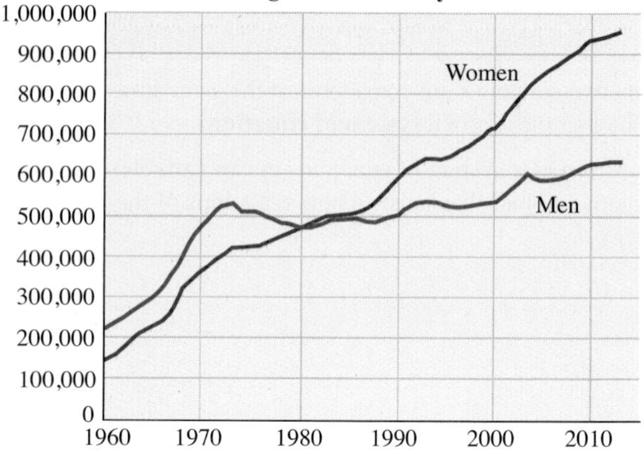

SECTION 4.2 Solving Systems of Equations by Substitution

DEFINITIONS AND CONCEPTS

To solve a system of equations in x and y by **substitution:**

1. Solve one of the equations for either x or y. If this is already done, go to step 2. (We call this equation the **substitution equation.**)
2. Substitute the expression for x (or for y) obtained in step 1 into the other equation and solve the equation.
3. Substitute the value of the variable found in step 2 into the substitution equation to find the value of the remaining variable.
4. Check the proposed solution in the equations of the original system.

With the substitution method, the objective is to use an appropriate substitution to obtain *one equation in one variable.*

If in step 2 the variable drops out and a false statement results, the system has **no solution.** If a true statement results, the system has **infinitely many solutions.**

EXAMPLES

Use substitution to solve the system: $\begin{cases} y = x + 7 \\ x + 2y = 5 \end{cases}$

Step 1: The first equation is already solved for y.

Step 2: Substitute $x + 7$ for y in the second equation.

$x + 2y = 5$

$x + 2(x + 7) = 5$ Substitute x + 7 for y.

$x + 2x + 14 = 5$ Distribute the multiplication by 2.

$3x + 14 = 5$ Combine like terms.

$3x = -9$ Subtract 14 from both sides.

$x = -3$ Divide both sides by 3. This is the x-value of the solution.

Step 3: $y = x + 7$ This is the substitution equation.

$y = -3 + 7$ Substitute −3 for x.

$y = 4$ This is the y-value of the solution.

Step 4: The following check verifies that the solution is $(-3, 4)$.

$y = x + 7$ The first equation.

$4 \stackrel{?}{=} -3 + 7$ Substitute.

$4 = 4$ True

$x + 2y = 5$ The second equation.

$-3 + 2(4) \stackrel{?}{=} 5$ Substitute.

$5 = 5$ True

REVIEW EXERCISES

Use the substitution method to solve each system.

9. $\begin{cases} y = 15 - 3x \\ 7y + 3x = 15 \end{cases}$ (5, 0)

10. $\begin{cases} x = y \\ 5x - 4y = 3 \end{cases}$ (3, 3)

11. $\begin{cases} 6x + 2y = 8 - y + x \\ 3x = 2 - y \end{cases}$ $\left(-\frac{1}{2}, \frac{7}{2}\right)$

12. $\begin{cases} r = 3s + 7 \\ r = 2s + 5 \end{cases}$ (1, −2)

13. $\begin{cases} 9x + 3y - 5 = 0 \\ 3x + y = \dfrac{5}{3} \end{cases}$ Infinitely many solutions

14. $\begin{cases} \dfrac{x}{2} + \dfrac{y}{2} = 11 \\ \dfrac{5x}{16} - \dfrac{3y}{16} = \dfrac{15}{8} \end{cases}$ (12, 10)

15. When solving a system using the substitution method, suppose you obtain the result $8 = 9$.
    a. How many solutions does the system have? No solution
    b. Describe the graph of the system. Two parallel lines
    c. What term is used to describe the system? Inconsistent system

16. Fill in the blank. With the substitution method, the objective is to use an appropriate substitution to obtain one equation in ___one___ variable.

SECTION 4.3 Solving Systems of Equations by Elimination (Addition)

DEFINITIONS AND CONCEPTS	EXAMPLES
To solve a system of equations in x and y using **elimination (addition):**	Use elimination to solve: $\begin{cases} 2x - 3y = 4 \\ 3x + y = -5 \end{cases}$
1. Write each equation in the standard $Ax + By = C$ form.	**Step 1:** Both equations are written in $Ax + By = C$ form.
2. Multiply one (or both) equations by nonzero quantities to make the coefficients of x (or y) opposites.	**Step 2:** Multiply the second equation by 3 so that the coefficients of y are opposites, −3 and 3.
3. Add the equations to eliminate the terms involving x (or y).	**Step 3:** $\begin{aligned} 2x - 3y &= 4 \\ 9x + 3y &= -15 \\ \hline 11x \quad &= -11 \end{aligned}$ In the middle column, $-3y + 3y = 0$. Add the like terms, column by column.
4. Solve the equation obtained in step 3.	**Step 4:** Solve for x. $11x = -11$; $x = -1$ Divide both sides by 11. This is the x-value of the solution.
5. Find the value of the other variable by substituting the value of the variable found in step 4 into any equation containing both variables.	**Step 5:** Find y. $3x + y = -5$ This is the second equation. $3(-1) + y = -5$ Substitute −1 for x. $y = -2$ This is the y-value of the solution.
6. Check the solution in the equations of the original system.	**Step 6:** The following check verifies that the solution is (−1, −2). $2x - 3y = 4$ First equation. $2(-1) - 3(-2) \stackrel{?}{=} 4$ Substitute. $-2 + 6 \stackrel{?}{=} 4$; $4 = 4$ True. $3x + y = -5$ Second equation. $3(-1) + (-2) \stackrel{?}{=} -5$ Substitute. $-3 - 2 \stackrel{?}{=} -5$; $-5 = -5$ True
With the elimination method, the basic objective is to obtain two equations whose sum will be one equation in one variable.	
If in step 3 both variables drop out and a false statement results, the system has **no solution.** If a true statement results, the system has **infinitely many solutions.**	

REVIEW EXERCISES

17. Write each equation of the system $\begin{cases} 4x + 2y - 7 = 0 \\ 3y = 5x + 6 \end{cases}$ in standard $Ax + By = C$ form.
$\begin{cases} 4x + 2y = 7 \\ 5x - 3y = -6 \end{cases}$

18. Fill in the blank. With the elimination method, the basic objective is to obtain two equations whose sum will be one equation in one variable.

Solve each system using the elimination (addition) method.

19. $\begin{cases} 2x + y = 1 \\ 5x - y = 20 \end{cases}$ $(3, -5)$

20. $\begin{cases} x + 8y = 7 \\ x - 4y = 1 \end{cases}$ $\left(3, \frac{1}{2}\right)$

21. $\begin{cases} 5a + b = 2 \\ 3a + 2b = 11 \end{cases}$ $(-1, 7)$

22. $\begin{cases} 11x + 3y = 27 \\ 8x + 4y = 36 \end{cases}$ $(0, 9)$

23. $\begin{cases} 9x + 3y = 15 \\ 3x = 5 - y \end{cases}$ Infinitely many solutions

24. $\begin{cases} 0.02x + 0.05y = 0 \\ 0.3x - 0.2y = -1.9 \end{cases}$ $(-5, 2)$

25. $\begin{cases} \frac{a}{4} - \frac{b}{3} = \frac{1}{12} \\ \frac{a}{2} - \frac{5b}{4} = \frac{7}{4} \end{cases}$ $(1, -1)$

26. $\begin{cases} -\frac{1}{4}x = 1 - \frac{2}{3}y \\ 6x - 18y = 5 - 2y \end{cases}$ No solution

For each system, determine which method, substitution or elimination (addition), would be easier to use to solve the system and explain why.

27. $\begin{cases} 6x + 2y = 5 \\ 3x - 3y = -4 \end{cases}$ Elimination; no variables have a coefficient of 1 or −1.

28. $\begin{cases} x = 5 - 7y \\ 3x - 3y = -4 \end{cases}$ Substitution; equation 1 is solved for x.

SECTION 4.4 Problem Solving Using Systems of Equations

DEFINITIONS AND CONCEPTS

We can solve many types of problems using a system of two linear equations in two variables:

- Geometry problems
- Number-value problems
- Interest problems
- Uniform motion problems
- Liquid and dry mixture problems

To solve problems involving two unknown quantities:

1. **Analyze** the facts of the problem. Make a table or a diagram if it is helpful.
2. **Assign** different variables to represent the two unknown quantities.
3. Translate the words of the problem to **form two equations** involving those variables.
4. **Solve** the system of equations using graphing, substitution, or elimination.
5. **State** the conclusion.
6. **Check** the results.

Two angles are said to be **complementary** if the sum of their measures is 90°. Two angles are said to be **supplementary** if the sum of their measures is 180°.

EXAMPLES

The difference of the measures of two supplementary angles is 40°. Find the measure of each angle.

Analyze Since the angles are supplementary, the sum of their measures is 180°. If we subtract the smaller angle from the larger, the result should be 40°.

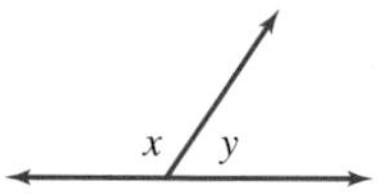

Assign Let x = the measure (in degrees) of the larger angle and y = the measure (in degrees) of the smaller angle.

Form $\begin{cases} x + y = 180 & \text{Their sum is 180°.} \\ x - y = 40 & \text{Their difference is 40°.} \end{cases}$

Solve If we add the equations, we get

$$\begin{aligned} x + y &= 180 \\ \underline{x - y} &\underline{= 40} \\ 2x \quad &= 220 \\ x &= 110 \end{aligned}$$

Divide both sides by 2. This is the measure of the larger angle.

We can use the first equation of the system to find y.

$x + y = 180$

$\mathbf{110} + y = 180$ Substitute 110 for x.

$y = 70$ Subtract 110 from both sides. This is the measure of the smaller angle.

State The angles measure 110° and 70°.

Check Angles with measures of 110° and 70° are supplementary (their sum is 180°) and their difference is 40°. The results check.

REVIEW EXERCISES

Write a system of two equations in two variables to solve each problem.

29. Elevations. The elevation of Las Vegas, Nevada, is 20 times greater than that of Baltimore, Maryland. The sum of their elevations is 2,100 feet. Find the elevation of each city. Las Vegas: 2,000 ft; Baltimore: 100 ft

30. Painting Equipment. When fully extended, a ladder is 35 feet in length. If the extension is 7 feet shorter than the base, how long is each part of the ladder? Base: 21 ft; extension: 14 ft

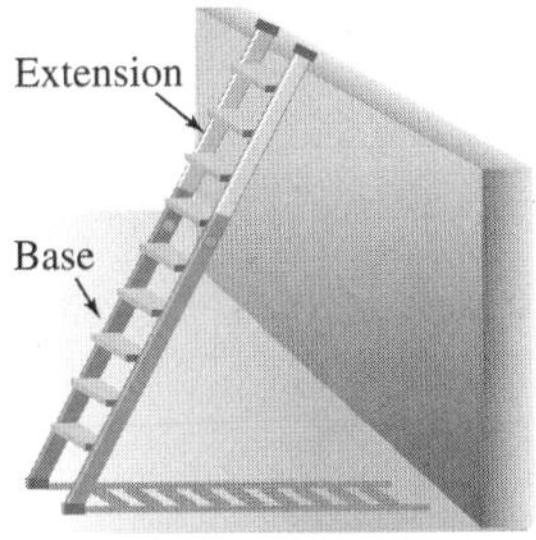

31. Geometry. Two angles are complementary. The measure of one is 15° more than twice the measure of the other. Find the measure of each angle. 65°, 25°

32. Crash Investigation. In an effort to protect evidence, investigators used 420 yards of yellow "Police Line—Do Not Cross" tape to seal off a large rectangular-shaped area around an airplane crash site. How much area will the investigators have to search if the width of the rectangle is three-fourths of the length? 10,800 yd^2

33. Complete each table.

a.

	Amount	· Strength =	Amount of pesticide
Weak	x	0.02	$0.02x$
Strong	y	0.09	$0.09y$
Mixture	100	0.08	0.08(100)

b.

	Rate	· Time =	Distance
With the wind	$s + w$	5	$5(s + w)$
Against the wind	$s - w$	7	$7(s - w)$

c.

	P ·	r	· t =	I
Mack Financial	x	0.11	1	$0.11x$
Union Savings	y	0.06	1	$0.06y$

d.

	Amount	· Price =	Total value
Caramel corn	x	4	$4x$
Peanuts	y	8	$8y$
Mixture	10	5	10(5)

34. Candy Store. A merchant wants to mix gummy worms worth \$6 per pound and gummy bears worth \$3 per pound to make 30 pounds of a mixture worth \$4.20 per pound. How many pounds of each type of candy should he use? 12 lb worms, 18 lb bears

35. Boating. It takes a motorboat 4 hours to travel 56 miles down a river, and 3 hours longer to make the return trip. Find the speed of the current. 3 mph

36. Shopping. Packages containing two bottles of contact lens cleaner and three bottles of soaking solution cost \$63.40, and packages containing three bottles of cleaner and two bottles of soaking solution cost \$69.60. Find the cost of a bottle of cleaner and a bottle of soaking solution. \$16.40, \$10.20

37. Investing. Carlos invested part of \$3,000 in a 10% certificate account and the rest in a 6% passbook account. The total annual interest from both accounts is \$270. How much did he invest at 6%? \$750

38. Antifreeze. How much of a 40% antifreeze solution must a mechanic mix with a 70% antifreeze solution if she needs 20 gallons of a 50% antifreeze solution? $13\frac{1}{3}$ gal 40%, $6\frac{2}{3}$ gal 70%

SECTION 4.5 Solving Systems of Linear Inequalities

DEFINITIONS AND CONCEPTS	EXAMPLES
A solution of a **system of linear inequalities** is an ordered pair that satisfies each inequality.	Graph the solutions of the system: $\begin{cases} y \le x + 1 \\ y > -1 \end{cases}$
To **solve a system of linear inequalities:** 1. Graph each inequality on the same coordinate system. 2. Use shading to highlight the intersection of the graphs. The points in this region are the solutions of the system. 3. As an informal check, pick a point from the region and verify that its coordinates satisfy each inequality of the original system.	**Step 1:** Graph each inequality on the same coordinate system as shown. **Step 2:** Use shading to highlight where the graphs intersect. 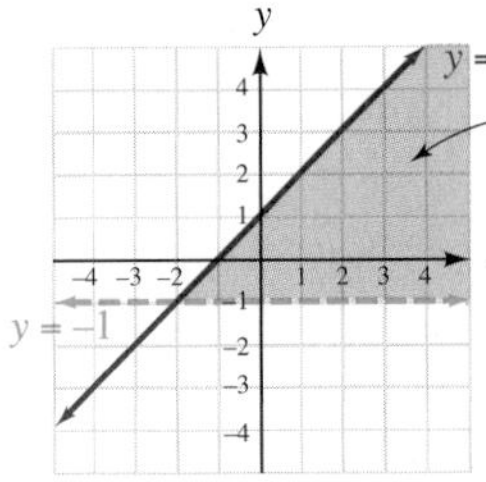 **Step 3:** Pick a point from the solution region such as (1, 0) and verify that it satisfies both inequalities.

REVIEW EXERCISES

Solve each system of inequalities. See AIE Appendix 3.

39. $\begin{cases} 5x + 3y < 15 \\ 3x - y > 3 \end{cases}$

40. $\begin{cases} 3y \le x \\ y > 3x \end{cases}$

41. $\begin{cases} x \le 0 \\ y < 0 \end{cases}$

42. $\begin{cases} y \ge x \\ y \le \frac{1}{3}x + 1 \\ x > -3 \end{cases}$

43. Use a check to determine whether each ordered pair is a solution of the system: $\begin{cases} x + 2y \le 3 \\ 2x - y > 1 \end{cases}$

a. $(5, -4)$ Yes **b.** $(-1, -3)$ No

44. Gift Shopping. A grandmother wants to spend at least $40 but no more than $60 on school clothes for her grandson. If T-shirts sell for $10 each and pants sell for $20 each, write a system of inequalities that describes the possible numbers of T-shirts x and pairs of pants y that she can buy. Graph the system and give two possible solutions.
$10x + 20y \ge 40$, $10x + 20y \le 60$; (3, 1): 3 shirts and 1 pair of pants; (1, 2): 1 shirt and 2 pairs of pants (answers may vary)

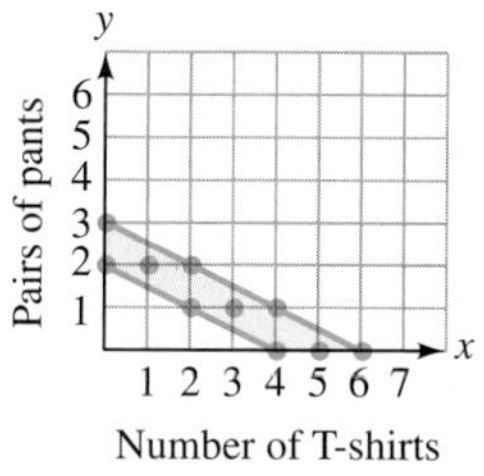

Teaching Tip: Because this Chapter Test is a comprehensive collection of problems that covers all of the topics discussed in Chapter 4, it is lengthy. If your students have time to complete it, that would be optimal. If, because of time constraints, they are unable to do so, assign an appropriate subset of problems that reflects the types of problems that the students will see on your exam.

4 CHAPTER TEST

Use a check to determine whether the ordered pair is a solution of the system.

1. $(5, 3)$, $\begin{cases} 3x + 2y = 21 \\ x + y = 8 \end{cases}$ Yes

2. $(-2, -1)$, $\begin{cases} 4x + y = -9 \\ 2x - 3y = -7 \end{cases}$ No

3. Fill in the blanks.

a. A <u>solution</u> of a system of linear equations is an ordered pair that satisfies each equation.

b. A system of equations that has at least one solution is called a <u>consistent</u> system.

c. A system of equations that has no solution is called an <u>inconsistent</u> system.

d. Equations with different graphs are called <u>independent</u> equations.

e. A system of <u>dependent</u> equations has an infinite number of solutions.

f. Two angles are said to be <u>supplementary</u> if the sum of their measures is 180°.

4. Energy. The graphs below show U.S. electricity generation from natural gas and nuclear sources from 1995 through 2009. What is the intersection point of the graphs? Explain its significance. (2005, 770); In 2005, the amount of electricity generated by natural gas and nuclear sources was the same, about 770 billion kilowatt hours each.

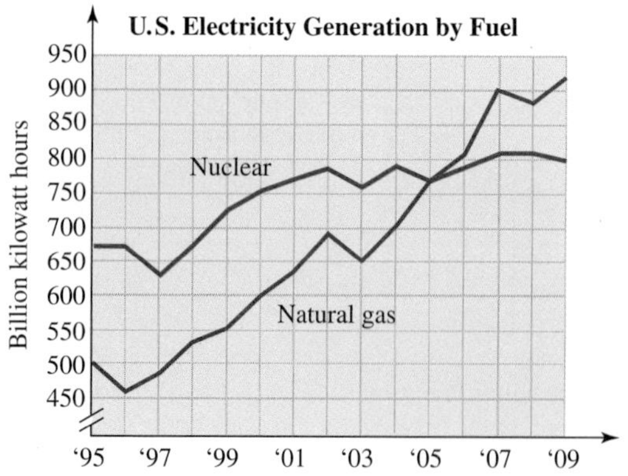

Solve each system by graphing. See AIE Appendix 3.

5. $\begin{cases} y = 2x - 1 \\ x - 2y = -4 \end{cases}$ (2, 3)

6. $\begin{cases} x + y = 5 \\ y = -x \end{cases}$ No solution

7. Find the slope and the y-intercept of the graph of each line in the system $\begin{cases} y = 4x - 10 \\ x - 2y = -16 \end{cases}$. Then, use that information to determine the *number of solutions* of the system. **Do not solve the system.**
Since the lines have different slopes, they will intersect at one point. The system has 1 solution.

8. How many solutions does the system of two linear equations graphed on the right have? Give three of the solutions. Infinitely many solutions; (0, 2), (3, 1), (6, 0) (answers may vary)

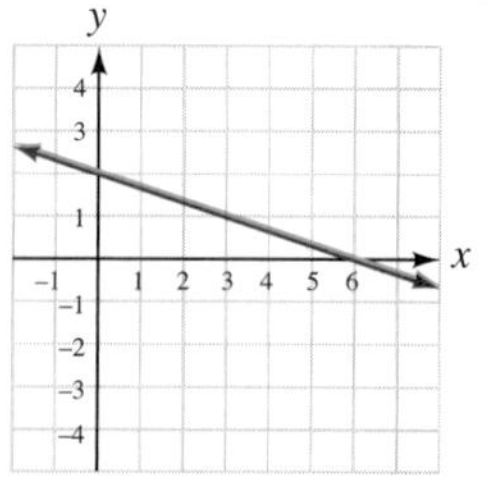

Solve each system by substitution.

9. $\begin{cases} y = x - 1 \\ 2x + y = -7 \end{cases}$ $(-2, -3)$

10. $\begin{cases} 3x + 6y = -15 \\ x + 2y = -5 \end{cases}$ Infinitely many solutions

Solve each system using elimination.

11. $\begin{cases} 3x - y = 2 \\ 2x + y = 8 \end{cases}$ (2, 4)

12. $\begin{cases} 4x + 3y = -3 \\ -3x = -4y + 21 \end{cases}$ $(-3, 3)$

Solve each system using substitution or elimination.

13. $\begin{cases} 3x - 5y - 16 = 0 \\ \frac{x}{2} - \frac{5}{6}y = \frac{1}{3} \end{cases}$ No solution

14. $\begin{cases} 3a + 4b = -7 \\ 2b - a = -1 \end{cases}$ $(-1, -1)$

15. $\begin{cases} y = 3x - 1 \\ y = 2x + 4 \end{cases}$ (5, 14)

16. $\begin{cases} 0.6c + 0.5d = 0 \\ 0.02c + 0.09d = 0 \end{cases}$ (0, 0)

17. $\begin{cases} a - 1 = 2b \\ 3a + 1 = -10b \end{cases}$ $\left(\frac{1}{2}, -\frac{1}{4}\right)$

18. $\begin{cases} \dfrac{x + 2}{6} = \dfrac{y + 3}{4} \\ \dfrac{x}{5} = \dfrac{3y - 3}{6} \end{cases}$ (10, 5)

19. $\begin{cases} 4(a - 2) + 5y = 19 \\ 3(a - 2) - y = 0 \end{cases}$ (3, 3)

20. $\begin{cases} 3x + 1 = 2x - 4y + 2 \\ 7x - y - 1 = 9y + 5x + 10 \end{cases}$ $\left(3, -\frac{1}{2}\right)$

Write a system of two equations in two variables to solve each problem.

21. **Child Care.** On a mother's 22-mile commute to work, she drops her daughter off at a child care center. The first part of the trip is 6 miles less than the second part. How long is each part of her morning commute? 1st part: 8 mi, 2nd part: 14 mi

22. **Vacationing.** It cost a family of 7 a total of \$219 for general admission tickets to the San Diego Zoo. How many adult tickets and how many child tickets were purchased? 3 adult tickets, 4 child tickets

23. **Financial Planning.** A woman invested some money at 8% and some at 9% annual simple interest. The interest for 1 year on the combined investment of \$10,000 was \$840. How much was invested at each rate? \$6,000 at 8%, \$4,000 at 9%

24. **Tailwinds/Headwinds.** Flying with a tailwind, a pilot flew an airplane 450 miles in 2.5 hours. Flying into a headwind, the return trip took 3 hours. Find the speed of the plane in calm air and the speed of the wind. The speed in calm air: 165 mph, speed of wind: 15 mph

25. **Tether Ball.** The angles shown in the illustration are complementary. The measure of the larger angle is 10° more than three times the measure of the smaller angle. Find the measure of each angle. Larger: 70°, smaller: 20°

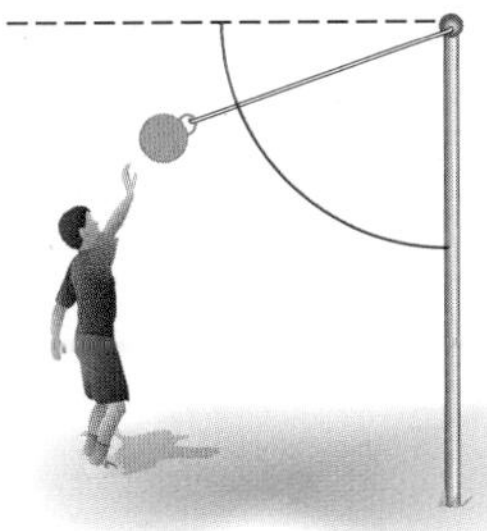

26. **Antifreeze.** How many pints of a 5% antifreeze solution and how many pints of a 20% antifreeze solution must be mixed to obtain 12 pints of a 15% solution? 5%: 4 pints, 20%: 8 pints

27. **Sunscreen.** A sunscreen selling for \$1.50 per ounce is to be combined with another sunscreen selling for \$0.80 per ounce. How many ounces of each are needed to make 10 ounces of a sunscreen mix that sells for \$1.01 per ounce? \$1.50 sunscreen: 3 oz, \$0.80 sunscreen: 7 oz

28. Use a check to determine whether (3, 1) is a solution of the system: $\begin{cases} y \le 2x - 1 \\ x + 3y > 6 \end{cases}$ No

Solve each system by graphing. See AIE Appendix 3.

29. $\begin{cases} 3x + 2y < 6 \\ y \ge x + 1 \end{cases}$

30. $\begin{cases} x - y < 3 \\ y \le 0 \\ x \ge 0 \end{cases}$

31. **Clothes Shopping.** This system of inequalities describes the number of \$20 shirts, x, and \$40 pairs of pants, y, a person can buy if he or she plans to spend not less than \$80 but not more than \$120. Graph the system. Then give three solutions. (1, 2), (2, 2), (3, 1) (answers may vary) See AIE Appendix 3.

$$\begin{cases} 20x + 40y \ge 80 \\ 20x + 40y \le 120 \end{cases}$$

32. Match each equation, inequality, or system with the graph of its solution.

a. $2x + y = 2$ iii

b. $2x + y \ge 2$ ii

c. $\begin{cases} 2x + y = 2 \\ 2x - y = 2 \end{cases}$ i

d. $\begin{cases} 2x + y \ge 2 \\ 2x - y \le 2 \end{cases}$ iv

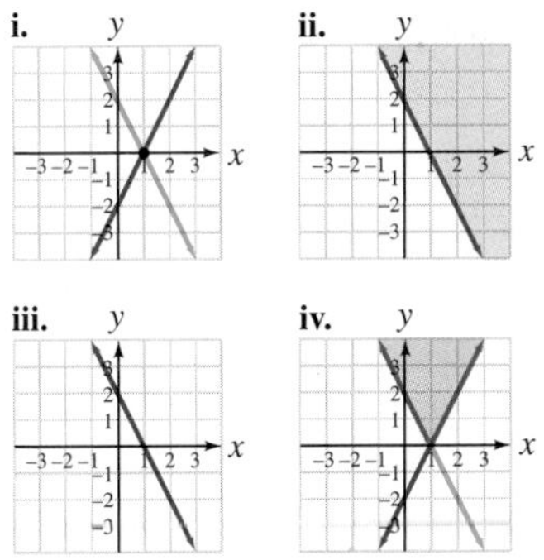

Group Project

WRITING APPLICATION PROBLEMS

Overview: In Section 4.4, you solved application problems by translating the words of the problem into a system of two equations. In this activity, you will reverse these steps.

Instructions: Form groups of 2 or 3 students. For each type of application, write a problem that could be solved using the given equations. If you need help getting started, refer to the specific problem types in the text. When finished writing the five applications, pick one problem and solve it completely.

A rectangle problem:

$$\begin{cases} 2l + 2w = 320 \\ l = w + 40 \end{cases}$$

A number-value problem:

$$\begin{cases} 5x + 2y = 23 \\ 3x + 7y = 37 \end{cases}$$

An interest problem:

$$\begin{cases} x + y = 75{,}000 \\ 0.03x + 0.05y = 2{,}750 \end{cases}$$

A with–against the wind problem:

$$\begin{cases} 2(x + y) = 600 \\ 3(x - y) = 600 \end{cases}$$

A liquid mixture problem:

$$\begin{cases} x + y = 36 \\ 0.50x + 0.20y = 0.30(36) \end{cases}$$

CUMULATIVE REVIEW Chapters 1–4

1. Fill in the blanks. The answer to an addition problem is called a sum. The answer to a subtraction problem is called a difference. The answer to a multiplication problem is called a product. The answer to a division problem is called a quotient. [Section 1.1]
2. Give the prime factorization of 100. [Section 1.2] $2^2 \cdot 5^2$
3. Divide: $\frac{3}{4} \div \frac{6}{5}$ [Section 1.2] $\frac{5}{8}$
4. Subtract: $\frac{7}{10} - \frac{1}{14}$ [Section 1.2] $\frac{22}{35}$
5. Is π a rational or irrational number? [Section 1.3] Irrational
6. Graph each member of the set on the number line. [Section 1.3] See AIE Appendix 3.

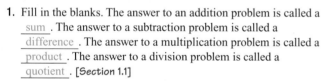

$$\left\{-2\frac{1}{4},\ \sqrt{2},\ -1.75,\ \frac{7}{2},\ 0.5\right\}$$

7. Write $\frac{2}{3}$ as a decimal. [Section 1.3] $0.\overline{6}$
8. What property of real numbers is illustrated? [Section 1.6]

$3(2x) = (3 \cdot 2)x$ Associative prop. of mult.

Evaluate each expression.

9. $-3^2 + |4^2 - 5^2|$ [Section 1.7] 0
10. $(4 - 5)^{20}$ [Section 1.7] 1
11. $\frac{-3 - (-7)}{2^2 - 3}$ [Section 1.7] 4
12. $12 - 2[1 - (-8 + 2)]$ [Section 1.7] -2
13. **Racing.** Suppose a driver has completed x laps of a 250-lap race. Write an expression for how many more laps he must make to finish the race. [Section 1.8] $250 - x$
14. What is the value of d dimes in cents? [Section 1.8] $10d$ cents

Simplify each expression.

15. $13r - 12r$ [Section 1.9] r
16. $27\left(\frac{2}{3}x\right)$ [Section 1.9] $18x$
17. $4(d - 3) - (d - 1)$ [Section 1.9] $3d - 11$
18. $(13c - 3)(-6)$ [Section 1.9] $-78c + 18$

Solve each equation. Check each result.

19. $3(x - 5) + 2 = 2x$ [Section 2.2] 13
20. $\frac{x - 5}{3} - 5 = 7$ [Section 2.2] 41
21. $\frac{2}{5}x + 1 = \frac{1}{3} + x$ [Section 2.2] $\frac{10}{9}$
22. $-\frac{5}{8}h = 15$ [Section 2.2] -24
23. **Gymnastics.** After the first day of registration, 119 children had been enrolled in a Gymboree class. That represented 85% of the available slots. Find the maximum number of children the center could enroll. [Section 2.3] 140 children

24. Greenhouse Gases. In 2008, the total U.S. greenhouse gas emissions were approximately 6,957 teragrams of carbon dioxide equivalent. Use the graph below to determine the greenhouse emissions that come from transportation sources. Round to the nearest teragram. [Section 2.3] 1,941 teragrams

U.S. Greenhouse Gas Emissions, by source

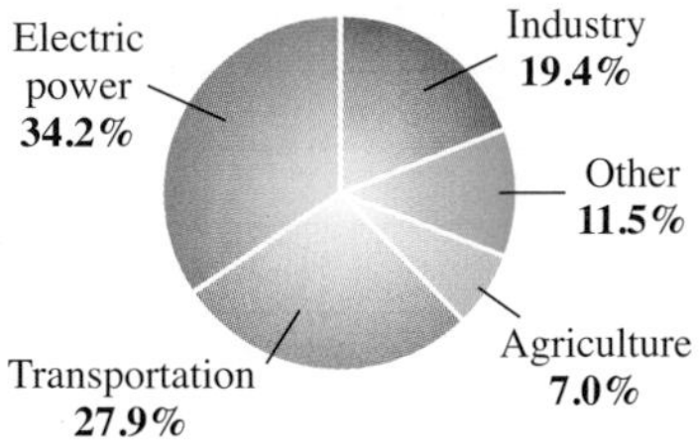

Source: U.S. Environmental Protection Agency

25. Solve $A = \frac{1}{2}h(b + B)$ for h. [Section 2.4] $h = \frac{2A}{b + B}$

26. Cancer. According to the National Lung Cancer Partnership, approximately 219,000 people are diagnosed with lung cancer in the U.S. each year, with 13,000 more of them being men than women. Approximately how many men and how many women are diagnosed with lung cancer each year? [Section 2.5] Men: 116,000; women: 103,000

27. Mixing Candy. The owner of a candy store wants to make a 30-pound mixture of two candies to sell for \$4 per pound. If red licorice bits sell for \$3.80 per pound and lemon gumdrops sell for \$4.40 per pound, how many pounds of each should be used? [Section 2.6] 20 lb of \$3.80 candy; 10 lb of \$4.40 candy

28. Solve $8(4 + x) > 10(6 + x)$. Write the solution set in interval notation and graph it. [Section 2.7] $(-\infty, -14)$

29. New York. In Manhattan, avenues run north and south, and streets run east and west to form a grid, as shown below. Give the location of the Chipotle Mexican Grill as an ordered pair of the form (avenue, street). [Section 3.1] (Ninth Avenue, 44th Street)

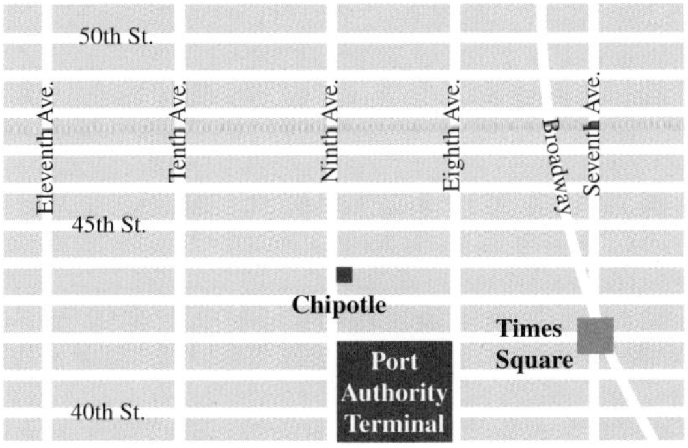

30. Perimeter. The eight vertices (corners) of a figure are $(-4, 6)$, $(4, 6)$, $(4, -3)$, $(2, -3)$, $(2, -5)$, $(-2, -5)$, $(-2, -3)$, and $(-4, -3)$. Find the perimeter of the figure. [Section 3.1] 38 units

31. In what quadrant does $(-3.5, 6)$ lie? [Section 3.1] II

32. Is $(-2, 8)$ a solution of $y = -2x + 3$? [Section 3.2] No

Graph each equation. See AIE Appendix 3.

33. $x = 4$ [Section 3.3]

34. $4x - 3y = 12$ [Section 3.3]

Find the slope of the line with the given properties.

35. Passing through $(-2, 4)$ and $(6, 8)$ [Section 3.4] $\frac{1}{2}$

36. A line that is horizontal [Section 3.4] 0

37. An equation of $2x - 3y = 12$ [Section 3.5] $\frac{2}{3}$

38. Are the graphs of the lines whose equations are given below parallel or perpendicular? [Section 3.5] Perpendicular

$$y = -\frac{3}{4}x + \frac{15}{4} \qquad 4x - 3y = 25$$

39. Find the slope and the x- and y-intercepts of the line graphed on the right. [Sections 3.3 and 3.4] Slope: 2; $(-1, 0)$, $(0, 2)$

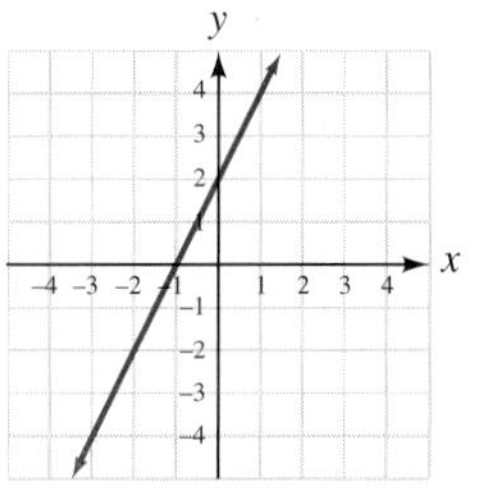

40. Newspapers. The line graph below approximates the percent of adults ages 25–34 who read a newspaper on a regular basis. Find the rate of change in newspaper readership for that age group. [Section 3.4] A decrease of 1.6% per year

Percent of those ages 25–34 who read a newspaper regularly

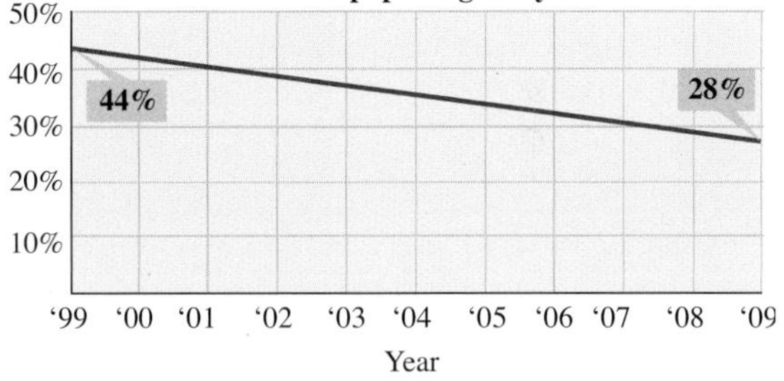

Source: The State of the News Media, 2010

Find an equation of the line with the following properties. Write the equation in slope–intercept form.

41. Slope $= \frac{2}{3}$, y-intercept $= (0, 5)$ [Section 3.5] $y = \frac{2}{3}x + 5$

42. Passing through $(-2, 4)$ and $(6, 10)$ [Section 3.6] $y = \frac{3}{4}x + \frac{11}{2}$

43. A horizontal line passing through $(2, 4)$ [Section 3.6] $y = 4$

44. Graph: $y < \frac{x}{3} - 1$ [Section 3.7] See AIE Appendix 3.

45. If $f(x) = -2x^2 - 3x^3$, find $f(-1)$. [Section 3.8] 1

46. Refer to the graph on the right. Is this the graph of a function? If it is, give the domain and range. If it is not, find ordered pairs that show a value of x that is assigned more than one value of y. [Section 3.8] No; $(1, 2)$, $(1, -2)$ (answers may vary)

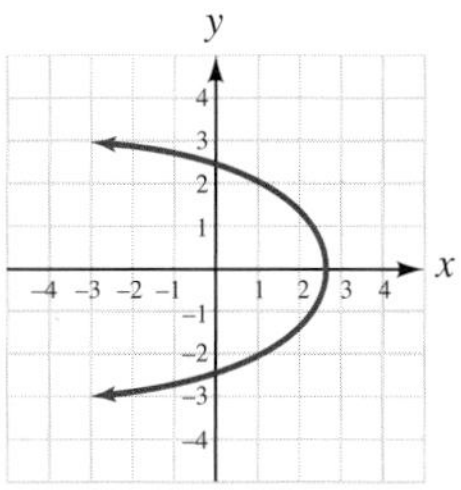

Solve each system by graphing. See AIE Appendix 3.

47. $\begin{cases} x + 4y = -2 \\ y = -x - 5 \end{cases}$ [Section 4.1] $(-6, 1)$

48. $\begin{cases} 2x - 3y < 0 \\ y > x - 1 \end{cases}$ [Section 4.5]

49. Solve $\begin{cases} x - 2y = 2 \\ 2x + 3y = 11 \end{cases}$ by substitution. [Section 4.2] $(4, 1)$

50. Solve $\begin{cases} \frac{3}{2}x - \frac{2}{3}y = 0 \\ \frac{3}{4}x + \frac{4}{3}y = \frac{5}{2} \end{cases}$ by elimination. [Section 4.3] $\left(\frac{2}{3}, \frac{3}{2}\right)$

Write a system of two equations in two variables to solve each problem.

51. Nutrition. The table shows per serving nutritional information for egg noodles and rice pilaf. How many servings of each food should be eaten to consume exactly 22 grams of protein and 21 grams of fat? [Section 4.4] Noodles: 2 servings, rice: 3 servings

	Protein (g)	Fat (g)
Egg noodles	5	3
Rice pilaf	4	5

52. Investment Clubs. Part of $8,000 was invested by an investment club at 10% interest and the rest at 12%. If the annual income from these investments is $900, how much was invested at each rate? [Section 4.4] $3,000 at 10%, $5,000 at 12%

5

Exponents and Polynomials

from Campus to Careers

Sound Engineering Technician

Today's digitally recorded music is crystal clear thanks to the talents of sound engineering technicians. They operate console mixing boards and microphones to record the music, voices, and sound effects that are so much a part of our media-filled lives. The job requires strong mathematical skills and an aptitude for working with electronic equipment. Sound technicians constantly work with numbers as they read meters and graphs, adjust dials and switches, and keep written logs.

Problem 125 in **Study Set 5.2, problem 89** in **Study Set 5.3,** and **problem 85** in **Study Set 5.4** involve situations that a sound engineering technician might encounter on the job. The mathematical concepts discussed in this chapter can be used to solve those problems.

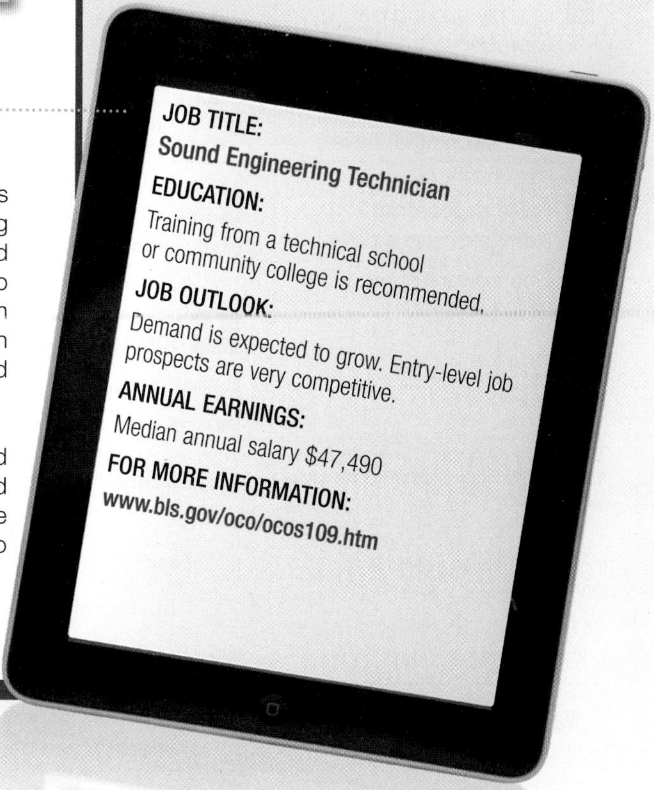

JOB TITLE:
Sound Engineering Technician
EDUCATION:
Training from a technical school or community college is recommended.
JOB OUTLOOK:
Demand is expected to grow. Entry-level job prospects are very competitive.
ANNUAL EARNINGS:
Median annual salary $47,490
FOR MORE INFORMATION:
www.bls.gov/oco/ocos109.htm

Study Skills Workshop ▶ Attendance

It's not uncommon for students' enthusiasm to lessen toward the middle of the term. Sometimes their effort and attendance begin to slip. Realize that missing even one class can have a great effect on your grade. Being tardy takes its toll as well. If you are just a few minutes late, or miss an entire class, you risk getting behind. So, keep the following tips in mind.

ARRIVE ON TIME, OR A LITTLE EARLY: When you arrive, get out your note-taking materials and homework. Identify any questions that you plan to ask your instructor once the class starts.

IF YOU MUST MISS A CLASS: Get a set of notes, the homework assignments, and any handouts that the instructor may have provided for the day(s) that you missed.

STUDY THE MATERIAL YOU MISSED: Take advantage of the online resources that are available with this textbook, such as video examples and problem-specific tutorials. Watch the explanations of the material from the section(s) that you missed.

Now Try This ▶

1. Plan ahead! List five possible situations that could cause you to be late to class or miss a class. (Some examples are parking/traffic delays, lack of a babysitter, oversleeping, or job responsibilities.) What can you do ahead of time so that these situations won't cause you to be tardy or absent?
2. Watch one section from the video series that accompanies this book. Take notes as you watch the explanations.

SECTION 5.1 Rules for Exponents

OBJECTIVES

1. Identify bases and exponents.
2. Multiply exponential expressions that have like bases.
3. Divide exponential expressions that have like bases.
4. Raise exponential expressions to a power.
5. Find powers of products and quotients.

ARE YOU READY? *Are You Ready? exercises available online at www.webassign.net/brookscole*

The following problems review some basic skills that are needed when working with exponents.

1. Evaluate: **a.** $5 + 5 + 5$ 15 **b.** $5 \cdot 5 \cdot 5$ 125
2. Evaluate: **a.** 2^6 64 **b.** $2 \cdot 6$ 12
3. Simplify: $\dfrac{x \cdot x \cdot x \cdot x}{x \cdot x \cdot x}$ x
4. Evaluate: $\dfrac{4^3}{4^2}$ 4

In this section, we will use the definition of exponent to develop some rules for simplifying expressions that contain exponents.

1 Identify Bases and Exponents.

Recall that an **exponent** indicates repeated multiplication. It indicates how many times the **base** is used as a factor. For example, 3^5 represents the product of five 3's.

$$3^5 = \underbrace{3 \cdot 3 \cdot 3 \cdot 3 \cdot 3}_{\text{5 factors of 3}}$$

(Exponent: 5; Base: 3)

In general, we have the following definition.

Teaching Tip: Remind your students that multiplication represents repeated addition:

$3 + 3 + 3 + 3 + 3 = 5 \cdot 3$

Exponents indicate repeated multiplication:

$3 \cdot 3 \cdot 3 \cdot 3 \cdot 3 = 3^5$

Natural-Number Exponents

A natural-number exponent tells how many times its base is to be used as a factor. For any number x and any natural number n,

$$x^n = \underbrace{x \cdot x \cdot x \cdot \cdots \cdot x}_{n\text{ factors of }x}$$

Expressions of the form x^n are called **exponential expressions.** The base of an exponential expression can be a number, a variable, or a combination of numbers and variables. Some examples are:

Notation

Bases that contain a − sign *must* be written within parentheses.

$(-2s)^3$ ← Exponent; Base: $-2s$

$10^5 = 10 \cdot 10 \cdot 10 \cdot 10 \cdot 10$ — The base is 10. The exponent is 5. Read as "10 to the fifth power" or simply as "10 to the fifth."

$y^2 = y \cdot y$ — The base is y. The exponent is 2. Read as "y squared."

$(-2s)^3 = (-2s)(-2s)(-2s)$ — The base is $-2s$. The exponent is 3. Read as "negative 2s raised to the third power" or "negative 2s cubed."

$-8^4 = -(8 \cdot 8 \cdot 8 \cdot 8)$ — Since the − sign is not written within parentheses, the base is 8. The exponent is 4. Read as "the opposite (or the negative) of 8 to the fourth power."

When an exponent is 1, it is usually not written. For example, $4 = 4^1$ and $x = x^1$.

EXAMPLE 1 Identify the base and the exponent in each expression: **a.** 9^5 **b.** $7a^3$ **c.** $(7a)^3$ **d.** $-t^{10}$

Strategy To identify the base and exponent, we will look for the form $\square^{\square}$.

Why The exponent is the small raised number to the right of the base.

Solution

a. In 9^5, the base is 9 and the exponent is 5.

b. $7a^3$ means $7 \cdot a^3$. Thus, the base is a, not $7a$. The exponent is 3.

c. Because of the parentheses in $(7a)^3$, the base is $7a$ and the exponent is 3.

d. Since the − symbol is not written within parentheses, the base in $-t^{10}$ is t and the exponent is 10.

Teaching Example 1 Identify the base and the exponent in each expression: **a.** 12^3 **b.** $7x^{10}$ **c.** -4^3 **d.** $(7x)^9$
Answers:
a. Base: 12; exponent: 3
b. Base: x; exponent: 10
c. Base: 4; exponent: 3
d. Base: $7x$; exponent: 9

Self Check 1 Identify the base and the exponent: **a.** 16^2 Base: 16; exponent: 2 **b.** $3y^4$ Base: y; exponent: 4 **c.** $(3y)^4$ Base: $3y$; exponent: 4 **d.** $-m^{15}$ Base: m; exponent: 15

Now Try Problems 13 and 17

EXAMPLE 2 Write as an exponential expression: **a.** $5 \cdot t \cdot t \cdot t$ **b.** $5t \cdot 5t \cdot 5t$ **c.** $\frac{p}{3} \cdot \frac{p}{3} \cdot \frac{p}{3} \cdot \frac{p}{3}$ **d.** $(a + 1)(a + 1)$

Teaching Tip: Remind your students that we read the notation $(a + 1)$ as "the quantity of a plus 1." Also remind them of the importance of writing the parentheses. Compare $(a + 1)^2$ to $a + 1^2$.

Strategy We will look for repeated factors and count the number of times each appears.

Why We can use an exponent to represent repeated multiplication.

Solution

a. There are three repeated factors of t in $5 \cdot t \cdot t \cdot t$. The expression can be written $5t^3$.

b. There are three repeated factors of $5t$ in $5t \cdot 5t \cdot 5t$. The expression can be written $(5t)^3$.

c. There are four repeated factors of $\frac{p}{3}$ in $\frac{p}{3} \cdot \frac{p}{3} \cdot \frac{p}{3} \cdot \frac{p}{3}$. The expression can be written $\left(\frac{p}{3}\right)^4$.

d. There are two repeated factors of $(a + 1)$. The expression can be written $(a + 1)^2$.

Teaching Example 2 Write as an exponential expression:
a. $3x \cdot 3x \cdot 3x \cdot 3x$
b. $3 \cdot x \cdot x \cdot x \cdot x$
c. $(y - 9)(y - 9)(y - 9)$
Answers:
a. $(3x)^4$ **b.** $3x^4$ **c.** $(y - 9)^3$

Self Check 2 Write as an exponential expression: **a.** $9 \cdot a \cdot a \cdot b \cdot b \cdot b \cdot b$ $9a^2b^4$ **b.** $9a \cdot 9a$ $(9a)^2$ **c.** $(x + y)(x + y)(x + y)(x + y)(x + y)$ $(x + y)^5$

Now Try Problems 19 and 21

2 Multiply Exponential Expressions That Have Like Bases.

To develop a rule for multiplying exponential expressions that have the same base, we consider the product $6^2 \cdot 6^3$. Since 6^2 means that 6 is to be used as a factor two times, and 6^3 means that 6 is to be used as a factor three times, we have

$$6^2 \cdot 6^3 = \overbrace{6 \cdot 6}^{\text{2 factors of 6}} \cdot \overbrace{6 \cdot 6 \cdot 6}^{\text{3 factors of 6}}$$

$$= \overbrace{6 \cdot 6 \cdot 6 \cdot 6 \cdot 6}^{\text{5 factors of 6}}$$

$$= 6^5$$

We can find this result quickly if we keep the common base of 6 and add the exponents on 6^2 and 6^3.

$$6^2 \cdot 6^3 = 6^{2+3} = 6^5$$

This example illustrates the following rule for exponents.

Product Rule for Exponents

To multiply exponential expressions that have the same base, keep the common base and add the exponents.

For any number x and any natural numbers m and n,

$$x^m \cdot x^n = x^{m+n}$$ Read as "x to the mth power times x to the nth power equals x to the m plus nth power."

EXAMPLE 3 Simplify: **a.** $9^5(9^6)$ **b.** $x^3 \cdot x^4$ **c.** y^2y^4y **d.** $(x + 2)^8(x + 2)^7$ **e.** $(c^2d^3)(c^4d^5)$

Strategy In each case, we want to write an equivalent expression using each base only once. We will use the product rule for exponents to do this.

Why The product rule for exponents is used to multiply exponential expressions that have the same base.

Solution

a. $9^5(9^6) = 9^{5+6} = 9^{11}$ Read as "9 to the fifth power times 9 to the sixth power." Keep the common base, 9, and add the exponents. Since 9^{11} is a very large number, we will leave the answer in this form. We won't evaluate it.

Caution

Don't make the mistake of multiplying the bases when using the product rule. Keep the same base.

$$9^5(9^6) \neq 81^{11}$$

b. $x^3 \cdot x^4 = x^{3+4} = x^7$ Keep the common base, x, and add the exponents.

c. $y^2y^4y = y^2y^4y^1$ Read as "y squared times y to the fourth power times y." Write y as y^1.

$= y^{2+4+1}$ Keep the common base, y, and add the exponents.

$= y^7$

Caution

Don't make the mistake of "distributing" an exponent over a sum (or difference). There is no such rule.

$$(x + 2)^{15} \neq x^{15} + 2^{15}$$

d. $(x + 2)^8(x + 2)^7 = (x + 2)^{8+7}$ Read as "the quantity of x + 2, raised to the eighth power, times the quantity of x + 2, raised to the seventh power." Keep the common base, x + 2, and add the exponents.

$= (x + 2)^{15}$

e. $(c^2d^3)(c^4d^5) = (c^2c^4)(d^3d^5)$ Read as "the quantity of c^2d^3 times the quantity of c^4d^5." Use the commutative and associative properties of multiplication to group like bases together.

$= (c^{2+4})(d^{3+5})$ Keep the common base, c, and add the exponents. Keep the common base, d, and add the exponents.

$= c^6d^8$

Teaching Example 3 Simplify:
a. $5^3(5^6)$ b. $a^3 \cdot a$ c. $y^2y^3y^5$
d. $(x^3y^2)(x^4y^3)$ e. $(a+b)^2(a+b)^4$
Answers:
a. 5^9 b. a^4 c. y^{10}
d. x^7y^5 e. $(a+b)^6$

Self Check 3 Simplify: **a.** $7^8(7^7)$ 7^{15} **b.** x^2x^3x x^6
c. $(y-1)^5(y-1)^5$ $(y-1)^{10}$ **d.** $(s^4t^3)(s^4t^4)$ s^8t^7

Now Try Problems 29 and 33

CAUTION We cannot use the product rule to simplify expressions like $3^2 \cdot 2^3$, where the bases are different. However, we can simplify this expression by doing the arithmetic:

$$3^2 \cdot 2^3 = 9 \cdot 8 = 72$$

EXAMPLE 4 **Geometry.** Find an expression that represents the area of the rectangle.

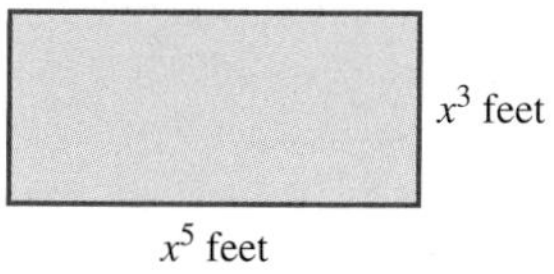

Strategy We will multiply the length of the rectangle by its width.

Why The area of a rectangle is equal to the product of its length and width.

Solution

$$\text{Area} = \textbf{length} \cdot \textbf{width} \quad \text{This is the formula for the area of a rectangle.}$$
$$= x^5 \cdot x^3 \quad \text{Substitute } x^5 \text{ for the length and } x^3 \text{ for the width.}$$
$$= x^{5+3} \quad \text{Use the product rule: Keep the common base, } x\text{, and add the exponents.}$$
$$= x^8$$

The area of the rectangle is x^8 square feet, which can be written as x^8 ft^2.

Teaching Tip: Have selected students read the following notation out loud and explain its meaning:
$m^5\text{ m}^2$ $y^{20}\text{ yd}^3$ $(c-m)^4\text{ cm}^2$

Teaching Example 4 Geometry. Find an expression that represents the area of the rectangle.

Answer: m^{14} yd^2

Self Check 4 **Geometry.** Find an expression that represents the area of a rectangle with length a^8 in. and width a^4 in. a^{12} in.^{2}

Now Try Problem 35

3 Divide Exponential Expressions That Have Like Bases.

To develop a rule for dividing exponential expressions that have the same base, we consider the quotient $\frac{4^5}{4^2}$, where the exponent in the numerator is greater than the exponent in the denominator. We can simplify this fraction by removing the common factors of 4 in the numerator and denominator:

$$\frac{4^5}{4^2} = \frac{4 \cdot 4 \cdot 4 \cdot 4 \cdot 4}{4 \cdot 4} = \frac{\overset{1}{\cancel{4}} \cdot \overset{1}{\cancel{4}} \cdot 4 \cdot 4 \cdot 4}{\underset{1}{\cancel{4}} \cdot \underset{1}{\cancel{4}}} = 4^3$$

We can find this result quickly if we keep the common base, 4, and subtract the exponents on 4^5 and 4^2.

$$\frac{4^5}{4^2} = 4^{5-2} = 4^3$$

This example illustrates another rule for exponents.

Quotient Rule for Exponents

To divide exponential expressions that have the same base, keep the common base and subtract the exponents.

For any nonzero number x and any natural numbers m and n, where $m > n$,

$$\frac{x^m}{x^n} = x^{m-n} \quad \text{Read as "x to the mth power divided by x to the nth power equals x to the m minus nth power."}$$

EXAMPLE 5 Simplify: **a.** $\frac{20^{16}}{20^9}$ **b.** $\frac{x^9}{x^3}$ **c.** $\frac{(7.5n)^{12}}{(7.5n)^{11}}$ **d.** $\frac{a^3b^8}{ab^5}$

Strategy In each case, we want to write an equivalent expression using each base only once. We will use the quotient rule for exponents to do this.

Why The quotient rule for exponents is used to divide exponential expressions that have the same base.

Solution

a. $\frac{20^{16}}{20^9} = 20^{16-9}$ Read as "20 to the sixteenth power divided by 20 to the ninth power." Keep the common base, 20, and subtract the exponents.

$= 20^7$ Since 20^7 is a very large number, we will leave the answer in this form. We won't evaluate it.

b. $\frac{x^9}{x^3} = x^{9-3}$ Keep the common base, x, and subtract the exponents.

$= x^6$

c. $\frac{(7.5n)^{12}}{(7.5n)^{11}} = (7.5n)^{12-11}$ Keep the common base, 7.5n, and subtract the exponents.

$= (7.5n)^1$

$= 7.5n$ Any number raised to the first power is simply that number.

d. $\frac{a^3b^8}{ab^5} = \frac{a^3}{a^1} \cdot \frac{b^8}{b^5}$ Group the common bases together. It is helpful to write a as a^1.

$= a^{3-1}b^{8-5}$ Keep the common base a and subtract the exponents. Keep the common base b and subtract the exponents.

$= a^2b^3$

Caution

Don't make the mistake of "removing the common bases" when using the quotient rule. Keep the same base.

$\frac{20^{16}}{20^9} = 1^7$ (crossed out)

The Language of Algebra

In this chapter, variables often appear in a denominator. In such cases, we will assume that the variables do not equal 0. That is, we will assume that **there are no divisions by 0**.

Teaching Example 5 Simplify: a. $\frac{13^{11}}{13^2}$ b. $\frac{x^8}{x^2}$ c. $\frac{y^4z^9}{yz^7}$ d. $\frac{(4.3a)^{20}}{(4.3a)^{11}}$
Answers:
a. 13^9 b. x^6 c. y^3z^2 d. $(4.3a)^9$

Self Check 5 Simplify: **a.** $\frac{55^{30}}{55^5}$ 55^{25} **b.** $\frac{a^5}{a^3}$ a^2 **c.** $\frac{(8.9t)^8}{(8.9t)^7}$ $8.9t$ **d.** $\frac{b^{15}c^4}{b^4c}$ $b^{11}c^3$

Now Try Problems 41 and 45

EXAMPLE 6 Simplify: $\frac{a^3a^5a^7}{a^4a}$

Strategy We want to write an equivalent expression using one base and one exponent. First, we will use the product rule to simplify the numerator and the denominator. Then, we will use the quotient rule to simplify that result.

Why The expression involves multiplication and division of exponential expressions that have the same base.

Solution We simplify the numerator and denominator separately and proceed as follows.

$\frac{a^3a^5a^7}{a^4a} = \frac{a^{15}}{a^5}$ In the numerator, keep the common base, a, and add the exponents. In the denominator, keep the common base, a, and add the exponents: 4 + 1 = 5.

$= a^{15-5}$ Keep the common base, a, and subtract the exponents.

$= a^{10}$

Success Tip

Sometimes, more than one rule for exponents is needed to simplify an expression.

Teaching Example 6 Simplify:
$\frac{x^5x^3x}{x^2x^3}$
Answer: x^4

Self Check 6 Simplify: $\frac{b^2b^6b}{b^4b^4}$ b

Now Try Problem 49

Recall that like terms are terms with exactly the same variables raised to exactly the same powers. To add or subtract exponential expressions, they must be like terms. To multiply or divide exponential expressions, only the bases need to be the same.

$x^5 + x^2$ These are not like terms; the exponents are different. We cannot add.

$x^2 + x^2 = 2x^2$ These are like terms; we can add. Recall that $x^2 = 1x^2$.

$x^5 \cdot x^2 = x^7$ The bases are the same; we can multiply.

$\frac{x^5}{x^2} = x^3$ The bases are the same; we can divide.

Teaching Tip: Compare the concepts of addition and subtraction of like terms with multiplication and division of exponential expressions with the same base.

4 Raise Exponential Expressions to a Power.

To develop another rule for exponents, we consider $(5^3)^4$. Here, an exponential expression, 5^3, is raised to a power. Since 5^3 is the base and 4 is the exponent, $(5^3)^4$ can be written as $5^3 \cdot 5^3 \cdot 5^3 \cdot 5^3$. Because each of the four factors of 5^3 contains three factors of 5, there are $4 \cdot 3$ or 12 factors of 5.

The Language of Algebra

An exponential expression raised to a power, such as $(5^3)^4$, is also called a **power of a power.**

$$(5^3)^4 = 5^3 \cdot 5^3 \cdot 5^3 \cdot 5^3 = \overbrace{\underbrace{5 \cdot 5 \cdot 5}_{5^3} \cdot \underbrace{5 \cdot 5 \cdot 5}_{5^3} \cdot \underbrace{5 \cdot 5 \cdot 5}_{5^3} \cdot \underbrace{5 \cdot 5 \cdot 5}_{5^3}}^{\text{12 factors of x}} = 5^{12}$$

We can find this result quickly if we keep the common base of 5 and multiply the exponents.

$$(5^3)^4 = 5^{3 \cdot 4} = 5^{12}$$

This example illustrates the following rule for exponents.

Power Rule for Exponents

To raise an exponential expression to a power, keep the base and multiply the exponents.

For any number x and any natural numbers m and n,

$(x^m)^n = x^{m \cdot n} = x^{mn}$ Read as "x to the mth power raised to the nth power equals x to the mnth power."

EXAMPLE 7 Simplify: **a.** $(2^3)^7$ **b.** $[(-6)^2]^5$ **c.** $(z^8)^8$

Strategy In each case, we want to write an equivalent expression using one base and one exponent. We will use the power rule for exponents to do this.

Why Each expression is a power of a power.

Solution

a. $(2^3)^7 = 2^{3 \cdot 7} = 2^{21}$ Read as "2 cubed raised to the seventh power." Keep the base, 2, and multiply the exponents.

b. $[(-6)^2]^5 = (-6)^{2 \cdot 5} = (-6)^{10}$ Read as "negative six squared raised to the fifth power." Keep the base, −6, and multiply the exponents. Since $(-6)^{10}$ is a very large number, we will leave the answer in this form.

c. $(z^8)^8 = z^{8 \cdot 8} = z^{64}$ Keep the base, z, and multiply the exponents.

Caution

Do not confuse the power and product rules for exponents:

$(2^3)^7 = 2^{3 \cdot 7} = 2^{21}$ Multiply exponents.

$2^3 \cdot 2^7 = 2^{3+7} = 2^{10}$ Add exponents.

Self Check 7 Simplify: **a.** $(4^6)^5$ 4^{30} **b.** $(y^5)^2$ y^{10}

Now Try Problems 53 and 55

Teaching Example 7 Simplify: **a.** $(7^3)^2$ **b.** $(a^4)^5$
Answers: **a.** 7^6 **b.** a^{20}

EXAMPLE 8 Simplify: **a.** $(x^2x^5)^8$ **b.** $(z^2)^4(z^3)^3$

Strategy In each case, we want to write an equivalent expression using one base and one exponent. We will use the product and power rules for exponents to do this.

Why The expressions involve multiplication of exponential expressions that have the same base and they involve powers of powers.

Solution **a.** $(x^2x^5)^8 = (x^7)^8$ Read as "the quantity of x squared times x to the fifth power, raised to the eighth power." Within the parentheses, keep the common base, x, and add the exponents.

$= x^{56}$ Keep the base, x, and multiply the exponents.

b. $(z^2)^4(z^3)^3 = z^8z^9$ For each power of z raised to a power, keep the base and multiply the exponents.

$= z^{17}$ Keep the common base, z, and add the exponents.

Teaching Example 8 Simplify:
a. $(r^2r^3)^4$ b. $(b^2)^5(b^3)^2$
Answers: a. r^{20} b. b^{16}

Self Check 8 Simplify: **a.** $(a^4a^3)^3$ a^{21} **b.** $(a^3)^3(a^4)^2$ a^{17}

Now Try Problems 59 and 63

5 Find Powers of Products and Quotients.

To develop more rules for exponents, we consider the expression $(2x)^3$, which is a *power of the product* of 2 and x, and the expression $\left(\frac{2}{x}\right)^3$, which is a *power of the quotient* of 2 and x.

$$\begin{aligned}(2x)^3 &= 2x \cdot 2x \cdot 2x \\ &= (2 \cdot 2 \cdot 2)(x \cdot x \cdot x) \\ &= 2^3x^3 \\ &= 8x^3\end{aligned}$$

$$\begin{aligned}\left(\frac{2}{x}\right)^3 &= \frac{2}{x} \cdot \frac{2}{x} \cdot \frac{2}{x} && \text{Assume } x \neq 0. \\ &= \frac{2 \cdot 2 \cdot 2}{x \cdot x \cdot x} && \text{Multiply the numerators. Multiply the denominators.} \\ &= \frac{2^3}{x^3} \\ &= \frac{8}{x^3} && \text{Evaluate: } 2^3 = 8.\end{aligned}$$

These examples illustrate the following rules for exponents.

Powers of a Product and a Quotient

To raise a product to a power, raise each factor of the product to that power. To raise a quotient to a power, raise the numerator and the denominator to that power.

For any numbers x and y, and any natural number n,

$$(xy)^n = x^ny^n \quad \text{and} \quad \left(\frac{x}{y}\right)^n = \frac{x^n}{y^n}, \quad \text{where } y \neq 0$$

EXAMPLE 9 Simplify: **a.** $(3c)^4$ **b.** $(x^2y^3)^5$ **c.** $\left(-\frac{1}{4}a^3b\right)^2$

Strategy In each case, we want to write the expression in an equivalent form in which each base is raised to a single power. We will use the power of a product rule for exponents to do this.

Why Within each set of parentheses is a product, and each of those products is raised to a power.

Solution

Caution

Never multiply an exponent and a base, or an exponent and a factor of the base!

$(3c)^4 = 12c^4$ (crossed out)

When simplifying expressions, *exponents interact only with other exponents* using the operations of addition, subtraction, or multiplication.

a. $(3c)^4 = 3^4c^4$ Raise each factor of the product 3c to the 4th power.

$= 81c^4$ Evaluate: $3^4 = 81$.

b. $(x^2y^3)^5 = (x^2)^5(y^3)^5$ Raise each factor of the product x^2y^3 to the 5th power.

$= x^{10}y^{15}$ For each power of a power, keep each base, x and y, and multiply the exponents.

c. $\left(-\frac{1}{4}a^3b\right)^2 = \left(-\frac{1}{4}\right)^2(a^3)^2b^2$ Raise each factor of the product $-\frac{1}{4}a^3b$ to the 2nd power.

$= \frac{1}{16}a^6b^2$ Evaluate: $\left(-\frac{1}{4}\right)^2 = \frac{1}{16}$. Keep the base a and multiply the exponents.

Teaching Example 9 Simplify:
a. $(5x)^3$ b. $(a^4b^3)^3$ c. $(-4a^5b^3)^2$
Answers:
a. $125x^3$ b. $a^{12}b^9$ c. $16a^{10}b^6$

Self Check 9 Simplify: **a.** $(2t)^4$ $16t^4$ **b.** $(c^3d^4)^6$ $c^{18}d^{24}$ **c.** $\left(-\frac{1}{3}ab^5\right)^3$ $-\frac{1}{27}a^3b^{15}$

Now Try Problems 67 and 73

EXAMPLE 10 Simplify: $\frac{(a^3b^4)^2}{ab^5}$

Strategy We want to write the expression in an equivalent form in which each base is raised to a single power. We will use the power of a product rule and the quotient rule for exponents to do this.

Why The expression involves a power of a product and it is the quotient of exponential expressions that have the same base.

Solution

$\frac{(a^3b^4)^2}{ab^5} = \frac{(a^3)^2(b^4)^2}{a^1b^5}$ In the numerator, raise each factor within the parentheses to the second power. In the denominator, it is helpful to write a as a^1.

$= \frac{a^6b^8}{a^1b^5}$ In the numerator, for each power of a power, keep each base, a and b, and multiply the exponents.

$= a^{6-1}b^{8-5}$ Keep each of the bases, a and b, and subtract the exponents.

$= a^5b^3$

Teaching Example 10 Simplify: $\frac{(6x^3y^4)^2}{x^2y^3}$
Answer: $36x^4y^5$

Self Check 10 Simplify: $\frac{(c^4d^5)^3}{c^2d^3}$ $c^{10}d^{12}$

Now Try Problem 75

EXAMPLE 11 Simplify: $\frac{(5b)^9}{(5b)^6}$

Strategy We want to write the expression in an equivalent form in which the base is raised to a single power. We will begin by using the quotient rule for exponents.

Why The expression involves division of exponential expressions that have the same base, 5b.

Solution

$\frac{(5b)^9}{(5b)^6} = (5b)^{9-6}$ Keep the common base, 5b, and subtract the exponents.

$= (5b)^3$

$= 5^3b^3$ Use the power of a product rule: Raise each factor within the parentheses to the 3rd power. Do not multiply the exponent 3 and the base 5.

$= 125b^3$ Evaluate 5^3.

Teaching Example 11 Simplify: $\frac{(3x)^5}{(3x)^2}$
Answer: $27x^3$

Self Check 11 Simplify: $\frac{(-2h)^{20}}{(-2h)^{14}}$ $64h^6$

Now Try Problem 79

EXAMPLE 12 Simplify: **a.** $\left(\frac{4}{k}\right)^3$ **b.** $\left(\frac{3x^2}{2y^3}\right)^5$

Strategy We want to write each expression in an equivalent form using each base raised to a single power. We will use the power of a quotient rule for exponents to do this.

Why Within each set of parentheses is a quotient, and each of those quotients is raised to a power.

Solution **a.** Since $\frac{4}{k}$ is the quotient of 4 and k, the expression $\left(\frac{4}{k}\right)^3$ is a power of a quotient.

$$\left(\frac{4}{k}\right)^3 = \frac{4^3}{k^3} = \frac{64}{k^3}$$ Raise the numerator and denominator to the 3rd power. Then evaluate: $4^3 = 64$. Do not multiply the exponent 3 and the base 4.

b. $$\left(\frac{3x^2}{2y^3}\right)^5 = \frac{(3x^2)^5}{(2y^3)^5}$$ Raise the numerator and the denominator to the 5th power.

$$= \frac{3^5(x^2)^5}{2^5(y^3)^5}$$ Raise each factor within parentheses to the 5th power. Do not multiply the base 3 and the exponent 5 or the base 2 and the exponent 5.

$$= \frac{243x^{10}}{32y^{15}}$$ Evaluate: $3^5 = 243$ and $2^5 = 32$. For each power of a power, keep the base and multiply the exponents.

Teaching Example 12 Simplify: a. $\left(\frac{a}{3}\right)^4$ b. $\left(\frac{5x^3}{3y^4}\right)^3$
Answers: a. $\frac{a^4}{81}$ b. $\frac{125x^9}{27y^{12}}$

Self Check 12 Simplify: **a.** $\left(\frac{x}{7}\right)^3$ $\frac{x^3}{343}$ **b.** $\left(\frac{2x^3}{3y^2}\right)^4$ $\frac{16x^{12}}{81y^8}$

Now Try Problems 83 and 85

Teaching Tip: Stress that these rules need to be memorized.

The following rules for exponents are used so often in this course, you need to memorize them.

Rules for Exponents

If m and n represent natural numbers and there are no divisions by zero, then

Exponent of 1	Product rule	Power rule
$x^1 = x$	$x^m x^n = x^{m+n}$	$(x^m)^n = x^{mn}$
Quotient rule	**Power of a product**	**Power of a quotient**
$\frac{x^m}{x^n} = x^{m-n}$	$(xy)^n = x^n y^n$	$\left(\frac{x}{y}\right)^n = \frac{x^n}{y^n}$

SECTION 5.1 STUDY SET

VOCABULARY

Fill in the blank.

1. Expressions such as x^4, 10^3, and $(5t)^2$ are called exponential expressions.

2. Match each expression below with the proper description on the next page.

$\frac{a^8}{a^2}$ $(a^4b^2)^5$ $\left(\frac{a^6}{a}\right)^3$ $(a^8)^4$ $a^5 \cdot a^3$

a. Product of exponential expressions with the same base $a^5 \cdot a^3$
b. Quotient of exponential expressions with the same base $\frac{a^8}{a^2}$
c. Power of an exponential expression $(a^8)^4$
d. Power of a product $(a^4b^2)^5$
e. Power of a quotient $\left(\frac{a^6}{a}\right)^3$

CONCEPTS

Fill in the blanks.

3. a. $(3x)^4 = 3x \cdot 3x \cdot 3x \cdot 3x$

b. $(-5y)(-5y)(-5y) = (-5y)^3$

▶ 4. a. $x = x^1$ b. $x^m x^n = x^{m+n}$

c. $(xy)^n = x^n y^n$ d. $(a^b)^c = a^{bc}$

e. $\frac{x^m}{x^n} = x^{m-n}$ f. $\left(\frac{a}{b}\right)^n = \frac{a^n}{b^n}$

5. To simplify each expression, determine whether you add, subtract, multiply, or divide the exponents.

a. $\frac{x^8}{x^2}$ Subtract b. $b^6 \cdot b^9$ Add

c. $(n^8)^4$ Multiply d. $(a^4b^2)^5$ Multiply

6. a. To simplify $(2y^3z^2)^4$, what factors within the parentheses must be raised to the fourth power? $2, y^3, z^2$

b. To simplify $\left(\frac{y^3}{z^2}\right)^4$, what two expressions must be raised to the fourth power? y^3 and z^2

Simplify each expression, if possible.

7. a. $x^2 + x^2$ $2x^2$ b. $x^2 \cdot x^2$ x^4
c. $x^2 + x$ Doesn't simplify d. $x^2 \cdot x$ x^3

8. a. $x^3 - x^2$ Doesn't simplify b. $\frac{x^3}{x^2}$ x

c. $4^2 \cdot 2^4$ 256 d. $\frac{x^3}{y^2}$ Doesn't simplify

NOTATION

Complete each solution to simplify each expression.

9. $(x^4x^2)^3 = (x^6)^3 = x^{18}$

▶ 10. $\frac{a^3a^4}{a^2} = \frac{a^7}{a^2} = a^{7-2} = a^5$

Fill in the blanks.

11. a. We read 9^4 as "nine to the fourth power."

b. We read $(a^2b^6)(a^4b^5)$ as "the quantity of a^2b^6 times the quantity of a^4b^5."

c. We read $(3^6)^9$ as "3 to the sixth power, raised to the ninth power."

▶ 12. a. We read n^2n^3n as "n squared times n cubed times n."

b. We read $\frac{x^7}{x^5}$ as "x to the seventh power divided by x to the fifth power."

c. We read $(b + 5)^6(b + 5)^8$ as "the quantity of $b + 5$, raised to the sixth power, times the quantity of $b + 5$, raised to the eighth power."

GUIDED PRACTICE

Identify the base and the exponent in each expression. **See Example 1.**

Look Alikes . . .

13. a. 4^3 4, 3 b. -4^3 4, 3 c. $(-4)^3$ −4, 3
▶ 14. a. x^5 x, 5 b. $-x^5$ x, 5 c. $(-x)^5$ $-x$, 5
15. a. $(-3x)^2$ $-3x$, 2 b. $-3x^2$ x, 2 c. $-(-3x)^2$ $-3x$, 2
▶ 16. a. $-\frac{1}{3}x^6$ x, 6 b. $\left(-\frac{1}{3}x\right)^6$ $-\frac{1}{3}x$, 6 c. $-\left(-\frac{1}{3}x\right)^6$ $-\frac{1}{3}x$, 6
17. a. $9m^{12}$ m, 12 b. $(9m)^{12}$ $9m$, 12 c. $-9m^{12}$ m, 12
▶ 18. a. $(y + 9)^4$ $y + 9$, 4 b. $y + 9^4$ 9, 4 c. $8(y + 9)^4$ $y + 9$, 4

Write each expression in an equivalent form using an exponent. **See Example 2.**

19. $4t \cdot 4t \cdot 4t \cdot 4t$ $(4t)^4$
▶ 20. $-5u(-5u)(-5u)(-5u)(-5u)$ $(-5u)^5$
21. $-4 \cdot t \cdot t \cdot t \cdot t \cdot t$ $-4t^5$ ▶ 22. $-5 \cdot u \cdot u \cdot u$ $-5u^3$
23. $\frac{t}{2} \cdot \frac{t}{2} \cdot \frac{t}{2}$ $\left(\frac{t}{2}\right)^3$ ▶ 24. $\frac{x}{c} \cdot \frac{x}{c} \cdot \frac{x}{c} \cdot \frac{x}{c}$ $\left(\frac{x}{c}\right)^4$
25. $(x - y)(x - y)$ $(x - y)^2$ ▶ 26. $(m + 4)(m + 4)$ $(m + 4)^2$

Use the product rule for exponents to simplify each expression. Write the results using exponents. **See Example 3.**

27. $5^3 \cdot 5^4$ 5^7 ▶ 28. $3^4 \cdot 3^6$ 3^{10}
▶ 29. bb^2b^3 b^6 30. aa^3a^5 a^9
31. $(y - 2)^5(y - 2)^2$ $(y - 2)^7$ ▶ 32. $(t + 1)^5(t + 1)^3$ $(t + 1)^8$
33. $(a^2b^3)(a^3b^3)$ a^5b^6 ▶ 34. $(u^3v^5)(u^4v^5)$ u^7v^{10}

Find an expression that represents the area or volume of each figure. Recall that the formula for the volume of a rectangular solid is V = length · width · height. **See Example 4.**

▶ 35.
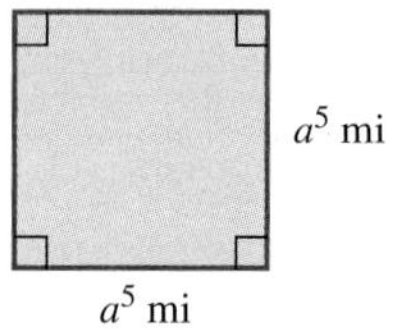

a^{10} mi^2

▶ 36.
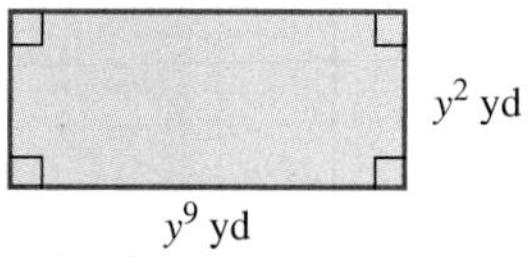

y^{11} yd^2

▶ 37.
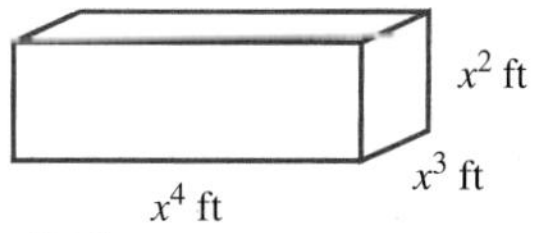

x^9 ft^3

▶ 38.
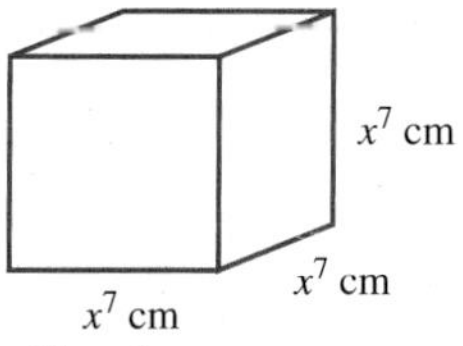

x^{21} cm^3

Use the quotient rule for exponents to simplify each expression. Write the results using exponents. **See Example 5.**

▶ 39. $\frac{8^{12}}{8^4}$ 8^8 40. $\frac{10^4}{10^2}$ 10^2
41. $\frac{x^{15}}{x^3}$ x^{12} ▶ 42. $\frac{y^6}{y^3}$ y^3
43. $\frac{(3.7p)^7}{(3.7p)^2}$ $(3.7p)^5$ ▶ 44. $\frac{(0.25y)^9}{(0.25y)^3}$ $(0.25y)^6$
▶ 45. $\frac{c^3d^7}{cd}$ c^2d^6 ▶ 46. $\frac{r^8s^9}{rs}$ r^7s^8

Use the product and quotient rules for exponents to simplify each expression. **See Example 6.**

47. $\frac{y^3y^4}{yy^2}$ y^4

48. $\frac{b^4b^5}{b^2b^3}$ b^4

49. $\frac{a^2a^3a^4}{a^8}$ a

50. $\frac{h^3h^6h}{h^9}$ h

Use the power rule for exponents to simplify each expression. Write the results using exponents. **See Example 7.**

51. $(3^2)^4$ 3^8

52. $(4^3)^3$ 4^9

53. $[(-4.3)^3]^8$ $(-4.3)^{24}$

54. $[(-1.7)^9]^8$ $(-1.7)^{72}$

55. $(m^{50})^{10}$ m^{500}

56. $(n^{25})^4$ n^{100}

57. $(y^5)^3$ y^{15}

58. $(b^3)^6$ b^{18}

Use the product and power rules for exponents to simplify each expression. **See Example 8.**

59. $(x^2x^3)^5$ x^{25}

60. $(y^3y^4)^4$ y^{28}

61. $(p^2p^3)^5$ p^{25}

62. $(r^3r^4)^2$ r^{14}

63. $(t^3)^4(t^2)^3$ t^{18}

64. $(b^2)^5(b^3)^2$ b^{16}

65. $(u^4)^2(u^3)^2$ u^{14}

66. $(v^5)^2(v^3)^4$ v^{22}

Use the power of a product rule for exponents to simplify each expression. **See Example 9.**

67. $(6a)^2$ $36a^2$

68. $(3b)^3$ $27b^3$

69. $(5y)^4$ $625y^4$

70. $(4t)^4$ $256t^4$

71. $(-2r^2s^3)^3$ $-8r^6s^9$

72. $(-2x^2y^4)^5$ $-32x^{10}y^{20}$

73. $\left(-\frac{1}{3}y^2z^4\right)^5$ $-\frac{1}{243}y^{10}z^{20}$

74. $\left(-\frac{1}{4}t^3u^8\right)^2$ $\frac{1}{16}t^6u^{16}$

Use rules for exponents to simplify each expression. **See Example 10.**

75. $\frac{(ab^2)^3}{a^2b^2}$ ab^4

76. $\frac{(m^3n^4)^3}{m^3n^6}$ m^6n^6

77. $\frac{(r^4s^3)^4}{r^3s^9}$ $r^{13}s^3$

78. $\frac{(x^2y^5)^5}{x^6y^2}$ x^4y^{23}

Use rules for exponents to simplify each expression. **See Example 11.**

79. $\frac{(6k)^7}{(6k)^4}$ $216k^3$

80. $\frac{(-3a)^{12}}{(-3a)^{10}}$ $9a^2$

81. $\frac{(3q)^5}{(3q)^3}$ $9q^2$

82. $\frac{(ab)^8}{(ab)^4}$ a^4b^4

Use the power of a quotient rule for exponents to simplify each expression. **See Example 12.**

83. $\left(\frac{a}{b}\right)^3$ $\frac{a^3}{b^3}$

84. $\left(\frac{r}{s}\right)^4$ $\frac{r^4}{s^4}$

85. $\left(\frac{8a^2}{11b^5}\right)^2$ $\frac{64a^4}{121b^{10}}$

86. $\left(\frac{7g^4}{6h^3}\right)^2$ $\frac{49g^8}{36h^6}$

TRY IT YOURSELF

Simplify each expression, if possible.

87. $\left(\frac{x^2}{y^3}\right)^5$ $\frac{x^{10}}{y^{15}}$

88. $\left(\frac{u^4}{v^2}\right)^6$ $\frac{u^{24}}{v^{12}}$

89. $y^3y^2y^4$ y^9

90. y^4yy^6 y^{11}

91. $\frac{15^9}{15^6}$ 15^3

92. $\frac{25^{13}}{25^7}$ 25^6

93. $\frac{t^5t^6t}{t^2t^3}$ t^7

94. $\frac{m^5m^{12}m}{m^7m^4}$ m^7

95. $\frac{(k-2)^{15}}{(k-2)}$ $(k-2)^{14}$

96. $\frac{(m+8)^{20}}{(m+8)}$ $(m+8)^{19}$

97. $cd^4 \cdot cd$ c^2d^5

98. $ab^3 \cdot ab^4$ a^2b^7

99. $\left(\frac{y^3y^5}{yy^2}\right)^3$ y^{15}

100. $\left(\frac{s^5s^6}{s^2s^2}\right)^4$ s^{28}

101. $\frac{s^2s^2s^2}{s^3s}$ s^2

102. $\frac{w^4w^4w^4}{w^2w}$ w^9

103. $(-6a^3b^2)^3$ $-216a^9b^6$

104. $(-10r^3s^2)^2$ $100r^6s^4$

105. $\left(\frac{3m^4}{2n^5}\right)^5$ $\frac{243m^{20}}{32n^{25}}$

106. $\left(\frac{2s^2}{3t^5}\right)^5$ $\frac{32s^{10}}{243t^{25}}$

107. $\frac{(a^2b^2)^{15}}{(ab)^9}$ $a^{21}b^{21}$

108. $\frac{(s^3t^3)^4}{(st)^2}$ $s^{10}t^{10}$

109. $(n^4n)^3(n^3)^6$ n^{33}

110. $(y^3y)^2(y^2)^2$ y^{12}

111. $\frac{(6h)^8}{(6h)^6}$ $36h^2$

112. $\frac{(-7r)^{10}}{(-7r)^8}$ $49r^2$

113. $\frac{x^4y^7}{xy^3}$ x^3y^4

114. $\frac{p^7q^{10}}{p^2q^7}$ p^5q^3

115. $\left(\frac{m}{3}\right)^4$ $\frac{m^4}{81}$

116. $\left(\frac{n}{5}\right)^3$ $\frac{n^3}{125}$

Look Alikes . . .

117. a. $a^3 \cdot a^3$ a^6 b. $(a^3)^3$ a^9 c. $a^3 + a^3$ $2a^3$

118. a. $(m^5)^7$ m^{35} b. $m^5 \cdot m^7$ m^{12} c. $m^5 - m^7$ Does not simplify

119. a. $b^3b^2b^4$ b^9 b. $(b^3b^2)^4$ b^{20} c. $\frac{b^3b^2}{b^4}$ b

120. a. $(2n^4n)^5$ $32n^{25}$ b. $2n^4n$ $2n^5$ c. $\left(\frac{2n^4}{n}\right)^5$ $32n^{15}$

APPLICATIONS

121. **Art History.** Leonardo da Vinci's drawing relating a human figure to a square and a circle is shown. Find an expression for the following:

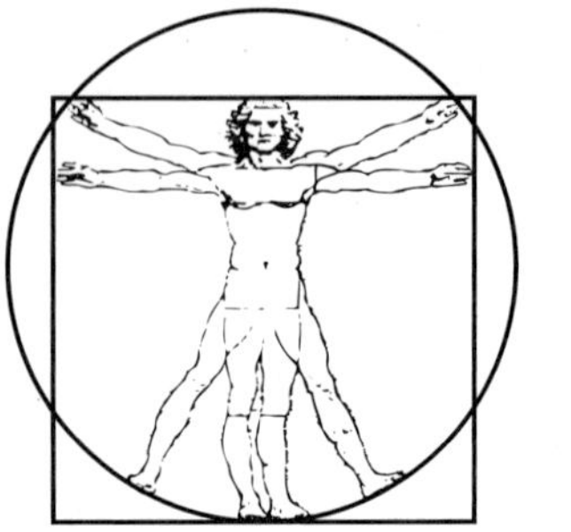

a. The area of the square if the man's height is $5x$ feet $25x^2$ ft^2

b. The area of the circle if the waist-to-feet distance is $3a$ feet. Leave π in your answer. $9a^2\pi$ ft^2

122. **Packaging.** A bowling ball fits tightly against all sides of a cardboard box that it is packaged in. Find expressions for the volume of the ball and box. Leave π in your answer. $36\pi x^3$ in.^{3}, $216x^3$ in.^{3}

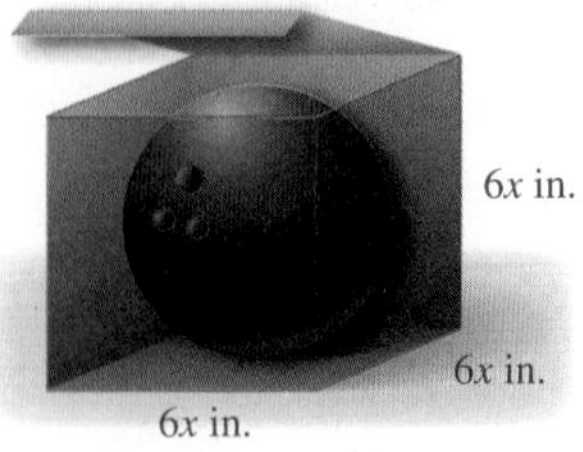

▶ **123. Childbirth.** Mr. and Mrs. Emory Harrison, of Johnson City, Tennessee, had 13 sons in a row during the 1940s and 1950s. The **probability** of a family of 13 children all being male is $\left(\frac{1}{2}\right)^{13}$. Evaluate this expression. $\frac{1}{8,192}$

124. Toys. A Super Ball is dropped from a height of 1 foot and always rebounds to four-fifths of its previous height. The rebound height of the ball after the third bounce is $\left(\frac{4}{5}\right)^3$ feet. Evaluate this expression. Is the third bounce more or less than $\frac{1}{2}$ foot high? $\frac{64}{125}$; slightly more (0.512 ft)

WRITING

▶ **125.** Explain the mistake in the following work.

a. ~~$2^3 \cdot 2^2 = 4^5 = 1,024$~~ b. $(5d^2)^3 = 15d^6$ (crossed out)

▶ **126.** Explain why we can simplify $x^4 \cdot x^5$, but cannot simplify $x^4 + x^5$.

REVIEW

Match each equation with its graph below.

127. $y = 2x - 1$ c

128. $y = 3x - 1$ a

▶ **129.** $y = 3$ d

▶ **130.** $x = 3$ b

a.
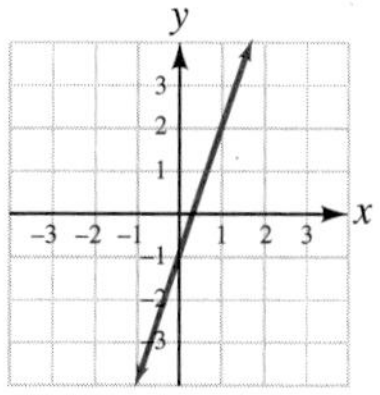

b.
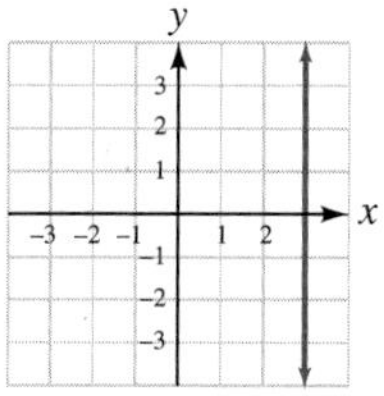

c.
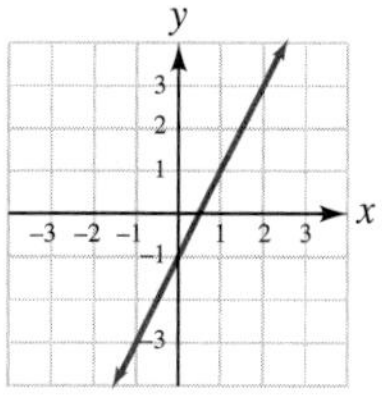

d.
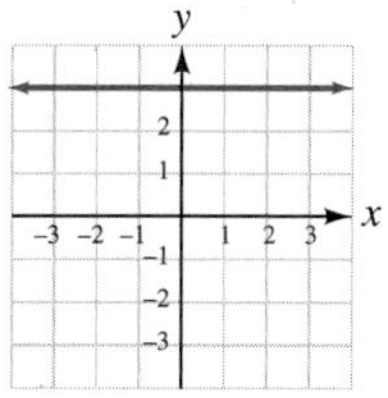

CHALLENGE PROBLEMS

▶ **131.** Simplify each expression. The variables represent natural numbers.

a. $x^{2m}x^{3m}$ x^{5m}

b. $(y^{5c})^4$ y^{20c}

c. $\frac{m^{8x}}{m^{4x}}$ m^{4x}

d. $(2a^{6y})^4$ $16a^{24y}$

132. Evaluate the following expression without using a calculator:

$$\frac{(108,642)^4}{(54,321)^4}$$ 16

SECTION 5.2

OBJECTIVES

1. Use the zero exponent rule.
2. Use the negative integer exponent rule.
3. Use exponent rules to change negative exponents in fractions to positive exponents.
4. Use all exponent rules to simplify expressions.

Zero and Negative Exponents

ARE YOU READY? *Are You Ready? exercises available online at www.webassign.net/brookscole*

The following problems review some basic skills that are needed when working with zero and negative exponents.

1. Simplify: $\frac{2 \cdot 2}{2 \cdot 2 \cdot 2}$ $\frac{1}{2}$

2. Simplify: $\frac{x \cdot x}{x \cdot x \cdot x \cdot x}$ $\frac{1}{x^2}$

3. Find the reciprocal: **a.** $\frac{1}{5}$ 5 **b.** w $\frac{1}{w}$

4. Identify the base and exponent: **a.** m^0 $m, 0$ **b.** 7^{-2} $7, -2$

We now extend the discussion of natural-number exponents to include exponents that are zero and exponents that are negative integers.

1 Use the Zero Exponent Rule.

To develop the definition of a zero exponent, we will simplify the expression $\frac{5^3}{5^3}$ in two ways and compare the results. First, we apply the quotient rule for exponents, where we subtract the equal exponents in the numerator and denominator. The result is 5^0, which is read as "5 to the zero power." In the second approach, we write 5^3 as $5 \cdot 5 \cdot 5$ and remove the common factors of 5 in the numerator and denominator. The result is 1.

$$\frac{5^3}{5^3} = 5^{3-3} = 5^0 \qquad \frac{5^3}{5^3} = \frac{\overset{1}{\cancel{5}} \cdot \overset{1}{\cancel{5}} \cdot \overset{1}{\cancel{5}}}{\underset{1}{\cancel{5}} \cdot \underset{1}{\cancel{5}} \cdot \underset{1}{\cancel{5}}} = 1$$

These results must be equal.

Since $\frac{5^3}{5^3} = 5^0$ and $\frac{5^3}{5^3} = 1$, we conclude that $5^0 = 1$. This observation illustrates the following definition.

Zero Exponents

Any nonzero base raised to the 0 power is 1.
For any nonzero real number x,

$x^0 = 1$ Read as "x to the zero power equals 1."

EXAMPLE 1 Simplify. Assume $a \neq 0$: **a.** $(-8)^0$ **b.** $\left(\frac{14}{15}\right)^0$ **c.** $(3a)^0$ **d.** $3a^0$

Strategy We note that each exponent is 0. To simplify the expressions, we will identify the base and use the zero-exponent rule.

Why If an expression contains a nonzero base raised to the 0 power, we can replace it with 1.

Solution

a. $(-8)^0 = 1$ Because the base is −8 and the exponent is 0.

b. $\left(\frac{14}{15}\right)^0 = 1$ Because the base is $\frac{14}{15}$ and the exponent is 0.

c. $(3a)^0 = 1$ Because of the parentheses, the base is 3a. The exponent is 0.

d. $3a^0 = 3 \cdot a^0$ Read as "3 times a to the zero power." Since there are no parentheses, the base is a, not 3a. The exponent is 0.

$= 3 \cdot 1$ Simplify: $a^0 = 1$.

$= 3$

The Language of Algebra

The zero exponent definition does not define 0^0. This expression is called an **indeterminate form,** which is beyond the scope of this book.

Teaching Example 1 Simplify each expression:
a. 30^0 b. $5y^0$ c. $(5y)^0$
Answers: a. 1 b. 5 c. 1

Self Check 1 Simplify each expression: **a.** $(0.75)^0$ 1 **b.** $-5c^0d$ $-5d$ **c.** $(5c)^0$ 1

Now Try Problems 13 and 17

2 Use the Negative Integer Exponent Rule.

The Language of Algebra

The **negative integers** are:
$-1, -2, -3, -4, -5, \ldots$

To develop the definition of a negative exponent, we will simplify $\frac{6^2}{6^5}$ in two ways and compare the results.

If we apply the quotient rule for exponents, where we subtract the greater exponent in the denominator from the lesser exponent in the numerator, we get 6^{-3}. In the second approach, we remove the two common factors of 6 to get $\frac{1}{6^3}$.

$$\frac{6^2}{6^5} = 6^{2-5} = 6^{-3} \qquad \frac{6^2}{6^5} = \frac{\overset{1}{\cancel{6}} \cdot \overset{1}{\cancel{6}}}{\underset{1}{\cancel{6}} \cdot \underset{1}{\cancel{6}} \cdot 6 \cdot 6 \cdot 6} = \frac{1}{6^3}$$

These must be equal.

Since $\frac{6^2}{6^5} = 6^{-3}$ and $\frac{6^2}{6^5} = \frac{1}{6^3}$, we conclude that $6^{-3} = \frac{1}{6^3}$. Note that 6^{-3} is equal to the reciprocal of 6^3. This observation illustrates the following definition.

Negative Exponents

For any nonzero real number x and any integer n,

$$x^{-n} = \frac{1}{x^n}$$ Read as "x to the negative nth power equals 1 over x to the nth power."

In words, x^{-n} is the reciprocal of x^n.

From the definition, we see that another way to write x^{-n} is to write its reciprocal and change the sign of the exponent. For example,

$$5^{-4} = \frac{1}{5^4}$$ Think of the reciprocal of 5^{-4}. Then change the sign of the exponent.

The definitions of a zero exponent and a negative integer exponent have been written in such a way as to remain consistent with the definition for a natural-number exponent that we learned earlier. This can be seen in the following list.

$2^4 = 16$
Divide by 2.
$2^3 = 8$
Divide by 2.
$2^2 = 4$
Divide by 2.
$2^1 = 2$

For natural number exponents, each time that we decrease the exponent by 1, the value of the exponential expression is divided by 2.

$2^0 = 1$ For the pattern to continue, we define $2^0 = 1$.

$2^{-1} = \frac{1}{2}$ For the pattern to continue, we define $2^{-1} = \frac{1}{2^1} = \frac{1}{2}$.

$2^{-2} = \frac{1}{4}$ For the pattern to continue, we define $2^{-2} = \frac{1}{2^2} = \frac{1}{4}$.

$2^{-3} = \frac{1}{8}$ For the pattern to continue, we define $2^{-3} = \frac{1}{2^3} = \frac{1}{8}$.

EXAMPLE 2 Express using positive exponents and simplify, if possible: **a.** 3^{-2} **b.** y^{-1} **c.** $(-2)^{-3}$ **d.** $5^{-2} - 10^{-2}$

Strategy Since each exponent is a negative number, we will use the negative exponent rule.

Why This rule enables us to write an exponential expression that has a negative exponent in an equivalent form using a positive exponent.

Solution

a. $3^{-2} = \frac{1}{3^2} = \frac{1}{9}$ Read as "3 to the negative second power." Write the reciprocal of 3^{-2} and change the sign of the exponent.

b. $y^{-1} = \frac{1}{y^1} = \frac{1}{y}$ Read as "y to the negative first power." Write the reciprocal of y^{-1} and change the sign of the exponent.

c. $(-2)^{-3} = \frac{1}{(-2)^3}$ Read as "−2 to the negative third power." Because of the parentheses, the base is −2. Write the reciprocal of $(-2)^{-3}$ and change the exponent.

$= -\frac{1}{8}$ Evaluate: $(-2)^3 = -8$. Write the − symbol in front of the fraction.

Caution

A negative exponent **does not,** itself, make the simplified expression negative. It indicates a reciprocal. Avoid these common mistakes:

~~$3^{-2} = -9$~~ ~~$3^{-2} = -\frac{1}{9}$~~

d. $5^{-2} - 10^{-2} = \frac{1}{5^2} - \frac{1}{10^2}$ Write the reciprocal of 5^{-2} and 10^{-2} and change the sign of each exponent.

$= \frac{1}{25} - \frac{1}{100}$ Evaluate: $5^2 = 25$ and $10^2 = 100$.

$= \frac{4}{100} - \frac{1}{100}$ Build $\frac{1}{25}$ to have a denominator of 100 so that the fractions can be subtracted: $\frac{1}{25} \cdot \frac{4}{4} = \frac{4}{100}$.

$= \frac{3}{100}$

Teaching Example 2 Express using positive exponents and simplify, if possible:
a. 10^{-3} b. a^{-12}
c. $(-2)^{-5}$ d. $3^{-2} - 6^{-2}$
Answers: a. $\frac{1}{1{,}000}$ b. $\frac{1}{a^{12}}$ c. $-\frac{1}{32}$
d. $\frac{1}{12}$

Self Check 2 Express using positive exponents and simplify, if possible:
a. 8^{-2} $\frac{1}{64}$ **b.** x^{-5} $\frac{1}{x^5}$ **c.** $(-3)^{-3}$ $-\frac{1}{27}$ **d.** $2^{-3} - 4^{-2}$ $\frac{1}{16}$

Now Try Problems 21, 25, and 29

EXAMPLE 3 Simplify. Do not use negative exponents in the answer. **a.** $9m^{-3}$ **b.** -5^{-2}

Strategy We note that each exponent is a negative number. We will identify the base for each negative exponent and then use the negative exponent rule.

Why This rule enables us to write an exponential expression that has a negative exponent in an equivalent form using a positive exponent.

Solution

a. $9m^{-3} = 9 \cdot m^{-3}$ Read as "9 times *m* to the negative third power." The base is *m*. The exponent is −3.

$= 9 \cdot \frac{1}{m^3}$ Write the reciprocal of m^{-3} and change the sign of the exponent. Since 9 is not part of the base, it is not part of the reciprocal.

$= \frac{9}{m^3}$ Multiply.

Teaching Tip: Remind your students that one key to simplifying exponential expressions successfully is this: When you see an exponent, always ask yourself "What is the base?"

b. $-5^{-2} = -1 \cdot 5^{-2}$ Read as "the opposite of 5 to the negative second power." The base is 5. The exponent is −2.

$= -1 \cdot \frac{1}{5^2}$ Write the reciprocal of 5^{-2} and change the sign of the exponent. Since −1 is not part of the base, it is not part of the reciprocal.

$= -\frac{1}{25}$ Evaluate 5^2 and multiply. The result is negative because of the − sign in front of -5^{-2}.

Teaching Example 3 Simplify. Do not use negative exponents in the answer.
a. $16t^{-4}$ b. -7^{-2}
Answers: a. $\frac{16}{t^4}$ b. $-\frac{1}{49}$

Self Check 3 Simplify. Do not use negative exponents in the answer. **a.** $12h^{-9}$ $\frac{12}{h^9}$
b. -2^{-4} $-\frac{1}{16}$

Now Try Problems 33 and 37

3 Use Exponent Rules to Change Negative Exponents in Fractions to Positive Exponents.

Negative exponents can appear in the numerator and/or the denominator of a fraction. To develop rules for such situations, we consider the following example.

$$\frac{a^{-4}}{b^{-3}} = \frac{\frac{1}{a^4}}{\frac{1}{b^3}} = \frac{1}{a^4} \div \frac{1}{b^3} = \frac{1}{a^4} \cdot \frac{b^3}{1} = \frac{b^3}{a^4}$$

Read $\frac{a^{-4}}{b^{-3}}$ as "*a* to the negative fourth power over (or divided by) *b* to the negative third power."

We can obtain this result in a simpler way. In $\frac{a^{-4}}{b^{-3}}$, we can move a^{-4} from the numerator to the denominator and change the sign of the exponent, and we can move b^{-3} from the denominator to the numerator and change the sign of the exponent.

The Language of Algebra

Factors of a numerator or denominator may be moved **across the fraction bar** if we change the sign of their exponent.

$$\frac{a^{-4}}{b^{-3}} = \frac{b^3}{a^4}$$

This example illustrates the following rules.

Changing from Negative to Positive Exponents

A factor can be moved from the denominator to the numerator or from the numerator to the denominator of a fraction if the sign of its exponent is changed.

For any nonzero real numbers x and y, and any integers m and n,

$$\frac{1}{x^{-n}} = x^n \quad \text{and} \quad \frac{x^{-m}}{y^{-n}} = \frac{y^n}{x^m}$$

These rules streamline the process when simplifying fractions involving negative exponents.

EXAMPLE 4 Simplify. Do not use negative exponents in the answer. **a.** $\frac{1}{d^{-10}}$ **b.** $\frac{2^{-3}}{3^{-4}}$ **c.** $\frac{-6s^{-2}}{t^{-9}}$

Strategy We will move any factors in the numerator that have a negative exponent to the denominator. Then we will move any factors in the denominator that have a negative exponent to the numerator.

Why In this process, the sign of a negative exponent changes to positive.

Solution

a. $\frac{1}{d^{-10}} = d^{10}$ Read as "1 over d to the negative tenth power." Move d^{-10} to the numerator and change the sign of the exponent.

b. $\frac{2^{-3}}{3^{-4}} = \frac{3^4}{2^3}$ Move 2^{-3} to the denominator and change the sign of the exponent. Move 3^{-4} to the numerator and change the sign of the exponent.

$= \frac{81}{8}$ Evaluate: $3^4 = 81$ and $2^3 = 8$.

c. $\frac{-6s^{-2}}{t^{-9}} = \frac{-6t^9}{s^2}$ Since $-6s^{-2}$ has no parentheses, s is the base. Move only s^{-2} to the denominator and change the sign of the exponent. Do not move -6. Move t^{-9} to the numerator and change the sign of the exponent.

Caution

A common error is to mistake the $-$ sign in -6 for a negative exponent. *It is not an exponent.* The -6 should not be moved to the denominator and have its sign changed.

$$\frac{-6s^{-2}}{t^{-9}} \neq \frac{t^9}{6s^2}$$

Teaching Example 4 Simplify. Do not use negative exponents in the answer.
a. $\frac{1}{d^{-9}}$ b. $\frac{8^{-2}}{9^{-2}}$ c. $\frac{-15r^{-6}}{s^{-10}}$
Answers: a. d^9 b. $\frac{81}{64}$ c. $-\frac{15s^{10}}{r^6}$

Self Check 4 Simplify. Do not use negative exponents in the answer.

a. $\frac{1}{w^{-5}}$ w^5 **b.** $\frac{5^{-2}}{4^{-3}}$ $\frac{64}{25}$ **c.** $\frac{-8h^{-6}}{a^{-7}}$ $-\frac{8a^7}{h^6}$

Now Try Problems 41, 45, and 47

When a fraction is raised to a negative power, we can use rules for exponents to change the sign of the exponent. For example, we see that

$$\left(\frac{a}{2}\right)^{-3} = \frac{a^{-3}}{2^{-3}} = \frac{2^3}{a^3} = \left(\frac{2}{a}\right)^3$$

The exponent is the opposite of -3.

The base is the reciprocal of $\frac{a}{2}$.

This process can be streamlined using the following rule.

Negative Exponents and Reciprocals

A fraction raised to a power is equal to the reciprocal of the fraction raised to the opposite power.

For any nonzero real numbers x and y, and any integer n,

$$\left(\frac{x}{y}\right)^{-n} = \left(\frac{y}{x}\right)^{n}$$

EXAMPLE 5 Simplify: $\left(\frac{4}{m}\right)^{-2}$

Strategy We want to write this fraction that is raised to a negative power in an equivalent form that involves a positive power. We will use the negative exponent and reciprocal rules to do this.

Why It is usually easier to simplify exponential expressions if the exponents are positive.

Solution

$$\left(\frac{4}{m}\right)^{-2} = \left(\frac{m}{4}\right)^{2}$$

Read as "the quantity of 4 over m, raised to the negative second power." The base is the fraction $\frac{4}{m}$ and the exponent is -2. Write the reciprocal of the base and change the sign of the exponent.

$$= \frac{m^2}{4^2}$$

Use the power of a quotient rule: Raise the numerator m and denominator 4 to the second power.

$$= \frac{m^2}{16}$$

Evaluate: $4^2 = 16$.

Teaching Example 5
Simplify: $\left(\frac{z}{15}\right)^{-2}$
Answer: $\frac{225}{z^2}$

Self Check 5 Simplify: $\left(\frac{c}{9}\right)^{-2}$ $\frac{81}{c^2}$

Now Try Problem 53

4 Use All Exponent Rules to Simplify Expressions.

The rules for exponents involving products, powers, and quotients are also true for zero and negative exponents.

Summary of Exponent Rules

If m and n represent integers and there are no divisions by zero, then

Product rule	*Power rule*	*Power of a product*
$x^m \cdot x^n = x^{m+n}$	$(x^m)^n = x^{mn}$	$(xy)^n = x^n y^n$
Quotient rule	*Power of a quotient*	*Exponents of 0 and 1*
$\frac{x^m}{x^n} = x^{m-n}$	$\left(\frac{x}{y}\right)^n = \frac{x^n}{y^n}$	$x^0 = 1$ and $x^1 = x$

Negative exponent

$$x^{-n} = \frac{1}{x^n}$$

Negative exponents appearing in fractions

$$\frac{1}{x^{-n}} = x^n \qquad \frac{x^{-m}}{y^{-n}} = \frac{y^n}{x^m} \qquad \left(\frac{x}{y}\right)^{-n} = \left(\frac{y}{x}\right)^{n}$$

Teaching Tip: Have students tell you why each of the following is not an acceptable "answer" when simplifying an exponential expression.

- $\frac{5m^2}{m}$
- $20x(y^2)^3$
- $m^{16}n^{-1}$

The rules for exponents are used to simplify expressions involving products, quotients, and powers. In general, an expression involving exponents is simplified when

- Each base occurs only once
- No powers are raised to powers
- There are no parentheses
- There are no negative or zero exponents

EXAMPLE 6 Simplify. Do not use negative exponents in the answer. **a.** $x^5 \cdot x^{-3}$ **b.** $\frac{x^3}{x^7}$ **c.** $(x^3)^{-2}$ **d.** $(2a^3b^{-5})^3$ **e.** $\left(\frac{3}{b^5}\right)^{-4}$

Strategy In each case, we want to write an equivalent expression such that each base is used only once and is raised to a positive power. We will use rules for exponents to do this.

Why These expressions are not in simplest form. In parts a and b, the base x occurs more than once. In parts a, c, d, and e, there is a negative exponent.

Solution

a. $x^5 \cdot x^{-3} = x^{5+(-3)} = x^2$ Use the product rule: Keep the base, x, and add exponents.

b. $\frac{x^3}{x^7} = x^{3-7}$ Use the quotient rule: Keep the base, x, and subtract the exponents.

$= x^{-4}$ Do the subtraction: $3 - 7 = -4$.

$= \frac{1}{x^4}$ Write the reciprocal of x^{-4} and change the sign of the exponent.

c. $(x^3)^{-2} = x^{-6}$ Use the power rule: Keep the base, x, and multiply exponents.

$= \frac{1}{x^6}$ Write the reciprocal of x^{-6} and change the sign of the exponent.

Success Tip

We can use the negative exponent rule to simplify $(x^3)^{-2}$ in an alternate way:

$$(x^3)^{-2} = \frac{1}{(x^3)^2} = \frac{1}{x^6}$$

d. $(2a^3b^{-5})^3 = 2^3(a^3)^3(b^{-5})^3$ Raise each factor of the product $2a^3b^{-5}$ to the 3rd power.

$= 8a^9b^{-15}$ Use the power rule: Multiply exponents.

$= \frac{8a^9}{b^{15}}$ Move b^{-15} to the denominator and change the sign of the exponent.

e. $\left(\frac{3}{b^5}\right)^{-4} = \left(\frac{b^5}{3}\right)^4$ The base is the fraction $\frac{3}{b^5}$ and its exponent is -4. Write the reciprocal of the base and change the sign of the exponent.

$= \frac{(b^5)^4}{(3)^4}$ Use the power of a quotient rule: Raise the numerator and denominator to the 4th power.

$= \frac{b^{20}}{81}$ Use the power rule: Keep the base, b, and multiply the exponents 5 and 4. Evaluate: $3^4 = 81$.

Teaching Example 6 Simplify. Do not use negative exponents in the answer.
a. $r^{15} \cdot r^{-4}$ **b.** $\frac{n^4}{n^9}$ **c.** $(b^3)^{-8}$ **d.** $(5g^3h^{-1})^3$ **e.** $\left(\frac{K^6}{6}\right)^{-3}$
Answers: **a.** r^{11} **b.** $\frac{1}{n^5}$ **c.** $\frac{1}{b^{24}}$ **d.** $\frac{125g^9}{h^3}$ **e.** $\frac{216}{K^{18}}$

Self Check 6 Simplify. Do not use negative exponents in the answer.
a. $t^8 \cdot t^{-4}$ t^4 **b.** $\frac{a^3}{a^8}$ $\frac{1}{a^5}$ **c.** $(n^4)^{-5}$ $\frac{1}{n^{20}}$
d. $(4c^2d^{-1})^3$ $\frac{64c^6}{d^3}$ **e.** $\left(\frac{c^4}{2}\right)^{-3}$ $\frac{8}{c^{12}}$

Now Try Problems 57, 61, 65, 69, and 71

EXAMPLE 7 Simplify. Do not use negative exponents in the answer.

a. $\dfrac{y^{-4}y^{-3}}{y^{-20}}$ **b.** $\dfrac{7^{-1}a^3b^4}{6^{-2}a^5b^2}$ **c.** $\left(\dfrac{x^{-3}y^2}{xy^{-3}}\right)^2$

Strategy In each case, we want to write an equivalent expression such that each base is used only once and is raised to a positive power.

Why These expressions are not in simplest form. The bases occur more than once and the expressions contain negative exponents.

Solution

a.

$\dfrac{y^{-4}y^{-3}}{y^{-20}} = \dfrac{y^{-7}}{y^{-20}}$ In the numerator, use the product rule: Keep the common base, y, and add exponents: $-4 + (-3) = -7$.

$= y^{-7-(-20)}$ Use the quotient rule: Keep the common base, y, and subtract exponents.

$= y^{13}$ Do the subtraction: $-7 - (-20) = -7 + 20 = 13$.

Alternate solution: To avoid working with negative numbers, we could move each factor across the fraction bar and change the sign of its exponent.

$$\frac{y^{-4}y^{-3}}{y^{-20}} = \frac{y^{20}}{y^4y^3} = \frac{y^{20}}{y^7} = y^{13}$$

Caution

We cannot use the alternate approach if the negative exponents occur in a sum or difference of terms. For example:

$$\frac{y^{-4} + y^{-3}}{y^{-20}} \neq \frac{y^{20}}{y^4 + y^3}$$

You will study this situation in more detail in your next algebra course.

b.

$\dfrac{7^{-1}a^3b^4}{6^{-2}a^5b^2} = \dfrac{6^2a^3b^4}{7^1a^5b^2}$ Move 7^{-1} to the denominator. Change the sign of the exponent. Move 6^{-2} to the numerator. Change the sign of the exponent.

$= \dfrac{36a^{3-5}b^{4-2}}{7}$ Use the quotient rule twice: Keep each base, a and b, and subtract exponents.

$= \dfrac{36a^{-2}b^2}{7}$ Do the subtractions: $3 - 5 = -2$ and $4 - 2 = 2$.

$= \dfrac{36b^2}{7a^2}$ Move a^{-2} to the denominator and change the sign of the exponent.

c.

$\left(\dfrac{x^{-3}y^2}{xy^{-3}}\right)^2 = [x^{-3-1}y^{2-(-3)}]^2$ Within the parentheses, use the quotient rule twice: Keep each base, x and y, and subtract exponents.

$= (x^{-4}y^5)^2$ Do the subtractions: $-3 - 1 = -4$ and $2 - (-3) = 5$.

$= x^{-8}y^{10}$ Raise each factor within the parentheses to the second power.

$= \dfrac{y^{10}}{x^8}$ Move x^{-8} to the denominator and change the sign of its exponent. y^{10} does not move.

Success Tip

Since the rules for exponents can be applied in different orders, there are several equally valid ways to simplify these expressions.

Alternate solution: To simplify the expression, we can begin on the "outside" by using the power of a quotient rule first.

$$\left(\frac{x^{-3}y^2}{xy^{-3}}\right)^2 = \frac{(x^{-3})^2(y^2)^2}{x^2(y^{-3})^2} = \frac{x^{-6}y^4}{x^2y^{-6}} = \frac{y^4y^6}{x^6x^2} = \frac{y^{10}}{x^8}$$

Self Check 7 Simplify. Do not use negative exponents in the answer.

a. $\dfrac{a^{-4}a^{-5}}{a^{-3}}$ $\frac{1}{a^6}$ **b.** $\dfrac{1^{-4}x^5y^3}{9^{-2}x^3y^6}$ $\frac{81x^2}{y^3}$ **c.** $\left(\dfrac{c^{-2}d^2}{c^4d^{-3}}\right)^3$ $\frac{d^{15}}{c^{18}}$

Now Try Problems 73, 77, and 79

Teaching Example 7 Simplify. Do not use negative exponents in the answer.

a. $\dfrac{x^{-2}x^{-4}}{x^{-1}}$ b. $\dfrac{1^{-8}a^6b^4}{3^{-4}a^2b^{14}}$

c. $\left(\dfrac{a^3b^{-2}}{a^2b^2}\right)^{-4}$

Answers:

a. $\dfrac{1}{x^5}$ b. $\dfrac{81a^4}{b^{10}}$ c. $\dfrac{b^{16}}{a^4}$

SECTION 5.2 STUDY SET

VOCABULARY

Fill in the blanks.

1. In the expression 5^{-1}, the exponent is a <u>negative</u> integer.
2. x^{-n} is the <u>reciprocal</u> of x^n.
3. We read a^0 as "a to the <u>zero</u> power."
4. We read 3^{-4} as "3 to the <u>negative</u> <u>fourth</u> power."

CONCEPTS

5. Complete the table.

Expression	Base	Exponent
4^{-2}	4	-2
$6x^{-5}$	x	-5
$\left(\frac{3}{y}\right)^{-8}$	$\frac{3}{y}$	-8
-7^{-1}	7	-1
$(-2)^{-3}$	-2	-3
$10a^0$	a	0

6. Complete each rule for exponents.
   a. $x^m \cdot x^n = x^{m+n}$
   b. $x^0 = 1$
   c. $(x^m)^n = x^{mn}$
   d. $(xy)^n = x^ny^n$
   e. $\left(\frac{x}{y}\right)^n = \frac{x^n}{y^n}$
   f. $x^{-n} = \frac{1}{x^n}$
   g. $\frac{1}{x^{-n}} = x^n$
   h. $\frac{x^{-m}}{y^{-n}} = \frac{y^n}{x^m}$
   i. $\frac{x^m}{x^n} = x^{m-n}$
   j. $\left(\frac{x}{y}\right)^{-n} = \left(\frac{y}{x}\right)^n$

Complete each table.

7.

x	3^x
2	9
1	3
0	1
-1	$\frac{1}{3}$
-2	$\frac{1}{9}$

8.

x	$(-9)^x$
2	81
1	-9
0	1
-1	$-\frac{1}{9}$
-2	$\frac{1}{81}$

9. Fill in the blanks.
   a. $2^{-3} = \frac{1}{2^3}$
   b. $\frac{1}{t^{-6}} = t^6$
   c. A factor can be moved from the denominator to the numerator or from the numerator to the denominator of a fraction if the <u>sign</u> of its exponent is changed.
   $$\frac{5^{-2}}{6^{-3}} = \frac{6^3}{5^2}$$
   d. A fraction raised to a power is equal to the <u>reciprocal</u> of the fraction raised to the opposite power.
   $$\left(\frac{3}{d}\right)^{-2} = \left(\frac{d}{3}\right)^2$$

10. Determine whether each statement is true or false.
    a. $6^{-2} = -36$ False
    b. $6^{-2} = \frac{1}{36}$ True
    c. $\frac{x^3}{y^{-2}} = \frac{y^2}{x^3}$ False
    d. $\frac{-6x^{-5}}{y^{-6}} = \frac{y^6}{6x^5}$ False

NOTATION

Complete each solution to simplify each expression.

11. $(y^5y^3)^{-5} = (y^8)^{-5} = y^{-40} = \frac{1}{y^{40}}$

12. $\left(\frac{a^2b}{a^{-3}b^3}\right)^3 = (a^{2-(-3)}b^{1-3})^3$
    $= (a^5b^{-2})^3$
    $= a^{15}b^{-6}$
    $= \frac{a^{15}}{b^6}$

GUIDED PRACTICE

Simplify each expression. See Example 1.

13. 7^0 1
14. 9^0 1
15. $\left(\frac{1}{4}\right)^0$ 1
16. $\left(\frac{3}{8}\right)^0$ 1
17. $2x^0$ 2
18. $8t^0$ 8
19. $\frac{5}{2x^0}$ $\frac{5}{2}$
20. $\frac{4}{3a^0}$ $\frac{4}{3}$

Express using positive exponents and simplify, if possible. See Example 2.

21. 2^{-2} $\frac{1}{4}$
22. 7^{-2} $\frac{1}{49}$
23. 6^{-1} $\frac{1}{6}$
24. 5^{-1} $\frac{1}{5}$
25. b^{-5} $\frac{1}{b^5}$
26. c^{-4} $\frac{1}{c^4}$
27. $(-5)^{-1}$ $-\frac{1}{5}$
28. $(-8)^{-1}$ $-\frac{1}{8}$
29. $2^{-2} + 4^{-1}$ $\frac{1}{2}$
30. $-9^{-1} + 9^{-2}$ $-\frac{8}{81}$
31. $9^0 - 9^{-1}$ $\frac{8}{9}$
32. $7^{-1} - 7^0$ $-\frac{6}{7}$

Simplify. Do not use negative exponents in the answer. See Example 3.

33. $15g^{-6}$ $\frac{15}{g^6}$
34. $16t^{-3}$ $\frac{16}{t^3}$
35. $5x^{-3}$ $\frac{5}{x^3}$
36. $27m^{-3}$ $\frac{27}{m^3}$
37. -3^{-3} $-\frac{1}{27}$
38. -6^{-3} $-\frac{1}{216}$
39. -8^{-2} $-\frac{1}{64}$
40. -4^{-2} $-\frac{1}{16}$

Simplify. Do not use negative exponents in the answer. See Example 4.

41. $\frac{1}{5^{-3}}$ 125
42. $\frac{1}{3^{-3}}$ 27
43. $\frac{8}{s^{-1}}$ $8s$
44. $\frac{6}{k^{-2}}$ $6k^2$

45. $\frac{2^{-4}}{3^{-1}}$ $\frac{3}{16}$

46. $\frac{7^{-2}}{2^{-3}}$ $\frac{8}{49}$

47. $\frac{-4d^{-1}}{p^{-10}}$ $-\frac{4p^{10}}{d}$

48. $\frac{-9m^{-1}}{n^{-30}}$ $-\frac{9n^{30}}{m}$

Simplify. See Example 5.

49. $\left(\frac{1}{6}\right)^{-2}$ 36

50. $\left(\frac{1}{7}\right)^{-2}$ 49

51. $\left(\frac{1}{2}\right)^{-3}$ 8

52. $\left(\frac{1}{5}\right)^{-3}$ 125

53. $\left(\frac{c}{d}\right)^{-8}$ $\frac{d^8}{c^8}$

54. $\left(\frac{a}{x}\right)^{-10}$ $\frac{x^{10}}{a^{10}}$

55. $\left(\frac{3}{m}\right)^{-4}$ $\frac{m^4}{81}$

56. $\left(\frac{2}{t}\right)^{-4}$ $\frac{t^4}{16}$

Simplify. Do not use negative exponents in the answer. See Example 6.

57. $y^8 \cdot y^{-2}$ y^6

58. $m^{10} \cdot m^{-6}$ m^4

59. $b^{-7} \cdot b^{14}$ b^7

60. $c^{-9} \cdot c^{14}$ c^5

61. $\frac{y^4}{y^5}$ $\frac{1}{y}$

62. $\frac{t^7}{t^{10}}$ $\frac{1}{t^3}$

63. $\frac{h^{-5}}{h^2}$ $\frac{1}{h^7}$

64. $\frac{y^{-3}}{y^4}$ $\frac{1}{y^7}$

65. $(x^4)^{-3}$ $\frac{1}{x^{12}}$

66. $(y^{-3})^2$ $\frac{1}{y^6}$

67. $(b^2)^{-4}$ $\frac{1}{b^8}$

68. $(n^3)^{-5}$ $\frac{1}{n^{15}}$

69. $(6s^4t^{-7})^2$ $\frac{36s^8}{t^{14}}$

70. $(11r^{10}s^{-3})^2$ $\frac{121r^{20}}{s^6}$

71. $\left(\frac{4}{x^3}\right)^{-3}$ $\frac{x^9}{64}$

72. $\left(\frac{2}{b^5}\right)^{-2}$ $\frac{b^{10}}{4}$

Simplify. Do not use negative exponents in the answer. See Example 7.

73. $\frac{y^{-3}}{y^{-4}y^{-2}}$ y^3

74. $\frac{x^{-12}}{x^{-3}x^{-4}}$ $\frac{1}{x^5}$

75. $\frac{a^{-5}a^{-9}}{a^{-8}}$ $\frac{1}{a^6}$

76. $\frac{b^{-2}b^{-3}}{b^{-9}}$ b^4

77. $\frac{2^{-1}a^4b^2}{3^{-2}a^2b^4}$ $\frac{9a^2}{2b^2}$

78. $\frac{6^{-2}b^9c^3}{5^{-3}b^4c^8}$ $\frac{125b^5}{36c^5}$

79. $\left(\frac{y^3z^{-2}}{y^{-4}z^3}\right)^2$ $\frac{y^{14}}{z^{10}}$

80. $\left(\frac{xy^3}{x^{-1}y^{-1}}\right)^3$ x^6y^{12}

TRY IT YOURSELF

Simplify. Do not use negative exponents in the answer. Assume that no variables are 0.

81. $\left(\frac{a^4}{2b}\right)^{-3}$ $\frac{8b^3}{a^{12}}$

82. $\left(\frac{n^8}{9m}\right)^{-2}$ $\frac{81m^2}{n^{16}}$

83. $\frac{r^{-50}}{r^{-70}}$ r^{20}

84. $\frac{m^{-30}}{m^{-40}}$ m^{10}

85. $\frac{a^{-5}}{b^{-2}}$ $\frac{b^2}{a^5}$

86. $\frac{r^{-6}}{s^{-1}}$ $\frac{s}{r^6}$

87. $(-10)^{-3}$ $-\frac{1}{1,000}$

88. $(-9)^{-2}$ $\frac{1}{81}$

89. $\left(\frac{a^2b^3}{ab^4}\right)^0$ 1

90. $\left(\frac{xyz}{x^2y}\right)^0$ 1

91. $\frac{9^{-2}s^6t}{4^{-3}s^4t^5}$ $\frac{64s^2}{81t^4}$

92. $\frac{2^{-5}m^{10}n^6}{5^{-2}m^6n^{10}}$ $\frac{25m^4}{32n^4}$

93. $(2u^{-2}v^5)^5$ $\frac{32v^{25}}{u^{10}}$

94. $(3w^{-8}x^3)^4$ $\frac{81x^{12}}{w^{32}}$

95. $\left(\frac{y^4}{3}\right)^{-2}$ $\frac{9}{y^8}$

96. $\left(\frac{p^3}{2}\right)^{-2}$ $\frac{4}{p^6}$

97. $-15x^0y$ $-15y$

98. $24g^0h^2$ $24h^2$

99. $\left(\frac{4}{h^{10}}\right)^{-2}$ $\frac{h^{20}}{16}$

100. $\left(\frac{x^4}{3}\right)^{-4}$ $\frac{81}{x^{16}}$

101. $x^{-3} \cdot x^{-3}$ $\frac{1}{x^6}$

102. $y^{-2} \cdot y^{-2}$ $\frac{1}{y^4}$

103. $\left(\frac{c^3d^{-4}}{c^{-1}d^5}\right)^3$ $\frac{c^{12}}{d^{27}}$

104. $\left(\frac{s^2t^{-8}}{s^{-9}t^2}\right)^4$ $\frac{s^{44}}{t^{40}}$

105. $15(-6x)^0$ 15

106. $4(-12y)^0$ 4

107. $\frac{2^{-2}g^{-2}h^{-3}}{9^{-1}h^{-3}}$ $\frac{9}{4g^2}$

108. $\frac{5^{-1}x^{-2}y^{-3}}{8^{-2}x^{-11}}$ $\frac{64x^9}{5y^3}$

109. $(5d^{-2})^3$ $\frac{125}{d^6}$

110. $(9s^{-6})^2$ $\frac{81}{s^{12}}$

111. $\left(\frac{x^2y^{-2}}{x^{-5}y^3}\right)^4$ $\frac{x^{28}}{y^{20}}$

112. $\left(\frac{r^4s^{-3}}{r^{-3}s^7}\right)^3$ $\frac{r^{21}}{s^{30}}$

113. $(2x^3y^{-2})^5$ $\frac{32x^{15}}{y^{10}}$

114. $(3u^{-2}v^3)^3$ $\frac{27v^9}{u^6}$

115. $\frac{t(t^{-2})^{-2}}{t^{-5}}$ t^{10}

116. $\frac{d(d^{-3})^{-3}}{d^{-7}}$ d^{17}

117. $\frac{-4s^{-5}}{t^{-2}}$ $-\frac{4t^2}{s^5}$

118. $\frac{-9k^{-8}}{m^{-2}}$ $-\frac{9m^2}{k^8}$

119. $(x^{-4}x^3)^3$ $\frac{1}{x^3}$

120. $(y^{-2}y)^3$ $\frac{1}{y^3}$

Look Alikes . . .

121. a. 8^{-1} $\frac{1}{8}$ b. -8^{-1} $-\frac{1}{8}$ c. $(-8)^{-1}$ $-\frac{1}{8}$ d. $-(-8)^{-1}$ $\frac{1}{8}$

122. a. 9^{-2} $\frac{1}{81}$ b. -9^{-2} $-\frac{1}{81}$ c. $(-9)^{-2}$ $\frac{1}{81}$ d. $-(-9)^{-2}$ $-\frac{1}{81}$

123. a. $4xy^{-2}$ $\frac{4x}{y^2}$ b. $(4xy)^{-2}$ $\frac{1}{16x^2y^2}$ c. $4x^{-2}y$ $\frac{4y}{x^2}$ d. $4^{-2}xy$ $\frac{xy}{16}$

124. a. $\left(\frac{r}{s}\right)^{-1}$ $\frac{s}{r}$ b. $\frac{r^{-1}}{s}$ $\frac{1}{rs}$ c. $\frac{r}{s^{-1}}$ rs d. $\frac{r^{-1}}{s^{-1}}$ $\frac{s}{r}$

APPLICATIONS

125. from **Campus to Careers**

Sound Engineering Technician

©iStockphoto.com/beetle8

The faintest sound that the typical human ear can detect (called the *threshold of hearing*) has an intensity of 10^{-12} units. A sound intensity of 10^4 units will cause an instant rupture of the eardrum. Consider the types of sounds in the table below. In the intensity column, write the most appropriate exponential expression from the following list. Each expression is used only once. (Source: physicsclassroom.com)

10^{-10} 10^{-6} 10^{-4} 10^{-1} 10^2

Lowest intensity **Greatest intensity**

Type of sound	Intensity
Front row rock concert	10^{-1}
Normal conversation	10^{-6}
Vacuum cleaner	10^{-4}
Military jet takeoff	10^2
Whisper	10^{-10}

▶ **126. Electronics.** The total resistance R of a certain circuit is given by

$$R = \left(\frac{1}{R_1} + \frac{1}{R_2}\right)^{-1} + R_3$$

Find R if $R_1 = 4$, $R_2 = 2$, and $R_3 = 1$. $\frac{7}{3}$

WRITING

▶ **127.** Explain how you would help a friend understand that 2^{-3} is not equal to -8.

128. Explain each error.

a. $\frac{-5x^{-2}}{y^{-2}} = \frac{y^2}{5x^2}$ (crossed out)

b. $4^{-2} = -\frac{1}{16}$ (crossed out)

REVIEW

Find the slope of the line that passes through the given points.

129. $(1, -4)$ and $(3, -7)$ $-\frac{3}{2}$

130. $(1, 3)$ and $(3, -1)$ -2

131. Write an equation of the line having slope $\frac{3}{4}$ and y-intercept -5. $y = \frac{3}{4}x - 5$

▶ **132.** Find an equation of the line that passes through $(4, 4)$ and $(-6, -6)$. Write the answer in slope–intercept form. $y = x$

CHALLENGE PROBLEMS

▶ **133.** Simplify each expression. Do not use negative exponents in the answer. The variable m represents a positive integer.

a. $r^{5m}r^{-6m}$ $\frac{1}{r^m}$

b. $\frac{x^{3m}}{x^{6m}}$ $\frac{1}{x^{3m}}$

134. Write an expression equivalent to $\left(\frac{2x^3y^7}{3z^5}\right)^9$ that involves only *negative* exponents. $\left(\frac{2^{-1}x^{-3}y^{-7}}{3^{-1}z^{-5}}\right)^{-9}$

SECTION 5.3

Scientific Notation

OBJECTIVES

1. Convert from scientific to standard notation.
2. Write numbers in scientific notation.
3. Perform calculations with scientific notation.

ARE YOU READY? *Are You Ready? exercises available online at www.webassign.net/brookscole*

The following problems review some basic skills that are needed when working with scientific notation.

1. Evaluate: 10^2 100

2. Multiply: $1{,}000 \cdot 4.528$ 4,528

3. Evaluate: 10^{-1} $\frac{1}{10}$

4. Multiply: $0.01 \cdot 6.22$ 0.0622

Scientists often deal with extremely large and small numbers. For example, the distance from the Earth to the sun is approximately 150,000,000 kilometers. The influenza virus, which causes flu symptoms of cough, sore throat, and headache, has a diameter of 0.00000256 inch.

©Sam D Cruz/Shutterstock.com

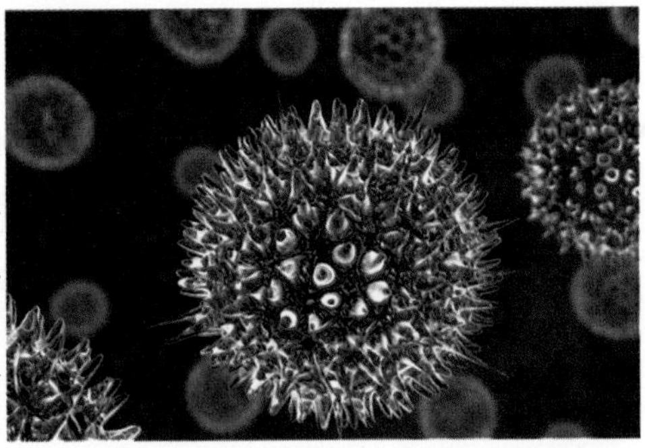
©iStockphoto.com/Björn Meyer

The numbers 150,000,000 and 0.00000256 are written in **standard notation,** which is also called **decimal notation.** Because they contain many zeros, they are difficult to read and cumbersome to work with in calculations. In this section, we will discuss a more convenient form in which we can write such numbers.

1 Convert from Scientific to Standard Notation.

Scientific notation provides a compact way of writing very large or very small numbers.

Scientific Notation

A positive number is written in **scientific notation** when it is written in the form $N \times 10^n$, where $1 \le N < 10$ and n is an integer.

Teaching Tip: Your students may be interested in knowing the names of some large powers of 10:

10^3 = one thousand
10^6 = one million
10^9 = one billion
10^{12} = one trillion
10^{15} = one quadrillion
10^{18} = one quintillion

To write numbers in scientific notation, you need to be familiar with **powers of 10,** like those listed in the table below.

Power of 10	10^4	10^3	10^2	10^1	10^0	10^{-1}	10^{-2}	10^{-3}	10^{-4}
Value	10,000	1,000	100	10	1	$\frac{1}{10} = 0.1$	$\frac{1}{100} = 0.01$	$\frac{1}{1,000} = 0.001$	$\frac{1}{10,000} = 0.0001$

Three examples of numbers written in scientific notation are shown below. Note that each of them is the product of a decimal number (between 1 and 10) and a power of 10.

An integer exponent

3.67×10^2 $\qquad$ 2.158×10^{-3} $\qquad$ 4.0×10^{57}

A decimal that is at least 1, but less than 10

Notation

The × sign for multiplication is usually used to write scientific notation. However, some books use a raised dot · instead.

A number written in scientific notation can be converted to standard notation by performing the indicated multiplication. For example, to convert 3.67×10^2, we recall that multiplying a decimal by 100 moves the decimal point 2 places to the right.

$$3.67 \times \mathbf{10^2} = 3.67 \times \mathbf{100} = 367.$$

To convert 2.158×10^{-3} to standard notation, we recall that dividing a decimal by 1,000 moves the decimal point 3 places to the left.

$$2.158 \times \mathbf{10^{-3}} = 2.158 \times \frac{\mathbf{1}}{\mathbf{10^3}} = 2.158 \times \frac{\mathbf{1}}{\mathbf{1,000}} = \frac{2.158}{\mathbf{1,000}} = 0.002158$$

In 3.67×10^2 and 2.158×10^{-3}, the exponent gives the number of decimal places that the decimal point moves, and the sign of the exponent indicates the direction in which it moves. Applying this observation to several other examples, we have

$5.32 \times 10^6 = 5320000.$ Move the decimal point 6 places to the right.

$1.95 \times 10^{-5} = 0.0000195$ Move the decimal point $|-5| = 5$ places to the left.

$9.7 \times 10^0 = 9.7$ There is no movement of the decimal point.

Success Tip

Since $10^0 = 1$, scientific notation involving 10^0 is easily simplified. For example,

$$9.7 \times \mathbf{10^0} = 9.7 \times \mathbf{1} = 9.7$$

The following procedure summarizes our observations.

Converting from Scientific to Standard Notation

1. If the exponent is positive, move the decimal point the same number of places to the right as the exponent.
2. If the exponent is negative, move the decimal point the same number of places to the left as the absolute value of the exponent.

EXAMPLE 1 Convert to standard notation: **a.** 3.467×10^5 **b.** 8.9×10^{-4}

Strategy In each case, we need to identify the exponent on the power of 10 and consider its sign.

Why The exponent gives the number of decimal places that we should move the decimal point. The sign of the exponent indicates whether it should be moved to the right or the left.

Solution

a. Since the exponent in 10^5 is 5, the decimal point moves 5 places to the right.

3 4 6 7 0 0. *To move 5 places to the right, two placeholder zeros must be written.*

Thus, $3.467 \times 10^5 = 346{,}700$.

b. Since the exponent in 10^{-4} is -4, the decimal point moves 4 places to the left.

0.0 0 0 8 9 *To move 4 places to the left, three placeholder zeros must be written.*

Thus, $8.9 \times 10^{-4} = 0.00089$.

The Language of Algebra

Standard notation is also called **decimal notation.**

Self Check 1 Convert to standard notation: **a.** 4.88×10^6 4,880,000 **b.** 9.8×10^{-3} 0.0098

Now Try Problems 13 and 17

Teaching Example 1 Convert to standard notation:
a. 9.03×10^7
b. 6.2×10^{-5}
Answers: a. 90,300,000 b. 0.000062

2 Write Numbers in Scientific Notation.

To write a number in scientific notation ($N \times 10^n$) we first determine N and then n.

EXAMPLE 2 Write each number in scientific notation: **a.** 150,000,000 **b.** 0.00000256 **c.** 432×10^5

Strategy We will write each number as the product of a number between 1 and 10 and a power of 10.

Why Numbers written in scientific notation have the form $N \times 10^n$.

Solution

a. We must write 150,000,000 (the distance in kilometers from the Earth to the sun) as the product of a number between 1 and 10 and a power of 10. We note that 1.5 lies between 1 and 10. To obtain 150,000,000, we must move the decimal point in 1.5 exactly 8 places to the right.

1.5 0 0 0 0 0 0 0

This will happen if we multiply 1.5 by 10^8. Therefore,

$150{,}000{,}000 = 1.5 \times 10^8$ *This is the distance (in kilometers) from the Earth to the sun.*

b. We must write 0.00000256 (the diameter in inches of a flu virus) as the product of a number between 1 and 10 and a power of 10. We note that 2.56 lies between 1 and 10. To obtain 0.00000256, the decimal point in 2.56 must be moved 6 places to the left.

0 0 0 0 0 0 2.56

This will happen if we multiply 2.56 by 10^{-6}. Therefore,

$0.00000256 = 2.56 \times 10^{-6}$ *This is the diameter (in inches) of a flu virus.*

Caution

At first glance, the following numbers might look like they are written in scientific notation, but they are not. In each case, the first factor is not a number between 1 and 10.

16.38×10^{24} 0.39×10^{-11}

Caution

Don't apply the negative exponent rule when writing numbers in scientific notation.

2.56×10^{-6} $2.56 \times \frac{1}{10^6}$

Calculators

Scientific notation

Calculators automatically change into scientific notation when a result is too large or too small to fit the answer display.

$453.46^5 = 1.917321395 \times 10^{13}$

1.917321395 13

$0.0005^{12} = 2.44140625 \times 10^{-40}$

2.44140625 -40

Teaching Example 2 Write in scientific notation:
a. 0.000042 **b.** 5,367,000,000
c. 783×10^8 **d.** 0.026×10^{-2}
Answers:
a. 4.2×10^{-5} **b.** 5.367×10^9
c. 7.83×10^{10} **d.** 2.6×10^{-4}

c. The number 432×10^5 is not written in scientific notation because 432 is not a number between 1 and 10. To write this number in scientific notation, we proceed as follows:

$432 \times 10^5 = \mathbf{4.32 \times 10^2} \times 10^5$ Write 432 in scientific notation.

$= 4.32 \times 10^7$ Use the product rule to find $10^2 \times 10^5$. Keep the base, 10, and add the exponents.

Written in scientific notation, 432×10^5 is 4.32×10^7.

Self Check 2 Write each number in scientific notation: **a.** 93,000,000 9.3×10^7 **b.** 0.00009055 9.055×10^{-5} **c.** 85×10^{-3} 8.5×10^{-2}

Now Try Problems 31, 35, and 55

The results from Example 2 illustrate the following forms to use when converting numbers from standard to scientific notation.

For real numbers between 0 and 1: $\square \times 10^{\text{negative integer}}$

For real numbers at least 1, but less than 10: $\square \times 10^0$

For real numbers greater than or equal to 10: $\square \times 10^{\text{positive integer}}$

3 Perform Calculations with Scientific Notation.

Another advantage of scientific notation becomes apparent when we evaluate products or quotients that involve very large or small numbers. If we express those numbers in scientific notation, we can use rules for exponents to make the calculations easier.

To multiply two numbers written in scientific notation, use the following rule:

$$(a \times 10^m)(b \times 10^n) = (a \cdot b) \times 10^{m+n}$$

EXAMPLE 3 **Astronomy.** Except for the sun, the nearest star visible to the naked eye from most parts of the United States is Sirius. Light from Sirius reaches Earth in about 70,000 hours. If light travels at approximately 670,000,000 mph, how far from Earth is Sirius?

Strategy We can use the formula $d = rt$ to find the distance from Sirius to Earth.

Why We know the *rate* at which light travels and the *time* it takes to travel from Sirius to the Earth.

Solution The rate at which light travels is 670,000,000 mph and the time it takes the light to travel from Sirius to Earth is 70,000 hr. To find the distance from Sirius to Earth, we proceed as follows:

NASA

$d = rt$ This is the formula for distance traveled.

$d = \mathbf{670,000,000(70,000)}$ Substitute 670,000,000 for r and 70,000 for t.

$= (6.7 \times 10^8)(7.0 \times 10^4)$ Write each number in scientific notation.

$= (6.7 \cdot 7.0) \times (\mathbf{10^8 \times 10^4})$ Group the decimals together and the powers of 10 together.

$= (6.7 \cdot 7.0) \times \mathbf{10^{8+4}}$ Use the product rule to find $10^8 \times 10^4$. Keep the base, 10, and add exponents.

$= \mathbf{46.9} \times 10^{12}$ Do the multiplication and the addition.

We note that 46.9 is not between 0 and 1, so 46.9×10^{12} is not written in scientific notation. To answer in scientific notation, we proceed as follows.

$= \mathbf{4.69 \times 10^1} \times 10^{12}$ — Write 46.9 in scientific notation as 4.69×10^1.

$= 4.69 \times 10^{13}$ — Use the product rule to find $10^1 \times 10^{12}$. Keep the base, 10, and add the exponents.

Sirius is approximately 4.69×10^{13} or 46,900,000,000,000 miles from Earth.

Teaching Example 3 Use scientific notation to evaluate: (0.0000024) (5,310,000,000,000)
Answer: 1.2744×10^7 or 12,744,000

Self Check 3 Use scientific notation to evaluate: (2,540,000,000,000)(0.00041)
1.0414×10^9

Now Try Problem 61

To divide two numbers written in scientific notation, use the following rule:

$$\frac{a \times 10^m}{b \times 10^n} = \frac{a}{b} \times 10^{m-n}$$

EXAMPLE 4 **Atoms.** As an example of how scientific notation is used in chemistry, we can approximate the weight (in grams) of one atom of the element uranium by evaluating the following expression.

$$\frac{2.4 \times 10^2}{6 \times 10^{23}}$$

Strategy To simplify, we will divide the numbers and powers of 10 separately.

Why We can then use the quotient rule for exponents to simplify the calculations.

Solution

$\frac{2.4 \times 10^2}{6 \times 10^{23}} = \frac{2.4}{6} \times \frac{10^2}{10^{23}}$ — Divide the decimals and the powers of 10 separately.

$= \frac{2.4}{6} \times 10^{\mathbf{2-23}}$ — For the powers of 10, use the quotient rule. Keep the base, 10, and subtract the exponents.

$= \mathbf{0.4} \times 10^{-21}$ — Divide the decimals. Subtract the exponents. The result is not in scientific notation form.

$= \mathbf{4 \times 10^{-1}} \times 10^{-21}$ — Write 0.4 in scientific notation as 4×10^{-1}.

$= 4 \times 10^{-22}$ — Use the product rule to find $10^{-1} \times 10^{-21}$. Keep the base, 10, and add the exponents.

One atom of uranium weighs 4×10^{-22} gram or 0.0000000000000000000004 g.

Calculators

Entering scientific notation
To evaluate the expression in Example 4 on a scientific calculator, we enter the numbers using the EE key:

2.4 EE 2 ÷ 6 EE 23 =

Teaching Example 4
Evaluate: $\frac{1.3 \times 10^{15}}{5.2 \times 10^{60}}$
Answer: 2.5×10^{-46}

Self Check 4 Find the approximate weight (in grams) of one atom of gold by evaluating: $\frac{1.98 \times 10^2}{6 \times 10^{23}}$ 3.3×10^{-22} g

Now Try Problem 65

SECTION 5.3 STUDY SET

VOCABULARY

Fill in the blanks.

1. 4.84×10^5 is written in scientific notation. 484,000 is written in standard notation.
2. 10^3, 10^{50}, and 10^{-4} are powers of 10.

CONCEPTS

Fill in the blanks.

3. When we multiply a decimal by 10^5, the decimal point moves 5 places to the right. When we multiply a decimal by 10^{-7}, the decimal point moves 7 places to the left.
4. Describe the procedure for converting a number from scientific notation to standard form.
   a. If the exponent on the base of 10 is positive, move the decimal point the same number of places to the right as the exponent.
   b. If the exponent on the base of 10 is negative, move the decimal point the same number of places to the left as the absolute value of the exponent.
5. a. When a real number greater than or equal to 10 is written in scientific notation, the exponent on 10 is a positive integer.
   b. When a real number between 0 and 1 is written in scientific notation, the exponent on 10 is a negative integer.
6. The arrows show the movement of a decimal point. By what power of 10 was each decimal multiplied?
   a. 0.0 0 0 0 0 0 0 5 5 6 10^{-7} b. 8, 0 4 1, 0 0 0, 0 0 0. 10^9

Fill in the blanks to write each number in scientific notation.

7. a. $7{,}700 = 7.7 \times 10^3$ b. $500{,}000 = 5.0 \times 10^5$
   c. $114{,}000{,}000 = 1.14 \times 10^8$
8. a. $0.0082 = 8.2 \times 10^{-3}$
   b. $0.0000001 = 1.0 \times 10^{-7}$
   c. $0.00003457 = 3.457 \times 10^{-5}$
9. Write each expression so that the decimal numbers are grouped together and the powers of ten are grouped together.
   a. $(5.1 \times 10^9)(1.5 \times 10^{22})$ $(5.1 \times 1.5)(10^9 \times 10^{22})$
   b. $\dfrac{8.8 \times 10^{30}}{2.2 \times 10^{19}}$ $\dfrac{8.8}{2.2} \times \dfrac{10^{30}}{10^{19}}$
10. Simplify each expression.
    a. $10^{24} \times 10^{33}$ 10^{57}
    b. $\dfrac{10^{50}}{10^{36}}$ 10^{14}
    c. $\dfrac{10^{15} \times 10^{27}}{10^{40}}$ 10^2

NOTATION

11. Fill in the blanks. A positive number is written in scientific notation when it is written in the form $N \times 10^n$, where $1 \le N < 10$ and n is an integer.
12. Express each power of 10 in fraction form and decimal form.
    a. 10^{-3} $\frac{1}{1{,}000}$, 0.001 b. 10^{-6} $\frac{1}{1{,}000{,}000}$, 0.000001

GUIDED PRACTICE

Convert each number to standard notation. **See Example 1.**

13. 2.3×10^2 230
14. 3.75×10^4 37,500
15. 8.12×10^5 812,000
16. 1.2×10^3 1,200
17. 1.15×10^{-3} 0.00115
18. 4.9×10^{-2} 0.049
19. 9.76×10^{-4} 0.000976
20. 7.63×10^{-5} 0.0000763
21. 6.001×10^6 6,001,000
22. 9.998×10^5 999,800
23. 2.718×10^0 2.718
24. 3.14×10^0 3.14
25. 6.789×10^{-2} 0.06789
26. 4.321×10^{-1} 0.4321
27. 2.0×10^{-5} 0.00002
28. 7.0×10^{-6} 0.000007

Write each number in scientific notation. **See Example 2.**

29. 23,000 2.3×10^4
30. 4,750 4.75×10^3
31. 1,700,000 1.7×10^6
32. 290,000 2.9×10^5
33. 0.062 6.2×10^{-2}
34. 0.00073 7.3×10^{-4}
35. 0.0000051 5.1×10^{-6}
36. 0.04 4.0×10^{-2}
37. 5,000,000,000 5.0×10^9
38. 7,000,000 7.0×10^6
39. 0.0000003 3.0×10^{-7}
40. 0.0001 1.0×10^{-4}
41. 909,000,000 9.09×10^8
42. 7,007,000,000 7.007×10^9
43. 0.0345 3.45×10^{-2}
44. 0.000000567 5.67×10^{-7}
45. 9 9.0×10^0
46. 2 2.0×10^0
47. 11 1.1×10^1
48. 55 5.5×10^1
49. 1,718,000,000,000,000,000 1.718×10^{18}
50. 44,180,000,000,000,000,000 4.418×10^{19}
51. 0.0000000000000123 1.23×10^{-14}
52. 0.0000000000000000555 5.55×10^{-17}
53. 73×10^4 7.3×10^5
54. 99×10^5 9.9×10^6
55. 201.8×10^{15} 2.018×10^{17}
56. 154.3×10^{17} 1.543×10^{19}
57. 0.073×10^{-3} 7.3×10^{-5}
58. 0.0017×10^{-4} 1.7×10^{-7}
59. 36.02×10^{-20} 3.602×10^{-19}
60. 56.29×10^{-30} 5.629×10^{-29}

Use scientific notation to perform the calculations. Give all answers in scientific notation and standard notation. **See Examples 3 and 4.**

61. $(3.4 \times 10^2)(2.1 \times 10^3)$ 7.14×10^5; 714,000
62. $(4.1 \times 10^{-3})(3.4 \times 10^4)$ 1.394×10^2; 139.4
63. $(8.4 \times 10^{-13})(4.8 \times 10^9)$ 4.032×10^{-3}; 0.004032
64. $(5.5 \times 10^{-15})(2.2 \times 10^{13})$ 1.21×10^{-1}; 0.121
65. $\dfrac{2.24 \times 10^4}{5.6 \times 10^7}$ 4.0×10^{-4}; 0.0004
66. $\dfrac{2.47 \times 10^5}{3.8 \times 10^{-5}}$ 6.5×10^9; 6,500,000,000
67. $\dfrac{9.3 \times 10^2}{3.1 \times 10^{-2}}$ 3.0×10^4; 30,000
68. $\dfrac{7.2 \times 10^6}{1.2 \times 10^8}$ 6.0×10^{-2}; 0.06
69. $\dfrac{0.00000129}{0.0003}$ 4.3×10^{-3}; 0.0043
70. $\dfrac{169{,}000{,}000{,}000}{26{,}000{,}000}$ 6.5×10^3; 6,500
71. (0.0000000056)(5,500,000) 3.08×10^{-2}; 0.0308
72. (0.000000061)(3,500,000,000) 2.135×10^2; 213.5

▶ 73. $\dfrac{96{,}000}{(12{,}000)(0.00004)}$ 2.0×10^5; 200,000

▶ 74. $\dfrac{(0.48)(14{,}400{,}000)}{96{,}000{,}000}$ 7.2×10^{-2}; 0.072

75. $\dfrac{2{,}475}{(132{,}000{,}000{,}000{,}000)(0.25)}$ 7.5×10^{-11}; 0.000000000075

▶ 76. $\dfrac{147{,}000{,}000{,}000{,}000}{(0.000049)(25)}$ 1.2×10^{17}; 120,000,000,000,000,000

Find each power.

77. $(456.4)^6$ $9.038030748 \times 10^{15}$

▶ 78. $(0.009)^{-6}$ $1.881676423 \times 10^{12}$

79. 225^{-5} $1.734152992 \times 10^{-12}$

80. $\left(\dfrac{1}{3}\right)^{-55}$ $1.74449211 \times 10^{26}$

APPLICATIONS

▶ 81. **Astronomy.** The distance from Earth to Alpha Centauri (the nearest star outside our solar system) is about 25,700,000,000,000 miles. Write this number in scientific notation. 2.57×10^{13} mi

82. **Water.** According to the U.S. Geological Survey, the total water supply of the world is 366,000,000,000,000,000,000 gallons. Write this number in scientific notation. 3.66×10^{20} gal

83. **Earth, Sun, Moon.** The surface area of Earth is 1.97×10^8 square miles, the surface area of the sun is 1.09×10^{17} square miles, and the surface area of the moon is 1.46×10^7 square miles. Convert each number to standard notation.
197,000,000 mi^2; 109,000,000,000,000,000 mi^2; 14,600,000 mi^2

▶ 84. **Atoms.** The number of atoms in 1 gram of iron is approximately 1.08×10^{22}. Convert this number to standard notation.
10,800,000,000,000,000,000,000

85. **Sand.** The mass of one grain of beach sand is approximately 0.00000000045 ounce. Write this number in scientific notation.
4.5×10^{-10} oz

▶ 86. **Molecules.** The mass of a water molecule is approximately 0.0000000000000000000000001056 ounce. Write this number in scientific notation.
1.056×10^{-24} oz

87. **Wavelengths.** Examples of the most common types of electromagnetic waves are given in the table in the next column. List the wavelengths in order from shortest to longest.
g, x, u, v, i, m, r

This distance between the two crests of the wave is called the wavelength.

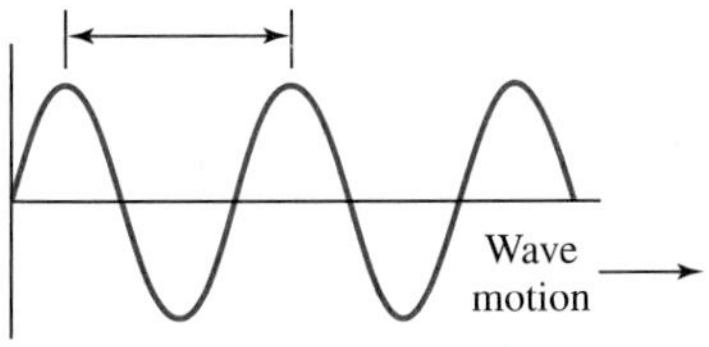

Type	Use	Wavelength (in meters)
visible light	lighting	9.3×10^{-6}
infrared	photography	3.7×10^{-5}
x-ray	medical	2.3×10^{-11}
radio wave	communication	3.0×10^{2}
gamma ray	treating cancer	8.9×10^{-14}
microwave	cooking	1.1×10^{-2}
ultraviolet	sun lamp	6.1×10^{-8}

▶ 88. **Exploration.** On July 4, 1997, the *Pathfinder,* carrying the rover vehicle called Sojourner, landed on Mars. The distance from Mars to Earth is approximately 3.5×10^7 miles. Use scientific notation to express this distance in feet. (*Hint:* 5,280 feet = 1 mile.)
1.848×10^{11} ft

▶ 89. **from Campus to Careers**
Sound Engineering Technician

© iStockphoto.com/beetle8

The speed of sound in air is approximately 3.3×10^4 centimeters per second. Use scientific notation to express this speed in kilometers per second. (*Hint:* 100 centimeters = 1 meter and 1,000 meters = 1 kilometer.) 3.3×10^{-1} km/sec

▶ 90. **Protons.** The mass of one proton is approximately 1.7×10^{-24} gram. Use scientific notation to express the mass of 1 million protons.
1.7×10^{-18} g

▶ 91. **Light Years.** One light year is about 5.87×10^{12} miles. Use scientific notation to express this distance in feet. (*Hint:* 5,280 feet = 1 mile.)
3.09936×10^{16} ft

▶ 92. **Oil.** As of 2009, Saudi Arabia was believed to have crude oil reserves of about 2.65×10^{11} barrels. A barrel contains 42 gallons of oil. Use scientific notation to express Saudi Arabia's oil reserves in gallons. (Source: British Petroleum)
1.113×10^{13} gal

93. **Insured Deposits.** As of June 2009, the total insured deposits in U.S. banks and savings and loans was approximately 7.6×10^{12} dollars. If this money was invested at a rate of 4% simple annual interest, how much would it earn in 1 year? Use scientific notation to express the answer. (Source: Federal Deposit Insurance Corporation.)
3.04×10^{11} dollars

▶ 94. **Currency.** As of December 2009, the number of \$20 bills in circulation was approximately 6.4×10^9. What was the total value of the currency? Express the answer in scientific notation and standard notation. (Source: The Federal Reserve.)
1.28×10^{11} dollars; \$128,000,000,000

95. Powers of 10. In the United States, we use Latin prefixes in front of "illion" to name extremely large numbers. Write each number in scientific notation.

One million: 1,000,000 1.0×10^6

One billion: 1,000,000,000 1.0×10^9

One trillion: 1,000,000,000,000 1.0×10^{12}

One quadrillion: 1,000,000,000,000,000 1.0×10^{15}

One quintillion: 1,000,000,000,000,000,000 1.0×10^{18}

96. Supercomputers. As of June 2009, the world's fastest computer was the Cray Jaguar, owned by the Oak Ridge National Laboratory in Tennessee. If it could make 1.75×10^{15} calculations in one second, how many could it make in one minute? Answer in scientific notation. (Source: datacenterknowledge.com) 1.05×10^{17} calculations

WRITING

97. In what situations would scientific notation be more convenient than standard notation?

98. To multiply a number by a power of 10, we move the decimal point. Which way, and how far? Explain.

99. 2.3×10^{-3} contains a negative sign but represents a positive number. Explain.

100. Explain why 237.8×10^8 is not written in scientific notation.

REVIEW

101. If $y = -1$, find the value of $-5y^{55}$. 5

102. What is the y-intercept of the graph of $y = -3x - 5$? $(0, -5)$

103. Counseling. At the end of her first year of practice, a family counselor had 75 clients. At the end of her second year, she had 105 clients. If a linear trend continues, write an equation that gives the number of clients c the counselor will have at the end of t years. $c = 30t + 45$

104. Is $(0, -5)$ a solution of $2x + 3y \geq -14$? No

CHALLENGE PROBLEMS

105. Consider 2.5×10^{-4}. Answer the following questions in scientific notation form.

a. What is its opposite? -2.5×10^{-4}

b. What is its reciprocal? 4.0×10^3

106. a. Write the numbers one million and one millionth in scientific notation. 1.0×10^6, 1.0×10^{-6}

b. By what number must we multiply one millionth to get one million? 10^{12}

SECTION 5.4 Polynomials

OBJECTIVES

1 Know the vocabulary for polynomials.

2 Evaluate polynomials.

3 Graph equations defined by polynomials.

ARE YOU READY? *Are You Ready? exercises available online at www.webassign.net/brookscole*

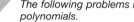

The following problems review some basic skills that are needed when working with polynomials.

1. How many terms does the expression $2x^2 - 5x + 8$ have? 3 terms
2. What is the coefficient of the term $6a^3$? 6
3. In $4b^3 + 5b^4 - 16b + b^2$, which term has the largest exponent? $5b^4$
4. For $y = x^3 + 1$, what is the value of y when $x = -3$? -26

In this section, we will discuss a special type of algebraic expression called a *polynomial.*

1 Know the Vocabulary for Polynomials.

Recall from Chapter 1 that a *term* is a product or quotient of numbers and/or variables. A single number or variable is also a term. Some examples of terms are:

$$14, \quad x, \quad -6y^3, \quad 9cd^2, \quad \text{and} \quad \frac{5}{y}$$

Polynomials

A **polynomial** is a single term or a sum of terms in which all variables have whole-number exponents and no variable appears in a denominator.

The Language of Algebra

The prefix **poly** means many. Some other words that begin with this prefix are *poly*gon, *poly*ester, and *poly*unsaturated.

Here are some examples of polynomials. Note that polynomials are expressions, not equations.

$$3x + 2, \quad 4y^2 - 2y - 3, \quad a^3 + 3a^2b + 3ab^2 + b^3, \quad \text{and} \quad -8xy^2z$$

Some examples of expressions that are *not polynomials* are:

$6x^3 + 4x^{-2}$ The variable x has an exponent that is not a whole number.

$y^2 + \frac{5}{y} + 1$ The variable y appears in the denominator.

The polynomial $3x + 2$ is the sum of two terms, $3x$ and 2, and we say it is a **polynomial in one variable, x.** A single number is called a **constant,** and so its last term, 2, is called the **constant term.**

A polynomial is defined as a single term or the sum of several terms. Since $4y^2 - 2y - 3$ can be written as the sum $4y^2 + (-2y) + (-3)$, it has three terms, $4y^2$, $-2y$, and -3. It is written in **descending powers** of y, because the exponents on y decrease from left to right. When a polynomial is written in descending powers, the first term, in this case $4y^2$, is called the **leading term.** The coefficient of the leading term, in this case 4, is called the **leading coefficient.**

A polynomial can have more than one variable. For example, $a^3 + 3a^2b + 3ab^3 + b^3$ is a **polynomial in two variables,** a and b. It has four terms and is written in descending powers of a and **ascending powers** of b. The polynomial $-8xy^2z$ is a polynomial in three variables, x, y, and z, and has only one term.

The Language of Algebra

The prefix **mono** means one; Jay Leno begins the *Tonight Show* with a monologue. The prefix **bi** means two, as in bicycle or binoculars. The prefix **tri** means three, as in triangle or the *Lord of the Rings Trilogy.*

Polynomials are classified according to the number of terms they have. A polynomial with exactly one term is called a **monomial;** exactly two terms, a **binomial;** and exactly three terms, a **trinomial.** Polynomials with four or more terms have no special names.

Polynomials			
Monomials	**Binomials**	**Trinomials**	**No special name**
$-6x$	$9u - 4$	$5t^2 + 4t + 3$	$a^4 + 2b^2 - 3b - 9$
$5.5x^3y^2$	$-29z^4 - z^2$	$27x^3 - 6x^2 - 2x$	$2.1 - 6.2t + 5.9t^2 + t^3$
11	$18a^2b + 4ab$	$\frac{1}{2}a^2 + 2ab + b^2$	$mn^2 - mn + m - n$

Polynomials and their terms can be described by the exponents on their variables.

Degree of a Term of a Polynomial

The **degree of a term** of a polynomial in one variable is the value of the exponent on the variable. If a polynomial is in more than one variable, the **degree of a term** is the sum of the exponents on the variables in that term. The **degree of a nonzero constant** is 0.

Think of the *degree of a term* as the number of variable factors in that term.

- $9x^6$ has degree **6**. x^6 represents 6 variable factors: $x \cdot x \cdot x \cdot x \cdot x \cdot x$.
- $-2a^4$ has degree **4**. a^4 represents 4 variable factors: $a \cdot a \cdot a \cdot a$.
- $47x^2y^{11}$ has degree **13**. Because $2 + 11 = 13$.
- 8 has degree **0** since it can be written as $8x^0$. There is no variable factor.

We determine the *degree of a polynomial* by considering the degrees of each of its terms.

Degree of a Polynomial

The **degree of a polynomial** is the same as the highest degree of any term of the polynomial.

EXAMPLE 1 Use the vocabulary of this section to describe each polynomial: **a.** $d^4 + 9d^2 - 16$ **b.** $\frac{1}{2}x^2 - x$ **c.** $-6y^{14} - 1.5y^9z^9 + 2.5y^8z^{10} + yz^{11}$

Teaching Tip: To help your students understand the vocabulary, use this example: Just as Coca-Cola, Sprite, and Pepsi are specific types of soft drinks, monomials, binomials, and trinomials are specific types of polynomials.

Strategy First, we will identify the variable(s) in the polynomial and determine whether it is written in ascending or descending powers. Then we will count the number of terms in the polynomial and determine the degree of each term.

Why The number of terms determines the type of polynomial. The highest degree of any term of the polynomial determines its degree.

Solution

a. $d^4 + 9d^2 - 16$ is a polynomial in one variable that is written in descending powers of d. If we write the subtraction as addition of the opposite, we see that it has 3 terms, d^4, $9d^2$, and -16, and is therefore a trinomial. The leading term is d^4 and the leading coefficient is 1. The highest degree of any of its terms is 4, so it is of degree 4.

$$d^4 + 9d^2 - 16 = \underset{\text{1st Term}}{d^4} + \underset{\text{2nd Term}}{9d^2} + \underset{\text{3rd Term}}{(-16)}$$

Term	Coefficient	Degree
d^4	1	4
$9d^2$	9	2
-16	-16	0

Degree of the polynomial: **4**

The Language of Algebra

To **descend** means to move from higher to lower. When we write $d^4 + 9d^2 - 16$ as $d^4 + 0d^3 + 9d^2 + 0d^1 - 16d^0$, we more clearly see the *descending* powers of d.

Success Tip

Recall that *terms* are separated by + symbols and that the *coefficient* of a term is the numerical factor of the term.

b. $\frac{1}{2}x^2 - x$ is a polynomial in one variable. It is written in descending powers of x. Since it has two terms, it is a binomial. The leading term is $\frac{1}{2}x^2$ and the leading coefficient is $\frac{1}{2}$. The highest degree of any of its terms is 2, so it is of degree 2.

Term	Coefficient	Degree
$\frac{1}{2}x^2$	$\frac{1}{2}$	**2**
$-x$	-1	1

Degree of the polynomial: **2**

c. $-6y^{14} - 1.5y^9z^9 + 2.5y^8z^{10} + yz^{11}$ is a polynomial in two variables, y and z. It is written in descending powers of y and ascending powers of z. It has 4 terms, and therefore has no special name. The leading term is $-6y^{14}$ and the leading coefficient is -6. The highest degree of any term is 18, so it is of degree 18.

Term	Coefficient	Degree
$-6y^{14}$	-6	14
$-1.5y^9z^9$	-1.5	**18**
$2.5y^8z^{10}$	2.5	**18**
yz^{11}	1	12

Degree of the polynomial: **18**

Self Check 1 Describe each polynomial: **a.** $x^2 + 4x - 16$ **b.** $-14s^5t + s^4t^3$

See AIE Appendix 3.

Now Try Problems 17 and 35

Teaching Example 1 Describe the polynomial: $n^7 - \frac{2}{3}n^3 + 20$
Answer: Terms: n^7, $-\frac{2}{3}n^3$, 20; coefficients: 1, $-\frac{2}{3}$, 20; trinomial; degree 7; leading term: n^7; leading coefficient: 1

2 Evaluate Polynomials.

A polynomial can have different values depending on the number that is substituted for its variable (or variables).

EXAMPLE 2 Evaluate $3x^2 + 4x - 5$ for $x = 0$ and $x = -2$.

Strategy We will substitute the given value for each x in the polynomial and follow the rules for the order of operations.

Why To *evaluate a polynomial* means to find its numerical value, once we know the value of its variable.

Solution

For $x = 0$:

$$3x^2 + 4x - 5 = 3(0)^2 + 4(0) - 5 = 3(0) + 4(0) - 5 = 0 + 0 - 5 = -5$$

For $x = -2$:

$$3x^2 + 4x - 5 = 3(-2)^2 + 4(-2) - 5 = 3(4) + 4(-2) - 5 = 12 + (-8) - 5 = -1$$

Caution

Recall that to evaluate $3(-2)^2$, the rules for the order of operations require that we find $(-2)^2$ first, and then multiply that result by 3.

Self Check 2 Evaluate $-x^3 + x - 2x + 3$ for $x = -3$. 33

Now Try Problem 55

Teaching Example 2 Evaluate $-x^2 + 5x - 1$ for $x = -3$.
Answer: -25

EXAMPLE 3 **Supermarket Displays.** The polynomial $\frac{1}{3}c^3 + \frac{1}{2}c^2 + \frac{1}{6}c$ gives the number of cans used in a display shaped like a square pyramid, having a square base formed by c cans per side. Find the number of cans used in the display.

Strategy We will evaluate the polynomial for $c = 4$.

Why From the illustration, we see that each side of the square base is formed by 4 cans.

Solution

$$\frac{1}{3}c^3 + \frac{1}{2}c^2 + \frac{1}{6}c = \frac{1}{3}(4)^3 + \frac{1}{2}(4)^2 + \frac{1}{6}(4) \quad \text{Substitute 4 for } c.$$

$$= \frac{1}{3}(64) + \frac{1}{2}(16) + \frac{1}{6}(4) \quad \text{Evaluate the exponential expressions first.}$$

$$= \frac{64}{3} + 8 + \frac{2}{3} \quad \text{Do the multiplication, and then simplify: } \tfrac{4}{6} = \tfrac{2}{3}.$$

$$= 30 \quad \text{Add the fractions: } \tfrac{64}{3} + \tfrac{2}{3} = \tfrac{66}{3} = 22.$$

There are 30 cans of soup in the display.

Self Check 3 Find the number of cans used in a display having a square base formed by 5 cans per side. 55 cans

Now Try Problem 81

Teaching Example 3 Find the number of cans used in a display having a square base formed by 7 cans per side.
Answer: 140 cans

In the following example, we evaluate a polynomial in two variables.

EXAMPLE 4 Evaluate $3p^2q - 4pq^2$ for $p = 2$ and $q = -3$.

Strategy We will substitute the given values for each p and q in the polynomial and follow the rules for the order of operations.

Why To evaluate a polynomial means to find its numerical value, once we know the value of its variables.

Solution

$$3p^2q - 4pq^2 = 3(2)^2(-3) - 4(2)(-3)^2 \quad \text{Substitute 2 for } p \text{ and } -3 \text{ for } q.$$

$$= 3(4)(-3) - 4(2)(9) \quad \text{Find the powers.}$$

$$= -36 - 72 \quad \text{Do the multiplication.}$$

$$= -108 \quad \text{Do the subtraction.}$$

Self Check 4 Evaluate $3a^3b^2 + 2a^2b$ for $a = 2$ and $b = -1$. 16

Now Try Problem 61

Teaching Example 4 Evaluate $2c^2d - 5cd^2$ for $c = 3$ and $d = -5$.
Answer: -465

3 Graph Equations Defined by Polynomials.

In Chapter 3, we graphed equations in two variables such as $y = x$ and $y = 2x - 3$ using the point-plotting method. Recall that these equations are called *linear equations* and that their graphs are straight lines. Note that the right side of the first two equations shown below is a polynomial of degree 1.

$y = x$ $\quad y = 2x - 3$ — The degree of each polynomial is 1.

$y = x^2$ — The degree of this polynomial is 2.

$y = x^3 + 1$ — The degree of this polynomial is 3.

We also can graph equations defined by polynomials with degrees greater than 1 such as $y = x^2$ and $y = x^3 + 1$ by plotting points.

EXAMPLE 5 Graph: $y = x^2$

Strategy We will find several solutions of the equation, plot them on a rectangular coordinate system, and then draw a smooth curve passing through the points.

Why To *graph* an equation in two variables means to make a drawing that represents all of its solutions.

Solution To find some solutions of this equation, we select several values of x that will make the calculations easy. Then we find each corresponding value of y. If $x = -3$, we substitute -3 for x in $y = x^2$ and find y.

$$y = x^2 = (-3)^2 = 9$$

Thus, $(-3, 9)$ is a solution. In a similar manner, we find the corresponding y-values for x-values of $-2, -1, 0, 1, 2,$ and 3. If we plot the ordered pairs listed in the table and join the points with a smooth curve, we get the graph shown below, which is called a **parabola.**

Success Tip

When constructing a table of solutions, it is wise to select some positive and negative integer-values for x, as well as 0.

$y = x^2$

x	y	(x, y)
-3	9	$(-3, 9)$
-2	4	$(-2, 4)$
-1	1	$(-1, 1)$
0	0	$(0, 0)$
1	1	$(1, 1)$
2	4	$(2, 4)$
3	9	$(3, 9)$

Select x. Find y. Plot (x, y).

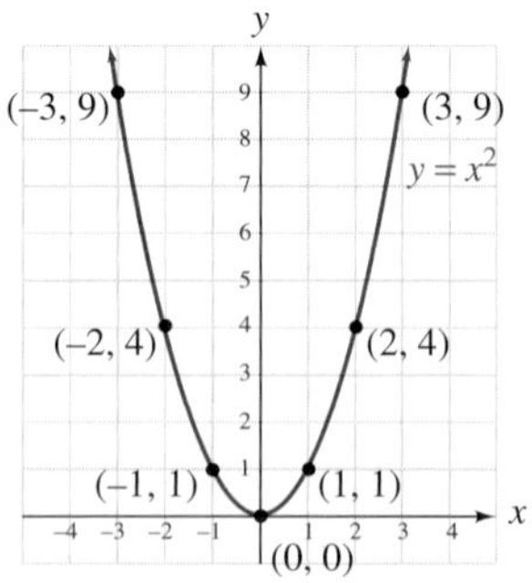

The Language of Algebra

The cup-like shape of a **parabola** has many real-life applications. The word *parabolic* (pronounced par • a • bol • ic) means having the form of a parabola. Did you know that a satellite TV dish is more formally known in the electronics industry as a *parabolic* dish?

Teaching Example 5 Graph: $y = x^2 + 3$
Answer:

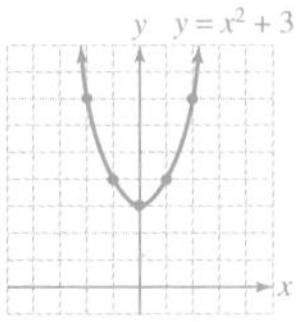

Self Check 5 Graph: $y = x^2 - 2$ See AIE Appendix 3.

Now Try Problem 69

Because the variable x is squared, $y = x^2$ is not a linear equation in two variables. We call it a **nonlinear equation.** In Examples 6 and 7, we will graph other nonlinear equations.

EXAMPLE 6 Graph: $y = -x^2 + 2$

Strategy We will find several solutions of the equation, plot them on a rectangular coordinate system, and then draw a smooth curve passing through the points.

Why To *graph* an equation in two variables means to make a drawing that represents all of its solutions.

Solution To make a table of solutions, we select x-values of $-3, -2, -1, 0, 1, 2,$ and 3 and find each corresponding y-value. For example, if $x = -3$, we have

Success Tip

To graph equations of lines, we must find at least two points on the line to draw its graph. Because the graphs of these equations are more complicated, more work is required. We must find a sufficient number of points on the graph so that its entire shape is revealed.

$y = -x^2 + 2$ This is the equation to graph.

$y = -(\mathbf{-3})^2 + 2$ Substitute -3 for x.

$y = -(9) + 2$ Evaluate the exponential expression first: $(-3)^2 = 9$.

$y = -7$ Do the addition: $-9 + 2 = -7$.

The ordered pair $(-3, -7)$ is a solution. Six other solutions appear in the table. After plotting each pair, we join the points with a smooth curve to obtain the graph, a parabola opening downward.

Teaching Tip: Tell your students that sometimes tables of solutions are written horizontally.

x				
y				

$y = -x^2 + 2$

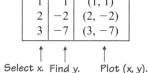

x	y	(x, y)
-3	-7	$(-3, -7)$
-2	-2	$(-2, -2)$
-1	1	$(-1, 1)$
0	2	$(0, 2)$
1	1	$(1, 1)$
2	-2	$(2, -2)$
3	-7	$(3, -7)$

↑ Select x. ↑ Find y. ↑ Plot (x, y).

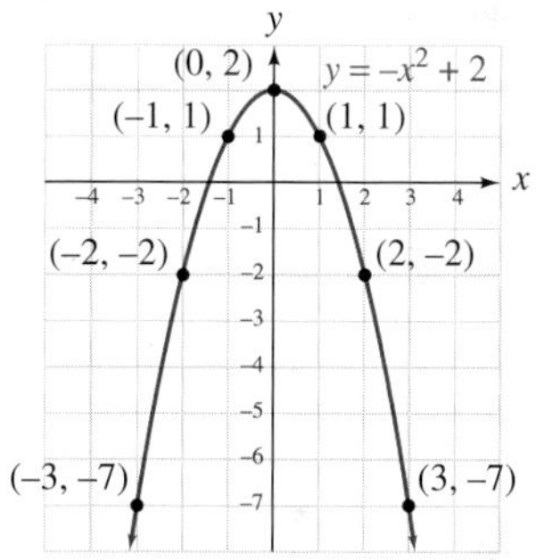

Teaching Example 6 Graph: $y = -x^2 + 1$
Answer:

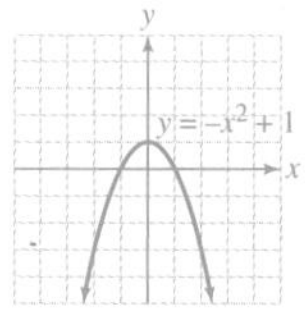

Self Check 6 Graph: $y = -x^2$ See AIE Appendix 3.

Now Try ▶ Problem 71

EXAMPLE 7 Graph: $y = x^3 + 1$

Strategy We will find several solutions of the equation, plot them on a rectangular coordinate system, and then draw a smooth curve passing through the points.

Why To *graph* an equation in two variables means to make a drawing that represents all of its solutions.

Solution If we let $x = -2$, we have

$y = x^3 + 1$ This is the equation to graph.

$y = (\mathbf{-2})^3 + 1$ Substitute -2 for x.

$y = -8 + 1$ Evaluate the exponential expression first: $(-2)^3 = -8$.

$y = -7$ Do the addition.

The ordered pair $(-2, -7)$ is a solution. This pair and others that satisfy the equation are listed in the table. Plotting the ordered pairs and joining the points with a smooth curve gives us the graph.

$y = x^3 + 1$

x	y	(x, y)
-2	-7	$(-2, -7)$
-1	0	$(-1, 0)$
0	1	$(0, 1)$
1	2	$(1, 2)$
2	9	$(2, 9)$

Select x. Find y. Plot (x, y).

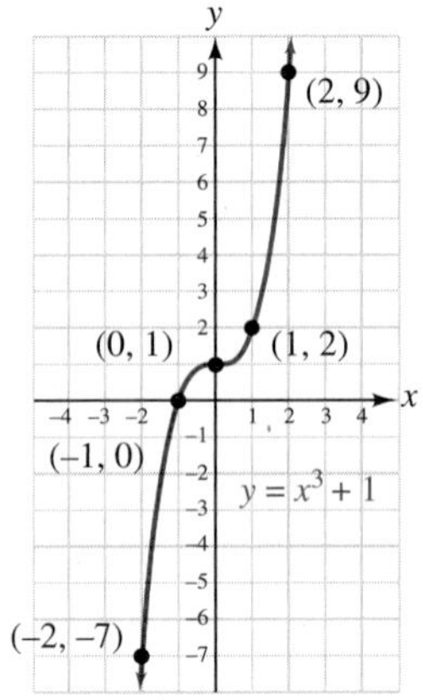

Teaching Example 7 Graph: $y = -x^3 + 1$
Answer:

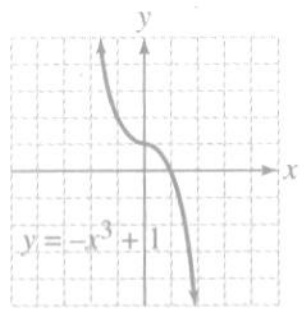

Self Check 7 Graph: $y = x^3 - 1$ See AIE Appendix 3.

Now Try Problem 75

SECTION 5.4 STUDY SET

VOCABULARY

Fill in the blanks.

1. A polynomial is a term or a sum of terms in which all variables have whole-number exponents and no variable appears in a denominator.
2. The terms of a polynomial are separated by + symbols.
3. $x^3 - 6x^2 + 9x - 2$ is a polynomial in one variable, and is written in descending powers of x, and $c^3 + 2c^2d - d^2$ is a polynomial in two variables and is written in ascending powers of d.
4. For the polynomial $6x^2 + 3x - 1$, the leading term is $6x^2$, and the leading coefficient is 6. The constant term is -1.
5. A monomial is a polynomial with exactly one term. A binomial is a polynomial with exactly two terms. A trinomial is a polynomial with exactly three terms.
6. The degree of the term $3x^7$ is 7 because x appears as a factor 7 times: $3 \cdot x \cdot x \cdot x \cdot x \cdot x \cdot x \cdot x$.
7. To evaluate the polynomial $x^2 - 2x + 1$ for $x = 6$, we substitute 6 for x and follow the rules for the order of operations.
8. The graph of $y = x^2$ is a cup-shaped curve called a parabola.

CONCEPTS

Determine whether each expression is a polynomial.

9. a. $x^3 - 5x^2 - 2$ Yes b. $x^{-4} - 5x$ No
 c. $x^2 - \dfrac{1}{2x} + 3$ No d. $x^3 - 1$ Yes
 e. $x^2 - y^2$ Yes f. $a^4 + a^3 + a^2 + a$ Yes

10. Fill in the blank so that the term has degree 5.
 a. $9x^{5}$ b. $-\dfrac{2}{3}xy^{4}$

Make a term-coefficient-degree table like that shown in Example 1 for each polynomial.

11. $8x^2 + x - 7$

Term	Coefficient	Degree
$8x^2$	8	2
x	1	1
-7	-7	0

Degree of the polynomial: 2

12. $y^4 - y^3 + 16y^2 + 3y$

Term	Coefficient	Degree
y^4	1	4
$-y^3$	-1	3
$16y^2$	16	2
$3y$	3	1

Degree of the polynomial: 4

13. $8a^6b^3 - 27ab$

Term	Coefficient	Degree
$8a^6b^3$	8	9
$-27ab$	-27	2

Degree of the polynomial: 9

14. $-1.2c^4 + 2.4c^2d^2 - 3.6d^4$

Term	Coefficient	Degree
$-1.2c^4$	-1.2	4
$2.4c^2d^2$	2.4	4
$-3.6d^4$	-3.6	4

Degree of the polynomial: 4

NOTATION

15. a. Write $x - 9 + 3x^2 + 5x^3$ in descending powers of x. $5x^3 + 3x^2 + x - 9$
    b. Write $-2xy + y^2 + x^2$ in ascending powers of y. $x^2 - 2xy + y^2$

16. Complete the solution. Evaluate $-2x^2 + 3x - 1$ for $x = -2$.

$$\begin{aligned} -2x^2 + 3x - 1 &= -2(-2)^2 + 3(-2) - 1 \\ &= -2(4) + 3(-2) - 1 \\ &= -8 + (-6) - 1 \\ &= -15 \end{aligned}$$

GUIDED PRACTICE

Classify each polynomial as a monomial, a binomial, a trinomial, or none of these. See Example 1.

17. $3x + 7$ Binomial
18. $3y - 5$ Binomial
19. $y^2 + 4y + 3$ Trinomial
20. $9xy$ Monomial
21. $\frac{3}{2}z^2$ Monomial
22. $\frac{3}{5}x^4 - \frac{2}{5}x^3 + \frac{3}{5}x - 1$ None of these
23. $t - 32$ Binomial
24. $12z^4$ Monomial
25. $s^2 - 23s + 31$ Trinomial
26. $2x^3 - 5x^2 + 6x - 3$ None of these
27. $6x^5 - x^4 - 3x^3 + 7$ None of these
28. x^3 Monomial
29. $3m^3n - 4m^2n^2 + mn - 1$ None of these
30. $4p^3q^2 + 7p^2q^3 + pq^4 - q^5$ None of these
31. $2a^2 - 3ab + b^2$ Trinomial
32. $a^3b - ab^3$ Binomial

Find the degree of each polynomial. See Example 1.

33. $3x^4$ 4th
34. $3x^5$ 5th
35. $-2x^2 + 3x + 1$ 2nd
36. $-5x^4 + 3x^2 - 3x$ 4th
37. $\frac{1}{3}x - 5$ 1st
38. $\frac{1}{2}y^3 + 4y^2$ 3rd
39. $-5r^2s^2 - r^3s + 3$ 4th
40. $4r^2s^3 - 5r^2s^8$ 10th
41. $x^{12} + 3x^2y^3$ 12th
42. $17ab^5 - 12a^3b$ 6th
43. 38 0th
44. -24 0th
45. $\frac{3}{2}m^7 - \frac{3}{4}m^{18}$ 18th
46. $\frac{7}{8}t^{10} - \frac{1}{8}t^{16}$ 16th
47. $5.5tw - 6.5t^2w - 7.5t^3$ 3rd
48. $0.4h + 0.6h^4c + 0.6h^5$ 5th

Evaluate each expression. See Example 2 and 3.

49. $x^2 - x + 1$ for
    a. $x = 2$ 3
    b. $x = -3$ 13
50. $x^2 - x + 7$ for
    a. $x = 6$ 37
    b. $x = -2$ 13
51. $4t^2 + 2t - 8$ for
    a. $t = -1$ −6
    b. $t = 0$ −8
52. $3s^2 - 2s + 8$ for
    a. $s = 1$ 9
    b. $s = 0$ 8
53. $\frac{1}{2}a^2 - \frac{1}{4}a$ for
    a. $a = 4$ 7
    b. $a = -8$ 34
54. $\frac{1}{3}b^2 - \frac{1}{9}b$ for
    a. $b = 9$ 26
    b. $b = -9$ 28
55. $-9.2x^2 + x - 1.4$ for
    a. $x = -1$ −11.6
    b. $x = -2$ −40.2
56. $-10.3x^2 - x + 6.5$ for
    a. $x = -1$ −2.8
    b. $x = -2$ −32.7
57. $x^3 + 3x^2 + 2x + 4$ for
    a. $x = 2$ 28
    b. $x = -2$ 4
58. $x^3 - 3x^2 - x + 9$ for
    a. $x = 3$ 6
    b. $x = -3$ −42
59. $y^4 - y^3 + y^2 + 2y - 1$ for
    a. $y = 1$ 2
    b. $y = -1$ 0
60. $-y^4 + y^3 + y^2 + y + 1$ for
    a. $y = 1$ 3
    b. $y = -1$ −1

Evaluate each polynomial for $a = -2$ and $b = 3$. See Example 4.

61. $6a^2b$ 72
62. $4ab^2$ −72
63. $a^3 + b^3$ 19
64. $a^3 - b^3$ −35
65. $a^2 + 5ab - b^2$ −35
66. $a^3 - 2ab + b^3$ 31
67. $5ab^3 - ab - b + 10$ −257
68. $-a^3b + ab - a - 21$ −1

Construct a table of solutions and then graph the equation. See Examples 5–7. See AIE Appendix 3.

69. $y = x^2 + 1$
70. $y = x^2 - 4$
71. $y = -x^2 - 2$
72. $y = -x^2 + 1$
73. $y = 2x^2 - 3$
74. $y = -2x^2 + 2$
75. $y = x^3 + 2$
76. $y = x^3 + 4$
77. $y = x^3 - 3$
78. $y = x^3 - 2$
79. $y = -x^3 - 1$
80. $y = -x^3$

APPLICATIONS

81. **Supermarkets.** A grocer plans to set up a pyramid-shaped display of cantaloupes like that shown in Example 3. If each side of the square base of the display is made of six cantaloupes, how many will be used in the display? 91 cantaloupes

82. **Packaging.** The polynomial $4x^3 - 44x^2 + 120x$ gives the volume (in cubic inches) of the resulting box when a square with sides x inches long is cut from each corner of a 10 in. × 12 in. piece of cardboard. Find the volume of a box if 3-inch squares are cut out. 72 in.^{3}

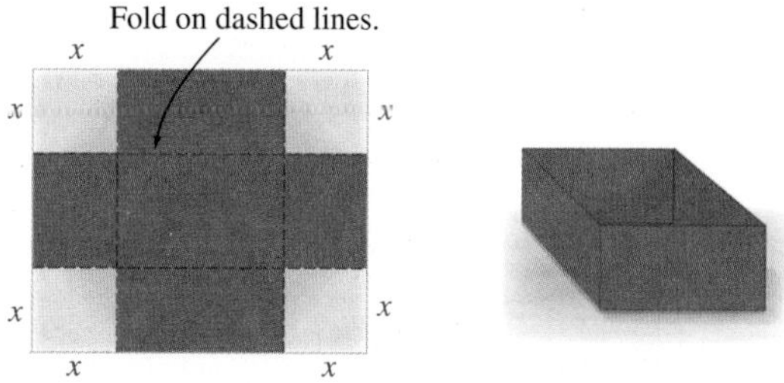

83. **Stopping Distance.** The number of feet that a car travels before stopping depends on the driver's reaction time and the braking distance, as shown in the illustration. For one driver, the stopping distance is given by the polynomial $0.04v^2 + 0.9v$ where v is the velocity of the car. Find the stopping distance when the driver is traveling at 30 mph. 63 ft

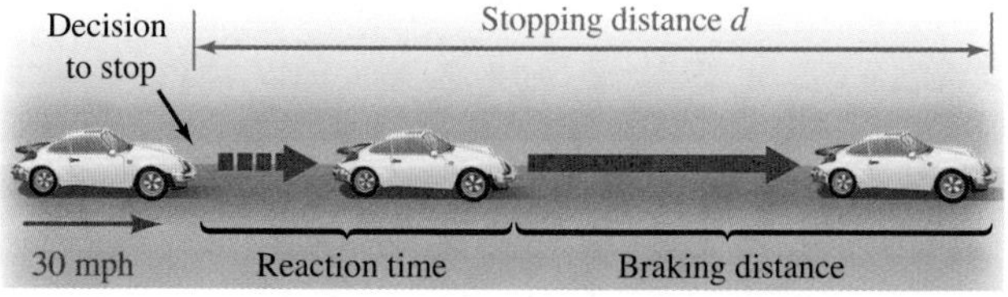

84. Suspension Bridges. The following polynomial $-0.0000001s^4 + 0.0066667s^2 + 400$ approximates the length of the cable between the two vertical towers of a bridge, where s is the sag in the cable (in feet). Estimate the length of the cable if the sag is 24.6 feet. About 404 ft

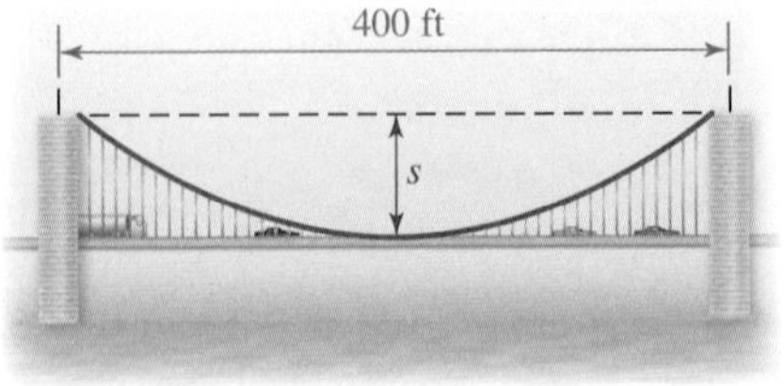

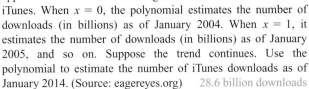

85. from **Campus to Careers**

Sound Engineering Technician

© iStockphoto.com/beetle8

Many people in the recording industry have been impressed by the success of Apple's iTunes Music Store. The polynomial $0.32x^2 - 0.36x + 0.21$ approximates the number of song downloads from iTunes. When $x = 0$, the polynomial estimates the number of downloads (in billions) as of January 2004. When $x = 1$, it estimates the number of downloads (in billions) as of January 2005, and so on. Suppose the trend continues. Use the polynomial to estimate the number of iTunes downloads as of January 2014. (Source: eagereyes.org) 28.6 billion downloads

86. Twitter. When $x = 1$, the polynomial $4.4x^2 + 36.2x + 42.5$ approximates the number of Tweets (in millions) on the social network Twitter for January, 2009. When $x = 2$, it approximates the number of Tweets (in millions) for the month of February, 2009, and so on. Use the polynomial to find the number of Tweets (in millions) for the month of October, 2009. (Source: pingdom.com) About 845 million Tweets

87. Science History. The Italian scientist Galileo Galilei (1564–1642) built an incline plane like that shown to study falling objects. As the ball rolled down, he measured the time it took the ball to travel different distances. Graph the data and then connect the points with a smooth curve. See AIE Appendix 3.

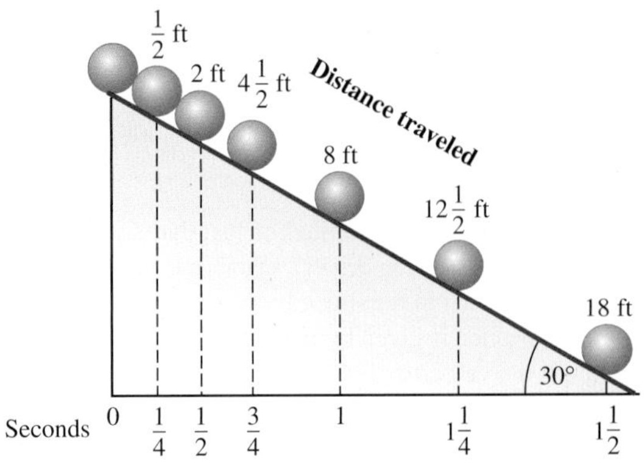

88. Dolphins. At a marine park, three trained dolphins jump in unison over an arching stream of water whose path can be described by the equation $y = -0.05x^2 + 2x$. Given the takeoff points for each dolphin, how high must each jump to clear the stream of water? 18.75 ft, 20 ft, 15 ft

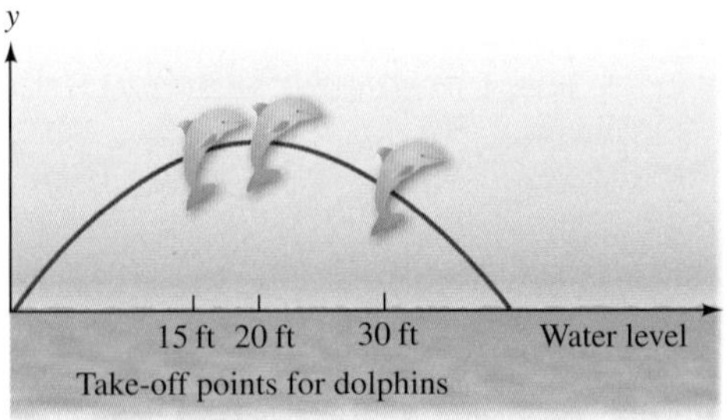

WRITING

89. Describe how to determine the degree of a polynomial.

90. List some words that contain the prefixes *mono, bi,* or *tri.*

91. To graph $y = x^2 - 4$, a table of solutions is constructed and a graph is drawn. Explain the error.

$y = x^2 - 4$

x	y	(x, y)
0	−4	(0, −4)
2	0	(2, 0)

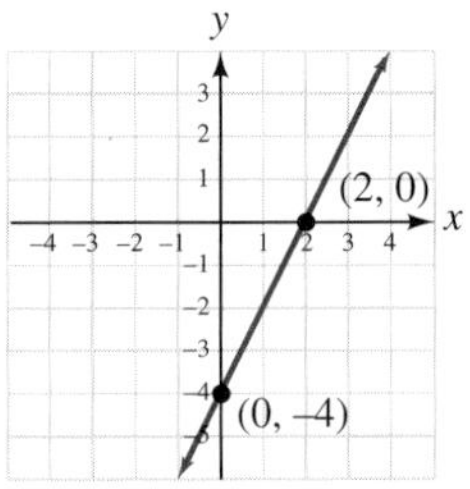

92. The expression $x + y$ is a binomial. Is xy also a binomial? Explain.

REVIEW

Solve each inequality. Write the solution set in interval notation and graph it. See AIE Appendix 3.

93. $-4(3y + 2) \le 28$ $[-3, \infty)$

94. $-5 < 3t + 4 \le 13$ $(-3, 3]$

Simplify each expression. Do not use negative exponents in the answer.

95. $(x^2x^4)^3$ x^{18}

96. $(a^2)^3(a^3)^2$ a^{12}

97. $\left(\frac{y^2y^5}{y^4}\right)^3$ y^9

98. $\left(\frac{2t^3}{t}\right)^{-4}$ $\frac{1}{16t^8}$

CHALLENGE PROBLEMS

99. Find a three-term polynomial of degree 2 whose value will be 1 when it is evaluated for $x = 2$. $x^2 - x - 1$ (Answers may vary)

100. Graph: $y = 2x^3 - 3x^2 - 11x + 6$ See AIE Appendix 3.

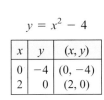

SECTION 5.5

OBJECTIVES

1 Simplify polynomials by combining like terms.
2 Add polynomials.
3 Subtract polynomials.

Adding and Subtracting Polynomials

ARE YOU READY? *Are You Ready? exercises available online at www.webassign.net/brookscole*

The following problems review some basic skills that are needed when adding and subtracting polynomials.

Combine like terms, if possible.

1. $8x + 9x$ $17x$

2. $2.5a^3 - 1.3a^3$ $1.2a^3$

3. $2t^2u - (-3t^2u)$ $5t^2u$

4. $8b^2 + 7b$ Does not simplify

5. Simplify: $-(4b^2 - 9b + 1)$ $-4b^2 + 9b - 1$

6. Subtract 10 from 2. -8

If we are to add (or subtract) objects, they must be similar. For example, we can add dollars to dollars and inches to inches, but we can't add dollars to inches. If you keep this concept in mind, then adding and subtracting polynomials will be easy. It simply involves combining like terms.

1 Simplify Polynomials by Combining Like Terms.

Recall that **like terms** have the same variables with the same exponents. Only the coefficients may differ.

Like terms	*Unlike terms*	
$-7x$ and $15x$	$-7x$ and $15a$	Different variables
$9.4y^3$ and $1.6y^3$	$9.4y^3$ and $1.6y^2$	Different exponents on the same variable
$\frac{1}{2}x^5y^2$ and $-\frac{1}{3}x^5y^2$	$\frac{1}{2}x^5y^2$ and $-\frac{1}{3}x^2y^5$	Different exponents on different variables

Teaching Tip: Remind students that *constant terms* are also like terms.

Also recall that to **combine like terms,** we combine their coefficients and keep the same variables with the same exponents. For example,

$$4y + 5y = (4 + 5)y = 9y \qquad 8x^2 - x^2 = (8 - 1)x^2 = 7x^2$$

The Language of Algebra

Simplifying the sum or difference of like terms is called **combining like terms.**

Polynomials with like terms can be simplified by combining like terms.

EXAMPLE 1 Simplify each polynomial by combining like terms:

a. $4x^4 + 81x^4$ **b.** $-0.3r - 0.4r + 0.6r$

c. $17x^2y^2 + 2x^2y - 6x^2y^2$ **d.** $\frac{3}{4}p^2 + \frac{1}{2}q^2 - 7 + \frac{1}{3}p^2 - \frac{5}{4}q^2 + 4$

Strategy We will use the distributive property in reverse to add (or subtract) the coefficients of the like terms. We will keep the same variables raised to the same powers.

Why To *combine like terms* means to add or subtract the like terms in an expression.

Solution

a. $4x^4 + 81x^4 = 85x^4$ Think: $(4 + 81)x^4 = 85x^4$.

b. $-0.3r - 0.4r + 0.6r = -0.1r$ Think: $(-0.3 - 0.4 + 0.6)r = -0.1r$.

c. The first and third terms are like terms.

$$\mathbf{17x^2y^2} + 2x^2y - \mathbf{6x^2y^2} = \mathbf{11x^2y^2} + 2x^2y$$ Think: $(17 - 6)x^2y^2 = 11x^2y^2$.

Caution

When combining like terms, simply combine their coefficients. Don't incorrectly add the exponents. The exponents on the variables *stay the same.*

d. $\frac{3}{4}p^2 + \frac{1}{2}q^2 - 7 + \frac{1}{3}p^2 - \frac{5}{4}q^2 + 4$

$= \left(\frac{3}{4} + \frac{1}{3}\right)p^2 + \left(\frac{1}{2} - \frac{5}{4}\right)q^2 - 7 + 4$ Combine like terms.

$= \left(\frac{9}{12} + \frac{4}{12}\right)p^2 + \left(\frac{2}{4} - \frac{5}{4}\right)q^2 - 7 + 4$ Build equivalent fractions: $\frac{3}{4} \cdot \frac{3}{3} = \frac{9}{12}, \frac{1}{3} \cdot \frac{4}{4} = \frac{4}{12}$, and $\frac{1}{2} \cdot \frac{2}{2} = \frac{2}{4}$.

$= \frac{13}{12}p^2 - \frac{3}{4}q^2 - 3$ Do the addition and the subtraction.

CAUTION Do not try to clear this expression of fractions by multiplying it by the LCD 12. That strategy works only when we multiply *both sides of an equation* by the LCD.

$$12\left(\frac{3}{4}p^2 + \frac{1}{2}q^2 - 7 + \frac{1}{3}p^2 - \frac{5}{4}q^2 + 4\right) \quad ✗$$

Teaching Example 1 Simplify each polynomial:
a. $4n^3 + 19n^3$
b. $-20a + 22a - a$
c. $2.9a^3t + 0.4a^3t - 0.7a^3t$
d. $\frac{3}{5}a^2 + \frac{4}{3}a - 7 + \frac{1}{2}a^2 - \frac{1}{6}a + 4$
Answers: a. $23n^3$ b. a c. $2.6a^3t$ d. $\frac{11}{10}a^2 + \frac{7}{6}a - 3$

Self Check 1 Simplify each polynomial: **a.** $6m^4 + 3m^4$ $9m^4$ **b.** $-19x + 21x - x$ x **c.** $1.7s^3t + 0.3s^2t - 0.6s^3t$ $1.1s^3t + 0.3s^2t$ **d.** $\frac{1}{8}c^5 + \frac{1}{3}d^5 - 9 + \frac{5}{4}c^5 - \frac{3}{5}d^5 + 1$ $\frac{11}{8}c^5 - \frac{4}{15}d^5 - 8$

Now Try Problems 13, 23, and 27

Notice that the terms of each solution in Example 1 are written in **descending powers** of one variable. When working with polynomials, answers are almost always presented in this form because it makes them easy to compare.

2 Add Polynomials.

When adding polynomials horizontally, each polynomial is usually enclosed within parentheses. For example,

$$(3x^2 + 6x + 7) + (2x - 5)$$

is the sum of a trinomial and a binomial. To find the sum, we reorder and regroup the terms using the commutative and associative properties of addition so that like terms are together.

$(3x^2 + 6x + 7) + (2x - 5) = 3x^2 + (\mathbf{6x + 2x}) + (\mathbf{7 - 5})$ The x-terms are together. The constant terms are together.

$= 3x^2 + \mathbf{8x} + \mathbf{2}$ Combine like terms.

This example illustrates the following rule.

Adding Polynomials To add polynomials, combine their like terms.

EXAMPLE 2 Add the polynomials: **a.** $(-6a^3 + 5a^2 - 7a + 9) + (4a^3 - 5a^2 - a - 8)$
b. $\left(\frac{1}{2}m^2 + \frac{2}{3}m + 1\right) + \left(\frac{3}{4}m^2 - \frac{7}{9}m - 4\right)$ **c.** $(16g^2 - h^2) + (4g^2 + 2gh + 10h^2)$

Strategy We will reorder and regroup to get the like terms together. Then we will combine like terms.

Why To add polynomials means to combine their like terms.

Solution a. $(-6a^3 + 5a^2 - 7a + 9) + (4a^3 - 5a^2 - a - 8)$

$= (-6a^3 + 4a^3) + (5a^2 - 5a^2) + (-7a - a) + (9 - 8)$ Group like terms together.

$= -2a^3 + 0a^2 + (-8a) + 1$ Combine like terms.

$= -2a^3 - 8a + 1$ It is not necessary to write $0a^2$.

Success Tip

Combine the like terms in order: a^3-terms first, a^2-terms second, a-terms third, and constants last. Then the answer will be in descending powers of a.

b. $\left(\frac{1}{2}m^2 + \frac{2}{3}m + 1\right) + \left(\frac{3}{4}m^2 - \frac{7}{9}m - 4\right)$

$= \left(\frac{1}{2}m^2 + \frac{3}{4}m^2\right) + \left(\frac{2}{3}m - \frac{7}{9}m\right) + (1 - 4)$ Group like terms together.

$= \left(\frac{2}{4}m^2 + \frac{3}{4}m^2\right) + \left(\frac{6}{9}m - \frac{7}{9}m\right) + (1 - 4)$ Build equivalent fractions: $\frac{1}{2} \cdot \frac{2}{2} = \frac{2}{4}$ and $\frac{2}{3} \cdot \frac{3}{3} = \frac{6}{9}$.

$= \frac{5}{4}m^2 - \frac{1}{9}m - 3$ Combine like terms.

c. $(16g^2 - h^2) + (4g^2 + 2gh + 10h^2)$

$= (16g^2 + 4g^2) + 2gh + (-h^2 + 10h^2)$ Group like terms together.

$= 20g^2 + 2gh + 9h^2$ Combine like terms.

Teaching Example 2 Add the polynomials:
a. $(3c^2 - 2c - 5) + (4c^2 - 7c + 3)$
b. $\left(\frac{5}{8}d^3 + \frac{2}{5}d + 9\right) + \left(\frac{1}{4}d^3 - \frac{11}{10}d - 11\right)$
c. $(8m^2 - 3mn - n^2) + (6m^2 - n^2)$
Answers:
a. $7c^2 - 9c - 2$ b. $\frac{7}{8}d^3 - \frac{7}{10}d - 2$
c. $14m^2 - 3mn - 2n^2$

Self Check 2 Add the polynomials:

a. $(2a^2 - a + 4) + (5a^2 + 6a - 5)$ $7a^2 + 5a - 1$

b. $\left(\frac{3}{2}b^3 + \frac{4}{5}b + 7\right) + \left(\frac{3}{4}b^3 - \frac{11}{10}b - 10\right)$ $\frac{9}{4}b^3 - \frac{3}{10}b - 3$

c. $(7x^2 - 2xy - y^2) + (4x^2 - y^2)$ $11x^2 - 2xy - 2y^2$

Now Try Problems 29, 31, and 35

EXAMPLE 3 Find a polynomial that represents the perimeter of the trapezoid.

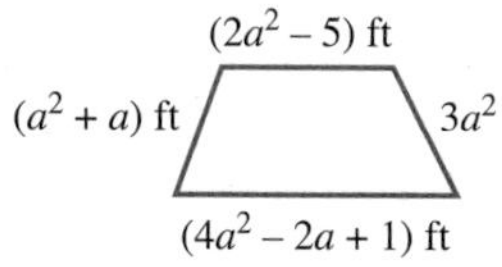

Strategy We will add the polynomials that represent the lengths of the sides of the trapezoid.

Why To find the perimeter of a figure, we find the distance around the figure by finding the sum of the lengths of its sides.

Solution To add the four polynomials that represent the lengths of the sides of the trapezoid, we combine their like terms.

The Language of Algebra

A **trapezoid** is a four-sided figure with exactly two sides parallel.

$(2a^2 - 5) + (a^2 + a) + (4a^2 - 2a + 1) + 3a^2$

$= (2a^2 + a^2 + 4a^2 + 3a^2) + (a - 2a) + (-5 + 1)$ Reorder and regroup terms.

$= 10a^2 - a - 4$ Combine like terms. Think: $2 + 1 + 4 + 3 = 10$, $1 - 2 = -1$, and $-5 + 1 = -4$.

The perimeter of the trapezoid is $(10a^2 - a - 4)$ ft.

Teaching Example 3 Find a polynomial that represents the perimeter of the figure.

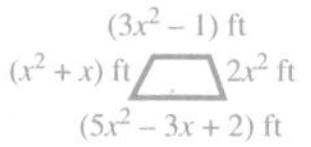

Answer: $(11x^2 - 2x + 1)$ ft

Self Check 3 Find a polynomial that represents the perimeter of a triangle with sides of length $(19h^2 + 11h - 2)$ in., $(4h^2 - 4h + 4)$ in., and $(6h^2 - 22h - 2)$ in. $(29h^2 - 15h)$ in.

Now Try Problem 39

Polynomials also can be added vertically by aligning like terms in columns.

EXAMPLE 4 Add $4x^2 - 3$ and $3x^2 - 8x + 8$ using vertical form.

Strategy First, we will write one polynomial underneath the other and draw a horizontal line beneath them. Then we will add the like terms, column by column, and write each result under the line.

Why *Vertical form* means to use an approach similar to that used in arithmetic to add two numbers.

Solution When performing vertical addition, any missing term may be written with a coefficient of 0. Since the first polynomial does not have an x-term, we insert a **placeholder** term $0x$ in the second column so that the constant terms line up in the third column.

Success Tip

In arithmetic, we use vertical form so that we add digits in like place-value columns. In polynomial addition, we combine the like terms in each column.

Hundreds, Tens, Ones

$$\begin{array}{r} 403 \\ +388 \\ \hline \end{array}$$

Teaching Example 4 Add $8c^2 + 2c - 1$ and $4c^2 + 5$ using vertical form.
Answer: $12c^2 + 2c + 4$

x^2-terms	x-terms	Constants	
$4x^2$	$+ 0x$	-3	In the x^2-column, find $4x^2 + 3x^2$.
$3x^2$	$-8x$	$+8$	In the x-column, find $0x + (-8x)$.
$7x^2$	$-8x$	$+5$	In the constant column, find $-3 + 8$.

The sum is $7x^2 - 8x + 5$.

Self Check 4 Add $4q^2 - 7$ and $2q^2 - 8q + 9$ using vertical form. $6q^2 - 8q + 2$

Now Try Problem 41

3 Subtract Polynomials.

Recall from Chapter 1 that we can use the distributive property to find the opposite of several terms enclosed within parentheses. For example, consider $-(2a^2 - a + 9)$.

$$-(2a^2 - a + 9) = \mathbf{-1}(2a^2 - a + 9)$$ Replace the − symbol in front of the parentheses with −1.

$$= -2a^2 + a - 9$$ Use the distributive property to remove parentheses.

This example illustrates the following method of subtracting polynomials.

Subtracting Polynomials To subtract two polynomials, change the signs of the terms of the polynomial being subtracted, drop the parentheses, and combine like terms.

EXAMPLE 5 Subtract the polynomials: **a.** $(3a^2 - 4a - 6) - (2a^2 - a + 9)$
b. $(-t^3u + 2t^2u - u + 1) - (-3t^2u - u + 8)$

Strategy In each case, we will change the signs of the terms of the polynomial being subtracted, drop the parentheses, and combine like terms.

Why This is the method for subtracting two polynomials.

Solution **a.** $(3a^2 - 4a - 6) - \mathbf{(2a^2 - a + 9)}$

$$= 3a^2 - 4a - 6 \mathbf{- 2a^2 + a - 9}$$ Change the sign of each term of $2a^2 - a + 9$ and drop the parentheses.

$$= a^2 - 3a - 15$$ Combine like terms.

Success Tip

After some practice, you will be able to reorder and regroup the terms of the polynomials in your head to combine them.

Teaching Example 5 Subtract the polynomials:
a. $(t^3 - 2t^2 + 6) - (2t^3 - t^2 - 9)$
b. $(r^2s - 3r + s - 8) - (7r + s - 3)$
Answers:
a. $-t^3 - t^2 + 15$
b. $r^2s - 10r - 5$

b. $(-t^3u + 2t^2u - u + 1) - (\mathbf{-3t^2u - u + 8})$

$= -t^3u + 2t^2u - u + 1 + \mathbf{3t^2u + u - 8}$ Change the sign of each term of $-3t^2u - u + 8$ and drop the parentheses.

$= -t^3u + 5t^2u - 7$ Combine like terms.

Self Check 5 Subtract the polynomials:
a. $(8a^3 - 5a^2 + 5) - (a^3 - a^2 - 7)$ $7a^3 - 4a^2 + 12$
b. $(x^2y - 2x + y - 2) - (6x + 9y - 2)$ $x^2y - 8x - 8y$

Now Try Problems 49 and 55

Polynomials can also be subtracted vertically by aligning like terms in columns.

EXAMPLE 6

a. Subtract $3x^2 - 2x + 3$ from $2x^2 + 4x - 1$ using vertical form.
b. Subtract $4x^3 - 6x^2 + x$ from $7x^3 - 2x$ using vertical form.

Strategy Since the first polynomial is to be subtracted from the second, we will write the first underneath the second, change the sign of each of its terms and add, column-by-column.

Why *Vertical form* means to arrange the like terms in columns.

Solution

a.

$$\begin{array}{r} 2x^2 + 4x - 1 \\ \underline{-(3x^2 - 2x + 3)} \end{array} \xrightarrow[\text{and add}]{\text{Change signs}} \begin{array}{r} 2x^2 + 4x - 1 \\ \underline{-3x^2 + 2x - 3} \\ -x^2 + 6x - 4 \end{array}$$

In the x^2-column, find $2x^2 + (-3x^2)$.
In the x-column, find $4x + 2x$.
In the constant column, find $-1 + (-3)$.

The difference is $-x^2 + 6x - 4$.

The Language of Algebra

Recall that we must be careful of the order when translating English phrases containing the words **subtract from.** For example, to subtract x from y, we write $y - x$.

Teaching Tip: Subtraction of polynomials in vertical form as seen here is used in long division of polynomials (Section 5.8).

b. Since $7x^3 - 2x$ is missing an x^2-term, we will insert the placeholder term $0x^2$ to ensure that like terms are in the same column.

$$\begin{array}{r} 7x^3 + \mathbf{0x^2} - 2x \\ \underline{-(4x^3 - 6x^2 + x)} \end{array} \xrightarrow[\text{and add}]{\text{Change signs}} \begin{array}{r} 7x^3 + 0x^2 - 2x \\ \underline{-4x^3 + 6x^2 - x} \\ 3x^3 + 6x^2 - 3x \end{array}$$

In the x^3-column, find $7x^3 + (-4x^3)$.
In the x^2-column, find $0x^2 + 6x^2$.
In the x-column, find $-2x + (-x)$.

The difference is $3x^3 + 6x^2 - 3x$.

Teaching Example 6 **a.** Subtract $3d^2 + 5d - 9$ from $12d^2 - 6d + 11$.
b. Subtract $y^3 - 9y^2 + y$ from $3y^3 - 10y$.
Answers: **a.** $9d^2 - 11d + 20$
b. $2y^3 + 9y^2 - 11y$

Self Check 6
a. Subtract $2p^2 + 2p - 8$ from $5p^2 - 6p + 7$ using vertical form. $3p^2 - 8p + 15$
b. Subtract $4m^3 - 6m^2 + 7m$ from $-8m^3 + 16m$. $-12m^3 + 6m^2 + 9m$

Now Try Problem 57

EXAMPLE 7

Subtract $1.2a^4 - 0.7a$ from the sum of $0.6a^4 + 1.5a$ and $0.4a^4 - 1.1a$.

Strategy First, we will translate the words of the problem into mathematical symbols. Then we will perform the indicated operations.

Why The words of the problem contain the key phrases *subtract from* and *sum.*

Solution

Since $1.2a^4 - 0.7a$ is to be subtracted from the sum, the order must be reversed when we translate to mathematical symbols.

Subtract $1.2a^4 - 0.7a$ from | the sum of $0.6a^4 + 1.5a$ and $0.4a^4 - 1.1a$.

$[(0.6a^4 + 1.5a) + (0.4a^4 - 1.1a)] - (\mathbf{1.2a^4 - 0.7a})$ Use brackets [ ] to enclose the sum.

Caution

A common error is to change the sign of only the first term of the polynomial being subtracted and forget to change the signs of its other terms.

Teaching Example 7 Subtract $-0.8p^2 - 0.9p$ from the sum of $0.4p^2 - 0.1p$ and $0.5p^2 + 0.2p$.
Answer: $1.7p^2 + p$

Next, we change the sign of each term within $(1.2a^4 - 0.7a)$ and drop the parentheses.

$$= 0.6a^4 + 1.5a + 0.4a^4 - 1.1a \mathbf{- 1.2a^4 + 0.7a}$$

$$= -0.2a^4 + 1.1a \quad \text{Combine like terms.}$$

Self Check 7 Subtract $-0.2q^2 - 0.2q$ from the sum of $0.1q^2 - 0.6q$ and $0.3q^2 + 0.1q$. $0.6q^2 - 0.3q$

Now Try Problem 65

EXAMPLE 8

Fireworks. Two firework shells are fired upward at the same time from different platforms. The height, after t seconds, of the first shell is $(-16t^2 + 160t + 3)$ feet. The height, after t seconds, of a higher-flying second shell is $(-16t^2 + 200t + 1)$ feet.

a. Find a polynomial that represents the difference in the heights of the shells.

b. In 5 seconds, the first shell reaches its peak and explodes. How much higher is the second shell at that time?

Strategy To find the difference in their heights, we will subtract the height of the first shell from the height of the higher-flying second shell.

Why The key word *difference* indicates that we should subtract the polynomials.

Solution

a. Since the height of the higher flying second shell is represented by $-16t^2 + 200t + 1$ and the height of the lower flying shell is represented by $-16t^2 + 160t + 3$, we can find their difference by performing the following subtraction.

$$(-16t^2 + 200t + 1) \mathbf{- (-16t^2 + 160t + 3)}$$

$$= -16t^2 + 200t + 1 \mathbf{+ 16t^2 - 160t - 3} \quad \text{Change the sign of each term of } -16t^2 + 160t + 3 \text{ and drop the parentheses.}$$

$$= 40t - 2 \quad \text{Combine like terms.}$$

The difference in the heights of the shells t seconds after being fired is $(40t - 2)$ feet.

b. To find the difference in their heights after 5 seconds, we will evaluate the polynomial found in part (a) at a value of 5 seconds. If we substitute 5 for t, we have

$$40t - 2 = 40(5) - 2 = 200 - 2 = 198$$

When the first shell explodes, the second shell will be 198 feet higher than the first shell.

Teaching Example 8 **Rescues.** Two warning flares are fired upward at the same time from different parts of a ship. The height of the first flare is $(-16t^2 + 115t + 25)$ feet and the height of the higher-traveling second flare is $(-16t^2 + 130t + 30)$ feet, after t seconds. Find a polynomial that represents the difference in the heights.
Answer: $(15t + 5)$ ft

Self Check 8 **Property values.** A real estate investor purchased two houses on the same day. The value of the first house, x years after its purchase, is given by \$$(2{,}500x + 95{,}000)$. The value of the second house, x years after its purchase, is given by \$$(4{,}500x + 125{,}000)$. Find a polynomial that represents the total value of the houses after x years.
\$$(7{,}000x + 220{,}000)$

Now Try Problem 109

SECTION 5.5 STUDY SET

VOCABULARY

Fill in the blanks.

▶ 1. $(b^3 - b^2 - 9b + 1) + (b^3 - b^2 - 9b + 1)$ is the sum of two ___polynomials___.

▶ 2. $(b^2 - 9b + 11) - (4b^2 - 14b)$ is the ___difference___ of a trinomial and a binomial.

▶ 3. ___Like___ terms have the same variables with the same exponents.

▶ 4. The polynomial $2t^4 + 3t^3 - 4t^2 + 5t - 6$ is written in ___descending___ powers of t.

CONCEPTS

Fill in the blanks.

5. To add polynomials, ___combine___ their like terms.

6. To subtract polynomials, ___change___ the signs of the terms of the polynomial being subtracted, drop parentheses, and combine like terms.

7. Simplify each polynomial, if possible.

a. $2x^2 + 3x^2$ $5x^2$

b. $15m^3 - m^3$ $14m^3$

c. $8a^3b - a^3b$ $7a^3b$

d. $6cd + 4c^2d$ $6cd + 4c^2d$

▶ 8. What is the result of the addition in the x-column? $-7x$

$$\begin{array}{r} 4x^2 + x - 12 \\ \underline{5x^2 - 8x + 23} \end{array}$$

▶ 9. Write without parentheses.

a. $-(5x^2 - 8x + 23)$
$-5x^2 + 8x - 23$

b. $-(-5y^4 + 3y^2 - 7)$
$5y^4 - 3y^2 + 7$

10. What is the result of the subtraction in the x-column? $-13x$

$$\begin{array}{rcr} 8x^2 - 7x - 1 & & 8x^2 - 7x - 1 \\ \underline{-(4x^2 + 6x - 9)} & \longrightarrow & \underline{-4x^2 - 6x + 9} \end{array}$$

NOTATION

Fill in the blanks to add (subtract) the polynomials.

11. $(6x^2 + 2x + 3) + (4x^2 - 7x + 1)$

$= (6x^2 + \boxed{4x^2}) + (\boxed{2x} - 7x) + (3 + \boxed{1})$

$= \boxed{10x^2} - 5x + \boxed{4}$

12. $(6x^2 + 2x + 3) - (4x^2 - 7x + 1)$

$= 6x^2 + 2x + 3 \boxed{-} 4x^2 \boxed{+} 7x - 1$

$= \boxed{2x^2} + 9x + \boxed{2}$

GUIDED PRACTICE

Simplify each polynomial and write it in descending powers of one variable. **See Example 1.**

▶ 13. $8t^2 + 4t^2$
$12t^2$

▶ 14. $15x^2 + 10x^2$
$25x^2$

15. $18x^2 - 19x + 2x^2$
$20x^2 - 19x$

16. $17y^2 - 22y - y^2$
$16y^2 - 22y$

▶ 17. $10x^2 - 8x + 9x - 9x^2$
$x^2 + x$

18. $-3y^2 - y - 6y^2 + 7y$
$-9y^2 + 6y$

▶ 19. $\frac{1}{5}x^2 - \frac{3}{8}x + \frac{2}{3}x^2 + \frac{1}{4}x$ $\frac{13}{15}x^2 - \frac{1}{8}x$

20. $\frac{6}{7}y^2 + \frac{1}{2}y - \frac{2}{3}y^2 + \frac{1}{5}y$ $\frac{4}{21}y^2 + \frac{7}{10}y$

21. $0.6x^3 + 0.8x^4 + 0.7x^3 + (-0.8x^4)$ $1.3x^3$

▶ 22. $1.9m^4 - 2.4m^6 - 3.7m^4 + 2.8m^6$ $0.4m^6 - 1.8m^4$

23. $\frac{1}{2}st + \frac{3}{2}st$ $2st$

24. $\frac{2}{5}at + \frac{1}{5}at$ $\frac{3}{5}at$

▶ 25. $-4ab + 4ab - ab$ $-ab$

▶ 26. $xy - 4xy - 2xy$ $-5xy$

27. $4x^2y + 5 - 6x^3y - 3x^2y + 2x^3y$ $-4x^3y + x^2y + 5$

▶ 28. $5b - 9ab^2 + 10a^3b - 8ab^2 - 9a^3b$ $a^3b - 17ab^2 + 5b$

Add the polynomials. **See Example 2.**

▶ 29. $(3q^2 - 5q + 7) + (2q^2 + q - 12)$ $5q^2 - 4q - 5$

30. $(2t^2 + 11t - 15) + (-5t^2 - 13t + 10)$ $-3t^2 - 2t - 5$

▶ 31. $\left(\frac{2}{3}y^3 + \frac{3}{4}y^2 + \frac{1}{2}\right) + \left(\frac{1}{3}y^3 + \frac{1}{5}y^2 - \frac{1}{6}\right)$ $y^3 + \frac{19}{20}y^2 + \frac{1}{3}$

▶ 32. $\left(\frac{1}{16}r^6 + \frac{1}{2}r^3 - \frac{11}{12}\right) + \left(\frac{9}{16}r^6 + \frac{9}{4}r^3 + \frac{1}{12}\right)$ $\frac{5}{8}r^6 + \frac{11}{4}r^3 - \frac{5}{6}$

33. $(0.3p + 2.1q) + (0.4p - 3q)$ $0.7p - 0.9q$

▶ 34. $(-0.3r - 5.2s) + (0.8r - 5.2s)$ $0.5r - 10.4s$

35. $(2x^2 + xy + 3y^2) + (5x^2 - y^2)$ $7x^2 + xy + 2y^2$

▶ 36. $(-4a^2 - ab + 15b^2) + (5a^2 - b^2)$ $a^2 - ab + 14b^2$

Find a polynomial that represents the perimeter of the figure. **See Example 3.**

▶ 37. $(3x^2 + 6x - 2)$ yd

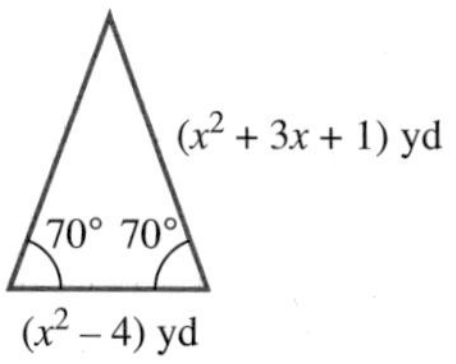

38. $(8y^5 - 7y^3)$ in.

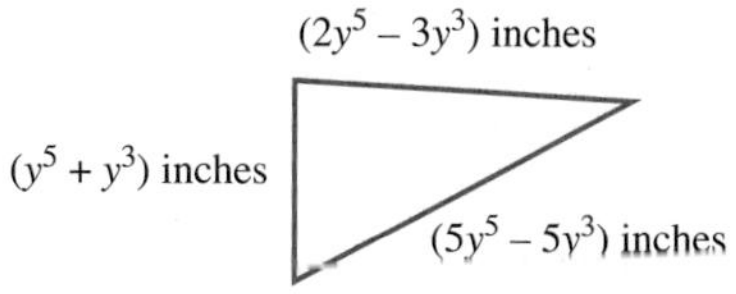

▶ 39. $(7x^2 + 5x + 6)$ mi

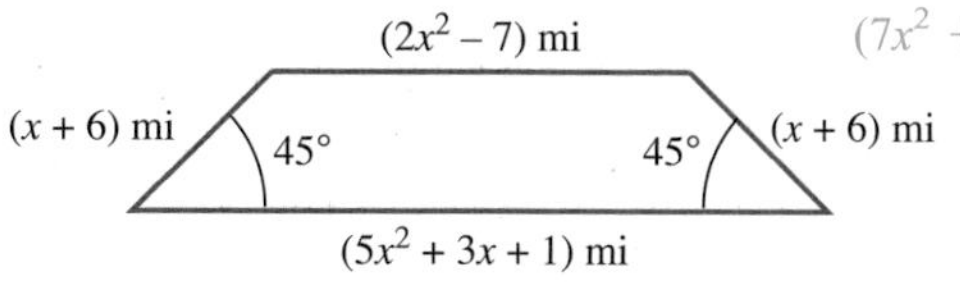

40. $(29a^2 - 10a - 1)$ ft

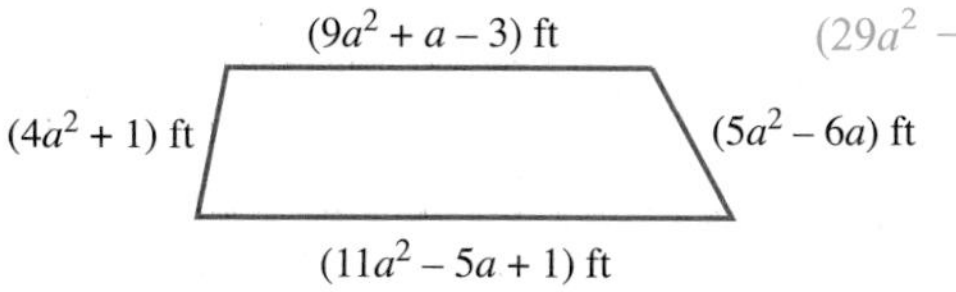

Use vertical form to add the polynomials. **See Example 4.**

41.
$$\begin{array}{r} 3x^2 + 4x + 5 \\ \underline{2x^2 - 3x + 6} \\ 5x^2 + x + 11 \end{array}$$

▶ 42.
$$\begin{array}{r} 6x^3 - 4x^2 + 7 \\ \underline{7x^3 + 9x^2 + 12} \\ 13x^3 + 5x^2 + 19 \end{array}$$

43. $\begin{array}{r} 6a^2 + 7a + 9 \\ \underline{-9a^2 \qquad - 2} \\ -3a^2 + 7a + 7 \end{array}$

44. $\begin{array}{r} -2c^2 - 3c - 5 \\ \underline{14c^2 \qquad - 1} \\ 12c^2 - 3c - 6 \end{array}$

45. $\begin{array}{r} z^3 + 6z^2 - 7z + 16 \\ \underline{9z^3 - 6z^2 + 8z - 18} \\ 10z^3 \qquad + z - 2 \end{array}$

46. $\begin{array}{r} 3x^3 + 4x^2 - 3x + 5 \\ \underline{3x^3 - 4x^2 - x - 7} \\ 6x^3 \qquad - 4x - 2 \end{array}$

47. $\begin{array}{r} -3x^3y^2 + 4x^2y - 4x + 9 \\ \underline{2x^3y^2 \qquad + 9x - 3} \\ -x^3y^2 + 4x^2y + 5x + 6 \end{array}$

48. $\begin{array}{r} 3x^2y^2 + 4xy + 25 \\ \underline{5x^2y^2 \qquad - 12} \\ 8x^2y^2 + 4xy + 13 \end{array}$

Subtract the polynomials. **See Example 5.**

49. $(3a^2 - 2a + 4) - (a^2 - 3a + 7)$ $2a^2 + a - 3$

50. $(2b^2 + 3b - 5) - (2b^2 - 4b - 9)$ $7b + 4$

51. $(-4h^3 + 5h^2 + 15) - (h^3 - 15)$ $-5h^3 + 5h^2 + 30$

52. $(-c^5 + 5c^4 - 12) - (2c^5 - c^4)$ $-3c^5 + 6c^4 - 12$

53. $\left(\frac{3}{8}s^8 - \frac{3}{4}s^7\right) - \left(\frac{1}{3}s^8 + \frac{1}{5}s^7\right)$ $\frac{1}{24}s^8 - \frac{19}{20}s^7$

54. $\left(\frac{5}{6}q^9 - \frac{4}{5}q^8\right) - \left(\frac{1}{4}q^9 + \frac{3}{8}q^8\right)$ $\frac{7}{12}q^9 - \frac{47}{40}q^8$

55. $(5ab + 2b^2) - (2 + ab + b^2)$ $b^2 + 4ab - 2$

56. $(mn + 8n^2) - (6 - 5mn + n^2)$ $7n^2 + 6mn - 6$

Use vertical form to subtract the polynomials. **See Example 6.**

57. Subtract $2x^2 - 2x + 3$ from $3x^2 + 4x + 5$.
$x^2 + 6x + 2$

58. Subtract $3y^2 - 6y + 7$ from $6y^2 + 4y + 13$.
$3y^2 + 10y + 6$

59. Subtract $(s^2 + 4s + 2)$ from $(5s^2 + 9)$.
$4s^2 - 4s + 7$

60. Subtract $(4p^2 - 4p - 40)$ from $(10p^2 - 30)$.
$6p^2 + 4p + 10$

61. Subtract $8a^3 + 8a^2 - 3a + 1$ from $17a^3 + 25a - 10$.
$9a^3 - 8a^2 + 28a - 11$

62. Subtract $m^3 + 20m^2 - 15m + 39$ from $-4m^3 - m + 22$.
$-5m^3 - 20m^2 + 14m - 17$

63. $\begin{array}{r} 0.8x^3 \qquad - 2.3x + 0.6 \\ \underline{-(0.2x^3 - 1.2x^2 - 3.6x + 0.9)} \\ 0.6x^3 + 1.2x^2 + 1.3x - 0.3 \end{array}$

64. $\begin{array}{r} 9.7y^3 \qquad + y + 1.1 \\ \underline{-(6.3y^3 - 4.4y^2 + 2.7y + 8.8)} \\ 3.4y^3 + 4.4y^2 - 1.7y - 7.7 \end{array}$

Perform the operations. **See Example 7.**

65. Subtract $(3x^2 + 4x - 7)$ from the sum of $(-2x^2 - 7x + 1)$ and $(-4x^2 + 8x - 7)$.
$-9x^2 - 3x + 1$

66. Subtract $(32x^2 - 17x + 45)$ from the sum of $(23x^2 - 12x - 7)$ and $(-11x^2 + 12x + 7)$.
$-20x^2 + 17x - 45$

67. Subtract $t^3 - 2t^2 + 2$ from the sum of $3t^3 + t^2$ and $-t^3 + 6t - 3$.
$t^3 + 3t^2 + 6t - 5$

68. Subtract $-3z^3 - 4z + 7$ from the sum of $2z^2 + 3z - 7$ and $-4z^3 - 2z - 3$.
$-z^3 + 2z^2 + 5z - 17$

TRY IT YOURSELF

Perform the operations.

69. $(9a^2 + 3a) - (2a - 4a^2)$ $13a^2 + a$

70. $(4b^2 + 3b) - (7b - b^2)$ $5b^2 - 4b$

71. Subtract $(-y^5 + 5y^4 - 1.2)$ from $(2y^5 - y^4)$.
$3y^5 - 6y^4 + 1.2$

72. Subtract $(-4w^3 + 5w^2 + 7.6)$ from $(w^3 - 15w^2)$.
$5w^3 - 20w^2 - 7.6$

73. $3r^4 - 4r + 7r^4$
$10r^4 - 4r$

74. $-2b^4 + 7b - 3b^4$
$-5b^4 + 7b$

75. $(0.03f^2 + 0.25f + 0.91) - (0.17f^2 - 1.18)$
$-0.14f^2 + 0.25f + 2.09$

76. $(0.05r^2 - 0.33r) - (0.48r^2 + 0.15r + 2.14)$
$-0.43r^2 - 0.48r - 2.14$

77. $\left(\frac{7}{8}r^4 + \frac{5}{9}r^2 - \frac{9}{4}\right) - \left(-\frac{3}{8}r^4 - \frac{2}{3}r^2 - \frac{1}{4}\right)$ $\frac{5}{4}r^4 + \frac{11}{9}r^2 - 2$

78. $\left(\frac{4}{5}t^4 - \frac{1}{3}t^2 + \frac{1}{2}\right) - \left(-\frac{1}{2}t^4 + \frac{3}{8}t^2 - \frac{1}{16}\right)$ $\frac{13}{10}t^4 - \frac{17}{24}t^2 + \frac{9}{16}$

79. $\begin{array}{r} 8c^2 - 4c - 5 \\ \underline{-(-c^2 + 2c + 9)} \\ 9c^2 - 6c - 14 \end{array}$

80. $\begin{array}{r} 3t^3 - 4t^2 - 3t + 5 \\ \underline{+11t^3 \qquad - 8t - 2} \\ 14t^3 - 4t^2 - 11t + 3 \end{array}$

81. $(12.1h^3 + 9.9h^2) + (7.3h^3 + 1.1h^2)$ $19.4h^3 + 11h^2$

82. $(5.7n^3 - 2.1n) + (-6.2n^3 - 3.9n)$ $-0.5n^3 - 6n$

83. $(20 - 4rt - 5r^2t) + (10 - 5rt)$ $-5r^2t - 9rt + 30$

84. $(5m^2 - 8m + 8) - (-20m^2 + m)$ $25m^2 - 9m + 8$

85. $(3x^2 - 3x - 2) + (3x^2 + 4x - 3)$ $6x^2 + x - 5$

86. $(4c^2 + 3c - 2) + (3c^2 + 4c + 2)$ $7c^2 + 7c$

87. $\frac{2}{3}d^2 - \frac{1}{4}c^2 + \frac{5}{6}c^2 - \frac{1}{2}cd + \frac{1}{3}d^2$ $\frac{7}{12}c^2 - \frac{1}{2}cd + d^2$

88. $\frac{3}{5}s^2 - \frac{2}{5}t^2 - \frac{1}{2}s^2 - \frac{7}{10}st - \frac{3}{10}st$ $\frac{1}{10}s^2 - st - \frac{2}{5}t^2$

89. $(3x + 7) + (4x - 3)$ $7x + 4$

90. $(2y - 3) + (4y + 7)$ $6y + 4$

91. Subtract $1.7t^2 - 1.1t$ from the sum of $-2.7t^2 + 2.1t - 1.7$ and $3.1t^2 - 2.5t + 2.3$.
$-1.3t^2 + 0.7t + 0.6$

92. Subtract $1.07x^2 - 2.07x$ from the sum of $1.04x^2 - 5.01$ and $1.33x - 1.9x^2 + 5.02$.
$-1.93x^2 + 3.4x + 0.01$

93. $-32u^3 - 16u^3$
$-48u^3$

94. $-25x^3 - 7x^3$
$-32x^3$

95. $(9d^2 + 6d) + (8d - 4d^2)$ $5d^2 + 14d$

96. $(2c^2 - 4c) + (8c - c^2)$ $c^2 + 4c$

97. $\begin{array}{r} 3x^3y^2 + 4x^2y + 7x + 12 \\ \underline{-(-4x^3y^2 + 6x^2y + 9x - 3)} \\ 7x^3y^2 - 2x^2y - 2x + 15 \end{array}$

98. $\begin{array}{r} -2x^2y^2 \qquad + 12y^2 \\ \underline{-(10x^2y^2 + 9xy - 24y^2)} \\ -12x^2y^2 - 9xy + 36y^2 \end{array}$

99. $(2x^2 - 3x + 1) - (4x^2 - 3x + 2) + (2x^2 + 3x + 2)$
$3x + 1$

100. $(-3z^2 - 4z + 7) + (2z^2 + 2z - 1) - (2z^2 - 3z + 7)$ $-3z^2 + z - 1$

101.
$$\begin{array}{r} 4x^3 + 4x^2 - 3x + 10 \\ +(5x^3 - 2x^2 - 4x - 4) \\ \hline 9x^3 + 2x^2 - 7x + 6 \end{array}$$

102.
$$\begin{array}{r} 7m^5 + m^3 + 9m^2 - m \\ -(8m^5 - 2m^3 + m^2 + m) \\ \hline -m^5 + 3m^3 + 8m^2 - 2m \end{array}$$

Look Alikes . . .

103. a. $(-8x^2 - 3x) + (-11x^2 + 6x + 10)$ $-19x^2 + 3x + 10$
b. $(-8x^2 - 3x) - (-11x^2 + 6x + 10)$ $3x^2 - 9x - 10$

104. a. $(10 - 2st - 3s^2t) + (4 - 6st)$ $-3s^2t - 8st + 14$
b. $(10 - 2st - 3s^2t) - (4 - 6st)$ $6 + 4st - 3s^2t$

APPLICATIONS

105. Greek Architecture.

a. Find a polynomial that represents the difference in the heights of the columns. $(x^2 - 8x + 12)$ ft

b. If the columns were stacked one atop the other, to what height would they reach? $(x^2 + 2x - 8)$ ft

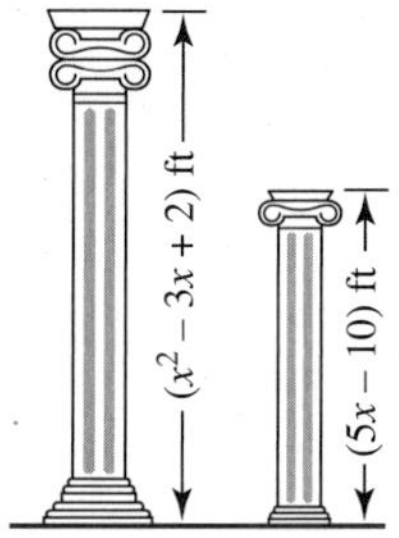

106. Jets. Find a polynomial that represents the length of the larger jet. $(11x - 12)$ ft

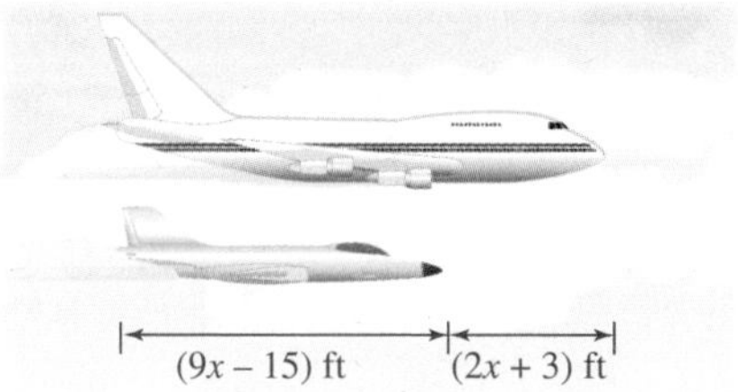

107. Piñatas. Find a polynomial that represents the length of the rope used to hold up the piñata. $(2a^2 + 6a + 5)$ in.

108. Reading Blueprints. Find a polynomial that represents

a. the difference in the length and width of the one-bedroom apartment. $(6x + 5)$ ft

b. the perimeter of the apartment. $(4x^2 + 26)$ ft

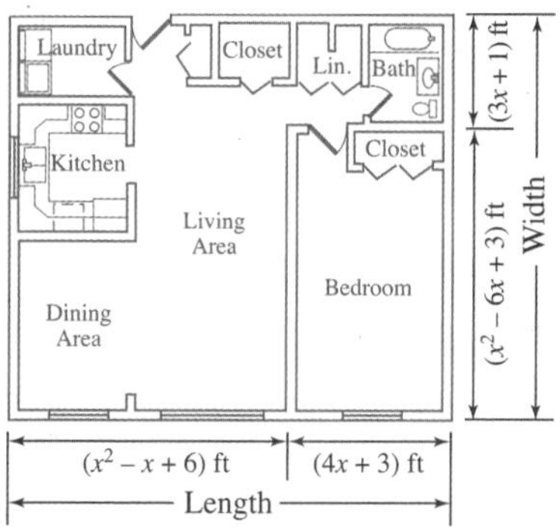

109. Naval Operations. Two warning flares are fired upward at the same time from different parts of a ship. The height of the first flare is $(-16t^2 + 128t + 20)$ feet and the height of the higher-traveling second flare is $(-16t^2 + 150t + 40)$ feet, after t seconds.

a. Find a polynomial that represents the difference in the heights of the flares. $(22t + 20)$ ft

b. In 4 seconds, the first flare reaches its peak, explodes, and lights up the sky. How much higher is the second flare at that time? 108 ft

110. Auto Mechanics. The length of a fan belt that wraps around three pulleys is $(3x^2 + 11x + 4.5\pi)$ in. Find a polynomial that represents the unknown length of a part of the belt shown in the illustration below. $(x^2 + 9x)$ in.

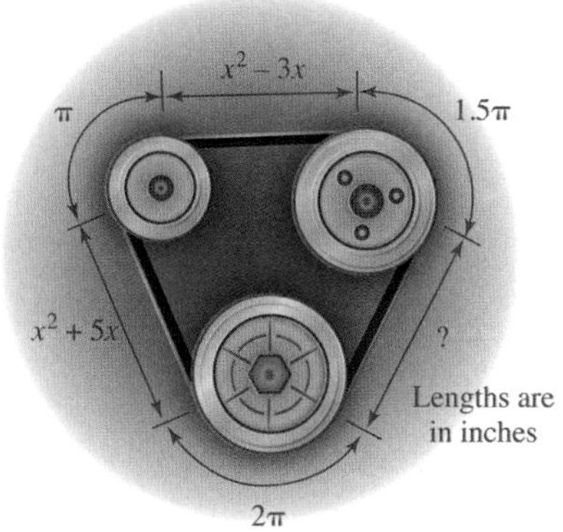

WRITING

111. How do you recognize like terms?

112. Explain why the vertical form used in algebra to add $2x^2 + 4x + 3$ and $5x^2 + 3x + 6$ is similar to the vertical form used in arithmetic to add 243 and 536.

113. Explain the error below.

$$7x^2y + 6x^2y = 13x^4y^2$$

114. Explain the error below.

$$(12x^2 - 4) - (3x^2 - 1) = 12x^2 - 4 - 3x^2 - 1$$
$$= 9x^2 - 5$$

115. A student was asked to simplify $\frac{1}{6}x^2 - 3 + \frac{2}{3}x^2$. Explain the error below:

$$6\left(\frac{1}{6}x^2 - 3 + \frac{2}{3}x^2\right) = x^2 - 18 + 4x^2$$
$$= 5x^2 - 18$$

116. Explain the error below.

Subtract $(2d^2 - d - 3)$ from $(d^2 - 9)$:

$$(2d^2 - d - 3) - (d^2 - 9) = d^2 - d + 6$$

REVIEW

117. What is the sum of the measures of the angles of a triangle? 180°

118. What are the formulas for the area of a circle and the area of a triangle? $A = \pi r^2$; $A = \frac{1}{2}bh$

119. Graph: $y = -\frac{1}{2}x + 2$

120. Graph: $2x + 3y = 9$

See AIE Appendix 3.

CHALLENGE PROBLEMS

121. What polynomial must be added to $2x^2 - x + 3$ so that the sum is $6x^2 - 7x - 8$? $4x^2 - 6x - 11$

122. Is the sum of two trinomials always a trinomial? Explain why or why not.
No; $(x^3 + x + 1) + (x^2 + x + 1) = x^3 + x^2 + 2x + 2$

SECTION 5.6 Multiplying Polynomials

OBJECTIVES

1. Multiply monomials.
2. Multiply a polynomial by a monomial.
3. Multiply binomials.
4. Multiply polynomials.

ARE YOU READY? *Are You Ready? exercises available online at www.webassign.net/brookscole*

The following problems review some basic skills that are needed when multiplying polynomials.

1. Multiply: $5 \cdot 10a$ $50a$
2. Multiply: $5(2x - 3)$ $10x - 15$
3. Multiply: $(4y + 8)7$ $28y + 56$
4. Multiply: $x^8 \cdot x^6$ x^{14}
5. Multiply: $-2(3)(5)$ -30
6. Simplify by combining like terms: $9x^2 + 6x - 3x + 2$ $9x^2 + 3x + 2$

We now discuss multiplying polynomials. We will begin with the simplest case—finding the product of two monomials.

1 Multiply Monomials.

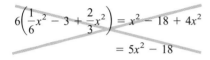

To find the product of two monomials, such as $8x^2$ and $3x^4$, we use the commutative and associative properties of multiplication to reorder and regroup the factors.

$$(8x^2)(3x^4) = (8 \cdot 3)(x^2 \cdot x^4)$$ Group the coefficients together and the variables together.
$$= 24x^6$$ Multiply: $8 \cdot 3 = 24$. Simplify: $x^2 \cdot x^4 = x^{2+4} = x^6$.

This example illustrates the following rule.

Multiplying Two Monomials

To multiply two monomials, multiply the numerical factors (the coefficients) and then multiply the variable factors.

EXAMPLE 1 Multiply: **a.** $6r \cdot r$ **b.** $3t^4(-2t^5)$ **c.** $\left(\frac{1}{3}a^2b^3\right)(21ab^2)$ **d.** $-4y^5z^2(2y^3z^3)(3yz)$

Strategy We will multiply the numerical factors and then multiply the variable factors.

Why The commutative and associative properties of multiplication enable us to reorder and regroup the factors.

Solution In each case, we will *multiply* the coefficients and *add* the exponents of the like bases.

a. $6r \cdot r = 6r^2$ Recall that $r = 1r$. Think: $6 \cdot 1 = 6$ and $r \cdot r = r^2$.

b. $(3t^4)(-2t^5) = -6t^9$ Think: $3(-2) = -6$ and $t^4 \cdot t^5 = t^{4+5} = t^9$.

c. $\left(\frac{1}{3}a^2b^3\right)(21ab^2) = 7a^3b^5$ Think: $\frac{1}{3} \cdot 21 = \frac{21}{3} = 7$, $a^2 \cdot a = a^3$, and $b^3 \cdot b^2 = b^5$.

d. $-4y^5z^2(2y^3z^3)(3yz) = -24y^9z^6$ Think: $-4(2)(3) = -24$, $y^5 \cdot y^3 \cdot y = y^9$, and $z^2 \cdot z^3 \cdot z = z^6$.

Caution

Only like terms can be added or subtracted.
Like terms:
$6r + r = 7r$
Unlike terms:
$3t^4 - 2t^5$ Does not simplify

However, like and unlike terms may be multiplied.
$6r \cdot r = 6r^2$
$(3t^4)(-2t^5) = -6t^9$

Teaching Example 1 Multiply:
a. $15x \cdot x$ **b.** $-9t^7(-5t^4)$
c. $(32a^{15})\left(\frac{1}{8}a^{11}\right)$
d. $(6c^2d^3)(-6c^3d^2)(cd)$
Answers: **a.** $15x^2$ **b.** $45t^{11}$
c. $4a^{26}$ **d.** $-36c^6d^6$

Self Check 1 Multiply: **a.** $18t \cdot t$ $18t^2$ **b.** $-10d^8(-6d^3)$ $60d^{11}$
c. $(16y^{12})\left(\frac{1}{4}y^2\right)$ $4y^{14}$ **d.** $(5a^3b^3)(-6a^3b^4)(ab)$ $-30a^7b^8$

Now Try Problems 13, 19, and 21

2 Multiply a Polynomial by a Monomial.

We can use the distributive property to find the product of a monomial and a binomial such as $5x$ and $2x + 4$:

$$5x(2x + 4) = 5x(2x) + 5x(4)$$ Read as "5x times the quantity of 2x plus 4." Distribute the multiplication by 5x.

$$= 10x^2 + 20x$$ Multiply the monomials.

This example suggests the following rule.

Multiplying Polynomials by Monomials

To multiply a monomial and a polynomial, multiply each term of the polynomial by the monomial.

EXAMPLE 2 Multiply: **a.** $3n^6(16n^{15} + n^{10})$ **b.** $3a^2(3a^2 - 5a + 2)$
c. $-2xz^3(6x^3z + x^2z^2 - xz^3 + 7z^4)$ **d.** $(-m^4 - 2.5)(4.1m^3)$

Strategy To find each product, we will multiply each term of the polynomial by the monomial.

Why We use the distributive property to multiply a monomial and a polynomial.

Solution **a.** Multiply each term of the binomial $16n^{15} + n^{10}$ by the monomial $3n^6$.

$$3n^6(16n^{15} + n^{10}) = 3n^6(16n^{15}) + 3n^6(n^{10})$$ Distribute the multiplication by $3n^6$.

$$= 48n^{21} + 3n^{16}$$ Multiply the monomials.

Teaching Tip: As you work through Examples 2–8, have your students predict how many monomial · monomial multiplications will have to be performed prior to working each problem.

b. Multiply each term of the trinomial $3a^2 - 5a + 2$ by the monomial $3a^2$.

$$3a^2(3a^2 - 5a + 2)$$

$$= 3a^2(3a^2) + 3a^2(-5a) + 3a^2(2)$$ Distribute the multiplication by $3a^2$.

$$= 9a^4 - 15a^3 + 6a^2$$ Multiply the monomials.

Success Tip

The rectangle below can be used to picture polynomial multiplication. The total area is $x(x + 2)$ and the sum of the two smaller areas is $x^2 + 2x$. Thus,

$$x(x + 2) = x^2 + 2x$$

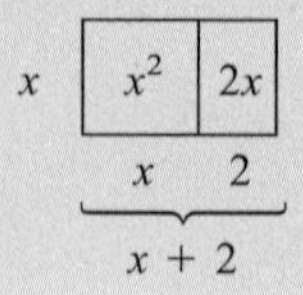

c. Multiply each term of $6x^3z + x^2z^2 - xz^3 + 7z^4$ by the monomial $-2xz^3$.

$$\mathbf{-2xz^3}(6x^3z + x^2z^2 - xz^3 + 7z^4)$$
$$= \mathbf{-2xz^3}(6x^3z) \mathbf{-2xz^3}(x^2z^2) \mathbf{-2xz^3}(-xz^3) \mathbf{-2xz^3}(7z^4)$$
$$= -12x^4z^4 - 2x^3z^5 + 2x^2z^6 - 14xz^7 \quad \text{Multiply the monomials.}$$

d. Multiply each term of the binomial $-m^4 - 2.5$ by the monomial $4.1m^3$.

$$(-m^4 - 2.5)(\mathbf{4.1m^3}) = -m^4(\mathbf{4.1m^3}) - 2.5(\mathbf{4.1m^3}) \quad \text{Distribute the multiplication by } 4.1m^3.$$
$$= -4.1m^7 - 10.25m^3 \quad \text{Multiply the monomials.}$$

Teaching Example 2 Multiply:
a. $27t^{18}(2t^5 + t^4)$
b. $8f^4(3f^2 - 4f - 7)$
c. $-9ab^2(6a^3b + a^2b^2 - ab^3 + 9b^4)$
d. $(-n^6 - 3.7)(5.2n^5)$
Answers:
a. $54t^{23} + 27t^{22}$
b. $24f^6 - 32f^5 - 56f^4$
c. $-54a^4b^3 - 9a^3b^4 + 9a^2b^5 - 81ab^6$
d. $-5.2b^{11} - 19.24n^5$

Self Check 2 Multiply: **a.** $22x(10x^5 + x^4)$ $220x^6 + 22x^5$
b. $5c^2(4c^3 - 9c - 8)$ $20c^5 - 45c^3 - 40c^2$
c. $-s^2t^2(-s^4t^2 + s^3t^3 - s^2t^4 + 7s)$ $s^6t^4 - s^5t^5 + s^4t^6 - 7s^3t^2$
d. $(w^7 - 2w)6w^5$ $6w^{12} - 12w^6$

Now Try Problems 27, 33, and 35

EXAMPLE 3 Find a polynomial that represents the area of the parallelogram.

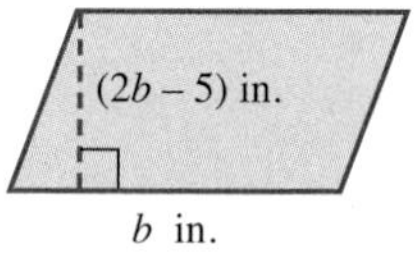

Strategy We will multiply the length of the base of the parallelogram by its height.

Why The area of a parallelogram is equal to the product of the length of its base and its height.

Solution

The Language of Algebra

A **parallelogram** is a four-sided figure whose opposite sides are parallel.

$$\text{Area} = \textbf{base} \cdot \textbf{height} \quad \text{This is the formula for the area of a parallelogram.}$$
$$= b(2b - 5) \quad b \text{ is the length of the base. Substitute } 2b - 5 \text{ for the height.}$$
$$= 2b^2 - 5b \quad \text{Distribute the multiplication by } b.$$

The area of the parallelogram is $(2b^2 - 5b)$ square inches, which can be written as $(2b^2 - 5b)$ in.^{2}.

Teaching Example 3 Find a polynomial that represents the area of the parallelogram.

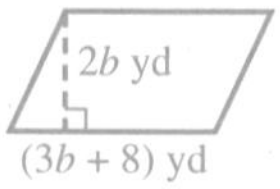

Answer: $(6b^2 + 16b)$ yd^2

Self Check 3 Find a polynomial that represents the area of a rectangle with length n^3 meters and width $(3n^2 + 2n - 9)$ meters. $(3n^5 + 2n^4 - 9n^3)$ m^2

Now Try Problem 37

3 Multiply Binomials.

The distributive property also can be used to multiply binomials. For example, to multiply $2a + 4$ and $3a + 5$, we think of $2a + 4$ as a single quantity and distribute it over each term of $3a + 5$.

$$(2a + 4)(3a + 5) = (2a + 4)3a + (2a + 4)5 \quad \text{Read as "the quantity of } 2a + 4 \text{ times the quantity of } 3a + 5\text{."}$$
$$= (2a + 4)\mathbf{3a} + (2a + 4)\mathbf{5}$$
$$= (2a)\mathbf{3a} + (4)\mathbf{3a} + (2a)\mathbf{5} + (4)\mathbf{5} \quad \text{Distribute the multiplication by } 3a \text{ and by } 5.$$
$$= 6a^2 + 12a + 10a + 20 \quad \text{Multiply the monomials.}$$
$$= 6a^2 + 22a + 20 \quad \text{Combine like terms.}$$

In the third line of the solution, notice that each term of $3a + 5$ has been multiplied by each term of $2a + 4$. This example suggests the following rule.

Multiplying Two Binomials

To multiply two binomials, multiply each term of one binomial by each term of the other binomial, and then combine like terms.

EXAMPLE 4 Multiply: $(5x - 8)(x + 1)$

Strategy To find the product, we will multiply $x + 1$ by $5x$ and by -8.

Why To multiply two binomials, each term of one binomial must be multiplied by each term of the other binomial.

Solution

Notation

We will write the answers to polynomial multiplication in descending powers.

$(5x - 8)(x + 1) = 5x(x + 1) - 8(x + 1)$ Read as "the quantity of $5x - 8$ times the quantity of $x + 1$." Multiply $x + 1$ by $5x$ and multiply $x + 1$ by -8.

$= 5x^2 + 5x - 8x - 8$ Distribute the multiplication by $5x$. Distribute the multiplication by -8.

$= 5x^2 - 3x - 8$ Combine like terms.

Teaching Example 4 Multiply: $(6a - 7)(a + 3)$
Answer: $6a^2 + 11a - 21$

Self Check 4 Multiply: $(9y + 3)(y - 4)$ $9y^2 - 33y - 12$

Now Try Problem 45

The Language of Algebra

An **acronym** is an abbreviation of several words in such a way that the abbreviation itself forms a word. The *acronym* FOIL helps us remember the order to follow when multiplying two binomials: First, Outer, Inner, Last.

We can use a shortcut method, called the **FOIL method,** to multiply binomials. FOIL is an acronym for **F**irst terms, **O**uter terms, **I**nner terms, **L**ast terms. The FOIL method is a form of the distributive property. To use the FOIL method to multiply $2a + 4$ by $3a + 5$, we

1. Multiply the **F**irst terms $2a$ and $3a$ to obtain $6a^2$,
2. Multiply the **O**uter terms $2a$ and 5 to obtain $10a$,
3. Multiply the **I**nner terms 4 and $3a$ to obtain $12a$, and
4. Multiply the **L**ast terms 4 and 5 to obtain 20.

Then we simplify the resulting polynomial by combining like terms, if possible.

Outer
First
F O I L

$(2a + 4)(3a + 5) = 2a(3a) + 2a(5) + 4(3a) + 4(5)$ Distribute $2a$ over $3a + 5$. Distribute 4 over $3a + 5$.

Inner Last

$= 6a^2 + 10a + 12a + 20$ Multiply the monomials.

$= 6a^2 + 22a + 20$ Combine like terms: $10a + 12a = 22a$.

EXAMPLE 5 Multiply: **a.** $(x + 5)(x + 7)$ **b.** $(3x + 4)(2x - 3)$

c. $\left(2r - \frac{1}{2}\right)\left(2r + \frac{5}{2}\right)$ **d.** $(3a^2 - 7b)(a^2 - b)$

Strategy We will use the FOIL method.

Why In each case we are to find the product of two binomials, and the FOIL method is a shortcut for multiplying two binomials.

Solution

Success Tip

The area of the large rectangle is given by $(x + 5)(x + 7)$. The sum of the areas of the smaller rectangles is $x^2 + 7x + 5x + 35$ or $x^2 + 12x + 35$. Thus,

$$(x + 5)(x + 7) = x^2 + 12x + 35$$

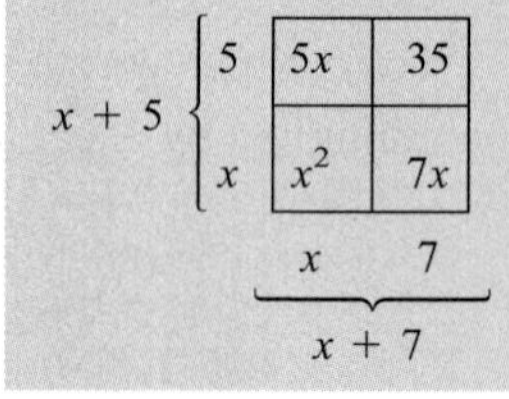

a.

$$\begin{aligned}(x + 5)(x + 7) &= \overset{F}{x(x)} + \overset{O}{x(7)} + \overset{I}{5(x)} + \overset{L}{5(7)} \\ &= x^2 + 7x + 5x + 35 && \text{Multiply the monomials.} \\ &= x^2 + 12x + 35 && \text{Combine like terms.}\end{aligned}$$

b.

$$\begin{aligned}(3x + 4)(2x - 3) &= \overset{F}{3x(2x)} + \overset{O}{3x(-3)} + \overset{I}{4(2x)} + \overset{L}{4(-3)} \\ &= 6x^2 - 9x + 8x - 12 && \text{Multiply the monomials.} \\ &= 6x^2 - x - 12 && \text{Combine like terms.}\end{aligned}$$

c.

$$\begin{aligned}\left(2r - \frac{1}{2}\right)\left(2r + \frac{5}{2}\right) &= \overset{F}{2r(2r)} + \overset{O}{2r\left(\frac{5}{2}\right)} - \overset{I}{\frac{1}{2}(2r)} - \overset{L}{\frac{1}{2}\left(\frac{5}{2}\right)} \\ &= 4r^2 + 5r - r - \frac{5}{4} && \text{Multiply the monomials.} \\ &= 4r^2 + 4r - \frac{5}{4} && \text{Combine like terms. We cannot clear the fraction because we began with an expression, not an equation.}\end{aligned}$$

d.

$$\begin{aligned}(3a^2 - 7b)(a^2 - b) &= \overset{F}{3a^2(a^2)} + \overset{O}{3a^2(-b)} - \overset{I}{7b(a^2)} - \overset{L}{7b(-b)} \\ &= 3a^4 - 3a^2b - 7a^2b + 7b^2 && \text{Multiply the monomials.} \\ &= 3a^4 - 10a^2b + 7b^2 && \text{Combine like terms.}\end{aligned}$$

Teaching Example 5 Multiply: **a.** $(x + 4)(x + 3)$ **b.** $(4x + 2)(5x - 6)$ **c.** $(3x + \frac{2}{3})(6x - \frac{1}{3})$ **d.** $(7m^2 - 4n)(m^2 - n)$
Answers: **a.** $x^2 + 7x + 12$ **b.** $20x^2 - 14x - 12$ **c.** $18x^2 + 3x - \frac{2}{9}$ **d.** $7m^4 - 11m^2n + 4n^2$

Self Check 5 Multiply: **a.** $(y + 3)(y + 1)$ $y^2 + 4y + 3$
b. $(2a - 1)(3a + 2)$ $6a^2 + a - 2$
c. $\left(4x - \frac{1}{2}\right)\left(4x + \frac{3}{4}\right)$ $16x^2 + x - \frac{3}{8}$
d. $(5y^3 - 2b)(2y^3 - 7b)$ $10y^6 - 39by^3 + 14b^2$

Now Try Problems 47, 51, and 55

4 Multiply Polynomials.

To develop a general rule for multiplying any two polynomials, we will find the product of $2x + 3$ and $3x^2 + 3x + 5$. In the solution, the distributive property is used four times.

$$\begin{aligned}(2x + 3)(3x^2 + 3x + 5) &= (2x + 3)3x^2 + (2x + 3)3x + (2x + 3)5 && \text{Distribute } (2x + 3). \\ &= (2x + 3)3x^2 + (2x + 3)3x + (2x + 3)5 \\ &= (2x)3x^2 + (3)3x^2 + (2x)3x + (3)3x + (2x)5 + (3)5 && \text{Distribute.} \\ &= 6x^3 + 9x^2 + 6x^2 + 9x + 10x + 15 && \text{Multiply the monomials.} \\ &= 6x^3 + 15x^2 + 19x + 15 && \text{Combine like terms.}\end{aligned}$$

Success Tip

In this section, you will see that every polynomial multiplication is basically a series of monomial multiplications. Then, if necessary, we combine like terms.

In the third line of the solution, note that each term of $3x^2 + 3x + 5$ has been multiplied by each term of $2x + 3$. This example suggests the following rule.

Multiplying Two Polynomials	To multiply two polynomials, multiply each term of one polynomial by each term of the other polynomial, and then combine like terms.

EXAMPLE 6 Multiply: $(7y + 3)(6y^2 - 8y + 1)$

Strategy We will multiply each term of the trinomial, $6y^2 - 8y + 1$, by each term of the binomial, $7y + 3$.

Why To multiply two polynomials, we must multiply each term of one polynomial by each term of the other polynomial.

Solution

$(7y + 3)(6y^2 - 8y + 1)$ Read as "the quantity of $7y + 3$ times the quantity of $6y^2 - 8y + 1$."

$= 7y(6y^2) + 7y(-8y) + 7y(1) + 3(6y^2) + 3(-8y) + 3(1)$

$= 42y^3 - 56y^2 + 7y + 18y^2 - 24y + 3$ Multiply the monomials.

$= 42y^3 - 38y^2 - 17y + 3$ Combine like terms.

Success Tip

The FOIL method cannot be applied here—only to products of two binomials.

Teaching Example 6 Multiply: $(3d - 1)(2d^2 - 5d - 3)$
Answer: $6d^3 - 17d^2 - 4d + 3$

Self Check 6 Multiply: $(3a^2 - 1)(2a^4 - a^2 - a)$ $6a^6 - 5a^4 - 3a^3 + a^2 + a$

Now Try Problem 59

It is often convenient to multiply polynomials using a vertical form similar to that used to multiply whole numbers.

EXAMPLE 7 Multiply using vertical form: **a.** $(3a^2 - 4a + 7)(2a + 5)$
b. $(6y^3 - 5y + 4)(-4y^2 - 3)$

Strategy First, we will write one polynomial underneath the other and draw a horizontal line beneath them. Then, we will multiply each term of the upper polynomial by each term of the lower polynomial.

Why *Vertical form* means to use an approach similar to that used in arithmetic to multiply two numbers.

Solution

a. Multiply:

$$\begin{array}{rrrrl} & 3a^2 & -\ 4a & +\ 7 & \\ & & 2a & +\ 5 & \\ \hline & \mathbf{15a^2} & \mathbf{-\ 20a} & \mathbf{+\ 35} & \text{Multiply } 3a^2 - 4a + 7 \text{ by } 5. \\ \mathbf{6a^3} & \mathbf{-\ 8a^2} & \mathbf{+\ 14a} & & \text{Multiply } 3a^2 - 4a + 7 \text{ by } 2a. \text{ Line up like terms.} \\ \hline 6a^3 & +\ 7a^2 & -\ 6a & +\ 35 & \text{In each column, combine like terms.} \end{array}$$

b. With this method, it is often necessary to leave a space for a missing term to align like terms vertically.

Multiply:

$$\begin{array}{rrrrrl} & 6y^3 & & -\ 5y & +\ 4 & \\ & & -4y^2 & & -\ 3 & \\ \hline & \mathbf{-18y^3} & & \mathbf{+\ 15y} & \mathbf{-\ 12} & \text{Multiply } 6y^3 - 5y + 4 \text{ by } -3. \\ \mathbf{-24y^5} & \mathbf{+\ 20y^3} & \mathbf{-\ 16y^2} & & & \text{Multiply } 6y^3 - 5y + 4 \text{ by } -4y^2. \\ \hline -24y^5 & +\ 2y^3 & -\ 16y^2 & +\ 15y & -\ 12 & \text{Leave a space for any missing powers of } y. \text{ In each column, combine like terms.} \end{array}$$

Success Tip

Multiplying two polynomials in vertical form is much like multiplying two numbers in arithmetic.

$$\begin{array}{r} 347 \\ \times\ 25 \\ \hline 1{,}735 \\ +694 \\ \hline 8{,}675 \end{array}$$

The Language of Algebra

The pair of polynomials written below the first horizontal line, $-18y^3 + 15y - 12$ and $-24y^5 + 20y^3 - 16y^2$, are called **partial products.**

Teaching Example 7 Multiply using vertical form:
a. $(5x - 3)(2x^2 + x - 4)$
b. $(4x^2 - 3x + 2)(-3x^2 + 4)$
Answers:
a. $10x^3 - x^2 - 23x + 12$
b. $-12x^4 + 9x^3 + 10x^2 - 12x + 8$

Self Check 7 Multiply using vertical form:
a. $(3x + 2)(2x^2 - 4x + 5)$ $6x^3 - 8x^2 + 7x + 10$
b. $(-2x^2 + 3)(2x^2 - 4x - 1)$ $-4x^4 + 8x^3 + 8x^2 - 12x - 3$

Now Try Problem 65

Multiplying Three Polynomials To multiply three polynomials, multiply any two of them, and then multiply that result by the third polynomial.

EXAMPLE 8 Multiply: $-3a(4a + 1)(a - 7)$

Strategy We will find the product of $4a + 1$ and $a - 7$ and then multiply that result by $-3a$.

Why It is wise to perform the most difficult multiplication first. (In this case, that would be the product of the two binomials). Save the simpler multiplication by $-3a$ for last.

Solution

$$-3a(4a + 1)(a - 7) = -3a(4a^2 - 28a + a - 7)$$ Multiply the two binomials.

$$= -3a(4a^2 - 27a - 7)$$ Combine like terms within the parentheses: $-28a + a = -27a$.

$$= -12a^3 + 81a^2 + 21a$$ Distribute the multiplication by $-3a$.

Teaching Example 8 Multiply: $-3x(x - 5)(2x + 1)$
Answer: $-6x^3 + 27x^2 + 15x$

Self Check 8 Multiply: $-2y(y + 3)(3y - 2)$ $-6y^3 - 14y^2 + 12y$

Now Try Problem 69

SECTION 5.6 STUDY SET

VOCABULARY

Fill in the blanks.

1. $(2x^3)(3x^4)$ is the product of two monomials and $(2a - 4)(3a + 5)$ is the product of two binomials.
2. We read $(x + 7)(2x - 3)$ as "the quantity of $x + 7$ times the quantity of $2x - 3$."
3. In the acronym FOIL, F stands for first terms, O for outer terms, I for inner terms, and L for last terms.
4. $(2a - 4)(3a^2 + 5a - 1)$ is the product of a binomial and a trinomial.

CONCEPTS

Fill in the blanks.

5. a. To multiply two polynomials, multiply each term of one polynomial by each term of the other polynomial, and then combine like terms.
   b. When multiplying three polynomials, we begin by multiplying any two of them, and then we multiply that result by the third polynomial.

6. Label each arrow using one of the letters F, O, I, or L. Then fill in the blanks.

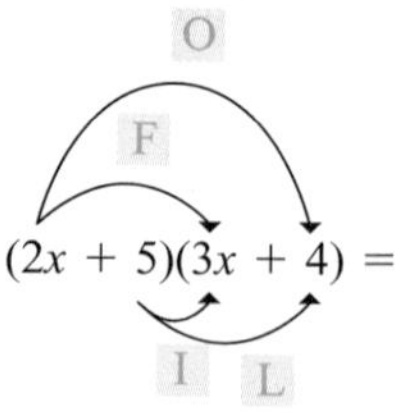

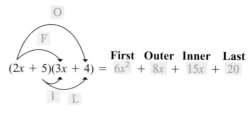

First Outer Inner Last
$(2x + 5)(3x + 4) = 6x^2 + 8x + 15x + 20$

7. Simplify each polynomial by combining like terms.
   a. $6x^2 - 8x + 9x - 12$ $6x^2 + x - 12$
   b. $5x^4 + 3ax^2 + 5ax^2 + 3a^2$ $5x^4 + 8ax^2 + 3a^2$
8. $(3a)(2a^2)$ can be classified as a monomial · monomial. Classify the following products by identifying the types of polynomial factors.
   a. $6x(x - 7)$ Monomial · binomial
   b. $(9a + 1)(5a - 3)$ Binomial · binomial
   c. $(c - d)(c^2 - c + d)$ Binomial · trinomial
   d. $6m(m^2 + 1)(m^2 - 1)$ Monomial · binomial · binomial

NOTATION

Complete each solution.

9. $(9n^3)(8n^2) = (9 \cdot 8)(n^3 \cdot n^2) = 72n^5$

10. $7x(3x^2 - 2x + 5) = 7x(3x^2) - 7x(2x) + 7x(5)$
$= 21x^3 - 14x^2 + 35x$

▶ 11. $(2x + 5)(3x - 2) = 2x(3x) - 2x(2) + 5(3x) - 5(2)$
$= 6x^2 - 4x + 15x - 10$
$= 6x^2 + 11x - 10$

▶ 12.
$$\begin{array}{r} 3x^2 + 4x - 2 \\ 2x + 3 \\ \hline 9x^2 + 12x - 6 \\ 6x^3 + 8x^2 - 4x \quad \\ \hline 6x^3 + 17x^2 + 8x - 6 \end{array}$$

GUIDED PRACTICE

Multiply. See Example 1.

13. $5m \cdot m$ $5m^2$
14. $4s \cdot s$ $4s^2$
15. $(3x^2)(4x^3)$ $12x^5$
▶ 16. $(-2a^3)(11a^2)$ $-22a^5$
17. $(1.2c^3)(5c^3)$ $6c^6$
▶ 18. $(2.5h^4)(2h^4)$ $5h^8$
▶ 19. $(3b^2)(-2b)(4b^3)$ $-24b^6$
20. $(3y)(7y^2)(-y^4)$ $-21y^7$
21. $(2x^2y^3)(4x^3y^2)$ $8x^5y^5$
▶ 22. $(-5x^3y^6)(2x^2y^2)$ $-10x^5y^8$
▶ 23. $(8a^5)\left(-\frac{1}{4}a^6\right)$ $-2a^{11}$
24. $\left(-\frac{2}{3}x^6\right)(9x^3)$ $-6x^9$

Multiply. See Example 2.

▶ 25. $3x(x + 4)$
$3x^2 + 12x$
26. $3a(a + 2)$
$3a^2 + 6a$
27. $-4t(t^2 - 7)$
$-4t^3 + 28t$
28. $-6s(s^2 - 3)$
$-6s^3 + 18s$
▶ 29. $-2x^3(3x^2 - x + 1)$
$-6x^5 + 2x^4 - 2x^3$
▶ 30. $-4b^3(2b^2 - 2b + 2)$
$-8b^5 + 8b^4 - 8b^3$
31. $\frac{5}{8}t^2(t^6 + 8t^2)$
$\frac{5}{8}t^8 + 5t^4$
▶ 32. $\frac{4}{9}a^2(9a^3 + a^2)$
$4a^5 + \frac{4}{9}a^4$
33. $-4x^2z(3x^2 + z^2 + xz - 1)$
$-12x^4z - 4x^2z^3 - 4x^3z^2 + 4x^2z$
▶ 34. $-3x^2y(x^2 + y^2 + xy - 1)$
$-3x^4y - 3x^2y^3 - 3x^3y^2 + 3x^2y$
35. $(x^2 - 12x)(6x^{12})$
$6x^{14} - 72x^{13}$
36. $(w^9 - 11w)(2w^7)$
$2w^{16} - 22w^8$

Find a polynomial that represents the area of the parallelogram or rectangle. See Example 3.

▶ 37.
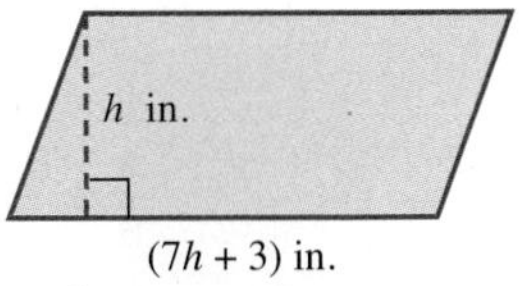

$(7h^2 + 3h)$ in.^{2}

38.
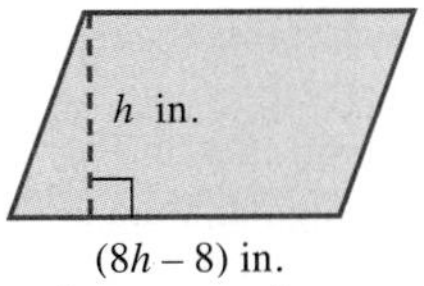

$(8h^2 - 8h)$ in.^{2}

39.
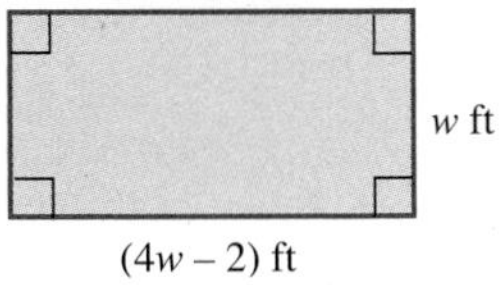

$(4w^2 - 2w)$ ft^2

▶ 40.
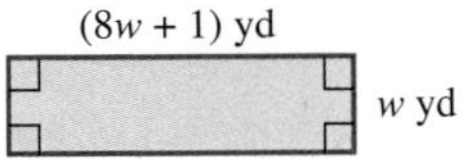

$(8w^2 + w)$ yd^2

Multiply. See Examples 4 and 5.

41. $(y + 3)(y + 5)$
$y^2 + 8y + 15$
▶ 42. $(a + 4)(a + 5)$
$a^2 + 9a + 20$
43. $(m + 6)(m - 9)$
$m^2 - 3m - 54$
44. $(n + 8)(n - 10)$
$n^2 - 2n - 80$
▶ 45. $(4y - 5)(y + 7)$
$4y^2 + 23y - 35$
▶ 46. $(3x - 4)(x + 5)$
$3x^2 + 11x - 20$
▶ 47. $(2x - 3)(6x - 5)$
$12x^2 - 28x + 15$
48. $(5x - 3)(2x - 3)$
$10x^2 - 21x + 9$
49. $(3.8y - 1)(2y - 1)$
$7.6y^2 - 5.8y + 1$
▶ 50. $(2.6x - 3)(2x - 1)$
$5.2x^2 - 8.6x + 3$
▶ 51. $\left(6m - \frac{2}{3}\right)\left(3m - \frac{4}{3}\right)$
$18m^2 - 10m + \frac{8}{9}$
52. $\left(8t - \frac{1}{2}\right)\left(4t - \frac{5}{2}\right)$
$32t^2 - 22t + \frac{5}{4}$
53. $(t^2 - 3)(t^2 - 4)$
$t^4 - 7t^2 + 12$
▶ 54. $(s^3 - 6)(s^3 - 8)$
$s^6 - 14s^3 + 48$
55. $(3a - 2b)(4a + b)$
$12a^2 - 5ab - 2b^2$
▶ 56. $(2t + 3s)(3t - s)$
$6t^2 + 7st - 3s^2$

Multiply. See Example 6.

57. $(x + 2)(x^2 - 2x + 3)$ $x^3 - x + 6$
▶ 58. $(x - 5)(x^2 + 2x - 3)$ $x^3 - 3x^2 - 13x + 15$
59. $(4t + 3)(t^2 + 2t + 3)$ $4t^3 + 11t^2 + 18t + 9$
▶ 60. $(3x + 1)(2x^2 - 3x + 1)$ $6x^3 - 7x^2 + 1$
▶ 61. $(x^2 + 6x + 7)(2x - 5)$ $2x^3 + 7x^2 - 16x - 35$
62. $(y^2 - 2y + 1)(4y + 8)$ $4y^3 - 12y + 8$
63. $(r^2 - r + 3)(r^2 - 4r - 5)$ $r^4 - 5r^3 + 2r^2 - 7r - 15$
▶ 64. $(w^2 + w - 9)(w^2 - w + 3)$ $w^4 - 7w^2 + 12w - 27$

Multiply using vertical form. See Example 7.

65.
$$\begin{array}{r} x^2 - 2x + 1 \\ x + 2 \\ \hline x^3 - 3x + 2 \end{array}$$
▶ 66.
$$\begin{array}{r} 5r^2 + r + 6 \\ 2r - 1 \\ \hline 10r^3 - 3r^2 + 11r - 6 \end{array}$$
67.
$$\begin{array}{r} 4x^2 + 3x - 4 \\ 3x + 2 \\ \hline 12x^3 + 17x^2 - 6x - 8 \end{array}$$
▶ 68.
$$\begin{array}{r} x^2 - x + 1 \\ x + 1 \\ \hline x^3 + 1 \end{array}$$

Multiply. See Example 8.

69. $4x(2x + 1)(x - 2)$ $8x^3 - 12x^2 - 8x$
70. $5a(3a - 2)(2a + 3)$ $30a^3 + 25a^2 - 30a$
71. $-3a(a + b)(a - b)$ $-3a^3 + 3ab^2$
▶ 72. $-2r(r + s)(r + s)$ $-2r^3 - 4r^2s - 2rs^2$
73. $(-2a^2)(-3a^3)(3a - 2)$ $18a^6 - 12a^5$
▶ 74. $(3x)(-2x^2)(x + 4)$ $-6x^4 - 24x^3$
75. $(x - 4)(x + 1)(x - 3)$ $x^3 - 6x^2 + 5x + 12$
▶ 76. $(x + 6)(x - 2)(x - 4)$ $x^3 - 28x + 48$

TRY IT YOURSELF

Multiply.

77. $(5x - 2)(6x - 1)$
$30x^2 - 17x + 2$
78. $(8x - 1)(3x - 7)$
$24x^2 - 59x + 7$
79. $(3x^2 + 4x - 7)(2x^2)$
$6x^4 + 8x^3 - 14x^2$
80. $(2y^2 - 7y - 8)(3y^3)$
$6y^5 - 21y^4 - 24y^3$
▶ 81. $2(t + 4)(t - 3)$
$2t^2 + 2t - 24$
82. $4(x + 7)(x - 6)$
$4x^2 + 4x - 168$

83. $\begin{array}{r} 2a^2 + 3a + 1 \\ \underline{3a^2 - 2a + 4} \end{array}$
$6a^4 + 5a^3 + 5a^2 + 10a + 4$

84. $\begin{array}{r} 3y^2 + 2y - 4 \\ \underline{2y^2 - 4y + 3} \end{array}$
$6y^4 - 8y^3 - 7y^2 + 22y - 12$

85. $(t + 2s)(9t - 3s)$
$9t^2 + 15st - 6s^2$

▶ 86. $(4t - u)(3t + u)$
$12t^2 + tu - u^2$

87. $\left(\frac{1}{2}a\right)(4a^4)(a^5)$ $2a^{10}$

▶ 88. $(12b)\left(\frac{7}{6}b\right)(b^4)$ $14b^6$

▶ 89. $\left(4a - \frac{5}{4}r\right)\left(4a + \frac{3}{4}r\right)$
$16a^2 - 2ar - \frac{15}{16}r^2$

90. $\left(5c - \frac{2}{5}t\right)\left(10c + \frac{1}{5}t\right)$
$50c^2 - 3ct - \frac{2}{25}t^2$

▶ 91. $(a + b)(a + b)$
$a^2 + 2ab + b^2$

92. $(m + n)(m + n)$
$m^2 + 2mn + n^2$

93. $(x + 6)(x^3 + 5x^2 - 4x - 4)$ $x^4 + 11x^3 + 26x^2 - 28x - 24$

94. $(x - 8)(x^3 - 4x^2 - 2x - 2)$ $x^4 - 12x^3 + 30x^2 + 14x + 16$

95. $9x^2(x^2 - 2x + 6)$
$9x^4 - 18x^3 + 54x^2$

▶ 96. $4y^2(y^2 + 5y - 10)$
$4y^4 + 20y^3 - 40y^2$

97. $4y(y + 3)(y + 7)$
$4y^3 + 40y^2 + 84y$

▶ 98. $2t(t + 8)(t + 10)$
$2t^3 + 36t^2 + 160t$

99. $0.3p^5(0.4p^4 - 6p^2)$
$0.12p^9 - 1.8p^7$

100. $0.5u^5(0.4u^6 - 0.5u^3)$
$0.2u^{11} - 0.25u^8$

101. $8.2pq(2pq - 3p + 5q)$
$16.4p^2q^2 - 24.6p^2q + 41pq^2$

▶ 102. $5.3ab(2ab + 6a - 3b)$
$10.6a^2b^2 + 31.8a^2b - 15.9ab^2$

103. $(-3x + y)(x^2 - 8xy + 16y^2)$ $-3x^3 + 25x^2y - 56xy^2 + 16y^3$

▶ 104. $(3x - y)(x^2 + 3xy - y^2)$ $3x^3 + 8x^2y - 6xy^2 + y^3$

Look Alikes . . .

Perform the indicated operations to simplify each expression, if possible.

105. a. $(x - 2) + (x^2 + 2x + 4)$
$x^2 + 3x + 2$
b. $(x - 2)(x^2 + 2x + 4)$
$x^3 - 8$

▶ 106. a. $(a + 3) + (a^2 - 3a + 9)$
$a^2 - 2a + 12$
b. $(a + 3)(a^2 - 3a + 9)$
$a^3 + 27$

107. a. $(6x^2z^5) - (-3xz^3)$
Does not simplify
b. $(6x^2z^5)(-3xz^3)$
$-18x^3z^8$

▶ 108. a. $(-5r^4t^2) - (2r^2t)$
Does not simplify
b. $(-5r^4t^2)(2r^2t)$
$-10r^6t^3$

109. a. $(2x^2 - x) - (3x^2 - 3x)$
$-x^2 + 2x$
b. $(2x^2 - x)(3x^2 - 3x)$
$6x^4 - 9x^3 + 3x^2$

110. a. $(4.9a - b) - (2a + b)$
$2.9a - 2b$
b. $(4.9a - b)(2a + b)$
$9.8a^2 + 2.9ab - b^2$

111. a. $3a + (4a - 1) + (6a + 2)$ $13a + 1$
b. $3a(4a - 1)(6a + 2)$ $72a^3 + 6a^2 - 6a$

▶ 112. a. $\left(\frac{1}{2}c - 3d\right) + \left(\frac{3}{4}c + d\right)$
$\frac{5}{4}c - 2d$
b. $\left(\frac{1}{2}c - 3d\right)\left(\frac{3}{4}c + d\right)$
$\frac{3}{8}c^2 - \frac{7}{4}cd - 3d^2$

APPLICATIONS

▶ 113. **Stamps.** Find a polynomial that represents the area of the stamp.
$(6x^2 + x - 1)\text{ cm}^2$

▶ 114. **Parking.** Find a polynomial that represents the total area of the van-accessible parking space and its access aisle. $(2x^2 + 20x)\text{ ft}^2$

▶ 115. **Sunglasses.** An ellipse is an oval-shaped curve. The area of an ellipse is approximately $0.785lw$, where l is its length and w is its width. Find a polynomial that represents the approximate area of one of the elliptical-shaped lenses of the sunglasses. $(0.785x^2 - 0.785)\text{ in.}^2$

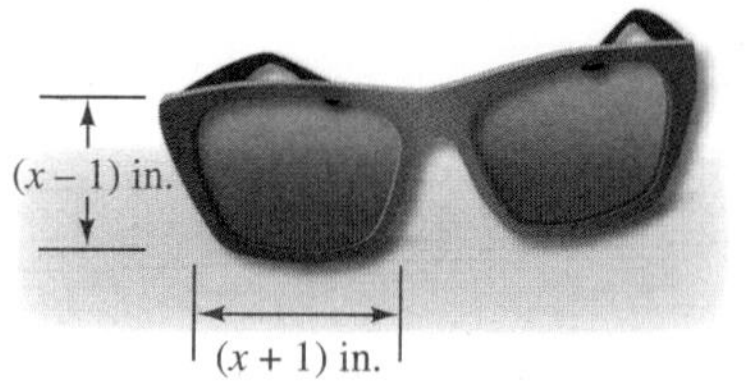

▶ 116. **Gardening.** Refer to the illustration below.

a. Find the area of the region planted with corn, tomatoes, beans, and carrots. Add your answers to find the total area of the garden.
$x^2\text{ ft}^2, 6x\text{ ft}^2, 5x\text{ ft}^2, 30\text{ ft}^2; (x^2 + 11x + 30)\text{ ft}^2$

b. Find the length and width of the garden. Multiply your answers to find its area.
$(x + 6)\text{ ft}, (x + 5)\text{ ft}; (x^2 + 11x + 30)\text{ ft}^2$

c. How do the answers from parts (a) and (b) for the area of the garden compare? They are the same.

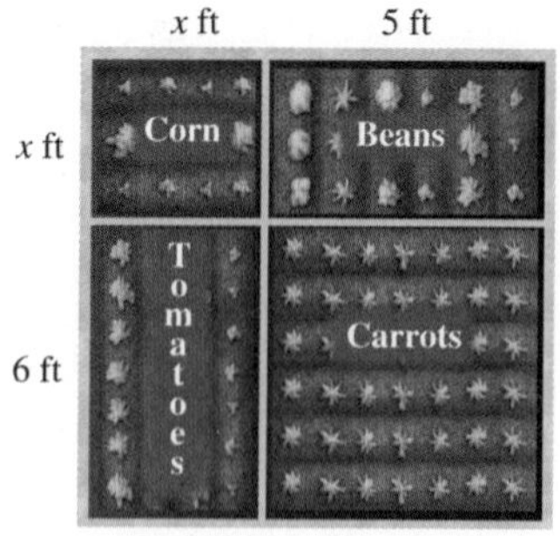

117. **Luggage.** Find a polynomial that represents the volume of the garment bag. (Recall that the formula for the volume of a rectangular solid is $V = lwh$.)
$(2x^3 - 4x^2 - 6x)\text{ in.}^3$

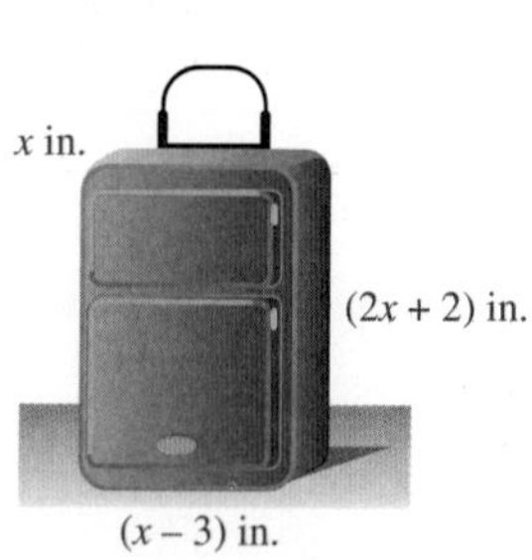

▶ **118. Baseball.** Find a polynomial that represents the volume within the batting cage. $(45x^3 + 12x^2 - 19x - 6)\text{ ft}^3$

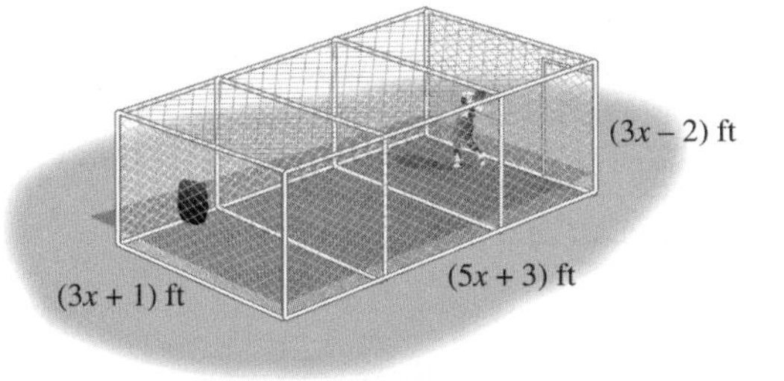

WRITING

119. Is the product of a monomial and a monomial always a monomial? Explain.

120. Explain this diagram.

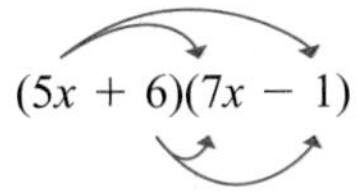

121. Explain why the FOIL method cannot be used to find $(3x + 2)(4x^2 - x + 10)$.

▶ **122.** Explain the error: ~~$(x + 3)(x - 2) = x^2 - 6$~~

123. Explain why the vertical form used in algebra to multiply $2x^2 + 3x + 1$ and $3x + 2$ is similar to the vertical form used in arithmetic to multiply 231 and 32.

124. Would the OLIF method give the same result as the FOIL method when multiplying two binomials? Explain why or why not.

REVIEW

▶ **125.** What is the slope of

a. Line 1? 1

b. Line 2? Undefined

c. Line 3? $-\frac{2}{3}$

▶ **d.** the x-axis? 0

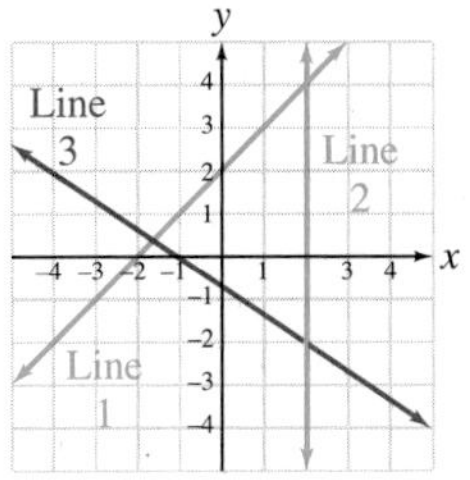

▶ **126. a.** What is the y-intercept of Line 1? $(0, 2)$

b. What is the x-intercept of Line 1? $(-2, 0)$

CHALLENGE PROBLEMS

127. a. Find each of the following products.

i. $(x - 1)(x + 1)$ $x^2 - 1$

ii. $(x - 1)(x^2 + x + 1)$ $x^3 - 1$

iii. $(x - 1)(x^3 + x^2 + x + 1)$ $x^4 - 1$

b. Write a product of two polynomials such that the result is $x^5 - 1$. $(x - 1)(x^4 + x^3 + x^2 + x + 1)$

▶ **128.** Solve: $(y - 1)(y + 6) = (y - 3)(y - 2) + 8$ 2

SECTION 5.7 Special Products

OBJECTIVES

1. Square a binomial.
2. Multiply the sum and difference of the same two terms.
3. Find higher powers of binomials.
4. Simplify expressions containing polynomial multiplication.

ARE YOU READY? *Are You Ready? exercises available online at www.webassign.net/brookscole*

The following problems review some basic skills that are needed when finding special products.

1. Simplify by combining like terms: $x^2 + 2x + 2x + 4$ $x^2 + 4x + 4$

2. In each expression, identify the base and the exponent:

a. $(x + 4)^2$ Base: $x + 4$; exponent: 2

b. $(x - 1)^3$ Base: $x - 1$; exponent: 3

3. Multiply: $(a + 5)(a + 5)$ $a^2 + 10a + 25$

4. Multiply: $(a + 5)(a - 5)$ $a^2 - 25$

Certain products of binomials, called **special products,** occur so often that it is worthwhile to learn their forms.

1 Square a Binomial.

To develop a rule to find the *square of a binomial sum,* we consider $(x + y)^2$. We can use the definition of exponent and the procedure for multiplying two binomials to find the product.

$$(x + y)^2 = (x + y)(x + y) \quad \text{In } (x + y)^2\text{, the base is } (x + y) \text{ and the exponent is 2.}$$
$$= x^2 + xy + xy + y^2 \quad \text{Multiply the binomials.}$$
$$= x^2 + 2xy + y^2 \quad \text{Combine like terms: } xy + xy = 1xy + 1xy = 2xy.$$

Success Tip

The illustration can be used to visualize a special product. The area of the large square is $(x + y)(x + y) = (x + y)^2$. The sum of the four smaller areas is $x^2 + xy + xy + y^2$ or $x^2 + 2xy + y^2$. Thus,

$$(x + y)^2 = x^2 + 2xy + y^2$$

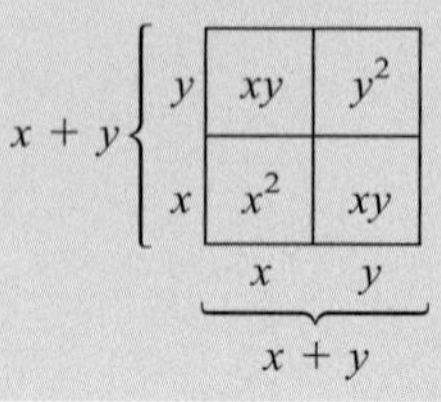

Teaching Tip: Review the difference between $(xy)^2$ and $(x + y)^2$. Show your students that exponents don't "distribute over addition or subtraction" using:
$(4 + 1)^2 \neq 4^2 + 1^2$
$(6 - 2)^2 \neq 6^2 - 2^2$

Note that the terms of the resulting trinomial are related to the terms of the binomial that was squared.

$$(x + y)^2 = x^2 + 2xy + y^2$$

- x^2: The square of the first term, *x*.
- $2xy$: Twice the product of the first and second terms, *x* and *y*.
- y^2: The square of the second term, *y*.

To develop a rule to find the *square of a binomial difference,* we consider $(x - y)^2$.

$$(x - y)^2 = (x - y)(x - y)$$ In $(x - y)^2$, the base is $(x - y)$ and the exponent is 2.

$$= x^2 - xy - xy + y^2$$ Multiply the binomials.

$$= x^2 - 2xy + y^2$$ Combine like terms: $-xy - xy = -1xy - 1xy = -2xy$.

Again, the terms of the resulting trinomial are related to the terms of the binomial that was squared.

$$(x - y)^2 = x^2 - 2xy + y^2$$

- x^2: The square of the first term, *x*.
- $-2xy$: Twice the product of the first and second terms, *x* and $-y$.
- y^2: The square of the second term, $-y$.

The observations from these two examples illustrate the following **special-product rules.**

Squaring a Binomial

The **square of a binomial** is a trinomial, such that:

- Its first term is the square of the first term of the binomial.
- Its last term is the square of the second term of the binomial.
- Its middle term is twice the product of both terms of the binomial.

$$(A + B)^2 = A^2 + 2AB + B^2 \qquad (A - B)^2 = A^2 - 2AB + B^2$$

EXAMPLE 1 Find each square: **a.** $(t + 9)^2$ **b.** $(8a - 5)^2$ **c.** $(d + 0.5)^2$ **d.** $\left(c^3 - \frac{7}{2}d\right)^2$

Strategy To find each square of a binomial, we will use one of the special-product rules.

Why This approach is faster than using the FOIL method.

Solution

a. $(t + 9)^2$ is the square of a binomial sum. The first term is t and the second term is 9.

$$(t + 9)^2 = \underbrace{t^2}_{\text{The square of the first term, } t.} + \underbrace{2(t)(9)}_{\text{Twice the product of both terms.}} + \underbrace{9^2}_{\text{The square of the second term, 9.}}$$

$$= t^2 + 18t + 81$$

The Language of Algebra

When squaring a binomial, the result is called a *perfect-square trinomial.*

$$(t + 9)^2 = \underbrace{t^2 + 18t + 81}_{\text{Perfect-square trinomial}}$$

b. $(8a - 5)^2$ is the square of a binomial difference. The first term is $8a$ and the second term is -5.

$$(8a - 5)^2 = \underbrace{(8a)^2}_{\text{The square of the first term, } 8a.} + \underbrace{2(8a)(-5)}_{\text{Twice the product of both terms.}} + \underbrace{(-5)^2}_{\text{The square of the second term, } -5.}$$

$$= 64a^2 - 80a + 25$$ Use the power of a product rule: $(8a)^2 = 8^2a^2 = 64a^2$.

Caution

The square of a binomial is a *trinomial.* A common error when squaring a binomial is to forget the middle term of the product.

$$(8a - 5)^2 \neq 64a^2 + 25$$

↑ Missing $-80a$

c. $(d + 0.5)^2$ is the square of a binomial sum. The first term is d and the second term is 0.5.

$$(d + 0.5)^2 = \underbrace{(d)^2}_{\text{The square of the first term, } d.} + \underbrace{2(d)(0.5)}_{\text{Twice the product of both terms.}} + \underbrace{(0.5)^2}_{\text{The square of the second term, 0.5.}}$$

$$= d^2 + d + 0.25$$

d. $\left(c^3 - \frac{7}{2}d\right)^2$ is the square of a binomial difference. The first term is c^3 and the second term is $-\frac{7}{2}d$.

$$\left(c^3 - \frac{7}{2}d\right)^2 = \underbrace{(c^3)^2}_{\text{The square of the first term, } c^3.} + \underbrace{2(c^3)\left(-\frac{7}{2}d\right)}_{\text{Twice the product of both terms.}} + \underbrace{\left(-\frac{7}{2}d\right)^2}_{\text{The square of the second term, } -\frac{7}{2}d.}$$

$$= c^6 - 7c^3d + \frac{49}{4}d^2$$ Use rules for exponents to find $(c^3)^2$ and $\left(-\frac{7}{2}d\right)^2$.

Teaching Example 1 Find each square: a. $(n + 4)^2$ b. $(9m - 3)^2$ c. $(z + 0.2)^2$ d. $(b^4 - \frac{3}{8}c)^2$
Answers:
a. $n^2 + 8n + 16$ b. $81m^2 - 54m + 9$
c. $z^2 + 0.4z + 0.04$
d. $b^8 - \frac{3}{4}b^4c + \frac{9}{64}c^2$

Self Check 1 Find each square: **a.** $(r + 6)^2$ $r^2 + 12r + 36$
b. $(7g - 2)^2$ $49g^2 - 28g + 4$ **c.** $(v + 0.8)^2$ $v^2 + 1.6v + 0.64$
d. $\left(w^4 - \frac{3}{2}y\right)^2$ $w^8 - 3w^4y + \frac{9}{4}y^2$

Now Try Problems 9, 15, and 23

2 Multiply the Sum and Difference of the Same Two Terms.

A final special product that occurs often has the form $(A + B)(A - B)$. In these products, one binomial is the sum of two terms and the other binomial is the difference of the same two terms. To develop a rule to find such products, consider the following multiplication:

$$(x + y)(x - y) = x^2 - xy + xy - y^2$$ Multiply the binomials.
$$= x^2 - y^2$$ Combine like terms: $-xy + xy = 0$.

Success Tip

We can use the FOIL method to find each of the special products discussed in this section. However, these forms occur so often, it is worthwhile to learn the special-product rules.

Note that when we combined like terms, we added opposites. This will always be the case for products of this type; the sum of the outer and inner products will be 0. The first and last products will be squares.

$$(x + y)(x - y) = x^2 - y^2$$

The square of the second term, y.
The square of the first term, x.

These observations suggest a third **special-product rule.**

Multiplying the Sum and Difference of Two Terms

The product of the sum of two terms and difference of the same two terms is the square of the first term minus the square of the second term.

$$(A + B)(A - B) = A^2 - B^2$$

EXAMPLE 2 Multiply: **a.** $(m + 2)(m - 2)$ **b.** $(3y + 4)(3y - 4)$ **c.** $\left(b - \frac{2}{3}\right)\left(b + \frac{2}{3}\right)$
d. $(t^4 - 6u)(t^4 + 6u)$

Strategy To find the product of each pair of binomials, we will use the special-product rule for the sum and difference of the same two terms.

Why This approach is faster than using the FOIL method.

Solution

The Language of Algebra

When multiplying the sum and difference of two terms, the result is called a *difference of two squares.*

$(m + 2)(m - 2) = \underbrace{m^2 - 4}_{\text{Difference of two squares}}$

a. $(m + 2)$ and $(m - 2)$ are the sum and difference of the same two terms, m and 2.

$$(m + 2)(m - 2) = \underbrace{m^2}_{\text{The square of the first term, } m.} - \underbrace{2^2}_{\text{The square of the second term, 2.}}$$

$$= m^2 - 4$$

b. $(3y + 4)$ and $(3y - 4)$ are the sum and difference of the same two terms, $3y$ and 4.

$$(3y + 4)(3y - 4) = \underbrace{(3y)^2}_{\text{The square of the first term, } 3y.} - \underbrace{4^2}_{\text{The square of the second term, 4.}}$$

$$= 9y^2 - 16$$

c. By the commutative property of multiplication, the special-product rule can be written with the factor containing the $-$ symbol first: $(A - B)(A + B) = A^2 - B^2$. Since $\left(b - \frac{2}{3}\right)$ and $\left(b + \frac{2}{3}\right)$ are the difference and sum of the same two terms, b and $\frac{2}{3}$, we have

$$\left(b - \frac{2}{3}\right)\left(b + \frac{2}{3}\right) = \underbrace{b^2}_{\text{The square of the first term, } b.} - \underbrace{\left(\frac{2}{3}\right)^2}_{\text{The square of the second term, } \frac{2}{3}.}$$

$$= b^2 - \frac{4}{9}$$

d. $(t^4 - 6u)$ and $(t^4 + 6u)$ are the difference and sum of the same two terms, t^4 and $6u$.

$$(t^4 - 6u)(t^4 + 6u) = \underbrace{(t^4)^2}_{\text{The square of the first term, } t^4.} - \underbrace{(6u)^2}_{\text{The square of the second term, } 6u.}$$

$$= t^8 - 36u^2$$

Teaching Example 2 Multiply:
a. $(d + 8)(d - 8)$
b. $(7t + 2)(7t - 2)$ c. $(r - \frac{4}{5})(r + \frac{4}{5})$
d. $(w^5 + 3z)(w^5 - 3z)$
Answers:
a. $d^2 - 64$ b. $49t^2 - 4$ c. $r^2 - \frac{16}{25}$
d. $w^{10} - 9z^2$

Self Check 2 Multiply: **a.** $(b + 4)(b - 4)$ $b^2 - 16$
b. $(5m + 9)(5m - 9)$ $25m^2 - 81$ **c.** $\left(s - \frac{3}{4}\right)\left(s + \frac{3}{4}\right)$ $s^2 - \frac{9}{16}$
d. $(c^3 + 2d)(c^3 - 2d)$ $c^6 - 4d^2$

Now Try Problems 25, 27, and 31

3 Find Higher Powers of Binomials.

When we find the third, fourth, or even higher powers of a binomial, we say that we are **expanding the binomial.** The special-product rules can be used in such cases. The result is an expression that has more terms than the original binomial.

EXAMPLE 3 Expand: $(x + 1)^3$

Strategy We will use a special-product rule to find the third power of $x + 1$.

Why Since $(x + 1)^3$ can be written as $(x + 1)(x + 1)^2$, we can use a special-product rule to find $(x + 1)^2$ quickly.

Solution

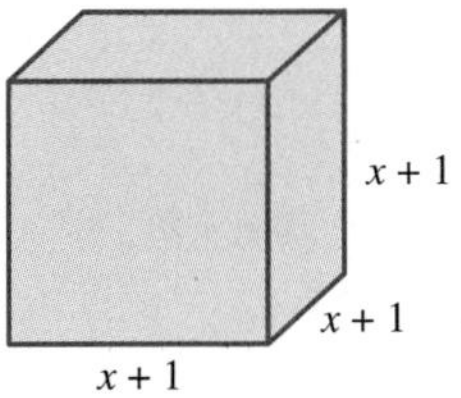

$(x + 1)^3 = (x + 1)(x + 1)^2$ Read as "the quantity of x + 1, cubed."

$= (x + 1)(x^2 + 2x + 1)$ Find $(x + 1)^2$ using the rule for the square of a sum.

$= (x + 1)(x^2 + 2x + 1)$ Multiply the binomial and the trinomial.

$= x(x^2) + x(2x) + x(1) + 1(x^2) + 1(2x) + 1(1)$ Multiply each term of $x^2 + 2x + 1$ by each term of $x + 1$.

$= x^3 + 2x^2 + x + x^2 + 2x + 1$ Multiply the monomials.

$= x^3 + 3x^2 + 3x + 1$ Combine like terms.

Notation

$(x + 1)^3$ represents the volume of a cube with sides of length $x + 1$ units.

Teaching Example 3
Expand: $(q - 2)^3$
Answer: $q^3 - 6q^2 + 12q - 8$

Self Check 3 Expand: $(n - 3)^3$ $n^3 - 9n^2 + 27n - 27$

Now Try Problem 37

4 Simplify Expressions Containing Polynomial Multiplication.

We can use a modified version of the order of operations rule to simplify expressions that involve polynomial addition, subtraction, multiplication, and raising to a power.

Order of Operations with Polynomials

1. If possible, simplify any polynomials within parentheses by combining like terms.
2. Square (or expand) all polynomials raised to powers using the FOIL method or a special-product rule.
3. Perform all polynomial multiplications using the distributive property, the FOIL method, or a special-product rule.
4. Perform all polynomial additions and subtractions by combining like terms.

EXAMPLE 4 Simplify each expression:

a. $-8(y^2 - 2y + 3) - 4(2y^2 + y - 6)$
b. $(x + 1)(x - 2) + 3x(x + 3)$
c. $(3y - 2)^2 - (y - 5)(y + 5)$

Strategy We will follow the rules for the order of operations to simplify each expression.

Why If we don't follow the correct order of operations, we can obtain different results that are not equivalent.

Solution **a.** The two polynomials within the parentheses do not simplify further and no polynomials are raised to a power. To perform the multiplication, we will use the distributive property twice. Then we will combine like terms.

$-8(y^2 - 2y + 3) - 4(2y^2 + y - 6) = -8y^2 + 16y - 24 - 8y^2 - 4y + 24$ Distribute.

$= -16y^2 + 12y$ Add and subtract to combine like terms.

b. The three polynomials within parentheses do not simplify further and no polynomials are raised to a power. To perform the multiplications, we use the FOIL method and the distributive property. Then we will combine like terms.

$$(x + 1)(x - 2) + 3x(x + 3)$$

$= x^2 - x - 2 + 3x^2 + 9x$ Use the FOIL method to find $(x + 1)(x - 2)$. Distribute the multiplication by $3x$.

$= 4x^2 + 8x - 2$ Combine like terms.

c. The three polynomials within parentheses do not simplify further. To square $3y - 2$, we use a special-product rule. To find the product of $(y - 5)(y + 5)$, we will use the special-product rule for the sum and difference of the same two terms. Then we will combine like terms.

$$(3y - 2)^2 - (y - 5)(y + 5)$$

$= 9y^2 - 12y + 4 - (y^2 - 25)$ Write $y^2 - 25$ within parentheses so that both terms are subtracted.

$= 9y^2 - 12y + 4 - y^2 + 25$ Change the sign of each term within $(y^2 - 25)$ and drop the parentheses.

$= 8y^2 - 12y + 29$ Combine like terms.

Teaching Example 4 Simplify each expression:
a. $-6(w^2 - 2w + 1) - 4(2w^2 + w - 5)$
b. $(a - 2)(a - 3) + 3a(a + 10)$
c. $(4d - 3)^2 - (d - 9)(d + 9)$
Answers:
a. $-14w^2 + 8w + 14$
b. $4a^2 + 25a + 6$
c. $15d^2 - 24d + 90$

Self Check 4 Simplify each expression: **a.** $2(a^2 - 3a) + 5(a^2 + 2a)$ $7a^2 + 4a$
b. $(x - 4)(x + 6) + 5x(2x - 1)$ $11x^2 - 3x - 24$
c. $(a + 9)(a - 9) - (2a - 4)^2$ $-3a^2 + 16a - 97$

Now Try Problems 45, 47, and 48

EXAMPLE 5

Find a polynomial that represents the area of the triangle.

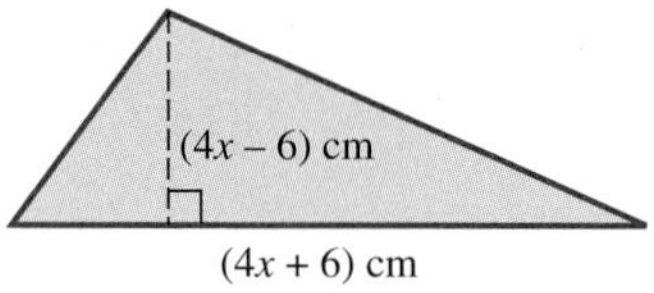

Strategy We will multiply one-half, the length of the base, and the height of the triangle.

Why The area of a triangle is equal to one-half the product of the length of its base and its height.

Solution We begin by substituting $4x + 6$ for the length of the base and $4x - 6$ for the height in the formula for the area of a triangle. It is wise to find $(4x + 6)(4x - 6)$ first, using a special-product rule, and then to multiply that result by $\frac{1}{2}$.

Success Tip

Recall that to multiply three polynomials, begin by multiplying any two of them. Then multiply that result by the third polynomial.

$\text{Area} = \frac{1}{2} \cdot \textbf{base} \cdot \textbf{height}$ This is the formula for the area of a triangle.

$= \frac{1}{2}(\mathbf{4x + 6})(\mathbf{4x - 6})$ Substitute.

$= \frac{1}{2}(16x^2 - 36)$ Use the special product rule for a sum and difference of two terms.

$= 8x^2 - 18$ Distribute the multiplication by $\frac{1}{2}$.

The area of the triangle is $(8x^2 - 18)$ square centimeters, which can be written as $(8x^2 - 18)\text{ cm}^2$.

Teaching Example 5 Find a polynomial that represents the area of the triangle.

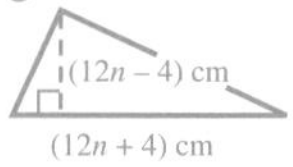

Answer: $(72n^2 - 8)\text{ cm}^2$

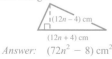

Self Check 5 Find a polynomial that represents the area of a triangle with height $(12a - 2)$ ft and the length of the base $(12a + 2)$ ft. $(72a^2 - 2)\text{ ft}^2$

Now Try Problem 53

SECTION 5.7 STUDY SET

VOCABULARY

Fill in the blanks.

1. Expressions of the form $(x + y)^2$, $(x - y)^2$, and $(x + y)(x - y)$ occur so frequently in algebra that they are called special products.
2. $(2x + 3)^2$ is the square of a binomial and $(a + 6)(a - 6)$ is the product of the sum and difference of the same two terms.

CONCEPTS

3. Fill in the blanks to describe each special product.

   a. $(x + y)^2 = x^2 + 2xy + y^2$
   - The square of the second term
   - Twice the product of the first and second terms
   - The square of the first term

   b. $(x + y)(x - y) = x^2 - y^2$
   - The square of the second term
   - The square of the first term

4. Consider the binomial $5x + 4$.

   a. What is the square of its first term? $25x^2$

   b. What is twice the product of its two terms? $40x$

   c. What is the square of its second term? 16

NOTATION

Complete each solution to find the product.

5. $(x + 4)^2 = x^2 + 2(x)(4) + 4^2$
   $= x^2 + 8x + 16$
6. $(6r - 1)^2 = (6r)^2 - 2(6r)(1) + (-1)^2$
   $= 36r^2 - 12r + 1$
7. $(s + 5)(s - 5) = s^2 - 5^2$
   $= s^2 - 25$
8. True or false: $(t + 7)(t - 7) = (t - 7)(t + 7)$? True

GUIDED PRACTICE

Find each product. See Example 1.

9. $(x + 1)^2$ $x^2 + 2x + 1$
10. $(y + 7)^2$ $y^2 + 14y + 49$
11. $(m - 6)^2$ $m^2 - 12m + 36$
12. $(b - 1)^2$ $b^2 - 2b + 1$
13. $(4x + 5)^2$ $16x^2 + 40x + 25$
14. $(6y + 3)^2$ $36y^2 + 36y + 9$
15. $(7m - 2)^2$ $49m^2 - 28m + 4$
16. $(9b - 2)^2$ $81b^2 - 36b + 4$
17. $(1 - 3y)^2$ $1 - 6y + 9y^2$
18. $(1 - 4a)^2$ $1 - 8a + 16a^2$
19. $(y + 0.9)^2$ $y^2 + 1.8y + 0.81$
20. $(d + 0.2)^2$ $d^2 + 0.4d + 0.04$
21. $(a^2 + b^2)^2$ $a^4 + 2a^2b^2 + b^4$
22. $(c^2 + d^2)^2$ $c^4 + 2c^2d^2 + d^4$
23. $\left(s + \frac{3}{4}\right)^2$ $s^2 + \frac{3}{2}s + \frac{9}{16}$
24. $\left(y - \frac{5}{3}\right)^2$ $y^2 - \frac{10}{3}y + \frac{25}{9}$

Find each product. See Example 2.

25. $(x + 3)(x - 3)$ $x^2 - 9$
26. $(y + 6)(y - 6)$ $y^2 - 36$
27. $(2p + 7)(2p - 7)$ $4p^2 - 49$
28. $(5t + 4)(5t - 4)$ $25t^2 - 16$
29. $(3n + 1)(3n - 1)$ $9n^2 - 1$
30. $(5a + 4)(5a - 4)$ $25a^2 - 16$
31. $\left(c + \frac{3}{4}\right)\left(c - \frac{3}{4}\right)$ $c^2 - \frac{9}{16}$
32. $\left(m + \frac{4}{5}\right)\left(m - \frac{4}{5}\right)$ $m^2 - \frac{16}{25}$
33. $(0.4 - 9m^2)(0.4 + 9m^2)$ $0.16 - 81m^4$
34. $(0.3 - 2c^2)(0.3 + 2c^2)$ $0.09 - 4c^4$
35. $(5 - 6g)(5 + 6g)$ $25 - 36g^2$
36. $(6 - c^2)(6 + c^2)$ $36 - c^4$

Expand each binomial. See Example 3.

37. $(x + 4)^3$ $x^3 + 12x^2 + 48x + 64$
38. $(y + 2)^3$ $y^3 + 6y^2 + 12y + 8$
39. $(n - 6)^3$ $n^3 - 18n^2 + 108n - 216$
40. $(m - 5)^3$ $m^3 - 15m^2 + 75m - 125$
41. $(2g - 3)^3$ $8g^3 - 36g^2 + 54g - 27$
42. $(3x - 2)^3$ $27x^3 - 54x^2 + 36x - 8$
43. $(a + b)^3$ $a^3 + 3a^2b + 3ab^2 + b^3$
44. $(c - d)^3$ $c^3 - 3c^2d + 3cd^2 - d^3$

Perform the operations. See Example 4.

45. $2(x^2 + 7x - 1) - 3(x^2 - 2x + 2)$ $-x^2 + 20x - 8$
46. $2t(t + 2) + (t - 1)(t + 9)$ $3t^2 + 12t - 9$
47. $(3x + 4)(2x - 2) - (2x + 1)(x + 3)$ $4x^2 - 5x - 11$
48. $(5a - 1)^2 - (a - 8)(a + 8)$ $24a^2 - 10a + 65$
49. $-5d(4d - 1)^2$ $-80d^3 + 40d^2 - 5d$
50. $-2h(7h - 2)^2$ $-98h^3 + 56h^2 - 8h$
51. $4d(d^2 + g^3)(d^2 - g^3)$ $4d^5 - 4dg^6$
52. $8y(x^2 + y^2)(x^2 - y^2)$ $8x^4y - 8y^5$

Find a polynomial that represents the area of the figure. Leave π in your answer. See Example 5.

53. 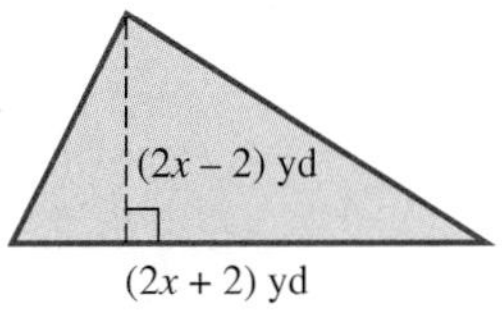

$(2x^2 - 2)\text{ yd}^2$

54. 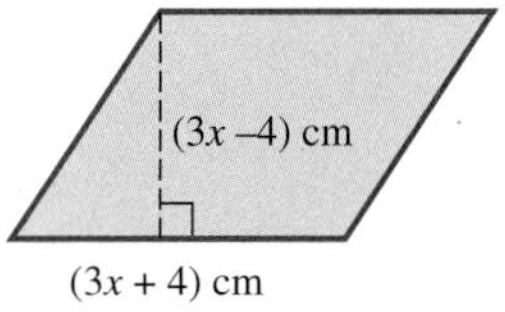

$(9x^2 - 16)\text{ cm}^2$

55. 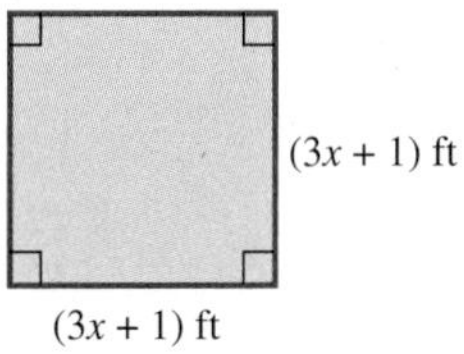

$(9x^2 + 6x + 1)\text{ ft}^2$

56. 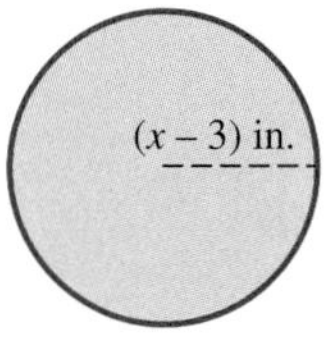

$(x^2 - 6x + 9)\pi\text{ in.}^2$

TRY IT YOURSELF

Perform the operations.

57. $(2v^3 - 8)^2$
$4v^6 - 32v^3 + 64$

58. $(8x^4 - 3)^2$
$64x^8 - 48x^4 + 9$

59. $3x(2x + 3)(2x + 3)$
$12x^3 + 36x^2 + 27x$

▶ 60. $4y(3y + 4)(3y + 4)$
$36y^3 + 96y^2 + 64y$

61. $(4f + 0.4)(4f - 0.4)$
$16f^2 - 0.16$

▶ 62. $(4t + 0.6)(4t - 0.6)$
$16t^2 - 0.36$

63. $(r^2 + 10s)^2$
$r^4 + 20r^2s + 100s^2$

64. $(m^2 + 8n)^2$
$m^4 + 16m^2n + 64n^2$

65. $2(x + 3) + 4(x - 2)$
$6x - 2$

66. $3(y - 4) - 5(y + 3)$
$-2y - 27$

67. $\left(d^4 + \frac{1}{4}\right)^2$
$d^8 + \frac{1}{2}d^4 + \frac{1}{16}$

▶ 68. $\left(q^6 + \frac{1}{3}\right)^2$
$q^{12} + \frac{2}{3}q^6 + \frac{1}{9}$

▶ 69. $(d + 7)(d - 7)$
$d^2 - 49$

▶ 70. $(t + 2)(t - 2)$
$t^2 - 4$

71. $(2a - 3b)^2$
$4a^2 - 12ab + 9b^2$

▶ 72. $(2x + 5y)^2$
$4x^2 + 20xy + 25y^2$

73. $(n + 6)(n - 6)$
$n^2 - 36$

74. $(a + 12)(a - 12)$
$a^2 - 144$

75. $(m + 10)^2 - (m - 8)^2$ $36m + 36$

76. $(5y - 1)^2 - (y + 7)(y - 7)$ $24y^2 - 10y + 50$

77. $(2m + n)^3$ $8m^3 + 12m^2n + 6mn^2 + n^3$

▶ 78. $(p - 2q)^3$ $p^3 - 6p^2q + 12pq^2 - 8q^3$

79. $\left(5m - \frac{6}{5}\right)^2$
$25m^2 - 12m + \frac{36}{25}$

80. $\left(6m - \frac{7}{6}\right)^2$
$36m^2 - 14m + \frac{49}{36}$

81. $(r^2 - s^2)^2$
$r^4 - 2r^2s^2 + s^4$

82. $(t^2 - u^2)^2$
$t^4 - 2t^2u^2 + u^4$

83. $(x - 2)^2$
$x^2 - 4x + 4$

84. $(a + 2)^2$
$a^2 + 4a + 4$

85. $(r + 2)^2$
$r^2 + 4r + 4$

86. $(n + 10)^2$
$n^2 + 20n + 100$

87. $(n - 2)^4$ $n^4 - 8n^3 + 24n^2 - 32n + 16$

▶ 88. $(c + d)^4$ $c^4 + 4c^3d + 6c^2d^2 + 4cd^3 + d^4$

89. $5(y^2 - 2y - 6) + 6(2y^2 + 2y - 5)$ $17y^2 + 2y - 60$

▶ 90. $(4b + 1)^2 - (b - 7)(b + 7)$ $15b^2 + 8b + 50$

91. $(3x - 2)^2 + (2x + 1)^2$ $13x^2 - 8x + 5$

▶ 92. $(4a - 3)^2 + (a + 6)^2$ $17a^2 - 12a + 45$

93. $(f - 8)^2$
$f^2 - 16f + 64$

94. $(w - 9)^2$
$w^2 - 18w + 81$

95. $\left(6b + \frac{1}{2}\right)\left(6b - \frac{1}{2}\right)$
$36b^2 - \frac{1}{4}$

96. $\left(4h + \frac{2}{3}\right)\left(4h - \frac{2}{3}\right)$
$16h^2 - \frac{4}{9}$

97. $3y(y + 2) + (y + 1)(y - 1)$ $4y^2 + 6y - 1$

▶ 98. $(x + y)(x - y) + x(x + y)$ $2x^2 + xy - y^2$

99. $(6 - 2d^3)^2$
$36 - 24d^3 + 4d^6$

▶ 100. $(6 - 5p^2)^2$
$36 - 60p^2 + 25p^4$

101. $(2e + 1)^3$ $8e^3 + 12e^2 + 6e + 1$

102. $(3m - 2n)^3$ $27m^3 - 54m^2n + 36mn^2 - 8n^3$

103. $(8x + 3)^2$
$64x^2 + 48x + 9$

▶ 104. $(4b - 8)^2$
$16b^2 - 64b + 64$

Look Alikes . . .

Perform the indicated operations.

105. a. $(xy)^2$ x^2y^2 b. $(x + y)^2$ $x^2 + 2xy + y^2$

▶ 106. a. $(cd)^2$ c^2d^2 b. $(c - d)^2$ $c^2 - 2cd + d^2$

107. a. $(2b^2d)^2$ $4b^4d^2$ b. $(2b^2 + d)^2$ $4b^4 + 4b^2d + d^2$

108. a. $(mn)^3$ m^3n^3 b. $(m + n)^3$ $m^3 + 3m^2n + 3mn^2 + n^3$

APPLICATIONS

▶ 109. **Playpens.** Find a polynomial that represents the area of the floor of the playpen. $(x^2 + 12x + 36)\text{ in.}^2$

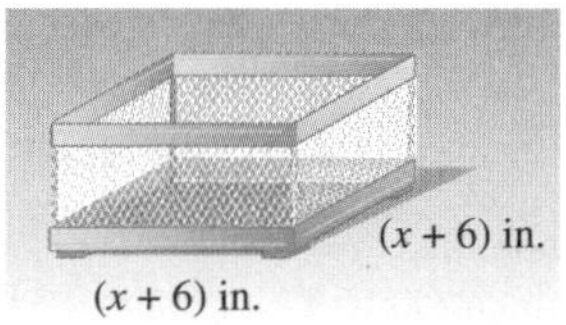

▶ 110. **Storage.** Find a polynomial that represents the volume of the cubicle. $(x^3 + 15x^2 + 75x + 125)\text{ in.}^3$

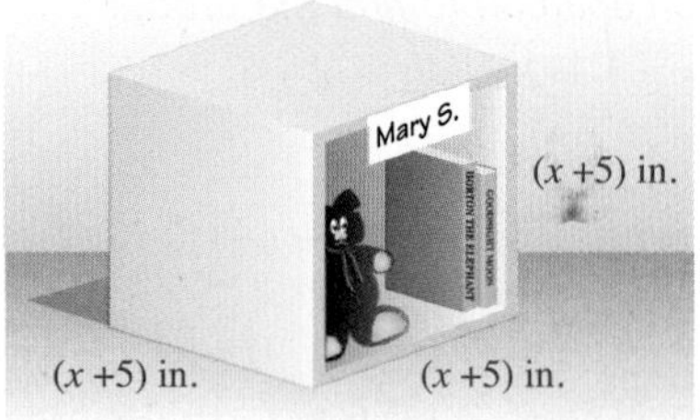

111. **Paper Towels.** The amount of space (volume) occupied by the paper on the roll of paper towels is given by the expression $\pi h(R + r)(R - r)$, where R is the outer radius and r is the inner radius. Perform the indicated multiplication.
$\pi hR^2 - \pi hr^2$

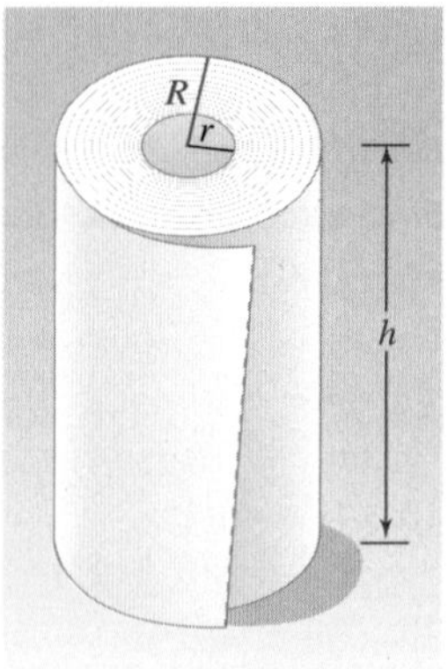

112. Signal Flags. Refer to the illustration below. Find a polynomial that represents the area in blue of the maritime signal flag for the letter *p*. The dimensions are given in centimeters. $(9x^2 + 13x + 2)\text{ cm}^2$

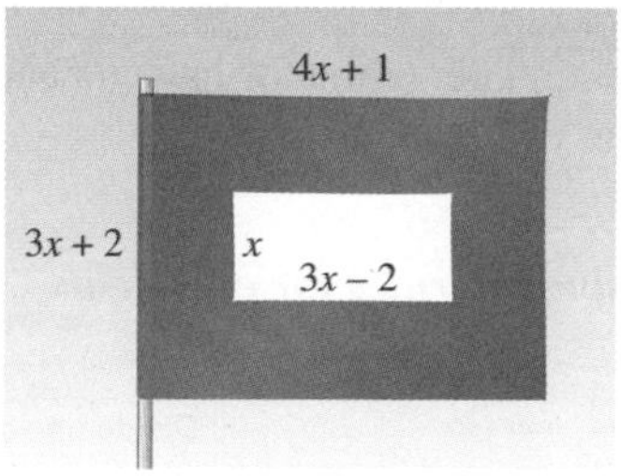

WRITING

113. What is a binomial? Explain how to square it.

114. Writing $(x + y)^2$ as $x^2 + y^2$ illustrates a common error. Explain.

115. We can find $(2x + 3)^2$ and $(5y - 6)^2$ using the FOIL method or using special product rules. Explain why the special product rules are faster.

116. **a.** Fill in the blanks: $(xy)^2$ is the <u>square</u> of a product and $(x + y)^2$ is the <u>square</u> of a sum.

b. Explain why $(xy)^2 \neq (x + y)^2$.

REVIEW

117. Simplify: $\frac{30}{36}$ $\frac{5}{6}$

118. Add: $\frac{5}{12} + \frac{1}{4}$ $\frac{2}{3}$

119. Multiply: $\frac{7}{8} \cdot \frac{3}{5}$ $\frac{21}{40}$

120. Divide: $\frac{1}{3} \div \frac{4}{5}$ $\frac{5}{12}$

CHALLENGE PROBLEMS

121. **a.** Find two binomials whose product is a binomial. $(x + 1)(x - 1) = x^2 - 1$

b. Find two binomials whose product is a trinomial. $(x + 1)(x + 1) = x^2 + 2x + 1$

c. Find two binomials whose product is a four-term polynomial. $(x + a)(x + b) = x^2 + bx + ax + ab$

122. A special-product rule can be used to find $31 \cdot 29$.

$$\begin{aligned} 31 \cdot 29 &= (30 + 1)(30 - 1) \\ &= 30^2 - 1^2 \\ &= 900 - 1 \\ &= 899 \end{aligned}$$

Use this method to find $52 \cdot 48$. 2,496

SECTION 5.8 Dividing Polynomials

OBJECTIVES

1. Divide a monomial by a monomial.
2. Divide a polynomial by a monomial.
3. Divide a polynomial by a polynomial.

ARE YOU READY? *Are You Ready? exercises available online at www.webassign.net/brookscole*

The following problems review some basic skills that are needed when dividing polynomials.

1. Simplify: **a.** $\frac{6}{15}$ $\frac{2}{5}$ **b.** $\frac{a^6}{a^4}$ a^2

2. Add: $\frac{a}{d} + \frac{b}{d}$ $\frac{a+b}{d}$

3. Divide: $24\overline{)864}$ 36

4. Subtract: $\begin{array}{r} 2x^3 + 8x^2 \\ \underline{-(2x^3 + 6x^2)} \end{array}$ $2x^2$

In this section, we will conclude our study of operations with polynomials by discussing division of polynomials. To begin, we consider the simplest case, the quotient of two monomials.

1 Divide a Monomial by a Monomial.

To divide monomials, we can use the method for simplifying fractions or a method that involves a rule for exponents.

EXAMPLE 1 Divide the monomials: **a.** $\frac{21x^5}{7x^2}$ **b.** $\frac{10r^6s}{6rs^3}$

Strategy We can use the rules for simplifying fractions and/or the quotient rule for exponents.

Why We need to make sure that the numerator and denominator have no common factors other than 1. If that is the case, then the fraction is in *simplest form*.

Solution

Success Tip

In this section, you will see that regardless of the number of terms involved, every polynomial division is a series of monomial divisions.

By simplifying fractions

a. $\frac{21x^5}{7x^2} = \frac{3 \cdot \overset{1}{\cancel{7}} \cdot \overset{1}{\cancel{x}} \cdot \overset{1}{\cancel{x}} \cdot x \cdot x \cdot x}{\underset{1}{\cancel{7}} \cdot \underset{1}{\cancel{x}} \cdot \underset{1}{\cancel{x}}}$

$= 3x^3$

b. $\frac{10r^6s}{6rs^3} = \frac{\overset{1}{\cancel{2}} \cdot 5 \cdot \overset{1}{\cancel{r}} \cdot r \cdot r \cdot r \cdot r \cdot r \cdot \overset{1}{\cancel{s}}}{\underset{1}{\cancel{2}} \cdot 3 \cdot \underset{1}{\cancel{r}} \cdot \underset{1}{\cancel{s}} \cdot s \cdot s}$

$= \frac{5r^5}{3s^2}$

Using the rules for exponents

$\frac{21x^5}{7x^2} = 3x^{5-2}$ Divide the coefficients.

$= 3x^3$ Subtract the exponents.

$\frac{10r^6s}{6rs^3} = \frac{5}{3}r^{6-1}s^{1-3}$ Simplify $\frac{10}{6}$.

$= \frac{5}{3}r^5s^{-2}$ Subtract exponents.

$= \frac{5r^5}{3s^2}$ Move s^{-2} to the denominator and change the sign of the exponent.

Teaching Example 1 Divide the monomials:
a. $\frac{60z^7}{10z^3}$ b. $\frac{9m^3n^9}{36m^5n^4}$
Answers: a. $6z^4$ b. $\frac{n^5}{4m^2}$

Self Check 1 Divide the monomials: **a.** $\frac{30y^4}{5y^2}$ $6y^2$ **b.** $\frac{8c^2d^6}{32c^5d^2}$ $\frac{d^4}{4c^3}$

Now Try Problems 15 and 21

2 Divide a Polynomial by a Monomial.

Recall that to add two fractions with the same denominator, we add their numerators and keep their common denominator.

$$\frac{a}{d} + \frac{b}{d} = \frac{a + b}{d}$$

We can use this rule in reverse to divide polynomials by monomials.

Dividing a Polynomial by a Monomial

To divide a polynomial by a monomial, divide each term of the polynomial by the monomial.

If A, B, and D represent monomials, where $D \neq 0$, then

$$\frac{A + B}{D} = \frac{A}{D} + \frac{B}{D}$$

EXAMPLE 2 Divide: **a.** $\frac{9x^2 + 3x}{3x}$ **b.** $\frac{12a^4b^3 - 18a^3b^2 + 2a^2}{6a^2b^2}$

Strategy We will divide each term of the polynomial in the numerator by the monomial in the denominator.

Why A fraction bar indicates division of the numerator by the denominator.

Solution **a.** Here, we have a binomial divided by a monomial.

$$\frac{9x^2 + 3x}{3x} = \frac{9x^2}{3x} + \frac{3x}{3x}$$ Divide each term of the numerator, $9x^2 + 3x$, by the denominator, $3x$.

$$= 3x^{2-1} + 1x^{1-1}$$ Do each monomial division. Divide the coefficients. Keep each base and subtract the exponents.

$$= 3x^1 + 1x^0$$

$$= 3x + 1$$ Recall that $x^0 = 1$.

Caution

A common error is to "remove" the $3x$'s in the expression incorrectly. Recall that only factors common to the numerator and denominator can be removed. Here, $3x$ is not a factor of the numerator—it is a *term*.

$$\frac{9x^2 + \overset{1}{\cancel{3x}}}{\underset{1}{\cancel{3x}}}$$

Check: We multiply the divisor, $3x$, and the quotient, $3x + 1$. The result should be the dividend, $9x^2 + 3x$.

$$3x(3x + 1) = 9x^2 + 3x$$ The answer checks.

b. Here, we have a trinomial divided by a monomial.

$$\frac{12a^4b^3 - 18a^3b^2 + 2a^2}{6a^2b^2} = \frac{12a^4b^3}{6a^2b^2} - \frac{18a^3b^2}{6a^2b^2} + \frac{2a^2}{6a^2b^2}$$ Divide each term of the numerator by the denominator, $6a^2b^2$.

$$= 2a^{4-2}b^{3-2} - 3a^{3-2}b^{2-2} + \frac{a^{2-2}}{3b^2}$$ Do each monomial division. Simplify: $\frac{2}{6} = \frac{1}{3}$.

$$= 2a^2b - 3a + \frac{1}{3b^2}$$ $b^{2-2} = b^0 = 1$ and $a^{2-2} = a^0 = 1$.

The Language of Algebra

The names of the parts of a division statement are

Dividend $\frac{9x^2 + 3x}{3x} = 3x + 1$ *Quotient*

Divisor

Recall that the variables in a polynomial must have whole-number exponents. Therefore, the result, $2a^2b - 3a + \frac{1}{3b^2}$, is not a polynomial because the last term has a variable in the denominator.

Check:

$$6a^2b^2\left(2a^2b - 3a + \frac{1}{3b^2}\right) = 12a^4b^3 - 18a^3b^2 + 2a^2$$ The answer checks.

Self Check 2 Divide: **a.** $\frac{50h^3 + 5h^2}{5h^2}$ $10h + 1$ **b.** $\frac{22s^5t^2 - s^4t^3 + 44s^2t^4}{11s^2t^2}$ $2s^3 - \frac{s^2t}{11} + 4t^2$

Now Try Problems 25, 29, and 33

Teaching Example 2 Divide: a. $\frac{35f^9 + 7f^3}{7f^3}$ b. $\frac{14p^3q + pq^2 - 2p}{2pq}$

Answers: a. $5f^6 + 1$ b. $7p^2 + \frac{q}{2} - \frac{1}{q}$

3 Divide a Polynomial by a Polynomial.

To divide a polynomial by a polynomial (other than a monomial), we use a method similar to long division in arithmetic.

EXAMPLE 3 Divide $x^2 + 5x + 6$ by $x + 2$.

Strategy We will use the long division method. The dividend is $x^2 + 5x + 6$ and the divisor is $x + 2$.

Why Since the divisor has more than one term, we must use the long division method to divide the polynomials.

Teaching Tip: To help them see each $\frac{\text{monomial}}{\text{monomial}}$ division that must be performed, have students circle the first terms of the divisor and the dividend.

Solution We write the division using a long division symbol $\overline{)\quad}$ and proceed as follows:

Step 1: $x + 2\overset{x}{\overline{)x^2 + 5x + 6}}$ Divide the first term of the dividend by the first term of the divisor: $\frac{x^2}{x} = x$. Write the result, x, above the long division symbol.

The Language of Algebra

Notice how the instruction translates:

Divide $x^2 + 5x + 6$ by $x + 2$

$x + 2\overline{)x^2 + 5x + 6}$

Success Tip

Notice that this method is much like that used for division of whole numbers.

$$\begin{array}{r} 13 \\ 12\overline{)156} \\ -12\downarrow \\ \hline 036 \\ -36 \\ \hline 0 \end{array}$$

Hundreds, Tens, Ones

Success Tip

The long division method aligns like terms vertically.

$$\begin{array}{r} x + 3 \\ x + 2\overline{)x^2 + 5x + 6} \\ -(x^2 + 2x) \\ \hline 3x + 6 \\ -(3x + 6) \\ \hline 0 \end{array}$$

x^2-terms, x-terms, Constants

Teaching Example 3 Divide $x^2 + 10x + 16$ by $x + 8$.
Answer: $x + 2$

Step 2:
$$\begin{array}{r} x \\ x + 2\overline{)x^2 + 5x + 6} \\ x^2 + 2x \\ \hline \end{array}$$

Multiply each term of the divisor by x. Write the result, $x^2 + 2x$, under $x^2 + 5x$, and draw a line. Be sure to align the like terms.

Step 3:
$$\begin{array}{r} x \\ x + 2\overline{)x^2 + 5x + 6} \\ -(x^2 + 2x) \downarrow \\ \hline 3x + 6 \end{array}$$

Write parentheses around $x^2 + 2x$ so that both of its terms are subtracted. Subtract $x^2 + 2x$ from $x^2 + 5x$. Work column by column: $x^2 - x^2 = 0$ and $5x - 2x = 3x$.
Bring down the next term, 6.

Step 4:
$$\begin{array}{r} x + 3 \\ x + 2\overline{)x^2 + 5x + 6} \\ -(x^2 + 2x) \\ \hline 3x + 6 \end{array}$$

Divide the first term of $3x + 6$ by the first term of the divisor: $\frac{3x}{x} = 3$. Write $+ 3$ above the long division symbol to form the second term of the quotient.

Step 5:
$$\begin{array}{r} x + 3 \\ x + 2\overline{)x^2 + 5x + 6} \\ -(x^2 + 2x) \\ \hline 3x + 6 \\ 3x + 6 \\ \hline \end{array}$$

Multiply each term of the divisor by 3. Write the result, $3x + 6$, under $3x + 6$ and draw a line. Be sure to align the like terms.

Step 6:
$$\begin{array}{r} x + 3 \\ x + 2\overline{)x^2 + 5x + 6} \\ -(x^2 + 2x) \\ \hline 3x + 6 \\ -(3x + 6) \\ \hline 0 \end{array}$$

Write parentheses around $3x + 6$ so that both of its terms are subtracted. Subtract $3x + 6$ from $3x + 6$. Work vertically: $3x - 3x = 0$ and $6 - 6 = 0$. There are no more terms to bring down.

This is the remainder.

The quotient is $x + 3$ and the remainder is 0.

Step 7: Check the result by verifying that $(x + 2)(x + 3)$ is $x^2 + 5x + 6$.

$$(x + 2)(x + 3) = x^2 + 3x + 2x + 6 \quad \text{Divisor} \cdot \text{quotient} = \text{dividend (if 0 remainder)}$$
$$= x^2 + 5x + 6 \quad \text{The result checks.}$$

Self Check 3 Divide $x^2 + 7x + 12$ by $x + 3$. $x + 4$

Now Try Problem 37

Dividing a Polynomial by a Polynomial

To divide a polynomial by a polynomial (other than a monomial) use long division. If there is a remainder, write the result in the form: quotient $+ \frac{\text{remainder}}{\text{divisor}}$.

The long division method used in algebra can have a remainder just as long division in arithmetic often does.

EXAMPLE 4 Divide: $(6x^2 - 7x - 2) \div (2x - 1)$

Strategy We will use the long division method. The dividend is $6x^2 - 7x - 2$ and the divisor is $2x - 1$.

Why Since the divisor has more than one term, we must use the long division method to divide the polynomials.

Solution

Caution

The long-division method is a series of four steps:

- Divide
- Multiply
- Subtract
- Bring down the next term

Use parentheses to avoid sign errors when subtracting.

Step 1:
$$\begin{array}{r} 3x \\ 2x - 1\overline{)6x^2 - 7x - 2} \end{array}$$

Divide the first term of the dividend by the first term of the divisor: $\frac{6x^2}{2x} = 3x$. Write the result, $3x$, above the long division symbol.

Step 2:
$$\begin{array}{r} 3x \\ 2x - 1\overline{)6x^2 - 7x - 2} \\ 6x^2 - 3x \end{array}$$

Multiply each term of the divisor by $3x$. Write the result, $6x^2 - 3x$, under $6x^2 - 7x$, and draw a line.

Step 3:
$$\begin{array}{r} 3x \\ 2x - 1\overline{)6x^2 - 7x - 2} \\ \underline{-(6x^2 - 3x)} \downarrow \\ -4x - 2 \end{array}$$

Write parentheses around $6x^2 - 3x$ so that both of its terms are subtracted. Subtract $6x^2 - 3x$ from $6x^2 - 7x$. Work vertically: $6x^2 - 6x^2 = 0$ and $-7x - (-3x) = -7x + 3x = -4x$. Bring down the next term, -2.

Step 4:
$$\begin{array}{r} 3x - 2 \\ 2x - 1\overline{)6x^2 - 7x - 2} \\ \underline{-(6x^2 - 3x)} \\ -4x - 2 \end{array}$$

Divide the first term of $-4x - 2$ by the first term of the divisor: $\frac{-4x}{2x} = -2$. Write -2 above the long division symbol to form the second term of the quotient.

Step 5:
$$\begin{array}{r} 3x - 2 \\ 2x - 1\overline{)6x^2 - 7x - 2} \\ \underline{-(6x^2 - 3x)} \\ -4x - 2 \\ \underline{-4x + 2} \end{array}$$

Multiply each term of the divisor by -2. Write the result, $-4x + 2$, under $-4x - 2$, and draw a line.

Success Tip

The long division method for polynomials continues until the degree of the remainder is less than the degree of the divisor. Here, the remainder, -4, has degree 0. The divisor, $2x - 1$, has degree 1. Therefore, the division ends.

Step 6:
$$\begin{array}{r} 3x - 2 \\ 2x - 1\overline{)6x^2 - 7x - 2} \\ \underline{-(6x^2 - 3x)} \\ -4x - 2 \\ \underline{-(-4x + 2)} \\ -4 \end{array}$$

Write $-4x + 2$ in parentheses so that both of its terms are subtracted. Subtract $-4x + 2$ from $-4x - 2$. Work vertically: $-4x - (-4x) = -4x + 4x = 0$ and $-2 - 2 = -4$. There are no more terms to bring down.

This is the remainder.

The quotient is $3x - 2$ and the remainder is -4. It is common to write the answer in *Quotient* + $\frac{\textit{remainder}}{\textit{divisor}}$ form as either

$$3x - 2 + \frac{-4}{2x - 1} \quad \text{or} \quad 3x - 2 - \frac{4}{2x - 1}$$

Step 7: We can check the result using the fact that for any division:

Divisor · quotient + remainder = dividend

$$(2x - 1)(3x - 2) + (-4) = 6x^2 - 4x - 3x + 2 + (-4)$$
$$= 6x^2 - 7x - 2 \quad \text{The result checks.}$$

Teaching Example 4 Divide: $(9x^2 + 9x + 7) \div (3x - 2)$. Check the result.
Answer: $3x + 5 + \frac{17}{3x - 2}$

Self Check 4 Divide: $(8x^2 + 6x - 3) \div (2x + 3)$. Check the result. $4x - 3 + \frac{6}{2x + 3}$

Now Try Problem 43

The division method works best when the terms of the divisor and the dividend are written in descending powers of the variable. If the powers in the dividend or divisor are not in descending order, we use the commutative property of addition to write them that way.

EXAMPLE 5 Divide $4x^2 + 2x^3 + 12 - 2x$ by $x + 3$.

Strategy We will write the dividend in descending powers of x and use the long division method.

Why It is easier to align like terms in columns when the powers of the variable are written in descending order.

Solution

$$\begin{array}{r}
2x^2 - 2x + 4 \\
x + 3\overline{)2x^3 + 4x^2 - 2x + 12} \\
-(2x^3 + 6x^2) \qquad\qquad \\
-2x^2 - 2x \qquad \\
-(-2x^2 - 6x) \qquad \\
4x + 12 \\
-(4x + 12) \\
0
\end{array}$$

The first division: $\frac{2x^3}{x} = 2x^2$.

The second division: $\frac{-2x^2}{x} = -2x$.

The third division: $\frac{4x}{x} = 4$.

Check: $(x + 3)(2x^2 - 2x + 4) = 2x^3 - 2x^2 + 4x + 6x^2 - 6x + 12$
$= 2x^3 + 4x^2 - 2x + 12$ The result checks.

Teaching Example 5 Divide: $(10x^3 + 12 + x^2 + 17x) \div (5x + 3)$
Answer: $2x^2 - x + 4$

Self Check 5 Divide $x^2 - 10x + 6x^3 + 4$ by $2x - 1$. $3x^2 + 2x - 4$

Now Try Problem 45

When we write the terms of a dividend in descending powers, we must determine whether some powers of the variable are missing. If any are missing, insert **placeholder terms** with a coefficient of 0 or leave blank spaces for them. This keeps like terms in the same column, which is necessary when performing the subtraction in vertical form.

EXAMPLE 6 Divide: $\frac{27x^3 + 1}{3x + 1}$

Strategy The divisor is $3x + 1$. The dividend, $27x^3 + 1$, does not have an x^2-term or an x-term. We will insert a $0x^2$ term and a $0x$ term as placeholders, and use the long division method.

Why We insert placeholder terms so that like terms will be aligned in the same column when we subtract.

Solution

$$\begin{array}{r}
9x^2 - 3x + 1 \\
3x + 1\overline{)27x^3 + \mathbf{0x^2} + \mathbf{0x} + 1} \\
-(27x^3 + 9x^2) \qquad\qquad \\
-9x^2 + \mathbf{0x} \qquad \\
-(-9x^2 - 3x) \qquad \\
3x + 1 \\
-(3x + 1) \\
0
\end{array}$$

The first division: $\frac{27x^3}{3x} = 9x^2$.

The second division: $\frac{-9x^2}{3x} = -3x$.

The third division: $\frac{3x}{3x} = 1$.

Check: $(3x + 1)(9x^2 - 3x + 1) = 27x^3 - 9x^2 + 3x + 9x^2 - 3x + 1$
$= 27x^3 + 1$ The result checks.

Teaching Example 6 Divide: $\frac{27x^3 - 1}{3x - 1}$
Answer: $9x^2 + 3x + 1$

Self Check 6 Divide: $\frac{x^2 - 9}{x - 3}$. Check the result. $x + 3$

Now Try Problem 49

EXAMPLE 7

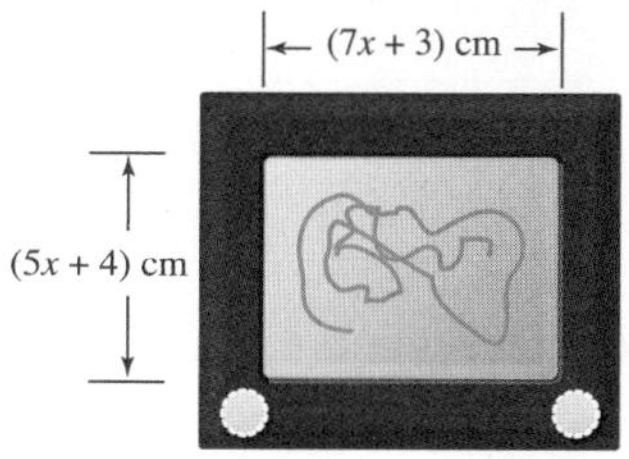

Toys. The area of an Etch A Sketch screen is represented by the polynomial $(35x^2 + 43x + 12)$ in.^{2}. If the width of the screen is $(5x + 4)$ inches, what polynomial represents its length?

Strategy We will find the length of the screen by dividing its area, $(35x^2 + 43x + 12)$ in.^{2}, by its width, $(5x + 4)$ inches.

Why Recall that the area of a rectangle is given by the formula $A = lw$. If we divide both sides of the formula by w, we see that $l = \frac{A}{w}$.

Solution

$$\text{Length} = \frac{\textbf{area}}{\textbf{width}}$$ This is the formula for the length of a rectangle.

$$= \frac{35x^2 + 43x + 12}{5x + 4}$$ Substitute $35x^2 + 43x + 12$ for the area and $5x + 4$ for the width.

To divide $35x^2 + 43x + 12$ by $5x + 4$, we use long division.

$$\begin{array}{r}
7x + 3 \\
5x + 4\overline{)35x^2 + 43x + 12} \\
-(35x^2 + 28x) \\
\hline
15x + 12 \\
-(15x + 12) \\
\hline
0
\end{array}$$

The first division: $\frac{35x^2}{5x} = 7x$.

The second division: $\frac{15x}{5x} = 3$.

The length of the Etch A Sketch screen is $(7x + 3)$ inches.

Teaching Example 7 Technology. The area of the rectangular display on a Blackberry Smartphone is represented by the polynomial $(48x^2 + 96x + 36)$ in.^{2}. If the width of the display is $(6x + 9)$ inches, what polynomial represents its length?
Answer: $(8x + 4)$ in.

Self Check 7 **Televisions.** The area of the rectangular screen of a plasma TV is represented by the polynomial $(36x^2 - 51x - 21)$ in^2. If the width of the screen is $(4x - 7)$ inches, what polynomial represents its length?
$(9x + 3)$ in.

Now Try Problem 103

SECTION 5.8 STUDY SET

VOCABULARY

Fill in the blanks.

1. The expression $\frac{18x^7}{9x^4}$ is a monomial divided by a monomial.

2. The expression $\frac{6x^3y - 4x^2y^2 + 8xy^3 - 2y^4}{2x^4}$ is a polynomial divided by a monomial.

3. The expression $\frac{x^2 - 8x + 12}{x - 6}$ is a trinomial divided by a binomial.

4. Divisor, Quotient, Dividend, Remainder

$$\begin{array}{r}
x - 2 \\
x - 6\overline{)x^2 - 8x - 4} \\
-(x^2 - 6x) \\
\hline
-2x - 4 \\
-(-2x + 12) \\
\hline
-16
\end{array}$$

(Divisor: $x - 6$; Quotient: $x - 2$; Dividend: $x^2 - 8x - 4$; Remainder: -16)

CONCEPTS

5. The long division method is a series of four steps that are repeated. Put them in the correct order:

subtract multiply bring down divide

Divide, multiply, subtract, bring down

6. In the following long divisions, find the answer to the subtraction that must be performed at this stage.

a. $\begin{array}{r} 3x \\ 2x + 1\overline{)6x^2 + 9x - 10} \\ -(6x^2 + 3x) \\ \hline 6x \end{array}$

b. $\begin{array}{r} x \\ x - 7\overline{)x^2 + 10x + 21} \\ -(x^2 - 7x) \\ \hline 17x \end{array}$

c. $\begin{array}{r} x \\ x - 7\overline{)x^2 - 9x - 6} \\ -(x^2 - 7x) \\ \hline -2x \end{array}$

d. $\begin{array}{r} x \\ x + 4\overline{)x^2 + 0x - 16} \\ -(x^2 + 4x) \\ \hline -4x \end{array}$

▶ Selected exercises available online at www.webassign.net/brookscole

7. Fill in the blanks: To check an answer of a long division, we use the fact that

Divisor · quotient + remainder = dividend

8. Check to see whether the following result of a long division is correct.

$$\frac{x^2 + 4x - 20}{x - 3} = x + 7 + \frac{1}{x - 3}$$ It is correct.

NOTATION

Complete each solution.

9. $\frac{28x^5 - x^3 + 5x^2}{7x^2} = \frac{28x^5}{7x^2} - \frac{x^3}{7x^2} + \frac{5x^2}{7x^2}$

$= 4x^{5-2} - \frac{x^{3-2}}{7} + \frac{5x^{2-2}}{7}$

$= 4x^3 - \frac{x}{7} + \frac{5}{7}$

10.
```
          x + 2
x + 2)x² + 4x + 5
     -(x² + 2x)
            2x + 5
          -(2x + 4)
                 1
```

11. Write the polynomial $2x^2 - 1 + 5x^4$ in descending powers of x and insert placeholders for each missing term. $5x^4 + 0x^3 + 2x^2 + 0x - 1$

▶ 12. True or false: $6x + 4 + \frac{-3}{x + 2} = 6x + 4 - \frac{3}{x + 2}$ True

GUIDED PRACTICE

Divide the monomials. See Example 1.

▶ 13. $\frac{x^5}{x^2}$ x^3

▶ 14. $\frac{a^{12}}{a^8}$ a^4

▶ 15. $\frac{12h^8}{9h^6}$ $\frac{4h^2}{3}$

16. $\frac{22b^9}{6b^6}$ $\frac{11b^3}{3}$

▶ 17. $\frac{-3d^4}{15d^8}$ $-\frac{1}{5d^4}$

18. $\frac{-4x^3}{16x^5}$ $-\frac{1}{4x^2}$

19. $\frac{10s^2}{s^3}$ $\frac{10}{s}$

20. $\frac{16y^3}{y^4}$ $\frac{16}{y}$

21. $\frac{8x^3y^2}{40xy^6}$ $\frac{x^2}{5y^4}$

22. $\frac{3y^3z}{18yz^6}$ $\frac{y^2}{6z^5}$

▶ 23. $\frac{-16r^3y^2}{-4r^2y^7}$ $\frac{4r}{y^5}$

24. $\frac{-35xz^6}{-7x^8z^2}$ $\frac{5z^4}{x^7}$

Divide the polynomial by the monomial. See Example 2.

25. $\frac{6x + 3}{3}$ $2x + 1$

26. $\frac{8x + 4}{4}$ $2x + 1$

27. $\frac{a - a^3 + a^4}{a^4}$ $\frac{1}{a^3} - \frac{1}{a} + 1$

▶ 28. $\frac{b^2 + b^3 - b^4}{b^4}$ $\frac{1}{b^2} + \frac{1}{b} - 1$

29. $\frac{6h^{12} + 48h^9}{24h^{10}}$ $\frac{h^2}{4} + \frac{2}{h}$

30. $\frac{4x^{14} - 36x^8}{36x^{12}}$ $\frac{x^2}{9} - \frac{1}{x^4}$

31. $\frac{9s^8 - 18s^5 + 12s^4}{3s^3}$ $3s^5 - 6s^2 + 4s$

▶ 32. $\frac{16b^{10} + 4b^6 - 20b^4}{4b^2}$ $4b^8 + b^4 - 5b^2$

▶ 33. $\frac{7c^5 + 21c^4 - 14c^3 - 35c}{7c^2}$ $c^3 + 3c^2 - 2c - \frac{5}{c}$

34. $\frac{12r^{15} - 48r^{12} + r^{10} - 18r^8}{6r^{10}}$ $2r^5 - 8r^2 + \frac{1}{6} - \frac{3}{r^2}$

35. $\frac{-25x^2y^3 + 30xy^2 - 5xy}{-5x^2y^2}$ $5y - \frac{6}{x} + \frac{1}{xy}$

▶ 36. $\frac{-30a^4b^4 - 15a^3b - 10a^2b^2}{-10a^2b^3}$ $3a^2b + \frac{3a}{2b^2} + \frac{1}{b}$

Perform each division. See Examples 3 and 4.

▶ 37. Divide $x^2 + 8x + 12$ by $x + 2$. $x + 6$

▶ 38. Divide $x^2 + 5x + 6$ by $x + 2$. $x + 3$

39. Divide $x^2 - 5x + 6$ by $x - 3$. $x - 2$

40. Divide $x^2 - 12x + 32$ by $x - 4$. $x - 8$

41. $\frac{2x^2 + 5x + 2}{2x + 3}$ $x + 1 + \frac{-1}{2x + 3}$

▶ 42. $\frac{3x^2 - 8x + 3}{3x - 2}$ $x - 2 + \frac{-1}{3x - 2}$

43. $\frac{6x^2 - 11x + 2}{3x - 1}$ $2x - 3 + \frac{-1}{3x - 1}$

▶ 44. $\frac{4x^2 + 6x - 1}{2x + 1}$ $2x + 2 + \frac{-3}{2x + 1}$

Perform each division. See Example 5.

45. $x + 2\overline{)3x + 2x^2 - 2}$ $2x - 1$

▶ 46. $x + 3\overline{)-x + 2x^2 - 21}$ $2x - 7$

47. $(3 + 11x + 10x^2) \div (5x + 3)$ $2x + 1$

▶ 48. $(6x + 1 + 9x^2) \div (3x + 1)$ $3x + 1$

Perform each division. See Example 6.

49. $(a^2 - 25) \div (a + 5)$ $a - 5$

▶ 50. $(b^2 - 36) \div (b + 6)$ $b - 6$

51. $(x^2 - 1) \div (x - 1)$ $x + 1$

52. $(x^2 - 9) \div (x + 3)$ $x - 3$

▶ 53. $\frac{4x^2 - 9}{2x + 3}$ $2x - 3$

▶ 54. $\frac{25x^2 - 16}{5x - 4}$ $5x + 4$

55. $\frac{81b^2 - 49}{9b - 7}$ $9b + 7$

56. $\frac{16t^2 - 121}{4t + 11}$ $4t - 11$

TRY IT YOURSELF

Perform each division.

57. Divide $y^2 + 13y + 13$ by $y + 1$. $y + 12 + \frac{1}{y + 1}$

▶ 58. Divide $z^2 + 7z + 14$ by $z + 3$. $z + 4 + \frac{2}{z + 3}$

59. $\frac{15a^8b^2 - 10a^2b^5}{5a^3b^2}$ $3a^5 - \frac{2b^3}{a}$

60. $\frac{9a^4b^3 - 16a^3b^4}{12a^2b}$ $\frac{3a^2b^2}{4} - \frac{4ab^3}{3}$

61. $3x + 2\overline{)2 + 7x + 6x^3 + 10x^2}$ $2x^2 + 2x + 1$

▶ 62. $3x - 2\overline{)4x - 4 + 6x^3 - x^2}$ $2x^2 + x + 2$

▶ 63. $\frac{8x^9 - 32x^6}{4x^4}$ $2x^5 - 8x^2$

▶ 64. $\frac{30y^8 + 40y^7}{10y^6}$ $3y^2 + 4y$

65. $\frac{6a^2 + 5a - 6}{2a + 3}$ $3a - 2$

66. $\frac{3b^2 - 5b + 2}{3b - 2}$ $b - 1$

67. $\frac{45m^{10}}{9m^5}$ $5m^5$

▶ 68. $\frac{24n^{12}}{8n^4}$ $3n^8$

69. $\frac{3b^2 + 11b + 6}{3b + 2}$ $b + 3$

▶ 70. $\frac{8a^2 + 2a - 3}{2a - 1}$ $4a + 3$

71. $2x - 7\overline{)-x - 21 + 2x^2}$ $x + 3$

72. $2x - 1\overline{)x - 2 + 6x^2}$ $3x + 2$

73. $\frac{x^3 + 1}{x + 1}$ $x^2 - x + 1$

74. $\frac{x^3 - 8}{x - 2}$ $x^2 + 2x + 4$

75. $\frac{-65rs^2}{15r^2s^5}$ $-\frac{13}{3rs^3}$

76. $\frac{112uz^4}{-42u^3z^8}$ $-\frac{8}{3u^2z^4}$

77. $\frac{-18w^6 - 9}{9w^4}$ $-2w^2 - \frac{1}{w^4}$

78. $\frac{-40f^4 + 16}{8f^3}$ $-5f + \frac{2}{f^3}$

79. $\frac{9m - 6}{m}$ $9 - \frac{6}{m}$

80. $\frac{10n - 6}{n}$ $10 - \frac{6}{n}$

81. $\frac{y^3 + y}{y - 2}$ $y^2 + 2y + 5 + \frac{10}{y - 2}$

82. $\frac{a^3 + a}{a + 3}$ $a^2 - 3a + 10 + \frac{-30}{a + 3}$

83. $\frac{5x^4 - 10x}{25x^3}$ $\frac{x}{5} - \frac{2}{5x^2}$

84. $\frac{24x^7 - 32x^2}{16x^3}$ $\frac{3x^4}{2} - \frac{2}{x}$

85. $3 + 4x\overline{)3 - 5x^2 - 2x + 4x^3}$ $x^2 - 2x + 1$

86. $2x + 3\overline{)7x^2 - 3 + 4x + 2x^3}$ $x^2 + 2x - 1$

87. $(x^2 + 6x + 15) \div (x + 5)$ $x + 1 + \frac{10}{x + 5}$

88. $(x^2 + 10x + 30) \div (x + 6)$ $x + 4 + \frac{6}{x + 6}$

89. $\frac{12x^3y^2 - 8x^2y - 4x}{4xy}$ $3x^2y - 2x - \frac{1}{y}$

90. $\frac{12a^2b^2 - 8a^2b - 4ab}{4ab}$ $3ab - 2a - 1$

91. $a - 5\overline{)a^2 - 17a + 64}$ $a - 12 + \frac{4}{a - 5}$

92. $b - 2\overline{)b^2 - 4b + 6}$ $b - 2 + \frac{2}{b - 2}$

93. $\frac{a^3 - 1}{a - 1}$ $a^2 + a + 1$

94. $\frac{y^3 + 8}{y + 2}$ $y^2 - 2y + 4$

95. $\frac{6x^3 + x^2 + 2x + 1}{3x - 1}$ $2x^2 + x + 1 + \frac{2}{3x - 1}$

96. $\frac{3y^3 - 4y^2 + 2y + 3}{y + 3}$ $3y^2 - 13y + 41 + \frac{-120}{y + 3}$

97. $\frac{8x^{17}y^{20}}{16x^{15}y^{30}}$ $\frac{x^2}{2y^{10}}$

98. $\frac{21a^{30}b^{15}}{14a^{40}b^{12}}$ $\frac{3b^3}{2a^{10}}$

99. $(6m^2 - m - 40) \div (2m + 5)$ $3m - 8$

100. $(12d^2 - 20d + 3) \div (6d - 1)$ $2d - 3$

Look Alikes . . .

Perform the indicated operations.

101. a. $\frac{16x^2 - 16x - 5}{4x}$ $4x - 4 - \frac{5}{4x}$
b. $\frac{16x^2 - 16x - 5}{4x + 1}$ $4x - 5$

102. a. $\frac{9x^3 + 3x^2 + 4x + 4}{3x}$ $3x^2 + x + \frac{4}{3} + \frac{4}{3x}$
b. $\frac{9x^3 + 3x^2 + 4x + 4}{3x + 2}$ $3x^2 - x + 2$

APPLICATIONS

103. **Furnace Filters.** The area of the rectangular-shaped furnace filter is $(x^2 - 2x - 24)$ square inches. What expression represents its height? $(x - 6)$ in.

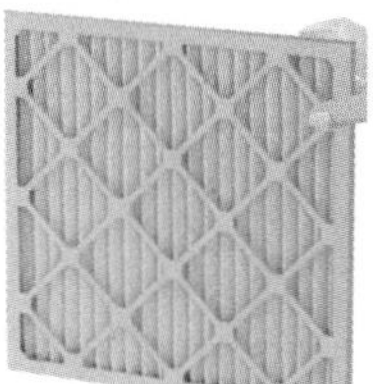

$(x + 4)$ in.

104. **Mini-Blinds.** The area covered by the mini-blinds is $(3x^3 - 6x)$ square feet. What expression represents the height of the blinds? $(x^2 - 2)$ ft

105. **Pool.** The rack shown in the illustration is used to set up the balls for a game of pool. If the perimeter of the rack, in inches, is given by the polynomial $6x^2 - 3x + 9$, what expression represents the approximate length of one side? $(2x^2 - x + 3)$ in.

106. **Communications.** Telephone poles were installed every $(2x - 3)$ feet along a stretch of railroad track $(8x^3 - 6x^2 + 5x - 21)$ feet long. What expression represents the number of poles that were used?
$(4x^2 + 3x + 7) + 1 = (4x^2 + 3x + 8)$ poles

WRITING

107. Explain how to check the following long division.

$$\begin{array}{r} x + 5 \\ 3x + 5\overline{)3x^2 + 20x - 5} \\ -(3x^2 + 5x) \\ 15x - 5 \\ -(15x + 25) \\ -30 \end{array}$$

108. Explain the error: $\frac{18x^2 + \cancel{6x}^{1}}{\cancel{6x}_{1}} = 18x^2 + 1$

109. How do you know when to stop the long division method when dividing polynomials?

110. When dividing $x^3 + 1$ by $x + 1$, why is it helpful to write $x^3 + 1$ as $x^3 + 0x^2 + 0x + 1$?

REVIEW

111. Write an equation of the line with slope $-\frac{11}{6}$ that passes through $(2, -6)$. Write the answer in slope–intercept form. $y = -\frac{11}{6}x - \frac{7}{3}$

112. Solve $S = 2\pi rh + 2\pi r^2$ for h. $h = \frac{S - 2\pi r^2}{2\pi r}$

CHALLENGE PROBLEMS

Perform each division.

113. $\dfrac{6a^3 - 17a^2b + 14ab^2 - 3b^3}{2a - 3b}$ $3a^2 - 4ab + b^2$

114. $(2x^4 + 3x^3 + 3x^2 - 5x - 3) \div (2x^2 - x - 1)$ $x^2 + 2x + 3$

115. $(x^6 + 2x^4 - 6x^2 - 9) \div (x^2 + 3)$ $x^4 - x^2 - 3$

116. $\dfrac{6x^{6m}y^{6n} + 15x^{4m}y^{7n} - 24x^{2m}y^{8n}}{3x^{2m}y^n}$ $2x^{4m}y^{5n} + 5x^{2m}y^{6n} - 8y^{7n}$

117. $\dfrac{a^8 + a^6 - 4a^4 + 5a^2 - 3}{a^4 + 2a^2 - 3}$ $a^4 - a^2 + 1$

▶ **118.** $\dfrac{-17x^2 + 5x + x^4 + 2}{4x + x^2 - 1}$ $x^2 - 4x + \dfrac{x + 2}{x^2 + 4x - 1}$

5 Summary & Review

SECTION 5.1 Rules for Exponents

DEFINITIONS AND CONCEPTS / EXAMPLES

An **exponent** indicates repeated multiplication. It tells how many times the **base** is to be used as a factor.

Exponent → $x^n = \underbrace{x \cdot x \cdot x \cdot \cdots \cdot x}_{n \text{ factors of } x}$ ← Base

Identify the base and the exponent in each expression.

$2^6 = 2 \cdot 2 \cdot 2 \cdot 2 \cdot 2 \cdot 2$ 2 is the base and 6 is the exponent.

$(-xy)^3 = (-xy)(-xy)(-xy)$ $-xy$ is the base and 3 is the exponent.

$5t^4 = 5 \cdot t \cdot t \cdot t \cdot t$ t is the base and 4 is the exponent.

$8^1 = 8$ 8 is the base and 1 is the exponent.

Rules for Exponents: If m and n represent integers and there are no divisions by 0, then

Product rule: $x^m x^n = x^{m+n}$

Quotient rule: $\dfrac{x^m}{x^n} = x^{m-n}$

Power rule: $(x^m)^n = x^{m \cdot n} = x^{mn}$

Power of a product rule: $(xy)^m = x^m y^m$

Power of a quotient rule: $\left(\dfrac{x}{y}\right)^n = \dfrac{x^n}{y^n}$

Simplify each expression:

$5^2 5^7 = 5^{2+7} = 5^9$

$\dfrac{t^7}{t^3} = t^{7-3} = t^4$

$(6^3)^7 = 6^{3 \cdot 7} = 6^{21}$

$(2p)^5 = 2^5 p^5 = 32p^5$

$\left(\dfrac{s}{4}\right)^4 = \dfrac{s^4}{4^4} = \dfrac{s^4}{256}$

REVIEW EXERCISES

1. Identify the base and the exponent in each expression.

a. n^{12} n, 12
b. $(2x)^6$ $2x$, 6
c. $3r^4$ r, 4
d. $(y - 7)^3$ $y - 7$, 3

2. Write each expression in an equivalent form using an exponent.

a. $m \cdot m \cdot m \cdot m \cdot m$ m^5
b. $-3 \cdot x \cdot x \cdot x \cdot x$ $-3x^4$
c. $(x + 8)(x + 8)$ $(x + 8)^2$
d. $\left(\frac{1}{2}pq\right)\left(\frac{1}{2}pq\right)\left(\frac{1}{2}pq\right)$ $\left(\frac{1}{2}pq\right)^3$

Simplify each expression. Assume there are no divisions by 0.

3. $7^4 \cdot 7^8$ 7^{12}

4. $mmnn$ m^2n^2

5. $(y^7)^3$ y^{21}

6. $(3x)^4$ $81x^4$

7. $\dfrac{b^{12}}{b^3}$ b^9

8. $-b^3b^4b^5$ $-b^{12}$

9. $(-16s^3)^2s^4$ $256s^{10}$

10. $(2.1x^2y)^2$ $4.41x^4y^2$

11. $[(-9)^3]^5$ $(-9)^{15}$

12. $(a^5)^3(a^2)^4$ a^{23}

13. $\left(\dfrac{1}{2}x^2x^3\right)^3$ $\frac{1}{8}x^{15}$

14. $\left(\dfrac{x^7}{3xy}\right)^2$ $\dfrac{x^{12}}{9y^2}$

15. $\dfrac{(m - 25)^{16}}{(m - 25)^4}$ $(m - 25)^{12}$

16. $\dfrac{(5y^2z^3)^3}{(yz)^5}$ $125yz^4$

17. $\dfrac{a^5a^4a^5}{a^2a}$ a^{11}

18. $\dfrac{(cd)^9}{(cd)^4}$ c^5d^5

Find an expression that represents the area or the volume of each figure, whichever is appropriate.

19. $64x^{12}$ in.^{3}

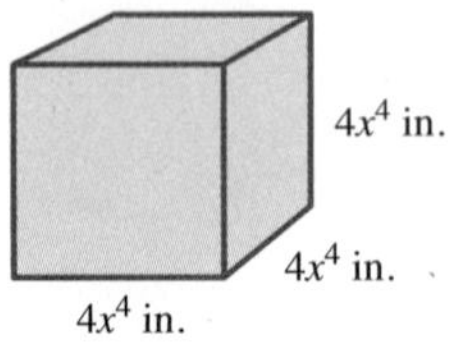

20. y^4 ft^2

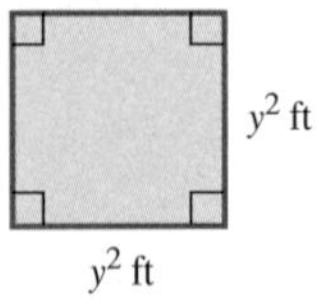

SECTION 5.2 Zero and Negative Exponents

DEFINITIONS AND CONCEPTS	EXAMPLES
Rules for exponents: For any nonzero real numbers x and y and any integers m and n,	Simplify each expression. Do not use negative exponents in the answer.
Zero exponent: $x^0 = 1$	$5^0 = 1$
Negative exponents: $x^{-n} = \frac{1}{x^n}$	$4^{-2} = \frac{1}{4^2} = \frac{1}{16}$ and $7c^{-6} = \frac{7}{c^6}$
Negative to positive rules: $\frac{1}{x^{-n}} = x^n \qquad \frac{x^{-m}}{y^{-n}} = \frac{y^n}{x^m}$	$\frac{1}{t^{-8}} = t^8$ and $\frac{2^{-4}}{x^{-6}} = \frac{x^6}{2^4} = \frac{x^6}{16}$
Negative exponents and reciprocals: $\left(\frac{x}{y}\right)^{-n} = \left(\frac{y}{x}\right)^n$	$\left(\frac{x}{10}\right)^{-3} = \left(\frac{10}{x}\right)^3 = \frac{10^3}{x^3} = \frac{1,000}{x^3}$

REVIEW EXERCISES

Simplify each expression. Do not use negative exponents in the answer.

21. x^0 1
22. $(3x^2y^2)^0$ 1
23. $3x^0$ 3
24. 10^{-3} $\frac{1}{1,000}$
25. -5^{-2} $-\frac{1}{25}$
26. $\frac{t^4}{t^{10}}$ $\frac{1}{t^6}$
27. $\frac{8}{x^{-5}}$ $8x^5$
28. $-6y^4y^{-5}$ $-\frac{6}{y}$
29. $\frac{7^{-2}}{2^{-3}}$ $\frac{8}{49}$
30. $(x^{-3}x^{-4})^{-2}$ x^{14}
31. $\left(\frac{-3r^4r^{-3}}{r^{-3}r^7}\right)^3$ $-\frac{27}{r^9}$
32. $\left(\frac{4z^4}{z^3}\right)^{-2}$ $\frac{1}{16z^2}$
33. $\frac{3^{-2}c^3d^3}{2^{-3}c^2d^8}$ $\frac{8c}{9d^5}$
34. $\frac{t^{-30}}{t^{-60}}$ t^{30}
35. $\frac{w(w^{-3})^{-4}}{w^{-9}}$ w^{22}
36. $\left(\frac{4}{f^4}\right)^{-10}$ $\frac{f^{40}}{4^{10}}$

SECTION 5.3 Scientific Notation

DEFINITIONS AND CONCEPTS	EXAMPLES
A positive number is written in **scientific notation** when it is written in the form $N \times 10^n$, where $1 \le N < 10$ and n is an integer.	Write each number in scientific notation. $32,500 = 3.25 \times 10^4$ (4 decimal places) and $0.0025 = 2.5 \times 10^{-3}$ (3 decimal places) Write each number in standard notation. $1.91 \times 10^5 = 191,000$ (5 decimal places) and $4.7 \times 10^{-6} = 0.0000047$ (6 decimal places)
Scientific notation provides an easier way to perform computations involving very large or very small numbers.	Use scientific notation to perform the calculation: $\frac{684,000,000}{456,000} = \frac{6.84 \times 10^8}{4.56 \times 10^5} = \frac{6.84}{4.56} \times \frac{10^8}{10^5} = 1.5 \times 10^3$ or 1,500

REVIEW EXERCISES

Write each number in scientific notation.

37. 720,000,000
7.2×10^8

38. 9,370,000,000,000,000
9.37×10^{15}

39. 0.00000000942
9.42×10^{-9}

40. 0.00013
1.3×10^{-4}

41. 0.018×10^{-2}
1.8×10^{-4}

42. 853×10^3
8.53×10^5

Write each number in standard notation.

43. 1.26×10^5
126,000

44. 3.919×10^{-8}
0.00000003919

45. 2.68×10^0 2.68

46. 5.76×10^1 57.6

Evaluate each expression by first writing each number in scientific notation. Express the result in scientific notation and standard notation.

47. $\dfrac{(0.000012)(0.000004)}{0.00000016}$
3.0×10^{-4}; 0.0003

48. $\dfrac{(4{,}800{,}000)(20{,}000{,}000)}{600{,}000}$
1.6×10^8; 160,000,000

49. World Population. As of January 2007, the world's population was estimated to be 6.57 billion. Write this number in standard notation and in scientific notation.
6,570,000,000; 6.57×10^9

50. Atoms. The illustration shows a cross section of an atom. How many nuclei (plural for nucleus), placed end to end, would it take to stretch across the atom? $1.0 \times 10^5 = 100{,}000$

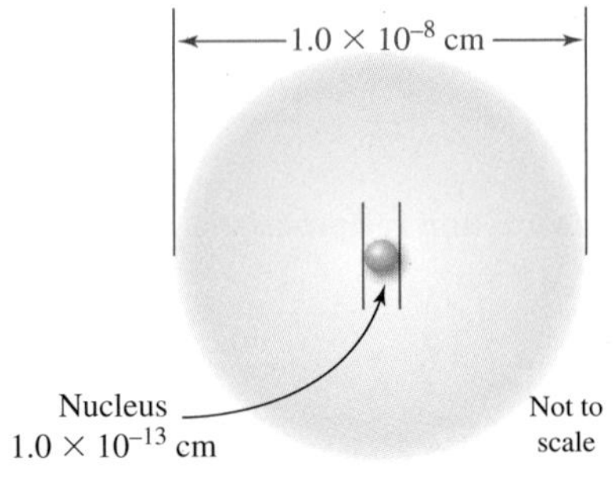

SECTION 5.4 Polynomials

DEFINITIONS AND CONCEPTS	EXAMPLES
A **polynomial** is a single term or a sum of terms in which all variables have whole-number exponents and no variable appears in a denominator.	Polynomials: 32, $-5x^2y^3$, $7p^3 - 14q^3$, $4m^2 + 5m - 12$ Not Polynomials: $y^2 - y^{-5}$, $4x^3 - \frac{7}{x} + 3x$
A polynomial with exactly one term is called a **monomial.** A polynomial with exactly two terms is called a **binomial.** A polynomial with exactly three terms is called a **trinomial.**	***Monomials***: $3x^2$, $-12m^3n^2$ ***Binomials***: $2y^3 + 3y$, $87t - 25$ ***Trinomials***: $3p^2 - 7p + 12$, $4p^2q^3 - 8p^2q^2 + 12p^2q$
The **coefficient of a term** is its numerical factor. The **degree of a term** of a polynomial in one variable is the value of the exponent on the variable. If a polynomial has more than one variable, the **degree of a term** is the sum of the exponents on the variables. The **degree of a nonzero constant** is 0.	***Term*** / ***Coefficient*** / ***Degree of the term*** $6a^7$ / 6 / 7 $-7.3x^5y^4$ / -7.3 / $5 + 4 = 9$ 32 / 32 / 0
The **degree of a polynomial** is equal to the highest degree of any term of the polynomial.	***Polynomial*** / ***Degree of the polynomial*** $7m^3 - 4m^2 + 5m - 12$ / 3 $\frac{1}{2}a^4b + \frac{3}{4}a^3b^2 - \frac{2}{3}a^2b^4$ / $2 + 4 = 6$
To **evaluate a polynomial** for a given value, substitute the value for the variable and follow the rules for the order of operations.	Evaluate $3x^2 - 4x + 2$ for $x = 2$. $3x^2 - 4x + 2 = 3(2)^2 - 4(2) + 2$ Substitute 2 for each x. $= 3(4) - 8 + 2$ Evaluate $(2)^2$ first. $= 6$

We can **graph equations defined by polynomials** such as $y = x^2 - 2$, $y = -x^2$, and $y = x^3 + 1$.

The graph of $y = x^2 - 2$ is a cup-shaped curve called a **parabola.**

Graph: $y = x^2 - 2$

Find several solutions of the equation, plot them on a rectangular coordinate system, and then draw a smooth curve passing through the points.

$y = x^2 - 2$

x	y	(x, y)
−2	2	(−2, 2)
−1	−1	(−1, −1)
0	−2	(0, −2)
1	−1	(1, −1)
2	2	(2, 2)

Select x-values. Find y-values. Plot points.

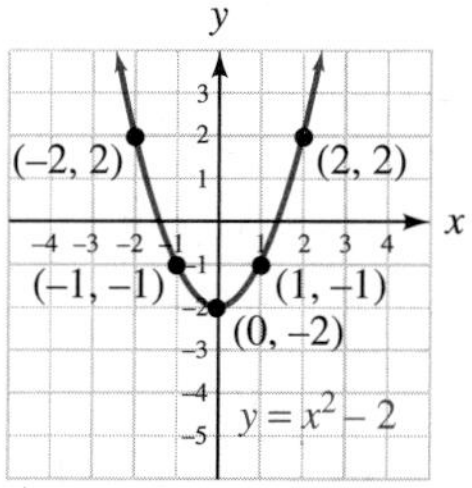

REVIEW EXERCISES

51. Consider the polynomial $3x^3 - x^2 + x + 10$.

a. How many terms does the polynomial have? 4

b. What is the lead term? $3x^3$

c. What is the coefficient of each term? 3, −1, 1, 10

d. What is the constant term? 10

52. Find the degree of each polynomial and classify it as a monomial, binomial, trinomial, or none of these.

a. $13x^7$ 7th, monomial

b. $-16a^2b$ 3rd, monomial

c. $5^3x + x^2$ 2nd, binomial

d. $-3x^5 + x - 1$ 5th, trinomial

e. $9xy^2 + 21x^3y^3$ 6th, binomial

f. $4s^4 - 3s^2 + 5s + 4$ 4th, none of these

53. Evaluate $-x^5 - 3x^4 + 3$ for $x = 0$ and $x = -2$. 3, −13

54. Diving. The number of inches that the woman deflects the diving board is given by the polynomial $0.1875x^2 - 0.0078125x^3$ where x is the number of feet that she stands from the front anchor point of the board. Find the amount of deflection if she stands on the end of the diving board, 8 feet from the anchor point. 8 in.

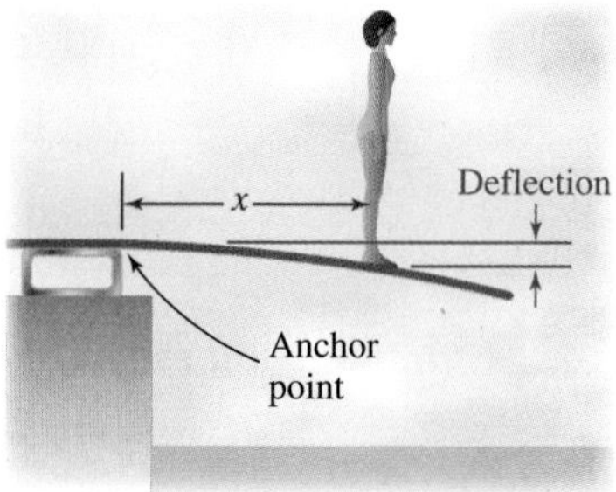

Construct a table of solutions like the one shown here and then graph the equation. See AIE Appendix 3.

55. $y = x^2$

x	−2	−1	0	1	2
y	4	1	0	1	4

56. $y = x^3 + 1$

x	−2	−1	0	1	2
y	−7	0	1	2	9

SECTION 5.5 Adding and Subtracting Polynomials

DEFINITIONS AND CONCEPTS	EXAMPLES
To **simplify a polynomial,** combine like terms.	Simplify: $3r^4 - 4r^3 + 7r^4 + 8r^2$ $= 10r^4 - 4r^3 + 8r^2$ Combine like terms. Think: $(3 + 7)r^4 = 10r^4$.
To **add polynomials,** combine their like terms.	Add: $(4x^2 + 9x + 4) + (3x^2 - 5x - 1)$ $= (4x^2 + 3x^2) + (9x - 5x) + (4 - 1)$ Group like terms. $= 7x^2 + 4x + 3$ Combine like terms.
To **subtract two polynomials,** change the signs of the terms of the polynomial being subtracted, drop the parentheses, and combine like terms.	Subtract: $(8a^3b - 4ab^2) - (-3a^3b + 9ab^2)$ $= 8a^3b - 4ab^2 + 3a^3b - 9ab^2$ Change the sign of each term of $-3a^3b + 9ab^2$ and drop the parentheses. $= 11a^3b - 13ab^2$ Combine like terms.

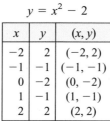

REVIEW EXERCISES

Simplify each polynomial and write the result in descending powers of one variable.

57. $6y^3 + 8y^4 + 7y^3 + (-8y^4)$ $13y^3$

58. $4a^2b + 5 - 6a^3b - 3a^2b + 2a^3b + 1$ $-4a^3b + a^2b + 6$

59. $\frac{5}{6}x^2 + \frac{1}{3}y^2 - \frac{1}{4}x^2 - \frac{3}{4}xy + \frac{2}{3}y^2$ $\frac{7}{12}x^2 - \frac{3}{4}xy + y^2$

60. $7.6c^5 - 2.1c^3 - 0.9c^5 + 8.1c^4$ $6.7c^5 + 8.1c^4 - 2.1c^3$

Perform the operations.

61. $(2r^6 + 14r^3) + (23r^6 - 5r^3 + 5r)$
$25r^6 + 9r^3 + 5r$

62. $(7.1a^2 + 2.2a - 5.8) - (3.4a^2 - 3.9a + 11.8)$
$3.7a^2 + 6.1a - 17.6$

63. $(3r^3s + r^2s^2 - 3rs^3 - 3s^4) + (r^3s - 8r^2s^2 - 4rs^3 + s^4)$
$4r^3s - 7r^2s^2 - 7rs^3 - 2s^4$

64. $\left(\frac{7}{8}m^4 - \frac{1}{5}m^3\right) - \left(\frac{1}{4}m^4 + \frac{1}{5}m^3\right) - \frac{3}{5}m^3$ $\frac{5}{8}m^4 - m^3$

65. Find the difference when $(-3z^3 - 4z + 7)$ is subtracted from the sum of $(2z^2 + 3z - 7)$ and $(-4z^3 - 2z - 3)$.
$-z^3 + 2z^2 + 5z - 17$

66. Gardening. Find a polynomial that represents the length of the wooden handle of the shovel. $(x^2 + x + 3)$ in.

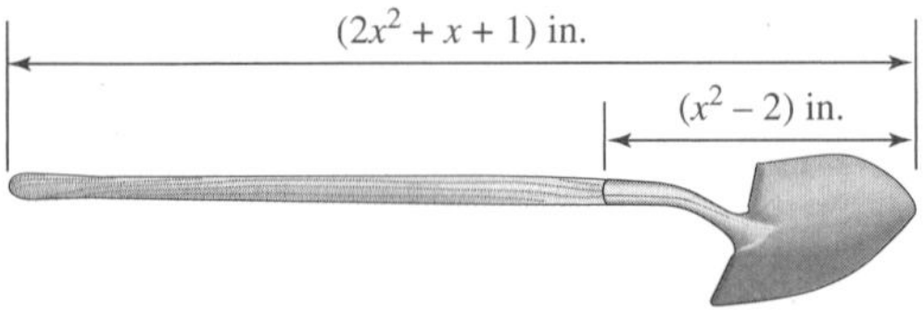

67. Add:

$$\begin{array}{r} 3x^2 + 5x + 2 \\ x^2 - 3x + 6 \\ \hline 4x^2 + 2x + 8 \end{array}$$

68. Subtract:

$$\begin{array}{r} 20x^3 \qquad\quad + 12x \\ -(12x^3 + 7x^2 - 7x) \\ \hline 8x^3 - 7x^2 + 19x \end{array}$$

SECTION 5.6 Multiplying Polynomials

DEFINITIONS AND CONCEPTS	EXAMPLES
To **multiply two monomials,** multiply the numerical factors (the coefficients) and then multiply the variable factors.	Multiply: $(5p^6)(2p^5) = (5 \cdot 2)(p^6 \cdot p^5)$ *Group the coefficients together and the variables together.* $= 10p^{11}$ *Think: $5 \cdot 2 = 10$ and $p^6 \cdot p^5 = p^{6+5} = p^{11}$.*
To **multiply a monomial and a polynomial,** multiply each term of the polynomial by the monomial.	Multiply: $3r^2(2r^4 + 7r^2 - 4)$ $= 3r^2(2r^4) + 3r^2(7r^2) + 3r^2(-4)$ *Distribute the multiplication by $3r^2$.* $= 6r^6 + 21r^4 - 12r^2$ *Multiply the monomials.*
To **multiply two binomials,** use the *FOIL method:* F: First O: Outer I: Inner L: Last	Multiply: $(3m + 4)(2m - 5) = \overset{F}{3m(2m)} + \overset{O}{3m(-5)} + \overset{I}{4(2m)} + \overset{L}{4(-5)}$ $= 6m^2 - 15m + 8m - 20$ *Multiply the monomials.* $= 6m^2 - 7m - 20$ *Combine like terms.*
To **multiply two polynomials,** multiply each term of one polynomial by each term of the other polynomial and then combine like terms.	Multiply: $(a - b)(6a^2 - 4ab + b^2)$ $= a(6a^2) + a(-4ab) + a(b^2) - b(6a^2) - b(-4ab) - b(b^2)$ $= 6a^3 - 4a^2b + ab^2 - 6a^2b + 4ab^2 - b^3$ *Multiply the monomials.* $= 6a^3 - 10a^2b + 5ab^2 - b^3$ *Combine like terms.*

When finding the **product of three polynomials,** begin by multiplying any two of them, and then multiply that result by the third polynomial.

Multiply: $-9x^4(x-1)(x-7) = -9x^4(x^2-7x-x+7)$ Multiply the two binomials.

$= -9x^4(x^2-8x+7)$ Combine like terms within the parentheses.

$= -9x^6+72x^5-63x^4$ Distribute the multiplication by $-9x^4$.

REVIEW EXERCISES

Multiply.

69. $(2x^2)(5x)$ $10x^3$
70. $(-6x^4z^3)(x^6z^2)$ $-6x^{10}z^5$
71. $5b^3 \cdot 6b^2 \cdot 4b^6$ $120b^{11}$
72. $\frac{2}{3}h^5(3h^9+12h^6)$ $2h^{14}+8h^{11}$
73. $3n^2(3n^2-5n+2)$ $9n^4-15n^3+6n^2$
74. $x^2y(y^2-xy)$ $x^2y^3-x^3y^2$
75. $2x(3x^4)(x+2)$ $6x^6+12x^5$
76. $-a^2b^2(-a^4b^2+a^3b^3-ab^4+7a)$ $a^6b^4-a^5b^5+a^3b^6-7a^3b^2$
77. $(x+3)(x+2)$ x^2+5x+6
78. $(2x+1)(x-1)$ $2x^2-x-1$
79. $(3t-3)(2t+2)$ $6t^2-6$
80. $(3n^4-5n^2)(2n^4-n^2)$ $6n^8-13n^6+5n^4$
81. $-a^5(a^2-b)(5a^2+b)$ $-5a^9+4a^7b+a^5b^2$
82. $6.6(a-1)(a+1)$ $6.6a^2-6.6$
83. $\left(3t-\frac{1}{3}\right)\left(6t+\frac{5}{3}\right)$ $18t^2+3t-\frac{5}{9}$
84. $(5.5-6b)(2-4b)$ $24b^2-34b+11$
85. $(2a-3)(4a^2+6a+9)$ $8a^3-27$
86. $(8x^2+x-2)(7x^2+x-1)$ $56x^4+15x^3-21x^2-3x+2$
87. Multiply using vertical form:
$$\begin{array}{r} 4x^2-2x+1 \\ 2x+1 \\ \hline \end{array}$$
$8x^3+1$
88. Refer to the illustration below. Find a polynomial that represents
    a. the perimeter of the base of the dishwasher. $(6x+10)$ in.
    b. the area of the base of the dishwasher. $(2x^2+11x-6)$ in.^{2}
    c. the volume occupied by the dishwasher. $(6x^3+33x^2-18x)$ in.^{3}

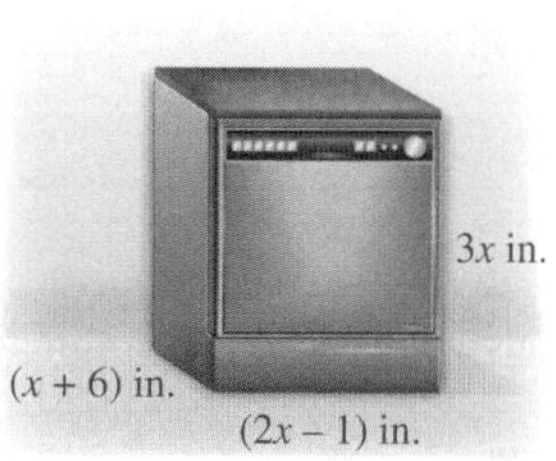

SECTION 5.7 Special Products

DEFINITIONS AND CONCEPTS	EXAMPLES

The following **special products** occur so often that it is worthwhile to learn their forms.

Square of a binomial:

$(A+B)^2 = A^2+2AB+B^2$

This is the square of a binomial sum.

$(A-B)^2 = A^2-2AB+B^2$

This is the square of a binomial difference.

Multiply: $(n+4)^2 = n^2 + 2(n)(4) + 4^2$

(The square of the first term, n. Twice the product of both terms. The square of the second term, 4.)

$= n^2+8n+16$

Multiply: $(5a-1)^2 = (5a)^2 + 2(5a)(-1) + (-1)^2$

(The square of the first term, $5a$. Twice the product of both terms. The square of the second term, -1.)

$= 25a^2-10a+1$

Multiplying the Sum and Difference of the Same Two Terms:

$(A+B)(A-B) = A^2-B^2$

Multiply: $(x+8)(x-8) = x^2 - 8^2$

(The square of the first term, x. The square of the second term, 8.)

$= x^2-64$

REVIEW EXERCISES

Find each product.

89. $(a - 3)^2$ $a^2 - 6a + 9$
90. $(m + 2)^3$ $m^3 + 6m^2 + 12m + 8$
91. $(x + 7)(x - 7)$ $x^2 - 49$
92. $(2x - 0.9)(2x + 0.9)$ $4x^2 - 0.81$
93. $(2y + 1)^2$ $4y^2 + 4y + 1$
94. $(y^2 + 1)(y^2 - 1)$ $y^4 - 1$
95. $(6r^2 + 10s)^2$ $36r^4 + 120r^2s + 100s^2$
96. $-(8a - 3c)^2$ $-64a^2 + 48ac - 9c^2$
97. $80s(r^2 + s^2)(r^2 - s^2)$ $80r^4s - 80s^5$
98. $4b(3b - 4)^2$ $36b^3 - 96b^2 + 64b$
99. $\left(t - \frac{3}{4}\right)^2$ $t^2 - \frac{3}{2}t + \frac{9}{16}$
100. $\left(x + \frac{4}{3}\right)^2$ $x^2 + \frac{8}{3}x + \frac{16}{9}$

Perform the operations.

101. $3(9x^2 + 3x + 7) - 2(11x^2 - 5x + 9)$ $5x^2 + 19x + 3$
102. $(5c - 1)^2 - (c + 6)(c - 6)$ $24c^2 - 10c + 37$

103. **Graphic Arts.** A Dr. Martin Luther King poster has his picture with a $\frac{1}{2}$-inch wide blue border around it. The length of the poster is $(x + 3)$ inches and the width is $(x - 1)$ inches. Find a polynomial that represents the area of the *picture* of Dr. King. $(x^2 - 4)$ in.^{2}

National Archives

104. Find a polynomial that represents the area of the triangle. $(50x^2 - 8)$ in.^{2}

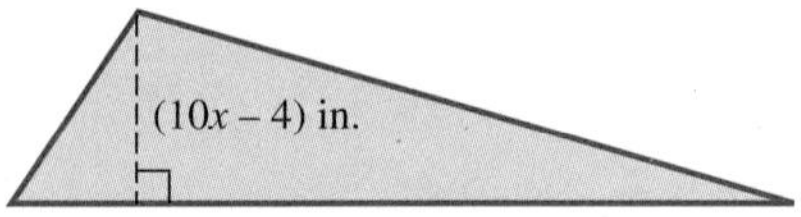

SECTION 5.8 Dividing Polynomials

DEFINITIONS AND CONCEPTS | **EXAMPLES**

To **divide monomials,** use the method for simplifying fractions and/or the quotient rule for exponents.

Divide the monomials:

$$\frac{8p^2q}{4pq^3} = \frac{2 \cdot \overset{1}{\cancel{4}} \cdot \overset{1}{\cancel{p}} \cdot p \cdot \overset{1}{\cancel{q}}}{\underset{1}{\cancel{4}} \cdot \underset{1}{\cancel{p}} \cdot \underset{1}{\cancel{q}} \cdot q \cdot q} = \frac{2p}{q^2}$$

or

$$\frac{8p^2q}{4pq^3} = \frac{8}{4}p^{2-1}q^{1-3}$$ Keep each base and subtract the exponents.

$$= 2p^1q^{-2}$$

$$= \frac{2p}{q^2}$$ Move q^{-2} to the denominator and change the sign of the exponent.

To **divide a polynomial by a monomial,** divide each term of the numerator by the denominator.

Divide: $$\frac{9a^2b^4 - 12a^3b^6}{\mathbf{18ab^5}} = \frac{9a^2b^4}{\mathbf{18ab^5}} - \frac{12a^3b^6}{\mathbf{18ab^5}}$$

$$= \frac{a}{2b} - \frac{2a^2b}{3}$$ Perform each monomial division.

Long division can be used to **divide a polynomial by a polynomial** (other than a monomial). The long division method is a series of four steps that are repeated: Divide, multiply, subtract, and bring down the next term.

When the division has a remainder, write the answer in the form: Quotient + $\frac{\text{remainder}}{\text{divisor}}$.

Divide $4x^2 - 4x + 5$ by $2x + 1$.

$$\begin{array}{r} \mathbf{2x - 3} \\ 2x + 1\overline{)4x^2 - 4x + 5} \\ -(4x^2 + 2x) \downarrow \\ \hline -6x + 5 \\ -(-6x - 3) \\ \hline 8 \end{array}$$

The first division: $\frac{4x^2}{2x} = 2x$.

The second division: $\frac{-6x}{2x} = -3$.

The remainder is 8.

The result is: $2x - 3 + \frac{8}{2x + 1}$

The long division method works best when the terms of the divisor and the dividend are written in **descending powers of the variable.** When the dividend has **missing terms,** insert such terms with a coefficient of 0, or leave a blank space.	Set up each long division. $\frac{5x + x^3 + 3 + 3x^2}{x + 1}$ — The terms of the dividend are not in descending powers of x. $x + 1\overline{)x^3 + 3x^2 + 5x + 3}$; $\frac{x^2 - 9}{x - 3}$ — The dividend is missing a term. $x - 3\overline{)x^2 + 0x - 9}$

REVIEW EXERCISES

Divide. Do not use negative exponents in the answer.

105. $\frac{16n^8}{8n^5}$ $2n^3$

106. $\frac{-14x^2y}{21xy^3}$ $-\frac{2x}{3y^2}$

107. $\frac{a^{15} - 24a^8}{6a^{12}}$ $\frac{a^3}{6} - \frac{4}{a^4}$

108. $\frac{15a^5b + ab^2 - 25b}{5a^2b}$ $3a^3 + \frac{b}{5a} - \frac{5}{a^2}$

109. $x - 1\overline{)x^2 - 6x + 5}$ $x - 5$

110. $\frac{2x^2 + 3 + 7x}{x + 3}$ $2x + 1$

111. $(15x^2 - 8x - 8) \div (3x + 2)$ $5x - 6 + \frac{4}{3x + 2}$

112. Divide $25y^2 - 9$ by $5y + 3$. $5y - 3$

113. $3x + 1\overline{)-13x - 4 + 9x^3}$ $3x^2 - x - 4$

114. $2x - 1\overline{)6x^3 + x^2 + 1}$ $3x^2 + 2x + 1 + \frac{2}{2x - 1}$

115. Use multiplication to show that $(3y^2 + 11y + 6) \div (y + 3)$ is $3y + 2$. $(y + 3)(3y + 2) = 3y^2 + 11y + 6$

116. **Bedding.** The area of a rectangular-shaped bed sheet is represented by the polynomial $(4x^3 + 12x^2 + x - 12)$ in.^{2}. If the width of the sheet is $(2x + 3)$ inches, find a polynomial that represents its length. $(2x^2 + 3x - 4)$ in.

Teaching Tip: Because this Chapter Test is a comprehensive collection of problems that covers all of the topics discussed in Chapter 5, it is lengthy. If your students have time to complete it, that would be optimal. If, because of time constraints, they are unable to do so, assign an appropriate subset of problems that reflects the types of problems that the students will see on your exam.

5 CHAPTER TEST

1. Fill in the blanks.
   a. In the expression y^{10}, the base is y and the exponent is 10.
   b. We call a polynomial with exactly one term a monomial, with exactly two terms a binomial, and with exactly three terms a trinomial.
   c. The degree of a term of a polynomial in one variable is the value of the exponent on the variable.
   d. $(x + y)^2$, $(x - y)^2$, and $(x + y)(x - y)$ are called special products.
2. Use exponents to rewrite $2xxxyyyy$. $2x^3y^4$

Simplify each expression. Do not use negative exponents in the answer.

3. $y^2(yy^3)$ y^6
4. $\left(\frac{1}{2}x^3\right)^5(x^2)^3$ $\frac{1}{32}x^{21}$
5. $3.5x^0$ 3.5
6. $2y^{-5}y^2$ $\frac{2}{y^3}$
7. 5^{-3} $\frac{1}{125}$
8. $\frac{(x + 1)^{15}}{(x + 1)^6}$ $(x + 1)^9$
9. $\frac{(y^{-5})^{-4}}{yy^{-2}}$ y^{21}
10. $\left(\frac{a^2b^{-1}}{4a^3b^{-2}}\right)^3$ $\frac{b^3}{64a^3}$
11. $\left(\frac{8}{m^6}\right)^{-2}$ $\frac{m^{12}}{64}$
12. $\frac{-6a}{b^{-9}}$ $-6ab^9$
13. Find an expression that represents the volume of a cube that has sides of length $10y^4$ inches. $1{,}000y^{12}$ in.^{3}
14. **Electricity.** One ampere (amp) corresponds to the flow of 6,250,000,000,000,000,000 electrons per second past any point in a direct current (DC) circuit. Write this number in scientific notation. 6.25×10^{18}
15. Write 9.3×10^{-5} in standard notation. 0.000093
16. Evaluate $(2.3 \times 10^{18})(4.0 \times 10^{-15})$. Write the answer in scientific notation and standard notation. 9.2×10^3; 9,200
17. Identify $x^4 + 8x^2 - 12$ as a monomial, binomial, or trinomial. Then complete the table. Trinomial

Term	Coefficient	Degree
x^4	1	4
$8x^2$	8	2
-12	-12	0

Degree of the polynomial: 4

18. Find the degree of the polynomial $3x^3y + 2x^2y^3 - 5xy^2 - 6y$. 5th degree
19. Complete the table of solutions for $y = x^2 + 2$ and then graph the equation. See AIE Appendix 3.

x	-2	-1	0	1	2
y	6	3	2	3	6

20. Free Fall. A visitor standing on the rim of the Grand Canyon drops a rock over the side. The distance (in feet) that the rock is from the canyon floor t seconds after being dropped is given by the polynomial $-16t^2 + 5{,}184$. Find the position of the rock 18 seconds after being dropped. Explain your answer. 0 ft; the rock hits the canyon floor 18 seconds after being dropped.

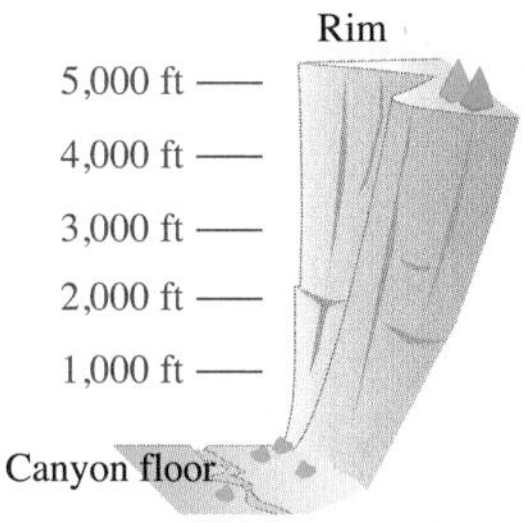

Simplify each polynomial.

21. $\frac{3}{5}x^2 + 6 + \frac{1}{4}x - 8 - \frac{1}{2}x^2 + \frac{1}{3}x$ $\quad \frac{1}{10}x^2 + \frac{7}{12}x - 2$

22. $4a^2b + 5 - 6a^3b - 3a^2b + 2a^3b$ $\quad -4a^3b + a^2b + 5$

Perform the operations.

23. $(12.1h^3 - 9.9h^2 + 9.5) + (7.3h^3 - 1.2h^2 - 10.1)$
$19.4h^3 - 11.1h^2 - 0.6$

24. Subtract $b^3c - 3bc + 12$ from the sum of $6b^3c - 3bc$ and $b^3c - 2bc$. $\quad 6b^3c - 2bc - 12$

25. Subtract:

$$\begin{array}{r} -5y^3 + 4y^2 \qquad\quad + 3 \\ \underline{-(-2y^3 - 14y^2 + 17y - 32)} \\ -3y^3 + 18y^2 - 17y + 35 \end{array}$$

26. Find a polynomial that represents the perimeter of the rectangle. $(10a^2 + 8a - 20)$ in.

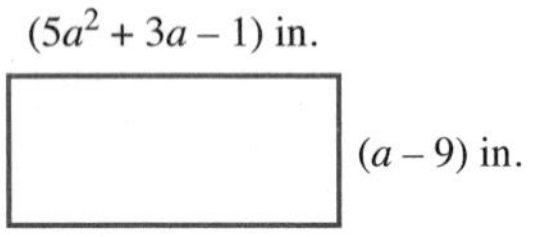

Multiply.

27. $(2x^3y^3)(5x^2y^8)$ $\quad 10x^5y^{11}$

28. $9b^3(8b^4)(-b)$ $\quad -72b^8$

29. $3y^2(y^2 - 2y + 3)$ $\quad 3y^4 - 6y^3 + 9y^2$

30. $0.6p^5(0.4p^6 - 0.9p^3)$ $\quad 0.24p^{11} - 0.54p^8$

31. $\frac{3}{4}s^3t^9(s^4t^8 + 16st)$ $\quad \frac{3}{4}s^7t^{17} + 12s^4t^{10}$

32. $(x - 5)(3x + 4)$ $\quad 3x^2 - 11x - 20$

33. $\left(6t + \frac{1}{2}\right)\left(2t - \frac{3}{2}\right)$ $\quad 12t^2 - 8t - \frac{3}{4}$

34. $(3.8m - 1)(2m - 1)$ $\quad 7.6m^2 - 5.8m + 1$

35. $(a^3 - 6)(a^3 + 7)$ $\quad a^6 + a^3 - 42$

36. $(2x - 3)(x^2 - 2x + 4)$ $\quad 2x^3 - 7x^2 + 14x - 12$

37. $(1 + 10c)(1 - 10c)$ $\quad 1 - 100c^2$

38. $(7b^3 - 3t)^2$ $\quad 49b^6 - 42b^3t + 9t^2$

39. $(2.2a)(a + 5)(a - 3)$ $\quad 2.2a^3 + 4.4a^2 - 33a$

40. Perform the operations: $(x + y)(x - y) + (x + y)^2$
$2x^2 + 2xy$

Divide.

41. $\frac{6a^2 - 12b^2}{24ab}$ $\quad \frac{a}{4b} - \frac{b}{2a}$

42. $\frac{x^2 + x - 6}{x + 3}$ $\quad x - 2$

43. $2x - 1\overline{)1 + x^2 + 6x^3}$ $\quad 3x^2 + 2x + 1 + \frac{2}{2x - 1}$

44. Find a polynomial that represents the width of a rectangle if its area is represented by the polynomial $(x^2 - 6x + 5)$ ft^2 and the length is $(x - 1)$ ft. $\quad (x - 5)$ ft

45. Use a check to determine whether $(5m^2 - 29m - 6) \div (5m + 1) = m - 6$.
Yes; $(5m + 1)(m - 6) = 5m^2 - 29m - 6$

46. Is $(a + b)^2 = a^2 + b^2$? Show why or why not.
No; $(a + b)^2 = a^2 + 2ab + b^2$

Group Project

BINOMIAL MULTIPLICATION AND THE AREA OF RECTANGLES

Overview: In this activity, rectangles are used to illustrate binomial multiplication.

Instructions: Form groups of 2 or 3 students. Study the figure. The area of the largest rectangle (outlined in blue) is given by $(x + 2)(x + 3)$. Its area is also the sum of the areas of the four smaller rectangles: $x^2 + 3x + 2x + 6$. Thus,

$$\begin{aligned}(x + 2)(x + 3) &= x^2 + 3x + 2x + 6 \\ &= x^2 + 5x + 6\end{aligned}$$

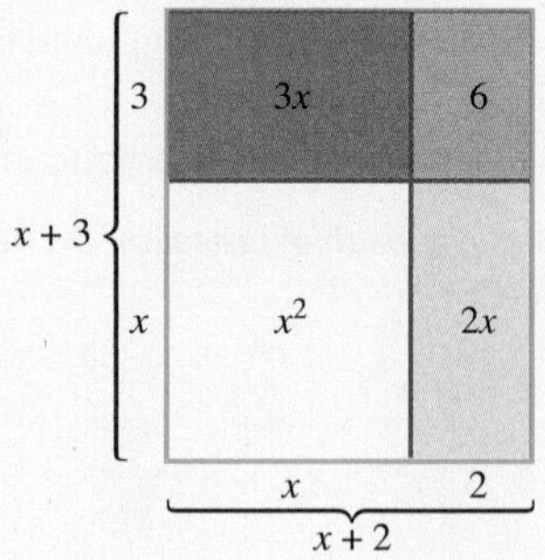

Draw three similar models to represent the following products.

1. $(x + 4)(x + 5)$ 2. $(x + 8)^2$ 3. $x(x + 6)$ 4. $(2x + 1)^2$

Determine the missing number so that each rectangle below has the given area.

5. Area: $x^2 + 9x + 14$

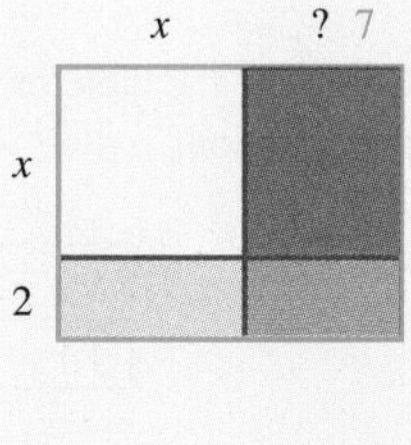

6. Area: $x^2 + 16x + 55$

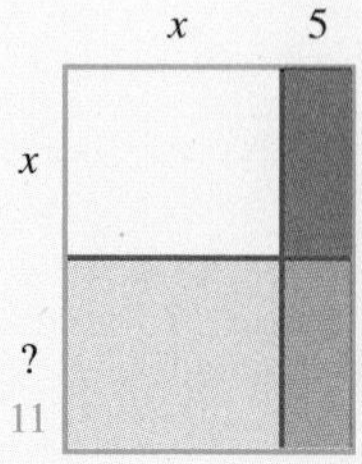

7. Draw a model that represents a rectangle with area $x^2 + 10x + 24$. Label it completely.

CUMULATIVE REVIEW Chapters 1–5

1. Use exponents to write the prime factorization of 270. [Section 1.2] $2 \cdot 3^3 \cdot 5$

2. **a.** Use the variables a and b to state the commutative property of addition. [Section 1.4] $a + b = b + a$

   b. Use the variables x, y, and z to state the associative property of multiplication. [Section 1.6] $(xy)z = x(yz)$

Evaluate each expression.

3. $3 - 4[-10 - 4(-5)]$ [Section 1.7] -37

4. $\dfrac{|-45|-2(-5) + 1^5}{2 \cdot 9 - 2^4}$ [Section 1.7] 28

Simplify each expression.

5. $27\left(\frac{2}{3}x\right)$ [Section 1.9] $18x$

6. $3x^2 + 2x^2 - 5x^2$ [Section 1.9] 0

Solve each equation.

7. $2 - (4x + 7) = 3 + 2(x + 2)$ [Section 2.2] -2

8. $\frac{2}{5}y + 3 = 9$ [Section 2.2] 15

9. **Candy Sales.** The circle graph shows how \$6.5 billion in seasonal candy sales for 2009 was spent. Find the candy sales for Halloween. [Section 2.3] \$2.21 billion

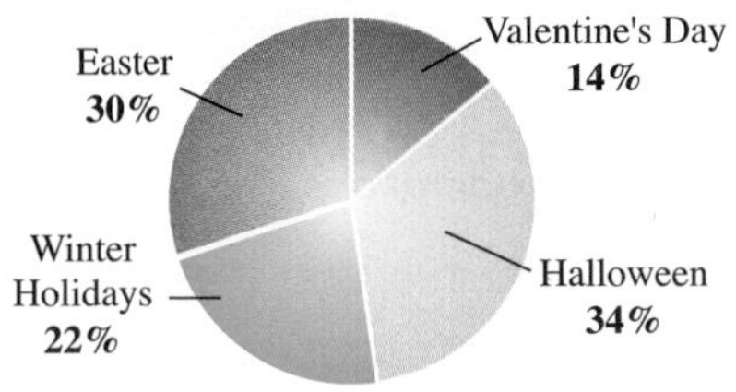

Source: National Confectioners Association

10. **Air Conditioning.** Find the volume of air contained in the duct. Round to the nearest tenth of a cubic foot. [Section 2.4] 1.2 ft^3

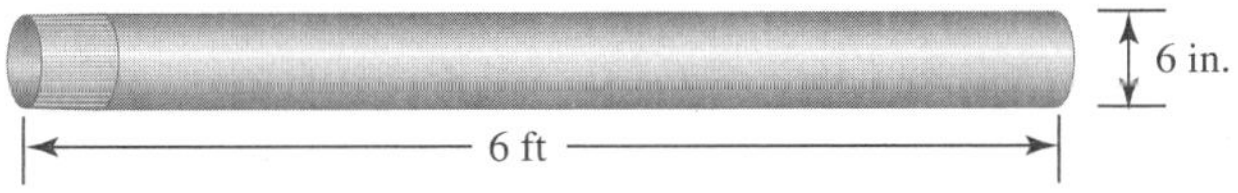

11. **Angle of Elevation.** Find x. [Section 2.5] 30°

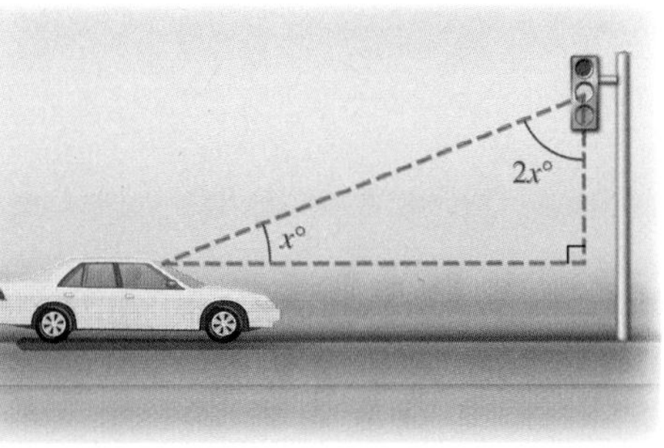

12. **Livestock Auction.** A farmer is going to sell one of her prize hogs at an auction and would like to make \$6,000 after paying a 4% commission to the auctioneer. For what selling price will the farmer make this amount of money? [Section 2.5] \$6,250

13. **Stock Market.** An investment club invested part of \$45,000 in a high-yield mutual fund that earned 12% annual simple interest. The remainder of the money was invested in Treasury bonds that earned 6.5% simple annual interest. The two investments earned \$4,300 in one year. How much was invested in each account? [Section 2.6] Mutual fund: \$25,000; bonds: \$20,000

14. Solve $-4x + 6 > 17$ and graph the solution set. Then describe the graph using interval notation. [Section 2.7] $\left(-\infty, -\frac{11}{4}\right)$ See AIE Appendix 3.

Graph each equation. See AIE Appendix 3.

15. $y = 3x$ [Section 3.2]

16. $x = -2$ [Section 3.3]

17. Find the slope of the line passing through $(6, -2)$ and $(-3, 2)$. [Section 3.4] $-\frac{4}{9}$

18. **Diabetes.** The graph below shows the number of people in the United States diagnosed with diabetes from 1996 to 2008. Find the rate of change in the number over this time span. [Section 3.4] 0.75 million people per year or $\frac{3}{4}$ million people per year

Number of People Diagnosed with Diabetes in U.S.

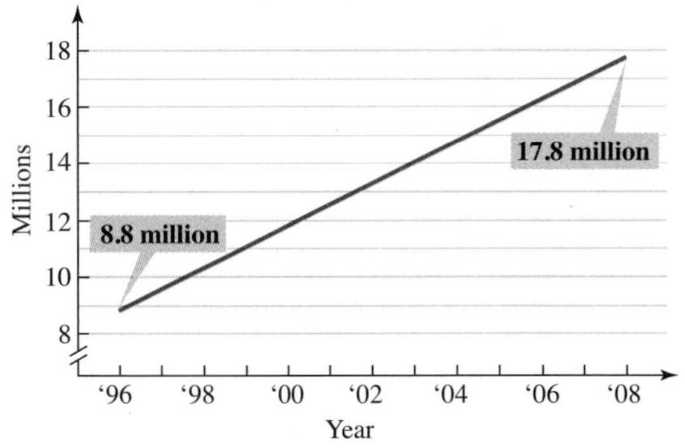

Source: U.S. Centers for Disease Control

19. Find the slope and y-intercept of the line. Then write the equation of the line. [Section 3.5] $m = 3$, $(0, -2)$; $y = 3x - 2$

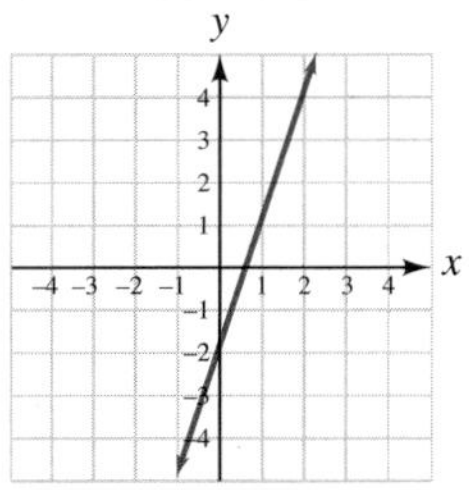

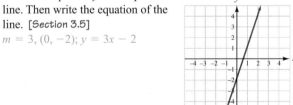

20. Without graphing, determine whether the graphs of $y = \frac{3}{2}x - 1$ and $2x + 3y = 10$ are parallel, perpendicular, or neither. [Section 3.5] Perpendicular

21. Write the equation of the line that passes through $(-2, 10)$ with slope -4. Write the result in slope–intercept form. [Section 3.6] $y = -4x + 2$

22. Is $(-2, 1)$ a solution of $2x - 3y \geq -6$? [Section 3.7] No

23. If $f(x) = 2x^2 + 3x - 9$, find $f(-5)$. [Section 3.8] 26

24. Determine whether the graph below is the graph of a function. If it is not, find ordered pairs that show a value of x that is assigned more than one value of y. [Section 3.8] Not a function; (1, 2), (1, −2); answers may vary.

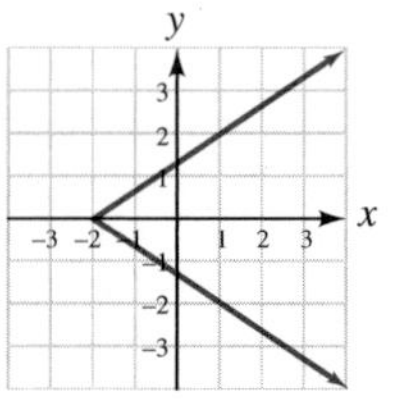

25. Is $\left(\frac{2}{3}, -1\right)$ a solution of the system $\begin{cases} y = -3x + 1 \\ 3x + 3y = -2 \end{cases}$?
[Section 4.1] No

26. Solve the system $\begin{cases} 3x + 2y = 14 \\ y = \frac{1}{4}x \end{cases}$ by graphing.
[Section 4.1] (4, 1) See AIE Appendix 3.

27. Solve the system $\begin{cases} 2b - 3a = 18 \\ a + 3b = 5 \end{cases}$ by substitution.
[Section 4.2] $(-4, 3)$

28. Solve the system $\begin{cases} 8s + 10t = 24 \\ 11s - 3t = -34 \end{cases}$ by elimination (addition).
[Section 4.3] $(-2, 4)$

29. Vacations. One-day passes to Universal Studios Hollywood cost a family of 5 (2 adults and 3 children) \$275. A family of 6 (3 adults and 3 children) paid \$336 for their one-day passes. Find the cost of an adult one-day pass and a child's one-day pass to Universal Studios.
[Section 4.4] Adult: \$61; child: \$51

30. Graph: $\begin{cases} y \leq 2x - 1 \\ x + 3y > 6 \end{cases}$
[Section 4.5] See AIE Appendix 3.

Simplify. Do not use negative exponents in the answer.

31. $(-3x^2y^4)^2$ [Section 5.1] $9x^4y^8$

32. $(v^5)^2(v^3)^4$ [Section 5.1] v^{22}

33. $ab^3c^4 \cdot ab^4c^2$ [Section 5.1] $a^2b^7c^6$

34. $\left(\frac{4t^3t^4t^5}{3t^2t^6}\right)^3$ [Section 5.1] $\frac{64t^{12}}{27}$

35. $(2y)^{-4}$ [Section 5.2] $\frac{1}{16y^4}$

36. $\frac{a^4b^0}{a^{-3}}$ [Section 5.2] a^7

37. -5^{-2} [Section 5.2] $-\frac{1}{25}$

38. $\left(\frac{a}{x}\right)^{-10}$ [Section 5.2] $\frac{x^{10}}{a^{10}}$

Write each number in scientific notation.

39. 615,000 [Section 5.3] 6.15×10^5

40. 0.0000013 [Section 5.3] 1.3×10^{-6}

41. Graph: $y = x^2$ [Section 5.4] See AIE Appendix 3.

42. Musical Instruments. The amount of deflection of the horizontal beam (in inches) is given by the polynomial $0.01875x^4 - 0.15x^3 + 1.2x$, where x is the distance (in feet) that the gong is hung from one end of the beam. Find the deflection if the gong is hung in the middle of the support.
[Section 5.4] 1.5 in.

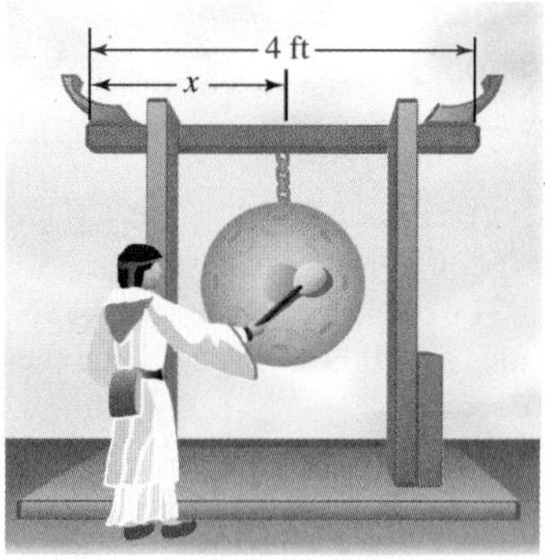

Perform the indicated operations.

43. $(4c^2 + 3c - 2) + (3c^2 + 4c + 2)$ [Section 5.5] $7c^2 + 7c$

44. Subtract:

$$\begin{array}{r} 17x^4 - 3x^2 - 65x - 12 \\ \underline{-(23x^4 + 14x^2 + 3x - 23)} \\ -6x^4 - 17x^2 - 68x + 11 \end{array}$$

[Section 5.5]

45. $(2t + 3s)(3t - s)$ [Section 5.6] $6t^2 + 7st - 3s^2$

46. $3x(2x + 3)^2$ [Section 5.7] $12x^3 + 36x^2 + 27x$

47. $5x + 3\overline{)11x + 10x^2 + 3}$ [Section 5.8] $2x + 1$

48. $\frac{2x - 32}{16x}$ [Section 5.8] $\frac{1}{8} - \frac{2}{x}$

Factoring and Quadratic Equations

6

from Campus to Careers

Elementary School Teacher

It has been said that a teacher takes a hand, opens a mind, and touches a heart. That is certainly true for the thousands of dedicated elementary school teachers across the country. Elementary school teachers use their training in mathematics in many ways. Besides teaching math on a daily basis, they calculate student grades, analyze test results, and order instructional materials and supplies. They use measurement and geometry for designing bulletin board displays and they construct detailed schedules so that the classroom time is used wisely.

Problem 129 in **Study Set 6.1, problem 103** in **Study Set 6.3,** and **problem 17** in **Study Set 6.8** involve situations that an elementary school teacher might encounter on the job. The mathematical concepts discussed in this chapter can be used to solve those problems.

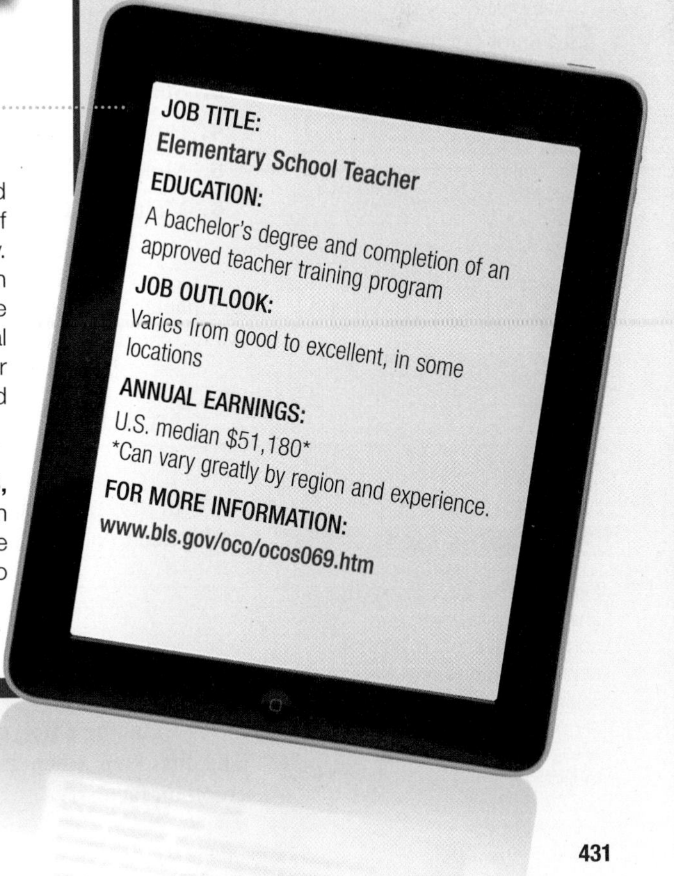

JOB TITLE:
Elementary School Teacher
EDUCATION:
A bachelor's degree and completion of an approved teacher training program
JOB OUTLOOK:
Varies from good to excellent, in some locations
ANNUAL EARNINGS:
U.S. median $51,180*
*Can vary greatly by region and experience.
FOR MORE INFORMATION:
www.bls.gov/oco/ocos069.htm

Study Skills Workshop ▶ Reading the Textbook

Reading an algebra textbook is different from reading a newspaper or a novel. Here are two ways that you should be reading this textbook.

SKIMMING FOR AN OVERVIEW: This is a quick way to look at material just before it is covered in class. It helps you become familiar with the new vocabulary and notation that will be used by your instructor in the lecture. It lays a foundation.

READING FOR UNDERSTANDING: This in-depth type of reading is done more slowly, with a pencil and paper at hand. Don't skip anything—every word counts! You should do just as much writing as you do reading. Highlight the important points and work each example. If you become confused, stop and reread the material until you understand it.

Now Try This ▶

Choose a section from this chapter and . . .

1. . . . quickly skim it. Write down any terms in bold face type and the titles of any properties, definitions, or strategies that are given in the colored boxes.
2. . . . work each *Self Check* problem. Your solutions should look like those in the *Examples.* Be sure to include your own "author notes" (the sentences in red to the right of each step of a solution).

SECTION 6.1 The Greatest Common Factor; Factoring by Grouping

OBJECTIVES

1. Find the greatest common factor of a list of terms.
2. Factor out the greatest common factor.
3. Factor by grouping.

ARE YOU READY? *Are You Ready? exercises available online at www.webassign.net/brookscole*

The following problems review some basic skills that are needed when factoring expressions.

1. Find the prime factorization of 54. $2 \cdot 3 \cdot 3 \cdot 3 = 2 \cdot 3^3$
2. Write $a \cdot a \cdot a \cdot a \cdot a \cdot a$ in an equivalent form using exponents. a^6
3. Simplify: $-(x - 7)$ $-x + 7$
4. How many terms does the expression $3x^3 + x^2 + 4x + 8$ have? 4
5. Multiply: $4d(d^3 + 2d)$ $4d^4 + 8d^2$
6. Multiply and simplify: $x^2(x + 1) - 3(x + 1)$ $x^3 + x^2 - 3x - 3$

In Chapter 5, we learned how to multiply polynomials. For example, to multiply $3x + 5$ by $4x$, we use the distributive property, as shown below.

$$\begin{aligned} 4x(3x + 5) &= 4x \cdot 3x + 4x \cdot 5 \\ &= 12x^2 + 20x \end{aligned}$$

In this section, we reverse the previous steps and determine what factors were multiplied to obtain $12x^2 + 20x$. We call that process *factoring the polynomial.*

Multiplication: Given the factors, we find a polynomial.

$$4x(3x + 5) = 12x^2 + 20x$$

Factoring: Given a polynomial, we find the factors.

To **factor a polynomial** means to express it as a product of two (or more) polynomials. The first step when factoring a polynomial is to determine whether its terms have any common factors.

Success Tip

On the game show *Jeopardy!,* answers are revealed and contestants respond with the appropriate questions. Factoring is similar. Answers to multiplications are given. You are to respond by telling what factors were multiplied.

1 Find the Greatest Common Factor of a List of Terms.

Teaching Tip: Now is a good time to remind your students about the difference between a *term* and a *factor* because they are two key concepts of the chapter.
$x + 7$ x is a term.
$7x$ x is a factor.

To determine whether two or more integers have common factors, it is helpful to write them as products of prime numbers. For example, the prime factorizations of 42 and 90 are given below.

$$42 = \mathbf{2} \cdot \mathbf{3} \cdot 7 \qquad 90 = \mathbf{2} \cdot \mathbf{3} \cdot 3 \cdot 5$$

The highlighting shows that 42 and 90 have one factor of 2 and one factor of 3 in common. To find their *greatest common factor* (*GCF*), we multiply the common factors: $2 \cdot 3 = 6$. Thus, the GCF of 42 and 90 is 6.

The Greatest Common Factor (GCF)

The **greatest common factor (GCF)** of a list of integers is the largest common factor of those integers. To find the greatest common factor of two (or more) integers:

1. Prime factor each number.
2. Identify the common prime factors.
3. The GCF is a product of all the common prime factors found in Step 2. If there are no common prime factors, the GCF is 1.

Recall from arithmetic that the factors of a number divide the number exactly, leaving no remainder. Therefore, the greatest common factor of two or more integers is the largest integer that divides each of them exactly.

EXAMPLE 1 Find the GCF of each list of numbers: **a.** 21 and 140 **b.** 24, 60, and 96 **c.** 9, 10, and 30

Teaching Tip: You may want to discuss divisibility rules at this time.

Strategy We will prime factor each number in the list. Then we will identify the common prime factors and find their product.

Why The product of the common prime factors is the GCF of the numbers in the list.

Solution

a. The prime factorization of each number is shown:

$$21 = 3 \cdot \mathbf{7}$$
$$140 = 2 \cdot 2 \cdot 5 \cdot \mathbf{7} \quad \text{This can be written as } 2^2 \cdot 5 \cdot 7.$$

$3\,|\underline{21}$ → 7; $2\,|\underline{140}$, $2\,|\underline{70}$, $5\,|\underline{35}$ → 7

Since the only prime factor common to 21 and 140 is 7, the GCF of 21 and 140 is 7.

Success Tip

Note that the GCF, 7, divides 21 and 140 exactly:

$$\frac{21}{7} = 3 \qquad \frac{140}{7} = 20$$

b. To find the GCF of three numbers, we proceed in a similar way by first finding the prime factorization of each number in the list.

$$24 = \mathbf{2} \cdot \mathbf{2} \cdot 2 \cdot \mathbf{3} \quad \text{This can be written as } 2^3 \cdot 3.$$
$$60 = \mathbf{2} \cdot \mathbf{2} \cdot \mathbf{3} \cdot 5 \quad \text{This can be written as } 2^2 \cdot 3 \cdot 5.$$
$$96 = \mathbf{2} \cdot \mathbf{2} \cdot 2 \cdot 2 \cdot 2 \cdot \mathbf{3} \quad \text{This can be written as } 2^5 \cdot 3.$$

$2\,|\underline{24}$, $2\,|\underline{12}$, $2\,|\underline{6}$ → 3; $2\,|\underline{60}$, $2\,|\underline{30}$, $3\,|\underline{15}$ → 5; $2\,|\underline{96}$, $2\,|\underline{48}$, $2\,|\underline{24}$, $2\,|\underline{12}$, $2\,|\underline{6}$ → 3

The highlighting shows that 24, 60, and 96 have two factors of 2 and one factor of 3 in common. The GCF of 24, 60, and 96 is the product of their common prime factors.

$$\text{GCF} = \mathbf{2} \cdot \mathbf{2} \cdot \mathbf{3} = 2^2 \cdot 3^1 = 12$$

Success Tip

The exponent on any factor in a GCF is the *smallest* exponent that appears on that factor in all of the numbers under consideration.

c. Since there are no prime factors common to 9, 10, and 30, their GCF is 1.

$$9 = 3 \cdot 3$$
$$10 = 2 \cdot 5$$
$$30 = 2 \cdot 3 \cdot 5$$

$3\,|\underline{9}$ → 3; $2\,|\underline{10}$ → 5; $2\,|\underline{30}$, $3\,|\underline{15}$ → 5

Teaching Example 1 Find the GCF of each list of numbers:
a. 28 and 70
b. 36, 54, and 90
c. 12, 15, and 98
Answers: a. 14 b. 18 c. 1

Self Check 1 Find the GCF of each list of numbers: **a.** 24 and 70 2 **b.** 22, 25, and 98 1 **c.** 45, 60, and 75 15

Now Try Problems 17 and 21

To find the greatest common factor of a list of terms, we can use the following approach.

Strategy for Finding the GCF

1. Write each coefficient as a product of prime factors.
2. Identify the numerical and variable factors common to each term.
3. Multiply the common numerical and variable factors identified in Step 2 to obtain the GCF. If there are no common factors, the GCF is 1.

EXAMPLE 2 Find the GCF of each list of terms: **a.** $12x^2$ and $20x$ **b.** $9a^5b^2$, $15a^4b^2$, and $90a^3b^3$

Strategy We will prime factor each coefficient of each term in the list. Then we will identify the numerical and variable factors common to each term and find their product.

Why The product of the common factors is the GCF of the terms in the list.

Solution

a. ***Step 1:*** We write each coefficient, 12 and 20, as a product of prime factors. Recall that an exponent, as in x^2, indicates repeated multiplication.

$$12x^2 = 2 \cdot 2 \cdot 3 \cdot x \cdot x \qquad \text{This can be written as } 2^2 \cdot 3 \cdot x^2.$$
$$20x = 2 \cdot 2 \cdot 5 \cdot x \qquad \text{This can be written as } 2^2 \cdot 5 \cdot x.$$

Step 2: There are two common factors of 2 and one common factor of x.

Step 3: We multiply the common factors, 2, 2, and x, to obtain the GCF.

$$\text{GCF} = 2 \cdot 2 \cdot x = 2^2 \cdot x = 4x$$

Success Tip

One way to identify common factors is to circle them:

$$12x^2 = 2 \cdot 2 \cdot 3 \cdot x \cdot x$$
$$20x = 2 \cdot 2 \cdot 5 \cdot x$$
$$\text{GCF} = 2 \cdot 2 \cdot x = 4x$$

b. ***Step 1:*** We write the coefficients, 9, 15, and 90, as products of primes. The exponents on the variables represent repeated multiplication.

$$9a^5b^2 = 3 \cdot 3 \cdot a \cdot a \cdot a \cdot a \cdot a \cdot b \cdot b \qquad \text{This can be written as } 3^2 \cdot a^5 \cdot b^2.$$
$$15a^4b^2 = 3 \cdot 5 \cdot a \cdot a \cdot a \cdot a \cdot b \cdot b \qquad \text{This can be written as } 3 \cdot 5 \cdot a^4 \cdot b^2.$$
$$90a^3b^3 = 2 \cdot 3 \cdot 3 \cdot 5 \cdot a \cdot a \cdot a \cdot b \cdot b \cdot b \qquad \text{This can be written as } 2 \cdot 3^2 \cdot 5 \cdot a^3 \cdot b^3.$$

Step 2: The highlighting shows one common factor of 3, three common factors of a, and two common factors of b.

Step 3: $\text{GCF} = 3 \cdot a \cdot a \cdot a \cdot b \cdot b = 3a^3b^2$

Success Tip

The exponent on any variable in a GCF is the *smallest* exponent that appears on that variable in all of the terms under consideration.

Teaching Example 2 Find the GCF of each list of terms:
a. $18y^2$ and $45y$
b. $12x^4y^6$, $18x^5y^4$, and $30x^3y^5$
Answers: a. $9y$ b. $6x^3y^4$

Self Check 2 Find the GCF of each list of terms: **a.** $33c$ and $22c^4$ $11c$ **b.** $42s^3t^2$, $63s^2t^4$, and $21s^3t^3$ $21s^2t^2$

Now Try Problems 29 and 35

2 Factor Out the Greatest Common Factor.

The concept of greatest common factor is used to factor polynomials. For example, to factor $12x^2 + 20x$, we note that there are two terms, $12x^2$ and $20x$. We previously determined that the GCF of $12x^2$ and $20x$ is $4x$. With this in mind, we write each term of $12x^2 + 20x$ as a product of the GCF and one other factor. Then we apply the distributive property in reverse: $ab + ac = a(b + c)$.

$$12x^2 + 20x = 4x \cdot 3x + 4x \cdot 5$$ Write $12x^2$ and $20x$ as the product of the GCF, $4x$, and one other factor.

$$= 4x(3x + 5)$$ Write an expression so that the multiplication by $4x$ distributes over the sum of the terms $3x$ and 5.

We have found that the factored form of $12x^2 + 20x$ is $4x(3x + 5)$. This process is called **factoring out the greatest common factor.**

The Language of Algebra

We have expressed a **sum of terms** as a **product of factors.**

$$12x^2 + 20x = 4x(3x + 5)$$

Sum of terms Factor Factor

EXAMPLE 3 Factor: **a.** $8m + 24$ **b.** $35a^3b^2 - 14a^2b^3$ **c.** $3x^4 - 5x^3 + x^2$

Strategy First, we will determine the GCF of the terms of the polynomial. Then we will write each term of the polynomial as the product of the GCF and one other factor.

Why We can then use the distributive property to factor out the GCF.

Solution

a. Since the GCF of $8m$ and 24 is 8, we write $8m$ and 24 as the product of 8 and one other factor.

$$8m + 24 = 8 \cdot m + 8 \cdot 3$$ This step can be done mentally.

$$= 8(m + 3)$$ Factor out the GCF, 8.

To check, we multiply: $8(m + 3) = 8 \cdot m + 8 \cdot 3 = 8m + 24$. Since we obtain the original polynomial, $8m + 24$, the factorization is correct.

Success Tip

Always verify a factorization by performing the indicated multiplication. The result should be the original polynomial.

CAUTION Remember to factor out the greatest common factor, not just a common factor. If we factored out 4 in the previous example, we would get

$$8m + 24 = 4(2m + 6)$$

However, the terms in red within parentheses have a common factor of 2, indicating that the factoring is not complete.

b. First, find the GCF of $35a^3b^2$ and $14a^2b^3$.

$$35a^3b^2 = 5 \cdot 7 \cdot a \cdot a \cdot a \cdot b \cdot b$$
$$14a^2b^3 = 2 \cdot 7 \cdot a \cdot a \cdot b \cdot b \cdot b$$

The GCF is $7a^2b^2$.

$5 \,\underline{|\, 35}$ → 7 $2 \,\underline{|\, 14}$ → 7

Now, we write $35a^3b^2$ and $14a^2b^3$ as the product of the GCF, $7a^2b^2$, and one other factor.

$$35a^3b^2 - 14a^2b^3 = 7a^2b^2 \cdot 5a - 7a^2b^2 \cdot 2b$$ $\frac{35a^3b^2}{7a^2b^2} = 5a$ and $\frac{14a^2b^3}{7a^2b^2} = 2b$.

$$= 7a^2b^2(5a - 2b)$$ Factor out the GCF, $7a^2b^2$.

We check by multiplying: $7a^2b^2(5a - 2b) = 35a^3b^2 - 14a^2b^3$.

Success Tip

With practice, you will learn how to factor out the GCF and write the correct terms within the parentheses in one step.

c. We factor out the GCF of the three terms, which is x^2.

$$3x^4 - 5x^3 + x^2 = x^2(3x^2) - x^2(5x) + x^2(1)$$ Write the last term, x^2, as $x^2(1)$.

$$= x^2(3x^2 - 5x + 1)$$ Factor out the GCF, x^2.

We check by multiplying: $x^2(3x^2 - 5x + 1) = 3x^4 - 5x^3 + x^2$.

Caution

If the GCF of the terms of a polynomial is the same as one of the terms, leave a 1 in place of that term when factoring out the GCF.

Self Check 3 Factor: **a.** $6f + 36$ $6(f + 6)$ **b.** $24s^2t^2 - 42s^3t$ $6s^2t(4t - 7s)$ **c.** $y^6 - 10y^4 - y^3$ $y^3(y^3 - 10y - 1)$

Now Try Problems 41, 49, and 51

Teaching Example 3
Factor: **a.** $9a + 27$
b. $30m^3n^2 - 45m^4n$
c. $r^7 - 20r^5 - r^4$
Answers: **a.** $9(a + 3)$
b. $15m^3n(2n - 3m)$
c. $r^4(r^3 - 20r - 1)$

When asked to factor a polynomial whose leading coefficient is negative, factor out the *opposite of the GCF.*

EXAMPLE 4 Factor -1 from each polynomial: **a.** $-a^3 + 2a^2 - 4$ **b.** $6 - x$

Strategy We will write each term of the polynomial as the product of -1 and one other factor.

Why We can then use the distributive property to factor out the -1.

Solution

a. $-a^3 + 2a^2 - 4 = (\mathbf{-1})a^3 + (\mathbf{-1})(-2a^2) + (\mathbf{-1})4$ This step can be done mentally.

$= \mathbf{-1}(a^3 - 2a^2 + 4)$ Factor out -1.

$= -(a^3 - 2a^2 + 4)$ The 1 need not be written.

We check by multiplying: $-(a^3 - 2a^2 + 4) = -a^3 + 2a^2 - 4$.

Success Tip

This result suggests a quick way to factor out -1. Simply change the sign of each term of $-a^3 + 2a^2 - 4$ and write a $-$ symbol in front of the parentheses.

b. First, we will write the terms of $6 - x$ in descending order to determine whether the leading coefficient is negative.

$6 - x = -x + 6$ Write in descending powers of x.

$= \mathbf{-1}(x) + (\mathbf{-1})(-6)$ This step can be done mentally.

$= \mathbf{-1}(x - 6)$ Factor out -1.

$= -(x - 6)$ The 1 need not be written.

Teaching Example 4 Factor -1 from each polynomial:
a. $-x^5 - 4x^3 + 15$ b. $27 - d$
Answers:
a. $-(x^5 + 4x^3 - 15)$
b. $-(d - 27)$

Self Check 4 Factor -1 from each polynomial: **a.** $-b^4 - 3b^2 + 2$ $-(b^4 + 3b^2 - 2)$
b. $9 - t$ $-(t - 9)$

Now Try Problems 57 and 61

EXAMPLE 5 Factor out the opposite of the GCF in $-20m + 30$.

Strategy First, we will determine the GCF of the terms of the polynomial. Then we will write each term of the polynomial as the product of the opposite of the GCF and one other factor.

Why We can then use the distributive property to factor out the opposite of the GCF.

Solution Since the GCF is 10, the opposite of the GCF is -10. We write each term of the polynomial as the product of -10 and another factor. Then we factor out -10.

$-20m + 30 = (\mathbf{-10})(2m) + (\mathbf{-10})(-3)$ This step can be done mentally.

$= \mathbf{-10}(2m - 3)$ Note that the leading coefficient of the polynomial within the parentheses is positive.

Success Tip

It is standard practice to factor in such a way that the leading coefficient of the polynomial within the parentheses is positive.

We check by multiplying: $-10(2m - 3) = -20m + 30$.

Teaching Example 5 Factor out the opposite of the GCF in $-30r + 24$.
Answer: $-6(5r - 4)$

Self Check 5 Factor out the opposite of the GCF in $-44c + 55$. $-11(4c - 5)$

Now Try Problem 67

EXAMPLE 6 Factor: $x(x + 4) + 3(x + 4)$

Strategy We will identify the terms of the expression and find their GCF.

Why We can then use the distributive property to factor out the GCF.

Solution The expression has two terms: $\underbrace{x(x + 4)}_{\text{The first term}} + \underbrace{3(x + 4)}_{\text{The second term}}$

The Language of Algebra

We say that the terms have the **common binomial factor** $x + 4$.

Teaching Tip: Warn students that a common error is to factor as $(x + 4)^2(x + 3)$. Multiply on each side to show that $(x + 4)^2(x + 3) \neq x(x + 4) + 3(x + 4)$.

Teaching Example 6 Factor: $3b(b - 5) + 11(b - 5)$
Answer: $(b - 5)(3b + 11)$

The GCF of the two terms is the binomial $x + 4$, which can be factored out.

$$x(x + 4) + 3(x + 4) = (x + 4)x + (x + 4)3$$ Write each term as the product of $(x + 4)$ and one other factor.

$$= (x + 4)(x + 3)$$ Factor out the common factor, $(x + 4)$. Caution: Don't write $(x + 4)^2$.

Self Check 6 Factor: $2y(y - 1) + 7(y - 1)$ $(y - 1)(2y + 7)$

Now Try Problem 73

3 Factor by Grouping.

Although the terms of many polynomials don't have a common factor, other than 1, it is possible to factor some of them by arranging their terms in convenient groups. This method is called **factoring by grouping.**

EXAMPLE 7 Factor by grouping: **a.** $2x^3 + x^2 + 12x + 6$ **b.** $5c - 5d + cd - d^2$

Strategy We will factor out a common factor from the first two terms and a common factor from the last two terms.

Why This often produces a common binomial factor that can then be factored out.

Solution **a.** Except for 1, there is no factor that is common to all four terms. However, the first two terms, $2x^3$ and x^2, have a common factor, x^2, and the last two terms, $12x$ and 6, have a common factor, 6.

$$2x^3 + x^2 \quad + \quad 12x + 6$$
$$\downarrow \qquad\qquad \downarrow$$
$$x^2(2x + 1) \quad + \quad 6(2x + 1)$$

We now see that $2x^3 + x^2$ and $12x + 6$ have a common binomial factor, $2x + 1$, which can be factored out.

$$2x^3 + x^2 + 12x + 6 = x^2(2x + 1) + 6(2x + 1)$$ Factor $2x^3 + x^2$ and $12x + 6$. Don't forget the blue + sign.

$$= (2x + 1)(x^2 + 6)$$ Factor out $2x + 1$. Caution: Don't write $(2x + 1)^2$.

Caution

Factoring by grouping can be attempted on any polynomial with four or more terms. However, not every such polynomial can be factored in this way.

We can check the factorization by multiplying. The result should be the original polynomial.

$$(2x + 1)(x^2 + 6) = 2x^3 + 12x + x^2 + 6$$
$$= 2x^3 + x^2 + 12x + 6$$ Rearrange the terms to get the original polynomial.

Caution

Don't think that $5(c - d) + d(c - d)$ is in factored form. It is a sum of two terms. To be in factored form, the result must be a product.

b. The first two terms have a common factor, 5, and the last two terms have a common factor, d.

$$5c - 5d + cd - d^2 = 5(c - d) + d(c - d)$$ Factor out 5 from $5c - 5d$ and d from $cd - d^2$. Don't forget the blue + sign.

$$= (c - d)(5 + d)$$ Factor out the common binomial factor, $c - d$. Caution: Don't write $(c - d)^2$.

We can check by multiplying:

$$(c - d)(5 + d) = 5c + cd - 5d - d^2$$
$$= 5c - 5d + cd - d^2$$

Teaching Example 7 Factor by grouping:
a. $t^3 + 3t^2 + 7t + 21$
b. $11a - 11b + ab - b^2$
Answers:
a. $(t + 3)(t^2 + 7)$
b. $(a - b)(11 + b)$

Self Check 7 Factor by grouping: **a.** $3n^3 + 2n^2 + 9n + 6$ $(3n + 2)(n^2 + 3)$
b. $7x - 7y + xy - y^2$ $(x - y)(7 + y)$

Now Try Problems 77 and 79

Factoring a Four-Termed Polynomial by Grouping

1. Group the terms of the polynomial so that the first two terms have a common factor and the last two terms have a common factor.
2. Factor out the common factor from each group.
3. Factor out the resulting common binomial factor. If there is no common binomial factor, regroup the terms of the polynomial and repeat steps 2 and 3.

By the multiplication property of 1, we know that 1 is a factor of every term. We can use this observation to factor certain polynomials by grouping.

EXAMPLE 8 Factor: **a.** $x^3 + 6x^2 + x + 6$ **b.** $x^2 - ax - x + a$

Strategy We will follow the steps for factoring a four-termed polynomial.

Why Since the terms of the polynomials do not have a common factor (other than 1), the only option is to attempt to factor these polynomials by grouping.

Solution **a.** The first two terms, x^3 and $6x^2$, have a common factor of x^2. The only common factor of the last two terms, x and 6, is 1.

$$x^3 + 6x^2 + x + 6 = x^2(x + 6) + 1(x + 6)$$ Factor out x^2 from $x^3 + 6x^2$. Factor out 1 from $x + 6$. Don't forget the blue + sign.

$$= (x + 6)(x^2 + 1)$$ Factor out the common binomial factor, $x + 6$. Caution: Don't write $(x + 6)^2$.

Check the factorization by multiplying.

Success Tip

When we factor out -1 from the last two terms,

$$x^2 - ax \; -x + a$$
$$= x(x - a) - 1(x - a)$$

the signs of those terms change within the parentheses. The binomials within both sets of parentheses are then identical.

b. Since x is a common factor of the first two terms, we can factor it out.

$$x^2 - ax - x + a = x(x - a) - x + a \quad \text{Factor out } x \text{ from } x^2 - ax.$$

When factoring four terms by grouping, if the coefficient of the 3rd term is negative, we often factor out a negative coefficient from the last two terms. If we factor -1 from $-x + a$, a common binomial factor $x - a$ appears within the second set of parentheses, which we can factor out.

$$x^2 - ax - x + a = x(x - a) - 1(x - a)$$

To factor out -1, change the sign of $-x$ and a, and write -1 in front of the parentheses.

$$= (x - a)(x - 1)$$

Factor out the common factor, $x - a$. Caution: Don't write $(x - a)^2$.

Check by multiplying.

Teaching Example 8 Factor:
a. $q^6 + 25q^5 + q + 25$
b. $y^2 - yz - y + z$
Answers:
a. $(q + 25)(q^5 + 1)$
b. $(y - z)(y - 1)$

Self Check 8 Factor: **a.** $a^5 + 11a^4 + a + 11$ $(a + 11)(a^4 + 1)$
b. $b^2 - bc - b + c$ $(b - c)(b - 1)$

Now Try Problems 85 and 87

EXAMPLE 9 Factor: $5x^3 - 8 + 10x^2 - 4x$

Strategy We will follow the steps for factoring by grouping.

Why Since the four terms of the polynomial do not have a common factor (other than 1), we will attempt to factor it by grouping.

Solution We note that the four terms of the polynomial do not have a common factor.

$$5x^3 - 8 + 10x^2 - 4x$$

No common factor (other than 1) for $5x^3 - 8$; GCF $= 2x$ for $10x^2 - 4x$. We cannot factor the polynomial in its current form.

We will write the polynomial in descending powers of x and attempt to factor by grouping again.

$$5x^3 - 8 + 10x^2 - 4x = 5x^3 + 10x^2 - 4x - 8$$
$$= 5x^2(x + 2) - 4(x + 2) \quad \text{Factor } 5x^2 \text{ from } 5x^3 + 10x^2 \text{ and } -4 \text{ from } -4x - 8.$$
$$= (x + 2)(5x^2 - 4) \quad \text{Factor out the GCF, } x + 2.$$

Success Tip

An equivalent factorization, $(5x^2 - 4)(x + 2)$, results if the terms are arranged as $5x^3 - 4x + 10x^2 - 8$ or as $5x^3 - 4x - 8 + 10x^2$.

Teaching Example 9 Factor: $f^3 - 16 + 2f^2 - 8f$
Answer: $(f + 2)(f^2 - 8)$

Self Check 9 Factor: $y^3 - 6 + 3y^2 - 2y$ $(y + 3)(y^2 - 2)$

Now Try Problem 93

The next example illustrates that when factoring a polynomial, we should *always look for a common factor first.*

EXAMPLE 10 Factor: $10k + 10m - 2km - 2m^2$

Strategy Since all four terms have a common factor of 2, we factor it out first. Then we will factor the resulting polynomial by grouping.

Why Factoring out the GCF first makes factoring by any method easier.

Solution After factoring out 2 from all four terms, notice that within the parentheses, the first two terms have a common factor of 5, and the last two terms have a common factor of $-m$.

$10k + 10m - 2km - 2m^2$

$= 2(5k + 5m - km - m^2)$ — Factor out the GCF, 2.

$= 2[5(k + m) - m(k + m)]$ — Factor out 5 from $5k + 5m$. Factor out $-m$ from $-km - m^2$. This causes the signs of $-km$ and $-m^2$ to change within the second set of parentheses. Enclose the factoring by grouping process within brackets [ ].

$= 2[(k + m)(5 - m)]$ — Factor out the common binomial factor, $k + m$.

$= 2(k + m)(5 - m)$ — Drop the unnecessary brackets.

Check by multiplying.

Success Tip

Here is a factoring guideline: If a new pair of parentheses is produced in the factoring process, always see if the resulting expression within those parentheses can be factored further.

Teaching Example 10 Factor: $15r + 15s - 3rs - 3s^2$
Answer: $3(r + s)(5 - s)$

Self Check 10 Factor: $4t + 4s + 4tz + 4sz$ $4(t + s)(1 + z)$

Now Try Problem 97

SECTION 6.1 STUDY SET

VOCABULARY

Fill in the blanks.

▶ 1. To factor a polynomial means to express it as a product of two (or more) polynomials.

▶ 2. GCF stands for greatest common factor. When we write $2x + 4$ as $2(x + 2)$, we say that we have factored out the GCF, 2.

▶ 3. To factor $m^3 + 3m^2 + 4m + 12$ by grouping, we begin by writing $m^2(m + 3) + 4(m + 3)$.

▶ 4. The terms $x(x - 1)$ and $4(x - 1)$ have the common binomial factor $x - 1$.

CONCEPTS

5. Complete each factorization.

a. $6x = 2 \cdot 3 \cdot x$

b. $35h^2 = 5 \cdot 7 \cdot h \cdot h$

c. $18y^3z = 2 \cdot 3 \cdot 3 \cdot y \cdot y \cdot y \cdot z$

▶ 6. a. Find the GCF of $30x^2$ and $105x^3$. $15x^2$

$30x^2 = 2 \cdot 3 \cdot 5 \cdot x \cdot x$

$105x^3 = 3 \cdot 5 \cdot 7 \cdot x \cdot x \cdot x$

b. Find the GCF of $12a^2b^2$, $15a^3b$, and $75a^4b^2$. $3a^2b$

$12a^2b^2 = 2 \cdot 2 \cdot 3 \cdot a \cdot a \cdot b \cdot b$

$15a^3b = 3 \cdot 5 \cdot a \cdot a \cdot a \cdot b$

$75a^4b^2 = 3 \cdot 5 \cdot 5 \cdot a \cdot a \cdot a \cdot a \cdot b \cdot b$

7. a. Write a binomial such that the GCF of its terms is 2.
$2x + 4$ (Answers may vary)

b. Write a trinomial such that the GCF of its terms is x.
$x^3 + x^2 + x$ (Answers may vary)

8. Check to determine whether each factorization is correct.

a. $9y^3 + 5y^2 - 15y = 3y(3y^2 + 2y - 5)$ No

b. $3s^3 + 2s^2 + 6s + 4 = (3s + 2)(s^2 + 2)$ Yes

Fill in the blanks to complete each factorization.

9. $2x^2 + 6x = \mathbf{2x} \cdot x + \mathbf{2x} \cdot 3$
$= 2x(x + 3)$

▶ 10. $3t^3 - t^2 + 15t - 5 = t^2(\mathbf{3t - 1}) + 5(\mathbf{3t - 1})$
$= (3t - 1)(t^2 + 5)$

11. Consider the polynomial $2k - 8 + hk - 4h$.

a. How many terms does the polynomial have? 4

b. Is there a common factor of all the terms, other than 1? No

c. What is the GCF of the first two terms and what is the GCF of the last two terms? 2; h

12. What is the first step in factoring $8y^2 - 16yz - 6y + 12z$?
Factor out the GCF, 2.

NOTATION

Complete each factorization.

13. $8m^2 - 32m + 16 = 8(m^2 - 4m + 2)$

14. $10a^4 - 15a^3 = 5a^3(2a - 3)$

15. $b^3 - 6b^2 + 2b - 12 = b^2(b - 6) + 2(b - 6)$
$= (b - 6)(b^2 + 2)$

16. $12 + 8n - 3m - 2mn = 4(3 + 2n) - m(3 + 2n)$
$= (3 + 2n)(4 - m)$

GUIDED PRACTICE

Find the GCF of each list of numbers. See Example 1.

17. 6, 10 — 2
▶ 18. 10, 15 — 5
19. 18, 24 — 6
▶ 20. 60, 72 — 12
▶ 21. 14, 21, 42 — 7
22. 16, 24, 48 — 8
23. 40, 32, 24 — 8
▶ 24. 28, 35, 21 — 7

Find the GCF of each list of terms. See Example 2.

25. m^4, m^3 — m^3
▶ 26. c^2, c^7 — c^2
27. $15x, 25$ — 5
▶ 28. $9a, 21$ — 3
29. $20c^2, 12c$ — $4c$
▶ 30. $18r, 27r^3$ — $9r$
31. $18a^4, 9a^3, 27a^3$ — $9a^3$
32. $33m^5, 22m^6, 11m^5$ — $11m^5$
▶ 33. $24a^2, 16a^3b, 40ab$ — $8a$
34. $12r^2, 15rs, 9r^2s^2$ — $3r$
▶ 35. $6m^4n, 12m^3n^2, 9m^3n^3$ — $3m^3n$
▶ 36. $15c^2d^4, 10c^2d, 40c^3d^3$ — $5c^2d$
37. $4(x+7), 9(x+7)$ — $x+7$
▶ 38. $2(y-1), 5(y-1)$ — $y-1$
39. $4(p-t), p(p-t)$ — $p-t$
▶ 40. $a(b+c), 3(b+c)$ — $b+c$

Factor out the GCF. See Example 3.

41. $3x+6$ — $3(x+2)$
▶ 42. $18x+24$ — $6(3x+4)$
43. $18m-9$ — $9(2m-1)$
▶ 44. $15s-35$ — $5(3s-7)$
45. d^2-7d — $d(d-7)$
▶ 46. a^2+9a — $a(a+9)$
47. $15c^3+25$ — $5(3c^3+5)$
▶ 48. $33h^4-22$ — $11(3h^4-2)$
49. $24a-16a^2$ — $8a(3-2a)$
▶ 50. $18r-30r^2$ — $6r(3-5r)$
▶ 51. $14x^2-7x-7$ — $7(2x^2-x-1)$
52. $27a^2-9a+9$ — $9(3a^2-a+1)$
53. $t^4+t^3+2t^2$ — $t^2(t^2+t+2)$
▶ 54. $b^4-b^3-3b^2$ — $b^2(b^2-b-3)$
55. $21x^2y^3+3xy^2$ — $3xy^2(7xy+1)$
▶ 56. $3x^2y^3-9x^4y^3$ — $3x^2y^3(1-3x^2)$

Factor out -1 *from each polynomial.* See Example 4.

57. $-a-b$ — $-(a+b)$
▶ 58. $-x-2y$ — $-(x+2y)$
59. $-x^2-x+16$ — $-(x^2+x-16)$
▶ 60. $-t^2-9t+1$ — $-(t^2+9t-1)$
61. $5-x$ — $-(-5+x)$ or $-(x-5)$
▶ 62. $10-m$ — $-(-10+m)$ or $-(m-10)$
63. $9-4a$ — $-(-9+4a)$ or $-(4a-9)$
▶ 64. $7-8b$ — $-(-7+8b)$ or $-(8b-7)$

Factor each polynomial by factoring out the opposite of the GCF. See Example 5.

65. $-3x^2-6x$ — $-3x(x+2)$
▶ 66. $-4a^2-6a$ — $-2a(2a+3)$
67. $-4a^2b+12a^3$ — $-4a^2(b-3a)$
▶ 68. $-25x^4+30x^2$ — $-5x^2(5x^2-6)$
69. $-24x^4-48x^3+36x^2$ — $-12x^2(2x^2+4x-3)$
▶ 70. $-28a^5-42a^4+14a^3$ — $-14a^3(2a^2+3a-1)$
71. $-4a^3b^2+14a^2b^2-10ab^2$ — $-2ab^2(2a^2-7a+5)$
▶ 72. $-30x^4y^3+24x^3y^2-60x^2y$ — $-6x^2y(5x^2y^2-4xy+10)$

Factor. See Example 6.

▶ 73. $y(x+2)+3(x+2)$ — $(x+2)(y+3)$
74. $r(t+v)+3(t+v)$ — $(t+v)(r+3)$
75. $m(p-q)-5(p-q)$ — $(p-q)(m-5)$
▶ 76. $ab(c-7)-12(c-7)$ — $(c-7)(ab-12)$

Factor by grouping. See Example 7.

77. $2x+2y+ax+ay$ — $(x+y)(2+a)$
▶ 78. $bx+bz+5x+5z$ — $(x+z)(b+5)$
79. $rs-ru+8sw-8uw$ — $(s-u)(r+8w)$
▶ 80. $12ab-4ac+3db-dc$ — $(3b-c)(4a+d)$
81. $7m^3-2m^2+14m-4$ — $(7m-2)(m^2+2)$
82. $9s^3-2s^2+36s-8$ — $(9s-2)(s^2+4)$
83. $5x^3-x^2+10x-2$ — $(5x-1)(x^2+2)$
▶ 84. $6a^3-a^2+18a-3$ — $(6a-1)(a^2+3)$

Factor by grouping. See Example 8.

▶ 85. $ab+ac+b+c$ — $(b+c)(a+1)$
86. $xy+3y^2+x+3y$ — $(x+3y)(y+1)$
87. $rs+4s^2-r-4s$ — $(r+4s)(s-1)$
▶ 88. $tx+tz-x-z$ — $(x+z)(t-1)$
▶ 89. $2ax+2bx-3a-3b$ — $(2x-3)(a+b)$
90. $rx+sx-ry-sy$ — $(r+s)(x-y)$
91. $mp-np-mq+nq$ — $(m-n)(p-q)$
▶ 92. $9p-9q-mp+mq$ — $(p-q)(9-m)$

Factor by grouping. See Example 9.

93. $5m^3+6+5m^2+6m$ — $(m+1)(5m^2+6)$
▶ 94. $4t^3+14+28t^2+2t$ — $(t+7)(4t^2+2)$
95. $y^3-12+3y-4y^2$ — $(y^2+3)(y-4)$
▶ 96. $h^3-8+h-8h^2$ — $(h^2+1)(h-8)$

Factor by grouping. Remember to factor out the GCF first. See Example 10.

97. $ax^3-2ax^2+5ax-10a$ — $a(x-2)(x^2+5)$
▶ 98. $x^3y^2-2x^2y^2+3xy^2-6y^2$ — $y^2(x-2)(x^2+3)$
▶ 99. $6x^3-6x^2+12x-12$ — $6(x^2+2)(x-1)$
100. $3x^3-6x^2+15x-30$ — $3(x^2+5)(x-2)$

TRY IT YOURSELF

Factor.

▶ 101. $h^2(14+r)+5(14+r)$ — $(14+r)(h^2+5)$
▶ 102. $x(y+9)-21(y+9)$ — $(y+9)(x-21)$
103. $22a^3-33a^2$ — $11a^2(2a-3)$
104. $39r^3+26r^2$ — $13r^2(3r+2)$
105. $ax+bx-a-b$ — $(a+b)(x-1)$
106. $2xy+y^2-2x-y$ — $(2x+y)(y-1)$
107. $15r^8-18r^6-30r^5$ — $3r^5(5r^3-6r-10)$
▶ 108. $24cm-12cn+16c$ — $4c(6m-3n+4)$
109. $27mp+9mq-9np-3nq$ — $3(3p+q)(3m-n)$
110. $4abc+4ac^2-2bc-2c^2$ — $2c(b+c)(2a-1)$
▶ 111. $-60p^2t^2-80pt^3$ — $-20pt^2(3p+4t)$
112. $-25x^5y^7+75x^3y^2$ — $-25x^3y^2(x^2y^5-3)$
113. $-2x+5$ — $-(2x-5)$
▶ 114. $-3x+8$ — $-(3x-8)$
▶ 115. $6x^2-2xy-15x+5y$ — $(3x-y)(2x-5)$
116. m^3+5m^2+m+5 — $(m+5)(m^2+1)$
▶ 117. $2x^3z-4x^2z+32xz-64z$ — $2z(x-2)(x^2+16)$
▶ 118. $4a^2b+12a^2-8ab-24a$ — $4a(b+3)(a-2)$
119. $12uvw^3-54uv^2w^2$ — $6uvw^2(2w-9v)$
▶ 120. $14xyz-16x^2y^2z$ — $2xyz(7-8xy)$

121. $x^3 + x^2 + x + 1$ $(x + 1)(x^2 + 1)$

122. $m^4 + m^3 + 2m + 2$ $(m + 1)(m^3 + 2)$

▶ **123.** $-3r + 2s - 3$ $-(3r - 2s + 3)$

124. $-6yz + 12xz + 5xy$ $-(6yz - 12xz - 5xy)$

Look Alikes . . .

125. **a.** $5t^3 + 6t^2 + 15t + 18$ $(5t + 6)(t^2 + 3)$ **b.** $3t^3 + 6t^2 + 15t + 18$ $3(t^3 + 2t^2 + 5t + 6)$

▶ **126.** **a.** $x^2 + xy + x + y$ $(x + y)(x + 1)$ **b.** $2x^2 + 2xy + 2x + 2y$ $2(x + y)(x + 1)$

APPLICATIONS

127. Geometry. The dimensions of the rectangle shown below can be found by factoring the polynomial that represents its area. Find the polynomials that represent the length and the width of the rectangle. $(x^2 + 5)$ ft; $(x + 4)$ ft

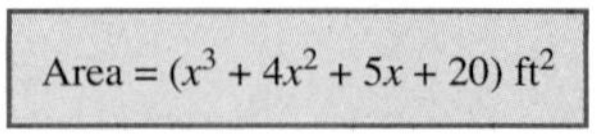

▶ **128. Front Doors.** Find a polynomial that represents the amount of surface on the front face of the door shown on the right that needs to be stained. Then factor the polynomial. $2x(15x + 16)$

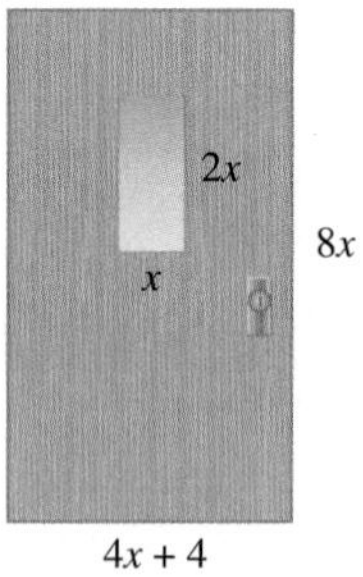

▶ **129.** ▶ from **Campus to Careers**

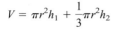

Elementary School Teacher

©iStockphoto.com/Neustockimages

A teacher is going to purchase a block of wax so that the children in her classroom can melt it down, add coloring, pour it into a mold, and make their own crayons. The amount of wax to make one crayon is given by the volume formula:

$$V = \pi r^2 h_1 + \frac{1}{3}\pi r^2 h_2$$

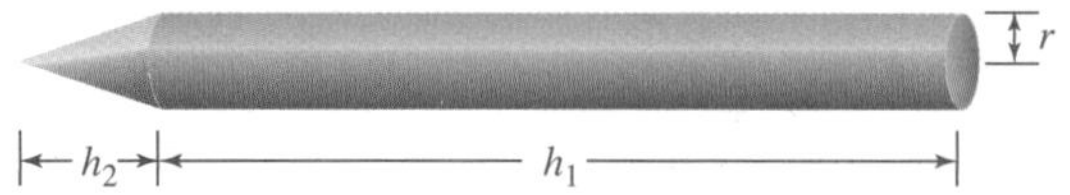

a. Rewrite the formula by factoring the expression on the right side. $V = \pi r^2(h_1 + \frac{1}{3}h_2)$

b. Use the formula to estimate what size block of wax (in cubic inches) she needs to purchase if there are 20 students in her class, and each student will get 5 crayons. The dimensions for the crayon molds they will use are: $h_1 = 1\frac{5}{6}$ in., $h_2 = \frac{1}{2}$ in., and $r = \frac{1}{4}$ in. Round up to the nearest cubic inch. A 40 in.^{3} block of wax is needed.

▶ **130. Interior Decorating.** The expression $\pi rs + \pi Rs$ can be used to find the amount of material needed to make the lamp shade shown. Factor the expression. $\pi s(r + R)$

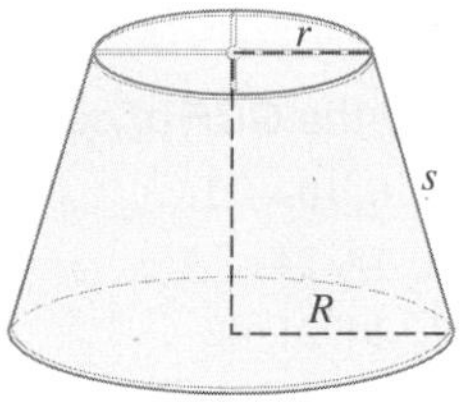

WRITING

▶ **131.** Explain how to find the GCF of $32a^3$ and $16a^2$.

132. Explain this diagram.

Multiplication ⟶

$$3x^2(5x^2 - 6x + 4) = 15x^4 - 18x^3 + 12x^2$$

⟵ Factoring

133. Explain how factorizations of polynomials are checked. Give an example.

134. Explain the error.

Factor out the GCF: ~~$30a^3 - 12a^2 = 6a(5a^2 - 2a)$~~

REVIEW

135. Insurance Costs. A college student's good grades earned her a student discount on her car insurance premium. What was the percent of decrease, to the nearest percent, if her annual premium was lowered from \$1,050 to \$925? 12%

▶ **136. Calculating Grades.** A student has test scores of 68%, 75%, and 79% in a government class. What must she score on the last exam to earn a B (80% or better) in the course? 98% or higher

CHALLENGE PROBLEMS

137. Factor: $6x^{4m}y^n + 21x^{3m}y^{2n} - 15x^{2m}y^{3n}$ $3x^{2m}y^n(2x^{2m} + 7x^my^n - 5y^{2n})$

▶ **138.** Factor $ab - b^2 - bc + ac - bc - c^2$ by grouping. $(a - b - c)(b + c)$

SECTION 6.2

Factoring Trinomials of the Form $x^2 + bx + c$

OBJECTIVES

1. Factor trinomials of the form $x^2 + bx + c$.
2. Factor trinomials of the form $x^2 + bx + c$ after factoring out the GCF.
3. Factor trinomials of the form $x^2 + bx + c$ using the grouping method.

ARE YOU READY? *Are You Ready? exercises available online at www.webassign.net/brookscole*

The following problems review some basic skills that are needed when factoring trinomials.

1. In $x^2 - 4x + 10$, what is the coefficient of the leading term? 1
2. Multiply: $(x + 8)(x - 1)$ $x^2 + 7x - 8$
3. What is the GCF of the terms of $2x^2 - 10x + 4$? 2
4. $x^2 - 4xy - 5y^2$ is a polynomial in how many variables? 2
5. Find two integers whose product is 8 and whose sum is 6. 2 and 4
6. Find two integers whose product is -15 and whose sum is 2. -3 and 5

In Chapter 5, we learned how to multiply binomials. For example, to multiply $x + 1$ and $x + 2$, we proceed as follows.

$$(x + 1)(x + 2) = x^2 + 2x + x + 2$$
$$= x^2 + 3x + 2$$

To *factor the trinomial* $x^2 + 3x + 2$, we will reverse the multiplication process and determine what factors were multiplied to obtain this result. Since the product of two binomials is often a trinomial, many trinomials factor into two binomials.

Multiplication: Given the binomial factors, we find a trinomial.

$$(x + 1)(x + 2) = x^2 + 3x + 2$$

Factoring: Given a trinomial, we find the binomial factors.

The Language of Algebra

Recall that when a polynomial in one variable is written in descending powers of that variable, the coefficient of the first term is called the **leading coefficient.**

To begin the discussion of trinomial factoring, we consider trinomials of the form $x^2 + bx + c$, such as

$$x^2 + 8x + 15, \quad y^2 - 13y + 12, \quad a^2 + a - 20, \quad \text{and} \quad z^2 - 20z - 21$$

In each case, the **leading coefficient**—the coefficient of the squared variable—is 1.

1 Factor Trinomials of the Form $x^2 + bx + c$.

To develop a method for factoring trinomials, we will find the product of $x + 6$ and $x + 4$ and make some observations about the result.

$$(x + 6)(x + 4) = \overset{F}{x \cdot x} + \overset{O}{x \cdot 4} + \overset{I}{6 \cdot x} + \overset{L}{6 \cdot 4} \quad \text{Use the FOIL method.}$$
$$= x^2 + 4x + 6x + 24$$
$$= x^2 + 10x + 24$$

First term: x^2; Middle term: $10x$; Last term: 24

The Language of Algebra

If a term of a trinomial is a number only, it is called a **constant term.**

$x^2 + 10x + 24$ — 24: Constant term

The result is a trinomial, where

- the first term, x^2, is the product of x and x
- the last term, 24, is the product of 6 and 4
- the coefficient of the middle term, 10, is the sum of 6 and 4

These observations suggest a strategy to use to factor trinomials that have 1 as the leading coefficient.

EXAMPLE 1 Factor: $x^2 + 8x + 15$

Strategy We will assume that $x^2 + 8x + 15$ is the product of two binomials and we will use a systematic method to find their terms.

Why Since the terms of $x^2 + 8x + 15$ do not have a common factor (other than 1), the only option available is to try to factor it as the product of two binomials.

Solution We represent the binomials using two sets of parentheses. Since the first term of the trinomial is x^2, we enter x and x as the first terms of its binomial factors.

$$x^2 + 8x + 15 = (x \square)(x \square)$$ Because $x \cdot x$ will give x^2.

Teaching Tip: You may want to illustrate the factoring process visually by examining the areas of four attached rectangles. Use an approach similar to that which was used to motivate binomial multiplication in Section 5.6.

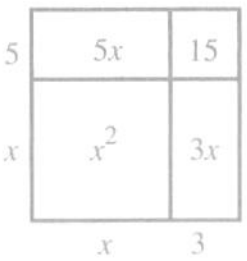

The second terms of the binomials must be two integers whose product is 15 and whose sum is 8. Since the integers must have a positive product and a positive sum, we consider only pairs of positive integer factors of 15. The only such pairs, $1 \cdot 15$ and $3 \cdot 5$, are listed in the table. Then we find the sum of each pair and enter each result in the table.

Positive factors of 15	Sum of the positive factors of 15
$1 \cdot 15 = 15$	$1 + 15 = 16$
$3 \cdot 5 = 15$	$3 + 5 = 8$

List all of the pairs of positive integers that multiply to give 15.

Add each pair of factors.

The second row of the table contains the correct pair of integers 3 and 5, whose product is 15 and whose sum is 8. To complete the factorization, we enter 3 and 5 as the second terms of the binomial factors.

$$x^2 + 8x + 15 = (x + 3)(x + 5)$$

We can check the factorization by multiplying:

$$(x + 3)(x + 5) = x^2 + 5x + 3x + 15$$
$$= x^2 + 8x + 15$$ This is the original trinomial.

Notation

By the commutative property of multiplication, the order of the binomial factors in a factorization does not matter. Thus, we can also write:

$$x^2 + 8x + 15 = (x + 5)(x + 3)$$

Teaching Example 1
Factor: $m^2 + 8m + 12$
Answer: $(m + 2)(m + 6)$

Self Check 1 Factor: $y^2 + 7y + 10$ $(y + 2)(y + 5)$

Now Try Problem 15

EXAMPLE 2 Factor: $y^2 - 13y + 12$

Strategy We will assume that $y^2 - 13y + 12$ is the product of two binomials and we will use a systematic method to find their terms.

Why Since the terms of $y^2 - 13y + 12$ do not have a common factor (other than 1), the only option available is to try to factor it as the product of two binomials.

Solution We represent the binomials using two sets of parentheses. Since the first term of the trinomial is y^2, the first term of each binomial factor must be y.

$$y^2 - 13y + 12 = (y \square)(y \square)$$ Because $y \cdot y$ will give y^2.

The second terms of the binomials must be two integers whose product is 12 and whose sum is −13. Since the integers must have a positive product and a negative sum, we only consider pairs of negative integer factors of 12. The possible pairs are listed in the table.

Negative factors of 12	Sum of the negative factors of 12
$-1(-12) = 12$	**$-1 + (-12) = -13$**
$-2(-6) = 12$	$-2 + (-6) = -8$
$-3(-4) = 12$	$-3 + (-4) = -7$

You can stop listing the factors after finding the correct combination.

The Language of Algebra

Make sure you understand this vocabulary: **Many trinomials factor as the product of two binomials.**

Trinomial: $y^2 - 13y + 12$ = Product of two binomials: $(y - 1)(y - 12)$

The first row of the table contains the correct pair of integers −1 and −12, whose product is 12 and whose sum is −13. To complete the factorization, we enter −1 and −12 as the second terms of the binomial factors.

$$y^2 - 13y + 12 = (y - \mathbf{1})(y - \mathbf{12})$$

We check the factorization by multiplying:

$$(y - 1)(y - 12) = y^2 - 12y - y + 12$$
$$= y^2 - 13y + 12 \quad \text{This is the original trinomial.}$$

Teaching Example 2 Factor: $t^2 - 17t + 16$
Answer: $(t - 1)(t - 16)$

Self Check 2 Factor: $p^2 - 6p + 8$ $(p - 2)(p - 4)$

Now Try Problem 19

EXAMPLE 3 Factor: $x^2 + x - 20$

Strategy We will assume that $x^2 + x - 20$ is the product of two binomials and we will use a systematic method to find their terms.

Why Since the terms of $x^2 + x - 20$ do not have a common factor (other than 1), the only option available is to try to factor it as the product of two binomials.

Solution We represent the binomials using two sets of parentheses. Since the first term of the trinomial is x^2, the first term of each binomial factor must be x.

$$x^2 + x - 20 = (x \square)(x \square) \quad \text{Because } x \cdot x \text{ will give } x^2.$$

To determine the second terms of the binomials, we must find two integers whose product is −20 and whose sum is 1. Because the integers must have a negative product, their signs must be different. The possible pairs are listed in the table.

Teaching Tip: Explain that some students find factoring easier if they carefully list the possible factor pairs using the organized approach shown in the table.

It is wise to follow an order when listing the factors in the table so that you don't skip the correct combination. Here, the first factors 1, 2, 4, 5, 10, and 20 are listed from least to greatest.

Factors of −20	Sum of the factors of −20
$1(-20) = -20$	$1 + (-20) = -19$
$2(-10) = -20$	$2 + (-10) = -8$
$4(-5) = -20$	$4 + (-5) = -1$
$5(-4) = -20$	**$5 + (-4) = 1$**
$10(-2) = -20$	$10 + (-2) = 8$
$20(-1) = -20$	$20 + (-1) = 19$

The fourth row of the table contains the correct pair of integers 5 and −4, whose product is −20 and whose sum is 1. To complete the factorization, we enter 5 and −4 as the second terms of the binomial factors.

$$x^2 + x - 20 = (x + 5)(x - 4)$$

Check the factorization by multiplying.

Teaching Example 3 Factor: $r^2 + 5r - 24$
Answer: $(r + 8)(r - 3)$

Self Check 3 Factor: $m^2 + m - 42$ $(m + 7)(m - 6)$

Now Try Problem 23

EXAMPLE 4 Factor: $z^2 - 4z - 21$

Strategy We will assume that $z^2 - 4z - 21$ is the product of two binomials and we will use a systematic method to find their terms.

Why Since the terms of $z^2 - 4z - 21$ do not have a common factor (other than 1), the only option available is to try to factor it as the product of two binomials.

Solution We represent the binomials using two sets of parentheses. Since the first term of the trinomial is z^2, the first term of each binomial factor must be z.

$$z^2 - 4z - 21 = (z \square)(z \square)$$ *Because $z \cdot z$ will give z^2.*

To determine the second terms of the binomials, we must find two integers whose product is −21 and whose sum is −4. Because the integers must have a negative product, their signs must be different. The possible pairs are listed in the table.

Success Tip

Show your students how to fix a factorization that is close, but not quite correct, quickly. For example, $(z + 7)(z - 3)$ does not produce the proper sign for the middle term: $(z + 7)(z - 3) = z^2 + 4z - 21$. However, a simple change of signs within the binomials does: $(z - 7)(z + 3) = z^2 - 4z - 21$.

Factors of −21	Sum of the factors of −21
$1(-21) = -21$	$1 + (-21) = -20$
$3(-7) = -21$	$3 + (-7) = -4$
$7(-3) = -21$	$7 + (-3) = 4$
$21(-1) = -21$	$21 + (-1) = 20$

The second row of the table contains the correct pair of integers 3 and −7, whose product is −21 and whose sum is −4. To complete the factorization, we enter 3 and −7 as the second terms of the binomial factors.

$$z^2 - 4z - 21 = (z + 3)(z - 7)$$

Check by multiplying.

Teaching Example 4 Factor: $h^2 - 7h - 44$
Answer: $(h + 4)(h - 11)$

Self Check 4 Factor: $q^2 - 2q - 24$ $(q + 4)(q - 6)$

Now Try Problem 29

The following guidelines are helpful when factoring trinomials.

Factoring Trinomials Whose Leading Coefficient Is 1

To factor a trinomial of the form $x^2 + bx + c$, find two numbers whose product is c and whose sum is b.

1. If c is positive, the numbers have the same sign.
2. If c is negative, the numbers have different signs.

Then write the trinomial as a product of two binomials. You can check by multiplying.

$$x^2 + bx + c = (x \square)(x \square)$$

The product of these numbers must be c and their sum must be b.

EXAMPLE 5 Factor: $-h^2 + 2h + 63$

Strategy We will factor out -1 and then factor the resulting trinomial.

Why It is easier to factor trinomials that have a positive leading coefficient.

Solution After factoring out -1, we factor the trinomial within the parentheses.

$$-h^2 + 2h + 63 = -1(h^2 - 2h - 63) \quad \text{Factor out } -1.$$
$$= -(h^2 - 2h - 63) \quad \text{The 1 need not be written.}$$
$$= -(h + 7)(h - 9) \quad \text{Factor } h^2 - 2h - 63.$$

Check:

$$-(h + 7)(h - 9) = -(h^2 - 9h + 7h - 63) \quad \text{Multiply the binomials first.}$$
$$= -(h^2 - 2h - 63) \quad \text{Combine like terms.}$$
$$= -h^2 + 2h + 63 \quad \text{Drop the } - \text{ sign and change the sign of every term within the parentheses.}$$

The result is the original trinomial.

Teaching Example 5
Factor: $-x^2 + 15x - 54$
Answer: $-(x - 6)(x - 9)$

Self Check 5 Factor: $-x^2 + 11x - 28$ $-(x - 4)(x - 7)$

Now Try Problem 33

We factor trinomials in two variables in a similar way.

EXAMPLE 6 Factor: $x^2 - 4xy - 5y^2$

Teaching Tip: If your course covers only factoring trinomials in one variable, you may want to skip this example.

Strategy We will assume that $x^2 - 4xy - 5y^2$ is the product of two binomials and we will use a systematic method to find their terms.

Why Since the terms of $x^2 - 4xy - 5y^2$ do not have a common factor (other than 1), the only option available is to try to factor it as the product of two binomials.

Solution We represent the binomials using two sets of parentheses. Since the first term of the trinomial is x^2, the first term of each binomial factor must be x. Since the third term contains y^2, the last term of each binomial factor must contain y. To complete the factorization, we need to determine the coefficient of each y-term.

$$x^2 - 4xy - 5y^2 = (x \square y)(x \square y) \quad \text{Because } x \cdot x \text{ will give } x^2 \text{ and } y \cdot y \text{ will give } y^2.$$

The Language of Algebra

The trinomial $x^2 - 4xy - 5y^2$ is in two variables, x and y. It is written in **descending** powers of x and **ascending** powers of y.

The coefficients of y must be two integers whose product is -5 and whose sum is -4. Such a pair is 1 and -5. Instead of writing the first factor as $(x + 1y)$, we write it as $(x + y)$, because $1y = y$.

$$x^2 - 4xy - 5y^2 = (x + y)(x - 5y)$$

Check: $(x + y)(x - 5y) = x^2 - 5xy + xy - 5y^2$

$= x^2 - 4xy - 5y^2$ This is the original trinomial.

Teaching Example 6
Factor: $m^2 + 3mn - 4n^2$
Answer: $(m + 4n)(m - n)$

Self Check 6 Factor: $s^2 + 6st - 7t^2$ $(s + 7t)(s - t)$

Now Try Problem 41

2 Factor Trinomials of the Form $x^2 + bx + c$ After Factoring Out the GCF.

If the terms of a trinomial have a common factor, it should be factored out first. The instruction "Factor" means for you to *factor the given expression completely.* A trinomial is **factored completely** when no factor can be factored further. Each factor of a completely factored expression will be prime.

EXAMPLE 7 Factor: $2x^4 + 26x^3 + 80x^2$

Strategy We will factor out the GCF, $2x^2$, first. Then we will factor the resulting trinomial.

Why The first step in factoring any polynomial is to factor out the GCF. Factoring out the GCF first makes factoring by any method easier.

Solution We begin by factoring out the GCF, $2x^2$, from $2x^4 + 26x^3 + 80x^2$.

$$2x^4 + 26x^3 + 80x^2 = 2x^2(x^2 + 13x + 40)$$

Next, we factor $x^2 + 13x + 40$. The integers 8 and 5 have a product of 40 and a sum of 13, so the completely factored form of the given trinomial is

$$2x^4 + 26x^3 + 80x^2 = 2x^2(x + 8)(x + 5)$$ The complete factorization must include $2x^2$.

Check: $2x^2(x + 8)(x + 5) = 2x^2(x^2 + 13x + 40)$

$= 2x^4 + 26x^3 + 80x^2$ This is the original trinomial.

Caution

When prime factoring 30, for example, you wouldn't stop here because 6 can be factored:

30
5 6

If a new pair of parentheses is produced in the factoring process, always see if the resulting expression within the parentheses can be factored further.

Self Check 7 Factor: $4m^5 + 8m^4 - 32m^3$ $4m^3(m + 4)(m - 2)$

Now Try Problem 47

Teaching Example 7
Factor: $3y^4 + 18y^3 - 48y^2$
Answer: $3y^2(y + 8)(y - 2)$

EXAMPLE 8 Factor: $-13g^2 + 36g + g^3$

Strategy We will write the terms of the trinomial in descending powers of g.

Why It is easier to factor a trinomial if its terms are written in descending powers of one variable.

Solution

$-13g^2 + 36g + g^3 = g^3 - 13g^2 + 36g$ Rearrange the terms.

$= g(g^2 - 13g + 36)$ Factor out the GCF, g.

$= g(g - 9)(g - 4)$ Factor the trinomial.

Check the factorization by multiplying.

Caution

For multistep factorizations, don't forget to write the GCF in the final factored form.

Teaching Example 8 Factor: $-63n + n^3 + 2n^2$
Answer: $n(n + 9)(n - 7)$

Self Check 8 Factor: $-12t + t^3 + 4t^2$ $t(t - 2)(t + 6)$

Now Try ▶ Problem 59

If a trinomial with integer coefficients cannot be factored using only integers, it is called a **prime trinomial.**

EXAMPLE 9 Factor: $x^2 + 2x + 3$

Strategy We will assume that $x^2 + 2x + 3$ is the product of two binomials and we will use a systematic method to find their terms.

Why Since the terms of $x^2 + 2x + 3$ do not have a common factor (other than 1), the only option available is to try to factor it as the product of two binomials.

Solution To factor the trinomial, we must find two integers whose product is 3 and whose sum is 2. The possible factorizations are shown in the table.

The Language of Algebra

When a trinomial is not factorable using only integers, we say it is **prime** and that it does not factor **over** the integers.

Factors of 3	Sum of the factors of 3
$1(3) = 3$	$1 + 3 = 4$
$-1(-3) = 3$	$-1 + (-3) = -4$

Since there are no two integers whose product is 3 and whose sum is 2, the trinomial $x^2 + 2x + 3$ cannot be factored and is a *prime trinomial.*

Teaching Example 9 Factor: $d^2 - 9d + 11$
Answer: Prime trinomial

Self Check 9 Factor: $x^2 - 4x + 6$ Prime trinomial

Now Try ▶ Problem 63

3 Factor Trinomials of the Form $x^2 + bx + c$ Using the Grouping Method.

Another way to factor trinomials is to write them as equivalent four-termed polynomials and factor by grouping. To factor $x^2 + 8x + 15$ using this method, we proceed as follows.

1. First, identify a as the coefficient of the x^2 term, b as the coefficient of the x-term, and c as the last (constant) term. In this section, we are only factoring trinomials that have a leading coefficient of 1. Thus, a will always be equal to 1.

$$\left.\begin{array}{c} ax^2 + bx + c \\ \downarrow \quad\ \downarrow \quad\ \downarrow \\ 1x^2 + 8x + 15 \end{array}\right\} a = 1, b = 8, \text{ and } c = 15$$

Then, find the product ac, called the **key number:** $ac = 1(15) = 15$.

2. Next, find two integers whose product is the key number, 15, and whose sum is $b = 8$. Since the integers must have a positive product and a positive sum, we consider only positive factors of 15.

Teaching Tip: Some instructors have their students show the thinking that leads to the correct pair of integers in a "diamond."

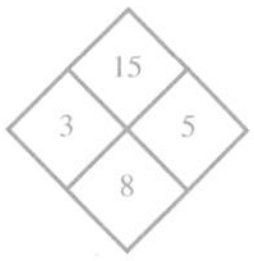

Key number = 15 $b = 8$

Positive factors of 15	Sum of the positive factors of 15
$1 \cdot 15 = 15$	$1 + 15 = 16$
$3 \cdot 5 = 15$	$3 + 5 = 8$

The second row of the table contains the correct pair of integers 3 and 5, whose product is the key number 15 and whose sum is $b = 8$.

3. Express the middle term, $8x$, of the trinomial as the *sum of two terms,* using the integers 3 and 5 found in step 2 as coefficients of the two terms.

$$x^2 + \mathbf{8x} + 15 = x^2 + \mathbf{3x} + \mathbf{5x} + 15 \quad \text{Express } 8x \text{ as } 3x + 5x.$$

4. Factor the equivalent four-term polynomial by grouping:

$$x^2 + 3x + 5x + 15 = x(x + 3) + 5(x + 3) \quad \text{Factor } x \text{ out of } x^2 + 3x \text{ and 5 out of } 5x + 15.$$

$$= (x + 3)(x + 5) \quad \text{Factor out } x + 3.$$

Check the factorization by multiplying.

The grouping method is an alternative to the method for factoring trinomials discussed earlier in this section. It is especially useful when the constant term, c, has many factors.

Factoring Trinomials of the Form $x^2 + bx + c$ Using Grouping

To factor a trinomial of the form $ax^2 + bx + c$, where $a = 1$:

1. Identify a, b, and c, and the key number, ac.
2. Find two integers whose product is the key number and whose sum is b.
3. Express the middle term, bx, as the sum (or difference) of two terms. Enter the two numbers found in step 2 as coefficients of x in the form shown below. Then factor the equivalent four-term polynomial by grouping.

$$x^2 + \square\, x + \square\, x + c$$

The product of these numbers must be ac, and their sum must be b.

4. Check the factorization using multiplication.

EXAMPLE 10 Factor by grouping: $x^2 + x - 20$

Strategy We will express the middle term, x, of the trinomial as the difference of two carefully chosen terms.

Why We want to produce an equivalent four-term polynomial that can be factored by grouping.

Solution Since $x^2 + x - 20 = \mathbf{1}x^2 + \mathbf{1}x - \mathbf{20}$, we identify a as **1**, b as **1**, c as **−20**, and the key number ac as $\mathbf{1(-20)} = -20$. We must find two integers whose product is -20 and whose sum is $b = \mathbf{1}$. Since the integers must have a negative product, their signs must be different.

Key number $= -20$ $\qquad$ $b = 1$

Factors of −20	Sum of the factors of −20
$1(-20) = -20$	$1 + (-20) = -19$
$2(-10) = -20$	$2 + (-10) = -8$
$4(-5) = -20$	$4 + (-5) = -1$
$\mathbf{5(-4)} = -20$	$\mathbf{5 + (-4)} = 1$
$10(-2) = -20$	$10 + (-2) = 8$
$20(-1) = -20$	$20 + (-1) = 19$

Success Tip

We could also express the middle term as $-4x + 5x$. We obtain the same binomial factors, but in reverse order.

$$x^2 - 4x + 5x - 20$$
$$= x(x - 4) + 5(x - 4)$$
$$= (x - 4)(x + 5)$$

The fourth row of the table contains the correct pair of integers 5 and -4, whose product is -20 and whose sum is 1. They serve as the coefficients of $5x$ and $-4x$, the two terms that we use to represent the middle term, x, of the trinomial.

$$x^2 + x - 20 = x^2 + 5x - 4x - 20 \quad \text{Express the middle term, } x\text{, as } 5x - 4x.$$
$$= x(x + 5) - 4(x + 5) \quad \text{Factor } x \text{ out of } x^2 + 5x \text{ and } -4 \text{ out of } -4x - 20.$$
$$= (x + 5)(x - 4) \quad \text{Factor out } x + 5.$$

Check the factorization by multiplying.

Teaching Example 10 Factor by grouping: $r^2 + 5r - 24$
Answer: $(r + 8)(r - 3)$

Self Check 10 Factor by grouping: $m^2 + m - 42$ $(m + 7)(m - 6)$

Now Try Problem 23

EXAMPLE 11 Factor by grouping: $x^2 - 4xy - 5y^2$

Teaching Tip: If your course covers only factoring trinomials in one variable, you may want to skip this example.

Strategy We will express the middle term, $-4xy$, of the trinomial as the sum of two carefully chosen terms.

Why We want to produce an equivalent four-term polynomial that can be factored by grouping.

Solution In $1x^2 - 4xy - 5y^2$, we identify a as 1, b as -4, c as -5, and the key number ac as $1(-5) = -5$. We must find two integers whose product is -5 and whose sum is $b = -4$. Such a pair is -5 and 1. They serve as the coefficients of $-5xy$ and $1xy$, the two terms that we use to represent the middle term, $-4xy$, of the trinomial.

Key number $= -5$ $b = -4$

Factors	Sum
$-5(1) = -5$	$-5 + 1 = -4$

$$x^2 - 4xy - 5y^2 = x^2 - 5xy + 1xy - 5y^2 \quad \text{Express the middle term, } -4xy\text{, as } -5xy + 1xy. \text{ (}1xy - 5xy \text{ could also be used.)}$$
$$= x(x - 5y) + y(x - 5y) \quad \text{Factor } x \text{ out of } x^2 - 5xy \text{ and } y \text{ out of } 1xy - 5y^2.$$
$$= (x - 5y)(x + y) \quad \text{Factor out } x - 5y.$$

Check the factorization by multiplying.

Teaching Example 11 Factor by grouping: $m^2 + 3mn - 4n^2$
Answer: $(m + 4n)(m - n)$

Self Check 11 Factor by grouping: $q^2 - 2qt - 24t^2$ $(q + 4t)(q - 6t)$

Now Try Problem 41

EXAMPLE 12 Factor: $2x^3 - 20x^2 + 18x$

Strategy We will factor out the GCF, $2x$, first. Then we will factor the resulting trinomial using the grouping method.

Why The first step in factoring any polynomial is to factor out the GCF.

Solution We begin by factoring out the GCF, $2x$, from $2x^3 - 20x^2 + 18x$.

$$2x^3 - 20x^2 + 18x = 2x(x^2 - 10x + 9)$$

To factor $x^2 - 10x + 9$ by grouping we identify a as 1, b as -10, and c as 9. We must find two integers whose product is the key number $ac = 1(9) = 9$ and whose sum is $b = -10$. Such a pair is -9 and -1.

Key number = 9 $b = -10$

Factors	Sum
$-9(-1) = 9$	$-9 + (-1) = -10$

$$x^2 - 10x + 9 = x^2 - 9x - 1x + 9$$ Express $-10x$ as $-9x - 1x$. ($-1x - 9x$ could also be used.)

$$= x(x - 9) - 1(x - 9)$$ Factor x out of $x^2 - 9x$ and -1 out of $-1x + 9$.

$$= (x - 9)(x - 1)$$ Factor out $x - 9$.

The complete factorization of the original trinomial is

$$2x^3 - 20x^2 + 18x = 2x(x - 9)(x - 1)$$ Don't forget to write the GCF, $2x$.

Check the factorization by multiplying.

Teaching Example 12 Factor: $4n^3 - 32n^2 + 28n$
Answer: $4n(n - 7)(n - 1)$

Self Check 12 Factor: $3m^3 - 27m^2 + 24m$ $3m(m - 8)(m - 1)$

Now Try Problem 47

SECTION 6.2 STUDY SET

VOCABULARY

Fill in the blanks.

1. The trinomial $x^2 - x - 12$ factors as the product of two binomials: $(x - 4)(x + 3)$.
2. A prime trinomial cannot be factored by using only integers.
3. The leading coefficient of $x^2 - 3x + 2$ is 1.
4. A trinomial is factored completely when no factor can be factored further.

CONCEPTS

Fill in the blanks.

5. a. Before attempting to factor a trinomial, be sure that it is written in descending powers of a variable.
   b. Before attempting to factor a trinomial into two binomials, always factor out any common factors first.
6. $x^2 + x - 56 = (x \square)(x \square)$

   The product of these numbers must be -56, and their sum must be 1.
7. $x^2 + 5x + 3$ cannot be factored because we cannot find two integers whose product is 3 and whose sum is 5.
8. Complete the following table.

Factors of 8	Sum of the factors of 8
1(8)	9
2(4)	6
−1(−8)	−9
−2(−4)	−6

9. Check to determine whether each factorization is correct.
   a. $x^2 - x - 20 = (x + 5)(x - 4)$ No
   b. $4a^2 + 12a - 16 = 4(a - 1)(a + 4)$ Yes
10. Find two integers whose
    a. product is 10 and whose sum is 7. 5, 2
    b. product is 8 and whose sum is −6. −2, −4
    c. product is −6 and whose sum is 1. 3, −2
    d. product is −9 and whose sum is −8. 1, −9
11. Consider a trinomial of the form $x^2 + bx + c$.
    a. If c is positive, what can be said about the two integers that should be chosen for the factorization?
       They are both positive or both negative.
    b. If c is negative, what can be said about the two integers that should be chosen for the factorization?
       One will be positive, the other negative.
12. Fill in each blank to explain how to factor $x^2 + 7x + 10$ by grouping.

    We express the middle term, $7x$, as the sum of two terms:
    $x^2 + 7x + 10 = x^2 + \square x + \square x + 10$

    The product of these numbers must be 10, and their sum must be 7.

NOTATION

13. To factor a trinomial, a student made a table and circled the correct pair of integers, as shown. Complete the factorization of the trinomial.

    $(x + 3)(x - 2)$

Factors	Sum
1(−6)	−5
2(−3)	−1
(3(−2)	1)
6(−1)	5

▶ 14. To factor a trinomial by grouping, a student made a table and circled the correct pair of integers, as shown. Enter the correct coefficients for the first stage in the factorization process.

$x^2 + 2x + 8x + 16$

Key number = 16

Factors	Sum
$1 \cdot 16$	17
$2 \cdot 8$	10
$4 \cdot 4$	8

GUIDED PRACTICE

***Factor.* See Example 1 or Objective 1.**

15. $x^2 + 3x + 2$ $(x + 2)(x + 1)$
▶ 16. $y^2 + 4y + 3$ $(y + 3)(y + 1)$
17. $z^2 + 7z + 12$ $(z + 4)(z + 3)$
▶ 18. $x^2 + 7x + 10$ $(x + 5)(x + 2)$

***Factor each trinomial.* See Example 2 or Example 10.**

19. $m^2 - 5m + 6$ $(m - 3)(m - 2)$
20. $n^2 - 7n + 10$ $(n - 5)(n - 2)$
21. $t^2 - 11t + 28$ $(t - 7)(t - 4)$
▶ 22. $c^2 - 9c + 8$ $(c - 8)(c - 1)$

***Factor.* See Example 3 or Example 10.**

▶ 23. $x^2 + 5x - 24$ $(x + 8)(x - 3)$
▶ 24. $u^2 + u - 42$ $(u + 7)(u - 6)$
25. $t^2 + 13t - 48$ $(t - 3)(t + 16)$
26. $m^2 + 2m - 48$ $(m - 6)(m + 8)$

***Factor each trinomial.* See Example 4 or Example 10.**

27. $a^2 - 6a - 16$ $(a - 8)(a + 2)$
28. $a^2 - 10a - 39$ $(a - 13)(a + 3)$
29. $b^2 - 9b - 36$ $(b - 12)(b + 3)$
▶ 30. $x^2 - 3x - 40$ $(x - 8)(x + 5)$

***Factor.* See Example 5.**

31. $-x^2 - 7x - 10$ $-(x + 5)(x + 2)$
▶ 32. $-x^2 + 9x - 20$ $-(x - 5)(x - 4)$
▶ 33. $-t^2 - t + 30$ $-(t + 6)(t - 5)$
34. $-t^2 - 15t + 34$ $-(t + 17)(t - 2)$
35. $-r^2 - 3r + 54$ $-(r + 9)(r - 6)$
▶ 36. $-d^2 - 2d + 63$ $-(d + 9)(d - 7)$
37. $-m^2 + 18m - 77$ $-(m - 7)(m - 11)$
38. $-n^2 + 14n - 33$ $-(n - 3)(n - 11)$

***Factor.* See Example 6 or Example 11.**

39. $a^2 + 4ab + 3b^2$ $(a + 3b)(a + b)$
▶ 40. $a^2 + 6ab + 5b^2$ $(a + b)(a + 5b)$
41. $x^2 - 6xy - 7y^2$ $(x - 7y)(x + y)$
▶ 42. $x^2 + 10xy - 11y^2$ $(x + 11y)(x - y)$
43. $r^2 + rs - 2s^2$ $(r + 2s)(r - s)$
▶ 44. $m^2 + mn - 6n^2$ $(m + 3n)(m - 2n)$
45. $a^2 - 5ab + 6b^2$ $(a - 3b)(a - 2b)$
▶ 46. $p^2 - 7pq + 10q^2$ $(p - 5q)(p - 2q)$

***Factor.* See Example 7 or Example 12.**

47. $2x^2 + 10x + 12$ $2(x + 3)(x + 2)$
▶ 48. $3y^2 - 21y + 18$ $3(y - 6)(y - 1)$
49. $6a^2 - 30a + 24$ $6(a - 4)(a - 1)$
▶ 50. $4b^2 + 12b - 16$ $4(b + 4)(b - 1)$
51. $5a^2 - 25a + 30$ $5(a - 3)(a - 2)$
▶ 52. $2b^2 - 20b + 18$ $2(b - 9)(b - 1)$
▶ 53. $-z^3 + 29z^2 - 100z$ $-z(z - 4)(z - 25)$
▶ 54. $-m^3 + m^2 + 56m$ $-m(m + 7)(m - 8)$

***Write each trinomial in descending powers of one variable and factor.* See Example 8.**

55. $80 - 24x + x^2$ $(x - 4)(x - 20)$
▶ 56. $y^2 + 100 + 25y$ $(y + 5)(y + 20)$
57. $10y + 9 + y^2$ $(y + 9)(y + 1)$
▶ 58. $x^2 - 13 - 12x$ $(x - 13)(x + 1)$
59. $r^3 - 16r + 6r^2$ $r(r - 2)(r + 8)$
▶ 60. $u^3 - 12u - u^2$ $u(u + 3)(u - 4)$
61. $4r^2x + r^3 + 3rx^2$ $r(r + 3x)(r + x)$
▶ 62. $a^3 + 5ab^2 + 6a^2b$ $a(a + b)(a + 5b)$

***Factor.* See Example 9.**

63. $u^2 + 10u + 15$ Prime
▶ 64. $v^2 + 9v + 15$ Prime
65. $r^2 + 2r - 4$ Prime
▶ 66. $r^2 - 9r - 12$ Prime

TRY IT YOURSELF

Choose the correct method from Section 6.1 or Section 6.2 to factor each of the following.

67. $5x + 15 + xy + 3y$ $(x + 3)(5 + y)$
68. $ab + b + 2a + 2$ $(a + 1)(b + 2)$
69. $26n^2 - 8n$ $2n(13n - 4)$
70. $40c^2 - 12c$ $4c(10c - 3)$
▶ 71. $a^2 - 4a - 5$ $(a - 5)(a + 1)$
▶ 72. $t^2 - 5t - 50$ $(t - 10)(t + 5)$
73. $-x^2 + 21x + 22$ $-(x - 22)(x + 1)$
▶ 74. $-r^2 + 14r - 45$ $-(r - 9)(r - 5)$
75. $4xy - 4x + 28y - 28$ $4(y - 1)(x + 7)$
76. $3xy - 3x + 15y - 15$ $3(y - 1)(x + 5)$
77. $12b^4 - 48b^3 - 36b^2$ $12b^2(b^2 - 4b - 3)$
▶ 78. $14n^5 - 42n^4 - 28n^3$ $14n^3(n^2 - 3n - 2)$
79. $r^2 - 9r + 18$ $(r - 3)(r - 6)$
80. $y^2 - 17y + 72$ $(y - 8)(y - 9)$
81. $-n^4 + 28n^3 + 60n^2$ $-n^2(n - 30)(n + 2)$
82. $-c^5 + 16c^4 + 80c^3$ $-c^3(c - 20)(c + 4)$
▶ 83. $x^2 + 4xy + 4y^2$ $(x + 2y)(x + 2y) = (x + 2y)^2$
84. $m^2 - 8mn + 16n^2$ $(m - 4n)(m - 4n) = (m - 4n)^2$
85. $a^2 - 4ab - 12b^2$ $(a - 6b)(a + 2b)$
▶ 86. $p^2 + pq - 6q^2$ $(p + 3q)(p - 2q)$
87. $4x^4 + 16x^3 + 16x^2$ $4x^2(x + 2)(x + 2) = 4x^2(x + 2)^2$
▶ 88. $3a^4 + 30a^3 + 75a^2$ $3a^2(a + 5)(a + 5) = 3a^2(a + 5)^2$
89. $a^2 - 46a + 45$ $(a - 45)(a - 1)$
90. $r^2 - 37r + 36$ $(r - 36)(r - 1)$
91. $r^2 - 2r + 4$ Prime
▶ 92. $m^2 + 3m - 20$ Prime
93. $t(x + 2) + 7(x + 2)$ $(x + 2)(t + 7)$
94. $r(t - v) + 10(t - v)$ $(t - v)(r + 10)$
95. $s^4 + 11s^3 - 26s^2$ $s^2(s + 13)(s - 2)$
96. $x^4 + 14x^3 + 45x^2$ $x^2(x + 5)(x + 9)$
97. $15s^3 + 75$ $15(s^3 + 5)$
98. $33g^4 - 99$ $33(g^4 - 3)$
99. $-13y + y^2 - 14$ $(y - 14)(y + 1)$
100. $-3a + a^2 + 2$ $(a - 2)(a - 1)$
101. $2x^2 - 12x + 16$ $2(x - 2)(x - 4)$
102. $6t^2 - 18t - 24$ $6(t - 4)(t + 1)$

Look Alikes . . .

103. a. $x^2 - 10x + 24$ $(x - 4)(x - 6)$ b. $x^2 - 10x - 24$ $(x - 12)(x + 2)$

▶ **104.** a. $x^2 - 5x + 6$ $(x - 2)(x - 3)$ b. $x^2 - 5x - 6$ $(x - 6)(x + 1)$

APPLICATIONS

▶ **105. Pets.** The cage shown on the right is used for transporting dogs. Its volume is $(x^3 + 12x^2 + 27x)$ in.^{3}. The dimensions of the cage can be found by factoring. If the cage is longer than it is tall and taller than it is wide, find the polynomials that represent its length, width, and height. $(x + 9)$ in., x in., $(x + 3)$ in.

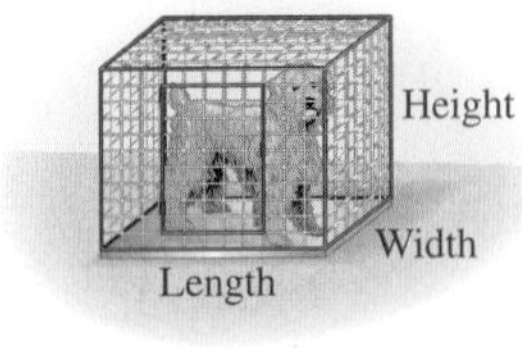

▶ **106. Photography.** A picture cube is a clever way to display 6 photographs in a small amount of space. Suppose the surface area of the entire cube is given by the polynomial $(6s^2 + 12s + 6)$ in.^{2}. Find the polynomial that represents the length of an edge of the cube. $(s + 1)$ in.

WRITING

▶ **107.** Explain what it means when we say that a trinomial is the product of two binomials. Give an example.

▶ **108.** Are $2x^2 - 12x + 16$ and $x^2 - 6x + 8$ factored in the same way? Explain.

▶ **109.** When factoring $x^2 - 2x - 3$, one student got $(x - 3)(x + 1)$, and another got $(x + 1)(x - 3)$. Are both answers acceptable? Explain.

110. In the partial solution shown below, a student began to factor the trinomial. Write a note to the student explaining his mistake.

Factor: $x^2 - 2x - 63$

$(x - \quad)(x - \quad)$

111. Explain the error in the following factorization.

$$x^3 + 8x^2 + 15x = x(x^2 + 8x + 15)$$
$$= (x + 3)(x + 5)$$

112. Explain why the factorization is not complete.

$$2y^2 - 12y + 16 = 2(y^2 - 6y + 8)$$

REVIEW

Simplify each expression. Write each answer without negative exponents.

113. $\dfrac{x^{12}x^{-7}}{x^3x^4}$ $\frac{1}{x^2}$

▶ **114.** $\dfrac{a^4a^{-2}}{a^2a^0}$ 1

115. $(x^{-3}x^{-2})^2$ $\frac{1}{x^{10}}$

116. $\left(\dfrac{18a^2b^3c^{-4}}{3a^{-1}b^2c}\right)^{-3}$ $\frac{c^{15}}{216a^9b^3}$

CHALLENGE PROBLEMS

Factor completely.

▶ **117.** $x^2 - \frac{6}{5}x + \frac{9}{25}$ $\left(x - \frac{3}{5}\right)^2$

118. $x^2 - 0.5x + 0.06$ $(x - 0.3)(x - 0.2)$

119. $x^{2m} - 12x^m - 45$ $(x^m + 3)(x^m - 15)$

120. $x^2(y + 1) - 3x(y + 1) - 70(y + 1)$ $(y + 1)(x + 7)(x - 10)$

121. Find all positive integer values of c that make $n^2 + 6n + c$ factorable. 5, 8, 9

▶ **122.** Find all integer values of b that make $x^2 + bx - 44$ factorable. $-43, 43, -7, 7, -20, 20$

SECTION 6.3 Factoring Trinomials of the Form $ax^2 + bx + c$

OBJECTIVES

1. Factor trinomials using the trial-and-check method.
2. Factor trinomials after factoring out the GCF.
3. Factor trinomials using the grouping method.

ARE YOU READY? *Are You Ready? exercises available online at www.webassign.net/brookscole*

The following problems review some basic skills that are needed when factoring trinomials.

1. In $3x^2 - x + 8$, what is the coefficient of the leading term? 3
2. Multiply: $5y \cdot y$ $5y^2$
3. Multiply: $(2x - 5)(3x + 1)$ $6x^2 - 13x - 5$
4. Find two integers whose product is -3. (There are two possible answers.) $1(-3)$ and $-1(3)$
5. Identify the coefficients of each term of $6x^2 - 3x + 7$. $6, -3, 7$
6. Factor by grouping: $10b^2 + 15b - 2b - 3$ $(2b + 3)(5b - 1)$

In this section, we will factor trinomials with leading coefficients other than 1, such as

$$2x^2 + 5x + 3, \qquad 6a^2 - 17a + 5, \qquad \text{and} \qquad 4b^2 + 8bc - 45c^2$$

We can use two methods to factor these trinomials. With the first method, we make educated guesses and then check them using multiplication. The correct factorization is determined through a process of elimination. The second method is an extension of factoring by grouping.

1 Factor Trinomials Using the Trial-and-Check Method.

EXAMPLE 1 Factor: $2x^2 + 5x + 3$

Strategy We will assume that $2x^2 + 5x + 3$ is the product of two binomials and we will use a systematic method to find their terms.

Why Since the terms of $2x^2 + 5x + 3$ do not have a common factor (other than 1), the only option available is to try to factor it as the product of two binomials.

Solution We represent the binomials using two sets of parentheses. Since the first term of the trinomial is $2x^2$, we enter $2x$ and x as the first terms of the binomial factors.

$$(2x \; \square)(x \; \square)$$ Because $2x \cdot x$ will give $2x^2$.

The Language of Algebra

To **interchange** means to put each in the place of the other. We create all of the possible factorizations by *interchanging* the second terms of the binomials.

$(2x + 1)(x + 3)$

$(2x + 3)(x + 1)$

The second terms of the binomials must be two integers whose product is 3. Since the coefficients of the terms of $2x^2 + 5x + 3$ are positive, we consider only pairs of positive integer factors of 3. Since there is just one such pair, $1 \cdot 3$, we can enter 1 and 3 as the second terms of the binomials, or we can reverse the order and enter 3 and 1.

$$(2x + 1)(x + 3) \qquad \text{or} \qquad (2x + 3)(x + 1)$$

The first possibility is incorrect, because when we find the outer and inner products and combine like terms, we obtain an incorrect middle term of $7x$.

Outer: $6x$

$(2x + 1)(x + 3)$ Multiply and add to find the middle term: $6x + x = 7x$. $7x$ is not the middle term we want.

Inner: x

The second possibility is correct, because it gives a middle term of $5x$.

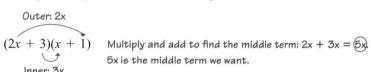

Outer: $2x$

$(2x + 3)(x + 1)$ Multiply and add to find the middle term: $2x + 3x = 5x$. $5x$ is the middle term we want.

Inner: $3x$

Thus, the factorization is:

$$2x^2 + 5x + 3 = (2x + 3)(x + 1)$$

Check the factorization by multiplying:

$$(2x + 3)(x + 1) = 2x^2 + 2x + 3x + 3$$
$$= 2x^2 + 5x + 3$$ This is the original trinomial.

Teaching Tip: You may want to illustrate the factoring process visually by examining the areas of four attached rectangles. Use an approach similar to that which was used to motivate binomial multiplication in Section 5.6.

1	$2x$	3
x	$2x^2$	$3x$
	$2x$	3

Teaching Example 1 Factor: $2x^2 + 9x + 7$
Answer: $(2x + 7)(x + 1)$

Self Check 1 Factor: $2x^2 + 5x + 2$ $(2x + 1)(x + 2)$

Now Try Problem 19

EXAMPLE 2 Factor: $6a^2 - 17a + 5$

Strategy We will assume that $6a^2 - 17a + 5$ is the product of two binomials and we will use a systematic method to find their terms.

Why Since the terms of $6a^2 - 17a + 5$ do not have a common factor (other than 1), the only option available is to try to factor it as the product of two binomials.

Solution We represent the binomials using two sets of parentheses. Since the first term is $6a^2$, the first terms of the factors must be $6a$ and a or $3a$ and $2a$.

$$(6a \;\square)(a \;\square) \quad \text{or} \quad (3a \;\square)(2a \;\square)$$ Because $6a \cdot a$ or $3a \cdot 2a$ will give $6a^2$.

The second terms of the binomials must be two integers whose product is 5. Since the last term of $6a^2 - 17a + 5$ is positive and the coefficient of the middle term is negative, we consider only negative integer factors of the last term. Since there is just one such pair, $-1(-5)$, we can enter -1 and -5, or we can reverse the order and enter -5 and -1 as second terms of the binomials.

$(6a - 1)(a - 5)$ (outer: $-30a$; inner: $-a$) $-30a - a = -31a$ Incorrect middle term

$(6a - 5)(a - 1)$ (outer: $-6a$; inner: $-5a$) $-6a - 5a = -11a$ Incorrect middle term

$(3a - 1)(2a - 5)$ (outer: $-15a$; inner: $-2a$) $-15a - 2a = -17a$ Correct middle term

$(3a - 5)(2a - 1)$ (outer: $-3a$; inner: $-10a$) $-3a - 10a = -13a$ Incorrect middle term

Only the possibility shown in blue gives the correct middle term of $-17a$. Thus,

$$6a^2 - 17a + 5 = (3a - 1)(2a - 5)$$

We check by multiplying: $(3a - 1)(2a - 5) = 6a^2 - 17a + 5$.

Teaching Example 2 Factor: $6a^2 - 11a + 3$
Answer: $(3a - 1)(2a - 3)$

Self Check 2 Factor: $6b^2 - 19b + 3$ $(6b - 1)(b - 3)$

Now Try Problem 27

EXAMPLE 3 Factor: $3y^2 - 7y - 6$

Strategy We will assume that $3y^2 - 7y - 6$ is the product of two binomials and we will use a systematic method to find their terms.

Why Since the terms of $3y^2 - 7y - 6$ do not have a common factor (other than 1), the only option available is to try to factor it as the product of two binomials.

Solution Since the first term is $3y^2$, the first terms of the binomial factors must be $3y$ and y.

$$(3y \;\square)(y \;\square)$$ Because $3y \cdot y$ will give $3y^2$.

The second terms of the binomials must be two integers whose product is -6. There are four such pairs: $1(-6)$, $-1(6)$, $2(-3)$, and $-2(3)$. When these pairs are entered, and then reversed as second terms of the binomials, there are eight possibilities to consider. Four of them can be discarded because they include a binomial whose terms have a common factor. If the terms of $3y^2 - 7y - 6$ do not have a common factor (other than 1), neither can any of its binomial factors.

Success Tip

If the terms of a trinomial do not have a common factor other than 1, the terms of each of its binomial factors will not have a common factor other than 1.

***For* 1 *and* −6:** $(3y + 1)(y - 6)$ (outer: $-18y$; inner: y) $-18y + y = -17y$ Incorrect middle term

or ~~$(3y - 6)(y + 1)$~~ See the Success Tip on the left. $3y - 6$ has a common factor of 3. This can't be the factorization.

Success Tip

Reversing the signs within the binomial factors reverses the sign of the middle term. For example, the factors 2 and -3 give the middle term $-7y$, while -2 and 3 give the middle term $7y$.

Success Tip

All eight possible factorizations are listed. In practice, you will often find the correct factorization without having to examine the entire list of possibilities.

***For* -1 *and* 6:** $(3y - 1)(y + 6)$ — outer: $18y$, inner: $-y$; $18y - y = 17y$ Incorrect middle term

or ~~$(3y + 6)(y - 1)$~~ — $3y + 6$ has a common factor of 3. This can't be the factorization.

***For* 2 *and* -3:** $(3y + 2)(y - 3)$ — outer: $-9y$, inner: $2y$; $-9y + 2y = -7y$ Correct middle term

or ~~$(3y - 3)(y + 2)$~~ — $3y - 3$ has a common factor of 3. This can't be the factorization.

***For* -2 *and* 3:** $(3y - 2)(y + 3)$ — outer: $9y$, inner: $-2y$; $9y - 2y = 7y$ Incorrect middle term

or ~~$(3y + 3)(y - 2)$~~ — $3y + 3$ has a common factor of 3. This can't be the factorization.

Only the possibility shown in green gives the correct middle term of $-7y$. Thus,

$$3y^2 - 7y - 6 = (3y + 2)(y - 3)$$

Check the factorization by multiplying.

Teaching Example 3 Factor: $5m^2 - 22m - 15$
Answer: $(5m + 3)(m - 5)$

Self Check 3 Factor: $5t^2 - 23t - 10$ $(5t + 2)(t - 5)$

Now Try Problem 35

EXAMPLE 4

Factor: $4b^2 + 8bc - 45c^2$

Teaching Tip: If your course covers only factoring trinomials in one variable, you may want to skip this example.

Strategy We will assume that $4b^2 + 8bc - 45c^2$ is the product of two binomials and we will use a systematic method to find their terms.

Why Since the terms of $4b^2 + 8bc - 45c^2$ do not have a common factor (other than 1), the only option available is to try to factor it as the product of two binomials.

Solution

Since the first term is $4b^2$, the first terms of the binomial factors must be $4b$ and b or $2b$ and $2b$. Since the last term contains c^2, the second terms of the binomial factors must contain c.

Notation

The trinomial is in two variables, b and c. It is written in descending powers of b and ascending powers of c.

$$(4b \,\square\, c)(b \,\square\, c) \quad \text{or} \quad (2b \,\square\, c)(2b \,\square\, c)$$

Because $4b \cdot b$ or $2b \cdot 2b$ gives $4b^2$, and because $c \cdot c$ gives c^2.

The coefficients of c must be two integers whose product is -45. Since the coefficient of the last term is negative, the signs of the integers must be different. If we pick factors of $4b$ and b for the first terms, and -1 and 45 for the coefficients of c, the multiplication gives an incorrect middle term of $179bc$.

$(4b - c)(b + 45c)$ — outer: $180bc$, inner: $-bc$; $180bc - bc = 179bc$ Incorrect middle term

Success Tip

When using the trial-and-check method to factor $ax^2 + bx + c$, here is one suggestion: If b is relatively small, test the factorizations that use the smaller factors of a and c first.

If we pick factors of $4b$ and b for the first terms, and 15 and -3 for the coefficients of c, the multiplication gives an incorrect middle term of $3bc$.

$-12bc$

$(4b + 15c)(b - 3c)$ $\quad -12bc + 15bc = 3bc$ Incorrect middle term

$15bc$

If we pick factors of $2b$ and $2b$ for the first terms, and -5 and 9 for the coefficients of c, we have

$18bc$

$(2b - 5c)(2b + 9c)$ $\quad 18bc - 10bc = 8bc$ Correct middle term

$-10bc$

which gives the correct middle term of $8bc$. Thus,

$$4b^2 + 8bc - 45c^2 = (2b - 5c)(2b + 9c)$$

Check the factorization by multiplying.

Teaching Example 4 Factor: $8x^2 + 2xy - 45y^2$
Answer: $(2x + 5y)(4x - 9y)$

Self Check 4 Factor: $4x^2 + 4xy - 3y^2$ $\quad (2x + 3y)(2x - y)$

Now Try Problem 43

Because guesswork is often necessary, it is difficult to give specific rules for factoring trinomials with leading coefficients other than 1. However, the following hints are helpful when using the **trial-and-check method.**

Factoring Trinomials with Leading Coefficients Other Than 1

To factor trinomials with leading coefficients other than 1:

1. Factor out any GCF (including -1 if that is necessary to make a positive in a trinomial of the form $ax^2 + bx + c$).
2. Write the trinomial as a product of two binomials. The coefficients of the first terms of each binomial factor must be factors of a, and the last terms must be factors of c.

   The product of these numbers must be a.

   $$ax^2 + bx + c = (\square x \;\; \square)(\square x \;\; \square)$$

   The product of these numbers must be c.

3. If c is positive, the signs within the binomial factors match the sign of b. If c is negative, the signs within the binomial factors are opposites.
4. Try combinations of the first terms and the second terms of the binomial factors until you find the one that gives the proper middle term in the trinomial. If no combination works, the trinomial is prime.
5. Check by multiplying.

2 Factor Trinomials After Factoring Out the GCF.

If the terms of a trinomial have a common factor, *the GCF (or the opposite of the GCF) should always be factored out first.*

EXAMPLE 5 Factor: $2x^2 - 8x^3 + 3x$

Strategy We will write the trinomial in descending powers of x and factor out the common factor , $-x$.

Why It is easier to factor trinomials that have a positive leading coefficient.

The Language of Algebra

When asked to *factor* a polynomial, that means we should *factor completely*.

Solution Write the trinomial in descending powers of x: $-8x^3 + 2x^2 + 3x$.

$$-8x^3 + 2x^2 + 3x = -x(8x^2 - 2x - 3) \quad \text{Factor out the opposite of the GCF, } -x.$$

We now factor $8x^2 - 2x - 3$. Its factorization has the form

$$(8x \;\square)(x \;\square) \quad \text{or} \quad (2x \;\square)(4x \;\square) \quad \text{Because } 8x \cdot x \text{ or } 4x \cdot 2x \text{ gives } 8x^2.$$

The second terms of the binomials must be two integers whose product is -3. There are two such pairs: $1(-3)$ and $-1(3)$. Since the coefficient of the middle term of the trinomial, $-2x$, is small, we pick the smaller factors of $8x^2$, which are $2x$ and $4x$, for the first terms and 1 and -3 for the second terms.

$$(2x + 1)(4x - 3) \qquad -6x + 4x = -2x$$

(outer product: $-6x$; inner product: $4x$)

This factorization gives the correct middle term of $-2x$. Thus,

$$8x^2 - 2x - 3 = (2x + 1)(4x - 3)$$

Caution

For multistep factorization, don't forget to write the GCF (or its opposite) in the final factored form.

We can now give the complete factorization of the original trinomial.

$$-8x^3 + 2x^2 + 3x = -x(8x^2 - 2x - 3)$$
$$= -x(2x + 1)(4x - 3)$$

Don't forget this factor. (pointing to $-x$)

Check the factorization by multiplying.

Teaching Example 5 Factor: $5x - 23x^2 - 10x^3$
Answer: $-x(2x + 5)(5x - 1)$

Self Check 5 Factor: $12y - 14y^3 + 22y^2$ $\quad -2y(7y + 3)(y - 2)$

Now Try Problem 53

3 Factor Trinomials Using the Grouping Method.

Another way to factor a trinomial of the form $ax^2 + bx + c$ is to write it as an equivalent four-termed polynomial and factor it by grouping. For example, to factor $2x^2 + 5x + 3$, we proceed as follows.

1. Identify the values of a, b, and c.

$$\left.\begin{matrix} ax^2 + bx + c \\ \downarrow \quad\; \downarrow \quad\; \downarrow \\ 2x^2 + 5x + 3 \end{matrix}\right\} a = 2,\ b = 5, \text{ and } c = 3$$

Then, find the product ac, called the **key number:** $ac = 2(3) = 6$.

2. Next, find two integers whose product is $ac = 6$ and whose sum is $b = 5$. Since the integers must have a positive product and a positive sum, we consider only positive factors of 6.

Key number = 6 $\qquad$ $b = 5$

Positive factors of 6	Sum of the positive factors of 6
$1 \cdot 6 = 6$	$1 + 6 = 7$
$2 \cdot 3 = 6$	$2 + 3 = 5$

The second row of the table contains the correct pair of integers 2 and 3, whose product is 6 and whose sum is 5.

3. Express the middle term, $5x$, of the trinomial as the *sum of two terms,* using the integers 2 and 3 found in step 2 as coefficients of the two terms.

$$2x^2 + 5x + 3 = 2x^2 + 2x + 3x + 3 \quad \text{Express } 5x \text{ as } 2x + 3x.$$

4. Factor the equivalent four-term polynomial by grouping:

$$2x^2 + 2x + 3x + 3 = 2x(x + 1) + 3(x + 1) \quad \text{Factor } 2x \text{ out of } 2x^2 + 2x \text{ and 3 out of } 3x + 3.$$
$$= (x + 1)(2x + 3) \quad \text{Factor out } x + 1.$$

Check by multiplying.

Factoring by grouping is especially useful when the leading coefficient, a, and the constant term, c, have many factors.

Factoring Trinomials by Grouping

To factor a trinomial by grouping:

1. Factor out any GCF (including -1 if that is necessary to make a positive in a trinomial of the form $ax^2 + bx + c$).
2. Identify a, b, and c, and find the key number ac.
3. Find two integers whose product is the key number and whose sum is b.
4. Express the middle term, bx, as the sum (or difference) of two terms. Enter the two numbers found in step 3 as coefficients of x in the form shown below. Then factor the equivalent four-term polynomial by grouping.

$$ax^2 + \square x + \square x + c$$

The product of these numbers must be ac and their sum must be b.

5. Check the factorization by multiplying.

EXAMPLE 6 Factor by grouping: $10x^2 + 13x - 3$

Strategy We will express the middle term, $13x$, of the trinomial as the sum of two carefully chosen terms.

Why We want to produce an equivalent four-term polynomial that can be factored by grouping.

Solution In $10x^2 + 13x - 3$, we have $a = 10$, $b = 13$, and $c = -3$. The key number is $ac = 10(-3) = -30$. We must find a factorization of -30 such that the sum of the factors is $b = 13$. The possible factor pairs are listed in the table. Since the factors must have a negative product, their signs must be different.

Key number $= -30$ $b = 13$

Factors of -30	Sum of the factors of -30
$1(-30) = -30$	$1 + (-30) = -29$
$2(-15) = -30$	$2 + (-15) = -13$
$3(-10) = -30$	$3 + (-10) = -7$
$5(-6) = -30$	$5 + (-6) = -1$
$6(-5) = -30$	$6 + (-5) = 1$
$10(-3) = -30$	$10 + (-3) = 7$
$\mathbf{15(-2)} = -30$	$\mathbf{15 + (-2)} = 13$
$30(-1) = -30$	$30 + (-1) = 29$

It is wise to follow an order when listing the factors in the table so that you don't skip the correct combination. Here, the first factors 1, 2, 3, 5, 6, 10, 15, and 30 are listed from least to greatest.

Notation

The middle term, $13x$, may be expressed as $15x - 2x$ or as $-2x + 15x$ when using factoring by grouping. The resulting factorizations will be equivalent.

The seventh row contains the correct pair of numbers 15 and -2, whose product is -30 and whose sum is 13. They serve as the coefficients of $15x$ and $-2x$, the two terms that we use to represent the middle term, $13x$, of the trinomial.

$$10x^2 + \mathbf{13x} - 3 = 10x^2 + \mathbf{15x - 2x} - 3 \qquad \text{Express } 13x \text{ as } 15x - 2x.$$

Finally, we factor by grouping.

$$10x^2 + 15x - 2x - 3 = 5x\mathbf{(2x + 3)} - 1\mathbf{(2x + 3)} \qquad \text{Factor out } 5x \text{ from } 10x^2 + 15x. \text{ Factor out } -1 \text{ from } -2x - 3.$$

$$= \mathbf{(2x + 3)}(5x - 1) \qquad \text{Factor out } 2x + 3.$$

So $10x^2 + 13x - 3 = (2x + 3)(5x - 1)$. Check the factorization by multiplying.

Teaching Example 6 Factor by grouping: $6x^2 + 7x - 5$
Answer: $(3x + 5)(2x - 1)$

Self Check 6 Factor by grouping: $15a^2 + 17a - 4$ $(3a + 4)(5a - 1)$

Now Try Problems 19, 27, and 35

EXAMPLE 7 Factor: $12x^5 - 17x^4 + 6x^3$

Strategy We will factor out the GCF, x^3, first. Then we will factor the resulting trinomial using the grouping method.

Why The first step in factoring any polynomial is to factor out the GCF.

Solution The GCF of the three terms of the trinomial is x^3.

$$12x^5 - 17x^4 + 6x^3 = x^3(12x^2 - 17x + 6)$$

To factor $12x^2 - 17x + 6$, we must find two integers whose product is $12(6) = 72$ and whose sum is -17. Two such numbers are -8 and -9. They serve as the coefficients of $-8x$ and $-9x$, the two terms that we use to represent the middle term, $-17x$, of the trinomial.

Key number $= 72$ $b = -17$

Factors	Sum
$-8(-9) = 72$	$-8 + (-9) = -17$

$$12x^2 - 17x + 6 = 12x^2 - 8x - 9x + 6$$ Express $-17x$ as $-8x - 9x$. ($-9x - 8x$ could also be used.)

$$= 4x(3x - 2) - 3(3x - 2)$$ Factor out $4x$ and factor out -3.

$$= (3x - 2)(4x - 3)$$ Factor out $3x - 2$.

The complete factorization of the original trinomial is

$$12x^5 - 17x^4 + 6x^3 = x^3(3x - 2)(4x - 3)$$ Don't forget to write the GCF, x^3.

Check the factorization by multiplying.

Teaching Example 7 Factor: $4m^4 - 15m^3 + 14m^2$
Answer: $m^2(4m - 7)(m - 2)$

Self Check 7 Factor: $21a^4 - 13a^3 + 2a^2$ $a^2(7a - 2)(3a - 1)$

Now Try Problem 53

SECTION 6.3 STUDY SET

VOCABULARY

Fill in the blanks.

1. The leading coefficient of $3x^2 - x - 12$ is 3.
2. Given $5y^2 + 16y + 3 = (5y + 1)(y + 3)$. We say that $5y^2 + 16y + 3$ factors as the product of two binomials.
3. The first terms of the binomial factors $(5y + 1)(y + 3)$ are $5y$ and y. The second terms of the binomial factors are 1 and 3.
4. To factor $2m^2 + 11m + 12$ by grouping, we write it as $2m^2 + 8m + 3m + 12$.

CONCEPTS

5. If $10x^2 - 27x + 5$ is to be factored as the product of two binomials, what are the possible *first terms* of the binomial factors? $10x$ and x, $5x$ and $2x$
6. Complete each sentence.

The product of these numbers must be 5.

$$5x^2 + 6x - 8 = (\square x\ \square)(\square x\ \square)$$

The product of these numbers must be -8.

7. a. Fill in the blanks. When factoring a trinomial, we write it in descending powers of the variable. Then we factor out any GCF (including -1 if that is necessary to make the leading coefficient positive).
   b. What is the GCF of the terms of $6s^4 + 33s^3 + 36s^2$? $3s^2$
   c. Factor out -1 from $-2d^2 + 19d - 8$. $-(2d^2 - 19d + 8)$
8. Check to determine whether $(3t - 1)(5t - 6)$ is the correct factorization of $15t^2 - 19t + 6$. No

A trinomial has been partially factored. Complete each statement that describes the type of integers we should consider for the blanks.

9. $5y^2 - 13y + 6 = (5y\ \square)(y\ \square)$
   Since the last term of the trinomial is positive and the middle term is negative, the integers must be negative factors of 6.
10. $5y^2 + 13y + 6 = (5y\ \square)(y\ \square)$
   Since the last term of the trinomial is positive and the middle term is positive, the integers must be positive factors of 6.
11. $5y^2 - 7y - 6 = (5y\ \square)(y\ \square)$
   Since the last term of the trinomial is negative, the signs of the integers will be different.
12. $5y^2 + 7y - 6 = (5y\ \square)(y\ \square)$
   Since the last term of the trinomial is negative, the signs of the integers will be different.
13. Complete the key number table.

Negative factors of 12	Sum of the negative factors of 12
$-1(-12)$	-13
$-2(-6)$	-8
$-3(-4)$	-7

14. Complete the sentence to explain how to factor $3x^2 + 16x + 5$ by grouping.

$$3x^2 + 16x + 5 = 3x^2 + \square x + \square x + 5$$

The product of these numbers must be 15 and their sum must be 16.

NOTATION

15. a. Suppose we wish to factor $12b^2 + 20b - 9$ by grouping. Identify a, b, and c. 12, 20, -9
   b. What is the key number, ac? -108

16. To factor $6x^2 + 13x + 6$ by grouping, a student made a table and circled the correct pair of integers, as shown. Enter the correct coefficients for the first stage in the factorization process.

$6x^2 + 4x + 9x + 6$

$ac = 36 \quad b = 13$

Factors	Sum
$1 \cdot 36$	37
$2 \cdot 18$	20
$3 \cdot 12$	15
$4 \cdot 9$	13
$6 \cdot 6$	12

Complete each step of the factorization of the trinomial by grouping.

17. $12t^2 + 17t + 6 = 12t^2 + 9t + 8t + 6$
$= 3t(4t + 3) + 2(4t + 3)$
$= (4t + 3)(3t + 2)$

▶ **18.** $35t^2 - 11t - 6 = 35t^2 + 10t - 21t - 6$
$= 5t(7t + 2) - 3(7t + 2)$
$= (7t + 2)(5t - 3)$

GUIDED PRACTICE

Factor. **See Example 1 or Example 6.**

19. $2x^2 + 3x + 1$ $(2x + 1)(x + 1)$

20. $3x^2 + 4x + 1$ $(3x + 1)(x + 1)$

21. $3a^2 + 10a + 3$ $(3a + 1)(a + 3)$

▶ **22.** $2b^2 + 7b + 3$ $(2b + 1)(b + 3)$

23. $5x^2 + 7x + 2$ $(5x + 2)(x + 1)$

▶ **24.** $7t^2 + 12t + 5$ $(7t + 5)(t + 1)$

25. $7x^2 + 18x + 11$ $(7x + 11)(x + 1)$

26. $5n^2 + 12n + 7$ $(5n + 7)(n + 1)$

Factor. **See Example 2 or Example 6.**

27. $4x^2 - 8x + 3$ $(2x - 3)(2x - 1)$

28. $4z^2 - 13z + 3$ $(z - 3)(4z - 1)$

▶ **29.** $8x^2 - 22x + 5$ $(4x - 1)(2x - 5)$

30. $15a^2 - 28a + 5$ $(5a - 1)(3a - 5)$

▶ **31.** $15t^2 - 26t + 7$ $(5t - 7)(3t - 1)$

▶ **32.** $10x^2 - 9x + 2$ $(5x - 2)(2x - 1)$

▶ **33.** $6y^2 - 13y + 2$ $(6y - 1)(y - 2)$

▶ **34.** $6y^2 - 43y + 7$ $(6y - 1)(y - 7)$

Factor. **See Example 3 or Example 6.**

35. $3x^2 - 2x - 21$ $(3x + 7)(x - 3)$

▶ **36.** $3u^2 - 44u - 15$ $(3u + 1)(u - 15)$

▶ **37.** $5m^2 - 7m - 6$ $(5m + 3)(m - 2)$

38. $5y^2 - 18y - 8$ $(5y + 2)(y - 4)$

39. $7y^2 + 55y - 8$ $(7y - 1)(y + 8)$

▶ **40.** $7x^2 + 33x - 10$ $(7x - 2)(x + 5)$

41. $11y^2 + 7y - 4$ $(11y - 4)(y + 1)$

▶ **42.** $13y^2 + 9y - 4$ $(13y - 4)(y + 1)$

Factor. **See Example 4.**

▶ **43.** $6r^2 + rs - 2s^2$ $(3r + 2s)(2r - s)$

▶ **44.** $3m^2 + 5mn + 2n^2$ $(3m + 2n)(m + n)$

▶ **45.** $4x^2 + 8xy + 3y^2$ $(2x + 3y)(2x + y)$

46. $4b^2 + 15bc - 4c^2$ $(4b - c)(b + 4c)$

▶ **47.** $8m^2 + 91mn + 33n^2$ $(8m + 3n)(m + 11n)$

48. $2m^2 + 17mn - 9n^2$ $(2m - n)(m + 9n)$

49. $15x^2 - xy - 6y^2$ $(5x + 3y)(3x - 2y)$

▶ **50.** $4a^2 - 15ab + 9b^2$ $(4a - 3b)(a - 3b)$

Factor. **See Example 5 or Example 7.**

▶ **51.** $-26x + 6x^2 - 20$ $2(3x + 2)(x - 5)$

▶ **52.** $-28 + 6a^2 - 2a$ $2(3a - 7)(a + 2)$

▶ **53.** $15a + 8a^3 - 26a^2$ $a(2a - 5)(4a - 3)$

54. $16r - 40r^2 + 25r^3$ $r(5r - 4)(5r - 4) = r(5r - 4)^2$

55. $2u^2 - 6v^2 - uv$ $(2u + 3v)(u - 2v)$

▶ **56.** $6a^2 + 6b^2 - 13ab$ $(3a - 2b)(2a - 3b)$

57. $36y^2 - 88y + 32$ $4(9y - 4)(y - 2)$

▶ **58.** $70a^2 - 95a + 30$ $5(7a - 6)(2a - 1)$

TRY IT YOURSELF

Factor. If an expression is prime, so indicate.

59. $6t^2 - 7t - 20$ $(2t - 5)(3t + 4)$

60. $6w^2 + 13w + 5$ $(2w + 1)(3w + 5)$

61. $15p^2 - 2pq - q^2$ $(3p - q)(5p + q)$

62. $8c^2 - 10cd + 3d^2$ $(4c - 3d)(2c - d)$

▶ **63.** $4t^2 - 16t + 7$ $(2t - 1)(2t - 7)$

64. $9x^2 - 32x + 15$ $(9x - 5)(x - 3)$

65. $130r^2 + 20r - 110$ $10(13r - 11)(r + 1)$

▶ **66.** $170h^2 - 210h - 260$ $10(17h + 13)(h - 2)$

▶ **67.** $8y^2 - 2y - 1$ $(4y + 1)(2y - 1)$

▶ **68.** $14y^2 + 11y + 2$ $(7y + 2)(2y + 1)$

69. $18x^2 + 31x - 10$ $(18x - 5)(x + 2)$

▶ **70.** $20y^2 - 93y - 35$ $(20y + 7)(y - 5)$

▶ **71.** $-y^3 - 13y^2 - 12y$ $-y(y + 12)(y + 1)$

▶ **72.** $-2xy^2 - 8xy + 24x$ $-2x(y + 6)(y - 2)$

▶ **73.** $10u^2 - 13u - 6$ Prime

▶ **74.** $8m^2 + 5m - 10$ Prime

75. $-6x^4 + 15x^3 + 9x^2$ $-3x^2(2x + 1)(x - 3)$

76. $-9y^4 - 3y^3 + 6y^2$ $-3y^2(3y - 2)(y + 1)$

77. $6p^2 + pq - q^2$ $(3p - q)(2p + q)$

78. $12m^2 - 11mn + 2n^2$ $(3m - 2n)(4m - n)$

79. $30r^5 + 63r^4 - 30r^3$ $3r^3(5r - 2)(2r + 5)$

▶ **80.** $6s^5 - 26s^4 - 20s^3$ $2s^3(3s + 2)(s - 5)$

▶ **81.** $16m^3n + 20m^2n^2 + 6mn^3$ $2mn(4m + 3n)(2m + n)$

82. $-28u^3v^3 + 26u^2v^4 - 6uv^5$ $-2uv^3(7u - 3v)(2u - v)$

83. $3x^2 + x + 6$ Prime

▶ **84.** $2u^2 + 3u + 25$ Prime

85. $-12y^2 - 12 + 25y$ $-(4y - 3)(3y - 4)$

▶ **86.** $-10t^2 + 1 + 3t$ $-(5t + 1)(2t - 1)$

Choose the correct method from Sections 6.1, 6.2, or 6.3 to factor each of the following.

87. $m^2 + 3m - 28$ $(m + 7)(m - 4)$

▶ **88.** $-b^2 - 5b + 24$ $-(b + 8)(b - 3)$

89. $6a^3 + 15a^2$ $3a^2(2a + 5)$

90. $9x^4 + 27x^6$ $9x^4(1 + 3x^2)$

91. $x^3 - 2x^2 + 5x - 10$ $(x - 2)(x^2 + 5)$

▶ **92.** $x^3 - x^2 + 2x - 2$ $(x - 1)(x^2 + 2)$

▶ **93.** $5y^2 + 3 - 8y$ $(5y - 3)(y - 1)$

94. $3t^2 + 7 - 10t$ $(3t - 7)(t - 1)$

95. $-2x^2 - 10x - 12$ $-2(x + 2)(x + 3)$

96. $4y^2 + 36y + 72$ $4(y + 3)(y + 6)$

97. $12x^3y^3 - 18x^2y^3 + 15x^2y^2$ $3x^2y^2(4xy - 6y + 5)$

▶ 98. $15c^2d^3 - 25c^3d^2 - 10c^4d^4$ $5c^2d^2(3d - 5c - 2c^2d^2)$

99. $a^2 - 7ab + 10b^2$ $(a - 5b)(a - 2b)$

100. $x^2 - 13xy + 12y^2$ $(x - 12y)(x - y)$

101. $9u^6 - 71u^5 - 8u^4$ $u^4(9u + 1)(u - 8)$

▶ 102. $25n^8 - 49n^7 - 2n^6$ $n^6(25n + 1)(n - 2)$

APPLICATIONS

▶ **103.** ▶ from **Campus to Careers**

Elementary School Teacher

©iStockphoto.com/Neustockimages

The area of a teacher's desktop is represented by the trinomial $(4x^2 + 20x - 11)$ in.^{2}. Factor it to find the polynomials that represent its length and width. $(2x + 11)$ in., $(2x - 1)$ in.

▶ **104. Storage.** The volume of an 8-foot-wide portable storage container is represented by the trinomial $(72x^2 + 120x - 400)$ ft^3. Its dimensions can be determined by factoring the trinomial. Find the polynomials that represent the height and the length of the container.
$(3x - 5)$ ft, $(3x + 10)$ ft

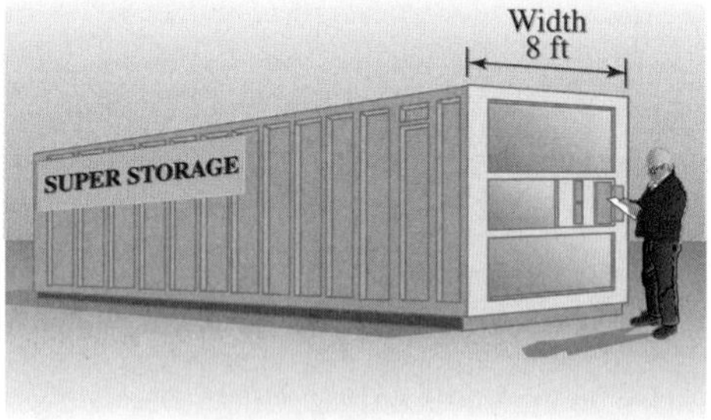

WRITING

105. Two students factor $2x^2 + 20x + 42$ and get two different answers: $(2x + 6)(x + 7)$ and $(x + 3)(2x + 14)$.
Do both answers check? Why don't they agree? Is either answer completely correct? Explain.

▶ 106. Why is the process of factoring $6x^2 - 5x - 6$ more complicated than the process of factoring $x^2 - 5x - 6$?

107. Suppose a factorization check of $(3x - 9)(5x + 7)$ gives a middle term $-24x$, but a middle term of $24x$ is actually needed. Explain how to quickly obtain the correct factorization.

108. Suppose we want to factor $2x^2 + 7x - 72$. Explain why $(2x - 1)(x + 72)$ is not a wise choice to try first.

REVIEW

Evaluate each expression.

109. -7^2 -49

▶ 110. $(-7)^2$ 49

111. 7^0 1

112. 7^{-2} $\frac{1}{49}$

113. $\frac{1}{7^{-2}}$ 49

114. $2 \cdot 7^2$ 98

CHALLENGE PROBLEMS

Factor.

115. $6a^{10} + 5a^5 - 21$ $(2a^5 - 3)(3a^5 + 7)$

116. $3x^4y^2 - 29x^2y + 56$ $(3x^2y - 8)(x^2y - 7)$

▶ 117. $8x^2(c^2 + c - 2) - 2x(c^2 + c - 2) - (c^2 + c - 2)$
$(c + 2)(c - 1)(4x + 1)(2x - 1)$

118. Find all integer values of b that make $2x^2 + bx - 5$ factorable. $-3, 3, -9, 9$

SECTION 6.4 Factoring Perfect-Square Trinomials and the Difference of Two Squares

OBJECTIVES

1. Recognize perfect-square trinomials.
2. Factor perfect-square trinomials.
3. Factor the difference of two squares.

ARE YOU READY? *Are You Ready? exercises available online at www.webassign.net/brookscole*

The following problems review some basic skills that are needed when factoring binomials and trinomials.

1. Multiply: $(3y + 2)(3y + 2)$ $9y^2 + 12y + 4$

2. Multiply: $(m + 9)(m - 9)$ $m^2 - 81$

3. Simplify: $(8d)^2$ $64d^2$

4. Translate to symbols: the sum of x^2 and 25 $x^2 + 25$

In this section, we will discuss a method that can be used to factor two types of trinomials, called *perfect-square trinomials.* We also develop techniques for factoring a type of binomial called the *difference of two squares.*

1 Recognize Perfect-Square Trinomials.

We have seen that the square of a binomial is a trinomial. We also have seen that the special-product rules shown below can be used to find the square of a sum and the square of a difference quickly. The terms of the resulting trinomial are related to the terms of the binomial that was squared.

Success Tip

To prepare for this section, it would be helpful to review Section 5.7 Special Products.

$$(A + B)^2 = A^2 + 2AB + B^2$$

This is the square of the first term of the binomial. — This is twice the product of A and B, or its opposite. — This is the square of the last term of the binomial.

$$(A - B)^2 = A^2 - 2AB + B^2$$

Trinomials that are squares of a binomial are called **perfect-square trinomials.** Some examples are

$y^2 + 6y + 9$ Because it is the square of $(y + 3)$: $(y + 3)^2 = y^2 + 6y + 9$

$t^2 - 14t + 49$ Because it is the square of $(t - 7)$: $(t - 7)^2 = t^2 - 14t + 49$

$4m^2 - 20m + 25$ Because it is the square of $(2m - 5)$: $(2m - 5)^2 = 4m^2 - 20m + 25$

EXAMPLE 1 Determine whether the following are perfect-square trinomials: **a.** $x^2 + 10x + 25$ **b.** $c^2 - 12c - 36$ **c.** $25y^2 - 30y + 9$ **d.** $4t^2 + 18t + 81$

Strategy We will compare each trinomial, term-by-term, to one of the special-product forms discussed in Section 5.7.

Why If a trinomial matches one of these forms, it is a perfect-square trinomial.

Solution **a.** To determine whether this is a perfect-square trinomial, we note that

$$x^2 + 10x + 25$$

The first term is the square of x. — The middle term is twice the product of x and 5: $2 \cdot x \cdot 5 = 10x$. — The last term is the square of 5.

Thus, $x^2 + 10x + 25$ is a perfect-square trinomial.

Teaching Tip: Remind students that the first and last terms of a perfect-square trinomial must be positive, because they are squares.

b. To determine whether this is a perfect-square trinomial, we note that

$c^2 - 12c - 36$ The last term, -36, is not the square of a real number.

Since the last term is negative, $c^2 - 12c - 36$ is not a perfect-square trinomial.

The Language of Algebra

The expressions $25y^2$ and 9 are called *perfect squares* because $25y^2 = (5y)^2$ and $9 = 3^2$.

c. To determine whether this is a perfect-square trinomial, we note that

$$25y^2 - 30y + 9$$

The first term is the square of 5y. — The middle term is the opposite of twice the product of 5y and 3: $-2(5y)(3) = -30y$. — The last term is the square of 3.

Thus, $25y^2 - 30y + 9$ is a perfect-square trinomial.

d. To determine whether this is a perfect-square trinomial, we note that

$$4t^2 + 18t + 81$$

The first term is the square of $2t$. The middle term is not twice the product of $2t$ and 9, because $2(2t)(9) = 36t$. The last term is the square of 9.

Thus, $4t^2 + 18t + 81$ is not a perfect-square trinomial.

Teaching Example 1 Determine whether the following are perfect-square trinomials:
a. $x^2 + 14x + 49$
b. $4x^2 - 20x + 25$
c. $x^2 - 10x - 25$
d. $9x^2 + 6x + 4$
Answers: a. Yes b. Yes c. No d. No

Self Check 1 Determine whether the following are perfect-square trinomials:
a. $y^2 + 4y + 4$ Yes **b.** $b^2 - 6b - 9$ No
c. $4z^2 + 4z + 4$ No **d.** $49x^2 - 28x + 16$ No

Now Try Problems 13 and 17

Recognizing Perfect-Square Trinomials

1. The first and last terms are squares of integers or monomials.
2. The middle term is twice the product of the expressions that are squared to produce the first and last terms, or its opposite.

2 Factor Perfect-Square Trinomials.

We can factor perfect-square trinomials using the methods previously discussed in Sections 6.2 and 6.3. However, in many cases, we can factor them more quickly by inspecting their terms and applying the special-product rules in reverse.

Factoring Perfect-Square Trinomials

Perfect-square trinomial		Square of a binomial
$A^2 + 2AB + B^2$	=	$(A + B)^2$
$A^2 - 2AB + B^2$	=	$(A - B)^2$

Each of these trinomials factors as the square of a binomial.

When factoring perfect-square trinomials, it is helpful to know the following perfect squares printed in red. The number 400, for example, is a perfect square, because $400 = 20^2$.

$1 = 1^2$	$25 = 5^2$	$81 = 9^2$	$169 = 13^2$	$289 = 17^2$
$4 = 2^2$	$36 = 6^2$	$100 = 10^2$	$196 = 14^2$	$324 = 18^2$
$9 = 3^2$	$49 = 7^2$	$121 = 11^2$	$225 = 15^2$	$361 = 19^2$
$16 = 4^2$	$64 = 8^2$	$144 = 12^2$	$256 = 16^2$	$400 = 20^2$

EXAMPLE 2 Factor: **a.** $x^2 + 20x + 100$ **b.** $9x^2 - 30xy + 25y^2$

Strategy The terms of each trinomial do not have a common factor (other than 1). We will determine whether each is a perfect-square trinomial.

Why If it is, we can factor it using a special-product rule in reverse.

Solution **a.** $x^2 + 20x + 100$ is a perfect-square trinomial, because:

- The first term x^2 is the square of x.
- The last term 100 is the square of 10.
- The middle term is twice the product of x and 10: $2(x)(10) = 20x$.

Success Tip

The sign of the middle term of a perfect-square trinomial is the same as the sign of the second term of the squared binomial.

Same sign

$A^2 + 2AB + B^2 = (A + B)^2$

$A^2 - 2AB + B^2 = (A - B)^2$

Same sign

Teaching Tip: You may want to illustrate the factoring process visually by examining the areas of four attached rectangles.

	x	10
x	x^2	$10x$
10	$10x$	100

To find the factorization, we match $x^2 + 20x + 100$ to the proper rule for factoring a perfect-square trinomial.

$$A^2 + 2\ A\ B + B^2 = (A + B)^2$$
$$\downarrow \quad \downarrow \quad \downarrow \quad \downarrow \quad \downarrow \quad \downarrow \quad \downarrow$$
$$x^2 + 20x + 10^2 = \mathbf{x}^2 + 2 \cdot \mathbf{x} \cdot \mathbf{10} + \mathbf{10}^2 = (\mathbf{x} + \mathbf{10})^2$$

Therefore, $x^2 + 20x + 10^2 = (x + 10)^2$. Check by finding $(x + 10)^2$.

b. $9x^2 - 30xy + 25y^2$ is a perfect-square trinomial, because:

- The first term $9x^2$ is the square of $\mathbf{3x}$: $(3x)^2 = 9x^2$.
- The last term $25y^2$ is the square of $\mathbf{5y}$: $(5y)^2 = 25y^2$.
- The middle term is the opposite of twice the product of $\mathbf{3x}$ and $\mathbf{5y}$: $-2(3x)(5y) = -30xy$.

We can use these observations to write the trinomial in one of the special-product forms that then leads to its factorization.

$$9x^2 - 30xy + 25y^2 = (\mathbf{3x})^2 - 2(\mathbf{3x})(\mathbf{5y}) + (\mathbf{5y})^2$$
$$= (\mathbf{3x} - \mathbf{5y})^2$$

Therefore, $9x^2 - 30xy + 25y^2 = (3x - 5y)^2$. Check by finding $(3x - 5y)^2$.

Teaching Example 2 Factor:
a. $x^2 - 22x + 121$
b. $x^2 + 16x + 64$
c. $4x^2 - 20xy + 25y^2$
Answers:
a. $(x - 11)^2$ **b.** $(x + 8)^2$
c. $(2x - 5y)^2$

Self Check 2 Factor: **a.** $x^2 + 18x + 81$ $(x + 9)^2$ **b.** $16x^2 - 8xy + y^2$ $(4x - y)^2$

Now Try Problems 21 and 29

EXAMPLE 3 Factor: $4a^3 - 4a^2 + a$

Strategy We will factor out the GCF, a, first. Then we will factor the resulting perfect-square trinomial using a special-product rule in reverse.

Why The first step in factoring any polynomial is to factor out the GCF.

Solution The terms of $4a^3 - 4a^2 + a$ have the common factor a, which should be factored out first. Within the parentheses, we recognize $4a^2 - 4a + 1$ as a perfect-square trinomial of the form $A^2 - 2AB + B^2$, and factor it as such.

$$4a^3 - 4a^2 + a = a(\mathbf{4a^2 - 4a + 1}) \quad \text{Factor out the GCF, } a.$$
$$= a(\mathbf{2a - 1})^2 \quad \text{Since } 4a^2 = (2a)^2,\ 1 = (1)^2, \text{ and } -4a = -2(2a)(1),\ 4a^2 - 4a + 1 \text{ can be factored as a perfect-square trinomial.}$$

Success Tip

Perfect-square trinomials, such as $4a^2 - 4a + 1$, also can be factored using the trial-and-check method or factoring by grouping. However, if you recognize that a trinomial is a perfect-square trinomial, the method of this section is often easier and faster.

Teaching Example 3 Factor: $9x^3 - 6x^2 + x$
Answer: $x(3x - 1)^2$

Self Check 3 Factor: $49x^3 - 14x^2 + x$ $x(7x - 1)^2$

Now Try Problem 33

3 Factor the Difference of Two Squares.

The Language of Algebra

The expression $A^2 - B^2$ is a **difference of two squares,** whereas $(A - B)^2$ is the **square of a difference.** They are not equivalent because $(A - B)^2 \neq A^2 - B^2$.

Recall the special-product rule for multiplying the sum and difference of the same two terms:

$$(A + B)(A - B) = A^2 - B^2$$

The binomial $A^2 - B^2$ is called a **difference of two squares,** because A^2 is the square of A and B^2 is the square of B. If we reverse this rule, we obtain a method for factoring a difference of two squares.

Factoring ⟶

$$A^2 - B^2 = (A + B)(A - B)$$

Teaching Tip: You may want to review this special form by showing all the steps to find:
$(x + 4)(x - 4)$ and $(5t + 3)(5t - 3)$

This pattern is easy to remember if we think of a difference of two squares as the square of a **F**irst quantity minus the square of a **L**ast quantity.

Factoring a Difference of Two Squares

To factor the square of a First quantity minus the square of a Last quantity, multiply the First plus the Last by the First minus the Last.

$$F^2 - L^2 = (F + L)(F - L)$$

To factor differences of two squares, it will be helpful if you recognize the perfect squares listed in red on page 466.

EXAMPLE 4 Factor: **a.** $x^2 - 9$ **b.** $16 - b^2$ **c.** $n^2 - 45$ **d.** $a^2 + 81$

Strategy The terms of each binomial do not have a common factor (other than 1). The only option available is to attempt to factor each as a difference of two squares.

Why If a binomial is a difference of two squares, we can factor it using a special-product rule in reverse.

Solution

Notation

By the commutative property of multiplication, the factors of a difference of two squares can be written in either order. For example, we can write:

$$x^2 - 9 = (x - 3)(x + 3)$$

a. $x^2 - 9$ is the difference of two squares because it can be written as $x^2 - 3^2$. We can match it to the rule for factoring a difference of two squares to find the factorization.

$$\mathbf{F}^2 - \mathbf{L}^2 = (\mathbf{F} + \mathbf{L})(\mathbf{F} - \mathbf{L})$$
$$\downarrow \quad \downarrow \qquad \downarrow \quad \downarrow \downarrow \quad \downarrow$$
$$x^2 - \mathbf{3}^2 = (x + \mathbf{3})(x - \mathbf{3}) \quad \text{9 is a perfect square: } 9 = 3^2.$$

Therefore, $x^2 - 9 = (x + 3)(x - 3)$.

Check by multiplying: $(x + 3)(x - 3) = x^2 - 9$.

b. $16 - b^2$ is the difference of two squares because $16 - b^2 = 4^2 - b^2$. Therefore,

$$16 - b^2 = (4 + b)(4 - b) \quad \text{16 is a perfect square: } 16 = 4^2.$$

Check by multiplying.

c. Since 45 is not a perfect square, $n^2 - 45$ cannot be factored using integers. It is *prime.*

The Language of Algebra

An expression of the form $A^2 + B^2$ is called the **sum of two squares,** whereas $(A + B)^2$ is the **square of a sum.** They are not equivalent because $(A + B)^2 \neq A^2 + B^2$.

d. $a^2 + 81$ can be written $a^2 + 9^2$, and is, therefore, the **sum of two squares.** We might attempt to factor $a^2 + 81$ as $(a + 9)(a + 9)$ or $(a - 9)(a - 9)$. However, the following checks show that neither product is $a^2 + 81$.

$$(a + 9)(a + 9) = a^2 + 18a + 81 \qquad (a - 9)(a - 9) = a^2 - 18a + 81$$

In general, **the sum of two squares (with no common factor other than 1) cannot be factored using real numbers.** Thus, $a^2 + 81$ is *prime.*

Teaching Example 4 Factor:
a. $a^2 - 144$ **b.** $25 - x^2$
c. $x^2 + 36$ **d.** $x^2 - 8$
Answers:
a. $(a + 12)(a - 12)$
b. $(5 + x)(5 - x)$
c. Prime **d.** Prime

Self Check 4 Factor: **a.** $c^2 - 4$ $(c + 2)(c - 2)$
b. $121 - t^2$ $(11 + t)(11 - t)$ **c.** $x^2 - 24$ Prime
d. $s^2 + 36$ Prime

Now Try Problems 37 and 45

Terms containing variables such as a^4, $25x^2$, and $4y^4$ are perfect squares, because they can be written as the square of a quantity. For example:

$$a^4 = (a^2)^2, \qquad 25x^2 = (5x)^2, \qquad \text{and} \qquad 4y^4 = (2y^2)^2$$

EXAMPLE 5 Factor: **a.** $25x^2 - 49$ **b.** $-121z^2 + 4y^4$

Strategy In each case, the terms of the binomial do not have a common factor (other than 1). To factor them, we will write each binomial in a form that clearly shows it is a difference of two squares.

Why We can then use a special-product rule in reverse to factor it.

Solution

a. We can write $25x^2 - 49$ in the form $(5x)^2 - 7^2$ and match it to the rule for factoring the difference of two squares:

Teaching Tip: Be sure to stress the following Success Tip with your class.

$$\begin{array}{ccccc} F^2 - L^2 & = & (F + L)(F - L) \\ \downarrow \quad \downarrow & & \downarrow \quad \downarrow \quad \downarrow \quad \downarrow \\ (5x)^2 - 7^2 & = & (5x + 7)(5x - 7) \end{array}$$

Therefore, $25x^2 - 49 = (5x + 7)(5x - 7)$. Check by multiplying.

Success Tip

Remember that a *difference of two squares* is a binomial. Each term is a square and the terms have different signs. The powers of the variables in the terms must be even.

b. If we reorder the terms, the resulting binomial, $4y^4 - 121z^2$ is obviously a difference of two squares. We can write $4y^4 - 121z^2$ in the form $(2y^2)^2 - (11z)^2$ and match it to the rule for factoring the difference of two squares:

$$\begin{array}{ccccc} F^2 - L^2 & = & (F + L)(F - L) \\ \downarrow \quad \downarrow & & \downarrow \quad \downarrow \quad \downarrow \quad \downarrow \\ (2y^2)^2 - (11z)^2 & = & (2y^2 + 11z)(2y^2 - 11z) \end{array}$$

Therefore, $-121z^2 + 4y^4 = 4y^4 - 121z^2 = (2y^2 + 11z)(2y^2 - 11z)$. Check by multiplying.

An alternate approach is first to factor out -1: $-(121z^2 - 4y^4)$. Then factor the difference of two squares within the parentheses to get an equivalent result: $-(11z + 2y^2)(11z - 2y^2)$.

Teaching Example 5 Factor: **a.** $36a^2 - 25$ **b.** $9x^4 - 49y^2$ **c.** $-81m^4 + 121$
Answers:
a. $(6a + 5)(6a - 5)$
b. $(3x^2 + 7y)(3x^2 - 7y)$
c. $(11 + 9m^2)(11 - 9m^2)$ or $-(9m^2 + 11)(9m^2 - 11)$

Self Check 5 Factor: **a.** $16y^2 - 9$ $(4y + 3)(4y - 3)$
b. $9m^2 - 64n^4$ $(3m + 8n^2)(3m - 8n^2)$ **c.** $-a^4 + 100$
$(10 + a^2)(10 - a^2)$ or $-(a^2 + 10)(a^2 - 10)$

Now Try Problems 49 and 53

EXAMPLE 6 Factor: $8x^2 - 8$

Strategy We will factor out the GCF, 8, first. Then we will factor the resulting difference of two squares.

Why The first step in factoring any polynomial is to factor out the GCF.

Solution

$$\begin{aligned} 8x^2 - 8 &= 8(x^2 - 1) && \text{The GCF is 8.} \\ &= 8(x + 1)(x - 1) && \text{Think of } x^2 - 1 \text{ as } x^2 - 1^2 \text{ and factor the difference of two squares.} \end{aligned}$$

Check:

$$\begin{aligned} 8(x + 1)(x - 1) &= 8(x^2 - 1) && \text{Multiply the binomials first.} \\ &= 8x^2 - 8 && \text{Distribute the multiplication by 8.} \end{aligned}$$

Teaching Example 6 Factor: $18x^2 - 32$
Answer: $2(3x + 4)(3x - 4)$

Self Check 6 Factor: $2p^2 - 200$ $2(p + 10)(p - 10)$

Now Try Problem 57

Sometimes we must factor a difference of two squares more than once to factor a polynomial completely.

EXAMPLE 7 Factor: $x^4 - 16$

Strategy The terms of $x^4 - 16$ do not have a common factor (other than 1). To factor this binomial, we will write it in a form that clearly shows it is a difference of two squares.

Why We can then use a special-product rule in reverse to factor it.

Solution

$$x^4 - 16 = (x^2)^2 - 4^2 \quad \text{Write } x^4 \text{ as } (x^2)^2 \text{ and 16 as } 4^2.$$
$$= (x^2 + 4)(x^2 - 4) \quad \text{Factor the difference of two squares.}$$
$$= (x^2 + 4)(x + 2)(x - 2) \quad \text{Factor } x^2 - 4\text{, which is itself a difference of two squares. The binomial } x^2 + 4 \text{ is a sum of two squares and does not factor further.}$$

Success Tip

Factoring a polynomial is complete when no factor can be factored further.

Teaching Example 7 Factor: $d^4 - 1$
Answer: $(d^2 + 1)(d + 1)(d - 1)$

Self Check 7 Factor: $a^4 - 81$ $(a^2 + 9)(a + 3)(a - 3)$

Now Try Problem 63

SECTION 6.4 STUDY SET

VOCABULARY

Fill in the blanks.

▶ 1. $x^2 + 6x + 9$ is a perfect-square trinomial because it is the square of the binomial $x + 3$.

▶ 2. The binomial $x^2 - 25$ is called a difference of two squares and it factors as $(x + 5)(x - 5)$. The binomial $x^2 + 25$ is a sum of two squares and since it does not factor using integers, it is prime.

CONCEPTS

Fill in the blanks.

3. Consider $25x^2 + 30x + 9$.
   a. The first term is the square of $5x$.
   b. The last term is the square of 3.
   c. The middle term is twice the product of $5x$ and 3.

4. Consider $49x^2 - 28xy + 4y^2$.
   a. The first term is the square of $7x$.
   b. The last term is the square of $2y$.
   c. The middle term is the opposite of twice the product of $7x$ and $2y$.

▶ 5. a. $x^2 + 2xy + y^2 = (x + y)^2$
   b. $x^2 - 2xy + y^2 = (x - y)^2$
   c. $x^2 - y^2 = (x + y)(x - y)$

6. a. $36x^2 = (6x)^2$ b. $100x^4 = (10x^2)^2$
   c. $4x^2 - 9 = (2x)^2 - (3)^2$

7. List the squares of the integers from 1 through 20.
   1, 4, 9, 16, 25, 36, 49, 64, 81, 100, 121, 144, 169, 196, 225, 256, 289, 324, 361, 400

8. Use multiplication to determine if each factorization is correct.
   a. $9y^2 - 12y + 4 = (3y - 2)^2$ Yes
   b. $n^2 - 16 = (n + 8)(n - 8)$ No

NOTATION

Complete each factorization.

9. $x^2 + 10x + 25 = (x + 5)^2$

10. $9b^2 - 12b + 4 = (3b - 2)^2$

▶ 11. $x^2 - 64 = (x + 8)(x - 8)$

12. $16t^2 - 49 = (4t + 7)(4t - 7)$

GUIDED PRACTICE

Determine whether each of the following is a perfect-square trinomial. See Example 1.

13. $x^2 + 18x + 81$ Yes
▶ 14. $x^2 + 14x + 49$ Yes
15. $y^2 + 2y + 4$ No
16. $y^2 + 4y + 16$ No
▶ 17. $9n^2 - 30n - 25$ No
18. $9a^2 - 48a - 64$ No
19. $4y^2 - 12y + 9$ Yes
▶ 20. $9y^2 - 30y + 25$ Yes

Factor. See Example 2.

21. $x^2 + 6x + 9$
   $(x + 3)^2$
▶ 22. $x^2 + 10x + 25$
   $(x + 5)^2$
▶ 23. $b^2 + 2b + 1$
   $(b + 1)^2$
▶ 24. $m^2 + 12m + 36$
   $(m + 6)^2$
25. $c^2 - 12c + 36$
   $(c - 6)^2$
▶ 26. $d^2 - 10d + 25$
   $(d - 5)^2$
▶ 27. $9 + 4x^2 + 12x$
   $(2x + 3)^2$
28. $121 + 4x^2 + 44x$
   $(2x + 11)^2$
29. $36m^2 + 60mn + 25n^2$
   $(6m + 5n)^2$
▶ 30. $25a^2 + 30ab + 9b^2$
   $(5a + 3b)^2$

▶ Selected exercises available online at www.webassign.net/brookscole

▶ **31.** $81x^2 - 72xy + 16y^2$
$(9x - 4y)^2$

32. $9x^2 - 48xy + 64y^2$
$(3x - 8y)^2$

***Factor.* See Example 3.**

33. $3u^2 - 18u + 27$
$3(u - 3)^2$

▶ **34.** $3v^2 - 42v + 147$
$3(v - 7)^2$

▶ **35.** $36x^3 + 12x^2 + x$
$x(6x + 1)^2$

▶ **36.** $4x^4 - 20x^3 + 25x^2$
$x^2(2x - 5)^2$

***Factor. If a polynomial can't be factored, write "prime."* See Example 4.**

37. $x^2 - 4$
$(x + 2)(x - 2)$

▶ **38.** $x^2 - 9$
$(x + 3)(x - 3)$

39. $x^2 - 16$
$(x + 4)(x - 4)$

▶ **40.** $x^2 - 25$
$(x + 5)(x - 5)$

▶ **41.** $36 - y^2$
$(6 + y)(6 - y)$

42. $49 - w^2$
$(7 + w)(7 - w)$

▶ **43.** $t^2 - 25$
$(t + 5)(t - 5)$

▶ **44.** $h^2 - 144$
$(h + 12)(h - 12)$

▶ **45.** $a^2 + b^2$
Prime

▶ **46.** $121a^2 + b^2$
Prime

▶ **47.** $y^2 - 63$
Prime

▶ **48.** $x^2 - 27$
Prime

***Factor.* See Example 5.**

49. $25t^2 - 64$
$(5t + 8)(5t - 8)$

▶ **50.** $49d^2 - 16$
$(7d + 4)(7d - 4)$

▶ **51.** $81y^2 - 1$
$(9y + 1)(9y - 1)$

52. $400z^2 - 1$
$(20z + 1)(20z - 1)$

▶ **53.** $9x^4 - y^2$
$(3x^2 + y)(3x^2 - y)$

54. $4x^2 - z^4$
$(2x + z^2)(2x - z^2)$

55. $-49d^4 + 16c^2$
$(4c + 7d^2)(4c - 7d^2)$ or $-(7d^2 + 4c)(7d^2 - 4c)$

▶ **56.** $-121b^4 + 36a^2$
$(6a + 11b^2)(6a - 11b^2)$ or $-(11b^2 + 6a)(11b^2 - 6a)$

***Factor.* See Example 6.**

▶ **57.** $8x^2 - 32y^2$
$8(x + 2y)(x - 2y)$

58. $2a^2 - 200b^2$
$2(a + 10b)(a - 10b)$

59. $63a^2 - 7$
$7(3a + 1)(3a - 1)$

▶ **60.** $20x^2 - 5$
$5(2x + 1)(2x - 1)$

***Factor.* See Example 7.**

61. $81 - s^4$
$(9 + s^2)(3 + s)(3 - s)$

▶ **62.** $y^4 - 625$
$(y^2 + 25)(y + 5)(y - 5)$

▶ **63.** $b^4 - 256$
$(b^2 + 16)(b + 4)(b - 4)$

▶ **64.** $m^4n^4 - 16$
$(m^2n^2 + 4)(mn + 2)(mn - 2)$

TRY IT YOURSELF

Factor.

▶ **65.** $a^4 - 144b^2$
$(a^2 + 12b)(a^2 - 12b)$

▶ **66.** $81y^4 - 100z^2$
$(9y^2 + 10z)(9y^2 - 10z)$

▶ **67.** $9x^2y^2 + 30xy + 25$
$(3xy + 5)^2$

68. $s^2t^2 - 20st + 100$
$(st - 10)^2$

69. $16t^4 - 16s^4$
$16(t^2 + s^2)(t + s)(t - s)$

▶ **70.** $2p^4 - 32q^4$
$2(p^2 + 4q^2)(p + 2q)(p - 2q)$

71. $t^2 - 20t + 100$
$(t - 10)^2$

▶ **72.** $r^2 + 24r + 144$
$(r + 12)^2$

73. $9y^2 - 24y + 16$
$(3y - 4)^2$

▶ **74.** $49z^2 - 14z + 1$
$(7z - 1)^2$

75. $z^2 - 64$
$(z + 8)(z - 8)$

76. $25 + B^2$
Prime

77. $25m^4 - 25$
$25(m^2 + 1)(m + 1)(m - 1)$

78. $9 - 9n^4$
$9(1 + n^2)(1 + n)(1 - n)$

79. $18a^5 + 84a^4b + 98a^3b^2$
$2a^3(3a + 7b)^2$

80. $32b^6 + 80b^5c + 50b^4c^2$
$2b^4(4b + 5c)^2$

81. $x^3 - 144x$
$x(x + 12)(x - 12)$

82. $g^3 - 121g$
$g(g + 11)(g - 11)$

83. $49t^2 - 28ts + 4s^2$
$(7t - 2s)^2$

84. $81p^2 - 36pq + 4q^2$
$(9p - 2q)^2$

85. $3m^4 - 3n^4$
$3(m^2 + n^2)(m + n)(m - n)$

86. $5a^4 - 80b^4$
$5(a^2 + 4b^2)(a + 2b)(a - 2b)$

87. $25m^2 + 70m + 49$
$(5m + 7)^2$

88. $25x^2 + 20x + 4$
$(5x + 2)^2$

89. $-100t^2 + 20t - 1$
$-(10t - 1)^2$

90. $-81r^2 - 18r - 1$
$-(9r + 1)^2$

91. $6x^4 - 6x^2y^2$
$6x^2(x + y)(x - y)$

▶ **92.** $4b^2y - 16c^2y$
$4y(b + 2c)(b - 2c)$

93. $100a^2 + 81$
Prime

▶ **94.** $25y^2 + 16$
Prime

95. $-169 + 25x^2$
$(5x + 13)(5x - 13)$ or $-(13 + 5x)(13 - 5x)$

▶ **96.** $-196 + 49f^2$
$(7f + 14)(7f - 14)$ or $-(14 + 7f)(14 - 7f)$

Choose the correct method from Section 6.1, Section 6.2, Section 6.3, or Section 6.4 to factor each of the following:

97. $x^2 + x - 42$
$(x + 7)(x - 6)$

▶ **98.** $rx - sx + r - s$
$(r - s)(x + 1)$

99. $x^2 - 9$
$(x + 3)(x - 3)$

100. $3a^2 - 4a - 4$
$(3a + 2)(a - 2)$

101. $24a^3b - 16a^2b$
$8a^2b(3a - 2)$

▶ **102.** $20ns^2 - 60nu + 100n$
$20n(s^2 - 3u + 5)$

103. $-2r^2 + 28r - 80$
$-2(r - 10)(r - 4)$

104. $10s - 39 + s^2$
$(s + 13)(s - 3)$

105. $x^3 + 3x^2 + 4x + 12$
$(x + 3)(x^2 + 4)$

▶ **106.** $2y^2 - 128z^2$
$2(y + 8z)(y - 8z)$

107. $4b^2 - 20b + 25$
$(2b - 5)^2$

108. $a^2 - 4ab - 12b^2$
$(a - 6b)(a + 2b)$

APPLICATIONS

▶ **109. Genetics.** The Hardy–Weinberg equation, one of the fundamental concepts in population genetics, is $p^2 + 2pq + q^2 = 1$, where p represents the frequency of a certain dominant gene and q represents the frequency of a certain recessive gene. Factor the left side of the equation.
$(p + q)^2$

▶ **110. Signal Flags.** The maritime signal flag for the letter X is shown. Find the polynomial that represents the area of the shaded region and express it in factored form.
$y^2 - 36 = (y + 6)(y - 6)$ square units

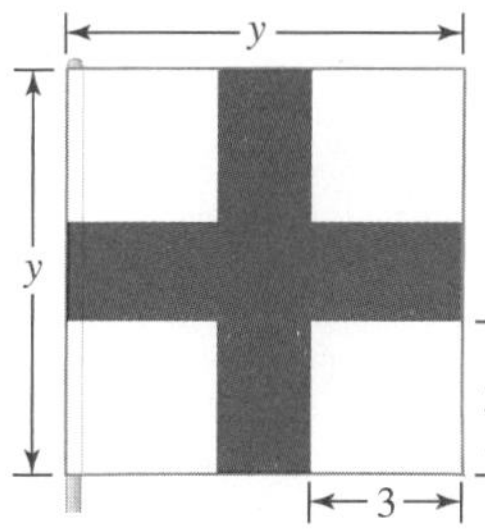

111. Physics. The illustration shows a time-sequence picture of a falling apple. Factor the expression, which gives the difference in the distance fallen by the apple during the time interval from t_1 to t_2 seconds.
$0.5g(t_1 + t_2)(t_1 - t_2)$

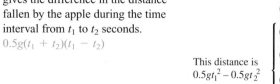

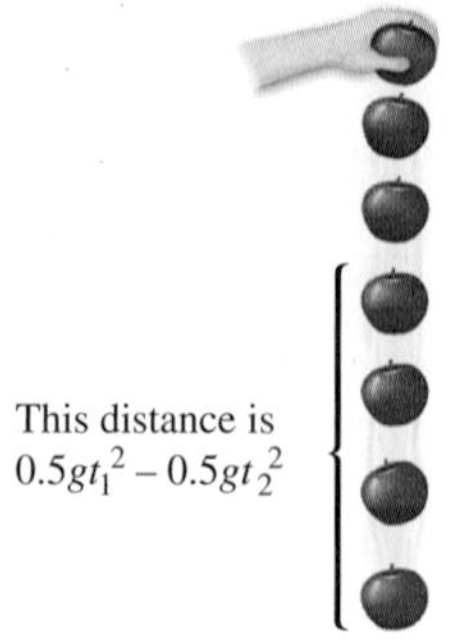

112. Darts. A circular dart board has a series of rings around a solid center, called the bullseye. To find the area of the outer black ring, we can use the formula $A = \pi R^2 - \pi r^2$. Factor the expression on the right side of the equation.
$\pi(R + r)(R - r)$

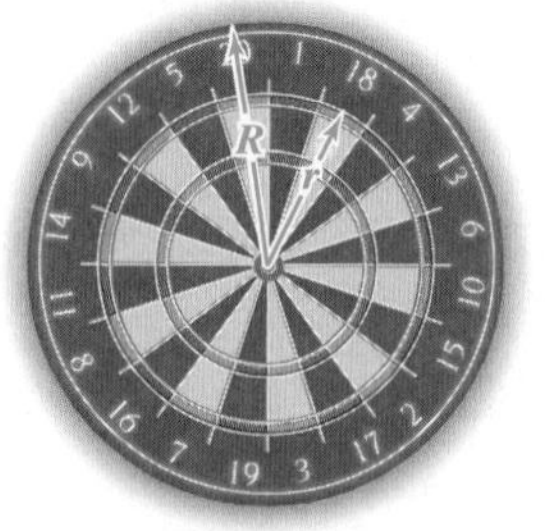

WRITING

113. When asked to factor $x^2 - 25$, one student wrote $(x + 5)(x - 5)$, and another student wrote $(x - 5)(x + 5)$. Are both answers correct? Explain.

114. Explain the error that was made in the following factorization:

$$x^2 - 100 = (x + 50)(x - 50)$$

115. Explain why the following factorization isn't complete.

$$x^4 - 625 = (x^2 + 25)(x^2 - 25)$$

116. Explain why $a^2 + 2a + 1$ is a perfect-square trinomial and why $a^2 + 4a + 1$ isn't a perfect-square trinomial.

REVIEW

Perform each division.

117. $\dfrac{-30c^2d^2 - 15c^2d - 10cd^2}{-10cd}$ $\quad 3cd + \frac{3c}{2} + d$

118. $2a - 1\overline{)a - 2 + 6a^2}$ $\quad 3a + 2$

CHALLENGE PROBLEMS

119. For what value of c does $80x^2 - c$ factor as $5(4x + 3)(4x - 3)$? 45

120. Find all values of b so that $0.16x^2 + bxy + 0.25y^2$ is a perfect-square trinomial. $-0.4, 0.4$

Factor completely.

121. $81x^6 + 36x^3y^2 + 4y^4$
$(9x^3 + 2y^2)^2$

122. $p^2 + p + \frac{1}{4}$
$\left(p + \frac{1}{2}\right)^2$

123. $c^2 + 1.6c + 0.64$
$(c + 0.8)^2$

124. $x^{2n} - y^{4n}$
$(x^n + y^{2n})(x^n - y^{2n})$

125. $(x + 5)^2 - y^2$
$(x + 5 + y)(x + 5 - y)$

126. $\frac{1}{2} - 2a^2$
$\frac{1}{2}(1 + 2a)(1 - 2a)$

127. $c^2 - \frac{1}{16}$
$\left(c + \frac{1}{4}\right)\left(c - \frac{1}{4}\right)$

128. $t^2 - \frac{9}{25}$
$\left(t + \frac{3}{5}\right)\left(t - \frac{3}{5}\right)$

SECTION 6.5 Factoring the Sum and Difference of Two Cubes

OBJECTIVES

1 Factor the sum and difference of two cubes.

ARE YOU READY? *Are You Ready? exercises available online at www.webassign.net/brookscole*

The following problems review some basic skills that are needed when factoring binomials and trinomials.

1. Multiply: $(x + 4)(x^2 - 4x + 16)$ $\quad x^3 + 64$
2. Multiply: $(2h - 1)(4h^2 + 2h + 1)$ $\quad 8h^3 - 1$
3. Evaluate: a. 3^3 27 b. 5^3 125
4. Explain why $x^2 - 2x + 4$ is prime. There are no two integers whose product is 4 and whose sum is -2.

In this section, we will discuss how to factor two types of binomials, called the *sum* and the *difference of two cubes.*

1 Factor the Sum and Difference of Two Cubes.

We have seen that the sum of two squares, such as $x^2 + 4$ or $25a^2 + 9b^2$, cannot be factored. However, the sum of two cubes and the difference of two cubes can be factored.

The sum of two cubes		***The difference of two cubes***	
x^3	$+ 8$	a^3	$- 64b^3$
This is x cubed.	This is 2 cubed: $2^3 = 8$.	This is a cubed.	This is 4b cubed: $(4b)^3 = 64b^3$.

The Language of Algebra

The expression $x^3 + y^3$ is a **sum of two cubes,** whereas $(x + y)^3$ is the **cube of a sum.** If you expand $(x + y)^3$, you will see that $(x + y)^3 \neq x^3 + y^3$.

To find rules for factoring the sum of two cubes and the difference of two cubes, we need to find the products shown below. Note that each term of the trinomial is multiplied by each term of the binomial.

$$(x + y)(x^2 - xy + y^2) = x^3 - \cancel{x^2y} + \cancel{xy^2} + \cancel{x^2y} - \cancel{xy^2} + y^3$$
$$= x^3 + y^3 \quad \text{Combine like terms: } -x^2y + x^2y = 0 \text{ and } xy^2 - xy^2 = 0.$$

$$(x - y)(x^2 + xy + y^2) = x^3 + \cancel{x^2y} + \cancel{xy^2} - \cancel{x^2y} - \cancel{xy^2} - y^3$$
$$= x^3 - y^3 \quad \text{Combine like terms.}$$

Teaching Tip: You may want to use vertical form to show the multiplication:

$$\begin{array}{r} x^2 + xy + y^2 \\ x \quad - y \\ \hline \end{array}$$

These results justify the rules for factoring the **sum** or **difference of two cubes.** They are easier to remember if we think of a sum (or a difference) of two cubes as the cube of a **F**irst quantity plus (or minus) the cube of the **L**ast quantity.

Factoring the Sum and Difference of Two Cubes

To factor the cube of a First quantity plus the cube of a Last quantity, multiply the First plus the Last by the First squared, minus the First times the Last, plus the Last squared.

$$F^3 + L^3 = (F + L)(F^2 - FL + L^2)$$

To factor the cube of a First quantity minus the cube of a Last quantity, multiply the First minus the Last by the First squared, plus the First times the Last, plus the Last squared.

$$F^3 - L^3 = (F - L)(F^2 + FL + L^2)$$

To factor the sum or difference of two cubes, it's helpful to know the cubes of integers from 1 to 10 shown in red below. The number 216, for example, is a **perfect cube,** because $216 = 6^3$.

$\mathbf{1} = 1^3$	$\mathbf{27} = 3^3$	$\mathbf{125} = 5^3$	$\mathbf{343} = 7^3$	$\mathbf{729} = 9^3$
$\mathbf{8} = 2^3$	$\mathbf{64} = 4^3$	$\mathbf{216} = 6^3$	$\mathbf{512} = 8^3$	$\mathbf{1{,}000} = 10^3$

EXAMPLE 1 Factor: $x^3 + 8$

Teaching Tip: Stress this wording: These factoring forms enable us to write certain *sums* and *differences* as equivalent *products.*

Strategy We will write $x^3 + 8$ in a form that clearly shows it is the sum of two cubes.

Why We can then use the rule for factoring the sum of two cubes.

Solution $x^3 + 8$ is the sum of two cubes because it can be written as $x^3 + 2^3$. We can match it to the rule for factoring the sum of two cubes to find its factorization.

$$\mathbf{F}^3 + \mathbf{L}^3 = (\mathbf{F} + \mathbf{L})(\mathbf{F}^2 - \mathbf{F}\ \mathbf{L} + \mathbf{L}^2)$$
$$x^3 + 2^3 = (x + 2)(x^2 - x \cdot 2 + 2^2)$$
$$= (x + 2)(x^2 - 2x + 4)$$

To write the trinomial factor:
- Square the first term of the binomial factor.
- Multiply the terms of the binomial factor.
- Square the last term of the binomial factor.

$x^2 - 2x + 4$ does not factor. It is prime.

Caution

A common error is to try to factor $x^2 - 2x + 4$. It is not a perfect square trinomial, because the middle term needs to be $-4x$. Furthermore, it cannot be factored by the methods of Section 6.2. It is prime.

Therefore, $x^3 + 8 = (x + 2)(x^2 - 2x + 4)$. We can check by multiplying.

$$(x + 2)(x^2 - 2x + 4) = x^3 - \cancel{2x^2} + \cancel{4x} + \cancel{2x^2} - \cancel{4x} + 8$$
$$= x^3 + 8 \quad \text{This is the original binomial.}$$

Teaching Example 1 Factor: $m^3 + 125$
Answer: $(m + 5)(m^2 - 5m + 25)$

Self Check 1 Factor: $h^3 + 27$ $(h + 3)(h^2 - 3h + 9)$

Now Try Problem 17

Terms containing variables such as a^6, $64b^3$, and $27m^6$ are also perfect cubes, because they can be written as the cube of a quantity:

$$a^6 = (a^2)^3, \quad 64b^3 = (4b)^3, \quad \text{and} \quad 27m^6 = (3m^2)^3$$

EXAMPLE 2 Factor: $a^3 - 64b^3$

Teaching Tip: Remind students that the *trinomial* in the factored form of a sum or difference of two cubes will always be prime.

Strategy We will write $a^3 - 64b^3$ in a form that clearly shows it is the difference of two cubes.

Why We can then use the rule for factoring the difference of two cubes.

Solution $a^3 - 64b^3$ is the difference of two cubes because it can be written as $a^3 - (4b)^3$. We can match it to the rule for factoring the difference of two cubes to find its factorization.

The Language of Algebra

An expression of the form $a^3 - 64b^3$ is called the **difference of two cubes,** whereas $(a - 64b)^3$ is the **cube of a difference.** They are not equivalent because $(a - 64b)^3 \neq a^3 - 64b^3$.

$$\mathrm{F}^3 - \mathrm{L}^3 = (\mathrm{F} - \mathrm{L})(\mathrm{F}^2 + \mathrm{F}\ \mathrm{L} + \mathrm{L}^2)$$
$$\downarrow$$
$$a^3 - (4b)^3 = (a - 4b)[a^2 + a \cdot 4b + (4b)^2]$$
$$= (a - 4b)(a^2 + 4ab + 16b^2)$$

To write the trinomial factor:
- Square the first term of the binomial factor.
- Multiply the terms of the binomial factor.
- Square the last term of the binomial factor.

$a^2 + 4ab + 16b^2$ does not factor.

Therefore, $a^3 - 64b^3 = (a - 4b)(a^2 + 4ab + 16b^2)$. Check by multiplying.

Teaching Example 2 Factor: $27x^3 - y^3$
Answer: $(3x - y)(9x^2 + 3xy + y^2)$

Self Check 2 Factor: $8c^3 - 1$ $(2c - 1)(4c^2 + 2c + 1)$

Now Try Problem 37

You should memorize the rules for factoring the sum and the difference of two cubes. Note that the right side of each rule has the form

(a binomial)(a trinomial)

and that there is a relationship between the signs that appear in these forms.

The Sum of Two Cubes

The same sign

$$F^3 + L^3 = (F + L)(F^2 - FL + L^2)$$

Opposite signs — Always plus

The Difference of Two Cubes

The same sign

$$F^3 - L^3 = (F - L)(F^2 + FL + L^2)$$

Opposite signs — Always plus

Success Tip

An easy way to remember the sign patterns within the binomial and trinomial of these factoring forms is with the word "SOAP."

Same, **O**pposite, **A**lways **P**lus

If the terms of a binomial have a common factor, the GCF (or the opposite of the GCF) should always be factored out first.

EXAMPLE 3 Factor: $-2t^5 + 250t^2$

Strategy We will factor out the common factor, $-2t^2$. We can then factor the resulting binomial as a difference of two cubes.

Why The first step in factoring any polynomial is to factor out the GCF, or its opposite.

Solution

$$-2t^5 + 250t^2 = -2t^2(t^3 - 125) \quad \text{Factor out the opposite of the GCF, } -2t^2.$$
$$= -2t^2(t - 5)(t^2 + 5t + 25) \quad \text{Factor } t^3 - 125.$$

Therefore, $-2t^5 + 250t^2 = -2t^2(t - 5)(t^2 + 5t + 25)$. Check by multiplying.

Teaching Example 3 Factor: $-3a^4 + 24a$
Answer: $-3a(a - 2)(a^2 + 2a + 4)$

Self Check 3 Factor: $4c^3 + 4d^3$ $4(c + d)(c^2 - cd + d^2)$

Now Try Problem 43

SECTION 6.5 STUDY SET

VOCABULARY

Fill in the blanks.

1. $x^3 + 27$ is the sum of two cubes and $a^3 - 125$ is the difference of two cubes.
2. The factorization of $x^3 + 8$ is $(x + 2)(x^2 - 2x + 4)$. The first factor is a binomial and the second is a trinomial.

CONCEPTS

Fill in the blanks.

3. a. $F^3 + L^3 = (F + L)(F^2 - FL + L^2)$
   b. $F^3 - L^3 = (F - L)(F^2 + FL + L^2)$
4. $m^3 + 64$
   This is m cubed. This is 4 cubed.
5. $216n^3 - 125$
   This is $6n$ cubed. This is 5 cubed.
6. a. $x^3 + 64y^3 = (x)^3 + (4y)^3$
   b. $8x^3 - 27 = (2x)^3 - (3)^3$
7. List the first ten positive integer cubes. 1, 8, 27, 64, 125, 216, 343, 512, 729, 1,000
8. $(x - 2)(x^2 + 2x + 4)$ is the factorization of what binomial? $x^3 - 8$
9. Use multiplication to determine if the factorization is correct.
   $b^3 + 27 = (b + 3)(b^2 + 3b + 9)$ No
10. The factorization of $y^3 + 27$ is $(y + 3)(y^2 - 3y + 9)$. Is this factored completely, or does $y^2 - 3y + 9$ factor further? It is factored completely.

NOTATION

Complete each factorization.

11. $a^3 + 8 = (a + 2)(a^2 - 2a + 4)$
12. $x^3 - 1 = (x - 1)(x^2 + x + 1)$
13. $b^3 + 27 = (b + 3)(b^2 - 3b + 9)$
14. $z^3 - 125 = (z - 5)(z^2 + 5z + 25)$

Give an example of each type of expression.

15. a. the sum of two cubes $x^3 + 8$ (Answers may vary)
    b. the cube of a sum $(x + 8)^3$
16. a. the difference of two cubes $y^3 - 27$ (Answers may vary)
    b. the cube of a difference $(y - 1)^3$

GUIDED PRACTICE

***Factor.* See Example 1.**

17. $y^3 + 125$ $(y + 5)(y^2 - 5y + 25)$
18. $b^3 + 216$ $(b + 6)(b^2 - 6b + 36)$
19. $a^3 + 64$ $(a + 4)(a^2 - 4a + 16)$
20. $n^3 + 1$ $(n + 1)(n^2 - n + 1)$
21. $n^3 + 512$ $(n + 8)(n^2 - 8n + 64)$
22. $t^3 + 729$ $(t + 9)(t^2 - 9t + 81)$
23. $8 + t^3$ $(2 + t)(4 - 2t + t^2)$
24. $27 + y^3$ $(3 + y)(9 - 3y + y^2)$
25. $a^3 + 1{,}000b^3$ $(a + 10b)(a^2 - 10ab + 100b^2)$
26. $8u^3 + w^3$ $(2u + w)(4u^2 - 2uw + w^2)$
27. $125c^3 + 27d^3$ $(5c + 3d)(25c^2 - 15cd + 9d^2)$
28. $64m^3 + 343n^3$ $(4m + 7n)(16m^2 - 28mn + 49n^2)$

***Factor.* See Example 2.**

29. $a^3 - 27$ $(a - 3)(a^2 + 3a + 9)$
30. $r^3 - 8$ $(r - 2)(r^2 + 2r + 4)$
31. $m^3 - 343$ $(m - 7)(m^2 + 7m + 49)$
32. $y^3 - 216$ $(y - 6)(y^2 + 6y + 36)$
33. $216 - v^3$ $(6 - v)(36 + 6v + v^2)$
34. $125 - t^3$ $(5 - t)(25 + 5t + t^2)$
35. $8s^3 - t^3$ $(2s - t)(4s^2 + 2st + t^2)$
36. $27a^3 - b^3$ $(3a - b)(9a^2 + 3ab + b^2)$
37. $1{,}000a^3 - w^3$ $(10a - w)(100a^2 + 10aw + w^2)$
38. $s^3 - 64t^3$ $(s - 4t)(s^2 + 4st + 16t^2)$
39. $64x^3 - 27y^3$ $(4x - 3y)(16x^2 + 12xy + 9y^2)$
40. $27x^3 - 1{,}000y^3$ $(3x - 10y)(9x^2 + 30xy + 100y^2)$

Factor. **See Example 3.**

41. $2x^3 + 2$
$2(x + 1)(x^2 - x + 1)$

42. $8y^3 + 8$
$8(y + 1)(y^2 - y + 1)$

43. $3d^3 + 81$
$3(d + 3)(d^2 - 3d + 9)$

44. $2x^3 + 54$
$2(x + 3)(x^2 - 3x + 9)$

45. $x^4 - 216x$
$x(x - 6)(x^2 + 6x + 36)$

46. $x^5 - 125x^2$
$x^2(x - 5)(x^2 + 5x + 25)$

47. $64m^3x - 8n^3x$ $8x(2m - n)(4m^2 + 2mn + n^2)$

48. $16r^4 - 128rs^3$ $16r(r - 2s)(r^2 + 2rs + 4s^2)$

TRY IT YOURSELF

Choose the correct method from Section 6.1 through Section 6.5 and factor completely.

49. $x^2 + 8x + 16$
$(x + 4)^2$

50. $64p^3 - 27$
$(4p - 3)(16p^2 + 12p + 9)$

51. $9r^2 - 16s^2$
$(3r + 4s)(3r - 4s)$

52. $-63 - 13x + 6x^2$
$(2x - 9)(3x + 7)$

53. $xy - ty + sx - st$
$(x - t)(y + s)$

54. $12p^2 + 14p - 6$
$2(3p - 1)(2p + 3)$

55. $4p^3 + 32q^3$
$4(p + 2q)(p^2 - 2pq + 4q^2)$

56. $56a^4 - 15a^3 + a^2$
$a^2(8a - 1)(7a - 1)$

57. $16c^3t^2 + 20c^2t^3 + 6ct^4$
$2ct^2(4c + 3t)(2c + t)$

58. $-t^2 - 9t + 1$
$-(t^2 + 9t - 1)$

59. $36e^4 - 36$
$36(e^2 + 1)(e + 1)(e - 1)$

60. $3(z + 4) - a(z + 4)$
$(z + 4)(3 - a)$

61. $35a^3b^2 - 14a^2b^3 + 14a^3b^3$
$7a^2b^2(5a - 2b + 2ab)$

62. $-y^2 - 15y + 34$
$-(y + 17)(y - 2)$

63. $36r^2 + 60rs + 25s^2$
$(6r + 5s)^2$

64. $16u^2 - 16$
$16(u + 1)(u - 1)$

Look Alikes . . .

65. a. $x^2 - 1$
$(x + 1)(x - 1)$

b. $x^3 - 1$
$(x - 1)(x^2 + x + 1)$

66. a. $x^2 - 64$
$(x + 8)(x - 8)$

b. $x^3 - 64$
$(x - 4)(x^2 + 4x + 16)$

67. a. $x^2 + 2x$
$x(x + 2)$

b. $x^2 + 2x + 1$
$(x + 1)^2$

68. a. $x^2 - 4x$
$x(x - 4)$

b. $x^2 - 4x + 4$
$(x - 2)^2$

APPLICATIONS

69. Mailing Breakables. Write a polynomial that describes the amount of space in the larger box that must be filled with styrofoam chips if the smaller box containing a glass tea cup is to be placed within the larger box for mailing. Then factor the polynomial.
$(1{,}000 - x^3)$ in.^{3}; $(10 - x)(100 + 10x + x^2)$

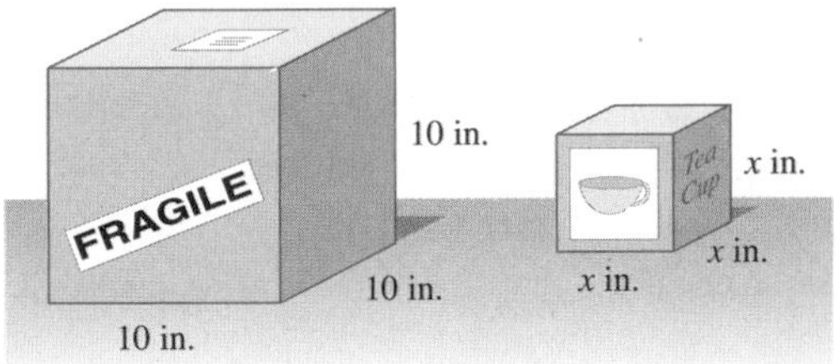

70. Melting Ice. In one hour, the block of ice shown below had melted to the size shown on the right. Write a polynomial that describes the volume of ice that melted away. Then factor the polynomial. $(729 - x^3)$ in.^{3}; $(9 - x)(81 + 9x + x^2)$

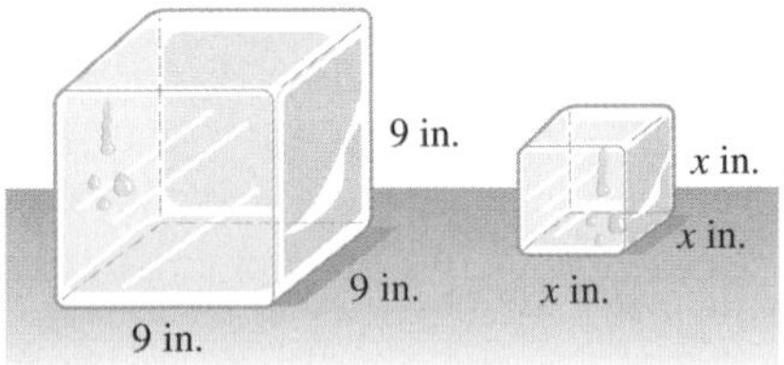

WRITING

71. Explain why $x^3 - 25$ is not a difference of two cubes.

72. Explain this diagram. Then draw a similar diagram for the difference of two cubes.

The same

$$F^3 + L^3 = (F + L)(F^2 - FL + L^2)$$

Opposite   Always plus

REVIEW

73. When expressed as a decimal, is $\frac{7}{9}$ a terminating or a repeating decimal? Repeating

74. Solve: $x + 20 = 4x - 1 + 2x$ $\frac{21}{5}$

75. Solve: $2x + 2 = \frac{2}{3}x - 2$ -3

76. Check to determine whether 4 is a solution of $3(m - 8) + 2m = 4 - (m + 2)$. It is not a solution.

CHALLENGE PROBLEMS

77. Consider: $x^6 - 1$. Write the binomial as a difference of two squares. Then factor.
$(x^3)^2 - 1^2$; $(x + 1)(x^2 - x + 1)(x - 1)(x^2 + x + 1)$

78. What binomial multiplied by $(a^2b^2 + 7ab + 49)$ produces a difference of two cubes? $(ab - 7)$

Factor.

79. $x^6 - y^9$ $(x^2 - y^3)(x^4 + x^2y^3 + y^6)$

80. $\frac{125}{8}s^3 + \frac{1}{27}t^3$ $\left(\frac{5}{2}s + \frac{1}{3}t\right)\left(\frac{25}{4}s^2 - \frac{5}{6}st + \frac{1}{9}t^2\right)$

81. $64x^{12} + y^{15}z^{18}$ $(4x^4 + y^5z^6)(16x^8 - 4x^4y^5z^6 + y^{10}z^{12})$

82. $x^{3m} - y^{3n}$ $(x^m - y^n)(x^{2m} + x^my^n + y^{2n})$

SECTION 6.6

OBJECTIVE

1 Use a general strategy for factoring polynomials.

A Factoring Strategy

ARE YOU READY? Are You Ready? exercises available online at www.webassign.net/brookscole

The following problems review some basic skills that are needed when factoring polynomials.

1. How many terms does each expression have?
   a. $x^4 - 5x^3 + x - 5$ 4
   b. $16x - 5x^2 + 6x^3$ 3
2. Do any of the factors in $(m^2 + 9)(m + 3)(m - 3)$ factor further? No
3. Multiply: $3n(n + 5)(n - 14)$ $3n^3 - 27n^2 - 210n$
4. What is the greatest common factor of the terms of $15c^3d^2 - 10cd^3$? $5cd^2$

The factoring methods discussed so far will be used in the remaining chapters to simplify expressions and solve equations. In such cases, we must determine the factoring method—it will not be specified. This section will give you practice in selecting the appropriate factoring method to use given a randomly chosen polynomial.

1 Use a General Strategy for Factoring Polynomials.

The following strategy is helpful when factoring polynomials.

Steps for Factoring a Polynomial

1. Is there a common factor? If so, factor out the GCF, or the opposite of the GCF so that the leading coefficient is positive. Remember to include it in your final answer.
2. How many terms does the polynomial have?
   If it has **two terms,** look for the following problem types:
   a. The difference of two squares
   b. The sum of two cubes
   c. The difference of two cubes
   If it has **three terms,** look for the following problem types:
   a. A perfect-square trinomial
   b. If the trinomial is not a perfect square, use the trial-and-check-method or the grouping method.
   If it has **four or more terms,** try to factor by grouping.
3. Can any factors be factored further? If so, factor them completely.
4. Does the factorization check? Check by multiplying.

EXAMPLE 1 Factor: $2x^4 - 162$

Strategy We will answer the four questions listed in the *Steps for Factoring a Polynomial.*

Why The answers to these questions help us determine which factoring techniques to use.

Solution ***Is there a common factor?*** Yes. Factor out the GCF, which is 2.

$$2x^4 - 162 = 2(x^4 - 81)$$

The Language of Algebra

Recall that **to factor a polynomial** means to express it as a product of two (or more) polynomials.

How many terms does it have? The polynomial within the parentheses, $x^4 - 81$, has two terms. It is a difference of two squares.

$$2x^4 - 162 = 2(x^4 - 81) \quad \text{Think of } x^4 - 81 \text{ as } (x^2)^2 - 9^2.$$
$$= 2(x^2 + 9)(x^2 - 9) \quad \text{Factor the difference of two squares.}$$

Is it factored completely? No. $x^2 - 9$ is also the difference of two squares and can be factored.

The Language of Algebra

Remember that the instruction to **factor** means to *factor completely.* A polynomial is *factored completely* when no factor can be factored further.

$$2x^4 - 162 = 2(x^4 - 81)$$
$$= 2(x^2 + 9)(\mathbf{x^2 - 9}) \quad \text{Think of } x^2 - 9 \text{ as } x^2 - 3^2.$$
$$= 2(x^2 + 9)(\mathbf{x + 3})(\mathbf{x - 3}) \quad x^2 + 9 \text{ is a sum of two squares and does not factor.}$$

Therefore, $2x^4 - 162 = 2(x^2 + 9)(x + 3)(x - 3)$.

Does it check? Yes.

$$2(x^2 + 9)(\mathbf{x + 3})(\mathbf{x - 3}) = 2(x^2 + 9)(\mathbf{x^2 - 9}) \quad \text{Multiply } (x + 3)(x - 3) \text{ first.}$$
$$= 2(x^4 - 81) \quad \text{Multiply } (x^2 + 9)(x^2 - 9).$$
$$= 2x^4 - 162 \quad \text{This is the original polynomial.}$$

Teaching Example 1 Factor: $3x^4 - 48$
Answer: $3(x^2 + 4)(x + 2)(x - 2)$

Self Check 1 Factor: $11a^6 - 11a^2$ $11a^2(a^2 + 1)(a + 1)(a - 1)$

Now Try Problem 21

EXAMPLE 2 Factor: $-4c^5d^2 - 12c^4d^3 - 9c^3d^4$

Strategy We will answer the four questions listed in the *Steps for Factoring a Polynomial.*

Why The answers to these questions help us determine which factoring techniques to use.

Solution ***Is there a common factor?*** Yes. Factor out the opposite of the GCF, $-c^3d^2$, so that the leading coefficient is positive.

$$-4c^5d^2 - 12c^4d^3 - 9c^3d^4 = -c^3d^2(4c^2 + 12cd + 9d^2)$$

Success Tip

Recall that if a pair of parentheses is produced in the factoring process, always see if the resulting expression within the parentheses can be factored further.

How many terms does it have? The polynomial within the parentheses has three terms. It is a perfect-square trinomial because $4c^2 = (2c)^2$, $9d^2 = (3d)^2$, and $12cd = 2 \cdot 2c \cdot 3d$.

$$-4c^5d^2 - 12c^4d^3 - 9c^3d^4 = -c^3d^2(\mathbf{4c^2 + 12cd + 9d^2})$$
$$= -c^3d^2(\mathbf{2c + 3d})^2$$

Is it factored completely? Yes. The binomial $2c + 3d$ does not factor further.

Therefore, $-4c^5d^2 - 12c^4d^3 - 9c^3d^4 = -c^3d^2(2c + 3d)^2$.

Does it check? Yes.

$$-c^3d^2(\mathbf{2c + 3d})^2 = -c^3d^2(\mathbf{4c^2 + 12cd + 9d^2}) \quad \text{Use a special-product rule.}$$
$$= -4c^5d^2 - 12c^4d^3 - 9c^3d^4 \quad \text{This is the original polynomial.}$$

Teaching Example 2 Factor: $-3x^3y^2 + 18x^2y^2 - 27xy^2$
Answer: $-3xy^2(x - 3)^2$

Self Check 2 Factor: $-32h^4 - 80h^3 - 50h^2$ $-2h^2(4h + 5)^2$

Now Try Problem 33

EXAMPLE 3 Factor: $y^4 - 3y^3 + y - 3$

Strategy We will answer the four questions listed in the *Steps for Factoring a Polynomial.*

Why The answers to these questions help us determine which factoring techniques to use.

Solution ***Is there a common factor?*** No. There is no common factor (other than 1).

How many terms does it have? Since the polynomial has four terms, we will try factoring by grouping.

$$y^4 - 3y^3 + y - 3 = y^3(y - 3) + 1(y - 3)$$ Factor out y^3 from $y^4 - 3y^3$. Factor out 1 from $y - 3$.

$$= (y - 3)(y^3 + 1)$$

Success Tip

Something as simple as counting the number of terms that a polynomial has is very important when determining how to factor it.

Is it factored completely? No. We can factor $y^3 + 1$ as a sum of two cubes.

$$y^4 - 3y^3 + y - 3 = y^3(y - 3) + 1(y - 3)$$

$$= (y - 3)(y^3 + 1)$$ Think of $y^3 + 1$ as $y^3 + 1^3$.

$$= (y - 3)(y + 1)(y^2 - y + 1)$$ $y^2 - y + 1$ does not factor further.

Therefore, $y^4 - 3y^3 + y - 3 = (y - 3)(y + 1)(y^2 - y + 1)$.

Does it check? Yes.

$$(y - 3)(y + 1)(y^2 - y + 1) = (y - 3)(y^3 + 1)$$ Multiply the last two factors.

$$= y^4 + y - 3y^3 - 3$$ Use the FOIL method.

$$= y^4 - 3y^3 + y - 3$$ This is the original polynomial.

Teaching Example 3 Factor: $x^4 - 6x^3 + x - 6$
Answer: $(x - 6)(x + 1)(x^2 - x + 1)$

Self Check 3 Factor: $b^4 + b^3 + 8b + 8$ $(b + 1)(b + 2)(b^2 - 2b + 4)$

Now Try Problem 37

EXAMPLE 4 Factor: $32n - 4n^2 + 4n^3$

Strategy We will answer the four questions listed in the *Steps for Factoring a Polynomial.*

Why The answers to these questions help us determine which factoring techniques to use.

Solution ***Is there a common factor?*** Yes. When we write the terms in descending powers of n, we see that the GCF is $4n$.

$$4n^3 - 4n^2 + 32n = 4n(n^2 - n + 8)$$

How many terms does it have? The polynomial within the parentheses has three terms. It is not a perfect-square trinomial because the last term, 8, is not a perfect square.

Negative factors of 8	Sum of the negative factors of 8
$-1(-8) = 8$	$-1 + (-8) = -9$
$-2(-4) = 8$	$-2 + (-8) = -10$

To factor the trinomial $n^2 - n + 8$, we must find two integers whose product is 8 and whose sum is -1. As we see in the table, there are no such integers. Thus, $n^2 - n + 8$ is prime.

Is it factored completely? Yes.

Therefore, $4n^3 - 4n^2 + 32n = 4n(n^2 - n + 8)$. Remember to write the GCF, $4n$, from the first step.

Does it check? Yes.

$$4n(n^2 - n + 8) = 4n^3 - 4n^2 + 32n$$ This is equivalent to the original polynomial.

Teaching Example 4 Factor: $15x^2 - 3x + 12x^3$
Answer: $3x(4x^2 + 5x - 1)$

Self Check 4 Factor: $6m^2 - 54m + 6m^3$ $6m(m^2 + m - 9)$

Now Try Problem 45

EXAMPLE 5 Factor: $3y^3 - 4y^2 - 4y$

Strategy We will answer the four questions listed in the *Steps for Factoring a Polynomial.*

Why The answers to these questions help us determine which factoring techniques to use.

Solution ***Is there a common factor?*** Yes. The GCF is y.

$$3y^3 - 4y^2 - 4y = y(3y^2 - 4y - 4)$$

The Language of Algebra

It is important to learn the names of all the factoring methods we have studied. Educational research has found that words play an important role in concept formation.

How many terms does it have? The polynomial within the parentheses has three terms. It is not a perfect-square trinomial because the first term, $3y^2$, is not a perfect square.

If we use grouping to factor $3y^2 - 4y - 4$, the key number is $ac = 3(-4) = -12$. We must find two integers whose product is -12 and whose sum is $b = -4$.

Key number $= -12$ $b = -4$

Factors of -12	Sum of the factors of -12
$2(-6) = -12$	$2 + (-6) = -4$

From the table, the correct pair is 2 and -6. They serve as the coefficients of $2y$ and $-6y$, the two terms that we use to represent the middle term, $-4y$, of the trinomial.

$3y^2 - \mathbf{4y} - 4 = 3y^2 + \mathbf{2y} - \mathbf{6y} - 4$ Express $-4y$ as $2y - 6y$.

$= y(3y + 2) - 2(3y + 2)$ Factor y from the first two terms and factor -2 from the last two terms.

$= (3y + 2)(y - 2)$ Factor out $3y + 2$.

The trinomial $3y^2 - 4y - 4$ factors as $(3y + 2)(y - 2)$.

Is it factored completely? Yes. Because $3y + 2$ and $y - 2$ do not factor.

Therefore, $3y^3 - 4y^2 - 4y = y(3y + 2)(y - 2)$. Remember to write the GCF, y, from the first step.

Does it check? Yes.

$y(3y + 2)(y - 2) = y(3y^2 - 4y - 4)$ Multiply the binomials.

$= 3y^3 - 4y^2 - 4y$ This is the original polynomial.

Teaching Example 5 Factor: $12x^3 + 14x^2 - 6x$
Answer: $2x(3x - 1)(2x + 3)$

Self Check 5 Factor: $6y^3 + 21y^2 - 12y$ $3y(2y - 1)(y + 4)$

Now Try Problem 67

SECTION 6.6 STUDY SET

VOCABULARY

Fill in the blanks.

1. To factor a polynomial means to express it as a product of two (or more) polynomials.
2. A polynomial is factored completely when each factor is prime.

CONCEPTS

For each of the following polynomials, which factoring method would you use first?

3. $2x^5y - 4x^3y$
Factor out the GCF
4. $9b^2 + 12y - 5$
Trinomial factoring
5. $x^2 + 18x + 81$
Perfect-square trinomial
6. $ax + ay - x - y$
Factoring by grouping
7. $x^3 + 27$
Sum of two cubes
8. $y^3 - 64$
Difference of two cubes
9. $m^2 + 3mn + 2n^2$
Trinomial factoring
10. $16 - 25z^2$
Difference of two squares
11. What is the first question that should be asked when using the strategy of this section to factor a polynomial? Is there a common factor?
12. Use multiplication to determine whether the factorization is correct. Yes

$$5c^3d^2 - 40c^2d^3 + 35cd^4 = 5cd^2(c - 7d)(c - d)$$

NOTATION

Complete each factorization.

13. $6m^3 - 28m^2 + 16m = 2m(3m^2 - 14m + 8)$
$= 2m(3m - 2)(m - 4)$
14. $2a^3 + 3a^2 - 2a - 3$
$= a^2(2a + 3) - 1(2a + 3)$
$= (2a + 3)(a^2 - 1)$
$= (2a + 3)(a + 1)(a - 1)$

TRY IT YOURSELF

The following is a list of random factoring problems. Factor each expression. If an expression is not factorable, write "prime." See Examples 1–5.

15. $2b^2 + 8b - 24$
$2(b + 6)(b - 2)$
16. $32 - 2t^4$
$2(4 + t^2)(2 + t)(2 - t)$
17. $8p^3q^7 + 4p^2q^3$
$4p^2q^3(2pq^4 + 1)$
18. $8m^2n^3 - 24mn^4$
$8mn^3(m - 3n)$
19. $2 + 24y + 40y^2$
$2(2y + 1)(10y + 1)$
20. $6r^2 + 3rs - 18s^2$
$3(2r - 3s)(r + 2s)$
21. $8x^4 - 8$
$8(x^2 + 1)(x + 1)(x - 1)$
22. $t - 90 + t^2$
$(t + 10)(t - 9)$
23. $14c - 147 + c^2$
$(c + 21)(c - 7)$
24. $ab^2 - 4a + 3b^2 - 12$
$(a + 3)(b + 2)(b - 2)$
25. $x^2 + 7x + 1$
Prime
26. $3a^3 + 24b^3$
$3(a + 2b)(a^2 - 2ab + 4b^2)$
27. $-2x^5 + 128x^2$
$-2x^2(x - 4)(x^2 + 4x + 16)$
28. $16 - 40z + 25z^2$
$(5z - 4)^2$
29. $a^2c + a^2d^2 + bc + bd^2$
$(c + d^2)(a^2 + b)$
30. $6t^4 + 14t^3 - 40t^2$
$2t^2(3t - 5)(t + 4)$
31. $-9x^2 + 6x - 1$
$-(3x - 1)^2$
32. $x^2y^2 - 2x^2 - y^2 + 2$
$(y^2 - 2)(x + 1)(x - 1)$
33. $-20m^3 - 100m^2 - 125m$
$-5m(2m + 5)^2$
34. $5x^3y^3z^4 + 25x^2y^4z^2 - 35x^3y^2z^5$
$5x^2y^2z^2(xyz^2 + 5y^2 - 7xz^3)$
35. $2c^2 - 5cd - 3d^2$
$(2c + d)(c - 3d)$
36. $125p^3 - 64y^3$
$(5p - 4y)(25p^2 + 20py + 16y^2)$
37. $p^4 - 2p^3 - 8p + 16$
$(p - 2)^2(p^2 + 2p + 4)$
38. $a^2 + 8a + 3$
Prime
39. $a^2(x - a) - b^2(x - a)$
$(x - a)(a + b)(a - b)$
40. $70p^4q^3 - 35p^4q^2 + 49p^5q^2$
$7p^4q^2(10q - 5 + 7p)$
41. $a^2b^2 - 144$
$(ab + 12)(ab - 12)$
42. $-16x^4y^2z + 24x^5y^3z^4 - 15x^2y^3z^7$
$-x^2y^2z(16x^2 - 24x^3yz^3 + 15yz^6)$
43. $2x^3 + 10x^2 + x + 5$
$(x + 5)(2x^2 + 1)$
44. $u^2 - 18u + 81$
$(u - 9)^2$
45. $8v^2 - 14v^3 + v^4$
$v^2(v^2 - 14v + 8)$
46. $28 - 3m - m^2$
$-(m + 7)(m - 4)$
47. $18a^2 - 6ab + 42ac - 14bc$
$2(3a - b)(3a + 7c)$
48. $81r^4 - 256$
$(9r^2 + 16)(3r + 4)(3r - 4)$
49. $8a^2x^3 - 2b^2x$
$2x(2ax + b)(2ax - b)$
50. $12x^2 + 14x - 6$
$2(3x - 1)(2x + 3)$
51. $6x^2 - 14x + 8$
$2(3x - 4)(x - 1)$
52. $12x^2 - 12$
$12(x + 1)(x - 1)$
53. $4x^2y^2 + 4xy^2 + y^2$
$y^2(2x + 1)^2$
54. $81r^4s^2 - 24rs^5$
$3rs^2(3r - 2s)(9r^2 + 6rs + 4s^2)$
55. $4m^5 + 500m^2$
$4m^2(m + 5)(m^2 - 5m + 25)$
56. $ae + bf + af + be$
$(a + b)(f + e)$
57. $a^3 - 24 - 4a + 6a^2$
$(a + 6)(a + 2)(a - 2)$
58. $6x^2 - x - 16$
Prime
59. $4x^2 + 9y^2$
Prime
60. $x^4y + 216xy^4$
$xy(x + 6y)(x^2 - 6xy + 36y^2)$
61. $16a^5 - 54a^2$
$2a^2(2a - 3)(4a^2 + 6a + 9)$
62. $25x^2 - 16y^2$
$(5x + 4y)(5x - 4y)$
63. $27x - 27y - 27z$
$27(x - y - z)$
64. $12x^2 + 52x + 35$
$(6x + 5)(2x + 7)$
65. $xy - ty + xs - ts$
$(x - t)(y + s)$
66. $bc + b + cd + d$
$(b + d)(c + 1)$
67. $35x^8 - 2x^7 - x^6$
$x^6(7x + 1)(5x - 1)$
68. $x^3 - 25$
Prime
69. $5(x - 2) + 10y(x - 2)$
$5(x - 2)(1 + 2y)$
70. $16x^2 - 40x^3 + 25x^4$
$x^2(5x - 4)^2$
71. $49p^2 + 28pq + 4q^2$
$(7p + 2q)^2$
72. $16d^2 - 56dz + 49z^2$
$(4d - 7z)^2$
73. $4t^2 + 36$
$4(t^2 + 9)$
74. $r^5 + 3r^3 + 2r^2 + 6$
$(r^2 + 3)(r^3 + 2)$
75. $m^2n^2 - 9m^2 + 3n^2 - 27$
$(n + 3)(n - 3)(m^2 + 3)$
76. $z^2 + 6yz^2 + 9y^2z^2$
$z^2(1 + 3y)^2$

WRITING

77. Which factoring method do you find the most difficult? Why?
78. What four questions make up the factoring strategy for polynomials discussed in this section?
79. What does it mean to factor a polynomial?
80. How is a factorization checked?

REVIEW See AIE Appendix 3.

81. Graph the real numbers $-3, 0, 2$, and $-\frac{3}{2}$ on a number line.
82. Graph the interval $(-2, 3]$ on a number line.
83. Graph: $y = \frac{1}{2}x + 1$
84. Graph: $y < 2 - 3x$

CHALLENGE PROBLEMS

Factor using rational numbers.

85. $x^6 - 4x^3 - 12$ $(x^3 + 2)(x^3 - 6)$
86. $x(x - y) - y(y - x)$ $(x - y)(x + y)$
87. $24 - x^3 + 8x^2 - 3x$ $(8 - x)(3 + x^2)$
88. $25b^2 + 14b + \frac{49}{25}$ $\left(5b + \frac{7}{5}\right)^2$
89. $x^9 + y^6$ $(x^3 + y^2)(x^6 - x^3y^2 + y^4)$
90. $\frac{1}{4} - \frac{u^2}{81}$ $\left(\frac{1}{2} + \frac{u}{9}\right)\left(\frac{1}{2} - \frac{u}{9}\right)$
91. $x^4 - 13x^2 + 36$ $(x + 2)(x - 2)(x + 3)(x - 3)$
92. $x^4 - 2x^2 - 8$ $(x + 2)(x - 2)(x^2 + 2)$
93. $x^2y^2 - 6xy - 16$ $(xy + 2)(xy - 8)$
94. $5x + 4y + 25x^2 - 16y^2$ $(5x + 4y)(1 + 5x - 4y)$

SECTION 6.7 Solving Quadratic Equations by Factoring

OBJECTIVES

1 Define quadratic equations.

2 Solve quadratic equations using the zero-factor property.

3 Solve third-degree equations by factoring.

ARE YOU READY? *Are You Ready? exercises available online at www.webassign.net/brookscole*

The following problems review some basic skills that are needed when solving quadratic equations.

1. Evaluate: $0 \cdot 5$ 0
2. Fill in the blank: $8 \cdot 0 = 0$
3. Solve: $x + 4 = 0$ -4
4. Solve: $8x = 0$ 0
5. Factor: $x^2 - x - 6$ $(x - 3)(x + 2)$
6. Factor: $3n^2 - n - 2$ $(3n + 2)(n - 1)$

The factoring methods that we have discussed have many applications in algebra. In this section, we will use factoring to solve *quadratic equations.* These equations are different from those that we solved in Chapter 2. They contain a term in which the variable is raised to the second power, such as x^2 or t^2.

1 Define Quadratic Equations.

In a linear, or first degree equation, such as $2x + 3 = 8$, the exponent on the variable is an unwritten 1. A quadratic, or second degree equation, has a term in which the exponent on the variable is 2, and has no other terms of higher degree.

Quadratic Equations

A **quadratic equation** is an equation that can be written in the **standard form**

$$ax^2 + bx + c = 0$$

where a, b, and c represent real numbers, and $a \neq 0$.

The Language of Algebra

Quadratic equations involve the square of a variable, not the 4th power as *quad* might suggest. Why is this? Because the origin of the word **quadratic** is the Latin word *quadratus,* meaning square.

Some examples of quadratic equations are

$$x^2 - 2x - 63 = 0, \quad x^2 - 25 = 0, \quad 6x^2 - 12x = 0, \quad \text{and} \quad 2x^2 + 3x = 2$$

The first three equations are in standard form. To write the fourth equation in standard form, we subtract 2 from both sides to get $2x^2 + 3x - 2 = 0$.

Even though it does not have an x-term, the equation $x^2 - 25 = 0$ is a quadratic equation because the definition allows b, the coefficient of x, to equal 0. And even though the equation $6x^2 - 12x = 0$ does not have a constant term, it is also a quadratic equation, because the definition allows c, the constant term, to equal 0. However, a quadratic equation must have a variable-squared term, because the definition requires that $a \neq 0$.

2 Solve Quadratic Equations Using the Zero-Factor Property.

To **solve a quadratic equation,** we find all values of the variable that make the equation true. The methods that we used to solve linear equations in Chapter 2 cannot be used to solve a quadratic equation, because we cannot isolate the variable on one side of the equation. However, we often can solve quadratic equations using factoring and the following property of real numbers.

The Zero-Factor Property

When the product of two real numbers is 0, at least one of them is 0.
If a and b represent real numbers, and

$$\text{if } ab = 0, \text{ then } a = 0 \text{ or } b = 0$$

EXAMPLE 1 Solve: $(4x - 1)(x + 6) = 0$

Strategy We will set $4x - 1$ equal to 0 and $x + 6$ equal to 0 and solve each equation.

Why If the product of $4x - 1$ and $x + 6$ is 0, then, by the zero-factor property, $4x - 1$ must equal 0, or $x + 6$ must equal 0.

Solution If $(4x - 1)(x + 6) = 0$ is to be a true statement, then either

$$4x - 1 = 0 \quad \text{or} \quad x + 6 = 0$$

Caution

It would not be helpful to multiply $(4x - 1)$ and $(x + 6)$. We want the left side of the equation to be in factored form so that we can use the zero-factor property.

Now we solve each of these linear equations using the methods from Chapter 2.

$$4x - 1 = 0 \qquad \text{or} \qquad x + 6 = 0$$

$$4x = 1 \quad \text{Add 1 to both sides.} \qquad\qquad x = -6 \quad \text{Subtract 6 from both sides.}$$

$$x = \frac{1}{4} \quad \text{Divide both sides by 4.}$$

The results must be checked separately to see whether each of them produces a true statement. We substitute $\frac{1}{4}$ and then -6 for x in the original equation and evaluate the left side.

The Language of Algebra

In the zero-factor property, the word **or** means one or the other or both. If the product of two numbers is 0, then one factor is 0, or the other factor is 0, or both factors can be 0.

Check $\frac{1}{4}$**:**

$$(4x - 1)(x + 6) = 0$$

$$\left[4\left(\frac{1}{4}\right) - 1\right]\left(\frac{1}{4} + 6\right) \stackrel{?}{=} 0$$

$$(1 - 1)\left(\frac{25}{4}\right) \stackrel{?}{=} 0$$

$$0\left(\frac{25}{4}\right) \stackrel{?}{=} 0 \quad \text{The factor } 4x - 1 \text{ is 0 when } x \text{ is } \tfrac{1}{4}.$$

$$0 = 0 \quad \text{True}$$

Check **−6:**

$$(4x - 1)(x + 6) = 0$$

$$[4(-6) - 1](-6 + 6) \stackrel{?}{=} 0$$

$$(-24 - 1)(0) \stackrel{?}{=} 0 \quad \text{The factor } x + 6 \text{ is 0 when } x \text{ is } -6.$$

$$-25(0) \stackrel{?}{=} 0$$

$$0 = 0 \quad \text{True}$$

The resulting true statements indicate that $(4x - 1)(x + 6) = 0$ has two solutions: $\frac{1}{4}$ and -6. Recall from Chapter 2 that the *solution set* of an equation is the set of all numbers that make the equation true. Thus, the solution set is $\left\{-6, \frac{1}{4}\right\}$.

Self Check 1 Solve: $(x - 12)(5x + 6) = 0$ $\quad 12, -\frac{6}{5}$

Now Try Problem 15

Teaching Example 1 Solve: $(x + 11)(7x - 4) = 0$

Answer: $-11, \frac{4}{7}$

In Example 1, the left side of $(4x - 1)(x + 6) = 0$ is in factored form and the right side is 0, so we can immediately use the zero-factor property. However, to solve many quadratic equations, we must factor before using the zero-factor property.

EXAMPLE 2 Solve: $x^2 - 2x - 63 = 0$

Strategy We will factor the trinomial on the left side of the equation and use the zero-factor property.

Why To use the zero-factor property, we need one side of the equation to be factored completely and the other side to be 0.

Solution

$x^2 - 2x - 63 = 0$ This is the equation to solve.

$(x + 7)(x - 9) = 0$ Factor the trinomial, $x^2 - 2x - 63$.

$x + 7 = 0$ or $x - 9 = 0$ Set each factor equal to 0.

$x = -7$ | $x = 9$ Solve each equation using the methods from Chapter 2.

Success Tip

When you see the word **solve** in this example, you probably think of steps from Chapter 2 such as combining like terms, distributing, or doing something to both sides. However, to solve this quadratic equation, we begin by factoring $x^2 - 2x - 63$.

To check the results, we substitute -7 and then 9 for x in the original equation and evaluate the left side.

***Check* -7:**	***Check* 9:**
$x^2 - 2x - 63 = 0$	$x^2 - 2x - 63 = 0$
$(-7)^2 - 2(-7) - 63 \stackrel{?}{=} 0$	$(9)^2 - 2(9) - 63 \stackrel{?}{=} 0$
$49 - (-14) - 63 \stackrel{?}{=} 0$	$81 - 18 - 63 \stackrel{?}{=} 0$
$63 - 63 \stackrel{?}{=} 0$	$63 - 63 \stackrel{?}{=} 0$
$0 = 0$ True	$0 = 0$ True

The solutions of $x^2 - 2x - 63 = 0$ are -7 and 9, and the solution set is $\{-7, 9\}$.

Teaching Example 2 Solve: $x^2 - 3x + 40 = 0$
Answer: $-5, 8$

Self Check 2 Solve: $x^2 + 5x + 6 = 0$ $-2, -3$

Now Try Problem 27

The previous examples suggest the following strategy to solve quadratic equations by factoring.

The Factoring Method for Solving a Quadratic Equation

1. Write the equation in standard form: $ax^2 + bx + c = 0$ or $0 = ax^2 + bx + c$.
2. Factor completely.
3. Use the zero-factor property to set each factor equal to 0.
4. Solve each resulting equation.
5. Check the results in the original equation.

With this method, we factor *expressions* to solve *equations.*

EXAMPLE 3 Solve: $x^2 - 25 = 0$

Teaching Tip: Remind students that there was no such "= 0" requirement when solving linear equations in one variable earlier. For example, consider:
$5x + 2 = 27$ and
$2 - 4(x + 1) = 16x$

Strategy We will factor the binomial on the left side of the equation and use the zero-factor property.

Why To use the zero-factor property, we need one side of the equation to be factored completely and the other side to be 0.

Solution We factor the difference of two squares on the left side of the equation and proceed as follows.

Notation

Although $x^2 - 25 = 0$ is missing a term involving x, it is a quadratic equation in standard $ax^2 + bx + c = 0$ form, where $a = 1$, $b = 0$, and $c = -25$.

$x^2 - 25 = 0$		This is the equation to solve.
$(x + 5)(x - 5) = 0$		Factor the difference of two squares, $x^2 - 25$.
$x + 5 = 0$ or	$x - 5 = 0$	Set each factor equal to 0.
$x = -5$	$x = 5$	Solve each equation.

Check each result by substituting it into the original equation.

***Check* −5:**	***Check* 5:**
$x^2 - 25 = 0$	$x^2 - 25 = 0$
$(-5)^2 - 25 \stackrel{?}{=} 0$	$5^2 - 25 \stackrel{?}{=} 0$
$25 - 25 \stackrel{?}{=} 0$	$25 - 25 \stackrel{?}{=} 0$
$0 = 0$ True	$0 = 0$ True

The solutions of $x^2 - 25 = 0$ are -5 and 5, and the solution set is $\{-5, 5\}$.

Teaching Example 3 Solve: $x^2 - 100 = 0$
Answer: -10, 10

Self Check 3 Solve: $x^2 - 49 = 0$ $-7, 7$

Now Try Problem 35

EXAMPLE 4 Solve: $6x^2 = 12x$

Strategy We will subtract $12x$ from both sides of the equation to get 0 on the right side. Then we will factor the resulting binomial and use the zero-factor property.

Why To use the zero-factor property, we need one side of the equation to be factored completely and the other side to be 0.

Solution The equation is not in standard form, $ax^2 + bx + c = 0$. To get 0 on the right side, we proceed as follows.

Caution

A creative, but incorrect, approach to solve $6x^2 = 12x$ is to divide both sides by $6x$.

$$\frac{6x^2}{6x} = \frac{12x}{6x}$$

You will obtain $x = 2$; however, you will lose the second solution, 0.

$6x^2 = 12x$	This is the equation to solve.
$6x^2 \mathbf{- 12x} = 12x \mathbf{- 12x}$	Use the subtraction property of equality to get 0 on the right side: Subtract 12x from both sides.
$6x^2 - 12x = 0$	Combine like terms: $12x - 12x = 0$. Although it is missing a constant term, this is a quadratic equation in standard $ax^2 + bx + c = 0$ form, where $a = 6$, $b = -12$, and $c = 0$.

To solve this equation, we factor the left side and proceed as follows.

$6x(x - 2) = 0$		Factor out the GCF, 6x.
$6x = 0$ or	$x - 2 = 0$	Set each factor equal to 0.
$x = \frac{0}{6}$	$x = 2$	Solve each equation using the methods from Chapter 2.
$x = 0$		

The solutions are 0 and 2 and the solution set is $\{0, 2\}$. Check each solution in the original equation, $6x^2 = 12x$.

Teaching Example 4 Solve: $15x^2 = 45x$
Answer: 0, 3

Self Check 4 Solve: $5x^2 = 25x$ 0, 5

Now Try Problem 47

EXAMPLE 5 Solve: $2x^2 - 2 = -3x$

Strategy We will add $3x$ to both sides of the equation to get 0 on the right side. Then we will factor the resulting trinomial and use the zero-factor property.

Why To use the zero-factor property, we need one side of the equation to be factored completely and the other side to be 0.

Solution The equation is not in standard form, $ax^2 + bx + c = 0$. To get 0 on the right side, we proceed as follows.

Teaching Tip: You may want to mention that to solve a quadratic equation in standard form at this stage of the course, it must be possible to factor the expression on the left side. If we can't factor that expression, we can't solve the equation. Explain that there will be more about this topic in a later chapter.

$2x^2 - 2 = -3x$ — This is the equation to solve.

$2x^2 + \mathbf{3x} - 2 = -3x + \mathbf{3x}$ — Use the addition property of equality to get 0 on the right side: Add 3x to both sides.

$2x^2 + 3x - 2 = 0$ — Combine like terms: $-3x + 3x = 0$. This equation is in standard form.

$(2x - 1)(x + 2) = 0$ — Factor the trinomial.

$2x - 1 = 0$ or $x + 2 = 0$ — Set each factor equal to 0.

$2x = 1$ | $x = -2$ — Solve each equation using the methods from Chapter 2.

$x = \frac{1}{2}$

The solutions are $\frac{1}{2}$ and -2 and the solution set is $\left\{-2, \frac{1}{2}\right\}$. Check each solution in the original equation, $2x^2 - 2 = -3x$.

Teaching Example 5 Solve: $2x^2 - 8 = -15x$
Answer: $-8, \frac{1}{2}$

Self Check 5 Solve: $3x^2 - 8 = -10x$ $\frac{2}{3}, -4$

Now Try Problem 51

Unlike linear equations, quadratic equations have two solutions. In some cases, however, the two solutions are the same number.

EXAMPLE 6 Solve: $x(9x - 12) = -4$

Strategy To write the equation in standard form, we will distribute the multiplication by x and add 4 to both sides. Then we will factor the resulting trinomial and use the zero-factor property.

Why To use the zero-factor property, we need one side of the equation to be factored completely and the other side to be 0.

Solution

$x(9x - 12) = -4$ — This is the equation to solve.

$9x^2 - 12x = -4$ — Distribute the multiplication by x.

$9x^2 - 12x + \mathbf{4} = -4 + \mathbf{4}$ — To get 0 on the right side, add 4 to both sides.

$9x^2 - 12x + 4 = 0$ — Combine like terms: $-4 + 4 = 0$. This equation is in standard form.

$(3x - 2)(3x - 2) = 0$ — Factor the trinomial, $9x^2 - 12x + 4$.

$3x - 2 = 0$ or $3x - 2 = 0$ — Set each factor equal to 0.

$3x = 2$ | $3x = 2$ — Solve each equation using the methods from Chapter 2.

$x = \frac{2}{3}$ | $x = \frac{2}{3}$

Caution

To use the zero-factor property, one side of the equation must be 0. In this example, it would be incorrect to set each factor equal to -4.

~~$x = -4$ or $9x - 12 = -4$~~

If the product of two numbers is -4, one of them does not have to be -4. For example, $2(-2) = -4$.

After solving both equations, we see that $\frac{2}{3}$ is a **repeated solution.** Thus, the solution set is $\left\{\frac{2}{3}\right\}$. Check this result by substituting it into the original equation.

Teaching Example 6 Solve: $3x(3x - 10) = -25$
Answer: $\frac{5}{3}$ repeated

Self Check 6 Solve: $x(4x + 12) = -9$ $-\frac{3}{2}$ repeated

Now Try Problem 59

3 Solve Third-Degree Equations by Factoring.

Some equations involving polynomials with degrees higher than 2 also can be solved by using the factoring method. In such cases, we use an extension of the zero-factor property: When the product of two *or more* real numbers is 0, at least one of them is 0.

EXAMPLE 7 Solve: $6x^3 + 12x = 17x^2$

Strategy This equation is not quadratic, because it contains a term involving x^3. However, we can solve it by using factoring. First we get 0 on the right side by subtracting $17x^2$ from both sides. Then we factor the polynomial on the left side and use an extension of the zero-factor property.

Why To use the zero-factor property, we need one side of the equation to be factored completely and the other side to be 0.

Solution

$$6x^3 + 12x = 17x^2$$ This is the equation to solve.

$$6x^3 - \mathbf{17x^2} + 12x = 17x^2 - \mathbf{17x^2}$$ To get 0 on the right side, subtract $17x^2$ from both sides.

$$6x^3 - 17x^2 + 12x = 0$$ Combine like terms: $17x^2 - 17x^2 = 0$.

$$x(6x^2 - 17x + 12) = 0$$ Factor out the GCF, x.

$$x(2x - 3)(3x - 4) = 0$$ Factor the trinomial, $6x^2 - 17x + 12$.

The Language of Algebra

Since the highest degree of any term in $6x^3 + 12x = 17x^2$ is 3, it is called a **third-degree** equation. Note that it has three solutions.

If $x(2x - 3)(3x - 4) = 0$, then at least one of the factors is equal to 0.

$x = 0$ or $2x - 3 = 0$ or $3x - 4 = 0$ Set each factor equal to 0.

$2x = 3$ | $3x = 4$ Solve each equation.

$x = \frac{3}{2}$ | $x = \frac{4}{3}$

Teaching Tip: You may want to ask students if they see a relationship between the exponent on the highest-degree term in an equation and the number of real solutions the equation has.

The solutions are 0, $\frac{3}{2}$, and $\frac{4}{3}$ and the solution set is $\{0, \frac{4}{3}, \frac{3}{2}\}$. Check each solution in the original equation, $6x^3 + 12x = 17x^2$.

Teaching Example 7 Solve: $6x^3 - 11x^2 = 7x$
Answer: $0, -\frac{1}{2}, \frac{7}{3}$

Self Check 7 Solve: $10x^3 + x^2 = 2x$ $0, \frac{2}{5}, -\frac{1}{2}$

Now Try Problem 63

SECTION 6.7 STUDY SET

VOCABULARY

Fill in the blanks.

1. $2x^2 + 3x - 1 = 0$ and $x^2 - 36 = 0$ are examples of quadratic equations.
2. $ax^2 + bx + c = 0$ is called the standard form of a quadratic equation.
3. The zero-factor property states that if the product of two numbers is 0, at least one of them is 0: If $ab = 0$, then $a = 0$ or $b = 0$.
4. Since the highest degree of any term in $x^3 - 5x^2 - 6x = 0$ is 3, it is called a third-degree equation.

Selected exercises available online at www.webassign.net/brookscole

CONCEPTS

▶ 5. Which of the following are quadratic equations?
   a. $x^2 + 2x - 10 = 0$ Yes
   b. $2x - 10 = 0$ No
   c. $x^2 = 15x$ Yes
   d. $x^3 + x^2 + 2x = 0$ No

6. Write each equation in the standard form $ax^2 + bx + c = 0$.
   a. $x^2 + 2x = 6$ $x^2 + 2x - 6 = 0$
   b. $x^2 = 5x$ $x^2 - 5x = 0$
   c. $3x(x - 8) = -9$ $3x^2 - 24x + 9 = 0$
   d. $4x^2 = 25$ $4x^2 - 25 = 0$

7. Set $5x + 4$ equal to 0 and solve for x. $-\frac{4}{5}$

▶ 8. What step should be performed first to solve $x^2 - 6x - 16 = 0$? Factor $x^2 - 6x - 16$

9. What step (or steps) should be performed first before factoring is used to solve each equation?
   a. $x^2 + 7x = -6$ Add 6 to both sides.
   b. $x(x + 7) = 3$ Distribute the multiplication by x and subtract 3 from both sides.

▶ 10. Check to determine whether the given number is a solution of the given quadratic equation.
   a. $x^2 - 4x = 0$; 4 Yes
   b. $x^2 - 2x - 7 = 0$; -2 No

NOTATION

Complete each solution to solve the equation.

11. $(x - 1)(x + 7) = 0$
   $x - 1 = 0$ or $x + 7 = 0$
   $x = 1$ | $x = -7$

▶ 12. $7y^2 + 14y = 0$
   $7y(y + 2) = 0$
   $7y = 0$ or $y + 2 = 0$
   $y = 0$ | $y = -2$

▶ 13. $p^2 - p - 6 = 0$
   $(p - 3)(p + 2) = 0$
   $p - 3 = 0$ or $p + 2 = 0$
   $p = 3$ | $p = -2$

14. $4y^2 - 25 = 0$
   $(2y + 5)(2y - 5) = 0$
   $2y + 5 = 0$ or $2y - 5 = 0$
   $2y = -5$ | $2y = 5$
   $y = -\frac{5}{2}$ | $y = \frac{5}{2}$

GUIDED PRACTICE

Solve each equation. See Example 1.

15. $(x - 3)(x - 2) = 0$ 3, 2
▶ 16. $(x + 2)(x + 3) = 0$ $-2, -3$
17. $(x + 7)(x - 7) = 0$ $-7, 7$
▶ 18. $(x - 8)(x + 8) = 0$ $8, -8$
19. $6x(2x - 5) = 0$ $0, \frac{5}{2}$
▶ 20. $5x(5x + 7) = 0$ $0, -\frac{7}{5}$
21. $-7a(3a + 10) = 0$ $0, -\frac{10}{3}$
▶ 22. $-6t(2t - 9) = 0$ $0, \frac{9}{2}$
23. $t(t - 6)(t + 8) = 0$ $0, 6, -8$
▶ 24. $n(n + 1)(n - 6) = 0$ $0, -1, 6$
▶ 25. $(x - 1)(x + 2)(x - 3) = 0$ $1, -2, 3$
▶ 26. $(x + 2)(x + 3)(x - 4) = 0$ $-2, -3, 4$

Solve each equation. See Example 2.

27. $x^2 - 13x + 12 = 0$ 12, 1
▶ 28. $x^2 + 7x + 6 = 0$ $-1, -6$
29. $x^2 - 4x - 21 = 0$ $-3, 7$
▶ 30. $x^2 + 2x - 15 = 0$ $3, -5$
31. $x^2 - 9x + 8 = 0$ 8, 1
32. $x^2 - 14x + 45 = 0$ 9, 5
33. $a^2 + 8a + 15 = 0$ $-3, -5$
▶ 34. $a^2 - 17a + 60 = 0$ 5, 12

Solve each equation. See Example 3.

35. $x^2 - 81 = 0$ $-9, 9$
▶ 36. $x^2 - 36 = 0$ $-6, 6$
▶ 37. $t^2 - 25 = 0$ $-5, 5$
38. $m^2 - 49 = 0$ $-7, 7$
39. $4x^2 - 1 = 0$ $-\frac{1}{2}, \frac{1}{2}$
▶ 40. $9y^2 - 1 = 0$ $-\frac{1}{3}, \frac{1}{3}$
41. $9y^2 - 49 = 0$ $-\frac{7}{3}, \frac{7}{3}$
▶ 42. $16z^2 - 25 = 0$ $-\frac{5}{4}, \frac{5}{4}$

Solve each equation. See Example 4.

43. $w^2 = 7w$ 0, 7
▶ 44. $x^2 = 5x$ 0, 5
45. $s^2 = 16s$ 0, 16
▶ 46. $p^2 = 20p$ 0, 20
▶ 47. $4y^2 = 12y$ 0, 3
48. $5m^2 = 15m$ 0, 3
▶ 49. $3x^2 = -8x$ $0, -\frac{8}{3}$
50. $3s^2 = -4s$ $0, -\frac{4}{3}$

Solve each equation. See Example 5.

51. $3x^2 + 5x = 2$ $-2, \frac{1}{3}$
▶ 52. $3x^2 + 14x = -8$ $-4, -\frac{2}{3}$
53. $2x^2 + x = 3$ $-\frac{3}{2}, 1$
▶ 54. $2x^2 - 5x = -2$ $\frac{1}{2}, 2$
55. $5x^2 + 1 = 6x$ $\frac{1}{5}, 1$
▶ 56. $6x^2 + 1 = 5x$ $\frac{1}{3}, \frac{1}{2}$
57. $2x^2 - 3x = 20$ $-\frac{5}{2}, 4$
▶ 58. $2x^2 - 3x = 14$ $\frac{7}{2}, -2$

Solve each equation. See Example 6.

59. $4r(r + 7) = -49$ $-\frac{7}{2}$ repeated
▶ 60. $5m(5m + 8) = -16$ $-\frac{4}{5}$ repeated
61. $9a(a - 3) = 3a - 25$ $\frac{5}{3}$ repeated
▶ 62. $3x(3x + 10) = 6x - 16$ $-\frac{4}{3}$ repeated

Solve each equation. See Example 7.

63. $x^3 + 3x^2 + 2x = 0$ $0, -1, -2$
▶ 64. $x^3 - 7x^2 + 10x = 0$ 0, 5, 2
▶ 65. $k^3 - 27k - 6k^2 = 0$ $0, 9, -3$
66. $j^3 - 22j - 9j^2 = 0$ $0, 11, -2$

TRY IT YOURSELF

Solve each equation.

▶ 67. $4x^2 = 81$ $-\frac{9}{2}, \frac{9}{2}$
68. $9y^2 = 64$ $-\frac{8}{3}, \frac{8}{3}$

69. $x^2 - 16x + 64 = 0$
8 repeated

70. $h^2 + 2h + 1 = 0$
-1 repeated

71. $(2s - 5)(s + 6) = 0$
$\frac{5}{2}, -6$

72. $h(3h - 4)(h + 1) = 0$
$0, \frac{4}{3}, -1$

73. $3b^2 - 30b = 6b - 60$
2, 10

74. $2m^2 - 8m = 2m - 12$
2, 3

75. $k^3 + k^2 - 20k = 0$
$0, -5, 4$

76. $n^3 - 6n^2 + 8n = 0$
0, 2, 4

77. $x^2 - 100 = 0$
$-10, 10$

78. $z^2 - 25 = 0$
$-5, 5$

79. $z(z - 7) = -12$
3, 4

80. $p(p + 1) = 6$
$2, -3$

81. $3y^2 - 14y - 5 = 0$
$-\frac{1}{3}, 5$

82. $4y^2 - 11y - 3 = 0$
$-\frac{1}{4}, 3$

83. $(x - 2)(x^2 - 8x + 7) = 0$
2, 7, 1

84. $(x - 1)(x^2 + 5x + 6) = 0$
$1, -2, -3$

85. $(n + 8)(n - 3) = -30$
$-3, -2$

86. $(2s + 5)(s + 1) = -1$
$-\frac{3}{2}, -2$

87. $x^3 - 6x^2 = -9x$
0, 3 repeated

88. $m^3 - 8m^2 + 16m = 0$
0, 4 repeated

89. $4a^2 + 1 = 8a + 1$
0, 2

90. $3b^2 - 6 = 12b - 6$
0, 4

91. $2b(6b + 13) = -12$
$-\frac{2}{3}, -\frac{3}{2}$

92. $5f(5f - 16) = -15$
$3, \frac{1}{5}$

93. $3a^3 + 4a^2 + a = 0$
$0, -1, -\frac{1}{3}$

94. $10b^3 - 15b^2 - 25b = 0$
$0, -1, \frac{5}{2}$

95. $2x^3 = 2x(x + 2)$
$0, -1, 2$

96. $x^3 + 7x^2 = x^2 - 9x$
$0, -3$ repeated

97. $-15x^2 + 2 + 7x = 0$
$\frac{2}{3}, -\frac{1}{5}$

98. $-8x^2 + 3 - 10x = 0$
$\frac{1}{4}, -\frac{3}{2}$

99. $4p^2 - 121 = 0$
$-\frac{11}{2}, \frac{11}{2}$

100. $q^2 - \frac{1}{4} = 0$
$-\frac{1}{2}, \frac{1}{2}$

101. $d(8d - 9) = -1$
$\frac{1}{8}, 1$

102. $6n^3 - 6n = 0$
$0, -1, 1$

Look Alikes . . .

Factor the expression in part a and solve the equation in part b.

103. a. $x^2 + 4x - 21$
$(x + 7)(x - 3)$
b. $x^2 + 4x - 21 = 0$
$-7, 3$

104. a. $4a^2 - 8a$
$4a(a - 2)$
b. $4a^2 - 8a = 0$
0, 2

105. a. $12n^2 - 5n - 2$
$(4n + 1)(3n - 2)$
b. $12n^2 - 5n - 2 = 0$
$-\frac{1}{4}, \frac{2}{3}$

106. a. $x^2 - 36$
$(x + 6)(x - 6)$
b. $x^2 - 36 = 0$
$-6, 6$

WRITING

107. Explain the zero-factor property.

108. Find the error in the following solution.

$$x(x + 1) = 6$$
$$x = 6 \quad \text{or} \quad x + 1 = 6$$
$$x = 5$$

The solutions are 6 and 5.

109. A student solved $x^2 - 5x + 6 = 0$ and obtained two solutions: 2 and 3. Explain the error in his check.

Check:
$$x^2 - 5x + 6 = 0$$
$$2^2 - 5(3) + 6 \stackrel{?}{=} 0$$
$$4 - 15 + 6 \stackrel{?}{=} 0$$
$$-5 = 0 \quad \text{False}$$

2 is not a solution. 3 is not a solution.

110. In this section, we solved quadratic equations by factoring. Did we always obtain two different solutions? Explain.

111. What is wrong with the step used to solve $x^2 = 2x$ shown below?

$$x^2 = 2x$$
$$\frac{x^2}{x} = \frac{2x}{x}$$
$$x = 2$$

The solution is 2.

112. Explain the error in the following solution.

Factor: $x^2 - 5x + 6$
$$(x - 2)(x - 3) = 0$$
$$x - 2 = 0 \quad \text{or} \quad x - 3 = 0$$
$$x = 2 \qquad\qquad x = 3$$

The solutions are 2 and 3.

REVIEW

113. **Exercise.** A doctor advises a patient to exercise at least 15 minutes but less than 30 minutes per day. Use a compound inequality to express the range of these times t in minutes. 15 min $\le t <$ 30 min

114. **Snacks.** A bag of peanuts is worth \$0.30 less than the same size bag of cashews. Equal amounts of peanuts and cashews are used to make 40 bags of a mixture that is worth \$1.05 per bag. How much is a bag of cashews worth? \$1.20

CHALLENGE PROBLEMS

Solve each equation.

115. $x^4 - 625 = 0$ $\quad -5, 5$

116. $2a^3 + a^2 - 32a - 16 = 0$ $\quad -4, -\frac{1}{2}, 4$

117. $(x - 3)^2 = 2x + 9$ $\quad 0, 8$

118. $(x + 3)^2 = (2x - 1)^2$ $\quad -\frac{2}{3}, 4$

SECTION 6.8

Applications of Quadratic Equations

OBJECTIVES

1. Solve problems involving geometric figures.
2. Solve problems involving consecutive integers.
3. Solve problems using the Pythagorean theorem.
4. Solve problems given the quadratic equation model.

ARE YOU READY? *Are You Ready? exercises available online at www.webassign.net/brookscole*

The following problems review some basic skills that are needed when solving application problems involving quadratic equations.

1. What is the formula for the area of a rectangle? $A = lw$

2. What is the formula for the area of a triangle? $A = \frac{1}{2}bh$

3. Let x = a number. Write an algebraic expression that represents 1 more than the number. $x + 1$

4. How many sides does a triangle have? 3

In Chapter 2, we solved mixture, investment, and uniform motion problems. To model those situations, we used *linear equations* in one variable. We will now consider situations that are modeled by *quadratic equations.*

1 Solve Problems Involving Geometric Figures.

We can use the six-step problem-solving strategy and the factoring method for solving quadratic equations to find the dimensions of certain figures, given their area.

EXAMPLE 1

Painting. In 2002, the pop art painting *The American Sweetheart,* by artist Robert Indiana, sold for \$614,500. The area of the rectangular painting is 32 square feet. Find the dimensions of the painting if it is twice as long as it is wide.

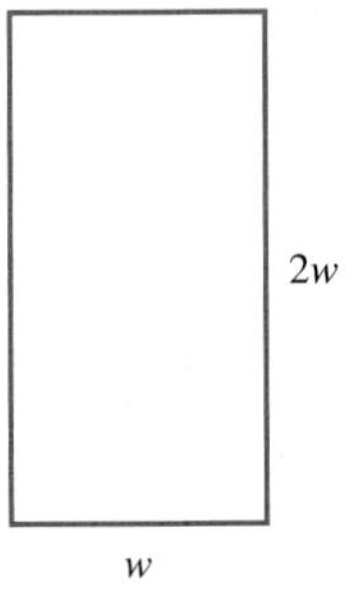

Analyze

- The area of the painting is 32 ft^2.
- The length is twice as long as the width.
- Find the length and width (the dimensions).

Assign Since the length is related to the width, let w = the width of the painting in feet. Then $2w$ = the length of the painting.

Form To form an equation, we use the formula for the area of a rectangle, $A = lw$, where $A = 32$.

The area of the rectangle	equals	the length	times	the width.
32	=	$2w$	$\cdot$	w

Solve

$32 = 2w \cdot w$	This is the equation to solve.
$32 = 2w^2$	Multiply $2w$ and w. This is a quadratic equation but it is not in standard form.
$0 = 2w^2 - 32$	To get 0 on the left side, subtract 32 from both sides.
$0 = 2(w^2 - 16)$	Factor out the GCF, 2.
$0 = 2(w + 4)(w - 4)$	Factor the difference of two squares, $w^2 - 16$.
$w + 4 = 0$ or $w - 4 = 0$	Since 2 cannot equal 0, discard that possibility. Set each factor that contains a variable equal to 0.
~~$w = -4$~~ $\quad$ $w = 4$	Solve each equation.

The Language of Algebra

When solving real-world application problems, we **discard** or **reject** any solutions of equations that do not make sense, such as a negative width of a painting.

Teaching Example 1 **Rugs.** A hall rug covers an area of 63 ft^2. Find the dimensions of the rug if it is seven times longer than it is wide.
Answer: Width: 3 ft; length: 21 ft

State The solutions of the equation are -4 and 4. Since w represents the width of the picture, and the width cannot be negative, we discard -4. Thus, the width of the picture is 4 feet and the length is $2 \cdot 4 = 8$ feet.

Check A rectangle with dimensions 4 feet by 8 feet has an area of 32 ft^2, and the length is twice the width. The answers check.

Self Check 1 **Geometry.** A rectangle has an area of 55 square meters. Its length is 1 meter more than twice its width. Find the dimensions of the rectangle. Width: 5 m; length: 11 m

Now Try Problem 13

EXAMPLE 2

Windmills. The height of a triangular canvas sail of a windmill is 1 foot less than twice the length of its base. If the sail has an area of 22.5 ft^2, find the length of the base and the height.

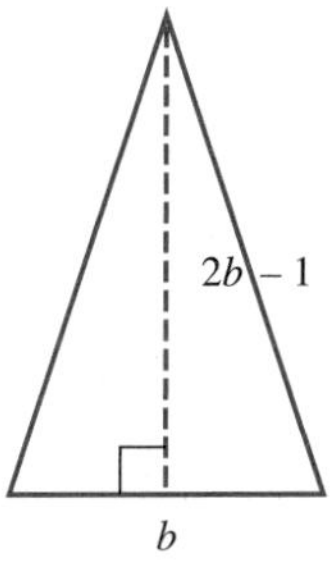

Analyze

- The height is 1 ft less than twice the length of the base.
- The area is 22.5 ft^2.
- Find the length of the base and the height.

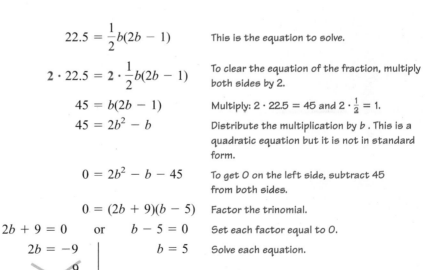

Assign Since the height is related to the length of the base, we let b = the length of the base of the sail in feet. Then $2b - 1$ = the height of the sail in feet.

Form To form an equation, we use the formula for the area of a triangle: $A = \frac{1}{2}bh$, where $A = 22.5$.

The area of the triangle	equals	one-half	times	the length of the base	times	the height.
22.5	=	$\frac{1}{2}$	$\cdot$	b	$\cdot$	$(2b - 1)$

Solve

$$22.5 = \frac{1}{2}b(2b - 1)$$ This is the equation to solve.

$$2 \cdot 22.5 = 2 \cdot \frac{1}{2}b(2b - 1)$$ To clear the equation of the fraction, multiply both sides by 2.

$$45 = b(2b - 1)$$ Multiply: $2 \cdot 22.5 = 45$ and $2 \cdot \frac{1}{2} = 1$.

$$45 = 2b^2 - b$$ Distribute the multiplication by b. This is a quadratic equation but it is not in standard form.

$$0 = 2b^2 - b - 45$$ To get 0 on the left side, subtract 45 from both sides.

$$0 = (2b + 9)(b - 5)$$ Factor the trinomial.

$$2b + 9 = 0 \quad \text{or} \quad b - 5 = 0$$ Set each factor equal to 0.

$$2b = -9 \qquad b = 5$$ Solve each equation.

$$b = -\frac{9}{2}$$ (crossed out)

Caution

A common error is to "distribute" the 2 incorrectly on the right side of the equation.

$$2 \cdot 22.5 = 2 \cdot \frac{1}{2}b(2b - 1)$$ (crossed out)

Teaching Tip: Warn students not to discard negative solutions unthinkingly if the instructions simply read "Solve: $2b^2 - b - 45$." Solutions are discarded when the value obtained does not make sense in a real-life situation.

State The solutions of the equation are $-\frac{9}{2}$ and 5. Since b represents the length of the base of the sail, and it cannot be negative, we discard $-\frac{9}{2}$. The length of the base is then 5 feet, and the height is $2(5) - 1 = 9$ feet.

Check A triangle with height 9 feet and base 5 feet has area $\frac{1}{2}(9)(5) = 22.5 \text{ ft}^2$, and the height is 1 foot less than twice the base. The answers check.

Teaching Example 2 Souvenirs. The height of a triangular pennant sold at a circus is 5 inches greater than the length of the base. If the area of the pennant is 42 in.^{2}, find the base and the height.
Answer: Base: 7 in.; height: 12 in.

Self Check 2 **Sailing.** The height of a triangular sail is 4 yards more than twice the length of the base. If the sail has an area of 15 yd^2, find the base and the height. Base: 3 yd; height: 10 yd

Now Try Problem 19

2 Solve Problems Involving Consecutive Integers.

Consecutive integers are integers that follow one another, such as 15 and 16. When solving consecutive integer problems, if we let $x =$ the first integer, then:

- two consecutive integers are x and $x + 1$
- two consecutive even integers are x and $x + 2$
- two consecutive odd integers are x and $x + 2$

EXAMPLE 3

Women's Tennis. In the 1998 Australian Open, sisters Venus and Serena Williams played against each other for the first time as professionals. Venus was victorious over her younger sister. At that time, their ages were consecutive integers whose product was 272. How old were Venus and Serena when they met in this match?

Analyze

- Venus is older than Serena.
- Their ages were consecutive integers.
- The product of their ages was 272.
- Find Venus' and Serena's ages when they played this match.

Assign Let $x =$ Serena's age when she played in the 1998 Australian Open. Since their ages were consecutive integers, and since Venus is older, we let $x + 1 =$ Venus' age.

Form The word *product* indicates multiplication.

Serena's age	times	Venus' age	was	272.
x	$\cdot$	$(x + 1)$	$=$	272

Solve

$$x(x + 1) = 272$$

$$x^2 + x = 272$$ Distribute the multiplication by x. This is a quadratic equation but it is not in standard form.

$$x^2 + x - 272 = 0$$ Subtract 272 from both sides to make the right side 0.

$$(x + 17)(x - 16) = 0$$ Factor $x^2 + x - 272$. Two numbers whose product is -272 and whose sum is 1 are 17 and -16.

$$x + 17 = 0 \quad \text{or} \quad x - 16 = 0$$ Set each factor equal to 0.

$$\cancel{x = -17} \quad\Big|\quad x = 16$$ Solve each equation.

State The solutions of the equation are -17 and 16. Since x represents Serena's age, and it cannot be negative, we discard -17. Thus, Serena Williams was 16 years old and Venus Williams was $16 + 1 = 17$ years old when they played against each other for the first time as professionals.

Success Tip

The prime factorization of 272 is helpful in determining that $272 = 17 \cdot 16$.

$$16\left\{\begin{array}{l} 2\,\underline{|272} \\ \;\;2\,\underline{|136} \\ \;\;\;\;2\,\underline{|68} \\ \;\;\;\;\;\;2\,\underline{|34} \end{array}\right.$$
17

Check Since 16 and 17 are consecutive integers, and since $16 \cdot 17 = 272$, the answers check.

Teaching Example 3 Classrooms. Two science classroom numbers are consecutive odd integers whose product is 255. Find the room numbers.
Answer: 15 and 17

Self Check 3 The product of two consecutive integers is 552. Find the integers. 23 and 24

Now Try Problem 23

3 Solve Problems Using the Pythagorean Theorem.

Caution

It doesn't matter which leg of the right triangle is labeled a and which is labeled b. However, the hypotenuse must be labeled c.

A **right triangle** is a triangle that has a 90° (right) angle. The longest side of a right triangle is the **hypotenuse,** which is the side opposite the right angle. The remaining two sides are the **legs** of the triangle. The **Pythagorean theorem** provides a formula relating the lengths of the three sides of a right triangle.

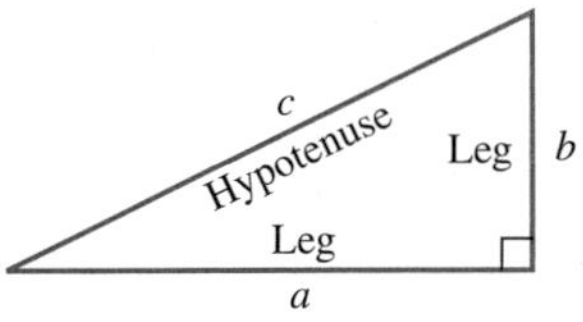

The Pythagorean Theorem

If a and b are the lengths of the legs of a right triangle and c is the length of the hypotenuse, then

$$a^2 + b^2 = c^2$$

In a right triangle, the sum of the squares of the lengths of the two legs is equal to the square of the length of the hypotenuse.

EXAMPLE 4

The Language of Algebra

A **theorem** is a mathematical statement that can be proved. The *Pythagorean theorem* is named after *Pythagoras*, a Greek mathematician who lived about 2,500 years ago. He is thought to have been the first to prove the theorem.

Right Triangles. The longer leg of a right triangle is 3 units longer than the shorter leg. If the hypotenuse is 6 units longer than the shorter leg, find the lengths of the sides of the triangle.

Analyze We begin by drawing a right triangle and labeling the legs and the hypotenuse.

Assign Let a = the length of the shorter leg. Then the length of the hypotenuse is $a + 6$ and the length of the longer leg is $a + 3$.

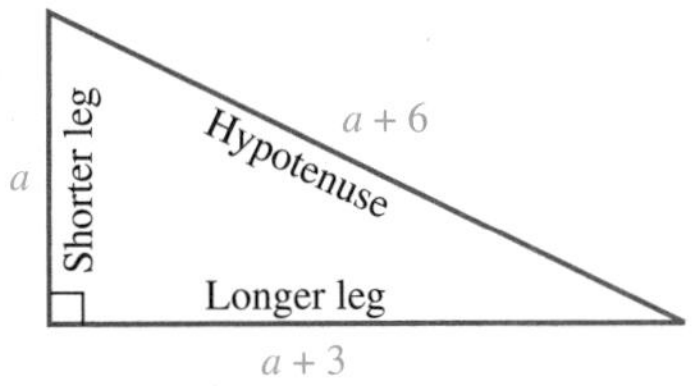

Form By the Pythagorean theorem, we have

The length of the shorter leg, squared	plus	the length of the longer leg, squared	equals	the length of the hypotenuse, squared
a^2	$+$	$(a+3)^2$	$=$	$(a+6)^2$

Solve

$$a^2 + (a + 3)^2 = (a + 6)^2$$

$a^2 + a^2 + 6a + 9 = a^2 + 12a + 36$ Find $(a + 3)^2$ and $(a + 6)^2$. Don't forget the middle terms.

$2a^2 + 6a + 9 = a^2 + 12a + 36$ On the left side: $a^2 + a^2 = 2a^2$. This is a quadratic equation but it is not in standard form.

$a^2 - 6a - 27 = 0$ To get 0 on the right side, subtract a^2, $12a$, and 36 from both sides. This is a quadratic equation.

$(a - 9)(a + 3) = 0$ Factor the trinomial.

$a - 9 = 0$ or $a + 3 = 0$ Set each factor equal to 0.

$a = 9$ | ~~$a = -3$~~ Solve each equation.

State Since a side cannot have a negative length, we discard the solution -3. Thus, the shorter leg is 9 units long, the hypotenuse is $9 + 6 = 15$ units long, and the longer leg is $9 + 3 = 12$ units long.

Check The longer leg, 12, is 3 units longer than the shorter leg, 9. The hypotenuse, 15, is 6 units longer than the shorter leg, 9, and the side lengths satisfy the Pythagorean theorem. So the results check.

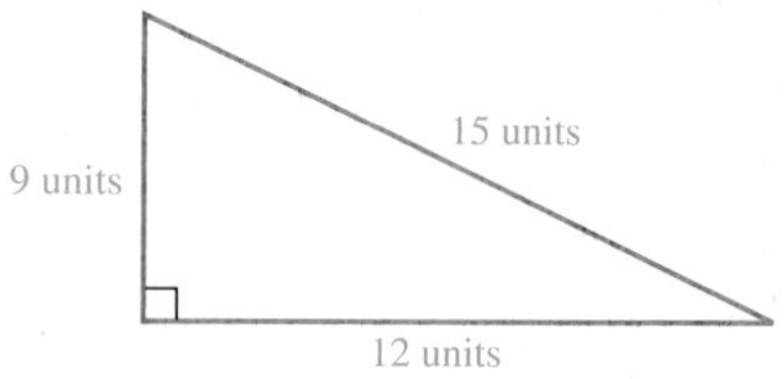

$$9^2 + 12^2 \stackrel{?}{=} 15^2$$
$$81 + 144 \stackrel{?}{=} 225$$
$$225 = 225$$

Teaching Example 4 Right Triangles. The longer leg of a right triangle is 7 cm longer than the shorter leg. If the hypotenuse is 8 cm longer than the shorter leg, find the lengths of the sides of the triangle.
Answer: 5 cm, 12 cm, and 13 cm

Self Check 4 **Right Triangles.** The longer leg of a right triangle is 7 inches longer than the shorter leg. If the hypotenuse is 9 inches longer than the shorter leg, find the lengths of the sides of the triangle. 8 in., 15 in., and 17 in.

Now Try Problem 31

4 Solve Problems Given the Quadratic Equation Model.

A quadratic equation can be used to describe the height of an object that is projected upward, such as a ball thrown into the air or an arrow shot into the sky.

EXAMPLE 5 **College Pranks.** A student uses rubber tubing to launch a water balloon from the roof of his dormitory. The height h (in feet) of the balloon, t seconds after being launched, is approximated by the formula $h = -16t^2 + 48t + 64$. After how many seconds will the balloon hit the ground?

Analyze When the water balloon hits the ground, its height will be 0 feet.

Assign To find the time that it takes for the balloon to hit the ground, we set h equal to 0, and solve the quadratic equation for t, the time.

Form

$h = -16t^2 + 48t + 64$

$0 = -16t^2 + 48t + 64$ Substitute 0 for the height, h. This is a quadratic equation.

Solve

$0 = -16t^2 + 48t + 64$ This is the equation to solve.

$0 = -16(t^2 - 3t - 4)$ Factor out the opposite of the GCF, -16.

$0 = -16(t + 1)(t - 4)$ Factor the trinomial.

$t + 1 = 0$ or $t - 4 = 0$ Since -16 cannot equal 0, discard that possibility. Set each factor that contains a variable equal to 0.

~~$t = -1$~~ $\quad$ $t = 4$ Solve each equation.

Success Tip

Note that the common factor, -16, divides -16, 48, and 64 exactly:

$$\frac{-16}{-16} = 1 \qquad \frac{48}{-16} = -3$$
$$\frac{64}{-16} = -4$$

State The equation has two solutions, -1 and 4. Since t represents time, and, in this case, time cannot be negative, we discard -1. The second solution, 4, indicates that the balloon hits the ground 4 seconds after being launched.

Check Check this result by substituting 4 for t in $h = -16t^2 + 48t + 64$. You should get $h = 0$.

Teaching Example 5 **Baseball.** A pitcher can throw a fast ball at 79 feet per second. If he throws the ball into the air with that velocity, its height h in feet, t seconds after being released, is approximated by $h = -16t^2 + 79t + 5$. After the ball is thrown, in how many seconds will it hit the ground?
Answer: 5 sec

Self Check 5 **Archery.** An arrow is shot into the air from a balcony. The height h (in feet) of the tip of the arrow, t seconds after being shot, is approximated by $h = -16t^2 + 77t + 15$. After how many seconds will the arrow hit the ground? 5 sec

Now Try Problem 37

SECTION 6.8 STUDY SET

VOCABULARY

Fill in the blanks.

1. Integers that follow one another, such as 6 and 7, are called consecutive integers.
2. A right triangle is a triangle that has a 90° angle.
3. The longest side of a right triangle is the hypotenuse. The remaining two sides are the legs of the triangle.
4. The Pythagorean theorem is a formula that relates the lengths of the three sides of a right triangle.

CONCEPTS

5. A rectangle has an area of 40 in.^{2}. The length is 3 inches longer than the width. Which rectangle below meets these conditions? ii.

i. 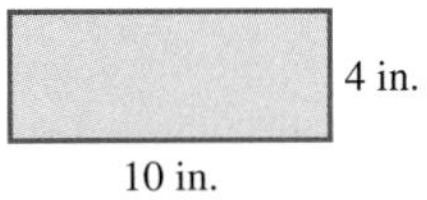

ii. 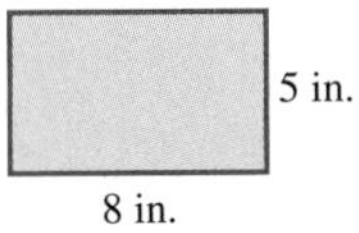

6. A triangle has an area of 15 ft^2. The height is 7 feet less than twice the length of the base. Which triangle below meets these conditions? i.

i. 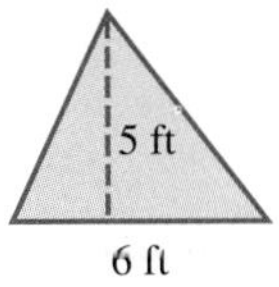

ii. 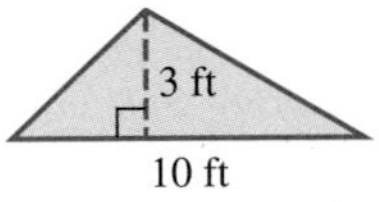

7. Multiply both sides of the equation by 2. ***Do not solve.***

$10 = \frac{1}{2}b(b + 5)$ $\quad 20 = b(b + 5)$

8. Fill in the blanks.
   a. If the length of the hypotenuse of a right triangle is c and the lengths of the other two legs are a and b, then $a^2 + b^2 = c^2$.
   b. In a right triangle, the sum of the squares of the lengths of the two legs is equal to the square of the length of the hypotenuse.

9. a. What kind of triangle is shown? A right triangle
   b. What are the lengths of the legs of the triangle? x ft; $(x + 1)$ ft
   c. How much longer is the hypotenuse than the shorter leg? 9 ft

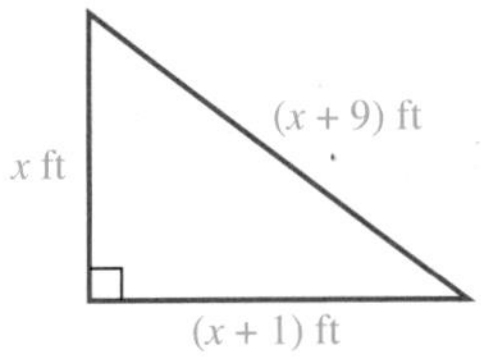

10. A ball is thrown into the air. Its height h in feet, t seconds after being released, is given by the formula $h = -16t^2 + 24t + 6$. When the ball hits the ground, what is the value of h? 0

NOTATION

Complete the solution to solve the equation.

11. $0 = -16t^2 + 32t + 48$

$0 = -16(t^2 - 2t - 3)$

$0 = -16(t - 3)(t + 1)$

$t - 3 = 0$ or $t + 1 = 0$

$t = 3$ $\quad$ $t = -1$

12. Fill in the blanks.
   a. Consecutive integers can be represented by x and $x + 1$.
   b. Consecutive odd integers can be represented by x and $x + 2$.
   c. Consecutive even integers can be represented by x and $x + 2$.

APPLICATIONS

Geometry Problems

13. **Flags.** The length of the flag of Australia is twice as long as the width. Find the dimensions of an Australian flag if its area is 18 ft^2. Width: 3 ft; length: 6 ft

14. **Billiards.** Pool tables are rectangular, and their length is twice the width. Find the dimensions of a pool table if it occupies 50 ft^2 of floor space. Width: 5 ft; length: 10 ft

15. X-Rays. A rectangular-shaped x-ray film has an area of 80 square inches. The length is 2 inches longer than the width. Find its width and length. 8 in., 10 in.

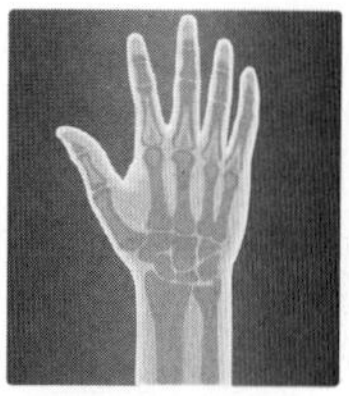

16. Insulation. The area of the rectangular slab of foam insulation in the illustration is 36 square meters. Find the dimensions of the slab. 4 m by 9 m

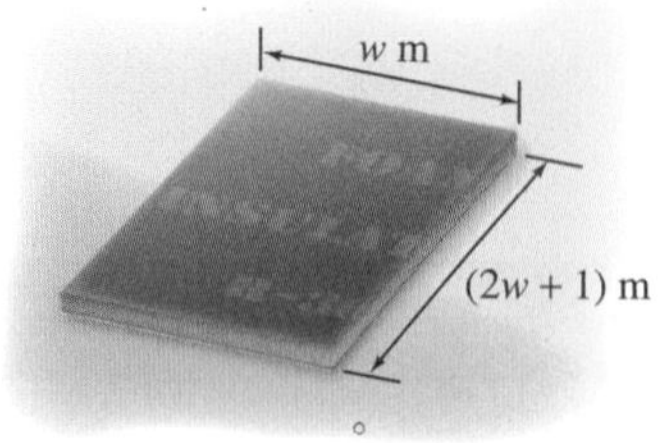

17. from Campus to Careers

Elementary School Teacher

©iStockphoto.com/Neustockimages

Suppose you are an elementary school teacher. You want to order a rectangular bulletin board to mount on a classroom wall that has an area of 90 square feet. Fire code requirements allow for no more than 30% of a classroom wall to be covered by a bulletin board. If the length of the board is to be three times as long as the width, what are the dimensions of the largest bulletin board that meets fire code? 3 ft by 9 ft

18. Tubing. Refer to the diagram below. A piece of cardboard in the shape of a parallelogram is twisted to form the tube. The parallelogram has an area of 60 square inches. If its height h is 7 inches more than the length of the base b, what is the length of the base? (*Hint:* The formula for the area of a parallelogram is $A = bh$.) 5 in.

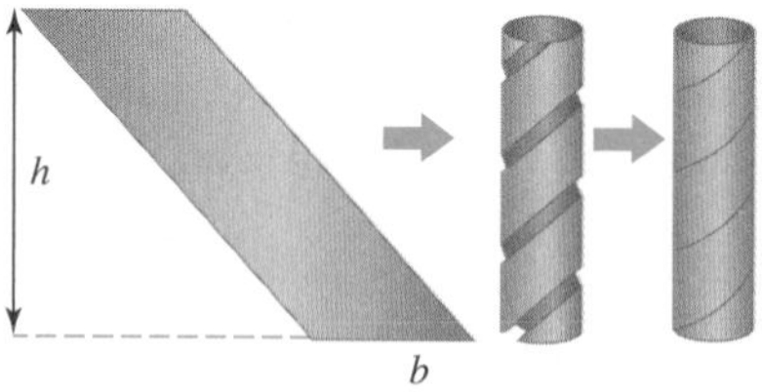

19. Jeans. The height of the triangular-shaped logo on a pair of jeans is 1 centimeter less than the length of its base. If the area of the logo is 15 square centimeters, find the length of the base and the height. Base: 6 cm; height: 5 cm

20. Shuffleboard. The area of the numbered triangle on a shuffleboard court is 27 ft^2. Its height is 3 feet more than the length of the base. Find the length of the base and the height. Base: 6 ft; height: 9 ft

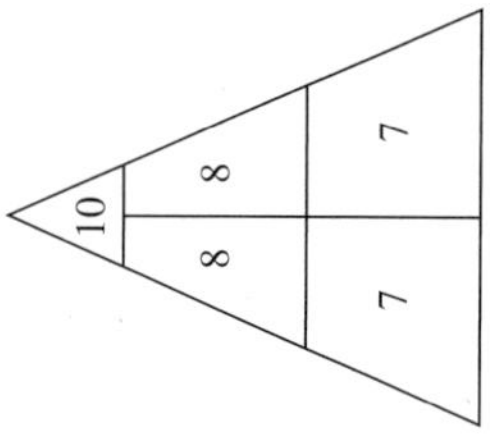

21. Sailboats. Refer to the diagram of a sail shown here. The length of the *luff* is 3 times longer than the length of the *foot* of the sail. Find the length of the foot and the length of the luff. Foot: 4 ft; luff: 12 feet

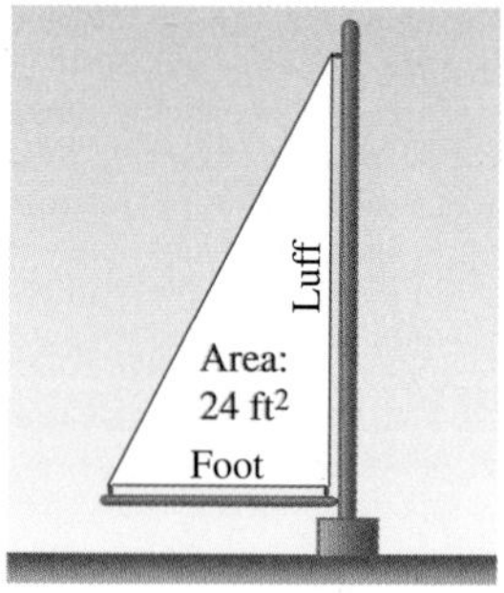

22. Designing Tents. The length of the base of the triangular sheet of canvas above the door of a tent is 2 feet more than twice its height. The area is 30 square feet. Find the height and the length of the base of the triangle. 5 ft, 12 ft

Consecutive Integer Problems

23. NASCAR. The car numbers of drivers Kasey Kahne and Scott Riggs are consecutive positive integers whose product is 90. If Kahne's car number is the smaller, what is the number of each car? Kahne: 9; Riggs: 10

24. Baseball. Catcher Thurman Munson and pitcher Whitey Ford are two of the sixteen New York Yankees who have had their uniform numbers retired. Their uniform numbers are consecutive integers whose product is 240. If Munson's was the smaller number, determine the uniform number of each player. Munson: 15; Ford: 16

25. Customer Service. At a pharmacy, customers take a ticket to reserve their turn for service. If the product of the ticket number now being served and the next ticket number to be served is 156, what number is now being served? 12

26. History. Delaware was the first state to enter the Union and Hawaii was the 50th. If we order the positions of entry for the rest of the states, we find that Kentucky entered the Union right after Vermont, and the product of their order-of-entry numbers is 210. Use the given information to complete these statements:

Kentucky was the 15 th state to enter the Union.

Vermont was the 14 th state to enter the Union.

27. Plotting Points. The x-coordinate and y-coordinate of a point in quadrant I are consecutive odd integers whose product is 143. The x-coordinate is less than the y-coordinate. Find the coordinates of the point. (11, 13)

28. Presidents. George Washington was born on 2-22-1732 (February 22, 1732). He died in 1799 at the age of 67. The month in which he died and the day of the month on which he died are consecutive even integers whose product is 168. When did Washington die? 12-14-1799

Pythagorean Theorem Problems

29. High-Ropes Adventures Courses. A builder of a high-ropes adventure course wants to secure a pole by attaching a support cable from the anchor stake 8 yards from its base to a point 6 yards up the pole. How long should the cable be? 10 yd

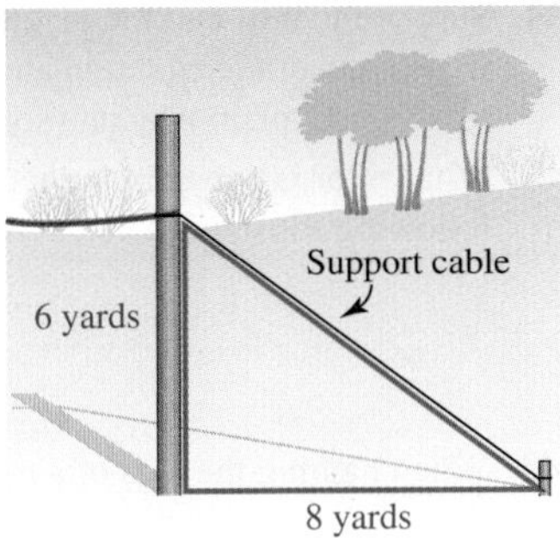

30. Wind Damage. A tree was blown over in a wind storm. Find x. Then find the height of the tree when it was standing upright. 6 ft, 16 ft

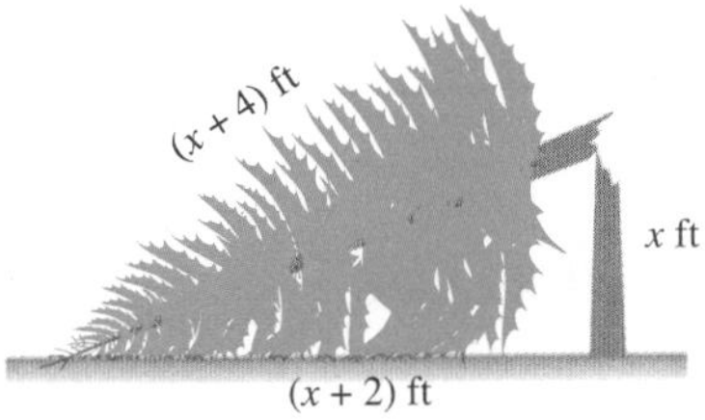

31. Moto X. Find x, the height of the landing ramp. 8 ft

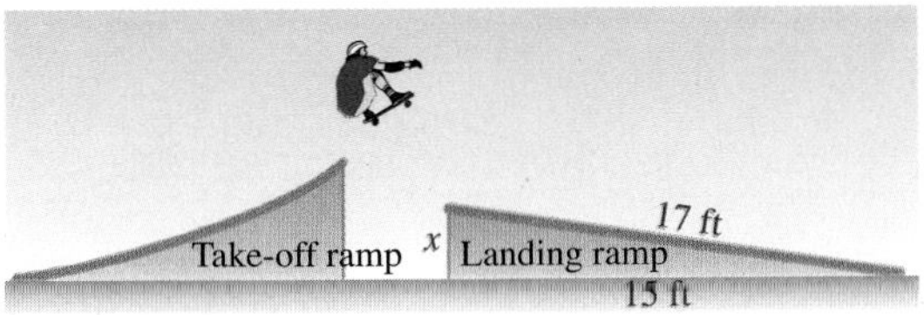

32. Gardening Tools. The dimensions (in millimeters) of the teeth of a pruning saw blade are given in the illustration. Find each length. 3 mm, 4 mm, 5 mm

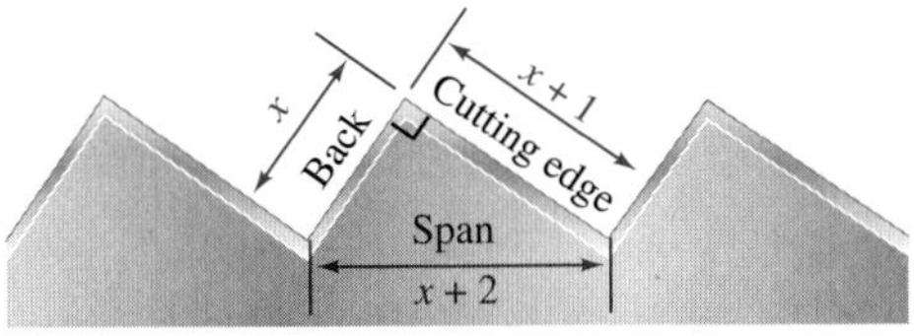

33. Boating. The inclined ramp of the boat launch shown in the next column is 8 meters longer than the rise of the ramp. The run is 7 meters longer than the rise. How long are the three sides of the ramp? 5 m, 12 m, 13 m

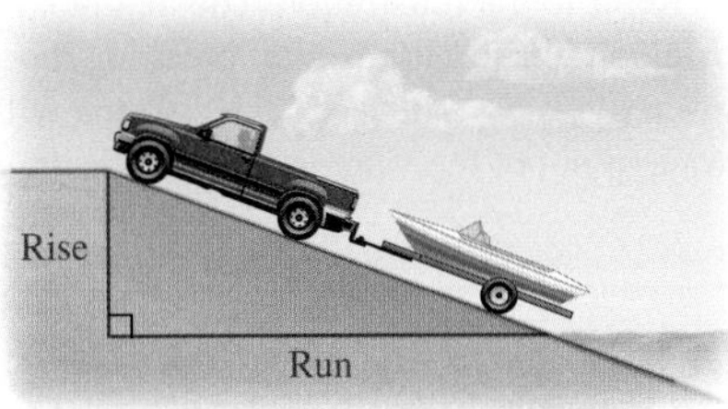

34. Car Repairs. To create some space to work under the front end of a car, a mechanic drives it up steel ramps. A ramp is 1 foot longer than the back, and the base is 2 feet longer than the back of the ramp. Find the length of each side of the ramp. 3 ft, 4 ft, 5 ft

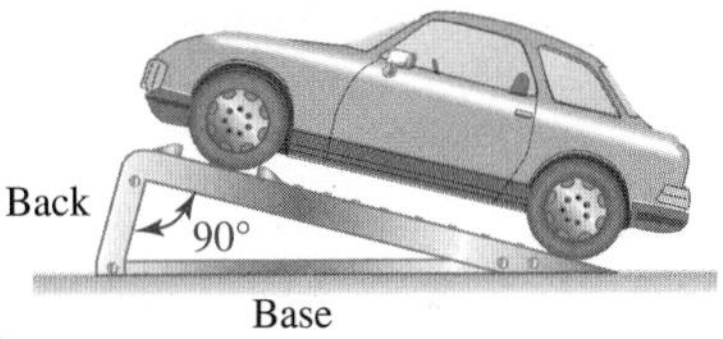

Quadratic Equation Model Problems

35. Thrill Rides. At the peak of a roller coaster ride, a rider's sunglasses fly off his head. The height h (in feet) of the glasses, t seconds after he loses them, is given by $h = -16t^2 + 64t + 80$. After how many seconds will the glasses hit the ground? (*Hint:* Factor out -16.) 5 sec

36. Parades. A celebrity on the top of a parade float is tossing pieces of candy to the people on the street below. The height h (in feet) of a piece of candy, t seconds after being thrown, is given by $h = -16t^2 + 16t + 32$. After how many seconds will the candy hit the ground? (*Hint:* Factor out -16.) 2 sec

37. Softball. A pitcher can throw a fastball underhand at 63 feet per second (about 45 mph). If she throws a ball into the air with that velocity, its height h in feet, t seconds after being released, is given by $h = -16t^2 + 63t + 4$. After the ball is thrown, in how many seconds will it hit the ground? (*Hint:* Factor out -1.) 4 sec

38. Officiating. Before a football game, a coin toss is used to determine which team will kick off. The height h (in feet) of a coin above the ground t seconds after being flipped up into the air is given by $h = -16t^2 + 22t + 3$. How long does a team captain have to call heads or tails if it must be done while the coin is in the air? (*Hint:* Factor out -1.) $\frac{3}{2} = 1.5$ sec

39. Dolphins. Refer to the illustration. The height h in feet reached by a dolphin t seconds after breaking the surface of the water is given by $h = -16t^2 + 32t$. How long will it take the dolphin to jump out of the water and touch the trainer's hand? 1 sec

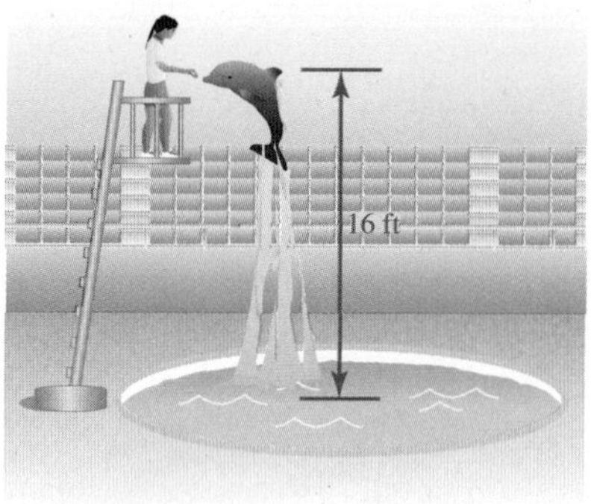

40. Exhibition Diving. In Acapulco, Mexico, men diving from a cliff to the water 64 feet below are quite a tourist attraction. A diver's height h above the water (in feet), t seconds after diving, is given by $h = -16t^2 + 64$. How long does a dive last? 2 sec

41. Choreography. For the finale of a musical, 36 dancers are to assemble in a triangular-shaped series of rows, where each row has one more dancer than the previous row. The illustration shows the beginning of such a formation. The relationship between the number of rows r and the number of dancers d is given by $d = \frac{1}{2}r(r + 1)$. Determine the number of rows in the formation. 8

42. Crafts. The illustration shows how a wall hanging can be created by stretching yarn from peg to peg across a wooden ring. The relationship between the number of pegs p placed evenly around the ring and the number of yarn segments s that criss-cross the ring is given by the formula $s = \frac{p(p - 3)}{2}$. How many pegs are needed if the designer wants 27 segments to criss-cross the ring? (*Hint:* Multiply both sides of the equation by 2.) 9

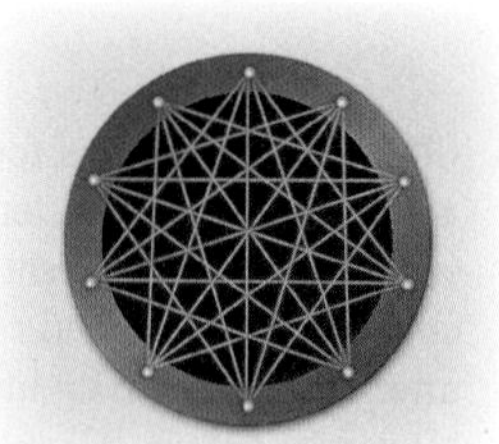

WRITING

43. A student was asked to solve the following problem: *The length of a rectangular room is 2 feet more than twice the width. If the area of the room is 60 square feet, find its dimensions.* Here is the student's solution:

    Since $10 \cdot 6 = 60$, the length of the room is 10 feet and the width is 6 feet.

    Explain why his solution is incorrect.

44. Suppose that to find the length of the base of a triangle, you write a quadratic equation and solve it to find $b = 6$ or $b = -8$. Explain why one solution should be discarded.

45. What error is apparent in the following illustration?

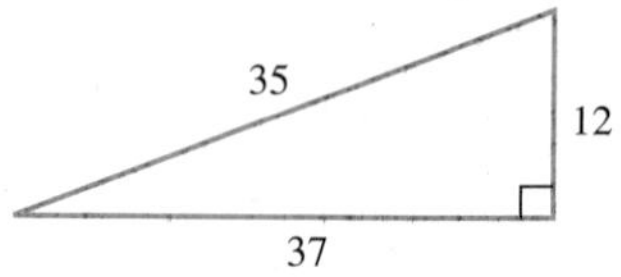

46. When naming the legs of a right triangle, explain why it doesn't matter which leg you label a and which leg you label b.

REVIEW

Find each special product.

47. $(5b - 2)^2$ $25b^2 - 20b + 4$

48. $(2a + 3)^2$ $4a^2 + 12a + 9$

49. $(s^2 + 4)^2$ $s^4 + 8s^2 + 16$

50. $(m^2 - 1)^2$ $m^4 - 2m^2 + 1$

51. $(9x + 6)(9x - 6)$ $81x^2 - 36$

52. $(5b + 2)(5b - 2)$ $25b^2 - 4$

CHALLENGE PROBLEMS

53. Pool Borders. The owners of a 10-meter-wide by 25-meter-long rectangular swimming pool want to surround the pool with a crushed-stone border of uniform width. They have enough stone to cover 74 square meters. How wide should they make the border? 1 m

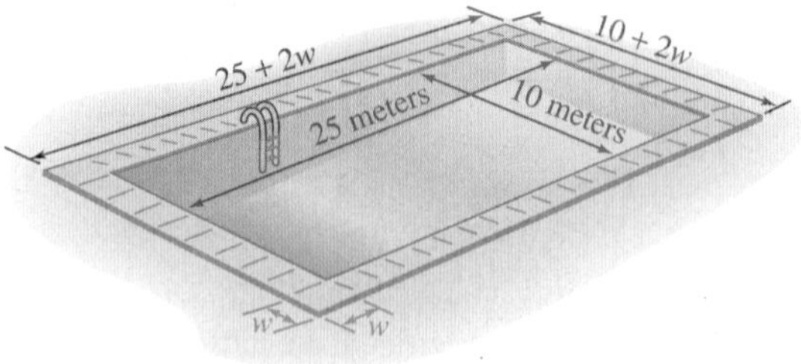

54. Find h. 12 in.

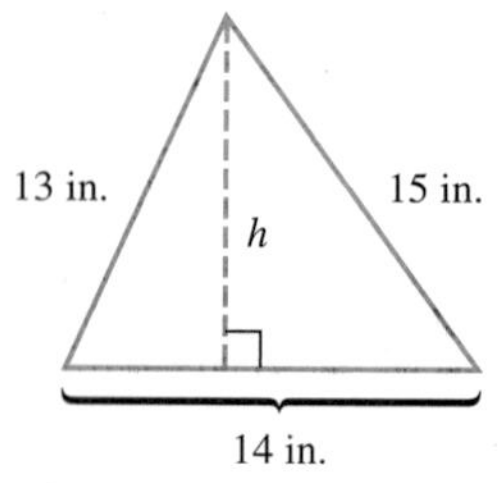

6 Summary & Review

SECTION 6.1 The Greatest Common Factor; Factoring by Grouping

DEFINITIONS AND CONCEPTS	EXAMPLES
Factoring is multiplication reversed. To **factor a polynomial** means to express it as a product of two (or more) polynomials.	Multiplication: Given the factors, we find a polynomial. ⟶ $2x(5x + 3) = 10x^2 + 6x$ ⟵ Factoring: Given a polynomial, we find the factors.
A natural number is in **prime-factored form** when it is written as the product of prime numbers.	The prime-factored form of 28 is $2 \cdot 2 \cdot 7 = 2^2 \cdot 7$.
To find the **greatest common factor, GCF,** of a list of terms **1.** Write each coefficient as a product of prime factors. **2.** Identify the numerical and variable factors common to each term. **3.** Multiply the common numerical and variable factors identified in step 2 to obtain the GCF. If there are no common factors, the GCF is 1.	Find the GCF of $35x^4$, $63x^3$, and $42x^2$. $\left.\begin{aligned} 35x^4 &= 5 \cdot 7 \cdot x \cdot x \cdot x \cdot x \\ 63x^3 &= 3 \cdot 3 \cdot 7 \cdot x \cdot x \cdot x \\ 42x^2 &= 2 \cdot 3 \cdot 7 \cdot x \cdot x \end{aligned}\right\}$ GCF $= 7 \cdot x \cdot x = 7x^2$
The first step of factoring a polynomial is to see whether the terms of the polynomial have a common factor. If they do, **factor out the GCF.**	Factor: $35x^4 + 63x^3 - 42x^2$ $= 7x^2(5x^2 + 9x - 6)$ Factor out the GCF, $7x^2$. Use multiplication to check the factorization: $7x^2(5x^2 + 9x - 6) = 35x^4 + 63x^3 - 42x^2$ This is the original polynomial.
If a polynomial has four terms, try **factoring by grouping.** **1.** Group the terms of the polynomial so that the first two terms have a common factor and the last two terms have a common factor. **2.** Factor out the common factor from each group. **3.** Factor out the resulting common binomial factor. If there is no common binomial factor, regroup the terms of the polynomial and repeat steps 2 and 3.	Factor: $ax - bx + ay - by$ $= x(a - b) + y(a - b)$ Factor out x from $ax - bx$ and y from $ay - by$. $= (a - b)(x + y)$ Factor out the common binomial factor, $(a - b)$.

REVIEW EXERCISES

Find the prime-factorization of each number.

1. 35 $5 \cdot 7$

2. 96 $2^5 \cdot 3$

Find the GCF of each list.

3. 28 and 35 7

4. $36a^4$, $54a^3$, and $126a^6$ $18a^3$

Factor.

5. $3x + 9y$ $3(x + 3y)$

6. $5ax^2 + 15a$ $5a(x^2 + 3)$

7. $7s^5 + 14s^3$ $7s^3(s^2 + 2)$

8. $\pi ab - \pi ac$ $\pi a(b - c)$

9. $24x^3 + 60x^2 - 48x$ $12x(2x^2 + 5x - 4)$

10. $x^5y^3z^2 + xy^5z^3 - xy^3z^2$ $xy^3z^2(x^4 + y^2z - 1)$

11. $-5ab^2 + 10a^2b - 15ab$ $-5ab(b - 2a + 3)$

12. $4(x - 2) - x(x - 2)$ $(x - 2)(4 - x)$

Factor out −1.

13. $-a - 7$ $-(a + 7)$

14. $-4t^2 + 3t - 1$ $-(4t^2 - 3t + 1)$

Factor.

15. $2c + 2d + ac + ad$ $(c + d)(2 + a)$

16. $3xy + 18x - 5y - 30$ $(y + 6)(3x - 5)$

17. $2a^3 + 2a^2 - a - 1$ $(a + 1)(2a^2 - 1)$

18. $4m^2n + 12m^2 - 8mn - 24m$ $4m(n + 3)(m - 2)$

SECTION 6.2 Factoring Trinomials of the Form $x^2 + bx + c$

DEFINITIONS AND CONCEPTS	EXAMPLES
Many trinomials factor as the product of two binomials. To **factor a trinomial** of the form $x^2 + bx + c$, whose **leading coefficient is 1,** find two integers whose product is c and whose sum is b. $x^2 + bx + c = (x \square)(x \square)$ The product of these numbers must be c and their sum must be b. Use the FOIL method to check the factorization.	Factor: $p^2 + 7p + 12$ $= (p \square)(p \square)$ $= (p + 3)(p + 4)$ Positive factors of 12 / Sum of positive factors of 12: $1 \cdot 12 = 12$ / $1 + 12 = 13$ $2 \cdot 6 = 12$ / $2 + 6 = 8$ $3 \cdot 4 = 12$ / $3 + 4 = 7$ ***Check:*** $(p + 3)(p + 4) = p^2 + 4p + 3p + 12$ $= p^2 + 7p + 12$
Before factoring a trinomial, write it in **descending powers** of one variable. Also, factor out -1 if that is necessary to make the **leading coefficient positive.**	Factor: $7q - q^2 - 6$ $= -q^2 + 7q - 6$ Write the terms in descending powers of q. $= -(q^2 - 7q + 6)$ Factor out -1. $= -(q - 1)(q - 6)$ Factor the trinomial.
If a trinomial cannot be factored using only integers, it is called a **prime trinomial.**	$t^2 + 2t - 5$ is a prime trinomial because there are no two integers whose product is -5 and whose sum is 2.
The GCF should always be factored out first. A trinomial is **factored completely** when no factor can be factored further. Use multiplication to check the factorization.	Factor: $3m^3 - 6m^2 - 24m$ $= 3m(m^2 - 2m - 8)$ Factor out the GCF, $3m$, first. $= 3m(m - 4)(m + 2)$ Factor the trinomial.
To factor a trinomial of the form $ax^2 + bx + c$ (where $a = 1$) by **grouping,** write it as a equivalent four-term polynomial: $x^2 + \square x + \square x + c$ The product of these numbers must be ac, and their sum must be b. Then factor the four-term polynomial by grouping. Use the FOIL method to check the factorization.	Factor by grouping: $p^2 + 7p + 12$ We must find two numbers whose product is $ac = 1(12) = 12$ and whose sum is $b = 7$. Two such numbers are 4 and 3. They serve as the coefficients of $4p$ and $3p$, the two terms that we use to represent the middle term, $7p$, of the trinomial. $p^2 + 7p + 12 = p^2 + 4p + 3p + 12$ Express $7p$ as $4p + 3p$. $= p(p + 4) + 3(p + 4)$ Factor p out of $p^2 + 4p$ and 3 out of $3p + 12$. $= (p + 4)(p + 3)$ Factor out $(p + 4)$.

REVIEW EXERCISES

19. What is the leading coefficient of $x^2 + 8x - 9$? 1

20. Complete the table.

Factors of 6	Sum of the factors of 6
1(6)	7
2(3)	5
−1(−6)	−7
−2(−3)	−5

Factor.

21. $x^2 + 2x - 24$
$(x + 6)(x - 4)$

22. $x^2 - 18x - 40$
$(x - 20)(x + 2)$

23. $x^2 - 14x + 45$
$(x - 5)(x - 9)$

24. $t^2 + 10t + 15$
Prime

25. $-y^2 + 15y - 56$
$-(y - 8)(y - 7)$

26. $10y + 9 + y^2$
$(y + 9)(y + 1)$

27. $c^2 + 3cd - 10d^2$
$(c + 5d)(c - 2d)$

28. $-3mn + m^2 + 2n^2$
$(m - 2n)(m - n)$

29. Explain how we can check to determine whether $(x - 4)(x + 5)$ is the factorization of $x^2 + x - 20$. Multiply

30. Explain why $x^2 + 7x + 11$ is prime. There are no two integers whose product is 11 and whose sum is 7.

Factor.

31. $5a^5 + 45a^4 - 50a^3$
$5a^3(a + 10)(a - 1)$

32. $-4x^2y - 4x^3 + 24xy^2$
$-4x(x + 3y)(x - 2y)$

SECTION 6.3 Factoring Trinomials of the Form $ax^2 + bx + c$

DEFINITIONS AND CONCEPTS

We can use the **trial-and-check method** to factor trinomials with **leading coefficients other than 1.** Write the trinomial as the product of two binomials and determine four integers.

The product of these numbers must be *a*.

$$ax^2 + bx + c = (\square x\ \square)(\square x\ \square)$$

The product of these numbers must be *c*.

Use the FOIL method to check the factorization to obtain the correct middle term.

EXAMPLES

Factor: $2x^2 - 5x - 12$

Since the first term is $2x^2$, the first terms of the binomial factors must be $2x$ and x.

$(2x\ \square)(x\ \square)$ Because $2x \cdot x$ will give $2x^2$

The second terms of the binomials must be two integers whose product is -12. There are six such pairs:

$12(-1)$, $6(-2)$, $4(-3)$, $\mathbf{3(-4)}$, $2(-6)$, and $1(-12)$

The pair in blue gives the correct middle term when we use the FOIL method to check:

Outer: $-8x$

$(2x + 3)(x - 4)$ Combine like terms: $-8x + 3x = -5x$. Correct middle term.

Inner: $3x$

Thus, $2x^2 - 5x - 12 = (2x + 3)(x - 4)$.

DEFINITIONS AND CONCEPTS

To factor $ax^2 + bx + c$ by **grouping,** write it as an equivalent four-term polynomial:

$$ax^2 + \square x + \square x + c$$

The product of these numbers must be *ac*, and their sum must be *b*.

Then factor the four-term polynomial by grouping.

Use the FOIL method to check your work.

EXAMPLES

Factor by grouping: $2x^2 - 5x - 12$

We must find two numbers whose product is $ac = 2(-12) = -24$ and whose sum is $b = -5$. Two such numbers are -8 and 3. They serve as the coefficients of $-8x$ and $3x$, the two terms that we use to represent the middle term, $-5x$, of the trinomial.

$$\begin{aligned} 2x^2 - 5x - 12 &= 2x^2 - 8x + 3x - 12 && \text{Express } -5x \text{ as } -8x + 3x. \\ &= 2x(x - 4) + 3(x - 4) && \text{Factor out } 2x \text{ and } 3. \\ &= (x - 4)(2x + 3) && \text{Factor out } (x - 4). \end{aligned}$$

REVIEW EXERCISES

Factor.

33. $2x^2 - 5x - 3$
$(2x + 1)(x - 3)$

34. $35y^2 + 11y - 10$
$(7y + 5)(5y - 2)$

35. $-3x^2 + 13x + 30$
$-(3x + 5)(x - 6)$

36. $-33p^2 - 6p + 18p^3$
$3p(6p + 1)(p - 2)$

37. $4b^2 - 17bc + 4c^2$
$(4b - c)(b - 4c)$

38. $7y^2 + 7y - 18$
Prime

39. Entertaining. The rectangular area occupied by a table setting is $(12x^2 - x - 1)$ square inches. Factor the polynomial to find the binomials that represent the length and width of the table setting. $(4x + 1)$ in., $(3x - 1)$ in.

40. In the following work, a student began to factor $5x^2 - 8x + 3$. Explain his mistake.

$(5x - \quad)(x + \quad)$

The signs of the second terms must be negative.

SECTION 6.4 Factoring Perfect-Square Trinomials and the Difference of Two Squares

DEFINITIONS AND CONCEPTS	EXAMPLES
Trinomials that are squares of a binomial are called **perfect-square trinomials.** We can factor perfect-square trinomials by applying the special-product rules in reverse. $A^2 + 2AB + B^2 = (A + B)^2$ $A^2 - 2AB + B^2 = (A - B)^2$	Factor: $g^2 + 8g + 16$ and $m^2 - 18mn + 81n^2$ We match each trinomial to a special-product form shown in the left column. $g^2 + 8g + 16 = g^2 + 2 \cdot 4 \cdot g + 4^2 = (g + 4)^2$ $m^2 - 18mn + 81n^2 = m^2 - 2 \cdot m \cdot 9n + (9n)^2 = (m - 9n)^2$
To factor the **difference of two squares,** use the rule $F^2 - L^2 = (F + L)(F - L)$ It will be helpful to review the table of **squares of integers** shown on page 466.	Factor: $25b^2 - 36$ $= (5b)^2 - 6^2$ This is a difference of two squares. $= (5b + 6)(5b - 6)$
In general, the **sum of two squares** (with no common factor other than 1) cannot be factored using real numbers.	$x^2 + 100$ and $36y^2 + 49$ are prime polynomials.

REVIEW EXERCISES

Factor.

41. $x^2 + 10x + 25$
$(x + 5)^2$
42. $9y^2 + 16 - 24y$
$(3y - 4)^2$
43. $-z^2 + 2z - 1$
$-(z - 1)^2$
44. $25a^2 + 20ab + 4b^2$
$(5a + 2b)^2$
45. $x^2 - 9$
$(x + 3)(x - 3)$
46. $49t^2 - 121y^2$
$(7t + 11y)(7t - 11y)$
47. $x^2y^2 - 400$
$(xy + 20)(xy - 20)$
48. $8at^2 - 32a$
$8a(t + 2)(t - 2)$
49. $c^4 - 256$
$(c^2 + 16)(c + 4)(c - 4)$
50. $h^2 + 36$
Prime

SECTION 6.5 Factoring the Sum and Difference of Two Cubes

DEFINITIONS AND CONCEPTS	EXAMPLES
To factor the **sum** and **difference of two cubes,** use the following rules. $F^3 + L^3 = (F + L)(F^2 - FL + L^2)$ $F^3 - L^3 = (F - L)(F^2 + FL + L^2)$ It will be helpful to review the table of **cubes of integers** shown on page 473.	Factor: $p^3 + 64$ and $125a^3 - 27b^3$ We match each binomial to a factoring rule shown in the left column. $p^3 + 64 = p^3 + 4^3$ This is a sum of two cubes. $= (p + 4)(p^2 - p \cdot 4 + 4^2)$ $= (p + 4)(p^2 - 4p + 16)$ $125a^3 - 27b^3 = (5a)^3 - (3b)^3$ This is a difference of two cubes. $= (5a - 3b)[(5a)^2 + 5a \cdot 3b + (3b)^2]$ $= (5a - 3b)(25a^2 + 15ab + 9b^2)$

REVIEW EXERCISES

Factor.

51. $b^3 + 1$ $(b + 1)(b^2 - b + 1)$
52. $x^3 - 216$ $(x - 6)(x^2 + 6x + 36)$
53. $p^3 + 125q^3$ $(p + 5q)(p^2 - 5pq + 25q^2)$
54. $16x^5 - 54x^2y^3$ $2x^2(2x - 3y)(4x^2 + 6xy + 9y^2)$

SECTION 6.6 A Factoring Strategy

DEFINITIONS AND CONCEPTS	EXAMPLES
To factor a random polynomial, use the **factoring strategy** discussed in Section 6.6 on page 477. Remember that the instruction to factor means to **factor completely.** A polynomial is factored completely when no factor can be factored further.	Factor: $a^5 + 8a^2 + 4a^3 + 32$ ***Is there a common factor?*** No. There is no common factor (other than 1). ***How many terms does it have?*** Since the polynomial has four terms, try factoring by grouping. $a^5 + 8a^2 + 4a^3 + 32 = a^2(a^3 + 8) + 4(a^3 + 8)$ Factor a^2 from $a^5 + 8a^2$ and 4 from $4a^3 + 32$. $= (a^3 + 8)(a^2 + 4)$ Factor out $a^3 + 8$. ***Is it factored completely?*** No. We can factor $a^3 + 8$ as a sum of two cubes. $a^5 + 8a^2 + 4a^3 + 32$ $= (a^3 + 8)(a^2 + 4)$ $a^2 + 4$ is prime. $= (a + 2)(a^2 - 2a + 4)(a^2 + 4)$ $a^2 - 2a + 4$ is prime. ***Does it check?*** Use multiplication to check.

REVIEW EXERCISES

Factor.

55. $14y^3 + 6y^4 - 40y^2$
$2y^2(3y - 5)(y + 4)$

56. $5s^2t + 5s^2u^2 + 5tv + 5u^2v$
$5(t + u^2)(s^2 + v)$

57. $j^4 - 16$
$(j^2 + 4)(j + 2)(j - 2)$

58. $-3j^3 - 24$
$-3(j + 2)(j^2 - 2j + 4)$

59. $400x + 400 - m^2x - m^2$
$(x + 1)(20 + m)(20 - m)$

60. $12w^4 - 36w^3 + 27w^2$
$3w^2(2w - 3)^2$

61. $2t^3 + 10$
$2(t^3 + 5)$

62. $121p^2 + 36q^2$
Prime

63. $x^2z + 64y^2z + 16xyz$
$z(x + 8y)^2$

64. $18c^3d^2 - 12c^3d - 24c^2d$
$6c^2d(3cd - 2c - 4)$

SECTION 6.7 Solving Quadratic Equations by Factoring

DEFINITIONS AND CONCEPTS	EXAMPLES
A **quadratic equation** is an equation that can be written in the **standard form** $ax^2 + bx + c = 0$, where a, b, and c are real numbers and $a \neq 0$.	Examples of quadratic equations are: $5x^2 + 25x = 0$, $4a^2 - 9 = 0$, and $y^2 - 13y = 6$
The Zero-Factor Property If the product of two (or more) numbers is 0, then at least one of the numbers is 0.	If $(x + 2)(x - 3) = 0$ then, $x + 2 = 0$ or $x - 3 = 0$.
To use the **factoring method to solve a quadratic equation:** **1.** Write the equation in standard form: $ax^2 + bx + c = 0$ or $0 = ax^2 + bx + c$ **2.** Factor completely. **3.** Use the *zero-factor property* to set each factor equal to 0. **4.** Solve each resulting equation. **5.** Check each result in the original equation.	Solve: $5x^2 + 25x = 0$ $5x(x + 5) = 0$ $5x = 0$ or $x + 5 = 0$ $x = 0$ | $x = -5$ The solutions are 0 and -5. The solution set is $\{0, -5\}$. Solve: $4a^2 - 9 = 0$ $(2a + 3)(2a - 3) = 0$ $2a + 3 = 0$ or $2a - 3 = 0$ $2a = -3$ | $2a = 3$ $a = -\frac{3}{2}$ | $a = \frac{3}{2}$ The solutions are $-\frac{3}{2}$ and $\frac{3}{2}$. The solution set is $\left\{-\frac{3}{2}, \frac{3}{2}\right\}$. Check each result in the original equation.

To use the zero-factor property to solve a quadratic equation, we need one side of the equation to be factored completely and the other side to be 0.	Solve: $5y^2 - 13y = 6$ This equation is not in standard form. $5y^2 - 13y - 6 = 6 - 6$ To get 0 on the right side, subtract 6 from both sides. $5y^2 - 13y - 6 = 0$ Do the subtraction. $(5y + 2)(y - 3) = 0$ Factor the trinomial. $5y + 2 = 0$ or $y - 3 = 0$ Set each factor equal to 0. $5y = -2$ | $y = 3$ Solve each equation. $y = -\frac{2}{5}$ The solutions are $-\frac{2}{5}$ and 3. Check each result in the original equation.

REVIEW EXERCISES

Solve each equation by factoring.

65. $8x(x - 6) = 0$ 0, 6

66. $(4x - 7)(x + 1) = 0$ $\frac{7}{4}, -1$

71. $2t^2 + 28t + 98 = 0$ -7 repeated

72. $2x - x^2 = -24$ 6, −4

67. $x^2 + 2x = 0$ 0, −2

68. $x^2 - 9 = 0$ −3, 3

73. $5a^2 - 6a + 1 = 0$ $1, \frac{1}{5}$

74. $2p^3 = 2p(p + 2)$ 0, −1, 2

69. $144x^2 - 25 = 0$ $-\frac{5}{12}, \frac{5}{12}$

70. $a^2 - 7a + 12 = 0$ 3, 4

SECTION 6.8 Applications of Quadratic Equations

DEFINITIONS AND CONCEPTS	EXAMPLES
To solve application problems, use the **six-step problem-solving strategy:** **1.** Analyze the problem **2.** Assign a variable **3.** Form an equation **4.** Solve the equation **5.** State the conclusion **6.** Check the result	Find two consecutive positive integers whose product is 72. **Analyze** *Consecutive integers* are integers that follow each other. The word *product* indicates multiplication. **Assign** Let $x =$ the smaller positive integer. Then $x + 1 =$ the larger integer. **Form** The smaller integer times the larger integer equals 72. $x \cdot (x + 1) = 72$ **Solve** $x(x + 1) = 72$ This is the equation to solve. $x^2 + x = 72$ Distribute the multiplication by x. This is a quadratic equation but it is not in standard form. $x^2 + x - 72 = 0$ To get 0 on the right side, subtract 72 from both sides. $(x + 9)(x - 8) = 0$ Factor the trinomial. $x + 9 = 0$ or $x - 8 = 0$ Set each factor equal to 0. $x = -9$ | $x = 8$ Solve each linear equation. **State** Since we are looking for positive integers, the solution −9 must be discarded. Thus, the smaller integer is 8 and the larger integer is $x + 1 = 9$. **Check** The integers 8 and 9 are consecutive positive integers and their product is 72.

The Pythagorean Theorem:

If a and b are the lengths of the legs of a right triangle and c is the length of the hypotenuse, then

$$a^2 + b^2 = c^2$$

To show that a triangle with sides of 5, 12, and 13 units is a right triangle, we verify that $5^2 + 12^2 = 13^2$.

$$5^2 + 12^2 \stackrel{?}{=} 13^2$$
$$25 + 144 \stackrel{?}{=} 169$$
$$169 = 169 \quad \text{True}$$

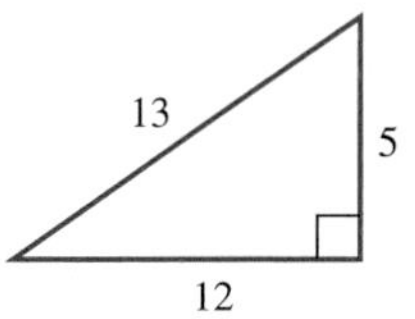

REVIEW EXERCISES

75. Sandpaper. A standard piece of sandpaper is 2 inches longer than it is wide. Find the dimensions of a piece of sandpaper if it has an area of 99 in^2. Width: 9 in.; length: 11 in.

76. Construction. The face of the triangular concrete panel has an area of 45 square meters, and its base is 3 meters longer than twice its height. Find the length of its base. 15 m

77. Fill in the blanks. If we let x = the first integer, then:
- two consecutive integers are x and $x + 1$
- two consecutive even integers are x and $x + 2$
- two consecutive odd integers are x and $x + 2$

78. Music Awards. The record for the most Grammy nominations in one year is held by Michael Jackson. Kanye West is currently in second place. The number of times Jackson and West were nominated are consecutive even integers whose product is 120. How many times was each artist nominated? (Source: *Wikipedia*) Jackson: 12 nominations; West: 10 nominations

79. Tightrope Walkers. A circus performer intends to walk up a taut cable shown in the illustration to a platform at the top of a pole. How high above the ground is the platform? 5 m

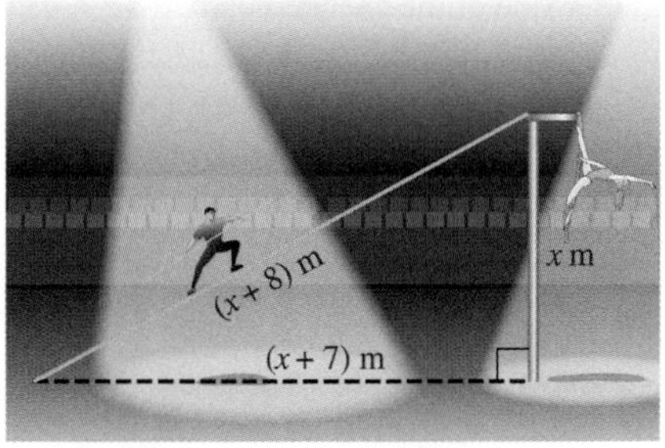

80. Ballooning. A hot-air balloonist accidentally dropped his camera overboard while traveling at a height of 1,600 ft. The height h in feet of the camera t seconds after being dropped is given by $h = -16t^2 + 1{,}600$. In how many seconds will the camera hit the ground? 10 sec

Teaching Tip: Because this Chapter Test is a comprehensive collection of problems that covers all of the topics discussed in Chapter 6, it is lengthy. If your students have time to complete it, that would be optimal. If, because of time constraints, they are unable to do so, assign an appropriate subset of problems that reflects the types of problems that the students will see on your exam.

6 CHAPTER TEST

1. Fill in the blanks.

a. The letters GCF stand for greatest common factor.

b. To factor a polynomial means to express it as a product of two (or more) polynomials.

c. The Pythagorean theorem provides a formula relating the lengths of the three sides of a right triangle.

d. $y^2 - 25$ is a difference of two squares.

e. The trinomial $x^2 + x - 6$ factors as the product of two binomials: $(x + 3)(x - 2)$.

2. a. Find the prime factorizations of 45 and 30. $45 = 3^2 \cdot 5$; $30 = 2 \cdot 3 \cdot 5$

b. Find the greatest common factor of $45x^4$ and $30x^3$. $15x^3$

Factor. If an expression cannot be factored, write "prime."

3. $4x + 16$ $\quad 4(x + 4)$

4. $q^2 - 81$ $\quad (q + 9)(q - 9)$

5. $30a^2b^3 - 20a^3b^2 + 5ab$ $\quad 5ab(6ab^2 - 4a^2b + 1)$

6. $x^2 + 9$ $\quad$ Prime

7. $2x(x + 1) + 3(x + 1)$ $\quad (x + 1)(2x + 3)$

8. $x^2 + 4x + 3$ $\quad (x + 3)(x + 1)$

9. $-x^2 + 9x + 22$ $\quad -(x - 11)(x + 2)$

10. $60x^2 - 32x^3 + x^4$ $\quad x^2(x - 30)(x - 2)$

11. $9a - 9b + ax - bx$ $\quad (a - b)(9 + x)$

12. $2a^2 + 5a - 12$ $\quad (2a - 3)(a + 4)$

13. $18x^2 + 60xy + 50y^2$ $\quad 2(3x + 5y)^2$

14. $x^3 + 8$ $\quad (x + 2)(x^2 - 2x + 4)$

15. $60m^8 - 45m^6$
$15m^6(4m^2 - 3)$

16. $3a^3 - 81$
$3(a - 3)(a^2 + 3a + 9)$

17. $16x^4 - 81$
$(4x^2 + 9)(2x + 3)(2x - 3)$

18. $a^3 + 5a^2 + a + 5$
$(a + 5)(a^2 + 1)$

19. $a^4 - 24 - 4a + 6a^3$
$(a + 6)(a^3 - 4)$

20. $3d - 4 + 10d^2$
$(5d + 4)(2d - 1)$

21. $8m^2 - 800$
$8(m + 10)(m - 10)$

22. $36n^2 - 84n + 49$
$(6n - 7)^2$

23. $8r^2 - 14r + 3$
$(4r - 1)(2r - 3)$

24. $t^2 - 6t + 10$
Prime

25. Checkers. The area of a square checkerboard is represented by $(25x^2 - 40x + 16)$ in.^{2}. Find the polynomial that represents the length of a side of the checkerboard. $(5x - 4)$ in.

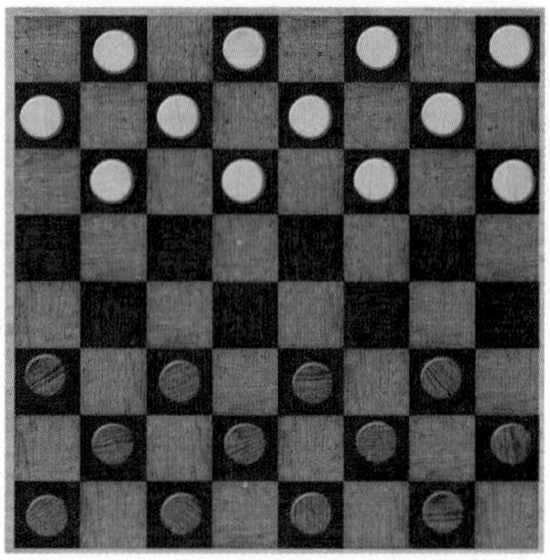
©iStockphoto.com/Philip Bonneau

26. Factor $x^2 - 3x - 54$. Show a check of your answer.
$(x - 9)(x + 6)$; $x^2 + 6x - 9x - 54 = x^2 - 3x - 54$

Solve each equation.

27. $(x + 3)(x - 2) = 0$
$-3, 2$

28. $x^2 - 25 = 0$
$-5, 5$

29. $36x^2 = 6x$
$0, \frac{1}{6}$

30. $x^2 + 6x = -9$
-3 repeated

31. $6x^2 + x - 1 = 0$
$\frac{1}{3}, -\frac{1}{2}$

32. $a(a - 7) = 18$
$9, -2$

33. $x^3 + 7x^2 = -6x$ $0, -1, -6$

34. Driving Safety. All cars have a blind spot where it is difficult for the driver to see a car behind and to the right. The area of the rectangular blind spot shown is 54 ft^2. Its length is 3 feet longer than its width. Find its dimensions. 6 ft by 9 ft

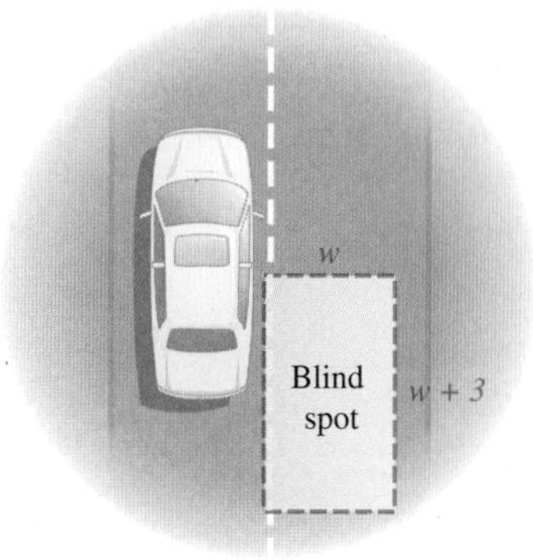

35. Rocketry. The height h, in feet, of a toy rocket t seconds after being launched is given by $h = -16t^2 + 80t$. After how many seconds will the rocket hit the ground? 5 sec

36. ATV's. The area of a triangular-shaped safety flag on an all-terrain vehicle is 33 in.^{2}. Its height is 1 inch less than twice the length of the base. Find the length of the base and the height of the flag. Base: 6 in.; height: 11 in.

37. Find two consecutive positive integers whose product is 156.
12, 13

38. Find the length of the hypotenuse of the right triangle shown. 10

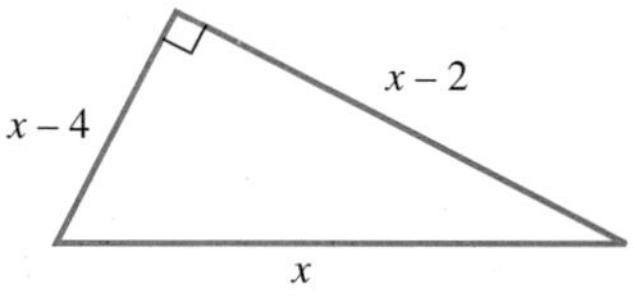

39. What is a quadratic equation? Give an example.
A quadratic equation is an equation that can be written in the form $ax^2 + bx + c = 0$; $x^2 - 2x + 1 = 0$. (Answers may vary.)

40. If the product of two numbers is 0, what conclusion can be drawn about the numbers? At least one of them is 0.

Group Project

FACTORING MODELS

Overview In this activity, you will construct geometric models to find factorizations of several trinomials.

Instructions Form groups of 2 or 3 students.

1. Copy and cut out each of the following figures. On each figure, write its area.

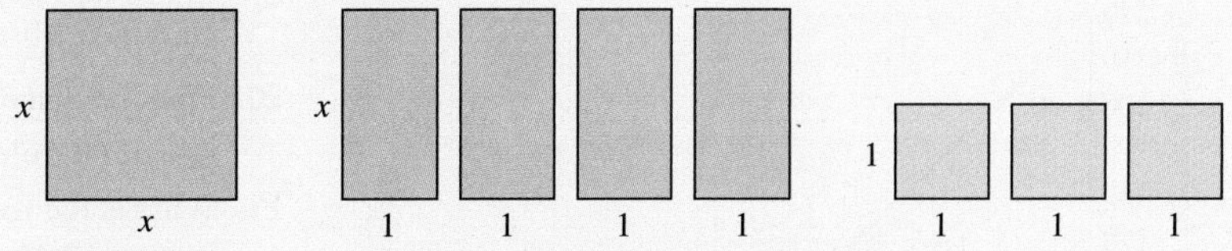

Write a trinomial that represents the *sum* of the areas of the eight figures by combining any like terms: _____ + _____ + _____

2. Now assemble the eight figures to form the large rectangle shown below.

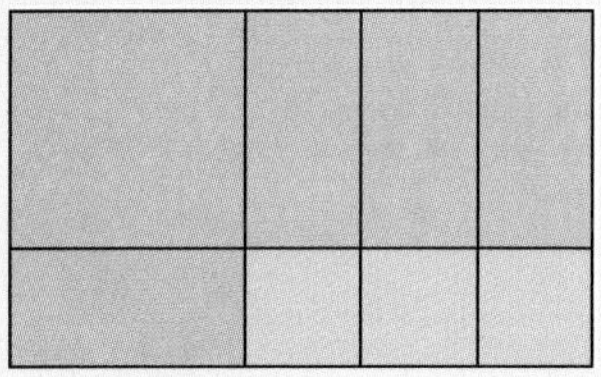

Write an expression that represents the *length* of the rectangle: _____ + _____

Write an expression that represents the *width* of the rectangle: _____ + _____

Express the area of the rectangle as the product of its length and width: (________)(________)

3. The set of figures used in step 1 and the set of figures used in step 2 are the same. Therefore, the expressions for the areas must be equal. Set your answers from steps 1 and 2 equal to find the factorization of the trinomial $x^2 + 4x + 3$.

________________ = ________________
Answer from step 1 Answer from step 2

4. Make a new model to find the factorization of $x^2 + 5x + 4$. (*Hint:* You will need to make one more 1-by-x figure and one more 1-by-1 figure.)

________________ = ________________

5. Make a new model to find the factorization of $2x^2 + 5x + 2$. (*Hint:* You will need to make one more x-by-x figure.)

________________ = ________________

CUMULATIVE REVIEW Chapters 1–6

1. **Heart Rates.** Refer to the graph. Determine the difference in the maximum heart beat rate for a 70-year-old as compared to someone half that age. [Section 1.1] About 35 beats/min difference

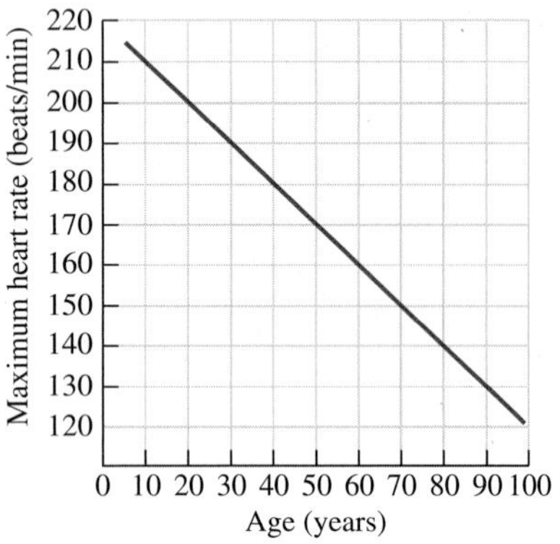

Based on data from *Cardiopulmonary Anatomy and Physiology: Essentials for Respiratory Care*, 2nd ed.

2. Find the prime factorization of 250. [Section 1.2] $2 \cdot 5^3$
3. Find the quotient: $\frac{16}{5} \div \frac{10}{3}$ [Section 1.2] $\frac{24}{25}$
4. Write $\frac{124}{125}$ as a decimal. [Section 1.3] 0.992
5. Determine whether each statement is true or false. [Section 1.2]
   a. Every integer is a whole number. False
   b. Every integer is a rational number. True
   c. π is a real number. True
6. Which division is undefined, $\frac{0}{5}$ or $\frac{5}{0}$? [Section 1.6] $\frac{5}{0}$

Evaluate each expression.

7. $3 + 2[-1 - 4(5)]$ [Section 1.7] -39
8. $\dfrac{|-25| - 2(-5)}{9 - 2^4}$ [Section 1.7] -5
9. What is -3 cubed? [Section 1.7] -27
10. What is the value of x twenty-dollar bills? [Section 1.8] $\$20x$
11. Evaluate $\frac{-x - a}{y - b}$ for $x = -2$, $y = 1$, $a = 5$, and $b = 2$. [Section 1.8] 3
12. Identify the coefficient of each term in the expression $8x^2 - x + 9$. [Section 1.8] 8, -1, 9

Simplify each expression.

13. $-8y^2 - 5y^2 + 6$ [Section 1.9] $-13y^2 + 6$
14. $3z + 2(y - z) + y$ [Section 1.9] $3y + z$

Solve each equation.

15. $-(3a + 1) + a = 2$ [Section 2.2] $-\frac{3}{2}$
16. $2 - (4x + 7) = 3 + 2(x + 2)$ [Section 2.2] -2
17. $\dfrac{3t - 21}{2} = t - 6$ [Section 2.2] 9
18. $-\dfrac{1}{3} - \dfrac{x}{5} = \dfrac{3}{2}$ [Section 2.2] $-\frac{55}{6}$
19. **Watermelons.** The heaviest watermelon on record weighed 270 pounds. If watermelon is 92% water by weight, what was its water weight? Round to the nearest pound. (Source: *Guinness World Records*) [Section 2.3] 248 lb
20. Find the distance traveled by a truck traveling for $5\frac{1}{2}$ hours at a rate of 60 miles per hour. [Section 2.4] 330 mi
21. What is the formula for simple interest? [Section 2.4] $I = Prt$
22. **Geometry Tools.** A compass is adjusted so that the distance between the pointed ends is 2 inches. Then a circle is drawn. What will the area of the circle be? Round to the nearest tenth of a square inch. [Section 2.4] 12.6 in.^{2}

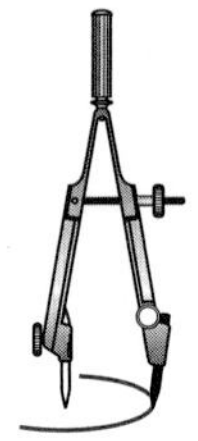

23. Solve $A = P + Prt$ for t. [Section 2.4] $t = \frac{A - P}{Pr}$
24. **History.** George Washington was the first president of the United States. John Adams was the second, and Thomas Jefferson was the third, and so on. Grover Cleveland was president two *different* times, as shown in the illustration. The sum of the numbers of Cleveland's presidencies is 46. Find these two numbers. [Section 2.5] 22nd president, 24th president

Grover Cleveland **Benjamin Harrison** **Grover Cleveland**

25. **Antique Shows.** A traveling antique show will be on the road for 17 weeks, visiting three cities. They will be in Los Angeles for 2 weeks longer than they will be in Las Vegas. Their stay in Dallas will be 1 week less than twice that in Las Vegas. How many weeks will they be in each city? [Section 2.5] Los Angeles: 6 wk; Las Vegas: 4 wk; Dallas: 7 wk
26. **Photographic Chemicals.** A photographer wishes to mix 6 liters of a 5% acetic acid solution with a 10% solution to get a 7% solution. How many liters of 10% solution must be added? [Section 2.6] 4 L
27. **Dried Fruits.** Dried apple slices cost \$4.60 per pound, and dried banana chips sell for \$3.40 per pound. How many pounds of each should be used to create a 10-pound mixture that sells for \$4 per pound? [Section 2.6] 5 lb apple slices, 5 lb banana chips
28. Solve: $-\frac{x}{2} + 4 > 5$. Write the solution set in interval notation and graph it. [Section 2.8] $(-\infty, -2)$ See AIE Appendix 3.
29. Is $(-2, 5)$ a solution of $3x + 2y = 4$? [Section 3.2] Yes
30. Graph: $y = 2x - 3$ [Section 3.2] See AIE Appendix 3.

31. Is the graph of $x = 3$ a vertical or horizontal line? [Section 3.3] Vertical line

32. If two lines are parallel, what can be said about their slopes? [Section 3.4] They are the same.

33. Encyclopedias. The graph below approximates the total number of articles on the English-language edition of *Wikipedia* for the years 2005 through 2010. Find the rate of change in the number of articles over that time span. [Section 3.4] (Source: Wikipedia) An increase of 536,000 articles per year

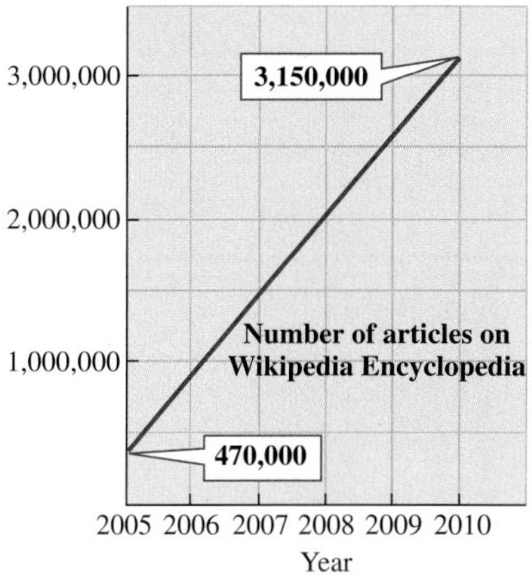

34. Find the slope and the y-intercept of the graph of $3x - 3y = 6$. [Section 3.5] $1; (0, -2)$

35. Find an equation of the line passing through $(-2, 5)$ and $(-3, -2)$. Write the equation in slope–intercept form. [Section 3.5] $y = 7x + 19$

36. Graph the line passing through $(-4, 1)$ that has slope -3. [Section 3.6] See AIE Appendix 3.

37. Graph: $8x + 4y \geq -24$ [Section 3.7] See AIE Appendix 3.

38. If $f(x) = 3x^2 - 2x + 1$, find $f(-2)$. [Section 3.8] 17

39. Is $\left(\frac{1}{2}, 1\right)$ a solution of the system $\begin{cases} 4x - y = 1 \\ 2x + y = 2 \end{cases}$? [Section 4.1]

Yes

40. Solve the system $\begin{cases} 3x - 2y = 6 \\ x - y = 1 \end{cases}$ by graphing. [Section 4.1]

$(4, 3)$ See AIE Appendix 3.

41. Solve the system $\begin{cases} y = -4x + 1 \\ 4x = y - 5 \end{cases}$ by substitution. [Section 4.2]

$\left(\frac{3}{4}, -2\right)$

42. Solve the system $\begin{cases} 5a + 3b = -8 \\ 2a + 9b = 2 \end{cases}$ by elimination (addition).

[Section 4.3] $\left(-2, \frac{2}{3}\right)$

43. Fundraising. A Rotary Club held a city-wide recycling drive. They collected a total of 14 tons of newspaper and cardboard that earned them \$356. They were paid \$31 per ton for the newspaper and \$18 per ton for the cardboard. How many tons of each did they collect? [Section 4.4] Newspaper: 8 tons, cardboard: 6 tons

44. Graph: $\begin{cases} 4x + 3y \geq 12 \\ y < 4 \end{cases}$ [Section 4.5] See AIE Appendix 3.

Simplify each expression. Write each answer without negative exponents.

45. $-y^2(4y^3)$ [Section 5.1] $-4y^5$

46. $\dfrac{(x^2y^5)^5}{(x^3y)^2}$ [Section 5.1] x^4y^{23}

47. $\dfrac{b^5}{b^{-2}}$ [Section 5.2] b^7

48. $2x^0$ [Section 5.1] 2

49. Write 0.00009011 in scientific notation. [Section 5.3] 9.011×10^{-5}

50. Write 1,700,000 in scientific notation. [Section 5.3] 1.7×10^6

51. Find the degree of $7y^3 + 4y^2 + y + 3$. [Section 5.4] 3

52. Graph: $y = x^3 + 2$ [Section 5.4] See AIE Appendix 3.

Perform the operations.

53. $(x^2 - 3x + 8) - (3x^2 + x + 3)$ [Section 5.5] $-2x^2 - 4x + 5$

54. $4b^3(2b^2 - 2b)$ [Section 5.6] $8b^5 - 8b^4$

55. $(3x - 2)(x + 4)$ [Section 5.6] $3x^2 + 10x - 8$

56. $(y - 6)^2$ [Section 5.7] $y^2 - 12y + 36$

57. $\dfrac{12a^2b^2 - 8a^2b - 4ab}{4ab}$ [Section 5.8] $3ab - 2a - 1$

58. $x - 3\overline{)2x^2 - 5x - 3}$ [Section 5.8] $2x + 1$

59. Playpens. Find an expression that represents the

a. perimeter of the playpen. [Section 5.5] $(4x + 8)$ in.

b. area of the floor of the playpen. [Section 5.6] $(x^2 + 4x + 3)$ in.^{2}

c. volume of the playpen. [Section 5.6] $(x^3 + 4x^2 + 3x)$ in.^{3}

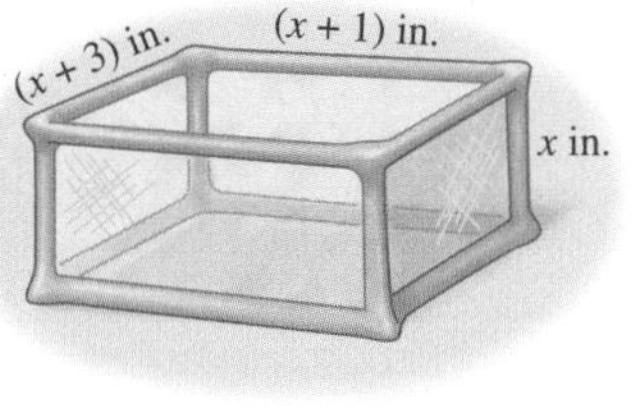

60. Find the GCF of $24x^5y^8$ and $54x^6y$. [Section 6.1] $6x^5y$

Factor.

61. $9b^3 - 27b^2$ [Section 6.1] $9b^2(b - 3)$

62. $ax + bx + ay + by$ [Section 6.1] $(x + y)(a + b)$

63. $u^2 - 3 + 2u$ [Section 6.2] $(u + 3)(u - 1)$

64. $10x^2 + x - 2$ [Section 6.3] $(2x + 1)(5x - 2)$

65. $4a^2 - 12a + 9$ [Section 6.4] $(2a - 3)^2$

66. $9z^2 - 1$ [Section 6.4] $(3z + 1)(3z - 1)$

67. $t^3 - 8$ [Section 6.5] $(t - 2)(t^2 + 2t + 4)$

68. $3a^2b^2 - 6a^2 - 3b^2 + 6$ [Section 6.6] $3(b^2 - 2)(a + 1)(a - 1)$

Solve each equation.

69. $15s^2 - 20s = 0$ [Section 6.7] $0, \frac{4}{3}$

70. $2x^2 - 5x = -2$ [Section 6.7] $\frac{1}{2}, 2$

71. $x^3 + 3x^2 + 2x = 0$ [Section 6.7] $0, -1, -2$

72. Camping. The rectangular-shaped cooking surface of a small camping stove is 108 in.^{2}. If its length is 3 inches longer than its width, what are its dimensions? [Section 6.8] 9 in. by 12 in.

Rational Expressions and Equations

7

from Campus to Careers

Recreation Director

People of all ages enjoy participating in activities, such as arts and crafts, camping, sports, and the performing arts. Recreation directors plan, organize, and oversee these activities in local playgrounds, camps, community centers, religious organizations, theme parks, and tourist attractions. The job of recreation director requires mathematical skills such as budgeting, scheduling, and forecasting trends.

Problem 33 in **Study Set 7.7** and **problem 77** in **Study Set 7.8** involve situations that a recreation director might encounter on the job. The mathematical concepts discussed in this chapter can be used to solve those problems.

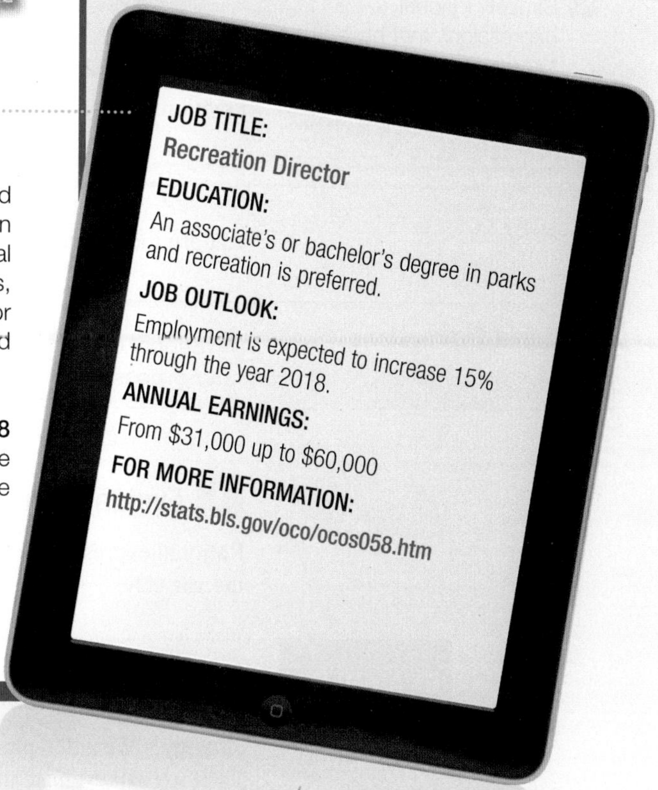

JOB TITLE:
Recreation Director
EDUCATION:
An associate's or bachelor's degree in parks and recreation is preferred.
JOB OUTLOOK:
Employment is expected to increase 15% through the year 2018.
ANNUAL EARNINGS:
From $31,000 up to $60,000
FOR MORE INFORMATION:
http://stats.bls.gov/oco/ocos058.htm

Study Skills Workshop ▶ Study Groups

Study groups give students an opportunity to ask their classmates questions, share ideas, compare lecture notes, and review for tests. If something like this interests you, here are some suggestions.

GROUP SIZE: A study group should be small—from 3 to 6 people is best.

TIME AND PLACE: You should meet regularly in a place where you can spread out and talk without disturbing others.

GROUND RULES: The study group will be more effective if, early on, you agree on some rules to follow.

Now Try This ▶

Would you like to begin a study group? If so, you need to answer the following questions.
Who will be in your group? Where will your group meet? How often will it meet? For how long will each session last? Will you have a group leader? What will be the leader's responsibilities? What will you try to accomplish each session? How will the members prepare for each meeting? Will you follow a set agenda each session? How will the members share contact information? When will you discuss ways to improve the study sessions?

SECTION 7.1 Simplifying Rational Expressions

OBJECTIVES

1. Evaluate rational expressions.
2. Find numbers that cause a rational expression to be undefined.
3. Simplify rational expressions.
4. Simplify rational expressions that have factors that are opposites.

ARE YOU READY? *Are You Ready? exercises available online at www.webassign.net/brookscole*

The following problems review some basic skills that are needed when simplifying rational expressions.

1. Evaluate: **a.** $\frac{0}{7}$ 0 **b.** $\frac{7}{0}$ Undefined

2. Simplify: **a.** $\frac{18}{24}$ $\frac{3}{4}$ **b.** $\frac{5}{45}$ $\frac{1}{9}$

3. Factor: $12x - 8$ $4(3x - 2)$

4. Factor: $a^2 - 16$ $(a + 4)(a - 4)$

5. Factor: $5x^2 - 23x - 10$ $(5x + 2)(x - 5)$

6. Evaluate: $\frac{6}{-6}$ -1

Fractions that are the quotient of two integers are *rational numbers.* Examples are $\frac{1}{2}$ and $\frac{9}{5}$. Fractions such as

$$\frac{3}{2y}, \qquad \frac{x}{x + 2}, \qquad \text{and} \qquad \frac{2a^2 - 8a}{a^2 - 6a + 8}$$

that are the quotient of two polynomials are called **rational expressions.**

Rational Expressions

A **rational expression** is an expression of the form $\frac{A}{B}$, where A and B are polynomials and B does not equal 0.

1 Evaluate Rational Expressions.

Rational expressions can have different values depending on the number that is substituted for the variable.

EXAMPLE 1 Evaluate $\frac{2x - 1}{x^2 + 1}$ for $x = -3$ and for $x = 0$.

Strategy We will replace each x in the rational expression with the given value of the variable. Then we will evaluate the numerator and denominator separately, and simplify, if possible.

Why Recall from Chapter 1 that to *evaluate an expression* means to find its numerical value, once we know the value of its variable.

Solution

For x = −3:

$$\frac{2x - 1}{x^2 + 1} = \frac{2(-3) - 1}{(-3)^2 + 1} = \frac{-6 - 1}{9 + 1} = -\frac{7}{10}$$

For x = 0:

$$\frac{2x - 1}{x^2 + 1} = \frac{2(0) - 1}{(0)^2 + 1} = \frac{0 - 1}{0 + 1} = -1$$

Teaching Example 1
Evaluate $\frac{2y + 9}{y^2 + 6}$ for $y = -1$ and for $y = 0$.

Answers: $1, \frac{3}{2}$

Self Check 1 Evaluate $\frac{2x - 1}{x^2 + 1}$ for $x = 7$ and for $x = -2$. $\frac{13}{50}, -1$

Now Try Problems 13 and 21

2 Find Numbers That Cause a Rational Expression to Be Undefined.

The fraction bar in a rational expression indicates division. Since division by 0 is undefined, we must make sure that the denominator of a rational expression is not equal to 0.

EXAMPLE 2 Find all real numbers for which the expression is undefined: **a.** $\frac{7x}{x - 5}$ **b.** $\frac{3x - 2}{x^2 - x - 6}$ **c.** $\frac{8}{x^2 + 1}$ **d.** $\frac{x + 9.4}{12}$

Strategy To find the real numbers for which each expression is undefined, we will find the values of the variable that make the *denominator* 0.

Why We don't need to examine the numerator of the rational expression; it can be any value, including 0. It's a denominator of 0 that makes a rational expression undefined, because a denominator of 0 indicates division by 0.

Solution

a. The denominator of $\frac{7x}{x - 5}$ will be 0 if we replace x with 5.

$$\frac{7x}{x - 5} = \frac{7(5)}{5 - 5} = \frac{35}{0}$$

Since $\frac{35}{0}$ is undefined, the expression $\frac{7x}{x - 5}$ is undefined for $x = 5$.

Teaching Tip: Remind students of the other division form that is often misunderstood: If the numerator of a fraction is 0 and the denominator is not, the fraction equals 0. For example, $\frac{0}{35} = 0$.

b. $\frac{3x - 2}{x^2 - x - 6}$ will be undefined for values of x that make the denominator 0. To find these values, we set $x^2 - x - 6$ equal to 0, and solve for x.

$x^2 - x - 6 = 0$ Set the denominator of $\frac{3x - 2}{x^2 - x - 6}$ equal to 0.

$(x - 3)(x + 2) = 0$ To solve the quadratic equation, factor the trinomial.

$x - 3 = 0$ or $x + 2 = 0$ Set each factor equal to 0.

$x = 3$ | $x = -2$ Solve each equation.

Since 3 and −2 make the denominator 0, the expression $\frac{3x - 2}{x^2 - x - 6}$ is undefined for $x = 3$ and $x = -2$. To check the results, we proceed as follows.

For x = 3:

$$\frac{3x - 2}{x^2 - x - 6} = \frac{3(3) - 2}{3^2 - 3 - 6} = \frac{9 - 2}{9 - 3 - 6} = \frac{7}{0}$$ This expression is undefined.

For x = −2:

$$\frac{3x - 2}{x^2 - x - 6} = \frac{3(-2) - 2}{(-2)^2 - (-2) - 6} = \frac{-6 - 2}{4 + 2 - 6} = \frac{-8}{0}$$ This expression is undefined.

The Language of Algebra

Another way that Example 2 could be phrased is: State the **restrictions** on the variable. For $\frac{3x - 2}{x^2 - x - 6}$, we can state the restrictions by writing $x \neq 3$ and $x \neq -2$.

c. No matter what real number is substituted for x, the denominator, $x^2 + 1$, will not be equal to 0. (A number squared plus 1 cannot equal 0.) Thus, no real numbers make $\frac{8}{x^2 + 1}$ undefined.

d. Since the denominator of $\frac{x + 9.4}{12}$ does not contain a variable, the denominator can never be equal to 0. Thus, no real numbers make the expression undefined.

Teaching Example 2
Find all real numbers for which the expression is undefined.
a. $\frac{3x}{x - 7}$ b. $\frac{x + 8}{x^2 - 7x + 6}$
c. $\frac{6 - x}{x^2 + 8}$ d. $\frac{x - 1.5}{16}$
Answers:
a. 7 b. 1, 6 c. None d. None

Self Check 2 Find all real numbers for which the expression is undefined:
a. $\frac{x}{x + 9}$ -9 **b.** $\frac{9x + 7}{x^2 - 25}$ $-5, 5$ **c.** $\frac{4 - x}{x^2 + 64}$ None **d.** $\frac{x + 3.8}{100}$ None

Now Try Problems 23, 27, 31, and 33

3 Simplify Rational Expressions.

In Section 1.2, we simplified fractions by removing factors equal to 1. For example, to simplify $\frac{6}{15}$, we factor 6 and 15, and then remove the factor $\frac{3}{3}$.

$$\frac{6}{15} = \frac{2 \cdot 3}{5 \cdot 3} = \frac{2}{5} \cdot \frac{3}{3} = \frac{2}{5} \cdot 1 = \frac{2}{5}$$

To streamline this process, we can replace $\frac{3}{3}$ in $\frac{2 \cdot 3}{5 \cdot 3}$ with the equivalent fraction $\frac{1}{1}$.

$$\frac{6}{15} = \frac{2 \cdot 3}{5 \cdot 3} = \frac{2 \cdot \overset{1}{\cancel{3}}}{5 \cdot \underset{1}{\cancel{3}}} = \frac{2}{5} \quad \text{We are removing } \frac{3}{3} = 1.$$

We can simplify rational expressions in a similar manner using a procedure that is based on the following property.

The Fundamental Property of Rational Expressions

If A, B, and C are polynomials, and B and C are not 0,

$$\frac{AC}{BC} = \frac{A}{B}$$

A rational expression is **simplified** if its numerator and denominator have no common factors other than 1. To simplify a rational expression, follow these steps.

Simplifying Rational Expressions

1. Factor the numerator and denominator completely to determine their common factors.
2. Remove factors equal to 1 by replacing each pair of factors common to the numerator and denominator with the equivalent fraction $\frac{1}{1}$.
3. Multiply the remaining factors in the numerator and in the denominator.

EXAMPLE 3 Simplify: $\frac{21x^3}{14x^2}$

Strategy We will write the numerator and denominator in factored form and then remove pairs of factors that are equal to 1.

Why The rational expression is simplified when the numerator and denominator have no common factor other than 1.

Solution

$$\frac{21x^3}{14x^2} = \frac{3 \cdot 7 \cdot x \cdot x \cdot x}{2 \cdot 7 \cdot x \cdot x}$$ To prepare to simplify the rational expression, factor the numerator and the denominator.

$$= \frac{3 \cdot \overset{1}{\cancel{7}} \cdot \overset{1}{\cancel{x}} \cdot \overset{1}{\cancel{x}} \cdot x}{2 \cdot \underset{1}{\cancel{7}} \cdot \underset{1}{\cancel{x}} \cdot \underset{1}{\cancel{x}}}$$ Simplify by replacing $\frac{7}{7}$ and $\frac{x}{x}$ with the equivalent fraction $\frac{1}{1}$. This removes the factor $\frac{7 \cdot x \cdot x}{7 \cdot x \cdot x}$, which is equal to 1.

$$= \frac{3x}{2}$$ Multiply the remaining factors in the numerator: $3 \cdot 1 \cdot 1 \cdot 1 \cdot x = 3x$. Multiply the remaining factors in the denominator: $2 \cdot 1 \cdot 1 \cdot 1 = 2$.

We say that $\frac{21x^3}{14x^2}$ simplifies to $\frac{3x}{2}$. Since $\frac{21x^3}{14x^2}$ is undefined for $x = 0$, the expressions $\frac{21x^3}{14x^2}$ and $\frac{3x}{2}$ are equal only if $x \neq 0$. That is,

$$\frac{21x^3}{14x^2} = \frac{3x}{2} \quad \text{provided } x \neq 0$$

Caution

Unless otherwise stated, we will now assume that the variables in rational expressions represent real numbers for which the denominator is not zero. Therefore, as we simplify rational expressions, warnings like the "provided $x \neq 0$" statement on the right are not necessary.

An alternate approach for Example 3 is to use rules for exponents to simplify rational expressions that are the quotient of two monomials.

$$\frac{21x^3}{14x^2} = \frac{3 \cdot \overset{1}{\cancel{7}} \cdot x^{3-2}}{2 \cdot \underset{1}{\cancel{7}}} = \frac{3x^1}{2} = \frac{3x}{2}$$ To divide exponential expressions with the same base, keep the base and subtract the exponents.

Teaching Example 3
Simplify: $\frac{20x^5}{15x^4}$
Answer: $\frac{4x}{3}$

Self Check 3 Simplify: $\frac{32a^3}{24a}$ $\frac{4a^2}{3}$

Now Try Problem 41

To simplify rational expressions, we often make use of the factoring methods discussed in Chapter 6.

EXAMPLE 4 Simplify: **a.** $\frac{30t - 6}{36}$ **b.** $\frac{x^2 + 13x + 12}{x^2 + 12x}$ **c.** $\frac{x^3 + x^2}{1 + x}$

Strategy We will begin by factoring the numerator and denominator. Then we will remove any factors common to the numerator and denominator.

Why We need to make sure that the numerator and denominator have no common factor other than 1. When this is the case, the rational expression is simplified.

Solution

a. $$\frac{30t - 6}{36} = \frac{6(5t - 1)}{6 \cdot 6}$$ To prepare to simplify the rational expression, factor the numerator: The GCF is 6. Factor the denominator.

$$= \frac{\overset{1}{\cancel{6}}(5t - 1)}{\underset{1}{\cancel{6}} \cdot 6}$$ Simplify by removing a factor equal to 1. Replace $\frac{6}{6}$ with $\frac{1}{1}$.

$$= \frac{5t - 1}{6}$$ Multiply the remaining factors in the numerator: $1 \cdot (5t - 1) = 5t - 1$. Multiply the remaining factors in the denominator: $1 \cdot 6 = 6$.

The Language of Algebra

Sometimes, a common factor of the numerator and denominator has two or more terms. Here, we remove the common **binomial factor,** $x + 12$.

b. $$\frac{x^2 + 13x + 12}{x^2 + 12x} = \frac{(x + 1)(x + 12)}{x(x + 12)}$$ To prepare to simplify the rational expression, factor the numerator. Factor the denominator: The GCF is x.

$$= \frac{(x + 1)\overset{1}{\cancel{(x + 12)}}}{x\underset{1}{\cancel{(x + 12)}}}$$ Simplify by replacing $\frac{x + 12}{x + 12}$ with the equivalent fraction $\frac{1}{1}$. This removes the factor $\frac{x + 12}{x + 12} = 1$.

$$= \frac{x + 1}{x}$$ This rational expression cannot be simplified further.

c. $\dfrac{x^3 + x^2}{1 + x} = \dfrac{x^2(x + 1)}{1 + x}$ To prepare to simplify the rational expression, factor the numerator. The GCF is x^2.

$= \dfrac{x^2\overset{1}{\cancel{(x + 1)}}}{\underset{1}{\cancel{1 + x}}}$ By the commutative property of addition, $x + 1 = 1 + x$. Simplify by removing a factor equal to 1. Replace $\frac{x+1}{1+x}$ with $\frac{1}{1}$.

$= \dfrac{x^2}{1}$ Multiply in the the numerator: $x^2 \cdot 1 = x^2$.

$= x^2$ Any number divided by 1 is itself.

Self Check 4 Simplify: **a.** $\dfrac{4t - 20}{12}$ $\dfrac{t-5}{3}$ **b.** $\dfrac{x^2 - x - 6}{x^2 - 3x}$ $\dfrac{x+2}{x}$ **c.** $\dfrac{2x^4 + 4x^3}{2 + x}$ $2x^3$

Now Try Problems 43 and 47

Teaching Example 4

Simplify: a. $\dfrac{21d - 28}{49}$ b. $\dfrac{x^2 + 11x + 30}{x^2 + 5x}$ c. $\dfrac{4x^2 + 32x}{8 + x}$

Answers:

a. $\dfrac{3d - 4}{7}$ b. $\dfrac{x + 6}{x}$ c. $4x$

When a rational expression is simplified, the result is an **equivalent expression.** In Example 4b, for instance, this means that $\frac{x^2 + 13x + 12}{x^2 + 12x}$ and $\frac{x+1}{x}$ have the same value for **all** values of x, except those that make either denominator 0. We can use that fact to perform an informal check of our work. If we let $x = 1$, for example, we see that the original rational expression and the simplified expression have the same value, 2.

Teaching Tip: Explain that this is an "informal" check. To be certain that the simplified expression is correct, it must have the same value as the original rational expression for all values of x that make either denominator 0. In this case, that would be -12 and 0.

The original expression	***The simplified expression***
$\dfrac{x^2 + 13x + 12}{x^2 + 12x} = \dfrac{(1)^2 + 13(1) + 12}{(1)^2 + 12(1)}$	$\dfrac{x + 1}{x} = \dfrac{1 + 1}{1}$
$= \dfrac{1 + 13 + 12}{1 + 12}$	$= \dfrac{2}{1}$
$= \dfrac{26}{13}$	$= 2$ Same result
$= 2$	

If the results are different, an error has been made, and the problem should be reworked.

EXAMPLE 5 Simplify: **a.** $\dfrac{3x^2 - 8x - 3}{2x^5 - 18x^3}$ **b.** $\dfrac{(x - y)^4}{x^2 - 2xy + y^2}$

Strategy We will begin by factoring the numerator and denominator using the methods discussed in Chapter 6. Then we will remove any factors common to the numerator and denominator.

Why We need to make sure that the numerator and denominator have no common factor other than 1. When this is the case, then the rational expression is simplified.

Solution **a.** $\dfrac{3x^2 - 8x - 3}{2x^5 - 18x^3} = \dfrac{(3x + 1)(x - 3)}{2x^3(x^2 - 9)}$ To prepare to simplify, factor the trinomial in the numerator. Factor the denominator: The GCF is $2x^3$.

$= \dfrac{(3x + 1)(x - 3)}{2x^3(x + 3)(x - 3)}$ In the denominator, factor the difference of two squares, $x^2 - 9$.

$= \dfrac{(3x + 1)\overset{1}{\cancel{(x - 3)}}}{2x^3(x + 3)\underset{1}{\cancel{(x - 3)}}}$ Simplify by replacing $\frac{x-3}{x-3}$ with the equivalent fraction $\frac{1}{1}$. This removes the factor $\frac{x-3}{x-3} = 1$.

$= \dfrac{3x + 1}{2x^3(x + 3)}$ It is not necessary to perform the multiplication $2x^3(x + 3)$ in the result. It is usually more convenient to leave the denominator in factored form.

Teaching Tip: Note that Example 5b involves expressions in two variables, x and y.

b. $\dfrac{(x-y)^4}{x^2-2xy+y^2} = \dfrac{(x-y)^4}{(x-y)^2}$ To prepare to simplify, factor the perfect-square trinomial $x^2 - 2xy + y^2$ in the denominator.

$= \dfrac{(x-y)(x-y)(x-y)(x-y)}{(x-y)(x-y)}$ Write the repeated multiplication indicated by each exponent.

$= \dfrac{\overset{1}{\cancel{(x-y)}}\overset{1}{\cancel{(x-y)}}(x-y)(x-y)}{\underset{1}{\cancel{(x-y)}}\underset{1}{\cancel{(x-y)}}}$ Simplify by replacing each $\frac{x-y}{x-y}$ with $\frac{1}{1}$.

$= (x-y)^2$ Use an exponent to write the repeated multiplication in the numerator. It is not necessary to find $(x-y)^2$. The result can be presented in factored form.

Teaching Example 5 Simplify:
a. $\dfrac{2x^2-x-15}{3x^3-27x}$ b. $\dfrac{(m+n)^4}{m^2+2mn+n^2}$
Answers: a. $\dfrac{2x+5}{3x(x+3)}$ b. $(m+n)^2$

Self Check 5 Simplify: **a.** $\dfrac{4x^2-4x-15}{8x^3-50x}$ $\dfrac{2x+3}{2x(2x+5)}$ **b.** $\dfrac{(a+3b)^5}{a^2+6ab+9b^2}$ $(a+3b)^3$

Now Try Problems 51 and 53

CAUTION When simplifying rational expressions, we can remove only factors common to the entire numerator and denominator. *It is incorrect to remove any terms common to the numerator and denominator.*

$\dfrac{\overset{1}{\cancel{x}}+1}{\underset{1}{\cancel{x}}}$ x is a term of $x + 1$.

$\dfrac{a^2-3a+\overset{1}{\cancel{2}}}{a+\underset{1}{\cancel{2}}}$ 2 is a term of $a^2 - 3a + 2$ and a term of $a + 2$.

$\dfrac{\overset{1}{\cancel{y^2}}-36}{\underset{1}{\cancel{y^2}}-y-7}$ y^2 is a term of $y^2 - 36$ and a term of $y^2 - y - 7$.

EXAMPLE 6 Simplify: $\dfrac{5(x+3)-5}{7(x+3)-7}$

Strategy We will begin by simplifying the numerator, $5(x + 3) - 5$, and the denominator, $7(x + 3) - 7$, separately. Then we will factor each result and remove any common factors.

Why We cannot remove $x + 3$ immediately because it is not a factor of the *entire* numerator and the *entire* denominator.

Solution

$\dfrac{5(x+3)-5}{7(x+3)-7} = \dfrac{5x+15-5}{7x+21-7}$ Use the distributive property in the numerator and in the denominator.

$= \dfrac{5x+10}{7x+14}$ Combine like terms: $15 - 5 = 10$ and $21 - 7 = 14$.

$= \dfrac{5(x+2)}{7(x+2)}$ To prepare to simplify, factor the numerator: The GCF is 5. Factor the denominator: The GCF is 7.

$= \dfrac{5\overset{1}{\cancel{(x+2)}}}{7\underset{1}{\cancel{(x+2)}}}$ Simplify by replacing $\frac{x+2}{x+2}$ with the equivalent fraction $\frac{1}{1}$. This removes the factor $\frac{x+2}{x+2} = 1$.

$= \dfrac{5}{7}$

The Language of Algebra

Some rational expressions cannot be simplified. For example, to attempt to simplify the following rational expression, we factor its numerator and denominator. Since there are no common factors, we say it **does not simplify** or that it is in **simplest form.**

$$\frac{x^2+x-2}{x^2+x} = \frac{(x+2)(x-1)}{x(x+1)}$$

Teaching Example 6
Simplify: $\dfrac{6(x+3)-6}{8(x+3)-8}$
Answer: $\frac{3}{4}$

Self Check 6 Simplify: $\dfrac{4(x-2)+4}{3(x-2)+3}$ $\frac{4}{3}$

Now Try Problem 55

4 Simplify Rational Expressions That Have Factors That Are Opposites.

If the terms of two polynomials are the same, except that they are opposite in sign, the polynomials are **opposites.** For example, the following pairs of polynomials are opposites.

$2a - 1$ and $1 - 2a$ — Compare terms: $2a$ and $-2a$; -1 and 1.

$-3x^2 - x + 5$ and $3x^2 + x - 5$ — Compare terms: $-3x^2$ and $3x^2$; $-x$ and x; 5 and -5.

Success Tip

When a difference is reversed, the original binomial and the resulting binomial are opposites. Here are some pairs of opposites:

$b - 11$ and $11 - b$

$x^2 - 4$ and $4 - x^2$

Notice that the sum of a polynomial and its opposite is 0.

$$(2a - 1) + (1 - 2a) = \mathbf{0} \qquad (-3x^2 - x + 5) + (3x^2 + x - 5) = \mathbf{0}$$

We have seen that the quotient of two real numbers that are opposites is always -1:

$$\frac{2}{-2} = -1 \qquad \frac{-78}{78} = -1 \qquad \frac{3.5}{-3.5} = -1$$

Likewise, the quotient of two polynomials that are opposites is always -1.

EXAMPLE 7 Simplify: $\dfrac{2a - 1}{1 - 2a}$

Strategy We will rearrange the terms of the numerator, $2a - 1$, and factor out -1.

Why This step is useful when the numerator and denominator contain factors that are opposites, such as $2a - 1$ and $1 - 2a$. It produces a common factor that can be removed.

Solution

$$\frac{2a - 1}{1 - 2a} = \frac{-1 + 2a}{1 - 2a}$$ Think of the numerator, $2a - 1$, as $2a + (-1)$. Then change the order of the terms: $2a + (-1) = -1 + 2a$.

$$= \frac{-1(1 - 2a)}{1 - 2a}$$ To prepare to simplify: factor out -1 from the two terms of the numerator.

$$= \frac{-1(\overset{1}{\cancel{1 - 2a}})}{\underset{1}{\cancel{1 - 2a}}}$$ Simplify by replacing $\frac{1 - 2a}{1 - 2a}$ with the equivalent fraction $\frac{1}{1}$. This removes the factor $\frac{1 - 2a}{1 - 2a} = 1$.

$$= \frac{-1}{1}$$ Multiply the remaining factors in the numerator.

$$= -1$$ Any number divided by 1 is itself.

Success Tip

The result would have been the same if, instead, we had factored out -1 from the denominator, $1 - 2a$.

Self Check 7 Simplify: $\frac{3p - 2}{2 - 3p}$ -1

Now Try Problem 59

Teaching Example 7
Simplify: $\dfrac{5x - 3}{3 - 5x}$
Answer: -1

In general, we have this fact.

The Quotient of Opposites The quotient of any nonzero polynomial and its opposite is -1.

Caution

Don't incorrectly apply the rule for opposites to a rational expression such as $\frac{x + 6}{6 + x}$. This is the quotient of a binomial and itself. The result is 1, not -1.

$$\frac{x + 6}{6 + x} = \frac{\overset{1}{\cancel{x + 6}}}{\underset{1}{\cancel{x + 6}}} = 1$$

For each of the following rational expressions, the numerator and denominator are opposites. Thus, each expression is equal to -1.

$$\frac{x - 6}{6 - x} = -1 \qquad \frac{2a - 9b}{9b - 2a} = -1 \qquad \frac{-3x^2 - x + 5}{3x^2 + x - 5} = -1$$

This fact can be used to simplify certain rational expressions by removing a factor equal to -1. If a factor of the numerator is the opposite of a factor of the denominator, we can replace them with the equivalent fraction $\frac{-1}{1}$, as shown in the following example.

EXAMPLE 8 Simplify: **a.** $\dfrac{y^2 - 1}{3 - 3y}$ **b.** $\dfrac{t + 8}{t - 8}$

Strategy We will begin by factoring the numerator and denominator. Then we look for common factors, or factors that are opposites, and remove them.

Why We need to make sure that the numerator and denominator have no common factor (or opposite factors) other than 1. When this is the case, then the rational expression is simplified.

Solution **a.** $\dfrac{y^2 - 1}{3 - 3y} = \dfrac{(y + 1)(y - 1)}{3(1 - y)}$ To prepare to simplify, factor the numerator, and factor the denominator.

$= \dfrac{(y + 1)\overset{-1}{\cancel{(y - 1)}}}{3\underset{1}{\cancel{(1 - y)}}}$ Since $y - 1$ and $1 - y$ are opposites, simplify by replacing $\frac{y - 1}{1 - y}$ with the equivalent fraction $\frac{-1}{1}$. This removes the factor $\frac{y - 1}{1 - y} = -1$.

$= \dfrac{-(y + 1)}{3}$ In the numerator, $-1 \cdot (y + 1)$ can be written as $-(y + 1)$. In the denominator, $3 \cdot 1 = 3$.

Teaching Tip: Caution students to write -1 only once (not twice) when they remove common factors of the numerator and denominator that are opposites.

Incorrect

$\dfrac{(y + 1)\overset{-1}{\cancel{(y - 1)}}}{3\underset{-1}{\cancel{(1 - y)}}}$

This result may be written in several other equivalent forms.

$\dfrac{-(y + 1)}{3} = -\dfrac{y + 1}{3}$ The $-$ symbol in $-(y + 1)$ can be written in the front of the fraction, and the parentheses can be dropped.

$\dfrac{-(y + 1)}{3} = \dfrac{-y - 1}{3}$ The $-$ symbol in $-(y + 1)$ represents a factor of -1. Distribute the multiplication by -1 in the numerator.

$\dfrac{-(y + 1)}{3} = \dfrac{y + 1}{-3}$ The $-$ symbol in $-(y + 1)$ can be applied to the denominator. However, we don't usually use this form.

Caution

A $-$ symbol in front of a fraction may be applied to the numerator or to the denominator, but not to both:

$$-\frac{y + 1}{3} \neq \frac{-(y + 1)}{-3}$$

b. The binomials $t + 8$ and $t - 8$ are not opposites because their first terms do not have opposite signs. Thus, $\frac{t + 8}{t - 8}$ does not simplify.

Teaching Example 8

Simplify: a. $\dfrac{a^2 - 9}{18 - 6a}$ b. $\dfrac{-19x + 3}{-19x - 3}$

Answers: a. $-\dfrac{a + 3}{6}$ b. Does not simplify

Self Check 8 Simplify: **a.** $\dfrac{m^2 - 100}{10m - m^2}$ $-\dfrac{m + 10}{m}$ **b.** $\dfrac{2x - 3}{2x + 3}$ Does not simplify

Now Try Problem 63

SECTION 7.1 STUDY SET

VOCABULARY

Fill in the blanks.

1. A quotient of two polynomials, such as $\frac{x^2 + x}{x^2 - 3x}$, is called a rational expression.
2. To simplify a rational expression, we remove common factors of the numerator and denominator.
3. Because of the division by 0, the expression $\frac{8}{0}$ is undefined.
4. The binomials $x - 15$ and $15 - x$ are called opposites, because their terms are the same, except that they are opposite in sign.

CONCEPTS

5. When we simplify $\frac{x^2 + 5x}{4x + 20}$, the result is $\frac{x}{4}$. These equivalent expressions have the same value for all real numbers, except $x = -5$. Show that they have the same value for $x = 1$. $\frac{6}{24} = \frac{1}{4}$
6. Determine whether each pair of polynomials are opposites. Write *yes* or *no.*
   a. $y + 7$ and $y - 7$ No
   b. $b - 20$ and $20 - b$ Yes
   c. $x^2 + 2x - 1$ and $-x^2 - 2x - 1$ No

7. Simplify each expression, if possible.

a. $\frac{x-8}{x-8}$ 1 b. $\frac{x-8}{8-x}$ -1

c. $\frac{x+8}{8+x}$ 1 d. $\frac{x+8}{x}$ Does not simplify

8. Simplify each expression.

a. $\frac{(x+2)(x-2)}{(x+1)(x+2)}$ $\frac{x-2}{x+1}$ b. $\frac{y(y-2)}{9(2-y)}$ $-\frac{y}{9}$

c. $\frac{(2m+7)(m-5)}{(2m+7)}$ $m-5$ d. $\frac{x \cdot x}{x \cdot x(x-30)}$ $\frac{1}{x-30}$

NOTATION

Complete the solution to simplify the rational expression.

9. $\frac{x^2+2x+1}{x^2+4x+3} = \frac{(x+1)(x+1)}{(x+3)(x+1)}$

$= \frac{(x+1)\overset{1}{\cancel{(x+1)}}}{(x+3)\underset{1}{\cancel{(x+1)}}}$

$= \frac{x+1}{x+3}$

10. In the following table, a student's answers to three homework problems are compared with the answers in the back of the book. Are the answers equivalent?

Student's Answer	Book's answer	Equivalent?
$\frac{-3}{x+3}$	$-\frac{3}{x+3}$	Yes
$\frac{-(x-4)}{6x+1}$	$\frac{-x+4}{6x+1}$	Yes
$\frac{x+7}{(x-4)(x+2)}$	$\frac{x+7}{(x+2)(x-4)}$	Yes
$\frac{x^2+6}{-x}$	$-\frac{x^2+6}{x}$	Yes

GUIDED PRACTICE

Evaluate each expression for x = 6. See Example 1.

11. $\frac{x-2}{x-5}$ 4

12. $\frac{3x-2}{x-2}$ 4

13. $\frac{x^2-4x-12}{x^2+x-2}$ 0

14. $\frac{x^2-36}{x^3-1}$ 0

15. $\frac{-x+1}{x^2-5x-6}$ Undefined

16. $\frac{-2x^2-3}{x-6}$ Undefined

Evaluate each expression for y = −3. See Example 1.

17. $\frac{y+5}{3y-2}$ $-\frac{2}{11}$

18. $\frac{2y+9}{y^2+25}$ $\frac{3}{34}$

19. $-\frac{y}{y^2-y+6}$ $\frac{1}{6}$

20. $-\frac{y^3}{3y^2+1}$ $\frac{27}{28}$

21. $\frac{y^2+9}{9-y^2}$ Undefined

22. $\frac{-y-11}{y^2+2y-3}$ Undefined

Find all real numbers for which the rational expression is undefined. See Example 2.

23. $\frac{15}{x-2}$ 2

24. $\frac{5x}{x+5}$ -5

25. $\frac{x+5}{8x}$ 0

26. $\frac{4x-1}{6x}$ 0

27. $\frac{15x+2}{x^2+6}$ None

28. $\frac{x^2-4x}{x^2+4}$ None

29. $\frac{x+1}{2x-1}$ $\frac{1}{2}$

30. $\frac{-6x}{3x-1}$ $\frac{1}{3}$

31. $\frac{x^2-6x}{9}$ None

32. $\frac{x^3-x^2}{15}$ None

33. $\frac{30x}{x^2-36}$ $-6, 6$

34. $\frac{2x-15}{x^2-49}$ $-7, 7$

35. $\frac{15}{x^2+x-2}$ $-2, 1$

36. $\frac{x-20}{x^2+2x-8}$ $-4, 2$

37. $\frac{16}{20-x}$ 20

38. $\frac{44}{57-x}$ 57

Simplify. See Example 3.

39. $\frac{45}{9a}$ $\frac{5}{a}$

40. $\frac{48}{16y}$ $\frac{3}{y}$

41. $\frac{6x^4}{4x^2}$ $\frac{3x^2}{2}$

42. $\frac{9x^3}{6x}$ $\frac{3x^2}{2}$

Simplify. See Example 4.

43. $\frac{6x+3}{9}$ $\frac{2x+1}{3}$

44. $\frac{4x+12}{16}$ $\frac{x+3}{4}$

45. $\frac{x+3}{3x+9}$ $\frac{1}{3}$

46. $\frac{2x-14}{x-7}$ 2

47. $\frac{x^2-4}{x^2-6x+8}$ $\frac{x+2}{x-4}$

48. $\frac{y^2-25}{y^2-3y-10}$ $\frac{y+5}{y+2}$

49. $\frac{x^2+5x+4}{x^2+4x}$ $\frac{x+1}{x}$

50. $\frac{x^2-10x+21}{x^2-3x}$ $\frac{x-7}{x}$

Simplify. See Example 5.

51. $\frac{m^2-2mn+n^2}{7m^2-7n^2}$ $\frac{m-n}{7(m+n)}$ or $\frac{m-n}{7m+7n}$

52. $\frac{11c^2-11d^2}{c^2-2cd+d^2}$ $\frac{11(c+d)}{c-d}$ or $\frac{11c+11d}{c-d}$

53. $\frac{4b^2+4b+1}{(2b+1)^3}$ $\frac{1}{2b+1}$

54. $\frac{9y^2-12y+4}{(3y-2)^3}$ $\frac{1}{3y-2}$

Simplify. See Example 6.

55. $\frac{10(c-3)+10}{3(c-3)+3}$ $\frac{10}{3}$

56. $\frac{6(d+3)-6}{7(d+3)-7}$ $\frac{6}{7}$

57. $\frac{6(x+3)-18}{3x-18}$ $\frac{2x}{x-6}$

58. $\frac{4(t-1)+4}{4t+4}$ $\frac{t}{t+1}$

Simplify. See Example 7.

59. $\frac{2x-7}{7-2x}$ -1

60. $\frac{18-d}{d-18}$ -1

61. $\frac{3-4t}{8t-6}$ $-\frac{1}{2}$

62. $\frac{5t-1}{3-15t}$ $-\frac{1}{3}$

Simplify. See Example 8.

63. $\frac{2-a}{a^2-a-2}$ $-\frac{1}{a+1}$

▶ 64. $\frac{4-b}{b^2-5b+4}$ $-\frac{1}{b-1}$

65. $\frac{25-5m}{m^2-25}$ $-\frac{5}{m+5}$

▶ 66. $\frac{36-6h}{h^2-36}$ $-\frac{6}{h+6}$

TRY IT YOURSELF

Simplify. If an expression cannot be simplified, write "Does not simplify."

67. $\frac{a^3-a^2}{a^4-a^3}$ $\frac{1}{a}$

68. $\frac{2c^4+2c^3}{4c^5+4c^4}$ $\frac{1}{2c}$

69. $\frac{4-x^2}{x^2-x-2}$ $-\frac{x+2}{x+1}$

▶ 70. $\frac{81-y^2}{y^2+10y+9}$ $-\frac{y-9}{y+1}$

71. $\frac{6x-30}{5-x}$ -6

72. $\frac{6t-42}{7-t}$ -6

73. $\frac{x^2+3x+2}{x^2+x-2}$ $\frac{x+1}{x-1}$

▶ 74. $\frac{x^2+x-6}{x^2-x-2}$ $\frac{x+3}{x+1}$

75. $\frac{15x^2y}{5xy^2}$ $\frac{3x}{y}$

▶ 76. $\frac{12xz}{4xz^2}$ $\frac{3}{z}$

77. $\frac{x^8+9x^7}{9+x}$ x^7

78. $\frac{x^9+50x^8}{50+x}$ x^8

▶ 79. $\frac{x(x-8)+16}{16-x^2}$ $\frac{4-x}{4+x}$ or $-\frac{x-4}{x+4}$

▶ 80. $\frac{x^2-3(2x-3)}{9-x^2}$ $\frac{3-x}{3+x}$ or $-\frac{x-3}{x+3}$

81. $\frac{4c+4d}{d+c}$ 4

82. $\frac{a+b}{5b+5a}$ $\frac{1}{5}$

83. $\frac{3x^2-27}{2x^2-5x-3}$ $\frac{3(x+3)}{2x+1}$ or $\frac{3x+9}{2x+1}$

▶ 84. $\frac{2x^2-8}{3x^2-5x-2}$ $\frac{2(x+2)}{3x+1}$ or $\frac{2x+4}{3x+1}$

85. $\frac{-3x^2+10x+77}{x^2-4x-21}$ $-\frac{3x+11}{x+3}$

▶ 86. $\frac{-2x^2+5x+3}{x^2+2x-15}$ $-\frac{2x+1}{x+5}$

87. $\frac{42c^3d}{18cd^3}$ $\frac{7c^2}{3d^2}$

88. $\frac{49m^4n^5}{35mn^6}$ $\frac{7m^3}{5n}$

▶ 89. $\frac{16a^2-1}{4a+4}$ Does not simplify

▶ 90. $\frac{25m^2-1}{5m+5}$ Does not simplify

91. $\frac{8u^2-2u-15}{4u^4+5u^3}$ $\frac{2u-3}{u^3}$

▶ 92. $\frac{6n^2-7n+2}{3n^3-2n^2}$ $\frac{2n-1}{n^2}$

93. $\frac{(2x+3)^4}{4x^2+12x+9}$ $(2x+3)^2$

94. $\frac{(3y-2)^5}{9y^2-12y+4}$ $(3y-2)^3$

▶ 95. $\frac{6a+3(a+2)+12}{a+2}$ 9

▶ 96. $\frac{2y+4(y-1)-2}{y-1}$ 6

97. $\frac{15x-3x^2}{25y-5xy}$ $\frac{3x}{5y}$

98. $\frac{18c-2c^2}{81d-9cd}$ $\frac{2c}{9d}$

▶ 99. $\frac{2x^2}{x+2}$ Does not simplify

100. $\frac{5y^2}{y+5}$ Does not simplify

101. $\frac{18+2x}{x^2-81}$ $\frac{2}{x-9}$

102. $\frac{12+6x}{x^2-4}$ $\frac{6}{x-2}$

APPLICATIONS

▶ 103. **Organ Pipes.** The number of vibrations n per second of an organ pipe is given by the formula $n=\frac{512}{L}$ where L is the length of the pipe in feet. How many times per second will a 6-foot pipe vibrate? $85\frac{1}{3}$

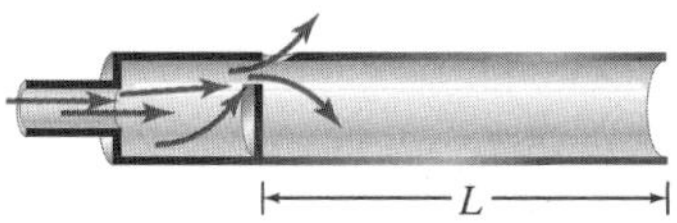

▶ 104. **Raising Turkeys.** The formula $T=\frac{2{,}000m}{m+1}$ gives the number T of turkeys on a poultry farm m months after the beginning of the year. How many turkeys will there be on the farm by the end of July? 1,750

105. **Medical Dosages.** The formula $c=\frac{4t}{t^2+1}$ gives the concentration c (in milligrams per liter) of a certain dosage of medication in a patient's bloodstream t hours after the medication is administered. Suppose the patient received the medication at noon. Find the concentration of medication in his blood at the following times later that afternoon. 2, 1.6, and 1.2 milligrams per liter

▶ 106. **Manufacturing.** If a company produces x child car seats, the average cost c (in dollars) to produce one car seat is given by the formula $c=\frac{50x+50{,}000}{x}$. Find the company's average production cost if 1,000 are produced. \$100 per car seat

WRITING

107. Explain why $\frac{x-7}{7-x}=-1$.

108. Explain why $\frac{x-3}{x+4}$ is undefined for $x=-4$ but defined for $x=3$.

109. Explain the error in the following work:

Simplify: $\frac{x}{x+2}=\frac{\overset{1}{\cancel{x}}}{\underset{1}{\cancel{x}}+2}$

$=\frac{1}{3}$

110. Explain the error in the following work:

Simplify: $\frac{30t-6}{36}=\frac{\overset{1}{\cancel{2}}\cdot\overset{1}{\cancel{3}}\cdot 5\cdot t-\overset{1}{\cancel{2}}\cdot\overset{1}{\cancel{3}}}{\underset{1}{\cancel{2}}\cdot\underset{1}{\cancel{2}}\cdot\underset{1}{\cancel{3}}\cdot\underset{1}{\cancel{3}}}$

$=5t-1$

▶ 111. Explain why there are no values for x for which $\frac{x-7}{x^2+49}$ is undefined.

112. Write a rational expression that is not defined for $x=5$. Then explain why that is so.

REVIEW

State each property using the variables a, b, and when necessary, c.

113. a. The associative property of addition $(a + b) + c = a + (b + c)$

b. The commutative property of multiplication $ab = ba$

114. a. The distributive property $a(b + c) = ab + ac$

b. The zero-factor property If $ab = 0$, then $a = 0$ or $b = 0$.

CHALLENGE PROBLEMS

Simplify.

115. $\dfrac{(x^2 + 2x + 1)(x^2 - 2x + 1)}{(x^2 - 1)^2}$ 1

▶ **116.** $\dfrac{2x^2 + 2x - 12}{x^3 + 3x^2 - 4x - 12}$ $\dfrac{2}{x + 2}$

117. $\dfrac{x^3 - 27}{x^3 - 9x}$ $\dfrac{x^2 + 3x + 9}{x(x + 3)}$

▶ **118.** $\dfrac{b^3 + a^3}{a^2 - ab + b^2}$ $b + a$

119. $\dfrac{m^3 + 64}{m^3 + 4m^2 + 3m + 12}$ $\dfrac{m^2 - 4m + 16}{m^2 + 3}$

120. $\dfrac{s^3 + s^2 - 6s - 6}{s^3 + 1}$ $\dfrac{s^2 - 6}{s^2 - s + 1}$

SECTION 7.2 Multiplying and Dividing Rational Expressions

OBJECTIVES

1. Multiply rational expressions.
2. Divide rational expressions.
3. Convert units of measurement.

ARE YOU READY? *Are You Ready? exercises available online at www.webassign.net/brookscole*

The following problems review some basic skills that are needed when multiplying and dividing rational expressions.

1. Multiply: $\dfrac{3}{4} \cdot \dfrac{1}{5}$ $\dfrac{3}{20}$

2. What is the reciprocal of $\dfrac{8}{9}$? $\dfrac{9}{8}$

3. Divide: $\dfrac{11}{16} \div \dfrac{7}{8}$ $\dfrac{11}{14}$

4. Factor: $x - x^2$ $x(1 - x)$

In this section, we will extend the rules for multiplying and dividing fractions to problems involving multiplication and division of rational expressions.

1 Multiply Rational Expressions.

Recall that to multiply fractions, we multiply their numerators and multiply their denominators. For example,

$$\frac{4}{7} \cdot \frac{3}{5} = \frac{4 \cdot 3}{7 \cdot 5} \quad \text{Multiply the numerators and multiply the denominators.}$$

$$= \frac{12}{35}$$

We use the same procedure to multiply rational expressions.

Multiplying Rational Expressions

To multiply two rational expressions, multiply their numerators and their denominators. Then, if possible, factor and simplify.

For any two rational expressions, $\frac{A}{B}$ and $\frac{C}{D}$,

$$\frac{A}{B} \cdot \frac{C}{D} = \frac{AC}{BD}$$

EXAMPLE 1 Multiply: **a.** $\frac{x+1}{x} \cdot \frac{9}{4x^2}$ **b.** $\frac{35x^3}{17y} \cdot \frac{y}{5x}$

Strategy To find the product, we will use the rule for multiplying rational expressions. In the process, we must be prepared to factor the numerators and denominators so that any common factors can be removed.

Why We want to give the result in simplified form, which requires that the numerator and denominator have no common factors other than 1.

Solution **a.** $\frac{x+1}{x} \cdot \frac{9}{4x^2} = \frac{9(x+1)}{4x^3}$ Multiply the numerators. Multiply the denominators.

Since the numerator and denominator do not share any common factors, $\frac{9(x+1)}{4x^3}$ cannot be simplified. We can leave the numerator in factored form, or we can distribute the multiplication by 9 and write the result as $\frac{9x+9}{4x^3}$.

Teaching Tip: Stress the importance of factoring first. Warn students that, if instead, they find the product of the numerators and the product of the denominators first, the resulting rational expression, $\frac{35x^3y}{85xy}$, is more difficult to simplify in that form.

b. $\frac{35x^3}{17y} \cdot \frac{y}{5x} = \frac{35x^3 \cdot y}{17y \cdot 5x}$ Multiply the numerators. Multiply the denominators.

It is obvious that the numerator and denominator of $\frac{35x^3 \cdot y}{17y \cdot 5x}$ have several common factors, such as 5, x, and y. These common factors become more apparent when we factor the numerator and denominator completely.

$$\frac{35x^3 \cdot y}{17y \cdot 5x} = \frac{5 \cdot 7 \cdot x \cdot x \cdot x \cdot y}{17 \cdot y \cdot 5 \cdot x}$$ To prepare to simplify, factor $35x^3$.

$$= \frac{\overset{1}{\cancel{5}} \cdot 7 \cdot \overset{1}{\cancel{x}} \cdot x \cdot x \cdot \overset{1}{\cancel{y}}}{17 \cdot \underset{1}{\cancel{y}} \cdot \underset{1}{\cancel{5}} \cdot \underset{1}{\cancel{x}}}$$ Simplify by replacing $\frac{5}{5}$, $\frac{x}{x}$, and $\frac{y}{y}$ with the equivalent fraction $\frac{1}{1}$. This removes the factor $\frac{5 \cdot x \cdot y}{5 \cdot x \cdot y} = 1$.

$$= \frac{7x^2}{17}$$ Multiply the remaining factors in the numerator. Multiply the remaining factors in the denominator.

Caution

When multiplying rational expressions, always write the result in simplest form by removing any factors common to the numerator and denominator.

Teaching Example 1
Multiply: a. $\frac{x-4}{x} \cdot \frac{10}{9x^2}$ b. $\frac{12a^3}{13b} \cdot \frac{b}{4a}$
Answers: a. $\frac{10(x-4)}{9x^3}$ b. $\frac{3a^2}{13}$

Self Check 1 Multiply: **a.** $\frac{a+7}{a} \cdot \frac{6}{5a^3}$ $\frac{6(a+7)}{5a^4}$ **b.** $\frac{a^4}{8b} \cdot \frac{24b}{11a^3}$ $\frac{3a}{11}$

Now Try Problems 13 and 17

EXAMPLE 2 Multiply: **a.** $\frac{x+3}{2x+4} \cdot \frac{6}{x^2-9}$ **b.** $\frac{8x^2-8x}{x^2+x-56} \cdot \frac{3x^2-22x+7}{x-x^2}$

Strategy To find the product, we will use the rule for multiplying rational expressions. In the process, we need to factor the monomials, binomials, or trinomials that are not prime, so that any common factors can be removed.

Why We want to give the result in simplified form, which requires that the numerator and denominator have no common factor other than 1.

Solution **a.** $\frac{x+3}{2x+4} \cdot \frac{6}{x^2-9} = \frac{(x+3)6}{(2x+4)(x^2-9)}$ Multiply the numerators and multiply the denominators.

$$= \frac{(x+3) \cdot 3 \cdot 2}{2(x+2)(x+3)(x-3)}$$ Factor 6. Factor out the GCF, 2, from $2x + 4$. Factor the difference of two squares, $x^2 - 9$.

$$= \frac{\overset{1}{\cancel{(x+3)}} \cdot 3 \cdot \overset{1}{\cancel{2}}}{\underset{1}{\cancel{2}}(x+2)\underset{1}{\cancel{(x+3)}}(x-3)}$$ Simplify by replacing $\frac{x+3}{x+3}$ and $\frac{2}{2}$ with $\frac{1}{1}$. This removes the factor $\frac{2 \cdot (x+3)}{2 \cdot (x+3)} = 1$.

$$= \frac{3}{(x+2)(x-3)}$$ Multiply the remaining factors in the numerator. Multiply the remaining factors in the denominator.

Notation

It is not necessary to multiply $(x + 2)(x - 3)$ in the denominator. When we add and subtract rational expressions in the next section, it is usually more convenient to leave the denominator in factored form as shown here.

b. $\dfrac{8x^2 - 8x}{x^2 + x - 56} \cdot \dfrac{3x^2 - 22x + 7}{x - x^2}$

$= \dfrac{(8x^2 - 8x)(3x^2 - 22x + 7)}{(x^2 + x - 56)(x - x^2)}$ Multiply the numerators and multiply the denominators.

$= \dfrac{8x(x - 1)(3x - 1)(x - 7)}{(x + 8)(x - 7)x(1 - x)}$ To prepare to simplify, factor all four polynomials.

$= \dfrac{8\overset{1}{\cancel{x}}\overset{-1}{\cancel{(x - 1)}}(3x - 1)\overset{1}{\cancel{(x - 7)}}}{(x + 8)\underset{1}{\cancel{(x - 7)}}\underset{1}{\cancel{x}}\underset{1}{\cancel{(1 - x)}}}$ Simplify. Since $x - 1$ and $1 - x$ are opposites, replace $\frac{x - 1}{1 - x}$ with $\frac{-1}{1}$. This removes the factor $\frac{x - 1}{1 - x} = -1$.

$= \dfrac{-8(3x - 1)}{x + 8}$ Multiply the remaining factors in the numerator. Multiply the remaining factors in the denominator.

The result can also be written as $-\dfrac{8(3x - 1)}{x + 8}$ or $-\dfrac{24x - 8}{x + 8}$.

Notation

We could distribute in the numerator and write the result as $\frac{-24x + 8}{x + 8}$. Check with your instructor to see which form of the result he or she prefers.

Teaching Example 2

Multiply: a. $\dfrac{n + 4}{2n + 14} \cdot \dfrac{18}{n^2 - 16}$

b. $\dfrac{11x^2 - 22x}{x^2 + 3x - 18} \cdot \dfrac{5x^2 - 16x + 3}{2x - x^2}$

Answers: a. $\dfrac{9}{(n + 7)(n - 4)}$ b. $-\dfrac{11(5x - 1)}{x + 6}$

Self Check 2 Multiply: **a.** $\dfrac{3n - 9}{3n + 2} \cdot \dfrac{9n^2 - 4}{6}$ $\dfrac{(n - 3)(3n - 2)}{2}$

b. $\dfrac{m^2 - 4m - 5}{2m - m^2} \cdot \dfrac{2m^2 - 4m}{3m^2 - 14m - 5}$ $-\dfrac{2(m + 1)}{3m + 1}$

Now Try Problems 21 and 29

EXAMPLE 3 Multiply: **a.** $63x\left(\dfrac{1}{7x}\right)$ **b.** $5a\left(\dfrac{3a - 1}{a}\right)$

Strategy We will write each of the monomials, $63x$ and $5a$, as rational expressions with denominator 1. (Remember, any number divided by 1 remains unchanged.) Then we will use the rule for multiplying rational expressions.

Why Writing $63x$ and $5a$ over 1 is helpful during the multiplication process when we multiply numerators and multiply denominators.

Solution

a. $63x\left(\dfrac{1}{7x}\right) = \dfrac{63x}{1}\left(\dfrac{1}{7x}\right)$ Write $63x$ as a fraction: $63x = \frac{63x}{1}$.

$= \dfrac{63x \cdot 1}{1 \cdot 7 \cdot x}$ Multiply the numerators and multiply the denominators.

$= \dfrac{9 \cdot \overset{1}{\cancel{7}} \cdot \overset{1}{\cancel{x}} \cdot 1}{1 \cdot \underset{1}{\cancel{7}} \cdot \underset{1}{\cancel{x}}}$ Write $63x$ in factored form as $9 \cdot 7 \cdot x$. Then simplify by removing a factor equal to 1: $\frac{7x}{7x}$.

$= 9$ Because $\frac{9}{1} = 9$

b. $5a\left(\dfrac{3a - 1}{a}\right) = \dfrac{5a}{1}\left(\dfrac{3a - 1}{a}\right)$ Write $5a$ as a fraction: $5a = \frac{5a}{1}$.

$= \dfrac{5\overset{1}{\cancel{a}}(3a - 1)}{1 \cdot \underset{1}{\cancel{a}}}$ Multiply the numerators and multiply the denominators. Then simplify by removing a factor equal to 1: $\frac{a}{a}$.

$= 5(3a - 1)$

$= 15a - 5$ Distribute the multiplication by 5.

Teaching Example 3

Multiply: a. $26x\left(\dfrac{1}{13x}\right)$

b. $3d\left(\dfrac{5d - 7}{d}\right)$

Answers: a. 2 b. $15d - 21$

Self Check 3 Multiply: **a.** $36b\left(\dfrac{1}{6b}\right)$ 6 **b.** $4x\left(\dfrac{x + 3}{x}\right)$ $4x + 12$

Now Try Problems 31 and 37

2 Divide Rational Expressions.

Recall that one number is the **reciprocal** of another if their product is 1. To find the reciprocal of a fraction, we invert its numerator and denominator. We have seen that to divide fractions, we multiply the first fraction by the reciprocal of the second fraction.

$$\frac{4}{7} \div \frac{3}{5} = \frac{4}{7} \cdot \frac{5}{3}$$ Invert $\frac{3}{5}$ and change the division to a multiplication.

$$= \frac{20}{21}$$ Multiply the numerators and multiply the denominators.

We use the same procedure to divide rational expressions.

Dividing Rational Expressions

To divide two rational expressions, multiply the first by the reciprocal of the second. Then, if possible, factor and simplify.

For any two rational expressions, $\frac{A}{B}$ and $\frac{C}{D}$, where $\frac{C}{D} \neq 0$,

$$\frac{A}{B} \div \frac{C}{D} = \frac{A}{B} \cdot \frac{D}{C} = \frac{AD}{BC}$$

EXAMPLE 4 Divide: **a.** $\frac{a}{13} \div \frac{17}{26}$ **b.** $\frac{9x}{35y} \div \frac{15x^2}{14}$

Teaching Tip: Before explaining Examples 4–6, have your students find the reciprocals of the expressions $\frac{15x^2}{14}$, $\frac{(x+1)^2}{6x-30}$, and $4y^2 - x^2$.

Strategy We will use the rule for dividing rational expressions. After multiplying by the reciprocal, we will factor the monomials that are not prime, and remove any common factors of the numerator and denominator.

Why We want to give the result in simplified form, which requires that the numerator and denominator have no common factor other than 1.

Solution

a. $$\frac{a}{13} \div \frac{17}{26} = \frac{a}{13} \cdot \frac{26}{17}$$ Multiply by the reciprocal of $\frac{17}{26}$.

$$= \frac{a \cdot 2 \cdot 13}{13 \cdot 17}$$ Multiply the numerators and multiply the denominators. Then, to prepare to simplify, factor 26 as $2 \cdot 13$.

$$= \frac{a \cdot 2 \cdot \overset{1}{\cancel{13}}}{\underset{1}{\cancel{13}} \cdot 17}$$ Simplify by removing common factors of the numerator and denominator.

$$= \frac{2a}{17}$$ Multiply the remaining factors in the numerator. Multiply the remaining factors in the denominator.

Caution

When dividing rational expressions, always write the result in simplest form, by removing any factors common to the numerator and denominator.

b. $$\frac{9x}{35y} \div \frac{15x^2}{14} = \frac{9x}{35y} \cdot \frac{14}{15x^2}$$ Multiply by the reciprocal of $\frac{15x^2}{14}$.

$$= \frac{3 \cdot 3 \cdot x \cdot 2 \cdot 7}{5 \cdot 7 \cdot y \cdot 3 \cdot 5 \cdot x \cdot x}$$ Multiply the numerators and multiply the denominators. Then, to prepare to simplify, factor 9, 35, 14, and $15x^2$.

$$= \frac{3 \cdot \overset{1}{\cancel{3}} \cdot \overset{1}{\cancel{x}} \cdot 2 \cdot \overset{1}{\cancel{7}}}{5 \cdot \underset{1}{\cancel{7}} \cdot y \cdot \underset{1}{\cancel{3}} \cdot 5 \cdot \underset{1}{\cancel{x}} \cdot x}$$ Simplify by removing factors equal to 1.

$$= \frac{6}{25xy}$$ Multiply the remaining factors in the numerator. Multiply the remaining factors in the denominator.

Teaching Tip: Caution students not to remove common factors of the numerator and denominator until the operation becomes multiplication.

Teaching Example 4

Divide: **a.** $\frac{x}{12} \div \frac{11}{24}$ **b.** $\frac{9x}{7y} \div \frac{18x}{35y^2}$

Answers: **a.** $\frac{2x}{11}$ **b.** $\frac{5y}{2}$

Self Check 4 Divide: $\frac{8a}{3b} \div \frac{16a^2}{9b^2}$ $\frac{3b}{2a}$

Now Try Problems 41 and 45

EXAMPLE 5 Divide: $\frac{x^2 + x}{3x - 15} \div \frac{(x + 1)^2}{6x - 30}$

Strategy To find the quotient, we will use the rule for dividing rational expressions. After multiplying by the reciprocal, we will factor the binomials that are not prime, and remove any common factors of the numerator and denominator.

Why We want to give the result in simplified form, which requires that the numerator and denominator have no common factor other than 1.

Solution

$$\frac{x^2 + x}{3x - 15} \div \frac{(x + 1)^2}{6x - 30}$$

$$= \frac{x^2 + x}{3x - 15} \cdot \frac{6x - 30}{(x + 1)^2}$$ Multiply by the reciprocal of $\frac{(x + 1)^2}{6x - 30}$.

$$= \frac{x(x + 1) \cdot 2 \cdot 3(x - 5)}{3(x - 5)(x + 1)(x + 1)}$$ Multiply the numerators and multiply the denominators. Then, to prepare to simplify, factor the binomials. Write $(x + 1)^2$ as repeated multiplication.

$$= \frac{x\overset{1}{\cancel{(x + 1)}} \cdot 2 \cdot \overset{1}{\cancel{3}}\overset{1}{\cancel{(x - 5)}}}{\underset{1}{\cancel{3}}\underset{1}{\cancel{(x - 5)}}\underset{1}{\cancel{(x + 1)}}(x + 1)}$$ Simplify by removing common factors of the numerator and denominator.

$$= \frac{2x}{x + 1}$$ Multiply the remaining factors in the numerator. Multiply the remaining factors in the denominator.

The Language of Algebra

To find the reciprocal of $\frac{(x + 1)^2}{6x - 30}$, we invert it. To **invert** means to turn upside down: $\frac{6x - 30}{(x + 1)^2}$. Some amusement park thrill rides have giant loops where the riders become inverted.

Teaching Example 5

Divide: $\frac{x^2 + 3x}{2x - 10} \div \frac{(x + 3)^2}{8x - 40}$

Answer: $\frac{4x}{x + 3}$

Self Check 5 Divide: $\frac{z^2 - 9}{z^2 + 4z + 3} \div \frac{z^2 - 3z}{(z + 1)^2}$ $\frac{z + 1}{z}$

Now Try Problem 51

EXAMPLE 6 Divide: $\frac{2x^2 - 3xy - 2y^2}{2x + y} \div (4y^2 - x^2)$

Strategy We begin by writing $4y^2 - x^2$ as a rational expression by inserting a denominator 1. Then we will use the rule for dividing rational expressions.

Why Writing $4y^2 - x^2$ over 1 is helpful when we invert its numerator and denominator to find its reciprocal.

Solution

$$\frac{2x^2 - 3xy - 2y^2}{2x + y} \div (4y^2 - x^2)$$

$$= \frac{2x^2 - 3xy - 2y^2}{2x + y} \div \frac{4y^2 - x^2}{1}$$ Write $4y^2 - x^2$ as a fraction with a denominator of 1.

$$= \frac{2x^2 - 3xy - 2y^2}{2x + y} \cdot \frac{1}{4y^2 - x^2}$$ Multiply by the reciprocal of $\frac{4y^2 - x^2}{1}$.

$$= \frac{(2x + y)(x - 2y) \cdot 1}{(2x + y)(2y + x)(2y - x)}$$ Multiply the numerators and denominators. Then, to simplify, factor $2x^2 - 3xy - 2y^2$ and $4y^2 - x^2$.

Teaching Tip: Continue to remind students that when the operation becomes multiplication, they are then to factor and remove any common factors of the numerator and denominator.

$$= \frac{\overset{1}{\cancel{(2x + y)}}\overset{-1}{\cancel{(x - 2y)}} \cdot 1}{\underset{1}{\cancel{(2x + y)}}(2y + x)\underset{1}{\cancel{(2y - x)}}}$$

Since $x - 2y$ and $2y - x$ are opposites, simplify by replacing $\frac{x - 2y}{2y - x}$ with $\frac{-1}{1}$.

$$= \frac{-1}{2y + x}$$

Multiply the remaining factors in the numerator. Multiply the remaining factors in the denominator.

$$= -\frac{1}{2y + x}$$

Write the − sign in front of the fraction.

Teaching Example 6
Divide:

$$\frac{3a^2 - 5ab - 2b^2}{3a + b} \div (4b^2 - a^2)$$

Answer: $-\frac{1}{2b + a}$

Self Check 6 Divide: $(b - a) \div \frac{a^2 - b^2}{a^2 + ab}$ $\quad -a$

Now Try Problem 63

3 Convert Units of Measurement.

Success Tip

Remember that unit conversion factors are equal to 1. Some examples are:

$$\frac{12 \text{ in.}}{1 \text{ ft}} = 1 \qquad \frac{60 \text{ min}}{1 \text{ hr}} = 1$$

We can use the concepts discussed in this section to make conversions from one unit of measure to another. *Unit conversion factors* play an important role in this process. A **unit conversion factor** is a fraction that has a value of 1. For example, we can use the fact that 1 square yard = 9 square feet to form two unit conversion factors:

$$\frac{1 \text{ yd}^2}{9 \text{ ft}^2} = 1$$ Read as "1 square yard per 9 square feet."

$$\frac{9 \text{ ft}^2}{1 \text{ yd}^2} = 1$$ Read as "9 square feet per 1 square yard."

Since a unit conversion factor is equal to 1, multiplying a measurement by a unit conversion factor does not change the measurement, it only changes the units of measure.

EXAMPLE 7

The Language of Algebra

The method we are using to convert from one unit of measurement to another is often referred to as **dimensional analysis.**

Carpeting. A roll of carpeting is 12 feet wide and 150 feet long. Find the number of square yards of carpeting on the roll.

Strategy We will begin by determining the number of square feet of carpeting on the roll. Then we will multiply that result by a unit conversion factor.

Why A properly chosen unit conversion factor can convert the number of square feet of carpeting on the roll to the number of square yards on the roll.

Solution

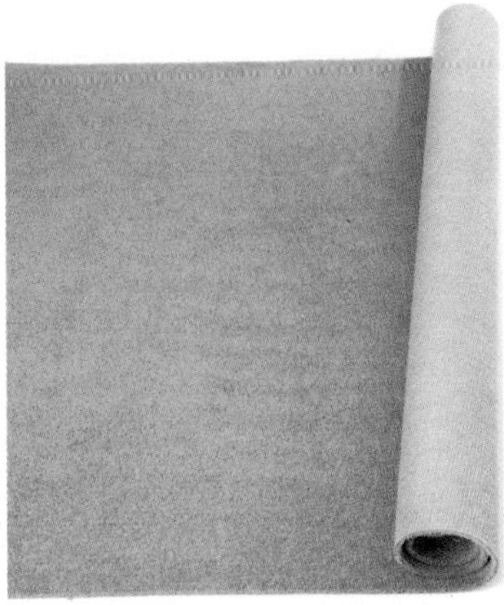

©iStockphoto.com/thumb

When unrolled, the carpeting forms a rectangular shape with an area of $12 \cdot 150 = 1{,}800$ square feet. We will multiply $1{,}800 \text{ ft}^2$ by a unit conversion factor such that the units of ft^2 are removed and the units of yd^2 are introduced. Since $1 \text{ yd}^2 = 9 \text{ ft}^2$, we will use $\frac{1\text{yd}^2}{9\text{ft}^2}$.

$$\frac{1{,}800 \text{ ft}^2}{1 \text{ roll}} = \frac{1{,}800 \text{ ft}^2}{1 \text{ roll}} \cdot \frac{\mathbf{1 \text{ yd}^2}}{\mathbf{9 \text{ ft}^2}}$$

Multiply by a unit conversion factor that relates yd^2 to ft^2.

$$= \frac{1{,}800 \cancel{\text{ft}^2}}{1 \text{ roll}} \cdot \frac{1 \text{ yd}^2}{9 \cancel{\text{ft}^2}}$$

Remove the units of ft^2 that are common to the numerator and denominator.

$$= \frac{200 \text{ yd}^2}{1 \text{ roll}}$$

Divide 1,800 by 9 to get 200.

There are 200 yd^2 of carpeting on the roll.

Teaching Example 7 Convert 630 square feet to square yards.
Answer: 70 yd^2

Self Check 7 Convert $5{,}400 \text{ ft}^2$ to square yards. $\quad 600 \text{ yd}^2$

Now Try Problem 67

EXAMPLE 8 **The Speed of Light.** The speed with which light moves through space is about 186,000 miles per second. Express this speed in miles per minute.

Strategy The speed of light can be expressed as $\frac{186{,}000 \text{ mi}}{1 \text{ sec}}$. We will multiply that fraction by a unit conversion factor.

Why A properly chosen unit conversion factor can convert the number of miles traveled per second to the number of miles traveled per minute.

Solution We will multiply $\frac{186{,}000 \text{ mi}}{1 \text{ sec}}$ by a unit conversion factor such that the units of seconds are removed and the units of minutes are introduced. Since 60 seconds = 1 minute, we will use $\frac{60 \text{ sec}}{1 \text{ min}}$.

$$\frac{186{,}000 \text{ mi}}{1 \text{ sec}} = \frac{186{,}000 \text{ mi}}{1 \text{ sec}} \cdot \frac{\mathbf{60 \text{ sec}}}{\mathbf{1 \text{ min}}}$$ Multiply by a unit conversion factor that relates seconds to minutes.

$$= \frac{186{,}000 \text{ mi}}{1 \cancel{\text{sec}}} \cdot \frac{60 \cancel{\text{sec}}}{1 \text{ min}}$$ Remove the units of seconds that are common to the numerator and denominator.

$$= \frac{11{,}160{,}000 \text{ mi}}{1 \text{ min}}$$ Multiply 186,000 and 60 to get 11,160,000.

The speed of light is about 11,160,000 miles per minute.

Success Tip

We can remove common units just as we remove factors that are common to the numerator and denominator when multiplying rational expressions.

Teaching Example 8 Birds. A hummingbird can beat its wings up to 200 times per second. Express this speed in beats per minute.
Answer: 12,000 beats per minute

Self Check 8 **Insects.** A mosquito beats it wings about 600 times per second. How many times is that per minute? 36,000 beats per minute

Now Try Problem 71

SECTION 7.2 STUDY SET

VOCABULARY

Fill in the blanks.

1. The reciprocal of $\frac{x^2 + 6x + 1}{10x}$ is $\frac{10x}{x^2 + 6x + 1}$.

2. A unit conversion factor is a fraction that is equal to 1, such as $\frac{3 \text{ ft}}{1 \text{ yd}}$.

CONCEPTS

Fill in the blanks.

3. a. To multiply rational expressions, multiply their numerators and multiply their denominators. To divide two rational expressions, multiply the first by the reciprocal of the second. In symbols,

b. $\frac{A}{B} \cdot \frac{C}{D} = \frac{AC}{BD}$ and $\frac{A}{B} \div \frac{C}{D} = \frac{A}{B} \cdot \frac{D}{C}$

Simplify each expression.

4. $\frac{(x + 7) \cdot 2 \cdot 5}{5(x + 1)(x + 7)(x - 9)}$ $\quad \frac{2}{(x + 1)(x - 9)}$

5. $\frac{y \cdot y \cdot y(15 - y)}{y(y - 15)(y + 1)}$ $\quad -\frac{y^2}{y + 1}$

6. a. Write $3x + 5$ in fractional form. $\frac{3x + 5}{1}$

b. What is the reciprocal of $18x$? $\frac{1}{18x}$

7. Find the product of the rational expression and its reciprocal. 1

$$\frac{3}{x + 2} \cdot \frac{x + 2}{3}$$

8. Use the fact that 1 tablespoon = 3 teaspoons to write two unit conversion factors. $\frac{1 \text{ tablespoon}}{3 \text{ teaspoons}}, \frac{3 \text{ teaspoons}}{1 \text{ tablespoon}}$

NOTATION

9. What units are common to the numerator and denominator of the following product? ft

$$\frac{45 \text{ ft}}{1} \cdot \frac{1 \text{ yd}}{3 \text{ ft}}$$

10. a. What fact is indicated by the unit conversion factor $\frac{1 \text{ day}}{24 \text{ hours}}$? 1 day = 24 hr

b. Fill in the blank: $\frac{1 \text{ day}}{24 \text{ hours}} = 1$.

GUIDED PRACTICE

Multiply, and then simplify, if possible. See Example 1.

11. $\frac{3}{7} \cdot \frac{y}{2}$ $\quad \frac{3y}{14}$

12. $\frac{2}{7} \cdot \frac{z}{3}$ $\quad \frac{2z}{21}$

13. $\frac{y + 2}{y} \cdot \frac{3}{y^2}$ $\quad \frac{3(y + 2)}{y^3}$ or $\frac{3y + 6}{y^3}$

14. $\frac{4}{a + 1} \cdot \frac{a}{7}$ $\quad \frac{4a}{7(a + 1)}$ or $\frac{4a}{7a + 7}$

15. $\frac{35n}{12} \cdot \frac{16}{7n^2}$ $\frac{20}{3n}$

16. $\frac{11m}{21} \cdot \frac{14}{55m^3}$ $\frac{2}{15m^2}$

17. $\frac{2x^2y}{3xy} \cdot \frac{3xy^2}{2}$ x^2y^2

▶ 18. $\frac{2x^2z}{z} \cdot \frac{5x}{z}$ $\frac{10x^3}{z}$

Multiply, and then simplify, if possible. **See Example 2.**

▶ 19. $\frac{x+5}{5} \cdot \frac{x}{x+5}$ $\frac{x}{5}$

20. $\frac{a-9}{9} \cdot \frac{8a}{a-9}$ $\frac{8a}{9}$

21. $\frac{2x+6}{x+3} \cdot \frac{3}{4x}$ $\frac{3}{2x}$

▶ 22. $\frac{3y-9}{y-3} \cdot \frac{y}{3y^2}$ $\frac{1}{y}$

▶ 23. $\frac{(x+1)^2}{x+2} \cdot \frac{x+2}{x+1}$ $x+1$

24. $\frac{(y-3)^2}{y-5} \cdot \frac{y-5}{y-3}$ $y-3$

▶ 25. $\frac{x^2-x}{x} \cdot \frac{3x-6}{3-3x}$ $-(x-2)$ or $-x+2$

▶ 26. $\frac{5z-10}{z+2} \cdot \frac{3}{6-3z}$ $-\frac{5}{z+2}$

27. $\frac{x^2+x-6}{5x} \cdot \frac{5x-10}{x+3}$ $\frac{(x-2)^2}{x}$

28. $\frac{z^2+4z-5}{5z-5} \cdot \frac{5z}{z+5}$ z

29. $\frac{m^2-2m-3}{2m+4} \cdot \frac{m^2-4}{m^2+3m+2}$ $\frac{(m-2)(m-3)}{2(m+2)}$

▶ 30. $\frac{p^2-p-6}{3p-9} \cdot \frac{2p^2-5p-3}{p^2-3p}$ $\frac{(p+2)(2p+1)}{3p}$

Multiply, and then simplify, if possible. **See Example 3.**

31. $7m\left(\frac{5}{m}\right)$ 35

▶ 32. $9p\left(\frac{10}{p}\right)$ 90

▶ 33. $15x\left(\frac{x+1}{5x}\right)$ $3x+3$

34. $30t\left(\frac{t-7}{10t}\right)$ $3t-21$

▶ 35. $12y\left(\frac{5y-8}{6y}\right)$ $10y-16$

36. $16x\left(\frac{3x+8}{4x}\right)$ $12x+32$

37. $24\left(\frac{3a-5}{2a}\right)$ $\frac{36a-60}{a}$

38. $28\left(\frac{8-3t}{4t}\right)$ $\frac{56-21t}{t}$

Divide, and then simplify, if possible. **See Example 4.**

39. $\frac{2}{y} \div \frac{4}{3}$ $\frac{3}{2y}$

▶ 40. $\frac{3}{a} \div \frac{9}{5}$ $\frac{5}{3a}$

41. $\frac{3a}{25} \div \frac{1}{5}$ $\frac{3a}{5}$

42. $\frac{3y}{8} \div \frac{3}{2}$ $\frac{y}{4}$

▶ 43. $\frac{x^3}{18y} \div \frac{x}{6y}$ $\frac{x^2}{3}$

44. $\frac{21x}{z^2} \div \frac{7x^3}{z^5}$ $\frac{3z^3}{x^2}$

45. $\frac{27p^4}{35q} \div \frac{9p}{21q}$ $\frac{9p^3}{5}$

46. $\frac{12}{25s^5} \div \frac{10}{15s^2}$ $\frac{18}{25s^3}$

Divide, and then simplify, if possible. **See Example 5.**

47. $\frac{9a-18}{28} \div \frac{9a^3}{35}$ $\frac{5(a-2)}{4a^3}$

▶ 48. $\frac{3x+6}{40} \div \frac{3x^2}{24}$ $\frac{3(x+2)}{5x^2}$

▶ 49. $\frac{x^2-4}{3x+6} \div \frac{2-x}{x+2}$ $-\frac{x+2}{3}$

50. $\frac{x^2-9}{5x+15} \div \frac{3-x}{x+3}$ $-\frac{x+3}{5}$

51. $\frac{x^2+7x}{5x-10} \div \frac{(x+7)^2}{15x-30}$ $\frac{3x}{x+7}$

▶ 52. $\frac{x^2-10x}{7x+7} \div \frac{(x-10)^2}{35x+35}$ $\frac{5x}{x-10}$

53. $\frac{m^2+m-20}{m} \div \frac{4-m}{m}$ $-(m+5)$ or $-m-5$

▶ 54. $\frac{n^2+4n-21}{n} \div \frac{3-n}{n}$ $-(n+7)$ or $-n-7$

55. $\frac{t^2+5t-14}{t} \div \frac{t-2}{t}$ $t+7$

56. $\frac{r^2+12r+11}{r} \div \frac{r+11}{r}$ $r+1$

57. $\frac{x^2-2x-35}{3x^2+27x} \div \frac{3x^2+17x+10}{18x^2+12x}$ $\frac{2(x-7)}{x+9}$

▶ 58. $\frac{x^2-x-6}{2x^2+9x+10} \div \frac{x^2-25}{2x^2+15x+25}$ $\frac{x-3}{x-5}$

Divide, and then simplify, if possible. **See Example 6.**

59. $\frac{x^2-1}{3x-3} \div (x+1)$ $\frac{1}{3}$

▶ 60. $\frac{x^2-16}{x-4} \div (3x+12)$ $\frac{1}{3}$

61. $\frac{n^2-10n+9}{n-9} \div (n-1)$ 1

62. $\frac{r^2-11r+18}{r-9} \div (r-2)$ 1

▶ 63. $\frac{2r-3s}{12} \div (4r^2-12rs+9s^2)$ $\frac{1}{12(2r-3s)}$

▶ 64. $\frac{3m+n}{18} \div (9m^2+6mn+n^2)$ $\frac{1}{18(3m+n)}$

▶ 65. $24n^2 \div \frac{18n^3}{n-1}$ $\frac{4(n-1)}{3n}$

66. $12m \div \frac{16m^2}{m+4}$ $\frac{3(m+4)}{4m}$

Complete each unit conversion. **See Examples 7 and 8.**

67. $\frac{150 \text{ yards}}{1} \cdot \frac{3 \text{ feet}}{1 \text{ yard}} = ?$ 450 ft

▶ 68. $\frac{60 \text{ inches}}{1} \cdot \frac{1 \text{ feet}}{12 \text{ inches}} = ?$ 5 ft

69. $\frac{6 \text{ pints}}{1} \cdot \frac{1 \text{gallon}}{8 \text{ pints}} = ?$ $\frac{3}{4}$ gal

70. $\frac{4 \text{ cups}}{1} \cdot \frac{1 \text{gallon}}{16 \text{ cups}} = ?$ $\frac{1}{4}$ gal

71. $\frac{30 \text{ miles}}{1 \text{ hour}} \cdot \frac{1 \text{ hour}}{60 \text{ minutes}} = ?$ $\frac{1}{2}$ mi per min

▶ 72. $\frac{300 \text{ meters}}{3 \text{ months}} \cdot \frac{12 \text{ months}}{1 \text{ year}} = ?$ 1,200 m per yr

▶ 73. $\frac{30 \text{ meters}}{1 \text{ seconds}} \cdot \frac{60 \text{ seconds}}{1 \text{ minute}} = ?$ 1,800 m per min

▶ 74. $\frac{288 \text{ inches}^2}{1 \text{ year}} \cdot \frac{1 \text{ feet}^2}{144 \text{ inches}^2} = ?$ 2 ft^2 per year

TRY IT YOURSELF

Perform the operations and simplify, if possible.

▶ 75. $\frac{b^2-5b+6}{b^2-10b+16} \div \frac{b^2+2b}{b^2-6b-16}$ $\frac{b-3}{b}$

76. $\frac{m^2+m-6}{m^2-6m+9} \div \frac{m^2-4}{m^2-9}$ $\frac{(m+3)^2}{(m-3)(m+2)}$

77. $\frac{5x+5}{25} \cdot \frac{5}{(x+1)^3}$ $\frac{1}{(x+1)^2}$

78. $\frac{7t-7}{28} \cdot \frac{4}{(t-1)^4}$ $\frac{1}{(t-1)^3}$

79. $\frac{6a^2}{a^2+6a+9} \cdot \frac{(a+3)^4}{4a^5}$ $\frac{3(a+3)^2}{2a^3}$

80. $\frac{9b^3}{b^2-8b+16} \cdot \frac{(b-4)^4}{15b^8}$ $\frac{3(b-4)^2}{5b^5}$

81. $\frac{36c^2-49d^2}{3d^3} \div \frac{12c+14d}{d^4}$ $\frac{d(6c-7d)}{6}$

82. $\frac{25y^2-16z^2}{2yz} \div \frac{10y-8z}{y^2}$ $\frac{y(5y+4z)}{4z}$

83. $10h\left(\frac{5h-3}{2h}\right)$ $25h-15$

84. $33r\left(\frac{5r+4}{11r}\right)$ $15r+12$

85. $\dfrac{n^2-9}{n^2-3n} \div \dfrac{n+3}{n^2-n}$ $n-1$

86. $\dfrac{b^2-b}{b+2} \div \dfrac{b^2-2b}{b^2-4}$ $b-1$

87. $\dfrac{10r^2s}{6rs^2} \cdot \dfrac{3r^3}{2rs}$ $\dfrac{5r^3}{2s^2}$

88. $\dfrac{3a^3b}{25cd^3} \cdot \dfrac{5cd^2}{6ab}$ $\dfrac{a^2}{10d}$

89. $\dfrac{7}{3p^3} \cdot \dfrac{p+2}{p}$ $\dfrac{7(p+2)}{3p^4}$ or $\dfrac{7p+14}{3p^4}$

90. $\dfrac{5t^2}{11} \cdot \dfrac{2t}{t-5}$ $\dfrac{10t^3}{11(t-5)}$ or $\dfrac{10t^3}{11t-55}$

▶ 91. $\dfrac{5x^2+13x-6}{x+3} \div \dfrac{5x^2-17x+6}{x-2}$ $\dfrac{x-2}{x-3}$

92. $\dfrac{3p^2+5p-2}{p^3+2p^2} \div \dfrac{6p^2+13p-5}{2p^3+5p^2}$ 1

▶ 93. $\dfrac{4x^2-12xy+9y^2}{x^3y^2} \cdot \dfrac{x^3y}{4x^2-9y^2}$ $\dfrac{2x-3y}{y(2x+3y)}$

94. $\dfrac{ab^4}{25a^2-16b^2} \cdot \dfrac{25a^2-40ab+16b^2}{a^2b^4}$ $\dfrac{5a-4b}{a(5a+4b)}$

95. $\dfrac{x-2}{x} \cdot \dfrac{2x}{2-x}$ -2

▶ 96. $\dfrac{y-3}{y} \cdot \dfrac{3y}{3-y}$ -3

Look Alikes . . .

97. a. $\dfrac{3x+6}{4} \cdot \dfrac{4x+8}{3}$ $(x+2)^2$
b. $\dfrac{3x+6}{4} \div \dfrac{4x+8}{3}$ $\dfrac{9}{16}$

▶ 98. a. $\dfrac{4a-8}{5} \cdot \dfrac{5a-10}{4}$ $(a-2)^2$
b. $\dfrac{4a-8}{5} \div \dfrac{5a-10}{4}$ $\dfrac{16}{25}$

99. a. $\dfrac{x^2-5x+6}{2x-4} \cdot \dfrac{2x-6}{x-2}$ $\dfrac{(x-3)^2}{x-2}$
b. $\dfrac{x^2-5x+6}{2x-4} \div \dfrac{2x-6}{x-2}$ $\dfrac{x-2}{4}$

100. a. $\dfrac{x^2+9x+20}{9x+36} \cdot \dfrac{9x+45}{x+4}$ $\dfrac{(x+5)^2}{x+4}$
b. $\dfrac{x^2+9x+20}{9x+36} \div \dfrac{9x+45}{x+4}$ $\dfrac{x+4}{81}$

APPLICATIONS

101. **Geometry.** Find the area of the rectangle. $\dfrac{x^2}{10}$ ft^2

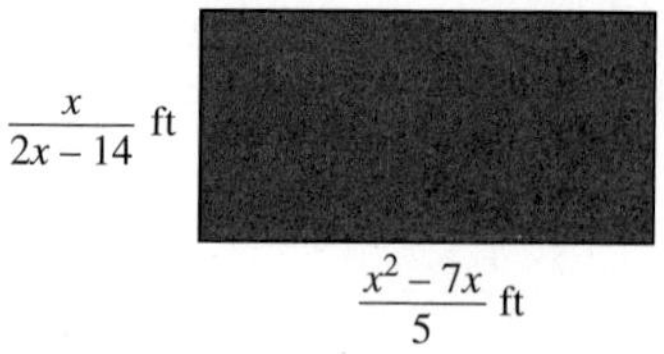

▶ 102. **Motion.** The table contains algebraic expressions for the rate an object travels and the time traveled at that rate. Complete the table.

Rate (mph)	Time (hr)	Distance (mi)
$\dfrac{k^2+k-6}{k-3}$	$\dfrac{k^2-9}{k^2-4}$	$\dfrac{(k+3)^2}{k+2}$

▶ 103. **Talking.** According to the *Sacramento Bee* newspaper, the number of words an average man speaks a day is about 12,000. How many words does an average man speak in 1 year? (*Hint:* 365 days = 1 year.) 4,380,000

▶ 104. **Classroom Space.** The recommended size of an elementary school classroom in the United States is approximately 900 square feet. Convert this to square yards. 100 yd^2

▶ 105. **Natural Light.** According to the University of Georgia School Design and Planning Laboratory, the basic classroom should have at least 72 square feet of windows for natural light. Convert this to square yards. 8 yd^2

▶ 106. **Trucking.** A cement truck holds 9 cubic yards of concrete. How many cubic feet of concrete does it hold? (*Hint:* 27 cubic feet = 1 cubic yard.) 243 ft^3

107. **Bears.** The maximum speed a grizzly bear can run is about 30 miles per hour. What is its maximum speed in miles per minute? $\frac{1}{2}$ mi per min

▶ 108. **Fuel Economy.** Use the information that follows to determine the miles per fluid ounce of gasoline for city and for highway driving for the Ford Ranger. (*Hint:* 1 gallon = 128 fluid ounces.) City: $\frac{1}{8}$ mi per oz; highway: $\frac{5}{32}$ mi per oz

2010 Ford Ranger
Fuel Economy

Fuel Type	Regular
MPG (city)	**16**
MPG (highway)	**20**

109. **TV Trivia.** On the comedy television series *Green Acres* (1965–1971), New York socialites Oliver Wendell Douglas (played by Eddie Albert) and his wife, Lisa Douglas (played by Eva Gabor), move from New York to purchase a 160-acre farm in Hooterville. Convert this to square miles. (*Hint:* 1 square mile = 640 acres.) $\frac{1}{4}$ mi^2

▶ 110. **Camping.** The capacity of backpacks is usually given in cubic inches. Convert a backpack capacity of 5,400 cubic inches to cubic feet. (*Hint:* 1 cubic foot = 1,728 cubic inches.) $3\frac{1}{8}$ ft^3

©Georgy Markov/Shutterstock.com

WRITING

111. Explain how to multiply rational expressions.

▶ 112. To divide rational expressions, you must first know how to multiply rational expressions. Explain why.

113. Explain why 60 miles per hour and 1 mile per minute are the same speed.

114. Explain why the unit conversion factor $\frac{1\text{ ft}}{12\text{ in.}}$ is equal to 1.

REVIEW

115. Hardware. A brace has a length that is 2 inches less than twice the width of the shelf that it supports. The brace is anchored to the wall 8 inches below the shelf. Find the width of the shelf and the length of the brace.
$w = 6$ in., $l = 10$ in.

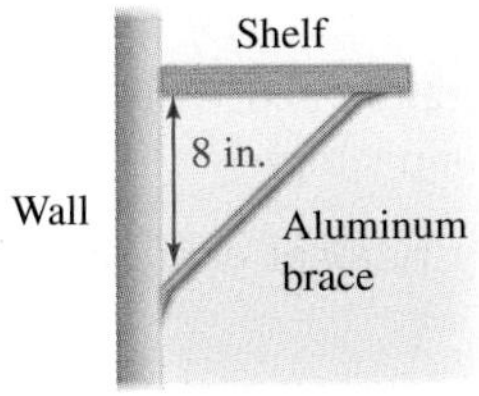

116. Solve $A = \frac{1}{2}h(b + d)$ for b. $b = \frac{2A - dh}{h}$ or $b = \frac{2A}{h} - d$

CHALLENGE PROBLEMS

Perform the operations. Simplify, if possible.

117. $\dfrac{c^3 - 2c^2 + 5c - 10}{c^2 - c - 2} \cdot \dfrac{c^3 + c^2 - 5c - 5}{c^4 - 25}$ 1

118. $\dfrac{x^3 - y^3}{x^3 + y^3} \div \dfrac{x^3 + x^2y + xy^2}{x^2y - xy^2 + y^3}$ $\dfrac{y(x - y)}{x(x + y)}$

119. $\dfrac{-x^3 + x^2 + 6x}{3x^3 + 21x^2} \div \left(\dfrac{2x + 4}{3x^2} \div \dfrac{2x + 14}{x^2 - 3x}\right)$ -1

120. $\dfrac{x^2 - y^2}{2x^4 - 2x^3} \div \left(\dfrac{x - y}{2x^2} \div \dfrac{x + y}{x^2 + 2xy + y^2}\right)$ $\dfrac{1}{x(x - 1)}$

SECTION 7.3

Adding and Subtracting with Like Denominators; Least Common Denominators

OBJECTIVES

1. Add and subtract rational expressions that have the same denominator.
2. Find the least common denominator.
3. Build rational expressions into equivalent expressions.

ARE YOU READY? *Are You Ready? exercises available online at www.webassign.net/brookscole*

The following problems review some basic skills that are needed when adding and subtracting rational expressions with like denominators.

1. Add: $\dfrac{5}{11} + \dfrac{3}{11}$ $\dfrac{8}{11}$

2. Subtract: $\dfrac{17}{21} - \dfrac{10}{21}$ $\dfrac{1}{3}$

3. Simplify: $x^2 + 8x - (3x - 1)$ $x^2 + 5x + 1$

4. Simplify: $\dfrac{x + 2}{x^2 - 5x - 14}$ $\dfrac{1}{x - 7}$

5. Factor completely: $24x^2$ $2 \cdot 2 \cdot 2 \cdot 3 \cdot x \cdot x$

6. Factor: $5x + 5$ $5(x + 1)$

In this section, we extend the rules for adding and subtracting fractions to problems involving addition and subtraction of rational expressions.

1 Add and Subtract Rational Expressions That Have the Same Denominator.

Recall from Chapter 1 that to add (or subtract) fractions that have the same denominator, we add (or subtract) their numerators and write the sum (or difference) over the common denominator. For example,

$$\frac{3}{7} + \frac{2}{7} = \frac{3 + 2}{7} = \frac{5}{7} \quad \text{and} \quad \frac{18}{25} - \frac{9}{25} = \frac{18 - 9}{25} = \frac{9}{25}$$

The Language of Algebra

We can describe $\frac{3}{7}$ and $\frac{2}{7}$ as having the **same** denominator, **common** denominators, or **like** denominators.

We use the same procedure to add and subtract rational expressions with like denominators.

Adding and Subtracting Rational Expressions That Have the Same Denominator

To add (or subtract) rational expressions that have same denominator, add (or subtract) their numerators and write the sum (or difference) over the common denominator. Then, if possible, factor and simplify.

If $\frac{A}{D}$ and $\frac{B}{D}$ are rational expressions,

$$\frac{A}{D} + \frac{B}{D} = \frac{A + B}{D} \quad \text{and} \quad \frac{A}{D} - \frac{B}{D} = \frac{A - B}{D}$$

EXAMPLE 1 Add: **a.** $\frac{x}{8} + \frac{3x}{8}$ **b.** $\frac{4s - 9}{9t} + \frac{7}{9t}$

Strategy We will add the numerators and write the sum over the common denominator. Then, if possible, we will factor and simplify.

Why This is the rule for adding rational expressions, such as these, that have the same denominator.

Solution **a.** The given rational expressions have the same denominator, 8.

$$\frac{x}{8} + \frac{3x}{8} = \frac{x + 3x}{8}$$

$= \frac{4x}{8}$ Combine like terms in the numerator: $x + 3x = 4x$. This result can be simplified.

$= \frac{\overset{1}{\cancel{4}} \cdot x}{2 \cdot \underset{1}{\cancel{4}}}$ Factor 8 as $4 \cdot 2$. Then simplify by removing a factor equal to 1.

$= \frac{x}{2}$

Caution

We *do not* add rational expressions by adding numerators and adding denominators!

$$\cancel{\frac{x}{8} + \frac{3x}{8} = \frac{4x}{16}}$$

The same caution applies when subtracting rational expressions.

b. The given rational expressions have the same denominator, $9t$.

$\frac{4s - 9}{9t} + \frac{7}{9t} = \frac{4s - 9 + 7}{9t}$ Add the numerators. Write the sum over the common denominator, $9t$.

$= \frac{4s - 2}{9t}$ Combine like terms in the numerator: $-9 + 7 = -2$.

To attempt to simplify the result, we factor the numerator to get $\frac{2(2s - 1)}{9t}$. Since the numerator and denominator do not have any common factors, $\frac{4s - 2}{9t}$ cannot be simplified. Thus,

$$\frac{4s - 9}{9t} + \frac{7}{9t} = \frac{4s - 2}{9t}$$

Notation

The numerator of the result may be written two ways:

Not factored	Factored
$\frac{4s - 2}{9t}$	$\frac{2(2s - 1)}{9t}$

Check with your instructor to see which form he or she prefers.

Teaching Example 1
Add: a. $\frac{x}{9} + \frac{5x}{9}$ b. $\frac{6m - 12}{3n} + \frac{4}{3n}$
Answers: a. $\frac{2x}{3}$
b. $\frac{6m - 8}{3n}$ or $\frac{2(3m - 4)}{3n}$

Self Check 1 Add: **a.** $\frac{2x}{15} + \frac{4x}{15}$ $\frac{2x}{5}$ **b.** $\frac{3m - 8}{23n} + \frac{2}{23n}$ $\frac{3m - 6}{23n}$ or $\frac{3(m - 2)}{23n}$

Now Try Problems 17 and 19

EXAMPLE 2 Add: **a.** $\frac{3x + 21}{5x + 10} + \frac{8x + 1}{5x + 10}$ **b.** $\frac{x^2 + 9x - 7}{2x(x - 6)} + \frac{x^2 - 9x}{(x - 6)2x}$

Strategy We will add the numerators and write the sum over the common denominator. Then, if possible, we will factor and simplify.

Why This is the rule for adding rational expressions that have the same denominator.

Solution **a.** $\frac{3x + 21}{5x + 10} + \frac{8x + 1}{5x + 10} = \frac{3x + 21 + 8x + 1}{5x + 10}$ Add the numerators. Write the sum over the common denominator, $5x + 10$.

$= \frac{11x + 22}{5x + 10}$ Combine like terms in the numerator: $3x + 8x = 11x$ and $21 + 1 = 22$.

$= \frac{11\overset{1}{\cancel{(x + 2)}}}{5\underset{1}{\cancel{(x + 2)}}}$ Factor the numerator: The GCF is 11. Factor the denominator: The GCF is 5. Then simplify by removing a factor equal to 1.

$= \frac{11}{5}$

Caution

When adding or subtracting rational expressions, always write the result in simplest form by removing any factors common to the numerator and denominator.

Caution

Don't make the mistake of removing the 2's in the result. This is incorrect because 2 is not a factor of the *entire* numerator.

$$\frac{\overset{1}{\cancel{2}}x^2 - 7}{\underset{1}{\cancel{2}}x(x - 6)}$$

b. By the commutative property of multiplication, $2x(x - 6) = (x - 6)2x$. Therefore, the denominators are the same. We add the numerators and write the sum over the common denominator.

$$\frac{x^2 + 9x - 7}{2x(x - 6)} + \frac{x^2 - 9x}{(x - 6)2x} = \frac{x^2 + 9x - 7 + x^2 - 9x}{2x(x - 6)}$$

$$= \frac{2x^2 - 7}{2x(x - 6)} \quad \text{Combine like terms in the numerator: } x^2 + x^2 = 2x^2 \text{ and } 9x - 9x = 0.$$

Since the numerator, $2x^2 - 7$, does not factor, $\frac{2x^2 - 7}{2x(x - 6)}$ is in simplest form.

Teaching Example 2

Add: **a.** $\frac{4x - 3}{7x - 14} + \frac{2x - 9}{7x - 14}$

b. $\frac{x^2 + 4x - 3}{3x(x - 8)} + \frac{x^2 - 7x}{(x - 8)3x}$

Answers: **a.** $\frac{6}{7}$ **b.** $\frac{2x^2 - 3x - 3}{3x(x - 8)}$

Self Check 2 Add: **a.** $\frac{m + 3}{3m - 9} + \frac{m - 9}{3m - 9}$ $\frac{2}{3}$ **b.** $\frac{c^2 - c}{(c - 1)(c + 2)} + \frac{c^2 - 10c}{(c + 2)(c - 1)}$ $\frac{2c^2 - 11c}{(c - 1)(c + 2)}$ or $\frac{c(2c - 11)}{(c - 1)(c + 2)}$

Now Try Problems 21 and 23

The method used to subtract rational expressions with like denominators is similar to the method used for adding such expressions.

EXAMPLE 3 Subtract: $\frac{x + 6}{x^2 + 4x - 5} - \frac{1}{x^2 + 4x - 5}$

Strategy We will subtract the numerators and write the sum over the common denominator. Then, if possible, we will factor and simplify.

Why This is the rule for subtracting rational expressions that have the same denominator.

Solution

$$\frac{x + 6}{x^2 + 4x - 5} - \frac{1}{x^2 + 4x - 5} = \frac{x + 6 - 1}{x^2 + 4x - 5} \quad \text{Subtract the numerators. Write the difference over the common denominator, } x^2 + 4x - 5.$$

$$= \frac{x + 5}{x^2 + 4x - 5} \quad \text{Combine like terms in the numerator: } 6 - 1 = 5.$$

$$= \frac{\overset{1}{\cancel{x + 5}}}{\underset{1}{(\cancel{x + 5})}(x - 1)} \quad \text{Factor the denominator. Then simplify by removing a factor equal to 1.}$$

$$= \frac{1}{x - 1}$$

Teaching Example 3

Subtract:

$\frac{y + 12}{y^2 + 3y - 54} - \frac{3}{y^2 + 3y - 54}$

Answer: $\frac{1}{y - 6}$

Self Check 3 Subtract: $\frac{n - 3}{n^2 - 16} - \frac{1}{n^2 - 16}$ $\frac{1}{n + 4}$

Now Try Problem 29

Be very careful when subtracting rational expressions. There is great potential for making sign errors.

EXAMPLE 4 Subtract: **a.** $\frac{x^2 + 10x}{x + 3} - \frac{4x - 9}{x + 3}$ **b.** $\frac{x^2}{(x + 7)(x - 8)} - \frac{-x^2 + 14x}{(x + 7)(x - 8)}$

Strategy We will use the rule for subtracting rational expressions that have the same denominators. In both cases, it is important to note that the numerator of the second fraction has *two* terms.

Why We must make sure that the entire numerator (not just the first term) of the second fraction is subtracted.

Solution **a.** To subtract the numerators, each term of $4x - 9$ must be subtracted from $x^2 + 10x$.

This − symbol applies to the entire numerator $4x - 9$.

This numerator is written within parentheses to make sure that we subtract both of its terms.

$$\frac{x^2 + 10x}{x + 3} - \frac{4x - 9}{x + 3} = \frac{x^2 + 10x - (4x - 9)}{x + 3}$$ Subtract the numerators. Write the difference over the common denominator.

$$= \frac{x^2 + 10x - 4x + 9}{x + 3}$$ In the numerator, use the distributive property: $-(4x - 9) = -1(4x - 9) = -4x + 9$.

$$= \frac{x^2 + 6x + 9}{x + 3}$$ Combine like terms in the numerator: $10x - 4x = 6x$.

$$= \frac{(x + 3)(x + 3)}{x + 3}$$ To see if the result simplifies, factor the numerator.

$$= \frac{\overset{1}{\cancel{(x + 3)}}(x + 3)}{\underset{1}{\cancel{x + 3}}}$$ Simplify by removing a factor equal to 1.

$$= x + 3$$

b. We subtract the numerators and write the difference over the common denominator.

$$\frac{x^2}{(x + 7)(x - 8)} - \frac{-x^2 + 14x}{(x + 7)(x - 8)} = \frac{x^2 - (-x^2 + 14x)}{(x + 7)(x - 8)}$$ Write the second numerator within parentheses.

$$= \frac{x^2 + x^2 - 14x}{(x + 7)(x - 8)}$$ Use the distributive property: $-(-x^2 + 14x) = x^2 - 14x$.

$$= \frac{2x^2 - 14x}{(x + 7)(x - 8)}$$ In the numerator, combine like terms: $x^2 + x^2 = 2x^2$.

In an attempt to simplify, we can factor $2x^2 - 14x$ as $2x(x - 7)$. However, the numerator and denominator have no common factors. The result is in simplest form.

Caution

Don't make this common error by forgetting to write $4x - 9$ within parentheses.

$$\frac{x^2 + 10x - 4x - 9}{x + 3}$$

Notation

A fraction bar is a grouping symbol. Think of parentheses around the terms of the numerator and the denominator.

$$\frac{4x - 9}{x + 3} = \frac{(4x - 9)}{(x + 3)}$$

Success Tip

$-(-x^2 + 14x)$ means $-1(-x^2 + 14x)$. The multiplication by -1 changes the sign of each term within the parentheses:

$$-(-x^2 + 14x) = x^2 - 14x$$

Teaching Example 4

Subtract: **a.** $\frac{x^2 + 26x}{x + 10} - \frac{6x - 100}{x + 10}$

b. $\frac{y^2}{(y + 4)(y - 9)} - \frac{-y^2 + 12y}{(y + 4)(y - 9)}$

Answers: **a.** $x + 10$

b. $\frac{2y^2 - 12y}{(y + 4)(y - 9)}$

Self Check 4 Subtract: **a.** $\frac{x^2 + 3x}{x - 1} - \frac{5x - 1}{x - 1}$ $x - 1$

b. $\frac{3y^2}{(y + 3)(y - 3)} - \frac{-3y^2 + y}{(y + 3)(y - 3)}$ $\frac{6y^2 - y}{(y + 3)(y - 3)}$

Now Try Problems 33 and 39

2 Find the Least Common Denominator.

We will now discuss two skills that are needed for adding and subtracting rational expressions that have unlike denominators. To begin, let's consider

$$\frac{11}{8x} + \frac{7}{18x^2}$$

To add these expressions, we must express them as equivalent expressions with a common denominator. The **least common denominator (LCD)** is usually the easiest one to use.

Finding the LCD

To find the LCD of a set of rational expressions:

1. Factor each denominator completely.
2. The LCD is a product that uses each different factor obtained in step 1 the greatest number of times it appears in any one factorization.

EXAMPLE 5 Find the LCD of each pair of rational expressions: **a.** $\frac{11}{8x}$ and $\frac{7}{18x^2}$ **b.** $\frac{20}{x}$ and $\frac{4x}{x-9}$

Teaching Tip: When finding an LCD for problems like these, students often incorrectly take each factor the *total* number of times it appears in both factorizations. Show them why that approach is wrong.

Strategy We will begin by factoring completely the denominator of each rational expression. Then we will form a product using each factor the greatest number of times it appears in any one factorization.

Why Since the LCD must contain the factors of each denominator, we need to write each denominator in factored form.

Solution

a. $8x = 2 \cdot 2 \cdot 2 \cdot x$ Prime factor 8.

$18x^2 = 2 \cdot 3 \cdot 3 \cdot x \cdot x$ Prime factor 18. Factor x^2.

$$\begin{array}{r|l} 2 & 8 \\ 2 & 4 \\ & 2 \end{array} \qquad \begin{array}{r|l} 2 & 18 \\ 3 & 9 \\ & 3 \end{array}$$

Success Tip

The factorizations can be written:

$$8x = 2^3 \cdot x$$
$$18x^2 = 2 \cdot 3^2 \cdot x^2$$

Note that the highest power of each factor is used to form the LCD.

$$\text{LCD} = 2^3 \cdot 3^2 \cdot x^2 = 72x^2$$

The factorizations of $8x$ and $18x^2$ contain the factors 2, 3, and x. The LCD of $\frac{11}{8x}$ and $\frac{7}{18x^2}$ should contain each factor of $8x$ and $18x^2$ the greatest number of times it appears in any one factorization.

The greatest number of times the factor 2 appears is three times.
The greatest number of times the factor 3 appears is twice.
The greatest number of times the factor x appears is twice.

$$\text{LCD} = 2 \cdot 2 \cdot 2 \cdot 3 \cdot 3 \cdot x \cdot x$$
$$= 72x^2$$

The LCD for $\frac{11}{8x}$ and $\frac{7}{18x^2}$ is $72x^2$.

Teaching Tip: Students often incorrectly list the LCD as only $x - 9$. Show them why that is wrong.

b. Since the denominators of $\frac{20}{x}$ and $\frac{4x}{x-9}$ are completely factored, the factor x appears once and the factor $x - 9$ appears once. Thus, the LCD is $x(x - 9)$.

Teaching Example 5
Find the LCD of each pair of rational expressions:
a. $\frac{7}{12a}$ and $\frac{13}{9a^2}$ b. $\frac{11}{y}$ and $\frac{4y}{y+5}$
Answers: a. $36a^2$ b. $y(y + 5)$

Self Check 5 Find the LCD of each pair of rational expressions:
a. $\frac{y+7}{6y^3}$ and $\frac{7}{75y}$ $150y^3$ **b.** $\frac{a-3}{a+3}$ and $\frac{21}{a}$ $a(a + 3)$

Now Try Problems 41 and 47

EXAMPLE 6 Find the LCD of each pair of rational expressions:
a. $\frac{x}{7x+7}$ and $\frac{x-2}{5x+5}$ **b.** $\frac{6-x}{x^2+8x+16}$ and $\frac{15x}{x^2-16}$

Strategy We will begin by factoring completely each binomial and trinomial in the denominators of the rational expressions. Then we will form a product using each factor the greatest number of times it appears in any one factorization.

Why Since the LCD must contain the factors of each denominator, we need to write each denominator in factored form.

Solution

a. Factor each denominator completely.

$7x + 7 = 7(x + 1)$ The GCF is 7.

$5x + 5 = 5(x + 1)$ The GCF is 5.

Notation

Rather than performing the multiplication, it is often better to leave an LCD in factored form:

$$\text{LCD} = 35(x + 1)$$

The factorizations of $7x + 7$ and $5x + 5$ contain the factors 7, 5, and $x + 1$. The LCD of $\frac{x}{7x+7}$ and $\frac{x-2}{5x+5}$ should contain each factor of $7x + 7$ and $5x + 5$ the greatest number of times it appears in any one factorization.

The greatest number of times the factor 7 appears is once.
The greatest number of times the factor 5 appears is once.
The greatest number of times the factor $x + 1$ appears is once.

$$\text{LCD} = 7 \cdot 5 \cdot (x + 1) = 35(x + 1)$$

b. Factor each denominator completely.

$$x^2 + 8x + 16 = (x + 4)(x + 4) \quad \text{Factor the trinomial.}$$
$$x^2 - 16 = (x + 4)(x - 4) \quad \text{Factor the difference of two squares.}$$

The factorizations of $x^2 + 8x + 16$ and $x^2 - 16$ contain the factors $x + 4$ and $x - 4$.

The greatest number of times the factor x + 4 appears is twice.
The greatest number of times the factor x − 4 appears is once.

$$\text{LCD} = (x + 4)(x + 4)(x - 4) = (x + 4)^2(x - 4)$$

Teaching Example 6
Find the LCD:
a. $\frac{a}{3a - 6}$ and $\frac{13}{5a - 10}$
b. $\frac{x + 5}{x^2 - 9}$ and $\frac{x - 7}{x^2 + 5x + 6}$
Answers: **a.** $15(a - 2)$ **b.** $(x + 3)(x - 3)(x + 2)$

Self Check 6 Find the LCD: **a.** $\frac{x^3}{x^2 - 6x}$ and $\frac{25x}{2x - 12}$ $2x(x - 6)$
b. $\frac{m + 1}{m^2 - 9}$ and $\frac{6m^2}{m^2 - 6m + 9}$ $(m + 3)(m - 3)^2$

Now Try Problems 51 and 57

3 Build Rational Expressions into Equivalent Expressions.

Recall from Chapter 1 that writing a fraction as an equivalent fraction with a larger denominator is called **building the fraction.** For example, to write $\frac{3}{5}$ as an equivalent fraction with a denominator of 35, we multiply it by 1 in the form of $\frac{7}{7}$:

$$\frac{3}{5} = \frac{3}{5} \cdot \frac{7}{7} = \frac{21}{35} \quad \begin{array}{l}\text{Multiply the numerators.}\\ \text{Multiply the denominators.}\end{array}$$

It is important to note that multiplying $\frac{3}{5}$ by $\frac{7}{7}$ changes its appearance but not its value, because we are multiplying it by 1.

To add and subtract rational expressions with different denominators, we must write them as equivalent expressions having a common denominator. To do so, we build rational expressions.

Building Rational Expressions

To build a rational expression, multiply it by 1 in the form of $\frac{c}{c}$, where c is any nonzero number or expression.

EXAMPLE 7 Write each rational expression as an equivalent expression with the indicated denominator:

a. $\frac{7}{15n}$, denominator $30n^3$ **b.** $\frac{6x}{x + 4}$, denominator $(x + 4)(x - 4)$

Teaching Tip: Show your students that the process of building a rational expression is the opposite of simplifying a rational expression.

Strategy We will begin by asking, "By what must we multiply the given denominator to get the required denominator?"

Why The answer to that question helps us determine the form of 1 to be used to build an equivalent rational expression.

Solution

a. We need to multiply the denominator of $\frac{7}{15n}$ by $2n^2$ to obtain a denominator of $30n^3$. It follows that $\frac{2n^2}{2n^2}$ is the form of 1 that should be used to build an equivalent expression.

$$\frac{7}{15n} = \frac{7}{15n} \cdot \frac{2n^2}{2n^2} \quad \text{Multiply the given rational expression by 1, in the form of } \frac{2n^2}{2n^2}.$$
$$= \frac{14n^2}{30n^3} \quad \begin{array}{l}\text{Multiply the numerators.}\\ \text{Multiply the denominators.}\end{array}$$

The Language of Algebra

We say that $\frac{7}{15n}$ and $\frac{14n^2}{30n^3}$ are **equivalent expressions** because they have the same value for all values of n, except those that make either denominator 0.

b. We need to multiply the denominator of $\frac{6x}{x+4}$ by $x - 4$ to obtain a denominator of $(x + 4)(x - 4)$. It follows that $\frac{x-4}{x-4}$ is the form of 1 that should be used to build an equivalent expression.

$$\frac{6x}{x+4} = \frac{6x}{x+4} \cdot \frac{x-4}{x-4}$$ Multiply the given rational expression by 1, in the form of $\frac{x-4}{x-4}$.

$$= \frac{6x(x-4)}{(x+4)(x-4)}$$ Multiply the numerators. Multiply the denominators.

$$= \frac{6x^2 - 24x}{(x+4)(x-4)}$$ In the numerator, distribute the multiplication by $6x$. Leave the denominator in factored form.

Teaching Tip: Stress that results should be written in this form when building rational expressions.

To get this answer, we multiplied the factors in the numerator to obtain a polynomial in unfactored form: $6x^2 - 24x$. However, we left the denominator in factored form. This approach is beneficial in the next section when we add and subtract rational expressions with unlike denominators.

Teaching Example 7
Write each rational expression as an equivalent expression with the indicated denominator:
a. $\frac{9}{16a}$, denominator $32a^3$
b. $\frac{15x}{x+1}$, denominator $(x + 1)(x - 1)$
Answers: a. $\frac{18a^2}{32a^3}$
b. $\frac{15x^2 - 15x}{(x+1)(x-1)}$

Self Check 7 Write each rational expression as an equivalent expression with the indicated denominator: **a.** $\frac{7}{20m^2}$, denominator $60m^3$ $\frac{21m}{60m^3}$
b. $\frac{2c}{c+1}$, denominator $(c + 1)(c + 3)$ $\frac{2c^2 + 6c}{(c+1)(c+3)}$

Now Try Problems 65 and 67

EXAMPLE 8 Write $\frac{x+1}{x^2+6x}$ as an equivalent expression with a denominator of $x(x + 6)(x + 2)$.

Strategy We will begin by factoring the denominator of $\frac{x+1}{x^2+6x}$. Then we will compare the factors of $x^2 + 6x$ to those of $x(x + 6)(x + 2)$.

Why This comparison will enable us to answer the question, "By what must we multiply $x^2 + 6x$ to obtain $x(x + 6)(x + 2)$?"

Solution We factor the denominator to determine what factors are missing.

$$\frac{x+1}{x^2+6x} = \frac{x+1}{x(x+6)}$$ Factor out the GCF, x, from $x^2 + 6x$.

It is now apparent that we need to multiply the denominator by $x + 2$ to obtain a denominator of $x(x + 6)(x + 2)$. It follows that $\frac{x+2}{x+2}$ is the form of 1 that should be used to build an equivalent expression.

Success Tip

When building rational expressions, write the numerator of the result as a polynomial in unfactored form. Write the denominator in factored form.

$$\frac{x+1}{x^2+6x} = \frac{x+1}{x(x+6)} \cdot \frac{x+2}{x+2}$$ Multiply the given rational expression by 1, in the form of $\frac{x+2}{x+2}$.

$$= \frac{(x+1)(x+2)}{x(x+6)(x+2)}$$ Multiply the numerators. Multiply the denominators.

$$= \frac{x^2 + 3x + 2}{x(x+6)(x+2)}$$ In the numerator, use the FOIL method to multiply $(x + 1)(x + 2)$. Leave the denominator in factored form.

Teaching Example 8
Write $\frac{x+4}{x^2-9x}$ as an equivalent expression with a denominator of $x(x - 9)(x + 1)$.
Answer: $\frac{x^2 + 5x + 4}{x(x-9)(x+1)}$

Self Check 8 Write $\frac{x-3}{x^2-4x}$ as an equivalent expression with a denominator of $x(x - 4)(x + 8)$. $\frac{x^2 + 5x - 24}{x(x-4)(x+8)}$

Now Try Problem 71

SECTION 7.3 STUDY SET

VOCABULARY

Fill in the blanks.

▶ 1. The rational expressions $\frac{7}{6n}$ and $\frac{n+1}{6n}$ have the common _denominator_ $6n$.

▶ 2. The _least_ common denominator of $\frac{x-8}{x+6}$ and $\frac{6-5x}{x}$ is $x(x+6)$.

▶ 3. To _build_ a rational expression, we multiply it by a form of 1. For example: $\frac{2}{n^2} \cdot \frac{8n}{8n} = \frac{16n}{8n^3}$

▶ 4. $\frac{16n}{8n^3}$ and $\frac{2}{n^2}$ are _equivalent_ expressions. They have the same value for all values of n, except for $n = 0$.

CONCEPTS

Fill in the blanks.

5. To add or subtract rational expressions that have the same denominator, add or subtract the _numerators_, and write the sum or difference over the common _denominator_. In symbols, $\frac{A}{D} + \frac{B}{D} = \frac{A+B}{D}$ and $\frac{A}{D} - \frac{B}{D} = \frac{A-B}{D}$.

6. When adding or subtracting rational expressions, always write the result in _simplest_ form by removing any factors common to the numerator and denominator.

▶ 7. The sum of two rational expressions is $\frac{4x+4}{5(x+1)}$. Factor the numerator and then simplify the result. $\frac{4}{5}$

8. Factor each denominator completely.

a. $\frac{17}{40x^2}$ $\quad 2 \cdot 2 \cdot 2 \cdot 5 \cdot x \cdot x$

b. $\frac{x+25}{2x^2-6x}$ $\quad 2x(x-3)$

9. Consider the following factorizations.

$$18x - 36 = 2 \cdot 3 \cdot 3 \cdot (x-2)$$
$$3x - 6 = 3(x-2)$$

a. What is the greatest number of times the factor 3 appears in any one factorization? Twice

b. What is the greatest number of times the factor $x - 2$ appears in any one factorization? Once

▶ 10. Fill in the blanks. To write $\frac{x}{x-9}$ as an equivalent rational expression with a denominator of $3x(x-9)$, we need to multiply the denominator by _$3x$_. It follows that $\frac{3x}{3x}$ is the form of 1 that should be used to build $\frac{x}{x-9}$.

NOTATION

Complete the solution.

11. $$\frac{6a-1}{4a+1} + \frac{2a+3}{4a+1} = \frac{6a-1+(2a+3)}{4a+1}$$
$$= \frac{8a+2}{4a+1}$$
$$= \frac{2(4a+1)}{4a+1}$$
$$= 2$$

12. The type of multiplication that is used to build rational expressions is shown below. Fill in the blanks.

a. $\frac{4x}{5} \cdot \frac{2}{2} = \frac{8x}{10}$

b. $\frac{3}{t} \cdot \frac{t-2}{t-2} = \frac{3t-6}{t(t-2)}$

c. $\frac{m+1}{m-3} \cdot \frac{m-5}{m-5} = \frac{m^2-4m-5}{(m-3)(m-5)}$

GUIDED PRACTICE

Add and simplify the result, if possible. See Example 1.

13. $\frac{9}{x} + \frac{2}{x}$ $\quad \frac{11}{x}$

▶ 14. $\frac{4}{s} + \frac{4}{s}$ $\quad \frac{8}{s}$

15. $\frac{x}{18} + \frac{5}{18}$ $\quad \frac{x+5}{18}$

▶ 16. $\frac{7}{10} + \frac{3y}{10}$ $\quad \frac{3y+7}{10}$

17. $\frac{a-5}{3a^3} + \frac{5}{3a^3}$ $\quad \frac{1}{3a^2}$

▶ 18. $\frac{b^3-8}{10b^4} + \frac{8}{10b^4}$ $\quad \frac{1}{10b}$

▶ 19. $\frac{x+3}{2y} + \frac{x+5}{2y}$ $\quad \frac{x+4}{y}$

▶ 20. $\frac{y+2}{10z} + \frac{y+4}{10z}$ $\quad \frac{y+3}{5z}$

Add and simplify the result, if possible. See Example 2.

▶ 21. $\frac{2}{r^2-3r-10} + \frac{r}{r^2-3r-10}$ $\quad \frac{1}{r-5}$

22. $\frac{1}{h^2-4h-5} + \frac{h}{h^2-4h-5}$ $\quad \frac{1}{h-5}$

23. $\frac{3x-5}{x-2} + \frac{6x-13}{x-2}$ $\quad 9$

▶ 24. $\frac{8x-7}{x+3} + \frac{2x+37}{x+3}$ $\quad 10$

Subtract and simplify the result, if possible. See Example 3.

25. $\frac{2x}{25} - \frac{x}{25}$ $\quad \frac{x}{25}$

▶ 26. $\frac{16c}{11} - \frac{4c}{11}$ $\quad \frac{12c}{11}$

27. $\frac{m-1}{6m^2} - \frac{5}{6m^2}$ $\quad \frac{m-6}{6m^2}$

▶ 28. $\frac{c+7}{4c^4} - \frac{3}{4c^4}$ $\quad \frac{c+4}{4c^4}$

29. $\frac{t}{t^2+t-2} - \frac{1}{t^2+t-2}$ $\quad \frac{1}{t+2}$

▶ 30. $\frac{r}{r^2-2r-3} - \frac{3}{r^2-2r-3}$ $\quad \frac{1}{r+1}$

▶ 31. $\frac{11w+6}{3w(w-9)} - \frac{11w}{3w(w-9)}$ $\quad \frac{2}{w(w-9)}$

32. $\frac{y+8}{2y(y-14)} - \frac{y}{2y(y-14)}$ $\quad \frac{4}{y(y-14)}$

Subtract and simplify the result, if possible. See Example 4.

▶ 33. $\frac{3y-2}{2y+6} - \frac{2y-5}{2y+6}$ $\quad \frac{1}{2}$

▶ 34. $\frac{5x+8}{3x+15} - \frac{3x-2}{3x+15}$ $\quad \frac{2}{3}$

▶ 35. $\frac{6x^2}{3x+2} - \frac{11x+10}{3x+2}$ $\quad 2x-5$

▶ 36. $\frac{8a^2}{2a+5} - \frac{4a^2+25}{2a+5}$ $\quad 2a-5$

37. $\frac{6x-5}{3xy} - \frac{3x-5}{3xy}$ $\quad \frac{1}{y}$

38. $\frac{7x+7}{5y} - \frac{2x+7}{5y}$ $\quad \frac{x}{y}$

39. $\frac{2-p}{p^2-p} - \frac{-p+2}{p^2-p}$ $\quad 0$

40. $\frac{2-7n}{n^2+5} - \frac{-7n+2}{n^2+5}$ $\quad 0$

Find the LCD of each pair of rational expressions. See Example 5.

41. $\frac{1}{2x}, \frac{9}{6x}$ $6x$

▶ 42. $\frac{4}{9y}, \frac{11}{3y}$ $9y$

▶ 43. $\frac{33}{15a^3}, \frac{9}{10a}$ $30a^3$

▶ 44. $\frac{m-21}{12m^4}, \frac{m+1}{18m}$ $36m^4$

▶ 45. $\frac{35}{3a^2b}, \frac{23}{a^2b^3}$ $3a^2b^3$

▶ 46. $\frac{27}{c^3d}, \frac{17}{2c^2d^3}$ $2c^3d^3$

▶ 47. $\frac{8}{c}, \frac{8-c}{c+2}$ $c(c+2)$

48. $\frac{d^2-5}{d+9}, \frac{d-3}{d}$ $d(d+9)$

Find the LCD of each pair of rational expressions. See Example 6.

49. $\frac{3x+1}{3x-3}, \frac{3x}{4x-4}$ $12(x-1)$

▶ 50. $\frac{b+1}{5b-10}, \frac{b}{6b-12}$ $30(b-2)$

▶ 51. $\frac{b-9}{4b+8}, \frac{b}{6}$ $12(b+2)$

52. $\frac{b^2-b}{10b-15}, \frac{11b}{10}$ $10(2b-3)$

53. $\frac{6-k}{2k+4}, \frac{11}{8k}$ $8k(k+2)$

▶ 54. $\frac{5m+6}{4m+12}, \frac{7}{6m}$ $12m(m+3)$

55. $\frac{-2x}{x^2-1}, \frac{5x}{x+1}$ $(x+1)(x-1)$

▶ 56. $\frac{7-y^2}{y^2-4}, \frac{y-49}{y+2}$ $(y+2)(y-2)$

57. $\frac{4x-5}{x^2-4x-5}, \frac{3x+1}{x^2-25}$ $(x+1)(x+5)(x-5)$

▶ 58. $\frac{44}{s^2-9}, \frac{s+9}{s^2-s-6}$ $(s+2)(s+3)(s-3)$

59. $\frac{5n^2-16}{2n^2+13n+20}, \frac{3n^2}{n^2+8n+16}$ $(2n+5)(n+4)^2$

▶ 60. $\frac{4y+25}{y^2+10y+25}, \frac{y^2-7}{2y^2+17y+35}$ $(2y+7)(y+5)^2$

Build each rational expression into an equivalent expression with the given denominator. See Example 7.

61. $\frac{5}{r}$; $10r$ $\frac{50}{10r}$

▶ 62. $\frac{4}{y}$; $7y$ $\frac{28}{7y}$

63. $\frac{8}{x}$; x^2y $\frac{8xy}{x^2y}$

▶ 64. $\frac{7}{y}$; xy^2 $\frac{7xy}{xy^2}$

65. $\frac{9}{4b}$; $12b^2$ $\frac{27b}{12b^2}$

66. $\frac{7}{6c}$; $30c^2$ $\frac{35c}{30c^2}$

67. $\frac{3x}{x+1}$; $(x+1)^2$ $\frac{3x^2+3x}{(x+1)^2}$

▶ 68. $\frac{5y}{y-2}$; $(y-2)^2$ $\frac{5y^2-10y}{(y-2)^2}$

Build each rational expression into an equivalent expression with the given denominator. See Example 8.

69. $\frac{x+9}{x^2+5x}$; $x^2(x+5)$ $\frac{x^2+9x}{x^2(x+5)}$

▶ 70. $\frac{a+11}{a^2+9a}$; $a^2(a+9)$ $\frac{a^2+11a}{a^2(a+9)}$

71. $\frac{t+5}{4t+8}$; $4(t+2)(t+9)$ $\frac{t^2+14t+45}{4(t+2)(t+9)}$

▶ 72. $\frac{x+7}{3x-15}$; $3(x-5)(x+10)$ $\frac{x^2+17x+70}{3(x-5)(x+10)}$

73. $\frac{y+3}{y^2-5y+6}$; $4y(y-2)(y-3)$ $\frac{4y^2+12y}{4y(y-2)(y-3)}$

▶ 74. $\frac{3x-4}{x^2+3x+2}$; $8x(x+1)(x+2)$ $\frac{24x^2-32x}{8x(x+1)(x+2)}$

75. $\frac{12-h}{h^2-81}$; $3(h+9)(h-9)$ $\frac{36-3h}{3(h+9)(h-9)}$

▶ 76. $\frac{m^2}{m^2-100}$; $9(m+10)(m-10)$ $\frac{9m^2}{9(m+10)(m-10)}$

TRY IT YOURSELF

Perform the operations. Then simplify, if possible.

▶ 77. $\frac{3t}{t^2-8t+7} - \frac{3}{t^2-8t+7}$ $\frac{3}{t-7}$

78. $\frac{10x}{x^2-2x+1} - \frac{10}{x^2-2x+1}$ $\frac{10}{x-1}$

79. $\frac{c}{c^2-d^2} - \frac{d}{c^2-d^2}$ $\frac{1}{c+d}$

▶ 80. $\frac{b}{b^2-4} - \frac{2}{b^2-4}$ $\frac{1}{b+2}$

81. $\frac{a^2+a}{4a^2-8a} + \frac{2a^2-7a}{4a^2-8a}$ $\frac{3}{4}$

▶ 82. $\frac{3b^2+16b}{6b^2+9b} + \frac{7b^2-b}{6b^2+9b}$ $\frac{5}{3}$

▶ 83. $\frac{17a}{2a+4} - \frac{7a}{2a+4}$ $\frac{5a}{a+2}$

84. $\frac{10b}{3b-18} - \frac{4b}{3b-18}$ $\frac{2b}{b-6}$

85. $\frac{8}{9-3x^2} - \frac{-6x+8}{9-3x^2}$ $\frac{2x}{3-x^2}$

86. $\frac{5}{10-5t^2} - \frac{-15t+5}{10-5t^2}$ $\frac{3t}{2-t^2}$

▶ 87. $\frac{11n}{(n+4)(n-2)} - \frac{4n-1}{(n-2)(n+4)}$ $\frac{7n+1}{(n+4)(n-2)}$

▶ 88. $\frac{1}{(t-1)(t+1)} - \frac{6-t}{(t+1)(t-1)}$ $\frac{t-5}{(t+1)(t-1)}$

▶ 89. $\frac{5r-27}{3r^2-9r} + \frac{4r}{3r^2-9r}$ $\frac{3}{r}$

90. $\frac{9a}{5a^2+25a} + \frac{a+50}{5a^2+25a}$ $\frac{2}{a}$

91. $\frac{11}{36y} + \frac{9}{36y}$ $\frac{5}{9y}$

92. $\frac{13}{24w} + \frac{17}{24w}$ $\frac{5}{4w}$

▶ 93. $\frac{-4x}{3x^2-7x+2} - \frac{-3x-2}{3x^2-7x+2}$ $-\frac{1}{3x-1}$

94. $\frac{-3c}{5c^2-16c+3} - \frac{-2c \quad 3}{5c^2-16c+3}$ $-\frac{1}{5c-1}$

95. $\frac{3x^2}{x+1} - \frac{-x+2}{x+1}$ $3x-2$

▶ 96. $\frac{8b^2}{3b-2} - \frac{-b^2+4}{3b-2}$ $3b+2$

Look Alikes . . .

97. a. $\frac{t}{12} + \frac{5t}{12}$ $\frac{t}{2}$ b. $\frac{t}{12} \cdot \frac{5t}{12}$ $\frac{5t^2}{144}$ c. $\frac{t}{12} \div \frac{5t}{12}$ $\frac{1}{5}$

▶ 98. a. $\frac{x}{9} + \frac{2x}{9}$ $\frac{x}{3}$ b. $\frac{x}{9} \cdot \frac{2x}{9}$ $\frac{2x^2}{81}$ c. $\frac{x}{9} \div \frac{2x}{9}$ $\frac{1}{2}$

99. a. $\frac{m+6}{5} - \frac{m+2}{5}$ $\frac{4}{5}$ b. $\frac{m+6}{5} \cdot \frac{m+2}{5}$ $\frac{m^2+8m+12}{25}$
c. $\frac{m+6}{5} \div \frac{m+2}{5}$ $\frac{m+6}{m+2}$

100. **a.** $\dfrac{2r+9}{2r} - \dfrac{1}{2r}$ $\quad \dfrac{r+4}{r}$ **b.** $\dfrac{2r+9}{2r} \cdot \dfrac{1}{2r}$ $\quad \dfrac{2r+9}{4r^2}$

c. $\dfrac{2r+9}{2r} \div \dfrac{1}{2r}$ $\quad 2r+9$

APPLICATIONS

101. Geometry. What is the difference of the length and width of the rectangle? $\dfrac{2x+6}{x+2}$ ft

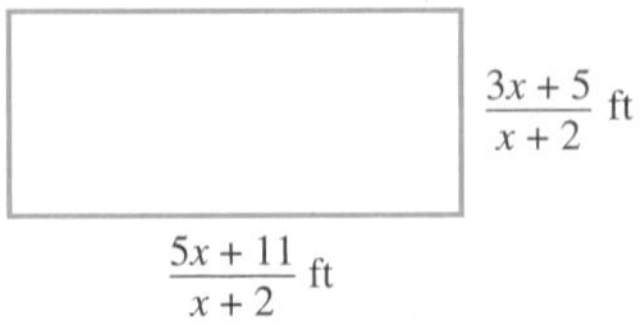

102. Geometry. What is the perimeter of the rectangle in Problem 101? 16 ft

WRITING

103. Explain how to add fractions with the same denominator.

104. Explain how to find a least common denominator.

105. Explain the error in the following solution:

$$\frac{2x+3}{x+5} - \frac{x+2}{x+5} = \frac{2x+3-x+2}{x+5} = \frac{x+5}{x+5} = 1$$

106. Explain the error in the following solution:

$$\frac{y+4}{y} - \frac{1}{y} = \frac{y+4-1}{y+y} = \frac{y+3}{2y}$$

107. **a.** Explain why the LCD of $\frac{5}{h^2}$ and $\frac{3}{h}$ is h^2 and not h^3.

b. Explain why the LCD of $\frac{1}{x}$ and $\frac{1}{x-9}$ is $x(x-9)$ and not $x-9$.

108. Explain how multiplication by 1 is used to build a rational expression. Give an example.

REVIEW

Give the formula for . . .

109. **a.** simple interest $\quad I = Prt$

b. the area of a triangle $\quad A = \frac{1}{2}bh$

c. the perimeter of a rectangle $\quad P = 2l + 2w$

110. **a.** the slope of a line $\quad m = \dfrac{y_2 - y_1}{x_2 - x_1}$

b. distance traveled $\quad d = rt$

c. the area of a circle $\quad A = \pi r^2$

CHALLENGE PROBLEMS

Perform the operations. Simplify the result, if possible.

111. $\dfrac{3xy}{x-y} - \dfrac{x(3y-x)}{x-y} - \dfrac{x(x-y)}{x-y}$ $\quad \dfrac{xy}{x-y}$

112. $\dfrac{9t^3 - 12t^2}{27t^3 - 64} - \dfrac{-3t+4}{27t^3-64}$ $\quad \dfrac{3t^2+1}{9t^2+12t+16}$

113. $\dfrac{2a^2+2}{a^3+8} + \dfrac{a^3+a}{a^3+8}$ $\quad \dfrac{a^2+1}{a^2-2a+4}$

114. Find the LCD of

$$\frac{2}{a^3+8}, \frac{a}{a^2-4}, \text{ and } \frac{2a+5}{a^3-8}.$$

$(a+2)(a-2)(a^2-2a+4)(a^2+2a+4)$

SECTION 7.4 Adding and Subtracting with Unlike Denominators

OBJECTIVES

1 Add and subtract rational expressions that have unlike denominators.

2 Add and subtract rational expressions that have denominators that are opposites.

ARE YOU READY? *Are You Ready? exercises available online at www.webassign.net/brookscole*

The following problems review some basic skills that are needed when adding and subtracting rational expressions with unlike denominators.

1. Write $\dfrac{5}{4a}$ as an equivalent expression with a denominator of $16a^2$. $\quad \dfrac{20a}{16a^2}$

2. Factor completely: $36x^3$ $\quad 2 \cdot 2 \cdot 3 \cdot 3 \cdot x \cdot x \cdot x$

3. Add: $\dfrac{6x}{9} + \dfrac{5x}{9}$ $\quad \dfrac{11x}{9}$

4. What is the opposite of $8 - t$? $\quad t - 8$

We have discussed a method for finding the least common denominator (LCD) of two rational expressions. We have also built rational expressions into equivalent expressions having a given denominator. We will now use these skills to add and subtract rational expressions with unlike denominators.

1 Add and Subtract Rational Expressions That Have Unlike Denominators.

The following steps summarize how to add or subtract rational expressions that have different denominators.

Adding and Subtracting Rational Expressions That Have Unlike Denominators

1. Find the LCD.
2. Rewrite each rational expression as an equivalent expression with the LCD as the denominator. To do so, build each fraction using a form of 1 that involves any factor(s) needed to obtain the LCD.
3. Add or subtract the numerators and write the sum or difference over the LCD.
4. Simplify the result, if possible.

EXAMPLE 1 Add: $\frac{9x}{7} + \frac{3x}{5}$

Strategy We will use the procedure for adding rational expressions that have unlike denominators. The first step is to determine the LCD.

Why If we are to add (or subtract) rational expressions, their denominators must be the same. Since the denominators of these rational expressions are different, we cannot add them in their present form.

$$\text{sevenths } \frac{9x}{7} + \frac{3x}{5} \text{ fifths}$$

Not the same number

Solution *Step 1:* The denominators are 7 and 5. The LCD is $7 \cdot 5 = 35$.

Step 2: We need to multiply the denominator of $\frac{9x}{7}$ by 5 and we need to multiply the denominator of $\frac{3x}{5}$ by 7 to obtain the LCD, 35. It follows that $\frac{5}{5}$ and $\frac{7}{7}$ are the forms of 1 that should be used to write the equivalent rational expressions.

$$\frac{9x}{7} + \frac{3x}{5} = \frac{9x}{7} \cdot \frac{5}{5} + \frac{3x}{5} \cdot \frac{7}{7}$$ Build the rational expressions so that each has a denominator of 35.

$$= \frac{45x}{35} + \frac{21x}{35}$$ Multiply the numerators: $9x \cdot 5 = 45x$ and $3x \cdot 7 = 21x$. Multiply the denominators. Now the denominators are like.

Step 3: $$= \frac{45x + 21x}{35}$$ Add the numerators. Write the sum over the common denominator, 35.

$$= \frac{66x}{35}$$ Combine like terms in the numerator: $45x + 21x = 66x$.

Step 4: Since 66 and 35 have no common factor other than 1, the result cannot be simplified.

Caution

In Step 2, don't simplify $\frac{45x}{35}$ and $\frac{21x}{35}$, because that will take you back to the original rational expressions and you will lose the common denominator.

Self Check 1 Add: $\frac{y}{2} + \frac{6y}{7}$ $\frac{19y}{14}$

Now Try Problem 13

Teaching Example 1
Add: $\frac{4a}{11} + \frac{2a}{3}$
Answer: $\frac{34a}{33}$

EXAMPLE 2 Subtract: $\frac{13}{18b^2} - \frac{1}{24b}$

Strategy We will use the procedure for subtracting rational expressions that have unlike denominators. The first step is to determine the LCD.

Why If we are to subtract rational expressions, their denominators must be the same. Since the denominators of these rational expressions are different, we cannot subtract them in their present form.

Solution *Step 1:* To find the LCD, we form a product that uses each different factor of $18b^2$ and $24b$ the greatest number of times it appears in any one factorization.

$$\left.\begin{aligned} 18b^2 &= 2 \cdot 3 \cdot 3 \cdot b \cdot b \\ 24b &= 2 \cdot 2 \cdot 2 \cdot 3 \cdot b \end{aligned}\right\} \text{LCD} = 2 \cdot 2 \cdot 2 \cdot 3 \cdot 3 \cdot b \cdot b = 72b^2$$

Notation

When checking your answers with those in the back of the book, remember that the results can often be presented in several equivalent forms. For example, this result can also be expressed as

$$\frac{-3b + 52}{72b^2}$$

Step 2: We need to multiply $18b^2$ by 4 to obtain $72b^2$, and $24b$ by $3b$ to obtain $72b^2$. It follows that we should use $\frac{4}{4}$ and $\frac{3b}{3b}$ to build the equivalent rational expressions.

$$\frac{13}{18b^2} - \frac{1}{24b} = \frac{13}{18b^2} \cdot \frac{4}{4} - \frac{1}{24b} \cdot \frac{3b}{3b}$$ Build the rational expressions so that each has a denominator of $72b^2$.

$$= \frac{52}{72b^2} - \frac{3b}{72b^2}$$ Multiply the numerators. Multiply the denominators. Now the denominators are like.

Step 3: $$= \frac{52 - 3b}{72b^2}$$ Subtract the numerators. Write the difference over the common denominator, $72b^2$.

Step 4: Since $52 - 3b$ does not factor, the result cannot be simplified.

Teaching Example 2

Subtract: $\frac{7}{36x^2} - \frac{11}{30x}$

Answer: $\frac{35 - 66x}{180x^2}$

Self Check 2 Subtract: $\frac{5}{21z^2} - \frac{3}{28z}$ $\frac{20 - 9z}{84z^2}$

Now Try Problem 21

EXAMPLE 3 Add: $\frac{3}{2x + 18} + \frac{27}{x^2 - 81}$

Strategy We use the procedure for adding rational expressions when the denominators are binomials. The first step is to find the LCD.

Why Since the denominators are different, we cannot add these rational expressions in their present form.

Solution After factoring the denominators, we see that the greatest number of times each of the factors 2, $x + 9$, and $x - 9$ appear in any one of the factorizations is once.

$$\left.\begin{aligned} 2x + 18 &= 2(x + 9) \\ x^2 - 81 &= (x + 9)(x - 9) \end{aligned}\right\} \text{LCD} = 2(x + 9)(x - 9)$$

Since we need to multiply $2(x + 9)$ by $x - 9$ to obtain the LCD and $(x + 9)(x - 9)$ by 2 to obtain the LCD, $\frac{x - 9}{x - 9}$ and $\frac{2}{2}$ are the forms of 1 to use to build the equivalent rational expressions.

$$\frac{3}{2x + 18} + \frac{27}{x^2 - 81} = \frac{3}{2(x + 9)} + \frac{27}{(x + 9)(x - 9)}$$

Write each denominator in factored form.

$$= \frac{3}{2(x + 9)} \cdot \frac{x - 9}{x - 9} + \frac{27}{(x + 9)(x - 9)} \cdot \frac{2}{2}$$

Build the expressions so that each has a denominator of $2(x + 9)(x - 9)$.

Multiply the numerators to prepare to combine like terms later.

Don't multiply the denominators. Leave them in factored form to possibly simplify the result later.

$$= \frac{3x - 27}{2(x + 9)(x - 9)} + \frac{54}{2(x + 9)(x - 9)}$$

Multiply: $3(x - 9) = 3x - 27$ and $27 \cdot 2 = 54$. Now the denominators are like.

Although it is not required, the factors of each denominator are written in the same order.

$$= \frac{3x - 27 + 54}{2(x + 9)(x - 9)}$$

Add the numerators. Write the sum over the common denominator, $2(x + 9)(x - 9)$.

$$= \frac{3x + 27}{2(x + 9)(x - 9)}$$

Combine like terms in the numerator: $-27 + 54 = 27$.

Caution

Always write the result in simplest form by removing any factors common to the numerator and denominator.

$$= \frac{3\overset{1}{\cancel{(x + 9)}}}{2\underset{1}{\cancel{(x + 9)}}(x - 9)}$$

Factor the numerator. Then simplify the expression by removing a factor equal to 1.

$$= \frac{3}{2(x - 9)}$$

This is the result in simplest form.

Teaching Example 3
Add: $\frac{5}{2n + 8} + \frac{20}{n^2 - 16}$
Answer: $\frac{5}{2(n - 4)}$

Self Check 3 Add: $\frac{2}{5x + 25} + \frac{4}{x^2 - 25}$ $\frac{2}{5(x - 5)}$

Now Try Problem 29

EXAMPLE 4 Subtract: $\frac{x}{x - 1} - \frac{x - 6}{x - 4}$

Strategy We use the same procedure for subtracting rational expressions when the denominators are binomials. The first step is to find the LCD.

Why Since the denominators are different, we cannot subtract these rational expressions in their present form.

Solution The denominators of $\frac{x}{x - 1}$ and $\frac{x - 6}{x - 4}$ are completely factored. The factor $x - 1$ appears once and the factor $x - 4$ appears once. Thus, the LCD $= (x - 1)(x - 4)$.

We need to multiply the first denominator by $x - 4$ to obtain the LCD and the second denominator by $x - 1$ to obtain the LCD. It follows that $\frac{x - 4}{x - 4}$ and $\frac{x - 1}{x - 1}$ are the forms of 1 to use to build the equivalent rational expressions.

$$\frac{x}{x - 1} - \frac{x - 6}{x - 4} = \frac{x}{x - 1} \cdot \frac{x - 4}{x - 4} - \frac{x - 6}{x - 4} \cdot \frac{x - 1}{x - 1}$$

Build the rational expressions so that each has a denominator of $(x - 1)(x - 4)$.

Multiply the numerators to prepare to combine like terms later.

Don't multiply the denominators. Leave them in factored form to possibly simplify later.

$$= \frac{x^2 - 4x}{(x - 1)(x - 4)} - \frac{x^2 - 7x + 6}{(x - 4)(x - 1)}$$

Multiply: $x(x - 4) = x^2 - 4x$.
Multiply: $(x - 6)(x - 1) = x^2 - 7x + 6$.
Now the denominators are like.

By the commutative property of multiplication, these are like denominators.

$$= \frac{x^2 - 4x - (x^2 - 7x + 6)}{(x - 1)(x - 4)}$$ Subtract the numerators. Remember the parentheses. Write the difference over the common denominator, $(x - 1)(x - 4)$.

$$= \frac{x^2 - 4x - x^2 + 7x - 6}{(x - 1)(x - 4)}$$ In the numerator, use the distributive property: $-(x^2 - 7x + 6) = -1(x^2 - 7x + 6) = -x^2 + 7x - 6$.

$$= \frac{3x - 6}{(x - 1)(x - 4)}$$ Combine like terms in the numerator: $x^2 - x^2 = 0$ and $-4x + 7x = 3x$.

Although, the numerator factors as $3(x - 2)$, the numerator and denominator do not have a common factor. Therefore, the result is in simplest form.

Teaching Example 4
Subtract: $\frac{x}{x + 3} - \frac{x + 1}{x + 7}$
Answer: $\frac{3x - 3}{(x + 3)(x + 7)}$

Self Check 4 Subtract: $\frac{x}{x + 9} - \frac{x - 7}{x + 8}$ $\frac{6x + 63}{(x + 9)(x + 8)}$

Now Try Problem 37

EXAMPLE 5 Subtract: $\frac{m}{m^2 + 5m + 6} - \frac{2}{m^2 + 3m + 2}$

Strategy We use the same procedure for subtracting rational expressions when the denominators are trinomials. The first step is to find the LCD.

Why Since the denominators are different, we cannot subtract these rational expressions in their present form.

Solution Factor each denominator and form the LCD.

$$\left.\begin{aligned} m^2 + 5m + 6 &= (m + 2)(m + 3) \\ m^2 + 3m + 2 &= (m + 2)(m + 1) \end{aligned}\right\} \text{LCD} = (m + 2)(m + 3)(m + 1)$$

Examining the factored forms, we see that the first denominator must be multiplied by $m + 1$, and the second must be multiplied by $m + 3$ to obtain the LCD. To build the expressions, we will use $\frac{m + 1}{m + 1}$ and $\frac{m + 3}{m + 3}$.

$$\frac{m}{m^2 + 5m + 6} - \frac{2}{m^2 + 3m + 2}$$

$$= \frac{m}{(m + 2)(m + 3)} - \frac{2}{(m + 2)(m + 1)}$$ Write each denominator in factored form.

$$= \frac{m}{(m + 2)(m + 3)} \cdot \frac{m + 1}{m + 1} - \frac{2}{(m + 2)(m + 1)} \cdot \frac{m + 3}{m + 3}$$ Build each expression so that it has a denominator of $(m + 2)(m + 3)(m + 1)$.

Multiply the numerators. → Leave the denominators in factored form. →

$$= \frac{m^2 + m}{(m + 2)(m + 3)(m + 1)} - \frac{2m + 6}{(m + 2)(m + 1)(m + 3)}$$ Multiply: $m(m + 1) = m^2 + m$ and $2(m + 3) = 2m + 6$. Now the denominators are like.

By the commutative property of multiplication, these are like denominators.

$$= \frac{m^2 + m - (2m + 6)}{(m + 2)(m + 3)(m + 1)}$$ Subtract the numerators. Remember the parentheses. Write the difference over the common denominator, $(m + 2)(m + 3)(m + 1)$.

$$= \frac{m^2 + m - 2m - 6}{(m + 2)(m + 3)(m + 1)}$$ Use the distributive property: $-(2m + 6) = -1(2m + 6) = -2m - 6$.

$$= \frac{m^2 - m - 6}{(m + 2)(m + 3)(m + 1)}$$ Combine like terms in the numerator: $m - 2m = -m$.

$$= \frac{(m-3)\overset{1}{\cancel{(m+2)}}}{\underset{1}{\cancel{(m+2)}}(m+3)(m+1)}$$ Factor the numerator and simplify the expression by removing a factor equal to 1.

$$= \frac{m-3}{(m+3)(m+1)}$$

Teaching Example 5
Subtract: $\frac{a}{a^2+9a+20} - \frac{4}{a^2+7a+12}$
Answer: $\frac{a-5}{(a+3)(a+5)}$

Self Check 5 Subtract: $\frac{b}{b^2-2b-8} - \frac{6}{b^2+b-20}$ $\frac{b+3}{(b+2)(b+5)}$

Now Try Problem 45

EXAMPLE 6 Add: $\frac{4b}{a-5} + b$

Strategy We will begin by writing the second addend, b, as $\frac{b}{1}$ and then find the LCD.

Why To add b to the rational expression, $\frac{4b}{a-5}$, we must rewrite b as a rational expression.

Solution The LCD of $\frac{4b}{a-5}$ and $\frac{b}{1}$ is $1(a-5)$, or simply $a-5$. Since we must multiply the denominator of $\frac{b}{1}$ by $a-5$ to obtain the LCD, we will use $\frac{a-5}{a-5}$ to write an equivalent rational expression.

$$\frac{4b}{a-5} + b = \frac{4b}{a-5} + \frac{b}{1}\cdot\frac{a-5}{a-5}$$ Build $\frac{b}{1}$ so that it has a denominator of $a-5$.

$$= \frac{4b}{a-5} + \frac{ab-5b}{a-5}$$ Multiply numerators: $b(a-5) = ab-5b$. Multiply denominators: $1(a-5) = a-5$. Now the denominators are like.

$$= \frac{4b+ab-5b}{a-5}$$ Add the numerators. Write the sum over the common denominator.

$$= \frac{ab-b}{a-5}$$ Combine like terms in the numerator: $4b-5b=-b$.

Although the numerator factors as $b(a-1)$, the numerator and denominator do not have a common factor. Therefore, the result is in simplest form.

Success Tip

Since the denominator of $\frac{4b}{a-5}$ is the LCD, we *do not* need to build it by multiplying it by a form of 1. The step below is unnecessary.

$$\cancel{\frac{4b}{a-5}\cdot\frac{1}{1}}$$

Teaching Example 6
Add: $\frac{9c}{b-3} + c$
Answer: $\frac{6c+bc}{b-3}$

Self Check 6 Add: $\frac{10y}{n+4} + y$ $\frac{14y+ny}{n+4}$

Now Try Problem 47

2 Add and Subtract Rational Expressions That Have Denominators That Are Opposites.

Recall that two polynomials are **opposites** if their terms are the same but they are opposite in sign. For example, $x-4$ and $4-x$ are opposites. If we multiply one of these binomials by -1, the subtraction is reversed, and the result is the other binomial.

$$-1(x-4) = -x+4$$
$$= 4-x$$ Write the expression with 4 first.

$$-1(4-x) = -4+x$$
$$= x-4$$ Write the expression with x first.

These results suggest that when a polynomial is multiplied by -1, the result is its opposite. This fact can be used when adding or subtracting rational expressions whose denominators are opposites.

Rational Expressions with Opposite Denominators

When adding or subtracting two rational expressions whose denominators are opposites, multiply either expression by 1 in the form of $\frac{-1}{-1}$ to obtain a common denominator.

EXAMPLE 7 Subtract: $\frac{x}{x-7} - \frac{1}{7-x}$

Strategy We note that the denominators are opposites. Either can serve as the LCD; we will choose $x - 7$. To obtain a common denominator, we will multiply $\frac{1}{7-x}$ by $\frac{-1}{-1}$.

Why When $7 - x$ is multiplied by -1, the subtraction is reversed, and the result is $x - 7$.

Solution We must multiply the denominator of $\frac{1}{7-x}$ by -1 to obtain the LCD. It follows that $\frac{-1}{-1}$ should be the form of 1 that is used to write an equivalent rational expression.

Success Tip

Either denominator can serve as the LCD. However, it is common to have a result whose denominator is written in descending powers of the variable. Therefore, we chose $x - 7$, as opposed to $7 - x$, as the LCD.

$$\frac{x}{x-7} - \frac{1}{7-x} = \frac{x}{x-7} - \frac{1}{7-x}\cdot\frac{-1}{-1}$$ Build $\frac{1}{7-x}$ so that it has a denominator of $x - 7$.

$$= \frac{x}{x-7} - \frac{-1}{-7+x}$$ Multiply the numerators: $1(-1) = -1$. Multiply the denominators.

$$= \frac{x}{x-7} - \frac{-1}{x-7}$$ Rewrite the second denominator: $-7 + x = x - 7$. Now the denominators are like.

$$= \frac{x-(-1)}{x-7}$$ Subtract the numerators. Remember the parentheses. Write the difference over the common denominator, $x - 7$.

$$= \frac{x+1}{x-7}$$ Do the subtraction in the numerator: $x - (-1) = x + 1$.

The result does not simplify.

Teaching Tip: Remind students that $\frac{-1}{-1}$ is a form of 1.

Self Check 7 Add: $\frac{n}{n-8} + \frac{12}{8-n}$ $\frac{n-12}{n-8}$

Now Try Problem 51

Teaching Example 7
Subtract: $\frac{11x}{x-2} - \frac{5}{2-x}$
Answer: $\frac{11x+5}{x-2}$

SECTION 7.4 STUDY SET

VOCABULARY

Fill in the blanks.

1. $\frac{x}{x-7}$ and $\frac{1}{x-7}$ have like denominators. $\frac{x+5}{x-7}$ and $\frac{4x}{x+7}$ have <u>unlike</u> denominators.
2. The polynomials $x - 3$ and $3 - x$ are <u>opposites</u>.

CONCEPTS

3. Write each denominator in factored form.
   a. $\frac{x+1}{20x^2}$ $2\cdot2\cdot5\cdot x\cdot x$
   b. $\frac{3x^2-4}{x^2+4x-12}$ $(x-2)(x+6)$
4. The factorizations of the denominators of two rational expressions are given. Find the LCD.
   a. $12a = 2\cdot2\cdot3\cdot a$ $36a^2$
   $18a^2 = 2\cdot3\cdot3\cdot a\cdot a$
   b. $x^2 - 36 = (x+6)(x-6)$ $3(x+6)(x-6)$
   $3x - 18 = 3(x-6)$

Selected exercises available online at www.webassign.net/brookscole

5. What is the LCD for $\frac{x-1}{x+6}$ and $\frac{1}{x+3}$? $(x+6)(x+3)$

6. The LCD for $\frac{1}{9n^2}$ and $\frac{37}{15n^3}$ is $3 \cdot 3 \cdot 5 \cdot n \cdot n \cdot n = 45n^3$. If we want to add these rational expressions, what form of 1 should be used

a. to build $\frac{1}{9n^2}$? $\frac{5n}{5n}$ b. to build $\frac{37}{15n^3}$? $\frac{3}{3}$

Fill in the blanks.

7. To build $\frac{x}{x+2}$ so that it has a denominator of $5(x+2)$, we multiply it by 1 in the form of $\frac{5}{5}$.

8. To build $\frac{8x}{2-x}$ so that it has a denominator of $x-2$, we multiply it by 1 in the form of $\frac{-1}{-1}$.

NOTATION

Complete the solution.

9. $\frac{2}{5} + \frac{7}{3x} = \frac{2}{5} \cdot \frac{3x}{3x} + \frac{7}{3x} \cdot \frac{5}{5}$

$= \frac{6x}{15x} + \frac{35}{15x}$

$= \frac{6x+35}{15x}$

▶ 10. Are the student's answers and the book's answers equivalent?

Student's answer	Book's answer	Equivalent?
$\frac{m^2+2m}{(m-1)(m-4)}$	$\frac{m^2+2m}{(m-4)(m-1)}$	Yes
$\frac{-5x^2-7}{4x(x+3)}$	$-\frac{5x^2-7}{4x(x+3)}$	No
$\frac{-2x}{x-y}$	$-\frac{2x}{x-y}$	Yes

GUIDED PRACTICE

Perform the operations. Simplify, if possible.
See Example 1.

▶ 11. $\frac{x}{3} + \frac{2x}{7}$ $\frac{13x}{21}$

12. $\frac{y}{4} + \frac{3y}{5}$ $\frac{17y}{20}$

13. $\frac{7a}{8} + \frac{4a}{5}$ $\frac{67a}{40}$

▶ 14. $\frac{5t}{6} + \frac{4t}{7}$ $\frac{59t}{42}$

Perform the operations. Simplify, if possible.
See Example 2.

15. $\frac{7}{m^2} - \frac{2}{m}$ $\frac{7-2m}{m^2}$

▶ 16. $\frac{6}{n^2} - \frac{2}{n}$ $\frac{6-2n}{n^2}$

17. $\frac{3}{5p^2} - \frac{5}{10p}$ $\frac{6-5p}{10p^2}$

▶ 18. $\frac{15}{16a} - \frac{3}{4a^2}$ $\frac{15a-12}{16a^2}$

19. $\frac{1}{6t} - \frac{11}{8t^3}$ $\frac{4t^2-33}{24t^3}$

20. $\frac{3}{10a} - \frac{13}{15a^3}$ $\frac{9a^2-26}{30a^3}$

▶ 21. $\frac{1}{6c^4} - \frac{8}{9c^2}$ $\frac{3-16c^2}{18c^4}$

▶ 22. $\frac{7}{8b^2} - \frac{5}{6b^3}$ $\frac{21b-20}{24b^3}$

Perform the operations. Simplify, if possible.
See Example 3.

23. $\frac{1}{2a+4} + \frac{5}{a^2-4}$ $\frac{a+8}{2(a+2)(a-2)}$

24. $\frac{5}{p^2-9} + \frac{2}{3p+9}$ $\frac{2p+9}{3(p+3)(p-3)}$

25. $\frac{2}{3a-2} + \frac{5}{9a^2-4}$ $\frac{6a+9}{(3a+2)(3a-2)}$

▶ 26. $\frac{2}{5b-3} + \frac{5}{25b^2-9}$ $\frac{10b+11}{(5b+3)(5b-3)}$

27. $\frac{4}{a+2} - \frac{7}{a^2+4a+4}$ $\frac{4a+1}{(a+2)^2}$

28. $\frac{9}{b^2-2b+1} - \frac{2}{b-1}$ $\frac{11-2b}{(b-1)^2}$

29. $\frac{6}{5m^2-5m} - \frac{3}{5m-5}$ $\frac{6-3m}{5m(m-1)}$

▶ 30. $\frac{9}{2c^2-2c} - \frac{5}{2c-2}$ $\frac{9-5c}{2c(c-1)}$

Perform the operations. Simplify, if possible.
See Example 4.

▶ 31. $\frac{9}{t+3} + \frac{8}{t+2}$ $\frac{17t+42}{(t+3)(t+2)}$

▶ 32. $\frac{2}{m-3} + \frac{7}{m-4}$ $\frac{9m-29}{(m-3)(m-4)}$

▶ 33. $\frac{3x}{2x-1} - \frac{2x}{2x+3}$ $\frac{2x^2+11x}{(2x-1)(2x+3)}$

34. $\frac{2y}{5y-1} - \frac{2y}{3y+2}$ $\frac{-4y^2+6y}{(5y-1)(3y+2)}$

35. $\frac{s+7}{s+3} - \frac{s-3}{s+7}$ $\frac{14s+58}{(s+3)(s+7)}$

▶ 36. $\frac{t+5}{t-5} - \frac{t-5}{t+5}$ $\frac{20t}{(t-5)(t+5)}$

▶ 37. $\frac{3m}{m-2} - \frac{m-3}{m+5}$ $\frac{2m^2+20m-6}{(m-2)(m+5)}$

▶ 38. $\frac{2x}{x+2} - \frac{x+1}{x-3}$ $\frac{x^2-9x-2}{(x-3)(x+2)}$

Perform the operations. Simplify, if possible.
See Example 5.

▶ 39. $\frac{4}{s^2+5s+4} + \frac{s}{s^2+2s+1}$ $\frac{s^2+8s+4}{(s+4)(s+1)(s+1)}$

▶ 40. $\frac{d}{d^2+6d+5} - \frac{3}{d^2+5d+4}$ $\frac{d^2+d-15}{(d+5)(d+1)(d+4)}$

41. $\frac{5}{x^2-9x+8} - \frac{3}{x^2-6x-16}$ $\frac{2x+13}{(x-8)(x-1)(x+2)}$

▶ 42. $\frac{3}{t^2+t-6} + \frac{1}{t^2+3t-10}$ $\frac{4t+18}{(t+5)(t-2)(t+3)}$

43. $\frac{2}{a^2+4a+3} + \frac{1}{a+3}$ $\frac{1}{a+1}$

▶ 44. $\frac{1}{c+6} + \frac{4}{c^2+8c+12}$ $\frac{1}{c+2}$

▶ 45. $\frac{8}{y^2-16} - \frac{7}{y^2-y-12}$ $\frac{1}{(y+3)(y+4)}$

46. $\frac{6}{s^2-9} - \frac{5}{s^2-s-6}$ $\frac{1}{(s+3)(s+2)}$

Perform the operations. Simplify, if possible.
See Example 6.

47. $\frac{9y}{x-4} + y$ $\frac{5y+xy}{x-4}$

▶ 48. $\frac{9n}{m+4} + n$ $\frac{13n+mn}{m+4}$

49. $\frac{8}{x} + z$ $\frac{8+xz}{x}$

▶ 50. $\frac{2}{y} + z$ $\frac{2+yz}{y}$

Perform the operations. Simplify, if possible.
See Example 7.

51. $\frac{7}{a-4} + \frac{5}{4-a}$ $\frac{2}{a-4}$

▶ 52. $\frac{4}{b-6} + \frac{b}{6-b}$ $\frac{4-b}{b-6}$

53. $\frac{c}{7c-d} - \frac{d}{d-7c}$ $\frac{c+d}{7c-d}$

▶ 54. $\frac{a}{5a-3b} - \frac{b}{3b-5a}$ $\frac{a+b}{5a-3b}$

TRY IT YOURSELF

Perform the operations and simplify, if possible.

55. $\frac{x-7}{x^2+4x-5} - \frac{x-9}{x^2+3x-10}$ $\frac{1}{(x-1)(x-2)}$

56. $\frac{r}{r^2+5r+6} - \frac{2}{r^2+3r+2}$ $\frac{r-3}{(r+3)(r+1)}$

57. $\frac{3d-3}{d-9} - \frac{3d}{9-d}$ $\frac{6d-3}{d-9}$

▶ 58. $\frac{2x+2}{x-2} - \frac{2x}{2-x}$ $\frac{4x+2}{x-2}$

▶ 59. $\frac{10}{x-1} + y$ $\frac{xy-y+10}{x-1}$

60. $\frac{3}{s-8} + t$ $\frac{st-8t+3}{s-8}$

61. $\frac{b}{b+1} - \frac{b-1}{b+2}$ $\frac{2b+1}{(b+1)(b+2)}$

62. $\frac{x}{x-2} - \frac{x+2}{x+3}$ $\frac{3x+4}{(x-2)(x+3)}$

▶ 63. $\frac{g}{g^2-4} + \frac{2}{4-g^2}$ $\frac{1}{g+2}$

▶ 64. $\frac{h}{h^2-49} + \frac{7}{49-h^2}$ $\frac{1}{h+7}$

▶ 65. $\frac{5y}{6} + \frac{5y}{3}$ $\frac{5y}{2}$

▶ 66. $\frac{4x}{3} + \frac{x}{6}$ $\frac{3x}{2}$

67. $\frac{1}{5x} + \frac{7x}{x+5}$ $\frac{35x^2+x+5}{5x(x+5)}$

▶ 68. $\frac{10h}{h-3} + \frac{7}{9h}$ $\frac{90h^2+7h-21}{9h(h-3)}$

▶ 69. $\frac{11}{5x} - \frac{5}{6x}$ $\frac{41}{30x}$

▶ 70. $\frac{5}{9y} - \frac{1}{4y}$ $\frac{11}{36y}$

71. $\frac{x}{x+1} + \frac{x-1}{x}$ $\frac{2x^2-1}{x(x+1)}$

▶ 72. $\frac{t-2}{t} + \frac{t}{t+3}$ $\frac{2t^2+t-6}{t(t+3)}$

73. $\frac{y}{y-1} - \frac{4}{1-y}$ $\frac{y+4}{y-1}$

74. $\frac{1}{t-7} - \frac{t}{7-t}$ $\frac{1+t}{t-7}$

75. $\frac{n}{5} - \frac{n-2}{15}$ $\frac{2n+2}{15}$

▶ 76. $\frac{m}{9} - \frac{m+1}{27}$ $\frac{2m-1}{27}$

77. $\frac{y+2}{5y^2} + \frac{y+4}{15y}$ $\frac{y^2+7y+6}{15y^2}$

78. $\frac{x+3}{x^2} + \frac{x+5}{2x}$ $\frac{x^2+7x+6}{2x^2}$

▶ 79. $\frac{x}{x-2} + \frac{4+2x}{x^2-4}$ $\frac{x+2}{x-2}$

80. $\frac{y}{y+3} - \frac{2y-6}{y^2-9}$ $\frac{y-2}{y+3}$

81. $b - \frac{3}{a^2}$ $\frac{a^2b-3}{a^2}$

▶ 82. $c - \frac{5}{3b}$ $\frac{3bc-5}{3b}$

83. $\frac{7}{3a} + \frac{1}{a-2}$ $\frac{10a-14}{3a(a-2)}$

84. $\frac{5}{9x} + \frac{4}{x+6}$ $\frac{41x+30}{9x(x+6)}$

▶ 85. $\frac{3}{x^2} + \frac{17}{x}$ $\frac{17x+3}{x^2}$

▶ 86. $\frac{7}{c} + \frac{14}{c^2}$ $\frac{7c+14}{c^2}$

87. $\frac{x+2}{x+1} - 5$
$\frac{-4x-3}{x+1}$ or $-\frac{4x+3}{x+1}$

88. $\frac{y+8}{y-8} - 4$
$\frac{-3y+40}{y-8}$ or $-\frac{3y-40}{y-8}$

▶ 89. $\frac{4b}{3} - \frac{5b}{12}$ $\frac{11b}{12}$

90. $\frac{21y}{12} - \frac{7y}{6}$ $\frac{7y}{12}$

Look Alikes . . .

91. a. $\frac{5}{2x} + \frac{4x}{15}$
$\frac{75+8x^2}{30x}$

b. $\frac{5}{2x} \cdot \frac{4x}{15}$
$\frac{2}{3}$

▶ 92. a. $\frac{2a+4}{3} - \frac{9}{a+2}$
$\frac{2a^2+8a-19}{3(a+2)}$

b. $\frac{2a+4}{3} \cdot \frac{9}{a+2}$
6

93. a. $\frac{t}{t-5} - \frac{t}{t^2-25}$
$\frac{t^2+4t}{(t-5)(t+5)}$

b. $\frac{t}{t-5} \div \frac{t}{t^2-25}$
$t+5$

▶ 94. a. $\frac{1}{m+2} - \frac{2}{m^2+4m+4}$
$\frac{m}{(m+2)^2}$

b. $\frac{1}{m+2} \div \frac{2}{m^2+4m+4}$
$\frac{m+2}{2}$

APPLICATIONS

95. Find the total height of the funnel. $\frac{20x+9}{6x^2}$ cm

▶ 96. **Funnels.** Refer to the illustration on the right. What is the difference between the diameter of the opening at the top of the funnel and the diameter of its spout?
$\frac{16x^2-3}{6x^3}$ cm

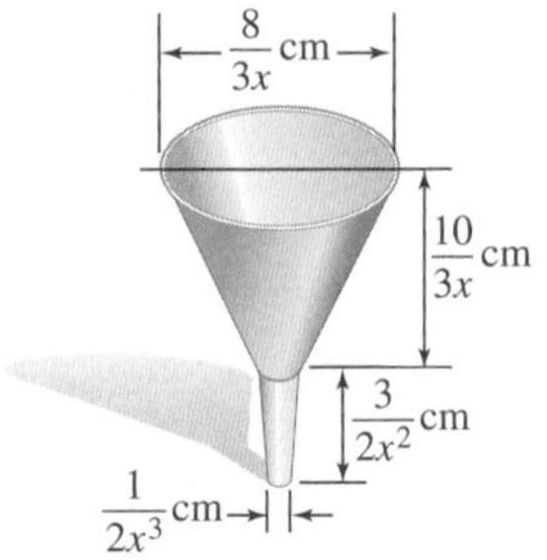

WRITING

97. Explain the error:

a. $\frac{3}{x} + \frac{8}{y} = \frac{3+8}{x+y} = \frac{11}{x+y}$ (crossed out)

b. $\frac{3}{x} + \frac{x}{3} = \frac{\cancel{3}^{1}}{\cancel{x}_{1}} + \frac{\cancel{x}^{1}}{\cancel{3}_{1}} = 1 + 1 = 2$ (crossed out)

▶ 98. Explain how to add two rational expressions with unlike denominators.

99. When will the LCD of two rational expressions be the product of the denominators of those rational expressions? Give an example.

100. Explain how multiplication by $\frac{-1}{-1}$ is used in this section.

REVIEW

101. Find the slope and y-intercept of the graph of $y = 8x + 2$.
8; (0, 2)

▶ 102. Find the slope and y-intercept of the graph of $3x + 4y = -36$.
$-\frac{3}{4}$; (0, −9)

103. What is the slope of the graph of $y = 2$?
0

104. Is the graph of the equation $x = 0$ the x-axis or the y-axis?
y-axis

CHALLENGE PROBLEMS

Perform the operations and simplify the result, if possible.

105. $\frac{a}{a-1} - \frac{2}{a+2} + \frac{3(a-2)}{a^2+a-2}$ $\frac{a+4}{a+2}$

▶ 106. $\frac{2x}{x^2-3x+2} + \frac{2x}{x-1} - \frac{x}{x-2}$ $\frac{x}{x-2}$

107. $\frac{1}{a+1} + \frac{a^2-7a+10}{2a^2-2a-4} \cdot \frac{2a^2-50}{a^2+10a+25}$ $\frac{a^2-9a+30}{(a+1)(a+5)}$

108. $1 - \frac{(x-2)^2}{(x+2)^2}$ $\frac{8x}{(x+2)^2}$

SECTION 7.5

OBJECTIVES

1. Simplify complex fractions using division.
2. Simplify complex fractions using the LCD.

Simplifying Complex Fractions

ARE YOU READY? *Are You Ready? exercises available online at www.webassign.net/brookscole*

The following problems review some basic skills that are needed when simplifying complex fractions.

1. What operation is indicated by the fraction bar in $\frac{56}{7}$? Division
2. Divide: $\frac{2}{9} \div \frac{4}{27}$ $\frac{3}{2}$
3. What is the LCD of $\frac{5}{2}$ and $\frac{3}{x}$? $2x$
4. Multiply: $10a\left(\frac{4}{5} - \frac{1}{a}\right)$ $8a - 10$

A **complex rational expression,** also called a **complex fraction,** is a rational expression whose numerator and/or denominator contains one or more rational expressions. The expression above the main fraction bar of a complex fraction is the numerator, and the expression below the main fraction bar is the denominator. Two examples of complex fractions are:

$$\frac{\frac{5x}{3}}{\frac{2x}{9}} \qquad \frac{\frac{1}{2} - \frac{1}{x}}{\frac{x}{3} + \frac{1}{5}}$$

← Numerator of complex fraction →
← Main fraction bar →
← Denominator of complex fraction →

In this section, we will discuss two methods for simplifying complex fractions. To **simplify a complex fraction** means to write it in the form $\frac{A}{B}$, where A and B are polynomials that have no common factors.

1 Simplify Complex Fractions Using Division.

One method for simplifying complex fractions uses the fact that the main fraction bar indicates division.

Simplifying Complex Fractions Method 1: Using Division

1. Add or subtract in the numerator and/or denominator so that the numerator is a single rational expression and the denominator is a single rational expression.
2. Perform the indicated division by multiplying the numerator of the complex fraction by the reciprocal of the denominator.
3. Simplify the result, if possible.

EXAMPLE 1

Simplify: $\frac{\frac{5x^2}{3}}{\frac{2x^3}{9}}$

Teaching Tip: Ask: "What is the rational expression in the numerator of the complex fraction?" and "What is the rational expression in the denominator of the complex fraction?"

Strategy We will perform the division indicated by the main fraction bar using the procedure for dividing rational expressions from Section 7.2.

Why We can skip the first step of method 1 and immediately divide because the numerator and the denominator of the complex fraction are already single rational expressions.

Solution

$$\frac{\frac{5x^2}{3}}{\frac{2x^3}{9}} = \frac{5x^2}{3} \div \frac{2x^3}{9}$$ Write the division indicated by the main fraction bar using a ÷ symbol.

$$= \frac{5x^2}{3} \cdot \frac{9}{2x^3}$$ To divide rational expressions, multiply the first by the reciprocal of the second.

$$= \frac{5x^2 \cdot 9}{3 \cdot 2x^3}$$ Multiply the numerators. Multiply the denominators.

$$= \frac{5 \cdot \overset{1}{\cancel{x}} \cdot \overset{1}{\cancel{x}} \cdot \overset{1}{\cancel{3}} \cdot 3}{\underset{1}{\cancel{3}} \cdot 2 \cdot \underset{1}{\cancel{x}} \cdot \underset{1}{\cancel{x}} \cdot x}$$ Factor 9 as 3 · 3. Then simplify by removing factors equal to 1.

$$= \frac{15}{2x}$$ Multiply the remaining factors in the numerator. Multiply the remaining factors in the denominator.

The Language of Algebra

The second step of this method could also be phrased: Perform the division by **inverting the denominator of the complex fraction and multiplying.**

Teaching Example 1 Simplify: $\frac{\frac{12x^2}{5}}{\frac{7x}{15}}$ *Answer:* $\frac{36x}{7}$

Self Check 1 Simplify: $\frac{\frac{7y^3}{8}}{\frac{21y^2}{20}}$ $\frac{5y}{6}$

Now Try Problem 17

In the next example, we must simplify the numerator and denominator of the complex fraction separately before the indicated division can be performed.

EXAMPLE 2 Simplify: $\frac{\frac{1}{2} - \frac{1}{x}}{\frac{x}{3} + \frac{1}{5}}$

Teaching Tip: Ask: "What is the rational expression in the numerator of the complex fraction?" and "What is the rational expression in the denominator of the complex fraction?"

Strategy We will simplify the expressions above and below the main fraction bar separately to write $\frac{1}{2} - \frac{1}{x}$ and $\frac{x}{3} + \frac{1}{5}$ as single rational expressions. Then we will perform the indicated division.

Why The numerator and the denominator of the complex fraction must be written as single rational expressions before dividing.

Solution To write the numerator as a single rational expression, we build $\frac{1}{2}$ and $\frac{1}{x}$ to have an LCD of $2x$, and then subtract. To write the denominator as a single rational expression, we build $\frac{x}{3}$ and $\frac{1}{5}$ to have an LCD of 15, and then add.

Success Tip

Notice that with this method, separate LCD's are found for the numerator and the denominator of the complex fraction.

$$\frac{\frac{1}{2} - \frac{1}{x}}{\frac{x}{3} + \frac{1}{5}} = \frac{\frac{1}{2} \cdot \frac{x}{x} - \frac{1}{x} \cdot \frac{2}{2}}{\frac{x}{3} \cdot \frac{5}{5} + \frac{1}{5} \cdot \frac{3}{3}}$$ ← The LCD for the numerator is 2x. Build each fraction so that each has a denominator of 2x. ← The LCD for the denominator is 15. Build each fraction so that each has a denominator of 15.

$$= \frac{\frac{x}{2x} - \frac{2}{2x}}{\frac{5x}{15} + \frac{3}{15}}$$ Perform each of the four multiplications shown in color.

$$= \frac{\frac{x - 2}{2x}}{\frac{5x + 3}{15}}$$ Subtract in the numerator and add in the denominator of the complex fraction.

Now that the numerator and the denominator of the complex fraction are single rational expressions, we perform the indicated division.

$$\frac{\frac{x-2}{2x}}{\frac{5x+3}{15}} = \frac{x-2}{2x} \div \frac{5x+3}{15}$$ Write the division indicated by the main fraction bar using a ÷ symbol.

$$= \frac{x-2}{2x} \cdot \frac{15}{5x+3}$$ Multiply by the reciprocal of $\frac{5x+3}{15}$.

$$= \frac{15(x-2)}{2x(5x+3)}$$ Multiply the numerators. Multiply the denominators. Since the numerator and denominator have no common factor, the result does not simplify.

$$= \frac{15x-30}{10x^2+6x}$$ Distribute 15 and distribute 2x.

Teaching Example 2 Simplify: $\frac{\frac{x}{5}-\frac{3}{x}}{\frac{1}{3}+\frac{2}{x}}$

Answer: $\frac{3x^2-45}{5x+30}$

Self Check 2 Simplify: $\frac{\frac{1}{3}+\frac{1}{x}}{\frac{x}{5}-\frac{1}{2}}$ $\frac{10x+30}{6x^2-15x}$

Now Try Problem 23

EXAMPLE 3 Simplify: $\frac{\frac{6}{x}+y}{\frac{6}{y}+x}$

Teaching Tip: Ask: "What is the rational expression in the numerator of the complex fraction?" and "What is the rational expression in the denominator of the complex fraction?"

Strategy We will simplify the expressions above and below the main fraction bar separately to write $\frac{6}{x}+y$ and $\frac{6}{y}+x$ as single rational expressions. Then we will perform the indicated division.

Why The numerator and the denominator of the complex fraction must be written as single rational expressions before dividing.

Solution To write $\frac{6}{x}+y$ as a single rational expression, we build y into a fraction with a denominator of x and add. To write $\frac{6}{y}+x$ as a single rational expression, we build x into a fraction with a denominator of y and add.

$$\frac{\frac{6}{x}+y}{\frac{6}{y}+x} = \frac{\frac{6}{x}+\frac{y}{1}\cdot\frac{x}{x}}{\frac{6}{y}+\frac{x}{1}\cdot\frac{y}{y}}$$

Write y as $\frac{y}{1}$. The LCD for the numerator is x. Build $\frac{y}{1}$ so that it has a denominator of x.

Write x as $\frac{x}{1}$. The LCD for the denominator is y. Build $\frac{x}{1}$ so that it has a denominator of y.

$$= \frac{\frac{6}{x}+\frac{xy}{x}}{\frac{6}{y}+\frac{xy}{y}}$$ Perform the two multiplications shown in color.

$$= \frac{\frac{6+xy}{x}}{\frac{6+xy}{y}}$$ Add in the numerator and in the denominator of the complex fraction.

Now that the numerator and the denominator of the complex fraction are single rational expressions, we can perform the division.

Success Tip

Simplifying using division (method 1) works well when a complex fraction is written, or can be easily written, as a quotient of two single rational expressions.

$$\frac{\frac{6+xy}{x}}{\frac{6+xy}{y}} = \frac{6+xy}{x} \div \frac{6+xy}{y}$$ Write the division indicated by the main fraction bar using a ÷ symbol.

$$= \frac{6+xy}{x} \cdot \frac{y}{6+xy}$$ Multiply by the reciprocal of $\frac{6+xy}{y}$.

$$= \frac{y(6+xy)}{x(6+xy)}$$ Multiply the numerators. Multiply the denominators.

$$= \frac{y\overset{1}{\cancel{(6+xy)}}}{x\underset{1}{\cancel{(6+xy)}}}$$ Simplify the result by removing a factor equal to 1.

$$= \frac{y}{x}$$

Teaching Example 3 Simplify: $\frac{\frac{8}{m}+n}{\frac{8}{n}+m}$ *Answer:* $\frac{n}{m}$

Self Check 3 Simplify: $\frac{\frac{2}{a}-b}{\frac{2}{b}-a}$ $\frac{b}{a}$

Now Try Problem 31

2 Simplify Complex Fractions Using the LCD.

A second method for simplifying complex fractions uses the concepts of LCD and multiplication by a form of 1. The multiplication by 1 produces a simpler, equivalent expression, which will not contain rational expressions in its numerator or denominator.

Simplifying Complex Fractions Method 2: Multiplying by the LCD

1. Find the LCD of all rational expressions within the complex fraction.
2. Multiply the complex fraction by 1 in the form $\frac{\text{LCD}}{\text{LCD}}$.
3. Perform the operations in the numerator and denominator. No rational expressions should remain within the complex fraction.
4. Simplify the result, if possible.

We will use method 2 to rework Example 2.

EXAMPLE 4 Simplify: $\frac{\frac{1}{2}-\frac{1}{x}}{\frac{x}{3}+\frac{1}{5}}$

Strategy Using method 1 to simplify this complex fraction, we worked with $\frac{1}{2}-\frac{1}{x}$ and $\frac{x}{3}+\frac{1}{5}$ separately. With method 2, we will use the LCD of *all four* rational expressions within the complex fraction.

Why Multiplying a complex fraction by 1 in the form of $\frac{\text{LCD}}{\text{LCD}}$ clears its numerator and denominator of fractions.

Solution The denominators of all the rational expressions within the complex fraction are 2, x, 3, and 5. Thus, their LCD is $2 \cdot x \cdot 3 \cdot 5 = 30x$.

We now multiply the complex fraction by a factor equal to 1, using the LCD: $\frac{30x}{30x} = 1$.

$$\frac{\frac{1}{2} - \frac{1}{x}}{\frac{x}{3} + \frac{1}{5}} = \frac{\frac{1}{2} - \frac{1}{x}}{\frac{x}{3} + \frac{1}{5}} \cdot \frac{\mathbf{30x}}{\mathbf{30x}}$$

$$= \frac{\left(\frac{1}{2} - \frac{1}{x}\right)\mathbf{30x}}{\left(\frac{x}{3} + \frac{1}{5}\right)\mathbf{30x}}$$ ← Multiply the numerators. ← Multiply the denominators.

$$= \frac{\frac{1}{2}(\mathbf{30x}) - \frac{1}{x}(\mathbf{30x})}{\frac{x}{3}(\mathbf{30x}) + \frac{1}{5}(\mathbf{30x})}$$ ← In the numerator, distribute the multiplication by $30x$. ← In the denominator, distribute the multiplication by $30x$.

$$= \frac{15x - 30}{10x^2 + 6x}$$ Perform each of the four multiplications by $30x$. Notice that no fractional expressions remain within the complex fraction.

To attempt to simplify the result, factor the numerator and denominator. Since they do not have a common factor, the result is in simplest form. This is the same result obtained in Example 2, which used method 1.

Success Tip

With method 2, each term of the numerator and each term of the denominator of the complex fraction is multiplied by the LCD. Arrows can be helpful in showing this.

$$\frac{\left(\frac{1}{2} - \frac{1}{x}\right)}{\left(\frac{x}{3} + \frac{1}{5}\right)} \cdot \frac{\mathbf{30x}}{\mathbf{30x}}$$

Success Tip

When simplifying a complex fraction, the same result will be obtained regardless of the method used. See Example 2.

Teaching Example 4 Use method 2 to simplify: $\frac{\frac{x}{5} - \frac{3}{x}}{\frac{1}{3} + \frac{2}{x}}$

Answer: $\frac{3x^2 - 45}{5x + 30}$

Self Check 4 Use method 2 to simplify: $\frac{\frac{1}{4} - \frac{1}{x}}{\frac{x}{5} + \frac{1}{3}}$ $\frac{15x - 60}{12x^2 + 20x}$

Now Try Problem 37

EXAMPLE 5 Simplify: $\frac{\frac{1}{8} - \frac{1}{y}}{\frac{8 - y}{4y^2}}$

Strategy Using method 1, we would work with $\frac{1}{8} - \frac{1}{y}$ and $\frac{8 - y}{4y^2}$ separately. With method 2, we use the LCD of all three rational expressions within the complex fraction.

Why Multiplying a complex fraction by 1 in the form of $\frac{\text{LCD}}{\text{LCD}}$ clears its numerator and denominator of fractions.

Solution The denominators of all the rational expressions within the complex fraction are 8, y, and $4y^2$. Therefore, the LCD is $8y^2$ and we multiply the complex fraction by a factor equal to 1, using the LCD: $\frac{8y^2}{8y^2} = 1$.

$$\frac{\frac{1}{8} - \frac{1}{y}}{\frac{8 - y}{4y^2}} = \frac{\frac{1}{8} - \frac{1}{y}}{\frac{8 - y}{4y^2}} \cdot \frac{\mathbf{8y^2}}{\mathbf{8y^2}}$$

Success Tip

Notice that with this method, all the rational expressions in the complex fraction are considered to find one universal LCD.

$$= \frac{\left(\frac{1}{8} - \frac{1}{y}\right)8y^2}{\left(\frac{8-y}{4y^2}\right)8y^2}$$ ← Multiply the numerators. ← Multiply the denominators.

$$= \frac{\frac{1}{8}(8y^2) - \frac{1}{y}(8y^2)}{\left(\frac{8-y}{4y^2}\right)(8y^2)}$$ In the numerator, distribute the multiplication by $8y^2$.

$$= \frac{y^2 - 8y}{(8-y)2}$$ Perform each of the three multiplications by $8y^2$.

$$= \frac{y\overset{-1}{\cancel{(y-8)}}}{\underset{1}{\cancel{(8-y)}}2}$$ In the numerator, factor out the GCF, y. Since $y - 8$ and $8 - y$ are opposites, simplify by replacing $\frac{y-8}{8-y}$ with $\frac{-1}{1}$.

$$= -\frac{y}{2}$$

The Language of Algebra

After multiplying a complex fraction by $\frac{\text{LCD}}{\text{LCD}}$ and performing the multiplications, the numerator and denominator of the complex fraction will be **cleared of fractions.**

Teaching Example 5 Simplify: $\dfrac{\frac{1}{12} + \frac{1}{x}}{\frac{6-x}{4x^2}}$

Answer: $\dfrac{x(x+12)}{3(6-x)}$

Self Check 5 Simplify: $\dfrac{\frac{10-n}{5n^2}}{\frac{1}{10} - \frac{1}{n}}$ $-\frac{2}{n}$

Now Try Problem 41

EXAMPLE 6 Simplify: $\dfrac{1}{1 + \frac{1}{x+1}}$

Strategy Although either method can be used, we will use method 2 to simplify this complex fraction.

Why Method 2 is often easier when the complex fraction contains a sum or difference.

Solution The only rational expression within the complex fraction has the denominator $x + 1$. Therefore, the LCD is $x + 1$. We multiply the complex fraction by a factor equal to 1, using the LCD: $\frac{x+1}{x+1} = 1$.

$$\frac{1}{1 + \frac{1}{x+1}} = \frac{1}{1 + \frac{1}{x+1}} \cdot \frac{x+1}{x+1}$$

$$= \frac{1(x+1)}{\left(1 + \frac{1}{x+1}\right)(x+1)}$$ Multiply the numerators. Multiply the denominators.

$$= \frac{1(x+1)}{1(x+1) + \frac{1}{x+1}(x+1)}$$ In the denominator, distribute the multiplication by $x + 1$.

$$= \frac{x+1}{x+1+1}$$ Perform each of the three multiplications by $x + 1$.

$$= \frac{x+1}{x+2}$$ Combine like terms in the denominator.

The result does not simplify.

Success Tip

Simplifying using the LCD (method 2) works well when the complex fraction has sums and/or differences in the numerator and/or denominator.

Teaching Example 6
Simplify: $\dfrac{1}{1 - \dfrac{1}{x-1}}$

Answer: $\dfrac{x-1}{x-2}$

Self Check 6 Simplify: $\dfrac{2}{\dfrac{1}{x+2} + 2}$ $\quad \dfrac{2x+4}{2x+5}$

Now Try Problem 47

SECTION 7.5 STUDY SET

VOCABULARY

Fill in the blanks.

1. The expression $\dfrac{\frac{2}{3} - \frac{1}{x}}{\frac{x-3}{4}}$ is called a _complex_ rational expression or a _complex_ fraction.
2. In a complex fraction, the numerator is above the _main_ fraction bar and the _denominator_ is below it.

CONCEPTS

Fill in the blanks.

3. Method 1: To simplify a complex fraction, write its numerator and denominator as _single_ rational expressions. Then perform the indicated _division_ by multiplying the numerator of the complex fraction by the _reciprocal_ of the denominator.
4. Method 2: To simplify a complex fraction, find the LCD of _all_ the rational expressions within the complex fraction. Multiply the complex fraction by 1 in the form $\frac{\text{LCD}}{\text{LCD}}$.
5. Consider: $\dfrac{\frac{x-3}{4}}{\frac{1}{12} - \frac{x}{6}}$
   a. What is the numerator of the complex fraction? Is it a single rational expression? $\frac{x-3}{4}$; yes
   b. What is the denominator of the complex fraction? Is it a single rational expression? $\frac{1}{12} - \frac{x}{6}$; no
6. Consider the complex fraction: $\dfrac{\frac{1}{y} - \frac{1}{3}}{\frac{5}{6} + \frac{1}{y}}$
   a. What is the LCD of all the rational expressions in the complex fraction? $6y$
   b. To simplify the complex fraction using method 2, it should be multiplied by what form of 1? $\frac{6y}{6y}$

NOTATION

Fill in the blanks.

7. $\dfrac{\frac{12}{y^2}}{\frac{4}{y^3}}$ means $\dfrac{12}{y^2}$ _÷_ $\dfrac{4}{y^3}$
8. $$\frac{\left(\frac{1}{5} - \frac{1}{a}\right)}{\left(\frac{a}{4} + \frac{2}{a}\right)} \cdot \frac{20a}{20a} = \frac{\frac{1}{5}(20a) - \frac{1}{a}(20a)}{\frac{a}{4}(20a) + \frac{2}{a}(20a)}$$
   $$= \frac{4a - 20}{5a^2 + 40}$$

GUIDED PRACTICE

***Simplify each complex fraction.* See Example 1.**

9. $\dfrac{\frac{2}{3}}{\frac{3}{4}}$ $\quad \frac{8}{9}$
10. $\dfrac{\frac{3}{5}}{\frac{2}{7}}$ $\quad \frac{21}{10}$
11. $\dfrac{\frac{x}{2}}{\frac{6}{5}}$ $\quad \frac{5x}{12}$
12. $\dfrac{\frac{9}{4}}{\frac{7}{x}}$ $\quad \frac{9x}{28}$
13. $\dfrac{\frac{x}{y}}{\frac{1}{x}}$ $\quad \frac{x^2}{y}$
14. $\dfrac{\frac{y}{x}}{\frac{x}{xy}}$ $\quad \frac{y^2}{x}$
15. $\dfrac{\frac{n}{8}}{\frac{1}{n^2}}$ $\quad \frac{n^3}{8}$
16. $\dfrac{\frac{1}{m}}{\frac{m^3}{15}}$ $\quad \frac{15}{m^4}$
17. $\dfrac{\frac{4a}{11}}{\frac{6a^4}{55}}$ $\quad \frac{10}{3a^3}$
18. $\dfrac{\frac{14}{15m}}{\frac{21}{25m^6}}$ $\quad \frac{10m^5}{9}$
19. $\dfrac{-\frac{x^4}{30}}{\frac{7x^2}{15}}$ $\quad -\frac{x^2}{14}$
20. $\dfrac{-\frac{5x^2}{24}}{\frac{x^5}{56}}$ $\quad -\frac{35}{3x^3}$

***Simplify each complex fraction.* See Examples 2 or 4.**

21. $\dfrac{\frac{1}{2} + \frac{3}{4}}{\frac{3}{2} + \frac{1}{4}}$ $\quad \frac{5}{7}$
22. $\dfrac{\frac{2}{3} - \frac{5}{2}}{\frac{2}{3} - \frac{3}{2}}$ $\quad \frac{11}{5}$
23. $\dfrac{\frac{1}{4} + \frac{1}{y}}{\frac{y}{3} - \frac{1}{2}}$ $\quad \frac{3y+12}{4y^2 - 6y}$
24. $\dfrac{\frac{2}{x} - \frac{1}{3}}{\frac{2}{3} + \frac{x}{5}}$ $\quad \frac{30 - 5x}{10x + 3x^2}$
25. $\dfrac{\frac{1}{y} - \frac{5}{2}}{\frac{3}{y}}$ $\quad \frac{2 - 5y}{6}$
26. $\dfrac{\frac{1}{6} - \frac{5}{s}}{\frac{2}{s}}$ $\quad \frac{s - 30}{12}$
27. $\dfrac{\frac{4}{c} - \frac{c}{6}}{\frac{2}{c}}$ $\quad \frac{24 - c^2}{12}$
28. $\dfrac{\frac{10}{n} - \frac{n}{4}}{\frac{8}{n}}$ $\quad \frac{40 - n^2}{32}$

***Simplify each complex fraction.* See Examples 3 or 5.**

29. $\dfrac{\frac{2}{3} + 1}{\frac{1}{3} + 1}$ $\quad \frac{5}{4}$
30. $\dfrac{\frac{3}{5} - 2}{\frac{2}{5} - 2}$ $\quad \frac{7}{8}$
31. $\dfrac{\frac{1}{x} - 3}{\frac{5}{x} + 2}$ $\quad \frac{1 - 3x}{5 + 2x}$
32. $\dfrac{\frac{1}{y} + 3}{\frac{3}{y} - 2}$ $\quad \frac{1 + 3y}{3 - 2y}$

▶ 33. $\dfrac{\frac{2}{x}+2}{\frac{4}{x}+2}$ $\quad \frac{1+x}{2+x}$

▶ 34. $\dfrac{\frac{3}{x}-3}{\frac{9}{x}-3}$ $\quad \frac{1-x}{3-x}$

▶ 35. $\dfrac{\frac{3y}{x}-y}{y-\frac{y}{x}}$ $\quad \frac{3-x}{x-1}$

▶ 36. $\dfrac{\frac{y}{x}+3y}{y+\frac{2y}{x}}$ $\quad \frac{3x+1}{x+2}$

Simplify each complex fraction. See Example 4.

37. $\dfrac{\frac{1}{6}-\frac{2}{x}}{\frac{1}{6}+\frac{1}{x}}$ $\quad \frac{x-12}{x+6}$

▶ 38. $\dfrac{\frac{3}{4}+\frac{1}{y}}{\frac{5}{6}-\frac{1}{y}}$ $\quad \frac{9y+12}{10y-12}$

39. $\dfrac{\frac{a}{7}-\frac{7}{a}}{\frac{1}{a}+\frac{1}{7}}$ $\quad a-7$

▶ 40. $\dfrac{\frac{t}{9}-\frac{9}{t}}{\frac{1}{t}+\frac{1}{9}}$ $\quad t-9$

Simplify each complex fraction. See Example 5.

41. $\dfrac{\frac{d^2}{4}+\frac{4d}{5}}{\frac{d+1}{2}}$ $\quad \frac{5d^2+16d}{10d+10}$

▶ 42. $\dfrac{\frac{d+2}{2}}{\frac{d}{3}-\frac{d}{4}}$ $\quad \frac{6d+12}{d}$

▶ 43. $\dfrac{\frac{2}{x}}{\frac{2}{y}-\frac{4}{x}}$ $\quad \frac{y}{x-2y}$

▶ 44. $\dfrac{\frac{2y}{3}}{\frac{2y}{3}-\frac{8}{y}}$ $\quad \frac{y^2}{y^2-12}$

Simplify each complex fraction. See Example 6.

▶ 45. $\dfrac{\frac{1}{x+1}}{1+\frac{1}{x+1}}$ $\quad \frac{1}{x+2}$

▶ 46. $\dfrac{\frac{1}{x-1}}{1-\frac{1}{x-1}}$ $\quad \frac{1}{x-2}$

47. $\dfrac{\frac{x}{x+2}}{\frac{x}{x+2}+x}$ $\quad \frac{1}{x+3}$

▶ 48. $\dfrac{\frac{2}{x-2}}{\frac{2}{x-2}-1}$ $\quad \frac{2}{4-x}$

TRY IT YOURSELF

Simplify each complex fraction.

▶ 49. $\dfrac{\frac{1}{p}+\frac{1}{q}}{\frac{1}{p}}$ $\quad \frac{q+p}{q}$

50. $\dfrac{\frac{m}{n}+1}{1-\frac{m}{n}}$ $\quad \frac{m+n}{n-m}$

51. $\dfrac{40x^2}{\frac{20x}{9}}$ $\quad 18x$

52. $\dfrac{18n^2}{\frac{6n}{13}}$ $\quad 39n$

53. $\dfrac{\frac{1}{c}+\frac{1}{2}}{\frac{1}{c^2}-\frac{1}{4}}$ $\quad \frac{2c}{2-c}$

54. $\dfrac{\frac{1}{m}-\frac{1}{n}}{\frac{m}{n}-\frac{n}{m}}$ $\quad -\frac{1}{m+n}$

55. $\dfrac{\frac{1}{r+1}+1}{\frac{3}{r-1}+1}$ $\quad \frac{r-1}{r+1}$

56. $\dfrac{5+\frac{1}{n+7}}{4-\frac{2}{n+7}}$ $\quad \frac{5n+36}{4n+26}$

▶ 57. $\dfrac{\frac{b^2-81}{18a^2}}{\frac{4b-36}{9a}}$ $\quad \frac{b+9}{8a}$

▶ 58. $\dfrac{\frac{8x-64}{y}}{\frac{x^2-64}{y^2}}$ $\quad \frac{8y}{x+8}$

59. $\dfrac{\frac{10x}{x-3}}{\frac{6}{x-3}}$ $\quad \frac{5x}{3}$

60. $\dfrac{\frac{18a}{a-4}}{\frac{12}{a-4}}$ $\quad \frac{3a}{2}$

61. $\dfrac{4-\frac{1}{8h}}{12+\frac{3}{4h}}$ $\quad \frac{32h-1}{96h+6}$

62. $\dfrac{12+\frac{1}{3b}}{12-\frac{1}{b^2}}$ $\quad \frac{36b^2+b}{36b^2-3}$

▶ 63. $\dfrac{\frac{m}{n}+\frac{n}{m}}{\frac{m}{n}-\frac{n}{m}}$ $\quad \frac{m^2+n^2}{m^2-n^2}$

64. $\dfrac{\frac{2a}{b}-\frac{b}{a}}{\frac{2a}{b}+\frac{b}{a}}$ $\quad \frac{2a^2-b^2}{2a^2+b^2}$

▶ 65. $\dfrac{\frac{2}{c^2}}{\frac{1}{c}+\frac{5}{4}}$ $\quad \frac{8}{4c+5c^2}$

▶ 66. $\dfrac{\frac{7}{s^2}}{\frac{1}{s}+\frac{10}{3}}$ $\quad \frac{21}{3s+10s^2}$

67. $\dfrac{\frac{4t-8}{t^2}}{\frac{8t-16}{t^5}}$ $\quad \frac{t^3}{2}$

68. $\dfrac{\frac{9m-27}{m^6}}{\frac{2m-6}{m^8}}$ $\quad \frac{9m^2}{2}$

▶ 69. $\dfrac{\frac{2}{s}-\frac{2}{s^2}}{\frac{4}{s^3}+\frac{4}{s^2}}$ $\quad \frac{s^2-s}{2+2s}$

▶ 70. $\dfrac{\frac{2}{x^3}-\frac{2}{x}}{\frac{4}{x}+\frac{8}{x^2}}$ $\quad \frac{1-x^2}{2x^2+4x}$

71. $\dfrac{1+\frac{6}{t}+\frac{8}{t^2}}{1+\frac{1}{t}-\frac{12}{t^2}}$ $\quad \frac{t+2}{t-3}$

72. $\dfrac{1-p+\frac{2}{p}}{\frac{6}{p^2}+\frac{1}{p}-1}$ $\quad \frac{p^3-p^2-2p}{p^2-p-6}$

▶ 73. $\dfrac{1}{\frac{1}{x}+\frac{1}{y}}$ $\quad \frac{xy}{y+x}$

▶ 74. $\dfrac{1}{\frac{b}{a}-\frac{a}{b}}$ $\quad \frac{ab}{b^2-a^2}$

75. $\dfrac{-\frac{25}{16x^2}}{\frac{15}{32x^5}}$ $\quad -\frac{10x^3}{3}$

76. $\dfrac{\frac{21}{8g^3}}{-\frac{35}{16g^8}}$ $\quad -\frac{6g^5}{5}$

▶ 77. $\dfrac{3+\frac{3}{x-1}}{3-\frac{3}{x-1}}$ $\quad \frac{x}{x-2}$

78. $\dfrac{2-\frac{2}{x+1}}{2+\frac{2}{x+1}}$ $\quad \frac{x}{x+2}$

79. $\dfrac{1-\frac{9}{d^2}}{2+\frac{6}{d}}$ $\quad \frac{d-3}{2d}$

80. $\dfrac{1-\frac{16}{a^2}}{\frac{12}{a}+3}$ $\quad \frac{a-4}{3a}$

▶ 81. $\dfrac{\frac{1}{a^2b}-\frac{5}{ab}}{\frac{3}{ab}-\frac{7}{ab^2}}$ $\quad \frac{b-5ab}{3ab-7a}$

82. $\dfrac{\frac{3}{ab^2}+\frac{6}{a^2b}}{\frac{6}{a}-\frac{9}{b^2}}$ $\quad \frac{a+2b}{2ab^2-3a^2}$

83. $\dfrac{m - \dfrac{1}{2m + 1}}{1 - \dfrac{m}{2m + 1}}$ $2m - 1$

84. $\dfrac{1 - \dfrac{r}{2r + 1}}{r - \dfrac{1}{2r + 1}}$ $\dfrac{1}{2r - 1}$

APPLICATIONS

85. **Slope.** We can use the slope formula shown below to find the slope of a line that passes through $\left(\frac{1}{2}, \frac{1}{3}\right)$ and $\left(\frac{3}{4}, \frac{5}{8}\right)$. Simplify the complex fraction to find m. $\frac{7}{6}$

$$m = \dfrac{\dfrac{5}{8} - \dfrac{1}{3}}{\dfrac{3}{4} - \dfrac{1}{2}}$$

86. **Pitching.** The earned run average (ERA) is a statistic that gives the average number of earned runs a pitcher allows. For a softball pitcher, this is based on a six-inning game. The formula for ERA is shown below. Simplify the complex fraction on the right side of the formula. $\text{ERA} = \dfrac{6 \cdot \text{earned runs}}{\text{innings pitched}}$

$$\text{ERA} = \dfrac{\text{earned runs}}{\dfrac{\text{innings pitched}}{6}}$$

87. **Electronics.** In electronic circuits, resistors are tiny components that limit the flow of an electric current. An important formula about two resistors in a circuit is shown below. Simplify the complex fraction on the right side of the formula. $\dfrac{R_1R_2}{R_2 + R_1}$

$$\text{Total resistance} = \dfrac{1}{\dfrac{1}{R_1} + \dfrac{1}{R_2}}$$

(Recall that R_1 is read as R sub one.)

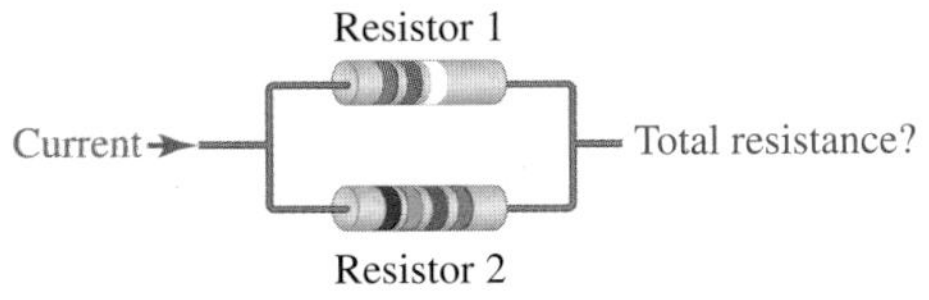

88. **Data Analysis.** Use the data in the table to find the average measurement for the three-trial experiment. $\frac{4k}{9}$

	Trial 1	Trial 2	Trial 3
Measurement	$\frac{k}{2}$	$\frac{k}{3}$	$\frac{k}{2}$

WRITING

89. What is a complex fraction? Give several examples.

90. Explain how to use method 1 to simplify: $\dfrac{1 + \frac{1}{x}}{3 - \frac{1}{x}}$

91. Explain how to use method 2 to simplify the expression in Problem 90.

92. a. List an advantage and a disadvantage of using method 1 to simplify a complex fraction.
    b. List an advantage and a disadvantage of using method 2 to simplify a complex fraction.

REVIEW

Simplify each expression. Write each answer without negative exponents.

93. $(8x)^0$ 1

94. $\left(-\dfrac{3r}{4r^3}\right)^4$ $\dfrac{81}{256r^8}$

95. $\left(\dfrac{4x^3}{5x^{-3}}\right)^{-2}$ $\dfrac{25}{16x^{12}}$

96. $\left(\dfrac{12xy^{-3}}{3x^{-2}y^2}\right)^{-2}$ $\dfrac{y^{10}}{16x^6}$

CHALLENGE PROBLEMS

Simplify.

97. $\dfrac{\dfrac{h}{h^2 + 3h + 2}}{\dfrac{4}{h + 2} - \dfrac{4}{h + 1}}$ $-\dfrac{h}{4}$

98. $\dfrac{\dfrac{2}{b^2 - 1} - \dfrac{3}{ab - a}}{\dfrac{3}{ab - a} - \dfrac{2}{b^2 - 1}}$ -1

99. $a + \dfrac{a}{1 + \dfrac{a}{a + 1}}$ $\dfrac{3a^2 + 2a}{2a + 1}$

100. $\dfrac{y^{-2} + 1}{y^{-2} - 1}$ $\dfrac{1 + y^2}{1 - y^2}$

SECTION 7.6 Solving Rational Equations

OBJECTIVES

1 Solve rational equations.

2 Solve for a specified variable in a formula.

ARE YOU READY? *Are You Ready? exercises available online at www.webassign.net/brookscole*

The following problems review some basic skills that are needed when solving rational equations.

1. What is the LCD of $\dfrac{7}{2x}$, $\dfrac{1}{9x}$, and $\dfrac{17}{6x}$? $18x$

2. Multiply: $5(x - 1)\left(\dfrac{x}{x - 1}\right)$ $5x$

3. Solve: $x^2 - x - 56 = 0$ $-7, 8$

4. Find all real numbers for which the rational expression $\dfrac{x + 1}{x + 8}$ is undefined. -8

In Chapter 2, we solved equations such as $\frac{1}{6}x + \frac{5}{2} = \frac{1}{3}$ by multiplying both sides by the LCD. With this approach, the equation that results is equivalent to the original equation, but easier to solve because it is cleared of fractions.

In this section, we will extend the fraction-clearing strategy to solve another type of equation, called a *rational equation.*

Rational Equations

A **rational equation** is an equation that contains one or more rational expressions.

Rational equations often have a variable in a denominator. Some examples of rational equations are:

$$\frac{2x}{3} = \frac{x}{6} + \frac{3}{2} \qquad \frac{2}{x} + \frac{1}{4} = \frac{5}{2x} \qquad \frac{11x}{x-5} = 6 + \frac{55}{x-5}$$

Teaching Tip: Remind students that to this point, they have learned how to solve two types of equations: *linear equations in one variable,* such as $6x - 18 = 3x$, and *quadratic equations,* such as $x^2 + 9x + 20 = 0$.

1 Solve Rational Equations.

To **solve a rational equation,** we find all the values of the variable that make the equation true. Any value of the variable that makes a denominator in a rational equation equal to 0 cannot be a solution of the equation. Such a number must be rejected, because division by 0 is undefined.

The goal when solving a rational equation is to use the multiplication property of equality to find an equivalent equation that we already know how to solve, such as a linear equation in one variable or a quadratic equation.

Strategy for Solving Rational Equations

1. Determine which numbers cannot be solutions of the equation.
2. Multiply both sides of the equation by the LCD of all rational expressions in the equation. This clears the equation of fractions.
3. Solve the resulting equation.
4. Check all possible solutions in the original equation.

EXAMPLE 1 Solve: $\frac{2x}{3} = \frac{x}{6} + \frac{3}{2}$

Teaching Tip: Point out that we are now working with equations (they contain an = symbol) and in previous sections, we were working with expressions (they do not contain an = symbol).

Strategy We will use the multiplication property of equality to clear this rational equation of fractions by multiplying both sides by the LCD.

Why Equations that contain only integers are usually easier to solve than equations that contain fractions.

Solution There are no restrictions on x, because no value of x ever makes a denominator 0. Since the denominators are 3, 6, and 2, we multiply both sides of the equation by the LCD, 6.

$$\frac{2x}{3} = \frac{x}{6} + \frac{3}{2}$$ This is the equation to solve.

Caution

Always enclose the left and right sides of an equation within parentheses when multiplying both sides by the LCD.

$$6\left(\frac{2x}{3}\right) = 6\left(\frac{x}{6} + \frac{3}{2}\right)$$ Multiply both sides of the equation by the LCD of $\frac{2x}{3}$, $\frac{x}{6}$, and $\frac{3}{2}$, which is 6.

$$6\left(\frac{2x}{3}\right) = 6\left(\frac{x}{6}\right) + 6\left(\frac{3}{2}\right)$$ Distribute the multiplication by 6.

$$2 \cdot \overset{1}{\cancel{3}}\left(\frac{2x}{\underset{1}{\cancel{3}}}\right) = \overset{1}{\cancel{6}}\left(\frac{x}{\underset{1}{\cancel{6}}}\right) + \overset{1}{\cancel{2}} \cdot 3\left(\frac{3}{\underset{1}{\cancel{2}}}\right)$$ Perform the three multiplications by 6 by first removing common factors of the numerator and denominator. Try to do this step in your head.

Notation

Here is an alternate way to remove common factors when multiplying by the LCD:

$$\overset{2}{\cancel{6}}\left(\frac{2x}{\underset{1}{\cancel{3}}}\right) \text{ and } \overset{3}{\cancel{6}}\left(\frac{3}{\underset{1}{\cancel{2}}}\right)$$

$$4x = x + 9$$ Simplify. Note that the fractions have been cleared. The result is a linear equation in one variable.

$$3x = 9$$ To eliminate x on the right side, subtract x from both sides.

$$x = 3$$ To undo the multiplication by 3, divide both sides by 3.

To check, we replace each x with 3 in the *original* equation.

$$\frac{2x}{3} = \frac{x}{6} + \frac{3}{2}$$

$$\frac{2(3)}{3} \stackrel{?}{=} \frac{3}{6} + \frac{3}{2}$$ Substitute 3 for x.

$$2 \stackrel{?}{=} \frac{1}{2} + \frac{3}{2}$$ Simplify: $\frac{2(\overset{1}{\cancel{3}})}{\underset{1}{\cancel{3}}} = 2$ and $\frac{3}{6} = \frac{1}{2}$.

$$2 = 2$$ Add: $\frac{1}{2} + \frac{3}{2} = \frac{4}{2} = 2$.

Since we obtain a true statement, 3 is the solution of $\frac{2x}{3} = \frac{x}{6} + \frac{3}{2}$. The solution set is $\{3\}$.

Teaching Example 1
Solve: $\frac{4x}{9} = \frac{x}{6} + \frac{5}{3}$
Answer: 6

Self Check 1 Solve: $\frac{3x}{5} = \frac{x}{2} + \frac{1}{10}$ 1

Now Try ▶ Problem 15

EXAMPLE 2 Solve: $\frac{2}{x} + \frac{1}{4} = \frac{5}{2x}$

Strategy This equation contains two rational expressions that have a variable in their denominators. We begin by asking, "What value(s) of x make either denominator 0?" Then we will clear the equation of fractions by multiplying both sides by the LCD.

Why If a number makes the denominator of a rational expression 0, that number cannot be a solution of the equation because division by 0 is undefined.

Solution If x is 0, the denominators of $\frac{2}{x}$ and $\frac{5}{2x}$ are 0 and the rational expressions would be undefined. Therefore, 0 cannot be a solution.

Since the denominators are x, 4, and $2x$, we multiply both sides of the equation by the LCD, $4x$, to clear the equation of fractions.

$$\frac{2}{x} + \frac{1}{4} = \frac{5}{2x}$$ This is the equation to solve.

$$4x\left(\frac{2}{x} + \frac{1}{4}\right) = 4x\left(\frac{5}{2x}\right)$$ Write each side of the equation within parentheses, and then multiply both sides by $4x$.

$$4x\left(\frac{2}{x}\right) + 4x\left(\frac{1}{4}\right) = 4x\left(\frac{5}{2x}\right)$$ On the left side, distribute the multiplication by $4x$.

$$4\overset{1}{\cancel{x}}\left(\frac{2}{\underset{1}{\cancel{x}}}\right) + \overset{1}{\cancel{4}}x\left(\frac{1}{\underset{1}{\cancel{4}}}\right) = 2 \cdot \overset{1}{\cancel{2}} \cdot \overset{1}{\cancel{x}}\left(\frac{5}{\underset{1}{\cancel{2}} \cdot \underset{1}{\cancel{x}}}\right)$$ On the right side, factor $4x$ as $2 \cdot 2 \cdot x$. Perform the three multiplications by $4x$ by first removing common factors of each numerator and denominator. Try to do this step in your head.

$$8 + x = 10$$ Simplify. Note that the fractions have been cleared. The result is a linear equation in one variable.

$$x = 2$$ To undo the addition of 8, subtract 8 from both sides.

The solution of $\frac{2}{x} + \frac{1}{4} = \frac{5}{2x}$ is 2. The solution set is $\{2\}$. Check by substituting 2 for each x in the *original* equation.

Caution

After multiplying both sides by the LCD and simplifying, the equation should not contain any fractions. If it does, check for an algebraic error, or perhaps your LCD is incorrect.

Success Tip

Don't confuse procedures. To *simplify the expression* $\frac{2}{x} + \frac{1}{4}$, we build each fraction and keep the LCD $4x$, add the numerators, and write the sum over the LCD. To *solve the equation* $\frac{2}{x} + \frac{1}{4} = \frac{5}{2x}$, we multiply both sides by the LCD $4x$ to eliminate the denominators.

Teaching Example 2
Solve: $\frac{3}{x} - \frac{1}{6} = \frac{5}{2x}$
Answer: 3

Self Check 2 Solve: $\frac{1}{6} + \frac{4}{3x} = \frac{5}{x}$ 22

Now Try Problem 25

EXAMPLE 3 Solve: $y - \frac{12}{y} = 4$

Strategy Since the only denominator is y, we will multiply both sides of the equation by y.

Why Multiplying both sides by y will clear the equation of the fraction, $\frac{12}{y}$.

Solution If y is 0, the denominator of $\frac{12}{y}$ is 0 and the fraction would be undefined. Therefore, 0 cannot be a solution.

$$y - \frac{12}{y} = 4$$ This is the equation to solve.

$$y\left(y - \frac{12}{y}\right) = y(4)$$ Write each side of the equation within parentheses and then multiply both sides by the LCD, y.

$$y(y) - y\left(\frac{12}{y}\right) = y(4)$$ Distribute the multiplication by y.

$$y^2 - 12 = 4y$$ Simplify: $\overset{1}{\cancel{y}}\left(\frac{12}{\underset{1}{\cancel{y}}}\right) = 12$. Note that the fraction has been cleared.

Note the y^2-term. The result is a quadratic equation.

Caution

By the multiplication property of equality, each term on both sides of the equation must be multiplied by the LCD. Here, it would be incorrect to multiply only the second term, $\frac{12}{y}$, by the LCD.

$$\cancel{y - y \cdot \frac{12}{y} = 4}$$

We can solve the resulting quadratic equation using the factoring method.

$$y^2 - 4y - 12 = 0$$ Subtract $4y$ from both sides to get 0 on the right side.

$$(y - 6)(y + 2) = 0$$ Factor the trinomial.

$$y - 6 = 0 \quad \text{or} \quad y + 2 = 0$$ Set each factor equal to 0.

$$y = 6 \quad | \quad y = -2$$ Solve each equation.

There are two possible solutions, 6 and −2, to check in the original equation.

Check y = 6:

$$y - \frac{12}{y} = 4$$

$$6 - \frac{12}{6} \stackrel{?}{=} 4$$

$$6 - 2 \stackrel{?}{=} 4$$

$$4 = 4 \quad \text{True}$$

Check y = −2:

$$y - \frac{12}{y} = 4$$ This is the original equation.

$$-2 - \frac{12}{-2} \stackrel{?}{=} 4$$

$$-2 - (-6) \stackrel{?}{=} 4$$

$$4 = 4 \quad \text{True}$$

Thus, the solutions of $y - \frac{12}{y} = 4$ are 6 and −2. The solution set is $\{-2, 6\}$.

Teaching Example 3
Solve: $x - \frac{63}{x} = 2$
Answer: −7, 9

Self Check 3 Solve: $x - \frac{24}{x} = -5$ 3, −8

Now Try Problem 31

EXAMPLE 4 Solve: $\frac{11x}{x-5} = 6 + \frac{55}{x-5}$

Strategy Since both denominators are $x - 5$, we multiply both sides by the LCD, $x - 5$.

Why This will clear the equation of fractions.

Solution If x is 5, the denominators of $\frac{11x}{x-5}$ and $\frac{55}{x-5}$ are 0, and the rational expressions are undefined. Therefore, 5 cannot be a solution of the equation.

Caution

Even if you do not make an arithmetic or algebraic error when solving a rational equation, a possible solution may not check.

$$\frac{11x}{x-5} = 6 + \frac{55}{x-5}$$ This is the equation to solve.

$$(x-5)\left(\frac{11x}{x-5}\right) = (x-5)\left(6 + \frac{55}{x-5}\right)$$ Write each side of the equation within parentheses and then multiply both sides by $x - 5$.

$$\overset{1}{\cancel{(x-5)}}\left(\frac{11x}{\underset{1}{\cancel{x-5}}}\right) = (x-5)6 + \overset{1}{\cancel{(x-5)}}\left(\frac{55}{\underset{1}{\cancel{x-5}}}\right)$$ Distribute the multiplication by $x - 5$. Remove the common binomial factor $(x - 5)$ of the numerator and denominator.

$$11x = (x-5)6 + 55$$ Simplify. Note that the fractions have been cleared. The result is a linear equation in one variable.

$$11x = 6x - 30 + 55$$ To solve the resulting equation, distribute the 6.

$$11x = 6x + 25$$ Combine like terms: $-30 + 55 = 25$.

$$5x = 25$$ To eliminate $6x$ on the right side, subtract $6x$ from both sides.

$$x = 5$$ To undo the multiplication by 5, divide both sides by 5.

Teaching Tip: For Example 4, show that when the possible solution 5 is substituted for x in the original equation, two undefined rational expressions appear in the check: $\frac{55}{0} = 6 + \frac{55}{0}$

Earlier, we determined that 5 makes both denominators in the original equation 0. Therefore, 5 cannot be a solution. Since 5 is the only possible solution, and it must be rejected, it follows that $\frac{11x}{x-5} = 6 + \frac{55}{x-5}$ has *no solution.* The solution set is written as $\{ \}$ or $\varnothing$.

When solving an equation, a possible solution that does not satisfy the original equation is called an **extraneous solution.** In this example, 5 is an extraneous solution.

The Language of Algebra

Extraneous means not a vital part. Mathematicians speak of *extraneous* solutions. Rock groups don't want *extraneous* sounds (like feedback) coming from their amplifiers. Artists erase *extraneous* marks on their sketches.

Teaching Example 4

Solve: $\frac{3n}{n-4} = 5 + \frac{12}{n-4}$

Answer: No solution, 4 is extraneous

Self Check 4 Solve $\frac{9x}{x-6} = 3 + \frac{54}{x-6}$, if possible. No solution, 6 is extraneous

Now Try Problem 41

EXAMPLE 5 Solve: $\frac{x+5}{x+3} + \frac{1}{x^2+2x-3} = 1$

Strategy We will multiply both sides by the LCD of the two rational expressions in the equation. But first, we must factor the second denominator.

Why To determine the restrictions on the variable and to find the LCD, we need to write $x^2 + 2x - 3$ in factored form.

Solution Since the trinomial $x^2 + 2x - 3$ factors as $(x + 3)(x - 1)$, we can write the given equation as:

$$\frac{x+5}{x+3} + \frac{1}{(x+3)(x-1)} = 1$$ If x is -3, the first denominator is 0. If x is -3 or 1, the second denominator is 0.

We see that -3 and 1 cannot be solutions of the equation, because they make rational expressions in the equation undefined.

Since the denominators are $x + 3$ and $(x + 3)(x - 1)$, we multiply both sides of the equation by the LCD, $(x + 3)(x - 1)$, to clear the fractions.

$$(x+3)(x-1)\left[\frac{x+5}{x+3} + \frac{1}{(x+3)(x-1)}\right] = (x+3)(x-1)[1]$$ Write each side within brackets [ ].

$$\overset{1}{\cancel{(x+3)}}(x-1)\frac{x+5}{\underset{1}{\cancel{x+3}}} + \overset{1}{\cancel{(x+3)}}\overset{1}{\cancel{(x-1)}}\frac{1}{\underset{1}{\cancel{(x+3)}}\underset{1}{\cancel{(x-1)}}} = (x+3)(x-1)[1]$$ Distribute the multiplication by $(x + 3)(x - 1)$ and remove common factors.

$$(x-1)(x+5) + 1 = (x+3)(x-1)$$ Simplify. The fractions are cleared.

The Language of Algebra

We say that the LCD is a **multiplier** that clears a rational equation of fractions.

To solve the resulting equation, we multiply the binomials on the left side and the right side, and proceed as follows.

$$x^2 + 4x - 5 + 1 = x^2 + 2x - 3$$ Find $(x - 1)(x + 5)$ and $(x + 3)(x - 1)$.

$$x^2 + 4x - 4 = x^2 + 2x - 3$$ Combine like terms: $-5 + 1 = -4$.

$$4x - 4 = 2x - 3$$ Subtract x^2 from both sides. The x^2-terms are eliminated. The result is a linear equation in one variable.

$$2x - 4 = -3$$ To eliminate $2x$ on the right side, subtract $2x$ from both sides.

$$2x = 1$$ To undo the subtraction of 4, add 4 to both sides.

$$x = \frac{1}{2}$$ To undo the multiplication by 2, divide both sides by 2.

A check will show that $\frac{1}{2}$ is the solution of the original equation.

Teaching Example 5
Solve: $\frac{x+1}{x+5} + \frac{1}{x^2+3x-10} = 1$
Answer: $\frac{9}{4}$

Self Check 5 Solve: $\frac{1}{x+3} + \frac{1}{x-3} = \frac{5}{x^2-9}$ $\frac{5}{2}$

Now Try Problem 45

2 Solve for a Specified Variable in a Formula.

Many formulas are expressed as rational equations. To solve such formulas for a specified variable, we use the same steps, in the same order, as we do when solving rational equations having only one variable.

EXAMPLE 6 **Determining a Child's Dosage.** The formula $C = \frac{AD}{A + 12}$ is called **Young's rule.** It is a way to find the approximate child's dose C of a medication, where A is the age of the child in years and D is the recommended dosage for an adult. Solve the formula for D.

Strategy As we have done in the previous examples, we will begin by multiplying both sides of the equation by the LCD to clear it of the fraction.

Why To isolate D on the right side of the equation, we must first isolate the term AD on that side. That calls for clearing the right side of the denominator $A + 12$.

Solution

$$C = \frac{AD}{A + 12}$$ This is Young's rule.

$$(A + 12)(C) = (A + 12)\left(\frac{AD}{A + 12}\right)$$ Write each side of the formula within parentheses, and then multiply both sides by the LCD, $A + 12$.

Success Tip

It is often helpful to circle the variable in the formula that you are solving for:

$$C = \frac{A\textcircled{D}}{A + 12}$$

$(A + 12)C = AD$ Simplify the right side: $(\cancel{A + 12})\left(\frac{AD}{\cancel{A + 12}}\right)$.

$AC + 12C = AD$ Distribute the multiplication by C.

$\frac{AC + 12C}{A} = \frac{AD}{A}$ To undo the multiplication by A on the right side and isolate D, divide both sides by A.

$\frac{AC + 12C}{A} = D$ Simplify the right side: $\frac{\cancel{A}D}{\cancel{A}} = D$.

Solving Young's rule for D, we have $D = \frac{AC + 12C}{A}$.

Teaching Example 6
Solve: $w = \frac{na}{c + b}$ for a.
Answer: $a = \frac{cw + bw}{n}$

Self Check 6 Solve $R = \frac{eS}{T - 10}$ for S. $S = \frac{RT - 10R}{e}$

Now Try Problem 49

EXAMPLE 7

©iStockphoto.com/Artem Zhushman

Photography. The design of a camera lens uses the formula $\frac{1}{f} = \frac{1}{p} + \frac{1}{q}$, where f is the focal length of the lens, p is the distance from the lens to the object, and q is the distance from the lens to the image. Solve the formula for q.

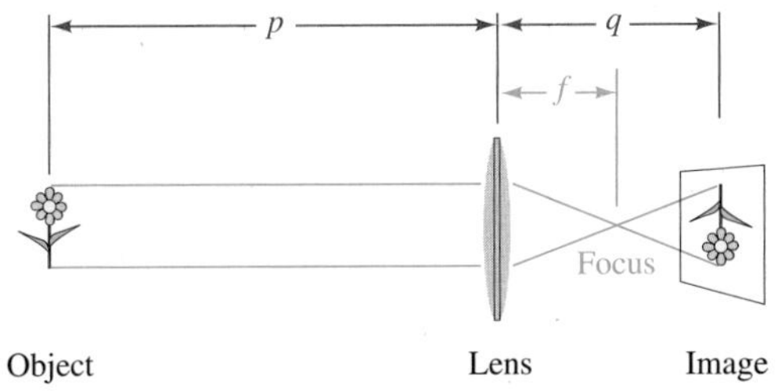

Strategy We will begin by multiplying both sides of the equation by the LCD.

Why It will be easier to isolate q if there are no fractions.

Solution

$\frac{1}{f} = \frac{1}{p} + \frac{1}{q}$ This is the given formula.

$fpq\left(\frac{1}{f}\right) = fpq\left(\frac{1}{p} + \frac{1}{q}\right)$ Write each side of the formula within parentheses and then multiply both sides by the LCD, fpq.

$\cancel{f}pq\left(\frac{1}{\cancel{f}}\right) = f\cancel{p}q\left(\frac{1}{\cancel{p}}\right) + fp\cancel{q}\left(\frac{1}{\cancel{q}}\right)$ Distribute the multiplication by fpq and then remove the common factors of each numerator and denominator.

$pq = fq + fp$ Simplify. Note that q is on both sides of the resulting equation.

Caution

A common error is to divide both sides of $pq = fq + fp$ by p to solve for q:

$$q = \frac{fq + fp}{p}$$ (crossed out)

The formula is not solved for q because q appears on *both sides* of the equation.

If we subtract fq from both sides, all terms that contain q will be on the left side.

$pq - fq = fp$ Subtract fq from both sides.

$q(p - f) = fp$ Factor out the GCF, q, from the two terms on the left side.

$\frac{q(p - f)}{p - f} = \frac{fp}{p - f}$ To undo the multiplication by $(p - f)$ and isolate q, divide both sides by $p - f$.

$q = \frac{fp}{p - f}$ Simplify the left side: $\frac{q(\cancel{p - f})}{\cancel{p - f}} = q$.

Teaching Tip: Stress the new equation-solving technique used here: Get all the terms involving q on the same side of the equation, and then factor out q. We use a concept from Chapter 6, factoring.

Solving the formula for q, we have $q = \frac{fp}{p - f}$.

Teaching Example 7
Solve $\frac{1}{R} = \frac{1}{S} + \frac{1}{T}$ for T.
Answer: $T = \frac{RS}{S - R}$

Self Check 7 Solve the formula in Example 7 for p. $p = \frac{fq}{q - f}$

Now Try Problem 51

SECTION 7.6 STUDY SET

VOCABULARY

Fill in the blanks.

1. Equations that contain one or more rational expressions, such as $\frac{x}{x+2} = 4 + \frac{10}{x+2}$, are called rational equations.
2. To solve a rational equation we find all the values of the variable that make the equation true.
3. To clear a rational equation of fractions, multiply both sides by the LCD of all rational expressions in the equation.
4. When solving a rational equation, if we obtain a number that does not satisfy the original equation, the number is called an extraneous solution.

CONCEPTS

5. Is 5 a solution of the given rational equation?
   a. $\frac{1}{x-1} = 1 - \frac{3}{x-1}$ Yes
   b. $\frac{x}{x-5} = 3 + \frac{5}{x-5}$ No
6. A student was asked to solve a rational equation. The first step of his solution is as follows:
   $$12x\left(\frac{5}{x} + \frac{2}{3}\right) = 12x\left(\frac{7}{4x}\right)$$
   a. What equation was he asked to solve? $\frac{5}{x} + \frac{2}{3} = \frac{7}{4x}$
   b. What LCD is used to clear the equation of fractions? $12x$
7. Consider the rational equation $\frac{x}{x-3} = \frac{1}{x} + \frac{2}{x-3}$.
   a. What values of x make a denominator 0? 3, 0
   b. What values of x make a rational expression undefined? 3, 0
   c. What numbers can't be solutions of the equation? 3, 0
8. A student solved a rational equation and found 8 to be a possible solution. When she checked 8, she obtained $\frac{3}{0} = \frac{1}{0} + \frac{2}{3}$. What conclusion can be drawn? 8 is an extraneous solution.

By what should both sides of the equation be multiplied to clear it of fractions?

9. a. $\frac{1}{y} = 20 - \frac{5}{y}$ y
   b. $\frac{x}{x^2-4} = \frac{4}{x-2}$ $(x+2)(x-2)$
10. a. $\frac{x}{5} = \frac{3x}{10} + \frac{7}{2x}$ $10x$
    b. $\frac{2x}{x-6} = 4 + \frac{1}{x-6}$ $x - 6$
11. Perform each multiplication.
    a. $4x\left(\frac{3}{4x}\right)$ 3
    b. $(x+6)(x-2)\left(\frac{3}{x-2}\right)$ $3x + 18$
12. Fill in the blanks.
    $$\underbrace{8x\left(\frac{3}{4x}\right)}_{6} = \underbrace{8x\left(\frac{1}{8x}\right)}_{1} + \underbrace{8x\left(\frac{5}{4}\right)}_{10x}$$

NOTATION

Complete the solution to solve the equation.

13. $$\frac{2}{a} + \frac{1}{2} = \frac{7}{2a}$$
    $$2a\left(\frac{2}{a} + \frac{1}{2}\right) = 2a\left(\frac{7}{2a}\right)$$
    $$2a\left(\frac{2}{a}\right) + 2a\left(\frac{1}{2}\right) = 2a\left(\frac{7}{2a}\right)$$
    $$4 + a = 7$$
    $$4 + a - 4 = 7 - 4$$
    $$a = 3$$
14. Can $5x\left(\frac{2}{x} + \frac{4}{5}\right)$ be written as $5x \cdot \frac{2}{x} + \frac{4}{5}$? Explain.
    No; $5x\left(\frac{2}{x}\right) + 5x\left(\frac{4}{5}\right) \neq 5x \cdot \frac{2}{x} + \frac{4}{5}$

GUIDED PRACTICE

***Solve each equation and check the result. If an equation has no solution, so indicate.* See Example 1.**

15. $\frac{2}{3} = \frac{1}{2} + \frac{x}{6}$ 1
16. $\frac{7}{4} = \frac{x}{8} + \frac{5}{2}$ -6
17. $\frac{s}{12} - \frac{s}{2} = \frac{5s}{4}$ 0
18. $\frac{n}{18} - \frac{n}{6} = \frac{4n}{3}$ 0
19. $\frac{x}{18} = \frac{1}{3} - \frac{x}{2}$ $\frac{3}{5}$
20. $\frac{x}{4} = \frac{1}{2} - \frac{3x}{20}$ $\frac{5}{4}$
21. $\frac{b}{4} + \frac{1}{2} = \frac{b}{3} - \frac{1}{4}$ 9
22. $\frac{n}{6} + \frac{2}{3} = \frac{n}{3} - \frac{1}{36}$ $\frac{25}{6}$

***Solve each equation and check the result. If an equation has no solution, so indicate.* See Example 2.**

23. $\frac{5}{3k} + \frac{1}{k} = -2$ $-\frac{4}{3}$
24. $\frac{3}{4h} + \frac{2}{h} = 1$ $\frac{11}{4}$
25. $\frac{1}{4} - \frac{5}{6} = \frac{1}{a}$ $-\frac{12}{7}$
26. $\frac{5}{9} - \frac{1}{3} = \frac{1}{b}$ $\frac{9}{2}$
27. $\frac{1}{8} + \frac{2}{b} - \frac{1}{12} = 0$ -48
28. $\frac{1}{14} + \frac{2}{n} - \frac{2}{21} = 0$ 84
29. $\frac{4}{5} - \frac{1}{10x} = \frac{7}{15}$ $\frac{3}{10}$
30. $\frac{5}{14} - \frac{1}{2x} = \frac{3}{7}$ -7

***Solve each equation and check the result. If an equation has no solution, so indicate.* See Example 3.**

31. $x + \frac{8}{x} = 6$ 2, 4
32. $z - \frac{16}{z} = 6$ 8, -2
33. $\frac{10}{t} - t = 3$ -5, 2
34. $\frac{7}{p} - p = -6$ -1, 7
35. $\frac{20}{c} + c = -9$ -4, -5
36. $d = 4 + \frac{21}{d}$ 7, -3
37. $4 + \frac{15}{p} = 3p$ 3, $-\frac{5}{3}$
38. $2x = 6 + \frac{8}{x}$ 4, -1

***Solve each equation and check the result. If an equation has no solution, so indicate.* See Example 4.**

39. $\frac{x}{x-5} = 3 + \frac{5}{x-5}$ No solution; 5 is extraneous

40. $\frac{3}{y-2} = \frac{3}{y-2} + 1$ No solution; 2 is extraneous

▶ 41. $\frac{a^2}{a+2} - a = \frac{4}{a+2}$ No solution; -2 is extraneous

▶ 42. $\frac{z^2}{z+1} + 2 = \frac{1}{z+1}$ No solution; -1 is extraneous

Solve each equation and check the result. If an equation has no solution, so indicate. **See Example 5.**

43. $\frac{x+6}{x+4} + \frac{1}{x^2+x-12} = 1$ $\frac{5}{2}$

▶ 44. $\frac{x+7}{x+2} + \frac{1}{x^2-3x-10} = 1$ $\frac{24}{5}$

45. $\frac{2x}{x^2+x-2} + \frac{2}{x+2} = 1$ 0, 3

▶ 46. $\frac{4x}{x^2+2x-3} + \frac{3}{x+3} = 1$ 0, 5

Solve each formula for the specified variable. **See Example 6.**

▶ 47. $h = \frac{2A}{b+d}$ for A $A = \frac{h(b+d)}{2}$

▶ 48. $T = \frac{3R}{M-n}$ for R $R = \frac{T(M-n)}{3}$

▶ 49. $I = \frac{E}{R+r}$ for r $r = \frac{E-IR}{I}$

50. $\frac{S}{k+h} = E$ for k $k = \frac{S-Eh}{E}$

51. $\frac{5}{x} - \frac{4}{y} = \frac{5}{z}$ for x $x = \frac{5yz}{5y+4z}$

52. $\frac{2}{c} + \frac{2}{d} = \frac{1}{h}$ for c $c = \frac{2dh}{d-2h}$

53. $\frac{1}{r} + \frac{1}{s} = \frac{1}{t}$ for r $r = \frac{st}{s-t}$

▶ 54. $\frac{1}{x} - \frac{1}{y} = \frac{1}{z}$ for x $x = \frac{yz}{y+z}$

Solve each formula for the specified variable. **See Example 7.**

▶ 55. $\frac{P}{n} = rt$ for P $P = nrt$

56. $\frac{F}{m} = a$ for F $F = ma$

57. $\frac{a}{b} = \frac{c}{d}$ for d $d = \frac{bc}{a}$

▶ 58. $\frac{pc}{s} = \frac{t}{r}$ for c $c = \frac{st}{pr}$

▶ 59. $\frac{1}{a} + \frac{1}{b} = 1$ for a $a = \frac{b}{b-1}$

▶ 60. $\frac{1}{a} - \frac{1}{b} = 1$ for b $b = \frac{a}{1-a}$

61. $F = \frac{L^2}{6d} + \frac{d}{2}$ for L^2 $L^2 = 6dF - 3d^2$

▶ 62. $H = \frac{J^3}{cd} - \frac{K^3}{d}$ for J^3 $J^3 = Hcd + cK^3$

TRY IT YOURSELF

Solve each equation and check the result. If an equation has no solution, so indicate.

63. $\frac{1}{3} + \frac{2}{x-3} = 1$ 6

▶ 64. $\frac{3}{5} + \frac{7}{x+2} = 2$ 3

65. $\frac{7}{q^2-q-2} + \frac{1}{q+1} = \frac{3}{q-2}$ 1

66. $\frac{3}{x-1} - \frac{1}{x+9} = \frac{18}{x^2+8x-9}$ -5

67. $\frac{2}{3-t} = \frac{-t}{t+3}$ $-1, 6$

68. $\frac{n}{n+1} = \frac{6}{n+7}$ $-3, 2$

69. $\frac{1}{8} + \frac{2}{y} = \frac{1}{y} + \frac{1}{10}$ -40

70. $\frac{7}{10} + \frac{4}{c} = \frac{1}{c} + \frac{11}{15}$ 90

71. $4 - \frac{8}{x+1} = \frac{8x}{x+1}$ No solution; -1 is extraneous

72. $\frac{x}{x-2} = \frac{2}{x-2} + 2$ No solution; 2 is extraneous

73. $\frac{5a}{a+1} - 4 = \frac{3}{a+1}$ 7

74. $\frac{4}{b-3} = \frac{b+5}{b-3} - 5$ 4

75. $\frac{2}{y+1} + 5 = \frac{12}{y+1}$ 1

76. $\frac{3}{p+6} - 2 = \frac{7}{p+6}$ -8

77. $\frac{3}{x+1} = \frac{x-2}{x+1} + \frac{x-2}{2}$ $-4, 3$

78. $\frac{2}{x-1} + \frac{x-2}{3} = \frac{4}{x-1}$ $4, -1$

79. $\frac{z-4}{z-3} = \frac{z+2}{z+1}$ 1

80. $\frac{a+2}{a+8} = \frac{a-3}{a-2}$ 4

81. $\frac{3}{x} + 2 = 3$ 3

▶ 82. $\frac{2}{x} + 9 = 11$ 1

83. $\frac{4}{y^2-4} = \frac{1}{y-2} + \frac{1}{y+2}$ No solution, 2 is extraneous

84. $\frac{2w}{w^2-9} = \frac{1}{w+3} - \frac{4}{w-3}$ No solution, -3 is extraneous

85. $\frac{3}{5d} + \frac{4}{3} = \frac{9}{10d}$ $\frac{9}{40}$

86. $\frac{2}{3d} + \frac{1}{4} = \frac{11}{6d}$ $\frac{14}{3}$

87. $\frac{n}{n^2-9} + \frac{n+8}{n+3} = \frac{n-8}{n-3}$ 0

▶ 88. $\frac{7}{x-5} = \frac{40}{x^2-25} + \frac{3}{x+5}$ $-\frac{5}{2}$

▶ 89. $\frac{3}{x-2} + \frac{1}{x} = \frac{6x+4}{x^2-2x}$ -3

90. $\frac{x}{x-1} - \frac{12}{x^2-x} = \frac{-1}{x-1}$ $3, -4$

▶ 91. $y + \frac{2}{3} = \frac{2y-12}{3y-9}$ 1, 2

▶ 92. $1 - \frac{3}{b} = \frac{-8b}{b^2+3b}$ $1, -9$

93. $\frac{a-1}{7} - \frac{a-2}{14} = \frac{1}{2}$ 7

94. $\frac{3x-1}{6} - \frac{x+3}{2} = \frac{3x+4}{3}$ -3

Look Alikes . . .

For each expression, perform the indicated operations and then simplify, if possible. Solve each equation and check the result.

95. a. $\frac{a}{3} + \frac{3}{5} + \frac{a}{15}$ $\frac{2a+3}{5}$ b. $\frac{a}{3} + \frac{3}{5} = \frac{a}{15}$ $-\frac{9}{4}$

▶ 96. a. $\frac{1}{6x} - \frac{2}{x-6}$ $\frac{-11x-6}{6x(x-6)}$ b. $\frac{1}{6x} = \frac{2}{x-6}$ $-\frac{6}{11}$

▶ 97. a. $\frac{x}{x-2} - \frac{1}{x-3}$ $\frac{x^2-4x+2}{(x-2)(x-3)}$ b. $\frac{x}{x-2} - \frac{1}{x-3} = 1$ 4

98. **a.** $\frac{u^2+1}{u^2-u} - \frac{u}{u-1}$ $\frac{1}{u(u-1)}$ **b.** $\frac{u^2+1}{u^2-u} - \frac{u}{u-1} = \frac{1}{u}$ 2

APPLICATIONS

99. Medicine. Radioactive tracers are used for diagnostic work in nuclear medicine. The ***effective half-life H*** of a radioactive material in an organism is given by the formula $H = \frac{RB}{R+B}$ where R is the radioactive half-life and B is the biological half-life of the tracer. Solve the formula for R. $R = \frac{HB}{B-H}$

100. Chemistry. Charles's law describes the relationship between the volume and temperature of a gas that is kept at a constant pressure. It can be expressed as $\frac{V_1}{V_2} = \frac{T_1}{T_2}$ where V_1 and V_2 are variables representing two different volumes, and T_1 and T_2 are variables representing two different temperatures. (Recall that the notation V_1 is read as *V sub one.*) Solve for V_2. $V_2 = \frac{V_1 T_2}{T_1}$

101. Electronics. Most electronic circuits require resistors to make them work properly. Resistors are components that limit current. An important formula about resistors in a circuit is $\frac{1}{r} = \frac{1}{r_1} + \frac{1}{r_2}$. Solve for r. $r = \frac{r_1 r_2}{r_2 + r_1}$

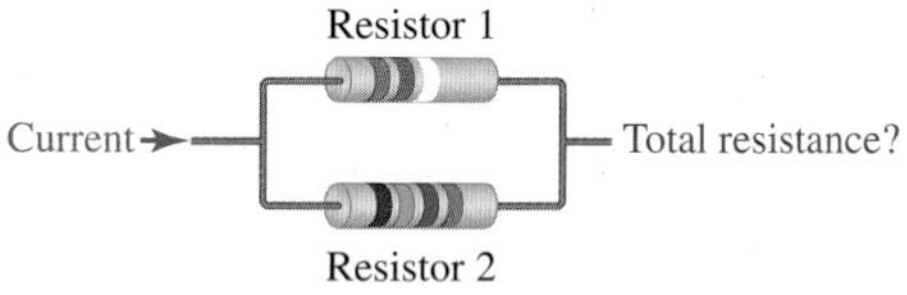

102. Mathematical Formulas. To quickly find the sum $\frac{1}{2} + \frac{1}{4} + \frac{1}{8} + \frac{1}{16} + \frac{1}{32} + \frac{1}{64} + \frac{1}{128}$, mathematicians use the formula $S = \frac{a(1-r^n)}{1-r}$. Solve the formula for a. $a = \frac{S(1-r)}{1-r^n}$

WRITING

103. Explain how the multiplication property of equality is used to solve rational equations. Give an example.

104. When solving rational equations, how do you know whether a solution is extraneous?

105. What is meant by clearing a rational equation of fractions? Give an example.

106. Explain the difference between the procedure used to simplify $\frac{1}{x} + \frac{1}{3}$ and the procedure used to solve $\frac{1}{x} + \frac{1}{3} = \frac{1}{2}$.

REVIEW

107. Uniforms. A cheerleading squad had their school mascot embroidered on the front of their uniform sweaters. They were charged \$18.50 per sweater plus a one time setup fee of \$75. If the project cost \$445, how many sweaters were embroidered? 20

108. Geometry. The vertex angle of an isosceles triangle is 46°. Find the measure of each base angle. 67°

CHALLENGE PROBLEMS

Solve each equation and check the result. If an equation has no solution, so indicate.

109. $\frac{x-4}{x-3} + \frac{x-2}{x-3} = x - 3$ 5; 3 is extraneous

110. $\frac{3}{x} = \frac{1 - \frac{1}{x}}{3 - \frac{7}{x}}$ 3, 7

111. $x^{-2} + 2x^{-1} + 1 = 0$ -1

112. Engines. A formula that is used in the design and testing of diesel engines is $E = 1 - \frac{T_4 - T_1}{a(T_3 - T_2)}$. Solve the formula for T_1. $T_1 = Ea(T_3 - T_2) - a(T_3 - T_2) + T_4$

SECTION 7.7 Problem Solving Using Rational Equations

OBJECTIVES

1 Solve number problems.

2 Solve uniform motion problems.

3 Solve shared-work problems.

4 Solve investment problems.

ARE YOU READY?

Are You Ready? exercises available online at www.webassign.net/brookscole

The following problems review some basic skills that are needed when using rational equations to solve application problems.

1. Solve the uniform motion formula $d = rt$ for t. $t = \frac{d}{r}$

2. Multiply: $\frac{1}{5} \cdot x$ $\frac{x}{5}$

3. What is the simple interest formula? $I = Prt$

4. What is the LCD for the fractions in the rational equation $\frac{x}{9} + \frac{x}{7} = 1$? 63

We will now use the six-step problem-solving strategy to solve application problems from a variety of areas, including banking, petroleum engineering, sports, and travel. In each case, we will use a rational equation to model the situation. We begin with an example in which we find an unknown number.

1 Solve Number Problems.

EXAMPLE 1 **Number Problem.** If the same number is added to both the numerator and the denominator of the fraction $\frac{3}{5}$, the result is $\frac{4}{5}$. Find the number.

Analyze

- Begin with the fraction $\frac{3}{5}$.
- Add the same number to the numerator and to the denominator.
- The result is $\frac{4}{5}$.
- Find the number.

Assign Let $n =$ the unknown number.

Form To form an equation, add the unknown number to the numerator and to the denominator of $\frac{3}{5}$. Then set the result equal to $\frac{4}{5}$.

$$\frac{3+n}{5+n} = \frac{4}{5}$$

Solve To solve this rational equation, we begin by clearing it of fractions.

$$\frac{3+n}{5+n} = \frac{4}{5}$$

$$5\overset{1}{(\cancel{5+n})}\left(\frac{3+n}{\underset{1}{\cancel{5+n}}}\right) = \overset{1}{\cancel{5}}(5+n)\left(\frac{4}{\underset{1}{\cancel{5}}}\right)$$ Multiply both sides by the LCD, $5(5 + n)$. Then remove common factors of the numerator and denominator.

$$5(3+n) = (5+n)4$$ Simplify. The fractions have been cleared.

$$15 + 5n = 20 + 4n$$ Distribute the multiplication by 5 and by 4.

$$15 + n = 20$$ To isolate the variable term on the left side, subtract $4n$ from both sides.

$$n = 5$$ To undo the addition of 15, subtract 15 from both sides.

State The number is 5.

Check When we add 5 to both the numerator and denominator of $\frac{3}{5}$, we get

$$\frac{3+5}{5+5} = \frac{8}{10} = \frac{4}{5}$$

The result checks.

Teaching Example 1 **Number Problems.** If the same number is added to both the numerator and denominator of the fraction $\frac{9}{11}$, the result is $\frac{10}{11}$. Find the number.
Answer: 11

Self Check 1 **Number Problem.** If the same number is added to both the numerator and denominator of the fraction $\frac{7}{9}$, the result is $\frac{8}{9}$. Find the number. 9

Now Try Problem 13

2 Solve Uniform Motion Problems.

The Language of Algebra

In uniform motion problems, the word **speed** is often used in place of the word **rate.** For example, we can say a car travels at a *rate* of 50 mph or its *speed* is 50 mph.

Recall that we use the distance formula $d = rt$ to solve motion problems. The relationship between distance, rate, and time can be expressed in another way by solving for t.

$$d = rt$$ Distance = rate · time.

$$\frac{d}{r} = \frac{rt}{r}$$ To undo the multiplication by r and isolate t, divide both sides by r.

$$\frac{d}{r} = t$$ Simplify the right side: $\frac{\overset{1}{\cancel{r}}t}{\underset{1}{\cancel{r}}} = t$.

$$t = \frac{d}{r}$$

This result suggests an alternate form of the distance formula, time $= \frac{\text{distance}}{\text{rate}}$, that is used to solve the next example.

EXAMPLE 2 **Runners.** A coach can run 10 miles in the same amount of time as his best student-athlete can run 12 miles. If the student runs 1 mile per hour (mph) faster than the coach, find the running speeds of the coach and the student.

©Exactostock/SuperStock

Analyze

- The coach runs 10 miles in the same time that the student runs 12 miles.
- The student runs 1 mph faster than the coach.
- Find the speed that each runs.

Assign Since the student's speed is 1 mph faster than the coach's, let $r =$ the speed that the coach can run. Then, $r + 1 =$ the speed that the student can run.

Form The expressions for the rates are entered in the Rate column of the table. The distances run by the coach and by the student are entered in the Distance column of the table.

Using $t = \frac{d}{r}$, we find that the time it takes the coach to run 10 miles, at a rate of r mph, is $\frac{10}{r}$ hours. Similarly, we find that the time it takes the student to run 12 miles, at a rate of $(r + 1)$ mph, is $\frac{12}{r+1}$ hours. These expressions are entered in the Time column of the table.

Success Tip

We could have let $r =$ the speed that the student can run. Then, since the coach runs slower than the student, $r - 1 =$ the speed of the coach.

	Rate	· Time	= Distance
Coach	r	$\frac{10}{r}$	10
Student	$r + 1$	$\frac{12}{r+1}$	12

Enter the information in these two columns first.

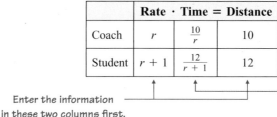

To get these entries, divide the distance by the rate to obtain an expression for the time: $t = \frac{d}{r}$.

The time it takes the coach to run 10 miles	is the same as	the time it takes the student to run 12 miles.
$\frac{10}{r}$	$=$	$\frac{12}{r+1}$

Solve To solve this rational equation, we begin by clearing it of fractions.

$$\frac{10}{r} = \frac{12}{r+1}$$

$$\overset{1}{\cancel{r}}(r + 1)\left(\frac{10}{\underset{1}{\cancel{r}}}\right) = r\overset{1}{\cancel{(r+1)}}\left(\frac{12}{\underset{1}{\cancel{r+1}}}\right)$$ Multiply both sides by the LCD, $r(r + 1)$. Then remove common factors of the numerator and denominator.

$$(r + 1)10 = 12r$$ Simplify. The fractions have been cleared.

$$10r + 10 = 12r$$ On the left side, distribute the multiplication by 10.

$$10 = 2r$$ To isolate the variable term on the right, subtract $10r$ from both sides.

$$5 = r$$ To undo the multiplication by 2, divide both sides by 2.

If $r = 5$, then $r + 1 = 6$.

State The coach's running speed is 5 mph and the student's running speed is 6 mph.

Check The coach will run 10 miles in $\frac{10 \text{ miles}}{5 \text{ mph}} = 2$ hours. The student will run 12 miles in $\frac{12 \text{ miles}}{6 \text{ mph}} = 2$ hours. The times are the same; the results check.

Caution

Check the result *in the words of the problem,* not by substituting it into the equation. Why? The equation may have been solved correctly, but the danger is that you may have formed it incorrectly.

Teaching Example 2 Cycling. Cyclists in the Tour de France can ride 50 miles in the same time the average cyclist can travel 24 miles. If the average cyclist's rate is 13 mph slower than the Tour de France competitors, how fast does the average cyclist ride?
Answer: 12 mph

Self Check 2 **Cycling.** A cyclist can ride 24 miles in the same amount of time that her friend can walk 8 miles. If the cyclist travels 8 mph faster than the walker, find the speed of the walker. 4 mph

Now Try Problem 23

3 Solve Shared-Work Problems.

Problems in which two or more people (or machines) work together to complete a job are called *shared-work problems.* To solve such problems, we must determine the **rate of work** for each person (or machine) involved. For example, suppose it takes you 4 hours to clean your house. Your rate of work can be expressed as $\frac{1}{4}$ of the job is completed per hour. If someone else takes 5 hours to clean the same house, they complete $\frac{1}{5}$ of the job per hour. In general, a rate of work can be determined in the following way.

Rate of Work

If a job can be completed in t units of time, the rate of work can be expressed as:

$\frac{1}{t}$ of the job is completed per unit of time.

To solve shared-work problems, we also must determine what fractional part of a job is completed. To do this, we use the formula

Work completed = rate of work · time worked or $W = rt$

EXAMPLE 3 **Payroll.** At the end of a pay period, it takes the president of a company 15 minutes to sign all of her employees' payroll checks. What fractional part of the job is completed if the president signs checks for 10 minutes?

Strategy We will begin by finding the president's check-signing rate. Then we can use the formula $W = rt$ to find the part of the job that is completed.

Why We know the time worked is 10 minutes. To use the work formula to find what part of the job is completed, we also need to know the president's work rate.

Solution If all of the checks can be signed in 15 minutes, the president's work rate is $\frac{1}{15}$ job per minute. Substituting into the work formula, we have

$$W = rt$$

$$= \frac{1}{15} \cdot 10 \quad \text{Substitute } \tfrac{1}{15} \text{ for } r\text{, the work rate, and 10 for } t\text{, the time worked.}$$

$$= \frac{10}{15} \quad \text{Do the multiplication. Think: } \tfrac{1}{15} \cdot \tfrac{10}{1} = \tfrac{10}{15}.$$

$$= \frac{2}{3} \quad \text{Simplify by removing the common factor of 5: } \tfrac{2 \cdot \cancel{5}^{1}}{3 \cdot \cancel{5}_{1}} = \tfrac{2}{3}.$$

In 10 minutes, the president will complete $\frac{2}{3}$ of the job of signing the payroll checks.

Teaching Example 3 Welding. It takes a welder 15 hours to complete a job. What part of the job is completed in 5 hours?
Answer: $\frac{1}{3}$ of the job

Self Check 3 **Farming.** It takes a farmer 8 days to harvest a wheat crop. What part of the job is completed in 6 days? $\frac{3}{4}$ of the job

Now Try Problem 5

EXAMPLE 4 **Filling a Tank.** An inlet pipe can fill an oil storage tank in 7 days, and a second inlet pipe can fill the same tank in 9 days. If both pipes are used, how long will it take to fill the tank?

Teaching Tip: Ask your students to estimate the time it will take to fill the tank if both pipes are used:
- Less than 7 days
- Between 7 and 9 days
- More than 9 days

Analyze

- The first pipe can fill the tank in 7 days.
- The second pipe can fill the tank in 9 days.
- How long will it take the two pipes, working together, to fill the tank?

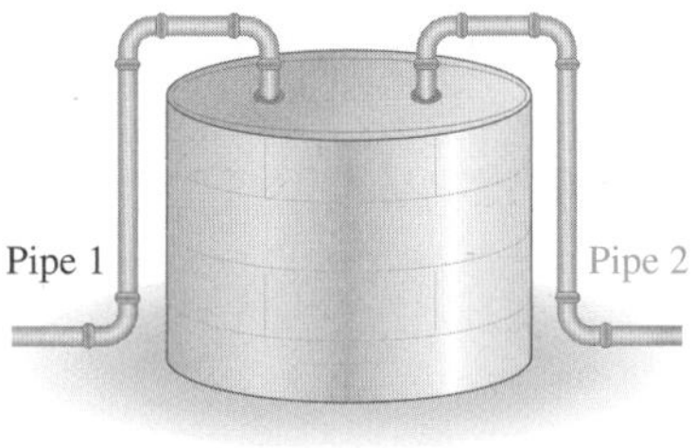

Assign Let $x =$ the number of days it will take to fill the tank if both pipes are used.

Form It is helpful to organize the facts of the problem in a table. Since the pipes will be open for the same amount of time as they fill the tank, enter x as the time worked for each pipe.

The first pipe can fill the tank in 7 days; its rate working alone is $\frac{1}{7}$ of the job per day. The second pipe can fill the tank in 9 days; its rate working alone is $\frac{1}{9}$ of the job per day. To determine the work completed by each pipe, multiply the rate by the time.

Success Tip

The tables and the formulas used to solve uniform motion problems and shared-work problems are similar:

Motion:	Work:
$d = rt$	$W = rt$

	Rate	· Time	= Work completed
1st pipe	$\frac{1}{7}$	x	$\frac{x}{7}$
2nd pipe	$\frac{1}{9}$	x	$\frac{x}{9}$

Think: $\frac{1}{7} \cdot \frac{x}{1} = \frac{x}{7}$.

Think: $\frac{1}{9} \cdot \frac{x}{1} = \frac{x}{9}$.

Enter this information first.

Multiply to get each of these entries: $W = rt$.

Teaching Tip: Students often have difficulty performing the simple multiplications such as $\frac{1}{7} \cdot x$ and $\frac{1}{9} \cdot x$ to fill in the last column of the table.

In shared-work problems, the number 1 represents one whole job completed. So we have

The part of job done by 1st pipe	plus	part of job done by 2nd pipe	equals	1 job completed.
$\frac{x}{7}$	$+$	$\frac{x}{9}$	$=$	1

Solve

$$\frac{x}{7} + \frac{x}{9} = 1$$ This is a rational equation.

$$63\left(\frac{x}{7} + \frac{x}{9}\right) = 63(1)$$ Clear the equation of fractions by multiplying both sides by the LCD, 63.

$$63\left(\frac{x}{7}\right) + 63\left(\frac{x}{9}\right) = 63$$ On the left side, distribute the multiplication by 63.

$$9x + 7x = 63$$ Simplify the left side: $\overset{1}{\cancel{7}} \cdot 9\left(\frac{x}{\underset{1}{\cancel{7}}}\right) = 9x$ and $7 \cdot \overset{1}{\cancel{9}}\left(\frac{x}{\underset{1}{\cancel{9}}}\right) = 7x$. The fractions have been cleared.

$$16x = 63$$ Combine like terms.

$$x = \frac{63}{16}$$ To undo the multiplication by 16 and isolate x, divide both sides by 16.

State If both pipes are used, it will take $\frac{63}{16}$ or $3\frac{15}{16}$ days to fill the tank.

Notation

For this application problem, the mixed-number form of the result, $3\frac{15}{16}$ days, is preferable to the improper fraction form, $\frac{63}{16}$ days.

Check To check, we use the work formula and multiply each rate by the time. In $\frac{63}{16}$ days, the first pipe fills $\frac{1}{7} \cdot \frac{63}{16} = \frac{9}{16}$ of the tank and the second pipe fills $\frac{1}{9} \cdot \frac{63}{16} = \frac{7}{16}$ of the tank. The sum of these efforts, $\frac{9}{16} + \frac{7}{16}$, is $\frac{16}{16}$ or 1 full tank. The result checks.

Teaching Example 4 Baking. A caterer can prepare cream puffs for an event in 4 hours. Her employee can prepare the same number of cream puffs in 7 hours. How long will it take them to make the dessert if they work together?
Answer: $2\frac{6}{11}$ hr

Self Check 4 **Mailing Flyers.** A school secretary can prepare a mass mailing of an informational flyer in 6 hours. A student worker would take 8 hours to prepare the mailing. How long will it take to prepare the mailing if they work together? $3\frac{3}{7}$ hr

Now Try Problem 37

Strategy for Solving Shared-Work Problems

Equations that model shared-work problems involving two people (or machines) have the form

$$\frac{x}{a} + \frac{x}{b} = 1$$

where x represents the time they work together on the job, and a and b represent the respective times each worker needs to complete the job alone.

Example 4 can be solved in a different way by considering the amount of work done by each pipe in 1 day. As before, if we let $x =$ the number of days it will take to fill the tank if both inlet pipes are used, then together, in 1 day, they will complete $\frac{1}{x}$ of the job. If we add what the first pipe can do in 1 day to what the second pipe can do in 1 day, the sum is what they can do together in 1 day.

What the first inlet pipe can do in 1 day	plus	what the second inlet pipe can do in 1 day	equals	what they can do together in 1 day.
$\frac{1}{7}$	$+$	$\frac{1}{9}$	$=$	$\frac{1}{x}$

To solve the equation, begin by clearing it of fractions.

$$\frac{1}{7} + \frac{1}{9} = \frac{1}{x}$$

$$63x\left(\frac{1}{7} + \frac{1}{9}\right) = 63x\left(\frac{1}{x}\right) \quad \text{Multiply both sides by the LCD, } 63x.$$

$$9x + 7x = 63 \quad \text{Distribute the multiplication by } 63x \text{ and simplify.}$$

$$16x = 63 \quad \text{Combine like terms.}$$

$$x = \frac{63}{16} \quad \text{To isolate } x\text{, divide both sides by 16.}$$

This is the same answer as the one obtained in Example 4.

4 Solve Investment Problems.

We have used the interest formula $I = Prt$ to solve investment problems. The relationships among interest, principal, rate, and time can be expressed in another way, by solving for P.

$$I = Prt \quad \text{Interest = principal · rate · time.}$$

$$\frac{I}{rt} = \frac{Prt}{rt} \quad \text{To undo the multiplication by } rt \text{ and isolate } P\text{, divide both sides by } rt.$$

$$\frac{I}{rt} = P \quad \text{Simplify the right side: } \frac{P\overset{1}{\cancel{r}}\overset{1}{\cancel{t}}}{\underset{1}{\cancel{r}}\underset{1}{\cancel{t}}} = P.$$

$$P = \frac{I}{rt} \quad \text{Reverse the sides of the equation so that } P \text{ is on the left.}$$

This alternate form of the interest formula, Principal $= \frac{\text{Interest}}{\text{rate} \cdot \text{time}}$, is used to solve the next example.

EXAMPLE 5 **Comparing Investments.** An amount of money invested for one year in bonds will earn \$120. At a bank, that same amount of money will only earn \$75 interest, because the interest rate paid by the bank is 3% less than that paid by the bonds. Find the rate of interest paid by each investment.

Analyze

- The investment in bonds earns \$120 in one year.
- The same amount of money, invested in a bank, earns \$75 in one year.
- The interest rate paid by the bank is 3% less than that paid by the bonds.
- Find the bond's rate of interest and the bank's rate of interest.

Caution

A common mistake is to forget to express 3% as the decimal 0.03 and incorrectly let $r - 3 =$ the bank's rate of interest.

Assign Since the interest rate paid by the bank is 3% less than that paid by the bonds, let $r =$ the bond's rate of interest, and $r - 0.03 =$ the bank's interest rate. (Recall that $3\% = 0.03$.)

Form If an investment earns \$120 interest in 1 year at some rate r, we can use $P = \frac{I}{rt}$ to find that the principal invested was $\frac{120}{r}$ dollars. Similarly, if another investment earns \$75 interest in 1 year at some rate $r - 0.03$, the principal invested was $\frac{75}{r - 0.03}$ dollars. We can organize the facts of the problem in a table.

	Principal ·	Rate ·	Time =	Interest
Bonds	$\frac{120}{r}$	r	1	120
Bank	$\frac{75}{r - 0.03}$	$r - 0.03$	1	75

Divide to get each of these entries: $P = \frac{I}{rt}$.

Enter this information first.

The amount invested in the bonds	equals	the amount invested in the bank.
$\frac{120}{r}$	$=$	$\frac{75}{r - 0.03}$

Solve

$$\frac{120}{r} = \frac{75}{r - 0.03}$$ This is a rational equation.

$$\overset{1}{\cancel{r}}(r - 0.03)\left(\frac{120}{\underset{1}{\cancel{r}}}\right) = r\overset{1}{\cancel{(r - 0.03)}}\left(\frac{75}{\underset{1}{\cancel{r - 0.03}}}\right)$$ Multiply both sides by the LCD, $r(r - 0.03)$. Then remove common factors of the numerator and denominator.

$$(r - 0.03)120 = 75r$$ Simplify. The fractions have been cleared.

$$120r - 3.6 = 75r$$ On the left side, distribute the multiplication by 120.

$$45r - 3.6 = 0$$ To isolate the variable term on the left side, subtract $75r$ from both sides.

$$45r = 3.6$$ To undo the subtraction of 3.6, add 3.6 to both sides.

$$r = 0.08$$ To undo the multiplication by 45 and isolate r, divide both sides by 45.

If $r = 0.08$, then the bank's interest rate is given by $r - 0.03 = 0.05$.

State The bonds pay 0.08, or 8%, interest. The bank's interest rate is 5%.

Check The amount invested at 8% that will earn \$120 interest in 1 year is $\frac{120}{(0.08)1} = \$1{,}500$. The amount invested at 5% that will earn \$75 interest in 1 year is $\frac{75}{(0.05)1} = \$1{,}500$. The amounts invested in the bonds and the bank are the same. The results check.

Teaching Example 5 **Banking.** At one bank, a sum of money invested for 1 year will earn \$24 interest. If invested in bonds, that money would earn \$72, because the interest rate paid by the bank is 4% less than that paid by the bonds. Find the rate of interest paid by each investment.
Answers: Bank: 2%; bonds: 6%

Self Check 5 **Comparing Investments.** An amount of money invested for one year in a certificate of deposit will earn \$210. The same amount of money in a savings account will earn \$70. If the certificate of deposit's interest rate is 2% more than the savings account's rate, find the interest rate of the savings account. 1%

Now Try Problem 41

SECTION 7.7 STUDY SET

VOCABULARY

Fill in the blanks.

1. In this section, problems that involve:
   - moving vehicles are called uniform _motion_ problems.
   - depositing money are called _investment_ problems.
   - people completing jobs are called shared-_work_ problems.
2. In the formula $W = rt$, the variable W stands for the _work_ completed, r is the _rate_, and t is the _time_.

CONCEPTS

3. Choose the equation that can be used to solve the following problem: *If the same number is added to the numerator and the denominator of the fraction $\frac{5}{8}$, the result is $\frac{2}{3}$. Find the number.* iii

   (i) $\frac{5}{8} + x = \frac{2}{3}$ (ii) $\frac{5+x}{8} = \frac{2}{3}$

   (iii) $\frac{5+x}{8+x} = \frac{2}{3}$ (iv) $\frac{5}{8} = \frac{2+x}{3+x}$

4. Fill in the blank: If a job can be completed in t hours, then the rate of work can be expressed as $\frac{1}{t}$ of the job is completed per hour.
5. **a.** It takes a night security officer 45 minutes to check each of the doors in an office building to make sure they are locked. What is the officer's rate of work? $\frac{1}{45}$ of the job per minute

   b. It takes an elementary school teacher 4 hours to make out the semester report cards. What part of the job does she complete in x hours? $\frac{x}{4}$
6. **Hospitals.** An experienced employee can sterilize an operating room in 3 hours. It takes a new employee 5 hours to sterilize the same room. Select the best estimate below of the time it will take them to sterilize the room if they work together. Less than 3 hours
   - Less than 3 hours
   - Between 3 and 5 hours
   - More than 5 hours
7. **a.** Solve $d = rt$ for t. $t = \frac{d}{r}$

   b. Solve $I = Prt$ for P. $P = \frac{I}{rt}$
8. Complete the table.

	r ·	t	$= d$
Snowmobile	r	$\frac{4}{r}$	4
4 × 4 truck	$r - 5$	$\frac{3}{r-5}$	3

9. Complete the table.

	Rate ·	Time =	Work completed
1st printer	$\frac{1}{15}$	x	$\frac{x}{15}$
2nd printer	$\frac{1}{8}$	x	$\frac{x}{8}$

10. Complete the table.

	P ·	r ·	t =	I
City savings bank	$\frac{50}{r}$	r	1	50
Credit union	$\frac{75}{r-0.02}$	$r - 0.02$	1	75

NOTATION

11. Write $\frac{55}{9}$ days using a mixed number. $6\frac{1}{9}$ days
12. **a.** Write 9% as a decimal. 0.09

    b. Write 0.035 as a percent. 3.5%

GUIDED PRACTICE

Solve each of these number problems. See Example 1.

13. If the same number is added to both the numerator and the denominator of $\frac{2}{5}$, the result is $\frac{2}{3}$. Find the number. 4
14. If the same number is subtracted from both the numerator and the denominator of $\frac{11}{13}$, the result is $\frac{3}{4}$. Find the number. 5
15. If the denominator of $\frac{3}{4}$ is increased by a number, and the numerator is doubled, the result is 1. Find the number. 2
16. If a number is added to the numerator of $\frac{7}{8}$, and the same number is subtracted from the denominator, the result is 2. Find the number. 3

17. If a number is added to the numerator of $\frac{3}{4}$, and twice as much is added to the denominator, the result is $\frac{4}{7}$. Find the number. 5

18. If a number is added to the numerator of $\frac{5}{7}$, and twice as much is subtracted from the denominator, the result is 8. Find the number. 3

19. The sum of a number and its reciprocal is $\frac{13}{6}$. Find the number. $\frac{2}{3}$ or $\frac{3}{2}$

20. The sum of the reciprocals of two consecutive even integers is $\frac{7}{24}$. Find each integer. 6 and 8

APPLICATIONS

21. Cooking. If the same number is added to both the numerator and the denominator of the amount of butter used in the following recipe for toffee, the result is the amount of brown sugar to be used. Find the number. 8

Toffee

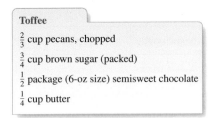

$\frac{2}{3}$ cup pecans, chopped

$\frac{3}{4}$ cup brown sugar (packed)

$\frac{1}{2}$ package (6-oz size) semisweet chocolate

$\frac{1}{4}$ cup butter

22. Tape Measures. If the same number is added to both the numerator and the denominator of the first measurement, the result is the second measurement. Find the number. 8

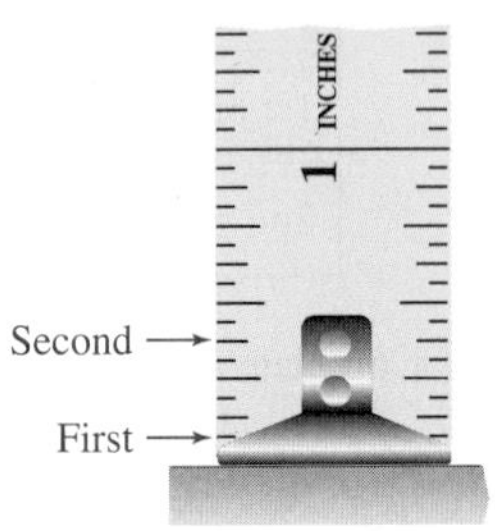

23. Tour De France. Maurice Garin of France won the first Tour de France bicycle road race in 1903. In 2005, American Lance Armstrong won his seventh consecutive Tour de France. Armstrong's average speed in 2005 was 10 mph faster than Garin's in 1903. In the time it took Garin to ride 80 miles, Armstrong could have ridden 130 miles. Find each cyclist's average speed. Garin: 16 mph; Armstrong: 26 mph

24. Physical Fitness. A woman can bicycle 28 miles in the same time as it takes her to walk 8 miles. She can ride 10 mph faster than she can walk. How fast can she walk? 4 mph

25. Packaging Fruit. The diagram below shows how apples are processed for market. Although the second conveyor belt is shorter, an apple spends the same amount of time on each belt because the second conveyor moves 1 foot per second slower than the first. Determine the speed of each conveyor belt. 1st: $1\frac{1}{2}$ ft per sec; 2nd: $\frac{1}{2}$ ft per sec

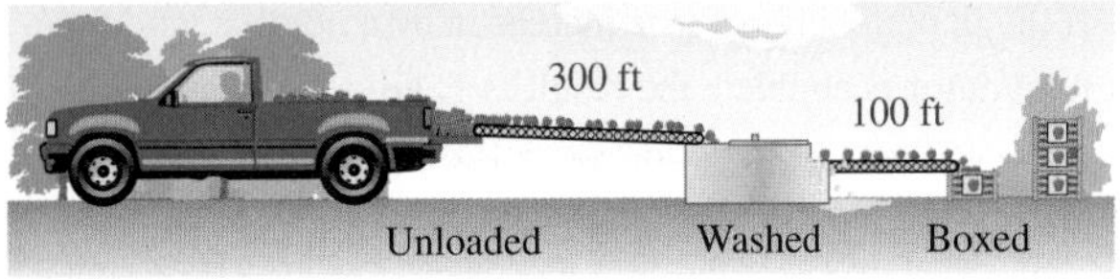

26. Comparing Travel. A plane can fly 300 miles in the same time as it takes a car to go 120 miles. If the car travels 90 mph slower than the plane, find the speed of the plane. 150 mph

27. Birds in Flight. Although flight speed is dependent upon the weather and the wind, in general, a Canada goose can fly about 10 mph faster than a great blue heron. In the same time that a Canada goose travels 120 miles, a great blue heron travels 80 miles. Find their flying speeds. Canada goose: 30 mph; great blue heron: 20 mph

28. Fast Cars. The top speed of a Dodge Charger SRT8 is 33 mph less than the top speed of a Chevrolet Corvette Z06. At their top speeds, a Corvette can travel 6 miles in the same time that a Charger can travel 5 miles. Find the top speed of each car. Charger: 165 mph; Corvette: 198 mph

29. Wind Speed. When a plane flies downwind, the wind pushes the plane so that its speed is the *sum* of the speed of the plane in still air and the speed of the wind. Traveling upwind, the wind pushes against the plane so that its speed is the *difference* of the speed of the plane in still air and the speed of the wind. Suppose a plane that travels 255 mph in still air can travel 300 miles downwind in the same time as it takes to travel 210 miles upwind. Complete the following table and find the speed of the wind, represented by x. 45 mph

	Rate ·	Time =	Distance
Downwind	$255 + x$	$\frac{300}{255 + x}$	300
Upwind	$255 - x$	$\frac{210}{255 - x}$	210

30. Boating. A boat that travels 18 mph in still water can travel 22 miles downstream in the same time as it takes to travel 14 miles upstream. Find the speed of the current in the river. (See problem 29.) 4 mph

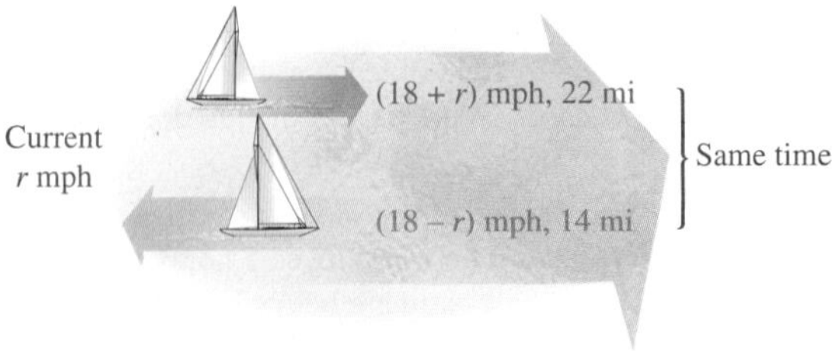

31. Roofing Houses. A homeowner estimates that it will take her 7 days to roof her house. A professional roofer estimates that he could roof the house in 4 days. How long will it take if the homeowner helps the roofer? $2\frac{6}{11}$ days

32. Holiday Decorating. One crew can put up holiday decorations in the mall in 8 hours. A second crew can put up the decorations in 10 hours. How long will it take if both crews work together to decorate the mall? $4\frac{4}{9}$ hr

33. **from Campus to Careers**

Recreation Director

©BananaStock/SuperStock

Suppose you are a recreation director at a summer camp. The water in the camp swimming pool was drained out for the winter and it is now time to refill the pool. One pipe can fill the empty pool in 12 hours and another can fill the empty pool in 18 hours. Suppose both pipes are opened at 8:00 A.M. and you have scheduled a swimming activity for 2:00 P.M. that day. Will the pool be filled by then? No; after the pipes are opened, the swimming is scheduled to take place in 6 hours. It takes 7.2 hr (7 hr 12 min) to fill the pool.

34. Groundskeeping. It takes a groundskeeper 45 minutes to prepare a softball field for a game. It takes his assistant 55 minutes to prepare the same field. How long will it take if they work together to prepare the field? 24.75 min $= 24\frac{3}{4}$ min

35. Filling a Pool. One inlet pipe can fill an empty pool in 4 hours, and a drain can empty the pool in 8 hours. How long will it take the pipe to fill the pool if the drain is left open? 8 hr

36. Sewage Treatment. A sludge pool is filled by two inlet pipes. One pipe can fill the pool in 15 days, and the other can fill it in 21 days. However, if no sewage is added, continuous waste removal will empty the pool in 36 days. How long will it take the two inlet pipes to fill an empty sludge pool? $11\frac{61}{109}$ days

37. Grading Papers. On average, it takes a teacher 30 minutes to grade a set of quizzes. It takes her teacher's aide twice as long to do the same grading. How long will it take if they work together to grade a set of quizzes? 20 min

38. Dog Kennels. It takes the owner/operator of a dog kennel 6 hours to clean all of the cages. It takes his assistant 2 hours more than that to clean the same cages. How long will it take if they work together? $3\frac{3}{7}$ hr

39. Printers. It takes a printer 6 hours to print the class schedules for all of the students enrolled in a community college. A faster printer can print the schedules in 4 hours. How long will it take the two printers working together to print $\frac{3}{4}$ of the class schedules? $1\frac{4}{5}$ hr $= 1.8$ hr

40. Office Work. In 5 hours, a secretary can address 100 envelopes. Another secretary can address 100 envelopes in 6 hours. How long would it take the secretaries, working together, to address 300 envelopes? (*Hint:* Think of addressing 300 envelopes as three 100-envelope jobs.) $8\frac{2}{11}$ hr

41. Comparing Investments. An amount of money invested for 1 year in tax-free bonds will earn \$300. In a certain credit union account, that same amount of money will only earn \$200 interest in a year, because the interest paid is 2% less than that paid by the bonds. Find the rate of interest paid by each investment. Credit union: 4%; bonds: 6%

42. Comparing Investments. An amount of money invested for 1 year in a savings account will earn \$1,500. That same amount of money invested in a mini-mall development will earn \$6,500 interest in a year, because the interest paid is 10% more than that paid by the savings account. Find the rate of interest paid by each investment. Sav. acct: 3%; mall: 13%

43. Comparing Investments. Two certificates of deposit (CDs) pay interest at rates that differ by 1%. Money invested for 1 year in the first CD earns \$175 interest. The same principal invested in the second CD earns \$200. Find the two rates of interest. 7% and 8%

44. Comparing Interest Rates. Two bond funds pay interest at rates that differ by 2%. Money invested for 1 year in the first fund earns \$315 interest. The same amount invested in the second fund earns \$385. Find the lower rate of interest. 9%

WRITING

45. In Example 4, one inlet pipe could fill an oil tank in 7 days, and another could fill the same tank in 9 days. We were asked to find how long it would take if both pipes were used. Explain why each of the following approaches is incorrect.

The time it would take to fill the tank

- is the *sum* of the lengths of time it takes each pipe to fill the tank: 7 days + 9 days = 16 days.
- is the *difference* in the lengths of time it takes each pipe to fill the tank: 9 days − 7 days = 2 days.
- is the *average* of the lengths of time it takes each pipe to fill the tank:

$$\frac{7 \text{ days} + 9 \text{ days}}{2} = \frac{16 \text{ days}}{2} = 8 \text{ days.}$$

46. Write a shared-work problem that can be modeled by the equation:

$$\frac{x}{3} + \frac{x}{4} = 1$$

REVIEW

47. Solve using substitution: $\begin{cases} x + y = 4 \\ y = 3x \end{cases}$ (1, 3)

48. Solve using elimination (addition): $\begin{cases} 5x - 4y = 19 \\ 3x + 2y = 7 \end{cases}$ (3, −1)

49. Use a check to determine whether $\frac{21}{5}$ is a solution of: $x + 20 = 4x - 1 + 2x$ Yes

50. Solve: $4x^2 + 8x = 0$ 0, −2

CHALLENGE PROBLEMS

51. River Tours. A river boat tour begins by going 60 miles upstream against a 5-mph current. There, the boat turns around and returns with the current. What still-water speed should the captain use to complete the tour in 5 hours? 25 mph

52. Travel Time. A company president flew 680 miles one way in the corporate jet, but returned in a smaller plane that could fly only half as fast. If the total travel time was 6 hours, find the speeds of the planes. 340 mph and 170 mph

53. Sales. A dealer bought some radios for a total of \$1,200. She gave away 6 radios as gifts, sold the rest for \$10 more than she paid for each radio, and broke even. How many radios did she buy? 30

54. Furnace Repairs. A repairman purchased several furnace-blower motors for a total cost of \$210. If his cost per motor had been \$5 less, he could have purchased one additional motor. How many motors did he buy at the regular rate? 6

SECTION 7.8 Proportions and Similar Triangles

OBJECTIVES

1. Write ratios and rates in simplest form.
2. Solve proportions.
3. Use proportions to solve problems.
4. Use proportions to solve problems involving similar triangles.

ARE YOU READY? *Are You Ready? exercises available online at www.webassign.net/brookscole*

The following problems review some basic skills that are needed when working with ratios and proportions.

1. Simplify: $\frac{42}{54}$ $\frac{7}{9}$

2. Multiply: **a.** $bd \cdot \frac{a}{d}$ ab **b.** $bd \cdot \frac{c}{d}$ bc

3. Solve: $16x = 136$ 8.5

4. Solve: $x^2 - 7x = 18$ $-2, 9$

In this section, we will discuss a problem-solving tool called a *proportion.* A proportion is a type of rational equation that involves two *ratios* or two *rates.*

1 Write Ratios and Rates in Simplest Form.

Ratios are used to compare two numbers or two quantities measured in the same units. Here are some examples.

- To prepare fuel for a lawnmower, gasoline is mixed with oil in a 50-to-1 ratio.
- In the stock market, winning stocks might outnumber losers by a ratio of 7 to 4.
- Gold is combined with other metals in the ratio of 14 to 10 to make 14-karat jewelry.

Ratios

A ratio is the quotient of two numbers or the quotient of two quantities that have the same units.

There are three ways to write a ratio: as a fraction, using the word *to,* or with a colon. For example, the comparison of the number of winning stocks to the number of losing stocks mentioned earlier can be written as

$\frac{7}{4}$, 7 to 4, or 7:4 Each of these forms can be read as "the ratio of 7 to 4."

EXAMPLE 1 Translate each phrase into a ratio written in fractional form: **a.** The ratio of 5 to 9 **b.** 12 ounces to 2 pounds

Strategy To translate, we need to identify the number (or quantity) before the word *to* and the number (or quantity) after it.

Why The number before the word *to* is the numerator of the ratio and the number after it is the denominator.

Solution

a. The ratio of 5 *to* 9 is written $\frac{5}{9}$. (5 → numerator; 9 → denominator)

b. To write a ratio of two quantities with the same units, we must express 2 pounds in terms of ounces. Since 1 pound = 16 ounces, 2 pounds = 32 ounces. The ratio of 12 ounces to 32 ounces can be simplified so that no units appear in the final form.

$$\frac{12 \text{ ounces}}{32 \text{ ounces}} = \frac{3 \cdot \overset{1}{\cancel{4}} \cancel{\text{ ounces}}}{\underset{1}{\cancel{4}} \cdot 8 \cancel{\text{ ounces}}} = \frac{3}{8}$$

Caution

A ratio that is the quotient of two quantities having the same units should be simplified so that no units appear in the final answer.

Teaching Example 1
Translate each phrase into a ratio written in fractional form:
a. The ratio of 11 to 27
b. 154 cm to 2 m

Answers: **a.** $\frac{11}{27}$ **b.** $\frac{77}{100}$

Self Check 1 Translate each phrase into a ratio written in fractional form: **a.** The ratio of 15 to 2 $\frac{15}{2}$ **b.** 12 hours to 2 days $\frac{1}{4}$

Now Try Problem 25

A quotient that compares quantities with different units is called a **rate.** For example, if the 495-mile drive from New Orleans to Dallas takes 9 hours, the average rate of speed is the quotient of the miles driven and the length of time the trip takes.

$$\text{Average rate of speed} = \frac{495 \text{ miles}}{9 \text{ hours}} = \frac{\overset{1}{\cancel{9}} \cdot 55 \text{ miles}}{\underset{1}{\cancel{9}} \cdot 1 \text{ hours}} = \frac{55 \text{ miles}}{1 \text{ hour}}$$

Rates A **rate** is a quotient of two quantities that have different units.

2 Solve Proportions.

If two ratios or two rates are equal, we say that they are *in proportion.*

Proportion A **proportion** is a mathematical statement that two ratios or two rates are equal.

The Language of Algebra

The word **proportion** implies a comparative relationship in size. For a picture to appear realistic, the artist must draw the shapes in the proper *proportion.* Remember the Y2K scare? The massive computer failures predicted by some experts were blown way out of *proportion.*

Some examples of proportions are:

$$\frac{1}{2} = \frac{3}{6} \qquad \frac{3 \text{ waiters}}{7 \text{ tables}} = \frac{9 \text{ waiters}}{21 \text{ tables}} \qquad \frac{a}{b} = \frac{c}{d}$$

- The proportion $\frac{1}{2} = \frac{3}{6}$ can be read as "1 is to 2 as 3 is to 6."
- The proportion $\frac{3 \text{ waiters}}{7 \text{ tables}} = \frac{9 \text{ waiters}}{21 \text{ tables}}$ can be read as "3 waiters is to 7 tables as 9 waiters is to 21 tables."
- The proportion $\frac{a}{b} = \frac{c}{d}$ can be read as "a is to b as c is to d."

Each of the four numbers in a proportion is called a **term.** The first and fourth terms are called the **extremes,** and the second and third terms are called the **means.**

First term → a, Second term → b, c ← Third term, d ← Fourth term

$$\frac{a}{b} = \frac{c}{d}$$

a and d are the extremes. b and c are the means.

For the proportion $\frac{a}{b} = \frac{c}{d}$, we can show that the product of the extremes, ad, is equal to the product of the means, bc, by multiplying both sides of the proportion by bd, and observing that $ad = bc$.

$$\frac{a}{b} = \frac{c}{d}$$

$$\overset{1}{\cancel{b}}d \cdot \frac{a}{\underset{1}{\cancel{b}}} = b\overset{1}{\cancel{d}} \cdot \frac{c}{\underset{1}{\cancel{d}}}$$ To clear the fractions, multiply both sides by the LCD, bd. Remove common factors of the numerator and denominator.

$$ad = bc$$ Simplify: $\frac{b}{b} = 1$ and $\frac{d}{d} = 1$.

Since $ad = bc$, the product of the extremes equals the product of the means.

The same products ad and bc can be found by multiplying diagonally in the proportion $\frac{a}{b} = \frac{c}{d}$. We call ad and bc **cross products.**

ad $\quad$ bc

$$\frac{a}{b} = \frac{c}{d}$$

The Fundamental Property of Proportions

In a proportion, the product of the extremes is equal to the product of the means.

If $\frac{a}{b} = \frac{c}{d}$, then $ad = bc$ and if $ad = bc$, then $\frac{a}{b} = \frac{c}{d}$.

EXAMPLE 2 Determine whether each equation is a proportion: **a.** $\frac{3}{7} = \frac{9}{21}$ **b.** $\frac{8}{3} = \frac{13}{5}$

Strategy We will check to see whether the product of the extremes is equal to the product of the means.

Why If the product of the extremes equals the product of the means, the equation is a proportion. If the cross products are not equal, the equation is not a proportion.

Solution

a. The product of the extremes is $3 \cdot 21 = 63$. The product of the means is $7 \cdot 9 = 63$. Since the cross products are equal, $\frac{3}{7} = \frac{9}{21}$ is a proportion.

$3 \cdot 21 = 63$ $\quad$ $7 \cdot 9 = 63$

$$\frac{3}{7} = \frac{9}{21}$$ Each cross product is 63.

b. The product of the extremes is $8 \cdot 5 = 40$. The product of the means is $3 \cdot 13 = 39$. Since the cross products are not equal, the equation is not a proportion: $\frac{8}{3} \neq \frac{13}{5}$.

$8 \cdot 5 = 40$ $\quad$ $3 \cdot 13 = 39$

$$\frac{8}{3} = \frac{13}{5}$$ One cross product is 40 and the other is 39.

Caution

We cannot remove common factors "across" an = symbol.

$$\frac{\overset{1}{\cancel{3}}}{7} = \frac{9}{\underset{7}{\cancel{21}}}$$

When this is done, the original proportion, $\frac{3}{7} = \frac{9}{21}$, which we found to be true, produces the false statement: $\frac{1}{7} = \frac{9}{7}$.

Teaching Example 2
Determine whether each equation is a proportion:
a. $\frac{7}{13} = \frac{5}{9}$ b. $\frac{5}{12} = \frac{20}{48}$
Answers: a. No b. Yes

Self Check 2 Determine whether the equation $\frac{6}{13} = \frac{24}{53}$ is a proportion. No

Now Try Problem 29

We have seen that a proportion contains four terms. If we know only three of the four terms of a proportion, we can use the fundamental property of proportions to find the value of the fourth term. This process is called **solving the proportion.**

EXAMPLE 3 Solve: $\frac{3}{2} = \frac{9}{x}$

Teaching Tip: Ask students how they would solve this equation using the method discussed in Section 7.6.

Strategy To solve for x, we will set the cross products equal.

Why This equation is a proportion, and in a proportion the product of the extremes equals the product of the means.

Solution If $x = 0$, the denominator of $\frac{9}{x}$ is 0 and the fraction would be undefined. Therefore, 0 cannot be a solution.

Caution

Remember that a cross product is the product of the means or extremes of a proportion. It would be incorrect to try to calculate cross products to solve $\frac{12}{18} = \frac{4}{x} + \frac{1}{2}$ because there is more than one term on the right side of the equation. It is *not* a proportion.

$\frac{3}{2} = \frac{9}{x}$ This is the given proportion. Since it is a type of rational equation, we can solve it by multiplying both sides by the LCD, 2x. However, it is often easier to solve a proportion using the cross products.

$3 \cdot x = 2 \cdot 9$ Find each cross product and set them equal.

$3x = 18$ Do the multiplication.

$\frac{3x}{3} = \frac{18}{3}$ To isolate x, divide both sides by 3.

$x = 6$ Do the division.

Check: To check the result, we substitute 6 for x in $\frac{3}{2} = \frac{9}{x}$ and find the cross products.

$3 \cdot 6 = 18$ $\quad$ $2 \cdot 9 = 18$

$\frac{3}{2} \stackrel{?}{=} \frac{9}{6}$ Each cross product is 18.

Since the cross products are equal, the solution of $\frac{3}{2} = \frac{9}{x}$ is 6. The solution set is $\{6\}$.

Teaching Example 3
Solve: $\frac{15}{6} = \frac{10}{x}$
Answer: 4

Self Check 3 Solve: $\frac{15}{x} = \frac{25}{40}$ 24

Now Try Problem 35

EXAMPLE 4 Solve: $\frac{a}{2} = \frac{4}{a - 2}$

Strategy To solve for a, we will set the cross products equal.

Why Since this equation is a proportion, the product of the means equals the product of the extremes.

Solution If $a = 2$, the denominator of $\frac{4}{a - 2}$ is 0 and the fraction would be undefined. Therefore, 2 cannot be a solution.

Success Tip

Since proportions are rational equations, they can also be solved by multiplying both sides by the LCD. Here, an alternate approach is to multiply both sides by $2(a - 2)$:

$\frac{a}{2} = \frac{4}{a - 2}$ This is the given proportion. Since it is a type of rational equation, we can solve it by multiplying both sides by the LCD, $2(a - 2)$, as shown in the margin. However, it is often easier to solve a proportion using the cross products.

$a(a - 2) = 2 \cdot 4$ Find each cross product and set them equal. Don't forget to write the parentheses.

$a^2 - 2a = 8$ On the left side, distribute the multiplication by a. This is a quadratic equation. On the right side, multiply.

$a^2 - 2a - 8 = 0$ To get 0 on the right side of the equation, subtract 8 from both sides.

$(a + 2)(a - 4) = 0$ Factor $a^2 - 2a - 8$.

$a + 2 = 0$ or $a - 4 = 0$ Set each factor equal to 0.

$a = -2$ | $a = 4$ Solve each equation.

The solutions are -2 and 4. Verify this using a check.

Teaching Example 4
Solve: $\frac{x}{5} = \frac{3}{x + 14}$
Answer: $-15, 1$

Self Check 4 Solve: $\frac{6}{c} = \frac{c - 1}{5}$ $-5, 6$

Now Try Problem 47

3 Use Proportions to Solve Problems.

We can use proportions to solve many problems. If we are given a ratio (or rate) comparing two quantities, the words of the problem can be translated into a proportion, and we can solve it to find the unknown.

EXAMPLE 5

iStockPhoto.com/Graeme Gilmour

Grocery Shopping. If 6 apples cost \$1.38, how much will 16 apples cost?

Analyze We know the cost of 6 apples; we are to find the cost of 16 apples.

Assign Let $c =$ the cost of 16 apples.

Form If we compare the number of apples to their cost, the two ratios must be equal.

*6 apples is **to** \$1.38 as 16 apples is **to** \$c.*

$$\begin{array}{r} \text{Number of apples} \rightarrow \\ \text{Cost} \rightarrow \end{array} \frac{6}{1.38} = \frac{16}{c} \begin{array}{l} \leftarrow \text{Number of apples} \\ \leftarrow \text{Cost} \end{array}$$

The units can be written outside the proportion.

Solve We drop the units, find each cross product, set them equal, and then solve the resulting equation for c.

$$6 \cdot c = 1.38(16)$$ In a proportion, the product of the extremes equals the product of the means.

$$6c = 22.08$$ Multiply: $1.38(16) = 22.08$.

$$\frac{6c}{6} = \frac{22.08}{6}$$ To undo the multiplication by 6 and isolate c, divide both sides by 6.

$$c = 3.68$$ Recall that c represents the cost of 16 apples.

State Sixteen apples will cost \$3.68.

Check We can use estimation to check the result. 16 apples are about 3 times as many as 6 apples, which cost \$1.38. If we multiply \$1.38 by 3, we get an estimate of the cost of 16 apples: $\$1.38 \cdot 3 = \$4.14$. The result, \$3.68, seems reasonable.

Self Check 5 **Concert Tickets.** If 9 tickets to a concert cost \$112.50, how much will 15 tickets cost? \$187.50

Now Try Problem 73

The Language of Algebra

Remember that the word **to** separates the numerator and denominator of a ratio. If the units are written outside the ratio, we can write 6 apples is *to* \$1.38 in fraction form as:

$$\frac{6}{1.38}$$

Teaching Tip: Stress that the units in the numerators of the two ratios must be the same and the units in the denominators of the two ratios must be the same.

Teaching Example 5 **Musicals.** If 6 tickets to a Broadway show cost \$345, how much do 11 tickets cost?
Answer: \$632.50

Be careful when solving problems using proportions. We must make sure that the units of both numerators are the same and the units of both denominators are the same. In Example 5, it would be incorrect to write

$$\begin{array}{r} \text{Cost of 6 apples} \rightarrow \\ \text{6 apples} \rightarrow \end{array} \frac{1.38}{6} = \frac{16}{c} \begin{array}{l} \leftarrow \text{16 apples} \\ \leftarrow \text{Cost of 16 apples} \end{array}$$

EXAMPLE 6

Carousel ratio
1 inch:160 inches

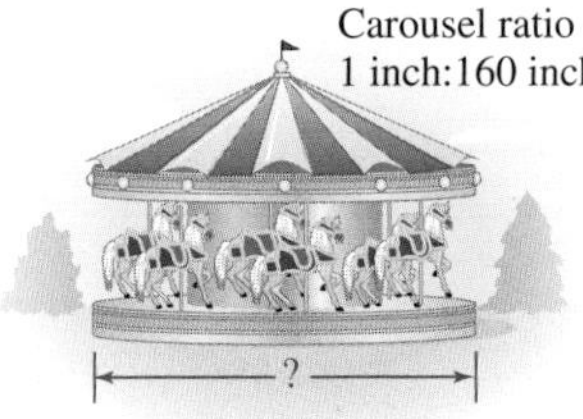

Miniatures. A **scale** is a ratio (or rate) that compares the size of a model, drawing, or map with the size of an actual object. The scale indicates that 1 inch on the model carousel is equivalent to 160 inches on the actual carousel. How wide should the model be if the actual carousel is 35 feet wide?

Analyze We are asked to determine the width of the miniature carousel if a ratio of 1 inch to 160 inches is used. We would like the width of the model to be given in inches, not feet, so we will express the 35-foot width of the actual carousel as $35 \cdot 12 = 420$ inches.

Assign Let $w =$ the width of the model.

The Language of Algebra

Architects, interior decorators, landscapers, and automotive engineers are a few of the professionals who construct **scale** drawings or *scale* models of the projects they are designing.

Form The ratios of the dimensions of the model to the corresponding dimensions of the actual carousel are equal.

*1 inch is **to** 160 inches as w inches is **to** 420 inches.*

$$\underset{\text{Actual size} \rightarrow}{\overset{\text{Model} \rightarrow}{}} \frac{1}{160} = \frac{w}{420} \underset{\leftarrow \text{Actual size}}{\overset{\leftarrow \text{Model}}{}}$$

Solve We drop the units, find each cross product, set them equal, and then solve the resulting equation for w.

$420 = 160w$ In a proportion, the product of the extremes is equal to the product of the means.

$\frac{420}{\mathbf{160}} = \frac{160w}{\mathbf{160}}$ To undo the multiplication by 160 and isolate w, divide both sides by 160.

$2.625 = w$ Recall that w represents the width of the model.

State The width of the miniature carousel should be 2.625 in., or $2\frac{5}{8}$ in.

Check A width of $2\frac{5}{8}$ in. is approximately 3 in. When we write the ratio of the model's approximate width to the width of the actual carousel, we get $\frac{3}{420} = \frac{1}{140}$, which is about $\frac{1}{160}$. The answer seems reasonable.

Teaching Example 6 **Model Railroads.** An N scale model railroad car is 4.75 inches long. If the N scale is 1 to 160, how long is the real railroad car, in inches and in feet?

Answer: 760 in., $63\frac{1}{3}$ ft

Self Check 6 **Blueprints.** The scale for a blueprint indicates that $\frac{1}{4}$ inch on the print is equivalent to 1 foot for the actual building. If the width of the building on the print is 7.5 inches, what is the width of the actual building? 30 ft

Now Try Problem 89

When shopping, *unit prices* can be used to compare costs of different sizes of the same brand to determine the best buy. The **unit price** gives the cost per unit, such as cost per ounce, cost per pound, or cost per sheet. We can find the unit price of an item using a proportion.

EXAMPLE 7 **Comparison Shopping.** Which size of toothpaste is the better buy?

$2.19

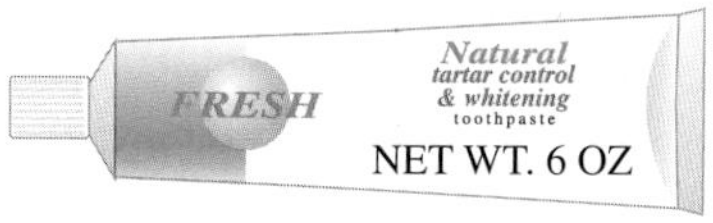

$2.79

Strategy We will find the unit price for each tube of toothpaste. Then we will identify which tube has the lower unit price.

Why The better buy is the tube that has the lower unit price.

Solution

The Language of Algebra

A **unit price** indicates the cost of 1 unit of an item, such as 1 ounce of bottled water or 1 pound of hamburger. In advanced mathematics, we study unit circles—circles that have a radius of 1 unit.

To find the unit price for each tube, we let $x =$ the price of 1 ounce of toothpaste. Then we set up and solve the following proportions.

For the 4-ounce tube:

$$\underset{\text{Ounces} \rightarrow}{\overset{\text{Price} \rightarrow}{}} \frac{2.19}{4} = \frac{x}{1} \underset{\leftarrow \text{Ounce}}{\overset{\leftarrow \text{Price}}{}}$$

$2.19 = 4x$

$\frac{2.19}{4} = x$

$0.55 \approx x$ The unit price is approximately $0.55.

For the 6-ounce tube:

$$\underset{\text{Ounces} \rightarrow}{\overset{\text{Price} \rightarrow}{}} \frac{2.79}{6} = \frac{x}{1} \underset{\leftarrow \text{Ounce}}{\overset{\leftarrow \text{Price}}{}}$$

$2.79 = 6x$

$\frac{2.79}{6} = x$

$0.47 \approx x$ The unit price is approximately $0.47.

Teaching Example 7 **Guitar Lessons.** A music store charges \$27 for a 30-minute guitar lesson and \$36 for a 45-minute lesson. Which is the better buy?
Answer: The 45-min lesson

The price of 1 ounce of toothpaste from the 4-ounce tube is about 55¢. The price for 1 ounce of toothpaste from the 6-ounce tube is about 47¢. Since the 6-ounce tube has the lower unit price, it is the better buy.

Self Check 7 **Comparison Shopping.** Which is the better buy: 3 pounds of hamburger for \$6.89 or 5 pounds for \$12.49? 3 lb for \$6.89

Now Try Problem 93

4 Use Proportions to Solve Problems Involving Similar Triangles.

If two angles of one triangle have the same measures as two angles of a second triangle, the triangles have the same shape. Triangles with the same shape, but not necessarily the same size, are called **similar triangles.** In the following figure, $\Delta ABC \sim \Delta DEF$. (Read the symbol $\sim$ as "is similar to.")

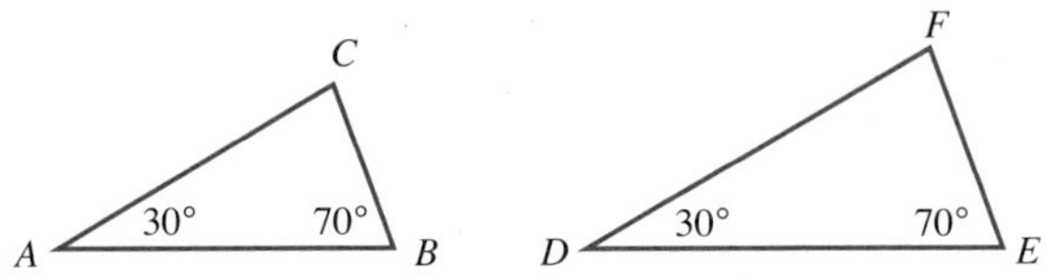

Property of Similar Triangles

If two triangles are **similar,** all pairs of corresponding sides are in proportion.

For the similar triangles previously shown, the following proportions are true.

$$\frac{AB}{DE} = \frac{BC}{EF}, \quad \frac{BC}{EF} = \frac{CA}{FD}, \quad \text{and} \quad \frac{CA}{FD} = \frac{AB}{DE}$$

Read AB as "the length of segment AB."

EXAMPLE 8 **Finding the Height of a Tree.** A tree casts a shadow 18 feet long at the same time as a woman 5 feet tall casts a shadow 1.5 feet long. Find the height of the tree.

Analyze The figure shows the similar triangles determined by the tree and its shadow and the woman and her shadow. Since the triangles are similar, the lengths of their corresponding sides are in proportion. We can use this fact to find the height of the tree.

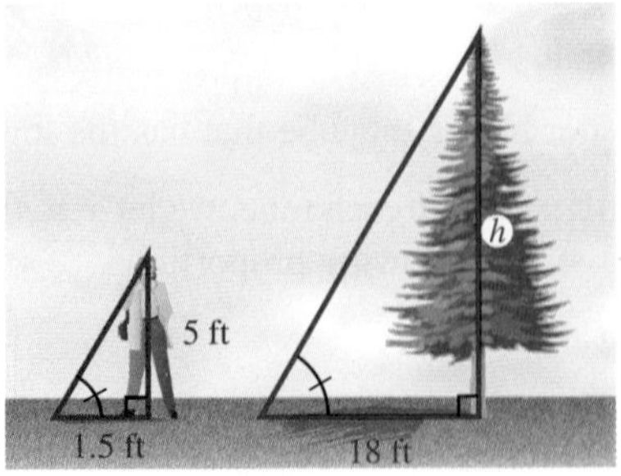

Each triangle has a right angle. Since the sun's rays strike the ground at the same angle, the angles highlighted with a tick mark have the same measure. Therefore, two angles of the smaller triangle have the same measures as two angles of the larger triangle; the triangles are similar.

Assign Let h = the height of the tree.

Form We can find h by solving the following proportion.

$$\frac{h}{5} = \frac{18}{1.5} \qquad \frac{\text{Height of the tree}}{\text{Height of the woman}} = \frac{\text{Length of shadow of the tree}}{\text{Length of shadow of the woman}}$$

Success Tip

Similar triangles do not have to be positioned the same. When they are placed differently, be careful to match their corresponding letters correctly. Here, $\Delta RST \sim \Delta MNO$.

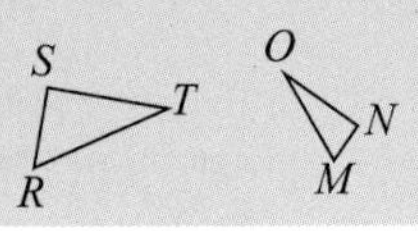

Success Tip

The quickest way to determine if two triangles are similar is to see if the measures of two angles of one triangle have the same measures as two angles of the other triangle.

Teaching Tip: Show how the sides of the similar triangles are compared:

$$\frac{\text{Right side large } \Delta}{\text{Right side small } \Delta} = \frac{\text{Base large } \Delta}{\text{Base small } \Delta}$$

Teaching Example 8 Finding Height. A light pole casts a shadow of 15 feet at the same time a man 6 feet tall casts a shadow of 2.5 feet. Find the height of the light pole.
Answer: 36 ft

Solve

$1.5h = 5(18)$ In a proportion, the product of the extremes equals the product of the means.

$1.5h = 90$ Multiply: $5(18) = 90$.

$\frac{1.5h}{1.5} = \frac{90}{1.5}$ To undo the multiplication by 1.5 and isolate h, divide both sides by 1.5.

$h = 60$ Do the decimal division, $1.5\overline{)90}$, to get 60.

State The tree is 60 feet tall.

Check $\frac{18}{1.5} = 12$ and $\frac{60}{5} = 12$. Since the ratios are the same, the result checks.

Self Check 8 **Shadows.** Find the height of the tree in Example 8 if the woman is 5 feet 6 inches tall and her shadow is 1.5 feet long. 66 ft

Now Try Problems 55 and 101

SECTION 7.8 STUDY SET

VOCABULARY

Fill in the blanks.

1. A ratio is the quotient of two numbers or the quotient of two quantities with the same units. A rate is a quotient of two quantities that have different units.
2. A proportion is a mathematical statement that two ratios or two rates are equal.
3. In $\frac{50}{3} = \frac{x}{9}$, the terms 50 and 9 are called the extremes and the terms 3 and x are called the means of the proportion.
4. The cross products for the proportion $\frac{5}{2} = \frac{6}{x}$ are $5x$ and 12.
5. Examples of unit prices are \$1.65 per gallon, 17¢ per day, and \$50 per foot.
6. Two triangles with the same shape, but not necessarily the same size, are called similar triangles.

CONCEPTS

7. Fill in the blanks: In a proportion, the product of the extremes is equal to the product of the means. In symbols,

$$\text{If } \frac{a}{b} = \frac{c}{d}, \text{ then } ad = bc.$$

8. Is 45 a solution of $\frac{5}{3} = \frac{75}{x}$? Yes

9. **Snacks.** In a sample of 25 bags of potato chips, 2 were found to be underweight. Complete the following proportion that could be used to find the number of underweight bags that would be expected in a shipment of 1,000 bags of potato chips.

$$\begin{matrix}\text{Number of bags} \rightarrow \\ \text{Number underweight} \rightarrow\end{matrix} \frac{25}{2} = \frac{1{,}000}{x} \begin{matrix}\leftarrow \text{Number of bags} \\ \leftarrow \text{Number underweight}\end{matrix}$$

10. **Miniatures.** A model of the Seattle Space Needle is to be made using a scale of 2 inches to 35 feet. Complete the following proportion to determine the height h of the model.

$$\frac{2}{35} = \frac{h}{605}$$

11. **Kleenex.** Complete the following proportion that can be used to find the unit price of facial tissue if a box of 85 tissues sells for \$2.19.

$$\begin{matrix}\text{Price} \rightarrow \\ \text{Number of sheets} \rightarrow\end{matrix} \frac{2.19}{85} = \frac{x}{1} \begin{matrix}\leftarrow \text{Price} \\ \leftarrow \text{Number of sheets}\end{matrix}$$

12. The two triangles shown in the following illustration are similar. Complete the proportion.

$$\frac{x}{3} = \frac{25}{10}$$

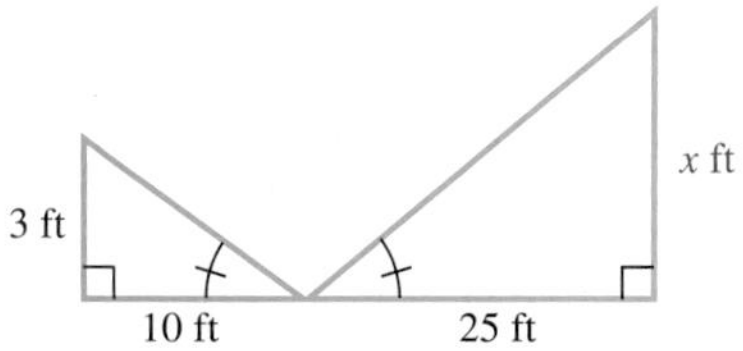

NOTATION

Complete the solution.

13. Solve for x: $\frac{12}{18} = \frac{x}{24}$

$$12 \cdot 24 = 18 \cdot x$$
$$288 = 18x$$
$$\frac{288}{18} = \frac{18x}{18}$$
$$16 = x$$

14. Write the ratio of 25 to 4 in two other forms. $\frac{25}{4}$, 25:4

▶ **15.** Fill in the blanks: The proportion $\frac{20}{1.6} = \frac{100}{8}$ can be read: 20 is to 1.6 as 100 is to 8.

16. Fill in the blank: We read $\Delta XYZ \sim \Delta MNO$ as: triangle XYZ is similar to triangle MNO.

GUIDED PRACTICE

Translate each ratio into a fraction in simplest form. **See Example 1.**

17. 4 boxes to 15 boxes $\frac{4}{15}$

▶ **18.** 2 miles to 9 miles $\frac{2}{9}$

19. 18 watts to 24 watts $\frac{3}{4}$

▶ **20.** 11 cans to 121 cans $\frac{1}{11}$

21. 30 days to 24 days $\frac{5}{4}$

22. 45 people to 30 people $\frac{3}{2}$

23. 90 minutes to 3 hours $\frac{1}{2}$

24. 20 inches to 2 feet $\frac{5}{6}$

▶ **25.** 8 quarts to 4 gallons $\frac{1}{2}$

26. 6 feet to 12 yards $\frac{1}{6}$

27. 6,000 feet to 1 mile (*Hint:* 1 mi = 5,280 ft) $\frac{25}{22}$

▶ **28.** 5 tons to 4,000 pounds (*Hint:* 1 ton = 2,000 lb) $\frac{5}{2}$

Determine whether each equation is a true proportion. **See Example 2.**

▶ **29.** $\frac{7}{3} = \frac{14}{6}$ Yes

▶ **30.** $\frac{7}{16} = \frac{3}{7}$ No

31. $\frac{5}{8} = \frac{12}{19.4}$ No

▶ **32.** $\frac{9}{32} = \frac{4.5}{16}$ Yes

Solve each proportion. **See Example 3.**

▶ **33.** $\frac{2}{3} = \frac{x}{6}$ 4

▶ **34.** $\frac{3}{6} = \frac{x}{8}$ 4

▶ **35.** $\frac{63}{g} = \frac{9}{2}$ 14

36. $\frac{27}{x} = \frac{9}{4}$ 12

37. $\frac{x+1}{5} = \frac{3}{15}$ 0

▶ **38.** $\frac{x-1}{7} = \frac{2}{21}$ $\frac{5}{3}$

39. $\frac{5-x}{17} = \frac{13}{34}$ $-\frac{3}{2}$

40. $\frac{4-x}{13} = \frac{11}{26}$ $-\frac{3}{2}$

41. $\frac{15}{7b+5} = \frac{5}{2b+1}$ -2

▶ **42.** $\frac{8}{3n+6} = \frac{16}{3n-3}$ -5

43. $\frac{8x}{3} = \frac{11x+9}{4}$ -27

44. $\frac{3x}{16} = \frac{x+2}{5}$ -32

Solve each proportion. **See Example 4.**

45. $\frac{2}{3x} = \frac{x}{6}$ 2, -2

▶ **46.** $\frac{y}{4} = \frac{4}{y}$ 4, -4

▶ **47.** $\frac{b-5}{3} = \frac{2}{b}$ 6, -1

48. $\frac{2}{q} = \frac{q-3}{2}$ 4, -1

49. $\frac{a-4}{a} = \frac{15}{a+4}$ -1, 16

▶ **50.** $\frac{s}{s-5} = \frac{s+5}{24}$ -1, 25

51. $\frac{t+3}{t+5} = \frac{-1}{2t}$ $-\frac{5}{2}$, -1

▶ **52.** $\frac{5h}{14h+3} = \frac{1}{h}$ $-\frac{1}{5}$, 3

Each pair of triangles is similar. Find the missing side length. **See Example 8.**

53. 15

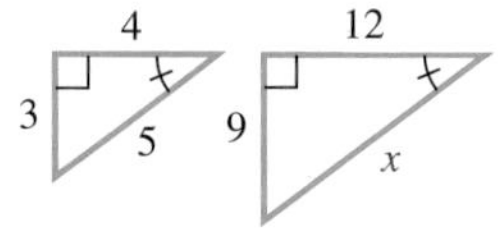

▶ **54.** 25

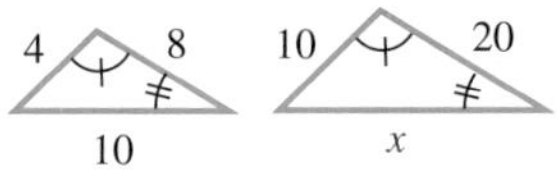

55. 8

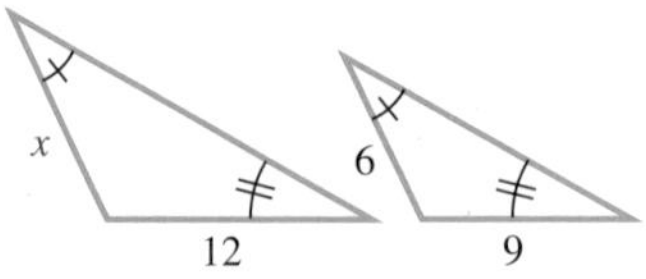

▶ **56.** 20

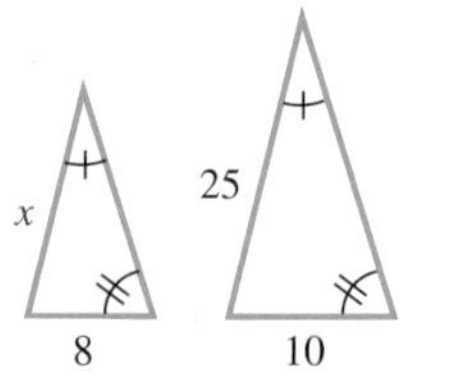

TRY IT YOURSELF

Solve each proportion.

▶ **57.** $\frac{x-1}{x+1} = \frac{2}{3x}$ $-\frac{1}{3}$, 2

▶ **58.** $\frac{2}{x+6} = \frac{-2x}{5}$ -5, -1

▶ **59.** $\frac{x+1}{4} = \frac{3x}{8}$ 2

60. $\frac{x-1}{9} = \frac{2x}{3}$ $-\frac{1}{5}$

61. $\frac{y-4}{y+1} = \frac{y+3}{y+6}$ $-\frac{27}{2}$

62. $\frac{r-6}{r-8} = \frac{r+1}{r-4}$ $\frac{32}{3}$

63. $\frac{c}{10} = \frac{10}{c}$ -10, 10

64. $\frac{-6}{r} = \frac{r}{-6}$ -6, 6

65. $\frac{m}{3} = \frac{4}{m+1}$ -4, 3

66. $\frac{n}{2} = \frac{5}{n+3}$ -5, 2

67. $\frac{3}{3b+4} = \frac{2}{5b-6}$ $\frac{26}{9}$

68. $\frac{2}{4d-1} = \frac{3}{2d+1}$ $\frac{5}{8}$

Look Alikes . . .

Solve each equation.

69. a. $-\frac{2}{5} = \frac{3}{4x}$ $-\frac{15}{8}$ b. $\frac{4}{x} - \frac{2}{5} = \frac{3}{4x}$ $\frac{65}{8}$

▶ **70.** a. $\frac{1}{4} = \frac{2}{3a}$ $\frac{8}{3}$ b. $\frac{5}{6a} + \frac{1}{4} = \frac{2}{3a}$ $-\frac{2}{3}$

71. a. $\frac{3}{a-1} = \frac{8}{a}$ $\frac{8}{5}$ b. $\frac{3}{a-1} + \frac{8}{a} = 3$ $\frac{2}{3}$, 4

72. a. $\frac{4}{3x} = \frac{1}{3}$ 4 b. $\frac{4}{3x} - \frac{1}{3} = x$ $-\frac{4}{3}$, 1

APPLICATIONS

▶ **73. Shopping for Clothes.** If shirts are on sale at two for \$25, how much do five shirts cost? \$62.50

▶ **74. Mixing Perfume.** A perfume is to be mixed in the ratio of 3 drops of pure essence to 7 drops of alcohol. How many drops of pure essence should be mixed with 56 drops of alcohol? 24

75. CPR. A first aid handbook states that when performing cardiopulmonary resuscitation on an adult, the ratio of chest compressions to breaths should be 30:2. If 210 compressions were administered to an adult patient, how many breaths should have been given? 14 breaths

76. Cooking. A recipe for wild rice soup follows. Find the amounts of chicken broth, rice, and flour needed to make 15 servings. $7\frac{1}{2}$ c, $1\frac{2}{3}$ c, 5 tbsp

Wild Rice Soup

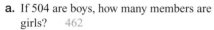

A sumptuous side dish with a nutty flavor

3 cups chicken broth	1 cup light cream
$\frac{2}{3}$ cup uncooked rice	2 tablespoons flour
$\frac{1}{4}$ cup sliced onions	$\frac{1}{8}$ teaspoon pepper
$\frac{1}{2}$ cup shredded carrots	Serves: 6

77. from **Campus to Careers**

Recreation Director

©BananaStock/SuperStock

A total of 966 boys and girls are members of a community recreation center.

a. If 504 are boys, how many members are girls? 462

b. Find the ratio of girls to boys who are members of the recreation center. $\frac{11}{12}$; 11:12

78. Gear Ratios. Write each ratio in two ways: as a fraction in simplest form and using a colon.

a. The number of teeth of the larger gear to the number of teeth of the smaller gear $\frac{3}{2}$, 3:2

b. The number of teeth of the smaller gear to the number of teeth of the larger gear $\frac{2}{3}$, 2:3

79. Computing a Paycheck. Billie earns \$412 for a 40-hour week. If she missed 10 hours of work last week, how much did she get paid? \$309

80. Waves. If the peak height to wavelength ratio is greater than 1:7, a wave becomes unstable and it breaks forward. (See the figure below.) What is the maximum height a wave with wavelength 637 feet can have before it breaks forward? 91 ft

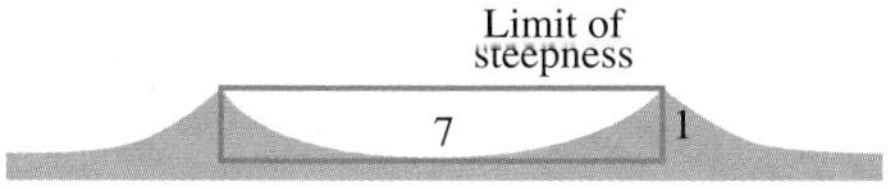

81. Twitter. According to a June, 2010 article in the Blog Herald, approximately 7,500 tweets are sent every 10 seconds. At this rate, about how many tweets are sent in one minute? 45,000 tweets

82. Engineering. A portion of a bridge is shown. Use the fact that $\frac{AB}{BC}$ is in proportion to $\frac{FE}{ED}$ to find FE. 6.4 ft

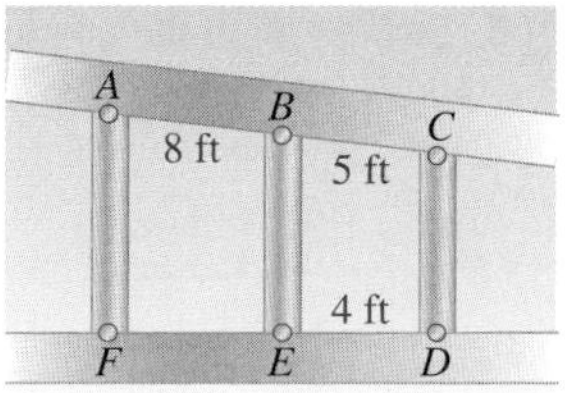

83. Nutrition. The table shows the nutritional facts about a 10-oz chocolate milkshake sold by a fast-food restaurant. Use the information to complete the table for the 16-oz shake. Round to the nearest unit when an answer is not exact.

	Calories	Fat (gm)	Protein (gm)
10-oz chocolate milkshake	355	8	9
16-oz chocolate milkshake	568	13	14

84. Photo Enlargements. The 3-by-5 photo is to be blown up to the larger size. Find x. $3\frac{3}{4}$ in.

85. Mixing Fuel. The instructions on a can of oil intended to be added to lawnmower gasoline are shown below. Are these instructions correct? (*Hint:* There are 128 ounces in 1 gallon.) Not exactly, but close

Recommended	Gasoline	Oil
50 to 1	6 gal	16 oz

86. Driver's Licenses. Of the 50 states, Alabama has one of the highest ratios of licensed drivers to residents. If the ratio is 399:500 and the population of Alabama is about 4,500,000, how many residents of that state have a driver's license? About 3,591,000

87. Capture–Release Method. To estimate the ground squirrel population on his acreage, a farmer trapped, tagged, and then released a dozen squirrels. Two weeks later, the farmer trapped 35 squirrels and noted that 3 were tagged. Use this information to estimate the number of ground squirrels on his acreage. 140

88. Concrete. A 2:3 concrete mix means that for every two parts of sand, three parts of gravel are used. How much sand should be used in a mix composed of 25 cubic feet of gravel? $16\frac{2}{3}$ ft^3

89. Model Railroads. An HO scale model railroad engine is 6 inches long. If the HO scale is 1 to 87, how long is a real engine, in inches? In feet? 522 in.; 43.5 ft

90. Model Railroads. An N scale model railroad caboose is 4.5 inches long. If the N scale is 1 to 160, how long is a real caboose, in inches? In feet? 720 in.; 60 ft

91. Blueprints. The scale for the drawing shown means that a $\frac{1}{4}$-inch length $\left(\frac{1}{4}''\right)$ on the drawing corresponds to an actual size of 1 foot (1′-0″). Suppose the length of the kitchen is $2\frac{1}{2}$ inches on the drawing. How long is the actual kitchen? 10 ft

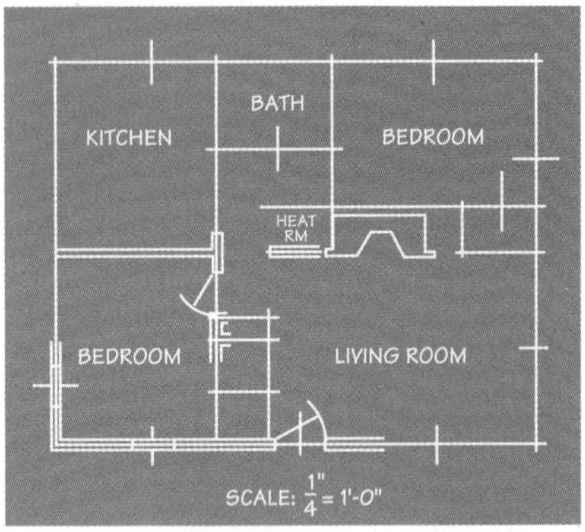

92. The Titanic. A 1:144 scale model of the *Titanic* is to be built. If the ship was 882 feet long, find the length of the model. 6.125 ft = $6\frac{1}{8}$ ft

***For each of the following purchases, determine the better buy.* See Example 7.**

93. Trumpet lessons: 45 minutes for \$25 or 60 minutes for \$35 45 min for \$25

94. Memory for a computer: 128 megabytes for \$26 or 512 megabytes for \$110 128 megabytes for \$26

95. Business cards: 100 for \$9.99 or 150 for \$12.99 150 for \$12.99

96. Dog food: 20 pounds for \$7.49 or 44 pounds for \$14.99 44 lb for \$14.99

97. Soft drinks: 6-pack for \$1.50 or a case (24 cans) for \$6.25 6-pack for \$1.50

98. Donuts: A dozen for \$6.24 or a baker's dozen (13) for \$6.65 A baker's dozen for \$6.65

99.

FAT-FREE PEACH YOGURT
4.79
Six 4-OZ CARTONS
FAT-FREE PEACH YOGURT
2.99
Four 4-OZ CARTONS

Four 4-oz cartons

100.

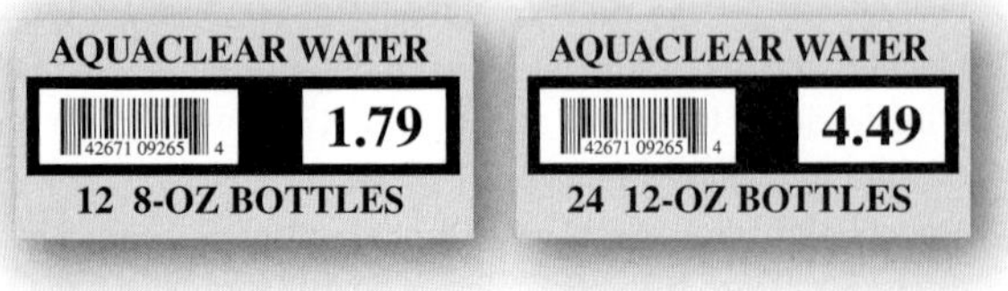

AQUACLEAR WATER
1.79
12 8-OZ BOTTLES
AQUACLEAR WATER
4.49
24 12-OZ BOTTLES

24 12-oz bottles

101. Height of a Tree. A tree casts a shadow of 26 feet at the same time as a 6-foot man casts a shadow of 4 feet. Find the height of the tree. 39 ft

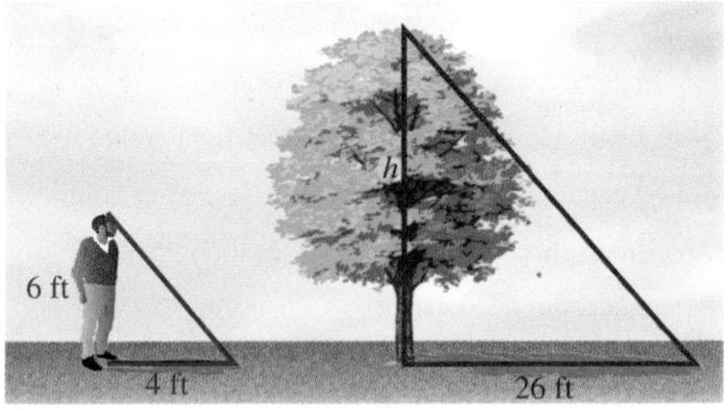

102. Height of a Building. A man places a mirror on the ground and sees the reflection of the top of a building, as shown. The two triangles in the illustration are similar. Find the height, h, of the building. 25 ft

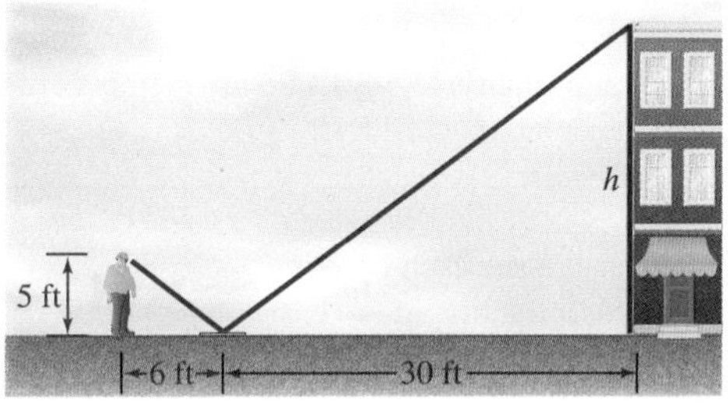

103. Surveying. To find the width of a river, a surveyor laid out the following similar triangles. Find w. $46\frac{7}{8}$ ft

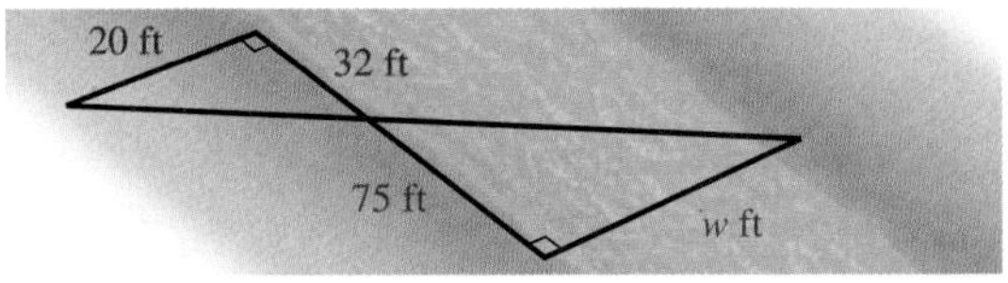

104. Flight Paths. An airplane ascends 100 feet as it flies a horizontal distance of 1,000 feet. How much altitude will it gain as it flies a horizontal distance of 1 mile? (*Hint:* 5,280 feet = 1 mile.) 528 ft

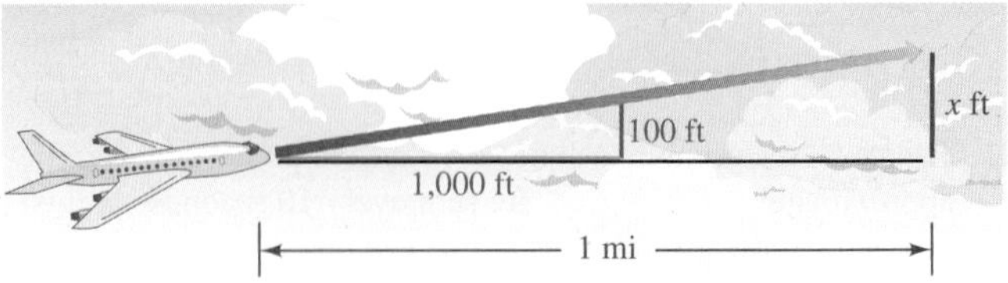

105. Slope. Find the unknown rise of the smaller slope triangle in the figure below. 8

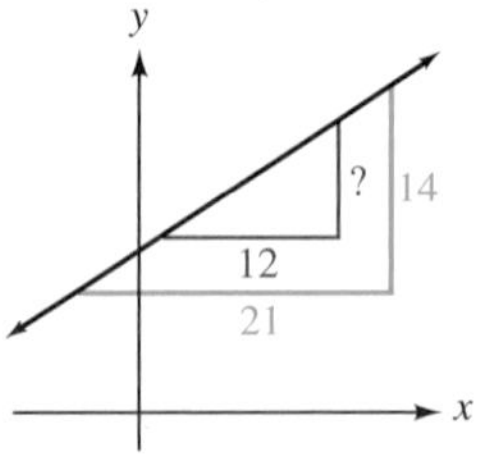

106. Washington, D.C. The Washington Monument casts a shadow of $166\frac{1}{2}$ feet at the same time as a 5-foot-tall tourist casts a shadow of $1\frac{1}{2}$ feet. Find the height of the monument. 555 ft

WRITING

107. Explain how to solve the equation $\frac{7}{6} = \frac{2}{x}$ and how to simplify the expression $\frac{7}{6} \cdot \frac{2}{x}$.

108. Explain why the concept of cross products cannot immediately be used to solve the equation:

$$\frac{x}{3} - \frac{3x}{4} = \frac{1}{12}$$

109. What are similar triangles?

▶ **110.** What is a unit price? Give an example.

REVIEW

111. Change $\frac{9}{10}$ to a percent. 90%

▶ **112.** Change $33\frac{1}{3}\%$ to a fraction. $\frac{1}{3}$

113. Find 30% of 1,600. 480

▶ **114. Shopping.** Maria bought a dress for 25% off the original price of \$98. How much did the dress cost? \$73.50

CHALLENGE PROBLEMS

▶ **115.** Suppose $\frac{a}{b} = \frac{c}{d}$. Write three other proportions using a, b, c, and d. $\frac{a}{c} = \frac{b}{d}$, $\frac{b}{a} = \frac{d}{c}$, $\frac{c}{a} = \frac{d}{b}$

116. a. Verify that $\frac{3}{5} = \frac{12}{20} = \frac{3 + 12}{5 + 20}$.

b. Is the following rule always true? Explain. Yes

$$\frac{a}{b} = \frac{c}{d} = \frac{a + c}{b + d}$$

7 Summary & Review

SECTION 7.1 Simplifying Rational Expressions

DEFINITIONS AND CONCEPTS	EXAMPLES
A **rational expression** is an expression of the form $\frac{A}{B}$, where A and B are polynomials and B does not equal 0.	Rational expressions: $\frac{8}{7t}$, $\frac{a}{a-3}$, and $\frac{4x^2 - 16x}{x^2 - 6x + 8}$
To **evaluate a rational expression,** we substitute the values of its variables and simplify.	Evaluate $\frac{3x+1}{x-2}$ for $x = 5$. $\frac{3x+1}{x-2} = \frac{3(5)+1}{5-2} = \frac{16}{3}$ Substitute 5 for x.
To find the real numbers for which a **rational expression is undefined,** find the values of the variable that make the denominator 0.	For which real numbers is $\frac{11}{2x-3}$ undefined? $2x - 3 = 0$ Set the denominator equal to 0 and solve for x. $2x = 3$ Add 3 to both sides. $x = \frac{3}{2}$ The expression is undefined for $x = \frac{3}{2}$.
To **simplify a rational expression:** 1. Factor the numerator and the denominator completely. 2. Remove factors equal to 1. 3. Multiply the remaining factors in the numerator and denominator.	Simplify: $\frac{x^2 - 4}{x^2 - 7x + 10} = \frac{(x+2)\overset{1}{\cancel{(x-2)}}}{(x-5)\underset{1}{\cancel{(x-2)}}}$ Factor and simplify. $= \frac{x+2}{x-5}$ Multiply the remaining factors in the numerator and the denominator.
The quotient of any nonzero expression and its **opposite** is -1.	$\frac{2t-3}{3-2t} = -1$ Because $2t - 3$ and $3 - 2t$ are opposites.

REVIEW EXERCISES

1. Find the values of x for which the rational expression $\frac{x-1}{x^2-16}$ is undefined. 4, −4

2. Evaluate $\frac{x^2-1}{x-5}$ for $x = -2$. $-\frac{3}{7}$

Simplify each rational expression, if possible. Assume that no denominators are zero.

3. $\frac{3x^2}{6x^3}$ $\frac{1}{2x}$

4. $\frac{5xy^2}{2x^2y^2}$ $\frac{5}{2x}$

5. $\frac{x^2}{x^2+x}$ $\frac{x}{x+1}$

6. $\frac{a^2-4}{a+2}$ $a-2$

7. $\frac{3p-2}{2-3p}$ -1

8. $\frac{8-x}{x^2-5x-24}$ $-\frac{1}{x+3}$

9. $\frac{2x^2-16x}{2x^2-18x+16}$ $\frac{x}{x-1}$

10. $\frac{x^2+x-2}{x^2-x-2}$ Does not simplify

11. $\frac{x^2-2xy+y^2}{(x-y)^3}$ $\frac{1}{x-y}$

12. $\frac{4(t+3)+8}{3(t+3)+6}$ $\frac{4}{3}$

13. Explain the error in the following work: $\frac{x+1}{x} = \frac{\overset{1}{\cancel{x}}+1}{\underset{1}{\cancel{x}}} = \frac{2}{1} = 2$. x is not a common factor of the numerator and the denominator; x is a term of the numerator.

14. **Dosages.** Cowling's rule is a formula that can be used to determine the dosage of a prescription medication for children. If C is the proper child's dosage, D is an adult dosage, and A is the child's age in years, then $C = \frac{D(A+1)}{24}$. Find the daily dosage of an antibiotic for an 11-year-old child if the adult daily dosage is 300 milligrams. 150 mg

SECTION 7.2 Multiplying and Dividing Rational Expressions

DEFINITIONS AND CONCEPTS	EXAMPLES
To **multiply rational expressions,** multiply their numerators and multiply their denominators. $\frac{A}{B} \cdot \frac{C}{D} = \frac{AC}{BD}$ Then simplify, if possible.	Multiply: $\frac{4b}{b+2} \cdot \frac{7}{b} = \frac{4b \cdot 7}{(b+2)b}$ Multiply the numerators. Multiply the denominators. $= \frac{4\overset{1}{\cancel{b}} \cdot 7}{(b+2)\underset{1}{\cancel{b}}}$ Simplify. $= \frac{28}{b+2}$ Multiply the remaining factors in the numerator and the denominator.
To find the **reciprocal** of a rational expression, invert its numerator and denominator.	The reciprocal of $\frac{c}{c+7}$ is $\frac{c+7}{c}$.
To **divide rational expressions,** multiply the first expression by the reciprocal of the second. $\frac{A}{B} \div \frac{C}{D} = \frac{A}{B} \cdot \frac{D}{C} = \frac{AD}{BC}$ Then simplify, if possible.	Divide: $\frac{t}{t+1} \div \frac{8}{t^2+t} = \frac{t}{t+1} \cdot \frac{t^2+t}{8}$ Multiply by the reciprocal. $= \frac{t \cdot t\overset{1}{\cancel{(t+1)}}}{\underset{1}{\cancel{(t+1)}}8}$ Factor and simplify. $= \frac{t^2}{8}$ Multiply the remaining factors in the numerator and the denominator.
A **unit conversion factor** is a fraction that has a value of 1.	$\frac{1\text{ yd}^2}{9\text{ ft}^2} = 1$ and $\frac{1\text{ mi}}{5{,}280\text{ ft}} = 1$

REVIEW EXERCISES

Multiply and simplify, if possible.

15. $\frac{3xy}{2x} \cdot \frac{4x}{2y^2}$ $\frac{3x}{y}$

16. $56x\left(\frac{12}{7x}\right)$ 96

17. $\frac{x^2 - 1}{x^2 + 2x} \cdot \frac{x}{x + 1}$ $\frac{x - 1}{x + 2}$

18. $\frac{x^2 + x}{3x - 15} \cdot \frac{6x - 30}{x^2 + 2x + 1}$ $\frac{2x}{x + 1}$

Divide and simplify, if possible.

19. $\frac{3x^2}{5x^2y} \div \frac{6x}{15xy^2}$ $\frac{3y}{2}$

20. $\frac{x^2 - x - 6}{1 - 2x} \div \frac{x^2 - 2x - 3}{2x^2 + x - 1}$ $-x - 2$

21. Determine whether the given fraction is a unit conversion factor.

a. $\frac{1 \text{ ft}}{12 \text{ in.}}$ Yes

b. $\frac{60 \text{ min}}{1 \text{ day}}$ No

c. $\frac{2{,}000 \text{ lb}}{1 \text{ ton}}$ Yes

d. $\frac{1 \text{ gal}}{4 \text{ qt}}$ Yes

22. **Traffic Signs.** Convert the speed limit on the sign from miles per hour to miles per minute. $\frac{1}{3}$ mile per minute

SECTION 7.3 Adding and Subtracting with Like Denominators; Least Common Denominators

DEFINITIONS AND CONCEPTS	EXAMPLES
To **add (or subtract) rational expressions** that have the same denominator, add (or subtract) their numerators and write the sum (or difference) over their common denominator. $\frac{A}{D} + \frac{B}{D} = \frac{A + B}{D}$ $\frac{A}{D} - \frac{B}{D} = \frac{A - B}{D}$ Then simplify, if possible.	Add: $\frac{2b}{3b - 9} + \frac{b}{3b - 9} = \frac{2b + b}{3b - 9}$ *Add the numerators and write the sum over the LCD, 3b − 9.* $= \frac{\overset{1}{\cancel{3}}b}{\underset{1}{\cancel{3}}(b - 3)}$ *Factor and simplify.* $= \frac{b}{b - 3}$ Subtract: $\frac{x + 1}{x} - \frac{x - 1}{x} = \frac{x + 1 - (x - 1)}{x}$ *Don't forget the parentheses.* $= \frac{x + 1 - x + 1}{x}$ $= \frac{2}{x}$ *Combine like terms.*
To find the **LCD** of several rational expressions, factor each denominator completely. Form a product using each different factor the greatest number of times it appears in any one factorization.	Find the LCD of $\frac{3}{x^3 - x^2}$ and $\frac{x}{x^2 - 1}$. $x^3 - x^2 = x \cdot x \cdot (x - 1)$ $x^2 - 1 = (x + 1)(x - 1)$ LCD $= x \cdot x \cdot (x - 1)(x + 1)$
To **build an equivalent rational expression,** multiply the given expression by 1 in the form of $\frac{c}{c}$ where $c \neq 0$.	$\frac{7}{4t} = \frac{7}{4t} \cdot \frac{3t}{3t}$ $= \frac{21t}{12t^2}$ and $\frac{x + 1}{x - 7} = \frac{x + 1}{x - 7} \cdot \frac{x - 1}{x - 1}$ $= \frac{(x + 1)(x - 1)}{(x - 7)(x - 1)}$ $= \frac{x^2 - 1}{x^2 - 8x + 7}$

REVIEW EXERCISES

Add or subtract and simplify, if possible.

23. $\dfrac{13}{15d} - \dfrac{8}{15d}$ $\quad \dfrac{1}{3d}$

24. $\dfrac{x}{x+y} + \dfrac{y}{x+y}$ $\quad 1$

25. $\dfrac{3x}{x-7} - \dfrac{x-2}{x-7}$ $\quad \dfrac{2x+2}{x-7}$

26. $\dfrac{a}{a^2-2a-8} + \dfrac{2}{a^2-2a-8}$ $\quad \dfrac{1}{a-4}$

Find the LCD of each pair of rational expressions.

27. $\dfrac{12}{x}, \dfrac{1}{9}$ $\quad 9x$

28. $\dfrac{1}{2x^3}, \dfrac{5}{8x}$ $\quad 8x^3$

29. $\dfrac{7}{m}, \dfrac{m+2}{m-8}$ $\quad m(m-8)$

30. $\dfrac{x}{5x+1}, \dfrac{5x}{5x-1}$ $\quad (5x+1)(5x-1)$

31. $\dfrac{6-a}{a^2-25}, \dfrac{a^2}{a-5}$ $\quad (a+5)(a-5)$

32. $\dfrac{4t+25}{t^2+10t+25}, \dfrac{t^2-7}{2t^2+17t+35}$ $\quad (2t+7)(t+5)^2$

Build each rational expression into an equivalent fraction having the denominator shown in red.

33. $\dfrac{9}{a}, \mathbf{7a}$ $\quad \dfrac{63}{7a}$

34. $\dfrac{2y+1}{x-9}, \mathbf{x(x-9)}$ $\quad \dfrac{2xy+x}{x(x-9)}$

35. $\dfrac{b+7}{3b-15}, \mathbf{6(b-5)}$ $\quad \dfrac{2b+14}{6(b-5)}$

36. $\dfrac{9r}{r^2+6r+5}, \mathbf{(r+1)(r-4)(r+5)}$ $\quad \dfrac{9r^2-36r}{(r+1)(r-4)(r+5)}$

SECTION 7.4 Adding and Subtracting with Unlike Denominators

DEFINITIONS AND CONCEPTS	EXAMPLES
To **add (or subtract) rational expressions** with unlike denominators: 1. Find the LCD. 2. Write each rational expression as an equivalent expression whose denominator is the LCD. 3. Add (or subtract) the numerators and write the sum (or difference) over the LCD. 4. Simplify the resulting rational expression if possible.	Add: $\dfrac{4x}{x} + \dfrac{2}{x-1} = \dfrac{4x}{x} \cdot \dfrac{x-1}{x-1} + \dfrac{2}{x-1} \cdot \dfrac{x}{x}$ The LCD is $x(x-1)$. Multiply → Don't multiply → $= \dfrac{4x(x-1)}{x(x-1)} + \dfrac{2x}{x(x-1)}$ Build so that each expression has a denominator of $x(x-1)$. $= \dfrac{4x^2-4x+2x}{x(x-1)}$ Distribute the multiplication by $4x$. $= \dfrac{4x^2-2x}{x(x-1)}$ Combine like terms. $= \dfrac{\overset{1}{\cancel{2x}}(2x-1)}{\underset{1}{\cancel{x}}(x-1)}$ Factor and simplify. $= \dfrac{2(2x-1)}{x-1}$
When a polynomial is multiplied by -1, the result is its opposite. This fact is used when adding or subtracting rational expressions whose **denominators are opposites.**	Add: $\dfrac{c}{c-4} + \dfrac{1}{4-c} = \dfrac{c}{c-4} + \dfrac{1}{4-c} \cdot \dfrac{-1}{-1}$ $= \dfrac{c}{c-4} + \dfrac{-1}{c-4}$ $-1(4-c) = c-4$ $= \dfrac{c-1}{c-4}$ Add the numerators. Write the sum over the LCD, $c-4$.

REVIEW EXERCISES

Add or subtract and simplify, if possible.

37. $\dfrac{1}{7} - \dfrac{1}{a}$ $\quad \dfrac{a-7}{7a}$

38. $\dfrac{x}{x-1} + \dfrac{1}{x}$ $\quad \dfrac{x^2+x-1}{x(x-1)}$

39. $\dfrac{2t+2}{t^2+2t+1} - \dfrac{1}{t+1}$ $\quad \dfrac{1}{t+1}$

40. $\dfrac{x+2}{2x} - \dfrac{2-x}{x^2}$ $\quad \dfrac{x^2+4x-4}{2x^2}$

41. $\dfrac{6}{b-1} - \dfrac{b}{1-b}$ $\quad \dfrac{b+6}{b-1}$

42. $\dfrac{8}{c} + 6$ $\quad \dfrac{6c+8}{c}$

43. $\dfrac{n+7}{n+3} - \dfrac{n-3}{n+7}$ $\quad \dfrac{14n+58}{(n+3)(n+7)}$

44. $\dfrac{4}{t+2} - \dfrac{7}{(t+2)^2}$ $\quad \dfrac{4t+1}{(t+2)^2}$

45. $\dfrac{6}{a^2-9} - \dfrac{5}{a^2-a-6}$ $\quad \dfrac{1}{(a+3)(a+2)}$

46. $\dfrac{2}{3y-6} + \dfrac{3}{4y+8}$ $\quad \dfrac{17y-2}{12(y-2)(y+2)}$

47. Working on a homework assignment, a student added two rational expressions and obtained $\dfrac{-5n^3-7}{3n(n+6)}$. The answer given in the back of the book was $-\dfrac{5n^3+7}{3n(n+6)}$. Are the answers equivalent? Yes

48. Digital Video Cameras. Find the perimeter and the area of the LED screen of the camera.
$\dfrac{14x+28}{(x+6)(x-1)}$ units, $\dfrac{12}{(x+6)(x-1)}$ square units

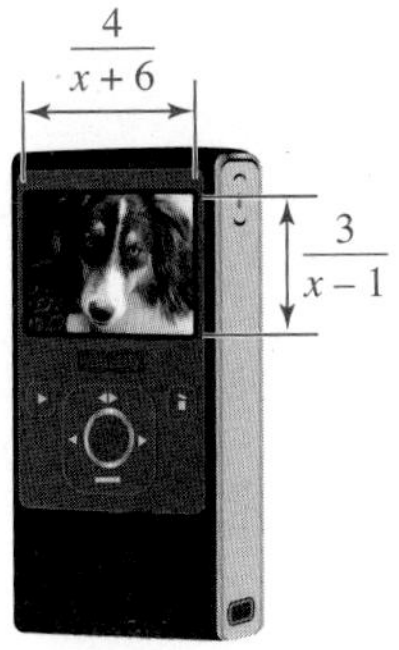

SECTION 7.5 Simplifying Complex Fractions

DEFINITIONS AND CONCEPTS	EXAMPLES
Complex fractions contain fractions in their numerators and/or their denominators.	Complex fractions: $\dfrac{\frac{2}{t}}{\frac{5}{4t}}$ and $\dfrac{\frac{3}{m}+\frac{m}{4}}{\frac{m}{2}}$
To **simplify a complex fraction:** **Method 1** Write the numerator and the denominator as single rational expressions and perform the indicated division.	Simplify: $\dfrac{\frac{3}{m}+\frac{m}{2}}{\frac{m}{4}} = \dfrac{\frac{3}{m}\cdot\frac{2}{2}+\frac{m}{2}\cdot\frac{m}{m}}{\frac{m}{4}}$ In the numerator, build to have an LCD of $2m$. $= \dfrac{\frac{6}{2m}+\frac{m^2}{2m}}{\frac{m}{4}}$ In the numerator, multiply the fractions. $= \dfrac{\frac{6+m^2}{2m}}{\frac{m}{4}}$ Add the fractions in the numerator. $= \dfrac{6+m^2}{2m} \div \dfrac{m}{4}$ The main fraction bar indicates division. $= \dfrac{(6+m^2)\cdot \overset{1}{\cancel{2}}\cdot 2}{\underset{1}{\cancel{2}}m\cdot m}$ Multiply by the reciprocal of $\frac{m}{4}$. Factor 4 and simplify. $= \dfrac{12+2m^2}{m^2}$ Distribute the multiplication by 2 in the numerator.

Method 2

Determine the LCD of all the rational expressions in the complex fraction and multiply the complex fraction by 1, written in the form $\frac{\text{LCD}}{\text{LCD}}$.

Simplify: $\dfrac{\frac{3}{m}+\frac{m}{2}}{\frac{m}{4}} = \dfrac{\frac{3}{m}+\frac{m}{2}}{\frac{m}{4}} \cdot \dfrac{4m}{4m}$ The LCD for all the rational expressions is $4m$.

$= \dfrac{\frac{3}{m}\cdot 4m + \frac{m}{2}\cdot 4m}{\frac{m}{4}\cdot 4m}$ In the numerator, distribute the multiplication by $4m$.

$= \dfrac{12+2m^2}{m^2}$ Perform each multiplication by $4m$.

REVIEW EXERCISES

Simplify each complex fraction.

49. $\dfrac{\frac{n^4}{30}}{\frac{7n}{15}}$ $\frac{n^3}{14}$

50. $\dfrac{\frac{r^2-81}{18s^2}}{\frac{4r-36}{9s}}$ $\frac{r+9}{8s}$

51. $\dfrac{\frac{1}{y}+1}{\frac{1}{y}-1}$ $\frac{1+y}{1-y}$

52. $\dfrac{\frac{7}{a^2}}{\frac{1}{a}+\frac{10}{3}}$ $\frac{21}{3a+10a^2}$

53. $\dfrac{\frac{2}{x-1}+\frac{x-1}{x+1}}{\frac{1}{x^2-1}}$ x^2+3

54. $\dfrac{\frac{1}{x^2y}-\frac{5}{xy}}{\frac{3}{xy}-\frac{7}{xy^2}}$ $\frac{y-5xy}{3xy-7x}$

SECTION 7.6 Solving Rational Equations

DEFINITIONS AND CONCEPTS	EXAMPLES

To **solve a rational equation** we use the multiplication property of equality to clear the equation of fractions. Use these steps:

1. Determine which numbers cannot be solutions.
2. Multiply both sides of the equation by the LCD of the rational expressions contained in the equation.
3. Solve the resulting equation.
4. Check all possible solutions in the *original* equation. A possible solution that does not satisfy the original equation is called an **extraneous solution.**

Solve: $\dfrac{y}{y-2} - 1 = \dfrac{1}{y}$ Since no denominators can be 0, $y \neq 2$ and $y \neq 0$.

$y(y-2)\left(\dfrac{y}{y-2} - 1\right) = y(y-2)\left(\dfrac{1}{y}\right)$ The LCD is $y(y-2)$.

$y(y-2)\left(\dfrac{y}{y-2}\right) - y(y-2)1 = y(y-2)\left(\dfrac{1}{y}\right)$ Simplify.

$y \cdot y - y(y-2) = (y-2)\cdot 1$

$y^2 - y^2 + 2y = y - 2$ Multiply.

$2y = y - 2$ Combine like terms.

$y = -2$ Solve for y.

REVIEW EXERCISES

Solve each equation and check the result. If an equation has no solution, so indicate.

55. $\dfrac{3}{x} = \dfrac{2}{x-1}$ 3

56. $\dfrac{a}{a-5} = 3 + \dfrac{5}{a-5}$ No solution; 5 is extraneous

57. $\dfrac{2}{3t} + \dfrac{1}{t} = \dfrac{5}{9}$ 3

58. $a = \dfrac{3a-50}{4a-24} - \dfrac{3}{4}$ 2, 4

59. $\dfrac{4}{x+2} - \dfrac{3}{x+3} = \dfrac{6}{x^2+5x+6}$ 0

60. $\dfrac{3}{x+1} - \dfrac{x-2}{2} = \dfrac{x-2}{x+1}$ −4, 3

61. **Engineering.** The efficiency E of a Carnot engine is given by the following formula. Solve it for T_1.

$$E = 1 - \frac{T_2}{T_1}$$ $T_1 = \frac{T_2}{1-E}$

62. Solve for y: $\dfrac{1}{x} = \dfrac{1}{y} + \dfrac{1}{z}$ $y = \frac{xz}{z-x}$

SECTION 7.7 Problem Solving Using Rational Equations

DEFINITIONS AND CONCEPTS	EXAMPLES
To solve application problems, follow these steps: **1.** Analyze the problem. **2.** Assign a variable. **3.** Form an equation. **4.** Solve the equation. **5.** State the conclusion. **6.** Check the result.	**Washing Cars.** Working alone, Carlos can wash the family SUV in 30 minutes. Victor, his brother, can wash the same SUV in 20 minutes working alone. How long will it take them if they wash the SUV together?

Rate of Work: If a job can be completed in t units of time, the rate of work can be expressed as $\frac{1}{t}$ of the job is completed per unit of time.

Shared-work problems:

Work completed = rate of work · time worked

Washing Cars. Working alone, Carlos can wash the family SUV in 30 minutes. Victor, his brother, can wash the same SUV in 20 minutes working alone. How long will it take them if they wash the SUV together?

Analyze It takes Carlos 30 minutes and it takes Victor 20 minutes. How long will it take working together?

Assign Let x = the number of minutes it will take Carlos and Victor, working together, to wash the SUV.

Form Enter the data in a table.

	Rate	· Time	= Work Completed
Carlos	$\frac{1}{30}$	x	$\frac{x}{30}$
Victor	$\frac{1}{20}$	x	$\frac{x}{20}$

The part of the job done by Carlos plus the part of the job done by Victor equals 1 job completed.

$$\frac{x}{30} + \frac{x}{20} = 1$$

Solve

$$60\left(\frac{x}{30} + \frac{x}{20}\right) = 60(1) \quad \text{Multiply both sides by the LCD, 60.}$$

$$60\left(\frac{x}{30}\right) + 60\left(\frac{x}{20}\right) = 60(1) \quad \text{On the left side, distribute the multiplication by 60.}$$

$$2x + 3x = 60 \quad \text{Perform each multiplication by 60.}$$

$$5x = 60 \quad \text{Combine like terms.}$$

$$x = \frac{60}{5} \quad \text{Divide both sides by 5.}$$

$$x = 12$$

State Working together, it will take Carlos and Victor 12 minutes to wash the family SUV.

Check In 12 minutes, Carlos will do $\frac{12}{30} = \frac{24}{60}$ of the job and Victor will do $\frac{12}{20} = \frac{36}{60}$ of the job. Together they will do $\frac{24}{60} + \frac{36}{60} = \frac{60}{60}$ or 1 whole job. The result checks.

DEFINITIONS AND CONCEPTS	EXAMPLES
Uniform motion problems: Time $= \dfrac{\text{distance}}{\text{rate}}$	See Example 2 in Section 7.7.
Investment problems: Principal $= \dfrac{\text{Interest}}{\text{rate} \cdot \text{time}}$	See Example 5 in Section 7.7.

REVIEW EXERCISES

63. Number Problems. If a number is subtracted from the denominator of $\frac{4}{5}$ and twice as much is added to the numerator, the result is 5. Find the number. 3

64. Exercise. A woman can bicycle 30 miles in the same time that it takes her to jog 10 miles. If she can ride 10 mph faster than she can jog, how fast can she jog? 5 mph

65. House Cleaning. A maid can clean a house in 4 hours. What is her rate of work? $\frac{1}{4}$ of the job per hr

66. House Painting. If a homeowner can paint a house in 14 days and a professional painter can paint it in 10 days, how long will it take if they work together? $5\frac{5}{6}$ days

67. Investments. In 1 year, a student earned \$100 interest on money she deposited at a savings and loan. She later learned that the money would have earned \$120 if she had deposited it at a credit union, because the credit union paid 1% more interest at the time. Find the rate she received from the savings and loan. 5%

68. Wind Speed. A plane flies 400 miles downwind in the same amount of time as it takes to travel 320 miles upwind. If the plane can fly at 360 mph in still air, find the velocity of the wind. 40 mph

SECTION 7.8 Proportions and Similar Triangles

DEFINITIONS AND CONCEPTS	EXAMPLES
A **ratio** is the quotient of two numbers or two quantities with the same units.	Ratios: $\frac{2}{3}$, $\frac{1}{50}$, and 2:3
A **rate** is the quotient of two quantities with different units.	Rates: $\frac{4\text{ oz}}{6\text{ lb}}$, $\frac{525\text{ mi}}{15\text{ hr}}$, and $\frac{\$1.95}{2\text{ lb}}$
A **proportion** is a statement that two ratios or two rates are equal. In the proportion $\frac{a}{b} = \frac{c}{d}$, a and d are the **extremes** and b and c are the **means.**	A proportion: $\frac{4}{9} = \frac{28}{63}$ Extremes: 4 and 63 Means: 9 and 28
In any proportion, the product of the extremes is equal to the product of the means. (The **cross products** are equal.)	A proportion: $\frac{4}{9} = \frac{28}{63}$ Cross product: $4 \cdot 63 = 252$ Cross product: $9 \cdot 28 = 252$
To **solve a proportion,** set the product of the extremes equal to the product of the means and solve the resulting equation.	Solve the proportion: $\frac{3}{2} = \frac{x}{10}$ $3 \cdot 10 = 2 \cdot x$ Set the cross products equal. $30 = 2x$ $15 = x$ Solve for x.
Triangles with the same shape but not necessarily the same size are called **similar triangles.** The lengths of the corresponding sides of two similar triangles are in proportion.	In these similar triangles: $\frac{a}{d} = \frac{b}{e} = \frac{c}{f}$ 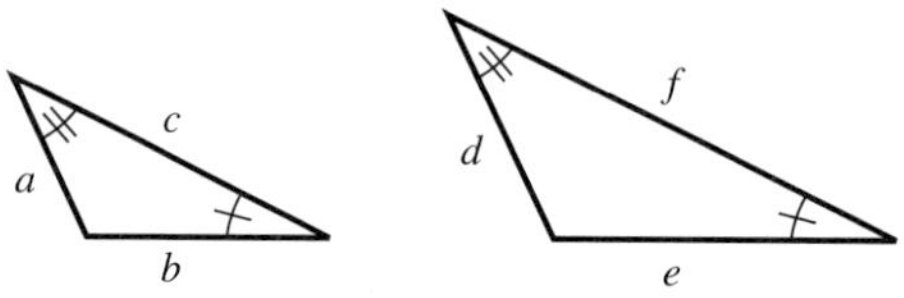
A **scale** is a ratio (or rate) that compares the size of a model to the size of an actual object.	See Example 6 in Section 7.8.
Unit prices can be used to compare costs of different sizes of the same brand to determine the best buy.	For the same item, a cost of $\frac{\$1.95}{1\text{ lb}}$ is a better buy than a cost of $\frac{\$1.99}{1\text{ lb}}$. See Example 7 in Section 7.8.

REVIEW EXERCISES

Determine whether each equation is a proportion.

69. $\frac{4}{7} = \frac{20}{34}$ No

70. $\frac{5}{7} = \frac{30}{42}$ Yes

Solve each proportion.

71. $\frac{3}{x} = \frac{6}{9}$ $\frac{9}{2}$

72. $\frac{x}{3} = \frac{x}{5}$ 0

73. $\frac{x-2}{5} = \frac{x}{7}$ 7

74. $\frac{2x}{x+4} = \frac{3}{x-1}$ $4, -\frac{3}{2}$

75. Dentistry. The diagram in the next column was displayed in a dentist's office. According to the diagram, if the dentist has 340 adult patients, how many will develop gum disease? 255

3 out of 4 adults will develop gum disease.

76. Utility Poles. A telephone pole casts a shadow 12 feet long at the same time that a man 6 feet tall casts a shadow of 3.6 feet. How tall is the pole? 20 ft

77. Porcelain Figurines. A model of a flutist, standing and playing at a music stand, was made using a 1/12th scale. If the scale model is 5.5 inches tall, how tall is the flutist? 5 ft 6 in.

78. Comparison Shopping. Which is the better buy for recordable compact discs: 150 for \$60 or 250 for \$98? 250 for \$98

7 CHAPTER TEST

1. Fill in the blanks.
   a. A quotient of two polynomials, such as $\frac{x+7}{x^2+2x}$, is called a rational expression.
   b. Two triangles with the same shape, but not necessarily the same size, are called similar triangles.
   c. A proportion is a mathematical statement that two ratios or two rates are equal.
   d. To build a rational expression, we multiply it by a form of 1. For example, $\frac{2}{5x} \cdot \frac{8}{8} = \frac{16}{40x}$.
   e. To simplify $\frac{x-3}{(x+3)(x-3)}$, we remove common factors of the numerator and denominator.

2. **Memory.** The formula $n = \frac{35+5d}{d}$ approximates the number of words n that a certain person can recall d days after memorizing a list of 50 words. How many words will the person remember in 1 week? 10 words

For what real numbers is each rational expression undefined?

3. $\frac{6x-9}{5x}$ 0

4. $\frac{x}{x^2+x-6}$ −3, 2

5. **The Internet.** A dial-up modem transmits up to 56K bits per second (K is an abbreviation for one thousand). Convert this to bits per minute. 3,360,000 or 3,360K bits per minute

6. Explain the error: $\frac{x+5}{5} = \frac{x+\overset{1}{\cancel{5}}}{\underset{1}{\cancel{5}}} = x+1$.

   5 is not a common factor of the numerator, and therefore cannot be removed. 5 is a term of the numerator.

Simplify each rational expression.

7. $\frac{48x^2y}{54xy^2}$ $\frac{8x}{9y}$

8. $\frac{7m-49}{7-m}$ −7

9. $\frac{2x^2-x-3}{4x^2-9}$ $\frac{x+1}{2x+3}$

10. $\frac{3(x+2)-3}{6x+5-(3x+2)}$ 1

Find the LCD of each pair of rational expressions.

11. $\frac{19}{3c^2d}, \frac{6}{c^2d^3}$ $3c^2d^3$

12. $\frac{4n+25}{n^2-4n-5}, \frac{6n}{n^2-25}$ $(n+1)(n+5)(n-5)$

Perform the operations. Simplify, if possible.

13. $\frac{12x^2y}{15xy} \cdot \frac{25y^2}{16x}$ $\frac{5y^2}{4}$

14. $\frac{x^2+3x+2}{3x+9} \cdot \frac{x+3}{x^2-4}$ $\frac{x+1}{3(x-2)}$

15. $\frac{x-x^2}{3x^2+6x} \div \frac{3x-3}{3x^3+6x^2}$ $-\frac{x^2}{3}$

16. $\frac{a^2-16}{a-4} \div (6a+24)$ $\frac{1}{6}$

17. $\frac{3y+7}{2y+3} - \frac{-3y-2}{2y+3}$ 3

18. $\frac{2n}{5m} - \frac{n}{2}$ $\frac{4n-5mn}{10m}$

19. $\frac{x+1}{x} + \frac{x-1}{x+1}$ $\frac{2x^2+x+1}{x(x+1)}$

20. $\frac{a+3}{a-1} - \frac{a+4}{1-a}$ $\frac{2a+7}{a-1}$

21. $\frac{9}{c-4} + c$ $\frac{c^2-4c+9}{c-4}$

22. $\frac{6}{t^2-9} - \frac{5}{t^2-t-6}$ $\frac{1}{(t+3)(t+2)}$

Simplify each complex fraction.

23. $\dfrac{\frac{3m-9}{8m}}{\frac{5m-15}{32}}$ $\frac{12}{5m}$

24. $\dfrac{\frac{3}{as^2}+\frac{6}{a^2s}}{\frac{6}{a}-\frac{9}{s^2}}$ $\frac{a+2s}{2as^2-3a^2}$

Solve each equation. If an equation has no solution, so indicate.

25. $\frac{1}{3} + \frac{4}{3y} = \frac{5}{y}$ 11

26. $\frac{9n}{n-6} = 3 + \frac{54}{n-6}$ No solution; 6 is extraneous

27. $\frac{7}{q^2-q-2} + \frac{1}{q+1} = \frac{3}{q-2}$ 1

28. $\frac{2}{3} = \frac{2c-12}{3c-9} - c$ 1, 2

29. $\frac{y}{y-1} = \frac{y-2}{y}$ $\frac{2}{3}$

30. $\frac{a}{a-3} + \frac{4}{a+3} = \frac{18}{a^2-9}$ −10; 3 is extraneous

31. Solve for B: $H = \frac{RB}{R+B}$ $B = \frac{HR}{R-H}$

32. Solve for s: $\frac{1}{r} + \frac{1}{s} = \frac{1}{t}$ $s = \frac{rt}{r-t}$

33. **Health Risks.** A medical newsletter states that a "healthy" waist-to-hip ratio for men is 19:20 or less. Does the patient shown in the illustration fall within the "healthy" range? Yes

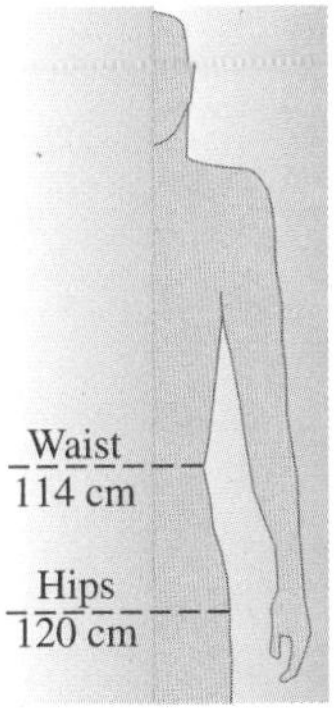

34. **Currency Exchange Rates.** Preparing for a visit to London, a New York resident exchanged 3,500 U.S. dollars for British pounds. (A pound is the basic monetary unit of Great Britain.) If the exchange rate was 100 U.S. dollars for 51 British pounds, how many British pounds did the traveler receive? 1,785

35. **TV Towers.** A television tower casts a shadow 114 feet long at the same time that a 6-foot-tall television reporter casts a shadow of 4 feet. Find the height of the tower. 171 ft

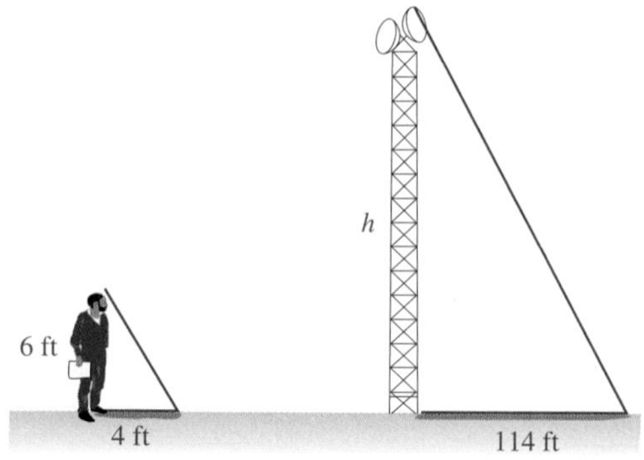

36. **Comparison Shopping.** Which is the better buy for fabric softener: 80 sheets for \$3.89 or 120 sheets for \$6.19? 80 sheets for \$3.89

37. **Cleaning Highways.** One highway worker can pick up all the trash on a strip of highway in 7 hours, and his helper can pick up the trash in 9 hours. How long will it take them if they work together? $3\frac{15}{16}$ hr

38. **Physical Fitness.** A man roller-blades at a rate 6 miles per hour faster than he jogs. In the same time it takes him to roller-blade 5 miles he can jog 2 miles. How fast does he jog? 4 mph

39. **Number Problem.** If a number is subtracted from the numerator of $\frac{5}{8}$ and twice as much is added to the denominator, the result is $\frac{1}{4}$. Find the number. 2

40. Explain the difference between the procedure used to simplify $\frac{1}{x} + \frac{1}{4}$ and the procedure used to solve $\frac{1}{x} + \frac{1}{4} = \frac{1}{2}$.

To simplify $\frac{1}{x} + \frac{1}{4}$, we build each fraction to have the LCD of $4x$. To solve $\frac{1}{x} + \frac{1}{4} = \frac{1}{2}$, we multiply both sides by the LCD $4x$ to eliminate the denominators and clear the equation of fractions.

Group Project

WHAT IS π?

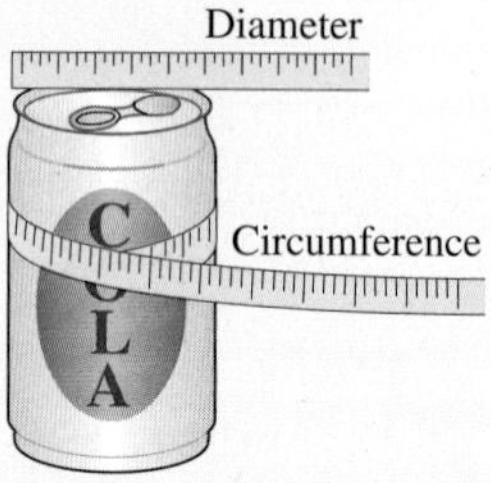

Overview In this activity, you will discover an important fact about the ratio of the circumference to the diameter of a circle.

Instructions Form groups of 2 or 3 students. With a piece of string or a cloth tape measure, find the circumference and the diameter of objects that are circular in shape. You can measure anything that is round: for example, a coin, the top of a can, a tire, or a wastepaper basket. Enter your results in a table, as shown below. Convert each measurement to a decimal, and then use a calculator to determine a decimal approximation of the ratio of the circumference C to diameter d.

Object	Circumference	Diameter	$\frac{C}{d}$ (approx.)
A quarter	$2\frac{15}{16}$ in. = 2.9375 in.	$\frac{15}{16}$ in. = 0.9375 in.	3.13333

Since early history, mathematicians have known that the ratio of the circumference to the diameter of a circle is the same for any size circle, approximately 3. Today, following centuries of study, we know that this ratio is exactly 3.141592653589. . . .

$$\frac{C}{d} = 3.141592653589\ldots$$

The Greek letter π (pi) is used to represent the ratio of circumference to diameter:

$$\pi = \frac{C}{d}, \quad \text{where } \pi = 3.141592653589\ldots$$

Are the ratios in your table numerically close to π? Give some reasons why they aren't exactly 3.141592653589 . . . in each case.

CUMULATIVE REVIEW Chapters 1–7

1. Determine whether each statement is true or false. [Section 1.3]
   a. Every integer is a whole number. False
   b. 0 is not a rational number. False
   c. π is an irrational number. True
   d. The set of integers is the set of whole numbers and their opposites. True
2. Insert the proper symbol, $<$ or $>$, in the blank to make a true statement.

   $|2 - 4|$ $<$ $-(-6)$ [Section 1.5]
3. Evaluate: $9^2 - 3[45 - 3(6 + 4)]$ [Section 1.7] 36
4. Find the average (mean) test score of a student in a history class with scores of 80, 73, 61, 73, and 98. [Section 1.7] 77
5. Simplify: $8(c + 7) - 2(c - 3)$ [Section 1.9] $6c + 62$
6. Solve: $\frac{4}{5}d = -4$ [Section 2.1] -5
7. Solve: $2 - 3(x - 5) = 4(x - 1)$ [Section 2.2] 3
8. **Grand King Size Beds.** Because Americans are taller compared to 100 years ago, bed manufacturers are making larger models. Find the percent of increase in sleeping area of the new grand king size bed compared to the standard king size. [Section 2.3] About 26%

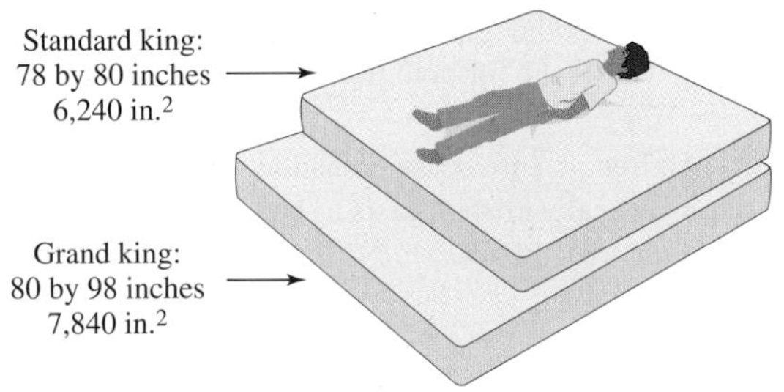

9. Solve $A - c = 2B + r$ for B. [Section 2.4] $B = \frac{A - c - r}{2}$
10. Change 40°C to degrees Fahrenheit. [Section 2.4] 104°F
11. Find the volume of a pyramid that has a square base, measuring 6 feet on a side, and whose height is 20 feet. [Section 2.4] 240 ft³
12. **Blending Tea.** One grade of tea (worth \$6.40 per pound) is to be mixed with another grade (worth \$4 per pound) to make 20 pounds of a mixture that will be worth \$5.44 per pound. How much of each grade of tea must be used? [Section 2.6] 12 lb of the \$6.40 tea and 8 lb of the \$4 tea
13. **Speed of a Plane.** Two planes are 6,000 miles apart and their speeds differ by 200 mph. If they travel toward each other and meet in 5 hours, find the speed of the slower plane. [Section 2.6] 500 mph
14. Solve $7x + 2 \geq 4x - 1$. Write the solution set in interval notation and graph it. [Section 2.7] $[-1, \infty)$, See AIE Appendix 3.
15. Graph: $y = 2x - 3$ [Section 3.2] See AIE Appendix 3.
16. Find the slope of the line that passes through $(-1, 3)$ and $(3, -1)$. [Section 3.4] -1
17. **Cutting Steel.** The graph shows the amount of wear (in millimeters) on a cutting blade for a given length of a cut (in meters). Find the rate of change in the length of the cutting blade. [Section 3.4] 0.008 mm/m

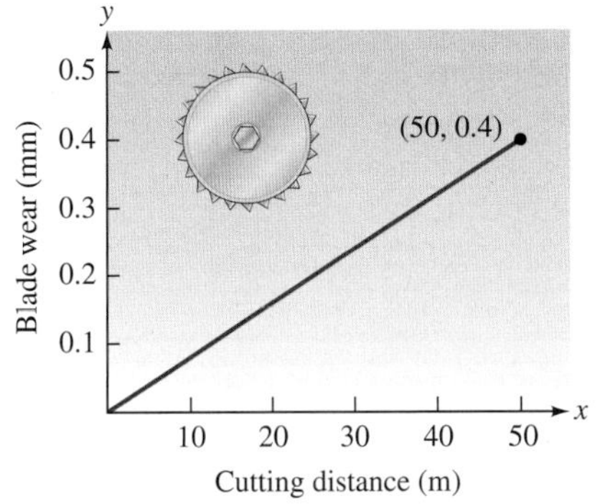

18. What is the slope of a line perpendicular to the line $y = -\frac{7}{8}x - 6$? [Section 3.5] $\frac{8}{7}$
19. Write an equation of the line that has slope 3 and passes through the point (1, 5). Write the answer in slope–intercept form. [Section 3.6] $y = 3x + 2$
20. Graph: $3x - 2y \leq 6$ [Section 3.7] See AIE Appendix 3.
21. If $f(x) = -3x^2 - 6x$, find $f(-2)$. [Section 3.8] 0
22. Fill in the blanks. The set of all possible input values for a function is called the domain and the set of all output values is called the range. [Section 3.8]
23. Solve the system $\begin{cases} x + y = 1 \\ y = x + 5 \end{cases}$ by graphing. [Section 4.1] $(-2, 3)$; See AIE Appendix 3.
24. Solve the system: $\begin{cases} x = 3y - 1 \\ 2x - 3y = 4 \end{cases}$ [Section 4.2] (5, 2)
25. Solve the system: $\begin{cases} 2x + 3y = -1 \\ 3x + 5y = -2 \end{cases}$ [Section 4.3] $(1, -1)$
26. **Poker.** After a night of cards, a poker player finished with some red chips (worth \$5 each) and some blue chips (worth \$10 each). He received \$190 when he cashed in the 23 chips. How many of each colored chip did he finish with? [Section 4.4] Red: 8, blue: 15

Simplify each expression. Write each answer without using negative exponents.

27. x^4x^3 [Section 5.1] x^7
28. $(x^2x^3)^5$ [Section 5.1] x^{25}
29. $\left(\frac{y^3y}{2yy^2}\right)^3$ [Section 5.1] $\frac{y^3}{8}$
30. $\left(\frac{-2a}{b}\right)^5$ [Section 5.1] $-\frac{32a^5}{b^5}$
31. $(a^{-2}b^3)^{-4}$ [Section 5.2] $\frac{a^8}{b^{12}}$
32. $\frac{9b^0b^3}{3b^{-3}b^4}$ [Section 5.2] $3b^2$
33. Write 290,000 in scientific notation. [Section 5.3] 2.9×10^5
34. What is the degree of the polynomial $5x^3 - 4x + 16$? [Section 5.4] 3

35. Graph: $y = -x^3$ [Section 5.4] See AIE Appendix 3.

36. Concentric Circles. The area of the ring between the two concentric circles of radius r and R is given by the formula

$$A = \pi(R + r)(R - r)$$

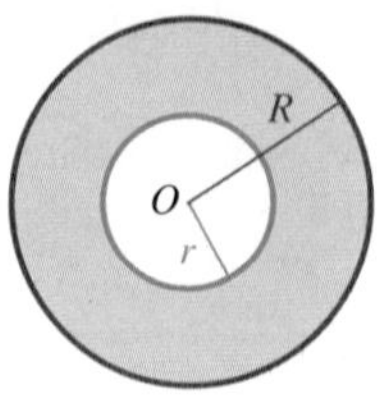

Do the multiplication on the right side of the equation. [Section 5.7] $A = \pi R^2 - \pi r^2$

Perform the operations.

37. $(3x^2 - 3x - 2) + (3x^2 + 4x - 3)$ [Section 5.5] $6x^2 + x - 5$

38. $\left(\frac{1}{16}t^3 + \frac{1}{2}t^2 - \frac{1}{6}t\right) - \left(\frac{9}{16}t^3 + \frac{9}{4}t^2 - \frac{1}{12}t\right)$ [Section 5.5] $-\frac{1}{2}t^3 - \frac{7}{4}t^2 - \frac{1}{12}t$

39. $(2x^2y^3)(3x^2y^2)$ [Section 5.6] $6x^4y^5$

40. $(2y - 5)(3y + 7)$ [Section 5.6] $6y^2 - y - 35$

41. $-4x^2z(3x^2 - z)$ [Section 5.6] $-12x^4z + 4x^2z^2$

42. $(3a - 4)^2$ [Section 5.7] $9a^2 - 24a + 16$

43. $\dfrac{6x + 9}{3}$ [Section 5.8] $2x + 3$

44. $2x + 3\overline{)2x^3 + 7x^2 + 4x - 3}$ [Section 5.8] $x^2 + 2x - 1$

Factor each polynomial completely.

45. $k^3t - 3k^2t$ [Section 6.1] $k^2t(k - 3)$

46. $2ab + 2ac + 3b + 3c$ [Section 6.1] $(b + c)(2a + 3)$

47. $u^2 - 18u + 81$ [Section 6.2] $(u - 9)^2$

48. $-r^2 + 2 + r$ [Section 6.2] $-(r - 2)(r + 1)$

49. $u^2 + 10u + 15$ [Section 6.2] Prime

50. $6x^2 - 63 - 13x$ [Section 6.3] $(2x - 9)(3x + 7)$

51. $2a^2 - 200b^2$ [Section 6.4] $2(a + 10b)(a - 10b)$

52. $b^3 + 125$ [Section 6.5] $(b + 5)(b^2 - 5b + 25)$

Solve each equation by factoring.

53. $5x^2 + x = 0$ [Section 6.7] $0, -\frac{1}{5}$

54. $6x^2 - 5x = -1$ [Section 6.7] $\frac{1}{3}, \frac{1}{2}$

55. Cooking. The electric griddle shown has a cooking surface of 160 square inches. Find the length and the width of the griddle. [Section 6.7] 16 in., 10 in.

56. For what values of x is the rational expression $\frac{3x^2}{x^2 - 25}$ undefined? [Section 7.1] $5, -5$

Perform the operations. Simplify, if possible.

57. $\dfrac{2x^2 - 8x}{x^2 - 6x + 8}$ [Section 7.1] $\frac{2x}{x - 2}$

58. $\dfrac{x^2 - 16}{4 - x} \div \dfrac{3x + 12}{x^3}$ [Section 7.2] $-\frac{x^3}{3}$

59. $\dfrac{8m^2}{2m + 5} - \dfrac{4m^2 + 25}{2m + 5}$ [Section 7.3] $2m - 5$

60. $\dfrac{4}{x - 3} + \dfrac{5}{3 - x}$ [Section 7.4] $-\frac{1}{x - 3}$

61. $\dfrac{m}{m^2 + 5m + 6} - \dfrac{2}{m^2 + 3m + 2}$ [Section 7.4] $\frac{m - 3}{(m + 3)(m + 1)}$

62. Simplify: $\dfrac{2 - \frac{2}{x + 1}}{2 + \frac{2}{x}}$ [Section 7.5] $\frac{x^2}{(x + 1)^2}$

Solve each equation.

63. $\dfrac{7}{5x} - \dfrac{1}{2} = \dfrac{5}{6x} + \dfrac{1}{3}$ [Section 7.6] $\frac{17}{25}$

64. $\dfrac{u}{u - 1} + \dfrac{1}{u} = \dfrac{u^2 + 1}{u^2 - u}$ [Section 7.6] 2

65. Draining a Tank. If one outlet pipe can drain a tank in 24 hours, and another pipe can drain the tank in 36 hours, how long will it take for both pipes to drain the tank? [Section 7.7] $14\frac{2}{5}$ hr

66. Height of a Tree. A tree casts a shadow of 29 feet at the same time as a vertical yardstick casts a shadow of 2.5 feet. Find the height of the tree. [Section 7.8] 34.8 ft

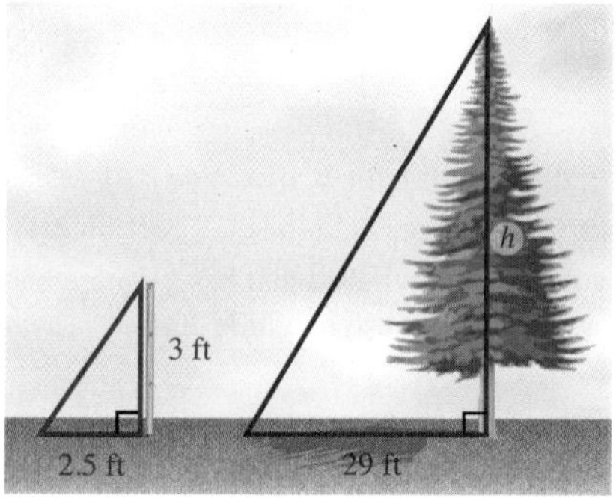

Transition to Intermediate Algebra

8

from Campus to Careers

Webmaster

If you use the Internet, then you have seen firsthand what webmasters do. They design and maintain websites for individuals and companies on the World Wide Web. The job of webmaster requires excellent computer and technical skills. A background in business, art, and design is also helpful. Since webmasters are often called on to be troubleshooters when technical difficulties arise, those considering entering the field are encouraged to study mathematics to strengthen their problem-solving abilities.

Problem 113 in Study Set 8.1, **problem 107** in Study Set 8.2, and **problem 33** in Study Set 8.9 involve situations that a webmaster might encounter on the job. The mathematical concepts discussed in this chapter can be used to solve those problems.

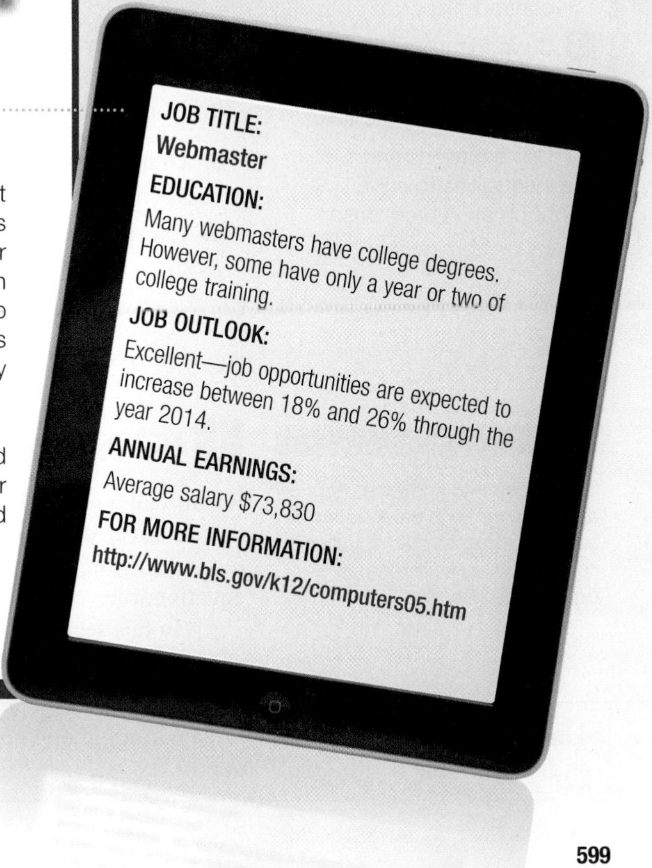

JOB TITLE:
Webmaster
EDUCATION:
Many webmasters have college degrees. However, some have only a year or two of college training.
JOB OUTLOOK:
Excellent—job opportunities are expected to increase between 18% and 26% through the year 2014.
ANNUAL EARNINGS:
Average salary $73,830
FOR MORE INFORMATION:
http://www.bls.gov/k12/computers05.htm

Study Skills Workshop ▶ Maintaining a Positive Attitude

Teaching Tip: This chapter is an excellent way to begin the second half of a combination Elementary and Intermediate Algebra course. Its purpose is three-fold:
1. It reviews some of the basic algebraic concepts studied in the first half of the book, such as equations, inequalities, factoring, rational expressions, graphing, and problem solving.
2. It expands on several of those concepts, such as factoring, functions, and graphing, with more in-depth study at the intermediate algebra level.
3. It offers students an opportunity to apply previously learned skills in new settings with topics such as compound inequalities, abolute value equations and inequalities, and variation.

Since a student's level of effort is significantly influenced by his or her attitude, you should strive to maintain a positive mental outlook for the entire term. Here are three suggestions to help you do that:

PERSONAL REMINDERS: From time to time, remind yourself of the ways in which you will benefit by passing the course.

DON'T DWELL ON THE NEGATIVE: Counterproductive feelings of stress and math anxiety often can be overcome with extra preparation, support services, and even relaxation techniques.

ACCOMPLISH GOALS AND EARN REWARDS: Reward yourself after studying, learning a difficult concept, or completing a homework assignment. The reward can be small, like listening to music, reading a novel, playing a sport, or spending some time with friends.

Now Try This

1. List six ways in which you will benefit by passing this course. For example, it will get you one step closer to a college degree or it will improve your problem-solving abilities.
2. List three ways in which you can to respond to feelings of stress or math anxiety, should they arise during the term.
3. List some simple ways that you can reward yourself when you complete one of the class goals that you set for yourself.

SECTION 8.1 Review of Solving Linear Equations, Formulas, and Linear Inequalities; Applications

OBJECTIVES

1. Use properties of equality to solve linear equations.
2. Identify identities and contradictions.
3. Solve formulas for a specified variable.
4. Solve linear inequalities.
5. Use equations and inequalities to solve application problems.

ARE YOU READY? *Are You Ready? exercises available online at www.webassign.net/brookscole*

The following problems review some basic skills that are needed when solving equations and inequalities.

1. Simplify: $5m - 2(8m - 4)$ $-11m + 8$
2. Multiply: $21\left(\frac{6}{7}n\right)$ $18n$
3. Multiply: $100 \cdot 0.27$ 27
4. True or false: $-15 \le -15$ True
5. Graph the set of real numbers greater than or equal to -2 on a number line. See AIE Appendix 3.
6. Express the fact that $6 > x$ using an $<$ symbol. $x < 6$

One of the most useful concepts in algebra is the equation. Writing and then solving an equation is a powerful problem-solving strategy.

1 Use Properties of Equality to Solve Linear Equations.

The Language of Algebra

It is important to know the difference between an **equation** and an **expression**. An equation contains an = symbol; an expression does not.

Recall that an **equation** is a statement indicating that two expressions are equal. A number that makes an equation true when substituted for the variable is called a **solution,** and it is said to *satisfy* the equation. The **solution set** of an equation is the set of all numbers that make the equation true.

In this section, we will solve *linear equations in one variable.*

Linear Equations in One Variable

A linear equation in one variable can be written in the form $ax + b = c$, where a, b, and c are real numbers, and $a \neq 0$.

Teaching Tip: Ask your students what the exponent is on the variable in a linear equation.

Success Tip

You may want to review the more detailed explanations of the properties of equality on pages 103, 105, 106, and 108.

Some examples of linear equations in one variable are

$$5a + 2 = 8, \qquad \frac{7}{3}y = -28, \qquad \text{and} \qquad 3(2x - 1) = 8x + 9 - 6x$$

We can solve a linear equation in one variable by using the following **properties of equality** to replace it with simpler **equivalent equations** that have the same solution set. We continue this process until the variable is isolated on one side of the = symbol.

Properties of Equality

1. Adding the same number to, or subtracting the same number from, both sides of an equation does not change its solution.
2. Multiplying or dividing both sides of an equation by the same nonzero number does not change its solution.

When solving equations, we should simplify the expressions that make up the left and right sides before applying any properties of equality. Often that involves using the **distributive property** to remove parentheses and/or **combining like terms.**

EXAMPLE 1 Solve: $3(2x - 1) = 8x + 9 - 6x$

Strategy We will use the distributive property on the left side of the equation and combine like terms on the right side.

Why It's best to simplify each side of an equation before using a property of equality. This makes it easier to determine the steps needed to get all the terms containing x on the same side of the equation.

Solution

$3(2x - 1) = 8x + 9 - 6x$	This is the equation to solve.
$6x - 3 = 2x + 9$	On the left side, distribute the multiplication by 3. On the right side, combine like terms: $8x - 6x = 2x$.
$6x - 3 - \mathbf{2x} = 2x + 9 - \mathbf{2x}$	To eliminate 2x from the right side, subtract 2x from both sides.
$4x - 3 = 9$	Combine like terms: $6x - 2x = 4x$ and $2x - 2x = 0$.
$4x - 3 + \mathbf{3} = 9 + \mathbf{3}$	To isolate the variable term 4x, undo the subtraction of 3 on the left side by adding 3 to both sides.
$4x = 12$	Combine like terms: $-3 + 3 = 0$.
$x = 3$	To isolate the variable x, undo the multiplication by 4 by dividing both sides by 4.

Success Tip

Since division by 4 is the same as multiplication by $\frac{1}{4}$, we can solve $4x = 12$ using the multiplication property of equality by multiplying both sides by the *multiplicative inverse* of 4, which is $\frac{1}{4}$.

$$\frac{1}{4} \cdot 4x = \frac{1}{4} \cdot 12$$
$$x = 3$$

Check: We substitute 3 for x in the original equation to see whether it satisfies the equation.

Evaluate the expression on the left side.

$$3(2x - 1) = 8x + 9 - 6x$$
$$3(2 \cdot \mathbf{3} - 1) \stackrel{?}{=} 8 \cdot \mathbf{3} + 9 - 6 \cdot \mathbf{3}$$
$$3(5) \stackrel{?}{=} 24 + 9 - 18$$
$$15 = 15 \quad \text{True}$$

Evaluate the expression on the right side.

The Language of Algebra

Read $\stackrel{?}{=}$ as "is possibly equal to."

Since the resulting statement $15 = 15$ is true, 3 satisfies the equation, and we say that 3 is a solution of $3(2x - 1) = 8x + 9 - 6x$. The solution set is $\{3\}$.

Teaching Example 1 Solve: $5(4x - 3) = 14x + 9 - 2x$
Answer: 3

Self Check 1 Solve: $2(3x - 2) = 10x - 13 - 7x$ -3

Now Try Problem 21

In general, we will follow these steps to solve linear equations in one variable. Not every step is needed to solve every equation.

Strategy for Solving Linear Equations in One Variable

1. **Clear the equation of fractions or decimals:** Multiply both sides by the LCD to clear fractions or multiply both sides by a power of 10 to clear decimals.
2. **Simplify each side of the equation:** Use the distributive property to remove parentheses and combine like terms on each side.
3. **Isolate the variable term on one side:** Add (or subtract) to get the variable term on one side of the equation and a number on the other using the addition (or subtraction) property of equality.
4. **Isolate the variable:** Multiply (or divide) to isolate the variable using the multiplication (or division) property of equality.
5. **Check the result:** Substitute the possible solution for the variable in the *original* equation to see if a true statement results.

EXAMPLE 2 Solve: $\frac{1}{3}(6x + 17) = \frac{3}{2}(x + 2) - 2$

Strategy We will follow the strategy for solving equations.

Why This is the most efficient way to solve a linear equation in one variable.

Solution *Step 1:* We can clear the equation of fractions by multiplying both sides by the least common denominator (LCD) of $\frac{1}{3}$ and $\frac{3}{2}$. The LCD of these fractions is the smallest number that can be divided by both 2 and 3 exactly. That number is 6.

Teaching Tip: It is wise to have students identify the number of terms on each side of the equation before multiplying both sides by the LCD.

$$\frac{1}{3}(6x + 17) = \frac{3}{2}(x + 2) - 2$$ This is the equation to solve.

$$6\left[\frac{1}{3}(6x + 17)\right] = 6\left[\frac{3}{2}(x + 2) - 2\right]$$ To eliminate the fractions, multiply both sides by the LCD, 6.

$$2(6x + 17) = \mathbf{6} \cdot \frac{3}{2}(x + 2) - \mathbf{6} \cdot 2$$ On the left side, perform the multiplication: $6 \cdot \frac{1}{3} = 2$. On the right side, distribute the multiplication by 6.

$$2(6x + 17) = 9(x + 2) - 12$$ Multiply on the right side: $6 \cdot \frac{3}{2} = 9$ and $6 \cdot 2 = 12$. Don't forget the parentheses shown in blue.

Success Tip

Before multiplying both sides of an equation by the LCD, frame the left side and frame the right side with parentheses or brackets.

Step 2: Simplify the expressions on each side of the equation.

$$12x + 34 = 9x + 18 - 12$$ Distribute the multiplication by 2 and the multiplication by 9.

$$12x + 34 = 9x + 6$$ Combine like terms: $18 - 12 = 6$.

Step 3: To get the variable term on the left side of the equation and the constant on the right side, subtract $9x$ and 34 from both sides.

Teaching Tip: Remind students that they may solve the equation so that the variable is isolated on either side of the equation.

$$12x + 34 \mathbf{- 9x - 34} = 9x + 6 \mathbf{- 9x - 34}$$

$$3x = -28$$ On each side, combine like terms.

Step 4: To isolate x, we undo the multiplication by 3 by dividing both sides by 3.

$$\frac{3x}{\mathbf{3}} = \frac{-28}{\mathbf{3}}$$

$$x = -\frac{28}{3}$$

Step 5: Verify that $-\frac{28}{3}$ is the solution by substituting it into the original equation and evaluating each side.

Teaching Example 2 Solve: $\frac{1}{5}(11x + 9) = \frac{4}{3}(x + 1) - 2$
Answer: $-\frac{37}{13}$

Self Check 2 Solve: $\frac{1}{3}(2x - 5) = \frac{3}{4}(5x + 1) + 2$ $-\frac{53}{37}$

Now Try Problem 29

2 Identify Identities and Contradictions.

The equations discussed so far are called **conditional equations.** For these equations, some real numbers satisfy the equation and others do not. Other equations are made true by *any* permissible replacement value for the variable. Such an equation is called an **identity.** Still other equations are false for *all* replacement values for the variable. We call such an equation a **contradiction.**

EXAMPLE 3 Solve: **a.** $-2(x - 1) - 4 = -4(1 + x) + 2x + 2$
b. $-6.2(-x - 1) - 4 = 4.2x - (-2x)$

Strategy In each case, we will follow the strategy for solving equations.

Why This is the most efficient way to solve a linear equation in one variable.

Solution

a. Since there are no fractions to clear, we will begin by using the distributive property to remove the parentheses on the left and right sides of the equation.

$-2(x - 1) - 4 = -4(1 + x) + 2x + 2$ This is the equation to solve.

$-2x + 2 - 4 = -4 - 4x + 2x + 2$ On each side, use the distributive property.

$-2x - 2 = -2x - 2$ On each side, combine like terms.

$-2x - 2 + \mathbf{2x} = -2x - 2 + \mathbf{2x}$ To attempt to isolate the variable on one side of the equation, add 2x to both sides.

$-2 = -2$ True

In the solution process, the terms involving x drop out. The resulting true statement indicates that the original equation is true for every permissible value of x. Therefore, *all real numbers* are solutions and this equation is an identity. Its solution set is written as {all real numbers} or using the symbol $\mathbb{R}$.

Success Tip

We know the given equation is an identity because we see in Step 3 that it is equivalent to the equation $-2x - 2 = -2x - 2$, which is true for all values of x.

Teaching Tip: A common mistake is to incorrectly "distribute" 10 over -6.2 and $(-x - 1)$.

b.

$-6.2(-x - 1) - 4 = 4.2x - (-2x)$ This is the equation to solve.

$-6.2(-x - 1) - 4 = 4.2x + 2x$ Simplify: $-(-2x) = 2x$.

$\mathbf{10}[-6.2(-x - 1) - 4] = \mathbf{10}(4.2x + 2x)$ To clear the decimals, multiply both sides by 10.

$\mathbf{10}(-6.2)(-x - 1) - \mathbf{10} \cdot 4 = \mathbf{10} \cdot 4.2x + \mathbf{10} \cdot 2x$ On each side, distribute the multiplication by 10.

$-62(-x - 1) - 40 = 42x + 20x$ Perform each multiplication by 10.

$62x + 62 - 40 = 42x + 20x$ On the left side, distribute -62.

$62x + 22 = 62x$ On each side, combine like terms.

$62x + 22 - \mathbf{62x} = 62x - \mathbf{62x}$ To attempt to isolate the variable on one side of the equation, subtract 62x from both sides.

$22 = 0$ False

In the solution process, the terms involving x drop out. The resulting false statement indicates that no value for x makes the original equation true. Therefore, this equation has *no solution* and it is a contradiction. Its solution set is the empty set, which is written as { } or using the symbol $\varnothing$.

The Language of Algebra

Contradiction is a form of the word *contradict,* meaning conflicting ideas. During a trial, evidence might be introduced that contradicts the testimony of a witness.

Teaching Example 3 Solve:
a. $-3(x+2)-5=x-7-4(x+1)$
b. $-0.5(x+7)-0.2=0.3x-0.8(x+1)$
Answers:
a. All real numbers, $\mathbb{R}$; identity
b. No solution, $\varnothing$; contradiction

Self Check 3 Solve: **a.** $3(a+1)-(20+a)=5(a-1)-3(a+4)$
All real numbers, $\mathbb{R}$; identity
b. $0.3(a+4)+0.2=0.2(a-1)+0.1a+1.9$
No solution, $\varnothing$; contradiction

Now Try Problems 31 and 35

3 Solve Formulas for a Specified Variable.

Real-world applications sometimes call for a formula solved for one variable to be solved for a different variable. **To solve a formula for a specified variable** means to isolate that variable on one side of the equal symbol, with all other variables and constants on the other side.

EXAMPLE 4 **Banking.** The formula $A = P + Prt$ gives the amount of money in an account at the end of a specific time t. A represents the amount, P the principal, and r the simple rate of interest. Solve the formula for t.

Strategy To solve for t, we treat it as if it were the only variable in the equation. To isolate t, we will use the same strategy that we used to solve linear equations in one variable.

Why We can solve this formula as if it were an equation in one variable because all the other variables, A, P, and r, are treated as if they were numbers (constants).

Solution

To solve for t, we will isolate it on this side of the equation.

$$A = P + Prt$$

$$A - P = Prt$$ To isolate the term involving t, subtract P from both sides. This step is done mentally.

$$\frac{A-P}{Pr} = \frac{Prt}{Pr}$$ To isolate t, divide both sides by Pr (or multiply both sides by $\frac{1}{Pr}$).

$$\frac{A-P}{Pr} = t$$ On the right side, remove the common factors: $\frac{\overset{1}{\cancel{P}}\overset{1}{\cancel{r}}t}{\underset{1}{\cancel{P}}\underset{1}{\cancel{r}}}$.

$$t = \frac{A-P}{Pr}$$ It is common practice to write the equation so that the specified variable, in this case t, is on the left side.

The Language of Algebra

We say the formula is **solved for t** because t is alone on one side of the equation and the other side does not contain t.

Teaching Example 4 Solve $T = mB + rst$ for s.
Answer: $s = \frac{T - mB}{rt}$

Self Check 4 Solve $A = P + Prt$ for r. $r = \frac{A-P}{Pt}$

Now Try Problem 39

EXAMPLE 5 **Geometry.** The formula for the area of a trapezoid is $A = \frac{1}{2}h(b_1 + b_2)$. Solve the formula for b_1.

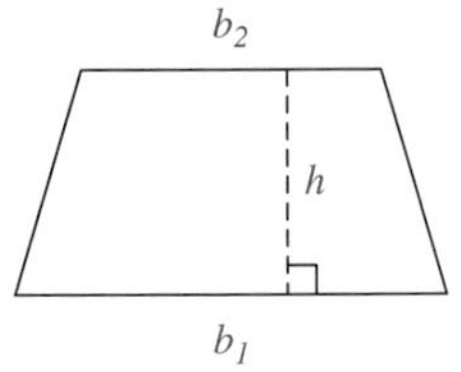

Trapezoid

Strategy To solve for b_1, we will treat it as if it were the only variable in the equation. To isolate b_1, we will use the same strategy that we used to solve linear equations in one variable.

Why We can solve the formula as if it were an equation in one variable because all the other variables, A, h, and b_2, are treated as if they were numbers (constants).

Solution

To solve for b_1, we will isolate b_1 on this side of the equation.

$A = \frac{1}{2}h(b_1 + b_2)$ — Read b_1 as "b-sub-one" and b_2 as "b-sub-two."

$2 \cdot A = 2 \cdot \frac{1}{2}h(b_1 + b_2)$ — Multiply both sides by 2 to clear the equation of the fraction.

$2A = h(b_1 + b_2)$ — Simplify each side of the equation.

$2A = hb_1 + hb_2$ — Distribute the multiplication by h.

$2A - hb_2 = hb_1$ — Subtract hb_2 from both sides to isolate the variable term hb_1 on the right side. This step is done mentally.

$\frac{2A - hb_2}{h} = \frac{hb_1}{h}$ — To isolate b_1, undo the multiplication by h by dividing both sides by h.

$\frac{2A - b_2h}{h} = b_1$ — On the right side, remove the common factor of h: $\frac{\overset{1}{\cancel{h}}b_1}{\underset{1}{\cancel{h}}}$.

$b_1 = \frac{2A - hb_2}{h}$ — Reverse the sides of the equation so that b_1 is on the left.

Teaching Tip: Remind students that 1 and 2 in b_1 and b_2 are subscripts and not exponents.

Caution

Do not try to simplify the result in the following way. It is incorrect because h is not a factor of the entire numerator.

$$b_1 = \frac{2A - \overset{1}{\cancel{h}}b_2}{\underset{1}{\cancel{h}}}$$

When solving formulas for a specified variable, there is often more than one way to express the result. In this case, we could perform the division by h on the right side term-by-term: $b_1 = \frac{2A}{h} - \frac{hb_2}{h}$. After removing the common factor of h in the numerator and denominator of the second fraction, we obtain the following equivalent form of the result: $b_1 = \frac{2A}{h} - b_2$.

Teaching Example 5 Solve $A = \frac{1}{2}h(b_1 + b_2)$ for b_2.

Answer: $b_2 = \frac{2A - hb_1}{h}$

Self Check 5 Solve $S = \frac{180(t - 2)}{7}$ for t. $t = \frac{7S + 360}{180}$

Now Try Problem 47

4 Solve Linear Inequalities.

Inequalities are statements indicating that two quantities are unequal and they contain one or more of the following symbols.

Inequality Symbols

$<$ is less than	$>$ is greater than	$\neq$ is not equal to
$\leq$ is less than or equal to	$\geq$ is greater than or equal to	

In this section, we will solve *linear inequalities in one variable*. A linear *inequality* in one variable is similar to a linear *equation* in one variable except that the equal symbol is replaced with an inequality symbol.

Linear Inequalities

A linear inequality in one variable (say, x) is any inequality that can be expressed in one of the following forms, where a, b, and c represent real numbers and $a \neq 0$.

$$ax + b < c \qquad ax + b \leq c \qquad ax + b > c \qquad ax + b \geq c$$

Some examples of linear inequalities are

$$3(2x - 9) < 9, \qquad -11x - 8.2 \leq x + 32.2, \qquad \text{and} \qquad 3x > 0$$

Success Tip

You may want to review the more detailed explanations of properties of inequality on pages 165 and 166.

To **solve a linear inequality** means to find all values that, when substituted for the variable, make the inequality true. The set of all solutions of an inequality is called its **solution set.** Most inequalities that we will solve have infinitely many solutions. To represent such solutions, we can use the graph of an **interval** on the number line and two special types of notation: **interval notation** and **set-builder notation.**

We use the following properties to solve inequalities.

Properties of Inequality

1. Adding the same number to, or subtracting the same number from, both sides of an inequality does not change the solutions.
2. Multiplying or dividing both sides of an inequality by the same positive number does not change the solutions.
3. If we multiply or divide both sides of an inequality by a negative number, the direction of the inequality symbol must be *reversed* for the inequalities to have the same solutions.

After applying one of these properties, the resulting inequality is equivalent to the original one. Like equivalent equations, **equivalent inequalities** have the same solution set.

EXAMPLE 6 Solve $3(2x - 9) < 9$. Graph the solution set and write it using interval notation.

Strategy We will use the properties of inequality and the same strategy for solving equations to isolate x on one side of the inequality.

Why Once we have obtained an equivalent inequality, with the variable isolated on one side, the solution set will be obvious.

Solution

$3(2x - 9) < 9$	This is the inequality to solve.
$6x - 27 < 9$	Distribute the multiplication by 3.
$6x < 36$	To isolate the variable term $6x$, undo the subtraction of 27 by adding 27 to both sides: $9 + 27 = 36$.
$x < 6$	To isolate x, undo the multiplication by 6 by dividing both sides by 6.

Teaching Tip: To prepare your students for the Look Alikes problems, ask them how to use this answer to solve

$3(2x - 9) \geq 9$

without having to do any work.

The solution set is the interval $(-\infty, 6)$, whose graph is shown. We also can write the solution set using set-builder notation: $\{x \mid x < 6\}$. We read this notation as "the set of all real numbers x such that x is less than 6."

Notation

When graphing solution sets on a number line, parentheses are used to exclude endpoints and brackets are used to include endpoints.

Since the solution set contains infinitely many real numbers, we cannot check all of them to see whether they satisfy the original inequality. However, as an informal check, we can pick one number in the graph, near the endpoint, such as 5, and see whether it satisfies the inequality. We also can pick one number not in the graph, but near the endpoint, such as 7, and see whether it fails to satisfy the inequality.

Check a value in the graph: $x = 5$	***Check a value not in the graph:*** $x = 7$
$3(2x - 9) < 9$	$3(2x - 9) < 9$
$3[2(5) - 9] \overset{?}{<} 9$	$3[2(7) - 9] \overset{?}{<} 9$
$3(10 - 9) \overset{?}{<} 9$	$3(14 - 9) \overset{?}{<} 9$
$3(1) \overset{?}{<} 9$	$3(5) \overset{?}{<} 9$
$3 < 9$ True	$15 < 9$ False

Since 5 satisfies $3(2x - 9) < 9$ and 7 does not, the solution set appears to be correct.

Teaching Example 6 Solve $4(2x - 3) > 4$. Graph the solution set and write it using interval notation.
Answer: $(2, \infty)$; $\{x \mid x > 2\}$

Self Check 6 Solve $2(3x + 2) > -44$. Graph the solution set and write it using interval notation. $(-8, \infty)$; see AIE Appendix 3.

Now Try Problem 53

EXAMPLE 7

Solve $-11x + 8.2 \leq x + 32.2$. Graph the solution set and write it using interval notation.

Teaching Tip: Select real numbers in the shaded and unshaded portions of the graph. Ask your students if the value, when substituted for x, makes the inequality true or false.

Strategy We will use the properties of inequality and the same strategy for solving equations to isolate x on one side of the inequality.

Why Once we have obtained an equivalent inequality, with the variable isolated on one side, the solution set will be obvious.

Solution

$-11x + 8.2 \leq x + 32.2$	This is the inequality to solve.
$-12x + 8.2 \leq 32.2$	To eliminate x from the right side, subtract x from both sides.
$-12x \leq 24$	To isolate the variable term $-12x$, undo the addition of 8.2 by subtracting 8.2 from both sides: $32.2 - 8.2 = 24$.
$\dfrac{-12x}{-12} \geq \dfrac{24}{-12}$	To isolate the x, undo the multiplication by -12 by dividing both sides by -12. Because we are dividing by a negative number, reverse the $\leq$ symbol.
$x \geq -2$	Do the division.

Success Tip

We must remember to reverse the inequality symbol every time we multiply or divide both sides by a negative number.

The solution set is $\{x \mid x \geq -2\}$ or the interval $[-2, \infty)$, whose graph is shown.

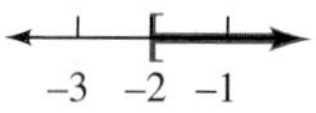

Teaching Example 7 Solve: $-8x - 3.6 \geq x + 32.4$. Graph the solution set and write it using interval notation.
Answer: $(-\infty, -4]$

Self Check 7 Solve: $-5x + 6 \leq x$. Graph the solution set and write it using interval notation. $[1, \infty)$; see AIE Appendix 3.

Now Try Problem 59

EXAMPLE 8

Solve $3a - 4 < 3(a + 5)$. Graph the solution set and write it using interval notation.

Strategy We will use the properties of inequality and the same strategy for solving equations to isolate a on one side of the inequality.

Why Once we have obtained an equivalent inequality, with the variable isolated on one side, the solution set is obvious.

Solution

$3a - 4 < 3(a + 5)$	This is the inequality to solve.
$3a - 4 < 3a + 15$	Distribute the multiplication by 3.
$3a - 4 - \mathbf{3a} < 3a + 15 - \mathbf{3a}$	To eliminate 3a on the right side, subtract 3a from both sides.
$-4 < 15$	True

Success Tip

If the terms involving the variable drop out when solving an inequality and the result is false, the solution set contains no elements and is denoted by ∅.

In the solution process, the terms involving a drop out. The resulting true statement indicates that the original inequality is true for every permissible value of a. Therefore, *all real numbers* are solutions and this inequality is an identity. Its solution set is the set of real numbers, written as $(-\infty, \infty)$ or $\mathbb{R}$, and its graph is as shown.

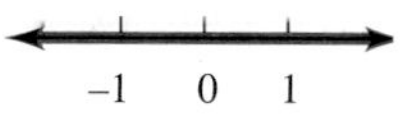

Teaching Example 8 Solve: $\dfrac{3a - 4}{-5} > \dfrac{3a + 15}{-5}$. Graph the solution set and write it using interval notation.
Answer: $(-\infty, \infty)$

Self Check 8 Solve $-8n + 10 \geq 1 - 2(4n - 2)$. Graph the solution set and write it using interval notation. $(-\infty, \infty)$; see AIE Appendix 3.

Now Try Problem 67

5 Use Equations and Inequalities to Solve Application Problems.

To become a good problem solver, you need a plan to follow such as the following six-step problem-solving strategy: **Analyze** the problem, **Assign** a variable, **Form** an equation (or inequality), **Solve** the equation (or inequality), **State** the conclusion, **Check** the result.

EXAMPLE 9

©oksana.perkins/Shutterstock.com

Travel Promotions. The price of a 7-day Alaskan cruise, normally \$2,752 per person, is reduced by \$1.75 per person for large groups traveling together. How large a group is needed for the price to be \$2,500 per person?

Analyze It is often helpful to consider some specific situations before attempting to form an equation. For a group of, say, 20 people, the price of the cruise would be reduced by 20(\$1.75), and each person would pay \$2,752 − 20(\$1.75). For a group of 30 people, the price of the cruise would be reduced by 30(\$1.75), and each person would pay \$2,752 − 30(\$1.75).

Assign We let x = the group size necessary for the price of the cruise to be \$2,500 per person.

Form We can form the following equation:

The price of the cruise	is	\$2,752	minus	the number of people in the group	times	\$1.75.
2,500	=	2,752	−	x	·	1.75

Solve

$$2{,}500 = 2{,}752 - 1.75x$$

$$2{,}500 - \mathbf{2{,}752} = 2{,}752 - 1.75x - \mathbf{2{,}752}$$ To isolate the variable term −1.75x, subtract 2,752 from both sides.

$$-252 = -1.75x$$ Do the subtraction.

$$\frac{-252}{\mathbf{-1.75}} = \frac{-1.75x}{\mathbf{-1.75}}$$ To isolate x, divide both sides by −1.75.

$$144 = x$$ Use a calculator to do the division. This is the required size of the group.

State If 144 people travel together, the price will be \$2,500 per person.

Check For 144 people, the cruise cost of \$2,752 will be reduced by 144(\$1.75) = \$252. If we subtract, \$2,752 − \$252 = \$2,500. The result checks.

Success Tip

You may want to review the six steps of the problem-solving strategy in detail. They are listed on page 142.

Teaching Example 9 Advertising. A trucking company had its logo embroidered on the front of baseball caps. The caps are normally \$20 each but will be reduced by \$0.05 per hat for large orders. How large an order is needed for the price per hat to be \$15? *Answer:* 100 hats

Self Check 9 **Broadway Shows.** Tickets to a Broadway show normally sell for \$90 per person. This price is reduced by \$0.50 per person for large groups. How large a group is needed for the price to be \$75 per person? 30 people

Now Try Problem 107

To determine whether to use an equation or an inequality to solve a problem, look for key words and phrases. For example, phrases like *does not exceed, is no more than,* and *is at least* translate into inequalities.

EXAMPLE 10

©DMSU/Shutterstock.com

Communication. A satellite phone is a mobile phone that sends and receives calls using satellites instead of cellular broadcasting towers. Satellite phone calls can be made from anywhere, such as the Sahara desert or the top of Mount Everest. If a satellite telephone company charges callers \$5.50 for the first three minutes and 88¢ for each additional minute, for how many minutes can a call last if the cost is not to exceed \$20?

Analyze We are given the rate at which a call is billed. Since the cost of a call is not to exceed \$20, the cost must be *less than or equal to* \$20. This phrase indicates that we should write an inequality to find how long a call can last.

Assign We will let x = the total number of minutes that a call can last.

Form If the call lasts x minutes, then the cost of a call will be \$5.50 for the first three minutes plus 88¢ times the number of additional minutes, where the number of *additional* minutes is $x - 3$ (the total number of minutes minus the first 3 minutes). With this information, we can form an inequality.

The cost of the first three minutes	plus	the cost of the additional minutes	is not to exceed	\$20.
5.50	+	0.88(x − 3)	≤	20

Solve First, we clear the inequality of decimals.

$$5.50 + 0.88(x - 3) \le 20 \quad \text{Write 88¢ as \$0.88.}$$

$$550 + 88(x - 3) \le 2{,}000 \quad \text{To eliminate the decimals, multiply both sides by 100.}$$

$$550 + 88x - 264 \le 2{,}000 \quad \text{Distribute the multiplication by 88.}$$

$$88x + 286 \le 2{,}000 \quad \text{Combine like terms: } 550 - 264 = 286.$$

$$88x \le 1{,}714 \quad \text{To isolate 88x, subtract 286 from both sides.}$$

$$x \le 19.4772727\ldots \quad \text{To isolate x, divide both sides by 88. Use a calculator.}$$

State Since the phone company doesn't bill for part of a minute, the longest time a call can last is 19 minutes. If a call lasts for $x = 19.4772727\ldots$ minutes, it will be charged as a 20-minute call, and the cost will be \$5.50 + \$0.88(17) = \$20.46.

Check If the call lasts 19 minutes, the cost will be \$5.50 + \$0.88(16) = \$19.58. This is less than \$20. The result checks.

Success Tip

You may want to review the phrases that translate to an inequality. They are listed on page 171.

Self Check 10 **Renting Cars.** Great Value Car Rental charges \$12 a day and \$0.15 per mile to rent a Ford Fusion. Be Thrifty Car Rental's daily charge for the same car is \$15 and \$0.12 per mile. If a businessman wants to rent the car for one day, for what range of miles driven is the Be Thrifty rental plan better? Be Thrifty's plan is better if the car is going to be driven more than 100 miles that day.

Now Try Problem 115

Teaching Example 10 County Fair Tickets. A county fair offers two ticket packages. With Package 1, admission is \$7.50 and carnival ride tickets cost \$1.25 each. With Package 2, admission is \$4.50 and carnival ride tickets cost \$1.75 each. How many carnival ride tickets would someone need to purchase for Package 1 to be the better deal?
Answer: Package 1 is a better deal if more than 6 carnival ride tickets are purchased.

SECTION 8.1 STUDY SET

VOCABULARY

Fill in the blanks.

1. An equation is a statement indicating that two expressions are equal.
2. $2x + 1 = 4$ is an example of a linear equation in one variable.
3. A number that makes an equation true when substituted for the variable is called a solution.
4. If two equations have the same solution set, they are called equivalent equations.
5. An equation that is made true by any permissible replacement value for the variable is called an identity.
6. An equation that is false for all replacement values for the variable is called a contradiction.
7. $<$, $>$, $\le$, and $\ge$ are inequality symbols.
8. To solve an inequality means to find all values of the variable that make the inequality true.

CONCEPTS

Fill in the blanks.

9. **a.** Adding the same number to, or subtracting the same number from, both sides of an equation does not change its solution.
   b. Multiplying or dividing both sides of an equation by the same nonzero number does not change its solution.
10. If we multiply both sides of an inequality by a negative number, the direction of the inequality must be reversed for the inequalities to have the same solutions.
11. Use a check to determine whether −5 is a solution of the following equation and inequality.
    a. $5(2x + 7) = 2x - 4$ No **b.** $3x + 6 \le -9$ Yes

▶ Selected exercises available online at www.webassign.net/brookscole

12. The solution set of a linear inequality in x is graphed on the right. Determine whether a true or false statement results when

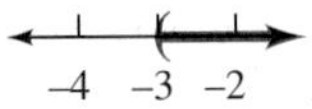

a. -4 is substituted for x. False
b. -3 is substituted for x. False
c. 0 is substituted for x. True

NOTATION

13. Match each interval with its graph.
a. $(-\infty, -1]$ iii b. $(-\infty, 1)$ i c. $[-1, \infty)$ ii

i.  ii.  iii. 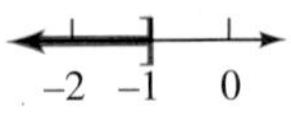

▶ 14. a. Suppose that when solving a linear equation, the variable drops out, and the result is $7 = -1$. What is the solution set?
{ } or $\varnothing$; The equation has no solutions.
b. Suppose that when solving a linear inequality, the variable drops out, and the result is $6 \le 10$. Write the solution set in interval notation and graph it.
$(-\infty, \infty)$; see AIE Appendix 3.

GUIDED PRACTICE

Solve each equation. Check the result. See Example 1.

15. $4x + 1 = 13$ 3
▶ 16. $4x - 8 = 16$ 6
17. $3(x + 1) = 15$ 4
▶ 18. $-2(x + 5) = 30$ -20
19. $2x + 6(2x + 3) = -10$ -2
20. $3(2y - 4) - 6 = 3y$ 6
▶ 21. $7(a + 2) = 11a + 17 - 7a$ 1
22. $5(5 - a) = 4a + 37 - 6a$ -4

Solve each equation. Check the result. See Example 2.

23. $\frac{1}{2}x - 4 = -1 + 2x$ -2
24. $2x + 3 = \frac{2}{3}x - 1$ -3
▶ 25. $\frac{x}{2} - \frac{x}{3} = 4$ 24
26. $\frac{x}{2} + \frac{x}{3} = 10$ 12
27. $\frac{1}{6}(x + 12) + 1 = \frac{x}{3}$ 18
▶ 28. $\frac{3}{2}(y + 4) = \frac{20 - y}{2}$ 2
29. $\frac{1}{5}(x + 6) = \frac{5}{8}(2x - 1) + 2$ $-\frac{1}{6}$
▶ 30. $\frac{2}{3}(x - 3) = \frac{1}{6}(7x + 29) + 3$ $-\frac{59}{3}$

Solve each equation. If an equation is an identity or a contradiction, so indicate. See Example 3.

▶ 31. $2x - 6 = -2x + 4(x - 2)$ No solution, $\varnothing$; contradiction
▶ 32. $-3x = -2x + 1 - (5 + x)$ No solution, $\varnothing$; contradiction
▶ 33. $2y + 1 = 5(0.2y + 1) - (4 - y)$ All real numbers, $\mathbb{R}$; identity
▶ 34. $4(2 - 3t) + 6t = -6t + 8$ All real numbers, $\mathbb{R}$; identity
35. $\frac{7}{2}(y - 1) + \frac{1}{2} = \frac{1}{2}(7y - 6)$ All real numbers, $\mathbb{R}$; identity
▶ 36. $2(x - 3) = \frac{3}{2}(x - 4) + \frac{x}{2}$ All real numbers, $\mathbb{R}$; identity
37. $0.3(x - 4) + 0.6 = -0.2(x + 4) + 0.5x$ No solution, $\varnothing$; contradiction
▶ 38. $0.5(y + 2) + 0.7 - 0.3y = 0.2(y + 9)$ No solution, $\varnothing$; contradiction

Solve each formula for the specified variable. See Examples 4 and 5.

39. $P = 2l + 2w$ for w $w = \frac{P - 2l}{2}$
▶ 40. $P = 2l + 2w$ for l $l = \frac{P - 2w}{2}$
▶ 41. $V = \frac{1}{3}Bh$ for B $B = \frac{3V}{h}$
▶ 42. $A = \frac{1}{2}bh$ for b $b = \frac{2A}{h}$
▶ 43. $T - W = ma$ for W $W = T - ma$
44. $G = U - TS + PV$ for S $S = \frac{U + PV - G}{T}$
45. $z = \frac{x - \mu}{\sigma}$ for x $x = z\sigma + \mu$
46. $P = L + \frac{s}{f}i$ for s $s = \frac{f(P - L)}{i}$
47. $S = \frac{n(a + l)}{2}$ for l $l = \frac{2S - na}{n}$ or $l = \frac{2S}{n} - a$
▶ 48. $h = 48t + \frac{1}{2}at^2$ for a $a = \frac{2h - 96t}{t^2}$ or $a = \frac{2(h - 48t)}{t^2}$
49. $l = a + (n - 1)d$ for d $d = \frac{l - a}{n - 1}$
50. $P = 2(w + h + l)$ for h $h = \frac{P - 2w - 2l}{2}$ or $h = \frac{P}{2} - w - l$

Solve each inequality. Graph the solution set and write it in interval notation. See Example 6. See AIE Appendix 3.

▶ 51. $5x - 3 > 7$ $(2, \infty)$
52. $7x - 9 < 5$ $(-\infty, 2)$
53. $9a + 11 \le 29$ $(-\infty, 2]$
▶ 54. $3b - 26 \ge 4$ $[10, \infty)$
▶ 55. $3(z - 2) \le 2(z + 7)$ $(-\infty, 20]$
▶ 56. $5(3 + z) > -3(z + 3)$ $(-3, \infty)$
57. $2x + 4 + 6x > 2 - 3x + 2$ $(0, \infty)$
▶ 58. $5x + 6 + 2x \ge 2 - x + 4$ $[0, \infty)$

Solve each inequality. Write the solution set in interval notation and then graph it. See Example 7. See AIE Appendix 3.

▶ 59. $-3x - 1 \le 5$ $[-2, \infty)$
60. $-2x + 6 \ge 16$ $(-\infty, -5]$
▶ 61. $-5t + 3 \le 5$ $\left[-\frac{2}{5}, \infty\right)$
62. $-9t + 6 \ge 16$ $\left(-\infty, -\frac{10}{9}\right]$
▶ 63. $-7y + 5 > -5y - 1$ $(-\infty, 3)$
▶ 64. $8 - 9y \ge -y$ $(-\infty, 1]$
▶ 65. $t + 1 - 3t \ge t - 20$ $(-\infty, 7]$
66. $a + 4 - 10a > a - 16$ $(-\infty, 2)$

Solve each inequality. Graph the solution set and write it in interval notation. See Example 8 and the Success Tip in the margin.

▶ 67. $2(5x - 6) > 4x - 15 + 6x$ $(-\infty, \infty)$; $\mathbb{R}$; see AIE Appendix 3.
68. $3(4x - 2) > 14x - 7 - 2x$ $(-\infty, \infty)$; $\mathbb{R}$; see AIE Appendix 3.
▶ 69. $\frac{3b + 7}{3} \le \frac{2b - 9}{2}$ No solution; $\varnothing$
70. $-\frac{5x}{4} > \frac{3 - 5x}{4}$ No solution; $\varnothing$

TRY IT YOURSELF

Solve each equation. If an equation is an identity or a contradiction, so indicate.

71. $2r - 5 = 1 - r$ 2
72. $5s - 13 = s - 1$ 3
73. $0.2(a - 5) - 0.1(3a + 1) = 0$ -11
▶ 74. $0.8(3a - 5) - 0.4(2a + 3) = 1.2$ 4

75. $-\frac{4}{5}s = 2$ $-\frac{5}{2}$

76. $-\frac{9}{8}s = 3$ $-\frac{8}{3}$

77. $\frac{1}{2}(3y + 2) - \frac{5}{8} = \frac{3}{4}y$ $-\frac{1}{2}$

78. $-\frac{3}{4}(4c - 3) + \frac{7}{8}c = \frac{19}{16}$ $\frac{1}{2}$

79. $8x + 3(2 - x) = 5x + 6$ All real numbers, $\mathbb{R}$; identity

80. $2(2a + 1) - 1 = 4a + 1$ All real numbers, $\mathbb{R}$; identity

▶ **81.** $12 + 3(x - 4) - 21 = 5[5 - 4(4 - x)]$ 2

82. $1 + 3[-2 + 6(4 - 2x)] = -(x + 3)$ 2

83. $\frac{3 + p}{3} - 4p = 1 - \frac{p + 7}{2}$ $\frac{21}{19}$

▶ **84.** $\frac{4 - t}{2} - \frac{3t}{5} = 2 + \frac{t + 1}{3}$ $-\frac{10}{43}$

85. $5x + 10 = 5x$ No solution, $\varnothing$; contradiction

86. $4(t - 2) - t = -(9 - 3t)$ No solution, $\varnothing$; contradiction

87. $0.06(a + 200) + 0.1a = 172$ 1,000

88. $0.03x + 0.05(6{,}000 - x) = 280$ 1,000

89. $-4[p - (3 - p)] = 3(6p - 2)$ $\frac{9}{13}$

90. $2[5(4 - a) + 2(a - 1)] = 3 - a$ $\frac{33}{5}$

Solve each inequality. Graph the solution set and write it in interval notation. See AIE Appendix 3.

91. $-3(a + 2) > 2(a + 1)$ $\left(-\infty, -\frac{8}{5}\right)$

92. $-4(y - 1) < y + 8$ $\left(-\frac{4}{5}, \infty\right)$

▶ **93.** $\frac{x - 7}{2} - \frac{x - 1}{5} \le -\frac{x}{4}$ $(-\infty, 6]$

▶ **94.** $\frac{3a + 1}{3} - \frac{4 - 3a}{5} \le -\frac{1}{15}$ $\left(-\infty, \frac{1}{4}\right]$

95. $5(2n + 2) - n > 3n - 3(1 - 2n)$ $(-\infty, \infty)$; $\mathbb{R}$

96. $-1 + 4(y - 1) + 2y \le \frac{1}{2}(12y - 30) + 15$ $(-\infty, \infty)$; $\mathbb{R}$

▶ **97.** $0.4x + 0.4 \le 0.1x + 0.85$ $(-\infty, 1.5]$

▶ **98.** $0.05 + 0.8x \le 0.5x - 0.7$ $(-\infty, -2.5]$

▶ **99.** $\frac{1}{2}y + 2 \ge \frac{1}{3}y - 4$ $[-36, \infty)$

▶ **100.** $\frac{1}{4}x - \frac{1}{3} \le x + 2$ $\left[-\frac{28}{9}, \infty\right)$

▶ **101.** $7 < \frac{5}{3}a - 3$ $(6, \infty)$

102. $5 > \frac{7}{2}a - 9$ $(-\infty, 4)$

Look Alikes . . .

Simplify each expression and solve each equation.

103. a. $\frac{1}{2}(6x + 8) - 10 - \frac{2}{3}(6x - 9)$ $-x$

b. $\frac{1}{2}(6x + 8) - 10 = -\frac{2}{3}(6x - 9)$ $\frac{12}{7}$

▶ **104. a.** $6.31w + 9.22 + 5(7.21w - 1.13)$ $42.36w + 3.57$

b. $6.31w + 9.22 = 5(7.21w - 1.13)$ 0.5

Solve the inequality in part a. Graph the solution set and write it in interval notation. Then use your answer to part a to determine the solution set for the inequality in part b. (No new work is necessary!) Graph the solution set and write it in interval notation. See AIE Appendix 3.

105. a. $12x - 33.16 \le 5.84$ $(-\infty, 3.25]$

b. $12x - 33.16 > 5.84$ $(3.25, \infty)$

▶ **106. a.** $-\frac{3}{4}x > -\frac{21}{32}$ $\left(-\infty, \frac{7}{8}\right)$

b. $-\frac{3}{4}x \le -\frac{21}{32}$ $\left[\frac{7}{8}, \infty\right)$

APPLICATIONS

107. Spring Tours. A group of junior high students will be touring Washington, D.C. Their chaperons will have the $1,810 cost of the tour reduced by $15.50 for each student they supervise. How many students will a chaperon have to supervise so that his or her cost to take the tour will be $1,500? 20

▶ **108. Machining.** Each pass through a lumber plane shaves off 0.015 inch of thickness from a board. How many times must a board, originally 0.875 inch thick, be run through the planer if a board of thickness 0.74 inch is desired? 9

109. Moving Expenses. To help move his furniture, a man rents a truck for $41.50 per day plus 35¢ per mile. If he has budgeted $150 for transportation expenses, how many miles will he be able to drive the truck if the move takes 1 day? 310 mi

110. Computing Salaries. A student working for a delivery company earns $57.50 per day plus $4.75 for each package she delivers. How many deliveries must she make each day to earn $200 a day? 30

▶ **111. Fencing Pens.** A man has 150 feet of fencing to build the two-part pen shown in the illustration. If one end is a square and the other a rectangle, find the outside dimensions of the pen. 20 ft by 45 ft

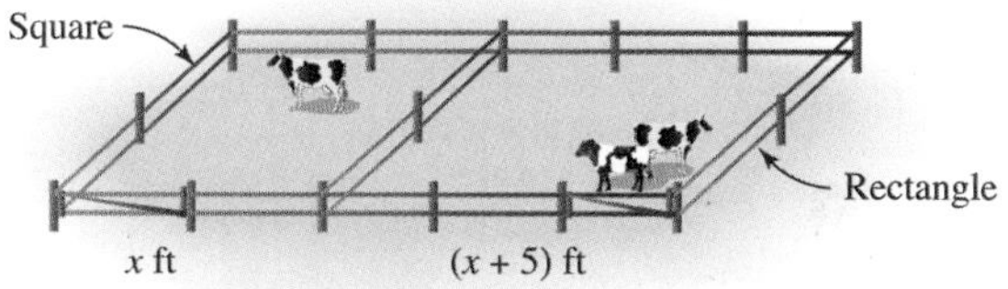

▶ **112. Fencing Pastures.** A farmer has 624 feet of fencing to enclose a pasture. Because a river runs along one side, fencing will be needed on only three sides. Find the dimensions of the pasture if its length is double its width. 156 ft by 312 ft

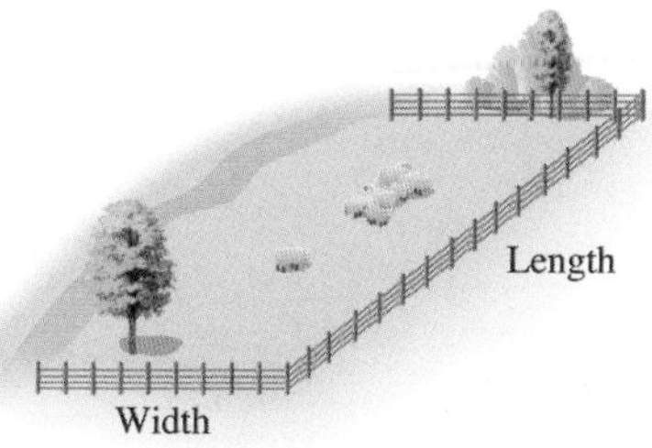

113. ▶ **from Campus to Careers**

Webmaster

©Ronen/Shutterstock.com

One of the most important duties that a webmaster has is to monitor the web traffic of the site that he or she oversees. Suppose a counter for a news organization website has recorded 650,568,999 *views* for April 1 through April 29. How many views does the site need on the last day of the month (April 30) to average 22,000,000 daily visitors? 9,431,001 views

114. Averaging Grades. A student has scores of 70, 77, and 85 on three government exams. What score does she need on a fourth exam to give her an average of 80 or better?
88 or higher

115. Fundraising. A school PTA wants to rent a dunking tank for its annual school fundraising carnival. The cost is \$85.00 for the first 3 hours and then \$19.50 for each additional hour or part thereof. How long can the tank be rented if up to \$185 is budgeted for this expense? 8 hr

116. Work Schedules. A student works two part-time jobs. He earns \$8 an hour for working at the college library and \$15 an hour for construction work. To save time for study, he limits his work to 25 hours a week. If he enjoys the work at the library more, how many hours can he work at the library and still earn at least \$300 a week? 10 hr

117. Scheduling Equipment. An excavating company charges \$300 an hour for the use of a backhoe and \$500 an hour for the use of a bulldozer. (Part of an hour counts as a full hour.) The company employs one operator for 40 hours per week to operate the machinery. If the company wants to bring in at least \$18,500 each week from equipment rental, how many hours per week can it schedule the operator to use a backhoe?
7 hr

118. Video Game Systems. A student who can afford to spend up to \$1,000 sees the ad shown in the illustration. If she decides to buy the video game system, find the greatest number of video games that she also can purchase. (Disregard sales tax.)
11

WRITING

119. Explain why the equation $x = x + 1$ doesn't have a real-number solution.

120. Explain what is wrong with the following statement:

When solving inequalities involving negative numbers, the direction of the inequality symbol must be reversed.

REVIEW

Simplify each expression. Write answers using only positive exponents.

121. $\left(\dfrac{t^3t^5t^{-6}}{t^2t^{-4}}\right)^{-3}$ $\dfrac{1}{t^{12}}$

122. $\left(\dfrac{a^{-2}b^3a^5b^{-2}}{a^6b^{-5}}\right)^{-4}$ $\dfrac{a^{12}}{b^{24}}$

CHALLENGE PROBLEMS

123. Find the value of k that makes 4 a solution of the following linear equation in x. -1

$$k + 3x - 6 = 3kx - k + 16$$

124. Solve: $0.75(x - 5) - \dfrac{4}{5} = \dfrac{1}{6}(3x + 1) + 3.2$ $\dfrac{95}{3}$

125. Consider the following "solution" of the inequality $\frac{1}{3} > \frac{1}{x}$ where it appears that the solution set is the interval $(3, \infty)$.

$$\frac{1}{3} > \frac{1}{x}$$

$$3x\left(\frac{1}{3}\right) > 3x\left(\frac{1}{x}\right)$$

$$x > 3$$

b. Multiplying both sides of an inequality by the same positive number does not change the solutions. In this case, however, both sides were multiplied by $3x$, an expression which, depending on the value of x, could be negative or 0.

a. Show that $x = -1$ makes the original inequality true.

b. If $x = -1$ makes the original inequality true, there must be an error in the "solution." Where is it?

126. Medical Plans. A college provides its employees with a choice of the two medical plans shown in the following table. For what size hospital bills is Plan 2 better for the employee than Plan 1? (*Hint:* The cost to the employee includes both the deductible payment and the employee's coinsurance payment.) Anything over \$900

Plan 1	Plan 2
Employee pays \$100	Employee pays \$200
Plan pays 70% of the rest	Plan pays 80% of the rest

SECTION 8.2

Functions

OBJECTIVES

1 Define relation, domain, and range.
2 Identify functions.
3 Use function notation.
4 Find the domain of a function.
5 Graph linear functions.
6 Write equations of linear functions.
7 Evaluate polynomial functions.

ARE YOU READY? *Are You Ready? exercises available online at www.webassign.net/brookscole*

The following problems review some basic skills that are needed when working with functions.

1. Which of the ordered pairs in the following set have the same x-coordinate: $\{(3, 5), (2, 9), (6, -7), (-1, 5), (3, 0), (1, 9)\}$? $(3, 5), (3, 0)$

2. Substitute 8 for x in $y = \frac{1}{2}x + 3$ and find y. 7

3. What are the slope and the y-intercept of the line described by the equation $y = 3x - 8$? 3, (0, −8)

4. What is the slope of a line perpendicular to the graph of the line that is described by the equation $y = \frac{2}{3}x + 1$? $-\frac{3}{2}$

The concept of a *function* is one of the most important ideas in all of mathematics. To introduce this topic, we will begin with a table that might be seen on television or printed in a newspaper.

1 Define Relation, Domain, and Range.

The following table shows the number of women serving in the U.S. House of Representatives for several recent sessions of Congress.

©Chris Howey/Shutterstock.com

Women in the U.S. House of Representatives

Session of Congress	105th	106th	107th	108th	109th	110th	111th
Number of Women Representatives	54	56	59	59	68	71	78

Source: womenincongress.house.gov

We can display the data in the table as a set of ordered pairs, where the **first component** represents the session of Congress and the **second component** represents the number of women representatives serving during that session:

$$\{(105, 54),\ (106, 56),\ (107, 59),\ (108, 59),\ (109, 68),\ (110, 71),\ (111, 78)\}$$

Sets of ordered pairs like this are called **relations.** The set of all first components is called the **domain of the relation,** and the set of all second components is called the **range of the relation.** A relation may consist of a finite (countable) number of ordered pairs or an infinite (unlimited) number of ordered pairs.

EXAMPLE 1 Find the domain and range of the relation: $\{(3, 2), (5, -7), (-8, 2), (9, 0)\}$

Teaching Tip: Remind your students that the members of a set are written within braces { }.

Strategy We will identify the first components and the second components of the ordered pairs.

Why The set of all first components is the domain of the relation, and the set of all second components is the range.

Solution The first components of the ordered pairs are highlighted in red, and the second components are highlighted in blue: $\{(\mathbf{3}, \mathbf{2}), (\mathbf{5}, \mathbf{-7}), (\mathbf{-8}, \mathbf{2}), (\mathbf{9}, \mathbf{0})\}$. When listing the elements of the domain and range, they are usually written in increasing order, and if a value is repeated, it is listed only once.

The domain of the relation is $\{\mathbf{-8, 3, 5, 9}\}$ and the range of the relation is $\{\mathbf{-7, 0, 2}\}$.

Teaching Example 1
Find the domain and range of the relation:
$\{(4, -9), (0, 5), (-2, 6), (0, 7)\}$
Answer:
D:$\{-2, 0, 4\}$, R:$\{-9, 5, 6, 7\}$

Self Check 1 Find the domain and range of the relation:
$\{(5, 6), (-12, 4), (8, 6), (-6, -6), (5, 4)\}$ D: $\{-12, -6, 5, 8\}$; R: $\{-6, 4, 6\}$

Now Try Problem 19

2 Identify Functions.

The relation in Example 1 was defined by a set of ordered pairs. Relations also can be defined using an **arrow** or **mapping diagram.** The data from the U.S. House of Representatives example is presented on the right in that form. Relations are also often defined using **two-column tables.**

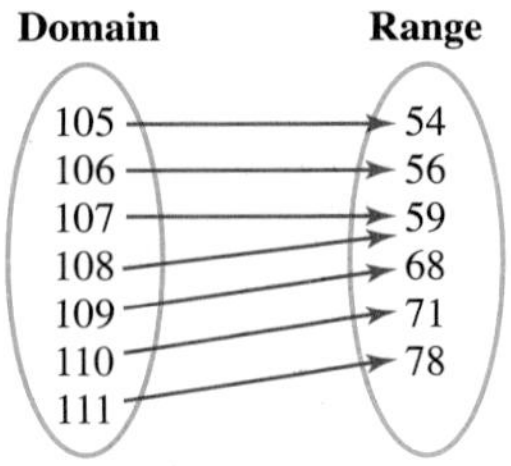

Notice that to each session of Congress, there corresponds exactly one number of women representatives. That is, to each member of the domain, there corresponds exactly one member of the range. Relations that have this characteristic are called *functions.*

Function, Domain, Range

A **function** is a set of ordered pairs (a relation) in which to each first component there corresponds exactly one second component. The set of first components is called the **domain of the function,** and the set of second components is called the **range of the function.**

Since we often will work with sets of ordered pairs of the form (x, y), it is helpful to define a function using the variables x and y.

y Is a Function of x

Given a relation in x and y, if to each value of x in the domain there corresponds exactly one value of y in the range, then y is said to be a function of x.

In the previous definition, since y depends on x, we call x the **independent variable** and y the **dependent variable.** The set of all possible values that can be used for the independent variable is the **domain** of the function, and the set of all values of the dependent variable is the **range** of the function.

EXAMPLE 2 Determine whether the relation defines y to be a function of x.

a.

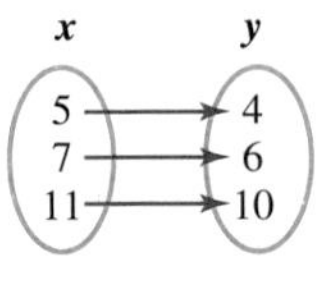

b.

x	y
8	2
1	4
8	3
9	9

c. $\{(-2, 3), (-1, 3), (0, 3), (1, 3)\}$

Teaching Tip: You may want to examine some real-life correspondences to determine whether they are functions:
- The students in your class and the vehicles that they drive
- The states in the union and number of U.S. senators each has
- The players on the L.A. Lakers and the players' uniform numbers

Strategy We will determine whether there is more than one value of y that corresponds to a single value of x.

Why If to any x-value there corresponds more than one y-value, then y is not a function of x.

Solution **a.** The arrow diagram defines a function because to each value of x there corresponds exactly one value of y.

- $5 \rightarrow 4$ To the x-value 5, there corresponds exactly one y-value, 4.
- $7 \rightarrow 6$ To the x-value 7, there corresponds exactly one y-value, 6.
- $11 \rightarrow 10$ To the x-value 11, there corresponds exactly one y-value, 10.

b. The table does not define a function, because to the x-value 8 there corresponds more than one y-value.

- In the first row, to the x-value 8, there corresponds the y-value 2.
- In the third row, to the same x-value 8, there corresponds a different y-value, 3.

When the correspondence in the table is written as a set of ordered pairs, it is apparent that the relation does not define a function:

The same x-value

$\{(\mathbf{8}, \mathbf{2}), (1, 4), (\mathbf{8}, \mathbf{3}), (9, 9)\}$ This is not a function.

Different y-values

c. Since to each value of x, there corresponds exactly one value of y, the set of ordered pairs defines y to be a function of x.

- $(-2, 3)$ To the x-value -2, there corresponds exactly one y-value, 3.
- $(-1, 3)$ To the x-value -1, there corresponds exactly one y-value, 3.
- $(0, 3)$ To the x-value 0, there corresponds exactly one y-value, 3.
- $(1, 3)$ To the x-value 1, there corresponds exactly one y-value, 3.

In this case, the same y-value, 3, corresponds to each x-value.

The results from parts (b) and (c) illustrate an important fact: *Two different ordered pairs of a function can have the same y-value, but they cannot have the same x-value.*

Success Tip

Every function is, by definition, a relation. However, not every relation is a function, as we see in part (b).

Teaching Example 2
Determine whether the relation defines y to be a function of x.

a.

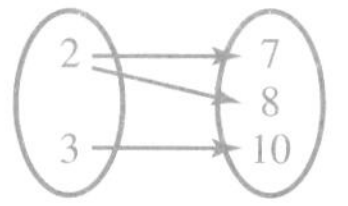

b.

x	y
5	3
2	9
7	9

c. $\{(-3, 4), (5, 12), (-3, 6)\}$

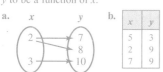

Answers: **a.** No; (2, 7), (2, 8) **b.** Yes **c.** No; $(-3, 4), (-3, 6)$

Self Check 2 Determine whether the relation defines y to be a function of x.

a.

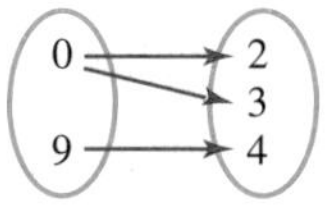

No; (0, 2), (0, 3)

b.

x	y
-1	-60
0	55
3	0

Yes

c. $\{(4, -1), (9, 2), (16, 15), (4, 4)\}$ No; $(4, -1), (4, 4)$

d. $\{(9,5), (10,5), (11,5), (12,5)\}$ Yes

Now Try Problems 23, 27, and 31

A function also can be defined by an equation. For example, $y = \frac{1}{2}x + 3$ sets up a rule in which to each value of x there corresponds exactly one value of y. To find the y-value (called an **output**) that corresponds to the x-value 4 (called an **input**), we substitute 4 for x and evaluate the right side of the equation.

$$y = \frac{1}{2}x + 3$$

$$y = \frac{1}{2}(4) + 3 \quad \text{Substitute 4 for x. The input is 4.}$$

$$y = 2 + 3 \quad \text{Do the multiplication: } \frac{1}{2}(4) = 2.$$

$$y = 5 \quad \text{Do the addition. This is the output.}$$

For the function defined by $y = \frac{1}{2}x + 3$, a y-value of 5 corresponds to an x-value of 4.

Not all equations define functions, as we will see in the next example.

The Language of Algebra

We also can think of a function as a rule or correspondence that **assigns** exactly one range value to each domain value.

EXAMPLE 3 Determine whether each equation defines y to be a function of x.

a. $y = 2x - 5$ **b.** $y^2 = x$

Strategy In each case, we will determine whether there is more than one value of y that corresponds to a single value of x.

Why If to any x-value there corresponds more than one y-value, then y is not a function of x.

Solution

a. To find the output value y that corresponds to an input value x, we multiply x by 2 and subtract 5. Since this arithmetic gives one result, to each value of x there corresponds exactly one value of y. Thus, $y = 2x - 5$ defines y to be a function of x.

b. The equation $y^2 = x$ does not define y to be a function of x, because more than one value of y corresponds to a single value of x. For example, if x is 16, then the equation becomes $y^2 = 16$ and y can be either 4 or -4. This is because $4^2 = 16$ and $(-4)^2 = 16$.

x	y
16	4
16	-4

Teaching Example 3
Determine whether each equation defines y to be a function of x.
a. $y = -6x + 1$ b. $|y| = x$
Answers: a. Yes
b. No; (1, 1), (1, −1)

Self Check 3 Determine whether each equation defines y to be a function of x.

a. $y = -x + 1$ Yes **b.** $\left|\frac{1}{2}y\right| = x$ No; (1, 2), (1, −2)

Now Try Problems 35 and 39

3 Use Function Notation.

A special notation is used to name functions that are defined by equations.

Function Notation The notation $y = f(x)$ indicates that y is a function of x.

If y is a function of x, the symbols y and $f(x)$ are interchangeable. In Example 3a, we saw that $y = 2x - 5$ defines y to be a function of x. To write this equation using function notation, we replace y with $f(x)$ to get $f(x) = 2x - 5$. This is read as "f of x is equal to $2x$ minus 5." In this notation, the parentheses do not indicate multiplication.

Caution

The symbol $f(x)$ denotes a function. It does not mean f times x.

This variable represents the input.

$$f(x) = 2x - 5$$

This is the name of the function.

This expression shows how to obtain an output from a given input.

Function notation provides a compact way of representing the output value that corresponds to some input value. For example, if $f(x) = 2x - 5$, the value that corresponds to $x = 6$ is represented by $f(6)$.

$f(x) = 2x - 5$

$f(6) = 2(6) - 5$ Substitute 6 for each x. (The input is 6.)

$= 12 - 5$ Do the multiplication.

$= 7$ Do the subtraction. This is the output.

The Language of Algebra

Another way to read $f(6) = 7$ is to say "the value of f at 6 is 7."

Thus, $f(6) = 7$.

We read this result as "f of 6 equals 7." It means that when $x = 6$, $f(x)$ or y is 7. The output 7 is called a **function value.** The input, 6, and output, 7, also can be written as an ordered pair of the form (6, 7).

To see why function notation is useful, we consider two sentences that ask you to do the same thing:

1. If $y = 2x - 5$, find the value of y when x is 6.
2. If $f(x) = 2x - 5$, find $f(6)$.

Statement 2, which uses $f(x)$ notation, is more compact.

The letter f used in the notation $f(x)$ represents the word *function.* However, other letters are often used to name functions. For example, in $g(x) = -5x + 15$, the name of the function is g and in $h(x) = x^3 - 7x + 9$, the name of the function is h.

Letters other than x can be used to represent the input of a function. Examples are $f(a) = 4a$, $g(t) = t^2 - 2t$, and $h(n) = -\frac{n^2 + 2}{2}$.

EXAMPLE 4 Let $f(x) = 4x + 3$ and $g(t) = t^2 - 2t$. Find: **a.** $f(3)$ **b.** $f(-1)$ **c.** $f(r + 1)$ **d.** $g(-2.4)$

Strategy We will substitute 3, -1, and $r + 1$ for each x in $f(x) = 4x + 3$ and evaluate the right side. We will substitute -2.4 for each t in $g(t) = t^2 - 2t$ and evaluate the right side.

Why Whatever appears within the parentheses in $f(\ )$ is to be substituted for each x in $f(x) = 4x + 3$. Whatever appears within the parentheses in $g(\ )$ is to be substituted for each t in $g(t) = t^2 - 2t$.

Solution

a. To find $f(3)$, we replace each x with 3.

$$\begin{aligned} f(x) &= 4x + 3 \\ f(\mathbf{3}) &= 4(\mathbf{3}) + 3 \\ &= 12 + 3 \\ &= 15 \end{aligned}$$

Thus, $f(3) = 15$ and the corresponding ordered pair is (3, 15).

b. To find $f(-1)$, we replace each x with -1:

$$\begin{aligned} f(x) &= 4x + 3 \\ f(\mathbf{-1}) &= 4(\mathbf{-1}) + 3 \\ &= -4 + 3 \\ &= -1 \end{aligned}$$

Thus, $f(-1) = -1$ and the corresponding ordered pair is $(-1, -1)$.

Notation

Note how function notation and ordered pair notation are related:

$$\begin{array}{cc} f(3) = & 15 \\ \uparrow & \uparrow \\ x & y \\ \downarrow & \downarrow \\ (3, & 15) \end{array}$$

c. To find $f(r + 1)$, we replace each x with $r + 1$:

$$\begin{aligned} f(x) &= 4x + 3 \\ f(\mathbf{r + 1}) &= 4(\mathbf{r + 1}) + 3 \\ &= 4r + 4 + 3 \\ &= 4r + 7 \end{aligned}$$

d. To find $g(-2.4)$, we replace each t with -2.4:

$$\begin{aligned} g(t) &= t^2 - 2t \qquad \text{Read as "}g\text{ of }t\text{."} \\ g(\mathbf{-2.4}) &= (\mathbf{-2.4})^2 - 2(\mathbf{-2.4}) \\ &= 5.76 + 4.8 \\ &= 10.56 \end{aligned}$$

The Language of Algebra

In part (d), if we use $g(t) = t^2 - 2t$ or $g(x) = x^2 - 2x$ or $g(a) = a^2 - 2a$, the results are the same: $g(-2.4) = 10.56$. The independent variable is really just a place holder, and for this reason, the letter that is used is often referred to as a **dummy variable.**

Self Check 4 If $f(x) = -2x - 1$ and $h(n) = -\frac{n^2 + 2n}{2}$, find: **a.** $f(2)$ -5 **b.** $f(-3)$ 5 **c.** $f(-t)$ $2t - 1$ **d.** $h(-0.6)$ 0.42

Now Try Problems 47, 67, and 71

Teaching Example 4
If $f(x) = -3x - 4$ and $s(t) = -t^2 - 3t$, find:
a. $f(0)$ **b.** $f(-3)$ **c.** $f(t - 2)$ **d.** $s(-10)$
Answers: **a.** -4 **b.** 5 **c.** $-3t + 2$ **d.** -70

In the next example, we are asked to find the input of a function when we are given the corresponding output.

EXAMPLE 5 Let $f(x) = 5x + 4$. For what value of x is $f(x) = -26$?

Strategy We will substitute -26 for $f(x)$ and solve for x.

Why In the equation, there are two unknowns, x and $f(x)$. If we replace $f(x)$ with -26, we can use our equation-solving skills to find x.

Teaching Tip: Point out that in Example 4, we were given an input and asked to find the corresponding output. In Example 5, it is just the reverse.

Solution

$f(x) = 5x + 4$ This is the given function.

$-26 = 5x + 4$ Substitute -26 for $f(x)$.

$-30 = 5x$ To isolate the variable term $5x$, subtract 4 from both sides.

$-6 = x$ To isolate x, divide both sides by 5.

We have found that $f(x) = -26$ when $x = -6$. To check this result, we can substitute -6 for x and verify that $f(-6) = -26$.

$f(x) = 5x + 4$ This is the given function.

$f(-6) = 5(-6) + 4$ Substitute -6 for x.

$= -30 + 4$ Do the multiplication.

$= -26$ This is the desired output.

Teaching Example 5
Let $f(x) = 16x - 20$. For what value of x is $f(x) = 76$?
Answer: 6

Self Check 5 Let $f(x) = 7x + 200$. For what value of x is $f(x) = 11$? -27

Now Try Problem 75

4 Find the Domain of a Function.

We can think of a function as a machine that takes some input x and turns it into some output $f(x)$, as shown in figure (a). The machine shown in figure (b) turns the input -6 into the output -11. The set of numbers that we put into the machine is the domain of the function, and the set of numbers that comes out is the range.

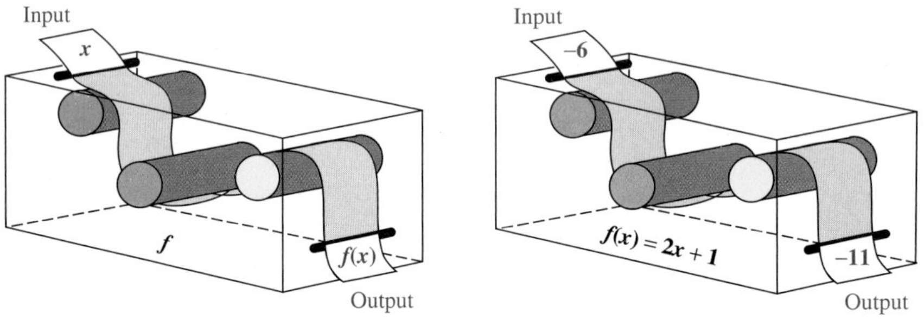

(a) (b)

Success Tip

To find the domain of a function defined by an equation, we must identify all the values of the input variable that produce a real-number output.

EXAMPLE 6 Find the domain of each function: **a.** $f(x) = 3x + 1$ **b.** $f(x) = \dfrac{x}{3x - 6}$

Strategy We will ask, "What values of x are permissible replacements for x in $3x + 1$ and $\dfrac{x}{3x - 6}$?"

Why These values of x form the domain of the function.

Solution

a. We will be able to evaluate $3x + 1$ for any value of x that is a real number. Thus, the domain of the function is *the set of real numbers,* which can be represented by the symbol $\mathbb{R}$.

b. Since division by 0 is undefined, we will not be able to evaluate $\frac{x}{3x - 6}$ for any number x that makes the denominator equal to 0. To find such x-values, we set the denominator equal to 0 and solve for x.

$3x - 6 = 0$

$3x = 6$ To isolate the variable term $3x$, add 6 to both sides.

$x = 2$ To isolate the variable, divide both sides by 3.

We have found that 2 must be excluded from the domain of the function because it makes the denominator 0. However, all other real numbers are permissible replacements for x. Thus, the domain of the function *is the set of all real numbers except* 2.

The Language of Algebra

The last two letters in the word *domain* help us remember that it is the set of all *inputs* of a function.

Success Tip

Notice that to determine the domain of $f(x) = \frac{x}{3x - 6}$, we don't have to examine the numerator. It can be any value, including 0.

Teaching Example 6
Find the domain of each function:
a. $f(x) = -5x + 3$
b. $f(x) = \dfrac{x}{4x + 36}$
Answers: a. The set of real numbers
b. The set of all real numbers except -9

Self Check 6 Find the domain of each function: **a.** $f(x) = 2x - 6$ The set of real numbers

b. $f(x) = \dfrac{2}{-4x - 12}$ The set of all real numbers except -3

Now Try Problem 79

5 Graph Linear Functions.

In this course, we will study several important "families" of functions. We will begin here with the most basic family, *linear functions.*

Linear Functions

A **linear function** is a function that can be defined by an equation of the form $f(x) = mx + b$, where m and b are real numbers. The graph of a linear function is a straight line with slope m and y-intercept $(0, b)$.

Some examples of linear functions are:

$$f(x) = \frac{1}{2}x + 3, \qquad f(x) = -6x - 10, \qquad \text{and} \qquad f(x) = x$$

The input-output pairs that a linear function such as $f(x) = \frac{1}{2}x + 3$ generates can be plotted on a rectangular coordinate system to get the **graph of the function.** Since the symbols y and $f(x)$ are interchangeable, we can graph $f(x) = \frac{1}{2}x + 3$ as we would $y = \frac{1}{2}x + 3$, using the methods of Sections 3.2, 3.3, and 3.5.

To use the **point-plotting method** to graph $f(x) = \frac{1}{2}x + 3$, we begin by constructing a table of function values. To make the table, we select several values for x, and find the corresponding values of $f(x)$. Then we plot the ordered pairs and draw a straight line through the points to get the graph of the function, as shown on the next page.

To use the **slope–intercept method** to graph $f(x) = \frac{1}{2}x + 3$, we begin by identifying m and b in the equation. Then we plot the y-intercept and use the slope to determine a second point on the line. Finally, we draw a line through the points to obtain a graph like that shown on the next page.

$$f(x) = \frac{1}{2}x + 3$$

Slope ↑ (under $\frac{1}{2}$) y-intercept: $(0, 3)$ ↑ (under 3)

To use the **intercept method** to graph $f(x) = \frac{1}{2}x + 3$, we begin by identifying b in the equation to determine the y-intercept. Then we let y, which in this case is $f(x)$, equal 0 and solve for x to find the x-intercept.

Find the y-intercept: Identify b

$$f(x) = \frac{1}{2}x + 3$$

The y-intercept is $(0, 3)$.

Find the x-intercept: Let f(x) = 0

$$f(x) = \frac{1}{2}x + 3$$
$$0 = \frac{1}{2}x + 3$$
$$-3 = \frac{1}{2}x$$
$$-6 = x$$

The x-intercept is $(-6, 0)$.

When we plot the intercepts $(0, 3)$ and $(-6, 0)$ and draw a straight line through them, we obtain a graph like that on the next page.

Notation

A table of function values is similar to a table of solutions, except that the second column is usually labeled $f(x)$ instead of y.

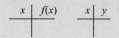

x	$f(x)$

x	y

$f(x) = \frac{1}{2}x + 3$

x	$f(x)$	
-2	2	→ $(-2, 2)$
0	3	→ $(0, 3)$
2	4	→ $(2, 4)$

↑ Select x. ↑ Find f(x). ↑ Plot.

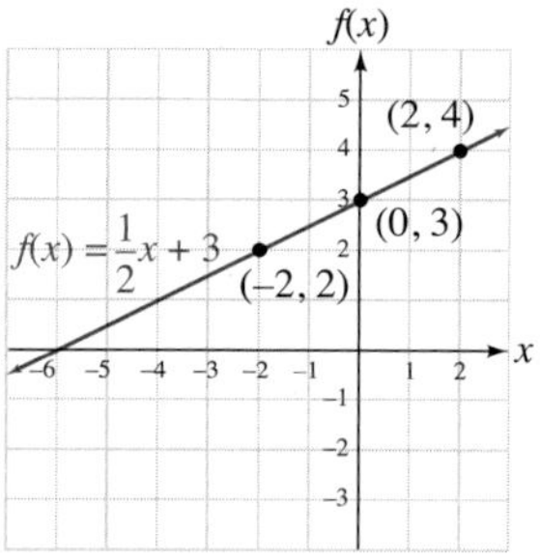

The vertical axis can be labeled $f(x)$ or y.

The most basic linear function is $f(x) = x$. It is called the **identity function** because it assigns each real number to itself. The graph of the identity function is a line with slope 1 and y-intercept (0, 0), as shown below in figure (a).

A linear function defined by $f(x) = b$ is called a **constant function,** because for any input x, the output is the constant b. The graph of a constant function is a horizontal line. The graph of $f(x) = 2$ is shown below in figure (b).

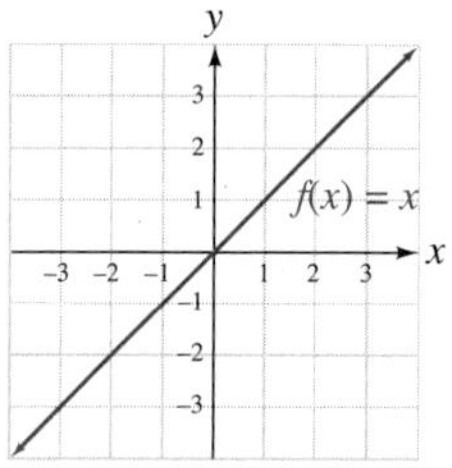

The identity function

(a)

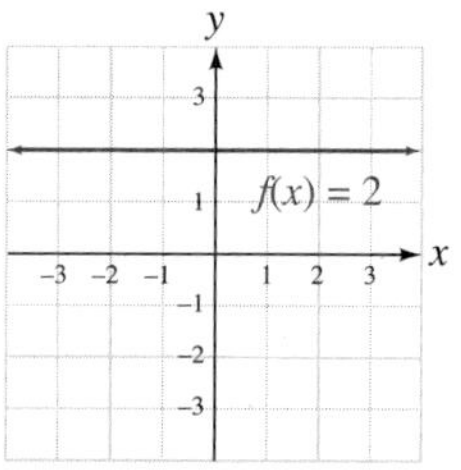

A constant function

(b)

6 Write Equations of Linear Functions.

The slope–intercept and point–slope forms for the equation of a line that we studied in Sections 3.5 and 3.6 can be adapted to write equations of linear functions.

EXAMPLE 7

a. Write an equation for the linear function whose graph has slope $-\frac{3}{4}$ and y-intercept (0, 2).

b. Write an equation for the linear function whose graph passes through (6, 4) with slope 5.

c. Write an equation for the linear function whose graph passes through $(-16, -12)$ and is perpendicular to the graph of $g(x) = -8x - 1$.

Strategy We will use either the slope–intercept or point–slope form to write each equation.

Why Writing equations of linear functions is similar to writing equations of linear equations in two variables, except that we replace y with $f(x)$.

Solution **a.** If the slope is $-\frac{3}{4}$ and the y-intercept is (0, 2), then $m = -\frac{3}{4}$ and $b = 2$.

$f(x) = mx + b$ — This is the slope–intercept form in which y is replaced with f(x).

$f(x) = -\frac{3}{4}x + 2$ — Substitute $-\frac{3}{4}$ for m and 2 for b.

b. If the slope is 5 and the graph passes through (6, 4), then $m = 5$ and $(x_1, y_1) = (6, 4)$.

$f(x) - y_1 = m(x - x_1)$ — This is the point–slope form in which y is replaced with f(x).

$f(x) - 4 = 5(x - 6)$ — Substitute 4 for y_1, 5 for m, and 6 for x_1.

$f(x) - 4 = 5x - 30$ — Distribute the multiplication by 5.

$f(x) = 5x - 26$ — To isolate f(x), add 4 to both sides.

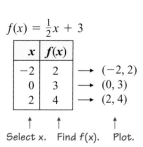

c. The slope of the line represented by $g(x) = -8x - 1$ is the coefficient of x: -8. Since the desired function is to have a graph that is perpendicular to the graph of $g(x) = -8x - 1$, its slope must be the negative reciprocal of -8, which is $\frac{1}{8}$.

$$f(x) - y_1 = m(x - x_1)$$ This is the point-slope form in which y is replaced with $f(x)$.

$$f(x) - (-12) = \frac{1}{8}[x - (-16)]$$ Substitute -12 for y_1, $\frac{1}{8}$ for m, and -16 for x_1.

$$f(x) + 12 = \frac{1}{8}(x + 16)$$ Simplify each side.

$$f(x) + 12 = \frac{1}{8}x + 2$$ Distribute the multiplication by $\frac{1}{8}$: $\frac{1}{8} \cdot 16 = 2$.

$$f(x) = \frac{1}{8}x - 10$$ To isolate $f(x)$, subtract 12 from both sides.

Teaching Example 7
a. Write an equation for the linear function whose graph has slope $-\frac{7}{8}$ and y-intercept (0, 14).
b. Write an equation for the linear function whose graph passes through (3, 1) with slope 9.
c. Write an equation for the linear function whose graph passes through $(-2, -11)$ and is perpendicular to the graph of $g(x) = -2x - 15$.
Answers: a. $f(x) = -\frac{7}{8}x + 14$
b. $f(x) = 9x - 26$
c. $f(x) = \frac{1}{2}x - 10$

Self Check 7 **a.** Write an equation for the linear function whose graph has slope $-\frac{1}{5}$ and y-intercept (0, 9). $f(x) = -\frac{1}{5}x + 9$ **b.** Write an equation for the linear function whose graph passes through (7, 1) with slope 2. $f(x) = 2x - 13$ **c.** Write an equation for the linear function whose graph passes through $(-20, -4)$ and is perpendicular to the graph of $g(x) = -10x$. $f(x) = \frac{1}{10}x - 2$

Now Try Problems 95, 97, and 103

EXAMPLE 8

The Language of Algebra

The phrase **constant rate** indicates that a linear function should be used to model the situation.

Alzheimer's Disease. The graph on the right is from a recent Alzheimer's Association report. It shows how the number of people in the United States with Alzheimer's disease is expected to increase steadily at a constant rate. The report estimated there were 4,800,000 people with Alzheimer's in 2005 and 5,100,000 with the disease in 2010. (Source: Medill Reports)
a. Write the linear function that models this situation. **b.** Use the function to estimate the number of people who will have Alzheimer's disease in 2025, assuming the trend continues.

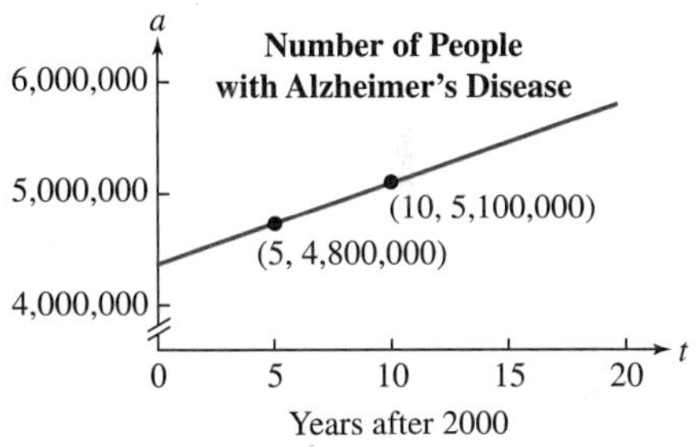

Strategy We will use the point–slope form to write a linear function.

Why We are given two points that lie on the graph of the function.

Solution Let a represent the approximate number of people in the U.S. with Alzheimer's disease and t represent the number of years after 2000. Two ordered pairs of the form (t, a) that lie on the graph of the function are:

- (5, 4,800,000) Since the year 2005 is 5 years after 2000, $t = 5$.
- (10, 5,100,00) Since the year 2010 is 10 years after 2000, $t = 10$.

Since we know two points that lie on the graph, we can write its equation. First, we find the slope of the line. If we write the slope formula using the variables t and a, and let $(t_1, a_1) =$ (5, 4,800,000) and $(t_2, a_2) =$ (10, 5,100,000), we have:

$$m = \frac{a_2 - a_1}{t_2 - t_1} = \frac{5{,}100{,}000 - 4{,}800{,}000}{10 - 5} = \frac{300{,}000}{5} = 60{,}000$$

The result indicates the number of people with Alzheimer's disease will increase at a rate of 60,000 per year.

a. To find the linear function, we substitute 60,000 for m, 5 for t_1, and 4,800,000 for a_1 in the point–slope form and simplify.

$$a(t) - a_1 = m(t - t_1)$$ This is the point–slope form using the variable t and replacing a with $a(t)$.

$$a(t) - 4{,}800{,}000 = 60{,}000(t - 5)$$ Substitute for m, t_1, and a_1.

$$a(t) - 4{,}800{,}000 = 60{,}000t - 300{,}000$$ Distribute the multiplication by 60,000.

$$a(t) = 60{,}000t + 4{,}500{,}000$$ To isolate $a(t)$, add 4,800,000 to both sides.

The approximate number of people with Alzheimer's disease t years after 2000 is given by $a(t) = 60{,}000t + 4{,}500{,}000$.

b. To estimate the number of people who will have Alzheimer's disease in 2025, which is 25 years after 2000, we find $a(25)$.

$$a(t) = 60{,}000t + 4{,}500{,}000$$ This is the linear function model.

$$a(25) = 60{,}000(25) + 4{,}500{,}000$$ Substitute 25 for t.

$$= 1{,}500{,}000 + 4{,}500{,}000$$ Do the multiplication.

$$= 6{,}000{,}000$$ Do the addition.

In 2025, there will be approximately 6,000,000 people in the U.S. with Alzheimer's disease.

Teaching Example 8 Fires. The dollar amount of fire damage from residential building fires in the United States can be modeled by a linear function. In 2004, there was \$6.4 billion dollars in damage. In 2008, there was \$8 billion dollars in damage. (Source: U.S. Fire Administration)
a. Let t be the number of years after 2003 and D be the dollar amount of fire damage (in billions). Write a linear function $D(t)$ to model the situation.
b. Predict the dollar amount of fire damage in 2020, if the trend continues.
Answers: **a.** $D(t) = 0.4t + 6.0$
b. \$12.8 billion

Self Check 8 **Energy.** The world's annual energy consumption can be modeled by a linear function. In 2004, the world consumed about 430 quadrillion Btu. By the year 2006, that number had increased to about 446 quadrillion Btu. (Source: Energy Information Administration)
a. Let t be the number of years after 2000 and E be the amount of energy (in quadrillion Btu). Write a linear function $E(t)$ to model the situation. $E(t) = 8t + 398$ **b.** Predict the world's energy consumption in 2030, if the trend continues. 638 quadrillion Btu

Now Try ▶ Problem 111

Using Your Calculator ▶ Evaluating Functions

We can use a graphing calculator to find function values. For example, suppose the linear function $f(c) = 12c - 18$ gives the daily income earned by a cosmetologist from serving c customers.

To find the income she earns for different numbers of customers, we first graph the income function $f(c) = 12c - 18$ as $y = 12x - 18$, using window settings of [0, 10] for x and [0, 100] for y to obtain figure (a). To find her income when she serves seven customers, we trace and move the cursor until the x-coordinate on the screen is nearly 7, as in figure (b). From the screen, we see that her income is about \$66.26.

To find her income when she serves nine customers, we trace and move the cursor until the x-coordinate is nearly 9, as in figure (c). From the screen, we see that her income is about \$90.51.

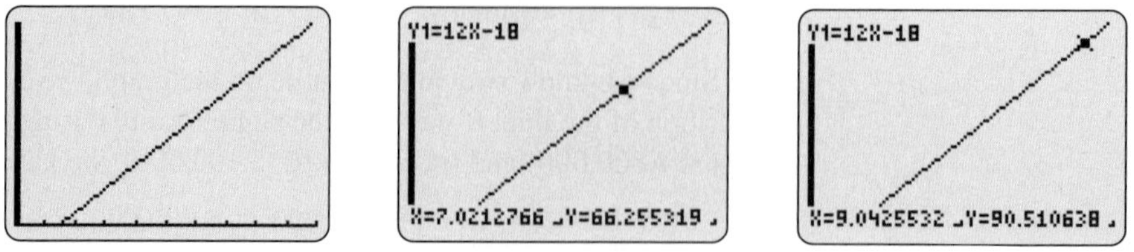

(a) (b) (c)

With some graphing calculator models, we can evaluate a function by entering function notation. To find $f(15)$, the income earned by the cosmetologist if she serves 15 customers, we use the following steps on a TI-84 Plus calculator.

With $f(c) = 12c - 18$ entered as $Y_1 = 12x - 18$, we call up the home screen by pressing 2nd QUIT. Then we enter VARS ▶ 1 ENTER. The symbolism Y_1 will be displayed. See figure (a). Next, we enter the input value 15 within parentheses, as shown in figure (b), and press ENTER. In figure (c) we see that $Y_1(15) = 162$. That is, $f(15) = 162$. The cosmetologist will earn \$162 if she serves 15 customers in one day.

(a) (b) (c)

7 Evaluate Polynomial Functions.

We have seen that linear functions are defined by equations of the form $f(x) = mx + b$. Some examples of linear functions are

$$f(x) = 8x - 10, \qquad g(x) = -\frac{1}{2}x + 1, \qquad \text{and} \qquad h(x) = 5x$$

In each case, the right side of the equation is a polynomial. For this reason, linear functions are members of a larger class of functions known as *polynomial functions.*

Polynomial Functions

A **polynomial function** is a function whose equation is defined by a polynomial in one variable.

Success Tip

Recall that a polynomial is a single term or the sum of terms in which all variables have whole-number exponents. No variable appears in the denominator.

Another example of a polynomial function is $f(x) = x^2 + 6x - 8$. This is a second-degree polynomial function, called a **quadratic function.** Quadratic functions are of the form $f(x) = ax^2 + bx + c$, where $a \neq 0$.

An example of a third-degree polynomial function is $f(x) = x^3 - 3x^2 - 9x + 2$. Third-degree polynomial functions, also called **cubic functions,** are of the form $f(x) = ax^3 + bx^2 + cx + d$, where $a \neq 0$.

Polynomial functions can be used to model many real-life situations. If we are given a polynomial function model, we can learn more about the situation by evaluating the function at specific values. To **evaluate a polynomial function** at a specific value, we replace the variable in the defining equation with that value, called the *input.* Then we simplify the resulting expression to find the *output.*

EXAMPLE 9

Packaging. To make boxes, a manufacturer cuts equal-sized squares from each corner of 10 in. × 12 in. pieces of cardboard and then folds up the sides. The polynomial function $f(x) = 4x^3 - 44x^2 + 120x$ gives the volume (in cubic inches) of the resulting box when a square with sides x inches long is cut from each corner. Find the volume of a box if 3-inch squares are cut out.

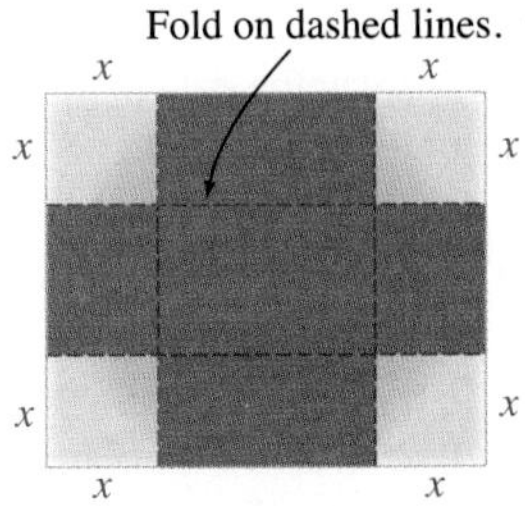

Strategy We will find $f(3)$.

Why The notation $f(3)$ represents the volume of the box when 3-inch squares are cut out of the corners of the piece of cardboard.

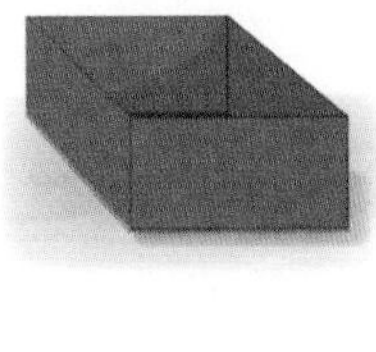

Solution

$f(x) = 4x^3 - 44x^2 + 120x$	This is the given function.
$f(3) = 4(3)^3 - 44(3)^2 + 120(3)$	Substitute 3 for each x. (The input is 3.)
$= 4(27) - 44(9) + 120(3)$	Evaluate the exponential expressions.
$= 108 - 396 + 360$	Do the multiplication.
$= 72$	The output is 72.

If 3-inch squares are cut out of the corners, the resulting box has a volume of 72 in.^{3}.

Teaching Example 9 In Example 9, find the volume of a box if 4-inch squares are cut out.
Answer: 32 in.^{3}

Self Check 9 In Example 9, find the volume of a box if 2-inch squares are cut out.
96 in.^{3}

Now Try Problem 119

SECTION 8.2 STUDY SET

VOCABULARY

Fill in the blanks.

▶ **1.** A set of ordered pairs is called a relation. The set of all first components of the ordered pairs is called the domain and the set of all second components is called the range.

▶ **2.** A function is a set of ordered pairs (a relation) in which to each first component there corresponds exactly one second component.

▶ **3.** Given a relation in x and y, if to each value of x in the domain there corresponds exactly one value of y in the range, y is said to be a function of x. We call x the independent variable and y the dependent variable.

▶ **4.** For a function, the set of all possible values that can be used for the independent variable is called the domain. The set of all values of the dependent variable is called the range.

▶ **5.** A linear function is a function that can be defined by an equation of the form $f(x) = mx + b$. A polynomial function is a function whose equation is defined by a polynomial in one variable.

▶ **6.** We call $f(x) = x$ the identity function because it assigns each real number to itself. We call $f(x) = 2$ a constant function, because for any input x, the output is always 2.

CONCEPTS

7. U.S. Recycling. The following table gives the approximate number of aluminum cans (in billions) collected each year for the years 2000–2006.

a. Display the data in the table as a relation, that is, as a set of ordered pairs. {(2000, 63), (2001, 56), (2002, 54), (2003, 50), (2004, 52), (2005, 51), (2006, 51)}

b. Find the domain and range of the relation.
D: {2000, 2001, 2002, 2003, 2004, 2005, 2006};
R: {50, 51, 52, 54, 56, 63}

c. Use an arrow diagram to show how members of the range correspond to members of the domain.
See AIE Appendix 3.

Year	2000	2001	2002	2003	2004	2005	2006
Billions of aluminum cans	63	56	54	50	52	51	51

Source: Aluminum Association of America, U.S. Dept. of Commerce

8. For the given input, what value will the function machine output? -120

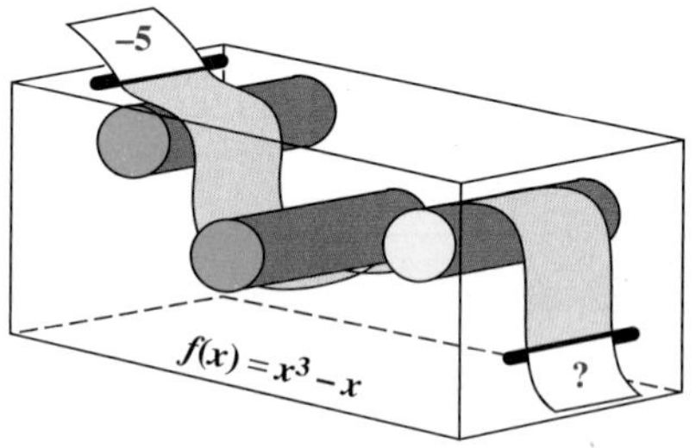

9. Explain why -4 is not in the domain of $f(x) = \frac{1}{x+4}$.
If $x = -4$, the denominator of $\frac{1}{x+4}$ is 0 and the fraction is undefined.

10. Consider the linear function $y = -\frac{4}{5}x + 3$.
a. What is the slope of its graph? $-\frac{4}{5}$
b. What is the y-intercept of its graph? $(0, 3)$

11. Consider the linear function $f(x) = -6x - 4$.
a. What is the y-intercept of its graph? $(0, -4)$
b. What is the x-intercept of its graph? $\left(-\frac{2}{3}, 0\right)$

12. The graphing calculator display shows a table of values for a function f. Fill in the blanks:

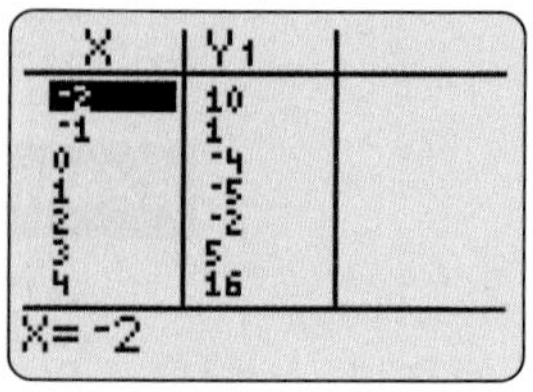

$f(-1) = 1$ $\quad f(3) = 5$

NOTATION

Fill in the blanks.

13. a. We read $f(x) = 5x - 6$ as "f of x is equal to $5x$ minus 6."
b. We read $g(t) = t + 9$ as "g of t is equal to t plus 9."

▶ Selected exercises available online at www.webassign.net/brookscole

14. This variable represents the input.

$$f(x) = 2x - 5$$

This is the name of the function. Use this expression to find the output.

15. Fill in the blank so that the following statements are equivalent:

- If $y = 5x + 1$, find the value of y when $x = 8$.
- If $f(x) = 5x + 1$, find $f(8)$.

▶ **16.** If $f(2) = 7$, the input 2 and the output 7 can be written as the ordered pair (2, 7).

17. When graphing $f(x) = -x + 5$, the vertical axis of the rectangular coordinate system can be labeled $f(x)$ or y.

18. To write the slope–intercept form $y = mx + b$ and the point–slope form $y - y_1 = m(x - x_1)$ using function notation, we simply replace y with $f(x)$.

GUIDED PRACTICE

Find the domain and range of each relation. **See Example 1.**

19. $\{(7, -1), (-1, -11), (-5, 3), (8, -6)\}$ D: $\{-5, -1, 7, 8\}$; R: $\{-11, -6, -1, 3\}$

▶ **20.** $\{(15, -3), (0, 0), (4, 6), (-3, -8)\}$ D: $\{-3, 0, 4, 15\}$; R: $\{-8, -3, 0, 6\}$

▶ **21.** $\{(0, 1), (-23, 35), (7, 1)\}$ D: $\{-23, 0, 7\}$; R: $\{1, 35\}$

22. $\{(1, -12), (-6, 8), (5, 8), (0, 0), (1, 4)\}$ D: $\{-6, 0, 1, 5\}$; R: $\{-12, 0, 4, 8\}$

Determine whether the relation defines y to be a function of x. If it does not, find two ordered pairs where more than one value of y corresponds to a single value of x. **See Example 2.**

▶ **23.**

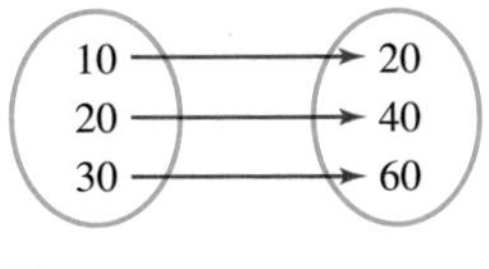

Yes

24.

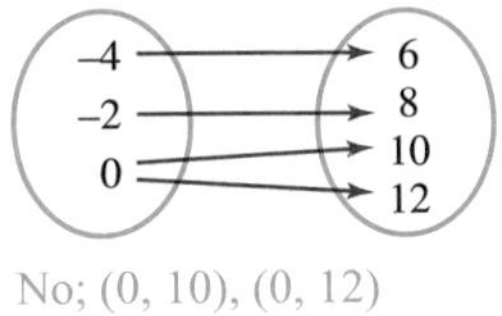

No; (0, 10), (0, 12)

▶ **25.**

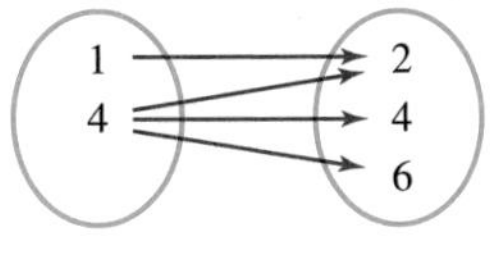

No; (4, 2), (4, 4), (4, 6)

▶ **26.**

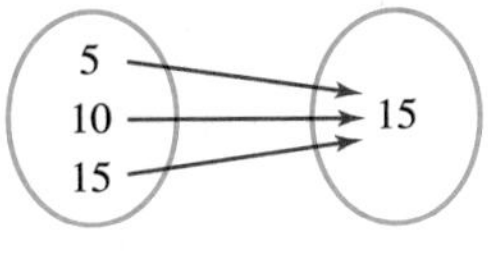

Yes

27. $\{(3, 4), (3, -4), (4, 3), (4, -3)\}$ No; (3, 4), (3, −4) or (4, 3), (4, −3)

▶ **28.** $\{(-1, 1), (-3, 1), (-5, 1), (-7, 1), (-9, 1)\}$ Yes

29. $\{(-2, 7), (-1, 10), (0, 13), (1, 16)\}$ Yes

▶ **30.** $\{(-2, 4), (-3, 8), (-3, 12), (-4, 16)\}$ No; (−3, 8), (−3, 12)

31.

x	y
1	7
2	15
3	23
4	16
5	8

Yes

▶ **32.**

x	y
30	2
30	4
30	6
30	8
30	10

No; (30, 2), (30, 4) (Answers may vary.)

▶ **33.**

x	y
−4	6
−1	0
0	−3
2	4
−1	2

No; (−1, 0), (−1, 2)

▶ **34.**

x	y
1	1
2	1
3	1
4	1

Yes

Determine whether each equation defines y to be a function of x. If it does not, find two ordered pairs where more than one value of y corresponds to a single value of x. **See Example 3.**

▶ **35.** $y = 2x + 3$ Yes

▶ **36.** $y = 4x - 1$ Yes

▶ **37.** $y = 4x^2$ Yes

▶ **38.** $y^2 = x$ No; (4, 2), (4, −2)

39. $y^4 = x$ No; (1, 1), (1, −1)

▶ **40.** $y = \frac{1}{x}$ Yes

41. $xy = 9$ Yes

▶ **42.** $y = |x|$ Yes

43. $y = \frac{1}{x^2}$ Yes

44. $x + 1 = |y|$ No; (1, 2), (1, −2)

▶ **45.** $x = |y|$ No; (1, 1), (1, −1)

▶ **46.** $xy = -4$ Yes

Find f(3) and f(−1). **See Example 4.**

47. $f(x) = 3x$ 9, −3

▶ **48.** $f(x) = -4x$ −12, 4

49. $f(x) = 2x - 3$ 3, −5

▶ **50.** $f(x) = 3x - 5$ 4, −8

Find g(2) and g(3). **See Example 4.**

51. $g(x) = x^2 - 10$ −6, −1

▶ **52.** $g(x) = x^2 - 2$ 2, 7

53. $g(x) = -x^3 + x$ −6, −24

54. $g(x) = x^3 - x$ 6, 24

55. $g(x) = (x + 1)^2$ 9, 16

▶ **56.** $g(x) = (x - 3)^2$ 1, 0

▶ **57.** $g(x) = 2x^2 - x + 1$ 7, 16

58. $g(x) = 5x^2 + 2x + 2$ 26, 53

Find h(5) and h(−2). **See Example 4.**

▶ **59.** $h(x) = |x| + 2$ 7, 4

60. $h(x) = |x| - 5$ 0, −3

▶ **61.** $h(x) = \frac{1}{x + 3}$ $\frac{1}{8}$, 1

▶ **62.** $h(x) = \frac{3}{x - 4}$ 3, $-\frac{1}{2}$

63. $h(x) = \frac{x}{x - 3}$ $\frac{5}{2}$, $\frac{2}{5}$

▶ **64.** $h(x) = \frac{x}{x^2 + 2}$ $\frac{5}{27}$, $-\frac{1}{3}$

65. $h(x) = \frac{x^2 + 2x - 35}{x^2 + 5x + 6}$ 0, undefined

▶ **66.** $h(x) = \frac{x^2 + x - 2}{x^2 - 5x}$ Undefined, 0

Complete each table. **See Example 4.**

67. $f(t) = |t - 2|$

t	$f(t)$
−1.7	3.7
0.9	1.1
5.4	3.4

▶ **68.** $f(r) = -2r^2 + 1$

Input	Output
−1.7	−4.78
0.9	−0.62
5.4	−57.32

69. $g(a) = a^3$

Input	Output
$-\frac{3}{4}$	$-\frac{27}{64}$
$\frac{1}{6}$	$\frac{1}{216}$
$\frac{5}{2}$	$\frac{125}{8}$

▶ **70.** $g(b) = 2\left(-b - \frac{1}{4}\right)$

b	$g(b)$
$-\frac{3}{4}$	1
$\frac{1}{6}$	$-\frac{5}{6}$
$\frac{5}{2}$	$-\frac{11}{2}$

Find g(w) and g(w + 1). **See Example 4.**

71. $g(x) = 2x$ $2w, 2w + 2$
72. $g(x) = -3x$ $-3w, -3w - 3$
73. $g(x) = 3x - 5$ $3w - 5, 3w - 2$
74. $g(x) = 2x - 7$ $2w - 7, 2w - 5$

Let f(x) = −2x + 5. For what value of x does function f have the given value? **See Example 5.**

75. $f(x) = 5$ 0
76. $f(x) = -7$ 6

Let $f(x) = \frac{3}{2}x - 2$. For what value of x does function f have the given value? **See Example 5.**

77. $f(x) = -\frac{1}{2}$ 1
78. $f(x) = \frac{2}{3}$ $\frac{16}{9}$

Find the domain of each function. **See Example 6.**

79. a. $h(x) = 3x + 6$ The set of real numbers
    b. $f(x) = \frac{1}{x - 4}$ The set of all real numbers except 4
80. a. $g(x) = |x - 7|$ The set of real numbers
    b. $f(x) = \frac{5}{x + 1}$ The set of all real numbers except −1
81. a. $f(x) = x^2$ The set of real numbers
    b. $s(x) = \frac{9}{2x + 1}$ The set of all real numbers except $-\frac{1}{2}$
82. a. $h(x) = x^3$ The set of real numbers
    b. $t(x) = \frac{15}{1 - 3x}$ The set of all real numbers except $\frac{1}{3}$

Graph each function. **See Objective 5.** See AIE Appendix 3.

83. $f(x) = 2x - 1$
84. $f(x) = -x + 2$
85. $f(x) = -\frac{3}{2}x - 3$
86. $f(x) = \frac{2}{3}x - 2$
87. $f(x) = x$
88. $f(x) = -x$
89. $f(x) = -4$
90. $f(x) = 2$
91. $g(x) = 0.75x$
92. $g(x) = -0.25x$
93. $s(x) = \frac{7}{8}x + 2$
94. $s(x) = -\frac{4}{5}x + 3$

Write an equation for a linear function whose graph has the given characteristics. **See Example 7.**

95. Slope 5, y-intercept $(0, -3)$ $f(x) = 5x - 3$
96. Slope 2, y-intercept $(0, 11)$ $f(x) = 2x + 11$
97. Slope $\frac{1}{5}$, passes through $(10, 1)$ $f(x) = \frac{1}{5}x - 1$
98. Slope $\frac{1}{4}$, passes through $(8, 1)$ $f(x) = \frac{1}{4}x - 1$
99. Passes through $(1, 7)$ and $(-2, 1)$ $f(x) = 2x + 5$
100. Passes through $(-2, 2)$ and $(2, -8)$ $f(x) = -\frac{5}{2}x - 3$
101. Passes through $(3, 0)$, parallel to the graph of $g(x) = -\frac{2}{3}x - 4$ $f(x) = -\frac{2}{3}x + 2$
102. Passes through $(2, 20)$, parallel to the graph of $g(x) = 8x + 1$ $f(x) = 8x + 4$
103. Passes through $(1, 2)$, perpendicular to the graph of $g(x) = -\frac{x}{6} + 1$ $f(x) = 6x - 4$
104. Passes through $(54, 0)$, perpendicular to the graph of $g(x) = 9x + 5$ $f(x) = -\frac{1}{9}x + 6$
105. Horizontal, passes through $(-8, 12)$ $f(x) = 12$
106. Horizontal, passes through $(9, -32)$ $f(x) = -32$

APPLICATIONS

107. from Campus to Careers

Webmaster

© Ronen/Shutterstock.com

According to data from *Netcraft Web Server Survey,* the total number of active websites available over the Internet (in millions) is approximated by the function $w(t) = 12.67t + 29.15$, where t is the number of years after 2005. Approximately how many active websites were there in 2011? 105.17 million websites

108. **Decongestants.** The temperature in degrees Celsius that is equivalent to a temperature in degrees Fahrenheit is given by the linear function $C(F) = \frac{5}{9}(F - 32)$. Refer to the label from a bottle of decongestant shown below. Use this function to find the low and high temperature extremes, in degrees Celsius, in which the bottle should be stored.
Between 20°C and 25°C

DIRECTIONS: Adults and children 12 years of age and over: Two teaspoons every 4 hours. DO NOT EXCEED 6 DOSES IN A 24-HOUR PERIOD. Store at a controlled room temperature between 68°F and 77°F.

109. **Concessionaires.** A baseball club pays a vendor $125 per game for selling bags of peanuts for $4.75 each.
    a. Write a linear function that describes the profit the vendor makes for the baseball club during a game if she sells b bags of peanuts. $p(b) = 4.75b - 125$
    b. Find the profit the baseball club will make if the vendor sells 110 bags of peanuts during a game. $397.50

110. **Home Construction.** In a proposal to some clients, a housing contractor listed the following costs:

Fees, permits, miscellaneous	$12,000
Construction, per square foot	$95

    a. Write a linear function that the clients could use to determine the cost of building a home having f square feet. $C(f) = 95f + 12{,}000$
    b. Find the cost to build a home having 1,950 square feet. $197,250

111. **Nurses.** The demand for full-time registered nurses in the United States can be modeled by a linear function. In 2005, approximately 2,175,500 nurses were needed. By the year 2015, that number is expected to increase to about 2,586,500. (Source: National Center for Health Workforce Analysis)
    a. Let t be the number of years after 2000 and N be the number of full-time registered nurses needed in the U.S. Write a linear function $N(t)$ to model the demand for nurses. $N(t) = 41{,}100t + 1{,}970{,}000$
    b. Use your answer to part a to predict the number of full-time registered nurses that will be needed in 2025, if the trend continues. 2,997,500 registered nurses

112. Wood Production. The total world wood production can be modeled by a linear function. In 1960, approximately 2,400 million cubic feet of wood were produced. Since then, the amount of increase has been approximately 25.5 million cubic feet per year. (Source: Earth Policy Institute)

a. Let t be the number of years after 1960 and W be the number of million cubic feet of wood produced. Write a linear function $W(t)$ to model the production of wood. $W(t) = 25.5t + 2{,}400$

b. Use your answer to part a to estimate how many million cubic feet of wood the world produced in 2010. 3,675 million cubic feet

113. Breathing Capacity. When fitness instructors prescribe exercise workouts for elderly patients, they must take into account age-related loss of lung function. Studies show that the percent of remaining breathing capacity for someone over 30 years old can be modeled by a linear function. (Source: alsearsmd.com)

a. At 35 years of age, approximately 90% of maximal breathing capacity remains and at 55 years of age, approximately 66% of maximal breathing capacity remains. Let a be the age of a patient and L be the percent of her maximal breathing capacity that remains. Write a linear function $L(a)$ to model this situation. $L(a) = -1.2a + 132$

b. Use your answer to part a to estimate the percent of maximal breathing capacity that remains in an 80-year-old. 36%

114. Chemical Reactions. When students mixed solutions of acetone and chloroform, they found that heat was generated. However, as time passed, the mixture cooled down. The graph shows data points of the form (time, temperature) taken by the students.

a. The linear function $T(t) = -\frac{t}{240} + 30$ models the relationship between the elapsed time t since the solutions were combined and the temperature $T(t)$ of the mixture. Graph the function. See AIE Appendix 3.

b. Predict the temperature of the mixture immediately after the two solutions are combined. 30°C

c. Is $T(180)$ more or less than the temperature recorded by the students for $t = 300$? More

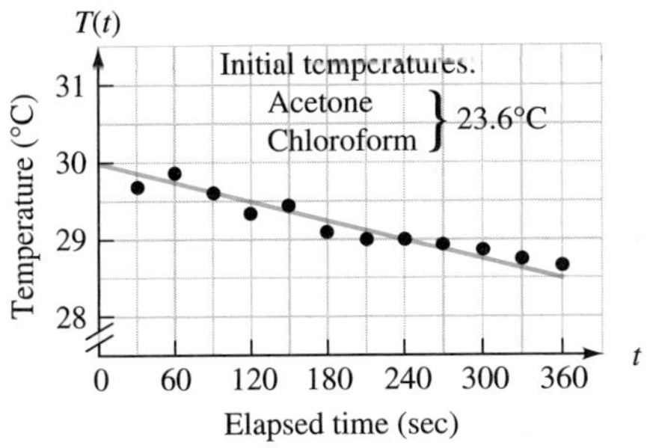

115. Taxes. The function

$$T(a) = 837.50 + 0.15(a - 8{,}375)$$

(where a is adjusted gross income) is a model of the instructions given on the first line of the following tax rate Schedule X.

a. Find $T(25{,}000)$ and interpret the result. \$3,331.25; the tax on an adjusted gross income of \$25,000 is \$ 3,331.25.

b. Write a function that models the second line on Schedule X. $T(a) = 4{,}681.25 + 0.25(a - 34{,}000)$

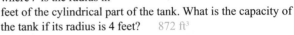

Schedule X–Use if your filing status is Single | 2010

If your adjusted gross income is: *Over—*	*But not over—*	Your tax is	*of the amount over—*
\$ 8,375	\$34,000	**\$ 837.50 + 15%**	**\$ 8,375**
\$34,000	\$82,400	**\$4,681.25 + 25%**	**\$34,000**

116. Storage Tanks. The volume $V(r)$ of the gasoline storage tank, in cubic feet, is given by the polynomial function $V(r) = 4.2r^3 + 37.7r^2$, where r is the radius in feet of the cylindrical part of the tank. What is the capacity of the tank if its radius is 4 feet? 872 ft³

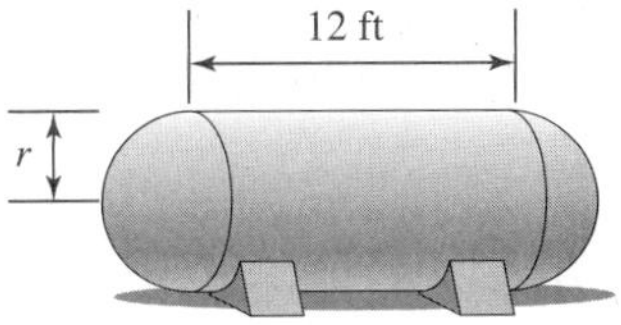

117. Roller Coasters. The polynomial function $f(x) = 0.001x^3 - 0.12x^2 + 3.6x + 10$ models the path of a portion of the track of a roller coaster. Use the function equation to find the height of the track for $x = 0, 20, 40$, and 60. 10 m, 42 m, 26 m, 10 m

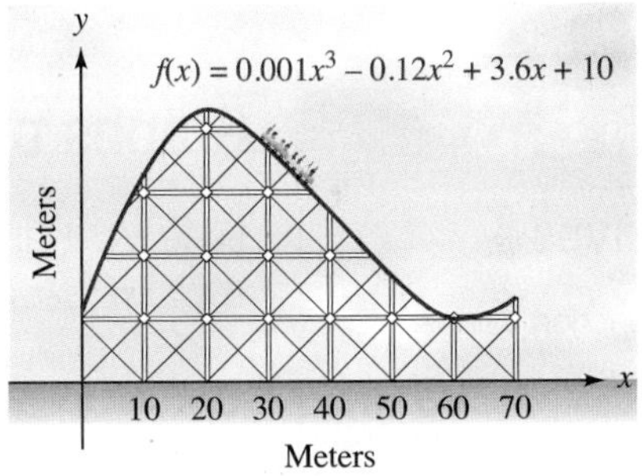

118. Customer Service. A software service hotline has found that on Mondays, the polynomial function $C(t) = -0.0625t^4 + t^3 - 6t^2 + 16t$ approximates the number of callers to the hotline at any one time. Here, t represents the time, in hours, since the hotline opened at 8:00 A.M. How many service technicians should be on duty on Mondays at noon if the company doesn't want any callers to the hotline waiting to be helped by a technician? 16

119. Rain Gutters. A rectangular sheet of metal will be used to make a rain gutter by bending up its sides, as shown. If the ends are covered, the capacity $f(x)$ of the gutter is a polynomial function of x: $f(x) = -240x^2 + 1{,}440x$. Find the capacity of the gutter if x is 3 inches. 2,160 in.³

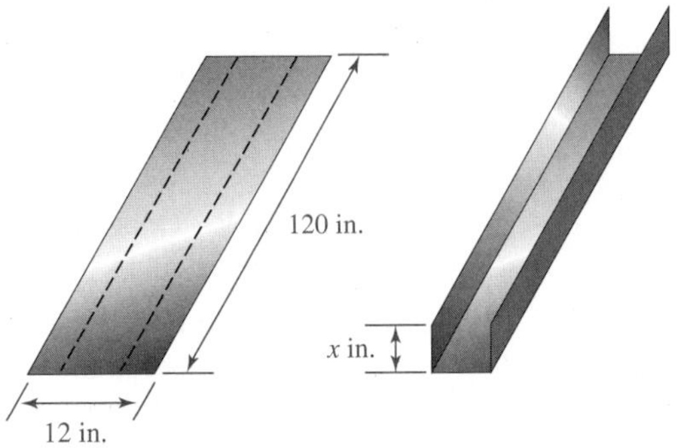

120. Stopping Distances. The number of feet that a car travels before stopping depends on the driver's reaction time and the braking distance. For one driver, the stopping distance $d(v)$, in feet, is given by the polynomial function $d(v) = 0.04v^2 + 0.9v$, where v is the velocity of the car in mph. Find the stopping distance at 60 mph. 198 ft

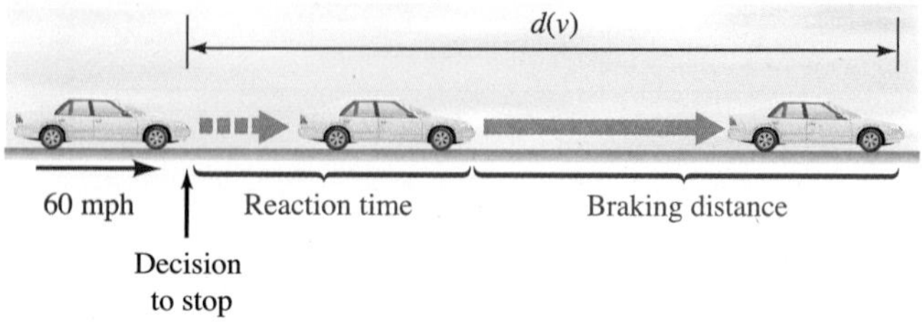

WRITING

121. Explain why 8 is not in the domain of the function $f(x) = \dfrac{5x - 7}{x - 8}$.

122. Explain why we can think of a function as a machine.

123. Consider the function defined by $y = 6x + 4$. Why do you think x is called the *independent* variable and y the *dependent* variable?

124. A website selling nutritional supplements contains the following sentence: "Health is a *function* of proper nutrition." Explain what this statement means.

REVIEW

Solve each equation. If the equation is an identity or a contradiction, so indicate.

125. $-2(t + 4) + 5t + 1 = 3(t - 4) + 7$ No solution, ∅; contradiction

126. $\frac{3}{2}(a - 4) = 2(a - 3) - \frac{a}{2}$ All real numbers, ℝ; identity

CHALLENGE PROBLEMS

127. Let $f(x) = 4x + 6$, function g be defined by $\{(4, 6), (6, 8), (8, 4)\}$, and function h be defined by the arrow diagram below. Find $\dfrac{f(8) + g(8)}{h(8)}$. 7

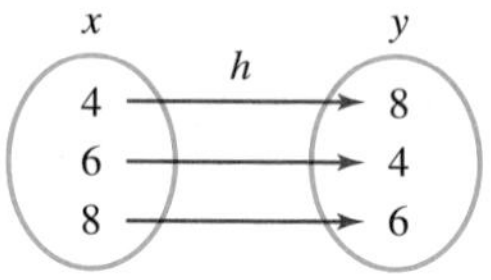

128. Find the domain of $f(x) = \dfrac{1}{5[9(x - 2) - 6(x - 3) + 3]}$.
The set of all real numbers except −1

SECTION 8.3 Graphs of Functions

OBJECTIVES

1. Find function values graphically.
2. Find the domain and range of a function graphically.
3. Graph nonlinear functions.
4. Translate graphs of functions.
5. Reflect graphs of functions.
6. Find function values and the domain and range of polynomial functions graphically.
7. Use the vertical line test.

ARE YOU READY? *Are You Ready? exercises available online at www.webassign.net/brookscole*

The following problems review some basic skills that are needed when graphing functions.

1. Fill in the blanks: $f(3) = 9$ corresponds to the ordered pair (3 , 9).
2. If $f(x) = x^2$, find $f(-1)$ and $f(1)$. 1, 1
3. If $f(x) = x^3$, find $f(-2)$ and $f(2)$. −8, 8
4. If $f(x) = |x|$, find $f(-4)$ and $f(4)$. 4, 4

Since a graph is often the best way to describe a function, we need to know how to construct and interpret their graphs.

1 Find Function Values Graphically.

From the graph of a function, we can determine function values. In general, the value of $f(a)$ is given by the y-coordinate of a point on the graph of function f with x-coordinate a.

EXAMPLE 1 Refer to the graph of function f in figure (a) on the next page.
a. Find $f(-3)$ **b.** Find the value of x for which $f(x) = -2$.

Strategy In each case, we will use the information provided by the function notation to locate a specific point on the graph and determine its x- and y-coordinates.

Why Once we locate the specific point, one of its coordinates will equal the value that we are asked to find.

Teaching Tip: Remind your students of the relationship between function notation and ordered pair notation
$f(-3) = 5$
↑ ↑
x y
↓ ↓
$(-3, 5)$

Solution **a.** To find $f(-3)$, we need to find the y-coordinate of the point on the graph of f whose x-coordinate is -3. If we draw a vertical line through -3 on the x-axis, as shown in figure (b), the line intersects the graph of f at $(-3, 5)$. Therefore, 5 corresponds to -3, and it follows that $f(-3) = 5$.

b. To find the input value x that has an output value $f(x) = -2$, we draw a horizontal line through -2 on the y-axis, as shown in figure (c) and note that it intersects the graph of f at $(4, -2)$. Since -2 corresponds to 4, it follows that $f(x) = -2$ if $x = 4$.

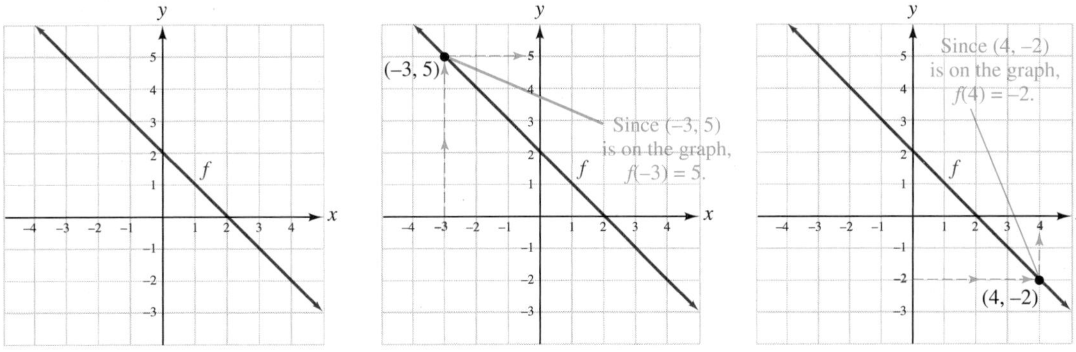

(a) (b) (c)

Teaching Example 1 Refer to the graph of function f in Example 1.
a. Find $f(-2)$
b. Find the value of x for which $f(x) = -1$
Answers: a. 4 b. 3

Self Check 1 Refer to the graph of function g on the right. **a.** Find $g(-3)$. -2 **b.** Find the x-value for which $g(x) = 2$. 2

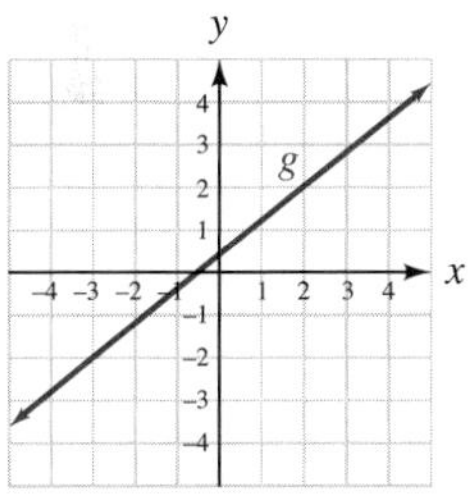

Now Try Problem 17

2 Find the Domain and Range of a Function Graphically.

We can find the domain and range of a function from its graph. For example, to find the domain of the linear function graphed in figure (a), we *project* the graph onto the x-axis. Because the graph of the function extends indefinitely to the left and to the right, the projection includes all the real numbers. Therefore, the domain of the function is the set of real numbers.

To find the range of the same linear function, we project the graph onto the y-axis, as shown in figure (b). Because the graph of the function extends indefinitely upward and downward, the projection includes all the real numbers. Therefore, the range of the function is the set of real numbers.

The Language of Algebra

Think of the **projection** of a graph on an axis as the "shadow" that the graph makes on the axis.

Project the graph onto the x-axis. Project the graph onto the y-axis.

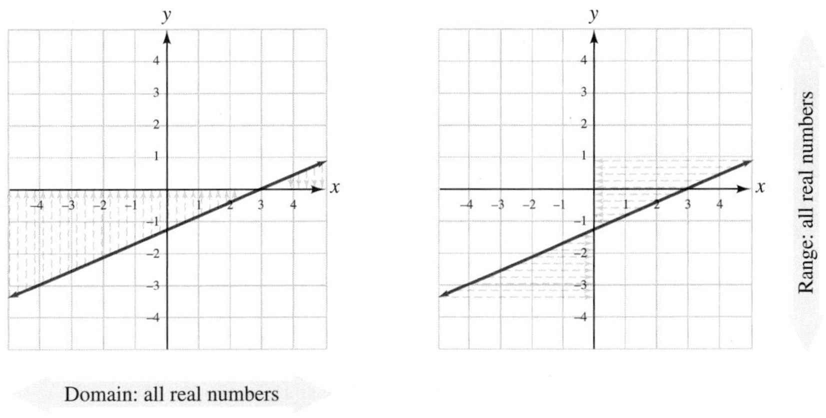

(a) (b)

3 Graph Nonlinear Functions.

We have seen that the graph of a linear function is a line. We will now consider several examples of **nonlinear functions** whose graphs are not lines. We will begin with $f(x) = x^2$, called the **squaring function.** We can graph this function using the **point-plotting method.**

EXAMPLE 2 Graph $f(x) = x^2$ and find its domain and range.

Strategy We will graph the function by creating a table of function values and plotting the corresponding ordered pairs.

Why After drawing a smooth curve through the plotted points, we will have the graph.

Solution To graph the function, we select several x-values and find the corresponding values of $f(x)$. For example, if we select -3 for x, we have

Teaching Tip: In the notation $f(x)$, the letter f is the name of the function— it is not a variable. Warn your students that they should not substitute numbers for f. They should only substitute values for x or $f(x)$.

$$f(x) = x^2 \quad \text{This is the function to graph.}$$
$$f(-3) = (-3)^2 \quad \text{Substitute } -3 \text{ for each } x.$$
$$= 9$$

Since $f(-3) = 9$, the ordered pair $(-3, 9)$ lies on the graph of f. In a similar manner, we find the corresponding values of $f(x)$ for six other x-values and list the ordered pairs in the table of values. Then we plot the points and draw a smooth curve through them to get the graph, called a **parabola.**

$f(x) = x^2$

x	$f(x)$	
−3	9	→ (−3, 9)
−2	4	→ (−2, 4)
−1	1	→ (−1, 1)
0	0	→ (0, 0)
1	1	→ (1, 1)
2	4	→ (2, 4)
3	9	→ (3, 9)

↑ Select x. ↑ Find f(x). ↑ Plot the point.

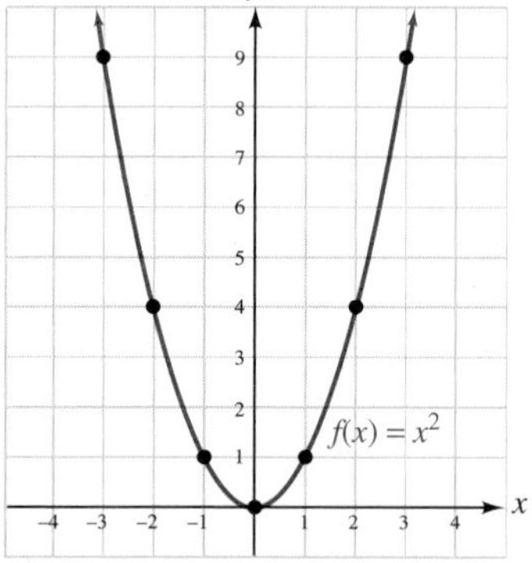

The Language of Algebra

The cuplike shape of a **parabola** has many real-life applications. For example, a satellite TV dish is called a **parabolic** dish.

Because the graph extends indefinitely to the left and to the right, the projection of the graph onto the x-axis includes all the real numbers. See figure (a). This means that the domain of the squaring function is the set of real numbers.

The Language of Algebra

The set of **nonnegative real numbers** is the set of real numbers greater than or equal to 0.

Project the graph onto the x-axis. Project the graph onto the y-axis.

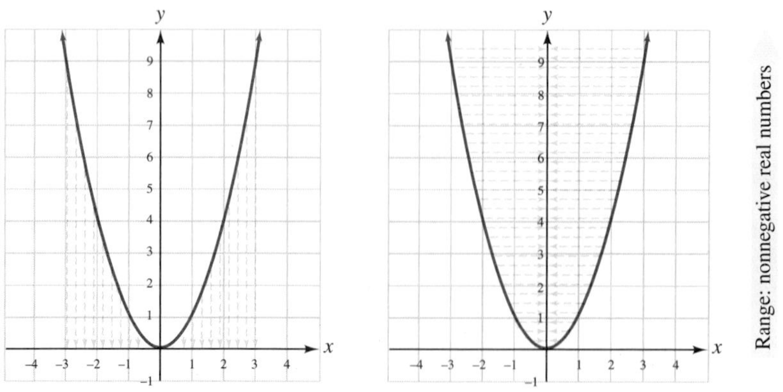

Domain: all real numbers

(a) (b)

Teaching Example 2 Graph $h(x) = x^2 + 3$ by plotting points. Then find its domain and range.
Answer:

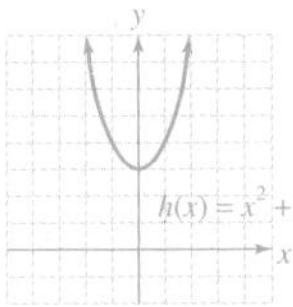

D: the set of all real numbers, R: the set of real numbers greater than or equal to 3.

Because the graph extends upward indefinitely from the point (0, 0), the projection of the graph on the y-axis includes only positive real numbers and 0. See figure (b) on the previous page. This means that the range of the squaring function is the set of nonnegative real numbers.

Self Check 2 Graph $g(x) = x^2 - 2$ by plotting points. Then find its domain and range. Compare the graph to the graph of $f(x) = x^2$.
D: the set of real numbers, R: the set of all real numbers greater than or equal to -2; the graph has the same shape, but is 2 units lower. See AIE Appendix 3.

Now Try Problem 29

Another important nonlinear function is $f(x) = x^3$, called the **cubing function.**

EXAMPLE 3

Graph $f(x) = x^3$ and find its domain and range.

Strategy We will graph the function by creating a table of function values and plotting the corresponding ordered pairs.

Why After drawing a smooth curve through the plotted points, we will have the graph.

Solution To graph the function, we select several values for x and find the corresponding values of $f(x)$. For example, if we select -2 for x, we have

$$f(x) = x^3 \quad \text{This is the function to graph.}$$
$$f(-2) = (-2)^3 \quad \text{Substitute } -2 \text{ for each } x.$$
$$= -8$$

Since $f(-2) = -8$, the ordered pair $(-2, -8)$ lies on the graph of f. In a similar manner, we find the corresponding values of $f(x)$ for four other x-values and list the ordered pairs in the table. Then we plot the points and draw a smooth curve through them to get the graph.

Success Tip

To graph a linear function, it is recommended that you find three points on the line to draw its graph. Because the graphs of nonlinear functions are more complicated, more work is required. You need to find a sufficient number of points on the graph so that its entire shape is revealed. At least five well-chosen input values are recommended.

$f(x) = x^3$

x	$f(x)$	
-2	-8	→ $(-2, -8)$
-1	-1	→ $(-1, -1)$
0	0	→ $(0, 0)$
1	1	→ $(1, 1)$
2	8	→ $(2, 8)$

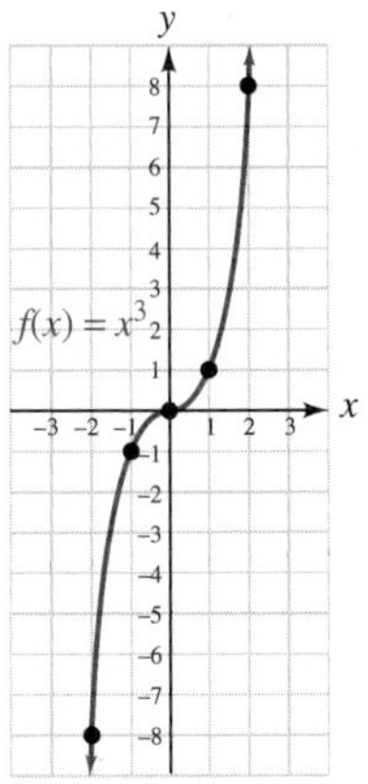

Because the graph of the function extends indefinitely to the left and to the right, the projection includes all the real numbers. Therefore, the domain of the cubing function is the set of real numbers.

Because the graph of the function extends indefinitely upward and downward, the projection includes all the real numbers. Therefore, the range of the cubing function is the set of real numbers.

Teaching Example 3 Graph $h(x) = x^3 - 2$ by plotting points. Then find the domain and range.
Answer:

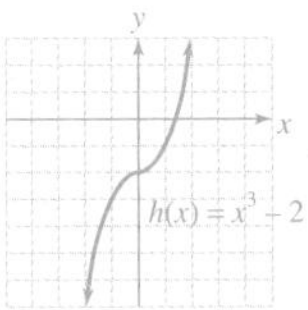

D: the set of all real numbers, R: the set of all real numbers

Self Check 3 Graph $g(x) = x^3 + 1$ by plotting points. Then find its domain and range. Compare the graph to the graph of $f(x) = x^3$. D: the set of real numbers, R: the set of all real numbers; the graph has the same shape, but is 1 unit higher. See AIE Appendix 3.

Now Try Problem 31

A third nonlinear function is $f(x) = |x|$, called the **absolute value function.**

EXAMPLE 4 Graph $f(x) = |x|$ and find its domain and range.

Strategy We will graph the function by creating a table of function values and plotting the corresponding ordered pairs.

Why After drawing straight lines through the plotted points, we will have the graph.

Solution To graph the function, we select several x-values and find the corresponding values for $f(x)$. For example, if we choose -3 for x, we have

$$f(x) = |x| \quad \text{This is the function to graph.}$$
$$f(-3) = |-3| \quad \text{Substitute } -3 \text{ for each } x.$$
$$= 3$$

Since $f(-3) = 3$, the ordered pair $(-3, 3)$ lies on the graph of f. In a similar manner, we find the corresponding values of $f(x)$ for six other x-values and list the ordered pairs in the table. Then we plot the points and connect them to get the following V-shaped graph.

$f(x) = |x|$

x	$f(x)$	
-3	3	→ $(-3, 3)$
-2	2	→ $(-2, 2)$
-1	1	→ $(-1, 1)$
0	0	→ $(0, 0)$
1	1	→ $(1, 1)$
2	2	→ $(2, 2)$
3	3	→ $(3, 3)$

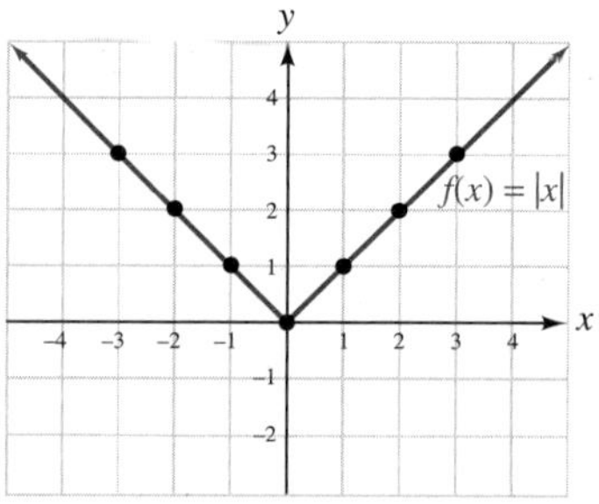

Success Tip

To determine the entire shape of the graph, several positive and negative values, along with 0, were selected as x-values when constructing the table of values.

Teaching Tip: Show your students that the shape of the graph of the absolute value function can be seen in its name:

Absolute Value function

Teaching Example 4 Graph $h(x) = |x - 1|$ by plotting points. Then find the domain and range.
Answer:

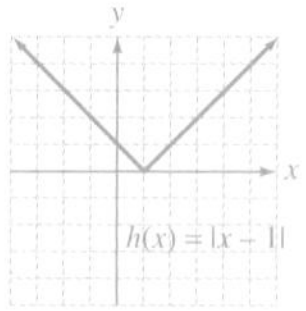

D: the set of all real numbers, R: the set of real numbers greater than or equal to 0

Because the graph extends indefinitely to the left and to the right, the projection of the graph onto the x-axis includes all the real numbers. Thus, the domain of the absolute value function is the set of real numbers.

Because the graph extends upward indefinitely from the point (0, 0), the projection of the graph on the y-axis includes only positive real numbers and 0. Thus, the range of the absolute value function is the set of nonnegative real numbers.

Self Check 4 Graph $g(x) = |x - 2|$ by plotting points. Then find its domain and range. Compare the graph to the graph of $f(x) = |x|$. D: the set of real numbers, R: the set of nonnegative real numbers; the graph has the same shape, but is 2 units to the right. See AIE Appendix 3.

Now Try ▶ Problem 33

Using Your Calculator ▶ Graphing Functions

We can graph nonlinear functions with a graphing calculator. For example, to graph $f(x) = x^2$ in a standard window of $[-10, 10]$ for x and $[-10, 10]$ for y, we first press Y =. Then we enter the function by typing x ^ 2 (or x followed by x^2), and press the GRAPH key. We will obtain the graph shown in figure (a).

To graph $f(x) = x^3$, we enter the function by typing x ^ 3 and then press the GRAPH key to obtain the graph in figure (b). To graph $f(x) = |x|$, we enter the function by selecting abs(from the NUM option within the MATH menu, typing x, and pressing the GRAPH key to obtain the graph in figure (c).

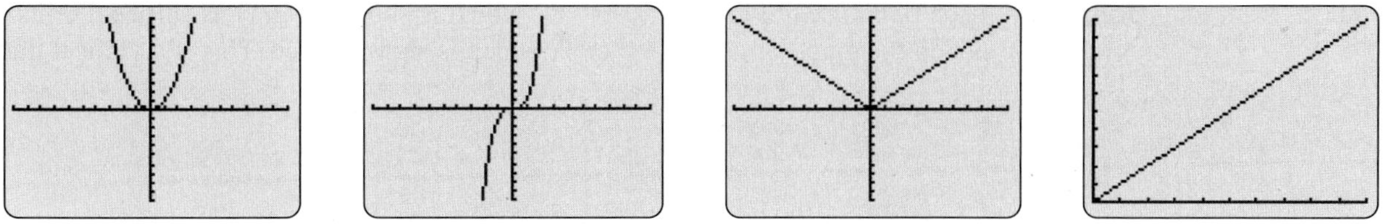

(a) (b) (c) (d)

When using a graphing calculator, we must be sure that the viewing window does not show a misleading graph. For example, if we graph $f(x) = |x|$ in the window [0, 10] for x and [0, 10] for y, we will obtain a misleading graph that looks like a line. See figure (d). This is not correct. The proper graph is the V-shaped graph shown in figure (c). One of the challenges of using graphing calculators is finding an appropriate viewing window.

4 Translate Graphs of Functions.

Examples 2, 3, and 4 and their Self Checks suggest that the graphs of different functions may be identical except for their positions in the coordinate plane. For example, the figure on the right shows the graph of $f(x) = x^2 + k$ for three different values of k. If $k = 0$, we get the graph of $f(x) = x^2$, shown in red. If $k = 3$, we get the graph of $f(x) = x^2 + 3$, shown in blue. Note that it is identical to the graph of $f(x) = x^2$ except that it is shifted 3 units upward. If $k = -4$, we get the graph of $f(x) = x^2 - 4$, shown in green. It is identical to the graph of $f(x) = x^2$ except that it is shifted 4 units downward. These shifts are called **vertical translations.**

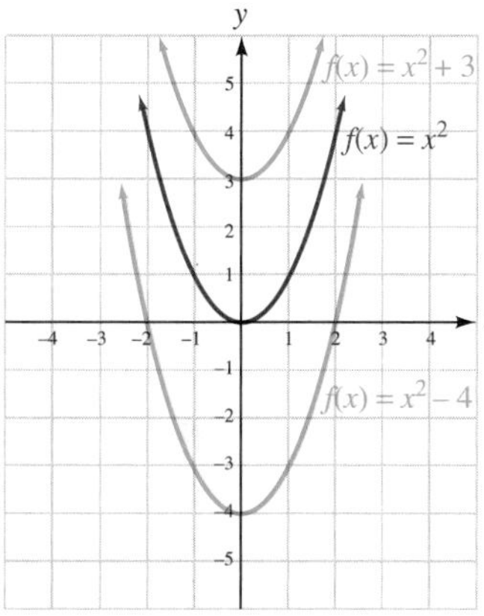

In general, we can make these observations.

Vertical Translations

If f is a function and k represents a positive number, then

- The graph of $y = f(x) + k$ is identical to the graph of $y = f(x)$ except that it is translated k units upward.
- The graph of $y = f(x) - k$ is identical to the graph of $y = f(x)$ except that it is translated k units downward.

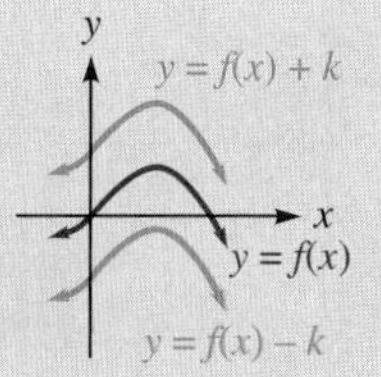

EXAMPLE 5 Graph: $g(x) = |x| + 2$

Strategy We will graph $g(x) = |x| + 2$ by translating (shifting) the graph of $f(x) = |x|$ upward 2 units.

Why The addition of 2 in $g(x) = |x| + \mathbf{2}$ causes a vertical shift of the graph of the absolute value function 2 units upward.

Solution Each point used to graph $f(x) = |x|$, which is shown in gray, is shifted 2 units upward to obtain the graph of $g(x) = |x| + 2$, which is shown in red. This is a vertical translation.

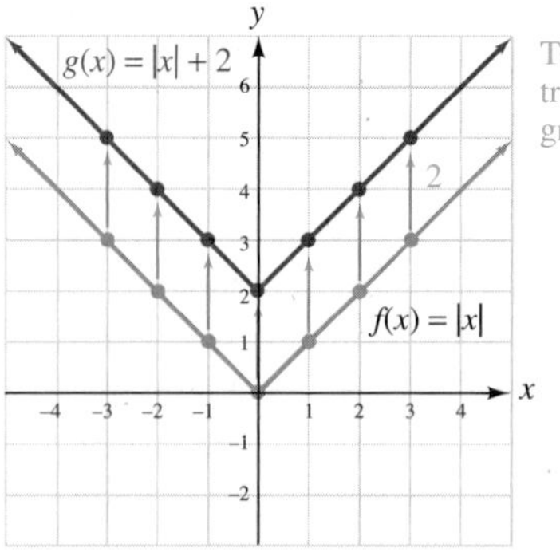

To graph $g(x) = |x| + 2$, translate each point on the graph of $f(x) = |x|$ up 2 units.

Teaching Example 5
Graph: $h(x) = |x| - 1$
Answer:

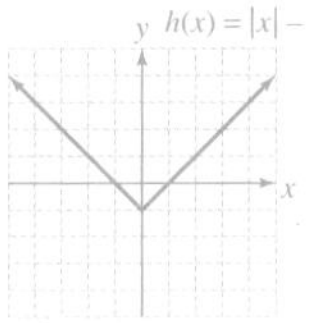

Self Check 5 Graph: $g(x) = |x| - 3$ See AIE Appendix 3.

Now Try ▶ Problem 37

The figure on the right shows the graph of $f(x) = (x + h)^2$ for three different values of h. If $h = 0$, we get the graph of $f(x) = x^2$, shown in red. The graph of $f(x) = (x - 3)^2$ shown in green is identical to the graph of $f(x) = x^2$ except that it is shifted 3 units to the right. The graph of $f(x) = (x + 2)^2$ shown in blue is identical to the graph of $f(x) = x^2$ except that it is shifted 2 units to the left. These shifts are called **horizontal translations.**

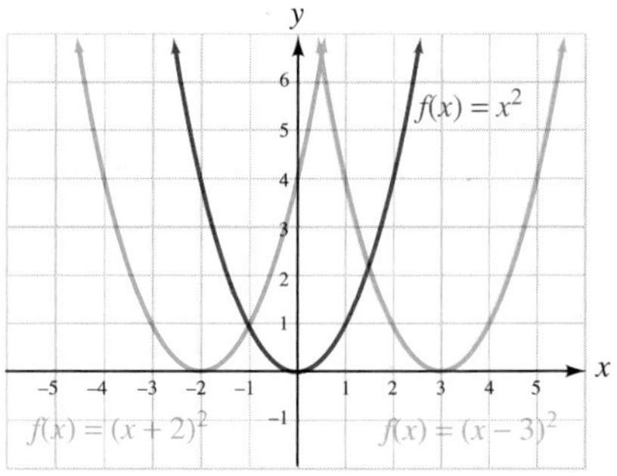

In general, we can make these observations.

Horizontal Translations

If f is a function and h is a positive number, then

- The graph of $y = f(x - h)$ is identical to the graph of $y = f(x)$ except that it is translated h units to the right.
- The graph of $y = f(x + h)$ is identical to the graph of $y = f(x)$ except that it is translated h units to the left.

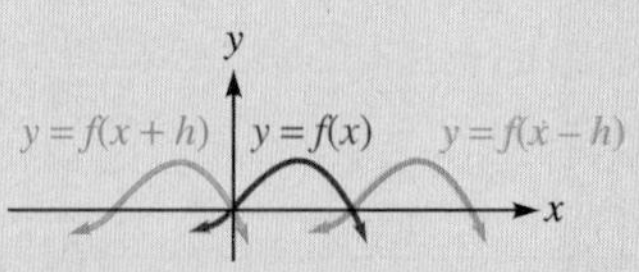

EXAMPLE 6 Graph: $g(x) = (x + 3)^3$

Success Tip

To determine the direction of the horizontal translation, find the value of x that makes the expression within the parentheses, in this case, $x + 3$, equal to 0. Since -3 makes $x + 3 = 0$, the translation is 3 units to the left.

Strategy We will graph $g(x) = (x + 3)^3$ by translating (shifting) the graph of $f(x) = x^3$ to the left 3 units.

Why The addition of 3 to x in $g(x) = (x + 3)^3$ causes a horizontal shift of the graph of the cubing function 3 units to the left.

To graph $g(x) = (x + 3)^3$, translate each point on the graph of $f(x) = x^3$ to the left 3 units.

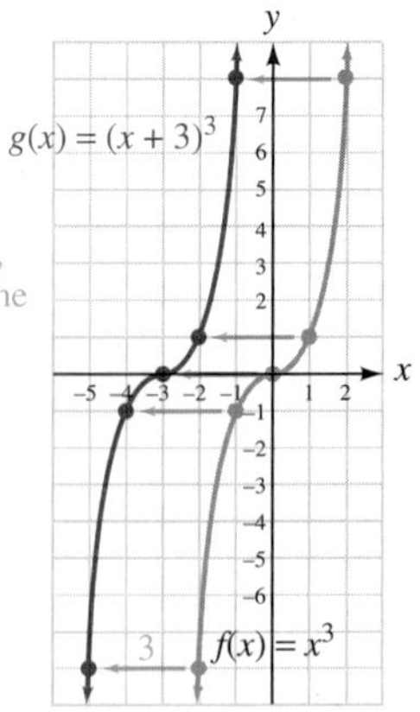

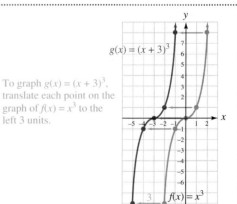

Solution Each point used to graph $f(x) = x^3$, which is shown in gray, is shifted 3 units to the left to obtain the graph of $g(x) = (x + 3)^3$, which is shown in red. This is a horizontal translation.

Teaching Example 6
Graph: $h(x) = (x - 1)^3$
Answer:

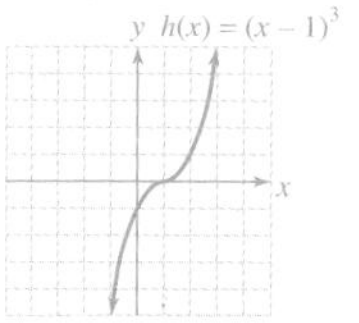

Self Check 6 Graph: $g(x) = (x - 2)^2$ See AIE Appendix 3.

Now Try Problem 47

The graphs of some functions involve horizontal and vertical translations.

EXAMPLE 7 Graph: $g(x) = (x - 5)^2 - 2$

Strategy To graph $g(x) = (x - 5)^2 - 2$, we will perform two translations by shifting the graph of $f(x) = x^2$ to the right 5 units and then 2 units downward.

Why The subtraction of 5 from x in $g(x) = (x - \mathbf{5})^2 - \mathbf{2}$ causes a horizontal shift of the graph of the squaring function 5 units to the right. The subtraction of 2 causes a vertical shift of the graph 2 units downward.

Solution Each point used to graph $f(x) = x^2$, which is shown in gray, is shifted 5 units to the right and 2 units downward to obtain the graph of $g(x) = (x - 5)^2 - 2$, which is shown in red. This is a horizontal and vertical translation.

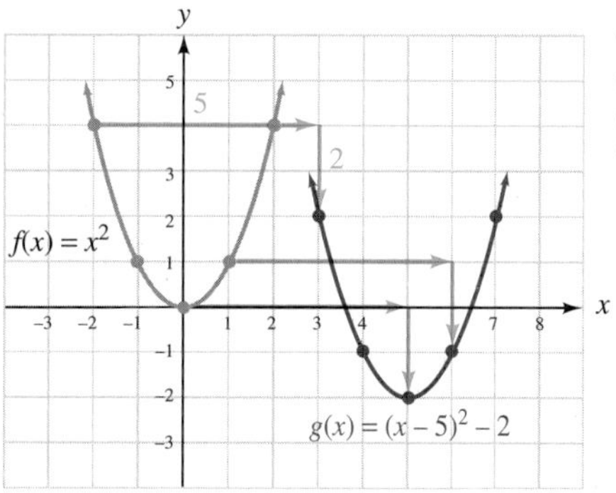

To graph $g(x) = (x - 5)^2 - 2$, translate each point on the graph of $f(x) = x^2$ to the right 5 units and then 2 units downward.

Teaching Example 7 Graph:
$f(x) = (x + 1)^3 - 2$
Answer:

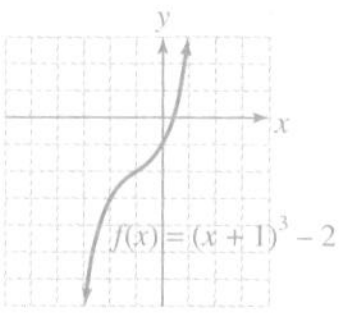

Self Check 7 Graph: $g(x) = |x + 2| - 3$ See AIE Appendix 3.

Now Try Problem 53

5 Reflect Graphs of Functions.

The following figure shows a table of values for $f(x) = x^2$ and for $g(x) = -x^2$. We note that for a given value of x, the corresponding y-value in the tables are opposites. When graphed, we see that the $-$ sign in $g(x) = -x^2$ has the effect of flipping the graph of $f(x) = x^2$ over the x-axis so that the parabola opens downward. We say that the graph of $g(x) = -x^2$ is a **reflection** of the graph of $f(x) = x^2$ about the x-axis.

$f(x) = x^2$

x	$f(x)$	
-2	4	$\rightarrow (-2, 4)$
-1	1	$\rightarrow (-1, 1)$
0	0	$\rightarrow (0, 0)$
1	1	$\rightarrow (1, 1)$
2	4	$\rightarrow (2, 4)$

$g(x) = -x^2$

x	$g(x)$	
-2	-4	$\rightarrow (-2, -4)$
-1	-1	$\rightarrow (-1, -1)$
0	0	$\rightarrow (0, 0)$
1	-1	$\rightarrow (1, -1)$
2	-4	$\rightarrow (2, -4)$

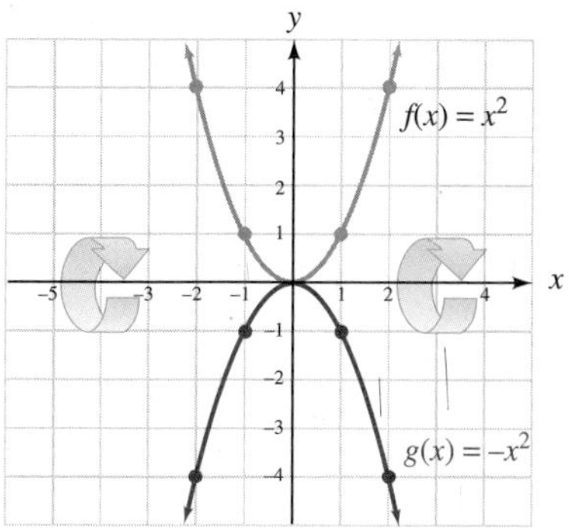

Reflection of a Graph The graph of $y = -f(x)$ is the graph of $y = f(x)$ reflected about the x-axis.

EXAMPLE 8 Graph: $g(x) = -x^3$

Strategy We will graph $g(x) = -x^3$ by reflecting the graph of $f(x) = x^3$ about the x-axis.

Why Because of the $-$ sign in $g(x) = -x^3$, the y-coordinate of each point on the graph of function g is the opposite of the y-coordinate of the corresponding point on the graph $f(x) = x^3$.

Solution To graph $g(x) = -x^3$, we use the graph of $f(x) = x^3$ from Example 3. First, we reflect the portion of the graph of $f(x) = x^3$ in quadrant I to quadrant IV, as shown. Then we reflect the portion of the graph of $f(x) = x^3$ in quadrant III to quadrant II.

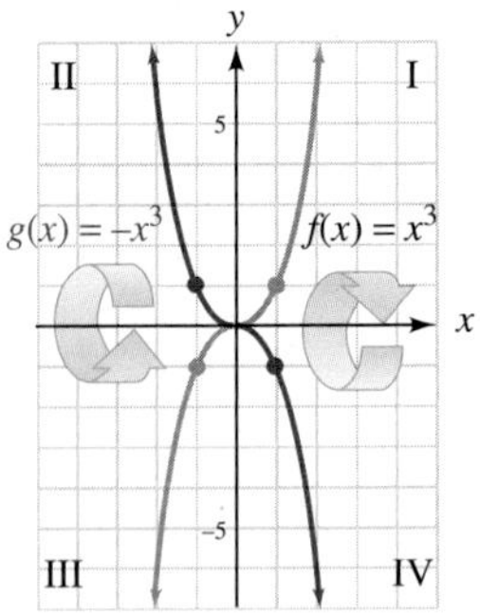

Teaching Example 8
Graph: $f(x) = -x^2 + 1$
Answer:

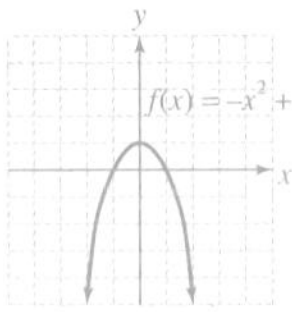

Self Check 8 Graph: $g(x) = -|x|$ See AIE Appendix 3.

Now Try Problem 63

6 Find Function Values and the Domain and Range of Polynomial Functions Graphically.

We have seen that the graphs of polynomial functions of degree 1, such as $f(x) = 4x - 1$ or $f(x) = \frac{1}{2}x + 3$, are straight lines and the graphs of polynomial functions of degree 2, such as $f(x) = x^2$ or $f(x) = x^2 - 2$, are parabolas. The graphs of polynomial functions of degree 3 or higher are often more complicated. Such graphs are always *smooth* and *continuous*. That is, they consist of only rounded curves with no sharp corners, and there are no breaks.

We can obtain important information about a polynomial function from its graph.

EXAMPLE 9 Refer to the graph of polynomial function $f(x) = x^3 - 3x^2 - 9x + 2$ in figure (a) on the next page. **a.** Find $f(2)$. **b.** Find any values of x for which $f(x) = -25$. **c.** Find the domain and range of f.

Strategy For parts a and b, we will use the information provided by the function notation to locate a specific point on the graph and determine its x- and y-coordinates. For part c, we will project the graph onto each axis.

Why Once we locate the specific point, one of its coordinates will equal the value that we are asked to find. Projecting the graph onto the x-axis gives the domain and projecting it onto the y-axis gives the range.

Solution

a. To find $f(2)$, we need to find the y-coordinate of the point on the graph of f whose x-coordinate is 2. If we draw a vertical line downward, from 2 on the x-axis, as shown in figure (b) on the next page, the line intersects the graph of f at $(2, -20)$. Therefore, -20 corresponds to 2, and it follows that $f(2) = -20$.

b. To find any input value x that has an output value $f(x) = -25$, we draw a horizontal line through -25 on the y-axis, as shown in figure (c) on the next page, and note that it intersects the graph of f at $(-3, -25)$ and $(3, -25)$. Therefore, $f(-3) = -25$ and $f(3) = -25$. It follows that the values of x for which $f(x) = -25$ are -3 and 3.

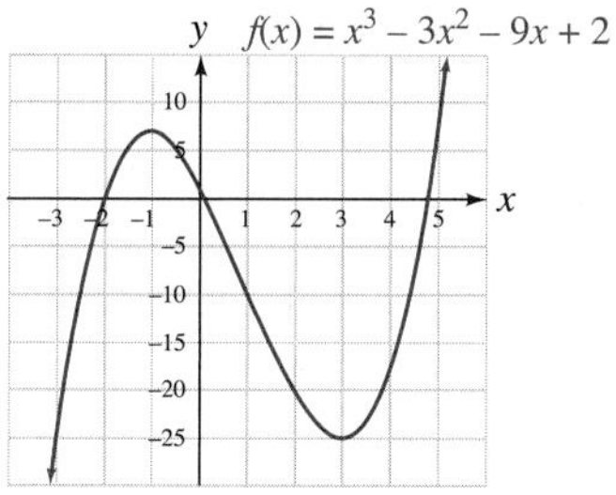

(a)

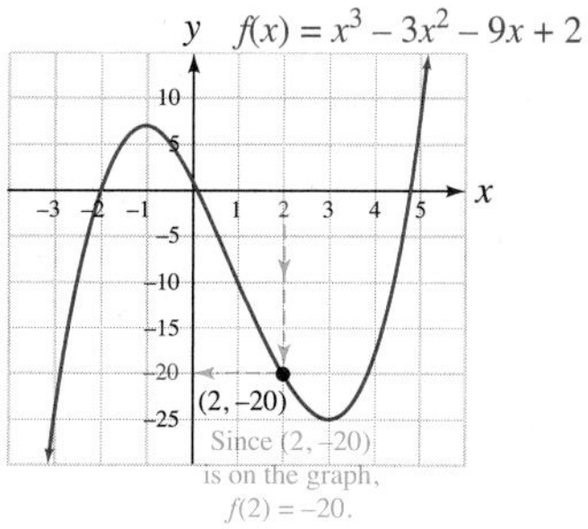

(b)

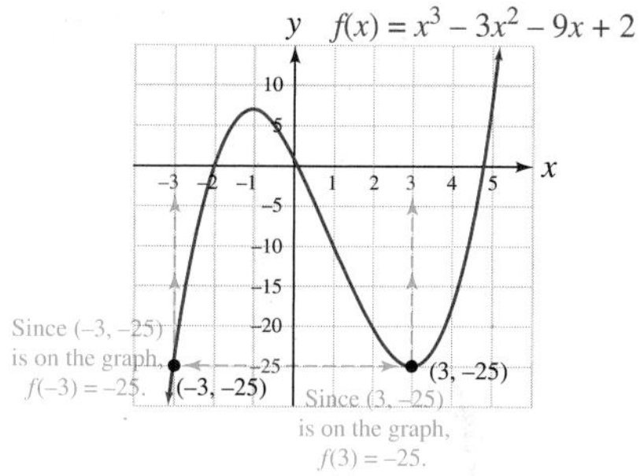

(c)

c. To find the domain of $f(x) = x^3 - 3x^2 - 9x + 2$, we project its graph onto the x-axis as shown in figure (d) below. Because the graph extends indefinitely to the left and right, the projection includes all real numbers. Therefore, the domain of the function is the set of real numbers, which can be written in interval notation as $(-\infty, \infty)$.

To determine the range of the same polynomial function, we project the graph onto the y-axis, as shown in figure (e) below. Because the graph of the function extends indefinitely upward and downward, the projection includes all real numbers. Therefore the range of the function is the set of real numbers, written $(-\infty, \infty)$.

Success Tip

The graphs of many polynomial functions of degree 3 and higher have "peaks" and "valleys" as shown here.

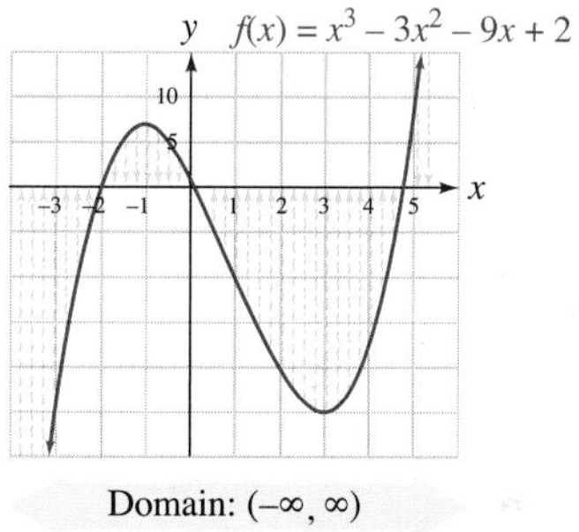

(d)

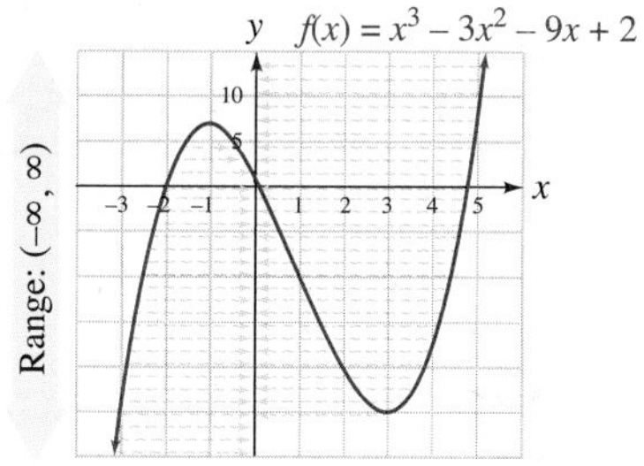

(e)

Teaching Example 9 Refer to the graph of function f in Example 9. Estimate each of the following:
a. $f(-1)$ **b.** Any x-values for which $f(x) = 7$
Answers: **a.** 7 **b.** $-1, 5$

Self Check 9 Refer to the graph of function f in Example 9. Estimate each of the following: **a.** $f(1)$ -10 **b.** Any x-values for which $f(x) = 0$ $-2, 0.2, 4.8$

Now Try Problem 65

7 Use the Vertical Line Test.

Some graphs define functions and some do not. If a vertical line intersects a graph more than once, the graph does not represent a function, because to one value of x there would correspond more than one value of y.

The Vertical Line Test If a vertical line intersects a graph in more than one point, the graph is not the graph of a function.

EXAMPLE 10 Determine whether the graph in figure (a) and figure (c) is the graph of a function.

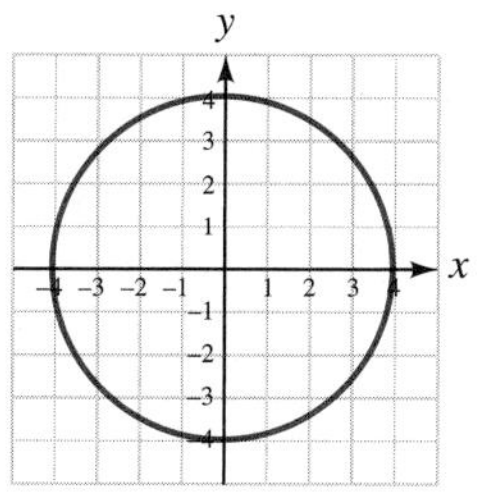

(a)

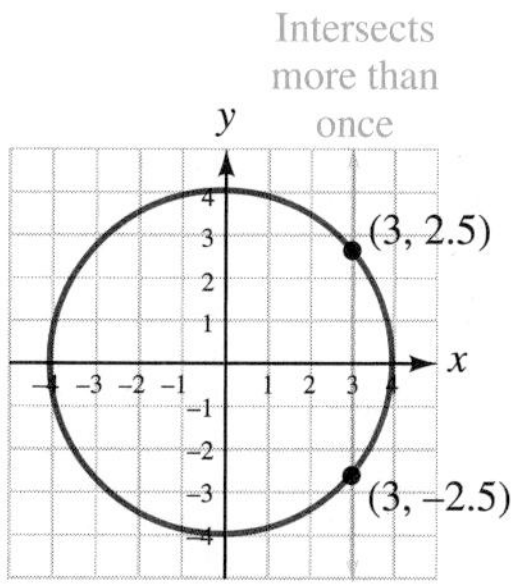

(b)

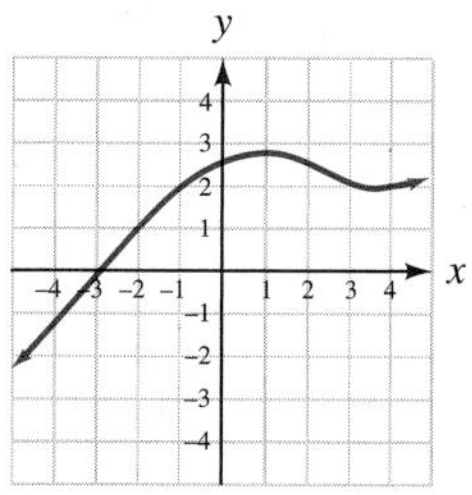

(c)

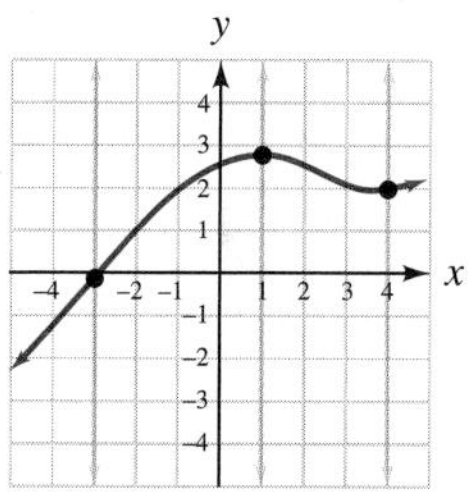

(d)

Strategy We will check to see whether any vertical lines intersect the graph more than once.

Why If any vertical line intersects the graph more than once, it is not the graph of a function.

Solution

Teaching Tip: Show your students that they can use the side of a credit card or a pencil to apply the vertical line test. Simply hold the card (or pencil) vertically and run it across the graph from left to right.

a. Refer to figure (b) on the previous page. The graph shown in red is not the graph of a function because a vertical line intersects the graph more than once. The points of intersection of the graph and the vertical line indicate that two values of y (2.5 and -2.5) correspond to the x-value 3.

x	y
3	2.5
3	-2.5

b. Refer to figure (d) on the previous page. The graph shown in red is the graph of a function, because no vertical line intersects the graph more than once.

Teaching Example 10 Determine whether the following graph is the graph of a function.

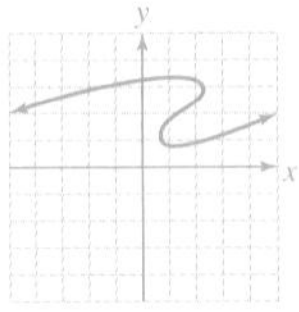

Answer: Not a function

Self Check 10 Determine whether the following graph is the graph of a function. Not a function

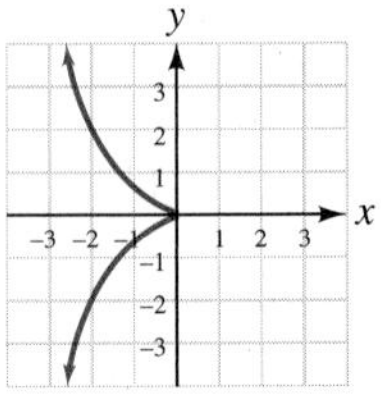

Now Try Problems 69 and 71

SECTION 8.3 STUDY SET

VOCABULARY

Fill in the blanks.

1. Functions whose graphs are not lines are called nonlinear functions.
2. The graph of $f(x) = x^2$ is a cuplike shape called a parabola.
3. The set of nonnegative real numbers is the set of real numbers greater than or equal to 0.
4. A shift of the graph of a function upward or downward is called a vertical translation.

CONCEPTS

5. Graph each basic function by plotting points and give its name. See AIE Appendix 3.
   a. $f(x) = x^2$ The squaring function
   b. $f(x) = x^3$ The cubing function
   c. $f(x) = |x|$ The absolute value function
6. Complete each sentence about finding function values graphically.
   a. To find $f(-3)$, we find the y-coordinate of the point on the graph whose x-coordinate is -3.
   b. To find the value of x for which $f(x) = -2$, we find the x-coordinate of the point(s) on the graph whose y-coordinate is -2.
   c. Suppose for a function f that $f(5) = 9$. The corresponding ordered pair that will be on the graph of the function is (5, 9).
7. Fill in the blank. The graph of $g(x) = -x^2$ is the reflection of the graph of $f(x) = x^2$ about the x-axis.
8. Fill in the blanks. The illustration shows the projection of the graph of function f on the y-axis. We see that the range of f is the set of real numbers less than or equal to 0.

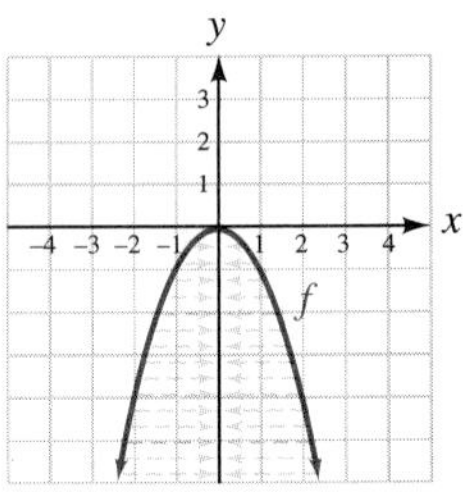

9. Consider the graph of the function f.
   a. Label each arrow in the illustration with the appropriate term: *domain* or *range.*
   b. Give the domain and range of f.
   D: the set of nonnegative real numbers; R: the set of all real numbers greater than or equal to 2

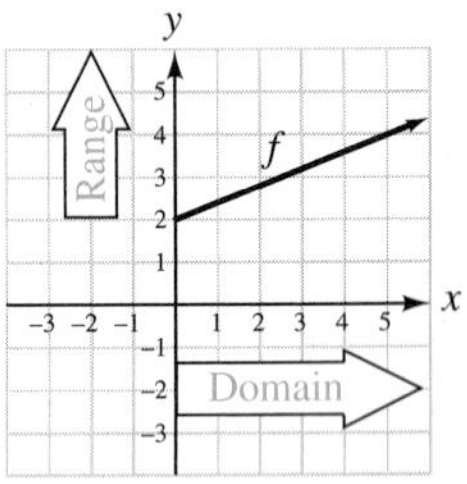

10. The graph of $f(x) = x^2 + k$ for three values of k is shown on the right. Find the value of k for
    a. the blue graph -3
    b. the red graph 1
    c. the green graph 2

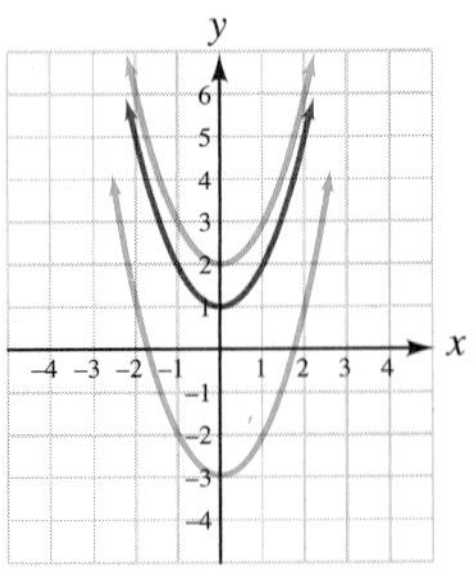

11. The graph of $f(x) = |x + h|$ for three values of h is shown. Find the value of h for
    a. the blue graph 4
    b. the red graph 0
    c. the green graph -2

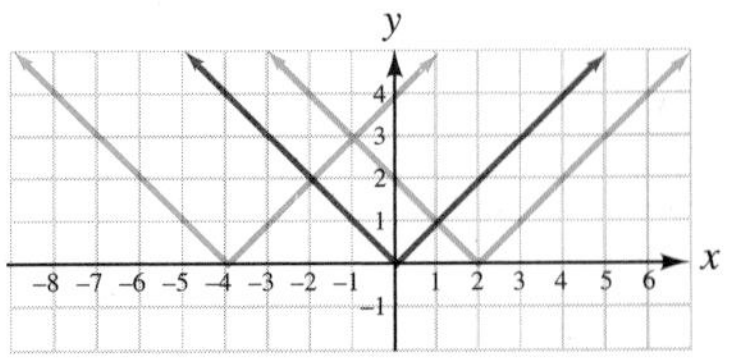

12. a. Translate each point plotted on the graph below to the left 5 units and then up 1 unit.

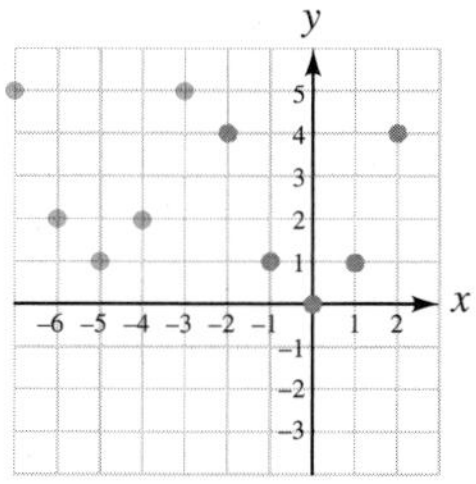

b. Translate each point plotted on the graph below to the right 4 units and then down 3 units.

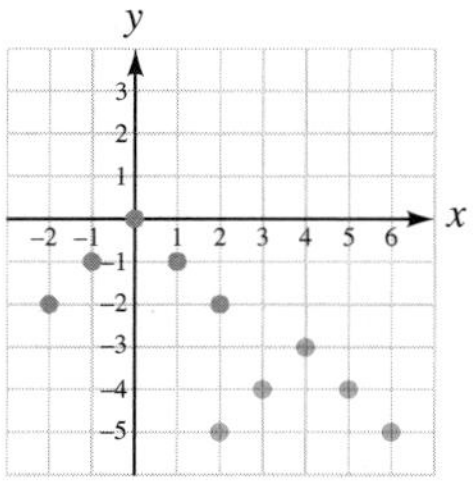

13. a. Give the coordinates of the points where the given vertical line intersects the graph in red. $(-2, 4), (-2, -4)$

b. Is this the graph of a function? Explain. No; to the x-value -2, there corresponds more than one y-value (4 and -4).

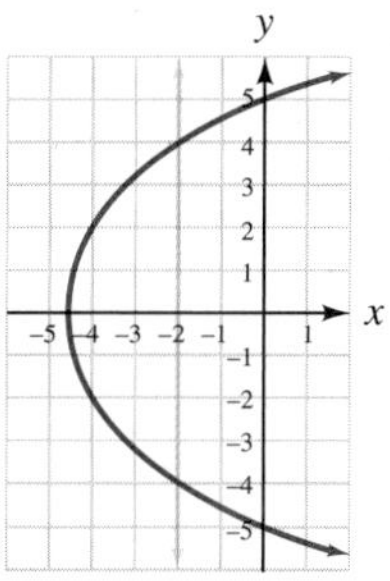

NOTATION

Fill in the blanks.

14. Fill in the blanks. The vertical line test: If a vertical line intersects a graph in more than one point, the graph is not the graph of a function.

15. a. The graph of $f(x) = (x + 4)^3$ is the same as the graph of $f(x) = x^3$ except that it is shifted 4 units to the left.

b. The graph of $f(x) = x^3 + 4$ is the same as the graph of $f(x) = x^3$ except that it is shifted 4 units up.

16. a. The graph of $f(x) = |x| - 5$ is the same as the graph of $f(x) = |x|$ except that it is shifted 5 units down.

b. The graph of $f(x) = |x - 5|$ is the same as the graph of $f(x) = |x|$ except that it is shifted 5 units to the right.

GUIDED PRACTICE

Refer to the given graph to find each value. **See Example 1.**

17. a. $f(-2)$ -4

b. $f(0)$ 0

c. The value of x for which $f(x) = 4$ 2

d. The value of x for which $f(x) = -2$ -1

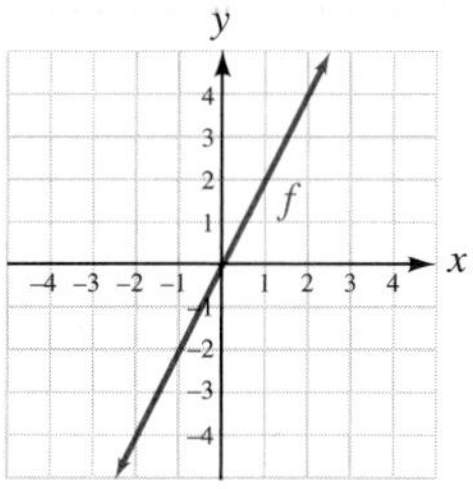

18. a. $s(-3)$ 2

b. $s(3)$ 4

c. The values of x for which $s(x) = 0$ $-1, 1, 5$

d. The values of x for which $s(x) = 3$ 2, 4

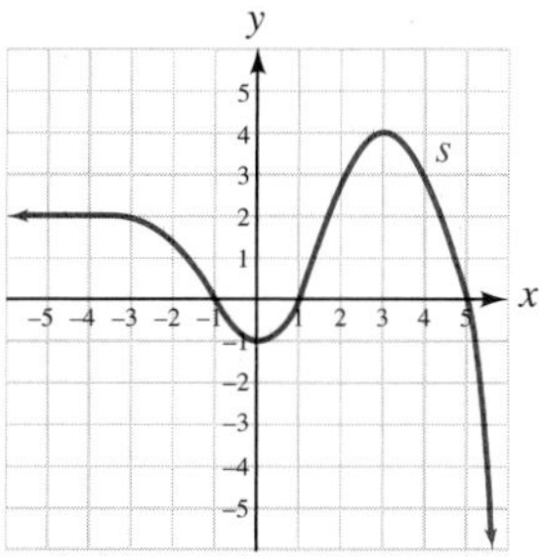

19. a. $g(-2)$ 2

b. $g(0)$ 2

c. The value of x for which $g(x) = 3$ -1

d. The values of x for which $g(x) = -1$ $-3, 1$

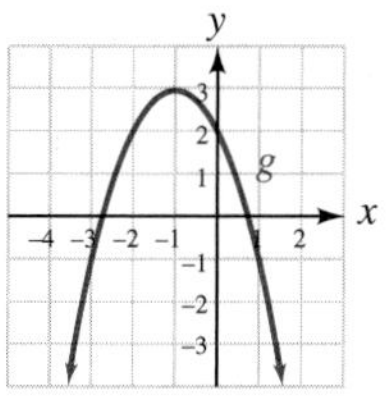

20. a. $h(-3)$ 4

b. $h(4)$ 3

c. The values of x for which $h(x) = 1$ 0, 2

d. The value of x for which $h(x) = 0$ 1

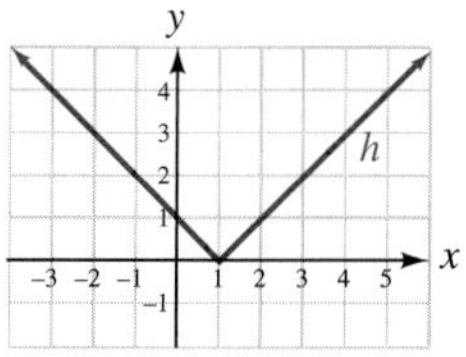

Find the domain and range of each function. **See Objective 2 and Example 2.**

21.

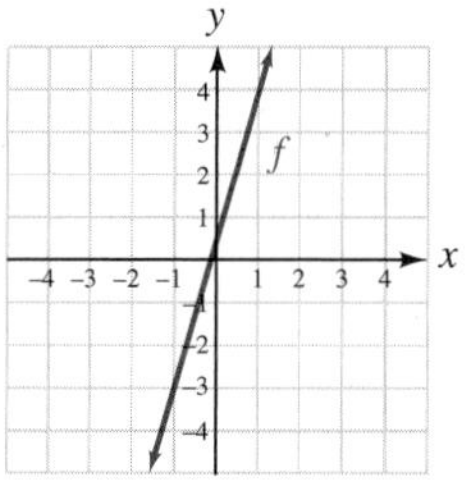

D: the set of real numbers; R: the set of real numbers

22.

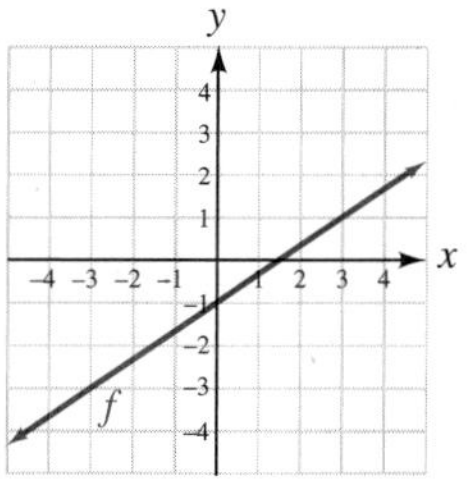

D: the set of real numbers; R: the set of real numbers

23.

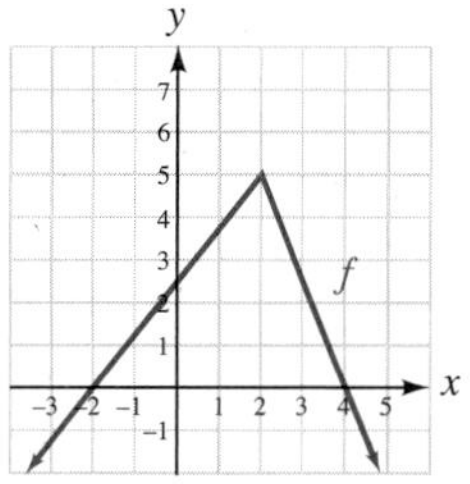

D: the set of real numbers; R: the set of real numbers less than or equal to 5

24.

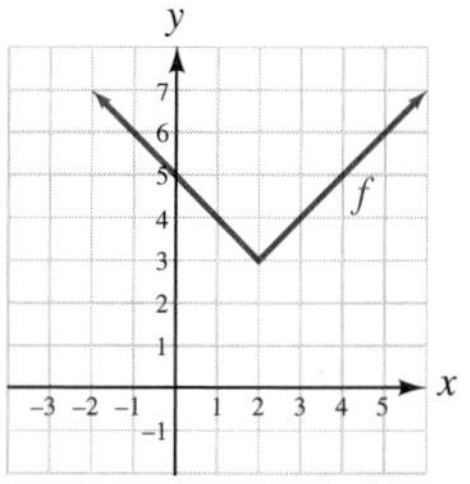

D: the set of real numbers; R: the set of real numbers greater than or equal to 3

25.

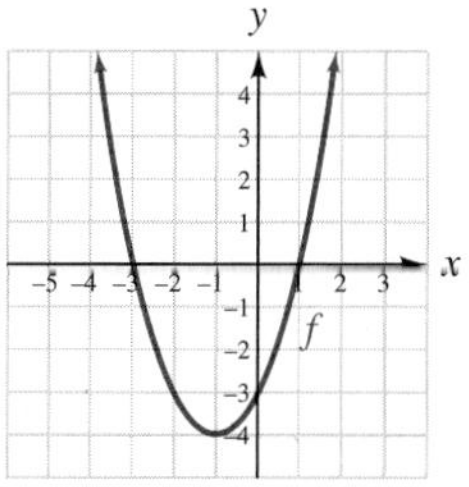

D: the set of real numbers; R: the set of real numbers greater than or equal to -4

26.

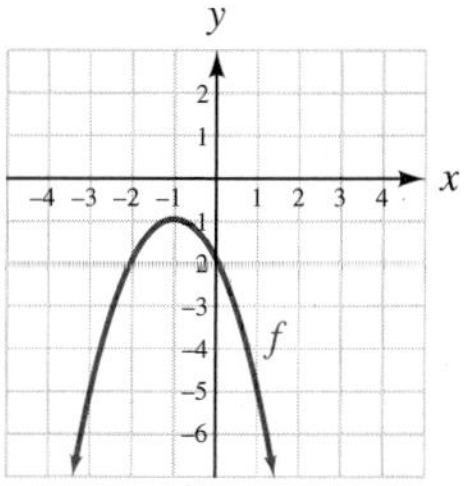

D: the set of real numbers; R: the set of real numbers less than or equal to -1

27.

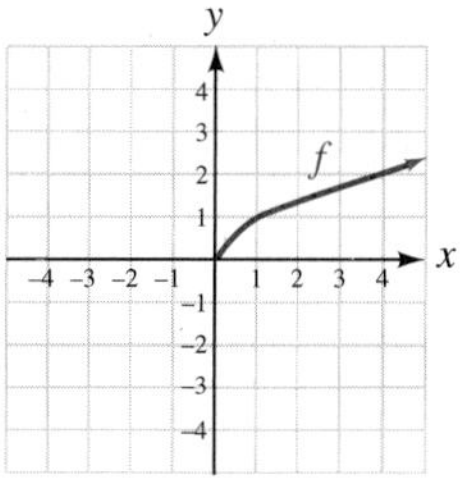

D: the set of nonnegative real numbers; R: the set of nonnegative real numbers

28.

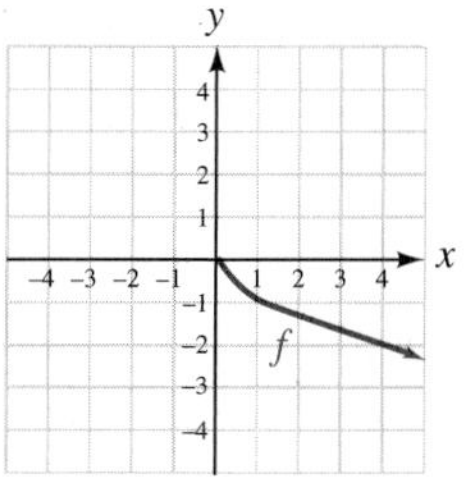

D: the set of nonnegative real numbers; R: the set of real numbers less than or equal to 0 or the set of nonpositive real numbers

Graph each function by creating a table of function values and plotting points. Give the domain and range of the function. **See Examples 2, 3, and 4.** See AIE Appendix 3.

▶ **29.** $f(x) = x^2 + 2$
D: the set of real numbers; R: the set of real numbers greater than or equal to 2

▶ **30.** $f(x) = x^2 - 4$
D: the set of real numbers; R: the set of real numbers greater than or equal to -4

▶ **31.** $f(x) = x^3 - 3$
D: the set of real numbers; R: the set of real numbers

▶ **32.** $f(x) = x^3 + 2$
D: the set of real numbers; R: the set of real numbers

▶ **33.** $f(x) = |x - 1|$
D: the set of real numbers; R: the set of nonnegative real numbers

▶ **34.** $f(x) = |x + 4|$
D: the set of real numbers; R: the set of nonnegative real numbers

35. $f(x) = (x + 4)^2$
D: the set of real numbers; R: the set of nonnegative real numbers

▶ **36.** $f(x) = (x - 1)^3$
D: the set of real numbers; R: the set of real numbers

For each of the following functions, first sketch the graph of its associated function, $f(x) = x^2$, $f(x) = x^3$, or $f(x) = |x|$. Then draw the graph of function g using a translation and give its domain and range. **See Examples 5 and 6.** See AIE Appendix 3.

▶ **37.** $g(x) = |x| - 2$ D: the set of real numbers; R: the set of real numbers greater than or equal to -2

▶ **38.** $g(x) = |x + 2|$ D: the set of real numbers; R: the set of nonnegative real numbers

▶ **39.** $g(x) = (x + 1)^3$ D: the set of real numbers; R: the set of real numbers.

▶ **40.** $g(x) = x^3 + 5$ D: the set of real numbers; R: the set of real numbers

▶ **41.** $g(x) = x^2 - 3$ D: the set of real numbers; R: the set of real numbers greater than or equal to -3

42. $g(x) = (x - 6)^2$ D: the set of real numbers; R: the set of nonnegative real numbers

43. $g(x) = (x - 4)^3$ D: the set of real numbers; R: the set of real numbers

▶ **44.** $g(x) = |x| + 1$ D: the set of real numbers; R: the set of real numbers greater than or equal to 1

45. $g(x) = x^3 + 4$ D: the set of real numbers; R: the set of real numbers

▶ **46.** $g(x) = x^2 - 5$ D: the set of real numbers; R: the set of real numbers greater than or equal to -5

47. $g(x) = (x + 4)^2$ D: the set of real numbers; R: the set of nonnegative real numbers

▶ **48.** $g(x) = (x - 1)^3$ D: the set of real numbers; R: the set of real numbers

For each of the following functions, first sketch the graph of its associated function, $f(x) = x^2$, $f(x) = x^3$, or $f(x) = |x|$. Then draw the graph of function g using translations and/or a reflection. **See Examples 7 and 8.** See AIE Appendix 3.

▶ **49.** $g(x) = |x - 2| - 1$

▶ **50.** $g(x) = (x + 2)^2 - 1$

51. $g(x) = (x + 1)^3 - 2$

▶ **52.** $g(x) = |x + 4| + 3$

53. $g(x) = (x - 2)^2 + 4$

▶ **54.** $g(x) = (x - 4)^2 + 3$

▶ **55.** $g(x) = |x + 3| + 5$

56. $g(x) = (x - 3)^2 - 2$

▶ **57.** $g(x) = -x^3$

58. $g(x) = -|x|$

▶ **59.** $g(x) = -x^2$

▶ **60.** $g(x) = -(x + 1)^2$

61. $g(x) = -|x + 5|$

▶ **62.** $g(x) = -(x + 4)^3$

63. $g(x) = -x^2 + 3$

▶ **64.** $g(x) = -|x| - 4$

Use the graph of the function to find each of the following. **See Example 9.**

65. Find:
a. $f(1)$ 15
b. $f(-3)$ 5
c. The values of x for which $f(x) = 0$ 0, -2, -4
d. The domain and range of f D: $(-\infty,\infty)$; R: $(-\infty,\infty)$

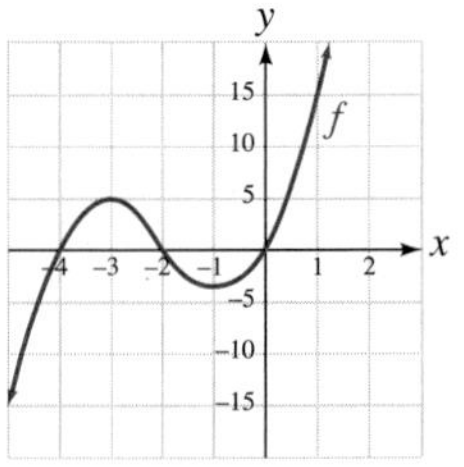

▶ **66.** Find:
a. $f(-1)$ -1
b. $f(0)$ 0
c. The values of x for which $f(x) = 3$ 1, -3
d. The domain and range of f D: $(-\infty, \infty)$; R: $[-1.8, \infty)$

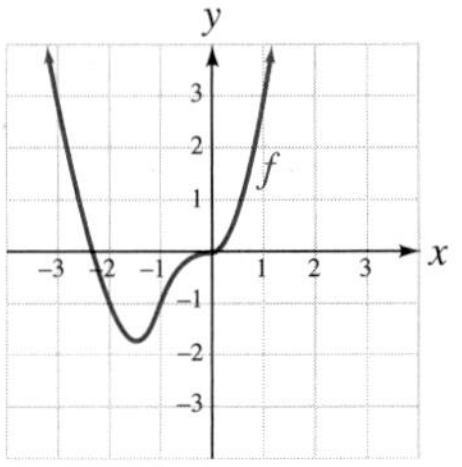

67. Find:
a. $g(1)$ 2
b. $g(-4)$ -5
c. The values of x for which $g(x) = 4$ $-2, 2$
d. The domain and range of g D: $(-\infty, \infty)$; R: $(-\infty, 4]$

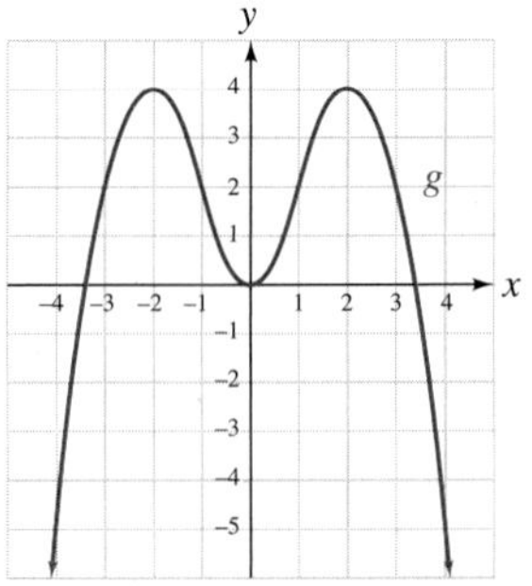

68. Find:
a. $h(0.5)$ -1
b. $h(-0.5)$ 0.5
c. The values of x for which $h(x) = -1.5$ -1.5, 2
d. The domain and range of h D: $(-\infty, \infty)$; R: $(-\infty, \infty)$

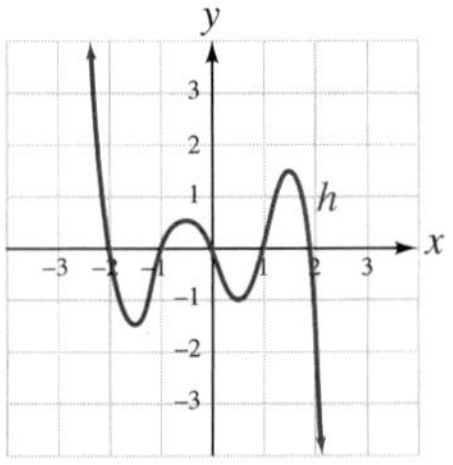

Determine whether each graph is the graph of a function. If it is not, find two ordered pairs where more than one value of y corresponds to a single value of x. **See Example 10.**

▶ **69.**

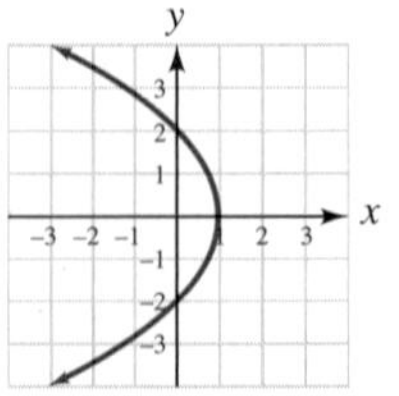

No, (0, 2), (0, -2)

▶ **70.**

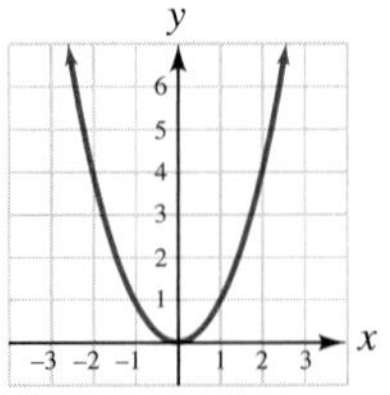

Yes

71.
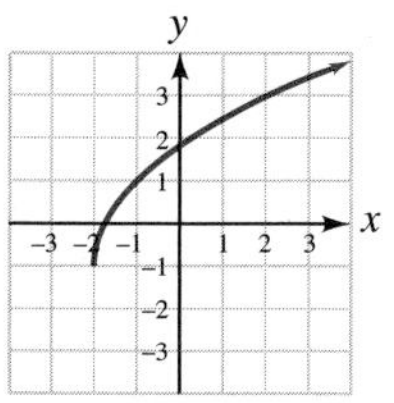

Yes

72.
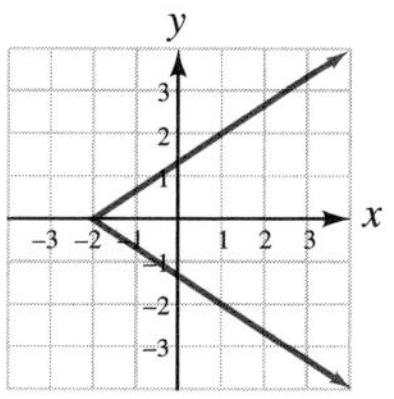

No, (1, 2), (1, −2)

73.
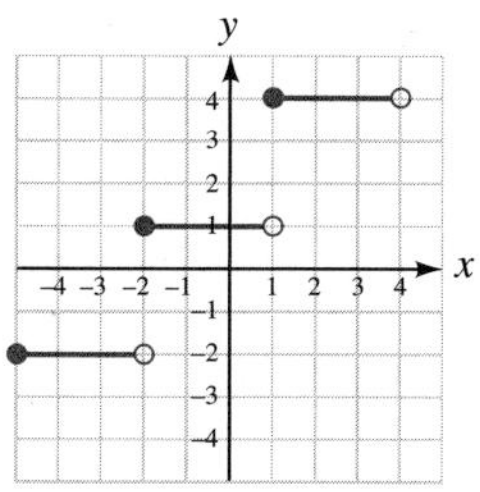

Yes

74.
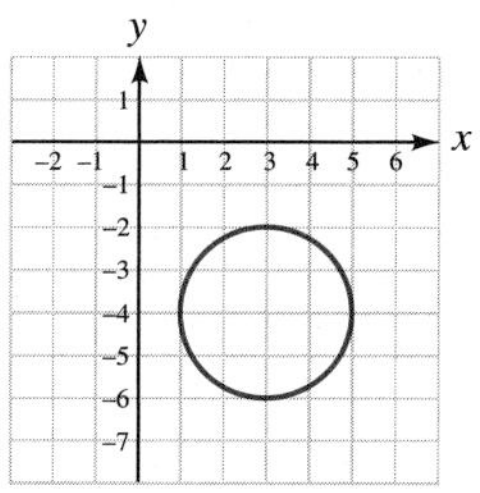

No, (3, −2), (3, −6)

75.
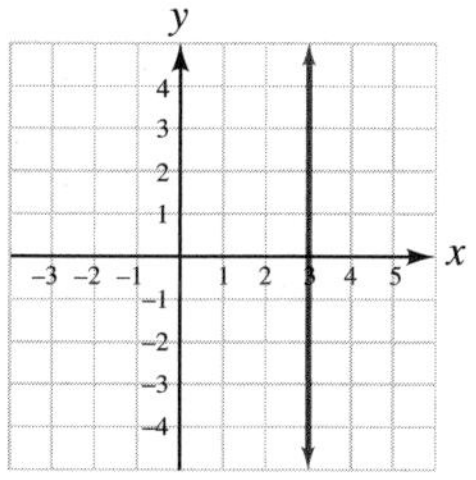

No, (3, 0), (3, 1)

76.
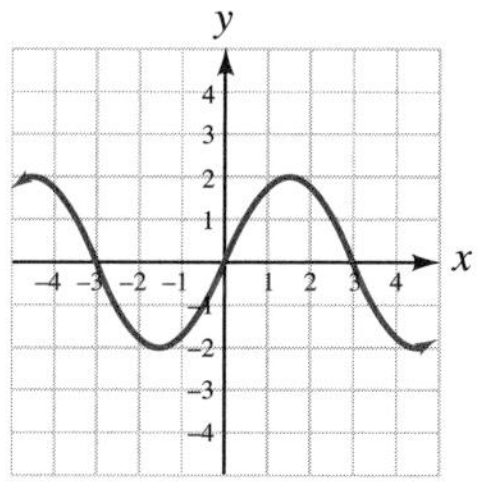

Yes

***Graph each function using window settings of* [−4, 4] *for x and* [−4, 4] *for y. The graph is not what it appears to be. Pick a better viewing window and find a better representation of the true graph.* See Using Your Calculator: Graphing Functions.** See AIE Appendix 3.

77. $f(x) = x^2 + 8$

78. $f(x) = x^3 - 8$

79. $f(x) = |x + 5|$

80. $f(x) = |x - 5|$

81. $f(x) = (x - 6)^2$

82. $f(x) = (x + 9)^2$

83. $f(x) = x^3 + 8$

84. $f(x) = x^3 - 12$

APPLICATIONS

85. **Optics.** See the illustration. The **law of reflection** states that the angle of reflection is equal to the angle of incidence. What function studied in this section models the path of the reflected light beam with an angle of incidence measuring 45°?
$f(x) = |x|$

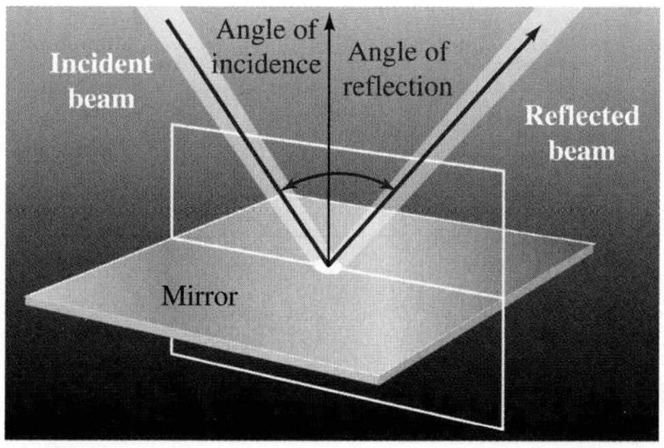

86. **Billiards.** In the illustration, a rectangular coordinate system has been superimposed over a billiard table. Write a function that models the path of the ball that is shown banking off of the cushion. $f(x) = -|x - 2| + 4$

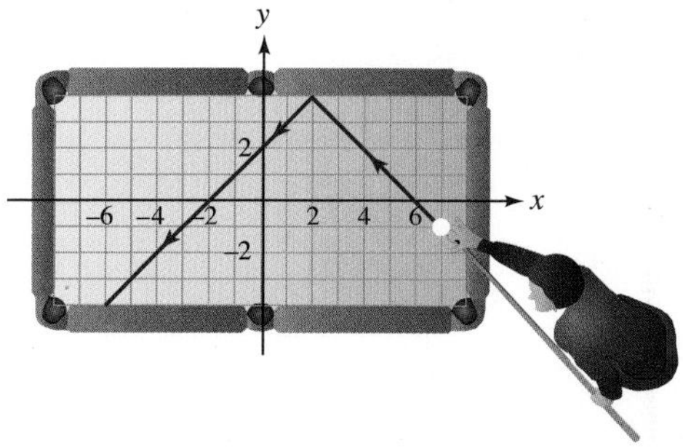

87. **Center of Gravity.** See the illustration. As a diver performs a $1\frac{1}{2}$-somersault in the tuck position, her center of gravity follows a path that can be described by a graph shape studied in this section. What graph shape is that?
A parabola

88 **Earth's Atmosphere.** The illustration below shows a graph of the temperatures of the atmosphere at various altitudes above Earth's surface. The temperature is expressed in degrees Kelvin, a scale widely used in scientific work.

a. Estimate the coordinates of three points on the graph that have an x-coordinate of 200.
(200, 25), (200, 90), (200, 105)

b. Explain why this is not the graph of a function.
It doesn't pass the vertical line test.

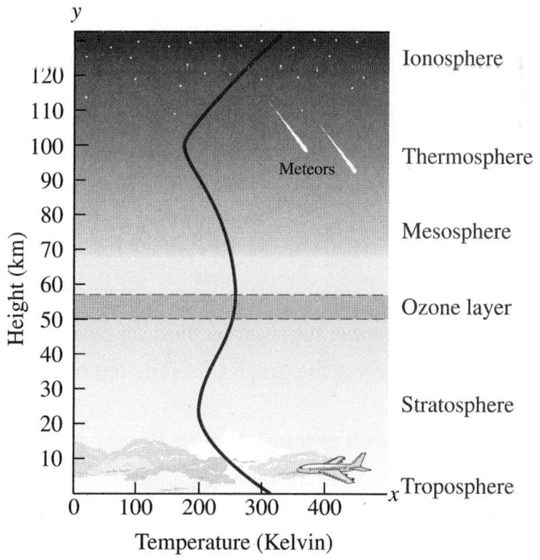

89. Labor Statistics. The polynomial function that is graphed below approximates the number of manufacturing jobs (in millions) in the United States, where x is the number of years after 2000. Use the graph to answer the following questions.

a. Estimate $J(9)$. Explain what the result means. $J(9) \approx 12.2$; in 2009, there were about 12.2 million manufacturing jobs in the U.S.

b. Estimate the value of x for which $J(x) = 14.5$. Explain what the result means. $x \approx 3$; in 2003, there were about 14.5 million manufacturing jobs in the U.S.

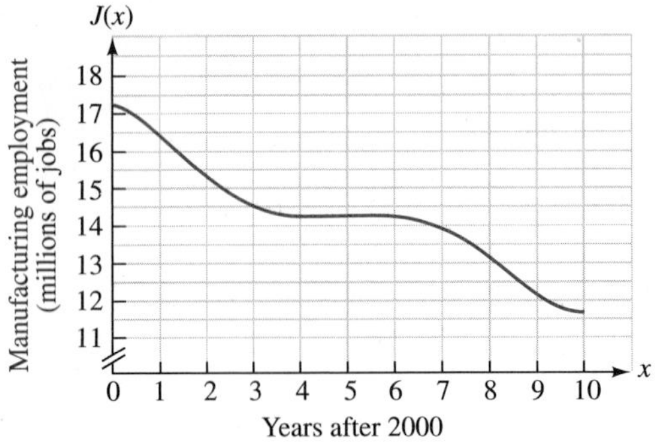

(Source: Bureau of Labor Statistics)

90. Transportation Engineering. The polynomial function A that is graphed below approximates the number of accidents per mile in one year on a 4-lane interstate, where x is the average daily traffic in number of vehicles. Use the graph to answer the following questions.

a. Estimate $A(20{,}000)$. Explain what the result means. $A(20{,}000) \approx 6.4$; There are about 6.4 accidents per mile if the average daily traffic is 20,000 vehicles.

b. Estimate the value of x for which $A(x) = 2$. Explain what the result means. $x \approx 9{,}000$; The average daily traffic is about 9,000 vehicles if there are 2 accidents per mile.

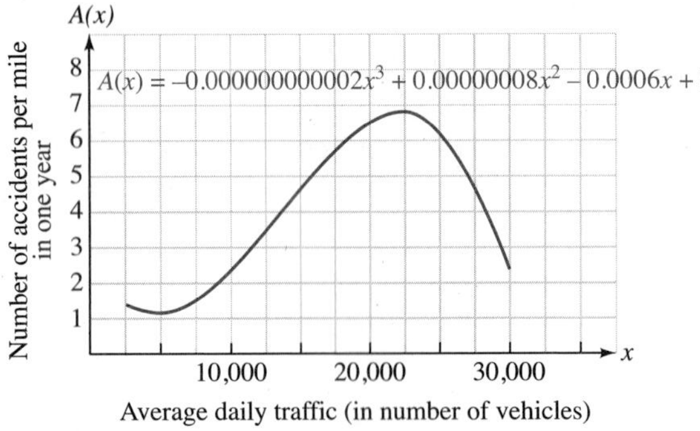

Source: Highway Safety Manual, Colorado Department of Transportation

WRITING

91. Explain how to graph a function by plotting points.

92. Explain how to *project* the graph of a function onto the x-axis. Give an example.

93. a. What does it mean to translate a graph vertically?

b. What does it mean to horizontally translate a graph?

c. What does it mean to reflect the graph of a function about the x-axis?

94. A student was asked to determine whether the graph shown below is the graph of a function. What is wrong with the following reasoning?

When I draw a vertical line through the graph, it intersects the graph only once. By the vertical line test, this is the graph of a function.

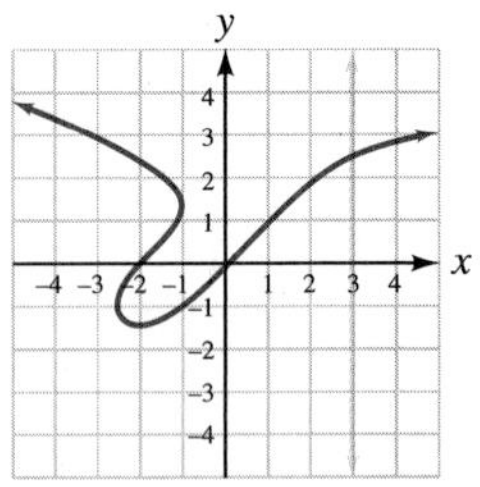

95. Explain why the graph of $g(x) = (x - 2)^2$ is two units to the right of the graph of $f(x) = x^2$.

96. Explain why the range of the polynomial function graphed below is not $(-\infty, \infty)$.

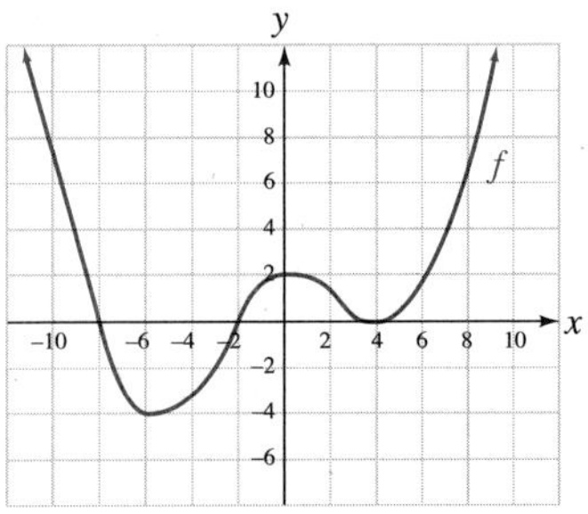

REVIEW

Solve each formula for the indicated variable.

97. $T - W = ma$ for W $W = T - ma$

98. $a + (n - 1)d = l$ for n $n = \frac{l - a + d}{d}$

99. $s = \frac{1}{2}gt^2 + vt$ for g $g = \frac{2(s - vt)}{t^2}$

100. $e = mc^2$ for m $m = \frac{e}{c^2}$

CHALLENGE PROBLEMS

Graph each function. See AIE Appendix 3.

101. $f(x) = \begin{cases} |x| & \text{for } x \geq 0 \\ x^3 & \text{for } x < 0 \end{cases}$

102. $f(x) = \begin{cases} x^2 & \text{for } x \geq 0 \\ |x| & \text{for } x < 0 \end{cases}$

Find the domain and range of each function.

103. a.

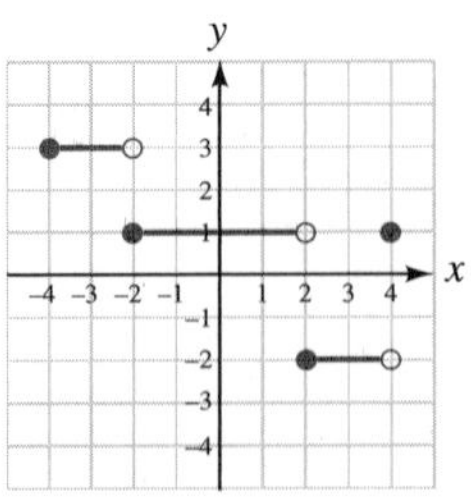

D: the set of real numbers from −4 to 4; R: $\{-2, 1, 3\}$

b.

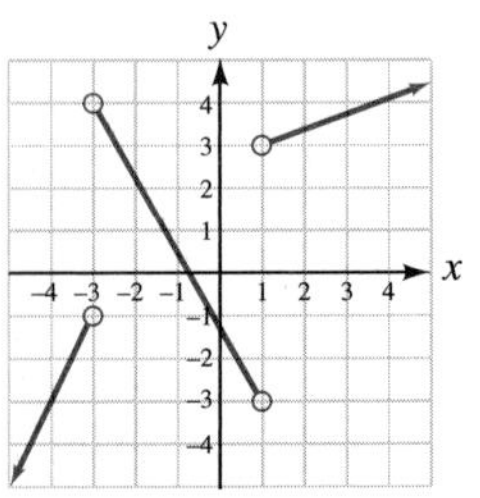

D: the set of real numbers except −3 and 1; R: the set of real numbers

104. Light. Light beams coming from a bulb are reflected outward by a parabolic mirror as parallel rays.

a. The cross-section of a parabolic mirror is given by the function $f(x) = x^2$ for the following values of x: $-0.7, -0.6, -0.5, -0.4, -0.3, -0.2, -0.1, 0, 0.1, 0.2, 0.3, 0.4, 0.5, 0.6, 0.7$. Sketch the parabolic mirror using the following grid.

b. From the lightbulb filament at (0, 0.25), draw a line segment representing a beam of light that strikes the mirror at (−0.4, 0.16) and then reflects outward, parallel to the y-axis.

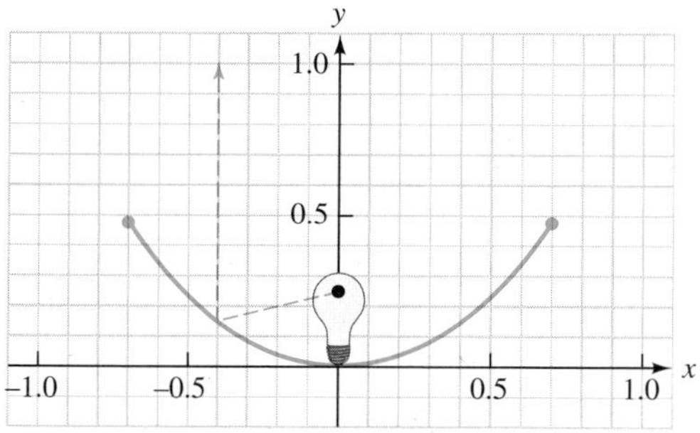

Complete the table and then graph the polynomial function. See AIE Appendix 3.

105. $f(x) = 2x^3 - 3x^2 - 11x + 6$

x	$f(x)$
−3	−42
−2	0
−1	12
0	6
1	−6
2	−12
3	0
4	42

D: $(-\infty, \infty)$; R: $(-\infty, \infty)$

106. $f(x) = -x^3 - x^2 + 6x$

x	$f(x)$
−4	24
−3	0
−2	−8
−1	−6
0	0
1	4
2	0
3	−18

D: $(-\infty, \infty)$; R: $(-\infty, \infty)$

SECTION 8.4 Solving Compound Inequalities

OBJECTIVES

1. Find the intersection and the union of two sets.
2. Solve compound inequalities containing the word *and.*
3. Solve double linear inequalities.
4. Solve compound inequalities containing the word *or.*

ARE YOU READY? *Are You Ready? exercises available online at www.webassign.net/brookscole*

The following problems review some basic skills that are needed to solve inequalities.

1. Let $A = \{-6, 1, 2, 3, 4\}$ and $B = \{0, 3, 4, 5, 6\}$. What numbers do sets A and B have in common? 3 and 4
2. Solve $6x + 8 + x \geq 5(x - 3) + 9$. Graph the solution set and write it in interval notation. $[-7, \infty)$; see AIE Appendix 3.
3. Consider the statements $x > 4$ and $x < 8$. Does the number 7 make both inequalities true? Yes
4. Graph the solution set for $x > 2$ and the solution set for $x \leq -1$ on the same number line. See AIE Appendix 3.

A label on a first-aid cream warns the user about the temperature at which the medication should be stored. A careful reading reveals that the storage instructions consist of two parts:

The storage temperature should be at least 59°F

and

the storage temperature should be at most 77°F

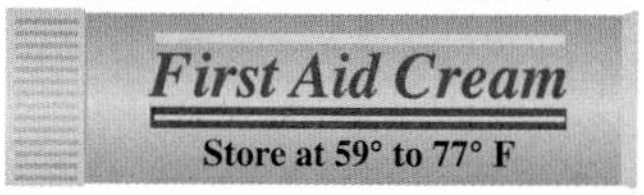

When the word *and* or the word *or* is used to connect pairs of inequalities, we call the statement a **compound inequality**. To solve compound inequalities, we need to know how to find the *intersection* and *union* of two sets.

1 Find the Intersection and the Union of Two Sets.

Just as operations such as addition and multiplication are performed on real numbers, operations also can be performed on sets. The operation of intersection of two sets produces a new third set that consists of all of the elements that the two given sets have in common.

The Intersection of Two Sets

The **intersection of set A and set B,** written $A \cap B$, is the set of all elements that are common to set A and set B.

The operation of union of two sets produces a third set that is a combination of all of the elements of the two given sets.

The Union of Two Sets

The **union of set A and set B,** written $A \cup B$, is the set of elements that belong to set A or set B or both.

Venn diagrams can be used to illustrate the intersection and union of sets. The area shown in purple in figure (a) represents $A \cap B$ and the area shown in both shades of red in figure (b) represents $A \cup B$.

The Language of Algebra

The **union** of two sets is the collection of elements that belong to either set. The concept is similar to that of a family *reunion,* which brings together the members of several families.

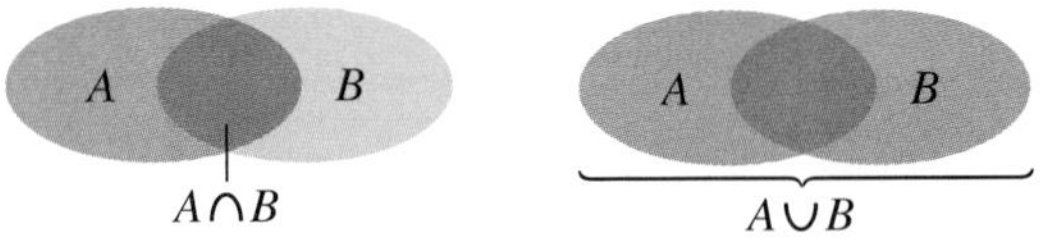

Read as "A intersect B." (a)

Read as "A union B." (b)

EXAMPLE 1 Let $A = \{0, 1, 2, 3, 4, 5, 6\}$ and $B = \{-4, -2, 0, 2, 4\}$.
a. Find $A \cap B$. **b.** Find $A \cup B$.

Strategy In part (a), we will find the elements that sets A and B have in common, and in part (b), we will find the elements that are in one set, or the other set, or both.

Why The symbol $\cap$ means intersection, and the symbol $\cup$ means union.

Solution

a. Since the numbers 0, 2, and 4 are common to both sets A and B, we have

$$A \cap B = \{0, 2, 4\}$$ Since the intersection is, itself, a set, braces are used.

b. Since the numbers in either or both sets are $-4, -2, 0, 1, 2, 3, 4, 5,$ and 6, we have

$$A \cup B = \{-4, -2, 0, 1, 2, 3, 4, 5, 6\}$$ 0, 2, and 4 are not listed twice.

The Language of Algebra

The **intersection** of two sets is the collection of elements that they have in common. When two streets cross, we call the area of pavement that they have in common an *intersection.*

Self Check 1 Let $C = \{8, 9, 10, 11\}$ and $D = \{3, 6, 9, 12, 15\}$.
a. Find $C \cap D$. $\{9\}$ **b.** Find $C \cup D$. $\{3, 6, 8, 9, 10, 11, 12, 15\}$

Now Try Problems 17 and 21

Teaching Example 1 Let $A = \{-5, -3, -1, 1, 3, 5\}$ and $B = \{0, 1, 2, 3\}$
a. Find $A \cap B$ **b.** Find $A \cup B$
Answers: **a.** $\{1, 3\}$
b. $\{-5, -3, -1, 0, 1, 2, 3, 5\}$

2 Solve Compound Inequalities Containing the Word *And.*

When two inequalities are joined with the word *and,* we call the statement a **compound inequality.** Some examples are

$$x \geq -3 \quad \text{and} \quad x \leq 6$$

$$\frac{x}{2} + 1 > 0 \quad \text{and} \quad 2x - 3 < 5$$

$$x + 3 \leq 2x - 1 \quad \text{and} \quad 3x - 2 < 5x - 4$$

The Language of Algebra

Compound means composed of two or more parts, as in *compound* inequalities, chemical *compounds,* and *compound* sentences.

Teaching Tip: Ask your students to think of some statements involving the word *and* such as: To get a license, you must pass the written test *and* the driving test.

The **solution set of a compound inequality containing the word *and*** includes all numbers that make both of the inequalities true. That is, it is the intersection of their solution sets. We can find the solution set of the compound inequality $x \geq -3$ and $x \leq 6$, for example, by graphing the solution sets of each inequality on the same number line and looking for the numbers common to both graphs.

In the following figure, the graph of the solution set of $x \geq -3$ is shown in red, and the graph of the solution set of $x \leq 6$ is shown in blue.

Preliminary work to determine the graph of the solution set

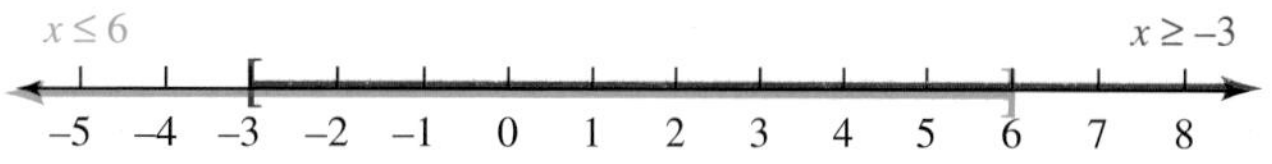

The figure below shows the graph of the solution of the compound inequality $x \geq -3$ and $x \leq 6$. The purple shaded interval, where the red and blue graphs intersect (or overlap), represents the real numbers that are common to the graphs of $x \geq -3$ and $x \leq 6$.

The graph of the solution set

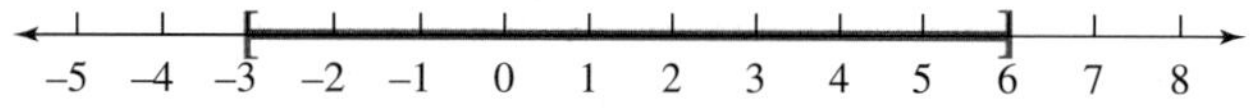

The solution set of $x \geq -3$ and $x \leq 6$ is the **bounded interval** $[-3, 6]$, where the brackets indicate that the endpoints, -3 and 6, are included. It represents all real numbers between -3 and 6, including -3 and 6. Intervals such as this, which contain both endpoints, are called **closed intervals.**

Since the solution set of $x \geq -3$ and $x \leq 6$ is the intersection of the solution sets of the two inequalities, we can write

$$[-3, \infty) \cap (-\infty, 6] = [-3, 6]$$

The solution set of the compound inequality $x \geq -3$ and $x \leq 6$ can be expressed in several ways:

Success Tip

Check with your instructor to see his or her preferred way (or ways) to present solutions of compound inequalities.

1. As a graph: 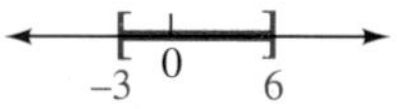

2. In words: all real numbers from -3 to 6
3. In interval notation: $[-3, 6]$
4. Using set-builder notation: $\{x \mid x \geq -3 \text{ and } x \leq 6\}$

EXAMPLE 2

Solve $\frac{x}{2} + 1 > 0$ and $2x - 3 < 5$. Graph the solution set and write it using interval notation and set-builder notation.

Strategy We will solve each inequality separately. Then we will graph the two solution sets on the same number line and determine their intersection.

Why The solution set of a compound inequality containing the word *and* is the intersection of the solution sets of the two inequalities.

Solution In each case, we can use properties of inequality to isolate the variable on one side of the inequality.

$\frac{x}{2} + 1 > 0$	and	$2x - 3 < 5$	This is the compound inequality to solve.
$\frac{x}{2} > -1$		$2x < 8$	
$x > -2$		$x < 4$	

Notation

When graphing on a number line, $(-2, 4)$ represents an *interval.* When graphing on a rectangular coordinate system, $(-2, 4)$ is an *ordered pair* that gives the coordinates of a point.

Next, we graph the solutions of each inequality on the same number line and determine their intersection.

Preliminary work to determine the graph of the solution set

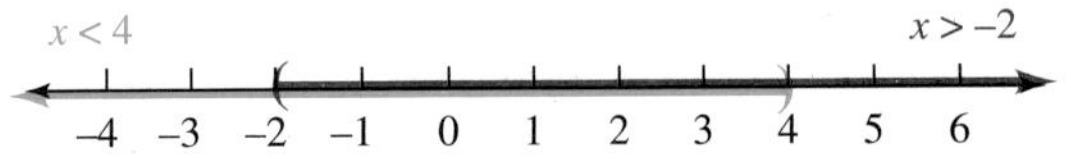

We see that the intersection of the graphs is the set of all real numbers between -2 and 4. The solution set of the compound inequality is the interval $(-2, 4)$, whose graph is shown below. This bounded interval, which does not include either endpoint, is called an **open interval.** Written using set-builder notation, the solution set is $\{x \mid x > -2 \text{ and } x < 4\}$.

The graph of the solution set

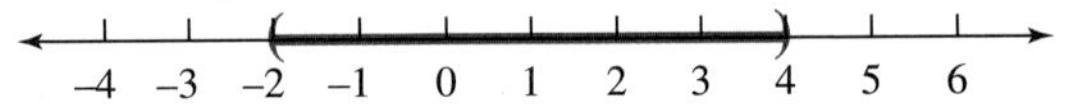

Teaching Example 2
Solve $2x + 3 \geq 5$ and $\frac{x}{3} - 2 < 1$. Graph the solution set and write it using interval notation.
Answer: $[1, 9)$

Self Check 2 Solve $3x > -18$ and $\frac{x}{5} - 1 \leq 1$. Graph the solution set and write it using interval notation. $(-6, 10]$; see AIE Appendix 3.

Now Try Problem 27

The solution of the compound inequality in the Self Check of Example 2 is the interval $(-6, 10]$. A bounded interval such as this, which includes only one endpoint, is called a **half-open interval.** The following chart shows the various types of bounded intervals, along with the inequalities and interval notation that describe them.

Intervals

Open intervals	The interval (a, b) includes all real numbers x such that $a < x < b$.	
Half-open intervals	The interval $[a, b)$ includes all real numbers x such that $a \leq x < b$.	
	The interval $(a, b]$ includes all real numbers x such that $a < x \leq b$.	
Closed intervals	The interval $[a, b]$ includes all real numbers x such that $a \leq x \leq b$.	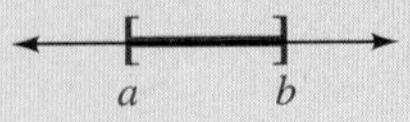

EXAMPLE 3 Solve $x + 3 \leq 2x - 1$ and $3x - 2 < 5x - 4$. Graph the solution set and write it using interval notation and set-builder notation.

Strategy We will solve each inequality separately. Then we will graph the two solution sets on the same number line and determine their intersection.

Why The solution set of a compound inequality containing the word *and* is the intersection of the solution sets of the two inequalities.

Solution In each case, we can use properties of inequality to isolate the variable on one side.

$x + 3 \leq 2x - 1$	and	$3x - 2 < 5x - 4$	This is the compound inequality to solve.
$4 \leq x$		$2 < 2x$	
$x \geq 4$		$1 < x$	
		$x > 1$	

The graph of $x \geq 4$ is shown below in red and the graph of $x > 1$ is shown below in blue.

Preliminary work to determine the graph of the solution set

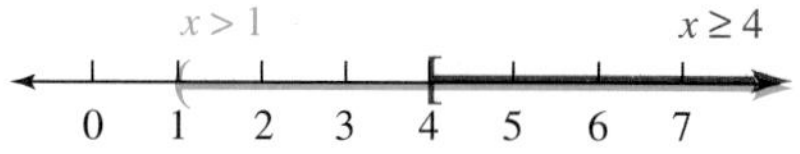

Only those values of x where $x \geq 4$ and $x > 1$ are in the solution set of the compound inequality. Since all numbers greater than or equal to 4 are also greater than 1, the solutions are the numbers x where $x \geq 4$. The solution set is the interval $[4, \infty)$, whose graph is shown below. Written using set-builder notation, the solution set is $\{x \mid x \geq 4\}$.

The graph of the solution set

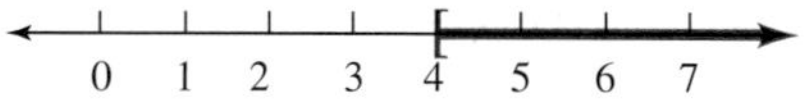

Teaching Example 3 Solve $3x + 4 \leq 2x + 6$ and $5x - 2 > 7x + 2$. Graph the solution set and write it using interval notation.
Answer: $(-\infty, -2)$

Self Check 3 Solve $2x + 3 < 4x + 2$ and $3x + 1 < 5x + 3$. Graph the solution set and write it using interval notation. $\left(\frac{1}{2}, \infty\right)$; see AIE Appendix 3.

Now Try Problem 29

EXAMPLE 4 Solve $x - 1 > -3$ and $2x < -8$, if possible.

Strategy We will solve each inequality separately. Then we will graph the two solution sets on the same number line and determine their intersection, if any.

Why The solution set of a compound inequality containing the word *and* is the intersection of the solution sets of the two inequalities.

Solution In each case, we can use properties of inequality to isolate the variable on one side.

$$x - 1 > -3 \quad \text{and} \quad 2x < -8 \qquad \text{This is the compound inequality to solve.}$$
$$x > -2 \qquad\qquad x < -4$$

The graphs of the solution sets shown below do not intersect. Since there are no numbers that make both parts of the original compound inequality true, $x - 1 > -3$ and $2x < -8$ has no solution.

Preliminary work to determine the graph of the solution set

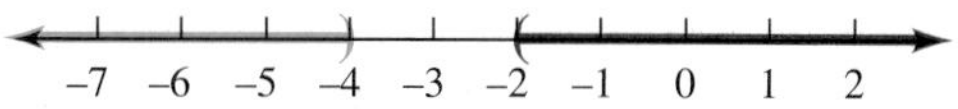

The solution set of the compound inequality is the empty set, which can be written as $\varnothing$. Since there is no solution, a graph is not needed.

Notation

The graphs of two linear inequalities can intersect at a single point, as shown below. The interval notation used to describe this point of intersection is $[3, 3]$.

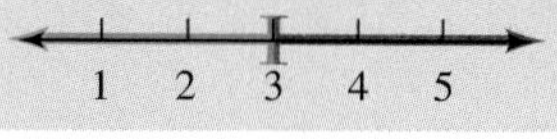

Teaching Example 4 Solve $6x - 3 < 6$ and $2x + 1 > 7$, if possible.
Answer: No solution; $\varnothing$

Self Check 4 Solve $2x - 3 < x - 2$ and $0 < x - 3.5$, if possible. No solution; $\varnothing$

Now Try Problem 31

3 Solve Double Linear Inequalities.

Inequalities that contain exactly two inequality symbols are called **double inequalities.** An example is

$$-3 \leq 2x + 5 < 7 \qquad \text{Read as "}-3\text{ is less than or equal to }2x + 5\text{ and }2x + 5\text{ is less than 7."}$$

Any double linear inequality can be written as a compound inequality containing the word *and.* In general, the following is true.

Double Linear Inequalities The compound inequality $c < x < d$ is equivalent to $c < x$ and $x < d$.

Thus, the double inequality $-3 \le 2x + 5 < 7$ is a shorter form for the compound inequality

$$-3 \le 2x + 5 \quad \text{and} \quad 2x + 5 < 7$$

As you will see in Example 5, it is easier to solve the double inequality because it enables us to solve both inequalities at once.

EXAMPLE 5 Solve $-3 \le 2x + 5 < 7$. Graph the solution set and write it using interval notation and set-builder notation.

Teaching Tip: If time allows, write Example 5 as the compound inequality $-3 \le 2x + 5$ and $2x + 5 < 7$. Solve it in this form and show that the same solution set results.

Strategy We will solve the double inequality by applying properties of inequality to *all three of its parts* to isolate x in the middle.

Why This double inequality $-3 \le 2x + 5 < 7$ means that $-3 \le 2x + 5$ and $2x + 5 < 7$. We can solve it more easily by leaving it in its original form.

Solution The objective is to isolate x in the middle of the double inequality.

$-3 \le 2x + 5 < 7$	This is the double inequality to solve.
$-3 - 5 \le 2x + 5 - 5 < 7 - 5$	To undo the addition of 5, subtract 5 from all three parts.
$-8 \le 2x < 2$	Perform the subtractions.
$\frac{-8}{2} \le \frac{2x}{2} < \frac{2}{2}$	To isolate x, undo the multiplication by 2 by dividing all three parts by 2.
$-4 \le x < 1$	Perform the divisions.

Notation

Note that the two inequality symbols in $-3 \le 2x + 5 < 7$ point in the same direction and point to the smaller number. And in $-4 \le x < 1$, the numbers -4 and 1 appear in the same order as they do on a number line.

The solution set of the double linear inequality is the half-open interval $[-4, 1)$, whose graph is shown below. Written using set-builder notation, the solution set is $\{x \mid -4 \le x < 1\}$.

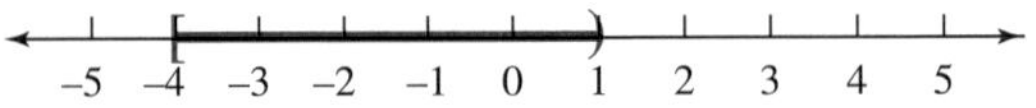

Self Check 5 Solve $-5 \le 3x - 8 \le 7$. Graph the solution set and write it using interval notation. $[1, 5]$; see AIE Appendix 3.

Now Try Problem 35

Teaching Example 5 Solve $2 < 3x - 1 \le 8$. Graph the solution set and write it using interval notation.
Answer: $(1, 3]$

CAUTION When multiplying or dividing all three parts of a double inequality by a negative number, don't forget to reverse the direction of both inequalities. As an example, we solve $-15 < -5x \le 25$.

$-15 < -5x \le 25$	
$\frac{-15}{-5} > \frac{-5x}{-5} \ge \frac{25}{-5}$	Divide all three parts by -5 to isolate x in the middle. Reverse both inequality signs.
$3 > x \ge -5$	Perform the divisions.
$-5 \le x < 3$	Write an equivalent double inequality with the smaller number, -5, on the left.

4 Solve Compound Inequalities Containing the Word *Or.*

A warning on the water temperature gauge of a commercial dishwasher cautions the operator to shut down the unit if

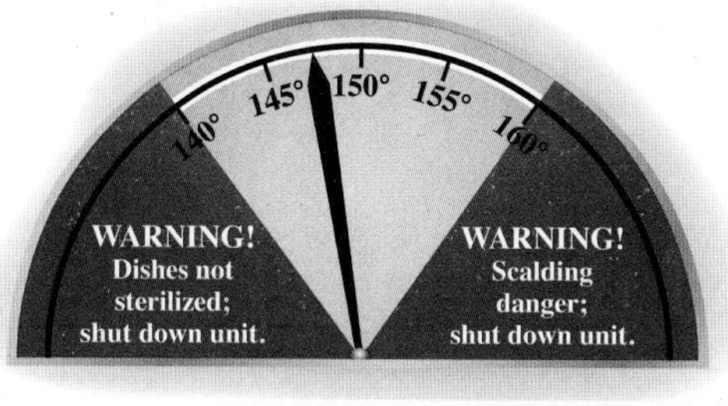

The water temperature goes below 140°

or

The water temperature goes above 160°

When two inequalities are joined with the word *or,* we also call the statement a compound inequality. Some examples are

$$x < 140 \quad \text{or} \quad x > 160$$

$$x \leq -3 \quad \text{or} \quad x \geq 2$$

$$\frac{x}{3} > \frac{2}{3} \quad \text{or} \quad -(x - 2) > 3$$

Caution

It is incorrect to write the statement $x \leq -3$ or $x \geq 2$ as the double inequality $2 \leq x \leq -3$, because that would imply that $2 \leq -3$, which is false.

The **solution set of a compound inequality containing the word *or*** includes all numbers that make one or the other or both inequalities true. That is, it is the union of their solution sets. We can find the solution set of $x \leq -3$ or $x \geq 2$, for example, by drawing the graphs of each inequality on the same number line.

In the following figure, the graph of the solution set of $x \leq -3$ is shown in red, and the graph of the solution set of $x \geq 2$ is shown in blue.

Preliminary work to determine the graph of the solution set

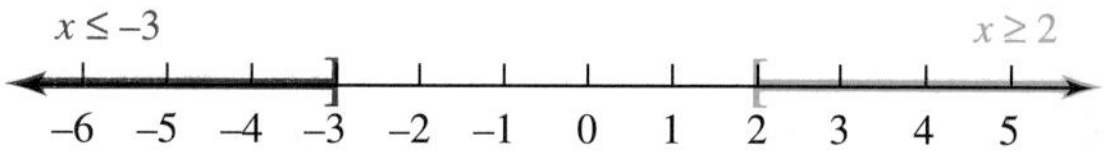

The figure below shows the graph of the solution set of $x \leq -3$ or $x \geq 2$. This graph is a union of the graph of $x \leq -3$ with the graph of $x \geq 2$.

The graph of the solution set

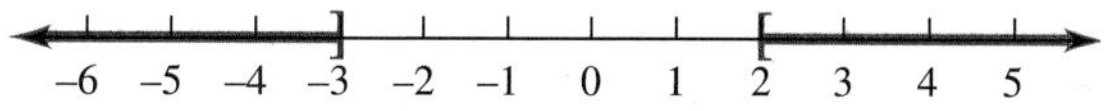

For the compound inequality $x \leq -3$ or $x \geq 2$, we can write the solution set as the union of two intervals:

$$(-\infty, -3] \cup [2, \infty)$$

We can express the solution set of the compound inequality $x \leq -3$ or $x \geq 2$ in several ways:

1. As a graph: 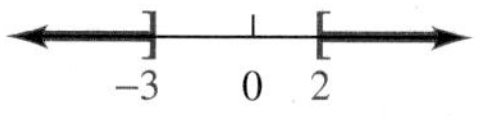

2. In words: all real numbers less than or equal to -3 *or* greater than or equal to 2
3. As the union of two intervals: $(-\infty, -3] \cup [2, \infty)$
4. Using set-builder notation: $\{x \mid x \leq -3 \text{ or } x \geq 2\}$

EXAMPLE 6

Solve $\frac{x}{3} > \frac{2}{3}$ or $-(x - 2) > 3$. Graph the solution set and write it using interval notation and set-builder notation.

Strategy We will solve each inequality separately. Then we will graph the two solution sets on the same number line to show their union.

Why The solution set of a compound inequality containing the word *or* is the union of the solution sets of the two inequalities.

Solution To solve each inequality, we proceed as follows:

$$\frac{x}{3} > \frac{2}{3} \quad \text{or} \quad -(x-2) > 3 \qquad \text{This is the compound inequality to solve.}$$

$$x > 2 \quad \Bigg| \quad \begin{aligned} -x + 2 &> 3 \\ -x &> 1 \\ x &< -1 \end{aligned}$$

Next, we graph the solutions of each inequality on the same number line and determine their union.

Preliminary work to determine the graph of the solution set

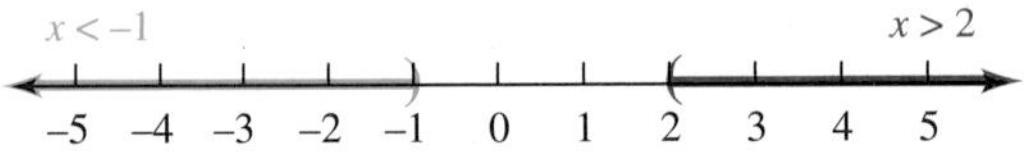

The union of the two solution sets consists of all real numbers less than -1 or greater than 2. The solution set of the compound inequality is the union of two intervals: $(-\infty, -1) \cup (2, \infty)$. Its graph appears below. Written using set-builder notation, the solution set is $\{x \mid x > 2 \text{ or } x < -1\}$.

The graph of the solution set

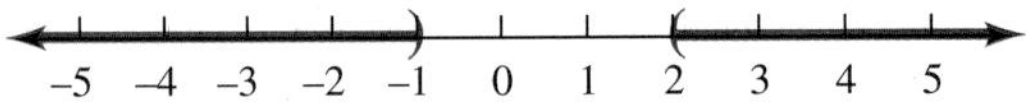

The Language of Algebra

The meaning of the word **or** in a compound inequality differs from our everyday use of the word. For example, when we say, "I will go shopping today *or* tomorrow," we mean that we will go one day or the other, but *not* both. With compound inequalities, *or* includes one possibility, or the other, or both.

Self Check 6 Solve $\frac{x}{2} > 2$ or $-3(x - 2) > 0$. Graph the solution set and write it using interval notation. $(-\infty, 2) \cup (4, \infty)$; see AIE Appendix 3.

Now Try Problem 41

Teaching Example 6 Solve $\frac{x}{6} \geq \frac{1}{2}$ or $-2(x + 1) > 4$. Graph the solution set and write it in interval notation.
Answer: $(-\infty, -3) \cup [3, \infty)$

EXAMPLE 7

Solve $x + 3 \geq -3$ or $-x > 0$. Graph the solution set and write it using interval notation and set-builder notation.

Strategy We will solve each inequality separately. Then we will graph the two solution sets on the same number line to show their union.

Why The solution set of a compound inequality containing the word *or* is the union of the solution sets of the two inequalities.

Solution To solve each inequality, we proceed as follows:

$$x + 3 \geq -3 \quad \text{or} \quad -x > 0 \qquad \text{This is the compound inequality to solve.}$$

$$x \geq -6 \quad | \quad x < 0$$

We graph the solution set of each inequality on the same number line and determine their union.

Preliminary work to determine the graph of the solution set

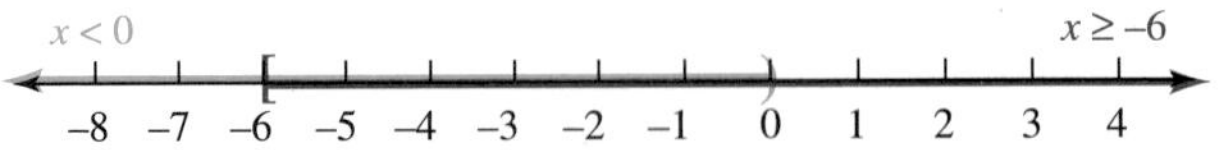

Since the entire number line is shaded, all real numbers satisfy the original compound inequality and the solution set is denoted as $(-\infty, \infty)$ or $\mathbb{R}$. Its graph is shown below. Written using set-builder notation, the solution set is $\{x \mid x \text{ is a real number}\}$.

The graph of the solution set

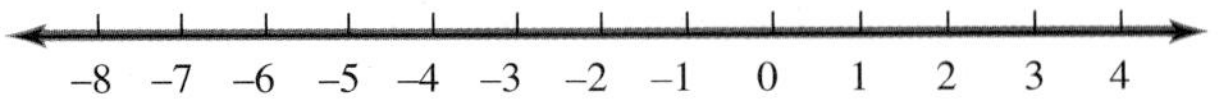

Teaching Tip: To prepare your students for the Look Alike problems, ask them how to use this answer to solve $x + 3 \geq -3$ and $-x > 0$ without having to do any new work.

Teaching Example 7 Solve $-3x \le 12$ or $5x + 1 < 11$. Graph the solution set and write it using interval notation.
Answer: $(-\infty, \infty)$

Self Check 7 Solve $x - 1 < 5$ or $-2x \le 10$. Graph the solution set and write it using interval notation. $(-\infty, \infty)$; see AIE Appendix 3.

Now Try Problem 43

Solving Compound Inequalities

1. Solve each inequality separately and graph their solution sets in different colors on the same number line.
2. If the inequalities are connected with the word *and*, find the intersection of the two solution sets. If the inequalities are connected with the word *or*, find the union of the two solution sets.
3. Write the solution set of the compound inequality using interval notation or set-builder notation, and graph it on a new number line.

SECTION 8.4 STUDY SET

VOCABULARY

Fill in the blanks.

1. The ___intersection___ of two sets is the set of elements that are common to both sets and the ___union___ of two sets is the set of elements that are in one set, or the other, or both.
2. $x \ge 3$ and $x < 4$ is a ___compound___ inequality.
3. $-6 < x + 1 \le 1$ is a ___double___ linear inequality.
4. (2, 8) is an example of an open ___interval___, [−4, 0] is an example of a ___closed___ interval, and (0, 9] is an example of a half-___open___ interval.

CONCEPTS

Fill in the blanks.

5. a. The solution set of a compound inequality containing the word *and* includes all numbers that make ___both___ inequalities true.
   b. The solution set of a compound inequality containing the word *or* includes all numbers that make ___one___, or the other, or ___both___ inequalities true.
6. The double inequality $4 < 3x + 5 \le 15$ is equivalent to $4 < 3x + 5$ ___and___ $3x + 5 \le 15$.
7. a. When solving a compound inequality containing the word *and*, the solution set is the ___intersection___ of the solution sets of the inequalities.
   b. When solving a compound inequality containing the word *or*, the solution set is the ___union___ of the solution sets of the inequalities.
8. When multiplying or dividing all three parts of a double inequality by a negative number, the direction of both inequality symbols must be ___reversed___.
9. Use a check to determine whether −3 is a solution of the compound inequality.
   a. $\frac{x}{3} + 1 \ge 0$ and $2x - 3 < -10$ No
   b. $2x \le 0$ or $-3x < -5$ Yes

10. Use a check to determine whether −3 is a solution of the double linear inequality.
   a. $-1 < -3x + 4 < 12$ No
   b. $-1 < -3x + 4 < 14$ Yes
11. Use interval notation, if possible, to describe the intersection of each pair of graphs.
   a.  [−2, 1)
   b. 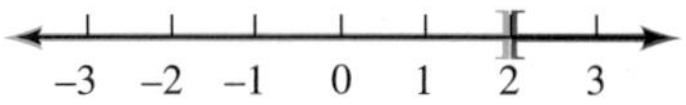 [2, 2]
   c. 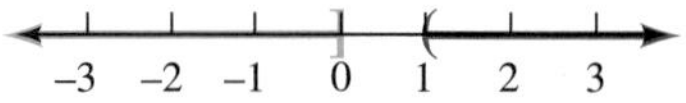 ∅
12. Use interval notation to describe the union of each pair of graphs.
   a. 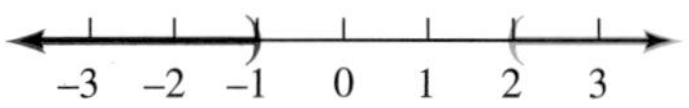 $(-\infty, -1) \cup (2, \infty)$
   b.  $(-\infty, \infty)$
   c.  $(-\infty, 0) \cup (0, \infty)$

NOTATION

13. Fill in the blanks: We read ∪ as ___union___ and ∩ as ___intersection___.
14. Match each interval with its corresponding graph.
   a. [2, 3) ii i. 
   b. (2, 3) iii ii. 
   c. [2, 3] i iii. 
15. What set is represented by the interval notation $(-\infty, \infty)$? Graph it. All real numbers; see AIE Appendix 3.
16. a. Graph: $(-\infty, 2) \cup [3, \infty)$ See AIE Appendix 3.
   b. Graph: $(-\infty, 3) \cap [-2, \infty)$

▶ Selected exercises available online at www.webassign.net/brookscole

GUIDED PRACTICE

Let $A = \{0, 1, 2, 3, 4, 5, 6\}$, $B = \{4, 6, 8, 10\}$, $C = \{-3, -1, 0, 1, 2\}$, ***and*** $D = \{-3, 1, 2, 5, 8\}$. ***Find each set.*** **See Example 1.**

17. $A \cap B$ $\{4, 6\}$
18. $A \cap D$ $\{1, 2, 5\}$
19. $C \cap D$ $\{-3, 1, 2\}$
20. $B \cap C$ $\varnothing$
21. $B \cup C$ $\{-3, -1, 0, 1, 2, 4, 6, 8, 10\}$
22. $A \cup C$ $\{-3, -1, 0, 1, 2, 3, 4, 5, 6\}$
23. $A \cup D$ $\{-3, 0, 1, 2, 3, 4, 5, 6, 8\}$
24. $C \cup D$ $\{-3, -1, 0, 1, 2, 5, 8\}$

Solve each compound inequality, if possible. Graph the solution set (if one exists) and write it using interval notation. **See Examples 2–4.** See AIE Appendix 3.

25. $x > -2$ and $x \le 5$ $(-2, 5]$
26. $x \le -4$ and $x \ge -7$ $[-7, -4]$
27. $2x - 1 > 3$ and $x + 8 \le 11$ $(2, 3]$
28. $5x - 3 \ge 2$ and $6 \ge 4x - 3$ $\left[1, \frac{9}{4}\right]$
29. $6x + 1 < 5x - 3$ and $\frac{x}{2} + 9 \le 6$ $(-\infty, -6]$
30. $\frac{2}{3}x + 1 > -9$ and $\frac{3}{4}x - 1 > -10$ $(-12, \infty)$
31. $x + 2 < -\frac{1}{3}x$ and $-6x < 9x$ No solution; $\varnothing$
32. $\frac{3}{2}x + \frac{1}{5} < 5$ and $2x + 1 > 9$ No solution; $\varnothing$

Solve each double inequality. Graph the solution set and write it using interval notation. **See Example 5.** See AIE Appendix 3.

33. $4 \le x + 3 \le 7$ $[1, 4]$
34. $-5.3 \le x - 2.3 \le -1.3$ $[-3, 1]$
35. $0.9 < 2x - 0.7 < 1.5$ $(0.8, 1.1)$
36. $7 < 3x - 2 < 25$ $(3, 9)$

Solve each compound inequality. Graph the solution set and write it using interval notation. **See Examples 6 and 7.** See AIE Appendix 3.

37. $x \le -2$ or $x > 6$ $(-\infty, -2] \cup (6, \infty)$
38. $x \ge -1$ or $x \le -3$ $(-\infty, -3] \cup [-1, \infty)$
39. $x - 3 < -4$ or $-x + 2 < 0$ $(-\infty, -1) \cup (2, \infty)$
40. $4x < -12$ or $\frac{x}{2} > 4$ $(-\infty, -3) \cup (8, \infty)$
41. $3x + 2 < 8$ or $2x - 3 > 11$ $(-\infty, 2) \cup (7, \infty)$
42. $3x + 4 < -2$ or $3x + 4 > 10$ $(-\infty, -2) \cup (2, \infty)$
43. $2x > x + 3$ or $\frac{x}{8} + 1 < \frac{13}{8}$ $(-\infty, \infty)$
44. $2(x + 2) < x - 11$ or $-\frac{x}{5} < 20$ $(-\infty, \infty)$

TRY IT YOURSELF

Solve each compound inequality, if possible. Graph the solution set (if one exists) and write it using interval notation. See AIE Appendix 3.

45. $-4(x + 2) \ge 12$ or $3x + 8 < 11$ $(-\infty, 1)$
46. $4.5x - 1 < -10$ or $6 - 2x \ge 12$ $(-\infty, -2)$
47. $2.2x < -19.8$ and $-4x < 40$ $(-10, -9)$
48. $\frac{1}{2}x \le 2$ and $0.75x \ge -6$ $[-8, 4]$
49. $-2 < -b + 3 < 5$ $(-2, 5)$
50. $2 < -t - 2 < 9$ $(-11, -4)$
51. $4.5x - 2 > 2.5$ or $\frac{1}{2}x \le 1$ $(-\infty, \infty)$
52. $0 < x$ or $3x - 5 > 4x - 7$ $(-\infty, \infty)$
53. $5(x - 2) \ge 0$ and $-3x < 9$ $[2, \infty)$
54. $x - 1 \le 2(x + 2)$ and $x \le 2x - 5$ $[5, \infty)$
55. $-x < -2x$ and $3x > 2x$ No solution; $\varnothing$
56. $-\frac{x}{4} > -2.5$ and $9x > 2(4x + 5)$ No solution; $\varnothing$
57. $-6 < -3(x - 4) \le 24$ $[-4, 6)$
58. $-4 \le -2(x + 8) < 8$ $(-12, -6]$
59. $2x + 1 \ge 5$ and $-3(x + 1) \ge -9$ $[2, 2]$
60. $2(-2) \le 3x - 1$ and $3x - 1 \le -1 - 3$ $[-1, -1]$
61. $\frac{4.5x - 12}{2} < x$ or $-15.3 > -3(x - 1.4)$ $(-\infty, 4.8) \cup (6.5, \infty)$
62. $y + 0.52 < 1.05y$ or $9.8 - 15y > -15.7$ $(-\infty, 1.7) \cup (10.4, \infty)$
63. $\frac{x}{0.7} + 5 > 4$ and $-4.8 \le \frac{3x}{-0.125}$ $(-0.7, 0.2]$
64. $5(x + 1) \le 4(x + 3)$ and $x + 12 < -3$ $(-\infty, -15)$
65. $-24 < \frac{3}{2}x - 6 \le -15$ $(-12, -6]$
66. $-4 > \frac{2}{3}x - 2 > -6$ $(-6, -3)$
67. $\frac{x}{3} - \frac{x}{4} > \frac{1}{6}$ or $\frac{x}{2} + \frac{2}{3} \le \frac{3}{4}$ $\left(-\infty, \frac{1}{6}\right] \cup (2, \infty)$
68. $\frac{a}{2} + \frac{7}{4} > 5$ or $\frac{3}{8} + \frac{a}{3} \le \frac{5}{12}$ $\left(-\infty, \frac{1}{8}\right] \cup \left(\frac{13}{2}, \infty\right)$
69. $0 \le \frac{4 - x}{3} \le 2$ $[-2, 4]$
70. $-2 \le \frac{5 - 3x}{2} \le 2$ $\left[\frac{1}{3}, 3\right]$
71. $x \le 6 - \frac{1}{2}x$ and $\frac{1}{2}x + 1 \ge 3$ $[4, 4]$
72. $3\left(x + \frac{2}{3}\right) \le -7$ and $2(x + 2) \ge -2$ $[-3, -3]$
73. $-6 < f(x) \le 0$ where $f(x) = 3x - 9$ $(1, 3]$
74. $-3 < f(x) < 7$ where $f(x) = \frac{2}{3}x - \frac{1}{3}$ $(-4, 11)$
75. Let $f(x) = 5x + 14$ and $g(x) = 2x + 8$. Find all values of x for which $f(x) > 29$ and $g(x) < 20$. $(3, 6)$
76. Let $f(x) = x - 2$. Find all values of x for which $f(x) > 5$ or $f(x) < -1$. $(-\infty, 1) \cup (7, \infty)$

Look Alikes . . .

Solve the inequality in part a. Graph the solution set and write it in interval notation. Then use your work from part a to determine the solution set for the compound inequality in part b. (No new work is necessary!) Graph the solution set and write it in interval notation. See AIE Appendix 3.

77. **a.** $3x - 2 \ge 4$ and $x + 6 \ge 12$ $[6, \infty)$
b. $3x - 2 \ge 4$ or $x + 6 \ge 12$ $[2, \infty)$

78. **a.** $x + 2 \le 10$ or $x - 3 \ge 2$ $(-\infty, \infty)$

b. $x + 2 \le 10$ and $x - 3 \ge 2$ $[5, 8]$

79. **a.** $2x + 1 \le 7$ and $3x + 5 \ge 23$ No solution; $\varnothing$

b. $2x + 1 \le 7$ or $3x + 5 \ge 23$ $(-\infty, 3] \cup [6, \infty)$

80. **a.** $7 \le 4x + 1 < 23$ $\left[\frac{3}{2}, \frac{11}{2}\right)$

b. $7 < 4x + 1 \le 23$ $\left(\frac{3}{2}, \frac{11}{2}\right]$

APPLICATIONS

81. **Baby Furniture.** Refer to the illustration. A company manufactures various sizes of play yard cribs having perimeters between 128 and 192 inches, inclusive.

a. Complete the double inequality that describes the range of the perimeters of the play yard shown.

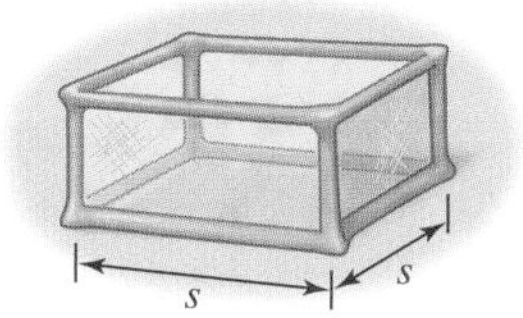

$128 \le 4s \le 192$

b. Solve the double inequality to find the range of the side lengths of the play yard. $32 \le s \le 48$, $[32, 48]$

82. **Trucking.** The distance that a truck can travel in 8 hours, at a constant rate of r mph, is given by $8r$. A trucker wants to travel at least 350 miles, and company regulations don't allow him to exceed 450 miles in one 8-hour shift.

a. Complete the double inequality that describes the mileage range of the truck.

$350 \le 8r \le 450$

b. Solve the double inequality to find the range of the average rate (speed) of the truck for the 8-hour trip. $43.75 \le r \le 56.25$, $[43.75, 56.25]$

83. **Thermostats.** During business hours, as shown in figure (a), the *Temp range* control in an office is set at 5. This means that the heater comes on when the room temperature gets 5 degrees below the *Temp setting* and the air conditioner comes on when the room temperature gets 5 degrees above the *Temp setting.*

a. Use interval notation to describe the temperature range when neither the heater nor the air conditioner will come on during business hours. $(67, 77)$

b. After business hours, the *Temp range* setting is changed to save energy. See figure (b). Use interval notation to describe the after-business-hours temperature range when neither the heater nor the air conditioner will come on. $(62, 82)$

Thermostat Thermostat

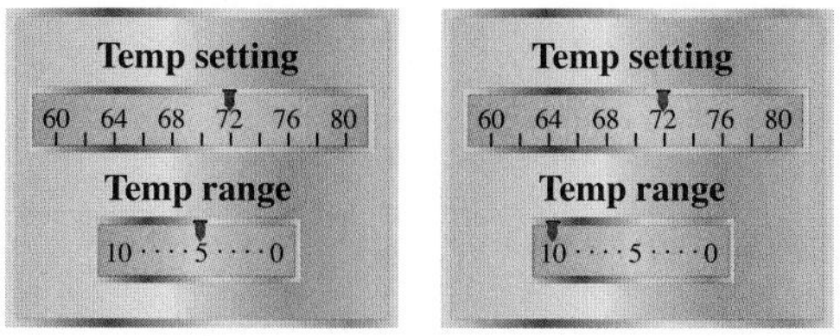

During business hours (a) After business hours (b)

84. **Treating Fevers.** Use the flow chart to determine what action should be taken for a 13-month-old child who has had a 99.8° temperature for 3 days and is not suffering any other symptoms. T represents the child's temperature, A the child's age in months, and S the number of hours the child has experienced the symptoms. See doctor today.

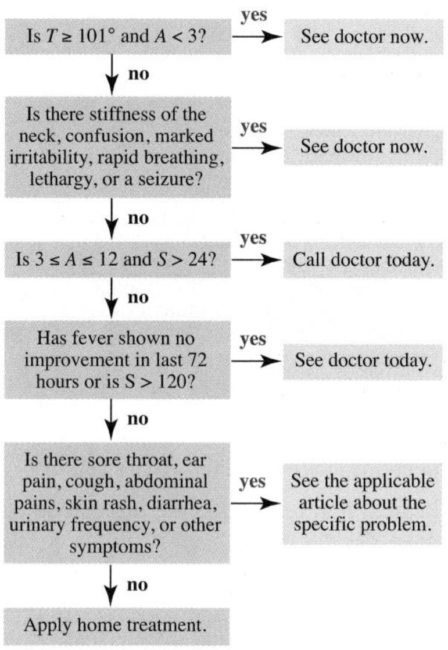

Based on information from *Take Care of Yourself* (Addison-Wesley, 1993)

85. **U.S. Health Care.** Refer to the following graph. Let P represent the percent of children covered by private insurance, M the percent covered by Medicaid, and N the percent not covered. For what years are the following true?

a. $P \ge 63$ and $M \ge 26$ 2004

b. $P > 60$ or $M \ge 29$ 2004, 2005, 2006, 2007

c. $M \ge 29$ and $N \le 10$ 2006, 2007

d. $M \ge 30$ or $N < 9.2$ 2007

U.S. Health Care Coverage for People Under 18 Years of Age (in percent)

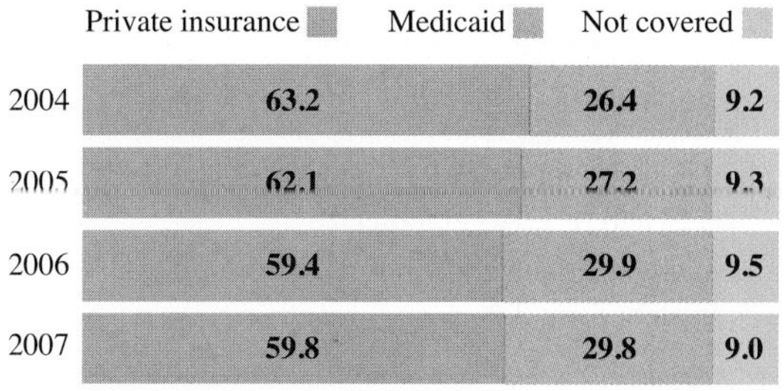

Source: U.S. Department of Health and Human Services

(The small percent of people covered each year by some nonstandard program is not shown.)

86. Polls. For each response to the poll question shown below, the *margin of error* is +/− (read as "plus or minus") 2.8%. This means that for the statistical methods used to do the polling, the actual response could be as much as 2.8 points more or 2.8 points less than shown. Use interval notation to describe the possible interval (in percent) for each response. All of them: $[16.2, 21.8]$; most of them: $[21.2, 26.8]$; some of them: $[26.2, 31.8]$; none of them: $[25.2, 30.8]$

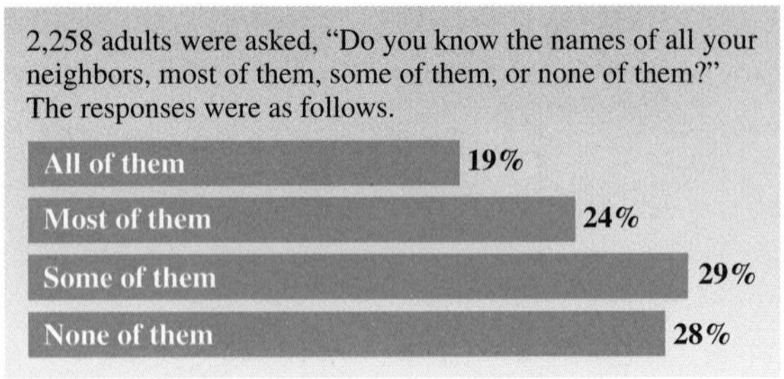

87. Street Intersections. Refer to figure (a) below.

a. Shade the area that represents the intersection of the two streets shown in the illustration.

b. Shade the area that represents the union of the two streets.

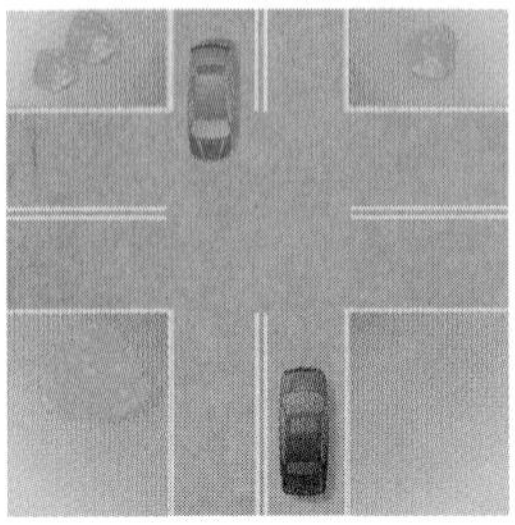

(a)

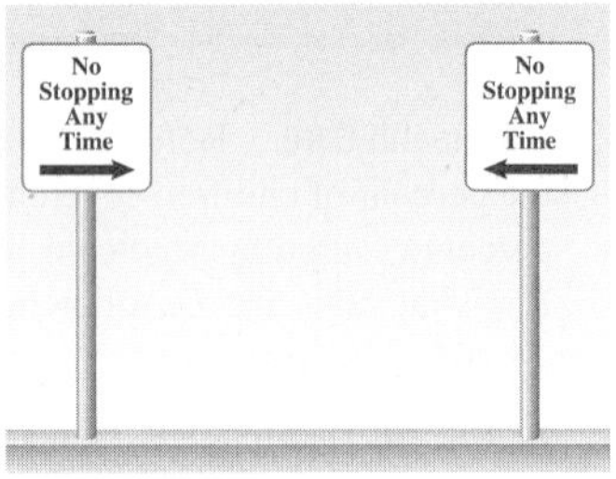

(b)

88. Traffic Signs. The pair of signs shown in figure (b) above are a real-life example of which concept discussed in this section? The no-stopping area is defined by the intersection of the arrows pointing in opposite directions.

WRITING

89. Explain how to find the union and how to find the intersection of $(-\infty, 5)$ and $(-2, \infty)$ graphically.

90. Explain why the double inequality $2 < x < 8$ can be written in the equivalent form $2 < x$ and $x < 8$.

91. Explain the meaning of the notation $(-1, 2)$ for each type of graph.

a. 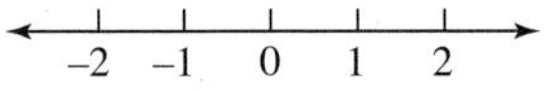

b. 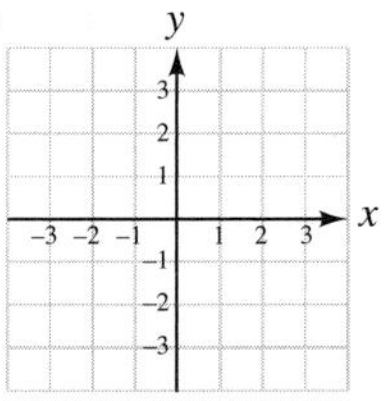

92. The meaning of the word *or* in a compound inequality differs from our everyday use of the word. Explain the difference.

93. Describe each set in words.

a. $(-3, 3)$ b. $[7, 12]$

c. $(-\infty, 5] \cup (6, \infty)$

94. What is incorrect about the double inequality $3 < -3x + 4 < -3$?

REVIEW

95. Airplanes. Together, a Delta B747 and a Delta B777 seat 681 passengers. If the B777 seats 125 less people than the B747, how many passengers does each seat? (Source: deltaskymag.com) B747: 403 passengers; B777: 278 passengers

96. Denzel. As of October 2010, Denzel Washington's three top domestic grossing films, *American Gangster, Remember the Titans,* and *The Pelican Brief,* had earned a total of \$346.7 million. If *American Gangster* earned \$14.5 million more than *Remember the Titans,* and if *Remember the Titans* earned \$14.9 million more than *The Pelican Brief,* how much did each film earn as of that date? (Source: boxofficemojo.com) *American Gangster:* \$130.2 million; *Titans:* \$115.7 million; *Pelican:* \$100.8 million

CHALLENGE PROBLEMS

Solve each compound inequality. Graph the solution set and write it in interval notation. See AIE Appendix 3.

97. $-5 < \dfrac{x+2}{-2} < 0$ or $2x + 10 \ge 30$ $(-2, 8) \cup [10, \infty)$

98. $-2 \le \dfrac{x-4}{3} \le 0$ and $\dfrac{x-5}{2} \ge -3$ $[-1, 4]$

99. $x - 12 < 4x < 2x + 16$ $(-4, 8)$

100. $6(x - 3) \le 3(3x + 2) \le 4(2x + 3)$ $[-8, 6]$

SECTION 8.5

OBJECTIVES

1 Solve equations of the form $|X| = k$.
2 Solve equations with two absolute values.
3 Solve inequalities of the form $|X| < k$.
4 Solve inequalities of the form $|X| > k$.

Solving Absolute Value Equations and Inequalities

ARE YOU READY? *Are You Ready? exercises available online at www.webassign.net/brookscole*

The following problems review some basic skills that are needed to solve absolute value equations and inequalities.

1. Find each absolute value.

a. $|12|$ 12 **b.** $|-7.5|$ 7.5

2. Tell whether each statement is true or false.

a. $|-3| \geq 2$ True **b.** $|-26| < -27$ False

3. Solve $3x + 6 \leq -3$ or $3x + 6 \geq 9$. Graph the solution set and write it using interval notation. $(-\infty, -3] \cup [1, \infty)$; see AIE Appendix 3.

4. Solve $-8 < 2x + 8 < 16$. Graph the solution set and write it using interval notation. $(-8, 4)$; see AIE Appendix 3.

Many quantities studied in mathematics, science, and engineering are expressed as positive numbers. To guarantee that a quantity is positive, we often use absolute value. In this section, we will consider equations and inequalities involving the absolute value of an algebraic expression. Some examples are

$$|3x - 2| = 5, \qquad |2x - 3| < 9, \qquad \text{and} \qquad \left|\frac{3 - x}{5}\right| \geq 6$$

To solve these *absolute value equations* and *inequalities,* we write and then solve equivalent compound equations and inequalities.

1 Solve Equations of the Form $|X| = k$.

Recall that the absolute value of a real number is its distance from 0 on a number line. To solve the **absolute value equation** $|x| = 5$, we must find all real numbers x whose distance from 0 on the number line is 5. There are two such numbers: 5 and -5. It follows that the solutions of $|x| = 5$ are 5 and -5 and the solution set is $\{5, -5\}$.

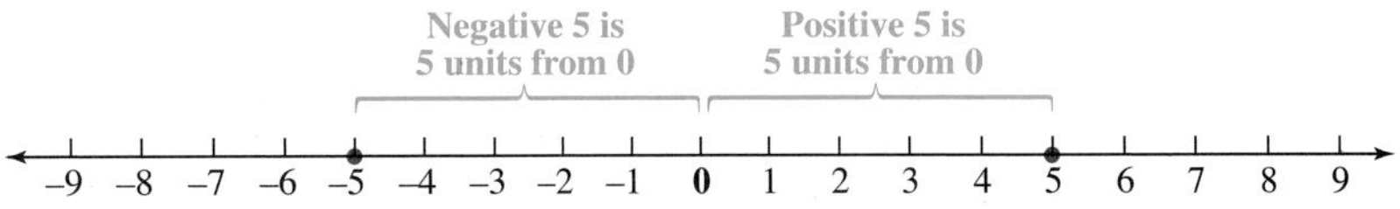

The results from this example suggest the following approach for solving absolute value equations.

Solving Absolute Value Equations

For any positive number k and any algebraic expression X:
To solve $|X| = k$, solve the equivalent compound equation

$$X = k \qquad \text{or} \qquad X = -k$$

Teaching Tip: Point out that the equivalent compound equation contains the word "or." Ask your students why the word "and" would not be correct.

The statement $X = k$ or $X = -k$ is called a **compound equation** because it consists of two equations joined with the word *or.*

EXAMPLE 1 Solve: **a.** $|s| = 0.003$ **b.** $|3x - 2| = 5$ **c.** $|10 - x| = -40$

Strategy To solve the first two equations, we will write and then solve an equivalent compound equation. We will solve the third equation by inspection.

Why All three of the equations are of the form $|X| = k$. However, the standard method for solving absolute value equations cannot be applied to $|10 - x| = -40$ because k is negative.

Solution

The Language of Algebra

When we say that the absolute value equation and a compound equation are **equivalent,** we mean that they have the same solution(s).

a. The absolute value equation $|s| = 0.003$ is equivalent to the compound equation

$$s = 0.003 \quad \text{or} \quad s = -0.003$$

Therefore, the solutions of $|s| = 0.003$ are 0.003 and -0.003, and the solution set is $\{0.003, -0.003\}$.

b. The equation-solving method used in part a can be extended to equations where the expression within absolute value bars is more complicated than a single variable. The absolute value equation $|3x - 2| = 5$ is equivalent to the compound equation

$$3x - 2 = 5 \quad \text{or} \quad 3x - 2 = -5$$

Caution

When you see the word *solve,* you probably think of steps such as combining like terms, distributing, or doing something to both sides. However, to solve absolute value equations of the form $|X| = k$, those familiar steps are not used until an equivalent compound equation is written.

Now we solve each equation for x:

$$\begin{array}{rcl|rcl} 3x - 2 &=& 5 \quad \text{or} & 3x - 2 &=& -5 \\ 3x &=& 7 & 3x &=& -3 \\ x &=& \frac{7}{3} & x &=& -1 \end{array}$$

The results must be checked separately to see whether each of them produces a true statement. We substitute $\frac{7}{3}$ for x and then -1 for x in the original equation.

Check: ***For*** $x = \frac{7}{3}$

$$|3x - 2| = 5$$
$$\left|3\left(\frac{7}{3}\right) - 2\right| \stackrel{?}{=} 5$$
$$|7 - 2| \stackrel{?}{=} 5$$
$$|5| \stackrel{?}{=} 5$$
$$5 = 5 \quad \text{True}$$

For $x = -1$

$$|3x - 2| = 5$$
$$|3(-1) - 2| \stackrel{?}{=} 5$$
$$|-3 - 2| \stackrel{?}{=} 5$$
$$|-5| \stackrel{?}{=} 5$$
$$5 = 5 \quad \text{True}$$

Teaching Tip: Ask your students why the instructions for Example 1 do not ask for the solution set to be written in interval notation. Also, ask them what the graphs of the solution sets for parts a and b would look like.

The resulting true statements indicate that the equation has two solutions: $\frac{7}{3}$ and -1. The solution set is $\left\{\frac{7}{3}, -1\right\}$.

c. Since an absolute value can never be negative, there are no real numbers x that make $|10 - x| = -40$ true. The equation has no solution and the solution set is $\varnothing$.

Teaching Example 1 Solve:
a. $|x| = \frac{3}{8}$ b. $|6 - 2x| = 12$
c. $\left|\frac{x + 1}{2}\right| = -4$
Answers: a. $\frac{3}{8}, -\frac{3}{8}$ b. $-3, 9$
c. No solution

Self Check 1 Solve: **a.** $|x| = \frac{1}{2}$ $\frac{1}{2}, -\frac{1}{2}$ **b.** $|2x - 3| = 7$ $5, -2$
c. $\left|\frac{x}{4} - 1\right| = -3$ No solution, $\varnothing$

Now Try Problems 19, 23, and 29

When solving absolute value equations (or inequalities), we must **isolate the absolute value expression on one side** before writing the equivalent compound statement.

EXAMPLE 2

Solve: $\left|\frac{2}{3}x + 3\right| + 4 = 10$

Strategy We will first isolate $\left|\frac{2}{3}x + 3\right|$ on the left side of the equation and then write and solve an equivalent compound equation.

Why After isolating the absolute value expression on the left, the resulting equation will have the desired form $|X| = k$.

Solution

$$\left|\frac{2}{3}x + 3\right| + 4 = 10 \quad \text{This is the equation to solve.}$$

$$\left|\frac{2}{3}x + 3\right| = 6 \quad \text{To isolate the absolute value expression, subtract 4 from both sides. The resulting equation is in the form } |X| = k.$$

With the absolute value now isolated, we can solve $\left|\frac{2}{3}x + 3\right| = 6$ by writing and solving an equivalent compound equation:

$$\frac{2}{3}x + 3 = 6 \quad \text{or} \quad \frac{2}{3}x + 3 = -6$$

Now we solve each equation for x:

$$\frac{2}{3}x + 3 = 6 \quad \text{or} \quad \frac{2}{3}x + 3 = -6$$

$$\frac{2}{3}x = 3 \qquad\qquad \frac{2}{3}x = -9$$

$$2x = 9 \qquad\qquad 2x = -27$$

$$x = \frac{9}{2} \qquad\qquad x = -\frac{27}{2}$$

Verify that both $\frac{9}{2}$ and $-\frac{27}{2}$ are solutions by substituting them into the original equation.

Caution

A common error when solving absolute value equations is to forget to isolate the absolute value expression first. Note:

$$\left|\frac{2}{3}x + 3\right| + 4 = 10$$

does not mean

$$\frac{2}{3}x + 3 + 4 = 10 \quad \text{or} \quad \frac{2}{3}x + 3 + 4 = -10$$

(crossed out)

Teaching Tip: Remind your students that the instruction "solve" means to find *all* of the solutions of the given equation. Here, for example, solving only $\frac{2}{3}x + 3 = 6$ is not sufficient.

Teaching Example 2
Solve: $\left|\frac{3x}{2} + 4\right| - 3 = -2$
Answer: $-\frac{10}{3}, -2$

Self Check 2 Solve: $|0.4x - 2| - 0.6 = 0.4$ 7.5, 2.5

Now Try Problem 33

EXAMPLE 3 Solve: $3\left|\frac{1}{2}x - 5\right| - 4 = -4$

Strategy We will first isolate $\left|\frac{1}{2}x - 5\right|$ on the left side of the equation and then write and solve an equivalent compound equation.

Why After isolating the absolute value expression, the resulting equation will have the desired form $|X| = k$.

Solution

$$3\left|\frac{1}{2}x - 5\right| - 4 = -4 \quad \text{This is the equation to solve.}$$

$$3\left|\frac{1}{2}x - 5\right| = 0 \quad \text{To isolate the absolute value expression } \left|\tfrac{1}{2}x - 5\right|, \text{ we first add 4 to both sides.}$$

$$\left|\frac{1}{2}x - 5\right| = 0 \quad \text{To complete the process to isolate the absolute value expression, divide both sides by 3. The resulting equation is in the form } |X| = k.$$

Since 0 is the only number whose absolute value is 0, the expression $\frac{1}{2}x - 5$ must be 0.

$$\frac{1}{2}x - 5 = 0 \quad \text{Set the expression within the absolute value bars equal to 0 and solve for x.}$$

$$\frac{1}{2}x = 5 \quad \text{To isolate the variable term } \tfrac{1}{2}x, \text{ add 5 to both sides.}$$

$$x = 10 \quad \text{To isolate x, multiply both sides by 2.}$$

The solution is 10 and the solution set is $\{10\}$. Verify that it satisfies the original equation.

Success Tip

To solve most absolute value equations, we must consider two cases. However, if an absolute value is equal to 0, we need only consider one: the case when the expression within the absolute value bars is equal to 0.

Teaching Tip: To summarize, ask: How many solutions does the equation $|X| = k$ have if

- k is positive
- k is negative
- k is 0

Teaching Example 3
Solve: $-2\left|\frac{1}{3}x + 5\right| + 6 = 6$
Answer: -15

Self Check 3 Solve: $-5\left|\frac{2}{3}x + 4\right| + 1 = 1$ -6

Now Try Problem 39

In Section 8.3 we discussed absolute value functions and their graphs. If we are given an output of an absolute value function, we can work in reverse to find the corresponding input(s).

EXAMPLE 4 Let $f(x) = |x + 4|$. For what value(s) of x is $f(x) = 20$?

Strategy We will substitute 20 for $f(x)$ and solve for x.

Why In the equation, there are two unknowns, x and $f(x)$. If we replace $f(x)$ with 20, we can solve the resulting absolute value equation for x.

Solution

$$f(x) = |x + 4| \quad \text{This is the given function.}$$

$$20 = |x + 4| \quad \text{Substitute 20 for } f(x).$$

$$|x + 4| = 20 \quad \text{Rewrite the equation so that the absolute value expression is on the left side.}$$

To solve $|x + 4| = 20$, we write and then solve an equivalent compound equation:

$$x + 4 = 20 \quad \text{or} \quad x + 4 = -20$$

Now we solve each equation for x:

$$x + 4 = 20 \quad \text{or} \quad x + 4 = -20$$
$$x = 16 \quad | \quad x = -24$$

The values of x for which $f(x) = 20$ are 16 and -24. To check, find $f(16)$ and $f(-24)$, and verify that the result is 20 in each case.

Teaching Example 4
Let $f(x) = |x - 7|$. For what value(s) of x is $f(x) = 6$?
Answer: 13, 1

Self Check 4 Let $f(x) = |x + 4|$. For what value(s) of x is $f(x) = 11$? 7, -15

Now Try Problem 71

2 Solve Equations with Two Absolute Values.

Equations can contain two absolute value expressions. To develop a strategy to solve them, consider the following four true statements.

$\|3\| = \|3\|$	or	$\|-3\| = \|-3\|$	or	$\|3\| = \|-3\|$	or	$\|-3\| = \|3\|$
The numbers are the same.		The numbers are the same.		The numbers are opposites.		The numbers are opposites.

These four possible cases are really just two cases: *Two absolute value expressions are equal when the expressions within the absolute value bars are equal to or opposites of each other.* This observation suggests the following approach for solving equations having two absolute value expressions.

Solving Equations with Two Absolute Values

For any algebraic expressions X and Y:

To solve $|X| = |Y|$, solve the compound equation $X = Y$ or $X = -Y$.

EXAMPLE 5 Solve: $|5x + 3| = |3x + 25|$

Strategy To solve this equation, we will write and then solve an equivalent compound equation.

Why We can use this approach because the equation is of the form $|X| = |Y|$.

Solution The equation $|5x + 3| = |3x + 25|$, with the two absolute value expressions, is equivalent to the following compound equation:

The expressions within the absolute value symbols are equal		The expressions within the absolute value symbols are opposites	
$5x + 3 = 3x + 25$	or	$5x + 3 = -(3x + 25)$	
$2x = 22$		$5x + 3 = -3x - 25$	*Solve each equation.*
$x = 11$		$8x = -28$	
		$x = -\frac{28}{8}$	
		$x = -\frac{7}{2}$	*Simplify the fraction.*

Verify that both solutions, 11 and $-\frac{7}{2}$, check by substituting them into the original equation. The solution set is $\left\{11, -\frac{7}{2}\right\}$.

Caution

Don't forget to use parentheses to write the opposite of expressions that have more than one term.

Expression	Opposite
$3x + 25$	$-(3x + 25)$

Teaching Example 5
Solve: $|4x + 2| = |x - 4|$
Answer: $-2, \frac{2}{5}$

Self Check 5 Solve: $|2x - 3| = |4x + 9|$ $-1, -6$

Now Try Problem 47

3 Solve Inequalities of the Form $|X| < k$.

To solve the **absolute value inequality** $|x| < 5$, we must find all real numbers x whose distance from 0 on the number line is less than 5. From the graph, we see that there are many such numbers. For example, -4.999, -3, -2.4, $-1\frac{7}{8}$, $-\frac{3}{4}$, 0, 1, 2.8, 3.001, and 4.999 all meet this requirement. We conclude that the solution set is all numbers between -5 and 5, which can be written in interval notation as $(-5, 5)$.

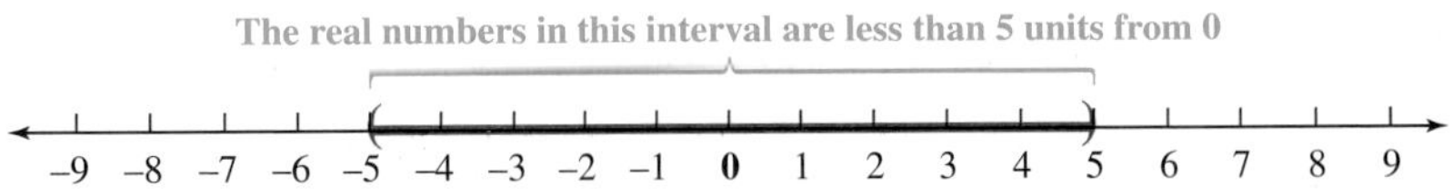
The real numbers in this interval are less than 5 units from 0
–9 –8 –7 –6 –5 –4 –3 –2 –1 0 1 2 3 4 5 6 7 8 9

Since x is between -5 and 5, it follows that $|x| < 5$ is equivalent to $-5 < x < 5$. This observation suggests the following approach for solving absolute value inequalities of the form $|X| < k$ and $|X| \le k$.

Solving $|X| < k$ and $|X| \le k$

For any positive number k and any algebraic expression X:

To solve $|X| < k$, solve the equivalent double inequality $-k < X < k$.

To solve $|X| \le k$, solve the equivalent double inequality $-k \le X \le k$.

EXAMPLE 6 Solve $|2x - 3| < 9$ and graph the solution set.

Strategy To solve this absolute value inequality, we will write and solve an equivalent double inequality.

Why We can use this approach because the inequality is of the form $|X| < k$, and k is positive.

Solution The absolute value inequality $|2x - 3| < 9$ is equivalent to the double inequality

$$-9 < 2x - 3 < 9$$

which we can solve for x:

Teaching Tip: After obtaining $-3 < x < 6$, ask your students which symbols (parentheses or brackets) should be used at the endpoints of the graph of the solution set.

$$-9 < 2x - 3 < 9$$

$$-6 < 2x < 12 \quad \text{To isolate the variable term } 2x\text{, add 3 to all three parts.}$$

$$-3 < x < 6 \quad \text{To isolate } x\text{, divide all parts by 2.}$$

Any number between -3 and 6 is in the solution set, which can be written as $\{x \mid -3 < x < 6\}$. This is the interval $(-3, 6)$; its graph is shown on the right.

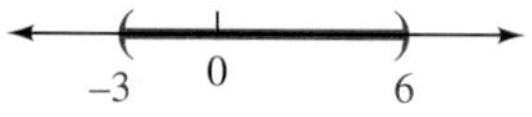

Teaching Example 6 Solve $|3x - 1| < 8$ and graph the solution set.
Answer: $\left(-\frac{7}{3}, 3\right)$

Self Check 6 Solve $|3x + 2| < 4$ and graph the solution set. $\left(-2, \frac{2}{3}\right)$; see AIE Appendix 3.

Now Try Problems 55 and 59

Because it is related to distance, absolute value can be used to describe the amount of error involved when measurements are taken.

EXAMPLE 7 **Tolerances.** When manufactured parts are inspected by a quality control engineer, they are classified as acceptable if each dimension falls within a given **tolerance range** of the dimensions listed on the blueprint. For the bracket shown in the margin, the distance between the two drilled holes is given as 2.900 inches. Because the tolerance is ± 0.015 inch, this distance can be as much as 0.015 inch longer or 0.015 inch shorter, and the part will be considered acceptable. The acceptable distance d between holes can be represented by the absolute value inequality $|d - 2.900| \leq 0.015$. Solve the inequality and explain the result.

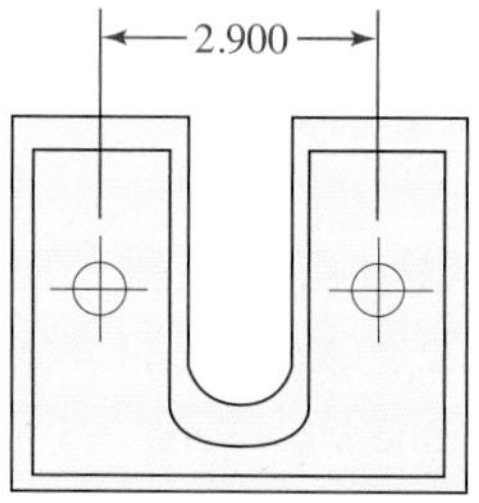

Unless otherwise specified, dimensions are in inches.
Tolerances ±0.015

Strategy To solve $|d - 2.900| \leq 0.015$, we will write and solve an equivalent double inequality.

Why We can use this approach because the inequality is of the form $|X| \leq k$, and k is positive.

Solution The absolute value inequality $|d - 2.900| \leq 0.015$ is equivalent to the double inequality

$$-0.015 \leq d - 2.900 \leq 0.015$$

which we can solve for d:

$$-0.015 \leq d - 2.900 \leq 0.015$$

$$2.885 \leq d \leq 2.915 \quad \text{To isolate } d\text{, add 2.900 to all three parts.}$$

The solution set is the interval [2.885, 2.915]. This means that the distance between the two holes should be between 2.885 and 2.915 inches, inclusive. If the distance is less than 2.885 inches or more than 2.915 inches, the part should be rejected.

Teaching Example 7 Tolerances. Refer to Example 7. Find the tolerance range if the tolerance is ±0.00015.
Answer: [2.89985, 2.90015]

Self Check 7 **Tolerances.** Refer to Example 7. Find the tolerance range if the tolerance is ±0.0015. [2.8985, 2.9015]

Now Try Problem 105

EXAMPLE 8 Solve: $|4x - 5| < -2$

Teaching Tip: Explain that if the problem was to solve $|4x - 5| \le 0$, we need only consider $|4x - 5| = 0$ because $|4x - 5|$ can never be less than 0.

Strategy We will solve this inequality by inspection.

Why The inequality $|4x - 5| < -2$ is of the form $|X| < k$. However, the standard method for solving such inequalities cannot be used because k (in this case, -2) is not positive.

Solution Since $|4x - 5|$ is always greater than or equal to 0 for any real number x, this absolute value inequality has no solution. The solution set is $\varnothing$.

Teaching Example 8 Solve: $|-5x + 4| < -10$
Answer: No solution, $\varnothing$

Self Check 8 Solve: $|6x + 24| < -51$ No solution, $\varnothing$

Now Try Problem 61

4 Solve Inequalities of the Form $|X| > k$.

To solve the absolute value inequality $|x| > 5$, we must find all real numbers x whose distance from 0 on the number line is greater than 5. From the following graph, we see that there are many such numbers. For example, -5.001, -6, -7.5, and $-8\frac{3}{8}$, as well as 5.001, 6.2, 7, 8, and $9\frac{1}{2}$ all meet this requirement. We conclude that the solution set is all numbers less than -5 or greater than 5, which can be written as the union of two intervals: $(-\infty, -5) \cup (5, \infty)$.

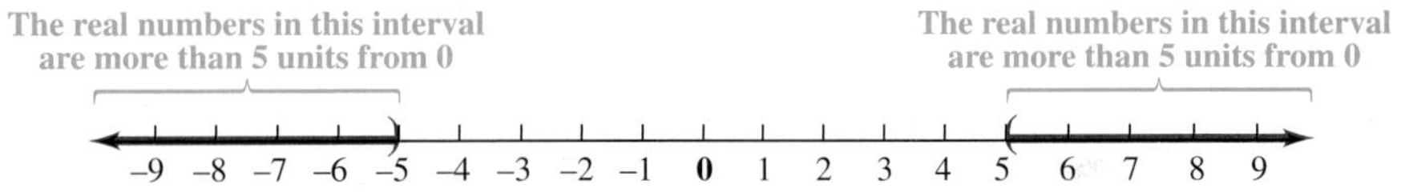
The real numbers in this interval are more than 5 units from 0
The real numbers in this interval are more than 5 units from 0
−9 −8 −7 −6 −5 −4 −3 −2 −1 0 1 2 3 4 5 6 7 8 9

Since x is less than -5 or greater than 5, it follows that $|x| > 5$ is equivalent to $x < -5$ or $x > 5$. This observation suggests the following approach for solving absolute value inequalities of the form $|X| > k$ and $|X| \ge k$.

Solving $|X| > k$ and $|X| \ge k$

For any positive number k and any algebraic expression X:

To solve $|X| > k$, solve the equivalent compound inequality $X < -k$ or $X > k$.

To solve $|X| \ge k$, solve the equivalent compound inequality $X \le -k$ or $X \ge k$.

EXAMPLE 9 Solve $\left|\frac{3 - x}{5}\right| \ge 6$ and graph the solution set.

Strategy To solve this absolute value inequality, we will write and solve an equivalent compound inequality.

Why We can use this approach because the inequality is of the form $|X| \ge k$, and k is positive.

Solution The absolute value inequality $\left|\frac{3 - x}{5}\right| \ge 6$ is equivalent to the compound inequality

$$\frac{3 - x}{5} \le -6 \quad \text{or} \quad \frac{3 - x}{5} \ge 6$$

Notation

In this case, the solution set cannot be represented by a single interval. The union symbol $\cup$ must be used to write the solution set in interval notation.

Teaching Tip: To prepare your students for the Look Alike problems, ask them how to use this answer to solve $\left|\frac{3-x}{5}\right| \le 6$ without having to do any new work.

Now we solve each inequality for x:

$$\frac{3-x}{5} \le -6 \quad \text{or} \quad \frac{3-x}{5} \ge 6$$

$3 - x \le -30$	$3 - x \ge 30$	To clear the fraction, multiply both sides by 5.
$-x \le -33$	$-x \ge 27$	To isolate the variable term $-x$, subtract 3 from both sides.
$x \ge 33$	$x \le -27$	To isolate x, divide both sides by -1 and reverse the direction of the inequality symbol.

The solution set is the union of two intervals: $(-\infty, -27] \cup [33, \infty)$. Using set-builder notation, the solution set is written as $\{x \mid x \le -27 \text{ or } x \ge 33\}$. Its graph appears on the right.

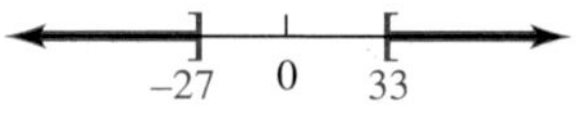

Teaching Example 9

Solve $\left|\frac{5-x}{2}\right| > 3$ and graph the solution set.

Answer: $(-\infty, -1) \cup (11, \infty)$

Self Check 9 Solve $\left|\frac{2-x}{4}\right| \ge 1$ and graph the solution set. $(-\infty, -2] \cup [6, \infty)$; see AIE Appendix 3.

Now Try Problems 63 and 65

EXAMPLE 10 Solve $6 < \left|\frac{2}{3}x - 2\right| - 3$ and graph the solution set.

Strategy We will first write the inequality in an equivalent form with the absolute value on the left side.

Why It's usually easier to solve an absolute value inequality if the absolute value appears on the left side of the inequality.

Solution

$$6 < \left|\frac{2}{3}x - 2\right| - 3 \quad \text{This is the inequality to solve.}$$

$$\left|\frac{2}{3}x - 2\right| - 3 > 6 \quad \text{Write the inequality with the absolute value on the left side.}$$

$$\left|\frac{2}{3}x - 2\right| > 9 \quad \text{Add 3 to both sides to isolate the absolute value expression.}$$

Success Tip

To write $6 < \left|\frac{2}{3}x - 2\right| - 3$ with the absolute value on the left, we used the following property:

If $a < b$, then $b > a$.

After isolating the absolute value expression on the left side, the resulting inequality has the form $|X| > k$. To solve this absolute value inequality, we write and solve an equivalent compound inequality:

$$\frac{2}{3}x - 2 < -9 \quad \text{or} \quad \frac{2}{3}x - 2 > 9$$

$\frac{2}{3}x < -7$	$\frac{2}{3}x > 11$	Add 2 to both sides.
$2x < -21$	$2x > 33$	Multiply both sides by 3.
$x < -\frac{21}{2}$	$x > \frac{33}{2}$	To isolate x, divide both sides by 2.

The solution set is the union of two intervals: $\left(-\infty, -\frac{21}{2}\right) \cup \left(\frac{33}{2}, \infty\right)$. Its graph appears on the right. Using set-builder notation, the solution set can be written as $\left\{x \mid x < -\frac{21}{2} \text{ or } x > \frac{33}{2}\right\}$.

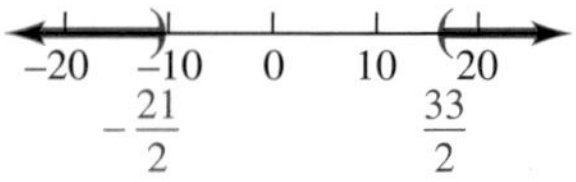

Teaching Example 10
Solve $7 \le \left|\frac{3}{5}x - 1\right| + 2$ and graph the solution set.

Answer: $\left(-\infty, -\frac{20}{3}\right] \cup [10,\infty)$

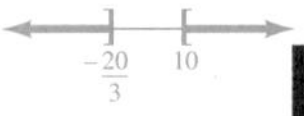

Self Check 10 Solve $3 < \left|\frac{3}{4}x + 2\right| - 1$ and graph the solution set.
$(-\infty, -8) \cup \left(\frac{8}{3}, \infty\right)$; see AIE Appendix 3.

Now Try Problem 67

EXAMPLE 11

Solve $\left|\frac{x}{8} - 1\right| \ge -4$ and graph the solution set.

Teaching Tip: Now would be a good time to review all of the special situations by working problem 15 in Study Set 8.5 as a class.

Strategy We will solve this inequality by inspection.

Why The inequality $\left|\frac{x}{8} - 1\right| \ge -4$ is of the form $|x| \ge k$. However, the standard method for solving such inequalities cannot be used because k is not positive.

Solution Since $\left|\frac{x}{8} - 1\right|$ is always greater than or equal to 0 for any real number x it will also be greater than or equal to -4 for any real number x. Therefore, this absolute value inequality is true for all real numbers. The solution set is the interval $(-\infty, \infty)$ or $\mathbb{R}$. Its graph appears on the right.

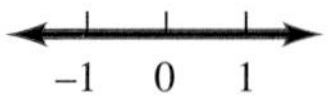

Teaching Example 11
Solve $\left|\frac{x}{0.9}\right| \ge -8.6$ and graph the solution set.

Answer: $(-\infty, \infty)$

Self Check 11 Solve $|-x - 9| > -0.5$ and graph the solution set. $(-\infty, \infty)$; see AIE Appendix 3.

Now Try Problem 69

The following summary shows how we can interpret absolute value in three ways. Assume $k > 0$.

Teaching Tip: Point out that the union of all three of the solution sets is the set of all real numbers.

Geometric description	***Graphic description***	***Algebraic description***
1. $|x| = k$ means that x is k units from 0 on the number line.	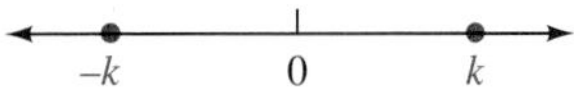	$|x| = k$ is equivalent to $x = k$ or $x = -k$.
2. $|x| < k$ means that x is less than k units from 0 on the number line.	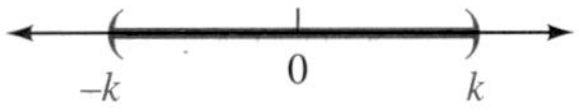	$|x| < k$ is equivalent to $-k < x < k$.
3. $|x| > k$ means that x is more than k units from 0 on the number line.	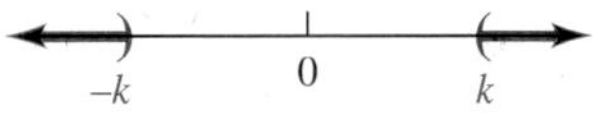	$|x| > k$ is equivalent to $x > k$ or $x < -k$.

Using Your Calculator ▶ Solving Absolute Value Equations and Inequalities

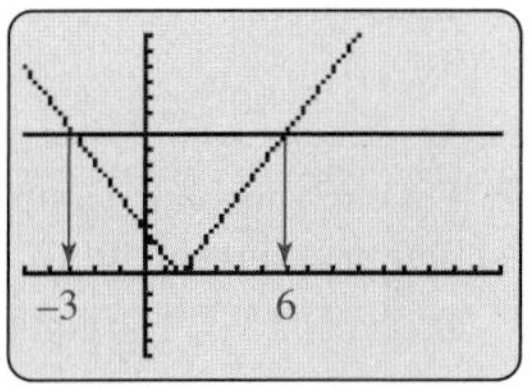

We can solve absolute value equations and inequalities with a graphing calculator. For example, to solve $|2x - 3| = 9$, we graph the equations $y = |2x - 3|$ and $y = 9$ on the same coordinate system, as shown in the figure. The equation $|2x - 3| = 9$ will be true for all x-coordinates of points that lie on *both* graphs. Using the TRACE or the INTERSECT feature, we can see that the graphs intersect at the points $(-3, 9)$ and $(6, 9)$. Thus, the solutions of the absolute value equation are -3 and 6.

The inequality $|2x - 3| < 9$ will be true for all x-coordinates of points that lie on the graph of $y = |2x - 3|$ and *below* the graph of $y = 9$. We see that these values of x are between -3 and 6. Thus, the solution set is the interval $(-3, 6)$.

The inequality $|2x - 3| > 9$ will be true for all x-coordinates of points that lie on the graph of $y = |2x - 3|$ and *above* the graph of $y = 9$. We see that these values of x are less than -3 or greater than 6. Thus, the solution set is the union of two intervals: $(-\infty, -3) \cup (6, \infty)$.

SECTION 8.5 STUDY SET

VOCABULARY

Fill in the blanks.

▶ 1. The absolute value of a number is its distance from 0 on a number line.

▶ 2. $|2x - 1| = 10$ is an absolute value equation and $|2x - 1| > 10$ is an absolute value inequality.

▶ 3. To isolate the absolute value in $|3 - x| - 4 = 5$, we add 4 to both sides.

▶ 4. When we say that the absolute value equation and a compound equation are equivalent, we mean that they have the same solution(s).

▶ 5. When two equations are joined by the word *or,* such as $x + 1 = 5$ or $x + 1 = -5$, we call the statement a compound equation.

▶ 6. $f(x) = |6x - 2|$ is called an absolute value function.

CONCEPTS

Fill in the blanks.

▶ 7. To solve absolute value equations and inequalities, we write and solve equivalent compound equations and inequalities.

8. Two absolute value expressions are equal when the expressions within the absolute value bars are equal to or opposites of each other.

9. Consider the following real numbers:
$-3, -2.01, -2, -1.99, -1, 0, 1, 1.99, 2, 2.01, 3$
   a. Which of them make $|x| = 2$ true? $-2, 2$
   b. Which of them make $|x| < 2$ true? $-1.99, -1, 0, 1, 1.99$
   c. Which of them make $|x| > 2$ true? $-3, -2.01, 2.01, 3$

10. Determine whether -3 is a solution of the given equation or inequality.
   a. $|x - 1| = 4$ Yes
   b. $|x - 1| > 4$ No
   c. $|x - 1| \le 4$ Yes
   d. $|5 - x| = |x + 12|$ No

11. For each absolute value equation, write an equivalent compound equation.
   a. $|x - 7| = 8$ is equivalent to

   $x - 7 = 8$ or $x - 7 = -8$

   b. $|x + 10| = |x - 3|$ is equivalent to

   $x + 10 = x - 3$ or $x + 10 = -(x - 3)$

▶ 12. For each absolute value inequality, write an equivalent compound inequality.
   a. $|x + 5| < 1$ is equivalent to

   $-1 < x + 5 < 1$

   b. $|x - 6| \ge 3$ is equivalent to

   $x - 6 \le -3$ or $x - 6 \ge 3$

13. For each absolute value equation or inequality, write an equivalent compound equation or inequality.
   a. $|x| = 8$ — $x = 8$ or $x = -8$
   b. $|x| \ge 8$ — $x \le -8$ or $x \ge 8$
   c. $|x| \le 8$ — $-8 \le x \le 8$
   d. $|5x - 1| = |x + 3|$ — $5x - 1 = x + 3$ or $5x - 1 = -(x + 3)$

14. Perform the necessary steps to isolate the absolute value expression on one side of the equation. *Do not solve.*
   a. $|3x + 2| - 7 = -5$ — $|3x + 2| = 2$
   b. $6 + 2|5x - 19| \le 40$ — $|5x - 19| \le 17$

15. Determine the solution set of each absolute value equation or inequality by inspection. (No work is necessary.) Your answer should be either *all real numbers* or *no solution.*
   a. $|7x + 6| = -8$ No solution
   b. $|7x + 6| \le -8$ No solution
   c. $|7x + 6| \ge -8$ All real numbers

16. Write the inequality $10 > |16x - 3|$ in an equivalent form with the absolute value expression on the left side.
$|16x - 3| < 10$

NOTATION

17. Match each equation or inequality with its graph.
   a. $|x| = 1$ ii
   b. $|x| > 1$ iii
   c. $|x| < 1$ i

i. 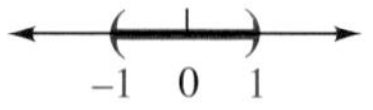

ii. 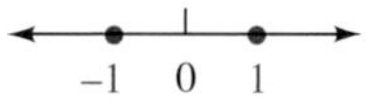

iii. 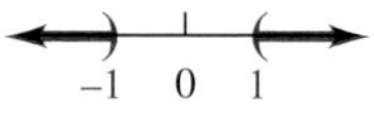

18. Describe the set graphed below using interval notation.
$(-\infty, -1) \cup (3, \infty)$

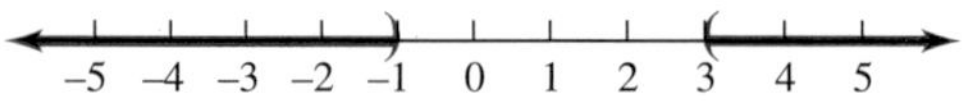

GUIDED PRACTICE

Solve each equation. See Example 1.

▶ 19. $|x| = 23$ $23, -23$

20. $|x| = 90$ $90, -90$

21. $|x - 5| = 8$ $13, -3$

▶ 22. $|x - 7| = 4$ $11, 3$

▶ 23. $|3x + 2| = 16$ $\frac{14}{3}, -6$

▶ 24. $|5x - 3| = 22$ $5, -\frac{19}{5}$

25. $\left|\frac{x}{5}\right| = 10$ $50, -50$

▶ 26. $\left|\frac{x}{7}\right| = 2$ $14, -14$

27. $|2x + 3.6| = 9.8$ $3.1, -6.7$

▶ 28. $|4x - 24.8| = 32.4$ $14.3, -1.9$

▶ 29. $\left|\frac{7}{2}x + 3\right| = -5$ No solution; $\varnothing$

▶ 30. $|x - 2.1| = -16.3$ No solution; $\varnothing$

Solve each equation. See Example 2.

▶ **31.** $|x - 3| - 19 = 3$
$25, -19$

32. $|x - 10| + 30 = 50$
$30, -10$

33. $|3x - 7| + 8 = 22$
$7, -\frac{7}{3}$

▶ **34.** $|6x - 3| + 7 = 28$
$4, -3$

▶ **35.** $|3 - 4x| + 1 = 6$
$2, -\frac{1}{2}$

▶ **36.** $|8 - 5x| - 8 = 10$
$\frac{26}{5}, -2$

▶ **37.** $\left|\frac{7}{8}x + 5\right| - 2 = 7$
$\frac{32}{7}, -16$

38. $\left|\frac{3}{4}x + 4\right| - 5 = 11$
$16, -\frac{80}{3}$

Solve each equation. See Example 3.

39. $\left|\frac{1}{5}x + 2\right| - 8 = -8$
-10

▶ **40.** $\left|\frac{1}{9}x + 4\right| + 25 = 25$
-36

41. $2|3x + 24| = 0$
-8

▶ **42.** $8\left|\frac{2x}{3} + 10\right| = 0$
-15

▶ **43.** $-5|2x - 9| + 14 = 14$
$\frac{9}{2}$

44. $-10|16x + 4| - 3 = -3$
$-\frac{1}{4}$

45. $6 - 3|10x + 5| = 6$
$-\frac{1}{2}$

▶ **46.** $15 - |12x + 12| = 15$
-1

Solve each equation. See Example 5.

47. $|5x - 12| = |4x - 16|$
$-4, \frac{28}{9}$

▶ **48.** $|4x - 7| = |3x - 21|$
$-14, 4$

▶ **49.** $|10x| = |x - 18|$
$-2, \frac{18}{11}$

50. $|6x| = |x + 45|$
$9, -\frac{45}{7}$

51. $|2 - x| = |3x + 2|$
$0, -2$

▶ **52.** $|4x + 3| = |9 - 2x|$
$1, -6$

53. $|5x - 7| = |4(x + 1)|$
$11, \frac{1}{3}$

▶ **54.** $|2x + 1| = |3(x + 1)|$
$-2, -\frac{4}{5}$

Solve each inequality. Graph the solution set and write it using interval notation. See Examples 6 and 8. See AIE Appendix 3.

▶ **55.** $|x| < 4$
$(-4, 4)$

56. $|x| < 9$
$(-9, 9)$

▶ **57.** $|x + 9| \le 12$
$[-21, 3]$

58. $|x - 8| \le 12$
$[-4, 20]$

▶ **59.** $|3x - 2| < 10$
$\left(-\frac{8}{3}, 4\right)$

▶ **60.** $|4 - 3x| \le 13$
$\left[-3, \frac{17}{3}\right]$

▶ **61.** $|5x - 12| < -5$
No solution; $\varnothing$

▶ **62.** $|3x + 2| \le -3$
No solution; $\varnothing$

Solve each inequality. Graph the solution set and write it using interval notation. See Examples 9–11. See AIE Appendix 3.

▶ **63.** $|x| > 3$
$(-\infty, -3) \cup (3, \infty)$

64. $|x| > 7$
$(-\infty, -7) \cup (7, \infty)$

▶ **65.** $|x - 12| > 24$
$(-\infty, -12) \cup (36, \infty)$

▶ **66.** $|x + 5| \ge 7$
$(-\infty, -12] \cup [2, \infty)$

67. $0 \le |5x - 1| - 2$
$\left(-\infty, -\frac{1}{5}\right] \cup \left[\frac{3}{5}, \infty\right)$

▶ **68.** $0 \le |6x - 3| - 5$
$\left(-\infty, -\frac{1}{3}\right] \cup \left[\frac{4}{3}, \infty\right)$

▶ **69.** $|4x + 3| \ge -5$
$(-\infty, \infty)$

70. $|7x + 2| \ge -8$
$(-\infty, \infty)$

See Examples 4, 6, and 9.

71. Let $f(x) = |x + 3|$. For what value(s) of x is $f(x) = 3$? $0, -6$

▶ **72.** Let $g(x) = |2 - x|$. For what value(s) of x is $g(x) = 2$? $0, 4$

73. Let $f(x) = |2(x - 1) + 4|$. For what value(s) of x is $f(x) < 4$? $(-3, 1)$

▶ **74.** Let $h(x) = \left|\frac{x}{5} - \frac{1}{2}\right|$. For what value(s) of x is $h(x) > \frac{9}{10}$?
$(-\infty, -2) \cup (7, \infty)$

TRY IT YOURSELF

Solve each equation and inequality. For the inequalities, graph the solution set and write it using interval notation. See AIE Appendix 3.

▶ **75.** $|3x + 2| + 1 > 15$
$\left(-\infty, -\frac{16}{3}\right) \cup (4, \infty)$

76. $|2x - 5| - 5 > 20$
$(-\infty, -10) \cup (15, \infty)$

77. $6\left|\frac{x - 2}{3}\right| \le 24$
$[-10, 14]$

78. $8\left|\frac{x - 2}{3}\right| > 32$
$(-\infty, -10) \cup (14, \infty)$

79. $-7 = 2 - |0.3x - 3|$
$40, -20$

80. $-1 = 1 - |0.1x + 8|$
$-60, -100$

▶ **81.** $|2 - 3x| \ge -8$
$(-\infty, \infty)$

▶ **82.** $|-1 - 2x| > 5$
$(-\infty, -3) \cup (2, \infty)$

▶ **83.** $|7x + 12| = |x - 6|$
$-3, -\frac{3}{4}$

84. $|8 - x| = |x + 2|$
3

85. $2 \ge 3|2 - 3x| + 2$
$\frac{2}{3}$

▶ **86.** $7 \ge |15x - 45| + 7$
3

87. $-14 = |x - 3|$
No solution; $\varnothing$

88. $-75 = |x + 4|$
No solution; $\varnothing$

89. $\frac{6}{5} = \left|\frac{3x}{5} + \frac{x}{2}\right|$
$\frac{12}{11}, -\frac{12}{11}$

▶ **90.** $\frac{11}{12} = \left|\frac{x}{3} - \frac{3x}{4}\right|$
$\frac{11}{5}, -\frac{11}{5}$

▶ **91.** $-|2x - 3| < -7$
$(-\infty, -2) \cup (5, \infty)$

▶ **92.** $-|3x + 1| < -8$
$(-\infty, -3) \cup \left(\frac{7}{3}, \infty\right)$

93. $|0.5x + 1| < -23$
No solution; $\varnothing$

94. $15 \ge 7 - |1.4x + 9|$
$(-\infty, \infty)$

Look Alikes . . .

95. **a.** $\frac{x}{10} - 1 = 1$ $\quad 20$
b. $\left|\frac{x}{10} - 1\right| = 1$ $\quad 20, 0$
c. $\frac{x}{10} - 1 > 1$
$(20, \infty)$
d. $\left|\frac{x}{10} - 1\right| > 1$
$(-\infty, 0) \cup (20, \infty)$

▶ **96.** **a.** $4x - 5 = 15$ $\quad 5$
b. $|4x - 5| = 15$ $\quad 5, -\frac{5}{2}$
c. $4x - 5 \le 15$ $\quad (-\infty, 5]$
d. $|4x - 5| \le 15$ $\quad \left[-\frac{5}{2}, 5\right]$

97. **a.** $0.9 - 0.3x = 8.4$
-25
b. $|0.9 - 0.3x| = 8.4$
$-25, 31$
c. $0.9 - 0.3x > 8.4$
$(-\infty, -25)$
d. $|0.9 - 0.3x| > 8.4$
$(-\infty, -25) \cup (31, \infty)$

98. **a.** $8(x - 4) = 6x - 44$ $\quad -6$
b. $|8(x - 4)| = |6x - 44|$ $\quad -6, \frac{38}{7}$

Solve the absolute value inequality in part a. Graph the solution set and write it in interval notation. Then use your work from part a to determine the solution set for the absolute value inequality in part b. (No new work is necessary!) Graph the solution set and write it in interval notation. See AIE Appendix 3.

99. **a.** $|8x - 40| \le 16$ $[3, 7]$ **b.** $|8x - 40| \ge 16$ $(-\infty, 3] \cup [7, \infty)$

100. **a.** $0 \le |14 - 27x|$ $(-\infty, \infty)$ **b.** $0 > |14 - 27x|$ No solution, $\varnothing$

101. **a.** $\left|\frac{4x - 4}{3}\right| - 1 > 11$ $(-\infty, -8) \cup (10, \infty)$

b. $\left|\frac{4x - 4}{3}\right| - 1 \le 11$ $[-8, 10]$

102. **a.** $\left|-\frac{1}{2}x - 3\right| + 2 > 7$ $(-\infty, -16) \cup (4, \infty)$

b. $\left|-\frac{1}{2}x - 3\right| + 2 < 7$ $(-16, 4)$

APPLICATIONS

103. Temperature Ranges. The temperatures on a sunny summer day satisfied the inequality $|t - 78°| \le 8°$, where t is a temperature in degrees Fahrenheit. Solve this inequality and express the range of temperatures as a double inequality. $70° \le t \le 86°$

104. Operating Temperatures. A car CD player has an operating temperature of $|t - 40°| < 80°$, where t is a temperature in degrees Fahrenheit. Solve the inequality and express this range of temperatures as an interval. $(-40°, 120°)$

105. Auto Mechanics. On most cars, the bottoms of the front wheels are closer together than the tops, creating a *camber angle.* This lessens road shock to the steering system. (See the illustration.) The specifications for a certain car state that the camber angle c of its wheels should be $0.6° \pm 0.5°$.

a. Express the range with an inequality containing absolute value symbols. $|c - 0.6°| \le 0.5°$

b. Solve the inequality and express this range of camber angles as an interval. $[0.1°, 1.1°]$

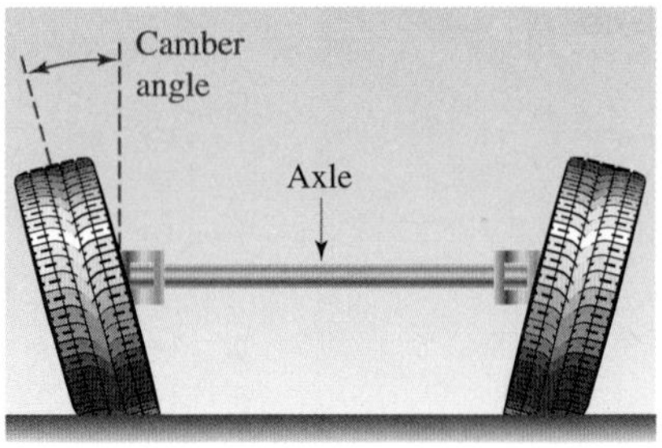

106. Steel Production. A sheet of steel is to be 0.250 inch thick with a tolerance of 0.025 inch.

a. Express this specification with an inequality containing absolute value symbols, using x to represent the thickness of a sheet of steel. $|x - 0.250| \le 0.025$

b. Solve the inequality and express the range of thickness as an interval. $[0.225, 0.275]$

107. Error Analysis. In a lab, students measured the percent of copper p in a sample of copper sulfate. The students know that copper sulfate is actually 25.46% copper by mass. They are to compare their results to the actual value and find the amount of *experimental error.* Which measurements shown in the illustration satisfy the absolute value inequality $|p - 25.46| \le 1.00$? 26.45%, 24.76%

Lab 4 Section A
Title:
"Percent copper (Cu) in copper sulfate ($CuSO_4 \cdot 5H_2O$)"

Results

	% Copper
Trial #1:	22.91%
Trial #2:	26.45%
Trial #3:	26.49%
Trial #4:	24.76%

108. Error Analysis. See Exercise 107. Which measurements satisfy the absolute value inequality $|p - 25.46| > 1.00$? 22.91%, 26.49%

WRITING

109. Explain the error.

Solve: $|x| + 2 = 6$

~~$x + 2 = 6$ or $x + 2 = -6$~~

~~$x = 4$ | $x = -8$~~

110. Explain why the equation $|x - 4| = -5$ has no solution.

111. Explain the differences between the solution sets of $|x| = 8$, $|x| < 8$, and $|x| > 8$.

112. Explain how to use the graph in the illustration to solve the following.

a. $|x - 2| = 3$

b. $|x - 2| \le 3$

c. $|x - 2| \ge 3$

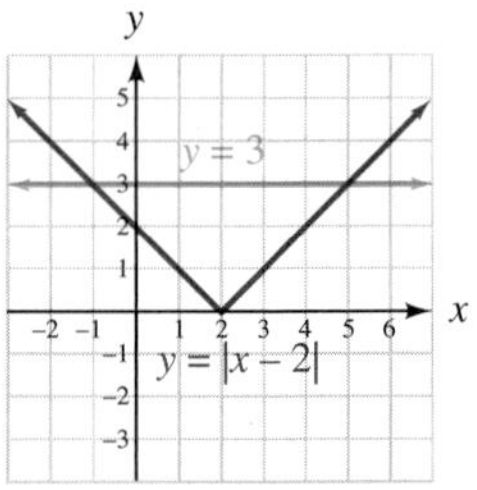

REVIEW

113. Flutes. When it is assembled, a flute is 29 inches long. The middle piece is 4 inches less than twice as long as the first piece. The last piece is two-thirds as long as the first piece. Find the length of each piece of the flute. First piece: 9 in.; middle piece: 14 in.; last piece: 6 in.

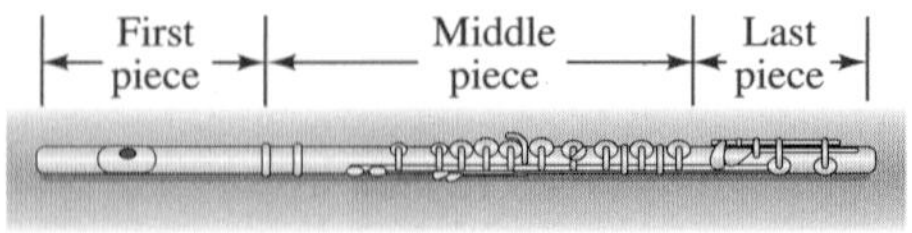

114. Commercials. For the typical "one-hour" prime-time television slot, the number of minutes of commercials is $\frac{3}{7}$ of the number of minutes of the actual program. Determine how many minutes of the program are shown in that one hour. 42 min

CHALLENGE PROBLEMS

115. **a.** For what values of k does $|x| + k = 0$ have exactly two solutions? $k < 0$

b. For what values of k does $|x| + k = 0$ have exactly one solution? $k = 0$

116. Solve: $2^{|2x-3|} = 64$ $\frac{9}{2}, -\frac{3}{2}$

Teaching Tip: This section reviews several of the factoring techniques from Chapter 6 and transitions students to more challenging factoring problems at the intermediate algebra level. The intermediate algebra topic of substitution also is introduced in this section.

SECTION 8.6

Review of Factoring Methods: GCF, Grouping, Trinomials

OBJECTIVES

1 Factor out the greatest common factor.
2 Factor by grouping.
3 Use factoring to solve formulas for a specified variable.
4 Factor trinomials.
5 Use substitution to factor trinomials.
6 Use the grouping method to factor trinomials.

ARE YOU READY? *Are You Ready? exercises available online at www.webassign.net/brookscole*

The following problems review some basic skills that are needed when factoring expressions.

1. Find the prime factorization of 108. $2 \cdot 2 \cdot 3 \cdot 3 \cdot 3 = 2^2 \cdot 3^3$
2. Multiply: $6b(b^3 + 2b + 4)$ $6b^4 + 12b^2 + 24b$
3. Multiply: $(x + 8)(x - 6)$ $x^2 + 2x - 48$
4. Multiply and simplify: $m(m + 2) - 5(m + 2)$ $m^2 - 3m - 10$
5. Find two integers whose product is 10 and whose sum is 7. 2 and 5
6. Find two integers whose product is -18 and whose sum is 3. -3 and 6

In Chapter 6, we discussed how to factor polynomials. In this section, we will review that material.

1 Factor Out the Greatest Common Factor.

Recall that *when we factor a polynomial, we write a sum of terms as a product of factors.* To perform the most basic type of factoring, we determine whether the terms of the given polynomial have any common factors. This process, called **factoring out the greatest common factor,** is based on the distributive property.

EXAMPLE 1 Factor: $3xy^2z^3 + 6xz^2 - 9xyz^4$

Strategy We will determine the GCF of the terms of the polynomial. Then we will write each term of the polynomial as the product of the GCF and one other factor.

Why We can then use the distributive property to factor out the GCF.

Solution We begin by factoring each term:

$$\left.\begin{aligned} 3xy^2z^3 &= 3 \cdot x \cdot y \cdot y \cdot z \cdot z \cdot z \\ 6xz^2 &= 2 \cdot 3 \cdot x \cdot z \cdot z \\ 9xyz^4 &= 3 \cdot 3 \cdot x \cdot y \cdot z \cdot z \cdot z \cdot z \end{aligned}\right\} \text{GCF} = 3 \cdot x \cdot z \cdot z = 3xz^2$$

Since each term has one factor of 3, one factor of x, and two factors of z, and there are no other common factors, $3xz^2$ is the greatest common factor of the three terms. We write each term as the product of the GCF, $3xz^2$, and one other factor and proceed as follows:

$$3xy^2z^3 + 6xz^2 - 9xyz^4 = 3xz^2 \cdot y^2z + 3xz^2 \cdot 2 - 3xz^2 \cdot 3yz^2$$ This can be done mentally.

$$= 3xz^2(y^2z + 2 - 3yz^2)$$ Factor out the GCF, $3xz^2$.

We can check the factorization using multiplication.

$$3xz^2(y^2z + 2 - 3yz^2) = 3xy^2z^3 + 6xz^2 - 9xyz^4$$ This is the original polynomial.

Self Check 1 Factor: $6a^2b^2 - 4ab^3 + 2ab^2$ $2ab^2(3a - 2b + 1)$

Now Try Problem 25

Success Tip

Always verify a factorization by performing the indicated multiplication. The result should be the original polynomial.

Teaching Example 1 Factor: $10x^4y^3z - 15xy^4z + 5xy^2z$
Answer: $5xy^2z(2x^3y - 3y^2 + 1)$

When asked to factor a polynomial whose leading coefficient is negative, we factor out the *opposite of the GCF.*

EXAMPLE 2 Factor out the opposite of the GCF from $-6u^2v^3 + 8u^3v^2$.

Strategy We will determine the GCF of the terms of the polynomial. Then we will write each term as the product of the opposite of the GCF and one other factor.

Why We can then use the distributive property to factor out the opposite of the GCF.

Solution Because the greatest common factor of the two terms is $2u^2v^2$, the opposite of the greatest common factor is $-2u^2v^2$. To factor out $-2u^2v^2$, we proceed as follows:

$$-6u^2v^3 + 8u^3v^2 = \mathbf{-2u^2v^2} \cdot 3v - (\mathbf{-2u^2v^2})4u \quad \text{This can be done mentally.}$$
$$= \mathbf{-2u^2v^2}(3v - 4u)$$

The leading coefficient of the polynomial within the parentheses is positive.

Success Tip

It is standard practice to factor in such a way that the leading coefficient of the polynomial within the parentheses is positive.

Teaching Example 2 Factor out the opposite of the GCF from $-20x^4y^3 + 10x^2y^4$.
Answer: $-10x^2y^3(2x^2 - y)$

Self Check 2 Factor out the opposite of the GCF: $-3p^3q + 6p^2q^2$ $\quad -3p^2q(p - 2q)$

Now Try Problem 35

A polynomial that cannot be factored is called a **prime polynomial** or an **irreducible polynomial.** For example, $9x + 16$ is a prime polynomial because its two terms, $9x$ and 16, have no common factors other than 1.

In the next example, we see that a common factor can have more than one term.

EXAMPLE 3 Factor: $a(x - y + z) - b(x - y + z) + 3(x - y + z)$

Strategy We will factor out the trinomial $x - y + z$ from each term.

Why $x - y + z$ is the GCF of each term of the given expression.

Solution

$$a(x - y + z) - b(x - y + z) + 3(x - y + z) = (x - y + z)(a - b + 3)$$

Teaching Example 3 Factor: $x(a + b - c) - y(a + b - c) + 5(a + b - c)$
Answer: $(a + b - c)(x - y + 5)$

Self Check 3 Factor: $c^2(y^2 + 1) + d^2(y^2 + 1)$ $\quad (y^2 + 1)(c^2 + d^2)$

Now Try Problem 41

2 Factor by Grouping.

Although the terms of many polynomials don't have a common factor, other than 1, it is possible to factor some of them by arranging their terms in convenient groups. This method is called **factoring by grouping.**

EXAMPLE 4 Factor: $2c - 2d + cd - d^2$

Strategy Since the four terms of the polynomial do not have a common factor (other than 1), we will attempt to factor the polynomial by grouping. We will factor out a common factor from the first two terms and from the last two terms.

Why This will produce a common binomial factor that can be factored out.

Solution If we group the terms as highlighted in blue, the first two terms have a common factor, 2, and the last two terms have a common factor, d. When we factor out the common factor from each group, a common binomial factor $c - d$ appears.

Success Tip

You may want to review the steps in the process of factoring by grouping on page 438.

$$2c - 2d + cd - d^2 = 2(c - d) + d(c - d)$$ Factor out 2 from $2c - 2d$ and d from $cd - d^2$. Don't forget the + sign.

$$= (c - d)(2 + d)$$ Factor out the common binomial factor, $c - d$.

We can check by multiplying:

$$(c - d)(2 + d) = 2c + cd - 2d - d^2$$

$$= 2c - 2d + cd - d^2$$ Rearrange the terms to get the original polynomial.

Teaching Example 4 Factor: $3a - 3b + ab - b^2$
Answer: $(a - b)(3 + b)$

Self Check 4 Factor: $7m - 7n + mn - n^2$ $(m - n)(7 + n)$

Now Try Problems 43 and 47

The instruction "Factor" means to factor the given expression completely. Each factor of a completely factored expression will be prime.

To factor a polynomial, it is often necessary to factor more than once. When factoring a polynomial, *always look for a common factor first.*

EXAMPLE 5 Factor: $3x^3y - 4x^2y^2 - 6x^2y + 8xy^2$

Strategy Since all four terms have a common factor of xy, we factor it out first. Then we will attempt to factor the resulting polynomial by grouping.

Why Factoring out the GCF first makes factoring by any method easier.

Solution We begin by factoring out the common factor xy.

$$3x^3y - 4x^2y^2 - 6x^2y + 8xy^2 = xy(3x^2 - 4xy - 6x + 8y)$$

We can now factor the resulting four-term polynomial $3x^2 - 4xy - 6x + 8y$ by grouping:

$$3x^3y - 4x^2y^2 - 6x^2y + 8xy^2$$
$$= xy(3x^2 - 4xy - 6x + 8y)$$
$$= xy[x(3x - 4y) - 2(3x - 4y)]$$ Brackets are needed to enclose the factoring by grouping steps. Factor x from $3x^2 - 4xy$ and -2 from $-6x + 8y$.

$$= xy(3x - 4y)(x - 2)$$ Factor out $3x - 4y$. The brackets are no longer needed.

Teaching Tip: Mention that it is helpful to write polynomials in descending powers of one variable before attempting to factor.

Because xy, $3x - 4y$, and $x - 2$ are prime, no further factoring can be done; the factorization is complete.

Teaching Example 5 Factor: $2ax^2 - 4bx^2 + 2axy - 4bxy$
Answer: $2x(x + y)(a - 2b)$

Self Check 5 Factor: $3a^3b + 3a^2b - 2a^2b^2 - 2ab^2$ $ab(3a - 2b)(a + 1)$

Now Try Problem 51

3 Use Factoring to Solve Formulas for a Specified Variable.

Factoring is often required to solve a formula for one of its variables.

EXAMPLE 6 **Electronics.** The formula $r_1r_2 = rr_2 + rr_1$ is used in electronics to relate the combined resistance, r, of two resistors wired in parallel. The variable r_1 represents the resistance of the first resistor, and the variable r_2 represents the resistance of the second. Solve for r_2.

Strategy To isolate r_2 on one side of the equation, we will get all the terms involving r_2 on the left side and all the terms not involving r_2 on the right side.

Why To *solve a formula for a specified variable* means to isolate that variable on one side of the equation, with all other variables and constants on the opposite side.

Solution

$r_1r_2 = rr_2 + rr_1$ — We want to isolate this variable on one side of the equation.

$r_1r_2 - rr_2 = rr_1$ — To eliminate rr_2 on the right side, subtract rr_2 from both sides.

$r_2(r_1 - r) = rr_1$ — On the left side, factor out the GCF r_2 from $r_1r_2 - rr_2$.

$\dfrac{r_2\overset{1}{\cancel{(r_1 - r)}}}{\underset{1}{\cancel{r_1 - r}}} = \dfrac{rr_1}{r_1 - r}$ — To isolate r_2 on the left side, divide both sides by $r_1 - r$.

$r_2 = \dfrac{rr_1}{r_1 - r}$ — Simplify the left side by removing the common factor $r_1 - r$ from the numerator and denominator.

Teaching Example 6 Solve $Mx = Dt + Cx$ for x.
Answer: $x = \dfrac{Dt}{M - C}$

Self Check 6 Solve $f_1f_2 = ff_1 + ff_2$ for f_1. $f_1 = \dfrac{ff_2}{f_2 - f}$

Now Try Problem 57

4 Factor Trinomials.

Recall that many trinomials factor as the product of two binomials.

EXAMPLE 7 Factor: $x^2 - 6x + 8$

Strategy We will assume that this trinomial is the product of two binomials. We must find the terms of the binomials.

Why Since the terms of $x^2 - 6x + 8$ do not have a common factor (other than 1), the only option is to try to factor it as the product of two binomials.

Solution We represent the binomials using two sets of parentheses. Since the first term of the trinomial is x^2, we enter x and x as the first terms of the binomial factors.

$$x^2 - 6x + 8 = (x\ \square)(x\ \square) \quad \text{Because } x \cdot x \text{ will give } x^2.$$

The second terms of the binomials must be two integers whose product is 8 and whose sum is -6. We list all possible integer-pair factors of 8 in the table.

Factors of 8	Sum of the factors of 8
$1(8) = 8$	$1 + 8 = 9$
$2(4) = 8$	$2 + 4 = 6$
$-1(-8) = 8$	$-1 + (-8) = -9$
$\mathbf{-2(-4) = 8}$	$\mathbf{-2 + (-4) = -6}$ ← This is the pair to choose.

The fourth row of the table contains the correct pair of integers -2 and -4, whose product is 8 and whose sum is -6. To complete the factorization, we enter -2 and -4 as the second terms of the binomial factors.

$$x^2 - 6x + 8 = (x - 2)(x - 4)$$

Success Tip

You may want to review the procedure for factoring trinomials whose leading coefficient is 1 on page 447.

The Language of Algebra

Make sure you understand the following vocabulary: *Many trinomials factor as the product of two binomials.*

$\underbrace{x^2 - 6x + 8}_{\text{Trinomial}} = \underbrace{(x - 2)(x - 4)}_{\text{Product of two binomials}}$

Check: We can verify the factorization by multiplication:

$$(x - 2)(x - 4) = x^2 - 4x - 2x + 8 \quad \text{Use the FOIL method.}$$
$$= x^2 - 6x + 8 \quad \text{This is the original trinomial.}$$

Teaching Example 7 Factor: $m^2 - 11m + 30$
Answer: $(m - 5)(m - 6)$

Self Check 7 Factor: $a^2 - 7a + 12$ $(a - 4)(a - 3)$

Now Try Problem 63

EXAMPLE 8 Factor: $2a^2 + 4ab - 30b^2$

Strategy We will factor out the GCF, 2, first. Then we will factor the resulting trinomial.

Why The first step in factoring any polynomial is to factor out the GCF. Factoring out the GCF first makes factoring by any method easier.

Solution Each term in this trinomial has a common factor of 2, which can be factored out.

$$2a^2 + 4ab - 30b^2 = 2(a^2 + 2ab - 15b^2)$$

Next, we factor $a^2 + 2ab - 15b^2$. Since the first term of the trinomial is a^2, the first term of each binomial factor must be a. Since the third term contains b^2, the last term of each binomial factor must contain b. To complete the factorization, we need to determine the coefficient of each b-term.

$$a^2 + 2ab - 15b^2 = (a \square b)(a \square b) \quad \text{Because } a \cdot a \text{ will give } a^2 \text{ and } b \cdot b \text{ will give } b^2.$$

The coefficients of b must be two integers whose product is -15 and whose sum is 2. We list the factors of -15 and find the pair whose sum is 2.

This is the pair to choose. ↓

$1(-15) \quad 3(-5) \quad 5(-3) \quad 15(-1)$

Caution

Be sure to include all factors in the final answer. Here, a common error is to forget to write the GCF, which is 2.

The only factorization where the sum of the factors is 2 (which is the coefficient of the middle term of $a^2 + \mathbf{2}ab - 15b^2$) is $5(-3)$. Thus,

$$2a^2 + 4ab - 30b^2 = 2(\mathbf{a^2 + 2ab - 15b^2})$$
$$= 2(\mathbf{a + 5b})(\mathbf{a - 3b})$$

Verify this result by multiplication.

Teaching Example 8 Factor: $3x^2 - 3xy - 60y^2$
Answer: $3(x + 4y)(x - 5y)$

Self Check 8 Factor: $3p^2 + 6pq - 24q^2$ $3(p + 4q)(p - 2q)$

Now Try Problem 67

There are more combinations of coefficients to consider when factoring trinomials with leading coefficients other than 1. Because it is not easy to give specific rules for factoring such trinomials, we will use a method called the **trial-and-check method.**

EXAMPLE 9 Factor: $3p^2 - 4p - 4$

Strategy We will assume that this trinomial is the product of two binomials. To find their terms, we will make educated guesses and then check them using multiplication.

Why Since the terms of the trinomial do not have a common factor (other than 1), the only option is to try to factor it as the product of two binomials.

Solution To factor the trinomial, we note that the first terms of the binomial factors must be $3p$ and p to give the first term of $3p^2$.

$$3p^2 - 4p - 4 = (3p\,\square)(p\,\square) \quad \text{Because } 3p \cdot p \text{ will give } 3p^2.$$

The second terms of the binomials must be two integers whose product is -4. There are three such pairs: $1(-4)$, $-1(4)$, and $-2(2)$. When these pairs are entered, and then reversed, as second terms of the binomials, there are six possibilities to consider.

Success Tip

You may want to review the procedure for factoring trinomials using the trial-and-check method on page 458.

For 1 and −4: $(3p + 1)(p - 4)$ or $(3p - 4)(p + 1)$

Outer: $-12p$, inner: p: $-12p + p = -11p$ | Outer: $3p$, inner: $-4p$: $3p + (-4p) = -p$

For −1 and 4: $(3p - 1)(p + 4)$ or $(3p + 4)(p - 1)$

Outer: $12p$, inner: $-p$: $12p + (-p) = 11p$ | Outer: $-3p$, inner: $4p$: $-3p + 4p = p$

For −2 and 2: $(3p - 2)(p + 2)$ or $(3p + 2)(p - 2)$

Outer: $6p$, inner: $-2p$: $6p + (-2p) = 4p$ | Outer: $-6p$, inner: $2p$: $-6p + 2p = -4p$

Notation

By the commutative property of multiplication, the factors of a trinomial can be written in either order. Thus, we also could write:

$3p^2 - 4p - 4 = (p - 2)(3p + 2)$

Of these possibilities, only the one in blue gives the required middle term of $-4p$. Thus,

$$3p^2 - 4p - 4 = (3p + 2)(p - 2)$$

Teaching Example 9 Factor: $3x^2 + 11x + 6$
Answer: $(3x + 2)(x + 3)$

Self Check 9 Factor: $4q^2 - 9q - 9$ $(4q + 3)(q - 3)$

Now Try Problem 75

EXAMPLE 10

Factor: $6y^3 + 13x^2y^3 + 6x^4y^3$

Strategy We write the expression in descending powers of x.

Why It is easier to factor a trinomial if its terms are written in descending powers of one variable.

Solution We write the expression in descending powers of x and factor out the greatest common factor, y^3.

Success Tip

Always write the terms of a trinomial in descending powers of one variable before attempting to factor it.

$$\begin{aligned} 6y^3 + 13x^2y^3 + 6x^4y^3 &= 6x^4y^3 + 13x^2y^3 + 6y^3 \\ &= y^3(6x^4 + 13x^2 + 6) \end{aligned}$$

To factor $6x^4 + 13x^2 + 6$, we examine its terms.

- Since the first term is $6x^4$, the first terms of the binomial factors must be either $2x^2$ and $3x^2$ or x^2 and $6x^2$.

$$6x^4 + 13x^2 + 6 = (2x^2\,\square)(3x^2\,\square) \text{ or } (x^2\,\square)(6x^2\,\square)$$

- Since the signs of the middle term and the last term of the trinomial are positive, the signs within each binomial factor will be positive.
- Since the product of the last terms of the binomial factors must be 6, we must find two numbers whose product is 6 that will lead to a middle term of $13x^2$.

After trying some combinations, we find the one that works.

$$6x^4y^3 + 13x^2y^3 + 6y^3 = y^3(6x^4 + 13x^2 + 6)$$
$$= y^3(2x^2 + 3)(3x^2 + 2)$$

Teaching Example 10 Factor: $4m^3n^4 - 11m^3n^2 + 6m^3$
Answer: $m^3(4n^2 - 3)(n^2 - 2)$

Self Check 10 Factor: $4b + 11a^2b + 6a^4b$ $b(2a^2 + 1)(3a^2 + 4)$

Now Try Problem 87

5 Use Substitution to Factor Trinomials.

For more complicated expressions, especially those involving a quantity within parentheses, a substitution sometimes helps to simplify the factoring process.

EXAMPLE 11 Factor: $(x + y)^2 + 7(x + y) + 12$

Strategy We will use a substitution where we will replace each expression $x + y$ with the variable z and factor the resulting trinomial.

Why The resulting trinomial will be easier to factor because it will be in only one variable, z.

Solution If we use the substitution $z = x + y$, we obtain

$$(x + y)^2 + 7(x + y) + 12 = z^2 + 7z + 12 \quad \text{Replace } x + y \text{ with } z.$$
$$= (z + 4)(z + 3) \quad \text{Factor the trinomial.}$$

To find the factorization of $(x + y)^2 + 7(x + y) + 12$, we substitute $x + y$ for each z in the expression $(z + 4)(z + 3)$.

$$(z + 4)(z + 3) = (x + y + 4)(x + y + 3)$$

Thus, $(x + y)^2 + 7(x + y) + 12 = (x + y + 4)(x + y + 3)$

Teaching Example 11 Factor: $(x + y)^2 - 2(x + y) - 8$
Answer: $(x + y - 4)(x + y + 2)$

Self Check 11 Factor: $(a + b)^2 - 3(a + b) - 10$ $(a + b + 2)(a + b - 5)$

Now Try Problem 95

6 Use the Grouping Method to Factor Trinomials.

Another way to factor trinomials is to write them as equivalent four-termed polynomials and factor by grouping.

EXAMPLE 12 Factor by grouping: **a.** $x^2 + 8x + 15$ **b.** $10x^2 + 13xy - 3y^2$

Strategy In each case, we will express the middle term of the trinomial as the sum of two terms.

Why We want to produce an equivalent four-termed polynomial that can be factored by grouping.

Success Tip

You may want to review the procedure for factoring trinomials using factoring by grouping (the key number method) on page 460.

Solution **a.** Since $x^2 + 8x + 15 = 1x^2 + 8x + 15$, we identify a as 1, b as 8, and c as 15. The **key number** is $ac = 1(15) = 15$. We must find two integers whose product is the key number 15 and whose sum is $b = 8$. Since the integers must have a positive product and a positive sum, we consider only positive factors of 15.

Key number = 15 $\quad b = 8$

Positive factors of 15	Sum of the factors of 15
$1 \cdot 15 = 15$	$1 + 15 = 16$
$\mathbf{3 \cdot 5 = 15}$	$\mathbf{3 + 5 = 8}$

The second row of the table contains the correct pair of integers 3 and 5, whose product is 15 and whose sum is 8.

We can express the middle term, $8x$, of the trinomial as the *sum of two terms,* using the integers 3 and 5 as coefficients of the two terms and factor the equivalent four-termed polynomial by grouping:

$$x^2 + 8x + 15 = x^2 + 3x + 5x + 15 \quad \text{Express } 8x \text{ as } 3x + 5x.$$

$$x^2 + 3x + 5x + 15 = x(x + 3) + 5(x + 3) \quad \text{Factor } x \text{ out of } x^2 + 3x \text{ and 5 out of } 5x + 15.$$

$$= (x + 3)(x + 5) \quad \text{Factor out the GCF, } x + 3.$$

Check the factorization by multiplying.

b. In $10x^2 + 13xy - 3y^2$, we have $a = 10$, $b = 13$, and $c = -3$. The key number is $ac = 10(-3) = -30$. We must find a factorization of -30 such that the sum of the factors is $b = 13$. Since the factors must have a negative product, their signs must be different. The possible factor pairs are listed in the table.

Key number = −30 $\quad b = 13$

Factors of −30	Sum of the factors of −30
$1(-30) = -30$	$1 + (-30) = -29$
$2(-15) = -30$	$2 + (-15) = -13$
$3(-10) = -30$	$3 + (-10) = -7$
$5(-6) = -30$	$5 + (-6) = -1$
$6(-5) = -30$	$6 + (-5) = 1$
$10(-3) = -30$	$10 + (-3) = 7$
$\mathbf{15(-2) = -30}$	$\mathbf{15 + (-2) = 13}$
$30(-1) = -30$	$30 + (-1) = 29$

The seventh row contains the correct pair of numbers 15 and -2, whose product is -30 and whose sum is 13. They serve as the coefficients of two terms, $15xy$ and $-2xy$, that we place between $10x^2$ and $-3y^2$.

Notation

The middle term, $13xy$, may be expressed as $15xy - 2xy$ or as $-2xy + 15xy$ when using factoring by grouping. The resulting factorizations will be equivalent.

$$10x^2 + 13xy - 3y^2 = 10x^2 + 15xy - 2xy - 3y^2 \quad \text{Express } 13xy \text{ as } 15xy - 2xy.$$

We factor the resulting four-term polynomial by grouping.

$$10x^2 + 15xy - 2xy - 3y^2 = 5x(2x + 3y) - y(2x + 3y) \quad \text{Factor out } 5x \text{ from } 10x^2 + 15xy. \text{ Factor out } -y \text{ from } -2xy - 3y^2.$$

$$= (2x + 3y)(5x - y) \quad \text{Factor out the GCF, } 2x + 3y.$$

Thus, $10x^2 + 13xy - 3y^2 = (2x + 3y)(5x - y)$. Check by multiplying.

Teaching Example 12 Factor by grouping:
a. $x^2 + 13x + 36$
b. $9c^2 - 12cd - 5d^2$
Answers: a. $(x + 4)(x + 9)$
b. $(3c - 5d)(3c + d)$

Self Check 12 Factor by grouping: **a.** $m^2 + 13m + 42$ $(m + 7)(m + 6)$
b. $15a^2 + 17ab - 4b^2$ $(3a + 4b)(5a - b)$

Now Try Problems 63 and 75

SECTION 8.6 STUDY SET

VOCABULARY

Fill in the blanks.

▶ **1.** When we write $2x + 4$ as $2(x + 2)$, we say that we have factored $2x + 4$.

▶ **2.** When we factor a polynomial, we write a sum of terms as a product of factors.

▶ **3.** The abbreviation GCF stands for greatest common factor.

▶ **4.** If a polynomial cannot be factored, it is called a prime polynomial or an irreducible polynomial.

▶ Selected exercises available online at www.webassign.net/brookscole

▶ 5. To factor $ab + 6a + 2b + 12$ by grouping, we begin by factoring out a from the first two terms and 2 from the last two terms.

▶ 6. The trinomial $4a^2 - 5a - 6$ is written in descending powers of a.

▶ 7. The leading coefficient of $x^2 - 3x + 2$ is 1, the coefficient of the middle term is -3, and the last term is 2.

▶ 8. The statement $x^2 - x - 12 = (x - 4)(x + 3)$ shows that $x^2 - x - 12$ factors into the product of two binomials.

CONCEPTS

▶ 9. The prime factorizations of three terms are shown here. Find their GCF. $6xy^2$

$2 \cdot 2 \cdot 3 \cdot x \cdot x \cdot y \cdot y \cdot y$

$2 \cdot 3 \cdot 3 \cdot x \cdot y \cdot y \cdot y \cdot y$

$2 \cdot 3 \cdot 3 \cdot 7 \cdot x \cdot x \cdot x \cdot y \cdot y$

10. Use multiplication to determine whether $(3t - 1)(5t - 6)$ is the correct factorization of $15t^2 - 19t + 6$. No

11. Complete the table.

Factors of 8	Sum of the factors of 8
$1(8) = 8$	9
$2(4) = 8$	6
$-1(-8) = 8$	-9
$-2(-4) = 8$	-6

12. Find two integers whose
   a. product is 10 and whose sum is 7. 5, 2
   b. product is 8 and whose sum is -6. $-2, -4$
   c. product is -6 and whose sum is 1. $3, -2$
   d. product is -9 and whose sum is -8. $1, -9$

13. Complete the key number table.

Key number = 12 $\qquad b = -7$

Negative factors of 12	Sum of the factors of 12
$-1(-12) = 12$	$-1 + (-12) = -13$
$-2(-6) = 12$	$-2 + (-6) = -8$
$-3(-4) = 12$	$-3 + (-4) = -7$

14. Use the substitution $x = a + b$ to rewrite the trinomial $6(a + b)^2 - 17(a + b) - 3$. $6x^2 - 17x - 3$

NOTATION

Complete each factorization.

15. $15c^3d^4 - 25c^2d^4 + 5c^3d^6 = 5c^2d^4(3c - 5 + cd^2)$

16. $x^3 - x^2 + 2x - 2 = x^2(x - 1) + 2(x - 1)$
$= (x - 1)(x^2 + 2)$

17. $6m^2 + 7m - 3 = (3m - 1)(2m + 3)$

18. $2y^2 + 10y + 12 = 2(y^2 + 5y + 6)$
$= 2(y + 3)(y + 2)$

GUIDED PRACTICE

***Factor each polynomial.* See Example 1.**

▶ 19. $2x^2 - 6x$
$2x(x - 3)$

20. $3y^3 + 3y^2$
$3y^2(y + 1)$

▶ 21. $15x^2y - 10x^2y^2$
$5x^2y(3 - 2y)$

22. $63x^3y^2 + 81x^2y^4$
$9x^2y^2(7x + 9y^2)$

23. $27z^3 + 12z^2 + 3z$
$3z(9z^2 + 4z + 1)$

▶ 24. $25t^6 - 10t^3 + 5t^2$
$5t^2(5t^4 - 2t + 1)$

25. $24s^3 - 12s^2t + 6st^2$
$6s(4s^2 - 2st + t^2)$

▶ 26. $18y^2z^2 + 12y^2z^3 - 24y^4z^3$
$6y^2z^2(3 + 2z - 4y^2z)$

27. $11x^3 - 12y$
Prime

28. $14s^3 + 15t^6$
Prime

▶ 29. $23a^2b^3 + 4x^3y^2$
Prime

30. $18p^3q^2 - 5t^5$
Prime

***Factor each polynomial by factoring out the opposite of the GCF.* See Example 2.**

31. $-8a - 16$
$-8(a + 2)$

32. $-6b - 30$
$-6(b + 5)$

▶ 33. $-6x^2 - 3xy$
$-3x(2x + y)$

▶ 34. $-15y^3 - 25y^2$
$-5y^2(3y + 5)$

▶ 35. $-18a^2b + 12ab^2$
$-6ab(3a - 2b)$

▶ 36. $-21t^5 + 28t^3$
$-7t^3(3t^2 - 4)$

37. $-8a^4c^8 + 28a^3c^8 - 20a^2c^9$
$-4a^2c^8(2a^2 - 7a + 5c)$

▶ 38. $-30x^{10}y + 24x^9y^2 - 60x^8y^2$
$-6x^8y(5x^2 - 4xy + 10y)$

***Factor.* See Example 3.**

39. $(x + y)u + (x + y)v$
$(x + y)(u + v)$

▶ 40. $4(x + y) + t(x + y)$
$(x + y)(4 + t)$

▶ 41. $5(a - b + c) - t(a - b + c)$ $\quad (a - b + c)(5 - t)$

42. $(a - b - c)r - (a - b - c)s$ $\quad (a - b - c)(r - s)$

***Factor by grouping.* See Example 4.**

▶ 43. $ax + bx + ay + by$
$(x + y)(a + b)$

44. $ar - br + as - bs$
$(r + s)(a - b)$

▶ 45. $x^2 + yx - x - y$
$(x + y)(x - 1)$

46. $d^2 + cd + c + d$
$(c + d)(1 + d)$

47. $t^3 - 3t^2 - 7t + 21$
$(t - 3)(t^2 - 7)$

▶ 48. $b^3 - 4b^2 - 3b + 12$
$(b - 4)(b^2 - 3)$

▶ 49. $a^2 - 4b + ab - 4a$
$(a + b)(a - 4)$

▶ 50. $3c - cd + 3d - c^2$
$(3 - c)(c + d)$

***Factor.* See Example 5.**

51. $6x^3 - 6x^2 + 12x - 12$ $\quad 6(x - 1)(x^2 + 2)$

▶ 52. $3x^3 - 6x^2 + 15x - 30$ $\quad 3(x - 2)(x^2 + 5)$

53. $28a^3b^3c + 14a^3c - 4b^3c - 2c$ $\quad 2c(2b^3 + 1)(7a^3 - 1)$

▶ 54. $12x^3z + 12xy^2z - 8x^2yz - 8y^3z$ $\quad 4z(x^2 + y^2)(3x - 2y)$

***Solve for the specified variable or expression.* See Example 6.**

55. $2g = ch + dh$ for h $\quad h = \frac{2g}{c + d}$

▶ 56. $d_1d_2 = fd_2 + fd_1$ for f $\quad f = \frac{d_1d_2}{d_2 + d_1}$

▶ 57. $r_1r_2 = rr_2 + rr_1$ for r_1 $\quad r_1 = \frac{rr_2}{r_2 - r}$

58. $rx - ty = by$ for y $\quad y = \frac{rx}{b + t}$

59. $b^2x^2 + a^2y^2 = a^2b^2$ for a^2 $\quad a^2 = \frac{b^2x^2}{b^2 - y^2}$

▶ 60. $b^2x^2 + a^2y^2 = a^2b^2$ for b^2 $\quad b^2 = \frac{a^2y^2}{a^2 - x^2}$

▶ 61. $Sn = (n - 2)180$ for n $\quad n = \frac{360}{180 - S}$

62. $S(1 - r) = a - lr$ for r $\quad r = \frac{S - a}{S - l}$

***Factor.* See Example 7 or 12.**

63. $x^2 - 5x + 6$
$(x - 3)(x - 2)$

64. $y^2 + 7y + 6$
$(y + 1)(y + 6)$

65. $x^2 + x - 30$
$(x + 6)(x - 5)$

66. $c^2 + 3c - 28$
$(c - 4)(c + 7)$

***Factor.* See Example 8.**

67. $3x^2 + 12xy - 63y^2$
$3(x + 7y)(x - 3y)$

68. $2y^2 + 4yz - 48z^2$
$2(y + 6z)(y - 4z)$

69. $6a^2 - 30ab + 24b^2$
$6(a - 4b)(a - b)$

70. $4b^2 + 12bc - 16c^2$
$4(b + 4c)(b - c)$

71. $n^4 - 28n^3t - 60n^2t^2$
$n^2(n - 30t)(n + 2t)$

72. $c^4 - 16c^3d - 80c^2d^2$
$c^2(c - 20d)(c + 4d)$

73. $-3x^2 + 15xy - 18y^2$
$-3(x - 3y)(x - 2y)$

74. $-2y^2 - 16yt + 40t^2$
$-2(y + 10t)(y - 2t)$

***Factor.* See Example 9 or 12.**

75. $5x^2 + 13x + 6$
$(5x + 3)(x + 2)$

76. $5x^2 + 18x + 9$
$(5x + 3)(x + 3)$

77. $7a^2 + 12a + 5$
$(7a + 5)(a + 1)$

78. $7a^2 + 36a + 5$
$(7a + 1)(a + 5)$

79. $11y^2 + 32y - 3$
$(11y - 1)(y + 3)$

80. $2y^2 - 9y - 18$
$(2y + 3)(y - 6)$

81. $8x^2 - 22x + 5$
$(4x - 1)(2x - 5)$

82. $4z^2 - 13z + 3$
$(z - 3)(4z - 1)$

83. $6y^2 - 13y + 6$
$(3y - 2)(2y - 3)$

84. $6x^2 - 11x + 3$
$(3x - 1)(2x - 3)$

85. $15b^2 + 4b - 4$
$(5b - 2)(3b + 2)$

86. $8a^2 + 6a - 9$
$(4a - 3)(2a + 3)$

***Factor each expression.* See Example 10.**

87. $30x^4 - 25x^2 - 20$
$5(3x^2 - 4)(2x^2 + 1)$

88. $14x^4 + 77x^2 + 84$
$7(2x^2 + 3)(x^2 + 4)$

89. $32x^4 - 96x^2 + 72$
$8(2x^2 - 3)^2$

90. $20a^4 + 60a^2 + 45$
$5(2a^2 + 3)^2$

91. $64h^5 - 4h + 24h^3$
$4h(8h^2 - 1)(2h^2 + 1)$

92. $9x^5 - 24x + 30x^3$
$3x(3x^2 - 2)(x^2 + 4)$

93. $-3a^4 - 5a^2b^2 - 2b^4$
$-(3a^2 + 2b^2)(a^2 + b^2)$

94. $-2x^4 + 3x^2y^2 + 5y^4$
$-(2x^2 - 5y^2)(x^2 + y^2)$

***Factor by using a substitution.* See Example 11.**

95. $(a + b)^2 - 2(a + b) - 24$ $(a + b + 4)(a + b - 6)$

96. $(x - y)^2 + 3(x - y) - 10$ $(x - y + 5)(x - y - 2)$

97. $(x + a)^2 + 2(x + a) + 1$ $(x + a + 1)^2$

98. $(a + b)^2 - 2(a + b) + 1$ $(a + b - 1)^2$

TRY IT YOURSELF

Factor completely. Factor out all common factors first including −1 if the first term is negative. If an expression is prime, so indicate.

99. $3(m + n + p) + x(m + n + p)$ $(m + n + p)(3 + x)$

100. $x(x - y - z) + y(x - y - z)$ $(x - y - z)(x + y)$

101. $-63u^3v^6 + 28u^2v^7 - 21u^3v^3$ $-7u^2v^3(9uv^3 - 4v^4 + 3u)$

102. $-56x^4y^3 - 72x^3y^4 + 80xy^2$ $-8xy^2(7x^3y + 9x^2y^2 - 10)$

103. $b^4x^2 - 12b^2x^2 + 35x^2$ $x^2(b^2 - 7)(b^2 - 5)$

104. $c^3x^4 + 11c^3x^2 - 42c^3$ $c^3(x^2 + 14)(x^2 - 3)$

105. $1 - n - m + mn$
$(1 - n)(1 - m)$

106. $a^2x^2 - 10 - 2x^2 + 5a^2$
$(a^2 - 2)(x^2 + 5)$

107. $-x^2 + 4xy + 21y^2$
$-(x + 3y)(x - 7y)$

108. $-a^2 - 4ab + 5b^2$
$-(a + 5b)(a - b)$

109. $a^2x + bx - a^2 - b$
$(a^2 + b)(x - 1)$

110. $x^2y - ax - xy + a$
$(xy - a)(x - 1)$

111. $4y^2 + 4y + 1$
$(2y + 1)^2$

112. $9x^2 + 6x + 1$
$(3x + 1)^2$

113. $b^2 + 8b + 18$
Prime

114. $x^2 + 4x - 28$
Prime

115. $13r + 3r^2 - 10$
$(3r - 2)(r + 5)$

116. $-r + 3r^2 - 10$
$(3r + 5)(r - 2)$

117. $y^3 - 12 + 3y - 4y^2$ $(y^2 + 3)(y - 4)$

118. $h^3 - 8 + h - 8h^2$ $(h^2 + 1)(h - 8)$

119. $2y^5 - 26y^3 + 60y$
$2y(y^2 - 10)(y^2 - 3)$

120. $2y^5 - 26y^3 + 84y$
$2y(y^2 - 7)(y^2 - 6)$

121. $14(q - r)^2 - 17(q - r) - 6$ $(7q - 7r + 2)(2q - 2r - 3)$

122. $8(h + s)^2 + 34(h + s) + 35$ $(4h + 4s + 7)(2h + 2s + 5)$

APPLICATIONS

123. **Crayons.** The amount of colored wax used to make the crayon shown in the illustration can be found by computing its volume using the formula

$$V = \pi r^2 h_1 + \frac{1}{3}\pi r^2 h_2$$

Factor the expression on the right side of this equation. $\pi r^2\left(h_1 + \frac{1}{3}h_2\right)$

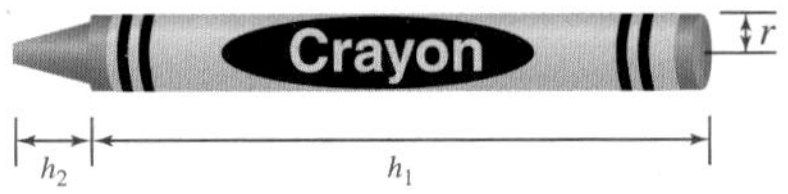

124. **Packaging.** The amount of cardboard needed to make the following cereal box can be found by finding the area A, which is given by the formula

$$A = 2wh + 4wl + 2lh$$

where w is the width, h the height, and l the length. Solve the equation for the width. $w = \frac{A - 2lh}{2h + 4l}$

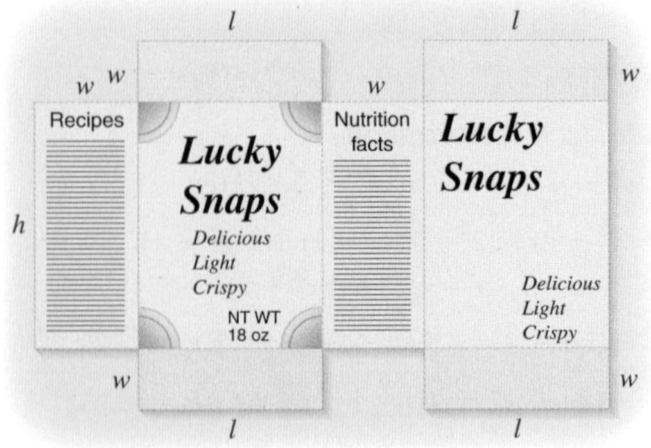

125. **Ice.** The surface area of the ice cube is represented by the expression $6x^2 + 36x + 54$. Use factoring to find the length of an edge of the cube. $x + 3$

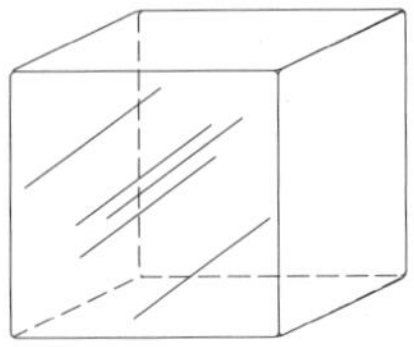

▶ **126. Checkers.** The area of the checkerboard is represented by the expression $25x^2 - 40x + 16$. Use factoring to find the length of each side. $5x - 4$

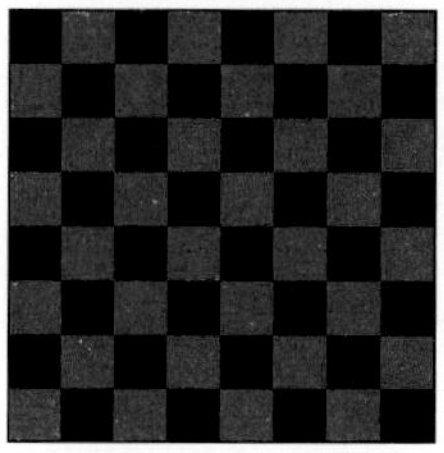

WRITING

▶ **127.** Explain the error in the following solution.

Solve for r_1: $r_1r_2 = rr_2 + rr_1$

$$\frac{r_1r_2}{r_2} = \frac{rr_2 + rr_1}{r_2}$$

$$r_1 = \frac{rr_2 + rr_1}{r_2}$$

128. Explain the error.

Factor: $2x^2 - 4x - 6 = (2x + 2)(x - 3)$

REVIEW

129. Use the graph to find:

a. $s(-5)$ 2

b. $s(4)$ 3

c. The values of x for which $s(x) = 0$ $-1, 1, 5$

d. The value of x for which $s(x) = 4$ 3

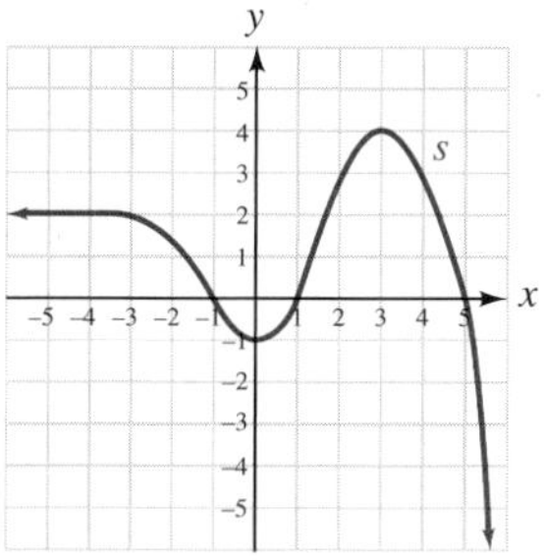

130. Determine whether each equation defines y to be a function of x. If it does not, find two ordered pairs where more than one value of y corresponds to a single value of x.

a. $x = |y|$ No; (1, 1), (1, −1)

b. $y = x^2$ Yes

c. $y^2 = x$ No; (1, 1), (1, −1)

CHALLENGE PROBLEMS

Factor out the specified factor.

131. t^{-3} from $t^5 + 4t^{-6}$ $t^{-3}(t^8 + 4t^{-3})$

▶ **132.** $7x^{-3n}$ from $21x^{6n} + 7x^{3n} + 14$ $7x^{-3n}(3x^{9n} + x^{6n} + 2x^{3n})$

Factor. Assume that n is a natural number.

133. $x^{2n} + 2x^n + 1$ $(x^n + 1)^2$

▶ **134.** $2a^{6n} - 3a^{3n} - 2$ $(2a^{3n} + 1)(a^{3n} - 2)$

135. $x^{4n} + 2x^{2n}y^{2n} + y^{4n}$ $(x^{2n} + y^{2n})^2$

136. $6x^{2n} + 7x^n - 3$ $(3x^n - 1)(2x^n + 3)$

SECTION 8.7

OBJECTIVES

1 Factor the difference of two squares.

2 Factor the sum and difference of two cubes.

Review of Factoring Methods: The Difference of Two Squares; the Sum and Difference of Two Cubes

ARE YOU READY? *Are You Ready? exercises available online at www.webassign.net/brookscole*

The following problems review some basic skills that are needed when factoring certain types of binomials.

1. Multiply: $(n + 9)(n - 9)$ $n^2 - 81$

2. Simplify: $(8d)^2$ $64d^2$

3. Evaluate: **a.** 3^3 27 **b.** 6^3 216

4. Simplify: $(4a)^3$ $64a^3$

5. Multiply: $(b + 4)(b^2 - 4b + 16)$ $b^3 + 64$

6. Multiply: $(2a - 1)(4a^2 + 2a + 1)$ $8a^3 - 1$

We will now review how to factor the difference of two squares and the sum and difference of two cubes.

1 Factor the Difference of Two Squares.

The Language of Algebra

The expression $A^2 - B^2$ is a **difference of two squares,** whereas $(A - B)^2$ is the **square of a difference.** They are not equivalent because $(A - B)^2 \neq A^2 - B^2$.

Recall the special-product rule for multiplying the sum and difference of the same two terms:

$$(A + B)(A - B) = A^2 - B^2$$

The binomial $A^2 - B^2$ is called a **difference of two squares,** because A^2 is the square of A and B^2 is the square of B. If we reverse this rule, we obtain a method for factoring a difference of two squares.

Factoring a Difference of Two Squares

To factor the square of a First quantity minus the square of a Last quantity, multiply the First plus the Last by the First minus the Last.

$$F^2 - L^2 = (F + L)(F - L)$$

Teaching Tip: The words "First" and "Last" help students verbalize this factorization rule. Educational research has found that verbalization promotes concept formation.

To factor the difference of two squares, it is helpful to know the first twenty **perfect-square integers.** The number 400, for example, is a perfect square, because $400 = 20^2$.

$1 = 1^2$	$25 = 5^2$	$81 = 9^2$	$169 = 13^2$	$289 = 17^2$
$4 = 2^2$	$36 = 6^2$	$100 = 10^2$	$196 = 14^2$	$324 = 18^2$
$9 = 3^2$	$49 = 7^2$	$121 = 11^2$	$225 = 15^2$	$361 = 19^2$
$16 = 4^2$	$64 = 8^2$	$144 = 12^2$	$256 = 16^2$	$400 = 20^2$

EXAMPLE 1 Factor: $49x^2 - 16$

Strategy The terms of this binomial do not have a common factor (other than 1). The only option is to attempt to factor it as a difference of two squares.

Why If a binomial is a difference of two squares, we can factor it using a special product.

Solution $49x^2 - 16$ is the difference of two squares because it can be written as $(7x)^2 - (4)^2$. We can match it to the rule for factoring a difference of two squares to find the factorization.

$$\begin{array}{c} \mathbf{F}^2 - \mathbf{L}^2 = (\mathbf{F} + \mathbf{L})(\mathbf{F} - \mathbf{L}) \\ \downarrow \quad \downarrow \quad \downarrow \quad \downarrow \; \downarrow \quad \downarrow \\ (7x)^2 - (4)^2 = (7x + 4)(7x - 4) \end{array}$$

Therefore, $49x^2 - 16 = (7x + 4)(7x - 4)$. We can verify this result using multiplication.

$$\begin{aligned} (7x + 4)(7x - 4) &= 49x^2 - 28x + 28x - 16 \\ &= 49x^2 - 16 \quad \text{This is the original binomial.} \end{aligned}$$

Success Tip

Always verify a factorization by doing the indicated multiplication. The result should be the original polynomial.

Teaching Example 1
Factor: $25x^2 - 36$
Answer: $(5x + 6)(5x - 6)$

Self Check 1 Factor: $81p^2 - 25$ $(9p + 5)(9p - 5)$

Now Try Problems 11 and 17

Expressions such as a^4 and x^6y^8 are also perfect squares, because they can be written as the square of another quantity:

$$a^4 = (a^2)^2, \qquad 81b^4 = (9b^2)^2, \qquad \text{and} \qquad x^6y^8 = (x^3y^4)^2$$

EXAMPLE 2 Factor: $64a^4 - 25b^2$

Strategy The terms of this binomial do not have a common factor (other than 1). The only option is to attempt to factor it as a difference of two squares.

Why If a binomial is a difference of two squares, we can factor it using a special-product rule.

Solution We can write $64a^4 - 25b^2$ in the form $(8a^2)^2 - (5b)^2$ and use the rule for factoring the difference of two squares.

$$\begin{array}{c} \mathbf{F}^2 - \mathbf{L}^2 = (\mathbf{F} + \mathbf{L})(\mathbf{F} - \mathbf{L}) \\ \downarrow \quad \downarrow \quad \downarrow \quad \downarrow \; \downarrow \quad \downarrow \\ (8a^2)^2 - (5b)^2 = (8a^2 + 5b)(8a^2 - 5b) \end{array}$$

Therefore, $64a^4 - 25b^2 = (8a^2 + 5b)(8a^2 - 5b)$.

Notation

By the commutative property of multiplication, this factorization could be written

$$(8a^2 - 5b)(8a^2 + 5b)$$

Teaching Example 2
Factor: $49x^2 - 81y^4$
Answer: $(7x + 9y^2)(7x - 9y^2)$

Self Check 2 Factor: $36r^4 - s^2$ $(6r^2 + s)(6r^2 - s)$

Now Try Problems 21 and 25

EXAMPLE 3 Factor: $x^4 - 1$

Strategy The terms of $x^4 - 1$ do not have a common factor (other than 1). To factor this binomial, we will write it in a form that shows it is a difference of two squares.

Why We can then use a special-product rule to factor it.

Solution Because the binomial is the difference of the squares of x^2 and 1, it factors into the sum of x^2 and 1 and the difference of x^2 and 1.

$$x^4 - 1 = (x^2)^2 - (1)^2$$
$$= (x^2 + 1)(x^2 - 1)$$

A prime polynomial (pointing to $x^2 + 1$)

The factor $x^2 + 1$ is the sum of two quantities and is prime. However, the factor $x^2 - 1$ is the difference of two squares and can be factored as $(x + 1)(x - 1)$. Thus,

$$x^4 - 1 = (x^2 + 1)(x^2 - 1)$$
$$= (x^2 + 1)(x + 1)(x - 1) \quad \text{Don't forget to write } (x^2 + 1).$$

Caution

The binomial $x^2 + 1$ is the **sum of two squares.** In general, after removing any common factor, a sum of two squares cannot be factored using real numbers.

Teaching Example 3
Factor: $x^4 - 16$
Answer: $(x^2 + 4)(x + 2)(x - 2)$

Self Check 3 Factor: $a^4 - 81$ $(a^2 + 9)(a + 3)(a - 3)$

Now Try Problem 29

At times, a **substitution** can be helpful in simplifying the factorization process.

EXAMPLE 4 Factor: $(x + y)^4 - z^4$

Strategy We will use a substitution to factor this difference of two squares.

Why For more complicated expressions, especially those involving a quantity within parentheses, a substitution often helps simplify the factoring process.

Solution If we use the substitution $a = x + y$, we obtain

$$(x + y)^4 - z^4 = a^4 - z^4 \quad \text{Replace } x + y \text{ with } a.$$
$$= (a^2 + z^2)(a^2 - z^2) \quad \text{Factor the difference of two squares.}$$
$$= (a^2 + z^2)(a + z)(a - z) \quad \text{Factor } a^2 - z^2\text{. } a^2 + z^2 \text{ is prime.}$$

To find the factorization of $(x + y)^4 - z^4$, we "reverse" substitute $x + y$ for each a in the expression $(a^2 + z^2)(a + z)(a - z)$.

$$(a^2 + z^2)(a + z)(a - z) = [(x + y)^2 + z^2](x + y + z)(x + y - z)$$

Thus, $(x + y)^4 - z^4 = [(x + y)^2 + z^2](x + y + z)(x + y - z)$.

If we square the binomial within the brackets, we have

$$(x + y)^4 - z^4 = [x^2 + 2xy + y^2 + z^2](x + y + z)(x + y - z)$$

Caution

A common error after making a substitution is to forget to "undo" or "reverse" it. In this example, if you are to factor an expression in x, y, and z, your answer must involve only x, y, and z. It should not contain the variable a.

Caution

When factoring a polynomial, be sure to factor it completely. Always check to see whether any of the factors of your result can be factored further.

Self Check 4 Factor: $(a - b)^4 - c^4$ $(a^2 - 2ab + b^2 + c^2)(a - b + c)(a - b - c)$

Now Try Problem 35

Teaching Example 4
Factor: $(x - y)^4 - z^4$
Answer:
$(x^2 - 2xy + y^2 + z^2)(x - y + z)(x - y - z)$

When possible, we always factor out a common factor before factoring the difference of two squares. The factoring process is easier when all common factors are factored out first.

EXAMPLE 5 Factor: $2x^4y - 32y$

Strategy We will factor out the GCF of $2y$ and factor the resulting difference of two squares.

Why The first step in factoring any polynomial is to factor out the GCF.

Solution

$2x^4y - 32y = 2y(x^4 - 16)$ Factor out the GCF, which is $2y$.

$= 2y(x^2 + 4)(x^2 - 4)$ Factor $x^4 - 16$.

$= 2y(x^2 + 4)(x + 2)(x - 2)$ Factor $x^2 - 4$. $x^2 + 4$ is prime.

Success Tip

Remember that a **difference of two squares** is a binomial. Each term is a square and the terms have different signs. The powers of the variables in the terms must be even.

Self Check 5 Factor: $3a^4 - 3$ $3(a^2 + 1)(a + 1)(a - 1)$

Now Try Problems 37 and 43

Teaching Example 5
Factor: $5x^4 - 405$
Answer: $5(x^2 + 9)(x + 3)(x - 3)$

To factor some expressions, we need to use some creative grouping to begin the process.

EXAMPLE 6 Factor: **a.** $x^2 - y^2 + x - y$ **b.** $x^2 + 6x + 9 - z^2$

Strategy The terms of each expression do not have a common factor (other than 1) and traditional factoring by grouping will not work. Instead, in part a, we will group only the first two terms of the polynomial and in part b we will group the first three terms.

Why Hopefully, those steps will produce equivalent expressions that can be factored.

Solution We group the first two terms, factor them as a difference of two squares, and look for a common factor.

a. $x^2 - y^2 + x - y = (x + y)(x - y) + (x - y)$ Factor $x^2 - y^2$. The terms of the resulting expression have a common binomial factor, $x - y$, that can be factored out later.

$= (x + y)(x - y) + 1(x - y)$ Factor out 1 from $x - y$.

$= (x - y)(x + y + 1)$ Factor out the GCF, $x - y$. Don't forget the 1.

b. We group the first three terms and factor that trinomial to get:

$x^2 + 6x + 9 - z^2 = (x + 3)(x + 3) - z^2$ $x^2 + 6x + 9$ is a perfect-square trinomial.

$= (x + 3)^2 - z^2$ Write $(x + 3)(x + 3)$ as $(x + 3)^2$. The expression that results is a difference of two squares.

$= (x + 3 + z)(x + 3 - z)$ Factor the difference of two squares.

Success Tip

We could use a substitution to factor $(x + 3)^2 - z^2$.

Self Check 6 Factor: **a.** $a^2 - b^2 + a + b$ $(a + b)(a - b + 1)$
b. $a^2 + 4a + 4 - b^2$ $(a + 2 + b)(a + 2 - b)$

Now Try Problems 45 and 49

Teaching Example 6
Factor:
a. $x^2 - 4 + x + 2$
b. $t^2 + 6t + 9 - r^2$
Answers: **a.** $(x + 2)(x - 2 + 1)$
b. $(t + 3 + r)(t + 3 - r)$

2 Factor the Sum and Difference of Two Cubes.

The number 64 is called a perfect cube, because $4^3 = 64$. To factor the sum or difference of two cubes, it is helpful to know the first ten **perfect-cube integers:**

$1 = 1^3$	$27 = 3^3$	$125 = 5^3$	$343 = 7^3$	$729 = 9^3$
$8 = 2^3$	$64 = 4^3$	$216 = 6^3$	$512 = 8^3$	$1{,}000 = 10^3$

Expressions such as b^6 and $64x^9y^{12}$ are also perfect cubes, because they can be written as the cube of another quantity:

$$b^6 = (b^2)^3 \qquad \text{and} \qquad 64x^9y^{12} = (4x^3y^4)^3$$

To find rules for factoring the sum of two cubes and the difference of two cubes, we need to find the products shown below. Note that each term of the trinomial is multiplied by each term of the binomial.

Teaching Tip: You may want to use vertical form to show the multiplication:

$$\begin{array}{r} x^2 - xy + y^2 \\ x + y \\ \hline \end{array}$$

$$\begin{aligned}(x + y)(x^2 - xy + y^2) &= x^3 - \cancel{x^2y} + \cancel{xy^2} + \cancel{x^2y} - \cancel{xy^2} + y^3 \\ &= x^3 + y^3 \end{aligned}$$

Combine like terms.

$$\begin{aligned}(x - y)(x^2 + xy + y^2) &= x^3 + \cancel{x^2y} + \cancel{xy^2} - \cancel{x^2y} - \cancel{xy^2} - y^3 \\ &= x^3 - y^3 \end{aligned}$$

Combine like terms.

We have found that

$$x^3 + y^3 = (x + y)(x^2 - xy + y^2) \qquad \text{and} \qquad x^3 - y^3 = (x - y)(x^2 + xy + y^2)$$

The Language of Algebra

The expression $x^3 + y^3$ is a **sum of two cubes,** whereas $(x + y)^3$ is the **cube of a sum.** If you expand $(x + y)^3$, you will see that they are not equivalent.

The binomial $x^3 + y^3$ is called the **sum of two cubes,** because x^3 represents the cube of x, y^3 represents the cube of y, and $x^3 + y^3$ represents the sum of these cubes. Similarly, $x^3 - y^3$ is called the **difference of two cubes.**

These results justify the rules for factoring the **sum and difference of two cubes.** They are easier to remember if we think of a sum (or a difference) of two cubes as the cube of a **F**irst quantity plus (or minus) the cube of the **L**ast quantity.

Factoring the Sum and Difference of Two Cubes

To factor the cube of a First quantity plus the cube of a Last quantity, multiply the First plus the Last by the First squared, minus the First times the Last, plus the Last squared.

$$F^3 + L^3 = (F + L)(F^2 - FL + L^2)$$

To factor the cube of a First quantity minus the cube of a Last quantity, multiply the First minus the Last by the First squared, plus the First times the Last, plus the Last squared.

$$F^3 - L^3 = (F - L)(F^2 + FL + L^2)$$

EXAMPLE 7 Factor: $a^3 + 8$

Strategy We will write the binomial in a form that shows it is the sum of two cubes.

Why We can then use the rule for factoring the sum of two cubes.

Solution Since $a^3 + 8$ can be written as $a^3 + 2^3$, it is a sum of two cubes, which factors as follows:

$$\mathbf{F^3 + L^3 = (F + L)(F^2 - FL + L^2)}$$
$$\downarrow \quad \downarrow \quad \downarrow \; \downarrow \; \downarrow \;\; \downarrow\downarrow \;\; \downarrow$$
$$\boldsymbol{a^3 + 2^3 = (a + 2)(a^2 - a2 + 2^2)}$$
$$= (a + 2)(a^2 - 2a + 4) \qquad a^2 - 2a + 4 \text{ does not factor, it is prime.}$$

Therefore, $a^3 + 8 = (a + 2)(a^2 - 2a + 4)$.

We can check by multiplying.

$$(a + 2)(a^2 - 2a + 4) = a^3 - 2a^2 + 4a + 2a^2 - 4a + 8$$
$$= a^3 + 8 \qquad \text{This is the original binomial.}$$

Caution

In Example 7, a common error is to try to factor $a^2 - 2a + 4$. It is not a perfect-square trinomial, because the middle term needs to be $-4a$. Furthermore, it cannot be factored by any of the methods of Section 6.2. It is prime.

Teaching Example 7
Factor: $x^3 + 125$
Answer: $(x + 5)(x^2 - 5x + 25)$

Self Check 7 Factor: $p^3 + 27$ $\quad (p + 3)(p^2 - 3p + 9)$

Now Try Problem 53

You should memorize the rules for factoring the sum and the difference of two cubes. Note that each has the form

(a binomial)(a trinomial)

and that there is a relationship between the signs that appear in these forms.

Success Tip

An easy way to remember the sign patterns within the binomial and trinomial of these factoring forms is with the word "SOAP."

Same, **O**pposite, **A**lways, **P**lus

The Sum of Cubes

The same sign

$$F^3 + L^3 = (F + L)(F^2 - FL + L^2)$$

Opposite signs   Always plus

The Difference of Cubes

The same sign

$$F^3 - L^3 = (F - L)(F^2 + FL + L^2)$$

Opposite signs   Always plus

EXAMPLE 8 Factor: $27a^3 - 64b^6$

Strategy We will write the binomial in a form that shows it is the difference of two cubes.

Why We can then use the rule for factoring the difference of two cubes.

Solution Since $27a^3 - 64b^6$ can be written as $(3a)^3 - (4b^2)^3$, it is a difference of two cubes, which factors as follows:

$$\mathbf{F^3 \;-\; L^3 \;= (\,F \;-\; L\,)(\,F^2 \;+\; F \;\; L \;+\; L^2)}$$
$$\downarrow \qquad \downarrow \qquad \downarrow \quad \downarrow \quad \downarrow \quad \downarrow \;\; \downarrow \quad \downarrow$$
$$\boldsymbol{(3a)^3 - (4b^2)^3 = (3a - 4b^2)[(3a)^2 + (3a)(4b^2) + (4b^2)^2]}$$
$$= (3a - 4b^2)(9a^2 + 12ab^2 + 16b^4)$$

Thus, $27a^3 - 64b^6 = (3a - 4b^2)(9a^2 + 12ab^2 + 16b^4)$. Multiply to check.

Teaching Tip: Remind students that the *trinomial* in the factored form of a sum or difference of two cubes will always be prime.

Teaching Example 8
Factor: $1{,}000x^3 - 27y^6$
Answer:
$(10x - 3y^2)(100x^2 + 30xy^2 + 9y^4)$

Self Check 8 Factor: $8c^6 - 125d^3$ $\quad (2c^2 - 5d)(4c^4 + 10c^2d + 25d^2)$

Now Try Problem 57

EXAMPLE 9 Factor: $a^3 - (c + d)^3$

Strategy To factor this expression, we will use the rule for factoring the difference of two cubes.

Why The terms a^3 and $(c + d)^3$ are perfect cubes.

Solution

$$\begin{array}{ccccccccc} F^3 & - & L^3 & = (F - & L &)(F^2 + F & L & + & L^2 \ ) \\ \downarrow & & \downarrow & \downarrow & \downarrow & \downarrow \quad \downarrow & \downarrow & & \downarrow \end{array}$$

$$a^3 - (c + d)^3 = [a - (c + d)][a^2 + a(c + d) + (c + d)^2]$$

Now we simplify the expressions within both sets of brackets. Thus,

$$a^3 - (c + d)^3 = (a - c - d)(a^2 + ac + ad + c^2 + 2cd + d^2)$$

Success Tip

We could also use the substitution $x = c + d$, factor $a^3 - x^3$ as $(a - x)(a^2 + ax + x^2)$, and then replace each x with $c + d$.

Teaching Example 9 Factor: $(x + y)^3 - b^3$ *Answer:* $(x + y - b)(x^2 + 2xy + y^2 + bx + by + b^2)$

Self Check 9 Factor: $(p + q)^3 - r^3$ $(p + q - r)(p^2 + 2pq + q^2 + pr + qr + r^2)$

Now Try Problem 61

EXAMPLE 10 Factor: $x^6 - 64$

Strategy This binomial is both the difference of two squares and the difference of two cubes. We will write it in a form that shows it is a difference of two squares to begin the factoring process.

Why It is easier to factor it as the difference of two squares first.

Solution

$$\begin{aligned} x^6 - 64 &= (x^3)^2 - 8^2 \\ &= (x^3 + 8)(x^3 - 8) \end{aligned}$$

Teaching Tip: Stress this wording: These factoring forms enable us to write certain *sums* and *differences* as equivalent *products*.

Each of these binomial factors can be factored further. The first is the sum of two cubes and the second is the difference of two cubes. Thus,

$$x^6 - 64 = (x + 2)(x^2 - 2x + 4)(x - 2)(x^2 + 2x + 4)$$

Teaching Example 10 Factor: $a^6 - b^6$ *Answer:* $(a - b)(a^2 + ab + b^2)(a + b)(a^2 - ab + b^2)$

Self Check 10 Factor: $1 - x^6$ $(1 + x)(1 - x + x^2)(1 - x)(1 + x + x^2)$

Now Try Problem 65

EXAMPLE 11 Factor: $2a^5 + 250a^2$

Strategy We will factor out the GCF $2a^2$ and then factor the resulting sum of two cubes.

Why The first step in factoring any polynomial is to factor out the GCF.

Solution We first factor out the common factor $2a^2$ to obtain

$$2a^5 + 250a^2 = 2a^2(a^3 + 125)$$

Then we factor $a^3 + 125$ as the sum of two cubes to obtain

$$2a^5 + 250a^2 = 2a^2(a + 5)(a^2 - 5a + 25) \qquad a^2 - 5a + 25 \text{ is prime.}$$

Teaching Example 11 Factor: $9n^6 + 9{,}000n^3$ *Answer:* $9n^3(n + 10)(n^2 - 10n + 100)$

Self Check 11 Factor: $3x^5 + 24x^2$ $3x^2(x + 2)(x^2 - 2x + 4)$

Now Try Problem 69

SECTION 8.7 STUDY SET

VOCABULARY

Fill in the blanks.

▶ 1. When the polynomial $4x^2 - 25$ is written as $(2x)^2 - (5)^2$, we see that it is the difference of two squares.

▶ 2. When the polynomial $8x^3 + 125$ is written as $(2x)^3 + (5)^3$, we see that it is the sum of two cubes.

▶ Selected exercises available online at www.webassign.net/brookscole

CONCEPTS

3. a. Write the first ten perfect-square integers.
   1, 4, 9, 16, 25, 36, 49, 64, 81, 100
   b. Write the first ten perfect-cube integers.
   1, 8, 27, 64, 125, 216, 343, 512, 729, 1,000
4. a. Use multiplication to verify that the sum of two squares $x^2 + 25$ does not factor as $(x + 5)(x + 5)$.
   $(x + 5)(x + 5) = x^2 + 10x + 25$
   b. Use multiplication to verify that the difference of two squares $x^2 - 25$ factors as $(x + 5)(x - 5)$.
   $(x + 5)(x - 5) = x^2 - 25$
5. Complete each factorization.
   a. $F^2 - L^2 = (F + L)(\ F - L\ )$
   b. $F^3 + L^3 = (F + L)(\ F^2 - FL + L^2\ )$
   c. $F^3 - L^3 = (F - L)(\ F^2 + FL + L^2\ )$
6. Factor each binomial.
   a. $5p^2 + 20$ $5(p^2 + 4)$
   b. $5p^2 - 20$ $5(p + 2)(p - 2)$
   c. $5p^3 + 20$ $5(p^3 + 4)$
   d. $5p^3 + 40$ $5(p + 2)(p^2 - 2p + 4)$

NOTATION

7. Give an example of each. Answers may vary.
   a. a difference of two squares $x^2 - 4$
   b. a square of a difference $(x - 4)^2$
   c. a sum of two squares $x^2 + 4$
   d. a sum of two cubes $x^3 + 8$
   e. a cube of a sum $(x + 8)^3$
8. Fill in the blanks.
   a. $36y^2 - 49m^4 = (\ 6y\ )^2 - (\ 7m^2\ )^2$
   b. $125h^3 - 27k^6 = (\ 5h\ )^3 - (\ 3k^2\ )^3$

GUIDED PRACTICE

Factor. See Example 1.

9. $x^2 - 16$ $(x + 4)(x - 4)$
10. $y^2 - 49$ $(y + 7)(y - 7)$
11. $9y^2 - 64$ $(3y + 8)(3y - 8)$
12. $16x^2 - 81$ $(4x + 9)(4x - 9)$
13. $144 - c^2$ $(12 + c)(12 - c)$
14. $25 - t^2$ $(5 + t)(5 - t)$
15. $100m^2 - 1$ $(10m + 1)(10m - 1)$
16. $144x^2 - 1$ $(12x + 1)(12x - 1)$
17. $81a^2 - 49b^2$ $(9a + 7b)(9a - 7b)$
18. $64r^2 - 121s^2$ $(8r + 11s)(8r - 11s)$
19. $x^2 + 25$ Prime
20. $a^2 + 36$ Prime

Factor each difference of two squares. See Example 2.

21. $9r^4 - 121s^2$ $(3r^2 + 11s)(3r^2 - 11s)$
22. $81a^4 - 16b^2$ $(9a^2 + 4b)(9a^2 - 4b)$
23. $16t^2 - 25w^4$ $(4t + 5w^2)(4t - 5w^2)$
24. $9r^2 - 25s^4$ $(3r + 5s^2)(3r - 5s^2)$
25. $100r^2s^4 - t^4$ $(10rs^2 + t^2)(10rs^2 - t^2)$
26. $400x^2z^4 - a^4$ $(20xz^2 + a^2)(20xz^2 - a^2)$
27. $36x^4y^2 - 49z^6$ $(6x^2y + 7z^3)(6x^2y - 7z^3)$
28. $4a^2b^4 - 9d^6$ $(2ab^2 + 3d^3)(2ab^2 - 3d^3)$

Factor completely. See Example 3.

29. $x^4 - y^4$ $(x^2 + y^2)(x + y)(x - y)$
30. $16n^4 - 1$ $(4n^2 + 1)(2n + 1)(2n - 1)$
31. $16a^4 - 81b^4$ $(4a^2 + 9b^2)(2a + 3b)(2a - 3b)$
32. $81m^4 - 256n^4$ $(9m^2 + 16n^2)(3m + 4n)(3m - 4n)$

Factor. See Example 4.

33. $(x + y)^2 - z^2$ $(x + y + z)(x + y - z)$
34. $a^2 - (b - c)^2$ $(a + b - c)(a - b + c)$
35. $(r - s)^2 - t^4$ $(r - s + t^2)(r - s - t^2)$
36. $(m + n)^2 - p^4$ $(m + n + p^2)(m + n - p^2)$

Factor each expression. Factor out any GCF first. See Example 5.

37. $2x^2 - 288$ $2(x + 12)(x - 12)$
38. $8x^2 - 72$ $8(x + 3)(x - 3)$
39. $3x^3 - 243x$ $3x(x + 9)(x - 9)$
40. $2x^3 - 32x$ $2x(x + 4)(x - 4)$
41. $5ab^4 - 5a$ $5a(b^2 + 1)(b + 1)(b - 1)$
42. $3ac^4 - 243a$ $3a(c^2 + 9)(c + 3)(c - 3)$
43. $64b - 4b^5$ $4b(4 + b^2)(2 + b)(2 - b)$
44. $1{,}250n - 2n^5$ $2n(25 + n^2)(5 + n)(5 - n)$

Factor by first grouping the appropriate terms. See Example 6.

45. $c^2 - d^2 + c + d$ $(c + d)(c - d + 1)$
46. $s^2 - t^2 + s - t$ $(s - t)(s + t + 1)$
47. $a^2 - b^2 + 2a - 2b$ $(a - b)(a + b + 2)$
48. $m^2 - n^2 + 3m + 3n$ $(m + n)(m - n + 3)$
49. $x^2 + 12x + 36 - y^2$ $(x + 6 + y)(x + 6 - y)$
50. $x^2 - 6x + 9 - 4y^2$ $(x - 3 + 2y)(x - 3 - 2y)$
51. $x^2 - 2x + 1 - 9z^2$ $(x - 1 + 3z)(x - 1 - 3z)$
52. $x^2 + 10x + 25 - 16z^2$ $(x + 5 + 4z)(x + 5 - 4z)$

Factor each sum of cubes. See Example 7.

53. $a^3 + 125$ $(a + 5)(a^2 - 5a + 25)$
54. $b^3 + 64$ $(b + 4)(b^2 - 4b + 16)$
55. $8r^3 + s^3$ $(2r + s)(4r^2 - 2rs + s^2)$
56. $27t^3 + u^3$ $(3t + u)(9t^2 - 3tu + u^2)$

Factor each difference of cubes. See Example 8.

57. $64t^6 - 27v^3$ $(4t^2 - 3v)(16t^4 + 12t^2v + 9v^2)$
58. $125m^3 - x^6$ $(5m - x^2)(25m^2 + 5mx^2 + x^4)$
59. $x^3 - 216y^6$ $(x - 6y^2)(x^2 + 6xy^2 + 36y^4)$
60. $8c^6 - 343w^3$ $(2c^2 - 7w)(4c^4 + 14c^2w + 49w^2)$

Factor. See Example 9.

61. $(a - b)^3 + 27$ $(a - b + 3)(a^2 - 2ab + b^2 - 3a + 3b + 9)$
62. $(b - c)^3 - 1{,}000$
    $(b - c - 10)(b^2 - 2bc + c^2 + 10b - 10c + 100)$
63. $64 - (a + b)^3$
    $(4 - a - b)(16 + 4a + 4b + a^2 + 2ab + b^2)$
64. $1 - (x + y)^3$
    $(1 - x - y)(1 + x + y + x^2 + 2xy + y^2)$

Factor each expression completely. Factor a difference of two squares first. See Example 10.

65. $x^6 - 1$ $(x + 1)(x^2 - x + 1)(x - 1)(x^2 + x + 1)$

▶ 66. $x^6 - y^6$ $(x + y)(x^2 - xy + y^2)(x - y)(x^2 + xy + y^2)$

67. $x^{12} - y^6$ $(x^2 + y)(x^4 - x^2y + y^2)(x^2 - y)(x^4 + x^2y + y^2)$

▶ 68. $a^{12} - 64$ $(a^2 + 2)(a^4 - 2a^2 + 4)(a^2 - 2)(a^4 + 2a^2 + 4)$

Factor each sum or difference of cubes. Factor out the GCF first. See Example 11.

▶ 69. $5x^3 + 625$ $5(x + 5)(x^2 - 5x + 25)$

▶ 70. $2x^3 - 128$ $2(x - 4)(x^2 + 4x + 16)$

▶ 71. $4x^5 - 256x^2$ $4x^2(x - 4)(x^2 + 4x + 16)$

▶ 72. $2x^6 + 54x^3$ $2x^3(x + 3)(x^2 - 3x + 9)$

TRY IT YOURSELF

Factor each expression.

▶ 73. $64a^3 - 125b^6$ $(4a - 5b^2)(16a^2 + 20ab^2 + 25b^4)$

74. $8x^6 - 27y^3$ $(2x^2 - 3y)(4x^4 + 6x^2y + 9y^2)$

75. $288b^2 - 2b^6$ $2b^2(12 + b^2)(12 - b^2)$

76. $98x - 2x^5$ $2x(7 + x^2)(7 - x^2)$

77. $x^2 - y^2 + 8x + 8y$ $(x + y)(x - y + 8)$

78. $5m - 5n + m^2 - n^2$ $(m - n)(5 + m + n)$

▶ 79. $x^9 + y^9$ $(x + y)(x^2 - xy + y^2)(x^6 - x^3y^3 + y^6)$

80. $x^6 + y^6$ $(x^2 + y^2)(x^4 - x^2y^2 + y^4)$

▶ 81. $144a^2t^2 - 169b^6$ $(12at + 13b^3)(12at - 13b^3)$

82. $25x^6 - 81y^2z^2$ $(5x^3 + 9yz)(5x^3 - 9yz)$

83. $100a^2 + 9b^2$ Prime

▶ 84. $25s^4 + 16t^2$ Prime

85. $81c^4d^4 - 16t^4$ $(9c^2d^2 + 4t^2)(3cd + 2t)(3cd - 2t)$

▶ 86. $256x^4 - 81y^4$ $(16x^2 + 9y^2)(4x + 3y)(4x - 3y)$

87. $128u^2v^3 - 2t^3u^2$ $2u^2(4v - t)(16v^2 + 4tv + t^2)$

88. $56rs^2t^3 + 7rs^2v^6$ $7rs^2(2t + v^2)(4t^2 - 2tv^2 + v^4)$

89. $y^2 - (2x - t)^2$ $(y + 2x - t)(y - 2x + t)$

90. $(15 - r)^2 - s^2$ $(15 - r + s)(15 - r - s)$

▶ 91. $x^2 + 20x + 100 - 9z^2$ $(x + 10 + 3z)(x + 10 - 3z)$

92. $49a^2 - b^2 - 14b - 49$ $(7a + b + 7)(7a - b - 7)$

▶ 93. $(c - d)^3 + 216$ $(c - d + 6)(c^2 - 2cd + d^2 - 6c + 6d + 36)$

94. $1 - (x + y)^3$ $(1 - x - y)(1 + x + y + x^2 + 2xy + y^2)$

95. $\frac{1}{36} - y^4$ $\left(\frac{1}{6} + y^2\right)\left(\frac{1}{6} - y^2\right)$

▶ 96. $\frac{4}{81} - m^4$ $\left(\frac{2}{9} + m^2\right)\left(\frac{2}{9} - m^2\right)$

97. $m^6 - 64$ $(m + 2)(m^2 - 2m + 4)(m - 2)(m^2 + 2m + 4)$

98. $1 - y^6$ $(1 + y)(1 - y + y^2)(1 - y)(1 + y + y^2)$

▶ 99. $(a + b)x^3 + 27(a + b)$ $(a + b)(x + 3)(x^2 - 3x + 9)$

100. $(c - d)r^3 - (c - d)s^3$ $(c - d)(r - s)(r^2 + rs + s^2)$

▶ 101. $x^9 - y^{12}z^{15}$ $(x^3 - y^4z^5)(x^6 + x^3y^4z^5 + y^8z^{10})$

▶ 102. $r^{12} + s^{18}t^{24}$ $(r^4 + s^6t^8)(r^8 - r^4s^6t^8 + s^{12}t^{16})$

Look Alikes . . .

103. a. $q^2 - 64$ $(q + 8)(q - 8)$ b. $q^3 - 64$ $(q - 4)(q^2 + 4q + 16)$

▶ 104. a. $a^2 - b^2$ $(a + b)(a - b)$ b. $a^3 - b^3$ $(a - b)(a^2 + ab + b^2)$

105. a. $d^2 - 25$ $(d + 5)(d - 5)$ b. $d^3 - 125$ $(d - 5)(d^2 + 5d + 25)$

106. a. $m^2 + 27$ Prime b. $m^3 + 27$ $(m + 3)(m^2 - 3m + 9)$

Factor the expression in part a. Then use your answer from part a to give the factorization of the expression in part b. (No new work is necessary!)

107. a. $a^6 - b^3$ $(a^2 - b)(a^4 + a^2b + b^2)$ b. $a^6 + b^3$ $(a^2 + b)(a^4 - a^2b + b^2)$

▶ 108. a. $c^3 - \frac{1}{8}$ $\left(c - \frac{1}{2}\right)\left(c^2 + \frac{1}{2}c + \frac{1}{4}\right)$ b. $c^3 + \frac{1}{8}$ $\left(c + \frac{1}{2}\right)\left(c^2 - \frac{1}{2}c + \frac{1}{4}\right)$

109. a. $125m^3 + 8n^3$ $(5m + 2n)(25m^2 - 10mn + 4n^2)$ b. $125m^3 - 8n^3$ $(5m - 2n)(25m^2 + 10mn + 4n^2)$

110. a. $w^3 + 0.001$ $(w + 0.1)(w^2 - 0.1w + 0.01)$ b. $w^3 - 0.001$ $(w - 0.1)(w^2 + 0.1w + 0.01)$

APPLICATIONS

▶ 111. **Candy.** To find the amount of chocolate used in the outer coating of the malted-milk ball shown, we can find the volume V of the chocolate shell using the formula

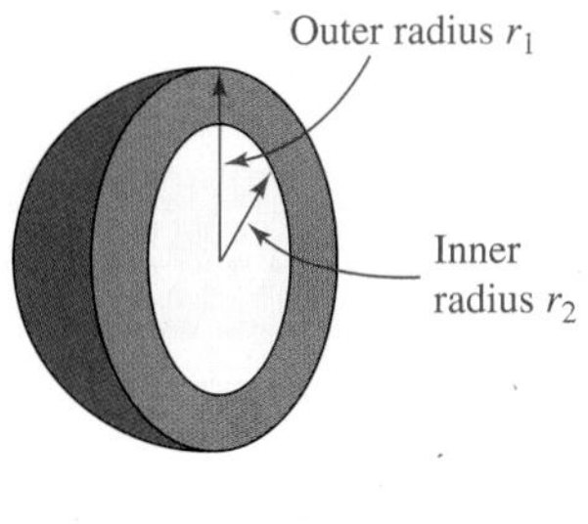

$$V = \frac{4}{3}\pi r_1^3 - \frac{4}{3}\pi r_2^3$$

Factor the expression on the right side of the formula.
$\frac{4}{3}\pi(r_1 - r_2)(r_1^2 + r_1r_2 + r_2^2)$

▶ 112. **Movie Stunts.** The function that gives the distance a stuntwoman is above the ground t seconds after she falls over the side of a 144-foot tall building is $h(t) = 144 - 16t^2$. Factor the right side. $16(3 + t)(3 - t)$

WRITING

▶ 113. Explain how the patterns used to factor the sum and difference of two cubes are similar and how they differ.

114. Explain why the factorization is not complete.

Factor: $1 - t^8 = (1 + t^4)(1 - t^4)$

115. Explain the error.

Factor: $4g^2 - 16 = (2g + 4)(2g - 4)$ (crossed out)

116. When asked to factor $81t^2 - 16$, one student answered $(9t - 4)(9t + 4)$, and another answered $(9t + 4)(9t - 4)$. Explain why both students are correct.

REVIEW

For each of the following purchases, determine the better buy.

117. Flute lessons: 45 minutes for \$25 or 1 hour for \$35.
45 minutes for \$25

118. Tissue paper: 15 sheets for \$1.39 or a dozen sheets for \$1.10.
A dozen sheets for \$1.10

CHALLENGE PROBLEMS

Factor. Assume all variables represent natural numbers.

119. $4x^{2n} - 9y^{2n}$ $(2x^n + 3y^n)(2x^n - 3y^n)$

120. $25 - x^{6n}$ $(5 + x^{3n})(5 - x^{3n})$

121. $a^{3b} - c^{3b}$ $(a^b - c^b)(a^{2b} + a^bc^b + c^{2b})$

122. $27x^{3n} + y^{3n}$ $(3x^n + y^n)(9x^{2n} - 3x^ny^n + y^{2n})$

123. Factor: $x^{32} - y^{32}$
$(x^{16} + y^{16})(x^8 + y^8)(x^4 + y^4)(x^2 + y^2)(x + y)(x - y)$

124. Find the error in this proof that $2 = 1$.

$$x = y$$
$$x^2 = xy$$
$$x^2 - y^2 = xy - y^2$$
$$(x + y)(x - y) = y(x - y)$$
$$\frac{(x + y)\cancel{(x - y)}}{\cancel{(x - y)}} = \frac{y\cancel{(x - y)}}{\cancel{(x - y)}}$$
Since $x = y$, there is a division by $x - y = 0$ in line 5.
$$x + y = y$$
$$y + y = y$$
$$2y = y$$
$$\frac{2y}{y} = \frac{y}{y}$$
$$2 = 1$$

SECTION 8.8 Review of Rational Expressions and Rational Equations; Rational Functions

OBJECTIVES

1. Evaluate rational functions.
2. Find the domain of a rational function.
3. Recognize the graphs of rational functions.
4. Simplify rational expressions.
5. Multiply and divide rational expressions.
6. Add and subtract rational expressions.
7. Solve rational equations.

ARE YOU READY?

Are You Ready? exercises available online at www.webassign.net/brookscole

The following problems review some basic skills that are needed when working with rational expressions.

1. Evaluate: **a.** $\frac{0}{10}$ 0 **b.** $\frac{10}{0}$ Undefined

2. Simplify: $\frac{36}{28}$ $\frac{9}{7}$

3. Simplify: $\frac{x(x + 5)(x - 7)}{4x(x - 5)(x - 7)}$ $\frac{x + 5}{4(x - 5)}$

4. What is the reciprocal of $\frac{2}{3x}$? $\frac{3x}{2}$

5. **a.** Multiply: $\frac{3}{8} \cdot \frac{1}{7}$ $\frac{3}{56}$

b. Divide: $\frac{21}{25} \div \frac{7}{15}$ $\frac{9}{5}$

6. **a.** Add: $\frac{5}{11} + \frac{3}{11}$ $\frac{8}{11}$

b. Subtract: $\frac{2}{3} - \frac{1}{5}$ $\frac{7}{15}$

Recall that rational expressions are algebraic fractions with polynomial numerators and denominators.

Rational Expressions

A **rational expression** is an expression of the form $\frac{A}{B}$, where A and B are polynomials and B does not equal 0.

Caution

Since division by 0 is undefined, the value of a polynomial in the denominator of a rational expression cannot be 0. For example, x cannot be 7 in the rational expression $\frac{3x}{x - 7}$, because the value of the denominator would be 0.

Some examples of rational expressions are

$$\frac{3x}{x - 7}, \quad \frac{8yz^4}{6y^2z^2}, \quad \frac{5m + n}{m^2 + 4mn + 4n^2}, \quad \text{and} \quad \frac{6a^2 - 13a + 6}{a^3 + 3a^2 + a - 2}$$

The rational expression $\frac{3x}{x - 7}$ is the quotient of the monomial $3x$ and the binomial $x - 7$. It is in one variable, x. The rational expression $\frac{5m + n}{m^2 + 4mn + 4n^2}$ is the quotient of the binomial $5m + n$ and the trinomial $m^2 + 4mn + 4n^2$. It is in two variables, m and n.

Rational expressions in one variable are used to define *rational functions.*

1 Evaluate Rational Functions.

We have previously studied linear and polynomial functions. In this section, we introduce another family of functions known as *rational functions*.

Rational Functions

A **rational function** is a function whose equation is defined by a rational expression in one variable, where the value of the polynomial in the denominator is never zero.

The Language of Algebra

Rational functions get their name from the fact that their defining equation contains a *ratio* (fraction) of two polynomials.

Teaching Tip: Have your students explain why the functions $f(x) = 3x - 9$ and $f(x) = x^3 + 2x^2 + 6x - 3$ are not rational functions.

Two examples of rational functions are

$$f(x) = \frac{1}{x + 8}$$ The rational expression $\frac{1}{x+8}$ that defines function f is in one variable, x.

$$s(t) = \frac{5t^2 + t}{t^2 + 6t - 1}$$ The rational expression $\frac{5t^2 + t}{t^2 + 6t - 1}$ that defines function s is in one variable, t.

Rational functions can be used to model many types of real-world situations.

EXAMPLE 1 **Internet Research.** The cost of subscribing to an online research network is \$6 a month plus \$1.50 per hour of access time. The rational function $c(n) = \frac{1.50n + 6}{n}$ gives the average hourly cost $c(n)$ of using the network for n hours a month. Find the average hourly cost for

a. a student who used the network for 2 hours in a month.

b. an instructor who used the network for 18 hours in a month.

©iStockphoto.com/quavondo

Strategy We will find $c(2)$ and $c(18)$.

Why The notation $c(2)$ represents the average hourly cost for using the network for 2 hours in a month and $c(18)$ represents the average hourly cost for using the network for 18 hours in a month.

Solution

a. To find the average hourly cost the student paid for 2 hours of access time in a month, we find $c(2)$.

$$c(2) = \frac{1.50(2) + 6}{2} = 4.5$$ Substitute 2 for n and evaluate the right side.

The student paid \$4.50 per hour to use the online research network for 2 hours.

b. To find the average hourly cost for 18 hours of access time in a month, we find $c(18)$.

$$c(18) = \frac{1.50(18) + 6}{18} = 1.833333333$$ Substitute 18 for n and evaluate the right side.

The instructor paid approximately \$1.83 per hour to use the online research network for 18 hours.

Teaching Example 1 Internet Research. Find the average hourly cost when the online research network in Example 1 is used for 20 hours in a month.
Answer: \$1.80 per hr

Self Check 1 **Internet Research.** Find the average hourly cost when the online research network in Example 1 is used for 100 hours in a month.
\$1.56 per hr

Now Try Problem 127

2 Find the Domain of a Rational Function.

Recall that the **domain** of a function is the set of all permissible input values for the variable. Since division by 0 is undefined, any input values that make the denominator 0 in a rational function must be excluded from the domain of the function.

EXAMPLE 2 Find the domain of the function: $f(x) = \dfrac{3x + 2}{x^2 + x - 6}$

Strategy We will set $x^2 + x - 6$ equal to 0 and solve for x.

Why We don't need to examine the numerator of the rational expression; it can be any value, including 0. The domain of the function includes all real numbers, except those that make the *denominator equal to* 0.

Teaching Tip: Remind students of the other division form that is often misunderstood: If the numerator of a fraction is 0 and the denominator is not, the fraction equals 0. For example, $\frac{0}{35} = 0$.

Solution

$$x^2 + x - 6 = 0 \quad \text{Set the denominator equal to 0.}$$
$$(x + 3)(x - 2) = 0 \quad \text{Factor the trinomial.}$$
$$x + 3 = 0 \quad \text{or} \quad x - 2 = 0 \quad \text{Set each factor equal to 0.}$$
$$x = -3 \quad | \quad x = 2 \quad \text{Solve each linear equation.}$$

Thus, the domain of the function is the set of all real numbers except -3 and 2. Using set-builder notation we can describe the domain as $\{x \mid x \text{ is a real number and } x \neq -3, x \neq 2\}$. In interval notation, the domain is $(-\infty, -3) \cup (-3, 2) \cup (2, \infty)$.

To check the answers, substitute -3 and 2 for x in $x^2 + x - 6$ and verify that the result is 0 for each.

The Language of Algebra

Two other ways that Example 2 could be phrased are:

- **State the restrictions** on the variable.
- Find the values of x for which the function is **undefined.**

Teaching Example 2 Find the domain of: $f(x) = \dfrac{4x + 9}{x^2 - x - 12}$
Answer: The set of all real numbers except -3 and 4: $(-\infty, -3) \cup (-3, 4) \cup (4, \infty)$; $\{x \mid x$ is a real number and $x \neq -3, x \neq 4\}$

Self Check 2 Find the domain of the function: $f(x) = \dfrac{x^2 + 1}{x^2 - 49}$ The domain is the set of all real numbers except -7 and 7: $(-\infty, -7) \cup (-7, 7) \cup (7, \infty)$.

Now Try Problem 31

3 Recognize the Graphs of Rational Functions.

The simplest of all rational functions, $f(x) = \frac{1}{x}$, is called the **reciprocal function.** We can use the point-plotting method to graph it. Because of the complex shape of the graph, we must select a large number of x-values, including positive and negative fractions, to determine how it looks. After plotting all of the ordered pairs, we draw a smooth curve through the points that form a "branch" on the left and a smooth curve through the points that form a "branch" on the right. Since 0 is not in the domain of the function, there is no point on the graph for $x = 0$.

$f(x) = \dfrac{1}{x}$

x	$f(x)$	
-4	$-\frac{1}{4}$	→ $\left(-4, -\frac{1}{4}\right)$
-3	$-\frac{1}{3}$	→ $\left(-3, -\frac{1}{3}\right)$
-2	$-\frac{1}{2}$	→ $\left(-2, -\frac{1}{2}\right)$
-1	-1	→ $(-1, -1)$
$-\frac{1}{2}$	-2	→ $\left(-\frac{1}{2}, -2\right)$
$-\frac{1}{3}$	-3	→ $\left(-\frac{1}{3}, -3\right)$
$-\frac{1}{4}$	-4	→ $\left(-\frac{1}{4}, -4\right)$
↑ Select x.	↑ Find the reciprocal.	↑ Plot the point.

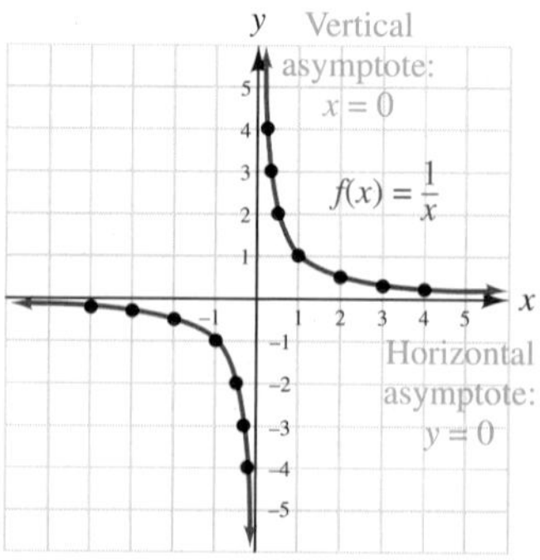

$f(x) = \dfrac{1}{x}$

x	$f(x)$	
4	$\frac{1}{4}$	→ $\left(4, \frac{1}{4}\right)$
3	$\frac{1}{3}$	→ $\left(3, \frac{1}{3}\right)$
2	$\frac{1}{2}$	→ $\left(2, \frac{1}{2}\right)$
1	1	→ $(1, 1)$
$\frac{1}{2}$	2	→ $\left(\frac{1}{2}, 2\right)$
$\frac{1}{3}$	3	→ $\left(\frac{1}{3}, 3\right)$
$\frac{1}{4}$	4	→ $\left(\frac{1}{4}, 4\right)$
↑ Select x.	↑ Find the reciprocal.	↑ Plot the point.

When a graph approaches a line, we call the line an **asymptote.** In this case, as x gets smaller and approaches 0, the graph of $f(x) = \frac{1}{x}$ approaches the y-axis. We say that the y-axis is a **vertical asymptote** of the graph.

On the far left and far right, the graph gets steadily closer to the x-axis. We say that the x-axis is a **horizontal asymptote** of the graph.

4 Simplify Rational Expressions.

To **simplify a rational expression** means to write it so that the numerator and denominator have no common factors other than 1.

Simplifying Rational Expressions

1. Factor the numerator and denominator completely to determine their common factors.
2. Remove factors equal to 1 by replacing each pair of factors common to the numerator and denominator with the equivalent fraction $\frac{1}{1}$.
3. Multiply the remaining factors in the numerator and in the denominator.

EXAMPLE 3 Simplify: $\dfrac{8yz^4}{6y^2z^2}$

Strategy We will begin by writing the numerator and denominator in factored form. Then we will remove any factors common to the numerator and denominator.

Why The rational expression is simplified when the numerator and denominator have no common factors other than 1.

Solution

$$\frac{8yz^4}{6y^2z^2} = \frac{2 \cdot 2 \cdot 2 \cdot y \cdot z \cdot z \cdot z \cdot z}{2 \cdot 3 \cdot y \cdot y \cdot z \cdot z}$$

To prepare to simplify the rational expression, factor $8yz^4$ and $6y^2z^2$ completely.

$$= \frac{\overset{1}{\cancel{2}} \cdot 2 \cdot 2 \cdot \overset{1}{\cancel{y}} \cdot \overset{1}{\cancel{z}} \cdot \overset{1}{\cancel{z}} \cdot z \cdot z}{\underset{1}{\cancel{2}} \cdot 3 \cdot \underset{1}{\cancel{y}} \cdot y \cdot \underset{1}{\cancel{z}} \cdot \underset{1}{\cancel{z}}}$$

Replace $\frac{2}{2}$, $\frac{y}{y}$, and $\frac{z}{z}$ with $\frac{1}{1}$. This removes the factor $\frac{2 \cdot y \cdot z \cdot z}{2 \cdot y \cdot z \cdot z} = 1$.

$$= \frac{4z^2}{3y}$$

Multiply the remaining factors in the numerator. Multiply the remaining factors in the denominator.

The Language of Algebra

When a rational expression is simplified, the result is an **equivalent expression.**

Caution

Unless otherwise stated, we will now assume that the variables in rational expressions represent real numbers for which the denominator is not zero. Therefore, as we simplify rational expressions, warnings like the "provided $y \neq 0$ and $z \neq 0$" statement in this example are not necessary.

Since $\frac{8yz^4}{6y^2z^2}$ is undefined for $y = 0$ and $z = 0$, the expressions $\frac{8yz^4}{6y^2z^2}$ and $\frac{4z^2}{3y}$ are equal only if $y \neq 0$ and $z \neq 0$. That is,

$$\frac{8yz^4}{6y^2z^2} = \frac{4z^2}{3y} \quad \text{provided } y \neq 0 \text{ and } z \neq 0.$$

An alternate approach is to use the rules for exponents to simplify the rational expressions that are the quotient of two monomials.

$$\frac{8yz^4}{6y^2z^2} = \frac{\overset{1}{\cancel{2}} \cdot 2 \cdot 2 \cdot y^{1-2}z^{4-2}}{\underset{1}{\cancel{2}} \cdot 3} = \frac{4y^{-1}z^2}{3} = \frac{4z^2}{3y}$$

To divide exponential expressions with the same base, keep the base and subtract the exponents.

Teaching Example 3 Simplify: $\dfrac{10kr^3}{25k^2r^2}$

Answer: $\dfrac{2r}{5k}$

Self Check 3 Simplify: $\dfrac{12a^4b^2}{20ab^4}$ $\quad \dfrac{3a^3}{5b^2}$

Now Try Problems 33 and 35

To simplify rational expressions, we often make use of the factoring methods discussed in Sections 8.6 and 8.7. We also often use the following property about polynomials and their opposites.

The Quotient of Opposites The quotient of any nonzero polynomials and its opposite is -1.

EXAMPLE 4 Simplify: **a.** $\dfrac{2x^2 + 11x + 12}{3x^2 + 11x - 4}$ **b.** $\dfrac{3x^2 - 10xy - 8y^2}{4y^2 - xy}$

Strategy We will begin by factoring the numerator and denominator completely. Then we will remove any factors common to the numerator and denominator.

Why We need to make sure that the numerator and denominator have no common factors other than 1. If that is the case, then the rational expression is simplified.

Solution

a. $\dfrac{2x^2 + 11x + 12}{3x^2 + 11x - 4} = \dfrac{(2x + 3)\overset{1}{\cancel{(x + 4)}}}{(3x - 1)\underset{1}{\cancel{(x + 4)}}}$ Factor the numerator and denominator. Remove a factor equal to 1: $\frac{x + 4}{x + 4} = 1$.

$= \dfrac{2x + 3}{3x - 1}$ This expression does not simplify further.

b. $\dfrac{3x^2 - 10xy - 8y^2}{4y^2 - xy} = \dfrac{(3x + 2y)\overset{-1}{\cancel{(x - 4y)}}}{y\underset{1}{\cancel{(4y - x)}}}$ Factor the numerator and denominator. Since $x - 4y$ and $4y - x$ are opposites, simplify by replacing $\frac{x - 4y}{4y - x}$ with the equivalent fraction $\frac{-1}{1} = -1$.

$= \dfrac{-(3x + 2y)}{y}$

This result can also be written as $-\dfrac{3x + 2y}{y}$ or $\dfrac{-3x - 2y}{y}$.

Caution

In Example 4, do not remove the x's in the result $\frac{2x + 3}{3x - 1}$. The x in the numerator is not a factor of the entire numerator. Likewise, the x in the denominator is not a factor of the entire denominator.

$$\frac{2\overset{1}{\cancel{x}} + 3}{3\underset{1}{\cancel{x}} - 1} = \frac{2 + 3}{3 - 1} = \frac{5}{2}$$

Caution

A $-$ symbol in front of a fraction may be placed in the numerator or the denominator, but not in both. For example,

$$-\frac{3x + 2y}{y} \neq \frac{-3x - 2y}{-y}$$

Self Check 4 Simplify: **a.** $\dfrac{2x^2 + 5x + 2}{3x^2 + 5x - 2}$ $\dfrac{2x + 1}{3x - 1}$

b. $\dfrac{2a^2 - 3ab - 9b^2}{3b^2 - ab}$ $-\dfrac{2a + 3b}{b}$ or $\dfrac{-2a - 3b}{b}$

Now Try Problems 39 and 43

Teaching Example 4 Simplify:
a. $\dfrac{3x^2 + 2x - 8}{3x^2 - x - 4}$
b. $\dfrac{x^2 - 13xy + 12y^2}{144y^2 - x^2}$
Answers: a. $\dfrac{x + 2}{x + 1}$
b. $-\dfrac{x - y}{x + 12y}$ or $\dfrac{y - x}{x + 12y}$

5 Multiply and Divide Rational Expressions.

Recall that to multiply two fractions, we multiply the numerators and multiply the denominators. We use the same strategy to multiply rational expressions.

Multiplying Rational Expressions To multiply rational expressions, multiply their numerators and their denominators. Then, if possible, factor and simplify.

For any two rational expressions, $\frac{A}{B}$ and $\frac{C}{D}$,

$$\frac{A}{B} \cdot \frac{C}{D} = \frac{AC}{BD}$$

EXAMPLE 5 Multiply: **a.** $\dfrac{x^2 - 6x + 9}{20x} \cdot \dfrac{5x^2}{x^2 - 9}$ **b.** $(2x - x^2) \cdot \dfrac{x}{x^2 - xb - 2x + 2b}$

Strategy To find the product, we will use the rule for multiplying rational expressions. In the process, we must be prepared to factor the numerators and denominators so that any common factors can be removed.

Why We want to give the result in simplified form.

Solution

a.

$$\frac{x^2 - 6x + 9}{20x} \cdot \frac{5x^2}{x^2 - 9} = \frac{(x^2 - 6x + 9)5x^2}{20x(x^2 - 9)}$$ Multiply the numerators. Multiply the denominators.

$$= \frac{(x-3)(x-3)5 \cdot x \cdot x}{4 \cdot 5 \cdot x(x+3)(x-3)}$$ To prepare to simplify, factor the numerator and factor the denominator.

$$= \frac{\overset{1}{\cancel{(x-3)}}(x-3)\overset{1}{\cancel{5}} \cdot \overset{1}{\cancel{x}} \cdot x}{4 \cdot \underset{1}{\cancel{5}} \cdot \underset{1}{\cancel{x}}(x+3)\underset{1}{\cancel{(x-3)}}}$$ Simplify by removing common factors of the numerator and denominator.

$$= \frac{x(x-3)}{4(x+3)}$$

Caution

When multiplying rational expressions, always write the result in simplest form by removing any factors common to the numerator and denominator.

We could distribute in the numerator and/or denominator and write the result as $\frac{x^2 - 3x}{4(x + 3)}$ or $\frac{x^2 - 3x}{4x + 12}$. Check with your instructor to see which form of the result he or she prefers.

b. Writing $2x - x^2$ as $\frac{2x - x^2}{1}$ is helpful during the multiplication process when we multiply numerators and multiply denominators.

$$(2x - x^2) \cdot \frac{x}{x^2 - xb - 2x + 2b}$$

$$= \frac{2x - x^2}{1} \cdot \frac{x}{x^2 - xb - 2x + 2b}$$ Write $2x - x^2$ as $\frac{2x - x^2}{1}$.

$$= \frac{(2x - x^2)x}{1(x^2 - xb - 2x + 2b)}$$ Multiply the numerators. Multiply the denominators.

$$= \frac{x(2-x)x}{1[x(x-b) - 2(x-b)]}$$ To prepare to simplify, factor out x in the numerator. In the denominator, begin factoring by grouping.

$$= \frac{x(2-x)x}{1(x-b)(x-2)}$$ In the denominator, complete the factoring by grouping. The brackets [ ] are no longer needed.

$$= \frac{x\overset{-1}{\cancel{(2-x)}}x}{1(x-b)\underset{1}{\cancel{(x-2)}}}$$ Simplify: The quotient of any nonzero quantity and its opposite is -1: $\frac{2-x}{x-2} = -1$.

$$= -\frac{x^2}{x-b}$$

Success Tip

We would obtain the same answer if we had factored the numerators and denominators first and simplified before we multiplied.

Self Check 5 Multiply: **a.** $\dfrac{a^2 + 6a + 9}{18a} \cdot \dfrac{3a^3}{a + 3}$ $\dfrac{a^2(a+3)}{6}$

b. $\dfrac{x^2 + 5x + 6}{4x + 8 - x^2 - 2x}(x^2 - 4x)$ $-x(x + 3)$

Now Try Problems 51 and 53

Teaching Example 5 Multiply:

a. $\dfrac{x^2 - 2x - 15}{x^2 - 25} \cdot \dfrac{x^2 + 9x + 20}{5x + 15}$

b. $(3x - x^2) \cdot \dfrac{x}{4x^3 - 12x^2 + 20x - 60}$

Answers: a. $\dfrac{x + 4}{5}$ b. $-\dfrac{x^2}{4(x^2 + 5)}$

Recall that to divide fractions, we multiply the first fraction by the reciprocal of the second fraction. We use the same strategy to divide rational expressions.

Dividing Rational Expressions

To divide two rational expressions, multiply the first by the reciprocal of the second. Then, if possible, factor and simplify.

For any two rational expressions, $\frac{A}{B}$ and $\frac{C}{D}$, where $\frac{C}{D} \neq 0$,

$$\frac{A}{B} \div \frac{C}{D} = \frac{A}{B} \cdot \frac{D}{C} = \frac{AD}{BC}$$

EXAMPLE 6 Divide: $\frac{x^3 + 8}{4x + 4} \div \frac{x^2 - 2x + 4}{2x^2 - 2}$

Strategy To find the quotient, we will use the rule for dividing rational expressions. After multiplying by the reciprocal, we will factor each polynomial that is not prime and remove any common factors of the numerator and denominator.

Why We want to give the result in simplified form.

Solution

$$\frac{x^3 + 8}{4x + 4} \div \frac{x^2 - 2x + 4}{2x^2 - 2}$$

$$= \frac{x^3 + 8}{4x + 4} \cdot \frac{2x^2 - 2}{x^2 - 2x + 4}$$ Multiply the first rational expression by the reciprocal of the second.

$$= \frac{(x^3 + 8)(2x^2 - 2)}{(4x + 4)(x^2 - 2x + 4)}$$ Multiply the numerators. Multiply the denominators.

$$= \frac{(x + 2)\overset{1}{\cancel{(x^2 - 2x + 4)}}\overset{1}{\cancel{2}}\overset{1}{\cancel{(x + 1)}}(x - 1)}{2 \cdot \underset{1}{\cancel{2}}\underset{1}{\cancel{(x + 1)}}\underset{1}{\cancel{(x^2 - 2x + 4)}}}$$ To prepare to simplify, factor completely. Note that $x^2 - 2x + 4$ is prime. Then simplify.

$$= \frac{(x + 2)(x - 1)}{2}$$ Multiply the remaining monomial factors in the numerator: $1 \cdot 1 \cdot 1 = 1$. Multiply the remaining factors in the denominator: $2 \cdot 1 \cdot 1 \cdot 1 = 2$.

Caution

When dividing rational expressions, always write the result in simplest form by removing any factors common to the numerator and denominator.

Self Check 6 Divide: $\frac{x^3 - 8}{9x - 9} \div \frac{x^2 + 2x + 4}{3x^2 - 3x}$ $\frac{x(x-2)}{3}$

Now Try Problem 61

Teaching Example 6
Divide: $\frac{a^3 - 125}{9a - 9} \div \frac{a^2 + 5a + 25}{3a^2 - 3}$
Answer: $\frac{(a - 5)(a + 1)}{3}$

6 Add and Subtract Rational Expressions.

To add (or subtract) fractions with like denominators, we add (or subtract) the numerators and keep the same denominator. We use the same strategy to add (or subtract) rational expressions with like denominators.

Adding and Subtracting Rational Expressions That Have the Same Denominator

To add (or subtract) rational expressions that have same denominator, add (or subtract) their numerators and write the sum (or difference) over the common denominator. Then, if possible, factor and simplify.

If $\frac{A}{D}$ and $\frac{B}{D}$ are rational expressions,

$$\frac{A}{D} + \frac{B}{D} = \frac{A + B}{D} \quad \text{and} \quad \frac{A}{D} - \frac{B}{D} = \frac{A - B}{D}$$

EXAMPLE 7 Add $\frac{a^2}{a^2 - 36} + \frac{6a}{a^2 - 36}$ and simplify the result.

Strategy We will add the numerators and write the sum over the common denominator.

Why This is the rule for adding rational expressions that have the *same* denominator.

Solution

$$\frac{a^2}{a^2 - 36} + \frac{6a}{a^2 - 36} = \frac{a^2 + 6a}{a^2 - 36}$$ Add the numerators. Write the sum over the common denominator, $a^2 - 36$.

$$= \frac{a(a + 6)}{(a + 6)(a - 6)}$$ To prepare to simplify, factor the numerator and denominator.

$$= \frac{a\overset{1}{\cancel{(a + 6)}}}{\underset{1}{\cancel{(a + 6)}}(a - 6)}$$ Simplify by removing the common factor of $a + 6$.

$$= \frac{a}{a - 6}$$

Caution

When adding or subtracting rational expressions, always write the result in simplest form by removing any factors common to the numerator and denominator.

Teaching Example 7

Add: $\frac{m^2}{m^2 - 64} + \frac{8m}{m^2 - 64}$

Answer: $\frac{m}{m - 8}$

Self Check 7 Add: $\frac{2b}{b^2 - 4} + \frac{b^2}{b^2 - 4}$ $\frac{b}{b - 2}$

Now Try Problem 65

To add or subtract rational expressions with unlike denominators, we build them to rational expressions with the same denominator.

Building Rational Expressions

To build a rational expression, multiply it by 1 in the form of $\frac{c}{c}$, where c is any nonzero number or expression.

When adding (or subtracting) rational expressions with unlike denominators, we will write the rational expressions with the smallest common denominator possible, called the **least** (or lowest) **common denominator (LCD).** To find the least common denominator of several rational expressions, we follow these steps.

Finding the LCD

1. Factor each denominator completely.
2. The LCD is a product that uses each different factor obtained in step 1 the greatest number of times it appears in any one factorization.

EXAMPLE 8 Add: $\frac{5a}{24b} + \frac{11a}{18b^2}$

Strategy We will find the LCD of these rational expressions. Then we will build the rational expressions so each one has the LCD as its denominator.

Why Since the denominators are different, we cannot add these rational expressions in their present form.

Solution To find the LCD, we write each denominator as the product of prime numbers and variables.

$$24b = 2 \cdot 2 \cdot 2 \cdot 3 \cdot b = 2^3 \cdot 3 \cdot b$$
$$18b^2 = 2 \cdot 3 \cdot 3 \cdot b \cdot b = 2 \cdot 3^2 \cdot b^2$$

Success Tip

Note that the highest power of each factor is used to form the LCD:

$$24b = 2^{③} \cdot 3 \cdot b$$
$$18b^2 = 2 \cdot 3^{②} \cdot b^{②}$$
$$\text{LCD} = 2^3 \cdot 3^2 \cdot b^2 = 72b^2$$

Then we form a product using each of these factors the greatest number of times it appears in any one factorization.

The greatest number of times the factor 2 appears is three times.
The greatest number of times the factor 3 appears is twice.
The greatest number of times the factor b appears is twice.

$$\text{LCD} = 2 \cdot 2 \cdot 2 \cdot 3 \cdot 3 \cdot b \cdot b$$
$$= 72b^2$$

We now multiply the numerator and denominator of each rational expression by whatever it takes to build their denominators to $72b^2$.

Success Tip

You may want to review the procedure for adding and subtracting rational expressions that have unlike denominators on page 541.

$$\frac{5a}{24b} + \frac{11a}{18b^2} = \frac{5a}{24b} \cdot \frac{3b}{3b} + \frac{11a}{18b^2} \cdot \frac{4}{4}$$ Build each rational expression by multiplying it by a form of 1.

$$= \frac{15ab}{72b^2} + \frac{44a}{72b^2}$$ Multiply the numerators. Multiply the denominators. Note that the expressions now have like denominators.

$$= \frac{15ab + 44a}{72b^2}$$ Add the numerators. Write the sum over the common denominator. The result does not simplify.

Teaching Example 8
Add: $\frac{13x}{15y^2} + \frac{11x}{25y}$
Answer: $\frac{65x + 33xy}{75y^2}$

Self Check 8 Add: $\frac{3}{28z^3} + \frac{5x}{21z}$ $\quad \frac{9 + 20xz^2}{84z^3}$

Now Try Problem 73

EXAMPLE 9

Subtract: $\frac{x + 1}{x^2 - 2x + 1} - \frac{x - 4}{x^2 - 1}$

Strategy We will factor each denominator, find the LCD, and build the rational expressions so each one has the LCD as its denominator.

Why Since the denominators are different, we cannot subtract these rational expressions in their present form.

Solution We factor each denominator to find the LCD:

$$x^2 - 2x + 1 = (x - 1)(x - 1) = (x - 1)^2$$
$$x^2 - 1 = (x + 1)(x - 1)$$

The greatest number of times $x - 1$ appears is twice. The greatest number of times $x + 1$ appears is once.

The LCD is $(x - 1)^2(x + 1)$ or $(x - 1)(x - 1)(x + 1)$.

We now write each rational expression with its denominator in factored form. Then we multiply each numerator and denominator by the missing factor, so that each rational expression has a denominator of $(x - 1)(x - 1)(x + 1)$.

Teaching Tip: Since a fraction bar is a grouping symbol, you may want to point out in step 3 of the solution that $\frac{x^2 - 5x + 4}{(x - 1)(x - 1)(x + 1)}$ can be written as $\frac{(x^2 - 5x + 4)}{(x - 1)(x - 1)(x + 1)}$.

$$\frac{x + 1}{x^2 - 2x + 1} - \frac{x - 4}{x^2 - 1}$$

$$= \frac{x + 1}{(x - 1)(x - 1)} - \frac{x - 4}{(x + 1)(x - 1)}$$ Write each denominator in factored form.

$$= \frac{x + 1}{(x - 1)(x - 1)} \cdot \frac{x + 1}{x + 1} - \frac{x - 4}{(x + 1)(x - 1)} \cdot \frac{x - 1}{x - 1}$$ Build each rational expression.

$$= \frac{x^2 + 2x + 1}{(x - 1)(x - 1)(x + 1)} - \frac{x^2 - 5x + 4}{(x - 1)(x - 1)(x + 1)}$$ Multiply the numerators. The denominators are now the same.

Multiply the numerators using the FOIL method to prepare to combine like terms.

Don't multiply the denominators. Leave them in factored form to possibly simplify the result later.

$$= \frac{x^2 + 2x + 1 - (x^2 - 5x + 4)}{(x - 1)(x - 1)(x + 1)}$$ Subtract the numerators. Write the difference over the common denominator. Don't forget the parentheses shown in blue.

$$= \frac{x^2 + 2x + 1 - x^2 + 5x - 4}{(x - 1)(x - 1)(x + 1)}$$ To subtract in the numerator, change the signs of x^2, $-5x$, and 4, and drop the parentheses.

$$= \frac{7x - 3}{(x - 1)(x - 1)(x + 1)}$$ Combine like terms in the numerator. The result does not simplify.

Teaching Example 9
Subtract: $\frac{x + 1}{x^2 + 5x + 6} - \frac{x - 3}{x^2 - 4}$
Answer: $\frac{-x + 7}{(x + 2)(x + 3)(x - 2)}$

Self Check 9 Subtract: $\frac{a + 2}{a^2 - 4a + 4} - \frac{a - 3}{a^2 - 4}$ $\frac{9a - 2}{(a - 2)(a - 2)(a + 2)}$

Now Try Problem 81

7 Solve Rational Equations.

If an equation contains one or more rational expressions, it is called a **rational equation.** Recall that we use a fraction-clearing strategy to solve rational equations. All possible solutions of a rational equation must be checked. Multiplying both sides of an equation by an expression that contains a variable can lead to **extraneous solutions,** which must be discarded.

EXAMPLE 10 Solve: $1 + \frac{8a}{a^2 + 3a} = \frac{3}{a}$

Strategy We will find the LCD of the rational expressions in the equation and multiply both sides by the LCD.

Why This will clear the equation of fractions.

Solution Since the binomial $a^2 + 3a$ factors as $a(a + 3)$, we can write the given equation as:

$$1 + \frac{8a}{a(a + 3)} = \frac{3}{a}$$ Factor $a^2 + 3a$.

Success Tip

You may want to review the procedure for solving rational equations on page 558.

We see that 0 and -3 cannot be solutions of the equation, because they make rational expressions in the equation undefined.

We can clear the equation of fractions by multiplying both sides by $a(a + 3)$, which is the LCD of the two rational expressions $\frac{8a}{a^2 + 3a}$ and $\frac{3}{a}$.

$$a(a + 3)\left[1 + \frac{8a}{a(a + 3)}\right] = a(a + 3)\left(\frac{3}{a}\right)$$ Multiply both sides by the LCD.

$$a(a + 3)1 + a(a + 3)\frac{8a}{a(a + 3)} = a(a + 3)\left(\frac{3}{a}\right)$$ On the left side, distribute the multiplication by $a(a + 3)$.

$$a(a + 3)1 + \frac{\overset{1}{\cancel{a}}\overset{1}{\cancel{(a + 3)}}\,8a}{\underset{1}{\cancel{a}}\underset{1}{\cancel{(a + 3)}}} = \overset{1}{\cancel{a}}(a + 3)\left(\frac{3}{\underset{1}{\cancel{a}}}\right)$$ Remove common factors of the numerator and denominator.

$$a^2 + 3a + 8a = 3a + 9$$ Simplify each side. The resulting quadratic equation does not contain any fractions.

$$a^2 + 8a - 9 = 0$$ To get 0 on the right side, subtract $3a$ and 9 from both sides.

$$(a + 9)(a - 1) = 0$$ Factor the left side.

$$a + 9 = 0 \quad \text{or} \quad a - 1 = 0$$ Set each factor equal to 0.

$$a = -9 \quad | \quad a = 1$$ Solve each equation.

The solutions are -9 and 1. Verify that both satisfy the original equation.

Caution

After multiplying both sides by the LCD and simplifying, the equation should not contain any fractions. If it does, check for an algebraic error, or perhaps your LCD is incorrect.

Teaching Tip: Point out the difference in the methods used to add (or subtract) rational expressions versus solving rational equations.

Teaching Example 10
Solve: $\frac{b}{4} = \frac{b-22}{5b-35} - \frac{1}{5}$
Answer: 3, 4

Self Check 10 Solve: $1 + \frac{2}{2b+1} = \frac{5}{2b^2+b}$ $-\frac{5}{2}, 1$

Now Try Problem 95

SECTION 8.8 STUDY SET

VOCABULARY

Fill in the blanks.

1. A quotient of two polynomials, such as $\frac{x^2+x}{x^2-3x}$, is called a _rational_ expression.
2. A _rational_ function, such as $f(x) = \frac{x-7}{x^2-x-6}$, is a function whose equation is defined by a rational expression in one variable.
3. The _domain_ of a function is the set of all permissible input values for the variable.
4. The rational function $f(x) = \frac{9x}{x-10}$ is _undefined_ for $x = 10$. In other words, there is a _restriction_ on the domain of the function: $x \neq 10$.
5. To _simplify_ a rational expression, we remove factors common to the numerator and denominator.
6. The quotient of _opposites_ is -1. For example, $\frac{x-8}{8-x} = -1$.
7. In the rational expression $\frac{(x+2)(3x-1)}{(x+2)(4x+2)}$, the binomial $x + 2$ is a common _factor_ of the numerator and the denominator.
8. The least _common_ _denominator_ of $\frac{x-8}{x+6}$ and $\frac{6-5x}{x}$ is $x(x+6)$.
9. To _build_ a rational expression, we multiply it by a form of 1. For example, $\frac{2}{n^2} \cdot \frac{8}{8} = \frac{16}{8n^2}$.
10. Equations that contain one or more rational expressions, such as $\frac{x}{x+2} = 4 + \frac{10}{x+1}$, are called _rational_ equations.

CONCEPTS

11. Let $f(x) = \frac{2x+1}{x^2+3x-4}$. Find
    a. $f(0)$ $-\frac{1}{4}$ b. $f(2)$ $\frac{5}{6}$ c. $f(1)$ Undefined
12. For what value(s) of x is each function undefined?
    a. $f(x) = \frac{x-7}{x}$ 0 b. $f(x) = \frac{x+1}{x-3}$ 3
    c. $f(x) = \frac{x^2-2}{x(x+8)}$ $-8, 0$ d. $f(x) = \frac{8x}{(x-1)(x+1)}$ $-1, 1$
13. The graph of rational function f is shown. Find each of the following.

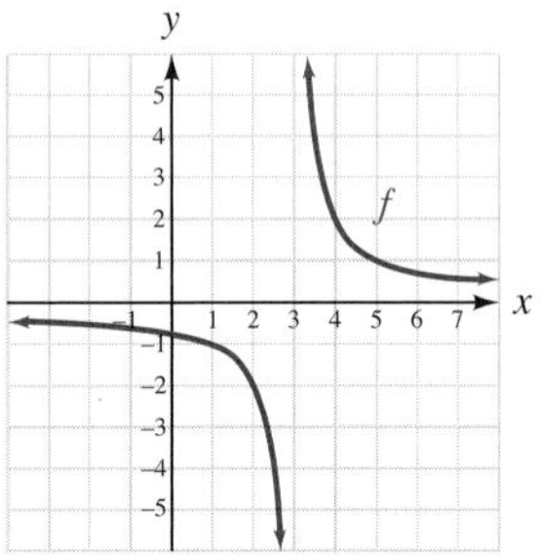

    a. $f(1)$ -1
    b. $f(4)$ 2
    c. The value of x for which $f(x) = -2$ 2
    d. The value of x for which $f(x) = 1$ 5

14. Match each function with the correct graph shown below.
    a. $f(x) = 2$ iii. b. $f(x) = x$ v. c. $f(x) = x^2$ i.
    d. $f(x) = x^3$ vi. e. $f(x) = |x|$ iv. f. $f(x) = \frac{1}{x}$ ii.

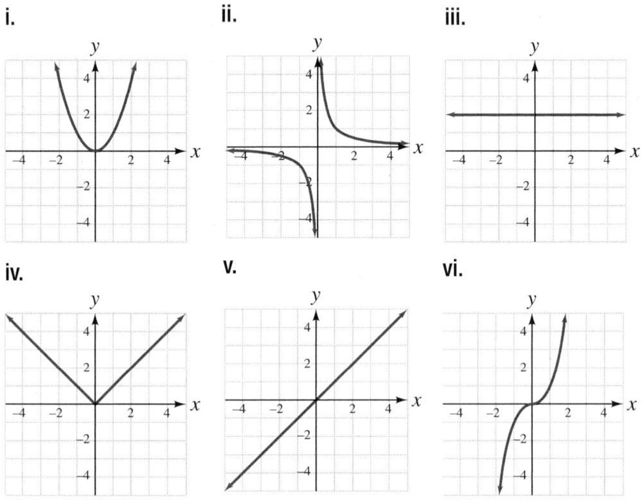

Fill in the blanks.

15. To multiply rational expressions, multiply their _numerators_ and multiply their _denominators_. To divide two rational expressions, multiply the first by the _reciprocal_ of the second.
16. To add or subtract rational expressions that have the same denominator, add or subtract the _numerators_ and write the sum or difference over the common _denominator_.
17. To find the least common denominator of several rational expressions, _factor_ each denominator completely. The LCD is a product that uses each different factor the _greatest_ number of times it appears in any one factorization.
18. The expression $4 - y$ must be multiplied by -1 to obtain $y - 4$.
19. Consider the following factorizations.

$$18x - 36 = 2 \cdot 3 \cdot 3 \cdot (x - 2)$$
$$3x^2 - 3x - 6 = 3(x - 2)(x + 1)$$

    a. What is the greatest number of times the factor 3 appears in any one factorization? Twice
    b. What is the greatest number of times the factor $x - 2$ appears in any one factorization? Once

20. The LCD for $\frac{2x+1}{x^2+5x+6}$ and $\frac{3x}{x^2-4}$ is

$$\text{LCD} = (x+2)(x+3)(x-2)$$

If we want to subtract these rational expressions, what form of 1 should be used

a. to build $\frac{2x+1}{x^2+5x+6}$? $\frac{x-2}{x-2}$

b. to build $\frac{3x}{x^2-4}$? $\frac{x+3}{x+3}$

21. To clear the following equations of fractions, by what should both sides be multiplied?

a. $\frac{1}{a} = \frac{1}{3} - \frac{2}{3a}$ $3a$

b. $\frac{2}{x-2} + \frac{10}{x+5} = \frac{2x}{x^2+3x-10}$ $(x-2)(x+5)$

22. Perform each multiplication.

a. $4x\left(\frac{3}{4x}\right)$ 3

b. $(x+6)(x-2)\left(\frac{3}{x-2}\right)$ $3x+18$

c. $8(x+4)\left[\frac{7x}{2(x+4)}\right]$ $28x$

d. $6(m-5)\left(\frac{7}{5-m}\right)$ -42

NOTATION

23. a. Write $5x^2 + 35x$ as a fraction. $\frac{5x^2+35x}{1}$

b. What is the reciprocal of $5x^2 + 35x$? $\frac{1}{5x^2+35x}$

24. What numbers are not included in each set of real numbers represented using interval notation?

a. $(-\infty, 4) \cup (4, \infty)$ 4

b. $(-\infty, -8) \cup (-8, 0) \cup (0, \infty)$ $-8, 0$

GUIDED PRACTICE

Find the domain of each rational function. Express your answer in words and using interval notation. See Example 2.

25. $f(x) = \frac{2}{x}$ All real numbers except 0; $(-\infty, 0) \cup (0, \infty)$

26. $f(x) = \frac{8}{x-1}$ All real numbers except 1; $(-\infty, 1) \cup (1, \infty)$

▶ 27. $f(x) = \frac{2x}{x+2}$
All real numbers except -2; $(-\infty, -2) \cup (-2, \infty)$

▶ 28. $f(x) = \frac{2x+1}{x^2-2x}$
All real numbers except 0 and 2; $(-\infty, 0) \cup (0, 2) \cup (2, \infty)$

29. $f(x) = \frac{3x-1}{x-x^2}$
All real numbers except 0 and 1; $(-\infty, 0) \cup (0, 1) \cup (1, \infty)$

▶ 30. $f(x) = \frac{x^2+36}{x^2-36}$ All real numbers except -6 and 6; $(-\infty, -6) \cup (-6, 6) \cup (6, \infty)$

▶ 31. $f(x) = \frac{x^2+3x+2}{x^2-x-56}$ All real numbers except -7 and 8; $(-\infty, -7) \cup (-7, 8) \cup (8, \infty)$

▶ 32. $f(x) = \frac{2x^2-3x-2}{x^2+2x-24}$ All real numbers except -6 and 4; $(-\infty, -6) \cup (-6, 4) \cup (4, \infty)$

Simplify each rational expression. See Example 3.

33. $\frac{15a^2}{25a^8}$ $\frac{3}{5a^6}$

34. $\frac{12x}{16x^7}$ $\frac{3}{4x^6}$

35. $\frac{24x^3y^4}{54x^4y^3}$ $\frac{4y}{9x}$

▶ 36. $\frac{15a^5b^4}{21a^2b^5}$ $\frac{5a^3}{7b}$

Simplify each rational expression. See Example 4.

37. $\frac{5x^2-10x}{x^2-4x+4}$ $\frac{5x}{x-2}$

38. $\frac{x^2+6x+9}{2x^2+6x}$ $\frac{x+3}{2x}$

▶ 39. $\frac{6x^2-7x-5}{2x^2+5x+2}$ $\frac{3x-5}{x+2}$

▶ 40. $\frac{6x^2+x-2}{8x^2+2x-3}$ $\frac{3x+2}{4x+3}$

41. $\frac{4-x^2}{x^2-x-2}$ $-\frac{x+2}{x+1}$

▶ 42. $\frac{x^2-2x-15}{25-x^2}$ $-\frac{x+3}{x+5}$

43. $\frac{p^3+p^2q-2pq^2}{pq^2+p^2q-2p^3}$ $-\frac{p+2q}{q+2p}$

▶ 44. $\frac{m^3-mn^2}{mn^2+m^2n-2m^3}$ $-\frac{m+n}{n+2m}$

Perform the operations and simplify, if possible. See Example 5.

45. $\frac{10a^2}{3b^4} \cdot \frac{12b^3}{5a^2}$ $\frac{8}{b}$

▶ 46. $\frac{16c^3}{5d^2} \cdot \frac{25d}{12c}$ $\frac{20c^2}{3d}$

47. $\frac{3p^2}{6p+24} \cdot \frac{p^2-16}{6p}$ $\frac{p(p-4)}{12}$

48. $\frac{y^2+6y+9}{15y} \cdot \frac{3y^2}{2y+6}$ $\frac{y(y+3)}{10}$

49. $\frac{x^2+2x+1}{9x} \cdot \frac{2x^2-2x}{2x^2-2}$ $\frac{x+1}{9}$

▶ 50. $\frac{a+6}{a^2-16} \cdot \frac{3a-12}{3a+18}$ $\frac{1}{a+4}$

51. $\frac{2x^2-x-3}{x^2-1} \cdot \frac{x^2+x-2}{2x^2+x-6}$ 1

▶ 52. $\frac{2p^2-5p-3}{p^2-9} \cdot \frac{2p^2+5p-3}{2p^2+5p+2}$ $\frac{2p-1}{p+2}$

53. $(6a-a^2) \cdot \frac{a^3}{a^3-6a^2+3a-18}$ $-\frac{a^4}{a^2+3}$

54. $(10n-n^2) \cdot \frac{n^6}{n^3-10n^2-2n+20}$ $-\frac{n^7}{(n^2-2)}$

▶ 55. $(x^2+x-2cx-2c) \cdot \frac{x^2+3x+2}{4c^2-x^2}$ $-\frac{(x+1)^2(x+2)}{2c+x}$

56. $(2ax-10x-a+5) \cdot \frac{x}{x-2x^2}$ $-a+5$

Perform the operations and simplify, if possible. See Example 6.

57. $\frac{m^2n}{4} \div \frac{mn^3}{6}$ $\frac{3m}{2n^2}$

58. $\frac{a^4b}{14} \div \frac{a^3b^2}{21}$ $\frac{3a}{2b}$

59. $\frac{x^2-16}{x^2-25} \div \frac{x+4}{x-5}$ $\frac{x-4}{x+5}$

60. $\frac{a^2-9}{a^2-49} \div \frac{a+3}{a+7}$ $\frac{a-3}{a-7}$

61. $\frac{5c+1}{6} \div \frac{125c^3+1}{6c+6}$ $\frac{c+1}{25c^2-5c+1}$

62. $\frac{6m-8}{9m^3} \div \frac{27m^3-64}{9m+9}$ $\frac{2(m+1)}{m^3(9m^2+12m+16)}$

▶ 63. $\frac{3n^2+5n-2}{12n^2-13n+3} \div \frac{n^2+3n+2}{4n^2+5n-6}$ $\frac{n+2}{n+1}$

64. $\dfrac{8y^2 - 14y - 15}{6y^2 - 11y - 10} \div \dfrac{4y^2 - 9y - 9}{3y^2 - 7y - 6}$ 1

Perform the operations and simplify, if possible. **See Example 7.**

65. $\dfrac{3x}{x^2 - 9} - \dfrac{9}{x^2 - 9}$ $\frac{3}{x+3}$

66. $\dfrac{9x}{x^2 - 1} - \dfrac{9}{x^2 - 1}$ $\frac{9}{x+1}$

67. $\dfrac{3y - 2}{2y + 6} - \dfrac{2y - 5}{2y + 6}$ $\frac{1}{2}$

68. $\dfrac{5x + 8}{3x + 15} - \dfrac{3x - 2}{3x + 15}$ $\frac{2}{3}$

Perform the operations and simplify, if possible. **See Example 8.**

69. $\dfrac{3}{4x} + \dfrac{2}{3x}$ $\frac{17}{12x}$

70. $\dfrac{2}{5a} + \dfrac{3}{2a}$ $\frac{19}{10a}$

71. $\dfrac{8}{9y^2} + \dfrac{1}{6y^4}$ $\frac{16y^2 + 3}{18y^4}$

72. $\dfrac{5}{6a^3} + \dfrac{7}{8a^2}$ $\frac{21a + 20}{24a^3}$

73. $\dfrac{3}{4ab^2} - \dfrac{5}{2a^2b}$ $\frac{3a - 10b}{4a^2b^2}$

74. $\dfrac{1}{5xy^3} - \dfrac{2}{15x^2y}$ $\frac{3x - 2y^2}{15x^2y^3}$

75. $\dfrac{y - 7}{y^2} - \dfrac{y + 7}{2y}$ $-\frac{y^2 + 5y + 14}{2y^2}$

76. $\dfrac{x + 5}{xy} - \dfrac{x - 1}{x^2y}$ $\frac{x^2 + 4x + 1}{x^2y}$

Perform the operations and simplify, if possible. **See Example 9.**

77. $\dfrac{3}{x + 2} + \dfrac{5}{x - 4}$ $\frac{8x - 2}{(x + 2)(x - 4)}$

78. $\dfrac{2}{a + 4} - \dfrac{6}{a + 3}$ $\frac{-2(2a + 9)}{(a + 4)(a + 3)}$

79. $\dfrac{x + 2}{x + 5} - \dfrac{x - 3}{x + 7}$ $\frac{7x + 29}{(x + 5)(x + 7)}$

80. $\dfrac{7}{x + 3} + \dfrac{4x}{x + 6}$ $\frac{4x^2 + 19x + 42}{(x + 3)(x + 6)}$

81. $\dfrac{x}{x^2 + 5x + 6} + \dfrac{x}{x^2 - 4}$ $\frac{2x^2 + x}{(x + 3)(x + 2)(x - 2)}$

82. $\dfrac{x}{x^2 + 2x + 1} + \dfrac{x}{x^2 - 1}$ $\frac{2x^2}{(x + 1)^2(x - 1)}$

83. $\dfrac{4}{x^2 - 2x - 3} - \dfrac{x}{3x^2 - 7x - 6}$ $\frac{-x^2 + 11x + 8}{(3x + 2)(x + 1)(x - 3)}$

84. $\dfrac{5}{x^2 + 5x - 6} - \dfrac{x}{2x^2 + 7x - 30}$ $\frac{-x^2 + 11x - 25}{(x + 6)(x - 1)(2x - 5)}$

85. $\dfrac{6}{5d^2 - 5d} - \dfrac{3}{5d - 5}$ $\frac{6 - 3d}{5d(d - 1)}$

86. $\dfrac{9}{2r^2 - 2r} - \dfrac{5}{2r - 2}$ $\frac{9 - 5r}{2r(r - 1)}$

87. $\dfrac{m}{m^2 + 9m + 20} - \dfrac{4}{m^2 + 7m + 12}$ $\frac{m - 5}{(m + 3)(m + 5)}$

88. $\dfrac{x + 3}{2x^2 - 5x + 2} - \dfrac{3x - 1}{x^2 - x - 2}$ $-\frac{5x + 1}{(2x - 1)(x + 1)}$

Solve each equation. **See Example 10.**

89. $\dfrac{3}{y} + \dfrac{7}{2y} = 13$ $\frac{1}{2}$

90. $\dfrac{2}{x} + \dfrac{1}{2} = \dfrac{7}{2x}$ 3

91. $\dfrac{2}{x} + \dfrac{1}{2} = \dfrac{9}{4x} - \dfrac{1}{2x}$ $-\frac{1}{2}$

92. $\dfrac{7}{5x} - \dfrac{1}{2} = \dfrac{5}{6x} + \dfrac{1}{3}$ $\frac{17}{25}$

93. $\dfrac{a}{2} = \dfrac{a - 6}{3a - 9} - \dfrac{1}{3}$ $1, 2$

94. $\dfrac{b}{5} = \dfrac{b - 14}{2b - 16} - \dfrac{1}{2}$ $3, 5$

95. $\dfrac{2}{5x - 5} + \dfrac{x - 2}{15} = \dfrac{4}{5x - 5}$ $4, -1$

96. $\dfrac{3}{2x + 4} = \dfrac{x - 2}{2} + \dfrac{x - 5}{2x + 4}$ $-4, 3$

TRY IT YOURSELF

Simplify each function. List any restrictions on the domain.

97. $f(x) = \dfrac{x^2 + 6x - 16}{x^2 - 4}$ $f(x) = \frac{x + 8}{x + 2}$, provided $x \neq -2, x \neq 2$

98. $f(x) = \dfrac{5x^2 + 50x}{x^5 + 10x^4}$ $f(x) = \frac{5}{x^3}$, provided $x \neq -10, x \neq 0$

99. $g(x) = \dfrac{x^3 + 64}{x^3 + 4x^2 + 3x + 12}$ $g(x) = \frac{x^2 - 4x + 16}{x^2 + 3}$, provided $x \neq -4$

100. $h(t) = \dfrac{t^3 - 5t^2 - 5t + 25}{t^3 - 125}$ $h(t) = \frac{t^2 - 5}{t^2 + 5t + 25}$, provided $t \neq 5$

In the following problems, simplify each expression by performing the indicated operations and solve each equation.

101. $\dfrac{p^3 - q^3}{q^2 - p^2} \cdot \dfrac{q^2 + pq}{p^3 + p^2q + pq^2}$ $-\frac{q}{p}$

102. $\dfrac{x^3 + y^3}{x^3 - y^3} \div \dfrac{x^2 - xy + y^2}{x^2 + xy + y^2}$ $\frac{x + y}{x - y}$

103. $\dfrac{t}{t^2 + 5t + 6} - \dfrac{2}{t^2 + 3t + 2}$ $\frac{t - 3}{(t + 3)(t + 1)}$

104. $\dfrac{2a}{a^2 - 2a - 8} + \dfrac{3}{a^2 - 5a + 4}$ $\frac{2a^2 + a + 6}{(a - 4)(a + 2)(a - 1)}$

105. $\dfrac{4}{m^2 - 9} + \dfrac{5}{m^2 - m - 12} = \dfrac{7}{m^2 - 7m + 12}$ 26

106. $\dfrac{34}{x^2} + \dfrac{13}{20x} = \dfrac{3}{2x}$ 40

107. $\dfrac{2}{x - 1} - \dfrac{2x}{x^2 - 1} - \dfrac{x}{x^2 + 2x + 1}$ $\frac{-x^2 + 3x + 2}{(x - 1)(x + 1)^2}$

108. $\dfrac{x - 2}{x^2 - 3x} + \dfrac{2x - 1}{x^2 + 3x} - \dfrac{2}{x^2 - 9}$ $\frac{3x + 1}{x(x + 3)}$

109. $(2x^2 - 15x + 25) \div \dfrac{2x^2 - 3x - 5}{x + 1}$ $x - 5$

110. $(x^2 - 6x + 9) \div \dfrac{x^2 - 9}{x + 3}$ $x - 3$

111. $\dfrac{y^3 - x^3}{2x^2 + 2xy + x + y} \cdot \dfrac{2x^2 - 5x - 3}{yx - 3y - x^2 + 3x}$ $\frac{y^2 + xy + x^2}{x + y}$

112. $\dfrac{ax + ay + bx + by}{x^3 - 27} \cdot \dfrac{x^2 + 3x + 9}{xc + xd + yc + yd}$ $\frac{a + b}{(x - 3)(c + d)}$

113. $\dfrac{24n^4}{16n^4 + 24n^3}$ $\frac{3n}{2n + 3}$

114. $\dfrac{18m^4}{36m^4 - 9m^3}$ $\frac{2m}{4m - 1}$

115. $1 + x - \dfrac{x}{x - 5}$ $\frac{x^2 - 5x - 5}{x - 5}$

116. $2 - x + \dfrac{3}{x - 9}$ $\frac{-x^2 + 11x - 15}{x - 9}$

117. $\dfrac{ax + by + ay + bx}{a^2 - b^2}$ $\frac{x + y}{a - b}$

118. $\dfrac{3x^2 - 3y^2}{x^2 + 2y + 2x + yx}$ $\frac{3(x - y)}{x + 2}$

119. $\dfrac{3}{s - 2} + \dfrac{s - 14}{2s^2 - 3s - 2} - \dfrac{4}{2s + 1} = 0$ 1

120. $\dfrac{1}{y^2 - 2y - 3} + \dfrac{1}{y^2 - 4y + 3} - \dfrac{1}{y^2 - 1} = 0$ -3

121. $\dfrac{5}{x + 4} + \dfrac{1}{x + 4} = x - 1$ $2, -5$

122. $\dfrac{5}{3x + 12} - \dfrac{1}{9} = \dfrac{x - 1}{3x}$ $-\frac{3}{2}, 2$

123. $\dfrac{5x}{x-3}+\dfrac{4x}{3-x}$ $\quad \dfrac{x}{x-3}$

124. $\dfrac{8x}{x-4}-\dfrac{10x}{4-x}$ $\quad \dfrac{18x}{x-4}$

APPLICATIONS

125. Environmental Cleanup. Suppose the cost (in dollars) of removing $p\%$ of the pollution in a river is given by the rational function

$$f(p)=\frac{50{,}000p}{100-p} \quad \text{where } 0 \le p < 100$$

Find the cost of removing each percent of pollution.

a. 50% $50,000 **b.** 80% $200,000

126. Directory Costs. The average (mean) cost for a service club to publish a directory of its members is given by the rational function

$$f(x)=\frac{1.25x+700}{x}$$

where x is the number of directories printed. Find the average cost per directory if

a. 500 directories are printed. $2.65

b. 2,000 directories are printed. $1.60

127. Utility Costs. An electric company charges $7.50 per month plus 9¢ for each kilowatt hour (kwh) of electricity used.

a. Find a linear function that gives the total cost of n kwh of electricity. (*Hint:* See Example 1.) $c(n)=0.09n+7.50$

b. Find a rational function that gives the average cost per kwh when using n kwh. $c(n)=\dfrac{0.09n+7.50}{n}$

c. Find the average cost per kwh when 775 kwh are used. About 10¢

128. Drafting. Among the tools used in drafting are $45°-45°-90°$ and $30°-60°-90°$ triangles. Find the perimeter of each triangle and express each result as a rational expression. $\dfrac{10r+20}{r}, \dfrac{3t+9}{t}$

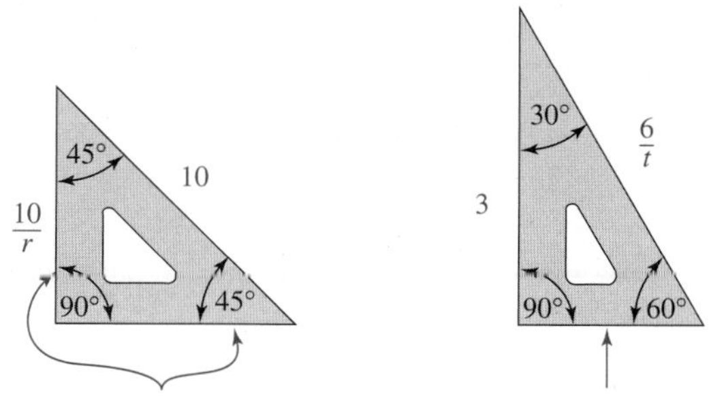

For a 45°–45°–90° triangle, these two sides are the same length.

For a 30°–60°–90° triangle, this side is half as long as the hypotenuse.

WRITING

129. A student compared his answer, $\frac{a-3b}{2b-a}$, with the answer, $\frac{3b-a}{a-2b}$, in the back of the text. Is the student's work correct? Explain.

130. Explain the error that is made in the following work:

$$\frac{3x^2+1}{3y}=\frac{\cancel{3}x^2+1}{\cancel{3}y}=\frac{x^2+1}{y}$$

131. Write some comments to the student who wrote the following solution, explaining his misunderstanding.

$$\text{Multiply: } \frac{1}{x}\cdot\frac{3}{2}=\frac{1}{x}\cdot\frac{2}{2}\cdot\frac{3}{2}\cdot\frac{x}{x}$$
$$=\frac{2}{2x}\cdot\frac{3x}{2x}$$
$$=\frac{6x}{4x^2}$$

132. Would you use the same approach to answer the following problems? Explain why or why not.

$$\text{Simplify: } \frac{x^2-10}{x^2-1}-\frac{3x}{x-1}-\frac{2x}{x+1}$$

$$\text{Solve: } \frac{x^2-10}{x^2-1}-\frac{3x}{x-1}=-\frac{2x}{x+1}$$

REVIEW

Solve each equation.

133. $-10|16h+4|-3=-3$ $\quad -\frac{1}{4}$

134. $-5|2a-9|+14=14$ $\quad \frac{9}{2}$

CHALLENGE PROBLEMS

135. Simplify: $\dfrac{a^6-64}{(a^2+2a+4)(a^2-2a+4)}$ $\quad (a-2)(a+2)$

136. Add: $x^{-1}+x^{-2}+x^{-3}+x^{-4}+x^{-5}$ $\quad \dfrac{x^4+x^3+x^2+x+1}{x^5}$

137. Simplify: $[(x^{-1}+1)^{-1}+1]^{-1}$ $\quad \dfrac{x+1}{2x+1}$

138. Find two rational expressions, each with denominator x^2+5x+6, such that their sum is $\dfrac{1}{x+2}$.

$\dfrac{x}{x^2+5x+6}, \dfrac{3}{x^2+5x+6}$

Graph each rational function. Show the vertical asymptote as a dashed line and label it. See AIE Appendix 3.

139. $f(x)=\dfrac{1}{x-1}$

140. $f(x)=\dfrac{1}{x+4}$

Perform the operations and simplify.

141. $\dfrac{6a^2-7a-3}{2a^2-2}\div\dfrac{4a^2-12a+9}{a^2-1}\cdot\dfrac{2a^2-a-3}{3a^2-2a-1}$ $\quad \dfrac{a+1}{2(a-1)}$

142. $\dfrac{2x^2-2x-4}{x^2+2x-8}\cdot\dfrac{3x^2+15x}{x+1}\div\dfrac{100-4x^2}{x^2-x-20}$ $\quad -\dfrac{3x}{2}$

143. $\dfrac{\dfrac{h}{h^2+3h+2}}{\dfrac{4}{h+2}-\dfrac{4}{h+1}}$ $\quad -\dfrac{h}{4}$

144. $\dfrac{\dfrac{2}{y-1}-\dfrac{2}{y}}{\dfrac{3}{y-1}-\dfrac{1}{1-y}}$ $\quad \dfrac{1}{2y}$

SECTION 8.9

OBJECTIVES

1. Solve problems involving direct variation.
2. Solve problems involving inverse variation.
3. Solve problems involving joint variation.
4. Solve problems involving combined variation.

Variation

ARE YOU READY? *Are You Ready? exercises available online at www.webassign.net/brookscole*

The following problems review some basic skills that are needed when working with variation.

1. Find y if $y = 45x$ and $x = 10$. 450

2. Find d if $d = \frac{90}{n}$ and $n = 18$. 5

3. Let $f = 0.0036Av^2$. Find f if $A = 500$ and $v = 30$. 1,620

4. Solve each equation:

a. $26 = 4k$ $\frac{2}{13}$ **b.** $80 = \frac{k}{2.5}$ 200

In this section, we introduce four types of *variation models,* each of which expresses a special relationship between two or more quantities. We will use these models to solve problems involving travel, lighting, geometry, and highway construction.

1 Solve Problems Involving Direct Variation.

To introduce direct variation, we consider the formula for the circumference of a circle

$$C = \pi D$$

Teaching Tip: You may want to point out that if we divide both sides of $y = kx$ by x, we obtain $\frac{y}{x} = k$. Thus, for the direct variation model, k is simply the quotient of one pair of values of x and y.

where C is the circumference, D is the diameter, and $\pi \approx 3.14159$. If we double the diameter of a circle, we determine another circle with a larger circumference C_1 such that

$$C_1 = \pi(2D) = 2\pi D = 2C$$

Thus, doubling the diameter results in doubling the circumference. Likewise, if we triple the diameter, we will triple the circumference.

In the formula, $C = \pi D$, we say that the variables C and D *vary directly,* or that they are *directly proportional.* This is because C is always found by multiplying D by a constant. In this example, the constant π is called the *constant of variation* or the *constant of proportionality.*

Direct Variation

The words **y varies directly as x** or **y is directly proportional to x** means that $y = kx$ for some nonzero constant k. The constant k is called the **constant of variation** or the **constant of proportionality.**

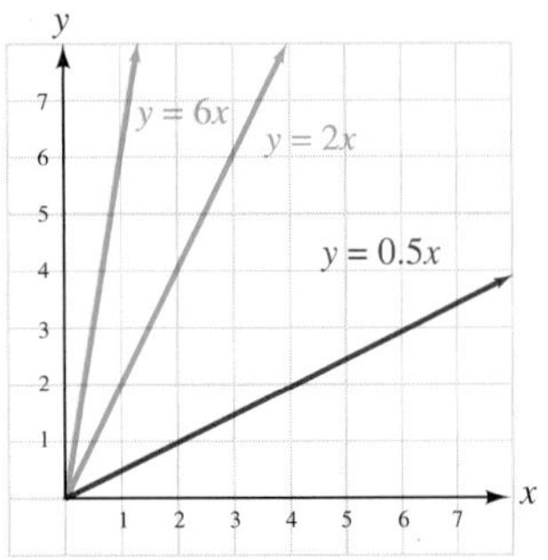

For any **direct variation** equation of the form $y = kx$, where $k > 0$: as x increases, y increases.

Since the formula for direct variation ($y = kx$) defines a linear function, its graph is always a line with a y-intercept at the origin. The graph of $y = kx$ where $x \geq 0$ appears on the left for three positive values of k.

One example of direct variation is Hooke's law from physics. Hooke's law states that the distance a spring will stretch varies directly as the force that is applied to it.

If d represents a distance and f represents a force, this verbal model of Hooke's law can be expressed as

$d = kf$ The direct variation model can be read as "d is directly proportional to f."

where k is the constant of variation. Suppose we know that a certain spring stretches 10 inches when a weight of 6 pounds is attached (see the figure). We can find k as follows:

$d = kf$

$10 = k(6)$ Substitute 10 for d and 6 for f.

$\frac{10}{6} = k$ To isolate k, divide both sides by 6.

$\frac{5}{3} = k$ Simplify the fraction. This is the constant of variation.

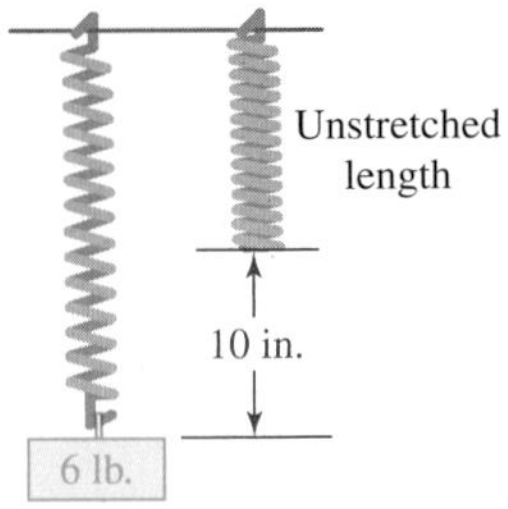

Success Tip

The value that we found for k is for this specific example. Another spring made out of a different type of steel will more than likely have a different value of k.

To find the force required to stretch the spring a distance of 35 inches, we can solve the equation $d = kf$ for f, with $d = 35$ and $k = \frac{5}{3}$.

$$d = kf \quad \text{This is the direct variation model.}$$

$$35 = \frac{5}{3}f \quad \text{Substitute 35 for } d \text{ and } \tfrac{5}{3} \text{ for } k.$$

$$105 = 5f \quad \text{Multiply both sides by 3.}$$

$$21 = f \quad \text{To isolate } f\text{, divide both sides by 5.}$$

Thus, the force required to stretch the spring a distance of 35 inches is 21 pounds.

We can use the following steps to solve variation problems.

Solving Variation Problems

To solve a variation problem:

1. Translate the verbal model into an equation.
2. Substitute the first set of values into the equation from step 1 to determine the value of k.
3. Substitute the value of k into the equation from step 1.
4. Substitute the remaining set of values into the equation from step 3 and solve for the unknown.

EXAMPLE 1

Currency Exchange. The currency calculator shown below converts from U.S. dollars to Russian rubles. When exchanging these currencies, the number of rubles received is directly proportional to the number of dollars to be exchanged. How many rubles will an exchange of $1,200 bring?

Strategy We will use a direct variation model to solve this problem.

Why The words *the number of rubles received is directly proportional to the number of dollars to be exchanged* indicate that this type of model should be used.

Solution

©Dimon/Shutterstock.com

Step 1: The verbal model can be represented by the equation

$$r = kd \quad \text{This is a direct variation model.}$$

where r is the number of rubles, k is the constant of variation, and d is the number of dollars.

Step 2: From the illustration, we see that an exchange of $500 brings 14,900 rubles. To find k, we substitute 500 for d and 14,900 for r, and then we solve for k.

$$r = kd$$

$$14{,}900 = k(500) \quad \text{Substitute for } r \text{ and } d.$$

$$29.8 = k \quad \text{To isolate } k\text{, divide both sides by 500. This is the constant of variation.}$$

Step 3: Now we substitute the value of k, 29.8, into the equation $r = kd$, to get

$$r = 29.8d$$

Step 4: To find how many rubles an exchange of $1,200 will bring, we substitute 1,200 for d in the direct variation model, and then we evaluate the right side.

$$r = 29.8d \quad \text{This is the ruble direct variation model.}$$

$$r = 29.8(1{,}200) \quad \text{Substitute for } d.$$

$$r = 35{,}760 \quad \text{Do the multiplication. The result is in rubles.}$$

An exchange of $1,200 will bring 35,760 rubles.

Teaching Example 1 Currency Exchange. When exchanging currencies, the number of euros received is directly proportional to the number of U.S. dollars to be exchanged. If \$500 converts to 370 euros, how many euros will be received if \$2,500 is exchanged? *Answer:* 1,850 euros

Self Check 1 **Currency Exchange.** When exchanging currencies, the number of British pounds received is directly proportional to the number of U.S. dollars to be exchanged. If \$800 converts to 392 pounds, how many pounds will be received if \$1,500 is exchanged? 735 British pounds

Now Try Problem 35

2 Solve Problems Involving Inverse Variation.

In the formula $w = \frac{12}{l}$, w gets smaller as l gets larger, and w gets larger as l gets smaller. Since these variables vary in opposite directions in a predictable way, we say that the variables *vary inversely*, or that they are *inversely proportional*. The constant 12 is the constant of variation.

Inverse Variation

The words **y varies inversely as x** or **y is inversely proportional to x** mean that $y = \frac{k}{x}$ for some nonzero constant k. The constant k is called the **constant of variation.**

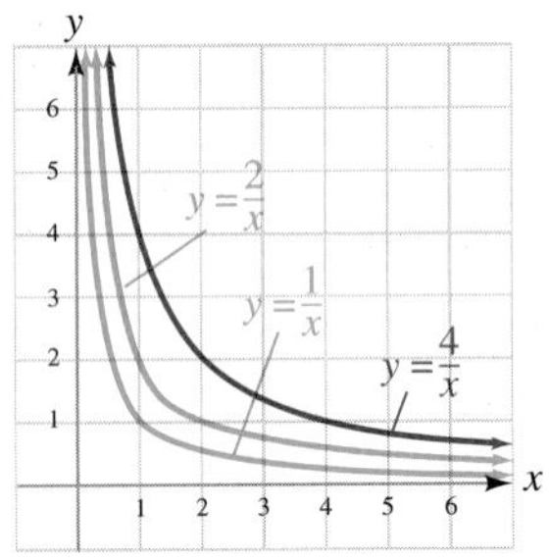

For any **inverse variation** equation of the form $y = \frac{k}{x}$, where $k > 0$: as x increases, y decreases.

The formula for inverse variation, $y = \frac{k}{x}$, defines a rational function whose graph will have the x- and y-axes as asymptotes. The graph of $y = \frac{k}{x}$ where $x > 0$ appears on the left for three positive values of k.

In an elevator, the amount of floor space per person varies inversely as the number of people in the elevator. If f represents the amount of floor space per person and n the number of people in the elevator, the relationship between f and n can be expressed by the equation:

$$f = \frac{k}{n}$$ This inverse variation model also can be read as "f is inversely proportional to n."

The figure on the right shows 6 people in an elevator; each has 8.25 square feet of floor space. To determine how much floor space each person would have if 15 people were in the elevator, we begin by determining k.

$$f = \frac{k}{n}$$ This is the inverse variation model.

$$8.25 = \frac{k}{6}$$ Substitute 8.25 for f and 6 for n.

$$k = 49.5$$ Multiply both sides by 6 to solve for k. This is the constant of variation.

To find the amount of floor space per person if 15 people are in the elevator, we proceed as follows:

Teaching Tip: You may want to point out that if we multiply both sides of $y = \frac{k}{x}$ by x, we get $xy = k$. Thus, for the inverse variation model, k is simply the product of one pair of values of x and y. (Assume $x \neq 0$.)

$$f = \frac{k}{n}$$ This is the inverse variation model.

$$f = \frac{49.5}{15}$$ Substitute 49.5 for k and 15 for n.

$$f = 3.3$$ Do the division.

If 15 people were in the elevator, each would have 3.3 square feet of floor space.

EXAMPLE 2 **Photography.** The intensity I of light received from a light source varies inversely as the square of the distance from the light source. If a photographer, 16 feet away from his subject, has a light meter reading of 4 foot-candles of luminance, what will the meter read if the photographer moves in for a close-up 4 feet away from the subject?

Strategy We will use the inverse variation model of the form $I = \frac{k}{d^2}$, where I represents the intensity and d^2 represents the square of the distance from the light source.

Why The words *intensity varies inversely as the square of the distance* indicate that this type of model should be used.

Solution

$I = \frac{k}{d^2}$ This inverse variation model also can be read as "I is inversely proportional to d^2."

$4 = \frac{k}{16^2}$ To find k, we substitute 4 for I and 16 for d and solve for k.

$4 = \frac{k}{256}$ Evaluate the denominator: $16^2 = 256$.

$1{,}024 = k$ To isolate k, multiply both sides by 256. This is the constant of variation.

To find the intensity when the photographer is 4 feet away from the subject, we substitute 4 for d and 1,024 for k and simplify.

$I = \frac{k}{d^2}$ This is the inverse variation model.

$I = \frac{1{,}024}{4^2}$ Substitute for k and d.

$I = 64$ Evaluate the right side.

The intensity at 4 feet is 64 foot-candles.

Success Tip

The constant of variation is usually positive, because most real-life applications involve only positive quantities. However, the definitions of *direct, inverse, joint,* and *combined variation* allow for a negative constant of variation.

Teaching Example 2 Photography. In Example 6, find the intensity when the photographer is 2 feet away from the subject.
Answer: 256 foot-candles

Self Check 2 **Photography.** Find the intensity when the photographer is 8 feet away from the subject. 16 foot-candles

Now Try Problem 39

3 Solve Problems Involving Joint Variation.

There are times when one variable varies as the product of several variables. For example, the area of a triangle varies directly with the product of its base and height:

$$A = \frac{1}{2}bh$$

Such variation is called *joint variation.*

Joint Variation

If one variable varies directly as the product of two or more variables, the relationship is called **joint variation.** If y varies jointly with x and z, then $y = kxz$. The nonzero constant k is called the **constant of variation.**

EXAMPLE 3 **Force of the Wind.** The force of the wind on a billboard varies jointly as the area of the billboard and the square of the wind velocity. When the wind is blowing at 20 mph, the force on a billboard 30 feet wide and 18 feet high is 972 pounds. Find the force on a billboard having an area of 300 square feet caused by a 40-mph wind.

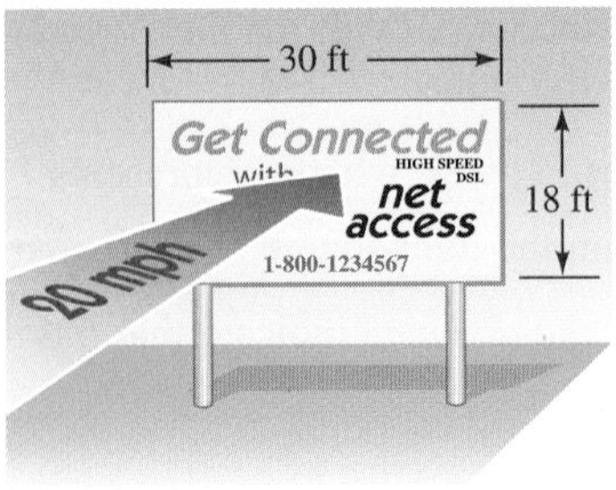

Strategy We will use the joint variation model $f = kAv^2$, where f represents the force of the wind, A represents the area of the billboard, and v^2 represents the square of the velocity of the wind.

Why The words *the force of the wind on a billboard varies jointly as the area of the billboard and the square of the wind velocity* indicate that this type of model should be used.

Solution

$f = kAv^2$ The joint variation model also can be read as "f is directly proportional to the product of A and v^2."

Since the billboard is 30 feet wide and 18 feet high, it has an area of $30 \cdot 18 = 540$ square feet. We can find k by substituting 972 for f, 540 for A, and 20 for v.

$$f = kAv^2$$
$$972 = k(540)(20)^2$$
$$972 = k(216{,}000) \quad \text{Evaluate: } (20)^2 = 400. \text{ Then do the multiplication.}$$
$$0.0045 = k \quad \text{Divide both sides by 216,000 to solve for } k. \text{ This is the constant of variation.}$$

To find the force exerted on a 300-square-foot billboard by a 40-mph wind, we use the formula $f = 0.0045Av^2$ and substitute 300 for A and 40 for v.

$$f = 0.0045Av^2 \quad \text{This is the joint variation model.}$$
$$f = 0.0045(300)(40)^2 \quad \text{Substitute for } A \text{ and } v.$$
$$f = 2{,}160 \quad \text{Evaluate the right side.}$$

The 40-mph wind exerts a force of 2,160 pounds on the billboard.

Teaching Example 3 Energy. Kinetic energy of an object varies jointly with its mass and the square of its velocity. A 25-gram mass moving at the rate of 30 centimeters per second has a kinetic energy of 11,250 dyne-centimeters. Find the kinetic energy of a 10-gram mass that is moving at 40 centimeters per second. *Answer:* 8,000 dyne-centimeters

Self Check 3 **Force of the Wind.** Refer to Example 3. Find the force of a 25-mph wind on a billboard having an area of 375 square feet. Approx. 1,055 lb

Now Try Problem 43

4 Solve Problems Involving Combined Variation.

Many applied problems involve a combination of direct and inverse variation. Such variation is called **combined variation.**

EXAMPLE 4 **Highway Construction.** The time it takes to build a highway varies directly as the length of the road, and inversely as the number of workers. If it takes 100 workers 4 weeks to build 2 miles of highway, how long will it take 80 workers to build 10 miles of highway?

Strategy We will use the combined variation model $t = \frac{kl}{w}$, where t represents the time in days, l represents the length of road built in miles, and w represents the number of workers.

Why The words *the time it takes to build a highway* ***varies directly*** *as the length of the road, and* ***inversely*** *as the number of workers* indicate that this type of model should be used.

Solution The relationship between these variables can be expressed by the equation

$$t = \frac{kl}{w}$$ This is a combined variation model.

$$4 = \frac{k(2)}{100}$$ Substitute 4 for t, 100 for w, and 2 for l to find k.

$$400 = 2k$$ Multiply both sides by 100.

$$200 = k$$ To isolate k, divide both sides by 2. This is the constant of variation.

We now substitute 80 for w, 10 for l, and 200 for k in the equation $t = \frac{kl}{w}$ and simplify:

$$t = \frac{kl}{w}$$ This is the combined variation model.

$$t = \frac{200(10)}{80}$$ Substitute for k, l, and w.

$$t = 25$$ Evaluate the right side.

It will take 25 weeks for 80 workers to build 10 miles of highway.

Teaching Example 4 Highway Construction. Refer to Example 4. How long will it take 100 workers to build 20 miles of highway?
Answer: 40 weeks

Self Check 4 **Highway Construction.** How long will it take 60 workers to build 6 miles of highway? 20 weeks

Now Try Problem 51

SECTION 8.9 STUDY SET

VOCABULARY

Fill in the blanks.

1. The equation $y = kx$ defines direct variation: As x increases, y increases.
2. The equation $y = \frac{k}{x}$ defines inverse variation: As x increases, y decreases.
3. The equation $y = kxz$ defines joint variation, and $y = \frac{kz}{x}$ defines combined variation.
4. The equation $y = \frac{kx}{z}$ means that y varies directly with x and inversely with z.

CONCEPTS

Determine whether direct or inverse variation applies and sketch a possible graph for the situation. (Hint: Refer to the graphs on pages 700 and 702.)

5. a.

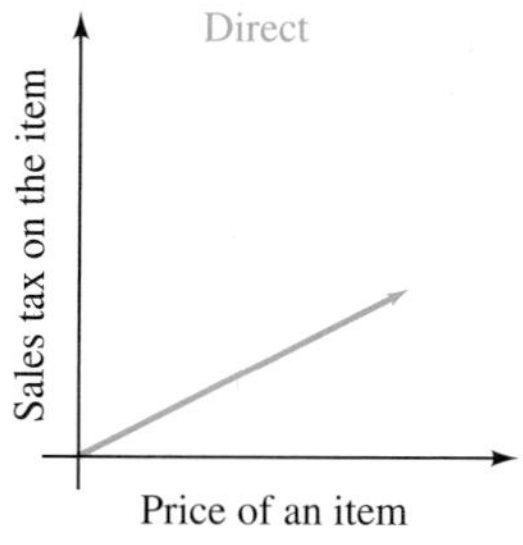

b.

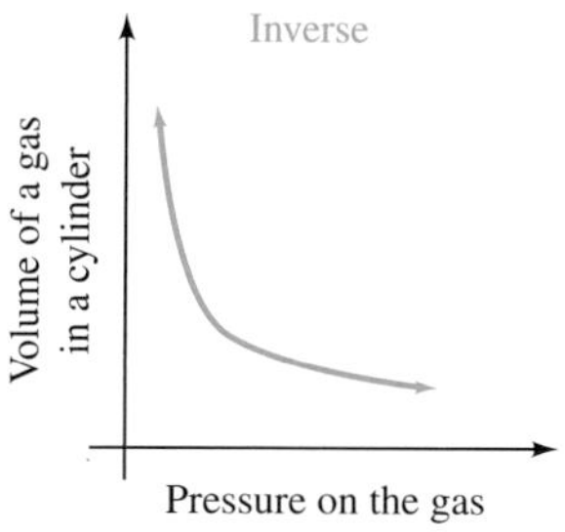

6. a.

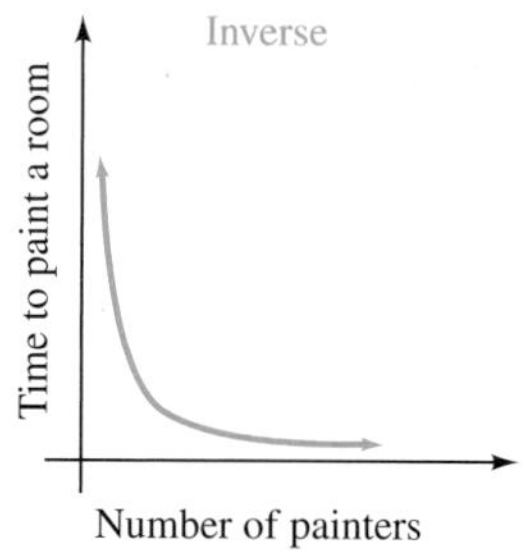

b.

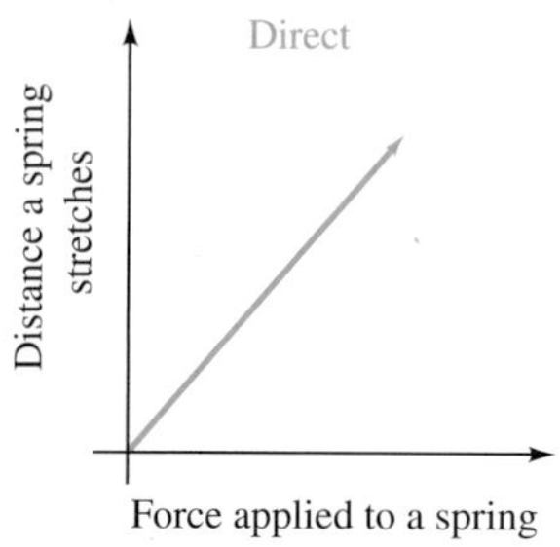

Tell whether each relationship suggests direct or inverse variation.

7. **Recycling.** The amount of money you receive and the number of aluminum cans you return Direct
8. **Karate.** The force needed to break a board and the length of the board Inverse
9. **Tools.** The force you must exert on the handle of a wrench to loosen a bolt and the length of the handle Inverse
10. **Anatomy.** The volume of blood pumped from your heart each minute and your pulse rate Direct
11. **Swimming Pools.** The amount of chlorine needed in a pool and the amount of water in the pool Direct
12. **Lightning.** The time it takes you to hear the lightning after a strike and your distance from the strike Direct
13. **Desserts.** The number of servings you can get from a wedding cake and the size of the piece that is served Inverse
14. **Remodeling.** The cost to remodel a house and the number of square feet to be added Direct

NOTATION

Complete each solution.

15. Determine whether the equation defines direct variation.

a. $y = kx$ Yes b. $y = k + x$ No

c. $y = \frac{k}{x}$ No d. $m = kc$ Yes

16. Determine whether each equation defines inverse variation.

a. $y = kx$ No b. $y = \frac{k}{x}$ Yes

c. $y = \frac{x}{k}$ No d. $d = \frac{k}{g}$ Yes

GUIDED PRACTICE

***Express each verbal model in symbols.* See Objectives 1 and 2.**

17. A varies directly as the square of p. $A = kp^2$

18. t varies directly as s. $t = ks$

19. z varies inversely as the cube of t. $z = \frac{k}{t^3}$

20. v varies inversely as the square of r. $v = \frac{k}{r^2}$

***Express each verbal model in symbols.* See Objectives 3 and 4.**

21. C varies jointly as x, y, and z. $C = kxyz$

22. d varies jointly as r and t. $d = krt$

23. P varies directly as the square of a and inversely as the cube of j. $P = \frac{ka^2}{j^3}$

24. M varies inversely as the cube of n and jointly as x and the square of z. $M = \frac{kxz^2}{n^3}$

***Express each variation model in words. In each equation, k is the constant of variation.* See Objectives 1 and 2.**

25. $r = kt$ r varies directly as t.

26. $A = kr^3$ A varies directly as the cube of r.

27. $b = \frac{k}{h}$ b varies inversely as h.

28. $d = \frac{k}{W^4}$ d varies inversely as the fourth power of W.

***Express each variation model in words. In each equation, k is the constant of variation.* See Objectives 3 and 4.**

29. $U = krs^2t$ U varies jointly as r, the square of s, and t.

30. $L = kmn$ L varies jointly as m and n.

31. $P = \frac{km}{n}$ P varies directly as m and inversely as n.

32. $R = \frac{kL}{d^2}$ R varies directly as L and inversely as d^2.

APPLICATIONS

33. from **Campus to Careers**

Webmaster

©Ronen/Shutterstock.com

The language of variation is often used to describe various aspects of the Internet and websites. Determine whether each statement, generally speaking, is true or false.

a. The dollar amount of sales that an Internet website receives is inversely proportional to the amount of Internet traffic that visits the website. False

b. The download time of an Internet website varies directly with the bandwidth being used. False

c. Search engines like Google place a value on a website that is directly proportional to the number of sites that link to it. True

34. **Pendulums.** The time it takes for one complete swing of a pendulum varies directly with the square root of the length L of the pendulum. Write an equation that models this type of variation.
$t = k\sqrt{L}$

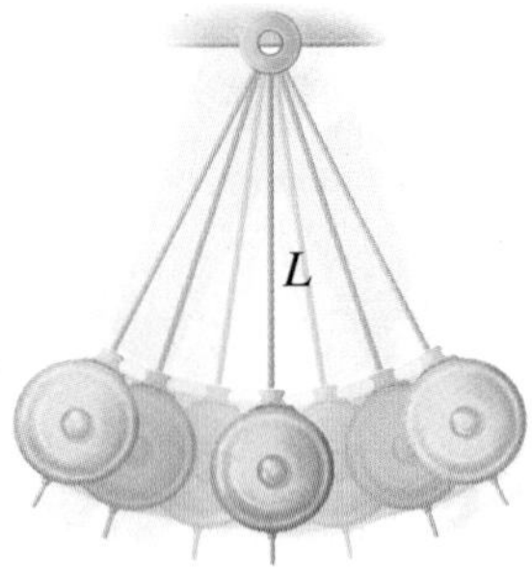

Solve each problem by writing a variation model.

35. **Gravity.** The force of gravity acting on an object varies directly as the mass of the object. The force on a mass of 5 kilograms is 49 newtons. What is the force acting on a mass of 12 kilograms? 117.6 newtons

36. **Free Fall.** An object in free fall travels a distance s that is directly proportional to the square of the time t. If an object falls 1,024 feet in 8 seconds, how far will it fall in 10 seconds? 1,600 ft

37. **Finding Distance.** The distance that a car can go varies directly as the number of gallons of gasoline it consumes. If a car can go 288 miles on 12 gallons of gasoline, how far can it go on a full tank of 18 gallons? 432 mi

38. **Braking.** Suppose the distance that a vehicle travels after its brakes have been applied varies directly as the square of the speed at which it was traveling. If the stopping distance for such a vehicle going 20 mph is 24 feet, what is the stopping distance for the vehicle traveling at 50 mph? 150 ft

39. **Farming.** The number of days that a given number of bushels of corn will last when feeding cattle varies inversely as the number of animals. If x bushels will feed 25 cows for 10 days, how long will the feed last for 10 cows? 25 days

40. Organ Pipes. The frequency of vibration of air in an organ pipe is inversely proportional to the length of the pipe. If a pipe 2 feet long vibrates 256 times per second, how many times per second will a 6-foot pipe vibrate? $85\frac{1}{3}$

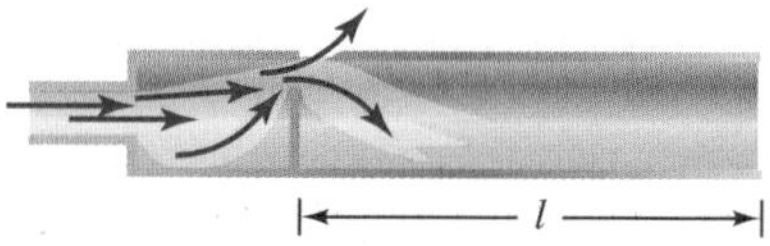

41. Gas Pressure. Under constant temperature, the volume occupied by a gas varies inversely to the pressure applied. If the gas occupies a volume of 20 cubic inches under a pressure of 6 pounds per square inch, find the volume when the gas is subjected to a pressure of 10 pounds per square inch. 12 in.^{3}

42. Real Estate. The following table shows the listing price for three homes in the same general locality. Write the variation model (direct or inverse) that describes the relationship between the listing price and the number of square feet of a house in this area. $P = 105f$

Number of square feet	Listing price
1,720	$180,600
1,205	$126,525
1,080	$113,400

43. Trucking Costs. The costs of a trucking company vary jointly as the number of trucks in service and the number of hours they are used. When 4 trucks are used for 6 hours each, the costs are $1,800. Find the costs of using 10 trucks, each for 12 hours. $9,000

44. Oil Storage. The number of gallons of oil that can be stored in a cylindrical tank varies jointly as the height of the tank and the square of the radius of its base. The constant of proportionality is 23.5. Find the number of gallons that can be stored in the cylindrical tank shown. 26,437.5 gal

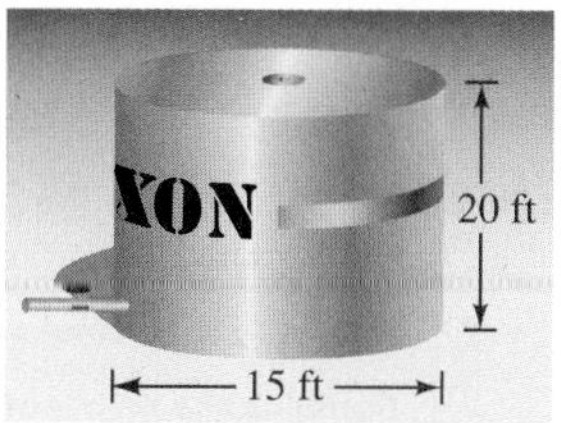

45. Electronics. The voltage (in volts) measured across a resistor is directly proportional to the current (in amperes) flowing through the resistor. The constant of variation is the **resistance** (in ohms). If 6 volts is measured across a resistor carrying a current of 2 amperes, find the resistance. 3 ohms

46. Electronics. The power (in watts) lost in a resistor (in the form of heat) varies directly as the square of the current (in amperes) passing through it. The constant of proportionality is the resistance (in ohms). What power is lost in a 5-ohm resistor carrying a 3-ampere current? 45 w

47. Structural Engineering. The deflection of a beam is inversely proportional to its width and the cube of its depth. If the deflection of a 4-inch-wide by 4-inch-deep beam is 1.1 inches, find the deflection of a 2-inch-wide by 8-inch-deep beam positioned as in figure (a) below. 0.275 in.

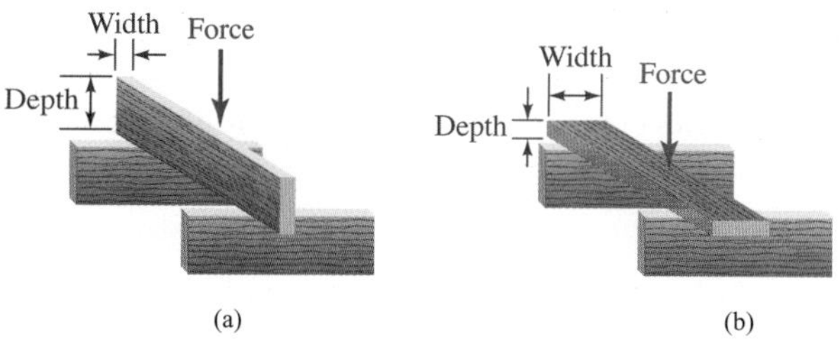

48. Structural Engineering. Find the deflection of the beam in Exercise 47 when the beam is positioned as in figure (b) above. 4.4 in.

49. Electronics. The resistance of a wire is directly proportional to the length of the wire and inversely proportional to the square of the diameter of the wire. If the resistance is 11.2 ohms in a 80-foot-long wire with diameter 0.01inch, what is the resistance in a 160-foot-long wire with diameter 0.04 inch? 1.4 ohms

50. Business Models. A businessman who sells widgets has found that the revenue from their sale varies directly as the advertising budget and inversely as the price. When $105,000 was spent on advertising and the widgets were priced at $19.95, the revenue from their sale was $200,000. How many widgets would he expect to sell at $17.50 each if $700,000 was spent on advertising? $1,520,000

51. Tension in a String. When playing with a Skip It toy, a child swings a weighted ball on the end of a string in a circular motion around one leg while jumping over the revolving string with the other leg. See the illustration. The tension T in the string is directly proportional to the square of the speed s of the ball and inversely proportional to the radius r of the circle. If the tension in the string is 6 pounds when the speed of the ball is 6 feet per second and the radius is 3 feet, find the tension when the speed is 8 feet per second and the radius is 2.5 feet. 12.8 lb

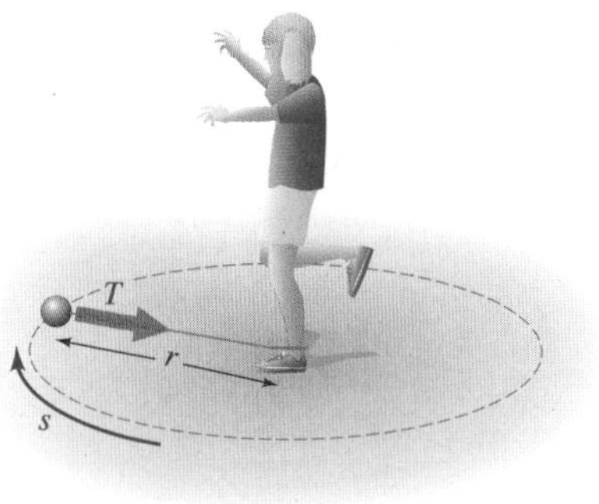

52. Gas Pressure. The pressure of a certain amount of gas is directly proportional to the temperature (measured on the Kelvin scale) and inversely proportional to the volume. A sample of gas at a pressure of 1 atmosphere occupies a volume of 1 cubic meter at a temperature of 273 Kelvin. When heated, the gas expands to twice its volume, but the pressure remains constant. To what temperature is it heated? 546 Kelvin

WRITING

53. Computer Printers. Is there a direct variation or an inverse variation between each pair of quantities? Explain. Draw a graph to support your answer.

a. The time it takes to print a term paper and the speed of the printer.

b. The time it takes to print a term paper and the length of the term paper.

54. Give examples of two quantities from everyday life that vary directly and two quantities that vary inversely.

REVIEW

Solve each compound inequality. Graph the solution set and write it using interval notation. See AIE Appendix 3.

55. $2a + 10 < 7a$ and $5a - 15 < 2a$ $(2, 5)$

56. $-22 \geq 5x - 2$ or $-3x - 4 > 2$ $(-\infty, -2)$

57. $\frac{x}{2} < \frac{x}{6} + 2$ or $0.1x + 0.1 > 1$ $(-\infty, 6) \cup (9, \infty)$

58. $\frac{7}{3}x + 2 \leq 16$ and $-8x \geq -48$ $(-\infty, 6]$

CHALLENGE PROBLEMS

59. As the cost of a purchase that is less than \$5 increases, the amount of change received from a five-dollar bill decreases. Is this inverse variation? Explain. No; change • cost is not a constant.

60. You've probably heard of Murphy's first law:

> If anything can go wrong, it will.

Another of Murphy's laws is:

> The chances of a piece of bread falling with the grape-jelly side down varies directly with the cost of the carpet.

Write one of your own witty sayings using the phrase *varies directly.*

8 Summary & Review

SECTION 8.1 Review of Solving Linear Equations, Formulas, and Linear Inequalities; Applications

DEFINITIONS AND CONCEPTS

Strategy for Solving Linear Equations in One Variable

1. Clear the equation of fractions or decimals.
2. Simplify each side of the equation by removing all sets of parentheses and combining like terms.
3. Isolate the variable term on one side of the equation.
4. Isolate the variable.
5. Check the result in the original equation.

EXAMPLES

Solve: $\frac{x-1}{6} + x = \frac{2}{3} - \frac{x+2}{6}$

$6\left(\frac{x-1}{6} + x\right) = 6\left(\frac{2}{3} - \frac{x+2}{6}\right)$ Multiply both sides by 6 to clear the fractions.

$6 \cdot \frac{x-1}{6} + 6 \cdot x = 6 \cdot \frac{2}{3} - 6 \cdot \frac{x+2}{6}$ Distribute 6 on both sides.

$x - 1 + 6x = 4 - (x + 2)$ Simplify. Don't forget the parentheses.

$x - 1 + 6x = 4 - x - 2$ Remove parentheses.

$7x - 1 = 2 - x$ Combine like terms on each side.

$7x - 1 + x = 2 - x + x$ To eliminate $-x$ on the right side, add x to both sides.

$8x - 1 = 2$ Combine like terms on each side.

$8x - 1 + 1 = 2 + 1$ To isolate the variable term $8x$, add 1 to both sides.

$8x = 3$ Simplify each side.

$\frac{8x}{8} = \frac{3}{8}$ Isolate the variable x by dividing both sides by 8.

$x = \frac{3}{8}$

The solution is $\frac{3}{8}$ and the solution set is $\left\{\frac{3}{8}\right\}$. Check this result to verify that it satisfies the *original* equation.

An equation that is satisfied by every number for which both sides are defined is called an **identity**.

When we solve $x + 5 + x = 2x + 5$, the variables drop out and we obtain a true statement $5 = 5$. All real numbers are solutions. The solution set is the set of real numbers written as $\mathbb{R}$.

A **contradiction** is an equation that is never true.

When we solve $y + 2 = y$, the variables drop out and we obtain a false statement $2 = 0$. The equation has no solutions. The solution set contains no elements and can be written as the **empty set** $\{\ \}$ or the **null set** $\varnothing$.

To **solve a formula for a specified variable** means to isolate that variable on one side of the equation, with all other variables and constants on the opposite side.

Solve $F = \dfrac{mMg}{r^2}$ for M.

$Fr^2 = mMg$ — To clear the fraction, multiply both sides by r^2.

$\dfrac{Fr^2}{mg} = M$ — To isolate M, divide both sides by mg.

$M = \dfrac{Fr^2}{mg}$ — Write M on the left side.

To **solve a linear inequality** in one variable, we use **properties of inequality** to find the values of its variable that make the inequality true.

If we multiply or divide both sides of an inequality by a negative number, the direction of the **inequality symbol must be reversed** for the inequalities to have the same solutions.

The set of all solutions of an inequality is called its **solution set.**

Solve: $-5x + 7 > 22$

$-5x + 7 - 7 > 22 - 7$ — Subtract 7 from both sides.

$-5x > 15$

$\dfrac{-5x}{-5} < \dfrac{15}{-5}$ — Divide both sides by -5 and reverse the direction of the inequality symbol.

$x < -3$

The solution set is:

Graph	Interval notation	Set-builder notation
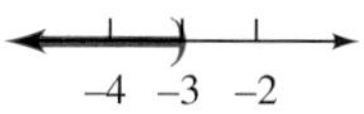	$(-\infty, -3)$	$\{x \mid x < -3\}$

REVIEW EXERCISES

Solve each equation. If an equation is an identity or a contradiction, so indicate.

1. $5x + 12 = 0$ $-\frac{12}{5}$
2. $-3x - 7 + x = 6x + 20 - 5x$ -9
3. $4(y - 1) = 28$ 8
4. $2 - 13(x - 1) = 4 - 6x$ $\frac{11}{7}$
5. $\frac{8}{3}(x - 5) = \frac{2}{5}(x - 4)$ $\frac{88}{17}$
6. $\frac{3y}{4} - 14 = -\frac{y}{3} - 1$ 12
7. $2x + 4 = 2(x + 3) - 2$ All real numbers, $\mathbb{R}$; identity
8. $3x - 2 - x = 2(x - 4)$ No solution, $\varnothing$; contradiction
9. $-\frac{5}{4}p = 10$ -8
10. $\frac{4t + 1}{3} - \frac{t + 5}{6} = \frac{t - 3}{6}$ 0

Solve each formula for the indicated variable.

11. $V = \pi r^2 h$ for h $h = \frac{V}{\pi r^2}$
12. $v = \frac{1}{6}ab(x + y)$ for x $x = \frac{6v - aby}{ab}$ or $x = \frac{6v}{ab} - y$

Solve each inequality. Give each solution set in interval notation and graph it.

13. $0.3x - 0.4 \ge 1.2 - 0.1x$ $[4, \infty)$

14. $\frac{7}{4}(x + 3) < \frac{3}{8}(x - 3)$ $\left(-\infty, -\frac{51}{11}\right)$

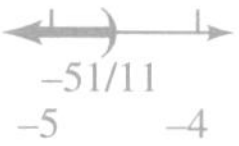

15. $-16 < -\frac{4}{5}x$ $(-\infty, 20)$

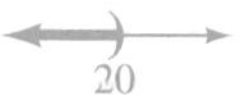

16. $5(2n + 2) - n > 3n - 3(1 - 2n)$ $(-\infty, \infty)$, $\mathbb{R}$

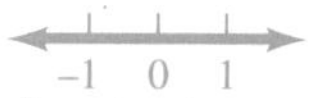

Use the six-step problem-solving strategy for each of the following application problems.

17. **Carpentry.** A carpenter wants to cut a 20-foot rafter so that one piece is 3 times as long as the other. Where should he cut the board? 5 ft from one end
18. **Geometry.** A rectangle is 4 meters longer than it is wide. If the perimeter of the rectangle is 28 meters, find its length and width. Length: 9 m, width: 5 m

SECTION 8.2 Functions

DEFINITIONS AND CONCEPTS | **EXAMPLES**

A **relation** is a set of ordered pairs. The set of first components is called the **domain** of the relation and the set of second components is called the **range.**

The relation $\{(2, 5), (7, -3), (4, 6)\}$ has domain $\{2, 4, 7\}$ and range $\{-3, 5, 6\}$.

A **function** is a set of ordered pairs (a relation) in which to each first component there corresponds exactly one second component.

Since we often work with sets of ordered pairs of the form (x, y), it is helpful to define a function using the variables x and y:

y* is a function of *x: Given a relation in x and y, if to each value of x in the domain there corresponds exactly one value of y in the range, then y is said to be a function of x. Since y depends on x, we call x the **independent variable** and y the **dependent variable.**

The relation $\{(2, 5), (7, -3), (4, 6)\}$ defines a function because to each first component there corresponds exactly one second component.

The relation $\{(-9, 1), (3, 8), (0, 0), (3, 24)\}$ *does not* define a function because to the first component 3 there corresponds two second components, 8 and 24.

The arrow diagram *does not* define y as a function of x because to the x-value 4 there corresponds more than one y-value: 2 and 6. The table defines a function and illustrates an important point: *Two different ordered pairs of a function can have the same y-value.*

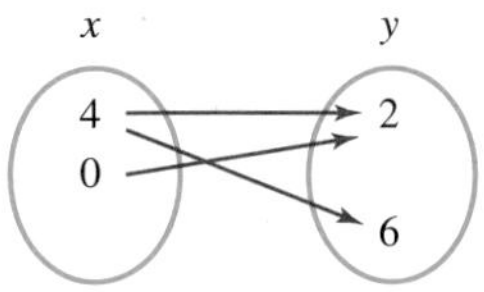

x	y
1	9
2	9
3	9

A function can be defined by an equation, however not all equations in two variables define functions.

The equation $y = 2x - 7$ defines y as a function of x because to each value of x there corresponds exactly one value of y.

The equation $x = |y|$ does not define y as a function of x because more than one value of y corresponds to a single value of x. If x is 2, for example, the equation becomes $2 = |y|$ and y can be either 2 or -2.

The **function notation** $y = f(x)$ indicates that the variable y is a function of x. It is read as "f of x."

Think of a function as a machine that takes some **input** x and turns it into some **output** $f(x)$, called a **function value.**

If $f(x) = 2x + 1$, find $f(-2)$ and $f(n + 1)$.

$$\begin{aligned} f(x) &= 2x + 1 \\ f(-2) &= 2(-2) + 1 \\ &= -4 + 1 \\ &= -3 \end{aligned}$$

Thus, $f(-2) = -3$.

$$\begin{aligned} f(x) &= 2x + 1 \\ f(n + 1) &= 2(n + 1) + 1 \\ &= 2n + 2 + 1 \\ &= 2n + 3 \end{aligned}$$

Thus, $f(n + 1) = 2n + 3$.

The input-output pairs that a function generates can be written as ordered pairs and plotted on a rectangular coordinate system to give the **graph of the function.**

A **linear function** is a function that can be defined by an equation of the form $f(x) = mx + b$. The graph of a linear function is a straight line.

Linear functions can be graphed using the **point-plotting method,** the **intercept method,** or the **slope–intercept method.**

To graph the linear function $f(x) = -4x - 2$, we make a table of values, plot the points, and draw the graph.

x	$f(x)$	
-1	2	→ $(-1, 2)$
0	-2	→ $(0, -2)$
1	-6	→ $(1, -6)$
↑ Select x.	↑ Find f(x).	↑ Plot the point.

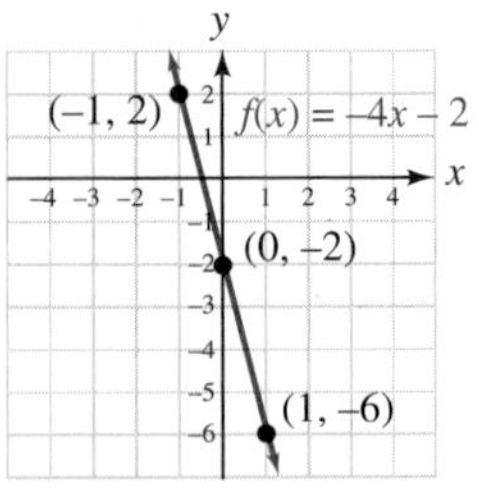

The **domain** of a function is the set of input values. The **range** is the set of output values.

To find the domain of a function defined by an equation, we must identify all the values of the input variable that produce a real-number output.

Find the domain of $f(x) = \frac{10}{x + 3}$.

The number -3 cannot be substituted for x, because that would make the denominator equal to 0. Since any real number except -3 can be substituted for x, the domain is *the set of all real numbers except* -3.

To write equations of linear functions, we can use:

slope–intercept form:
$f(x) = mx + b$

point–slope form:
$f(x) - y_1 = m(x - x_1)$

Write an equation for the linear function whose graph has slope -4 and passes through $(1, 8)$.

$f(x) - y_1 = m(x - x_1)$ This is point–slope form.

$f(x) - 8 = -4(x - 1)$ Substitute for m, y_1, and x_1.

$f(x) - 8 = -4x + 4$ Distribute.

$f(x) = -4x + 12$ Solve for $f(x)$.

A **polynomial function** is a function whose equation is defined by a polynomial in one variable.

To **evaluate a polynomial function,** we replace the variable in the defining equation with its value, called the **input.** Then we simplify to find the **output.**

Let $f(x) = x^3 - 3x^2 - 9x + 2$. Find $f(2)$.

$f(x) = x^3 - 3x^2 - 9x + 2$

$f(2) = (2)^3 - 3(2)^2 - 9(2) + 2$ Substitute 2 for x.

$= 8 - 3(4) - 18 + 2$

$= -20$ The output is -20.

REVIEW EXERCISES

19. Find the domain and the range of the relation: $\{(-4, 0), (5, 16), (2, -2), (-1, -2)\}$
D: $\{-4, -1, 2, 5\}$; R: $\{-2, 0, 16\}$

20. Fill in the blanks.

a. A _function_ is a set of ordered pairs (a relation) in which to each first component there corresponds exactly one second component.

b. Given a relation in x and y, if to each value of x in the domain there corresponds exactly one value of y in the range, y is said to be a _function_ of x. We call x the independent _variable_ and y the _dependent_ variable.

Determine whether the relation defines y as a function of x. If it does not, find two ordered pairs where more than one value of y corresponds to a single value of x.

21. a.

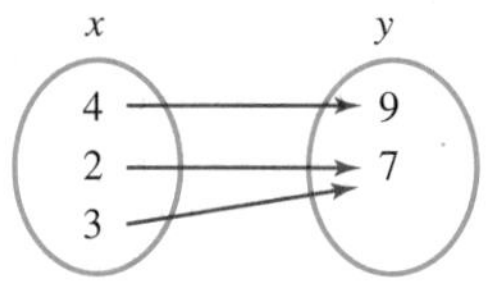

Yes

b.

x	y
-1	8
0	5
4	1
-1	9

No; $(-1, 8), (-1, 9)$

c. $\{(14, 6), (-1, 14), (6, 0), (-3, 8)\}$ Yes

22. For the given input, what value will the function machine output?
0

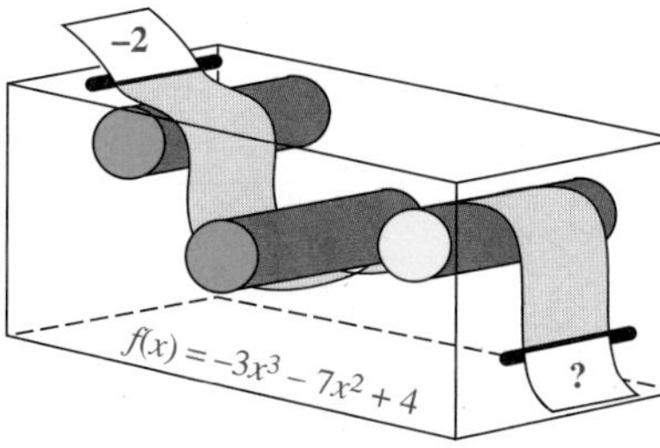

Determine whether each equation defines y as a function of x. If it does not, find two ordered pairs where more than one value of y corresponds to a single value of x.

23. $y = 6x - 4$ Yes

24. $y = 4 - x^2$ Yes

25. $y^2 = x$
No; $(25, 5), (25, -5)$

26. $|y| = x + 1$
No; $(3, 4), (3, -4)$

Let $f(x) = 3x + 2$ and $g(a) = \frac{a^2 - 4a + 4}{2}$. Find each function value.

27. $f(-3)$ -7

28. $g(8)$ 18

29. $g(-2)$ 8

30. $f(t + 2)$ $3t + 8$

31. Let $f(x) = -5x + 7$. For what value of x is $f(x) = -8$? 3

32. Let $g(t) = \frac{3}{4}t - 1$. For what value of t is $g(t) = 0$? $\frac{4}{3}$

Find the domain of each function.

33. $f(x) = 4x - 1$ The set of real numbers

34. $s(t) = t^2 + 1$ The set of real numbers

35. $h(x) = \dfrac{4}{2 - x}$ The set of all real numbers except 2

36. $g(b) = \dfrac{12}{5b + 25}$ The set of all real numbers except -5

37. What are the slope and the y-intercept of the graph of $g(x) = -2x - 16$? $-2, (0, -16)$

38. Graph: $f(x) = \frac{2}{3}x - 2$ See AIE Appendix 3.

Write an equation for a linear function whose graph has the given characteristics.

39. Slope: $\dfrac{9}{10}$, y-intercept: $\left(0, \dfrac{7}{8}\right)$ $f(x) = \frac{9}{10}x + \frac{7}{8}$

40. Slope: $\dfrac{1}{5}$, passes through $(10, 1)$ $f(x) = \frac{1}{5}x - 1$

41. Horizontal, passes through $(-8, 12)$ $f(x) = 12$

42. Passes through $(2, 5)$, parallel to the graph of $g(x) = 4x - 7$
$f(x) = 4x - 3$

43. Passes through $(-6, 3)$, perpendicular to the graph of $g(x) = -3x - 12$ $f(x) = \frac{1}{3}x + 5$

44. Electricity. The relationship between the electrical resistance R of a coil of wire and its temperature t can be modeled with a linear function.

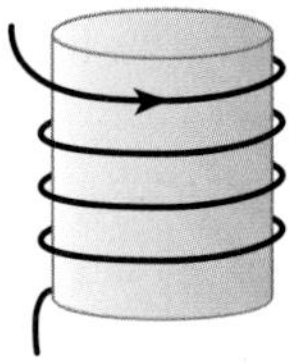

a. Use the following data in the table to write a function that describes this relationship. $R(t) = 0.02t + 5.05$

b. Use the answer to part (a) to predict the resistance if the temperature of the coil of wire is 100° Celsius. 7.05 milliohms

***t* (in degrees Celsius)**	10	30
***R* (in milliohms)**	5.25	5.65

45. Squirt Guns. The volume of the reservoir on top of the squirt gun is given by the polynomial function $V(r) = 4.19r^3 + 25.13r^2$, where r is the radius in inches. Find $V(2)$ to the nearest cubic inch. 134 in.³

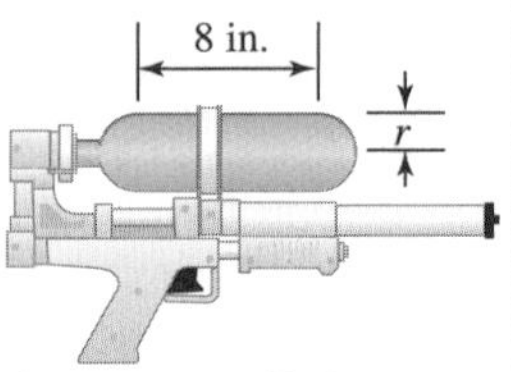

46. Calculus. In the advanced mathematics course called Calculus, an important polynomial function is:

$$f(x) = 1 + x + \frac{x^2}{2} + \frac{x^3}{6} + \frac{x^4}{24}$$

Find $f(1)$. $\frac{65}{24} = 2\frac{17}{24}$

SECTION 8.3 Graphs of Functions

DEFINITIONS AND CONCEPTS	EXAMPLES
From the graph of a function, we can determine function values. For the example, the value of $f(2)$ is given by the y-coordinate of a point on the graph of f with x-coordinate 2.	To find $f(2)$ from the graph, draw a vertical line through 2 on the x-axis. It intersects the graph at (2, 3). Therefore, 3 corresponds to 2 and it follows that $f(2) = 3$. 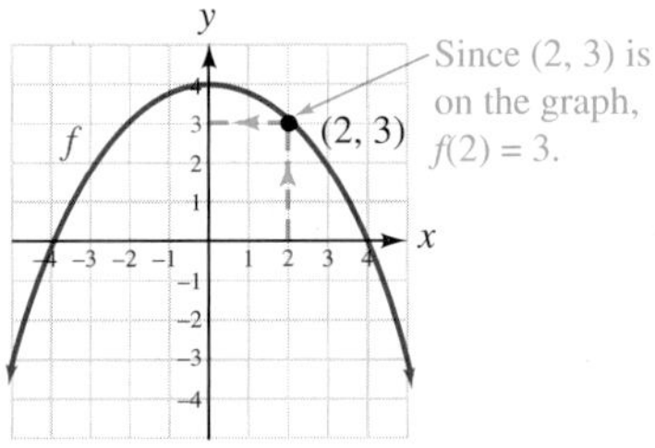
These three basic functions are used so often in algebra that you should memorize their names and their graphs. They are called **nonlinear functions** because their graphs are not lines.	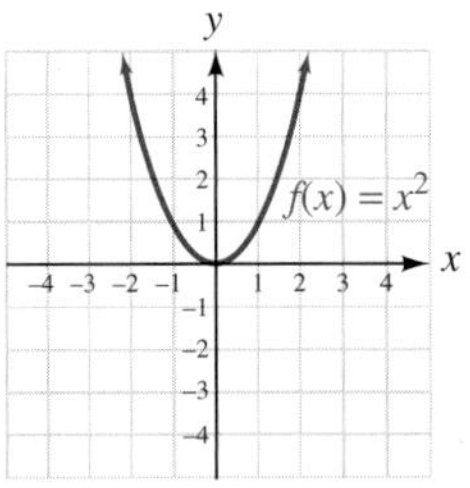 The squaring function 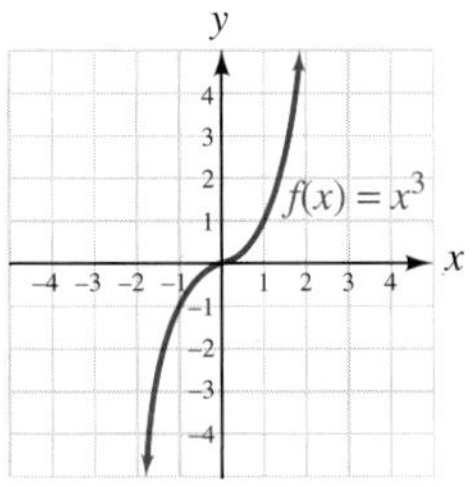 The cubing function 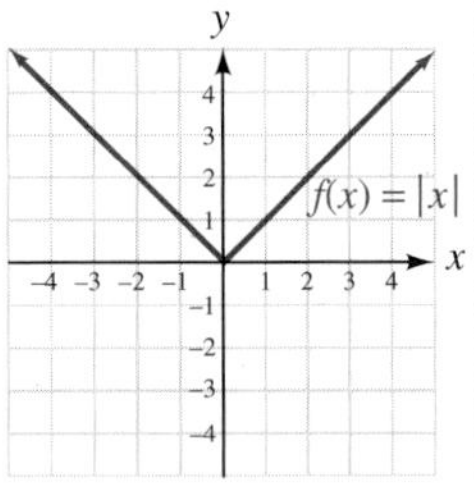 The absolute value function
We can find the domain and range of a function from its graph. The **domain of a function** is the projection of its graph onto the x-axis. **The range of a function** is the projection of its graph onto the y-axis.	Find the domain and range of function f. 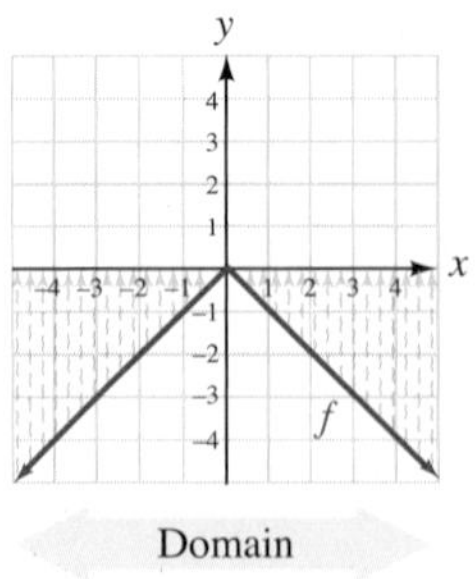 D: The set of real numbers 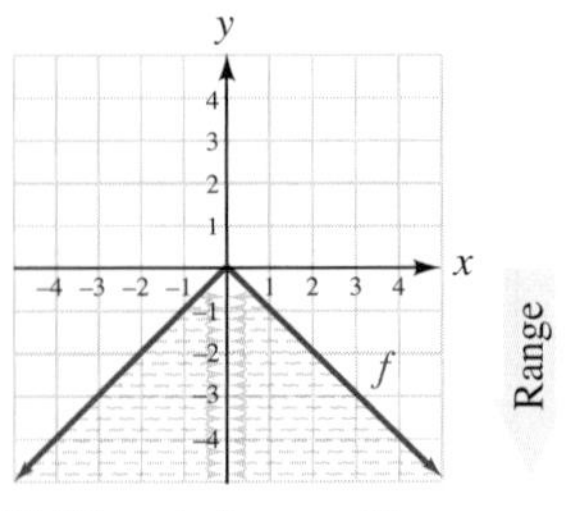 R: The set of nonpositive real numbers

A **vertical translation** shifts a graph upward or downward. A **horizontal translation** shifts a graph left or right. A **reflection** "flips" a graph about the x-axis.

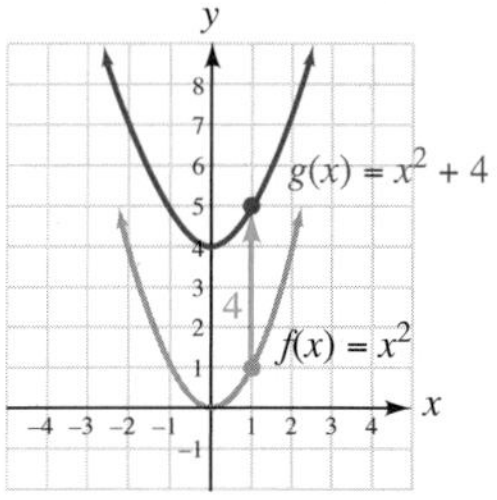

To graph $g(x) = x^2 + 4$, translate each point on the graph of $f(x) = x^2$ up 4 units.

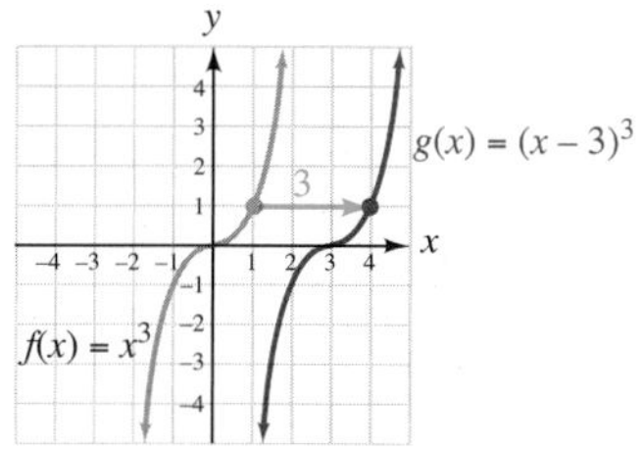

To graph $g(x) = (x - 3)^3$, translate each point on the graph of $f(x) = x^3$ to the right 3 units.

The **vertical line test:** If a vertical line intersects a graph in more than one point, the graph is not the graph of a function.

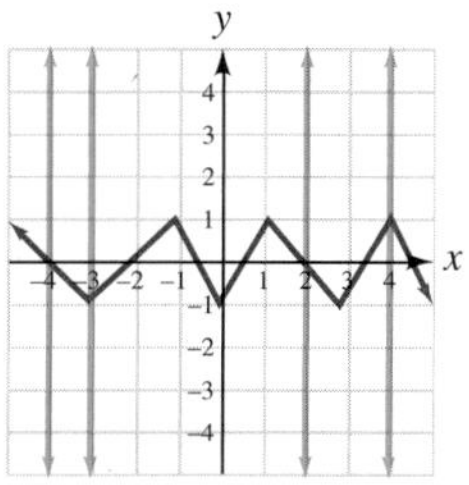

The graph of a function

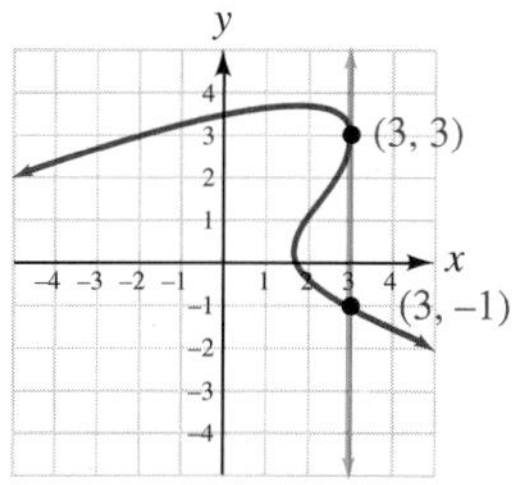

Not the graph of a function

REVIEW EXERCISES

47. Use the graph to find each value.

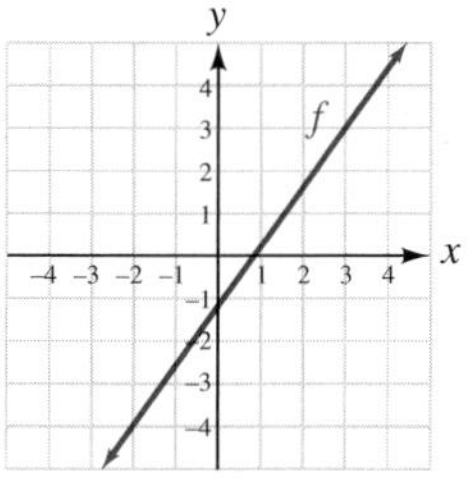

a. $f(-2)$ -4

b. $f(3)$ 3

c. The value of x for which $f(x) = 0$. 1

48. Use the graph to find each value.

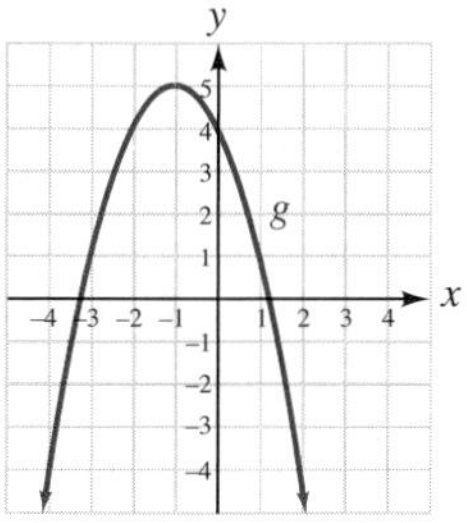

a. $g(0)$ 4

b. $g(-3)$ 1

c. The value(s) of x for which $g(x) = -4$. $-4, 2$

49. Use the graph of function f to find each of the following.

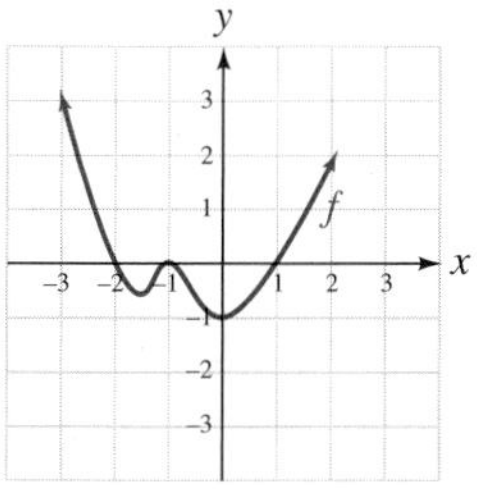

a. $f(0)$ -1

b. The values of x for which $f(x) = 0$ $-2, -1, 1$

c. Write the domain and range of f in interval notation. D: $(-\infty, \infty)$; R: $[-1, \infty)$

50. What are the nonnegative real numbers? The real numbers greater than or equal to 0

Give the domain and range of each function graphed below.

51.

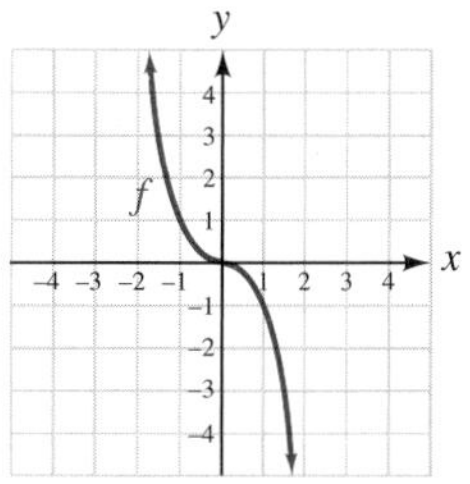

D: the set of real numbers; R: the set of real numbers

52.

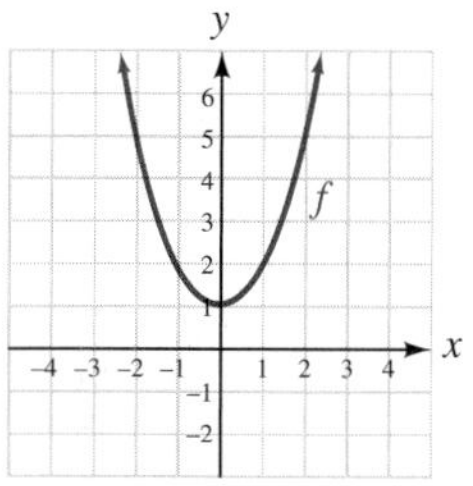

D: the set of real numbers; R: the set of real numbers greater than or equal to 1

53. Graph $f(x) = |x + 2|$ by creating a table of function values and plotting points. Give the domain and range of the function. D: the set of real numbers; R: the set of nonnegative real numbers See AIE Appendix 3.

54. Fill in the blanks.

a. The graph of $f(x) = x^2 + 6$ is the same as the graph of $f(x) = x^2$ except that it is shifted 6 units up.

b. The graph of $f(x) = (x + 6)^2$ is the same as the graph of $f(x) = x^2$ except that it is shifted 6 units to the left.

For each of the following functions, first sketch the graph of its associated function, $f(x) = x^2$, $f(x) = x^3$, or $f(x) = |x|$. Then draw the graph of function g using a translation and/or a reflection and give its domain and range. See AIE Appendix 3.

55. $g(x) = x^2 - 3$ D: the set of real numbers; R: the set of real numbers greater than or equal to -3

56. $g(x) = |x - 4|$ D: the set of real numbers; R: the set of nonnegative real numbers

57. $g(x) = (x - 2)^3 + 1$
D: the set of real numbers; R: the set of real numbers

58. $g(x) = -x^3$
D: the set of real numbers; R: the set of real numbers

Determine whether each graph is the graph of a function. If it is not, find two ordered pairs where more than one value of y corresponds to a single value of x.

59.

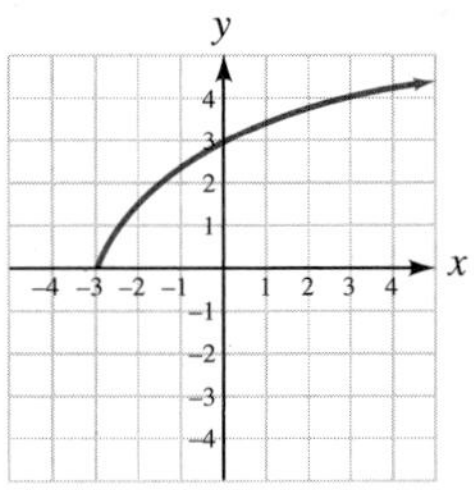

Function

60.

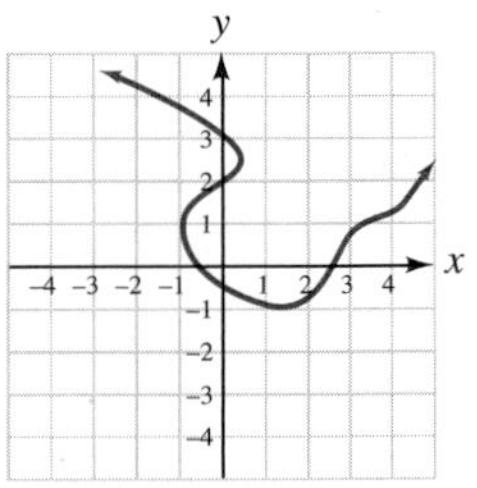

Not a function; (0, 2), (0,3)

SECTION 8.4 Solving Compound Inequalities

DEFINITIONS AND CONCEPTS	EXAMPLES
The **intersection** of two sets A and B, written $A \cap B$, is the set of all elements that are common to set A and set B. The **union** of two sets A and B, written $A \cup B$, is the set of all elements that are in set A, set B, or both.	Let $A = \{-2, 0, 3, 5\}$ and $B = \{-3, 0, 5, 7\}$. $A \cap B = \{0, 5\}$ The intersection contains the elements that the sets have in common. $A \cup B = \{-3, -2, 0, 3, 5, 7\}$ The union contains the elements that are in one or the other set, or both.
When the word *and* or the word *or* is used to connect pairs of inequalities, we call the statement a **compound inequality.** The solution set of a **compound inequality containing the word *and*** includes all numbers that make both of the inequalities true. That is, it is the intersection of their solution sets.	Solve: $2x - 1 \le 5$ and $5x + 1 > 4$ We solve each inequality separately. Then we graph the two solution sets (one in red, the other in blue) on the same number line and determine their intersection. $2x - 1 \le 5$ and $5x + 1 > 4$ $2x \le 6$ | $5x > 3$ $x \le 3$ | $x > \frac{3}{5}$ ***Preliminary Work:*** 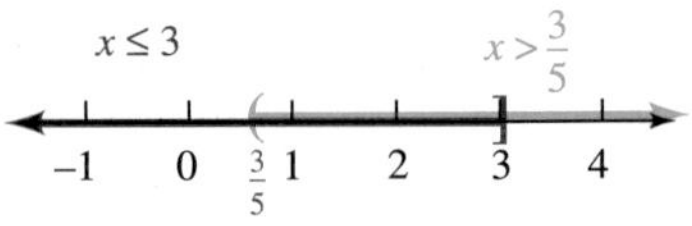 The purple-shaded interval shown below is where the red and blue graphs overlap. Thus, the solution set is: 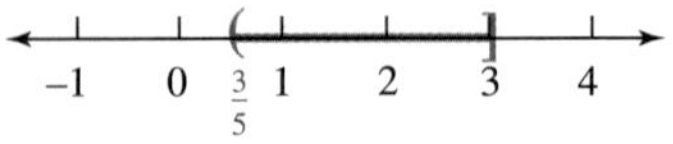 Interval notation: $\left(\frac{3}{5}, 3\right]$
Inequalities that contain exactly two inequality symbols are called **double inequalities.** Any double linear inequality can be written as a compound inequality containing the word *and.* For example: $c < x < d$ is equivalent to $c < x$ and $x < d$	Solve: $-7 \le 3x - 1 < 5$ We apply properties of inequality to *all three of its parts* to isolate x in the middle. $-7 \le 3x - 1 < 5$ $-7 + 1 \le 3x - 1 + 1 < 5 + 1$ Add 1 to all three parts. $-6 \le 3x < 6$ $\frac{-6}{3} \le \frac{3x}{3} < \frac{6}{3}$ Divide each part by 3. $-2 \le x < 2$ The solution set is: 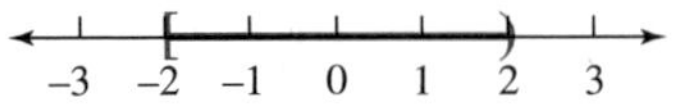 Interval notation: $[-2, 2)$

The solution set of a **compound inequality containing the word *or*** includes all numbers that make one or the other, or both, inequalities true. That is, it is the union of their solution sets.

Solve: $2x - 1 > 5$ or $-(5x - 7) \geq 2$

We solve each inequality separately. Then we graph the two solution sets on the same number line to show their union.

$$\begin{array}{rc|c} 2x - 1 > 5 & \text{or} & -(5x - 7) \geq 2 \\ 2x > 6 & & -5x + 7 \geq 2 \\ x > 3 & & -5x \geq -5 \\ & & x \leq 1 \end{array}$$

The solution set is:

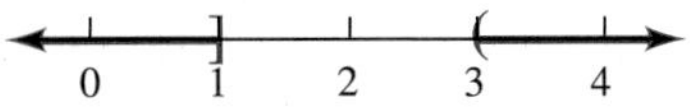

Interval notation: $(-\infty, 1] \cup (3, \infty)$

This is the union of two intervals.

REVIEW EXERCISES

Let $A = \{-6, -3, 0, 3, 6\}$ and $B = \{-5, -3, 3, 8\}$.

61. Find $A \cap B$. $\{-3, 3\}$

62. Find $A \cup B$. $\{-6, -5, -3, 0, 3, 6, 8\}$

Check to determine whether −4 is a solution of the compound inequality.

63. $x < 0$ and $x > -5$ Yes

64. $x + 3 < -3x - 1$ and $4x - 3 > 3x$ No

Graph each set. See AIE Appendix 3.

65. $(-3, 3) \cup [1, 6]$

66. $(-\infty, 2] \cap [1, 4)$

Solve each compound inequality. Graph the solution set and write it using interval notation. See AIE Appendix 3.

67. $-2x > 8$ and $x + 4 \geq -6$ $[-10, -4)$

68. $5(x + 2) \leq 4(x + 1)$ and $11 + x < 0$ $(-\infty, -11)$

69. $\frac{2}{5}x - 2 < -\frac{4}{5}$ and $-\frac{x}{3} < -1$ No solution; $\varnothing$

70. $4\left(x - \frac{1}{4}\right) \leq 3x - 1$ and $x \geq 0$ $[0, 0]$

Solve each double inequality. Graph the solution set and write it using interval notation. See AIE Appendix 3.

71. $3 < 3x + 4 < 10$ $\left(-\frac{1}{3}, 2\right)$

72. $-2 \leq \frac{5 - x}{2} \leq 2$ $[1, 9]$

Check to determine whether −4 is a solution of the compound inequality.

73. $x < 1.6$ or $x > -3.9$ Yes

74. $x + 1 < 2x - 1$ or $4x - 3 > 3x$ No

Solve each compound inequality. Graph the solution set and write it using interval notation. See AIE Appendix 3.

75. $x + 1 < -4$ or $x - 4 > 0$ $(-\infty, -5) \cup (4, \infty)$

76. $\frac{x}{2} + 3 > -2$ or $4 - x > 4$ $(-\infty, \infty)$

77. Rugs. A manufacturer makes a line of decorator rugs that are 4 feet wide and of varying lengths x (in feet). The floor area covered by the rugs ranges from 17 ft^2 to 25 ft^2. Write and then solve a double linear inequality to find the range of the lengths of the rugs. $17 \leq 4x \leq 25$, $4.25 \text{ ft} \leq x \leq 6.25 \text{ ft}$, $[4.25, 6.25]$

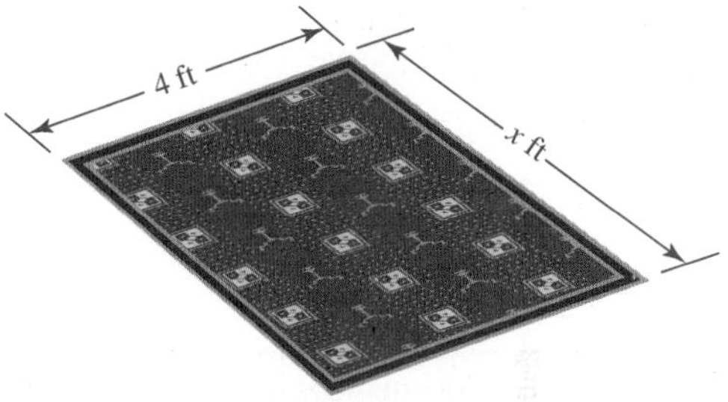

78. Match each word in Column I with *two* associated items in Column II.

Column I		***Column II***
a. or	ii, iv	**i.** $\cap$
		ii. $\cup$
b. and	i, iii	**iii.** intersection
		iv. union

79. Let $f(x) = \frac{5}{4}x - 140$. Find all values of x for which $f(x) < -48$ or $f(x) \geq 32$. $(-\infty, 73.6) \cup [137.6, \infty)$; see AIE Appendix 3.

80. Let $f(x) = 3x - 5$. Find all values of x for which $-4 \geq f(x) > -12$.

$\left(-\frac{7}{3}, \frac{1}{3}\right]$; see AIE Appendix 3.

SECTION 8.5 Solving Absolute Value Equations and Inequalities

DEFINITIONS AND CONCEPTS	EXAMPLES
To **solve absolute value equations** of the form $\lvert X\rvert = k$, where $k > 0$, solve the equivalent **compound equation** $X = k \quad \text{or} \quad X = -k$ If k is negative, then $\lvert X\rvert = k$ has no solution. Recall that **equivalent equations** have the same solutions.	Solve: $\lvert 2x + 1\rvert = 7$ This absolute value equation is equivalent to the following compound equation: $2x + 1 = 7$ or $2x + 1 = -7$ $2x = 6 \quad \mid \quad 2x = -8$ $x = 3 \quad \mid \quad x = -4$ This equation has two solutions: 3 and -4. The solution set is $\{-4, 3\}$.
Since absolute value expresses distance, the absolute value of a number is always positive or zero, but never negative.	Solve: $\lvert 4x - 5\rvert = -3$ Since an absolute value can never be negative, there are no real numbers x that make $\lvert 4x - 5\rvert = -3$ true. The equation has no solution and the solution set is $\varnothing$.
To **solve absolute value equations** of the form $\lvert X\rvert = \lvert Y\rvert$, solve the compound equation $X = Y \quad \text{or} \quad X = -Y$ The expressions within the absolute value symbols are equal, or they are opposites.	Solve: $\lvert 3x - 2\rvert = \lvert 2x + 4\rvert$ This equation is equivalent to the following compound equation: $3x - 2 = 2x + 4$ or $3x - 2 = -(2x + 4)$ $x - 2 = 4 \quad \mid \quad 3x - 2 = -2x - 4$ $x = 6 \quad \mid \quad 5x - 2 = -4$ $5x = -2$ $x = -\frac{2}{5}$ This equation has two solutions: 6 and $-\frac{2}{5}$. The solution set is $\left\{-\frac{2}{5}, 6\right\}$.
To **solve absolute value inequalities** of the form $\lvert X\rvert < k$, where $k > 0$, solve the equivalent double inequality $-k < X < k$. Use a similar approach to solve $\lvert X\rvert \le k$. When solving absolute value equations or inequalities, **isolate the absolute value expression** on one side **before** writing the equivalent compound statement.	Solve: $\lvert 4x - 3\rvert - 2 < 7$ $\lvert 4x - 3\rvert - 2 + 2 < 7 + 2$ To isolate the absolute value, add 2 to both sides. $\lvert 4x - 3\rvert < 9$ The resulting inequality is equivalent to the following double inequality: $-9 < 4x - 3 < 9$ $-6 < 4x < 12$ Add 3 to all three parts. $-\frac{3}{2} < x < 3$ Divide each part by 4 and simplify. The solution set is: 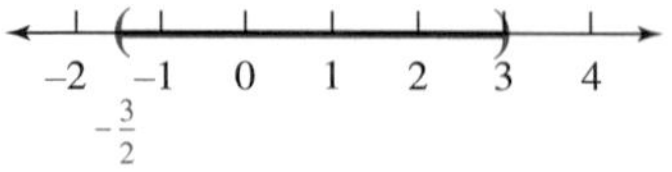 Interval notation: $\left(-\frac{3}{2}, 3\right)$ Set builder: $\left\{x \mid -\frac{3}{2} < x < 3\right\}$

To **solve absolute value inequalities** of the form $|X| \geq k$, where $k > 0$, solve the equivalent compound inequality $X \leq -k$ or $X \geq k$.

Use a similar approach to solve $|X| > k$.

Solve: $|3x + 1| \geq 7$

This inequality is equivalent to the following compound inequality:

$$3x + 1 \leq -7 \quad \text{or} \quad 3x + 1 \geq 7$$
$$3x \leq -8 \quad | \quad 3x \geq 6$$
$$x \leq -\frac{8}{3} \quad | \quad x \geq 2$$

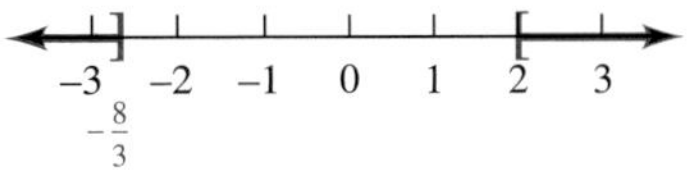

Interval notation: $\left(-\infty, -\frac{8}{3}\right] \cup [2, \infty)$

This is the union of two intervals.

Set builder: $\left\{x \mid x \leq -\frac{8}{3} \text{ or } x > 2\right\}$

REVIEW EXERCISES

Solve each absolute value equation.

81. $|4x| = 8$ $2, -2$

82. $2|3x + 1| - 1 = 19$ $3, -\frac{11}{3}$

83. $\left|\frac{3}{2}x - 4\right| - 10 = -1$ $\frac{26}{3}, -\frac{10}{3}$

84. $\left|\frac{2 - x}{3}\right| = -4$ No solution; $\varnothing$

85. $|-4(2x - 6)| = 0$ 3

86. $\left|\frac{3}{8} + \frac{x}{3}\right| = \frac{5}{12}$ $\frac{1}{8}, -\frac{19}{8}$

87. $|3x + 2| = |2x - 3|$ $\frac{1}{5}, -5$

88. $\left|\frac{2(1 - x) + 1}{2}\right| = \left|\frac{3x - 2}{3}\right|$ $\frac{13}{12}$

Solve each absolute value inequality. Graph the solution set and write it using interval notation. See AIE Appendix 3.

89. $|x| \leq 3$ $[-3, 3]$

90. $|2x + 7| < 3$ $(-5, -2)$

91. $2|5 - 3x| \leq 28$ $\left[-3, \frac{19}{3}\right]$

92. $\left|\frac{2}{3}x + 14\right| + 6 < 6$ No solution; $\varnothing$

93. $|x| > 1$ $(-\infty, -1) \cup (1, \infty)$

94. $\left|\frac{1 - 5x}{3}\right| \geq 7$ $(-\infty, -4] \cup \left[\frac{22}{5}, \infty\right)$

95. $|3x - 8| - 4 > 0$ $\left(-\infty, \frac{4}{3}\right) \cup (4, \infty)$

96. $\left|\frac{3}{2}x - 14\right| \geq 0$ $(-\infty, \infty)$, $\mathbb{R}$

97. **Produce.** Before packing, freshly picked tomatoes are weighed on the scale shown. Tomatoes having a weight w (in ounces) that falls within the highlighted range are sold to grocery stores.

   a. Complete the following absolute value inequality that expresses the acceptable weight range: $|w - \ 8\ | \leq 2$

   b. Solve the inequality from part (a) and express the acceptable weight range using interval notation. $[6, 10]$

98. Let $f(x) = \frac{1}{3}|6x| - 1$. For what value(s) of x is $f(x) = 5$? $3, -3$

99. Let $f(x) = 2|3(x + 4)| + 1.5$. Find all values of x for which $f(x) = 25.5$. $0, -8$

100. Let $f(x) = |7 - x|$. Find all values of x for which $f(x) < 5$. $(2, 12)$; see AIE Appendix 3.

101. Explain why $|0.04x - 8.8| < -2$ has no solution. Since $|0.04x - 8.8|$ is always greater than or equal to 0 for any real number x, it can never be less than -2. This absolute value inequality has no solution.

102. Explain why the solution set of $\left|\frac{3x}{50} + \frac{1}{45}\right| \geq -\frac{4}{5}$ is the set of all real numbers.
Since $\left|\frac{3x}{50} + \frac{1}{45}\right|$ is always greater than or equal to 0 for any real number x, it will always be greater than or equal to $-\frac{4}{5}$. This absolute value inequality is true for all real numbers.

SECTION 8.6 Review of Factoring Methods: GCF, Grouping, Trinomials

DEFINITIONS AND CONCEPTS	EXAMPLES
The first step of factoring a polynomial is to see whether the terms of the polynomial have a common factor. If they do, **factor out the GCF.**	Factor: $14a^4 + 35a^3 - 56a^2 = 7a^2(2a^2 + 5a - 8)$ Factor out the GCF, $7a^2$. Use multiplication to check the factorization: $7a^2(2a^2 + 5a - 8) = 14a^4 + 35a^3 - 56a^2$ This is the original polynomial.

If an expression has four or more terms, try to factor the expression by **grouping.**

Factor: $ax - 2x + 3a - 6$

$$ax - 2x + 3a - 6 = (ax - 2x) + (3a - 6)$$ Group the terms.

$$= x(a - 2) + 3(a - 2)$$ Factor x from $ax - 2x$. Factor 3 from $3a - 6$.

$$= (a - 2)(x + 3)$$ Factor out the GCF, $a - 2$.

Many trinomials factor as the product of two binomials. To **factor a trinomial** of the form $x^2 + bx + c$, whose **leading coefficient is 1,** find two integers whose product is c and whose sum is b.

Factor: $p^2 + 14p + 45$

We must find two integers whose product is 45 and whose sum is 14. Since $5 \cdot 9 = 45$ and $5 + 9 = 14$, two such numbers are 5 and 9, and we have

$$p^2 + 14p + 45 = (p + 5)(p + 9)$$

Check: $(p + 5)(p + 9) = p^2 + 9p + 5p + 45 = p^2 + 14p + 45$

We can use the **trial-and-check method** to factor trinomials with **leading coefficients other than 1.** Write the trinomial as the product of two binomials and determine four integers.

Factor: $2x^2 - 5x - 12$

Since the first term is $2x^2$, the first terms of the binomial factors must be $2x$ and x.

$$(2x \square)(x \square)$$ Because $2x \cdot x$ will give $2x^2$

The second terms of the binomials must be two integers whose product is -12. There are six such pairs:

$$1(-12),\quad 2(-6),\quad 3(-4),\quad 4(-3),\quad 6(-2),\quad \text{and}\quad 12(-1)$$

The pair in blue gives the correct middle term, $-5x$, when placed in the binomial factors shown above. We use the FOIL method to check:

Outer: $-8x$

$$(2x + 3)(x - 4)$$ Combine like terms: $-8x + 3x = -5x$.

Inner: $3x$

Thus, $2x^2 - 5x - 12 = (2x + 3)(x - 4)$.

To factor $ax^2 + bx + c$ by **grouping,** write it as an equivalent four-term polynomial:

$$ax^2 + \square x + \square x + c$$

The product of these numbers must be ac, and their sum must be b.

Then factor the four-term polynomial by grouping. Use the FOIL method to check.

Factor by grouping: $2x^2 - 5x - 12$

We must find two integers whose product is $ac = 2(-12) = -24$ and whose sum is $b = -5$. Two such numbers are -8 and 3. They serve as the coefficients of $-8x$ and $3x$, the two terms that we use to represent the middle term, $-5x$, of the trinomial.

$$2x^2 - 5x - 12 = 2x^2 - 8x + 3x - 12$$ Express $-5x$ as $-8x + 3x$.

$$= 2x(x - 4) + 3(x - 4)$$

$$= (x - 4)(2x + 3)$$ Factor out the GCF, $(x - 4)$.

REVIEW EXERCISES

Factor, if possible.

103. $z^2 - 11z + 30$ $(z - 5)(z - 6)$

104. $x^4 + 4x^2 + x^2y + 4y$ $(x^2 + 4)(x^2 + y)$

105. $4a^2 - 5a + 1$ $(4a - 1)(a - 1)$

106. $27x^3y^3z^3 + 81x^4y^5z^2 - 90x^2y^3z^7$ $9x^2y^3z^2(3xz + 9x^2y^2 - 10z^5)$

107. $15b^2 + 4b - 4$ $(5b - 2)(3b + 2)$

108. $-x^2 - 3x + 28$ $-(x + 7)(x - 4)$

109. $15x^2 - 57xy - 12y^2$ $3(5x + y)(x - 4y)$

110. $w^8 - w^4 - 90$ $(w^4 - 10)(w^4 + 9)$

111. $r^2y - ar - ry + a + r - 1$ $(ry - a + 1)(r - 1)$

112. $49a^6 + 84a^3b^2 + 36b^4$ $(7a^3 + 6b^2)^2$

113. $3b^2 + 2b + 1$ Prime

114. $2a^4 + 4a^3 - 6a^2$ $2a^2(a + 3)(a - 1)$

115. Use a substitution to factor: $(s + t)^2 - 2(s + t) + 1$ $(s + t - 1)^2$

116. Solve $m_1m_2 = mm_2 + mm_1$ for m_1. $m_1 = \frac{mm_2}{m_2 - m}$

SECTION 8.7 Review of Factoring Methods: The Difference of Two Squares; the Sum and Difference of Two Cubes

DEFINITIONS AND CONCEPTS	EXAMPLES
The **difference of two squares:** To factor the square of a First quantity minus the square of a Last quantity, multiply the First plus the Last by the First minus the Last. $F^2 - L^2 = (F + L)(F - L)$	Factor: $x^2y^2 - 100$ $x^2y^2 - 100 = (xy)^2 - 10^2$ This is a difference of two squares. $= (xy + 10)(xy - 10)$
In general, the **sum of two squares** (with no common factor other than 1) cannot be factored using real numbers.	$x^2 + 100$ and $36y^2 + 49z^4$ are prime polynomials.
The **sum of two cubes:** To factor the cube of a First quantity plus the cube of a Last quantity, multiply the First plus the Last by the First squared, minus the First times the Last, plus the Last squared. $F^3 + L^3 = (F + L)(F^2 - FL + L^2)$	Factor: $y^3 + 27z^6$ $y^3 + 27z^6 = y^3 + (3z^2)^3$ This is a sum of two cubes. $= (y + 3z^2)[y^2 - y \cdot 3z^2 + (3z^2)^2]$ $= (y + 3z^2)(y^2 - 3yz^2 + 9z^4)$
The **difference of two cubes:** To factor the cube of a First quantity minus the cube of a Last quantity, multiply the First minus the Last by the First squared, plus the First times the Last, plus the Last squared. $F^3 - L^3 = (F - L)(F^2 + FL + L^2)$	Factor: $125s^3 - 64$ $125s^3 - 64 = (5s)^3 - 4^3$ This is a difference of two cubes. $= (5s - 4)[(5s)^2 + 5s \cdot 4 + 4^2]$ $= (5s - 4)(25s^2 + 20s + 16)$

REVIEW EXERCISES

Factor, if possible.

117. $z^2 - 16$ $(z + 4)(z - 4)$

118. $x^2y^4 - 64z^6$ $(xy^2 + 8z^3)(xy^2 - 8z^3)$

119. $a^2b^2 + c^2$ Prime

120. $c^2 - (a + b)^2$ $(c + a + b)(c - a - b)$

121. $32a^4c - 162b^4c$ $2c(4a^2 + 9b^2)(2a + 3b)(2a - 3b)$

122. $k^2 + 2k + 1 - 9m^2$ $(k + 1 + 3m)(k + 1 - 3m)$

123. $m^2 - n^2 - m - n$ $(m + n)(m - n - 1)$

124. $t^3 + 64$ $(t + 4)(t^2 - 4t + 16)$

125. $8a^3 - 125b^9$ $(2a - 5b^3)(4a^2 + 10ab^3 + 25b^6)$

126. Spanish Roof Tile. The amount of clay used to make a roof tile is given by

$$V = \frac{\pi}{2}r_1^2h - \frac{\pi}{2}r_2^2h$$

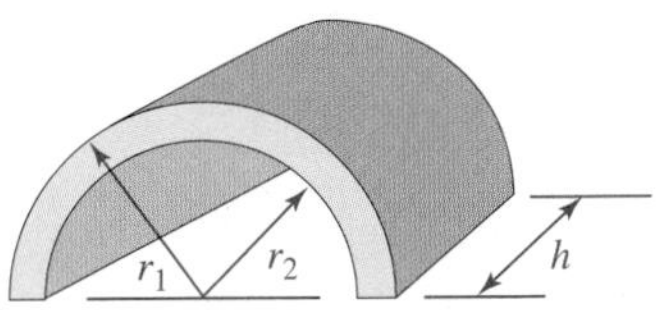

Factor the right side of the formula completely. $\frac{\pi}{2}h(r_1 + r_2)(r_1 - r_2)$

SECTION 8.8 Review of Rational Expressions and Rational Equations; Rational Functions

DEFINITIONS AND CONCEPTS	EXAMPLES
A **rational expression** is an expression of the form $\frac{A}{B}$, where A and B are polynomials and B does not equal 0.	Rational expressions: $\frac{3x^2}{xy}$, $\frac{5b - 15}{b^2 - 25}$, and $\frac{a + 2}{a^2 - 3a - 4}$
A **rational function** is a function whose equation is defined by a rational expression in one variable.	Rational functions: $f(x) = \frac{6x}{x - 2}$ and $f(n) = \frac{n + 3}{n^3 + 2n - 9}$

Since division by 0 is undefined, any values that make the denominator 0 in a rational function must be excluded from the **domain** of the function.

When finding the domain of a rational function, we *don't need to examine the numerator* of the expression; it can be any value, including 0.

Find the domain of the rational function: $f(x) = \frac{x+3}{x^2-4}$

$x^2 - 4 = 0$ Set the denominator equal to 0.

$(x+2)(x-2) = 0$ Factor the difference of two squares.

$x + 2 = 0$ or $x - 2 = 0$ Set each factor equal to 0.

$x = -2$ | $x = 2$ Solve each equation.

The domain of the function is the set of all real numbers except -2 and 2. In interval notation, the domain is $(-\infty, -2) \cup (-2, 2) \cup (2, \infty)$.

To simplify a rational expression:

1. Factor the numerator and denominator completely.
2. Remove factors equal to 1 by replacing each pair of factors common to the numerator and denominator with the equivalent fraction $\frac{1}{1}$.
3. Multiply the remaining factors in the numerator and in the denominator.

Simplify: $\frac{x^2-4}{2x+4} = \frac{\overset{1}{\cancel{(x+2)}}(x-2)}{2\underset{1}{\cancel{(x+2)}}} = \frac{x-2}{2}$

Simplify: $\frac{2a^3-5a^2-12a}{2a^3-11a^2+12a} = \frac{\overset{1}{\cancel{a}}(2a+3)\overset{1}{\cancel{(a-4)}}}{\underset{1}{\cancel{a}}(2a-3)\underset{1}{\cancel{(a-4)}}} = \frac{2a+3}{2a-3}$

To **multiply rational expressions,** multiply the numerators and multiply the denominators.

$$\frac{A}{B} \cdot \frac{C}{D} = \frac{AC}{BD}$$

Then simplify, if possible.

Multiply, and then simplify, if possible.

$\frac{x^2-4}{x+3} \cdot \frac{3x+9}{x+2} = \frac{(x^2-4)(3x+9)}{(x+3)(x+2)}$ Multiply the numerators. Multiply the denominators.

$= \frac{\overset{1}{\cancel{(x+2)}}(x-2) \cdot 3 \cdot \overset{1}{\cancel{(x+3)}}}{\underset{1}{\cancel{(x+3)}}\underset{1}{\cancel{(x+2)}}}$ Factor completely and then simplify.

$= 3(x-2)$

To **divide rational expressions,** multiply the first by the reciprocal of the second.

$$\frac{A}{B} \div \frac{C}{D} = \frac{A}{B} \cdot \frac{D}{C} = \frac{AD}{BC}$$

Then simplify, if possible.

Divide, and then simplify, if possible.

$\frac{x^2+4x+3}{x^2+3x} \div \frac{3}{x} = \frac{x^2+4x+3}{x^2+3x} \cdot \frac{x}{3}$ Multiply the first rational expression by the reciprocal of the second.

$= \frac{(x^2+4x+3) \cdot x}{(x^2+3x) \cdot 3}$ Multiply the numerators. Multiply the denominators.

$= \frac{(x+1)\overset{1}{\cancel{(x+3)}} \cdot \overset{1}{\cancel{x}}}{\underset{1}{\cancel{x}}\underset{1}{\cancel{(x+3)}} \cdot 3}$ Factor completely and then simplify.

$= \frac{x+1}{3}$ Multiply the remaining factors in the numerator. Multiply the remaining factors in the denominator.

To **add (or subtract) two rational expressions with like denominators,** add (or subtract) the numerators and keep the common denominator. Then, if possibe, factor and simplify.

Add:

$\frac{x^2-26}{x-5} + \frac{1}{x-5} = \frac{x^2-26+1}{x-5}$ Add the numerators. Write the sum over the common denominator, $x - 5$.

$= \frac{x^2-25}{x-5}$ Combine like terms.

$= \frac{(x+5)\overset{1}{\cancel{(x-5)}}}{\underset{1}{\cancel{x-5}}}$ To simplify the result, factor the numerator and remove the factor common to the numerator and denominator.

$= x + 5$

To **add or subtract rational expressions with unlike denominators,** find the LCD and express each rational expression with a denominator that is the LCD. Add (or subtract) the resulting fractions and simplify the result, if possible.

To build a rational expression, multiply it by 1 in the form of $\frac{c}{c}$, where c is any nonzero number or expression.

Subtract:

$$\frac{2x}{x+5} - \frac{1}{x} = \frac{2x}{x+5} \cdot \frac{x}{x} - \frac{1}{x} \cdot \frac{x+5}{x+5}$$ Build each rational expression to have the LCD of $x(x + 5)$.

$$= \frac{2x^2}{x(x+5)} - \frac{x+5}{x(x+5)}$$ Multiply the numerators. Multiply the denominators.

$$= \frac{2x^2 - (x+5)}{x(x+5)}$$ Subtract the numerators. Write the difference over the common denominator.

$$= \frac{2x^2 - x - 5}{x(x+5)}$$ The result does not simplify.

To **solve a rational equation:**

1. Factor all denominators.
2. Determine which numbers cannot be solutions of the equation.
3. Multiply both sides of the equation by the LCD of all rational expressions in the equation.
4. Use the distributive property to remove parentheses, remove any factors equal to 1, and write the result in simplified form.
5. Solve the resulting equation.
6. Check all possible solutions in the original equation.

All possible solutions of a rational equation must be checked. Multiplying both sides of an equation by an expression that contains a variable can lead to **extraneous solutions**, which must be discarded.

Solve: $\frac{3}{2} + \frac{1}{a-4} = \frac{5}{2a-8}$

If we factor the last denominator, the equation can be written as:

$$\frac{3}{2} + \frac{1}{a-4} = \frac{5}{2(a-4)}$$

We see that 4 cannot be a solution of the equation, because it makes at least one of the rational expressions in the equation undefined.

We can clear the equation of fractions by multiplying both sides by $2(a - 4)$, which is the LCD of the three rational expressions.

$$2(a-4)\left(\frac{3}{2} + \frac{1}{a-4}\right) = 2(a-4)\left[\frac{5}{2(a-4)}\right]$$ Multiply both sides by the LCD.

$$2(a-4)\left(\frac{3}{2}\right) + 2(a-4)\left(\frac{1}{a-4}\right) = 2(a-4)\left[\frac{5}{2(a-4)}\right]$$ Distribute.

$$\frac{\overset{1}{\cancel{2}}(a-4)}{1}\left(\frac{3}{\underset{1}{\cancel{2}}}\right) + 2\overset{1}{\cancel{(a-4)}}\left(\frac{1}{\underset{1}{\cancel{a-4}}}\right) = \overset{1}{\cancel{2}}\overset{1}{\cancel{(a-4)}}\left[\frac{5}{\underset{1}{\cancel{2}}\underset{1}{\cancel{(a-4)}}}\right]$$ Remove common factors.

$(a - 4)3 + 2 = 5$ Simplify.

$3a - 12 + 2 = 5$ Distribute.

$3a - 10 = 5$ Combine like terms.

$3a = 15$

$a = 5$

The solution is 5. Verify that it satisfies the original equation.

REVIEW EXERCISES

127. Use the graph of function f to find each of the following:

a. $f(2)$ 1

b. $f(-1)$ $-\frac{1}{2}$

c. The value of x for which $f(x) = -1$ 0

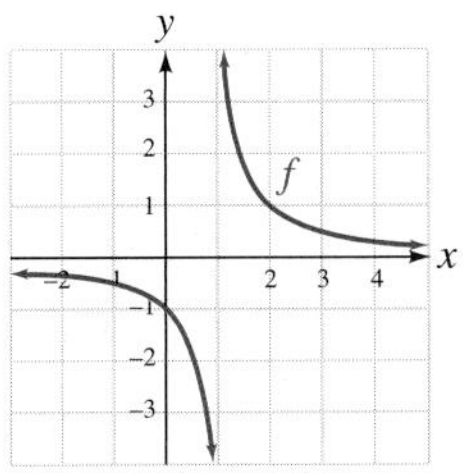

128. Find the domain of the rational function $f(x) = \frac{2x^2 + 8x}{x^2 + 2x - 24}$. Express your answer in words and using interval notation. The domain is the set of all real numbers except −6 and 4: $(-\infty, -6) \cup (-6, 4) \cup (4, \infty)$.

129. Injections. A hospital patient was given an injection for pain. The rational function

$$n(t) = \frac{28t}{t^2 + 1}$$

gives the number of milligrams n of pain medication per liter in the patient's bloodstream, where t is the number of hours since the injection. Find $n(3)$ and explain its meaning. $n(3) = 8.4$; Three hours after the injection, the concentration of pain medication in the patient's bloodstream was 8.4 milligrams per liter.

130. Let $f(x) = \frac{4x^2 - 6x + 7}{3x^2 - 27}$. Find $f(3)$. Undefined

Simplify each rational expression.

131. $\dfrac{62x^2y}{144xy^2}$ $\dfrac{31x}{72y}$

132. $\dfrac{2m-2n}{n-m}$ -2

Perform the operations and simplify, if possible.

133. $\dfrac{3x^3y^4}{c^2d}\cdot\dfrac{c^3d^2}{21x^5y^4}$ $\dfrac{cd}{7x^2}$

134. $\dfrac{2a^2-5a-3}{a^2-9}\div\dfrac{2a^2+5a+2}{2a^2+5a-3}$ $\dfrac{2a-1}{a+2}$

135. $\dfrac{m^2+3m+9}{m^2+mp+mr+pr}\div\dfrac{m^3-27}{am+ar+bm+br}$ $\dfrac{a+b}{(m+p)(m-3)}$

136. $\dfrac{x^3+3x^2+2x}{2x^2-2x-12}\cdot\dfrac{3x^2-3x}{x^3-3x^2-4x}\div\dfrac{x^2+3x+2}{2x^2-4x-16}$ $\dfrac{3x(x-1)}{(x-3)(x+1)}$

137. $\dfrac{d^2}{c^3-d^3}+\dfrac{c^2+cd}{c^3-d^3}$ $\dfrac{1}{c-d}$

138. $\dfrac{4}{t-3}+\dfrac{6}{3-t}$ $-\dfrac{2}{t-3}$

139. $\dfrac{5x}{14z^2}+\dfrac{y^2}{16z}$ $\dfrac{40x+7y^2z}{112z^2}$

140. $\dfrac{4}{3xy-6y}-\dfrac{4}{10-5x}$ $\dfrac{12y+20}{15y(x-2)}$

141. $\dfrac{y+7}{y+3}-\dfrac{y-3}{y+7}$ $\dfrac{14y+58}{(y+3)(y+7)}$

142. $\dfrac{2x}{x+1}+\dfrac{3x}{x+2}+\dfrac{4x}{x^2+3x+2}$ $\dfrac{5x^2+11x}{(x+1)(x+2)}$

Solve each equation. If a solution is extraneous, so indicate.

143. $\dfrac{4}{x}-\dfrac{1}{10}=\dfrac{7}{2x}$ 5

144. $\dfrac{3}{y}-\dfrac{2}{y+1}=\dfrac{1}{2}$ $-2, 3$

145. $\dfrac{2}{3x+15}-\dfrac{1}{18}=\dfrac{1}{3x+12}$ $-1, -2$

146. $\dfrac{3}{x+2}=\dfrac{1}{2-x}+\dfrac{2}{x^2-4}$ $\dfrac{3}{2}$

147. $\dfrac{x+3}{x-5}+\dfrac{2x^2+6}{x^2-7x+10}=\dfrac{3x}{x-2}$ 0

148. $\dfrac{5a}{a-3}-7=\dfrac{15}{a-3}$ No solution; 3 is extraneous

SECTION 8.9 Variation

DEFINITIONS AND CONCEPTS	EXAMPLES
The words ***y varies directly as x*** or ***y is directly proportional to x*** mean that $y = kx$ for some nonzero constant k, called the **constant of variation.** The words ***y varies inversely as x*** or ***y is inversely proportional to x*** mean that $y = \frac{k}{x}$ for some nonzero constant k.	The distance d that a spring stretches ***varies directly*** as the force f attached to the spring: $d = kf$. If the voltage in an electric circuit is kept constant, the current I ***varies inversely*** as the resistance R: $I = \dfrac{k}{R}$
Strategy for solving variation problems 1. Translate the verbal model into an equation. 2. Substitute the first set of values into the equation from step 1 to determine the value of k. 3. Substitute the value of k into the equation from step 1. 4. Substitute the remaining set of values into the equation from step 3 and solve for the unknown.	Suppose d varies inversely as h. If $d = 5$ when $h = 4$, find d when $h = 10$. 1. The words *d varies inversely as h* translate to $d = \frac{k}{h}$. 2. If we substitute 5 for d and 4 for h, we have $5 = \dfrac{k}{4}$ $20 = k$ To find k, multiply both sides by 4. This is the constant of variation. 3. Since $k = 20$, the inverse variation equation is $d = \frac{20}{h}$. 4. To answer the final question, we substitute 10 for h in the inverse variation model: $d = \dfrac{20}{10} = 2$
Joint variation: One variable varies as the product of several variables. For example, $y = kxz$ (k is a constant).	The number of gallons g of oil that can be stored in a cylindrical tank ***varies jointly*** as the height h of the tank and the square of the radius r of its base: $g = khr^2$.

Combined variation: A combination of direct and inverse variation. For example, $y = \frac{kx}{z}$ (k is a constant)	The gravitational force F between two objects with masses m_1 and m_2 ***varies directly*** as the product of their masses and ***inversely*** as the square of the distance d between them: $F = \frac{km_1m_2}{d^2}$

REVIEW EXERCISES

149. Property Tax. The property tax in a certain county varies directly as assessed valuation. If a tax of \$1,575 is charged on a single-family home assessed at \$90,000, determine the property tax on an apartment complex assessed at \$312,000. \$5,460

150. Electricity. For a fixed voltage, the current in an electrical circuit varies inversely as the resistance in the circuit. If a certain circuit has a current of $2\frac{1}{2}$ amps when the resistance is 150 ohms, find the current in the circuit when the resistance is doubled. 1.25 amps

151. Assume that y varies jointly with x and z. Find the constant of variation if $x = 24$ when $y = 3$ and $z = 4$. $\frac{1}{32}$

152. Hurricane Winds. The wind force on a vertical surface varies jointly as the area of the surface and the square of the wind's velocity. If a 10-mph wind exerts a force of 1.98 pounds on the sign shown in the next column, find the force on the sign when the wind is blowing at 80 mph. 126.72 lb

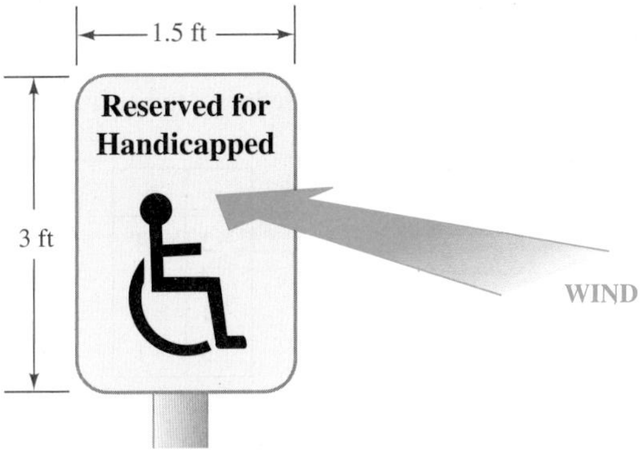

153. Does the graph on the right show direct or inverse variation? Inverse variation

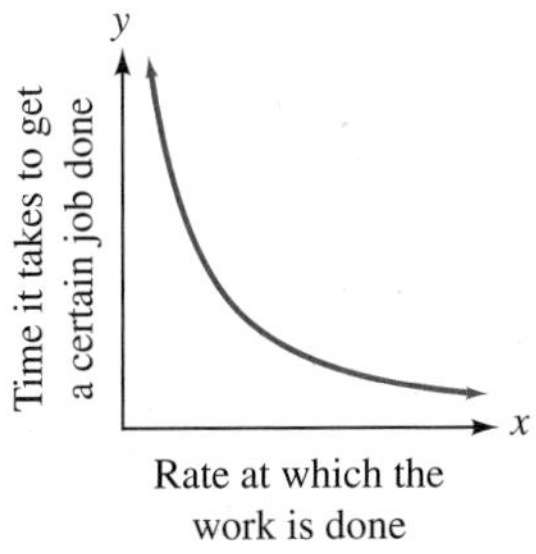

154. Assume that x_1 varies directly with the third power of t and inversely with x_2. Find the constant of variation if $x_1 = 1.6$ when $t = 8$ and $x_2 = 64$. 0.2

Teaching Tip: Because this Chapter Test is a comprehensive collection of problems that covers all of the topics discussed in Chapter 8, it is lengthy. If your students have time to complete it, that would be optimal. If, because of time constraints, they are unable to do so, assign an appropriate subset of problems that reflects the types of problems that the students will see on your exam.

8 Chapter Test

1. Fill in the blanks.

a. To <u>solve</u> an equation means to find all of the values of the variable that make the equation true.

b. $<$, $>$, $\leq$, and $\geq$ are <u>inequality</u> symbols.

c. The statement $x^2 - x - 12 = (x - 4)(x + 3)$ shows that the trinomial $x^2 - x - 12$ <u>factors</u> as the product of two binomials.

d. The <u>reciprocal</u> of $\frac{x+1}{x-7}$ is $\frac{x-7}{x+1}$.

e. Given a relation in x and y, if to each value of x in the domain there corresponds exactly one value of y in the range, y is said to be a <u>function</u> of x.

f. For a function, the set of all possible values that can be used for the independent variable is called the <u>domain</u>. The set of all values of the dependent variable is called the <u>range</u>.

2. Use a check to determine whether 6.7 is a solution of $1.6y + (-3) = y + 1.02$. Yes

Solve each equation.

3. $t + 18 = 5t - 3 + t$ $\quad \frac{21}{5}$

4. $\frac{2}{3}(2s + 2) = \frac{1}{6}(5s + 29) - 4$ $\quad -1$

5. $6 - (x - 3) - 5x = 3[1 - 2(x + 2)]$
No solution, $\varnothing$; contradiction

6. Solve $y - y_1 = m(x - x_1)$ for x_1. $\quad x_1 = \frac{y_1 + mx - y}{m}$

7. Hand Tools. With each pass that a craftsman makes with a sander over a piece of fiberglass, he removes 0.03125 inch of thickness. If the fiberglass was originally 0.9375 inch thick, how many passes are needed to obtain the desired thickness of 0.6875 inch? 8

8. Averaging Grades. Use the information from the gradebook to determine what score Karen Nelson-Sims needs on the fifth exam so that her exam average exceeds 80. More than 78

Sociology 101 8:00-10:00 pm MW	Exam 1	Exam 2	Exam 3	Exam 4	Exam 5
Nelson-Sims, Karen	70	79	85	88	

9. Determine whether the relation defines y to be a function of x. If it does not, explain why.

a.

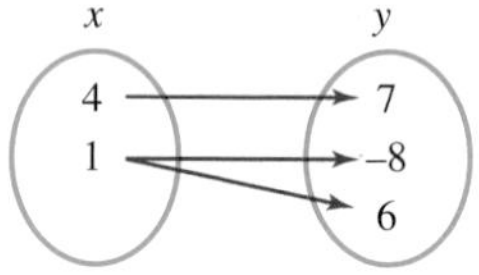

No; (1, −8), (1, 6)

b.

x	y
−3	4
4	−3
1	5
5	1

Yes

c. $\{(0, \frac{1}{2}), (-10, \frac{1}{2}), (-20, \frac{1}{2}), (-30, \frac{1}{2})\}$ Yes

d. $|y| = x$ No; (4, −4), (4, 4)

10. Find the domain of $f(x) = \frac{15}{12 - 2x}$. The set of all real numbers except 6

11. Let $f(x) = -\frac{4}{5}x - 12$. For what value of x is $f(x) = 4$? −20

12. Determine the slope and y-intercept of the graph of $f(x) = 8x - 9$. 8, (0, −9)

13. Write an equation for the linear function whose graph passes through (2, 0) and is perpendicular to the graph of $g(x) = \frac{4}{5}x + \frac{1}{9}$. $f(x) = -\frac{5}{4}x + \frac{5}{2}$

14. Vehicle Performance. The average fleet-wide performance of light-duty trucks has made constant improvement since 1980. For example, the time for the average 1980 model light-duty truck to accelerate from 0 to 60 mph was about 14.5 seconds. For 2005 models, that time was only 10.5 seconds. (Source: United States EPA)

a. Let m be the model year of a light-duty truck and T be the time in seconds for it to accelerate from 0 to 60 mph. Write a linear function $T(m)$ to model the situation. $T(m) = -0.16m + 331.3$

b. Use your answer to part a to predict the time it will take the average light-duty truck to accelerate from 0 to 60 mph in 2015, if the trend continues. 8.9 sec

Let $f(x) = 3x + 1$ and $g(t) = t^2 - 2t + 1$. Find each value.

15. $f(3)$ 10

16. $g(-6)$ 49

17. $g\left(\frac{1}{4}\right)$ $\frac{9}{16}$

18. $f(r + 8)$ $3r + 25$

19. Boating. The height (in feet) of a warning flare from the surface of the ocean t seconds after being shot into the air is approximated by the polynomial function $h(t) = -16t^2 + 80t + 10$. What is the height of the flare 2.5 seconds after being fired? 110 ft

20. Use the graph of function f to find each of the following.

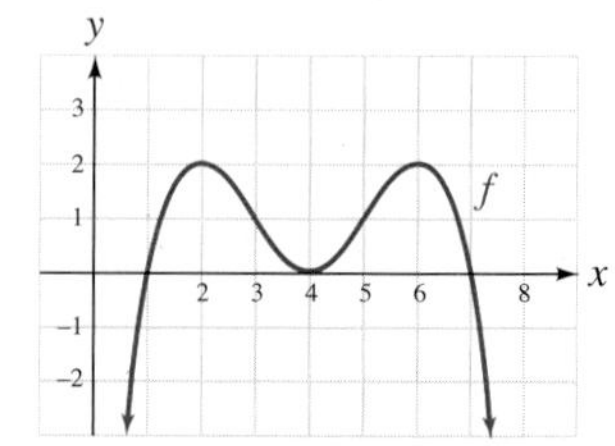

a. $f(4)$ 0

b. The values of x for which $f(x) = 2$. 2, 6

c. The domain and range of f. D: $(-\infty, \infty)$; R: $(-\infty, 2]$

Determine whether each graph is the graph of a function. If it is not, explain why.

21.

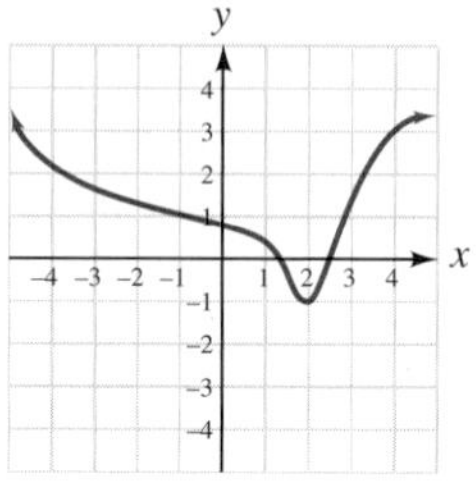

Function

22.

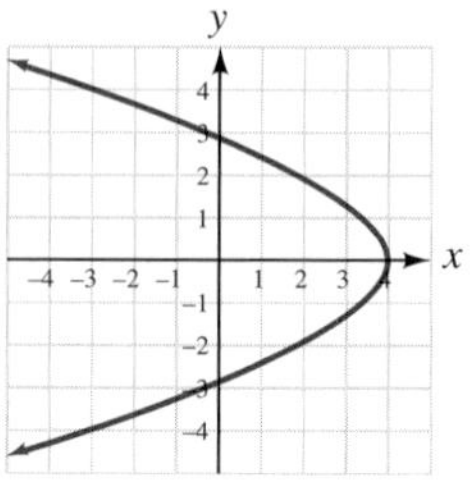

Not a function; (2, 2), (2, −2)

23. Graph $f(x) = |x| + 3$ by creating a table of function values and plotting points. Give the domain and range of the function. D: the set of real numbers; R: the set of real numbers greater than or equal to 3; see AIE Appendix 3.

24. Draw the graph of $g(x) = (x - 4)^3 + 1$ using a translation by first sketching the graph of its associated function. Then give the domain and range of function g. D: the set of real numbers; R: the set of real numbers; see AIE Appendix 3.

Solve each inequality. Write the solution set in interval notation and graph it. See AIE Appendix 3.

25. $-2(2x + 3) \geq 14$ $(-\infty, -5]$

26. $-2 < \frac{x - 4}{3} < 4$ $(-2, 16)$

27. $3x \geq -2x + 5$ and $7 \geq 4x - 2$ $\left[1, \frac{9}{4}\right]$

28. $3x < -9$ or $-\frac{x}{4} < -2$ $(-\infty, -3) \cup (8, \infty)$

29. $|2x - 4| > 22$ $(-\infty, -9) \cup (13, \infty)$

30. $2|3(x - 2)| \leq 4$ $\left[\frac{4}{3}, \frac{8}{3}\right]$

Solve each equation.

31. $|2x + 3| - 19 = 0$ 8, −11

32. $|3x + 4| = |x + 12|$ 4, −4

Factor.

33. $12a^3b^2c - 3a^2b^2c^2 + 6abc^3$ $3abc(4a^2b - abc + 2c^2)$

34. $4y^4 - 64$ $4(y^2 + 4)(y + 2)(y - 2)$

35. $b^3 + 125$ $(b + 5)(b^2 - 5b + 25)$

36. $6u^2 + 9u - 6$ $3(u + 2)(2u - 1)$

37. $ax - xy + ay - y^2$ $(a - y)(x + y)$

38. $25m^8 - 60m^4n + 36n^2$ $(5m^4 - 6n)^2$

39. $144b^2 + 25$ Prime

40. $x^2 + 6x + 9 - y^2$ $(x + 3 + y)(x + 3 - y)$

41. $64a^3 - 125b^6$ $(4a - 5b^2)(16a^2 + 20ab^2 + 25b^4)$

42. $(x - y)^2 + 3(x - y) - 10$ $(x - y + 5)(x - y - 2)$

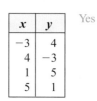

43. Refer to the graph of function f on the right. Find

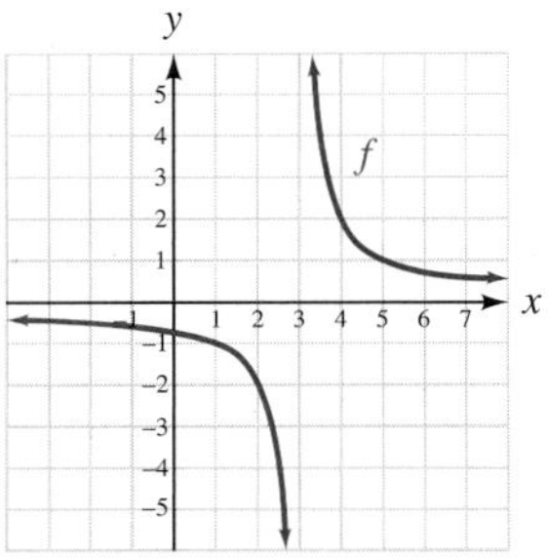

a. $f(5)$ 1

b. $f(-1)$ $-\frac{1}{2}$

c. $f(1)$ -1

d. The value of x for which $f(x) = -2$ 2

44. Wildlife. The number of prairie dogs p living in a given area of land is approximated by the rational function

$$p(t) = \frac{200t}{t + 1}$$

where t is the time, in months, since the first of the year. Find the number of prairie dogs in that area by the end of July. $p(7) = 175$

45. Are the expressions $\frac{x - 4}{x^2 + x - 18}$ and $-\frac{x + 4}{x^2 + x - 18}$ equivalent? No

46. Find the domain of the rational function $f(x) = \frac{x^2 + 6x + 5}{x - x^2}$. Express it using words and interval notation. The set of all real numbers except 0 and 1; $(-\infty, 0) \cup (0, 1) \cup (1, \infty)$

Simplify each rational expression.

47. $\dfrac{3y - 6z}{2z - y}$ -3

48. $\dfrac{2x^2 + 7xy + 3y^2}{4xy + 12y^2}$ $\frac{2x + y}{4y}$

Perform the operations and simplify, if possible.

49. $\dfrac{x^3 + y^3}{4} \div \dfrac{x^2 - xy + y^2}{2x + 2y}$ $\frac{(x + y)^2}{2}$

50. $\dfrac{xu + 2u + 3x + 6}{u^2 - 9} \cdot \dfrac{13u - 39}{x^2 + 3x + 2}$ $\frac{13}{x + 1}$

51. $\dfrac{-3t + 4}{t^2 + t - 20} + \dfrac{5t + 6}{t^2 + t - 20}$ $\frac{2}{t - 4}$

52. $\dfrac{a + 3}{a^2 - a - 2} - \dfrac{a - 4}{a^2 - 2a - 3}$ $\frac{6a - 17}{(a + 1)(a - 2)(a - 3)}$

Solve each equation.

53. $\dfrac{34}{a^2} = \dfrac{3}{2a} - \dfrac{13}{20a}$ 40

54. $\dfrac{u - 2}{u - 3} + 3 = u + \dfrac{u - 4}{3 - u}$ 5; 3 is extraneous.

55. $\dfrac{3}{x - 2} = \dfrac{x + 3}{2x}$ 6, -1

56. $\dfrac{4}{x^2 - 9} = \dfrac{7}{x^2 - 7x + 12} - \dfrac{5}{x^2 - x - 12}$ 26

57. Assume that y varies directly with x. If $x = 30$ when $y = 4$, find y when $x = 9$. $\frac{6}{5}$

58. Sound. Sound intensity (loudness) varies inversely as the square of the distance from the source. If a rock band has a sound intensity of 100 decibels 30 feet away from the amplifier, find the sound intensity 60 feet away from the amplifier. 25 decibels

Group Project

VENN DIAGRAMS

Overview: In this activity, we will discuss several of the fundamental concepts of what is known as *set theory.*

Instructions: *Venn diagrams* are a convenient way to visualize relationships between sets and operations on sets. They were invented by the English mathematician John Venn (1834–1923). To draw a Venn diagram, we begin with a large rectangle, called the *universal set.* Ovals or circles are then drawn in the interior of the rectangle to represent subsets of the universal set.

Form groups of 2 or 3 students. Study the following figures, which illustrate three set operations: union, intersection, and complement.

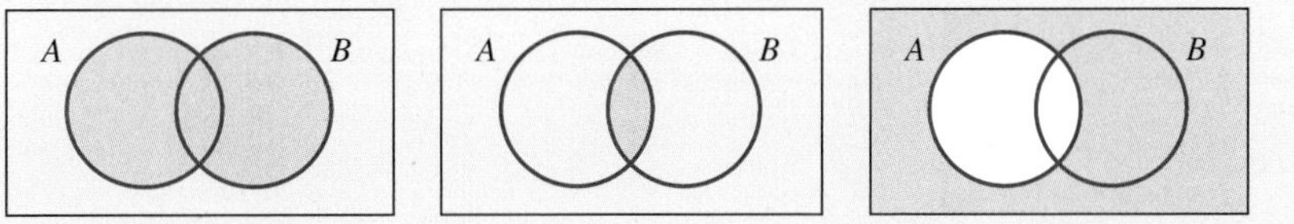

$A \cup B$ The shaded region is the *union* of set A and set B.

$A \cap B$ The shaded region is the *intersection* of set A and set B.

$\overline{A}$ The shaded region is the complement of set A.

For each of the following exercises, sketch the following blank Venn diagram and then shade the indicated region.

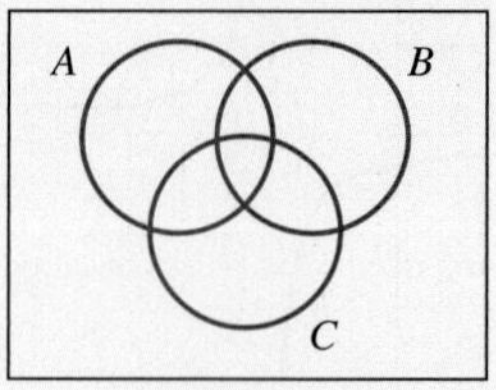

1. $A \cup B$
2. $A \cap B$
3. $A \cap C$
4. $A \cup C$
5. $A \cup B \cup C$
6. $A \cap B \cap C$
7. $(B \cup C) \cap A$
8. $C \cup (A \cap B)$
9. $\overline{A}$
10. $\overline{B} \cup \overline{C}$
11. $\overline{B} \cap \overline{C}$
12. $\overline{A \cup B}$

Radical Expressions and Equations

9

from Campus to Careers

General Contractor

The growing popularity of remodeling has created a boom for general contractors. If it's an additional bedroom you need or a makeover of a dated kitchen or bathroom, they can provide design and construction expertise, as well as knowledge of local building code requirements. From the planning stages of a project through its completion, general contractors use mathematics every step of the way.

Problem 141 in **Study Set 9.2, problem 113** in **Study Set 9.3,** and **problem 109** in **Study Set 9.5** involve situations that a general contractor encounters on the job. The mathematical concepts in this chapter can be used to solve those problems.

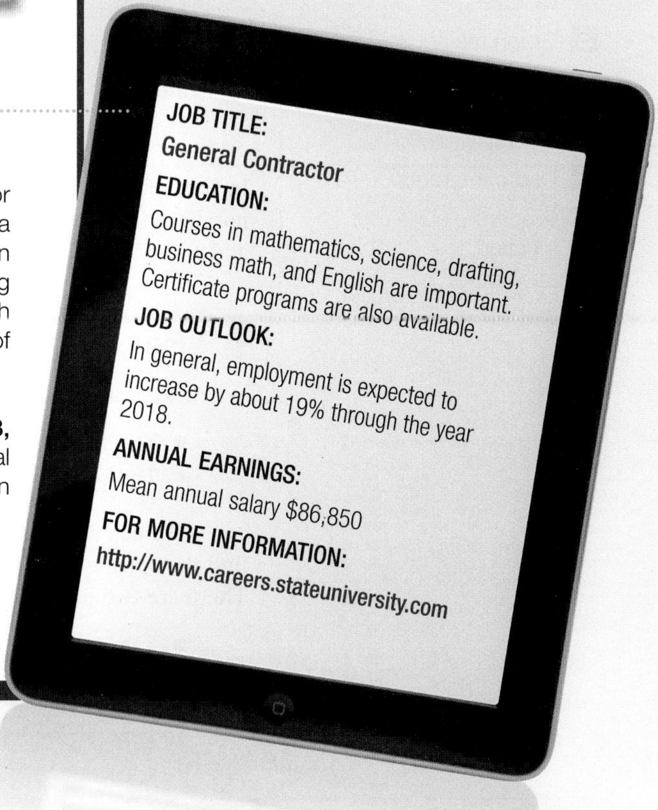

JOB TITLE:
General Contractor
EDUCATION:
Courses in mathematics, science, drafting, business math, and English are important. Certificate programs are also available.
JOB OUTLOOK:
In general, employment is expected to increase by about 19% through the year 2018.
ANNUAL EARNINGS:
Mean annual salary $86,850
FOR MORE INFORMATION:
http://www.careers.stateuniversity.com

Study Skills Workshop ▶ Don't Just Memorize

Many Students attempt to learn algebra by rote memorization. Unfortunately, as the term progresses, they find that does not work. When they encounter problem types slightly different from those that they have memorized, they experience great difficulty. Remember, memorization only provides a superficial grasp of the concepts. When learning a new algebraic procedure, it is most important that you:

UNDERSTAND "WHY": Be able to exaplin the purpose for each stp in the procedure and why they are applied in that order.

UNDERSTAND "WHEN": Be able to explain what types of problems are solved using the procedure and what types are not.

Now Try This ▶

1. Choose five problems in the Guided Practice section of a Study Set to solve. Write a *Strategy* and *Why* statement for each solution in your own words.
2. Select a procedure that is introduced in this chapter and explain when it should be used. Write a *Caution* statement that warns of a possible pitfall when using it. Give an example of an application problem that can be solved using the procedure.

SECTION 9.1

OBJECTIVES

1. Find square roots.
2. Find square roots of expressions containing variables.
3. Graph the square root function.
4. Evaluate radical functions.
5. Find cube roots.
6. Graph the cube root function.
7. Find *n*th roots.

Radical Expressions and Radical Functions

ARE YOU READY? *Are You Ready? exercises available online at www.webassign.net/brookscole*

The following problems review some basic skills that are needed when working with radical expressions and radical functions.

1. Evaluate: **a.** 15^2 225 **b.** 4^3 64

2. Evaluate: **a.** $\left(\frac{7}{5}\right)^2$ $\frac{49}{25}$ **b.** $(0.2)^2$ 0.04

3. Evaluate: **a.** $(-6)^3$ -216 **b.** 3^4 81

4. Simplify: **a.** $a^4 \cdot a^4$ a^8 **b.** $x^3 \cdot x^3 \cdot x^3$ x^9

5. Multiply: $(x + 8)(x + 8)$ $x^2 + 16x + 64$

6. Let $f(x) = |2x - 3|$. Find $f(4)$. 5

In this section, we will reverse the squaring process and learn how to find *square roots* of numbers. Then we will generalize the concept of root and consider cube roots, fourth roots, and so on. We will also discuss a new family of functions, called *radical functions.*

1 Find Square Roots.

When we raise a number to the second power, we are squaring it, or finding its **square.**

- The square of 5 is 25 because $5^2 = 25$.
- The square of -5 is 25, because $(-5)^2 = 25$.

We can reverse the squaring process to find **square roots** of numbers. For example, to find the square roots of 25, we ask ourselves "What number, when squared, is equal to 25?" There are two possible answers.

- 5 is a square root of 25, because $5^2 = 25$.
- -5 is also a square root of 25, because $(-5)^2 = 25$.

In general, we have the following definition.

Square Root of *a* — The number b is a **square root** of the number a if $b^2 = a$.

The Language of Algebra

Precise vocabulary must be used when discussing square roots. For example, 3 is **a** square root of 9 and -3 is also **a** square root of 9. However, when we speak of **the** square root of a number, we mean only the *nonnegative* square root of that number. Thus, **the** square root of 9 is 3, which is written as $\sqrt{9} = 3$.

Teaching Tip: Stress that it is incorrect to say that $\sqrt{9} = 3$ or -3.

Every positive number has two square roots, one positive and one negative. For example, the two square roots of 9 are 3 and -3, and the two square roots of 144 are 12 and -12. The number 0 is the only real number with exactly one square root. In fact, it is its own square root, because $0^2 = 0$.

A **radical symbol** $\sqrt{}$ represents the **positive** or **principal square root** of a number. Since 3 is the positive square root of 9, we can write

$\sqrt{9} = 3$ Read as "the square root of 9."

The symbol $-\sqrt{}$ represents the **negative square root** of a number. It is the opposite of the principal square root. Since -12 is the negative square root of 144, we can write

$-\sqrt{144} = -12$ Read as "the negative square root of 144 is -12" or "the opposite of the square root of 144 is -12."

Square Root Notation

If a is a positive real number,

1. $\sqrt{a}$ represents the **positive** or **principal square root** of a. It is the positive number we square to get a.
2. $-\sqrt{a}$ represents the **negative square root** of a. It is the opposite of the principal square root of a: $-\sqrt{a} = -1 \cdot \sqrt{a}$.
3. The principal square root of 0 is 0: $\sqrt{0} = 0$.

The Language of Algebra

The word **radical** comes from the Latin word *radix,* meaning root. The radical symbol has evolved over the years from ℞ in the 1300s, to √ in the 1500s, to the familiar $\sqrt{}$ in the 1600s.

The number or variable expression under a radical symbol is called the **radicand.** Together, the radical symbol and radicand are called a **radical.** An algebraic expression containing a radical is called a **radical expression.**

Radical symbol ↘ $\sqrt{81}$ ← Radicand

Radical

Read $\sqrt{81}$ as "the square root of 81" or as "radical 81."

Some more examples of radical expressions are:

$$\sqrt{100}, \quad \sqrt{2} + 3, \quad \sqrt{x^2}, \quad \text{and} \quad \sqrt{\frac{a-1}{49b^2}}$$

To evaluate (find the value of) square root radical expressions, you need to quickly recognize each of the natural-number **perfect squares** shown below in red.

1 $= 1^2$	**25** $= 5^2$	**81** $= 9^2$	**169** $= 13^2$	**289** $= 17^2$
4 $= 2^2$	**36** $= 6^2$	**100** $= 10^2$	**196** $= 14^2$	**324** $= 18^2$
9 $= 3^2$	**49** $= 7^2$	**121** $= 11^2$	**225** $= 15^2$	**361** $= 19^2$
16 $= 4^2$	**64** $= 8^2$	**144** $= 12^2$	**256** $= 16^2$	**400** $= 20^2$

EXAMPLE 1 Evaluate: **a.** $\sqrt{81}$ **b.** $-\sqrt{225}$ **c.** $\sqrt{\frac{49}{4}}$ **d.** $\sqrt{0.36}$

Strategy In each case, we will determine what positive number, when squared, produces the radicand.

Why The symbol $\sqrt{}$ indicates that the positive square root of the number written under it should be found.

Solution **a.** $\sqrt{81} = 9$ Because $9^2 = 81$ **b.** $-\sqrt{225} = -15$ Because $-\sqrt{225} = -1 \cdot \sqrt{225}$

c. $\sqrt{\frac{49}{4}} = \frac{7}{2}$ Because $\left(\frac{7}{2}\right)^2 = \frac{49}{4}$ **d.** $\sqrt{0.36} = 0.6$ Because $(0.6)^2 = 0.36$

Teaching Example 1 Evaluate each square root:
a. $\sqrt{121}$ b. $-\sqrt{25}$
c. $\sqrt{\frac{169}{225}}$ d. $-\sqrt{0.36}$
Answers:
a. 11 b. −5 c. $\frac{13}{15}$ d. −0.6

Self Check 1 Evaluate: **a.** $\sqrt{64}$ 8 **b.** $-\sqrt{1}$ −1 **c.** $\sqrt{\frac{1}{16}}$ $\frac{1}{4}$

d. $\sqrt{0.09}$ 0.3

Now Try Problems 23, 27, and 29

A number, such as 81, 225, $\frac{49}{4}$, and 0.36, that is the square of some rational number, is called a **perfect square.** In Example 1, we saw that the square root of a perfect square is a rational number.

If a positive number is not a perfect square, its square root is irrational. For example, $\sqrt{83}$ is an irrational number because 83 is not a perfect square. Since $\sqrt{83}$ is irrational, its decimal representation is nonterminating and nonrepeating. We can find a rational-number approximation of $\sqrt{83}$ using the square root key $\boxed{\sqrt{\ }}$ on a calculator or from the table of square roots found in Appendix I at the back of the book.

$$\sqrt{83} \approx 9.110433579 \quad \text{or} \quad \sqrt{83} \approx 9.11 \quad \text{Round to the nearest hundredth.}$$

Success Tip

Estimation is helpful when approximating square roots. For example, $\sqrt{83}$ must be a number between 9 and 10, because $\sqrt{81} < \sqrt{83} < \sqrt{100}$.

Teaching Tip: Use a calculator to show that $(9.11)^2 = 82.9921$. Then ask your students why the result is not exactly 83.

CAUTION Square roots of negative numbers are not real numbers. For example, $\sqrt{-9}$ is not a real number, because no real number squared equals −9. Square roots of negative numbers come from a set called the **imaginary numbers,** which we will discuss later in this chapter. If we attempt to evaluate $\sqrt{-9}$ using a calculator, we will get an error message.

Caution

Although they look similar, these radical expressions have very different meanings.

$$-\sqrt{9} = -3$$

$\sqrt{-9}$ is not a real number.

Scientific calculator

Graphing calculator

We summarize three important facts about square roots as follows.

Square Roots

1. If a is a perfect square, then $\sqrt{a}$ is rational.
2. If a is a positive number that is not a perfect square, then $\sqrt{a}$ is irrational.
3. If a is a negative number, then $\sqrt{a}$ is not a real number.

2 Find Square Roots of Expressions Containing Variables.

If $x \neq 0$, the positive number x^2 has x and $-x$ for its two square roots. To denote the positive square root of $\sqrt{x^2}$, we must know whether x is positive or negative.

If x is positive, we can write

$$\sqrt{x^2} = x \quad \sqrt{x^2} \text{ represents the positive square root of } x^2\text{, which is } x.$$

If x is negative, then $-x$ is positive and we can write

$$\sqrt{x^2} = -x \quad \sqrt{x^2} \text{ represents the positive square root of } x^2\text{, which is } -x.$$

Teaching Tip: Some students struggle with this definition for negative values of x. Here is an example to help explain:

$$\sqrt{(-6)^2} = \sqrt{36} = 6$$

↑ x ↑ opposite of x

If we don't know whether x is positive or negative, we can use absolute value symbols to ensure that $\sqrt{x^2}$ is not negative.

Simplifying $\sqrt{x^2}$

For any real number x,

$$\sqrt{x^2} = |x| \quad \text{The principal square root of x is equal to the absolute value of x.}$$

We use this definition to *simplify* square root radical expressions.

EXAMPLE 2 Simplify. Assume all variables are unrestricted. **a.** $\sqrt{16x^2}$ **b.** $\sqrt{t^2 + 2t + 1}$ **c.** $\sqrt{m^6}$ **d.** $\sqrt{49r^8}$

Strategy In each case, we will determine what positive expression, when squared, produces the radicand.

Why The symbol $\sqrt{}$ indicates that the positive square root of the expression written under it should be found.

Solution If x, t, m, and r can be any real number, we have

a. $\sqrt{16x^2} = |4x|$ Because $(4x)^2 = 16x^2$. Since 4x could be negative (for example, when $x = -5$, the expression 4x is -20), absolute value symbols are needed to ensure that the result is not negative.

$= 4|x|$ Since 4 is a positive constant in the product 4x, we can write it outside the absolute value symbols.

b. $\sqrt{t^2 + 2t + 1} = \sqrt{(t+1)^2}$ Factor the radicand: $t^2 + 2t + 1 = (t + 1)^2$.

$= |t + 1|$ Since $t + 1$ can be negative (for example, when $t = -5$, the expression $t + 1$ is -4), absolute value symbols are needed to ensure that the result is not negative.

c. $\sqrt{m^6} = |m^3|$ Because $(m^3)^2 = m^6$. Since m^3 can be negative (for example, when $m = -3$, the expression m^3 is -27), absolute value symbols are needed to ensure that the result is not negative.

d. $\sqrt{49r^8} = 7r^4$ Because $(7r^4)^2 = 49r^8$. Since r^4 is never negative for any value of r, no absolute value symbols are needed.

The Language of Algebra

The phrase **"assume all variables are unrestricted"** in the instructions for Example 2 reminds us that negative numbers are permissible replacement values for the variables. Therefore, absolute value symbols may be necessary to ensure the answer is not negative.

Self Check 2 Simplify. Assume all variables are unrestricted. **a.** $\sqrt{25a^2}$ $5|a|$ **b.** $\sqrt{b^{14}}$ $|b^7|$ **c.** $\sqrt{x^2 - 18x + 81}$ $|x - 9|$ **d.** $\sqrt{100n^8}$ $10n^4$

Now Try Problems 39, 41, 45, and 47

Teaching Example 2 Simplify. Assume all variables are unrestricted.
a. $\sqrt{81x^2}$ b. $\sqrt{36a^4}$ c. $\sqrt{x^2 + 14x + 49}$ d. $\sqrt{400y^{10}}$
Answers:
a. $9|x|$ b. $6a^2$ c. $|x + 7|$ d. $20|y^5|$

If we know that x, t, and m are positive in Example 2, we don't need to use absolute value symbols in the answers. For example,

$\sqrt{16x^2} = 4x$ If x is positive, 4x is positive.

$\sqrt{m^6} = m^3$ If m is positive, m^3 is positive.

$\sqrt{t^2 + 2t + 1} = t + 1$ If t is positive, $t + 1$ is positive.

The Language of Algebra

If we are told **"All variables represent positive real numbers,"** then absolute value symbols are not needed when simplifying square root radical expressions.

3 Graph the Square Root Function.

Since there is one principal square root for every nonnegative real number x, the equation $f(x) = \sqrt{x}$ determines a function, called a **square root function.** Square root functions belong to a larger family of functions known as **radical functions.**

EXAMPLE 3 Graph $f(x) = \sqrt{x}$ and find the domain and range of the function.

Strategy We will graph the function by creating a table of function values and plotting the corresponding ordered pairs.

Why After drawing a smooth curve through the plotted points, we will have the graph.

Solution To graph the function, we select several values for x that are perfect squares, such as 0, 1, 4, and 9, and find the corresponding values of $f(x)$. We begin with $x = 0$, since 0 is the smallest input for which $\sqrt{x}$ is defined.

$$f(x) = \sqrt{x} \quad f(0) = \sqrt{0} = 0 \qquad f(x) = \sqrt{x} \quad f(1) = \sqrt{1} = 1 \qquad f(x) = \sqrt{x} \quad f(4) = \sqrt{4} = 2 \qquad f(x) = \sqrt{x} \quad f(9) = \sqrt{9} = 3$$

We also can select values of x that are not perfect squares when creating a table of function values. For example, if $x = 6$, then $f(6) = \sqrt{6}$, and it follows that the point $(2, \sqrt{6})$ is on the graph. To help locate this point on the coordinate system, it is helpful to approximate: $\sqrt{6} \approx 2.45$.

We enter each value of x and its corresponding value of $f(x)$ in the table below. After plotting the ordered pairs, we draw a smooth curve through the points to get the graph shown in figure (a). Since the equation defines a function, its graph passes the vertical line test.

To find the domain of the function graphically, we project the graph onto the x-axis, as shown in figure (b). Because the graph begins at (0, 0) and extends indefinitely to the right, the projection includes 0 and all positive real numbers. Thus, the domain of $f(x) = \sqrt{x}$ is the set of nonnegative real numbers, which can be written in interval notation as $[0, \infty)$.

To find the range of the function graphically, we project the graph onto the y-axis, as shown in figure (b). Because the graph of the function begins at (0, 0) and extends indefinitely upward, the projection includes all nonnegative real numbers. Therefore, the range of the function is $[0, \infty)$.

The Language of Algebra

Together, 0 and the positive real numbers are called the **nonnegative real numbers.**

Teaching Tip: Two questions to ask the class are:

- Why wasn't $x = -1$ selected when creating the table of function values?
- If we want to include another entry in the table, what is the next natural-number perfect square to select for x?

$f(x) = \sqrt{x}$

x	$f(x)$	
0	0	→ (0, 0)
1	1	→ (1, 1)
4	2	→ (4, 2)
6	$\sqrt{6}$	→ $(6, \sqrt{6})$
9	3	→ (9, 3)
16	4	→ (16, 4)

↑ Select x ↑ Find $f(x)$. ↑ Plot the point.

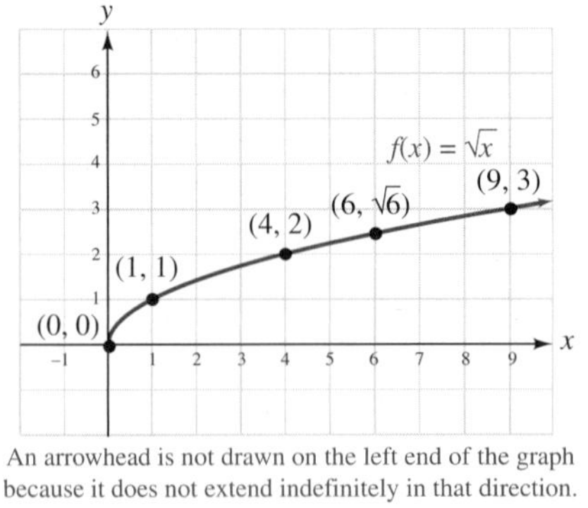

An arrowhead is not drawn on the left end of the graph because it does not extend indefinitely in that direction.

(a)

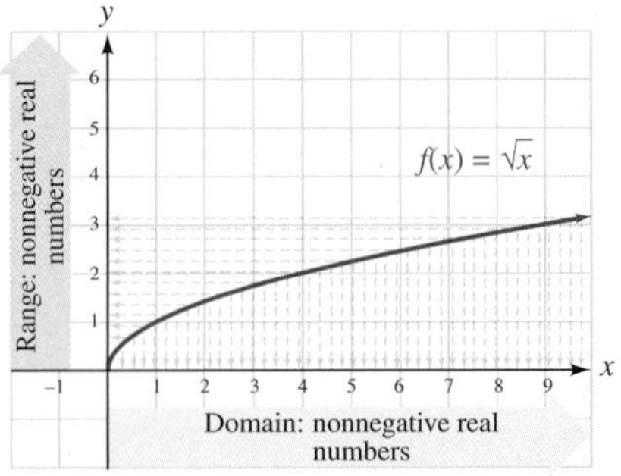

(b)

Teaching Example 3 Graph $f(x) = \sqrt{x} - 3$ and find its domain and range.

Answer:

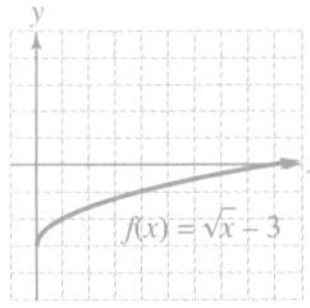

D: $[0, \infty)$, R: $[-3, \infty)$

Self Check 3 Graph: $g(x) = \sqrt{x} + 2$. Then give its domain and range and compare it with the graph of $f(x) = \sqrt{x}$. D: $[0, \infty)$; R: $[2, \infty)$; the graph is 2 units higher; see AIE Appendix 3.

Now Try Problem 51

EXAMPLE 4 Let $g(x) = \sqrt{x + 3}$. **a.** Find its domain algebraically. **b.** Graph the function. **c.** Find its range.

Strategy We will determine the domain algebraically by finding all the values of x for which $x + 3 \geq 0$.

Why Since the expression $\sqrt{x + 3}$ is not a real number when $x + 3$ is negative, we must require that $x + 3 \geq 0$.

Solution

Success Tip

We solved linear inequalities in one variable, such as $x + 3 \geq 0$, in Section 2.7.

a. To determine the domain of the function, we solve the following inequality:

$x + 3 \geq 0$ Because we cannot find the square root of a negative number

$x \geq -3$ To solve for x, subtract 3 from both sides.

The x-inputs must be real numbers greater than or equal to -3. Thus, the domain of $g(x) = \sqrt{x + 3}$ is the interval $[-3, \infty)$.

b. To graph the function, we construct a table of function values. We begin by selecting $x = -3$, since -3 is the smallest input for which $\sqrt{x + 3}$ is defined. Then we let $x = -2$, 1, and 6 and list each corresponding function value $g(x)$ in the table.

$$g(x) = \sqrt{x + 3} \quad g(-3) = \sqrt{-3 + 3} = \sqrt{0} = 0$$

$$g(x) = \sqrt{x + 3} \quad g(-2) = \sqrt{-2 + 3} = \sqrt{1} = 1$$

$$g(x) = \sqrt{x + 3} \quad g(1) = \sqrt{1 + 3} = \sqrt{4} = 2$$

$$g(x) = \sqrt{x + 3} \quad g(6) = \sqrt{6 + 3} = \sqrt{9} = 3$$

After plotting the ordered pairs, we draw a smooth curve through the points to get the graph shown in figure (a). In figure (b), we see that the graph of $g(x) = \sqrt{x + 3}$ is the graph of $f(x) = \sqrt{x}$, translated 3 units to the left.

$g(x) = \sqrt{x + 3}$

x	$g(x)$	
-3	0	→ $(-3, 0)$
-2	1	→ $(-2, 1)$
1	2	→ $(1, 2)$
6	3	→ $(6, 3)$

↑ Select values of x that make x + 3 a perfect square.

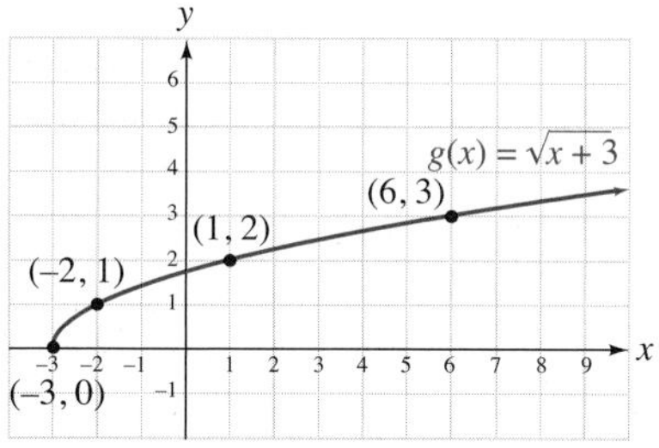

(a)

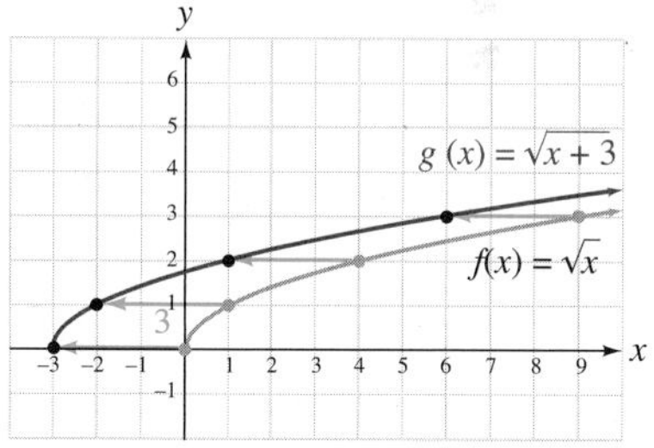

(b)

c. From the graph in figure (a), we see that the range of $g(x) = \sqrt{x + 3}$ is $[0, \infty)$.

Self Check 4 Let $h(x) = \sqrt{x - 2}$. **a.** Find its domain algebraically. $[2, \infty)$ **b.** Graph the function. See AIE Appendix 3. **c.** Find its range. $[0, \infty)$

Now Try Problems 53 and 55

Teaching Example 4 Let $h(x) = \sqrt{x - 1}$.
a. Find its domain algebraically.
b. Graph the function.
c. Find its range.
Answers:
a. $[1, \infty)$
b.

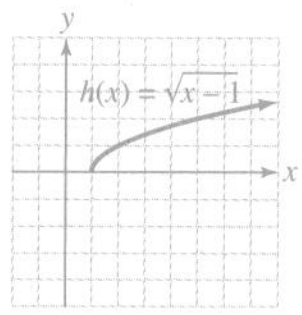

c. $[0, \infty)$

4 Evaluate Radical Functions.

Radical functions can be used to model certain real-life situations that exhibit growth that eventually levels off. A calculator is often helpful when evaluating a radical function for a given value of the input variable.

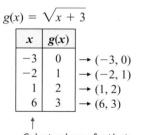

EXAMPLE 5 **Pendulums.** The **period of a pendulum** is the time required for the pendulum to swing back and forth to complete one cycle. The period (in seconds) is a function of the pendulum's length L (in feet) and is given by $f(L) = 2\pi\sqrt{\frac{L}{32}}$. Find the period of the 5-foot-long pendulum of a clock. Round the result to the nearest tenth.

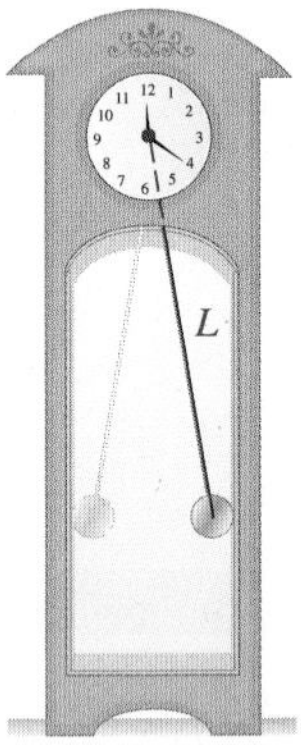

Strategy To find the period of the pendulum we will find $f(5)$.

Why The notation $f(5)$ represents the period (in seconds) of a pendulum whose length L is 5 feet.

Solution

$$f(L) = 2\pi\sqrt{\frac{L}{32}}$$

$$f(5) = 2\pi\sqrt{\frac{5}{32}}$$ Substitute 5 for L.

$$\approx 2.483647066$$ Evaluate the radical expression. Use a calculator to find an approximation. $2\pi\sqrt{\frac{5}{32}}$ means $2 \cdot \pi \cdot \sqrt{\frac{5}{32}}$.

The period is approximately 2.5 seconds.

Teaching Example 5 Pendulums. Find the period of a pendulum that is 4 feet long. Round to the nearest tenth. *Answer:* 2.2 sec

Self Check 5 **Pendulums.** Find the period of a pendulum that is 3 feet long. Round the result to the nearest hundredth. 1.92 sec

Now Try Problem 123

Using Your Calculator ▶ Graphing and Evaluating a Square Root Function

We can graph radical functions on a graphing calculator. To graph $f(x) = \sqrt{x}$ from Example 3, we press Y = and enter the right side of the equation. If the window settings for x and y are $[-1, 9]$, we get the graph shown in figure (a) when we press GRAPH. The function $g(x) = \sqrt{x + 3}$ from Example 4 is graphed in a similar way with window settings of $[-4, 10]$ for x and $[-1, 9]$ for y. See figure (b).

To answer Example 5 with a graphing calculator, we graph the function $f(x) = 2\pi\sqrt{\frac{x}{32}}$. We then trace and move the cursor toward an x-value of 5 until we see the coordinates shown in figure (c). The pendulum's period is given by the y-value shown on the screen. By zooming in, we can get better results.

After entering $Y_1 = 2\pi\sqrt{\frac{x}{32}}$, we can also use the TABLE mode to find $f(5)$. See figure (d).

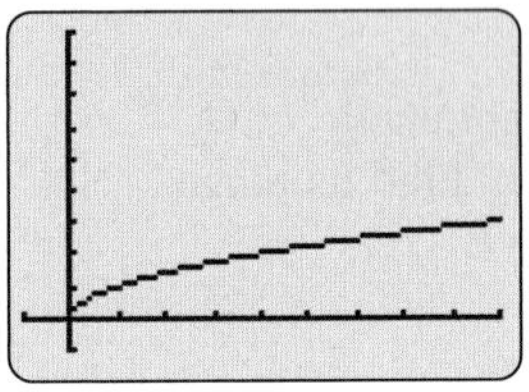
(a)

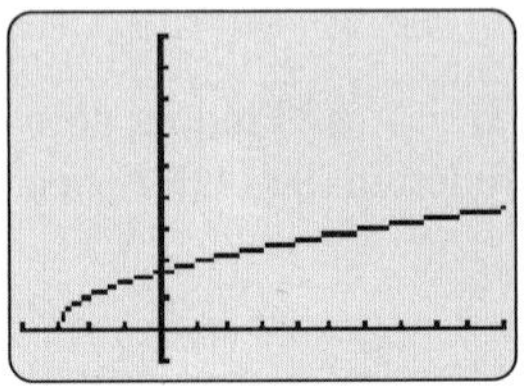
(b)

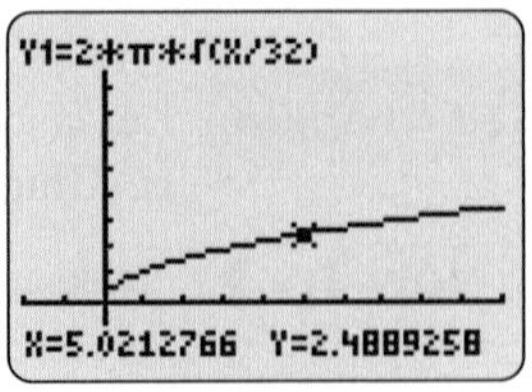

(c)

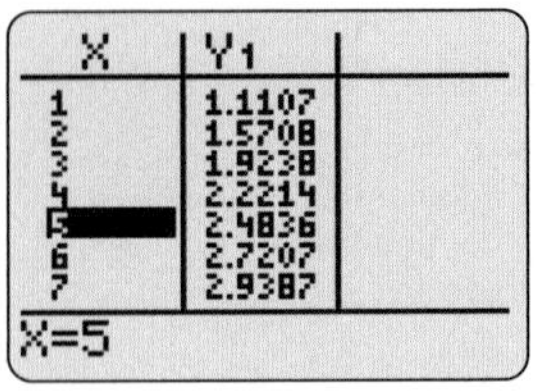

(d)

5 Find Cube Roots.

When we raise a number to the third power, we are cubing it, or finding its **cube.** We can reverse the cubing process to find **cube roots** of numbers. To find the cube root of 8, we ask "What number, when cubed, is equal to 8?" It follows that 2 is a cube root of 8, because $2^3 = 8$.

In general, we have this definition.

Cube Root of a

The number b is a **cube root** of the real number a if $b^3 = a$.

All real numbers have one real cube root. A positive number has a positive cube root, a negative number has a negative cube root, and the cube root of 0 is 0.

Cube Root Notation

The **cube root of a** is written as $\sqrt[3]{a}$. By definition,

$$\sqrt[3]{a} = b \quad \text{if} \quad b^3 = a$$

Notation

For the square root symbol $\sqrt{\ }$, the unwritten index is understood to be 2.

$$\sqrt{a} = \sqrt[2]{a}$$

Earlier, we determined that the cube root of 8 is 2. In symbols, we can write: $\sqrt[3]{8} = 2$. The number 3 is called the **index,** 8 is called the **radicand,** and the entire expression is called a **radical.**

Index
↓
$\sqrt[3]{8}$ ← Radicand   *Read as "the cube root of 8."*
Radical

A number, such as 125, $\frac{1}{64}$, -27, and -8, that is the cube of some rational number is called a **perfect cube.**

To simplify cube root radical expressions, you need to quickly recognize each of the following natural-number perfect cubes shown in blue.

$1 = 1^3$	$27 = 3^3$	$125 = 5^3$	$343 = 7^3$	$729 = 9^3$
$8 = 2^3$	$64 = 4^3$	$216 = 6^3$	$512 = 8^3$	$1{,}000 = 10^3$

The following property is also used to simplify cube root radical expressions.

Simplifying $\sqrt[3]{x^3}$

For any real number x,

$$\sqrt[3]{x^3} = x$$

EXAMPLE 6

Simplify: **a.** $\sqrt[3]{125}$ **b.** $\sqrt[3]{-\frac{1}{64}}$ **c.** $\sqrt[3]{27x^3}$ **d.** $\sqrt[3]{-8a^9b^6}$

Strategy In each case, we will determine what number or expression, when cubed, produces the radicand.

Why The symbol $\sqrt[3]{\ }$ indicates that the cube root of the number written under it should be found.

Solution

a. $\sqrt[3]{125} = 5$ *Because $5^3 = 5 \cdot 5 \cdot 5 = 125$*

b. $\sqrt[3]{-\frac{1}{64}} = -\frac{1}{4}$ *Because $\left(-\frac{1}{4}\right)^3 = \left(-\frac{1}{4}\right)\left(-\frac{1}{4}\right)\left(-\frac{1}{4}\right) = -\frac{1}{64}$*

c. $\sqrt[3]{27x^3} = 3x$ *Because $(3x)^3 = (3x)(3x)(3x) = 27x^3$. No absolute value symbols are needed.*

d. $\sqrt[3]{-8a^9b^6} = -2a^3b^2$ *Because $(-2a^3b^2)^3 = (-2a^3b^2)(-2a^3b^2)(-2a^3b^2) = -8a^9b^6$.*

Success Tip

Since every real number has exactly one real cube root, absolute value symbols are not to be used when simplifying cube roots.

Teaching Example 6 Simplify:

a. $\sqrt[3]{64}$ b. $\sqrt[3]{\frac{8}{27}}$ c. $\sqrt[3]{-216a^3}$ d. $\sqrt[3]{125m^6n^9}$

Answers:

a. 4 b. $\frac{2}{3}$ c. $-6a$ d. $5m^2n^3$

Self Check 6 Simplify: **a.** $\sqrt[3]{64}$ 4 **b.** $\sqrt[3]{-\frac{1}{1{,}000}}$ $-\frac{1}{10}$ **c.** $\sqrt[3]{-125a^3}$ $-5a$ **d.** $\sqrt[3]{27m^6n^9}$ $3m^2n^3$

Now Try Problems 73, 75, and 81

6 Graph the Cube Root Function.

Since there is one cube root for every real number x, the equation $f(x) = \sqrt[3]{x}$ defines a function, called the **cube root function.** Like square root functions, cube root functions belong to the family of radical functions.

EXAMPLE 7 Consider $f(x) = \sqrt[3]{x}$. **a.** Graph the function. **b.** Find its domain and range. **c.** Graph: $g(x) = \sqrt[3]{x} - 2$

Teaching Tip: In part b, compare square roots to cube roots. Point out that the radicand of a cube root can be negative and the result can be negative as well.

Strategy We will graph the function by creating a table of function values and plotting the corresponding ordered pairs.

Why After drawing a smooth curve through the plotted points, we will have the graph. The answers to parts (b) and (c) can then be determined from the graph.

Solution

Teaching Tip: Two questions to ask the class are:
- If we want to include another entry in the table, what is the next positive perfect cube to select for x?
- Why would that number make the graphing process more cumbersome?

a. To graph the function, we select several values for x, that are perfect cubes, such as -8, -1, 0, 1, and 8, and find the corresponding values of $f(x)$. The results are entered in the table below.

$f(x) = \sqrt[3]{x}$	$f(x) = \sqrt[3]{x}$	$f(x) = \sqrt[3]{x}$	$f(x) = \sqrt[3]{x}$	$f(x) = \sqrt[3]{x}$
$f(-8) = \sqrt[3]{-8}$	$f(-1) = \sqrt[3]{-1}$	$f(0) = \sqrt[3]{0}$	$f(1) = \sqrt[3]{1}$	$f(8) = \sqrt[3]{8}$
$= -2$	$= -1$	$= 0$	$= 1$	$= 2$

After plotting the ordered pairs, we draw a smooth curve through the points to get the graph shown in figure (a).

$f(x) = \sqrt[3]{x}$

x	$f(x)$	
-8	-2	→ $(-8, -2)$
-1	-1	→ $(-1, -1)$
0	0	→ $(0, 0)$
1	1	→ $(1, 1)$
8	2	→ $(8, 2)$

↑ Select values of x that are perfect cubes.

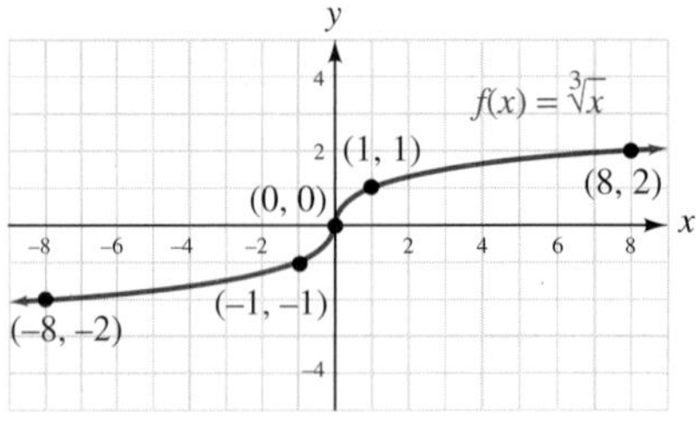

(a)

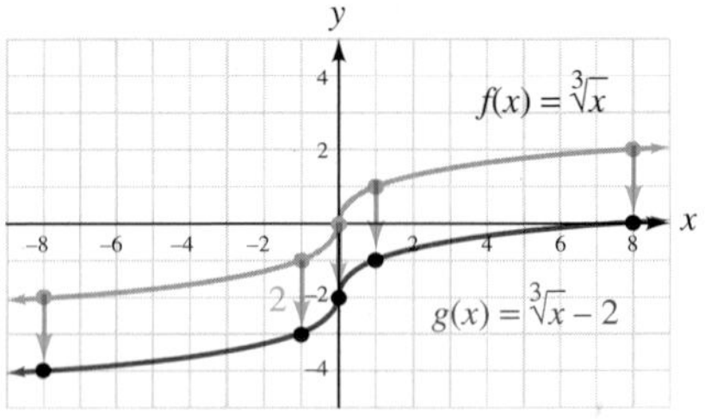

(b)

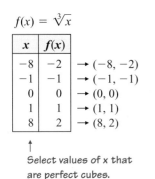

b. If we project the graph in figure (a) onto the x- and y-axes, we see that the domain and the range of the function are the set of real numbers. Thus, the domain is $(-\infty, \infty)$ and the range is $(-\infty, \infty)$.

c. Refer to figure (b). The graph of $g(x) = \sqrt[3]{x} - 2$ is the graph of $f(x) = \sqrt[3]{x}$, translated 2 units downward.

Teaching Example 7 Consider $f(x) = \sqrt[3]{x} + 3$.
a. Graph the function.
b. Find its domain and range.
Answers:
a.

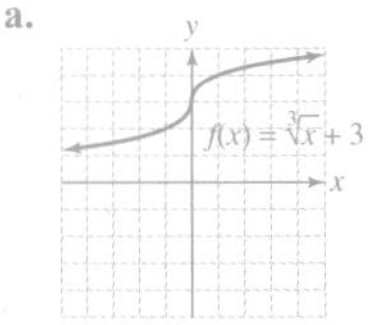

b. D: $(-\infty, \infty)$, R: $(-\infty, \infty)$

Self Check 7 Consider $f(x) = \sqrt[3]{x} + 1$. **a.** Graph the function. **b.** Find its domain and range. D: $(-\infty, \infty)$; R: $(-\infty, \infty)$; see AIE Appendix 3.

Now Try Problem 83

7 Find *n*th Roots.

Just as there are square roots and cube roots, there are fourth roots, fifth roots, sixth roots, and so on. In general, we have the following definition.

nth Roots of a

The ***n*th root of *a*** is written as $\sqrt[n]{a}$, and

$$\sqrt[n]{a} = b \quad \text{if} \quad b^n = a$$

The number n is called the **index** (or **order**) of the radical. If n is an even natural number, a must be positive or zero, and b must be positive.

Success Tip

When simplifying radical expressions, it is helpful to recognize the following perfect fourth and fifth powers:

$16 = 2^4$	$32 = 2^5$
$81 = 3^4$	$243 = 3^5$
$256 = 4^4$	$1{,}024 = 4^5$
$625 = 5^4$	

Teaching Tip: If $x > 0$, explain why absolute value symbols are not needed when simplifying these two even roots.

When n is an odd natural number, the expression $\sqrt[n]{x}$, where $n > 1$, represents an **odd root.** Since every real number has just one real *n*th root when n is odd, we don't need absolute value symbols when finding odd roots. For example,

$\sqrt[5]{243} = 3$ Because $3^5 = 243$

$\sqrt[5]{-32x^5} = -2x$ Because $(-2x)^5 = -32x^5$

When n is an even natural number, the expression $\sqrt[n]{x}$, where $x > 0$, represents an **even root.** In this case, there will be one positive and one negative real *n*th root. For example, the real sixth roots of 729 are 3 and -3, because $3^6 = 729$ and $(-3)^6 = 729$. When finding even roots, we can use absolute value symbols to guarantee that the *n*th root is positive.

$\sqrt[4]{(-3)^4} = |-3| = 3$ We also could simplify this as follows: $\sqrt[4]{(-3)^4} = \sqrt[4]{81} = 3$.

$\sqrt[6]{729x^6} = |3x| = 3|x|$ Because $(3x)^6 = 729x^6$. The absolute value symbols ensure that the result is positive.

In general, we have the following rules.

Simplifying $\sqrt[n]{x^n}$

If x is a real number and $n > 1$, then

If n is an odd natural number, $\sqrt[n]{x^n} = x$.

If n is an even natural number, $\sqrt[n]{x^n} = |x|$.

EXAMPLE 8 Evaluate: **a.** $\sqrt[4]{625}$ **b.** $\sqrt[4]{-1}$ **c.** $\sqrt[5]{-32}$ **d.** $\sqrt[6]{\frac{1}{64}}$

Strategy In each case, we will determine what number, when raised to the fourth, fifth, or sixth power, produces the radicand.

Why The symbols $\sqrt[4]{}$, $\sqrt[5]{}$, and $\sqrt[6]{}$ indicate that the fourth, fifth, or sixth root of the number written under it should be found.

Solution

a. $\sqrt[4]{625} = 5$, because $5^4 = 625$. Read $\sqrt[4]{625}$ as "the fourth root of 625."

b. $\sqrt[4]{-1}$ is not a real number. An even root of a negative number is not a real number.

c. $\sqrt[5]{-32} = -2$, because $(-2)^5 = -32$. Read $\sqrt[5]{-32}$ as "the fifth root of -32."

d. $\sqrt[6]{\frac{1}{64}} = \frac{1}{2}$, because $\left(\frac{1}{2}\right)^6 = \frac{1}{64}$. Read $\sqrt[6]{\frac{1}{64}}$ as "the sixth root of $\frac{1}{64}$."

Caution

When n is even $(n > 1)$ and $x < 0$, $\sqrt[n]{x}$ is not a real number. For example, $\sqrt[4]{-81}$ is not a real number, because no real number raised to the fourth power is -81.

Teaching Example 8 Evaluate:
a. $\sqrt[5]{\frac{1}{32}}$ b. $\sqrt[4]{-625}$
c. $\sqrt[5]{-243}$ d. $-\sqrt[6]{64}$
Answers:
a. $\frac{1}{2}$ b. Not a real number c. -3
d. -2

Self Check 8 Evaluate: **a.** $\sqrt[4]{\frac{1}{81}}$ $\frac{1}{3}$ **b.** $\sqrt[5]{10^5}$ 10 **c.** $\sqrt[6]{-64}$ Not a real number

Now Try Problems 87, 91, and 93

Using Your Calculator ▶ Finding Roots

The square root key $\boxed{\sqrt{\ }}$ on a reverse entry scientific calculator can be used to evaluate square roots. To evaluate roots with an index greater than 2, we can use the root key $\boxed{\sqrt[x]{y}}$. For example, the function $r(V) = \sqrt[3]{\frac{3V}{4\pi}}$ gives the radius of a sphere with volume V. To find the radius of the spherical propane tank shown on the left, we substitute 113 for V to get:

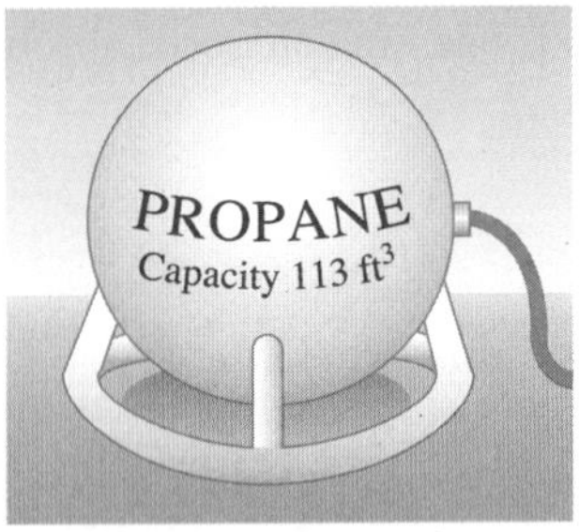

$$r(113) = \sqrt[3]{\frac{3(113)}{4\pi}}$$

To evaluate a root, we enter the radicand and press the root key $\boxed{\sqrt[x]{y}}$ followed by the index of the radical, which in this case is 3.

3 $\boxed{\times}$ 113 $\boxed{\div}$ $\boxed{(}$ 4 $\boxed{\times}$ $\boxed{\pi}$ $\boxed{)}$ $\boxed{=}$ $\boxed{\text{2nd}}$ $\boxed{\sqrt[x]{y}}$ 3 $\boxed{=}$ 2.999139118

To evaluate the cube root of $\frac{3(113)}{4\pi}$ using a direct entry calculator, we enter:

3 $\boxed{\text{2nd}}$ $\boxed{\sqrt[x]{y}}$ $\boxed{(}$ 3 $\boxed{\times}$ 113 $\boxed{\div}$ $\boxed{(}$ 4 $\boxed{\times}$ $\boxed{\pi}$ $\boxed{)}$ $\boxed{)}$ $\boxed{\text{ENTER}}$

To evaluate the cube root of $\frac{3(113)}{4\pi}$ with a graphing calculator, we enter:

$\boxed{\text{MATH}}$ 4 $\boxed{(}$ 3 $\boxed{\times}$ 113 $\boxed{)}$ $\boxed{\div}$ $\boxed{(}$ 4 $\boxed{\times}$ $\boxed{\text{2nd}}$ $\boxed{\pi}$ $\boxed{)}$ $\boxed{)}$ $\boxed{\text{ENTER}}$

```
³√((3*113)/(4*π)
)
            2.999139118
```

If we round the result to the nearest foot, we see that the radius of the propane tank is about 3 feet.

EXAMPLE 9 Simplify each radical expression. Assume that x can be any real number.
a. $\sqrt[5]{x^5}$ **b.** $\sqrt[4]{16x^4}$ **c.** $\sqrt[6]{(x+4)^6}$ **d.** $\sqrt[4]{81x^8}$

Strategy When the index n is odd, we will determine what expression, when raised to the nth power, produces the radicand. When the index n is even, we will determine what positive expression, when raised to the nth power, produces the radicand.

Why This is the definition of nth root.

Solution

a. $\sqrt[5]{x^5} = x$ Since n is odd, absolute value symbols aren't needed.

b. $\sqrt[4]{16x^4} = |2x| = 2|x|$ Since n is even and x can be negative, absolute value symbols are needed to ensure that the result is positive.

c. $\sqrt[6]{(x+4)^6} = |x+4|$ Absolute value symbols are needed to ensure that the result is positive.

d. $\sqrt[4]{81x^8} = 3x^2$ Because $(3x^2)^4 = 81x^8$. Since $3x^2 \geq 0$ for any value of x, no absolute value symbols are needed.

The Language of Algebra

Another way to say that x can be any real number is to say that **the variable is unrestricted.**

Teaching Example 9 Simplify each expression. Assume all variables are unrestricted.
a. $\sqrt[5]{32a^5}$ b. $\sqrt[4]{x^8}$
c. $\sqrt[6]{64x^6}$ d. $\sqrt[4]{(x+1)^4}$
Answers:
a. $2a$ b. x^2
c. $2|x|$ d. $|x+1|$

Self Check 9 Simplify. Assume all variables are unrestricted.
a. $\sqrt[6]{x^6}$ $|x|$ **b.** $\sqrt[5]{(a+5)^5}$ $a+5$ **c.** $\sqrt[4]{16a^8}$ $2a^2$

Now Try Problems 95, 97, 99, and 101

If we know that x is positive in parts (b) and (c) of Example 9, we don't need to use absolute value symbols. For example, if $x > 0$, then

$\sqrt[4]{16x^4} = 2x$ If x is positive, 2x is positive.

$\sqrt[6]{(x+4)^6} = x + 4$ If x is positive, x + 4 is positive.

We now summarize the definitions concerning nth roots.

Summary of the Definitions of $\sqrt[n]{x}$

If n is a natural number greater than 1 and x is a real number,

If $x > 0$, then $\sqrt[n]{x}$ is the positive number such that $\left(\sqrt[n]{x}\right)^n = x$.

If $x = 0$, then $\sqrt[n]{x} = 0$.

If $x < 0$ $\begin{cases} \text{and } n \text{ is odd, then } \sqrt[n]{x} \text{ is the negative number such that } \left(\sqrt[n]{x}\right)^n = x. \\ \text{and } n \text{ is even, then } \sqrt[n]{x} \text{ is not a real number.} \end{cases}$

SECTION 9.1 STUDY SET

VOCABULARY

Fill in the blanks.

1. $5x^2$ is the square root of $25x^4$ because $(5x^2)^2 = 25x^4$. The cube root of 216 is 6 because $6^3 = 216$.
2. The symbol $\sqrt{\ }$ is called a radical symbol or a square root symbol.
3. A radical symbol $\sqrt{\ }$ represents the positive or principal square root of a number.
4. The number 4 has two square roots, -2 and 2. When we speak of *the* square root of 4, we mean only the nonnegative square root of 4, which is 2.
5. The number 100 has two square roots. The positive or principal square root of 100 is 10.
6. In the expression $\sqrt[3]{27x^6}$, the index is 3 and $27x^6$ is the radicand.
7. When we write $\sqrt{b^4} = b^2$, we say that we have simplified the radical expression.
8. When n is an odd number, $\sqrt[n]{x}$ represents an odd root. When n is an even number, $\sqrt[n]{x}$ represents an even root.
9. $f(x) = \sqrt{x}$ and $g(x) = \sqrt[3]{x}$ are radical functions.
10. Together, 0 and the positive real numbers are called the nonnegative real numbers.

CONCEPTS

Fill in the blanks.

11. b is a square root of a if $b^2 = a$.
12. $\sqrt{0} = 0$ and $\sqrt[3]{0} = 0$.
13. $\sqrt{-4}$ is not a real number, because no real number squared equals -4.
14. $\sqrt[3]{x} = y$ if $y^3 = x$.
15. $\sqrt{x^2} = |x|$ and $\sqrt[3]{x^3} = x$
16. a. The graph of $g(x) = \sqrt{x} + 3$ is the graph of $f(x) = \sqrt{x}$ translated 3 units up.

    b. The graph of $g(x) = \sqrt{x + 5}$ is the graph of $f(x) = \sqrt{x}$ translated 5 units to the left.
17. The graph of a square root function f is shown below. Find each of the following, if possible.

    a. $f(11)$ 3

    b. $f(2)$ 0

    c. $f(-1)$ Undefined

    d. The value of x for which $f(x) = 2$ 6

    e. The value of x for which $f(x) = -1$ None

    f. The domain and range of f (Use interval notation.) D: $[2, \infty)$, R: $[0, \infty)$

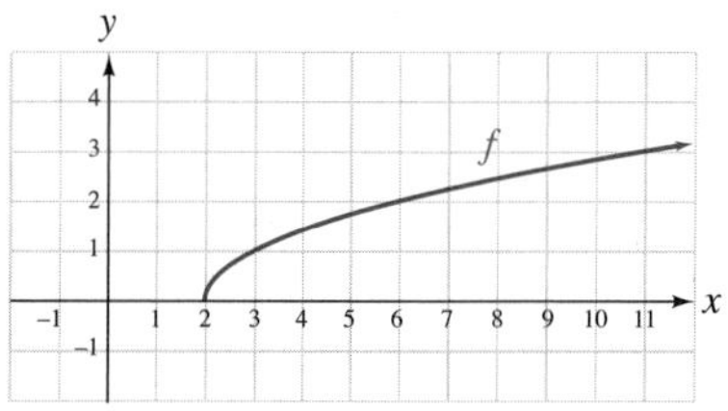

18. Refer to the graph in problem 17. Estimate each of the following function values.

    a. $f(4)$ 1.4

    b. $f(10)$ 2.8

19. The graph of a cube root function f is shown below. Find each of the following.

a. $f(-8)$ -5 **b.** $f(0)$ -3

c. The value of x for which $f(x) = -2$ 1

d. The domain and range of f (Use interval notation.) D: $(-\infty, \infty)$, R: $(-\infty, \infty)$

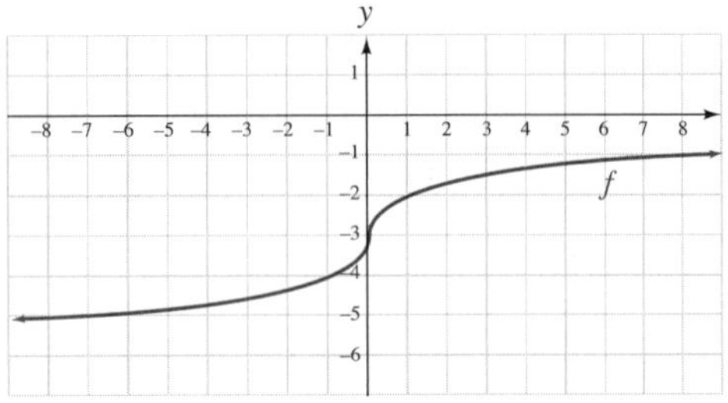

20. Match each function with the correct graph shown below.

a. $f(x) = x - 1$ ii. **b.** $f(x) = \sqrt{x} - 1$ iii.

c. $f(x) = x^2 - 1$ v. **d.** $f(x) = \sqrt[3]{x} - 1$ i.

e. $f(x) = |x| - 1$ vi. **f.** $f(x) = \dfrac{1}{x - 1}$ iv.

i. 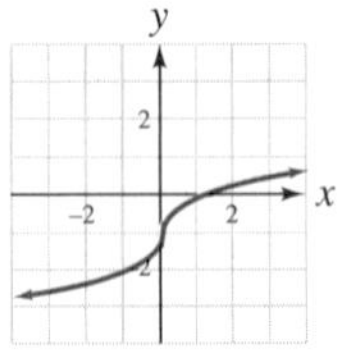

ii. 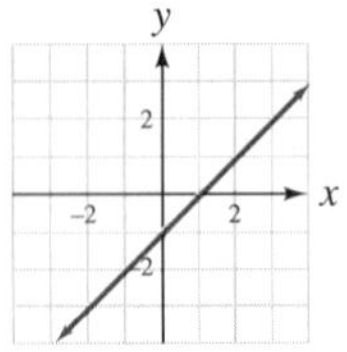

iii. 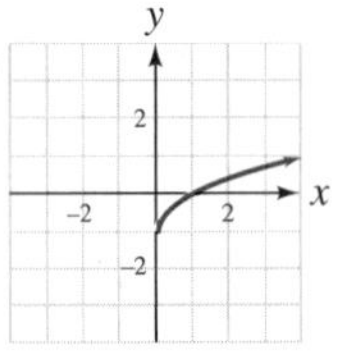

iv. 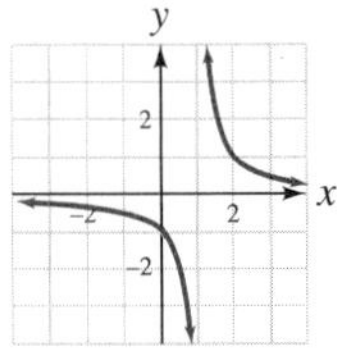

v. 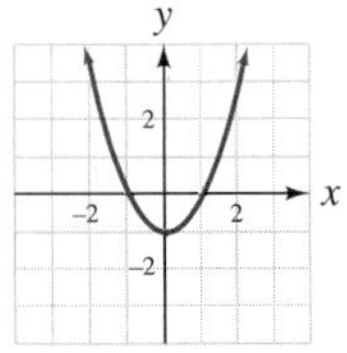

vi. 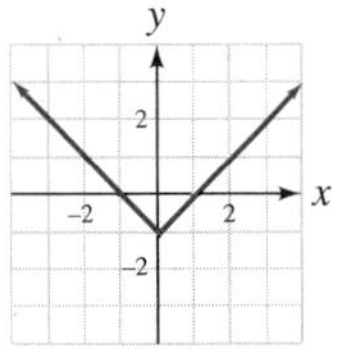

NOTATION

Translate each sentence into mathematical symbols.

▶ **21. a.** The square root of x squared is the absolute value of x.
$\sqrt{x^2} = |x|$

b. The cube root of x cubed is x. $\sqrt[3]{x^3} = x$

c. The fifth root of negative thirty-two is negative two.
$\sqrt[5]{-32} = -2$

22. a. f of x equals the square root of the quantity x minus five.
$f(x) = \sqrt{x - 5}$

b. g of x equals the cube root of x squared.
$g(x) = \sqrt[3]{x^2}$

GUIDED PRACTICE

Evaluate each square root without using a calculator. See Objective 1 and Example 1.

▶ **23.** $\sqrt{100}$ 10 **24.** $\sqrt{49}$ 7 ▶ **25.** $-\sqrt{64}$ -8 **26.** $-\sqrt{1}$ -1

▶ **27.** $\sqrt{\dfrac{1}{9}}$ $\frac{1}{3}$ **28.** $\sqrt{\dfrac{4}{25}}$ $\frac{2}{5}$ **29.** $\sqrt{0.25}$ 0.5 ▶ **30.** $\sqrt{0.16}$ 0.4

▶ **31.** $\sqrt{-81}$ Not real **32.** $-\sqrt{-49}$ Not real ▶ **33.** $\sqrt{121}$ 11 **34.** $\sqrt{144}$ 12

Use a calculator to find each square root. Give each answer to four decimal places. See Objective 1.

35. $\sqrt{12}$ 3.4641 ▶ **36.** $\sqrt{340}$ 18.4391

37. $\sqrt{679.25}$ 26.0624 ▶ **38.** $\sqrt{0.0063}$ 0.0794

Simplify each expression. Assume that all variables are unrestricted and use absolute value symbols when necessary. See Example 2.

▶ **39.** $\sqrt{4x^2}$ $2|x|$ **40.** $\sqrt{64t^2}$ $8|t|$

41. $\sqrt{81h^4}$ $9h^2$ ▶ **42.** $\sqrt{36y^4}$ $6y^2$

43. $\sqrt{36s^6}$ $6|s^3|$ ▶ **44.** $\sqrt{9y^6}$ $3|y^3|$

45. $\sqrt{144m^8}$ $12m^4$ **46.** $\sqrt{4n^8}$ $2n^4$

47. $\sqrt{y^2 - 2y + 1}$ $|y - 1|$ ▶ **48.** $\sqrt{b^2 - 14b + 49}$ $|b - 7|$

49. $\sqrt{a^4 + 6a^2 + 9}$ $a^2 + 3$ ▶ **50.** $\sqrt{x^4 + 10x^2 + 25}$ $x^2 + 5$

Complete each table and then graph the function. Give the domain and range. See Examples 3 and 4.
See AIE Appendix 3.

51. $f(x) = -\sqrt{x}$

x	y
0	0
1	-1
4	-2
9	-3
16	-4

D: $[0, \infty)$; R: $(-\infty, 0]$

▶ **52.** $f(x) = \sqrt{x} + 2$

x	y
0	2
1	3
4	4
9	5
16	6

D: $[0, \infty)$; R: $[2, \infty)$

Graph each function. Give the domain and range. See Examples 3 and 4. See AIE Appendix 3.

53. $f(x) = \sqrt{x + 4}$ D: $[-4, \infty)$; R: $[0, \infty)$ ▶ **54.** $f(x) = \sqrt{x} - 1$ D: $[0, \infty)$; R: $[-1, \infty)$

Find the domain of each function. See Example 4.

55. $f(x) = \sqrt{x + 6}$ D: $[-6, \infty)$ ▶ **56.** $g(x) = \sqrt{x + 12}$ D: $[-12, \infty)$

57. $g(x) = \sqrt{8 - 2x}$ D: $(-\infty, 4]$ ▶ **58.** $h(x) = \sqrt{35 - 5x}$ D: $(-\infty, 7]$

59. $s(t) = \sqrt{9t - 4}$ D: $\left[\frac{4}{9}, \infty\right)$ **60.** $T(a) = \sqrt{3a + 17}$ D: $\left[-\frac{17}{3}, \infty\right)$

61. $c(x) = \sqrt{0.5x - 20}$ D: $[40, \infty)$ ▶ **62.** $H(b) = \sqrt{0.4b - 36}$ D: $[90, \infty)$

Find each function value, if possible. Do not use a calculator. See Example 5.

▶ **63.** $f(x) = \sqrt{3x + 1}$
a. $f(8)$ 5 **b.** $f(-2)$ Undefined

▶ **64.** $h(t) = \sqrt{t^2 + t - 3}$
a. $h(-4)$ 3 **b.** $h(-1)$ Undefined

▶ **65.** $g(x) = \sqrt[3]{x - 4}$
a. $g(12)$ 2 **b.** $g(-23)$ -3

▶ **66.** $s(a) = -\sqrt[3]{32s}$
a. $s(-2)$ 4 **b.** $s(2)$ -4

Use a calculator to find each function value. Round to the nearest ten-thousandth. See Example 5 and *Using Your Calculator.*

67. $f(x) = \sqrt{x^2 + 1}$
 a. $f(4)$ 4.1231 b. $f(2.35)$ 2.5539

▶ 68. $g(x) = \sqrt{7 - 4x}$
 a. $g(-\pi)$ 4.4234 b. $g(0.5)$ 2.2361

69. $g(x) = \sqrt[3]{x^2 + 1}$
 a. $g(6)$ 3.3322 b. $g(21.57)$ 7.7543

▶ 70. $h(t) = \sqrt[3]{2.1t + 11}$
 a. $h(-0.4)$ 2.1659 b. $h(15)$ 3.4898

Simplify each cube root. See Example 6.

71. $\sqrt[3]{1}$ 1
▶ 72. $\sqrt[3]{8}$ 2
▶ 73. $\sqrt[3]{-125}$ −5
74. $\sqrt[3]{-27}$ −3
75. $\sqrt[3]{\frac{8}{27}}$ $\frac{2}{3}$
▶ 76. $\sqrt[3]{\frac{125}{64}}$ $\frac{5}{4}$
▶ 77. $\sqrt[3]{64}$ 4
78. $\sqrt[3]{1{,}000}$ 10
▶ 79. $\sqrt[3]{-216a^3}$ $-6a$
▶ 80. $\sqrt[3]{-512x^3}$ $-8x$
81. $\sqrt[3]{-1{,}000p^6q^3}$ $-10p^2q$
▶ 82. $\sqrt[3]{-343a^6b^3}$ $-7a^2b$

Complete each table and then graph the function. Give the domain and range. See Example 7. See AIE Appendix 3.

▶ 83. $f(x) = \sqrt[3]{x} - 3$

x	y
−8	−5
−1	−4
0	−3
1	−2
8	−1

D: $(-\infty, \infty)$; R: $(-\infty, \infty)$

▶ 84. $f(x) = -\sqrt[3]{x}$

x	y
−8	2
−1	1
0	0
1	−1
8	−2

D: $(-\infty, \infty)$; R: $(-\infty, \infty)$

Graph each function. Give the domain and range. See Example 7. See AIE Appendix 3.

85. $f(x) = \sqrt[3]{x - 3}$
 D: $(-\infty, \infty)$; R: $(-\infty, \infty)$

▶ 86. $f(x) = \sqrt[3]{x} + 3$
 D: $(-\infty, \infty)$; R: $(-\infty, \infty)$

Evaluate each radical expression, if possible, without using a calculator. See Example 8.

87. $\sqrt[4]{81}$ 3
▶ 88. $\sqrt[6]{64}$ 2
89. $-\sqrt[5]{243}$ −3
▶ 90. $-\sqrt[4]{625}$ −5
▶ 91. $\sqrt[6]{-256}$ Not real
▶ 92. $\sqrt[6]{-729}$ Not real
93. $\sqrt[5]{-\frac{1}{32}}$ $-\frac{1}{2}$
▶ 94. $\sqrt[5]{-\frac{243}{32}}$ $-\frac{3}{2}$

Simplify each radical expression. Assume all variables are unrestricted. See Example 9.

▶ 95. $\sqrt[5]{32a^5}$ $2a$
▶ 96. $\sqrt[5]{-32x^5}$ $-2x$
97. $\sqrt[4]{81a^4}$ $3|a|$
▶ 98. $\sqrt[8]{t^8}$ $|t|$
99. $\sqrt[6]{k^{12}}$ k^2
▶ 100. $\sqrt[6]{64b^6}$ $2|b|$
101. $\sqrt[4]{(m + 4)^8}$ $(m + 4)^2$
▶ 102. $\sqrt[4]{(x - 7)^8}$ $(x - 7)^2$

TRY IT YOURSELF

Simplify each radical expression, if possible. Assume all variables are unrestricted.

103. $\sqrt[3]{64s^9t^6}$ $4s^3t^2$
104. $\sqrt[3]{1{,}000a^6b^6}$ $10a^2b^2$
105. $-\sqrt{49b^8}$ $-7b^4$
106. $-\sqrt{144t^4}$ $-12t^2$
107. $-\sqrt[5]{-\frac{1}{32}}$ $\frac{1}{2}$
108. $-\sqrt[5]{-243}$ 3
▶ 109. $\sqrt[3]{-125m^6}$ $-5m^2$
110. $\sqrt[3]{-216z^9}$ $-6z^3$
111. $\sqrt{400m^{16}n^2}$ $20m^8|n|$
112. $\sqrt{169p^4q^2}$ $13p^2|q|$
113. $\sqrt[6]{64a^6b^6}$ $2|ab|$
114 $\sqrt[6]{(x + 4)^6}$ $|x + 4|$
115. $\sqrt[4]{-81}$ Not real
116. $\sqrt[6]{-1}$ Not real
117. $\sqrt{n^2 + 12n + 36}$
 $|n + 6|$
118. $\sqrt{s^2 - 20s + 100}$
 $|s - 10|$

Look Alikes . . .

119. a. $\sqrt{64}$ 8 b. $\sqrt[3]{64}$ 4
▶ 120. a. $\sqrt{-64}$ Not real b. $\sqrt[3]{-64}$ −4
121. a. $\sqrt{81}$ 9 b. $\sqrt[4]{81}$ 3
122. a. $\sqrt{16}$ 4 b. $\sqrt[4]{16}$ 2

APPLICATIONS

Use a calculator to solve each problem. Round answers to the nearest tenth.

123. **Embroidery.** The radius r of a circle is given by the formula $r = \sqrt{\frac{A}{\pi}}$, where A is its area. Find the diameter of the embroidery hoop if there are 38.5 in.^{2} of stretched fabric on which to embroider. 7.0 in.

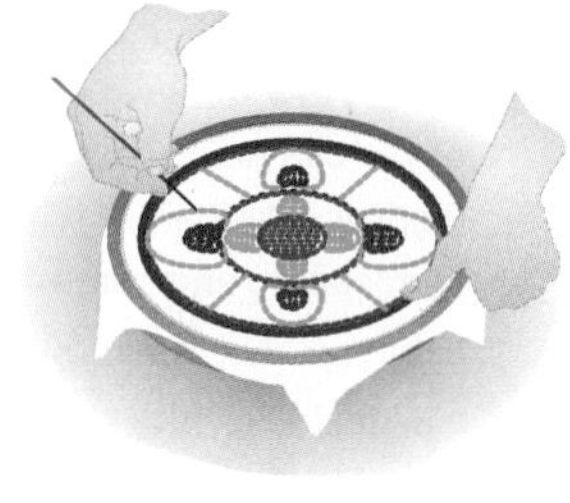

124. **Pendulums.** Find the period of a pendulum with length 1 foot. *See Example 5.* 1.1 sec

125. **Shoelaces.** The formula
$$S = 2\left[H + L + (p - 1)\sqrt{H^2 + V^2}\right]$$ can be used to calculate the correct shoelace length for the criss-cross lacing pattern shown in the illustration, where p represents the number of *pairs* of eyelets. Find the correct shoelace length if H(horizontal distance) = 50 mm, L (length of end) = 250 mm, and V(vertical distance) = 20 mm. Round to the nearest tenth. (*Source:* Ian's Shoelace Site at www.fieggen.com) 1,138.5 mm

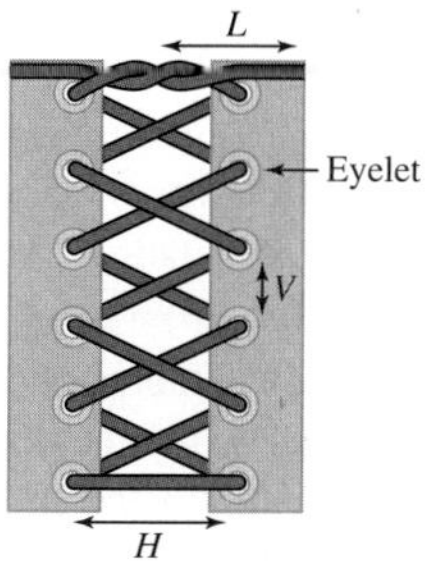

126. Baseball. The length of a diagonal of a square is given by the function $d(s) = \sqrt{2s^2}$, where s is the length of a side of the square. Find the distance from home plate to second base on a softball diamond and on a baseball diamond. Round to the nearest tenth. The illustration gives the dimensions of each type of infield. 84.9 ft, 127.3 ft

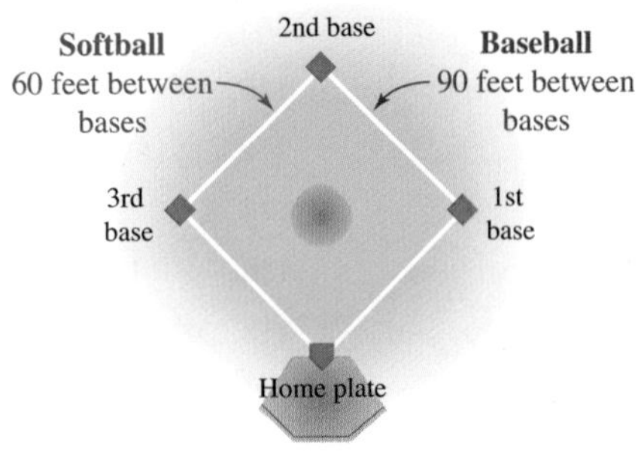

127. Pulse Rates. The approximate pulse rate (in beats per minute) of an adult who is t inches tall is given by the function $p(t) = \frac{590}{\sqrt{t}}$. The *Guinness Book of World Records 2008* lists Leonid Stadnyk of Ukraine as the tallest living man, at 8 feet, 5.5 inches. Find his approximate pulse rate as predicted by the function. About 58.6 beats/min

128. The Grand Canyon. The time t (in seconds) that it takes for an object to fall a distance of s feet is given by the formula $t = \frac{\sqrt{s}}{4}$. In some places, the Grand Canyon is one mile (5,280 feet) deep. How long would it take a stone dropped over the edge of the canyon to hit bottom? 18.2 sec

129. Biology. Scientists will place five rats inside a clear plastic hemisphere and control the environment to study the rats' behavior. The function $d(V) = \sqrt[3]{12\left(\frac{V}{\pi}\right)}$ gives the diameter of a hemisphere with volume V. Use the function to determine the diameter of the base of the hemisphere, if each rat requires 125 cubic feet of living space. 13.4 ft

130. Aquariums. The function $s(g) = \sqrt[3]{\frac{g}{7.5}}$ determines how long (in feet) an edge of a cube-shaped tank must be if it is to hold g gallons of water. What dimensions should a cube-shaped aquarium have if it is to hold 1,250 gallons of water? 5.5 ft × 5.5 ft × 5.5 ft

131. Collectibles. The *effective rate of interest r* earned by an investment is given by the formula $r = \sqrt[n]{\frac{A}{P}} - 1$, where P is the initial investment that grows to value A after n years. Determine the effective rate of interest earned by a collector on a Lladró porcelain figurine purchased for \$800 and sold for \$950 five years later. 3.5%

132. Law Enforcement. The graphs of the two radical functions shown in the illustration in the next column can be used to estimate the speed (in mph) of a car involved in an accident. Suppose a police accident report listed skid marks to be 220 feet long but failed to give the road conditions. Estimate the possible speeds the car was traveling prior to the brakes being applied. Dry: about 72 mph; wet: about 47 mph

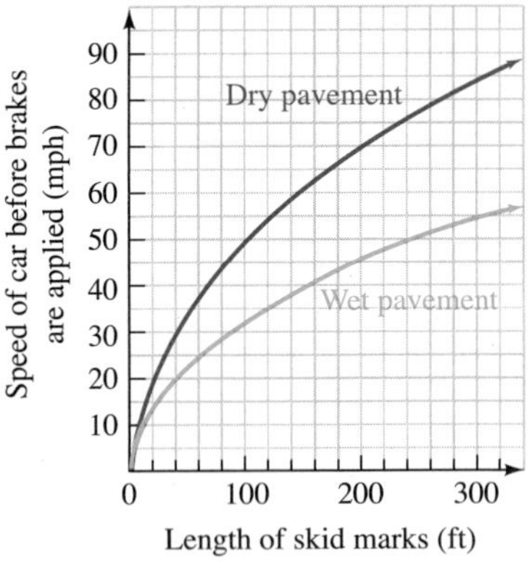

WRITING

133. If x is any real number, that is, if x is unrestricted, then $\sqrt{x^2} = x$ is not correct. Explain why.

134. Explain why $\sqrt{36}$ is just 6, and not also -6.

135. Explain what is wrong with the graph in the illustration if it is supposed to be the graph of $f(x) = \sqrt{x}$.

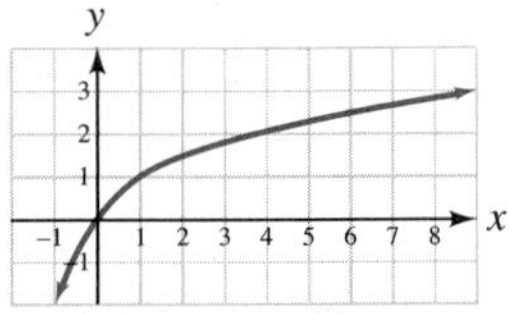

136. Explain how to estimate the domain and range of the radical function whose graph is shown here.

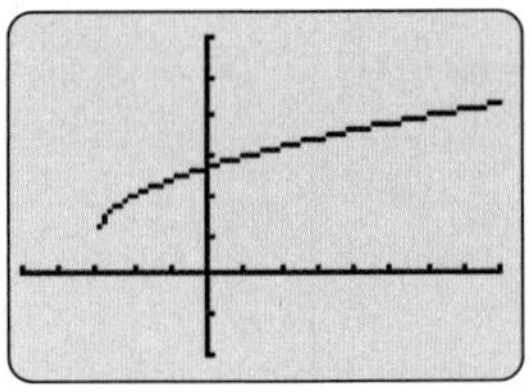

REVIEW

Perform the operations and simplify when possible.

137. $\dfrac{x^2 - 3xy - 4y^2}{x^2 + cx - 2yx - 2cy} \div \dfrac{x^2 - 2xy - 3y^2}{x^2 + cx - 4yx - 4cy}$ $\frac{(x-4y)^2}{(x-2y)(x-3y)}$

138. $\dfrac{2x + 3}{3x - 1} - \dfrac{x - 4}{2x + 1}$ $\frac{x^2 + 21x - 1}{(3x - 1)(2x + 1)}$

CHALLENGE PROBLEMS

139. Graph $f(x) = -\sqrt{x - 2} + 3$ and find the domain and range. D: $[2, \infty)$; R: $(-\infty, 3]$; see AIE Appendix 3.

140. Simplify $\sqrt{9a^{16} + 12a^8b^{25} + 4b^{50}}$ and assume that $a > 0$ and $b > 0$. $3a^8 + 2b^{25}$

SECTION 9.2

OBJECTIVES

1. Simplify expressions of the form $a^{1/n}$.
2. Simplify expressions of the form $a^{m/n}$.
3. Convert between radicals and rational exponents.
4. Simplify expressions with negative rational exponents.
5. Use rules for exponents to simplify expressions.
6. Simplify radical expressions.

Rational Exponents

ARE YOU READY? *Are You Ready? exercises available online at www.webassign.net/brookscole*

The following problems review some basic skills that are needed when working with rational exponents.

1. Evaluate: **a.** $\sqrt{64}$ 8 **b.** $\sqrt[3]{-64}$ -4

2. Evaluate: **a.** $\sqrt[4]{81}$ 3 **b.** $\sqrt[5]{\frac{1}{32}}$ $\frac{1}{2}$

3. Simplify: **a.** $\sqrt{(x+5)^2}$ $|x+5|$ **b.** $\sqrt[3]{27a^6}$ $3a^2$

4. Evaluate: $\left(\sqrt{36}\right)^3$ 216

5. Simplify: 7^{-2} $\frac{1}{49}$

6. Simplify: $\frac{x^5x^7}{x^3}$ x^9

In this section, we will extend the definition of exponent to include rational (fractional) exponents. We will see how expressions such as $9^{1/2}$, $\left(\frac{1}{16}\right)^{3/4}$, and $(-32x^5)^{-2/5}$ can be simplified by writing them in an equivalent radical form using two new rules for exponents.

1 Simplify Expressions of the Form $a^{1/n}$.

The Language of Algebra

Rational exponents are also called **fractional exponents.**

It is possible to raise numbers to fractional powers. To give meaning to rational exponents, we first consider $\sqrt{7}$. Because $\sqrt{7}$ is the positive number whose square is 7, we have

$$\left(\sqrt{7}\right)^2 = 7$$

We now consider the notation $7^{1/2}$, which is read as "7 to the one-half power." If rational exponents are to follow the same rules as integer exponents, the square of $7^{1/2}$ must be 7, because

$$(7^{1/2})^2 = 7^{1/2 \cdot 2}$$ Keep the base and multiply the exponents.

$$= 7^1$$ Do the multiplication: $\frac{1}{2} \cdot 2 = 1$.

$$= 7$$

Teaching Tip: As you introduce the rules for exponents, have selected students read each one out loud to the class.

Since the square of $7^{1/2}$ and the square of $\sqrt{7}$ are both equal to 7, we define $7^{1/2}$ to be $\sqrt{7}$. Similarly,

$$7^{1/3} = \sqrt[3]{7}, \quad 7^{1/4} = \sqrt[4]{7}, \quad \text{and} \quad 7^{1/5} = \sqrt[5]{7}$$

In general, we have the following definition.

The Definition of $x^{1/n}$

A rational exponent of $\frac{1}{n}$ indicates the nth root of its base.

If n represents a positive integer greater than 1 and $\sqrt[n]{x}$ represents a real number,

$$x^{1/n} = \sqrt[n]{x}$$ Read as "x to the $\frac{1}{n}$ power equals the nth root of x."

We can use this definition to simplify exponential expressions that have rational exponents with a numerator of 1. For example, to simplify $8^{1/3}$, we write it as an equivalent expression in radical form and proceed as follows:

Index

$$8^{1/3} = \sqrt[3]{8} = 2$$

Radicand

The base of the exponential expression, 8, is the radicand of the radical expression. The denominator of the fractional exponent, 3, is the index of the radical.

Thus, $8^{1/3} = 2$.

EXAMPLE 1 Evaluate: **a.** $9^{1/2}$ **b.** $(-64)^{1/3}$ **c.** $16^{1/4}$ **d.** $-\left(\frac{1}{32}\right)^{1/5}$

Strategy First, we will identify the base and the exponent of the exponential expression. Then we will write the expression in an equivalent radical form using the rule for rational exponents $x^{1/n} = \sqrt[n]{x}$.

Why We can then use the methods from Section 7.1 to evaluate the resulting square root, cube root, fourth root, and fifth root.

Solution

a. $9^{1/2} = \sqrt{9} = 3$ — The base is 9 and the exponent is $\frac{1}{2}$. Because the denominator of the exponent $\frac{1}{2}$ is 2, find the square root of the base, 9.

b. $(-64)^{1/3} = \sqrt[3]{-64} = -4$ — Read as "-64 to the one-third power." Because the denominator of the exponent $\frac{1}{3}$ is 3, find the cube root of the base, -64.

c. $16^{1/4} = \sqrt[4]{16} = 2$ — Because the denominator of the exponent $\frac{1}{4}$ is 4, find the fourth root of the base, 16.

d. $-\left(\frac{1}{32}\right)^{1/5} = -\sqrt[5]{\frac{1}{32}} = -\frac{1}{2}$ — Read as "the opposite of the one-fifth power of $\frac{1}{32}$." Because the denominator of the exponent $\frac{1}{5}$ is 5, find the fifth root of the base, $\frac{1}{32}$.

Success Tip

One key to successfully simplifying the exponential expressions in this section is this: When you see an exponent, always ask yourself, "What is the base?"

Teaching Example 1 Evaluate:
a. $25^{1/2}$ b. $\left(\frac{81}{16}\right)^{1/4}$ c. $-(32)^{1/5}$
d. $(-125)^{1/3}$
Answers:
a. 5 b. $\frac{3}{2}$ c. -2 d. -5

Self Check 1 Evaluate: **a.** $16^{1/2}$ 4 **b.** $\left(-\frac{27}{8}\right)^{1/3}$ $-\frac{3}{2}$ **c.** $-(81)^{1/4}$ -3 **d.** $1^{1/5}$ 1

Now Try Problems 19, 23, 25, and 27

As with radicals, when n is an *odd natural number* in the expression $x^{1/n}$, where $n > 1$, there is exactly one real nth root, and we don't need to use absolute value symbols.

When n is an *even natural number,* there are two nth roots. Since we want the expression $x^{1/n}$ to represent the positive nth root, we often must use absolute value symbols to ensure that the simplified result is positive. Thus, if n is even,

$$(x^n)^{1/n} = |x|$$

When n is even and x is negative, the expression $x^{1/n}$ is not a real number.

EXAMPLE 2 Simplify. Assume that the variables are unrestricted. **a.** $(-27x^3)^{1/3}$ **b.** $(256a^8)^{1/8}$ **c.** $[(y + 4)^2]^{1/2}$ **d.** $(25b^4)^{1/2}$ **e.** $(-256)^{1/4}$

Strategy We will write each exponential expression in an equivalent radical form using the rule for rational exponents $x^{1/n} = \sqrt[n]{x}$.

Why We can then use the methods of Section 7.1 to simplify the resulting radical expression.

Solution

a. $(-27x^3)^{1/3} = \sqrt[3]{-27x^3} = -3x$ — Read as "the quantity $-27x^3$ raised to the one-third power." Because $(-3x)^3 = -27x^3$. Since n is odd, no absolute value symbols are needed.

b. $(256a^8)^{1/8} = \sqrt[8]{256a^8} = 2|a|$ — Because $(2a)^8 = 256a^8$. Since n is even and a can be any real number, $2a$ can be negative. Thus, absolute value symbols are needed.

c. $\left[(y + 4)^2\right]^{1/2} = \sqrt{(y + 4)^2} = |y + 4|$ — Because $|y + 4|^2 = (y + 4)^2$. Since n is even and y can be any real number, $y + 4$ can be negative. Thus, absolute value symbols are needed.

The Language of Algebra

The phrase **"assume that the variables are unrestricted"** in the instructions for Example 2 reminds us that negative numbers are permissible replacement values for the variables. Therefore, absolute value symbols may be necessary to ensure the answer is not negative.

d. $(25b^4)^{1/2} = \sqrt{25b^4} = 5b^2$ Because $(5b^2)^2 = 25b^4$. Since $b^2 \geq 0$, no absolute value symbols are needed.

e. $(-256)^{1/4} = \sqrt[4]{-256}$, which is not a real number. Because no real number raised to the 4th power is -256.

Teaching Example 2 Simplify. Assume that the variables are unrestricted.
a. $(-27m^3)^{1/3}$ b. $(81b^4)^{1/4}$ c. $(x^8)^{1/2}$ d. $(-1)^{1/6}$
Answers:
a. $-3m$ b. $3|b|$ c. x^4 d. Not a real number

Self Check 2 Simplify. Assume that the variables are unrestricted.
a. $(-8n^3)^{1/3}$ $-2n$ **b.** $(625a^4)^{1/4}$ $5|a|$ **c.** $(b^4)^{1/2}$ b^2 **d.** $(-64)^{1/6}$ Not a real number

Now Try Problems 29, 35, 37, and 39

Teaching Tip: Stress that if *all variables represent positive real numbers* then absolute value symbols are not needed when simplifying even-root radical expressions.

If we were told that the variables represent positive real numbers in parts (b) and (c) of Example 2, the absolute value symbols in the answers would not be needed.

$(256a^8)^{1/8} = 2a$ If a represents a positive real number, then $2a$ is positive.

$[(y + 4)^2]^{1/2} = y + 4$ If y represents a positive real number, then $y + 4$ is positive.

We summarize the cases as follows.

Summary of the Definitions of $x^{1/n}$

If n is a natural number greater than 1 and x is a real number,

If $x > 0$, then $x^{1/n}$ is the real number such that $(x^{1/n})^n = x$.

If $x = 0$, then $x^{1/n} = 0$.

If $x < 0$ $\begin{cases} \text{and } n \text{ is odd, then } x^{1/n} \text{ is the negative number such that } (x^{1/n})^n = x. \\ \text{and } n \text{ is even, then } x^{1/n} \text{ is not a real number.} \end{cases}$

2 Simplify Expressions of the Form $a^{m/n}$.

We can extend the definition of $x^{1/n}$ to include fractional exponents with numerators other than 1. For example, since $8^{2/3}$ can be written as $(8^{1/3})^2$, we have

$8^{2/3} = (8^{1/3})^2$ Read $8^{2/3}$ as "8 to the two-thirds power."

$= (\sqrt[3]{8})^2$ Write $8^{1/3}$ in radical form.

$= 2^2$ Find the cube root first: $\sqrt[3]{8} = 2$.

$= 4$ Then find the power.

Thus, we can simplify $8^{2/3}$ by finding the second power of the cube root of 8.

The numerator of the rational exponent is the power.

$$8^{2/3} = (\sqrt[3]{8})^2$$

The base of the exponential expression is the radicand.

The denominator of the rational exponent is the index of the radical.

We also can simplify $8^{2/3}$ by taking the cube root of 8 squared.

$8^{2/3} = (8^2)^{1/3}$

$= 64^{1/3}$ Find the power first: $8^2 = 64$.

$= \sqrt[3]{64}$ Write $64^{1/3}$ in radical form.

$= 4$ Now find the cube root.

In general, we have the following definition.

The Definition of $x^{m/n}$

If m and n represent positive integers ($n \neq 1$) and $\sqrt[n]{x}$ represents a real number,

$$x^{m/n} = \left(\sqrt[n]{x}\right)^m \quad \text{and} \quad x^{m/n} = \sqrt[n]{x^m}$$

We read the first definition given above as "x to the m divided by n power equals the nth root of x, raised to the mth power."

Because of the previous definition, we can interpret $x^{m/n}$ in two ways:

1. $x^{m/n}$ means the nth root of the mth power of x.
2. $x^{m/n}$ means the mth power of the nth root of x.

We can use this definition to evaluate exponential expressions that have rational exponents with a numerator that is not 1. To avoid large numbers, we usually find the root of the base first and then calculate the power using the rule $x^{m/n} = \left(\sqrt[n]{x}\right)^m$.

EXAMPLE 3 Evaluate: **a.** $32^{2/5}$ **b.** $81^{3/4}$ **c.** $(-64)^{2/3}$ **d.** $-\left(\frac{1}{25}\right)^{3/2}$

Strategy First, we will identify the base and the exponent of the exponential expression. Then we will write the expression in an equivalent radical form using the rule for rational exponents $x^{m/n} = \left(\sqrt[n]{x}\right)^m$.

Why We know how to evaluate square roots, cube roots, fourth roots, and fifth roots.

Solution

a. To evaluate $32^{2/5}$, we write it in an equivalent radical form. The denominator of the rational exponent is the same as the index of the corresponding radical. The numerator of the rational exponent indicates the power to which the radical base is raised.

Power
Root

$$32^{2/5} = \left(\sqrt[5]{32}\right)^2 = (2)^2 = 4$$

Read as "32 to the two-fifths power." Because the exponent is 2/5, find the fifth root of the base, 32, to get 2. Then find the second power of 2.

Power
Root

b. $81^{3/4} = \left(\sqrt[4]{81}\right)^3 = (3)^3 = 27$

Read as "81 to the three-fourths power." Because the exponent is 3/4, find the fourth root of the base, 81, to get 3. Then find the third power of 3.

c. For $(-64)^{2/3}$, the base is -64.

Power
Root

$$(-64)^{2/3} = (\sqrt[3]{-64})^2 = (-4)^2 = 16$$

Read as "-64 to the two-thirds power." Because the exponent is 2/3, find the cube root of the base, -64, to get -4. Then find the second power of -4.

d. For $-\left(\frac{1}{25}\right)^{3/2}$, the base is $\frac{1}{25}$, not $-\frac{1}{25}$.

Power
Root

$$-\left(\frac{1}{25}\right)^{3/2} = -\left(\sqrt[2]{\frac{1}{25}}\right)^3 = -\left(\frac{1}{5}\right)^3 = -\frac{1}{125}$$

Read as "the opposite of the three-halves power of $\frac{1}{25}$." Because the exponent is 3/2, find the square root of the base, $\frac{1}{25}$, to get $\frac{1}{5}$. Then find the third power of $\frac{1}{5}$.

Caution

We also can evaluate $x^{m/n}$ using $\sqrt[n]{x^m}$. However, the resulting radicand is often extremely large. For example,

$$\begin{aligned} 81^{3/4} &= \sqrt[4]{81^3} \\ &= \sqrt[4]{531{,}441} \\ &= 27 \end{aligned}$$

Teaching Example 3 Evaluate:
a. $25^{3/2}$ b. $16^{3/4}$
c. $(-27)^{4/3}$ d. $-\left(\frac{1}{8}\right)^{2/3}$
Answers:
a. 125 b. 8 c. 81 d. $-\frac{1}{4}$

Self Check 3 Evaluate: **a.** $16^{3/2}$ 64 **b.** $125^{4/3}$ 625 **c.** $(-216)^{2/3}$ 36
d. $-\left(\frac{1}{32}\right)^{4/5}$ $-\frac{1}{16}$

Now Try Problems 41, 45, and 47

EXAMPLE 4 Simplify. All variables represent positive real numbers. **a.** $(36m^4)^{3/2}$ **b.** $(-8x^3)^{4/3}$ **c.** $-(x^5y^5)^{2/5}$

Strategy We will write each exponential expression in an equivalent radical form using the rule for rational exponents $x^{m/n} = \left(\sqrt[n]{x}\right)^m$.

Why We can then use the methods of Section 9.1 to simplify the resulting radical expression.

Solution

Power, Root

a. $(36m^4)^{3/2} = \left(\sqrt[2]{36m^4}\right)^3 = (6m^2)^3 = 216m^6$

Read as "the quantity of $36m^4$ raised to the three-halves power." Because the exponent is 3/2, find the square root of the base, $36m^4$, to get $6m^2$. Then find the third power of $6m^2$.

Power, Root

b. $(-8x^3)^{4/3} = \left(\sqrt[3]{-8x^3}\right)^4 = (-2x)^4 = 16x^4$

Read as "the quantity of $-8x^3$ raised to the four-thirds power." Because the exponent is 4/3, find the cube root of the base, $-8x^3$, to get $-2x$. Then find the fourth power of $-2x$.

Power, Root

c. $-(x^5y^5)^{2/5} = -\left(\sqrt[5]{x^5y^5}\right)^2 = -(xy)^2 = -x^2y^2$

Read as "the opposite of the two-fifths power of the quantity x^5y^5." Because the exponent is 2/5, find the fifth root of the base, x^5y^5, to get xy. Then find the second power of xy.

Teaching Example 4 Simplify. All variables represent positive real numbers.
a. $(9y^4)^{3/2}$ b. $(-8a^3b^3)^{2/3}$
c. $-(32r^{15})^{3/5}$
Answers:
a. $27y^6$ b. $-2a^2b^2$ c. $-8r^9$

Self Check 4 Simplify. All variables represent positive real numbers. **a.** $(4c^4)^{3/2}$ $8c^6$ **b.** $(-27m^3n^3)^{2/3}$ $9m^2n^2$ **c.** $-(32a^{10})^{3/5}$ $-8a^6$

Now Try Problems 49 and 51

Using Your Calculator ▶ Rational Exponents

We can evaluate expressions containing rational exponents using the exponential key [y^x] or [x^y] on a scientific calculator. For example, to evaluate $10^{2/3}$, we enter

10 [y^x] [( ] 2 [÷] 3 [ )] [=] 4.641588834

Note that parentheses were used when entering the power. Without them, the calculator would interpret the entry as $10^2 \div 3$.

To evaluate the exponential expression using a direct entry or graphing calculator, we use the [^] key, which raises a base to a power. Again, we use parentheses when entering the power.

10 [^] [( ] 2 [÷] 3 [ )] [ENTER]

```
10^(2/3)
        4.641588834
```

To the nearest hundredth, $10^{2/3} \approx 4.64$.

3 Convert Between Radicals and Rational Exponents.

We can use the rules for rational exponents to convert expressions from radical form to exponential form, and vice versa.

EXAMPLE 5 Write $\sqrt{5xyz}$ as an exponential expression with a rational exponent.

Strategy We will use the first rule for rational exponents in reverse: $\sqrt[n]{x} = x^{1/n}$.

Why We are given a radical expression and we want to write an equivalent exponential expression.

Solution The radicand is $5xyz$, so the base of the exponential expression is $5xyz$. The index of the radical is an understood 2, so the denominator of the fractional exponent is 2.

$$\sqrt{5xyz} = (5xyz)^{1/2} \quad \text{Recall: } \sqrt[2]{5xyz} = \sqrt{5xyz}.$$

Teaching Example 5 Write $\sqrt[11]{9xy^2z}$ as an exponential expression with a rational exponent.
Answer: $(9xy^2z)^{1/11}$

Self Check 5 Write $\sqrt[6]{7ab}$ as an exponential expression with a rational exponent. $(7ab)^{1/6}$

Now Try Problem 57

Rational exponents appear in formulas used in many disciplines, such as science and engineering.

EXAMPLE 6 **Satellites.** The formula $r = \left(\frac{GMP^2}{4\pi^2}\right)^{1/3}$ gives the orbital radius (in meters) of a satellite circling Earth, where G and M are constants and P is the time in seconds for the satellite to make one complete revolution. Write the formula using a radical.

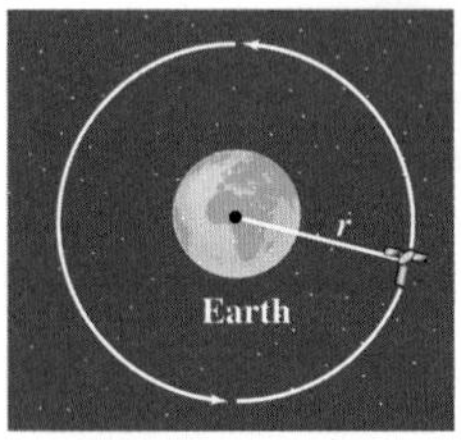

Strategy We will use the first rule for rational exponents: $x^{1/n} = \sqrt[n]{x}$.

Why We are given an exponential expression involving a rational exponent with a numerator of 1 and we want to write an equivalent radical expression.

Solution The fractional exponent $\frac{1}{3}$, with a numerator of 1 and a denominator of 3, indicates that we are to find the cube root of the base of the exponential expression. So we have

$$r = \sqrt[3]{\frac{GMP^2}{4\pi^2}}$$

Teaching Example 6 Statistics. The formula $s = \left[\frac{\Sigma(x - \bar{x})^2}{n - 1}\right]^{1/2}$ gives the sample standard deviation of a set of data. Write the formula using a radical.
Answer: $s = \sqrt{\frac{\Sigma(x - \bar{x})^2}{n - 1}}$

Self Check 6 **Statistics.** The formula $\sigma = \left(\frac{\Sigma(x - \mu)^2}{N}\right)^{1/2}$ gives the population standard deviation. Write the formula using a radical. $\sigma = \sqrt{\frac{\Sigma(x - \mu)^2}{N}}$

Now Try Problem 61

4 Simplify Expressions with Negative Rational Exponents.

To be consistent with the definition of negative integer exponents, we define $x^{-m/n}$ as follows.

Definition of $x^{-m/n}$

If m and n are positive integers, $\frac{m}{n}$ is in simplified form, and $x^{1/n}$ is a real number, then

$$x^{-m/n} = \frac{1}{x^{m/n}} \quad \text{and} \quad \frac{1}{x^{-m/n}} = x^{m/n} \quad (x \neq 0)$$

Teaching Tip: Now is a good time to review how to simplify expressions involving negative exponents, such as 9^{-2} and $(-6)^{-3}$.

From the definition, we see that another way to write $x^{-m/n}$ is to write its reciprocal and change the sign of the exponent.

EXAMPLE 7 Simplify. Assume that x can represent any nonzero real number.

a. $64^{-1/2}$ **b.** $(-16)^{-5/4}$ **c.** $-625^{-3/4}$ **d.** $(-32x^5)^{-2/5}$ **e.** $\dfrac{1}{25^{-3/2}}$

Strategy We will use one of the rules $x^{-m/n} = \dfrac{1}{x^{m/n}}$ or $\dfrac{1}{x^{-m/n}} = x^{m/n}$ to write the reciprocal of each exponential expression and change the exponent's sign to positive.

Why If we can produce an equivalent expression having a positive rational exponent, we can use the methods of this section to simplify it.

Solution

Reciprocal

a. $64^{-1/2} = \dfrac{1}{64^{1/2}} = \dfrac{1}{\sqrt{64}} = \dfrac{1}{8}$ Read as "64 to the negative one-half power." Because the exponent is negative, write the reciprocal of $64^{-1/2}$, and change the sign of the exponent.

Change sign

Caution

A negative exponent does not, itself, indicate a negative number. For example,

$$64^{-1/2} = \frac{1}{8}$$

b. $(-16)^{-5/4}$ is read as "-16 to the negative five-fourths power." It is not a real number because $(-16)^{1/4}$ is not a real number.

c. In $-625^{-3/4}$, the base is 625.

$$-625^{-3/4} = -\frac{1}{625^{3/4}} = -\frac{1}{(\sqrt[4]{625})^3} = -\frac{1}{(5)^3} = -\frac{1}{125}$$

d. This is read as "the quantity of $-32x^5$, raised to the negative two-fifths power."

$$(-32x^5)^{-2/5} = \frac{1}{(-32x^5)^{2/5}} = \frac{1}{(\sqrt[5]{-32x^5})^2} = \frac{1}{(-2x)^2} = \frac{1}{4x^2}$$

Caution

A base of 0 raised to a negative power is undefined. For example, $0^{-2} = \frac{1}{0^2}$ is undefined because we cannot divide by 0.

e. This is read as "1 over 25 to the negative three-halves power."

$$\frac{1}{25^{-3/2}} = 25^{3/2} = (\sqrt{25})^3 = (5)^3 = 125$$

Because the exponent is negative, write the reciprocal of $\frac{1}{25^{-3/2}}$, and change the sign of the exponent.

Teaching Example 7 Simplify. Assume that x can represent any nonzero real number.
a. $16^{-1/2}$ b. $(-8)^{2/3}$
c. $(27x^3)^{-2/3}$ d. $-125^{-2/3}$
e. $\dfrac{1}{8^{-2/3}}$
Answers:
a. $\frac{1}{4}$ b. 4 c. $\frac{1}{9x^2}$ d. $-\frac{1}{25}$ e. 4

Self Check 7 Simplify. Assume that a can represent any nonzero real number.
a. $9^{-1/2}$ $\frac{1}{3}$ **b.** $(36)^{-3/2}$ $\frac{1}{216}$ **c.** $(-27a^3)^{-2/3}$ $\frac{1}{9a^2}$ **d.** $-\frac{1}{81^{-3/4}}$ -27

Now Try Problems 65, 69, and 75

5 Use Rules for Exponents to Simplify Expressions.

We can use the rules for exponents to simplify many expressions with fractional exponents. If all variables represent positive real numbers, absolute value symbols are not needed.

EXAMPLE 8 Simplify. All variables represent positive real numbers. Write all answers using positive exponents only. **a.** $5^{2/7} \cdot 5^{3/7}$ **b.** $(11^{2/7})^3$ **c.** $(a^{2/3}b^{1/2})^6$ **d.** $\dfrac{a^{8/3}a^{1/3}}{a^2}$

Strategy We will use the product, power, and quotient rules for exponents to simplify each expression.

Why The familiar rules for exponents discussed in Chapter 5 are valid for rational exponents.

Solution

a. $5^{2/7} \cdot 5^{3/7} = 5^{2/7+3/7}$ Use the rule $x^m x^n = x^{m+n}$. Do not multiply the bases.

$= 5^{5/7}$ Add the exponents: $\frac{2}{7} + \frac{3}{7} = \frac{5}{7}$.

b. $(11^{2/7})^3 = 11^{(2/7)(3)}$ Use the rule $(x^m)^n = x^{mn}$.

$= 11^{6/7}$ Multiply the exponents: $\frac{2}{7}(3) = \frac{6}{7}$.

c. $(a^{2/3}b^{1/2})^6 = (a^{2/3})^6(b^{1/2})^6$ Use the rule $(xy)^n = x^n y^n$.

$= a^{12/3}b^{6/2}$ Use the rule $(x^m)^n = x^{mn}$ twice. Multiply the exponents.

$= a^4b^3$ Simplify the exponents.

d. $\dfrac{a^{8/3}a^{1/3}}{a^2} = a^{8/3+1/3-2}$ Use the rules $x^m x^n = x^{m+n}$ and $\frac{x^m}{x^n} = x^{m-n}$.

$= a^{8/3+1/3-6/3}$ To establish an LCD, write -2 as $-\frac{6}{3}$.

$= a^{3/3}$ Simplify: $\frac{8}{3} + \frac{1}{3} - \frac{6}{3} = \frac{3}{3}$.

$= a$ Simplify: $\frac{3}{3} = 1$.

Success Tip

Now is a good time to review the rules for exponents on page 364 in Chapter 5. These rules are not only valid when m and n are integers, but when m and n are rational numbers, as well.

Teaching Example 8 Simplify. All variables represent positive numbers.
a. $7^{2/9} \cdot 7^{5/9}$ b. $(7^{1/5})^4$
c. $(x^{1/8}y^{2/3})^{24}$ d. $\dfrac{x^{1/5}x^{7/5}}{x^{3/5}}$
Answers:
a. $7^{7/9}$ b. $7^{4/5}$
c. x^3y^{16} d. x

Self Check 8 Simplify. All variables represent positive real numbers.
a. $2^{1/5} \cdot 2^{2/5}$ $2^{3/5}$ **b.** $(12^{1/3})^4$ $12^{4/3}$ **c.** $(x^{1/3}y^{3/2})^6$ x^2y^9 **d.** $\dfrac{x^{5/3}x^{2/3}}{x^{1/3}}$ x^2

Now Try Problems 77, 81, and 85

EXAMPLE 9 Perform each multiplication and simplify when possible. Assume all variables represent positive real numbers. Write all answers using positive exponents only.

a. $a^{4/5}(a^{1/5} + a^{3/5})$ **b.** $x^{1/2}(x^{-1/2} - x^{1/2})$

Strategy We will use the distributive property and multiply each term within the parentheses by the term outside the parentheses.

Why The first expression has the form $a(b + c)$ and the second has the form $a(b - c)$.

Solution

a. $a^{4/5}(a^{1/5} + a^{3/5}) = a^{4/5}a^{1/5} + a^{4/5}a^{3/5}$ Use the distributive property.

$= a^{4/5+1/5} + a^{4/5+3/5}$ Use the rule $x^m x^n = x^{m+n}$.

$= a^{5/5} + a^{7/5}$ Add the exponents.

$= a + a^{7/5}$ We cannot add these terms because they are not like terms.

b. $x^{1/2}(x^{-1/2} - x^{1/2}) = x^{1/2}x^{-1/2} - x^{1/2}x^{1/2}$ Use the distributive property.

$= x^{1/2+(-1/2)} - x^{1/2+1/2}$ Use the rule $x^m x^n = x^{m+n}$.

$= x^0 - x^1$ Add the exponents.

$= 1 - x$ Simplify: $x^0 = 1$.

Caution

A common error when distributing $a^{4/5}$ over the terms $a^{1/5}$ and $a^{3/5}$ is to multiply the exponents incorrectly :

$a^{4/5}(a^{1/5} + a^{3/5}) = a^{4/25} + a^{12/25}$

Teaching Example 9 Simplify. All variables represent positive numbers. Write all answers using positive exponents only.
a. $x^{1/4}(x^{1/4} - x^{3/4})$
b. $x^{5/8}(x^{3/8} + x^{-5/8})$
Answers: a. $x^{1/2} - x$ b. $x + 1$

Self Check 9 Simplify: $t^{5/8}(t^{3/8} - t^{-5/8})$. Assume t represents a positive real number. $t - 1$

Now Try Problem 91

6 Simplify Radical Expressions.

We can simplify many radical expressions by using the following steps.

Using Rational Exponents to Simplify Radicals

1. Change the radical expression into an exponential expression.
2. Simplify the rational exponents.
3. Change the exponential expression back into a radical.

EXAMPLE 10 Simplify: **a.** $\sqrt[4]{3^2}$ **b.** $\sqrt[8]{x^6}$ **c.** $\sqrt[9]{27x^6y^3}$ **d.** $\sqrt[5]{\sqrt[3]{t}}$

Strategy We will write each radical expression as an equivalent exponential expression and use rules for exponents to simplify it. Then we will change that result back into a radical.

Why When the given expression is written in an equivalent exponential form, we can use rules for exponents and our arithmetic skills with fractions to simplify the exponents.

Solution

a.

$\sqrt[4]{3^2} = 3^{2/4}$ Use the rule $\sqrt[n]{x^m} = x^{m/n}$.

$= 3^{1/2}$ Simplify the fractional exponent: $\frac{2}{4} = \frac{1}{2}$.

$= \sqrt{3}$ Change back to radical form.

b.

$\sqrt[8]{x^6} = x^{6/8}$ Use the rule $\sqrt[n]{x^m} = x^{m/n}$.

$= x^{3/4}$ Simplify the fractional exponent: $\frac{6}{8} = \frac{3}{4}$.

$= (x^3)^{1/4}$ Write $\frac{3}{4}$ as $3\left(\frac{1}{4}\right)$.

$= \sqrt[4]{x^3}$ Change back to radical form.

c.

$\sqrt[9]{27x^6y^3} = (3^3x^6y^3)^{1/9}$ Write 27 as 3^3 and change the radical to an exponential expression.

$= 3^{3/9}x^{6/9}y^{3/9}$ Raise each factor to the $\frac{1}{9}$ power by multiplying the fractional exponents.

$= 3^{1/3}x^{2/3}y^{1/3}$ Simplify each fractional exponent.

$= (3x^2y)^{1/3}$ Use the rule $(xy)^n = x^ny^n$.

$= \sqrt[3]{3x^2y}$ Change back to radical form.

d.

$\sqrt[5]{\sqrt[3]{t}} = \sqrt[5]{t^{1/3}}$ Change the radical $\sqrt[3]{t}$ to exponential notation.

$= (t^{1/3})^{1/5}$ Change the radical $\sqrt[5]{t^{1/3}}$ to exponential notation.

$= t^{1/15}$ Use the rule $(x^m)^n = x^{mn}$. Multiply: $\frac{1}{3} \cdot \frac{1}{5} = \frac{1}{15}$.

$= \sqrt[15]{t}$ Change back to radical form.

Teaching Example 10 Simplify:
a. $\sqrt[9]{2^3}$ b. $\sqrt[9]{x^6}$ c. $\sqrt[21]{x^7y^{14}}$
d. $\sqrt[6]{\sqrt[3]{x}}$
Answers:
a. $\sqrt[3]{2}$ b. $\sqrt[3]{x^2}$ c. $\sqrt[3]{xy^2}$ d. $\sqrt[18]{x}$

Self Check 10 Simplify: **a.** $\sqrt[6]{3^3}$ $\sqrt{3}$ **b.** $\sqrt[4]{49x^2y^2}$ $\sqrt{7xy}$ **c.** $\sqrt[3]{\sqrt[4]{m}}$ $\sqrt[12]{m}$

Now Try Problems 93, 99, and 101

SECTION 9.2 STUDY SET

VOCABULARY

Fill in the blanks.

1. The expressions $4^{1/2}$ and $(-8)^{-2/3}$ have rational (or fractional) exponents.
2. We read $16^{3/2}$ as "16 to the three-halves power."
3. We read $27^{-1/3}$ as "27 to the negative one-third power."
4. We read $(-64a^5)^{4/5}$ as "the quantity of $-64a^5$, raised to the four-fifths power."
5. In the radical expression $\sqrt[4]{16x^8}$, 4 is the index, and $16x^8$ is the radicand.
6. $32^{4/5}$ means the fourth power of the fifth root of 32.

CONCEPTS

7. Complete the table by writing the given expression in the alternate form. Also give the base and exponent for the exponential form.

Radical form	Exponential form	Base	Exponent
$\sqrt[5]{25}$	$25^{1/5}$	25	$\frac{1}{5}$
$(\sqrt[3]{-27})^2$	$(-27)^{2/3}$	-27	$\frac{2}{3}$
$(\sqrt[4]{16})^{-3}$	$16^{-3/4}$	16	$-\frac{3}{4}$
$(\sqrt{81})^3$	$81^{3/2}$	81	$\frac{3}{2}$
$-\sqrt{\frac{9}{64}}$	$-\left(\frac{9}{64}\right)^{1/2}$	$\frac{9}{64}$	$\frac{1}{2}$

8. In your own words, explain the three rules for rational exponents illustrated in the diagrams below.

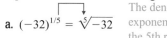

a. $(-32)^{1/5} = \sqrt[5]{-32}$ The denominator of the rational exponent indicates that we are to find the 5th root of the base.

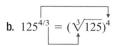

b. $125^{4/3} = (\sqrt[3]{125})^4$ The denominator of the rational exponent indicates that we are to find the cube root of the base. The numerator indicates that we are to find the 4th power of the cube root.

c. $8^{-1/3} = \dfrac{1}{8^{1/3}}$ Write the reciprocal of $8^{-1/3}$ and change the sign of the exponent.

▶ 9. Graph the following real numbers on a number line.

$$\left\{8^{2/3},\ (-125)^{1/3},\ -16^{-1/4},\ 4^{3/2},\ -\left(\frac{9}{100}\right)^{-1/2}\right\}$$

See AIE Appendix 3.

▶ 10. a. Evaluate $25^{3/2}$ by writing it in the form $(25^{1/2})^3$. 125

b. Evaluate $25^{3/2}$ by writing it in the form $(25^3)^{1/2}$. 125

c. Which way was easier? The first way

Complete each rule for exponents.

11. $x^{1/n} = \sqrt[n]{x}$

12. $x^{m/n} = (\sqrt[n]{x})^m = \sqrt[n]{x^m}$

13. $x^{-m/n} = \dfrac{1}{x^{m/n}}$

▶ 14. $\dfrac{1}{x^{-m/n}} = x^{m/n}$

NOTATION

Complete each solution.

15. Simplify:

$$(100a^4)^{3/2} = (\sqrt{100a^4})^3 = (10a^2)^3 = 1{,}000a^6$$

▶ 16. Simplify:

$$(m^{1/3}n^{1/2})^6 = (m^{1/3})^6(n^{1/2})^6 = m^{6/3}n^{6/2} = m^2n^3$$

GUIDED PRACTICE

Evaluate each expression. See Example 1.

▶ 17. $125^{1/3}$ 5

18. $8^{1/3}$ 2

▶ 19. $81^{1/4}$ 3

▶ 20. $625^{1/4}$ 5

21. $32^{1/5}$ 2

22. $0^{1/5}$ 0

▶ 23. $(-216)^{1/3}$ −6

24. $(-1{,}000)^{1/3}$ −10

▶ 25. $-16^{1/4}$ −2

26. $-125^{1/3}$ −5

27. $\left(\frac{1}{4}\right)^{1/2}$ $\frac{1}{2}$

▶ 28. $\left(\frac{1}{16}\right)^{1/2}$ $\frac{1}{4}$

Simplify each expression. Assume that the variables can be any real number, and use absolute value symbols when necessary. See Example 2.

29. $(4x^4)^{1/2}$ $2x^2$

▶ 30. $(25a^8)^{1/2}$ $5a^4$

31. $(x^2)^{1/2}$ $|x|$

▶ 32. $(x^3)^{1/3}$ x

▶ 33. $(-64p^8)^{1/2}$ Not real

34. $(-16q^4)^{1/2}$ Not real

35. $(-27n^9)^{1/3}$ $-3n^3$

▶ 36. $(-64t^9)^{1/3}$ $-4t^3$

37. $(-64x^8)^{1/8}$ Not real

▶ 38. $(243x^{10})^{1/5}$ $3x^2$

39. $[(x+1)^6]^{1/6}$ $|x+1|$

▶ 40. $[(x+5)^8]^{1/8}$ $|x+5|$

Evaluate each expression. See Example 3.

41. $36^{3/2}$ 216

▶ 42. $27^{2/3}$ 9

43. $16^{3/4}$ 8

▶ 44. $-100^{3/2}$ −1,000

45. $\left(-\frac{1}{216}\right)^{2/3}$ $\frac{1}{36}$

▶ 46. $\left(\frac{4}{9}\right)^{3/2}$ $\frac{8}{27}$

47. $-4^{5/2}$ −32

▶ 48. $(-125)^{4/3}$ 625

Simplify each expression. All variables represent positive real numbers. See Example 4.

▶ 49. $(25x^4)^{3/2}$ $125x^6$

50. $(27a^3b^3)^{2/3}$ $9a^2b^2$

51. $(-8x^6y^3)^{2/3}$ $4x^4y^2$

▶ 52. $(-32x^{10}y^5)^{4/5}$ $16x^8y^4$

▶ 53. $(81x^4y^8)^{3/4}$ $27x^3y^6$

54. $\left(\frac{1}{16}x^8y^4\right)^{3/4}$ $\frac{1}{8}x^6y^3$

55. $-\left(\frac{x^5}{32}\right)^{4/5}$ $-\frac{x^4}{16}$

▶ 56. $-\left(\frac{27}{64y^6}\right)^{2/3}$ $-\frac{9}{16y^4}$

Change each radical to an exponential expression. See Example 5.

▶ 57. $\sqrt[5]{8abc}$ $(8abc)^{1/5}$

▶ 58. $\sqrt[7]{7p^2q}$ $(7p^2q)^{1/7}$

59. $\sqrt[3]{a^2-b^2}$ $(a^2-b^2)^{1/3}$

▶ 60. $\sqrt{x^2+y^2}$ $(x^2+y^2)^{1/2}$

Change each exponential expression to a radical. See Example 6.

61. $(6x^3y)^{1/4}$ $\sqrt[4]{6x^3y}$

▶ 62. $(7a^2b^2)^{1/5}$ $\sqrt[5]{7a^2b^2}$

63. $(2s^2-t^2)^{1/2}$ $\sqrt{2s^2-t^2}$

▶ 64. $(x^3+y^3)^{1/3}$ $\sqrt[3]{x^3+y^3}$

Simplify each expression. All variables represent positive real numbers. See Example 7.

▶ 65. $4^{-1/2}$ $\frac{1}{2}$

66. $49^{-1/2}$ $\frac{1}{7}$

67. $125^{-1/3}$ $\frac{1}{5}$

▶ 68. $8^{-1/3}$ $\frac{1}{2}$

69. $-(1{,}000y^3)^{-2/3}$ $-\frac{1}{100y^2}$

▶ 70. $-(81c^4)^{-3/2}$ $-\frac{1}{729c^6}$

▶ 71. $\left(-\frac{27}{8}\right)^{-4/3}$ $\frac{16}{81}$

72. $\left(\frac{25}{49}\right)^{-3/2}$ $\frac{343}{125}$

73. $\left(\frac{16}{81y^4}\right)^{-3/4}$ $\frac{27y^3}{8}$

▶ 74. $\left(-\frac{8x^3}{27}\right)^{-1/3}$ $-\frac{3}{2x}$

▶ 75. $\dfrac{1}{9^{-5/2}}$ 243

76. $\dfrac{1}{16^{-5/2}}$ 1,024

Simplify each expression. Write the answers without negative exponents. All variables represent positive real numbers. See Example 8.

77. $9^{3/7} \cdot 9^{2/7}$ $9^{5/7}$

▶ 78. $4^{2/5} \cdot 4^{2/5}$ $4^{4/5}$

▶ 79. $6^{-2/3}6^{-4/3}$ $\frac{1}{36}$

80. $5^{1/3}5^{-5/3}$ $\frac{1}{5^{4/3}}$

81. $(m^{2/3}m^{1/3})^6$ m^6

▶ 82. $(b^{3/5}b^{2/5})^8$ b^8

83. $(a^{1/2}b^{1/3})^{3/2}$ $a^{3/4}b^{1/2}$

▶ 84. $(mn^{-2/3})^{-3/5}$ $\frac{n^{2/5}}{m^{3/5}}$

85. $\dfrac{3^{4/3}3^{1/3}}{3^{2/3}}$ 3

▶ 86. $\dfrac{2^{5/6}2^{1/3}}{2^{1/2}}$ $2^{2/3}$

87. $\dfrac{a^{3/4}a^{3/4}}{a^{1/2}}$ a

▶ 88. $\dfrac{b^{4/5}b^{4/5}}{b^{3/5}}$ b

Perform the multiplications. All variables represent positive real numbers. See Example 9.

89. $y^{1/3}(y^{2/3}+y^{5/3})$ $y+y^2$

▶ 90. $y^{2/5}(y^{-2/5}+y^{3/5})$ $1+y$

91. $x^{3/5}(x^{7/5}-x^{-3/5}+1)$ $x^2-1+x^{3/5}$

▶ 92. $x^{4/3}(x^{2/3}+3x^{5/3}-4)$ $x^2+3x^3-4x^{4/3}$

Use rational exponents to simplify each radical. All variables represent positive real numbers. See Example 10.

93. $\sqrt[4]{5^2}$ $\sqrt{5}$

▶ 94. $\sqrt[6]{7^3}$ $\sqrt{7}$

95. $\sqrt[9]{11^3}$ $\sqrt[3]{11}$

▶ 96. $\sqrt[12]{13^4}$ $\sqrt[3]{13}$

97. $\sqrt[6]{p^3}$ $\sqrt{p}$

▶ 98. $\sqrt[8]{q^2}$ $\sqrt[4]{q}$

99. $\sqrt[10]{x^2y^2}$ $\sqrt[5]{xy}$

▶ 100. $\sqrt[6]{x^2y^2}$ $\sqrt[3]{xy}$

101. $\sqrt[9]{\sqrt{c}}$ $\sqrt[18]{c}$

▶ 102. $\sqrt[4]{\sqrt{x}}$ $\sqrt[8]{x}$

▶ 103. $\sqrt[5]{\sqrt[3]{7m}}$ $\sqrt[15]{7m}$

104. $\sqrt[3]{\sqrt[4]{21x}}$ $\sqrt[12]{21x}$

Use a calculator to evaluate each expression. Round to the nearest hundredth. **See Using Your Calculator: Rational Exponents.**

105. $15^{1/3}$ 2.47

▶ 106. $(50.5)^{1/4}$ 2.67

107. $(1.045)^{2/5}$ 1.02

108. $(-1{,}000)^{3/5}$ −63.10

TRY IT YOURSELF

Simplify each expression. All variables represent positive real numbers.

109. $(25y^2)^{1/2}$ $5y$

▶ 110. $(-27x^3)^{1/3}$ $-3x$

111. $-\left(\dfrac{a^4}{81}\right)^{3/4}$ $-\dfrac{a^3}{27}$

▶ 112. $-\left(\dfrac{b^8}{625}\right)^{3/4}$ $-\dfrac{b^6}{125}$

▶ 113. $16^{-3/2}$ $\dfrac{1}{64}$

114 $(16)^{-5/4}$ $\dfrac{1}{32}$

115. $\dfrac{p^{8/5}p^{7/5}}{p^2}$ p

116. $\dfrac{c^{2/3}c^{2/3}}{c^{1/3}}$ c

117. $(-27x^6)^{-1/3}$ $-\dfrac{1}{3x^2}$

118. $(16a^4)^{-1/2}$ $\dfrac{1}{4a^2}$

119. $\dfrac{1}{32^{-1/5}}$ 2

120. $\dfrac{1}{64^{-1/6}}$ 2

121. $n^{1/5}(n^{2/5} - n^{-1/5})$ $n^{3/5} - 1$

▶ 122. $t^{4/3}(t^{5/3} + t^{-4/3})$ $t^3 + 1$

123. $\dfrac{1}{4^{-5/2}}$ 32

▶ 124. $\dfrac{1}{100^{-5/2}}$ 100,000

125. $(m^4)^{1/2}$ m^2

126. $(a^4)^{1/4}$ a

127. $\sqrt[4]{25b^2}$ $\sqrt{5b}$

128. $\sqrt[9]{8x^6}$ $\sqrt[3]{2x^2}$

129. $(16x^4)^{1/4}$ $2x$

▶ 130. $(-x^4)^{1/4}$ Not real

131. $-(8a^3b^6)^{-2/3}$ $-\dfrac{1}{4a^2b^4}$

132. $-(25s^4t^6)^{-3/2}$ $-\dfrac{1}{125s^6t^9}$

Look Alikes . . .

133. a. $-125^{2/3}$ −25 b. $(-125)^{2/3}$ 25

c. $-125^{-2/3}$ $-\dfrac{1}{25}$ d. $\dfrac{1}{(-125)^{-2/3}}$ 25

▶ 134. a. $81^{1/4}$ 3 b. $81^{-1/4}$ $\dfrac{1}{3}$

c. $-81^{1/4}$ −3 d. $\dfrac{1}{81^{-1/4}}$ 3

135. a. $(64a^4)^{1/2}$ $8a^2$ b. $(64a^4)^{-1/2}$ $\dfrac{1}{8a^2}$

c. $-(64a^4)^{1/2}$ $-8a^2$ d. $\dfrac{1}{(64a^4)^{1/2}}$ $\dfrac{1}{8a^2}$

136. a. $r^{1/3} \cdot r^{1/5}$ $r^{8/15}$ b. $(r^{1/3})^{1/5}$ $r^{1/15}$

c. $r^{1/3} \cdot r^{-1/5}$ $r^{2/15}$ d. $(r^{-1/3})^{-1/5}$ $r^{1/15}$

APPLICATIONS

▶ 137. **Ballistic Pendulums.** The formula $v = \dfrac{m + M}{m}(2gh)^{1/2}$ gives the velocity (in ft/sec) of a bullet with weight m fired into a block with weight M, that raises the height of the block h feet after the collision. See the illustration in the next column. The letter g represents a constant, 32. Find the velocity of the bullet to the nearest ft/sec. 736 ft/sec

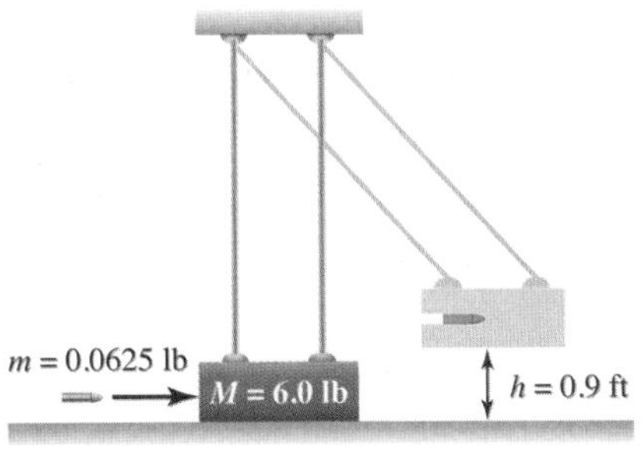

▶ 138. **Geography.** The formula $A = [s(s - a)(s - b)(s - c)]^{1/2}$ gives the area of a triangle with sides of length a, b, and c, where s is one-half of the perimeter. Estimate the area of Virginia (to the nearest square mile) using the data given in the illustration. 40,700 mi^2

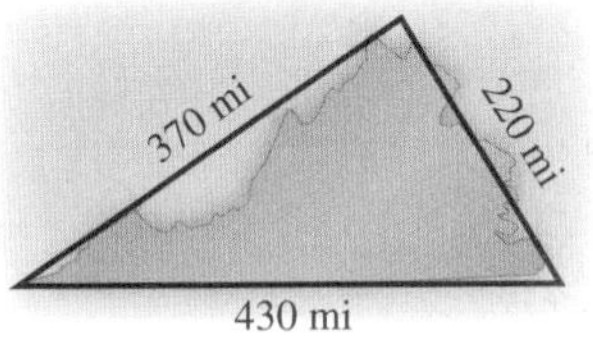

▶ 139. **Relativity.** One concept of relativity theory is that an object moving past an observer at a speed near the speed of light appears to have a larger mass because of its motion. If the mass of the object is m_0 when the object is at rest relative to the observer, its mass m will be given by the formula $m = m_0\left(1 - \frac{v^2}{c^2}\right)^{-1/2}$ when it is moving with speed v (in miles per second) past the observer. The variable c is the speed of light, 186,000 mi/sec. If a proton with a rest mass of 1 unit is accelerated by a nuclear accelerator to a speed of 160,000 mi/sec, what mass will the technicians observe it to have? Round to the nearest hundredth. 1.96 units

▶ 140. **Logging.** The width w and height h of the strongest rectangular beam that can be cut from a cylindrical log of radius a are given by $w = \frac{2a}{3}\left(3^{1/2}\right)$ and $h = a\left(\frac{8}{3}\right)^{1/2}$. Find the width, height, and cross-sectional area of the strongest beam that can be cut from a log with *diameter* 4 feet. Round to the nearest hundredth. 2.31 ft, 3.27 ft, 7.55 ft^2

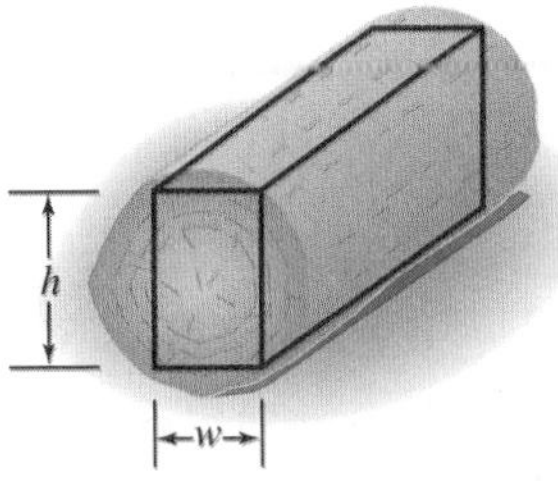

141. **from Campus to Careers**

General Contractor

©Goodluz/Shutterstock.com

The length L of the longest board that can be carried horizontally around the right-angle corner of two intersecting hallways is given by the formula $L = (a^{2/3} + b^{2/3})^{3/2}$, where a and b represent the widths of the hallways. Find the longest shelf that a carpenter can carry around the corner if $a = 40$ in. and $b = 64$ in. Give your result in inches and in feet. In each case, round to the nearest tenth. 145.8 in. or 12.1 ft

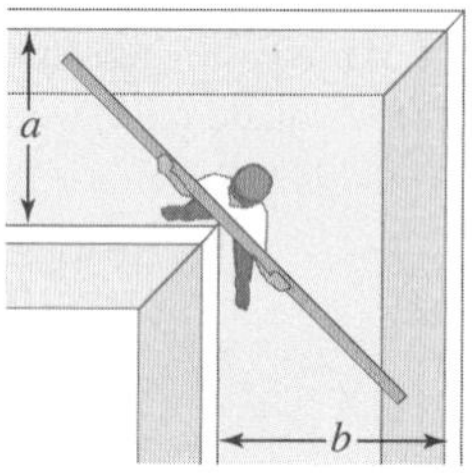

142. Cubicles. The area of the base of a cube is given by the function $A(V) = V^{2/3}$, where V is the volume of the cube. In a preschool room, 18 children's cubicles like the one shown are placed on the floor around the room. Estimate how much floor space is lost to the cubicles. Give your answer in square inches and in square feet. 4,608 in.^{2}, 32 ft^2

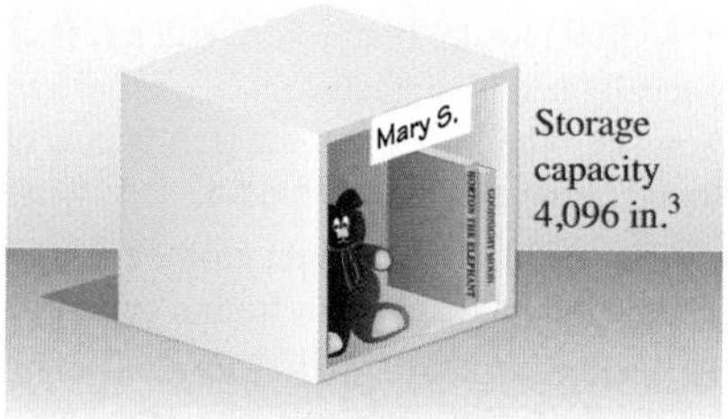

WRITING

143. What is a rational exponent? Give some examples.

144. Explain how the root key $\boxed{\sqrt[x]{y}}$ on a scientific calculator can be used in combination with other keys to evaluate the expression $16^{3/4}$.

REVIEW

145. Commuting Time. The time it takes a car to travel a certain distance varies inversely with its rate of speed. If a certain trip takes 3 hours at 50 miles per hour, how long will the trip take at 60 miles per hour? $2\frac{1}{2}$ hr

146. Bankruptcy. After filing for bankruptcy, a company was able to pay its creditors only 15 cents on the dollar. If the company owed a lumberyard \$9,712, how much could the lumberyard expect to be paid? \$1,456.80

CHALLENGE PROBLEMS

147. The fraction $\frac{2}{4}$ is equal to $\frac{1}{2}$. Is $16^{2/4}$ equal to $16^{1/2}$? Explain. Yes

148. Explain how would you evaluate an expression with a mixed-number exponent. For example, what is $8^{1\frac{1}{3}}$? What is $25^{2\frac{1}{2}}$? 16; 3,125

SECTION 9.3 Simplifying and Combining Radical Expressions

OBJECTIVES

1. Use the product rule to simplify radical expressions.
2. Use prime factorization to simplify radical expressions.
3. Use the quotient rule to simplify radical expressions.
4. Add and subtract radical expressions.

ARE YOU READY? *Are You Ready? exercises available online at www.webassign.net/brookscole*

The following problems review some basic skills that are needed when adding and subtracting radical expressions.

1. Complete each factorization:
a. $28 = 4 \cdot 7$ **b.** $54 = 27 \cdot 2$

2. Simplify:
a. $\sqrt[3]{-8}$ -2 **b.** $\sqrt[4]{16}$ 2

3. Multiply: $3 \cdot 3 \cdot 3 \cdot a^4 \cdot b^8$ $27a^4b^8$

4. Which of the following are like terms? $7x, 9x$
$7x \quad 3x^2 \quad 9x \quad 2x^3$

5. Combine like terms:
$15m^2 + 5m - m^2 - 6m$ $14m^2 - m$

6. Simplify: $\dfrac{\sqrt{25}}{\sqrt{36}}$ $\frac{5}{6}$

In algebra, it is often helpful to replace an expression with a simpler equivalent expression. This is certainly true when working with radicals. In most cases, radical expressions should be written in simplified form. We use two rules for radicals to do this.

1 Use the Product Rule to Simplify Radical Expressions.

To introduce the product rule for radicals, we will find $\sqrt{4 \cdot 25}$ and $\sqrt{4}\sqrt{25}$, and compare the results.

Square root of a product	***Product of square roots***
$\sqrt{4 \cdot 25} = \sqrt{100}$	$\sqrt{4}\sqrt{25} = 2 \cdot 5$
$= 10$	$= 10$

In each case, the answer is 10. Thus, $\sqrt{4 \cdot 25} = \sqrt{4}\sqrt{25}$.

Similarly, we will find $\sqrt[3]{8 \cdot 27}$ and $\sqrt[3]{8}\sqrt[3]{27}$ and compare the results.

Cube root of a product	***Product of cube roots***
$\sqrt[3]{8 \cdot 27} = \sqrt[3]{216}$	$\sqrt[3]{8}\sqrt[3]{27} = 2 \cdot 3$
$= 6$	$= 6$

In each case, the answer is 6. Thus, $\sqrt[3]{8 \cdot 27} = \sqrt[3]{8}\sqrt[3]{27}$. These results illustrate the *product rule for radicals.*

Notation

The products $\sqrt{4}\sqrt{25}$ and $\sqrt[3]{8}\sqrt[3]{27}$ also can be written using a raised dot:

$\sqrt{4} \cdot \sqrt{25}$ $\quad$ $\sqrt[3]{8} \cdot \sqrt[3]{27}$

The Product Rule for Radicals

The nth root of the product of two numbers is equal to the product of their nth roots. If $\sqrt[n]{a}$ and $\sqrt[n]{b}$ are real numbers,

$$\sqrt[n]{ab} = \sqrt[n]{a}\sqrt[n]{b}$$

Read as "the nth root of a times b equals the nth root of a times the nth root of b."

CAUTION The product rule for radicals applies to the nth root of a product. There is no such property for sums or differences. For example,

$$\sqrt{9 + 4} \neq \sqrt{9} + \sqrt{4} \qquad \sqrt{9 - 4} \neq \sqrt{9} - \sqrt{4}$$
$$\sqrt{13} \neq 3 + 2 \qquad \sqrt{5} \neq 3 - 2$$
$$\sqrt{13} \neq 5 \qquad \sqrt{5} \neq 1$$

Thus, $\sqrt{a + b} \neq \sqrt{a} + \sqrt{b}$ and $\sqrt{a - b} \neq \sqrt{a} - \sqrt{b}$.

The product rule for radicals can be used to simplify radical expressions. When a radical expression is written in **simplified form,** each of the following is true.

Simplified Form of a Radical Expression

1. Each factor in the radicand is to a power that is less than the index of the radical.
2. The radicand contains no fractions or negative numbers.
3. No radicals appear in the denominator of a fraction.

To simplify radical expressions, we must often factor the radicand using two natural-number factors. Since one factor should be a perfect square, perfect cube, perfect-fourth power, and so/on, it is helpful to memorize the following lists.

Teaching Tip: Stress the importance of memorizing the lists of numbers shown in color. Look for the color coding (red for squares, blue for cubes, green for fourth powers, purple for fifth powers) in the following worked examples.

Perfect squares: **1, 4, 9, 16, 25, 36, 49, 64, 81, 100, 121, 144, 169, 196, 225, . . .**

Perfect cubes: **1, 8, 27, 64, 125, 216, 343, 512, 729, 1,000, . . .**

Perfect-fourth powers: **1, 16, 81, 256, 625, . . .**

Perfect-fifth powers: **1, 32, 243, 1,024, . . .**

EXAMPLE 1 Simplify: **a.** $\sqrt{12}$ **b.** $\sqrt{98}$ **c.** $5\sqrt[3]{54}$ **d.** $-\sqrt[4]{48}$

Strategy We will factor each radicand into two factors, one of which is a perfect square, perfect cube, or perfect-fourth power, depending on the index of the radical. Then we can use the product rule for radicals to simplify the expression.

Why Factoring the radicand in this way leads to a square root, cube root, or fourth root of a perfect square, perfect cube, or perfect-fourth power that we can easily simplify.

Solution

The Language of Algebra

The instructions **simplify** and **approximate** do not mean the same thing.

Simplify: $\sqrt{12} = 2\sqrt{3}$ (exact)

Approximate:

$\sqrt{12} \approx 3.464$ (not exact)

When referrring to a square root, the instruction *simplify* means to remove any perfect-square factors from the radicand.

a. To simplify $\sqrt{12}$, we first factor 12 so that one factor is the largest perfect square that divides 12. Since 4 is the largest perfect-square factor of 12, we write 12 as $4 \cdot 3$, use the product rule for radicals, and simplify.

$\sqrt{12} = \sqrt{4 \cdot 3}$ Write 12 as $12 = 4 \cdot 3$.

Write the perfect-square factor first.

$= \sqrt{4}\sqrt{3}$ The square root of a product is equal to the product of the square roots.

$= 2\sqrt{3}$ Evaluate $\sqrt{4}$. Read as "2 times the square root of 3" or as "2 radical 3."

We say that $2\sqrt{3}$ is the simplified form of $\sqrt{12}$. Note that $\sqrt{12}$ and $2\sqrt{3}$ are just different representations of the same number. When we compare calculator approximations of each, that fact seems reasonable.

$$\sqrt{12} \approx 3.464101615 \qquad 2\sqrt{3} \approx 3.464101615$$

b. The largest perfect-square factor of 98 is 49. Thus,

$\sqrt{98} = \sqrt{49 \cdot 2}$ Write 98 in factored form: $98 = 49 \cdot 2$.

$= \sqrt{49}\sqrt{2}$ The square root of a product is equal to the product of the square roots: $\sqrt{49 \cdot 2} = \sqrt{49}\sqrt{2}$.

$= 7\sqrt{2}$ Evaluate $\sqrt{49}$.

c. Since the largest perfect-cube factor of 54 is 27, we have

The Language of Algebra

In Example 1, **a radical of a product** is written as **a product of radicals:**

$$\sqrt[n]{ab} = \sqrt[n]{a}\sqrt[n]{b}$$

$5\sqrt[3]{54} = 5\sqrt[3]{27 \cdot 2}$ Write 54 as $27 \cdot 2$.

$= 5\sqrt[3]{27}\sqrt[3]{2}$ The cube root of a product is equal to the product of the cube roots: $\sqrt[3]{27 \cdot 2} = \sqrt[3]{27}\sqrt[3]{2}$.

$= 5 \cdot 3\sqrt[3]{2}$ Evaluate $\sqrt[3]{27}$.

$= 15\sqrt[3]{2}$ Multiply: $5 \cdot 3 = 15$.

d. The largest perfect-fourth power factor of 48 is 16. Thus,

$-\sqrt[4]{48} = -\sqrt[4]{16 \cdot 3}$ Write 48 as $16 \cdot 3$.

$= -\sqrt[4]{16}\sqrt[4]{3}$ The fourth root of a product is equal to the product of the fourth roots: $\sqrt[4]{16 \cdot 3} = \sqrt[4]{16} \cdot \sqrt[4]{3}$.

$= -2\sqrt[4]{3}$ Evaluate $\sqrt[4]{16}$.

Teaching Example 1 Simplify:
a. $\sqrt{27}$ b. $10\sqrt{75}$
c. $\sqrt[3]{72}$ d. $-\sqrt[4]{80}$
Answers:
a. $3\sqrt{3}$ b. $50\sqrt{3}$ c. $2\sqrt[3]{9}$
d. $-2\sqrt[4]{5}$

Self Check 1 Simplify: **a.** $\sqrt{18}$ $3\sqrt{2}$ **b.** $7\sqrt[3]{24}$ $14\sqrt[3]{3}$ **c.** $\sqrt[4]{32}$ $2\sqrt[4]{2}$ **d.** $\sqrt[5]{128}$ $2\sqrt[5]{4}$

Now Try Problems 13, 17, and 19

Variable expressions also can be perfect squares, perfect cubes, perfect-fourth powers, and so on. For example,

Perfect squares: $x^2, x^4, x^6, x^8, x^{10}, \ldots$

Perfect cubes: $x^3, x^6, x^9, x^{12}, x^{15}, \ldots$

Perfect-fourth powers: $x^4, x^8, x^{12}, x^{16}, x^{20}, \ldots$

Perfect-fifth powers: $x^5, x^{10}, x^{15}, x^{20}, x^{25}, \ldots$

EXAMPLE 2 Simplify. All variables represent positive real numbers.

a. $\sqrt{m^9}$ **b.** $10\sqrt{128a^5}$ **c.** $\sqrt[3]{-24x^5}$ **d.** $\sqrt[5]{a^9b^5}$

Strategy We will factor each radicand into two factors, one of which is a perfect *n*th power.

Why We can then apply the rule *the nth root of a product is the product of the nth roots* to simplify the radical expression.

Solution **a.** The largest perfect-square factor of m^9 is m^8.

$$\sqrt{m^9} = \sqrt{m^8 \cdot m} \quad \text{Write } m^9 \text{ in factored form as } m^8 \cdot m.$$
$$= \sqrt{m^8}\sqrt{m} \quad \text{Use the product rule for radicals.}$$
$$= m^4\sqrt{m} \quad \text{Simplify } \sqrt{m^8}.$$

Caution

$\sqrt{m^9} \neq m^3$ because $(m^3)^2 \neq m^9$. By the power rule for exponents, $(m^3)^2 = m^6$. Similarly, $\sqrt{y^{25}} \neq y^5$ because $(y^5)^2 \neq y^{25}$.

b. Since the largest perfect-square factor of 128 is 64 and the largest perfect-square factor of a^5 is a^4, the largest perfect-square factor of $128a^5$ is $64a^4$. We write $128a^5$ as $64a^4 \cdot 2a$ and proceed as follows:

$$10\sqrt{128a^5} = 10\sqrt{64a^4 \cdot 2a} \quad \text{Write } 128a^5 \text{ in factored form as } 64a^4 \cdot 2a.$$
$$= 10\sqrt{64a^4}\sqrt{2a} \quad \text{Use the product rule for radicals.}$$
$$= 10 \cdot 8a^2\sqrt{2a} \quad \text{Simplify } \sqrt{64a^4}.$$
$$= 80a^2\sqrt{2a} \quad \text{Multiply: } 10 \cdot 8 = 80.$$

c. We write $-24x^5$ as $-8x^3 \cdot 3x^2$ and proceed as follows:

$$\sqrt[3]{-24x^5} = \sqrt[3]{-8x^3 \cdot 3x^2} \quad 8x^3 \text{ is the largest perfect-cube factor of } 24x^5. \text{ Since the radicand is negative, we factor it using } -8x^3.$$
$$= \sqrt[3]{-8x^3}\sqrt[3]{3x^2} \quad \text{Use the product rule for radicals.}$$
$$= -2x\sqrt[3]{3x^2} \quad \text{Simplify } \sqrt[3]{-8x^3}.$$

Caution

$\sqrt[3]{-24x^5} = -2x\sqrt[3]{3x^2}$

↑

In simplified form, the radicand should not have any perfect-cube factors.

d. The largest perfect-fifth power factor of a^9 is a^5, and b^5 is a perfect-fifth power.

$$\sqrt[5]{a^9b^5} = \sqrt[5]{a^5b^5 \cdot a^4} \quad a^5b^5 \text{ is the largest perfect-fifth power factor of } a^9b^5.$$
$$= \sqrt[5]{a^5b^5}\sqrt[5]{a^4} \quad \text{Use the product rule for radicals.}$$
$$= ab\sqrt[5]{a^4} \quad \text{Simplify } \sqrt[5]{a^5b^5}.$$

Teaching Example 2 Simplify. All variables represent positive real numbers.
a. $\sqrt{b^{11}}$ b. $\sqrt{125x^7}$
c. $9\sqrt[3]{-54x^{11}}$ d. $\sqrt[4]{a^8b^{19}}$
Answers:
a. $b^5\sqrt{b}$ b. $5x^3\sqrt{5x}$
c. $-27x^3\sqrt[3]{2x^2}$ d. $a^2b^4\sqrt[4]{b^3}$

Self Check 2 Simplify. All variables represent positive real numbers.
a. $6\sqrt{98b^3}$ $42b\sqrt{2b}$ **b.** $\sqrt[3]{-54y^5}$ $-3y\sqrt[3]{2y^2}$ **c.** $\sqrt[4]{t^8u^{15}}$ $t^2u^3\sqrt[4]{u^3}$

Now Try Problems 21, 25, and 27

2 Use Prime Factorization to Simplify Radical Expressions.

When simplifying radical expressions, prime factorization can be helpful in determining how to factor the radicand.

EXAMPLE 3 Simplify. All variables represent positive real numbers.

a. $\sqrt{150}$ **b.** $\sqrt[3]{297b^4}$ **c.** $\sqrt[4]{224s^8t^7}$

Strategy In each case, the way to factor the radicand is not obvious. Another approach is to prime-factor the coefficient of the radicand and look for groups of like factors.

Why Identifying groups of like factors of the radicand leads to a factorization of the radicand that can be easily simplified.

Solution

a. $\sqrt{150} = \sqrt{2 \cdot 3 \cdot 5 \cdot 5}$ Write 150 in prime-factored form.

$= \sqrt{2 \cdot 3}\sqrt{5 \cdot 5}$ Group the pair of like factors together and use the product rule for radicals.

$= \sqrt{2 \cdot 3}\sqrt{5^2}$ Write $5 \cdot 5$ as 5^2.

$= \sqrt{6} \cdot 5$ Evaluate $\sqrt{5^2}$.

$= 5\sqrt{6}$ Write the factor 5 first.

$$\begin{array}{r|l} 2 & 150 \\ 3 & 75 \\ 5 & 25 \\ & 5 \end{array}$$

Caution

To simplify $\sqrt{150}$, we could have factored 150 as $10 \cdot 15$. However, neither of the resulting radical expressions is the square root of a perfect square.

$$\sqrt{150} = \sqrt{10}\sqrt{15}$$

Teaching Tip: You may want to show your students calculator approximations:

$\sqrt{150} \approx 12.24744871$

$5\sqrt{6} \approx 12.24744871$

b. $\sqrt[3]{297b^4} = \sqrt[3]{3 \cdot 3 \cdot 3 \cdot 11 \cdot b^3 \cdot b}$ Write 297 in prime-factored form. The largest perfect-cube factor of b^4 is b^3.

$= \sqrt[3]{3 \cdot 3 \cdot 3 \cdot b^3}\sqrt[3]{11b}$ Group the three like factors of 3 together and use the product rule for radicals.

$= \sqrt[3]{3^3b^3}\sqrt[3]{11b}$ Write $3 \cdot 3 \cdot 3$ as 3^3.

$= 3b\sqrt[3]{11b}$ Simplify $\sqrt[3]{3^3b^3}$.

$$\begin{array}{r|l} 3 & 297 \\ 3 & 99 \\ 3 & 33 \\ & 11 \end{array}$$

Teaching Tip: Remind students that to avoid negative radicands for even roots, we are assuming that any variables within the radical symbol do not represent negative real numbers.

c. $\sqrt[4]{224s^8t^7} = \sqrt[4]{2 \cdot 2 \cdot 2 \cdot 2 \cdot 2 \cdot 7 \cdot s^8 \cdot t^4 \cdot t^3}$ Write 224 in prime-factored form. The largest perfect-fourth power factor of t^7 is t^4.

$= \sqrt[4]{2 \cdot 2 \cdot 2 \cdot 2 \cdot s^8 \cdot t^4}\sqrt[4]{2 \cdot 7 \cdot t^3}$ Group the four like factors of 2 together and use the product rule for radicals.

$= \sqrt[4]{2^4s^8t^4}\sqrt[4]{2 \cdot 7 \cdot t^3}$ Write $2 \cdot 2 \cdot 2 \cdot 2$ as 2^4.

$= 2s^2t\sqrt[4]{14t^3}$ Simplify $\sqrt[4]{2^4s^8t^4}$.

$$\begin{array}{r|l} 2 & 224 \\ 2 & 112 \\ 2 & 56 \\ 2 & 28 \\ 2 & 14 \\ & 7 \end{array}$$

Teaching Example 3 Simplify. All variables represent positive real numbers.
a. $\sqrt{175}$ b. $\sqrt[3]{-384x^5}$
c. $\sqrt[4]{162x^5y^{11}}$
Answers:
a. $5\sqrt{7}$ b. $-4x\sqrt[3]{6x^2}$
c. $3xy^2\sqrt[4]{2xy^3}$

Self Check 3 Simplify. All variables represent positive real numbers.

a. $\sqrt{275}$ $5\sqrt{11}$ **b.** $\sqrt[3]{189c^4d^3}$ $3cd\sqrt[3]{7c}$ **c.** $\sqrt[4]{1{,}250x^8y^6}$ $5x^2y\sqrt[4]{2y^2}$

Now Try Problems 29 and 35

3 Use the Quotient Rule to Simplify Radical Expressions.

To introduce the quotient rule for radicals, we will find $\sqrt{\frac{100}{4}}$ and $\frac{\sqrt{100}}{\sqrt{4}}$ and compare the results.

Square root of a quotient	***Quotient of square roots***
$\sqrt{\frac{100}{4}} = \sqrt{25}$	$\frac{\sqrt{100}}{\sqrt{4}} = \frac{10}{2}$
$= 5$	$= 5$

Since the answer is 5 in each case, $\sqrt{\frac{100}{4}} = \frac{\sqrt{100}}{\sqrt{4}}$.

Similarly, we will find $\sqrt[3]{\frac{64}{8}}$ and $\frac{\sqrt[3]{64}}{\sqrt[3]{8}}$, and compare the results.

Cube root of a quotient	***Quotient of cube roots***
$\sqrt[3]{\frac{64}{8}} = \sqrt[3]{8}$	$\frac{\sqrt[3]{64}}{\sqrt[3]{8}} = \frac{4}{2}$
$= 2$	$= 2$

Since the answer is 2 in each case, $\sqrt[3]{\frac{64}{8}} = \frac{\sqrt[3]{64}}{\sqrt[3]{8}}$. These results illustrate the *quotient rule for radicals.*

The Quotient Rule for Radicals

The nth root of the quotient of two numbers is equal to the quotient of their nth roots. If $\sqrt[n]{a}$ and $\sqrt[n]{b}$ are real numbers, then

$$\sqrt[n]{\frac{a}{b}} = \frac{\sqrt[n]{a}}{\sqrt[n]{b}} \quad (b \neq 0)$$

Read as "the nth root of a divided by b equals the nth root of a divided by the nth root of b."

In the following examples, variables appear in the denominators of radical expressions. To avoid undefined situations, we will assume ***all variables represent positive real numbers.***

EXAMPLE 4 Simplify each expression: **a.** $\sqrt{\frac{7}{64}}$ **b.** $\sqrt{\frac{15}{49x^2}}$ **c.** $\sqrt[3]{\frac{10x^2}{27y^6}}$

Strategy In each case, the radical is not in simplified form because the radicand contains a fraction. To write each of these expressions in simplified form, we will use the quotient rule for radicals.

Why Writing these expressions in $\frac{\sqrt[n]{a}}{\sqrt[n]{b}}$ form leads to square roots of perfect squares and cube roots of perfect cubes that we can easily simplify.

Solution **a.** We can use the quotient rule for radicals to simplify each expression.

$$\sqrt{\frac{7}{64}} = \frac{\sqrt{7}}{\sqrt{64}}$$ *The square root of a quotient is equal to the quotient of the square roots.*

$$= \frac{\sqrt{7}}{8}$$ *Evaluate $\sqrt{64}$.*

b. $$\sqrt{\frac{15}{49x^2}} = \frac{\sqrt{15}}{\sqrt{49x^2}}$$ *The square root of a quotient is equal to the quotient of the square roots.*

$$= \frac{\sqrt{15}}{7x}$$ *Simplify the denominator: $\sqrt{49x^2} = 7x$.*

c. $$\sqrt[3]{\frac{10x^2}{27y^6}} = \frac{\sqrt[3]{10x^2}}{\sqrt[3]{27y^6}}$$ *The cube root of a quotient is equal to the quotient of the cube roots.*

$$= \frac{\sqrt[3]{10x^2}}{3y^2}$$ *Simplify the denominator.*

The Language of Algebra

In Example 4, a **radical of a quotient** is written as a **quotient of radicals:**

$$\sqrt[n]{\frac{a}{b}} = \frac{\sqrt[n]{a}}{\sqrt[n]{b}}$$

Teaching Example 4 Simplify: a. $\sqrt{\frac{13}{36}}$ b. $\sqrt{\frac{22}{81x^2}}$ c. $\sqrt[3]{\frac{5a^2}{8b^9}}$
Answers: a. $\frac{\sqrt{13}}{6}$ b. $\frac{\sqrt{22}}{9x}$ c. $\frac{\sqrt[3]{5a^2}}{2b^3}$

Self Check 4 Simplify: **a.** $\sqrt[3]{\frac{25}{27}}$ $\frac{\sqrt[3]{25}}{3}$ **b.** $\sqrt{\frac{11}{36a^2}}$ $\frac{\sqrt{11}}{6a}$ **c.** $\sqrt[4]{\frac{a^3}{625y^{12}}}$ $\frac{\sqrt[4]{a^3}}{5y^3}$

Now Try Problems 37, 39, and 43

EXAMPLE 5 Simplify: **a.** $\frac{\sqrt{45xy^2}}{\sqrt{5x}}$ **b.** $\frac{\sqrt[3]{-432x^5}}{\sqrt[3]{8x}}$

Strategy We will use the quotient rule for radicals in reverse: $\frac{\sqrt[n]{a}}{\sqrt[n]{b}} = \sqrt[n]{\frac{a}{b}}$

Why When the radicands are written under a single radical symbol, the result is a rational expression. Our hope is that the rational expression can be simplified.

Solution

a. We can write the quotient of the square roots as the square root of a quotient.

$$\frac{\sqrt{45xy^2}}{\sqrt{5x}} = \sqrt{\frac{45xy^2}{5x}}$$ Use the quotient rule for radicals. Note that the resulting radicand is a rational expression.

$$= \sqrt{9y^2}$$ Simplify the radicand: $\frac{45xy^2}{5x} = \frac{\overset{1}{\cancel{5}} \cdot 9 \cdot \overset{1}{\cancel{x}} \cdot y^2}{\underset{1}{\cancel{5}} \cdot \underset{1}{\cancel{x}}} = 9y^2$.

$$= 3y$$ Simplify the radical.

b. We can write the quotient of the cube roots as the cube root of a quotient.

$$\frac{\sqrt[3]{-432x^5}}{\sqrt[3]{8x}} = \sqrt[3]{-\frac{432x^5}{8x}}$$ Use the quotient rule for radicals. Note that the resulting radicand is a rational expression.

$$= \sqrt[3]{-54x^4}$$ Simplify the radicand: $-\frac{432x^5}{8x} = -54x^4$.

$$= \sqrt[3]{-27x^3 \cdot 2x}$$ $27x^3$ is the largest perfect cube that divides $54x^4$.

$$= \sqrt[3]{-27x^3}\sqrt[3]{2x}$$ Use the product rule for radicals.

$$= -3x\sqrt[3]{2x}$$ Simplify: $\sqrt[3]{-27x^3} = -3x$.

Self Check 5 Simplify: **a.** $\frac{\sqrt{50ab^2}}{\sqrt{2a}}$ $5b$ **b.** $\frac{\sqrt[3]{-2{,}000x^5y^3}}{\sqrt[3]{2x}}$ $-10xy\sqrt[3]{x}$

Now Try Problems 47 and 51

The Language of Algebra

In Example 5, a **quotient of radicals** is written as a **radical of a quotient.**

$$\frac{\sqrt[n]{a}}{\sqrt[n]{b}} = \sqrt[n]{\frac{a}{b}}$$

Success Tip

In parts a and b, we use skills learned in Chapter 7 to simplify the **rational expressions** in each radicand: $\frac{45xy^2}{5x}$ and $-\frac{432x^5}{8x}$.

Teaching Example 5 Simplify: a. $\frac{\sqrt{98x^2y}}{\sqrt{2y}}$ b. $\frac{\sqrt[3]{-375x^{11}}}{\sqrt[3]{3x^2}}$

Answers: a. $7x$ b. $-5x^3$

4 Add and Subtract Radical Expressions.

Radical expressions with the same index and the same radicand are called **like** or **similar radicals.** For example, $3\sqrt{2}$ and $2\sqrt{2}$ are like radicals. However,

- $3\sqrt{5}$ and $4\sqrt{2}$ are not like radicals, because the radicands are different.
- $3\sqrt[4]{5}$ and $2\sqrt[3]{5}$ are not like radicals, because the indices are different.

For an expression with two or more radical terms, we should attempt to combine like radicals, if possible. For example, to simplify the expression $3\sqrt{2} + 2\sqrt{2}$, we use the distributive property to factor out $\sqrt{2}$ and simplify.

$$3\sqrt{2} + 2\sqrt{2} = (3 + 2)\sqrt{2}$$ Factor out $\sqrt{2}$.

$$= 5\sqrt{2}$$ Do the addition.

Radicals with the same index but different radicands often can be written as like radicals. For example, to simplify the expression $\sqrt{75} - \sqrt{27}$, we simplify both radicals first and then combine the like radicals.

$$\sqrt{75} - \sqrt{27} = \sqrt{25 \cdot 3} - \sqrt{9 \cdot 3}$$ Write 75 and 27 in factored form.

$$= \sqrt{25}\sqrt{3} - \sqrt{9}\sqrt{3}$$ Use the product rule for radicals.

$$= 5\sqrt{3} - 3\sqrt{3}$$ Evaluate $\sqrt{25}$ and $\sqrt{9}$.

$$= (5 - 3)\sqrt{3}$$ Factor out $\sqrt{3}$.

$$= 2\sqrt{3}$$ Do the subtraction.

As the previous examples suggest, we can add or subtract radicals as follows.

Success Tip

Combining like radicals is similar to combining like terms.

$$3\sqrt{2} + 2\sqrt{2} = 5\sqrt{2}$$

$$3x + 2x = 5x$$

Adding and Subtracting Radicals

To add or subtract radicals, simplify each radical, if possible, and combine like radicals.

EXAMPLE 6 Simplify: **a.** $18 + 2\sqrt{12} - 3\sqrt{48} + 9$ **b.** $\sqrt[3]{16} + \sqrt[3]{54} - \sqrt[3]{24}$

Strategy Since the radicals in each part are unlike radicals, we cannot add or subtract them in their current form. However, we will simplify the radicals and hope that like radicals result.

Why Like radicals can be combined.

Solution **a.** We begin by simplifying each radical expression:

$$18 + 2\sqrt{12} - 3\sqrt{48} + 9 = 18 + 2\sqrt{4 \cdot 3} - 3\sqrt{16 \cdot 3} + 9$$ Factor the radicands, 12 and 48.

$$= 18 + 2\sqrt{4}\sqrt{3} - 3\sqrt{16}\sqrt{3} + 9$$ Use the product rule for radicals.

$$= 18 + 2(2)\sqrt{3} - 3(4)\sqrt{3} + 9$$ Evaluate $\sqrt{4}$ and $\sqrt{16}$.

$$= 18 + \mathbf{4}\sqrt{3} - \mathbf{12}\sqrt{3} + 9$$ Both radicals now have the same index, 2, and radicand, 3.

$$= (\mathbf{4} - \mathbf{12})\sqrt{3} + 27$$ Add the constants: $18 + 9 = 27$. Combine like radicals.

$$= -8\sqrt{3} + 27$$ Do the subtraction.

Teaching Tip: In Section 9.4, students will have to simplify expressions such as $18 + 4\sqrt{3} - 12\sqrt{3} + 9$ after multiplying two radical expressions.

b. We begin by simplifying each radical expression:

$$\sqrt[3]{16} + \sqrt[3]{54} - \sqrt[3]{24} = \sqrt[3]{8 \cdot 2} + \sqrt[3]{27 \cdot 2} - \sqrt[3]{8 \cdot 3}$$ Factor the radicands.

$$= \sqrt[3]{8}\sqrt[3]{2} + \sqrt[3]{27}\sqrt[3]{2} - \sqrt[3]{8}\sqrt[3]{3}$$ Use the product rule.

$$= \mathbf{2}\sqrt[3]{\mathbf{2}} + \mathbf{3}\sqrt[3]{\mathbf{2}} - 2\sqrt[3]{3}$$ Evaluate $\sqrt[3]{8}$ and $\sqrt[3]{27}$.

$$= (\mathbf{2} + \mathbf{3})\sqrt[3]{2} - 2\sqrt[3]{3}$$ Combine the first two radical expressions because they have the same index and radicand.

$$= 5\sqrt[3]{2} - 2\sqrt[3]{3}$$ Do the addition.

CAUTION Even though the expressions $5\sqrt[3]{2}$ and $2\sqrt[3]{3}$ have the same index, we cannot combine them, because their radicands are different. Neither can we combine radical expressions having the same radicand but a different index. For example, the expression $\sqrt[3]{2} + \sqrt[4]{2}$ cannot be simplified.

Teaching Example 6 Simplify:
a. $10 + 5\sqrt{50} - 3\sqrt{98} + 12$
b. $\sqrt[3]{40} + 2\sqrt[3]{135} - \sqrt[3]{250}$
Answers:
a. $4\sqrt{2} + 22$
b. $8\sqrt[3]{5} - 5\sqrt[3]{2}$

Self Check 6 Simplify: **a.** $14 + 3\sqrt{75} - 2\sqrt{12} + 2\sqrt{48} - 8$ $19\sqrt{3} + 6$
b. $\sqrt[3]{24} - \sqrt[3]{16} + \sqrt[3]{54}$ $2\sqrt[3]{3} + \sqrt[3]{2}$

Now Try Problems 57 and 63

EXAMPLE 7 Simplify: $\sqrt[3]{16x^4} + 4\sqrt[3]{54x^4} - x\sqrt[3]{-128x}$

Strategy Since the radicals are unlike radicals, we cannot add or subtract them in their current form. However, we will simplify the radicals and hope that like radicals result.

Why Like radicals can be combined.

Solution We begin by simplifying each radical expression.

$$\sqrt[3]{16x^4} + 4\sqrt[3]{54x^4} - x\sqrt[3]{-128x}$$

$$= \sqrt[3]{8x^3 \cdot 2x} + 4\sqrt[3]{27x^3 \cdot 2x} - x\sqrt[3]{-64 \cdot 2x}$$

$$= \sqrt[3]{\mathbf{8x^3}}\sqrt[3]{2x} + 4\sqrt[3]{\mathbf{27x^3}}\sqrt[3]{2x} - x\sqrt[3]{\mathbf{-64}}\sqrt[3]{2x}$$

$$= \mathbf{2x}\sqrt[3]{2x} + 4 \cdot \mathbf{3x}\sqrt[3]{2x} + \mathbf{x \cdot 4}\sqrt[3]{2x}$$ All three radicals have the same index and radicand.

$$= 2x\sqrt[3]{2x} + 12x\sqrt[3]{2x} + 4x\sqrt[3]{2x}$$ Combine like radicals.

$$= (2x + 12x + 4x)\sqrt[3]{2x}$$ Multiply: $4 \cdot 3x = 12x$ and $x \cdot 4 = 4x$.

$$= 18x\sqrt[3]{2x}$$ Within the parentheses, combine like terms: $2x + 12x + 4x = 18x$.

Teaching Tip: Ask your students at what stage of the solution they can tell that there will be like radicals to combine. Hopefully, they say the first line, where all three radicands involve a factor of $2x$.

Teaching Example 7 Simplify: $5\sqrt[4]{48x^7} - 2x\sqrt[4]{3x^3} + x\sqrt[4]{243x^3}$
Answer: $11x\sqrt[4]{3x^3}$

Self Check 7 Simplify: $\sqrt{32x^3} + 4\sqrt{50x^3} - x\sqrt{18x}$ $21x\sqrt{2x}$

Now Try Problems 67 and 69

SECTION 9.3 STUDY SET

VOCABULARY

Fill in the blanks.

1. Radical expressions such as $\sqrt[3]{4}$ and $6\sqrt[3]{4}$ with the same index and the same radicand are called like radicals.
2. Numbers such as 1, 4, 9, 16, 25, and 36 are called perfect squares. Numbers such as 1, 8, 27, 64, and 125 are called perfect cubes. Numbers such as 1, 16, 81, 256, and 625 are called perfect-fourth powers.
3. The largest perfect-square factor of 27 is 9. The largest perfect-cube factor of 16 is 8.
4. To simplify $\sqrt{24}$ means to write it as $2\sqrt{6}$.

CONCEPTS

Fill in the blanks.

5. The product rule for radicals: $\sqrt[n]{ab} = \sqrt[n]{a}\sqrt[n]{b}$. In words, the nth root of the product of two numbers is equal to the product of their nth roots.
6. The quotient rule for radicals: $\sqrt[n]{\frac{a}{b}} = \frac{\sqrt[n]{a}}{\sqrt[n]{b}}$. In words, the nth root of the quotient of two numbers is equal to the quotient of their nth roots.
7. Consider the expressions $\sqrt{4 \cdot 5}$ and $\sqrt{4}\sqrt{5}$. Which expression is
   a. the square root of a product? $\sqrt{4 \cdot 5}$
   b. the product of square roots? $\sqrt{4}\sqrt{5}$
   c. How are these two expressions related? $\sqrt{4 \cdot 5} = \sqrt{4}\sqrt{5}$
8. Consider $\frac{\sqrt[3]{a}}{\sqrt[3]{x^2}}$ and $\sqrt[3]{\frac{a}{x^2}}$. Which expression is
   a. the cube root of a quotient? $\sqrt[3]{\frac{a}{x^2}}$
   b. the quotient of cube roots? $\frac{\sqrt[3]{a}}{\sqrt[3]{x^2}}$
   c. How are these two expressions related? $\sqrt[3]{\frac{a}{x^2}} = \frac{\sqrt[3]{a}}{\sqrt[3]{x^2}}$
9. a. Write two radical expressions that have the same radicand but a different index. Can the expressions be added?
   $\sqrt{5}$, $\sqrt[3]{5}$ (Answers may vary); no
   b. Write two radical expressions that have the same index but a different radicand. Can the expressions be added?
   $\sqrt{5}$, $\sqrt{6}$ (Answers may vary); no
10. Fill in the blanks.
   a. $5\sqrt{6} + 3\sqrt{6} = (5 + 3)\sqrt{6} = 8\sqrt{6}$
   b. $9\sqrt[3]{n} - 2\sqrt[3]{n} = (9 - 2)\sqrt[3]{n} = 7\sqrt[3]{n}$

NOTATION

Complete each solution.

11. Simplify:
$$\sqrt[3]{32k^4} = \sqrt[3]{8k^3 \cdot 4k} = \sqrt[3]{8k^3}\sqrt[3]{4k} = 2k\sqrt[3]{4k}$$

12. Simplify:
$$\frac{\sqrt{80s^2t^4}}{\sqrt{5s^2}} = \sqrt{\frac{80s^2t^4}{5s^2}} = \sqrt{16t^4} = 4t^2$$

GUIDED PRACTICE

Simplify each expression. **See Example 1.**

13. $\sqrt{50}$ $5\sqrt{2}$
14. $\sqrt{28}$ $2\sqrt{7}$
15. $8\sqrt{45}$ $24\sqrt{5}$
16. $9\sqrt{54}$ $27\sqrt{6}$
17. $\sqrt[3]{32}$ $2\sqrt[3]{4}$
18. $\sqrt[3]{40}$ $2\sqrt[3]{5}$
19. $\sqrt[4]{48}$ $2\sqrt[4]{3}$
20. $\sqrt[4]{32}$ $2\sqrt[4]{2}$

Simplify each radical expression. All variables represent positive real numbers. **See Example 2.**

21. $\sqrt{75a^2}$ $5a\sqrt{3}$
22. $\sqrt{50x^2}$ $5x\sqrt{2}$
23. $\sqrt{128a^3b^5}$ $8ab^2\sqrt{2ab}$
24. $\sqrt{75b^8c}$ $5b^4\sqrt{3c}$
25. $2\sqrt[3]{-54x^6}$ $-6x^2\sqrt[3]{2}$
26. $4\sqrt[3]{-81a^3}$ $-12a\sqrt[3]{3}$
27. $\sqrt[4]{32x^{12}y^4}$ $2x^3y\sqrt[4]{2}$
28. $\sqrt[5]{64x^{10}y^5}$ $2x^2y\sqrt[5]{2}$

Simplify each radical expression. All variables represent positive real numbers. **See Example 3.**

29. $\sqrt{242}$ $11\sqrt{2}$
30. $\sqrt{363}$ $11\sqrt{3}$
31. $\sqrt{112a^3}$ $4a\sqrt{7a}$
32. $\sqrt{147a^5}$ $7a^2\sqrt{3a}$
33. $-\sqrt[5]{96a^4}$ $-2\sqrt[5]{3a^4}$
34. $-\sqrt[7]{256t^6}$ $-2\sqrt[7]{2t^6}$
35. $\sqrt[3]{405x^{12}y^4}$ $3x^4y\sqrt[3]{15y}$
36. $\sqrt[3]{280a^5b^6}$ $2ab^2\sqrt[3]{35a^2}$

Simplify each radical expression. All variables represent positive real numbers. **See Example 4.**

37. $\sqrt{\frac{11}{9}}$ $\frac{\sqrt{11}}{3}$
38. $\sqrt{\frac{3}{4}}$ $\frac{\sqrt{3}}{2}$
39. $\sqrt[4]{\frac{3}{625}}$ $\frac{\sqrt[4]{3}}{5}$
40. $\sqrt[5]{\frac{2}{243}}$ $\frac{\sqrt[5]{2}}{3}$
41. $\sqrt[5]{\frac{3x^{10}}{32}}$ $\frac{x^2\sqrt[5]{3}}{2}$
42. $\sqrt[6]{\frac{5y^{12}}{64}}$ $\frac{y^2\sqrt[6]{5}}{2}$
43. $\sqrt{\frac{z^2}{16x^2}}$ $\frac{z}{4x}$
44. $\sqrt{\frac{b^4}{64a^8}}$ $\frac{b^2}{8a^4}$

Selected exercises available online at www.webassign.net/brookscole

Simplify each expression. All variables represent positive real numbers. **See Example 5.**

45. $\frac{\sqrt{500}}{\sqrt{5}}$ 10

▶ 46. $\frac{\sqrt{128}}{\sqrt{2}}$ 8

47. $\frac{\sqrt{98x^3}}{\sqrt{2x}}$ $7x$

▶ 48. $\frac{\sqrt{75y^5}}{\sqrt{3y}}$ $5y^2$

49. $\frac{\sqrt[3]{48x^7}}{\sqrt[3]{6x}}$ $2x^2$

▶ 50. $\frac{\sqrt[3]{64y^8}}{\sqrt[3]{8y^2}}$ $2y^2$

51. $\frac{\sqrt[3]{189a^5}}{\sqrt[3]{7a}}$ $3a\sqrt[3]{a}$

▶ 52. $\frac{\sqrt[3]{243x^8}}{\sqrt[3]{9x}}$ $3x^2\sqrt[3]{x}$

Simplify by combining like radicals. **See Objective 4 and Example 6.**

53. $5\sqrt{7} + 3\sqrt{7}$ $8\sqrt{7}$

▶ 54. $11\sqrt{3} + 2\sqrt{3}$ $13\sqrt{3}$

55. $20\sqrt[3]{4} - 15\sqrt[3]{4}$ $5\sqrt[3]{4}$

▶ 56. $30\sqrt[3]{6} - 10\sqrt[3]{6}$ $20\sqrt[3]{6}$

▶ 57. $4 + \sqrt{8} + \sqrt{2} + 8$ $12 + 3\sqrt{2}$

58. $9 + \sqrt{45} + \sqrt{20} + 16$ $25 + 5\sqrt{5}$

▶ 59. $\sqrt{98} - \sqrt{50} - \sqrt{72}$ $-4\sqrt{2}$

▶ 60. $\sqrt{20} + \sqrt{125} - \sqrt{80}$ $3\sqrt{5}$

▶ 61. $8 + \sqrt[3]{32} - \sqrt[3]{108} - 7$ $1 - \sqrt[3]{4}$

62. $12 + \sqrt[3]{80} - \sqrt[3]{10{,}000} + 4$ $16 - 8\sqrt[3]{10}$

63. $14\sqrt[4]{32} - 15\sqrt[4]{2}$ $13\sqrt[4]{2}$

▶ 64. $23\sqrt[4]{3} + \sqrt[4]{48}$ $25\sqrt[4]{3}$

Simplify by combining like radicals. All variables represent positive real numbers. **See Example 7.**

▶ 65. $4\sqrt{2x} + 6\sqrt{2x}$ $10\sqrt{2x}$

66. $6\sqrt[3]{5y} + 3\sqrt[3]{5y}$ $9\sqrt[3]{5y}$

67. $\sqrt{18t} + \sqrt{300t} - \sqrt{243t}$ $3\sqrt{2t} + \sqrt{3t}$

▶ 68. $\sqrt{80m} - \sqrt{128m} + \sqrt{288m}$ $4\sqrt{5m} + 4\sqrt{2m}$

▶ 69. $2\sqrt[3]{16} - \sqrt[3]{54} - 3\sqrt[3]{128}$ $-11\sqrt[3]{2}$

70. $\sqrt[3]{250} - 4\sqrt[3]{5} + \sqrt[3]{16}$ $7\sqrt[3]{2} - 4\sqrt[3]{5}$

71. $\sqrt[4]{32} + 5\sqrt[4]{2} - \sqrt[4]{162}$ $4\sqrt[4]{2}$

▶ 72. $\sqrt[4]{48} - \sqrt[4]{243} - \sqrt[4]{768}$ $-5\sqrt[4]{3}$

TRY IT YOURSELF

Simplify each expression, if possible. All variables represent positive real numbers.

73. $\sqrt[6]{m^{11}}$ $m\sqrt[6]{m^5}$

74. $\sqrt[5]{n^{13}}$ $n^2\sqrt[5]{n}$

75. $2\sqrt[3]{64a} + 2\sqrt[3]{8a}$ $12\sqrt[3]{a}$

▶ 76. $3\sqrt[4]{x^4y} - 2\sqrt[4]{x^4y}$ $x\sqrt[4]{y}$

77. $\sqrt{8y^7} + \sqrt{32y^7} - \sqrt{2y^7}$ $5y^3\sqrt{2y}$

▶ 78. $\sqrt{y^5} - \sqrt{9y^5} - \sqrt{25y^5}$ $-7y^2\sqrt{y}$

79. $\sqrt{32b}$ $4\sqrt{2b}$

80. $\sqrt{80c}$ $4\sqrt{5c}$

81. $\sqrt{\frac{125n^5}{64n}}$ $\frac{5n^2\sqrt{5}}{8}$

82. $\sqrt{\frac{72q^7}{25q^3}}$ $\frac{6q^2\sqrt{2}}{5}$

83. $2\sqrt[3]{125} - 5\sqrt[3]{64}$ -10

84. $3\sqrt[3]{27} + 12\sqrt[3]{216}$ 81

85. $\sqrt{300xy}$ $10\sqrt{3xy}$

▶ 86. $\sqrt{200x^2y}$ $10x\sqrt{2y}$

87. $\sqrt[4]{\frac{5x}{16z^4}}$ $\frac{\sqrt[4]{5x}}{2z}$

88. $\sqrt[3]{\frac{11a^2}{125b^6}}$ $\frac{\sqrt[3]{11a^2}}{5b^2}$

89. $8\sqrt[5]{7a^2} - 7\sqrt[5]{7a^2}$ $\sqrt[5]{7a^2}$

90. $10\sqrt[6]{12xy} - \sqrt[6]{12xy}$ $9\sqrt[6]{12xy}$

91. $\sqrt[5]{x^6y^2} + \sqrt[5]{32x^6y^2} + \sqrt[5]{x^6y^2}$ $4x\sqrt[5]{xy^2}$

92. $\sqrt[3]{xy^4} + \sqrt[3]{8xy^4} - \sqrt[3]{27xy^4}$ 0

93. $\sqrt[4]{208m^4n}$ $2m\sqrt[4]{13n}$

▶ 94. $\sqrt[4]{128p^8q^3}$ $2p^2\sqrt[4]{8q^3}$

95. $\sqrt[3]{\frac{a^7}{64a}}$ $\frac{a^2}{4}$

96. $\sqrt[3]{\frac{b^3c^8}{125c^5}}$ $\frac{bc}{5}$

97. $\sqrt[3]{\frac{7}{64}}$ $\frac{\sqrt[3]{7}}{4}$

98. $\sqrt[3]{\frac{4}{125}}$ $\frac{\sqrt[3]{4}}{5}$

99. $\sqrt{80} + \sqrt{45} - \sqrt{27}$ $7\sqrt{5} - 3\sqrt{3}$

100. $\sqrt{63} + \sqrt{72} - \sqrt{28}$ $\sqrt{7} + 6\sqrt{2}$

101. $\sqrt[5]{64t^{11}}$ $2t^2\sqrt[5]{2t}$

▶ 102. $\sqrt[5]{243r^{22}}$ $3r^4\sqrt[5]{r^2}$

103. $\sqrt[3]{24x} + \sqrt[3]{3x}$ $3\sqrt[3]{3x}$

▶ 104. $\sqrt[3]{16y} + \sqrt[3]{128y}$ $6\sqrt[3]{2y}$

Look Alikes . . .

105. a. $\sqrt{20} + \sqrt{20}$ $4\sqrt{5}$ b. $\sqrt{21} + \sqrt{21}$ $2\sqrt{21}$

▶ 106. a. $\sqrt{2} + \sqrt{18}$ $4\sqrt{2}$ b. $\sqrt{2} + \sqrt{19}$ Cannot be simplified

107. a. $\sqrt{9x^2} - \sqrt{25x^2} + \sqrt{16x^2}$ $2x$
b. $\sqrt{9x^3} - \sqrt{25x^3} + \sqrt{16x^3}$ $2x\sqrt{x}$

108. a. $2\sqrt{5} - \sqrt[3]{5} + 4\sqrt{5} - 6\sqrt[3]{5}$ $6\sqrt{5} - 7\sqrt[3]{5}$
b. $2\sqrt[3]{5} - \sqrt[3]{5} + 4\sqrt[3]{5} - 6\sqrt[3]{5}$ $-\sqrt[3]{5}$

109. a. $3\sqrt{16} + \sqrt{54}$ $12 + 3\sqrt{6}$ b. $3\sqrt[3]{16} + \sqrt[3]{54}$ $9\sqrt[3]{2}$

110. a. $\sqrt[3]{27} - 5\sqrt[3]{8}$ -7 b. $\sqrt{27} - 5\sqrt{8}$ $3\sqrt{3} - 10\sqrt{2}$

111. a. $24\sqrt[5]{6x} + 16\sqrt[5]{6x}$ $40\sqrt[5]{6x}$ b. $24\sqrt[4]{6x} + 16\sqrt[4]{6x}$ $40\sqrt[4]{6x}$

▶ 112. a. $x\sqrt[3]{64x^6} - x\sqrt[3]{x^6}$ $3x^3$ b. $x\sqrt{64x^6} - x\sqrt{x^6}$ $7x^4$

APPLICATIONS

First give the exact answer, expressed as a simplified radical expression. Then give an approximation, rounded to the nearest tenth.

▶ **113.** from **Campus to Careers**

General Contractor

©Goodluz/Shutterstock.com

Structural engineers have determined that two additional supports (shown in red) need to be added to strengthen the truss shown below. Find the length L of a support using the formula

$$L = \sqrt{\frac{b^2}{2} + \frac{c^2}{2} - \frac{a^2}{4}}$$ $3\sqrt{14}$ ft; 11.2 ft

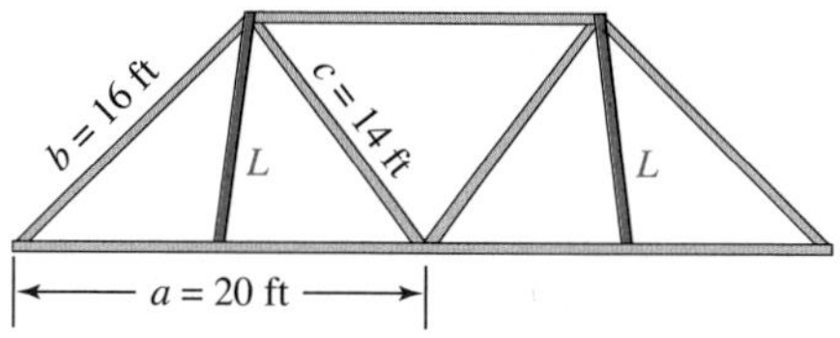

114. Unbrellas. The surface area of a cone is given by the formula $S = \pi r\sqrt{r^2 + h^2}$, where r is the radius of the base and h is its height. Use this formula to find the number of square feet of waterproof cloth used to make the umbrella shown. $8\pi\sqrt{5}$ ft^2; 56.2 ft^2

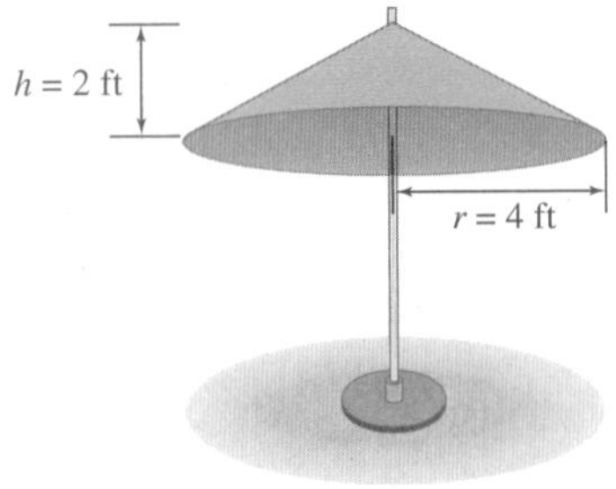

115. Blow Dryers. The current I (in amps), the power P (in watts), and the resistance R (in ohms) are related by the formula $I = \sqrt{\frac{P}{R}}$. What current is needed for a 1,200-watt hair dryer if the resistance is 16 ohms?
$5\sqrt{3}$ amps; 8.7 amps

116. Communications Satellites. Engineers have determined that a spherical communications satellite needs to have a capacity of 565.2 cubic feet to house all of its operating systems. The volume V of a sphere is related to its radius r by the formula $r = \sqrt[3]{\frac{3V}{4\pi}}$. What radius must the satellite have to meet the engineer's specification? Use 3.14 as an approximation of π.
$3\sqrt[3]{5}$ ft; 5.1 ft

117. Ductwork. The following pattern is laid out on a sheet of galvanized tin. Then it is cut out and bent on the dashed lines to make an air conditioning duct connection. Find the total length of the cut that must be made. (All measurements are in inches.)
$(26\sqrt{5} + 10\sqrt{3})$ in.; 75.5 in.

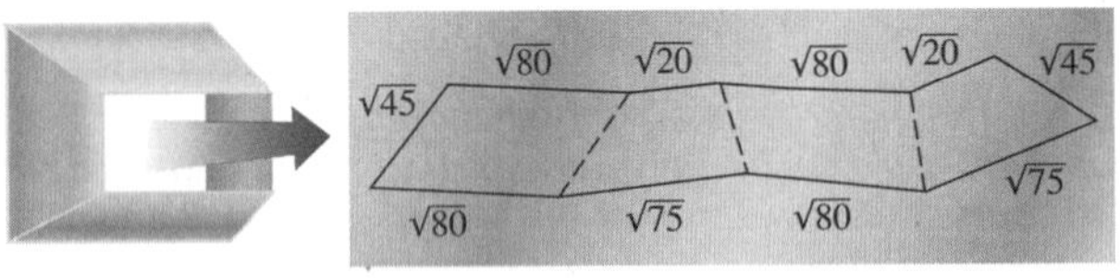

118. Outdoor Cooking. The diameter of a circle is given by the function $d(A) = 2\sqrt{\frac{A}{\pi}}$, where A is the area of the circle. Find the difference between the diameters of the barbecue grills. $6\sqrt{3}$ in.; 10.4 in.

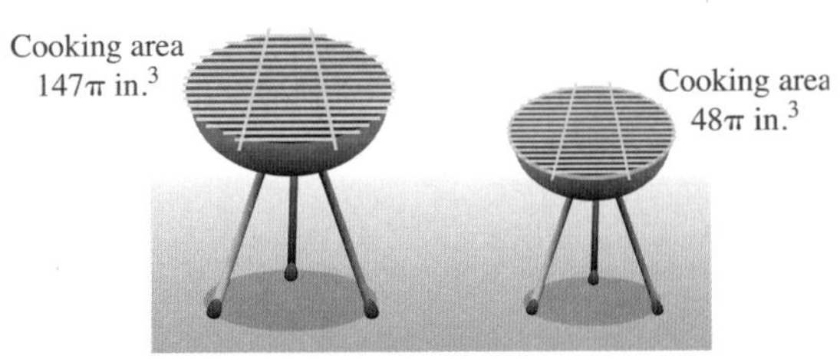

WRITING

119. Explain why each expression is not in simplified form.

a. $\sqrt[3]{9x^4}$ b. $\sqrt{\frac{24m}{25}}$ c. $\frac{\sqrt[4]{c^3}}{\sqrt[4]{16}}$

120. How are the procedures used to simplify $3x + 4x$ and $3\sqrt{x} + 4\sqrt{x}$ similar?

121. Explain the mistake in the student's solution shown below.
Simplify: $\sqrt[3]{54}$

$$\begin{aligned}\sqrt[3]{54} &= \sqrt[3]{27 + 27}\\ &= \sqrt[3]{27} + \sqrt[3]{27}\\ &= 3 + 3\\ &= 6\end{aligned}$$

122. Explain how the graphs of $Y_1 = 3\sqrt{24x} + \sqrt{54x}$ (on the left) and $Y_1 = 9\sqrt{6x}$ (on the right) can be used to verify the simplification $3\sqrt{24x} + \sqrt{54x} = 9\sqrt{6x}$. In each graph, settings of $[-5, 20]$ for x and $[-5, 100]$ for y were used.

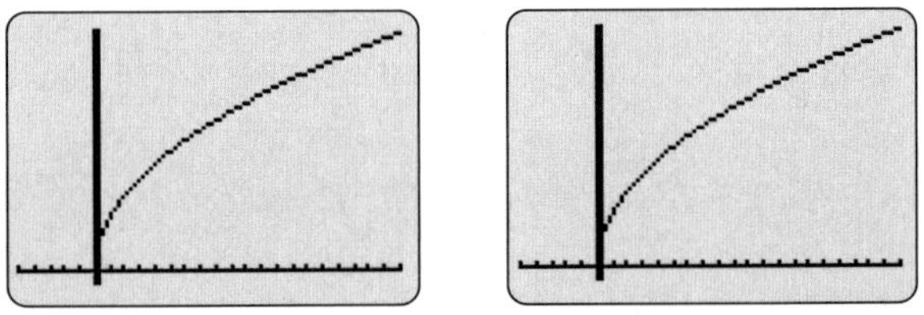

REVIEW

Perform each operation.

123. $3x^2y^3(-5x^3y^{-4})$ $-\frac{15x^5}{y}$

124. $(2x^2 - 9x - 5) \cdot \frac{x}{2x^2 + x}$ $x - 5$

125. $2p - 5\overline{)6p^2 - 7p - 25}$ $3p + 4 - \frac{5}{2p - 5}$

126. $\frac{xy}{\frac{1}{x} - \frac{1}{y}}$ $\frac{x^2y^2}{y - x}$

CHALLENGE PROBLEMS

Simplify each expression. All variables represent positive real numbers.

127. $\frac{\sqrt{24}}{3} + \frac{\sqrt{6}}{5}$ $\frac{13\sqrt{6}}{15}$

128. $\sqrt{3} + \sqrt{3^2} + \sqrt{3^3} + \sqrt{3^4} + \sqrt{3^5}$ $12 + 13\sqrt{3}$

129. $\sqrt[3]{\frac{3b}{8}} - 9\sqrt[3]{3b}$ $-\frac{17\sqrt[3]{3b}}{2}$

130. $\frac{\sqrt[4]{32}}{3y} - \frac{3\sqrt[4]{2}}{7y}$ $\frac{5\sqrt[4]{2}}{21y}$

131. $\sqrt{25x + 25} - \sqrt{x + 1}$ $4\sqrt{x + 1}$

132. $\sqrt[3]{216m^4} + \sqrt[3]{125m}$ $(6m + 5)\sqrt[3]{m}$

SECTION 9.4

OBJECTIVES

1 Multiply radical expressions.
2 Find powers of radical expressions.
3 Rationalize denominators.
4 Rationalize denominators that have two terms.
5 Rationalize numerators.

Multiplying and Dividing Radical Expressions

ARE YOU READY? *Are You Ready? exercises available online at www.webassign.net/brookscole*

The following problems review some basic skills that are needed when multiplying and dividing radical expressions.

Perform the indicated operations and simplify, if possible.

1. $(5a^6)^2$ $25a^{12}$

2. $9t^3(6t^3 + 2t^2)$ $54t^6 + 18t^5$

3. $(2x + 3)(x - 1)$ $2x^2 + x - 3$

4. $7 - 3\sqrt{14} + \sqrt{14} - 6$ $1 - 2\sqrt{14}$

5. $(x + 10)(x - 10)$ $x^2 - 100$

6. Build an equivalent fraction for $\frac{2}{3a}$ with a denominator of $27a$. $\frac{18}{27a}$

In this section, we will discuss the methods we can use to multiply and divide radical expressions.

1 Multiply Radical Expressions.

We have used the *product rule for radicals* to write radical expressions in simplified form. We also can use this rule to multiply radical expressions that have the same index.

The Product Rule for Radicals

The product of the nth roots of two nonnegative numbers is equal to the nth root of the product of those numbers.

If $\sqrt[n]{a}$ and $\sqrt[n]{b}$ are real numbers,

$$\sqrt[n]{a} \cdot \sqrt[n]{b} = \sqrt[n]{a \cdot b}$$

EXAMPLE 1 Multiply and then simplify: **a.** $\sqrt{5}\sqrt{10}$ **b.** $3\sqrt{6}(2\sqrt{3})$ **c.** $-2\sqrt[3]{7x} \cdot 6\sqrt[3]{49x^2}$

Strategy In each expression, we will use the product rule for radicals to multiply factors of the form $\sqrt[n]{a}$ and $\sqrt[n]{b}$.

Why The product rule for radicals is used to multiply radicals that have the same index.

Solution

a.
$\sqrt{5}\sqrt{10} = \sqrt{5 \cdot 10}$ Use the product rule for radicals.
$= \sqrt{50}$ Multiply under the radical. Note that $\sqrt{50}$ can be simplified.
$= \sqrt{25 \cdot 2}$ Prepare to simplify: factor 50.
$= 5\sqrt{2}$ Simplify: $\sqrt{25 \cdot 2} = \sqrt{25}\sqrt{2} = 5\sqrt{2}$.

Success Tip

Here we use the product rule for radicals in two ways: to *multiply* and to *simplify* the result.

b. We use the commutative and associative properties of multiplication to multiply the integer factors and the radicals separately. Then we simplify any radicals in the product, if possible.

$3\sqrt{6}(2\sqrt{3}) = 3(2)\sqrt{6}\sqrt{3}$ Multiply the integer factors, 3 and 2, and multiply the radicals.
$= 6\sqrt{18}$ Use the product rule for radicals.
$= 6\sqrt{9}\sqrt{2}$ Simplify: $\sqrt{18} = \sqrt{9 \cdot 2} = \sqrt{9}\sqrt{2}$.
$= 6(3)\sqrt{2}$ Evaluate: $\sqrt{9} = 3$.
$= 18\sqrt{2}$ Multiply 6 and 3 to get 18.

Notation

Multiplication of radicals can be shown in several ways:

$\sqrt{5} \cdot \sqrt{10} = \sqrt{5}(\sqrt{10}) = \sqrt{5}\sqrt{10}$

c. $-2\sqrt[3]{7x} \cdot 6\sqrt[3]{49x^2} = -2(6)\sqrt[3]{7x}\sqrt[3]{49x^2}$ Write the integer factors together and the radicals together.

$= -12\sqrt[3]{7x \cdot 49x^2}$ Multiply the integer factors, −2 and 6, and multiply the radicals.

$= -12\sqrt[3]{7x \cdot 7^2x^2}$ Write 49 as 7^2.

$= -12\sqrt[3]{7^3x^3}$ Prepare to simplify: write $7x \cdot 7^2x^2$ as 7^3x^3.

$= -12(7x)$ Simplify: $\sqrt[3]{7^3x^3} = 7x$.

$= -84x$ Multiply.

Caution

Note that to multiply radical expressions, they must have the same index.

$$\sqrt[n]{a} \cdot \sqrt[n]{b} = \sqrt[n]{a \cdot b}$$

Teaching Tip: To explain how to find the products in parts b and c, you may want to compare them to how we find $3x \cdot 2x$ and $-2y \cdot 6y$.

Teaching Example 1 Multiply and then simplify:
a. $\sqrt{5}\sqrt{15}$ **b.** $5\sqrt{8}(7\sqrt{3})$
c. $3\sqrt[3]{25x} \cdot \sqrt[3]{15x^2}$
Answers:
a. $5\sqrt{3}$ **b.** $70\sqrt{6}$ **c.** $15x\sqrt[3]{3}$

Self Check 1 Multiply and then simplify: **a.** $\sqrt{7}\sqrt{14}$ $7\sqrt{2}$
b. $-2\sqrt[3]{2}(5\sqrt[3]{12})$ $-20\sqrt[3]{3}$ **c.** $\sqrt[4]{4x^3} \cdot 9\sqrt[4]{8x^2}$ $18x\sqrt[4]{2x}$

Now Try Problems 15, 23, and 25

Recall that to multiply a polynomial by a monomial, we use the distributive property. We use the same technique to multiply a radical expression that has two or more terms by a radical expression that has only one term.

EXAMPLE 2 Multiply and then simplify: $3\sqrt{3}(4\sqrt{8} - 5\sqrt{10})$

Strategy We will use the distributive property and multiply each term within the parentheses by the term outside the parentheses.

Why The given expression has the form $a(b - c)$.

Solution

$3\sqrt{3}(4\sqrt{8} - 5\sqrt{10})$

$= 3\sqrt{3} \cdot 4\sqrt{8} - 3\sqrt{3} \cdot 5\sqrt{10}$ Distribute the multiplication by $3\sqrt{3}$.

$= 12\sqrt{24} - 15\sqrt{30}$ Multiply the integer factors and use the product rule to multiply the radicals.

$= 12\sqrt{4}\sqrt{6} - 15\sqrt{30}$ Simplify: $\sqrt{24} = \sqrt{4 \cdot 6} = \sqrt{4}\sqrt{6}$.

$= 12(2)\sqrt{6} - 15\sqrt{30}$ Evaluate: $\sqrt{4} = 2$.

$= 24\sqrt{6} - 15\sqrt{30}$ Multiply 12 and 2 to get 24.

Teaching Tip: To explain how to find this product, you may want to compare it to how we find $3a(4a^3 - 5a^2)$.

Teaching Example 2 Multiply and then simplify:
$2\sqrt{7}(4\sqrt{14} - 5\sqrt{21})$
Answer:
$56\sqrt{2} - 70\sqrt{3}$

Self Check 2 Multiply and then simplify: $4\sqrt{2}(3\sqrt{5} - 2\sqrt{8})$ $12\sqrt{10} - 32$

Now Try Problems 31 and 33

Recall that to multiply two binomials, we multiply each term of one binomial by each term of the other binomial and simplify. We multiply two radical expressions, each having two terms, in the same way.

EXAMPLE 3 Multiply and then simplify: **a.** $(\sqrt{7} + \sqrt{2})(\sqrt{7} - 9\sqrt{2})$
b. $(\sqrt[3]{x^2} - 4\sqrt[3]{5})(\sqrt[3]{x} + \sqrt[3]{2})$

Strategy As with binomials, we will multiply each term within the first set of parentheses by each term within the second set of parentheses.

Why This is an application of the FOIL method for multiplying binomials.

Solution a. $(\sqrt{7} + \sqrt{2})(\sqrt{7} - 9\sqrt{2})$

Teaching Tip: To explain how to find these products, you may want to compare them to how we find $(x + 2y)(x - 9y)$.

$= \sqrt{7}\sqrt{7} - 9\sqrt{7}\sqrt{2} + \sqrt{2}\sqrt{7} - 9\sqrt{2}\sqrt{2}$ (F, O, I, L) Use the FOIL method.

$= 7 - 9\sqrt{14} + \sqrt{14} - 9(2)$ Perform each multiplication.

$= 7 - 8\sqrt{14} - 18$ Combine like radicals: $-9\sqrt{14} + \sqrt{14} = -8\sqrt{14}$.

$= -11 - 8\sqrt{14}$ Combine like terms: $7 - 18 = -11$.

Caution

A common error is to "simplify" incorrectly by subtracting:

$-11 - 8\sqrt{14} \neq -19\sqrt{14}$

b. $(\sqrt[3]{x^2} - 4\sqrt[3]{5})(\sqrt[3]{x} + \sqrt[3]{2})$

$= \sqrt[3]{x^2}\sqrt[3]{x} + \sqrt[3]{x^2}\sqrt[3]{2} - 4\sqrt[3]{5}\sqrt[3]{x} - 4\sqrt[3]{5}\sqrt[3]{2}$ Use the FOIL method.

$= \sqrt[3]{x^3} + \sqrt[3]{2x^2} - 4\sqrt[3]{5x} - 4\sqrt[3]{10}$ Perform each multiplication.

$= x + \sqrt[3]{2x^2} - 4\sqrt[3]{5x} - 4\sqrt[3]{10}$ Simplify the first term. There are no like radicals or terms to combine.

Teaching Example 3
Multiply and then simplify:
a. $(\sqrt{11} - 3\sqrt{2})(4\sqrt{11} - \sqrt{2})$
b. $(\sqrt[3]{n^2} - 2\sqrt[3]{5})(\sqrt[3]{n} + \sqrt[3]{3})$
Answers:
a. $50 - 13\sqrt{22}$
b. $n + \sqrt[3]{3n^2} - 2\sqrt[3]{5n} - 2\sqrt[3]{15}$

Self Check 3 Multiply and then simplify:

a. $(\sqrt{5} + 2\sqrt{3})(\sqrt{5} - \sqrt{3})$ $-1 + \sqrt{15}$

b. $(\sqrt[3]{a} + 9\sqrt[3]{2})(\sqrt[3]{a^2} - \sqrt[3]{3})$ $a - \sqrt[3]{3a} + 9\sqrt[3]{2a^2} - 9\sqrt[3]{6}$

Now Try Problems 37 and 41

2 Find Powers of Radical Expressions.

To find the power of a radical expression, such as $(\sqrt{5})^2$ or $(\sqrt[3]{2})^3$, we can use the definition of exponent and the product rule for radicals.

$(\sqrt{5})^2 = \sqrt{5}\sqrt{5}$	$(\sqrt[3]{2})^3 = \sqrt[3]{2} \cdot \sqrt[3]{2} \cdot \sqrt[3]{2}$
$= \sqrt{25}$	$= \sqrt[3]{8}$
$= 5$	$= 2$

These results illustrate the following property of radicals.

The *n*th Power of the *n*th Root

If $\sqrt[n]{a}$ is a real number,

$$(\sqrt[n]{a})^n = a$$

EXAMPLE 4 Find: a. $(\sqrt{5})^2$ b. $(2\sqrt[3]{7x^2})^3$ c. $(8 - \sqrt{3})^2$ d. $(\sqrt{m+1} + 2)^2$

Strategy In part (a), we will use the definition of square root. In part (b), we will use a power rule for exponents. In parts (c) and (d), we will use the FOIL method.

Why Part (a) is the square of a square root, part (b) has the form $(xy)^n$, and part (c) has the form $(x + y)^2$.

Solution a. $(\sqrt{5})^2 = 5$ Because the square of the square root of 5 is 5

b. We can use the power of a product rule for exponents to find $(2\sqrt[3]{7x^2})^3$.

$(2\sqrt[3]{7x^2})^3 = 2^3(\sqrt[3]{7x^2})^3$ Raise each factor of $2\sqrt[3]{7x^2}$ to the 3rd power.

$= 8(7x^2)$ Evaluate: $2^3 = 8$. Use $(\sqrt[n]{a})^n = a$.

$= 56x^2$ Multiply: $8 \cdot 7 = 56$.

Success Tip

Since $(8 - \sqrt{3})^2$ has the form $(x - y)^2$, we could also use a special-product rule to find this square of a difference quickly.

c. $(8 - \sqrt{3})^2 = (8 - \sqrt{3})(8 - \sqrt{3})$ Write the base $8 - \sqrt{3}$ as a factor twice.

$= 64 - 8\sqrt{3} - 8\sqrt{3} + \sqrt{3}\sqrt{3}$ Use the FOIL method.

$= 64 - 16\sqrt{3} + 3$ Combine like radicals: $-8\sqrt{3} - 8\sqrt{3} = -16\sqrt{3}$.

$= 67 - 16\sqrt{3}$ Combine like terms: $64 + 3 = 67$.

Caution

A common error when squaring a two-term expression is to simply square the first term and square the last term, and forget the middle term of the product.

$(\sqrt{m+1} + 2)^2 = (\sqrt{m+1})^2 + 2^2$
$= m + 1 + 4$

d. We can use the FOIL method to find the product.

$$(\sqrt{m+1} + 2)^2 = (\sqrt{m+1} + 2)(\sqrt{m+1} + 2)$$
$$= (\sqrt{m+1})^2 + 2\sqrt{m+1} + 2\sqrt{m+1} + 2 \cdot 2$$
$$= m + 1 + 2\sqrt{m+1} + 2\sqrt{m+1} + 4 \quad \text{Use } (\sqrt[n]{a})^n = a.$$
$$= m + 4\sqrt{m+1} + 5 \quad \text{Combine like terms.}$$

Teaching Example 4 Find:
a. $(\sqrt{13})^2$ b. $(4\sqrt[3]{2x^2})^3$
c. $(\sqrt{7} + 3)^2$ d. $(\sqrt{r-1} - 9)^2$
Answers:
a. 13 b. $128x^2$ c. $7 + 6\sqrt{7} + 16$
d. $r - 18\sqrt{r-1} + 80$

Self Check 4 Find: **a.** $(\sqrt{11})^2$ 11 **b.** $(3\sqrt[3]{4y})^3$ $108y$ **c.** $(\sqrt{3} + 4)^2$ $19 + 8\sqrt{3}$
d. $(\sqrt{x-8} - 5)^2$ $x - 10\sqrt{x-8} + 17$

Now Try Problems 43, 49, and 53

3 Rationalize Denominators.

We have seen that when a radical expression is written in simplified form, each of the following statements is true.

Simplified Form of a Radical Expression

1. Each factor in the radicand is to a power that is less than the index of the radical.
2. The radicand contains no fractions or negative numbers.
3. No radicals appear in the denominator of a fraction.

The Language of Algebra

Since $\sqrt{3}$ is an irrational number, $\frac{\sqrt{5}}{\sqrt{3}}$ has an **irrational denominator.** Since 3 is a rational number, $\frac{\sqrt{15}}{3}$ has a **rational denominator**.

Success Tip

As an informal check, we can use a calculator to approximate the value of each expression.

$$\frac{\sqrt{5}}{\sqrt{3}} \approx 1.290994449$$
$$\frac{\sqrt{15}}{3} \approx 1.290994449$$

We now consider radical expressions that do not satisfy requirements 2 or 3. We will introduce an algebraic technique, called **rationalizing the denominator,** that is used to write such expressions in an equivalent simplified form. In this process, we multiply the expression by a form of 1 and use the fact that $\sqrt[n]{a^n} = a$.

As an example, let's consider the following expression:

$\frac{\sqrt{5}}{\sqrt{3}}$ This radical expression is not in simplified form, because a radical appears in the denominator. It doesn't satisfy requirement 3 listed above.

We want to find a fraction equivalent to $\frac{\sqrt{5}}{\sqrt{3}}$ that does not have a radical in its denominator. If we multiply $\frac{\sqrt{5}}{\sqrt{3}}$ by $\frac{\sqrt{3}}{\sqrt{3}}$, the denominator becomes $\sqrt{3} \cdot \sqrt{3} = 3$, a rational number.

$\frac{\sqrt{5}}{\sqrt{3}} = \frac{\sqrt{5}}{\sqrt{3}} \cdot \frac{\sqrt{3}}{\sqrt{3}}$ To build an equivalent fraction, multiply by a form of 1: $\frac{\sqrt{3}}{\sqrt{3}} = 1$.

$= \frac{\sqrt{15}}{3}$ Multiply the numerators: $\sqrt{5} \cdot \sqrt{3} = \sqrt{15}$. Multiply the denominators: $\sqrt{3} \cdot \sqrt{3} = (\sqrt{3})^2 = 3$. The denominator is now a rational number, 3.

Thus, $\frac{\sqrt{5}}{\sqrt{3}} = \frac{\sqrt{15}}{3}$. These equivalent fractions represent the same number, but have different forms. Since there is no radical in the denominator, and $\sqrt{15}$ is in simplest form, the expression $\frac{\sqrt{15}}{3}$ is in simplified form. We say that we have *rationalized the denominator* of $\frac{\sqrt{5}}{\sqrt{3}}$.

EXAMPLE 5 Rationalize the denominator: **a.** $\sqrt{\frac{20}{7}}$ **b.** $\frac{4}{\sqrt[3]{2}}$

Strategy We look at each denominator and ask, "By what must we multiply it to obtain a rational number?" Then we will multiply each expression by a carefully chosen form of 1.

Why We want to produce an equivalent expression that does not have a radical in its denominator.

Solution **a.** This radical expression is not in simplified form, because the radicand contains a fraction. (It doesn't satisfy requirement 2.) We begin by writing the square root of the quotient as the quotient of two square roots:

$$\sqrt{\frac{20}{7}} = \frac{\sqrt{20}}{\sqrt{7}}$$ Use the division property of radicals: $\sqrt[n]{\frac{a}{b}} = \frac{\sqrt[n]{a}}{\sqrt[n]{b}}$.

To rationalize the denominator, we proceed as follows:

$$\frac{\sqrt{20}}{\sqrt{7}} = \frac{\sqrt{20}}{\sqrt{7}} \cdot \frac{\sqrt{7}}{\sqrt{7}}$$ To build an equivalent fraction, multiply by $\frac{\sqrt{7}}{\sqrt{7}} = 1$.

$$= \frac{\sqrt{140}}{7}$$ Multiply the numerators. Multiply the denominators: $\sqrt{7} \cdot \sqrt{7} = (\sqrt{7})^2 = 7$. The denominator is now a rational number, 7.

$$= \frac{2\sqrt{35}}{7}$$ Simplify: $\sqrt{140} = \sqrt{4 \cdot 35} = \sqrt{4}\sqrt{35} = 2\sqrt{35}$.

Caution

Do not attempt to remove a common factor of 7 from the numerator and denominator of $\frac{2\sqrt{35}}{7}$. The numerator, $2\sqrt{35}$, does not have a factor of 7.

$$\frac{2\sqrt{35}}{7} = \frac{2 \cdot \sqrt{5 \cdot 7}}{7}$$

Teaching Tip: Ask your students to explain why the value of the given fraction does not change in the rationalizing process.

b. This expression is not in simplified form because a radical appears in the denominator of a fraction. (It doesn't satisfy requirement 3.) Here, we must rationalize a denominator that is a cube root. We multiply the numerator and the denominator by a number that will give a perfect cube under the radical. Since $2 \cdot 4 = 8$ is a perfect cube, $\sqrt[3]{4}$ is such a number.

$$\frac{4}{\sqrt[3]{2}} = \frac{4}{\sqrt[3]{2}} \cdot \frac{\sqrt[3]{4}}{\sqrt[3]{4}}$$ To build an equivalent fraction, multiply by $\frac{\sqrt[3]{4}}{\sqrt[3]{4}} = 1$.

$$= \frac{4\sqrt[3]{4}}{\sqrt[3]{8}}$$ Multiply the numerators. Multiply the denominators. This radicand is now a perfect cube.

$$= \frac{4\sqrt[3]{4}}{2}$$ Evaluate the denominator: $\sqrt[3]{8} = 2$. The denominator is now a rational number, 2.

$$= 2\sqrt[3]{4}$$ Simplify the fraction: $\frac{4\sqrt[3]{4}}{2} = \frac{\overset{1}{\cancel{2}} \cdot 2\sqrt[3]{4}}{\underset{1}{\cancel{2}}} = 2\sqrt[3]{4}$.

Caution

Multiplying $\frac{4}{\sqrt[3]{2}}$ by $\frac{\sqrt[3]{2}}{\sqrt[3]{2}}$ does not rationalize the denominator.

$$\frac{4}{\sqrt[3]{2}} \cdot \frac{\sqrt[3]{2}}{\sqrt[3]{2}} = \frac{4\sqrt[3]{2}}{\sqrt[3]{4}}$$

↑

Since 4 is not a perfect cube, this radical does not simplify.

An alternate way to rationalize the denominator is to use the equivalent form $\frac{\sqrt[3]{2^2}}{\sqrt[3]{2^2}}$ instead of $\frac{\sqrt[3]{4}}{\sqrt[3]{4}}$ to build an equivalent fraction:

$$\frac{4}{\sqrt[3]{2}} \cdot \frac{\sqrt[3]{2^2}}{\sqrt[3]{2^2}} = \frac{4\sqrt[3]{2^2}}{\sqrt[3]{2^3}}$$ Multiply the numerators. Multiply the denominators. In the denominator, the radicand is now a perfect cube because $2 \cdot 2^2 = 2^3$.

$$= \frac{4\sqrt[3]{4}}{2}$$ In the numerator, $2^2 = 4$. Simplify in the denominator, $\sqrt[3]{2^3} = 2$.

$$= 2\sqrt[3]{4}$$ Simplify the fraction: $\frac{4\sqrt[3]{4}}{2} = \frac{\overset{1}{\cancel{2}} \cdot 2\sqrt[3]{4}}{\underset{1}{\cancel{2}}} = 2\sqrt[3]{4}$.

Teaching Example 5
Rationalize the denominator:
a. $\sqrt{\frac{12}{11}}$ b. $\frac{7}{\sqrt[3]{3}}$
Answers:
a. $\frac{2\sqrt{33}}{11}$ b. $\frac{7\sqrt[3]{9}}{3}$

Self Check 5 Rationalize the denominator: **a.** $\sqrt{\frac{8}{5}}$ $\frac{2\sqrt{10}}{5}$ **b.** $\frac{5}{\sqrt[3]{9}}$ $\frac{5\sqrt[3]{3}}{3}$

Now Try Problems 57, 59, and 63

EXAMPLE 6 Rationalize the denominator: $\frac{\sqrt{5xy^2}}{\sqrt{xy^3}}$

Strategy We will begin by using the quotient rule for radicals in reverse: $\frac{\sqrt[n]{a}}{\sqrt[n]{b}} = \sqrt[n]{\frac{a}{b}}$

Why When the radicands are written under a single radical symbol, the result is a rational expression. Our hope is that the rational expression can be simplified, which could possibly make rationalizing the denominator easier.

Solution There are two methods we can use to rationalize the denominator. In each method, we simplify the rational expression $\frac{5xy^2}{xy^3}$ that appears in the radicand first.

Caution

We will assume that all of the variables appearing in the following examples represent positive numbers.

***Method* 1**

$$\frac{\sqrt{5xy^2}}{\sqrt{xy^3}} = \sqrt{\frac{5xy^2}{xy^3}}$$

$$= \sqrt{\frac{5}{y}} \quad \text{Simplify the radicand.}$$

$$= \frac{\sqrt{5}}{\sqrt{y}} \quad \text{Use the quotient rule.}$$

$$= \frac{\sqrt{5}}{\sqrt{y}} \cdot \frac{\sqrt{y}}{\sqrt{y}} \quad \text{Multiply outside the radical.}$$

$$= \frac{\sqrt{5y}}{y} \quad \text{Multiply the numerators and the denominators.}$$

***Method* 2**

$$\frac{\sqrt{5xy^2}}{\sqrt{xy^3}} = \sqrt{\frac{5xy^2}{xy^3}}$$

$$= \sqrt{\frac{5}{y}} \quad \text{Simplify the radicand.}$$

$$= \sqrt{\frac{5}{y} \cdot \frac{y}{y}} \quad \text{Multiply within the radical.}$$

$$= \frac{\sqrt{5y}}{\sqrt{y^2}} \quad \text{Use the quotient rule.}$$

$$= \frac{\sqrt{5y}}{y} \quad \text{Simplify the denominator.}$$

Teaching Example 6
Rationalize the denominator: $\frac{\sqrt{6a^2b^5}}{\sqrt{3a^3b}}$
Answer:
$\frac{b^2\sqrt{2a}}{a}$

Self Check 6 Rationalize the denominator: $\frac{\sqrt{4ab^3}}{\sqrt{2a^2b^2}}$ $\frac{\sqrt{2ab}}{a}$

Now Try Problems 67 and 73

EXAMPLE 7 Rationalize the denominator: $\frac{11}{\sqrt{20q^5}}$

Strategy We will simplify the radical expression in the denominator before rationalizing the denominator.

Why We could begin by multiplying $\frac{11}{\sqrt{20q^5}}$ by $\frac{\sqrt{20q^5}}{\sqrt{20q^5}}$. However, to work with smaller numbers and simpler radical expressions, it is easier if we simplify $\sqrt{20q^5}$ first, and then rationalize the denominator.

Solution

$$\frac{11}{\sqrt{20q^5}} = \frac{11}{\sqrt{4q^4 \cdot 5q}} \quad \text{To prepare to simplify } \sqrt{20q^5}\text{, factor } 20q^5 \text{ as } 4q^4 \cdot 5q.$$

$$= \frac{11}{2q^2\sqrt{5q}} \quad \text{Simplify: } \sqrt{4q^4 \cdot 5q} = \sqrt{4q^4}\sqrt{5q} = 2q^2\sqrt{5q}.$$

Success Tip

We usually simplify a radical expression before rationalizing the denominator.

$$= \frac{11}{2q^2\sqrt{5q}} \cdot \frac{\sqrt{5q}}{\sqrt{5q}}$$ To rationalize the denominator, multiply by $\frac{\sqrt{5q}}{\sqrt{5q}} = 1$.

$$= \frac{11\sqrt{5q}}{2q^2(5q)}$$ Multiply the numerators. Multiply the denominators: $\sqrt{5q} \cdot \sqrt{5q} = \left(\sqrt{5q}\right)^2 = 5q$.

$$= \frac{11\sqrt{5q}}{10q^3}$$ Multiply in the denominator: $2 \cdot 5 = 10$ and $q^2 \cdot q = q^3$.

Teaching Example 7 Rationalize the denominator: $\frac{5}{\sqrt{24x^5}}$

Answer: $\frac{5\sqrt{6x}}{12x^3}$

Self Check 7 Rationalize the denominator: $\frac{7}{\sqrt{18c^3}}$ $\frac{7\sqrt{2c}}{6c^2}$

Now Try Problems 75 and 81

EXAMPLE 8 Rationalize each denominator: **a.** $\frac{5}{\sqrt[3]{6n^2}}$ **b.** $\frac{\sqrt[4]{2}}{\sqrt[4]{3a}}$

Strategy In part (a), we will examine the radicand in the denominator and ask, "By what must we multiply it to obtain a perfect cube?" In part (b), we will examine the radicand in the denominator and ask, "By what must we multiply it to obtain a perfect-fourth power?"

Why The answers to those questions will determine what form of 1 we use to rationalize each denominator.

Solution

a. To rationalize the denominator $\sqrt[3]{6n^2}$, we need the radicand to be a perfect cube. Since $6n^2 = 6 \cdot n \cdot n$, the radicand needs two more factors of 6 and one more factor of n. It follows that we should multiply the given expression by $\frac{\sqrt[3]{6^2n}}{\sqrt[3]{6^2n}}$.

$$\frac{5}{\sqrt[3]{6n^2}} = \frac{5}{\sqrt[3]{6n^2}} \cdot \frac{\sqrt[3]{6^2n}}{\sqrt[3]{6^2n}}$$ Multiply by a form of 1 to rationalize the denominator.

$$= \frac{5\sqrt[3]{36n}}{\sqrt[3]{6^3n^3}}$$ Multiply the numerators. Multiply the denominators. ← This radicand is now a perfect cube.

$$= \frac{5\sqrt[3]{36n}}{6n}$$ Simplify the denominator: $\sqrt[3]{6^3n^3} = 6n$.

Success Tip

An alternate form of 1 that can be used to rationalize the denominator is $\frac{\sqrt[3]{36n}}{\sqrt[3]{36n}}$.

b. To rationalize the denominator $\sqrt[4]{3a}$, we need the radicand to be a perfect-fourth power. Since $3a = 3 \cdot a$, the radicand needs three more factors of 3 and three more factors of a. It follows that we should multiply the given expression by $\frac{\sqrt[4]{3^3a^3}}{\sqrt[4]{3^3a^3}}$.

$$\frac{\sqrt[4]{2}}{\sqrt[4]{3a}} = \frac{\sqrt[4]{2}}{\sqrt[4]{3a}} \cdot \frac{\sqrt[4]{3^3a^3}}{\sqrt[4]{3^3a^3}}$$ Multiply by a form of 1 to rationalize the denominator.

$$= \frac{\sqrt[4]{54a^3}}{\sqrt[4]{3^4a^4}}$$ ← Multiply the numerators: $2 \cdot 27 = 54$. Multiply the denominators. This radicand is now a perfect-fourth power.

$$= \frac{\sqrt[4]{54a^3}}{3a}$$ Simplify the denominator: $\sqrt[4]{3^4a^4} = 3a$.

Success Tip

An alternate form of 1 that can be used to rationalize the denominator is $\frac{\sqrt[4]{27a^3}}{\sqrt[4]{27a^3}}$.

Teaching Example 8 Rationalize each denominator:

a. $\frac{\sqrt[3]{2}}{\sqrt[3]{5a^2}}$ b. $\frac{\sqrt[4]{7}}{\sqrt[4]{4t}}$

Answers:

a. $\frac{\sqrt[3]{50a}}{5a}$ b. $\frac{\sqrt[4]{28t^3}}{2t}$

Self Check 8 Rationalize each denominator: **a.** $\frac{27}{\sqrt[3]{100a}}$ $\frac{27\sqrt[3]{10a^2}}{10a}$

b. $\frac{\sqrt[4]{3}}{\sqrt[4]{4y^2}}$ $\frac{\sqrt[4]{12y^2}}{2y}$

Now Try Problems 83 and 87

4 Rationalize Denominators That Have Two Terms.

So far, we have rationalized denominators that have only one term. We will now discuss a method to rationalize denominators that have two terms.

One-termed denominators			*Two-termed denominators*	
$\frac{\sqrt{5}}{\sqrt{3}}$	$\frac{11}{\sqrt{20q^5}}$	$\frac{4}{\sqrt[3]{2}}$	$\frac{1}{\sqrt{2}+1}$	$\frac{\sqrt{x}+\sqrt{2}}{\sqrt{x}-\sqrt{2}}$

To rationalize the denominator of $\frac{1}{\sqrt{2}+1}$, for example, we multiply the numerator and denominator by $\sqrt{2}-1$, because the product $(\sqrt{2}+1)(\sqrt{2}-1)$ contains no radicals.

$$(\sqrt{2}+1)(\sqrt{2}-1) = (\sqrt{2})^2 - (1)^2 \quad \text{Use a special-product rule.}$$
$$= 2 - 1$$
$$= 1$$

Radical expressions that involve the sum and difference of the same two terms, such as $\sqrt{2}+1$ and $\sqrt{2}-1$, are called **conjugates.**

EXAMPLE 9 Rationalize the denominator: **a.** $\frac{1}{\sqrt{2}+1}$ **b.** $\frac{\sqrt{x}+\sqrt{2}}{\sqrt{x}-\sqrt{2}}$

Strategy In each part, we will rationalize the denominator by multiplying the numerator and the denominator by the conjugate of the denominator.

Why Multiplying each denominator by its conjugate will produce a new denominator that does not contain radicals.

Solution **a.** To find a fraction equivalent to $\frac{1}{\sqrt{2}+1}$ that does not have a radical in its denominator, we multiply $\frac{1}{\sqrt{2}+1}$ by a form of 1 that uses the conjugate of $\sqrt{2}+1$.

$$\frac{1}{\sqrt{2}+1} = \frac{1}{\sqrt{2}+1} \cdot \frac{\sqrt{2}-1}{\sqrt{2}-1}$$

$$= \frac{\sqrt{2}-1}{(\sqrt{2})^2-(1)^2} \quad \text{Multiply the numerators. Multiply the denominators using a special-product rule.}$$

$$= \frac{\sqrt{2}-1}{2-1} \quad \text{In the denominator, } (\sqrt{2})^2 = 2.$$

$$= \frac{\sqrt{2}-1}{1}$$

$$= \sqrt{2}-1$$

Success Tip

In the denominator, we can use the special-product rule $(A+B)(A-B) = A^2 - B^2$ to multiply $\sqrt{2}+1$ and $\sqrt{2}-1$ quickly.

b. We multiply the numerator and denominator by $\sqrt{x}+\sqrt{2}$, which is the conjugate of $\sqrt{x}-\sqrt{2}$, and simplify.

$$\frac{\sqrt{x}+\sqrt{2}}{\sqrt{x}-\sqrt{2}} = \frac{\sqrt{x}+\sqrt{2}}{\sqrt{x}-\sqrt{2}} \cdot \frac{\sqrt{x}+\sqrt{2}}{\sqrt{x}+\sqrt{2}}$$

$$= \frac{x+\sqrt{2x}+\sqrt{2x}+2}{(\sqrt{x})^2-(\sqrt{2})^2} \quad \text{Multiply the numerators. Multiply the denominators using a special-product formula.}$$

$$= \frac{x+\sqrt{2x}+\sqrt{2x}+2}{x-2} \quad \text{In the denominator, } (\sqrt{x})^2 = x \text{ and } (\sqrt{2})^2 = 2.$$

$$= \frac{x+2\sqrt{2x}+2}{x-2} \quad \text{In the numerator, combine like radicals: } \sqrt{2x}+\sqrt{2x} = 2\sqrt{2x}.$$

Teaching Example 9 Rationalize the denominator:

a. $\frac{3}{\sqrt{5}-1}$ b. $\frac{\sqrt{x}-\sqrt{3}}{\sqrt{x}+\sqrt{3}}$

Answers:

a. $\frac{3\sqrt{5}+3}{4}$ b. $\frac{x-2\sqrt{3x}+3}{x-3}$

Self Check 9 Rationalize the denominator: $\frac{\sqrt{x}-\sqrt{2}}{\sqrt{x}+\sqrt{2}}$ $\frac{x-2\sqrt{2x}+2}{x-2}$

Now Try Problems 91 and 97

5 Rationalize Numerators.

In some advanced mathematics courses, like calculus, we sometimes have to rationalize a numerator by multiplying the numerator and denominator of the fraction by the conjugate of the numerator.

EXAMPLE 10 Rationalize the numerator: $\frac{\sqrt{x}-3}{\sqrt{x}}$

Strategy To rationalize the numerator, we will multiply the numerator and the denominator by the conjugate of the numerator.

Why After rationalizing the numerator, we can simplify the expression. Although the result will not be in simplified form, this nonsimplified form is often desirable in calculus.

Solution We multiply the numerator and denominator by $\sqrt{x}+3$, which is the conjugate of the numerator.

$$\frac{\sqrt{x}-3}{\sqrt{x}} = \frac{\sqrt{x}-3}{\sqrt{x}} \cdot \frac{\sqrt{x}+3}{\sqrt{x}+3}$$ Multiply by a form of 1 to rationalize the numerator.

$$= \frac{\left(\sqrt{x}\right)^2 - 3^2}{x+3\sqrt{x}}$$ Multiply the numerators using a special-product rule. Multiply the denominators using the distributive property.

$$= \frac{x-9}{x+3\sqrt{x}}$$ In the numerator, $\left(\sqrt{x}\right)^2 = x$ and $3^2 = 9$.

Teaching Example 10 Rationalize the numerator: $\frac{\sqrt{x}+5}{\sqrt{x}}$

Answer: $\frac{x-25}{x-5\sqrt{x}}$

Self Check 10 Rationalize the numerator: $\frac{\sqrt{x}+3}{\sqrt{x}}$ $\frac{x-9}{x-3\sqrt{x}}$

Now Try Problem 101

SECTION 9.4 STUDY SET

VOCABULARY

Fill in the blanks.

1. In this section, we used the product rule for radicals in reverse: $\sqrt[n]{a} \cdot \sqrt[n]{b} = \sqrt[n]{ab}$.
2. To multiply $2\sqrt{5}\left(3\sqrt{8}+\sqrt{3}\right)$, use the distributive property.
3. To rationalize the denominator of $\frac{4}{\sqrt{5}}$, we multiply the fraction by $\frac{\sqrt{5}}{\sqrt{5}}$.
4. The denominator of the fraction $\frac{4}{\sqrt{5}}$ is an irrational number.
5. To obtain a perfect -cube radicand in the denominator of $\frac{\sqrt[3]{7}}{\sqrt[3]{5n}}$, we multiply the fraction by $\frac{\sqrt[3]{25n^2}}{\sqrt[3]{25n^2}}$.
6. The conjugate of $\sqrt{x}+1$ is $\sqrt{x}-1$.

CONCEPTS

7. Tell why each of the following expressions is not in simplified radical form. Then simplify it. Finally, use a calculator to approximate its value.

	Why isn't it in simplified form?	Simplified form	Approximation
$\frac{3}{\sqrt{2}}$	A radical appears in the denominator.	$\frac{3\sqrt{2}}{2}$	2.121320344
$\frac{\sqrt{18}}{2}$	There is a perfect-square factor in the radicand: 9	$\frac{3\sqrt{2}}{2}$	2.121320344
$\sqrt{\frac{9}{2}}$	The radicand contains a fraction.	$\frac{3\sqrt{2}}{2}$	2.121320344

8. Fill in the blank: To rationalize the denominator of $\frac{3}{\sqrt{2}}$, we multiply it by $\frac{\sqrt{2}}{\sqrt{2}}$, which is a form of 1.

9. Fill in the blanks to complete this special product:

$$(5 - \sqrt{x})^2 = (5)^2 - 2(5)(\sqrt{x}) + (\sqrt{x})^2$$
$$= 25 - 10\sqrt{x} + x$$

10. Fill in the blanks to complete this special product:

$$(\sqrt{7} + 2)(\sqrt{7} - 2) = (\sqrt{7})^2 - (2)^2$$
$$= 7 - 4$$
$$= 3$$

11. Perform each operation, if possible.

a. $4\sqrt{6} + 2\sqrt{6}$ $6\sqrt{6}$

b. $4\sqrt{6}(2\sqrt{6})$ 48

c. $3\sqrt{2} - 2\sqrt{3}$ Can't be simplified

d. $3\sqrt{2}(-2\sqrt{3})$ $-6\sqrt{6}$

▶ 12. Perform each operation, if possible.

a. $5 + 6\sqrt[3]{6}$ Can't be simplified

b. $5(6\sqrt[3]{6})$ $30\sqrt[3]{6}$

c. $\frac{30\sqrt[3]{15}}{5}$ $6\sqrt[3]{15}$

d. $\frac{\sqrt[3]{15}}{5}$ Can't be simplified

NOTATION

Fill in the blanks.

13. Multiply:

$$5\sqrt{8} \cdot 7\sqrt{6} = 5(7)\sqrt{8}\sqrt{6}$$
$$= 35\sqrt{48}$$
$$= 35\sqrt{16 \cdot 3}$$
$$= 35(4)\sqrt{3}$$
$$= 140\sqrt{3}$$

▶ 14. Rationalize the denominator:

$$\frac{9}{\sqrt[3]{4a^2}} = \frac{9}{\sqrt[3]{4a^2}} \cdot \frac{\sqrt[3]{2a}}{\sqrt[3]{2a}}$$
$$= \frac{9\sqrt[3]{2a}}{\sqrt[3]{8a^3}}$$
$$= \frac{9\sqrt[3]{2a}}{2a}$$

GUIDED PRACTICE

Multiply and simplify. All variables represent positive real numbers. See Example 1.

15. $\sqrt{3}\sqrt{15}$ $3\sqrt{5}$

▶ 16. $\sqrt{5}\sqrt{15}$ $5\sqrt{3}$

▶ 17. $2\sqrt{3}\sqrt{6}$ $6\sqrt{2}$

18. $-3\sqrt{11}\sqrt{33}$ $-33\sqrt{3}$

▶ 19. $(3\sqrt[3]{9})(2\sqrt[3]{3})$ 18

20. $(2\sqrt[3]{16})(-\sqrt[3]{4})$ -8

21. $\sqrt[3]{2} \cdot \sqrt[3]{12}$ $2\sqrt[3]{3}$

▶ 22. $\sqrt[3]{3} \cdot \sqrt[3]{18}$ $3\sqrt[3]{2}$

▶ 23. $6\sqrt{ab^3}(8\sqrt{ab})$ $48ab^2$

24. $3\sqrt{8x}(2\sqrt{2x^3y})$ $24x^2\sqrt{y}$

25. $\sqrt[4]{5a^3}\sqrt[4]{125a^2}$ $5a\sqrt[4]{a}$

▶ 26. $\sqrt[4]{2r^3}\sqrt[4]{8r^2}$ $2r\sqrt[4]{r}$

Multiply and simplify. All variables represent positive real numbers. See Example 2.

27. $3\sqrt{5}(4 - \sqrt{5})$ $12\sqrt{5} - 15$

▶ 28. $2\sqrt{7}(3 - \sqrt{7})$ $6\sqrt{7} - 14$

▶ 29. $\sqrt{2}(4\sqrt{6} + 2\sqrt{7})$ $8\sqrt{3} + 2\sqrt{14}$

▶ 30. $-\sqrt{3}(\sqrt{7} - \sqrt{15})$ $-\sqrt{21} + 3\sqrt{5}$

▶ 31. $-2\sqrt{5x}(4\sqrt{2x} - 3\sqrt{3})$ $-8x\sqrt{10} + 6\sqrt{15x}$

▶ 32. $3\sqrt{7t}(2\sqrt{7t} + 3\sqrt{3t^2})$ $42t + 9t\sqrt{21t}$

33. $\sqrt[3]{2}(4\sqrt[3]{4} + \sqrt[3]{12})$ $8 + 2\sqrt[3]{3}$

▶ 34. $\sqrt[3]{3}(2\sqrt[3]{9} + \sqrt[3]{18})$ $6 + 3\sqrt[3]{2}$

Multiply and simplify. All variables represent positive real numbers. See Example 3.

▶ 35. $(\sqrt{2} + 1)(\sqrt{2} - 3)$ $-1 - 2\sqrt{2}$

▶ 36. $(2\sqrt{3} + 1)(\sqrt{3} - 1)$ $5 - \sqrt{3}$

37. $(\sqrt{3x} - \sqrt{2y})(\sqrt{3x} + \sqrt{2y})$ $3x - 2y$

▶ 38. $(\sqrt{3m} + \sqrt{2n})(\sqrt{3m} - \sqrt{2n})$ $3m - 2n$

39. $(2\sqrt[3]{4} - 3\sqrt[3]{2})(3\sqrt[3]{4} + 2\sqrt[3]{10})$ $12\sqrt[3]{2} + 8\sqrt[3]{5} - 18 - 6\sqrt[3]{20}$

▶ 40. $(4\sqrt[3]{9} - 3\sqrt[3]{3})(4\sqrt[3]{3} + 2\sqrt[3]{6})$ $48 + 24\sqrt[3]{2} - 12\sqrt[3]{9} - 6\sqrt[3]{18}$

41. $(\sqrt[3]{5z} + \sqrt[3]{3})(\sqrt[3]{5z} + 2\sqrt[3]{3})$ $\sqrt[3]{25z^2} + 3\sqrt[3]{15z} + 2\sqrt[3]{9}$

▶ 42. $(\sqrt[3]{3p} - 2\sqrt[3]{2})(\sqrt[3]{3p} + \sqrt[3]{2})$ $\sqrt[3]{9p^2} - \sqrt[3]{6p} - 2\sqrt[3]{4}$

Square or cube each quantity and simplify the result. See Example 4.

43. $(\sqrt{7})^2$ 7

▶ 44. $(\sqrt{11})^2$ 11

45. $(\sqrt[3]{12})^3$ 12

▶ 46. $(\sqrt[3]{9})^3$ 9

▶ 47. $(3\sqrt{2})^2$ 18

48. $(2\sqrt{5})^2$ 20

49. $\left(-2\sqrt[3]{2x^2}\right)^3$ $-16x^2$

▶ 50. $\left(-3\sqrt[3]{10y^3}\right)^3$ $-270y^3$

51. $(6-\sqrt{3})^2$ $39-12\sqrt{3}$

▶ 52. $(9-\sqrt{11})^2$ $92-18\sqrt{11}$

53. $(\sqrt{3x}+\sqrt{3})^2$ $3x+6\sqrt{x}+3$

▶ 54. $(\sqrt{5x}-\sqrt{3})^2$ $5x-2\sqrt{15x}+3$

Rationalize each denominator. **See Example 5.**

▶ 55. $\sqrt{\dfrac{2}{7}}$ $\dfrac{\sqrt{14}}{7}$

56. $\sqrt{\dfrac{5}{3}}$ $\dfrac{\sqrt{15}}{3}$

57. $\sqrt{\dfrac{8}{3}}$ $\dfrac{2\sqrt{6}}{3}$

▶ 58. $\sqrt{\dfrac{8}{7}}$ $\dfrac{2\sqrt{14}}{7}$

59. $\dfrac{4}{\sqrt{6}}$ $\dfrac{2\sqrt{6}}{3}$

▶ 60. $\dfrac{8}{\sqrt{10}}$ $\dfrac{4\sqrt{10}}{5}$

▶ 61. $\dfrac{1}{\sqrt[3]{2}}$ $\dfrac{\sqrt[3]{4}}{2}$

▶ 62. $\dfrac{2}{\sqrt[3]{6}}$ $\dfrac{\sqrt[3]{36}}{3}$

▶ 63. $\dfrac{3}{\sqrt[3]{9}}$ $\sqrt[3]{3}$

▶ 64. $\dfrac{2}{\sqrt[3]{a}}$ $\dfrac{2\sqrt[3]{a^2}}{a}$

65. $\dfrac{1}{\sqrt[4]{8}}$ $\dfrac{\sqrt[4]{2}}{2}$

▶ 66. $\dfrac{1}{\sqrt[5]{2}}$ $\dfrac{\sqrt[5]{16}}{2}$

Rationalize each denominator. All variables represent positive real numbers. **See Example 6.**

67. $\dfrac{\sqrt{10y^2}}{\sqrt{2y^3}}$ $\dfrac{\sqrt{5y}}{y}$

▶ 68. $\dfrac{\sqrt{15b^2}}{\sqrt{5b^3}}$ $\dfrac{\sqrt{3b}}{b}$

69. $\dfrac{\sqrt{48x^2}}{\sqrt{8x^2y}}$ $\dfrac{\sqrt{6y}}{y}$

▶ 70. $\dfrac{\sqrt{9xy}}{\sqrt{3x^2y}}$ $\dfrac{\sqrt{3x}}{x}$

71. $\dfrac{\sqrt[3]{12t^3}}{\sqrt[3]{54t^2}}$ $\dfrac{\sqrt[3]{6t}}{3}$

▶ 72. $\dfrac{\sqrt[3]{15m^4}}{\sqrt[3]{12m^3}}$ $\dfrac{\sqrt[3]{10m}}{2}$

73. $\dfrac{\sqrt[3]{4a^6}}{\sqrt[3]{2a^5b}}$ $\dfrac{\sqrt[3]{2ab^2}}{b}$

▶ 74. $\dfrac{\sqrt[3]{9x^5y^4}}{\sqrt[3]{3x^5y^5}}$ $\dfrac{\sqrt[3]{3y^2}}{y}$

Rationalize each denominator. All variables represent positive real numbers. **See Example 7.**

75. $\dfrac{23}{\sqrt{50p^5}}$ $\dfrac{23\sqrt{2p}}{10p^3}$

▶ 76. $\dfrac{11}{\sqrt{75s^5}}$ $\dfrac{11\sqrt{3s}}{15s^3}$

77. $\dfrac{7}{\sqrt{24b^3}}$ $\dfrac{7\sqrt{6b}}{12b^2}$

78. $\dfrac{13}{\sqrt{32n^3}}$ $\dfrac{13\sqrt{2n}}{8n^2}$

Rationalize each denominator. All variables represent positive real numbers. **See Example 8.**

79. $\sqrt[3]{\dfrac{5}{16}}$ $\dfrac{\sqrt[3]{20}}{4}$

▶ 80. $\sqrt[3]{\dfrac{2}{81}}$ $\dfrac{\sqrt[3]{18}}{9}$

81. $\sqrt[3]{\dfrac{4}{81}}$ $\dfrac{\sqrt[3]{36}}{9}$

82. $\sqrt[3]{\dfrac{7}{16}}$ $\dfrac{\sqrt[3]{28}}{4}$

83. $\dfrac{19}{\sqrt[3]{5c^2}}$ $\dfrac{19\sqrt[3]{25c}}{5c}$

84. $\dfrac{1}{\sqrt[3]{4m^2}}$ $\dfrac{\sqrt[3]{2m}}{2m}$

85. $\dfrac{\sqrt[3]{3}}{\sqrt[3]{2r}}$ $\dfrac{\sqrt[3]{12r^2}}{2r}$

▶ 86. $\dfrac{\sqrt[3]{7}}{\sqrt[3]{100s}}$ $\dfrac{\sqrt[3]{70s^2}}{10s}$

87. $\dfrac{\sqrt[4]{2}}{\sqrt[4]{3t^2}}$ $\dfrac{\sqrt[4]{54t^2}}{3t}$

▶ 88. $\dfrac{\sqrt[4]{3}}{\sqrt[4]{5b^3}}$ $\dfrac{\sqrt[4]{375b}}{5b}$

89. $\dfrac{25}{\sqrt[4]{8a}}$ $\dfrac{25\sqrt[4]{2a^3}}{2a}$

▶ 90. $\dfrac{4}{\sqrt[4]{9t}}$ $\dfrac{\sqrt[4]{9t^3}}{3t}$

Rationalize each denominator. All variables represent positive real numbers. **See Example 9.**

91. $\dfrac{\sqrt{2}}{\sqrt{5}+3}$ $\dfrac{3\sqrt{2}-\sqrt{10}}{4}$

92. $\dfrac{\sqrt{3}}{\sqrt{3}-2}$ $-3-2\sqrt{3}$

93. $\dfrac{2}{\sqrt{x}+1}$ $\dfrac{2(\sqrt{x}-1)}{x-1}$ or $\dfrac{2\sqrt{x}-2}{x-1}$

▶ 94. $\dfrac{3}{\sqrt{x}-2}$ $\dfrac{3(\sqrt{x}+2)}{x-4}$ or $\dfrac{3\sqrt{x}+6}{x-4}$

▶ 95. $\dfrac{\sqrt{7}-\sqrt{2}}{\sqrt{2}+\sqrt{7}}$ $\dfrac{9-2\sqrt{14}}{5}$

▶ 96. $\dfrac{\sqrt{3}+\sqrt{2}}{\sqrt{3}-\sqrt{2}}$ $5+2\sqrt{6}$

97. $\dfrac{\sqrt{x}-\sqrt{y}}{\sqrt{x}+\sqrt{y}}$ $\dfrac{x-2\sqrt{xy}+y}{x-y}$

▶ 98. $\dfrac{\sqrt{x}+\sqrt{y}}{\sqrt{x}-\sqrt{y}}$ $\dfrac{x+2\sqrt{xy}+y}{x-y}$

Rationalize each numerator. All variables represent positive real numbers. **See Example 10.**

99. $\dfrac{\sqrt{x}+3}{x}$ $\dfrac{x-9}{x(\sqrt{x}-3)}$

▶ 100. $\dfrac{2+\sqrt{x}}{5x}$ $\dfrac{4-x}{5x(2-\sqrt{x})}$

101. $\dfrac{\sqrt{x}+\sqrt{y}}{\sqrt{x}}$ $\dfrac{x-y}{\sqrt{x}(\sqrt{x}-\sqrt{y})}$

▶ 102. $\dfrac{\sqrt{x}-\sqrt{y}}{\sqrt{x}+\sqrt{y}}$ $\dfrac{x-y}{x+2\sqrt{xy}+y}$

TRY IT YOURSELF

The following problems involve addition, subtraction, and multiplication of radical expressions, as well as rationalizing the denominator. Perform the operations and simplify, if possible. All variables represent positive real numbers.

103. $\sqrt{x}(\sqrt{14x}+\sqrt{2})$ $x\sqrt{14}+\sqrt{2x}$

104. $2\sqrt[3]{16}-3\sqrt[3]{128}-\sqrt[3]{54}$ $-11\sqrt[3]{2}$

105. $\dfrac{3\sqrt{2}-5\sqrt{3}}{2\sqrt{3}-3\sqrt{2}}$ $\dfrac{3\sqrt{6}+4}{2}$

106. $\dfrac{3\sqrt{6}+5\sqrt{5}}{2\sqrt{5}-3\sqrt{6}}$ $-\dfrac{104+21\sqrt{30}}{34}$

▶ 107. $\left(10\sqrt[3]{2x}\right)^3$ $2{,}000x$

108. $\dfrac{\sqrt{3}}{\sqrt{98x^2}}$ $\dfrac{\sqrt{6}}{14x}$

109. $-4\sqrt[3]{5r^2s}\left(5\sqrt[3]{2r}\right)$ $-20r\sqrt[3]{10s}$

110. $-\sqrt[3]{3xy^2}\left(-\sqrt[3]{9x^3}\right)$ $3x\sqrt[3]{xy^2}$

111. $(3p+\sqrt{5})^2$ $9p^2+6p\sqrt{5}+5$

112. $\sqrt{288t}+\sqrt{80t}-\sqrt{128t}$ $4\sqrt{5t}+4\sqrt{2t}$

113. $\sqrt{\dfrac{72m^8}{25m^3}}$ $\dfrac{6m^2\sqrt{2m}}{5}$

114. $(\sqrt{14x}+\sqrt{3})(\sqrt{14x}-\sqrt{3})$ $14x-3$

115. $\sqrt[4]{3n^2}\sqrt[4]{27n^3}$ $3n\sqrt[4]{n}$

116. $\dfrac{\sqrt{y}-2}{\sqrt{y}+3}$ $\dfrac{y-5\sqrt{y}+6}{y-9}$

117. $\dfrac{\sqrt[3]{x}}{\sqrt[3]{9}}$ $\dfrac{\sqrt[3]{3x}}{3}$

118. $\sqrt[5]{\dfrac{2}{243}}$ $\dfrac{\sqrt[5]{2}}{3}$

119. $\left(3\sqrt{2r}-2\right)^2$ $18r - 12\sqrt{2r} + 4$

120. $\left(2\sqrt{3t}+5\right)^2$ $12t + 20\sqrt{3t} + 25$

121. $\sqrt{x(x+3)}\sqrt{x^3(x+3)}$ $x^2(x+3)$

122. $\sqrt{y^2(x+y)}\sqrt{(x+y)^3}$ $y(x+y)^2$

123. $\dfrac{2z-1}{\sqrt{2z}-1}$ (*Hint:* Do not perform the multiplication of the numerators.) $\sqrt{2z}+1$

124. $\dfrac{3t-1}{\sqrt{3t}+1}$ (*Hint:* Do not perform the multiplication of the numerators.) $\sqrt{3t}-1$

Look Alikes . . .

125. **a.** $\left(3\sqrt{a}\right)^2$ $9a$ **b.** $\left(3+\sqrt{a}\right)^2$ $9 + 6\sqrt{a} + a$

126. **a.** $\left(9\sqrt{x-5}\right)^2$ $81(x-5) = 81x - 405$ **b.** $\left(9+\sqrt{x-5}\right)^2$ $76 + 18\sqrt{x-5} + x$

127. **a.** $\left(\sqrt{m-6}\right)^2$ $m - 6$ **b.** $\left(\sqrt{m}-6\right)^2$ $m - 12\sqrt{m} + 36$

128. **a.** $\dfrac{1}{\sqrt{xy}}$ $\dfrac{\sqrt{xy}}{xy}$ **b.** $\dfrac{1}{\sqrt{x}+\sqrt{y}}$ $\dfrac{\sqrt{x}-\sqrt{y}}{x-y}$

APPLICATIONS

129. Statistics. An example of a normal distribution curve, or *bell-shaped* curve, is shown. A fraction that is part of the equation that models this curve is $\dfrac{1}{\sigma\sqrt{2\pi}}$, where σ is a letter from the Greek alphabet. Rationalize the denominator of the fraction. $\dfrac{\sqrt{2\pi}}{2\pi\sigma}$

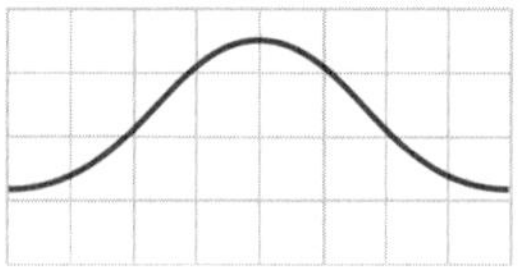

130. Analytical Geometry. The length of the perpendicular segment drawn from $(-2, 2)$ to the line with equation $2x - 4y = 4$ is given by

$$L = \frac{|2(-2) + (-4)(2) + (-4)|}{\sqrt{(2)^2 + (-4)^2}}$$

Find L. Express the result in simplified radical form. Then give an approximation to the nearest tenth. $\dfrac{8\sqrt{5}}{5} \approx 3.6$

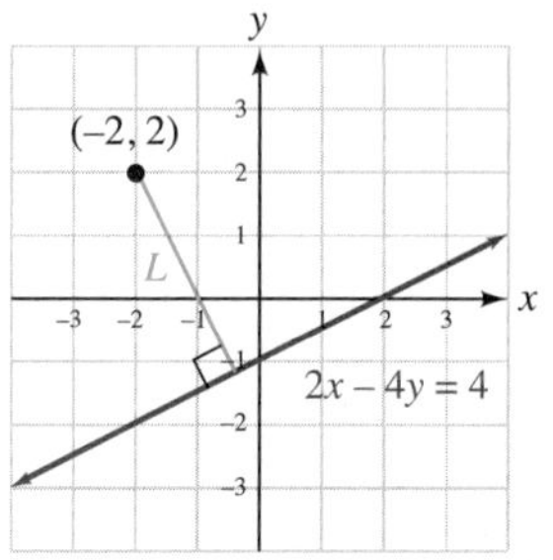

131. Trigonometry. In trigonometry, we often must find the ratio of the lengths of two sides of right triangles. Use the information in the illustration to find the ratio

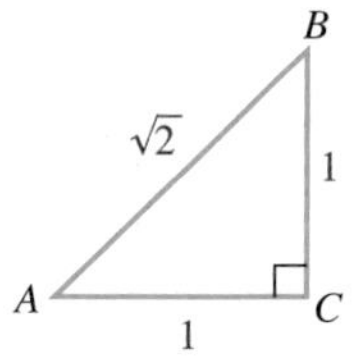

$$\frac{\text{length of side } AC}{\text{length of side } AB}$$

Write the result in simplified radical form. $\dfrac{\sqrt{2}}{2}$

132. Engineering. Refer to the illustration below that shows a block connected to two walls by springs. A measure of how fast the block will oscillate when the spring system is set in motion is given by the formula $\omega = \sqrt{\dfrac{k_1 + k_2}{m}}$ where k_1 and k_2 indicate the stiffness of the springs and m is the mass of the block. Rationalize the right side and restate the formula. $\omega = \dfrac{\sqrt{(k_1 + k_2)m}}{m}$

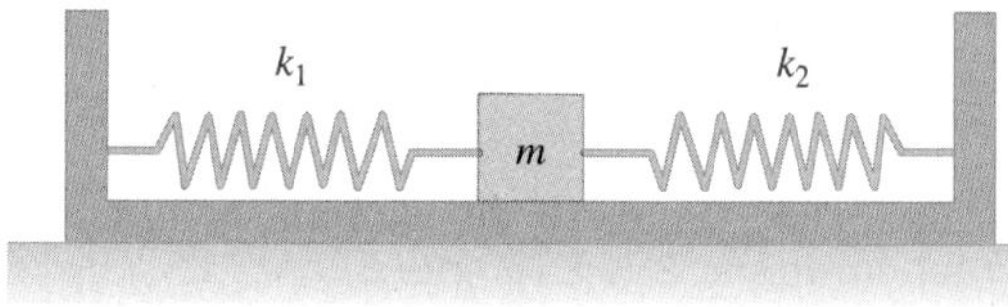

WRITING

133. Consider $\dfrac{\sqrt{3}}{\sqrt{7}} = \dfrac{\sqrt{3}}{\sqrt{7}} \cdot \dfrac{\sqrt{7}}{\sqrt{7}}$. Explain why the expressions on the left side and the right side of the equation are equal.

134. To rationalize the denominator of $\dfrac{\sqrt[4]{12}}{\sqrt[4]{3}}$, why wouldn't we multiply the numerator and denominator by $\dfrac{\sqrt[4]{3}}{\sqrt[4]{3}}$?

135. Explain why $\dfrac{\sqrt[3]{12}}{\sqrt[3]{5}}$ is not in simplified form.

136. Explain why $\sqrt{\dfrac{3a}{11k}}$ is not in simplified form.

137. Explain why $\sqrt{m} \cdot \sqrt{m} = m$ but $\sqrt[3]{m} \cdot \sqrt[3]{m} \neq m$. Assume that m represents a positive number.

138. Explain why the product of $\sqrt{m} + 3$ and $\sqrt{m} - 3$ does not contain a radical.

REVIEW

Solve each equation.

139. $\dfrac{8}{b-2} + \dfrac{3}{2-b} = -\dfrac{1}{b}$ $\dfrac{1}{3}$

140. $\dfrac{2}{x-2} + \dfrac{1}{x+1} = \dfrac{1}{(x+1)(x-2)}$ $\dfrac{1}{3}$

CHALLENGE PROBLEMS

141. Multiply: $\sqrt{2} \cdot \sqrt[3]{2}$. (*Hint:* Keep in mind two things. The indices (plural for *index*) must be the same to use the product rule for radicals, and radical expressions can be written using rational exponents.) $\sqrt[6]{32}$

142. Show that $\dfrac{\sqrt[3]{a^2} + \sqrt[3]{a}\sqrt[3]{b} + \sqrt[3]{b^2}}{\sqrt[3]{a^2} + \sqrt[3]{a}\sqrt[3]{b} + \sqrt[3]{b^2}}$ can be used to rationalize the denominator of $\dfrac{1}{\sqrt[3]{a} - \sqrt[3]{b}}$.

SECTION 9.5

Solving Radical Equations

OBJECTIVES

1 Solve equations containing one radical.

2 Solve equations containing two radicals.

3 Solve formulas containing radicals.

ARE YOU READY? *Are You Ready? exercises available online at www.webassign.net/brookscole*

The following problems review some basic skills that are needed when solving radical equations.

1. Simplify: **a.** $\left(\sqrt{x-1}\right)^2$ $x - 1$
   b. $\left(\sqrt[3]{x^3+7}\right)^3$ $x^3 + 7$
2. Simplify: $\left(3\sqrt{2x+5}\right)^2$ $18x + 45$
3. Expand: $(x-4)^2$ $x^2 - 8x + 16$
4. Solve: $x^2 - 6x - 27 = 0$ $-3, 9$
5. Expand: $(x+2)^3$ $x^3 + 6x^2 + 12x + 8$
6. Multiply: $\left(2-\sqrt{x}\right)^2$ $4 - 4\sqrt{x} + x$

When we solve equations containing fractions, we clear them of the fractions by multiplying both sides by the LCD. To solve equations containing radical expressions, we take a similar approach. The first step is to clear them of the radicals by raising both sides to a power.

1 Solve Equations Containing One Radical.

Radical equations contain at least one radical expression with a variable in the radicand. Some examples are

$$\sqrt{x+3} = 4 \qquad \sqrt[3]{x^3+7} = x + 1 \qquad \sqrt{x} + \sqrt{x+2} = 2$$

To **solve a radical equation,** we find all the values of the variable that make the equation true. The goal when solving a radical equation is to use the following **power rule** to find an equivalent equation that we already know how to solve, such as a linear equation in one variable or a quadratic equation.

The Power Rule for Solving Radical Equations

If we raise two equal quantities to the same power, the results are equal quantities. If x, y, and n are real numbers and $x = y$, then

$$x^n = y^n \quad \text{for any exponent } n.$$

If both sides of an equation are raised to the same power, all solutions of the original equation are also solutions of the new equation. However, the resulting equation might not be equivalent to the original equation. For example, if we square both sides of the equation $x = 3$ with a solution set of $\{3\}$, we obtain the equation $x^2 = 9$ with a solution set of $\{3, -3\}$.

The equations $x = 3$ and $x^2 = 9$ are not equivalent, because they have different solution sets. The solution -3 of $x^2 = 9$ does not satisfy the equation $x = 3$. Since raising both sides of an equation to the same power can produce an equation with proposed solutions that don't satisfy the original equation, **we must always check each proposed solution in the original equation** and discard any **extraneous solutions.**

Caution

A similar warning about checking proposed solutions was given in Chapter 7 when we solved rational equations, such as:

$$\frac{11x}{x-5} = 6 + \frac{55}{x-5}$$

When we use the power rule to solve square root radical equations, it produces expressions of the form $\left(\sqrt{a}\right)^2$. We have seen that when this expression is simplified, the radical symbol is removed.

The Square of a Square Root

For any nonnegative real number a, $\left(\sqrt{a}\right)^2 = a$.

Here are some examples of the square of a square root. Notice how squaring such an expression *removes the square root symbol.*

$$\left(\sqrt{4n}\right)^2 = 4n, \quad \left(\sqrt{x-3}\right)^2 = x - 3, \quad \text{and} \quad \left(\sqrt{5a+8}\right)^2 = 5a + 8$$

EXAMPLE 1 Solve: $\sqrt{x+3} = 4$

Teaching Tip: Point out that we are now working with *equations* and, in previous sections, we were working with *expressions*.

Strategy We will use the power rule and square both sides of the equation.

Why Squaring both sides will produce, on the left side, the expression $\left(\sqrt{x+3}\right)^2$ that simplifies to $x + 3$. This step clears the equation of the radical.

Solution

The Language of Algebra

When we square both sides of an equation, we are **raising both sides to the second power.**

$\sqrt{x+3} = 4$ This is the equation to solve.

$\left(\sqrt{x+3}\right)^2 = (4)^2$ To clear the equation of the square root, square both sides.

$x + 3 = 16$ Perform the operations on each side.

$x = 13$ Solve the resulting linear equation by subtracting 3 from both sides.

We must check the proposed solution 13 to see whether it satisfies the original equation.

The Language of Algebra

Proposed solutions are also called **potential** or **possible** solutions.

Evaluate the left side. Do not square both sides when checking!

Check:

$\sqrt{x+3} = 4$ This is the original equation.

$\sqrt{13+3} \stackrel{?}{=} 4$ Substitute 13 for x.

$\sqrt{16} \stackrel{?}{=} 4$

$4 = 4$ True

Since 13 satisfies the original equation, it is the solution. The solution set is $\{13\}$.

Teaching Example 1 Solve: $\sqrt{x-6} = 5$
Answer: 31

Self Check 1 Solve: $\sqrt{a-2} = 3$ 11

Now Try Problems 15 and 19

The method used in Example 1 to solve a radical equation containing a square root can be generalized, as follows.

Solving an Equation Containing Radicals

1. Isolate a radical term on one side of the equation.
2. Raise both sides of the equation to the power that is the same as the index of the radical.
3. If it still contains a radical, go back to step 1. If it does not contain a radical, solve the resulting equation.
4. Check the proposed solutions in the original equation.

EXAMPLE 2 **Free Fall.** The distance d in feet that an object will fall in t seconds is given by the formula $t = \sqrt{\frac{d}{16}}$. If the designers of the amusement park attraction want the riders to experience 3 seconds of vertical free fall, what length of vertical drop is needed?

Strategy We will begin by substituting 3 for the time t in the formula.

Why We can then solve the resulting radical equation in one variable to find the unknown distance d.

Solution

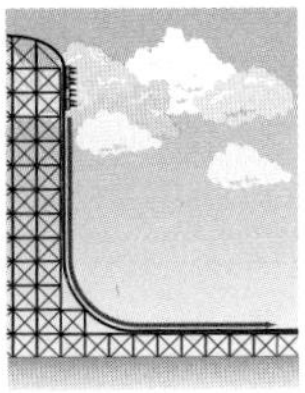

$$t = \sqrt{\frac{d}{16}}$$ This is the given formula.

$$3 = \sqrt{\frac{d}{16}}$$ Substitute 3 for *t*. Here the radical is isolated on the right side.

$$(3)^2 = \left(\sqrt{\frac{d}{16}}\right)^2$$ To clear the equation of the square root, square both sides.

$$9 = \frac{d}{16}$$ Perform the operations on each side.

$$144 = d$$ Solve the resulting equation by multiplying both sides by 16.

The amount of vertical drop needs to be 144 feet.

Caution

When using the power rule, don't forget to raise both sides to the same power. For this example, a common error would be to write

$$3 = \left(\sqrt{\frac{d}{16}}\right)^2$$

Self Check 2 **Free Fall.** How long a vertical drop is needed if the riders are to free fall for 3.5 seconds? 196 ft

Now Try Problem 105

Teaching Example 2 **Free Fall.** In Example 2, how long a vertical drop is needed if the riders are to free fall for 4 seconds?
Answer: 256 ft

When solving radical equations, always isolate a radical term before using the power rule.

EXAMPLE 3 Solve: $\sqrt{3x + 1} + 1 = x$

Strategy Since 1 is outside the square root symbol, there are two terms on the left side of the equation. To isolate the radical term, we will subtract 1 from both sides.

Why This will put the equation in a form in which we can square both sides to clear the radical.

Solution

$$\sqrt{3x + 1} + 1 = x$$ This is the equation to solve.

$$\sqrt{3x + 1} = x - 1$$ To isolate the radical on the left side, subtract 1 from both sides.

$$\left(\sqrt{3x + 1}\right)^2 = (x - 1)^2$$ Square both sides to eliminate the square root. Don't forget the parentheses.

$$3x + 1 = x^2 - 2x + 1$$ On the right side, use a special-product formula to square the binomial: $(x - 1)^2 = x^2 - 2x + 1$. The resulting equation is quadratic.

$$0 = x^2 - 5x$$ To get 0 on the left side, subtract 3x and 1 from both sides.

$$0 = x(x - 5)$$ Factor out the GCF, *x*.

$$x = 0 \quad \text{or} \quad x - 5 = 0$$ Set each factor equal to 0.

$$x = 0 \quad | \quad x = 5$$

Caution

If both sides of the equation in Example 3 are squared, as is, the radical is not cleared. In fact, the equation becomes more complicated.

$$\left(\sqrt{3x + 1} + 1\right)^2 = (x)^2$$
$$3x + 1 + 2\sqrt{3x + 1} + 1 = x^2$$

Teaching Tip: In this section, students often must square binomials. Remind them how to use the special-product formulas to find the results quickly.

We must check each proposed solution to see whether it satisfies the original equation.

***This is the check for* 0:**	***This is the check for* 5:**	
$\sqrt{3x + 1} + 1 = x$	$\sqrt{3x + 1} + 1 = x$	This is the original equation.
$\sqrt{3(0) + 1} + 1 \stackrel{?}{=} 0$	$\sqrt{3(5) + 1} + 1 \stackrel{?}{=} 5$	
$\sqrt{1} + 1 \stackrel{?}{=} 0$	$\sqrt{16} + 1 \stackrel{?}{=} 5$	
$2 = 0$ False	$5 = 5$ True	

The proposed solution 0 does not check; it must be discarded. Since the only solution is 5, the solution set is $\{5\}$.

Success Tip

Even if you are certain that no algebraic mistakes were made when solving a radical equation, you must still check your solutions. Raising both sides to an even power can introduce extraneous solutions that must be discarded.

Teaching Example 3 Solve: $\sqrt{2x + 4} + 2 = x$
Answer: 6, 0 is extraneous

Self Check 3 Solve: $\sqrt{4x + 1} + 1 = x$ 6, 0 is extraneous

Now Try ▶ Problems 23 and 27

Using Your Calculator ▶ Solving Radical Equations

To find solutions for $\sqrt{3x + 1} + 1 = x$ with a graphing calculator, we graph the functions $f(x) = \sqrt{3x + 1} + 1$ and $g(x) = x$, as in figure (a). We then trace to find the approximate x-coordinate of their intersection point, as in figure (b). After repeated zooms, we will see that $x = 5$.

We also can use the INTERSECT feature to approximate the point of intersection of the graphs. See figure (c). The intersection point of **(5**, 5), with x-coordinate **5**, implies that 5 is a solution of the radical equation.

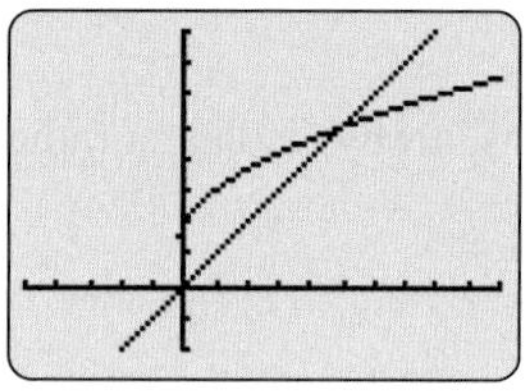

(a)

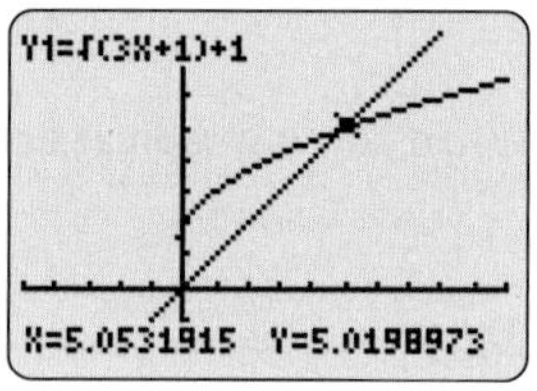

(b)

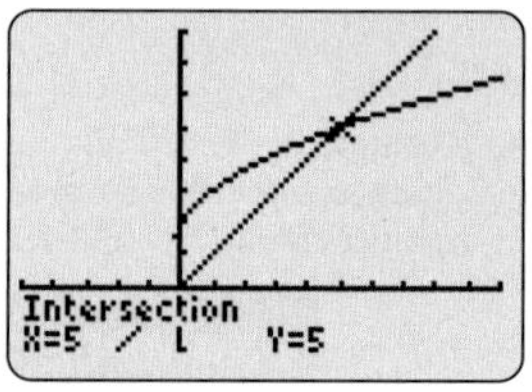

(c)

EXAMPLE 4 Solve: $\sqrt{3x} + 8 = 2$

Strategy Since 8 is outside the square root symbol, there are two terms on the left side of the equation. To isolate the radical, we will subtract 8 from both sides.

Why This will put the equation in a form in which we can square both sides to clear the radical.

Solution

$\sqrt{3x} + 8 = 2$ This is the equation to solve.

$\sqrt{3x} = -6$ To isolate the radical on the left side, subtract 8 from both sides.

$\left(\sqrt{3x}\right)^2 = (-6)^2$ Square both sides to eliminate the square root.

$3x = 36$ Perform the operations on each side.

$x = 12$ To solve the resulting linear equation, divide both sides by 3.

We check the proposed solution 12 in the original equation.

$\sqrt{3x} + 8 = 2$

$\sqrt{3(\mathbf{12})} + 8 \stackrel{?}{=} 2$ Substitute 12 for x.

$\sqrt{36} + 8 \stackrel{?}{=} 2$

$6 + 8 \stackrel{?}{=} 2$

$14 = 2$ False

Since 12 does not satisfy the original equation, it is extraneous. The equation $\sqrt{3x} + 8 = 2$ has no solution. The solution set is $\varnothing$.

Success Tip

After isolating the radical, we obtained the equation $\sqrt{3x} = -6$. Since the principal square root of a number cannot be negative, we immediately know that $\sqrt{3x} + 8 = 2$ has no solution.

Teaching Example 4 Solve: $\sqrt{9m} + 50 = 41$
Answer: 9 is extraneous, no solution, $\varnothing$

Self Check 4 Solve: $\sqrt{a - 9} + 3 = 0$ 18 is extraneous, no solution, $\varnothing$

Now Try ▶ Problem 31

Recall the following property of radicals from Section 9.4.

The *n*th Power of the *n*th Root

If $\sqrt[n]{a}$ is a real number,

$$\left(\sqrt[n]{a}\right)^n = a$$

Here are some examples of the cube of a cube root, and the fourth power of a fourth root. Notice how raising each expression to the appropriate power *removes the radical symbol.*

$$\left(\sqrt[3]{5x}\right)^3 = 5x \qquad \left(\sqrt[3]{2a - 11}\right)^3 = 2a - 11 \qquad \left(\sqrt[4]{6 - y}\right)^4 = 6 - y$$

The power rule and the *n*th power of the *n*th root property can be used in combination to solve radical equations that involve cube roots, fourth roots, fifth roots, and so on.

EXAMPLE 5 Solve: $(x^3 + 7)^{1/3} = x + 1$

Strategy We will use the definition of a fractional exponent with numerator 1 to rewrite the given equation as an equivalent radical equation.

Why Then we can use the equation-solving strategy of this section to find the solution(s).

Teaching Tip: Show that Example 4 could be written as:

$$(3x)^{1/2} + 8 = 2$$

Solution Recall from Section 9.2 that $x^{1/n} = \sqrt[n]{x}$. Thus, we can replace $(x^3 + 7)^{1/3}$ on the left side of the given equation with $\sqrt[3]{x^3 + 7}$.

Success Tip

Be careful cubing the binomial $x + 1$.

$$(x + 1)^3 \neq x^3 + 1^3$$

$(x^3 + 7)^{1/3} = x + 1$	This is the equation to solve.
$\sqrt[3]{x^3 + 7} = x + 1$	Write $(x^3 + 7)^{1/3}$ using radical notation.
$\left(\sqrt[3]{x^3 + 7}\right)^3 = (x + 1)^3$	Cube both sides to eliminate the cube root. Don't forget the parentheses.
$x^3 + 7 = x^3 + 3x^2 + 3x + 1$	Perform the operations on each side. On the right: $(x + 1)^3 = (x + 1)(x + 1)^2$.
$0 = 3x^2 + 3x - 6$	To get 0 on the left side, subtract x^3 and 7 from both sides. This is a quadratic equation.
$0 = x^2 + x - 2$	Divide both sides by 3.
$0 = (x + 2)(x - 1)$	Factor the trinomial.
$x + 2 = 0$ or $x - 1 = 0$	Set each factor equal to 0.
$x = -2$ | $x = 1$	Solve each linear equation.

Success Tip

Raising both sides of an equation to an even power may not produce an equivalent equation. Raising both sides to an odd power does produce an equivalent equation and, in such cases, there will be no extraneous solutions. However, it is still recommended that all proposed solutions be checked regardless of the type of root involved.

We check each proposed solution, −2 and 1, to see whether they satisfy the original equation.

Check:

$(x^3 + 7)^{1/3} = x + 1$	$(x^3 + 7)^{1/3} = x + 1$
$\sqrt[3]{x^3 + 7} = x + 1$	$\sqrt[3]{x^3 + 7} = x + 1$
$\sqrt[3]{(-2)^3 + 7} \stackrel{?}{=} -2 + 1$	$\sqrt[3]{1^3 + 7} \stackrel{?}{=} 1 + 1$
$\sqrt[3]{-8 + 7} \stackrel{?}{=} -1$	$\sqrt[3]{1 + 7} \stackrel{?}{=} 2$
$\sqrt[3]{-1} \stackrel{?}{=} -1$	$\sqrt[3]{8} \stackrel{?}{=} 2$
$-1 = -1$ True	$2 = 2$ True

Both −2 and 1 satisfy the original equation. Thus, the solution set is $\{-2, 1\}$.

Teaching Example 5 Solve: $(x^3 + 27)^{1/3} = x + 3$
Answer: 0, −3

Self Check 5 Solve: $(x^3 + 8)^{1/3} = x + 2$ 0, −2

Now Try Problems 35 and 39

EXAMPLE 6 Let $f(x) = \sqrt[4]{2x + 1}$. For what value(s) of x is $f(x) = 5$?

Strategy We will substitute 5 for $f(x)$ and solve the equation $5 = \sqrt[4]{2x + 1}$. To do so, we will raise both sides of the equation to the fourth power.

Why Raising both sides to the fourth power will produce, on the right side, the expression $\left(\sqrt[4]{2x + 1}\right)^4$ that simplifies to $2x + 1$. This step clears the equation of the radical.

Solution To find the value(s) for which $f(x) = 5$, we substitute 5 for $f(x)$ and solve for x.

$$f(x) = \sqrt[4]{2x + 1}$$
$$5 = \sqrt[4]{2x + 1} \quad \text{This is the equation to solve.}$$

Since the equation contains a fourth root, we raise both sides to the fourth power to solve for x.

$(5)^4 = \left(\sqrt[4]{2x + 1}\right)^4$	Use the power rule to clear the radical.
$625 = 2x + 1$	Perform the operations on each side.
$624 = 2x$	To solve the resulting equation, subtract 1 from both sides.
$312 = x$	Divide both sides by 2.

If $x = 312$, then $f(x) = 5$. Verify this by evaluating $f(312)$ using a calculator, if necessary.

Teaching Example 6 Let $f(x) = \sqrt[3]{4x - 4}$. For what value(s) of x is $f(x) = 2$?
Answer: 3

Self Check 6 Let $g(x) = \sqrt[5]{10x + 1}$. For what value(s) of x is $g(x) = 1$? 0

Now Try Problem 43

2 Solve Equations Containing Two Radicals.

To solve an equation containing two radicals, we want to have one radical on the left side and one radical on the right side.

EXAMPLE 7 Solve: $\sqrt{5x + 9} = 2\sqrt{3x + 4}$

Strategy We will square both sides to clear the equation of both radicals.

Why We can square both sides immediately since each radical is isolated on one side of the equation.

Solution

$\sqrt{5x + 9} = 2\sqrt{3x + 4}$	This is the equation to solve.
$\left(\sqrt{5x + 9}\right)^2 = \left(2\sqrt{3x + 4}\right)^2$	Square both sides to eliminate the radicals.
$5x + 9 = 2^2\left(\sqrt{3x + 4}\right)^2$	Simplify on the left. On the right, raise each factor of the product $2\sqrt{3x + 4}$ to the second power.
$5x + 9 = 4(3x + 4)$	Perform the operations on the right.
$5x + 9 = 12x + 16$	To solve the resulting linear equation, distribute the multiplication by 4.
$-7 = 7x$	Subtract 5x and 16 from both sides.
$-1 = x$	Divide both sides by 7.

Caution

When finding $\left(2\sqrt{3x + 4}\right)^2$, remember to square both of the factors, 2 and $\sqrt{3x + 4}$, to get:

$$2^2\left(\sqrt{3x + 4}\right)^2$$

This is an application of the power of a product rule for exponents:

$$(xy)^n = x^n y^n$$

We check the solution by substituting -1 for x in the original equation.

$\sqrt{5x + 9} = 2\sqrt{3x + 4}$	
$\sqrt{5(-1) + 9} \stackrel{?}{=} 2\sqrt{3(-1) + 4}$	Substitute -1 for x.
$\sqrt{4} \stackrel{?}{=} 2\sqrt{1}$	
$2 = 2$	True

The solution is -1 and the solution set is $\{-1\}$.

Teaching Example 7 Solve: $\sqrt{3x+21} = 3\sqrt{x-5}$
Answer: 11

Self Check 7 Solve: $\sqrt{x-4} = 2\sqrt{x-16}$ 20

Now Try Problem 49

When more than one radical appears in an equation, we often must use the power rule more than once.

EXAMPLE 8 Solve: $\sqrt{x} + \sqrt{x+2} = 2$

Strategy We will isolate $\sqrt{x+2}$ on the left side of the equation and square both sides to eliminate it. After simplifying the resulting equation, we will isolate the remaining radical term and square both sides a second time to eliminate it.

Why Each time that we square both sides, we are able to clear the equation of one radical.

Solution

$\sqrt{x} + \sqrt{x+2} = 2$ This is the equation to solve.

$\sqrt{x+2} = 2 - \sqrt{x}$ To isolate $\sqrt{x+2}$, subtract $\sqrt{x}$ from both sides.

$\left(\sqrt{x+2}\right)^2 = \left(2 - \sqrt{x}\right)^2$ Square both sides to eliminate the square root on the left side.

$x + 2 = \left(2 - \sqrt{x}\right)^2$ Perform the operation on the left side.

Success Tip

Note in this example that we isolate the more complicated radical term, $\sqrt{x+2}$. As a result, we work with the less complicated radical term $\sqrt{x}$ when applying the special-product rule.

Caution

When finding $\left(2 - \sqrt{x}\right)^2$, it is like squaring a binomial. Remember to use a special-product rule (or FOIL). **Do not just square the first term and the last term.**

$$\left(2 - \sqrt{x}\right)^2 \neq 4 + x$$

To square the expression $2 - \sqrt{x}$ on the right side, we can use a special-product rule:

$$x + 2 = \underbrace{2^2}_{\text{Square the first term, 2.}} - \underbrace{2(2)\left(\sqrt{x}\right)}_{\text{Twice the product of both terms}} + \underbrace{\left(\sqrt{x}\right)^2}_{\text{Square the last term, }\sqrt{x}.}$$

$$x + 2 = 4 - 4\sqrt{x} + x$$

Since the equation still contains a radical, we need to square both sides again. Before doing that, we must isolate the radical on one side.

$2 = 4 - 4\sqrt{x}$ Subtract x from both sides.

$-2 = -4\sqrt{x}$ To isolate the radical term $-4\sqrt{x}$, subtract 4 from both sides.

$\frac{1}{2} = \sqrt{x}$ To isolate the radical, divide both sides by -4.

$\left(\frac{1}{2}\right)^2 = \left(\sqrt{x}\right)^2$ To eliminate the radical, square both sides again.

$\frac{1}{4} = x$ Perform the operations on each side.

Check:

$\sqrt{x} + \sqrt{x+2} = 2$ This is the original equation.

$\sqrt{\frac{1}{4}} + \sqrt{\frac{1}{4} + 2} \stackrel{?}{=} 2$ Substitute $\frac{1}{4}$ for x.

$\frac{1}{2} + \sqrt{\frac{9}{4}} \stackrel{?}{=} 2$ Think of 2 as $\frac{8}{4}$ and add: $\frac{1}{4} + \frac{8}{4} = \frac{9}{4}$.

$\frac{1}{2} + \frac{3}{2} \stackrel{?}{=} 2$ Evaluate $\sqrt{\frac{9}{4}}$.

$2 = 2$ True

The result $\frac{1}{4}$ checks. The solution set is $\left\{\frac{1}{4}\right\}$.

Teaching Example 8 Solve: $\sqrt{9x+16}+\sqrt{9x}=8$
Answer: 1

Self Check 8 Solve: $\sqrt{a}+\sqrt{a+3}=3$ 1

Now Try Problems 55 and 59

Using Your Calculator ▶ Solving Radical Equations

To find solutions for $\sqrt{x}+\sqrt{x+2}=4$ (an equation similar to Example 8) with a graphing calculator, we graph the functions $f(x)=\sqrt{x}+\sqrt{x+2}$ and $g(x)=4$. We then trace to find an approximation of the x-coordinate of their intersection point, as in figure (a). From the figure, we can see that $x\approx 2.98$. We can zoom to get better results.

Figure (b) shows that the INTERSECT feature gives the approximate coordinates of the point of intersection of the two graphs as (3.06, 4). Therefore, an approximate solution of the radical equation is 3.06. Check its reasonableness.

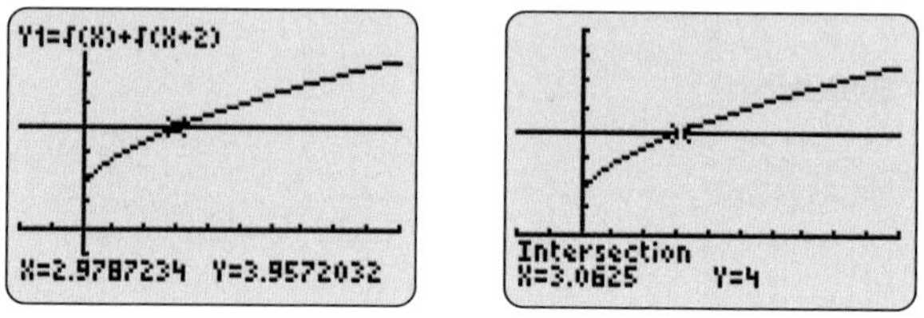

(a) (b)

3 Solve Formulas Containing Radicals.

To *solve a formula for a variable* means to isolate that variable on one side of the equation, with all other quantities on the other side.

EXAMPLE 9 **Depreciation Rates.** Some office equipment that is now worth V dollars originally cost C dollars 3 years ago. The rate r at which it has depreciated is given by $r=1-\sqrt[3]{\frac{V}{C}}$. Solve the formula for C.

Strategy To isolate the radical, we will subtract 1 from both sides. We can then eliminate the radical by cubing both sides.

Why Cubing both sides will produce, on the right, the expression $\left(\sqrt[3]{\frac{V}{C}}\right)^3$ that simplifies to $\frac{V}{C}$. This step clears the equation of the radical.

Solution We begin by isolating the cube root on the right side of the equation.

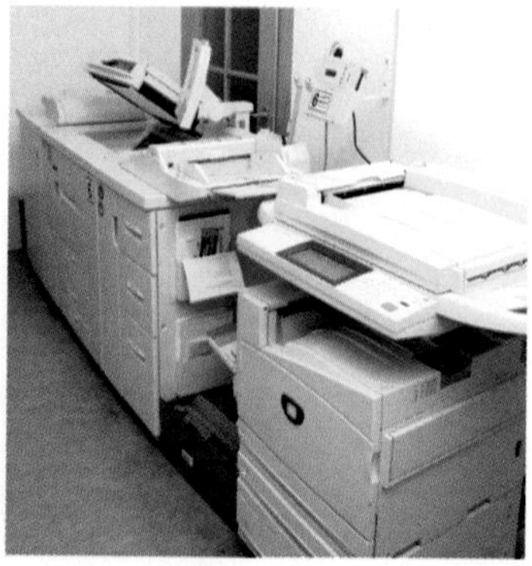
iStockphoto.com/wakila

$$r=1-\sqrt[3]{\frac{V}{C}}$$ This is the depreciation model.

$$r-1=-\sqrt[3]{\frac{V}{C}}$$ Subtract 1 from both sides to isolate the radical.

$$(r-1)^3=\left(-\sqrt[3]{\frac{V}{C}}\right)^3$$ To eliminate the radical, cube both sides.

$$(r-1)^3=-\frac{V}{C}$$ Simplify the right side.

$$C(r-1)^3=-V$$ To clear the equation of the fraction, multiply both sides by C.

$$C=-\frac{V}{(r-1)^3}$$ To isolate C, divide both sides by $(r-1)^3$.

Teaching Example 9 Solve $N = \frac{1}{2\pi}\sqrt{\frac{a}{r}}$ for r.
Answer: $r = \frac{a}{4\pi^2 N^2}$

Self Check 9 **Statistics.** A formula used in statistics to determine the size of a sample to obtain a desired degree of accuracy is $E = z_0\sqrt{\frac{pq}{n}}$. Solve the formula for n. $n = \frac{z_0^2 pq}{E^2}$

Now Try Problem 67

SECTION 9.5 STUDY SET

VOCABULARY

Fill in the blanks.

1. Equations such as $\sqrt{x + 4} - 4 = 5$ and $\sqrt[3]{x + 1} = 12$ are called radical equations.
2. To solve a radical equation, we find all the values of the variable that make the equation true.
3. When we square both sides of a radical equation, we say we are raising both sides to the second power.
4. When solving equations containing radicals, first we isolate one radical expression on one side of the equation.
5. Proposed solutions of a radical equation that don't satisfy it are called extraneous solutions.
6. To check a proposed solution means to substitute it into the original equation and see whether a true statement results.

CONCEPTS

7. Fill in the blanks.
   a. The power rule for solving radical equations states that if x, y, and n are real numbers and $x = y$, then
   $x^n = y^n$
   b. If $\sqrt[n]{a}$ is a real number, then $\left(\sqrt[n]{a}\right)^n = a$.
8. Determine whether 6 is a solution of each radical equation.
   a. $\sqrt{x + 3} = x - 3$ Yes b. $\sqrt[3]{5x - 3} + 9 = x$ No
9. What is the first step in solving each equation?
   a. $\sqrt{x + 11} = 5$ Square both sides.
   b. $\sqrt[3]{5x + 4} + 3 = 30$ Subtract 3 from both sides.
   c. $\sqrt{x + 8} - \sqrt{2x} = 1$ Add $\sqrt{2x}$ to both sides.
10. Simplify each expression.
    a. $\left(\sqrt{x}\right)^2$ x b. $\left(\sqrt{x - 5}\right)^2$ $x - 5$
    c. $\left(\sqrt[3]{4x - 8}\right)^3$ $4x - 8$ d. $\left(\sqrt[4]{8x}\right)^4$ $8x$
    e. $\left(4\sqrt{2x}\right)^2$ $32x$ f. $\left(3\sqrt[3]{x + 1}\right)^3$ $27x + 27$
11. Find: $\left(\sqrt{x} - 3\right)^2$ $x - 6\sqrt{x} + 9$
12. Find: $\left(\sqrt{5x + 2} - 4\right)^2$ $5x - 8\sqrt{5x + 2} + 18$

NOTATION

Complete each solution.

13. Solve: $\sqrt{3x + 3} - 1 = 5$
$$\sqrt{3x + 3} = 6$$
$$\left(\sqrt{3x + 3}\right)^2 = (6)^2$$
$$3x + 3 = 36$$
$$3x = 33$$
$$x = 11$$
Does the proposed solution check? Yes
14. Fill in the blanks. Write each radical equation using a rational exponent.
    a. $\sqrt{x + 10} + 5 = 15$ can be written $(x + 10)^{1/2} + 5 = 15$
    b. $\sqrt[3]{2t + 4} = t - 1$ can be written $(2t + 4)^{1/3} = t - 1$

GUIDED PRACTICE

Solve each equation. See Example 1.

15. $\sqrt{a - 3} = 1$ 4
16. $\sqrt{x - 10} = 1$ 11
17. $\sqrt{4x + 5} = 5$ 5
18. $\sqrt{5x - 6} = 2$ 2
19. $\sqrt{6x + 13} = 7$ 6
20. $\sqrt{6x + 1} = 5$ 4
21. $\sqrt{\frac{1}{3}x - 2} = 8$ 198
22. $\sqrt{\frac{1}{2}x + 3} = 6$ 66

Solve each equation. Write all proposed solutions. Cross out those that are extraneous. See Example 3.

23. $\sqrt{2x + 11} + 2 = x$ 7, ~~−1~~
24. $\sqrt{2a - 3} + 3 = a$ 6, ~~2~~
25. $\sqrt{2r - 3} + 9 = r$ 14, ~~6~~
26. $\sqrt{-x + 2} + 2 = x$ 2, ~~1~~
27. $\sqrt{3t + 7} - t = 1$ 3, ~~−2~~
28. $\sqrt{t + 3} - t = 1$ 1, ~~−2~~
29. $\sqrt{9 - a} - a = 3$ 0, ~~−7~~
30. $\sqrt{4 - a} - a = 2$ 0, ~~−5~~

Solve each equation. Write all proposed solutions. Cross out those that are extraneous. See Example 4.

31. $\sqrt{5x} + 10 = 8$
~~$\frac{4}{5}$~~, no solution
32. $\sqrt{3x} + 5 = 2$
~~3~~, no solution
33. $\sqrt{5 - x} + 10 = 9$
~~4~~, no solution
34. $1 = 2 + \sqrt{4x + 75}$
~~$-\frac{37}{2}$~~, no solution

Solve each equation. See Example 5.

35. $\sqrt[3]{7n - 1} = 3$
4
36. $\sqrt[3]{12m + 4} = 4$
5
37. $\sqrt[3]{x^3 - 7} = x - 1$
2, −1
38. $\sqrt[3]{b^3 - 63} = b - 3$
4, −1

39. $(m^3 + 26)^{1/3} = m + 2$
$1, -3$

▶ 40. $(x^3 + 56)^{1/3} = x + 2$
$2, -4$

41. $(5r + 14)^{1/3} = 4$
10

▶ 42. $(2b + 29)^{1/3} = 3$
-1

See Example 6.

43. Let $f(x) = \sqrt[4]{3x + 1}$. For what value(s) of x is $f(x) = 4$?
85

▶ 44. Let $f(x) = \sqrt{2x^2 - 7x}$. For what value(s) of x is $f(x) = 2$?
$-\frac{1}{2}, 4$

45. Let $f(x) = \sqrt[3]{3x - 6}$. For what value(s) of x is $f(x) = -3$?
-7

▶ 46. Let $f(x) = \sqrt[5]{4x - 4}$. For what value(s) of x is $f(x) = -2$?
-7

Solve each equation. See Example 7.

47. $\sqrt{3x + 12} = \sqrt{5x - 12}$
12

▶ 48. $\sqrt{m + 4} = \sqrt{2m - 5}$
9

49. $2\sqrt{4x + 1} = \sqrt{x + 4}$
0

▶ 50. $\sqrt{6 - 2x} = 4\sqrt{x - 3}$
3

51. $\sqrt{6t + 9} = 3\sqrt{t}$
3

▶ 52. $\sqrt{12x + 24} = 6\sqrt{x}$
1

53. $(34x + 26)^{1/3} = 4(x - 1)^{1/3}$
3

▶ 54. $(a^2 + 2a)^{1/3} = 2(a - 1)^{1/3}$
$2, 4$

Solve each equation. Write all proposed solutions. Cross out those that are extraneous. See Example 8.

55. $\sqrt{x - 5} + \sqrt{x} = 5$
9

56. $\sqrt{x - 7} + \sqrt{x} = 7$
16

57. $\sqrt{z + 3} - \sqrt{z} = 1$
1

▶ 58. $\sqrt{x + 12} + \sqrt{x} = 6$
4

▶ 59. $3 = \sqrt{y + 4} - \sqrt{y + 7}$
~~-3~~, no solution

▶ 60. $3 = \sqrt{u - 3} - \sqrt{u}$
~~4~~, no solution

▶ 61. $2 = \sqrt{2u + 7} - \sqrt{u}$
$1, 9$

▶ 62. $1 = \sqrt{4s + 5} - \sqrt{2s + 2}$
$-1, 1$

Solve each equation for the specified variable or expression. See Example 9.

▶ 63. $v = \sqrt{2gh}$ for h
$h = \frac{v^2}{2g}$

▶ 64. $d = 1.4\sqrt{h}$ for h
$h = \frac{d^2}{1.96}$

65. $T = 2\pi\sqrt{\frac{l}{32}}$ for l
$l = \frac{8T^2}{\pi^2}$

▶ 66. $d = \sqrt[3]{\frac{12V}{\pi}}$ for V
$V = \frac{\pi d^3}{12}$

67. $r = \sqrt[3]{\frac{A}{P}} - 1$ for A
$A = P(r + 1)^3$

▶ 68. $r = \sqrt[3]{\frac{A}{P}} - 1$ for P
$P = \frac{A}{(r + 1)^3}$

69. $L_A = L_B\sqrt{1 - \frac{v^2}{c^2}}$ for v^2
$v^2 = c^2\left(1 - \frac{L_A{}^2}{L_B{}^2}\right)$

▶ 70. $R_1 = \sqrt{\frac{A}{\pi} - R_2{}^2}$ for A
$A = \pi R_1{}^2 + \pi R_2{}^2$

TRY IT YOURSELF

Solve each equation. Write all proposed solutions. Cross out those that are extraneous.

71. $2\sqrt{x} = \sqrt{5x - 16}$
16

72. $3\sqrt{x} = \sqrt{3x + 54}$
9

73. $\sqrt{x + 5} + \sqrt{x - 3} = 4$
4

▶ 74. $\sqrt{b + 7} - \sqrt{b - 5} = 2$
9

75. $n = (n^3 + n^2 - 1)^{1/3}$
$-1, 1$

76. $(m^4 + m^2 - 25)^{1/4} = m$
~~-5~~, 5

▶ 77. $\sqrt{y + 2} + y = 4$
2, ~~7~~

▶ 78. $\sqrt{22y + 86} - y = 9$
$5, -1$

79. $\sqrt[3]{x + 8} = -2$
-16

▶ 80. $\sqrt[3]{x + 4} = -1$
-5

▶ 81. $2 = \sqrt{x + 5} - \sqrt{x} + 1$
4

82. $4 = \sqrt{x + 8} - \sqrt{x} + 2$
1

83. $x = \frac{\sqrt{12x - 5}}{2}$ $\frac{5}{2}, \frac{1}{2}$

84. $x = \frac{\sqrt{16x - 12}}{2}$ $1, 3$

85. $(n^2 + 6n + 3)^{1/2} = (n^2 - 6n - 3)^{1/2}$ $-\frac{1}{2}$

86. $(m^2 - 12m - 3)^{1/2} = (m^2 + 12m + 3)^{1/2}$ $-\frac{1}{4}$

87. $\sqrt{x - 5} - \sqrt{x + 3} = 4$
~~6~~, no solution

▶ 88. $\sqrt{x + 8} - \sqrt{x - 4} = -2$
~~8~~, no solution

89. $\sqrt[4]{10y + 6} = 2\sqrt[4]{y}$
1

90. $\sqrt[4]{21a + 39} = 3\sqrt[4]{a - 1}$
2

91. $\sqrt{-5x + 24} = 6 - x$
$4, 3$

▶ 92. $-s - 3 = 2\sqrt{5 - s}$
-11, ~~1~~

93. $\sqrt{2x} + 5 = 1$
~~8~~, no solution

94. $\sqrt{3x} + 10 = 1$
~~27~~, no solution

▶ 95. $\sqrt{6x + 2} - \sqrt{5x + 3} = 0$
1

96. $\sqrt{5x + 2} - \sqrt{x + 10} = 0$
2

97. Let $f(x) = \sqrt{x + 16}$ and $g(x) = 7 - \sqrt{x + 9}$. Find all values of x for which $f(x) = g(x)$. 0

98. Let $s(t) = \sqrt{t + 8}$ and $h(t) = 6 - \sqrt{t - 4}$. Find all values of t for which $s(t) = h(t)$. 8

99. Let $f(x) = \sqrt[4]{x + 8} - \sqrt[4]{2x}$. Find all values of x for which $f(x) = 0$. 8

▶ 100. Let $h(a) = \sqrt[4]{a + 11} - \sqrt[4]{2a + 6}$. Find all values of a for which $h(a) = 0$. 5

Look Alikes . . .

101. a. $3\sqrt{5n - 9} = \sqrt{5n}$
$\frac{81}{40}$

b. $3 + \sqrt{5n - 9} = \sqrt{5n}$
$\frac{9}{5}$

▶ 102. a. $\sqrt{8 + a} = 2\sqrt{a}$
$\frac{8}{3}$

b. $\sqrt{8 + a} = 2 + \sqrt{a}$
1

103. a. $\sqrt{2x} - 10 = 0$
50

b. $\sqrt{2x} + 10 = 0$
~~50~~, no solution

104. a. $x^{1/2} + 6 = 8$
4

b. $x^{1/4} + 6 = 8$
16

APPLICATIONS

▶ 105. **Highway Design.** A curved road will accommodate traffic traveling s mph if the radius of the curve is r feet, according to the formula $s = 3\sqrt{r}$. If engineers expect 40-mph traffic, what radius should they specify? Give the result to the nearest foot. 178 ft

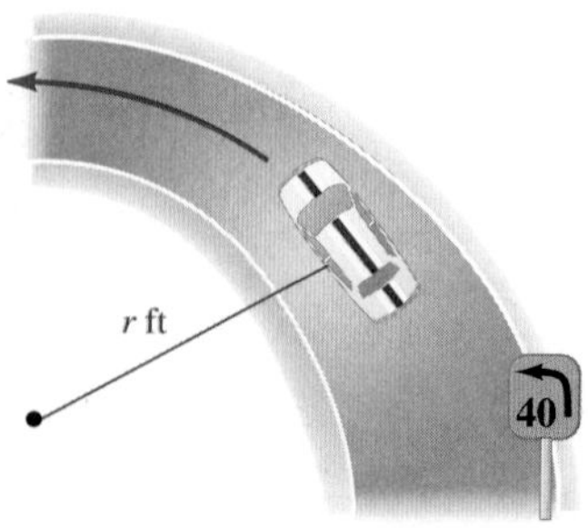

106. Forestry. The taller a lookout tower, the farther an observer can see. That distance d (called the *horizon distance,* measured in miles) is related to the height h of the observer (measured in feet) by the formula $d = 1.22\sqrt{h}$. How tall must a lookout tower be to see the edge of the forest, 25 miles away? (Round to the nearest foot.) 420 ft

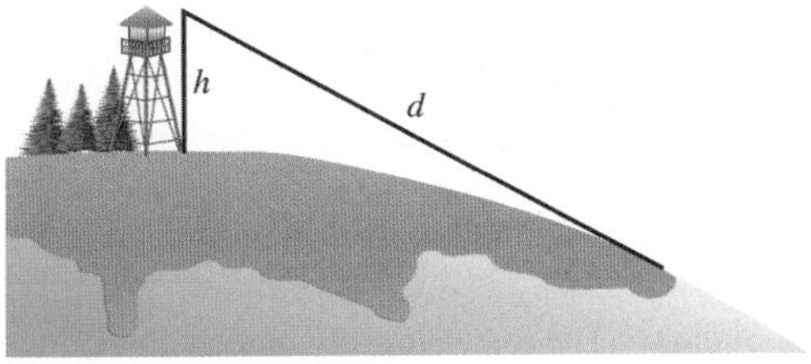

107. Wind Power. The power generated by a windmill is related to the velocity of the wind by the formula $v = \sqrt[3]{\frac{P}{0.02}}$ where P is the power (in watts) and v is the velocity of the wind (in mph). Find how much power the windmill is generating when the wind is 29 mph. About 488 watts

108. Diamonds. The *effective rate of interest r* earned by an investment is given by the formula $r = \sqrt[n]{\frac{A}{P}} - 1$ where P is the initial investment that grows to value A after n years. If a diamond buyer got \$4,000 for a 1.73-carat diamond that he had purchased 4 years earlier, and earned an annual rate of return of 6.5% on the investment, what did he originally pay for the diamond? About \$3,109

109. from **Campus to Careers**

General Contractor

©Goodluz/Shutterstock.com

During construction, carpenters often brace walls as shown in the illustration, where the length L of the brace is given by the formula $L = \sqrt{f^2 + h^2}$. If a carpenter nails a 10-ft brace to the wall 6 feet above the floor, how far from the base of the wall should he nail the brace to the floor? 8 ft

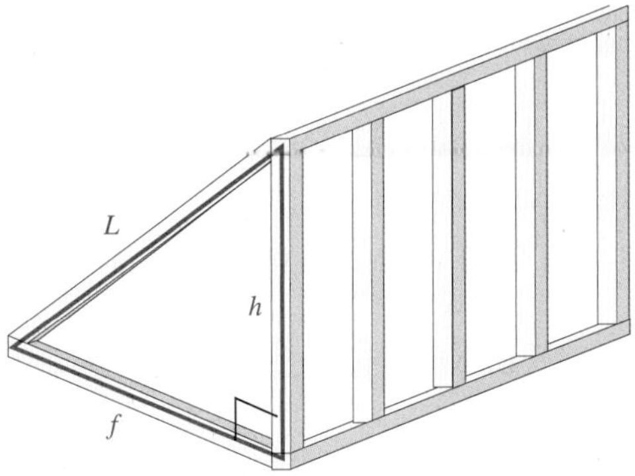

110. Theater Productions. The ropes, pulleys, and sandbags shown in the illustration are part of a mechanical system used to raise and lower scenery for a stage play. For the scenery to be in the proper position, the following formula must apply: $w_2 = \sqrt{w_1{}^2 + w_3{}^2}$. If $w_2 = 12.5$ lb and $w_3 = 7.5$ lb, find w_1. 10 lb

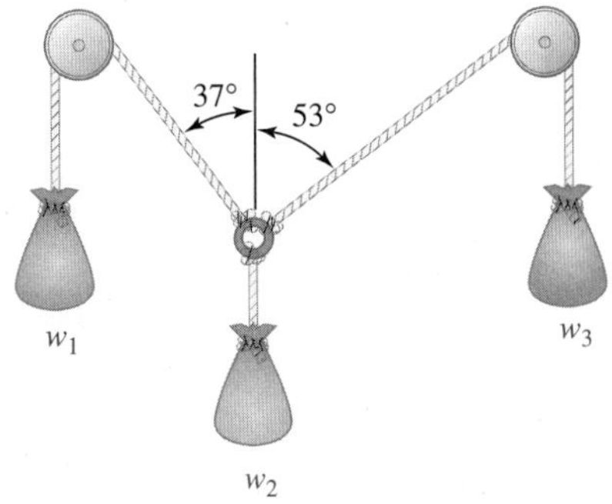

111. Supply and Demand. The number of wrenches that will be produced at a given price can be predicted by the formula $s = \sqrt{5x}$, where s is the supply (in thousands) and x is the price (in dollars). The demand d for wrenches can be predicted by the formula $d = \sqrt{100 - 3x^2}$. Find the equilibrium price—that is, find the price at which supply will equal demand. \$5

112. Supply and Demand. The number of mirrors that will be produced at a given price can be predicted by the formula $s = \sqrt{23x}$, where s is the supply (in thousands) and x is the price (in dollars). The demand d for mirrors can be predicted by the formula $d = \sqrt{312 - 2x^2}$. Find the equilibrium price—that is, find the price at which supply will equal demand. \$8

WRITING

113. What is wrong with the work shown below?

Solve: $\sqrt[3]{x + 1} - 3 = 8$

$$\sqrt[3]{x + 1} = 11$$
$$\left(\sqrt[3]{x + 1}\right)^3 = 11$$
$$x + 1 = 11$$
$$x = 10$$

114. The first step of a student's solution is shown below. What is a better way to begin the solution?

Solve: $\sqrt{x} + \sqrt{x + 22} = 12$

$$\left(\sqrt{x} + \sqrt{x + 22}\right)^2 = 12^2$$

115. Explain the error in the following work.

Solve: $\sqrt{2y + 1} = \sqrt{y + 7} + 3$

$$\left(\sqrt{2y + 1}\right)^2 = \left(\sqrt{y + 7} + 3\right)^2$$
$$2y + 1 = y + 7 + 9$$

116. Explain why it is immediately apparent that $\sqrt{8x - 7} = -2$ has no solution.

117. To solve the equation $\sqrt{2x + 7} = \sqrt{x}$ we need only square both sides once. To solve the equation $\sqrt{2x + 7} = \sqrt{x} + 2$ we have to square both sides twice. Why does the second equation require more work?

118. Explain how to solve $\sqrt{x-2}+2=4$ using the graphs below. 6

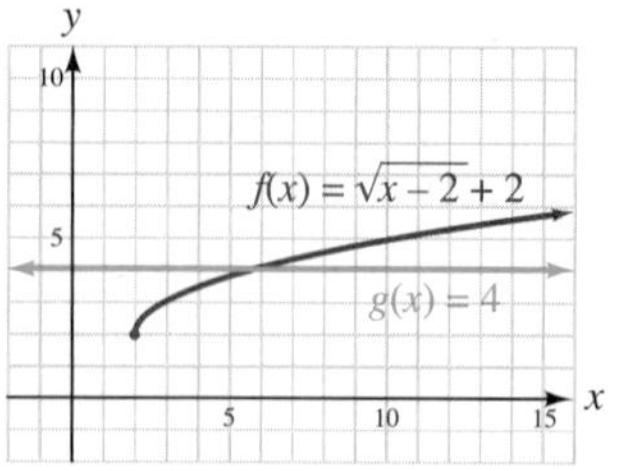

119. Explain how the table can be used to solve $\sqrt{4x-3}-2=\sqrt{2x-5}$ if $Y_1=\sqrt{4x-3}-2$ and $Y_2=\sqrt{2x-5}$.

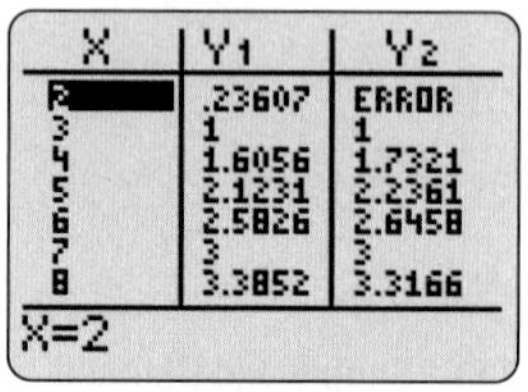

X	Y1	Y2
2	.23607	ERROR
3	1	1
4	1.6056	1.7321
5	2.1231	2.2361
6	2.5826	2.6458
7	3	3
8	3.3852	3.3166

X=2

120. Explain how to use the graph of $f(x)=\sqrt[3]{x-0.5}-1$, shown in the illustration, to approximate the solution of $\sqrt[3]{x-0.5}=1$.

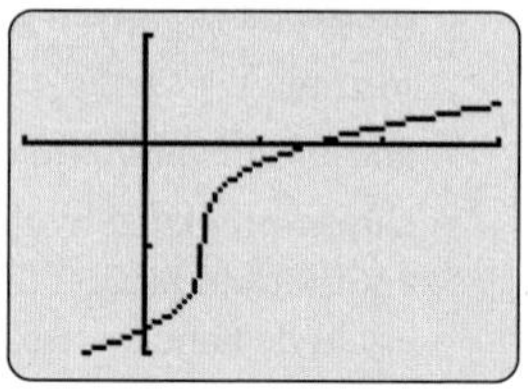

REVIEW

121. Lighting. The intensity of light from a lightbulb varies inversely as the square of the distance from the bulb. If you are 5 feet away from a bulb and the intensity is 40 foot-candles, what will the intensity be if you move 20 feet away from the bulb? 2.5 foot-candles

122. Property Tax. The property tax in a certain county varies directly as assessed valuation. If a tax of \$1,575 is charged on a single-family home assessed at \$90,000, determine the property tax on an apartment complex assessed at \$312,000. \$5,460

123. Typesetting. If 12-point type is 0.166044 inch tall, how tall is 30-point type? 0.41511 in.

▶ **124. Guitar Strings.** The frequency of vibration of a string varies directly as the square root of the tension and inversely as the length of the string. Suppose a string 2.5 feet long, under a tension of 16 pounds, vibrates 25 times per second. Find k, the constant of proportionality. 15.625

CHALLENGE PROBLEMS

Solve each equation. Write all proposed solutions. Cross out those that are extraneous.

125. $\sqrt[4]{x}=\sqrt{\frac{x}{4}}$ 0, 16

▶ **126.** $\sqrt[3]{2x}=\sqrt{x}$ A repeated solution of 0, 4

127. $\sqrt{x+2}+\sqrt{2x}=\sqrt{18-x}$ 2, ~~16~~

▶ **128.** $\sqrt{8-x}-\sqrt{3x-8}=\sqrt{x-4}$ 4, ~~$\frac{68}{13}$~~

129. $\sqrt{2\sqrt{x+1}}=\sqrt{16-4x}$ 3, ~~$\frac{21}{4}$~~

130. $(2x-1)^{2/3}=x^{1/3}$ 1, $\frac{1}{4}$

SECTION 9.6 Geometric Applications of Radicals

OBJECTIVES

1. Use the Pythagorean theorem to solve problems.
2. Solve problems involving 45°–45°–90° triangles.
3. Solve problems involving 30°–60°–90° triangles.
4. Use the distance formula to solve problems.
5. Find the midpoint of a line segment.

ARE YOU READY? *Are You Ready? exercises available online at www.webassign.net/brookscole*

The following problems review some basic skills that are needed when working with special triangles.

1. What is the sum of the measures of the angles of any triangle? 180°

2. Simplify: $\sqrt{3a^2}$ $a\sqrt{3}$

3. Rationalize the denominator: $\frac{25}{\sqrt{3}}$ $\frac{25\sqrt{3}}{3}$

4. Evaluate: $\sqrt{(9-4)^2+(-1-11)^2}$ 13

5. Approximate to the nearest hundredth: $2\sqrt{3}$ 3.46

6. Multiply: $2\cdot\frac{7\sqrt{3}}{3}$ $\frac{14\sqrt{3}}{3}$

We will now consider applications of square roots in geometry. Then we will find the distance between two points on a rectangular coordinate system, using a formula that contains a square root. We begin by considering an important theorem about right triangles.

1 Use the Pythagorean Theorem to Solve Problems.

If we know the lengths of two legs of a right triangle, we can find the length of the **hypotenuse** (the side opposite the 90° angle) by using the **Pythagorean theorem.**

The Pythagorean Theorem

If a and b are the lengths of two legs of a right triangle and c is the length of the hypotenuse,

$$a^2 + b^2 = c^2$$

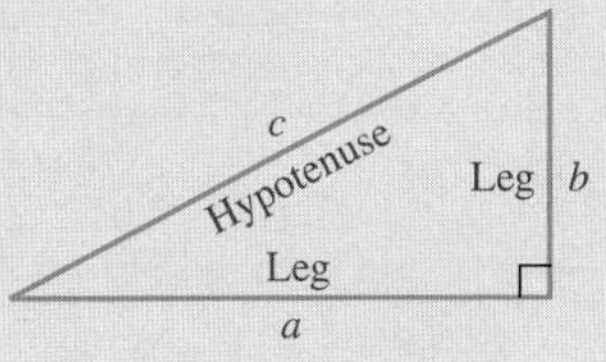

In words, the Pythagorean theorem is expressed as follows:

In a right triangle, the sum of the squares of the lengths of two legs is equal to the square of the length of the hypotenuse.

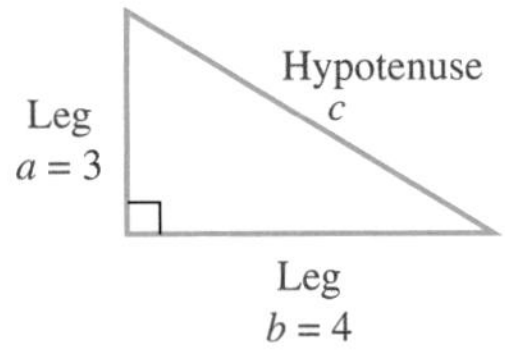

Suppose the right triangle shown in the margin has legs of length 3 and 4 units. To find the length of the hypotenuse, we use the **Pythagorean equation.**

$$\mathbf{a}^2 + \mathbf{b}^2 = c^2$$

$$\mathbf{3}^2 + \mathbf{4}^2 = c^2 \quad \text{Substitute 3 for } a \text{ and 4 for } b.$$

$$9 + 16 = c^2$$

$$25 = c^2$$

To find c, we ask "What number, when squared, is equal to 25?" There are two such numbers: the positive square root of 25 and the negative square root of 25. Since c represents the length of the hypotenuse, and it cannot be negative, it follows that c is the positive square root of 25.

$$\sqrt{25} = c \quad \text{Recall that a radical symbol } \sqrt{\ } \text{ is used to represent the positive, or principal, square root of a number.}$$

$$5 = c$$

The length of the hypotenuse is 5 units.

The Language of Algebra

A **theorem** is a mathematical statement that can be proved. The Pythagorean theorem is named after Pythagoras, a Greek mathematician who lived about 2,500 years ago. He is thought to have been the first to prove the theorem.

EXAMPLE 1 **Firefighting.** To fight a fire, the forestry department plans to clear a rectangular firebreak around the fire, as shown in the illustration on the left. Crews are equipped with mobile communications that have a 3,000-yard range. Can crews at points A and B remain in radio contact?

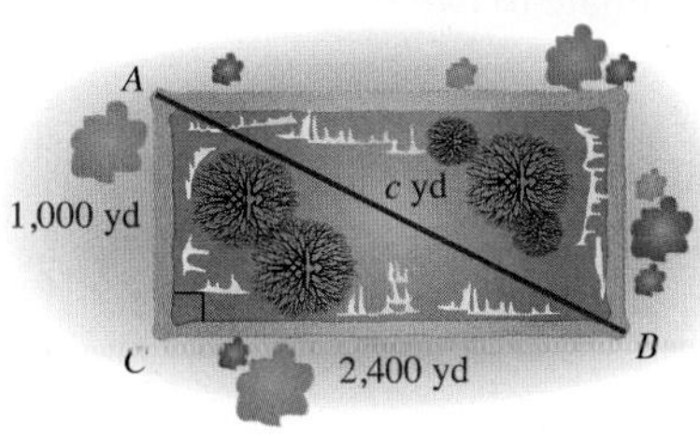

Strategy We will use the Pythagorean theorem to find the distance between points A and B.

Why If this distance is less than 3,000 yards, they can communicate. If it is greater than 3,000 yards, they cannot communicate.

Solution The line segments connecting points A, B, and C form a right triangle. To find the distance c from point A to point B, we can use the Pythagorean theorem, substituting 2,400 for a and 1,000 for b and solving for c.

$$\mathbf{a}^2 + \mathbf{b}^2 = c^2 \quad \text{This is the Pythagorean equation.}$$

$$\mathbf{2{,}400}^2 + \mathbf{1{,}000}^2 = c^2 \quad \text{Substitute for } a \text{ and } b.$$

$$5{,}760{,}000 + 1{,}000{,}000 = c^2 \quad \text{Evaluate each exponential expression.}$$

$$6{,}760{,}000 = c^2 \quad \text{Do the addition.}$$

$$\sqrt{6{,}760{,}000} = c \quad \text{If } c^2 = 6{,}760{,}000 \text{, then } c \text{ must be a square root of 6,760,000. Because } c \text{ represents a length, it must be the positive square root of 6,760,000.}$$

$$2{,}600 = c \quad \text{Use a calculator to find the square root.}$$

The two crews are 2,600 yards apart. Because this distance is less than the 3,000-yard range of the radios, they can communicate by radio.

Caution

When using the Pythagorean equation $a^2 + b^2 = c^2$, we can let a represent the length of either leg of the right triangle. We then let b represent the length of the other leg. The variable c must always represent the length of the hypotenuse.

Teaching Example 1 Firefighting. In Example 1, can the crews communicate if $b = 2{,}000$ yards?
Answer: No

Self Check 1 **Firefighting.** In Example 1, can the crews communicate if $b = 1{,}500$ yards? Yes

Now Try Problems 15 and 55

2 Solve Problems Involving 45°–45°–90° Triangles.

An **isosceles right triangle** is a right triangle with two legs of equal length. Isosceles right triangles have angle measures of 45°, 45°, and 90°. If we know the length of one leg of an isosceles right triangle, we can use the Pythagorean theorem to find the length of the hypotenuse. Since the triangle shown in the margin is a right triangle, we have

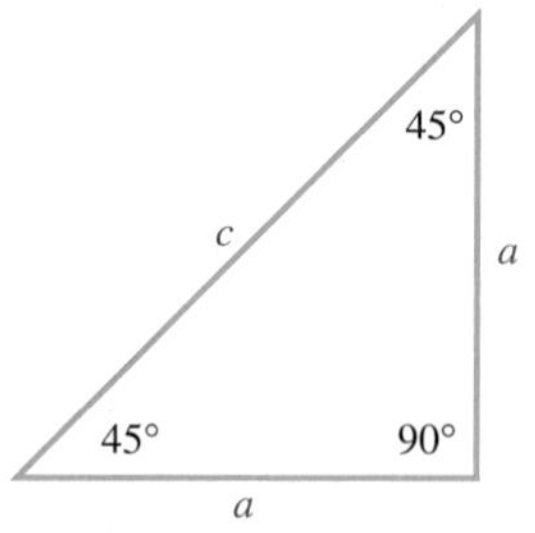

$c^2 = a^2 + b^2$ This is the Pythagorean equation.

$c^2 = a^2 + a^2$ Both legs are a units long, so replace b with a.

$c^2 = 2a^2$ Combine like terms.

$c = \sqrt{2a^2}$ If $c^2 = 2a^2$, then c must be a square root of $2a^2$. Because c represents a length, it must be the positive square root of $2a^2$.

$c = a\sqrt{2}$ Simplify the radical: $\sqrt{2a^2} = \sqrt{2}\sqrt{a^2} = \sqrt{2}a = a\sqrt{2}$.

Thus, **in an isosceles right triangle, the length of the hypotenuse is $\sqrt{2}$ times the length of one leg.**

EXAMPLE 2 If one leg of an isosceles right triangle is 10 feet long, find the exact length of the hypotenuse. Then approximate the length to two decimal places.

Strategy We will multiply the length of the known leg by $\sqrt{2}$.

Why The length of the hypotenuse of an isosceles right triangle is $\sqrt{2}$ times the length of one leg.

Solution Since the length of the hypotenuse is the length of a leg times $\sqrt{2}$, we have

$$c = 10\sqrt{2}$$

The exact length of the hypotenuse is $10\sqrt{2}$ feet. If we approximate to two decimal places, the length is 14.14 feet.

Teaching Example 2 Find the exact length of the hypotenuse of an isosceles right triangle if one leg is 14 cm. Then approximate the length to two decimal places.
Answer: $14\sqrt{2}$ cm ≈ 19.80 cm

Self Check 2 Find the exact length of the hypotenuse of an isosceles right triangle if one leg is 12 meters long. Then approximate the length to two decimal places. $12\sqrt{2}$ m ≈ 16.97 m

Now Try Problem 23

EXAMPLE 3 Find the exact length of each leg of the isosceles right triangle shown in the margin. Then approximate the lengths to two decimal places.

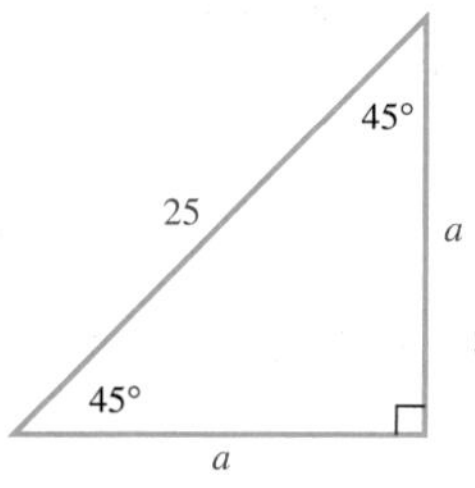

Strategy We will find the length of each leg of the triangle by substituting 25 for c in the formula $c = a\sqrt{2}$ and solving for a.

Why The variable a represents the unknown length of each leg of the triangle.

Solution

$c = a\sqrt{2}$ This is the formula for the length of the hypotenuse of an isosceles right triangle.

$25 = a\sqrt{2}$ Substitute 25 for c, the length of the hypotenuse.

$\frac{25}{\sqrt{2}} = \frac{a\sqrt{2}}{\sqrt{2}}$ To isolate a, undo the multiplication by $\sqrt{2}$ by dividing both sides by $\sqrt{2}$.

$$\frac{25}{\sqrt{2}} = a \qquad \text{Simplify the right side.}$$

$$a = \frac{25}{\sqrt{2}} \qquad \text{Reverse the sides of the equation so that } a \text{ is on the left.}$$

$$a = \frac{25}{\sqrt{2}} \cdot \frac{\sqrt{2}}{\sqrt{2}} \qquad \text{Rationalize the denominator.}$$

$$a = \frac{25\sqrt{2}}{2} \qquad \text{Simplify the denominator: } \sqrt{2} \cdot \sqrt{2} = 2.$$

The exact length of each leg is $\frac{25\sqrt{2}}{2}$ units. If we approximate to two decimal places, the length is 17.68 units.

Teaching Example 3 Find the exact length of each leg of an isosceles right triangle if the hypotenuse is 18 meters. Then approximate the lengths to two decimal places.
Answer: $9\sqrt{2}$ m $\approx$ 12.73 m

Self Check 3 Find the exact length of each leg of an isosceles right triangle if the length of the hypotenuse is 9 inches. Then approximate the lengths to two decimal places. $\frac{9\sqrt{2}}{2}$ in. $\approx$ 6.36 in.

Now Try Problem 27

3 Solve Problems Involving 30°–60°–90° Triangles.

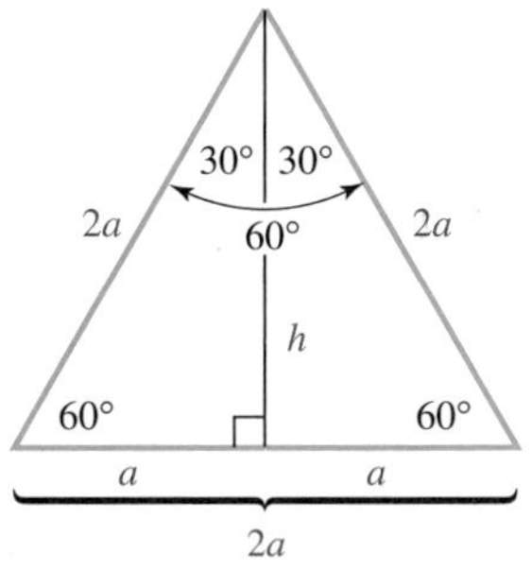

From geometry, we know that an **equilateral triangle** is a triangle with three sides of equal length and three 60° angles. Each side of the equilateral triangle shown in the margin is $2a$ units long. If an **altitude** (height) is drawn to its base, the altitude divides the base into two segments of equal length and divides the equilateral triangle into two 30°–60°–90° triangles. From the figure, we can see that the shorter leg of each 30°–60°–90° triangle (the side *opposite* the 30° angle) is a units long. Thus,

The length of the hypotenuse of a 30°–60°–90° triangle is twice as long as the shorter leg.

We can discover another relationship between the legs of a 30°–60°–90° triangle if we find the length of the altitude h in the figure. We begin by applying the Pythagorean theorem to one of the 30°–60°–90° triangles.

Teaching Tip: Mention that 30°–60°–90° and 45°–45°–90° triangles play an important role in an advanced mathematics course called trigonometry. If any of your students are planning on taking trigonometry in the future, the special relationships between the side lengths discussed in this section should be memorized.

$$a^2 + b^2 = c^2 \qquad \text{This is the Pythagorean equation.}$$

$$a^2 + h^2 = (2a)^2 \qquad \text{The altitude is } h \text{ units long, so replace } b \text{ with } h. \text{ The hypotenuse is } 2a \text{ units long, so replace } c \text{ with } 2a.$$

$$a^2 + h^2 = 4a^2 \qquad (2a)^2 = (2a)(2a) = 4a^2.$$

$$h^2 = 3a^2 \qquad \text{Subtract } a^2 \text{ from both sides.}$$

$$h = \sqrt{3a^2} \qquad \text{If } h^2 = 3a^2, \text{ then } h \text{ must be the positive square root of } 3a^2.$$

$$h = a\sqrt{3} \qquad \text{Simplify the radical: } \sqrt{3a^2} = \sqrt{3}\sqrt{a^2} = a\sqrt{3}.$$

We see that the altitude—the longer leg of the 30°–60°–90° triangle—is $\sqrt{3}$ times as long as the shorter leg. Thus,

The length of the longer leg of a 30°–60°–90° triangle is $\sqrt{3}$ times the length of the shorter leg.

EXAMPLE 4 Find the length of the hypotenuse and the length of the longer leg of the 30°–60°–90° triangle shown in the margin.

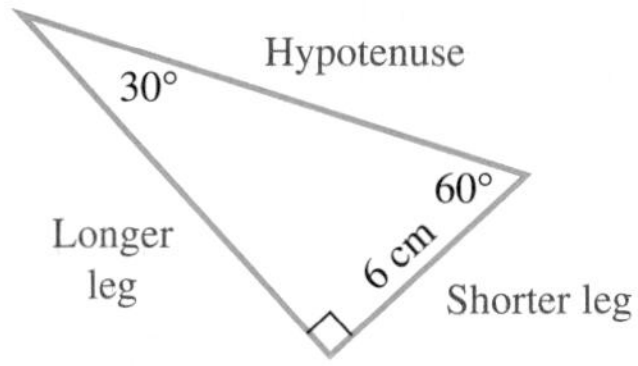

Strategy To find the length of the hypotenuse, we will multiply the length of the shorter leg by 2. To find the length of the longer leg, we will multiply the length of the shorter leg by $\sqrt{3}$.

Why These side-length relationships are true for any 30°–60°–90° triangle.

Solution Since the length of the hypotenuse of a 30°–60°–90° triangle is twice as long as the shorter leg, and the length of the shorter leg is 6 cm, the hypotenuse is $2 \cdot 6 = 12$ cm.

Since the length of the longer leg is $\sqrt{3}$ times the length of the shorter leg, and the length of the shorter leg is 6 cm, the longer leg is $6\sqrt{3}$ cm (about 10.39 cm).

Teaching Example 4 Find the length of the hypotenuse and the longer leg of a 30°–60°–90° triangle if the shorter leg is 10 cm long.
Answer: 20 cm, $10\sqrt{3}$ cm

Self Check 4 Find the length of the hypotenuse and the longer leg of a 30°–60°–90° triangle if the shorter leg is 8 centimeters long. 16 cm, $8\sqrt{3}$ cm

Now Try Problem 31

EXAMPLE 5 Find the length of the hypotenuse and the length of the shorter leg of the 30°–60°–90° triangle shown in the margin. Then approximate the lengths to two decimal places.

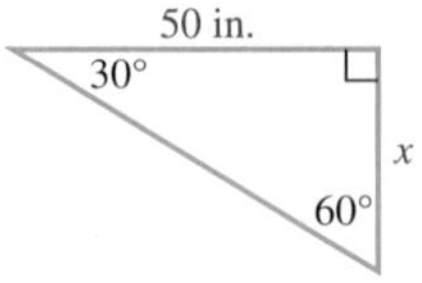

Strategy We will find the length of the shorter leg first.

Why Once we know the length of the shorter leg, we can multiply it by 2 to find the length of the hypotenuse.

Solution If we let x = the length in inches of the shorter leg of the triangle, we can form an equation by translating the following statement:

Success Tip

In a 30°–60°–90° triangle, the side opposite the 30° angle is the shorter leg, the side opposite the 60° angle is the longer leg, and the hypotenuse is opposite the 90° angle.

The length of the longer leg of a 30°–60°–90° triangle	is	$\sqrt{3}$	times	the length of the shorter leg.
50	=	$\sqrt{3}$	$\cdot$	x

To find the length of the shorter leg, we solve the equation for x.

$$50 = \sqrt{3}x$$

$$\frac{50}{\sqrt{3}} = \frac{\sqrt{3}x}{\sqrt{3}} \quad \text{To isolate } x\text{, divide both sides by } \sqrt{3}.$$

$$\frac{50}{\sqrt{3}} = x$$

The length of the shorter leg is exactly $\frac{50}{\sqrt{3}}$ inches. To write this number in simplified radical form, we rationalize the denominator.

$$\frac{50}{\sqrt{3}} \cdot \frac{\sqrt{3}}{\sqrt{3}} = \frac{50\sqrt{3}}{3}$$

Thus, the length of the shorter leg is exactly $\frac{50\sqrt{3}}{3}$ inches (about 28.87 inches).

Since the length of the hypotenuse of a 30°–60°–90° triangle is twice as long as the shorter leg, the hypotenuse is $2 \cdot \frac{50\sqrt{3}}{3} = \frac{100\sqrt{3}}{3}$ inches (about 57.74 inches).

Teaching Example 5 Find the length of the shorter leg and the length of the hypotenuse of a 30°–60°–90° triangle if the longer leg is 16 cm long. Then approximate the lengths to two decimal places.
Answer: $\frac{16\sqrt{3}}{3}$ cm ≈ 9.24 cm, $\frac{32\sqrt{3}}{3}$ cm ≈ 18.48 cm

Self Check 5 Find the length of the hypotenuse and the shorter leg of a 30°–60°–90° triangle if the longer leg is 15 feet long. Then approximate the lengths to two decimal places. $10\sqrt{3}$ in. ≈ 17.32 in., $5\sqrt{3}$ in. ≈ 8.66 in.

Now Try Problem 35

EXAMPLE 6 **Stretching Exercises.** A doctor prescribed the exercise shown in figure (a) on the next page. The patient was instructed to raise his leg to an angle of 60° and hold the position for 10 seconds. If the patient's leg is 36 inches long, how high off the floor will his foot be when his leg is held at the proper angle?

Strategy This situation is modeled by a 30°–60°–90° triangle. We will begin by finding the length of the shorter leg.

Why Once we know the length of the shorter leg, we can easily find the length of the longer leg, which represents the distance the patient's foot is off the ground.

Solution In figure (b), we see right triangle ABC, which models the situation. Since the length of the hypotenuse is twice as long as the side opposite the 30° angle, side AC is half as long as the hypotenuse. Since the hypotenuse is given to be 36 inches long, side AC must be 18 inches long.

Since the length of the longer leg (the leg opposite the 60° angle) is $\sqrt{3}$ times the length of the shorter leg (side AC), side BC is $18\sqrt{3}$, or about 31 inches long. So the patient's foot will be about 31 inches from the floor when his leg is in the proper position.

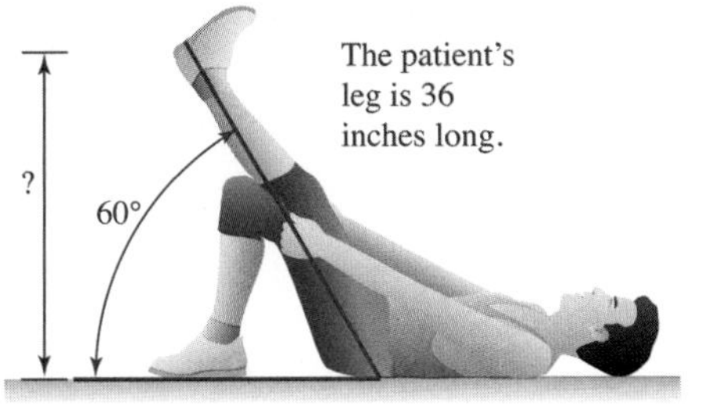

(a)

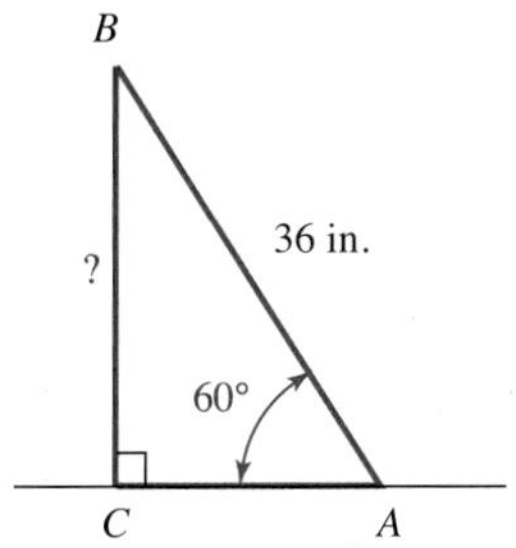

(b)

Teaching Example 6 Stretching Exercises. Repeat Example 6 for a patient whose leg is 30 inches long.
Answer: $15\sqrt{3}$ in. ≈ 26 in.

Self Check 6 **Stretching Exercises.** Refer to Example 6. The doctor prescribed the same exercise for a patient whose leg is 29 inches long. Find how high off the floor this patient's foot is when her leg is held at the proper angle.

$\frac{29\sqrt{3}}{2}$ in. ≈ 25 in.

Now Try Problems 39 and 61

4 Use the Distance Formula to Solve Problems.

With the **distance formula,** we can find the distance between any two points graphed on a rectangular coordinate system. To find the distance d between points $P(x_1, y_1)$ and $Q(x_2, y_2)$ shown in the figure on the right, we construct the right triangle PRQ. The distance between P and R is $|x_2 - x_1|$, and the distance between R and Q is $|y_2 - y_1|$ We apply the Pythagorean theorem to the right triangle PRQ to get

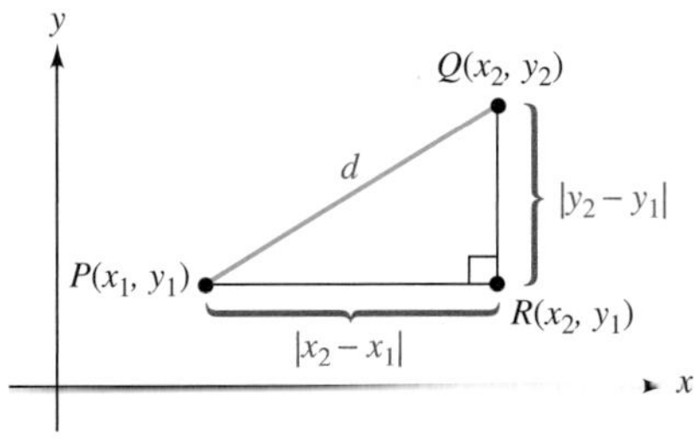

$$d^2 = |x_2 - x_1|^2 + |y_2 - y_1|^2$$
$$= (x_2 - x_1)^2 + (y_2 - y_1)^2 \quad \text{Because } |x_2 - x_1|^2 = (x_2 - x_1)^2 \text{ and } |y_2 - y_1|^2 = (y_2 - y_1)^2$$

Because d represents the distance between two points, it must be equal to the positive square root of $(x_2 - x_1)^2 + (y_2 - y_1)^2$.

$$d = \sqrt{(x_2 - x_1)^2 + (y_2 - y_1)^2}$$

We call this result the *distance formula.*

Distance Formula The distance d between two points with coordinates (x_1, y_1) and (x_2, y_2) is given by

$$d = \sqrt{(x_2 - x_1)^2 + (y_2 - y_1)^2}$$

EXAMPLE 7 Find the distance between the points: **a.** $(-2, 3)$ and $(4, -5)$ **b.** $(27, 46)$ and $(33, 50)$

Strategy We will use the distance formula.

Why We know the x- and y-coordinates of both points.

Solution **a.** To find the distance, we can use the distance formula by substituting 4 for x_2, -2 for x_1, -5 for y_2, and 3 for y_1.

$d = \sqrt{(x_2 - x_1)^2 + (y_2 - y_1)^2}$ This is the distance formula.

$= \sqrt{[4 - (-2)]^2 + (-5 - 3)^2}$ Substitute for x_1, x_2, y_1, and y_2.

$= \sqrt{(4 + 2)^2 + (-5 - 3)^2}$ Simplify within the parentheses.

$= \sqrt{6^2 + (-8)^2}$ Do the addition and subtraction.

$= \sqrt{36 + 64}$ Evaluate each exponential expression.

$= \sqrt{100}$ Do the addition.

$= 10$ Evaluate the square root.

The distance between the points is 10 units.

b. $d = \sqrt{(x_2 - x_1)^2 + (y_2 - y_1)^2}$ This is the distance formula.

$d = \sqrt{(33 - 27)^2 + (50 - 46)^2}$ Substitute 33 for x_2, 27 for x_1, 50 for y_2, and 46 for y_1.

$= \sqrt{6^2 + 4^2}$ Do the subtraction.

$= \sqrt{52}$ Evaluate: $6^2 + 4^2 = 36 + 16 = 52$.

$= 2\sqrt{13}$ Simplify: $\sqrt{52} = \sqrt{4 \cdot 13} = 2\sqrt{13}$.

The distance between the points is exactly $2\sqrt{13}$ units, which is about 7.21 units.

Teaching Example 7 Find the distance between the points: **a.** $(-5, 2)$ and $(4, 14)$ **b.** $(88, 20)$ and $(80, 24)$
Answers:
a. 15 units **b.** $4\sqrt{5} \approx 8.94$ units

Self Check 7 Find the distance between the points: **a.** $(-2, -2)$ and $(3, 10)$ 13 units **b.** $(55, 29)$ and $(61, 32)$ $3\sqrt{5} \approx 6.71$ units

Now Try Problems 47 and 49

5 Find the Midpoint of a Line Segment.

If point M in the figure on the right lies midway between point P and point Q, it is called the **midpoint** of line segment PQ. We call the points P and Q, the **endpoints** of the segment.

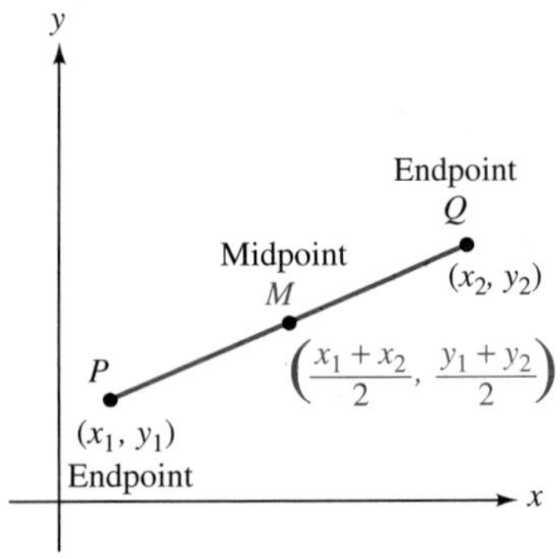

To distinguish between the coordinates of the endpoints of a line segment, we can use *subscript notation.* In the figure, the point P with coordinates (x_1, y_1) is read as "point P with coordinates x sub 1 and y sub 1," and the point Q with coordinates (x_2, y_2) is read as "point Q with coordinates x sub 2 and y sub 2."

To find the coordinates of point M, we find the average of the x-coordinates and the average of the y-coordinates of points P and Q. Using subscript notation, we can write the midpoint formula in the following way.

The Midpoint Formula

The **midpoint** of a line segment with endpoints (x_1, y_1) and (x_2, y_2) is the point with coordinates

$$\left(\frac{x_1 + x_2}{2}, \frac{y_1 + y_2}{2}\right)$$

EXAMPLE 8 Find the midpoint of the line segment with endpoints $(-2, 5)$ and $(4, -2)$.

Teaching Tip: Point out that it doesn't matter which of the given endpoints is labeled (x_1, y_1) and which is labeled (x_2, y_2).

Strategy To find the coordinates of the midpoint, we find the average of the x-coordinates and the average of the y-coordinates of the endpoints.

Why This is what is called for by the expressions $\frac{x_1 + x_2}{2}$ and $\frac{y_1 + y_2}{2}$ of the midpoint formula.

Solution We can let $(x_1, y_1) = (-2, 5)$ and $(x_2, y_2) = (4, -2)$. After substituting these values into the expressions for the x- and y-coordinates in the midpoint formula, we evaluate each expression to find the coordinates of the midpoint.

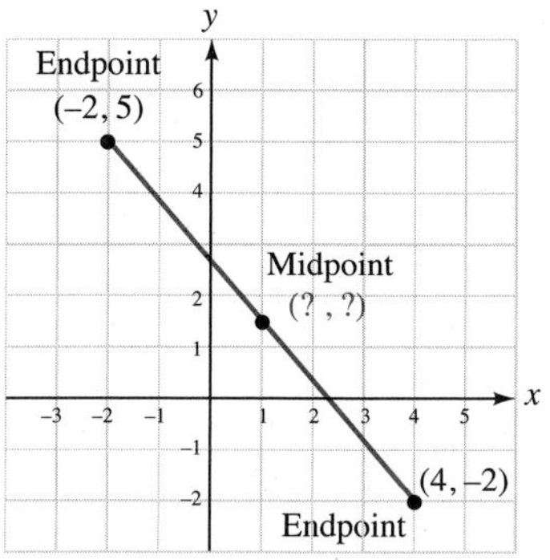

Find the x-coordinate of the midpoint:

$$\frac{x_1 + x_2}{2} = \frac{-2 + 4}{2} = \frac{2}{2} = 1$$

Find the y-coordinate of the midpoint:

$$\frac{y_1 + y_2}{2} = \frac{5 + (-2)}{2} = \frac{3}{2}$$

Thus, the midpoint is $\left(1, \frac{3}{2}\right)$.

Teaching Example 8 Find the midpoint of the line segment with endpoints $(7, -3)$ and $(5, -7)$.
Answer: $(6, -5)$

Self Check 8 Find the midpoint of the line segment with endpoints $(-1, 8)$ and $(5, 2)$. $(2, 5)$

Now Try Problem 57

EXAMPLE 9 The midpoint of the line segment joining $(-5, -3)$ and a point Q is the point $(-1, 2)$. Find the coordinates of point Q.

Strategy As in Example 8, we will use the midpoint formula to find the unknown coordinates. However, this time, we need to find x_2 and y_2.

Why We want to find the coordinates of one of the endpoints.

Solution We can let $(x_1, y_1) = (-5, -3)$ and $(x_M, y_M) = (-1, 2)$, where x_M represents the x-coordinate and y_M represents the y-coordinate of the midpoint. To find the coordinates of point Q, we substitute for x_1, x_M, y_1, and y_M in the expressions for the coordinates in the midpoint formula and solve the resulting equations for x_2 and y_2.

Read x_M as "x sub M" and y_M as "y sub M."

$$x_M = \frac{x_1 + x_2}{2}$$

$$-1 = \frac{-5 + x_2}{2} \quad \text{Substitute.}$$

$$-2 = -5 + x_2 \quad \text{Multiply both sides by 2.}$$

$$3 = x_2 \quad \text{Add 5 to both sides.}$$

$$y_M = \frac{y_1 + y_2}{2}$$

$$2 = \frac{-3 + y_2}{2} \quad \text{Substitute.}$$

$$4 = -3 + y_2 \quad \text{Multiply both sides by 2.}$$

$$7 = y_2 \quad \text{Add 3 to both sides.}$$

Since $x_2 = 3$ and $y_2 = 7$, the coordinates of point Q are $(3, 7)$.

Teaching Example 9 The midpoint of a line segment joining $(5, 1)$ and a point P is the point $(-3, 7)$. Find the coordinates of point P.
Answer: $(-11, 13)$

Self Check 9 The midpoint of the line segment joining $(-5, -3)$ and a point P is the point $(-2, 5)$. Find the coordinates of point P. $(1, 13)$

Now Try Problem 63

SECTION 9.6 STUDY SET

VOCABULARY

Fill in the blanks.

1. In a right triangle, the side opposite the 90° angle is called the hypotenuse.
2. An isosceles right triangle is a right triangle with two legs of equal length.
3. The Pythagorean theorem states that in a right triangle, the sum of the squares of the lengths of the two legs is equal to the square of the hypotenuse.
4. An equilateral triangle has three sides of equal length and three 60° angles.

CONCEPTS

Fill in the blanks.

5. If a and b are the lengths of the legs of a right triangle and c is the length of the hypotenuse, then $a^2 + b^2 = c^2$. This is called the Pythagorean equation.
6. In any right triangle, the square of the hypotenuse is equal to the sum of the squares of the two legs.
7. In an isosceles right triangle, the length of the hypotenuse is $\sqrt{2}$ times the length of one leg.
8. The shorter leg of a 30°–60°–90° triangle is half as long as the hypotenuse.
9. The length of the longer leg of a 30°–60°–90° triangle is $\sqrt{3}$ times the length of the shorter leg.
10. In a 30°–60°–90° triangle, the shorter leg is opposite the 30° angle, and the longer leg is opposite the 60° angle.
11. The formula to find the distance between points (x_1, y_1) and (x_2, y_2) is $d = \sqrt{(x_2-x_1)^2 + (y_2-y_1)^2}$.
12. Solve for c, where c represents the length of the hypotenuse of a right triangle. Simplify the result, if possible.
   a. $c^2 = 64$ 8
   b. $c^2 = 15$ $\sqrt{15}$
   c. $c^2 = 24$ $2\sqrt{6}$

NOTATION

Complete each solution.

13. Evaluate. Approximate to two decimal places.

$$\sqrt{(-1-3)^2 + [2-(-4)]^2} = \sqrt{(-4)^2 + [6]^2}$$
$$= \sqrt{52}$$
$$= \sqrt{4 \cdot 13}$$
$$= 2\sqrt{13}$$
$$\approx 7.21$$

14. Solve $8^2 + 4^2 = c^2$ and assume $c > 0$. Approximate to two decimal places.

$$64 + 16 = c^2$$
$$80 = c^2$$
$$\sqrt{80} = c$$
$$\sqrt{16 \cdot 5} = c$$
$$4\sqrt{5} = c$$
$$c \approx 8.94$$

GUIDED PRACTICE

The lengths of two sides of the right triangle ABC are given. Find the length of the missing side. **See Example 1.**

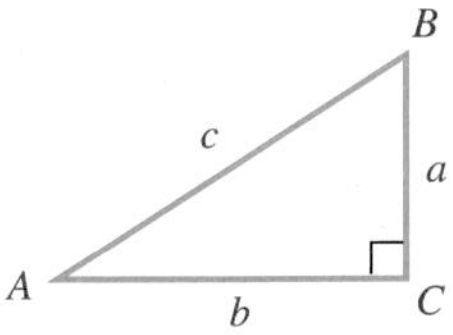

15. $a = 6$ ft and $b = 8$ ft 10 ft
16. $a = 5$ in. and $b = 12$ in. 13 in.
17. $a = 8$ ft and $b = 15$ ft 17 ft
18. $a = 24$ yd and $b = 7$ yd 25 yd
19. $b = 9$ ft and $c = 41$ ft 40 ft
20. $b = 18$ m and $c = 82$ m 80 m
21. $a = 10$ cm and $c = 26$ cm 24 cm
22. $a = 14$ in. and $c = 50$ in. 48 in.

Find the missing side lengths in each triangle. Give the exact answer and then an approximation to two decimal places when appropriate. **See Example 2.**

23.

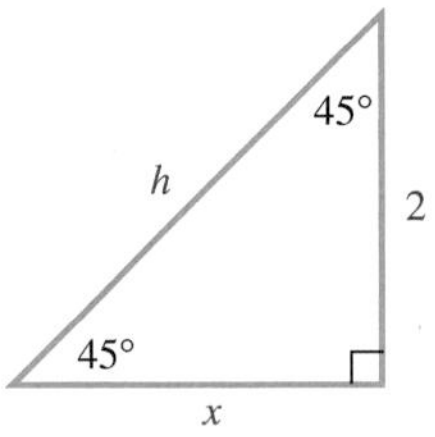

$x = 2$, $h = 2\sqrt{2} \approx 2.83$

24.

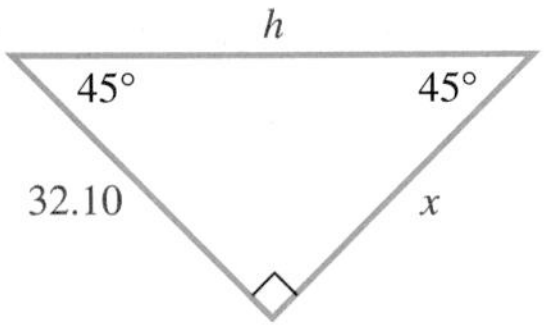

$x = 32.10$, $h = 32.10\sqrt{2} \approx 45.40$

25. One leg of an isosceles right triangle is 3.2 feet long. Find the length of its hypotenuse. Give the exact answer and then an approximation to two decimal places. $3.2\sqrt{2}$ ft ≈ 4.53 ft
26. One side of a square is $5\frac{1}{2}$ in. long. Find the length of its diagonal. Give the exact answer and then an approximation to two decimal places. $5.5\sqrt{2}$ in. ≈ 7.78 in.

Find the missing side lengths in each triangle. Give the exact answer and then an approximation to two decimal places. **See Example 3.**

27.

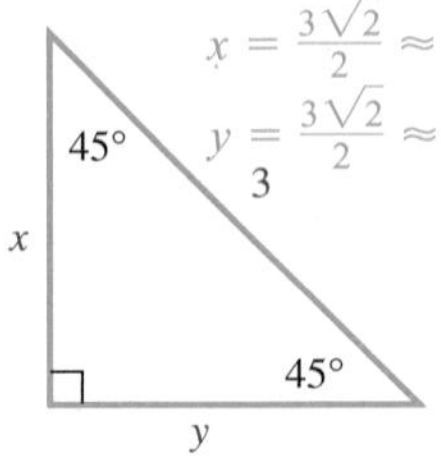

$x = \frac{3\sqrt{2}}{2} \approx 2.12$, $y = \frac{3\sqrt{2}}{2} \approx 2.12$

28.

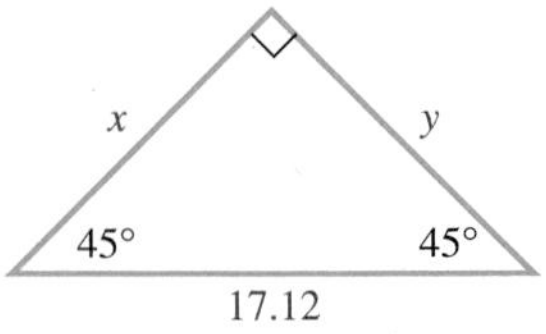

$x = \frac{17.12\sqrt{2}}{2} \approx 12.11$, $y = \frac{17.12\sqrt{2}}{2} \approx 12.11$

29. Photographs. The diagonal of a square photograph measures 10 inches. Find the length of one of its sides. Give the exact answer and then an approximation to two decimal places. $5\sqrt{2}$ in. ≈ 7.07 in.

30. Parking Lots. The diagonal of a square parking lot is approximately 1,414 feet long.

a. Find the length of one side of the parking lot. Round to the nearest foot. 1,000 ft

b. Find the approximate area of the parking lot. 1,000,000 ft^2

Find the missing side lengths in each triangle. Give the exact answer and then an approximation to two decimal places, when appropriate. **See Example 4.**

31.

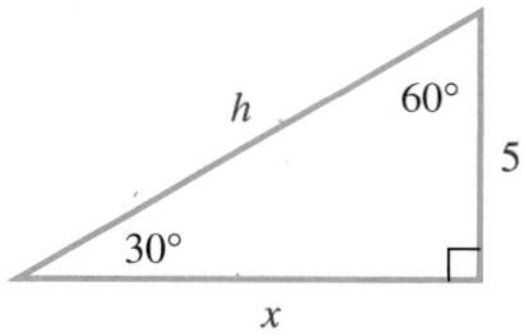

32.

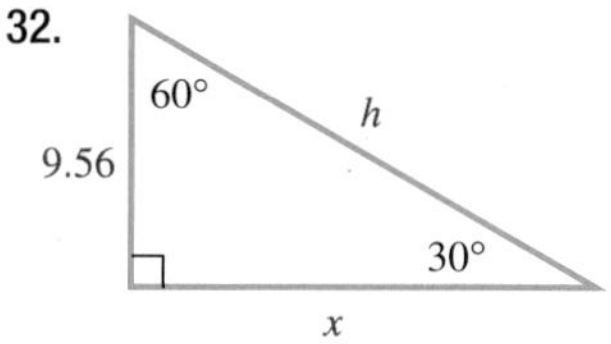

31. $x = 5\sqrt{3} \approx 8.66, h = 10$ 32. $x = 9.56\sqrt{3} \approx 16.56, h = 19.12$

33. In a 30°–60°–90° triangle, the length of the leg opposite the 30° angle is 75 cm. Find the length of the leg opposite the 60° angle and the length of the hypotenuse. Give the exact answer and then an approximation to two decimal places, when appropriate. $75\sqrt{3}$ cm ≈ 129.90 cm, 150 cm

34. In a 30°–60°–90° triangle, the length of the shorter leg is $5\sqrt{2}$ inches. Find the length of the hypotenuse and the length of the longer leg. Give the exact answer and then an approximation to two decimal places.
$10\sqrt{2}$ in. ≈ 14.14 in., $5\sqrt{6}$ in. ≈ 12.25 in.

Find the missing lengths in each triangle. Give the exact answer and then an approximation to two decimal places. **See Example 5.**

35.

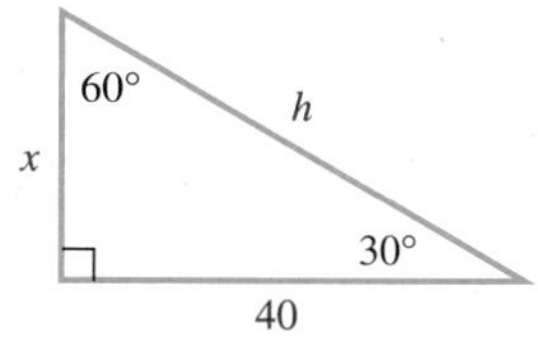

36.

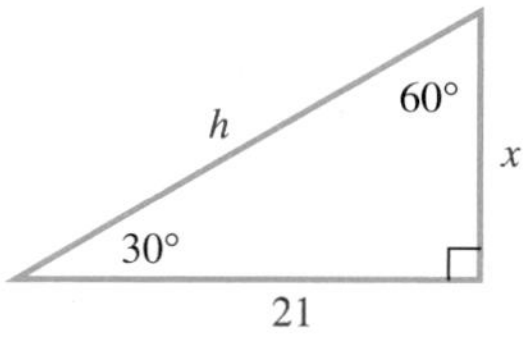

35. $x = \frac{40\sqrt{3}}{3} \approx 23.09$, $h = \frac{80\sqrt{3}}{3} \approx 46.19$ 36. $x = 7\sqrt{3} \approx 12.12$, $h = 14\sqrt{3} \approx 24.25$

37. In a 30°–60°–90° right triangle, the length of the leg opposite the 60° angle is 55 millimeters. Find the length of the leg opposite the 30° angle and the length of the hypotenuse. Give the exact answer and then an approximation to two decimal places. $\frac{55\sqrt{3}}{3}$ mm ≈ 31.75 mm, $\frac{110\sqrt{3}}{3}$ mm ≈ 63.51 mm

38. In a 30°–60°–90° right triangle, the length of the longer leg is 24 yards. Find the length of the hypotenuse and the length of the shorter leg. Give the exact answer and then an approximation to two decimal places.
$16\sqrt{3}$ yd ≈ 27.71 yd, $8\sqrt{3}$ yd ≈ 13.86 yd

Find the missing lengths in each triangle. Give the exact answer and then an approximation to two decimal places, when appropriate. **See Example 6.**

39.

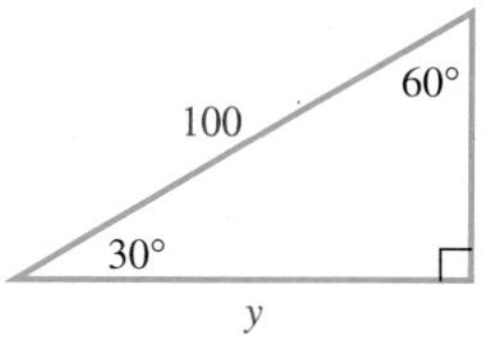

40.

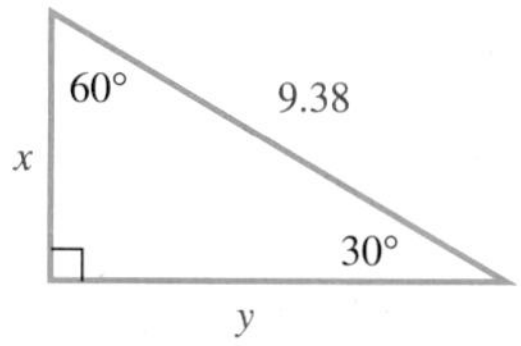

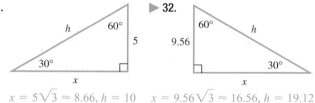

39. $x = 50, y = 50\sqrt{3} \approx 86.60$ 40. $x = 4.69, y = 4.69\sqrt{3} \approx 8.12$

41. In a 30°–60°–90° right triangle, the length of the hypotenuse is 1.5 feet. To the nearest hundredth, find the length of the shorter leg and the length of the longer leg. Give the exact answer and then an approximation to two decimal places, when appropriate.
0.75 ft, $0.75\sqrt{3}$ ft ≈ 1.30 ft

42. In a 30°–60°–90° right triangle, the length of the hypotenuse is $12\sqrt{3}$ inches. Find the length of the leg opposite the 30° angle and the length of the leg opposite the 60° angle. Give the exact answer and then an approximation to two decimal places, when appropriate. $6\sqrt{3}$ in. ≈ 10.39 in., 18 in.

Find the exact distance between each pair of points. **See Example 7.**

43. (0, 0), (3, −4) 5
44. (0, 0), (−12, 16) 20
45. (−2, −8), (3, 4) 13
46. (−5, −2), (7, 3) 13
47. (6, 8), (12, 16) 10
48. (10, 4), (2, −2) 10
49. (−2, 1), (3, 4) $\sqrt{34}$
50. (2, −3), (4, −8) $\sqrt{29}$
51. (−1, −6), (3, −4) $2\sqrt{5}$
52. (−3, 5), (−5, −5) $2\sqrt{26}$
53. (−2, −1), (−5, 8) $3\sqrt{10}$
54. (4, 7), (−4, −5) $4\sqrt{13}$

Find the midpoint of the line segment with the given endpoints. **See Example 8.**

55. (0, 0), (6, 8) (3, 4)
56. (10, 12), (0, 0) (5, 6)
57. (6, 8), (12, 16) (9, 12)
58. (10, 4), (2, −2) (6, 1)
59. (−2, −8), (3, −8) $\left(\frac{1}{2}, -8\right)$
60. (−5, −2), (7, 3) $\left(1, \frac{1}{2}\right)$
61. (7, 1), (−10, 4) $\left(-\frac{3}{2}, \frac{5}{2}\right)$
62. (−4, −3), (4, −8) $\left(0, -\frac{11}{2}\right)$

Solve each problem. **See Example 9.**

63. If (−2, 3) is the midpoint of segment PQ and the coordinates of P are (−8, 5), find the coordinates of Q. (4, 1)

64. If (6, −5) is the midpoint of segment PQ and the coordinates of Q are (−5, −8), find the coordinates of P. (17, −2)

65. If (−7, −3) is the midpoint of segment QP and the coordinates of Q are (6, −3), find the coordinates of P. (−20, −3)

66. If $\left(\frac{1}{2}, -2\right)$ is the midpoint of segment QP and the coordinates of P are $\left(-\frac{5}{2}, 5\right)$, find the coordinates of Q. $\left(\frac{7}{2}, -9\right)$

APPLICATIONS

67. Soccer. The allowable length of a rectangular soccer field used for international adult matches can be from 100 to 110 meters and the width can be from 64 to 75 meters.

a. Find the length of the diagonal of the field that has the minimum allowable length and minimum allowable width. Give an approximation to two decimal places. 118.73 m

b. Find the length of the diagonal of the field that has the maximum allowable length and maximum allowable width. Give the exact answer and an approximation to two decimal places. 133.14 m

68. Cubes. Find the exact length of the diagonal (in blue) of one of the *faces* of the cube shown here. $7\sqrt{2}$ cm

69. Cubes. Find the exact length of the diagonal (in green) of the cube shown here. $7\sqrt{3}$ cm

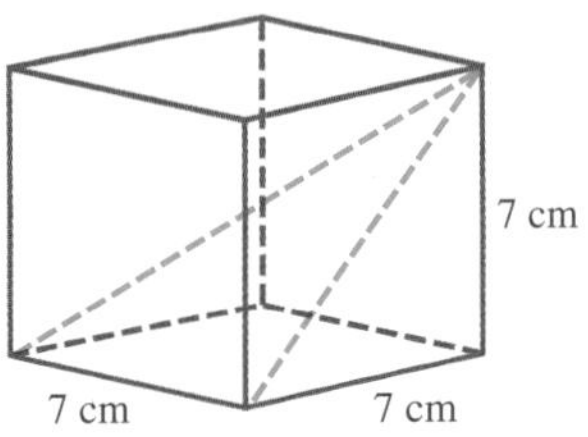

70. Geometry. Use the distance formula to show that a triangle with vertices $(-2, 4)$, $(2, 8)$, and $(6, 4)$ is isosceles.

71. Washington, D.C. The square in the map shows the 100-square-mile site selected by George Washington in 1790 to serve as a permanent capital for the United States. In 1847, the part of the district lying on the west bank of the Potomac was returned to Virginia. Find the exact coordinates of each corner of the original square that outlined the District of Columbia and approximations to two decimal places.
$(5\sqrt{2}, 0), (0, 5\sqrt{2}), (-5\sqrt{2}, 0), (0, -5\sqrt{2})$;
$(7.07, 0), (0, 7.07), (-7.07, 0), (0, -7.07)$

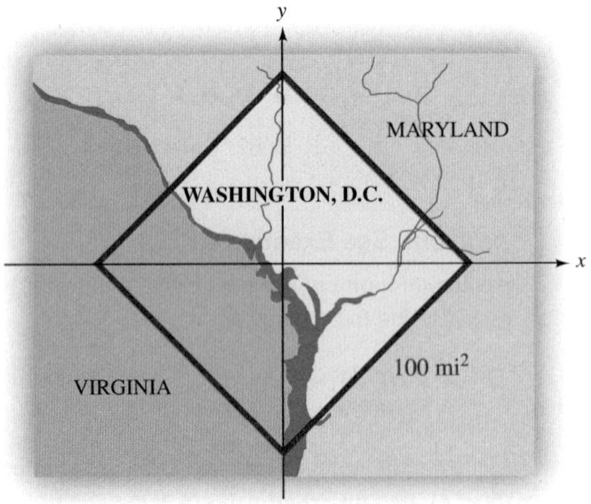

72. Paper Airplanes. The illustration gives the directions for making a paper airplane from a square piece of paper with sides 8 inches long. Find the length l of the plane when it is completed. Give the exact answer and an approximation to two decimal places. $8\sqrt{2}$ in. ≈ 11.31 in.

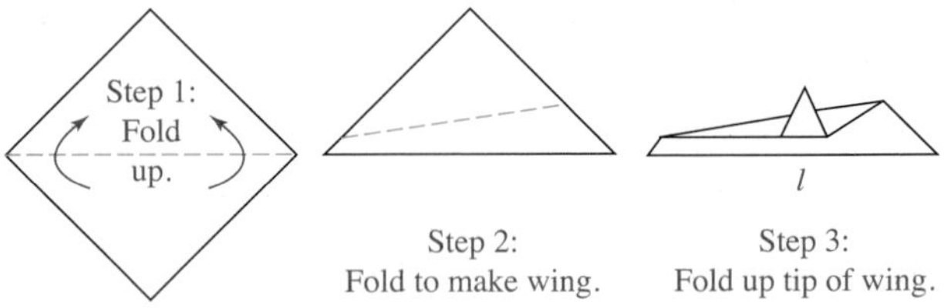

73. Hardware. The sides of a regular hexagonal nut are 10 millimeters long. Find the height h of the nut. Give the exact answer and an approximation to two decimal places.
$10\sqrt{3}$ mm ≈ 17.32 mm

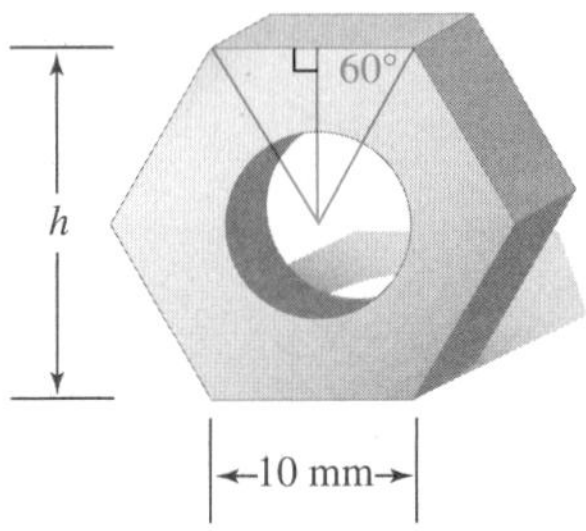

74. Ironing Boards. Find the height h of the ironing board shown in the illustration. Give the exact answer and an approximation to two decimal places.
$20\sqrt{3}$ in. ≈ 34.64 in.

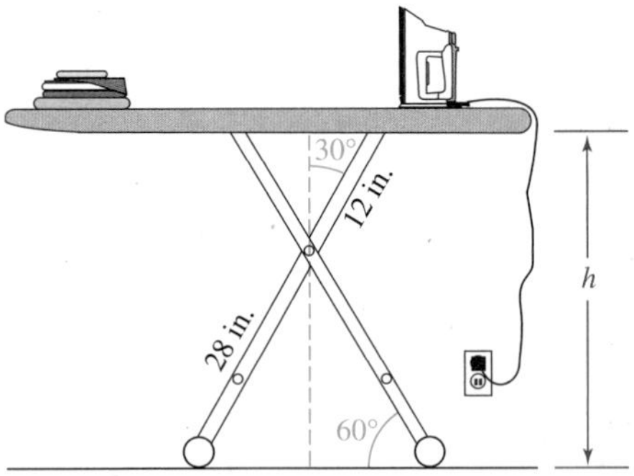

75. Baseball. A baseball diamond is a square, 90 feet on a side. If the third baseman fields a ground ball 10 feet directly behind third base, how far must he throw the ball to throw a runner out at first base? Give the exact answer and an approximation to two decimal places. $10\sqrt{181}$ ft ≈ 134.54 ft

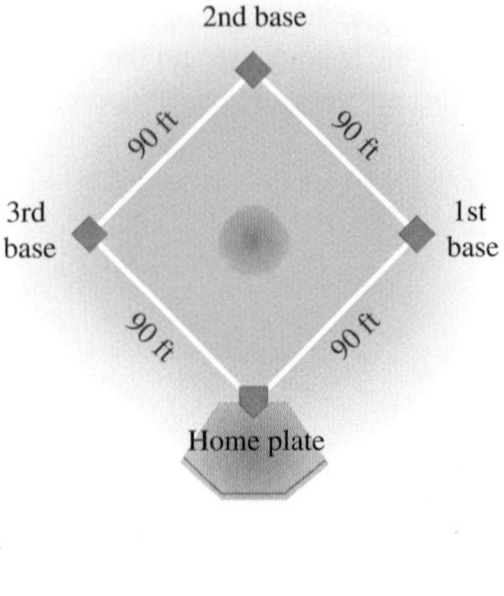

76. Baseball. A shortstop fields a grounder at a point one-third of the way from second base to third base. How far will he have to throw the ball to make an out at first base? Give the exact answer and an approximation to two decimal places.
$30\sqrt{10}$ ft ≈ 94.87 ft

77. Clotheslines. A pair of damp jeans are hung on a clothesline to dry. They pull the center down 1 foot. By how much is the line stretched? Give an approximation to two decimal places.
About 0.13 ft

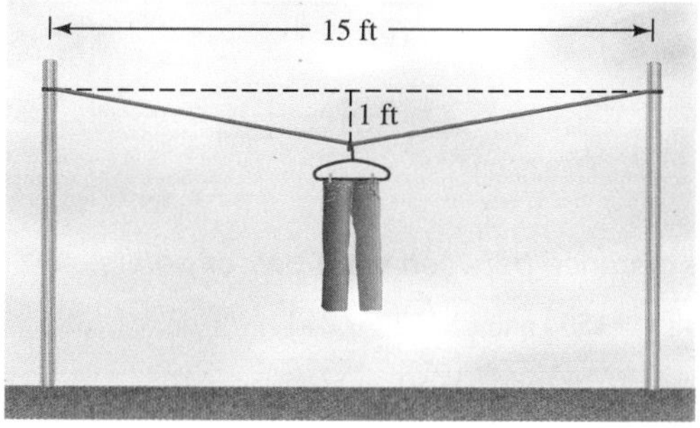

78. Firefighting. The base of the 37-foot ladder is 9 feet from the wall. Will the top reach a window ledge that is 35 feet above the ground? Verify your result. Yes

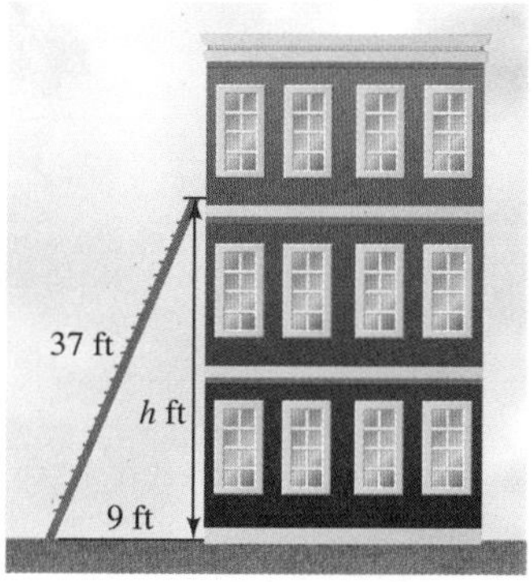

79. Art History. A figure displaying some of the characteristics of Egyptian art is shown in the illustration. Use the distance formula to find the following dimensions of the drawing. Round your answers to two decimal places.

a. From the foot to the eye 21.21 units

b. From the belt to the hand holding the staff 8.25 units

c. From the shoulder to the symbol held in the hand
13.00 units

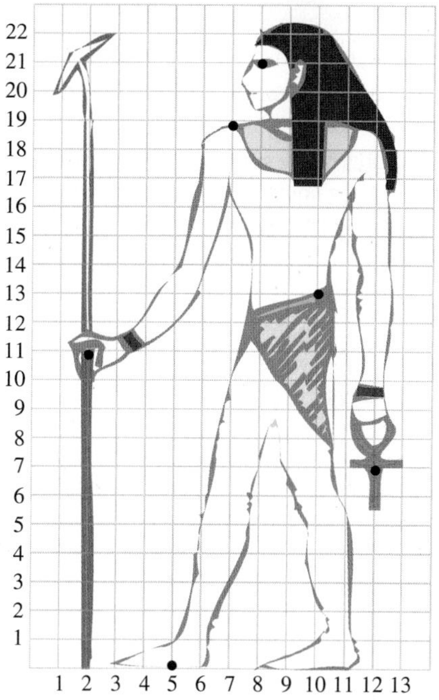

80. Packaging. The diagonal d of a rectangular box with dimensions $a \times b \times c$ is given by $d = \sqrt{a^2 + b^2 + c^2}$. Will the umbrella fit in the shipping carton in the illustration? Verify your result. Not quite

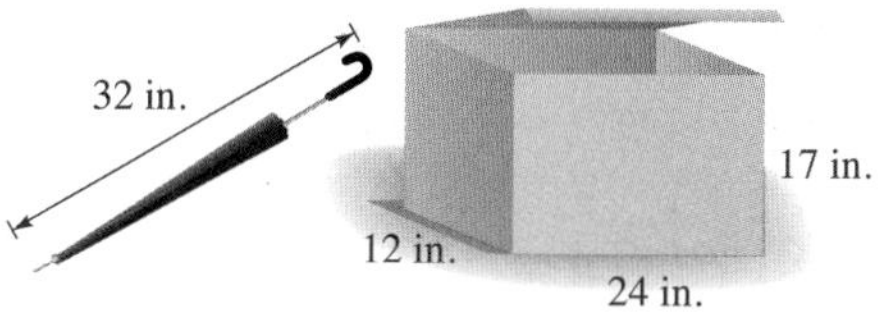

81. Packaging. An archaeologist wants to ship a 34-inch femur bone. Will it fit in a 4-inch-tall box that has a square base with sides 24 inches long? (See Exercise 68.) Verify your result. Yes

82. Telephone. The telephone cable in the illustration runs from A to B to C to D. How much cable is required to run from A to D directly? 173 yd

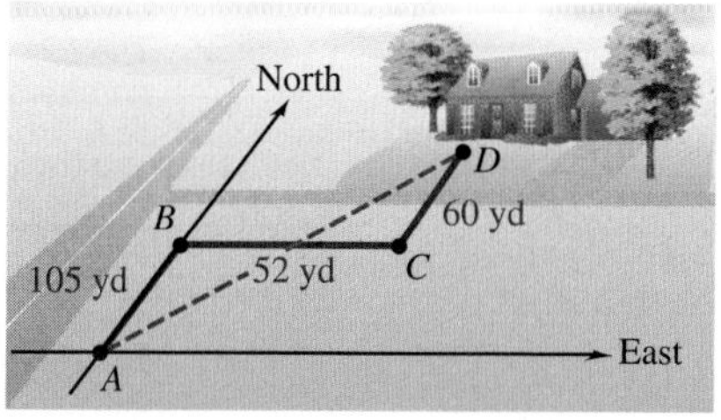

WRITING

83. State the Pythagorean theorem in words.

▶ 84. List the facts that you learned about special right triangles in this section.

▶ 85. When the lengths of the sides of a certain triangle are substituted into the equation of the Pythagorean theorem, the result is a false statement. Explain why.

$$a^2 + b^2 = c^2$$
$$2^2 + 4^2 = 5^2$$
$$4 + 16 = 25$$
$$20 = 25 \quad \text{False}$$

86. Explain how the distance formula and the Pythagorean theorem can be used to show that a triangle with vertices (2, 3), (−3, 4), and (1, −2) is a right triangle.

REVIEW

87. **Discount Buying.** A repairman purchased some washing-machine motors for a total of \$224. When the unit cost decreased by \$4, he was able to buy one extra motor for the same total price. How many motors did he buy originally? 7

▶ 88. **Aviation.** An airplane can fly 650 miles with the wind in the same amount of time as it can fly 475 miles against the wind. If the wind speed is 40 mph, find the speed of the plane in still air. About 257 mph

CHALLENGE PROBLEMS

89. Find the length of the diagonal of the cube shown in figure (a) below. $a\sqrt{3}$ in.

▶ 90. Show that the length of the diagonal of the rectangular solid shown in figure (b) below is $\sqrt{a^2 + b^2 + c^2}$ cm.

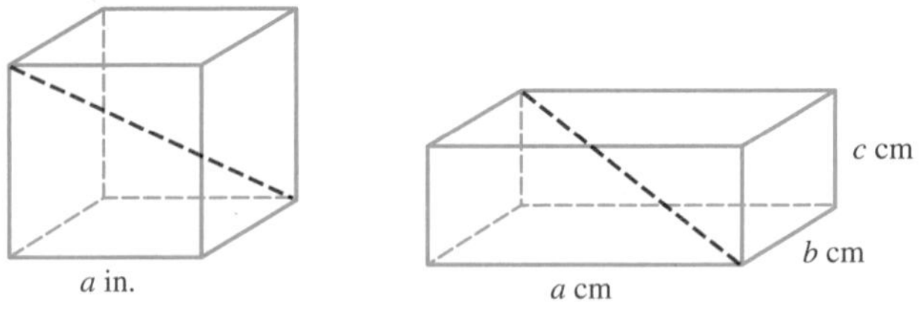

(a) (b)

Find the distance between each pair of points.

91. $(\sqrt{48}, \sqrt{150})$ and $(\sqrt{12}, \sqrt{24})$ $\sqrt{66}$

▶ 92. $(\sqrt{8}, -\sqrt{20})$ and $(\sqrt{50}, -\sqrt{45})$ $\sqrt{23}$

SECTION 9.7 Complex Numbers

OBJECTIVES

1 Express square roots of negative numbers in terms of *i*.

2 Write complex numbers in the form *a* + *bi*.

3 Add and subtract complex numbers.

4 Multiply complex numbers.

5 Divide complex numbers.

6 Perform operations involving powers of *i*.

ARE YOU READY? *Are You Ready? exercises available online at www.webassign.net/brookscole*

The following problems review some basic skills that are needed when working with complex numbers.

1. Explain why $\sqrt{-16}$ is not a real number. No real number squared is equal to −16.

2. Simplify: $(x^2 + 8x) + (5x^2 - 10x)$ $6x^2 - 2x$

3. Multiply: $(2n + 3)(7n - 1)$ $14n^2 + 19n - 3$

4. Simplify: $\sqrt{63}$ $3\sqrt{7}$

5. Rationalize the denominator: $\dfrac{7}{\sqrt{x} + 4}$ $\dfrac{7(\sqrt{x} - 4)}{x - 16}$

6. Divide and give the remainder: $4\overline{)87}$ 21 R3

Recall that the square root of a negative number is not a real number. However, an expanded number system, called the *complex number system,* gives meaning to square roots of negative numbers, such as $\sqrt{-9}$ and $\sqrt{-25}$. To define complex numbers, we use a number that is denoted by the letter *i*.

1 Express Square Roots of Negative Numbers in Terms of *i*.

Teaching Tip: Now would be a good time to review the diagram of the real number system on page 26.

Some equations do not have real-number solutions. For example, $x^2 = -1$ has no real-number solutions because the square of a real number is never negative. To provide a solution to this equation, mathematicians have defined the number i so that $i^2 = -1$.

The Number *i*

The imaginary number i is defined as

$$i = \sqrt{-1}$$

From the definition, it follows that $i^2 = -1$.

This definition enables us to write the square root of any negative number in terms of i.

We can use extensions of the product and quotient rules for radicals to write the square root of a negative number as the product of a real number and i.

EXAMPLE 1 Write each expression in terms of i: **a.** $\sqrt{-9}$ **b.** $\sqrt{-7}$ **c.** $-\sqrt{-18}$ **d.** $\sqrt{-\frac{24}{49}}$

Strategy We will write each radicand as the product of -1 and a positive number. Then we will apply the appropriate rules for radicals.

Why We want our work to produce a factor of $\sqrt{-1}$ so that we can replace it with i.

Solution After factoring the radicand, we use an extension of the product rule for radicals.

a. $\sqrt{-9} = \sqrt{-1 \cdot 9} = \sqrt{-1}\sqrt{9} = i \cdot 3 = 3i$ Replace $\sqrt{-1}$ with i.

b. $\sqrt{-7} = \sqrt{-1 \cdot 7} = \sqrt{-1}\sqrt{7} = i\sqrt{7}$ or $\sqrt{7}i$ Replace $\sqrt{-1}$ with i.

c. $-\sqrt{-18} = -\sqrt{-1 \cdot 9 \cdot 2} = -\sqrt{-1}\sqrt{9}\sqrt{2} = -i \cdot 3 \cdot \sqrt{2} = -3i\sqrt{2}$ or $-3\sqrt{2}i$

d. After factoring the radicand, use an extension of the product and quotient rules for radicals.

$$\sqrt{-\frac{24}{49}} = \sqrt{-1 \cdot \frac{24}{49}} = \frac{\sqrt{-1 \cdot 24}}{\sqrt{49}} = \frac{\sqrt{-1}\sqrt{4}\sqrt{6}}{\sqrt{49}} = \frac{2i\sqrt{6}}{7} \quad \text{or} \quad \frac{2\sqrt{6}}{7}i$$

Notation

Since it is easy to confuse $\sqrt{b}i$ with $\sqrt{bi}$, we usually write i first so that it is clear that the i is not under the radical symbol. However, both $i\sqrt{b}$ and $\sqrt{b}i$ are correct. For example: $i\sqrt{7} = \sqrt{7}i$.

Self Check 1 Write each expression in terms of i: **a.** $\sqrt{-25}$ $5i$ **b.** $-\sqrt{-19}$ $-i\sqrt{19}$ **c.** $\sqrt{-45}$ $3i\sqrt{5}$ **d.** $\sqrt{-\frac{50}{81}}$ $\frac{5\sqrt{2}}{9}i$

Now Try Problems 19, 21, and 27

Teaching Example 1 Write each expression in terms of i:
a. $\sqrt{-49}$ b. $\sqrt{-19}$
c. $-\sqrt{-12}$ d. $\sqrt{-\frac{18}{121}}$
Answers:
a. $7i$ b. $i\sqrt{19}$ c. $-2i\sqrt{3}$
d. $\frac{3\sqrt{2}}{11}i$

The results from Example 1 illustrate a rule for simplifying square roots of negative numbers.

Square Root of a Negative Number

For any positive real number b,

$$\sqrt{-b} = i\sqrt{b}$$

To justify this rule, we use the fact that $\sqrt{-1} = i$.

$$\begin{aligned}\sqrt{-b} &= \sqrt{-1 \cdot b} \\ &= \sqrt{-1}\sqrt{b} \\ &= i\sqrt{b}\end{aligned}$$

2 Write Complex Numbers in the Form $a + bi$.

The imaginary number i is used to define *complex numbers.*

Complex Numbers

A **complex number** is any number that can be written in the form $a + bi$, where a and b are real numbers and $i = \sqrt{-1}$.

Complex numbers of the form $a + bi$, where $b \neq 0$, are also called **imaginary numbers.***

*Some textbooks define imaginary numbers as complex numbers with $a = 0$ and $b \neq 0$.

Notation

It is acceptable to use $a - bi$ as a substitute for the form $a + bi$. For example:

$$6 - 9i = 6 + (-9)i$$

For a complex number written in the **standard form** $a + bi$, we call a the **real part** and b the **imaginary part.** Some examples of complex numbers written in standard form are

$2 + 11i$	$6 - 9i$	$-\frac{1}{2} + 0i$	$0 + i\sqrt{3}$
$a = 2, b = 11$	$a = 6, b = -9$	$a = -\frac{1}{2}, b = 0$	$a = 0, b = \sqrt{3}$

Two complex numbers $a + bi$ and $c + di$ are equal if and only if $a = c$ and $b = d$. Thus, $0.5 + 0.9i = \frac{1}{2} + \frac{9}{10}i$ because $0.5 = \frac{1}{2}$ and $0.9 = \frac{9}{10}$.

EXAMPLE 2 Write each number in the form $a + bi$: **a.** 6 **b.** $\sqrt{-64}$ **c.** $-2 + \sqrt{-63}$

Strategy We will determine a, the real part, and we will simplify the radical (if necessary) to determine the bi part.

Why We can put the two parts together to produce the desired $a + bi$ form.

Solution

a. $6 = 6 + 0i$ The real part a is 6. The imaginary part b is 0.

b. $\sqrt{-64} = 0 + 8i$ The real part a is 0. Simplify: $\sqrt{-64} = \sqrt{-1}\sqrt{64} = 8i$. Thus, b is 8.

c. $-2 + \sqrt{-63} = -2 + 3i\sqrt{7}$ The real part a is -2. Simplify: $\sqrt{-63} = \sqrt{-1}\sqrt{63} = \sqrt{-1}\sqrt{9}\sqrt{7} = 3i\sqrt{7}$. Thus, $b = 3\sqrt{7}$.

Caution

Use i only when working with the *square root* of a negative number. It does not apply to cube roots.

$\sqrt{-64} = 8i$ and $\sqrt[3]{-64} = -4$

Teaching Example 2 Write each number in the form $a + bi$:
a. 23 **b.** $\sqrt{-144}$ **c.** $5 + \sqrt{-18}$
Answers:
a. $23 + 0i$ **b.** $0 + 12i$
c. $5 + 3i\sqrt{2}$

Self Check 2 Write each number in the form $a + bi$: **a.** -18 $-18 + 0i$ **b.** $\sqrt{-36}$ $0 + 6i$ **c.** $1 + \sqrt{-24}$ $1 + 2i\sqrt{6}$

Now Try Problems 29 and 33

Teaching Tip: Have students classify each of the examples of real numbers as either rational or irrational.

The following illustration shows the relationship between the real numbers, the imaginary numbers, and the complex numbers.

The complex numbers: $a + bi$

The real numbers: $a + bi$ where $b = 0$				The imaginary numbers: $a + bi$ where $b \neq 0$		
-6	$\frac{5}{16}$	-1.75	π	$9 + 7i$	$-2i$	$\frac{1}{4} - \frac{3}{4}i$
$48 + 0i$	0	$-\sqrt{10}$	$-\frac{7}{2}$	$0.56i$	$\sqrt{-10}$	$6 + i\sqrt{3}$
$\sqrt[3]{-10}$	$4 + \pi$	$15\frac{5}{8}$	$\sqrt[4]{99}$	$25i\sqrt{5}$	$\frac{7}{8}i$	$\sqrt{3} + 3i\sqrt{2}$

Success Tip

Just as real numbers are either rational or irrational, but not both, complex numbers are either real or imaginary, but not both.

3 Add and Subtract Complex Numbers.

Adding and subtracting complex numbers is similar to adding and subtracting polynomials.

Addition and Subtraction of Complex Numbers	1. To add complex numbers, add their real parts and add their imaginary parts. 2. To subtract complex numbers, add the opposite of the complex number being subtracted.

EXAMPLE 3 Perform each operation. Write the answers in the form $a + bi$. **a.** $(8 + 4i) + (12 + 8i)$ **b.** $(-6 + 4i) - (3 + 2i)$ **c.** $\left(7 - \sqrt{-16}\right) + \left(9 + \sqrt{-4}\right)$

Strategy To add the complex numbers, we will add their real parts and add their imaginary parts. To subtract the complex numbers, we will add the opposite of the complex number to be subtracted.

Why We perform the indicated operations as if the complex numbers were polynomials with i as a variable.

Solution

a. $(8 + 4i) + (12 + 8i) = (8 + 12) + (4 + 8)i$ — The sum of the real parts; The sum of the imaginary parts

$= 20 + 12i$

b. $(-6 + 4i) - (3 + 2i) = (-6 + 4i) + (-3 - 2i)$ — Add the opposite. To find the opposite, change the sign of each term of $3 + 2i$.

$= [-6 + (-3)] + [4 + (-2)]i$ — Add the real parts. Add the imaginary parts.

$= -9 + 2i$

c. $\left(7 - \sqrt{-16}\right) + \left(9 + \sqrt{-4}\right)$

$= (7 - 4i) + (9 + 2i)$ — Write $\sqrt{-16}$ and $\sqrt{-4}$ in terms of i.

$= (7 + 9) + (-4 + 2)i$ — Add the real parts. Add the imaginary parts.

$= 16 - 2i$ — Write $16 + (-2i)$ in the form $16 - 2i$.

The Language of Algebra

For years, mathematicians thought numbers like $\sqrt{-4}$ and $\sqrt{-16}$ were useless. In the 17th century, French mathematician René Descartes (1596–1650) called them **imaginary numbers.** Today they have important uses such as describing alternating electric current.

Success Tip

Always express square roots of negative numbers in terms of i before performing any operations.

Self Check 3 Perform the operations. Write the answers in the form $a + bi$.
a. $(3 - 5i) + (-2 + 7i)$ $1 + 2i$
b. $\left(3 - \sqrt{-25}\right) - \left(-2 + \sqrt{-49}\right)$ $5 - 12i$

Now Try Problems 37 and 43

Teaching Example 3 Perform each operation. Write the answers in the form $a + bi$.
a. $(4 - 14i) + (12 + 56i)$
b. $(22 - 11i) - (5 + 2i)$
c. $\left(9 - \sqrt{-49}\right) + \left(2 - \sqrt{-64}\right)$
Answers:
a. $16 + 42i$ b. $17 - 13i$
c. $11 - 15i$

4 Multiply Complex Numbers.

Since imaginary numbers are not real numbers, some properties of real numbers do not apply to imaginary numbers. For example, we cannot use the product rule for radicals to multiply two imaginary numbers.

Teaching Tip: Have your students turn to page 765 and read the product rule for radicals. Ask why $\sqrt{-4} \cdot \sqrt{-9}$ does not satisfy the "If" portion of this rule.

CAUTION If a and b are both negative, then $\sqrt{a}\sqrt{b} \neq \sqrt{ab}$. For example, if $a = -4$ and $b = -9$,

~~$\sqrt{-4} \cdot \sqrt{-9} = \sqrt{-4(-9)} = \sqrt{36} = 6$~~ $\sqrt{-4}\sqrt{-9} = 2i(3i) = 6i^2 = 6(-1) = -6$

EXAMPLE 4 Multiply: $\sqrt{-2} \cdot \sqrt{-20}$

Strategy To multiply the imaginary numbers, first we will write $\sqrt{-2}$ and $\sqrt{-20}$ in $i\sqrt{b}$ form. Then we will use the product rule for radicals.

Why We cannot use the product rule for radicals immediately because it does not apply when both radicands are negative.

Solution

$$\sqrt{-2} \cdot \sqrt{-20} = \left(i\sqrt{2}\right)\left(2i\sqrt{5}\right) \quad \text{Simplify: } \sqrt{-20} = i\sqrt{20} = 2i\sqrt{5}.$$

$$= 2i^2\sqrt{2 \cdot 5} \quad \text{Multiply: } i \cdot 2i = 2i^2\text{. Use the product rule for radicals.}$$

$$= 2i^2\sqrt{10}$$

$$= 2(-1)\sqrt{10} \quad \text{Replace } i^2 \text{ with } -1.$$

$$= -2\sqrt{10} \quad \text{Multiply.}$$

Success Tip

The algebraic methods we have used with polynomials and rational expressions (such as combining like terms, distributive property, FOIL, and building fractions) are also used with complex numbers. However, when you encounter i^2 in your work, remember to replace it with -1.

Self Check 4 Multiply: $\sqrt{-3} \cdot \sqrt{-32}$ $-4\sqrt{6}$

Now Try Problem 47

Teaching Example 4 Multiply: $\sqrt{-5} \cdot \sqrt{-45}$
Answer: -15

Multiplying complex numbers is similar to multiplying polynomials.

EXAMPLE 5 Multiply. Write the answers in the form $a + bi$. **a.** $6(2 + 9i)$ **b.** $-5i(4 - 8i)$

Strategy We will use the distributive property to find the products.

Why We perform the indicated operations as if the complex numbers were polynomials with i as a variable.

Solution

a. $6(2 + 9i) = 6(2) + 6(9i)$ Use the distributive property.
$= 12 + 54i$ Perform each multiplication.

b. $-5i(4 - 8i) = -5i(4) - (-5i)8i$ Use the distributive property.
$= -20i + 40i^2$ Perform each multiplication.
$= -20i + 40(-1)$ Replace i^2 with -1.
$= -20i - 40$ Multiply.
$= -40 - 20i$ Write the real part, -40, as the first term.

Caution

A common mistake is to replace i with -1. Remember, $i \neq -1$. By definition, $i = \sqrt{-1}$ and $i^2 = -1$.

Teaching Example 5 Multiply. Write the answers in the form $a + bi$.
a. $-3(4 - 5i)$ **b.** $6i(2 - 5i)$
Answers: **a.** $-12 + 15i$ **b.** $30 + 12i$

Self Check 5 Multiply. Write the answers in the form $a + bi$.
a. $-2(-9 - i)$ $18 + 2i$ **b.** $10i(7 + 4i)$ $-40 + 70i$

Now Try Problems 49 and 55

EXAMPLE 6 Multiply. Write the answers in the form $a + bi$. **a.** $(2 + 3i)(3 - 2i)$
b. $(-4 + 2i)(2 + i)$

Strategy We will use the FOIL method to multiply the two complex numbers.

Why We perform the indicated operations as if the complex numbers were binomials with i as a variable.

Solution

Success Tip

i is not a variable, but you can think of it as one when adding, subtracting, and multiplying. For example:

$$-4i + 9i = 5i$$
$$6i - 2i = 4i$$
$$i \cdot i = i^2$$

Remember that the expression i^2 simplifies to -1.

F O I L

a. $(2 + 3i)(3 - 2i) = 6 - 4i + 9i - 6i^2$ Use the FOIL method.

$= 6 + 5i - 6(-1)$ Combine the imaginary terms: $-4i + 9i = 5i$. Replace i^2 with -1.

$= 6 + 5i + 6$ Simplify the last term.

$= 12 + 5i$ Combine like terms.

b. $(-4 + 2i)(2 + i) = -8 - 4i + 4i + 2i^2$ Use the FOIL method.

$= -8 + 0i + 2(-1)$ Combine like terms: $-4i + 4i = 0i$. Replace i^2 with -1.

$= -8 + 0i - 2$ Multiply.

$= -10 + 0i$ Combine like terms.

Teaching Example 6 Multiply. Write the answers in the form $a + bi$.
a. $(3 - 5i)(-2 + 6i)$
b. $\left(4 + \sqrt{-25}\right)\left(1 - \sqrt{-4}\right)$
Answers: **a.** $24 + 28i$ **b.** $14 - 3i$

Self Check 6 Multiply. Write the answers in the form $a + bi$. $(-2 + 3i)(3 - 2i)$
$0 + 13i$

Now Try Problem 59

5 Divide Complex Numbers.

Before we can discuss division of complex numbers, we must introduce an important fact about *complex conjugates*.

Complex Conjugates The complex numbers $a + bi$ and $a - bi$ are called **complex conjugates.**

For example,

- $7 + 4i$ and $7 - 4i$ are complex conjugates.
- $5 - i$ and $5 + i$ are complex conjugates.
- $-6i$ and $6i$ are complex conjugates, because $-6i = 0 - 6i$ and $6i = 0 + 6i$.

The Language of Algebra

Recall that the word **conjugate** was used earlier when we rationalized the denominators of radical expressions such as

$$\frac{5}{\sqrt{6} - 1}$$

In general, the product of the complex number $a + bi$ and its complex conjugate $a - bi$ is the real number $a^2 + b^2$, as the following work shows:

$(a + bi)(a - bi) = a^2 - abi + abi - b^2i^2$ Use the FOIL method.

$= a^2 - b^2(-1)$ $-abi + abi = 0$. Replace i^2 with -1.

$= a^2 + b^2$

EXAMPLE 7 Find the product of $3 + 5i$ and its complex conjugate.

Strategy The complex conjugate of $3 + 5i$ is $3 - 5i$. We will find their product by using the FOIL method.

Why We perform the indicated operations as if the complex numbers were binomials with i as a variable.

Solution We can find the product as follows:

$$\begin{aligned}(3 + 5i)(3 - 5i) &= 9 - 15i + 15i - 25i^2 && \text{Use the FOIL method.}\\ &= 9 - 25i^2 && \text{Combine like terms: } -15i + 15i = 0.\\ &= 9 - 25(-1) && \text{Replace } i^2 \text{ with } -1.\\ &= 9 + 25\\ &= 34\end{aligned}$$

The product of $3 + 5i$ and its complex conjugate $3 - 5i$ is the real number 34.

Teaching Example 7 Find the product of $-2 + 7i$ and its complex conjugate.
Answer: 53

Self Check 7 Multiply: $(2 + 3i)(2 - 3i)$ 13

Now Try Problem 65

Recall that to divide *radical expressions,* we rationalized the denominator. We will use a similar approach to divide complex numbers. To divide two complex numbers when the divisor has two terms, we use the following strategy.

Division of Complex Numbers

To divide complex numbers, multiply the numerator and denominator by the complex conjugate of the denominator.

EXAMPLE 8 Divide. Write the answers in the form $a + bi$. **a.** $\dfrac{3}{6 + i}$ **b.** $\dfrac{1 + 2i}{3 - 4i}$

Strategy We will build each fraction by multiplying it by a form of 1 that uses the conjugate of the denominator.

Why This step produces a *real number* in the denominator so that the result then can be written in the form $a + bi$.

Solution **a.** We want to build a fraction equivalent to $\frac{3}{6 + i}$ that does not have i in the denominator. To make the denominator, $6 + i$, a real number, we need to multiply it by its complex conjugate, $6 - i$. It follows that $\frac{6 - i}{6 - i}$ should be the form of 1 that is used to build $\frac{3}{6 + i}$.

$$\begin{aligned}\frac{3}{6 + i} &= \frac{3}{6 + i} \cdot \frac{6 - i}{6 - i} && \text{To build an equivalent fraction, multiply by } \tfrac{6 - i}{6 - i} = 1.\\ &= \frac{18 - 3i}{36 - 6i + 6i - i^2} && \text{To multiply the numerators, distribute the multiplication by 3. Use the FOIL method to multiply the denominators.}\\ &= \frac{18 - 3i}{36 - (-1)} && \text{Combine like terms: } -6i + 6i = 0. \text{ Replace } i^2 \text{ with } -1. \text{ Note that the denominator no longer contains } i.\\ &= \frac{18 - 3i}{37} && \text{Simplify the denominator. This notation represents the difference of two fractions that have the common denominator 37: } \tfrac{18}{37} \text{ and } \tfrac{3i}{37}.\\ &= \frac{18}{37} - \frac{3}{37}i && \text{Write the complex number in the form } a + bi.\end{aligned}$$

Success Tip

When multiplying the denominator and its complex conjugate, we can use the special-product rule for the sum and difference of the same two terms to simplify the calculations:

$$(3 - 4i)(3 + 4i) = 3^2 - (4i)^2 = 9 - 16i^2$$

Caution

Remember, we can remove only factors common to the entire numerator and denominator. Don't make the mistake of removing a common term, as shown below:

$$\frac{\overset{1}{\cancel{-5}} + 10i}{\underset{5}{\cancel{25}}}$$

b. We can make the denominator of $\frac{1 + 2i}{3 - 4i}$ a real number by multiplying it by the complex conjugate of $3 - 4i$, which is $3 + 4i$. It follows that $\frac{3 + 4i}{3 + 4i}$ should be the form of 1 that is used to build $\frac{1 + 2i}{3 - 4i}$.

$$\frac{1 + 2i}{3 - 4i} = \frac{1 + 2i}{3 - 4i} \cdot \frac{3 + 4i}{3 + 4i}$$ To build an equivalent fraction, multiply by $\frac{3 + 4i}{3 + 4i} = 1$.

$$= \frac{3 + 4i + 6i + 8i^2}{9 + 12i - 12i - 16i^2}$$ Use the FOIL method to multiply the numerators and the denominators.

$$= \frac{3 + 10i + 8(-1)}{9 - 16(-1)}$$ Combine like terms in the numerator and denominator. Replace i^2 with -1. The denominator is now a real number.

$$= \frac{3 + 10i - 8}{9 + 16}$$ Simplify the numerator and denominator.

$$= \frac{-5 + 10i}{25}$$ Combine like terms in the numerator and denominator.

$$= \frac{\overset{1}{\cancel{5}}(-1 + 2i)}{\underset{1}{\cancel{5}} \cdot 5}$$ Factor out 5 in the numerator and remove the common factor of 5.

$$= \frac{-1 + 2i}{5}$$ Simplify. This notation represents the sum of two fractions that have the common denominator 5.

$$= -\frac{1}{5} + \frac{2}{5}i$$ Write the complex number in the form $a + bi$.

Teaching Example 8 Divide. Write the answer in the form $a + bi$.
a. $\frac{5}{3 + 2i}$ b. $\frac{3 - 2i}{2 - i}$
Answers:
a. $\frac{15}{13} - \frac{10}{13}i$ b. $\frac{8}{5} - \frac{1}{5}i$

Self Check 8 Divide. Write the answers in the form $a + bi$. **a.** $\frac{6}{5 + 2i}$ $\frac{30}{29} - \frac{12}{29}i$
b. $\frac{2 - 4i}{5 - 3i}$ $\frac{11}{17} - \frac{7}{17}i$

Now Try Problems 71 and 77

EXAMPLE 9 Divide and write the answer in the form $a + bi$: $\frac{4 + \sqrt{-16}}{2 + \sqrt{-4}}$

Strategy We will begin by writing $\sqrt{-16}$ and $\sqrt{-4}$ in $i\sqrt{b}$ form.

Why To perform any operations, the numerator and denominator should be written in the form $a + bi$.

Solution

$$\frac{4 + \sqrt{-16}}{2 + \sqrt{-4}} = \frac{4 + 4i}{2 + 2i}$$ Simplify: $\sqrt{-16} = \sqrt{-1}\sqrt{16} = 4i$ and $\sqrt{-4} = \sqrt{-1}\sqrt{4} = 2i$.

$$= \frac{2\overset{1}{\cancel{(2 + 2i)}}}{\underset{1}{\cancel{2 + 2i}}}$$ Factor out 2 in the numerator and remove the common factor of $2 + 2i$.

$$= 2$$

$$= 2 + 0i$$ Write 2 in the form $a + bi$.

Teaching Example 9 Divide and write the answer in the form $a + bi$:
$\frac{2 + \sqrt{-25}}{6 + \sqrt{-225}}$
Answer: $\frac{1}{3} + 0i$

Self Check 9 Divide and write the answer in the form $a + bi$: $\frac{21 - \sqrt{-49}}{3 - \sqrt{-1}}$ $7 + 0i$

Now Try Problem 81

EXAMPLE 10 Divide and write the result in the form $a + bi$: $\dfrac{7}{2i}$

Strategy We will use $\frac{-2i}{-2i}$ as the form of 1 to build $\frac{7}{2i}$.

Why Since the denominator $2i$ can be expressed as $0 + 2i$, its complex conjugate is $0 - 2i$. However, instead of building with $\frac{0 - 2i}{0 - 2i}$, we will drop the zeros and just use $\frac{-2i}{-2i}$.

Solution

$$\frac{7}{2i} = \frac{7}{2i} \cdot \frac{-2i}{-2i}$$ To build an equivalent fraction, multiply by $\frac{-2i}{-2i} = 1$.

$$= \frac{-14i}{-4i^2}$$ Multiply the numerators and multiply the denominators.

$$= \frac{-14i}{-4(-1)}$$ Replace i^2 with -1. The denominator is now a real number.

$$= \frac{-14i}{4}$$ Simplify the denominator.

$$= -\frac{7i}{2}$$ Simplify the fraction: $-\frac{\overset{1}{\cancel{2}} \cdot 7i}{\underset{1}{\cancel{2}} \cdot 2}$.

$$= 0 - \frac{7}{2}i$$ Write in the form $a + bi$.

Self Check 10 Divide and write the result in the form $a + bi$: $\frac{3}{4i}$ $\quad 0 - \frac{3}{4}i$

Now Try Problem 85

Success Tip

In this example, the denominator of $\frac{7}{2i}$ is of the form bi. In such cases, we can eliminate i in the denominator by simply multiplying by $\frac{i}{i}$.

$$\frac{7}{2i} = \frac{7}{2i} \cdot \frac{i}{i} = \frac{7i}{2i^2} = -\frac{7}{2}i$$

Teaching Example 10 Divide and write the result in the form $a + bi$: $\frac{11}{6i}$

Answer: $0 - \frac{11}{6}i$

6 Perform Operations Involving Powers of i.

The powers of i produce an interesting pattern:

$$i = \sqrt{-1} = i \qquad i^5 = i^4 i = 1i = i$$
$$i^2 = \left(\sqrt{-1}\right)^2 = -1 \qquad i^6 = i^4 i^2 = 1(-1) = -1$$
$$i^3 = i^2 i = -1i = -i \qquad i^7 = i^4 i^3 = 1(-i) = -i$$
$$i^4 = i^2 i^2 = (-1)(-1) = 1 \qquad i^8 = i^4 i^4 = (1)(1) = 1$$

The pattern continues: $i, -1, -i, 1, \ldots$.

Larger powers of i can be simplified by using the fact that $i^4 = 1$. For example, to simplify i^{29}, we note that 29 divided by 4 gives a quotient of 7 and a remainder of 1. Thus, $29 = 4 \cdot 7 + 1$ and

$$i^{29} = i^{4 \cdot 7 + 1}$$ $4 \cdot 7 = 28$.

$$= (i^4)^7 \cdot i^1$$ Use the rules for exponents $x^{m \cdot n} = (x^m)^n$ and $x^{m+n} = x^m \cdot x^n$.

$$= 1^7 \cdot i$$ Simplify: $i^4 = 1$.

$$= i$$ Simplify: $1 \cdot i = i$.

The result of this example illustrates the following fact.

Success Tip

Note that the powers of i cycle through four possible outcomes:

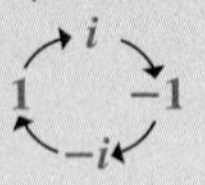

Teaching Tip: Stress that there are only four possible results when simplifying a power of i.

Powers of i

If n is a natural number that has a remainder of R when divided by 4, then

$$i^n = i^R$$

EXAMPLE 11 Simplify: **a.** i^{55} **b.** i^{98}

Strategy We will examine the remainder when we divide the exponents 55 and 98 by 4.

Why The remainder determines the power to which i is raised in the simplified form.

Solution

a. We divide 55 by 4 and get a remainder of 3. Therefore,

$$i^{55} = i^3 = -i \qquad \begin{array}{r} 13\text{R }3 \\ 4\overline{)55} \\ \underline{-4} \\ 15 \\ \underline{-12} \\ 3 \end{array}$$

b. We divide 98 by 4 and get a remainder of 2. Therefore,

$$i^{98} = i^2 = -1 \qquad \begin{array}{r} 24\text{R }2 \\ 4\overline{)98} \\ \underline{-8} \\ 18 \\ \underline{-16} \\ 2 \end{array}$$

Success Tip

If we divide the natural number exponent n of a power of i by 4, the remainder indicates the simplified form of i^n:

$R = 1: \; i$
$R = 2: \; -1$
$R = 3: \; -i$
$R = 0: \; 1$

Self Check 11 Simplify: **a.** i^{62} -1 **b.** i^{105} i

Now Try Problem 91

Teaching Example 11 Simplify:
a. i^{33} **b.** i^{52} **c.** i^{63}
Answers: **a.** i **b.** 1 **c.** $-i$

SECTION 9.7 STUDY SET

VOCABULARY

Fill in the blanks.

1. The imaginary number i is defined as $i = \sqrt{-1}$. We call i^{25} a power of i.
2. A complex number is any number that can be written in the form $a + bi$, where a and b are real numbers and $i = \sqrt{-1}$.
3. For the complex number $2 + 5i$, we call 2 the real part and 5 the imaginary part.
4. $6 + 3i$ and $6 - 3i$ are called complex conjugates.

CONCEPTS

Fill in the blanks.

5. **a.** $i = \sqrt{-1}$ **b.** $i^2 = -1$
   c. $i^3 = -i$ **d.** $i^4 = 1$
   e. In general, the powers of i cycle through four possible outcomes.
6. Simplify:

$$\sqrt{-36} = \sqrt{-1 \cdot 36} = \sqrt{-1}\sqrt{36} = 6i$$

7. **a.** To add (or subtract) complex numbers, add (or subtract) their real parts and add (or subtract) their imaginary parts.
   b. To multiply two complex numbers, such as $(2 + 3i)(3 + 5i)$, we can use the FOIL method.
8. To divide $6 + 7i$ by $1 - 8i$, we multiply $\frac{6 + 7i}{1 - 8i}$ by 1 in the form of $\frac{1 + 8i}{1 + 8i}$.
9. Give the complex conjugate of each number.
   a. $2 - 3i$ $2 + 3i$ **b.** 2 $2 - 0i$ **c.** $-3i$ $0 + 3i$
10. Factor each numerator. Then remove the factor common to the numerator and denominator. Write the result in the form $a + bi$.
    a. $\frac{3 + 6i}{3}$ $1 + 2i$ **b.** $\frac{15 + 25i}{10}$ $\frac{3}{2} + \frac{5}{2}i$
11. Complete the illustration. Label the *real numbers,* the *imaginary numbers,* the *complex numbers,* the *rational numbers,* and the *irrational numbers.*

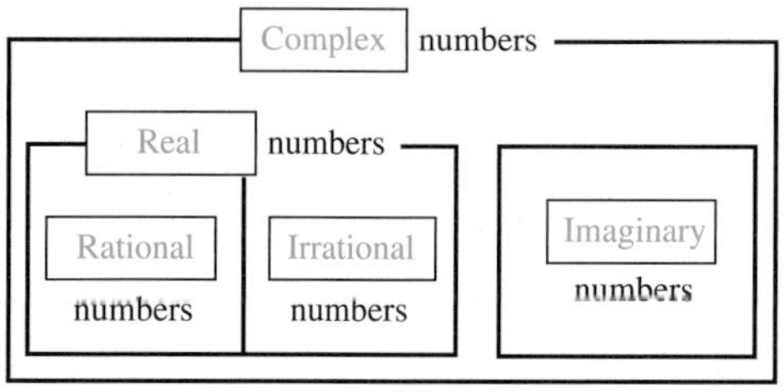

12. Determine whether each statement is true or false.
    a. Every complex number is a real number. False
    b. Every real number is a complex number. True
    c. i is a real number. False
    d. The square root of a negative number is an imaginary number. True
    e. The product of a complex number and its complex conjugate is always a real number. True

NOTATION

Complete each solution.

13. $(3 + 2i)(3 - i) = 9 - 3i + 6i - 2i^2$
$= 9 + 3i + 2$
$= 11 + 3i$

▶ 14. $\frac{3}{2-i} = \frac{3}{2-i} \cdot \frac{2+i}{2+i}$
$= \frac{6 + 3i}{4 - i^2}$
$= \frac{6+3i}{5}$
$= \frac{6}{5} + \frac{3}{5}i$

15. Determine whether each statement is true or false.
a. $\sqrt{6i} = i\sqrt{6}$ True
b. $\sqrt{8i} = \sqrt{8}i$ False
c. $\sqrt{-25} = -\sqrt{25}$ False
d. $-i = i$ False

▶ 16. Write each number in the form $a + bi$.
a. $\frac{9 + 11i}{4}$ $\frac{9}{4} + \frac{11}{4}i$
b. $\frac{1 - i}{18}$ $\frac{1}{18} - \frac{1}{18}i$

GUIDED PRACTICE

Express each number in terms of i. See Example 1.

▶ 17. $\sqrt{-9}$ $3i$
18. $\sqrt{-4}$ $2i$
19. $\sqrt{-7}$ $\sqrt{7}i$ or $i\sqrt{7}$
▶ 20. $\sqrt{-11}$ $\sqrt{11}i$ or $i\sqrt{11}$
▶ 21. $\sqrt{-24}$ $2\sqrt{6}i$ or $2i\sqrt{6}$
▶ 22. $\sqrt{-28}$ $2\sqrt{7}i$ or $2i\sqrt{7}$
23. $-\sqrt{-72}$ $-6\sqrt{2}i$ or $-6i\sqrt{2}$
▶ 24. $-\sqrt{-24}$ $-2\sqrt{6}i$ or $-2i\sqrt{6}$
25. $5\sqrt{-81}$ $45i$
▶ 26. $6\sqrt{-49}$ $42i$
▶ 27. $\sqrt{-\frac{25}{9}}$ $\frac{5}{3}i$
▶ 28. $-\sqrt{-\frac{121}{144}}$ $-\frac{11}{12}i$

Write each number in the form a + bi. See Example 2.

29. a. 5 $5 + 0i$
b. $\sqrt{-49}$ $0 + 7i$
▶ 30. a. -43 $-43 + 0i$
b. $\sqrt{-169}$ $0 + 13i$
31. a. $1 + \sqrt{-25}$ $1 + 5i$
b. $-3 + \sqrt{-8}$ $-3 + 2i\sqrt{2}$
▶ 32. a. $21 + \sqrt{-16}$ $21 + 4i$
b. $-9 + \sqrt{-12}$ $-9 + 2i\sqrt{3}$
▶ 33. a. $76 - \sqrt{-54}$ $76 - 3i\sqrt{6}$
b. $-7 + \sqrt{-19}$ $-7 + i\sqrt{19}$
34. a. $88 - \sqrt{-98}$ $88 - 7i\sqrt{2}$
b. $-2 + \sqrt{-35}$ $-2 + i\sqrt{35}$
35. a. $-6 - \sqrt{-9}$ $-6 - 3i$
b. $3 + \sqrt{-6}$ $3 + i\sqrt{6}$
▶ 36. a. $-45 - \sqrt{-81}$ $-45 - 9i$
b. $8 + \sqrt{-7}$ $8 + i\sqrt{7}$

Perform the operations. Write all answers in the form a + bi. See Example 3.

▶ 37. $(3 + 4i) + (5 - 6i)$ $8 - 2i$
38. $(8 + 3i) + (-7 - 2i)$ $1 + i$
39. $(6 - i) + (9 + 3i)$ $15 + 2i$
▶ 40. $(5 + 3i) - (6 - 9i)$ $-1 + 12i$
▶ 41. $(7 - 3i) - (4 + 2i)$ $3 - 5i$
▶ 42. $(5 - 4i) - (3 + 2i)$ $2 - 6i$
▶ 43. $\left(8 + \sqrt{-25}\right) - \left(7 + \sqrt{-4}\right)$ $1 + 3i$
▶ 44. $\left(-7 + \sqrt{-81}\right) - \left(-2 - \sqrt{-64}\right)$ $-5 + 17i$

Multiply. See Example 4.

45. $\sqrt{-1} \cdot \sqrt{-36}$ -6
▶ 46. $\sqrt{-9} \cdot \sqrt{-100}$ -30
▶ 47. $\sqrt{-2}\sqrt{-12}$ $-2\sqrt{6}$
▶ 48. $\sqrt{-3}\sqrt{-45}$ $-3\sqrt{15}$

Multiply. Write all answers in the form a + bi. See Example 5.

49. $3(2 - 9i)$ $6 - 27i$
▶ 50. $-4(3 + 4i)$ $-12 - 16i$
51. $7(5 - 4i)$ $35 - 28i$
52. $-5(3 + 2i)$ $-15 - 10i$
53. $2i(7 - 3i)$ $6 + 14i$
54. $i(8 + 2i)$ $-2 + 8i$
▶ 55. $-5i(5 - 5i)$ $-25 - 25i$
56. $2i(7 + 2i)$ $-4 + 14i$

Multiply. Write all answers in the form a + bi. See Example 6.

57. $(2 + i)(3 - i)$ $7 + i$
▶ 58. $(4 - i)(2 + i)$ $9 + 2i$
▶ 59. $(3 - 2i)(2 + 3i)$ $12 + 5i$
60. $(3 - i)(2 + 3i)$ $9 + 7i$
61. $\left(4 + \sqrt{-1}\right)\left(3 - \sqrt{-1}\right)$ $13 - i$
▶ 62. $\left(1 - \sqrt{-25}\right)\left(1 - \sqrt{-16}\right)$ $-19 - 9i$
63. $(2 + i)^2$ $3 + 4i$
▶ 64. $(3 - 2i)^2$ $5 - 12i$

Find the product of the given complex number and its conjugate. See Example 7.

65. $2 + 6i$ 40
▶ 66. $5 + 2i$ 29
67. $-4 - 7i$ 65
▶ 68. $-10 - 9i$ 181

Divide. Write all answers in the form a + bi. See Example 8.

▶ 69. $\frac{9}{5 + i}$ $\frac{45}{26} - \frac{9}{26}i$
▶ 70. $\frac{4}{2 - i}$ $\frac{8}{5} + \frac{4}{5}i$
71. $\frac{11i}{4 - 7i}$ $-\frac{77}{65} + \frac{44}{65}i$
▶ 72. $\frac{2i}{3 + 8i}$ $\frac{16}{73} + \frac{6}{73}i$
73. $\frac{3 - 2i}{4 - i}$ $\frac{14}{17} - \frac{5}{17}i$
74. $\frac{6 - i}{2 + i}$ $\frac{11}{5} - \frac{8}{5}i$
75. $\frac{7 + 4i}{2 - 5i}$ $-\frac{6}{29} + \frac{43}{29}i$
▶ 76. $\frac{2 + 3i}{2 - 3i}$ $-\frac{5}{13} + \frac{12}{13}i$
▶ 77. $\frac{7 + 3i}{4 - 2i}$ $\frac{11}{10} + \frac{13}{10}i$
78. $\frac{5 - 3i}{4 + 2i}$ $\frac{7}{10} - \frac{11}{10}i$
79. $\frac{1 - 3i}{3 + i}$ $0 - i$
▶ 80. $\frac{3 + 5i}{1 - i}$ $-1 + 4i$

Divide. Write all answers in the form a + bi. See Example 9. (*Hint:* Factor the numerator.)

81. $\frac{8 + \sqrt{-144}}{2 + \sqrt{-9}}$ $4 + 0i$
▶ 82. $\frac{3 + \sqrt{-36}}{1 + \sqrt{-4}}$ $3 + 0i$
83. $\frac{-4 - \sqrt{-4}}{2 + \sqrt{-1}}$ $-2 + 0i$
▶ 84. $\frac{-5 - \sqrt{-25}}{1 + \sqrt{-1}}$ $-5 + 0i$

Divide. Write all answers in the form a + bi. See Example 10.

85. $\frac{5}{3i}$ $0 - \frac{5}{3}i$
▶ 86. $\frac{3}{8i}$ $0 - \frac{3}{8}i$
87. $-\frac{2}{7i}$ $0 + \frac{2}{7}i$
▶ 88. $-\frac{8}{5i}$ $0 + \frac{8}{5}i$

Simplify each expression. See Example 11.

89. i^{21} i 90. i^{19} $-i$ 91. i^{27} $-i$ 92. i^{22} -1

93. i^{100} 1 94. i^{97} i 95. i^{42} -1 96. i^{200} 1

TRY IT YOURSELF

Perform the operations. Write all answers in the form a + bi.

97. $(3 - i) - (-1 + 10i)$ $4 - 11i$

98. $(14 + 4i) - (-9 - i)$ $23 + 5i$

99. $(2 - \sqrt{-16})(3 + \sqrt{-4})$ $14 - 8i$

100. $(3 - \sqrt{-4})(4 - \sqrt{-9})$ $6 - 17i$

101. $(-6 - 9i) + (4 + 3i)$ $-2 - 6i$

102. $(-3 + 11i) + (-1 - 6i)$ $-4 + 5i$

103. $\frac{-2i}{3 + 2i}$ $-\frac{4}{13} - \frac{6}{13}i$

104. $\frac{-4i}{2 - 6i}$ $\frac{3}{5} - \frac{1}{5}i$

105. $6i(2 - 3i)$ $18 + 12i$

106. $-9i(4 - 6i)$ $-54 - 36i$

107. $\frac{4}{5i^{35}}$ $0 + \frac{4}{5}i$

108. $\frac{3}{2i^{17}}$ $0 - \frac{3}{2}i$

109. $(2 + i\sqrt{2})(3 - i\sqrt{2})$ $8 + \sqrt{2}i$ or $8 + i\sqrt{2}$

110. $(5 + i\sqrt{3})(2 - i\sqrt{3})$ $13 - 3\sqrt{3}i$ or $13 - 3i\sqrt{3}$

111. $\frac{5 + 9i}{1 - i}$ $-2 + 7i$

112. $\frac{5 - i}{3 + 2i}$ $1 - i$

113. $(4 - 8i)^2$ $-48 - 64i$

114. $(7 - 3i)^2$ $40 - 42i$

115. $\frac{\sqrt{5} - i\sqrt{3}}{\sqrt{5} + i\sqrt{3}}$ $\frac{1}{4} - \frac{\sqrt{15}}{4}i$

116. $\frac{\sqrt{3} + i\sqrt{2}}{\sqrt{3} - i\sqrt{2}}$ $\frac{1}{5} + \frac{2\sqrt{6}}{5}i$

Look Alikes . . .

117. a. $\sqrt{-8}$ $2i\sqrt{2}$ b. $\sqrt[3]{-8}$ -2

118. a. $(3 + i) + (2 + 4i)$ $5 + 5i$ b. $(3 + i)(2 + 4i)$ $2 + 14i$

119. a. $(2i)^2$ $-4 + 0i$ b. $(2 + i)^2$ $3 + 4i$

120. a. $\sqrt{-9}\sqrt{-16}$ $-12 + 0i$ b. $\sqrt{9}\sqrt{16}$ $12 + 0i$

APPLICATIONS

121. **Fractals.** Complex numbers are fundamental in the creation of the intricate geometric shape shown below, called a *fractal.* The process of creating this image is based on the following sequence of steps, which begins by picking any complex number, which we will call z.

1. Square z, and then add that result to z.
2. Square the result from step 1, and then add it to z.
3. Square the result from step 2, and then add it to z.

If we begin with the complex number i, what is the result after performing steps 1, 2, and 3? $-1 + i$

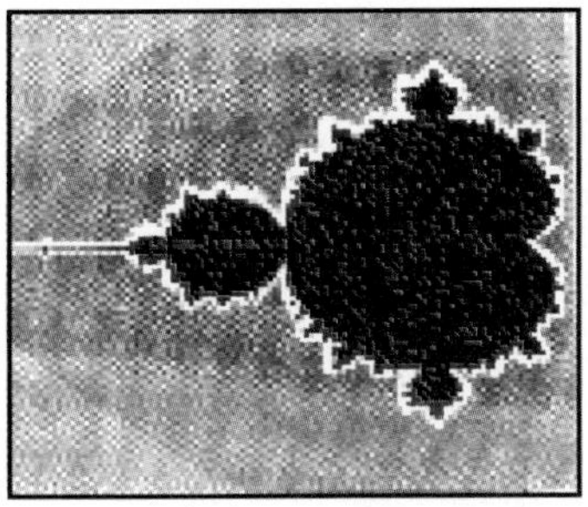

122. **Electronics.** The impedance Z in an AC (alternating current) circuit is a measure of how much the circuit impedes (hinders) the flow of current through it. The impedance is related to the voltage V and the current I by the formula $V = IZ$. If a circuit has a current of $(0.5 + 2.0i)$ amps and an impedance of $(0.4 - 3.0i)$ ohms, find the voltage. $6.2 - 0.7i$

WRITING

123. What is an imaginary number? What is a complex number?

124. The method used to divide complex numbers is similar to the method used to divide radical expressions. Explain why. Give an example.

125. Explain the error. Then find the correct result.

a. Add: $\sqrt{-16} + \sqrt{-9} = \sqrt{-25}$

b. Multiply: $\sqrt{-2}\sqrt{-3} = \sqrt{-2(-3)} = \sqrt{6}$

126. Determine whether the pair of complex numbers are equal. Explain your reasoning.

a. $4 - \frac{2}{5}i, \frac{8}{2} - 0.4i$ b. $0.25 + 0.7i, \frac{1}{4} + \frac{7}{10}i$

REVIEW

127. **Wind Speeds.** A plane that can fly 200 mph in still air makes a 330-mile flight with a tail wind and returns, flying into the same wind. Find the speed of the wind if the total flying time is $3\frac{1}{3}$ hours. 20 mph

128. **Finding Rates.** A student drove a distance of 135 miles at an average speed of 50 mph. How much faster would she have to drive on the return trip to save 30 minutes of driving time? About 11.4 mph faster

CHALLENGE PROBLEMS

129. Simplify: $(i^{349})^{-i^{456}}$ $-i$

130. Simplify $(2 + 3i)^{-2}$ and write the result in the form $a + bi$. $-\frac{5}{169} - \frac{12}{169}i$

9 Summary & Review

SECTION 9.1 Radical Expressions and Radical Functions

DEFINITIONS AND CONCEPTS	EXAMPLES
The number b is a **square root of a** if $b^2 = a$. Every positive real number has two square roots.	7 is a square root of 49 because $7^2 = 49$. -7 is also a square root of 49 because $(-7)^2 = 49$.
A **radical symbol** $\sqrt{\ }$ represents the **positive** or **principal square root** of a number. For any real number x, $\sqrt{x^2} = \|x\|$ The symbol $-\sqrt{\ }$ represents the **negative square root** of a number. Review the list of perfect squares on page 729.	$\sqrt{25} = 5$ because $5^2 = 25$. $\sqrt{36x^2} = \|6x\| = 6\|x\|$ because $(6x)^2 = 36x^2$. Since x could be negative, absolute value symbols ensure that the result is not negative. $\sqrt{\frac{r^8}{100}} = \frac{r^4}{10}$ because $\left(\frac{r^4}{10}\right)^2 = \frac{r^8}{100}$. Since $\frac{r^4}{10}$ cannot be negative, no absolute value symbols are needed. $-\sqrt{81} = -9$ and $\sqrt{-81}$ is not a real number.
A function of the form $f(x) = \sqrt{x}$ is called a **square root function.** Review its graph on page 732.	Let $f(x) = \sqrt{x - 2}$. Find $f(38)$. $f(38) = \sqrt{38 - 2} = \sqrt{36} = 6$
The **domain of a square root function** is the set of all real numbers for which the radicand is nonnegative. To find the domain, set the radicand greater than or equal to 0 and solve for the variable.	Find the domain of $g(x) = \sqrt{2x - 4}$. Since $\sqrt{2x - 4}$ is not a real number when $2x - 4$ is negative, we must require that $2x - 4 \geq 0$ Because we cannot find the square root of a negative number $2x \geq 4$ Solve for x. $x \geq 2$ The domain of g is $[2, \infty)$.
The **cube root** of x is denoted as $\sqrt[3]{x}$ and is defined as $\sqrt[3]{x} = y$ if $y^3 = x$ Review the list of perfect cubes on page 735.	$\sqrt[3]{8} = 2$ because $2^3 = 8$. $\sqrt[3]{-64} = -4$ because $(-4)^3 = -64$. $\sqrt[3]{-27a^6} = -3a^2$ because $(-3a^2)^3 = -27a^6$.
A function of the form $f(x) = \sqrt[3]{x}$ is called a **cube root function.** Review its graph on page 736.	Let $f(x) = \sqrt[3]{x + 4}$. Find $f(23)$. $f(23) = \sqrt[3]{23 + 4} = \sqrt[3]{27} = 3$
The ***n*th root of *x*** is denoted as $\sqrt[n]{x}$. Index → $\sqrt[4]{16a^8}$ ← Radicand If x is a real number and $n > 1$, then: If n is an odd natural number, $\sqrt[n]{x^n} = x$. If n is an even natural number, $\sqrt[n]{x^n} = \|x\|$.	$\sqrt[4]{81x^4} = \|3x\| = 3\|x\|$ because $(3x)^4 = 81x^4$. $\sqrt[4]{-16}$ Not a real number $\sqrt[5]{32a^{10}} = 2a^2$ because $(2a^2)^5 = 32a^{10}$. No absolute value symbols are needed. $\sqrt[6]{(m-4)^6} = \|m - 4\|$ because $(m-4)^6 = (m-4)^6$.

REVIEW EXERCISES

1. The graph of a square root function f is shown here. Find each of the following:

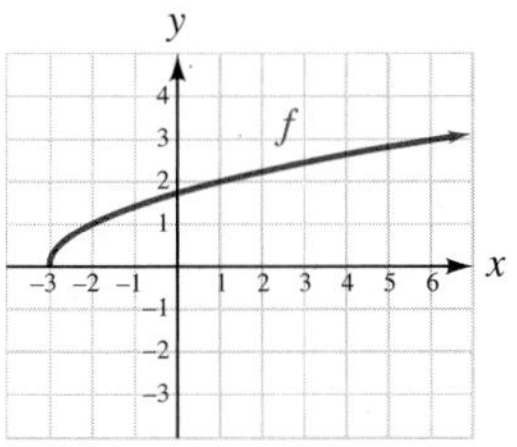

   a. $f(1)$ 2 b. $f(-3)$ 0
   c. The value of x for which $f(x) = 3$ 6
   d. The domain and range of f
   D: $[-3, \infty)$, R: $[0, \infty)$

2. a. Simplify $\sqrt{100a^2}$. Assume that the variable a is unrestricted. $10|a|$
   b. Simplify $\sqrt{100a^2}$. Assume that the variable a is a positive real number. $10a$

Simplify each expression. Assume that all variables are unrestricted and use absolute value symbols when necessary.

3. $\sqrt{49}$ 7
4. $-\sqrt{121}$ -11
5. $\sqrt{\frac{225}{49}}$ $\frac{15}{7}$
6. $\sqrt{-4}$ Not a real number
7. $\sqrt{100a^{12}}$ $10a^6$
8. $\sqrt{25x^2}$ $5|x|$
9. $\sqrt{x^8}$ x^4
10. $\sqrt{x^2 + 4x + 4}$ $|x + 2|$
11. $\sqrt[3]{-27}$ -3
12. $-\sqrt[3]{216}$ -6
13. $\sqrt[3]{64x^6y^3}$ $4x^2y$
14. $\sqrt[3]{\frac{x^9}{125}}$ $\frac{x^3}{5}$
15. $\sqrt[6]{64}$ 2
16. $\sqrt[5]{-32}$ -2
17. $\sqrt[4]{256x^8y^4}$ $4x^2|y|$
18. $\sqrt[5]{(x + 1)^5}$ $x + 1$
19. $-\sqrt[4]{\frac{1}{16}}$ $-\frac{1}{2}$
20. $\sqrt[4]{-81}$ Not a real number
21. $\sqrt[6]{-1}$ Not a real number
22. $\sqrt[5]{0}$ 0
23. Find the domain of $f(x) = \sqrt{3x + 15}$. $[-5, \infty)$
24. **Geometry.** The side of a square with area A square feet is given by the function $s(A) = \sqrt{A}$. Find the length of one side of a square that has an area of 169 ft^2. 13 ft
25. **Cubes.** The total surface area of a cube is related to its volume V by the function $A(V) = 6\sqrt[3]{V^2}$. Find the surface area of a cube with a volume of 8 cm^3. 24 cm^2
26. Let $g(x) = \sqrt[3]{x^2 + 9}$. Use a calculator fo find $g(-1.9)$ to four decimal places. 2.3276

Graph each function. Find the domain and range. See AIE Appendix 3.

27. $f(x) = \sqrt{x}$
    D: $[0, \infty)$; R: $[0, \infty)$
28. $f(x) = \sqrt[3]{x}$
    D: $(-\infty, \infty)$; R: $(-\infty, \infty)$
29. $f(x) = \sqrt{x + 2}$
    D: $[-2, \infty)$; R: $[0, \infty)$
30. $f(x) = -\sqrt[3]{x} + 3$
    D: $(-\infty, \infty)$; R: $(-\infty, \infty)$

SECTION 9.2 Rational Exponents

DEFINITIONS AND CONCEPTS	EXAMPLES
To simplify exponential expressions involving **rational (fractional) exponents,** use the following rules to write the expressions in an equivalent radical form. $x^{1/n} = \sqrt[n]{x}$ $x^{m/n} = \left(\sqrt[n]{x}\right)^m = \sqrt[n]{x^m}$	Simplify. All variables represent positive real numbers. $25^{1/2} = \sqrt{25} = 5$ $\left(\frac{256}{d^4}\right)^{1/4} = \sqrt[4]{\frac{256}{d^4}} = \frac{4}{d}$ $8^{2/3} = \left(\sqrt[3]{8}\right)^2 = (2)^2 = 4$ $(t^{10})^{6/5} = \left(\sqrt[5]{t^{10}}\right)^6 = (t^2)^6 = t^{12}$
To be consistent with the definition of negative integer exponents, we define $x^{-m/n}$ as follows. $x^{-m/n} = \frac{1}{x^{m/n}}$ $\frac{1}{x^{-m/n}} = x^{m/n}$	Simplify. All variables represent positive real numbers. $(125)^{-2/3} = \frac{1}{(125)^{2/3}} = \frac{1}{\left(\sqrt[3]{125}\right)^2} = \frac{1}{25}$ $\frac{1}{(-32x^5)^{-3/5}} = (-32x^5)^{3/5} = \left(\sqrt[5]{-32x^5}\right)^3 = -8x^3$
The **rules for exponents** can be used to simplify expressions with rational (fractional) exponents.	Simplify: $\frac{p^{5/3}p^{8/3}}{p^4} = p^{5/3+8/3-4} = p^{5/3+8/3-12/3} = p^{1/3}$
We can write certain radical expressions as an equivalent exponential expression and use rules for exponents to simplify it. Then we can change that result back into a radical.	Simplify: $\sqrt[4]{9} = \sqrt[4]{3^2} = 3^{2/4} = 3^{1/2} = \sqrt{3}$

REVIEW EXERCISES

Write each expression in radical form.

31. $t^{1/2}$ $\sqrt{t}$

32. $(5xy^3)^{1/4}$ $\sqrt[4]{5xy^3}$

Simplify each expression, if possible. Assume that all variables represent positive real numbers.

33. $25^{1/2}$ 5

34. $-36^{1/2}$ -6

35. $(-36)^{1/2}$ Not a real number

36. $1^{1/5}$ 1

37. $\left(\frac{9}{x^2}\right)^{1/2}$ $\frac{3}{x}$

38. $(-8)^{1/3}$ -2

39. $625^{1/4}$ 5

40. $(81c^4d^4)^{1/4}$ $3cd$

41. $9^{3/2}$ 27

42. $8^{-2/3}$ $\frac{1}{4}$

43. $-49^{5/2}$ $-16{,}807$

44. $\frac{1}{100^{-1/2}}$ 10

45. $\left(\frac{4}{9}\right)^{-3/2}$ $\frac{27}{8}$

46. $\frac{1}{25^{5/2}}$ $\frac{1}{3{,}125}$

47. $(25x^2y^4)^{3/2}$ $125x^3y^6$

48. $(8u^6v^3)^{-2/3}$ $\frac{1}{4u^4v^2}$

Perform the operations. Write answers without negative exponents. Assume that all variables represent positive real numbers.

49. $5^{1/4} \cdot 5^{1/2}$ $5^{3/4}$

50. $a^{3/7}a^{-2/7}$ $a^{1/7}$

51. $(k^{4/5})^{10}$ k^8

52. $\frac{3^{5/6}3^{1/3}}{3^{1/2}}$ $3^{2/3}$

Perform the multiplications. Assume all variables represent positive real numbers.

53. $u^{1/2}(u^{1/2} - u^{-1/2})$ $u - 1$

54. $v^{2/3}(v^{1/3} + v^{4/3})$ $v + v^2$

Use rational exponents to simplify each radical. All variables represent positive real numbers.

55. $\sqrt[4]{a^2}$ $\sqrt{a}$

56. $\sqrt[3]{\sqrt{c}}$ $\sqrt[6]{c}$

57. **Visibility.** The distance d in miles a person in an airplane can see to the horizon on a clear day is given by the formula $d = 1.22a^{1/2}$, where a is the altitude of the plane in feet. Find d. 183 mi

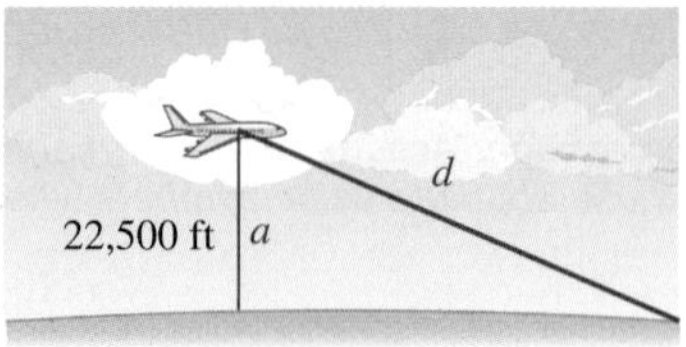

58. Substitute the x- and y-coordinates of each point labeled in the graph into the equation

$$x^{2/3} + y^{2/3} = 32$$

Show that each one satisfies the equation. Two true statements result: $32 = 32$.

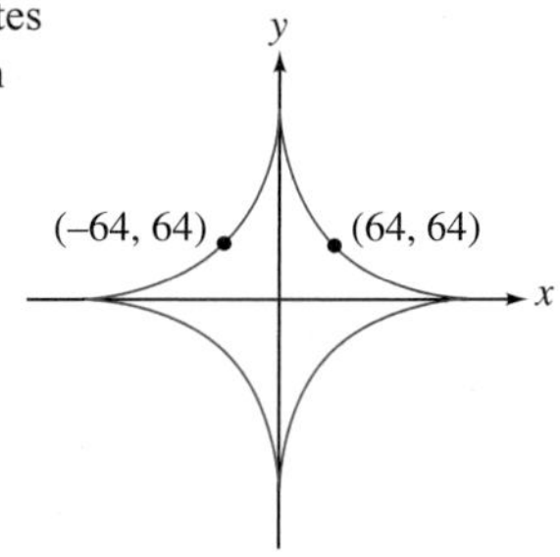

SECTION 9.3 Simplifying and Combining Radical Expressions

DEFINITIONS AND CONCEPTS	EXAMPLES
Product rule for radicals: $\sqrt[n]{ab} = \sqrt[n]{a}\sqrt[n]{b}$ The product rule for radicals can be used to simplify radical expressions.	Simplify: $\sqrt{98} = \sqrt{49 \cdot 2}$ Write 98 as the product of its greatest perfect-square factor and one other factor. $= \sqrt{49}\sqrt{2}$ The square root of a product is equal to the product of the square roots. $= 7\sqrt{2}$ Evaluate $\sqrt{49}$.
Simplified form of a radical: 1. Except for 1, the radicand has no perfect-square factors. 2. No fraction appears in the radicand. 3. No radical appears in the denominator. Review the lists of perfect squares, cubes, fourth powers, and fifth powers on pages 755 and 756.	Simplify: $\sqrt[3]{16x^4} = \sqrt[3]{8x^3 \cdot 2x}$ Write $16x^4$ as the product of its greatest perfect-cube factor and one other factor. $= \sqrt[3]{8x^3}\sqrt[3]{2x}$ The cube root of a product is equal to the product of the cube roots. $= 2x\sqrt[3]{2x}$ Simplify $\sqrt[3]{8x^3}$.

Quotient rule for radicals: $\sqrt[n]{\frac{a}{b}} = \frac{\sqrt[n]{a}}{\sqrt[n]{b}}$	Simplify: $\sqrt{\frac{10}{25x^4}} = \frac{\sqrt{10}}{\sqrt{25x^4}} = \frac{\sqrt{10}}{5x^2}$ Simplify: $\sqrt[3]{\frac{16y^3}{125a^3}} = \frac{\sqrt[3]{16y^3}}{\sqrt[3]{125a^3}} = \frac{\sqrt[3]{8y^3}\sqrt[3]{2}}{5a} = \frac{2y\sqrt[3]{2}}{5a}$
Radical expressions with the same index and radicand are called **like radicals.** Like radicals can be combined by addition and subtraction. To **combine like radicals,** we use the distributive property in reverse.	Add: $3\sqrt{6} + 5\sqrt{6} = (3 + 5)\sqrt{6} = 8\sqrt{6}$ Subtract: $8\sqrt[4]{2y} - 9\sqrt[4]{2y} = (8 - 9)\sqrt[4]{2y} = -\sqrt[4]{2y}$
If a sum or difference involves unlike radicals, make sure that each one is written in simplified form. After doing so, like radicals may result that can be combined.	Simplify: $\sqrt[3]{54a^4} - \sqrt[3]{16a^4} = \sqrt[3]{27a^3 \cdot 2a} - \sqrt[3]{8a^3 \cdot 2a}$ $= \sqrt[3]{27a^3}\sqrt[3]{2a} - \sqrt[3]{8a^3}\sqrt[3]{2a}$ $= 3a\sqrt[3]{2a} - 2a\sqrt[3]{2a}$ $= (3a - 2a)\sqrt[3]{2a}$ $= a\sqrt[3]{2a}$

REVIEW EXERCISES

Simplify each expression. All variables represent positive real numbers.

59. $\sqrt{80}$ $4\sqrt{5}$
60. $\sqrt[3]{54}$ $3\sqrt[3]{2}$
61. $\sqrt[4]{160}$ $2\sqrt[4]{10}$
62. $\sqrt[5]{-96}$ $-2\sqrt[5]{3}$
63. $\sqrt{8x^5}$ $2x^2\sqrt{2x}$
64. $\sqrt[4]{r^{17}}$ $r^4\sqrt[4]{r}$
65. $\sqrt[3]{-27j^7k}$ $-3j^2\sqrt[3]{jk}$
66. $\sqrt[3]{-16x^5y^4}$ $-2xy\sqrt[3]{2x^2y}$
67. $\sqrt{\frac{m}{144n^{12}}}$ $\frac{\sqrt{m}}{12n^6}$
68. $\sqrt{\frac{17xy}{64a^4}}$ $\frac{\sqrt{17xy}}{8a^2}$
69. $\frac{\sqrt[5]{64x^8}}{\sqrt[5]{2x^3}}$ $2x$
70. $\frac{\sqrt[5]{243x^{16}}}{\sqrt[5]{x}}$ $3x^3$

Simplify and combine like radicals. All variables represent positive real numbers.

71. $\sqrt{2} + 2\sqrt{2}$ $3\sqrt{2}$
72. $6\sqrt{20} - \sqrt{5}$ $11\sqrt{5}$
73. $2\sqrt[3]{3} - \sqrt[3]{24}$ 0
74. $-\sqrt[4]{32a^5} - 2\sqrt[4]{162a^5}$ $-8a\sqrt[4]{2a}$
75. $2x\sqrt{8} + 2\sqrt{200x^2} + \sqrt{50x^2}$ $29x\sqrt{2}$
76. $\sqrt[3]{54x^3} - 3\sqrt[3]{16x^3} + 4\sqrt[3]{128x^3}$ $13x\sqrt[3]{2}$
77. $2\sqrt[4]{32t^3} - 8\sqrt[4]{6t^3} + 5\sqrt[4]{2t^3}$ $9\sqrt[4]{2t^3} - 8\sqrt[4]{6t^3}$
78. $10\sqrt[4]{16x^9} - 8x^2\sqrt[4]{x} + 5\sqrt[4]{x^5}$ $12x^2\sqrt[4]{x} + 5x\sqrt[4]{x}$
79. Explain the error in each simplification.
    a. $2\sqrt{5x} + 3\sqrt{5x} = 5\sqrt{10x}$
    b. $30 + 30\sqrt[4]{2} = 60\sqrt[4]{2}$
    c. $7\sqrt[3]{y^2} - 5\sqrt[3]{y^2} = 2$
    d. $6\sqrt{11ab} - 3\sqrt{5ab} = 3\sqrt{6ab}$
80. **Sewing.** A corner of fabric is folded over to form a collar and stitched down as shown. From the dimensions given in the figure, determine the exact number of inches of stitching that must be made. Then give an approximation to one decimal place. (All measurements are in inches.) $(6\sqrt{2} + 2\sqrt{10})$ in., 14.8 in.

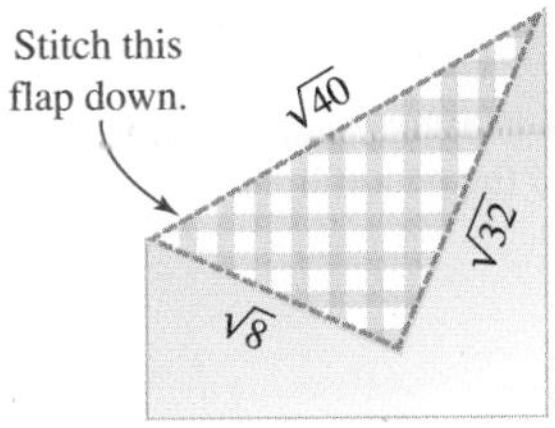

SECTION 9.4 Multiplying and Dividing Radical Expressions

DEFINITIONS AND CONCEPTS	EXAMPLES
We can use the product rule for radicals to **multiply radical expressions** that have the same index: $\sqrt[n]{a}\sqrt[n]{b} = \sqrt[n]{ab}$ provided $\sqrt[n]{a}$ and $\sqrt[n]{b}$ are real numbers.	Multiply and then simplify, if possible: $\sqrt{6}\sqrt{8} = \sqrt{6 \cdot 8} = \sqrt{48} = \sqrt{16 \cdot 3} = \sqrt{16}\sqrt{3} = 4\sqrt{3}$ $\sqrt[3]{9x^4}\sqrt[3]{3x^2} = \sqrt[3]{9x^4 \cdot 3x^2} = \sqrt[3]{27x^6} = 3x^2$
We can use the **distributive property** to multiply a radical expression with two or more terms by a radical expression with one term.	Multiply and then simplify, if possible: $2\sqrt{3}(4\sqrt{5} - 5\sqrt{2}) = 2\sqrt{3} \cdot 4\sqrt{5} - 2\sqrt{3} \cdot 5\sqrt{2}$ $= 2 \cdot 4\sqrt{3 \cdot 5} - 2 \cdot 5\sqrt{3 \cdot 2}$ $= 8\sqrt{15} - 10\sqrt{6}$
We can use the **FOIL method** to multiply a radical expression with two terms by another radical expression with two terms.	Multiply and then simplify, if possible: F O I L $(\sqrt[3]{x} - \sqrt[3]{3})(\sqrt[3]{x} + \sqrt[3]{9}) = \sqrt[3]{x}\sqrt[3]{x} + \sqrt[3]{x}\sqrt[3]{9} - \sqrt[3]{3}\sqrt[3]{x} - \sqrt[3]{3}\sqrt[3]{9}$ $= \sqrt[3]{x^2} + \sqrt[3]{9x} - \sqrt[3]{3x} - \sqrt[3]{27}$ $= \sqrt[3]{x^2} + \sqrt[3]{9x} - \sqrt[3]{3x} - 3$
If a radical appears in a denominator of a fraction, or if a radicand contains a fraction, we can write the radical in simplest form by **rationalizing the denominator.** To rationalize a denominator, we multiply the given expression by a carefully chosen form of 1.	Rationalize the denominator: $\sqrt{\frac{10}{3}} = \frac{\sqrt{10}}{\sqrt{3}} \cdot \frac{\sqrt{3}}{\sqrt{3}}$ $= \frac{\sqrt{30}}{3}$ $\sqrt[3]{\frac{5}{2p^2}} = \frac{\sqrt[3]{5}}{\sqrt[3]{2p^2}} \cdot \frac{\sqrt[3]{4p}}{\sqrt[3]{4p}}$ $= \frac{\sqrt[3]{20p}}{\sqrt[3]{8p^3}}$ $= \frac{\sqrt[3]{20p}}{2p}$
Radical expressions that involve the sum and difference of the same two terms are called **conjugates.**	Conjugates: $\sqrt{2x} + 3$ and $\sqrt{2x} - 3$
To **rationalize a two-termed denominator** of a fraction, multiply the numerator and the denominator by the **conjugate** of the denominator.	Rationalize the denominator: $\frac{\sqrt{x} - 2}{\sqrt{x} + 2} = \frac{\sqrt{x} - 2}{\sqrt{x} + 2} \cdot \frac{\sqrt{x} - 2}{\sqrt{x} - 2}$ $= \frac{\sqrt{x}\sqrt{x} - 2\sqrt{x} - 2\sqrt{x} + 4}{\sqrt{x}\sqrt{x} - 2\sqrt{x} + 2\sqrt{x} - 4}$ $= \frac{x - 4\sqrt{x} + 4}{x - 4}$

REVIEW EXERCISES

Simplify each expression. All variables represent positive real numbers.

81. $\sqrt{7}\sqrt{7}$
7

82. $(2\sqrt{5})(3\sqrt{2})$
$6\sqrt{10}$

83. $(-2\sqrt{8})^2$
32

84. $2\sqrt{6}\sqrt{15}$
$6\sqrt{10}$

85. $\sqrt{9x}\sqrt{x}$
$3x$

86. $(\sqrt[3]{x+1})^3$
$x + 1$

87. $-\sqrt[3]{2x^2}\sqrt[3]{4x^8}$
$-2x^3\sqrt[3]{x}$

88. $\sqrt[5]{9} \cdot \sqrt[5]{27}$
3

89. $3\sqrt{7t}(2\sqrt{7t} + 3\sqrt{3t^2})$
$42t + 9t\sqrt{21t}$

90. $-\sqrt[4]{4x^5y^{11}}\sqrt[4]{8x^9y^3}$
$-2x^3y^3\sqrt[4]{2x^2y^2}$

91. $(\sqrt{3b} + \sqrt{3})^2$
$3b + 6\sqrt{b} + 3$

92. $(\sqrt[3]{3p} - 2\sqrt[3]{2})(\sqrt[3]{3p} + \sqrt[3]{2})$
$\sqrt[3]{9p^2} - \sqrt[3]{6p} - 2\sqrt[3]{4}$

Rationalize each denominator. All variables represent positive real numbers.

93. $\dfrac{10}{\sqrt{3}}$ $\quad \dfrac{10\sqrt{3}}{3}$

94. $\sqrt{\dfrac{3}{5xy}}$ $\quad \dfrac{\sqrt{15xy}}{5xy}$

95. $\dfrac{\sqrt[3]{6u}}{\sqrt[3]{u^5}}$ $\quad \dfrac{\sqrt[3]{6u^2}}{u^2}$

96. $\dfrac{\sqrt[4]{a}}{\sqrt[4]{3b^2}}$ $\quad \dfrac{\sqrt[4]{27ab^2}}{3b}$

97. $\dfrac{2}{\sqrt{2} - 1}$
$2(\sqrt{2} + 1)$ or $2\sqrt{2} + 2$

98. $\dfrac{4\sqrt{x} - 2\sqrt{z}}{\sqrt{z} + 4\sqrt{x}}$
$\dfrac{12\sqrt{xz} - 16x - 2z}{z - 16x}$

99. Rationalize the numerator: $\dfrac{\sqrt{a} - \sqrt{b}}{\sqrt{a}}$ $\quad \dfrac{a - b}{a + \sqrt{ab}}$

100. **Volume.** The formula relating the radius r of a sphere and its volume V is $r = \sqrt[3]{\frac{3V}{4\pi}}$. Write the radical in simplest form.
$r = \dfrac{\sqrt[3]{6\pi^2 V}}{2\pi}$

SECTION 9.5 Solving Radical Equations

DEFINITIONS AND CONCEPTS	EXAMPLES
We can use the **power rule** to solve radical equations. If $x = y$, then $x^n = y^n$. To **solve equations containing radicals:** 1. Isolate one radical expression on one side of the equation. 2. Raise both sides of the equation to the power that is the same as the index. 3. If it still contains a radical, go back to step 1. If it does not contain a radical, solve the resulting equation. 4. Check the proposed solutions in the original equation to eliminate **extraneous** solutions.	Solve each equation. $\sqrt{2x-2} + 1 = x$ $\sqrt{2x-2} = x - 1$ $(\sqrt{2x-2})^2 = (x-1)^2$ $2x - 2 = x^2 - 2x + 1$ $0 = x^2 - 4x + 3$ $0 = (x-3)(x-1)$ $x - 3 = 0$ or $x - 1 = 0$ $x = 3$ | $x = 1$ The solutions are 3 and 1. Verify that each satisfies the original equation. $\sqrt[3]{x+2} = 3$ $(\sqrt[3]{x+2})^3 = 3^3$ $x + 2 = 27$ $x = 25$ The solution is 25. Verify that it satisfies the original equation.
When **more than one radical** appears in an equation, we often must use the power rule more than once to solve the equation. The **special-product rules** are often helpful when squaring expressions containing two terms, such as $(5 - \sqrt{x})^2$. Warning: The result is not $25 + x$.	Solve: $\sqrt{x} + \sqrt{x+5} = 5$ $\sqrt{x+5} = 5 - \sqrt{x}$ — To isolate one radical, subtract $\sqrt{x}$ from both sides. $(\sqrt{x+5})^2 = (5 - \sqrt{x})^2$ — To eliminate one radical, square both sides. $x + 5 = 25 - 10\sqrt{x} + x$ — Perform the operations on each side. $-20 = -10\sqrt{x}$ — To isolate the radical term, subtract 25 and x from both sides. $2 = \sqrt{x}$ — To isolate the radical, divide both sides by −10. $(2)^2 = (\sqrt{x})^2$ — To eliminate the radical, square both sides again. $4 = x$ The solution is 4. Verify that it satisfies the original equation.

REVIEW EXERCISES

Solve each equation. Write all proposed solutions. Cross out those that are extraneous.

101. $\sqrt{7x - 10} - 1 = 11$
22

102. $u = \sqrt{25u - 144}$
16, 9

103. $2\sqrt{y - 3} = \sqrt{2y + 1}$
$\frac{13}{2}$

104. $\sqrt{z + 1} + \sqrt{z} = 2$
$\frac{9}{16}$

105. $\sqrt[3]{x^3 + 56} - 2 = x$
2, −4

106. $a = \sqrt{a^2 + 5a - 35}$
7

107. $(x + 2)^{1/2} - (4 - x)^{1/2} = 0$
1

108. $\sqrt{b^2 + b} = \sqrt{3 - b^2}$
$-\frac{3}{2}, 1$

109. $\sqrt[4]{8x - 8} + 2 = 0$ ~~3~~, no solution

110. $\sqrt{2m + 4} - \sqrt{m + 3} = 1$ 6, ~~−2~~

111. Let $f(x) = \sqrt{5x + 1}$ and $g(x) = x + 1$. For what values of x is $f(x) = g(x)$? 0, 3

112. Let $f(x) = \sqrt{2x^2 - 7x}$. For what value(s) of x is $f(x) = 2$? $-\frac{1}{2}, 4$

Solve each equation for the specified variable.

113. $r = \sqrt{\frac{A}{P}} - 1$ for P
$P = \frac{A}{(r + 1)^2}$

114. $h = \sqrt[3]{\frac{12I}{b}}$ for I
$I = \frac{h^3 b}{12}$

SECTION 9.6 Geometric Applications of Radicals

DEFINITIONS AND CONCEPTS | **EXAMPLES**

The Pythagorean theorem:

If a and b are the lengths of the **legs** of a right triangle and c is the length of the **hypotenuse,** then $a^2 + b^2 = c^2$.

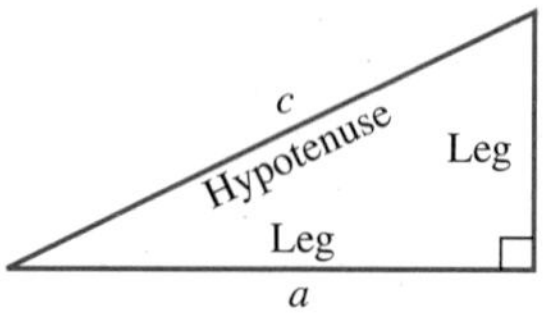

Find the length of the third side of the right triangle.

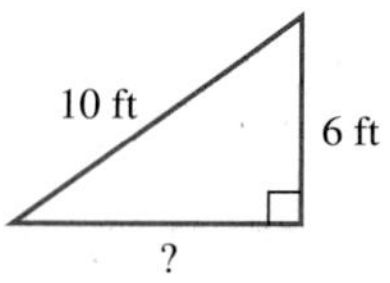

$a^2 + b^2 = c^2$ This is the Pythagorean equation.

$6^2 + b^2 = 10^2$ Substitute 6 for a and 10 for c.

$36 + b^2 = 100$

$b^2 = 64$ To isolate b^2, subtract 36 from both sides.

$b = \sqrt{64}$ Since b must be positive, find the positive square root of 64.

$b = 8$

The length of the third side of the triangle is 8 ft.

In an **isosceles right triangle,** the length of the hypotenuse is $\sqrt{2}$ times the length of one leg.

If the length of one leg of an isosceles right triangle is 7 feet, the length of the hypotenuse is $7\sqrt{2}$ feet.

Isosceles right triangle

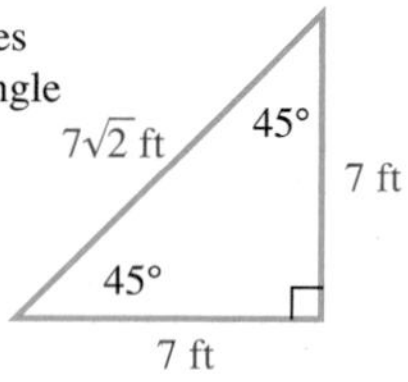

The hypotenuse of a **30°–60°–90° triangle** is twice as long as the shorter leg (the leg opposite the 30° angle.) The length of the longer leg (the leg opposite the 60° angle) is $\sqrt{3}$ times the length of the shorter leg.

If the shorter leg of a 30°–60°–90° triangle is 9 inches long:

- The length of the hypotenuse is $2 \cdot 9 = 18$ inches.
- The length of the longer leg is $\sqrt{3} \cdot 9 = 9\sqrt{3}$ inches.

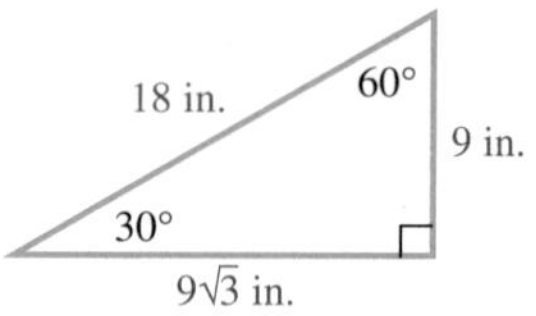

30°–60°–90° Triangle

The distance d between two points with coordinates (x_1, y_1) and (x_2, y_2) is given by **the distance formula:**

$$d = \sqrt{(x_2 - x_1)^2 + (y_2 - y_1)^2}$$

The distance between points $(-2, 3)$ and $(1, 7)$ is:

$$\begin{aligned} d &= \sqrt{(x_2 - x_1)^2 + (y_2 - y_1)^2} \\ &= \sqrt{[1 - (-2)]^2 + (7 - 3)^2} \\ &= \sqrt{3^2 + 4^2} \\ &= \sqrt{9 + 16} \\ &= \sqrt{25} \\ &= 5 \end{aligned}$$

The **midpoint** of a line segment with endpoints (x_1, y_1) and (x_2, y_2) is the point with coordinates

$$\left(\frac{x_1 + x_2}{2}, \frac{y_1 + y_2}{2}\right)$$

Find the midpoint of the segment joining $(-3, 7)$ and $(5, -8)$.

We let $(x_1, y_1) = (-3, 7)$ and $(x_2, y_2) = (5, -8)$ and substitute the coordinates into the midpoint formula.

$$\left(\frac{x_1 + x_2}{2}, \frac{y_1 + y_2}{2}\right) = \left(\frac{-3 + 5}{2}, \frac{7 + (-8)}{2}\right)$$

$$= \left(1, -\frac{1}{2}\right) \quad \text{This is the midpoint.}$$

REVIEW EXERCISES

115. Carpentry. The gable end of the roof shown below is divided in half by a vertical brace, 8 feet in height. Find the length of the roof line. 17 ft

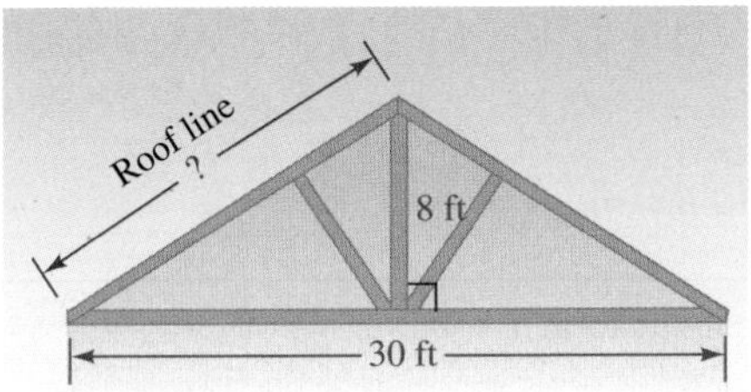

116. Sailing. A technique called *tacking* allows a sailboat to make progress into the wind. A sailboat follows the course shown below. Find d, the distance the boat advances into the wind after tacking. 88 yd

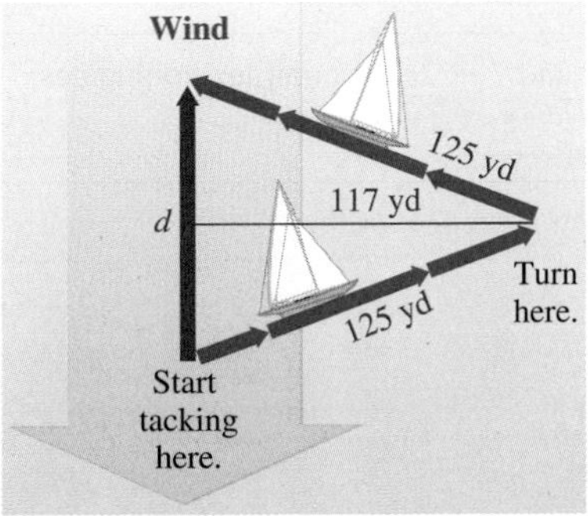

For problems 117–122, give the exact answer and then an approximation to two decimal places, when appropriate.

117. Find the length of the hypotenuse of an isosceles right triangle if the length of one leg is 7 meters. $7\sqrt{2}$ m ≈ 9.90 m

118. The length of the hypotenuse of an isosceles right triangle is 15 yards. Find the length of one leg of the triangle. $\frac{15\sqrt{2}}{2}$ yd ≈ 10.61 yd

119. The length of the hypotenuse of a 30°–60°–90° triangle is 12 centimeters. Find the length of each leg. Shorter leg: 6 cm; longer leg: $6\sqrt{3}$ cm ≈ 10.39 cm

120. In a 30°–60°–90° triangle, the length of the longer leg is 60 feet. Find the length of the hypotenuse and the length of the shorter leg. $40\sqrt{3}$ ft ≈ 69.28 ft; $20\sqrt{3}$ ft ≈ 34.64 ft

121. Find x and y.

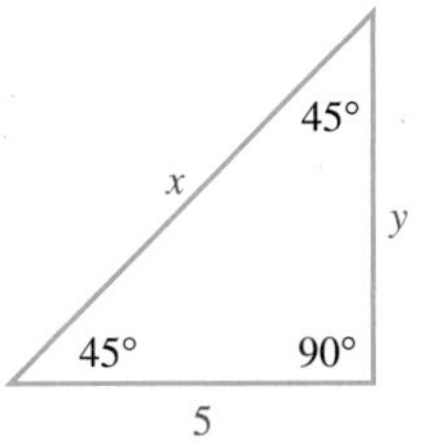

$x = 5\sqrt{2} \approx 7.07, y = 5$

122. Find x and y.

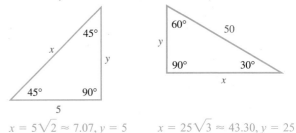

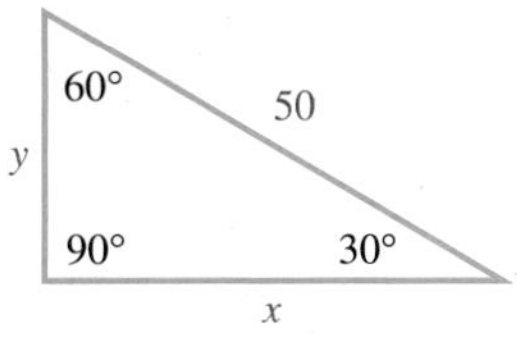

$x = 25\sqrt{3} \approx 43.30, y = 25$

Find the distance between the points.

123. $(1, 3)$ and $(6, -9)$ 13

124. $(-4, 6)$ and $(-2, 8)$ $2\sqrt{2}$

125. Find the midpoint of the segment joining $(8, -2)$ and $(6, -4)$. $(7, -3)$

126. If $(6, 1)$ is the midpoint of segment PQ and the coordinates of P are $(10, 4)$, find the coordinates of Q. $(2, -2)$

SECTION 9.7 Complex Numbers

DEFINITIONS AND CONCEPTS	EXAMPLES
The **imaginary number** i is defined as $i = \sqrt{-1}$. From the definition, it follows that $i^2 = -1$.	Write each expression in terms of i: $\sqrt{-81} = \sqrt{-1 \cdot 81} = \sqrt{-1}\sqrt{81} = i \cdot 9 = 9i$; $\sqrt{-24} = \sqrt{-1 \cdot 24} = \sqrt{-1}\sqrt{24} = i\sqrt{4}\sqrt{6} = 2i\sqrt{6}$ or $2\sqrt{6}i$

A **complex number** is any number that can be written in the form $a + bi$, where a and b are real numbers and $i = \sqrt{-1}$. We call a the **real part** and b the **imaginary part.**	Complex numbers: $5 + 3i$ — 5 is the real part a and 3 is the imaginary part b. $16 = 16 + 0i$ — 16 is the real part a and 0 is the imaginary part b. $9i = 0 + 9i$ — 0 is the real part a and 9 is the imaginary part b.
Adding and subtracting complex numbers is similar to adding and subtracting polynomials. To **add two complex numbers,** add their real parts and add their imaginary parts. To **subtract two complex numbers,** add the opposite of the complex number being subtracted.	Add. Write the answer in the form $a + bi$. $(7 - 5i) + (3 + 9i) = (7 + 3) + (-5 + 9)i$ — Add the real parts. Add the imaginary parts. $= 10 + 4i$ Subtract. Write the answer in the form $a + bi$. $(8 - i) - (-1 + 6i) = (8 - i) + (1 - 6i)$ — Add the opposite of $-1 + 6i$. $= (8 + 1) + [-1 + (-6)]i$ — Add the real parts. Add the imaginary parts. $= 9 - 7i$
Multiplying complex numbers is similar to multiplying polynomials.	Multiply. Write the answers in the form $a + bi$. $3i(6 - 4i) = 18i - 12i^2$ $= 18i - 12(-1)$ $= 18i + 12$ $= 12 + 18i$ $(4 + 7i)(2 - i) = 8 - 4i + 14i - 7i^2$ $= 8 + 10i - 7(-1)$ $= 8 + 10i + 7$ $= 15 + 10i$
The complex numbers $a + bi$ and $a - bi$ are called **complex conjugates.**	The complex numbers $7 - 2i$ and $7 + 2i$ are complex conjugates.
To **divide complex numbers,** multiply the numerator and denominator by the complex conjugate of the denominator. The process is similar to rationalizing denominators.	Divide. Write the answers in the form $a + bi$. $\frac{3}{1+i} \cdot \frac{1-i}{1-i} = \frac{3(1-i)}{1 - i + i - i^2}$ $= \frac{3(1-i)}{1-(-1)}$ $= \frac{3-3i}{2}$ $= \frac{3}{2} - \frac{3}{2}i$ $\frac{6+i}{2-i} \cdot \frac{2+i}{2+i} = \frac{12 + 6i + 2i + i^2}{4 + 2i - 2i - i^2}$ $= \frac{12 + 8i + (-1)}{4 - (-1)}$ $= \frac{11 + 8i}{5}$ $= \frac{11}{5} + \frac{8}{5}i$
The **powers of i** cycle through four possible outcomes: i, -1, $-i$, and 1. $i^1 = i = i^5 = i^9 = \ldots$ R = 1 $i^2 = -1 = i^6 = i^{10} = \ldots$ R = 2 $i^3 = -i = i^7 = i^{11} = \ldots$ R = 3 $i^4 = 1 = i^8 = i^{12} = \ldots$ R = 0	Simplify: i^{66} We divide 66 by 4 to get a remainder of **2**. The remainder determines the power to which i is raised in the simplified form. Thus, $i^{66} = i^2 = -1$. $66 \div 4 = 16$ R2: $4\overline{)66}$, -4, 26, -24, **2**

REVIEW EXERCISES

Write each expression in terms of i.

127. $\sqrt{-25}$ $5i$

128. $\sqrt{-18}$ $3i\sqrt{2}$

129. $-\sqrt{-6}$ $-i\sqrt{6}$

130. $\sqrt{-\frac{9}{64}}$ $\frac{3}{8}i$

131. Complete the diagram.

Complex numbers	
Real numbers	Imaginary numbers

132. Determine whether each statement is true or false.

a. Every real number is a complex number. True

b. $3 - 4i$ is a complex number. True

c. $\sqrt{-4}$ is a real number. False

d. i is a real number. False

Give the complex conjugate of each number.

133. a. $3 + 6i$ $3 - 6i$

b. $19i$ $0 - 19i$

134. a. $-1 - 7i$ $-1 + 7i$

b. $-i$ $0 + i$

Perform the operations. Write all answers in the form a + bi.

135. $(3 + 4i) + (5 - 6i)$ $8 - 2i$

136. $\left(7 - \sqrt{-9}\right) - \left(4 + \sqrt{-4}\right)$ $3 - 5i$

137. $3i(2 - i)$ $3 + 6i$

138. $(2 - 7i)(-3 + 4i)$ $22 + 29i$

139. $\sqrt{-3} \cdot \sqrt{-9}$ $-3\sqrt{3} + 0i$

140. $(9i)^2$ $-81 + 0i$

141. $\frac{5 + 14i}{2 + 3i}$ $4 + i$

142. $\frac{3}{11i}$ $0 - \frac{3}{11}i$

Simplify each expression.

143. i^{42} -1

144. i^{97} i

Teaching Tip: Because this Chapter Test is a comprehensive collection of problems that covers all of the topics discussed in Chapter 9, it is lengthy. If your students have time to complete it, that would be optimal. If, because of time constraints, they are unable to do so, assign an appropriate subset of problems that reflects the types of problems that the students will see on your exam.

9 Chapter Test

1. Fill in the blanks.

a. The symbol $\sqrt{\ }$ is called a radical symbol.

b. The imaginary number i is defined as $i = \sqrt{-1}$.

c. Squaring both sides of an equation can introduce extraneous solutions.

d. An isosceles right triangle is a right triangle with two legs of equal length.

e. To rationalize the denominator of $\frac{4}{\sqrt{5}}$, we multiply the fraction by $\frac{\sqrt{5}}{\sqrt{5}}$.

f. A complex number is any number that can be written in the form $a + bi$, where a and b are real numbers and $i = \sqrt{-1}$.

2. a. State the product rule for radicals.
If $\sqrt[n]{a}$ and $\sqrt[n]{b}$ are real numbers, $\sqrt[n]{ab} = \sqrt[n]{a}\sqrt[n]{b}$.

b. State the quotient rule for radicals.
If $\sqrt[n]{a}$ and $\sqrt[n]{b}$ are real numbers, then $\sqrt[n]{\frac{a}{b}} = \frac{\sqrt[n]{a}}{\sqrt[n]{b}}$, $(b \neq 0)$.

c. Explain why $\sqrt[4]{-16}$ is not a real number. No real number raised to the fourth power is -16.

3. Graph $f(x) = \sqrt{x - 1}$. Find the domain and range of the function. D: $[1, \infty)$; R: $[0, \infty)$; See AIE Appendix 3.

4. Diving. Refer to the illustration in the next column. The velocity v of an object in feet per second after it has fallen a distance of d feet is approximated by the function $v(d) = \sqrt{64.4d}$. Olympic diving platforms are 10 meters tall (approximately 32.8 feet). Estimate the velocity at which a diver hits the water from this height. Round to the nearest foot per second. 46 ft/sec

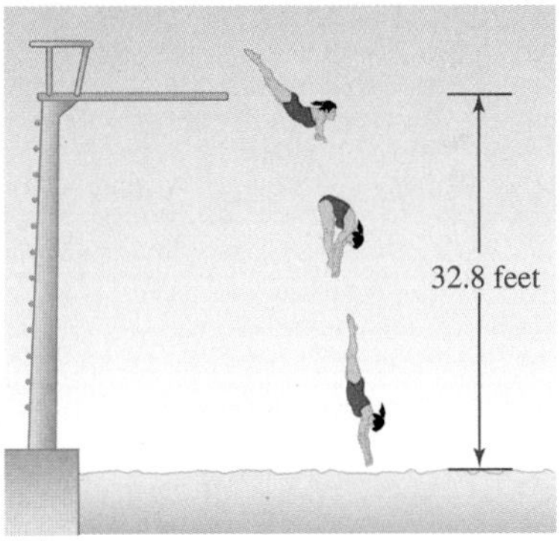

5. Use the graph of function f below to find each of the following.

a. $f(-1)$ -1

b. $f(8)$ 2

c. The value of x for which $f(x) = 1$ 1

d. The domain and range of f D: $(-\infty, \infty)$; R: $(-\infty, \infty)$

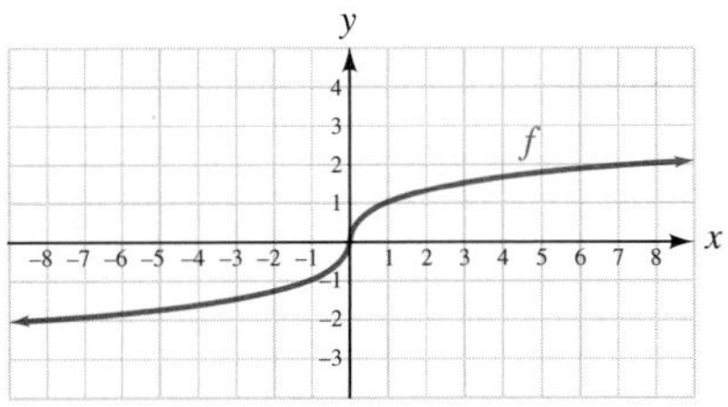

6. Find the domain of $f(x) = \sqrt{10x + 50}$. $[-5, \infty)$

Simplify each expression. All variables represent positive real numbers. Write answers without using negative exponents.

7. $(49x^4)^{1/2}$ $7x^2$
8. $-27^{2/3}$ -9
9. $36^{-3/2}$ $\frac{1}{216}$
10. $\left(-\frac{8}{125n^6}\right)^{-2/3}$ $\frac{25n^4}{4}$
11. $\frac{2^{5/3}2^{1/6}}{2^{1/2}}$ $2^{4/3}$
12. $(a^{2/3})^{1/6}$ $a^{1/9}$

Simplify each expression. The variables are unrestricted.

13. $\sqrt{x^2}$ $|x|$
14. $\sqrt{y^2 - 10y + 25}$ $|y - 5|$

Simplify each expression. All variables represent positive real numbers.

15. $\sqrt[3]{-64x^3y^6}$ $-4xy^2$
16. $\sqrt{\frac{4a^2}{9}}$ $\frac{2}{3}a$
17. $\sqrt[5]{(t + 8)^5}$ $t + 8$
18. $\sqrt{540x^3y^5}$ $6xy^2\sqrt{15xy}$
19. $\frac{\sqrt[3]{24x^{15}y^4}}{\sqrt[3]{y}}$ $2x^5y\sqrt[3]{3}$
20. $\sqrt[4]{32}$ $2\sqrt[4]{2}$

Perform the operations and simplify. All variables represent positive real numbers.

21. $2\sqrt{48y^5} - 3y\sqrt{12y^3}$ $2y^2\sqrt{3y}$
22. $2\sqrt[3]{40} - \sqrt[3]{5{,}000} + 4\sqrt[3]{625}$ $14\sqrt[3]{5}$
23. $\sqrt[4]{243z^{13}} + z\sqrt[4]{48z^9}$ $5z^3\sqrt[4]{3z}$
24. $-2\sqrt{xy}\left(3\sqrt{x} + \sqrt{xy^3}\right)$ $-6x\sqrt{y} - 2xy^2$
25. $\left(3\sqrt{2} + \sqrt{3}\right)\left(2\sqrt{2} - 3\sqrt{3}\right)$ $3 - 7\sqrt{6}$
26. $\left(\sqrt[3]{2a} + 9\right)^2$ $\sqrt[3]{4a^2} + 18\sqrt[3]{2a} + 81$
27. $\frac{8}{\sqrt{10}}$ $\frac{4\sqrt{10}}{5}$
28. $\frac{\sqrt{x} + \sqrt{y}}{\sqrt{x} - \sqrt{y}}$ $\frac{x + 2\sqrt{xy} + y}{x - y}$
29. $\sqrt[3]{\frac{9}{4a}}$ $\frac{\sqrt[3]{18a^2}}{2a}$
30. Rationalize the numerator: $\frac{\sqrt{5} + 3}{-4\sqrt{2}}$

$\frac{1}{\sqrt{2}\left(\sqrt{5} - 3\right)} = \frac{1}{\sqrt{10} - 3\sqrt{2}}$

Solve each equation. Write all proposed solutions. Cross out those that are extraneous.

31. $4\sqrt{x} = \sqrt{x + 1}$ $\frac{1}{15}$
32. $\sqrt[3]{6n + 4} - 4 = 0$ 10
33. $1 = \sqrt{u - 3} + \sqrt{u}$ ~~4~~, no solution
34. $(2m^2 - 9)^{1/2} = m$ 3, ~~−3~~
35. $\sqrt{t - 2} - t + 2 = 0$ 2, 3
36. $\sqrt{x - 8} + 10 = 0$ ~~108~~, no solution
37. Let $f(x) = \sqrt[4]{15 - x}$ and $g(x) = \sqrt[4]{13 - 2x}$. Find all values of x for which $f(x) = g(x)$. -2
38. Solve $r = \sqrt[3]{\frac{GMt^2}{4\pi^2}}$ for G. $G = \frac{4\pi^2r^3}{Mt^2}$

Find the missing side lengths in each triangle. Give the exact answer and then an approximation to two decimal places, when appropriate.

39.

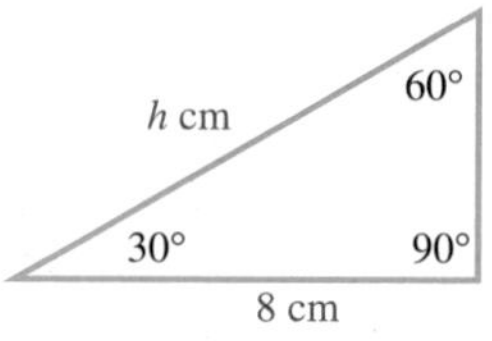

$x = \frac{8\sqrt{3}}{3}$ cm ≈ 4.62 cm;
$h = \frac{16\sqrt{3}}{3}$ cm ≈ 9.24 cm

40.

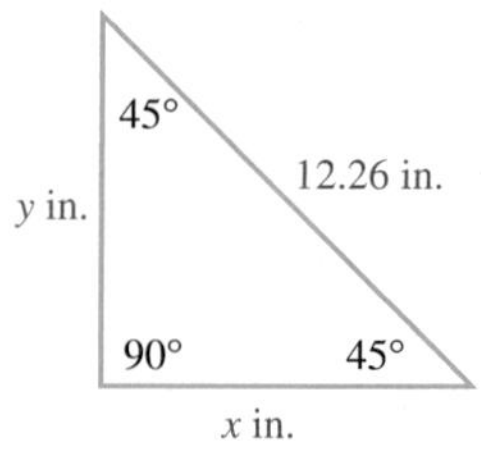

$x = 6.13\sqrt{2}$ in. ≈ 8.67 in.;
$y = 6.13\sqrt{2}$ in. ≈ 8.67 in.

41. Find the distance between $(-2, 5)$ and $(22, 12)$. 25
42. Find the coordinates of the midpoint of the line segment joining $(-2, -5)$ and $(7, -11)$. $\left(\frac{5}{2}, -8\right)$
43. If $(3, 0)$ is the midpoint of segment PQ and the coordinates of P are $(-1, -3)$, find the coordinates of Q. $(7, 3)$
44. **Shipping Crates.** The diagonal brace on the shipping crate in the illustration is 53 inches. Find the height h of the crate. 28 in.

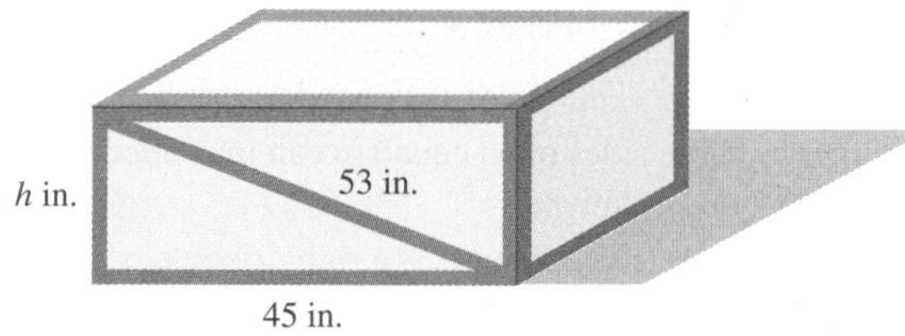

45. Express $\sqrt{-45}$ in terms of i. $3i\sqrt{5}$
46. Simplify: i^{106} -1

Perform the operations. Write all answers in the form a + bi.

47. $(9 + 4i) + (-13 + 7i)$ $-4 + 11i$
48. $\left(3 - \sqrt{-9}\right) - \left(-1 + \sqrt{-16}\right)$ $4 - 7i$
49. $15i(3 - 5i)$ $75 + 45i$
50. $(8 + 10i)(-7 - i)$ $-46 - 78i$
51. $\frac{1}{i\sqrt{2}}$ $0 - \frac{\sqrt{2}}{2}i$
52. $\frac{2 + i}{3 - i}$ $\frac{1}{2} + \frac{1}{2}i$

Group Project

A Spiral of Roots

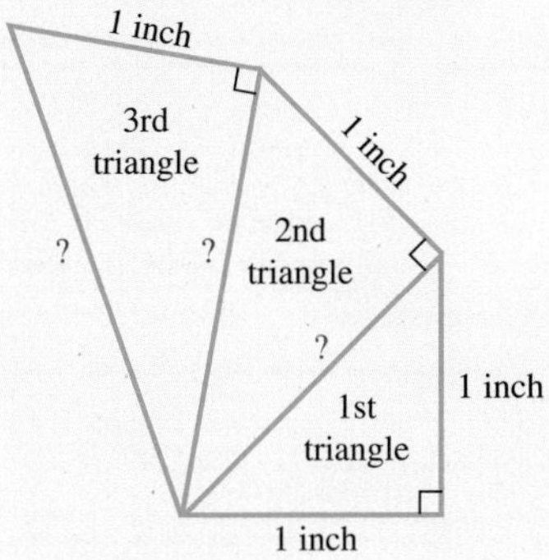

Overview: In this activity, you will create a visual representation of a collection of square roots.

Instructions: Form groups of 2 or 3 students. You will need a piece of unlined paper, a protractor, a ruler, and a pencil. Begin by drawing an isosceles right triangle with legs of length 1 inch in the middle of the paper. (See the illustration.) Use the Pythagorean theorem to determine the length of the hypotenuse. Draw a second right triangle using the hypotenuse of the first right triangle as one leg. Draw its second leg with length 1 inch. Find the length of the hypotenuse of the second triangle.

Continue creating right triangles, using the previous hypotenuse as one leg and drawing a new second leg of length 1 inch each time. Calculate the length of each resulting hypotenuse. When the figure begins to spiral onto itself, you may stop the process. Make a list of the lengths of each hypotenuse. What pattern do you see?

Graphing in Three Dimensions

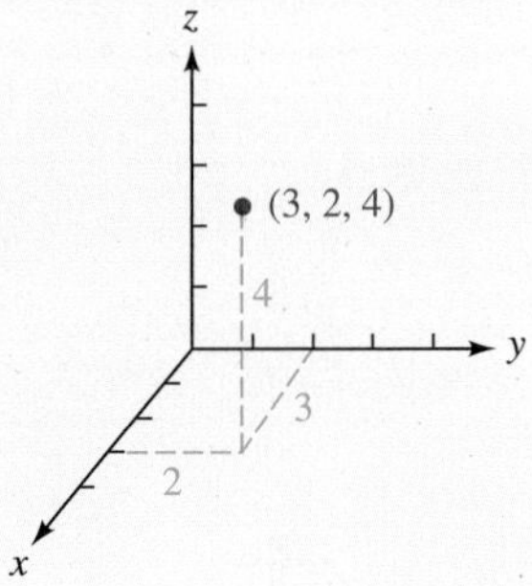

Overview: In this activity, you will find the distance between two points that lie in three-dimensional space.

Instructions: Form groups of 2 or 3 students. In a three-dimensional Cartesian coordinate system, the positive x-axis is horizontal and pointing toward the viewer (out of the page), the positive y-axis is also horizontal and pointing to the right, and the positive z-axis is vertical, pointing up. A point is located by plotting an ordered triple of numbers (x, y, z). In the illustration, the point $(3, 2, 4)$ is plotted.

In three dimensions, the distance formula is

$$d = \sqrt{(x_2 - x_1)^2 + (y_2 - y_1)^2 + (z_2 - z_1)^2}$$

1. Copy the illustration shown. Then plot the point $(1, 4, 3)$. Use the distance formula to find the distance between these two points.
2. Draw another three-dimensional coordinate system and plot the points $(-3, 3, -4)$ and $(2, -3, 2)$. Use the distance formula to find the distance between these two points.

10

Quadratic Equations, Functions, and Inequalities

from Campus to Careers

Police Patrol Officer

The responsibilities of a police patrol officer are extremely broad. Quite often, he or she must make a split-second decision while under enormous pressure. One internet vocational website cautions anyone considering such a career to "pay attention in your mathematics and science classes. Those classes help sharpen your ability to think things through and solve problems—an important part of police work."

Problem 93 in **Study Set 10.2**, **problem 87** in **Study Set 10.3**, and **problem 89** in **Study Set 10.4** involve situations that a police patrol officer encounters on the job. The mathematical concepts discussed in this chapter can be used to solve those problems.

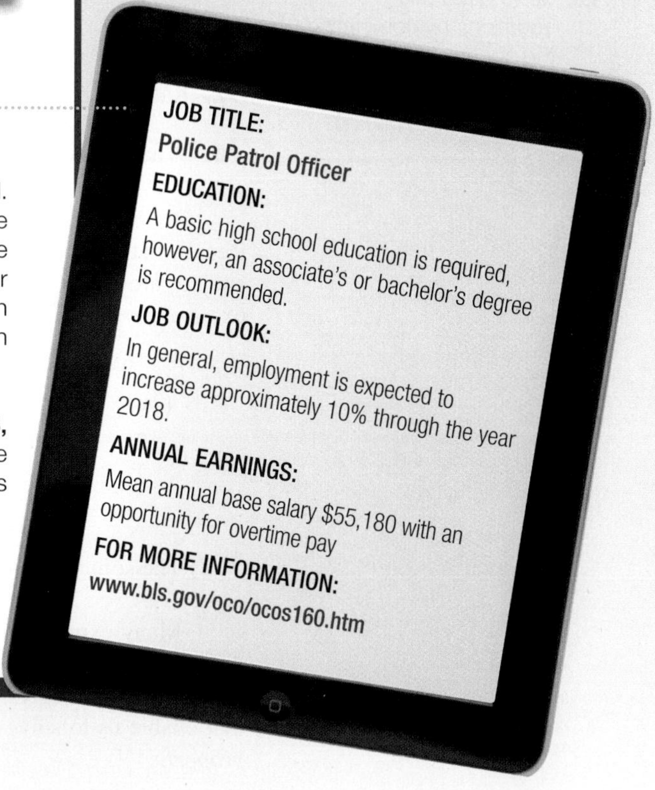

JOB TITLE:
Police Patrol Officer
EDUCATION:
A basic high school education is required, however, an associate's or bachelor's degree is recommended.
JOB OUTLOOK:
In general, employment is expected to increase approximately 10% through the year 2018.
ANNUAL EARNINGS:
Mean annual base salary $55,180 with an opportunity for overtime pay
FOR MORE INFORMATION:
www.bls.gov/oco/ocos160.htm

Study Skills Workshop ▶ Organizing Your Notebook

If you're like most students, your algebra notebook could probably use some attention at this stage of the course. You will definitely appreciate a well-organized notebook when it comes time to study for the final exam. Here are some suggestions to put it in tip-top shape.

ORGANIZE YOUR NOTEBOOK INTO SECTIONS: Create a separate section in the notebook for each chapter (or unit of study) that your class has covered this term.

ORGANIZE THE PAPERS WITHIN EACH SECTION: One recommended order is to begin each section with your class notes, followed by your completed homework assignments, then any study sheets or handouts, and, finally, all graded quizzes and tests.

Now Try This

1. Organize your algebra notebook using the guidelines given above.
2. Write a Table of Contents to place at the beginning of your notebook. List each chapter (or unit of study) and include the dates over which the material was covered.
3. Compare your completed notebook with those of other students in your class. Have you overlooked any important items that would be useful when studying for the final exam?

SECTION 10.1 The Square Root Property and Completing the Square

OBJECTIVES

1. Use the square root property to solve quadratic equations.
2. Solve quadratic equations by completing the square.

ARE YOU READY? *Are You Ready? exercises available online at www.webassign.net/brookscole*

The following problems review some basic skills that are needed when using the square root property and completing the square.

1. Simplify: **a.** $\sqrt{28}$ $2\sqrt{7}$ **b.** $\sqrt{-36}$ $6i$
2. Rationalize: $\sqrt{\frac{5}{6}}$ $\frac{\sqrt{30}}{6}$
3. What is one-half of 9? $\frac{9}{2}$
4. Find the square of $\frac{7}{2}$. $\frac{49}{4}$
5. Fill in the blank: $1 + \frac{25}{144} = \frac{169}{144}$
6. Factor: $x^2 - 8x + 16$ $(x - 4)^2$

Recall that a *quadratic equation* is an equation of the form $ax^2 + bx + c = 0$, where a, b, and c are real numbers and $a \neq 0$. We have solved quadratic equations, such as $6x^2 - 7x - 3 = 0$, using factoring and the zero-factor property as shown below.

$$6x^2 - 7x - 3 = 0$$

$$(2x - 3)(3x + 1) = 0 \quad \text{Factor the trinomial.}$$

$$2x - 3 = 0 \quad \text{or} \quad 3x + 1 = 0 \quad \text{Set each factor equal to 0.}$$

$$x = \frac{3}{2} \quad \Big| \quad x = -\frac{1}{3} \quad \text{Solve each linear equation.}$$

The solutions are $-\frac{1}{3}$ and $\frac{3}{2}$.

Many expressions do not factor as easily as $6x^2 - 7x - 3$. For example, it would be difficult to solve $2x^2 + 4x + 1 = 0$ by factoring, because $2x^2 + 4x + 1$ cannot be factored by using only integers. In this section and the next, we will develop more general methods that enable us to solve any quadratic equation. Those methods are based on the *square root property*.

1 Use the Square Root Property to Solve Quadratic Equations.

To develop general methods for solving quadratic equations, we first consider the equation $x^2 = c$. If $c \geq 0$, we can find the real solutions of $x^2 = c$ as follows:

$$x^2 = c$$

$$x^2 - c = 0 \quad \text{Subtract } c \text{ from both sides.}$$

$$x^2 - (\sqrt{c})^2 = 0 \quad \text{Replace } c \text{ with } (\sqrt{c})^2, \text{ since } c = (\sqrt{c})^2.$$

$$(x + \sqrt{c})(x - \sqrt{c}) = 0 \quad \text{Factor the difference of two squares.}$$

$$x + \sqrt{c} = 0 \quad \text{or} \quad x - \sqrt{c} = 0 \quad \text{Set each factor equal to 0.}$$

$$x = -\sqrt{c} \quad \Big| \quad x = \sqrt{c} \quad \text{Solve each linear equation.}$$

The solutions of $x^2 = c$ are $\sqrt{c}$ and $-\sqrt{c}$.

The Square Root Property

For any nonnegative real number c, if $x^2 = c$, then

$$x = \sqrt{c} \quad \text{or} \quad x = -\sqrt{c}$$

The Language of Algebra

The ± symbol is often seen in political polls. A candidate with 48% (±4%) support could be between 48 + 4 = 52% and 48 − 4 = 44%

We can write the conclusion of the square root property in a more compact form, called **double-sign notation.**

$x = \pm\sqrt{c}$ This is formally read as "x equals the positive or negative square root of c." However, it is often read more informally as "x equals plus or minus the square root of c."

EXAMPLE 1 Solve: $x^2 - 12 = 0$

Strategy Since $x^2 - 12$ does not factor as a difference of two integer squares, we must take an alternate approach. We will add 12 to both sides of the equation and use the square root property to solve for x.

Why After adding 12 to both sides, the resulting equivalent equation will have the desired form $x^2 = c$.

Teaching Tip: Ask your students how they would solve the equation if it were $x^2 - 16 = 0$ instead.

Solution

$$x^2 - 12 = 0 \quad \text{This is the equation to solve. It is a quadratic equation that is missing an } x\text{-term.}$$

$$x^2 = 12 \quad \text{To isolate } x^2 \text{ on the left side, add 12 to both sides.}$$

$$x = \sqrt{12} \quad \text{or} \quad x = -\sqrt{12} \quad \text{Use the square root property.}$$

$$x = 2\sqrt{3} \quad \Big| \quad x = -2\sqrt{3} \quad \text{Simplify: } \sqrt{12} = \sqrt{4}\sqrt{3} = 2\sqrt{3}.$$

Check:

$$x^2 - 12 = 0 \qquad\qquad x^2 - 12 = 0$$

$$(2\sqrt{3})^2 - 12 \stackrel{?}{=} 0 \qquad\qquad (-2\sqrt{3})^2 - 12 \stackrel{?}{=} 0$$

$$12 - 12 \stackrel{?}{=} 0 \qquad\qquad 12 - 12 \stackrel{?}{=} 0$$

$$0 = 0 \quad \text{True} \qquad\qquad 0 = 0 \quad \text{True}$$

The exact solutions are $2\sqrt{3}$ and $-2\sqrt{3}$, which can be written using double-sign notation as $\pm 2\sqrt{3}$. The solution set is $\{-2\sqrt{3}, 2\sqrt{3}\}$. We can use a calculator to approximate the solutions. To the nearest hundredth, they are ± 3.46.

Teaching Tip: Ask a student to explain the difference between an exact solution and an approximate solution.

Teaching Example 1 Solve: $x^2 - 75 = 0$
Answer: $\pm 5\sqrt{3}$

Self Check 1 Solve: $x^2 - 18 = 0$ $\pm 3\sqrt{2}$

Now Try Problems 15 and 19

EXAMPLE 2 **Phonograph Records.** Before compact discs, music was recorded on thin vinyl discs. The discs used for long-playing records had a surface area of about 111 square inches per side. Find the radius of a long-playing record to the nearst tenth of an inch.

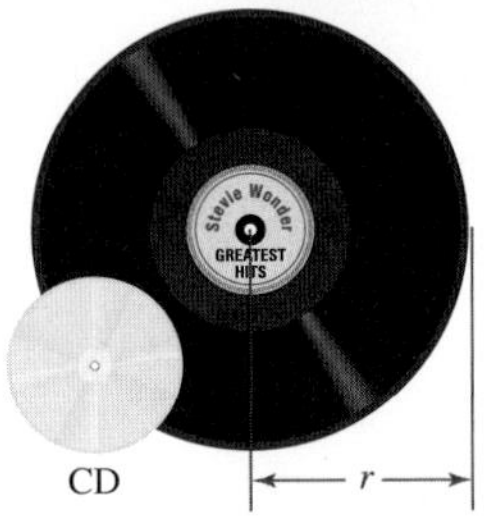

Strategy The area A of a circle with radius r is given by the formula $A = \pi r^2$. We will find the radius of a record by substituting 111 for A and dividing both sides by π. Then we will use the square root property to solve for r.

Why After substituting 111 for A and dividing both sides by π, the resulting equivalent equation will have the desired form $r^2 = c$.

Solution

$$A = \pi r^2 \quad \text{This is the formula for the area of a circle.}$$

$$\mathbf{111} = \pi r^2 \quad \text{Substitute 111 for A.}$$

$$\frac{111}{\pi} = r^2 \quad \text{To undo the multiplication by } \pi\text{, divide both sides by } \pi.$$

$$r = \sqrt{\frac{111}{\pi}} \quad \text{or} \quad r = -\sqrt{\frac{111}{\pi}}$$

Use the square root property. Since the radius of the record cannot be negative, discard the second solution.

The radius of a long-playing record is $\sqrt{\frac{111}{\pi}}$ inches—to the nearest tenth, 5.9 inches.

Teaching Example 2 Compact Discs. The surface area of a compact disc (CD) is approximately 15.9 square inches. What is the radius of a CD? Round to the nearest hundredth.
Answer: $\sqrt{\frac{15.9}{\pi}}$ in. ≈ 2.25 in.

Self Check 2 **Restaurant Seating.** A restaurant requires 14 square feet of floor area for one of their round tables. Find the radius of a table. Round to the nearest hundredth. $\sqrt{\frac{14}{\pi}}$ ft ≈ 2.11 ft

Now Try Problems 35 and 103

Some quadratic equations have solutions that are not real numbers.

EXAMPLE 3 Solve: $4x^2 + 25 = 0$

Strategy We will subtract 25 from both sides of the equation and divide both sides by 4. Then we will use the square root property to solve for x.

Why After subtracting 25 from both sides and dividing both sides by 4, the resulting equivalent equation will have the desired form $x^2 =$ a constant.

Solution

Teaching Tip: You may want to review complex numbers before explaining this example. For example, ask your students to simplify $\sqrt{-25}$.

$$4x^2 + 25 = 0 \quad \text{This is the equation to solve.}$$

$$x^2 = -\frac{25}{4} \quad \text{To isolate } x^2\text{, subtract 25 from both sides and divide both sides by 4.}$$

$$x = \pm\sqrt{-\frac{25}{4}} \quad \text{Use the square root property and write the result using double-sign notation.}$$

$$x = \pm\frac{5}{2}i \quad \text{Simplify the radical: } \sqrt{-\frac{25}{4}} = \sqrt{-1\cdot\frac{25}{4}} = \sqrt{-1}\frac{\sqrt{25}}{\sqrt{4}} = \frac{5}{2}i$$

Since the solutions are $\frac{5}{2}i$ and $-\frac{5}{2}i$, the solution set is $\left\{\frac{5}{2}i, -\frac{5}{2}i\right\}$.

Check:

$$4x^2 + 25 = 0 \qquad\qquad 4x^2 + 25 = 0$$

$$4\left(\frac{5}{2}i\right)^2 + 25 \stackrel{?}{=} 0 \qquad\qquad 4\left(-\frac{5}{2}i\right)^2 + 25 \stackrel{?}{=} 0$$

$$4\left(\frac{25}{4}\right)i^2 + 25 \stackrel{?}{=} 0 \qquad\qquad 4\left(\frac{25}{4}\right)i^2 + 25 \stackrel{?}{=} 0$$

$$25(-1) + 25 \stackrel{?}{=} 0 \qquad\qquad 25(-1) + 25 \stackrel{?}{=} 0$$

$$0 = 0 \quad \text{True} \qquad\qquad 0 = 0 \quad \text{True}$$

Teaching Example 3 Solve: $36x^2 + 25 = 0$
Answer: $\pm\frac{5}{6}i$

Self Check 3 Solve: $16x^2 + 49 = 0$ $\pm\frac{7}{4}i$

Now Try ▶ Problem 27

We can extend the square root property to solve equations that involve the square of a binomial and a constant.

EXAMPLE 4 Use the square root property to solve $(x - 1)^2 = 16$.

Strategy Instead of a variable squared on the left side, we have a quantity squared. We still use the square root property to solve the equation.

Why We want to eliminate the square on the binomial, so that we can eventually isolate the variable on one side of the equation.

Solution

$(x - 1)^2 = 16$ This is the equation to solve.

$x - 1 = \pm\sqrt{16}$ Use the square root property and write the result using double-sign notation.

$x - 1 = \pm 4$ Simplify: $\sqrt{16} = 4$.

$x = 1 \pm 4$ To isolate x, add 1 to both sides. It is standard practice to write the addition of the 1 in front of the ± symbol. Read as "one plus or minus four."

$x = 1 + 4$ or $x = 1 - 4$ To find one solution, use +. To find the other, use −.

$x = 5$ | $x = -3$ Add (subtract).

Verify that 5 and -3 satisfy the original equation.

Caution

It might be tempting to square the binomial on the left side of $(x - 1)^2 = 16$. However, that causes unnecessary, additional steps to solve the equation another way.

$$(x - 1)^2 = 16$$

$$\cancel{x^2 - 2x + 1 = 16}$$

Teaching Example 4 Solve: $(x - 7)^2 = 25$
Answer: 2, 12

Self Check 4 Use the square root property to solve $(x + 2)^2 = 9$. 1, −5

Now Try ▶ Problems 31 and 33

EXAMPLE 5 Let $f(x) = (3x + 8)^2$. For what value(s) of x is $f(x) = 6$?

Strategy We will substitute 6 for $f(x)$ and solve for x.

Why In the equation, there are two unknowns, x and $f(x)$. If we replace $f(x)$ with 6, we can use equation-solving techniques to find x.

Solution

$f(x) = (3x + 8)^2$	This is the given function.
$6 = (3x + 8)^2$	Substitute 6 for $f(x)$.
$\pm\sqrt{6} = 3x + 8$	Use the square root property and write the result using double-sign notation.
$-8 \pm \sqrt{6} = 3x$	To isolate the variable term $3x$, subtract 8 from both sides. It is standard practice to write the subtraction of -8 in front of the $\pm$.
$\dfrac{-8 \pm \sqrt{6}}{3} = x$	To isolate x, undo the multiplication by 3 by dividing both sides by 3.

We have found that $f(x) = 6$ when $x = \frac{-8 - \sqrt{6}}{3}$ or $x = \frac{-8 + \sqrt{6}}{3}$. To check, verify that $f\left(\frac{-8 - \sqrt{6}}{3}\right) = 6$ and $f\left(\frac{-8 + \sqrt{6}}{3}\right) = 6$.

Teaching Tip: Mention that $\frac{-8 - \sqrt{6}}{3}$ and $\frac{-8 + \sqrt{6}}{3}$ are the most complicated function input values that we have seen so far. Point out that they are, indeed, real numbers. Have selected students approximate each value using a calculator.

Self Check 5 Let $f(x) = (4x + 11)^2$. For what values of x is $f(x) = 3$? $\frac{-11 \pm \sqrt{3}}{4}$

Now Try Problem 35

Teaching Example 5 Let $f(x) = (5x + 23)^2$. For what values of x is $f(x) = 7$?

Answer: $\frac{-23 \pm \sqrt{7}}{5}$

2 Solve Quadratic Equations by Completing the Square.

When the polynomial in a quadratic equation doesn't factor easily, we can solve the equation by **completing the square.** This method is based on the following perfect-square trinomials (with leading coefficients of 1) and their factored forms:

$$x^2 + 2bx + b^2 = (x + b)^2 \quad \text{and} \quad x^2 - 2bx + b^2 = (x - b)^2$$

In each of these perfect-square trinomials, the third term is the square of one-half of the coefficient of x.

- In $x^2 + \mathbf{2b}x + b^2$, the coefficient of x is $\mathbf{2b}$. If we find $\frac{1}{2} \cdot \mathbf{2b}$, which is b, and square it, we get the third term, b^2.
- In $x^2 - \mathbf{2b}x + b^2$, the coefficient of x is $\mathbf{-2b}$. If we find $\frac{1}{2}(\mathbf{-2b})$, which is $-b$, and square it, we get the third term: $(-b)^2 = b^2$.

The Language of Algebra

Recall that trinomials that are the square of a binomial are called **perfect-square trinomials.**

We can use these observations to change certain binomials into perfect-square trinomials using a 3-step process. For example, to change $x^2 + 12x$ into a perfect-square trinomial, we find one-half of the coefficient of x, square the result, and add the square to $x^2 + 12x$.

$$x^2 + \mathbf{12}x + \square$$

Step 1: Find one-half of the coefficient of x. $\frac{1}{2} \cdot \mathbf{12} = 6$

Step 2: Square the result. $6^2 = \mathbf{36}$

Step 3: Add the square to the binomial.

We obtain the perfect-square trinomial $x^2 + 12x + 36$, which factors as $(x + 6)^2$. By adding 36 to $x^2 + 12x$, we say that we have *completed the square on* $x^2 + 12x$.

Completing the Square

To complete the square on $x^2 + bx$, add the square of one-half of the coefficient of x:

$$x^2 + bx + \left(\frac{1}{2}b\right)^2$$

EXAMPLE 6 Complete the square and factor the resulting perfect-square trinomial: **a.** $x^2 + 10x$ **b.** $x^2 - 11x$ **c.** $x^2 + \frac{7}{3}x$

Teaching Tip: After you finish discussing part b, ask your students why fractions were involved in the solution to part b, but they were not involved in the solution to part a.

Strategy We will add the square of one-half of the coefficient of x to the given binomial.

Why Adding such a term will change the binomial into a perfect-square trinomial that will factor.

Solution

a. To make $x^2 + 10x$ a perfect-square trinomial, we find one-half of 10, square it, and add the result to $x^2 + 10x$.

$$x^2 + 10x + \mathbf{25} \quad \tfrac{1}{2} \cdot 10 = 5 \text{ and } 5^2 = 25. \text{ Add 25 to the given binomial.}$$

This trinomial factors as $(x + 5)^2$. To check, we square $x + 5$ and verify that the result is $x^2 + 10x + 25$.

b. To make $x^2 - 11x$ a perfect-square trinomial, we find one-half of -11, square it, and add the result to $x^2 - 11x$.

$$x^2 - 11x + \mathbf{\frac{121}{4}} \quad \tfrac{1}{2}(-11) = -\tfrac{11}{2} \text{ and } \left(-\tfrac{11}{2}\right)^2 = \tfrac{121}{4}. \text{ Add } \tfrac{121}{4} \text{ to the given binomial.}$$

This trinomial factors as $\left(x - \frac{11}{2}\right)^2$. Check using multiplication or using a special-product rule.

c. To make $x^2 + \frac{7}{3}x$ a perfect-square trinomial, we find one-half of $\frac{7}{3}$, square it, and add the result to $x^2 + \frac{7}{3}x$.

$$x^2 + \frac{7}{3}x + \mathbf{\frac{49}{36}} \quad \text{To complete the square: } \tfrac{1}{2}\left(\tfrac{7}{3}\right) = \tfrac{7}{6} \text{ and } \left(\tfrac{7}{6}\right)^2 = \tfrac{49}{36}. \text{ Add } \tfrac{49}{36} \text{ to the given binomial.}$$

The trinomial factors as $\left(x + \frac{7}{6}\right)^2$.

Caution

Realize that when we complete the square on a binomial, we are not writing an equivalent trinomial expression. Since the result is a completely different polynomial, it would be incorrect to use an = symbol between the two.

$x^2 + 10x = x^2 + 10x + 25$ (crossed out)

The Language of Algebra

When we add $\frac{121}{4}$ to $x^2 - 11x$, we say we have **completed the square** on $x^2 - 11x$. For that reason, binomials such as $x^2 - 11x$ are often called **incomplete squares.**

Self Check 6 Complete the square on $a^2 - 5a$ and factor the resulting trinomial. $a^2 - 5a + \frac{25}{4} = \left(a - \frac{5}{2}\right)^2$

Now Try Problems 39 and 41

Teaching Example 6 Complete the square and factor the resulting trinomial:
a. $y^2 - 8y$ **b.** $y^2 + 3y$ **c.** $x^2 + \frac{7}{5}x$
Answers:
a. $y^2 - 8y + 16 = (y - 4)^2$
b. $y^2 + 3y + \frac{9}{4} = \left(y + \frac{3}{2}\right)^2$
c. $x^2 + \frac{7}{5}x + \frac{49}{100} = \left(x + \frac{7}{10}\right)^2$

To solve an equation of the form $ax^2 + bx + c = 0$ by completing the square, we use the following steps.

Completing the Square to Solve a Quadratic Equation in *x*

1. If the coefficient of x^2 is 1, go to step 2. If it is not, make it 1 by dividing both sides of the equation by the coefficient of x^2.
2. Get all variable terms on one side of the equation and constants on the other side.
3. Complete the square by finding one-half of the coefficient of x, squaring the result, and adding the square to both sides of the equation.
4. Factor the perfect-square trinomial as the square of a binomial.
5. Solve the resulting equation using the square root property.
6. Check your answers in the original equation.

EXAMPLE 7 Solve by completing the square: $x^2 - 8x - 5 = 0$. Approximate the solutions to the nearest hundredth.

Strategy We will use the addition property of equality and add 5 to both sides. Then we will complete the square to solve for x.

Why To prepare to complete the square, we need to isolate the variable terms, x^2 and $-8x$, on the left side of the equation and the constant term on the right side.

Solution

$x^2 - 8x - 5 = 0$ This is the equation to solve.

$x^2 - 8x = 5$ Add 5 to both sides so that the constant term is on the right side.

The Language of Algebra

For $x^2 - 8x - 5$, we call x^2 the **leading term,** the **first term,** or the **quadratic term.** We call $-8x$ the **second term** or the **linear term.** Together, x^2 and $-8x$ are called **variable terms.** We call -5 the **constant term** or the **third term.**

In $x^2 - 8x$, the coefficient of x is $\mathbf{-8}$. One-half of $\mathbf{-8}$ is -4, and $(-4)^2 = 16$. If we add 16 to $x^2 - 8x$, it becomes a perfect-square trinomial.

$x^2 - 8x + \mathbf{16} = 5 + \mathbf{16}$ To complete the square on the left side, add 16 to both sides.

$(x - 4)^2 = 21$ Factor the perfect-square trinomial. Add on the right side.

$x - 4 = \pm\sqrt{21}$ By the square root property, $x - 4 = \sqrt{21}$ or $x - 4 = -\sqrt{21}$. Use double-sign notation to show this.

$x = 4 \pm \sqrt{21}$ To isolate x, add 4 to both sides. It is common practice to write the 4 in front of the $\pm$ symbol.

The exact solutions of $x^2 - 8x - 5 = 0$ are two irrational numbers, $4 + \sqrt{21}$ and $4 - \sqrt{21}$.

We also can approximate each solution by using the decimal approximation of $\sqrt{21}$, which is 4.582575695:

$$4 + \sqrt{21} \approx 4 + 4.582575695 \approx 8.58 \qquad 4 - \sqrt{21} \approx 4 - 4.582575695 \approx -0.58$$

To the nearest hundredth, the solutions are 8.58 and -0.58. These approximations can be used to check the exact solutions informally by substituting each of them into the original equation.

Teaching Tip: Explain that since the solutions of this quadratic equation are irrational numbers, it cannot be solved by factoring. The reason the solutions are irrational is because the radicand, 21, is not a perfect square.

Self Check 7 Solve by completing the square: $x^2 - 10x - 4 = 0$. Approximate the solutions to the nearest hundredth. $5 \pm \sqrt{29}$; 10.39, -0.39

Now Try Problem 47

Teaching Example 7 Solve by completing the square. Approximate the solutions to the nearest hundredth. $x^2 - 12x - 3 = 0$
Answer: $6 \pm \sqrt{39}$; 12.24, -0.24

If the coefficient of the squared variable (called the **leading coefficient**) of a quadratic equation is not 1, we must make it 1 before we can complete the square.

EXAMPLE 8 Solve $6x^2 + 5x - 6 = 0$ by completing the square.

Strategy We will begin by dividing both sides of the equation by 6.

Why This will create a leading coefficient that is 1 so that we can proceed to complete the square to solve the equation.

Solution ***Step 1:*** To make the coefficient of x^2 equal to 1, we divide both sides of the equation by 6.

$6x^2 + 5x - 6 = 0$ This is the equation to solve.

$\frac{6x^2}{6} + \frac{5x}{6} - \frac{6}{6} = \frac{0}{6}$ Divide both sides by 6, term-by-term.

$x^2 + \frac{5}{6}x - 1 = 0$ Simplify. The coefficient of x^2 is now 1.

Success Tip

When solving a quadratic equation using the factoring method, one side of the equation must be 0. When we complete the square, we are not concerned with that requirement.

Step 2: To have the constant term on one side of the equation and the variable terms on the other, add 1 to both sides.

$$x^2 + \frac{5}{6}x \qquad = 1$$

Some students find it helpful to leave additional space here to prepare for completing the square.

Step 3: The coefficient of x is $\frac{5}{6}$, one-half of $\frac{5}{6}$ is $\frac{5}{12}$, and $\left(\frac{5}{12}\right)^2 = \frac{25}{144}$. To complete the square, we add $\frac{25}{144}$ to both sides.

$$x^2 + \frac{5}{6}x + \mathbf{\frac{25}{144}} = 1 + \mathbf{\frac{25}{144}}$$

$$x^2 + \frac{5}{6}x + \frac{25}{144} = \frac{169}{144}$$ On the right side, add: $1 + \frac{25}{144} = \frac{144}{144} + \frac{25}{144} = \frac{169}{144}$.

Step 4: Factor the left side of the equation.

$$\left(x + \frac{5}{12}\right)^2 = \frac{169}{144}$$ $x^2 + \frac{5}{6}x + \frac{25}{144}$ is a perfect-square trinomial.

Caution

When using the square root property to solve an equation, always write the $\pm$ symbol. If you forget, you will lose one of the solutions.

Step 5: We can solve the resulting equation by using the square root property.

$$x + \frac{5}{12} = \pm\sqrt{\frac{169}{144}}$$ Don't forget to write the $\pm$ symbol.

$$x + \frac{5}{12} = \pm\frac{13}{12}$$ Simplify: $\pm\sqrt{\frac{169}{144}} = \pm\frac{13}{12}$.

$$x = -\frac{5}{12} \pm \frac{13}{12}$$ To isolate x, subtract $\frac{5}{12}$ from both sides. These are like terms that can be added and subtracted.

Success Tip

Since the solutions of $6x^2 + 5x - 6 = 0$ are rational numbers, that indicates it can also be solved by factoring:

$$(3x - 2)(2x + 3) = 0$$
$$x = \frac{2}{3} \quad \text{or} \quad x = -\frac{3}{2}$$

To find the first solution, we evaluate the expression using the + symbol. To find the second solution, we evaluate the expression using the − symbol.

$$x = -\frac{5}{12} + \frac{13}{12} \quad \text{or} \quad x = -\frac{5}{12} - \frac{13}{12}$$

$$x = \frac{8}{12} \qquad\qquad x = -\frac{18}{12}$$ Add (subtract) the fractions.

$$x = \frac{2}{3} \qquad\qquad x = -\frac{3}{2}$$ Simplify each fraction. The solutions are two rational numbers.

Step 6: Check each solution, $\frac{2}{3}$ and $-\frac{3}{2}$, in the original equation.

Teaching Example 8 Solve by completing the square: $3a^2 + 2a - 1 = 0$

Answer: $-1, \frac{1}{3}$

Self Check 8 Solve by completing the square: $3x^2 + 2x - 8 = 0$ $\frac{4}{3}, -2$

Now Try Problem 55

EXAMPLE 9 Solve: $2x^2 + 4x + 1 = 0$. Approximate the solutions to the nearest hundredth.

Strategy We will follow the steps for solving a quadratic equation by completing the square.

Why Since the trinomial $2x^2 + 4x + 1$ cannot be factored using only integers, solving the equation by completing the square is our only option at this time.

Caution

A common error is to add a constant to one side of an equation to complete the square and forget to add it to the other side.

Solution

$2x^2 + 4x + 1 = 0$	This is the equation to solve.
$x^2 + 2x + \frac{1}{2} = 0$	Divide both sides by 2 to make the coefficient of x^2 equal to 1.
$x^2 + 2x = -\frac{1}{2}$	Subtract $\frac{1}{2}$ from both sides so that the constant term is on the right side.
$x^2 + 2x + 1 = -\frac{1}{2} + 1$	To complete the square, square one-half of the coefficient of x and add it to both sides.
$(x + 1)^2 = \frac{1}{2}$	Factor on the left side and do the addition on the right side.
$x + 1 = \pm\sqrt{\frac{1}{2}}$	Use the square root property. Don't forget the $\pm$ symbol.
$x = -1 \pm \sqrt{\frac{1}{2}}$	To isolate x, subtract 1 from both sides.

To write $\sqrt{\frac{1}{2}}$ in simplified radical form, we use the quotient rule for radicals and then rationalize the denominator.

$$x = -1 + \frac{\sqrt{2}}{2} \quad \text{or} \quad x = -1 - \frac{\sqrt{2}}{2}$$

Rationalize: $\sqrt{\frac{1}{2}} = \frac{\sqrt{1}}{\sqrt{2}} = \frac{1 \cdot \sqrt{2}}{\sqrt{2}\sqrt{2}} = \frac{\sqrt{2}}{2}$.

We can express each solution in an alternate form if we write -1 as a fraction with a denominator of 2.

Caution

Recall that to simplify a fraction, we remove common factors of the numerator and denominator. Since -2 is a term of the numerator of $\frac{-2+\sqrt{2}}{2}$, no further simplification of this expression can be made.

$$x = -\frac{2}{2} + \frac{\sqrt{2}}{2} \quad \text{or} \quad x = -\frac{2}{2} - \frac{\sqrt{2}}{2}$$

Write -1 as $-\frac{2}{2}$.

$$x = \frac{-2 + \sqrt{2}}{2} \quad \Bigg| \quad x = \frac{-2 - \sqrt{2}}{2}$$

Add (subtract) the numerators and keep the common denominator 2.

The exact solutions are $\frac{-2+\sqrt{2}}{2}$ and $\frac{-2-\sqrt{2}}{2}$, or simply, $\frac{-2\pm\sqrt{2}}{2}$. We can use a calculator to approximate them. To the nearest hundredth, they are -0.29 and -1.71.

Teaching Example 9 Solve: $2x^2 - 12x + 3 = 0$. Approximate the solutions to the nearest hundredth.

Answer: $\frac{6 \pm \sqrt{30}}{2}$; 5.74, 0.26

Self Check 9 Solve: $3x^2 + 6x + 1 = 0$. Approximate the solutions to the nearest hundredth. $\frac{-3 \pm \sqrt{6}}{3}$; -0.18, -1.82

Now Try Problem 59

Using Your Calculator ▶ Checking Solutions of Quadratic Equations

We can use a graphing calculator to check the solutions of the quadratic equation $2x^2 + 4x + 1 = 0$ found in Example 9. After entering $Y_1 = 2x^2 + 4x + 1$, we call up the home screen by pressing [2nd] [QUIT]. Then we press [VARS], arrow to Y-VARS, press [ENTER], and enter 1 to get the display shown in figure (a). We evaluate $2x^2 + 4x + 1$ for $x = \frac{-2+\sqrt{2}}{2}$ by entering the solution using function notation, as shown in figure (b). When [ENTER] is pressed, the result of 0 is confirmation that $x = \frac{-2+\sqrt{2}}{2}$ is a solution of the equation.

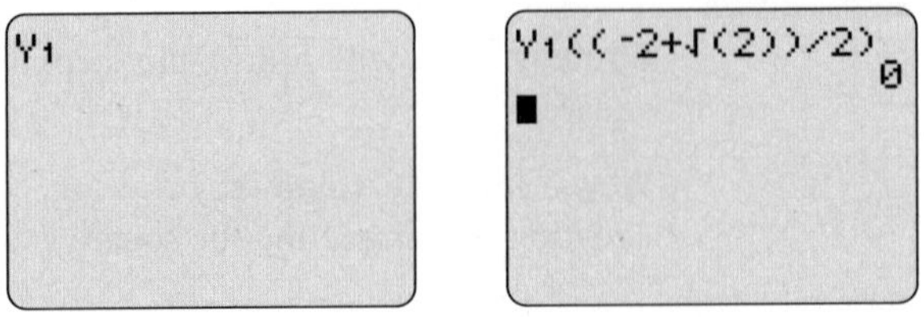

(a) (b)

The solutions of some quadratic equations are complex numbers that contain i.

EXAMPLE 10 Solve: **a.** $x^2 - 4x + 16 = 0$ **b.** $x^2 + \frac{2}{3}x + 6 = 0$

Strategy We will follow the steps for solving a quadratic equation by completing the square.

Why Since the trinomials $x^2 - 4x + 16$ and $x^2 + \frac{2}{3}x + 6 = 0$ cannot be factored using only integers, solving the equation by completing the square is our only option.

Solution

a.

$x^2 - 4x + 16 = 0$ This is the equation to solve.

$x^2 - 4x = -16$ Subtract 16 from both sides so that the constant term is on the right side.

$x^2 - 4x + \mathbf{4} = -16 + \mathbf{4}$ Complete the square: $\frac{1}{2}(-4) = -2$ and $(-2)^2 = 4$. Add 4 to both sides.

$(x - 2)^2 = -12$ Factor the left side. On the right side, add.

$x - 2 = \pm\sqrt{-12}$ Use the square root property. Don't forget the ± symbol.

$x = 2 \pm \sqrt{-12}$ To isolate x, add 2 to both sides.

$x = 2 \pm 2i\sqrt{3}$ Simplify the radical: $\sqrt{-12} = \sqrt{-1 \cdot 4 \cdot 3} = \sqrt{-1}\sqrt{4}\sqrt{3} = 2i\sqrt{3}$.

There are two complex number solutions involving i: $2 + 2i\sqrt{3}$ and $2 - 2i\sqrt{3}$. Verify this by checking them in the original equation.

b.

$x^2 + \frac{2}{3}x + 6 = 0$ This is the equation to solve.

$x^2 + \frac{2}{3}x = -6$ Subtract 6 from both sides so that the constant term is on the right side.

$x^2 + \frac{2}{3}x + \mathbf{\frac{1}{9}} = \mathbf{\frac{1}{9}} - 6$ Complete the square: $\frac{1}{2} \cdot \frac{2}{3} = \frac{1}{3}$ and $\left(\frac{1}{3}\right)^2 = \frac{1}{9}$. Add $\frac{1}{9}$ to both sides.

$\left(x + \frac{1}{3}\right)^2 = \frac{1}{9} - \frac{6}{1} \cdot \frac{9}{9}$ Factor the left side. On the right side, prepare to subtract. Write 6 as $\frac{6}{1}$ and multiply by $\frac{9}{9}$ to build an equivalent fraction with denominator 9.

$\left(x + \frac{1}{3}\right)^2 = \frac{1}{9} - \frac{54}{9}$ On the right side, multiply the numerators. Multiply the denominators.

$\left(x + \frac{1}{3}\right)^2 = -\frac{53}{9}$ On the right side, subtract the fractions.

$x + \frac{1}{3} = \pm\sqrt{-\frac{53}{9}}$ Use the square root property. Don't forget the ± symbol.

$x = -\frac{1}{3} \pm \sqrt{-\frac{53}{9}}$ To isolate x, subtract $\frac{1}{3}$ from both sides.

$x = -\frac{1}{3} \pm \frac{\sqrt{53}}{3}i$ Simplify the radical: $\sqrt{-\frac{53}{9}} = \sqrt{-1}\frac{\sqrt{53}}{\sqrt{9}} = \frac{\sqrt{53}}{3}i$.

The solutions are $-\frac{1}{3} + \frac{\sqrt{53}}{3}i$ and $-\frac{1}{3} - \frac{\sqrt{53}}{3}i$.

Success Tip

Example 8 can be solved by factoring. However, Examples 7, 9, and 10 cannot. This illustrates an important fact: completing the square can be used to solve any quadratic equation.

Success Tip

Notice how we multiply $\frac{6}{1}$ by a form of 1 to build an equivalent fraction with a denominator 9:

$$\frac{6}{1} \cdot \frac{9}{9}$$

Notation

The solutions are written in complex number form $a + bi$. They also could be written as

$$\frac{-1 \pm i\sqrt{53}}{3}$$

Self Check 10 Solve: **a.** $x^2 - 6x + 27 = 0$ $3 \pm 3i\sqrt{2}$ **b.** $x^2 + \frac{2}{5}x + 3 = 0$ $-\frac{1}{5} \pm \frac{i\sqrt{74}}{5}$

Now Try Problems 63 and 67

Teaching Example 10 Solve:

a. $x^2 - 4x + 54 = 0$

b. $x^2 + \frac{2}{7}x + 2 = 0$

Answers: **a.** $2 \pm 5i\sqrt{2}$

b. $-\frac{1}{7} \pm \frac{i\sqrt{97}}{7}$

SECTION 10.1 STUDY SET

VOCABULARY

Fill in the blanks.

1. An equation of the form $ax^2 + bx + c = 0$, where $a \neq 0$, is called a quadratic equation.
2. $x^2 + 6x + 9$ is called a perfect-square trinomial because it factors as $(x + 3)^2$.
3. When we add 16 to $x^2 + 8x$, we say that we have completed the square on $x^2 + 8x$.
4. The leading coefficient of $5x^2 - 2x + 7$ is 5 and the constant term is 7.

CONCEPTS

Fill in the blanks.

5. For any nonnegative number c, if $x^2 = c$, then $x = \sqrt{c}$ or $x = -\sqrt{c}$.
6. To complete the square on $x^2 + 10x$, add the square of one-half of the coefficient of x.
7. Find one-half of the given number and square the result.
   a. 12 36
   b. -5 $\frac{25}{4}$
8. Fill in the blanks to factor the perfect-square trinomial.
   a. $x^2 + 8x + 16 = (x + 4)^2$
   b. $x^2 - 9x + \frac{81}{4} = \left(x - \frac{9}{2}\right)^2$
9. What is the first step to solve each equation by completing the square? **Do not solve.**
   a. $x^2 + 9x + 7 = 0$ Subtract 7 from both sides.
   b. $4x^2 + 5x - 16 = 0$ Divide both sides by 4.
10. Divide both sides of the equation by the proper number to make the coefficient of x^2 equal to 1. **Do not solve.**
   a. $2x^2 - 3x + 6 = 0$ $x^2 - \frac{3}{2}x + 3 = 0$
   b. $3x^2 + 6x - 4 = 0$ $x^2 + 2x - \frac{4}{3} = 0$
11. Use a check to determine whether $-2 + \sqrt{2}$ is a solution of $x^2 + 4x + 2 = 0$. It is a solution.
12. Determine whether each statement is true or false.
   a. Any quadratic equation can be solved by the factoring method. False
   b. Any quadratic equation can be solved by completing the square. True

NOTATION

13. We read $8 \pm \sqrt{3}$ as "eight plus or minus the square root of 3."
14. When solving a quadratic equation, a student obtains $x = \frac{-5 \pm \sqrt{7}}{3}$.
   a. How many solutions are represented by this notation? List them. 2; $\frac{-5 + \sqrt{7}}{3}$, $\frac{-5 - \sqrt{7}}{3}$
   b. Approximate the solutions to the nearest hundredth. $-0.78, -2.55$

GUIDED PRACTICE

Use the square root property to solve each equation. See Example 1.

15. $t^2 - 11 = 0$ $\pm\sqrt{11}$
16. $w^2 - 47 = 0$ $\pm\sqrt{47}$
17. $x^2 - 35 = 0$ $\pm\sqrt{35}$
18. $x^2 - 101 = 0$ $\pm\sqrt{101}$
19. $z^2 - 50 = 0$ $\pm 5\sqrt{2}$
20. $u^2 - 24 = 0$ $\pm 2\sqrt{6}$
21. $3x^2 - 16 = 0$ $\pm\frac{4\sqrt{3}}{3}$
22. $5x^2 - 49 = 0$ $\pm\frac{7\sqrt{5}}{5}$

Use the square root property to solve each equation. See Example 3.

23. $p^2 = -16$ $\pm 4i$
24. $q^2 = -25$ $\pm 5i$
25. $a^2 + 8 = 0$ $\pm 2i\sqrt{2}$
26. $m^2 + 18 = 0$ $\pm 3i\sqrt{2}$
27. $4m^2 + 81 = 0$ $\pm\frac{9}{2}i$
28. $9n^2 + 121 = 0$ $\pm\frac{11}{3}i$
29. $6b^2 + 144 = 0$ $\pm 2i\sqrt{6}$
30. $9n^2 + 288 = 0$ $\pm 4i\sqrt{2}$

Use the square root property to solve each equation. See Example 4.

31. $(x + 5)^2 = 9$ $-8, -2$
32. $(x - 1)^2 = 4$ $3, -1$
33. $(t + 4)^2 = 16$ $0, -8$
34. $(s - 7)^2 = 9$ $4, 10$

See Example 5.

35. Let $f(x) = (x + 5)^2$. For what value(s) of x is $f(x) = 3$? $-5 \pm \sqrt{3}$
36. Let $f(x) = (x + 3)^2$. For what value(s) of x is $f(x) = 7$? $-3 \pm \sqrt{7}$
37. Let $g(a) = (7a - 2)^2$. For what value(s) of a is $g(a) = 8$? $\frac{2 \pm 2\sqrt{2}}{7}$
38. Let $h(c) = (11c - 2)^2$. For what value(s) of c is $h(c) = 12$? $\frac{2 \pm 2\sqrt{3}}{11}$

Complete the square and factor the resulting perfect-square trinomial. See Example 6.

39. $x^2 + 24x$ $x^2 + 24x + 144 = (x + 12)^2$
40. $y^2 - 18y$ $y^2 - 18y + 81 = (y - 9)^2$
41. $a^2 - 7a$ $a^2 - 7a + \frac{49}{4} = \left(a - \frac{7}{2}\right)^2$
42. $b^2 + 11b$ $b^2 + 11b + \frac{121}{4} = \left(b + \frac{11}{2}\right)^2$
43. $x^2 + \frac{2}{3}x$ $x^2 + \frac{2}{3}x + \frac{1}{9} = \left(x + \frac{1}{3}\right)^2$
44. $x^2 + \frac{2}{5}x$ $x^2 + \frac{2}{5}x + \frac{1}{25} = \left(x + \frac{1}{5}\right)^2$
45. $m^2 - \frac{5}{6}m$ $m^2 - \frac{5}{6}m + \frac{25}{144} = \left(m - \frac{5}{12}\right)^2$
46. $n^2 - \frac{5}{3}n$ $n^2 - \frac{5}{3}n + \frac{25}{36} = \left(n - \frac{5}{6}\right)^2$

Use completing the square to solve each equation. Approximate each solution to the nearest hundredth. See Example 7.

47. $x^2 - 4x - 2 = 0$ $2 \pm \sqrt{6}$; 4.45, -0.45
48. $x^2 - 6x - 4 = 0$ $3 \pm \sqrt{13}$; 6.61, -0.61
49. $x^2 - 12x + 1 = 0$ $6 \pm \sqrt{35}$; 11.92, 0.08
50. $x^2 - 18x + 2 = 0$ $9 \pm \sqrt{79}$; 17.89, 0.11
51. $t^2 + 20t + 25 = 0$ $-10 \pm 5\sqrt{3}$; $-1.34, -18.66$
52. $t^2 + 22t + 93 = 0$ $-11 \pm 2\sqrt{7}$; $-5.71, -16.29$

53. $t^2 + 16t - 16 = 0$
$-8 \pm 4\sqrt{5}$; 0.94, −16.94

54. $t^2 + 14t - 7 = 0$
$-7 \pm 2\sqrt{14}$; 0.48, −14.48

Use completing the square to solve each equation. See Example 8.

55. $2x^2 - x - 1 = 0$
$-\frac{1}{2}, 1$

▶ 56. $2x^2 - 5x + 2 = 0$
$\frac{1}{2}, 2$

▶ 57. $12t^2 - 5t - 3 = 0$
$\frac{3}{4}, -\frac{1}{3}$

58. $5m^2 + 13m - 6 = 0$
$\frac{2}{5}, -3$

Use completing the square to solve each equation. Approximate each solution to the nearest hundredth. See Example 9.

59. $3x^2 - 12x + 1 = 0$
$\frac{6 \pm \sqrt{33}}{3}$; 3.91, 0.09

60. $6x^2 - 12x + 1 = 0$
$\frac{6 \pm \sqrt{30}}{6}$; 1.91, 0.09

▶ 61. $2x^2 + 5x - 2 = 0$
$\frac{-5 \pm \sqrt{41}}{4}$; 0.35, −2.85

▶ 62. $2x^2 - 8x + 5 = 0$
$\frac{4 \pm \sqrt{6}}{2}$; 3.22, 0.78

Use completing the square to solve each equation. See Example 10.

63. $p^2 + 2p + 2 = 0$
$-1 \pm i$

▶ 64. $x^2 - 6x + 10 = 0$
$3 \pm i$

65. $y^2 + 8y + 18 = 0$
$-4 \pm i\sqrt{2}$

▶ 66. $n^2 + 10n + 28 = 0$
$-5 \pm i\sqrt{3}$

67. $x^2 + \frac{2}{3}x + 7 = 0$
$-\frac{1}{3} \pm \frac{i\sqrt{62}}{3}$

68. $x^2 + \frac{2}{5}x + 6 = 0$
$-\frac{1}{5} \pm \frac{i\sqrt{149}}{5}$

69. $a^2 - \frac{1}{2}a + 1 = 0$
$\frac{1}{4} \pm \frac{i\sqrt{15}}{4}$

70. $b^2 - \frac{1}{4}b + 1 = 0$
$\frac{1}{8} \pm \frac{3i\sqrt{7}}{8}$

TRY IT YOURSELF

Solve each equation. Approximate the solutions to the nearest hundredth when appropriate.

71. $(3x - 1)^2 = 25$
$2, -\frac{4}{3}$

▶ 72. $(5x - 2)^2 = 64$
$2, -\frac{6}{5}$

▶ 73. $3x^2 - 6x = 1$
$\frac{3 \pm 2\sqrt{3}}{3}$; 2.15, −0.15

▶ 74. $2x^2 - 6x = -3$
$\frac{3 \pm \sqrt{3}}{2}$; 2.37, 0.63

▶ 75. $x^2 + 8x + 6 = 0$
$-4 \pm \sqrt{10}$; −7.16, −0.84

▶ 76. $x^2 + 6x + 4 = 0$
$-3 \pm \sqrt{5}$; −5.24, −0.76

77. $6x^2 + 72 = 0$
$\pm 2i\sqrt{3}$

78. $5x^2 + 40 = 0$
$\pm 2i\sqrt{2}$

▶ 79. $x^2 - 2x = 17$
$1 \pm 3\sqrt{2}$; 5.24, −3.24

80. $x^2 + 10x = 7$
$-5 \pm 4\sqrt{2}$; 0.66, −10.66

81. $m^2 - 7m + 3 = 0$
$\frac{7 \pm \sqrt{37}}{2}$; 6.54, 0.46

82. $m^2 - 5m + 3 = 0$
$\frac{5 \pm \sqrt{13}}{2}$; 4.30, 0.70

83. $7h^2 = 35$
$\pm\sqrt{5}$; ±2.24

84. $9n^2 = 99$
$\pm\sqrt{11}$; ±3.32

85. $\frac{7x + 1}{5} = -x^2$
$\frac{-7 \pm \sqrt{29}}{10}$; −0.16, −1.24

▶ 86. $\frac{3}{8}x^2 = \frac{1}{8} - x$
$\frac{-4 \pm \sqrt{19}}{3}$; 0.12, −2.79

87. $t^2 + t + 3 = 0$
$-\frac{1}{2} \pm \frac{\sqrt{11}}{2}i$

88. $b^2 - b + 5 = 0$
$\frac{1}{2} \pm \frac{\sqrt{19}}{2}i$

89. $(8x + 5)^2 = 24$
$\frac{-5 \pm 2\sqrt{6}}{8}$; −1.24, −0.01

90. $(3y - 2)^2 = 18$
$\frac{2 \pm 3\sqrt{2}}{3}$; 2.08, −0.75

91. $r^2 - 6r - 27 = 0$
$-3, 9$

92. $s^2 - 6s - 40 = 0$
$-4, 10$

▶ 93. $4p^2 + 2p + 3 = 0$
$-\frac{1}{4} \pm \frac{\sqrt{11}}{4}i$

94. $3m^2 - 2m + 3 = 0$
$\frac{1}{3} \pm \frac{2\sqrt{2}}{3}i$

Look Alikes . . .

95. a. $x^2 - 24 = 0$
$\pm 2\sqrt{6}$; ±4.90

b. $x^2 + 24 = 0$
$\pm 2i\sqrt{6}$

▶ 96. a. $x^2 + 7x - 8 = 0$
$-8, 1$

b. $x^2 + 7x - 9 = 0$
$\frac{-7 \pm \sqrt{85}}{2}$; 1.11, −8.11

97. a. $2m^2 - 8m = 0$
$0, 4$

b. $2m^2 - 8m = 1$
$\frac{4 \pm 3\sqrt{2}}{2}$; 4.12, −0.12

98. a. $a^2 + a - 7 = 0$
$\frac{-1 \pm \sqrt{29}}{2}$; 2.19, −3.19

b. $a^2 - a - 7 = 0$
$\frac{1 \pm \sqrt{29}}{2}$; 3.19, −2.19

99. a. $x^2 - 4x + 20 = 0$
$2 \pm 4i$

b. $x^2 - 4x - 20 = 0$
$2 \pm 2\sqrt{6}$; 6.90, −2.90

100. a. $a^2 = 6a - 3$
$3 \pm \sqrt{6}$; 5.45, 0.55

b. $n^2 = 6n + 3$
$3 \pm 2\sqrt{3}$; 6.46, −0.46

101. a. $2r^2 - 4r + 3 = 0$
$\frac{2 \pm i\sqrt{2}}{2}$

b. $2r^2 - 4r - 3 = 0$
$\frac{2 \pm \sqrt{10}}{2}$; 2.58, −0.58

▶ 102. a. $5y^2 + 15y + 12 = 0$
$\frac{-15 \pm i\sqrt{15}}{10}$

b. $5y^2 + 15y - 12 = 0$
$\frac{-15 \pm \sqrt{465}}{10}$; 0.66, −3.66

APPLICATIONS

▶ 103. **Movie Stunts.** According to the *Guinness Book of World Records,* stuntman Dan Koko fell a distance of 312 feet into an airbag after jumping from the Vegas World Hotel and Casino. The distance d in feet traveled by a free-falling object in t seconds is given by the formula $d = 16t^2$. To the nearest tenth of a second, how long did the fall last? 4.4 sec

▶ 104. **Geography.** The surface area S of a sphere is given by the formula $S = 4\pi r^2$, where r is the radius of the sphere. An almanac lists the surface area of the Earth as 196,938,800 square miles. Assuming the Earth to be spherical, what is its radius to the nearest mile? 3,959 mi

▶ 105. **Accidents.** The height h (in feet) of an object that is dropped from a height of s feet is given by the formula $h = s - 16t^2$, where t is the time the object has been falling. A 5-foot-tall woman on a sidewalk looks directly overhead and sees a window washer drop a bottle from four stories up. How long does she have to get out of the way? Round to the nearest tenth. (A story is 12 feet.) 1.6 sec

106. **Flags.** In 1912, an order by President Taft fixed the width and length of the U.S. flag in the ratio 1 to 1.9. If 100 square feet of cloth are to be used to make a U.S. flag, estimate its dimensions to the nearest $\frac{1}{4}$ foot. Width: $7\frac{1}{4}$ ft; length: $13\frac{3}{4}$ ft

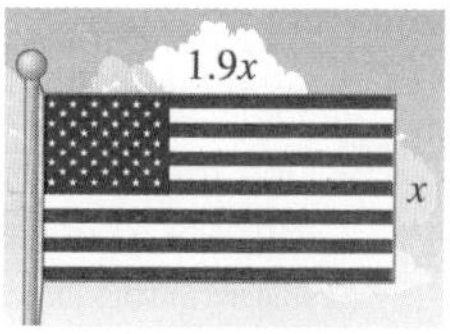

107. Automobile Engines. As the piston shown moves upward, it pushes a cylinder of a gasoline/air mixture that is ignited by the spark plug. The formula that gives the volume of a cylinder is $V = \pi r^2 h$, where r is the radius and h the height. Find the radius of the piston (to the nearest hundredth of an inch) if it displaces 47.75 cubic inches of gasoline/air mixture as it moves from its lowest to its highest point. 1.70 in.

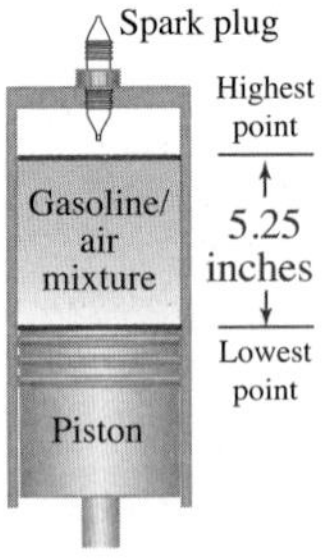

108. Investments. If P dollars are deposited in an account that pays an annual rate of interest r, then in n years, the amount of money A in the account is given by the formula $A = P(1 + r)^n$. A savings account was opened on January 3, 2006, with a deposit of $10,000 and closed on January 2, 2008, with an ending balance of $11,772.25. Find the rate of interest. 8.5%

109. Physics. Albert Einstein discovered a connection between energy and mass. This relationship (*energy equals mass times the velocity of light squared*) is expressed in the equation $E = mc^2$. Solve for c. $c = \frac{\sqrt{Em}}{m}$

110. Right Triangles. The Pythagorean theorem relates the lengths of the sides in a right triangle: $a^2 + b^2 = c^2$, where a and b represent the lengths of the legs and c represents the length of the hypotenuse. Solve for b. $b = \sqrt{c^2 - a^2}$

WRITING

111. Give an example of a perfect-square trinomial. Why do you think the word "perfect" is used to describe it?

112. Explain why completing the square on $x^2 + 5x$ is more difficult than completing the square on $x^2 + 4x$.

113. Explain the error in the work shown below.

a. $\frac{4 \pm \sqrt{3}}{8} = \frac{\cancel{4}^{1} \pm \sqrt{3}}{\cancel{4}_{1} \cdot 2} = \frac{1 \pm \sqrt{3}}{2}$

b. $\frac{1 \pm \sqrt{5}}{5} = \frac{1 \pm \sqrt{\cancel{5}}^{1}}{\cancel{5}_{1}} = \frac{1 \pm 1}{1}$

114. Explain the steps involved in expressing $8 \pm \frac{\sqrt{15}}{2}$ as a single fraction with denominator 2.

REVIEW

Simplify each expression. All variables represent positive real numbers.

115. $\sqrt[3]{40a^3b^6}$ $2ab^2\sqrt[3]{5}$

116. $\sqrt[8]{x^{24}}$ x^3

117. $\sqrt[4]{\frac{16}{625}}$ $\frac{2}{5}$

118. $\sqrt{175a^2b^3}$ $5ab\sqrt{7b}$

CHALLENGE PROBLEMS

119. What number must be added to $x^2 + \sqrt{3}x$ to make a perfect-square trinomial? $\frac{3}{4}$

120. Solve $x^2 + \sqrt{3}x - \frac{1}{4} = 0$ by completing the square.
$-\frac{\sqrt{3}}{2} \pm 1 = \frac{-\sqrt{3} \pm 2}{2}$

SECTION 10.2 The Quadratic Formula

OBJECTIVES

1. Derive the quadratic formula.
2. Solve quadratic equations using the quadratic formula.
3. Write equivalent equations to make quadratic formula calculations easier.
4. Use the quadratic formula to solve application problems.

ARE YOU READY? *Are You Ready? exercises available online at www.webassign.net/brookscole*

The following problems review some basic skills that are needed when solving quadratic equations using the quadratic formula.

1. Evaluate: $\sqrt{5^2 - 4(4)(-6)}$ 11

2. Simplify: $\sqrt{45}$ $3\sqrt{5}$

3. How many terms does $2x^2 - x + 7$ have? What is the coefficient of each term? 3; 2, −1, 7

4. Evaluate: $\frac{-5 \pm 11}{8}$ $\frac{3}{4}, -2$

5. Classify $\sqrt{-28}$ as a rational, irrational, or not a real number. Not a real number

6. Approximate $\frac{6 + \sqrt{3}}{2}$ to the nearest hundredth. 3.87

We can solve quadratic equations by completing the square, but that method is often lengthy. In this section, we will develop a formula, called the *quadratic formula,* that enables us to solve quadratic equations with less effort.

1 Derive the Quadratic Formula.

To develop a formula that will produce the solutions of any given quadratic equation, we start with a quadratic equation in **standard form**, $ax^2 + bx + c = 0$, where $a > 0$. We can solve for x by completing the square.

Teaching Tip: You may want to derive the quadratic formula this way:
$ax^2 + bx + c = 0$
$ax^2 + bx = -c$
$4a^2x^2 + 4abx = -4ac$
$4a^2x^2 + 4abx + b^2 = b^2 - 4ac$
$(2ax + b)^2 = b^2 - 4ac$
$2ax + b = \pm\sqrt{b^2 - 4ac}$
$2ax = -b \pm \sqrt{b^2 - 4ac}$
$x = \frac{-b \pm \sqrt{b^2 - 4ac}}{2a}$

$$ax^2 + bx + c = 0$$

$$\frac{ax^2}{a} + \frac{bx}{a} + \frac{c}{a} = \frac{0}{a}$$ Divide both sides by a so that the coefficient of x^2 is 1.

$$x^2 + \frac{b}{a}x + \frac{c}{a} = 0$$ Simplify: $\frac{ax^2}{a} = x^2$. Write $\frac{bx}{a}$ as $\frac{b}{a}x$

$$x^2 + \frac{b}{a}x = -\frac{c}{a}$$ Subtract $\frac{c}{a}$ from both sides so that only the variable terms are on the left side of the equation and the constant is on the right side.

We can complete the square on $x^2 + \frac{b}{a}x$ by adding the square of one-half of the coefficient of x. Since the coefficient of x is $\frac{b}{a}$, we have $\frac{1}{2} \cdot \frac{b}{a} = \frac{b}{2a}$ and $\left(\frac{b}{2a}\right)^2 = \frac{b^2}{4a^2}$.

$$x^2 + \frac{b}{a}x + \frac{b^2}{4a^2} = -\frac{c}{a} + \frac{b^2}{4a^2}$$ To complete the square, add $\frac{b^2}{4a^2}$ to both sides.

$$x^2 + \frac{b}{a}x + \frac{b^2}{4a^2} = -\frac{4ac}{4aa} + \frac{b^2}{4a^2}$$ On the right side, build $-\frac{c}{a}$ by multiplying it by $\frac{4a}{4a}$. Now the fractions on that side have the common denominator $4a^2$.

$$\left(x + \frac{b}{2a}\right)^2 = \frac{b^2 - 4ac}{4a^2}$$ On the left side, factor the perfect-square trinomial. On the right side, add the fractions. In the numerator, write $-4ac + b^2$ as $b^2 - 4ac$.

$$x + \frac{b}{2a} = \pm\sqrt{\frac{b^2 - 4ac}{4a^2}}$$ Use the square root property.

$$x + \frac{b}{2a} = \pm\frac{\sqrt{b^2 - 4ac}}{\sqrt{4a^2}}$$ On the right side, the square root of a quotient is the quotient of square roots.

$$x + \frac{b}{2a} = \pm\frac{\sqrt{b^2 - 4ac}}{2a}$$ On the right side, simplify the denominator. Since $a > 0$, $\sqrt{4a^2} = 2a$.

$$x = -\frac{b}{2a} \pm \frac{\sqrt{b^2 - 4ac}}{2a}$$ To isolate x, subtract $\frac{b}{2a}$ from both sides.

$$x = \frac{-b \pm \sqrt{b^2 - 4ac}}{2a}$$ Combine the fractions. Write the sum (and difference) over the common denominator $2a$.

The Language of Algebra

To **derive** means to obtain by reasoning. To *derive* the quadratic formula means to solve $ax^2 + bx + c = 0$ for x, using the series of steps shown here, to obtain

$$x = \frac{-b \pm \sqrt{b^2 - 4ac}}{2a}$$

This result is called the **quadratic formula.** To develop this formula, we assumed that a was positive. If a is negative, similar steps are used, and we obtain the same result. This formula is very useful and should be memorized.

The Quadratic Formula

The solutions of $ax^2 + bx + c = 0$, with $a \neq 0$, are given by

$$x = \frac{-b \pm \sqrt{b^2 - 4ac}}{2a}$$

Read as "x equals the opposite of b plus or minus the square root of b squared minus 4ac, all over 2a."

The quadratic formula is a compact way of representing two solutions:

$$x = \frac{-b + \sqrt{b^2 - 4ac}}{2a} \quad \text{or} \quad x = \frac{-b - \sqrt{b^2 - 4ac}}{2a}$$

2 Solve Quadratic Equations Using the Quadratic Formula.

In the next example, we will use the quadratic formula to solve a quadratic equation.

EXAMPLE 1 Solve $2x^2 - 5x - 3 = 0$ by using the quadratic formula.

Teaching Tip: Point out that the leading coefficient does not have to be 1 to use the quadratic formula.

Strategy We will begin by comparing $2x^2 - 5x - 3 = 0$ to the standard form $ax^2 + bx + c = 0$.

Why To use the quadratic formula, we need to identify the values of a, b, and c.

Solution

$$\begin{array}{l} 2x^2 - 5x - 3 = 0 \\ \uparrow \quad\ \ \uparrow \quad\ \uparrow \\ ax^2 + bx + c = 0 \end{array}$$ This is the equation to solve.

Caution

Make sure to include the correct sign when determining a, b, and c. In this example, $b = -5$ and $c = -3$.

We see that $a = 2$, $b = -5$, and $c = -3$. To find the solutions of the equation, we substitute these values into the quadratic formula and evaluate the right side.

$$x = \frac{-b \pm \sqrt{b^2 - 4ac}}{2a}$$ This is the quadratic formula.

$$x = \frac{-(-5) \pm \sqrt{(-5)^2 - 4(2)(-3)}}{2(2)}$$ Substitute 2 for a, -5 for b, and -3 for c.

$$x = \frac{5 \pm \sqrt{25 - (-24)}}{4}$$ Simplify: $-(-5) = 5$. Evaluate the power and multiply within the radical. Multiply in the denominator.

$$x = \frac{5 \pm \sqrt{49}}{4}$$ Simplify within the radical.

$$x = \frac{5 \pm 7}{4}$$ Evaluate the radical: $\sqrt{49} = 7$.

Caution

When writing the quadratic formula, be careful to draw the fraction bar so that it includes the entire numerator. Do not write

$$x = -b \pm \frac{\sqrt{b^2 - 4ac}}{2a}$$

or

$$x = -b \pm \sqrt{\frac{b^2 - 4ac}{2a}}$$

To find the first solution, we evaluate the expression using the + symbol. To find the second solution, we evaluate the expression using the − symbol.

$$x = \frac{5 + 7}{4} \quad \text{or} \quad x = \frac{5 - 7}{4}$$

$$x = \frac{12}{4} \qquad\qquad x = \frac{-2}{4}$$

$$x = 3 \qquad\qquad x = -\frac{1}{2}$$

Teaching Tip: Explain that since the solutions of this quadratic equation are rational numbers, the equation also can be solved by factoring.

The solutions are 3 and $-\frac{1}{2}$ and the solution set is $\{3, -\frac{1}{2}\}$. Check each solution in the original equation.

Self Check 1 Solve $4x^2 - 7x - 2 = 0$ by using the quadratic formula. $2, -\frac{1}{4}$

Teaching Example 1 Solve $3x^2 - 11x - 4 = 0$ by using the quadratic formula.
Answer: $4, -\frac{1}{3}$

Now Try Problem 13

To solve a quadratic equation in x using the quadratic formula, we follow these steps.

Solving a Quadratic Equation in *x* Using the Quadratic Formula

1. Write the equation in standard form: $ax^2 + bx + c = 0$.
2. Identify a, b, and c.
3. Substitute the values for a, b, and c in the quadratic formula and evaluate the right side to obtain the solutions.

$$x = \frac{-b \pm \sqrt{b^2 - 4ac}}{2a}$$

EXAMPLE 2 Solve: $2x^2 = -4x - 1$

Teaching Tip: Before solving, ask your students how the given equation's form is different from that in Example 1.

Strategy We will write the equation in standard form $ax^2 + bx + c = 0$. Then we will identify the values of a, b, and c, and substitute these values into the quadratic formula.

Why The quadratic equation must be in standard form to identify the values of a, b, and c.

Solution To write the equation in standard form, we need to have all nonzero terms on the left side and 0 on the right side.

$$2x^2 = -4x - 1 \quad \text{This is the equation to solve.}$$

$$2x^2 + 4x + 1 = 0 \quad \text{To get 0 on the right side, add 4x and 1 to both sides.}$$

In the resulting equivalent equation, $a = 2$, $b = 4$, and $c = 1$.

$$x = \frac{-b \pm \sqrt{b^2 - 4ac}}{2a} \quad \text{This is the quadratic formula.}$$

$$x = \frac{-4 \pm \sqrt{4^2 - 4(2)(1)}}{2(2)} \quad \text{Substitute 2 for } a\text{, 4 for } b\text{, and 1 for } c.$$

$$x = \frac{-4 \pm \sqrt{16 - 8}}{4} \quad \text{Evaluate the expression within the radical. Multiply in the denominator.}$$

$$x = \frac{-4 \pm \sqrt{8}}{4} \quad \text{Do the subtraction within the radical.}$$

$$x = \frac{-4 \pm 2\sqrt{2}}{4} \quad \text{Simplify the radical: } \sqrt{8} = \sqrt{4 \cdot 2} = 2\sqrt{2}.$$

We can write the solutions in simpler form by factoring out 2 from the two terms in the numerator and removing the common factor of 2 in the numerator and denominator.

$$x = \frac{-4 \pm 2\sqrt{2}}{4} = \frac{2(-2 \pm \sqrt{2})}{4} = \frac{\overset{1}{\cancel{2}}(-2 \pm \sqrt{2})}{\underset{1}{\cancel{2}} \cdot 2} = \frac{-2 \pm \sqrt{2}}{2} \quad \text{Factor first.}$$

The two irrational solutions are $\frac{-2 + \sqrt{2}}{2}$ and $\frac{-2 - \sqrt{2}}{2}$ and the solution set is $\left\{\frac{-2 + \sqrt{2}}{2}, \frac{-2 - \sqrt{2}}{2}\right\}$. We can approximate the solutions using a calculator. To the nearest hundredth, they are -0.29 and -1.71.

Success Tip

Perhaps you noticed that Example 1 could be solved by factoring. However, that is not the case for Example 2. These observations illustrate an important fact: The quadratic formula can be used to solve ***any*** quadratic equation.

Teaching Tip: Explain that since the solutions of this quadratic equation are irrational numbers, it cannot be solved by factoring. The reason the solutions are irrational is because the radicand, 8, is not a perfect square.

Notation

The solutions also can be written:

$$-\frac{2}{2} \pm \frac{\sqrt{2}}{2} = -1 \pm \frac{\sqrt{2}}{2}$$

Self Check 2 Solve $3x^2 = 2x + 3$. Approximate the solutions to the nearest hundredth. $\frac{1 \pm \sqrt{10}}{3}$; $-0.72, 1.39$

Now Try Problem 25

Teaching Example 2 Solve $4x^2 = 2x + 1$. Approximate the solutions to the nearest hundredth.

Answer: $\frac{1 \pm \sqrt{5}}{4}$; $0.81, -0.31$

EXAMPLE 3 Solve: $m^2 + m = -1$

Strategy We will write the equation in standard form $am^2 + bm + c = 0$. Then we will identify the values of a, b, and c, and substitute these values into the quadratic formula.

Why The quadratic equation must be in standard form to identify the values of a, b, and c.

Solution To write $m^2 + m = -1$ in standard form, we add 1 to both sides, to get

$m^2 + m + 1 = 0$ This is the equation to solve.

In the resulting equivalent equation, $a = 1$, $b = 1$, and $c = 1$:

$m = \frac{-b \pm \sqrt{b^2 - 4ac}}{2a}$ This is the quadratic formula.

$m = \frac{-1 \pm \sqrt{1^2 - 4(1)(1)}}{2(1)}$ Substitute 1 for *a*, 1 for *b*, and 1 for *c*.

$m = \frac{-1 \pm \sqrt{1 - 4}}{2}$ Evaluate the expression within the radical. Multiply in the denominator.

$m = \frac{-1 \pm \sqrt{-3}}{2}$ Do the subtraction within the radical.

$m = \frac{-1 \pm i\sqrt{3}}{2}$ Simplify the radical: $\sqrt{-3} = \sqrt{-1 \cdot 3} = \sqrt{-1}\sqrt{3} = i\sqrt{3}$.

The solutions are two complex numbers involving i: $-\frac{1}{2} + \frac{\sqrt{3}}{2}i$ and $-\frac{1}{2} - \frac{\sqrt{3}}{2}i$ and the solution set is $\left\{-\frac{1}{2} + \frac{\sqrt{3}}{2}i, -\frac{1}{2} - \frac{\sqrt{3}}{2}i\right\}$.

Caution

Since the variable in the equation is m, not x, we must change the variable in the quadratic formula to reflect this:

$$m = \frac{-b \pm \sqrt{b^2 - 4ac}}{2a}$$

Teaching Tip: Ask your students to point out the first step in the solution process in which they could see that the solutions of this equation would involve i.

Notation

The solutions are written in complex number form $a + bi$. They also could be written as

$$\frac{-1 \pm i\sqrt{3}}{2}$$

Self Check 3 Solve: $a^2 + 3a = -5$ $-\frac{3}{2} \pm \frac{\sqrt{11}}{2}i$

Now Try Problem 29

Teaching Example 3 Solve: $n^2 - n = -2$

Answer: $\frac{1}{2} \pm \frac{\sqrt{7}}{2}i$

3 Write Equivalent Equations to Make Quadratic Formula Calculations Easier.

When solving a quadratic equation by the quadratic formula, we often can simplify the calculations by solving a simpler, but equivalent equation.

EXAMPLE 4 For each equation below, write an equivalent equation so that the quadratic formula calculations will be simpler.

a. $-2x^2 + 4x - 1 = 0$ **b.** $x^2 + \frac{4}{5}x - \frac{1}{3} = 0$

c. $20x^2 - 60x - 40 = 0$ **d.** $0.03x^2 - 0.04x - 0.01 = 0$

Strategy We will multiply both sides of each equation by a carefully chosen number.

Why In each case, the objective is to find an equivalent equation whose values of a, b, and c are easier to work with than those of the given equation.

Solution **a.** It is often easier to solve a quadratic equation using the quadratic formula if a is positive. If we multiply (or divide) both sides of $-2x^2 + 4x - 1 = 0$ by -1, we obtain an equivalent equation with $a > 0$.

$-2x^2 + 4x - 1 = 0$ Here, $a = -2$.

$-1(-2x^2 + 4x - 1) = -1(0)$ Don't forget to multiply each term by -1.

$2x^2 - 4x + 1 = 0$ Now $a = 2$.

Success Tip

Unlike completing the square, the quadratic formula does not require the leading coefficient to be 1.

b. For $x^2 + \frac{4}{5}x - \frac{1}{3} = 0$, two coefficients are fractions: $b = \frac{4}{5}$ and $c = -\frac{1}{3}$. We can multiply both sides of the equation by their least common denominator, 15, to obtain an equivalent equation having coefficients that are integers.

$$x^2 + \frac{4}{5}x - \frac{1}{3} = 0$$ Here, $a = 1$, $b = \frac{4}{5}$, and $c = -\frac{1}{3}$.

$$\mathbf{15}\left(x^2 + \frac{4}{5}x - \frac{1}{3}\right) = \mathbf{15}(0)$$

$$15x^2 + 12x - 5 = 0$$ On the left side, distribute the multiplication by 15. Now $a = 15$, $b = 12$, and $c = -5$.

The Language of Algebra

Recall that **equivalent equations** have the same solutions.

c. For $20x^2 - 60x - 40 = 0$, the coefficients 20, -60, and -40 have a common factor of 20. If we divide both sides of the equation by their GCF, we obtain an equivalent equation having smaller coefficients.

$$20x^2 - 60x - 40 = 0$$ Here, $a = 20$, $b = -60$, and $c = -40$.

$$\frac{20x^2}{\mathbf{20}} - \frac{60x}{\mathbf{20}} - \frac{40}{\mathbf{20}} = \frac{0}{\mathbf{20}}$$ The division by 20 is done term-by-term.

$$x^2 - 3x - 2 = 0$$ Now $a = 1$, $b = -3$, and $c = -2$.

d. For $0.03x^2 - 0.04x - 0.01 = 0$, all three coefficients are decimals. We can multiply both sides of the equation by 100 to obtain an equivalent equation having coefficients that are integers.

$$0.03x^2 - 0.04x - 0.01 = 0$$ Here, $a = 0.03$, $b = -0.04$, and $c = -0.01$.

$$\mathbf{100}(0.03x^2 - 0.04x - 0.01) = \mathbf{100}(0)$$

$$3x^2 - 4x - 1 = 0$$ On the left side, distribute 100. Now $a = 3$, $b = -4$, and $c = -1$.

Teaching Example 4 For each equation, write an equivalent equation so that the quadratic formula calculations will be simpler:
a. $\frac{1}{2}x^2 - \frac{3}{5}x + \frac{7}{10} = 0$
b. $-5x^2 + 3x - 9 = 0$
c. $75x^2 - 25x + 50 = 0$
d. $0.06n^2 - 0.09n - 0.02 = 0$
Answers: a. $5x^2 - 6x + 7 = 0$
b. $5x^2 - 3x + 9 = 0$
c. $3x^2 - x + 2 = 0$
d. $6n^2 - 9n - 2 = 0$

Self Check 4 For each equation, write an equivalent equation so that the quadratic formula calculations will be simpler.

a. $-6x^2 + 7x - 9 = 0$ $6x^2 - 7x + 9 = 0$

b. $\frac{1}{3}x^2 - \frac{2}{3}x - \frac{5}{6} = 0$ $2x^2 - 4x - 5 = 0$

c. $44x^2 + 66x - 99 = 0$ $4x^2 + 6x - 9 = 0$

d. $0.08x^2 - 0.07x - 0.02 = 0$ $8x^2 - 7x - 2 = 0$

Now Try Problems 37 and 39

4 Use the Quadratic Formula to Solve Application Problems.

A variety of real-world applications can be modeled by quadratic equations. However, such equations are often difficult or even impossible to solve using the factoring method. In those cases, we can use the quadratic formula to solve the equation.

EXAMPLE 5 **Shortcuts.** Instead of using the hallways, students are wearing a path through a planted quad area to walk 195 feet directly from the classrooms to the cafeteria. If the length of the hallway from the office to the cafeteria is 105 feet longer than the hallway from the office to the classrooms, how much walking are the students saving by taking the shortcut?

Analyze The two hallways and the shortcut form a right triangle with a hypotenuse 195 feet long. We will use the Pythagorean theorem to solve this problem.

Assign If we let $x =$ the length (in feet) of the hallway from the classrooms to the office, then the length of the hallway from the office to the cafeteria is $(x + 105)$ feet.

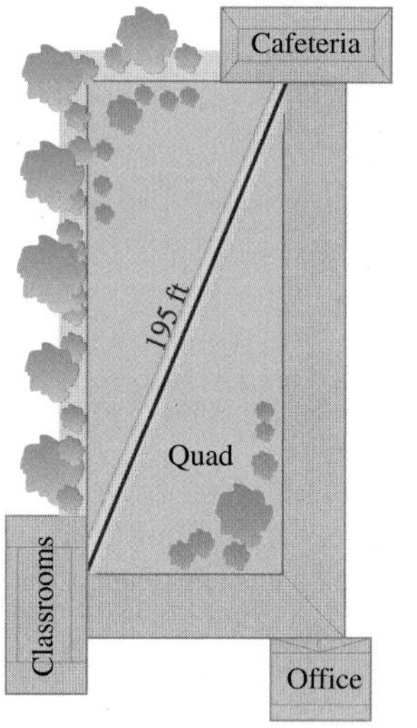

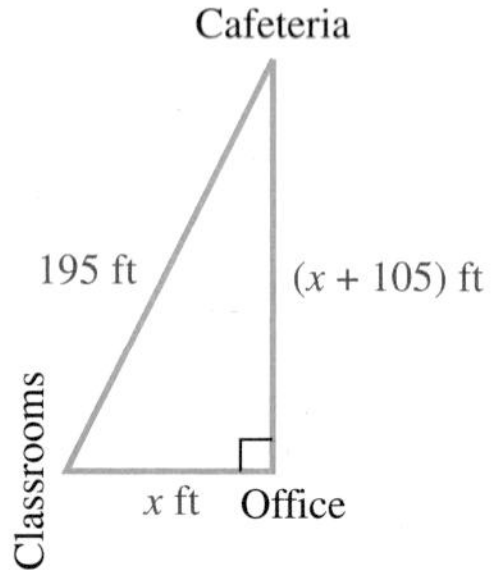

Form Substituting the lengths into the Pythagorean equation, we have

$a^2 + b^2 = c^2$	This is the Pythagorean equation.
$x^2 + (x + 105)^2 = 195^2$	Substitute x for a, (x + 105) for b, and 195 for c.
$x^2 + x^2 + 105x + 105x + 11{,}025 = 38{,}025$	Find $(x + 105)^2$.
$2x^2 + 210x + 11{,}025 = 38{,}025$	Combine like terms.
$2x^2 + 210x - 27{,}000 = 0$	To get 0 on the right side, subtract 38,025 from both sides. This is a quadratic equation.
$x^2 + 105x - 13{,}500 = 0$	The coefficients have a common factor of 2: Divide both sides by 2.

Solve To solve $x^2 + 105x - 13{,}500 = 0$, we will use the quadratic formula with $a = 1$, $b = 105$, and $c = -13{,}500$.

$x = \dfrac{-b \pm \sqrt{b^2 - 4ac}}{2a}$	This is the quadratic formula.
$x = \dfrac{-105 \pm \sqrt{105^2 - 4(1)(-13{,}500)}}{2(1)}$	Substitute for a, b, and c.
$x = \dfrac{-105 \pm \sqrt{65{,}025}}{2}$	Evaluate within the radical: $105^2 - 4(1)(-13{,}500) = 11{,}025 + 54{,}000 = 65{,}025$. Multiply in the denominator.
$x = \dfrac{-105 \pm 255}{2}$	Use a calculator: $\sqrt{65{,}025} = 255$.
$x = \dfrac{150}{2}$ or $x = \dfrac{-360}{2}$	Add: $-105 + 255 = 150$. Subtract: $-105 - 255 = -360$.
$x = 75$ | ~~$x = -180$~~	Do the division. Since the length of the hallway can't be negative, discard the solution -180.

State The length of the hallway from the classrooms to the office is 75 feet. The length of the hallway from the office to the cafeteria is $75 + 105 = 180$ feet. Instead of using the hallways, a distance of $75 + 180 = 255$ feet, the students are taking the 195-foot shortcut to the cafeteria, a savings of $(255 - 195)$, or 60 feet.

Check The length of the 180-foot hallway is 105 feet longer than the length of the 75-foot hallway. The sum of the squares of the lengths of the hallways is $75^2 + 180^2 = 38{,}025$. This equals the square of the length of the 195-foot shortcut: $195^2 = 38{,}025$. The result checks.

Teaching Example 5 Right Triangles. The hypotenuse of a right triangle is 40 in. long. The longer leg is 8 in. longer than the shorter leg. Find the lengths of the legs of the triangle. *Answer:* 24 in., 32 in.

Self Check 5 **Right Triangles.** The hypotenuse of a right triangle is 41 in. long. The longer leg is 31 inches longer than the shorter leg. Find the lengths of the legs of the triangle. 9 in., 40 in.

Now Try Problem 81

EXAMPLE 6 **Mass Transit.** A bus company has 4,000 passengers daily, each currently paying a 75¢ fare. For each 15¢ fare increase, the company estimates that it will lose 50 passengers. If the company needs to bring in \$6,570 per day to stay in business, what fare must be charged to produce this amount of revenue?

©Krivosheev Vitaly/Shutterstock.com

Analyze To understand how a fare increase affects the number of passengers, let's consider what happens if there are two fare increases. We organize the data in a table. The fares are expressed in terms of dollars.

Number of increases	New fare	Number of passengers
One \$0.15 increase	\$0.75 + \$0.15(1) = \$0.90	4,000 − 50(1) = 3,950
Two \$0.15 increases	\$0.75 + \$0.15(2) = \$1.05	4,000 − 50(2) = 3,900

In general, the new fare will be the old fare (\$0.75) plus the number of fare increases times \$0.15. The number of passengers who will pay the new fare is 4,000 minus 50 times the number of \$0.15 fare increases.

Assign If we let $x =$ the number of \$0.15 fare increases necessary to bring in \$6,570 daily, then $\$(0.75 + 0.15x)$ is the fare that must be charged. The number of passengers who will pay this fare is $4{,}000 - 50x$.

Form We can now form an equation, using the words of the problem.

The bus fare	times	the number of passengers who will pay that fare	equals	\$6,570.
$(0.75 + 0.15x)$	$\cdot$	$(4{,}000 - 50x)$	$=$	$6{,}570$

Solve

$$(0.75 + 0.15x)(4{,}000 - 50x) = 6{,}570$$

$$3{,}000 - 37.5x + 600x - 7.5x^2 = 6{,}570$$ Multiply the binomials on the left side.

$$-7.5x^2 + 562.5x + 3{,}000 = 6{,}570$$ Combine like terms: $-37.5x + 600x = 562.5x$.

$$-7.5x^2 + 562.5x - 3{,}570 = 0$$ To get 0 on the right side, subtract 6,570 from both sides. This is a quadratic equation.

$$7.5x^2 - 562.5x + 3{,}570 = 0$$ Multiply both sides by −1 so that the value of a, 7.5, is positive.

To solve this equation, we will use the quadratic formula.

$$x = \frac{-b \pm \sqrt{b^2 - 4ac}}{2a}$$ This is the quadratic formula.

$$x = \frac{-(-562.5) \pm \sqrt{(-562.5)^2 - 4(7.5)(3{,}570)}}{2(7.5)}$$ Substitute 7.5 for a, −562.5 for b, and 3,570 for c.

$$x = \frac{562.5 \pm \sqrt{209{,}306.25}}{15}$$ Evaluate within the radical: $(-562.5)^2 - 4(7.5)(3{,}570) = 316{,}406.25 - 107.100 = 209{,}306.25$. Multiply in the denominator.

$$x = \frac{562.5 \pm 457.5}{15}$$ Use a calculator: $\sqrt{209{,}306.25} = 457.5$.

$$x = \frac{1{,}020}{15} \quad \text{or} \quad x = \frac{105}{15}$$ Add: $562.5 + 457.5 = 1{,}020$. Subtract: $562.5 - 457.5 = 105$.

$$x = 68 \quad | \quad x = 7$$ Do the division.

State If there are 7 fifteen-cent increases in the fare, the new fare will be \$0.75 + \$0.15(7) = \$1.80. If there are 68 fifteen-cent increases in the fare, the new fare will be \$0.75 + \$0.15(68) = \$10.95. Although this fare would bring in the necessary revenue, a \$10.95 bus fare is unreasonable, so we discard it.

Check A fare of \$1.80 will be paid by $[4{,}000 - 50(7)] = 3{,}650$ bus riders. The amount of revenue brought in would be $\$1.80(3{,}650) = \$6{,}570$. The result checks.

Teaching Example 6 Ticket Prices. A jazz group on tour has been drawing average crowds of 500 people. It is projected that for every \$1 increase in the \$12 ticket price, the average attendance will decrease by 50. At what ticket price will nightly receipts be \$5,600?
Answer: \$14

Self Check 6 **Airport Shuttles.** A bus company shuttles 1,120 passengers daily between Rockford, Illinois, and O'Hare Airport. The current one-way fare is \$10. For each 25¢ increase in the fare, the company predicts that it will lose 48 passengers. What increase in fare will produce daily revenue of \$10, 208? \$1

Now Try Problem 91

EXAMPLE 7 **Lawyers.** The number of lawyers in the United States is approximated by the function $N(x) = 45x^2 + 21{,}000x + 560{,}000$, where $N(x)$ is the number of lawyers and x is the number of years after 1980. In what year does this model indicate that the United States had one million lawyers? (Based on data from the American Bar Association)

Strategy We will substitute 1,000,000 for $N(x)$ in the equation and solve for x.

Why The value of x will give the number of years after 1980 that the United States had 1,000,000 lawyers.

Solution

©Fuse/JupiterImages

$$N(x) = 45x^2 + 21{,}000x + 560{,}000$$ This is the quadratic function model.

$$1{,}000{,}000 = 45x^2 + 21{,}000x + 560{,}000$$ Replace N(x) with 1,000,000.

$$0 = 45x^2 + 21{,}000x - 440{,}000$$ To get 0 on the left side, subtract 1,000,000 from both sides.

We can simplify the calculations by dividing both sides of the equation by 5, which is the greatest common factor of 45, 21,000, and 440,000.

$$9x^2 + 4{,}200x - 88{,}000 = 0$$ Divide both sides by 5.

We solve this equation using the quadratic formula.

$$x = \frac{-b \pm \sqrt{b^2 - 4ac}}{2a}$$

$$x = \frac{-4{,}200 \pm \sqrt{(4{,}200)^2 - 4(9)(-88{,}000)}}{2(9)}$$ Substitute 9 for a, 4,200 for b, and −88,000 for c.

$$x = \frac{-4{,}200 \pm \sqrt{20{,}808{,}000}}{18}$$ Evaluate the expression within the radical. Multiply in the denominator.

$$x \approx \frac{362}{18} \quad \text{or} \quad x \approx \frac{-8{,}762}{18}$$ Use a calculator to evaluate each numerator.

$$x \approx 20.1 \quad \quad x \approx -486.8$$ Do the division. Since the model is defined for only positive values of x, we discard the second solution.

In 20.1 years after 1980, or in early 2000, the model predicts that the United States had approximately 1,000,000 lawyers.

Teaching Example 7 Lawyers. See Example 7. In what year does the model indicate that the United States had 800,000 lawyers?
Answer: Early 1991

Self Check 7 **Lawyers.** See Example 7. In what year does the model indicate that the United States had three-quarters of a million lawyers? Late 1988

Now Try Problem 95

EXAMPLE 8

Graduation Announcements. To create the announcement shown, a graphic artist must follow two design requirements.

- A border of uniform width should surround the text.
- Equal areas should be devoted to the text and to the border.

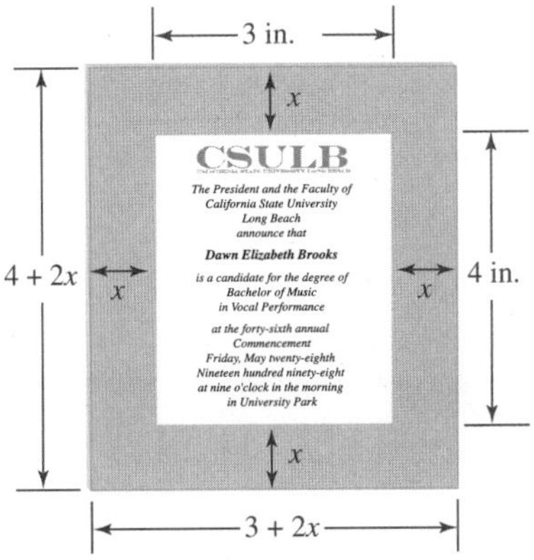

To meet these requirements, how wide should the border be?

Analyze The text occupies $4 \cdot 3 = 12$ in.^{2} of space. The border must also have an area of 12 in.^{2}.

Assign If we let x = the width of the border in inches, the length of the announcement is $(4 + 2x)$ inches and the width is $(3 + 2x)$ inches.

Form We can now form an equation. Recall that the area of a rectangle is the product of its length and width.

The area of the announcement	minus	the area of the text	equals	the area of the border.
$(4 + 2x)(3 + 2x)$	$-$	12	$=$	12

Solve

$$(4 + 2x)(3 + 2x) - 12 = 12$$

$$12 + 8x + 6x + 4x^2 - 12 = 12 \quad \text{On the left side, multiply the binomials.}$$

$$4x^2 + 14x = 12 \quad \text{Combine like terms. This is a quadratic equation.}$$

$$4x^2 + 14x - 12 = 0 \quad \text{To get 0 on the right side, subtract 12 from both sides.}$$

$$2x^2 + 7x - 6 = 0 \quad \text{The coefficients have a common factor of 2. Divide both sides by 2.}$$

To solve this equation, we will use the quadratic formula with $a = 2$, $b = 7$, and $c = -6$.

$$x = \frac{-b \pm \sqrt{b^2 - 4ac}}{2a}$$

$$x = \frac{-7 \pm \sqrt{7^2 - 4(2)(-6)}}{2(2)} \quad \text{Substitute for } a, b, \text{ and } c.$$

$$x = \frac{-7 \pm \sqrt{97}}{4} \quad \text{Evaluate within the radical: } 7^2 - 4(2)(-6) = 49 + 48 = 97. \text{ Multiply in the denominator.}$$

$$x \approx \frac{-7 + \sqrt{97}}{4} \quad \text{or} \quad x \approx \frac{-7 - \sqrt{97}}{4} \quad \text{These are the two exact solutions.}$$

$x \approx 0.71$ | ~~$x \approx -4.21$~~

State The width of the border should be about 0.71 inch. (We discard the solution $\frac{-7 - \sqrt{97}}{4}$ since it is negative.)

Check If the border is 0.71 inch wide, the announcement has an area of about $5.42 \cdot 4.42 \approx 23.96$ in.^{2}. If we subtract the area of the text from the area of the announcement, we get $23.96 - 12 = 11.96$ in.^{2}. This represents the area of the border, which was to be 12 in.^{2}. The result seems reasonable.

Teaching Example 8 Graduation Announcements. See Example 8. Find the width of the border if the text occupies an area that is 4 inches by 5 inches.

Answer: $\frac{-9 + \sqrt{161}}{4}$ in. ≈ 0.92 in.

Self Check 8 **Graduation Announcements.** See Example 8. Find the width of the border if the text occupies an area 3 inches by 5 inches.

$\frac{-4 + \sqrt{31}}{2}$ in. ≈ 0.78 in.

Now Try Problem 97

SECTION 10.2 STUDY SET

VOCABULARY

Fill in the blanks.

1. The standard form of a quadratic equation is $ax^2 + bx + c = 0$.
2. $x = \frac{-b \pm \sqrt{b^2 - 4ac}}{2a}$ is called the quadratic formula.

CONCEPTS

3. Write each quadratic equation in standard form.
   a. $x^2 + 2x = -5$ $\quad x^2 + 2x + 5 = 0$
   b. $3x^2 = -2x + 1$ $\quad 3x^2 + 2x - 1 = 0$
4. For each quadratic equation, find the values of a, b, and c.
   a. $x^2 + 5x + 6 = 0$ $\quad 1, 5, 6$
   b. $8x^2 - x = 10$ $\quad 8, -1, -10$
5. Determine whether each statement is true or false.
   a. Any quadratic equation can be solved by using the quadratic formula. True
   b. Any quadratic equation can be solved by completing the square. True
   c. Any quadratic equation can be solved by factoring using integers. False
6. What is wrong with the beginning of the solution shown below?

   Solve: $x^2 - 3x = 2$

   $a = 1 \quad b = -3 \quad c = 2$

   The equation wasn't written in standard form first: $c = -2$
7. Evaluate each expression.
   a. $\frac{-2 \pm \sqrt{2^2 - 4(1)(-8)}}{2(1)}$ $\quad 2, -4$
   b. $\frac{-(-1) \pm \sqrt{(-1)^2 - 4(2)(-4)}}{2(2)}$ $\quad \frac{1 \pm \sqrt{33}}{4}$
8. A student used the quadratic formula to solve a quadratic equation and obtained $x = \frac{-2 \pm \sqrt{3}}{2}$.
   a. How many solutions does the equation have? What are they exactly? $\quad 2; \frac{-2 + \sqrt{3}}{2}, \frac{-2 - \sqrt{3}}{2}$
   b. Graph the solutions on a number line. See AIE Appendix 3.
9. Simplify each of the following.
   a. $\frac{3 \pm 6\sqrt{2}}{3}$ $\quad 1 \pm 2\sqrt{2}$
   b. $\frac{-12 \pm 4\sqrt{7}}{8}$ $\quad \frac{-3 \pm \sqrt{7}}{2}$
10. a. Write an expression that represents the width of the larger rectangle shown in red. $(2 + 2x)$ in.
    b. Write an expression that represents the length of the larger rectangle shown in red. $(5 + 2x)$ in.

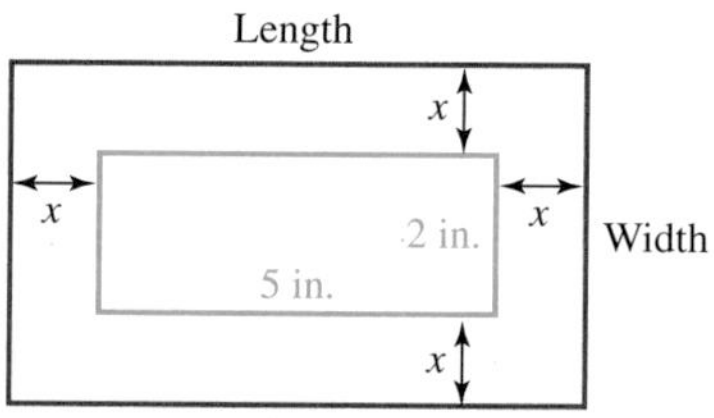

NOTATION

11. On a quiz, students were asked to write the quadratic formula. What is wrong with each answer shown below?
    a. $x = -b \pm \frac{\sqrt{b^2 - 4ac}}{2a}$ The fraction bar wasn't drawn under both parts of the numerator.
    b. $x = \frac{-b\sqrt{b^2 - 4ac}}{2a}$ A $\pm$ sign wasn't written between $-b$ and the radical.
12. When reading $\frac{-b \pm \sqrt{b^2 - 4ac}}{2a}$, we say, "The opposite of b, plus or minus the square root of b squared minus 4 times a times c, all over (divided by) $2a$."

GUIDED PRACTICE

Use the quadratic formula to solve each equation. See Example 1.

13. $x^2 - 3x + 2 = 0$ $\quad 1, 2$
14. $x^2 + 3x + 2 = 0$ $\quad -1, -2$
15. $x^2 + 12x = -36$ $\quad$ A repeated solution of -6
16. $x^2 - 18x + 81 = 0$ $\quad$ A repeated solution of 9
17. $2x^2 + x - 3 = 0$ $\quad -\frac{3}{2}, 1$
18. $6x^2 - x - 1 = 0$ $\quad \frac{1}{2}, -\frac{1}{3}$
19. $12t^2 - 5t - 2 = 0$ $\quad \frac{2}{3}, -\frac{1}{4}$
20. $12z^2 + 5z - 3 = 0$ $\quad \frac{1}{3}, -\frac{3}{4}$

Solve each equation. Approximate the solutions to the nearest hundredth. See Example 2.

21. $x^2 = x + 7$ $\quad \frac{1 \pm \sqrt{29}}{2}; 3.19, -2.19$
22. $t^2 = t + 4$ $\quad \frac{1 \pm \sqrt{17}}{2}; 2.56, -1.56$
23. $5x^2 + 5x = -1$ $\quad \frac{-5 \pm \sqrt{5}}{10}; -0.28, -0.72$
24. $2x^2 + 7x = -1$ $\quad \frac{-7 \pm \sqrt{41}}{4}; -0.15, -3.35$

25. $3y^2 + 1 = -6y$ $\frac{-3 \pm \sqrt{6}}{3}$; $-0.18, -1.82$

▶ 26. $4w^2 + 1 = -6w$ $\frac{-3 \pm \sqrt{5}}{4}$; $-0.19, -1.31$

▶ 27. $4m^2 = 4m + 19$ $\frac{1 \pm 2\sqrt{5}}{2}$; $2.74, -1.74$

28. $3y^2 = 12y - 4$ $\frac{6 \pm 2\sqrt{6}}{3}$; $3.63, 0.37$

Solve each equation. See Example 3.

29. $2x^2 + x + 1 = 0$ $-\frac{1}{4} \pm \frac{\sqrt{7}}{4}i$

▶ 30. $2x^2 + 3x + 5 = 0$ $-\frac{3}{4} \pm \frac{\sqrt{31}}{4}i$

▶ 31. $3x^2 - 2x + 1 = 0$ $\frac{1}{3} \pm \frac{\sqrt{2}}{3}i$

▶ 32. $3x^2 - 2x + 5 = 0$ $\frac{1}{3} \pm \frac{\sqrt{14}}{3}i$

33. $x^2 - 2x + 2 = 0$ $1 \pm i$

▶ 34. $x^2 - 4x + 8 = 0$ $2 \pm 2i$

▶ 35. $4a^2 + 4a + 5 = 0$ $-\frac{1}{2} \pm i$

▶ 36. $4b^2 + 4b + 17 = 0$ $-\frac{1}{2} \pm 2i$

For each equation, write an equivalent quadratic equation that will be easier to solve. Do not solve the equation. See Example 4.

37. a. $-5x^2 + 9x - 2 = 0$ $5x^2 - 9x + 2 = 0$
b. $1.6t^2 + 2.4t - 0.9 = 0$ $16t^2 + 24t - 9 = 0$

38. a. $\frac{1}{8}x^2 + \frac{1}{2}x - \frac{3}{4} = 0$ $x^2 + 4x - 6 = 0$
b. $33y^2 + 99y - 66 = 0$ $y^2 + 3y - 2 = 0$

▶ 39. a. $45x^2 + 30x - 15 = 0$ $3x^2 + 2x - 1 = 0$
b. $\frac{1}{3}m^2 - \frac{1}{2}m - \frac{1}{3} = 0$ $2m^2 - 3m - 2 = 0$

▶ 40. a. $0.6t^2 - 0.1t - 0.2 = 0$ $6t^2 - t - 2 = 0$
b. $-a^2 - 15a + 12 = 0$ $a^2 + 15a - 12 = 0$

TRY IT YOURSELF

Solve each equation. Approximate the solutions to the nearest hundredth when appropriate.

41. $x^2 - \frac{14}{15}x = \frac{8}{15}$ $\frac{4}{3}, -\frac{2}{5}$

▶ 42. $x^2 = -\frac{5}{4}x + \frac{3}{2}$ $\frac{3}{4}, -2$

43. $3x^2 - 4x = -2$ $\frac{2}{3} \pm \frac{\sqrt{2}}{3}i$

44. $2x^2 + 3x = -3$ $-\frac{3}{4} \pm \frac{\sqrt{15}}{4}i$

45. $-16y^2 - 8y + 3 = 0$ $\frac{1}{4}, -\frac{3}{4}$

46. $-16x^2 - 16x - 3 = 0$ $-\frac{1}{4}, -\frac{3}{4}$

▶ 47. $2x^2 - 3x - 1 = 0$ $\frac{3 \pm \sqrt{17}}{4}$; $1.78, -0.28$

48. $3x^2 - 9x - 2 = 0$ $\frac{9 \pm \sqrt{105}}{6}$; $3.21, -0.21$

▶ 49. $-x^2 + 10x = 18$ $5 \pm \sqrt{7}$; $2.35, 7.65$

▶ 50. $-x^2 - 6x - 2 = 0$ $-3 \pm \sqrt{7}$; $-5.65, -0.35$

51. $x(x - 6) = 391$ $23, -17$

52. $x(x - 27) = 280$ $35, -8$

53. $x^2 + 5x - 5 = 0$ $\frac{-5 \pm 3\sqrt{5}}{2}$; $0.85, -5.85$

54. $x^2 - 3x - 27 = 0$ $\frac{3 \pm 3\sqrt{13}}{2}$; $6.91, -3.91$

55. $9h^2 - 6h + 7 = 0$ $\frac{1}{3} \pm \frac{\sqrt{6}}{3}i$

▶ 56. $5x^2 = 2x - 1$ $\frac{1}{5} \pm \frac{2}{5}i$

57. $50x^2 + 30x - 10 = 0$ $\frac{-3 \pm \sqrt{29}}{10}$; $0.24, -0.84$

▶ 58. $120b^2 + 120b - 40 = 0$ $\frac{-3 \pm \sqrt{21}}{6}$; $0.26, -1.26$

▶ 59. $0.6x^2 + 0.03 - 0.4x = 0$ $\frac{10 \pm \sqrt{55}}{30}$; $0.58, 0.09$

▶ 60. $2x^2 + 0.1x = 0.04$ $\frac{-1 \pm \sqrt{33}}{40}$; $0.12, -0.17$

61. $\frac{1}{8}x^2 - \frac{1}{2}x + 1 = 0$ $2 \pm 2i$

62. $\frac{1}{2}x^2 + 3x + \frac{13}{2} = 0$ $-3 \pm 2i$

63. $\frac{a^2}{10} - \frac{3a}{5} + \frac{7}{5} = 0$ $3 \pm i\sqrt{5}$

64. $\frac{c^2}{4} + c + \frac{11}{4} = 0$ $-2 \pm i\sqrt{7}$

▶ 65. $\frac{x^2}{2} + \frac{5}{2}x = -1$ $\frac{-5 \pm \sqrt{17}}{2}$; $-0.44, -4.56$

▶ 66. $\frac{x^2}{8} - \frac{x}{4} = \frac{1}{2}$ $1 \pm \sqrt{5}$; $3.24, -1.24$

67. $900x^2 - 8{,}100x = 1{,}800$ $\frac{9 \pm \sqrt{89}}{2}$; $9.22, -0.22$

68. $14x^2 - 21x = 49$ $\frac{3 \pm \sqrt{65}}{4}$; $2.77, -1.27$

69. $\frac{1}{4}x^2 - \frac{1}{6}x - \frac{1}{6} = 0$ $\frac{1 \pm \sqrt{7}}{3}$; $1.22, -0.55$

▶ 70. $81x^2 + 12x - 80 = 0$ $\frac{-2 \pm 2\sqrt{181}}{27}$; $0.92, -1.07$

71. Let $f(x) = 0.7x^2 - 3.5x$ For what value(s) of x is $f(x) = 25$? $\frac{35 \pm 5\sqrt{329}}{14}$; $8.98, -3.98$

▶ 72. Let $g(x) = 4.5x^2 + 0.2x$. For what value(s) of x is $g(x) = 3.75$? $\frac{-2 \pm \sqrt{6.754}}{90}$; $-0.94, 0.89$

Look Alikes . . .

73. a. $a^2 + 4a - 7 = 0$ $-2 \pm \sqrt{11}$; $1.32, -5.32$
b. $a^2 - 4a - 7 = 0$ $2 \pm \sqrt{11}$; $5.32, -1.32$

▶ 74. a. $n^2 + 6n - 2 = 0$ $-3 \pm \sqrt{11}$; $0.32, -6.32$
b. $n^2 + 6n + 2 = 0$ $-3 \pm \sqrt{7}$; -0.35; -5.65

75. a. $(x + 2)(x - 4) = 16$ $6, -4$
b. $(x + 2)(x - 4) = -16$ $1 \pm i\sqrt{7}$

76. a. $5n^2 - 14n - 3 = 0$ $3, -\frac{1}{5}$
b. $-5n^2 - 14n - 3 = 0$ $\frac{-7 \pm \sqrt{34}}{5}$; $-0.23, -2.57$

▶ 77. a. $x^2 - 42x + 441 = 0$ A repeated solution of 21
b. $x^2 + 42x + 441 = 0$ A repeated solution of -21

78. a. $0.3y^2 + 0.6y + 0.5 = 0$ $\frac{-3 \pm i\sqrt{6}}{3}$
b. $0.003y^2 + 0.006y + 0.005 = 0$ $\frac{-3 \pm i\sqrt{6}}{3}$

APPLICATIONS

79. **Crosswalks.** Refer to the illustration below. Instead of using the Main Street and First Avenue crosswalks to get from Nordstrom to Best Buy, a shopper uses the diagonal crosswalk to walk 97 feet directly from one corner to the other. If the length of the Main Street crosswalk is 7 feet longer than the First Avenue crosswalk, how much walking does the shopper save by using the diagonal crosswalk? 40 ft

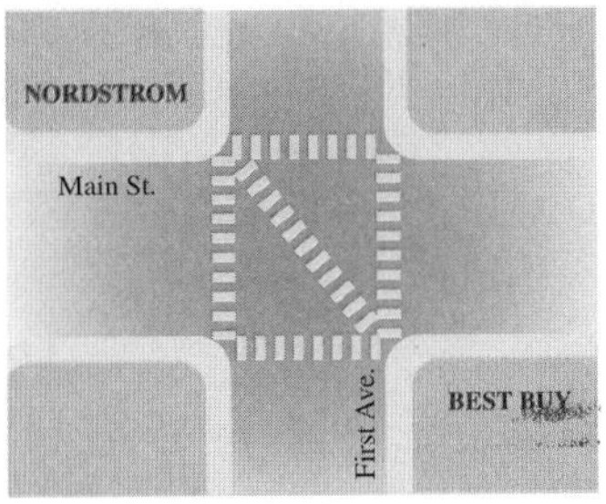

80. Badminton. The person who wrote the instructions for setting up the badminton net shown below forgot to give the specific dimensions for securing the pole. How long is the support string? 34 in.

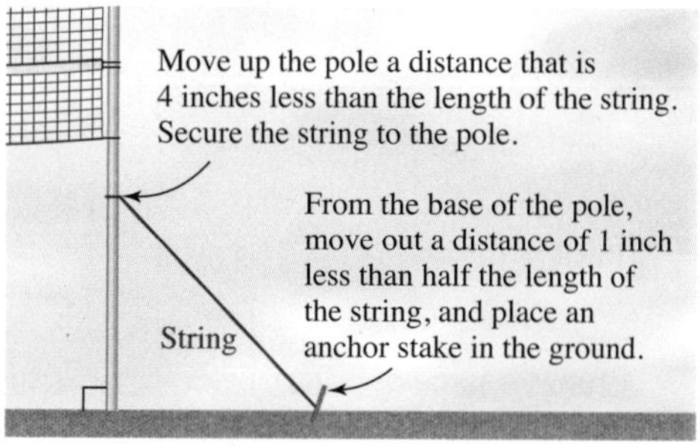
Move up the pole a distance that is
4 inches less than the length of the string.
Secure the string to the pole.
From the base of the pole,
move out a distance of 1 inch
less than half the length of
the string, and place an
anchor stake in the ground.
String

81. Right Triangles. The hypotenuse of a right triangle is 2.5 units long. The longer leg is 1.7 units longer than the shorter leg. Find the lengths of the legs of the triangle. 0.7 units, 2.4 units

82. Televisions. The screen size of a television is measured diagonally from one corner to the opposite corner. In 2007, Sharp developed the world's largest TV screen to date—a 108-inch flat-panel liquid crystal display (LCD). Find the width and the height of the rectangular screen if the width were 41 inches greater than the height. Round to the nearest inch. Height: 53 in., width: 94 in.

83. IMAX Screens. The largest permanent movie screen is in the Panasonic Imax theater at Darling Harbor, Sydney, Australia. The rectangular screen has an area of 11,349 square feet. Find the dimensions of the screen if it is 20 feet longer than it is wide. 97 ft by 117 ft

84. World's Largest LED Screen. A huge suspended LED screen is the centerpiece of The Place, a popular mall in Beijing, China. Find the length and width of the rectangular screen if the length is 10 meters more than 8 times its width, and the viewable area is 7,500 square meters. Length: 250 m; width: 30 m

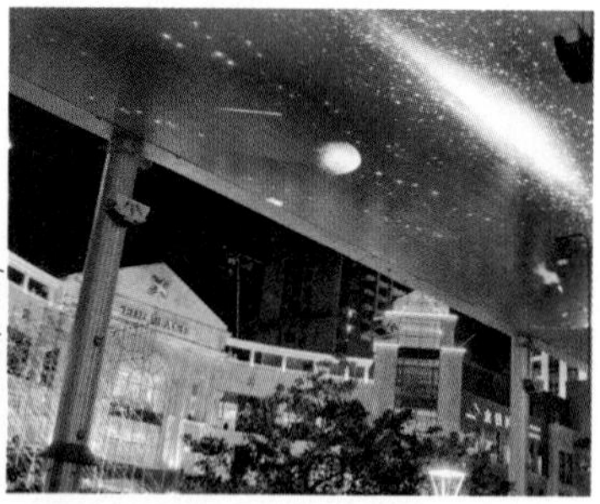
©China Photos/Alamy

85. Parks. Central Park is one of New York's best-known landmarks. Rectangular in shape, its length is 5 times its width. When measured in miles, its perimeter numerically exceeds its area by 4.75. Find the dimensions of Central Park if we know that its width is less than 1 mile. About 0.5 mi by 2.5 mi

86. History. One of the important cities of the ancient world was Babylon. Greek historians wrote that the city was square. Measured in miles, its area numerically exceeded its perimeter by about 124. Find its dimensions. (Round to the nearest tenth.) 13.3 mi by 13.3 mi

87. Polygons. A five-sided polygon, called a *pentagon,* has 5 diagonals. The number of diagonals d of a polygon of n sides is given by the formula $d = \frac{n(n-3)}{2}$. Find the number of sides of a polygon if it has 275 diagonals. 25 sides

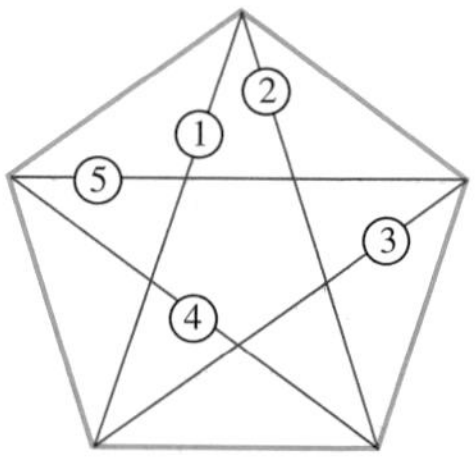
1
2
3
4
5

88. Metal Fabrication. A box with no top is to be made by cutting a 2-inch square from each corner of a square sheet of metal. After bending up the sides, the volume of the box is to be 220 cubic inches. Find the length of a side of the square sheet of metal that should be used in the construction of the box. Round to the nearest hundredth. 14.49 in.

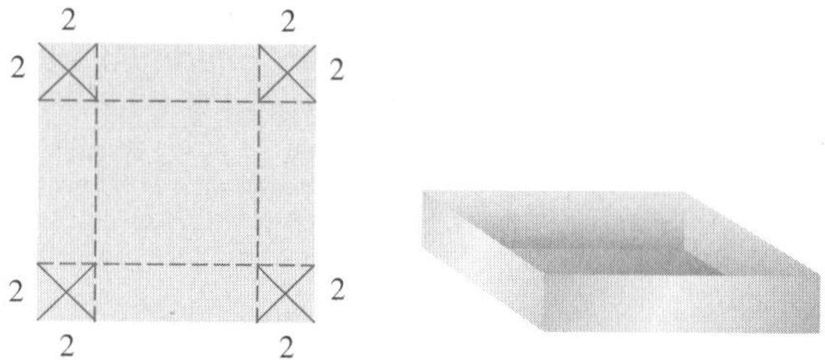
2
2
2
2
2
2
2
2

89. Dances. Tickets to a school dance cost \$4 and the projected attendance is 300 people. It is further projected that for every 10¢ increase in ticket price, the average attendance will decrease by 5. At what ticket price will the receipts from the dance be \$1,248? \$4.80 or \$5.20

90. Ticket Sales. A carnival usually sells three thousand 75¢ ride tickets on a Saturday. For each 15¢ increase in price, management estimates that 80 fewer tickets will be sold. What increase in ticket price will produce \$2,982 of revenue on Saturday? 30¢

91. Magazine Sales. The *Gazette's* profit is \$20 per year for each of its 3,000 subscribers. Management estimates that the profit per subscriber will increase by 1¢ for each additional subscriber over the current 3,000. How many subscribers will bring a total profit of \$120,000? 4,000 subscribers

92. Investment Rates. A woman invests \$1,000 in a fund for which interest is compounded annually at a rate r. After one year, she deposits an additional \$2,000. After two years, the balance in the account is $\$1{,}000(1 + r)^2 + \$2{,}000(1 + r)$. If this amount is \$3,368.10, find r. 9%

93. from **Campus to Careers**

Police Patrol Officer

© Robert E Daemmrich/Getty Images

The number of female police officers (sworn) in the United States is approximated by the function $f(t) = -24t^2 + 1{,}534t + 72{,}065$, where t is the number of years after 2000. In what year does the model indicate that the number of female officers reached 75,000? (Source: *Crime in the United States, 2000–2010*) 2002

94. Shopping Centers. The number of shopping centers in the United States is approximated by the function $s(t) = 48t^2 + 581t + 77{,}383$, where t is the number of years after 1990. In what year does the model indicate that the number of shopping centers reached 100,000? (Source: U.S. Census Bureau) About mid-year 2006

▶ **95. Picture Framing.** The matting around the picture has a uniform width. How wide is the matting if its area equals the area of the picture? Round to the nearest hundredth of an inch. 0.92 in.

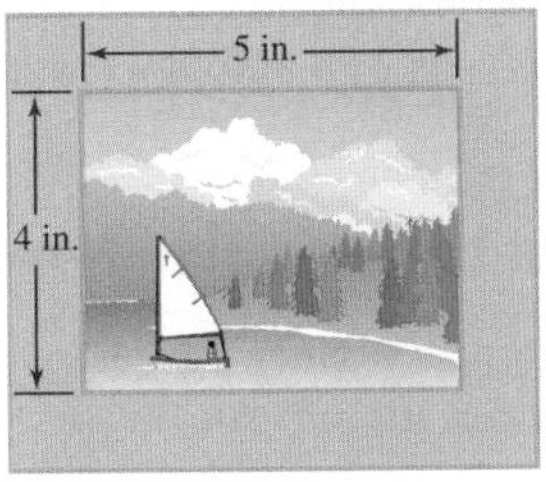

▶ **96. Swimming Pools.** In the advertisement shown, how wide will the free concrete decking be if a uniform width is constructed around the perimeter of the pool? Round to the nearest hundredth of a yard. (*Hint:* Note the difference in units. Convert the dimensions of the pool to yards.) 0.80 yd

SAHARA POOL & SPA
SUMMER SPECIAL
This 18 ft x 30 ft pool: only $28,500
Buy now and receive 28 square yards of concrete decking FREE!

97. Dimensions of Rectangle. A rectangle is 4 feet longer than it is wide, and its area is 20 square feet. Find its dimensions to the nearest tenth of a foot. 2.9 ft, 6.9 ft

▶ **98. Dimensions of a Triangle.** The height of a triangle is 4 meters longer than twice its base. Find the base and height if the area of the triangle is 10 square meters. Round to the nearest hundredth of a meter. 2.32 m, 8.63 m

WRITING

99. Explain why the quadratic formula, in most cases, is easier to use to solve a quadratic equation than is the method of completing the square.

▶ **100.** On an exam, a student was asked to solve the equation $-4w^2 - 6w - 1 = 0$. Her first step was to multiply both sides of the equation by -1. She then used the quadratic formula to solve $4w^2 + 6w + 1 = 0$ instead. Is this a valid approach? Explain.

REVIEW

Change each radical to an exponential expression.

101. $\sqrt{n}$ $n^{1/2}$

▶ **102.** $\sqrt[7]{8r^2s}$ $(8r^2s)^{1/7}$

▶ **103.** $\sqrt[4]{3b}$ $(3b)^{1/4}$

104. $3\sqrt[3]{c^2 - d^2}$ $3(c^2 - d^2)^{1/3}$

Write each expression in radical form.

105. $t^{1/3}$ $\sqrt[3]{t}$

106. $(3m^2n^2)^{1/5}$ $\sqrt[5]{3m^2n^2}$

107. $(3t)^{1/4}$ $\sqrt[4]{3t}$

108. $(c^2 + d^2)^{1/2}$ $\sqrt{c^2 + d^2}$

CHALLENGE PROBLEMS

All of the equations we have solved so far have had rational-number coefficients. However, the quadratic formula can be used to solve quadratic equations with irrational or even imaginary coefficients. Solve each equation.

109. $x^2 + 2\sqrt{2}x - 6 = 0$ $\sqrt{2}, -3\sqrt{2}$

110. $\sqrt{2}x^2 + x - \sqrt{2} = 0$ $-\sqrt{2}, \frac{\sqrt{2}}{2}$

▶ **111.** $x^2 - 3ix - 2 = 0$ $i, 2i$

112. $100ix^2 + 300x - 200i = 0$ $i, 2i$

SECTION 10.3

OBJECTIVES

1. Use the discriminant to determine number and type of solutions.
2. Solve equations that are quadratic in form.
3. Solve application problems using quadratic equations.

The Discriminant and Equations That Can Be Written in Quadratic Form

ARE YOU READY? *Are You Ready? exercises available online at www.webassign.net/brookscole*

The following problems review some basic skills that are needed when working with discriminants and equations that can be written in quadratic form.

1. Find the value of the expression within the radical symbol only:
$$\frac{6 \pm \sqrt{(-6)^2 - 4(3)(-4)}}{6}$$ 84

2. Fill in the blanks: **a.** $x^4 = (x^2)^2$ **b.** $(x^{1/3})^2 = x^{2/3}$

3. Fill in the blanks: **a.** $(\sqrt{x})^2 = x$ **b.** $15a^{-2} = \frac{15}{a^2}$

4. Solve: $x^2 = -4$ $\pm 2i$

5. Solve: **a.** $\sqrt{x} = 3$ 9 **b.** $\sqrt[3]{x} = -\frac{1}{2}$ $-\frac{1}{8}$

6. Solve: $\frac{1}{a} = \frac{1}{5}$ 5

We have seen that solutions of the quadratic equation $ax^2 + bx + c = 0$ with $a \neq 0$ are given by the formula

$$x = \frac{-b \pm \sqrt{b^2 - 4ac}}{2a}$$

In this section, we will examine the radicand within the quadratic formula to distinguish or "*discriminate*" among the three types of solutions—rational, irrational, or imaginary.

1 Use the Discriminant to Determine Number and Type of Solutions.

The expression $b^2 - 4ac$ that appears under the radical symbol in the quadratic formula is called the **discriminant.** The discriminant can be used to predict what kind of solutions a quadratic equation has without solving it.

The Discriminant

For a quadratic equation of the form $ax^2 + bx + c = 0$ with rational-number coefficients and $a \neq 0$, the expression $b^2 - 4ac$ is called the **discriminant** and can be used to determine the number and type of the solutions of the equation.

Discriminant: $b^2 - 4ac$	*Number and type of solutions*
Positive	Two different real numbers
0	One repeated solution, a rational number
Negative	Two different imaginary numbers that are complex conjugates

Discriminant: $b^2 - 4ac$	*Number and type of solutions*
A perfect square	Two different rational numbers
Positive and not a perfect square	Two different irrational numbers

EXAMPLE 1 Determine the number and type of solutions for each equation:

a. $x^2 + x + 1 = 0$ **b.** $3x^2 + 5x + 2 = 0$

Teaching Tip: Point out that the discriminant does not include a radical symbol.

Strategy We will identify the values of a, b, and c in each equation. Then we will use those values to compute $b^2 - 4ac$, the discriminant.

Why Once we know whether the discriminant is positive, 0, or negative, and whether it is a perfect square, we can determine the number and type of the solutions of the equation.

Solution **a.** For $x^2 + x + 1 = 0$, the discriminant is:

$$\begin{aligned} b^2 - 4ac &= 1^2 - 4(1)(1) && \text{Substitute: } a = 1, b = 1, \text{ and } c = 1. \\ &= -3 && \text{The result is a negative number.} \end{aligned}$$

Since $b^2 - 4ac < 0$, the solutions of $x^2 + x + 1 = 0$ are two different imaginary numbers that are complex conjugates.

b. For $3x^2 + 5x + 2 = 0$, the discriminant is:

$$\begin{aligned} b^2 - 4ac &= 5^2 - 4(3)(2) && \text{Substitute: } a = 3, b = 5, \text{ and } c = 2. \\ &= 25 - 24 \\ &= 1 && \text{The result is a positive number.} \end{aligned}$$

Since $b^2 - 4ac > 0$ and $b^2 - 4ac$ is a perfect square, the solutions of $3x^2 + 5x + 2 = 0$ are two different rational numbers.

Success Tip

The discriminant also can be used to determine factorability. The trinomial $ax^2 + bx + c$ with integer coefficients and $a \neq 0$ will factor as the product of two binomials with integer coefficients if the value of $b^2 - 4ac$ is a perfect square. If $b^2 - 4ac = 0$, the factors will be the same.

Teaching Example 1 Determine the number and type of solutions for:
a. $x^2 - x - 1 = 0$
b. $9x^2 - 6x + 1 = 0$
Answers:
a. Two different irrational numbers
b. One repeated solution, a rational number

Self Check 1 Determine the number and type of solutions for:
a. $x^2 + x - 1 = 0$ Two different irrational numbers
b. $3x^2 + 4x + 2 = 0$ Two different imaginary numbers that are complex conjugates

Now Try Problems 11, 13, and 15

2 Solve Equations That Are Quadratic in Form.

We have discussed four methods that are used to solve quadratic equations. The table below shows some advantages and disadvantages of each method.

Method	Advantages	Disadvantages	Examples
Factoring and the zero-factor property	It can be very fast. When each factor is set equal to 0, the resulting equations are usually easy to solve.	Some polynomials may be difficult to factor and others impossible.	$x^2 - 2x - 24 = 0$ $4a^2 - a = 0$
Square root property	It is the fastest way to solve equations of the form $ax^2 = n$ (n = a number) or $(ax + b)^2 = n$.	It applies only to equations that are in these forms.	$x^2 = 27$ $(2y + 3)^2 = 25$
Completing the square*	It can be used to solve any quadratic equation. It works well with equations of the form $x^2 + bx = n$, where b is even.	It involves more steps than the other methods. The algebra can be cumbersome if the leading coefficient is not 1 or if b is odd or a fraction.	$x^2 + 4x = -1$ $t^2 - 14t - 9 = 0$
Quadratic formula	It can be used to solve any quadratic equation.	It involves several calculations in which sign errors can be made. Often the result must be simplified.	$x^2 + 3x - 33 = 0$ $4s^2 - 10s + 5 = 0$

*The quadratic formula is just a condensed version of completing the square and is usually easier to use. However, you need to know how to complete the square because it is used in more advanced mathematics courses.

Success Tip

The solutions of a quadratic equation can always be found using the quadratic formula. The solutions cannot always be found using factoring.

To determine the most efficient method for a given equation, we can use the following strategy.

Strategy for Solving Quadratic Equations

1. See whether the equation is in a form such that the **square root method** is easily applied.
2. If step 1 does not apply, write the equation in $ax^2 + bx + c = 0$ form.
3. See whether the equation can be solved using the **factoring method.**
4. If you can't factor, solve the equation by the **quadratic formula. Completing the square** can also be used.

Many nonquadratic equations can be written in quadratic form ($ax^2 + bx + c = 0$) and solved using the techniques discussed in previous sections. For example, a careful inspection of the equation $x^4 - 5x^2 + 4 = 0$ leads to the following observations:

$$x^4 - 5x^2 + 4 = 0$$

The leading term, x^4, is the square of the expression x^2 in the middle term: $x^4 = (x^2)^2$.

The last term is a constant.

Equations that contain an expression, the same expression squared, and a constant term are said to be **quadratic in form**. One method used to solve such equations is to make a substitution.

EXAMPLE 2 Solve: $x^4 - 3x^2 - 4 = 0$

Teaching Tip: Some students may recognize that factoring could be used:

$x^4 - 3x^2 - 4 = 0$
$(x^2 - 4)(x^2 + 1) = 0$
$(x + 2)(x - 2)(x^2 + 1) = 0$

Strategy Since the leading term, x^4, is the square of the expression x^2 in the middle term, we will substitute y for x^2.

Why Our hope is that such a substitution will produce an equation that we can solve using one of the methods previously discussed.

Solution If we write x^4 as $(x^2)^2$, the equation takes the form

$$(x^2)^2 - 3x^2 - 4 = 0$$

and it is said to be **quadratic in x^2**. We can solve this equation by letting $y = x^2$.

$$y^2 - 3y - 4 = 0 \quad \text{Replace each } x^2 \text{ with } y.$$

Notation

The choice of the letter y for the substitution is arbitrary. We could just as well let $b = x^2$.

We can solve the resulting quadratic equation by factoring.

$(y - 4)(y + 1) = 0$ Factor $y^2 - 3y - 4$.
$y - 4 = 0$ or $y + 1 = 0$ Set each factor equal to 0.
$y = 4$ | $y = -1$ Solve for y.

Caution

If you are solving an equation in x, you can't answer with values of y. Remember to reverse (undo) any substitutions, and solve for the variable in the original equation.

These are not the solutions for x. To find x, we reverse the substitution by replacing each y with x^2 and proceed as follows:

$x^2 = 4$ or $x^2 = -1$ Undo the substitution. Substitute x^2 for y.
$x = \pm\sqrt{4}$ | $x = \pm\sqrt{-1}$ Use the square root property. Don't forget $\pm$.
$x = \pm 2$ | $x = \pm i$ Simplify each radical.

This equation has four solutions: 2, -2, i, and $-i$. Check each one in the original equation. In general, a fourth-degree polynomial equation, such as this, can have up to four distinct solutions.

Teaching Example 2 Solve: $x^4 + 15x^2 - 16 = 0$
Answer: $1, -1, 4i, -4i$

Self Check 2 Solve: $x^4 - 5x^2 - 36 = 0$ $3, -3, 2i, -2i$

Now Try Problem 23

EXAMPLE 3 Solve: $x - 7\sqrt{x} + 12 = 0$

Strategy Since the leading term, x, is the square of the expression $\sqrt{x}$ in the middle term, we will substitute y for $\sqrt{x}$.

Teaching Tip: Some students may recognize that this is a radical equation that could be solved by isolating the radical and squaring both sides:

$x - 7\sqrt{x} + 12 = 0$
$x + 12 = 7\sqrt{x}$
$(x + 12)^2 = (7\sqrt{x})^2$

Why Our hope is that such a substitution will produce an equation that we can solve using one of the methods previously discussed.

Solution We examine the leading term and the middle term.

The leading term, x, is the square of the expression $\sqrt{x}$ in middle term: $x = \left(\sqrt{x}\right)^2$.

$$x - 7\sqrt{x} + 12 = 0$$

The last term is a constant.

If we write x as $\left(\sqrt{x}\right)^2$, the equation takes the form

$$\left(\sqrt{x}\right)^2 - 7\sqrt{x} + 12 = 0$$

The Language of Algebra

Equations such as $x - 7\sqrt{x} + 12 = 0$ that are quadratic in form are also said to be **reducible** to a quadratic.

and it is said to be **quadratic in $\sqrt{x}$**. We can solve this equation by letting $y = \sqrt{x}$ and factoring.

$y^2 - 7y + 12 = 0$ Replace each $\sqrt{x}$ with y. This is a quadratic equation.
$(y - 3)(y - 4) = 0$ Factor the trinomial.
$y - 3 = 0$ or $y - 4 = 0$ Set each factor equal to 0.
$y = 3$ | $y = 4$ Solve for y.

Success Tip

Recall from Chapter 9 that we must check each proposed solution of a radical equation and eliminate any extraneous solutions.

To find x, we reverse the substitution and replace each y with $\sqrt{x}$. Then we solve the resulting radical equations by squaring both sides.

$$\sqrt{x} = 3 \quad \text{or} \quad \sqrt{x} = 4 \qquad \text{Undo the substitution.}$$
$$x = 9 \quad \Big| \quad x = 16$$

The solutions are 9 and 16. Check each solution in the original equation.

Teaching Example 3 Solve: $x - 12\sqrt{x} + 35 = 0$
Answer: 49, 25

Self Check 3 Solve: $x + \sqrt{x} - 6 = 0$ 4, -9 does not check

Now Try Problem 27

EXAMPLE 4 Solve: $2m^{2/3} - 2 = 3m^{1/3}$

Strategy We will write the equation in descending powers of m and look for a possible substitution to make.

Why Our hope is that a substitution will produce an equation that we can solve using one of the methods previously discussed.

Solution After writing the equation in descending powers of m, we see that

$$2m^{2/3} - 3m^{1/3} - 2 = 0$$

is **quadratic in $m^{1/3}$**, because $m^{2/3} = (m^{1/3})^2$. We will use the substitution $y = m^{1/3}$ to write this equation in quadratic form.

The Language of Algebra

Recall that in $2m^{2/3}$, we call 2/3 a **rational** or **fractional exponent.**

$$2m^{2/3} - 3m^{1/3} - 2 = 0 \qquad \text{This is the equation to solve.}$$
$$2(m^{1/3})^2 - 3m^{1/3} - 2 = 0 \qquad \text{Write } m^{2/3} \text{ as } (m^{1/3})^2.$$
$$2y^2 - 3y - 2 = 0 \qquad \text{Substitute: Replace each } m^{1/3} \text{ with } y. \text{ This is a quadratic equation.}$$
$$(2y + 1)(y - 2) = 0 \qquad \text{Factor } 2y^2 - 3y - 2.$$
$$2y + 1 = 0 \quad \text{or} \quad y - 2 = 0 \qquad \text{Set each factor equal to 0.}$$
$$y = -\frac{1}{2} \quad \Big| \quad y = 2 \qquad \text{Solve for } y.$$

To find m, we reverse the substitution and replace each y with $m^{1/3}$. Then we solve the resulting equations by cubing both sides.

Success Tip

We could write $m^{1/3} = -\frac{1}{2}$ in the equivalent radical form $\sqrt[3]{m} = -\frac{1}{2}$ and solve in the same way—by cubing both sides.

$$m^{1/3} = -\frac{1}{2} \quad \text{or} \quad m^{1/3} = 2 \qquad \text{Undo the substitution.}$$
$$(m^{1/3})^3 = \left(-\frac{1}{2}\right)^3 \quad \Big| \quad (m^{1/3})^3 = (2)^3 \qquad \text{Recall that } m^{1/3} = \sqrt[3]{m}. \text{ To solve for } m, \text{ cube both sides.}$$
$$m = -\frac{1}{8} \quad \Big| \quad m = 8 \qquad \text{Find each power.}$$

The solutions are $-\frac{1}{8}$ and 8. Check each solution in the original equation.

Teaching Example 4 Solve: $n^{2/5} = n^{1/5} + 2$
Answer: 32, -1

Self Check 4 Solve: $a^{2/3} = -3a^{1/3} + 10$ $-125, 8$

Now Try Problem 31

EXAMPLE 5 Solve: $(4t + 2)^2 - 30(4t + 2) + 224 = 0$

Strategy Since the leading term, $(4t + 2)^2$, is the square of the expression $4t + 2$ in the middle term, we will substitute y for $4t + 2$.

Why Our hope is that such a substitution will produce an equation that we can solve using one of the methods previously discussed.

Solution This equation is **quadratic in 4t + 2**. If we make the substitution $y = 4t + 2$, we obtain

$$y^2 - 30y + 224 = 0$$

which can be solved by using the quadratic formula.

$$y = \frac{-b \pm \sqrt{b^2 - 4ac}}{2a}$$

$$y = \frac{-(-30) \pm \sqrt{(-30)^2 - 4(1)(224)}}{2(1)}$$ Substitute 1 for a, -30 for b, and 224 for c.

$$y = \frac{30 \pm \sqrt{4}}{2}$$ Evaluate within the radical.

$$y = \frac{30 \pm 2}{2}$$ Evaluate the radical: $\sqrt{4} = 2$.

$y = 16$ or $y = 14$ Evaluate: $\frac{30+2}{2} = 16$ and $\frac{30-2}{2} = 14$

To find t, we reverse the substitution and replace y with $4t + 2$. Then we solve for t.

$4t + 2 = 16$	or	$4t + 2 = 14$	Undo the substitution.
$4t = 14$		$4t = 12$	Isolate the variable term, $4t$.
$t = 3.5$		$t = 3$	Solve for t.

Verify that 3.5 and 3 satisfy the original equation.

Teaching Example 5 Solve: $(2a - 3)^2 + 4(2a - 3) = 12$
Answer: $-\frac{3}{2}, \frac{5}{2}$

Self Check 5 Solve: $(n + 3)^2 - 6(n + 3) = -8$ $-1, 1$

Now Try Problem 35

EXAMPLE 6

Solve: $15a^{-2} - 8a^{-1} + 1 = 0$

Strategy We will write the equation with positive exponents and look for a possible substitution to make.

Why Our hope is that a substitution will produce an equation that we can solve using one of the methods previously discussed.

Solution When we write the terms $15a^{-2}$ and $-8a^{-1}$ using positive exponents, we see that this equation is **quadratic in** $\frac{1}{a}$.

Success Tip

We could also solve this equation by multiplying both sides by the LCD, a^2.

$$\frac{15}{a^2} - \frac{8}{a} + 1 = 0$$ Think of this equation as $15 \cdot \left(\frac{1}{a}\right)^2 - 8 \cdot \frac{1}{a} + 1 = 0$. Note that 0 is not a possible solution.

If we let $y = \frac{1}{a}$, the resulting quadratic equation can be solved by factoring.

$15y^2 - 8y + 1 = 0$ Substitute y^2 for $\frac{1}{a^2}$ and y for $\frac{1}{a}$. This is a quadratic equation.

$(5y - 1)(3y - 1) = 0$ Factor $15y^2 - 8y + 1$.

$5y - 1 = 0$	or	$3y - 1 = 0$	Set each factor equal to 0.
$y = \frac{1}{5}$		$y = \frac{1}{3}$	Solve for y.

Success Tip

Recall that we use cross products to solve proportions.

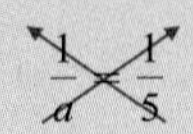

$5 = a$

To find a, we reverse the substitution and replace each y with $\frac{1}{a}$. Then we proceed as follows:

$$\frac{1}{a} = \frac{1}{5} \quad \text{or} \quad \frac{1}{a} = \frac{1}{3} \qquad \text{Undo the substitution.}$$

$$5 = a \qquad\qquad 3 = a \qquad \text{Solve the proportions by finding the cross products.}$$

The solutions are 5 and 3. Check each solution in the original equation.

Teaching Example 6
Solve: $2n^{-2} - 7n^{-1} + 3 = 0$

Answer: $2, \frac{1}{3}$

Self Check 6 Solve: $28c^{-2} - 3c^{-1} - 1 = 0$ $-7, 4$

Now Try Problem 41

3 Solve Application Problems Using Quadratic Equations.

EXAMPLE 7

Electronic Temperature Control

Water Temp

Household Appliances. The illustration shows a temperature control on a washing machine. When the *warm* setting is selected, both the hot and cold water pipes open to fill the tub in 2 minutes 15 seconds. When the *cold* setting is chosen, the tub fills 45 seconds faster than when the *hot* setting is used. How long does it take to fill the washing machine if the *hot* setting is used?

Analyze It is helpful to organize the facts of this shared-work problem in a table. We note that the hot and cold water inlets will be open for the same time: 2 minutes 15 seconds, or 135 seconds.

Assign Let $x =$ the number of seconds it takes to fill the tub with hot water. Since the cold water inlet fills the tub in 45 seconds less time, $x - 45 =$ the number of seconds it takes to fill the tub with cold water.

Form To determine the work completed by each inlet, multiply the rate by the time.

	Rate ·	Time =	Work completed
Hot water	$\frac{1}{x}$	135	$\frac{135}{x}$
Cold water	$\frac{1}{x - 45}$	135	$\frac{135}{x - 45}$

Enter this information first. Multiply to get each of these entries: $W = rt$.

Success Tip

An alternate way to form an equation is to note that what the hot water inlet can do in 1 second plus what the cold water inlet can do in 1 second equals what they can do together in 1 second:

$$\frac{1}{x} + \frac{1}{x - 45} = \frac{1}{135}$$

In shared-work problems, 1 represents one whole job completed. So we have

The fraction of tub filled with hot water	plus	the fraction of the tub filled with cold water	equals	1 tub filled.
$\frac{135}{x}$	$+$	$\frac{135}{x - 45}$	$=$	1

Solve

$$\frac{135}{x} + \frac{135}{x - 45} = 1$$ This is a rational equation.

$$x(x - 45)\left(\frac{135}{x} + \frac{135}{x - 45}\right) = x(x - 45)(1)$$ Multiply both sides by the LCD $x(x - 45)$ to clear the equation of fractions.

$$x(x - 45)\frac{135}{x} + x(x - 45)\frac{135}{x - 45} = x(x - 45)(1)$$ Distribute $x(x - 45)$.

$$135(x - 45) + 135x = x(x - 45)$$ Simplify each side.

$$135x - 6{,}075 + 135x = x^2 - 45x$$ Distribute 135 and x.

$$270x - 6{,}075 = x^2 - 45x$$ Combine like terms.

$$0 = x^2 - 315x + 6{,}075$$ Get 0 on the left side.

To solve for x, we will use the quadratic formula: $a = 1$, $b = -315$, and $c = 6{,}075$.

$$x = \frac{-b \pm \sqrt{b^2 - 4ac}}{2a}$$

$$x = \frac{-(-315) \pm \sqrt{(-315)^2 - 4(1)(6{,}075)}}{2(1)}$$ Substitute 1 for a, -315 for b, and 6,075 for c.

$$x = \frac{315 \pm \sqrt{74{,}925}}{2}$$ Evaluate the expression within the radical.

$x \approx 294$ or ~~$x \approx 21$~~ Use a calculator to find each solution.

State We can discard the solution of 21 seconds, because this would imply that the cold water inlet fills the tub in a negative number of seconds $(21 - 45 = -24)$. Therefore, the hot water inlet fills the washing machine tub in about 294 seconds, which is 4 minutes 54 seconds.

Check Use estimation to check the result. The work completed by the hot water inlet is $\frac{135}{294} \approx 0.46$ and the work completed by the cold water inlet is $\frac{135}{294 - 45} \approx 0.54$. Since $0.46 + 0.54 \approx 1$, the result seems reasonable.

Teaching Tip: Ask your students to point out the first step in the solution process in which they could see that they are working with a quadratic equation.

Self Check 7 **Mowing Lawns.** Carly can mow a lawn in 1 hour less time than her friend Lindsey. Together they can finish the job in 5 hours. How long would it take Carly if she worked alone? About 9.5 hr

Now Try ▶ Problem 87

Teaching Example 7 Water Storage Tanks. Two pipes are used to fill a water storage tank. The first pipe can fill the tank in 4 hours, and the two pipes together can fill the tank in 2 hours less time than the second pipe alone. How long would it take for the two pipes together to fill the tank? *Answer:* 2 hr

SECTION 10.3 STUDY SET

VOCABULARY

Fill in the blanks.

▶ 1. For the quadratic equation $ax^2 + bx + c = 0$, the discriminant is $b^2 - 4ac$.

▶ 2. We can solve $x - 2\sqrt{x} - 8 = 0$ by making the substitution $y = \sqrt{x}$.

CONCEPTS

Consider the quadratic equation $ax^2 - bx + c = 0$, where a, b, and c represent rational numbers, and fill in the blanks.

3. If $b^2 - 4ac < 0$, the solutions of the equation are two different imaginary numbers that are complex conjugates.

▶ 4. If $b^2 - 4ac = 0$, the equation has one repeated rational-number solution.

5. If $b^2 - 4ac$ is a perfect square, the solutions of the equation are two different rational numbers.

▶ 6. If $b^2 - 4ac$ is positive and not a perfect square, the solutions of the equation are two different irrational numbers.

7. For each equation, determine the substitution that should be made to write the equation in quadratic form.

a. $x^4 - 12x^2 + 27 = 0$ Let $y = x^2$

b. $x - 13\sqrt{x} + 40 = 0$ Let $y = \sqrt{x}$

c. $x^{2/3} + 2x^{1/3} - 3 = 0$ Let $y = x^{1/3}$

d. $x^{-2} - x^{-1} - 30 = 0$ Let $y = \frac{1}{x}$

e. $(x + 1)^2 - (x + 1) - 6 = 0$ Let $y = x + 1$

8. Fill in the blanks.

a. $x^4 = (x^2)^2$ b. $x = (\sqrt{x})^2$

c. $x^{2/3} = (x^{1/3})^2$ d. $\frac{1}{x^2} = \left(\frac{1}{x}\right)^2$

NOTATION

Complete the solution.

9. To find the type of solutions for the equation $x^2 + 5x + 6 = 0$, we calculate the discriminant.

$$b^2 - 4ac = 5^2 - 4(1)(6)$$
$$= 25 - 24$$
$$= 1$$

Since a, b, and c are rational numbers and the value of the discriminant is a perfect square, the solutions are two different rational numbers.

10. Fill in the blanks to write each equation in quadratic form.

a. $x^4 - 2x^2 - 15 = 0 \rightarrow (x^2)^2 - 2x^2 - 15 = 0$

b. $x - 2\sqrt{x} + 3 = 0 \rightarrow (\sqrt{x})^2 - 2\sqrt{x} + 3 = 0$

c. $8m^{2/3} - 10m^{1/3} - 3 = 0 \rightarrow 8(m^{1/3})^2 - 10m^{1/3} - 3 = 0$

GUIDED PRACTICE

Use the discriminant to determine the number and type of solutions for each equation. Do not solve. **See Example 1.**

11. $4x^2 - 4x + 1 = 0$
One repeated rational-number solution

▶ 12. $6x^2 - 5x - 6 = 0$
Two different rational-number solutions

13. $5x^2 + x + 2 = 0$
Two imaginary-number solutions (complex conjugates)

▶ 14. $3x^2 + 10x - 2 = 0$
Two different irrational-number solutions

15. $2x^2 = 4x - 1$
Two different irrational-number solutions

▶ 16. $9x^2 = 12x - 4$
One repeated rational-number solution

17. $x(2x - 3) = 20$
Two different rational-number solutions

▶ 18. $x(x - 3) = -10$
Two imaginary-number solutions (complex conjugates)

19. $3x^2 - 10 = 0$
Two different irrational-number solutions

20. $5x^2 - 24 = 0$
Two different irrational-number solutions

21. $x^2 - \frac{14}{15}x = \frac{8}{15}$
Two different rational-number solutions

▶ 22. $x^2 = -\frac{5}{4}x + \frac{3}{2}$
Two different rational-number solutions

Solve each equation. **See Example 2.**

▶ 23. $x^4 - 17x^2 + 16 = 0$
$-1, 1, -4, 4$

24. $x^4 - 10x^2 + 9 = 0$
$3, -3, 1, -1$

▶ 25. $x^4 + 5x^2 - 36 = 0$
$2, -2, 3i, -3i$

26. $x^4 - 15x^2 - 16 = 0$
$4, -4, i, -i$

Solve each equation. **See Example 3.**

27. $x - 13\sqrt{x} + 40 = 0$
$25, 64$

▶ 28. $x - 9\sqrt{x} + 18 = 0$
$9, 36$

▶ 29. $2x + \sqrt{x} - 3 = 0$
1, $\frac{9}{4}$ is extraneous

30. $2x - \sqrt{x} - 1 = 0$
1, $\frac{1}{4}$ is extraneous

Solve each equation. **See Example 4.**

▶ 31. $a^{2/3} - 2a^{1/3} = 3$
$-1, 27$

▶ 32. $r^{2/3} + 4r^{1/3} = 5$
$-125, 1$

▶ 33. $x^{2/3} + 2x^{1/3} - 8 = 0$
$-64, 8$

34. $x^{2/3} - 7x^{1/3} + 12 = 0$
$64, 27$

Solve each equation. **See Example 5.**

▶ 35. $(c + 1)^2 - 4(c + 1) + 3 = 0$ $0, 2$

▶ 36. $(a - 5)^2 - 4(a - 5) - 21 = 0$ $2, 12$

▶ 37. $2(2x + 1)^2 - 7(2x + 1) + 6 = 0$ $\frac{1}{4}, \frac{1}{2}$

▶ 38. $3(2 - x)^2 + 10(2 - x) - 8 = 0$ $6, \frac{4}{3}$

Solve each equation. **See Example 6.**

39. $m^{-2} + m^{-1} - 6 = 0$
$-\frac{1}{3}, \frac{1}{2}$

▶ 40. $t^{-2} + t^{-1} - 42 = 0$
$-\frac{1}{7}, \frac{1}{6}$

▶ 41. $8x^{-2} - 10x^{-1} - 3 = 0$
$-4, \frac{2}{3}$

▶ 42. $2x^{-2} - 5x^{-1} - 3 = 0$
$-2, \frac{1}{3}$

Solve each equation. **See Example 7.**

43. $1 - \frac{5}{x} = \frac{10}{x^2}$ $\frac{5 \pm \sqrt{65}}{2}$

▶ 44. $1 - \frac{3}{x} = \frac{5}{x^2}$ $\frac{3 \pm \sqrt{29}}{2}$

▶ 45. $\frac{1}{2} + \frac{1}{b} = \frac{1}{b - 7}$ $\frac{7 \pm \sqrt{105}}{2}$

46. $\frac{1}{4} - \frac{1}{n} = \frac{1}{n + 3}$ $\frac{5 \pm \sqrt{73}}{2}$

TRY IT YOURSELF

Solve each equation.

47. $2x - \sqrt{x} = 3$
$\frac{9}{4}$, 1 is extraneous

▶ 48. $3x + 4\sqrt{x} = 4$
$\frac{4}{9}$, 4 is extraneous

49. $x^{-2} + 2x^{-1} - 3 = 0$
$-\frac{1}{3}, 1$

50. $x^{-2} + 2x^{-1} - 8 = 0$
$-\frac{1}{4}, \frac{1}{2}$

51. $x^4 + 19x^2 + 18 = 0$
$-i, i, -3i\sqrt{2}, 3i\sqrt{2}$

▶ 52. $t^4 + 4t^2 - 5 = 0$
$-1, 1, -i\sqrt{5}, i\sqrt{5}$

▶ 53. $(k - 7)^2 + 6(k - 7) + 10 = 0$ $4 \pm i$

54. $(d + 9)^2 - 4(d + 9) + 8 = 0$ $-7 \pm 2i$

55. $\frac{2}{x - 1} + \frac{1}{x + 1} = 3$
$\frac{3 \pm \sqrt{57}}{6}$

56. $\frac{3}{x - 2} - \frac{1}{x + 2} = 5$
$\frac{1 \pm \sqrt{141}}{5}$

▶ 57. $x - 6x^{1/2} = -8$
$16, 4$

▶ 58. $x - 5x^{1/2} + 4 = 0$
$16, 1$

59. $(y^2 - 9)^2 + 2(y^2 - 9) - 99 = 0$ $\pm i\sqrt{2}, \pm 3\sqrt{2}$

60. $(a^2 - 4)^2 - 4(a^2 - 4) - 32 = 0$
A repeated solution of $0, -2\sqrt{3}, 2\sqrt{3}$

▶ **61.** $x^{-4} - 2x^{-2} + 1 = 0$
Repeated solutions of 1 and −1

62. $4x^{-4} + 1 = 5x^{-2}$
$1, -1, 2, -2$

63. $t^4 + 3t^2 = 28$
$2, -2, i\sqrt{7}, -i\sqrt{7}$

▶ **64.** $3h^4 + h^2 - 2 = 0$
$\frac{\sqrt{6}}{3}, -\frac{\sqrt{6}}{3}, i, -i$

65. $2x^{2/5} - 5x^{1/5} = -3$
$\frac{243}{32}, 1$

66. $2x^{2/5} + 3x^{1/5} = -1$
$-\frac{1}{32}, -1$

▶ **67.** $9\left(\frac{3m+2}{m}\right)^2 - 30\left(\frac{3m+2}{m}\right) + 25 = 0$
A repeated solution of $-\frac{3}{2}$

68. $4\left(\frac{c-7}{c}\right)^2 - 12\left(\frac{c-7}{c}\right) + 9 = 0$
A repeated solution of −14

69. $\frac{3}{a-1} = 1 - \frac{2}{a}$
$3 \pm \sqrt{7}$

70. $1 + \frac{4}{x} = \frac{3}{x^2}$
$-2 \pm \sqrt{7}$

71. $\left(8 - \sqrt{a}\right)^2 + 6\left(8 - \sqrt{a}\right) - 7 = 0$ 49, 225

72. $\left(10 - \sqrt{t}\right)^2 - 4\left(10 - \sqrt{t}\right) - 45 = 0$ 1, 225

73. $x + \frac{2}{x-2} = 0$
$1 \pm i$

▶ **74.** $x + \frac{x+5}{x-3} = 0$
$1 \pm 2i$

75. $3x + 5\sqrt{x} + 2 = 0$
No solution

▶ **76.** $3x - 4\sqrt{x} + 1 = 0$
$1, \frac{1}{9}$

▶ **77.** $x^4 - 6x^2 + 5 = 0$
$1, -1, \sqrt{5}, -\sqrt{5}$

78. $2x^4 - 26x^2 + 24 = 0$
$1, -1, 2\sqrt{3}, -2\sqrt{3}$

79. $8(t+1)^{-2} - 30(t+1)^{-1} + 7 = 0$ $-\frac{5}{7}, 3$

80. $2(s-2)^{-2} + 3(s-2)^{-1} - 5 = 0$ $\frac{8}{5}, 3$

81. $\frac{1}{x+2} + \frac{24}{x+3} = 13$
$-1, -\frac{27}{13}$

▶ **82.** $\frac{3}{x} + \frac{4}{x+1} = 2$
$3, -\frac{1}{2}$

Look Alikes . . .

83. **a.** $y^{2/3} + y^{1/3} - 20 = 0$
$64, -125$

b. $y^{-2} + y^{-1} - 20 = 0$
$\frac{1}{4}, -\frac{1}{5}$

▶ **84.** **a.** $x + 6\sqrt{x} - 16 = 0$
4, 64 does not check

b. $x^{2/3} + 6x^{1/3} - 16 = 0$
$8, -512$

85. **a.** $\frac{1}{x} = \frac{2x}{x+1}$
$1, -\frac{1}{2}$

b. $\frac{1}{x} + \frac{2x}{x+1} = 1$
$\pm i$

86. **a.** $(m^2 - 1)^2 - (m^2 - 1) = 2$
$0, \pm\sqrt{3}$

b. $m^4 - m^2 = 2$
$\pm\sqrt{2}, \pm i$

APPLICATIONS

▶ **87.** ▶ **from Campus to Careers**

Police Patrol Officer

Crowd Control. After a sporting event at a stadium, police have found that a public parking lot can be emptied in 60 minutes if both the east and west exits are opened. If just the east exit is used, it takes 40 minutes longer to clear the lot than it does if just the west exit is opened. How long does it take to clear the parking lot if every car must use the west exit? Round to the nearest minute.
103 min

▶ **88. Paper Routes.** When a father, in a car, and his son, on a bicycle, work together to distribute the morning newspaper, it takes them 35 minutes to complete the route. Working alone, it takes the son 25 minutes longer than the father. To the nearest minute, how long does it take the son to cover the route on his bicycle? 85 min

89. Assembly Lines. A newly manufactured product traveled 300 feet on a high-speed conveyor belt at a rate of r feet per second. It could have traveled the 300 feet in 3 seconds less time if the speed of the conveyor belt was increased by 5 feet per second. Find r. 20 feet per second

▶ **90. Bicycling.** Tina bicycles 160 miles at the rate of r mph. The same trip would have taken 2 hours longer if she had decreased her speed by 4 mph. Find r. 20 mph

▶ **91. Architecture.** A **golden rectangle** is one of the most visually appealing of all geometric forms. The Parthenon, built by the Greeks in the 5th century B.C., fits into a golden rectangle if its ruined triangular pediment is included. See the illustration. In a golden rectangle, the length l and width w must satisfy the equation $\frac{l}{w} = \frac{w}{l-w}$. If a rectangular billboard is to have a width of 20 feet, what should its length be so that it is a golden rectangle? Round to the nearest tenth. 32.4 ft

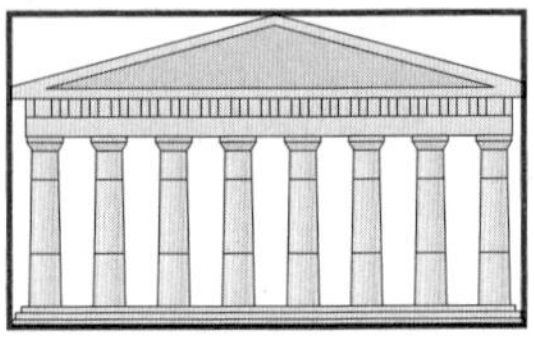

92. Door Designs. An architect needs to determine the height h of the window shown in the illustration. The radius r, the width w, and the height h of the circular-shaped window are related by the formula $r = \frac{4h^2 + w^2}{8h}$. If w is to be 34 inches and r is to be 18 inches, find h to the nearest tenth of an inch. 12.1 in.

WRITING

93. Describe how to predict what type of solutions the equation $3x^2 - 4x + 5 = 0$ will have.

▶ **94.** What **error** is made in the following solution?

Solve: $x^4 - 12x^2 + 27 = 0$

$y^2 - 12y + 27 = 0$ Let $y = x^2$.

$(y - 9)(y - 3) = 0$

$y - 9 = 0$ or $y - 3 = 0$

$y = 9$ | $y = 3$

The solutions of $x^4 - 12x^2 + 27 = 0$ are 9 and 3.

REVIEW

95. Write an equation of the vertical line that passes through (3, 4). $x = 3$

▶ **96.** Find an equation for a linear function whose graph passes through (−1, −6) and (−2, −1). $y = -5x - 11$

▶ **97.** Write an equation of the line with slope $\frac{2}{3}$ that passes through the origin. $y = \frac{2}{3}x$

98. Find an equation of the line that passes through $(2, -3)$ and is perpendicular to the line whose equation is $y = \frac{x}{5} + 6$. Write the equation in slope–intercept form. $y = -5x + 7$

CHALLENGE PROBLEMS

99. Find the real-number solutions of $x^4 - 3x^2 - 2 = 0$. Rationalize the denominators of the solutions.
$\pm\frac{\sqrt{6 + 2\sqrt{17}}}{2}$

100. Solve: $x^6 + 17x^3 + 16 = 0$ $\quad -2\sqrt[3]{2}, -1$

101. Solve: $x^3 - x^2 + 16x - 16 = 0$ $\quad 1, \pm 4i$

102. Solve: $\sqrt{x^2 + 10} = 4\sqrt{x}$ $\quad 8 \pm 3\sqrt{6}$

SECTION 10.4 Quadratic Functions and Their Graphs

OBJECTIVES

1. Graph functions of the form $f(x) = ax^2$ and $f(x) = ax^2 + k$.
2. Graph functions of the form $f(x) = a(x - h)^2$ and $f(x) = a(x - h)^2 + k$.
3. Graph functions of the form $f(x) = ax^2 + bx + c$ by completing the square.
4. Find the vertex using $-\frac{b}{2a}$.
5. Determine minimum and maximum values.
6. Solve quadratic equations graphically.

ARE YOU READY? *Are You Ready? exercises available online at www.webassign.net/brookscole*

The following problems review some basic skills that are needed when graphing quadratic functions.

1. Graph: $f(x) = x^2$
See AIE Appendix 3.

2. Let $f(x) = 2(x - 3)^2 - 4$. Find $f(4)$.
-2

3. Complete the square on $x^2 + 8x$. Then factor the resulting trinomial.
$x^2 + 8x + 16 = (x + 4)^2$

4. Complete the square on $x^2 + x$. Then factor the resulting trinomial.
$x^2 + x + \frac{1}{4} = \left(x + \frac{1}{2}\right)^2$

5. Solve: $x^2 + 6x + 9 = 0$
A repeated solution of -3

6. Solve: $-2x^2 - 8x - 8 = 0$
A repeated solution of -2

7. Evaluate $-\frac{b}{2a}$ for $a = -2$ and $b = -20$. $\quad -5$

8. Is the graph of $x = -1$ a horizontal or a vertical line? A vertical line

In this section, we will discuss methods for graphing *quadratic functions.*

Quadratic Functions

A **quadratic function** is a second-degree polynomial function that can be written in the form

$$f(x) = ax^2 + bx + c$$

where a, b, and c are real numbers and $a \neq 0$.

Quadratic functions often are written in another form, called **standard form,**

$$f(x) = a(x - h)^2 + k$$

where a, h, and k are real numbers and $a \neq 0$. This form is useful because a, h, and k give us important information about the graph of the function. To develop a strategy for graphing quadratic functions written in standard form, we will begin by considering the simplest case, $f(x) = ax^2$.

Notation

Since $y = f(x)$, quadratic functions can also be written as $y = ax^2 + bx + c$ and $y = a(x - h)^2 + k$.

1 Graph Functions of the Form $f(x) = ax^2$ and $f(x) = ax^2 + k$.

One way to graph quadratic functions is to plot points.

EXAMPLE 1 Graph: **a.** $f(x) = x^2$ **b.** $g(x) = 3x^2$ **c.** $s(x) = \frac{1}{3}x^2$

Teaching Tip: Remind your students that y and $f(x)$ are interchangable.

Strategy We can make a table of values for each function, plot each point, and connect them with a smooth curve.

Why At this time, this method is our only option.

Solution After graphing each curve, we see that the graph of $g(x) = 3x^2$ is narrower than the graph of $f(x) = x^2$, and the graph of $s(x) = \frac{1}{3}x^2$ is wider than the graph of $f(x) = x^2$. For $f(x) = ax^2$, the smaller the value of $|a|$, the wider the graph.

$f(x) = x^2$

x	$f(x)$
−2	4
−1	1
0	0
1	1
2	4

$g(x) = 3x^2$

x	$g(x)$
−2	12
−1	3
0	0
1	3
2	12

$s(x) = \frac{1}{3}x^2$

x	$s(x)$
−2	$\frac{4}{3}$
−1	$\frac{1}{3}$
0	0
1	$\frac{1}{3}$
2	$\frac{4}{3}$

The values of $g(x)$ increase faster than the values of $f(x)$, making its graph steeper.

The values of $s(x)$ increase more slowly than the values of $f(x)$, making its graph flatter.

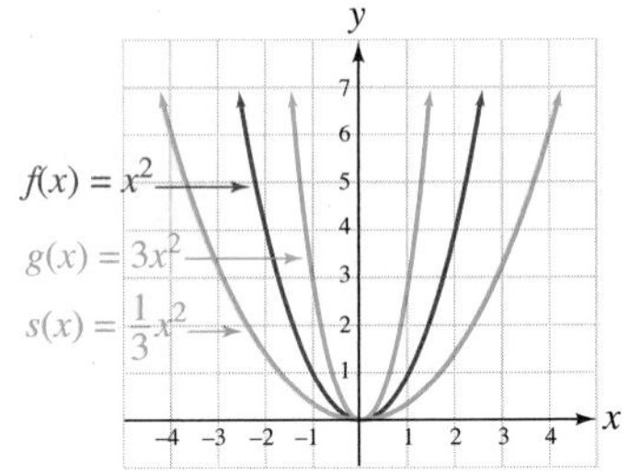

Teaching Example 1 Graph: $g(x) = \frac{3}{2}x^2$

Answer:

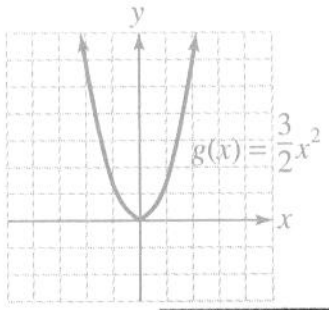

Self Check 1 Graph: $g(x) = \frac{2}{3}x^2$. See AIE Appendix 3.

Now Try Problem 15

EXAMPLE 2 Graph: $f(x) = -3x^2$

Strategy We make a table of values for the function, plot each point, and connect them with a smooth curve.

Why At this time, this method is our only option.

Solution After graphing the curve, we see that it opens downward and has the same shape as the graph of $g(x) = 3x^2$ that was graphed in Example 1.

Teaching Tip: You may also want to ask your students for the domain and range of the function.

$f(x) = -3x^2$

x	$f(x)$	
−2	−12	→ (−2, −12)
−1	−3	→ (−1, −3)
0	0	→ (0, 0)
1	−3	→ (1, −3)
2	−12	→ (2, −12)

This axis could also be labeled $f(x)$.

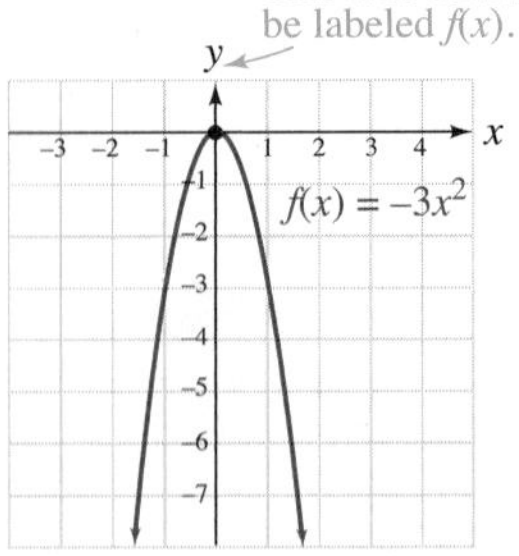

Teaching Example 2 Graph: $f(x) = -\frac{3}{2}x^2$

Answer:

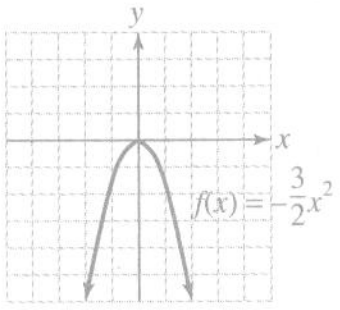

Self Check 2 Graph: $f(x) = -\frac{1}{3}x^2$ See AIE Appendix 3.

Now Try Problem 17

The graphs of functions of the form $f(x) = ax^2$ are **parabolas.** The lowest point on a parabola that opens upward, or the highest point on a parabola that opens downward, is called the **vertex** of the parabola. The vertical line, called an **axis of symmetry,** that passes through the vertex divides the parabola into two congruent halves. If we fold the paper along the axis of symmetry, the two sides of the parabola will match.

The Language of Algebra

An axis of symmetry divides a parabola into two matching sides. The sides are said to be **mirror images** of each other.

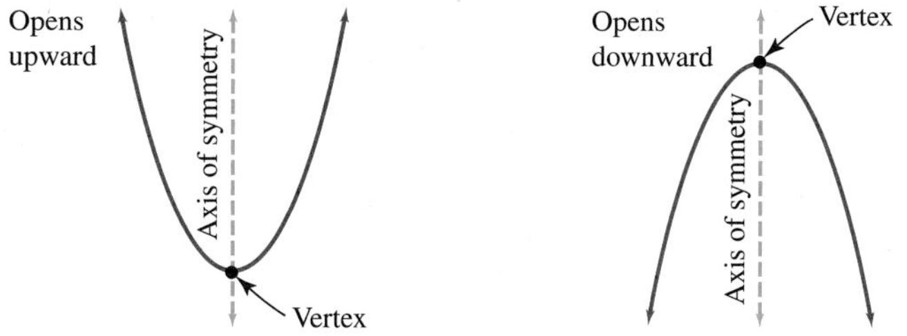

The results from Examples 1 and 2 confirm the following facts.

The Graph of $f(x) = ax^2$

The graph of $f(x) = ax^2$ is a parabola opening upward when $a > 0$ and downward when $a < 0$, with vertex at the point $(0, 0)$ and axis of symmetry the line $x = 0$.

EXAMPLE 3 Graph: **a.** $f(x) = 2x^2$ **b.** $g(x) = 2x^2 + 3$ **c.** $s(x) = 2x^2 - 3$

Strategy We make a table of values for each function, plot each point, and connect them with a smooth curve.

Why At this time, this method is our only option.

Solution After graphing the curves, we see that the graph of $g(x) = 2x^2 + 3$ is identical to the graph of $f(x) = 2x^2$, except that it has been translated 3 units upward. The graph of $s(x) = 2x^2 - 3$ is identical to the graph of $f(x) = 2x^2$, except that it has been translated 3 units downward. In each case, the axis of symmetry is the line $x = 0$.

$f(x) = 2x^2$

x	$f(x)$
−2	8
−1	2
0	0
1	2
2	8

$g(x) = 2x^2 + 3$

x	$g(x)$
−2	11
−1	5
0	3
1	5
2	11

$s(x) = 2x^2 - 3$

x	$s(x)$
−2	5
−1	−1
0	−3
1	−1
2	5

For each x-value, g(x) is 3 more than f(x).

For each x-value, s(x) is 3 less than f(x).

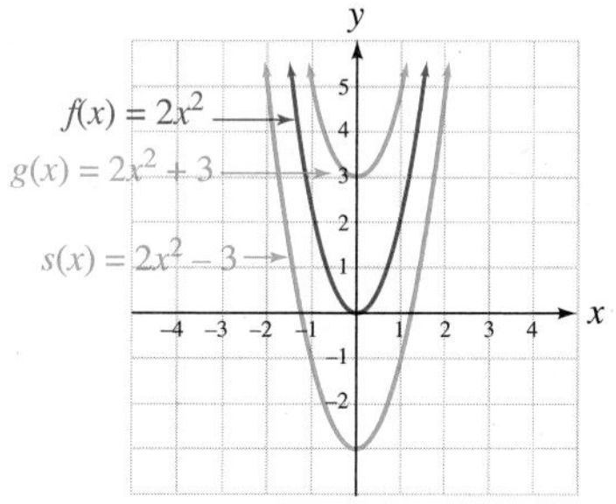

Teaching Example 3 Graph: $f(x) = 2x^2 - 1$
Answer:

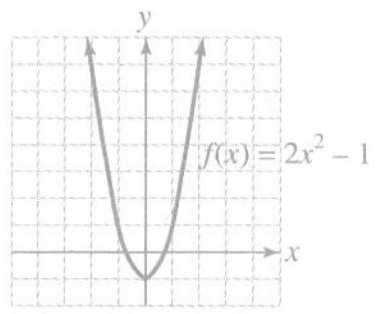

Self Check 3 Graph: $f(x) = 2x^2 + 1$ See AIE Appendix 3.

Now Try Problem 19

The results of Example 3 confirm the following facts.

The Graph of $f(x) = ax^2 + k$

The graph of $f(x) = ax^2 + k$ is a parabola having the same shape as $f(x) = ax^2$ but translated k units upward if k is positive and $|k|$ units downward if k is negative. The vertex is at the point $(0, k)$, and the axis of symmetry is the line $x = 0$.

2 Graph Functions of the Form $f(x) = a(x - h)^2$ and $f(x) = a(x - h)^2 + k$.

EXAMPLE 4 Graph: **a.** $f(x) = 2x^2$ **b.** $g(x) = 2(x - 3)^2$ **c.** $s(x) = 2(x + 3)^2$

Strategy We make a table of values for each function, plot each point, and connect them with a smooth curve.

Why At this time, this method is our only option.

Solution We note that the graph of $g(x) = 2(x - 3)^2$ below is identical to the graph of $f(x) = 2x^2$, except that it has been translated 3 units to the right. The graph of $s(x) = 2(x + 3)^2$ is identical to the graph of $f(x) = 2x^2$, except that it has been translated 3 units to the left.

$f(x) = 2x^2$

x	$f(x)$
−2	8
−1	2
0	0
1	2
2	8

$g(x) = 2(x - 3)^2$

x	$g(x)$
1	8
2	2
3	0
4	2
5	8

$s(x) = 2(x + 3)^2$

x	$s(x)$
−5	8
−4	2
−3	0
−2	2
−1	8

When an x-value is increased by 3, the function's outputs are the same.

When an x-value is decreased by 3, the function's outputs are the same.

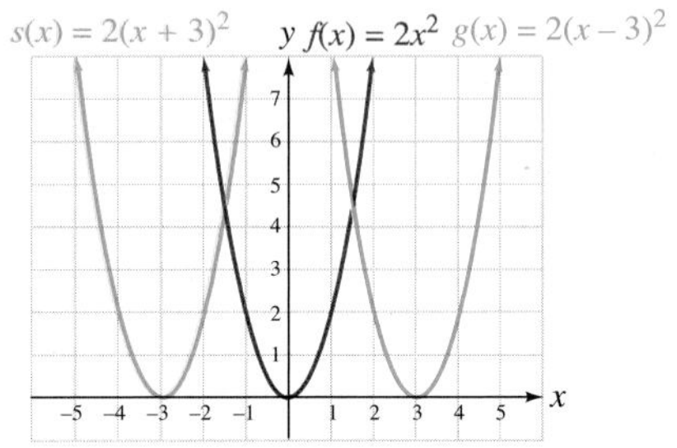

Teaching Example 4 Graph: $g(x) = 2(x - 1)^2$
Answer:

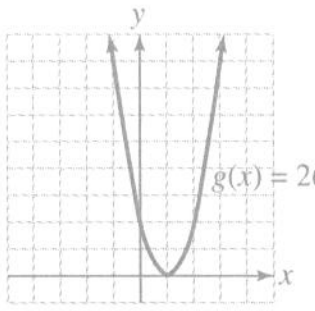

Self Check 4 Graph: $g(x) = 2(x + 1)^2$ See AIE Appendix 3.

Now Try Problem 21

The results of Example 4 confirm the following facts.

> **The Graph of $f(x) = a(x - h)^2$**
>
> The graph of $f(x) = a(x - h)^2$ is a parabola having the same shape as $f(x) = ax^2$ but translated h units to the right if h is positive and $|h|$ units to the left if h is negative. The vertex is at the point $(h, 0)$, and the axis of symmetry is the line $x = h$.

The results of Examples 1–4 suggest a general strategy for graphing quadratic functions that are written in the form $f(x) = a(x - h)^2 + k$.

> **Graphing a Quadratic Function in Standard Form**
>
> The graph of the quadratic function $f(x) = a(x - h)^2 + k$, where $a \neq 0$, is a parabola with vertex at (h, k). The axis of symmetry is the line $x = h$. The parabola opens upward when $a > 0$ and downward when $a < 0$.

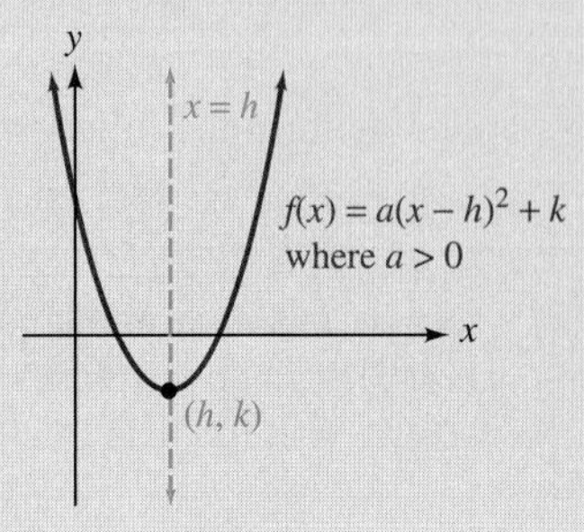

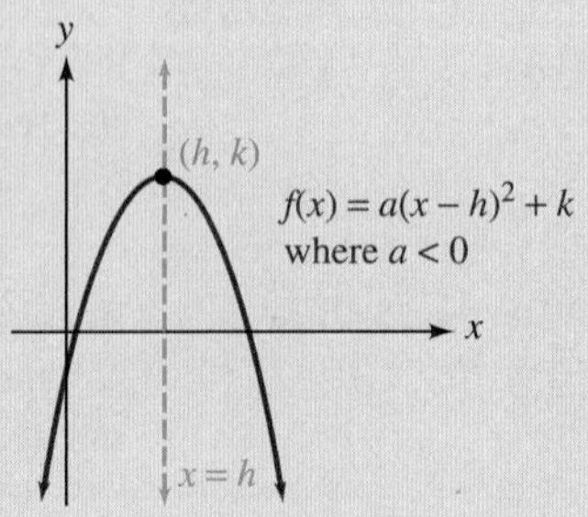

EXAMPLE 5 Graph: $f(x) = 2(x - 3)^2 - 4$. Label the vertex and draw the axis of symmetry.

Strategy We will determine whether the graph opens upward or downward and find its vertex and axis of symmetry. Then we will plot some points and complete the graph.

Why This method will be more efficient than plotting many points.

Solution The graph of $f(x) = 2(x - 3)^2 - 4$ is identical to the graph of $g(x) = 2(x - 3)^2$, except that it has been translated 4 units downward. The graph of $g(x) = 2(x - 3)^2$ is identical to the graph of $s(x) = 2x^2$, except that it has been translated 3 units to the right.

We can learn more about the graph of $f(x) = 2(x - 3)^2 - 4$ by determining a, h, and k.

Success Tip

The most important point to find when graphing a quadratic function is the vertex.

$$\left.\begin{array}{l} f(x) = \mathbf{2}(x - \mathbf{3})^2 - 4 \\ \quad\ \uparrow \qquad\ \uparrow \qquad \uparrow \\ f(x) = \boldsymbol{a}(x - \boldsymbol{h})^2 + k \end{array}\right\} a = 2,\ h = 3, \text{ and } k = -4$$

Upward/downward: Since $a = 2$ and $2 > 0$, the parabola opens upward.

Vertex: The vertex of the parabola is $(h, k) = (3, -4)$, as shown below.

Axis of symmetry: Since $h = 3$, the axis of symmetry is the line $x = 3$, as shown below.

Plotting points: We can construct a table of values to determine several points on the parabola. Since the x-coordinate of the vertex is 3, we choose the x-values of 4 and 5, find $f(4)$ and $f(5)$, and record the results in a table. Then we plot $(4, -2)$ and $(5, 4)$, and use symmetry to locate two other points on the parabola: $(2, -2)$ and $(1, 4)$. Finally, we draw a smooth curve through the points to get the graph.

$f(x) = 2(x - 3)^2 - 4$

x	$f(x)$	
4	-2	→ $(4, -2)$
5	4	→ $(5, 4)$

↑ The x-coordinate of the vertex is 3. Choose values for x close to 3 and on the same side of the axis of symmetry.

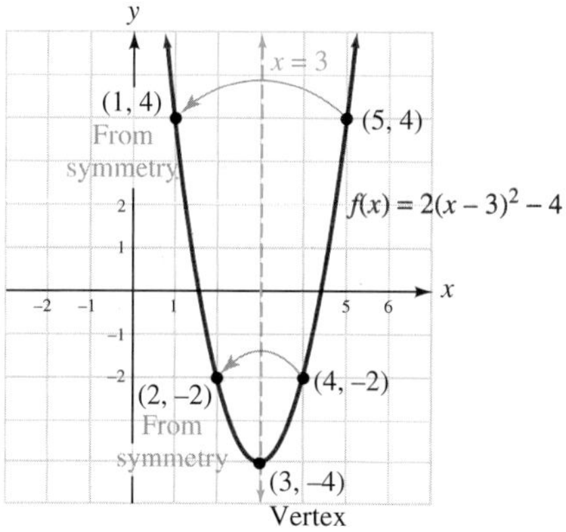

Success Tip

When graphing, remember that any point on a parabola to the right of the axis of symmetry yields a second point to the left of the axis of symmetry, and vice versa. Think of it as two-for-one.

Self Check 5 Graph: $f(x) = 2(x - 1)^2 - 2$. Label the vertex and draw the axis of symmetry. See AIE Appendix 3.

Now Try Problems 25 and 31

Teaching Example 5 Graph $f(x) = 2(x + 1)^2 - 3$. Label the vertex and draw the axis of symmetry.
Answer:

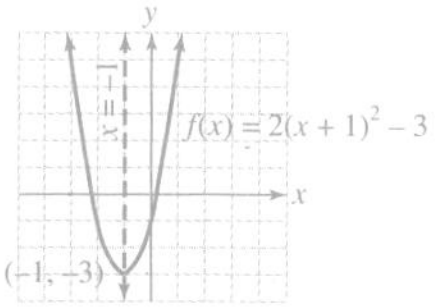

3 Graph Functions of the Form $f(x) = ax^2 + bx + c$ by Completing the Square.

To graph functions of the form $f(x) = ax^2 + bx + c$, we can complete the square to write the function in standard form $f(x) = a(x - h)^2 + k$.

EXAMPLE 6 Determine the vertex and the axis of symmetry of the graph of $f(x) = x^2 + 8x + 21$. Will the graph open upward or downward?

Strategy To find the vertex and the axis of symmetry, we will complete the square on x and write the equation of the function in standard form.

Why Once the equation is written in standard form, we can determine the values of a, h, and k. The coordinates of the vertex will be (h, k) and the equation of the axis of symmetry will be $x = h$. The graph will open upward if $a > 0$ or downward if $a < 0$.

Solution To determine the vertex and the axis of symmetry of the graph, we complete the square on the right side so we can write the function in $f(x) = a(x - h)^2 + k$ form.

$$f(x) = x^2 + 8x + 21$$

$$f(x) = (x^2 + 8x \quad) + 21 \quad \text{Prepare to complete the square on } x \text{ by writing parentheses around } x^2 + 8x.$$

To complete the square on $x^2 + 8x$, we note that one-half of the coefficient of x is $\frac{1}{2} \cdot 8 = 4$, and $4^2 = 16$. If we add 16 to $x^2 + 8x$, we obtain a perfect-square trinomial within the parentheses. Since this step adds 16 to the right side, we must also subtract 16 from the right side so that it remains in an equivalent form.

Add 16 to the right side. ↓ ↓ Subtract 16 from the right side to counteract the addition of 16.

$$f(x) = (x^2 + 8x + \mathbf{16}) + 21 - \mathbf{16}$$

$$f(x) = (x + 4)^2 + 5 \quad \text{Factor } x^2 + 8x + 16 \text{ and combine like terms.}$$

Success Tip

When a number is added to and that same number is subtracted from one side of an equation, the value of that side of the equation remains the same.

The function is now written in standard form, and we can determine a, h, and k.

The standard form requires a minus symbol here.

$$f(x) = \left[(x - (-4)\right]^2 + 5 \quad \text{Write } x + 4 \text{ as } x - (-4) \text{ to determine } h.$$

$a = 1$ $h = -4$ $k = 5$

Teaching Tip: Remind your students that h is the number that follows a subtraction symbol.

The vertex is $(h, k) = (-4, 5)$ and the axis of symmetry is the line $x = -4$. Since $a = 1$ and $1 > 0$, the parabola opens upward.

Self Check 6 Determine the vertex and the axis of symmetry of the graph of $f(x) = x^2 + 4x + 10$. Will the graph open upward or downward?
$(-2, 6)$; $x = -2$; opens upward

Now Try Problem 39

Teaching Example 6 Determine the vertex and the axis of symmetry of the graph of $f(x) = x^2 + 6x + 15$. Will the graph open upward or downward? *Answer:* $(-3, 6)$; $x = -3$; opens upward

EXAMPLE 7 Graph: $f(x) = 2x^2 - 4x - 1$

Strategy We will complete the square on x and write the equation of the function in standard form, $f(x) = a(x - h)^2 + k$.

Why When the equation is in standard form, we can identify the values of a, h, and k from the equation. This information will help us sketch the graph.

Solution Recall that to complete the square on $2x^2 - 4x$, the coefficient of x^2 must be equal to 1. Therefore, we factor 2 from $2x^2 - 4x$.

$$f(x) = 2x^2 - 4x - 1$$

$$f(x) = 2(x^2 - 2x \quad) - 1$$

To complete the square on $x^2 - 2x$, we note that one-half of the coefficient of x is $\frac{1}{2}(-2) = -1$, and $(-1)^2 = 1$. If we add 1 to $x^2 - 2x$, we obtain a perfect-square trinomial within the parentheses. Since this step adds 2 to the right side, we must also subtract 2 from the right side so that it remains in an equivalent form.

By the distributive property, when 1 is added to the expression within the parentheses, $2 \cdot 1 = 2$ is added to the right side.

Subtract 2 to counteract the addition of 2 shown in red.

$$f(x) = 2(x^2 - 2x + 1) - 1 - 2$$

$$f(x) = 2(x - 1)^2 - 3 \qquad \text{Factor } x^2 - 2x + 1 \text{ and combine like terms.}$$

We see that $a = 2$, $h = 1$, and $k = -3$. Thus, the vertex is at the point $(1, -3)$, and the axis of symmetry is $x = 1$. Since $a = 2$ and $2 > 0$, the parabola opens upward. We plot the vertex and axis of symmetry as shown below.

Finally, we construct a table of values, plot the points, use symmetry to plot the corresponding points, and then draw the graph.

$f(x) = 2x^2 - 4x - 1$
or
$f(x) = 2(x - 1)^2 - 3$

x	$f(x)$	
2	−1	→ (2, −1)
3	5	→ (3, 5)

↑ The x-coordinate of the vertex is 1. Choose values for x close to 1 and on the same side of the axis of symmetry.

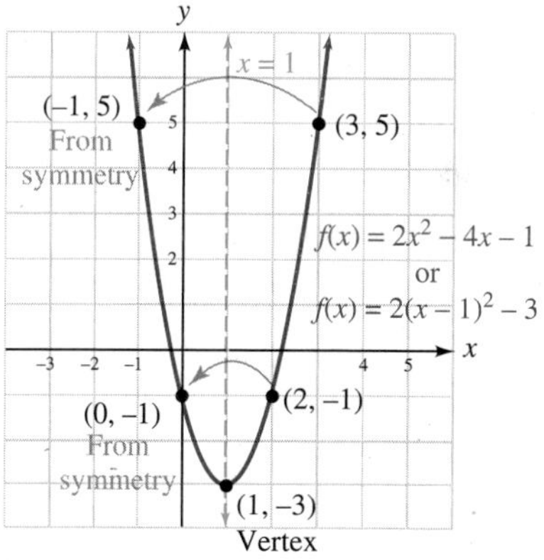

Success Tip

To find additional points on the graph, select values of x that are close to the x-coordinate of the vertex.

Teaching Example 7 Graph: $f(x) = 2x^2 - 4x + 5$
Answer:

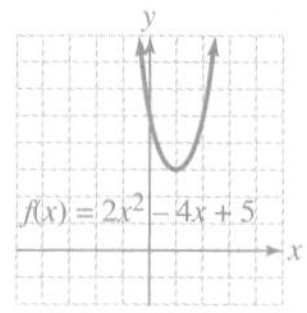

Self Check 7 Graph: $f(x) = 3x^2 - 12x + 8$. See AIE Appendix 3.

Now Try ▶ Problem 45

4 Find the Vertex Using $-\dfrac{b}{2a}$.

Because of symmetry, if a parabola has two x-intercepts, the x-coordinate of the vertex is exactly midway between them. We can use this fact to derive a formula to find the vertex of a parabola.

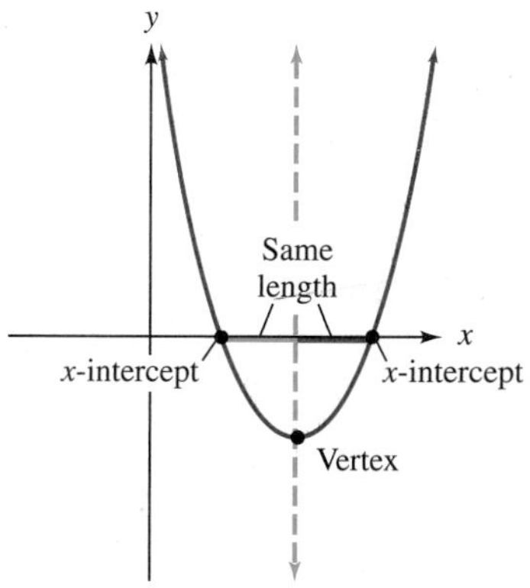

In general, if a parabola has two x-intercepts, they can be found by solving $0 = ax^2 + bx + c$ for x. We can use the quadratic formula to find the solutions. They are

$$x = \frac{-b - \sqrt{b^2 - 4ac}}{2a} \quad \text{and} \quad x = \frac{-b + \sqrt{b^2 - 4ac}}{2a}$$

Thus, the parabola's x-intercepts are $\left(\frac{-b - \sqrt{b^2 - 4ac}}{2a}, 0\right)$ and $\left(\frac{-b + \sqrt{b^2 - 4ac}}{2a}, 0\right)$.

Since the x-value of the vertex of a parabola is halfway between the two x-intercepts, we can find this value by finding the average, or $\frac{1}{2}$ of the sum of the x-coordinates of the x-intercepts.

$$x = \frac{1}{2}\left(\frac{-b - \sqrt{b^2 - 4ac}}{2a} + \frac{-b + \sqrt{b^2 - 4ac}}{2a}\right)$$

$$x = \frac{1}{2}\left(\frac{-b - \sqrt{b^2 - 4ac} + (-b) + \sqrt{b^2 - 4ac}}{2a}\right)$$ Add the numerators and keep the common denominator.

$$x = \frac{1}{2}\left(\frac{-2b}{2a}\right)$$ Combine like terms: $-b + (-b) = -2b$ and $-\sqrt{b^2 - 4ac} + \sqrt{b^2 - 4ac} = 0$.

$$x = -\frac{b}{2a}$$ Remove the common factor of 2 in the numerator and denominator and simplify.

This result is true even if the graph has no x-intercepts.

Formula for the Vertex of a Parabola

The vertex of the graph of the quadratic function $f(x) = ax^2 + bx + c$ is

$$\left(-\frac{b}{2a}, f\left(-\frac{b}{2a}\right)\right)$$

and the axis of symmetry of the parabola is the line $x = -\frac{b}{2a}$.

EXAMPLE 8 Find the vertex of the graph of $f(x) = 2x^2 - 4x - 1$.

Strategy We will determine the values of a and b and substitute into the formula for the vertex of a parabola.

Why It is easier to find the coordinates of the vertex using the formula than it is to complete the square on $2x^2 - 4x - 1$.

Solution The function is written in $f(x) = ax^2 + bx + c$ form, where $a = 2$ and $b = -4$. To find the vertex of its graph, we calculate

$$-\frac{b}{2a} = -\frac{-4}{2(2)} = -\frac{-4}{4} = 1$$ This is the x-coordinate of the vertex.

$$f\left(-\frac{b}{2a}\right) = f(1) = 2(1)^2 - 4(1) - 1 = -3$$ This is the y-coordinate of the vertex.

The vertex is the point $(1, -3)$. This agrees with the result we obtained in Example 7 by completing the square.

Self Check 8 Find the vertex of the graph of $f(x) = 3x^2 - 12x + 8$. $(2, -4)$

Now Try Problem 55

Success Tip

We can find the vertex of the graph of a quadratic function by completing the square or by using the formula.

Teaching Example 8 Find the vertex of the graph of $f(x) = 3x^2 - 6x + 1$. *Answer:* $(1, -2)$

Using Your Calculator ▶ Finding the Vertex

We can use a graphing calculator to graph the function $f(x) = 2x^2 + 6x - 3$ and find the coordinates of the vertex and the axis of symmetry of the parabola. If we enter the function, we will obtain the graph shown in figure (a) on the next page.

We then trace to move the cursor to the lowest point on the graph, as shown in figure (b). By zooming in, we can see that the vertex is the point $(-1.5, -7.5)$, or $\left(-\frac{3}{2}, -\frac{15}{2}\right)$, and that the line $x = -\frac{3}{2}$ is the axis of symmetry.

Some calculators have an fmin or fmax feature that also can be used to find the vertex.

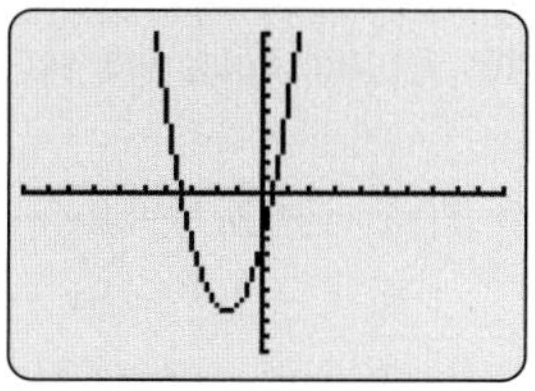

(a)

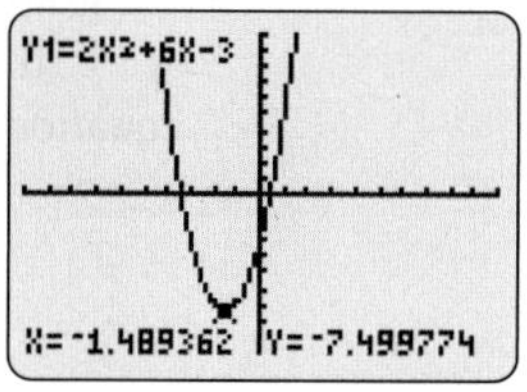

(b)

We can determine much about the graph of $f(x) = ax^2 + bx + c$ from the coefficients a, b, and c. This information is summarized as follows:

Graphing a Quadratic Function $f(x) = ax^2 + bx + c$

- Determine whether the parabola opens upward or downward by finding the value of a.
- The x-coordinate of the vertex of the parabola is $x = -\frac{b}{2a}$.
- To find the y-coordinate of the vertex, substitute $-\frac{b}{2a}$ for x and find $f\left(-\frac{b}{2a}\right)$.
- The axis of symmetry is the vertical line passing through the vertex.
- The y-intercept is determined by the value of $f(x)$ when $x = 0$: the y-intercept is $(0, c)$.
- The x-intercepts (if any) are determined by the values of x that make $f(x) = 0$. To find them, solve the quadratic equation $ax^2 + bx + c = 0$.

EXAMPLE 9 Graph: $f(x) = -2x^2 - 8x - 8$

Strategy We will follow the steps for graphing a quadratic function.

Why This is the most efficient way to graph a general quadratic function.

Solution ***Step 1: Determine whether the parabola opens upward or downward.*** The function is in the form $f(x) = ax^2 + bx + c$, with $a = -2$, $b = -8$, and $c = -8$. Since $a < 0$, the parabola opens downward.

Step 2: Find the vertex and draw the axis of symmetry. To find the coordinates of the vertex, we calculate

$$x = -\frac{b}{2a}$$

$$x = -\frac{-8}{2(-2)} \quad \text{Substitute } -2 \text{ for } a \text{ and } -8 \text{ for } b.$$

$$= -2 \quad \text{This is the } x\text{-coordinate of the vertex.}$$

$$f\left(-\frac{b}{2a}\right) = f(-2)$$

$$= -2(-2)^2 - 8(-2) - 8$$

$$= -8 + 16 - 8$$

$$= 0 \quad \text{This is the } y\text{-coordinate of the vertex.}$$

The vertex of the parabola is the point $(-2, 0)$. This point is in blue on the graph. The axis of symmetry is the line $x = -2$.

Success Tip

An easy way to remember the vertex formula is to note that $x = \frac{-b}{2a}$ is part of the quadratic formula:

$$x = \frac{-b \pm \sqrt{b^2 - 4ac}}{2a}$$

Step 3: ***Find the x- and y-intercepts.*** Since $c = -8$, the y-intercept of the parabola is $(0, -8)$. The point $(-4, -8)$, two units to the left of the axis of symmetry, must also be on the graph. We plot both points in black on the graph.

To find the x-intercepts, we set $f(x)$ equal to 0 and solve the resulting quadratic equation.

$$f(x) = -2x^2 - 8x - 8 \quad \text{This is the function to graph.}$$
$$0 = -2x^2 - 8x - 8 \quad \text{Set } f(x) = 0.$$
$$0 = x^2 + 4x + 4 \quad \text{Divide both sides by } -2.$$
$$0 = (x + 2)(x + 2) \quad \text{Factor the trinomial.}$$
$$x + 2 = 0 \quad \text{or} \quad x + 2 = 0 \quad \text{Set each factor equal to 0.}$$
$$x = -2 \quad \Big| \quad x = -2$$

Since the solutions are the same, the graph has only one x-intercept: $(-2, 0)$. This point is the vertex of the parabola and has already been plotted.

Step 4: ***Plot another point.*** Finally, we find another point on the parabola. If $x = -3$, then $f(-3) = -2$. We plot $(-3, -2)$ and use symmetry to determine that $(-1, -2)$ is also on the graph. Both points are in green.

Step 5: ***Draw a smooth curve through the points.***

$f(x) = -2x^2 - 8x - 8$

x	$f(x)$	
−3	−2	→ (−3, −2)

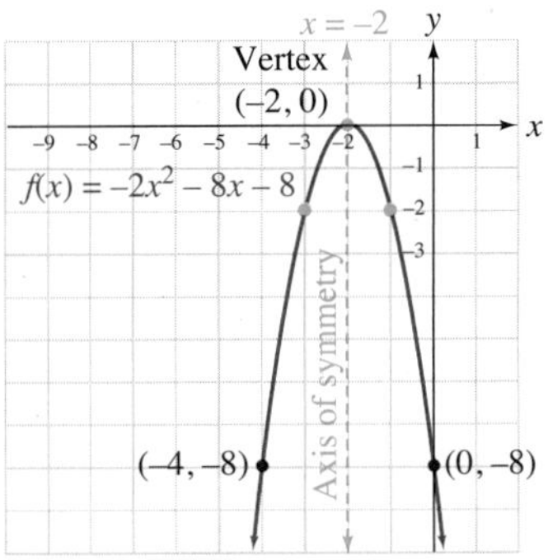

Teaching Example 9 Graph: $f(x) = -2x^2 - 4x + 1$
Answer:

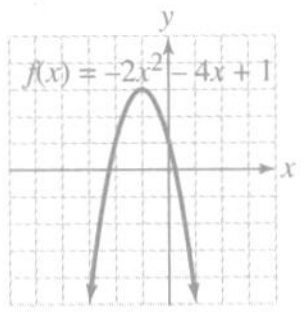

Self Check 9 Graph: $f(x) = -2x^2 + 12x - 16$ See AIE Appendix 3.

Now Try ▶ **Problems 59 and 69**

5 Determine Minimum and Maximum Values.

It is often useful to know the smallest or largest possible value a quantity can assume. For example, companies try to minimize their costs and maximize their profits. If the quantity is expressed by a quadratic function, the y-coordinate of the vertex of the graph of the function gives its minimum or maximum value.

EXAMPLE 10

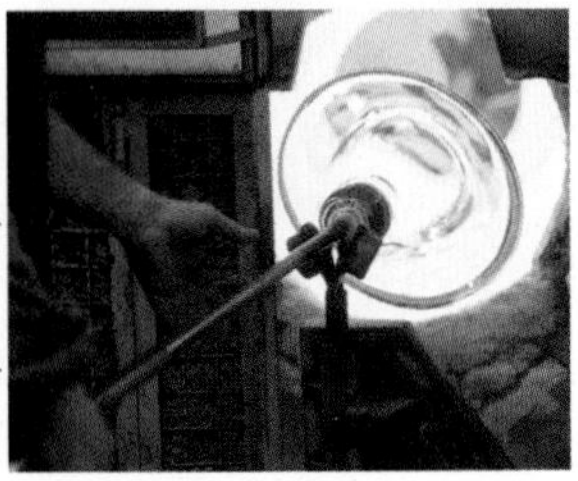

Minimizing Costs. A glassworks that makes lead crystal vases has daily production costs given by the function $C(x) = 0.2x^2 - 10x + 650$, where x is the number of vases made each day. How many vases should be produced to minimize the per-day costs? What will the costs be?

Strategy We will find the vertex of the graph of the quadratic function.

Why The x-coordinate of the vertex indicates the number of vases to make to keep costs at a minimum, and the y-coordinate indicates the minimum cost.

Solution The graph of $C(x) = 0.2x^2 - 10x + 650$ is a parabola opening upward. The vertex is the lowest point on the graph. To find the vertex, we calculate

$$-\frac{b}{2a} = -\frac{-10}{2(0.2)} \quad b = -10 \text{ and } a = 0.2.$$
$$= -\frac{-10}{0.4}$$
$$= 25$$

$$f\left(-\frac{b}{2a}\right) = f(25)$$
$$= 0.2(25)^2 - 10(25) + 650$$
$$= 525$$

The Language of Algebra

We say that 25 is the value for which the function $C(x) = 0.2x^2 - 10x + 650$ is a minimum.

The vertex is (25, 525), and it indicates that the costs are a minimum of \$525 when 25 vases are made daily.

To solve this problem with a graphing calculator, we graph the function $C(x) = 0.2x^2 - 10x + 650$. By using TRACE and ZOOM, we can locate the vertex of the graph. The coordinates of the vertex indicate that the minimum cost is \$525 when the number of vases produced is 25.

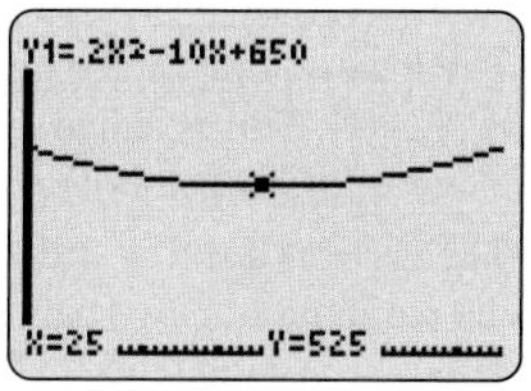

Teaching Example 10 **Profits.** The profit function for a company is $P(x) = -0.0025x^2 + 120x - 125{,}000$, where x is the number of units sold. What sales level will give a maximum profit?
Answer: 24,000 units

Self Check 10 **Minimizing Costs.** A manufacturing company has a daily production cost of $c(x) = 0.25x^2 - 10x + 800$, where x is the number of items produced and $c(x)$ is the cost. How many items should be produced to minimize the per-day cost? What is the minimum cost? 20, \$700

Now Try Problem 85

6 Solve Quadratic Equations Graphically.

When solving quadratic equations graphically, we must consider three possibilities. If the graph of the associated quadratic function has two x-intercepts, the quadratic equation has two real-number solutions. Figure (a) shows an example of this. If the graph has one x-intercept, as shown in figure (b), the equation has one repeated real-number solution. Finally, if the graph does not have an x-intercept, as shown in figure (c), the equation does not have any real-number solutions.

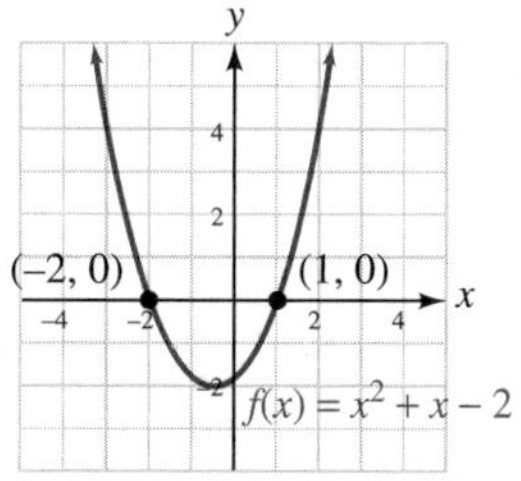

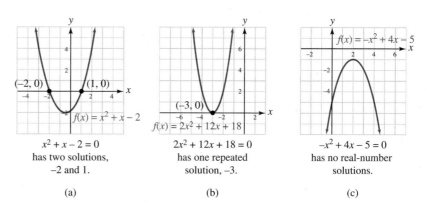

$x^2 + x - 2 = 0$
has two solutions, −2 and 1.

(a)

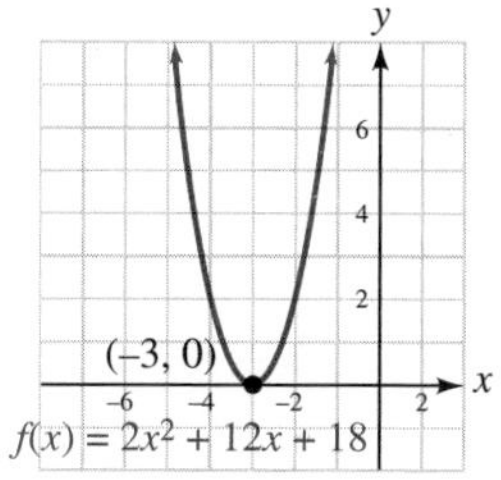

$2x^2 + 12x + 18 = 0$
has one repeated solution, −3.

(b)

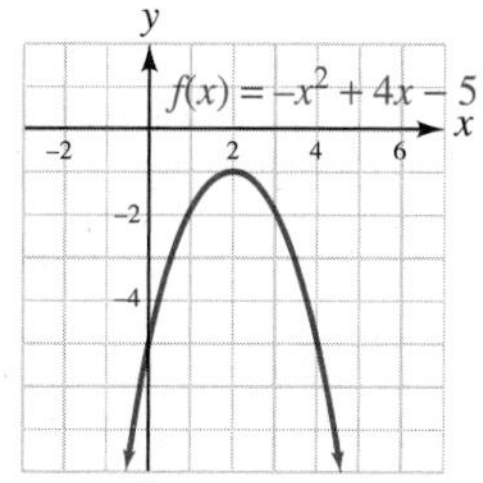

$-x^2 + 4x - 5 = 0$
has no real-number solutions.

(c)

Success Tip

Note that the solutions of each quadratic equation are given by the x-coordinate of the x-intercept of each respective graph.

Using Your Calculator ▶ Solving Quadratic Equations Graphically

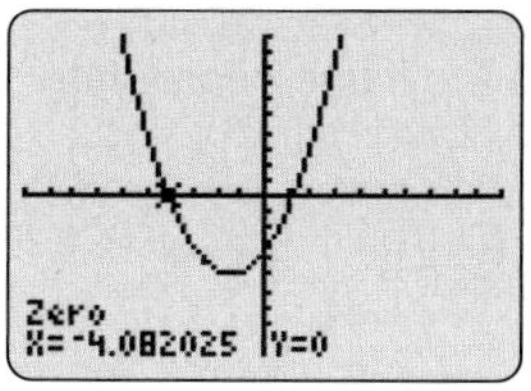

We can use a graphing calculator to find approximate solutions of quadratic equations. For example, the solutions of $0.7x^2 + 2x - 3.5 = 0$ are the numbers x that will make $y = 0$ in the quadratic function $f(x) = 0.7x^2 + 2x - 3.5$. To approximate these numbers, we graph the quadratic function and read the x-intercepts from the graph using the ZERO feature. (The ZERO feature can be found by pressing 2nd, CALC, and then 2.) In the figure, we see that the x-coordinate of the left-most x-intercept of the graph is given as -4.082025. This means that an approximate solution of the equation is -4.08. To find the positive x-intercept, we use similar steps.

SECTION 10.4 STUDY SET

VOCABULARY

Refer to the graph. Fill in the blanks.

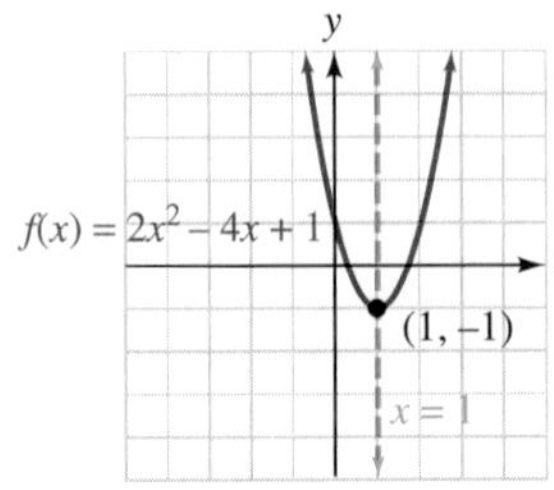

1. $f(x) = 2x^2 - 4x + 1$ is called a quadratic function. Its graph is a cup-shaped figure called a parabola.
2. The lowest point on the graph is $(1, -1)$. This is called the vertex of the parabola.
3. The vertical line $x = 1$ divides the parabola into two halves. This line is called the axis of symmetry.
4. $f(x) = a(x - h)^2 + k$ is called the standard form of the equation of a quadratic function.

CONCEPTS

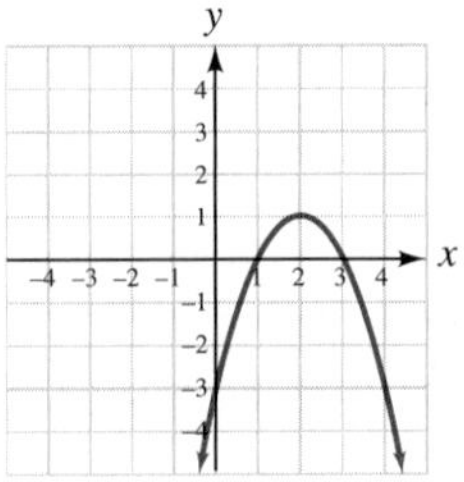

5. Refer to the graph.
   a. What are the x-intercepts of the graph? $(1, 0), (3, 0)$
   b. What is the y-intercept of the graph? $(0, -3)$
   c. What is the vertex? $(2, 1)$
   d. What is the axis of symmetry? $x = 2$
   e. What are the domain and the range of the function? Domain: $(-\infty, \infty)$; range: $(-\infty, 1]$

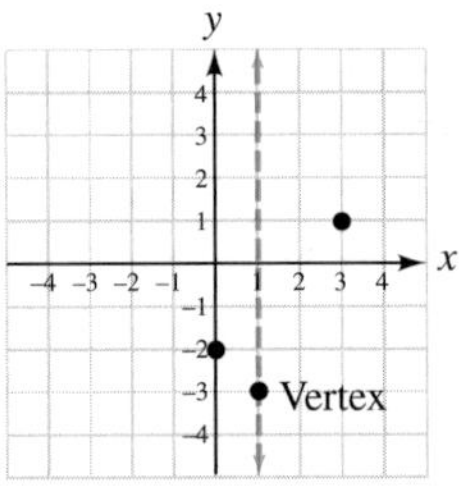

6. The vertex of a parabola is at $(1, -3)$, its y-intercept is $(0, -2)$, and it passes through the point $(3, 1)$, as shown in the illustration. Use the axis of symmetry shown in blue to help determine two other points on the parabola. $(-1, 1)$, $(2, -2)$

7. Draw the graph of a quadratic function using the given facts about its graph. See AIE Appendix 3.
   - Opens upward
   - y-intercept: $(0, -3)$
   - Vertex: $(-1, -4)$
   - x-intercepts: $(-3, 0), (1, 0)$
   -

x	$f(x)$
2	5

8. For $f(x) = -x^2 + 6x - 7$, the value of $-\frac{b}{2a}$ is 3. Find the y-coordinate of the vertex of the graph of this function. 2

9. Fill in the blanks.
   a. To complete the square on the right side of $f(x) = 2x^2 + 12x + 11$, what should be factored from the first two terms?

   $$f(x) = 2(x^2 + 6x) + 11$$

   b. To complete the square on $x^2 + 6x$ shown below, what should be added within the parentheses and what should be subtracted outside the parentheses?

   $$f(x) = 2(x^2 + 6x + 9) + 11 - 18$$

10. Fill in the blanks. To complete the square on $x^2 + 4x$ shown below, what should be added within the parentheses and what should be added outside the parentheses?

   $$f(x) = -5(x^2 + 4x + 4) + 7 + 20$$

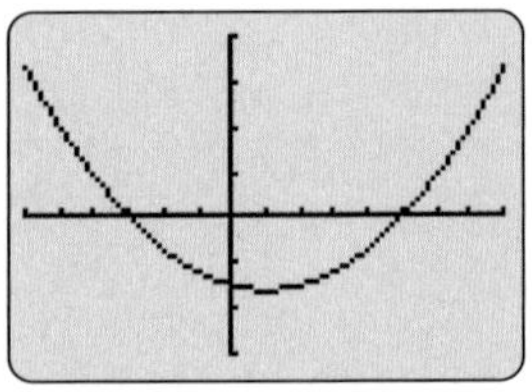

11. Use the graph of $f(x) = \frac{1}{10}x^2 - \frac{1}{5}x - \frac{3}{2}$, shown here, to estimate the solutions of the equation $\frac{1}{10}x^2 - \frac{1}{5}x - \frac{3}{2} = 0$. $-3, 5$

12. Three quadratic equations are to be solved graphically. The graphs of their associated quadratic functions are shown here. Determine which graph indicates that the equation has
   a. two real solutions. ii
   b. one repeated real solution. iii
   c. no real solutions. i

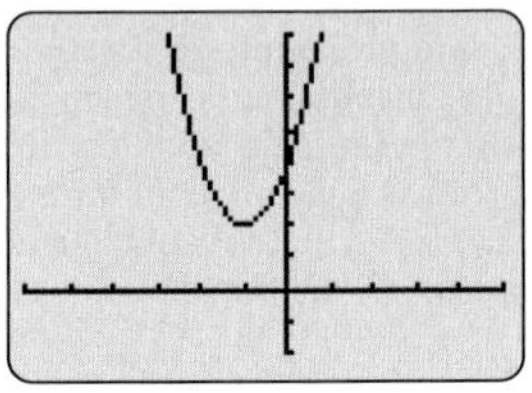

(i)

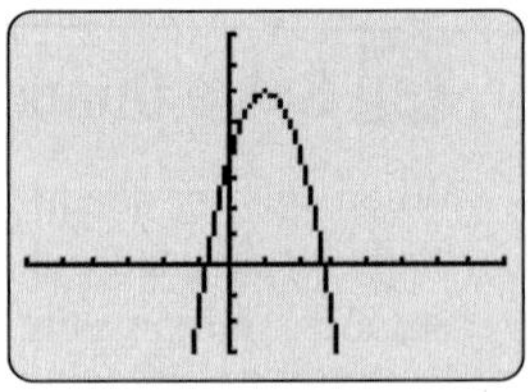

(ii)

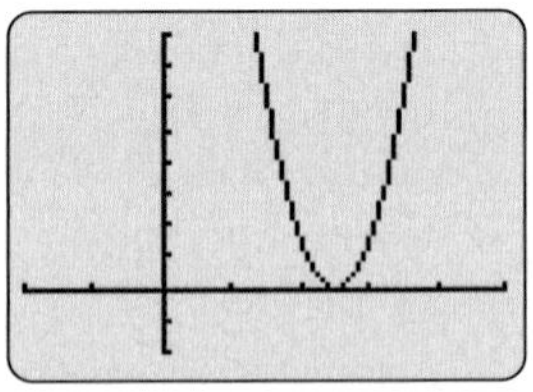

(iii)

NOTATION

13. The function $f(x) = 2(x + 1)^2 + 6$ is written in the form $f(x) = a(x - h)^2 + k$. Is $h = -1$ or is $h = 1$? Explain. $h = -1$; $f(x) = 2[x - (-1)]^2 + 6$
14. Consider the function $f(x) = 2x^2 + 4x - 8$.
   a. What are a, b, and c? $2, 4, -8$
   b. Find $-\frac{b}{2a}$. -1

GUIDED PRACTICE

Graph each group of functions on the same coordinate system. See Example 1. See AIE Appendix 3.

15. $f(x) = x^2,\ g(x) = 2x^2,\ s(x) = \frac{1}{2}x^2$
16. $f(x) = x^2,\ g(x) = 4x^2,\ s(x) = \frac{1}{4}x^2$

Graph each pair of functions on the same coordinate system. See Example 2. See AIE Appendix 3.

17. $f(x) = 2x^2,\ g(x) = -2x^2$
18. $f(x) = \frac{1}{2}x^2,\ g(x) = -\frac{1}{2}x^2$

***Graph each group of functions on the same coordinate system.* See Example 3.** See AIE Appendix 3.

19. $f(x) = 4x^2,\ g(x) = 4x^2 + 3,\ s(x) = 4x^2 - 2$

▶ **20.** $f(x) = \frac{1}{3}x^2,\ g(x) = \frac{1}{3}x^2 + 4,\ s(x) = \frac{1}{3}x^2 - 3$

***Graph each group of functions on the same coordinate system and describe how the graphs are similar and how they are different.* See Example 4.** See AIE Appendix 3.

▶ **21.** $f(x) = 3x^2,\ g(x) = 3(x + 2)^2,\ s(x) = 3(x - 3)^2$

▶ **22.** $f(x) = \frac{1}{2}x^2,\ g(x) = \frac{1}{2}(x + 3)^2,\ s(x) = \frac{1}{2}(x - 2)^2$

***Find the vertex and the axis of symmetry of the graph of each function. Do not graph the function, but determine whether the graph will open upward or downward.* See Example 5.**

▶ **23.** $f(x) = (x - 1)^2 + 2$ $(1, 2)$; $x = 1$; upward

▶ **24.** $f(x) = 2(x - 2)^2 - 1$ $(2, -1)$; $x = 2$; upward

▶ **25.** $f(x) = -2(x + 3)^2 - 4$ $(-3, -4)$; $x = -3$; downward

▶ **26.** $f(x) = -3(x + 1)^2 + 3$ $(-1, 3)$; $x = -1$; downward

27. $f(x) = -0.5(x - 7.5)^2 + 8.5$ $(7.5, 8.5)$; $x = 7.5$; downward

▶ **28.** $f(x) = -\frac{3}{2}\left(x + \frac{1}{4}\right)^2 + \frac{7}{8}$ $\left(-\frac{1}{4}, \frac{7}{8}\right)$; $x = -\frac{1}{4}$; downward

▶ **29.** $f(x) = 2x^2 - 4$ $(0, -4)$; $x = 0$; upward

30. $f(x) = 3x^2 - 3$ $(0, -3)$, $x = 0$; upward

***Determine the vertex and the axis of symmetry of the graph of each function. Then plot several points and complete the graph.* See Example 5.** See AIE Appendix 3.

31. $f(x) = (x - 3)^2 + 2$
$(3, 2), x = 3$

▶ **32.** $f(x) = (x + 1)^2 - 2$
$(-1, -2), x = -1$

▶ **33.** $f(x) = -(x - 2)^2$
$(2, 0), x = 2$

▶ **34.** $f(x) = -(x + 2)^2$
$(-2, 0), x = -2$

35. $f(x) = -2(x + 3)^2 + 4$
$(-3, 4), x = -3$

▶ **36.** $f(x) = -2(x - 2)^2 - 4$
$(2, -4), x = 2$

▶ **37.** $f(x) = \frac{1}{2}(x + 1)^2 - 3$
$(-1, -3), x = -1$

38. $f(x) = \frac{1}{3}(x - 1)^2 + 2$
$(1, 2), x = 1$

***Determine the vertex and the axis of symmetry of the graph of each function. Will the graph open upward or downward?* See Example 6.**

▶ **39.** $f(x) = x^2 + 4x + 5$ $(-2, 1)$; $x = -2$; upward

▶ **40.** $f(x) = x^2 - 4x - 1$ $(2, -5)$; $x = 2$; upward

▶ **41.** $f(x) = -x^2 + 6x - 15$ $(3, -6)$; $x = 3$; downward

42. $f(x) = -x^2 - 6x + 3$ $(-3, 12)$; $x = -3$; downward

***Complete the square to write each function in $f(x) = a(x - h)^2 + k$ form. Determine the vertex and the axis of symmetry of the graph of the function. Then plot several points and complete the graph.* See Examples 6 and 7.** See AIE Appendix 3.

43. $f(x) = x^2 + 2x - 3$
$f(x) = (x + 1)^2 - 4; (-1, -4), x = -1$

▶ **44.** $f(x) = x^2 + 6x + 5$
$f(x) = (x + 3)^2 - 4; (-3, -4), x = -3$

45. $f(x) = 4x^2 + 24x + 37$
$f(x) = 4(x + 3)^2 + 1; (-3, 1), x = -3$

▶ **46.** $f(x) = 3x^2 - 12x + 10$
$f(x) = 3(x - 2)^2 - 2; (2, -2), x = 2$

47. $f(x) = x^2 + x - 6$
$f(x) = \left(x + \frac{1}{2}\right)^2 - \frac{25}{4}; \left(-\frac{1}{2}, -\frac{25}{4}\right), x = -\frac{1}{2}$

▶ **48.** $f(x) = x^2 - x - 6$
$f(x) = \left(x - \frac{1}{2}\right)^2 - \frac{25}{4}; \left(\frac{1}{2}, -\frac{25}{4}\right), x = \frac{1}{2}$

49. $f(x) = -4x^2 + 16x - 10$
$f(x) = -4(x - 2)^2 + 6; (2, 6), x = 2$

▶ **50.** $f(x) = -2x^2 + 4x + 3$
$f(x) = -2(x - 1)^2 + 5; (1, 5), x = 1$

51. $f(x) = 2x^2 + 8x + 6$
$f(x) = 2(x + 2)^2 - 2; (-2, -2), x = -2$

▶ **52.** $f(x) = 3x^2 - 12x + 9$
$f(x) = 3(x - 2)^2 - 3; (2, -3), x = 2$

53. $f(x) = -x^2 - 8x - 17$
$f(x) = -(x + 4)^2 - 1; (-4, -1), x = -4$

▶ **54.** $f(x) = -x^2 + 6x - 8$
$f(x) = -(x - 3)^2 + 1; (3, 1), x = 3$

***Use the vertex formula to find the vertex of the graph of each function.* See Example 8.**

55. $f(x) = x^2 + 2x - 5$
$(-1, -6)$

▶ **56.** $f(x) = -x^2 + 4x - 5$
$(2, -1)$

57. $f(x) = 2x^2 - 3x + 4$
$\left(\frac{3}{4}, \frac{23}{8}\right)$

58. $f(x) = 2x^2 - 7x - 4$
$\left(\frac{7}{4}, -\frac{81}{8}\right)$

***Find the x- and y-intercepts of the graph of the quadratic function.* See Example 9.**

▶ **59.** $f(x) = x^2 - 2x - 35$
$(-5, 0), (7, 0); (0, -35)$

▶ **60.** $f(x) = -x^2 - 10x - 21$
$(-3, 0), (-7, 0); (0, -21)$

▶ **61.** $f(x) = -2x^2 + 4x$
$(0, 0), (2, 0); (0, 0)$

▶ **62.** $f(x) = 3x^2 + 6x - 9$
$(-3, 0), (1, 0); (0, -9)$

***Determine the coordinates of the vertex of the graph of each function using the vertex formula. Then determine the x- and y-intercepts of the graph. Finally, plot several points and complete the graph.* See Example 9.** See AIE Appendix 3.

▶ **63.** $f(x) = x^2 + 4x + 4$ $(-2, 0); (-2, 0); (0, 4)$

64. $f(x) = x^2 - 6x + 9$ $(3, 0); (3, 0); (0, 9)$

65. $f(x) = -x^2 + 2x - 1$ $(1, 0); (1, 0); (0, -1)$

▶ **66.** $f(x) = -x^2 - 2x - 1$ $(-1, 0); (-1, 0); (0, -1)$

67. $f(x) = x^2 - 2x$ $(1, -1); (0, 0), (2, 0); (0, 0)$

▶ **68.** $f(x) = x^2 + x$ $\left(-\frac{1}{2}, -\frac{1}{4}\right); (-1, 0), (0, 0); (0, 0)$

▶ **69.** $f(x) = 2x^2 - 8x + 6$ $(2, -2); (1, 0), (3, 0); (0, 6)$

▶ **70.** $f(x) = 3x^2 - 12x + 12$ $(2, 0); (2, 0); (0, 12)$

71. $f(x) = -6x^2 - 12x - 8$ $(-1, -2)$; no x-intercept; $(0, -8)$

▶ **72.** $f(x) = -2x^2 + 8x - 10$ $(2, -2)$; no x-intercept; $(0, -10)$

73. $f(x) = 4x^2 - 12x + 9$ $\left(\frac{3}{2}, 0\right); \left(\frac{3}{2}, 0\right); (0, 9)$

▶ **74.** $f(x) = 4x^2 + 4x - 3$ $\left(-\frac{1}{2}, -4\right); \left(-\frac{3}{2}, 0\right), \left(\frac{1}{2}, 0\right); (0, -3)$

***Use a graphing calculator to find the coordinates of the vertex of the graph of each quadratic function. Round to the nearest hundredth.* See Using Your Calculator: Finding the Vertex.**

75. $f(x) = 2x^2 - x + 1$
$(0.25, 0.88)$

76. $f(x) = x^2 + 5x - 6$
$(-2.50, -12.25)$

77. $f(x) = -x^2 + x + 7$
(0.50, 7.25)

78. $f(x) = 2x^2 - 3x + 2$
(0.75, 0.88)

Use a graphing calculator to solve each equation. If an answer is not exact, round to the nearest hundredth. **See Using Your Calculator: Solving Quadratic Equations Graphically.**

79. $x^2 + x - 6 = 0$
2, −3

80. $2x^2 - 5x - 3 = 0$
3, −0.5

81. $0.5x^2 - 0.7x - 3 = 0$
−1.85, 3.25

82. $2x^2 - 0.5x - 2 = 0$
−0.88, 1.13

APPLICATIONS

83. **Crossword Puzzles.** Darken the appropriate squares to the right of the dashed red line so that the puzzle has symmetry with respect to that line.

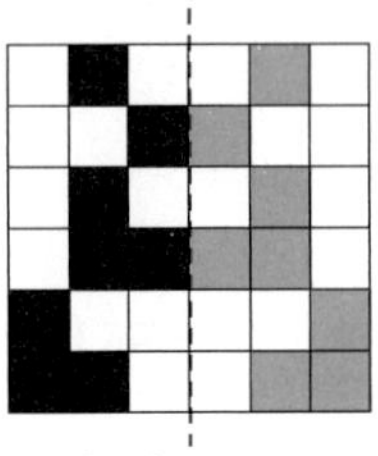

84. **Graphic Arts.** Draw an axis of symmetry over the letter shown here.

85. **Operating Costs.** The cost C in dollars of operating a certain concrete-cutting machine is related to the number of minutes n the machine is run by the function $C(n) = 2.2n^2 - 66n + 655$. For what number of minutes is the cost of running the machine a minimum? What is the minimum cost? 15 min, \$160

86. **Water Usage.** The height (in feet) of the water level in a reservoir over a 1-year period is modeled by the function $H(t) = 3.3(t - 9)^2 + 14$ where $t = 1$ represents January, $t = 2$ represents February, and so on. How low did the water level get that year, and when did it reach the low mark? 14 ft, September

87. **Fireworks.** A fireworks shell is shot straight up with an initial velocity of 120 feet per second. Its height s in feet after t seconds is approximated by the equation $s = 120t - 16t^2$. If the shell is designed to explode when it reaches its maximum height, how long after being fired, and at what height, will the fireworks appear in the sky? 3.75 sec, 225 ft

88. **Projectiles.** A ball is thrown straight upward from the top of a building with an initial velocity of 32 feet per second. The equation $s = -16t^2 + 32t + 48$ gives the height s of the ball in feet t seconds after it is thrown. Find the maximum height reached by the ball and the time it takes for the ball to hit the ground. 64 ft, 3 sec

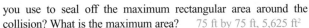

89. from **Campus to Careers**

Police Patrol Officer

Suppose you are a police patrol officer and you have a 300-foot-long roll of yellow "DO NOT CROSS" barricade tape to seal off an automobile accident, as shown in the illustration. What dimensions should you use to seal off the maximum rectangular area around the collision? What is the maximum area? 75 ft by 75 ft, 5,625 ft²

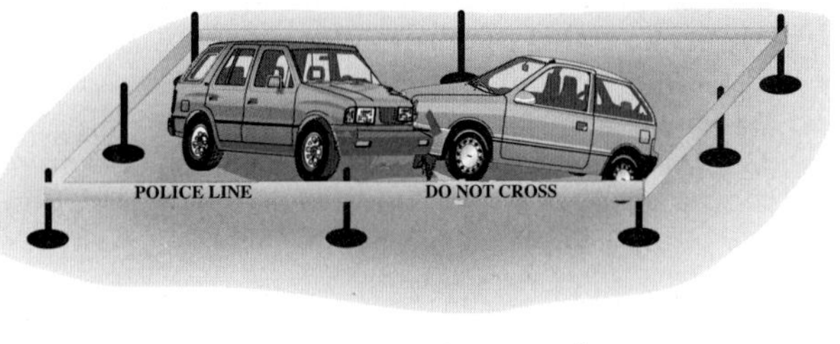

90. **Ranching.** See the illustration. A farmer wants to fence in three sides of a rectangular field with 1,000 feet of fencing. The other side of the rectangle will be a river. If the enclosed area is to be maximum, find the dimensions of the field. 250 ft by 500 ft

91. **Military History.** The function $N(x) = -0.0534x^2 + 0.337x + 0.97$ gives the number of active-duty military personnel in the United States Army (in millions) for the years 1965–1972, where $x = 0$ corresponds to 1965, $x = 1$ corresponds to 1966, and so on. For this period, when was the army's personnel strength level at its highest, and what was it? Historically, can you explain why?
1968, about 1.5 million; the U.S. involvement in the war in Vietnam was at its peak

92. **School Enrollment.** After peaking in 1970, school enrollment in the United States fell during the 1970s and 1980s. The total annual enrollment (in millions) in U.S. elementary and secondary schools for the years 1975–1996 is given by the model $E(x) = 0.058x^2 - 1.162x + 50.604$ where $x = 0$ corresponds to 1975, $x = 1$ corresponds to 1976, and so on. For this period, when was enrollment the lowest? What was the enrollment? 1985, about 44.8 million

93. **Maximizing Revenue.** The revenue R received for selling x stereos is given by the formula $R = -\frac{x^2}{5} + 80x - 1{,}000$. How many stereos must be sold to obtain the maximum revenue? Find the maximum revenue. 200, \$7,000

94. Maximizing Revenue. When priced at \$30 each, a toy has annual sales of 4,000 units. The manufacturer estimates that each \$1 increase in price will decrease sales by 100 units. Find the unit price that will maximize total revenue. (*Hint:* Total revenue = price · the number of units sold.) \$35

WRITING

95. Use the example of a stream of water from a drinking fountain to explain the concepts of the vertex and the axis of symmetry of a parabola. Draw a picture.

96. What are some quantities that are good to maximize? What are some quantities that are good to minimize?

97. A mirror is held against the y-axis of the graph of a quadratic function. What fact about parabolas does this illustrate?

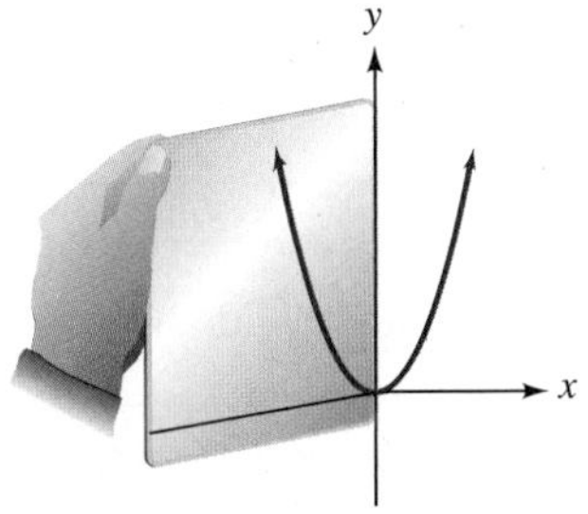

98. The vertex of a quadratic function $f(x) = ax^2 + bx + c$ is given by the formula $\left(-\frac{b}{2a}, f\left(-\frac{b}{2a}\right)\right)$. Explain what is meant by the notation $f\left(-\frac{b}{2a}\right)$.

99. A table of values for $f(x) = 2x^2 - 4x + 3$ is shown. Explain why it appears that the vertex of the graph of f is the point (1, 1).

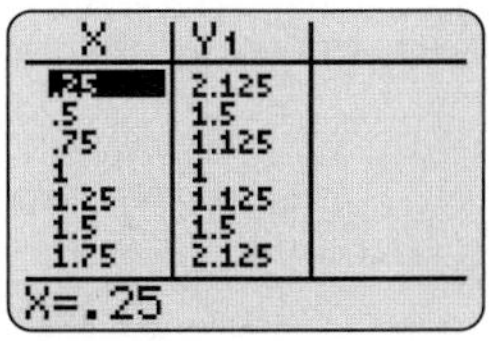

100. The illustration shows the graph of the quadratic function $f(x) = -4x^2 + 12x$ with domain [0, 3]. Explain how the value of $f(x)$ changes as the value of x increases from 0 to 3.

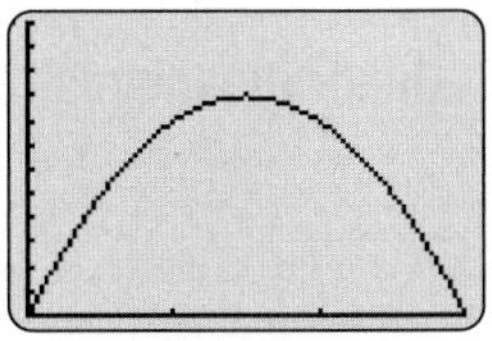

REVIEW

Simplify each expression. Assume all variables represent positive numbers.

101. $\dfrac{\sqrt{3}}{\sqrt{50}}$ $\quad\dfrac{\sqrt{6}}{10}$

102. $\dfrac{3}{\sqrt[3]{9}}$ $\quad\sqrt[3]{3}$

103. $3\left(\sqrt{5b} - \sqrt{3}\right)^2$ $\quad 15b - 6\sqrt{15b} + 9$

104. $-2\sqrt{5b}\left(4\sqrt{2b} - 3\sqrt{3}\right)$ $\quad -8b\sqrt{10} + 6\sqrt{15b}$

CHALLENGE PROBLEMS

105. Find a number between 0 and 1 such that the difference of the number and its square is a maximum. $\frac{1}{2}$

106. Determine a quadratic function whose graph has x-intercepts of (2, 0) and (−4, 0).
$f(x) = x^2 + 2x - 8$; answers may vary.

SECTION 10.5 Quadratic and Other Nonlinear Inequalities

OBJECTIVES

1. Solve quadratic inequalities.
2. Solve rational inequalities.
3. Graph nonlinear inequalities in two variables.

ARE YOU READY? *Are You Ready? exercises available online at www.webassign.net/brookscole*

The following problems review some basic skills that are needed when working with quadratic and nonlinear inequalities?

1. Does $x = -2$ satisfy $x^2 + x - 6 < 0$? Yes

2. Solve: $x^2 - 5x - 50 = 0$ $\quad -5, 10$

3. Graph the set of real numbers between −3 and 2 on a number line. See AIE Appendix 3.

4. Graph the set of real numbers greater than 1or less than or equal to −2 on a number line. See AIE Appendix 3.

5. What values of x make the denominator of $\dfrac{x+1}{x^2-9}$ equal to 0? $\quad -3, 3$

6. Graph: $y = -x^2 + 4$ See AIE Appendix 3.

We have previously solved *linear* inequalities in one variable such as $2x + 3 > 8$ and $6x - 7 < 4x - 9$. To find their solution sets, we used properties of inequalities to isolate the variable on one side of the inequality.

In this section, we will solve *quadratic* inequalities in one variable such as $x^2 + x - 6 < 0$ and $x^2 + 4x \geq 5$. We will use an interval testing method on the number line to determine their solution sets.

1 Solve Quadratic Inequalities.

Recall that a quadratic equation can be written in the form $ax^2 + bx + c = 0$. If we replace the $=$ symbol with an inequality symbol, we have a quadratic inequality.

Quadratic Inequalities

A **quadratic inequality** can be written in one of the standard forms

$$ax^2 + bx + c < 0 \quad ax^2 + bx + c > 0 \quad ax^2 + bx + c \le 0 \quad ax^2 + bx + c \ge 0$$

where a, b, and c are real numbers and $a \neq 0$.

To solve a quadratic inequality in one variable, we will use the following steps to find the values of the variable that make the inequality true.

Solving Quadratic Inequalities

1. Write the inequality in standard form and solve its related quadratic equation.
2. Locate the solutions (called **critical numbers**) of the related quadratic equation on a number line.
3. Test each interval on the number line created in step 2 by choosing a test value from the interval and determining whether it satisfies the inequality. The solution set includes the interval(s) whose test value makes the inequality true.
4. Determine whether the endpoints of the intervals are included in the solution set.

EXAMPLE 1 Solve: $x^2 + x - 6 < 0$

Strategy We will solve the related quadratic equation $x^2 + x - 6 = 0$ by factoring to determine the critical numbers. These critical numbers will separate the number line into intervals.

Why We can test each interval to see whether numbers in the interval are in the solution set of the inequality.

Solution The expression $x^2 + x - 6$ can be positive, negative, or 0, depending on what value is substituted for x. Solutions of the inequality are x-values that make $x^2 + x - 6$ less than 0. To find them, we will follow the steps for solving quadratic inequalities.

Step 1: Solve the related quadratic equation. For the quadratic inequality $x^2 + x - 6 < 0$, the related quadratic *equation* is $x^2 + x - 6 = 0$.

$$x^2 + x - 6 = 0$$

$$(x + 3)(x - 2) = 0 \qquad \text{Factor the trinomial.}$$

$$x + 3 = 0 \quad \text{or} \quad x - 2 = 0 \qquad \text{Set each factor equal to 0.}$$

$$x = -3 \quad | \quad x = 2 \qquad \text{Solve each equation.}$$

The solutions of $x^2 + x - 6 = 0$ are -3 and 2. These solutions are the critical numbers.

Step 2: Locate the critical numbers on a number line. When we highlight -3 and 2 on a number line, they separate the number line into three intervals:

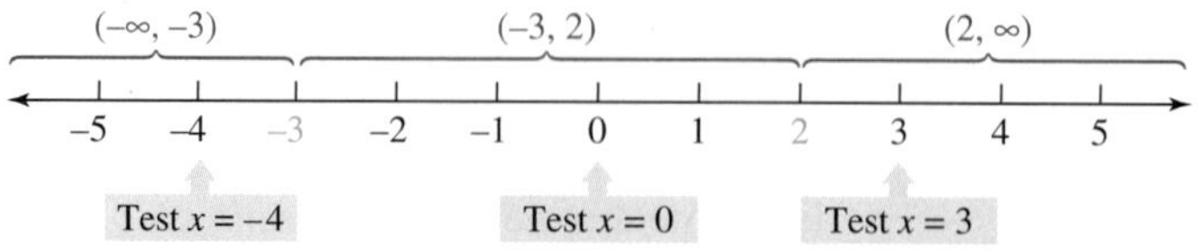

The Language of Algebra

We say that the critical numbers **partition** the real-number line into test intervals. Interior decorators use freestanding screens to *partition* off parts of a room.

Step 3: ***Test each interval.*** To determine whether the numbers in $(-\infty, -3)$ are solutions of the inequality, we choose a number from that interval, substitute it for x, and see whether it satisfies $x^2 + x - 6 < 0$. *If one number in that interval satisfies the inequality, all numbers in that interval will satisfy the inequality.*

If we choose -4 from $(-\infty, -3)$, we have:

$$x^2 + x - 6 < 0 \quad \text{This is the original inequality.}$$
$$(-4)^2 + (-4) - 6 \stackrel{?}{<} 0 \quad \text{Substitute } -4 \text{ for } x.$$
$$16 + (-4) - 6 \stackrel{?}{<} 0$$
$$6 < 0 \quad \text{False}$$

Since -4 does not satisfy the inequality, none of the numbers in $(-\infty, -3)$ are solutions.

To test the second interval, $(-3, 2)$, we choose $x = 0$.

$$x^2 + x - 6 < 0 \quad \text{This is the original inequality.}$$
$$0^2 + 0 - 6 \stackrel{?}{<} 0 \quad \text{Substitute 0 for } x.$$
$$-6 < 0 \quad \text{True}$$

Since 0 satisfies the inequality, all of the numbers in $(-3, 2)$ are solutions.

To test the third interval, $(2, \infty)$, we choose $x = 3$.

$$x^2 + x - 6 < 0 \quad \text{This is the original inequality.}$$
$$3^2 + 3 - 6 \stackrel{?}{<} 0 \quad \text{Substitute 3 for } x.$$
$$9 + 3 - 6 \stackrel{?}{<} 0$$
$$6 < 0 \quad \text{False}$$

Since 3 does not satisfy the inequality, none of the numbers in $(2, \infty)$ are solutions.

Step 4: ***Are the endpoints included?*** From the interval testing, we see that only numbers from $(-3, 2)$ satisfy $x^2 + x - 6 < 0$. The endpoints -3 and 2 are not included in the solution set because they do not satisfy the inequality. (Recall that -3 and 2 make $x^2 + x - 6$ equal to 0.) The solution set is the interval $(-3, 2)$ as graphed on the right.

Success Tip

If a quadratic inequality contains $\le$ or $\ge$, the endpoints of the intervals are included in the solution set. If the inequality contains $<$ or $>$, they are not.

Teaching Example 1 Solve: $x^2 - x - 12 < 0$
Answer: $(-3, 4)$

Self Check 1 Solve: $x^2 + x - 12 < 0$ $(-4, 3)$; see AIE Appendix 3.

Now Try ▶ Problem 15

EXAMPLE 2 Solve: $x^2 + 4x \ge 5$

Strategy This inequality is not in standard form because it does not have 0 on the right side. We will write it in standard form and solve its related quadratic equation to find any critical numbers. These critical numbers will separate the number line into intervals.

Why We can then test each interval to see whether numbers in the interval are in the solution set of the inequality.

Solution To get 0 on the right side, we subtract 5 from both sides.

$$x^2 + 4x \ge 5 \quad \text{This is the inequality to solve.}$$
$$x^2 + 4x - 5 \ge 0 \quad \text{Write the inequality in the equivalent form } ax^2 + bx + c \ge 0.$$

We can solve the related quadratic *equation* $x^2 + 4x - 5 = 0$ by factoring.

$$x^2 + 4x - 5 = 0$$
$$(x + 5)(x - 1) = 0 \quad \text{Factor the trinomial.}$$
$$x + 5 = 0 \quad \text{or} \quad x - 1 = 0 \quad \text{Set each factor equal to 0.}$$
$$x = -5 \quad \Big| \quad x = 1$$

Success Tip

When choosing a test value from an interval, pick a convenient number that makes the calculations easy. When applicable, 0 is an obvious choice.

The critical numbers -5 and 1 separate the number line into three intervals. We pick a test value from each interval to see whether it satisfies $x^2 + 4x - 5 \geq 0$.

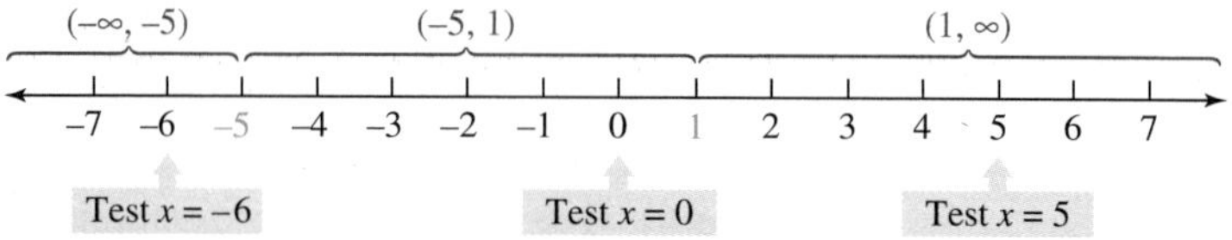

$$x^2 + 4x - 5 \geq 0 \qquad x^2 + 4x - 5 \geq 0 \qquad x^2 + 4x - 5 \geq 0$$

$$(-6)^2 + 4(-6) - 5 \stackrel{?}{\geq} 0 \qquad 0^2 + 4(0) - 5 \stackrel{?}{\geq} 0 \qquad 5^2 + 4(5) - 5 \stackrel{?}{\geq} 0$$

$$7 \geq 0 \quad \text{True} \qquad -5 \geq 0 \quad \text{False} \qquad 40 \geq 0 \quad \text{True}$$

The numbers in the intervals $(-\infty, -5)$ and $(1, \infty)$ satisfy the inequality. Since the endpoints -5 and 1 also satisfy $x^2 + 4x - 5 \geq 0$, they are included in the solution set. (Recall that -5 and 1 make $x^2 + 4x - 5$ equal to 0.) Thus, the solution set is the union of two intervals: $(-\infty, -5] \cup [1, \infty)$. The graph of the solution set is shown on the right.

Teaching Example 2 Solve: $x^2 + 8x \geq -15$
Answer: $(-\infty, -5] \cup [-3, \infty)$

Self Check 2 Solve: $x^2 + 3x \geq 40$ $(-\infty, -8] \cup [5, \infty)$; see AIE Appendix 3.

Now Try Problem 19

2 Solve Rational Inequalities.

Rational inequalities in one variable such as $\frac{9}{x} < 8$ and $\frac{x^2 + x - 2}{x - 4} \geq 0$ can also be solved using the interval testing method.

Solving Rational Inequalities

1. Write the inequality in standard form with a single quotient on the left side and 0 on the right side. Then solve its related rational equation.
2. Set the denominator equal to zero and solve that equation.
3. Locate the solutions (called **critical numbers**) found in steps 1 and 2 on a number line.
4. Test each interval on the number line created in step 3 by choosing a test value from the interval and determining whether it satisfies the inequality. The solution set includes the interval(s) whose test value makes the inequality true.
5. Determine whether the endpoints of the intervals are included in the solution set. Exclude any values that make the denominator 0.

EXAMPLE 3 Solve: $\frac{9}{x} < 8$

Strategy This rational inequality is not in standard form because it does not have 0 on the right side. We will write it in standard form and solve its related rational equation to find any critical numbers. These critical numbers will separate the number line into intervals.

Why We can test each interval to see whether numbers in the interval are in the solution set of the inequality.

Solution To get 0 on the right side, we subtract 8 from both sides. We then find a common denominator to write the left side as a single quotient.

Caution

When solving rational inequalities such as $\frac{9}{x} < 8$, a common error is to multiply both sides by x to clear it of the fraction. However, we don't know whether x is positive or negative, so we don't know whether or not to reverse the inequality symbol.

$$\frac{9}{x} < 8 \quad \text{This is the inequality to solve.}$$

$$\frac{9}{x} - 8 < 0 \quad \text{Subtract 8 from both sides.}$$

$$\frac{9}{x} - 8 \cdot \frac{x}{x} < 0 \quad \text{To write the left side as a single quotient, build 8 to a fraction with denominator } x.$$

$$\frac{9}{x} - \frac{8x}{x} < 0$$

$$\frac{9 - 8x}{x} < 0 \quad \text{Subtract the numerators and keep the common denominator, } x.$$

Now we solve the related rational equation.

Success Tip

Recall that if a fraction is equal to 0, its numerator must be 0.

$$\frac{9 - 8x}{x} = 0$$

$$9 - 8x = 0 \quad \text{If } x \neq 0\text{, we can clear the equation of the fraction by multiplying both sides by } x.$$

$$-8x = -9 \quad \text{To isolate the variable term } -8x\text{, subtract 9 from both sides.}$$

$$x = \frac{9}{8} \quad \text{Solve for } x\text{. This is a critical number.}$$

If we set the denominator of $\frac{9 - 8x}{x}$ equal to 0, we obtain a second critical number, $x = 0$. When graphed, the critical numbers 0 and $\frac{9}{8}$ separate the number line into three intervals. We pick a test value from each interval to see whether it satisfies $\frac{9 - 8x}{x} < 0$.

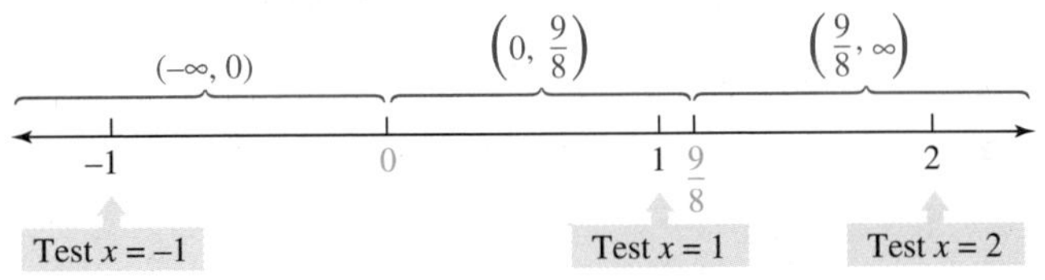

$\frac{9 - 8x}{x} < 0$	$\frac{9 - 8x}{x} < 0$	$\frac{9 - 8x}{x} < 0$
$\frac{9 - 8(-1)}{-1} \stackrel{?}{<} 0$	$\frac{9 - 8(1)}{1} \stackrel{?}{<} 0$	$\frac{9 - 8(2)}{2} \stackrel{?}{<} 0$
$-17 < 0$ True	$1 < 0$ False	$-\frac{7}{2} < 0$ True

The numbers in the intervals $(-\infty, 0)$ and $\left(\frac{9}{8}, \infty\right)$ satisfy the inequality. We do not include the endpoint 0 in the solution set, because it makes the denominator of the original inequality 0. Neither do we include $\frac{9}{8}$, because it does not satisfy $\frac{9 - 8x}{x} < 0$. (Recall that $\frac{9}{8}$ makes $\frac{9 - 8x}{x}$ equal to 0.) Thus, the solution set is the union of two intervals: $(-\infty, 0) \cup \left(\frac{9}{8}, \infty\right)$. Its graph is shown on the right.

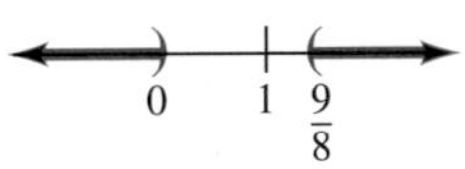

Self Check 3 Solve: $\frac{3}{x} < 5$ $(-\infty, 0) \cup \left(\frac{3}{5}, \infty\right)$; see AIE Appendix 3.

Now Try ▶ **Problem 23**

Teaching Example 3 Solve: $\frac{12}{x} < 3$

Answer: $(-\infty, 0) \cup (4, \infty)$

EXAMPLE 4 Solve: $\dfrac{x^2 + x - 2}{x - 4} \geq 0$

Strategy This inequality is in standard form. We will solve its related rational equation to find any critical numbers. These critical numbers will separate the number line into intervals.

Why We can test each interval to see whether numbers in the interval are in the solution set of the inequality.

Solution To solve the related rational equation, we proceed as follows:

$$\frac{x^2 + x - 2}{x - 4} = 0$$

$$x^2 + x - 2 = 0 \qquad \text{If } x \neq 4\text{, we can clear the equation of the fraction by multiplying both sides by } x - 4.$$

$$(x + 2)(x - 1) = 0 \qquad \text{Factor the trinomial.}$$

$$x + 2 = 0 \quad \text{or} \quad x - 1 = 0 \qquad \text{Set each factor equal to 0.}$$

$$x = -2 \quad \Big| \quad x = 1 \qquad \text{These are critical numbers.}$$

Caution

We cannot multiply both sides of the inequality by $x - 4$ to clear it of fractions. We don't know if $x - 4$ is positive or negative, so we don't know whether or not to reverse the inequality symbol.

If we set the denominator of $\frac{x^2 + x - 2}{x - 4}$ equal to 0, we see that $x = 4$ is also a critical number. When graphed, the critical numbers, -2, 1, and 4, separate the number line into four intervals. We pick a test value from each interval to see whether it satisfies $\frac{x^2 + x - 2}{x - 4} \geq 0$.

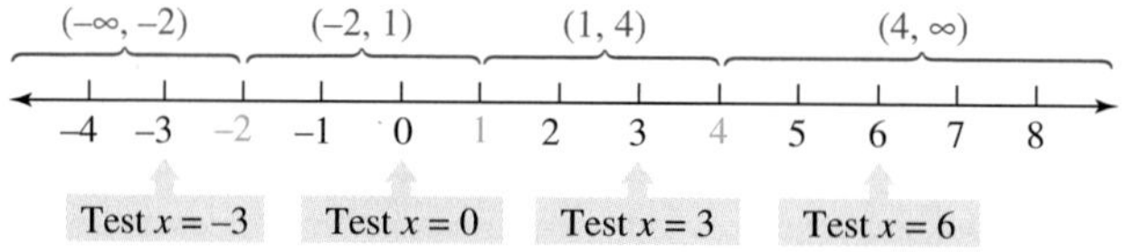

$\dfrac{(-3)^2 + (-3) - 2}{-3 - 4} \overset{?}{\geq} 0$	$\dfrac{0^2 + 0 - 2}{0 - 4} \overset{?}{\geq} 0$	$\dfrac{3^2 + 3 - 2}{3 - 4} \overset{?}{\geq} 0$	$\dfrac{6^2 + 6 - 2}{6 - 4} \overset{?}{\geq} 0$
$-\dfrac{4}{7} \geq 0$	$\dfrac{1}{2} \geq 0$	$-10 \geq 0$	$20 \geq 0$
False	True	False	True

Teaching Tip: Point out that we never include any values of the variable that make the denominator of a rational expression equal to 0 in the solution set.

The numbers in the intervals $(-2, 1)$ and $(4, \infty)$ satisfy the inequality. We include the endpoints -2 and 1 in the solution set because they satisfy the inequality. We do not include 4 because it makes the denominator of the fraction 0. Thus, the solution set is the union of two intervals $[-2, 1] \cup (4, \infty)$, as graphed on the right.

Teaching Example 4 Solve: $\dfrac{x^2 - 2x - 8}{x + 3} \leq 0$
Answer: $(-\infty, -3) \cup [-2, 4]$

Self Check 4 Solve: $\dfrac{x + 2}{x^2 - 2x - 3} \geq 0$ $[-2, -1) \cup (3, \infty)$; see AIE Appendix 3.

Now Try Problem 27

EXAMPLE 5 Solve: $\dfrac{3}{x - 1} < \dfrac{2}{x}$

Strategy We will subtract $\frac{2}{x}$ from both sides to get 0 on the right side and solve the resulting related rational equation to find any critical numbers. These critical numbers will separate the number line into intervals.

Why We can test each interval to see whether numbers in the interval are in the solution set of the inequality.

Solution

$$\frac{3}{x-1} < \frac{2}{x}$$ This is the inequality to solve.

$$\frac{3}{x-1} - \frac{2}{x} < 0$$ Subtract $\frac{2}{x}$ from both sides.

$$\frac{3}{x-1} \cdot \frac{x}{x} - \frac{2}{x} \cdot \frac{x-1}{x-1} < 0$$ To get a single quotient on the left side, build each rational expression to have the common denominator $x(x - 1)$.

$$\frac{3x - 2x + 2}{x(x-1)} < 0$$ Subtract the numerators and keep the common denominator.

$$\frac{x+2}{x(x-1)} < 0$$ Combine like terms.

Teaching Tip: Stress that the interval testing method requires that the right side of the inequality be 0 and the left side be in factored form.

The only solution of the related rational equation $\frac{x+2}{x(x-1)} = 0$ is -2. Thus, -2 is a critical number. When we set the denominator equal to 0 and solve $x(x - 1) = 0$, we find two more critical numbers, 0 and 1. These three critical numbers create four intervals to test.

Success Tip

When the endpoints of an interval are consecutive integers, such as with the third interval (0, 1), we cannot choose an integer as a test value. For these cases, choose a fraction or decimal that lies within the interval.

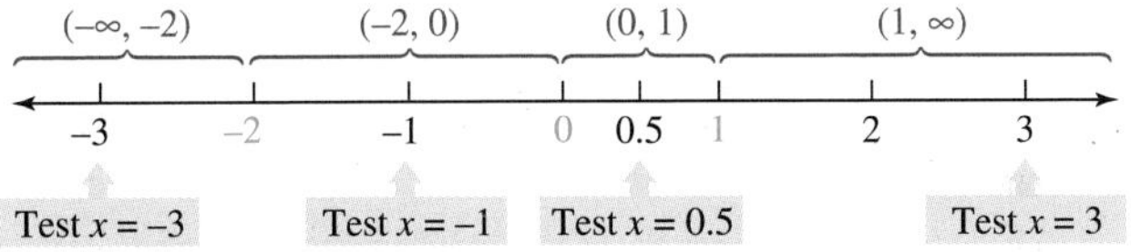

$\frac{-3+2}{-3(-3-1)} \overset{?}{<} 0$	$\frac{-1+2}{-1(-1-1)} \overset{?}{<} 0$	$\frac{0.5+2}{0.5(0.5-1)} \overset{?}{<} 0$	$\frac{3+2}{3(3-1)} \overset{?}{<} 0$
$\frac{-1}{-3(-4)} \overset{?}{<} 0$	$\frac{1}{-1(-2)} \overset{?}{<} 0$	$\frac{2.5}{0.5(-0.5)} \overset{?}{<} 0$	$\frac{5}{3(2)} \overset{?}{<} 0$
$-\frac{1}{12} < 0$	$\frac{1}{2} < 0$	$-10 < 0$	$\frac{5}{6} < 0$
True	False	True	False

The numbers 0 and 1 are not included in the solution set because they make the denominator 0, and the number -2 is not included because it does not satisfy the inequality. The solution set is the union of two intervals $(-\infty, -2) \cup (0, 1)$, as graphed on the right.

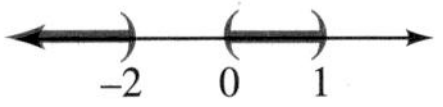

Teaching Example 5 Solve: $\frac{4}{x-2} > \frac{5}{x}$
Answer: $(-\infty, 0) \cup (2, 10)$

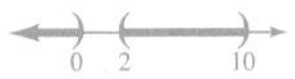

Self Check 5 Solve: $\frac{2}{x+1} > \frac{1}{x}$ $(-1, 0) \cup (1, \infty)$; see AIE Appendix 3.

Now Try ▶ Problem 31

Using Your Calculator ▶ Solving Inequalities Graphically

We can solve $x^2 + 4x \geq 5$ (Example 2) graphically by writing the inequality as $x^2 + 4x - 5 \geq 0$ and graphing the quadratic function $f(x) = x^2 + 4x - 5$, as shown in figure (a) on the next page. The solution set of the inequality will be those values of x for which the graph lies on or above the x-axis. We can trace to determine that this is the union of two intervals: $(-\infty, -5] \cup [1, \infty)$.

To solve $\frac{3}{x-1} < \frac{2}{x}$ (Example 5) graphically, we first write the inequality in the form $\frac{x+2}{x(x-1)} < 0$ and then graph the rational function $f(x) = \frac{x+2}{x(x-1)}$, as shown in figure (b) on the next page. The solution of the inequality will be those values of x for which the graph lies below the axis.

We can trace to see that the graph is below the x-axis when x is less than -2. Since we cannot see the graph in the interval $0 < x < 1$, we redraw the graph using window settings of $[-1, 2]$ for x and $[-25, 10]$ for y, as shown in figure (c).

Now we see that the graph is below the x-axis in the interval $(0, 1)$. Thus, the solution set of the inequality is the union of the two intervals: $(-\infty, -2) \cup (0, 1)$.

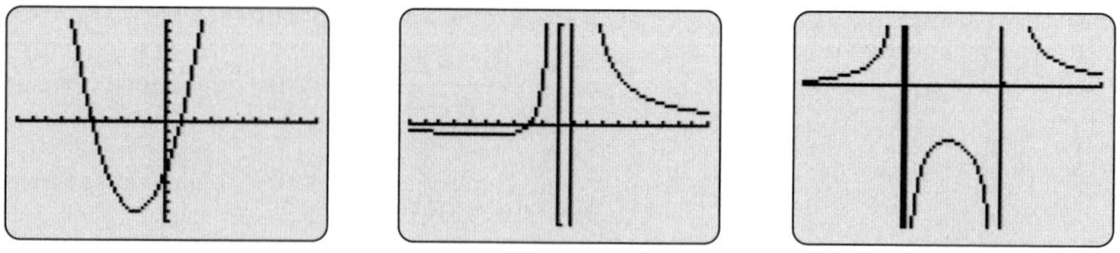

(a) (b) (c)

3 Graph Nonlinear Inequalities in Two Variables.

We have previously graphed linear inequalities in two variables such as $y > 3x + 2$ and $2x - 3y \le 6$ using the following steps.

Graphing Inequalities in Two Variables

1. Graph the related equation to find the boundary line of the region. If the inequality allows equality (the symbol is either $\le$ or $\ge$), draw the boundary as a solid line. If equality is not allowed ($<$ or $>$), draw the boundary as a dashed line.
2. Pick a test point that is on one side of the boundary line. (Use the origin if possible.) Replace x and y in the original inequality with the coordinates of that point. If the inequality is satisfied, shade the side that contains that point. If the inequality is not satisfied, shade the other side of the boundary.

We use the same procedure to graph **nonlinear inequalities** in two variables.

EXAMPLE 6 Graph: $y < -x^2 + 4$

Strategy We will graph the related equation $y = -x^2 + 4$ to establish a boundary parabola. Then we will determine which side of the boundary parabola represents the solution set of the inequality.

Why To *graph a nonlinear inequality* in two variables means to draw a "picture" of the ordered pairs (x, y) that make the inequality true.

Solution The graph of the boundary $y = -x^2 + 4$ is a parabola opening downward, with vertex at $(0, 4)$ and axis of symmetry $x = 0$ (the y-axis). Since the inequality contains an $<$ symbol and equality is not allowed, we draw the parabola using a dashed curve.

To determine which region to shade, we pick the test point $(0, 0)$ and substitute its coordinates into the inequality. We shade the region containing $(0, 0)$ because its coordinates satisfy $y < -x^2 + 4$.

Graph the boundary

$y = -x^2 + 4$

Compare to $y = a(x - h)^2 + k$

$a = -1$: Opens downward

$h = 0$ and $k = 4$: Vertex $(0, 4)$

Axis of symmetry $x = 0$

x	y
1	3
2	0

Shading: Use the test point (0, 0)

$y < -x^2 + 4$

$0 \overset{?}{<} -0^2 + 4$

$0 < 4$ True

Since $0 < 4$ is true, $(0, 0)$ is a solution of $y < -x^2 + 4$.

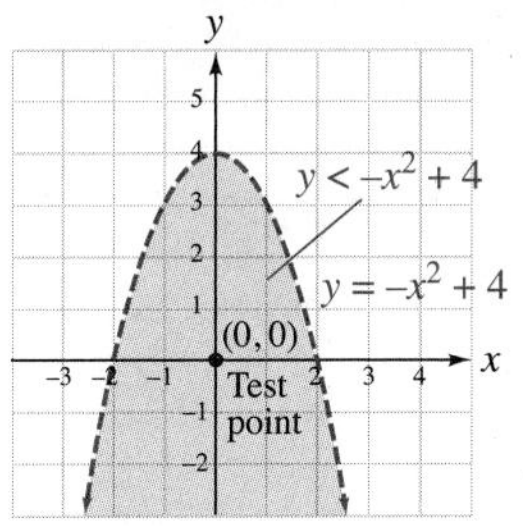

Teaching Example 6 Graph: $y > x^2 - 4$
Answer:

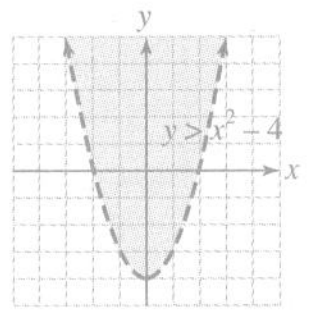

Self Check 6 Graph: $y \geq -x^2 + 4$ See AIE Appendix 3.

Now Try Problem 35

EXAMPLE 7 Graph: $x \leq |y|$

Strategy We will graph the related equation $x = |y|$ to establish a boundary. Then we will determine which side of the boundary represents the solution set of the inequality.

Why To *graph a nonlinear inequality* in two variables means to draw a "picture" of the ordered pairs (x, y) that make the inequality true.

Solution To graph the boundary, $x = |y|$, we construct a table of solutions, as shown in figure (a). In figure (b), the boundary is a solid line because the inequality contains a $\leq$ symbol and equality is permitted. Since the origin is on the graph, we cannot use it as a test point. However, any other point, such as $(1, 0)$, will do. We substitute 1 for x and 0 for y into the inequality to get

$x \leq |y|$ This is the inequality to graph.

$1 \overset{?}{\leq} |0|$ Substitute.

$1 \leq 0$ False

Since $1 \leq 0$ is a false statement, the point $(1, 0)$ does not satisfy the inequality and is not part of the graph. Thus, the graph of $x \leq |y|$ is to the left of the boundary.

The complete graph is shown in figure (c).

$x = |y|$

x	y
0	0
1	1
1	−1
2	2
2	−2

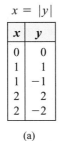

(a)

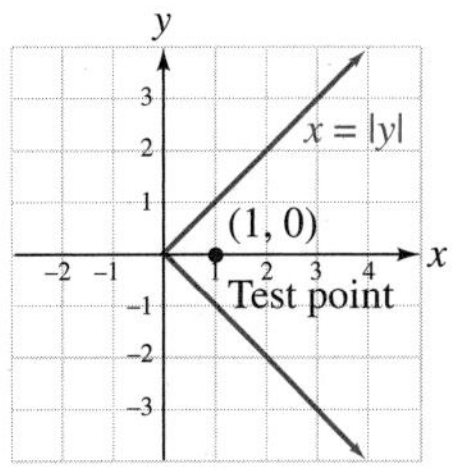

(b)

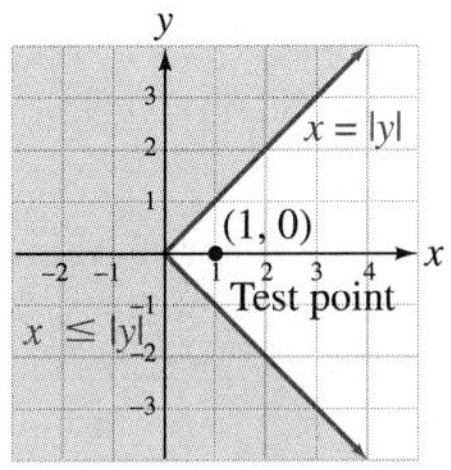

(c)

Teaching Example 7 Graph: $-x > |y|$
Answer:

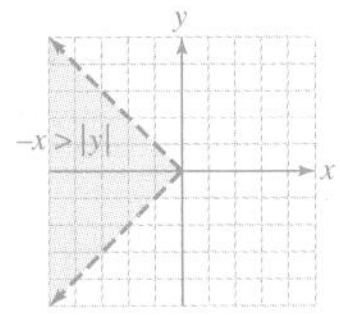

Self Check 7 Graph: $x \geq -|y|$ See AIE Appendix 3.

Now Try Problem 39

SECTION 10.5 STUDY SET

VOCABULARY

Fill in the blanks.

1. $x^2 + 3x - 18 < 0$ is an example of a quadratic inequality in one variable.
2. $\frac{x - 1}{x^2 - x - 20} \le 0$ is an example of a rational inequality in one variable.
3. $y \le x^2 - 4x + 3$ is an example of a nonlinear inequality in two variables.
4. The set of real numbers greater than 3 can be represented using the interval notation $(3, \infty)$.

CONCEPTS

5. The critical numbers of a quadratic inequality are highlighted in red on the number line shown below. Use interval notation to represent each interval that must be tested to solve the inequality. $(-\infty, -1), (-1, 4), (4, \infty)$

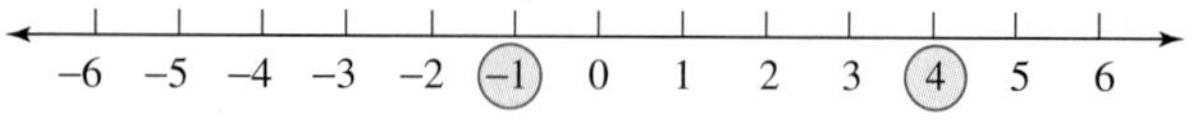

6. Graph each of the following solution sets. See AIE Appendix 3.
   a. $(-2, 4)$
   b. $(-\infty, -2) \cup (3, 5]$
7. The graph of the solution set of a rational inequality in one variable is shown. Determine whether each of the following numbers is a solution of the inequality.

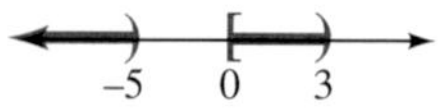

   a. -10 Yes
   b. -5 No
   c. 0 Yes
   d. 4 No
8. What are the critical numbers for each inequality?
   a. $x^2 - 2x - 48 \ge 0$ $-6, 8$
   b. $\frac{x - 3}{x(x + 4)} > 0$ $-4, 0, 3$
9. a. The results after interval testing for a quadratic inequality containing a $>$ symbol are shown below. (The critical numbers are highlighted in red.) What is the solution set? $(-3, 2)$

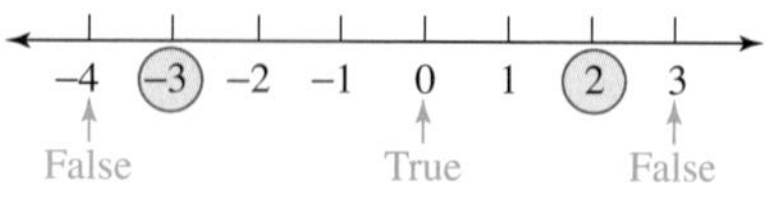

   b. The results after interval testing for a quadratic inequality containing a $\le$ symbol are shown below. (The critical numbers are highlighted in red.) What is the solution set? $(-\infty, -1] \cup [1, \infty)$

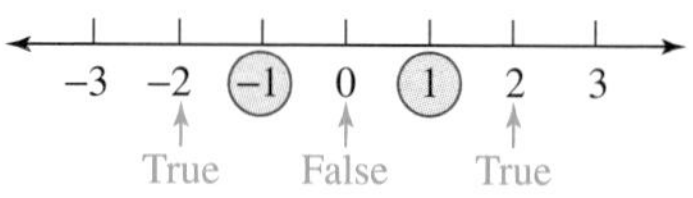

10. Fill in the blank to complete this important fact about the interval testing method discussed in this section: *If one number in an interval satisfies the inequality,* all *numbers in that interval will satisfy the inequality.*

11. a. When graphing the solution of $y \le x^2 + 2x + 1$, should the boundary be solid or dashed? Solid
    b. Does the test point (0, 0) satisfy the inequality? Yes
12. a. Estimate the solution of $x^2 - x - 6 > 0$ using the graph of $y = x^2 - x - 6$ shown in figure (a) below. $(-\infty, -2) \cup (3, \infty)$
    b. Estimate the solution of $\frac{x - 3}{x} \le 0$ using the graph of $y = \frac{x - 3}{x}$ shown in figure (b) below. $(0, 3]$

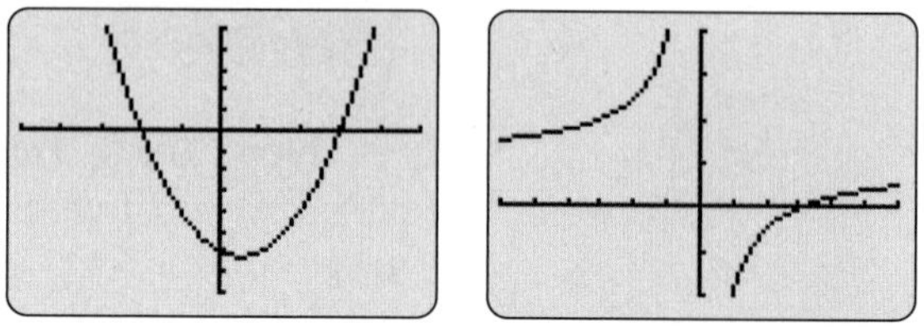

(a) (b)

NOTATION

13. Write the quadratic inequality $x^2 - 6x \ge 7$ in standard form. $x^2 - 6x - 7 \ge 0$
14. The solution set of a rational inequality consists of the intervals $(-1, 4]$ and $(7, \infty)$. When writing the solution set, what symbol is used between the two intervals? $\cup$

GUIDED PRACTICE

Solve each inequality. Write the solution set in interval notation and graph it. See Example 1. See AIE Appendix 3.

15. $x^2 - 5x + 4 < 0$ $(1, 4)$
16. $x^2 + 2x - 8 < 0$ $(-4, 2)$
17. $x^2 - 8x + 15 > 0$ $(-\infty, 3) \cup (5, \infty)$
18. $x^2 - 3x - 4 > 0$ $(-\infty, -1) \cup (4, \infty)$

Solve each inequality. Write the solution set in interval notation and graph it. See Example 2. See AIE Appendix 3.

19. $x^2 - x \ge 42$ $(-\infty, -6] \cup [7, \infty)$
20. $x^2 - x \ge 72$ $(-\infty, -8] \cup [9, \infty)$
21. $x^2 + x \le 12$ $[-4, 3]$
22. $x^2 - 8x \le -15$ $[3, 5]$

Solve each inequality. Write the solution set in interval notation and graph it. See Example 3. See AIE Appendix 3.

23. $\frac{1}{x} < 2$ $(-\infty, 0) \cup \left(\frac{1}{2}, \infty\right)$
24. $\frac{1}{x} < 3$ $(-\infty, 0) \cup \left(\frac{1}{3}, \infty\right)$
25. $\frac{5}{x} \ge -3$ $\left(-\infty, -\frac{5}{3}\right] \cup (0, \infty)$
26. $\frac{4}{x} \ge 8$ $\left(0, \frac{1}{2}\right]$

Solve each inequality. Write the solution set in interval notation and graph it. See Example 4. See AIE Appendix 3.

27. $\frac{x^2 - x - 12}{x - 1} < 0$ $(-\infty, -3) \cup (1, 4)$
28. $\frac{x^2 + x - 6}{x - 4} \ge 0$ $[-3, 2] \cup (4, \infty)$
29. $\frac{6x^2 - 5x + 1}{2x + 1} \ge 0$ $\left(-\frac{1}{2}, \frac{1}{3}\right] \cup \left[\frac{1}{2}, \infty\right)$
30. $\frac{6x^2 + 11x + 3}{3x - 1} < 0$ $\left(-\infty, -\frac{3}{2}\right) \cup \left(-\frac{1}{3}, \frac{1}{3}\right)$

Solve each inequality. Write the solution set in interval notation and graph it. **See Example 5.** See AIE Appendix 3.

31. $\frac{3}{x-2} < \frac{4}{x}$ $(0, 2) \cup (8, \infty)$

32. $\frac{-6}{x+1} \geq \frac{1}{x}$ $(-\infty, -1) \cup \left[-\frac{1}{7}, 0\right)$

33. $\frac{7}{x-3} \geq \frac{2}{x+4}$ $\left[-\frac{34}{5}, -4\right) \cup (3, \infty)$

34. $\frac{-5}{x-4} < \frac{3}{x+1}$ $\left(-1, \frac{7}{8}\right) \cup (4, \infty)$

Graph each inequality. **See Example 6.** See AIE Appendix 3.

35. $y < x^2 + 1$

36. $y > x^2 - 3$

37. $y \leq x^2 + 5x + 6$

38. $y \geq x^2 + 5x + 4$

Graph each inequality. **See Example 7.** See AIE Appendix 3.

39. $y < |x + 4|$

40. $y \leq |x - 3|$

41. $y \geq -|x| + 2$

42. $y > |x| - 2$

Use a graphing calculator to solve each inequality. Write the solution set in interval notation. **See Using Your Calculator: Solving Inequalities Graphically.**

43. $x^2 - 2x - 3 < 0$ $(-1, 3)$

44. $x^2 + x - 6 > 0$ $(-\infty, -3) \cup (2, \infty)$

45. $\frac{x+3}{x-2} > 0$ $(-\infty, -3) \cup (2, \infty)$

46. $\frac{3}{x} < 2$ $(-\infty, 0) \cup \left(\frac{3}{2}, \infty\right)$

TRY IT YOURSELF

Solve each inequality. Write the solution set in interval notation and graph it. See AIE Appendix 3.

47. $\frac{x}{x+4} \leq \frac{1}{x+1}$ $(-4, -2] \cup (-1, 2]$

48. $\frac{x}{x+9} \geq \frac{1}{x+1}$ $(-\infty, -9) \cup [-3, -1) \cup [3, \infty)$

49. $x^2 \geq 9$ $(-\infty, -3] \cup [3, \infty)$

50. $x^2 \geq 16$ $(-\infty, -4] \cup [4, \infty)$

51. $x^2 + 6x \geq -9$ $(-\infty, \infty)$

52. $x^2 + 8x < -16$ No solutions

53. $\frac{x^2 + x - 2}{x - 3} > 0$ $(-2, 1) \cup (3, \infty)$

54. $\frac{x-2}{x^2-1} > 0$ $(-1, 1) \cup (2, \infty)$

55. $2x^2 - 50 < 0$ $(-5, 5)$

56. $3x^2 - 243 < 0$ $(-9, 9)$

57. $\frac{2x-3}{3x+1} < 0$ $\left(-\frac{1}{3}, \frac{3}{2}\right)$

58. $\frac{x-5}{x+1} < 0$ $(-1, 5)$

59. $x^2 - 6x + 9 < 0$ No solutions

60. $x^2 + 4x + 4 > 0$ $(-\infty, -2) \cup (-2, \infty)$

61. $\frac{5}{x+1} > \frac{3}{x-4}$ $(-1, 4) \cup \left(\frac{23}{2}, \infty\right)$

62. $\frac{3}{x-2} \leq -\frac{2}{x+3}$ $(-\infty, -3) \cup [-1, 2)$

APPLICATIONS

63. **Bridges.** If an x-axis is superimposed over the roadway of the Golden Gate Bridge, with the origin at the center of the bridge, the length L in feet of a vertical support cable can be approximated by the formula

$$L = \frac{1}{9{,}000}x^2 + 5$$

For the Golden Gate Bridge, $-2{,}100 < x < 2{,}100$. For what intervals along the x-axis are the vertical cables more than 95 feet long? $(-2{,}100, -900) \cup (900, 2{,}100)$

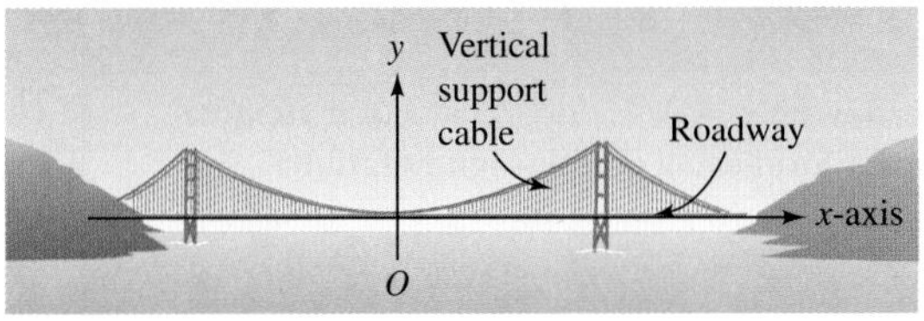

64. **Malls.** The number of people n in a mall is modeled by the formula

$$n = -100x^2 + 1{,}200x$$

where x is the number of hours since the mall opened. If the mall opened at 9 A.M., when were there 2,000 or more people in it? Between 11 A.M. and 7 P.M., inclusive

WRITING

65. How are critical numbers used when solving a quadratic inequality in one variable?

66. Explain how to graph $y \geq x^2$.

67. The graph of $f(x) = x^2 - 3x + 4$ is shown here. Explain why the quadratic inequality $x^2 - 3x + 4 < 0$ has no solution.

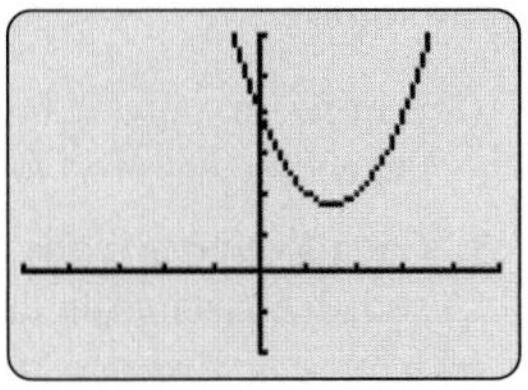

68. Describe the following solution set of a rational inequality in words: $(-\infty, 4] \cup (6, 7)$.

REVIEW

Translate each statement into an equation.

69. x varies directly as y. $x = ky$

70. y varies inversely as t. $y = \frac{k}{t}$

71. t varies jointly as x and y. $t = kxy$

72. d varies directly as t and inversely as u^2. $d = \frac{kt}{u^2}$

CHALLENGE PROBLEMS

73. **a.** Solve: $x^2 - x - 12 > 0$ $(-\infty, -3) \cup (4, \infty)$

b. Find a rational inequality in one variable that has the same solution set as the quadratic inequality in part (a). $\frac{x-4}{x+3} > 0$ (Answers may vary.)

74. **a.** Solve: $\frac{1}{x} < 1$ $(-\infty, 0) \cup (1, \infty)$

b. Now incorrectly "solve" $\frac{1}{x} < 1$ by multiplying both sides by x to clear it of the fraction. What part of the solution set is not obtained with this incorrect approach? $(-\infty, 0)$

10 Summary & Review

SECTION 10.1 The Square Root Property and Completing the Square

DEFINITIONS AND CONCEPTS	EXAMPLES
We can use the **square root property** to solve equations of the form $x^2 = c$, where $c > 0$. The two solutions are $x = \sqrt{c}$ or $x = -\sqrt{c}$ We can write $x = \sqrt{c}$ or $x = -\sqrt{c}$ in more compact form using **double-sign notation:** $x = \pm\sqrt{c}$	Solve: $x^2 = 24$ $x = \sqrt{24}$ or $x = -\sqrt{24}$ Use the square root property. $x = \pm\sqrt{24}$ Use double-sign notation. $x = \pm 2\sqrt{6}$ Simplify $\sqrt{24}$. The solutions are $2\sqrt{6}$ and $-2\sqrt{6}$. Verify this using a check. Solve: $(x - 3)^2 = -81$ $x - 3 = \pm\sqrt{-81}$ Use the square root property and double-sign notation. $x = 3 \pm \sqrt{-81}$ To isolate x, add 3 to both sides. $x = 3 \pm 9i$ Simplify the radical expression. The solutions are $3 + 9i$ and $3 - 9i$. Verify this using a check.
To **complete the square** on $x^2 + bx$, add the square of one-half of the coefficient of x. $x^2 + bx + \left(\frac{1}{2}b\right)^2$	Complete the square on $x^2 + 8x$ and factor the resulting perfect-square trinomial. $x^2 + 8x + \mathbf{16}$ The coefficient of x is 8. To complete the square: $\frac{1}{2} \cdot 8 = 4$ and $4^2 = 16$. Add 16 to the binomial. Now we factor: $x^2 + 8x + 16 = (x + 4)^2$
To **solve a quadratic equation in x by completing the square:** **1.** If necessary, divide both sides of the equation by the coefficient of x^2 to make its coefficient 1. **2.** Get all variable terms on one side of the equation and all constants on the other side. **3.** Complete the square. **4.** Factor the perfect-square trinomial. **5.** Solve the resulting equation by using the square root property. **6.** Check your answers in the original equation.	Solve: $3x^2 - 12x + 6 = 0$ $\frac{3x^2}{3} - \frac{12x}{3} + \frac{6}{3} = \frac{0}{3}$ To make the leading coefficient 1, divide both sides by 3, term-by-term. $x^2 - 4x + 2 = 0$ Do the division. $x^2 - 4x = -2$ Subtract 2 from both sides so that the constant term, −2, is on the right side. $x^2 - 4x + \mathbf{4} = -2 + \mathbf{4}$ The coefficient of x is −4. To complete the square: $\frac{1}{2}(-4) = -2$ and $(-2)^2 = 4$. Add 4 to both sides. $(x - 2)^2 = 2$ Factor the perfect-square trinomial on the left side. Add on the right side. $x - 2 = \pm\sqrt{2}$ Use the square root property. $x = 2 \pm \sqrt{2}$ To isolate x, add 2 to both sides. The solutions are $2 + \sqrt{2}$ and $2 - \sqrt{2}$. Verify this using a check.

REVIEW EXERCISES

Solve each equation by factoring.

1. $x^2 + 9x + 20 = 0$ $-5, -4$ **2.** $6x^2 + 17x + 5 = 0$ $-\frac{1}{3}, -\frac{5}{2}$

Solve each equation using the square root property. Approximate the solutions to the nearest hundredth when appropriate.

3. $x^2 = 28$ $\pm 2\sqrt{7}; \pm 5.29$ **4.** $(t + 2)^2 = 36$ $4, -8$

5. $a^2 + 25 = 0$ $\pm 5i$ **6.** $5x^2 - 49 = 0$ $\pm\frac{7\sqrt{5}}{5}; \pm 3.13$

7. Solve $A = \pi r^2$ for r. Assume all variables represent positive numbers. Express the result in simplified radical form. $r = \frac{\sqrt{\pi A}}{\pi}$

8. Complete the square on $x^2 - x$ and then factor the resulting perfect-square trinomial. $x^2 - x + \frac{1}{4} = \left(x - \frac{1}{2}\right)^2$

Solve each equation by completing the square. Approximate the solutions to the nearest hundredth when appropriate.

9. $x^2 + 6x + 8 = 0$ $-4, -2$ **10.** $2x^2 - 6x + 3 = 0$ $\frac{3 \pm \sqrt{3}}{2}; 2.37, 0.63$

11. $6a^2 - 12a = -1$

$\frac{6 \pm \sqrt{30}}{6}$; 1.91, 0.09

12. $x^2 - 2x = -13$

$1 \pm 2i\sqrt{3}$

13. Let $f(x) = x^2$. Find all values of x for which $f(x) = 32$. $\pm 4\sqrt{2}$

14. Let $g(x) = (7x - 51)^2$. Find all values of x for which $g(x) = 11$.

$\frac{51 \pm \sqrt{11}}{7}$

15. Explain why completing the square on $x^2 + 7x$ is more difficult than completing the square on $x^2 + 6x$. Because 7 is an odd number and not divisible by 2, the calculations involved in completing the square on $x^2 + 7x$ create fractions. The calculations involved in completing the square on $x^2 + 6x$ do not.

16. Happy New Year. As part of a New Year's Eve celebration, a huge ball is to be dropped from the top of a 605-foot-tall building at the proper moment so that it strikes the ground at exactly 12:00 midnight. The distance d in feet traveled by a free-falling object in t seconds is given by the formula $d = 16t^2$. To the nearest second, when should the ball be dropped from the building? 6 seconds before midnight

SECTION 10.2 The Quadratic Formula

DEFINITIONS AND CONCEPTS	EXAMPLES
To **solve a quadratic equation in x using the quadratic formula:** **1.** Write the equation in standard form: $ax^2 + bx + c = 0$ **2.** Identify a, b, and c. **3.** Substitute the values for a, b, and c in the quadratic formula $x = \frac{-b \pm \sqrt{b^2 - 4ac}}{2a}$ and evaluate the right side to obtain the solutions.	Solve: $3x^2 - 2x - 2 = 0$ Here, $a = 3$, $b = -2$, and $c = -2$. $x = \frac{-b \pm \sqrt{b^2 - 4ac}}{2a}$ This is the quadratic formula. $x = \frac{-(-2) \pm \sqrt{(-2)^2 - 4(3)(-2)}}{2(3)}$ Substitute 3 for a, -2 for b, and -2 for c. $x = \frac{2 \pm \sqrt{28}}{6}$ Evaluate the expression within the radical. $x = \frac{2 \pm 2\sqrt{7}}{6}$ Simplify the radical: $\sqrt{28} = \sqrt{4}\sqrt{7} = 2\sqrt{7}$. $x = \frac{\overset{1}{\cancel{2}}(1 \pm \sqrt{7})}{\underset{1}{\cancel{2}} \cdot 3}$ Factor out the GCF, 2, from the two terms in the numerator. In the denominator, factor 6 as $2 \cdot 3$. Then remove the common factor, 2. $x = \frac{1 \pm \sqrt{7}}{3}$ The exact solutions are $\frac{1 + \sqrt{7}}{3}$ and $\frac{1 - \sqrt{7}}{3}$. We can use a calculator to approximate them. To the nearest hundredth, they are 1.22 and -0.55.
When solving a quadratic equation using the quadratic formula, we can often **simplify the calculations** by solving an equivalent equation that does not involve fractions or decimals, and whose leading coefficient is positive.	**Before solving . . .** / **do this . . .** / **to get this** $-3x^2 + 5x - 1 = 0$ / Multiply both sides by -1. / $3x^2 - 5x + 1 = 0$ $x^2 + \frac{7}{8}x - \frac{1}{2} = 0$ / Multiply both sides by 8. / $8x^2 + 7x - 4 = 0$ $60x^2 - 40x + 90 = 0$ / Divide both sides by 10. / $6x^2 - 4x + 9 = 0$ $0.05x^2 + 0.16x + 0.71 = 0$ / Multiply both sides by 100. / $5x^2 + 16x + 71 = 0$

REVIEW EXERCISES

Solve each equation using the quadratic formula. Approximate the solutions to the nearest hundredth when appropriate.

17. $2x^2 + 13x = 7$
$\frac{1}{2}, -7$

18. $-x^2 + 10x - 18 = 0$
$5 \pm \sqrt{7}$; 7.65, 2.35

19. $x^2 - 10x = 0$
0, 10

20. $3y^2 = 26y - 2$
$\frac{13 \pm \sqrt{163}}{3}$; 8.59, 0.08

21. $\frac{1}{3}p^2 + \frac{1}{2}p + \frac{1}{2} = 0$
$-\frac{3}{4} \pm \frac{\sqrt{15}}{4}i$

22. $3{,}000t^2 - 4{,}000t = -2{,}000$
$\frac{2}{3} \pm \frac{\sqrt{2}}{3}i$

23. $0.5x^2 + 0.3x - 0.1 = 0$
$\frac{-3 \pm \sqrt{29}}{10}$; 0.24, −0.84

24. $x^2 - 3x - 27 = 0$
$\frac{3 \pm 3\sqrt{13}}{2}$; 6.91; −3.91

25. Let $h(x) = x^2 + 3x - 7$. Find all values of x for which $h(x) = 1$.
$\frac{-3 \pm \sqrt{41}}{2}$

26. Let $T(x) = -4x^2 - 2x$. Find all values of x for which $T(x) = -3$.
$\frac{-1 \pm \sqrt{13}}{4}$

27. Explain the error: $\frac{2 \pm \sqrt{7}}{2} = \frac{\overset{1}{\cancel{2}} \pm \sqrt{7}}{\underset{1}{\cancel{2}}}$

2 is not a factor of the numerator—it is a term. Only common factors of the numerator and denominator can be removed.

28. **a.** Write an expression that represents the width of the larger rectangle shown in red. $(2 + 2x)$ ft

b. Write an expression that represents the length of the larger rectangle shown in red. $(6 + 2x)$ ft

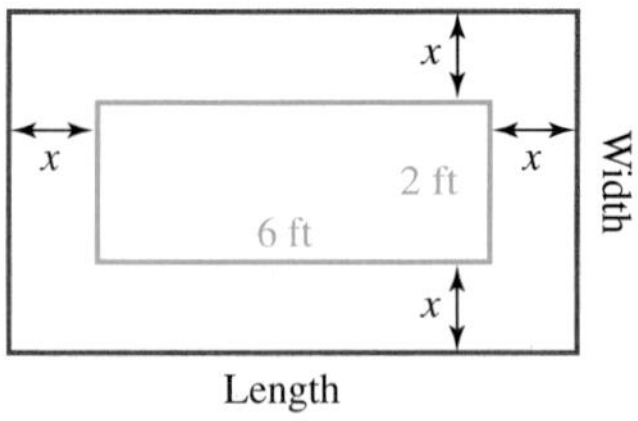

29. **Posters.** The specifications for a poster of Cesar Chavez call for a 615-square-inch photograph to be surrounded by a green border. The borders on the top and bottom of the poster are to be twice as wide as those on the sides. Find the width of each border. Sides: 1.25 in. wide; top/bottom: 2.5 in. wide

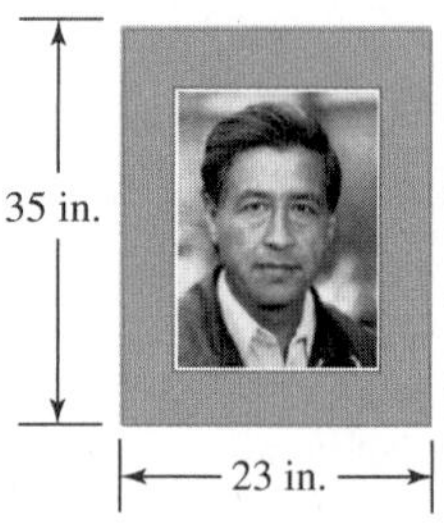

30. **Tutoring.** A private tutoring company charges \$20 for a 1-hour session. Currently, 300 students are tutored each week. Since the company is losing money, the owner has decided to increase the price. For each 50¢ increase, she estimates that 5 fewer students will participate. If the company needs to bring in \$6,240 per week to stay in business, what price must be charged for a 1-hour tutoring session to produce this amount of revenue? \$24 or \$26

31. **Acrobats.** To begin his routine on a trapeze, an acrobat is catapulted upward as shown in the illustration. His distance d (in feet) from the arena floor during this maneuver is given by the function $d(t) = -16t^2 + 40t + 5$, where t is the time in seconds since being launched. If the trapeze bar is 25 feet in the air, at what two times will he be able to grab it? Round to the nearest tenth. 0.7 sec, 1.8 sec

32. **Triangles.** The length of the longer leg of a right triangle exceeds the length of the shorter leg by 23 inches and the length of the hypotenuse is 65 inches. Find the length of each leg of the triangle. 33 in., 56 in.

SECTION 10.3 The Discriminant and Equations That Can Be Written in Quadratic Form

DEFINITIONS AND CONCEPTS	EXAMPLES
The **discriminant** predicts the type of solutions of $ax^2 + bx + c = 0$, where a, b, and c are rational numbers and $a \neq 0$. Review the table on page 852.	In the quadratic equation $2x^2 - 5x - 3 = 0$, we have $a = 2$, $b = -5$, and $c = -3$. So the value of the discriminant is $b^2 - 4ac = (-5)^2 - 4(2)(-3) = 25 + 24 = 49$ Since the value of the discriminant is positive and a perfect square, the equation $2x^2 - 5x - 3 = 0$ has two different rational-number solutions.

Equations that contain an expression, the same expression squared, and a constant term are said to be **quadratic in form.** One method used to solve such equations is to make a **substitution.** The leading term, x, is the square of the expression $\sqrt{x}$ in the middle term: $x = (\sqrt{x})^2$. $x - 9\sqrt{x} + 20 = 0$ The last term is a constant.	Solve: $x^{2/3} - 6x^{1/3} + 5 = 0$ The equation can be written in quadratic form: $(x^{1/3})^2 - 6x^{1/3} + 5 = 0$ Because $x^{2/3} = (x^{1/3})^2$ We substitute y for $x^{1/3}$ and use factoring to solve the resulting quadratic equation $y^2 - 6y + 5 = 0$. $(y - 1)(y - 5) = 0$ Let $y = x^{1/3}$. $y = 1$ or $y = 5$ Now we reverse (undo) the substitution $y = x^{1/3}$ and solve for x. $x^{1/3} = 1$ or $x^{1/3} = 5$ $(x^{1/3})^3 = (1)^3$ $(x^{1/3})^3 = (5)^3$ Recall: $x^{1/3} = \sqrt[3]{x}$. $x = 1$ $x = 125$ The solutions are 1 and 125. Check both in the original equation.

REVIEW EXERCISES

Use the discriminant to determine the number and type of solutions for each equation.

33. $3x^2 + 4x - 3 = 0$ Two different irrational-number solutions

34. $4x^2 - 5x + 7 = 0$ Two imaginary-number solutions that are complex conjugates

35. $3x^2 - 4x + \frac{4}{3} = 0$ One repeated solution, a rational number

36. $m(2m - 3) = 20$ Two different rational-number solutions

Solve each equation.

37. $x - 13\sqrt{x} + 12 = 0$ 1, 144

38. $a^{2/3} + a^{1/3} - 6 = 0$ 8, -27

39. $3x^4 + x^2 - 2 = 0$ $i, -i, \frac{\sqrt{6}}{3}, -\frac{\sqrt{6}}{3}$

40. $\frac{6}{x + 2} + \frac{6}{x + 1} = 5$ $1, -\frac{8}{5}$

41. $(x - 7)^2 + 6(x - 7) + 10 = 0$ $4 \pm i$

42. $m^{-4} - 2m^{-2} + 1 = 0$ Repeated solutions of -1 and 1

43. $4\left(\frac{x + 1}{x}\right)^2 + 12\left(\frac{x + 1}{x}\right) + 9 = 0$ A repeated solution of $-\frac{2}{5}$

44. $2m^{2/5} - 5m^{1/5} + 2 = 0$ $\frac{1}{32}, 32$

45. Weekly Chores. Working together, two sisters can do the yard work at their house in 45 minutes. When the older girl does it all herself, she can complete the job in 20 minutes less time than it takes the younger girl working alone. How long does it take the older girl to do the yard work? About 81 min

46. Road Trips. A woman drives her automobile 150 miles at a rate of r mph. She could have gone the same distance in 2 hours less time if she had increased her speed by 20 mph. Find r. 30 mph

SECTION 10.4 Quadratic Functions and Their Graphs

DEFINITIONS AND CONCEPTS	EXAMPLES
A **quadratic function** is a second-degree polynomial function of the form $f(x) = ax^2 + bx + c$	Quadratic functions: $f(x) = 2x^2 - 3x + 5$, $g(x) = -x^2 + 4x$, and $s(x) = \frac{1}{4}x^2 - 10$

The graph of the quadratic function $f(x) = a(x - h)^2 + k$ where $a \neq 0$ is a **parabola** with **vertex** at (h, k). The **axis of symmetry** is the line $x = h$. The parabola opens upward when $a > 0$ and downward when $a < 0$.

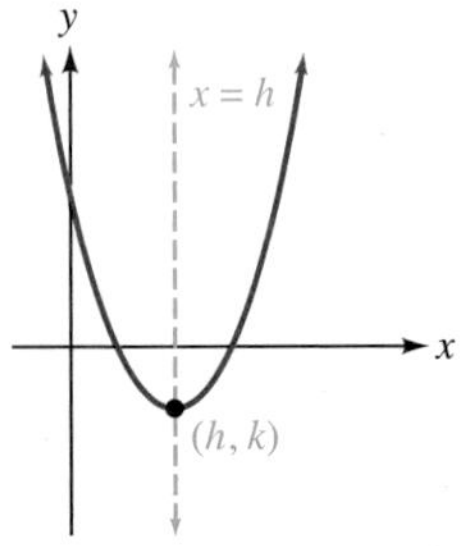

$f(x) = a(x - h)^2 + k$ where $a > 0$

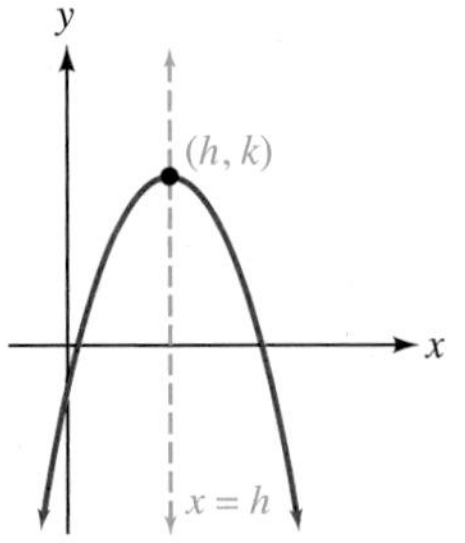

$f(x) = a(x - h)^2 + k$ where $a < 0$

Graph: $f(x) = 2(x + 1)^2 - 8$

$$f(x) = \mathbf{2}[x - (\mathbf{-1})]^2 - \mathbf{8}$$
$$\uparrow \quad \uparrow \quad \uparrow$$
$$f(x) = \mathbf{a}(x - \mathbf{h})^2 + \mathbf{k}$$

We see that $a = 2$, $h = -1$, and $k = -8$. The graph is a parabola with vertex $(h, k) = (-1, -8)$ and axis of symmetry $x = -1$. Since a is positive, the parabola opens upward.

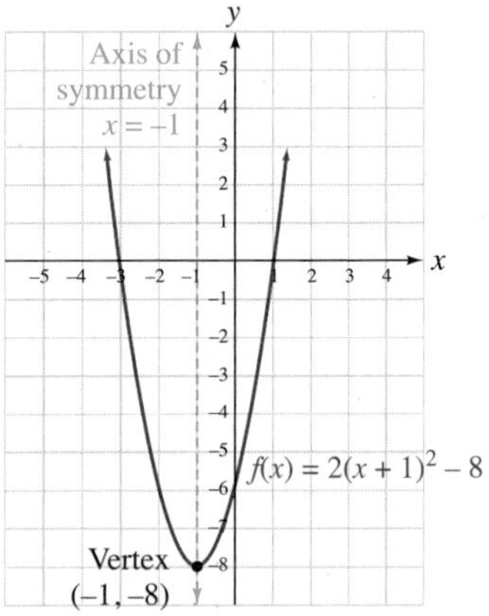

The **vertex** of the graph of $f(x) = ax^2 + bx + c$ is

$$\left(-\frac{b}{2a}, f\left(-\frac{b}{2a}\right)\right)$$

and the axis of symmetry is the line

$$x = -\frac{b}{2a}$$

The y-coordinate of the vertex of the graph of a quadratic function gives the **minimum or maximum value** of the function.

Graph: $f(x) = -x^2 + 3x + 4$

Here, $a = -1$, $b = 3$, and $c = 4$.

- Since $a < 0$, the graph opens downward.
- The x-coordinate of the vertex of the graph is:

$$-\frac{b}{2a} = -\frac{3}{2(-1)} = \frac{3}{2}$$

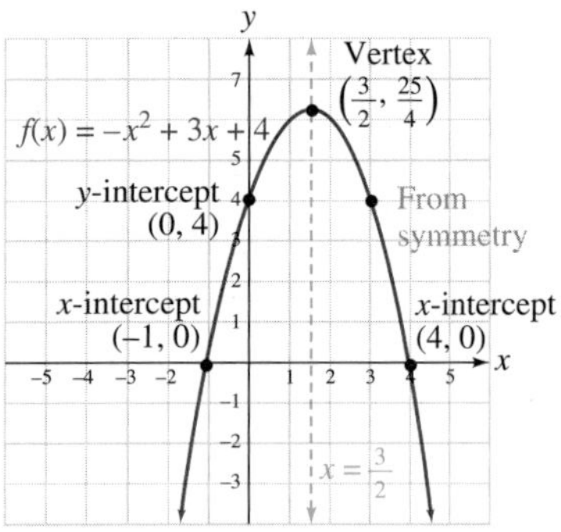

To find the y-coordinate of the vertex, we substitute $\frac{3}{2}$ for x in the function.

$$f\left(\frac{3}{2}\right) = -\left(\frac{3}{2}\right)^2 + 3\left(\frac{3}{2}\right) + 4 = \frac{25}{4}$$

The vertex of the parabola is $\left(\frac{3}{2}, \frac{25}{4}\right)$.

The **y-intercept** is determined by the value of $f(x)$ when $x = 0$: the y-intercept is $(0, c)$.

To find the **x-intercepts**, let $f(x) = 0$ and solve $ax^2 + bx + c = 0$.

- The y-intercept is the value of the function when $x = 0$. Since $f(0) = 4$, the y-intercept is $(0, 4)$.
- To find the x-intercepts, we solve:

$$-x^2 + 3x + 4 = 0 \quad \text{Let } f(x) = 0.$$
$$x^2 - 3x - 4 = 0 \quad \text{Multiply both sides by } -1.$$
$$(x + 1)(x - 4) = 0 \quad \text{Factor.}$$
$$x = -1 \quad \text{or} \quad x = 4$$

The x-intercepts are $(-1, 0)$ and $(4, 0)$.

REVIEW EXERCISES

47. Hospitals. The annual number of in-patient admissions to U.S. community hospitals for the years 1980–2008 can be modeled by the quadratic function $A(x) = 0.03x^2 - 0.88x + 37.3$, where $A(x)$ is the number of admissions in millions and x is the number of years after 1980. Use the function to estimate the number of in-patient admissions for the year 2005. Round to the nearest tenth of one million. (Source: American Hospital Association) 34.1 million

48. Fill in the blanks. The graph of the quadratic function $f(x) = a(x - h)^2 + k$ is a parabola with vertex at (h , k). The axis of symmetry is the line $x = h$. The parabola opens upward when $a > 0$ and downward when $a < 0$.

Graph each pair of functions on the same coordinate system. See AIE Appendix 3.

49. $f(x) = 2x^2$, $g(x) = 2x^2 - 3$

50. $f(x) = -\frac{1}{4}x^2$, $g(x) = -\frac{1}{4}(x + 2)^2$

51. Find the vertex and the axis of symmetry of the graph of $f(x) = -2(x - 1)^2 + 4$. Then plot several points and complete the graph. $(1, 4)$, $x = 1$; See AIE Appendix 3.

52. Complete the square to write $f(x) = 4x^2 + 16x + 9$ in the form $f(x) = a(x - h)^2 + k$. Determine the vertex and the axis of symmetry of the graph. Then plot several points and complete the graph. $f(x) = 4(x + 2)^2 - 7$; $(-2, -7)$, $x = -2$; see AIE Appendix 3.

53. Find the vertex of the graph of $f(x) = -2x^2 + 4x - 8$ using the vertex formula. $(1, -6)$

54. First determine the coordinates of the vertex and the axis of symmetry of the graph of $f(x) = x^2 + x - 2$ using the vertex formula. Then determine the x- and y-intercepts of the graph. Finally, plot several points and complete the graph. $\left(-\frac{1}{2}, -\frac{9}{4}\right)$, $x = -\frac{1}{2}$; $(-2, 0)$, $(1, 0)$; $(0, -2)$; see AIE Appendix 3.

55. Farming. The number of farms in the United States for the years 1870–1970 is approximated by

$$N(x) = -1{,}526x^2 + 155{,}652x + 2{,}500{,}200$$

where $x = 0$ represents 1870, $x = 1$ represents 1871, and so on. For this period, when was the number of U.S. farms a maximum? How many farms were there? 1921; 6,469,326

56. Estimate the solutions of $-3x^2 - 5x + 2 = 0$ from the graph of $f(x) = -3x^2 - 5x + 2$, shown here. $-2, \frac{1}{3}$

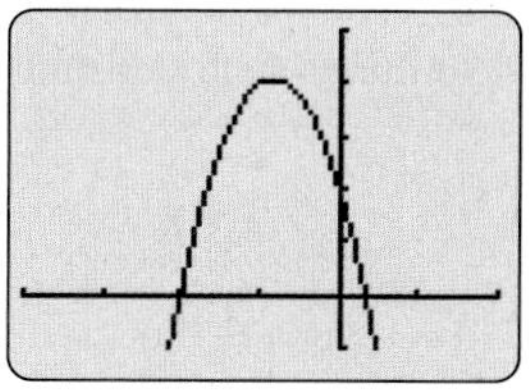

SECTION 10.5 Quadratic and Other Nonlinear Inequalities

DEFINITIONS AND CONCEPTS	EXAMPLES
To solve a quadratic inequality, get 0 on the right side and solve the related quadratic equation. Then locate the **critical numbers** on a number line, test each interval, and check the endpoints.	To solve $x^2 - x - 6 \geq 0$, we solve the related quadratic equation:

$$x^2 - x - 6 = 0$$

$(x - 3)(x + 2) = 0$ *Factor.*

$x = 3$ or $x = -2$ *These are the critical numbers that divide the number line into three intervals.*

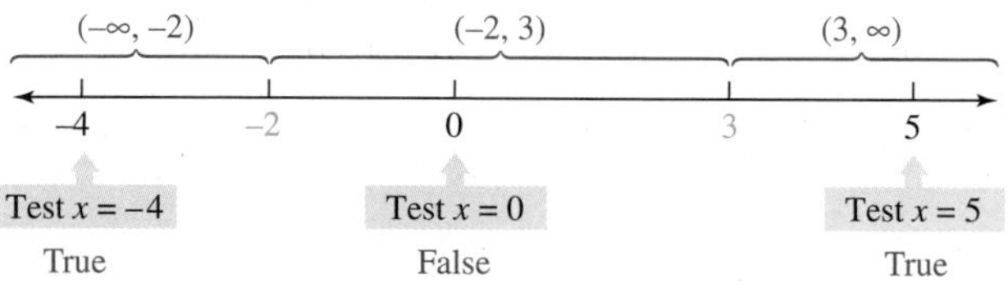

After testing each interval and noting that 3 and -2 satisfy the inequality, we see that the solution set is $(-\infty, -2] \cup [3, \infty)$.

To solve a rational inequality, get 0 on the right side and solve the related rational equation. Then locate the **critical numbers** (including any values that make the denominator 0) on a number line, test each interval, and check the endpoints.

To solve $\frac{x+1}{x-4} < 0$, we solve the related rational equation $\frac{x+1}{x-4} = 0$ to obtain the solution $x = -1$, which is a critical number. Another critical number is $x = 4$, the value that makes the denominator 0. These critical numbers divide the number line into three intervals.

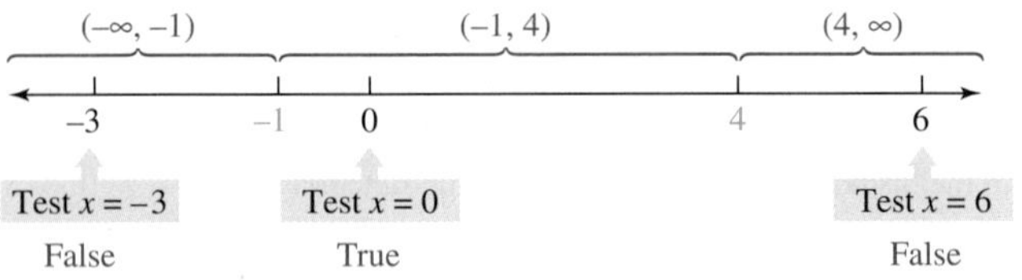

After testing each interval and noting that -1 and 4 do not satisfy the inequality, we see that the solution set is the interval $(-1, 4)$.

To graph a nonlinear inequality in two variables, first graph the boundary. Then use a test point to determine which side of the boundary to shade.

This is the graph of $y \le x^2 + 5x + 4$.

Since the inequality contains the symbol $\le$, and equality is allowed, we draw the parabola determined by $y = x^2 + 5x + 4$ using a **solid curve.**

We shade the region containing the test point (0, 0) because its coordinates satisfy $y \le x^2 + 5x + 4$.

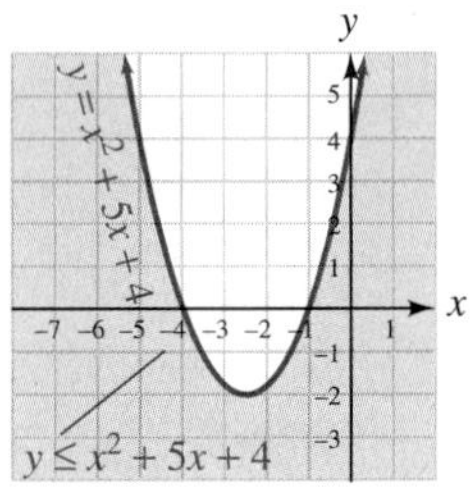

REVIEW EXERCISES

Solve each inequality. Write the solution set in interval notation and graph it. See AIE Appendix 3.

57. $x^2 + 2x - 35 > 0$
$(-\infty, -7) \cup (5, \infty)$

58. $x^2 \le 81$
$[-9, 9]$

59. $\frac{3}{x} \le 5$
$(-\infty, 0) \cup \left[\frac{3}{5}, \infty\right)$

60. $\frac{2x^2 - x - 28}{x - 1} > 0$
$\left(-\frac{7}{2}, 1\right) \cup (4, \infty)$

61. Estimate the solution set of $3x^2 + 10x - 8 \le 0$ from the graph of $f(x) = 3x^2 + 10x - 8$ shown in figure (a). $\left[-4, \frac{2}{3}\right]$

62. Estimate the solution set of $\frac{x-1}{x} > 0$ from the graph of $f(x) = \frac{x-1}{x}$ shown in figure (b). $(-\infty, 0) \cup (1, \infty)$

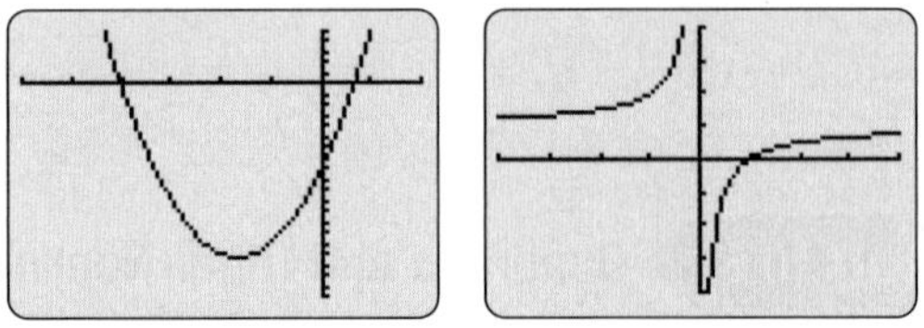

(a) (b)

Graph each inequality. See AIE Appendix 3.

63. $y < \frac{1}{2}x^2 - 1$

64. $y \ge -|x|$

Teaching Tip: Because this Chapter Test is a comprehensive collection of problems that covers all of the topics discussed in Chapter 10, it is lengthy. If your students have time to complete it, that would be optimal. If, because of time constraints, they are unable to do so, assign an appropriate subset of problems that reflects the types of problems that the students will see on your exam.

10 Chapter Test

1. Fill in the blanks.

a. An equation of the form $ax^2 + bx + c = 0$, where $a \ne 0$, is called a quadratic equation.

b. When we add 81 to $x^2 + 18x$, we say that we have completed the square on $x^2 + 18x$.

c. The lowest point on a parabola that opens upward, or the highest point on a parabola that opens downward, is called the vertex of the parabola.

d. $\frac{x-5}{x^2 - x - 56} > 0$ is an example of a rational inequality in one variable.

e. $y \le x^2 - 4x + 3$ is an example of a nonlinear inequality in two variables.

2. Solve $x^2 - 63 = 0$ using the square root property. Approximate the solutions to the nearest hundredth. $\pm 3\sqrt{7}$; ± 7.94

Solve each equation using the square root property.

3. $(a + 7)^2 = 50$
$-7 \pm 5\sqrt{2}$

4. $m^2 + 4 = 0$
$\pm 2i$

5. Add a number to make $x^2 + 11x$ a perfect-square trinomial. Then factor the result. $x^2 + 11x + \frac{121}{4} = \left(x + \frac{11}{2}\right)^2$

6. Solve $x^2 + 3x - 2 = 0$ by completing the square. Approximate the solutions to the nearest hundredth.

$\frac{-3 \pm \sqrt{17}}{2}$; $-3.56, 0.56$

7. Solve $2x^2 + 8x + 12 = 0$ by completing the square.
$-2 \pm i\sqrt{2}$

8. Let $f(x) = (3x - 2)^2$. Find all values of x for which $f(x) = 18$.
$\frac{2 \pm 3\sqrt{2}}{3}$

Use the quadratic formula to solve each equation. Approximate each solution to the nearest hundredth, when appropriate.

9. $4x^2 + 4x - 1 = 0$ $\frac{-1 \pm \sqrt{2}}{2}$; $-1.21, 0.21$

10. $\frac{1}{8}t^2 - \frac{1}{4}t = \frac{1}{2}$ $1 \pm \sqrt{5}$; $-1.24, 3.24$

11. $-t^2 + 4t - 13 = 0$
$2 \pm 3i$

12. $0.01x^2 = -0.08x - 0.15$
$-5, -3$

13. $m^2 - 94m = -2{,}209$ A repeated solution of 47

14. Let $g(x) = 3x^2 - 20$. Find all values of x for which $g(x) = 10$.
$\pm\sqrt{10}$

Solve each equation by any method. Approximate each solution to the nearest hundredth, when appropriate.

15. $2y - 3\sqrt{y} + 1 = 0$
$1, \frac{1}{4}$

16. $3 = m^{-2} - 2m^{-1}$
$-1, \frac{1}{3}$

17. $x^4 - x^2 - 12 = 0$ $2, -2, i\sqrt{3}, -i\sqrt{3}$

18. $4\left(\frac{x+2}{3x}\right)^2 - 4\left(\frac{x+2}{3x}\right) - 3 = 0$ $-\frac{4}{5}, \frac{4}{7}$

19. $\frac{1}{n+2} = \frac{1}{3} - \frac{1}{n}$ $2 \pm \sqrt{10}$; $5.16, -1.16$

20. $5a^{2/3} + 11a^{1/3} = -2$ $-8, -\frac{1}{125}$

21. $10x(x + 1) = -3$ $-\frac{1}{2} \pm \frac{\sqrt{5}}{10}i$

22. $a^3 - 3a^2 + 8a - 24 = 0$ $3, \pm 2i\sqrt{2}$

23. Solve $E = mc^2$ for c. Assume that all variables represent positive numbers. Express any radical in simplified form. $c = \frac{\sqrt{Em}}{m}$

24. Use the discriminant to determine the number and type of solutions for each equation.
   a. $3x^2 + 5x + 17 = 0$ Two different imaginary-number solutions that are complex conjugates
   b. $9m^2 - 12m = -4$ One repeated solution, a rational number

25. **Tablecloths.** In 1990, Sportex of Highland, Illinois, made what was at the time the world's longest tablecloth. Find the dimensions of the rectangular tablecloth if it covered an area of 6,759 square feet and its length was 8 feet more than 332 times its width. 4.5 ft by 1,502 ft

26. **Cooking.** Working together, a chef and his assistant can make a pastry dessert in 25 minutes. When the chef makes it himself, it takes him 8 minutes less time than it takes his assistant working alone. How long does it take the chef to make the dessert? About 46 min

27. **Drawing.** An artist uses four equal-sized right triangles to block out a perspective drawing of an old hotel. See the illustration in the next column. For each triangle, the leg on the horizontal line is 14 inches longer than the leg on the center line. The length of each hypotenuse is 26 inches. On the center line of the drawing, what is the length of the segment extending from the ground to the top of the building? 20 in.

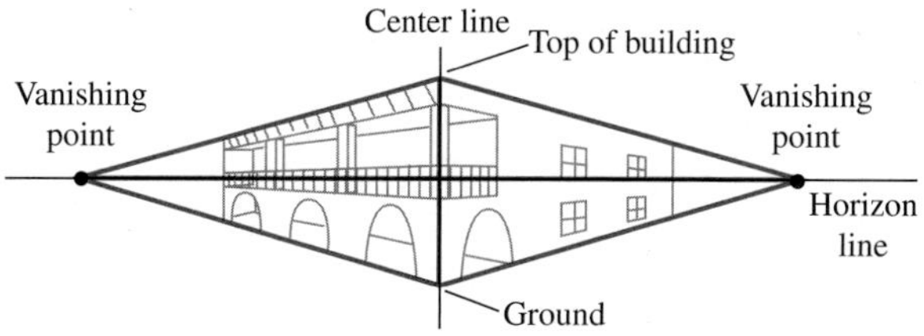

28. **Home Decorating.** A woman is going to put a cork border of uniform width around an 18-inch by 24-inch wall picture, as shown here. If the area covered by the cork is to be the same as the surface area of the picture, how wide should the border be? Round to the nearest tenth of an inch.
4.3 in.

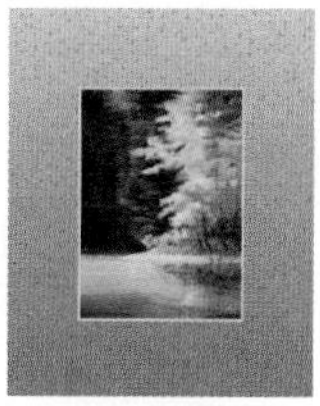

29. **ER Rooms.** The number of hospital emergency departments in the United States is approximated by the function $E(t) = 2.29t^2 - 75.72t + 5{,}206.95$, where t is the number of years after 1990. In what year does the model indicate that the number of hospital emergency departments was 5,000? (Source: American Hospital Association) 1993

30. **Anthropology.** Anthropologists refer to the shape of the human jaw as a *parabolic dental arcade.* Which function is the best mathematical model of the parabola shown in the illustration?
iii

   i. $f(x) = -\frac{3}{8}(x - 4)^2 + 6$
   ii. $f(x) = -\frac{3}{8}(x - 6)^2 + 4$
   iii. $f(x) = -\frac{3}{8}x^2 + 6$
   iv. $f(x) = \frac{3}{8}x^2 + 6$

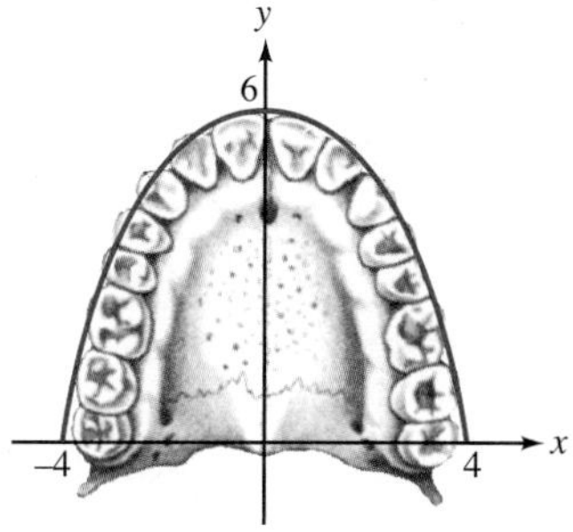

31. Find the vertex and the axis of symmetry of the graph of $f(x) = -3(x - 1)^2 + 2$. Then plot several points and complete the graph. $(1, 2)$, $x = 1$; see AIE Appendix 3.

32. Complete the square to write the function $f(x) = 5x^2 + 10x - 1$ in the form $f(x) = a(x - h)^2 + k$. Determine the vertex and the axis of symmetry of the graph. Then plot several points and complete the graph.
$f(x) = 5(x + 1)^2 - 6$; $(-1, -6)$, $x = -1$; see AIE Appendix 3.

33. First determine the coordinates of the vertex and the axis of symmetry of the graph of $f(x) = 2x^2 + x - 1$ using the vertex formula. Then determine the x- and y-intercepts of the graph. Finally, plot several points and complete the graph.
$\left(-\frac{1}{4}, -\frac{9}{8}\right)$, $x = -\frac{1}{4}$, $(-1, 0)$, $\left(\frac{1}{2}, 0\right)$; $(0, -1)$; see AIE Appendix 3.

34. Distress Signals. A flare is fired directly upward into the air from a boat that is experiencing engine problems. The height of the flare (in feet) above the water, t seconds after being fired, is given by the function $h(t) = -16t^2 + 112t + 15$. If the flare is designed to explode when it reaches its highest point, at what height will this occur? 211 ft

Solve each inequality. Write the solution set in interval notation and then graph it. See AIE Appendix 3.

35. $x^2 - 2x > 8$ $(-\infty, -2) \cup (4, \infty)$

36. $\dfrac{x-2}{x+3} \le 0$ $(-3, 2]$

37. Climate. The average monthly temperature in San Antonio, Texas, is approximated by the function $T(m) = -1.1m^2 + 15.3m + 29.5$, where m is the number of the month (January = 1, February = 2, and so on). Use the function to approximate the average monthly temperature in July. (Source: cityrating.com) 82.7°F

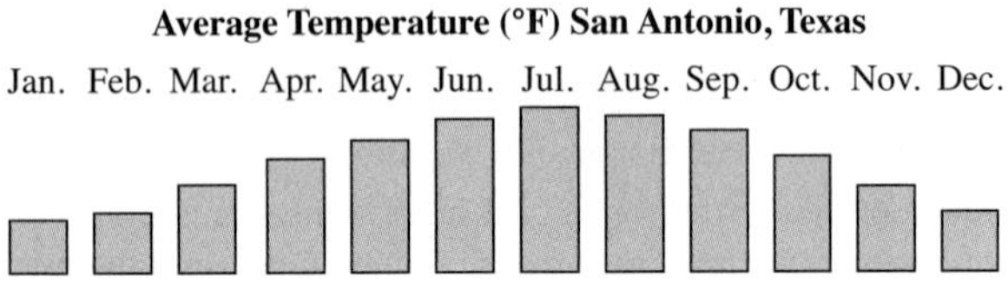

38. Graph: $y \le -x^2 + 3$ See AIE Appendix 3.

39. The graph of a quadratic function of the form $f(x) = ax^2 + bx + c$ is shown. Estimate the solutions of the corresponding quadratic equation $ax^2 + bx + c = 0$. $-2, 3$

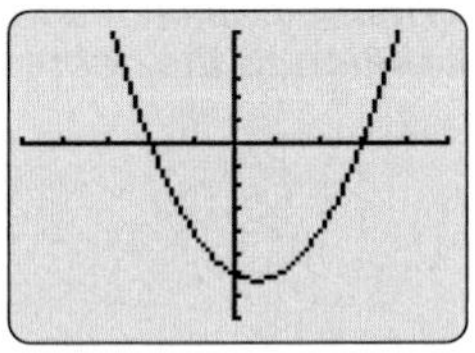

40. See Exercise 39. Estimate the solution of the quadratic inequality $ax^2 + bx + c \le 0$. $[-2, 3]$

Group Project

Picture Framing

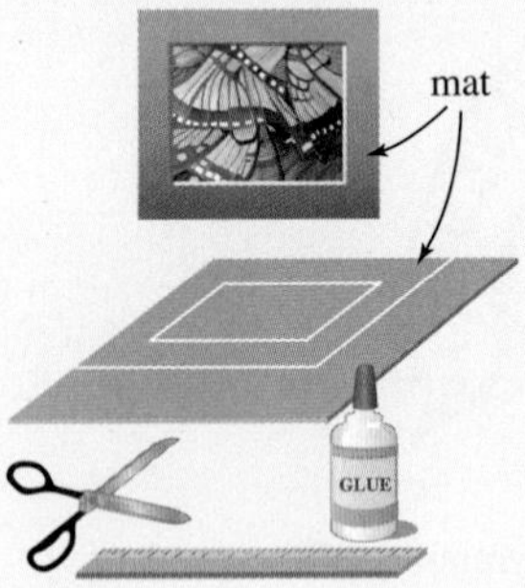

Overview: When framing pictures, mats are often used to enhance the images and give them a sense of depth. In this activity, you will use the quadratic formula to design the matting for several pictures.

Instructions: Form groups of 3 students. Each person in your group is to bring a picture to class. You can use a picture from a magazine or newspaper, a picture postcard, or a photograph that is no larger than 5 in. × 7 in. You will also need a pair of scissors, a ruler, glue, and three pieces of construction paper (12 in. × 18 in.).

Select one of the pictures and find its area. A mat of *uniform* width is to be placed around the picture. The area of the mat should equal the area of the picture. To determine the proper width of the matting, follow the steps of Example 8 in Section 10.2. Once you have determined the proper width, cut out the mat from the construction paper and glue it to the picture.

Then, choose another picture and find its area. Determine the uniform width that a matting should have so that its area is double that of the picture. Cut out the proper-size matting from the construction paper and glue it to the second picture.

Finally, find the area of the third picture and determine the uniform width that a matting should have so that its area is one-half that of the picture. Cut out the proper-size matting from the construction paper and glue it to the third picture.

Is one size matting more visually appealing than another? Discuss this among the members of your group.

CUMULATIVE REVIEW Chapters 1–10

1. Determine whether each statement is true or false. [Section 1.3]
   a. All whole numbers are integers. True
   b. π is a rational number. False
   c. A real number is either rational or irrational. True

2. Evaluate: $\dfrac{-3(3+2)^2-(-5)}{17-|-22|}$ [Section 1.7] 14

3. Simplify: $-9(3a-9)-7(2a-7)$ [Section 1.9] $-41a+130$

4. Use a check to determine whether -11 is a solution of $\dfrac{3(b+2)}{2}=\dfrac{4b-10}{4}$ [Section 2.2] It is a solution.

5. Solve: $2-(4x+7)=3+2(x+2)$ [Section 2.2] -2

6. Solve: $2(6n+5)=4(3n+2)+2$ [Section 2.2] All real numbers

7. **Backpacks.** Pediatricians advise that children should not carry more than 20% of their own body weight in a backpack. According to this warning, how much weight can a fifth-grade girl who weighs 85 pounds safely carry in her backpack? [Section 2.3] 17 lb

8. **Surface Area.** The total surface area A of a box with dimensions l, w, and h is given by the formula $A=2lw+2wh+2lh$. If $A=202$ square inches, $l=9$ inches, and $w=5$ inches, find h. [Section 2.4] 4 in.

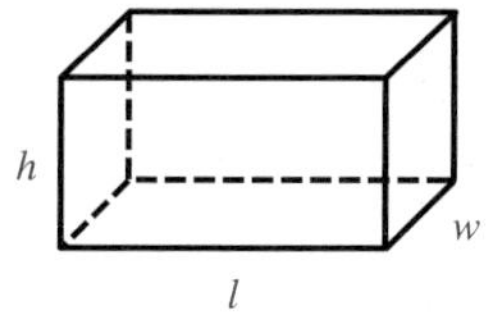

9. **Geometry.** If the vertex angle of an isosceles triangle is 53°, find the measure of each base angle. [Section 2.5] 63.5°

10. **Search and Rescue.** Two search and rescue teams leave base at the same time, looking for a lost boy. The first team, on foot, heads north at 2 mph and the other, on horseback, south at 4 mph. How long will it take them to search a distance of 21 miles between them? [Section 2.6] 3.5 hr

11. **Pharmacist.** How many liters of a 1% glucose solution should a pharmacist mix with 2 liters of a 5% glucose solution to obtain a 2% glucose solution? [Section 2.6] 6 L

12. **Blending Coffee.** A store sells regular coffee for \$8 a pound and gourmet coffee for \$14 a pound. Using 40 pounds of the gourmet coffee, the owner makes a blend to put on sale for \$10 a pound. How many pounds of regular coffee should he use? [Section 2.6] 80 lb

13. Solve: $3-3x\ge 6+x$. Graph the solution set. Then describe the graph using interval notation. [Section 2.7] $\left(-\infty,-\frac{3}{4}\right]$ See AIE Appendix 3.

14. Check to determine whether $(-6,-7)$ is a solution of $4x-3y=-4$. [Section 3.1] No

Graph each equation. See AIE Appendix 3.

15. $y=\dfrac{1}{2}x$ [Section 3.2]

16. $3x-4y=12$ [Section 3.3]

17. $x=5$ [Section 3.3]

18. $y=2x^2-3$ [Section 5.4]

19. Find the x-intercept and the y-intercept of the graph of the linear equation $5x-3y=6$. [Section 3.3] x-intercept: $\left(\frac{6}{5},0\right)$; y-intercept: $(0,-2)$

20. **Skype.** The line graph below shows the approximate growth in the number of subscribers to Skype, a software application that allows users to make video calls over the Internet. Find the rate of change in the number of subscribers to Skype during the years 2007–2010 by finding the slope of the line. [Section 3.4] An increase of 114 million subscribers per year

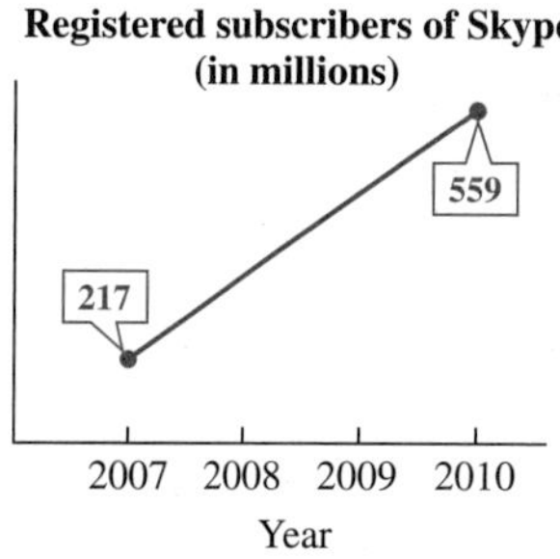

Source: businessinsider.com

21. What is the slope of the line defined by each equation? [Section 3.5]
    a. $y=3x-7$ 3
    b. $5x-6y=13$ $\frac{5}{6}$

22. What is the slope of the line whose equation is $y=-3$? [Section 3.5] 0

Find an equation of the line with the given properties. Write the equation in slope–intercept form.

23. Slope 3, passes through $(-2,-4)$ [Section 3.5] $y=3x+2$

24. Parallel to the graph of $2x+3y=6$ and passes through $(0,-2)$ [Section 3.6] $y=-\frac{2}{3}x-2$

25. If $f(x)=2x^2-3x+1$, find $f(-3)$. [Section 3.8] 28

26. **Boating.** The graph shows the vertical distance from a point on the tip of a propeller to the centerline as the propeller spins. Is this the graph of a function? [Section 3.8] Yes

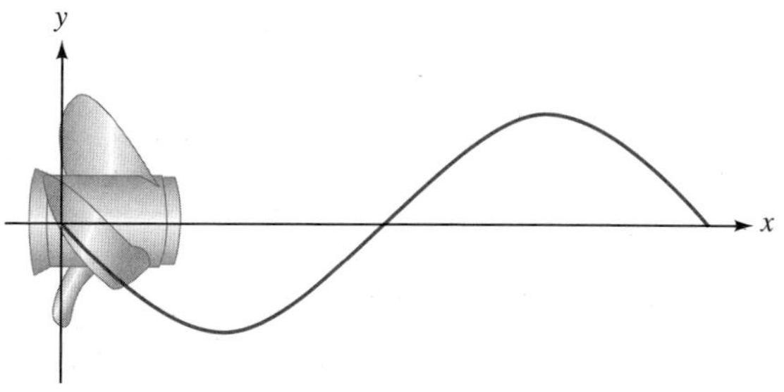

27. Use a check to determine whether $\left(2,\frac{1}{3}\right)$ is a solution of $\begin{cases}x-3y=1\\-2x+6=-6y\end{cases}$ [Section 4.1] It is not a solution.

28. Solve the system $\begin{cases}x+y=4\\y=x+6\end{cases}$ by graphing. [Section 4.1] $(-1,5)$ See AIE Appendix 3.

Solve each system of equations.

29. $\begin{cases} x = y + 4 \\ 2x + y = 5 \end{cases}$ [Section 4.2] $(3, -1)$

30. $\begin{cases} 3s + 4t = 5 \\ 2s - 3t = -8 \end{cases}$ [Section 4.3] $(-1, 2)$

31. Financial Planning. In investing \$6,000 of a couple's money, a financial planner put some of it into a savings account paying 6% annual interest. The rest was invested in a riskier mini-mall development plan paying 12% annually. The combined interest earned for the first year was \$540. How much money was invested at each rate? Use two variables to solve this problem. [Section 4.4] 6%: \$3,000; 12%: \$3,000

32. Graph: $\begin{cases} 3x + 2y \ge 6 \\ x + 3y \le 6 \end{cases}$ [Section 4.5] See AIE Appendix 3.

Simplify. Use only positive exponents in your answers.

33. $(x^5)^2(x^7)^3$ [Section 5.1] x^{31}

34. $\left(\dfrac{a^3b}{c^4}\right)^5$ [Section 5.1] $\dfrac{a^{15}b^5}{c^{20}}$

35. $4^{-3} \cdot 4^{-2} \cdot 4^5$ [Section 5.2] 1

36. $\left(\dfrac{2a^2b^3c^{-4}}{5a^{-2}b^{-1}c^3}\right)^{-3}$ [Section 5.2] $\dfrac{125c^{21}}{8a^{12}b^{12}}$

37. $5n^0$ 5

38. $\dfrac{4b^{-4}}{(a^{-2})^{-5}}$ $\dfrac{4}{a^{10}b^4}$

39. Write each number in scientific notation and perform the operations. Give the answer in scientific notation and in standard notation:

$$\frac{(1{,}280{,}000{,}000)(2{,}700{,}000)}{240{,}000}$$ [Section 5.3]

1.44×10^{10}; 14,400,000,000

40. Simplify: $\frac{3}{5}s^2 - \frac{2}{5}t^2 - \frac{1}{2}s^2 - \frac{7}{10}st - \frac{3}{10}st$ [Section 5.5] $\frac{1}{10}s^2 - st - \frac{2}{5}t^2$

Perform the indicated operations.

41. $(-8.9t^3 - 2.4t) - (2.1t^3 + 0.8t^2 - t)$ [Section 5.5] $-11t^3 - 0.8t^2 - 1.4t$

42. $(2a - b)(4a^2 + 2ab + b^2)$ [Section 5.6] $8a^3 - b^3$

43. $(-3t + 2s)(2t - 3s)$ [Section 5.6] $-6t^2 + 13st - 6s^2$

44. $(4b - 8)^2$ [Section 5.7] $16b^2 - 64b + 64$

45. $\left(6b + \dfrac{1}{2}\right)\left(6b - \dfrac{1}{2}\right)$ [Section 5.7] $36b^2 - \frac{1}{4}$

46. $x + 2\overline{)2x^2 + 3x - 2}$ [Section 5.8] $2x - 1$

Factor each expression.

47. $12uvw^3 - 18uv^2w^2$ [Section 6.1] $6uvw^2(2w - 3v)$

48. $x^2 + 4y - xy - 4x$ [Section 6.1] $(x - y)(x - 4)$

49. $x^2 + 7x + 10$ [Section 6.2] $(x + 5)(x + 2)$

50. $6 + 3x^2 + x$ [Section 6.3] Prime

51. $6a^2 - 7a - 20$ [Section 6.3] $(3a + 4)(2a - 5)$

52. $30a^4 - 4a^3 - 16a^2$ [Section 6.3] $2a^2(3a + 2)(5a - 4)$

53. $49s^6 - 84s^3n^2 + 36n^4$ [Section 6.4] $(7s^3 - 6n^2)^2$

54. $x^4 - 16y^4$ [Section 6.4] $(x^2 + 4y^2)(x + 2y)(x - 2y)$

55. $x^3 - 64$ [Section 6.4] $(x - 4)(x^2 + 4x + 16)$

56. $8x^6 + 125y^3$ [Section 6.5] $(2x^2 + 5y)(4x^4 - 10x^2y + 25y^2)$

Solve each equation.

57. $x^2 + 3x + 2 = 0$ [Section 6.7] $-1, -2$

58. $5x^2 = 10x$ [Section 6.7] $0, 2$

59. $6x^2 - x = 2$ [Section 6.7] $\frac{2}{3}, -\frac{1}{2}$

60. $a^2 - 25 = 0$ [Section 6.7] $-5, 5$

61. $(m + 4)(2m + 3) - 22 = 10m$ [Section 6.7] $2, -\frac{5}{2}$

62. $6a^3 - 2a = a^2$ [Section 6.7] $0, \frac{2}{3}, -\frac{1}{2}$

63. Children's Stickers. A rectangular-shaped sticker has an area of 20 cm^2. The width is 1 cm shorter than the length. Find the length of the sticker. [Section 6.8] 5 cm

64. Simplify: $\dfrac{x^2 + 2x + 1}{x^2 - 1}$ [Section 7.1] $\dfrac{x + 1}{x - 1}$

Perform the operations. Simplify, if possible.

65. $\dfrac{p^2 - p - 6}{3p - 9} \div \dfrac{p^2 + 6p + 9}{p^2 - 9}$ [Section 7.2] $\dfrac{(p - 3)(p + 2)}{3(p + 3)}$

66. $\dfrac{12x^2}{7 - x} \cdot \dfrac{x - 7}{20x^3}$ [Section 7.2] $-\dfrac{3}{5x}$

67. $\dfrac{13}{15a} - \dfrac{8}{15a}$ [Section 7.3] $\dfrac{1}{3a}$

68. $\dfrac{x + 2}{x + 5} - \dfrac{x - 3}{x + 7}$ [Section 7.4] $\dfrac{7x + 29}{(x + 5)(x + 7)}$

69. $\dfrac{1}{6b^4} - \dfrac{8}{9b^2}$ [Section 7.4] $\dfrac{3 - 16b^2}{18b^4}$

70. $\dfrac{\frac{1}{x} + \frac{1}{y}}{\frac{1}{x} - \frac{1}{y}}$ [Section 7.5] $\dfrac{y + x}{y - x}$

71. Solve: $\dfrac{7}{a^2 - a - 2} + \dfrac{1}{a + 1} = \dfrac{3}{a - 2}$ [Section 7.6] 1

72. Solve: $\dfrac{x - 4}{x - 3} + \dfrac{x - 2}{x - 3} = x - 3$ [Section 7.6] 5; 3 is extraneous

73. Solve for R: $\dfrac{1}{R} = \dfrac{1}{R_1} + \dfrac{1}{R_2} + \dfrac{1}{R_3}$ [Section 7.6] $R = \dfrac{R_1R_2R_3}{R_2R_3 + R_1R_3 + R_1R_2}$

74. Filling a Pool. An inlet pipe can fill an empty swimming pool in 5 hours, and another inlet pipe can fill the pool in 4 hours. How long will it take both pipes to fill the pool? [Section 7.7] $2\frac{2}{9}$ hr

75. Online Sales. A company found that, on average, it made 9 online sales transactions for every 500 hits on its Internet Web site. If the company's Web site had 360,000 hits in one year, how many sales transactions did it have that year? [Section 7.8] 6,480

76. The triangles shown below are similar. Find x. [Section 7.8] 20

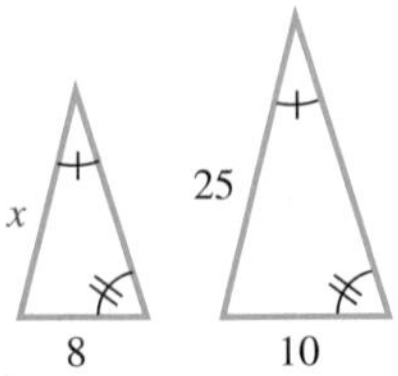

77. Solve: $\frac{1}{6}(a + 12) + 1 = \frac{a}{3}$ [Section 8.1] 18

78. Solve: $2x + 1 = 5(0.2x + 1) - (4 - x)$ [Section 8.1] All real numbers, $\mathbb{R}$; identity

79. **Grades.** A student has scores of 75, 83, 91, and 68 on four algebra tests. What score does she need on a fifth test to give her an average of 80 or better? [Section 8.1] 83 or better

80. Fill in the blank: Given a relation in x and y, if to each value of x in the domain there corresponds exactly one value of y in the range, then y is said to be a _function_ of x. [Section 8.2]

Refer to the following graph of function f.

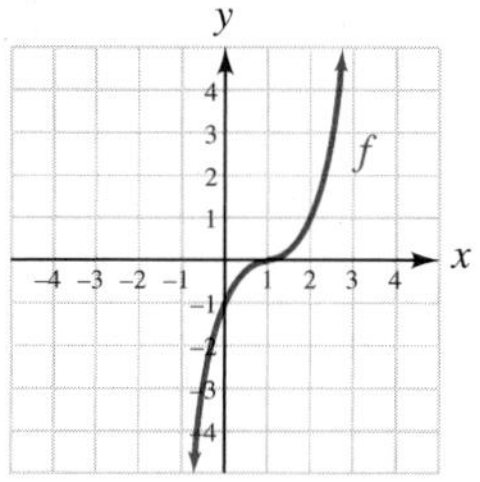

81. a. Find $f(1)$. [Section 8.2] 0

b. Find the value of x for which $f(x) = 1$. [Section 8.2] 2

82. Find the domain and range of the function. [Section 8.3] D: The set of real numbers R: The set of real numbers

83. Find the domain and range of the relation $\{(1, -12), (-6, 7), (5, 8), (0, 7), (0, 4)\}$. Is it a function? [Section 8.2] D: $\{-6, 0, 1, 5\}$; R: $\{-12, 4, 7, 8\}$; no

84. Does $y^2 = x$ define y as a function of x? If it does not, find two ordered pairs where more than one value of y corresponds to a single value of x. [Section 8.2] No; (4, 2), (4, −2)

85. Let $h(x) = -\frac{1}{5}x - 12$. For what value of x is $h(x) = 0$? [Section 8.2] −60

86. Let $f(x) = 3x^2 + 2$ and $g(x) = -2x - 1$. Find each function value. [Section 8.2]

a. $g(-2)$ 3

b. $f(-r)$ $3r^2 + 2$

87. Find the domain of the function $f(x) = \frac{5}{x + 1}$. [Section 8.2] The set of all real numbers except −1

88. What are the slope and y-intercept of the graph of the linear function $f(x) = 6x + 15$? [Section 8.2] 6, (0, 15)

89. Write an equation for the linear function whose graph is horizontal and passes through $\left(\frac{3}{2}, -\frac{7}{8}\right)$. [Section 8.2] $f(x) = -\frac{7}{8}$

90. **Driving.** The total number of miles driven each day in the United States on all roads by all vehicles since the year 1980 can be modeled by a linear function. In 1990, approximately 5.76 billion vehicle miles were driven daily. In 2005, approximately 8.31 billion vehicle miles were driven daily. (Source: FHWA Highway Statistics) [Section 8.2]

a. Let t be the number of years after 1980 and M be the number of billion vehicle miles driven daily. Write a linear function $M(t)$. $M(t) = 0.17t + 4.06$

b. Use your answer to part a to estimate how many billion vehicle miles will be driven daily in 2020. 10.86 billion vehicle miles

91. Graph $f(x) = -x^2 + 1$ by plotting points. Then give the domain and range of the function. [Section 8.3] D: the set of real numbers; R: the set of real numbers less than or equal to 1. See AIE Appendix 3.

92. First sketch the graph of the basic function associated with $g(x) = |x - 3| - 4$. Then draw the graph of function g using a translation. Give the domain and range of function g. [Section 8.3] D: the set of real numbers; R: the set of real numbers greater than or equal to −4. See AIE Appendix 3.

Solve each equation.

93. $2|4x - 3| + 1 = 19$ [Section 8.5] $3, -\frac{3}{2}$

94. $|2x - 1| = |3x + 4|$ [Section 8.5] $-5, -\frac{3}{5}$

Solve each inequality. Write the solution set in interval notation and then graph it. See AIE Appendix 3.

95. $5(-2x + 2) > 20 - x$ [Section 8.1] $\left(-\infty, -\frac{10}{9}\right)$

96. $3x + 4 < -2$ or $3x + 4 > 10$ [Section 8.4] $(-\infty, -2) \cup (2, \infty)$

97. $5x - 3 \ge 2$ and $6 \ge 4x - 3$ [Section 8.4] $\left[1, \frac{9}{4}\right]$

98. $|2x - 5| \ge 25$ [Section 8.5] $(-\infty, -10] \cup [15, \infty)$

99. $|3x - 2| \le 4$ [Section 8.5] $\left[-\frac{2}{3}, 2\right]$

100. $5|4 - x| + 6 < 1$ [Section 8.5] No solution

101. Use the substitution to factor $(x + y)^2 + 7(x + y) + 12$. [Section 8.6] $(x + y + 4)(x + y + 3)$

102. Solve $d_1d_2 - fd_2 = fd_1$ for f. [Section 8.6] $f = \frac{d_1d_2}{d_2 + d_1}$

103. Factor $x^6 - 1$ by factoring it as a difference of two squares first. [Section 8.7] $(x + 1)(x^2 - x + 1)(x - 1)(x^2 + x + 1)$

104. Find the domain of the function $f(x) = \frac{2x + 1}{x^2 - 2x}$. Express your answer in words and using interval notation. [Section 8.8] All real numbers except 0 and 2; $(-\infty, 0) \cup (0, 2) \cup (2, \infty)$

105. Match each function with the correct graph shown below.

a. $f(x) = x$ iii. [Section 8.2]

b. $f(x) = x^2$ i. [Section 8.3]

c. $f(x) = x^3$ iv. [Section 8.3]

d. $f(x) = \frac{1}{x}$ ii. [Section 8.8]

i.

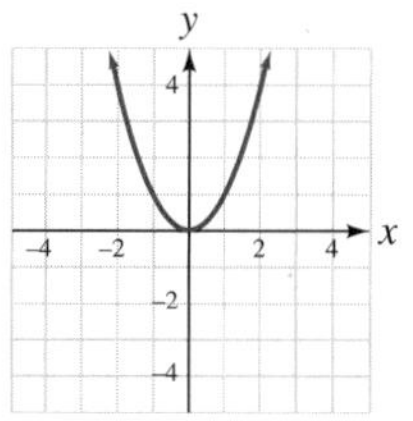

ii.

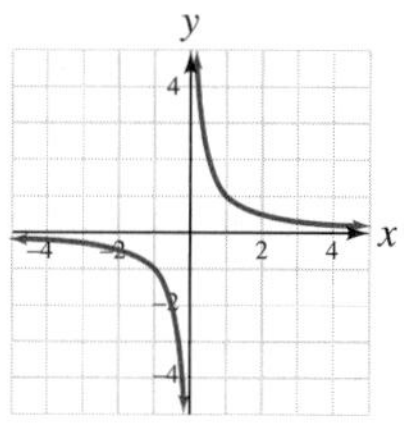

iii.

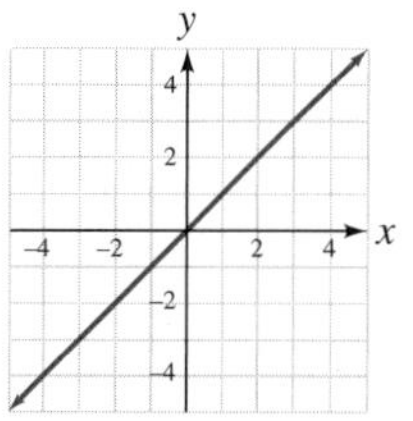

iv.

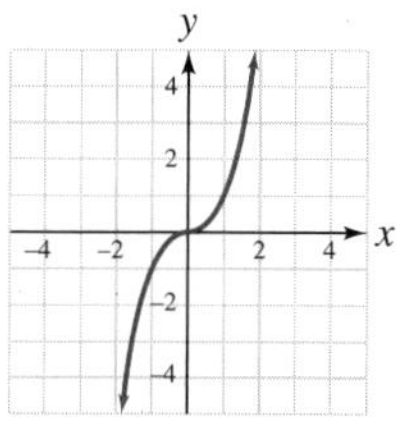

106. **Deliveries.** The costs of a delivery company vary jointly with the number of trucks in service and the number of hours they are used. When 8 trucks are used for 12 hours each, the costs are \$3,600. Find the costs of using 20 trucks, each for 12 hours. [Section 8.9] \$9,000

107. Graph the function $f(x) = \sqrt{x - 2}$ and give its domain and range. [Section 9.1] D: $[2, \infty)$; R: $[0, \infty)$; see AIE Appendix 3.

108. Find the domain of the function $f(x) = \sqrt{0.4x - 36}$. [Section 9.1] D: $[90, \infty)$

Simplify each expression. All variables represent real numbers.

109. $\sqrt[3]{-27x^3}$ [Section 9.1] $-3x$

110. $\sqrt{48t^3}$ [Section 9.1] $4t\sqrt{3t}$

111. $64^{-2/3}$ [Section 9.2] $\frac{1}{16}$

112. $\frac{x^{5/3}x^{1/2}}{x^{3/4}}$ [Section 9.2] $x^{17/12}$

113. $-3\sqrt[4]{32} - 2\sqrt[4]{162} + 5\sqrt[4]{48}$ [Section 9.3] $-12\sqrt[4]{2} + 10\sqrt[4]{3}$

114. $3\sqrt{2}\left(2\sqrt{3} - 4\sqrt{12}\right)$ [Section 9.4] $-18\sqrt{6}$

115. $\frac{\sqrt{x}+2}{\sqrt{x}-1}$ [Section 9.4] $\frac{x+3\sqrt{x}+2}{x-1}$

116. $\frac{5}{\sqrt[3]{x}}$ [Section 9.4] $\frac{5\sqrt[3]{x^2}}{x}$

Solve each equation.

117. $5\sqrt{x+2} = x + 8$ [Section 9.5] 2, 7

118. $\sqrt{x} + \sqrt{x+2} = 2$ [Section 9.5] $\frac{1}{4}$

119. Let $f(x) = \sqrt[3]{x^2 + 2x}$ and $g(x) = 2\sqrt[3]{x-1}$. Find all values of x for which $f(x) = g(x)$. [Section 9.5] 2, 4

120. **a.** Find the length of the hypotenuse of the right triangle in figure (a). [Section 9.6] $3\sqrt{2}$ in.

b. Find the length of the hypotenuse of the right triangle in figure (b). [Section 9.6] $2\sqrt{3}$ in.

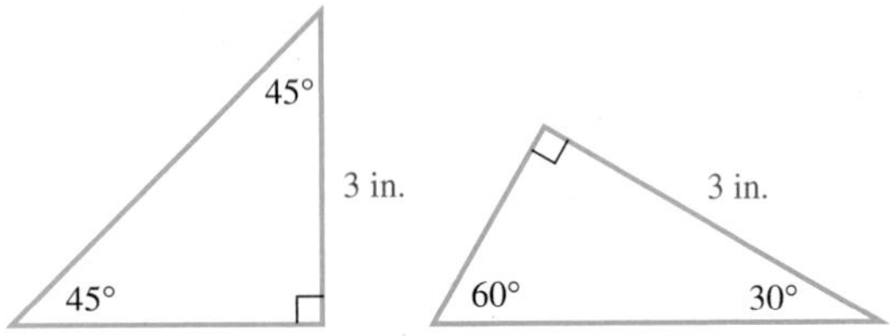

(a) (b)

121. **a.** Find the distance between $(-2, 6)$ and $(4, 14)$. [Section 9.6] 10

b. Find the midpoint of the line segment with endpoints $(7, 1)$ and $(-10, 4)$. [Section 9.6] $\left(-\frac{3}{2}, \frac{5}{2}\right)$

122. Simplify: i^{43} [Section 9.7] $-i$

Perform the indicated operations. Write each result in a + bi form.

123. $\left(-7 + \sqrt{-81}\right) - \left(-2 - \sqrt{-64}\right)$ [Section 9.7] $-5 + 17i$

124. $\frac{5}{3 - i}$ [Section 9.7] $\frac{3}{2} + \frac{1}{2}i$

125. $(2 + i)^2$ [Section 9.7] $3 + 4i$

126. $\frac{-4}{6i^7}$ [Section 9.7] $0 - \frac{2}{3}i$

Solve each equation.

127. $x^2 = 28$ [Section 10.1] $\pm 2\sqrt{7}$

128. $(x - 19)^2 = -5$ [Section 10.1] $19 \pm i\sqrt{5}$

129. Use the method of completing the square to solve $2x^2 - 6x + 3 = 0$. [Section 10.1] $\frac{3 \pm \sqrt{3}}{2}$

130. Use the quadratic formula to solve $a^2 - \frac{2}{5}a = -\frac{1}{5}$. [Section 10.2] $\frac{1}{5} \pm \frac{2}{5}i$

131. **Community Gardens.** Residents of a community can work their own 16-ft × 24-ft plot of city-owned land if they agree to the following conditions:

- The area of the garden cannot exceed 180 square feet.
- A path of uniform width must be maintained around the garden.

Find the dimensions of the largest possible garden. [Section 10.2] 10 ft by 18 ft

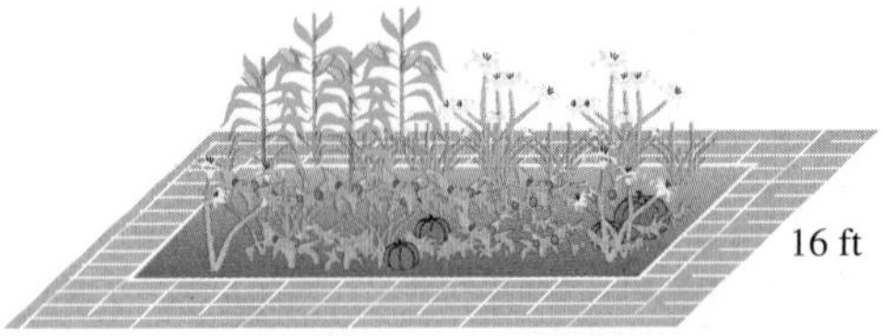

132. **Sidewalks.** A 170-meter-long sidewalk from the mathematics building M to the student center C is shown in red in the illustration. However, students prefer to walk directly from M to C, across a lawn. How long are the two segments of the existing sidewalk? [Section 10.2] 50 m and 120 m

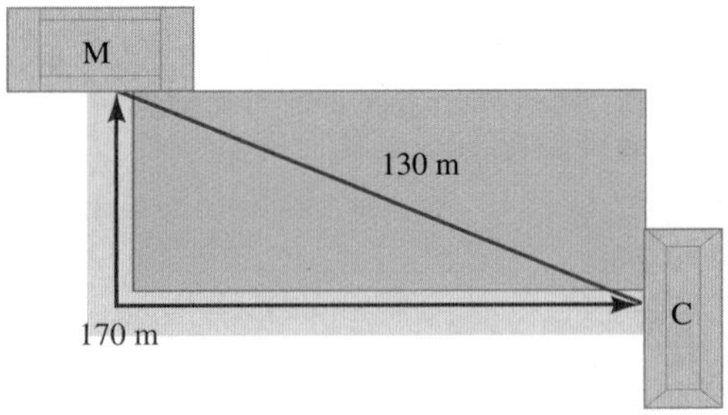

Solve each equation.

133. $t^{2/3} - t^{1/3} = 6$ [Section 10.3] $-8, 27$

134. $x^{-4} - 2x^{-2} + 1 = 0$ [Section 10.3] Repeated solutions of -1 and 1

135. First determine the vertex and the axis of symmetry of the graph of $f(x) = -x^2 - 4x$ using the vertex formula. Then determine the x- and y-intercepts of the graph. Finally, plot several points and complete the graph. [Section 10.4] $(-2, 4)$, $x = -2$; $(-4, 0)$, $(0, 0)$; $(0, 0)$; see AIE Appendix 3.

Solve each inequality. Write the solution set in interval notation and then graph it. See AIE Appendix 3.

136. $x^2 - 81 < 0$ [Section 10.5] $(-9, 9)$

137. $\frac{1}{x+1} \geq \frac{x}{x+4}$ [Section 10.5] $(-4, -2] \cup (-1, 2]$

138. **a.** The graph of $f(x) = 16x^2 + 24x + 9$ is shown below. Estimate the solution(s) of $16x^2 + 24x + 9 = 0$. [Section 10.5] $-\frac{3}{4}$

b. Use the graph to determine the solution of $16x^2 + 24x + 9 < 0$. [Section 10.5] No solution

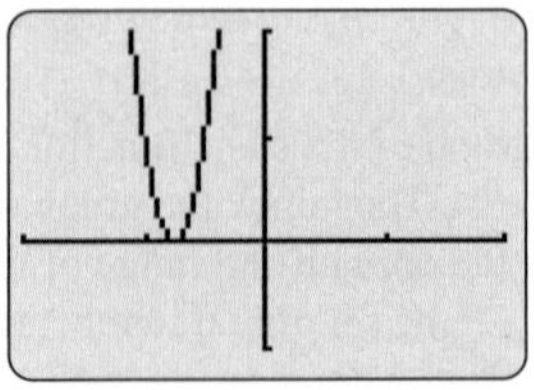

Exponential and Logarithmic Functions

11

from Campus to Careers

Social Worker

For those with a desire to help improve other people's lives, social work is one career option to consider. Social workers offer guidance and counseling to people in crisis. They must be critical thinkers—able to use their logic and reasoning to brainstorm alternate solutions to problems faced by their clients. Social workers use their mathematical skills to construct family budgets, plan personnel schedules, gather and interpret data, and comprehend the statistical methods used in research studies.

Problem 49 in **Study Set 11.3**, **problem 115** in **Study Set 11.4**, and **problem 87** in **Study Set 11.5** involve situations that a social worker encounters on the job. The mathematical concepts discussed in this chapter can be used to solve those problems.

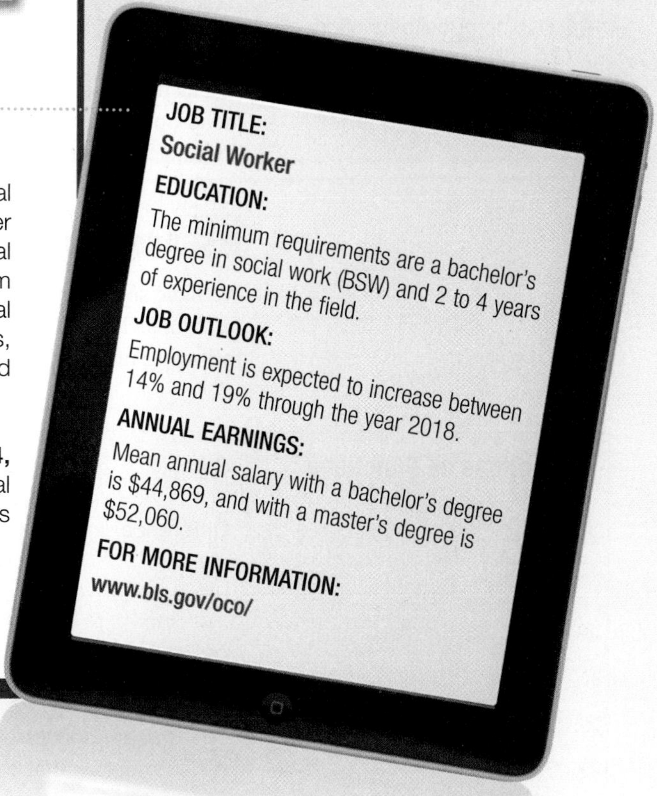

JOB TITLE:
Social Worker
EDUCATION:
The minimum requirements are a bachelor's degree in social work (BSW) and 2 to 4 years of experience in the field.
JOB OUTLOOK:
Employment is expected to increase between 14% and 19% through the year 2018.
ANNUAL EARNINGS:
Mean annual salary with a bachelor's degree is $44,869, and with a master's degree is $52,060.
FOR MORE INFORMATION:
www.bls.gov/oco/

Study Skills Workshop ▶ Participating in Class

One of the keys to success in algebra is to learn as much as you can in class. To get the most out of class meetings, and to make them more enjoyable, you should participate in the following ways:

ASK QUESTIONS: During class, clear up any questions that may arise from your homework assignments or from your instructor's lectures. Also, pay close attention when other students ask questions. You never know when you might face the same difficulty.

ANSWER QUESTIONS: Many instructors direct questions to the class while lecturing. Take advantage of this opportunity to increase your knowledge by attempting to answer all such questions from your instructor.

INTERACT WITH CLASSMATES: Before class begins and after class ends, regularly discuss the material that you are studying with fellow classmates.

Now Try This

1. List the reasons why you do not feel comfortable asking questions in class.
2. While working on your next homework assignment, write down any questions that occur to you so that you will not forget to ask them in class.
3. Exchange a written question about a homework problem with a classmate. See if you can answer each other's question.

SECTION 11.1 Algebra and Composition of Functions

OBJECTIVES

1. Add, subtract, multiply, and divide functions.
2. Find the composition of functions.
3. Use graphs to evaluate functions.
4. Use composite functions to solve problems.

ARE YOU READY? *Are You Ready? exercises available online at www.webassign.net/brookscole*

The following problems review some basic skills that are needed when performing the algebra of functions.

1. Add: $(7x^2 + 5x - 12) + (2x^2 - 3x + 4)$ $9x^2 + 2x - 8$
2. Subtract: $(8x^3 - 4x^2) - (6x^3 - 3x^2)$ $2x^3 - x^2$
3. Multiply: $(8x - 5)(9x^2 - 4)$ $72x^3 - 45x^2 - 32x + 20$
4. Divide: $(x^2 - x - 6) \div (x - 3)$ $x + 2$

Just as it is possible to perform arithmetic operations on real numbers, it is possible to perform those operations on functions. We call the process of adding, subtracting, multiplying, and dividing functions the **algebra of functions.**

1 Add, Subtract, Multiply, and Divide Functions.

We have seen that the sum, difference, product, and quotient of two functions are themselves functions. The new functions that result from such operations can be represented using the following notation.

Operations on Functions

Sum: $(f + g)(x) = f(x) + g(x)$ Read as "f plus g of x equals f of x plus g of x."

Difference: $(f - g)(x) = f(x) - g(x)$ Read as "f minus g of x equals f of x minus g of x."

Product: $(f \cdot g)(x) = f(x)g(x)$ Read as "f times g of x equals f of x times g of x."

Quotient: $(f/g)(x) = \dfrac{f(x)}{g(x)}$ where $g(x) \neq 0$ Read as "f divided by g of x equals f of x divided by g of x."

The domain of each sum, difference, and product function shown above is the set of real numbers x that are in the domains of both f and g. The domain of the quotient function is the set of real numbers x that are the domains of f and g, excluding any values of x where $g(x) = 0$.

EXAMPLE 1 Let $f(x) = 2x^2 + 1$ and $g(x) = 5x - 3$. Find each function and give its domain:
a. $f + g$ **b.** $f - g$ **c.** $f \cdot g$ **d.** f/g

Strategy We will add, subtract, multiply, and divide the functions as if they were binomials.

Why We add because of the plus symbol in $f + g$, we subtract because of the minus symbol in $f - g$, we multiply because of the raised dot in $f \cdot g$, and we divide because of the fraction bar in f/g.

Solution

Teaching Tip: Before beginning your explanation of this example, ask your students to tell you the domain of f and the domain of g.

a. $(f + g)(x) = f(x) + g(x)$ — This is the definition of a sum function.
$= (2x^2 + 1) + (5x - 3)$ — Replace $f(x)$ with $2x^2 + 1$ and $g(x)$ with $5x - 3$.
$= 2x^2 + 5x - 2$ — Drop the parentheses and combine like terms.

The domain of $f + g$ is the set of real numbers that are in the domain of both f and g. Since the domain of both f and g is $(-\infty, \infty)$, the domain of $f + g$ is the $(-\infty, \infty)$.

b. $(f - g)(x) = f(x) - g(x)$ — This is the definition of a difference function.
$= (2x^2 + 1) - (5x - 3)$ — Replace $f(x)$ with $2x^2 + 1$ and $g(x)$ with $5x - 3$.
$= 2x^2 + 1 - 5x + 3$ — Change the sign of each term of $5x - 3$ and drop the parentheses.
$= 2x^2 - 5x + 4$ — Combine like terms.

Since the domain of both f and g is $(-\infty, \infty)$, the domain of $f - g$ is $(-\infty, \infty)$.

Notation

The parentheses around $\frac{3}{5}$ in the notation $\left(-\infty, \frac{3}{5}\right) \cup \left(\frac{3}{5}, \infty\right)$ indicate that $\frac{3}{5}$ is not included in the domain of f/g. The domain could also be shown graphically:

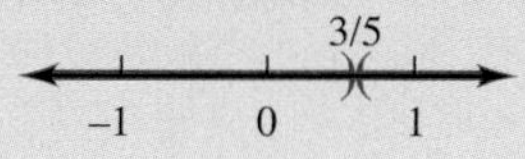

c. $(f \cdot g)(x) = f(x) \cdot g(x)$ — This is the definition of a product function.
$= (2x^2 + 1)(5x - 3)$ — Replace $f(x)$ with $2x^2 + 1$ and $g(x)$ with $5x - 3$.
$= 10x^3 - 6x^2 + 5x - 3$ — Multiply the binomials. There are no like terms.

The domain of $f \cdot g$ is the set of real numbers that are in the domain of both f and g. Since the domain of both f and g is $(-\infty, \infty)$, the domain of $f \cdot g$ is $(-\infty, \infty)$.

d. $(f/g)(x) = \dfrac{f(x)}{g(x)}$ — This is the definition of a quotient function.
$= \dfrac{2x^2 + 1}{5x - 3}$ — Replace $f(x)$ with $2x^2 + 1$ and $g(x)$ with $5x - 3$. The result does not simplify.

Since the denominator of the fraction cannot be 0, it follows that this function is undefined if $5x - 3 = 0$. If we solve for x, we see that x cannot be $\frac{3}{5}$. Thus, the domain of f/g is the union of two intervals: $\left(-\infty, \frac{3}{5}\right) \cup \left(\frac{3}{5}, \infty\right)$.

Teaching Example 1
Let $f(x) = 5x^2 + 4$ and $g(x) = 7x - 2$. Find each function and give its domain:
a. $f + g$ b. $f - g$ c. $f \cdot g$ d. f/g
Answers:
a. $(f + g)(x) = 5x^2 + 7x + 2$; D: $(-\infty, \infty)$
b. $(f - g)(x) = 5x^2 - 7x + 6$; D: $(-\infty, \infty)$
c. $(f \cdot g)(x) = 35x^3 - 10x^2 + 28x - 8$; D: $(-\infty, \infty)$
d. $(f/g)(x) = \frac{5x^2 + 4}{7x - 2}$; D: $\left(-\infty, \frac{2}{7}\right) \cup \left(\frac{2}{7}, \infty\right)$

Self Check 1 Let $f(x) = 3x - 2$ and $g(x) = 2x^2 + 3x$. Find each function and give its domain:
a. $f + g$ $(f + g)(x) = 2x^2 + 6x - 2$; D: $(-\infty, \infty)$
b. $f - g$ $(f - g)(x) = -2x^2 - 2$; D: $(-\infty, \infty)$
c. $f \cdot g$ $(f \cdot g)(x) = 6x^3 + 5x^2 - 6x$; D: $(-\infty, \infty)$
d. f/g $(f/g)(x) = \frac{3x - 2}{2x^2 + 3x}$; D: $\left(-\infty, -\frac{3}{2}\right) \cup \left(-\frac{3}{2}, 0\right) \cup (0, \infty)$

Now Try Problems 13, 15, 17, and 19

EXAMPLE 2 Use the results from Example 1 to find: **a.** $(f + g)(-3)$ **b.** $(f - g)(6)$ **c.** $(f \cdot g)(0)$ **d.** $(f/g)(10)$

Strategy We will substitute the given values within the second set of parentheses for each x in the sum, difference, product, and quotient functions found in Example 1. Then we will evaluate the right side of each equation.

Why The number that is within the second set of parentheses is the input of the function.

Solution

a. $(f + g)(x) = 2x^2 + 5x - 2$ This is the sum function found in Example 1, part a.

$(f + g)(\mathbf{-3}) = 2(\mathbf{-3})^2 + 5(\mathbf{-3}) - 2$ Substitute -3 for each x.

$= 18 + (-15) - 2$ Evaluate the right side.

$= 1$

b. $(f - g)(x) = 2x^2 - 5x + 4$ This is the difference function found in Example 1, part b.

$(f - g)(\mathbf{6}) = 2(\mathbf{6})^2 - 5(\mathbf{6}) + 4$ Substitute 6 for each x.

$= 72 - 30 + 4$ Evaluate the right side.

$= 46$

c. $(f \cdot g)(x) = 10x^3 - 6x^2 + 5x - 3$ This is the product function found in Example 1, part c.

$(f \cdot g)(\mathbf{0}) = 10(\mathbf{0})^3 - 6(\mathbf{0})^2 + 5(\mathbf{0}) - 3$ Substitute 0 for each x.

$= -3$ Evaluate the right side.

d. $(f/g)(x) = \dfrac{2x^2 + 1}{5x - 3}$ This is the quotient function found in Example 1, part d.

$(f/g)(\mathbf{10}) = \dfrac{2(\mathbf{10})^2 + 1}{5(\mathbf{10}) - 3}$ Substitute 10 for each x.

$= \dfrac{2(100) + 1}{50 - 3}$ Evaluate the right side.

$= \dfrac{201}{47}$

Teaching Example 2 Use the results from Teaching Example 1 to find:
a. $(f + g)(4)$ b. $(f - g)(-1)$
c. $(f \cdot g)(0)$ d. $(f/g)(2)$

Answers: a. 110 b. 18 c. -8
d. 2

Self Check 2 Use the results from Self Check 1 to find: **a.** $(f + g)(5)$ 78 **b.** $(f - g)(-9)$ -164 **c.** $(f \cdot g)(2)$ 56 **d.** $(f/g)(0)$ Undefined

Now Try Problems 29 and 31

There is a relationship that can be seen between the graphs of two functions and the graph of their sum (or difference) function. For example, in the following illustration, the graph of $f + g$, which gives the total number of elementary and secondary students in the United States, can be found by adding the graph of f, which gives the number of secondary students, to the graph of g, which gives the number of elementary students. For any given x-value, we simply add the two corresponding y-values to get the graph of the sum function $f + g$.

Actual and Projected Enrollment in U. S. Public and Private Elementary and Secondary Schools

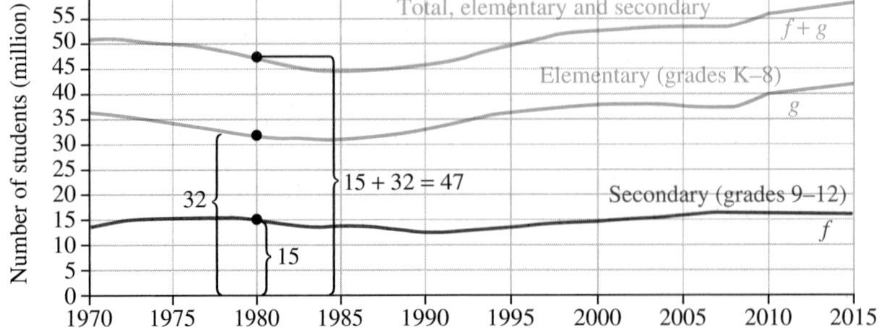

Source: U.S. Department of Education

To find the total enrollment in 1980, add the elementary enrollment of 32 million to the secondary enrollment of 15 million to get 47 million.

2 Find the Composition of Functions.

We have seen that a function can be represented by a machine: We put in a number from the domain, and a number from the range comes out. For example, if we put the number 2 into the machine shown on the right, the number $f(2) = 8$ comes out. In general, if we put x into the machine, the value $f(x)$ comes out.

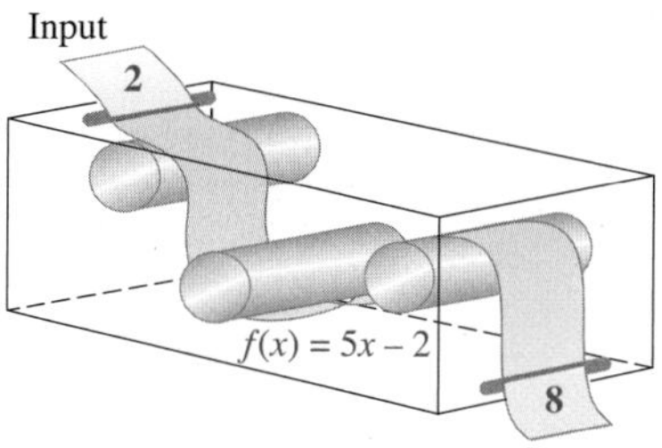

Often one quantity is a function of a second quantity that depends, in turn, on a third quantity. For example, the cost of a car trip is a function of the gasoline consumed. The amount of gasoline consumed, in turn, is a function of the number of miles driven. Such chains of dependence can be analyzed mathematically as **compositions of functions.**

The Language of Algebra

A **composite** is something that is formed by bringing together distinct parts. For example, a police artist creates a *composite* sketch from eyewitness descriptions of the suspect.

Suppose that $y = f(x)$ and $y = g(x)$ define two functions. Any number x in the domain of g will produce the corresponding value $g(x)$ in the range of g. If $g(x)$ is in the domain of function f, then $g(x)$ can be substituted into f, and a corresponding value $f(g(x))$ will be determined. Because of the **nested parentheses** in $f(g(x))$, we read it as "f of g of x." This two-step process defines a new function, called a **composite function,** denoted by $f \circ g$. (This is read as "f composed with g" or "the composition of f and g" or "f circle g.")

The function machines shown below illustrate the composition $f \circ g$. When we put a number into the function g, a value $g(x)$ comes out. The value $g(x)$ then goes into function f, which transforms $g(x)$ into $f(g(x))$. If the function machines for g and f were connected to make a single machine, that machine would be named $f \circ g$.

The Language of Algebra

$f(\underbrace{g(x)})$

We call these **nested parentheses.**

Teaching Tip: The illustration below can also be used to show the "flow" of the composition function.

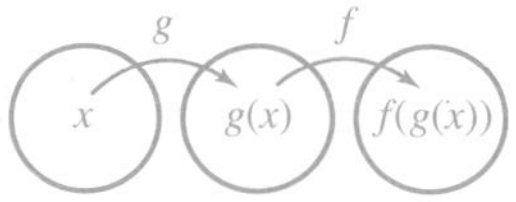

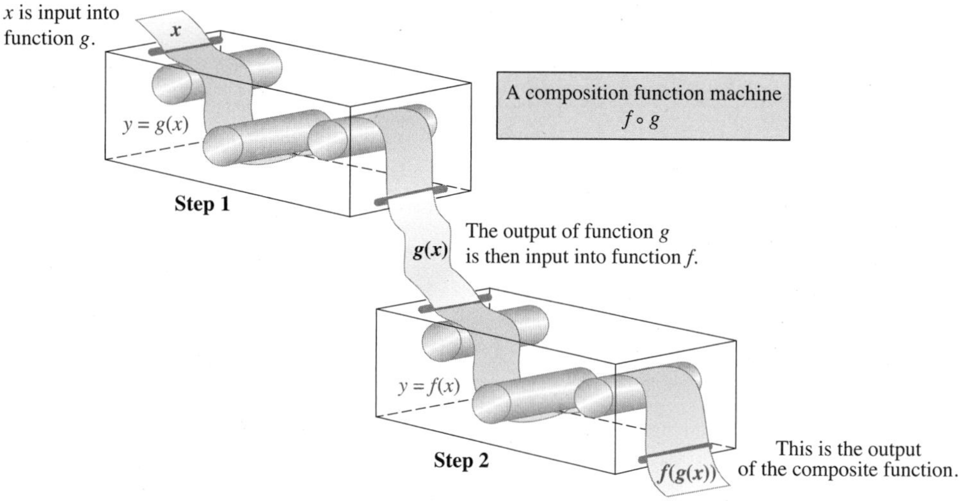

To be in the domain of the composite function $f \circ g$, a number x has to be in the domain of g and the output of g must be in the domain of f. Thus, the domain of $f \circ g$ consists of those numbers x that are in the domain of g, and for which $g(x)$ is in the domain of f.

Composite Functions

The composite function $f \circ g$ is defined by

$$(f \circ g)(x) = f(g(x))$$

If $f(x) = 4x$ and $g(x) = 3x + 2$, to find $f \circ g$ and $g \circ f$, we proceed as follows.

$$\begin{aligned}(f \circ g)(x) &= f(g(x)) \\ &= f(3x + 2) \\ &= 4(3x + 2) \\ &= 12x + 8\end{aligned} \qquad \begin{aligned}(g \circ f)(x) &= g(f(x)) \\ &= g(4x) \\ &= 3(4x) + 2 \\ &= 12x + 2\end{aligned}$$

Different results

Caution

$(f \circ g)(x)$ — Composition

does not mean

$(f \cdot g)(x)$ — Multiplication

The different results illustrate that the composition of functions is not commutative. Usually, we will find that $(f \circ g)(x) \neq (g \circ f)(x)$.

EXAMPLE 3 Let $f(x) = 2x + 1$ and $g(x) = x - 4$. Find: **a.** $(f \circ g)(9)$ **b.** $(f \circ g)(x)$ **c.** $(g \circ f)(-2)$

Strategy In part (a), we will find $f(g(9))$. In part (b), we will find $f(g(x))$. In part (c), we will find $g(f(-2))$.

Why To evaluate a composite function written with the circle $\circ$ notation, we rewrite it using nested parentheses.

Solution **a.** $(f \circ g)(9)$ means $f(g(9))$, where the function g is applied first and function f is applied second.

$$(f \circ g)(9) = f(\underbrace{g(9)}_{\text{Apply first}})$$

In figure (a) on the next page, we see that function g receives the number 9, subtracts 4, and releases the number $g(9) = 5$. Then 5 goes into the f function, which doubles 5 and adds 1. The final result, 11, is the output of the composite function $f \circ g$.

Read as "f composed with g of 9." ↓ $(f \circ g)(9)$; Read as "f of g of 9." ↓ $f(g(9))$

$(f \circ g)(9) = f(g(9))$	Change from $\circ$ notation to nested parentheses notation.
$= f(5)$	Evaluate: $g(9) = 9 - 4 = 5$.
$= 2(5) + 1$	Evaluate $f(5)$ using $f(x) = 2x + 1$.
$= 11$	

Thus, $(f \circ g)(9) = 11$.

b. $(f \circ g)(x)$ means $f(g(x))$. In figure (a) on the next page, function g receives the number x, subtracts 4, and releases the number $x - 4$. Then $x - 4$ goes into the f function, which doubles $x - 4$ and adds 1. The final result, $2x - 7$, is the output of the composite function $f \circ g$.

Read as "f composed with g of x." ↓ $(f \circ g)(x)$; Read as "f of g of x." ↓ $f(g(x))$

$(f \circ g)(x) = f(g(x))$	Change from $\circ$ notation to nested parentheses notation.
$= f(x - 4)$	We are given $g(x) = x - 4$. Replace $g(x)$ with $x - 4$.
$= 2(x - 4) + 1$	Find $f(x - 4)$ using $f(x) = 2x + 1$.
$= 2x - 8 + 1$	Distribute the multiplication by 2.
$= 2x - 7$	Combine like terms.

Thus, $(f \circ g)(x) = 2x - 7$.

c. $(g \circ f)(-2)$ means $g(f(-2))$. In figure (b) on the next page, function f receives the number -2, doubles it and adds 1, and releases -3 into the g function. Function g subtracts 4 from -3 and outputs a final result of -7. Thus,

Read as "g composed with f of −2." ↓ $(g \circ f)(-2)$; Read as "g of f of −2." ↓ $g(f(-2))$

$(g \circ f)(-2) = g(f(-2))$	Change from $\circ$ notation to nested parentheses notation.
$= g(-3)$	Evaluate $f(-2)$ using $f(x) = 2x + 1$.
$= -3 - 4$	Evaluate $g(-3)$ using $g(x) = x - 4$.
$= -7$	Do the subtraction.

Thus, $(g \circ f)(-2) = -7$.

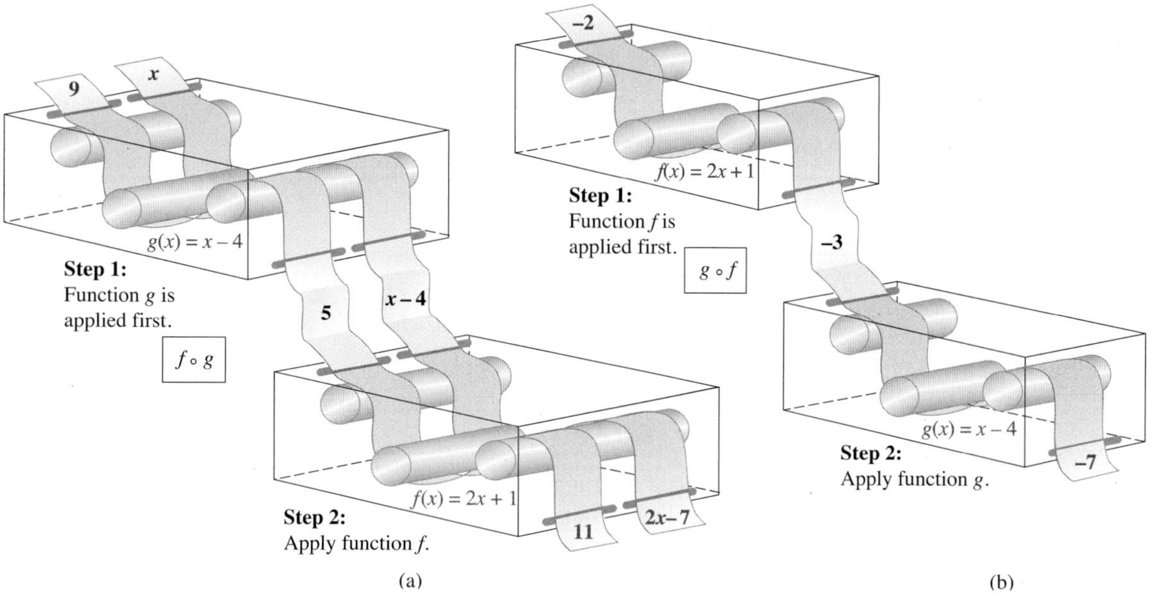

Teaching Example 3 Let $f(x) = 5x + 2$ and $g(x) = 2x - 1$. Find: **a.** $(f \circ g)(3)$ **b.** $(f \circ g)(x) = (f \circ g)(x)$ **c.** $(g \circ f)(-2)$
Answers: **a.** 27 **b.** $10x - 3$ **c.** -17

Self Check 3 Let $f(x) = x^3$ and $g(x) = 6 - x$. Find: **a.** $(f \circ g)(8)$ -8 **b.** $(g \circ f)(1)$ 5 **c.** $(g \circ f)(x)$ $(g \circ f)(x) = 6 - x^3$

Now Try Problems 37, 39, and 43

3 Use Graphs to Evaluate Functions.

EXAMPLE 4 Refer to the graphs of functions f and g on the left to find each of the following.
a. $(f + g)(-4)$ **b.** $(f \cdot g)(2)$ **c.** $(f \circ g)(-3)$

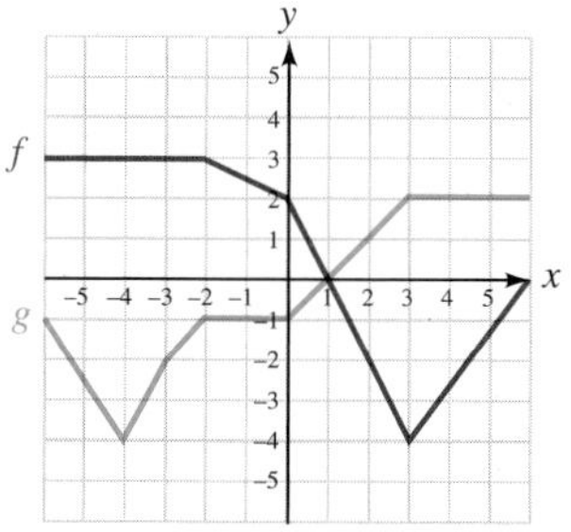

Strategy We will express the sum, product, and composite functions using the functions from which they are formed.

Why We can evaluate sum, product, and composite functions at a given x-value by evaluating each function from which they are formed at that x-value.

Solution

a. The value of $f(-4)$ is found by looking in quadrant II of the graph and the value of $g(-4)$ by looking in quadrant III.

$$\begin{aligned}(f + g)(-4) &= \mathbf{f(-4)} + \mathbf{g(-4)} \\ &= \mathbf{3} + (\mathbf{-4}) \\ &= -1\end{aligned}$$

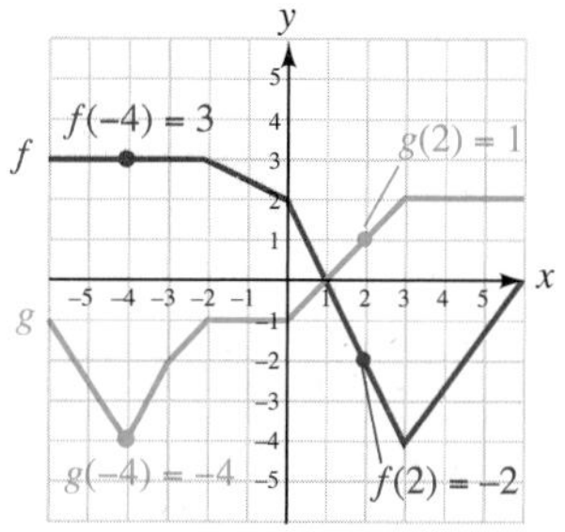

b. The value of $f(2)$ is found by looking in quadrant IV of the graph and the value of $g(2)$ by looking in quadrant I.

$$\begin{aligned}(f \cdot g)(2) &= \mathbf{f(2)} \cdot \mathbf{g(2)} \\ &= \mathbf{-2} \cdot \mathbf{1} \\ &= -2\end{aligned}$$

c. The value of $g(-3)$ is found by looking in quadrant III of the graph and the value of $f(-2)$ by looking in quadrant II.

$$\begin{aligned}(f \circ g)(-3) &= f(\mathbf{g(-3)}) \quad \text{Change from } \circ \text{ notation to nested parentheses notation.} \\ &= f(\mathbf{-2}) \\ &= 3\end{aligned}$$

Teaching Example 4 Refer to the graph in Example 4 to find each of the following:
a. $(f + g)(2)$ b. $(f \cdot g)(-4)$
c. $(g \circ f)(-3)$
Answers: a. -1 b. -12 c. 2

Self Check 4 Refer to the graph in Example 4 to find each of the following.

a. $(f - g)(3)$ -6 **b.** $\left(\frac{f}{g}\right)(-2)$ -3 **c.** $(g \circ f)(3)$ -4

Now Try Problems 45 and 47

EXAMPLE 5 If $h(x) = (4x - 6)^2$, find f and g such that $h(x) = (f \circ g)(x)$.

Strategy We will determine the order in which we would evaluate function h for a given value of x.

Why If we can see the evaluation as a two-step process, it will help us determine the unknown functions f and g.

Solution If we were to evaluate function h for a given value of x, we would first find $4x - 6$, and then we would square that result. This suggests that the first function that operates in the composition, function g, should receive an input x and produce $4x - 6$. Thus, $g(x) = 4x - 6$. The second function that operates in the composition, function f, should receive an input and square it. Thus, $f(x) = x^2$.

We can check these results by forming the composition function.

$$(f \circ g)(x) = f(g(x)) \quad \text{Change from} \circ \text{notation to nested parentheses.}$$
$$= f(4x - 6) \quad \text{Replace } g(x) \text{ with } 4x - 6.$$
$$= (4x - 6)^2 \quad \text{Find } f(4x - 6) \text{ using } f(x) = x^2.$$

The composition of functions $f(x) = x^2$ and $g(x) = 4x - 6$ does indeed produce $h(x) = (4x - 6)^2$. It is important to note that there are many other possibilities for f and g, but the ones that we found here are the most obvious.

Teaching Example 5 If $h(x) = (7 - 4x)^3$, find f and g such that $h(x) = (f \circ g)(x)$.
Answer: $f(x) = x^3$, $g(x) = 7 - 4x$

Self Check 5 If $h(x) = \sqrt{x + 15}$, find f and g such that $h(x) = (f \circ g)(x)$.
$f(x) = \sqrt{x}$, $g(x) = x + 15$

Now Try Problems 49 and 51

4 Use Composite Functions to Solve Problems.

EXAMPLE 6 **Biological Research.** A specimen is stored in refrigeration at a temperature of 15° Fahrenheit. Biologists remove the specimen and warm it at a controlled rate of 3°F per hour. Express its Celsius temperature as a function of the time t since it was removed from refrigeration.

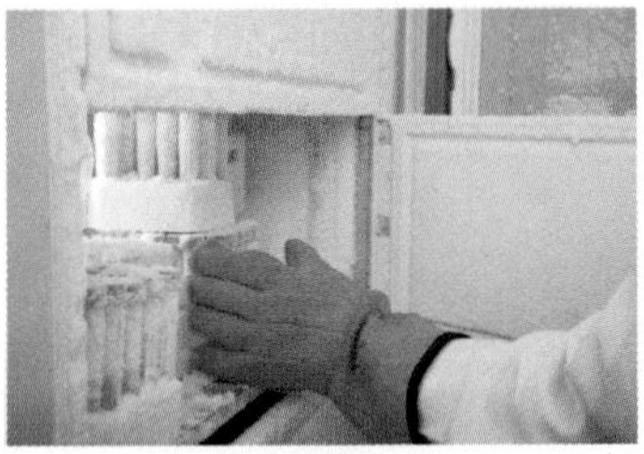

Strategy We will express the Fahrenheit temperature of the specimen as a function of the time t since it was removed from refrigeration. Then we will express the Celsius temperature of the specimen as a function of its Fahrenheit temperature and find the composition of the two functions.

Why The Celsius temperature of the specimen is a function of its Fahrenheit temperature. Its Fahrenheit temperature is a function of the time since it was removed from refrigeration. This chain of dependence suggests that we write a composition of functions.

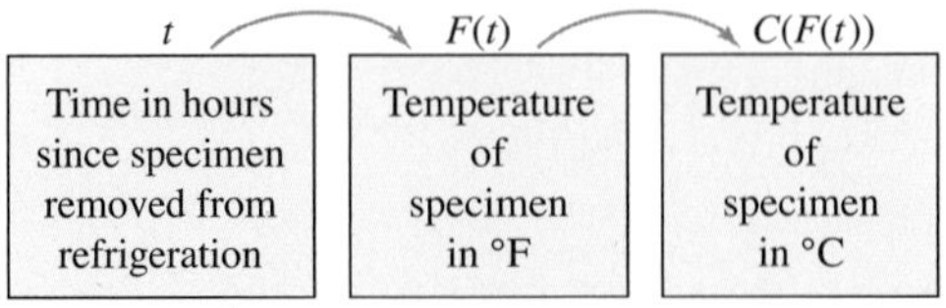

Solution The temperature of the specimen is 15°F when the time $t = 0$. Because it warms at a rate of 3°F per hour, its initial temperature of 15°F increases by $3t$°F in t hours. The Fahrenheit temperature at time t of the specimen is given by the function

$$F(t) = 3t + 15$$

The Celsius temperature C is a function of this Fahrenheit temperature F, given by the function

$$C(F) = \frac{5}{9}(F - 32)$$

To express the specimen's Celsius temperature as a function of *time,* we find the composite function $(C \circ F)(t)$.

$$(C \circ F)(t) = C(F(t))$$ Change from $\circ$ notation to nested parentheses notation.

$$= C(3t + 15)$$ Substitute $3t + 15$ for $F(t)$.

$$= \frac{5}{9}\left(3t + 15 - 32\right)$$ Find $C(3t + 15)$ by substituting $3t + 15$ for F in $C(F) = \frac{5}{9}(F - 32)$.

$$= \frac{5}{9}(3t - 17)$$ Subtract within the parentheses: $15 - 32 = -17$.

$$= \frac{15}{9}t - \frac{85}{9}$$ Distribute the multiplication by $\frac{5}{9}$.

$$= \frac{5}{3}t - \frac{85}{9}$$ Simplify $\frac{15}{9}$: $\frac{\overset{1}{\cancel{3}} \cdot 5}{\underset{1}{\cancel{3}} \cdot 3} = \frac{5}{3}$.

The composite function, $C(t) = \frac{5}{3}t - \frac{85}{9}$, gives the temperature of the specimen in degrees Celsius t hours after it is removed from refrigeration.

Teaching Example 6 Weather Forecasting. A changing weather pattern is bringing in warmer weather for the next 24 hours. The temperature is now 50° Fahrenheit and is expected to increase 6° every 5 hours. Write a composite function that expresses the Celsius temperature as a function of the number of hours from now.

Answer: $C(t) = \frac{2}{3}t + 10$

Self Check 6 **Weather Forecasting.** A low-pressure area is bringing in colder weather for the next 12 hours. The temperature is now 86° Fahrenheit and is expected to fall 3° every 2 hours. Write a composite function that expresses the Celsius temperature as a function of the number of hours from now. $C(t) = -\frac{5}{6}t + 30$

Now Try Problem 87

SECTION 11.1 STUDY SET

VOCABULARY

Fill in the blanks.

1. The __sum__ of f and g, denoted as $f + g$, is defined by $(f + g)(x) =$ __$f(x) + g(x)$__ and the __difference__ of f and g, denoted as $f - g$, is defined by $(f - g)(x) =$ __$f(x) - g(x)$__.
2. The __product__ of f and g, denoted as $f \cdot g$, is defined by $(f \cdot g)(x) =$ __$f(x)g(x)$__ and the __quotient__ of f and g, denoted as f/g, is defined by $(f/g)(x) = \frac{f(x)}{g(x)}$.
3. The __domain__ of the function $f + g$ is the set of real numbers x that are in the domain of both f and g.
4. The __composite__ function $f \circ g$ is defined by $(f \circ g)(x) =$ __$f(g(x))$__.
5. When we write $(f \circ g)(x)$ as $f(g(x))$, we have changed from $\circ$ notation to __nested__ parentheses notation.
6. When reading the notation $f(g(x))$, we say "f __of__ g __of__ x."

CONCEPTS

7. Fill in the blanks.
   a. $(f \circ g)(3) = f($__$g(3)$__$)$
   b. To find $f(g(3))$, we first find __$g(3)$__ and then substitute that value for x in $f(x)$.
8. a. If $f(x) = 3x + 1$ and $g(x) = 1 - 2x$, find $f(g(3))$ and $g(f(3))$. $-14, -19$
   b. Is the composition of functions commutative? No

9. Fill in the three blanks in the drawing of the function machines that show how to compute $g(f(-2))$.

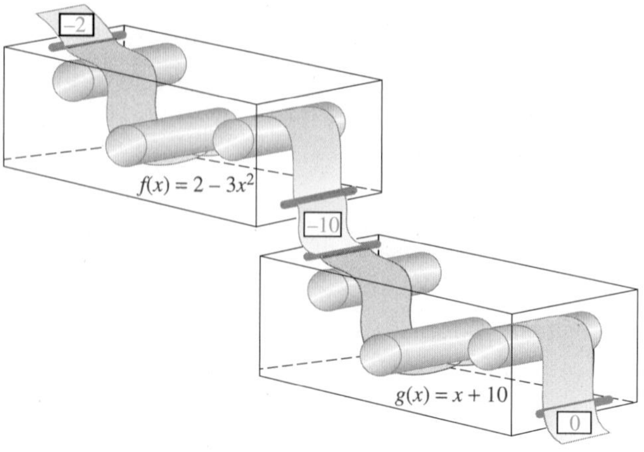

10. If $f(x) = x^2 + 3$ and $g(x) = x - 4$, then $(f/g)(x) = \dfrac{x^2 + 3}{x - 4}$.

a. What value of x makes $g(x) = 0$? 4

b. Fill in the blank: The domain of f/g is $(-\infty, 4) \cup (4, \infty)$.

NOTATION

Complete each solution.

▶ 11. Let $f(x) = 3x - 1$ and $g(x) = 2x + 3$. Find $f \cdot g$.

$$\begin{aligned}(f \cdot g)(x) &= f(x) \cdot g(x)\\ &= (3x - 1)(2x + 3)\\ &= 6x^2 + 9x - 2x - 3\\ (f \cdot g)(x) &= 6x^2 + 7x - 3\end{aligned}$$

▶ 12. Let $f(x) = 3x - 1$ and $g(x) = 2x + 3$. Find $f \circ g$.

$$\begin{aligned}(f \circ g)(x) &= f(g(x))\\ &= f(2x + 3)\\ &= 3(2x + 3) - 1\\ &= 6x + 9 - 1\\ (f \circ g)(x) &= 6x + 8\end{aligned}$$

GUIDED PRACTICE

Let $f(x) = 2x + 1$ and $g(x) = x - 3$. Find each function and give its domain. See Example 1.

▶ 13. $f + g$ $(f + g)(x) = 3x - 2, (-\infty, \infty)$

▶ 14. $f - g$ $(f - g)(x) = x + 4, (-\infty, \infty)$

15. $g - f$ $(g - f)(x) = -x - 4, (-\infty, \infty)$

16. $g + f$ $(g + f)(x) = 3x - 2, (-\infty, \infty)$

▶ 17. $f \cdot g$ $(f \cdot g)(x) = 2x^2 - 5x - 3, (-\infty, \infty)$

▶ 18. f/g $(f/g)(x) = \frac{2x + 1}{x - 3}, (-\infty, 3) \cup (3, \infty)$

19. g/f $(g/f)(x) = \frac{x - 3}{2x + 1}, \left(-\infty, -\frac{1}{2}\right) \cup \left(-\frac{1}{2}, \infty\right)$

20. $g \cdot f$ $(g \cdot f)(x) = 2x^2 - 5x - 3, (-\infty, \infty)$

Let $f(x) = 3x$ and $g(x) = 4x$. Find each function and give its domain. See Example 1.

21. $f + g$ $(f + g)(x) = 7x, (-\infty, \infty)$

▶ 22. $f - g$ $(f - g)(x) = -x, (-\infty, \infty)$

23. $g - f$ $(g - f)(x) = x, (-\infty, \infty)$

▶ 24. $g + f$ $(g + f)(x) = 7x, (-\infty, \infty)$

▶ 25. $f \cdot g$ $(f \cdot g)(x) = 12x^2, (-\infty, \infty)$

▶ 26. f/g $(f/g)(x) = \frac{3}{4}, (-\infty, 0) \cup (0, \infty)$

27. g/f $(g/f)(x) = \frac{4}{3}, (-\infty, 0) \cup (0, \infty)$

28. $g \cdot f$ $(g \cdot f)(x) = 12x^2, (-\infty, \infty)$

Let $f(x) = 2x - 5$ and $g(x) = x + 1$. Find each of the following function values. See Example 2.

29. $(f + g)(8)$ 20

▶ 30. $(f - g)(-4)$ -10

31. $(f \cdot g)(0)$ -5

▶ 32. $(f/g)(2)$ $-\frac{1}{3}$

Let $s(x) = 3 - x$ and $t(x) = x^2 - x - 6$. Find each function value. See Example 2.

33. $(s \cdot t)(-2)$ 0

▶ 34. $(s + t)(3)$ 0

▶ 35. $(s/t)(1)$ $-\frac{1}{3}$

36. $(s - t)(12)$ -135

Let $f(x) = 2x + 1$ and $g(x) = x^2 - 1$. Find each of the following. See Example 3.

▶ 37. $(f \circ g)(2)$ 7

▶ 38. $(g \circ f)(2)$ 24

39. $(g \circ f)(-3)$ 24

▶ 40. $(f \circ g)(-3)$ 17

41. $(f \circ g)\left(\frac{1}{2}\right)$ $-\frac{1}{2}$

▶ 42. $(g \circ f)\left(\frac{1}{3}\right)$ $\frac{16}{9}$

43. $(g \circ f)(2x)$ $(g \circ f)(2x) = 16x^2 + 8x$

▶ 44. $(f \circ g)(2x)$ $(f \circ g)(2x) = 8x^2 - 1$

Refer to graphs at the right. Find each function value. See Example 4.

▶ 45. a. $(f + g)(-5)$ 0

b. $(f - g)(3)$ 6

c. $(f \cdot g)(-3)$ -3

▶ 46. a. $(f/g)(0)$ -2

b. $(f \circ g)(3)$ 1

c. $(g \circ f)(2)$ -3

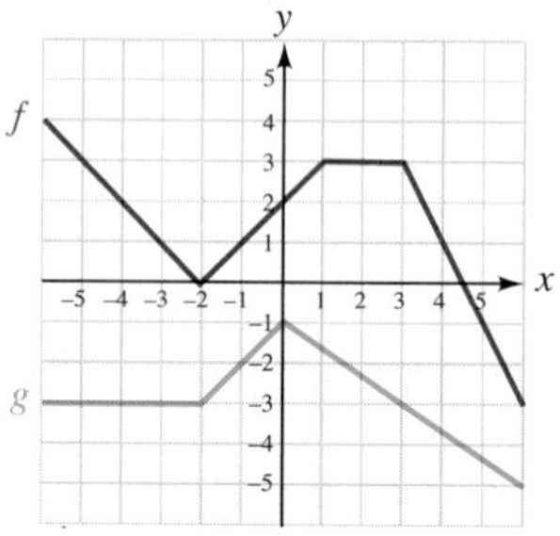

Refer to graphs at the right. Find each function value. See Example 4.

47. a. $(g + f)(2)$ -1

b. $(g - f)(-5)$ 6

c. $(g \cdot f)(1)$ -6

48. a. $(g/f)(-6)$ -1

b. $(g \circ f)(4)$ -3

c. $(f \circ g)(6)$ About $\frac{1}{2}$

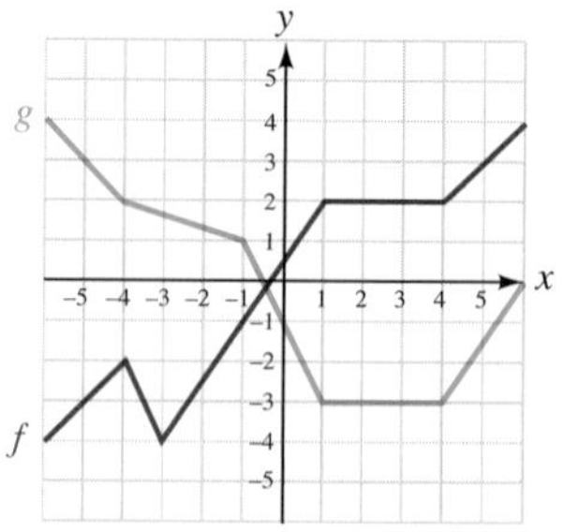

Find $f(x)$ and $g(x)$ such that $h(x) = (f \circ g)(x)$. Answers may vary. See Example 5.

49. $h(x) = (x + 15)^2$
$f(x) = x^2$; $g(x) = x + 15$

▶ 50. $h(x) = (x - 9)^3$
$f(x) = x^3$; $g(x) = x - 9$

51. $h(x) = x^5 + 9$
$f(x) = x + 9$; $g(x) = x^5$

▶ 52. $h(x) = x^6 - 100$
$f(x) = x - 100$; $g(x) = x^6$

53. $h(x) = \sqrt{16x - 1}$
$f(x) = \sqrt{x}$; $g(x) = 16x - 1$

▶ 54. $h(x) = \sqrt[3]{10 - x}$
$f(x) = \sqrt[3]{x}$; $g(x) = 10 - x$

55. $h(x) = \dfrac{1}{x - 4}$
$f(x) = \frac{1}{x}$; $g(x) = x - 4$

▶ 56. $h(x) = \dfrac{1}{3x - 16}$
$f(x) = \frac{1}{x}$; $g(x) = 3x - 16$

TRY IT YOURSELF

Let $f(x) = 3x - 2$ and $g(x) = x^2 + x$. Find each of the following.

57. $(f \circ g)(4)$ 58
58. $(g \circ f)(4)$ 110
59. $(g \circ f)(-3)$ 110
60. $(f \circ g)(-3)$ 16
61. $(g \circ f)(0)$ 2
62. $(f \circ g)(0)$ -2
63. $(g \circ f)(x)$ $(g \circ f)(x) = 9x^2 - 9x + 2$
64. $(f \circ g)(x)$ $(f \circ g)(x) = 3x^2 + 3x - 2$

Let $f(x) = 3x - 2$ and $g(x) = 2x^2 + 1$. Find each function and give its domain.

65. $f - g$ $(f - g)(x) = -2x^2 + 3x - 3, (-\infty, \infty)$
66. $f + g$ $(f + g)(x) = 2x^2 + 3x - 1, (-\infty, \infty)$
67. f/g $(f/g)(x) = \frac{3x - 2}{2x^2 + 1}, (-\infty, \infty)$
68. $f \cdot g$ $(f \cdot g)(x) = 6x^3 - 4x^2 + 3x - 2, (-\infty, \infty)$

Let $f(x) = \frac{1}{x}$ and $g(x) = \frac{1}{x^2}$. Find each of the following.

69. $(g \circ f)\left(\frac{1}{3}\right)$ $\frac{1}{9}$
70. $(g \circ f)\left(\frac{1}{10}\right)$ $\frac{1}{100}$
71. $(g \circ f)(8x)$ $(g \circ f)(8x) = 64x^2$
72. $(f \circ g)(5x)$ $(f \circ g)(5x) = 25x^2$

Let $f(x) = x^2 - 1$ and $g(x) = x^2 - 4$. Find each function and give its domain.

73. $f - g$ $(f - g)(x) = 3, (-\infty, \infty)$
74. $f + g$ $(f + g)(x) = 2x^2 - 5, (-\infty, \infty)$
75. g/f $(g/f)(x) = \frac{x^2 - 4}{x^2 - 1}, (-\infty, -1) \cup (-1, 1) \cup (1, \infty)$
76. $g \cdot f$ $(g \cdot f)(x) = x^4 - 5x^2 + 4, (-\infty, \infty)$

Let $h(t) = \sqrt{t + 3}$ and $k(t) = t - 5$. Find each of the following.

77. $(h \circ k)(18)$ 4
78. $(h \circ k)(11)$ 3
79. $(k \circ h)(22)$ 0
80. $(k \circ h)(-2)$ -4

81. Use the tables of values for functions f and g to find each of the following.

a. $(f + g)(1)$ 7 b. $(f - g)(5)$ 8
c. $(f \cdot g)(1)$ 12 d. $(g/f)(5)$ 0

x	$f(x)$
1	3
5	8

x	$g(x)$
1	4
5	0

82. Use the table of values for functions f and g to find each of the following.

a. $(f \circ g)(1)$ 5 b. $(g \circ f)(2)$ -3

x	$f(x)$
2	5
4	7

x	$g(x)$
1	2
5	-3

83. If $f(x) = x + 1$ and $g(x) = 2x - 5$, show that $(f \circ g)(x) \neq (g \circ f)(x)$.

84. If $f(x) = x^2 + 1$ and $g(x) = 3x^2 - 2$, show that $(f \circ g)(x) \neq (g \circ f)(x)$.

APPLICATIONS

85. **SAT Scores.** The graph of function m in the next column gives the average score on the mathematics portion of the SAT college entrance exam, the graph of function r gives the average score on the critical reading portion, and x represents the number of years since 2000.

a. Find $(m + r)(4)$ and explain what information about SAT scores it gives. 1,026; in 2004, the average combined math and reading score was 1,026.

b. Find $(m - r)(4)$ and explain what information about SAT scores it gives. 10; in 2004, the average difference in the math and reading scores was 10.

c. Find: $(m + r)(9)$ 1,016

d. Find: $(m - r)(9)$ 14

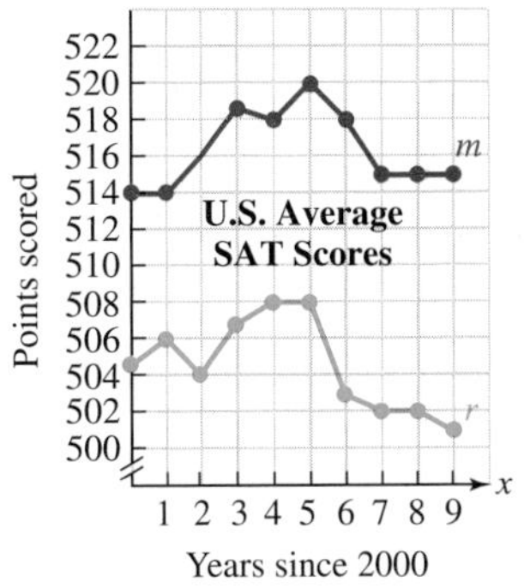

Source: National Center for Education Statistics

86. **Bachelor's Degrees.** The graph of function m below gives the number of bachelor's degrees awarded to men, and the graph of function w gives the number of bachelor's degrees awarded to women in the U. S. for the years 1990 through 2008.

a. Estimate $(w + m)(2004)$ and explain what information about bachelor's degrees it gives. In 2004, the total number of degrees awarded was about 1,400,000.

b. Estimate $(w - m)(2004)$ and explain what information about bachelor's degrees it gives. In 2004, the difference in the number of degrees awarded to women and men was about 200,000.

c. Estimate: $(w + m)(1994)$ 1,150,000

d. Estimate: $(w - m)(2000)$ 180,000

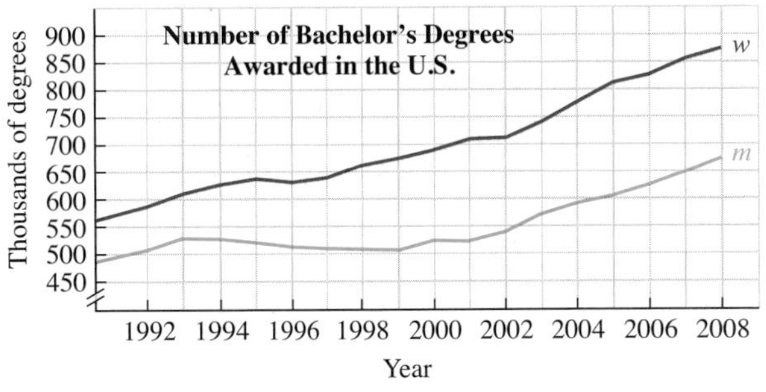

Source: National Center for Education Statistics

87. **Metallurgy.** A molten alloy must be cooled slowly to control crystallization. When removed from the furnace, its temperature is 2,700°F, and it will be cooled at 200° per hour. Write a composition function that expresses the Celsius temperature as a function of the number of hours t since cooling began. (*Hint*: $C(F) = \frac{5}{9}(F - 32)$.) $C(t) = \frac{5}{9}(2{,}668 - 200t)$

88. **Weather Forecasting.** A high-pressure area promises increasingly warmer weather for the next 48 hours. The temperature is now 34° Celsius and is expected to rise 1° every 6 hours. Write a composition function that expresses the Fahrenheit temperature as a function of the number of hours from now. (*Hint*: $F(C) = \frac{9}{5}C + 32$.) $F(t) = \frac{3}{10}t + \frac{466}{5}$

89. Vacation Mileage Costs.

a. Use the following graphs to determine the cost of the gasoline consumed if a family drove 500 miles on a summer vacation. About \$75

b. Write a composition function that expresses the cost of the gasoline consumed on the vacation as a function of the miles driven. $C(m) = \frac{3m}{20} = 0.15m$

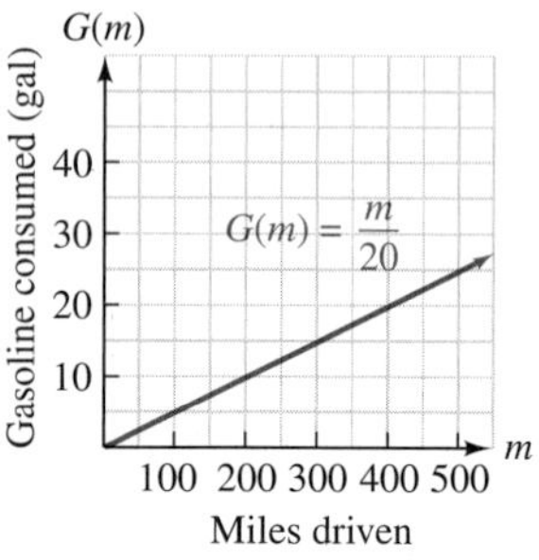

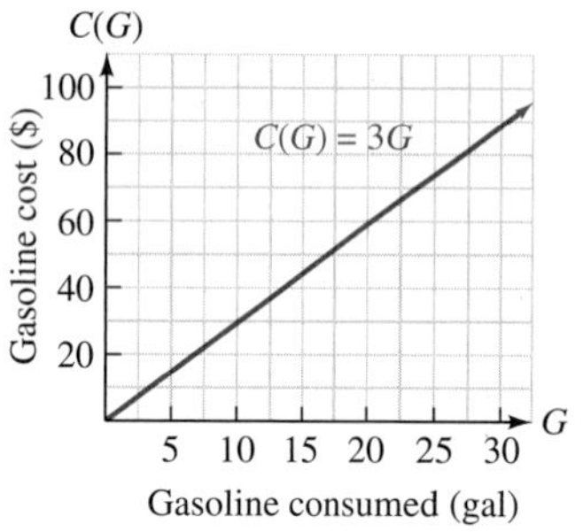

90. Halloween Costumes. The tables on the back of a pattern package can be used to determine the number of yards of material needed to make a rabbit costume for a child.

a. How many yards of material are needed if the child's chest measures 29 inches? $3\frac{3}{4}$ yd

b. In this exercise, one quantity is a function of a second quantity that depends, in turn, on a third quantity. Explain this dependence. The amount of material needed depends on the size of the pattern used. The size of the pattern to use depends on the chest measurement of the child.

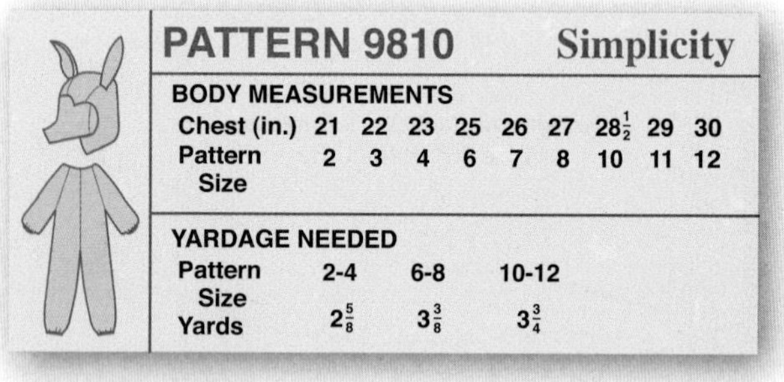
PATTERN 9810 Simplicity

BODY MEASUREMENTS

Chest (in.)	21	22	23	25	26	27	$28\frac{1}{2}$	29	30
Pattern Size	2	3	4	6	7	8	10	11	12

YARDAGE NEEDED

Pattern Size	2-4	6-8	10-12
Yards	$2\frac{5}{8}$	$3\frac{3}{8}$	$3\frac{3}{4}$

WRITING

91. Exercise 89 illustrates a chain of dependence between the cost of the gasoline, the gasoline consumed, and the miles driven. Describe another chain of dependence that could be represented by a composition function.

92. In this section, what operations are performed on functions? Give an example of each.

93. Write out in words how to say each of the following:

$$(f \circ g)(2) \qquad g(f(-8))$$

94. If $Y_1 = f(x)$ and $Y_2 = g(x)$, explain how to use the following tables to find $g(f(2))$.

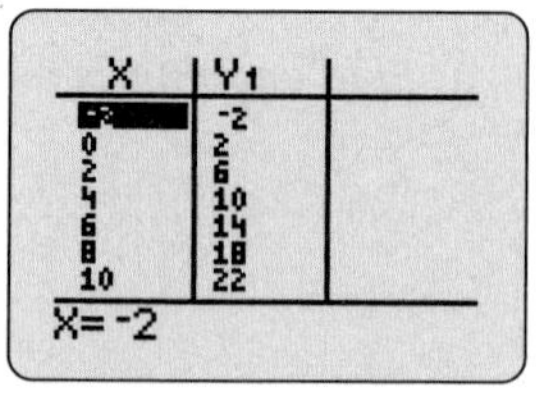

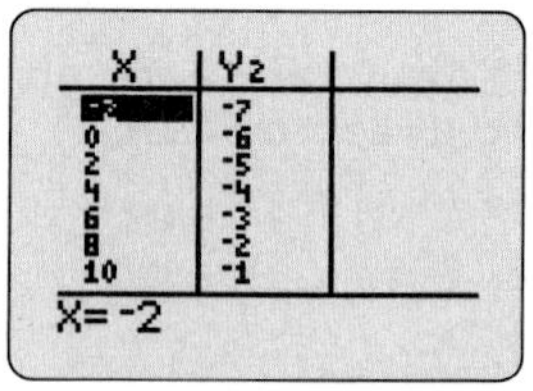

REVIEW

Simplify each complex fraction.

95. $\dfrac{\dfrac{ac - ad - c + d}{a^3 - 1}}{\dfrac{c^2 - 2cd + d^2}{a^2 + a + 1}}$ $\quad \frac{1}{c - d}$

96. $\dfrac{2 + \dfrac{1}{x^2 - 1}}{1 + \dfrac{1}{x - 1}}$ $\quad \frac{2x^2 - 1}{x^2 + x}$

CHALLENGE PROBLEMS

Fill in the blanks.

97. If $f(x) = x^2$ and $g(x) = 2x + 5$, then $(f \circ g)(x) = 4x^2 + 20x + 25$.

98. If $f(x) = \sqrt{3x}$ and $g(x) = x^4 + 7$, then $(g \circ f)(x) = 9x^2 + 7$.

Refer to the following graphs of functions f and g.

99. Graph the sum function $f + g$ on the given coordinate system.

100. Graph the difference function $f - g$ on the given coordinate system.

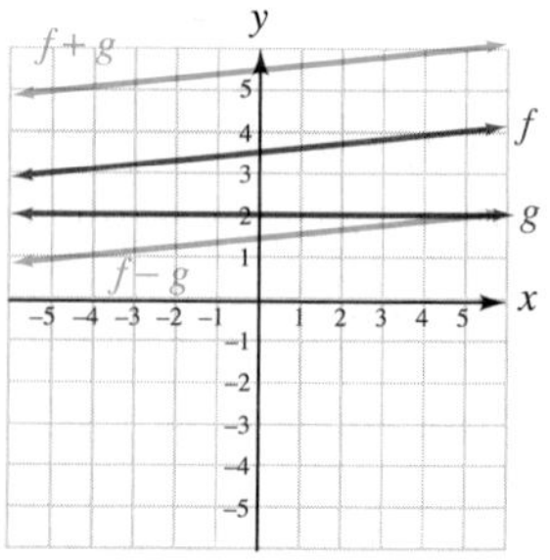

SECTION 11.2 Inverse Functions

OBJECTIVES

1. Determine whether a function is a one-to-one function.
2. Use the horizontal line test to determine whether a function is one-to-one.
3. Find the equation of the inverse of a function.
4. Find the composition of a function and its inverse.
5. Graph a function and its inverse.

ARE YOU READY? *Are You Ready? exercises available online at www.webassign.net/brookscole*

The following problems review some basic skills that are needed when working with inverse functions.

1. What are the domain and the range of the function $\{(-2, 8), (3, -3), (5, 10), (9, 1)\}$? D: $\{-2, 3, 5, 9\}$, R: $\{-3, 1, 8, 10\}$
2. Is a parabola that opens upward the graph of a function? Yes
3. Fill in the blank: If y is a <u>function</u> of x, the symbols y and $f(x)$ are interchangeable.
4. Let $f(x) = 2x + 6$. Find $f(8x)$. $16x + 6$

In the previous section, we created new functions from given functions by using the operations of arithmetic and composition. Another way to create new functions is to find the *inverse of a function.*

1 Determine Whether a Function Is a One-to-One Function.

In figure (a) below, the arrow diagram defines a function f. If we reverse the arrows as shown in figure (b), we obtain a new correspondence where the range of f becomes the domain of the new correspondence, and the domain of f becomes the range. The new correspondence is a function because to each member of the domain, there corresponds exactly one member of the range. We call this new correspondence the **inverse** of f, or f inverse.

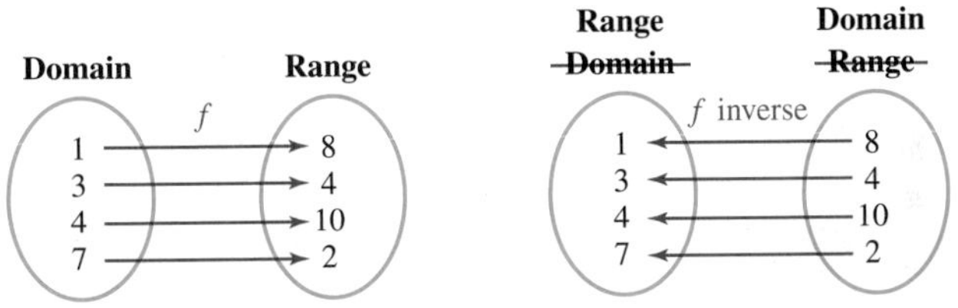

(a) (b)

This reversing process does not always produce a function. For example, if we reverse the arrows in function g defined by the diagram in figure (a) below, the resulting correspondence shown in figure (b) is not a function. This is because to the number 2 in the domain, there corresponds two members of the range: 8 and 4.

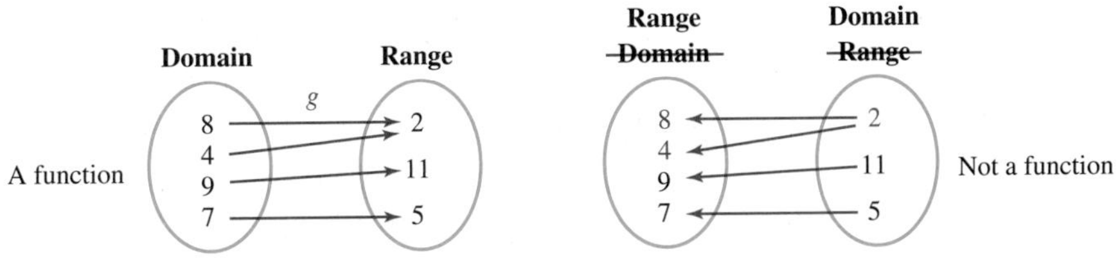

(a) (b)

The question that arises is, "What must be true of an original function to guarantee that the reversing process produces a function?" The answer is: *The original function must be one-to-one.*

We have seen that in a function, each input determines exactly one output. For some functions, different inputs determine different outputs, as in figure (a) below. For other functions, different inputs might determine the *same* output, as in figure (b). When a function has the property that different inputs determine different outputs, as in figure (a), we say the function is *one-to-one.*

Teaching Tip: Stress that a function is either one-to-one or not one-to-one. Draw a horizontal line on the classroom board. First ask if it is the graph of a function. Then ask if it is a one-to-one function.

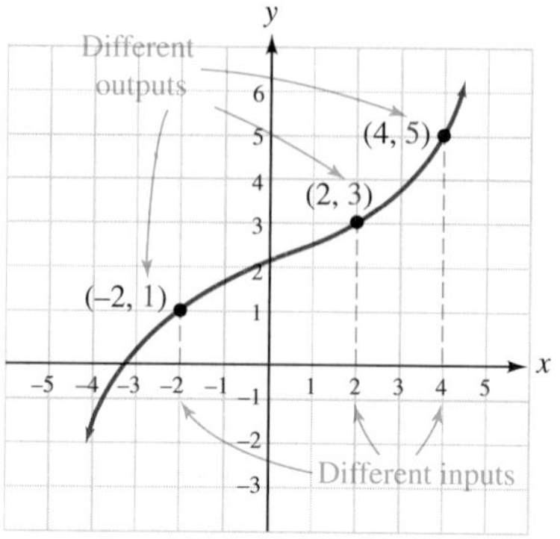

A one-to-one function

(a)

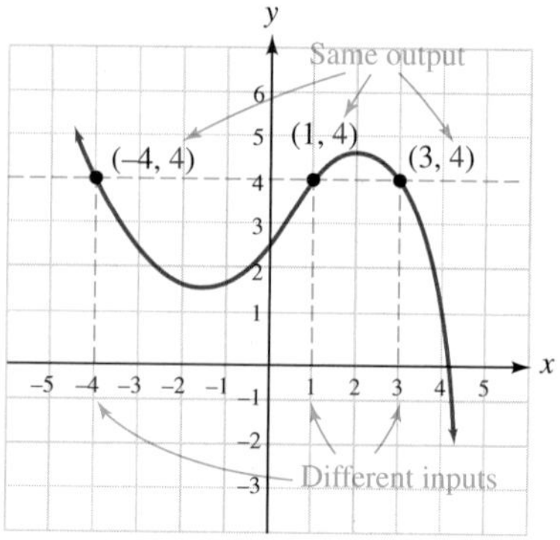

Not a one-to-one function

(b)

One-to-One Functions	A function is called a **one-to-one function** if different inputs determine different outputs.

EXAMPLE 1 Determine whether each function is one-to-one. **a.** $f(x) = x^2$ **b.** $f(x) = x^3$

Strategy We will determine whether different inputs have different outputs.

Why If different inputs have different outputs, the function is one-to-one. If different inputs have the same output, the function is not one-to-one.

Solution **a.** Since two different inputs, -3 and 3, have the same output 9, $f(x) = x^2$ is not a one-to-one function.

Success Tip

Example 1 illustrates that not every function is one-to-one.

$f(-3) = (-3)^2 = 9$ and $f(3) = 3^2 = 9$

x	$f(x)$
-3	9
3	9

The output 9 does not correspond to exactly one input.

b. Since different numbers have different cubes, each input of $f(x) = x^3$ determines a different output. This function is one-to-one.

Teaching Example 1
Determine whether $f(x) = |x|$ is one-to-one.
Answer:
No, $f(-2) = |-2| = 2$ and $f(2) = |2| = 2$

Self Check 1 Determine whether each function is one-to-one. If not, find an output that corresponds to more than one input. **a.** $f(x) = 2x + 3$ Yes **b.** $f(x) = x^4$ No, $(-1, 1)$, $(1, 1)$

Now Try Problems 19 and 21

2 Use the Horizontal Line Test to Determine Whether a Function Is One-to-One.

To determine whether a function is one-to-one, it is often easier to view its graph rather than its defining equation. If two (or more) points on the graph of a function have the same y-coordinate, the function is not one-to-one. This observation suggests the following **horizontal line test.**

The Horizontal Line Test	A function is one-to-one if each horizontal line that intersects its graph does so exactly once.

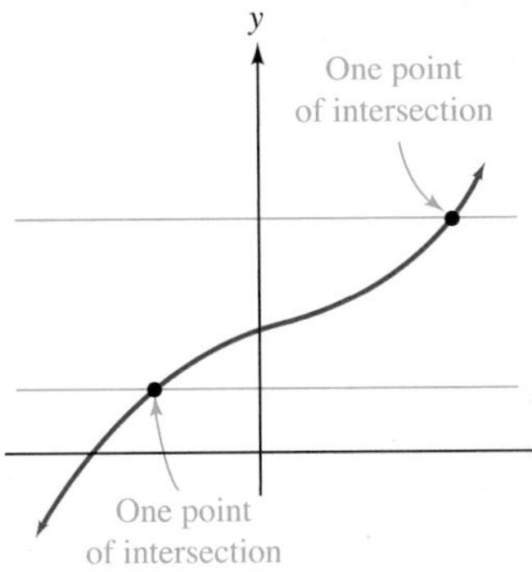

A one-to-one function

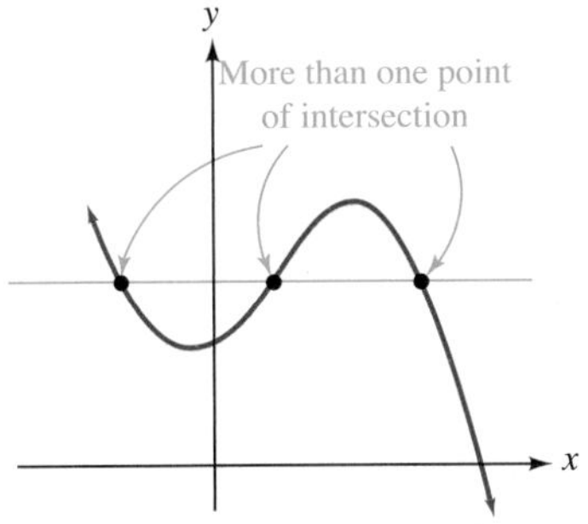

Not a one-to-one function

EXAMPLE 2 Use the horizontal line test to determine whether the following graphs of functions represent one-to-one functions.

Strategy We will draw horizontal lines through the graph of the function and see how many times each line intersects the graph.

Why If each horizontal line intersects the graph of the function exactly once, the graph represents a one-to-one function. If any horizontal line intersects the graph of the function more than once, the graph does not represent a one-to-one function.

Solution

a. Because every horizontal line that intersects the graph of $f(x) = -\frac{3}{4}x - 2$ in figure (a) does so exactly once, the graph represents a one-to-one function. We simply say, the function $f(x) = -\frac{3}{4}x - 2$ is one-to-one.

b. Refer to figure (b). Because we can draw a horizontal line that intersects the graph of $f(x) = x^2 - 4$ twice, the graph does not represent a one-to-one function. We simply say, the function $f(x) = x^2 - 4$ is not one-to-one.

c. Because every horizontal line that intersects the graph of $f(x) = \sqrt{x}$ in figure (c) does so exactly once, the graph represents a one-to-one function. We simply say, the function $f(x) = \sqrt{x}$ is one-to-one.

Success Tip

Recall that we use the **vertical line test** to determine whether a graph represents a function. We use the **horizontal line test** to determine whether the function that is graphed is one-to-one.

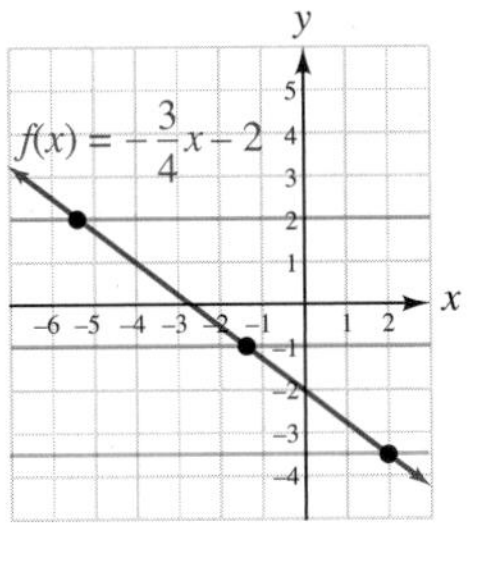

(a)

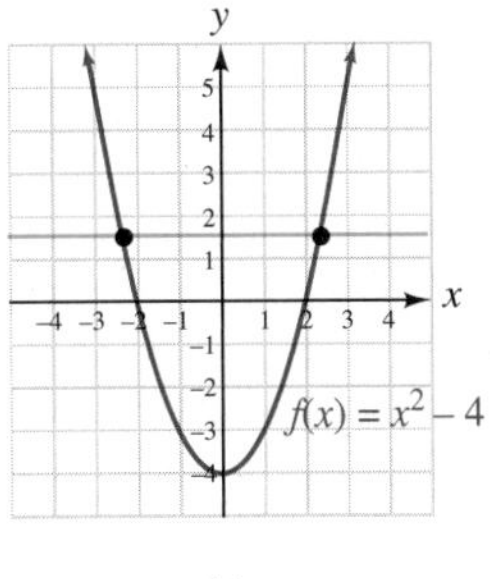

(b)

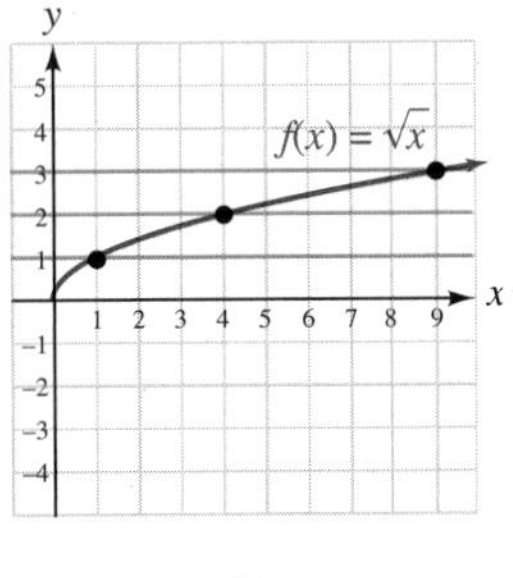

(c)

Teaching Example 2
Use the horizontal line test to determine whether the following graphs represent one-to-one functions.

a.

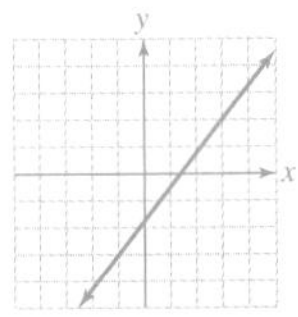

b.

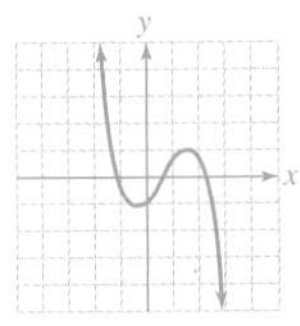

Answers: **a.** Yes **b.** No

Self Check 2 Use the horizontal line test to determine whether the graph represents a one-to-one function.

a. No

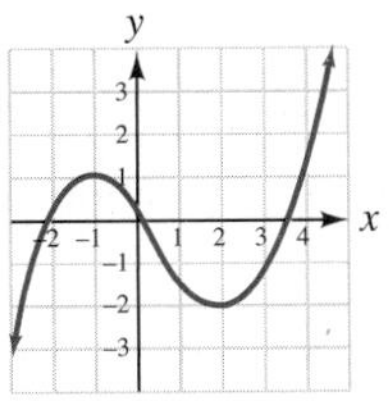

b. Yes

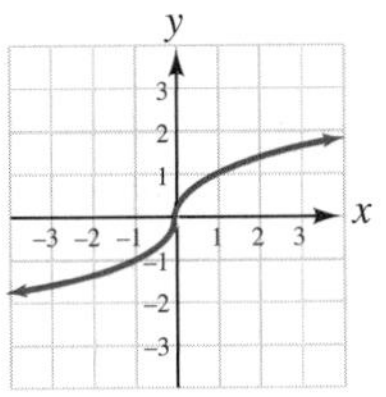

Now Try Problems 27 and 29

3 Find the Equation of the Inverse of a Function.

If f is the one-to-one function defined by the arrow diagram in figure (a), it turns the number 1 into 10, 2 into 20, and 3 into 30. The ordered pairs that define f can be listed in a table. Since the inverse of f must turn 10 back into 1, 20 back into 2, and 30 back into 3, it consists of the ordered pairs shown in the table in figure (b).

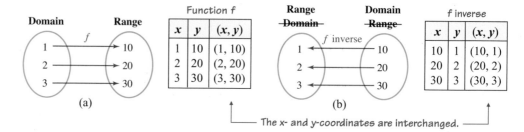

Teaching Tip: You may want to point out the similarity in the words *inverse* and *reverse*.

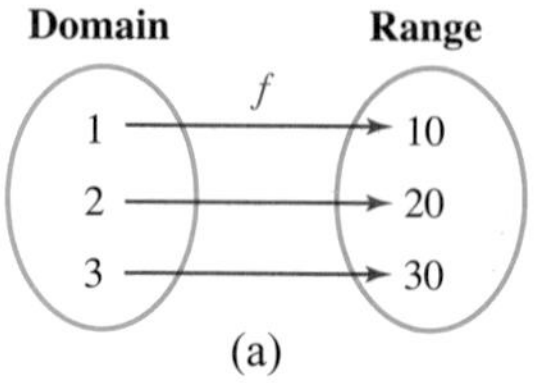

(a)

Function f

x	y	(x, y)
1	10	(1, 10)
2	20	(2, 20)
3	30	(3, 30)

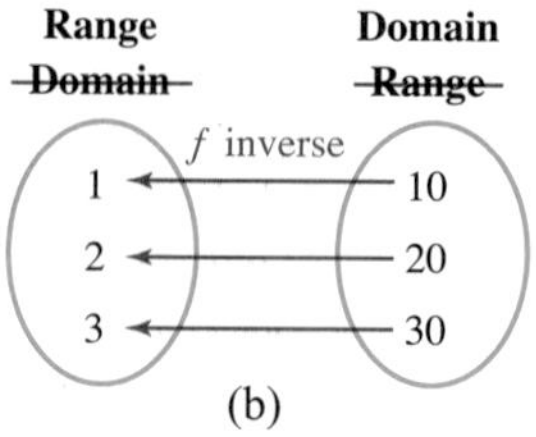

(b)

f inverse

x	y	(x, y)
10	1	(10, 1)
20	2	(20, 2)
30	3	(30, 3)

The x- and y-coordinates are interchanged.

Caution

The -1 in the notation $f^{-1}(x)$ is not an exponent:

$$f^{-1}(x) \neq \frac{1}{f(x)}$$

We note that the domain of f and the range of its inverse is $\{1, 2, 3\}$. The range of f and the domain of its inverse is $\{10, 20, 30\}$.

This example suggests that to form the inverse of a function f, we simply interchange the coordinates of each ordered pair that determines f. When the inverse of a function is also a function, we call it ***f* inverse** and denote it with the symbol f^{-1}. The symbol $f^{-1}(x)$ is read as "the inverse of $f(x)$" or "f inverse of x."

The Inverse of a Function

If f is a one-to-one function consisting of ordered pairs of the form (x, y), **the inverse of** f, denoted f^{-1}, is the one-to-one function consisting of all ordered pairs of the form (y, x).

When a one-to-one function is defined by an equation, we use the following method to find the equation of its inverse.

Finding the Equation of the Inverse of a Function

If a function is one-to-one, we find its inverse as follows:

1. If the function is written using function notation, replace $f(x)$ with y.
2. Interchange the variables x and y.
3. Solve the resulting equation for y.
4. Substitute $f^{-1}(x)$ for y.

EXAMPLE 3 Determine whether each function is one-to-one. If so, find the equation of its inverse.
a. $f(x) = 4x + 2$ **b.** $f(x) = x^3$

Teaching Tip: Before beginning the explanation of part a, point out that function f takes an input, multiplies it by 4, and then adds 2 to that result. Ask your students how they would expect the inverse of f to operate on an input. Ask them what operations f inverse would involve, and in what order they would be performed.

Strategy We will determine whether each function is one-to-one. If it is, we can find the equation of its inverse by replacing $f(x)$ with y, interchanging x and y, and solving for y.

Why The reason for interchanging the variables is this: If a one-to-one function takes an input x into an output y, by definition, its inverse function has the reverse effect.

Solution

a. We recognize $f(x) = 4x + 2$ as a linear function whose graph is a straight line with slope 4 and y-intercept $(0, 2)$. Since such a graph would pass the horizontal line test, we conclude that f is one-to-one.

Success Tip

Recall that $f(x) = 4x + 2$ is a linear function. Every linear function, except those of the form $f(x) = c$, where c is a constant, is one-to-one.

To find the inverse of function f, we proceed as follows:

$f(x) = 4x + 2$	Function f multiplies all inputs by 4 and then adds 2.
$y = 4x + 2$	Replace $f(x)$ with y.
$\mathbf{x} = 4\mathbf{y} + 2$	Interchange the variables x and y.
$x - 2 = 4y$	To isolate the term $4y$, subtract 2 from both sides.
$\frac{x-2}{4} = y$	To solve for y, divide both sides by 4.
$y = \frac{x-2}{4}$	Write the equation with y on the left side.

To denote that this equation is the inverse of function f, we replace y with $f^{-1}(x)$.

$$\mathbf{f^{-1}(x)} = \frac{x-2}{4} \quad \text{Function } f^{-1} \text{ subtracts 2 from each input and then divides by 4.}$$

As an informal check, we see below that if $x = 1$, function f produces an output of 6. And if $x = 6$, function f^{-1} produces an output of 1.

$f(1) = 4(1) + 2$	$f^{-1}(6) = \frac{6-2}{4}$
$= 4 + 2$	$= \frac{4}{4}$
$= 6$	$= 1$
Ordered pair: (1, 6)	Ordered pair: (6, 1)

The coordinates are interchanged.

Success Tip

Notice that f^{-1} operates on an input using the inverse operations of f in the reverse order.

Teaching Tip: Before beginning the explanation of part b, point out that function f takes an input and cubes it. Ask your students how they would expect the inverse of f to operate on an input.

b. The graph of $f(x) = x^3$ is shown on the right. Since such a graph would pass the horizontal line test, we conclude that f is a one-to-one function.

To find its inverse, we proceed as follows:

$f(x) = x^3$	Function f cubes all inputs.
$y = x^3$	Replace $f(x)$ with y.
$\mathbf{x} = \mathbf{y}^3$	Interchange the variables x and y.
$\sqrt[3]{x} = y$	To solve for y, take the cube root of both sides.
$y = \sqrt[3]{x}$	Write the equation with y on the left side.

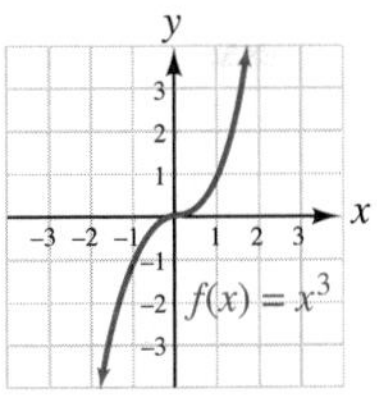

Caution

Only one-to-one functions have inverse functions.

Replacing y with $f^{-1}(x)$, we have

$$f^{-1}(x) = \sqrt[3]{x} \quad \text{Function } f^{-1} \text{ finds the cube root of each input.}$$

As an informal check, let $x = 4$ and determine whether f and f^{-1} produce ordered pairs whose coordinates are reversed.

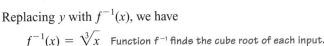

Self Check 3 Determine whether each function is one-to-one. If it is, find the equation of its inverse. **a.** $f(x) = -5x - 3$ $f^{-1}(x) = \frac{-x-3}{5}$ **b.** $f(x) = x^5$ $f^{-1}(x) = \sqrt[5]{x}$

Now Try Problems 35 and 45

Teaching Example 3 Determine whether each function is one-to-one. If it is, find the equation of its inverse.
a. $f(x) = 6x - 4$ **b.** $f(x) = x^7$

Answers: **a.** $f^{-1}(x) = \frac{x+4}{6}$
b. $f^{-1}(x) = \sqrt[7]{x}$

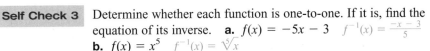

4 Find the Composition of a Function and Its Inverse.

To emphasize a relationship between a function and its inverse, we substitute some number x, such as $x = 3$, into the function $f(x) = 4x + 2$ of Example 3(a). The corresponding value of y that is produced is

$$f(3) = 4(3) + 2 = \mathbf{14} \quad f \text{ determines the ordered pair } (3, 14).$$

If we substitute 14 into the inverse function, $f^{-1}(x) = \frac{x-2}{4}$, the corresponding value of y that is produced is

$$f^{-1}(14) = \frac{14-2}{4} = 3 \quad f^{-1} \text{ determines the ordered pair } (14, 3).$$

Thus, the function f turns 3 into 14, and the inverse function f^{-1} turns 14 back into 3.

In general, the composition of a function and its inverse function is the identity function, $f(x) = x$, such that any input x has the output x. This fact can be stated symbolically as follows.

The Composition of Inverse Functions

For any one-to-one function f and its inverse, f^{-1},

$$(f \circ f^{-1})(x) = x \quad \text{and} \quad (f^{-1} \circ f)(x) = x$$

We can use this property to determine whether two functions are inverses.

EXAMPLE 4 Show that $f(x) = 4x + 2$ and $f^{-1}(x) = \frac{x-2}{4}$ are inverses.

Strategy We will find the composition of $f(x)$ and $f^{-1}(x)$ in both directions and show that the result is x.

Why Only when the result of the composition is x in both directions are the functions inverses.

Solution To show that $f(x) = 4x + 2$ and $f^{-1}(x) = \frac{x-2}{4}$ are inverses, we must show that for each composition, an input of x gives an output of x.

$$\begin{aligned}(f \circ f^{-1})(x) &= f(f^{-1}(x)) \\ &= f\left(\frac{x-2}{4}\right) \\ &= 4\left(\frac{x-2}{4}\right) + 2 \\ &= x - 2 + 2 \\ &= x\end{aligned}$$

$$\begin{aligned}(f^{-1} \circ f)(x) &= f^{-1}(f(x)) \\ &= f^{-1}(4x + 2) \\ &= \frac{4x + 2 - 2}{4} \\ &= \frac{4x}{4} \\ &= x\end{aligned}$$

Because $(f \circ f^{-1})(x) = x$ and $(f^{-1} \circ f)(x) = x$, the functions are inverses.

Self Check 4 Show that $f(x) = x - 4$ and $g(x) = x + 4$ are inverses.

Now Try Problem 55

Teaching Example 4 Show that $f(x) = 6x - 4$ and $f^{-1}(x) = \frac{x+4}{6}$ are inverses.
Answer: $(f \circ f^{-1})(x) = x$ and $(f^{-1} \circ f)(x) = x$

Success Tip

Recall that the line $y = x$ passes through points whose x- and y-coordinates are equal: $(-1, -1)$, $(0, 0)$, $(1, 1)$, $(2, 2)$, and so on.

5 Graph a Function and Its Inverse.

If a point (a, b) is on the graph of function f, it follows that the point (b, a) is on the graph of f^{-1}, and vice versa. There is a geometric relationship between a pair of points whose coordinates are interchanged. For example, in the graph, we see that the line segment between $(1, 3)$ and $(3, 1)$ is perpendicular to and cut in half by the line $y = x$. We say that $(1, 3)$ and $(3, 1)$ are mirror images of each other with respect to $y = x$.

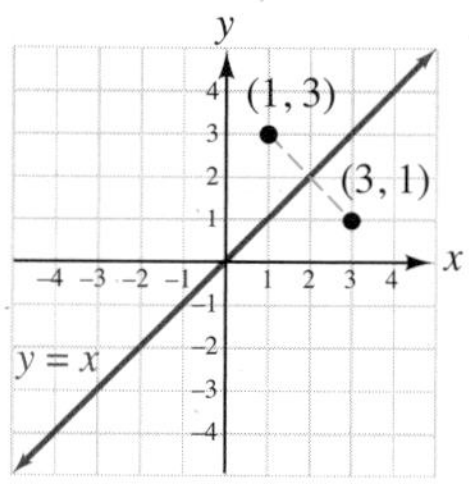

Since each point on the graph of f^{-1} is a mirror image of a point on the graph of f, and vice versa, the graphs of f and f^{-1} must be mirror images of each other with respect to $y = x$.

EXAMPLE 5 Find the equation of the inverse of $f(x) = -\frac{3}{2}x + 3$. Then graph f and its inverse on one coordinate system.

Strategy We will determine whether the function has an inverse. If so, we will replace $f(x)$ with y, interchange x and y, and solve for y to obtain the equation of the inverse.

Why The reason for interchanging the variables is this: If a one-to-one function takes an input x into an output y, by definition, its inverse function has the reverse effect.

Solution Since $f(x) = -\frac{3}{2}x + 3$ is a linear function, it is one-to-one and has an inverse. To find the inverse function, we replace $f(x)$ with y, and interchange x and y to obtain

$$x = -\frac{3}{2}y + 3$$

Then we solve for y to get

$$x - 3 = -\frac{3}{2}y \quad \text{Subtract 3 from both sides.}$$

$$-\frac{2}{3}x + 2 = y \quad \text{To isolate } y\text{, multiply both sides by } -\tfrac{2}{3}.$$

When we replace y with $f^{-1}(x)$, we have $f^{-1}(x) = -\frac{2}{3}x + 2$.

To graph f, we construct a table of values, plot points, and draw the graph in red as shown below. To graph f^{-1}, we don't need to do any calculations to construct a table of values. We can simply interchange the coordinates of the ordered pairs in the table for f and use them to graph f^{-1}. The result is the graph in blue shown below. Because the functions are inverses of each other, their graphs are **mirror images** about the line $y = x$.

Success Tip

If the graphs of f and f^{-1} intersect, it will always be on the line $y = x$.

$f(x) = -\frac{3}{2}x + 3$

x	$f(x)$	
0	3	→ (0, 3)
2	0	→ (2, 0)
4	−3	→ (4, −3)

$f^{-1}(x) = -\frac{2}{3}x + 2$

x	$f^{-1}(x)$	
3	0	→ (3, 0)
0	2	→ (0, 2)
−3	4	→ (−3, 4)

Simply interchange the coordinates to graph f^{-1}.

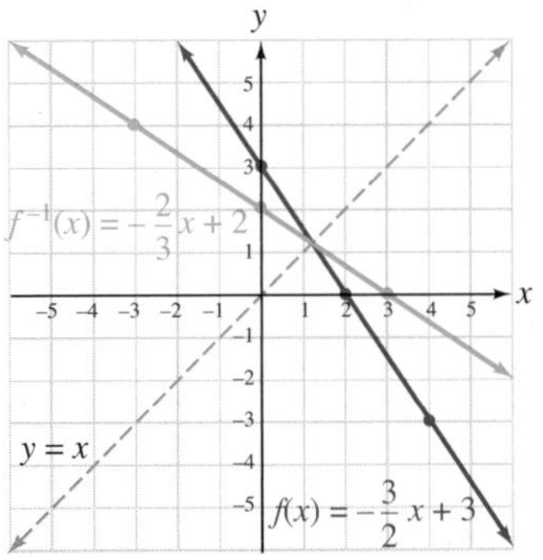

Self Check 5 Find the inverse of $f(x) = \frac{2}{3}x - 2$. Then graph the function and its inverse on one coordinate system. $f^{-1}(x) = \frac{3}{2}x + 3$; see AIE Appendix 3.

Now Try Problem 59

Teaching Example 5
Find the inverse function of the one-to-one function $f(x) = \frac{1}{2}x + 1$. Graph the function, its inverse, and $y = x$ on one coordinate system.

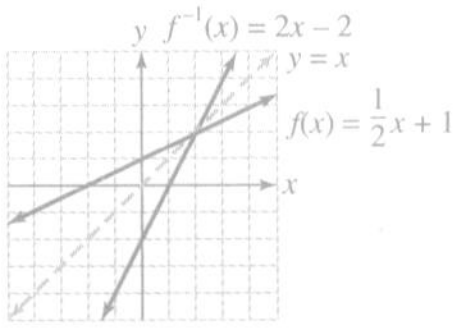

Answer: $f^{-1}(x) = 2x - 2$

Using Your Calculator ▶ Graphing the Inverse of a Function

We can use a graphing calculator to check the result found in Example 5. First, we enter $f(x) = -\frac{3}{2}x + 3$ and then enter what we believe to be the inverse function, $f^{-1}(x) = -\frac{2}{3}x + 2$, as well as the equation $y = x$. See figure (a) on the next page. Before graphing, we adjust the display so that the graphing grid will be composed of squares. The line of symmetry $y = x$ is then at a 45° angle to the positive x-axis.

In figure (b), it appears that the two graphs are symmetric about the line $y = x$. Although it is not definitive, this visual check does help to validate the result of Example 5.

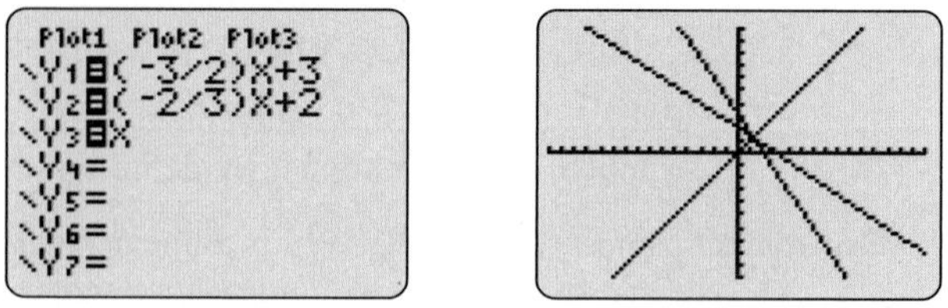

(a) (b)

EXAMPLE 6 Graph the inverse of function f shown in figure (a).

Strategy We will find the coordinates of several points on the graph of f in figure (a). After interchanging the coordinates of these points, we will plot them as shown in figure (b).

Why The reason for interchanging the coordinates is this: If (a, b) is a point on the graph of a one-to-one function, then the point (b, a) is on the graph of its inverse.

Solution In figure (a), we see that the points $(-5, -3)$, $(-2, -1)$, $(0, 2)$, $(3, 3)$, $(5, 4)$, and $(7, 5)$ lie on the graph of function f. To graph the inverse, we interchange their coordinates, and plot them in blue, as shown in figure (b). Then we graph the line $y = x$ and use symmetry to draw a smooth curve through those points to get the graph of f^{-1}.

The Language of Algebra

We also can say that the graphs of f and f^{-1} are **reflections** of each other about the line $y = x$, or they are **symmetric about** $y = x$.

Teaching Example 6 Graph the inverse of function f.

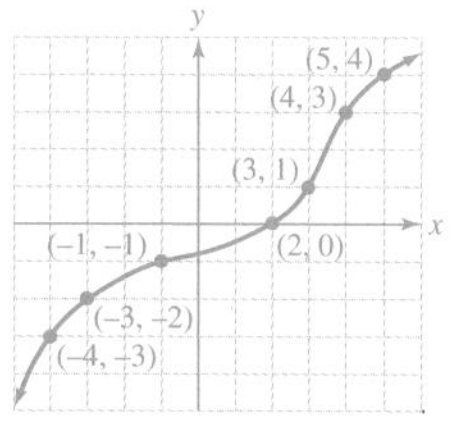

Answer:

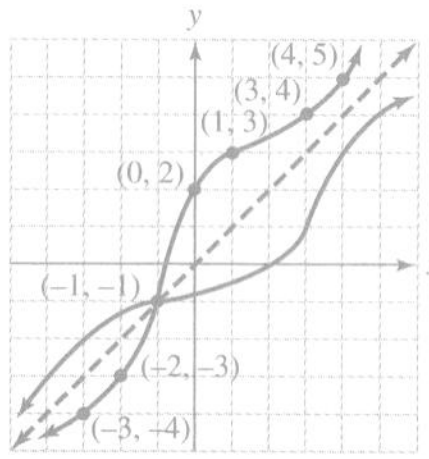

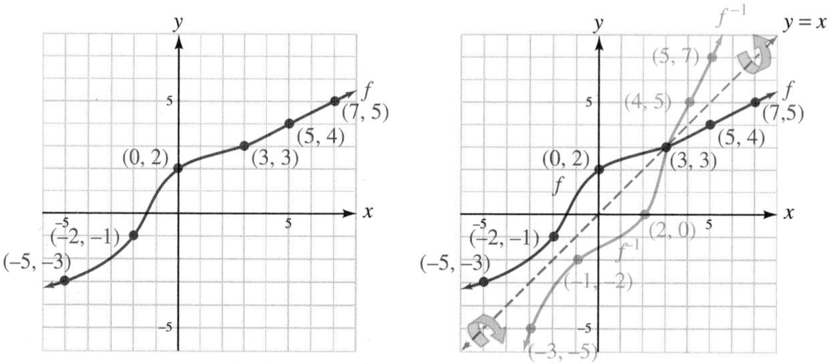

(a) (b)

Now Try Problem 65

SECTION 11.2 STUDY SET

VOCABULARY

Fill in the blanks.

1. A function is called a <u>one-to-one</u> function if different inputs determine different outputs.
2. The <u>horizontal</u> line test can be used to determine whether the graph of a function represents a one-to-one function.
3. The functions f and f^{-1} are <u>inverses</u>.

4. The graphs of a function and its inverse are <u>mirror</u> images of each other with respect to $y = x$. We also say that their graphs are <u>symmetric</u> with respect to the line $y = x$.

CONCEPTS

Fill in the blanks.

5. If any horizontal line that intersects the graph of a function does so more than once, the function is not <u>one-to-one</u>.
6. To find the inverse of the function $f(x) = 2x - 3$, we begin by replacing $f(x)$ with y, and then we <u>interchange</u> x and y.

▶ 7. If f is a one-to-one function, the domain of f is the range of f^{-1}, and the range of f is the domain of f^{-1}.

8. If a function turns an input of 2 into an output of 5, the inverse function will turn an input of 5 into the output 2.

9. If f is a one-to-one function, and if $f(1) = 6$, then $f^{-1}(6) = 1$.

▶ 10. If the point $(9, -4)$ is on the graph of the one-to-one function f, then the point $(-4, 9)$ is on the graph of f^{-1}.

11. a. Is the correspondence defined by the arrow diagram in figure (a) below a one-to-one function? No

b. Is the correspondence defined by the table in figure (b) below a one-to-one function? No

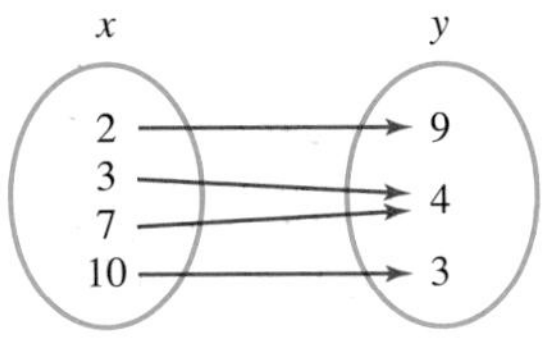

(a)

x	$f(x)$
-2	4
-1	1
0	0
2	4
3	9

(b)

12. Is the inverse of a one-to-one function always a function? Yes

▶ 13. Use the table of values of the one-to-one function f to complete a table of values for f^{-1}.

x	$f(x)$
-4	-2
0	0
8	4

x	$f^{-1}(x)$
-2	-4
0	0
4	8

14. Redraw the graph of function f. Then graph f^{-1} and the axis of symmetry on the same coordinate system.

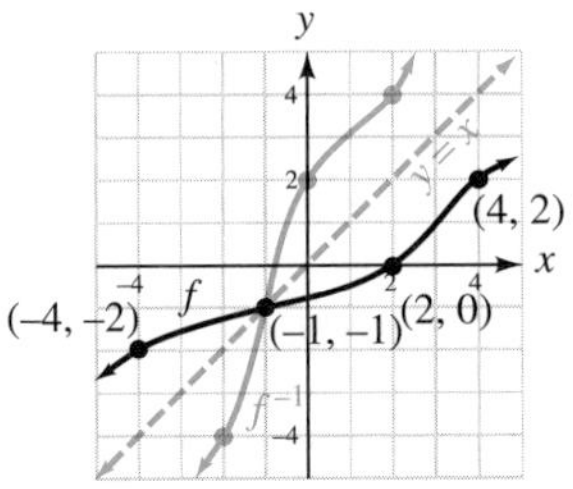

NOTATION

Complete each solution.

15. Find the inverse of $f(x) = 2x - 3$.

$$y = 2x - 3$$
$$x = 2y - 3$$
$$x + 3 = 2y$$
$$\frac{x+3}{2} = y$$

The inverse of $f(x) = 2x - 3$ is $f^{-1}(x) = \frac{x+3}{2}$.

16. Find the inverse of $f(x) = \sqrt[3]{x} + 2$.

$$y = \sqrt[3]{x} + 2$$
$$x = \sqrt[3]{y} + 2$$
$$x - 2 = \sqrt[3]{y}$$
$$(x - 2)^3 = y$$

The inverse of $f(x) = \sqrt[3]{x} + 2$ is $f^{-1}(x) = (x - 2)^3$.

17. The symbol f^{-1} is read as "the inverse of f" or "f inverse."

▶ 18. Explain the difference in the meaning of the -1 in the notation $f^{-1}(x)$ as compared with a^{-1}. In $f^{-1}(x)$, the -1 denotes the inverse function. In a^{-1}, the -1 is a negative exponent and $a^{-1} = \frac{1}{a}$.

GUIDED PRACTICE

***Determine whether each function is one-to-one.* See Example 1.**

▶ 19. $f(x) = 2x$ Yes

▶ 20. $f(x) = |x|$ No

21. $f(x) = x^4$ No

▶ 22. $f(x) = x^3 + 1$ Yes

▶ 23. $f(x) = -x^2 + 3x$ No

24. $f(x) = \frac{2}{3}x + 8$ Yes

▶ 25. $\{(1, 1), (2, 1), (3, 1), (4, 1)\}$ No

▶ 26. $\{(3, 2), (2, 1), (1,0)\}$ Yes

***Each graph represents a function. Use the horizontal line test to determine whether the function is one-to-one.* See Example 2.**

▶ 27.

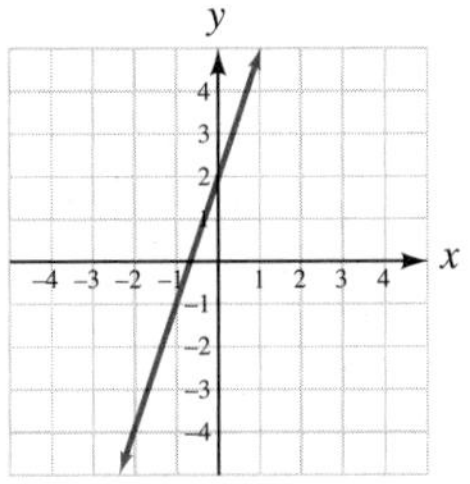

One-to-one

28.

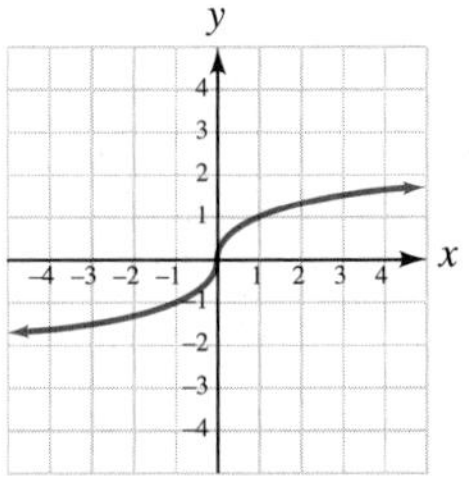

One-to-one

▶ 29.

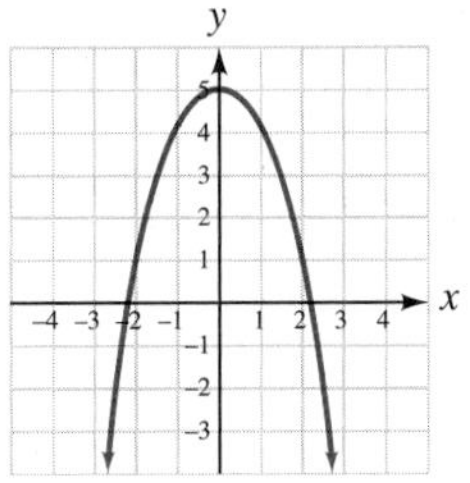

Not one-to-one

▶ 30.

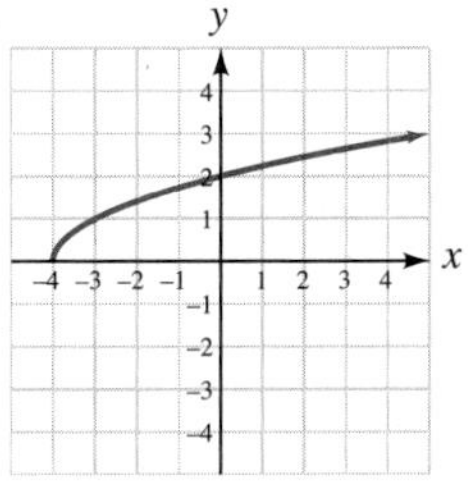

One-to-one

31.

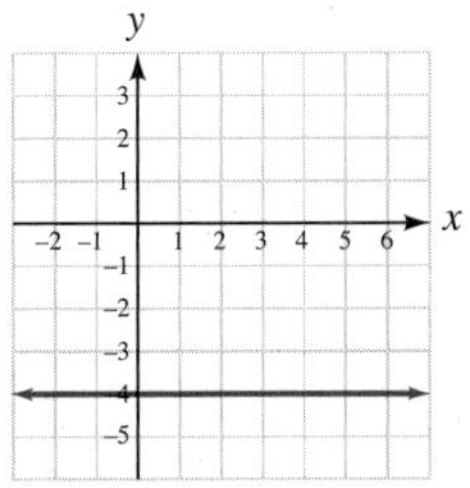

Not one-to-one

▶ 32.

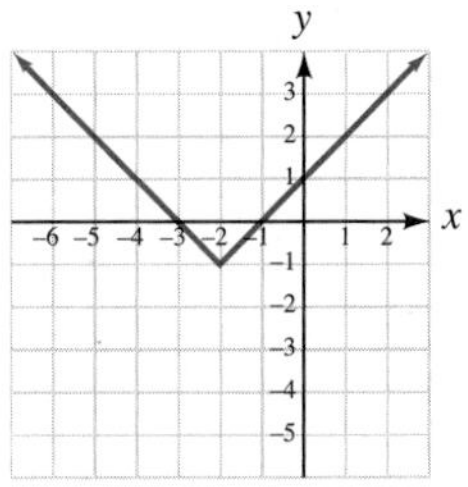

Not one-to-one

33.

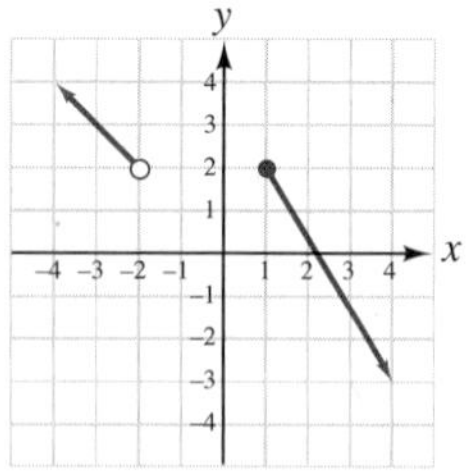

One-to-one

▶ 34.

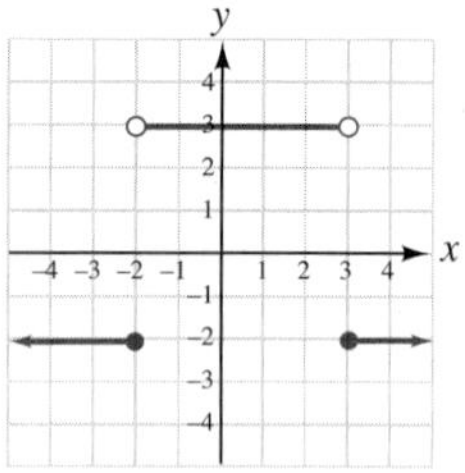

Not one-to-one

Each of the following functions is one-to-one. Find the inverse of each function and express it using $f^{-1}(x)$ notation. **See Example 3.**

35. $f(x) = 2x + 4$
$f^{-1}(x) = \frac{x-4}{2}$

▶ 36. $f(x) = 5x - 1$
$f^{-1}(x) = \frac{1}{5}x + \frac{1}{5}$

37. $f(x) = \frac{x}{5} + \frac{4}{5}$
$f^{-1}(x) = 5x - 4$

▶ 38. $f(x) = \frac{x}{3} - \frac{1}{3}$
$f^{-1}(x) = 3x + 1$

39. $f(x) = \frac{x-4}{5}$
$f^{-1}(x) = 5x + 4$

▶ 40. $f(x) = \frac{2x+6}{3}$
$f^{-1}(x) = \frac{3}{2}x - 3$

41. $f(x) = \frac{2}{x-3}$
$f^{-1}(x) = \frac{2}{x} + 3$

▶ 42. $f(x) = \frac{3}{x+1}$
$f^{-1}(x) = \frac{3}{x} - 1$

▶ 43. $f(x) = \frac{4}{x}$
$f^{-1}(x) = \frac{4}{x}$

44. $f(x) = \frac{1}{x}$
$f^{-1}(x) = \frac{1}{x}$

45. $f(x) = x^3 + 8$
$f^{-1}(x) = \sqrt[3]{x-8}$

▶ 46. $f(x) = x^3 - 4$
$f^{-1}(x) = \sqrt[3]{x+4}$

47. $f(x) = \sqrt[3]{x}$
$f^{-1}(x) = x^3$

▶ 48. $f(x) = \sqrt[3]{x-5}$
$f^{-1}(x) = x^3 + 5$

49. $f(x) = (x + 10)^3$
$f^{-1}(x) = \sqrt[3]{x} - 10$

50. $f(x) = (x - 9)^3$
$f^{-1}(x) = \sqrt[3]{x} + 9$

▶ 51. $f(x) = 2x^3 - 3$
$f^{-1}(x) = \sqrt[3]{\frac{x+3}{2}}$

▶ 52. $f(x) = \frac{3}{x^3} - 1$
$f^{-1}(x) = \sqrt[3]{\frac{3}{x+1}}$

53. $f(x) = \frac{x^7}{2}$
$f^{-1}(x) = \sqrt[7]{2x}$

▶ 54. $f(x) = \frac{x^9}{4}$
$f^{-1}(x) = \sqrt[9]{4x}$

Show that each pair of functions are inverses. **See Example 4.**

▶ 55. $f(x) = 2x + 9$, $f^{-1}(x) = \frac{x-9}{2}$

▶ 56. $f(x) = 5x - 1$, $f^{-1}(x) = \frac{x+1}{5}$

57. $f(x) = \frac{2}{x-3}$, $f^{-1}(x) = \frac{2}{x} + 3$

▶ 58. $f(x) = \sqrt[3]{x-6}$, $f^{-1}(x) = x^3 + 6$

Find the inverse of each function. Then graph the function and its inverse on one coordinate system. Show the line of symmetry on the graph. **See Examples 5 and 6.** See AIE Appendix 3.

59. $f(x) = 2x$
$f^{-1}(x) = \frac{1}{2}x$

▶ 60. $f(x) = -3x$
$f^{-1}(x) = -\frac{1}{3}x$

▶ 61. $f(x) = 4x + 3$
$f^{-1}(x) = \frac{x-3}{4}$

▶ 62. $f(x) = \frac{x}{3} + \frac{1}{3}$
$f^{-1}(x) = 3x - 1$

63. $f(x) = -\frac{2}{3}x + 3$
$f^{-1}(x) = -\frac{3}{2}x + \frac{9}{2}$

▶ 64. $f(x) = -\frac{1}{3}x + \frac{4}{3}$
$f^{-1}(x) = -3x + 4$

65. $f(x) = x^3$
$f^{-1}(x) = \sqrt[3]{x}$

▶ 66. $f(x) = x^3 + 1$
$f^{-1}(x) = \sqrt[3]{x-1}$

▶ 67. $f(x) = x^2 - 1 \ (x \geq 0)$
$f^{-1}(x) = \sqrt{x+1}$

▶ 68. $f(x) = x^2 + 1 \ (x \geq 0)$
$f^{-1}(x) = \sqrt{x-1}$

APPLICATIONS

▶ 69. **Interpersonal Relationships.** Feelings of anxiety in a relationship can increase or decrease, depending on what is going on in the relationship. The graph shows how a person's anxiety might vary as a relationship develops over time.

a. Is this the graph of a function? Is its inverse a function? Yes; no

b. Does each anxiety level correspond to exactly one point in time? Use the dashed lined labeled *Maximum threshold* to explain. No. Twice during this period, the person's anxiety level was at the maximum threshold value.

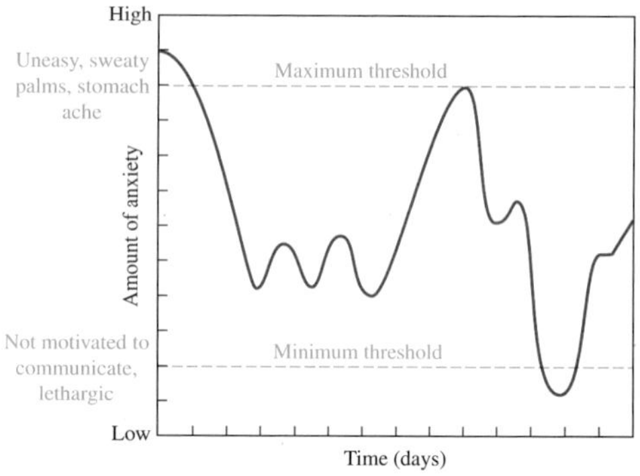

Source: Gudykunst, *Building Bridges: Interpersonal Skills for a Changing World* (Houghton Mifflin, 1994)

▶ 70. **Lighting Levels.** The ability of the eye to see detail increases as the level of illumination increases. This relationship can be modeled by a function E, whose graph is shown here.

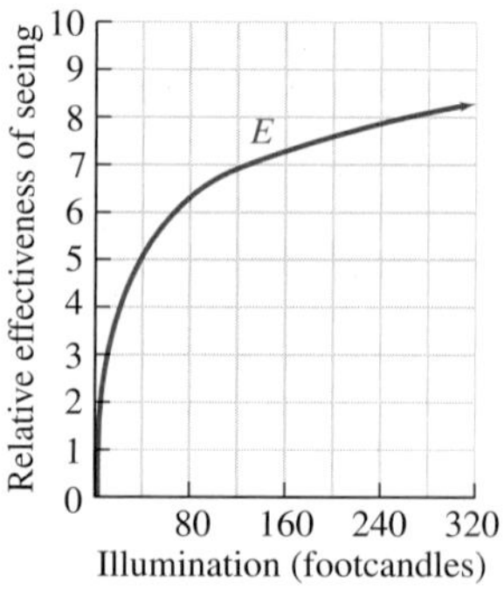

Based on information from *World Book Encyclopedia*

a. From the graph, determine $E(240)$. 7.9

b. Is function E one-to-one? Does E have an inverse? Yes; yes

c. If the effectiveness of seeing in an office is 7, what is the illumination in the office? How can this question be asked using inverse function notation? 120 footcandles; What is $E^{-1}(7)$?

WRITING

71. In your own words, what is a one-to-one function?

72. Two functions are graphed on the square grid on the right along with the line $y = x$. Explain why the functions cannot be inverses of each other.

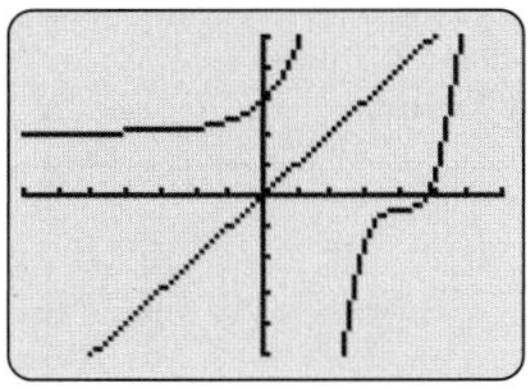

73. Explain how the graph of a one-to-one function can be used to draw the graph of its inverse function.

74. a. Explain the purpose of the vertical line test.
    b. Explain the purpose of the horizontal line test.

75. In the illustration, a function f and its inverse f^{-1} have been graphed on the same coordinate system. Explain what concept can be demonstrated by folding the graph paper on the dashed line.

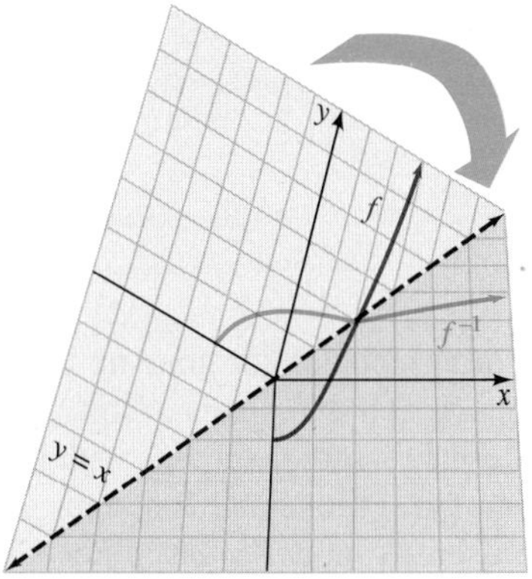

76. Write in words how to read the notation.
    a. $f^{-1}(x) = \frac{1}{2}x - 3$
    b. $(f \circ f^{-1})(x) = x$

REVIEW

Simplify. Write the result in the form a + bi.

77. $3 - \sqrt{-64}$ $3 - 8i$

78. $(2 - 3i) + (4 + 5i)$ $6 + 2i$

79. $(3 + 4i)(2 - 3i)$ $18 - i$

80. $\frac{6 + 7i}{3 - 4i}$ $-\frac{2}{5} + \frac{9}{5}i$

81. $(6 - 8i)^2$ $-28 - 96i$

82. i^{100} $1 + 0i$

CHALLENGE PROBLEMS

83. Find the inverse of $f(x) = \frac{x + 1}{x - 1}$. $f^{-1}(x) = \frac{x+1}{x-1}$

84. Using the functions of Exercise 83, show that $(f \circ f^{-1})(x) = x$ and $(f^{-1} \circ f)(x) = x$.

85. A table of values for a function f is shown in figure (a). A table of values for f^{-1} is shown in figure (b). Use the tables to find $f^{-1}(f(4))$ and $f(f^{-1}(2))$.
$f(4) = 10$ and $f^{-1}(10) = 4$; $f^{-1}(2) = 0$ and $f(0) = 2$

X	Y1
-2	-2
0	2
2	6
4	10
6	14
8	18
10	22

X=-2

(a)

X	Y1
-2	-2
0	-1
2	0
4	1
6	2
8	3
10	4

X=-2

(b)

86. a. The graph of a one-to-one function lies entirely in quadrant I. In what quadrant does the graph of its inverse lie? Quadrant I
    b. The graph of a one-to-one function lies entirely in quadrant II. In what quadrant does the graph of its inverse lie? Quadrant IV
    c. The graph of a one-to-one function lies entirely in quadrant III. In what quadrant does the graph of its inverse lie? Quadrant III
    d. The graph of a one-to-one function lies entirely in quadrant IV. In what quadrant does the graph of its inverse lie? Quadrant II

SECTION 11.3 Exponential Functions

OBJECTIVES

1. Define exponential functions.
2. Graph exponential functions.
3. Use exponential functions in applications involving growth or decay.

ARE YOU READY? *Are You Ready? exercises available online at www.webassign.net/brookscole*

The following problems review some basic skills that are needed when working with exponential functions.

1. Simplify: **a.** $3^4 \cdot 3^8$ 3^{12} **b.** $(3^4)^8$ 3^{32}
2. Evaluate: **a.** 2^3 8 **b.** 2^0 1 **c.** 2^{-2} $\frac{1}{4}$
3. Evaluate: **a.** $\left(\frac{1}{3}\right)^2$ $\frac{1}{9}$ **b.** $\left(\frac{1}{3}\right)^0$ 1 **c.** $\left(\frac{1}{3}\right)^{-3}$ 27
4. Fill in the blanks: The graph of $f(x) = x$ is a line and the graph of $f(x) = x^2$ is a parabola.

In previous chapters, we have discussed linear functions, polynomial functions, rational functions, and radical functions. We now begin a study of a new family of functions known as *exponential functions.*

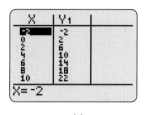

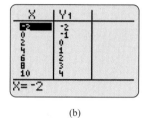

As an example, consider the graph in figure (a) below, which models the soaring popularity of the social network website Twitter in recent years. The rapidly rising red curve is the graph of an exponential function.

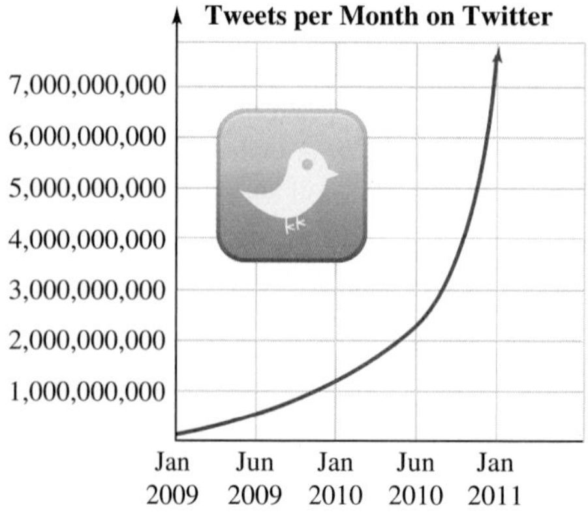

Source: pingdom.com
©iStockphoto.com/scottdunlap

(a)

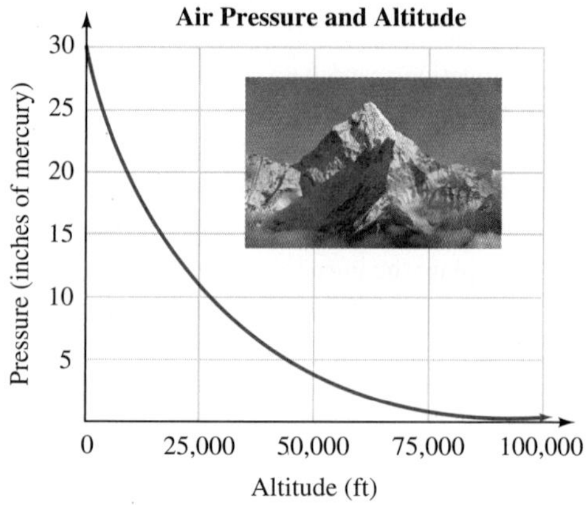

©iStockphoto.com/isoft

(b)

If you have ever climbed a high mountain or gone up in an airplane that does not have a pressurized cabin, you have probably felt the effects of low air pressure. The graph in figure (b) above shows how the atmospheric pressure decreases with increasing altitude. The rapidly falling red curve is also the graph of an exponential function.

Exponential functions are used to model many other situations, such as population growth, the spread of an epidemic, the temperature of a heated object as it cools, and radioactive decay.

1 Define Exponential Functions.

In this course, we have evaluated exponential expressions with integer exponents, such as 2^5, 7^0, and 5^{-1}, and rational number exponents, such as $9^{1/2}$, $64^{3/4}$, and $8^{-2/3}$. To define exponential functions, we also must be able to evaluate exponential expressions that have irrational exponents. For example, let's consider the expression

Teaching Tip: Remind students that an irrational number is a nonterminating, nonrepeating decimal. You may want to give other examples of irrational exponents such as: $9^{\sqrt{7}}$, 4^{π}, and $\left(\frac{1}{2}\right)^{3\sqrt{5}}$.

$5^{\sqrt{2}}$ where $\sqrt{2}$ is the irrational number 1.414213562 . . .

We can successively approximate $5^{\sqrt{2}}$ using the following *rational* powers:

$5^{1.4}, \quad 5^{1.41}, \quad 5^{1.414}, \quad 5^{1.4142}, \quad 5^{1.41421}, \ldots$

Using concepts from advanced mathematics, it can be shown that there is exactly one number that these powers approach. We define $5^{\sqrt{2}}$ to be that number. This process can be used to approximate $5^{\sqrt{2}}$ to as many decimal places as desired. Any other positive irrational exponent can be defined in the same manner, and negative irrational exponents can be defined using reciprocals.

This discussion leads us to the following conclusion: If b is positive, the exponential expression b^x has meaning and can be evaluated for any real number exponent x. Furthermore, it can be shown that all of the familiar rules of exponents are also true for irrational exponents.

Using Your Calculator ▶ Evaluating Exponential Expressions

We can use a calculator to obtain a very good approximation of an exponential expression with an irrational exponent. To find the value of $5^{\sqrt{2}}$ with a reverse entry scientific calculator, we enter:

5 [y^x] 2 [$\sqrt{\ }$] = 9.738517742

With a direct entry graphing calculator, we enter:

5 [^] [2nd] [$\sqrt{\ }$] 2 [)] [ENTER]

```
5^√(2)
          9.738517742
```

Teaching Tip: Ask your students what shape the graph of $f(x) = 1^x$ would have. Then ask what choices of x would make $h(x) = (-4)^x$ undefined.

If $b > 0$ and $b \neq 1$, the function $f(x) = b^x$ is called an **exponential function.** Since x can be any real number, its domain is the set of real numbers, which can be written as $(-\infty, \infty)$.

Because b is positive, the value of $f(x)$ is positive, and the range is the set of positive numbers, which can be written as $(0, \infty)$.

Since $b \neq 1$, an exponential function cannot be the constant function $f(x) = 1^x$, in which $f(x) = 1$ for every real number x.

Exponential Functions

An exponential function with base b is defined by the equations

$$f(x) = b^x \quad \text{or} \quad y = b^x$$

where $b > 0$, $b \neq 1$, and x is a real number. The domain of $f(x) = b^x$ is the interval $(-\infty, \infty)$, and the range is the interval $(0, \infty)$.

Exponential functions have a *constant base* and a *variable exponent.* Some examples of exponential functions are:

$f(x) = 2^x$	$g(x) = \left(\frac{1}{3}\right)^x$	$s(x) = 4^{x+1}$	$P(t) = (1.45)^{-3.04t}$
The base is 2.	The base is $\frac{1}{3}$.	The base is 4.	The base is 1.45.

In the third example, we see that the exponent of an exponential function doesn't have to be just x. It can be a variable expression, such as $x + 1$. In the fourth example, we see that the base can be a positive decimal, and the input variable can be a letter other than x.

The following functions are not exponential functions.

$f(x) = x^2$ $\quad$ $g(x) = x^{1/3}$	$h(x) = (-4)^x$	$f(x) = 1^x$
These have a variable base and a constant exponent.	The base cannot be negative.	The base cannot be 1.

2 Graph Exponential Functions.

Since the domain and range of $f(x) = b^x$ are sets of real numbers, we can graph exponential functions on a rectangular coordinate system. To do this, we will use the familiar **point-plotting method.**

EXAMPLE 1 Graph: $f(x) = 2^x$

Strategy We will graph the function by creating a table of function values and plotting the corresponding ordered pairs.

Why After drawing a smooth curve through the plotted points, we will have the graph.

Solution To graph $f(x) = 2^x$, we select several values for x and find the corresponding values of $f(x)$. If x is -3, and if x is -2, we have:

Caution

We have previously graphed the linear function $f(x) = 2x$ and the squaring function $f(x) = x^2$. For the exponential function $f(x) = 2^x$, note that the variable is in the exponent.

$$f(x) = 2^x$$
$$f(-3) = 2^{-3} \quad \text{Substitute } -3 \text{ for } x.$$
$$= \frac{1}{2^3}$$
$$= \frac{1}{8}$$

$$f(x) = 2^x$$
$$f(-2) = 2^{-2} \quad \text{Substitute } -2 \text{ for } x.$$
$$= \frac{1}{2^2}$$
$$= \frac{1}{4}$$

The points $\left(-3, \frac{1}{8}\right)$ and $\left(-2, \frac{1}{4}\right)$ are on the graph of $f(x) = 2^x$. In a similar way, we find the corresponding values of $f(x)$ for x values of -1, 0, 1, 2, 3, and 4 and list them in a table. Then we plot the ordered pairs and draw a smooth curve through them, as shown below. Notice with the ordered pairs, as the value of x increases, the value of y also increases, but very rapidly as compared with x. This graph is an example of **exponential growth.**

Teaching Tip: To illustrate just how dramatically $f(x) = 2^x$ increases, point out that when $x = 15$, the value of $f(x)$ is about 33,000.

$f(x) = 2^x$

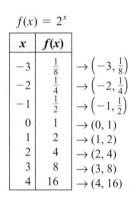

x	$f(x)$	
-3	$\frac{1}{8}$	$\to \left(-3, \frac{1}{8}\right)$
-2	$\frac{1}{4}$	$\to \left(-2, \frac{1}{4}\right)$
-1	$\frac{1}{2}$	$\to \left(-1, \frac{1}{2}\right)$
0	1	$\to (0, 1)$
1	2	$\to (1, 2)$
2	4	$\to (2, 4)$
3	8	$\to (3, 8)$
4	16	$\to (4, 16)$

Because of the variable exponent in their equations, the graphs of exponential functions rise or fall sharply. When graphing them, make sure you plot enough points to show this.

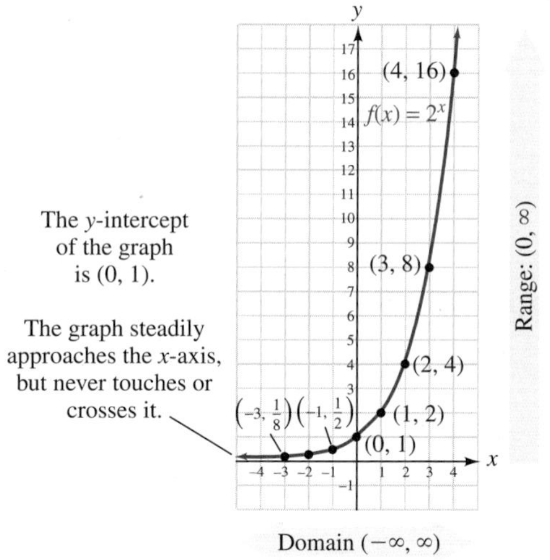

By the vertical line test, we see that this is indeed the graph of a function. Because the graph extends indefinitely to the left and right, the projection of the graph onto the x-axis includes all real numbers. Thus, the domain of $f(x) = 2^x$ is $(-\infty, \infty)$.

Because the projection of the graph onto the y-axis covers only the positive portion of that axis, the range of the function is $(0, \infty)$. Since the graph passes the horizontal line test, the function is one-to-one.

Note that as x decreases, the values of $f(x)$ decrease and approach 0. Thus, the x-axis is a **horizontal asymptote** of the graph. The graph does not have an x-intercept, the y-intercept is $(0, 1)$, and the graph passes through the point $(1, 2)$.

Teaching Example 1 Graph: $f(x) = 3^x$
Answer:

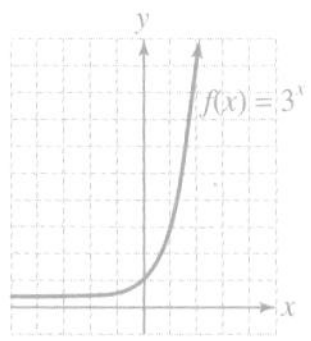

Self Check 1 Graph: $g(x) = 4^x$ See AIE Appendix 3.

Now Try Problem 19

Exponential functions can have a base that is a real number between 0 and 1.

EXAMPLE 2 Graph: $f(x) = \left(\frac{1}{3}\right)^x$

Strategy We will graph the function by creating a table of function values and plotting the corresponding ordered pairs.

Why After drawing a smooth curve through the plotted points, we will have the graph.

Solution If $x = -2$ and if $x = -1$, we have

$$f(x) = \left(\frac{1}{3}\right)^x \qquad f(x) = \left(\frac{1}{3}\right)^x$$

$$f(-2) = \left(\frac{1}{3}\right)^{-2} \qquad f(-1) = \left(\frac{1}{3}\right)^{-1}$$

$$= \left(\frac{3}{1}\right)^2 \quad \text{Recall: } \left(\frac{x}{y}\right)^{-n} = \left(\frac{y}{x}\right)^n. \qquad = \left(\frac{3}{1}\right)^1$$

$$= 9 \qquad = 3$$

Success Tip

Since $\frac{1}{3} = 3^{-1}$, and since the rules for exponents hold for any real number, we can rewrite this function as:

$$f(x) = \left(\frac{1}{3}\right)^x = (3^{-1})^x = 3^{-x}$$

The points $(-2, 9)$ and $(-1, 3)$ are on the graph of $f(x) = \left(\frac{1}{3}\right)^x$. In a similar way, we find the corresponding values of $f(x)$ for $x = 0$, 1, and 2 and list them in a table. Then we plot the ordered pairs and draw a smooth curve through them, as shown below. Notice with the ordered pairs, as the value of x increases, the value of y decreases very rapidly as compared with x. This graph is an example of **exponential decay.**

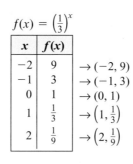

Teaching Tip: Ask why -3 was not selected to be used as a value of x in the table.

$f(x) = \left(\frac{1}{3}\right)^x$

x	$f(x)$	
-2	9	$\to (-2, 9)$
-1	3	$\to (-1, 3)$
0	1	$\to (0, 1)$
1	$\frac{1}{3}$	$\to \left(1, \frac{1}{3}\right)$
2	$\frac{1}{9}$	$\to \left(2, \frac{1}{9}\right)$

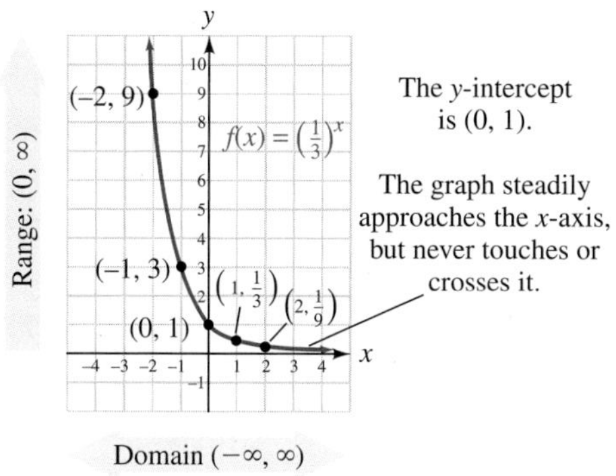

The graph passes the vertical line test, and so it is indeed the graph of a function. Because the graph extends indefinitely to the left and right, the projection of the graph onto the x-axis includes all real numbers. Thus, the domain of $f(x) = \left(\frac{1}{3}\right)^x$ is $(-\infty, \infty)$. Because the projection of the graph onto the y-axis covers only the positive portion of that axis, the range of the function is $(0, \infty)$.

Note that as x increases, the values of $f(x)$ decrease and approach 0. Thus, the x-axis is a horizontal asymptote of the graph. The graph does not have an x-intercept, the y-intercept is $(0, 1)$, and the graph passes through the point $\left(1, \frac{1}{3}\right)$.

Teaching Example 2 Graph: $f(x) = \left(\dfrac{1}{4}\right)^x$

Answer:

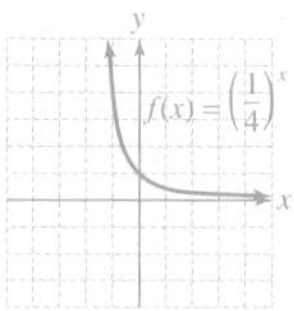

Self Check 2 Graph: $g(x) = \left(\frac{1}{2}\right)^x$ See AIE Appendix 3.

Now Try Problem 23

In Example 1 (where $b = 2$), the values of y increase as the values of x increase. Since the graph rises as we move to the right, we call the function an *increasing function.* When $b > 1$, the larger the value of b, the steeper the curve, as shown in figure (a) below.

In Example 2 $\left(\text{where } b = \frac{1}{3}\right)$, the values of y decrease as the values of x increase. Since the graph drops as we move to the right, we call the function a *decreasing function.* When $0 < b < 1$, the smaller the value of b, the steeper the curve, as shown in figure (b) below.

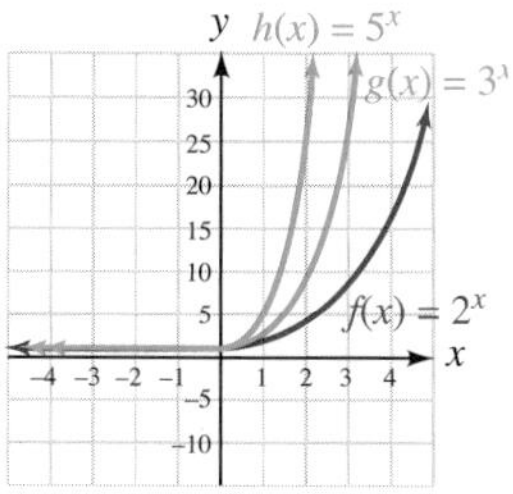

(a)

The bases of these exponential functions are 2, 3, and 5. Notice that the largest base, 5, has the steepest graph.

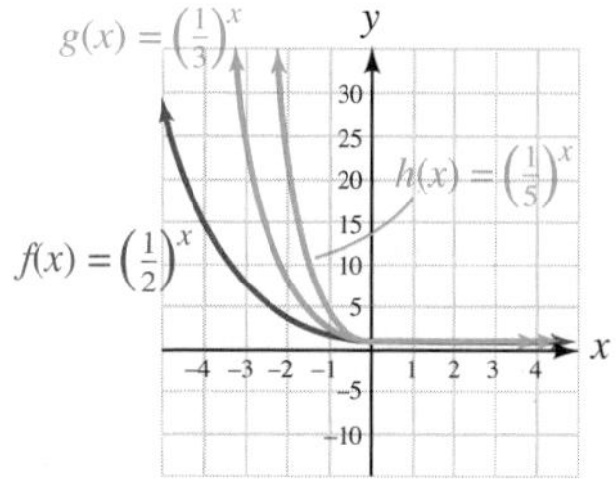

(b)

The bases of these exponential functions are $\frac{1}{2}$, $\frac{1}{3}$, and $\frac{1}{5}$, which are numbers between 0 and 1. Notice that the smallest base, $\frac{1}{5}$, has the steepest graph.

Examples 1 and 2 illustrate the following properties of exponential functions.

Properties of Exponential Functions

1. The domain of the exponential function $f(x) = b^x$ is the interval $(-\infty, \infty)$ and the range is the interval $(0, \infty)$.
2. The graph has a y-intercept of $(0, 1)$.
3. The x-axis is an asymptote of the graph.
4. The graph of $f(x) = b^x$ passes through the point $(1, b)$.
5. Exponential functions are one-to-one.
6. If $b > 1$, then $f(x) = b^x$ is an **increasing function.** If $0 < b < 1$, then $f(x) = b^x$ is a **decreasing function.**

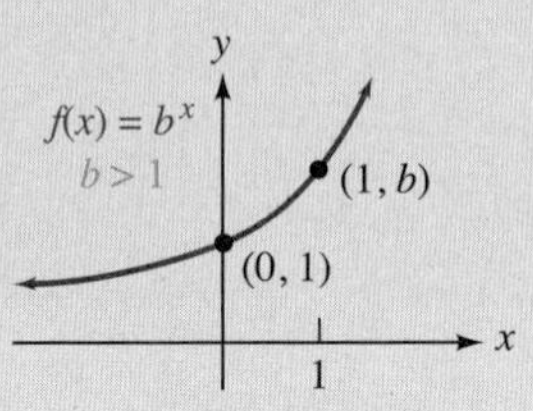

Increasing function

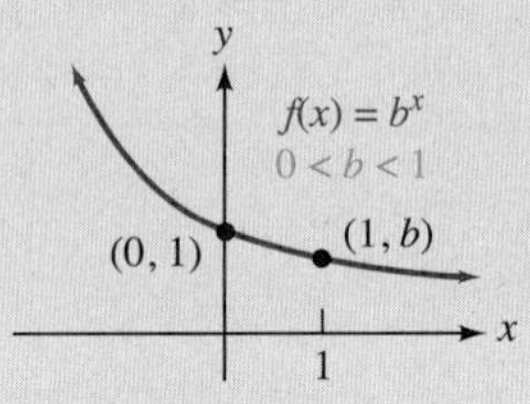

Decreasing function

Using Your Calculator ▶ Graphing Exponential Functions

To use a graphing calculator to graph $f(x) = \left(\frac{2}{3}\right)^x$ and $g(x) = \left(\frac{3}{2}\right)^x$, we enter the right sides of the equations after the symbols $Y_1 =$ and $Y_2 =$. The screen will show the following equations.

$Y_1 = (2/3)$^X
$Y_2 = (3/2)$^X

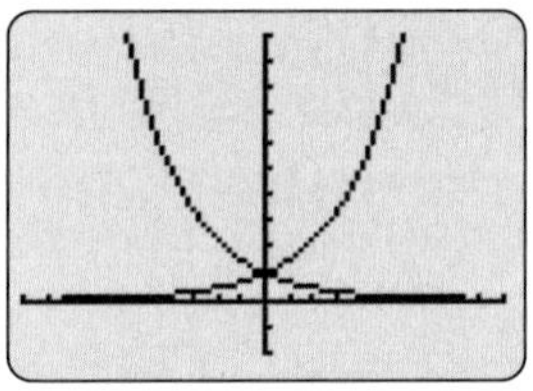

If we press GRAPH, we will obtain the display shown.

We note that the graph of $f(x) = \left(\frac{2}{3}\right)^x$ passes through (0, 1). Since $\frac{2}{3} < 1$, the function is decreasing. The graph of $g(x) = \left(\frac{3}{2}\right)^x$ also passes through (0, 1). Since $\frac{3}{2} > 1$, the function is increasing. Since both graphs pass the horizontal line test, each function is one-to-one.

The graphs of many exponential functions are horizontal and vertical translations of basic graphs.

EXAMPLE 3 Graph each function by using a translation:

a. $g(x) = 2^x - 4$ **b.** $g(x) = \left(\frac{1}{3}\right)^{x+3}$

Strategy We will graph $g(x) = 2^x - 4$ by translating the graph of $f(x) = 2^x$ downward 4 units. We will graph $g(x) = \left(\frac{1}{3}\right)^{x+3}$ by translating the graph of $f(x) = \left(\frac{1}{3}\right)^x$ to the left 3 units.

Why The subtraction of 4 in $g(x) = 2^x - 4$ causes a vertical shift of the graph of the base-2 exponential function 4 units downward. The addition of 3 to x in $g(x) = \left(\frac{1}{3}\right)^{x+3}$ causes a horizontal shift of the graph of the base-$\frac{1}{3}$ exponential function 3 units to the left.

Solution

Teaching Tip: Ask your students for the function that would translate the graph of $f(x) = 2^x$ upward 4 units.

Answer: $f(x) = 2^x + 4$

a. The graph of $g(x) = 2^x - 4$ will be the same shape as the graph of $f(x) = 2^x$. We call this a **vertical translation.** To graph $g(x) = 2^x - 4$, simply translate each point on the graph of $f(x) = 2^x$ down 4 units. See figure (a) below.

b. The graph of $g(x) = \left(\frac{1}{3}\right)^{x+3}$ will be the same shape as the graph of $f(x) = \left(\frac{1}{3}\right)^x$. We call this a **horizontal translation**. To graph $g(x) = \left(\frac{1}{3}\right)^{x+3}$, simply translate each point on the graph of $f(x) = \left(\frac{1}{3}\right)^x$ to the left 3 units. See figure (b) below.

Teaching Tip: Ask your students for the function that would translate the graph of $f(x) = \left(\frac{1}{3}\right)^x$ to the right 3 units.

Answer: $f(x) = \left(\frac{1}{3}\right)^{x-3}$

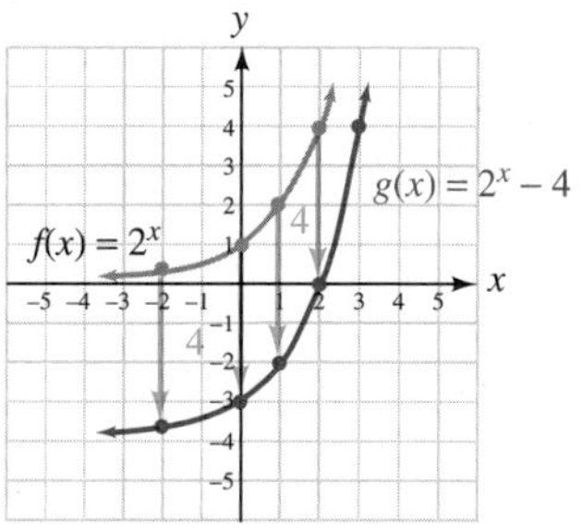

(a)

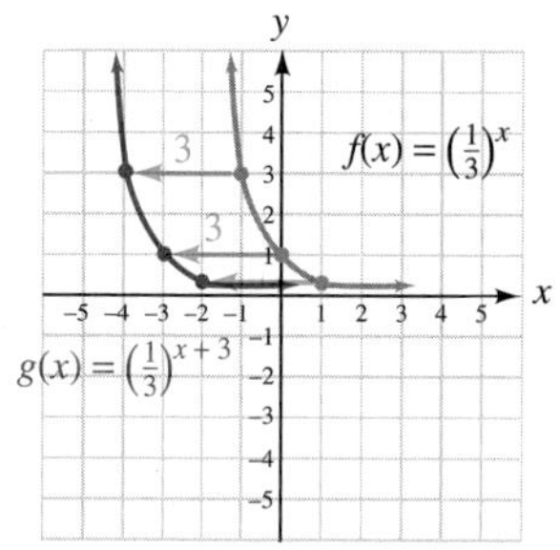

(b)

Teaching Example 3 Graph the function $g(x) = 3^x - 2$ by using a translation.

Answer:

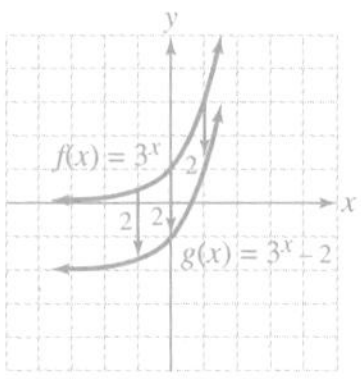

Self Check 3 Graph each function by using a translation:

a. $g(x) = \left(\frac{1}{4}\right)^x + 2$ **b.** $g(x) = 4^{x-3}$ See AIE Appendix 3.

Now Try Problems 27 and 29

3 Use Exponential Functions in Applications Involving Growth or Decay.

Many real-world situations can be modeled by exponential functions that describe how a quantity grows or decays over time. Some examples of this include the studies of populations, bacteria, heat transfer, radioactive substances, drug concentrations, and financial accounts. Two examples of such functions are:

Exponential growth:

$c(t) = 5(1.034)^t$

A constant — A base greater than 1

Exponential decay:

$f(n) = 650(0.94)^n$

A constant — A base between 0 and 1

EXAMPLE 4

©jason aron/Shutterstock.com

Professional Baseball Salaries. The exponential function $s(t) = 650{,}000(1.09)^t$ approximates the average annual salary of a major league baseball player, where t is the number of years after 1990. (Source: Baseball Almanac) **a.** Graph the function. **b.** Use the function to determine the average annual salary in 2020, if the current trend continues.

Strategy For part a, we will graph the function by creating a table of function values and plotting the resulting ordered pairs. For part b, we will find $s(30)$.

Why After drawing a smooth curve through the plotted points, we will have the graph. Since the year 2020 is 30 years after 1990, $t = 30$.

Solution **a.** The function values for $t = 0$ and $t = 5$ are calculated as follows:

The Language of Algebra

The word **exponential** is used in many settings to describe rapid growth. For example, we hear that the processing power of computers is growing *exponentially.*

$t = 0$ (*the year 1990*)	$t = 5$ (*the year 1995*)
$s(t) = 650{,}000(1.09)^t$	$s(t) = 650{,}000(1.09)^t$
$s(0) = 650{,}000(1.09)^0$	$s(5) = 650{,}000(1.09)^5$
$= 650{,}000(1)$	$\approx 1{,}000{,}106$ Use a calculator.
$= 650{,}000$	

To approximate $s(5)$, use the keystrokes 650000 [×] 1.09 [y^x] 5 [=] on a scientific calculator and 650000 [×] 1.09 [^] 5 [ENTER] on a graphing calculator.

In a similar way, we find the corresponding values of $s(t)$ for t-values of 10, 15, and 20 and list them in a table. Then we plot the ordered pairs and draw a smooth curve through them to get the graph shown here.

Teaching Tip: Ask your students what part of the function equation indicates that it describes exponential growth.

t	$s(t)$
0	650,000
5	1,000,106
10	1,538,786
15	2,367,614
20	3,642,867

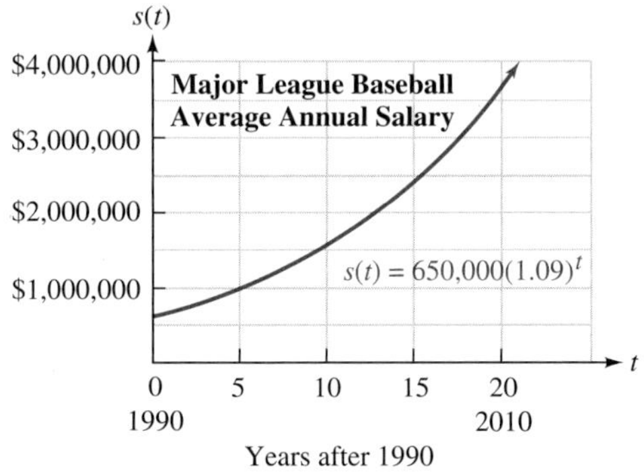

b. To estimate the average annual major league baseball salary in 2020, which is 30 years after 1990, we find $s(30)$.

$$s(t) = 650{,}000(1.09)^t \quad \text{This is the exponential growth model.}$$
$$s(30) = 650{,}000(1.09)^{30} \quad \text{Substitute 30 for } t.$$
$$\approx 8{,}623{,}991 \quad \text{Use a calculator.}$$

If the trend continues, in 2020, the average annual salary will be approximately \$8,623,991.

Teaching Example 4 Salaries. Use the function in Example 4 to determine the average annual salary in 2016, if the current trend continues.
Answer: \$6,109,453

Self Check 4 **Salaries.** Use the function in Example 4 to determine the average annual salary in 2018, if the current trend continues. \$7,258,641

Now Try Problem 43

Using Your Calculator ▶ Graphing Exponential Functions

To use a graphing calculator to graph the exponential function $s(t) = 170{,}000(1.12)^t$, we enter the right side of the equation after the symbol $Y_1 =$ and replace the variable t with x. The display will show the equation

$$Y_1 = 170000(1.12\text{^}X)$$

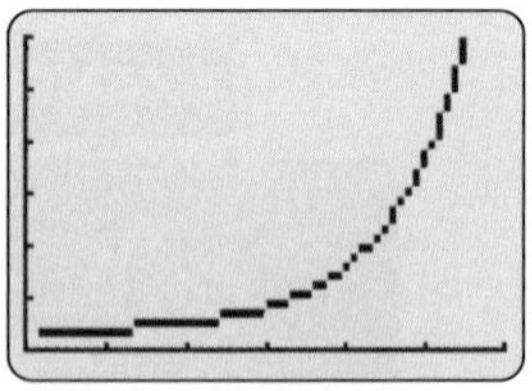

With window settings [0, 30] for x and Xscale = 5 and [0, 3000000] for y and Yscale = 500000, we obtain the display shown when we press [GRAPH].

Up to this point, the financial application problems that we have solved involved **simple interest,** which is calculated using the formula $I = Prt$. However, most savings accounts and investments pay compound interest rather than simple interest. **Compound interest** is paid more than once a year on the principal and *previously earned interest.* The following compound interest formula is a useful application of exponential functions.

Formula for Compound Interest

If \$$P$ is deposited in an account and interest is paid k times a year at an annual rate r, the amount A in the account after t years is given by

$$A = P\left(1 + \frac{r}{k}\right)^{kt}$$

EXAMPLE 5 **Educational Savings Plan.** To save for college, parents of a newborn child invest \$12,000 in a mutual fund at 10% interest, compounded quarterly.

a. Find a function for the amount in the account after t years.

b. If the quarterly interest paid is continually reinvested, how much money will be in the account when the child is 18 years old?

Strategy To write a function for the amount in the account after t years, we will substitute the given values for P, r, and k into the compound interest formula.

Why The resulting equation will involve only two variables, A and t. Then we can write that equation using function notation.

Solution

a. When we substitute 12,000 for P, 0.10 for r, and 4 for k in the formula for compound interest, the resulting formula involves only two variables, A and t.

$$A = P\left(1 + \frac{r}{k}\right)^{kt}$$ This is the compound interest model.

$$A = \mathbf{12{,}000}\left(1 + \frac{\mathbf{0.10}}{\mathbf{4}}\right)^{\mathbf{4}t}$$ Since the interest is compounded quarterly, $k = 4$. Express $r = 10\%$ as a decimal.

The Language of Algebra

The following words indicate the number of times that interest is paid by a financial institution in one year.

Annually:	1 time
Semiannually:	2 times
Quarterly:	4 times
Monthly:	12 times
Daily:	365 times

Since the value of A depends on the value of t, we can express this relationship using function notation.

$$A(t) = 12{,}000\left(1 + \frac{0.10}{4}\right)^{4t}$$

$$A(t) = 12{,}000(1 + 0.025)^{4t}$$ Evaluate within the parentheses: $\frac{0.10}{4} = 0.025$.

$$A(t) = 12{,}000(1.025)^{4t}$$ The base of this exponential function is 1.025.

b. To find how much money will be in the account when the child is 18 years old, we need to find $A(18)$.

$$A(t) = 12{,}000(1.025)^{4(t)}$$ This is the exponential growth model.

$$A(\mathbf{18}) = 12{,}000(1.025)^{4(\mathbf{18})}$$ Substitute 18 for t.

$$= 12{,}000(1.025)^{72}$$ Evaluate the exponent: $4(18) = 72$.

$$\approx 71{,}006.74$$ Use a scientific calculator and press these keys: 12000 × 1.025 y^x 72 =.

When the child is 18 years old, the account will contain \$71,006.74.

Teaching Example 5 Savings plans. In Example 5, how much money would be in the account after 18 years if the parents initially invested $10,000?
Answer: $59,172.28

Self Check 5 **Savings Plans.** In Example 5, how much money would be in the account after 18 years if the parents initially invested $20,000? $118,344.56

Now Try Problem 53

Using Your Calculator ▶ Solving Investment Problems

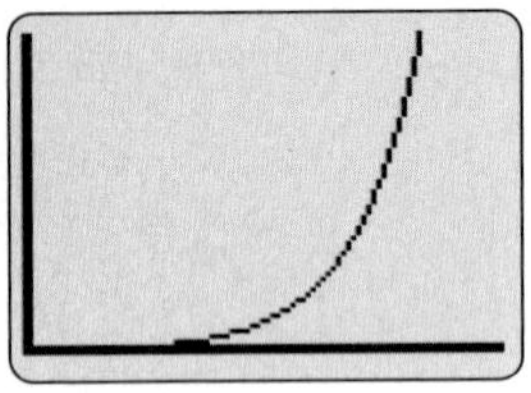

Suppose $1 is deposited in an account earning 6% annual interest, compounded monthly. To use a graphing calculator to estimate how much will be in the account in 100 years, we can substitute 1 for P, 0.06 for r, and 12 for k in the formula and simplify.

$$A = P\left(1 + \frac{r}{k}\right)^{kt} = \mathbf{1}\left(1 + \frac{\mathbf{0.06}}{\mathbf{12}}\right)^{\mathbf{12}t} = (1.005)^{12t}$$

We now graph the function $A(t) = (1.005)^{12t}$ using window settings of [0, 120] and [0, 400] with Xscale = 1 and Yscale = 1 to obtain the graph shown. We can then trace and zoom to estimate that $1 grows to be approximately $397 in 100 years. From the graph, we can see that the money grows slowly in the early years and rapidly in the later years.

Examples 4 and 5 are applications illustrating exponential growth. In the next example, we see an application of exponential decay.

EXAMPLE 6 **Medications.** The most common way people take medications is orally (by mouth). Medications, when swallowed, travel from the stomach or small intestine into the bloodstream, but they are eventually eliminated from the body by the kidneys and the liver. If a patient takes a 250-milligram dose of an antibiotic, the function $A(t) = 250(0.58)^t$ approximates the amount of medication (in milligrams) left in the patient's bloodstream t hours after it is taken.

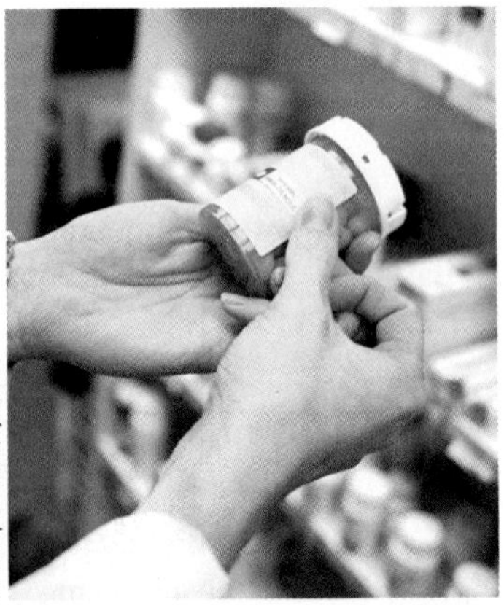

©iStockphoto.com/sjlocke

a. Graph the function.

b. Use the function to determine the amount of medication in the patient's bloodstream 10 hours after taking the dose.

Strategy For part a, we will graph the function by creating a table of function values and plotting the resulting ordered pairs. For part b, we will find $A(10)$.

Why After drawing a smooth curve through the plotted points, we will have the graph. Since the variable t represents the time since taking the dose, $t = 10$.

Solution

a. The function values for $t = 0$ and $t = 2$ are calculated as follows:

$t = 0$:

$$A(t) = 250(0.58)^t$$
$$A(\mathbf{0}) = 250(0.58)^{\mathbf{0}}$$
$$= 250(1)$$
$$= 250$$

$t = 2$:

$$A(t) = 250(0.58)^t$$
$$A(\mathbf{2}) = 250(0.58)^{\mathbf{2}}$$
$$\approx 84.1 \quad \text{Use a calculator.}$$

Teaching Tip: Ask your students what part of the function equation indicates that it describes exponential decay.

Success Tip

$A(t) = 250(0.58)^t$ is a decreasing function because the base, 0.58, is such that $0 < 0.58 < 1$.

In a similar way, we find the corresponding values of $A(t)$ for t-values of 4 and 6, and list them in a table. Then we plot the ordered pairs and draw a smooth curve through them to get the graph shown below.

t	$A(t)$
0	250
2	84.1
4	28.3
6	9.5

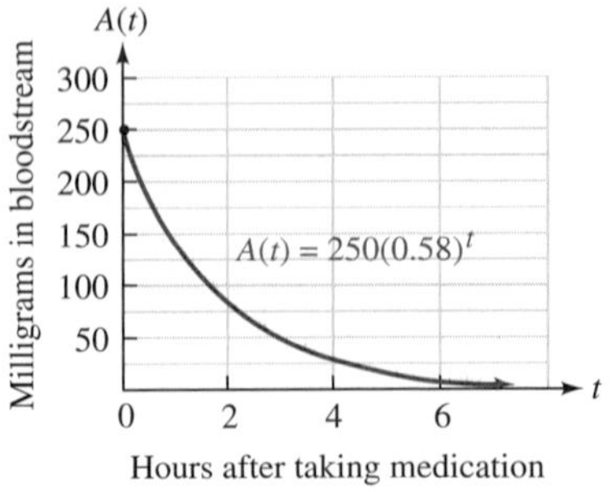

b. To estimate the amount of medication in the patient's bloodstream 10 hours after taking the dose, we find $A(10)$.

$$A(t) = 250(0.58)^t \quad \text{This is the exponential decay model.}$$
$$A(10) = 250(0.58)^{10} \quad \text{Substitute 10 for } t.$$
$$\approx 1.1 \quad \text{Use a calculator.}$$

In 10 hours, there will be approximately 1.1 milligrams of medication in the patient's bloodstream.

Teaching Example 6 Medications. If a patient takes a 50-milligram dose of an antihistamine, the function $A(t) = 50(0.72)^t$ approximates the amount of medication (in milligrams) left in the patient's bloodstream t hours after it is taken. Determine the amount of medication in the patient's bloodstream 8 hours after taking the dose.
Answer: About 3.6 milligrams

Self Check 6 **Medications.** Use the function in Example 6 to determine the amount of medication in the patient's bloodstream $8\frac{1}{2}$ hours after taking the dose. About 2.4 milligrams

Now Try Problem 44

SECTION 11.3 STUDY SET

VOCABULARY

Fill in the blanks.

1. $f(x) = 2^x$ and $f(x) = \left(\frac{1}{4}\right)^x$ are examples of exponential functions.
2. Exponential functions have a constant base and a variable exponent .
3. The graph of $f(x) = 3^x$ approaches, but never touches, the negative portion of the x-axis. Thus, the x-axis is an asymptote of the graph.
4. Compound interest is paid on the principal and previously earned interest.

CONCEPTS

5. Refer to the graph shown at the right.

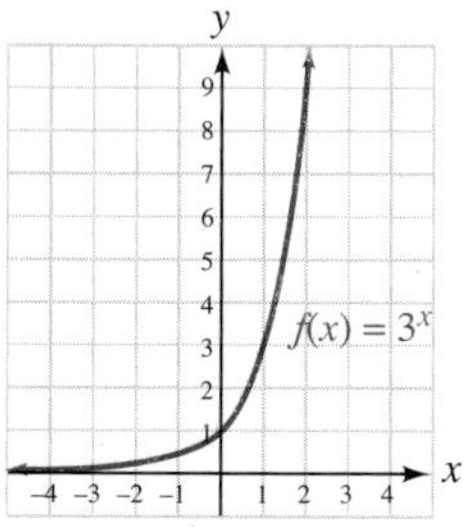

   a. What type of function is $f(x) = 3^x$? Exponential
   b. What is the domain of the function? $(-\infty, \infty)$
   c. What is the range of the function? $(0, \infty)$
   d. What is the y-intercept of the graph? What is the x-intercept of the graph? (0, 1); none
   e. Is the function one-to-one? Yes
   f. What is an asymptote of the graph? The x-axis ($y = 0$)
   g. Is f an increasing or a decreasing function? Increasing
   h. The graph passes through the point $(1, y)$. What is y? 3

6. Which of the following functions are exponential functions? c, e

   a. $f(x) = x^2$ b. $g(x) = 4x$ c. $h(x) = 8^x$
   d. $s(x) = \frac{1}{x}$ e. $T(x) = (0.92)^{x+1}$ f. $r(x) = x^3$
   g. $P(x) = \sqrt{x}$ h. $d(x) = |x|$

7. Evaluate each expression without a calculator.

   a. 3^{-2} $\frac{1}{9}$ b. $\left(\frac{1}{2}\right)^4$ $\frac{1}{16}$ c. $\left(\frac{1}{5}\right)^{-2}$ 25

8. Evaluate each expression using a calculator. Round to the nearest tenth.

   a. $20{,}000(1.036)^{52}$ 125,816.4 b. $92(0.88)^6$ 42.7

9. Match each function with its graph shown below.

   a. $f(x) = x^2$ iii. b. $f(x) = 2^x$ iv.
   c. $f(x) = 2$ ii. d. $f(x) = 2x$ i.

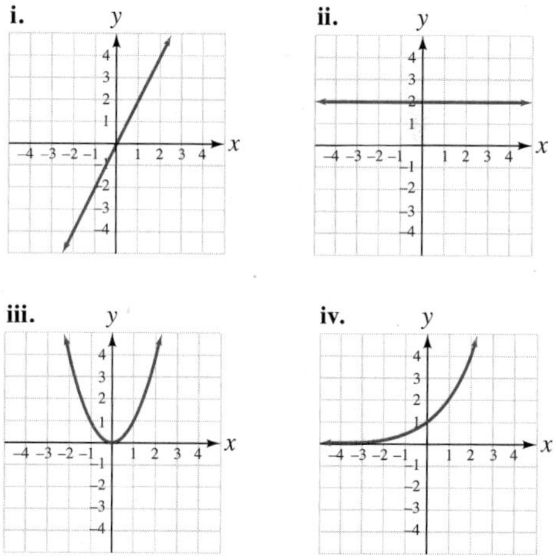

10. a. Two exponential functions of the form $f(x) = b^x$ are graphed in figure (a) below. Which function has the larger base b, the one graphed in red or the one graphed in blue? Red

    b. Two exponential functions of the form $f(x) = b^x$ are graphed in figure (b) below. Which function has the smaller base b? Red

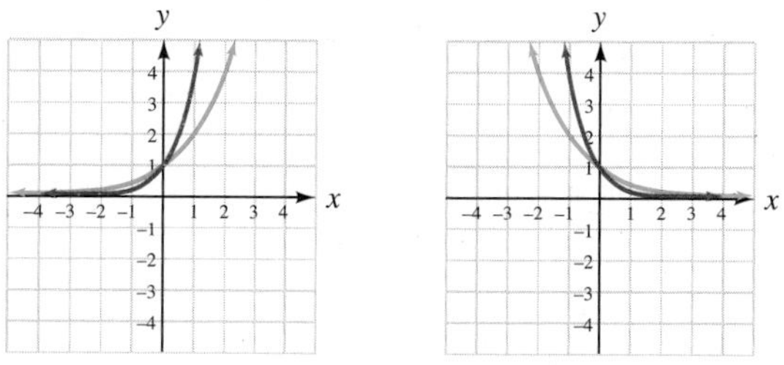

(a) (b)

11. Determine the domain and range of each exponential function graphed below.

a.

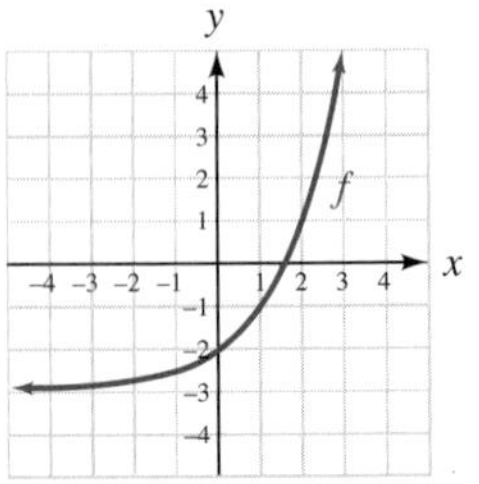

D: $(-\infty, \infty)$, R: $(-3, \infty)$

b.

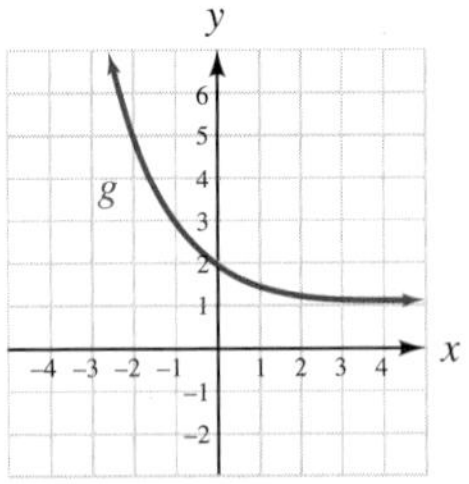

D: $(-\infty, \infty)$, R: $(1, \infty)$

12. Fill in the blanks.

a. The graph of $g(x) = 4^x + 3$ is similar to the graph of $f(x) = 4^x$, but it is translated 3 units upward.

b. The graph of $g(x) = 4^{x-3}$ is similar to the graph of $f(x) = 4^x$, but it is translated 3 units to the right.

13. Complete the table of function values shown here.

$f(x) = 5^x$

x	$f(x)$
-3	$\frac{1}{125}$
-2	$\frac{1}{25}$
-1	$\frac{1}{5}$
0	1
1	5
2	25
3	125

14. Determine whether each of the following functions model exponential growth or exponential decay.

a. $D(t) = 150(0.44)^t$ Decay b. $H(t) = 15{,}000(1.03)^t$ Growth

15. Match each situation to the exponential graph that best models it.

a. The number of cell phone subscribers in the world over the past 5 years ii.

b. The level of caffeine in the bloodstream after drinking a cup of coffee i.

c. The amount of money in a bank account earning interest compounded quarterly ii.

d. The number of rabbits in a population with a high birth rate ii.

e. The amount of water in a shirt that was just washed and hung on a clothesline to dry i.

i.

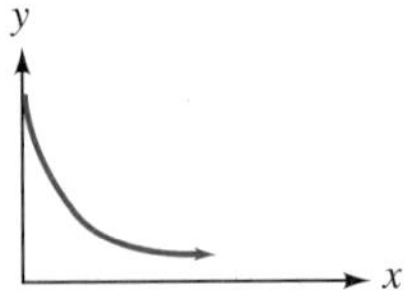

ii.

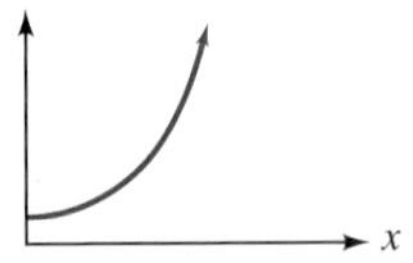

16. What formula is used to determine the amount of money in a savings account earning compound interest?

$A = P\left(1 + \frac{r}{k}\right)^{kt}$

NOTATION

17. For an exponential function of the form $f(x) = b^x$, what are the restrictions on b? $b > 0, b \neq 1$

18. In $A(t) = 16{,}000\left(1 + \frac{0.05}{365}\right)^{365t}$, what is the base and what is the exponent? $1 + \frac{0.05}{365}$, $365t$

GUIDED PRACTICE

***Graph each function.* See Examples 1 and 2.** See AIE Appendix 3.

19. $f(x) = 3^x$

20. $f(x) = 6^x$

21. $f(x) = 5^x$

22. $f(x) = 7^x$

23. $f(x) = \left(\frac{1}{4}\right)^x$

24. $f(x) = \left(\frac{1}{5}\right)^x$

25. $f(x) = \left(\frac{1}{6}\right)^x$

26. $f(x) = \left(\frac{1}{8}\right)^x$

***Graph each function by plotting points or using a translation.* See Example 4.** See AIE Appendix 3.

27. $g(x) = 3^x - 2$

28. $g(x) = 2^x + 1$

29. $g(x) = 2^{x+1}$

30. $g(x) = 3^{x-1}$

31. $g(x) = 4^{x-1} + 2$

32. $g(x) = 4^{x+1} - 2$

33. $g(x) = -2^x$

34. $g(x) = -3^x$

***Use a graphing calculator to graph each function. Determine whether the function is an increasing or a decreasing function.* See Using Your Calculator: Graphing Exponential Functions.** See AIE Appendix 3.

35. $f(x) = \frac{1}{2}(3^{x/2})$ Inc

36. $f(x) = -3(2^{x/3})$ Dec

37. $f(x) = 2(3^{-x/2})$ Dec

38. $f(x) = -\frac{1}{4}(2^{-x/2})$ Inc

APPLICATIONS

39. **CO_2 Concentration.** The exponential growth model below illustrates the rise in atmospheric carbon dioxide from 1744 to 2006. The historical data (shown with blue points) comes from ice cores, and modern data (shown with red points) was collected from the Mauna Loa Observatory in Hawaii.

a. Estimate the atmospheric carbon dioxide concentrations in 1800, 1900, and 2000. 280 ppm, 295 ppm, 370 ppm

b. In approximately what year did the concentration surpass 325 parts per million? About 1970

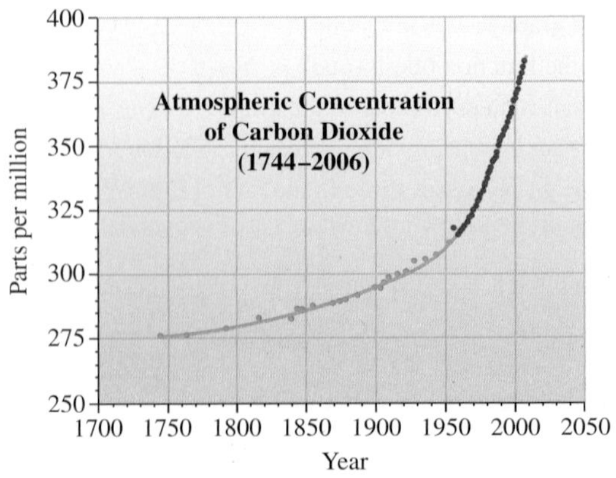

Source: www.eoearth.org

40. Global Warming. The following graph from the United States Environmental Protection Agency shows the projected sea level changes due to anticipated global warming.

a. What type of function does it appear could be used to model the sea level change? Exponential

b. When were the earliest instrumental records of sea level change made? 1870

c. For the year 2100, what is the upper-end projection for sea level change? What is the lower-end projection? About 480 mm, about 210 mm

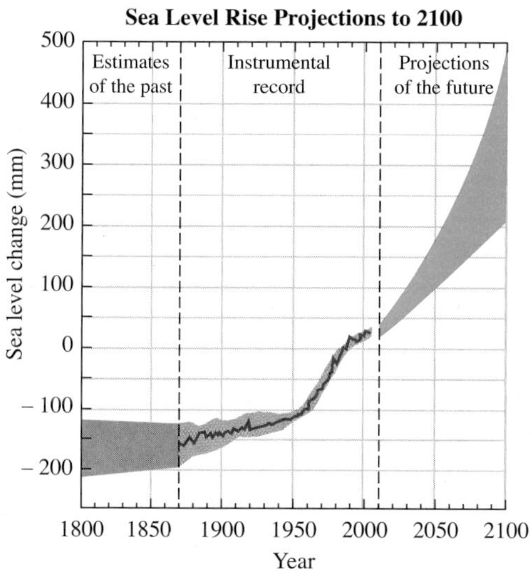

Source: United States Environmental Protection Agency

41. Value of a Car. The graph shows how the value of the average car depreciates as a percent of its original value over a 10-year period. It also shows the yearly maintenance costs as a percent of the car's value.

a. When is the car worth half of its purchase price? At the end of the 2nd year

b. When is the car worth a quarter of its purchase price? At the end of the 4th year

c. When do the average yearly maintenance costs surpass the value of the car? During the 7th year

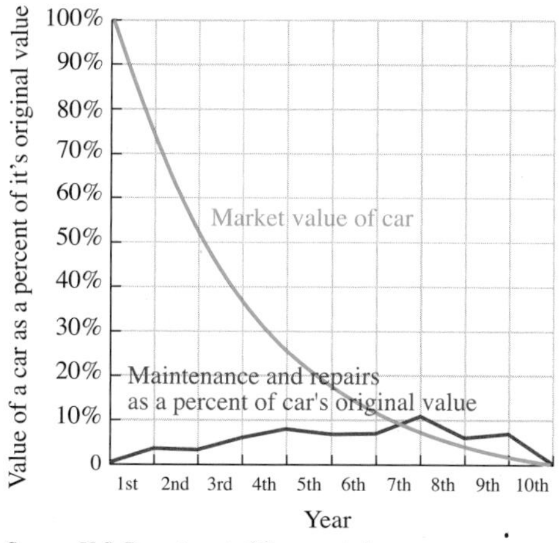

Source: U.S. Department of Transportation

42. Diving. *Bottom time* is the time a scuba diver spends descending plus the actual time spent at a certain depth. Graph the bottom time limits given in the table as ordered pairs of the form (depth, bottom time).Then draw a smooth curve through the points. See AIE Appendix 3.

Depth (ft)	Bottom time limit (min)
30	no limit
35	310
40	200
50	100
60	60
70	50

Depth (ft)	Bottom time limit (min)
80	40
90	30
100	25
110	20
120	15
130	10

43. Computer Viruses. Suppose the number of computers infected by the spread of a virus through an e-mail is described by the exponential function $c(t) = 5(1.034)^t$, where t is the number of minutes since the first infected e-mail was opened.

a. Graph the function. Scale the t-axis from 0 to 400, in units of 50. Scale the $c(t)$-axis from 0 to 800,000 in units of 100,000. See AIE Appendix 3.

b. Use the function to determine the number of infected computers in 8 hours, which is 480 minutes. About 46,650,000 computers.

44. Salvage Value. A small business purchased a computer for \$5,000. The value (in dollars) of the computer, t years after its purchase, is given by the exponential function $v(t) = 5{,}000(0.75)^t$.

a. Graph the function. Scale the t-axis from 0 to 10 in units of 2. Scale the $v(t)$-axis from 0 to 6,000 in units of 1,000. See AIE Appendix 3.

b. Use the function to determine the value of the computer 12 years after it is purchased. About \$158

45. Guitars. The frets on the neck of a guitar are placed so that pressing a string against them determines the strings' vibrating length. The exponential function $f(n) = 650(0.94)^n$ gives the vibrating length (in millimeters) of a string on a certain guitar for the fret number n. Find the length of the vibrating string when a guitarist holds down a string at the 7th fret. About 422 mm

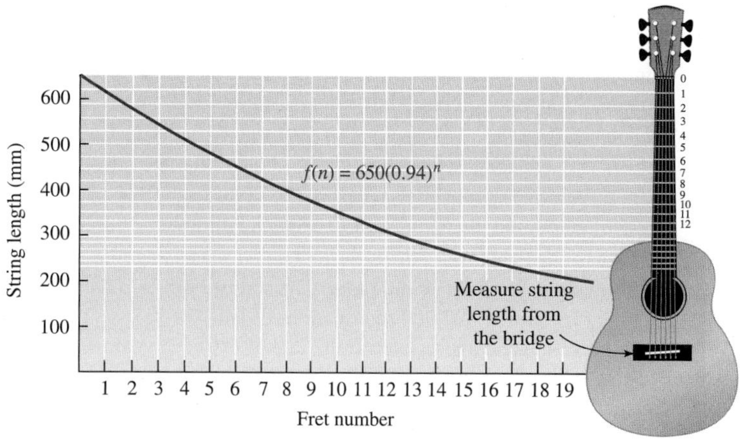

46. Bacterial Cultures. A colony of 6 million bacteria was determined to be growing in the culture medium shown in illustration (a). If the population P of bacteria after t hours is given by the function $P(t) = 6{,}000{,}000(2.3)^t$, find the population in the culture later in the day using the information given in illustration (b). 167,904,600

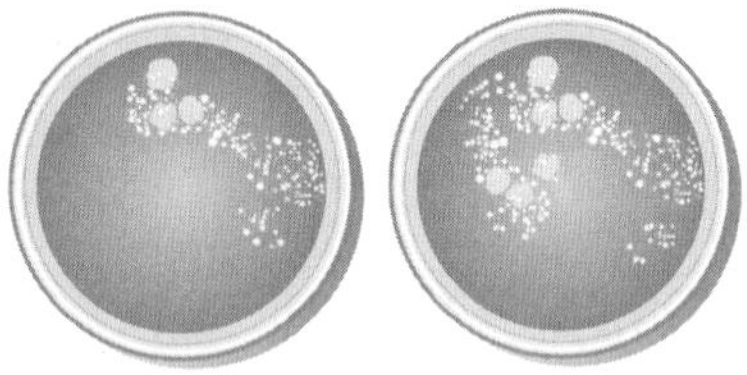

Date: 8-16-08 Time: 6:00 AM (a) | Date: 8-16-08 Time: 10:00 AM (b)

47. Radioactive Decay. Five hundred grams of a radioactive material decays according to the formula $A = 500\left(\frac{2}{3}\right)^t$, where t is measured in years. Find the amount present in 10 years. Round to the nearest one-tenth of a gram. 8.7 gm

48. Discharging a Battery. The charge remaining in a battery decreases as the battery discharges. The charge C (in coulombs) after t days is given by the function $C(t) = 0.0003(0.7)^t$. Find the charge after 5 days. 0.000050421 coulombs

49. from **Campus to Careers**

Social Worker

©iStockphoto/Chris Schmidt

The function $P(t) = 35.8(1.06)^t$ approximates the number of people (in millions) in the United States living in poverty, where t is the number of years after 2006. Use the function to complete the table below. Round to the nearest tenth. (Source: U.S. Census Bureau)

Year	2006	2007	2008	2009
Number in poverty (in millions)	35.8	37.9	40.2	42.6

50. Publishing Books. The function $W(f) = 4{,}066(0.8753)^f$ approximates the number of words that can be typeset on a standard page using the Times Roman font size f. Find the number of words that can be typeset on a page using the font size 12. (Source: writersservices.com) About 822 words

51. Population Growth. The population of North Rivers is decreasing exponentially according to the formula $P = 3{,}745(0.93)^t$, where t is measured in years from the present date. Find the population in 6 years, 9 months. 2,295

52. The Louisiana Purchase. In 1803, the United States negotiated the Louisiana Purchase with France. The country doubled its territory by adding 827,000 square miles of land for $15 million. If the land appreciated at the rate of 6% each year, what would one square mile of land be worth in 2005? About $2,346,230

In Exercises 53–58, assume that there are no deposits or withdrawals.

53. Compound Interest. An initial deposit of $10,000 earns 8% interest, compounded quarterly. How much will be in the account after 10 years? $22,080.40

54. Compound Interest. An initial deposit of $10,000 earns 8% interest, compounded monthly. How much will be in the account after 10 years? $22,196.40

55. Comparing Interest Rates. How much more interest could $1,000 earn in 5 years, compounded quarterly, if the annual interest rate were $5\frac{1}{2}$% instead of 5%? $32.03

56. Comparing Savings Plans. Which institution in the ads provides the better investment? Fidelity

Fidelity Savings & Loan	Union Trust
Earn 5.25% compounded monthly	Money Market Account paying 5.35% compounded annually

57. Compound Interest. If $1 had been invested on July 4, 1776, at 5% interest, compounded annually, what would it be worth on July 4, 2076? $2,273,996.13

58. Frequency of Compounding. $10,000 is invested in each of two accounts, both paying 6% annual interest. In the first account, interest compounds quarterly, and in the second account, interest compounds daily. Find the difference between the accounts after 20 years. $291.27

WRITING

59. If world population is increasing exponentially, why is there cause for concern?

60. How do the graphs of $f(x) = 3^x$ and $g(x) = \left(\frac{1}{3}\right)^x$ differ? How are they similar?

61. A snowball rolling downhill grows *exponentially* with time. Explain what this means. Sketch a simple graph that models the situation.

62. Explain why the change in temperature of a cup of hot coffee left unattended on a kitchen table is an example of exponential decay.

63. Let $f(x) = \left(\frac{1}{5}\right)^x$. Explain why we can rewrite the function equation as $f(x) = 5^{-x}$.

64. Explain why the graph of $f(x) = 3^x$ gets closer and closer to the x-axis as the values of x decrease. Does the graph ever cross the x-axis? Explain why or why not.

65. Describe the graphs of $f(x) = x^2$ and $g(x) = 2^x$ in words.

66. Write a paragraph explaining the concept that is illustrated in the graph.

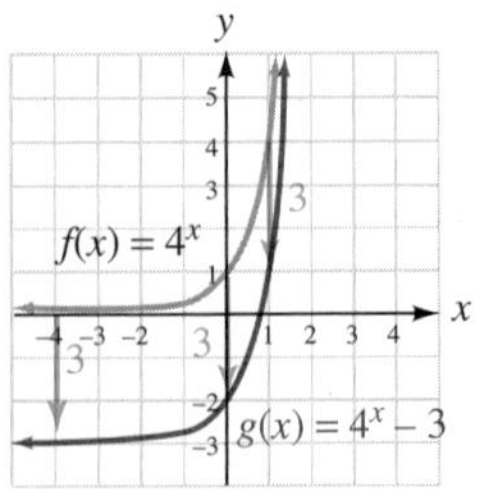

67. In the definition of the exponential function, b could not be negative. Why?

68. How does exponential growth differ from linear growth? Give an example.

REVIEW

In Exercises 69–72, refer to the illustration below in which lines r and s are parallel.

69. Find x. 40

70. Find the measure of $\angle 1$. 60°

71. Find the measure of $\angle 2$. 120°

72. Find the measure of $\angle 3$. 60°

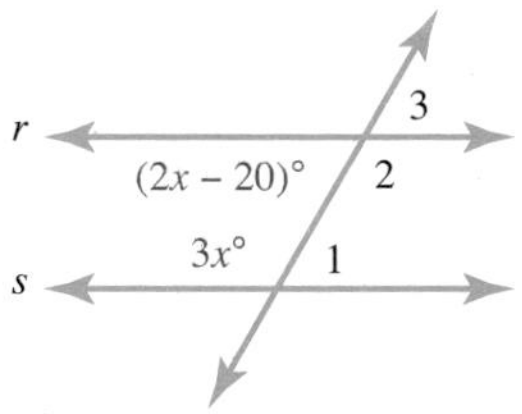

CHALLENGE PROBLEMS

73. Graph $f(x) = 3^x$. Then use the graph to estimate the value of $3^{1.5}$. About 5.2

74. Graph $y = x^{1/2}$ and $y = \left(\frac{1}{2}\right)^x$ on the same set of coordinate axes. Estimate the coordinates of any point(s) that the graphs have in common. See AIE Appendix 3; about $\left(\frac{1}{2}, \frac{2}{3}\right)$

75. Find the value of b that would cause the graph of $f(x) = b^x$ to look like the graph on the right. $\frac{1}{50}$

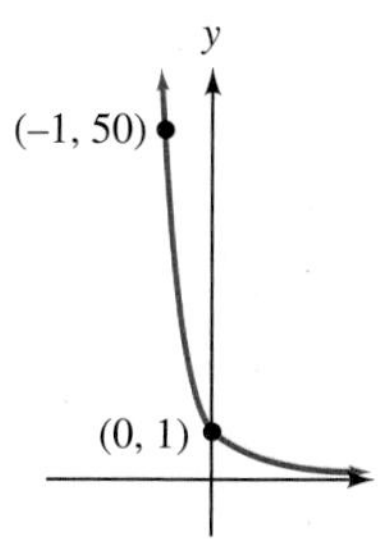

Simplify each expression. Write answers using positive exponents.

76. **a.** $\left(2^{\sqrt{3}}\right)^{\sqrt{3}}$ 8 **b.** $7^{\sqrt{3}}7^{\sqrt{12}}$ $7^{3\sqrt{3}}$

c. $\dfrac{5^{6\sqrt{2}}}{5^{4\sqrt{2}}}$ $5^{2\sqrt{2}}$ **d.** $5^{-\sqrt{5}}$ $\dfrac{1}{5^{\sqrt{5}}}$

77. Graph: $f(x) = 2^{|x|}$ See AIE Appendix 3.

78. Graph the inverse of $f(x) = 3^x$. See AIE Appendix 3.

SECTION 11.4 Logarithmic Functions

OBJECTIVES

1. Define logarithm.
2. Write logarithmic equations as exponential equations.
3. Write exponential equations as logarithmic equations.
4. Evaluate logarithmic expressions.
5. Graph logarithmic functions.
6. Use logarithmic formulas and functions in applications.

ARE YOU READY? *Are You Ready? exercises available online at www.webassign.net/brookscole*

The following problems review some basic skills that are needed when working with logarithmic functions.

1. A table of values for a one-to-one function f is shown below. Complete the table of values for f^{-1}.

x	$f(x)$
0	1
1	2
2	4

x	$f^{-1}(x)$
1	0
2	1
4	2

2. Fill in the blanks: **a.** $5^{-3} = \dfrac{1}{125}$ **b.** $3^{3} = 27$

3. Fill in the blanks: **a.** $4^{1} = 4$ **b.** $7^{1/2} = \sqrt{7}$

4. Evaluate: **a.** 10^3 1,000 **b.** 10^{-2} $\dfrac{1}{100}$

In this section, we will discuss inverses of exponential functions. These functions are called *logarithmic functions,* and they can be used to solve problems from fields such as electronics, seismology (the study of earthquakes), and business.

1 Define Logarithm.

The graph of the exponential function $f(x) = 2^x$ is shown in red below. Since it passes the horizontal line test, it is a one-to-one function and has an inverse. To graph f^{-1}, we interchange the coordinates of the ordered pairs in the table, plot those points, and draw a smooth curve through them, as shown in blue. As expected, the graphs of f and f^{-1} are symmetric with respect to the line $y = x$.

$f(x) = 2^x$

To graph f^{-1}, interchange each pair of coordinates.

x	$f(x)$		
-3	$\frac{1}{8}$	$\rightarrow\left(-3, \frac{1}{8}\right)$	$\left(\frac{1}{8}, -3\right)$
-2	$\frac{1}{4}$	$\rightarrow\left(-2, \frac{1}{4}\right)$	$\left(\frac{1}{4}, -2\right)$
-1	$\frac{1}{2}$	$\rightarrow\left(-1, \frac{1}{2}\right)$	$\left(\frac{1}{2}, -1\right)$
0	1	$\rightarrow(0, 1)$	$(1, 0)$
1	2	$\rightarrow(1, 2)$	$(2, 1)$
2	4	$\rightarrow(2, 4)$	$(4, 2)$
3	8	$\rightarrow(3, 8)$	$(8, 3)$

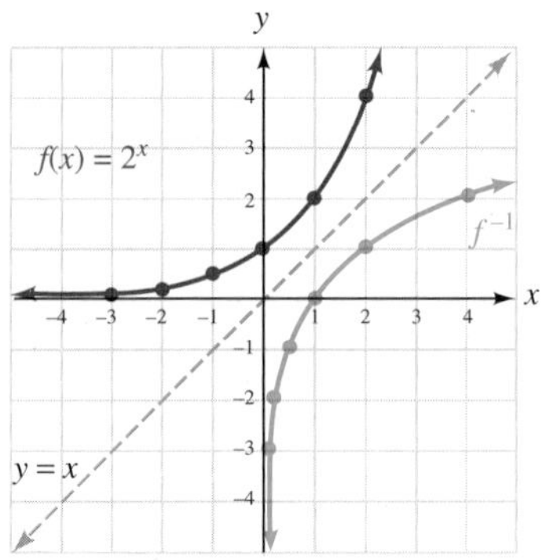

To write an equation for the inverse of $f(x) = 2^x$, we proceed as follows:

$$f(x) = 2^x$$

$$y = 2^x \quad \text{Replace } f(x) \text{ with } y.$$

$$x = 2^y \quad \text{Interchange the variables } x \text{ and } y.$$

Teaching Tip: Immediately stress that, like $f(x)$ function notation, $\log_2 x$ does not mean to multiply $\log_2$ times x.

We cannot solve the equation for y because we have not discussed methods for solving equations with a variable exponent. However, we can translate the relationship $x = 2^y$ into words:

$y =$ the power to which we raise 2 to get x

If we substitute the notation $f^{-1}(x)$ for y, we see that

$f^{-1}(x) =$ the power to which we raise 2 to get x

The Language of Algebra

The abbreviation log is used for the word **logarithm.** A *logarithm* is an exponent.

If we define the symbol $\log_2 x$ to mean *the power to which we raise* 2 *to get* x, we can write the equation for the inverse as

$f^{-1}(x) = \log_2 x$ Read $\log_2 x$ as "the logarithm, base 2, of x" or "log, base 2, of x."

We have found that the inverse of the exponential function $f(x) = 2^x$ is $f^{-1}(x) = \log_2 x$. To find the inverse of exponential functions with other bases, such as $f(x) = 3^x$ and $f(x) = 10^x$, we define *logarithm* in the following way.

Definition of Logarithm

For all positive numbers b, where $b \neq 1$, and all positive numbers x,

$y = \log_b x$ is equivalent to $x = b^y$

Success Tip

Here are two examples of inverses of other exponential functions:

$f(x) = 3^x \quad f^{-1}(x) = \log_3 x$

$g(x) = 10^x \quad g^{-1}(x) = \log_{10} x$

This definition guarantees that any pair (x, y) that satisfies the logarithmic equation $y = \log_b x$ also satisfies the exponential equation $x = b^y$. Because of this relationship, a statement written in logarithmic form can be written in an equivalent exponential form, and vice versa. They are just different ways of expressing the same thing. The following diagram will help you remember the respective positions of the exponent and base in each form.

Exponent

$y = \log_{\boldsymbol{b}} x$ means $x = \boldsymbol{b}^y$

Base

2 Write Logarithmic Equations as Exponential Equations.

The following table shows the relationship between logarithmic and exponential notation. We need to be able to work in both directions.

Success Tip

Study the relationship between exponential and logarithmic statements carefully. Your success with the material in the rest of this chapter depends greatly on your understanding of this definition.

Logarithmic equation		*Exponential equation*
$\log_2 8 = 3$		$2^3 = 8$
$\log_3 81 = 4$	means	$3^4 = 81$
$\log_4 4 = 1$		$4^1 = 4$
$\log_5 \frac{1}{125} = -3$		$5^{-3} = \frac{1}{125}$

EXAMPLE 1 Write each logarithmic equation as an exponential equation:

a. $\log_4 64 = 3$ **b.** $\log_7 \sqrt{7} = \frac{1}{2}$ **c.** $\log_6 \frac{1}{36} = -2$

Strategy To write an equivalent exponential equation, we will determine which number will serve as the base and which will serve as the exponent.

Why We can then use the definition of logarithm to move from one form to the other: $\log_b x = y$ is equivalent to $x = b^y$.

Solution

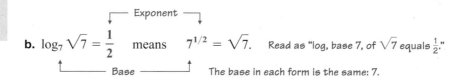

a. $\log_4 64 = 3$ means $4^3 = 64$. (Exponent: 3; Base: 4) Read as "log, base 4, of 64 equals 3." The base in each form is the same: 4.

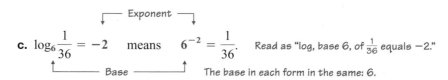

Caution

Recall that $f(x)$ is not read as $f \cdot x$. Similarly, $\log_4 64$ is not read as $\log_4 \cdot 64$.

b. $\log_7 \sqrt{7} = \frac{1}{2}$ means $7^{1/2} = \sqrt{7}$. (Exponent: $\frac{1}{2}$; Base: 7) Read as "log, base 7, of $\sqrt{7}$ equals $\frac{1}{2}$." The base in each form is the same: 7.

c. $\log_6 \frac{1}{36} = -2$ means $6^{-2} = \frac{1}{36}$. (Exponent: −2; Base: 6) Read as "log, base 6, of $\frac{1}{36}$ equals −2." The base in each form in the same: 6.

Self Check 1 Write $\log_2 128 = 7$ as an exponential equation. $2^7 = 128$

Now Try Problems 23, 27, and 35

Teaching Example 1 Write each logarithmic equation as an exponential equation:
a. $\log_5 25 = 2$ b. $\log_{2/3} \frac{8}{27} = 3$
c. $\log_2 \frac{1}{8} = -3$
Answers: a. $5^2 = 25$ b. $\left(\frac{2}{3}\right)^3 = \frac{8}{27}$
c. $2^{-3} = \frac{1}{8}$

3 Write Exponential Equations as Logarithmic Equations.

EXAMPLE 2 Write each exponential equation as a logarithmic equation:

a. $8^0 = 1$ **b.** $6^{1/3} = \sqrt[3]{6}$ **c.** $\left(\frac{1}{4}\right)^2 = \frac{1}{16}$

Strategy To write an equivalent logarithmic equation, we will determine which number will serve as the base and where we will place the exponent.

Why We can then use the definition of logarithm to move from one form to the other: $x = b^y$ is equivalent to $\log_b x = y$.

Solution

a. $8^0 = 1$ means $\log_8 1 = 0$ *In each form, the base is 8.* (Exponent: 0; Base: 8)

The Language of Algebra

Just as 8 is called the **base** in the exponential form $8^0 = 1$, we call 8 the **base** in the logarithmic form $\log_8 1 = 0$.

b. $6^{1/3} = \sqrt[3]{6}$ means $\log_6 \sqrt[3]{6} = \frac{1}{3}$ *In each form, the base is 6.* (Exponent: $\frac{1}{3}$; Base: 6)

c. $\left(\frac{1}{4}\right)^2 = \frac{1}{16}$ means $\log_{1/4} \frac{1}{16} = 2$ *In each form, the base is $\frac{1}{4}$.* (Exponent: 2; Base: $\frac{1}{4}$)

Teaching Example 2 Write each exponential equation as a logarithmic equation:
a. $3^{1/2} = \sqrt{3}$ b. $\left(\frac{1}{5}\right)^{-1} = 5$
c. $2^0 = 1$
Answers:
a. $\log_3 \sqrt{3} = \frac{1}{2}$ b. $\log_{1/5} 5 = -1$
c. $\log_2 1 = 0$

Self Check 2 Write $9^{-1} = \frac{1}{9}$ as a logarithmic equation. $\log_9 \frac{1}{9} = -1$

Now Try Problems 39, 43, and 49

Certain logarithmic equations can be solved by writing them as exponential equations.

EXAMPLE 3 Solve each equation for x: **a.** $\log_x 25 = 2$ **b.** $\log_3 x = -3$ **c.** $\log_{1/2} \frac{1}{16} = x$

Strategy To solve each logarithmic equation, we will instead write and solve an equivalent exponential equation.

Why The resulting exponential equation is easier to solve because the variable term is often isolated on one side.

Solution

a. Since $\log_x 25 = 2$ is equivalent to $x^2 = 25$, we can solve $x^2 = 25$ to find x. (Exponent: 2; Base: x)

The Language of Algebra

Recall that **equivalent equations** have the same solutions.

$$x^2 = 25$$
$$x = \pm\sqrt{25} \quad \text{Use the square root property.}$$
$$x = \pm 5$$

In the expression $\log_x 25$, the base of the logarithm is x. Because the base must be positive, we discard -5 and we have

$$x = 5$$

To check the solution of 5, verify that $\log_5 25 = 2$. The solution set is written as $\{5\}$.

b. Since $\log_3 x = -3$ is equivalent to $3^{-3} = x$, we can instead solve $3^{-3} = x$ to find x. (Exponent: -3; Base: 3)

$$3^{-3} = x$$
$$\frac{1}{3^3} = x$$
$$x = \frac{1}{27}$$

To check the solution of $\frac{1}{27}$, verify that $\log_3 \frac{1}{27} = -3$. The solution set is $\left\{\frac{1}{27}\right\}$.

Success Tip

To solve this equation, we note that if the bases are equal, the exponents must be equal.

$$\left(\frac{1}{2}\right)^x = \left(\frac{1}{2}\right)^4$$

Exponent

c. Since $\log_{1/2} \frac{1}{16} = x$ is equivalent to $\left(\frac{1}{2}\right)^x = \frac{1}{16}$, we can instead solve $\left(\frac{1}{2}\right)^x = \frac{1}{16}$ to find x.

Base

$$\left(\frac{1}{2}\right)^x = \frac{1}{16}$$

$$\left(\frac{1}{2}\right)^x = \left(\frac{1}{2}\right)^4 \quad \text{Write } \tfrac{1}{16} \text{ as a power of } \tfrac{1}{2} \text{ to match the bases: } \tfrac{1}{2} \cdot \tfrac{1}{2} \cdot \tfrac{1}{2} \cdot \tfrac{1}{2} = \tfrac{1}{16}.$$

$$x = 4 \quad \text{Since the bases are the same, and since exponential functions are one-to-one, the exponents must be equal.}$$

To check the solution of 4, verify that $\log_{1/2} \frac{1}{16} = 4$. The solution set is $\{4\}$.

Teaching Example 3 Solve for x:
a. $\log_x 9 = 2$ **b.** $\log_2 x = 4$
c. $\log_{1/7} \frac{1}{49} = x$
Answers: **a.** 3 **b.** 16 **c.** 2

Self Check 3 Solve each equation for x: **a.** $\log_x 49 = 2$ 7 **b.** $\log_{1/3} x = 2$ $\frac{1}{9}$
c. $\log_6 216 = x$ 3

Now Try Problems 51, 53, and 55

4 Evaluate Logarithmic Expressions.

In the previous examples, we have seen that the logarithm of a number is an exponent. In fact,

$\log_b x$ is the exponent to which b is raised to get x.

Translating this statement into symbols, we have

$$b^{\log_b x} = x$$

EXAMPLE 4 Evaluate each expression: **a.** $\log_8 64$ **b.** $\log_3 \frac{1}{3}$ **c.** $\log_4 2$

Strategy After identifying the base, we will ask "To what power must the base be raised to get the other number?"

Why That power is the value of the logarithmic expression.

Solution

a. $\log_8 64 = 2$ Ask: "To what power must we raise 8 to get 64?" Since $8^2 = 64$, the answer is the 2nd power.

An alternate approach for this evaluation problem is to let $\log_8 64 = x$. When we write the equivalent exponential equation $8^x = 64$, it is easy to see that $x = 2$.

b. $\log_3 \frac{1}{3} = -1$ Ask: "To what power must we raise 3 to get $\frac{1}{3}$?" Since $3^{-1} = \frac{1}{3}$, the answer is the -1 power.

We could also let $\log_3 \frac{1}{3} = x$. The equivalent exponential equation is $3^x = \frac{1}{3}$. Thus, x must be -1.

c. $\log_4 2 = \frac{1}{2}$ Ask: "To what power must we raise 4 to get 2?" Since $\sqrt{4} = 4^{1/2} = 2$, the answer is the $\frac{1}{2}$ power.

We could also let $\log_4 2 = x$. Then the equivalent exponential equation is $4^x = 2$. Thus, x must be $\frac{1}{2}$.

Teaching Example 4 Evaluate each expression:
a. $\log_9 81$ **b.** $\log_{11} \frac{1}{121}$ **c.** $\log_8 2$
Answers:
a. 2 **b.** -2 **c.** $\frac{1}{3}$

Self Check 4 Evaluate each expression: **a.** $\log_9 81$ 2 **b.** $\log_4 \frac{1}{16}$ -2 **c.** $\log_9 3$ $\frac{1}{2}$

Now Try Problems 75 and 77

The Language of Algebra

London professor Henry Briggs (1561–1630) and Scottish lord John Napier (1550–1617) are credited with developing the concept of **common logarithms.** Their tables of logarithms were useful tools at that time for those performing large calculations.

In many applications, base-10 logarithms (also called **common logarithms**) are used. When the base b is not indicated in the notation $\log x$, we assume that $b = 10$:

$$\log x \quad \text{means} \quad \log_{10} x$$

The table below shows the relationship between base-10 logarithmic notation and exponential notation.

Logarithmic form		*Exponential form*
$\log 100 = 2$	means	$10^2 = 100$ Read log 100 as "log of 100."
$\log \frac{1}{10} = -1$		$10^{-1} = \frac{1}{10}$
$\log 1 = 0$		$10^0 = 1$

In general, we have

$$\log 10^x = x$$

EXAMPLE 5 Evaluate each expression: **a.** $\log 1{,}000$ **b.** $\log \frac{1}{100}$ **c.** $\log 10$ **d.** $\log(-10)$

Strategy After identifying the base, we will ask "To what power must 10 be raised to get the other number?"

Why That power is the value of the logarithmic expression.

Solution

a. $\log 1{,}000 = 3$ Ask: "To what power must we raise 10 to get 1,000?" Since $10^3 = 1{,}000$, the answer is: the 3rd power.

b. $\log \frac{1}{100} = -2$ Ask: "To what power must we raise 10 to get $\frac{1}{100}$?" Since $10^{-2} = \frac{1}{100}$, the answer is: the -2 power.

c. $\log 10 = 1$ Ask: "To what power must we raise 10 to get 10?" Since $10^1 = 10$, the answer is: the 1st power.

d. To find $\log(-10)$, we must find a power of 10 such that $10^? = -10$. There is no such number. Thus, $\log(-10)$ is undefined.

Teaching Tip: Remind your students that to evaluate logarithmic expressions, they can always let that expression equal x. Then they can write an equivalent exponential expression and solve it for x. However, with some practice, that approach probably will not be necessary.

Teaching Example 5 Evaluate each expression:
a. $\log 100$ b. $\log \frac{1}{10{,}000}$
c. $\log(-100)$
Answers:
a. 2 b. -4 c. Undefined

Self Check 5 Evaluate each expression: **a.** $\log 10{,}000$ 4 **b.** $\log \frac{1}{1{,}000}$ -3 **c.** $\log 0$ Undefined

Now Try Problems 79 and 81

Teaching Tip: Before using a calculator, see if your students can locate the value of log 2.34 between two integers: $0 < \log 2.34 < 1$.

Many logarithmic expressions cannot be evaluated by inspection. For example, to find log 2.34, we ask, "To what power must we raise 10 to get 2.34?" This answer isn't obvious. In such cases, we use a calculator.

Using Your Calculator ▶ Evaluating Logarithms

To find log 2.34 with a scientific calculator we enter

2.34 [LOG] .369215857

On some calculators, the [10^x] key also serves as the [LOG] key when [2nd] or [SHIFT] is pressed. This is because $f(x) = 10^x$ and $f(x) = \log x$ are inverses.

To use a graphing calculator, we enter

[LOG] 2.34 [)] [ENTER]

```
log(2.34)
       .369215857
```

To four decimal places, $\log 2.34 = 0.3692$. This means, $10^{0.3692} \approx 2.34$.

If we attempt to evaluate logarithmic expressions such as log 0, or the logarithm of a negative number, such as $\log(-5)$, an error message like the following will be displayed.

Error	ERR:DOMAIN 1:QUIT 2:Go to	ERR:NONREAL ANS 1:QUIT 2:Go to

EXAMPLE 6 Solve $\log x = 0.3568$ and round to four decimal places.

Strategy To solve this logarithmic equation, we will instead write and solve an equivalent exponential equation.

Why The resulting exponential equation is easier to solve because the variable term is isolated on one side.

Solution The equation $\log x = 0.3568$ means $\log_{10} x = 0.3568$, which is equivalent to $\mathbf{10^{0.3568}} = x$. Since we cannot determine $10^{0.3568}$ by inspection, we will use a calculator to find an approximate solution. We enter

10 [yˣ] .3568 [=]

The display reads [2.274049951]. To four decimal places,

$x = 2.2740$

If your calculator has a [10^x] key, enter .3568 and press it to get the same result. The solution is 2.2740. To check, use your calculator to verify that $\log 2.2740 \approx 0.3568$.

Teaching Example 6 Solve $\log x = 0.7429$ and round to four decimal places.
Answer: 5.5322

Self Check 6 Solve $\log x = 1.87737$ and round to four decimal places. 75.3998

Now Try Problem 91

5 Graph Logarithmic Functions.

Because an exponential function defined by $f(x) = b^x$ is one-to-one, it has an inverse function that is defined by $x = b^y$. When we write $x = b^y$ in the equivalent form $y = \log_b x$, the result is called a *logarithmic function.*

Logarithmic Functions

If $b > 0$ and $b \neq 1$, the **logarithmic function with base b** is defined by the equations

$$f(x) = \log_b x \quad \text{or} \quad y = \log_b x$$

The domain of $f(x) = \log_b x$ is the interval $(0, \infty)$ and the range is the interval $(-\infty, \infty)$.

Caution

Since the domain of the logarithmic function is the set of positive real numbers, it is impossible to find the logarithm of 0 or the logarithm of a negative number. For example, $\log_2(-4)$ and $\log_2 0$ are undefined.

Since every logarithmic function is the inverse of a one-to-one exponential function, logarithmic functions are one-to-one.

We can plot points to graph logarithmic functions. For example, to graph $f(x) = \log_2 x$, we construct a table of function values, plot the resulting ordered pairs, and draw a smooth curve through the points to get the graph, as shown in figure (a) on the next page. To graph $f(x) = \log_{1/2} x$, we use the same method, as shown in figure (b).

By the vertical line test, we see that each graph is indeed the graph of a function. Because in each case the projection of the graph onto the x-axis covers only the positive portion of that axis, the domain of each function is $(0, \infty)$. Because the graphs extend indefinitely upward and downward, the projection of the graphs onto the y-axis includes all real numbers. Thus, the range of each function is $(-\infty, \infty)$.

$f(x) = \log_2 x$

x	$f(x)$	
$\frac{1}{4}$	-2	$\rightarrow \left(\frac{1}{4}, -2\right)$
$\frac{1}{2}$	-1	$\rightarrow \left(\frac{1}{2}, -1\right)$
1	0	$\rightarrow (1, 0)$
2	1	$\rightarrow (2, 1)$
4	2	$\rightarrow (4, 2)$
8	3	$\rightarrow (8, 3)$

↑

Because the base of the function is 2, choose values for x that are integer powers of 2.

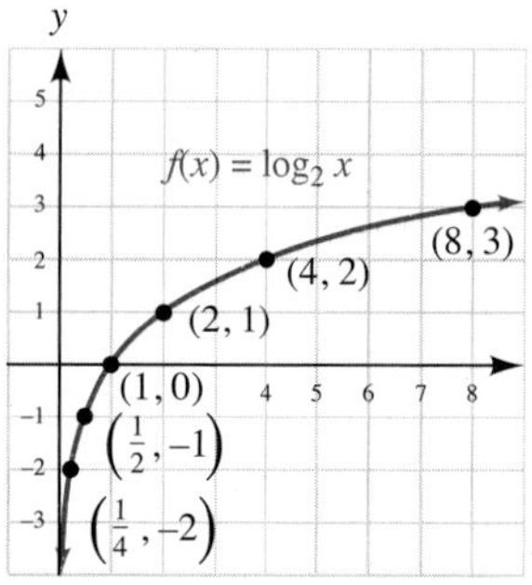

(a)

$f(x) = \log_{1/2} x$

x	$f(x)$	
$\frac{1}{4}$	2	$\rightarrow \left(\frac{1}{4}, 2\right)$
$\frac{1}{2}$	1	$\rightarrow \left(\frac{1}{2}, 1\right)$
1	0	$\rightarrow (1, 0)$
2	-1	$\rightarrow (2, -1)$
4	-2	$\rightarrow (4, -2)$
8	-3	$\rightarrow (8, -3)$

↑

Because the base of the function is $\frac{1}{2}$, choose values for x that are integer powers of $\frac{1}{2}$.

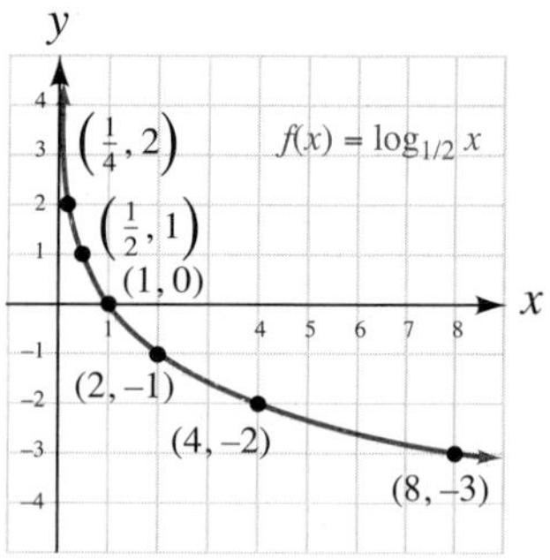

(b)

The graphs of all logarithmic functions are similar to those shown below. If $b > 1$, the logarithmic function is increasing, as in figure (a). If $0 < b < 1$, the logarithmic function is decreasing, as in figure (b).

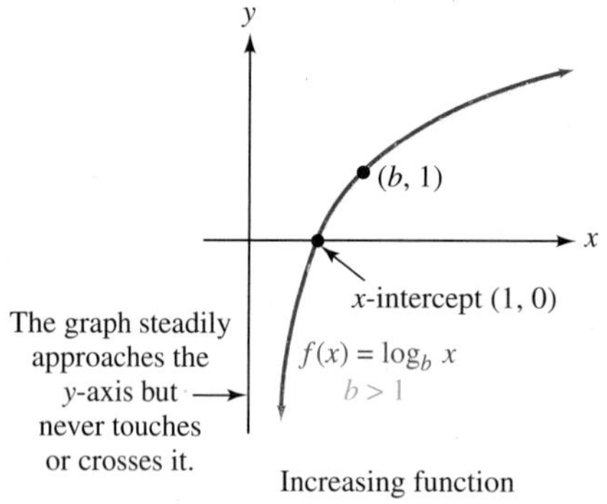

Increasing function
(a)

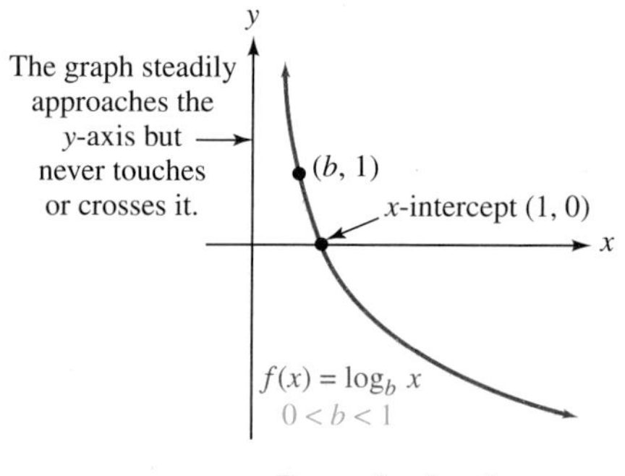

Decreasing function
(b)

Properties of Logarithmic Functions

The graph of $f(x) = \log_b x$ (or $y = \log_b x$) has the following properties.

1. It passes through the point $(1, 0)$.
2. It passes through the point $(b, 1)$.
3. The y-axis (the line $x = 0$) is an asymptote.
4. The domain is the interval $(0, \infty)$ and the range is the interval $(-\infty, \infty)$.

The exponential and logarithmic functions are inverses of each other, so their graphs have symmetry about the line $y = x$. The graphs of $f(x) = \log_b x$ and $g(x) = b^x$ are shown in figure (a) when $b > 1$ and in figure (b) when $0 < b < 1$.

Notation

If it is helpful, imagine parentheses around the input value in the logarithmic function notation.

$f(8) = 3$

Function name — Input — Output

$\log_2 8 = 3$

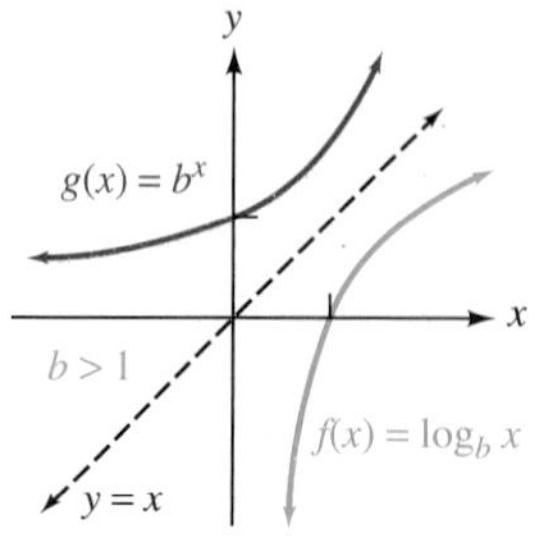

(a)

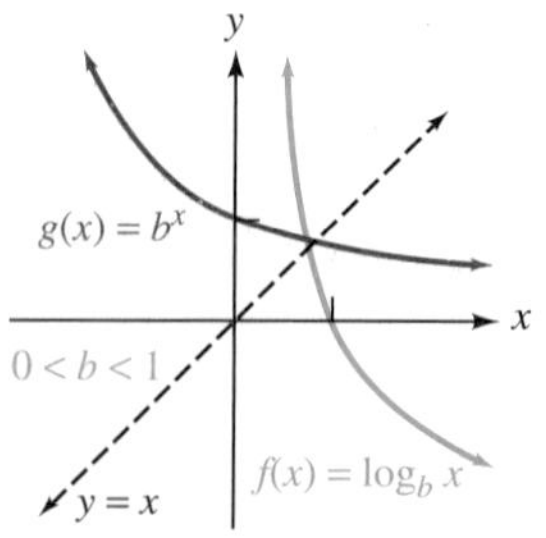

(b)

The graphs of many functions involving logarithms are translations of the basic logarithmic graphs.

EXAMPLE 7 Graph each function by using a translation: **a.** $g(x) = 3 + \log_2 x$ **b.** $g(x) = \log_{1/2}(x - 1)$

Notation

Since $y = f(x)$, we can write

$$f(x) = \log_2 x$$
as
$$y = \log_2 x$$

Strategy We will graph $g(x) = 3 + \log_2 x$ by translating the graph of $f(x) = \log_2 x$ upward 3 units. We will graph $g(x) = \log_{1/2}(x - 1)$ by translating the graph of $f(x) = \log_{1/2} x$ to the right 1 unit.

Why The addition of 3 in $g(x) = \mathbf{3} + \log_2 x$ causes a vertical shift of the graph of the base-2 logarithmic function 3 units upward. The subtraction of 1 from x in $g(x) = \log_{1/2}(x - \mathbf{1})$ causes a horizontal shift of the graph of the base-$\frac{1}{2}$ logarithmic function 1 unit to the right.

Solution

a. The graph of $g(x) = 3 + \log_2 x$ will be the same shape as the graph of $f(x) = \log_2 x$. We simply translate each point on the graph of $f(x) = \log_2 x$ up 3 units. See figure (a) below.

b. The graph of $g(x) = \log_{1/2}(x - 1)$ will be the same shape as the graph of $f(x) = \log_{1/2} x$. We simply translate each point on the graph of $f(x) = \log_{1/2} x$ to the right 1 unit. See figure (b) below.

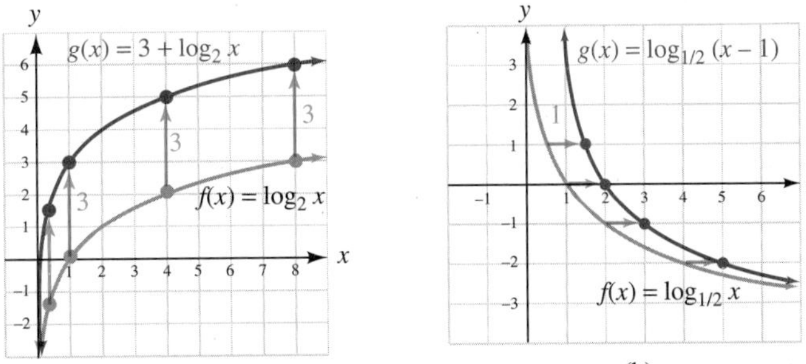

(a) (b)

Teaching Example 7 Graph each function by using a translation:
a. $g(x) = -2 + \log_2 x$
b. $g(x) = \log_2 (x + 1)$
Answers:
a.

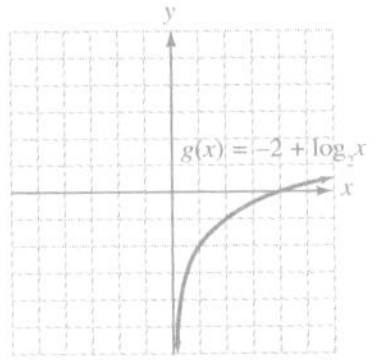

b.

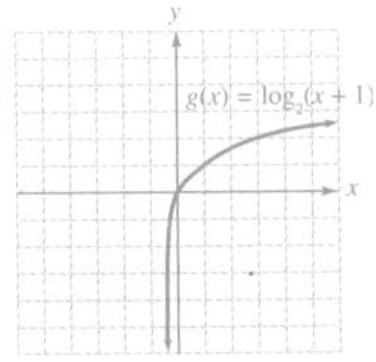

Self Check 7 Graph each function by using a translation: **a.** $g(x) = (\log_3 x) - 2$ **b.** $g(x) = \log_{1/3}(x + 2)$ See AIE Appendix 3.

Now Try Problems 103 and 105

To graph more complicated logarithmic functions, a graphing calculator is a useful tool.

Using Your Calculator ▶ Graphing Logarithmic Functions

To use a calculator to graph the logarithmic function $f(x) = -2 + \log_{10}\frac{x}{2}$, we enter the right side of the equation after the symbol $Y_1 =$. The display will show the equation

$$Y_1 = -2 + \log(X/2)$$

If we use window settings of $[-1, 5]$ for x and $[-4, 1]$ for y and press the GRAPH key, we will obtain the graph shown.

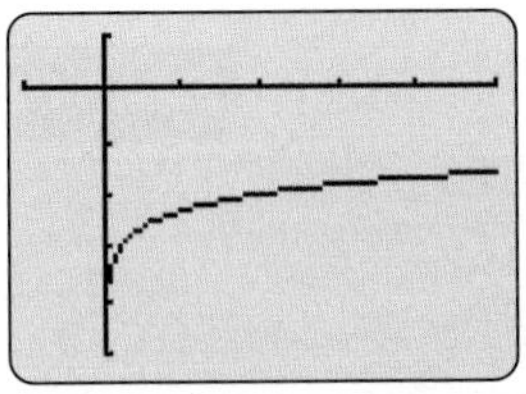

6 Use Logarithmic Formulas and Functions in Applications.

Logarithmic functions, like exponential functions, can be used to model certain types of growth and decay. Logarithms are especially useful when measuring a quantity that varies over a very large range of values, such as the intensity of earthquakes, the loudness of sounds, and the brightness of stars.

The following common logarithm formula is used in electrical engineering to express the gain (or loss) of an electronic device such as an amplifier as it takes an input signal and produces an output signal. The unit of gain (or loss) in such cases is called the **decibel,** which is abbreviated **dB.**

Decibel Voltage Gain

If E_O is the output voltage of a device and E_I is the input voltage, the decibel voltage gain of the device (dB gain) is given by

$$\text{dB gain} = 20 \log \frac{E_O}{E_I}$$

EXAMPLE 8

©iStockphoto.com/PeterAlbrektsen

dB Gain. If the input to an amplifier is 0.5 volt and the output is 40 volts, find the decibel voltage gain of the amplifier.

Strategy We will substitute into the formula for dB gain and evaluate the right side using a calculator.

Why We can use this formula to find the dB gain because we are given the input voltage E_I and the output voltage E_O.

Solution We can find the decibel voltage gain by substituting 0.5 for E_I and 40 for E_O into the formula for dB gain:

Notation

$20 \log \frac{E_O}{E_I}$ means $20 \cdot \log \frac{E_O}{E_I}$

$$\text{dB gain} = 20 \log \frac{E_O}{E_I}$$ Read as "20 times the log of E sub O divided by E sub I."

$$\text{dB gain} = 20 \log \frac{40}{0.5}$$ Substitute the input voltage 0.5 for E_I and the output voltage 40 for E_O.

$$= 20 \log 80$$ Divide: $\frac{40}{0.5} = 80$.

$$\approx 38$$ Use a scientific calculator and press: 20 × 80 LOG =.

The amplifier provides a 38-decibel voltage gain.

Teaching Example 8 dB Gain. If the input to an amplifier is 0.7 volt and the output is 40 volts, find the decibel voltage gain of the amplifier.
Answer: About 35 dB

Self Check 8 **dB Gain.** If the input to an amplifier is 0.6 volt and the output is 40 volts, find the decibel voltage gain of the amplifier. About 36 dB

Now Try Problem 111

EXAMPLE 9

©Coral Coolahan/Shutterstock.com

Stocking Lakes. To create the proper environmental balance, 250 hybrid bluegill were introduced into a lake by the local Fish and Game Department. Department biologists found that the number of bluegill in the lake could be approximated by the logarithmic function $f(t) = 250 + 400 \log (t + 1)$, where t is the number of years since the lake was stocked. Find the bluegill population in the lake after 5 years.

Strategy We will find $f(5)$.

Why Since the variable t represents the time since the lake was stocked, $t = 5$.

Solution

$$f(t) = 250 + 400 \log (t + 1)$$ This is the logarithmic growth model.

$$f(5) = 250 + 400 \log (5 + 1)$$ Substitute 5 for t.

$$= 250 + 400 \log 6$$ Do the addition within the parentheses.

$$\approx 561$$ Use a scientific calculator and press: 250 + 400 × 6 LOG =.

There were approximately 561 bluegill in the lake 5 years after it was stocked.

Teaching Example 9 Stocking Lakes. Zoologists found that the number of bass in a lake could be approximated by the logarithmic function $f(t) = 500 + 1{,}200 \log (t + 1)$, where t is the number of years since the lake was stocked. Find the bass population in the lake after 4 years.
Answer: About 1,339 bass

Self Check 9 **Stocking Ponds.** A rancher stocked a pond on his property with 50 catfish. He was told that with the proper care, the catfish population could be approximated by the logarithmic function $f(t) = 50 + 22 \log (t + 1)$, where t is the number of years since the pond was stocked. Find the number of catfish he can expect in the pond in 3 years. About 63 catfish

Now Try Problem 117

SECTION 11.4 STUDY SET

VOCABULARY

Fill in the blanks.

1. $f(x) = \log_2 x$ and $g(x) = \log x$ are examples of logarithmic functions.
2. Base-10 logarithms are called common logarithms.
3. The graph of $f(x) = \log_2 x$ approaches, but never touches, the negative portion of the y-axis. Thus the y- axis is an asymptote of the graph.
4. $\log_x 81 = 4$ is equivalent to $x^4 = 81$.

CONCEPTS

5. Refer to the graph on the right.

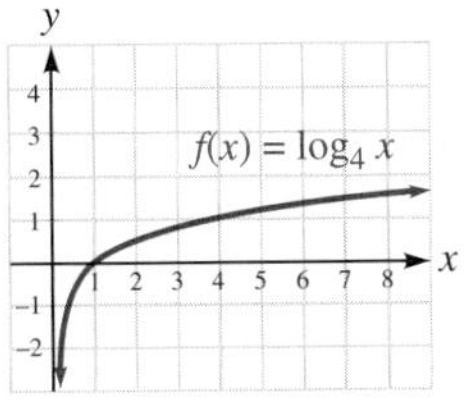

   a. What type of function is $f(x) = \log_4 x$? Logarithmic
   b. What is the domain of the function? What is the range of the function? D: $(0, \infty)$, R: $(-\infty, \infty)$
   c. What is the y-intercept of the graph? What is the x-intercept of the graph? None, $(1, 0)$
   d. Is f a one-to-one function? Yes
   e. What is an asymptote of the graph? The y-axis $(x = 0)$
   f. Is f an increasing or a decreasing function? Increasing
   g. The graph passes through the point $(4, y)$. What is y? 1
6. Determine the domain and range of each logarithmic function graphed below.

   a.

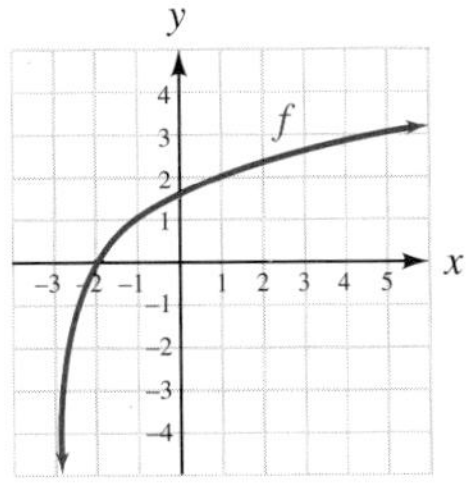

   D: $(-3, \infty)$, R: $(-\infty, \infty)$

   b.

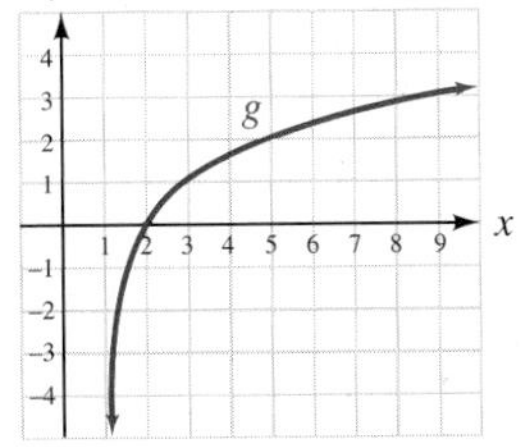

   D: $(1, \infty)$, R: $(-\infty, \infty)$

7. Match each function with its graph shown below.
   a. $f(x) = x^3$ ii.
   b. $f(x) = \log_3 x$ iii.
   c. $f(x) = \sqrt[3]{x}$ i.
   d. $f(x) = 3x$ iv.

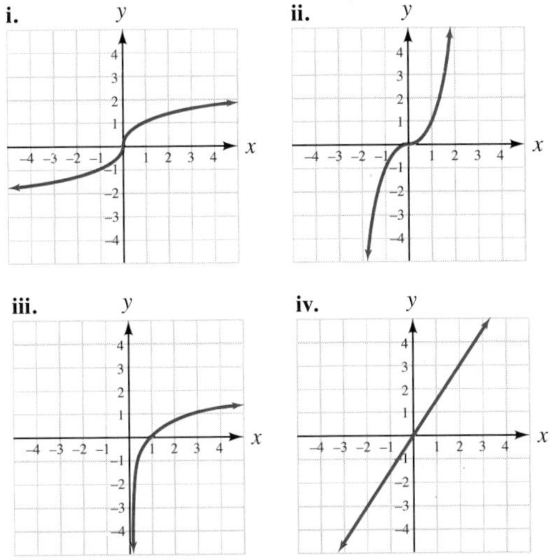

8. a. An exponential function is graphed below. Graph its inverse and the axis of symmetry on the same coordinate system.
   b. What type of function is the inverse function? Logarithmic

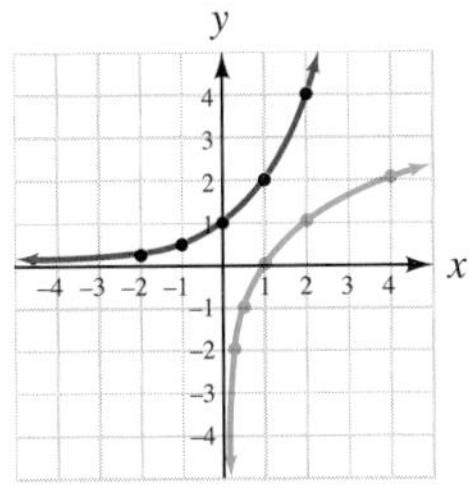

Fill in the blanks.

9. $\log_6 36 = 2$ means $6^2 = 36$.
10. $\log x = -2$ is equivalent to $10^{-2} = x$.
11. $\log_b x$ is the exponent to which b is raised to get x.
12. The functions $f(x) = \log_{10} x$ and $f(x) = 10^x$ are inverse functions.
13. The inverse of an exponential function is called a logarithmic function.

14. Fill in the blanks.

a. The graph of $g(x) = 4 + \log x$ is similar to the graph of $f(x) = \log x$, but it is translated 4 units upward.

b. The graph of $g(x) = \log_4 (x + 2)$ is similar to the graph of $f(x) = \log_4 x$, but it is translated 2 units to the left.

Complete the table of values.

15. $f(x) = \log x$

x	$f(x)$
100	2
$\frac{1}{100}$	-2

▶ **16.** $f(x) = \log_5 x$

x	$f(x)$
25	2
$\frac{1}{25}$	-2

▶ **17.** $f(x) = \log_6 x$

Input	Output
6	1
-6	Undefined
0	Undefined

18. $f(x) = \log_8 x$

Input	Output
8	1
-8	Undefined
0	Undefined

19. a. Use a calculator to complete the table of values for $f(x) = \log x$. Round to the nearest hundredth.

b. Graph $f(x) = \log x$. Note that the units on the x- and y-axes are different.

x	$f(x)$
0.5	-0.30
1	0
2	0.30
4	0.60
6	0.78
8	0.90
10	1

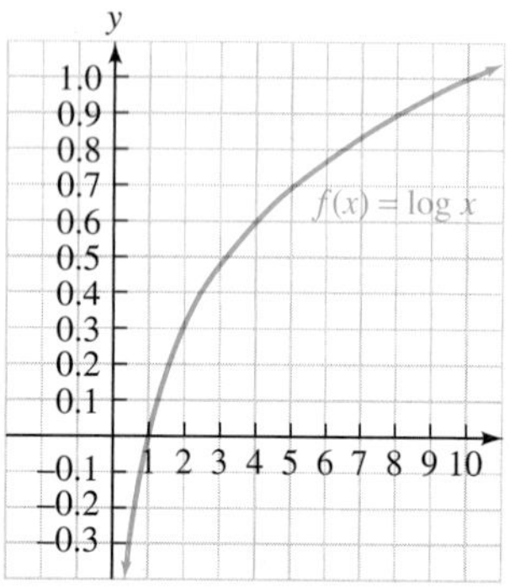

20. For each function, determine its inverse, $f^{-1}(x)$.

a. $f(x) = 10^x$ $\quad f^{-1}(x) = \log x$

b. $f(x) = 3^x$ $\quad f^{-1}(x) = \log_3 x$

c. $f(x) = \log x$ $\quad f^{-1}(x) = 10^x$

d. $f(x) = \log_2 x$ $\quad f^{-1}(x) = 2^x$

NOTATION

Fill in the blanks.

▶ **21. a.** $\log x = \log_{10} x$ **b.** $\log_{10} 10^x = x$

▶ **22. a.** We read $\log_5 25$ as "log, base 5, of 25."

b. We read $\log x$ as "log of x."

GUIDED PRACTICE

Write each logarithmic equation as an exponential equation. **See Example 1. Do not solve.**

▶ **23.** $\log_3 81 = 4$ $\quad 3^4 = 81$

24. $\log_7 7 = 1$ $\quad 7^1 = 7$

25. $\log_{10} 10 = 1$ $\quad 10^1 = 10$

▶ **26.** $\log_{10} 100 = 2$ $\quad 10^2 = 100$

27. $\log_4 \frac{1}{64} = -3$ $\quad 4^{-3} = \frac{1}{64}$

▶ **28.** $\log_6 \frac{1}{36} = -2$ $\quad 6^{-2} = \frac{1}{36}$

▶ **29.** $\log_5 \sqrt{5} = \frac{1}{2}$ $\quad 5^{1/2} = \sqrt{5}$

30. $\log_8 \sqrt[3]{8} = \frac{1}{3}$ $\quad 8^{1/3} = \sqrt[3]{8}$

31. $\log 0.1 = -1$ $\quad 10^{-1} = 0.1$

▶ **32.** $\log 0.01 = -2$ $\quad 10^{-2} = 0.01$

33. $x = \log_8 64$ $\quad 8^x = 64$

34. $x = \log_9 81$ $\quad 9^x = 81$

35. $t = \log_b T_1$ $\quad b^t = T_1$

▶ **36.** $n = \log_b R_1$ $\quad b^n = R_1$

37. $\log_n C = -42$ $\quad n^{-42} = C$

38. $\log_m P = 101$ $\quad m^{101} = P$

Write each exponential equation as a logarithmic equation. **See Example 2.**

39. $8^2 = 64$ $\quad \log_8 64 = 2$

▶ **40.** $10^3 = 1{,}000$ $\quad \log_{10} 1{,}000 = 3$

▶ **41.** $4^{-2} = \frac{1}{16}$ $\quad \log_4 \frac{1}{16} = -2$

42. $3^{-4} = \frac{1}{81}$ $\quad \log_3 \frac{1}{81} = -4$

▶ **43.** $\left(\frac{1}{2}\right)^{-5} = 32$ $\quad \log_{1/2} 32 = -5$

▶ **44.** $\left(\frac{1}{3}\right)^{-3} = 27$ $\quad \log_{1/3} 27 = -3$

45. $x^y = z$ $\quad \log_x z = y$

▶ **46.** $m^n = p$ $\quad \log_m p = n$

47. $y^t = 8.6$ $\quad \log_y 8.6 = t$

▶ **48.** $b^r = \frac{2}{3}$ $\quad \log_b \frac{2}{3} = r$

49. $7^{4.3} = B + 1$ $\quad \log_7 (B + 1) = 4.3$

▶ **50.** $12^{-2.6} = N + 1$ $\quad \log_{12} (N + 1) = -2.6$

Solve for x. **See Example 3.**

51. $\log_x 81 = 2$ $\quad 9$

▶ **52.** $\log_x 9 = 2$ $\quad 3$

▶ **53.** $\log_8 x = 2$ $\quad 64$

54. $\log_7 x = 0$ $\quad 1$

55. $\log_5 125 = x$ $\quad 3$

56. $\log_4 16 = x$ $\quad 2$

57. $\log_5 x = -2$ $\quad \frac{1}{25}$

▶ **58.** $\log_3 x = -4$ $\quad \frac{1}{81}$

▶ **59.** $\log_{36} x = -\frac{1}{2}$ $\quad \frac{1}{6}$

▶ **60** $\log_{27} x = -\frac{1}{3}$ $\quad \frac{1}{3}$

61. $\log_x 0.01 = -2$ $\quad 10$

▶ **62.** $\log_x 0.001 = -3$ $\quad 10$

▶ **63.** $\log_{27} 9 = x$ $\quad \frac{2}{3}$

▶ **64.** $\log_{12} x = 0$ $\quad 1$

▶ **65.** $\log_x 5^3 = 3$ $\quad 5$

66. $\log_x 5 = 1$ $\quad 5$

67. $\log_{100} x = \frac{3}{2}$ $\quad 1{,}000$

68. $\log_x \frac{1}{1{,}000} = -\frac{3}{2}$ $\quad 100$

69. $\log_x \frac{1}{64} = -3$ $\quad 4$

▶ **70.** $\log_x \frac{1}{100} = -2$ $\quad 10$

71. $\log_8 x = 0$ $\quad 1$

72. $\log_4 8 = x$ $\quad \frac{3}{2}$

▶ **73.** $\log_x \frac{\sqrt{3}}{3} = \frac{1}{2}$ $\quad \frac{1}{3}$

▶ **74.** $\log_x \frac{9}{4} = 2$ $\quad \frac{3}{2}$

Evaluate each logarithmic expression. **See Examples 4 and 5.**

▶ **75.** $\log_2 8$ $\quad 3$

76. $\log_3 9$ $\quad 2$

77. $\log_4 16$ $\quad 2$

▶ **78.** $\log_6 216$ $\quad 3$

79. $\log 1{,}000{,}000$ $\quad 6$

▶ **80** $\log 100{,}000$ $\quad 5$

▶ **81.** $\log \frac{1}{10}$ $\quad -1$

▶ **82.** $\log \frac{1}{10{,}000}$ $\quad -4$

▶ 83. $\log_{1/2} \frac{1}{32}$ 5

84. $\log_{1/3} \frac{1}{81}$ 4

85. $\log_9 3$ $\frac{1}{2}$

▶ 86. $\log_{125} 5$ $\frac{1}{3}$

Use a calculator to find each value. Give answers to four decimal places. **See Using Your Calculator: Evaluating Logarithms.**

87. $\log 3.25$ 0.5119

▶ 88. $\log 0.57$ −0.2441

89. $\log 0.00467$ −2.3307

▶ 90. $\log 375.876$ 2.5750

Use a calculator to solve each equation. Round answers to four decimal places. **See Example 6.**

91. $\log x = 3.7813$ 6,043.6597

92. $\log x = 2.8945$ 784.3321

93. $\log x = -0.7630$ 0.1726

94. $\log x = -1.3587$ 0.0438

▶ 95. $\log x = -0.5$ 0.3162

96. $\log x = -0.926$ 0.1186

97. $\log x = -1.71$ 0.0195

98. $\log x = 1.4023$ 25.2522

Graph each function. Determine whether each function is an increasing or a decreasing function. **See Objective 5.** See AIE Appendix 3.

▶ 99. $f(x) = \log_3 x$ Increasing

▶ 100. $f(x) = \log_{1/3} x$ Decreasing

▶ 101. $y = \log_{1/2} x$ Decreasing

102. $y = \log_4 x$ Increasing

Graph each function by plotting points or by using a translation. (The basic logarithmic functions graphed in Exercises 99–102 will be helpful.) **See Example 7.** See AIE Appendix 3.

▶ 103. $f(x) = 3 + \log_3 x$

104. $f(x) = (\log_{1/3} x) - 1$

105. $y = \log_{1/2} (x - 2)$

▶ 106. $y = \log_4 (x + 2)$

Graph each pair of inverse functions on the same coordinate system. Draw the axis of symmetry. **See Objective 1.** See AIE Appendix 3.

▶ 107. $f(x) = 6^x$, $f^{-1}(x) = \log_6 x$

108. $f(x) = 3^x$, $f^{-1}(x) = \log_3 x$

▶ 109. $f(x) = 5^x$, $f^{-1}(x) = \log_5 x$

110. $f(x) = 8^x$, $f^{-1}(x) = \log_8 x$

APPLICATIONS

▶ 111. **dB Gain.** Find the dB gain of the amplifier shown below. Round to the nearest tenth. 49.5 dB

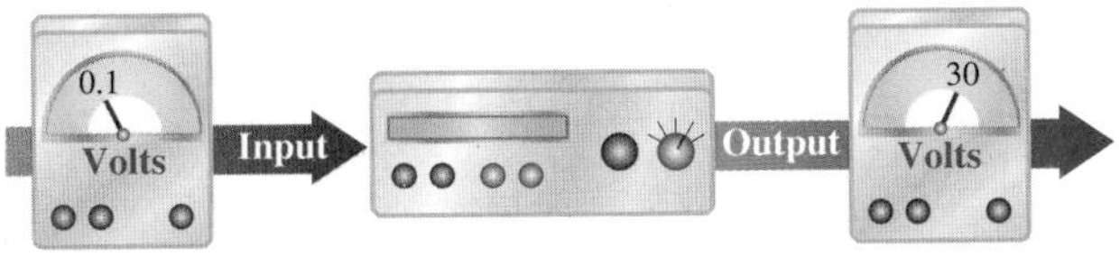

▶ 112. **Output Voltage.** Find the dB gain of an amplifier if the output voltage is 2.8 volts when the input voltage is 0.05 volt. Round to the nearest dB. 35 dB

▶ 113. **Earthquakes.** Refer to the illustration in the next column. Common logarithms are used to measure the intensity of earthquakes. If R is the intensity of an earthquake on the **Richter scale,** A is the amplitude (measured in micrometers) of the ground motion and P is the period (the time of one oscillation of the Earth's surface measured in seconds), then $R = \log \frac{A}{P}$. If an earthquake has amplitude of 5,000 micrometers and a period of 0.2 second, what is its measure on the Richter scale? Round to the nearest tenth. 4.4

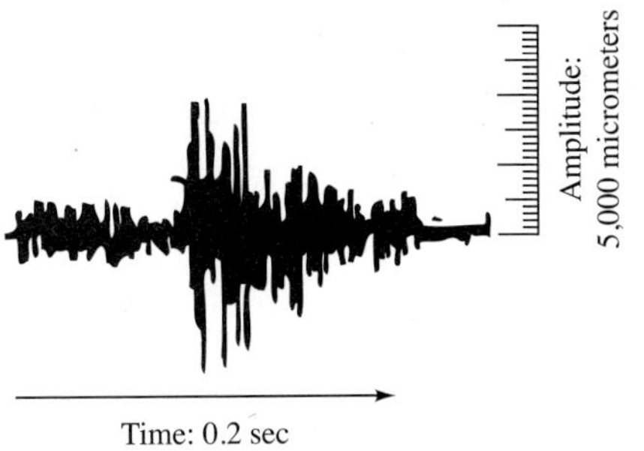

114. **Earthquakes.** If an earthquake has amplitude of 95,000 micrometers and a period of $\frac{1}{4}$ second, what is its measure on the Richter scale? See problem 113. Round to the nearest tenth. 5.6

115. **from Campus to Careers**

Social Worker

The logarithmic function $c(m) = 500 \log (m + 1)$ approximates the total number of cases handled one year by the staff at a county department of social services. The variable m represents the month of the year, where January is 1, February is 2, and so on. Find the total number of cases handled by the department that year by the end of June. About 423 cases

116. **Advertising.** The dollar amount of sales of a certain new product is approximated by the logarithmic function $D(m) = 20{,}000 + 100{,}000 \log (15m + 1)$, where m is the number of minutes of advertising of the product that is shown on cable television. Find the total sales if 30 minutes of cable television advertising time is purchased. About $285,418 of sales

117. **Stocking Lakes.** A farmer stocked a lake on her property with 75 sunfish. She was told that with the proper oversight, the sunfish population could be approximated by the logarithmic function $f(t) = 75 + 45 \log (t + 1)$, where t is the number of years since the lake was stocked. Find the number of sunfish she can expect in the lake in $2\frac{1}{2}$ years. About 99 sunfish

118. **Zoology.** A trap-and-release program run by zoologists found that the ground squirrel population in a wilderness area could be estimated by the logarithmic function $s(t) = 800 + 600 \log (50t + 1)$, where t is the number of months after the program started. Find the ground squirrel population 3 years after the program began. (*Hint*: Be careful to use the correct units in your solution.) About 2,753 ground squirrels

119. **Children's Height.** The logarithmic function $h(A) = 29 + 48.8 \log(A + 1)$ gives the percent of the adult height a male child A years old has attained. If a boy is 9 years old, what percent of his adult height will he have reached? 77.8%

▶ **120. Depreciation.** In business, equipment is often depreciated using the double declining-balance method. In this method, a piece of equipment with a life expectancy of N years, costing $\$C$, will depreciate to a value of $\$V$ in n years, where n is given by the formula

$$n = \frac{\log V - \log C}{\log\left(1 - \frac{2}{N}\right)}$$

A computer that cost $37,000 has a life expectancy of 5 years. If it has depreciated to a value of $8,000, how old is it? 3 yr old

▶ **121. Investing.** If $\$P$ is invested at the end of each year in an annuity earning annual interest at a rate r, the amount in the account will be $\$A$ after n years, where

$$n = \frac{\log\left(\frac{Ar}{P} + 1\right)}{\log(1 + r)}$$

If $1,000 is invested each year in an annuity earning 12% annual interest, how long will it take for the account to be worth $20,000? Round to the nearest tenth of a year. 10.8 yr

▶ **122. Growth of Money.** If $5,000 is invested each year in an annuity earning 8% annual interest, how long will it take for the account to be worth $50,000? (See Exercise 121.) Round to the nearest tenth of a year. 7.6 yr

WRITING

123. Explain the mathematical relationship between $f(x) = \log x$ and $g(x) = 10^x$.

▶ **124.** Explain why it is impossible to find the logarithm of a negative number.

125. A table of solutions for $f(x) = \log x$ is shown here. As x decreases and gets close to 0, what happens to the values of $f(x)$?

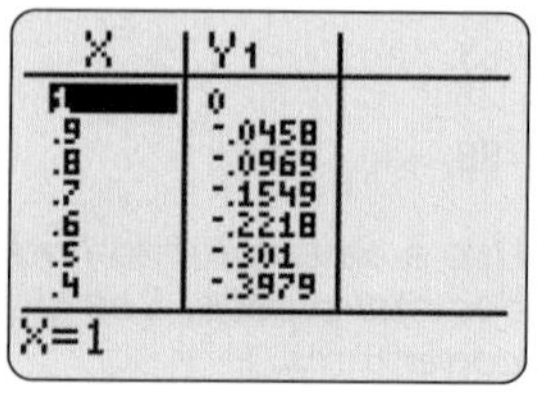

X	Y1
1	0
.9	-.0458
.8	-.0969
.7	-.1549
.6	-.2218
.5	-.301
.4	-.3979

X=1

126. What question should be asked when evaluating the expression $\log_4 16$?

REVIEW

Solve each equation.

127. $\sqrt[3]{6x + 4} = 4$ 10

▶ **128.** $\sqrt{3x + 4} = \sqrt{7x + 2}$ $\frac{1}{2}$

129. $\sqrt{a + 1} - 1 = 3a$ 0; $-\frac{5}{9}$ is extraneous

▶ **130.** $3 - \sqrt{t - 3} = \sqrt{t}$ 4

CHALLENGE PROBLEMS

131. Without graphing, determine the domain of the function $f(x) = \log_5 (x^2 - 1)$. Express the result in interval notation. $(-\infty, -1) \cup (1, \infty)$

▶ **132.** Evaluate: $\log_6(\log_5(\log_4 1{,}024))$ 0

▶ **133. Earthquakes.** In 1985, Mexico City experienced an earthquake of magnitude 8.1 on the Richter scale. In 1989, the San Francisco Bay area was rocked by an earthquake measuring 7.1. By what factor must the amplitude of an earthquake change to increase its severity by 1 point on the Richter scale? (Assume that the period remains constant.) A factor of 10

134. Graph: $f(x) = \log_2|x|$ See AIE Appendix 3.

SECTION 11.5 Base-*e* Exponential and Logarithmic Functions

OBJECTIVES

1. Define the natural exponential function.
2. Graph the natural exponential function.
3. Use base-e exponential formulas and functions in applications.
4. Define base-e logarithms.
5. Evaluate natural logarithmic expressions.
6. Graph the natural logarithmic function.
7. Use base-e logarithmic formulas and functions in applications.

ARE YOU READY? *Are You Ready? exercises available online at www.webassign.net/brookscole*

The following problems review some basic skills that are needed when working with base-e exponential and logarithmic functions.

1. Evaluate $\left(1 + \frac{1}{n}\right)^n$ for $n = 2$. 2.25 **2.** Evaluate: $(2.718)^0$ 1

3. Round 2.718281828459 to the nearest tenth. 2.7

4. Use a calculator to evaluate $55(2.718)^{0.4}$. Round to the nearest hundredth. 82.05

5. Fill in the blank: $\log_3 x = 6$ is equivalent to $3^6 = x$.

6. Fill in the blank: If (3, 8) is on the graph of a one-to-one function f, then the point (8, 3) is on the graph of f^{-1}.

Any positive real number not equal to 1 can be used as a base of an exponential or a logarithmic function. However, some bases are used more often than others. Exponential and logarithmic functions that have many applications are ones whose base is an irrational number represented by the letter *e*.

1 Define the Natural Exponential Function.

n	$\left(1 + \frac{1}{n}\right)^n$
1	2
2	2.25
4	2.44140625 ...
12	2.61303529 ...
365	2.71456748 ...
1,000	2.71692393 ...
100,000	2.71826830 ...
1,000,000	2.71828137 ...

A scientific calculator was used to evaluate the expression.

The number called e is defined to be the value that $\left(1 + \frac{1}{n}\right)^n$ approaches as n gets larger and larger. The table in the margin shows the value of that expression as n increases from 1 to 1,000,000. It can be shown that as n approaches infinity, the value of $\left(1 + \frac{1}{n}\right)^n$ approaches:

$$e = \mathbf{2.718281828459} \ldots$$

We give this number a letter-name because it makes communication easier. Reciting its first thirteen digits every time we refer to it would be overwhelming. So, we simply call it by the name "e."

Like π, the number e is an irrational number. That means its decimal representation is nonterminating and nonrepeating. Rounded to four decimal places, $e \approx 2.7183$.

Of all possible bases for an exponential function, e is the most convenient for problems involving growth or decay. Since these situations occur often in natural settings, we call $f(x) = e^x$ the *natural exponential function.*

The Natural Exponential Function

The function defined by $f(x) = e^x$ is the **natural exponential function** (or the **base-e exponential function**) where $e = 2.71828\ldots$. The domain of $f(x) = e^x$ is the interval $(-\infty, \infty)$. The range is the interval $(0, \infty)$.

The e^x key on a calculator is used to find powers of e.

Using Your Calculator ▶ Finding powers of e

To find the value of e^5 with a reverse entry scientific calculator, we press:

5 e^x 148.4131591

On some calculators, the e^x key also serves as the LN key. (Later in this section we will see why.) To activate the e^x key, we must begin by pressing 2nd or SHIFT.

With a direct entry graphing calculator, we press:

2nd e^x 5) ENTER e^(5) 148.4131591

Since powers of e are irrational numbers (nonterminating, nonrepeating decimals), we often round such answers. For example, to the nearest hundredth, $e^5 \approx 148.41$

2 Graph the Natural Exponential Function.

To graph $f(x) = e^x$, we construct a table of function values by choosing several values for x and finding the corresponding values of $f(x)$. For example, if $x = -2$, we have

$f(x) = e^x$

$f(-2) = e^{-2}$ Substitute −2 for each x.

$= 0.135335283 \ldots$ Use a calculator. On a scientific calculator, press: 2 +/− 2nd e^x.

≈ 0.1 Round to the nearest tenth.

We enter $(-2, 0.1)$ in the table on the next page. Similarly, we find $f(-1)$, $f(0)$, $f(1)$, and $f(2)$, enter each result in the table, and plot the ordered pairs. We draw a smooth curve through the points to get the graph.

From the graph, we can verify that the domain of $f(x) = e^x$ is the interval $(-\infty, \infty)$ and the range is the interval $(0, \infty)$. Since the graph passes the horizontal line test, the function is one-to-one.

Notation

Swiss born Leonhard Euler (1707–1783) is said to have published more than any mathematician in history. Through his work, the symbol e came into common use.

Note that as x decreases, the values of $f(x)$ decrease and approach 0. Thus, the x-axis is an asymptote of the graph. The graph does not have an x-intercept, the y-intercept is (0, 1), and the graph passes through the point $(1, e)$.

$f(x) = e^x$

x	$f(x)$	
-2	$\frac{1}{e^2} \approx 0.1$	$\rightarrow (-2, 0.1)$
-1	$\frac{1}{e^1} \approx 0.4$	$\rightarrow (-1, 0.4)$
0	$e^0 = 1$	$\rightarrow (0, 1)$
1	$e^1 \approx 2.7$	$\rightarrow (1, 2.7)$
2	$e^2 \approx 7.4$	$\rightarrow (2, 7.4)$

The outputs can be found using the e^x key on a calculator. In such cases, round to the nearest tenth to make point-plotting easier.

Teaching Tip: Remind your students that e is not a variable, it is a constant.

The graph steadily approaches the x-axis, but never touches or crosses it.

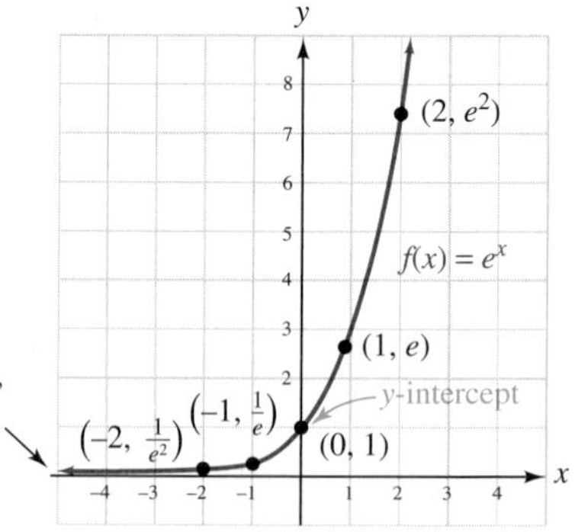

The graph of the natural exponential function can be translated horizontally and vertically, as shown below.

Success Tip

Recall that exponential functions have a constant base and a variable exponent. For the natural exponential function $f(x) = e^x$, the constant base is e.

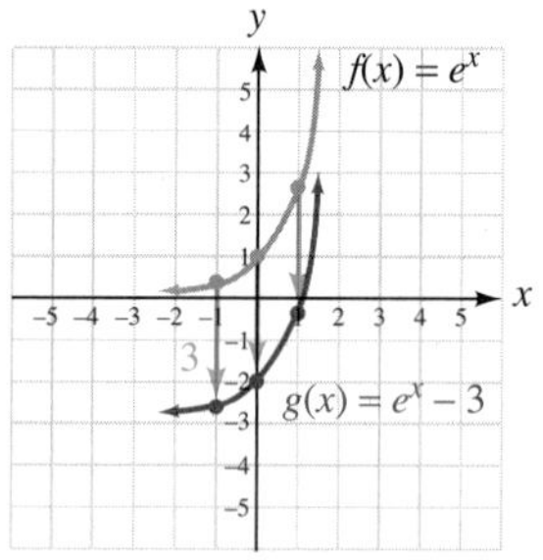

To graph $g(x) = e^x - 3$, translate each point on the graph of $f(x) = e^x$ down 3 units.

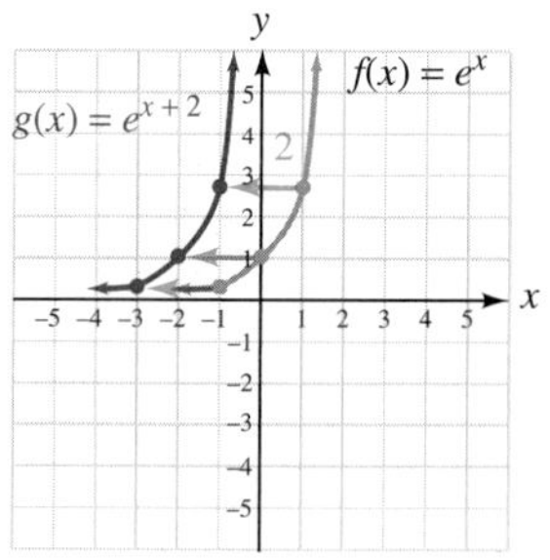

To graph $g(x) = e^{x+2}$, translate each point on the graph of $f(x) = e^x$ to the left 2 units.

We can illustrate the effects of vertical and horizontal translations of the natural exponential function by using a graphing calculator.

Using Your Calculator ▶ Graphing Base *e* (Natural) Exponential Functions

Figure (a) shows the calculator graphs of $f(x) = e^x$, $g(x) = e^x + 5$, and $h(x) = e^x - 3$. To graph them, we enter the right sides of the equations after $Y_1 =$, $Y_2 =$, and $Y_3 =$. The display will show:

$Y_1 = e^\wedge(X) \qquad Y_2 = e^\wedge(X) + 5 \qquad Y_3 = e^\wedge(X) - 3$

The graph of $g(x) = e^x + 5$ is 5 units above the graph of $f(x) = e^x$ and the graph of $h(x) = e^x - 3$ is 3 units below the graph of $f(x) = e^x$.

Figure (b) shows the calculator graphs of $f(x) = e^x$, $g(x) = e^{x+5}$, and $h(x) = e^{x-3}$. The graph of $g(x) = e^{x+5}$ is 5 units to the left of the graph of $f(x) = e^x$ and the graph of $h(x) = e^{x-3}$ is 3 units to the right of the graph of $f(x) = e^x$.

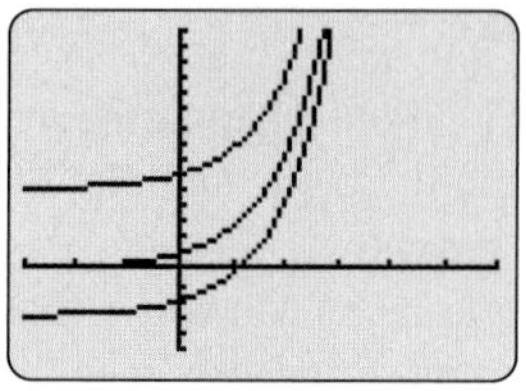

(a)

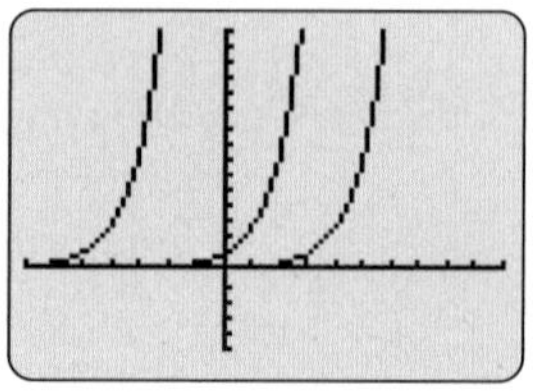

(b)

To graph complicated natural exponential functions, point-plotting can be tedious. In such cases, we will use a graphing calculator. For example, the figure shows the calculator graph of $f(x) = 3e^{-x/2}$. To graph this function, we enter the right side of the equation after the symbol $Y_1 =$. The display will show the equation

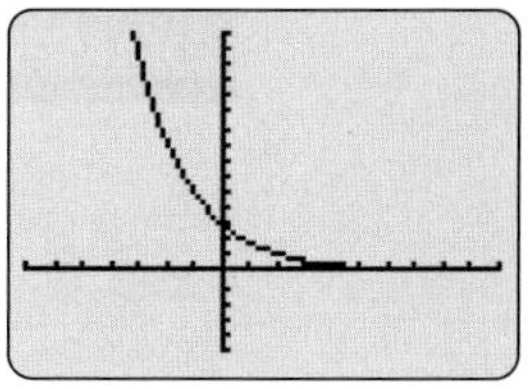

$$Y_1 = 3(e\text{^}(-X/2))$$

3 Use Base-*e* Exponential Formulas and Functions in Applications.

The following formula involving *e* provides a mathematical model for growth and decay applications. It is used in many areas, such as continuous compound interest, population trends, heat transfer, radioactivity, and learning retention.

Formula for Exponential Growth/Decay

If a quantity P increases or decreases at an annual rate r, compounded continuously, the amount A after t years is given by

$A = Pe^{rt}$ Read as "A equals P times e to the rt power."

For the formula above, if the time is measured in years, r is called the **annual growth rate.** If r is negative, the growth represents a decrease.

For a given quantity P, say 10,000, and a given rate r, say 5%, we can write the formula for exponential growth using function notation:

$$A(t) = 10{,}000e^{0.05t}$$

Recall that if a bank pays interest twice a year, we say the interest is compounded *semiannually.* If it pays interest four times a year, we say the interest is paid *quarterly.* If it pays interest continuously (infinitely many times a year), we say that the interest is **compounded continuously.**

EXAMPLE 1 **Investing.** If \$25,000 accumulates interest at an annual rate of 8%, compounded continuously, find the balance in the account in 50 years.

Strategy We will substitute 25,000 for P, 0.08 for r, and 50 for t in the formula $A = Pe^{rt}$ and calculate the value of A.

Why The words *compounded continuously* indicate that we should use the base-*e* exponential growth/decay formula.

Solution

Teaching Tip: As a comparison, you may want to use $A = P\left(1 + \frac{r}{k}\right)^{kt}$ to calculate the account balance in Example 1 if the interest is compounded quarterly or daily.

$A = Pe^{rt}$ This is the formula for continuous compound interest.

$A = 25{,}000e^{0.08(50)}$ Write 8% as 0.08. Substitute for P, r, and t.

$= 25{,}000e^{4}$ Evaluate the exponent: 0.08(50) = 4.

$\approx 1{,}364{,}953.751$ Use a calculator. On a scientific calculator, press: 25000 × 4 2nd e^x =.

$\approx 1{,}364{,}953.75$ Round to the nearest hundredth.

In 50 years, the balance will be \$1,364,953.75—more than a million dollars.

Teaching Example 1 Investing. In Example 1, find the balance in 55 years. *Answer:* \$2,036,271.72

Self Check 1 **Investing.** In Example 1, find the balance in 60 years. \$3,037,760.44

Now Try Problems 33 and 73

EXAMPLE 2 **City Planning.** The population of a city is currently 15,000, but economic conditions are causing the population to decrease 3% each year. If this trend continues, find the population in 30 years.

Strategy We will substitute 15,000 for P, -0.03 for r, and 30 for t in the formula $A = Pe^{rt}$ and calculate the value of A.

Why Since the population is decreasing 3% each year, the annual growth rate is -3%, or -0.03.

Solution

Success Tip

For quantities that are decreasing, remember to enter a negative value for r, the annual rate, in the formula $A = Pe^{rt}$.

$A = Pe^{rt}$ This is the model for population growth/decay.

$A = \mathbf{15{,}000}e^{\mathbf{-0.03(30)}}$ Substitute for P, r, and t.

$= 15{,}000e^{-0.9}$ Evaluate the exponent: $-0.03(30) = -0.9$.

$\approx 6{,}098.544896$ Use a calculator. On a scientific calculator, press: 15000 × .9 +/− 2nd e^x =.

$\approx 6{,}099$ Round to the nearest whole number.

In 30 years, the expected population will be 6,099.

Teaching Example 2 City Planning. In Example 2, find the population in 40 years.
Answer: 4,518

Self Check 2 **City Planning.** In Example 2, find the population in 50 years. 3,347

Now Try Problems 37 and 81

EXAMPLE 3 **Baking.** A mother takes a cake out of the oven and sets it on a rack to cool. The function $T(t) = 68 + 220e^{-0.18t}$ gives the cake's temperature in degrees Fahrenheit after it has cooled for t minutes. If her children will be home from school in 20 minutes, will the cake have cooled enough for the children to eat it? (Assume that 80°F, or cooler, would be a comfortable eating temperature.)

©RoJo Images/Shutterstock.com

Strategy We will substitute 20 for t in the function $T(t) = 68 + 220e^{-0.18t}$.

Why The variable t represents the number of minutes the cake has cooled.

Solution When the children arrive home, the cake will have cooled for 20 minutes. To find the temperature of the cake at that time, we need to find $T(20)$.

$T(t) = 68 + 220e^{-0.18t}$ This is the cooling model.

$T(\mathbf{20}) = 68 + 220e^{-0.18(\mathbf{20})}$ Substitute 20 for t.

$= 68 + 220e^{-3.6}$ Evaluate the exponent: $-0.18(20) = -3.6$.

≈ 74.01121894 Use a calculator. On a scientific calculator, press: 68 + 220 × 3.6 +/− 2nd e^x =.

≈ 74.0 Round to the nearest tenth.

When the children return home, the temperature of the cake will be about 74°, and it can be eaten.

Teaching Example 3 Baking. In Example 3, find the temperature of the cake 15 minutes after it is removed from the oven. Round to the nearest tenth.
Answer: 82.8°

Self Check 3 **Baking.** In Example 3, find the temperature of the cake 10 minutes after it is removed from the oven. Round to the nearest tenth. 104.4°

Now Try Problem 91

4 Define Base-*e* Logarithms.

Of all possible bases for a logarithmic function, e is the most convenient for problems involving growth or decay. Since these situations occur often in natural settings, base-e

Teaching Tip: You may want to tell your students that natural logarithmic functions play a very important role in advanced mathematics classes, such as calculus.

Caution

Because of the font used to print the natural log of x, some students initially misread the notation as In x. In handwriting, ln x should look like *ℓn x*.

logarithms are called **natural logarithms** or **Napierian logarithms** after John Napier (1550–1617). They are usually written as $\ln x$ rather than $\log_e x$:

$\ln x$ means $\log_e x$ Read ln x letter-by-letter as "ℓ … n … of x."

In general, the logarithm of a number is an exponent. For natural logarithms,

ln x is the exponent to which e is raised to get x.

Translating this statement into symbols, we have

$$e^{\ln x} = x$$

5 Evaluate Natural Logarithmic Expressions.

EXAMPLE 4 Evaluate each natural logarithmic expression:

a. $\ln e$ **b.** $\ln \frac{1}{e^2}$ **c.** $\ln 1$ **d.** $\ln \sqrt{e}$

Strategy Since the base is e in each case, we will ask "To what power must e be raised to get the given number?"

Why That power is the value of the logarithmic expression.

Solution

a. $\ln e = 1$ Ask: "To what power must we raise e to get e?" Since $e^1 = e$, the answer is: the 1st power.

b. $\ln \frac{1}{e^2} = -2$ Ask: "To what power must we raise e to get $\frac{1}{e^2}$?" Since $e^{-2} = \frac{1}{e^2}$, the answer is: the -2 power.

c. $\ln 1 = 0$ Ask: "To what power must we raise e to get 1?" Since $e^0 = 1$, the answer is: the 0 power.

d. $\ln \sqrt{e} = \frac{1}{2}$ Ask: "To what power must we raise e to get $\sqrt{e}$?" Since $e^{1/2} = \sqrt{e}$, the answer is: the $\frac{1}{2}$ power.

Success Tip

Every natural logarithmic equation has a corresponding natural exponential equation. For example:

$$\ln \frac{1}{e^2} = -2 \quad \text{means} \quad e^{-2} = \frac{1}{e^2}$$

and

$$\ln \sqrt{e} = \frac{1}{2} \quad \text{means} \quad e^{1/2} = \sqrt{e}$$

Self Check 4 Evaluate each expression: **a.** $\ln e^3$ 3 **b.** $\ln \frac{1}{e}$ -1 **c.** $\ln \sqrt[3]{e}$ $\frac{1}{3}$

Now Try Problems 41, 45, and 47

Teaching Example 4 Evaluate each natural logarithmic expression:

a. $\ln e^5$ **b.** $\ln \sqrt[5]{e}$ **c.** $\ln \frac{1}{e^5}$

Answers:

a. 5 **b.** $\frac{1}{5}$ **c.** -5

Many natural logarithmic expressions are not as easy to evaluate as those in the previous example. For example, to find ln 2.34, we ask, "To what power must we raise e to get 2.34?" The answer isn't obvious. In such cases, we use a calculator.

Using Your Calculator ▸ **Evaluating Base-*e* (Natural) Logarithms**

To find ln 2.34 with a reverse entry scientific calculator, we press:

2.34 [LN] `.850150929`

On some calculators, the [e^x] key also serves as the [LN] key when [2nd] or [SHIFT] is pressed. We will see why this is so later in this section.

To use a direct entry graphing calculator, we press:

[LN] 2.34 [)] [ENTER]

```
ln(2.34)
         .8501509294
```

To four decimal places, $\ln 2.34 \approx 0.8502$. This means that $e^{0.8502} \approx 2.34$.

If we attempt to evaluate logarithmic expressions such as ln 0, or the logarithm of a negative number, such as ln (−5), then one of the following error statements will be displayed.

Error

```
ERR:DOMAIN
1:QUIT
2:Go to
```

```
ERR:NONREAL ANS
1:QUIT
2:Go to
```

Certain natural logarithmic equations can be solved by writing them as equivalent natural exponential equations.

EXAMPLE 5 Solve each equation: **a.** $\ln x = 1.335$ and **b.** $\ln x = -5.5$. Give each result to four decimal places.

Strategy To solve these logarithmic equations, we will instead write and solve equivalent exponential equations.

Why The resulting exponential equations are easier to solve because the variable term is isolated on one side.

Solution **a.** Since the base of the natural logarithmic function is e:

Success Tip

Recall that equivalent equations have the same solutions.

$\ln x = 1.335$ is equivalent to $e^{1.335} = x$ (Exponent: 1.335; Base: e)

To use a reverse entry scientific calculator to find x, press:

1.335 [2nd] [e^x] 3.799995946

To four decimal places, $x = 3.8000$. Thus, the solution is 3.8000. To check, use your calculator to verify that $\ln 3.8000 \approx 1.335$.

b. The logarithmic equation $\ln x = -5.5$ is equivalent to the exponential equation $e^{-5.5} = x$. To use a reverse entry scientific calculator to find x, press:

5.5 [+/−] [2nd] [e^x] 0.004086771

To four decimal places, $x = 0.0041$. Thus, the solution is 0.0041. To check, use your calculator to verify that $\ln 0.0041 \approx -5.5$.

Teaching Example 5 Solve each equation. Give each result to four decimal places.
a. $\ln x = 2.049$ b. $\ln x = -1.722$
Answers: a. 7.7601 b. 0.1787

Self Check 5 Solve each equation. Give each result to four decimal places.
a. $\ln x = 1.9344$ 6.9199 **b.** $-3 = \ln x$ 0.0498

Now Try Problems 61 and 65

6 Graph the Natural Logarithmic Function.

Because the natural exponential function defined by $f(x) = e^x$ is one-to-one, it has an inverse function that is defined by $x = e^y$. When we write $x = e^y$ in the equivalent form $y = \ln x$, the result is called the *natural logarithmic function.*

The Natural Logarithmic Function

The natural logarithmic function with base e is defined by the equations

$$f(x) = \ln x \text{ or } y = \ln x, \text{ where } \ln x = \log_e x$$

The domain of $f(x) = \ln x$ is the interval $(0, \infty)$, and the range is the interval $(-\infty, \infty)$.

Since the natural logarithmic function is the inverse of the one-to-one natural exponential function, the natural logarithmic function is one-to-one.

To graph $f(x) = \ln x$, we can construct a table of function values, plot the resulting ordered pairs, and draw a smooth curve through the points to get the graph shown in figure (a). Figure (b) shows the calculator graph of $f(x) = \ln x$. We see that the domain of the natural logarithmic function is $(0, \infty)$ and the range is $(-\infty, \infty)$.

$f(x) = \ln x$

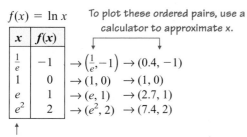

To plot these ordered pairs, use a calculator to approximate x.

x	$f(x)$		
$\frac{1}{e}$	-1	$\to \left(\frac{1}{e}, -1\right)$	$\to (0.4, -1)$
1	0	$\to (1, 0)$	$\to (1, 0)$
e	1	$\to (e, 1)$	$\to (2.7, 1)$
e^2	2	$\to (e^2, 2)$	$\to (7.4, 2)$

Since the base of the natural logarithmic function is e, choose x-values that are integer powers of e.

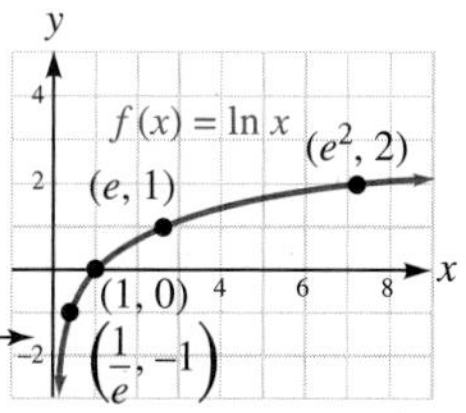

(a)

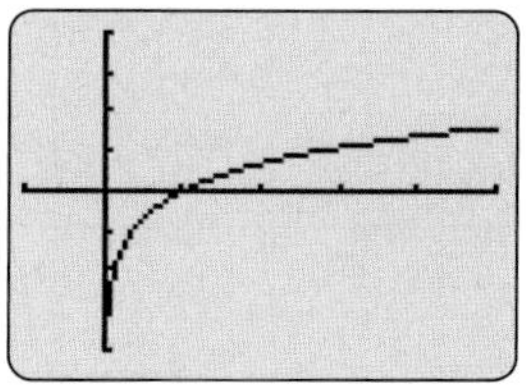
(b)

Notation

If it is helpful, imagine parentheses around the input value in the natural logarithmic function notation.

$f(e) = 1$
Function name — Input — Output
$\ln (e) = 1$

The natural exponential function and the natural logarithm function are inverse functions. The figure shows that their graphs are symmetric to the line $y = x$.

You may want to confirm that these functions are indeed inverses using the composition of inverses test in Section 11.2.

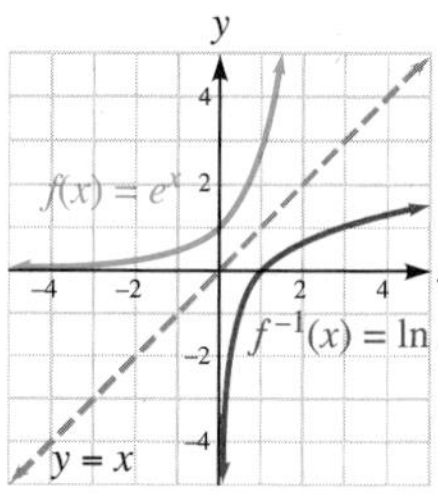

Using Your Calculator ▶ Graphing Base-*e* Logarithmic Functions

Many graphs of logarithmic functions involve translations of the graph of $f(x) = \ln x$. For example, the figure below shows calculator graphs of the functions $f(x) = \ln x$, $g(x) = (\ln x) + 2$, and $h(x) = (\ln x) - 3$.

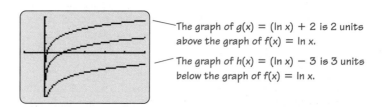

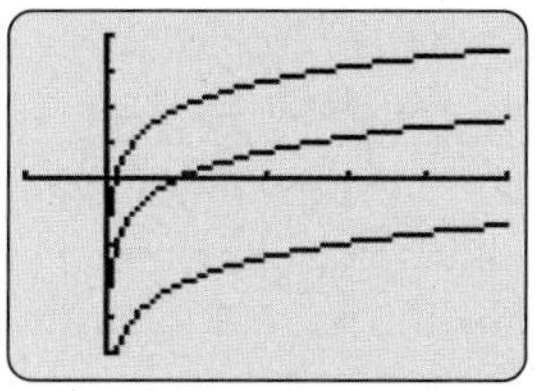

The graph of g(x) = (ln x) + 2 is 2 units above the graph of f(x) = ln x.
The graph of h(x) = (ln x) − 3 is 3 units below the graph of f(x) = ln x.

The next figure shows the calculator graph of the functions $f(x) = \ln x$, $g(x) = \ln (x - 2)$, and $h(x) = \ln (x + 3)$.

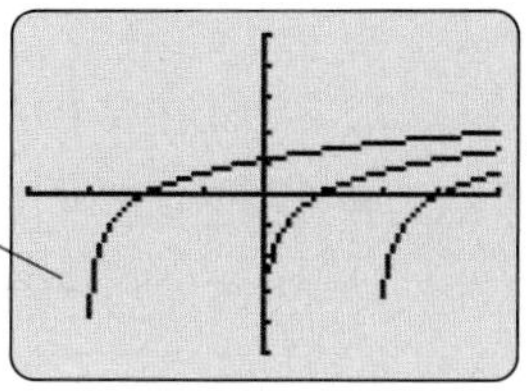

The graph of h(x) = ln (x + 3) is 3 units to the left of the graph of f(x) = ln x.
The graph of g(x) = ln (x − 2) is 2 units to the right of the graph of f(x) = ln x.

7 Use Base-*e* Logarithmic Formulas and Functions in Applications.

If a population grows exponentially at a certain annual rate, the time required for the population to double is called the **doubling time.** It is given by the following formula.

Formula for Doubling Time

If r is the annual rate, compounded continuously, and t is the time required for a population to double, then

$$t = \frac{\ln 2}{r}$$

EXAMPLE 6

Doubling Time. The population of the Earth is growing at the approximate rate of 1.133% per year. If this rate continues, how long will it take for the population to double? (Source: CIA World Fact Book, 2009 data)

Strategy We will substitute 1.133% expressed as a decimal for r in the formula for doubling time and evaluate the right side using a calculator.

Why We can use this formula because we are given the annual rate of continuous compounding.

Solution Since the population is growing at the rate of 1.133% per year, we substitute 0.0133 for r in the formula for doubling time and simplify.

$$t = \frac{\ln 2}{r}$$ Don't forget to substitute the decimal form of 1.133%, which is 0.0133, for r.

$$t = \frac{\ln 2}{\mathbf{0.0133}}$$

$$\approx 52.11632937$$ Use a calculator. On a scientific calculator, press: 2 LN ÷ .0033 =.

$$\approx 52$$ Round to the nearest year.

At the current growth rate, the population of the Earth will double in about 52 years.

Teaching Example 6 Doubling Time. See Example 6. If the population's annual growth rate could be reduced to 1% per year, what would be the doubling time?
Answer: About 69 years

Self Check 6 **Doubling Time.** If the annual growth rate of the Earth's population could be reduced to 1.1% per year, what would be the doubling time?

About 63 years

Now Try Problem 93

EXAMPLE 7

The Pace of Life. A study by psychologists M. H. Bornstein and H. G. Bornstein found that the average walking speed s, in feet per second, of a pedestrian in a city of population p, is approximated by the natural logarithmic function $s(p) = 0.05 + 0.37 \ln p$. According to this study, what is the average walking speed of a pedestrian in Detroit if its population is 910,000? (Source: *Nature*, Volume 259, Feb. 19, 1976 and infoplease.com)

Strategy We will find $s(910{,}000)$.

Why Since the variable p represents the population of the city, $p = 910{,}000$.

Solution

$$s(p) = 0.05 + 0.37 \ln p$$ This is the natural logarithmic model.

$$s(\mathbf{910{,}000}) = 0.05 + 0.37 \ln \mathbf{910{,}000}$$ Substitute 910,000 for p.

$$\approx 5.126843955$$ Use a calculator. On a scientific calculator, press: 0.05 + 0.37 × 910000 LN =.

$$\approx 5.1$$ Round to the nearest tenth.

The average walking speed of a pedestrian in Detroit is about 5.1 feet per second.

Teaching Tip: You may want to ask your students why the larger the population, the greater the walking speed. The authors of this study reasoned that the greater walking speed of people in larger cities was in response to "stimulatory overload."

Teaching Example 7 The Pace of Life. Use the natural logarithmic function from Example 7 to find the average walking speed of a pedestrian in Mesa, Arizona if its population is 462,500.
Answer: About 4.9 feet per second

Self Check 7 **The Pace of Life.** Use the natural logarithmic function from Example 7 to find the average walking speed of a pedestrian in Dallas if its population is 1,300,000. About 5.3 feet per second

Now Try Problem 101

SECTION 11.5 STUDY SET

VOCABULARY

Fill in the blanks.

1. $f(x) = e^x$ is called the natural exponential function. The base is e.
2. $f(x) = \ln x$ is called the natural logarithmic function. The base is e.
3. If a bank pays interest infinitely many times a year, we say that the interest is compounded continuously.
4. Like π, the number e is an irrational number. Its decimal representation is nonterminating and nonrepeating.

CONCEPTS

5. Refer to the graph on the right.

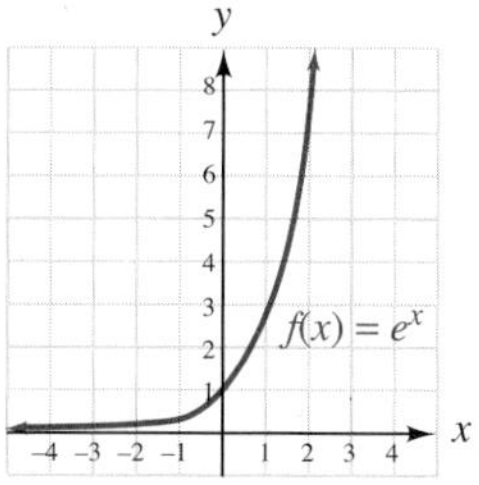

   a. What is the name of the function $f(x) = e^x$? The natural exponential function
   b. What is the domain of the function? What is the range of the function? D: $(-\infty, \infty)$; R: $(0, \infty)$
   c. What is the y-intercept of the graph? What is the x-intercept of the graph? $(0, 1)$, none
   d. Is the function one-to-one? Yes
   e. What is an asymptote of the graph? The x-axis ($y = 0$)
   f. Is f an increasing or a decreasing function? Increasing
   g. The graph passes through the point $(1, y)$. What is y? e

6. a. Use a calculator to complete the table of values in the next column for $f(x) = \ln x$. Round to the nearest hundredth.
   b. Graph $f(x) = \ln x$. Note that the units on the x- and y-axes are different.
   c. What are the domain and range of the function? D: $(0, \infty)$; R: $(-\infty, \infty)$
   d. What is the x-intercept of the graph? What is the y-intercept? $(1, 0)$, none
   e. What is an asymptote of the graph? The y-axis ($x = 0$)
   f. Is f increasing or decreasing? Increasing
   g. Is the function one-to-one? Yes

x	$f(x)$
0.5	−0.69
1	0
2	0.69
4	1.39
6	1.79
8	2.08
10	2.30

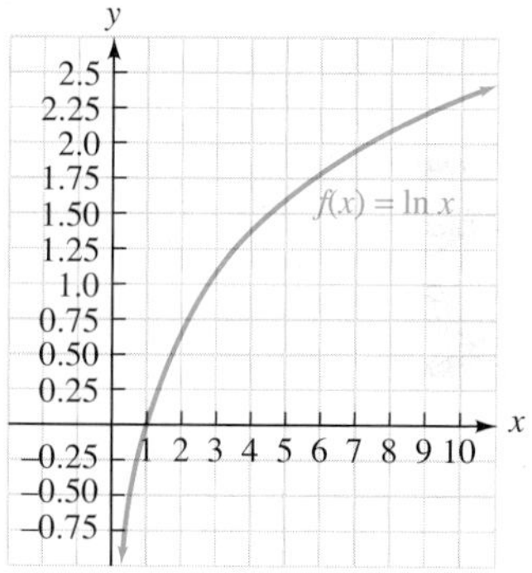

Fill in the blanks.

7. $e = 2.718281828459\ldots$
8. To two decimal places, the value of e is 2.72.
9. If n gets larger and larger, the value of $\left(1 + \frac{1}{n}\right)^n$ approaches the value of e.
10. The formula for exponential growth/decay is $A = Pe^{rt}$.
11. To find $\ln e^2$, we ask, "To what power must we raise e to get e^2?" Since the answer is the 2nd power, $\ln e^2 = 2$.
12. The logarithmic equation $\ln x = 1.5318$ is equivalent to the exponential equation $e^{1.5318} = x$.
13. Graph each irrational number on the number line: $\{\pi, e, \sqrt{2}\}$.

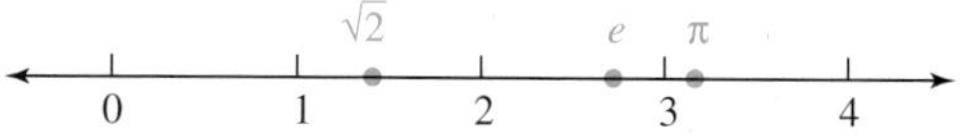

14. Complete the table of values. Use a calculator when necessary and round to the nearest hundredth.

x	−2	−1	0	1	2
e^x	0.14	0.37	1	2.72	7.39

15. a. The function $f(x) = e^x$ is graphed in figure (a) below and the TRACE feature is used. What is the y-coordinate of the point on the graph having an x-coordinate of 1? What is the symbol that represents this number? 2.7182818 . . .; e

b. Figure (b) below shows a table of values for $f(x) = e^x$. As x decreases, what happens to the values of $f(x)$ listed in the Y_1 column? Will the value of $f(x)$ ever be 0 or negative? They decrease; no

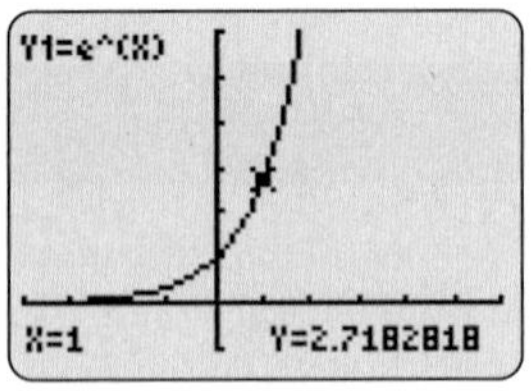

(a)

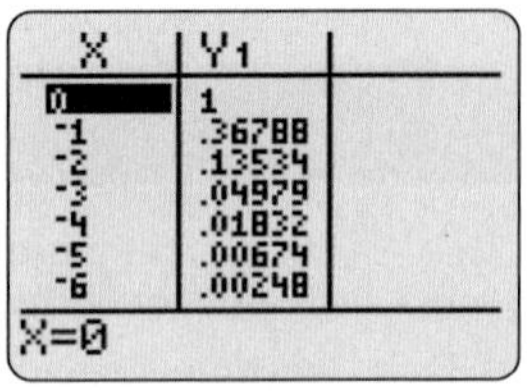

(b)

16. a. The illustration shows the graph of $f(x) = \ln x$, as well as a vertical translation of that graph. Using the notation $g(x)$ for the translation, write the defining equation for that function. $g(x) = 2 + \ln x$

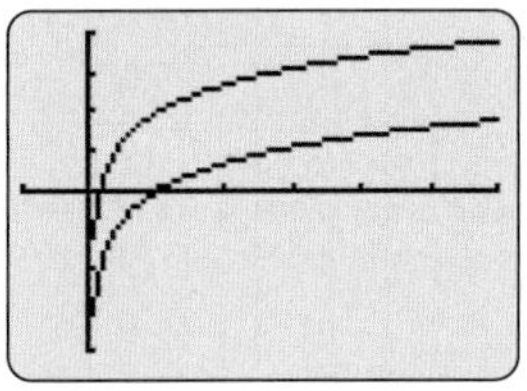

b. In the illustration, $f(x) = \ln x$ was graphed, and the TRACE feature was used. What is the x-coordinate of the point on the graph having a y-coordinate of 1? What is the name given this number? 2.7182818 . . . ; e

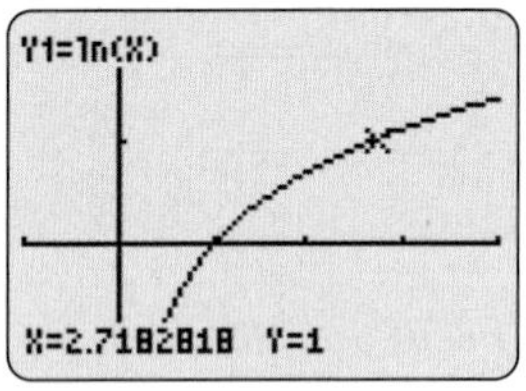

17. What is the inverse of the natural logarithmic function $f(x) = \ln x$? $f^{-1}(x) = e^x$

18. Let $f(x) = 75 + 3{,}570 \ln x$. Find $f(28.1)$. Round to the nearest tenth. 11,983.7

NOTATION

Find A using the formula $A = Pe^{rt}$ given the following values of P, r, and t. Round to the nearest tenth.

19. $P = 1{,}000$, $r = 0.09$, and $t = 10$

$A = 1{,}000\, e^{(0.09)(10)}$

$= 1{,}000e^{0.9}$

$\approx 2{,}459.6$ Use a calculator.

20. $P = 50{,}000$, $r = -0.12$, and $t = 50$

$A = 50{,}000e^{(-0.12)(50)}$

$= 50{,}000e^{-6}$

≈ 123.9 Use a calculator.

Fill in the blanks.

21. We read $\ln x$ letter-by-letter as "ℓ … n … of x."

22. a. $\ln 2$ means $\log_e 2$.

b. $\log 2$ means $\log_{10} 2$.

23. To evaluate a base-10 logarithm with a calculator, use the LOG key. To evaluate the base-e logarithm, use the LN key.

24. If a population grows exponentially at a rate r, the time it will take the population to double is given by the formula $t = \frac{\ln 2}{r}$.

GUIDED PRACTICE

Graph each function. **See Objective 2.** See AIE Appendix 3.

25. $f(x) = e^x$ **26.** $f(x) = -e^x$

27. $f(x) = e^x + 1$ **28.** $f(x) = e^x - 2$

29. $y = e^{x+3}$ **30.** $y = e^{x-5}$

31. $f(x) = 2e^x$ **32.** $f(x) = \frac{1}{2}e^x$

Find A using the formula $A = Pe^{rt}$ given the following values of P, r, and t. Round to the nearest hundredth. **See Example 1.**

33. $P = 5{,}000$, $r = 8\%$, $t = 20$ years 24,765.16

34. $P = 15{,}000$, $r = 6\%$, $t = 40$ years 165,347.65

35. $P = 20{,}000$, $r = 10.5\%$, $t = 50$ years 3,811,325.37

36. $P = 25{,}000$, $r = 6.5\%$, $t = 100$ years 16,628,540.83

Find A using the formula $A = Pe^{rt}$ given the following values of P, r, and t. Round to the nearest hundredth. **See Example 2.**

37. $P = 15{,}895$, $r = -2\%$, $t = 16$ years 11,542.14

38. $P = 33{,}999$, $r = -4\%$, $t = 21$ years 14,677.73

39. $P = 565$, $r = -0.5\%$, $t = 8$ years 542.85

40. $P = 110$, $r = -0.25\%$, $t = 9$ years 107.55

Evaluate each expression without using a calculator. **See Example 4.**

41. $\ln e^5$ 5 **42.** $\ln e^2$ 2

43. $\ln e^6$ 6 **44.** $\ln e^4$ 4

45. $\ln \frac{1}{e}$ -1 **46.** $\ln \frac{1}{e^3}$ -3

47. $\ln \sqrt[4]{e}$ $\frac{1}{4}$ **48.** $\ln \sqrt[5]{e}$ $\frac{1}{5}$

49. $\ln \sqrt[3]{e^2}$ $\frac{2}{3}$ **50.** $\ln \sqrt[4]{e^3}$ $\frac{3}{4}$

51. $\ln e^{-7}$ -7 **52.** $\ln e^{-10}$ -10

Use a calculator to evaluate each expression, if possible. Express all answers to four decimal places. **See Using Your Calculator: Evaluating Base-e (Natural) Logarithms.**

53. $\ln 35.15$ 3.5596 **54.** $\ln 0.675$ -0.3930

55. $\ln 0.00465$ -5.3709 **56.** $\ln 378.96$ 5.9374

57. $\ln 1.72$ 0.5423 **58.** $\ln 2.7$ 0.9933

59. $\ln(-0.1)$ Undefined **60.** $\ln(-10)$ Undefined

Solve each equation. Express all answers to four decimal places. **See Example 5.**

61. $\ln x = 1.4023$ 4.0645 **62.** $\ln x = 2.6490$ 14.1399

63. $\ln x = 4.24$ 69.4079 **64.** $\ln x = 0.926$ 2.5244

65. $\ln x = -3.71$ 0.0245 **66.** $\ln x = -0.28$ 0.7558

67. $\ln x = 1.001$ 2.7210 **68.** $\ln x = -0.001$ 0.9990

Use a graphing calculator to graph each function. **See Objective 2. See Using Your Calculator: Graph Base-e Logarithmic Functions.** See AIE Appendix 3.

69. $f(x) = \ln\left(\frac{1}{2}x\right)$ **70.** $f(x) = \ln x^2$

71. $f(x) = \ln(-x)$ **72.** $f(x) = \ln(3x)$

APPLICATIONS

In Exercises 73–78, assume that there are no deposits or withdrawals.

73. **Continuous Compound Interest.** An initial investment of \$5,000 earns 8.2% interest, compounded continuously. What will the investment be worth in 12 years? \$13,375.68

74. **Continuous Compound Interest.** An initial investment of \$2,000 earns 8% interest, compounded continuously. What will the investment be worth in 15 years? \$6,640.23

75. **Comparison of Compounding Methods.** An initial deposit of \$5,000 grows at an annual rate of 8.5% for 5 years. Compare the final balances resulting from annual compounding and continuous compounding. \$7,518.28 from annual compounding, \$7,647.95 from continuous compounding

76. **Comparison of Compounding Methods.** An initial deposit of \$30,000 grows at an annual rate of 8% for 20 years. Compare the final balances resulting from annual compounding and continuous compounding. \$139,828.71 from annual compounding, \$148,590.97 from continuous compounding

77. **Determining the Initial Deposit.** An account now contains \$11,180 and has been accumulating interest at 7% annual interest, compounded continuously, for 7 years. Find the initial deposit. \$6,849.16

78. **Determining the Previous Balance.** An account now contains \$3,610 and has been accumulating interest at 8% annual interest, compounded continuously. How much was in the account 4 years ago? \$2,621.40

79. **The 20th Century.** The exponential function $A(t) = 123e^{0.0117t}$ approximates the population of the United States (in millions), where t is the number of years after 1930. Use the function to estimate the U.S. population for these important dates:
   - 1937 The Golden Gate Bridge is completed 133 million
   - 1941 The United States enters World War II 140 million
   - 1955 Rosa Parks refuses to give up her seat on a Montgomery, Alabama, bus 165 million
   - 1969 Astronaut Neil Armstrong walks on the moon 194 million
   - 1974 President Nixon resigns 206 million
   - 1986 The *Challenger* space shuttle explodes 237 million
   - 1997 *The Simpsons* becomes the longest running cartoon television series in history 269 million

80. **World Population Growth.** The population of Earth is approximately 6.8 billion people and is growing at an annual rate of 1.133%. Use the exponential growth model to predict the world population in 30 years. About 9.6 billion

81. **Highs and Lows.** Kuwait, located at the head of the Persian Gulf, has one of the greatest population growth rates in the world. Bulgaria, in southeastern Europe, has one of the smallest. Use an exponential growth/decay model to complete the table.

Country	Population 2010	Annual growth rate	Estimated population 2025
Kuwait	2,789,132	3.501%	4,715,620
Bulgaria	7,148,785	−0.768%	6,370,911

Source: CIA World Factbook

82. **Bed Bugs.** If not checked, the population of a colony of bed bugs will grow exponentially at a rate of 65% per week. If a colony currently has 50 bed bugs, how many will there be in 6 weeks? About 2,470 bed bugs

83. **Epidemics.** The spread of hoof-and-mouth disease through a herd of cattle can be modeled by the function $P(t) = 2e^{0.27t}$ (t is in days). If a rancher does not quickly treat the two cows that now have the disease, how many cattle will have the disease in 12 days? About 51

84. **Oceanography.** The width w (in millimeters) of successive growth spirals of the sea shell *Catapulus voluto,* shown below, is given by the exponential function $w(n) = 1.54e^{0.503n}$ where n is the spiral number. Find the width, to the nearest tenth of a millimeter, of the sixth spiral. 31.5 mm

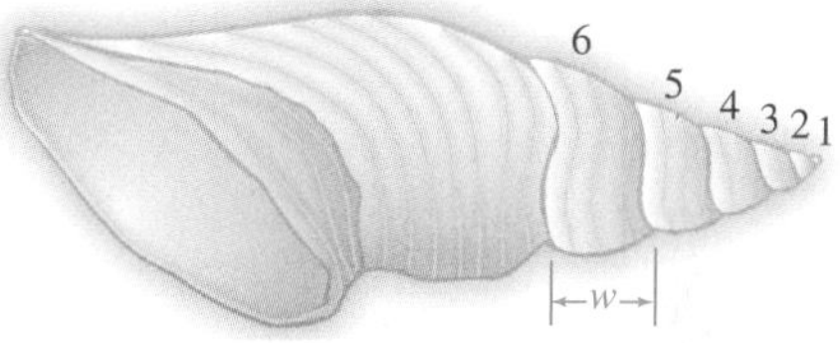

85. **Ants.** Shortly after an explorer ant discovers a food source, a recruitment process begins in which numerous additional ants travel to the source. The number of ants at the source grows exponentially according to the function $a(t) = 1.36\left(\frac{e}{2.5}\right)^t$, where t is the number of minutes since the explorer discovered the food. How many ants will be at the source in 40 minutes? About 39

86. **Half-Life of a Drug.** The quantity of a prescription drug in the bloodstream of a patient t hours after it is administered can be modeled by an exponential function. (See the graph.) From the graph determine the time it takes to eliminate half of the initial dose from the body. 12 hr

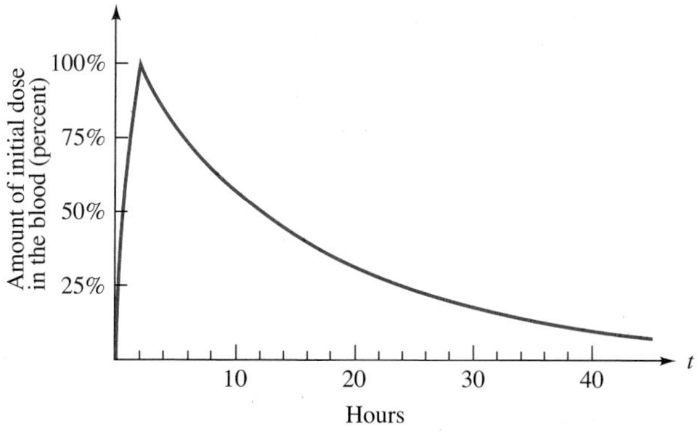

87. from **Campus to Careers**

Social Worker

Social workers often use occupational test results when counseling their clients about employment options. The "learning curve" below shows that as a factory trainee assembled more chairs, the assembly time per chair generally decreased. If company standards required an average assembly time of 10 minutes or less, how many chairs did the trainee have to assemble before meeting company standards? (Notice that the graph is a model of exponential decay.) 14 chairs

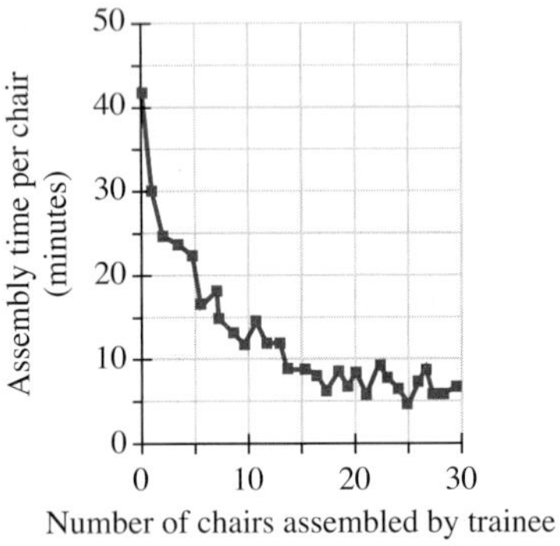

88. Ozone Concentrations. A ***Dobson*** unit is the most basic measure used in ozone research. Roughly 300 Dobson units are equivalent to the height of 2 pennies stacked on top of each other. Suppose the ozone layer thickness (in Dobsons) over a certain city is modeled by the function $A(t) = 300e^{-0.0011t}$, where t is the number of years after 1990. Estimate how thick the ozone layer will be in 2015. About 292 Dobsons

89. Disinfectants. The exponential function $A(t) = 2{,}000{,}000e^{-0.588t}$ approximates the number of germs on a table top, t minutes after disinfectant was sprayed on it. Estimate the germ count on the table 5 minutes after it is sprayed. 105,731 germs

90. Medicine. The concentration of a certain prescription drug in an organ after t minutes is modeled by the function $f(t) = 0.08\left(1 - e^{-0.1t}\right)$ where $f(t)$ is the concentration at time t. Find the concentration of the drug at 30 minutes. About 0.076

91. Sky Diving. Before the parachute opens, a skydiver's velocity in meters per second is modeled by the function $f(t) = 50\left(1 - e^{-0.2t}\right)$ where $f(t)$ is the velocity at time t. Find the velocity after 20 seconds of free fall. About 49 meters per second

92. Free Fall. After t seconds a certain falling object has a velocity in meters per second given by the function $f(t) = 50\left(1 - e^{-0.3t}\right)$. Which is falling faster after 2 seconds—the object or the skydiver in Exercise 91? The object

93. The Tarheel State. The 4.3% annual population growth rate for the Raleigh-Cary metropolitan area in North Carolina is one of the largest of any metropolitan area in the United States. If its growth rate remains constant, how long will it take for its population to double? (Source: U.S. Bureau of the Census) About 16 years

94. The Big Easy. New Orleans has steadily won back some of the population it lost in the wake of Hurricane Katrina in 2005. If the current 8.2% annual increase in population remains constant, how long will it take for its population to double? (Source: money.cnn.com) About 8.5 years

95. The Equality State. In 2009, the state with the fastest annual population growth rate was Wyoming. If the 2.13% annual increase in population remains constant, what is the first full year that the population of Wyoming will be double what it was in 2009? (Source: U.S. Bureau of the Census) The year 2042

96. Doubling Money. How long will it take \$1,000 to double if it is invested at an annual rate of 5% compounded continuously? About 13.9 yr

97. Population Growth. A population growing continuously at an annual rate r will triple in a time t given by the formula $t = \frac{\ln 3}{r}$. How long will it take the population of a town to triple if it is growing at the rate of 12% per year? About 9.2 yr

98. Tripling Money. Find the length of time for \$25,000 to triple when it is invested at 6% annual interest, compounded continuously. See Exercise 97. About 18.3 yr

99. Forensic Medicine. To estimate the number of hours t that a murder victim had been dead, a coroner used the formula $t = \frac{1}{0.25} \ln \frac{98.6 - T_s}{82 - T_s}$ where T_s is the temperature of the surroundings where the body was found. If the crime took place in an apartment where the thermostat was set at 70°F, approximately how long ago did the murder occur? About 3.5 hr

100. Making Jello. After the contents of a package of JELL-O are combined with boiling water, the mixture is placed in a refrigerator whose temperature remains a constant 42°F. Estimate the number of hours t that it will take for the JELL-O to cool to 50°F using the formula $t = -\frac{1}{0.9} \ln \frac{50 - T_r}{200 - T_r}$ where T_r is the temperature of the refrigerator. About 3.3 hr

101. Cross Country Skiing. The function $H(s) = -47.73 + 107.38 \ln s$ approximates the heart rate (in beats/minute) for an Olympic-class cross country skier traveling at s miles per hour, where $s > 5$ mph. Find the heart rate of a skier traveling at a rate of 7.5 miles per hour. (Source: btc.ontana.edu/Olympics/physiology) About 169 beats per minute

102. Strength Loss. After participating in an eight-week weight training program, a college student was selected to be part of a study to see how much strength he would lose if he discontinued working out. The function $M_B(w) = 225 - 14 \ln (4w + 1)$ approximates his maximum bench press (in pounds) w weeks after stopping weight training. What was his maximum bench press:

a. At the end of the eight-week training course? About 225 lb

b. 6 weeks after stopping the weight training? About 180 lb

103. The Pace of Life. According to the study discussed in Example 7, how much faster does the average pedestrian in New York City (population 8,392,000) walk than the average pedestrian in Atlanta (population 541,000)? (Source: infoplease.com) About 1 foot per second faster (5.9 ft/sec – 4.9 ft/sec)

104. Maturity Levels. The function $P(a) = 41.0 + 20.4 \ln a$ approximates the percent of adult height attained by an early-maturing girl of age a years, for $1 \le a \le 18$. The function $P(a) = 37.5 + 20.2 \ln a$ does the same for a late-maturing girl. Find the difference in percent of their adult height for both maturity types on their 10th birthday. (Source: Growth, Maturation, and Physical Activity, Human Kinetic Books, Robert Malina) About 4% difference (88% – 84%)

WRITING

105. Explain why the graph of $y = e^x - 5$ is five units below the graph of $y = e^x$.

106. A feature article in a newspaper stated that the sport of snowboarding was growing *exponentially.* Explain what the author of the article meant by that.

107. As of 2007, the population growth rate for Russia was -0.37% annually. What are some of the consequences for a country that has a negative population growth?

108. What is e?

109. Explain the difference between the functions $f(x) = \log x$ and $g(x) = \ln x$.

110. How are the functions $f(x) = \ln x$ and $g(x) = e^x$ related?

111. Explain why $\ln e = 1$.

112. Why is $f(x) = \ln x$ called the natural logarithmic function?

113. A table of values for $f(x) = \ln x$ is shown in figure (a) below. Explain why ERROR appears in the Y_1 column for the first three entries.

114. The graphs of $f(x) = \ln x$, $g(x) = e^x$, and $y = x$ are shown in figure (b) below. Describe the relationship between the graphs in words.

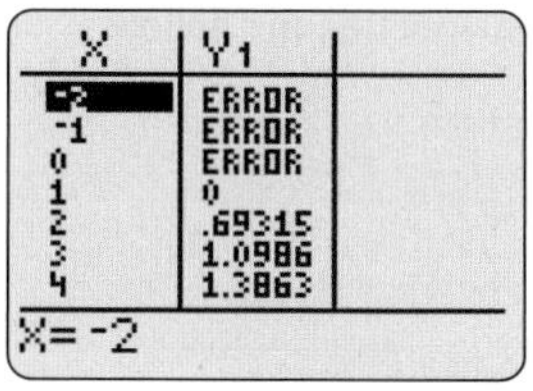

(a)

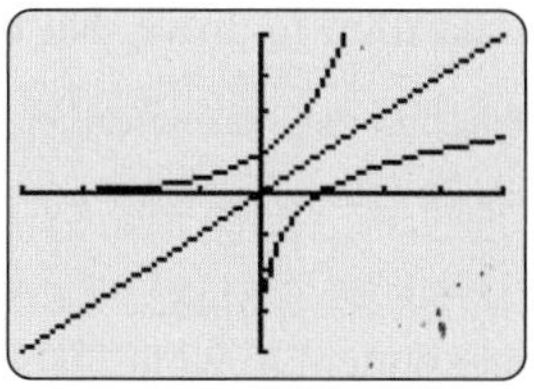
(b)

REVIEW

Simplify each expression. Assume that all variables represent positive numbers.

115. $\sqrt{240x^5}$ $4x^2\sqrt{15x}$

116. $\sqrt[3]{-125x^5y^4}$ $-5xy\sqrt[3]{x^2y}$

117. $4\sqrt{48y^3} - 3y\sqrt{12y}$ $10y\sqrt{3y}$

118. $\sqrt[4]{48z^5} + \sqrt[4]{768z^5}$ $6z\sqrt[4]{3z}$

CHALLENGE PROBLEMS

119. Without using a calculator, determine whether the statement $e^e > e^3$ is true or false. Explain your reasoning. False

120. Graph the function defined by the equation $f(x) = \dfrac{e^x + e^{-x}}{2}$ from $x = -2$ to $x = 2$. The graph will look like a parabola, but it is not. The graph, called a **catenary,** is important in the design of power distribution networks, because it represents the shape of a uniform flexible cable whose ends are suspended from the same height. See AIE Appendix 3.

121. If $e^{t+5} = ke^t$, find k. $k = e^5$

122. If $e^{5t} = k^t$, find k. $k = e^5$

123. Use the formula $P = P_0e^{rt}$ to verify that P will be twice P_0 when $t = \frac{\ln 2}{r}$.

124. Use the formula $P = P_0e^{rt}$ to verify that P will be three times as large as P_0 when $t = \frac{\ln 3}{r}$.

125. Use a graphing calculator to graph the function $f(x) = \frac{1}{1 + e^{-2x}}$. Describe its graph in words. See AIE Appendix 3.

126. Food Shortages. Suppose that a country with a population of 1,000 people is growing according to the formula $P = 1{,}000e^{0.02t}$ where t is in years. Furthermore, assume that the food supply F, measured in adequate food per day per person, is growing linearly according to the formula $F = 30.625t + 2{,}000$ (t is time in years). Use a graphing calculator to determine in how many years the population will outstrip the food supply. In 72 yr

SECTION 11.6

OBJECTIVES

1. Use the four basic properties of logarithms.
2. Use the product rule for logarithms.
3. Use the quotient rule for logarithms.
4. Use the power rule for logarithms.
5. Write logarithmic expressions as a single logarithm.
6. Use the change-of-base formula.
7. Use properties of logarithms to solve application problems.

Properties of Logarithms

ARE YOU READY? *Are You Ready? exercises available online at www.webassign.net/brookscole*

The following problems review some basic skills that are needed when working with properties of logarithms.

1. Evaluate: $\log_2 8 + \log_2 1$ 3
2. Evaluate: $\log 10{,}000 - \log 10$ 3
3. Evaluate: $9 \log_3 \frac{1}{3}$ −9
4. Evaluate: $\frac{\log_7 49}{\log_7 7}$ 2
5. **a.** Write $\sqrt{x}$ using a fractional exponent. $x^{1/2}$
   b. Write $(x - 2)^{1/2}$ using radical notation. $\sqrt{x-2}$
6. Use a calculator to find ln 5. Round to four decimal places. 1.6094

Since a logarithm is an exponent, we would expect there to be properties of logarithms just as there are properties of exponents. In this section, we will introduce seven properties of logarithms and use them to simplify and expand logarithmic expressions.

1 Use the Four Basic Properties of Logarithms.

The first four properties of logarithms follow directly from the definition of logarithm.

Properties of Logarithms For all positive numbers b, where $b \neq 1$,

1. $\log_b 1 = 0$ 2. $\log_b b = 1$ 3. $\log_b b^x = x$ 4. $b^{\log_b x} = x \quad (x > 0)$

We can use the definition of logarithm to prove that these properties are true.

1. $\log_b 1 = 0$, because $b^0 = 1$. *Read as "the log base b of 1 equals 0."*
2. $\log_b b = 1$, because $b^1 = b$. *Read as "the log base b of b equals 1."*
3. $\log_b b^x = x$, because $b^x = b^x$. *Read as "the log base b of b to the x power equals x."*
4. $b^{\log_b x} = x$, because $\log_b x$ is the exponent to which b is raised to get x.
   Read as "b raised to the log base b of x power equals x."

Properties 3 and 4 also indicate that the composition of the exponential and logarithmic functions (in both directions) is the identity function. This is expected, because the exponential and logarithmic functions are inverse functions.

EXAMPLE 1 Simplify: **a.** $\log_5 1$ **b.** $\log_3 3$ **c.** $\ln e^3$ **d.** $6^{\log_6 7}$

Strategy We will compare each logarithmic expression to the left side of the previous four properties of logarithms.

Why When we get a match, the property will provide the answer.

Solution
a. By property 1, $\log_5 1 = 0$, because $5^0 = 1$.
b. By property 2, $\log_3 \mathbf{3} = 1$, because $3^1 = 3$.
c. By property 3, $\ln e^3 = 3$, because $e^3 = e^3$.
d. By property 4, $\mathbf{6}^{\log_6 7} = 7$, because $\log_6 7$ is the power to which 6 is raised to get 7.

Teaching Example 1 Simplify:
a. $\log_\pi \pi$ b. $\log_7 7^9$ c. $6^{\log_6 5}$
d. $\log_3 1$
Answers:
a. 1 b. 9 c. 5 d. 0

Self Check 1 Simplify: **a.** $\log_4 1$ 0 **b.** $\log_4 4$ 1 **c.** $\log_2 2^4$ 4 **d.** $5^{\log_5 2}$ 2

Now Try Problems 19, 21, 23, and 27

2 Use the Product Rule for Logarithms.

The next property of logarithms is related to the product rule for exponents: $x^m \cdot x^n = x^{m+n}$.

The Product Rule for Logarithms

The logarithm of a product equals the sum of the logarithms of the factors. For all positive real numbers M, N, and b, where $b \neq 1$,

$$\log_b MN = \log_b M + \log_b N$$ Read as "the log base b of M times N equals the log base b of M plus the log base b of N."

Teaching Tip: Have selected students read each of the logarithm rules as you introduce them.

As we apply properties of logarithms to rewrite expressions, we assume that all variables represent positive numbers.

EXAMPLE 2 Write each expression as a sum of logarithms. Then simplify, if possible. **a.** $\log_2(2 \cdot 7)$ **b.** $\log 100x$ **c.** $\log_5 125yz$

Strategy In each case, we will use the product rule for logarithms.

Why We use the product rule because each of the logarithmic expressions has the form $\log_b MN$.

Solution

Teaching Tip: Some students may be inclined to "distribute" $\log_2$ incorrectly to simplify $\log_2(2 \cdot 7)$. Point out that $\log_2$ is not, itself, a real number. It is the name of a type of logarithmic function.

a. To avoid any confusion, the product $2 \cdot 7$ is written within parentheses.

$\log_2(2 \cdot 7) = \log_2 2 + \log_2 7$ Read as "the log base 2 of 2 times 7." The log of the product $2 \cdot 7$ is the sum of the logs of its two factors.

$= 1 + \log_2 7$ Simplify: By property 2, $\log_2 2 = 1$.

b. Recall that $100x$ means $100 \cdot x$.

$\log 100x = \log 100 + \log x$ The log of the product $100x$ is the sum of the logs of its two factors.

$= 2 + \log x$ Simplify: By property 3, $\log 100 = \log 10^2 = 2$.

c. The product $125yz$ has three factors: $125 \cdot y \cdot z$.

$\log_5 125yz = \log_5 125 + \log_5 y + \log_5 z$ The log of the product $125yz$ is the sum of the logs of its three factors.

$= 3 + \log_5 y + \log_5 z$ Simplify: By property 3, $\log_5 125 = \log_5 5^3 = 3$.

Success Tip

Your eyes should go immediately to the "input" of each log function, and then ask:

$\log 100x$

↑

"What are the factors?"

↓

$\log_5 125yz$

Teaching Example 2 Write each expression as the sum of logarithms. Then simplify, if possible.
a. $\log_5(5 \cdot 3)$ **b.** $\log(10{,}000x)$
c. $\log_9 81mn$
Answers:
a. $1 + \log_5 3$ **b.** $4 + \log x$
c. $2 + \log_9 m + \log_9 n$

Self Check 2 Write each expression as the sum of logarithms. Then simplify, if possible. **a.** $\log_3(3 \cdot 4)$ $1 + \log_3 4$ **b.** $\log 1{,}000y$ $3 + \log y$ **c.** $\log_5 25cd$ $2 + \log_5 c + \log_5 d$

Now Try Problems 31 and 35

PROOF To prove the product rule for logarithms, we let $x = \log_b M$, $y = \log_b N$, and use the definition of logarithm to write each equation in exponential form.

$$M = b^x \text{ and } N = b^y$$

Then $MN = b^x b^y$, and a property of exponents gives

$$MN = b^{x+y}$$ Keep the base and add the exponents: $b^x b^y = b^{x+y}$.

We write this exponential equation in logarithmic form as

$$\log_b MN = x + y$$

Substituting the values of x and y completes the proof.

$$\log_b MN = \log_b M + \log_b N$$ This is the product rule for logarithms.

Caution

The log of a sum **does not equal** the sum of the logs. The log of a difference **does not equal** the difference of the logs.

CAUTION By the product rule, the logarithm of a *product* is equal to the *sum* of the logarithms. The logarithm of a sum or a difference usually does not simplify. **Do not incorrectly apply the distributive property in such cases.** In general,

$$\log_b (M + N) \neq \log_b M + \log_b N \quad \text{and} \quad \log_b (M - N) \neq \log_b M - \log_b N$$

For example,

$$\log_2 (2 + 7) \neq \log_2 2 + \log_2 7 \quad \text{and} \quad \log (100 - y) \neq \log 100 - \log y$$

Using Your Calculator ▶ Verifying Properties of Logarithms

We can use a calculator to illustrate the product rule for logarithms by showing that

$$\log (3.7 \cdot 15.9) = \log 3.7 + \log 15.9$$

We calculate the left and right sides of the equation separately and compare the results. To use a scientific calculator to find log (3.7 · 15.9), we enter

3.7 [×] 15.9 [=] [LOG] `1.769598848`

To find log 3.7 + log 15.9, we enter

3.7 [LOG] [+] 15.9 [LOG] [=] `1.769598848`

Since the screen displays for the left and right sides are equal, that would suggest that the equation log (3.7 · 15.9) = log 3.7 + log 15.9 is true.

3 Use the Quotient Rule for Logarithms.

The next property of logarithms is related to the quotient rule for exponents: $\frac{x^m}{x^n} = x^{m-n}$.

The Quotient Rule for Logarithms

The logarithm of a quotient equals the difference of the logarithms of the numerator and denominator. For all positive real numbers M, N, and b, where $b \neq 1$,

$$\log_b \frac{M}{N} = \log_b M - \log_b N$$

Read as "the log base b of M divided by N equals the log base b of M minus the log base b of N."

The proof of the quotient rule for logarithms is similar to the proof for the product rule for logarithms.

EXAMPLE 3 Write each expression as a difference of logarithms. Then simplify, if possible.

a. $\ln \frac{10}{7}$ **b.** $\log_4 \frac{x}{64}$

Teaching Tip: You may want to use a calculator to illustrate the quotient rule.
$\ln \frac{10}{7} \approx 0.356674944$
$\ln 10 - \ln 7 \approx 0.356674944$

Strategy In both cases, we will apply the quotient rule for logarithms.

Why We use the quotient rule because each of the logarithmic expressions has the form $\log_b \frac{M}{N}$.

Solution **a.** $\ln \frac{10}{7} = \ln 10 - \ln 7$ *Recall that $\ln \frac{10}{7}$ is a natural logarithm and means $\log_e \frac{10}{7}$. The log of the quotient $\frac{10}{7}$ is the difference of the logs of its numerator and denominator.*

b. $\log_4 \frac{x}{64} = \log_4 x - \log_4 64$ The log of the quotient $\frac{x}{64}$ is the difference of the logs of its numerator and denominator.

$= \log_4 x - 3$ Simplify: $\log_4 64 = \log_4 4^3 = 3$.

Teaching Example 3 Write each expression as a difference of logarithms. Then simplify, if possible.
a. $\ln \frac{5}{2}$ **b.** $\log \frac{100}{x}$
Answers:
a. $\ln 5 - \ln 2$ **b.** $2 - \log x$

Self Check 3 Write each expression as a difference of logarithms. Then simplify, if possible. **a.** $\log_6 \frac{6}{5}$ $1 - \log_6 5$ **b.** $\ln \frac{y}{100}$ $\ln y - \ln 100$

Now Try Problem 39

CAUTION By the quotient rule, the logarithm of a *quotient* is equal to the *difference* of the logarithms. The logarithm of a quotient is not the quotient of the logarithms:

$$\log_b \frac{M}{N} \neq \frac{\log_b M}{\log_b N}$$

For example,

$$\ln \frac{10}{7} \neq \frac{\ln 10}{\ln 7} \quad \text{and} \quad \log_4 \frac{x}{64} \neq \frac{\log_4 x}{\log_4 64}$$

Using a calculator, we have:
$\ln \frac{10}{7} \approx 0.356674944$
$\frac{\ln 10}{\ln 7} \approx 1.183294662$

In the next example, the product and quotient rules for logarithms are used in combination to rewrite an expression.

EXAMPLE 4 Write $\log \frac{xy}{10z}$ as the sum and/or difference of logarithms of a single quantity. Then simplify, if possible.

Strategy We will use the quotient rule for logarithms and then the product rule.

Why We use the quotient rule because $\log \frac{xy}{10z}$ has the form $\log_b \frac{M}{N}$. We later use the product rule because the numerator and denominator of $\frac{xy}{10z}$ contain products.

Solution We begin by applying the quotient rule for logarithms.

$\log \frac{xy}{10z} = \log xy - \log 10z$ The log of a quotient is the difference of the logs.

$= \log x + \log y - (\log 10 + \log z)$ The log of a product is the sum of the logs. Write parentheses here so that both terms of the sum $\log 10 + \log z$ are subtracted.

$= \log x + \log y - \log 10 - \log z$ Change the sign of each term of $\log 10 + \log z$ and drop the parentheses.

$= \log x + \log y - 1 - \log z$ Simplify: $\log 10 = 1$.

Success Tip

Your eyes should go immediately to the "input" of the log function, and identify any quotients and/or products.

$\log \frac{xy}{10z}$ ↑

Teaching Example 4 Write $\log \frac{xy}{100z}$ as the sum and/or difference of logarithms of a single quantity. Then simplify, if possible.
Answer:
$\log x + \log y - 2 - \log z$

Self Check 4 Write $\log_b \frac{x}{yz}$ as the sum and/or difference of logarithms of a single quantity. Then simplify, if possible. $\log_b x - \log_b y - \log_b z$

Now Try Problem 45

4 Use the Power Rule for Logarithms.

The next property of logarithms is related to the power rule for exponents: $(x^m)^n = x^{mn}$.

The Power Rule for Logarithms

The logarithm of a number raised to a power equals the power times the logarithm of the number.

For all positive real numbers M and b, where $b \neq 1$, and any real number p,

$$\log_b M^p = p \log_b M$$ Read as "the log base b of M to the p power equals p times the log base b of M."

EXAMPLE 5 Write each logarithm without an exponent or a square root: **a.** $\log_5 6^2$ **b.** $\log \sqrt{10}$

Strategy In each case, we will use the power rule for logarithms.

Why We use the power rule because $\log_5 6^2$ has the form $\log_b M^p$, as will $\log \sqrt{10}$ if we write $\sqrt{10}$ as $10^{1/2}$.

Solution

Teaching Tip: Show how the product rule can be used to simplify $\log_5 6^2$, but more work is involved:

$\log_5 6^2 = \log_5 (6 \cdot 6)$
$= \log_5 6 + \log_5 6$
$= 2 \log_5 6$

a. $\log_5 6^2 = 2 \log_5 6$ The log of a power is equal to the power times the log. Write the exponent 2 in front of $\log_5 6$.

b. $\log \sqrt{10} = \log 10^{1/2}$ Write $\sqrt{10}$ using a fractional exponent: $\sqrt{10} = (10)^{1/2}$.

$= \frac{1}{2} \log 10$ The log of a power is equal to the power times the log. Write the exponent $\frac{1}{2}$ in front of log 10.

$= \frac{1}{2}$ Simplify: $\log 10 = 1$.

Teaching Example 5 Write each logarithm without an exponent or fourth root:
a. $\log_3 7^4$ **b.** $\log \sqrt[4]{5}$ **c.** $\log x^9$
Answers:
a. $4 \log_3 7$ **b.** $\frac{1}{4} \log 5$ **c.** $9 \log x$

Self Check 5 Write each logarithm without an exponent or a cube root:

a. $\ln x^4$ $4 \ln x$ **b.** $\log_2 \sqrt[3]{3}$ $\frac{1}{3} \log_2 3$

Now Try Problems 51 and 53

PROOF To prove the power rule, we let $x = \log_b M$, write the expression in exponential form, and raise both sides to the pth power:

$M = b^x$

$(M)^p = (b^x)^p$ Raise both sides to the pth power.

$M^p = b^{px}$ Keep the base and multiply the exponents.

Using the definition of logarithms gives

$$\log_b M^p = px$$

Substituting $\log_b M$ for x completes the proof.

$$\log_b M^p = p \log_b M$$ This is the power rule.

It is often necessary to use more than one rule for logarithms to **expand a logarithmic expression.**

EXAMPLE 6 Write each expression as the sum and/or difference of logarithms of a single quantity:

a. $\log_b x^2y^3z$ **b.** $\ln \frac{y^3\sqrt[4]{x}}{z}$

Strategy In part (a), we will use the product rule and the power rules for logarithms. In part (b), we will use the quotient rule, the product rule, and the power rule for logarithms.

Why In part (a), we first use the product rule because the expression has the form $\log_b MN$. In part (b), we first use the quotient rule because the expression has the form $\log_b \frac{M}{N}$.

Solution

The Language of Algebra

In Examples 2, 3, 4, and 6, we use properties of logarithms to **expand** single logarithmic expressions. For example:

$$\log_b MN \xrightarrow{\text{Expand}} \log_b M + \log_b N$$

a. The expression $\log_b x^2y^3z$ is the logarithm of a product.

$$\log_b x^2y^3z = \log_b x^2 + \log_b y^3 + \log_b z$$ The log of a product is the sum of the logs.

$$= 2\log_b x + 3\log_b y + \log_b z$$ The log of a power is the power times the log.

b. The expression $\ln \frac{y^3\sqrt[4]{x}}{z}$ is the logarithm of a quotient.

$$\ln \frac{y^3\sqrt[4]{x}}{z} = \ln y^3\sqrt[4]{x} - \ln z$$ This is a natural log expression. The log of a quotient is the difference of the logs.

$$= \ln y^3 + \ln \sqrt[4]{x} - \ln z$$ The log of a product is the sum of the logs.

$$= \ln y^3 + \ln x^{1/4} - \ln z$$ Write $\sqrt[4]{x}$ as $x^{1/4}$.

$$= 3\ln y + \frac{1}{4}\ln x - \ln z$$ The log of a power is the power times the log.

Success Tip

Your eyes should go immediately to the "input" of the log function, and identify any quotients, products, powers, or roots.

$$\ln \frac{y^3\sqrt{x}}{z}$$
↑

Self Check 6 Expand: $\log \sqrt[4]{\frac{x^3y}{z}}$ $\frac{3}{4}\log x + \frac{1}{4}\log y - \frac{1}{4}\log z$

Now Try Problems 59 and 61

Teaching Example 6 Write each expression as the sum and/or difference of logarithms of a single quantity:

a. $\ln \frac{5\sqrt[3]{x}}{y^2}$ **b.** $\log_2 \sqrt[3]{\frac{x^2y}{z}}$

Answers:

a. $\ln 5 + \frac{1}{3}\ln x - 2\ln y$

b. $\frac{2}{3}\log_2 x + \frac{1}{3}\log_2 y - \frac{1}{3}\log_2 z$

5 Write Logarithmic Expressions as a Single Logarithm.

We also can use properties of logarithms in reverse to **condense logarithmic expressions** having two or more terms.

EXAMPLE 7 Write each logarithmic expression as one logarithm:

a. $16\log_8 x + \frac{1}{3}\log_8 y$ **b.** $\frac{1}{2}\log_b (x-2) - \log_b y + 3\log_b z$

Strategy In part (a), we will use the power rule and product rule for logarithms in reverse. In part (b), we will use the power rule, the quotient rule, and the product rule for logarithms in reverse.

Why We use the power rule because we see expressions of the form $p\log_b M$. The $+$ symbol between logarithmic terms suggests that we use the product rule and the $-$ symbol between such terms suggests that we use the quotient rule.

Solution

a. We begin by using the power rule on both terms of the expression.

$$16\log_8 x + \frac{1}{3}\log_8 y = \log_8 x^{16} + \log_8 y^{1/3}$$ A power times a log is the log of the power.

$$= \log_8 (x^{16} \cdot y^{1/3})$$ The sum of two logs is the log of the product.

$$= \log_8 x^{16}y^{1/3}$$

$$= \log_8 x^{16}\sqrt[3]{y}$$ Write $y^{1/3}$ using radical notation: $\sqrt[3]{y}$.

Success Tip

When condensing logarithmic expressions, you should first identify the coefficients of any logarithmic terms, and then apply the power rule in reverse.

$$16\log_8 x + \frac{1}{3}\log_8 y$$
↑ ↑

The Language of Algebra

In these examples, we use properties of logarithms in "reverse" to **condense** the given expression into a single logarithmic expression. For example:

Condense

$\log_b MN = \log_b M + \log_b N$

b. The first and third terms of this expression can be rewritten using the power rule of logarithms. Note that the base of each logarithm is b. We do not need to know the value of b to apply properties of logarithms.

$$\frac{1}{2}\log_b (x-2) - \log_b y + 3\log_b z$$

$= \log_b (x-2)^{1/2} - \log_b y + \log_b z^3$ — A power times a log is the log of the power.

$= \log_b \frac{(x-2)^{1/2}}{y} + \log_b z^3$ — The difference of two logs is the log of the quotient.

$= \log_b \frac{\sqrt{x-2}}{y} + \log_b z^3$ — Write $(x-2)^{1/2}$ using a radical notation: $\sqrt{x-2}$.

$= \log_b\left(\frac{\sqrt{x-2}}{y} \cdot z^3\right)$ — The sum of two logs is the log of the product.

$= \log_b \frac{z^3\sqrt{x-2}}{y}$

Teaching Example 7 Write the expression as one logarithm:
$\frac{2}{3}\log_b x - 5\log_b y + 3\log_b (x+4)$
Answer:
$\log_b \frac{(x+4)^3\sqrt[3]{x^2}}{y^5}$

Self Check 7 Write the expression as one logarithm:
$2\log_a x + \frac{1}{2}\log_a y - 2\log_a (x-y)$ $\quad \log_a \frac{x^2\sqrt{y}}{(x-y)^2}$

Now Try Problems 75 and 79

The properties of logarithms can be used when working with numerical values.

EXAMPLE 8 If $\log 2 \approx 0.3010$ and $\log 3 \approx 0.4771$, use properties of logarithms to find approximations for: **a.** $\log 6$ **b.** $\log 18$

Strategy We will express 6 and 18 using factors of 2 and 3 and then use properties of logarithms to simplify each resulting expression.

Why We express 6 and 18 using factors of 2 and 3 because we are given values of log 2 and log 3.

Solution

a. $\log 6 = \log (2 \cdot 3)$ — Write 6 using the factors 2 and 3.

$= \log 2 + \log 3$ — The log of a product is the sum of the logs.

$\approx 0.3010 + 0.4771$ — Substitute the given values for log 2 and log 3. We must now use an ≈ symbol. Do the addition.

≈ 0.7781

b. $\log 18 = \log (2 \cdot 3^2)$ — Write 18 using the factors 2 and 3.

$= \log 2 + \log 3^2$ — The log of a product is the sum of the logs.

$= \log 2 + 2\log 3$ — The log of a power is the power times the log. Write the exponent 2 in front of log 3.

$\approx 0.3010 + 2(0.4771)$ — Substitute the given value for log 2 and log 3. We must now use an ≈ symbol.

≈ 1.2552 — Evaluate the expression.

Teaching Example 8 If $\log_b 5 \approx 1.462$ and $\log_b 7 \approx 1.768$, use properties of logarithms to find approximations for:

a. $\log_b 35$ **b.** $\log_b \frac{7}{5}$ **c.** $\log_b 49$

Answers:
a. 3.230 **b.** 0.306 **c.** 3.536

Self Check 8 See Example 8. Find approximations for each logarithm.
a. $\log 1.5$ 0.1761 **b.** $\log 0.75$ −0.1249

Now Try Problems 87 and 89

We summarize the properties of logarithms as follows.

Properties of Logarithms

If b, M, and N are positive real numbers, $b \neq 1$, and p is any real number,

1. $\log_b 1 = 0$
2. $\log_b b = 1$
3. $\log_b b^x = x$
4. $b^{\log_b x} = x$
5. $\log_b MN = \log_b M + \log_b N$
6. $\log_b \frac{M}{N} = \log_b M - \log_b N$
7. $\log_b M^p = p \log_b M$

6 Use the Change-of-Base Formula.

Most calculators can find common logarithms (base 10) and natural logarithms (base e). If we need to find a logarithm with some other base, we can use the following **change-of-base formula.**

Change-of-Base Formula

For any logarithmic bases a and b, and any positive real number x,

$$\log_b x = \frac{\log_a x}{\log_a b}$$ This formula converts a logarithm of one base to a ratio of logarithms of a different base.

We can use any positive number other than 1 for base a in the change-of-base formula. However, we usually use 10 or e because of the capabilities of a standard calculator.

EXAMPLE 9 Find: $\log_3 5$

Strategy To evaluate this base-3 logarithm, we will substitute into the change-of-base formula.

Why We assume that the reader does not have a calculator that evaluates base-3 logarithms (at least not directly). Thus, the only alternative is to change the base.

Solution To find $\log_3 5$, we substitute 3 for b, 10 for a, and 5 for x in the change-of-base formula and simplify:

$$\log_b x = \frac{\log_a x}{\log_a b}$$ This is the change-of-base formula.

$$\log_3 5 = \frac{\log_{10} 5}{\log_{10} 3}$$ The old base is 3. The new base we want to introduce is 10. Substitute: $b = 3$, $x = 5$, and $a = 10$.

$$\approx 1.464973521$$ Approximate. On a scientific calculator, enter: 5 LOG ÷ 3 LOG = .

To four decimal places, $\log_3 5 = 1.4650$. To check this result, use a calculator to verify that $3^{1.4650} \approx 5$.

Caution

Don't misapply the quotient rule: $\frac{\log_{10} 5}{\log_{10} 3}$ means $\log_{10} 5 \div \log_{10} 3$. It is the expression $\log_{10} \frac{5}{3}$ that means $\log_{10} 5 - \log_{10} 3$.

Caution

Wait until the final calculation has been made to round. Don't round any values when performing intermediate calculations. That could make the final result incorrect because of a build-up of rounding errors.

We also can use the natural logarithm function (base e) in the change-of-base formula to find a base-3 logarithm.

$\log_b x = \frac{\log_a x}{\log_a b}$ This is the change-of-base formula.

$\log_3 5 = \frac{\log_e 5}{\log_e 3}$ The old base is 3. The new base we want to introduce is e. Substitute: $b = 3$, $x = 5$, and $a = e$.

$\log_3 5 = \frac{\ln 5}{\ln 3}$ Write $\log_e 5$ as $\ln 5$ and $\log_e 3$ as $\ln 3$.

≈ 1.464973521 Approximate. On a scientific calculator, enter: 5 LN ÷ 3 LN = .

We obtain the same result.

Teaching Example 9 Find $\log_9 12$ to four decimal places.
Answer: 1.1309

Self Check 9 Find $\log_5 3$ to four decimal places. 0.6826

Now Try Problem 95

PROOF To prove the change-of-base formula, we begin with the equation $\log_b x = y$.

$y = \log_b x$

$x = b^y$ Change the equation from logarithmic to exponential form.

$\log_a x = \log_a b^y$ Take the base-a logarithm of both sides.

$\log_a x = y \log_a b$ The log of a power is the power times the log.

$y = \frac{\log_a x}{\log_a b}$ Divide both sides by $\log_a b$.

$\log_b x = \frac{\log_a x}{\log_a b}$ Refer to the first equation, $y = \log_b x$ and substitute $\log_b x$ for y. This is the change-of-base formula.

7 Use Properties of Logarithms to Solve Application Problems.

In chemistry, common logarithms are used to express how basic or acidic a solution is. The more acidic a solution, the greater the concentration of hydrogen ions. (A **hydrogen ion** is a positively charged hydrogen atom missing its electron.) The concentration of hydrogen ions in a solution is commonly measured using the **pH scale.** The pH of a solution is defined as follows.

pH of a Solution

If $[H^+]$ is the hydrogen ion concentration in gram-ions per liter, then

$$\text{pH} = -\log[H^+]$$

EXAMPLE 10 **pH Meters.** One of the most accurate ways to measure pH is with a probe and meter. What reading should the meter give for lemon juice if it has a hydrogen ion concentration $[H^+]$ of approximately 6.2×10^{-3} gram-ions per liter?

Strategy We will substitute into the formula for pH and use the power rule for logarithms to simplify the right side.

Why After substituting 6.2×10^{-3} for $[H^+]$ in $-\log[H^+]$, the resulting expression will have the form $\log_b M^p$.

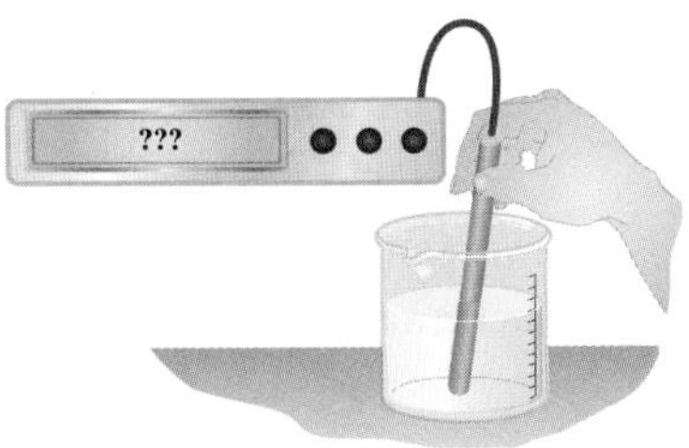

Solution Since lemon juice has approximately 6.2×10^{-3} gram-ions per liter, its pH is

$$\text{pH} = -\log[H^+]$$ This is the formula for pH. Read as "the opposite of the log of the hydrogen ion concentration."

$$\text{pH} = -\log(\mathbf{6.2 \times 10^{-3}})$$ Substitute 6.2×10^{-3} for $[H^+]$.

$$= -(\log 6.2 + \log 10^{-3})$$ The log of a product is the sum of the logs.

$$= -[\log 6.2 + (-3)\,\mathbf{\log 10}]$$ The log of a power is the power times the log.

$$= -[\log 6.2 + (-3) \cdot \mathbf{1}]$$ Evaluate: $\log 10 = 1$.

$$\approx 2.207608311$$ Use a calculator.

$$\approx 2.2$$ Round to the nearest tenth.

The meter should give a reading of approximately 2.2.

Teaching Example 10 pH Meters. The hydrogen ion concentration range for a freshwater aquarium has a high-end value of 1.6×10^{-7}. Find the pH level that corresponds to this value. *Answer:* 6.8

Self Check 10 **pH Meters.** The hydrogen ion concentration range for a freshwater aquarium has a low-end value of 2.5×10^{-8}. Find the pH level that corresponds to this value. 7.6

Now Try Problem 107

SECTION 11.6 STUDY SET

VOCABULARY

Fill in the blanks.

1. The logarithm of a product, such as $\log_3 4x$, equals the sum of the logarithms of the factors.
2. The logarithm of a quotient, such as $\log_2 \frac{5}{x}$, equals the difference of the logarithms of the numerator and denominator.
3. The logarithm of a number to a power, such as $\log_4 5^3$, equals the power times the logarithm of the number.
4. The change-of-base formula converts a logarithm of one base to a ratio of logarithms of a different base.

CONCEPTS

Fill in the blanks. In problem 6, also give the name of each rule.

5. a. $\log_b 1 = 0$ b. $\log_b b = 1$
   c. $\log_b b^x = x$ d. $b^{\log_b x} = x$
6. a. $\log_b MN = \log_b M + \log_b N$ Product rule
   b. $\log_b \frac{M}{N} = \log_b M - \log_b N$ Quotient rule
   c. $\log_b M^p = p \log_b M$ Power rule
   d. $\log_b x = \frac{\log_a x}{\log_a b}$ Change-of-base rule

Use a calculator to verify that each equation is true. See Using Your Calculator: Verifying Properties of Logarithms.

7. $\log(2.5 \cdot 3.7) = \log 2.5 + \log 3.7$
8. $\ln(2.25)^4 = 4 \ln 2.25$
9. $\ln \frac{11.3}{6.1} = \ln 11.3 - \ln 6.1$
10. $\log \sqrt{24.3} = \frac{1}{2} \log 24.3$

Match each expression with an equivalent expression from the list on the right.

11. $\log_3 10$ c
12. $\log_3 \frac{10}{11}$ d
13. $\log_3 10^{11}$ b
14. $\log_3 11$ a

a. $\frac{\log 11}{\log 3}$
b. $11 \log_3 10$
c. $\log_3 5 + \log_3 2$
d. $\log_3 10 - \log_3 11$

NOTATION

Complete each solution.

15. $\log_8 8a^3 = \log_8 8 + \log_8 a^3$
   $= \log_8 8 + 3 \log_8 a$
   $= 1 + 3 \log_8 a$
16. $\log \frac{r}{st} = \log r - \log(st)$
   $= \log r - (\log s + \log t)$
   $= \log r - \log s - \log t$

17. True or False? True

$\log 10{,}000x = \log(10{,}000x)$

18. Fill in the blanks:

a. $y^{1/3} = \sqrt[3]{y}$ b. $\sqrt[5]{x} = x^{1/5}$

GUIDED PRACTICE

In this Study Set, assume that all variables represent positive numbers and $b \neq 1$.

Evaluate each expression. **See Example 1.**

19. $\log_6 1$ 0
20. $\log_9 9$ 1
21. $\log_4 4^7$ 7
22. $\ln e^8$ 8
23. $5^{\log_5 10}$ 10
24. $8^{\log_8 10}$ 10
25. $\log_5 5^2$ 2
26. $\log_4 4^2$ 2
27. $\ln e$ 1
28. $\log_7 1$ 0
29. $\log_3 3^7$ 7
30. $5^{\log_5 8}$ 8

Write each logarithm as a sum. Then simplify, if possible. **See Example 2.**

31. $\log_2(4 \cdot 5)$ $2 + \log_2 5$
32. $\log_3(27 \cdot 5)$ $3 + \log_3 5$
33. $\log 25y$ $\log 25 + \log y$
34. $\log xy$ $\log x + \log y$
35. $\log 100pq$ $2 + \log p + \log q$
36. $\log 1{,}000rs$ $3 + \log r + \log s$
37. $\log 5xyz$ $\log 5 + \log x + \log y + \log z$
38. $\log 10abc$ $1 + \log a + \log b + \log c$

Write each logarithm as a difference. Then simplify, if possible. **See Example 3.**

39. $\log \frac{100}{9}$ $2 - \log 9$
40. $\ln \frac{27}{e}$ $\ln 27 - 1$
41. $\log_6 \frac{x}{36}$ $\log_6 x - 2$
42. $\log_8 \frac{y}{8}$ $\log_8 y - 1$

Write each logarithm as the sum and/or difference of logarithms of a single quantity. Then simplify, if possible. **See Example 4.**

43. $\log \frac{7c}{2}$ $\log 7 + \log c - \log 2$
44. $\log \frac{9t}{4}$ $\log 9 + \log t - \log 4$
45. $\log \frac{10x}{y}$ $1 + \log x - \log y$
46. $\log_2 \frac{ab}{4}$ $\log_2 a + \log_2 b - 2$
47. $\ln \frac{exy}{z}$ $1 + \ln x + \ln y - \ln z$
48. $\ln \frac{5p}{e}$ $\ln 5 + \ln p - 1$
49. $\log_8 \frac{1}{8m}$ $-1 - \log_8 m$
50. $\log_6 \frac{1}{36r}$ $-2 - \log_6 r$

Write each logarithm without an exponent or a radical symbol. Then simplify, if possible. **See Example 5.**

51. $\ln y^7$ $7 \ln y$
52. $\ln z^9$ $9 \ln z$
53. $\log \sqrt{5}$ $\frac{1}{2}\log 5$
54. $\log \sqrt[3]{7}$ $\frac{1}{3}\log 7$
55. $\log e^{-3}$ $-3 \log e$
56. $\log e^{-1}$ $-\log e$
57. $\log_7 (\sqrt[5]{100})^3$ $\frac{3}{5}\log_7 100$
58. $\log_3 (\sqrt{10})^5$ $\frac{5}{2}\log_3 10$

Write each logarithm as the sum and/or difference of logarithms of a single quantity. Then simplify, if possible. **See Example 6.**

59. $\log xyz^2$ $\log x + \log y + 2\log z$
60. $\log 4xz^2$ $\log 4 + \log x + 2\log z$
61. $\log_2 \frac{2\sqrt[3]{x}}{y}$ $1 + \frac{1}{3}\log_2 x - \log_2 y$
62. $\log_3 \frac{\sqrt[4]{x}}{yz}$ $\frac{1}{4}\log_3 x - \log_3 y - \log_3 z$
63. $\log x^3y^2$ $3\log x + 2\log y$
64. $\log xy^2z^3$ $\log x + 2\log y + 3\log z$
65. $\log_b \sqrt{xy}$ $\frac{1}{2}\log_b x + \frac{1}{2}\log_b y$
66. $\log_b x^3\sqrt{y}$ $3\log_b x + \frac{1}{2}\log_b y$
67. $\log_a \frac{\sqrt[3]{x}}{\sqrt[4]{yz}}$ $\frac{1}{3}\log_a x - \frac{1}{4}\log_a y - \frac{1}{4}\log_a z$
68. $\log_b \sqrt[4]{\frac{x^3y^2}{z^4}}$ $\frac{3}{4}\log_b x + \frac{1}{2}\log_b y - \log_b z$
69. $\ln x^{20}\sqrt{z}$ $20\ln x + \frac{1}{2}\ln z$
70. $\ln \sqrt{xy}$ $\frac{1}{2}(\ln x + \ln y)$
71. $\log_5 \left(\frac{1}{t^3}\right)^d$ $-3d\log_5 t$
72. $\log_6 \left(\frac{1}{x^4}\right)^t$ $-4t\log_6 x$
73. $\ln \sqrt{ex}$ $\frac{1}{2} + \frac{1}{2}\ln x$
74. $\ln \sqrt[3]{e^2x}$ $\frac{2}{3} + \frac{1}{3}\ln x$

Write each logarithmic expression as one logarithm. **See Example 7.**

75. $\log_2(x+1) + 9\log_2 x$ $\log_2 x^9(x+1)$
76. $2\log x + \frac{1}{2}\log y$ $\log x^2\sqrt{y}$
77. $\log_3 x + \log_3(x+2) - \log_3 8$ $\log_3 \frac{x(x+2)}{8}$
78. $-2\log x - 3\log y + \log z$ $\log \frac{z}{x^2y^3}$
79. $-3\log_b x - 2\log_b y + \frac{1}{2}\log_b z$ $\log_b \frac{\sqrt{z}}{x^3y^2}$
80. $3\log_b(x+1) - 2\log_b(x+2) + \log_b x$ $\log_b \frac{x(x+1)^3}{(x+2)^2}$
81. $\frac{1}{3}[\log_b(M^2-9) - \log_b(M+3)]$ $\log_b \sqrt[3]{M-3}$
82. $\frac{1}{4}[\log_r(n^2-16) - \log_r(n-4)]$ $\log_r \sqrt[4]{n+4}$
83. $\ln\left(\frac{x}{z} + x\right) - \ln\left(\frac{y}{z} + y\right)$ $\ln \frac{\frac{x}{z}+x}{\frac{y}{z}+y} = \ln \frac{x}{y}$
84. $\ln(xy + y^2) - \ln(xz + yz) + \ln z$ $\ln y$
85. $\frac{1}{2}\log_6(x^2+1) - \log_6(x^2+2)$ $\log_6 \frac{\sqrt{x^2+1}}{x^2+2}$
86. $\frac{1}{2}\log_8(x^2+5) - \log_8(x^2+5)$ $\log_8 \frac{\sqrt{x^2+5}}{x^2+5}$

Assume that $\log 4 \approx 0.6021$, $\log 7 \approx 0.8451$, and $\log 9 \approx 0.9542$. Use these values to evaluate each logarithm. **See Example 8.**

87. $\log_b 28$ 1.4472
88. $\log_b \frac{7}{4}$ 0.2430
89. $\log_b \frac{4}{63}$ −1.1972
90. $\log_b 36$ 1.5563

91. $\log_b \frac{63}{4}$ 1.1972
▶ 92. $\log_b 2.25$ 0.3521
93. $\log_b 64$ 1.8063
▶ 94. $\log_b 49$ 1.6902

Use the change-of-base formula to find each logarithm to four decimal places. See Example 9.

95. $\log_3 7$ 1.7712
▶ 96. $\log_7 3$ 0.5646
▶ 97. $\log_{1/3} 3$ −1.0000
▶ 98. $\log_{1/2} 6$ −2.5850
▶ 99. $\log_3 8$ 1.8928
100. $\log_5 10$ 1.4307
101. $\log_{\sqrt{2}} \sqrt{5}$ 2.3219
▶ 102. $\log_{\pi} e$ 0.8736

Look Alikes . . .
Which pair of expressions in each list are equivalent?

103. a. $\log (9 \cdot 3)$ b. $\log 9 \cdot \log 3$ c. $\log 9 + \log 3$ a, c
▶ 104. a. $\log_6 \frac{7}{9}$ b. $\frac{\log_6 7}{\log_6 9}$ c. $\log_6 7 - \log_6 9$ a, c
105. a. $\log_2 11^4$ b. $4 \log_2 11$ c. $(\log_2 11)^4$ a, b
106. a. $\ln \sqrt{t}$ b. $\sqrt{\ln t}$ c. $\frac{1}{2} \ln t$ a, c

APPLICATIONS

▶ 107. **pH of a Solution.** Find the pH of a solution with a hydrogen ion concentration of 1.7×10^{-5} gram-ions per liter. About 4.8

▶ 108. **pH of Pickles.** The hydrogen ion concentration of sour pickles is 6.31×10^{-4}. Find the pH. About 3.2

109. **Formulas.** Use properties of logarithms to write the right side of each formula in an equivalent condensed form.
a. From sound engineering: $B = 10(\log I - \log I_0)$ $B = \log \left(\frac{I}{I_0}\right)^{10}$
b. From medicine: $T = \frac{1}{k}(\ln C_2 - \ln C_1)$ $T = \ln \sqrt[k]{\frac{C_2}{C_1}}$

110. **Doubling Time.** The formula $t = \frac{\ln 2}{r}$ gives the time t for a population to double, where r is the annual rate of continuous compounding. Write the formula in an equivalent form so that it involves a common logarithm, not a natural logarithm.
$T = \frac{\log 2}{r \log e}$

WRITING

111. Explain the difference between a logarithm of a product and the product of logarithms.
▶ 112. How can the LOG key on a calculator be used to find $\log_2 7$?

Explain why each statement is false.

113. ~~$\log AB = (\log A)(\log B)$~~
114. ~~$\log (A + B) = \log A + \log B$~~
115. ~~$\log_b (A - B) = \frac{\log_b A}{\log_b B}$~~
116. ~~$\frac{\log_b A}{\log_b B} = \log_b A - \log_b B$~~
117. Explain the meaning of the arrow:
$\log 7^{15}$
118. When is the change-of-base formula helpful?

REVIEW

Consider the line that passes through P(−2, 3) and Q(4, −4).

119. Find the slope of line PQ. $-\frac{7}{6}$
▶ 120. Find the distance between P and Q. $\sqrt{85}$
121. Find the midpoint of line segment PQ. $\left(1, -\frac{1}{2}\right)$
122. Write the equation in slope–intercept form of line PQ.
$y = -\frac{7}{6}x + \frac{2}{3}$

CHALLENGE PROBLEMS

123. Explain why $e^{\ln x} = x$.
▶ 124. If $\log_b 3x = 1 + \log_b x$, find b. 3
125. Show that $\log_{b^2} x = \frac{1}{2} \log_b x$.
126. Show that $e^{x \ln a} = a^x$.

SECTION 11.7 Exponential and Logarithmic Equations

OBJECTIVES

1 Solve exponential equations.
2 Solve logarithmic equations.
3 Use exponential and logarithmic equations to solve application problems.

ARE YOU READY? *Are You Ready? exercises available online at www.webassign.net/brookscole*

The following problems review some basic skills that are needed when solving exponential and logarithmic equations.

1. Fill in the blanks: **a.** $16 = 2^{\square}$ 4 **b.** $\frac{1}{125} = 5^{\square}$ −3
2. Find $\frac{\log 12}{\log 9}$. Round to four decimal places. 1.1309
3. Use the power rule for logarithms: $\log_3 8^x$ $x \log_3 8$
4. Evaluate: $\ln e$ 1
5. Use a property of logarithms to write each expression as a single logarithm:
a. $\log_2 5 + \log_2 x$ $\log_2 5x$
b. $\ln 10 - \ln (2t + 1)$ $\ln \frac{10}{2t + 1}$
6. Evaluate: $\log (-6)$ Undefined

Teaching Tip: Ask selected students for an example of each of the types of equations they have solved in earlier chapters.

In earlier chapters, we have solved linear equations, absolute value equations, quadratic equations, rational equations, and radical equations. In this section, we will solve two new types of equations: exponential equations and logarithmic equations.

1 Solve Exponential Equations.

An **exponential equation** contains a variable in one of its exponents. Some examples of exponential equations are

$$3^{x+1} = 81, \qquad 6^{x-3} = 2^x, \qquad \text{and} \qquad e^{0.9t} = 8$$

If both sides of an exponential equation can be expressed as a power of the same base, we can use the following property to solve it.

Exponent Property of Equality

If two exponential expressions with the same base are equal, their exponents are equal. For any real number b, where $b \neq -1, 0,$ or 1,

$b^x = b^y$ is equivalent to $x = y$

EXAMPLE 1 Solve: $3^{x+1} = 81$

Strategy We will express the right side of the equation as a power of 3.

Why If each side of the equation is expressed as a power of the *same* base (in this case, 3), we can use the exponent property of equality to set the exponents equal and solve for x.

Solution

$3^{x+1} = 81$	This is the equation to solve.
$3^{x+1} = 3^4$	Write 81 as a power of 3: $81 = 3^4$.
$x + 1 = 4$	If two exponential expressions with the same base are equal, their exponents are equal.
$x = 3$	Solve for x by subtracting 1 from both sides.

The solution is 3 and the solution set is $\{3\}$. To check this result, we substitute 3 for x in the original equation.

Check:

$$3^{x+1} = 81$$
$$3^{3+1} \stackrel{?}{=} 81$$
$$3^4 \stackrel{?}{=} 81$$
$$81 = 81 \quad \text{True}$$

Teaching Example 1 Solve: $4^{x-2} = 64$
Answer: 5

Self Check 1 Solve: $5^{3x-4} = 25$ 2

Now Try Problem 21

EXAMPLE 2 Solve: $2^{x^2+2x} = \frac{1}{2}$

Strategy We will express the right side of the equation as a power of 2.

Why If each side of the equation is expressed as a power of the same base (in this case, 2), we can use the exponent property of equality to set the exponents equal and solve for x.

Solution

$2^{x^2+2x} = \frac{1}{2}$	This is the equation to solve.
$2^{x^2+2x} = 2^{-1}$	Write $\frac{1}{2}$ as a power of 2: $\frac{1}{2} = 2^{-1}$.
$x^2 + 2x = -1$	If two exponential expressions with the same base are equal, their exponents are equal.
$x^2 + 2x + 1 = 0$	Add 1 to both sides.
$(x + 1)(x + 1) = 0$	Factor the trinomial.
$x + 1 = 0$ or $x + 1 = 0$	Set each factor equal to 0.
$x = -1$ | $x = -1$	Solve each linear equation.

We see that the two solutions are the same. Thus, -1 is a repeated solution and the solution set is $\{-1\}$. Verify that -1 satisfies the original equation.

Teaching Example 2 Solve: $5^{x^2-4x} = \frac{1}{125}$
Answer: 1, 3

Self Check 2 Solve: $3^{x^2-2x} = \frac{1}{3}$ A repeated solution of 1

Now Try ▶ Problem 25

Using Your Calculator ▶ Solving Exponential Equations Graphically

To use a graphing calculator to approximate the solutions of $2^{x^2+2x} = \frac{1}{2}$ (see Example 2), we can subtract $\frac{1}{2}$ from both sides of the equation to get $2^{x^2+2x} - \frac{1}{2} = 0$ and graph the corresponding function $f(x) = 2^{x^2+2x} - \frac{1}{2}$ as shown in figure (a).

The solutions of $2^{x^2+2x} - \frac{1}{2} = 0$ are the x-coordinates of the x-intercepts of the graph of $f(x) = 2^{x^2+2x} - \frac{1}{2}$. Using the ZERO feature, we see in figure (a) that the graph has only one x-intercept, $(-1, 0)$. Therefore, -1 is the only solution of $2^{x^2+2x} - \frac{1}{2} = 0$.

We also can solve $2^{x^2+2x} = \frac{1}{2}$ using the INTERSECT feature found on most graphing calculators. After graphing $Y_1 = 2^{x^2+2x}$ and $Y_2 = \frac{1}{2}$, we select INTERSECT, which approximates the coordinates of the point of intersection of the two graphs. From the display shown in figure (b), we can conclude that the solution is -1. Verify this by checking.

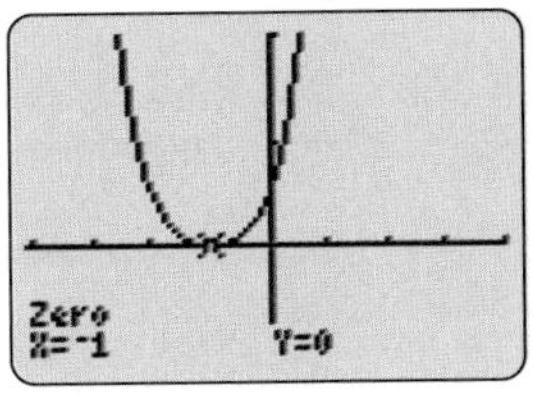

(a)

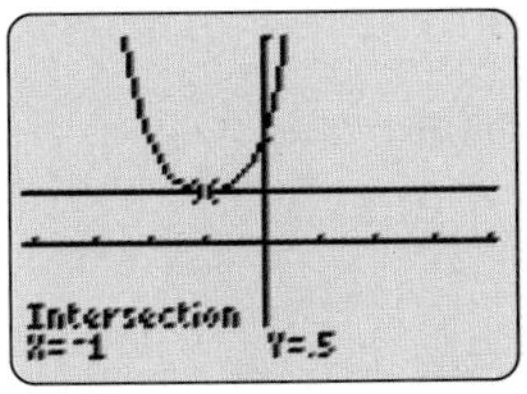

(b)

When it is difficult or impossible to write each side of an exponential equation as a power of the same base, we often can use the following property of logarithms to solve the equation.

Logarithm Property of Equality

If two positive numbers are equal, the logarithms base-b of the numbers are equal. For any positive number b, where $b \neq 1$, and positive numbers x and y,

$\log_b x = \log_b y$ is equivalent to $x = y$

EXAMPLE 3 Solve: $3^x = 5$

Strategy We will take the base-10 logarithm on both sides of the equation.

Why We can then use the power rule of logarithms to move the variable x from its current position as an exponent to a position as a factor.

Solution

The Language of Algebra

Some instructors phrase the first step of the solution process as "taking the logarithm *of* both sides."

Unlike Example 1, where we solved $3^{x+1} = 81$, it is not possible to write each side of $3^x = 5$ as an integer power of the same base 3. Instead, we use the logarithm property of equality and *take the logarithm on each side* to solve the equation. Although any base logarithm can be chosen, the calculations with a calculator are usually simplest if we use a common or natural logarithm.

$3^x = 5$ — This is the equation to solve.

$\log 3^x = \log 5$ — Take the common logarithm on each side.

$x \log 3 = \log 5$ — The log of a power is the power times the log: $\log 3^x = x \log 3$. The variable x is now a factor of $x \log 3$ and not an exponent.

$\frac{x \log 3}{\log 3} = \frac{\log 5}{\log 3}$ — $x \log 3$ means $x \cdot \log 3$. To isolate x, undo the multiplication by $\log 3$ by dividing both sides by $\log 3$.

$\frac{x \overset{1}{\cancel{\log 3}}}{\underset{1}{\cancel{\log 3}}} = \frac{\log 5}{\log 3}$ — Simplify the left side by removing the common factor of $\log 3$ from the numerator and denominator.

$x = \frac{\log 5}{\log 3}$ — This is the exact solution.

$x \approx 1.464973521$ — Approximate. On a reverse-entry scientific calculator, press: 5 LOG ÷ 3 LOG = .

Success Tip

The power rule of logarithms provides a way of **moving the variable x from its position in an exponent to a position as a factor** of $x \log 3$.

$$\log 3^x = \log 5$$
$$x \log 3 = \log 5$$

Teaching Tip: After applying the power rule, some students will have difficulty solving for x. Explain that $\log 3$ is a real number, and that $x \log 3$ means $x \cdot \log 3$. Thus, to isolate x, we must divide both sides by $\log 3$.

The exact solution is $\frac{\log 5}{\log 3}$ and the solution set is $\left\{\frac{\log 5}{\log 3}\right\}$. Rounded to four decimal places, an approximate solution is 1.4650.

We also can take the natural logarithm on each side of the equation to solve for x.

$3^x = 5$

$\ln 3^x = \ln 5$ — Take the natural logarithm on each side.

$x \ln 3 = \ln 5$ — Use the power rule of logarithms: $\ln 3^x = x \ln 3$.

$\frac{x \ln 3}{\ln 3} = \frac{\ln 5}{\ln 3}$ — $x \ln 3$ means $x \cdot \ln 3$. To isolate x, undo the multiplication by $\ln 3$ by dividing both sides by $\ln 3$.

$\frac{x \overset{1}{\cancel{\ln 3}}}{\underset{1}{\cancel{\ln 3}}} = \frac{\ln 5}{\ln 3}$ — Simplify the left side by removing the common factor of $\ln 3$ from the numerator and denominator.

$x = \frac{\ln 5}{\ln 3}$ — This is the exact solution.

$x \approx 1.464973521$ — Approximate. On a scientific calculator, press: 5 LN ÷ 3 LN = .

Caution

Don't misuse the quotient rule:

$\frac{\ln 5}{\ln 3}$ means $\ln 5 \div \ln 3$

It is the expression $\ln \frac{5}{3}$ that means $\ln 5 - \ln 3$.

The result is the same using the natural logarithm. To check the approximate solution, we substitute 1.4650 for x in 3^x and see if $3^{1.4650}$ is approximately 5.

Check:

$$3^x = 5$$
$$3^{1.4650} \stackrel{?}{=} 5$$
$$5.000145454 \approx 5$$

On a scientific calculator, press: 3 y^x 1.4650 = .

Self Check 3 Solve: $5^x = 4$ $\quad \frac{\log 4}{\log 5} \approx 0.8614$

Now Try Problem 29

Teaching Example 3 Solve: $6^x = 8$

Answer: $\frac{\log 8}{\log 6} \approx 1.1606$

EXAMPLE 4 Solve: $6^{x-3} = 2^x$

Strategy We will take the common logarithm on both sides of the equation.

Why We can then use the power rule of logarithms to move the expressions $x - 3$ and x from their current positions as exponents to positions as factors.

Solution

$6^{x-3} = 2^x$	This is the equation to solve.
$\log 6^{x-3} = \log 2^x$	Take the common logarithm on each side.
$(x - 3)\log 6 = x \log 2$	The log of a power is the power times the log. The expression x − 3 is now a factor of (x − 3) log 6 and not an exponent.
$x \log 6 - 3 \log 6 = x \log 2$	Distribute the multiplication by log 6.
$x \log 6 - x \log 2 = 3 \log 6$	To get the terms involving x on the left side, add 3 log 6 and subtract x log 2 on both sides.
$x(\log 6 - \log 2) = 3 \log 6$	Factor out x on the left side.
$\frac{x \overset{1}{\cancel{(\log 6 - \log 2)}}}{\underset{1}{\cancel{\log 6 - \log 2}}} = \frac{3 \log 6}{\log 6 - \log 2}$	To isolate x, undo the multiplication by log 6 − log 2 by dividing both side by log 6 − log 2. Then simplify the left side.
$x = \frac{3 \log 6}{\log 6 - \log 2}$	This is the exact solution.
$x \approx 4.892789261$	Approximate. On a reverse-entry scientific calculator, press: 3 × 6 LOG ÷ (6 LOG − 2 LOG) =.

The Language of Algebra

$\frac{3 \log 6}{\log 6 - \log 2}$ is the **exact** solution of $6^{x-3} = 2^x$. An **approximate** solution is 4.8928.

The solution is $\frac{3 \log 6}{\log 6 - \log 2}$ and the solution set is $\left\{\frac{3 \log 6}{\log 6 - \log 2}\right\}$. To four decimal places, an approximate solution is 4.8928. To check, we substitute 4.8928 for each x in $6^{x-3} = 2^x$. The resulting values on the left and right sides of the equation should be approximately equal.

Teaching Example 4 Solve: $3^{x+4} = 2^x$
Answer: $\frac{-4 \log 3}{\log 3 - \log 2} \approx -10.8380$

Self Check 4 Solve: $5^{x-2} = 3^x$ $\frac{2 \log 5}{\log 5 - \log 3} \approx 6.3013$

Now Try Problem 33

When an exponential equation involves an exponential expression with base e, it is easiest to take the natural logarithm on both sides to solve the equation.

EXAMPLE 5 Solve: $e^{0.9t} = 10$

Strategy We will take the natural (base-e) logarithm on both sides of the equation.

Why We can then use the power rule of logarithms to move the expression $0.9t$ from its current position as an exponent to a position as a factor.

Solution The exponential expression on the left side has base e. In such cases, the calculations are easier when we take the natural logarithm of each side.

$e^{0.9t} = 10$	This is the equation to solve.
$\ln e^{0.9t} = \ln 10$	Take the natural logarithm on each side.
$0.9t \ln e = \ln 10$	Use the power rule of logarithms: $\ln e^{0.9t} = 0.9t \ln e$. The expression 0.9t is now a factor of 0.9t ln e and not an exponent.

Success Tip

When we take the natural logarithm of both sides, that conveniently produces the factor $\ln e$, which is equal to 1, on the left side of the equation.

$0.9t \cdot \mathbf{1} = \ln 10$	Simplify: $\ln e = 1$.
$0.9t = \ln 10$	Simplify the left side.
$t = \dfrac{\ln 10}{0.9}$	To isolate t, undo the multiplication by 0.9 by dividing both sides by 0.9.
$t \approx 2.558427881$	Approximate. On a reverse-entry scientific calculator, press: 10 LN ÷ .9 =.

The exact solution is $\frac{\ln 10}{0.9}$. To four decimal places, an approximate solution is 2.5584. Verify this by using a calculator to show that $e^{0.9(2.5584)} \approx 10$.

Teaching Example 5 Solve: $e^{3.5t} = 42$
Answer: $\frac{\ln 42}{3.5} \approx 1.0679$

Self Check 5 Solve: $e^{2.1t} = 35$ $\frac{\ln 35}{2.1} \approx 1.6930$

Now Try Problem 37

Strategy for Solving Exponential Equations

1. Isolate one of the exponential expressions in the equation.
2. If both sides of the equation can be written as exponential expressions with the same base, do so. Then set the exponents equal and solve the resulting equation.
3. If step 2 is difficult or impossible, take the common or natural logarithm on both sides. Use the power rule of logarithms to write the variable exponent as a factor, and then solve the resulting equation.
4. Check the results in the original equation.

2 Solve Logarithmic Equations.

A **logarithmic equation** is an equation with a logarithmic expression that contains a variable. Some examples of logarithmic equations are

$$\log 5x = 3, \qquad \log(3x + 2) = \log(2x - 3), \qquad \text{and} \qquad \log_2 7 - \log_2 x = 5$$

Some logarithmic equations can be solved by rewriting them in equivalent exponential form.

EXAMPLE 6 Solve: $\log 5x = 3$

Strategy Recall that $\log 5x = \log_{10} 5x$. To solve $\log 5x = 3$, we will instead write and solve an equivalent base-10 exponential equation.

Why The resulting exponential equation is easier to solve because the variable term is isolated on one side.

Solution

$\log 5x = 3$	This is the equation to solve.
$\log_{10} 5x = 3$	The base of the logarithm is 10.
$10^3 = 5x$	Write the equivalent base-10 exponential equation.
$1{,}000 = 5x$	Simplify: $10^3 = 1{,}000$.
$200 = x$	To isolate x, divide both sides by 5.

The solution is 200 and the solution set is $\{200\}$.

Check:

$\log 5x = 3$	This is the original equation.
$\log 5(\mathbf{200}) \stackrel{?}{=} 3$	Substitute 200 for x.
$\log 1{,}000 \stackrel{?}{=} 3$	Multiply $5(200) = 1{,}000$.
$3 = 3$	Evaluate: $\log 1{,}000 = \log 10^3 = 3$.

Caution

Always check your solutions to a logarithmic equation to identify any extraneous solutions.

Teaching Example 6 Solve: $\log 25x = 2$
Answer: 4

Self Check 6 Solve: $\log_2 (x - 3) = -1$ $\frac{7}{2}$

Now Try ▶ Problem 41

There is a possibility of obtaining **extraneous solutions** when solving logarithmic equations. Always discard any possible solutions that produce the logarithm of a negative number or the logarithm of 0 in the original equation.

EXAMPLE 7 Solve: $\log(3x + 2) = \log(2x - 3)$

Strategy We will use the logarithmic property of equality to see that $3x + 2 = 2x - 3$.

Why We can use the logarithm property of equality because the given equation, $\log(3x + 2) = \log(2x - 3)$, has the form $\log_b x = \log_b y$.

Solution

$\log(3x + 2) = \log(2x - 3)$ This is the equation to solve.

$3x + 2 = 2x - 3$ If the logarithms of two numbers are equal, the numbers are equal.

$x + 2 = -3$ Subtract 2x from both sides.

$x = -5$ To isolate x, subtract 2 from both sides.

Check:

$\log(3x + 2) = \log(2x - 3)$ This is the original equation.

$\log[3(-5) + 2] \stackrel{?}{=} \log[2(-5) - 3]$ Substitute −5 for x.

~~$\log(-13) \stackrel{?}{=} \log(-13)$~~ Evaluate within brackets. Recall that $\log(-13)$ is undefined.

Since the logarithm of a negative number does not exist, the proposed solution of −5 must be discarded. This equation has no solution. Its solution set is ∅.

Caution

Don't make this error of trying to "distribute" log:

~~$\log(3x + 2)$~~

The notation log is not a number, it is the name of a function and cannot be distributed.

Teaching Example 7 Solve: $\log(7x + 1) = \log(6x - 5)$
Answer: No solution

Self Check 7 Solve: $\log(5x + 14) = \log(7x - 2)$ 8

Now Try ▶ Problem 49

In Examples 8 and 9, we will use the product and quotient rules of logarithms to "condense" one side of the equation first, before solving for the variable.

EXAMPLE 8 Solve: $\log x + \log(x - 3) = 1$

Strategy We will use the product rule for logarithms in reverse: The sum of two logarithms is equal to the logarithm of a product. Then we will write and solve an equivalent exponential equation.

Why We use the product rule of logarithms because the left side of the equation, $\log x + \log(x - 3)$, has the form $\log_b M + \log_b N$.

Solution

$\log x + \log(x - 3) = 1$ This is the equation to solve.

$\log x(x - 3) = 1$ On the left side, use the product rule of logarithms.

$\log_{10} x(x - 3) = 1$ The base of the logarithm is 10.

$x(x - 3) = 10^1$ Write the equivalent base-10 exponential equation.

$x^2 - 3x - 10 = 0$ Distribute the multiplication by x, and then subtract 10 from both sides.

$(x + 2)(x - 5) = 0$ Factor the trinomial.

$x + 2 = 0$ or $x - 5 = 0$ Set each factor equal to 0.

$x = -2$ | $x = 5$

Success Tip

The objective is to use the product rule to "condense" the left side of the equation. We want to write an equivalent equation in which the variable x appears in only a single logarithmic expression.

Caution

Don't automatically discard proposed solutions that are negative. Only discard proposed solutions that produce undefined logarithms in the original equation.

Check: The number -2 is not a solution because it does not satisfy the equation (a negative number does not have a logarithm). We will check the other result, 5.

$$\log x + \log(x-3) = 1 \quad \text{This is the original equation.}$$
$$\log 5 + \log(5-3) \stackrel{?}{=} 1 \quad \text{Substitute 5 for } x.$$
$$\log 5 + \log 2 \stackrel{?}{=} 1 \quad \text{Do the subtraction within the parentheses.}$$
$$\log 10 \stackrel{?}{=} 1 \quad \text{Use the product rule of logarithms: } \log 5 + \log 2 = \log(5 \cdot 2) = \log 10.$$
$$1 = 1 \quad \text{Evaluate: } \log 10 = 1.$$

Since 5 satisfies the equation, it is the solution.

Teaching Example 8 Solve: $\log x + \log(x - 21) = 2$
Answer: 25

Self Check 8 Solve: $\log x + \log(x + 3) = 1$ 2

Now Try Problem 53

Using Your Calculator ▶ Solving Logarithmic Equations Graphically

To use a graphing calculator to approximate the solutions of the logarithmic equation $\log x + \log(x - 3) = 1$ (see Example 8), we can subtract 1 from both sides of the equation to get $\log x + \log(x - 3) - 1 = 0$ and graph the corresponding function $f(x) = \log x + \log(x - 3) - 1$ as shown in figure (a). Since the solution of the equation is the x-value that makes $f(x) = 0$, the solution is the x-coordinate of the x-intercept of the graph. We can use the ZERO feature to find that this x-value is 5.

We also can solve $\log x + \log(x - 3) = 1$ using the INTERSECT feature. After graphing $Y_1 = \log x + \log(x - 3)$ and $Y_2 = 1$, we select INTERSECT, which approximates the coordinates of the point of intersection of the two graphs. From the display shown in figure (b), we can conclude that the solution is 5.

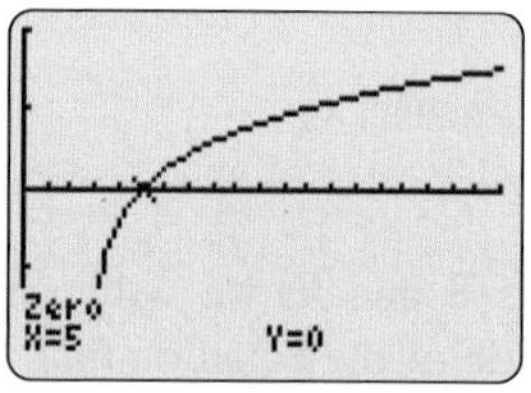

(a)

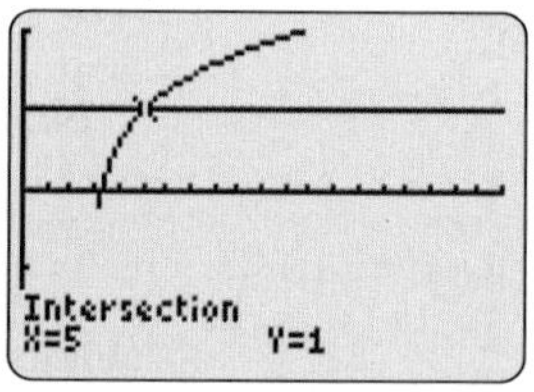

(b)

EXAMPLE 9 Solve: $\log_2 7 - \log_2 x = 5$

Strategy We will use the quotient rule for logarithms in reverse: The difference of two logarithms is equal to the logarithm of a quotient. Then we will write and solve an equivalent exponential equation.

Why We use the quotient rule for logarithms because the left side of the equation, $\log_2 7 - \log_2 x$, has the form $\log_b M - \log_b N$.

Solution

$$\log_2 7 - \log_2 x = 5 \quad \text{This is the equation to solve.}$$
$$\log_2 \frac{7}{x} = 5 \quad \text{On the left side, use the quotient rule for logarithms.}$$
$$\frac{7}{x} = 2^5 \quad \text{Write the equivalent base-2 exponential equation.}$$
$$\frac{7}{x} = 32 \quad \text{Evaluate: } 2^5 = 32.$$

Success Tip

The objective is to use the quotient rule to "condense" the left side of the equation. We want to write an equivalent equation in which the variable x appears in only a single logarithmic expression.

$$7 = 32x \quad \text{To clear the equation of the fraction, multiply both sides by } x.$$

$$\frac{7}{32} = x \quad \text{To isolate } x\text{, divide both sides by 32.}$$

The solution is $\frac{7}{32}$. Verify that it satisfies the original equation.

Teaching Example 9 Solve: $\log_2 11 - \log_2 x = 10$
Answer: $\frac{11}{1,024}$

Self Check 9 Solve: $\log_2 9 - \log_2 x = 4$ $\frac{9}{16}$

Now Try ▶ Problem 57

3 Use Exponential and Logarithmic Equations to Solve Application Problems.

Recall from Section 11.6 that a **hydrogen ion** $[H^+]$ is the positively charged nucleus of a hydrogen atom, without its electron. The concentration of hydrogen ions in a solution is commonly measured using the **pH scale.** The pH of a solution is defined as follows.

pH of a Solution

If $[H^+]$ is the hydrogen ion concentration in gram-ions per liter, then

$$\text{pH} = -\log[H^+]$$

EXAMPLE 10

©iStockphoto.com/Reniw-Imagery

Hydrogen Ion Concentration. Find the hydrogen ion concentration of seawater if its pH is 8.5.

Strategy To find the hydrogen ion concentration, we will substitute 8.5 for pH in the formula $\text{pH} = -\log[H^+]$ and solve the resulting equation for $[H^+]$.

Why After substituting for pH, the resulting logarithmic equation can be solved by solving an equivalent exponential equation.

Solution

Notation

In this logarithmic equation, the variable is the symbol $[H^+]$.

$$\textbf{pH} = -\log[H^+] \quad \text{This is the formula for pH.}$$

$$\textbf{8.5} = -\log[H^+] \quad \text{Substitute 8.5 for pH.}$$

$$-8.5 = \log[H^+] \quad \text{Multiply both sides by } -1.$$

$$-8.5 = \log_{\mathbf{10}}[H^+] \quad \text{The base of the logarithm is 10.}$$

$$[H^+] = \mathbf{10}^{-8.5} \quad \text{Write the equivalent base-10 exponential equation.}$$

$$[H^+] \approx 0.000000003 \quad \text{Approximate. On a reverse-entry scientific calculator, press:}$$

10 y^x 8.5 +/− =.

We can write the result using scientific notation:

$$[H^+] \approx 3.0 \times 10^{-9} \text{ gram-ions per liter}$$

Teaching Example 10 Hydrogen Ion Concentration. Find the hydrogen ion concentration of pickles if their pH is 3.2.
Answer: About 6.31×10^{-4} gram-ions per liter

Self Check 10 **Hydrogen Ion Concentration.** Find the hydrogen ion concentration of a solution with a pH value of 4.8. About 1.58×10^{-5} gram-ions per liter

Now Try ▶ Problem 101

Experiments have determined the time it takes for half of a sample of a radioactive material to decompose. This time is a constant, called the material's **half-life.**

When living organisms die, the oxygen–carbon dioxide cycle common to all living things ceases, and carbon-14, a radioactive isotope with a half-life of 5,700 years, is no longer absorbed. By measuring the amount of carbon-14 present in an ancient object, archaeologists can estimate the object's age by using the radioactive decay formula.

Radioactive Decay Formula

If A is the amount of radioactive material present at time t, A_0 was the amount present at $t = 0$, and h is the material's half-life, then

$$A = A_0 2^{-t/h}$$

EXAMPLE 11 **Carbon-14 Dating.** How old is a piece of wood that retains only one-third of its original carbon-14 content?

Strategy If A_0 is the original carbon-14 content, then today's content $A = \frac{1}{3}A_0$. We will substitute $\frac{A_0}{3}$ for A and 5,700 for h in the radioactive decay formula and solve for t.

Why The value of t is the estimated age of the piece of wood.

Solution To find the time t when $A = \frac{1}{3}A_0$, we substitute $\frac{A_0}{3}$ for A and 5,700 for h in the radioactive decay formula and solve for t:

$$A = A_0 2^{-t/h}$$ This is the radioactive decay model.

$$\frac{A_0}{3} = A_0 2^{-t/5,700}$$ The half-life of carbon-14 is 5,700 years.

$$1 = 3(2^{-t/5,700})$$ Divide both sides by A_0 and multiply both sides by 3.

$$\log 1 = \log 3(2^{-t/5,700})$$ Take the common logarithm on both sides.

$$0 = \log 3 + \log 2^{-t/5,700}$$ $\log 1 = 0$, and use the product rule for logarithms.

$$-\log 3 = -\frac{t}{5,700}\log 2$$ Subtract log 3 from both sides and use the power rule of logarithms.

$$5,700\left(\frac{\log 3}{\log 2}\right) = t$$ Multiply both sides by $-\frac{5,700}{\log 2}$.

$$t \approx 9,034.286254$$ Approximate. On a reverse-entry scientific calculator, press: 5700 × 3 LOG ÷ 2 LOG =.

The piece of wood is approximately 9,000 years old.

Notation

The initial amount of radioactive material is represented by A_0, and it is read as "A sub 0."

Teaching Example 11 Carbon-14 Dating. In Example 11, how old is a statue that retains 60% of its original carbon-14 content?
Answer: About 4,200 years

Self Check 11 **Carbon-14 Dating.** How old is a piece of wood that retains 25% of its original carbon-14 content? About 11,400 yr

Now Try Problem 103

When there is sufficient food and space available, populations of living organisms tend to increase exponentially according to the following growth model.

Exponential Growth Model

If P is the population at some time t, P_0 is the initial population at $t = 0$, and k depends on the rate of growth, then

$$P = P_0 e^{kt}$$

EXAMPLE 12 **Population Growth.** The bacteria in a laboratory culture increased from an initial population of 500 to 1,500 in 3 hours. How long will it take for the population to reach 10,000?

Strategy We will substitute 500 for P_0, 1,500 for P, and 3 for t in the exponential growth model and solve for k.

Why Once we know the value of k, we can substitute 10,000 for P, 500 for P_0, and the value of k in the exponential growth model and solve for the time t.

Solution

$$P = P_0e^{kt} \quad \text{This is the population growth formula.}$$
$$1{,}500 = 500(e^{k3}) \quad \text{Substitute 1,500 for } P\text{, 500 for } P_0\text{, and 3 for } t.$$
$$3 = e^{3k} \quad \text{Divide both sides by 500.}$$
$$3k = \ln 3 \quad \text{Write the equivalent base-}e\text{ logarithmic equation.}$$
$$k = \frac{\ln 3}{3} \quad \text{Divide both sides by 3.}$$

Notation

The initial population of bacteria is represented by P_0, and it is read as "P sub 0."

To find when the population will reach 10,000, we substitute 10,000 for P, 500 for P_0, and $\frac{\ln 3}{3}$ for k in the growth model and solve for t:

$$P = P_0e^{kt}$$
$$10{,}000 = 500e^{[(\ln 3)/3]t}$$
$$20 = e^{[(\ln 3)/3]t} \quad \text{Divide both sides by 500.}$$
$$\left(\frac{\ln 3}{3}\right)t = \ln 20 \quad \text{Write the equivalent base-}e\text{ logarithmic equation.}$$
$$t = \frac{3 \ln 20}{\ln 3} \quad \text{To isolate } t\text{, multiply both sides by the reciprocal of } \tfrac{\ln 3}{3}\text{, which is } \tfrac{3}{\ln 3}.$$
$$\approx 8.180499084 \quad \text{Approximate. On a reverse-entry scientific calculator, press: 3 × 20 LN ÷ 3 LN =.}$$

The culture will reach 10,000 bacteria in about 8 hours.

Teaching Tip: You may want to point out that once the values of P_0 and k are substituted into the growth model formula, it can be written as the growth model function $P(t) = 500e^{[(\ln 3)/3]t}$.

Teaching Example 12 Population Growth. In Example 12, how long will it take the population to reach 15,000?
Answer: About 9 hours

Self Check 12 **Population Growth.** In Example 12, how long will it take the population to reach 20,000? About 10 hr

Now Try Problem 115

SECTION 11.7 STUDY SET

VOCABULARY

Fill in the blanks.

1. An equation with a positive constant base and a variable in its exponent, such as $3^{2x} = 8$, is called an exponential equation.
2. An equation with a logarithmic expression that contains a variable, such as $\log_5 (2x - 3) = \log_5 (x + 4)$, is a logarithmic equation.

CONCEPTS

Fill in the blanks.

3. a. If two exponential expressions with the same base are equal, their exponents are equal.

   $b^x = b^y$ is equivalent to $x = y$.

   b. If the logarithms base-b of two numbers are equal, the numbers are equal.

   $\log_b x = \log_b y$ is equivalent to $x = y$.

4. The right side of the exponential equation $5^{x-3} = 125$ can be written as a power of 5.
5. If $6^{4x} = 6^{-2}$, then $4x = -2$.
6. a. Write the equivalent base-10 exponential equation for $\log (x + 1) = 2$. $10^2 = x + 1$

   b. Write the equivalent base-e exponential equation for $\ln (x + 1) = 2$. $e^2 = x + 1$

Fill in the blanks.

7. To solve $5^x = 2$, we can take the logarithm of both sides of the equation to get $\log 5^x = \log 2$.
8. a. For $5^x = 2$, the power rule for logarithms provides a way of moving the variable x from its position as an exponent to a position as a factor.

   b. If the power rule for logarithms is used on the left side of the equation $\log 5^x = 2$, the resulting equation is $x \log 5 = 2$.

9. If $e^{x+2} = 4$, then $\ln e^{x+2} = \ln 4$.
10. Perform a check to determine whether -2 is a solution of $5^{2x+3} = \frac{1}{5}$. It is a solution.
11. Perform a check to determine whether 4 is a solution of $\log_5 (x + 1) = 2$. Not a solution

12. Use a calculator to determine whether 2.5646 is an approximate solution of $2^{2x+1} = 70$. Yes

13. a. How do we solve $x \ln 3 = \ln 5$ for x? Divide both sides by ln 3.
    b. What is the exact solution? $\frac{\ln 5}{\ln 3}$
    c. What is an approximate solution to four decimal places? 1.4650

14. Use a property of logarithms to condense the left side of each equation to a single logarithm. **Do not solve.**
    a. $\log_5 x + \log_5 \cdot (4x - 1) = 1$ $\log_5 x(4x - 1) = 1$
    b. $\log_3 4x - \log_3 7 = 2$ $\log_3 \frac{4x}{7} = 2$

15. a. Find $\frac{\log 8}{\log 5}$. Round to four decimal places. 1.2920
    b. Find $\frac{3 \ln 12}{\ln 4 - \ln 2}$. Round to four decimal places. 10.7549

▶ 16. Does $\frac{\log 7}{\log 3} = \log 7 - \log 3$? No

17. Complete each formula.
    a. $\text{pH} = -\log\,[H^+]$
    b. Radioactive decay: $A = A_0 2^{-t/h}$
    c. Population growth: $P = P_0 e^{kt}$

18. Use the graphs below to estimate the solution of each equation.
    a. $2^x = 3^{-x+3}$ About 1.8
    b. $3 \log (x - 1) = 2 \log x$ About 4.6

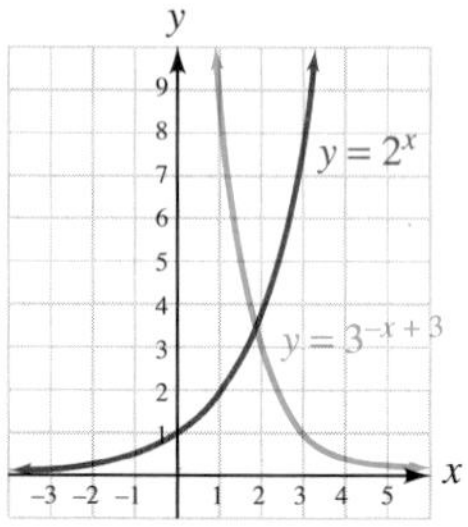

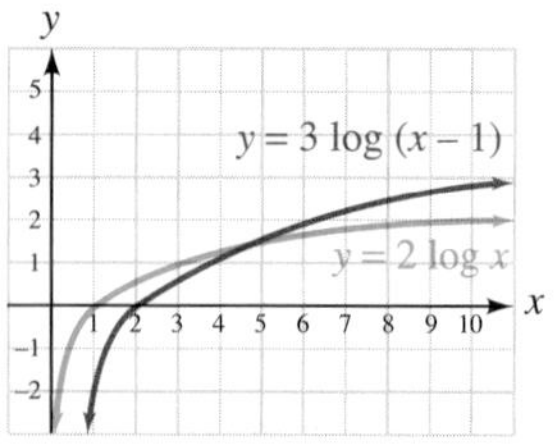

NOTATION

Complete each solution.

19. Solve: $2^x = 7$

$$\log 2^x = \log 7$$
$$x \log 2 = \log 7$$
$$x = \frac{\log 7}{\log 2}$$
$$x \approx 2.8074$$

▶ 20. Solve: $\log_2 (2x - 3) = \log_2 (x + 4)$

$$2x - 3 = x + 4$$
$$x = 7$$

GUIDED PRACTICE

Solve each equation. See Example 1.

21. $6^{x-2} = 36$ 4

▶ 22. $3^{x+1} = 27$ 2

▶ 23. $5^{4x} = \frac{1}{125}$ $-\frac{3}{4} = -0.75$

24. $8^{-2x+1} = \frac{1}{64}$ $\frac{3}{2} = 1.5$

Solve each equation. See Example 2.

▶ 25. $2^{x^2-2x} = 8$ 3, −1

▶ 26. $3^{x^2-3x} = 81$ 4, −1

▶ 27. $3^{x^2+4x} = \frac{1}{81}$ A repeated solution of −2

28. $7^{x^2+3x} = \frac{1}{49}$ −2, −1

Solve each equation. Give the exact solution and an approximation to four decimal places. See Example 3.

▶ 29. $4^x = 5$ $\frac{\log 5}{\log 4} \approx 1.1610$

30. $7^x = 12$ $\frac{\log 12}{\log 7} \approx 1.2770$

▶ 31. $13^{x-1} = 2$ $\frac{\log 2 + \log 13}{\log 13} \approx 1.2702$

▶ 32. $5^{x+1} = 3$ $\frac{\log 3 - \log 5}{\log 5} \approx -0.3174$

Solve each equation. Give the exact solution and an approximation to four decimal places. See Example 4.

▶ 33. $2^{x+1} = 3^x$ $\frac{\log 2}{\log 3 - \log 2} \approx 1.7095$

34. $6^x = 7^{x-4}$ $\frac{4 \log 7}{\log 7 - \log 6} \approx 50.4937$

▶ 35. $5^{x-3} = 3^{2x}$ $\frac{3 \log 5}{\log 5 - 2 \log 3} \approx -8.2144$

36. $8^{3x} = 9^{x+1}$ $\frac{\log 9}{3 \log 8 - \log 9} \approx 0.5437$

Solve each equation. Give the exact solution and an approximation to four decimal places. See Example 5.

37. $e^{2.9x} = 4.5$ $\frac{\ln 4.5}{2.9} \approx 0.5186$

▶ 38. $e^{3.3t} = 9.1$ $\frac{\ln 9.1}{3.3} \approx 0.6692$

▶ 39. $e^{-0.2t} = 14.2$ $-\frac{\ln 14.2}{0.2} \approx -13.2662$

40. $e^{-0.7x} = 6.2$ $-\frac{\ln 6.2}{0.7} \approx -2.6065$

Solve each equation. See Example 6.

41. $\log 2x = 4$ 5,000

▶ 42. $\log 5x = 4$ 2,000

43. $\log_3 (x - 3) = 2$ 12

▶ 44. $\log_4 (2x - 1) = 3$ 32.5

45. $\log (7 - x) = 2$ −93

▶ 46. $\log (2 - x) = 3$ −998

47. $\log \frac{1}{8}x = -2$ 0.08

▶ 48. $\log \frac{1}{5}x = -3$ 0.005

Solve each equation. See Example 7.

49. $\log (3 - 2x) = \log (x + 24)$ −7

▶ 50. $\log (3x + 5) = \log (2x + 6)$ 1

51. $\ln (3x + 1) = \ln (x + 7)$ 3

▶ 52. $\ln (x^2 + 4x) = \ln (x^2 + 16)$ 4

Solve each equation. See Example 8.

▶ 53. $\log x + \log (x - 48) = 2$ 50

▶ 54. $\log x + \log (x + 9) = 1$ 1

▶ 55. $\log_5 (4x - 1) + \log_5 x = 1$ $\frac{5}{4} = 1.25$

▶ 56. $\log_2 (x - 7) + \log_2 x = 3$ 8

Solve each equation. See Example 9.

57. $\log 5 - \log x = 1$ 0.5

▶ 58. $\log 11 - \log x = 2$ 0.11

▶ 59. $\log_3 4x - \log_3 7 = 2$ 15.75

60. $\log_2 5x - \log_2 3 = 4$ 9.6

TRY IT YOURSELF

Solve each equation. Give the exact solution and, when appropriate, an approximation to four decimal places.

▶ 61. $\log 2x = \log 4$ 2

▶ 62. $\log 3x = \log 9$ 3

63. $\ln x = 1$ $e \approx 2.7183$

64. $\ln x = 5$ $e^5 \approx 148.4132$

65. $7^{x^2} = 10$ $\pm\sqrt{\frac{\log 10}{\log 7}} \approx \pm 1.0878$

66. $8^{x^2} = 11$ $\pm\sqrt{\frac{\log 11}{\log 8}} \approx \pm 1.0738$

67. $\log (x + 90) + \log x = 3$ 10

▶ 68. $\log (x - 90) + \log x = 3$ 100

69. $3^{x-6} = 81$ 10

70. $5^{x+4} = 125$ −1

71. $\log \frac{4x + 1}{2x + 9} = 0$ 4

▶ 72. $\log \frac{2 - 5x}{2(x + 8)} = 0$ −2

73. $15 = 9^{x+2}$ $\frac{\log 15 - 2 \log 9}{\log 9} \approx -0.7675$

74. $29 = 5^{x-6}$ $\frac{\log 29 + 6 \log 5}{\log 5} \approx 8.0922$

75. $\log x^2 = 2$ $10, -10$

76. $\log x^3 = 3$ 10

77. $\log (x - 6) - \log (x - 2) = \log \dfrac{5}{x}$ 10

78. $\log (3 - 2x) - \log (x + 9) = 0$ -2

79. $\log_3 x = \log_3 \left(\dfrac{1}{x}\right) + 4$ 9

80. $\log_5 (7 + x) + \log_5 (8 - x) - \log_5 2 = 2$ $3, -2$

81. $2 \log_2 x = 3 + \log_2 (x - 2)$ A repeated solution of 4

82. $2 \log_3 x - \log_3 (x - 4) = 2 + \log_3 2$ $6, 12$

83. $\log (7y + 1) = 2 \log (y + 3) - \log 2$ $1, 7$

84. $2 \log (y + 2) = \log (y + 2) - \log 12$ $-\frac{23}{12}$

85. $e^{3x} = 9$ $\frac{\ln 9}{3} \approx 0.7324$

86. $e^{4x} = 60$ $\frac{\ln 60}{4} \approx 1.0236$

87. $\dfrac{\log (5x + 6)}{2} = \log x$ 6

88. $\dfrac{1}{2} \log (4x + 5) = \log x$ 5

Look Alikes . . .

89. a. $\log 5x = 1.7$ $\frac{10^{1.7}}{5} \approx 10.0237$
b. $\ln 5x = 1.7$ $\frac{e^{1.7}}{5} \approx 1.0948$

90. a. $\log_2 (x^2 - x) = 1$ $-1, 2$
b. $\log_6 (x^2 - x) = 1$ $-2, 3$

91. a. $4^{3x-5} = 90$ $\frac{\log 90 + 5 \log 4}{3 \log 4} \approx 2.7486$
b. $e^{3x-5} = 90$ $\frac{5 + \ln 90}{3} \approx 3.1666$

92. a. $\log x + 2 \log x = \log 8$ 2
b. $\log x - 2 \log x = \log 8$ $\frac{1}{8}$

93. a. $\log_2 (x + 5) - \log_2 4x = \log_2 x$ $\frac{5}{4}$
b. $\ln (x + 5) - \ln 4x = \ln x$ $\frac{5}{4}$

94. a. $5^{9x-1} = 125$ $\frac{4}{9}$
b. $5^{9x-1} = 124$ $\frac{\log 124 + \log 5}{9 \log 5} \approx 0.4439$

95. a. $\left(\dfrac{2}{3}\right)^{6-x} = \dfrac{8}{27}$ 3
b. $\left(\dfrac{2}{3}\right)^{6-x} = \dfrac{16}{81}$ 2

96. a. $\log x - \log (x + 7) = -1$ $\frac{7}{9}$
b. $\log x - \log (x + 7) = 1$ No solution

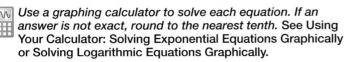

Use a graphing calculator to solve each equation. If an answer is not exact, round to the nearest tenth. **See Using Your Calculator: Solving Exponential Equations Graphically or Solving Logarithmic Equations Graphically.**

97. $2^{x+1} = 7$ 1.8

98. $3^x - 10 = 3^{-x}$ 2.1

99. $\log x + \log (x - 15) = 2$ 20

100. $\ln (2x + 5) - \ln 3 = \ln (x - 1)$ 8

APPLICATIONS

101. **Hydrogen Ion Concentration.** Find the hydrogen ion concentration of a saturated solution of calcium hydroxide whose pH is 13.2. About 6.3×10^{-14} gram-ions per liter

102. **Aquariums.** The safe pH range for a freshwater aquarium is shown on the scale in the next column. Find the corresponding hydrogen ion concentration. Approximate range: From 2.5×10^{-8} to 1.58×10^{-7}

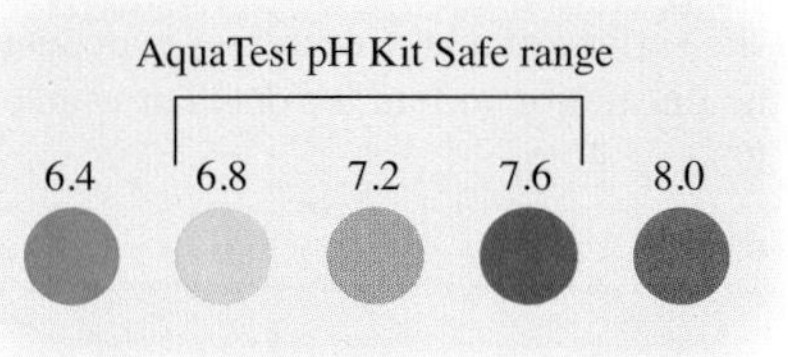

103. **Tritium Decay.** The half-life of tritium is 12.4 years. How long will it take for 25% of a sample of tritium to decompose? About 5.1 yr

104. **Radioactive Decay.** In 2 years, 20% of a radioactive element decays. Find its half-life. About 6.2 yr

105. **Thorium Decay.** An isotope of thorium, written as ^{227}Th, has a half-life of 18.4 days. How long will it take for 80% of the sample to decompose? About 42.7 days

106. **Lead Decay.** An isotope of lead, written as ^{201}Pb, has a half-life of 8.4 hours. How many hours ago was there 30% more of the substance? About 3.2 hr

107. **Carbon-14 Dating.** A bone fragment analyzed by archaeologists contains 60% of the carbon-14 that it is assumed to have had initially. How old is it? About 4,200 yr

108. **Carbon-14 Dating.** Only 10% of the carbon-14 in a small wooden bowl remains. How old is the bowl? About 19,000 yr

109. **Compound Interest.** If \$500 is deposited in an account paying 8.5% annual interest, compounded semiannually, how long will it take for the account to increase to \$800? About 5.6 yr

110. **Continuous Compund Interest.** In Exercise 109, how long will it take if the interest is compounded continuously? About 5.5 yr

111. **Compound Interest.** If \$1,300 is deposited in a savings account paying 9% interest, compounded quarterly, how long will it take the account to increase to \$2,100? About 5.4 yr

112. **Compound Interest.** A sum of \$5,000 deposited in an account grows to \$7,000 in 5 years. Assuming annual compounding, what interest rate is being paid? 6.96%

113. **Rule of Seventy.** A rule of thumb for finding how long it takes an investment to double is called the **rule of seventy.** To apply the rule, divide 70 by the interest rate written as a percent. At 5%, an investment takes $\frac{70}{5} = 14$ years to double. At 7%, it takes $\frac{70}{7} = 10$ years. Explain why this formula works. Because $\ln 2 \approx 0.7$

114. **Bacterial Growth.** A bacterial culture grows according to the function $P(t) = P_0 a^t$. If it takes 5 days for the culture to triple in size, how long will it take to double in size? About 3.2 days

115. **Rodent Control.** The rodent population in a city is currently estimated at 30,000. If it is expected to double every 5 years, when will the population reach 1 million? About 25.3 yr

116. **Population Growth.** The population of a city is expected to triple every 15 years. When can the city planners expect the present population of 140 persons to double? About 9.5 yr

117. **Bacterial Culture.** A bacteria culture doubles in size every 24 hours. By how much will it have increased in 36 hours? 2.828 times larger

118. Oceanography. The intensity I of a light a distance x meters beneath the surface of a lake decreases exponentially. Use the data in the illustration to find the depth at which the intensity will be 20%. 27 m

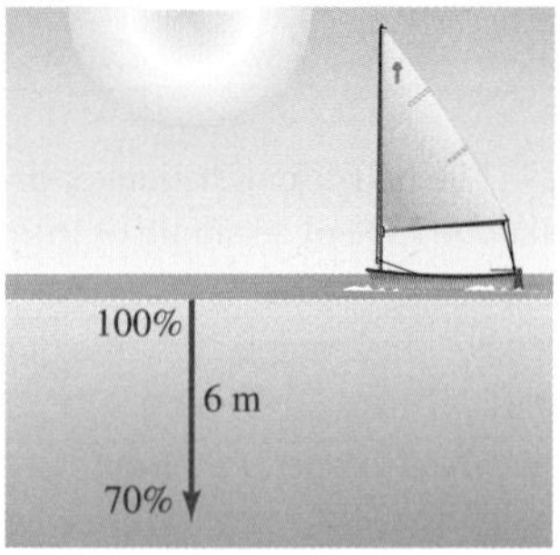

119. Newton's Law of Cooling. Water initially at 100°C is left to cool in a room at temperature 60°C. After 3 minutes, the water temperature is 90°. The water temperature T is a function of time t given by the formula $T = 60 + 40e^{kt}$. Find k. $\frac{1}{3}\ln 0.75 \approx -0.0959$

120. Newton's Law of Cooling. Refer to Exercise 119 and find the time for the water temperature to reach 70°C. About 14.5 min

WRITING

121. Explain how to solve the equation $2^{x+1} = 31$.

122. Explain how to solve the equation $2^{x+1} = 32$.

123. Write a justification for each step of the solution.

$15^x = 9$ This is the equation to solve.

$\log 15^x = \log 9$ ____________________.

$x \log 15 = \log 9$ ____________________.

$x = \dfrac{\log 9}{\log 15}$ ____________________.

124. What is meant by the term *half-life*?

REVIEW

125. Find the length of leg AC. $\sqrt{137}$ in.

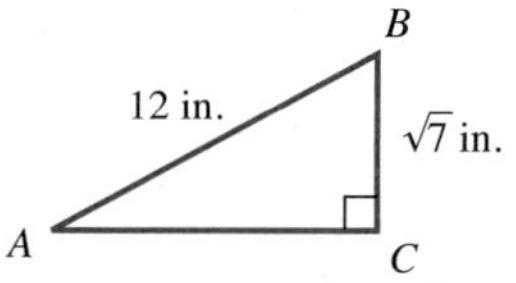

126. Dosages. The amount of medicine a patient should take is often proportional to his or her weight. If a patient weighing 83 kilograms needs 150 milligrams of medicine, how much will be needed by a person weighing 99.6 kilograms? 180 mg

CHALLENGE PROBLEMS

Solve each equation.

127. $\log_3 x + \log_3 (x + 2) = 2$ $-1 + \sqrt{10} \approx 2.1623$

128. $x^{\log x} = 10{,}000$ $100, \frac{1}{100}$

129. $\dfrac{\log_2 (6x - 8)}{\log_2 x} = 2$ 2, 4

130. $\dfrac{\log (3x - 4)}{\log x} = 2$ No solution

11 Summary & Review

SECTION 11.1 Algebra and Composition of Functions

DEFINITIONS AND CONCEPTS	EXAMPLES
Just as it is possible to perform arithmetic operations on real numbers, it is possible to perform those operations on functions. The **sum, difference, product,** and **quotient functions** are defined as: $(f + g)(x) = f(x) + g(x)$ $(f - g)(x) = f(x) - g(x)$ $(f \cdot g)(x) = f(x)g(x)$ $(f/g)(x) = \dfrac{f(x)}{g(x)}$, with $g(x) \neq 0$	Let $f(x) = 2x + 1$ and $g(x) = x^2$. $(f + g)(x) = f(x) + g(x) = 2x + 1 + x^2 = x^2 + 2x + 1$ $(f - g)(x) = f(x) - g(x) = 2x + 1 - x^2 = -x^2 + 2x + 1$ $(f \cdot g)(x) = f(x) \cdot g(x) = (2x + 1)x^2 = 2x^3 + x^2$ $(f/g)(x) = \dfrac{f(x)}{g(x)} = \dfrac{2x + 1}{x^2}$

Often one quantity is a function of a second quantity that depends, in turn, on a third quantity. Such chains of dependence can be modeled by a **composition of functions.** Read as "f of g of x." $(f \circ g)(x) = f(g(x))$ Read as "f composed with g of x."	Let $f(x) = 4x - 9$ and $g(x) = x^3$. Find $(f \circ g)(2)$ and $(f \circ g)(x)$. $(f \circ g)(2) = f(g(2))$ Change to nested parentheses notation. $= f(8)$ Evaluate: $g(2) = 2^3 = 8$. $= 4(8) - 9$ Evaluate $f(8)$ using $f(x) = 4x - 9$. $= 23$ $(f \circ g)(x) = f(g(x)) = f(x^3) = 4x^3 - 9$

REVIEW EXERCISES

Let f(x) = 2x and g(x) = x + 1. Find each function and give its domain.

1. $f + g$ $(f + g)(x) = 3x + 1, (-\infty, \infty)$
2. $f - g$ $(f - g)(x) = x - 1, (-\infty, \infty)$
3. $f \cdot g$ $(f \cdot g)(x) = 2x^2 + 2x, (-\infty, \infty)$
4. f/g $(f/g)(x) = \frac{2x}{x + 1}, (-\infty, -1) \cup (-1, \infty)$

Let $f(x) = x^2 + 2$ and $g(x) = 2x + 1$. Find each of the following.

5. $(f \circ g)(-1)$ 3
6. $(g \circ f)(0)$ 5
7. $(f \circ g)(x)$ $(f \circ g)(x) = 4x^2 + 4x + 3$
8. $(g \circ f)(x)$ $(g \circ f)(x) = 2x^2 + 5$

9. Use the graphs of functions f and g to find each of the following.
   a. $(f + g)(2)$ 0
   b. $(f \cdot g)(-4)$ −8
   c. $(f \circ g)(4)$ 0
   d. $(g \circ f)(6)$ −1

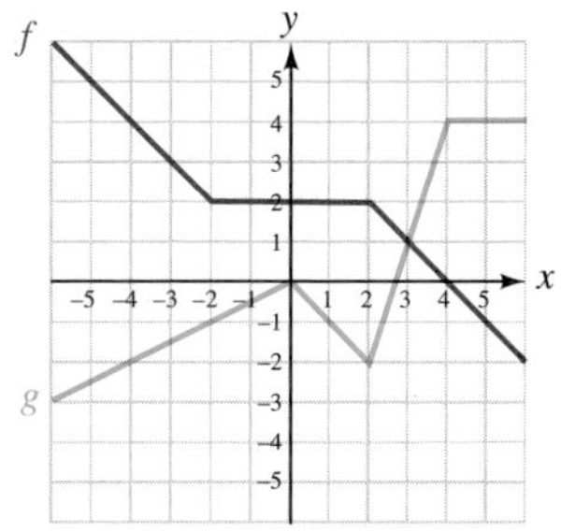

10. **Mileage Costs.** The function $f(m) = \frac{m}{8}$ gives the number of gallons of fuel consumed if a bus travels m miles. The function $C(f) = 3.25f$ gives the cost (in dollars) of f gallons of fuel. Write a composition function that expresses the cost of the fuel consumed as a function of the number of miles driven.
    $C(m) = \frac{3.25m}{8}$

SECTION 11.2 Inverse Functions

DEFINITIONS AND CONCEPTS	EXAMPLES
A function is called a **one-to-one function** if different inputs determine different outputs.	The function $f(x) = 3x - 5$ is a one-to-one function because different inputs have different outputs. Since two different inputs, -2 and 2, have the same output 16, the function $f(x) = x^4$ is not one-to-one.
Horizontal line test: A function is one-to-one if every horizontal line intersects the graph of the function at most once.	The function $f(x) = \|x\| - 2$ is not a one-to-one function because we can draw a horizontal line that intersects its graph twice. 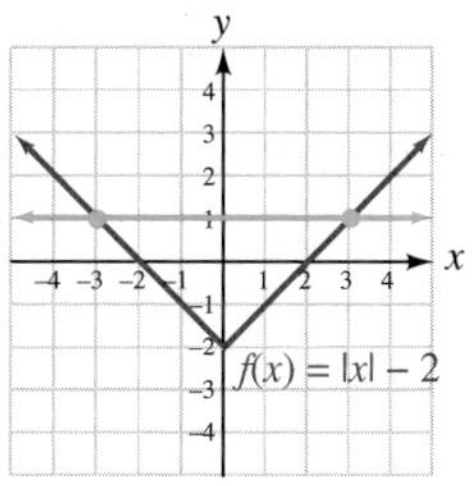
To find the inverse of a function, replace $f(x)$ with y, interchange the variables x and y, solve for y, and then replace y with $f^{-1}(x)$.	Find the inverse of the one-to-one function $f(x) = 2x + 1$. $f(x) = 2x + 1$ $y = 2x + 1$ Replace f(x) with y. $x = 2y + 1$ Interchange the variables x and y. $\frac{x - 1}{2} = y$ Solve for y. $f^{-1}(x) = \frac{x - 1}{2}$ Replace y with $f^{-1}(x)$.

If a point (a, b) is on the graph of function f, it follows that the point (b, a) is on the graph of f^{-1}, and vice versa.

The graph of a function and its inverse are **symmetric about the line $y = x$.**

The graphs of $f(x) = 2x + 1$ and $f^{-1}(x) = \frac{x-1}{2}$ are symmetric about the line $y = x$ as shown in the illustration.

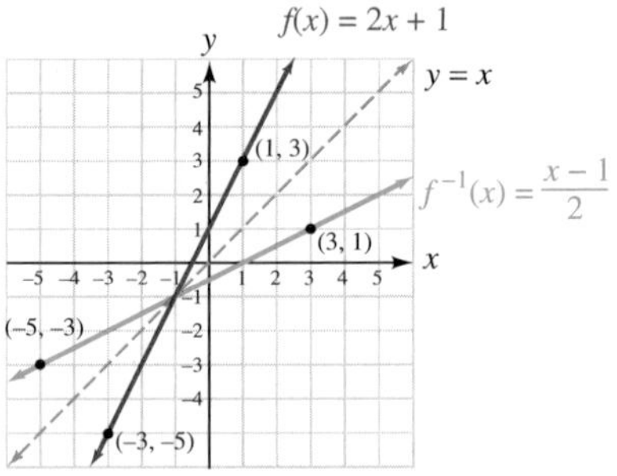

For any one-to-one function f and its inverse, f^{-1},

$$(f \circ f^{-1})(x) = x \quad \text{and} \quad (f^{-1} \circ f)(x) = x$$

The composition of $f(x) = 2x + 1$ and its inverse $f^{-1}(x) = \frac{x-1}{2}$ is the identity function $f(x) = x$.

$$(f \circ f^{-1})(x) = f(f^{-1}(x)) = f\left(\frac{x-1}{2}\right) = 2\left(\frac{x-1}{2}\right) + 1 = x - 1 + 1 = x$$

$$(f^{-1} \circ f)(x) = f^{-1}(f(x)) = f^{-1}(2x + 1) = \frac{2x + 1 - 1}{2} = \frac{2x}{2} = x$$

REVIEW EXERCISES

In Exercises 11–16, determine whether the function is one-to-one.

11. $f(x) = x^2 + 3$ No

12. $f(x) = \frac{1}{3}x - 8$ Yes

13. $\{(3, 4), (5, 10), (10, -1), (6, 6)\}$ Yes

14. No

x	$f(x)$
0	−5
2	10
4	−5
6	15

15.

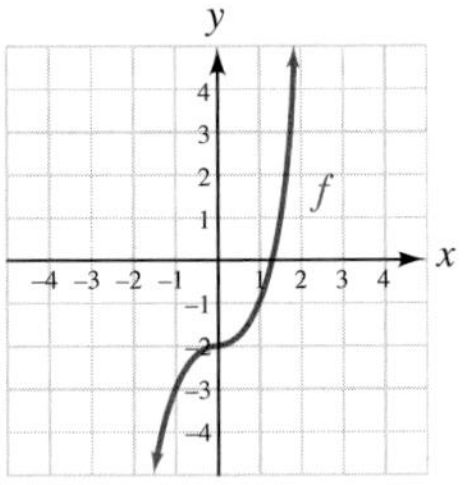

Yes

16.

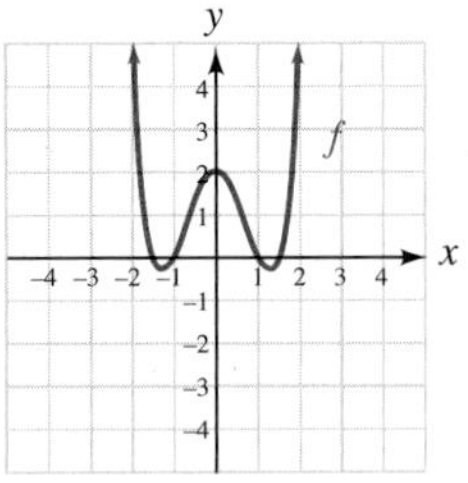

No

17. Use the table of values of the one-to-one function f to complete a table of values for f^{-1}.

x	$f(x)$
−6	−6
−1	−3
7	12
20	3

x	$f^{-1}(x)$
−6	−6
−3	−1
12	7
3	20

18. Given the graph of function f, graph f^{-1} on the same coordinate axes. Label the axis of symmetry.

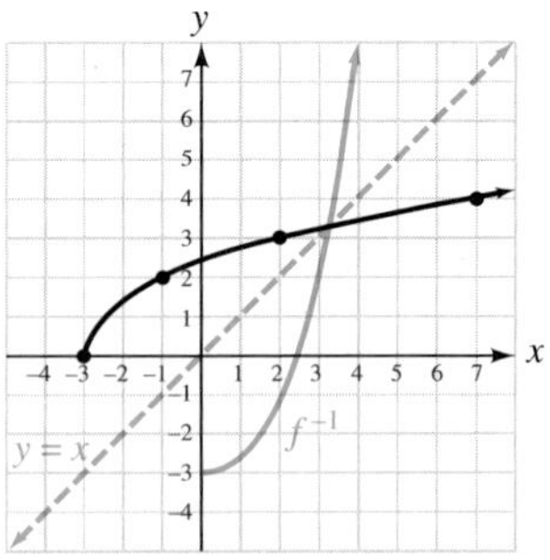

Find the inverse of each function.

19. $f(x) = 6x - 3$ $f^{-1}(x) = \frac{x+3}{6}$

20. $f(x) = \frac{4}{x-1}$ $f^{-1}(x) = \frac{4}{x} + 1$

21. $f(x) = (x + 2)^3$ $f^{-1}(x) = \sqrt[3]{x} - 2$

22. $f(x) = \frac{x}{6} - \frac{1}{6}$ $f^{-1}(x) = 6x + 1$

23. Find the inverse of $f(x) = \sqrt[3]{x - 1}$. Then graph the function and its inverse on one coordinate system. Show the axis of symmetry on the graph. $f^{-1}(x) = x^3 + 1$; see AIE Appendix 3.

24. Use composition to show that $f(x) = 5 - 4x$ and $f^{-1}(x) = -\frac{x-5}{4}$ are inverse functions.

SECTION 11.3 Exponential Functions

DEFINITIONS AND CONCEPTS	EXAMPLES

Exponential functions have a *constant base* and a *variable exponent* and are defined by the equation

$$f(x) = b^x, \quad \text{with } b > 0,\ b \neq 1$$

Properties of an exponential function $f(x) = b^x$:

The **domain** is the interval $(-\infty, \infty)$.

The **range** is the interval $(0, \infty)$.

Its graph has a **y-intercept** of (0, 1).

The x-axis is an **asymptote** of its graph.

The graph **passes through** the point $(1, b)$.

If $b > 1$, then $f(x) = b^x$ is an **increasing function.**

If $0 < b < 1$, then $f(x) = b^x$ is a **decreasing function.**

The graphs of $f(x) = 2^x$ and $g(x) = \left(\frac{1}{2}\right)^x$ are shown below.

$g(x) = \left(\frac{1}{2}\right)^x$

x	$g(x)$
−3	8
−2	4
−1	2
0	1
1	$\frac{1}{2}$
2	$\frac{1}{4}$

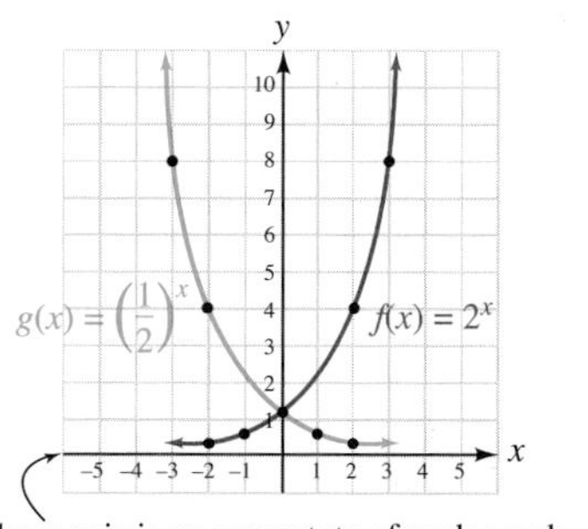

The x-axis is an asymptote of each graph.

$f(x) = 2^x$

x	$f(x)$
−2	$\frac{1}{4}$
−1	$\frac{1}{2}$
0	1
1	2
2	4
3	8

Since the base 2 is greater than 1, the function $f(x) = 2^x$ is an increasing function.

Since the base $\frac{1}{2}$ is such that $0 < \frac{1}{2} < 1$, the function $g(x) = \left(\frac{1}{2}\right)^x$ is a decreasing function.

The graphs of exponential functions can be **translated** horizontally and vertically.

$h(x) = 2^{x+3}$ The graph of $f(x) = 2^x$ is moved 3 units to the left.

$s(x) = \left(\frac{1}{2}\right)^x - 4$ The graph of $g(x) = \left(\frac{1}{2}\right)^x$ is moved 4 units downward.

Exponential functions are used to model many situations, such as population **growth,** the spread of an epidemic, the temperature of a heated object as it cools, and radioactive **decay.**

Exponential functions are suitable models for describing **compound interest.**

If $\$P$ is the deposit, and interest is paid k times a year at an annual rate r, the amount A in the account after t years is given by

$$A = P\left(1 + \frac{r}{k}\right)^{kt}$$

If $15,000 is deposited in an account paying an annual interest rate of 7.5%, compounded monthly, how much will be in the account in 60 years?

$$A(t) = 15{,}000\left(1 + \frac{0.075}{12}\right)^{12t}$$ To write the formula in function notation, substitute for P, r, and k.

$$A(60) = 15{,}000\left(1 + \frac{0.075}{12}\right)^{12(60)}$$ Substitute 60 for the time t.

$$= 15{,}000\left(1 + \frac{0.075}{12}\right)^{720}$$ Evaluate the exponent: $12(60) = 720$.

$$\approx 1{,}331{,}479.52$$ Use a calculator with an exponential key: y^x or ^.

In 60 years, the account will contain about $1,331,479.52.

REVIEW EXERCISES

25. a. Which of the following are exponential functions?

$f(x) = 2x$ $g(x) = x^2$ $h(a) = \sqrt{a}$ $n(x) = 2^x$,

$n(x) = 2^x$ $t(x) = \frac{1}{x}$ $s(t) = 1.08^t$ $s(t) = 1.08^t$

b. Use a calculator to find $0.9(1.42)^{14}$. Round to four decimal places. 121.9774

26. Determine whether each application is an example of exponential growth or decay.

a.

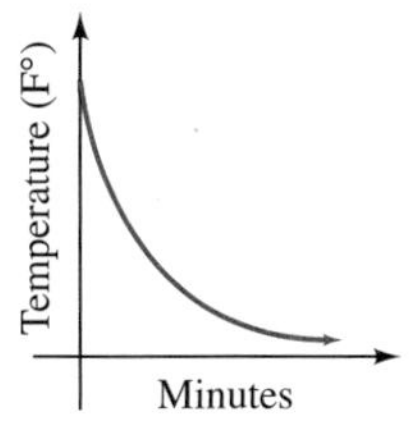

Exponential decay

b.

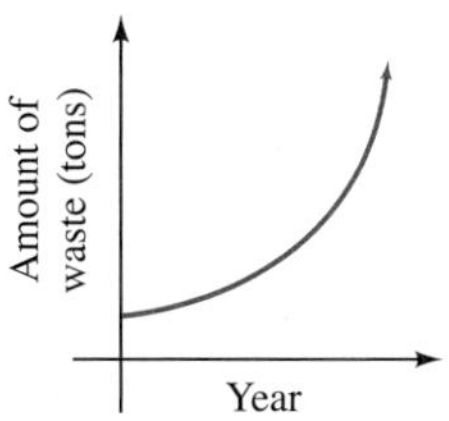

Exponential growth

Graph each function and give the domain and the range. Label the y-intercept. See AIE Appendix 3.

27. $f(x) = 3^x$ D: $(-\infty, \infty)$; R: $(0, \infty)$

28. $f(x) = \left(\frac{1}{3}\right)^x$ D: $(-\infty, \infty)$; R: $(0, \infty)$

29. $f(x) = \left(\frac{1}{2}\right)^x - 2$ D: $(-\infty, \infty)$; R: $(-2, \infty)$

30. $f(x) = 3^{x-1}$ D: $(-\infty, \infty)$; R: $(0, \infty)$

31. In Exercise 30, what is the asymptote of the graph of $f(x) = 3^{x-1}$? The x-axis ($y = 0$)

32. Coal Production. The function $c(t) = 128{,}000\,(1.08)^t$ approximates the number of tons of coal produced in the United States for the years 1800–1910, where t is the number of years after 1800. How many tons of coal did the U.S. produce in:

a. 1800? 128,000 tons

b. 1900? About 281,569,441 tons

33. Compound Interest. How much will \$10,500 become if it earns 9% annual interest, compounded quarterly, for 60 years? About \$2,189,703.45

34. Depreciation. The value (in dollars) of a certain model car is given by the function $V(t) = 12{,}000(10^{-0.155t})$, where t is the number of years from the present. Find the value of the car in 5 years. About \$2,015

SECTION 11.4 Logarithmic Functions

DEFINITIONS AND CONCEPTS	EXAMPLES
Definition of logarithm: If $b > 0$, $b \neq 1$, and x is positive, then $y = \log_b x$ is equivalent to $x = b^y$ (Exponent: y; Base: b)	***Logarithmic form*** ***Exponential form*** $\log_5 125 = 3$ means $5^3 = 125$ (Exponent: 3; Base: 5) $\log_2 \frac{1}{16} = -4$ means $2^{-4} = \frac{1}{16}$ (Exponent: −4; Base: 2)
$\log_b x$ is the exponent to which b is raised to get x. $b^{\log_b x} = x$	To evaluate $\log_4 16$ we ask: "To what power must we raise 4 to get 16?" Since $4^2 = 16$, the answer is: the 2nd power. Thus, $\log_4 16 = \mathbf{2}$.
We cannot find the logarithm of 0 or a negative number.	$\log_4 \mathbf{0}$ is undefined. $\log_8 (\mathbf{-64})$ is undefined.
For calculation purposes and in many applications, we use base-10 logarithms, called **common logarithms.** $\log x$ means $\log_{10} x$	$\log 1{,}000{,}000 = \mathbf{6}$ because $10^6 = 1{,}000{,}000$. $\log \frac{1}{1{,}000} = \mathbf{-3}$ because $10^{-3} = \frac{1}{1{,}000}$.
If $b > 0$ and $b \neq 1$, the **logarithmic function with base *b*** is defined by $\boldsymbol{f(x) = \log_b x}$. The domain is $(0, \infty)$ and the range is $(-\infty, \infty)$. If $b > 1$, then $f(x) = \log_b x$ is an increasing function. If $0 < b < 1$, then $f(x) = \log_b x$ is a decreasing function. The graphs of logarithmic functions can be **translated** horizontally and vertically.	The graph of the logarithmic function $f(x) = \log_2 x$ is shown at the right. From the graph, we see that $f(x) = \log_2 x$ is an increasing function. To graph $g(x) = \log_2 (x - 3)$, move the graph of $f(x) = \log_2 x$ to the right 3 units. 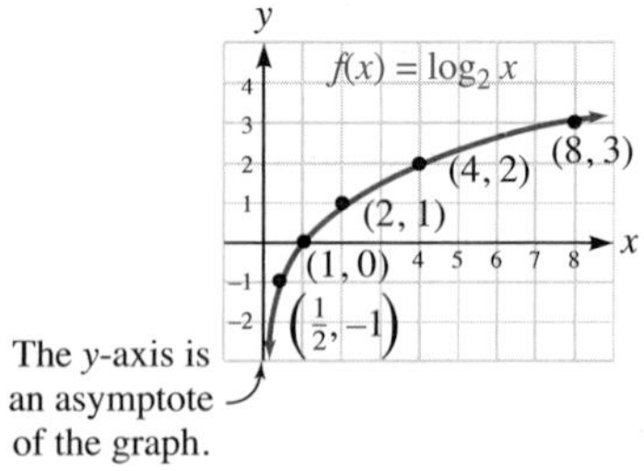
The exponential function $f(x) = b^x$ and the logarithmic function $f(x) = \log_b x$ are inverses of each other.	$f(x) = 3^x$ and $f^{-1}(x) = \log_3 x$ are inverses of each other. Their graphs are symmetric about the line $y = x$. Similarly, $f(x) = 10^x$ and $f^{-1}(x) = \log x$ are inverses. 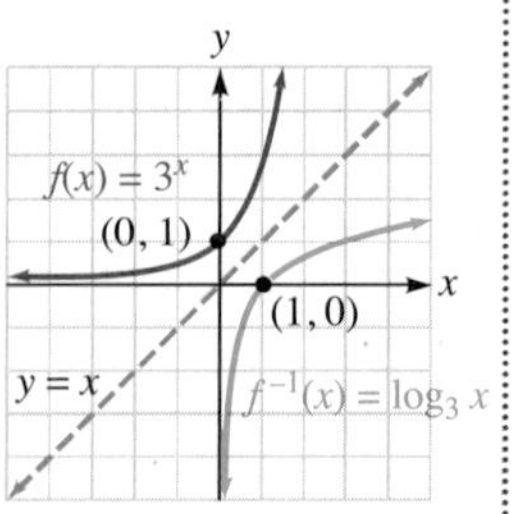

Logarithmic functions, like exponential functions, can be used to **model** certain types of growth and decay.

Decibel voltage gain: dB gain $= 20 \log \frac{E_O}{E_I}$

The Richter scale: $R = \log \frac{A}{P}$

The input to an amplifier is 0.4 volt and the output is 30 volts. Find the dB gain.

$$\text{dB gain} = 20 \log \frac{E_O}{E_I}$$

$$= 20 \log \frac{30}{0.4} \quad \text{Substitute 30 for } E_O \text{ and 0.4 for } E_I.$$

$$\approx 37.50122527 \quad \text{Use a calculator with a LOG key.}$$

The dB gain is about 38 decibels.

REVIEW EXERCISES

35. Give the domain and range of $f(x) = \log x$. D: $(0, \infty)$; R: $(-\infty, \infty)$

36. Explain why a student got an error message when she used a calculator to evaluate log 0. Since there is no real number such that $10^? = 0$, log 0 is undefined.

37. Write the statement $\log_4 64 = 3$ in exponential form. $4^3 = 64$

38. Write the statement $7^{-1} = \frac{1}{7}$ in logarithmic form. $\log_7 \frac{1}{7} = -1$

Evaluate.

39. $\log_3 9$ 2

40 $\log_9 \frac{1}{81}$ -2

41. $\log_{1/2} 1$ 0

42. $\log_5 (-25)$ Undefined

43. $\log_6 \sqrt{6}$ $\frac{1}{2}$

44. $\log 1{,}000$ 3

Solve for x.

45. $\log_2 x = 5$ 32

46. $\log_3 x = -4$ $\frac{1}{81}$

47. $\log_x 16 = 2$ 4

48. $\log_x \frac{1}{100} = -2$ 10

49. $\log_9 3 = x$ $\frac{1}{2}$

50. $\log_{27} 3 = x$ $\frac{1}{3}$

Use a calculator to find the value of x to four decimal places.

51. $\log 4.51 = x$ 0.6542

52. $\log x = 1.43$ 26.9153

Graph each function and its inverse on the same coordinate system. Draw the axis of symmetry. See AIE Appendix 3.

53. $f(x) = \log_4 x$ and $g(x) = 4^x$

54. $f(x) = \log_{1/3} x$ and $g(x) = \left(\frac{1}{3}\right)^x$

Graph each function. Label the x-intercept.
See AIE Appendix 3.

55. $f(x) = \log (x - 2)$

56. $f(x) = 3 + \log x$

57. Electrical Engineering. Find the dB gain of an amplifier with an output of 18 volts and an input of 0.04 volt. About 53 dB

58. Earthquakes. An earthquake had a period of 0.3 second and an amplitude of 7,500 micrometers. Use the formula $R = \log \frac{A}{P}$ to find its measure on the Richter scale. About 4.4

59. Organ Pipes. The design for a set of brass pipes for a church organ is shown below. The function $h(n) = 52 + 25 \log n$ approximates the height (in centimeters) of the pipe with number n. Find the height of

a. Pipe 1 52 cm

b. Pipe 8 About 74.6 cm

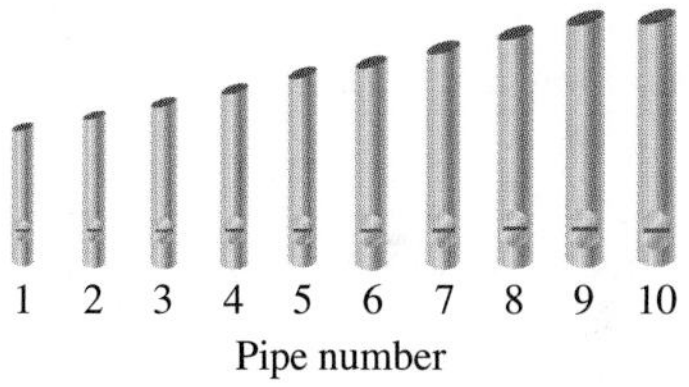

60. Girls' Heights. The function $P(A) = 61.8 + 34.9 \log(A - 4)$ approximates the percent of the adult height a female child A years old has attained, where $5 \leq A \leq 15$. What percent of her adult height will a girl have reached the day she first becomes a teenager? About 95%

SECTION 11.5 Base-*e* Exponential and Logarithmic Functions

DEFINITIONS AND CONCEPTS	EXAMPLES

Of all possible bases for an exponential function, ***e*** is the most convenient for problems involving growth or decay.

$$e = 2.718281828459 \ldots$$

As n approaches infinity, the value of $\left(1 + \frac{1}{n}\right)^n$ approaches e.

The function defined by $\boldsymbol{f(x) = e^x}$ is called the **natural exponential function.** It is called this because it can be applied to many natural settings.

From the graph, we see that the domain of the natural exponential function is $(-\infty, \infty)$ and the range is $(0, \infty)$.

The graph of $f(x) = e^x$ can be **translated** horizontally and vertically.

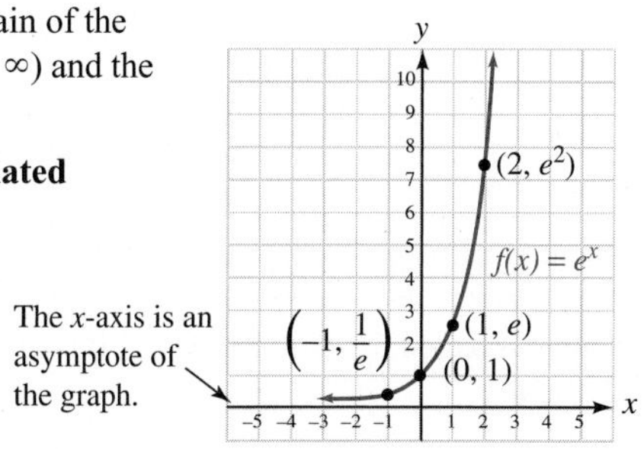

Exponential growth/decay: If a quantity increases or decreases at an annual rate r, **compounded continuously,** the amount A after t years is given by

$$A = Pe^{rt}$$

If r is negative, the amount decreases.

If interest is paid coutinuously (infinitely many times a year), we say that the interest is **compounded continuously.**

If \$30,000 accumulates interest at an annual rate of 9%, compounded continuously, find the amount in the account after 25 years.

$A = Pe^{rt}$ This is the formula for continuous compound interest.

$= \mathbf{30{,}000}e^{\mathbf{0.09 \cdot 25}}$ Substitute 30,000 for P, 0.09 for r, and 25 for t.

$= 30{,}000e^{2.25}$ Evaluate the exponent: $0.09 \cdot 25 = 2.25$

$\approx 284{,}632.08$ Use a calculator with an e^x key.

In 30 years, the account will contain \$284,632.08.

Suppose the population of a city of 50,000 people is decreasing exponentially according to the function $P(t) = 50{,}000e^{-0.003t}$, where t is measured in years from the present date. Find the expected population of the city in 20 years.

$P(t) = 50{,}000e^{-0.003t}$ Since r is negative, this is the exponential decay model.

$P(\mathbf{20}) = 50{,}000e^{-0.003(\mathbf{20})}$ Substitute 20 for t.

$= 50{,}000e^{-0.06}$ Evaluate the exponent: $-0.003\,(20) = -0.06$.

$\approx 47{,}088$ Use a calculator with an e^x key.

After 20 years, the expected population will be about 47,088 people.

Of all possible bases for a logarithmic function, e is the most convenient for problems involving growth or decay. Since these situations occur often in natural settings, base-e logarithms are called **natural logarithms:**

$\ln x$ means $\log_e x$

To evaluate $\ln \frac{1}{e^4}$ we ask: "To what power must we raise e to get $\frac{1}{e^4}$?" Since $e^{-4} = \frac{1}{e^4}$, the answer is: the -4th power. Thus,

$$\ln \frac{1}{e^4} = -4$$

$\ln \mathbf{0}$ and $\ln(\mathbf{-5})$ are undefined.

$\ln x$ is the exponent to which e is raised to get x.

$\ln 9 \approx \mathbf{2.1972}$ means $e^{\mathbf{2.1972}} \approx 9$

The **natural logarithmic function** with base e is defined by

$$\mathbf{f(x) = \ln x}$$

The domain is the interval $(0, \infty)$ and the range is the interval $(-\infty, \infty)$.

The graph of the natural logarithmic function $f(x) = \ln x$ is shown on the right.

From the graph, we see that $f(x) = \ln x$ is an increasing function.

The graph of $f(x) = \ln x$ can be translated horizontally and vertically.

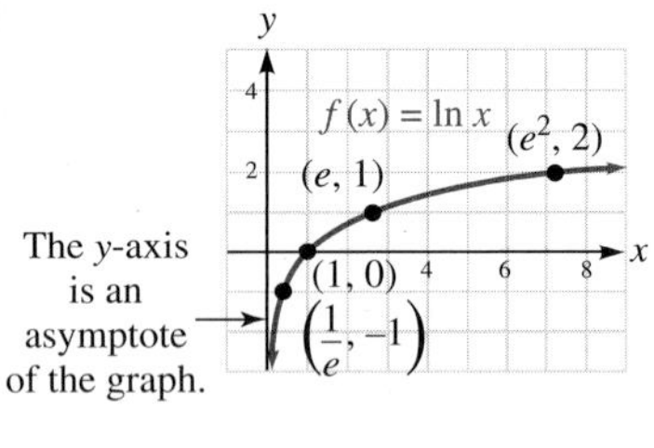

The natural exponential function $f(x) = e^x$ and the natural logarithmic function $f^{-1}(x) = \ln x$ are **inverses** of each other.

The graphs are symmetric about the line $y = x$.

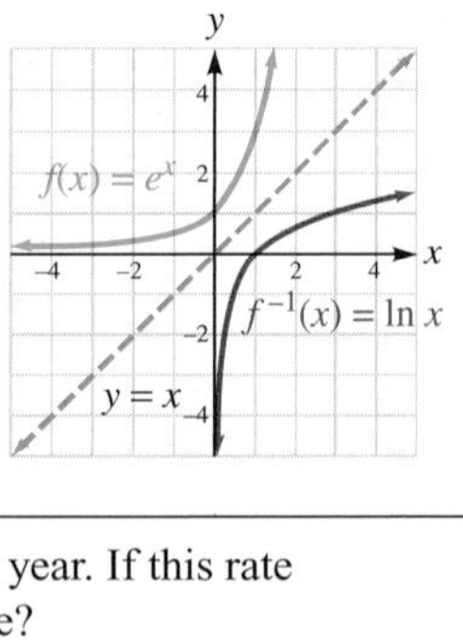

If a population grows exponentially at a certain annual rate r, the time required for the population to double is called the **doubling time.** It is given by the formula:

$$t = \frac{\ln 2}{r}$$

The population of a town is growing at a rate of 3% per year. If this rate continues, how long will it take the population to double?

We substitute 0.03 for r and use a calculator to perform the calculation.

$t = \frac{\ln 2}{r} = \frac{\ln 2}{\mathbf{0.03}} \approx 23.10490602$ Use a calculator with an LN key.

The population will double in about 23.1 years.

REVIEW EXERCISES

61. a. Approximate e to the nearest hundredth. $e \approx 2.72$

b. Fill in the blanks:

$\ln 15 \approx 2.7081$ means $e^{2.7081} \approx 15$

62. Use a calculator to find $16.4 + 252.7e^{-0.76(9)}$. Round to four decimal places. 16.6704

Graph each function, and give the domain and the range. See AIE Appendix 3.

63. $f(x) = e^x + 1$ D: $(-\infty, \infty)$; R: $(1, \infty)$

64 $f(x) = e^{x-3}$ D: $(-\infty, \infty)$; R: $(0, \infty)$

65. **Interest Compounded Continuously.** If \$10,500 accumulates interest at an annual rate of 9%, compounded continuously, how much will be in the account in 60 years? About \$2,324,767.37

66. **The Sooner State.** In 2009, Oklahoma had the largest *gross domestic product* growth rate of all fifty states: 6.6%. (The GDP, as it is called, is the value of all goods and services produced within a state.) If the 2009 GDP totaled \$142.5 billion, predict Oklahoma's GDP in 2015, assuming the growth rate remains the same. (Source: huffingtonpost.com) About \$211.7 billion

67. **Mortgage Rates.** There was the housing boom in the 1980s as the baby boomers (those born from 1946 through 1964) bought their homes. The average annual interest rate in percent on a 30-year fixed-rate home mortgage for the years 1980–1996 can be approximated by the function $r(t) = 13.9e^{-0.035t}$, where t is the number of years since 1980. To the nearest hundredth of a percent, what does this model predict was the 30-year fixed rate in 1980? In 1985? In 1990? 13.9%, 11.67%, 9.80%

68. **Medical Tests.** A radioactive dye is injected into a patient as part of a test to detect heart disease. The amount of dye remaining in his bloodstream t hours after the injection is given by the function $f(t) = 10e^{-0.27t}$. How can you determine from the function that the amount of dye in the bloodstream is decreasing? The exponent on the base e is negative.

Evaluate each expression. Do not use a calculator.

69. $\ln e$ 1

70. $\ln e^2$ 2

71. $\ln \frac{1}{e^5}$ -5

72. $\ln \sqrt{e}$ $\frac{1}{2}$

73. $\ln(-e)$ Undefined

74. $\ln 0$ Undefined

75. $\ln 1$ 0

76. $\ln e^{-7}$ -7

Use a calculator to evaluate each expression. Express all answers to four decimal places.

77. $\ln 452$ 6.1137

78. $\ln 0.85$ -0.1625

Solve each equation. Express all answers to four decimal places.

79. $\ln x = 2.336$ 10.3398

80. $\ln x = -8.8$ 0.0002

81. Explain the difference between the functions $f(x) = \log x$ and $g(x) = \ln x$. They have different bases: $\log x = \log_{10} x$ and $\ln x = \log_e x$.

82. What function is the inverse of $f(x) = \ln x$? $f^{-1}(x) = e^x$

Graph each function. See AIE Appendix 3.

83. $f(x) = 1 + \ln x$

84. $f(x) = \ln(x + 1)$

85. **Population Growth.** How long will it take the population of Mexico to double if the growth rate is currently about 1.118%? (Source: CIA World Fact Book) About 62 yr

86. **Botany.** The height (in inches) of a certain plant is approximated by the function $H(a) = 13 + 20.03 \ln a$, where a is its age in years. How tall will it be when it is 19 years old? About 72 in. (6 ft)

SECTION 11.6 Properties of Logarithms

DEFINITIONS AND CONCEPTS	EXAMPLES
Properties of logarithms: If M, N, and b are positive real numbers, $b \neq 1$	Apply a property of logarithms and then simplify, if possible.
1. $\log_b 1 = 0$ 2. $\log_b b = 1$ 3. $\log_b b^x = x$ 4. $b^{\log_b x} = x$	1. $\log_3 1 = 0$ 2. $\log_7 7 = 1$ 3. $\log_5 5^3 = 3$ 4. $9^{\log_9 10} = 10$
5. ***Product rule for logarithms:*** $\log_b MN = \log_b M + \log_b N$	5. $\log_2(6 \cdot 8) = \log_2 6 + \log_2 8 = \log_2 6 + 3$
6. ***Quotient rule for logarithms:*** $\log_b \frac{M}{N} = \log_b M - \log_b N$	6. $\log_3 \frac{81}{x} = \log_3 81 - \log_3 x = 4 - \log_3 x$
7. ***Power rule for logarithms:*** $\log_b M^p = p \log_b M$	7. $\log_8 7^3 = 3 \log_8 7$ and $\log \sqrt[4]{x} = \log x^{\frac{1}{4}} = \frac{1}{4} \log x$

Properties of logarithms can be used to **expand** logarithmic expressions.

Write $\log_3 (x^2y^3)$ as the sum and/or difference of logarithms of a single quantity.

$$\log_3 (x^2y^3) = \log_3 x^2 + \log_3 y^3 \quad \text{The log of a product is the sum of the logs.}$$
$$= 2\log_3 x + 3\log_3 y \quad \text{The log of a power is the power times the log.}$$

Properties of logarithms can be used to **condense** certain logarithmic expressions.

To condense, apply the power rule first to make the coefficients of logarithms 1. Then use the product and quotient rules.

Write $3 \ln x - \frac{1}{2} \ln y$ as a single logarithm.

$$3 \ln x - \frac{1}{2} \ln y = \ln x^3 - \ln y^{1/2} \quad \text{A power times a log is the log of the power.}$$
$$= \ln \frac{x^3}{y^{1/2}} \quad \text{The difference of two logs is the log of the quotient.}$$
$$= \ln \frac{x^3}{\sqrt{y}} \quad \text{Write } y^{1/2} \text{ as } \sqrt{y}.$$

If we need to find a logarithm with some base other than 10 or e, we can use a conversion formula.

Change-of-base formula:

$$\log_b x = \frac{\log_a x}{\log_a b}$$

Find $\log_7 6$ to four decimal places.

$$\log_7 6 = \frac{\log 6}{\log 7} \approx 0.920782221 \quad \text{Change to the ratio of base-10 logarithms.}$$

To four decimal places, $\log_7 6 = 0.9208$. To check, verify that $7^{0.9208}$ is approximately 6.

In chemistry, common logarithms are used to express the acidity of solutions using pH.

pH scale: $\text{pH} = -\log[\text{H}^+]$

where the symbol $[\text{H}^+]$ represents the hydrogen ion concentration in gram-ions per liter.

Find the pH of a liquid with a hydrogen ion concentration of 10^{-8} gram-ions per liter.

$$\text{pH} = -\log [\text{H}^+] \quad \text{This is the pH formula.}$$
$$\text{pH} = -\log 10^{-8} \quad \text{Substitute } 10^{-8} \text{ for } [\text{H}^+].$$
$$= -(-8) \log 10 \quad \text{The log of a power is the power times the log.}$$
$$= 8 \quad \text{Simplify: } \log 10 = 1.$$

REVIEW EXERCISES

Simplify each expression.

87. $\log_2 1$ 0

88. $\log_9 9$ 1

89. $\log 10^3$ 3

90. $7^{\log_7 4}$ 4

Write each logarithm as the sum and/or difference of logarithms of a single quantity. Then simplify, if possible.

91. $\log_3 27x$ $3 + \log_3 x$

92. $\log \frac{100}{x}$ $2 - \log x$

93. $\log_5 \sqrt{27}$ $\frac{1}{2} \log_5 27$

94. $\log_b 10ab$ $\log_b 10 + \log_b a + 1$

Write each logarithm as the sum and/or difference of logarithms of a single quantity.

95. $\log_b \frac{x^2y^3}{z}$ $2\log_b x + 3\log_b y - \log_b z$

96. $\ln \sqrt{\frac{x}{yz^2}}$ $\frac{1}{2}\ln x - \frac{1}{2}\ln y - \ln z$

Write each logarithmic expression as one logarithm.

97. $3\log_2 x - 5\log_2 y + 7\log_2 z$ $\log_2 \frac{x^3z^7}{y^5}$

98. $-3\log_b y - 7\log_b z + \frac{1}{2}\log_b (x + 2)$ $\log_b \frac{\sqrt{x+2}}{y^3z^7}$

99. $\log_b (a^2 - 25) - \log_b (a + 5)$ $\log_b(a - 5)$

100. $3\log_8 x + 4\log_8 x$ $\log_8 x^7$

Assume that $\log_b 5 = 1.1609$ and $\log_b 8 = 1.5000$ and find each value to four decimal places.

101. $\log_b 40$ 2.6609

102. $\log_b 64$ 3.0000

103. Find $\log_5 17$ to four decimal places. 1.7604

104. pH of Grapefruit. Find the pH of grapefruit juice if its hydrogen ion concentration is 7.9×10^{-4} gram-ions per liter. Round to the nearest tenth. 3.1

SECTION 11.7 Exponential and Logarithmic Equations

DEFINITIONS AND CONCEPTS	EXAMPLES
An **exponential equation** contains a variable in one of its exponents. Two examples are: $6^{x-3} = 9$ and $e^{-2.5t} = 56$ If both sides of an exponential equation can be expressed as a power of the same base, we can use the following property to solve it: $b^x = b^y$ is equivalent to $x = y$	Solve: $3^{x+2} = 27$ $3^{x+2} = 3^3$ Express the right side of the equation as a power of 3: $27 = 3^3$. $x + 2 = 3$ If two exponential expressions with the same base are equal, their exponents are equal. $x = 1$ The solution is 1. Check it in the original equation.
When it is difficult to write each side of an exponential equation as a power of the same base, **take the logarithm on each side.** With this method, we often obtain **exact solutions** involving logarithms that we can **approximate.**	Solve: $4^x = 7$. Give an approximate solution to four decimal places. $\log 4^x = \log 7$ Take the base-10 logarithm on both sides of the equation. $x \log 4 = \log 7$ The log of a power is the power times the log. $x = \frac{\log 7}{\log 4}$ To isolate x, divide both sides by log 4. This is the exact solution. $x \approx 1.4037$ Use a calculator with a LOG key. To four decimal places, the approximate solution is 1.4037. To check it, we substitute 1.4037 for x in $4^x = 7$ and use a calculator to evaluate the left side: $4^{1.4037} \approx 7$
To solve an exponential equation that contains a base-e exponential expression, take the natural logarithm on both sides of the equation.	Solve: $e^{5t} = 11$ $\ln e^{5t} = \ln 11$ Take the base-e logarithm on both sides. $5t \ln e = \ln 11$ The log of a power is the power times the log. $5t \cdot 1 = \ln 11$ Simplify: $\ln e = 1$. $5t = \ln 11$ Simplify the left side. $t = \frac{\ln 11}{5}$ This is the exact solution. An approximate solution to four decimal places is 0.4796.
A **logarithmic equation** is an equation containing a variable in a logarithmic expression. Two examples are: $2 \log_3 x + \log_3 5x = 4$ and $\ln(x + 1) = \ln(3x - 4)$ Certain logarithmic equations can be solved using the following property: $\log_b x = \log_b y$ is equivalent to $x = y$	Solve: $\log(4x - 3) = \log(2x + 7)$ $4x - 3 = 2x + 7$ If the logarithms of two numbers are equal, the numbers are equal. $2x = 10$ $x = 5$ The solution is 5. Check it in the original equation.
To solve some logarithmic equations, we write and solve an equivalent exponential equation.	Solve: $\log_4(x + 1) = 2$ $x + 1 = 4^2$ Write the equivalent base-4 exponential equation. $x + 1 = 16$ $x = 15$ The solution is 15. Check it in the original equation.

To solve some logarithmic equations, we first apply properties of logarithms, such as:

- The product rule
- The quotient rule
- The power rule

Solve: $\log_2 (x + 14) - \log_2 x = 3$

$$\log_2 \frac{x + 14}{x} = 3$$ On the left side, use the quotient rule for logarithms.

$$\frac{x + 14}{x} = 2^3$$ Write the equivalent base-2 exponential equation.

$$\frac{x + 14}{x} = 8$$ Evaluate: $2^3 = 8$.

$$x + 14 = 8x$$ Multiply both sides by x.

$$14 = 7x$$ Subtract x from both sides.

$$2 = x$$ Divide both sides by 7.

The solution is 2. Check by substituting it into the original equation.

When there is sufficient food and space available, populations of living organisms tend to increase exponentially according to the following **growth model.**

Population growth: $P = P_0e^{kt}$

Find the number of bacteria in a culture of 1,000 bacteria if they are allowed to reproduce for 5 hours. Assume $k = \frac{\ln 3}{3}$.

$$P = P_0e^{kt}$$ This is the population growth model.

$$= 1{,}000e^{\frac{\ln 3}{3} \cdot 5}$$ Substitute for P_0, k, and t.

$$\approx 6{,}240$$ Use a calculator with LN and e^x keys.

In 5 hours, there will be approximately 6,240 bacteria.

REVIEW EXERCISES

Solve each equation. Give the exact solution and an approximate solution to four decimal places, when appropriate.

105. $5^{x+6} = 25$ -4

106. $2^{x^2+4x} = \frac{1}{8}$ $-3, -1$

107. $3^x = 7$ $\frac{\log 7}{\log 3} \approx 1.7712$

108. $2^x = 3^{x-4}$ $\frac{4 \log 3}{\log 3 - \log 2} \approx 10.8380$

109. $e^x = 7$ $\ln 7 \approx 1.9459$

110. $e^{-0.4t} = 25$ $-\frac{\ln 25}{0.4} \approx -8.0472$

Solve each equation.

111. $\left(\frac{2}{5}\right)^{3x-4} = \frac{8}{125}$ $\frac{7}{3}$

112. $9^{x^2} = 33$ $\pm\sqrt{\frac{\log 33}{\log 9}} \approx \pm 1.2615$

113. $\log (x - 4) = 2$ 104

114. $\ln (2x - 3) = \ln 15$ 9

115. $\log x + \log (29 - x) = 2$ 25, 4

116. $\log_2 x + \log_2 (x - 2) = 3$ 4

117. $\frac{\log (7x - 12)}{\log x} = 2$ 4, 3

118. $\log_2 (x + 2) + \log_2 (x - 1) = 2$ 2

119. $\log x + \log (x - 5) = \log 6$ 6

120. $\log 3 - \log (x - 1) = -1$ 31

121. Evaluate both sides of the statement $\frac{\log 8}{\log 15} \neq \log 8 - \log 15$ to show that the sides are indeed not equal.
$0.76787 \neq -0.27300$

122. Carbon-14 Dating. A wooden statue found in Egypt has a carbon-14 content that is two-thirds of that found in living wood. If the half-life of carbon-14 is 5,700 years, how old is the statue? About 3,300 yr

123. Ants. The number of ants in a colony is estimated to be 800. If the ant population is expected to triple every 14 days, how long will it take for the population to reach one million?
About 91 days

124. The approximate coordinates of the points of intersection of the graphs of $f(x) = \log x$ and $g(x) = 1 - \log (7 - x)$ are shown in parts (a) and (b) of the illustration. Use the graphs to estimate the solutions of the logarithmic equation $\log x = 1 - \log (7 - x)$. Then check your answers. 2, 5

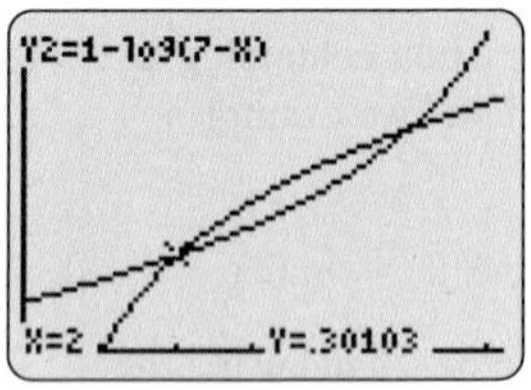

(a)

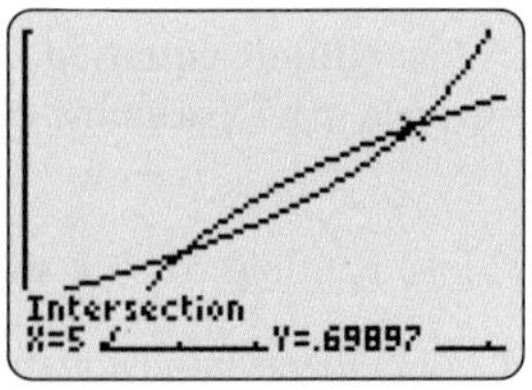

(b)

11 Chapter Test

1. Fill in the blanks.
   a. A composite function is denoted by $f \circ g$.
   b. $f(x) = e^x$ is the natural exponential function.
   c. In continuous compound interest, the number of compoundings is infinitely large.
   d. The functions $f(x) = \log_{10} x$ and $f(x) = 10^x$ are inverse functions.
   e. $f(x) = \log_4 x$ is a logarithmic function.
2. Write out in words how to say each of the following:
   a. $(f \circ g)(x)$ f composed with g of x
   b. $g(f(8))$ g of f of eight
   c. $f^{-1}(x)$ f inverse of x

Let $f(x) = x + 9$ and $g(x) = 4x^2 - 3x + 2$. Find each function and give its domain.

3. $f + g$ $(f + g)(x) = 4x^2 - 2x + 11$, $(-\infty, \infty)$
4. g/f $(g/f)(x) = \frac{4x^2 - 3x + 2}{x + 9}$, $(-\infty, -9) \cup (-9, \infty)$

Let $f(x) = 2x^2 + 3$ and $g(x) = 4x - 8$. Find each composition.

5. $(g \circ f)(-3)$ 76
6. $(f \circ g)(x)$ $32x^2 - 128x + 131$

Use the tables of values for functions f and g to find each of the following.

7. a. $(f \cdot g)(9)$ -16
   b. $(f \circ g)(-3)$ 17

x	$f(x)$
9	-1
10	17

x	$g(x)$
-3	10
9	16

8. Refer to the graphs of functions f and g below to find each of the following.
   a. $(g/f)(-4)$ 0
   b. $(f \circ g)(1)$ 3
   c. $(f + g)(2)$ 1
   d. $(f \cdot g)(0)$ -2
   e. $(g - f)(1)$ -5

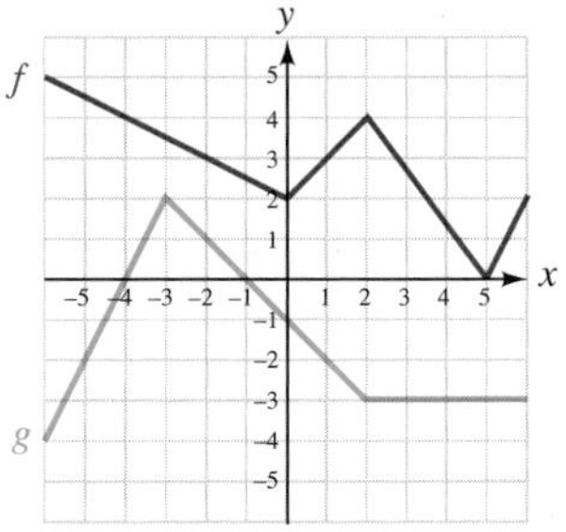

9. Determine whether each function is one-to-one.
   a. $f(x) = |x|$ No
   b. {(1,7), (7,1), (2,8), (8,2)} Yes
   c. Yes
   d. No

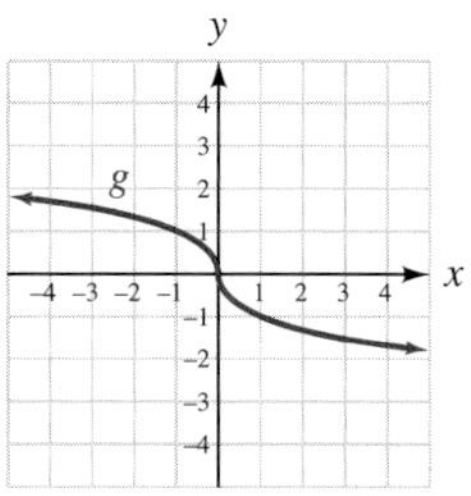

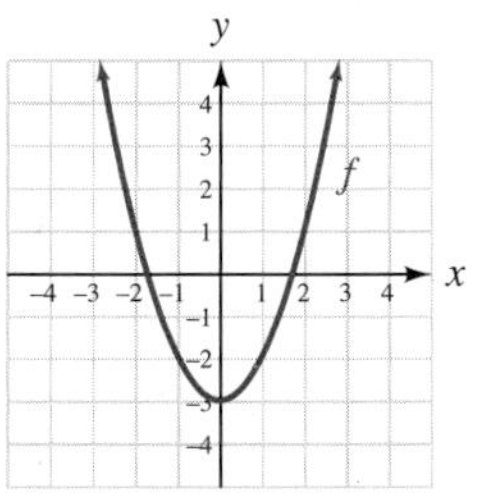

10. Find the inverse of $f(x) = -\frac{1}{3}x$ and then graph f and its inverse on the same coordinate axes. Label the axis of symmetry. $f^{-1}(x) = -3x$; see AIE Appendix 3.
11. Determine whether $f(x) = \frac{1}{3}x + 2$ is a one-to-one function. If it is, find its inverse. One-to-one function; $f^{-1}(x) = 3x - 6$
12. Find the inverse of $f(x) = (x - 15)^3$. $f^{-1}(x) = \sqrt[3]{x} + 15$
13. Use composition to show that $f(x) = 4x + 4$ and $f^{-1}(x) = \frac{x - 4}{4}$ are inverse functions.
14. Consider the following graph of the function f.
    a. Is f a one-to-one function? Yes
    b. Is its inverse a function? Yes
    c. What is $f^{-1}(260)$? What information does it give? 80; when the temperature of the tire tread is 260°, the vehicle is traveling 80 mph

Relationship Between Car Speed and Tire Temperature

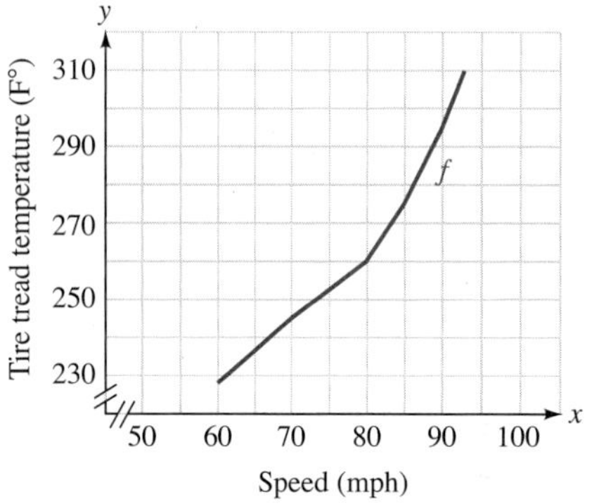

Graph each function and give the domain and the range. See AIE Appendix 3.

15. $f(x) = 2^x + 1$ D: $(-\infty, \infty)$; R: $(1, \infty)$
16. $f(x) = 3^{-x}$ D: $(-\infty, \infty)$; R: $(0, \infty)$
17. **Radioactive Decay.** A radioactive material decays according to the formula $A = A_0(2)^{-t}$. How much of a 3-gram sample will be left in 6 years? $\frac{3}{64}$ g = 0.046875 g
18. **Compound Interest.** An initial deposit of \$1,000 earns 6% interest, compounded twice a year. How much will be in the account in one year? About \$1,060.90
19. a. Graph $f(x) = e^x$. Label the y-intercept and the asymptote of the graph. See AIE Appendix 3.
    b. Give the domain and range. D: $(-\infty, \infty)$; R: $(0, \infty)$
    c. What is the inverse of $f(x) = e^x$? $f^{-1}(x) = \ln x$
20. **Population Growth.** As of July 2010, the population of India was estimated to be 1,173,108,018, with an annual growth rate of 1.376%. If the growth rate remains the same, how large will the population be in July, 2020? Round to the nearest thousand. (Source: CIA World Fact Book) 1,346,161,000

21. **Biology.** Human growth hormone, known as HGH, is produced by the pituitary gland in the brain and released into the blood stream. It stimulates growth and cell production. After the age of 20, levels of HGH in the body decrease dramatically, as shown in the graph. Use the given function to approximate the amount of HGH produced per day by a person 55 years old. About 88 micrograms per day

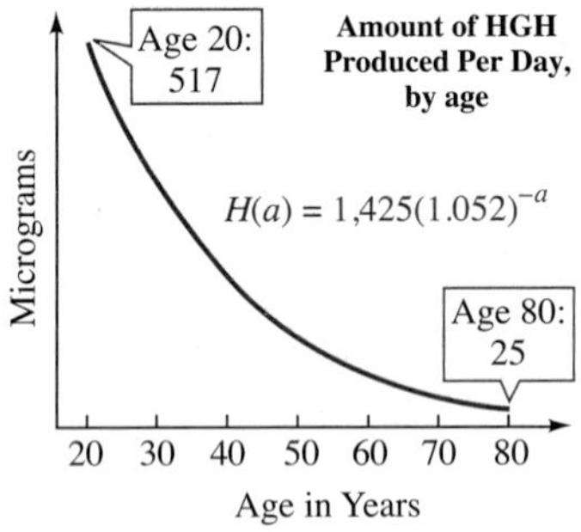

Source: Aging Well, James L.Holly, MD

22. Write the statement $\log_6 \frac{1}{36} = -2$ in exponential form. $6^{-2} = \frac{1}{36}$

23. **a.** What are the domain and range of the function $f(x) = \log x$? D: $(0, \infty)$; R: $(-\infty, \infty)$

b. What is the inverse of $f(x) = \log x$? $f^{-1}(x) = 10^x$

24. **Botany.** Which phrase best describes the relationship between the rate of photosynthesis in plants and light intensity that is graphed below: linear growth, exponential growth, or logarithmic growth? Logarithmic growth

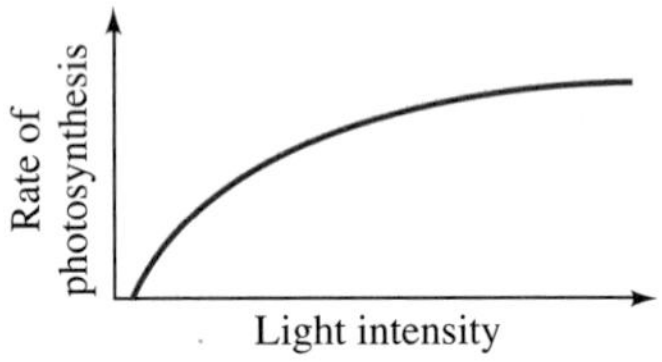

Evaluate each logarithmic expression, if possible.

25. $\log_5 25$ 2

26. $\log_9 \frac{1}{81}$ -2

27. $\log(-100)$ Undefined

28. $\ln \frac{1}{e^6}$ -6

29. $\log_4 2$ $\frac{1}{2}$

30. $\log_{1/3} 1$ 0

Solve for x.

31. $\log_x 32 = 5$ 2

32. $\log_8 x = \frac{4}{3}$ 16

33. $\log_3 x = -3$ $\frac{1}{27}$

34. $\ln x = 1$ e

Graph each function. See AIE Appendix 3.

35. $f(x) = -\log_3 x$

36. $f(x) = \ln x$

37. **Chemistry.** Find the pH of a solution with a hydrogen ion concentration of 3.7×10^{-7} gram-ions per liter. (*Hint:* $\text{pH} = -\log[\text{H}^+]$.) About 6.4

38. **Electronics.** Find the dB gain of an amplifier when $E_O = 60$ volts and $E_I = 0.3$ volt. *Hint:* dB gain $= 20 \log \frac{E_O}{E_I}$. About 46 dB

39. Use a calculator to find x to four decimal places: $\log x = -1.06$ 0.0871

40. Use the change-of-base formula to find $\log_7 3$ to four decimal places. 0.5646

41. Write the expression $\log_b a^2bc^3$ as the sum and/or difference of logarithms of a single quantity. Then simplify, if possible. $2 \log_b a + 1 + 3 \log_b c$

42. Write the expression $\frac{1}{2} \ln(a + 2) + \ln b - 3 \ln c$ as a logarithm of a single quantity. $\ln \frac{b\sqrt{a + 2}}{c^3}$

Solve each equation. Give the exact solution and an approximate solution to four decimal places, when appropriate.

43. $5^x = 3$ $\frac{\log 3}{\log 5} \approx 0.6826$

44. $3^{x-1} = 27$ 4

45. $\left(\frac{3}{2}\right)^{6x+2} = \frac{27}{8}$ $\frac{1}{6}$

46. $e^{0.08t} = 4$ $\frac{\ln 4}{0.08} \approx 17.3287$

47. $2 \log x = \log 25$ 5

48. $\log_2(x + 2) - \log_2(x - 5) = 3$ 6

49. $\ln(5x + 2) = \ln(2x + 5)$ 1

50. $\log x + \log(x - 9) = 1$ 10

51. The illustration shows the graphs of $y = \frac{1}{2} \ln(x - 1)$ and $y = \ln 2$ and the approximate coordinates of their point of intersection. Estimate the solution of the logarithmic equation $\frac{1}{2} \ln(x - 1) = \ln 2$. Then check the result. 5

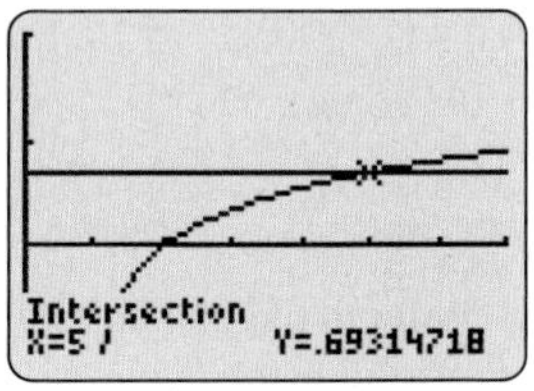

52. **Insects.** The number of insects attracted to a bright light is currently 5. If the number is expected to quadruple every 6 minutes, how long will it take for the number to reach 500? About 20 min

Group Project

The Number *e*

Overview: In this activity, you will use a calculator to find progressively more accurate approximations of *e*.

Instructions: Form groups of two students. Each student will need a scientific calculator.

Begin by finding an approximation of *e* using the e^x key on your calculator. Copy the table shown below, and write the number displayed on the calculator screen at the top of the table.

The value of *e* can be calculated to any degree of accuracy by adding the terms of the following pattern:

$$e = 1 + 1 + \frac{1}{2} + \frac{1}{2 \cdot 3} + \frac{1}{2 \cdot 3 \cdot 4} + \frac{1}{2 \cdot 3 \cdot 4 \cdot 5} + \cdots$$

The more terms that are added, the closer the sum will be to *e*.

You are to add as many terms as necessary until you obtain a sum that matches the value of *e*, given by the e^x key on your calculator. Work together as a team. One member of the group should compute the fractional form of the term to be added. (See the middle column of the table.) The other member should take that information and calculate the cumulative sum. (See the right column of the table.)

How many terms must be added so that the cumulative sum approximation and the e^x key approximation match in each decimal place?

Approximation of *e* found using the e^x key: $e \approx$ ____________________

Number of terms in the sum	Term (expressed as a fraction)	Cumulative sum (an approximation of *e*)
1	1	1
2	1	2
3	$\frac{1}{2}$	2.5
4	$\frac{1}{2 \cdot 3} = \frac{1}{6}$	2.666666667
⋮	⋮	⋮

More on Systems of Equations

12

from Campus to Careers

Fashion Designer

Fashion designers help create the billions of clothing articles, shoes, and accessories purchased every year by consumers. Fashion design relies heavily on mathematical skills, including knowledge of lines, angles, curves, and measurement. Designers also use mathematics in the manufacturing and marketing parts of the industry as they calculate labor costs and determine the markups and markdowns involved in retail pricing.

Problems 91 and **99** in **Study Set 12.1** and **problem 11** in **Study Set 12.3** involve situations that a fashion designer might encounter on the job. The mathematical concepts discussed in this chapter can be used to solve those problems.

JOB TITLE:
Fashion Designer
EDUCATION:
Employers usually seek designers with a 2-year or 4-year degree who are knowledgeable about textiles, fabrics, ornaments, and fashion trends.
JOB OUTLOOK:
The best opportunities will be in designing clothing sold in department stores and retail chains.
ANNUAL EARNINGS:
The average annual income is $74,100.
FOR MORE INFORMATION:
www.bls.gov/oco/ocos291.htm

Teaching Tip: This chapter begins with a review of Chapter 4: Systems of Linear Equations and Inequalities. Then there is more in-depth study of systems at the intermediate algebra level.

Study Skills Workshop ▶ Preparing for a Final Exam

Final exams can be stressful for many students because the number of topics to study can seem overwhelming. Here are some suggestions to help reduce the stress and prepare you for the test.

GET ORGANIZED: Gather all of your notes, study sheets, homework assignments, and especially all of your returned tests to review.

TALK WITH YOUR INSTRUCTOR: Ask your instructor to list the topics that may appear on the final and those that won't be covered.

MANAGE YOUR TIME: Adjust your daily schedule 1 week before the final so that it includes extended periods of study time.

Now Try This

1. Review your old tests. Make a list of the test problems that you are still unsure about and see a tutor or your instructor to get help.
2. Make a practice final exam that includes one or more of each type of problem that may appear on the test.
3. Make a detailed study plan. Determine when, where, and what you will study each day for 1 week before the final.

SECTION 12.1

OBJECTIVES

1. Determine whether an ordered pair is a solution of a system.
2. Solve systems of linear equations by graphing.
3. Use graphing to identify inconsistent systems and dependent equations.
4. Solve systems of linear equations by substitution.
5. Solve systems of linear equations by the elimination (addition) method.
6. Use substitution and elimination (addition) to identify inconsistent systems and dependent equations.
7. Solve application problems using systems of equations.

Solving Systems of Equations in Two Variables; Applications

ARE YOU READY? *Are You Ready? exercises available online at www.webassign.net/brookscole*

The following problems review some basic skills that are needed when solving systems of linear equations in two variables.

1. Is $(-1, 4)$ a solution of $y = 3x - 1$? Not a solution
2. Use the slope and the y-intercept to graph $y = -4x + 2$. See AIE Appendix 3.
3. Graph $3x + 4y = 12$ by finding the x- and y-intercepts. See AIE Appendix 3.
4. Solve $5x - y = -4$ for y. $y = 5x + 4$
5. Substitute 4 for x in $y = -2x - 1$ and find y. -9
6. Multiply both sides of the equation $7x - y = 9$ by -4. $-28x + 4y = -36$

In this section, we will review graphical and algebraic methods for solving systems of two linear equations in two variables.

1 Determine Whether an Ordered Pair Is a Solution of a System.

When two equations with the same variables are considered simultaneously (at the same time), we say that they form a **system of equations.** We will use a left brace { when writing a system of equations. An example is

$$\begin{cases} 2x + 5y = -1 \\ x - y = -4 \end{cases}$$

Read as "the system of equations $2x + 5y = -1$ and $x - y = -4$."

A **solution of a system** of equations in two variables is an ordered pair that satisfies both equations of the system.

EXAMPLE 1 Determine whether $(-3, 1)$ is a solution of each system of equations.

a. $\begin{cases} 2x + 5y = -1 \\ x - y = -4 \end{cases}$ **b.** $\begin{cases} 5y = 2 - x \\ y = 3x \end{cases}$

Teaching Tip: If the desks in your classroom are arranged in rows and columns, you can use that configuration to introduce this section.
1. Have all the students in a given row raise their right hands.
2. Have all the students in a given column raise their left hands.
3. Ask: "Who raised both hands?"

Strategy We will substitute the x- and y-coordinates of $(-3, 1)$ for the corresponding variables in both equations of the system.

Why If both equations are satisfied (made true) by the x- and y-coordinates, the ordered pair is a solution of the system.

Solution

The Language of Algebra

A system of equations is two (or more) equations that are considered **simultaneously**—at the same time. On June 3, 2007, in Kansas City, more than 1,680 guitarists set a world record for the most people playing the same song *simultaneously.* The song was Deep Purple's *Smoke on the Water.*

a. To determine whether $(-3, 1)$ is a solution, we substitute -3 for x and 1 for y in each equation.

Check:

$$\begin{aligned} 2x + 5y &= -1 && \text{First equation} \\ 2(-3) + 5(1) &\stackrel{?}{=} -1 \\ -6 + 5 &\stackrel{?}{=} -1 \\ -1 &= -1 && \text{True} \end{aligned}$$

$$\begin{aligned} x - y &= -4 && \text{Second equation} \\ -3 - 1 &\stackrel{?}{=} -4 \\ -4 &= -4 && \text{True} \end{aligned}$$

Since $(-3, 1)$ satisfies both equations, it is a solution of the system.

b. We substitute -3 for x and 1 for y in each equation in the second system.

Check:

$$\begin{aligned} 5y &= 2 - x && \text{First equation} \\ 5(1) &\stackrel{?}{=} 2 - (-3) \\ 5 &\stackrel{?}{=} 2 + 3 \\ 5 &= 5 && \text{True} \end{aligned}$$

$$\begin{aligned} y &= 3x && \text{Second equation} \\ 1 &\stackrel{?}{=} 3(-3) \\ 1 &= -9 && \text{False} \end{aligned}$$

The Language of Algebra

We say that $(-3, 1)$ **satisfies** both equations because it makes both equations true. To *satisfy* means to make content, as in *satisfy* your thirst or a *satisfied* customer.

Although $(-3, 1)$ satisfies the first equation, it does not satisfy the second. Because it does not satisfy both equations, $(-3, 1)$ is *not a solution* of the system.

Self Check 1 Determine whether $(6, -2)$ is a solution of $\begin{cases} x - 2y = 10 \\ y = 3x - 20 \end{cases}$. It is a solution.

Now Try Problem 13

Teaching Tip: Use this analogy: Both the bride and groom must say "I do" during a wedding ceremony or there is no marriage. Similarly, both ordered pairs must satisfy the equations.

Teaching Example 1

Determine whether $(4, -3)$ is a solution of the system: $\begin{cases} 3x - 2y = 18 \\ 4y = 3x \end{cases}$

Answer: Not a solution

2 Solve Systems of Linear Equations by Graphing.

To **solve a system** of equations means to find all of the solutions of the system. One way to solve a system of linear equations in two variables is to graph each equation and find where the graphs intersect.

The Graphing Method

1. Carefully graph each equation on the same rectangular coordinate system.
2. If the lines intersect, determine the coordinates of the point of intersection of the graphs. That ordered pair is the solution of the system.
3. If the graphs have no point in common, the system has no solution.
4. Check the proposed solution in each equation of the original system.

A system of two linear equations can have exactly one solution, no solution, or infinitely many solutions. When a system of equations (as in Example 2) has at least one solution, the system is called a **consistent system.**

EXAMPLE 2 Solve the system by graphing: $\begin{cases} x + 2y = 4 \\ 2x - y = 3 \end{cases}$

Strategy We will graph both equations on the same coordinate system.

Why The graph of a linear equation is a picture of its solutions. If both equations are graphed on the same coordinate system, we can see whether they have any common solutions.

Solution The intercept method is a convenient way to graph equations such as $x + 2y = 4$ and $2x - y = 3$, because they are in standard $Ax + By = C$ form.

Success Tip

Since accuracy is crucial when using the graphing method to solve a system:

- Use graph paper.
- Use a sharp pencil.
- Use a straightedge.

$x + 2y = 4$

x	y	(x, y)
4	0	(4, 0)
0	2	(0, 2)
−2	3	(−2, 3)

$2x - y = 3$

x	y	(x, y)
$\frac{3}{2}$	0	$\left(\frac{3}{2}, 0\right)$
0	−3	(0, −3)
−1	−5	(−1, −5)

Success Tip

When determining the coordinates of a point of intersection from a graph, realize that they are simply estimates. Only after algebraically checking a proposed solution can we be sure that it is an actual solution.

Although infinitely many ordered pairs (x, y) satisfy $x + 2y = 4$, and infinitely many ordered pairs (x, y) satisfy $2x - y = 3$, only the coordinates of the point where the graphs intersect satisfy both equations. From the graph, it appears that the intersection point has coordinates (2, 1). To verify that it is the solution, we substitute 2 for x and 1 for y in both equations and show that (2, 1) satisfies each one.

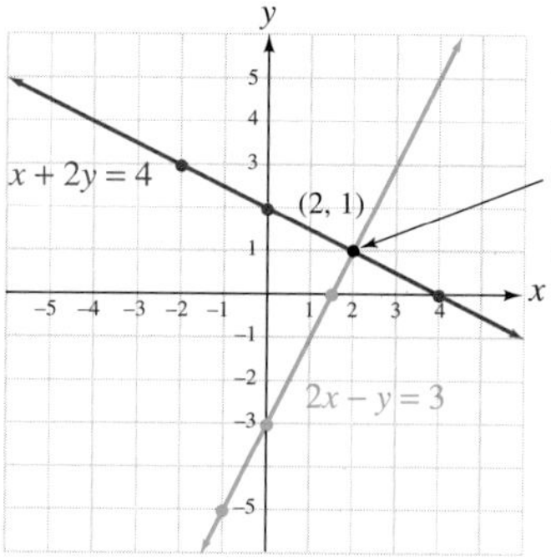

The point of intersection gives the solution of the system.

Check:

$$\begin{aligned} x + 2y &= 4 && \text{First equation} \\ 2 + 2(\mathbf{1}) &\stackrel{?}{=} 4 \\ 2 + 2 &\stackrel{?}{=} 4 \\ 4 &= 4 && \text{True} \end{aligned}$$

$$\begin{aligned} 2x - y &= 3 && \text{Second equation} \\ 2(\mathbf{2}) - \mathbf{1} &\stackrel{?}{=} 3 \\ 4 - 1 &\stackrel{?}{=} 3 \\ 3 &= 3 && \text{True} \end{aligned}$$

Since (2, 1) makes both equations true, it is the solution of the system. The solution set is $\{(2, 1)\}$.

Teaching Example 2
Solve the system by graphing:
$\begin{cases} x - 3y = 5 \\ 2x + y = 3 \end{cases}$
Answer: (2, −1)

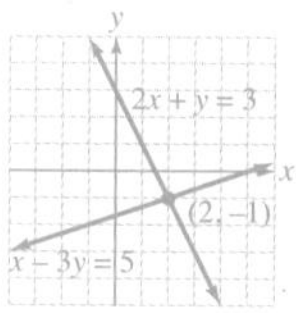

Self Check 2 Solve the system by graphing: $\begin{cases} x - 3y = -5 \\ 2x + y = 4 \end{cases}$ (1, 2); see AIE Appendix 3.

Now Try Problem 21

3 Use Graphing to Identify Inconsistent Systems and Dependent Equations.

When a system has no solution (as in Example 3), it is called an **inconsistent system.**

EXAMPLE 3 Solve the system $\begin{cases} 2x + 3y = 6 \\ 4x + 6y = 24 \end{cases}$ by graphing, if possible.

Strategy We will graph both equations on the same coordinate system.

Why If both equations are graphed on the same coordinate system, we can see whether they have any common solutions.

Solution Using the intercept method, we graph both equations on one set of coordinate axes, as shown on the right.

$2x + 3y = 6$

x	y	(x, y)
3	0	(3, 0)
0	2	(0, 2)
−3	4	(−3, 4)

$4x + 6y = 24$

x	y	(x, y)
6	0	(6, 0)
0	4	(0, 4)
−3	6	(−3, 6)

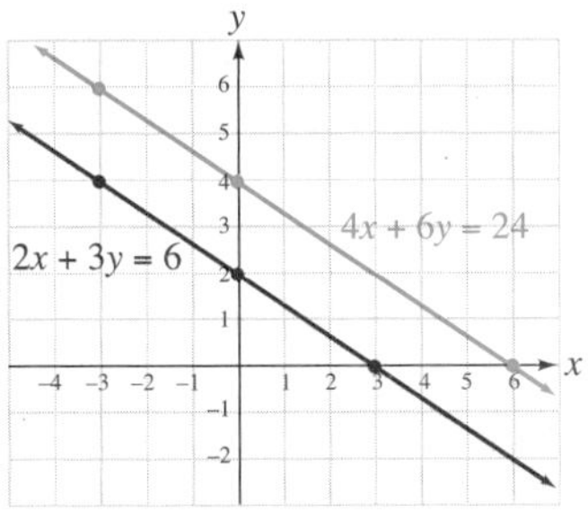

Parallel lines — no solution

Teaching Tip: Ask your students to describe the graph of the system:
1. if the lines have the same slope.
2. if the lines have different slopes.

Caution

A common error is to graph the parallel lines, but forget to answer with the words

no solution

In this example, the graphs are parallel, because the slopes of the two lines are equal and they have different y-intercepts. We can see that the slope of each line is $-\frac{2}{3}$ by writing the equations in slope–intercept form. To do that, we solve each for y.

$$2x + 3y = 6 \quad \text{First equation}$$
$$3y = -2x + 6$$
$$y = -\frac{2}{3}x + 2 \quad \text{Divide both sides by 3 and simplify.}$$

$$4x + 6y = 24 \quad \text{Second equation}$$
$$6y = -4x + 24$$
$$y = -\frac{2}{3}x + 4 \quad \text{Divide both sides by 6 and simplify.}$$

Teaching Example 3
Solve the system $\begin{cases} 2y - x = 8 \\ x - 2y = 2 \end{cases}$ by graphing, if possible.
Answer: No solutions; $\varnothing$

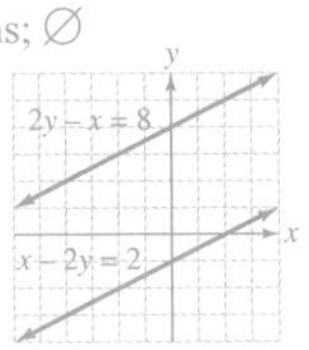

Because the lines are parallel, there is no point of intersection. Such a system has *no solution* and it is called an **inconsistent system.** The solution set is the empty set, which is written $\varnothing$.

Self Check 3 Solve the system $\begin{cases} 3y - 2x = 6 \\ 2x - 3y = 6 \end{cases}$ by graphing, if possible.
No solution; $\varnothing$; see AIE Appendix 3.

Now Try Problem 23

When the equations of a system have different graphs (as in Examples 2 and 3), the equations are called **independent equations.**

EXAMPLE 4 Solve the system by graphing: $\begin{cases} y = \frac{1}{2}x + 2 \\ 2x + 8 = 4y \end{cases}$

Strategy We will graph both equations on the same coordinate system.

Why If both equations are graphed on the same coordinate system, we can see whether they have any common solutions.

Solution We graph each equation on one set of coordinate axes, as shown below.

The Language of Algebra

Here the graphs of the lines **coincide.** That is, they occupy the same location. To illustrate this concept, think of a clock. At noon and midnight, the hands of the clock *coincide.*

Graph using the slope and y-intercept.

$$y = \frac{1}{2}x + 2$$

$m = \frac{1}{2}$

y-intercept: (0, 2)

Graph using the intercept method.

$2x + 8 = 4y$

x	y	(x, y)
−4	0	(−4, 0)
0	2	(0, 2)
2	3	(2, 3)

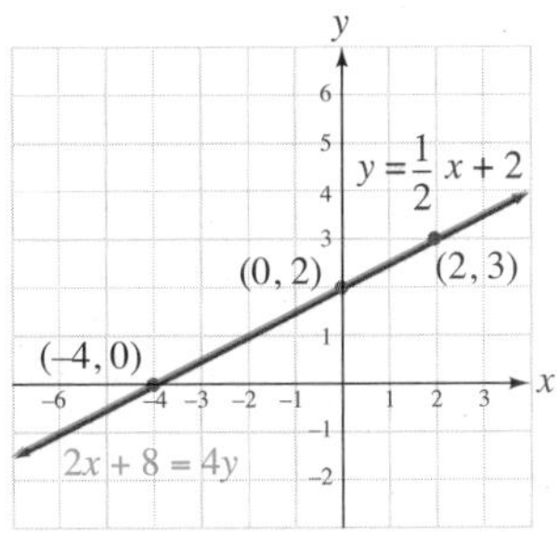

The same line — infinitely many solutions

The graphs appear to be identical. We can verify this by writing the second equation in slope–intercept form and observing that it is the same as the first equation.

$y = \frac{1}{2}x + 2$ First equation

$2x + 8 = 4y$ Second equation

$\frac{2x}{4} + \frac{8}{4} = \frac{4y}{4}$ Divide both sides by 4.

$\frac{1}{2}x + 2 = y$

We see that the equations of the system are equivalent. Because $y = \frac{1}{2}x + 2$ and $2x + 8 = 4y$ are different forms of the same equation, they are called **dependent equations.**

Since the graphs are the same line, they have infinitely many points in common. All of the points that lie on the common line are solutions because the coordinates of each of those points satisfy both equations of the system. In cases like this, we say that there are *infinitely many solutions.* The solution set can be written using **set-builder notation** as

$\left\{(x, y) \mid y = \frac{1}{2}x + 2\right\}$ Read as, "the set of all ordered pairs (x, y), such that $y = \frac{1}{2}x + 2$."

Teaching Tip: Remind your students that equations with the same solutions are called *equivalent equations.*

We also can express the solution set using the second equation of the system in the set-builder notation: $\{(x, y) \mid 2x + 8 = 4y\}$.

Some instructors prefer that the set-builder notation use an equation in standard form with coefficients that are integers having no common factor other than 1. Such an equation that is equivalent to $y = \frac{1}{2}x + 2$ and $2x + 8 = 4y$ is $x - 2y = -4$. The set-builder notation solution for this example could, therefore, be written as $\{(x, y) \mid x - 2y = -4\}$.

From the graph, it appears that three of the infinitely many solutions are $(-4, 0)$, $(0, 2)$, and $(2, 3)$. Check each of them to verify that both equations of the system are satisfied.

Caution

A common error is to graph the identical lines, but forget to answer with the words *infinitely many solutions* and the set-builder notation.

Teaching Example 4

Solve the system by graphing:
$\begin{cases} 2x - 5y = 10 \\ y = \frac{2}{5}x - 2 \end{cases}$

Answer: $\{(x, y) \mid 2x - 5y = 10\}$ or $\{(x, y) \mid y = \frac{2}{5}x - 2\}$; infinitely many solutions

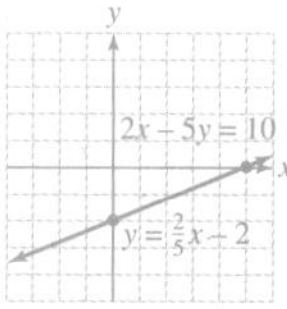

Self Check 4 Solve the system by graphing: $\begin{cases} 2x - y = 4 \\ y = 2x - 4 \end{cases}$

$\{(x, y) \mid 2x - y = 4\}$; infinitely many solutions; see AIE Appendix 3.

Now Try Problem 25

We now summarize the possibilities that can occur when two linear equations, each with two variables, are graphed.

Solving a System of Equations by the Graphing Method

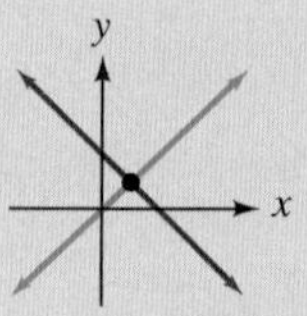

If the lines are different and intersect, the equations are independent, and the system is consistent. **One solution exists.** It is the point of intersection.

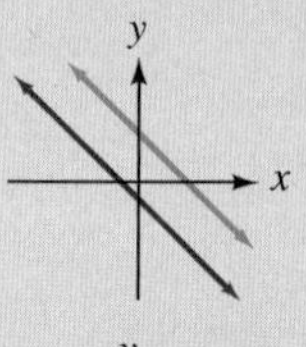

If the lines are different and parallel, the equations are independent, and the system is inconsistent. **No solution exists.**

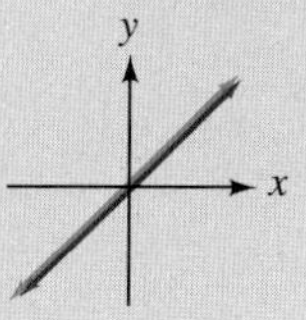

If the lines are identical, the equations are dependent, and the system is consistent. **Infinitely many solutions exist.** Any point on the line is a solution.

If each equation in one system is equivalent to a corresponding equation in another system, the systems are called **equivalent systems.**

EXAMPLE 5 Solve the system by graphing: $\begin{cases} \frac{3}{2}x - y = \frac{5}{2} \\ \frac{1}{8}y = 1 - \frac{x}{4} \end{cases}$

Strategy We will use the multiplication property of equality to clear both equations of fractions and solve the resulting equivalent system by graphing.

Why It is usually easier to solve systems of equations that do not contain fractions.

Solution We multiply both sides of $\frac{3}{2}x - y = \frac{5}{2}$ by 2 to eliminate the fractions and obtain the equation $3x - 2y = 5$. We multiply both sides of $\frac{1}{8}y = 1 - \frac{x}{4}$ by 8 to eliminate the fractions and obtain the equation $y = 8 - 2x$.

The original system ***An equivalent system***

$$\begin{cases} \frac{3}{2}x - y = \frac{5}{2} \xrightarrow{\text{Multiply by 2}} 2\left(\frac{3}{2}x - y\right) = 2\left(\frac{5}{2}\right) \xrightarrow{\text{Simplify}} 3x - 2y = 5 \\ \frac{1}{8}y = 1 - \frac{x}{4} \xrightarrow[\text{Multiply by 8}]{} 8\left(\frac{1}{8}y\right) = 8\left(1 - \frac{x}{4}\right) \xrightarrow[\text{Simplify}]{} y = 8 - 2x \end{cases}$$

Since the new system is equivalent to the original system, they have the same solution. If we graph the equations of the new system, it appears that the point where the lines intersect is (3, 2).

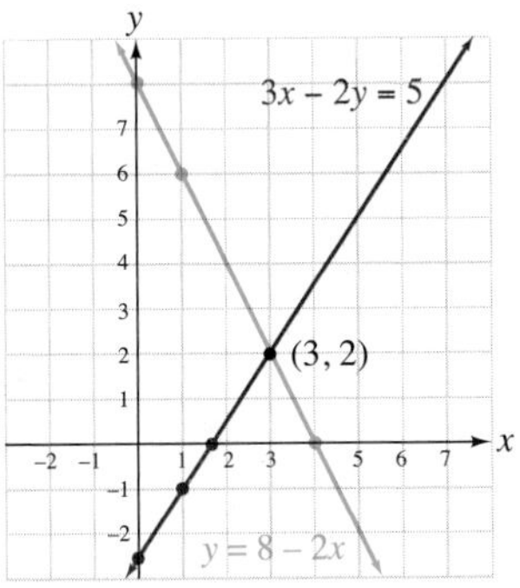

Graph using the intercept method.

$3x - 2y = 5$

x	y	(x, y)
$\frac{5}{3}$	0	$\left(\frac{5}{3}, 0\right)$
0	$-\frac{5}{2}$	$\left(0, -\frac{5}{2}\right)$
1	-1	$(1, -1)$

Graph using the slope and y-intercept.

$y = 8 - 2x$

Slope = −2

y-intercept: (0, 8)

Caution

When checking the solution of a system of equations, always substitute the values of the variables into the original equations.

To verify that (3, 2) is the solution, we substitute 3 for x and 2 for y in each equation of the original system.

Check:

$\frac{3}{2}x - y = \frac{5}{2}$ First equation

$\frac{3}{2}(3) - 2 \stackrel{?}{=} \frac{5}{2}$

$\frac{9}{2} - \frac{4}{2} \stackrel{?}{=} \frac{5}{2}$

$\frac{5}{2} = \frac{5}{2}$ True

$\frac{1}{8}y = 1 - \frac{x}{4}$ Second equation

$\frac{1}{8}(2) \stackrel{?}{=} 1 - \frac{3}{4}$

$\frac{2}{8} \stackrel{?}{=} \frac{1}{4}$

$\frac{1}{4} = \frac{1}{4}$ True

Teaching Example 5
Solve the system by graphing:
$\begin{cases} x = -\frac{2}{3}y \\ \frac{x}{2} - \frac{y}{6} = -\frac{3}{2} \end{cases}$

Answer: (−2, 3)

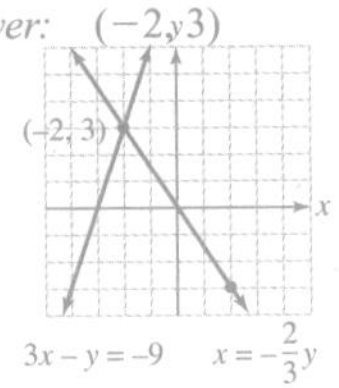

Self Check 5 Solve the system by graphing: $\begin{cases} \frac{1}{2}x + \frac{1}{2}y = -1 \\ \frac{1}{3}x - \frac{1}{2}y = -4 \end{cases}$

(−6, 4); see AIE Appendix 3.

Now Try ▶ Problem 29

Using Your Calculator ▶ Solving Systems by Graphing

The TRACE and INTERSECT features found on most graphing calculators enable us to get very good approximations of solutions of systems of two linear equations. To illustrate this, consider the following system. To graph the equations, we must first solve each equation for y.

$$\text{To solve the system } \begin{cases} 3x + 2y = 12 \\ 2x - 3y = 12 \end{cases} \quad \begin{matrix} \xrightarrow{\text{Solve for } y} \\ \xrightarrow{\text{Solve for } y} \end{matrix} \quad \begin{cases} y = -\frac{3}{2}x + 6 \\ y = \frac{2}{3}x - 4 \end{cases}$$

Next, we press [Y =] and enter the right side of each equation of the equivalent system after the symbols Y_1 and Y_2, as shown in figure (a) below. Then we press [GRAPH]. If we use window settings of $[-10, 10]$ for x and for y, the graphs of the equations will look like those in figure (b). If we zoom in on the intersection point of the two lines and trace, we will get an approximate solution like the one shown in figure (c). To get better results, we can do more zooms. We would then find that, to the nearest hundredth, the solution is $(4.63, -0.94)$. Verify that this is reasonable.

A more efficient method for finding the intersection point of two lines uses the INTERSECT feature. With this feature, the cursor automatically highlights the intersection point, and the x- and y-coordinates are displayed. To locate INTERSECT, press [2nd], [CALC], 5, followed by [ENTER]. The result is a graph similar to figure (d). The display shows the approximate coordinates of the point of intersection.

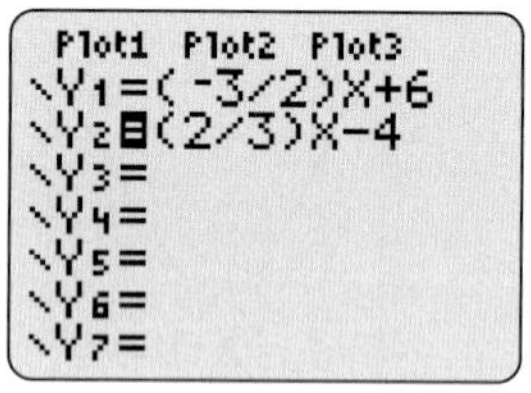

(a)

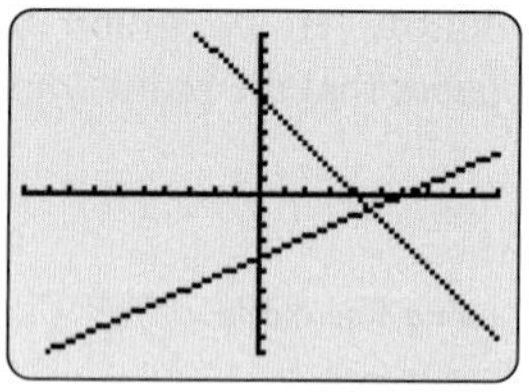
(b)

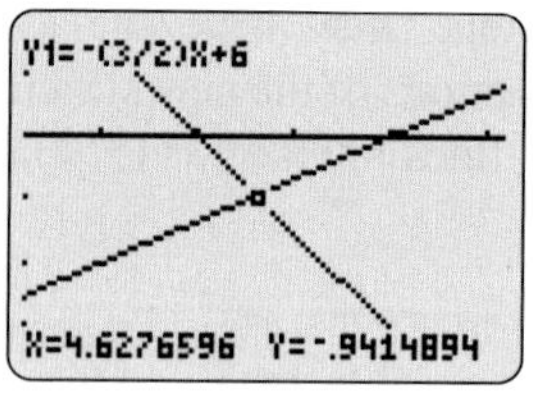

(c)

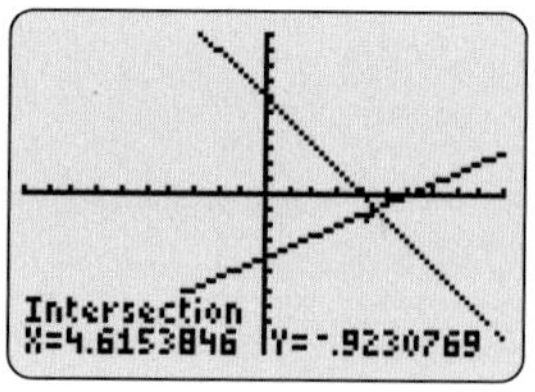

(d)

4 Solve Systems of Linear Equations by Substitution.

The graphing method enables us to visualize the process of solving systems of equations. However, it can be difficult to determine the exact coordinates of the point of intersection. We now review an algebraic method that we can use to find the exact solutions of systems of equations.

The **substitution method** works well for solving systems where one equation is solved, or can be easily solved, for one of the variables. To solve a system of two linear equations in x and y by the substitution method, we can follow these steps.

The Substitution Method

1. Solve one of the equations for either x or y—preferably a variable with a coefficient of 1 or -1. If this is already done, go to step 2. (We call this equation the **substitution equation.**)
2. Substitute the expression for x or for y obtained in step 1 into the other equation and solve that equation.
3. Substitute the value of the variable found in step 2 into the substitution equation to find the value of the remaining variable.
4. Check the proposed solution in each equation of the original system. Write the solution as an ordered pair.

EXAMPLE 6

Solve the system by substitution: $\begin{cases} 4x + y = 13 \\ -2x + 3y = -17 \end{cases}$

Teaching Tip: Before beginning the solution, have the students tell you the coefficients of each of the variable terms in both equations.

Strategy We will use the substitution method. Since the system does not contain an equation solved for x or y, we must choose an equation and solve it for x or y. It is easiest to solve for y in the first equation, because y has a coefficient of 1.

Why Solving $4x + y = 13$ for x or solving $-2x + 3y = -17$ for x or y would involve working with cumbersome fractions.

Solution

The Language of Algebra

Since substitution involves algebra and not graphing, it is called an **algebraic** method for solving a system.

Step 1: We solve the first equation for y, because y has a coefficient of 1.

$$4x + y = 13$$

$$4x + y - \mathbf{4x} = \mathbf{-4x} + 13 \quad \text{To isolate y, subtract 4x from both sides.}$$

$$y = -4x + 13 \quad \text{This is the substitution equation. It could also be written: } y = 13 - 4x.$$

Because y and $-4x + 13$ are equal, we can substitute $-4x + 13$ for y in the second equation of the system.

$$y = \mathbf{-4x + 13} \qquad -2x + 3\mathbf{y} = -17$$

Success Tip

Throughout the course, we have been substituting numbers for variables. With this method, we substitute a *variable expression for a variable.* The objective is to use an appropriate substitution to obtain *one* equation in *one* variable.

Step 2: We then substitute $-4x + 13$ for y in the second equation to eliminate the variable y from that equation. The result will be an equation containing only one variable, x.

$$-2x + 3y = -17 \quad \text{This is the second equation of the system.}$$

$$-2x + 3(\mathbf{-4x + 13}) = -17 \quad \text{Substitute } -4x + 13 \text{ for y. Write the parentheses so that the multiplication by 3 is distributed over both terms of } -4x + 13.$$

$$-2x - 12x + 39 = -17 \quad \text{Distribute the multiplication by 3.}$$

$$-14x + 39 = -17 \quad \text{Combine like terms.}$$

$$-14x = -56 \quad \text{Subtract 39 from both sides.}$$

$$x = 4 \quad \text{To solve for x, divide both sides by } -14. \text{ This is the x-value of the solution.}$$

The Language of Algebra

The phrase **back-substitute** can also be used to describe step 3 of the substitution method. To find y, we *back-substitute* 4 for x in the equation $y = -4x + 13$.

Step 3: To find y, we substitute 4 for x in the substitution equation and evaluate the right side.

$$y = -4x + 13 \quad \text{This is the substitution equation.}$$

$$y = -4(\mathbf{4}) + 13$$

$$y = -16 + 13 \quad \text{Multiply: } -4(4) = -16.$$

$$y = -3 \quad \text{This is the y-value of the solution.}$$

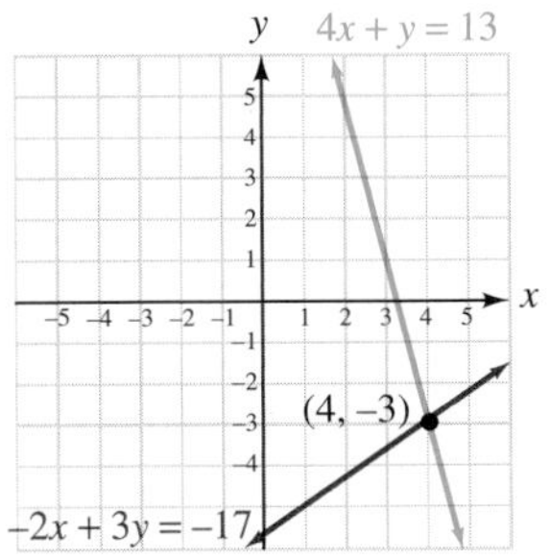

Step 4: To verify that $(4, -3)$ satisfies both equations, we substitute 4 for x and -3 for y into each equation of the original system and simplify.

Check:

$$4x + y = 13 \quad \text{First equation}$$
$$4(\mathbf{4}) + (\mathbf{-3}) \stackrel{?}{=} 13$$
$$16 - 3 \stackrel{?}{=} 13$$
$$13 = 13 \quad \text{True}$$

$$-2x + 3y = -17 \quad \text{Second equation}$$
$$-2(\mathbf{4}) + 3(\mathbf{-3}) \stackrel{?}{=} -17$$
$$-8 - 9 \stackrel{?}{=} -17$$
$$-17 = -17 \quad \text{True}$$

Since $(4, -3)$ satisfies both equations of the system, it is the solution of the system. The solution set is $\{(4, -3)\}$. The graphs of the equations of the system help to verify this—they appear to intersect at $(4, -3)$, as shown on the left.

Teaching Example 6
Solve the system by substitution:
$\begin{cases} 2x + 5y = -3 \\ 3x - y = 4 \end{cases}$
Answer: $(1, -1)$

Self Check 6 Solve the system by substitution: $\begin{cases} x + 3y = 9 \\ 2x - y = -10 \end{cases}$ $(-3, 4)$

Now Try Problem 35

5 Solve Systems of Linear Equations by the Elimination (Addition) Method.

Recall that with the **elimination (addition) method,** we combine the equations of the system in a way that will eliminate terms involving one of the variables.

The Elimination (Addition) Method

1. Write both equations of the system in standard form: $Ax + By = C$.
2. If necessary, multiply one or both of the equations by a nonzero number chosen to make the coefficients of x (or the coefficients of y) opposites.
3. Add the equations to eliminate the terms involving x (or y).
4. Solve the equation resulting from step 3.
5. Find the value of the remaining variable by substituting the solution found in step 4 into any equation containing both variables. Or, repeat steps 2–4 to eliminate the other variable.
6. Check the proposed solution in each equation of the original system. Write the solution as an ordered pair.

EXAMPLE 7 Solve: $\begin{cases} \frac{4}{3}x + \frac{1}{2}y = -\frac{2}{3} \\ 0.3x + 0.4y = 1 \end{cases}$

Strategy We will find an equivalent system without fractions or decimals and use the elimination method to solve it.

Why It's usually easier to solve a system of equations that involves only integers.

Solution *Step 1:* To clear the first equation of the fractions, we multiply both sides by 6. To clear the second equation of decimals, we multiply both sides by 10.

Teaching Tip: Add the equations to show that neither variable gets eliminated:

$$\begin{array}{r} 8x + 3y = -4 \\ 3x + 4y = 10 \\ \hline 11x + 7y = 6 \end{array}$$

$$\begin{cases} \frac{4}{3}x + \frac{1}{2}y = -\frac{2}{3} \\ 0.3x + 0.4y = 1 \end{cases} \begin{array}{c} \xrightarrow{\text{Multiply by 6}} \\ \xrightarrow[\text{Multiply by 10}]{} \end{array} \begin{array}{c} 6\left(\frac{4}{3}x + \frac{1}{2}y\right) = 6\left(-\frac{2}{3}\right) \\ \mathbf{10}(0.3x + 0.4y) = \mathbf{10}(1) \end{array} \begin{array}{c} \xrightarrow{\text{Simplify}} \\ \xrightarrow[\text{Simplify}]{} \end{array} \begin{cases} 8x + 3y = -4 \\ 3x + 4y = 10 \end{cases}$$

Step 2: To make the y-terms drop out when we add the equations, we multiply both sides of $8x + 3y = -4$ by 4 and both sides of $3x + 4y = 10$ by -3 to get

$$\begin{cases} 8x + 3y = -4 \\ 3x + 4y = 10 \end{cases} \begin{array}{c} \xrightarrow{\text{Multiply by 4}} \\ \xrightarrow[\text{Multiply by } -3]{} \end{array} \begin{cases} 32x + 12y = -16 \\ -9x - 12y = -30 \end{cases}$$

Success Tip

The basic objective of the elimination method is to obtain two equations whose sum will be *one* equation in *one* variable.

Step 3: When these equations are added, the y-terms drop out.

$$\begin{array}{rcl} 32x + \mathbf{12y} &=& -16 \\ -9x - \mathbf{12y} &=& -30 \\ \hline 23x \qquad &=& -46 \end{array}$$

Add the like terms, column by column: $32x + (-9x) = 23x$, $12y + (-12y) = 0$, and $-16 + (-30) = -46$.

Step 4: We solve the resulting equation to find x.

$$23x = -46$$

$$x = -2$$ To solve for x, divide both sides by 23. This is the x-value of the solution.

Teaching Tip: As you explain, purposefully make an error. Multiply only the left sides of the equations by the required numbers. See if any students catch the mistake.

Step 5: To find y, we can substitute -2 for x in either of the equations of the original system or either of the equations of the equivalent system. It appears the calculations will be the simplest if we use $3x + 4y = 10$.

$$3x + 4y = 10 \quad \text{This is the second equation of the equivalent system.}$$
$$3(-2) + 4y = 10 \quad \text{Substitute } -2 \text{ for } x.$$
$$-6 + 4y = 10 \quad \text{Simplify.}$$
$$4y = 16 \quad \text{To isolate the variable term, add 6 to both sides.}$$
$$y = 4 \quad \text{To solve for } y \text{, divide both sides by 4. This is the } y\text{-value of the solution.}$$

Step 6: The solution is $(-2, 4)$ and the solution set is $\{(-2, 4)\}$. Verify that the solution checks using the original equations.

Teaching Example 7
Solve the system:
$$\begin{cases} \frac{3}{16}x + \frac{1}{4}y = -\frac{3}{2} \\ 0.5x + 7.2 = -1.2y \end{cases}$$
Answer: $(0, -6)$

Self Check 7 Solve the system: $\begin{cases} \frac{2}{3}x - \frac{2}{5}y = 10 \\ 0.3x + 0.4y = -4.2 \end{cases}$ $(6, -15)$

Now Try ▶ Problem 49

6 Use Substitution and Elimination (Addition) to Identify Inconsistent Systems and Dependent Equations.

We have solved inconsistent systems and systems of dependent equations by graphing. We also can solve these systems using the substitution and elimination methods.

EXAMPLE 8 Solve the system: $\begin{cases} y = 2x + 4 \\ 8x - 4y = 7 \end{cases}$

Strategy We will use the substitution method to solve this system.

Why The substitution method works well when one of the equations of the system (in this case, $y = 2x + 4$) is solved for a variable.

Solution Since the first equation is solved for y, we will use the substitution method.

$$y = \mathbf{2x + 4} \quad \text{This is the substitution equation.}$$
$$8x - 4y = 7 \quad \text{This is the second equation of the system.}$$
$$8x - 4(\mathbf{2x + 4}) = 7 \quad \text{Substitute } 2x + 4 \text{ for } y.$$

Now we can try to solve this equation for x:

$$8x - 8x - 16 = 7 \quad \text{Distribute the multiplication by } -4.$$
$$-16 = 7 \quad \text{Simplify the left side: } 8x - 8x = 0.$$

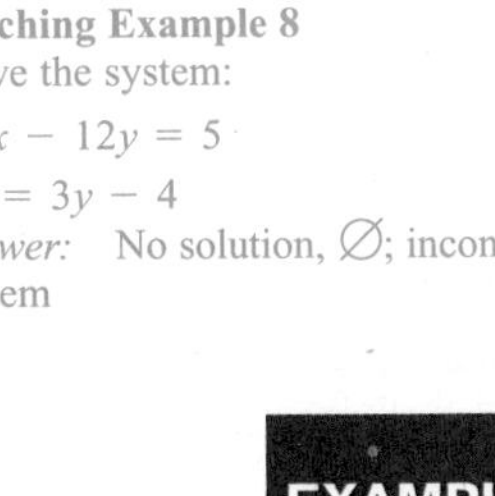

Here, the terms involving x drop out, and we get $-16 = 7$. This false statement indicates that the system has *no solution* and is, therefore, inconsistent. The solution set is $\varnothing$. The graphs of the equations of the system help to verify this—they appear to be parallel lines, as shown on the left.

Teaching Example 8
Solve the system:
$$\begin{cases} 4x - 12y = 5 \\ x = 3y - 4 \end{cases}$$
Answer: No solution, $\varnothing$; inconsistent system

Self Check 8 Solve the system: $\begin{cases} x = -2.5y + 8 \\ y = -0.4x + 2 \end{cases}$

Now Try ▶ Problem 51 No solution, $\varnothing$; inconsistent system

EXAMPLE 9 Solve: $\begin{cases} 2(2x + 3y) = 12 \\ -2x = 3y - 6 \end{cases}$

Strategy We will write each equation in standard (general) form $Ax + By = C$ and use the elimination (addition) method to solve the resulting equivalent system.

Why Since no variable has a coefficient of 1 or -1, it would be difficult to solve this system using substitution.

Solution To write the first equation in standard (general) form, we use the distributive property. To write the second equation in standard form, we subtract $3y$ from both sides.

The first equation	***The second equation***
$2(2x + 3y) = 12$	$-2x = 3y - 6$
$4x + 6y = 12$	$-2x - 3y = -6$

We now copy $4x + 6y = 12$ and multiply both sides of $-2x - 3y = -6$ by 2 to get

$$\begin{array}{r} 4x + 6y = \quad 12 \\ \underline{-4x - 6y = -12} \\ 0 = 0 \end{array}$$

When we add like terms, column by column, the result is $0x + 0y = 0$, which simplifies to $0 = 0$.

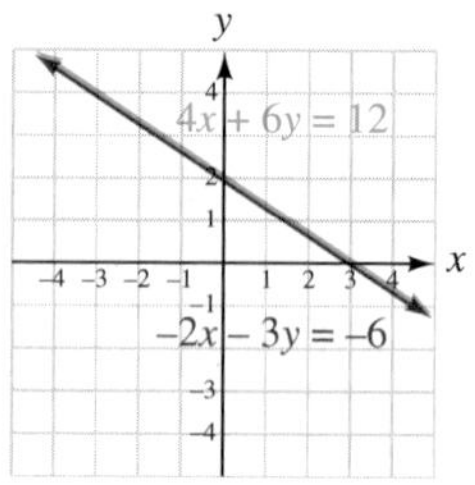

Here, both the x- and y-terms drop out. The resulting true statement $0 = 0$ indicates that the equations are dependent and that the system has an *infinitely many solutions.* The solution set is written using set-builder notation as $\{(x, y) \mid 4x + 6y = 12\}$ and is read as "the set of all ordered pairs (x, y) such that $4x + 6y = 12$."

Note that the equations of the system are dependent equations, because when the second equation is multiplied by -2, it becomes the first equation. The graphs of these equations are, therefore, the same line. To find some of the infinitely many solutions of the system, we can substitute 0, 3, and -3 for x in either equation to obtain (0, 2), (3, 0), and $(-3, 4)$.

Teaching Example 9 Solve the system:
$\begin{cases} y = 7x - 2 \\ 2(7x - y) = 4 \end{cases}$
Answer: $\{(x, y) \mid 7x - y = 2\}$, infinitely many solutions; dependent equations

Self Check 9 Solve the system: $\begin{cases} 2x = 5(y + 2) \\ -4x + 10y = -20 \end{cases}$

Infinitely many solutions, $\{(x, y) \mid 2x - 5y = 10\}$; dependent equations

Now Try Problem 53

Examples 8 and 9 illustrate the following facts.

Inconsistent Systems and Dependent Equations

When solving a system of two linear equations in two variables using substitution or elimination (addition):

1. If the variables drop out and a true statement (identity) is obtained, the system has an infinite number of solutions. The equations are dependent and the system is consistent.
2. If the variables drop out and a false statement (contradiction) is obtained, the system has no solution and is inconsistent.

7 Solve Application Problems Using Systems of Equations.

We have solved applied problems involving two unknown quantities by modeling the situation with an equation in one variable. It's often easier to solve such problems using a two-variable approach. We write two equations in two variables to model the situation, and then we use the methods of this section to solve the system formed by the pair of equations.

EXAMPLE 10 **Wedding Pictures.** A professional photographer offers two different packages for wedding pictures. Use the information in the illustration on the next page to determine the cost of one 8×10-inch photograph and the cost of one 5×7-inch photograph.

Analyze

- Eight 8×10 and twelve 5×7 pictures cost \$133.
- Six 8×10 and twenty-two 5×7 pictures cost \$168.
- Find the cost of one 8×10 photograph and the cost of one 5×7 photograph.

Assign Let $x =$ the cost of one 8×10 photograph (in dollars), and let $y =$ the cost of one 5×7 photograph (in dollars).

Form We can use the fact that **Number · value = total value** to construct tables that model the cost of each package.

Package 1

Size of photo	Number	· Value	= Total value
8×10	8	x	$8x$
5×7	12	y	$12y$

Total: \$133

↑ One equation comes from this column.

Package 2

Size of photo	Number	· Value	= Total value
8×10	6	x	$6x$
5×7	22	y	$22y$

Total: \$168

↑ The second equation comes from this column.

From the *Total value column* of the first table:

The cost of eight 8×10 photographs	plus	the cost of twelve 5×7 photographs	is	the cost of the first package.
$8x$	$+$	$12y$	$=$	133

From the *Total value column* of the second table:

The cost of six 8×10 photographs	plus	the cost of twenty-two 5×7 photographs	is	the cost of the second package.
$6x$	$+$	$22y$	$=$	168

The resulting system is: $\begin{cases} 8x + 12y = 133 & \textbf{(1)} \\ 6x + 22y = 168 & \textbf{(2)} \end{cases}$

Success Tip

In this section, to solve the application problems that involve two unknowns, we:
- Assign **two** variables
- Form **two** equations
- Solve a system of **two** eqns.
- State **two** results
- Perform **two** checks

Solve We will use elimination to solve this system. To make the x-terms drop out, we multiply both sides of equation 1 by 3. Then we multiply both sides of equation 2 by -4, add the resulting equations, and solve for y:

$$24x + 36y = 399 \quad \text{This is } 3(8x + 12y) = 3(133).$$
$$\underline{-24x - 88y = -672} \quad \text{This is } -4(6x + 22y) = -4(168).$$
$$-52y = -273 \quad \text{Add the terms, column by column. The x-terms drop out.}$$
$$y = 5.25 \quad \text{Divide both sides by } -52. \text{ This is the cost of one } 5 \times 7 \text{ photograph.}$$

To find x, we substitute 5.25 for y in equation 1 and solve for x:

$$8x + 12y = 133$$
$$8x + 12(\mathbf{5.25}) = 133 \quad \text{Substitute 5.25 for y.}$$
$$8x + 63 = 133 \quad \text{Do the multiplication.}$$
$$8x = 70 \quad \text{Subtract 63 from both sides.}$$
$$x = 8.75 \quad \text{Divide both sides by 8. This is the cost of one } 8 \times 10 \text{ photograph.}$$

Caution

In this problem we are to find two unknowns, the cost of an 8×10 photo and the cost of a 5×7 photo. Remember to give both in the *State* step of the solution.

State The cost of one 8×10 photo is \$8.75, and the cost of one 5×7 photo is \$5.25.

Check If the first package contains eight 8×10 and twelve 5×7 photographs, the value of the package is $8(\$8.75) + 12(\$5.25) = \$70 + \$63 = \$133$. If the second package contains six 8×10 and twenty-two 5×7 photographs, the value of the package is $6(\$8.75) + 22(\$5.25) = \$52.50 + \$115.50 = \$168$. The results check.

Teaching Example 10 Team Photos. A soccer picture package that includes one team photo and 8 trading cards costs \$15.00. Another package that includes two team photos and 12 trading cards costs \$25.00. Find the cost of one team photo and one trading card.
Answer: Team photo: \$5.00, trading card: \$1.25

Self Check 10 **Theater Snacks.** At a movie theater, 1 large popcorn and 2 medium drinks cost \$10.50. Another package offers 1 large popcorn and 3 medium drinks for \$13.00. Find the cost of one large popcorn and 1 medium drink. Popcorn: \$5.50, medium drink: \$2.50

Now Try Problem 97

SECTION 12.1 STUDY SET

VOCABULARY

Fill in the blanks.

1. $\begin{cases} x - y = -1 \\ 2x - y = 1 \end{cases}$ is called a <u>system</u> of equations.
2. Because the ordered pair (2, 3) satisfies both equations in problem 1, it is a <u>solution</u> of the system of equations.
3. When the graphs of the equations of a system are identical lines, the equations are called dependent and the system has <u>infinitely</u> many solutions.
4. In the equation $x + 3y = -1$, the x-term has an understood <u>coefficient</u> of 1.

CONCEPTS

5. Refer to the illustration. Determine whether a true or a false statement is obtained when the coordinates of
   a. point A are substituted into the equation for line l_1 True
   b. point B are substituted into the equation for line l_1 False
   c. point C are substituted into the equation for line l_1 True
   d. point C are substituted into the equation for line l_2 True

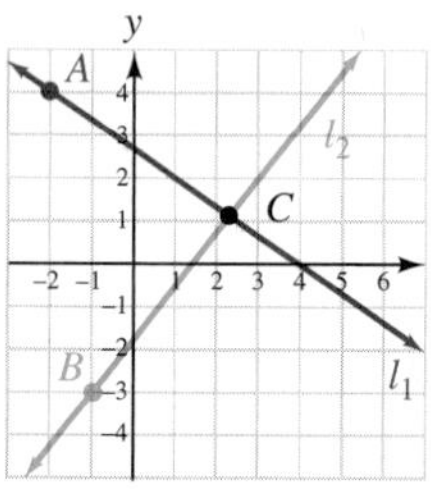

6. Refer to the illustration.
   a. How many ordered pairs satisfy the equation $3x + y = 3$? Name three. Infinitely many; $(-1, 6)$, $(0, 3)$, $(1, 0)$
   b. How many ordered pairs satisfy the equation $\frac{2}{3}x - y = -3$? Name three. Infinitely many; $(-3, 1)$, $(0, 3)$, $(3, 5)$
   c. How many ordered pairs satisfy both equations? Name it or them. One; (0, 3)

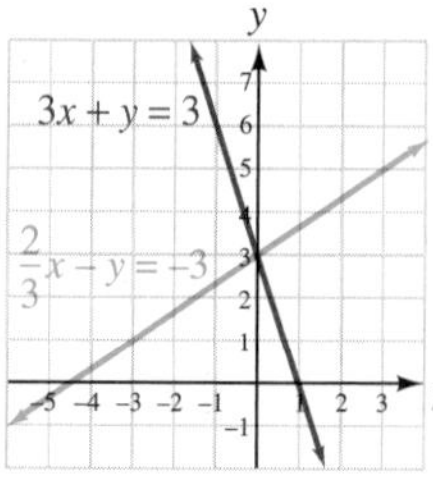

7. If the system $\begin{cases} 4x - 3y = 7 \\ 3x - 2y = 6 \end{cases}$ is to be solved using the elimination method, by what constant should each equation be multiplied if
   a. the x-terms are to drop out? 3; −4 (Answers may vary)
   b. the y-terms are to drop out? 2; −3 (Answers may vary)

8. Consider the system: $\begin{cases} \frac{2}{3}x - \frac{y}{6} = \frac{16}{9} \\ 0.03x + 0.02y = 0.03 \end{cases}$
   a. What step should be performed to clear the first equation of fractions? Multiply both sides by 18.
   b. What step should be performed to clear the second equation of decimals? Multiply both sides by 100.

NOTATION

Complete each solution.

9. Solve: $\begin{cases} y = 3x - 7 \\ x + y = 5 \end{cases}$

$$x + (3x - 7) = 5$$
$$x + 3x - 7 = 5$$
$$4x - 7 = 5$$
$$4x = 12$$
$$x = 3$$
$$y = 3x - 7$$
$$y = 3(3) - 7$$
$$y = 2$$

The solution is (3, 2).

10. Solve: $\begin{cases} 6x + 2y = 0 \\ x - 2y = 7 \end{cases}$

$$6x + 2y = 0$$
$$\underline{x - 2y = 7}$$
$$7x = 7$$
$$x = 1$$
$$x - 2y = 7$$
$$1 - 2y = 7$$
$$-2y = 6$$
$$y = -3$$

The solution is (1, −3).

GUIDED PRACTICE

Use a check to determine whether the ordered pair is a solution of the system of equations. See Example 1.

11. $(-4, 3)$; $\begin{cases} 4x - y = -19 \\ 3x + 2y = -6 \end{cases}$ Yes
12. $(-1, 2)$; $\begin{cases} 3x - y = -5 \\ x - y = -4 \end{cases}$ No
13. $(2, -3)$; $\begin{cases} y + 2 = \frac{1}{2}x \\ 3x + 2y = 0 \end{cases}$ No
14. $(1, 2)$; $\begin{cases} 2x - y = 0 \\ y = \frac{1}{2}x + \frac{3}{2} \end{cases}$ Yes
15. $\left(\frac{1}{2}, \frac{1}{3}\right)$; $\begin{cases} 2x + 3y = 2 \\ 4x - 9y = 1 \end{cases}$ No
16. $\left(-\frac{3}{4}, \frac{2}{3}\right)$; $\begin{cases} 4x + 3y = -1 \\ 4x - 3y = -5 \end{cases}$ Yes
17. $(-0.2, 0.5)$; $\begin{cases} 2x + 5y = 2.1 \\ 5x + y = -0.5 \end{cases}$ Yes
18. $(2.1, -3.2)$; $\begin{cases} x + y = -1.1 \\ 2x - 3y = 13.8 \end{cases}$ Yes

Solve each system by graphing. **See Example 2.** See AIE Appendix 3.

▶ 19. $\begin{cases} x + y = 6 \\ x - y = 2 \end{cases}$ (4, 2)

20. $\begin{cases} x - y = 4 \\ 2x + y = 5 \end{cases}$ (3, −1)

▶ 21. $\begin{cases} y = -2x + 1 \\ x - 2y = -7 \end{cases}$ (−1, 3)

▶ 22. $\begin{cases} 3x - y = -3 \\ y = -2x - 7 \end{cases}$ (−2, −3)

Solve each system by graphing, if possible. If a system is inconsistent or if the equations are dependent, state this. **See Examples 3 and 4.** See AIE Appendix 3.

23. $\begin{cases} 3x - 3y = 4 \\ x - y = 4 \end{cases}$
No solution, ∅; inconsistent system

▶ 24. $\begin{cases} 5x + 2y = 6 \\ -10x - 4y = -12 \end{cases}$
Infinitely many solutions; $\{(x, y) \mid 5x + 2y = 6\}$; dependent equations

▶ 25. $\begin{cases} x = 3 - 2y \\ 2x + 4y = 6 \end{cases}$
Infinitely many solutions; $\{(x, y) \mid x = 3 - 2y\}$ or $\{(x, y) \mid x + 2y = 3\}$; dependent equations

▶ 26. $\begin{cases} 3x = 5 - 2y \\ 3x + 2y = 7 \end{cases}$
No solution, ∅; inconsistent system

Solve each system by graphing. **See Example 5.** See AIE Appendix 3.

27. $\begin{cases} \frac{1}{6}x = \frac{1}{3}y + \frac{1}{2} \\ y = x \end{cases}$ (−3, −3)

▶ 28. $\begin{cases} x = y + 3 \\ \frac{1}{4}x - \frac{1}{6}y = \frac{1}{3} \end{cases}$ (−2, −5)

▶ 29. $\begin{cases} \frac{1}{3}x - \frac{7}{6}y = \frac{1}{2} \\ \frac{1}{5}y = \frac{1}{3}x + \frac{7}{15} \end{cases}$ (−2, −1)

30. $\begin{cases} \frac{3}{5}x + \frac{1}{4}y = -\frac{11}{10} \\ \frac{1}{8}x = \frac{13}{24} + \frac{1}{3}y \end{cases}$ (−1, −2)

Use a graphing calculator to solve each system. Give all answers to the nearest hundredth. **See Using Your Calculator: Solving Systems by Graphing.**

31. $\begin{cases} y = 3.2x - 1.5 \\ y = -2.7x - 3.7 \end{cases}$
(−0.37, −2.69)

▶ 32. $\begin{cases} y = -0.45x + 5 \\ y = 5.55x - 13.7 \end{cases}$
(3.12, 3.60)

33. $\begin{cases} 1.7x + 2.3y = 3.2 \\ y = 0.25x + 8.95 \end{cases}$
(−7.64, 7.04)

▶ 34. $\begin{cases} 2.75x = 12.9y - 3.79 \\ 7.1x - y = 35.76 \end{cases}$
(5.24, 1.41)

Solve each system by substitution. **See Example 6.**

35. $\begin{cases} y = 3x \\ x + y = 8 \end{cases}$ (2, 6)

▶ 36. $\begin{cases} y = x + 2 \\ x + 2y = 16 \end{cases}$ (4, 6)

37. $\begin{cases} x = 2 + y \\ 2x + y = 13 \end{cases}$ (5, 3)

▶ 38. $\begin{cases} x = -5 + y \\ 3x - 2y = -7 \end{cases}$ (3, 8)

▶ 39. $\begin{cases} x + 2y = 6 \\ 3x - y = -10 \end{cases}$ (−2, 4)

▶ 40. $\begin{cases} 2x - y = -21 \\ 4x + 5y = 7 \end{cases}$ (−7, 7)

41. $\begin{cases} 5x + 3y = -26 \\ 3x + y = -14 \end{cases}$ (−4, −2)

42. $\begin{cases} 3x + 5y = 4 \\ 5x + y = 14 \end{cases}$ (3, −1)

Solve each system by elimination (addition). **See Example 7.**

▶ 43. $\begin{cases} x - y = 7 \\ x + y = 11 \end{cases}$ (9, 2)

44. $\begin{cases} a + b = 5 \\ a - b = 11 \end{cases}$ (8, −3)

45. $\begin{cases} 2s + 3t = -8 \\ 2s - 3t = -8 \end{cases}$ (−4, 0)

▶ 46. $\begin{cases} x + 2y = -21 \\ x - 2y = 11 \end{cases}$ (−5, −8)

▶ 47. $\begin{cases} 3x + 4y = -24 \\ 5x + 12y = -72 \end{cases}$ (0, −6)

▶ 48. $\begin{cases} 5x + 2y = 11 \\ 7x + 6y = 9 \end{cases}$ (3, −2)

49. $\begin{cases} \frac{5}{6}x + \frac{1}{2}y = 12 \\ 0.3x + 0.5y = 5.6 \end{cases}$ (12, 4)

50. $\begin{cases} \frac{1}{3}x + \frac{1}{2}y = \frac{31}{6} \\ 0.3x + 0.2y = 3.9 \end{cases}$ (11, 3)

Solve each system, if possible. If a system is inconsistent or if the equations are dependent, state this. **See Examples 8 and 9.**

51. $\begin{cases} 6x + 3y = 18 \\ y = -2x + 5 \end{cases}$
No solution, ∅; inconsistent system

▶ 52. $\begin{cases} 8x - 4y = 16 \\ 2x - 4 = y \end{cases}$
Infinitely many solutions; $\{(x, y) \mid 2x - 4 = y\}$; dependent equations

53. $\begin{cases} 3x - y = 5 \\ 21x = 7(y + 5) \end{cases}$
Infinitely many solutions; $\{(x, y) \mid 3x - y = 5\}$; dependent equations

▶ 54. $\begin{cases} 4x + 8y = 15 \\ x = 2(2 - y) \end{cases}$
No solution, ∅; inconsistent system

TRY IT YOURSELF

Solve each system by graphing, if possible. If a system is inconsistent or if the equations are dependent, state this. (Hint: Several coordinates of points of intersection are fractions.) See AIE Appendix 3.

▶ 55. $\begin{cases} 4x - 3y = 5 \\ y = -2x \end{cases}$
$\left(\frac{1}{2}, -1\right)$

▶ 56. $\begin{cases} 2x + 2y = -1 \\ 3x + 4y = 0 \end{cases}$
$\left(-2, \frac{3}{2}\right)$

57. $\begin{cases} y = -\frac{5}{2}x + \frac{1}{2} \\ 2x - \frac{3}{2}y = 5 \end{cases}$
(1, −2)

58. $\begin{cases} \frac{5}{2}x + 3y = 6 \\ y = -\frac{5}{6}x + 2 \end{cases}$
Infinitely many solutions, $\left\{(x, y) \mid y = -\frac{5}{6}x + 2\right\}$ or $\{(x, y) \mid 5x + 6y = 12\}$; dependent equations

59. $\begin{cases} x = \frac{11 - 2y}{3} \\ y = \frac{11 - 6x}{4} \end{cases}$
No solution, ∅; inconsistent system

▶ 60. $\begin{cases} x = \frac{1 - 3y}{4} \\ y = \frac{12 + 3x}{2} \end{cases}$
(−2, 3)

61. $\begin{cases} x = 13 - 4y \\ 3x = 4 + 2y \end{cases}$
$\left(3, \frac{5}{2}\right)$

▶ 62. $\begin{cases} 3x = 7 - 2y \\ 2x = 2 + 4y \end{cases}$
$\left(2, \frac{1}{2}\right)$

▶ 63. $\begin{cases} x = 2 \\ y = -\frac{1}{2}x + 2 \end{cases}$
(2, 1)

▶ 64. $\begin{cases} y = -2 \\ y = \frac{2}{3}x - \frac{4}{3} \end{cases}$
(−1, −2)

▶ 65. $\begin{cases} x + 3y = 6 \\ y = -\frac{1}{3}x + 2 \end{cases}$
Infinitely many solutions; $\{(x, y) \mid x + 3y = 6\}$; dependent equations

▶ 66. $\begin{cases} 2x - y = -4 \\ 2y = 4x - 6 \end{cases}$
No solution, ∅; inconsistent system

Solve each system by any method. If a system is inconsistent or if the equations are dependent, so indicate.

67. $\begin{cases} 2x + 3y = 8 \\ 3x - 2y = -1 \end{cases}$

(1, 2)

68. $\begin{cases} x = \frac{3}{2}y + 5 \\ 2x - 3y = 8 \end{cases}$

No solution, ∅; dependent equations

▶ 69. $\begin{cases} 4(x - 2) = -9y \\ 2(x - 3y) = -3 \end{cases}$

$\left(\frac{1}{2}, \frac{2}{3}\right)$

▶ 70. $\begin{cases} 2(2x + 3y) = 5 \\ 8x = 3(1 + 3y) \end{cases}$

$\left(\frac{3}{4}, \frac{1}{3}\right)$

▶ 71. $\begin{cases} 0.3a + 0.1b = 0.5 \\ \frac{4}{3}a + \frac{1}{3}b = 3 \end{cases}$

(4, −7)

72. $\begin{cases} 0.9p + 0.2q = 1.2 \\ \frac{2}{3}p + \frac{1}{9}q = 1 \end{cases}$

(2, −3)

73. $\begin{cases} \frac{x}{2} + \frac{y}{2} = 6 \\ \frac{x}{3} + \frac{y}{3} = 4 \end{cases}$

Infinitely many solutions; $\{(x, y) \mid x + y = 12\}$; dependent equations

74. $\begin{cases} \frac{x}{2} - \frac{y}{3} = -4 \\ \frac{x}{2} + \frac{y}{9} = 0 \end{cases}$

(−2, 9)

75. $\begin{cases} x = \frac{2}{3}y \\ y = 4x + 5 \end{cases}$

(−2, −3)

76. $\begin{cases} 5x - 2y = 19 \\ y = \frac{1 - 3x}{4} \end{cases}$

(3, −2)

77. $\begin{cases} 3x - 4y = 9 \\ x + 2y = 8 \end{cases}$

$\left(5, \frac{3}{2}\right)$

78. $\begin{cases} 3x - 2y = -10 \\ 6x + 5y = 25 \end{cases}$

(0, 5)

79. $\begin{cases} x - \frac{4y}{5} = 4 \\ \frac{y}{3} = \frac{x}{2} - \frac{5}{2} \end{cases}$

$\left(10, \frac{15}{2}\right)$

80. $\begin{cases} 3x - 2y = \frac{9}{2} \\ \frac{x}{2} - \frac{3}{4} = 2y \end{cases}$

$\left(\frac{3}{2}, 0\right)$

81. $\begin{cases} \frac{2}{3}x - \frac{1}{4}y = -8 \\ 0.5x - 0.375y = -9 \end{cases}$

(−6, 16)

82. $\begin{cases} 0.5x + 0.5y = 6 \\ \frac{x}{2} - \frac{y}{2} = -2 \end{cases}$

(4, 8)

83. $\begin{cases} \frac{3}{2}p + \frac{1}{3}q = 2 \\ \frac{2}{3}p + \frac{1}{9}q = 1 \end{cases}$

(2, −3)

84. $\begin{cases} a + \frac{b}{3} = \frac{5}{3} \\ \frac{a + b}{3} = 3 - a \end{cases}$

(4, −7)

▶ 85. $\begin{cases} \frac{m - n}{5} + \frac{m + n}{2} = 6 \\ \frac{m - n}{2} - \frac{m + n}{4} = 3 \end{cases}$

(9, −1)

▶ 86. $\begin{cases} \frac{r - 2}{5} + \frac{s + 3}{2} = 5 \\ \frac{r + 3}{2} + \frac{s - 2}{3} = 6 \end{cases}$

(7, 5)

Solve each system. To do so, substitute a for $\frac{1}{x}$ and b for $\frac{1}{y}$ and solve for a and b. Then find x and y using the fact that $a = \frac{1}{x}$ and $b = \frac{1}{y}$.

▶ 87. $\begin{cases} \frac{1}{x} + \frac{1}{y} = \frac{5}{6} \\ \frac{1}{x} - \frac{1}{y} = \frac{1}{6} \end{cases}$

(2, 3)

▶ 88. $\begin{cases} \frac{1}{x} + \frac{1}{y} = \frac{9}{20} \\ \frac{1}{x} - \frac{1}{y} = \frac{1}{20} \end{cases}$

(4, 5)

89. $\begin{cases} \frac{1}{x} + \frac{2}{y} = -1 \\ \frac{2}{x} - \frac{1}{y} = -7 \end{cases}$

$\left(-\frac{1}{3}, 1\right)$

90. $\begin{cases} \frac{3}{x} - \frac{2}{y} = -30 \\ \frac{2}{x} - \frac{3}{y} = -30 \end{cases}$

$\left(-\frac{1}{6}, \frac{1}{6}\right)$

APPLICATIONS

91. ▶ **from Campus to Careers**

Fashion Designer

One of the line graphs below gives the percent share of the U.S. footwear market for shoes produced in the United States. The other line gives the percent share of the U.S. footwear market for imports.

a. Estimate the coordinates of the point of intersection of the graphs. (1978, 50%)

b. What important percent-of-the-market information does your answer to part a give? In 1978, the percent share of the U.S. footwear market for shoes produced in the United States and imports was the same: 50%.

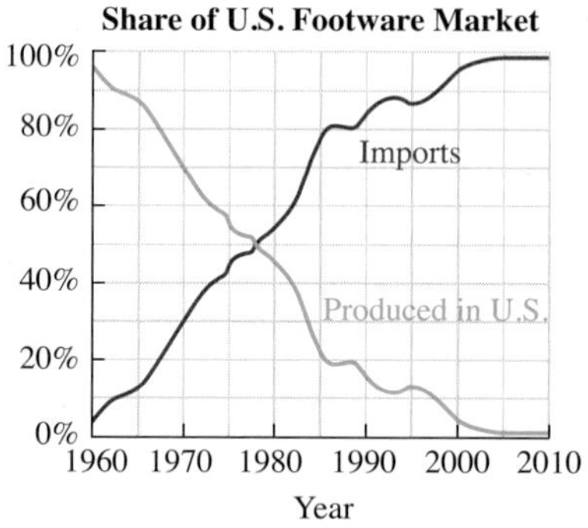

Source: Americal Apparel and Footwear Association

92. **The Internet.**

The graph on the next page shows the growing importance of the Internet in the daily lives of Americans. Determine when the time spent on the following activities was the same. Approximately how many hours per year were spent on each?

a. Internet and reading magazines 2001: 125 hr

b. Internet and reading newspapers 2006: 180 hr

c. Internet and reading books 2000: 105 hr

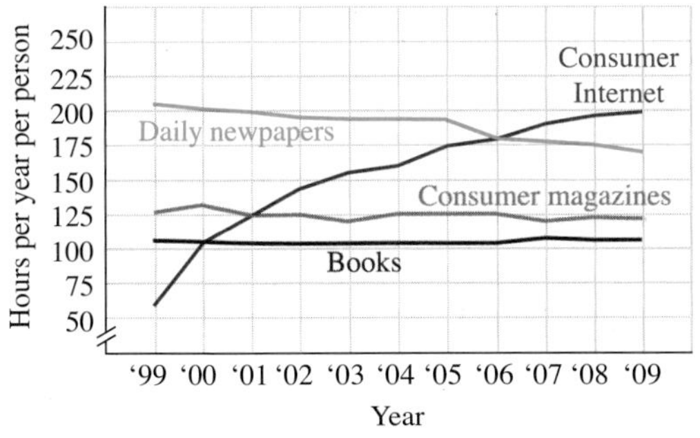

Source: Vernois Suhler Stevenson

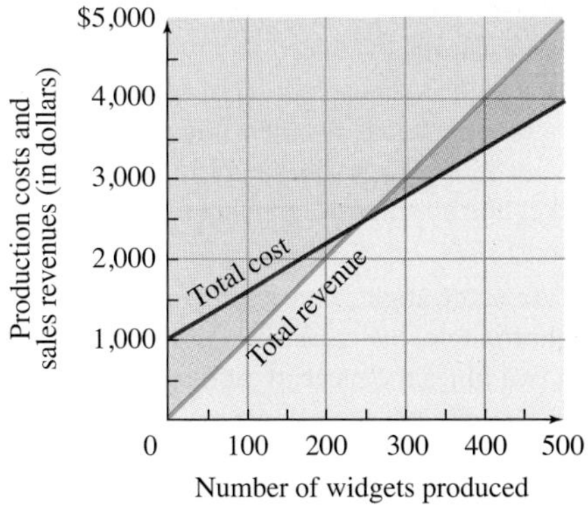

93. Law of Supply and Demand. The demand function, graphed below, describes the relationship between the price x of a certain camera and the demand for the camera.

a. The supply function, $S(x) = \frac{25}{4}x - 525$, describes the relationship between the price x of the camera and the number of cameras the manufacturer is willing to supply. Graph this function in the illustration.

b. For what price will the supply of cameras equal the demand? $140

c. As the price of the camera is increased, what happens to supply and what happens to demand? Supply increases and demand decreases.

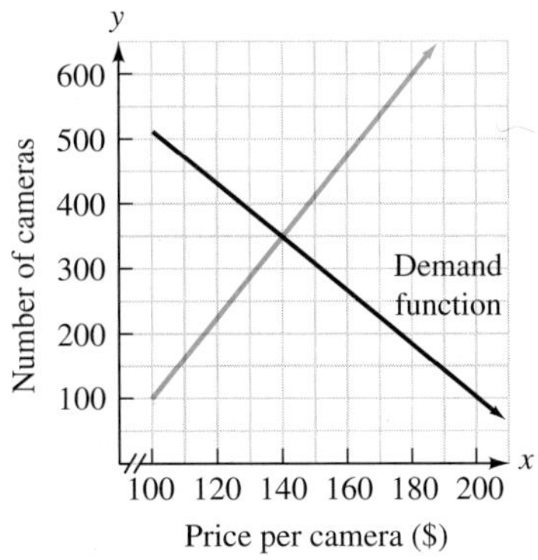

94. Cost and Revenue. The function $C(x) = 200x + 400$ gives the cost for a college to offer x sections of an introductory class in CPR (cardiopulmonary resuscitation). The function $R(x) = 280x$ gives the amount of revenue the college brings in when offering x sections of CPR.

a. Find the *break-even point* (where cost = revenue) by graphing each function on the same coordinate system. (5, 1,400); see AIE Appendix 3.

b. How many sections does the college need to offer to make a profit on the CPR training course? More than 5

95. Business. Estimate the break-even point (where cost = revenue) on the graph in the next column. Explain why it is called the *break-even point.* (250, 2,500); if the company makes 250 widgets, the cost to make them and the revenue obtained from their sale will be equal: $2,500.

96. Navigation. The paths of two ships are tracked on the same coordinate system. One ship is following a path described by the equation $2x + 3y = 6$, and the other is following a path described by the equation $y = \frac{2}{3}x - 3$.

a. Is there a possibility of a collision? Yes

b. What are the coordinates of the danger point? $(3.75, -0.5)$

c. Is a collision a certainty? No

In Exercises 97–112, write a system of two equations in two variables to solve each problem.

97. Ticket Sales. The ticket prices for a Halloween haunted house were $5 for adults and $3 for children. On a day when a total of 390 tickets were purchased, the receipts were $1,470. How many of each type of ticket were sold? Adults: 150, children: 240

98. Advertising. Use the information in the ad to find the cost of a 15-second and the cost of a 30-second radio commercial on radio station KLIZ. 15 sec: $475, 30 sec: $800

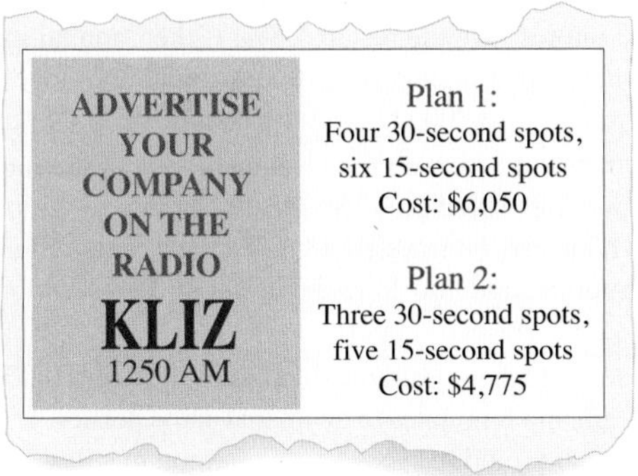

ADVERTISE YOUR COMPANY ON THE RADIO
KLIZ
1250 AM
Plan 1:
Four 30-second spots,
six 15-second spots
Cost: $6,050
Plan 2:
Three 30-second spots,
five 15-second spots
Cost: $4,775

99. from **Campus to Careers**

Fashion Designer

©iStockphoto.com/CandyBox Photography

In 2009, there was a combined total of 4,046 Gap and Aéropostale clothing stores worldwide. The number of Gap stores was $3\frac{1}{4}$ times more than the number of Aéropostale stores. How many Gap stores and how many Aéropostale stores were there that year? (Source: wikinvest.com) Gap: 3,094 stores; Aéropostale: 952 stores

100. Summer Concerts. According to *StubHub.com,* in 2009, two tickets to a Jonas Brothers concert and two tickets to an Elton John concert cost, on average, a total of \$560. At those prices, four tickets to see the Jonas Brothers and two tickets to see Elton John cost \$806. What was the average cost of a Jonas Brothers ticket and an Elton John ticket in 2009? Jonas Brothers, \$123; Elton John, \$157

101. Geometry. An acute angle is an angle with measure less than 90°. In a right triangle, the measure of one acute angle is 15° greater than two times the measure of the other acute angle. Find the measure of each acute angle. 65°, 25°

102. New York City. The triangular-shaped Flatiron Building in Manhattan has a perimeter of 499 feet at its base. It is bordered on each side by a street. The 5th Avenue front of the building is 198 feet long. The Broadway front is 43 feet more than twice as long as the East 22nd Street front. Find the length of the Broadway front and East 22nd Street front. (Source: New York Public Library) Broadway: 215 ft, East 22nd Street: 86 ft

103. Investment Clubs. Part of \$8,000 was invested by an investment club at 10% interest and the rest at 12%. If the annual income from these investments is \$900, how much was invested at each rate? \$3,000 at 10%, \$5,000 at 12%

104. Retirement Income. A retired couple invested part of \$12,000 at 6% interest and the rest at 7.5%. If their annual income from these investments is \$810, how much was invested at each rate? \$6,000 at 6%, \$6,000 at 7.5%

105. Snowmobiling. A man rode a snowmobile at the rate of 20 mph and then skied cross country at the rate of 4 mph. During the 6-hour trip, he traveled 48 miles. How long did he snowmobile, and how long did he ski? Snowmobile: 1.5 hr, ski: 4.5 hr

106. Salmon. It takes a salmon 40 minutes to swim 10,000 feet upstream and 8 minutes to swim that same portion of a river downstream. Find the speed of the salmon in still water and the speed of the current. Salmon: 750 ft/min; current: 500 ft/min

107. Production Planning. A manufacturer builds racing bikes and mountain bikes, with the per unit manufacturing costs shown in the table. The company has budgeted \$26,150 for materials and \$31,800 for labor. How many bicycles of each type can be built? 85 racing bikes, 120 mountain bikes

Model	Cost of materials	Cost of labor
Racing	\$110	\$120
Mountain	\$140	\$180

108. Farming. A farmer keeps some animals on a strict diet. Each animal is to receive 15 grams of protein and 7.5 grams of carbohydrates. The farmer uses two food mixes, with nutrients as shown in the table. How many grams of each mix should be used to provide the correct nutrients for each animal? 50 g of A, 60 g of B

Mix	Protein	Carbohydrates
Mix A	12%	9%
Mix B	15%	5%

109. Cosmotology. A beauty shop specializing in permanents has fixed costs of \$2,101.20 per month. The owner estimates that the cost for each permanent is \$23.60, which covers labor, chemicals, and electricity. If her shop can give as many permanents as she wants at a price of \$44 each, how many must be given each month for her to break even? 103

110. Mixing Candy. How many pounds of each candy shown in the illustration must be mixed to obtain 60 pounds of candy that would be worth \$4 per pound? Gummy bears: 45 lb; jelly beans: 15 lb

111. Dermatology. Tests of an antibacterial face-wash cream showed that a mixture containing 0.3% Triclosan (active ingredient) gave the best results. How many grams of cream from each tube should be used to make an equal-size tube of the 0.3% cream? 148 g of the 0.2%, 37 g of the 0.7%

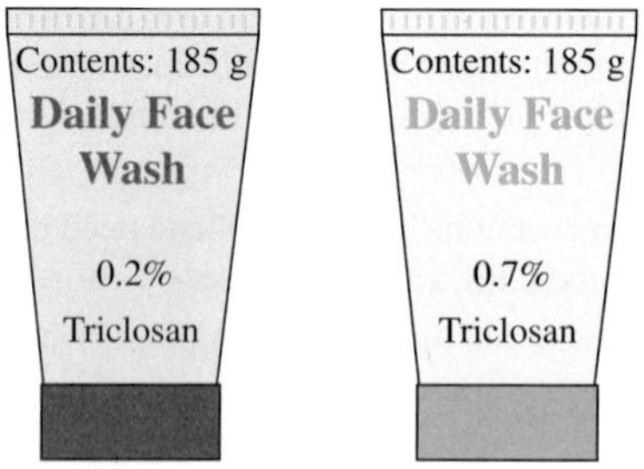

112. Mixing Solutions. How many ounces of the two alcohol solutions in the illustration must be mixed to obtain 100 ounces of a 12.2% solution? 40 oz of 8% solution, 60 oz of 15% solution

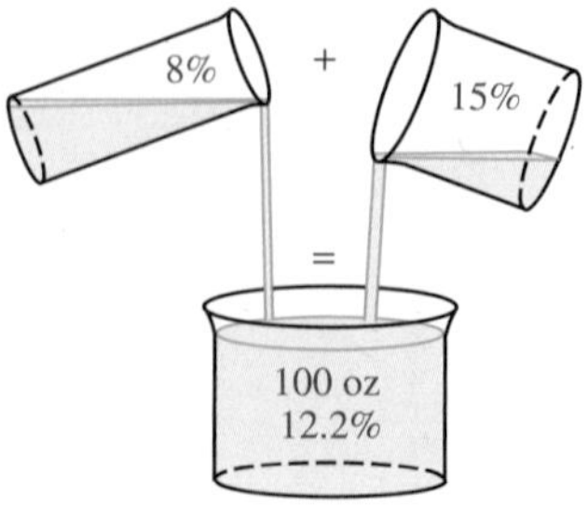

WRITING

113. Which method would you use to solve the system? Explain.

$$\begin{cases} y - 1 = 3x \\ 3x + 2y = 12 \end{cases}$$

114. Which method would you use to solve the system? Explain.

$$\begin{cases} 2x + 4y = 9 \\ 3x - 5y = 20 \end{cases}$$

115. When solving a system, what advantages are there with the substitution and elimination methods compared with the graphing method?

116. When using the elimination (addition) method, how can you tell whether

a. a system of linear equations has no solution?

b. a system of linear equations has infinitely many solutions?

REVIEW

Solve each formula for the specified variable.

117. $\frac{V_2}{V_1} = \frac{P_1}{P_2}$ for P_1 $\quad P_1 = \frac{P_2V_2}{V_1}$

118. $\frac{1}{r} = \frac{1}{r_1} + \frac{1}{r_2}$ for r $\quad r = \frac{r_1r_2}{r_2 + r_1}$

119. $S = \frac{a - lr}{1 - r}$ for r $\quad r = \frac{S - a}{S - l}$

120. $P = \frac{Q_1}{Q_2 - Q_1}$ for Q_1 $\quad Q_1 = \frac{PQ_2}{1 + P}$

CHALLENGE PROBLEMS

121. If the solution of the system $\begin{cases} Ax + By = -2 \\ Bx - Ay = -26 \end{cases}$ is $(-3, 5)$, find the values of A and B. A is 4, B is 2

122. Solve $\begin{cases} 2ab - 3cd = 1 \\ 3ab - 2cd = 1 \end{cases}$ and assume that b and d are constants. $\left(\frac{1}{5b}, -\frac{1}{5d}\right)$

SECTION 12.2 Solving Systems of Equations in Three Variables

OBJECTIVES

1. Determine whether an ordered triple is a solution of a system.
2. Solve systems of three linear equations in three variables.
3. Solve systems of equations with missing variable terms.
4. Identify inconsistent systems and dependent equations.

ARE YOU READY? *Are You Ready? exercises available online at www.webassign.net/brookscole*

The following problems review some basic skills that are needed when solving systems of equations in three variables.

1. What is the coefficient of each term on the left side of the equation $x - 4y + 5z = 2$? $\quad$ 1, −4, 5

2. For $6x + y + 2z = 36$, if $x = 5$ and $y = -2$, what is z? $\quad$ 4

3. Write the equation $7x + 5y = z + 9$ so that all three variable terms are on the left side. $\quad 7x + 5y - z = 9$

4. Solve the system by elimination: $\begin{cases} 5x + 2y = -5 \\ -6x - 2y = 10 \end{cases}$ $\quad (-5, 10)$

In previous sections, we solved systems of linear equations in two variables. We will now extend this discussion to consider systems of linear equations in *three* variables.

1 Determine Whether an Ordered Triple Is a Solution of a System.

The equation $x - 5y + 7z = 10$, where each variable is raised to the first power, is an example of a linear equation in three variables. In general, we have the following definition.

Standard Form

A **linear equation in three variables** is an equation that can be written in the form

$$Ax + By + Cz = D$$

where A, B, C, and D are real numbers and A, B, and C are not all 0.

A solution of a linear equation in three variables is an **ordered triple** of numbers of the form (x, y, z) whose coordinates satisfy the equation. For example, $(2, 0, 1)$ is a solution of $x + y + z = 3$ because a true statement results when we substitute 2 for x, 0 for y, and 1 for z: $2 + 0 + 1 = 3$.

A **solution of a system of three linear equations** in three variables is an ordered triple that satisfies each equation of the system.

EXAMPLE 1 Determine whether $(-4, 2, 5)$ is a solution of the system:

$$\begin{cases} 2x + 3y + 4z = 18 \\ 3x + 4y + z = 1 \\ x + y + 3z = 13 \end{cases}$$

Strategy We will substitute the x-, y-, and z-coordinates of $(-4, 2, 5)$ for the corresponding variables in each equation of the system.

Why If each equation is satisfied by the x-, y-, and z-coordinates, the ordered triple is a solution of the system.

Solution We substitute -4 for x, 2 for y, and 5 for z in each equation.

The first equation	*The second equation*	*The third equation*
$2x + 3y + 4z = 18$	$3x + 4y + z = 1$	$x + y + 3z = 13$
$2(-4) + 3(2) + 4(5) \stackrel{?}{=} 18$	$3(-4) + 4(2) + 5 \stackrel{?}{=} 1$	$-4 + 2 + 3(5) \stackrel{?}{=} 13$
$-8 + 6 + 20 \stackrel{?}{=} 18$	$-12 + 8 + 5 \stackrel{?}{=} 1$	$-4 + 2 + 15 \stackrel{?}{=} 13$
$18 = 18$ True	$1 = 1$ True	$13 = 13$ True

Since $(-4, 2, 5)$ satisfies each equation, it is a solution of the system.

Teaching Example 1
Determine whether $(-2, 3, -1)$ is a solution of the system:

$$\begin{cases} 3x - y + 2z = -11 \\ x + 2y - 3z = 7 \\ 3x + 2y + z = 1 \end{cases}$$

Answer: Not a solution

Self Check 1 Is $(6, -3, 1)$ a solution of: $\begin{cases} x - y + z = 10 \\ x + 4y - z = -7 \\ 3x - y + 4z = 24 \end{cases}$ Not a solution

Now Try Problem 11

The Language of Algebra

Recall that when a system of equations has at least one solution, the system is called a **consistent** system, and if a system has no solution, the system is called **inconsistent.**

The graph of an equation of the form $Ax + By + Cz = D$ can be drawn on a coordinate system with three axes. The graph of such an equation is a flat surface called a **plane.** A system of three linear equations with three variables is consistent or inconsistent, depending on how the three planes corresponding to the three equations intersect. The following illustration shows some of the possibilities. Just as in the case of two variables, a system of three linear equations in three variables can have exactly one solution, no solution, or infinitely many solutions, as shown below.

A consistent system | A consistent system | Three types of inconsistent systems

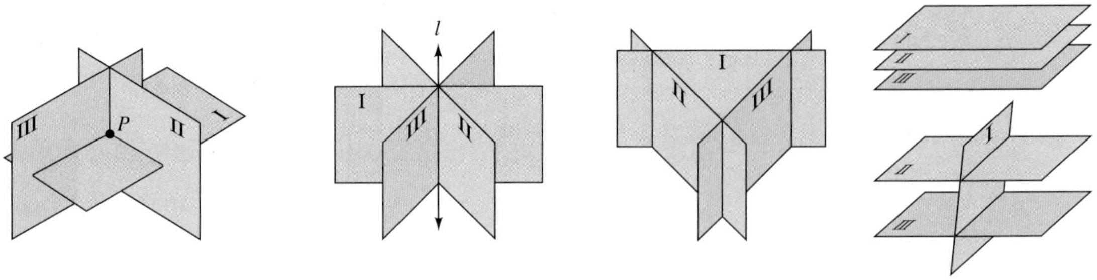

(a) The three planes intersect at a single point: The system has **one solution**.

(b) The three planes have a line l in common. The system has **infinitely many solutions**.

(c) The three planes have no point in common to all three. The system has **no solutions**.

2 Solve Systems of Three Linear Equations in Three Variables.

Teaching Tip: You can use the walls, floor, and ceiling of your classroom to illustrate three planes and how they intersect or are parallel.

To **solve a system of three linear equations** in three variables means to find all of the solutions of the system. Solving such a system by graphing is not practical because it requires a three-dimensional coordinate system.

The substitution method is useful to solve systems of three equations where one or more equations have only two variables. However, the best way to solve systems of three linear equations in three variables is usually the elimination method.

Solving a System of Three Linear Equations by Elimination

1. Write each equation in standard form $Ax + By + Cz = D$ and clear any decimals or fractions.
2. Pick any two equations and eliminate a variable.
3. Pick a different pair of equations and eliminate the same variable as in step 1.
4. Solve the resulting pair of two equations in two variables.
5. To find the value of the third variable, substitute the values of the two variables found in step 4 into any equation containing all three variables and solve the equation.
6. Check the proposed solution in all three of the original equations. Write the solution as an ordered triple.

EXAMPLE 2

Solve the system: $\begin{cases} 2x + y + 4z = 12 \\ x + 2y + 2z = 9 \\ 3x - 3y - 2z = 1 \end{cases}$

Strategy Since the coefficients of the z-terms are opposites in the second and third equations, we will add the left and right sides of those equations to eliminate z. Then we will choose another pair of equations and eliminate z again.

Why The result will be a system of two equations in x and y that we can solve by elimination.

Solution

Step 1: We can skip step 1 because each equation is written in standard form and there are no fractions or decimals to clear. We will number each equation and move to step 2.

$$\begin{array}{l} (1) \\ (2) \\ (3) \end{array} \begin{cases} 2x + y + 4z = 12 \\ x + 2y + 2z = 9 \\ 3x - 3y - 2z = 1 \end{cases}$$

Notation

We number the equations (1), (2), and (3) to help describe how the system is solved using the elimination method.

Step 2: If we pick equations 2 and 3 and add them, the variable z is eliminated.

$$\begin{array}{llr} (2) & x + 2y + 2z = & 9 \\ (3) & \underline{3x - 3y - 2z =} & \underline{1} \\ (4) & 4x - y \qquad = & 10 \end{array}$$ This equation does not contain z.

Teaching Tip: In step 3, stress the importance of eliminating the same variable as that eliminated in step 2.

Step 3: We now pick a different pair of equations (equations 1 and 3) and eliminate z again. If each side of equation 3 is multiplied by 2, and the resulting equation is added to equation 1, z is eliminated.

$$\begin{array}{llr} (1) & 2x + y + 4z = & 12 \\ & \underline{6x - 6y - 4z =} & \underline{2} \\ (5) & 8x - 5y \qquad = & 14 \end{array}$$

This is $2(3x - 3y - 2z) = 2(1)$.
This equation does not contain z.

Success Tip

With this method, we use elimination to reduce a system of three equations in three variables to a system of two equations in two variables.

Step 4: Equations 4 and 5 form a system of two equations in x and y.

$$\begin{array}{l} (4) \\ (5) \end{array} \begin{cases} 4x - y = 10 \\ 8x - 5y = 14 \end{cases}$$

To solve this system, we multiply equation 4 by -5 and add the resulting equation to equation 5 to eliminate y.

$$\begin{array}{llr} & -20x + 5y = & -50 \\ (5) & \underline{8x - 5y =} & \underline{14} \\ & -12x \qquad = & -36 \end{array}$$ This is $-5(4x - y) = -5(10)$.

$x = 3$ Divide both sides by -12. This is the x-value of the solution.

To find y, we substitute 3 for x in any equation containing x and y (such as equation 5) and solve for y:

(5) $8x - 5y = 14$

$8(3) - 5y = 14$ Substitute 3 for x.

$24 - 5y = 14$ Simplify.

$-5y = -10$ Subtract 24 from both sides.

$y = 2$ Divide both sides by -5. This is the y-value of the solution.

Step 5: To find z, we substitute 3 for x and 2 for y in any equation containing x, y, and z (such as equation 1) and solve for z:

(1) $2x + y + 4z = 12$

$2(3) + 2 + 4z = 12$ Substitute 3 for x and 2 for y.

$8 + 4z = 12$ Simplify.

$4z = 4$ Subtract 8 from both sides.

$z = 1$ Divide both sides by 4. This is the z-value of the solution.

Teaching Tip: Remind your students that the coordinates of the ordered triple solution are written in alphabetical order.

Step 6: To verify that the solution is (3, 2, 1), we substitute 3 for x, 2 for y, and 1 for z in the three equations of the original system. The solution set is written as $\{(3, 2, 1)\}$. Since this system has a solution, it is a consistent system.

Teaching Example 2
Solve the system:
$$\begin{cases} 2x - 3y + 2z = -7 \\ x + 4y - z = 10 \\ 3x + 2y + z = 4 \end{cases}$$
Answer: (2, 1, −4)

Self Check 2 Solve the system: $\begin{cases} 2x - 3y + 2z = -7 \\ x + 4y - z = 10 \\ 3x + 2y + z = 4 \end{cases}$ (2, 1, −4)

Now Try Problem 15

3 Solve Systems of Equations with Missing Variable Terms.

When one or more of the equations of a system is missing a variable term, the elimination of a variable that is normally performed in step 2 of the solution process can be skipped.

EXAMPLE 3 Solve the system: $\begin{cases} 3x = 6 - 2y + z \\ -y - 2z = -8 - x \\ x = 1 - 2z \end{cases}$

Strategy Since the third equation does not contain the variable y, we will work with the first and second equations to obtain another equation that does not contain y.

Why Then we can use the elimination method to solve the resulting system of two equations in x and z.

Solution ***Step 1:*** We use the addition property of equality to write each equation in the standard form $Ax + By + Cz = D$ and number each equation.

(1) $3x + 2y - z = 6$ Add 2y and subtract z from both sides of $3x = 6 - 2y + z$.

(2) $x - y - 2z = -8$ Add x to both sides of $-y - 2z = -8 - x$.

(3) $x + 2z = 1$ Add 2z to both sides of $x = 1 - 2z$.

Step 2: Since equation 3 does not have a y-term, we can skip to step 3, where we will find another equation that does not contain a y-term.

Step 3: If each side of equation 2 is multiplied by 2 and the resulting equation is added to equation 1, y is eliminated.

$$\begin{array}{lrl} (1) & 3x + 2y - z = & 6 \\ & 2x - 2y - 4z = & -16 \quad \text{This is } 2(x - y - 2z) = 2(-8). \\ \hline (4) & 5x - 5z = & -10 \end{array}$$

Step 4: Equations 3 and 4 form a system of two equations in x and z:

$$\begin{array}{l} (3) \\ (4) \end{array} \begin{cases} x + 2z = 1 \\ 5x - 5z = -10 \end{cases}$$

Success Tip

We don't have to find the values of the variables in alphabetical order. In step 2, choose the variable that is the easiest to eliminate. In this example, the value of z is found first.

To solve this system, we multiply equation 3 by -5 and add the resulting equation to equation 4 to eliminate x:

$$\begin{array}{lrl} & -5x - 10z = -5 & \text{This is } -5(x + 2z) = -5(1). \\ (4) & 5x - 5z = -10 & \\ \hline & -15z = -15 & \\ & z = 1 & \text{Divide both sides by } -15. \text{ This is the } z\text{-value of the solution.} \end{array}$$

To find x, we substitute 1 for z in equation 3.

$$\begin{array}{lrl} (3) & x + 2z = 1 & \\ & x + 2(\mathbf{1}) = 1 & \text{Substitute 1 for } z. \\ & x + 2 = 1 & \text{Multiply.} \\ & x = -1 & \text{Subtract 2 from both sides.} \end{array}$$

Step 5: To find y, we substitute -1 for x and 1 for z in equation 1:

$$\begin{array}{lrl} (1) & 3x + 2y - z = 6 & \\ & 3(\mathbf{-1}) + 2y - \mathbf{1} = 6 & \text{Substitute } -1 \text{ for } x \text{ and 1 for } z. \\ & -3 + 2y - 1 = 6 & \text{Multiply.} \\ & 2y = 10 & \text{Simplify and add 4 to both sides.} \\ & y = 5 & \text{Divide both sides by 2.} \end{array}$$

The solution of the system is $(-1, 5, 1)$ and the solution set is $\{(-1, 5, 1)\}$.

Step 6: Check the proposed solution in all three of the original equations.

Self Check 3 Solve the system: $\begin{cases} x + 2y = 1 + z \\ 2x = 3 + y - z \\ x + z = 3 \end{cases}$ (1, 1, 2)

Now Try Problem 23

EXAMPLE 4 Solve the system: $\begin{cases} x - y + 4z = -30 & (1) \\ x + 2y = 200 & (2) \\ y + z = 30 & (3) \end{cases}$

Strategy Since the second equation does not contain the variable z, we will work with the first and third equations to obtain another equation that does not contain z.

Why Then we can use the elimination method to solve the resulting system of two equations in x and y.

Solution ***Step 1:*** Each of the equations is written in standard form, however, two of the equations are missing variable terms.

Step 2: Since equation 2 does not have a z-term, we can skip to step 3, where we will find another equation that does not contain a z-term.

Step 3: If each side of equation 3 is multiplied by -4 and the resulting equation is added to the equation 1, z is eliminated.

$$\begin{array}{lrl} (1) & x - y + 4z = -30 & \\ & -4y - 4z = -120 & \text{This is } -4(y + z) = -4(30). \\ \hline (4) & x - 5y \quad = -150 & \end{array}$$

Step 4: Equations 2 and 4 form a system of two equations in x and y.

$$\begin{array}{l} (2) \\ (4) \end{array} \begin{cases} x + 2y = 200 \\ x - 5y = -150 \end{cases}$$

To solve this system, we multiply equation 4 by -1 and add the resulting equation to equation 2 to eliminate x.

$$\begin{array}{lrl} (2) & x + 2y = 200 & \\ & -x + 5y = 150 & \text{This is } -1(x - 5y) = -1(-150). \\ \hline & 7y = 350 & \\ & y = 50 & \text{To find } y \text{, divide both sides by 7.} \end{array}$$

We then find the value of x to complete step 4 and the value of z (step 5) in the same way as in Examples 2 and 3.

Step 6: Use a check to verify that the solution is $(100, 50, -20)$.

Teaching Example 4
Solve the system:
$$\begin{cases} x - y + 3z = 70 \\ x + 6y = 440 \\ y + z = 10 \end{cases}$$
Answer: $(200, 40, -30)$

Self Check 4 Solve the system: $\begin{cases} x + y + 3z = 35 \\ x + 3y = -20 \\ 2y + z = -35 \end{cases}$ $(40, -20, 5)$

Now Try ▶ **Problem 27**

In the following example, we will solve the same system from Example 4 using a different approach.

EXAMPLE 5 Use substitution to solve the system: $\begin{cases} x - y + 4z = -30 \\ x + 2y = 200 \\ y + z = 30 \end{cases}$

Strategy We will solve the second equation for x and the third equation for z. This creates two substitution equations. Then we will substitute the results for x and for z in the first equation.

Why These substitutions will produce one equation in the variable y.

Solution The method we will use is much like solving systems of two equations by substitution. However, we must find two substitution equations, instead of one.

Teaching Tip: Ask your students which method they prefer for solving this system: the elimination method used in Example 4 or the substitution used in Example 5.

$$\begin{cases} x - y + 4z = -30 \\ x + 2y = 200 \\ y + z = 30 \end{cases} \quad \xrightarrow[\text{Solve for } z]{\text{Solve for } x} \quad \begin{cases} x - y + 4z = -30 & (1) \\ x = \mathbf{200 - 2y} & (2) \\ z = \mathbf{30 - y} & (3) \end{cases}$$

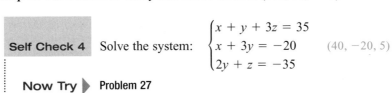

Since the variable x is isolated in equation 2, we will substitute $200 - 2y$ for x in equation 1, and since the variable z is isolated in equation 3, we will substitute $30 - y$ for z in equation 1. These substitutions will eliminate x and z from equation 1, leaving an equation in one variable, y.

Success Tip

Just like with systems of two equations, the objective is to use the appropriate substitutions to obtain one equation in one variable.

$x - y + 4z = -30$	This is equation 1.
$\mathbf{200 - 2y} - y + 4(\mathbf{30 - y}) = -30$	Substitute $200 - 2y$ for x and $30 - y$ for z.
$200 - 2y - y + 120 - 4y = -30$	Distribute the multiplication by 4.
$320 - 7y = -30$	On the left side, combine like terms.
$-7y = -350$	Subtract 320 from both sides.
$y = 50$	To solve for y, divide both sides by -7.

As expected, this is the same value for y that we obtained using elimination in Example 4. We can substitute 50 for y in equation 2 to find that $x = 100$ and 50 for y in equation 3 to find that $z = -20$. Using elimination or substitution, we find that the solution is $(100, 50, -20)$.

Teaching Example 5
Use substitution to solve the system:
$$\begin{cases} x - y + 3z = 70 \\ x + 6y = 440 \\ y + z = 10 \end{cases}$$
Answer: $(200, 40, -30)$

Self Check 5 Use substitution to solve the system: $\begin{cases} x + y - 4z = -54 \\ x - y = -6 \\ 3y + z = 12 \end{cases}$ $(-6, 0, 12)$

Now Try Problem 31

4 Identify Inconsistent Systems and Dependent Equations.

We have seen that a system of three linear equations in three variables represents three planes. If the planes have no common point of intersection, the system is said to be **inconsistent** with no solution. Illustrations of these types of inconsistent systems are shown in figure c on page 1020.

EXAMPLE 6 Solve the system: $\begin{cases} 2a + b - 3c = -3 & \textbf{(1)} \\ 3a - 2b + 4c = 2 & \textbf{(2)} \\ 4a + 2b - 6c = -7 & \textbf{(3)} \end{cases}$

Strategy Since the coefficients of the b-terms are opposites in the second and third equations, we will add the left and right sides of those equations to eliminate b. Then we will choose another pair of equations and eliminate b again.

Why The result will be a system of two equations in a and c that we can attempt to solve by elimination.

Solution We add equations 2 and 3 of the system to eliminate b.

$$\begin{array}{ll} \textbf{(2)} & 3a - 2b + 4c = 2 \\ \textbf{(3)} & \underline{4a + 2b - 6c = -7} \\ \textbf{(4)} & 7a \qquad\quad - 2c = -5 \end{array}$$

We can multiply both sides of equation 1 by 2 and add the resulting equation to equation 2 to eliminate b again:

$$\begin{array}{ll} & 4a + 2b - 6c = -6 \quad \text{This is } 2(2a + b - 3c) = 2(-3). \\ \textbf{(2)} & \underline{3a - 2b + 4c = 2} \\ \textbf{(5)} & 7a \qquad\quad - 2c = -4 \end{array}$$

Equations 4 and 5 form a system in a and c.

$$\begin{array}{ll} \textbf{(4)} & \begin{cases} 7a - 2c = -5 \\ 7a - 2c = -4 \end{cases} \\ \textbf{(5)} & \end{array}$$

Teaching Tip: After determining that this system has no solution, you may want to refer to the illustrations of inconsistent systems on page 1020.

Success Tip

If you obtain a false statement at any time in the solution process, you need not proceed. Such an outcome indicates that the system has no solution.

If we multiply both sides of equation 5 by -1 and add the result to equation 4, the terms involving a and c are both eliminated.

$$\begin{array}{lrl} \textbf{(4)} & 7a - 2c = -5 & \\ & -7a + 2c = 4 & \text{This is } -1(7a - 2c) = -1(-4). \\ \hline & 0 = -1 & \end{array}$$

Note that in the solution process, all three variables have been eliminated. The false statement $0 = -1$ indicates that the system has *no solution* and is, therefore, inconsistent. The solution set is $\varnothing$.

Teaching Example 6
Solve the system:
$$\begin{cases} 4a + 2b + c = 1 \\ -9a + b - c = 3 \\ 14a - 4b + c = -2 \end{cases}$$
Answer: No solution, $\varnothing$; inconsistent system

Self Check 6 Solve the system: $\begin{cases} 2a + b - 3c = 8 \\ 3a - 2b + 4c = 10 \\ 4a + 2b - 6c = -5 \end{cases}$ No solution, $\varnothing$; inconsistent system

Now Try Problems 33 and 35

When the equations in a system of *two* equations with *two* variables are dependent, the system has infinitely many solutions. In such cases, we classify the system as consistent. This is not always true for systems of three equations with three variables. In fact, a system can have dependent equations* and still be inconsistent. The following illustration shows the different possibilities.

A consistent system

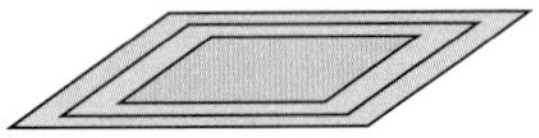

When three planes coincide, the equations are dependent, and there are **infinitely many solutions**.

(a)

A consistent system

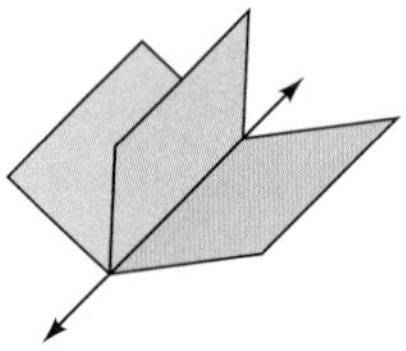

When three planes intersect in a common line, the equations are dependent, and there are **infinitely many solutions**.

(b)

An inconsistent system

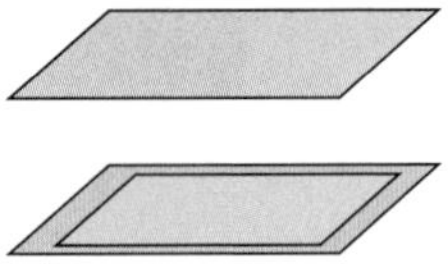

When two planes coincide and are parallel to a third plane, the system is inconsistent, and there are **no solutions**.

(c)

*A set of equations is dependent if at least one equation can be expressed as a sum of multiples of other equations in that set.

EXAMPLE 7 Solve the system: $\begin{cases} \frac{4}{5}x - y + z = \frac{53}{5} \\ x - 2y - z = 8 \\ 0.2x - 0.3y + 0.1z = 2.3 \end{cases}$

Strategy We will find an equivalent system without fractions or decimals and use elimination to solve it.

Why It's easier to solve a system of equations that involves only integers.

Solution To clear equation 1 of fractions, we multiply both sides by the LCD of the fractions, which is 5. To clear equation 3 of decimals, we multiply both sides by 10.

$$\begin{array}{ll} \textbf{(1)} & \frac{4}{5}x - y + z = \frac{53}{5} \\ \textbf{(2)} & x - 2y - z = 8 \\ \textbf{(3)} & 0.2x - 0.3y + 0.1z = 2.3 \end{array} \quad \begin{array}{c} \xrightarrow{\text{Multiply by 5}} \\ \xrightarrow{\text{Unchanged}} \\ \xrightarrow[\text{Multiply by 10}]{} \end{array} \quad \begin{array}{ll} 4x - 5y + 5z = 53 & \textbf{(4)} \\ x - 2y - z = 8 & \textbf{(2)} \\ 2x - 3y + z = 23 & \textbf{(5)} \end{array}$$

If each side of equation 2 is multiplied by 5 and the resulting equation is added to equation 4, the variable z is eliminated.

$$\begin{array}{lrl} \textbf{(4)} & 4x - 5y + 5z = 53 & \\ & \underline{5x - 10y - 5z = 40} & \text{This is } 5(x - 2y - z) = 5(8). \\ \textbf{(6)} & 9x - 15y = 93 & \end{array}$$

If we add equations 2 and 5, the variable z is eliminated again.

$$\begin{array}{lr} \textbf{(2)} & x - 2y - z = 8 \\ \textbf{(5)} & \underline{2x - 3y + z = 23} \\ \textbf{(7)} & 3x - 5y = 31 \end{array}$$

Equations 6 and 7 form a system in x and y. If each side of equation 7 is multiplied by -3 and the resulting equation is added to equation 6, both x and y are eliminated.

$$\begin{array}{lrl} \textbf{(6)} & 9x - 15y = 93 & \\ & \underline{-9x + 15y = -93} & \text{This is } -3(3x - 5y) = -3(31). \\ & 0 = 0 & \end{array}$$

Note that in the solution process, all three variables are eliminated. The resulting true statement, $0 = 0$, indicates that we are working with a set of dependent equations and that the system has an infinite number of solutions.

Success Tip

If you obtain a true statement at any time in the solution process, you need not proceed. Such an outcome indicates that the system contains dependent equations.

Teaching Example 7
Solve the system:
$$\begin{cases} 4x - 5y + 5z = 53 \\ x - 2y - z = 8 \\ 2x - 3y + z = 23 \end{cases}$$
Answer: Infinitely many solutions; dependent equations

Self Check 7 Solve the system: $\begin{cases} x - 2y - z = 1 \\ x + \frac{4}{3}y + z = \frac{5}{3} \\ 0.02x + 0.01y + 0.01z = 0.03 \end{cases}$ Infinitely many solutions; dependent equations

Now Try Problems 33 and 35

SECTION 12.2 STUDY SET

VOCABULARY

Fill in the blanks.

1. $\begin{cases} 2x + y - 3z = 0 \\ 3x - y + 4z = 5 \\ 4x + 2y - 6z = 0 \end{cases}$ is called a _system_ of three linear equations in three variables. Each equation is written in _standard_ $Ax + By + Cz = D$ form.
2. If the first two equations of the system in Exercise 1 are added, the variable y is _eliminated_.
3. Solutions of a system of three equations in three variables, x, y, and z, are written in the form (x, y, z) and are called ordered _triples_.
4. The graph of the equation $2x + 3y + 4z = 5$ is a flat surface called a _plane_.
5. When three planes coincide, the equations of the system are _dependent_, and there are infinitely many solutions.
6. When three planes intersect in a line, the system will have _infinitely_ many solutions.

CONCEPTS

7. For each graph of a system of three equations, determine whether the solution set contains one solution, infinitely many solutions, or no solution.

a.

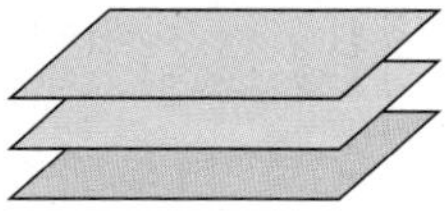

No solution

b.

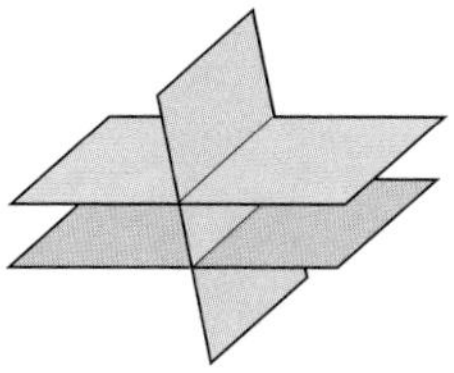

No solution

8. Consider the system: $\begin{cases} \textbf{(1)} & -2x + y + 4z = 3 \\ \textbf{(2)} & x - y + 2z = 1 \\ \textbf{(3)} & x + y - 3z = 2 \end{cases}$
   a. What is the result if equation 1 and equation 2 are added? $-x + 6z = 4$
   b. What is the result if equation 2 and equation 3 are added? $2x - z = 3$
   c. What variable was eliminated in the steps performed in parts (a) and (b)? y

NOTATION

9. For the following system, clear the equations of any fractions or decimals and write each equation in $Ax + By + Cz = D$ form.

$$\begin{cases} x + y = 3 - 4z \\ 0.7x - 0.2y + 0.8z = 1.5 \\ \frac{x}{2} + \frac{y}{3} - \frac{z}{6} = \frac{2}{3} \end{cases} \longrightarrow \begin{cases} x + y + 4z = 3 \\ 7x - 2y + 8z = 15 \\ 3x + 2y - z = 4 \end{cases}$$

10. What is the purpose of the numbers shown in red in front of the equations below?

$$\begin{matrix} (1) \\ (2) \\ (3) \end{matrix} \begin{cases} x + y - z = 6 \\ 2x - y + z = 3 \\ 5x + 3y - z = -2 \end{cases}$$

Numbering the equations helps us describe how the system is solved using the elimination method.

GUIDED PRACTICE

Use a check to determine whether the ordered triple is a solution of the system. See Example 1.

11. (2, 1, 1) $\begin{cases} x - y + z = 2 \\ 2x + y - z = 4 \\ 2x - 3y + z = 2 \end{cases}$ Yes

12. (−3, 2, −1) $\begin{cases} 3x + y - z = -6 \\ 2x + 2y + 3z = -1 \\ x + y + 2z = 1 \end{cases}$ No

13. (6, −7, −5) $\begin{cases} 3x - 2y - z = 37 \\ x - 3y = 27 \\ 2x + 7y + 2z = -48 \end{cases}$ No

14. (−4, 0, 9) $\begin{cases} x + 2y - 3z = -31 \\ 2x + 6z = 46 \\ 3x - y = -12 \end{cases}$ Yes

Solve each system. See Example 2.

15. $\begin{cases} x + y + z = 4 \\ 2x + y - z = 1 \\ 2x - 3y + z = 1 \end{cases}$ (1, 1, 2)

16. $\begin{cases} x + y + z = 4 \\ x - y + z = 2 \\ x - y - 2z = -1 \end{cases}$ (2, 1, 1)

17. $\begin{cases} 3x + 2y - 5z = 3 \\ 4x - 2y - 3z = -10 \\ 5x - 2y - 2z = -11 \end{cases}$ (−1, 3, 0)

18. $\begin{cases} 5x + 4y + 2z = -2 \\ 3x + 4y - 3z = -27 \\ 2x - 4y - 7z = -23 \end{cases}$ (0, −3, 5)

19. $\begin{cases} 2x + 6y + 3z = 9 \\ 5x - 3y - 5z = 3 \\ 4x + 3y + 2z = 15 \end{cases}$ (3, −1, 3)

20. $\begin{cases} 4x - 3y + 5z = 23 \\ 2x - 5y - 3z = 13 \\ -4x - 6y + 7z = 7 \end{cases}$ (3, −2, 1)

21. $\begin{cases} 4x - 5y - 8z = -52 \\ 2x - 3y - 4z = -26 \\ 3x + 7y + 8z = 31 \end{cases}$ (−3, 0, 5)

22. $\begin{cases} 2x + 6y + 3z = -20 \\ 5x - 3y - 5z = 47 \\ 4x + 3y + 2z = 4 \end{cases}$ (5, −4, −2)

Solve each system. See Example 3.

23. $\begin{cases} 3x + 3z = 6 - 4y \\ 7x - 5z = 46 + 2y \\ 4x = 31 - z \end{cases}$ (7, −6, 3)

24. $\begin{cases} 5x + 6z = 4y - 21 \\ 9x + 2y = 3z - 47 \\ 3x + y = -19 \end{cases}$ (−5, −4, −2)

25. $\begin{cases} 2x + z = -2 + y \\ 8x - 3y = -2 \\ 6x - 2y + 3z = -4 \end{cases}$ $\left(\frac{1}{2}, 2, -1\right)$

26. $\begin{cases} 3y + z = -1 \\ -x + 2z = -9 + 6y \\ 9y + 3z = -9 + 2x \end{cases}$ $\left(3, \frac{1}{3}, -2\right)$

Solve each system using elimination. See Example 4.

27. $\begin{cases} x + y + 3z = 35 \\ -x - 3y = 20 \\ 2y + z = -35 \end{cases}$ (40, −20, 5)

28. $\begin{cases} x + 2y + 3z = 11 \\ 5x - y = 13 \\ 2x - 3z = -11 \end{cases}$ (2, −3, 5)

29. $\begin{cases} 3x + 2y - z = 7 \\ 6x - 3y = -2 \\ 3y - 2z = 8 \end{cases}$ $\left(\frac{2}{3}, 2, -1\right)$

30. $\begin{cases} 2x + y = 4 \\ -x - 2y + 8z = 7 \\ -y + 4z = 5 \end{cases}$ $\left(3, -2, \frac{3}{4}\right)$

Solve each system using substitution. See Example 5.

31. $\begin{cases} r + s - 3t = 21 \\ r + 4s = 9 \\ 5s + t = -4 \end{cases}$ (9, 0, −4)

32. $\begin{cases} r - s + 6t = 12 \\ r + 6s = -28 \\ 7s + t = -26 \end{cases}$ (−4, −4, 2)

33. $\begin{cases} x - 8z = -30 \\ 3x + y - 4z = 5 \\ y + 7z = 30 \end{cases}$ (10, −5, 5)

34. $\begin{cases} x + 6z = -36 \\ 5x + 3y - 2z = -20 \\ y + 4z = -20 \end{cases}$ (−6, 0, −5)

Solve each system. If a system is inconsistent or if the equations are dependent, state this. See Examples 6 and 7.

35. $\begin{cases} 7a + 9b - 2c = -5 \\ 5a + 14b - c = -11 \\ 2a - 5b - c = 3 \end{cases}$ No solution, ∅; inconsistent system

36. $\begin{cases} 3x + 4y + z = 10 \\ x - 2y + z = -3 \\ 2x + y + z = 5 \end{cases}$ No solution, ∅; inconsistent system

37. $\begin{cases} 7x - y - z = 10 \\ x - 3y + z = 2 \\ x + 2y - z = 1 \end{cases}$ Inf. many sol.; dep. eq.

38. $\begin{cases} 2a - b + c = 6 \\ -5a - 2b - 4c = -30 \\ a + b + c = 8 \end{cases}$ Inf. many sol.; dep. eq.

TRY IT YOURSELF

Solve each system, if possible. If a system is inconsistent or if the equations are dependent, state this.

39. $\begin{cases} 2a + 3b - 2c = 18 \\ 5a - 6b + c = 21 \\ 4b - 2c - 6 = 0 \end{cases}$ (8, 4, 5)

40. $\begin{cases} r - s + t = 4 \\ r + 2s - t = -1 \\ r + s - 3t = -2 \end{cases}$ (2, −1, 1)

41. $\begin{cases} 2x + 2y - z = 2 \\ x + 3z - 24 = 0 \\ y = 7 - 4z \end{cases}$ (12, −9, 4)

42. $\begin{cases} r - 3t = -11 \\ r + s + t = 13 \\ s - 4t = -12 \end{cases}$ (2.5, 6, 4.5)

43. $\begin{cases} b + 2c = 7 - a \\ a + c = 2(4 - b) \\ 2a + b + c = 9 \end{cases}$ (3, 2, 1)

44. $\begin{cases} 0.02a = 0.02 - 0.03b - 0.01c \\ 4a + 6b + 2c - 5 = 0 \\ a + c = 3 + 2b \end{cases}$ No solution, ∅; inconsistent system

45. $\begin{cases} 2x + y - z = 1 \\ x + 2y + 2z = 2 \\ 4x + 5y + 3z = 3 \end{cases}$ No solution, ∅; inconsistent system

46. $\begin{cases} 2x + 2y + 3z = 10 \\ 3x + y - z = 0 \\ x + y + 2z = 6 \end{cases}$ (0, 2, 2)

▶ **47.** $\begin{cases} 0.4x + 0.3z = 0.4 \\ 2y - 6z = -1 \\ 4(2x + y) = 9 - 3z \end{cases}$
$\left(\frac{3}{4}, \frac{1}{2}, \frac{1}{3}\right)$

▶ **48.** $\begin{cases} a + b + c = 180 \\ \frac{a}{4} + \frac{b}{2} + \frac{c}{3} = 60 \\ 2b + 3c - 330 = 0 \end{cases}$
(60, 30, 90)

49. $\begin{cases} r + s + 4t = 3 \\ 3r + 7t = 0 \\ 3s + 5t = 0 \end{cases}$
No solution, $\varnothing$; inconsistent system

50. $\begin{cases} x - y = 3 \\ 2x - y + z = 1 \\ x + z = -2 \end{cases}$
Infinitely many solutions; dependent equations

51. $\begin{cases} 0.5a + 0.3b = 2.2 \\ 1.2c - 8.5b = -24.4 \\ 3.3c + 1.3a = 29 \end{cases}$
(2, 4, 8)

▶ **52.** $\begin{cases} 4a - 3b = 1 \\ 6a - 8c = 1 \\ 2b - 4c = 0 \end{cases}$
$\left(-\frac{1}{2}, -1, -\frac{1}{2}\right)$

▶ **53.** $\begin{cases} 2x + 3y = 6 - 4z \\ 2x = 3y + 4z - 4 \\ 4x + 6y + 8z = 12 \end{cases}$
Inf. many sol.; dep. eq.

54. $\begin{cases} -x + 5y - 7z = 0 \\ 4x + y - z = 0 \\ x + y - 4z = 0 \end{cases}$
(0, 0, 0)

55. $\begin{cases} a + b = 2 + c \\ a = 3 + b - c \\ -a + b + c - 4 = 0 \end{cases}$
(2.5, 3, 3.5)

▶ **56.** $\begin{cases} 0.1x - 0.3y + 0.4z = 0.2 \\ 2x + y + 2z = 3 \\ 4x - 5y + 10z = 7 \end{cases}$
Inf. many sol.; dep. eq.

▶ **57.** $\begin{cases} x + \frac{1}{3}y + z = 13 \\ \frac{1}{2}x - y + \frac{1}{3}z = -2 \\ x + \frac{1}{2}y - \frac{1}{3}z = 2 \end{cases}$
(2, 6, 9)

58. $\begin{cases} x - \frac{1}{5}y - z = 9 \\ \frac{1}{4}x + \frac{1}{5}y - \frac{1}{2}z = 5 \\ 2x + y + \frac{1}{6}z = 12 \end{cases}$
(4, 5, −6)

APPLICATIONS

59. Graphs of Systems. Explain how each of the following pictures is an example of the graph of a system of three equations. Then describe the solution, if there is any.

a.

Infinitely many solutions, all lying on the line running down the binding

b.

3 parallel planes (shelves); no solution

c.

Each pair of planes (cards) intersect; no solution

d.

3 planes (faces of die) intersect at a corner; 1 solution

▶ **60. Zoology.** An X-ray of a mouse revealed a cancerous tumor located at the intersection of the coronal, sagittal, and transverse planes. From this description, would you expect the tumor to be at the base of the tail, on the back, in the stomach, on the tip of the right ear, or in the mouth of the mouse? In the stomach

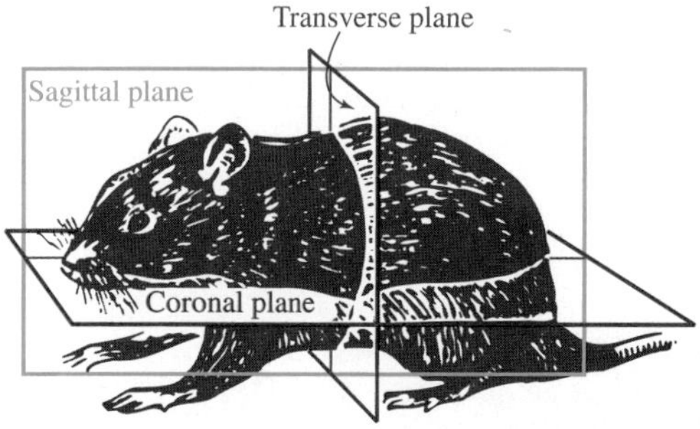

▶ **61. NBA Records.** The three highest one-game point totals by one player in a National Basketball Association game are shown below. Solve the following system to find x, y, and z. 100, 81, 78

$$\begin{cases} x + y + z = 259 \\ x - y = 19 \\ x - z = 22 \end{cases}$$

Pts	Player, team	Date
x	Wilt Chamberlain, Philadelphia	3/2/1962
y	Kobe Bryant, Los Angeles	1/22/2006
z	Wilt Chamberlain, Philadelphia	12/8/1961

62. Bicycle Frames. The angle measures of the triangular part of the bicycle frame shown can be found by solving the following system. Find x, y, and z. 45, 75, 60

$$\begin{cases} x + y + z = 180 \\ x + y = 120 \\ y + z = 135 \end{cases}$$

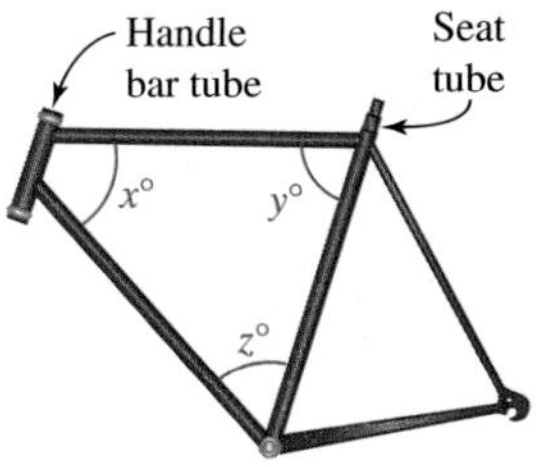

WRITING

63. Explain how a system of three equations in three variables can be reduced to a system of two equations in two variables.

▶ **64.** What makes a system of three equations with three variables inconsistent?

65. What does the graph of a linear equation in three variables such as $2x - 3y + 9z = 10$ look like?

66. What situation discussed in this section looks like two walls of a room and the floor meeting in a corner?

REVIEW

Graph each of the basic functions. See AIE Appendix 3.

67. $f(x) = |x|$

68. $g(x) = x^2$

69. $h(x) = x^3$

▶ **70.** $S(x) = x$

CHALLENGE PROBLEMS

Solve each system.

71. $\begin{cases} w + x + y + z = 3 \\ w - x + y + z = 1 \\ w + x - y + z = 1 \\ w + x + y - z = 3 \end{cases}$

(1, 1, 1, 0)

72. $\begin{cases} \frac{1}{x} + \frac{1}{y} + \frac{1}{z} = 3 \\ \frac{2}{x} + \frac{1}{y} - \frac{1}{z} = 0 \\ \frac{1}{x} - \frac{2}{y} + \frac{4}{z} = 21 \end{cases}$

$\left(\frac{1}{3}, -\frac{1}{3}, \frac{1}{3}\right)$

73. $\begin{cases} 4a + b + 2c - 3d = -16 \\ 3a - 3b + c - 4d = -20 \\ a - 2b - 5c - d = 4 \\ 5a + 4b + 3c - d = -10 \end{cases}$

(−1, 1, −2, 3)

74. $\begin{cases} a + c + 2d = -4 \\ b - 2c = 1 \\ a + 2b - c = -2 \\ 2a + b + 3c - 2d = -4 \end{cases}$

(−1, −1, −1, −1)

SECTION 12.3 Problem Solving Using Systems of Three Equations

OBJECTIVES

1. Assign variables to three unknowns.
2. Use systems to solve curve-fitting problems.

ARE YOU READY? *Are You Ready? exercises available online at www.webassign.net/brookscole*

The following problems review some basic skills that are needed when solving application problems using systems of three equations.

1. Translate to mathematical symbols: 75 more than the number n. $n + 75$
2. Write an algebraic expression that represents the value of x coats, if each coat has a value of \$150. $\$150x$
3. Write an algebraic expression that represents the value (in cents) of n nickels. $5n$¢
4. Solve the system $\begin{cases} x - y + 6z = 12 \\ x + 6y = -28 \\ 7y + z = -26 \end{cases}$ by substitution. (−4, −4, 2)

In this section, we will solve application problems involving three unknown quantities.

1 Assign Variables to Three Unknowns.

The six step problem-solving strategy that was used to solve application problems involving two unknowns in Section 4.4 can be extended to situations involving three unknowns. In the Assign step, we assign *three* variables to represent *three* unknown quantities. In the Form and Solve steps, we form a system of three equations in three variables and solve it using the methods of the previous section.

EXAMPLE 1

Teaching Tip: Have your students turn to page 314 to review the six-step problem solving strategy in Section 4.4 for application problems involving two unknowns.

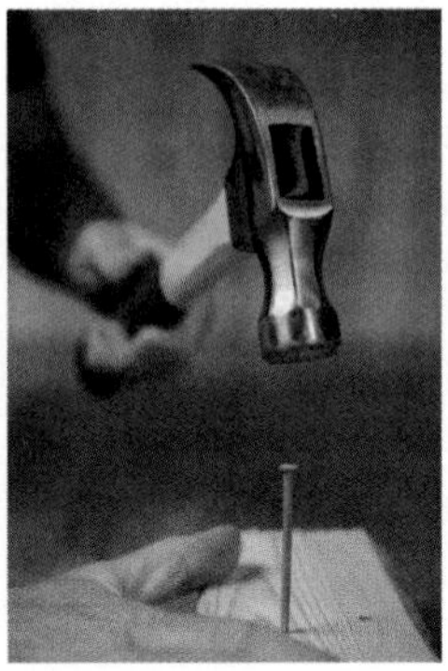

©iStockphoto.com/Jim Jurica

Tool Manufacturing. A company makes three types of hammers, which are marketed as "good," "better," and "best." The cost of manufacturing each type of hammer is \$4, \$6, and \$7, respectively, and the hammers sell for \$6, \$9, and \$12. Each day, the cost of manufacturing 100 hammers is \$520, and the daily revenue from their sale is \$810. How many hammers of each type are manufactured per day?

Analyze We need to find how many of each type of hammer are manufactured daily. Since there are three unknowns, we must write three equations to find them.

Assign Let x = the number of good hammers
y = the number of better hammers
z = the number of best hammers

Form We can organize the facts of the problem in a table, as shown below. To fill in the "Cost to manufacture" column, we multiply the number of good hammers, x, by the cost to manufacture a good hammer, \$4, to find that it costs $\$4 \cdot x$, or $\$4x$, to manufacture the good hammers. Similarly, we use multiplication to find that the manufacturing cost for y better hammers is $\$6y$ and for z best hammers is $\$7z$. The total manufacturing cost of \$520 is entered at the bottom of that column.

To complete the "Revenue received" column, we again use multiplication. The revenue received from the sale of x good hammers, sold for \$6 per hammer, is $\$6 \cdot x$, or $\$6x$. Similarly, the revenue received from the sale of y better hammers is $\$9y$, and the revenue received from the sale of z best hammers is $\$12z$. The total revenue received of \$810 is entered at the bottom of that column.

We can use the facts of the problem to write three equations.

Type of hammer	Number	Cost to manufacture	Revenue received
Good	x	$4x$	$6x$
Better	y	$6y$	$9y$
Best	z	$7z$	$12z$
	Total: 100	Total: \$520	Total: \$810
	↑ One equation comes from this column.	↑ A second equation comes from this column.	↑ A third equation comes from this column.

The resulting system is:
$$\begin{cases} x + y + z = 100 & \textbf{(1)} \\ 4x + 6y + 7z = 520 & \textbf{(2)} \\ 6x + 9y + 12z = 810 & \textbf{(3)} \end{cases}$$

Solve We will use the elimination method to solve this system of three equations in three variables. If we multiply equation 1 by -7 and add the result to equation 2, we get

$$\begin{array}{lrl} & -7x - 7y - 7z = -700 & \text{This is } -7(x + y + z) = -7(100). \\ \textbf{(2)} & 4x + 6y + 7z = 520 & \\ \hline \textbf{(4)} & -3x - y \quad = -180 & \end{array}$$

If we multiply equation 1 by -12 and add the result to equation 3, we get

$$\begin{array}{lrl} & -12x - 12y - 12z = -1{,}200 & \text{This is } -12(x + y + z) = -12(100). \\ \textbf{(3)} & 6x + 9y + 12z = 810 & \\ \hline \textbf{(5)} & -6x - 3y \quad = -390 & \end{array}$$

We can multiply equation 4 by -3 and add it to equation 5 to eliminate y.

$$\begin{array}{lrl} & 9x + 3y = 540 & \text{This is } -3(-3x - y) = -3(-180). \\ \textbf{(5)} & -6x - 3y = -390 & \\ \hline & 3x \quad = 150 & \\ & x = 50 & \text{To solve for x, divide both sides by 3.} \\ & & \text{This is the number of good hammers manufactured.} \end{array}$$

To find y, we substitute 50 for x in equation 4:

$$\begin{aligned} -3x - y &= -180 & \\ -3(\mathbf{50}) - y &= -180 & \text{Substitute 50 for x.} \\ -150 - y &= -180 & \\ -y &= -30 & \text{Add 150 to both sides.} \\ y &= 30 & \text{To solve for y, divide both sides by } -1. \\ & & \text{This is the number of better hammers manufactured.} \end{aligned}$$

To find z, we substitute 50 for x and 30 for y in equation 1:

$$\begin{aligned} x + y + z &= 100 & \\ \mathbf{50} + \mathbf{30} + z &= 100 & \\ z &= 20 & \text{To solve for z, subtract 80 from both sides.} \\ & & \text{This is the number of best hammers manufactured.} \end{aligned}$$

Success Tip

If three variables are used to represent three unknowns, then three equations must be formed, and a three-part check must be performed to verify the results.

State Each day, the company manufactures 50 good hammers, 30 better hammers, and 20 best hammers.

Check If the company manufactures **50** good hammers, **30** better hammers, and **20** best hammers each day, that is a total of **50** + **30** + **20** = 100 hammers. The cost of manufacturing the three types of hammers is \$4(**50**) + \$6(**30**) + \$7(**20**) = \$200 + \$180 + \$140 or \$520. The revenue from the sale of the hammers is \$6(**50**) + \$9(**30**) + \$12(**20**) = \$300 + \$270 + \$240 or \$810. The results check.

Teaching Example 1 Bakeries. At a local bakery, 3 price levels of pies are available. Cream pies sell for \$10 per pie, fruit pies sell for \$12 per pie, and specialty pies sell for \$15 per pie. The cost of producing the pies is \$4.00, \$4.50, and \$5.00, respectively. Each day, the revenue from the 50 pies sold is \$585 and the cost to make them is \$220. How many pies of each type are made?
Answer: Cream pies: 15, fruit pies: 30, specialty pies: 5

Self Check 2 **Computer Storage.** A manufacturer of memory cards makes 1-GB, 2-GB, and 4-GB storage size cards. The cost of manufaturing each is \$2, \$3, and \$5, respectively. Each day the cost of manufacturing 500 cards is \$1,500. The cards sell for \$15, \$20, and \$30, respectively, with a daily revenue of \$10,000. How many memory cards of each type are manufactured? 1-GB: 200 cards; 2-GB: 200 cards; 4-GB: 100 cards

Now Try Problem 7

EXAMPLE 2 **The Olympics.** The three countries that won the most medals in the 2008 summer Olympic games were the United States, China, and Russia, in that order. Together they won a total of 282 medals, with the U.S. medal count 10 more than China's and China's medal count 28 more than Russia's. Find the number of medals won by each country. (Source: sportsillustrated.cnn.com)

Analyze We need to find how many medals the United States, China, and Russia won. Since there are three unknowns, we must write three equations to find them.

Assign Let x = the number of medals won by the United States
y = the number of medals won by China
z = the number of medals won by Russia

Form We can use the facts of the problem to write three equations.

The number of medals won by the United States	plus	the number of medals won by China	plus	the number of medals won by Russia	was	282.
x	$+$	y	$+$	z	$=$	282

The United States' medal count	was	10 more than	China's medal count.
x	$=$	y $+$	10

China's medal count	was	28 more than	Russia's medal count.
y	$=$	z $+$	28

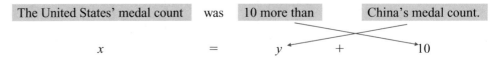

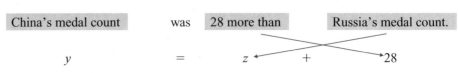

Solve We can use substitution to solve the resulting system of three equations. If we solve equation 3 for z, then the resulting equation 4 and equation 2 can serve as substitution equations.

$$\begin{array}{ll} (1) \\ (2) \\ (3) \end{array} \begin{cases} x + y + z = 282 \\ x = y + 10 \\ y = z + 28 \end{cases} \quad \begin{array}{l} \text{Unchanged} \rightarrow \\ \text{Unchanged} \rightarrow \\ \text{Solve for } z \rightarrow \end{array} \begin{cases} x + y + z = 282 \\ x = y + 10 \\ z = y - 28 \end{cases} \begin{array}{l} (1) \\ (2) \\ (4) \end{array}$$

When we substitute for x and z in equation 1, we obtain an equation in one variable, y.

$x + y + z = 282$	This is equation 1.
$\mathbf{y + 10} + y + \mathbf{y - 28} = 282$	Substitute $y + 10$ for x and $y - 28$ for z.
$3y - 18 = 282$	On the left side, combine like terms.
$3y = 300$	Add 18 to both sides.
$y = 100$	To solve for y, divide both sides by 3. This is the number of medals won by China.

To find x, we substitute 100 for y in equation 2. To find z, we substitute 100 for y in equation 4.

$x = y + 10$	This is equation 2.	$z = y - 28$	This is equation 4.
$x = \mathbf{100} + 10$		$z = \mathbf{100} - 28$	
$x = 110$	This is the U.S. medal count.	$z = 72$	Russia's medal count

State In the 2008 summer Olympics, the United States won 110 medals, China won 100, and Russia won 72.

Check The sum of 110 + 100 + 72 is 282. Furthermore, 110 is 10 more than 100, and 100 is 28 more than 72. The results check.

Teaching Example 2 The Olympics. The three countries that won the most medals in the 2010 winter Olympics were the United States, Germany, and Canada, in that order. Together they won a total of 93 medals, with the U.S. medal count 7 more than Germany's, and Germany's medal count 4 more than Canada's. Find the number of medals won by each country. (Source: *Vancouver Examiner*)
Answer: United States: 37 medals; Germany: 30 medals; Canada: 26 medals

Self Check 2 **The Winter Olympics.** The three countries that are at the top of the list for the most all-time medal wins are Norway, the United States, and Austria, in that order. Together they have won a total of 618 medals, with Norway's medal count 70 more than the U.S., and the U.S. medal count 31 more than Austria's. Find the number of medals won by each country. (Source: nationmaster.com) Norway: 263 medals; United States: 193 medals; Austria: 162 medals

Now Try Problems 33 and 35

2 Use Systems to Solve Curve-Fitting Problems.

The process of determining an equation whose graph contains given points is called **curve fitting.**

EXAMPLE 3 The equation of a parabola opening upward or downward is of the form $y = ax^2 + bx + c$. Find the equation of the parabola graphed on the left by determining the values of a, b, and c.

Strategy We will substitute the x- and y-coordinates of three points that lie on the graph into the equation $y = ax^2 + bx + c$. This will produce a system of three equations in three variables that we can solve to find a, b, and c.

Why Once we know a, b, and c, we can write the equation.

Solution Since the parabola passes through the points $(-1, 5)$, $(1, 1)$, and $(2, 2)$, each pair of coordinates must satisfy the equation $y = ax^2 + bx + c$. If we substitute each pair into $y = ax^2 + bx + c$, we will get a system of three equations in three variables.

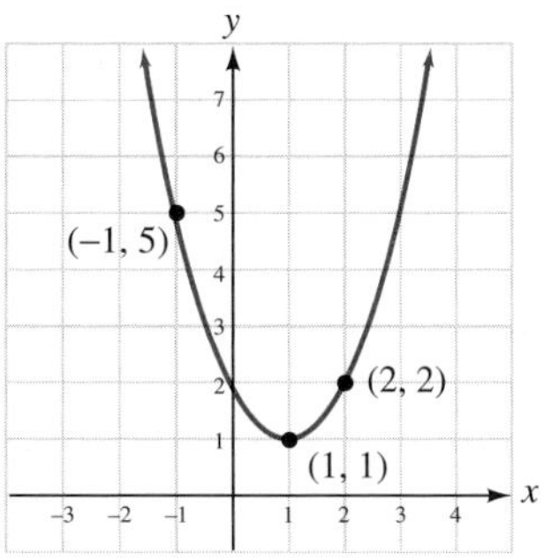

***Substitute* (−1, 5)**	***Substitute* (1, 1)**	***Substitute* (2, 2)**
$y = ax^2 + bx + c$	$y = ax^2 + bx + c$	$y = ax^2 + bx + c$
$5 = a(-1)^2 + b(-1) + c$	$1 = a(1)^2 + b(1) + c$	$2 = a(2)^2 + b(2) + c$
$5 = a - b + c$	$1 = a + b + c$	$2 = 4a + 2b + c$
This is equation 1.	This is equation 2.	This is equation 3.

Success Tip

If a point lies on the graph of an equation, it is a solution of the equation, and the coordinates of the point satisfy the equation.

The three equations above give the system, which we can solve to find a, b, and c.

$$\begin{cases} a - b + c = 5 & (1) \\ a + b + c = 1 & (2) \\ 4a + 2b + c = 2 & (3) \end{cases}$$

If we add equations 1 and 2, we obtain

$$\begin{array}{rl} & a - b + c = 5 \\ & a + b + c = 1 \\ \hline (4) & 2a \quad + 2c = 6 \end{array}$$

If we multiply equation 1 by 2 and add the result to equation 3, we get

$$\begin{array}{rl} & 2a - 2b + 2c = 10 \\ (3) & 4a + 2b + c = 2 \\ \hline (5) & 6a \quad + 3c = 12 \end{array}$$

We can then divide both sides of equation 4 by 2 to get equation 6 and divide both sides of equation 5 by 3 to get equation 7. We now have the system

$$\begin{cases} a + c = 3 & (6) \\ 2a + c = 4 & (7) \end{cases}$$

To eliminate c, we multiply equation 6 by -1 and add the result to equation 7. We get

$$\begin{array}{rl} -a - c = -3 & \text{This is } -1(a + c) = -1(3). \\ 2a + c = 4 & \\ \hline a \quad = 1 & \end{array}$$

To find c, we can substitute 1 for a in equation 6 and find that $c = 2$. To find b, we can substitute 1 for a and 2 for c in equation 2 and find that $b = -2$.

After we substitute these values of a, b, and c into the equation $y = ax^2 + bx + c$, we have the equation of the parabola.

$$y = ax^2 + bx + c$$
$$y = 1x^2 - 2x + 2$$
$$y = x^2 - 2x + 2 \quad \text{This is the equation of the parabola graphed above.}$$

Teaching Example 3
Find the equation of the parabola, $y = ax^2 + bx + c$, that passes through $(-2, 0)$, $(-1, 1)$, and $(1, -3)$.
Answer: $y = -x^2 - 2x$

Self Check 3 Find the equation of the parabola, $y = ax^2 + bx + c$, that passes through $(1, 6)$, $(-4, 1)$, and $(-3, -2)$. $y = x^2 + 4x + 1$

Now Try ▶ Problem 27

SECTION 12.3 STUDY SET

VOCABULARY

Fill in the blanks.

1. If a point lies on the graph of an equation, it is a solution of the equation, and the coordinates of the point _satisfy_ the equation.
2. The process of determining an equation whose graph contains given points is called curve _fitting_.

CONCEPTS

Write a system of three equations in three variables that models the situation. Do not solve the system.

3. **Desserts.** A bakery makes three kinds of pies: chocolate cream, which sells for \$5; apple, which sells for \$6; and cherry, which sells for \$7. The cost to make the pies is \$2, \$3, and \$4, respectively. Let x = the number of chocolate cream pies made daily, y = the number of apple pies made daily, and z = the number of cherry pies made daily.
   - Each day, the bakery makes 50 pies.
   - Each day, the revenue from the sale of the pies is \$295.
   - Each day, the cost to make the pies is \$145.

   $$\begin{cases} x + y + z = 50 \\ 5x + 6y + 7z = 295 \\ 2x + 3y + 4z = 145 \end{cases}$$

4. **Fast Foods.** Let x = the number of calories in a Big Mac hamburger, y = the number of calories in a small order of French fries, and z = the number of calories in a medium Coca-Cola.
   - The total number of calories in a Big Mac hamburger, a small order of French fries, and a medium Coke is 1,000.
   - The number of calories in a Big Mac is 260 more than in a small order of French fries.
   - The number of calories in a small order of French fries is 40 more than in a medium Coke. (Source: McDonald's USA)

$$\begin{cases} x + y + z = 1{,}000 \\ x = y + 260 \\ y = z + 40 \end{cases}$$

5. What equation results when the coordinates of the point $(2, -3)$ are substituted into $y = ax^2 + bx + c$? $-3 = 4a + 2b + c$
6. The equation $y = 5x^2 - 6x + 1$ is written in the form $y = ax^2 + bx + c$. What are a, b, and c? 5, −6, 1

APPLICATIONS

7. **Making Statues.** An artist makes three types of ceramic statues (large, medium, and small) at a monthly cost of \$650 for 180 statues. The manufacturing costs for the three types are \$5, \$4, and \$3. If the statues sell for \$20, \$12, and \$9, respectively, how many of each type should be made to produce \$2,100 in monthly revenue? 30 large, 50 medium, 100 small
8. **Puppets.** A toy company makes a total of 500 puppets in three sizes during a production run. The small puppets cost \$5 to make and sell for \$8 each, the standard-size puppets cost \$10 to make and sell for \$16 each, and the super-size puppets cost \$15 to make and sell for \$25. The total cost to make the puppets is \$4,750 and the revenue from their sale is \$7,700. How many small, standard, and super-size puppets are made during a production run? Small: 150, standard: 250, super-size: 100
9. **Nutrition.** A dietician is to design a meal using Foods A, B, and C that will provide a patient with exactly 14 grams of fat, 13 grams of carbohydrates, and 9 grams of protein.
   - Each ounce of Food A contains 2 grams of fat, 3 grams of carbohydrates, and 2 grams of protein.
   - Each ounce of Food B contains 3 grams of fat, 2 grams of carbohydrates, and 1 gram of protein.
   - Each ounce of Food C contains 1 gram of fat, 1 gram of carbohydrates, and 2 grams of protein.

   a. Complete the following table and then form a system of three equations that could be used to determine how many ounces of each food should be used in the meal.

Name of food	Number of ounces used	Grams of fat	Grams of carbohydrates	Grams of protein
A	a	$2a$	$3a$	$2a$
B	b	$3b$	$2b$	b
C	c	c	c	$2c$
		Total: 14	Total: 13	Total: 9

   b. Solve the system from part a. Food A: 2 oz, Food B: 3 oz, Food C: 1 oz

$$\begin{cases} 2a + 3b + c = 14 \\ 3a + 2b + c = 13 \\ 2a + b + 2c = 9 \end{cases}$$

10. **Nutritional Planning.** One ounce of each of three foods has the vitamin and mineral content shown in the table. How many ounces of each must be used to provide exactly 22 milligrams (mg) of niacin, 12 mg of zinc, and 20 mg of vitamin C? Food A: 2 oz, Food B: 4 oz, Food C: 6 oz

Milligrams per ounce in each food type

Food	Niacin	Zinc	Vitamin C
A	1 mg	1 mg	2 mg
B	2 mg	1 mg	1 mg
C	2 mg	1 mg	2 mg

11. from **Campus to Careers**

Fashion Designer

©iStockphoto.com/CandyBox Photography

A clothing manufacturer makes coats, shirts, and slacks. The time required for cutting, sewing, and packaging each item is shown in the table. How many of each should be made to use all available labor hours? 120 coats, 200 shirts, 150 slacks

	Coats	Shirts	Slacks	Time available
Cutting	20 min	15 min	10 min	115 hr
Sewing	60 min	30 min	24 min	280 hr
Packaging	5 min	12 min	6 min	65 hr

12. **Sculpting.** A wood sculptor carves three types of statues with a chainsaw. The number of hours required for carving, sanding, and painting a totem pole, a bear, and a deer are shown in the table. How many of each should be produced to use all available labor hours? 3 poles, 2 bears, 4 deer

	Totem pole	Bear	Deer	Time available
Carving	2 hr	2 hr	1 hr	14 hr
Sanding	1 hr	2 hr	2 hr	15 hr
Painting	3 hr	2 hr	2 hr	21 hr

13. **NFL Records.** Jerry Rice, who played the majority of his career with the San Francisco 49ers and the Oakland Raiders, holds the all-time record for touchdown (TD) passes caught. Here are some interesting facts about this feat.
    - He caught 30 more TD passes from Steve Young than he did from Joe Montana.
    - He caught 39 more TD passes from Joe Montana than he did from Rich Gannon.
    - He caught a total of 156 TD passes from Young, Montana, and Gannon.

    Determine the number of touchdown passes Rice has caught from Young, from Montana, and from Gannon. Young: 85 TD passes, Montana: 55 TD passes, Gannon: 16 TD passes
14. **Hot Dogs.** In 10 minutes, the top three finishers in the 2010 Nathan's Hot Dog Eating Contest consumed a total of 136 hot dogs. The winner, Joey Chestnut, ate 9 more hot dogs than the runner-up, Tim Janus. Pat Bertoletti finished a distant third, 8 hot dogs behind Janus. How many hot dogs did each person eat? (Source: nathansfamous.com) Chestnut: 54, Janus: 45, Bertoletti: 37

15. Earth's Atmosphere. Use the information in the circle graph to determine what percent of Earth's atmosphere is nitrogen, is oxygen, and is other gases. Nitrogen: 78%, oxygen: 21%, other gases: 1%

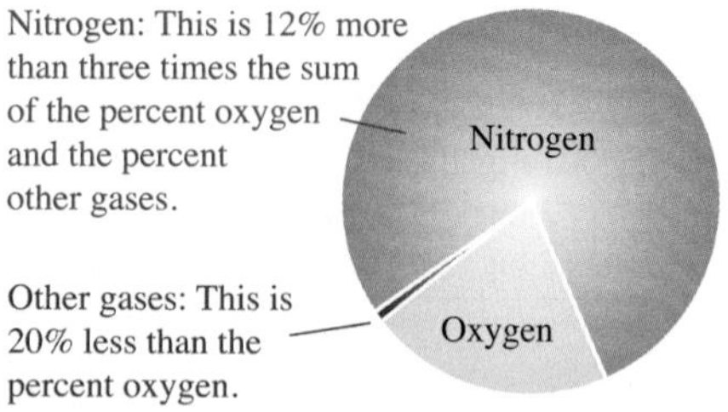

16. Deceased Celebrities. Between October 2009 and October 2010, the estates of Michael Jackson, Elvis Presley, and J.R.R. Tolkien (author of *Lord of the Rings*) earned a total of \$385 million. Together, the Presley and Tolkien estates earned \$165 million less than the Jackson estate. The Jackson estate earned 5.5 times as much as the Tolkien estate. Use this information to label each bar on the graph below. (Source: Forbes.com) Jackson: \$275 million, Presley: \$60 million, Tolkien: \$50 million

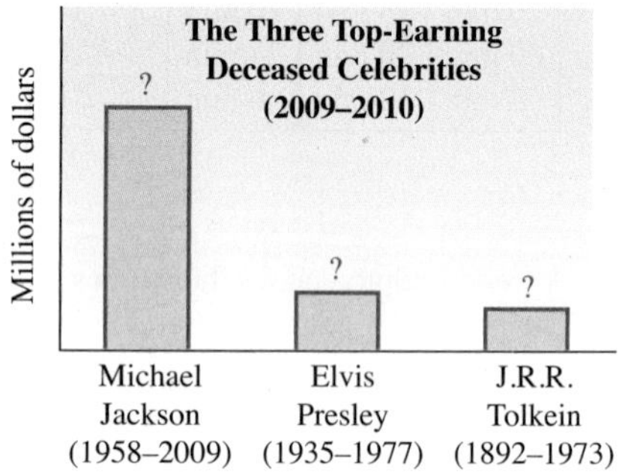

17. Triangles. The sum of the measures of the angles of any triangle is 180°. In ΔABC, $\angle A$ measures 100° less than the sum of the measures of $\angle B$ and $\angle C$, and the measure of $\angle C$ is 40° less than twice the measure of $\angle B$. Find the measure of each angle of the triangle. $\angle A$: 40°, $\angle B$: 60°, $\angle C$: 80°

18. Quadrilaterals. A quadrilateral is a four-sided polygon. The sum of the measures of the angles of any quadrilateral is 360°. In the illustration, the measures of $\angle A$ and $\angle B$ are the same. The measure of $\angle C$ is 20° greater than the measure of $\angle A$, and the measure of $\angle D$ is 60° less than $\angle B$. Find the measure of $\angle A$, $\angle B$, $\angle C$, and $\angle D$.
$\angle A$: 100°, $\angle B$: 100°, $\angle C$: 120°, $\angle D$: 40°

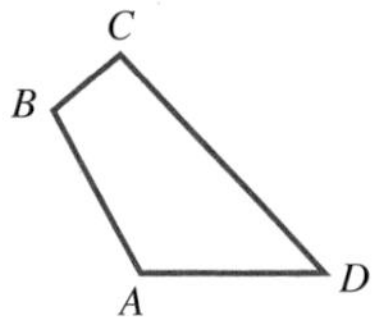

19. TV History. *X-Files, Will & Grace,* and *Seinfeld* are three of the most popular television shows of all time. The total number of episodes of these three shows is 575. There are 21 more episodes of *X-Files* than *Seinfeld,* and the difference between the number of episodes of *Will & Grace* and *Seinfeld* is 14. Find the number of episodes of each show.
X-Files: 201 episodes, *Will & Grace:* 194 episodes, *Seinfeld:* 180 episodes

20. Traffic Lights. At a traffic light, one cycle through green-yellow-red lasts for 80 seconds. The green light is on eight times longer than the yellow light, and the red light is on eleven times longer than the yellow light. For how long is each colored light on during one cycle? Green: 32 sec, yellow: 4 sec, red: 44 sec

21. Ice Skating. Three circles are traced out by a figure skater during her performance as shown below. If the centers of the circles are the given distances apart (10 yd, 14 yd, and 18 yd), find the radius of each circle. (*Hint:* Each red line segment is composed of two radii. Label the radii r_1, r_2, *and* r_3.) r_1: 3 yd, r_2: 7 yd, r_3: 11 yd

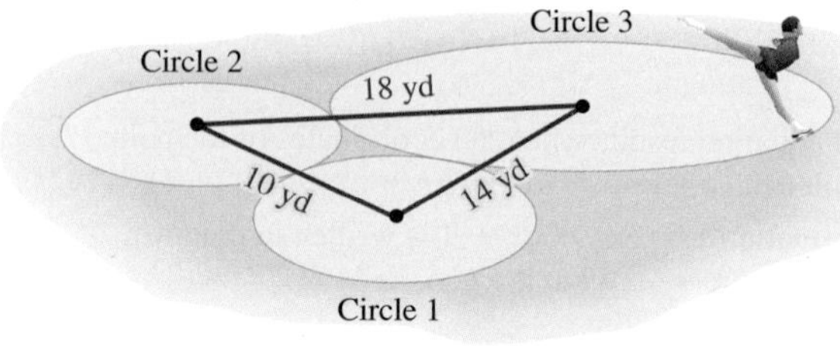

22. NBA Centers. Together, Shaquille O'Neil (Boston Celtics) and Dwight Howard (Orlando Magic) weigh 590 pounds. Together O'Neil and Yao Ming (Houston Rockets) weigh 635 pounds. Together, Howard and Yao Ming weigh 575 pounds. Find the weight of each center. O'Neil: 325 lb, Howard: 265 lb, Yao Ming: 310 lb

23. Potpourri. The owner of a home decorating shop wants to mix dried rose petals selling for \$6 per pound, dried lavender selling for \$5 per pound, and buckwheat hulls selling for \$4 per pound to get 10 pounds of a mixture that would sell for \$5.50 per pound. She wants to use twice as many pounds of rose petals as lavender. How many pounds of each should she use? 6 lb rose petals, 3 lb lavender, 1 lb buckwheat hulls

24. Mixing Nuts. The owner of a candy store wants to mix some peanuts worth \$3 per pound, some cashews worth \$9 per pound, and some Brazil nuts worth \$9 per pound to get 50 pounds of a mixture that will sell for \$6 per pound. She uses 15 fewer pounds of cashews than peanuts. How many pounds of each did she use? 25 lb peanuts, 10 lb cashews, 15 lb Brazil nuts

25. Piggy Banks. When a child breaks open her piggy bank, she finds a total of 64 coins, consisting of nickels, dimes, and quarters. The total value of the coins is \$6. If the nickels were dimes, and the dimes were nickels, the value of the coins would be \$5. How many nickels, dimes, and quarters were in the piggy bank? Nickels: 20, dimes: 40, quarters: 4

26. Theater Seating. The illustration shows the cash receipts and the ticket prices from two sold-out Sunday performances of a play. Find the number of seats in each of the three sections of the 800-seat theater. Founder's circle: 100 tickets, box seats: 300 tickets, promenade: 400 tickets

Sunday Ticket Receipts

Matinee	\$13,000
Evening	\$23,000

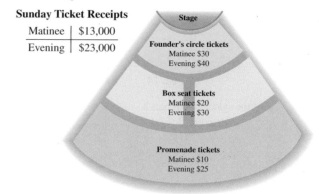

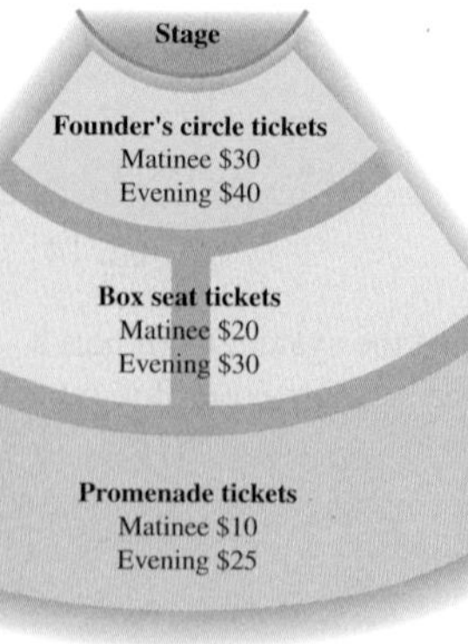

27. Astronomy. Comets have elliptical orbits, but the orbits of some comets are so large that they look much like a parabola. Find an equation of the form $y = ax^2 + bx + c$ for the parabola that closely describes the orbit of the comet shown in the illustration. $y = \frac{1}{2}x^2 - 2x - 1$

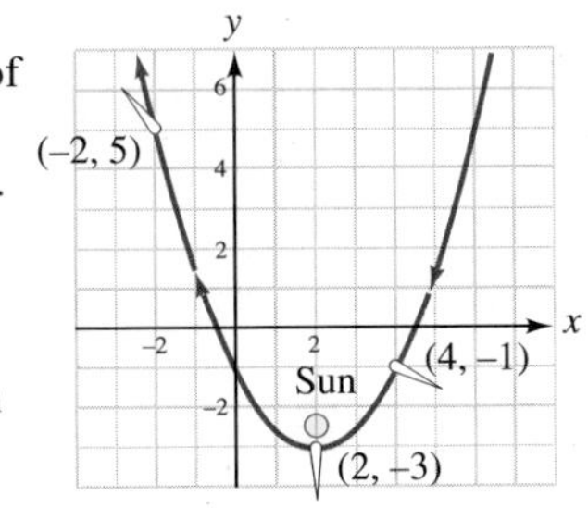

28. Curve Fitting. Find an equation of the form $y = ax^2 + bx + c$ for the parabola shown in the illustration. $y = -x^2 + x - 1$

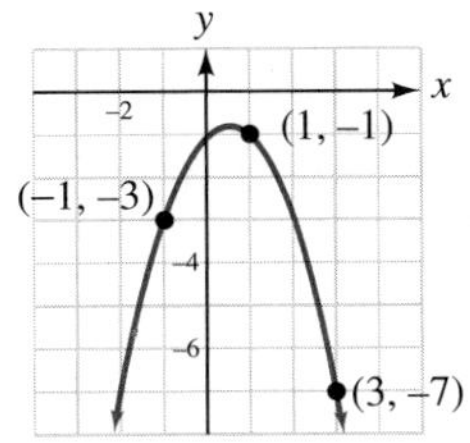

29. Walkways. A circular sidewalk is to be constructed in a city park. The walk is to pass by three particular areas of the park, as shown in the illustration. If an equation of a circle is of the form $x^2 + y^2 + Cx + Dy + E = 0$, find an equation that describes the path of the sidewalk by determining C, D, and E. $x^2 + y^2 - 2x - 2y - 2 = 0$

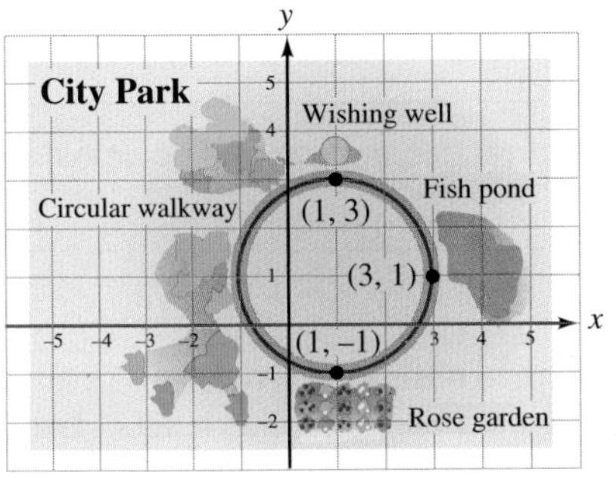

30. Curve Fitting. The equation of a circle is of the form $x^2 + y^2 + Cx + Dy + E = 0$. Find an equation of the circle shown in the illustration by determining C, D, and E. $x^2 - 6x + y^2 = 0$

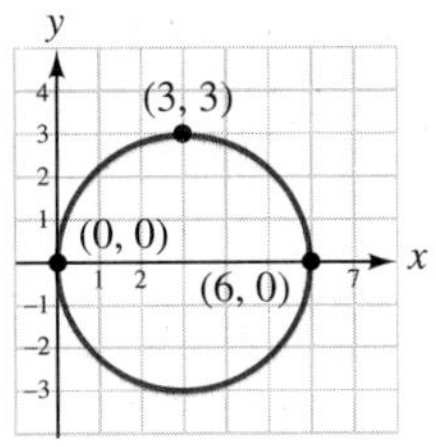

WRITING

31. Explain why the following problem does not give enough information to answer the question: The sum of three integers is 48. If the first integer is doubled, the sum is 60. Find the integers.

32. Write an application problem that can be solved using a system of three equations in three variables.

REVIEW

Determine whether each equation defines y to be a function of x. If it does not, find two ordered pairs where more than one value of y corresponds to a single value of x.

33. $y = \frac{1}{x}$ Yes

34. $y^4 = x$ No; (1, 1), (1, −1)

35. $xy = 9$ Yes

36. $y = |x|$ Yes

37. $x + 1 = |y|$ No; (1, 2), (1, −2)

38. $y = \frac{1}{x^2}$ Yes

39. $y^2 = x$ No; (4, 2), (4, −2)

40. $x = |y|$ No; (1, 1), (1, −1)

CHALLENGE PROBLEMS

41. Digits Problems. The sum of the digits of a three-digit number is 8. Twice the hundreds digit plus the tens digit is equal to the ones digit. If the digits of the number are reversed, the new number is 82 more than twice the original number. What is the three-digit number? 215

42. Purchasing Pets. A pet store owner spent \$100 to buy 100 animals. He bought at least one iguana, one guinea pig, and one mouse, but no other kinds of animals. If an iguana cost \$10.00, a guinea pig cost \$3.00, and a mouse cost \$0.50, how many of each did he buy? 5 iguana, 1 guinea pig, 94 mice

SECTION 12.4

OBJECTIVES

1. Define a matrix and determine its order.
2. Write the augmented matrix for a system.
3. Perform elementary row operations on matrices.
4. Use matrices to solve a system of two equations.
5. Use matrices to solve a system of three equations.
6. Use matrices to identify inconsistent systems and dependent equations.

Solving Systems of Equations Using Matrices

ARE YOU READY? *Are You Ready? exercises available online at www.webassign.net/brookscole*

The following problems review some basic skills that are needed when solving systems of equations using matrices.

1. Consider the system $\begin{cases} 3x - 7y = 14 \\ 5x - y = -9 \end{cases}$. What are the coefficients of the variable terms on the left side of:

a. the first equation? 3, −7 **b.** the second equation? 5, −1

2. Multiply −2 times 4 and add the result to 9. What is the answer? 1

3. Multiply −18 by $-\frac{1}{18}$. What is the result? 1

4. Multiply both sides of $2x - 6y = 5$ by −2. What is the result? $-4x + 12y = -10$

In this section, we will discuss another way to solve systems of linear equations. This technique uses a mathematical tool called a *matrix* in a series of steps that are based on the elimination (addition) method.

1 Define a Matrix and Determine Its Order.

Another way to solve systems of equations involves rectangular arrays of numbers called *matrices* (plural of matrix).

Matrices

A **matrix** is any rectangular array of numbers arranged in rows and columns, written within brackets.

The Language of Algebra

An **array** is an orderly arrangement. For example, a jewelry store might display an impressive *array* of gemstones.

The **rows** of a matrix are horizontal and the **columns** are vertical. A pair of brackets is used to write a matrix. Matrices are often named using capital letters. Two examples of matrices are:

$$A = \begin{bmatrix} 1 & -3 & 8 \\ 2 & 5 & -1 \end{bmatrix} \begin{matrix} \leftarrow \text{Row 1} \\ \leftarrow \text{Row 2} \end{matrix} \qquad B = \begin{bmatrix} 1 & 4 & -2 & -4 \\ 6 & -2 & 6 & 1 \\ 3 & 8 & -3 & 12 \end{bmatrix} \begin{matrix} \leftarrow \text{Row 1} \\ \leftarrow \text{Row 2} \\ \leftarrow \text{Row 3} \end{matrix}$$

(A: ↑ Column 1, ↑ Column 2, ↑ Column 3; B: ↑ Column 1, ↑ Column 2, ↑ Column 3, ↑ Column 4)

Teaching Tip: The rows and columns of student desks in a typical classroom can serve as an example of a matrix.

Each number in a matrix is called an **element** or an **entry** of the matrix. A matrix with m rows and n columns has **order** $m \times n$, which is read as "m by n." Because matrix A has two rows and three columns, its order is 2×3. The order of matrix B is 3×4 because it has three rows and four columns.

2 Write the Augmented Matrix for a System.

Caution

The equations of a system must be written in standard form before the corresponding augmented matrix can be written.

To show how to use matrices to solve systems of linear equations, consider the following system (written in standard form) and the corresponding **augmented matrix** to its right.

A system of two linear equations $\qquad$ ***Its augmented matrix***

$$\begin{cases} x - y = 4 \\ 2x + y = 5 \end{cases} \qquad \left[\begin{array}{cc|c} 1 & -1 & 4 \\ 2 & 1 & 5 \end{array}\right]$$

Each row of the augmented matrix represents one equation of the system. The first two columns of the augmented matrix are determined by the coefficients of x and y in the equations of the system. The last column is determined by the constants in the equations. A dashed vertical bar is written in place of the equal symbols.

$$\left[\begin{array}{cc:c} 1 & -1 & 4 \\ 2 & 1 & 5 \end{array}\right]$$

This row represents the equation $x - y = 4$.
This row represents the equation $2x + y = 5$.

Coefficients of x — Coefficients of y — Constants

EXAMPLE 1 Represent each system using an augmented matrix:

a. $\begin{cases} 3x + y = 11 \\ x - 8y = 0 \end{cases}$ **b.** $\begin{cases} 2a + b - 3c = -3 \\ 9a + 4c = 2 \\ a - b - 6c = -7 \end{cases}$

Strategy We will write the coefficients of the variables and the constants from each equation in rows to form a matrix. The coefficients are written to the left of a vertical dashed line and constants to the right.

Why In an augmented matrix, each row represents one equation of the system.

Solution Since the equations of each system are written in standard form, we can easily write the corresponding augmented matrices.

a. $\begin{cases} 3x + y = 11 \\ x - 8y = 0 \end{cases} \leftrightarrow \left[\begin{array}{cc:c} 3 & 1 & 11 \\ 1 & -8 & 0 \end{array}\right]$

b. $\begin{cases} 2a + b - 3c = -3 \\ 9a + 4c = 2 \\ a - b - 6c = -7 \end{cases} \leftrightarrow \left[\begin{array}{ccc:c} 2 & 1 & -3 & -3 \\ 9 & 0 & 4 & 2 \\ 1 & -1 & -6 & -7 \end{array}\right]$

In the second row, 0 is entered as the coefficient of the missing b-term.

Teaching Example 1
Represent each system using an augmented matrix:

a. $\begin{cases} 5a - 7b = 11 \\ 3a + 5b = 2 \end{cases}$

b. $\begin{cases} 2a + b - 3c = 1 \\ a - 2b + c = 7 \\ 3a + b - 4c = 2 \end{cases}$

Answers:

a. $\left[\begin{array}{cc:c} 5 & -7 & 11 \\ 3 & 5 & 2 \end{array}\right]$

b. $\left[\begin{array}{ccc:c} 2 & 1 & -3 & 1 \\ 1 & -2 & 1 & 7 \\ 3 & 1 & -4 & 2 \end{array}\right]$

Self Check 1 Represent each system using an augmented matrix:

a. $\begin{cases} 2x - 4y = 9 \\ 5x - y = -2 \end{cases}$ **b.** $\begin{cases} a + b - c = -4 \\ -2b + 7c = 0 \\ 10a + 8b - 4c = 5 \end{cases}$

a. $\left[\begin{array}{cc:c} 2 & -4 & 9 \\ 5 & -1 & -2 \end{array}\right]$

b. $\left[\begin{array}{ccc:c} 1 & 1 & -1 & -4 \\ 0 & -2 & 7 & 0 \\ 10 & 8 & -4 & 5 \end{array}\right]$

Now Try Problem 13

3 Perform Elementary Row Operations on Matrices.

The Language of Algebra

The word **echelon** (pronounced esh · e · lon) means a steplike formation in which each unit is slightly to the left or right of the one preceding it, such as geese flying in *echelon*.

To solve a system of linear equations using matrices, we transform the entries to the left of the dashed line in the augmented matrix into an equivalent matrix that has 1's down its **main diagonal** and 0's in all the remaining entries directly above and below this diagonal. A matrix written in this form is said to be in **reduced row-echelon form.** We can easily determine the solution of the associated system of equations when an augmented matrix is written in this form.

Reduced row-echelon form

$$\left[\begin{array}{cc:c} 1 & 0 & a \\ 0 & 1 & b \end{array}\right] \qquad \left[\begin{array}{ccc:c} 1 & 0 & 0 & a \\ 0 & 1 & 0 & b \\ 0 & 0 & 1 & c \end{array}\right]$$

Main diagonal — Main diagonal

a, b, and c represent real numbers.

To write an augmented matrix in reduced row-echelon form, we use three operations called **elementary row operations.**

Elementary Row Operations

Type 1: Any two rows of a matrix can be interchanged.
Type 2: Any row of a matrix can be multiplied by a nonzero constant.
Type 3: Any row of a matrix can be changed by adding a nonzero constant multiple of another row to it.

The Language of Algebra

Two matrices are **equivalent** if they represent systems that have the same solution set.

None of these row operations affect the solution of a given system of equations. The changes to the augmented matrix produce equivalent matrices that correspond to systems with the same solution.

- A type 1 row operation corresponds to interchanging two equations of the system.
- A type 2 row operation corresponds to multiplying both sides of an equation by a nonzero constant.
- A type 3 row operation corresponds to adding a nonzero multiple of one equation to another.

EXAMPLE 2 Perform the following elementary row operations.

$$A = \left[\begin{array}{cc|c} 2 & 4 & -3 \\ 1 & -8 & 0 \end{array}\right] \quad B = \left[\begin{array}{cc|c} 1 & -1 & 2 \\ 4 & -8 & 0 \end{array}\right] \quad C = \left[\begin{array}{ccc|c} 2 & 1 & -8 & 4 \\ 0 & 1 & 4 & -2 \\ 0 & 0 & -6 & 24 \end{array}\right]$$

a. Type 1: Interchange rows 1 and 2 of matrix A.

b. Type 2: Multiply row 3 of matrix C by $-\frac{1}{6}$.

c. Type 3: To the numbers in row 2 of matrix B, add the results of multiplying each number in row 1 by -4.

Strategy We will perform elementary row operations on each matrix as if we were performing those operations on the equations of a system.

Why The rows of an augmented matrix correspond to the equations of a system.

Solution **a.** Interchanging rows 1 and 2 of matrix $A = \left[\begin{array}{cc|c} 2 & 4 & -3 \\ 1 & -8 & 0 \end{array}\right]$ gives $\left[\begin{array}{cc|c} 1 & -8 & 0 \\ 2 & 4 & -3 \end{array}\right]$.

We can represent the instruction to interchange rows 1 and 2 with the symbol $R_1 \leftrightarrow R_2$.

b. We multiply each number in row 3 of matrix $C = \left[\begin{array}{ccc|c} 2 & 1 & -8 & 4 \\ 0 & 1 & 4 & -2 \\ 0 & 0 & -6 & 24 \end{array}\right]$ by $-\frac{1}{6}$. Note that rows 1 and 2 remain unchanged.

$$\left[\begin{array}{ccc|c} 2 & 1 & -8 & 4 \\ 0 & 1 & 4 & -2 \\ 0 & 0 & 1 & -4 \end{array}\right]$$

We can represent the instruction to multiply the third row by $-\frac{1}{6}$ with the symbol $-\frac{1}{6}R_3$.

c. We multiply each number from the first row of matrix $B = \left[\begin{array}{cc|c} 1 & -1 & 2 \\ 4 & -8 & 0 \end{array}\right]$ by -4 to get

$$-4 \quad 4 \quad -8$$ *This is $-4R_1$.*

We then add these numbers to the entries in row 2 of matrix B. (Note that row 1 remains unchanged.)

$$\left[\begin{array}{cc|c} 1 & -1 & 2 \\ 4 + (-4) & -8 + 4 & 0 + (-8) \end{array}\right]$$

This procedure is represented by $-4R_1 + R_2$, which means "Multiply row 1 by -4 and add the result to row 2."

After simplifying the bottom row, we have the matrix $\left[\begin{array}{cc|c} 1 & -1 & 2 \\ 0 & -4 & -8 \end{array}\right]$.

Success Tip

This elementary row operation corresponds to the following elimination method step for solving a system:

$$\begin{aligned} 4x - 8y &= 0 \\ -4x + 4y &= -8 \\ \hline -4y &= -8 \end{aligned}$$

Teaching Example 2
Refer to Example 2.

a. Interchange rows 1 and 2 of matrix C.

b. Multiply row 2 of matrix B by $\frac{1}{4}$.

c. To the numbers in row 1 of matrix A, add the results of multiplying each number in row 2 by -2.

Answers:

a. $\left[\begin{array}{ccc|c} 0 & 1 & 4 & -2 \\ 2 & 1 & -8 & 4 \\ 0 & 0 & -6 & 24 \end{array}\right]$

b. $\left[\begin{array}{cc|c} 1 & -1 & 2 \\ 1 & -2 & 0 \end{array}\right]$

c. $\left[\begin{array}{cc|c} 0 & 20 & -3 \\ 1 & -8 & 0 \end{array}\right]$

Self Check 2 Use the matrices of Example 2 and perform the following:

a. Interchange the rows of matrix B.

b. To the numbers in row 2 of matrix B, add the results of multiplying each number in row 1 by -4.

c. Multiply row 1 of matrix C by $\frac{1}{2}$.

a. $\left[\begin{array}{cc|c} 4 & -8 & 0 \\ 1 & -1 & 2 \end{array}\right]$ b. $\left[\begin{array}{cc|c} 1 & -1 & 2 \\ 0 & -4 & -8 \end{array}\right]$

c. $\left[\begin{array}{ccc|c} 1 & \frac{1}{2} & -4 & 2 \\ 0 & 1 & 4 & -2 \\ 0 & 0 & -6 & 24 \end{array}\right]$

Now Try Problems 17, 19, and 23

4 Use Matrices to Solve a System of Two Equations.

We can solve a system of two linear equations using a series of elementary row operations on the augmented matrix to produce a simpler, equivalent matrix. This process is called **Gauss-Jordan elimination.**

EXAMPLE 3 Use Gauss-Jordan elimination to solve the system: $\begin{cases} 2x + y = 5 \\ x - y = 4 \end{cases}$

Strategy We will represent the system with an augmented matrix and use a series of elementary-row operations to produce an equivalent matrix in reduced row-echelon form.

Why The values of the variables x and y that solve the original system of equations will be in the last column of the reduced row-echelon form matrix.

Solution We can represent the system with the following augmented matrix:

$$\left[\begin{array}{cc|c} 2 & 1 & 5 \\ 1 & -1 & 4 \end{array}\right]$$

The Language of Algebra

This matrix method of solving a system of linear equations is called **Gauss-Jordan elimination**, after the German mathematicians Carl Friedrich Gauss (1777–1855) and Wilhelm Jordan (1842–1899).

First, we want to get a 1 in the top row of the first column where the shaded 2 is. This can be done by applying a type 1 row operation and interchanging rows 1 and 2.

$$\left[\begin{array}{cc|c} \mathbf{1} & -1 & 4 \\ \mathbf{2} & 1 & 5 \end{array}\right] \quad R_1 \leftrightarrow R_2$$

To get a 0 in the first column where the shaded 2 is, we use a type 3 row operation and multiply each entry in row 1 by -2 to get

$$\mathbf{-2} \qquad \mathbf{2} \qquad \mathbf{-8}$$

and add these numbers to the entries in row 2.

$$\left[\begin{array}{cc|c} 1 & -1 & 4 \\ 2 + (-2) & 1 + 2 & 5 + (-8) \end{array}\right] \quad \text{This is } -2R_1 + R_2.$$

After simplifying the bottom row, we have

$$\left[\begin{array}{cc|c} 1 & -1 & 4 \\ \mathbf{0} & 3 & -3 \end{array}\right]$$

To get a 1 in the bottom row of the second column where the shaded 3 is, we use a type 2 row operation and multiply row 2 by $\frac{1}{3}$.

$$\left[\begin{array}{cc|c} 1 & -1 & 4 \\ 0 & \mathbf{1} & -1 \end{array}\right] \quad \tfrac{1}{3}R_2$$

To get a 0 in the first row where the shaded -1 is, we use a type 3 row operation and add to those numbers the entries in row 2 to get

$$\left[\begin{array}{cc|c} 1 + \mathbf{0} & -1 + \mathbf{1} & 4 + \mathbf{(-1)} \\ 0 & 1 & -1 \end{array}\right] \quad \text{This is } R_1 + R_2.$$

Success Tip

Follow this order in getting 1's and 0's in the proper positions of the augmented matrix.

Step 1: $\left[\begin{array}{cc|c} 1 & \square & \square \\ \square & \square & \square \end{array}\right]$

Step 2: $\left[\begin{array}{cc|c} 1 & \square & \square \\ 0 & \square & \square \end{array}\right]$

Step 3: $\left[\begin{array}{cc|c} 1 & \square & \square \\ 0 & 1 & \square \end{array}\right]$

Step 4: $\left[\begin{array}{cc|c} 1 & 0 & \square \\ 0 & 1 & \square \end{array}\right]$

After simplifying the top row, we obtain an equivalent matrix in the desired reduced row-echelon form.

$$\left[\begin{array}{cc|c} 1 & 0 & 3 \\ 0 & 1 & -1 \end{array}\right]$$

This augmented matrix represents a simpler system of equations that is equivalent to the original system:

$$\begin{cases} 1x + 0y = 3 \\ 0x + 1y = -1 \end{cases}$$

Writing the equations without the coefficients, and dropping the $0y$ and $0x$ terms, we have

$$\begin{cases} x = 3 \\ y = -1 \end{cases}$$

Thus, the solution of the given system is $(3, -1)$ and the solution set is $\{(3, -1)\}$. Verify that this ordered pair satisfies both equations in the original system.

Teaching Example 3
Use Gauss-Jordan elimination to solve the system:
$$\begin{cases} 3x + 4y = 11 \\ x - 3y = -5 \end{cases}$$
Answer: $(1, 2)$

Self Check 3 Use Gauss-Jordan elimination to solve the system: $\begin{cases} 2x + y = -9 \\ x - 3y = 13 \end{cases}$

Now Try ▶ **Problem 27**

$(-2, -5)$

When a system of linear equations has one solution, we can use the following steps to solve it.

Solving Systems of Linear Equations Using Gauss-Jordan Elimination

1. Write an augmented matrix for the system.
2. Use elementary row operations to transform the augmented matrix into a matrix in reduced row-echelon form with 1's down its main diagonal and 0's directly above and below the 1's.
3. When step 2 is complete, write the resulting equivalent system to find the solution.
4. Check the proposed solution in the equations of the original system.

5 Use Matrices to Solve a System of Three Equations.

To show how to use Gauss-Jordan elimination to solve systems of three linear equations in three variables, consider the following system (written in standard form) and the corresponding augmented matrix to its right.

A system of three linear equations	***Its augmented matrix***
$\begin{cases} 3x + y + 5z = 8 \\ 2x + 3y - z = 6 \\ x + 2y + 2z = 10 \end{cases}$	$\left[\begin{array}{cc\|cc} 3 & 1 & 5 & 8 \\ 2 & 3 & -1 & 6 \\ 1 & 2 & 2 & 10 \end{array}\right]$

To solve the 3×3 system of equations, we transform the augmented matrix into a matrix in reduced row-echelon form with 1's down its main diagonal and 0's directly above and below its main diagonal.

$$\left[\begin{array}{ccc|c} 1 & 0 & 0 & a \\ 0 & 1 & 0 & b \\ 0 & 0 & 1 & c \end{array}\right]$$

Main diagonal

a, b, and c represent real numbers.

EXAMPLE 4 Use Gauss-Jordan elimination to solve the system:

$$\begin{cases} 3x + y + 5z = 8 \\ 2x + 3y - z = 6 \\ x + 2y + 2z = 10 \end{cases}$$

Strategy We will represent the system with an augmented matrix and use a series of elementary row operations to produce an equivalent matrix in reduced row-echelon form.

Why The values of the variables x, y, and z that solve the original system of equations will be in the last column of the reduced row-echelon form matrix.

Solution This system can be represented by the augmented matrix

$$\left[\begin{array}{ccc|c} 3 & 1 & 5 & 8 \\ 2 & 3 & -1 & 6 \\ 1 & 2 & 2 & 10 \end{array}\right]$$

Success Tip

Follow this order in getting 1's and 0's in the proper positions of the augmented matrix.

Step 1: $\left[\begin{array}{ccc|c} 1 & \blacksquare & \blacksquare & \blacksquare \\ \blacksquare & \blacksquare & \blacksquare & \blacksquare \\ \blacksquare & \blacksquare & \blacksquare & \blacksquare \end{array}\right]$

Step 2: $\left[\begin{array}{ccc|c} 1 & \blacksquare & \blacksquare & \blacksquare \\ 0 & \blacksquare & \blacksquare & \blacksquare \\ 0 & \blacksquare & \blacksquare & \blacksquare \end{array}\right]$

Step 3: $\left[\begin{array}{ccc|c} 1 & \blacksquare & \blacksquare & \blacksquare \\ 0 & 1 & \blacksquare & \blacksquare \\ 0 & \blacksquare & \blacksquare & \blacksquare \end{array}\right]$

Step 4: $\left[\begin{array}{ccc|c} 1 & 0 & \blacksquare & \blacksquare \\ 0 & 1 & \blacksquare & \blacksquare \\ 0 & 0 & \blacksquare & \blacksquare \end{array}\right]$

Step 5: $\left[\begin{array}{ccc|c} 1 & 0 & \blacksquare & \blacksquare \\ 0 & 1 & \blacksquare & \blacksquare \\ 0 & 0 & 1 & \blacksquare \end{array}\right]$

Step 6: $\left[\begin{array}{ccc|c} 1 & 0 & 0 & \blacksquare \\ 0 & 1 & 0 & \blacksquare \\ 0 & 0 & 1 & \blacksquare \end{array}\right]$

To get a 1 in the first column where the shaded 3 is, we perform a type 1 row operation by interchanging rows 1 and 3.

$$\left[\begin{array}{ccc|c} 1 & 2 & 2 & 10 \\ 2 & 3 & -1 & 6 \\ 3 & 1 & 5 & 8 \end{array}\right] \quad R_1 \leftrightarrow R_3$$

To get a 0 where the shaded 2 is, we perform a type 3 row operation by multiplying each entry in row 1 by -2 to get: **−2 −4 −4 −20**. Then we add these numbers to the entries in row 2. And to get a 0 where the shaded 3 is, we perform another type 3 row operation by multiplying the entries in row 1 by -3 and adding the results to row 3.

$$\left[\begin{array}{ccc|c} 1 & 2 & 2 & 10 \\ 0 & -1 & -5 & -14 \\ 0 & -5 & -1 & -22 \end{array}\right] \quad \begin{array}{l} \\ \text{This is } -2R_1 + R_2. \\ -3R_1 + R_3 \end{array}$$

To get a 1 where the shaded -1 is, we perform a type 2 row operation by multiplying row 2 by -1.

$$\left[\begin{array}{ccc|c} 1 & 2 & 2 & 10 \\ 0 & 1 & 5 & 14 \\ 0 & -5 & -1 & -22 \end{array}\right] \quad -1R_2$$

To get a 0 where the shaded 2 is, we perform a type 3 row operation by multiplying the entries in row 2 by -2 and adding the results to row 1. And to get a 0 where the shaded -5 is, we multiply the entries in row 2 by 5 and add the results to row 3.

$$\left[\begin{array}{ccc|c} 1 & 0 & -8 & -18 \\ 0 & 1 & 5 & 14 \\ 0 & 0 & 24 & 48 \end{array}\right] \quad \begin{array}{l} -2R_2 + R_1 \\ \\ 5R_2 + R_3 \end{array}$$

To get a 1 where the shaded 24 is, we perform a type 2 row operation by multiplying row 3 by $\frac{1}{24}$.

$$\left[\begin{array}{ccc|c} 1 & 0 & -8 & -18 \\ 0 & 1 & 5 & 14 \\ 0 & 0 & 1 & 2 \end{array}\right] \quad \frac{1}{24}R_3$$

To get a 0 where the shaded 5 is, we multiply the entries in row 3 by -5 and add the results to row 2. And to get a 0 where the shaded -8 is, we multiply the entries in row 3 by 8 and add the results to row 1.

$$\left[\begin{array}{ccc|c} 1 & 0 & 0 & -2 \\ 0 & 1 & 0 & 4 \\ 0 & 0 & 1 & 2 \end{array}\right] \quad \begin{array}{l} 8R_3 + R_1 \\ -5R_3 + R_2 \end{array}$$

This is the desired reduced row-echelon form.

This augmented matrix represents a simpler system of equations that is equivalent to the original system:

$$\begin{cases} 1x + 0y + 0z = -2 \\ 0x + 1y + 0z = 4 \\ 0x + 0y + 1z = 2 \end{cases} \quad \text{which can be written without the coefficients of 0 and 1 as} \quad \begin{cases} x = -2 \\ y = 4 \\ z = 2 \end{cases}$$

Thus, the solution of the given system is $(-2, 4, 2)$ and the solution set is $\{(-2, 4, 2)\}$. Verify that this ordered triple satisfies each equation of the original system.

Teaching Example 4
Use Gauss-Jordan elimination to solve the system:

$$\begin{cases} 2x + 3y - z = -8 \\ x - y - z = -2 \\ -4x + 3y + z = 6 \end{cases}$$

Answer: $(-2, -1, 1)$

Self Check 4 Use Gauss-Jordan elimination to solve: $\begin{cases} 2x - y + z = 5 \\ x + y - z = -2 \\ -x + 2y + 2z = 1 \end{cases}$ $(1, -1, 2)$

Now Try Problem 29

6 Use Matrices to Identify Inconsistent Systems and Dependent Equations.

In the next example, we will see how to recognize inconsistent systems and systems of dependent equations when matrices are used to solve them.

EXAMPLE 5 Use Gauss-Jordan elimination to solve the system:

a. $\begin{cases} x + y = -1 \\ -3x - 3y = -5 \end{cases}$ **b.** $\begin{cases} 2x - y = 4 \\ -6x + 3y = -12 \end{cases}$

Strategy We will represent the system with an augmented matrix and attempt to use Gauss-Jordan elimination to produce an equivalent matrix in reduced row-echelon form.

Why It is easy to recognize an inconsistent system or dependent equations during that process.

Solution **a.** The system $\begin{cases} x + y = -1 \\ -3x - 3y = -5 \end{cases}$ can be represented by the augmented matrix

$$\left[\begin{array}{cc|c} 1 & 1 & -1 \\ -3 & -3 & -5 \end{array}\right]$$

Since the matrix has a 1 in the top row of the first column, we proceed to get a 0 where the shaded -3 is by multiplying row 1 by 3 and adding the results to row 2.

$$\left[\begin{array}{cc|c} 1 & 1 & -1 \\ 0 & 0 & -8 \end{array}\right] \quad 3R_1 + R_2$$

The second row of this augmented matrix represents the equation $0 = -8$. This false statement indicates that the system is inconsistent and has no solution. The solution set is $\varnothing$.

b. The system $\begin{cases} 2x - y = 4 \\ -6x + 3y = -12 \end{cases}$ can be represented by the augmented matrix

$$\left[\begin{array}{cc|c} 2 & -1 & 4 \\ -6 & 3 & -12 \end{array}\right]$$

To get a 1 where the shaded 2 is, we perform a type 2 row operation by multiplying row 1 by $\frac{1}{2}$.

$$\left[\begin{array}{cc|c} 1 & -\frac{1}{2} & 2 \\ -6 & 3 & -12 \end{array}\right] \quad \tfrac{1}{2}R_1$$

To get a 0 where the shaded -6 is, we perform a type 3 row operation by multiplying the entries in row 1 by 6 and adding the results to the entries in row 2.

$$\left[\begin{array}{cc|c} 1 & -\frac{1}{2} & 2 \\ 0 & 0 & 0 \end{array}\right] \quad 6R_1 + R_2$$

The second row of this augmented matrix represents the equation $0 = 0$. This true statement indicates that the equations are dependent and that the system has infinitely many solutions. The solution set is $\{(x, y) \mid 2x - y = 4\}$.

Teaching Example 5
Use Gauss-Jordan elimination to solve the system:

a. $\begin{cases} x + 3y = 4 \\ -2x - 6y = 16 \end{cases}$

b. $\begin{cases} 2a + 5b = 1 \\ -4a - 10b = -2 \end{cases}$

Answers: **a.** No solution, $\varnothing$; inconsistent system
b. $\{(a, b) \mid 2a + 5b = 1\}$, infinitely many solutions; dependent equations

Self Check 5 Use Gauss-Jordan elimination to solve the system:

a. $\begin{cases} 4x - 8y = 9 \\ x - 2y = -5 \end{cases}$ No solution, $\varnothing$; inconsistent system

b. $\begin{cases} x - 3y = 6 \\ -4x + 12y = -24 \end{cases}$ $\{(x, y) \mid x - 3y = 6\}$, infinitely many solutions; dependent equations

Now Try ▶ Problems 33 and 35

Using Your Calculator ▶ Augmented Matrices

Systems of linear equations like those found in this section can be solved on a graphing calculator by entering their corresponding augmented matrices. For example, the augmented matrix for the system in Example 4 is shown below in figure (a). Note that it has been defined to be matrix "A." See your owner's manual for the specific instructions on how to enter the elements of a matrix.

We then use the reduced row-echelon matrix command (on a TI-84, locate **B:rref** by pressing MATRIX, arrowing to MATH, and scrolling down) to put the matrix in reduced row-echelon form, as shown in figure (b). The solution of the given system, $(-2, 4, 2)$, can be read from the last column of the display.

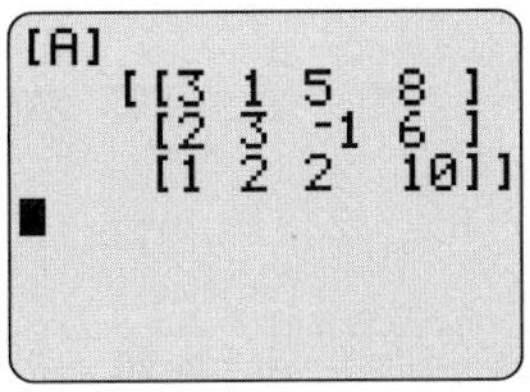

(a)

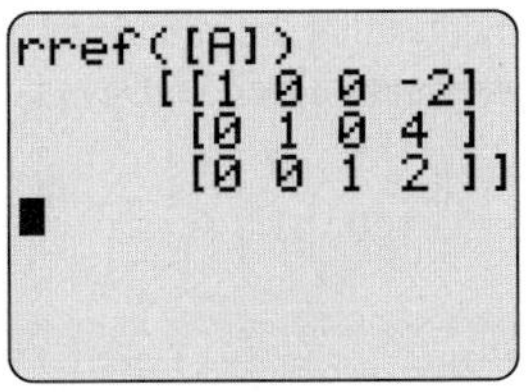

(b)

SECTION 12.4 STUDY SET

VOCABULARY

Fill in the blanks.

▶ **1.** A matrix is a rectangular array of numbers written within brackets.

▶ **2.** Each number in a matrix is called an element or entry of the matrix.

▶ **3.** If the order of a matrix is 3×4, it has 3 rows and 4 columns. We read 3×4 as "3 by 4."

▶ **4.** A matrix that represents the equations of a system is called an augmented matrix.

▶ **5.** Elementary row operations can be used on an augmented matrix to produce a simpler equivalent matrix that gives the solution of a system. This process is called Gauss - Jordan elimination.

▶ **6.** The matrix $\left[\begin{array}{cc:c} 1 & 0 & -2 \\ 0 & 1 & 4 \end{array}\right]$, with 1's down its main diagonal and 0's directly above and below it, is in reduced row- echelon form.

CONCEPTS

7. For each matrix, determine the number of rows and the number of columns.

a. $\left[\begin{array}{cc:c} 4 & 6 & -1 \\ 1 & 9 & -3 \end{array}\right]$ 2×3 **b.** $\left[\begin{array}{ccc:c} 1 & -2 & 3 & 1 \\ 0 & 1 & 6 & 4 \\ 0 & 0 & 1 & \frac{1}{3} \end{array}\right]$ 3×4

8. Fill in the blanks to complete each elementary row operation:

a. Type 1: Any two rows of a matrix can be interchanged.

b. Type 2: Any row of a matrix can be multiplied by a nonzero constant.

c. Type 3: Any row of a matrix can be changed by adding a nonzero constant multiple of another row to it.

▶ **9.** Gauss-Jordan elimination was used to solve a system. The final augmented matrix is shown. Fill in the blanks.

a. $\left[\begin{array}{cc:c} 1 & 0 & 10 \\ 0 & 1 & 6 \end{array}\right]$ represents $\begin{cases} x = 10 \\ y = 6 \end{cases}$

The solution of the system is (10, 6).

b. $\left[\begin{array}{ccc:c} 1 & 0 & 0 & -16 \\ 0 & 1 & 0 & 8 \\ 0 & 0 & 1 & 4 \end{array}\right]$ represents $\begin{cases} x = -16 \\ y = 8 \\ z = 4 \end{cases}$

The solution of the system is (−16, 8, 4).

▶ **10. a.** Which matrix shown below indicates that its corresponding system of equations has no solution? iii.

b. Which matrix indicates that the equations of its corresponding system are dependent? i.

i. $\left[\begin{array}{cc:c} 1 & 2 & -4 \\ 0 & 0 & 0 \end{array}\right]$ **ii.** $\left[\begin{array}{cc:c} 1 & 0 & 6 \\ 0 & 1 & 0 \end{array}\right]$ **iii.** $\left[\begin{array}{cc:c} 1 & 2 & -4 \\ 0 & 0 & 2 \end{array}\right]$

NOTATION

▶ **11.** Explain what each symbolism means.

a. $R_1 \leftrightarrow R_2$ Interchange rows 1 and 2

b. $\frac{1}{2}R_1$ Multiply row 1 by $\frac{1}{2}$

c. $6R_2 + R_3$ Add row 3 to 6 times row 2

12. Complete the solution to solve the system using matrices.

$\begin{cases} 4x - y = 14 \\ x + y = 6 \end{cases}$

$\left[\begin{array}{cc:c} 4 & -1 & 14 \\ 1 & 1 & 6 \end{array}\right]$ This is the augmented matrix.

$\left[\begin{array}{cc:c} 1 & 1 & 6 \\ 4 & -1 & 14 \end{array}\right]$ $R_1 \leftrightarrow R_2$

$\left[\begin{array}{cc:c} 1 & 1 & 6 \\ 0 & -5 & -10 \end{array}\right]$ $-4R_1 + R_2$

$\left[\begin{array}{cc:c} 1 & 1 & 6 \\ 0 & 1 & 2 \end{array}\right]$ $-\frac{1}{5}R_2$

$\left[\begin{array}{cc:c} 1 & 0 & 4 \\ 0 & 1 & 2 \end{array}\right]$ $-R_2 + R_1$

This augmented matrix represents the equivalent system

$\begin{cases} x = 4 \\ y = 2 \end{cases}$

The solution is (4, 2).

GUIDED PRACTICE

***Represent each system using an augmented matrix.* See Example 1.**

▶ **13.** $\begin{cases} x + 2y = 6 \\ 3x - y = -10 \end{cases}$ $\left[\begin{array}{cc:c} 1 & 2 & 6 \\ 3 & -1 & -10 \end{array}\right]$

▶ **14.** $\begin{cases} x + y + z = 4 \\ 2x + y - z = 1 \\ 2x - 3y = 1 \end{cases}$ $\left[\begin{array}{ccc:c} 1 & 1 & 1 & 4 \\ 2 & 1 & -1 & 1 \\ 2 & -3 & 0 & 1 \end{array}\right]$

For each augmented matrix, give the system of equations that it represents.

▶ **15.** $\left[\begin{array}{cc:c} 1 & 6 & 7 \\ 0 & 1 & 4 \end{array}\right]$ $\begin{cases} x + 6y = 7 \\ y = 4 \end{cases}$

▶ **16.** $\left[\begin{array}{ccc:c} 1 & -2 & 9 & 1 \\ 0 & 1 & 4 & 0 \\ 0 & 0 & 1 & -7 \end{array}\right]$ $\begin{cases} x - 2y + 9z = 1 \\ y + 4z = 0 \\ z = -7 \end{cases}$

Perform each of the following elementary row operations on the augmented matrix $\left[\begin{array}{cc:c} -3 & 1 & -6 \\ 1 & -4 & 4 \end{array}\right]$**. See Example 2.**

▶ **17.** $R_1 \leftrightarrow R_2$ $\left[\begin{array}{cc:c} 1 & -4 & 4 \\ -3 & 1 & -6 \end{array}\right]$

18. $5R_2$ $\left[\begin{array}{cc:c} -3 & 1 & -6 \\ 5 & -20 & 20 \end{array}\right]$

19. $-\frac{1}{3}R_1$ $\left[\begin{array}{cc:c} 1 & -\frac{1}{3} & 2 \\ 1 & -4 & 4 \end{array}\right]$

▶ **20.** $3R_2 + R_1$ $\left[\begin{array}{cc:c} 0 & -11 & 6 \\ 1 & -4 & 4 \end{array}\right]$

Perform each of the following elementary row operations on the augmented matrix $\left[\begin{array}{ccc|c} 3 & 6 & -9 & 0 \\ 1 & 5 & -2 & 1 \\ -2 & 2 & -2 & 5 \end{array}\right]$. **See Example 2.**

21. $R_2 \leftrightarrow R_3$ $\left[\begin{array}{ccc|c} 3 & 6 & -9 & 0 \\ -2 & 2 & -2 & 5 \\ 1 & 5 & -2 & 1 \end{array}\right]$

▶ **22.** $-\frac{1}{2}R_3$ $\left[\begin{array}{ccc|c} 3 & 6 & -9 & 0 \\ 1 & 5 & -2 & 1 \\ 1 & -1 & 1 & -\frac{5}{2} \end{array}\right]$

▶ **23.** $-R_1 + R_2$ $\left[\begin{array}{ccc|c} 3 & 6 & -9 & 0 \\ -2 & -1 & 7 & 1 \\ -2 & 2 & -2 & 5 \end{array}\right]$

24. $2R_2 + R_3$ $\left[\begin{array}{ccc|c} 3 & 6 & -9 & 0 \\ 1 & 5 & -2 & 1 \\ 0 & 12 & -6 & 7 \end{array}\right]$

Use matrices to solve each system of equations. **See Example 3.**

▶ **25.** $\begin{cases} x + y = 2 \\ x - y = 0 \end{cases}$ (1, 1)

▶ **26.** $\begin{cases} x + y = 3 \\ x - y = -1 \end{cases}$ (1, 2)

▶ **27.** $\begin{cases} 2x + y = 1 \\ x + 2y = -4 \end{cases}$ (2, −3)

▶ **28.** $\begin{cases} 5x - 4y = 10 \\ x - 7y = 2 \end{cases}$ (2, 0)

Use matrices to solve each system of equations. **See Example 4.**

▶ **29.** $\begin{cases} x + y + z = 6 \\ x + 2y + z = 8 \\ x + y + 2z = 7 \end{cases}$ (3, 2, 1)

▶ **30.** $\begin{cases} x + y + z = 6 \\ x + 2y + z = 8 \\ x + y + 2z = 9 \end{cases}$ (1, 2, 3)

▶ **31.** $\begin{cases} 3x + y - 3z = 5 \\ x - 2y + 4z = 10 \\ x + y + z = 13 \end{cases}$ (4, 5, 4)

▶ **32.** $\begin{cases} 2x + y - 3z = -1 \\ 3x - 2y - z = -5 \\ x - 3y - 2z = -12 \end{cases}$ (1, 3, 2)

Use matrices to solve each system of equations. If the equations of a system are dependent or if a system is inconsistent, state this. **See Example 5.**

▶ **33.** $\begin{cases} x - 3y = 9 \\ -2x + 6y = 18 \end{cases}$ No solution, ∅; inconsistent system

▶ **34.** $\begin{cases} -6x + 12y = 10 \\ 2x - 4y = 8 \end{cases}$ No solution, ∅; inconsistent system

▶ **35.** $\begin{cases} -4x - 4y = -12 \\ x + y = 3 \end{cases}$ $\{(x, y) \mid x + y = 3\}$, infinitely many solutions; dependent equations

36. $\begin{cases} 5x - 15y = 10 \\ x - 3y = 2 \end{cases}$ $\{(x, y) \mid x - 3y = 2\}$, infinitely many solutions; dependent equations

TRY IT YOURSELF

Use matrices to solve each system of equations. If the equations of a system are dependent or if a system is inconsistent, state this.

▶ **37.** $\begin{cases} 2x + 3y - z = -8 \\ x - y - z = -2 \\ -4x + 3y + z = 6 \end{cases}$ (−2, −1, 1)

▶ **38.** $\begin{cases} 2a + b + 3c = 3 \\ -2a - b + c = 5 \\ 4a - 2b + 2c = 2 \end{cases}$ (−1, −1, 2)

▶ **39.** $\begin{cases} 2x - y = -1 \\ x - 2y = 1 \end{cases}$ (−1, −1)

▶ **40.** $\begin{cases} 2x - y = 0 \\ x + y = 3 \end{cases}$ (1, 2)

41. $\begin{cases} 3x + 4y = -12 \\ 9x - 2y = 6 \end{cases}$ (0, −3)

▶ **42.** $\begin{cases} 2x - 3y = 16 \\ -4x + y = -22 \end{cases}$ (5, −2)

43. $\begin{cases} 2x + y - z = 1 \\ x + 2y + 2z = 2 \\ 4x + 5y + 3z = 3 \end{cases}$ No solution, ∅; inconsistent system

44. $\begin{cases} x - y = 1 \\ 2x - z = 0 \\ 2y - z = -2 \end{cases}$ Dependent equations

45. $\begin{cases} 8x - 2y = 4 \\ 4x - y = 2 \end{cases}$ $\{(x, y) \mid 4x - y = 2\}$, infinitely many solutions; dependent equations

46. $\begin{cases} 9x - 3y = 6 \\ 3x - y = 8 \end{cases}$ No solution, ∅; inconsistent system

47. $\begin{cases} 2x + y - 2z = 6 \\ 4x - y + z = -1 \\ 6x - 2y + 3z = -5 \end{cases}$ $\left(\frac{1}{2}, 1, -2\right)$

▶ **48.** $\begin{cases} 2x - 3y + 3z = 14 \\ 3x + 3y - z = 2 \\ -2x + 6y + 5z = 9 \end{cases}$ $\left(2, -\frac{1}{3}, 3\right)$

▶ **49.** $\begin{cases} 6x + y - z = -2 \\ x + 2y + z = 5 \\ 5y - z = 2 \end{cases}$ (0, 1, 3)

▶ **50.** $\begin{cases} 2x + 3y - 2z = 18 \\ 5x - 6y + z = 21 \\ 4y - 2z = 6 \end{cases}$ (8, 4, 5)

51. $\begin{cases} 5x + 3y = 4 \\ 3y - 4z = 4 \\ x + z = 1 \end{cases}$ (−4, 8, 5)

52. $\begin{cases} y + 2z = -2 \\ x + y = 1 \\ 2x - z = 0 \end{cases}$ (−1, 2, −2)

APPLICATIONS

53. Digital Photography. A digital camera stores the black and white photograph shown on the right as a 512×512 matrix. Each element of the matrix corresponds to a small dot of grey scale shading, called a *pixel,* in the picture. How many elements does a 512×512 matrix have? 262,144

▶ **54. Digital Imaging.** A scanner stores a black and white photograph as a matrix that has a total of 307,200 elements. If the matrix has 480 rows, how many columns does it have? 640

55. Complementary Angles. The following system can be used to find the angle measures (in degrees) of the two complementary angles shown in the illustration below. Solve the system using matrices to find the measure of each angle. (*Hint:* Each equation must be in standard $Ax + By = C$ form.) 22°; 68°

$\begin{cases} x + y = 90 \\ y = x + 46 \end{cases}$

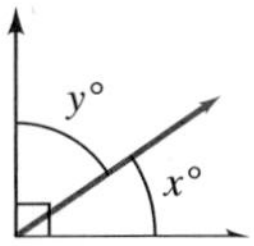

56. Supplementary Angles. The following system can be used to find the angle measures (in degrees) of the two supplementary angles shown in the illustration below. Solve the system using matrices to find the measure of each angle. (*Hint:* Each equation must be in standard $Ax + By = C$ form.) 97°, 83°

$$\begin{cases} x + y = 180 \\ y = x - 14 \end{cases}$$

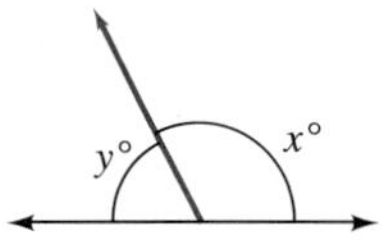

57. Triangles. The following system can be used to find the measures (in degrees) of $\angle A$, $\angle B$, and $\angle C$ shown in the illustration below. Solve the system using matrices to find the measure of each angle of the triangle. (*Hint:* Each equation must be in standard $Ax + By = C$ form.) $\angle A$: 40°, $\angle B$: 65°, $\angle C$: 75°

$$\begin{cases} A + B + C = 180 \\ B = A + 25 \\ C = 2A - 5 \end{cases}$$

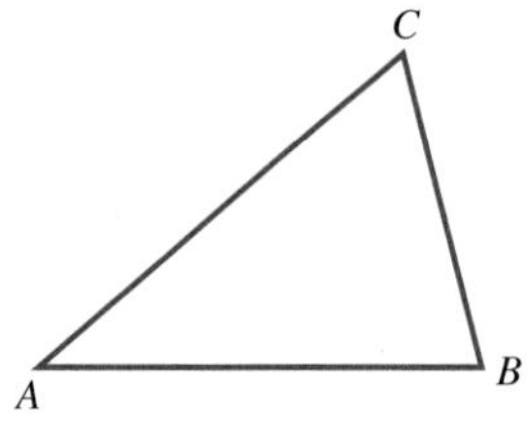

58. Triangles. The following system can be used to find the measures (in degrees) of $\angle A$, $\angle B$, and $\angle C$ shown in the illustration below. Solve the system using matrices to find the measure of each angle of the triangle. (*Hint:* Each equation must be in standard $Ax + By = C$ form.) $\angle A$: 50°, $\angle B$: 60°, $\angle C$: 70°

$$\begin{cases} A + B + C = 180 \\ A = B - 10 \\ B = C - 10 \end{cases}$$

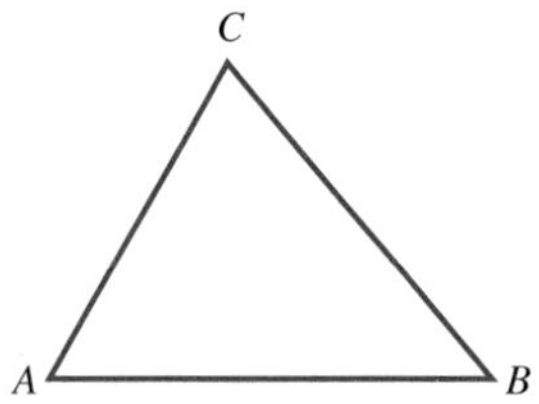

WRITING

59. For the system $\begin{cases} 2x - 3y = 5 \\ 4x + 8 = y \end{cases}$, explain what is wrong with writing its corresponding augmented matrix as $\left[\begin{array}{rr|r} 2 & -3 & 5 \\ 4 & 8 & 1 \end{array}\right]$. How should it be written?

60. Explain what is meant by the phrase *reduced row-echelon form*. Give an example of how it was used in this section.

61. Explain how a type 3 row operation is similar to the elimination (addition) method of solving a system of equations.

62. If the system represented by the augmented matrix at the right has no solution, what do you know about k? Explain your answer.
$$\left[\begin{array}{rrr|r} 1 & 1 & 0 & 1 \\ 0 & 0 & 1 & 2 \\ 0 & 0 & 0 & k \end{array}\right]$$

REVIEW

63. What is the formula used to find the slope of a line, given two points on the line? $m = \frac{y_2 - y_1}{x_2 - x_1}$ $(x_2 \neq x_1)$

64. What is the form of the equation of a horizontal line? Of a vertical line? $y = b$, $x = a$

65. What is the point–slope form of the equation of a line? $y - y_1 = m(x - x_1)$

66. What is the slope–intercept form of the equation of a line? $y = mx + b$

CHALLENGE PROBLEMS

Use matrices to solve the system.

67. $\begin{cases} x^2 + y^2 + z^2 = 14 \\ 2x^2 + 3y^2 - 2z^2 = -7 \\ x^2 - 5y^2 + z^2 = 8 \end{cases}$ $(\pm 2, \pm 1, \pm 3)$

68. $\begin{cases} w + x + y + z = 0 \\ w - 2x + y - 3z = -3 \\ 2w + 3x + y - 2z = -1 \\ 2w - 2x - 2y + z = -12 \end{cases}$ $(-3, 1, 2, 0)$

SECTION 12.5 Solving Systems of Equations Using Determinants

OBJECTIVES

1. Evaluate 2×2 and 3×3 determinants.
2. Use Cramer's rule to solve systems of two equations.
3. Use Cramer's rule to solve systems of three equations.

ARE YOU READY? *Are You Ready? exercises available online at www.webassign.net/brookscole*

The following problems review some basic skills that are needed when solving systems of equations using determinants.

1. How many rows and columns does the matrix $\begin{bmatrix} 2 & -4 & 9 \\ 1 & 0 & 4 \\ -6 & 5 & 11 \end{bmatrix}$ have? 3 rows, 3 columns

2. What numbers lie on the main diagonal of the matrix $\begin{bmatrix} 3 & 5 \\ 6 & -1 \end{bmatrix}$? 3, −1

3. Evaluate: $6(-8) - (-2)(-3)$ −54

4. Evaluate: $6[3 - (-1)] - 9[7 - (-8)] + 4(-2 - 1)$ −123

In this section, we will discuss another method for solving systems of linear equations. With this method, called *Cramer's rule,* we work with combinations of the coefficients and the constants of the equations written as *determinants.*

1 Evaluate 2 × 2 and 3 × 3 Determinants.

An idea related to the concept of matrix is the **determinant.** A determinant is a number that is associated with a **square matrix,** a matrix that has the same number of rows and columns. For any square matrix A, the symbol $|A|$ represents the determinant of A. To write a determinant, we put the elements of a square matrix between two vertical lines.

Notation

Brackets are used to write matrices and vertical bars are used to write determinants.

$$\begin{bmatrix} 2 & 1 \\ 8 & -3 \end{bmatrix} \qquad \begin{vmatrix} 7 & -6 \\ 4 & 0 \end{vmatrix}$$

Matrix Determinant

$$\begin{bmatrix} 3 & 2 \\ 6 & 9 \end{bmatrix} \qquad \begin{vmatrix} 3 & 2 \\ 6 & 9 \end{vmatrix} \qquad \begin{bmatrix} 9 & 3 & -2 \\ 0 & 1 & 8 \\ 1 & 2 & 4 \end{bmatrix} \qquad \begin{vmatrix} 9 & 3 & -2 \\ 0 & 1 & 8 \\ 1 & 2 & 4 \end{vmatrix}$$

Matrix Determinant Matrix Determinant

Like matrices, determinants are classified according to the number of rows and columns they contain. The determinant above, on the left, is a 2 × 2 determinant. The other is a 3 × 3 determinant.

The determinant of a 2 × 2 matrix is the number that is equal to the product of the numbers on the main diagonal minus the product of the numbers on the other diagonal.

$$\begin{vmatrix} a & b \\ c & d \end{vmatrix} \qquad \begin{vmatrix} a & b \\ c & d \end{vmatrix}$$

Main diagonal Other diagonal

Value of a 2 × 2 Determinant

If a, b, c, and d are numbers, the **determinant** of the matrix $\begin{bmatrix} a & b \\ c & d \end{bmatrix}$ is

$$\begin{vmatrix} a & b \\ c & d \end{vmatrix} = ad - bc$$

EXAMPLE 1 Evaluate each determinant: **a.** $\begin{vmatrix} 3 & 2 \\ 6 & 9 \end{vmatrix}$ **b.** $\begin{vmatrix} -20 & 1 \\ -8 & 4 \end{vmatrix}$

Strategy We will find the product of the numbers on the main diagonal and the product of the numbers along the other diagonal and subtract the results.

Why The value of a determinant of the form $\begin{vmatrix} a & b \\ c & d \end{vmatrix}$ is $ad - bc$.

Solution To evaluate these 2 × 2 determinants, we proceed as follows:

a. This is always minus.

$$\begin{vmatrix} 3 & 2 \\ 6 & 9 \end{vmatrix} = 3(9) - 2(6) = 27 - 12 = 15$$

b. $$\begin{vmatrix} -20 & 1 \\ -8 & 4 \end{vmatrix} = -20(4) - 1(-8) = -80 - (-8) = -80 + 8 = -72$$

Caution

Although the symbols used to write a determinant are a pair of vertical lines just like the vertical lines used for absolute value, the two concepts are unrelated. In part (b), notice that the value of the determinant is negative.

Self Check 1 Evaluate: $\begin{vmatrix} 4 & -3 \\ 2 & 1 \end{vmatrix}$ 10

Now Try Problems 15 and 17

Teaching Example 1

Evaluate: $\begin{vmatrix} -2 & 3 \\ -1 & 4 \end{vmatrix}$

Answer: −5

A 3×3 determinant is evaluated by **expanding by minors.** The following definition shows how we can evaluate a 3×3 determinant by expanding by minors along the first row.

Value of a 3 × 3 Determinant

$$\begin{vmatrix} a_1 & b_1 & c_1 \\ a_2 & b_2 & c_2 \\ a_3 & b_3 & c_3 \end{vmatrix} = a_1 \underbrace{\begin{vmatrix} b_2 & c_2 \\ b_3 & c_3 \end{vmatrix}}_{\text{Minor of } a_1} - b_1 \underbrace{\begin{vmatrix} a_2 & c_2 \\ a_3 & c_3 \end{vmatrix}}_{\text{Minor of } b_1} + c_1 \underbrace{\begin{vmatrix} a_2 & b_2 \\ a_3 & b_3 \end{vmatrix}}_{\text{Minor of } c_1}$$

The Language of Algebra

The 2 × 2 determinants in the definition box above are called **minors** of the elements in the 3 × 3 determinant.

To find the minor of a_1, we cross out the elements of the determinant that are in the same row and column as a_1:

$$\begin{vmatrix} a_1 & b_1 & c_1 \\ a_2 & b_2 & c_2 \\ a_3 & b_3 & c_3 \end{vmatrix} \quad \text{The minor of } a_1 \text{ is } \begin{vmatrix} b_2 & c_2 \\ b_3 & c_3 \end{vmatrix}.$$

To find the minor of b_1, we cross out the elements of the determinant that are in the same row and column as b_1:

$$\begin{vmatrix} a_1 & b_1 & c_1 \\ a_2 & b_2 & c_2 \\ a_3 & b_3 & c_3 \end{vmatrix} \quad \text{The minor of } b_1 \text{ is } \begin{vmatrix} a_2 & c_2 \\ a_3 & c_3 \end{vmatrix}.$$

To find the minor of c_1, we cross out the elements of the determinant that are in the same row and column as c_1:

$$\begin{vmatrix} a_1 & b_1 & c_1 \\ a_2 & b_2 & c_2 \\ a_3 & b_3 & c_3 \end{vmatrix} \quad \text{The minor of } c_1 \text{ is } \begin{vmatrix} a_2 & b_2 \\ a_3 & b_3 \end{vmatrix}.$$

EXAMPLE 2 Evaluate the determinant: $\begin{vmatrix} 1 & 3 & -2 \\ 2 & 0 & 3 \\ 1 & 2 & 3 \end{vmatrix}$

Strategy We will expand the determinant along the first row using the numbers in the first row and their corresponding minors.

Why We can then evaluate the resulting 2×2 determinants and simplify.

Solution To evaluate this 3×3 determinant, we can use the first row and expand the determinant by minors:

Teaching Tip: Tell your students that the value of some determinants is positive, the value of others is negative, and some have value 0.

$$\begin{vmatrix} 1 & 3 & -2 \\ 2 & 0 & 3 \\ 1 & 2 & 3 \end{vmatrix} = 1 \underbrace{\begin{vmatrix} 0 & 3 \\ 2 & 3 \end{vmatrix}}_{\text{Minor of } 1} - 3 \underbrace{\begin{vmatrix} 2 & 3 \\ 1 & 3 \end{vmatrix}}_{\text{Minor of } 3} + (-2) \underbrace{\begin{vmatrix} 2 & 0 \\ 1 & 2 \end{vmatrix}}_{\text{Minor of } -2}$$

$= 1(0 - 6) - 3(6 - 3) - 2(4 - 0)$ Evaluate each 2 × 2 determinant.

$= 1(-6) - 3(3) - 2(4)$ Do each subtraction within parentheses.

$= -6 - 9 - 8$ Do each multiplication.

$= -23$ Do the subtraction.

Teaching Example 2
Evaluate:

$$\begin{vmatrix} -2 & 3 & 2 \\ -1 & 4 & 5 \\ 1 & 7 & 6 \end{vmatrix}$$

Answer: 33

Self Check 2 Evaluate: $\begin{vmatrix} 2 & 3 & -1 \\ 0 & 2 & 4 \\ -2 & 5 & 6 \end{vmatrix}$ −44

Now Try Problem 27

We can evaluate a 3 × 3 determinant by expanding it by the minors of any row or column. We will get the same value. To determine the signs between the terms of the expansion of a 3 × 3 determinant, we use the following array of signs.

Array of Signs for a 3 × 3 Determinant

$$\begin{matrix} + & - & + \\ - & + & - \\ + & - & + \end{matrix}$$

This array of signs is often called the checkerboard pattern.

To remember the sign pattern, note that there is a + sign in the upper left position and that the signs alternate for all of the positions that follow.

EXAMPLE 3 Evaluate the determinant $\begin{vmatrix} 1 & 3 & -2 \\ 2 & 0 & 3 \\ 1 & 2 & 3 \end{vmatrix}$ by expanding by the minors of the middle column. (This is the determinant of Example 2.)

Strategy We will expand the determinant using the numbers in the middle column and their corresponding minors. We will use the sign pattern − + − between the terms of the expansion.

Why We can then evaluate the resulting 2 × 2 determinants and simplify.

Solution To evaluate the determinant, we proceed as follows:

$$\begin{vmatrix} 1 & 3 & -2 \\ 2 & 0 & 3 \\ 1 & 2 & 3 \end{vmatrix} = -3\underbrace{\begin{vmatrix} 2 & 3 \\ 1 & 3 \end{vmatrix}}_{\text{Minor of 3}} + 0\underbrace{\begin{vmatrix} 1 & -2 \\ 1 & 3 \end{vmatrix}}_{\text{Minor of 0}} - 2\underbrace{\begin{vmatrix} 1 & -2 \\ 2 & 3 \end{vmatrix}}_{\text{Minor of 2}}$$

Use the middle column of the checkerboard pattern: $\begin{matrix} + & - & + \\ - & + & - \\ + & - & + \end{matrix}$

$= -3(6 - 3) + 0 - 2[3 - (-4)]$ Evaluate each 2 × 2 determinant.
$= -3(3) + 0 - 2(7)$ Do each subtraction within parentheses.
$= -9 + 0 - 14$ Do each multiplication.
$= -23$ Add and subtract.

As expected, we get the same value as in Example 2.

Success Tip

When evaluating a determinant, expanding along a row or column that contains 0's can simplify the calculations.

Teaching Example 3
Evaluate $\begin{vmatrix} 1 & 4 & -3 \\ 2 & -1 & 6 \\ 5 & 0 & 9 \end{vmatrix}$ by expanding by the minors of the last row.
Answer: 24

Self Check 3 Evaluate $\begin{vmatrix} 2 & 3 & -1 \\ 0 & 2 & 4 \\ -2 & 5 & 6 \end{vmatrix}$ by expanding by the minors of the first column. −44

Now Try Problem 31

Using Your Calculator ▶ Evaluating Determinants

It is possible to use a graphing calculator to evaluate determinants. For example, to evaluate the determinant in Example 3, we first enter the matrix by pressing the MATRIX key, selecting EDIT, and pressing the ENTER key. Next, we enter the dimensions and the elements of the matrix to get figure (a). We then press 2nd QUIT to clear the screen, press MATRIX, select MATH, and press 1 to get figure (b). We then press MATRIX, select NAMES, press 1, and press) and ENTER to get the value of the determinant. Figure (c) shows that the value of the determinant is -23.

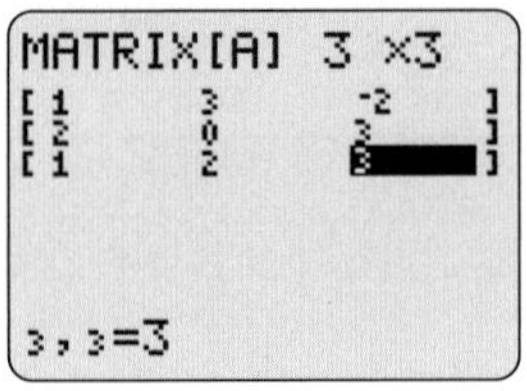

(a)

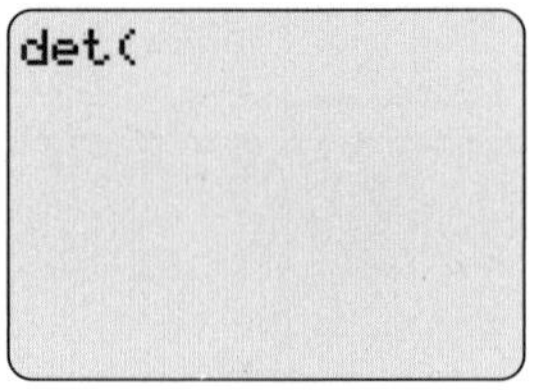

(b)

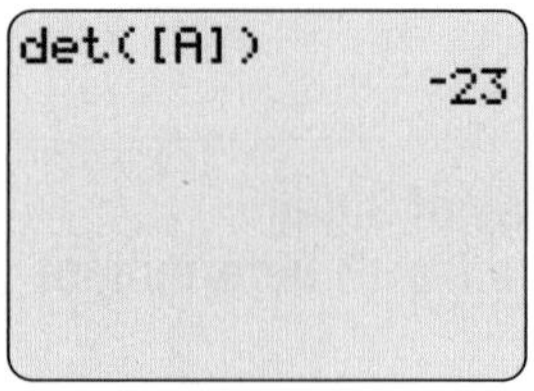

(c)

2 Use Cramer's Rule to Solve Systems of Two Equations.

The method of using determinants to solve systems of linear equations is called **Cramer's rule,** named after the 18th-century Swiss mathematician Gabriel Cramer. To develop Cramer's rule, we consider the system

$$\begin{cases} ax + by = e \\ cx + dy = f \end{cases}$$

where x and y are variables and a, b, c, d, e, and f are constants.

If we multiply both sides of the first equation by d and multiply both sides of the second equation by $-b$, we can add the equations and eliminate y:

$$\begin{aligned} adx + bdy &= ed & \text{This is } d(ax + by) = d(e). \\ -bcx - bdy &= -bf & \text{This is } -b(cx + dy) = -b(f). \\ \hline adx - bcx \quad &= ed - bf \end{aligned}$$

To solve for x, we use the distributive property to write $adx - bcx$ as $(ad - bc)x$ on the left side and divide each side by $ad - bc$:

$$(ad - bc)x = ed - bf$$

$$x = \frac{ed - bf}{ad - bc} \quad \text{where } ad - bc \neq 0$$

We can find y in a similar way. After eliminating the variable x, we get

$$y = \frac{af - ec}{ad - bc} \quad \text{where } ad - bc \neq 0$$

Note that the denominator for both x and y is

$$\begin{vmatrix} a & b \\ c & d \end{vmatrix} = ad - bc$$

The numerators can be expressed as determinants also:

$$x = \frac{ed - bf}{ad - bc} = \frac{\begin{vmatrix} e & b \\ f & d \end{vmatrix}}{\begin{vmatrix} a & b \\ c & d \end{vmatrix}} \quad \text{and} \quad y = \frac{af - ec}{ad - bc} = \frac{\begin{vmatrix} a & e \\ c & f \end{vmatrix}}{\begin{vmatrix} a & b \\ c & d \end{vmatrix}}$$

Cramer's Rule for Systems of Two Equations in Two Variables

The solution of the system $\begin{cases} ax + by = e \\ cx + dy = f \end{cases}$ is given by $x = \frac{D_x}{D}$ and $y = \frac{D_y}{D}$, where

$$D = \begin{vmatrix} a & b \\ c & d \end{vmatrix}$$ Use only the coefficients of the variables.

$$D_x = \begin{vmatrix} e & b \\ f & d \end{vmatrix}$$ Replace the x-term coefficients with the constants.

$$D_y = \begin{vmatrix} a & e \\ c & f \end{vmatrix}$$ Replace the y-term coefficients with the constants.

If every determinant is 0, the system is consistent, but the equations are dependent.

If $D = 0$ and D_x or D_y is nonzero, the system is inconsistent. If $D \neq 0$, the system is consistent, and the equations are independent.

The following observations are helpful when memorizing the three determinants of Cramer's rule.

- The denominator determinant, D, is formed by using the coefficients a, b, c, and d of the variables in the equations.

$$D = \begin{vmatrix} a & b \\ c & d \end{vmatrix}$$

x-term coefficients (first column) y-term coefficients (second column)

- The numerator determinants, D_x and D_y, are the same as the denominator determinant, D, except that the column of coefficients of the variable for which we are solving is replaced with the column of constants e and f.

$$D_x = \begin{vmatrix} e & b \\ f & d \end{vmatrix} \qquad D_y = \begin{vmatrix} a & e \\ c & f \end{vmatrix}$$

Replace the x-term coefficients with the constants. Replace the y-term coefficients with the constants.

EXAMPLE 4 Use Cramer's rule to solve the system: $\begin{cases} 4x - 3y = 6 \\ -2x + 5y = 4 \end{cases}$

Strategy We will evaluate three determinants, D, D_x, and D_y.

Why The x-value of the solution of the system is the quotient of D_x and D and the y-value of the solution is the quotient of two determinants, D_y and D.

Solution By Cramer's rule, $x = \frac{D_x}{D}$ and $y = \frac{D_y}{D}$. It is best to find D first because if $D = 0$, the system is inconsistent or has dependent equations. The denominator determinant D is made up of the coefficients of x and y:

$$D = \begin{vmatrix} 4 & -3 \\ -2 & 5 \end{vmatrix} = 4(5) - (-3)(-2) = 20 - 6 = 14$$

We form the numerator determinant D_x from D by replacing its first column (the coefficients of x) with the column of constants (**6** and **4**).

$$D_x = \begin{vmatrix} \mathbf{6} & -3 \\ \mathbf{4} & 5 \end{vmatrix} = 6(5) - (-3)(4) = 30 + 12 = 42$$

Success Tip

To find D, D_x, and D_y, we must evaluate three 2×2 determinants. The values of D_x and D_y cannot be obtained from the calculations used to find D.

Success Tip

We can now solve systems of two linear equations in two variables using:

- graphing
- substitution
- elimination (addition)
- matrices
- Cramer's rule

We form the numerator determinant D_y from D by replacing the second column (the coefficients of y) with the column of constants (6 and 4).

$$D_y = \begin{vmatrix} 4 & 6 \\ -2 & 4 \end{vmatrix} = 4(4) - 6(-2) = 16 + 12 = 28$$

Thus, $x = \frac{D_x}{D} = \frac{42}{14} = 3$ and $y = \frac{D_y}{D} = \frac{28}{14} = 2$. The solution is (3, 2) and the solution set is $\{(3, 2)\}$. Verify that (3, 2) satisfies both equations of the given system.

Teaching Example 4
Use Cramer's rule to solve:
$\begin{cases} 5x - 2y = 6 \\ 3x + 4y = -2 \end{cases}$
Answer: $\left(\frac{10}{13}, -\frac{14}{13}\right)$

Self Check 4 Use Cramer's rule to solve the system: $\begin{cases} 2x - 3y = -16 \\ 3x + 5y = 14 \end{cases}$ (−2, 4)

Now Try Problem 39

In the next example, we will see how to recognize inconsistent systems when Cramer's rule is used to solve them.

EXAMPLE 5 Use Cramer's rule to solve: $\begin{cases} 7x = 8 - 4y \\ 2y = 3 - \frac{7}{2}x \end{cases}$

Strategy We will evaluate three determinants, D, D_x, and D_y.

Why The x-value of the solution of the system is the quotient of D_x and D, and the y-value of the solution is the quotient of two determinants, D_y and D.

Solution Before we can form the required determinants, the equations of the system must be written in standard $Ax + By = C$ form.

Caution

The equations of a system must be written in standard form before the determinants of Cramer's rule can be constructed.

$$\begin{cases} 7x = 8 - 4y \\ 2y = 3 - \frac{7}{2}x \end{cases} \xrightarrow[\text{Multiply both sides by 2 and add } 7x \text{ to both sides.}]{\text{Add } 4y \text{ to both sides.}} \begin{cases} 7x + 4y = 8 \\ 7x + 4y = 6 \end{cases}$$

When we attempt to use Cramer's rule to solve this system for x, we obtain

$$D = \begin{vmatrix} 7 & 4 \\ 7 & 4 \end{vmatrix} = 28 - 28 = 0 \quad \text{and} \quad D_x = \begin{vmatrix} 8 & 4 \\ 6 & 4 \end{vmatrix} = 32 - 24 = 8$$

Success Tip

If any two rows or any two columns of a determinant are identical, the value of the determinant is 0.

Thus,

$$x = \frac{8}{0}, \text{ which is undefined.}$$

Since the denominator determinant D is 0 and the numerator determinant D_x is not 0, the system is inconsistent. It has no solution and the solution set is $\varnothing$.

We can see directly from the system that it is inconsistent. For any values of x and y, it is impossible that 7 times x plus 4 times y could be both 8 and 6.

Teaching Example 5
Use Cramer's rule to solve:
$\begin{cases} 2x = 7 + y \\ y = 2x + 5 \end{cases}$
Answer: No solution, $\varnothing$; inconsistent system

Self Check 5 Use Cramer's rule to solve the system: $\begin{cases} 3x = 8 - 4y \\ y = \frac{5}{2} - \frac{3}{4}x \end{cases}$

No solution, $\varnothing$; inconsistent system

Now Try Problem 47

3 Use Cramer's Rule to Solve Systems of Three Equations.

Cramer's rule can be extended to solve systems of three linear equations with three variables.

Cramer's Rule for Three Equations in Three Variables

The solution of the system $\begin{cases} ax + by + cz = j \\ dx + ey + fz = k \\ gx + hy + iz = l \end{cases}$ is given by

$$x = \frac{D_x}{D}, \quad y = \frac{D_y}{D}, \quad \text{and} \quad z = \frac{D_z}{D}, \quad \text{where}$$

$$D = \begin{vmatrix} a & b & c \\ d & e & f \\ g & h & i \end{vmatrix}$$ Use only the coefficients of the variables.

$$D_x = \begin{vmatrix} j & b & c \\ k & e & f \\ l & h & i \end{vmatrix}$$ Replace the x-term coefficients with the constants.

$$D_y = \begin{vmatrix} a & j & c \\ d & k & f \\ g & l & i \end{vmatrix}$$ Replace the y-term coefficients with the constants.

$$D_z = \begin{vmatrix} a & b & j \\ d & e & k \\ g & h & l \end{vmatrix}$$ Replace the z-term coefficients with the constants.

If every determinant is 0, the system is consistent, but the equations are dependent.

If $D = 0$ and D_x or D_y or D_z is nonzero, the system is inconsistent. If $D \neq 0$, the system is consistent, and the equations are independent.

EXAMPLE 6 Use Cramer's rule to solve the system: $\begin{cases} 2x + y + 4z = 12 \\ x + 2y + 2z = 9 \\ 3x - 3y - 2z = 1 \end{cases}$

Strategy We will evaluate four determinants, D, D_x, D_y, and D_z.

Why The x-value of the solution of the system is the quotient of D_x and D, the y-value of the solution is the quotient of two determinants, D_y and D, and the z-value of the solution is the quotient of two determinants, D_z and D.

Solution The denominator determinant D is the determinant formed by the coefficients of the variables. The numerator determinants, D_x, D_y, and D_z, are formed by replacing the coefficients of the variable being solved for by the column of constants. We will evaluate each determinant by expanding by minors about the first row:

$$D = \begin{vmatrix} 2 & 1 & 4 \\ 1 & 2 & 2 \\ 3 & -3 & -2 \end{vmatrix} = 2\begin{vmatrix} 2 & 2 \\ -3 & -2 \end{vmatrix} - 1\begin{vmatrix} 1 & 2 \\ 3 & -2 \end{vmatrix} + 4\begin{vmatrix} 1 & 2 \\ 3 & -3 \end{vmatrix} = 2(2) - 1(-8) + 4(-9) = -24$$

$$D_x = \begin{vmatrix} 12 & 1 & 4 \\ 9 & 2 & 2 \\ 1 & -3 & -2 \end{vmatrix} = 12\begin{vmatrix} 2 & 2 \\ -3 & -2 \end{vmatrix} - 1\begin{vmatrix} 9 & 2 \\ 1 & -2 \end{vmatrix} + 4\begin{vmatrix} 9 & 2 \\ 1 & -3 \end{vmatrix} = 12(2) - 1(-20) + 4(-29) = -72$$

$$D_y = \begin{vmatrix} 2 & 12 & 4 \\ 1 & 9 & 2 \\ 3 & 1 & -2 \end{vmatrix} = 2\begin{vmatrix} 9 & 2 \\ 1 & -2 \end{vmatrix} - 12\begin{vmatrix} 1 & 2 \\ 3 & -2 \end{vmatrix} + 4\begin{vmatrix} 1 & 9 \\ 3 & 1 \end{vmatrix} = 2(-20) - 12(-8) + 4(-26) = -48$$

$$D_z = \begin{vmatrix} 2 & 1 & 12 \\ 1 & 2 & 9 \\ 3 & -3 & 1 \end{vmatrix} = 2\begin{vmatrix} 2 & 9 \\ -3 & 1 \end{vmatrix} - 1\begin{vmatrix} 1 & 9 \\ 3 & 1 \end{vmatrix} + 12\begin{vmatrix} 1 & 2 \\ 3 & -3 \end{vmatrix} = 2(29) - 1(-26) + 12(-9) = -24$$

Success Tip

To find D, D_x, D_y, and D_z, we must evaluate four 3×3 determinants. The values of D_x, D_y, and D_z cannot be obtained from the calculations used to find D.

Thus,

$$x = \frac{D_x}{D} = \frac{-72}{-24} = 3$$

$$y = \frac{D_y}{D} = \frac{-48}{-24} = 2$$

$$z = \frac{D_z}{D} = \frac{-24}{-24} = 1$$

The solution is (3, 2, 1) and the solution set is {(3, 2, 1)}. Verify that (3, 2, 1) satisfies the three original equations.

Teaching Example 6
Use Cramer's rule to solve:
$$\begin{cases} 6x + y - z = -2 \\ 2x + 5y - z = 2 \\ x + 2y + z = 5 \end{cases}$$
Answer: (0, 1, 3)

Self Check 6 Use Cramer's rule to solve the system: $\begin{cases} x + y + 2z = 6 \\ 2x - y + z = 9 \\ x + y - 2z = -6 \end{cases}$ (2, −2, 3)

Now Try Problem 51

SECTION 12.5 STUDY SET

VOCABULARY

Fill in the blanks.

1. $\begin{vmatrix} 4 & 9 \\ -6 & 1 \end{vmatrix}$ is a determinant. The numbers 4 and 1 lie along its main diagonal.

2. A determinant is number that is associated with a square matrix.

3. The minor of b_1 in $\begin{vmatrix} a_1 & b_1 & c_1 \\ a_2 & b_2 & c_2 \\ a_3 & b_3 & c_3 \end{vmatrix}$ is $\begin{vmatrix} a_2 & c_2 \\ a_3 & c_3 \end{vmatrix}$.

4. Cramer's rule uses determinants to solve systems of linear equations.

CONCEPTS

Fill in the blanks.

5. $\begin{vmatrix} a & b \\ c & d \end{vmatrix} = ad - bc$

6. To find the minor of 5, we cross out the elements of the determinant that are in the same row and column as 5.

$$\begin{vmatrix} 3 & 5 & 1 \\ 6 & -2 & 2 \\ 8 & -1 & 4 \end{vmatrix}$$

7. In evaluating the determinant below, about what row or column was it expanded? The third column

$$\begin{vmatrix} 5 & 1 & -1 \\ 8 & 7 & 4 \\ 9 & 7 & 6 \end{vmatrix} = -1\begin{vmatrix} 8 & 7 \\ 9 & 7 \end{vmatrix} - 4\begin{vmatrix} 5 & 1 \\ 9 & 7 \end{vmatrix} + 6\begin{vmatrix} 5 & 1 \\ 8 & 7 \end{vmatrix}$$

8. What is the denominator determinant D for the system $\begin{cases} 3x + 4y = 7 \\ 2x - 3y = 5 \end{cases}$? $\begin{vmatrix} 3 & 4 \\ 2 & -3 \end{vmatrix}$

9. What is the denominator determinant D for the system $\begin{cases} x + 2y = -8 \\ 3x + y - z = -2 \\ 8x + 4y - z = 6 \end{cases}$? $\begin{vmatrix} 1 & 2 & 0 \\ 3 & 1 & -1 \\ 8 & 4 & -1 \end{vmatrix}$

10. For the system $\begin{cases} 3x + 2y = 1 \\ 4x - y = 3 \end{cases}$, $D_x = -7$, $D_y = 5$, and $D = -11$. Find the solution of the system. $\left(\frac{7}{11}, -\frac{5}{11}\right)$

11. For the system $\begin{cases} 2x + 3y - z = -8 \\ x - y - z = -2 \\ -4x + 3y + z = 6 \end{cases}$, $D_x = -28$, $D_y = -14$, $D_z = 14$, and $D = 14$. Find the solution of the system. (−2, −1, 1)

12. Fill in the blank. If the denominator determinant D for a system of equations is 0, the equations of the system are dependent or the system is inconsistent.

NOTATION

Complete the evaluation of each determinant.

13. $\begin{vmatrix} 5 & -2 \\ -2 & 6 \end{vmatrix} = 5(6) - (-2)(-2)$
$= 30 - 4$
$= 26$

▶ **14.** $\begin{vmatrix} 2 & 1 & 3 \\ 3 & 4 & 2 \\ 1 & 5 & 3 \end{vmatrix} = 2\begin{vmatrix} 4 & 2 \\ 5 & 3 \end{vmatrix} - 1\begin{vmatrix} 3 & 2 \\ 1 & 3 \end{vmatrix} + 3\begin{vmatrix} 3 & 4 \\ 1 & 5 \end{vmatrix}$

$= 2(12 - 10) - 1(9 - 2) + 3(15 - 4)$

$= 2(2) - 1(7) + 3(11)$

$= 4 - 7 + 33$

$= 30$

GUIDED PRACTICE

Evaluate each determinant. **See Example 1.**

15. $\begin{vmatrix} 2 & 3 \\ 2 & 5 \end{vmatrix}$ 4

▶ **16.** $\begin{vmatrix} 3 & 2 \\ 2 & 4 \end{vmatrix}$ 8

▶ **17.** $\begin{vmatrix} -9 & 7 \\ 4 & -2 \end{vmatrix}$ −10

▶ **18.** $\begin{vmatrix} -1 & 2 \\ 3 & -4 \end{vmatrix}$ −2

19. $\begin{vmatrix} 5 & 20 \\ 10 & 6 \end{vmatrix}$ −170

▶ **20.** $\begin{vmatrix} 10 & 15 \\ 15 & 5 \end{vmatrix}$ −175

21. $\begin{vmatrix} -6 & -2 \\ 15 & 4 \end{vmatrix}$ 6

▶ **22.** $\begin{vmatrix} 3 & -2 \\ 12 & -8 \end{vmatrix}$ 0

23. $\begin{vmatrix} -9 & -1 \\ -10 & -5 \end{vmatrix}$ 35

▶ **24.** $\begin{vmatrix} -7 & -7 \\ -6 & -4 \end{vmatrix}$ −14

25. $\begin{vmatrix} 8 & 8 \\ -9 & -9 \end{vmatrix}$ 0

▶ **26.** $\begin{vmatrix} 20 & -3 \\ 20 & -3 \end{vmatrix}$ 0

Evaluate each determinant. **See Examples 2 and 3.**

27. $\begin{vmatrix} 3 & 2 & 1 \\ 4 & 1 & 2 \\ 5 & 3 & 1 \end{vmatrix}$ 4

▶ **28.** $\begin{vmatrix} 6 & 2 & 3 \\ 1 & 5 & 4 \\ 2 & 3 & 5 \end{vmatrix}$ 63

▶ **29.** $\begin{vmatrix} 1 & -2 & 3 \\ -2 & 1 & 1 \\ -3 & -2 & 1 \end{vmatrix}$ 26

▶ **30.** $\begin{vmatrix} 1 & 1 & 2 \\ 2 & 1 & -2 \\ 3 & 1 & 3 \end{vmatrix}$ −9

31. $\begin{vmatrix} -2 & 5 & 1 \\ 0 & 3 & 4 \\ -1 & 2 & 6 \end{vmatrix}$ −37

▶ **32.** $\begin{vmatrix} 4 & -1 & 2 \\ 6 & -1 & 0 \\ 1 & -3 & 4 \end{vmatrix}$ −26

33. $\begin{vmatrix} 1 & -4 & 1 \\ 3 & 0 & -2 \\ 3 & 1 & -2 \end{vmatrix}$ 5

▶ **34.** $\begin{vmatrix} 8 & -3 & 1 \\ 1 & 0 & 2 \\ 3 & -9 & 4 \end{vmatrix}$ 129

▶ **35.** $\begin{vmatrix} 1 & 2 & 1 \\ -3 & 7 & 3 \\ -4 & 3 & -5 \end{vmatrix}$ −79

▶ **36.** $\begin{vmatrix} 1 & 4 & 7 \\ 2 & 5 & 8 \\ 3 & 6 & 9 \end{vmatrix}$ 0

37. $\begin{vmatrix} 1 & 2 & 0 \\ 0 & 1 & 2 \\ 0 & 0 & 1 \end{vmatrix}$ 1

▶ **38.** $\begin{vmatrix} 1 & 0 & 1 \\ 0 & 1 & 0 \\ 1 & 1 & 1 \end{vmatrix}$ 0

Use Cramer's rule to solve each system of equations. **See Example 4.**

▶ **39.** $\begin{cases} x + y = 6 \\ x - y = 2 \end{cases}$ (4, 2)

▶ **40.** $\begin{cases} x - y = 4 \\ 2x + y = 5 \end{cases}$ (3, −1)

41. $\begin{cases} x + 2y = -21 \\ x - 2y = 11 \end{cases}$ (−5, −8)

▶ **42.** $\begin{cases} 5x + 2y = 11 \\ 7x + 6y = 9 \end{cases}$ (3, −2)

43. $\begin{cases} 3x - 4y = 9 \\ x + 2y = 8 \end{cases}$ $\left(5, \frac{3}{2}\right)$

▶ **44.** $\begin{cases} 2x + 2y = -1 \\ 3x + 4y = 0 \end{cases}$ $\left(-2, \frac{3}{2}\right)$

45. $\begin{cases} 2x + 3y = 31 \\ 3x + 2y = 39 \end{cases}$ (11, 3)

▶ **46.** $\begin{cases} 5x + 3y = 72 \\ 3x + 5y = 56 \end{cases}$ (12, 4)

Use Cramer's rule to solve each system of equations. If a system is inconsistent or if the equations are dependent, so indicate. **See Example 5.**

▶ **47.** $\begin{cases} 3x + 2y = 11 \\ 6x + 4y = 11 \end{cases}$ No sol., ∅; incons. syst.

▶ **48.** $\begin{cases} 5x - 4y = 20 \\ 10x - 8y = 30 \end{cases}$ No sol., ∅; incons. syst.

▶ **49.** $\begin{cases} \frac{5}{6}x = 2 - y \\ 10x + 12y = 24 \end{cases}$ $\{(x, y) \mid 5x = 12 - 6y\}$, inf. many sol.; dep. eq.

▶ **50.** $\begin{cases} 16x - 8y = 32 \\ x - 2 = \frac{y}{2} \end{cases}$ $\{(x, y) \mid 2x - 4 = y\}$, inf. many sol.; dep. eq.

Use Cramer's rule to solve each system of equations. **See Example 6.**

▶ **51.** $\begin{cases} x + y + z = 4 \\ x + y - z = 0 \\ x - y + z = 2 \end{cases}$ (1, 1, 2)

52. $\begin{cases} x + y + z = 4 \\ x - y + z = 2 \\ x - y - z = 0 \end{cases}$ (2, 1, 1)

▶ **53.** $\begin{cases} 3x + 2y - z = -8 \\ 2x - y + 7z = 10 \\ 2x + 2y - 3z = -10 \end{cases}$ (−2, 0, 2)

▶ **54.** $\begin{cases} x + 2y + 2z = 10 \\ 2x + y + 2z = 9 \\ 2x + 2y + z = 1 \end{cases}$ (−2, −1, 7)

TRY IT YOURSELF

Use Cramer's rule to solve each system of equations. If a system is inconsistent or if the equations are dependent, so indicate.

▶ **55.** $\begin{cases} 2x + y + z = 5 \\ x - 2y + 3z = 10 \\ x + y - 4z = -3 \end{cases}$ (3, −2, 1)

▶ **56.** $\begin{cases} x + y + 2z = 7 \\ x + 2y + z = 8 \\ 2x + y + z = 9 \end{cases}$ (3, 2, 1)

57. $\begin{cases} y = \frac{-2x + 1}{3} \\ 3x - 2y = 8 \end{cases}$ (2, −1)

▶ **58.** $\begin{cases} 2x + 3y = -1 \\ x = \frac{y - 9}{4} \end{cases}$ (−2, 1)

59. $\begin{cases} 4x - 3y = 1 \\ 6x - 8z = 1 \\ 2y - 4z = 0 \end{cases}$ $\left(-\frac{1}{2}, -1, -\frac{1}{2}\right)$

60. $\begin{cases} 4x + 3z = 4 \\ 2y - 6z = -1 \\ 8x + 4y + 3z = 9 \end{cases}$ $\left(\frac{3}{4}, \frac{1}{2}, \frac{1}{3}\right)$

61. $\begin{cases} 2x + y - z - 1 = 0 \\ x + 2y + 2z - 2 = 0 \\ 4x + 5y + 3z - 3 = 0 \end{cases}$ No sol., ∅; incons. syst.

▶ **62.** $\begin{cases} 2x - y + 4z + 2 = 0 \\ 5x + 8y + 7z = -8 \\ x + 3y + z + 3 = 0 \end{cases}$ No sol., ∅; incons. syst.

63. $\begin{cases} 3x - 16 = 5y \\ -3x + 5y - 33 = 0 \end{cases}$ No sol., ∅; incons. syst.

64. $\begin{cases} 2x + 5y - 13 = 0 \\ -2x + 13 = 5y \end{cases}$ $\{(x, y) \mid 2x + 5y = 13\}$, dep. eqns.; inf. many sol.

65. $\begin{cases} x + y = 1 \\ \frac{1}{2}y + z = \frac{5}{2} \\ x - z = -3 \end{cases}$ (−2, 3, 1)

▶ **66.** $\begin{cases} \frac{1}{2}x + y + z + \frac{3}{2} = 0 \\ x + \frac{1}{2}y + z - \frac{1}{2} = 0 \\ x + y + \frac{1}{2}z + \frac{1}{2} = 0 \end{cases}$ $\left(\frac{9}{5}, -\frac{11}{5}, -\frac{1}{5}\right)$

67. $\begin{cases} 2x + 3y = 0 \\ 4x - 6y = -4 \end{cases}$
$\left(-\frac{1}{2}, \frac{1}{3}\right)$

68. $\begin{cases} 4x - 3y = -1 \\ 8x + 3y = 4 \end{cases}$
$\left(\frac{1}{4}, \frac{2}{3}\right)$

69. $\begin{cases} 2x + 3y + 4z = 6 \\ 2x - 3y - 4z = -4 \\ 4x + 6y + 8z = 12 \end{cases}$
Infinitely many solutions; dependent equations

70. $\begin{cases} x - 3y + 4z - 2 = 0 \\ 2x + y + 2z - 3 = 0 \\ 4x - 5y + 10z - 7 = 0 \end{cases}$
Infinitely many solutions; dependent equations

***Use a calculator with matrix capabilities. Evaluate each determinant. See Using Your Calculator:* Evaluating Determinants.**

71. $\begin{vmatrix} 25 & -36 & 44 \\ -11 & 21 & 54 \\ 37 & -31 & 19 \end{vmatrix}$
$-46{,}811$

72. $\begin{vmatrix} 13 & -27 & 62 \\ -38 & 27 & -52 \\ 10 & -300 & 42 \end{vmatrix}$
$472{,}950$

73. $\begin{vmatrix} -280 & 191 & -356 \\ -211 & -102 & -422 \\ 400 & -213 & -333 \end{vmatrix}$
$-60{,}527{,}941$

74. $\begin{vmatrix} 4.1 & 2.2 & -3.3 \\ 2.7 & -5.9 & 6.8 \\ 2.3 & 5.3 & 0.6 \end{vmatrix}$
-223.438

APPLICATIONS

75. Signaling. Solve the following system using Cramer's rule to find the measures (in degrees) of the angles shown in the illustration. (*Hint:* Each equation must be written in standard $Ax + By = C$ form.) $50°, 80°$

$$\begin{cases} 2x + y = 180 \\ y = x + 30 \end{cases}$$

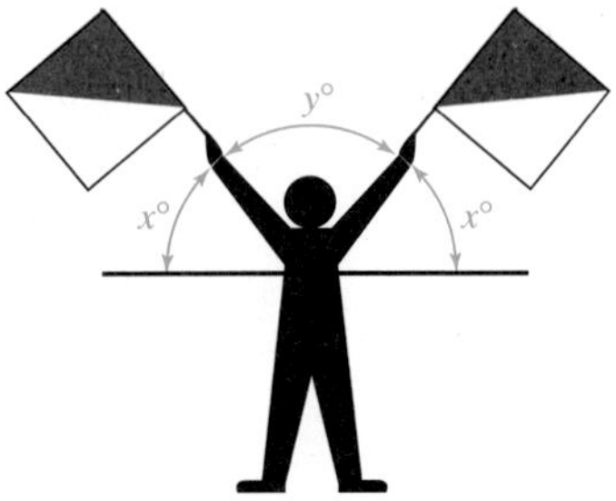

76. Investing. Solve the following system using Cramer's rule to find the amount invested in each stock. (*Hint:* Each equation must be written in standard $Ax + By + Cz = D$ form.)

$$\begin{cases} x + y + z = 20{,}000 \\ 0.10x + 0.05y + 0.06z = 1{,}320 \\ x = \frac{1}{3}(y + z) \end{cases}$$

HiTech: \$5,000, SaveTel: \$8,000, OilCo: \$7,000

Stock	Amount Invested	Rate of return
HiTech	x	10%
SaveTel	y	5%
OilCo	z	6%

WRITING

77. Explain the difference between a matrix and a determinant. Give an example of each.

78. When evaluating $\begin{vmatrix} 4 & -1 & 2 \\ 6 & -1 & 0 \\ 1 & -3 & 4 \end{vmatrix}$, why is it helpful to expand by the minors of the numbers in the third column?

79. Explain how to find the minor of an element of a determinant.

80. Explain how to find x when solving a system of three linear equations in x, y, and z by Cramer's rule. Use the words *coefficients* and *constants* in your explanation.

81. Explain how the following checkerboard pattern is used when evaluating a 3×3 determinant.

+	−	+
−	+	−
+	−	+

82. Briefly describe each of the five methods of this chapter that can be used to solve a system of two linear equations in two variables.

REVIEW

83. Are the lines described by $y = 2x - 7$ and $x - 2y = 7$ perpendicular? No

84. Are the lines described by $y = 2x - 7$ and $2x - y = 10$ parallel? Yes

85. How are the graphs of $f(x) = x^2$ and $g(x) = x^2 - 2$ related? The graph of function g is 2 units below the graph of function f.

86. Is the graph of a circle the graph of a function? No

87. The graph of a line passes through $(0, -3)$. Is this the x-intercept or the y-intercept of the line? y-intercept

88. What is the name of the function $f(x) = |x|$? The absolute value function

89. For the function $y = 2x^2 + 6x + 1$, what is the independent variable and what is the dependent variable? x; y

90. If $f(x) = x^3 - x$, what is $f(-1)$? 0

CHALLENGE PROBLEMS

91. Show that $\begin{vmatrix} x & y & 1 \\ -2 & 3 & 1 \\ 3 & 5 & 1 \end{vmatrix} = 0$ represents the equation of the line passing through $(-2, 3)$ and $(3, 5)$.

92. Show that $\frac{1}{2}\begin{vmatrix} 0 & 0 & 1 \\ 3 & 0 & 1 \\ 0 & 4 & 1 \end{vmatrix}$ represents the area of the triangle with vertices at $(0, 0)$, $(3, 0)$, and $(0, 4)$.

12 Summary & Review

SECTION 12.1 Solving Systems of Equations in Two Variables; Applications

DEFINITIONS AND CONCEPTS

When two equations are considered at the same time, we say that they form a **system of equations.**

A **solution of a system** of equations in two variables is an ordered pair that satisfies both equations of the system.

EXAMPLES

The ordered pair $(2, -3)$ is a solution of the system $\begin{cases} x + y = -1 \\ x - 2y = 8 \end{cases}$ because its coordinates, $x = 2$ and $y = -3$, satisfy both equations.

$$\begin{aligned} x + y &= -1 && \text{First equation.} \\ 2 + (-3) &\stackrel{?}{=} -1 \\ -1 &= -1 && \text{True} \end{aligned} \qquad \begin{aligned} x - 2y &= 8 && \text{Second equation.} \\ 2 - 2(-3) &\stackrel{?}{=} 8 \\ 8 &= 8 && \text{True} \end{aligned}$$

DEFINITIONS AND CONCEPTS

To **solve a system graphically:**

1. Graph each equation on the same rectangular coordinate system.
2. Determine the coordinates of the point where the graphs intersect. That ordered pair is the solution.
3. If the graphs have no point in common, the system has no solution.
4. Check the proposed solution in each equation of the original system.

EXAMPLES

To solve the system $\begin{cases} y = -x - 1 \\ x - 2y = 8 \end{cases}$ by graphing, we graph each equation as shown on the right.

The graphs appear to intersect at the point $(2, -3)$. The check shown above verifies that $(2, -3)$ is the solution of the system.

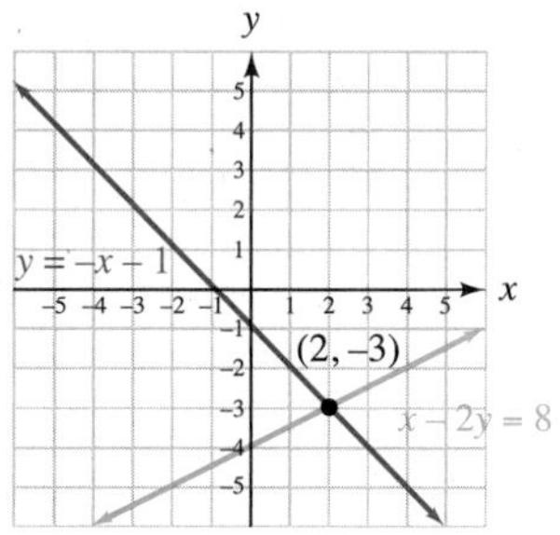

DEFINITIONS AND CONCEPTS

A system of equations that has at least one solution is called a **consistent system.** If the graphs are parallel lines, the system has no solution, and it is called an **inconsistent system.**

Equations with different graphs are called **independent equations.** If the graphs are the same line, the system has infinitely many solutions. The equations are called **dependent equations.**

EXAMPLES

Since the system shown above has a solution, it is a *consistent system.* Since the graphs are different, the equations are *independent.*

Consistent system

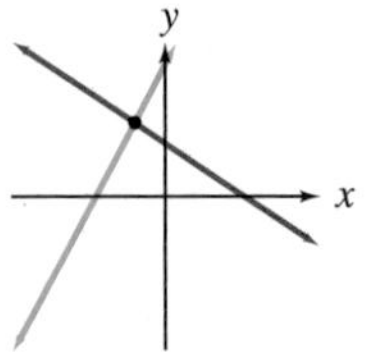

Inconsistent system

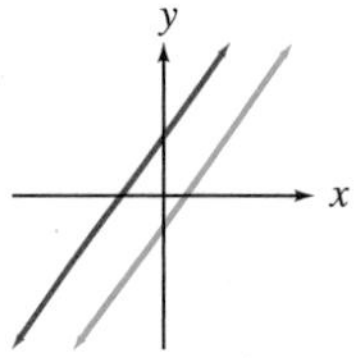

Dependent equations

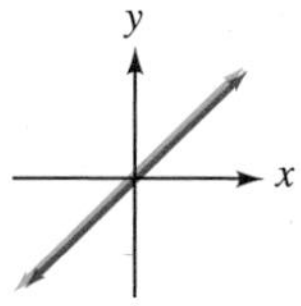

DEFINITIONS AND CONCEPTS

To solve a system of two linear equations in x and y by the **substitution method:**

1. Solve one equation for either x or y. This is called the **substitution equation.**
2. Substitute the resulting expression for that variable into the other equation and solve it.
3. Substitute the value of the variable found in step 2 into the substitution equation and solve that equation.
4. Check the proposed solution in each of the original equations. Write the solution as an ordered pair.

If in step 2 the variable drops out and a false statement results, the system has **no solution.** If a true statement results, the system has **infinitely many solutions** and we can use **set-builder notation** to write the solution set.

EXAMPLES

Use substitution to solve the system: $\begin{cases} x + y = -1 \\ x - 2y = 8 \end{cases}$

Step 1: We will solve $x + y = -1$ for y.

$$y = -x - 1 \quad \text{This is the substitution equation.}$$

Step 2: We substitute $-x - 1$ for y in the second equation and solve for x.

$$\begin{aligned} x - 2(-x - 1) &= 8 \\ x + 2x + 2 &= 8 && \text{Distribute.} \\ 3x + 2 &= 8 && \text{Combine like terms.} \\ 3x &= 6 && \text{Subtract 2 from both sides.} \\ x &= 2 && \text{Divide both sides by 3.} \end{aligned}$$

Step 3: We substitute 2 for x in the substitution equation and solve for y.

$$\begin{aligned} y &= -x - 1 \\ y &= -2 - 1 \\ y &= -3 \end{aligned}$$

Step 4: The solution is $(2, -3)$. Verify this by checking it in each of the original equations.

To solve a system of two linear equations in x and y by the **elimination (addition) method:**

1. Write both equations of the system in standard form: $Ax + By = C$.
2. If necessary, multiply one or both of the equations by a nonzero number chosen to make the coefficients of x (or the coefficients of y) opposites.
3. Add the equations to eliminate the terms involving x (or y).
4. Solve the equation resulting from step 3.
5. Find the value of the remaining variable by substituting the solution found in step 4 into any equation containing both variables. Or, repeat steps 2–4 to eliminate the other variable.
6. Check the proposed solution in each equation of the original system. Write the solution as an ordered pair.

If in step 3 both variables drop out and a false statement results, the system has **no solution.** If a true statement results, the system has **infinitely many solutions** and we can use **set-builder notation** to write the solution set.

Use elimination to solve the system: $\begin{cases} x + y = -1 \\ x - 2y = 8 \end{cases}$

Step 1: Since both equations are in standard form, we move to step 2.

Step 2: We can multiply both sides of the first equation by 2 to get the coefficients of y to be opposites.

$$\begin{matrix} (1) \\ (2) \end{matrix} \begin{cases} x + y = -1 \\ x - 2y = 8 \end{cases} \begin{matrix} \xrightarrow{\text{Multiply by 2}} \\ \xrightarrow[\text{Unchanged}]{} \end{matrix} \begin{cases} 2x + 2y = -2 \\ x - 2y = 8 \end{cases} \begin{matrix} (3) \\ (2) \end{matrix}$$

Step 3: We add equations 3 and 2 to eliminate y.

$$\begin{aligned} 2x + 2y &= -2 \\ x - 2y &= 8 \quad \text{Add like terms, column-by-column.} \\ \hline 3x \quad &= 6 \end{aligned}$$

Step 4: Since the resulting equation has only one variable, we can solve it for x.

$$3x = 6$$
$$x = 2 \quad \text{Divide both sides by 3.}$$

Step 5: To find y, we can substitute 2 for x in equation 1.

$$x + y = -1$$
$$2 + y = -1$$
$$y = -3 \quad \text{Subtract 2 from both sides.}$$

Step 6: The solution is $(2, -3)$. Verify this by checking it in each of the original equations.

To solve problems involving two unknown quantities, we can use the following **problem-solving strategy:**

1. **Analyze** the facts of the problem.
2. **Assign** different variables to represent the two unknown quantities.
3. **Form** two equations involving those variables.
4. **Solve** the system of equations by graphing, substitution, elimination, using matrices, or using Cramer's rule.
5. **State** the conclusion.
6. **Check** the result in the words of the problem.

Retirement Income. A retired office manager invested \$10,000 in two accounts, one paying 5% annual interest and the other 6%. If the interest earned for the first year was \$540, how much did she invest at each rate?

Analyze A total of \$10,000 is invested at two different rates for 1 year. The total interest earned was \$540.

Assign Let x = the number of dollars invested at 5% and y = the number of dollars invested at 6%.

Form We can use the formula $I = Prt$ to determine that the interest earned on the 5% investment is \$0.05x and the interest earned on the 6% investment is \$0.06y. This information is shown in the table.

	P ·	r ·	t =	I
First account	x	0.05	1	$0.05x$
Second account	y	0.06	1	$0.06y$
Total	\$10,000			\$540

One equation comes from this column. (↑ P column) A second equation comes from this column. (↑ I column)

The resulting system is: $\begin{cases} x + y = 10{,}000 \\ 0.05x + 0.06y = 540 \end{cases}$

Solve To solve by elimination we can multiply both sides of the first equation by -6, multiply the second equation by 100, and add the equations.

$$\begin{aligned} -6x - 6y &= -60{,}000 && \text{This is } -6(x + y) = -6(10{,}000). \\ 5x + 6y &= 54{,}000 && \text{This is } 100(0.05x + 0.06y) = 100(540). \\ \hline -x &= -6{,}000 \\ x &= 6{,}000 \end{aligned}$$

To find y, we substitute 6,000 for x in the first equation of the system.

$$\begin{aligned} x + y &= 10{,}000 \\ \mathbf{6{,}000} + y &= 10{,}000 \\ y &= 4{,}000 \end{aligned}$$

State The amount invested at 5% was \$6,000 and the amount invested at 6% was \$4,000.

Check The sum of \$6,000 and \$4,000 is \$10,000.

\$6,000 invested at 5% for 1 year would earn \$6,000 · 0.05 = \$300.
\$4,000 invested at 6% for 1 year would earn \$4,000 · 0.06 = \$240.
The total interest earned is \$300 + \$240 = \$540. The results check.

REVIEW EXERCISES

Use a check to determine whether the ordered pair is a solution of the system of equations.

1. $\left(-1, \frac{1}{2}\right)$, $\begin{cases} x + 2y = 0 \\ x + 4y = 1 \end{cases}$ Yes

2. $(13, 23)$, $\begin{cases} 3a - 2b + 7 = 0 \\ -2a + b = -4 \end{cases}$ No

3. See the illustration.

a. Give three points that satisfy the equation $2x + y = 5$. (1, 3), (2, 1), (4, −3) (Answers may vary)

b. Give three points that satisfy the equation $y = x - 4$. (0, −4), (2, −2), (4, 0) (Answers may vary)

c. Find the solution of: $\begin{cases} 2x + y = 5 \\ y = x - 4 \end{cases}$

(3, −1)

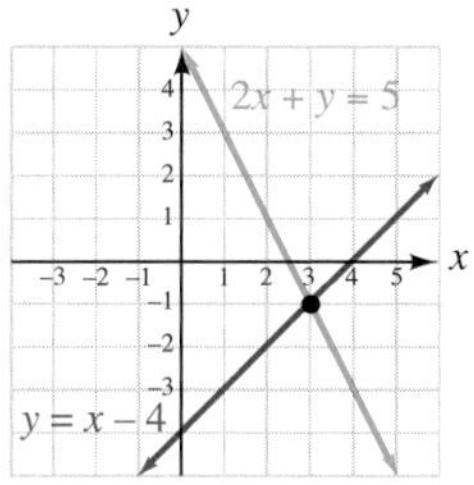

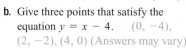

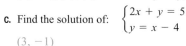

4. Video Viewing. The graph below by the Diffusion Group (TDG) predicts American video viewing habits in the future. In what year does the graph predict that the weekly amount of time spent viewing live broadcast television will be the same as that spent viewing Internet video? Approximately how many hours of weekly viewing of each type will that be? 2019, about 15 hr

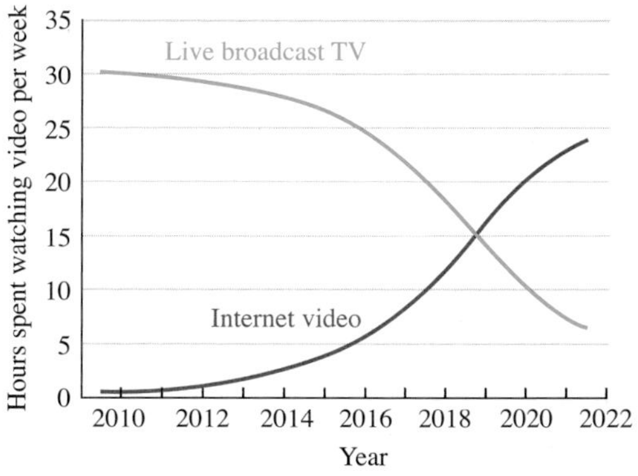

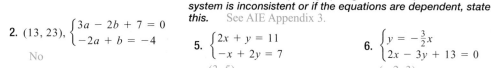

Solve each system by the graphing method, if possible. If a system is inconsistent or if the equations are dependent, state this. See AIE Appendix 3.

5. $\begin{cases} 2x + y = 11 \\ -x + 2y = 7 \end{cases}$ (3, 5)

6. $\begin{cases} y = -\frac{3}{2}x \\ 2x - 3y + 13 = 0 \end{cases}$ (−2, 3)

7. $\begin{cases} \frac{1}{2}x + \frac{1}{3}y = 2 \\ y = 6 - \frac{3}{2}x \end{cases}$ $\{(x, y) \mid 3x + 2y = 12\}$, infinitely many solutions; dependent equations

8. $\begin{cases} \frac{x}{3} - \frac{y}{2} = 1 \\ 6x - 9y = 3 \end{cases}$ No solution, ∅; inconsistent system

Solve each system using the substitution method.

9. $\begin{cases} x = y - 4 \\ 2x + 3y = 7 \end{cases}$ (−1, 3)

10. $\begin{cases} y = 2x + 5 \\ 3x - 5y = -4 \end{cases}$ (−3, −1)

Solve each system using the elimination (addition) method.

11. $\begin{cases} x - 2y = 11 \\ x + 2y = -21 \end{cases}$ (−5, −8)

12. $\begin{cases} 4a + 5b = -9 \\ 6a = 3b - 3 \end{cases}$ (−1, −1)

Solve each system by any method, if possible. If a system is inconsistent or if the equations are dependent, state this.

13. $\begin{cases} \dfrac{1}{2}a - \dfrac{3}{8}b = -9 \\ \dfrac{2}{3}a - \dfrac{1}{4}b = -8 \end{cases}$ (−6, 16)

14. $\begin{cases} y = \dfrac{x - 3}{2} \\ x = \dfrac{2y + 7}{2} \end{cases}$ $\left(4, \frac{1}{2}\right)$

15. $\begin{cases} 4x = 8y + 5 \\ 8x = 1 - 2y \end{cases}$ $\left(\frac{1}{4}, -\frac{1}{2}\right)$

16. $\begin{cases} x + 3y = -2 \\ -2(x + 3y) = 4 \end{cases}$ $\{(x, y) \mid x + 3y = -2\}$, inf. many sol.; dep. eq.

17. $\begin{cases} 0.07x = 0.05 + 0.09y \\ 7x - 9y = 8 \end{cases}$ No solution, ∅; inconsistent system

18. $\begin{cases} 0.1x + 0.2y = 1.1 \\ 2x - y = 2 \end{cases}$ (3, 4)

19. Estimate the solution of the system $\begin{cases} y = -\frac{2}{3}x \\ 2x - 3y = -4 \end{cases}$ from the graphs in the illustration. Then solve the system algebraically. $(-1, 0.7)$ (Answers may vary); $\left(-1, \frac{2}{3}\right)$

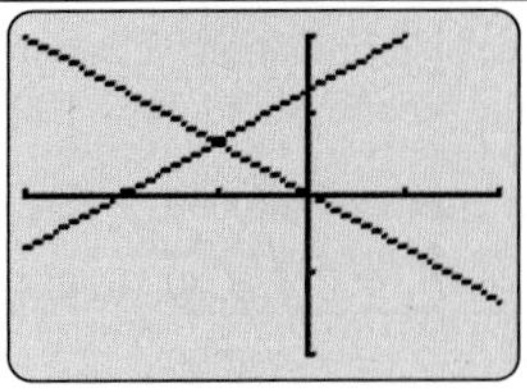

20. Give one advantage and one disadvantage for each of the following methods for solving a system of two linear equations in two variables.

a. Graphing method
b. Substitution method
c. Elimination method

Write a system of two equations in two variables to solve each problem.

21. Library Cards. Residents of a city are charged \$3 for a library card; nonresidents are charged \$7. On a day when a total of 11 library cards were purchased, the receipts were \$41. How many resident and nonresident library cards were sold? 9 cards, 2 cards

22. Maps. Refer to the illustration in the next column. The distance between Austin and Houston is 4 miles less than twice the distance between Austin and San Antonio. The round trip from Houston to Austin to San Antonio and back to Houston is 442 miles. Determine the mileages between Austin and Houston and between Austin and San Antonio. Austin–Houston: 162 mi; Austin–San Antonio: 83 mi

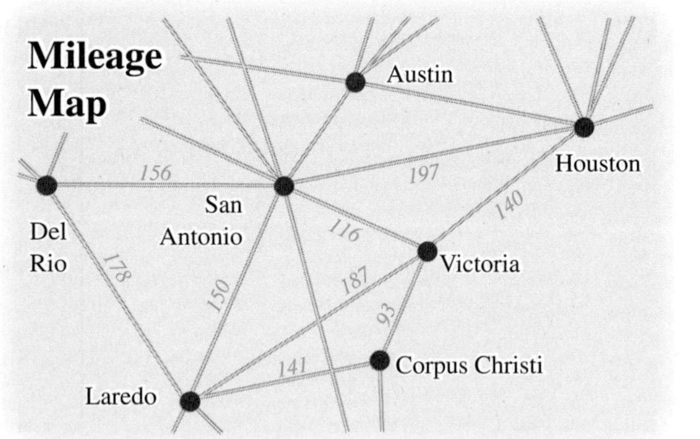

23. Riverboats. A Mississippi riverboat travels 30 miles downstream in three hours and then makes the return trip upstream in five hours. Find the speed of the riverboat in still water and the speed of the current. 8 mph, 2 mph

24. Mixing Solutions. How many fluid ounces of 6% sucrose solution must be mixed with 18% sucrose solution to make 750 ounces of a 10% sucrose solution? 500 oz of 6%; 250 oz of 18%

25. Investing. One year, a couple invested a total of \$10,000 in two projects. The first investment, a mini-mall, made a 6% profit. The other investment, a skateboard park, made a 12% profit. If their investments made \$960, how much was invested at each rate? \$4,000 at 6%; \$6,000 at 12%

26. Cooking. Two teaspoons and five tablespoons is a total of 85 milliliters of liquid. Five teaspoons and two tablespoons is a total of 55 milliliters of liquid. Find the number of milliliters in one teaspoon and the number of milliliters in one tablespoon. Teaspoon: 5 mL; tablespoon: 15 mL

SECTION 12.2 Solving Systems of Equations In Three Variables

DEFINITIONS AND CONCEPTS	EXAMPLES
The graph of an equation of the form $Ax + By + Cz = D$ is a flat surface called a **plane**. A **solution of a system of three linear equations** in three variables is an **ordered triple** that satisfies each equation of the system.	The ordered triple $(4, 0, -3)$ is a solution of $\begin{cases} x + y - z = 7 \\ x - y + z = 1 \\ 2x + y + z = 5 \end{cases}$ because its coordinates, $x = 4$, $y = 0$, and $z = -3$, satisfy each equation: $x + y - z = 7$; $4 + 0 - (-3) \stackrel{?}{=} 7$; $7 = 7$ True $x - y + z = 1$; $4 - 0 + (-3) \stackrel{?}{=} 1$; $1 = 1$ True $2x + y + z = 5$; $2(4) + 0 + (-3) \stackrel{?}{=} 5$; $5 = 5$ True

To **solve a system of three linear equations** by elimination:

1. Write each equation in standard form $Ax + By + Cz = D$ and clear any decimals or fractions.
2. Pick any two equations and eliminate a variable.
3. Pick a different pair of equations and eliminate the same variable as in step 2.
4. Solve the resulting pair of two equations in two variables.
5. To find the value of the third variable, substitute the values of the two variables found in step 4 into any equation containing all three variables and solve the equation.
6. Check the proposed solution in all three of the original equations. Write the solution as an ordered triple.

If at any time in the elimination process the variables drop out and a false statement results, the system has **no solution.** If a true statement results, the system has **infinitely many solutions.**

Solve the system: $\begin{cases} x + 2y - z = 1 & \textbf{(1)} \\ 2x - y + z = 6 & \textbf{(2)} \\ x + 3y - z = 2 & \textbf{(3)} \end{cases}$

Step 1: Each equation is written in standard form.

Step 2: To eliminate z, we add equations 1 and 2.

$$\begin{array}{ll} \textbf{(1)} & x + 2y - z = 1 \\ \textbf{(2)} & \underline{2x - y + z = 6} \\ \textbf{(4)} & 3x + y \quad\;\; = 7 \end{array}$$

Step 3: To eliminate z again, we add equations 2 and 3.

$$\begin{array}{ll} \textbf{(2)} & 2x - y + z = 6 \\ \textbf{(3)} & \underline{x + 3y - z = 2} \\ \textbf{(5)} & 3x + 2y \quad\;\; = 8 \end{array}$$

Step 4: Equations 4 and 5 form a system of two equations in x and y. To solve this system, we multiply equation 4 by -1 and add the resulting equation to equation 5 to eliminate x.

$$\begin{array}{lll} & -3x - y = -7 & \text{This is } -1(3x + y) = -1(7). \\ \textbf{(5)} & \underline{3x + 2y = 8} & \\ & y = 1 & \end{array}$$

To find x, we substitute 1 for y in any equation containing x and y (such as equation 4) and solve for x:

$$\begin{array}{ll} 3x + y = 7 & \text{This is equation 4.} \\ 3x + \mathbf{1} = 7 & \text{Substitute 1 for y.} \\ x = 2 & \text{Solve for x.} \end{array}$$

Step 5: To find z, we substitute 2 for x and 1 for y in any equation containing x, y, and z (such as equation 2) and solve for z:

$$\begin{array}{ll} 2x - y + z = 6 & \text{This is equation 2.} \\ 2(\mathbf{2}) - \mathbf{1} + z = 6 & \text{Substitute for x and y.} \\ 4 - 1 + z = 6 & \\ z = 3 & \text{Solve for z.} \end{array}$$

Step 6: The solution is (2, 1, 3). Verify this by checking it in each of the original equations.

REVIEW EXERCISES

27. Use a check to determine whether (2, −1, 1) is a solution of the system: $\begin{cases} x - y + z = 4 \\ x + 2y - z = -1 \\ x + y - 3z = -1 \end{cases}$ Not a solution

28. A system of three linear equations in three variables is graphed on the right. Does the system have a solution? If so, how many solutions does it have? Yes; one solution

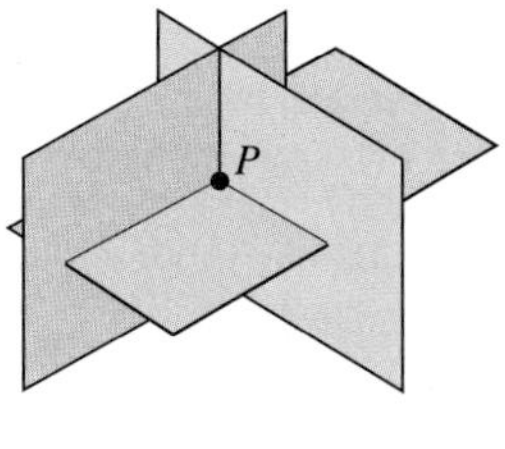

Solve each system, if possible. If a system is inconsistent or if the equations are dependent, state this.

29. $\begin{cases} x - 2y + 3z = -7 \\ -x + 3y + 2z = -8 \\ 2x - y - z = 7 \end{cases}$ (2, 0, −3)

30. $\begin{cases} x + y + z = 4 \\ x - 2y - z = 1 \\ 2x - y - 2z = -1 \end{cases}$ (2, −1, 3)

31. $\begin{cases} x + y - z = -3 \\ x + z = 2 \\ 2x - y = 3 - 2z \end{cases}$ (−1, 1, 3)

32. $\begin{cases} b - 4c = 2 \\ a - b + 2c = 1 \\ 2a - 2b = -2 - 5c \end{cases}$ (−5, −14, −4)

33. $\begin{cases} x + 2z = 10 \\ 3x + 2y - 3z = 8 \\ y + 4z = 6 \end{cases}$ (6, −2, 2)

34. $\begin{cases} x + 3y + z = 14 \\ x - 5y = -19 \\ 3y + z = 13 \end{cases}$ (1, 4, 1)

35. $\begin{cases} 2x + 3y + z = -5 \\ -x + 2y - z = -6 \\ 3x + y + 2z = 4 \end{cases}$ No solution, ∅; inconsistent system

36. $\begin{cases} 3x + 3y + 6z = -6 \\ -x - y - 2z = 2 \\ 2x + 2y + 4z = -4 \end{cases}$ Infinitely many solutions; dependent equations

SECTION 12.3 Problem Solving Using Systems of Three Equations

DEFINITIONS AND CONCEPTS	EXAMPLES

Problems that involve **three unknown quantities** can be solved using a strategy similar to that for solving problems involving two unknowns.

Batteries. A hardware store sells three types of batteries: AA size for \$1 each, C size for \$1.50 each, and D size for \$2.00 each. One Saturday, the store sold 25 batteries for a total of \$34. If the number of C batteries that were sold was four less than the number of AA batteries that were sold, how many of each size battery were sold?

Analyze To find the three unknowns we will write a system of three equations in three variables.

Assign Let A = the number of AA batteries sold, C = the number of C batteries sold, and D = the number of D batteries sold.

Form The information can be shown in a table.

Battery	Number ·	Value =	Total value
AA size	A	1	A
C size	C	1.50	1.50C
D size	D	2	2D
Total	25		\$34

↑ One equation comes from this column. (Number) ↑ Another equation comes from this column. (Total value)

$$\begin{cases} A + C + D = 25 \\ 1A + 1.50C + 2D = 34 \\ C = A - 4 \end{cases}$$

The total number of batteries sold was 25.
The total value of the batteries sold was \$34.
The number of C batteries sold was 4 less than AA batteries sold.

If we multiply the second equation by 10 to clear the decimal and write the third equation in standard form, we have the system:

$$\begin{array}{l} (1) \\ (2) \\ (3) \end{array} \begin{cases} A + C + D = 25 \\ 10A + 15C + 20D = 340 \\ -A + C = -4 \end{cases}$$

Elimination, substitution, matrices, or Cramer's rule can be used to solve the system of equations.

Solve Since equation 3 does not contain a D-term, we will find another equation that does not contain a D-term. If each side of equation 1 is multiplied by -20 and the resulting equation is added to the equation 2, D is eliminated, and we obtain

$$\begin{array}{lrcl} & -20A - 20C - 20D &=& -500 \\ (2) & 10A + 15C + 20D &=& 340 \\ \hline (4) & -10A - 5C &=& -160 \end{array}$$

This is $-20(A + C + D) = -20(25)$.

Equations 3 and 4 form a system of two equations in A and C that can be solved in the usual manner. (The remaining work is left to the reader.)

$$\begin{cases} -10A - 5C = -160 \\ -A + C = -4 \end{cases}$$

State There were 12 AA batteries, 8 C batteries, and 5 D batteries sold.

Check Verify that these results are correct by checking them in the words of the problem.

REVIEW EXERCISES

37. Teddy Bears. A toy company produces three sizes of teddy bears. Each day, the total cost to produce the bears is \$850, the total time needed to stuff them is 480 minutes, and the total time needed to sew them is 1,260 minutes. Use the information in the table to determine how many of each type of teddy bear are produced daily. Small: 50; medium: 60; large: 40

Size of teddy bear	Production cost	Stuffing time	Sewing time
Small	\$3	2 min	6 min
Medium	\$5	3 min	8 min
Large	\$10	5 min	12 min

38. Financial Planning. A financial planner invested \$22,000 in three accounts, paying 5%, 6%, and 7% annual interest. She invested \$2,000 more at 6% than at 5%. If the total interest earned in one year was \$1,370, how much was invested at each rate? \$5,000 at 5%, \$7,000 at 6%, and \$10,000 at 7%

39. Ballistics. The path of a thrown object is a parabola with an equation of $y = ax^2 + bx + c$. The parabola passes through the points $(0, 0)$, $(8, 12)$, and $(12, 15)$. Find a, b, and c.
$a = -\frac{1}{16}, b = 2, c = 0$

40. Veterinary Medicine. The daily requirements of a balanced diet for an animal are shown in the nutritional pyramid. The number of grams per cup of nutrients in three food mixes are shown in the table. How many cups of each mix should be used to meet the daily requirements for protein, carbohydrates, and essential fatty acids in the animal's diet? 2 cups mix A, 1 cup mix B, 1 cup mix C

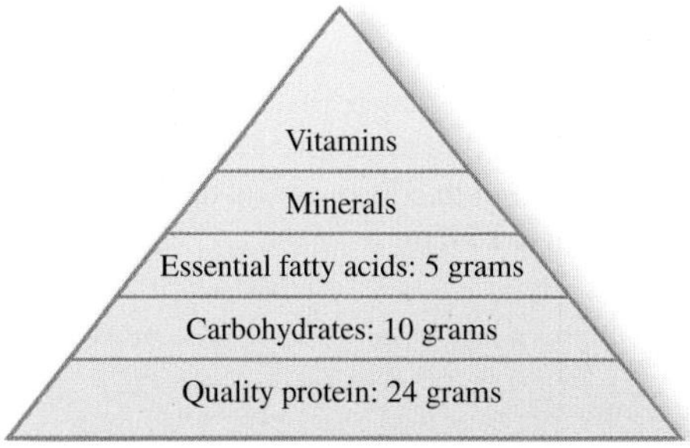

	Grams per cup		
	Protein	Carbohydrates	Fatty acids
Mix A	5	2	1
Mix B	6	3	2
Mix C	8	3	1

SECTION 12.4 Solving Systems of Equations Using Matrices

DEFINITIONS AND CONCEPTS	EXAMPLES
A **matrix** is a rectangular array of numbers. Each number in a matrix is called an **element** or an **entry** of the matrix. A matrix with m rows and n columns has **order** $m \times n$.	A 2×3 matrix: $\begin{bmatrix} 2 & -7 & 5 \\ -3 & 4 & 1 \end{bmatrix}$ A 3×3 matrix: $\begin{bmatrix} 5 & -3 & 12 \\ 4 & 7 & -5 \\ 1 & -4 & 2 \end{bmatrix}$
A system of linear equations can be represented by an **augmented matrix.** Each row of the augmented matrix represents one equation of the system.	The system of equations $\begin{cases} 3x + 5y = 12 \\ 2x - 7y = -5 \end{cases}$ can be represented by the augmented matrix $\left[\begin{array}{cc\|c} 3 & 5 & 12 \\ 2 & -7 & -5 \end{array}\right]$.

Systems of linear equations can be solved by using Gauss-Jordan elimination and **elementary row operations:**

1. Any two rows can be interchanged.
2. Any row can be multiplied by a nonzero constant.
3. Any row can be changed by adding a nonzero constant multiple of another row to it.

To solve a system of two linear equations in two unknowns using matrices, we transform the augmented matrix into an equivalent matrix that has 1's down its main diagonal and 0's directly above and below the 1's. A matrix written in this form is said to be in **reduced row-echelon form.**

$$\left[\begin{array}{cc|c} 1 & 0 & a \\ 0 & 1 & b \end{array}\right]$$ a and b represent real numbers.

Main diagonal

Matrices also can be used to solve systems of three linear equations containing three variables.

Solve the system using matrices: $\begin{cases} 2x - 3y = 0 \\ x + 2y = 7 \end{cases}$

This system is represented by the augmented matrix

$$\left[\begin{array}{cc|c} 2 & -3 & 0 \\ 1 & 2 & 7 \end{array}\right]$$

We can write this matrix in reduced row-echelon form by performing the following elementary row operations

$$\left[\begin{array}{cc|c} 1 & 2 & 7 \\ 2 & -3 & 0 \end{array}\right]$$ Interchange row 1 and row 2. In symbols: $R_1 \leftrightarrow R_2$.

$$\left[\begin{array}{cc|c} 1 & 2 & 7 \\ 0 & -7 & -14 \end{array}\right]$$ Multiply row 1 by -2 and add to row 2: $-2R_1 + R_2$.

$$\left[\begin{array}{cc|c} 1 & 2 & 7 \\ 0 & 1 & 2 \end{array}\right]$$ Multiply row 2 by $-\frac{1}{7}$. In symbols: $-\frac{1}{7}R_2$.

$$\left[\begin{array}{cc|c} 1 & 0 & 3 \\ 0 & 1 & 2 \end{array}\right]$$ Multiply row 2 by -2 and add to row 1: $-2R_2 + R_1$.

This augmented matrix represents the equivalent system: $\begin{cases} x = 3 \\ y = 2 \end{cases}$

Thus, the solution of the given system is (3, 2). The solution set is {(3, 2)}.

REVIEW EXERCISES

Represent each system of equations using an augmented matrix.

41. $\begin{cases} 5x + 4y = 3 \\ x - y = -3 \end{cases}$ $\left[\begin{array}{cc|c} 5 & 4 & 3 \\ 1 & -1 & -3 \end{array}\right]$

42. $\begin{cases} x + 2y + 3z = 6 \\ x - 3y - z = 4 \\ 6x + y - 2z = -1 \end{cases}$ $\left[\begin{array}{ccc|c} 1 & 2 & 3 & 6 \\ 1 & -3 & -1 & 4 \\ 6 & 1 & -2 & -1 \end{array}\right]$

43. Perform each of the following elementary row operations on the augmented matrix $\left[\begin{array}{cc|c} 6 & 12 & -6 \\ 1 & 3 & -2 \end{array}\right]$.

a. $R_1 \leftrightarrow R_2$ $\left[\begin{array}{cc|c} 1 & 3 & -2 \\ 6 & 12 & -6 \end{array}\right]$

b. $\frac{1}{6}R_1$ $\left[\begin{array}{cc|c} 1 & 2 & -1 \\ 1 & 3 & -2 \end{array}\right]$

c. $-6R_2 + R_1$ $\left[\begin{array}{cc|c} 0 & -6 & 6 \\ 1 & 3 & -2 \end{array}\right]$

44. Perform each of the following elementary row operations on the augmented matrix: $\left[\begin{array}{ccc|c} 2 & -1 & 1 & 3 \\ 1 & 1 & 0 & -1 \\ 3 & -1 & -2 & 7 \end{array}\right]$

a. $R_1 \leftrightarrow R_2$ $\left[\begin{array}{ccc|c} 1 & 1 & 0 & -1 \\ 2 & -1 & 1 & 3 \\ 3 & -1 & -2 & 7 \end{array}\right]$

b. $3R_2$ $\left[\begin{array}{ccc|c} 2 & -1 & 1 & 3 \\ 3 & 3 & 0 & -3 \\ 3 & -1 & -2 & 7 \end{array}\right]$

c. $-2R_2 + R_1$ $\left[\begin{array}{ccc|c} 0 & -3 & 1 & 5 \\ 1 & 1 & 0 & -1 \\ 3 & -1 & -2 & 7 \end{array}\right]$

Solve each system using matrices, if possible. If a system is inconsistent or if the equations are dependent, state this.

45. $\begin{cases} x - y = 4 \\ 3x + 7y = -18 \end{cases}$

(1, −3)

46. $\begin{cases} x + 2y - 3z = 5 \\ x + y + z = 0 \\ 3x + 4y + 2z = -1 \end{cases}$

(5, −3, −2)

47. $\begin{cases} 16x - 8y = 32 \\ -2x + y = -4 \end{cases}$

$\{(x, y) \mid -2x + y = -4\}$, inf. many sol.; dep. eq.

48. $\begin{cases} x + 2y - z = 4 \\ x + 3y + 4z = 1 \\ 2x + 4y - 2z = 3 \end{cases}$

No solution, ∅; inconsistent system

SECTION 12.5 Solving Systems of Equations Using Determinants

DEFINITIONS AND CONCEPTS	EXAMPLES
A determinant is a number that is associated with a **square matrix,** a matrix that has the same number of rows and columns.	A 2×2 determinant: $\begin{vmatrix} 3 & -3 \\ -4 & 5 \end{vmatrix}$ A 3×3 determinant: $\begin{vmatrix} 3 & 8 & 3 \\ 7 & 2 & 2 \\ 1 & 5 & 1 \end{vmatrix}$
To **evaluate** a 2×2 determinant: $\begin{vmatrix} a & b \\ c & d \end{vmatrix} = ad - bc$	Evaluate: $\begin{vmatrix} 3 & -3 \\ -4 & 5 \end{vmatrix} = 3(5) - (-3)(-4) = 15 - 12 = 3$
To evaluate a 3×3 determinant, we **expand it by minors** along any row or column.	Evaluate: $\begin{vmatrix} 3 & 8 & 3 \\ 7 & 2 & 2 \\ 1 & 5 & 1 \end{vmatrix} = 3\underbrace{\begin{vmatrix} 2 & 2 \\ 5 & 1 \end{vmatrix}}_{\text{Minor of 3}} - 8\underbrace{\begin{vmatrix} 7 & 2 \\ 1 & 1 \end{vmatrix}}_{\text{Minor of 8}} + 3\underbrace{\begin{vmatrix} 7 & 2 \\ 1 & 5 \end{vmatrix}}_{\text{Minor of 3}}$ $= 3(-8) - 8(5) + 3(33)$ $= -24 - 40 + 99$ $= 35$
Cramer's rule can be used to solve systems of two linear equations in two variables. Cramer's rule can be extended to solve systems of **three linear equations** with three variables.	Use Cramer's rule to solve: $\begin{cases} 2x - 3y = 0 \\ x + 2y = 7 \end{cases}$ The denominator determinant is D: $\begin{vmatrix} 2 & -3 \\ 1 & 2 \end{vmatrix} = 4 - (-3) = 7$ The numerator determinant for x is D_x: $\begin{vmatrix} 0 & -3 \\ 7 & 2 \end{vmatrix} = 0 - (-3)(7) = \mathbf{21}$ The numerator determinant for y is D_y: $\begin{vmatrix} 2 & 0 \\ 1 & 7 \end{vmatrix} = 14 - 0 = 14$ Thus, we have: $x = \frac{D_x}{D} = \frac{\mathbf{21}}{7} = 3$ and $y = \frac{D_y}{D} = \frac{14}{7} = 2$ Thus, the solution of the given system is (3, 2). The solution set is {(3, 2)}.

REVIEW EXERCISES

Evaluate each determinant.

49. $\begin{vmatrix} 2 & 3 \\ -4 & 3 \end{vmatrix}$ 18

50. $\begin{vmatrix} -3 & -4 \\ 5 & -6 \end{vmatrix}$ 38

51. $\begin{vmatrix} -1 & 2 & -1 \\ 2 & -1 & 3 \\ 1 & -2 & 2 \end{vmatrix}$ −3

52. $\begin{vmatrix} 3 & -2 & 2 \\ 1 & -2 & -2 \\ 2 & 1 & -1 \end{vmatrix}$ 28

Use Cramer's rule to solve each system, if possible. If a system is inconsistent or if the equations are dependent, state this.

53. $\begin{cases} 3x + 4y = 10 \\ 2x - 3y = 1 \end{cases}$ (2, 1)

54. $\begin{cases} -6x - 4y = -6 \\ 3x + 2y = 5 \end{cases}$ No solution, ∅; inconsistent system

55. $\begin{cases} x + 2y + z = 0 \\ 2x + y + z = 3 \\ x + y + 2z = 5 \end{cases}$ (1, −2, 3)

56. $\begin{cases} 2x + 3y + z = 2 \\ x + 3y + 2z = 7 \\ x - y - z = -7 \end{cases}$ (−3, 2, 2)

 Teaching Tip: Because this Chapter Test is a comprehensive collection of problems that covers all of the topics discussed in Chapter 12, it is lengthy. If your students have time to complete it, that would be optimal. If, because of time constraints, they are unable to do so, assign an appropriate subset of problems that reflects the types of problems that the students will see on your exam.

12 Chapter Test

1. Fill in the blanks.

   a. $\begin{cases} 2x - 7y = 1 \\ 4x - y = -8 \end{cases}$ is called a _system_ of linear equations.

   b. The matrix $\begin{bmatrix} -10 & 3 \\ 4 & 9 \end{bmatrix}$ has 2 _rows_ and 2 _columns_.

   c. Solutions of a system of three equations in three variables, x, y, and z, are written in the form (x, y, z) and are called ordered _triples_.

   d. The graph of the equation $2x + 3y + 4z = 5$ is a flat surface called a _plane_.

   e. A _matrix_ is a rectangular array of numbers written within brackets.

2. Solve the system by graphing: $\begin{cases} 2x + y = 5 \\ y = 2x - 3 \end{cases}$ (2, 1)

   See AIE Appendix 3.

3. Use a check to determine whether $\left(-\frac{1}{2}, -\frac{2}{3}\right)$ is a solution of the system:

   $\begin{cases} 10x - 12y = 3 \\ 18x - 15y = 1 \end{cases}$ It is a solution.

4. **Politics.** Explain the importance of the point of intersection of the graphs shown below. The point of intersection is (6/06, 44%). Governor Schwarzenegger's job approval and disapproval ratings were the same in June of 2006, approximately 44%.

Governor Arnold Schwarzenegger's Job Performance

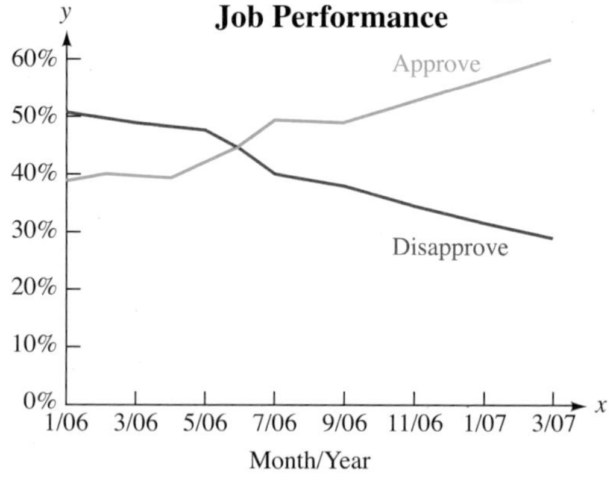

5. Use the vocabulary of this chapter to describe each system of two linear equations in two variables graphed below. Does the system have a solution (or solutions)?

   a.

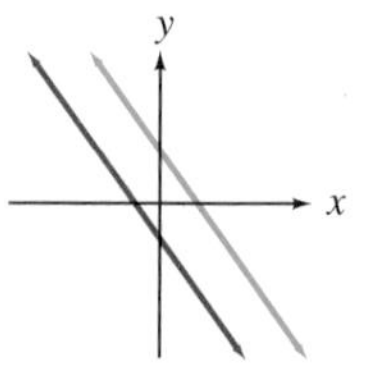

   Inconsistent system; no solution, ∅

   b.

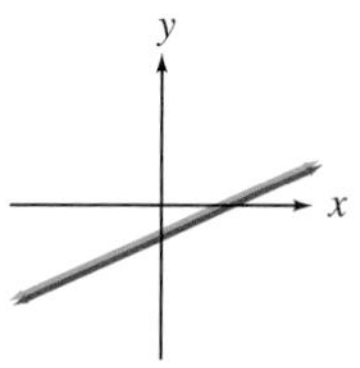

   Dependent equations; infinitely many solutions

6. Use the graphs in the illustration to solve $3(x - 2) - 2(-2 + x) = 1$. 3

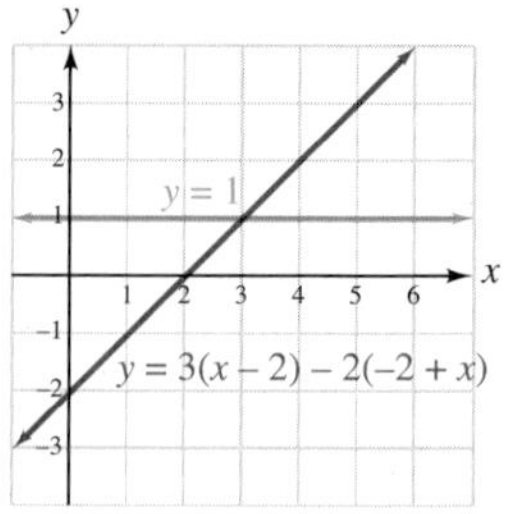

7. Use substitution to solve the system: $\begin{cases} 2x - 4y = 14 \\ x + 2y = 7 \end{cases}$ (7, 0)

8. Use elimination (addition) to solve the system:

   $\begin{cases} 2c + 3d = -5 \\ 3c - 2d = 12 \end{cases}$ (2, −3)

Solve each system by any method, if possible. If a system is inconsistent or if the equations are dependent, state this.

9. $\begin{cases} 3(x + y) = x - 3 \\ -y = \dfrac{2x + 3}{3} \end{cases}$ $\{(x, y) \mid 2x + 3y = -3\}$, infinitely many solutions; dependent equations

10. $\begin{cases} 0.6x + 0.5y = 1.2 \\ x - \dfrac{4}{9}y + \dfrac{5}{9} = 0 \end{cases}$ $\left(\frac{1}{3}, 2\right)$

Write a system of two equations in two variables to solve each problem.

11. **Traffic Signs.** In the sign, find x and y, if y is 15 more than x. 55, 70

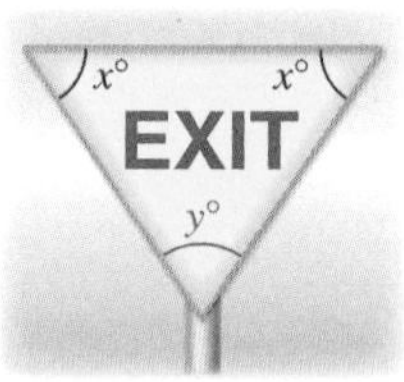

12. **Antifreeze.** How much of a 40% antifreeze solution must a mechanic mix with an 80% antifreeze solution if 20 gallons of a 50% antifreeze solution are needed? 15 gal of 40%; 5 gal of 80%

13. **Avalanches.** For the 2009–2010 snow season, the total number of avalanche fatalities in the United States and Canada was 48. If the number in the United States was three times greater than the number in Canada, how many avalanche fatalities were there in each country? (Source: Avalanche.org) Canada: 12; United States: 36

14. **Airport Walkways.** A man walks at a steady pace as he steps onto a moving walkway. It takes him 40 seconds to reach the end, 320 feet away. If he walks at the same rate against the flow of the walkway, it would take him 80 seconds to reach the end. Find his rate of walking and the rate of the moving walkway. Walking: 6 ft per sec; moving walkway: 2 ft per sec

15. **Student Loans.** A student took out two loans to pay for $8,500 of college expenses. One of the loans was at 2.5% simple annual interest and the other at 4% simple annual interest. After one year, he owed a total of $265 in interest. What was the amount of each loan? 2.5% loan: $5,000; 4% loan: $3,500

16. **Gourmet Fruit.** A fruit pack that contains 6 Royal Riviera pears and 4 Honey Crisp apples costs $25.50. Another pack with 4 pears and 10 apples costs $31.30. Find the cost of one pear and the cost of one apple. Pear: $2.95; apple: $1.95

17. Determine whether $\left(-1, -\frac{1}{2}, 5\right)$ is a solution of:

$$\begin{cases} x - 2y + z = 5 \\ 2x + 4y = -4 \\ -6y + 4z = 22 \end{cases}$$ No

18. A system of three linear equations in three variables is graphed on the right. Does the system have a solution? If so, how many solutions does it have? No solutions

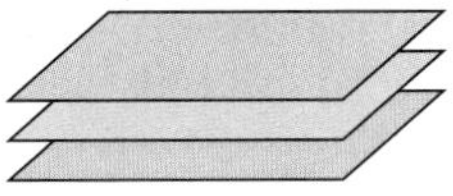

19. Solve the system: $\begin{cases} x + y + z = 4 \\ x + y - z = 6 \\ 2x - 3y + z = -1 \end{cases}$ $(3, 2, -1)$

20. Solve the system: $\begin{cases} z - 2y = 1 \\ x + y + z = 1 \\ x + 5y = 4 \end{cases}$ $(-6, 2, 5)$

21. **Movie Tickets.** The receipts for one showing of a movie were $410 for an audience of 100 people. The ticket prices are given in the table. If twice as many children's tickets as general admission tickets were purchased, how many of each type of ticket were sold? Children: 60, general admission: 30, seniors: 10

Ticket prices	
Children	$3
General admission	$6
Seniors	$5

22. **Weight Training.** Refer to the following illustrations. Determine the weight of the bar, the weight of one large plate, and the weight of one small plate. Bar: 45 lb; large plate: 45 lb; small plate: 10 lb

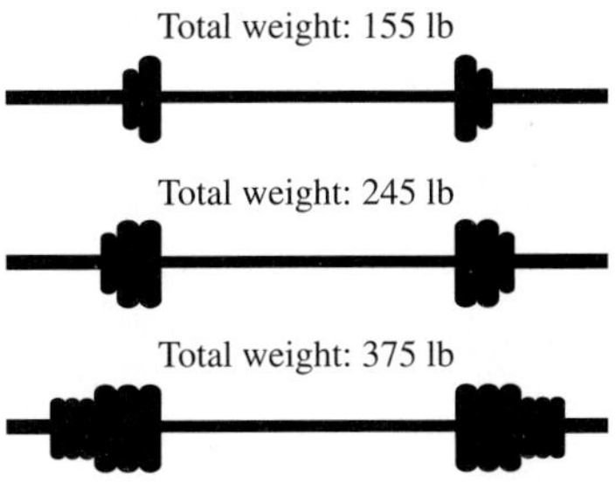

23. Let $A = \left[\begin{array}{cc|c} 1 & 7 & -3 \\ 3 & -1 & 13 \end{array}\right]$. Write the matrix obtained when the elementary row operations $-3R_1 + R_2$ are performed on matrix A. $\left[\begin{array}{cc|c} 1 & 7 & -3 \\ 0 & -22 & 22 \end{array}\right]$

Use matrices to solve each system, if possible. If a system is inconsistent or if the equations are dependent, state this.

24. $\begin{cases} x + y = 4 \\ 2x - y = 2 \end{cases}$ $(2, 2)$

25. $\begin{cases} x - 3y + 2z = 1 \\ x - 2y + 3z = 5 \\ 2x - 6y + 4z = 3 \end{cases}$ No solution, $\varnothing$; inconsistent system

26. $\begin{cases} a + 2b + 2c = 10 \\ 2a + 3b - c = 6 \\ 3a + b + 5c = 8 \end{cases}$ $(-2, 4, 2)$

Evaluate each determinant.

27. $\begin{vmatrix} 2 & -3 \\ -4 & 5 \end{vmatrix}$ -2

28. $\begin{vmatrix} 1 & 2 & 0 \\ 2 & 0 & 3 \\ 1 & -2 & 2 \end{vmatrix}$ 4

29. Use Cramer's rule to solve the system:

$\begin{cases} x - y = -6 \\ 3x + y = -6 \end{cases}$ $(-3, 3)$

30. Solve the following system for z only, using Cramer's rule.

$\begin{cases} x + y + z = 4 \\ x + y - z = 6 \\ 2x - 3y + z = -1 \end{cases}$ -1

Group Project

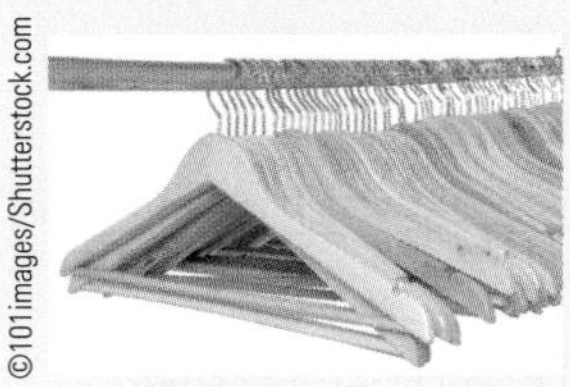

Overview: In this activity, you are to interpret a graph that contains a break point and submit your observations in writing in the form of a financial report.

Instructions: Form groups of 2 or 3 students. Suppose you are a financial analyst for a coat hanger company. The setup cost of a machine that makes wooden coat hangers is $400. After setup, it costs $1.50 to make each hanger (the unit cost). Management is considering the purchase of a new machine that can manufacture the same type of coat hanger at a cost of $1.25 per hanger. If the setup cost of the new machine is $500, find the number of coat hangers that the company would need to manufacture to make the cost the same using either machine. This is called the **break point.**

Then write a brief report that could be given to company managers, explaining their options concerning the purchase of the new machine. Under what conditions should they keep the machine currently in use? Under what conditions should they buy the new machine?

Overview: In this activity, you will explore the advantages and disadvantages of several methods for solving a system of linear equations.

Instructions: Form groups of 5 students. Have each member of your group solve the system

$$\begin{cases} x - y = 4 \\ 2x + y = 5 \end{cases}$$

in a different way. The methods to use are graphing, substitution, elimination, matrices, and Cramer's rule. Have each person briefly explain to the group his or her method of solution. After everyone has presented a solution, discuss the advantages and drawbacks of each method. Then rank the five methods, from most desirable to least desirable.

CUMULATIVE REVIEW Chapters 1–12

1. Determine whether each statement is true or false. [Section 1.3]
   a. Every rational number can be written as a ratio of two integers. True
   b. The set of real numbers corresponds to all points on the number line. True
   c. The whole numbers and their opposites form the set of integers. True
2. **The Stock Market.** In the graph below, positive numbers represent new cash *inflow* into the U.S. stock market. Negative numbers represent cash *outflow* from the market. Was there a net inflow or outflow over the 9-year period from 2000 to 2009? What was it? [Section 1.4] There was a net inflow of $328 billion.

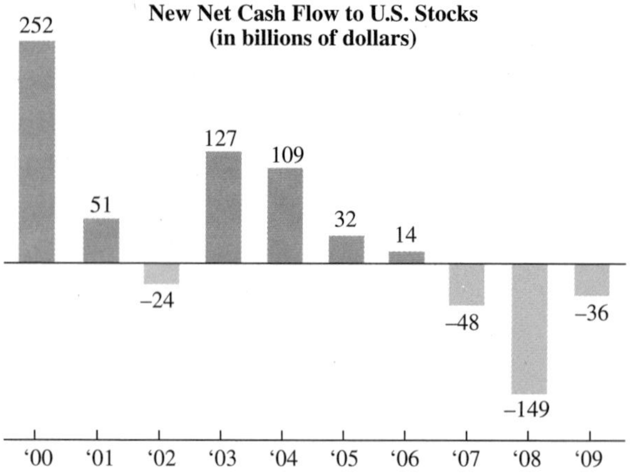

Source: Investment Company Institute

3. Evaluate: $-4 + 2[-7 - 3(-9)]$ [Section 1.7] 36
4. Evaluate: $\left|\frac{4}{5} \cdot 10 - 12\right|$ [Section 1.7] 4
5. Evaluate $(x - a)^2 + (y - b)^2$ for $x = -2, y = 1, a = 5$, and $b = -3$. [Section 1.8] 65
6. Simplify: $3p - 6(p - 9) + p$ [Section 1.9] $-2p + 54$
7. Solve $\frac{5}{6}k = 10$ and check the result. [Section 2.2] 12
8. Solve $-(3a + 1) + a = 2$ and check the result. [Section 2.2] $-\frac{3}{2}$
9. **Loose Change.** The Coinstar machines that are in many grocery stores count unsorted coins and print out a voucher that can be exchanged for cash at the checkout stand. However, to use this service, a processing fee is charged. If a boy turned in a jar of coins worth $60 and received a voucher for $54.12, what was the processing fee (expressed as a percent) charged by Coinstar? [Section 2.3] 9.8%
10. Solve $T = 2r + 2t$ for r. [Section 2.4] $r = \frac{T - 2t}{2}$
11. **Selling a Home.** At what price should a home be listed if the owner wants to make $330,000 on its sale after paying a 4% real estate commission? [Section 2.5] $343,750
12. **Business Loans.** Last year, a women's professional organization made two small-business loans totaling $28,000 to young women beginning their own businesses. The money was lent at 7% and 10% simple interest rates. If the annual income the organization received from these loans was $2,560, what was each loan amount? [Section 2.6] $8,000 at 7%, $20,000 at 10%
13. Solve $5x + 7 < 2x + 1$ and graph the solution set. Then use interval notation to describe the solution. [Section 2.7] $(-\infty, -2)$; ←) → −2
14. Check to determine whether $(-5, -3)$ is a solution of $2x - 3y = -1$. [Section 3.1] Yes

Graph each equation or inequality. See AIE Appendix 3.

15. $y = -x + 2$ [Section 3.2]
16. $2y - 2x = 6$ [Section 3.3]
17. $y = -3$ [Section 3.3]
18. $y < 3x$ [Section 3.7]
19. Find the slope of the line passing through $(-2, -2)$ and $(-12, -8)$. [Section 3.4] $\frac{3}{5}$

20. **TV News.** The line graph in red below approximates the evening news viewership on all networks for the years 1995–2009. Find the rate of decrease over this period of time. [Section 3.4] A decrease of 900,000 viewers per year

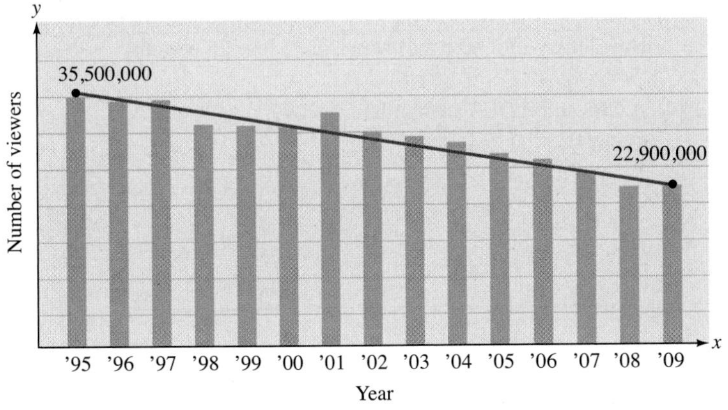

Source: The State of the News Media, 2010

21. What is the slope of the line defined by $4x + 5y = 6$? [Section 3.5] $-\frac{4}{5}$

22. Write the equation of the line whose graph has slope -2 and y-intercept $(0, 1)$. [Section 3.5] $y = -2x + 1$

23. Are the graphs of $y = 4x + 9$ and $x + 4y = -10$ parallel, perpendicular, or neither? [Section 3.5] Perpendicular

24. Write the equation of the line whose graph has slope $\frac{1}{4}$ and passes through the point $(8, 1)$. Write the equation in slope–intercept form. [Section 3.6] $y = \frac{1}{4}x - 1$

25. Graph the line passing through $(-2, -1)$ and having slope $\frac{4}{3}$. [Section 3.6] See AIE Appendix 3.

26. If $f(x) = 3x^2 + 3x - 8$, find $f(-1)$. [Section 3.8] -8

27. Find the domain and range of the relation: $\{(1, 8), (4, -3), (-4, 2), (5, 8)\}$ [Section 3.8] Domain: $\{-4, 1, 4, 5\}$; range: $\{-3, 2, 8\}$

28. Is a function shown? [Section 3.8]

a.

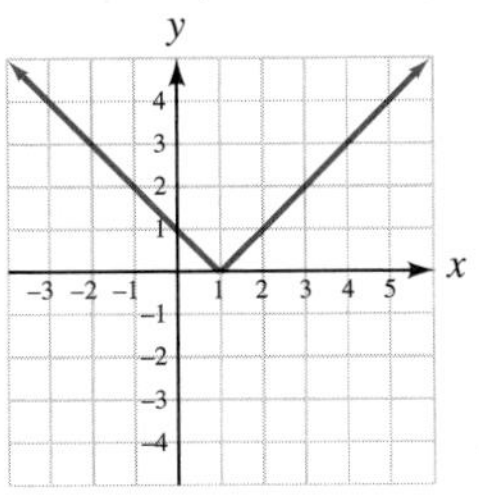

Yes

b.

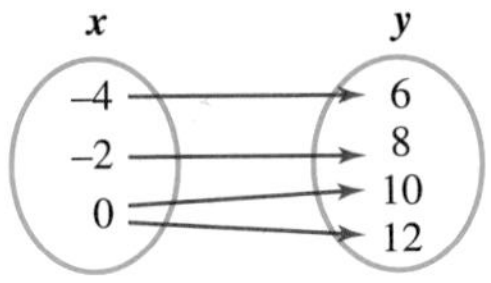

No; (0, 10), (0, 12)

29. Solve using the graphing method. [Section 4.1]

$$\begin{cases} x + y = 1 \\ y = x + 5 \end{cases}$$

$(-2, 3)$; see AIE Appendix 3.

30. Solve using the substitution method.

$$\begin{cases} y = 2x + 5 \\ x + 2y = -5 \end{cases}$$

[Section 4.2] $(-3, -1)$

31. Solve using the elimination (addition) method.

$$\begin{cases} \frac{3}{5}s + \frac{4}{5}t = 1 \\ -\frac{1}{4}s + \frac{3}{8}t = 1 \end{cases}$$

[Section 4.3] $(-1, 2)$

32. **Aviation.** With the wind, a plane can fly 3,000 miles in 5 hours. Against the wind, the trip takes 6 hours. Find the airspeed of the plane (the speed in still air). Use two variables to solve this problem. [Section 4.4] 550 mph

33. **Mixing Candy.** How many pounds of each candy must be mixed to obtain 48 pounds of candy that would be worth \$4.50 per pound? Use two variables to solve this problem. [Section 4.4] 36 lb of hard candy, 12 lb soft candy

34. Solve the system of linear inequalities.

$$\begin{cases} 3x + 4y > -7 \\ 2x - 3y \geq 1 \end{cases}$$

[Section 4.5] See AIE Appendix 3.

Simplify each expression. Write each answer without using parentheses or negative exponents.

35. $y^3(y^2y^4)$ [Section 5.1] y^9

36. $\left(\frac{b^2}{3a}\right)^3$ [Section 5.1] $\frac{b^6}{27a^3}$

37. $\frac{10a^4a^{-2}}{5a^2a^0}$ [Section 5.2] 2

38. $\left(\frac{21x^{-2}y^2z^{-2}}{7x^3y^{-1}}\right)^{-2}$ [Section 5.2] $\frac{x^{10}z^4}{9y^6}$

39. **Five-Card Poker.** The odds against being dealt the hand shown are about 2.6×10^6 to 1. Express 2.6×10^6 using standard notation. [Section 5.3] 2,600,000

40. Write 0.00073 in scientific notation. [Section 5.3] 7.3×10^{-4}

41. Graph: $y = x^3 - 2$ [Section 5.4] See AIE Appendix 3.

42. Write a polynomial that represents the perimeter of the rectangle. [Section 5.5] $6x^3 + 4x$

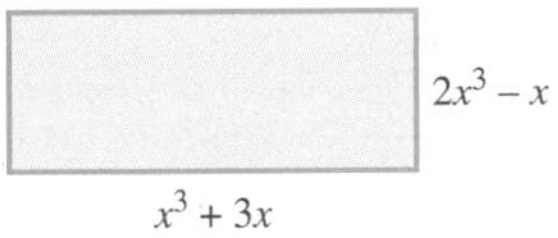

Perform the operations.

43. $4(4x^3 + 2x^2 - 3x - 8) - 5(2x^3 - 3x + 8)$ [Section 5.5] $6x^3 + 8x^2 + 3x - 72$

44. $(-2a^3)(3a^2)$ [Section 5.6] $-6a^5$

45. $(2b - 1)(3b + 4)$ [Section 5.6] $6b^2 + 5b - 4$

46. $(3x + y)(2x^2 - 3xy + y^2)$ [Section 5.6] $6x^3 - 7x^2y + y^3$

47. $(2x + 5y)^2$ [Section 5.7] $4x^2 + 20xy + 25y^2$

48. $(9m^2 - 1)(9m^2 + 1)$ [Section 5.7] $81m^4 - 1$

49. $\dfrac{12a^3b - 9a^2b^2 + 3ab}{6a^2b}$ [Section 5.8] $2a - \frac{3}{2}b + \frac{1}{2a}$

50. $x - 3\overline{)2x^2 - 3 - 5x}$ [Section 5.8] $2x + 1$

Factor each expression completely.

51. $6a^2 - 12a^3b + 36ab$ [Section 6.1] $6a(a - 2a^2b + 6b)$

52. $2x + 2y + ax + ay$ [Section 6.1] $(x + y)(2 + a)$

53. $x^2 - 6x - 16$ [Section 6.2] $(x + 2)(x - 8)$

54. $30y^5 + 63y^4 - 30y^3$ [Section 6.3] $3y^3(5y - 2)(2y + 5)$

55. $t^4 - 16$ [Section 6.4] $(t^2 + 4)(t + 2)(t - 2)$

56. $b^3 + 125$ [Section 6.5] $(b + 5)(b^2 - 5b + 25)$

Solve each equation by factoring.

57. $3x^2 + 8x = 0$ [Section 6.7] $0, -\frac{8}{3}$

58. $15x^2 - 2 = 7x$ [Section 6.7] $\frac{2}{3}, -\frac{1}{5}$

59. **Geometry.** The triangle shown has an area of 22.5 square inches. Find its height. [Section 6.8] 5 in.

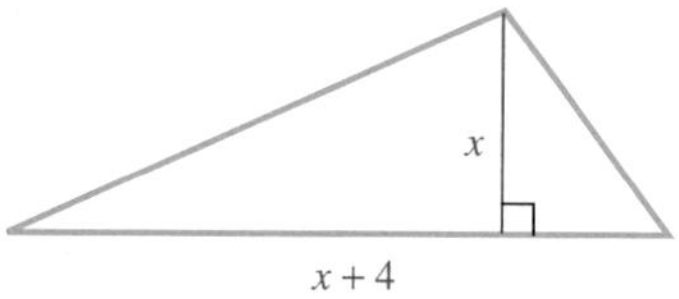

60. For what value is $\frac{x}{x + 8}$ undefined? [Section 7.1] -8

Simplify each expression.

61. $\dfrac{3x^2 - 27}{x^2 + 3x - 18}$ [Section 7.1] $\frac{3(x + 3)}{x + 6}$

62. $\dfrac{a - 15}{15 - a}$ [Section 7.1] -1

Perform the operations and simplify when possible.

63. $\dfrac{x^2 - x - 6}{2x^2 + 9x + 10} \div \dfrac{x^2 - 25}{2x^2 + 15x + 25}$ [Section 7.2] $\frac{x - 3}{x - 5}$

64. $\dfrac{1}{s^2 - 4s - 5} + \dfrac{s}{s^2 - 4s - 5}$ [Section 7.3] $\frac{1}{s - 5}$

65. $\dfrac{x + 5}{xy} - \dfrac{x - 1}{x^2y}$ [Section 7.4] $\frac{x^2 + 4x + 1}{x^2y}$

66. $\dfrac{x}{x - 2} + \dfrac{3x}{x^2 - 4}$ [Section 7.4] $\frac{x^2 + 5x}{x^2 - 4}$

Simplify each complex fraction.

67. $\dfrac{\frac{9m - 27}{m^6}}{\frac{2m - 6}{m^8}}$ [Section 7.5] $\frac{9m^2}{2}$

68. $\dfrac{\frac{5}{y} + \frac{4}{y + 1}}{\frac{4}{y} - \frac{5}{y + 1}}$ [Section 7.5] $\frac{9y + 5}{4 - y}$

Solve each equation.

69. $\dfrac{2p}{3} - \dfrac{1}{p} = \dfrac{2p - 1}{3}$ [Section 7.6] 3

70. $\dfrac{7}{q^2 - q - 2} + \dfrac{1}{q + 1} = \dfrac{3}{q - 2}$ [Section 7.6] 1

71. Solve the formula $\frac{1}{a} + \frac{1}{b} = 1$ for a. [Section 7.6] $a = \frac{b}{b - 1}$

72. **Roofing.** A homeowner estimates that it will take him 7 days to roof his house. A professional roofer estimates that he can roof the house in 4 days. How long will it take if the homeowner helps the roofer? [Section 7.7] $2\frac{6}{11}$ days

73. **Losing Weight.** If a person cuts his or her daily calorie intake by 100, it will take 350 days for that person to lose 10 pounds. How long will it take for the person to lose 25 pounds? [Section 7.8] 875 days

74. ΔABC and ΔDEF are similar triangles. Find x. [Section 7.8] 39

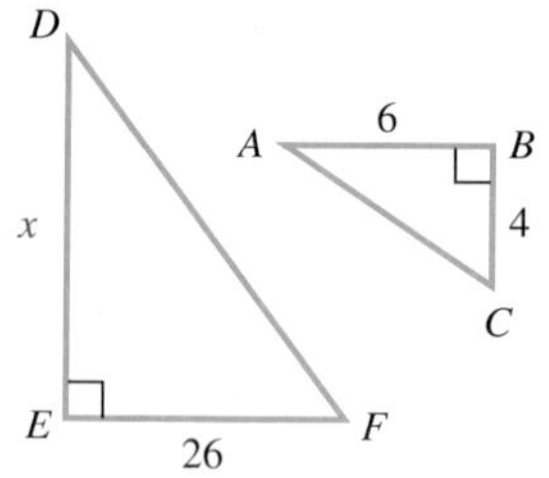

75. Solve: $\dfrac{8}{5} - \dfrac{x + 9}{2} = \dfrac{x + 2}{5} - 4x$ [Section 8.1] 1

76. Solve: $-6.2(-a - 1) - 4 = 4.2a - (-2a)$ [Section 8.1] No solution

77. Solve $A = \frac{1}{2}h(b_1 + b_2)$ for b_1. [Section 8.1] $b_1 = \frac{2A - hb_2}{h}$

78. **Spring Tours.** A group of junior high students will be touring Washington, D.C. Their chaperons will have the \$1,810 cost of the tour reduced by \$15.50 for each student they personally supervise. How many students will a chaperon have to supervise so that his or her cost to take the tour will be \$1,500? [Section 8.1] 20 students

79. Solve $2(5x - 6) > 4x - 15 + 6x$. Graph the solution set and write it in interval notation. [Section 8.1] $(-\infty, \infty)$; $\mathbb{R}$

80. **National Parks.** The number of visitors to U.S. national parks can be approximated by the expression $-100{,}000t + 3{,}600{,}000$, where t is the number of years after 1990. Use an inequality to determine those years for which the number of visitors fell below 2,400,000. (Source: National Park Service Stats) [Section 8.1] Years after 2002

81. If $f(x) = -x^2 - \frac{x}{2}$, find each of the following.

a. $f(10)$ [Section 8.2] -105

b. $f(r)$ [Section 8.2] $-r^2 - \frac{r}{2}$

82. **Railroad Safety.** There has been a downward trend in the number of highway–railroad-crossing accidents in the United States since 1990. For example, the number of such accidents in 1995 was about 4,500 and the number in 2005 was about 2,800. (Source: Federal Railroad Administration) [Section 8.2]

a. Let t be the number of years since 1990 and A be the number of highway–railroad-crossing accidents in the U.S. Write a linear function $A(t)$ to model the situation. $A(t) = -170t + 5{,}350$

b. Use your answer to part a to predict the number of highway–railroad-crossing accidents in 2015, if the trend continues. 1,100 accidents

83. Refer to the graph of function f on the right. Find

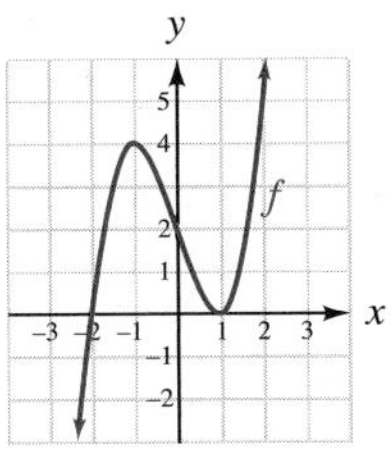

a. $f(-1)$ 4

b. $f(2)$ 5

c. The values of x for which $f(x) = 0$.
$-2, 1$

d. Give the domain and range of f using interval notation.
[Section 8.2] Domain: $(-\infty, \infty)$; range: $(-\infty, \infty)$

84. **Physics.** An object shot from the ground straight up into the air with an initial velocity of 144 feet per second will reach a height of h feet after t seconds, according to the function $h(t) = -16t^2 + 144t$. Find the height of the object 3 seconds after being shot. [Section 8.2] 288 ft

85. a. Does the table define y as a function of x?
[Section 8.2] Yes

x	y
-2	5
-1	2
0	2
5	5

b. Determine whether the graph on the right is the graph of a function. Explain why or why not.
[Section 8.3] No. It does not pass the vertical line test.
$(-2, 6), (-2, 2)$

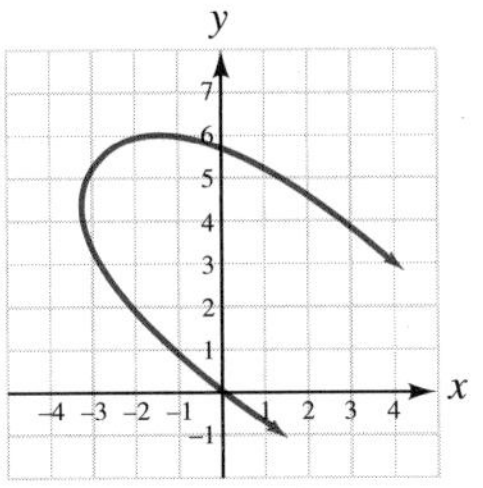

86. a. Graph the function $f(x) = x^2$. Then draw the graph of the associated function $g(x) = (x + 4)^2$ using a translation. Give the domain and range of function g. [Section 8.3]
D: the set of real numbers; R: the set of real numbers greater than or equal to 0. See AIE Appendix 3.

b. The graph of function g on the right is a vertical translation of the graph of $f(x) = |x|$. Write the equation that describes function g. [Section 8.3]
$g(x) = |x| + 3$

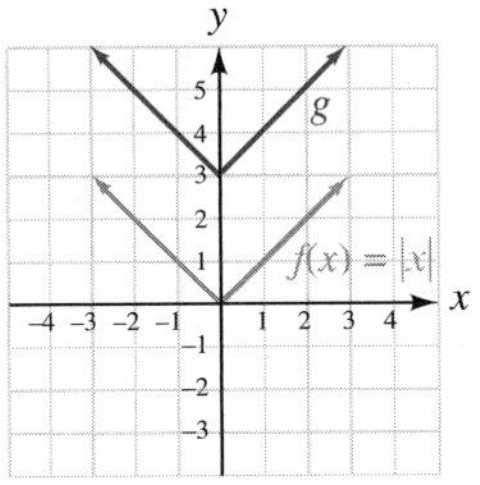

Solve each inequality. Write the solution set in interval notation and graph it. See AIE Appendix 3.

87. $\left|\frac{x - 2}{3}\right| - 4 \le 0$ [Section 8.5] $[-10, 14]$

88. $3x + 2 < 8$ or $2x - 3 > 11$ [Section 8.4]
$(-\infty, 2) \cup (7, \infty)$

Factor completely.

89. $x^2 + 4x + 4 - y^2$ [Section 8.6] $(x + 2 + y)(x + 2 - y)$

90. $b^4 - 17b^2 + 16$ [Section 8.7]
$(b + 1)(b - 1)(b + 4)(b - 4)$

91. Find the domain of the function $f(x) = \frac{5x - 10}{x^3 - 3x}$. Express your answer in words and using interval notation. [Section 8.8]
All real numbers except 0 and 3; $(-\infty, 0) \cup (0, 3) \cup (3, \infty)$

Perform the indicated operation and then simplify, if possible.

92. $(10n - n^2) \cdot \frac{n^6}{n^4 - 10n^3 - 2n^2 + 20n}$ [Section 8.8] $-\frac{n^6}{n^2 - 2}$

93. Simplify: $\frac{2x^2y + xy - 6y}{3x^2y + 5xy - 2y}$ [Section 8.8] $\frac{2x - 3}{3x - 1}$

94. Graph: $f(x) = \frac{1}{x}$ [Section 8.8] See AIE Appendix 3.

95. Suppose w varies directly as x. If $w = 1.2$ when $x = 4$, find w when $x = 30$. [Section 8.9] 9

96. **Gears.** The speed of a gear varies inversely with the number of teeth. If a gear with 10 teeth makes 3 revolutions per second, how many revolutions per second will a gear with 25 teeth make? [Section 8.9] 1.2 revolutions per second

97. Graph the function $f(x) = \sqrt{x - 2}$ and give its domain and range. [Section 9.1] D: $[2, \infty)$; R: $[0, \infty)$; see AIE Appendix 3.

98. **Oceanography.** The speed of a deep-water wave (in meters per second) is given by the function $C(L) = 1.25\sqrt{L}$, where L is the wave length (in meters). Find the speed of a wave if each adjacent crest is 6.5 meters apart. Round to the nearest tenth.
[Section 9.1] About 3.2 meters per sec

99. Simplify $\sqrt{4x^2}$ and assume that the variable is unrestricted.
[Section 9.1] $2|x|$

100. Simplify: $(-8)^{-4/3}$ [Section 9.2] $\frac{1}{16}$

Simplify each expression. All variables represent positive numbers.

101. $\sqrt{100a^6b^4}$ [Section 9.3]
$10a^3b^2$

102. $\sqrt[4]{16x^7y^4}$ [Section 9.3]
$2xy\sqrt[4]{x^3}$

103. $3\sqrt{24} + \sqrt{54}$ [Section 9.3] $9\sqrt{6}$

104. $-3\sqrt[4]{32} - 2\sqrt[4]{162} + 5\sqrt[4]{48}$ [Section 9.3]
$-12\sqrt[4]{2} + 10\sqrt[4]{3}$

105. $\sqrt{\frac{72x^3}{y^2}}$ [Section 9.4] $\frac{6x\sqrt{2x}}{y}$

106. $\sqrt[3]{\frac{27m^3}{8n^6}}$ [Section 9.4] $\frac{3m}{2n^2}$

Rationalize the denominator. All variables represent positive real numbers.

107. $\frac{2}{\sqrt[3]{a}}$ [Section 9.4]
$\frac{2\sqrt[3]{a^2}}{a}$

108. $\frac{\sqrt{x} - \sqrt{y}}{\sqrt{x} + \sqrt{y}}$ [Section 9.4]
$\frac{x - 2\sqrt{xy} + y}{x - y}$

109. Solve: $2 + \sqrt{u} = \sqrt{2u + 7}$ [Section 9.5] 1, 9

110. **Storage Cubes.** The diagonal distance across the face of each of the stacking cubes is 15 inches. What is the height of the entire storage arrangement? Round to the nearest tenth of an inch.
[Section 9.6] 21.2 in.

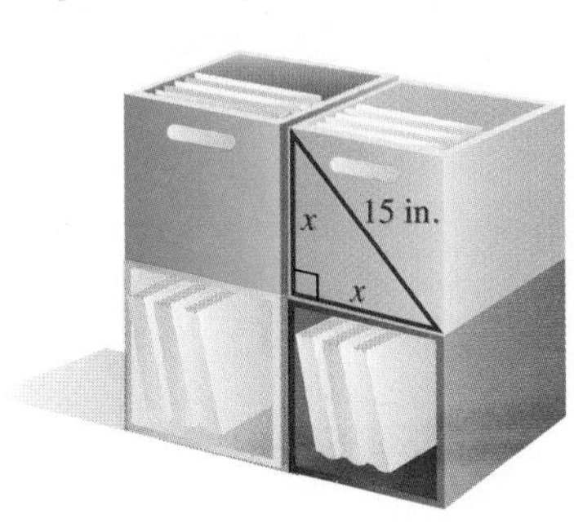

Write each expression in terms of i.

111. $\sqrt{-49}$ [Section 9.7] $7i$

112. $\sqrt{-54}$ [Section 9.7] $3i\sqrt{6}$

Perform the operations. Express each answer in the form $a + bi$.

113. a. $(2 + 3i) - (1 - 2i)$ [Section 9.7] $1 + 5i$

b. $(7 - 4i) + (9 + 2i)$ [Section 9.7] $16 - 2i$

114. a. $(3 - 2i)(4 - 3i)$ [Section 9.7] $6 - 17i$

b. $\frac{3 - i}{2 + i}$ [Section 9.7] $1 - i$

115. Solve $x^2 + 8x + 12 = 0$ by completing the square. [Section 10.1] $-2, -6$

116. Solve $4x^2 - x - 2 = 0$ using the quadratic formula. Give the exact solutions, and then approximate each to the nearest hundredth. [Section 10.2] $\frac{1 \pm \sqrt{33}}{8}$; $-0.59, 0.84$

Solve each equation. Express the solutions in the form $a + bi$.

117. $x^2 + 16 = 0$ [Section 10.2] $0 \pm 4i$

118. $x^2 - 4x = -5$ [Section 10.2] $2 \pm i$

119. Solve: $a^{2/3} + a^{1/3} - 6 = 0$ [Section 10.3] $8, -27$

120. Graph the quadratic function $y = 2x^2 + 8x + 6$. Find the vertex, the x- and y-intercepts, and the axis of symmetry of the graph. [Section 10.4] Vertex: $(-2, -2)$; $(-1, 0)$, $(-3, 0)$; $(0, 6)$; $x = -2$. See AIE Appendix 3.

121. Let $f(x) = 3x - 2$ and $g(x) = x^2 + x$. Find $(f \circ g)(-3)$. [Section 11.1] 16

122. Find the inverse of $f(x) = -\frac{3}{2}x + 3$. [Section 11.2] $f^{-1}(x) = -\frac{2}{3}x + 2$

Graph each function. Determine the domain and range. See AIE Appendix 3.

123. $f(x) = 5^x$ [Section 11.3] D: $(-\infty, \infty)$; R: $(0, \infty)$

124. $f(x) = \ln x$ [Section 11.5] D: $(0, \infty)$; R: $(-\infty, \infty)$

125. Give an approximate value of e to the nearest tenth. [Section 11.5] 2.7

126. World Population Growth. The population of the Earth is approximately 6.9 billion people and is growing at an annual rate of 1.092%. Use the exponential growth model to estimate the world population in 20 years. [Section 11.5] About 8.6 billion

Find each value of x.

127. $\log_x 5 = 1$ [Section 11.4] 5

128. $\log_8 x = 2$ [Section 11.4] 64

Evaluate each expression.

129. $\log_9 \frac{1}{81}$ [Section 11.4] -2

130. $\ln e$ [Section 11.5] 1

131. Write $\ln \frac{y^3\sqrt{x}}{z}$ as the sum and/or difference of logarithms of a single quantity. Then simplify, if possible. [Section 11.6] $3 \ln y + \frac{1}{2}\ln x - \ln z$

132. Write $2 \log x - 3 \log y + \log z$ as one logarithm. [Section 11.6] $\log \frac{x^2z}{y^3}$

Solve each equation. Give answers to four decimal places when necessary.

133. $5^{x-3} = 3^{2x}$ [Section 11.7] -8.2144

134. $\log (x + 90) = 3 - \log x$ [Section 11.7] 10

135. Solve: $\begin{cases} x - y + z = 4 \\ x + 2y - z = -1 \\ x + y - 3z = -2 \end{cases}$ [Section 12.2] $(2, -1, 1)$

136. Triangles. The sum of the measures of the angles of any triangle is 180°. In ΔDEF, $\angle D$ measures 100° less than the sum of the measures of $\angle E$ and $\angle F$, and the measure of $\angle F$ is 40° less than twice the measure of $\angle E$. Find the measure of each angle of the triangle. [Section 12.3] $\angle D = 40°$, $\angle E = 60°$, $\angle F = 80°$

137. Solve by using matrices: $\begin{cases} 2x + y = 1 \\ x + 2y = -4 \end{cases}$ [Section 12.4] $(2, -3)$

138. Solve by using Cramer's rule: $\begin{cases} 3x - 4y = 9 \\ x + 2y = 8 \end{cases}$ [Section 12.5] $\left(5, \frac{3}{2}\right)$

Conic Sections; More Graphing

13

from Campus to Careers

Civil Engineer

Civil engineers design roads, buildings, airports, tunnels, dams, bridges, or water supply and sewage systems. They often must manage people as well as projects. A civil engineer may oversee a construction site or be a city engineer. Others may work in design, construction, research, and teaching. During the planning stages, they make detailed drawings and graphs of the project. Because construction projects are often publicly funded, they make budgets, submit bid proposals, and perform cost analysis studies to make sure that public funding is spent wisely.

Problem 89 in **Study Set 13.1, problem 57** in **Study Set 13.2,** and **problem 63** in **Study Set 13.3** involve situations that a civil engineer might encounter on the job. The mathematical concepts discussed in this chapter can be used to solve those problems.

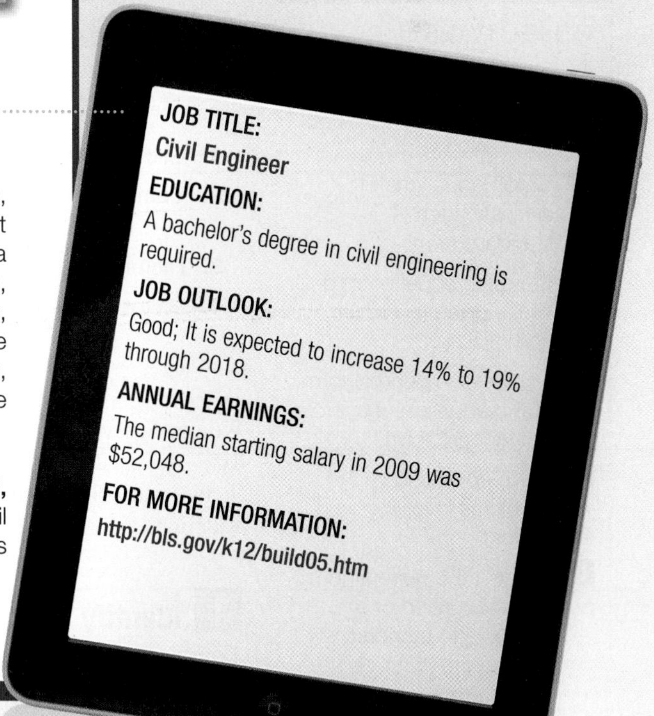
JOB TITLE:
Civil Engineer
EDUCATION:
A bachelor's degree in civil engineering is required.
JOB OUTLOOK:
Good; It is expected to increase 14% to 19% through 2018.
ANNUAL EARNINGS:
The median starting salary in 2009 was $52,048.
FOR MORE INFORMATION:
http://bls.gov/k12/build05.htm

Study Skills Workshop ▸ Preparing for Your Next Math Course

Before moving on to a new mathematics course, it's worthwhile to take some time to reflect on your effort and performance in this course.

Now Try This

As this course draws to a close, here are some questions to ask yourself.

1. How was my attendance?
2. Was I organized? Did I have the right materials?
3. Did I follow a regular schedule?
4. Did I pay attention in class and take good notes?
5. Did I spend the appropriate amount of time on homework?
6. How did I prepare for tests? Did I have a test-taking strategy?
7. Was I part of a study group? If not, why not? If so, was it worthwhile?
8. Did I ever seek extra help from a tutor or from my instructor?
9. In what topics was I the strongest? In what topics was I the weakest?
10. If I had it to do over, would I do anything differently?

SECTION 13.1 The Circle and the Parabola

OBJECTIVES

1. Identify conic sections and some of their applications.
2. Graph equations of circles written in standard form.
3. Write the equation of a circle, given its center and radius.
4. Convert the general form of the equation of a circle to standard form.
5. Solve application problems involving circles.
6. Convert the general form of the equation of a parabola to standard form to graph it.

ARE YOU READY? *Are You Ready? exercises available online at www.webassign.net/brookscole*

The following problems review some basic skills that are needed when working with circles and parabolas.

1. Use a special-product formula to find $(x - 8)^2$. $x^2 - 16x + 64$
2. Complete the square on $x^2 - 6x$ and factor the resulting trinomial. $x^2 - 6x + 9 = (x - 3)^2$
3. Factor: $y^2 + 10y + 25$ $(y + 5)^2$
4. Factor out -3 from the terms of the expression $-3y^2 - 12y$. $-3(y^2 + 4y)$

We have previously graphed first-degree equations in two variables such as $y = 3x + 8$ and $4x - 3y = 12$. Their graphs are lines. In this section, we will graph second-degree equations in two variables such as $x^2 + y^2 = 25$ and $x = -3y^2 - 12y - 13$. The graphs of these equations are *conic sections.*

1 Identify Conic Sections and Some of Their Applications.

The curves formed by the intersection of a plane with an infinite right-circular cone are called **conic sections.** Those curves have four basic shapes, called **circles, parabolas, ellipses,** and **hyperbolas,** as shown on the next page.

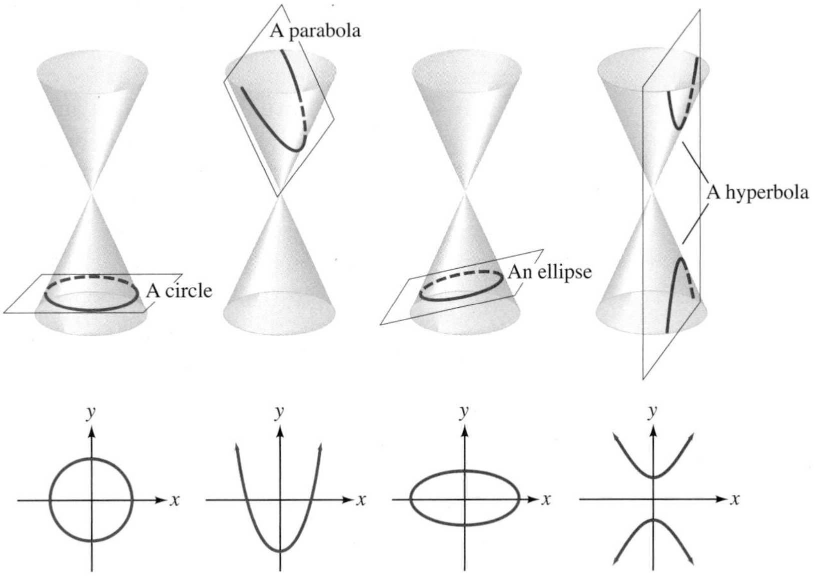

Conic sections have many applications. For example, everyone is familiar with circular wheels and gears, pizza cutters, and hula hoops.

The Language of Algebra

Conic sections are often simply called **conics.**

Parabolas can be rotated to generate dish-shaped surfaces called **paraboloids.** Any light or sound placed at the **focus** of a paraboloid is reflected outward in parallel paths. This property makes parabolic surfaces ideal for flashlight and headlight reflectors. It also makes parabolic surfaces good antennas, because signals captured by such antennas are concentrated at the focus. Parabolic mirrors are capable of concentrating the rays of the sun at a single point, thereby generating tremendous heat. This property is used in the design of solar furnaces.

Any object thrown upward and outward travels in a parabolic path. An example of this is a stream of water flowing from a drinking fountain. In architecture, many arches are parabolic in shape, because this gives them strength. Cables that support suspension bridges hang in the shape of a parabola.

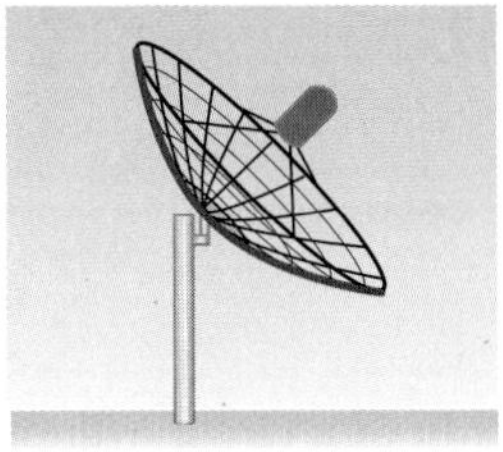
Radar dish

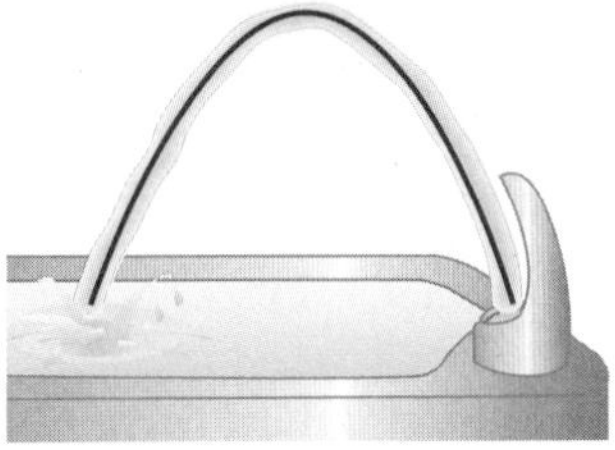
Stream of water

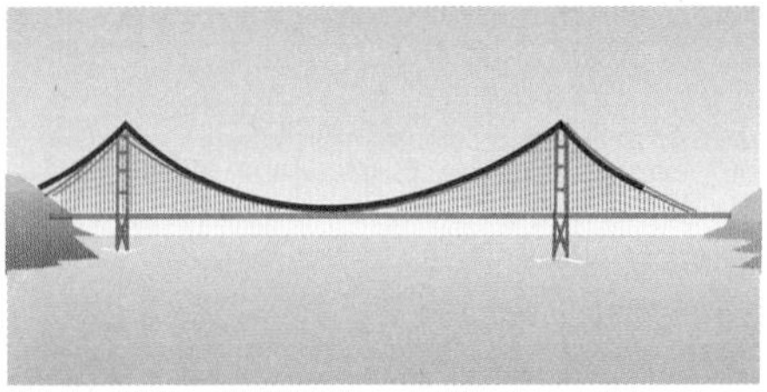
Support cables

Ellipses have optical and acoustical properties that are useful in architecture and engineering. Many arches are portions of an ellipse, because the shape is pleasing to the eye. The planets and many comets have elliptical orbits. Certain gears have elliptical shapes to provide nonuniform motion.

Arches

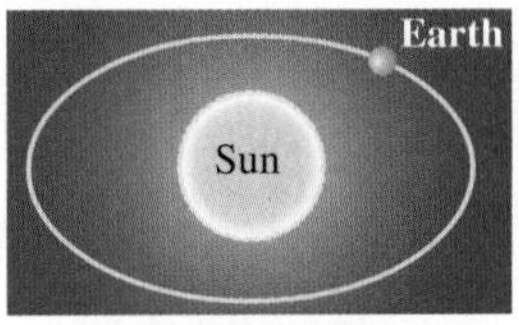

Earth's orbit

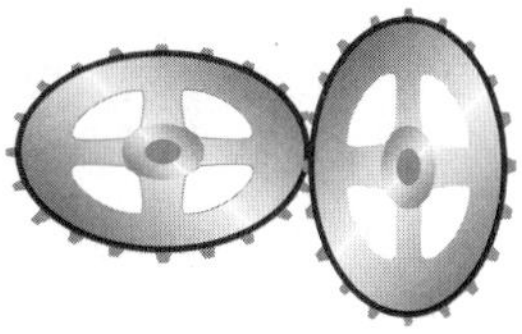
Gears

Hyperbolas serve as the basis of a navigational system known as LORAN (LOng RAnge Navigation). They also are used to find the source of a distress signal, are the basis for the design of hypoid gears, and describe the orbits of some comets.

A sonic shock wave created by a jet aircraft has the shape of a cone. In level flight, the sound wave intersects the ground as one branch of a hyperbola, as shown below. People in different places along the curve on the ground hear and feel the sonic boom at the same time.

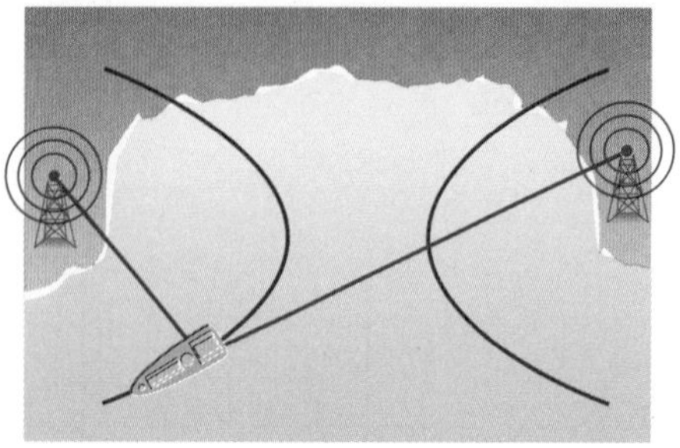
Navigation

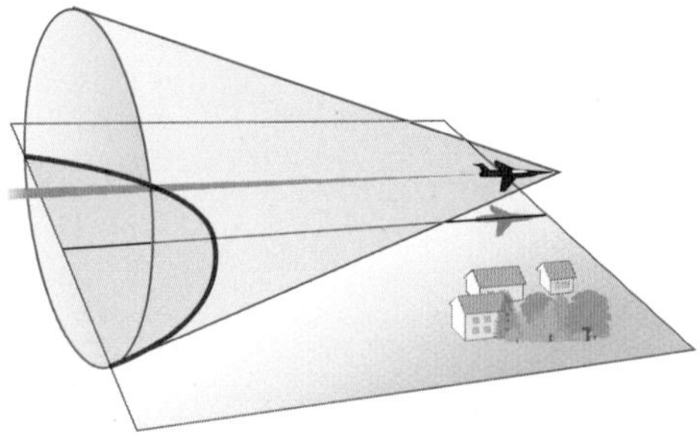
Sonic boom

2 Graph Equations of Circles Written in Standard Form.

Every conic section can be represented by a second-degree equation in x and y. To find the equation of a circle, we use the following definition.

Definition of a Circle

A **circle** is the set of all points in a plane that are a fixed distance from a fixed point called its **center.** The fixed distance is called the **radius** of the circle.

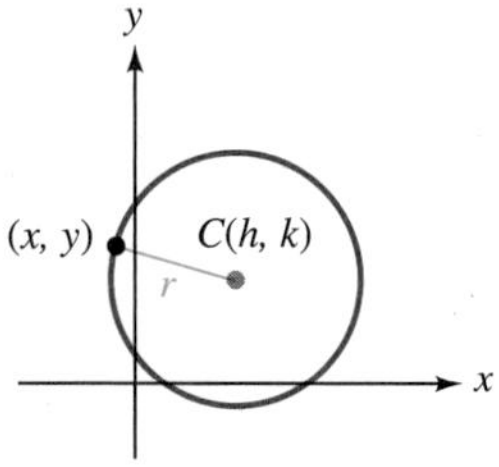

If we let (h, k) be the center of a circle and (x, y) be some point on a circle that is graphed on a rectangular coordinate system, the distance from (h, k) to (x, y) is the radius r of the circle. We can use the distance formula to find r.

$$r = \sqrt{(x - h)^2 + (y - k)^2}$$

We can square both sides to eliminate the radical and obtain

$$r^2 = (x - h)^2 + (y - k)^2$$

This result is called the **standard form of the equation of a circle** with radius r and center at (h, k).

Equation of a Circle

The **standard form of the equation of a circle** with radius r and center at (h, k) is

$$(x - h)^2 + (y - k)^2 = r^2$$

EXAMPLE 1 Find the center and the radius of each circle and then graph it:
a. $(x - 4)^2 + (y - 1)^2 = 9$ **b.** $x^2 + y^2 = 25$ **c.** $(x + 3)^2 + y^2 = 12$

Strategy We will compare each equation to the standard form of the equation of a circle, $(x - h)^2 + (y - k)^2 = r^2$, and identify h, k, and r.

Why The center of the circle is the point with coordinates (h, k) and the radius of the circle is r.

Solution

Success Tip

A compass can be used to draw a circle. The distance between the compass point and pencil should be set to equal the radius.

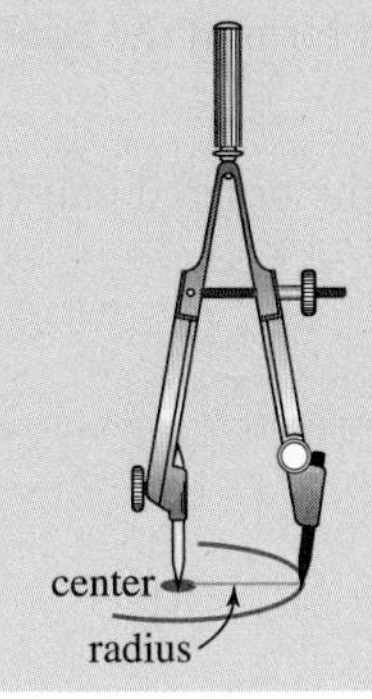

a. The color highlighting shows how to compare the given equation to the standard form to find h, k, and r.

$$(x - 4)^2 + (y - 1)^2 = 9$$
$$(x - h)^2 + (y - k)^2 = r^2$$

$h = 4$, $k = 1$, and $r^2 = 9$. Since the radius of a circle must be positive, $r = 3$.

The center of the circle is $(h, k) = (4, 1)$ and the radius is 3.

To plot four points on the circle, we move up, down, left, and right 3 units from the center, as shown in figure (a). Then we draw a circle through the points to get the graph of $(x - 4)^2 + (y - 1)^2 = 9$, as shown in figure (b).

The center (4, 1) is not part of the graph of a circle; it only helps us sketch the graph.

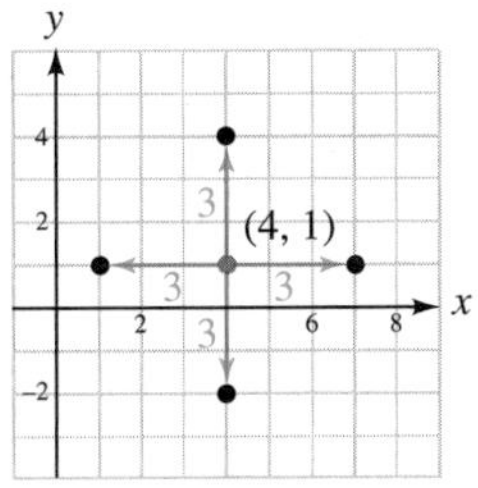

(a)

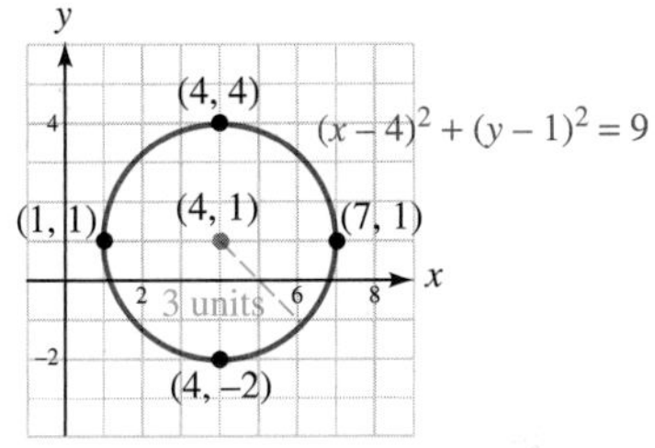

(b)

Success Tip

An equation of the form

$$x^2 + y^2 = r^2$$

has a graph that is a circle with radius r and center at (0, 0).

b. To find h and k, we will write $x^2 + y^2 = 25$ in the following way:

$$(x - 0)^2 + (y - 0)^2 = 25$$
$$(x - h)^2 + (y - k)^2 = r^2$$

$h = 0$, $k = 0$, and $r^2 = 25$. Since the radius must be positive, $r = 5$.

The center of the circle is at (0, 0) and the radius is 5.

To plot four points on the circle, we move up, down, left, and right 5 units from the center. Then we draw a circle through the points to get the graph of $x^2 + y^2 = 25$, as shown.

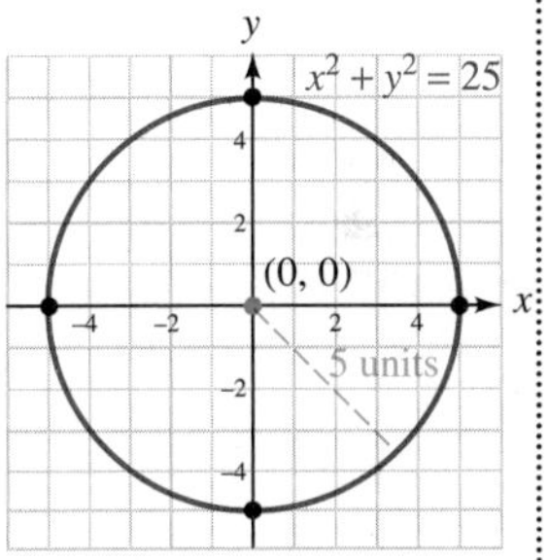

Success Tip

When the equation of a circle is written in standard form, h and k are the numbers within the parentheses that *follow* the − (subtraction) symbols.

c. To find h, we will write $x + 3$ as $x - (-3)$.

Standard form requires a minus symbol here.

$$[x - (-3)]^2 + (y - 0)^2 = 12$$
$$(x - h)^2 + (y - k)^2 = r^2$$

$h = -3$, $k = 0$, and $r^2 = 12$.

Since $r^2 = 12$, we have

$$r = \pm\sqrt{12} = \pm 2\sqrt{3} \quad \text{Use the square root property.}$$

Since the radius can't be negative, $r = 2\sqrt{3}$. The center of the circle is at $(-3, 0)$ and the radius is $2\sqrt{3}$.

To plot four points on the circle, we move up, down, left, and right $2\sqrt{3} \approx 3.5$ units from the center. We then draw a circle through the points to get the graph of $(x + 3)^2 + y^2 = 12$, as shown on the left.

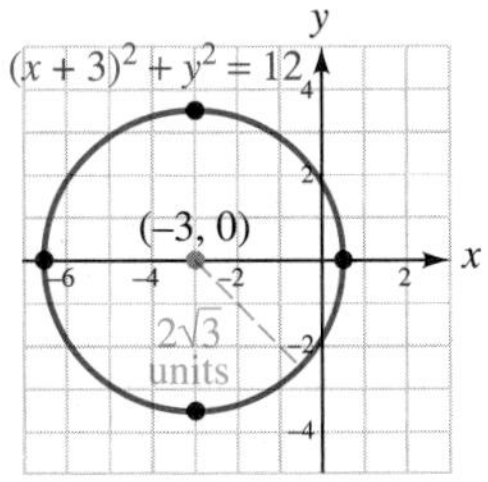

Teaching Example 1 Find the center and the radius of each circle and then graph it:
a. $(x - 1)^2 + (y - 2)^2 = 36$
b. $x^2 + y^2 = 18$
c. $x^2 + (y + 2)^2 = 9$
Answers: **a.** $(1, 2)$, $r = 6$
b. $(0, 0)$, $r = 3\sqrt{2}$ **c.** $(0, -2)$, $r = 3$

Self Check 1 Find the center and the radius of each circle and then graph it:
a. $(x - 3)^2 + (y + 4)^2 = 4$ $(3, -4)$, $r = 2$ **b.** $x^2 + y^2 = 8$
$(0, 0)$, $r = 2\sqrt{2} \approx 2.8$; see AIE Appendix 3.

Now Try Problems 15, 19, and 21

3 Write the Equation of a Circle, Given Its Center and Radius.

Because a circle is determined by its center and radius, that information is all we need to know to write its equation.

EXAMPLE 2 Write the equation of the circle with radius 9 and center at $(6, -5)$.

Strategy We substitute 9 for r, 6 for h, and -5 for k in the standard form of the equation of a circle, $(x - h)^2 + (y - k)^2 = r^2$.

Why When writing the standard form, the center is represented by the ordered pair (h, k) and the radius as r.

Solution

$(x - h)^2 + (y - k)^2 = r^2$ This is the standard form.

$(x - 6)^2 + [y - (-5)]^2 = 9^2$ Substitute 6 for h, -5 for k, and 9 for r.

$(x - 6)^2 + (y + 5)^2 = 9^2$ Write $y - (-5)$ as $y + 5$.

If we express 9^2 as 81, we can write the equation as $(x - 6)^2 + (y + 5)^2 = 81$.

Notation

Standard form can be written

$(x - 6)^2 + (y + 5)^2 = 9^2$
or
$(x - 6)^2 + (y + 5)^2 = 81$

Teaching Example 2 Write the equation of the circle with radius 5 and center at $(-3, -2)$.
Answer: $(x + 3)^2 + (y + 2)^2 = 25$

Self Check 2 Write the equation of the circle with radius 10 and center at $(-7, 1)$.
$(x + 7)^2 + (y - 1)^2 = 100$

Now Try Problems 23, 27, and 31

4 Convert the General Form of the Equation of a Circle to Standard Form.

In Example 2, the result was written in standard form: $(x - 6)^2 + (y + 5)^2 = 81$. If we square $x - 6$ and $y + 5$, we obtain a different form for the equation of the circle.

$(x - 6)^2 + (y + 5)^2 = 9^2$

$x^2 - 12x + 36 + y^2 + 10y + 25 = 81$ Square each binomial.

$x^2 - 12x + y^2 + 10y - 20 = 0$ Subtract 81 from both sides. Combine like terms.

$x^2 + y^2 - 12x + 10y - 20 = 0$ Rearrange the terms, writing the squared terms first.

This result is written in the *general form of the equation of a circle.*

Success Tip

This example illustrates an important fact: The equation of a circle contains both x^2 and y^2 terms on the same side of the equation with equal coefficients.

Equation of a Circle

The general form of the equation of a circle is

$$x^2 + y^2 + Dx + Ey + F = 0$$

We can convert from the general form to the standard form of the equation of a circle by completing the square.

EXAMPLE 3 Write the equation $x^2 + y^2 - 4x + 2y - 11 = 0$ in standard form and graph it.

Strategy We will rearrange the terms to write the equation in the form $x^2 - 4x + y^2 + 2y = 11$ and complete the square on x and y.

Why Standard form contains the expressions $(x - h)^2$ and $(y - k)^2$. We can obtain a perfect-square trinomial that factors as $(x - 2)^2$ by completing the square on $x^2 - 4x$. We can complete the square on $y^2 + 2y$ to obtain an expression of the form $(y + 1)^2$.

Solution To write the equation in standard form, we complete the square twice.

Teaching Tip: Ask if this is the graph of a function.

$$x^2 + y^2 - 4x + 2y - 11 = 0$$

$$x^2 - 4x \quad + y^2 + 2y \quad = 11$$ Write the x-terms together, the y-terms together, and add 11 to both sides.

To complete the square on $x^2 - 4x$, we note that $\frac{1}{2}(-4) = -2$ and $(-2)^2 = \mathbf{4}$. To complete the square on $y^2 + 2y$, we note that $\frac{1}{2}(2) = 1$ and $1^2 = \mathbf{1}$. We add **4** and **1** to both sides of the equation.

$$x^2 - 4x + \mathbf{4} + y^2 + 2y + \mathbf{1} = 11 + \mathbf{4} + \mathbf{1}$$

$$(x - 2)^2 + (y + 1)^2 = 16$$ Factor $x^2 - 4x + 4$ and $y^2 + 2y + 1$.

$x^2 + y^2 - 4x + 2y - 11 = 0$
or
$(x - 2)^2 + (y + 1)^2 = 16$

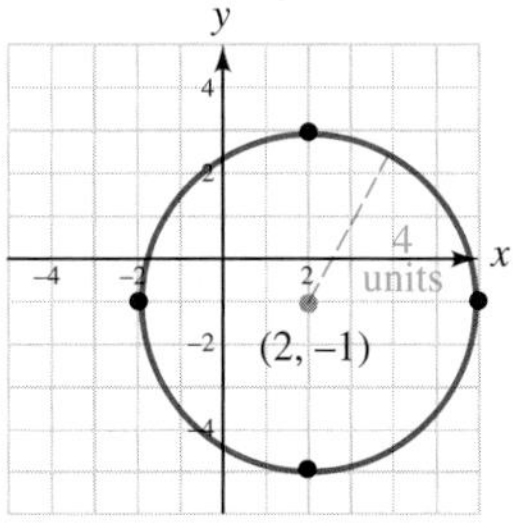

The equation also can be written as $(x - 2)^2 + (y + 1)^2 = 4^2$.

We can determine the circle's center and radius by comparing this equation to the standard form of the equation of a circle, $(x - h)^2 + (y - k)^2 = r^2$. We see that $h = 2$, $k = -1$, and $r = 4$. We can use the center, $(h, k) = (2, -1)$ and the radius $r = 4$, to graph the circle as shown on the left.

Teaching Example 3 Write the equation $x^2 + y^2 + 2x - 12y = -28$ in standard form and graph it.
Answer: $(x + 1)^2 + (y - 6)^2 = 9$

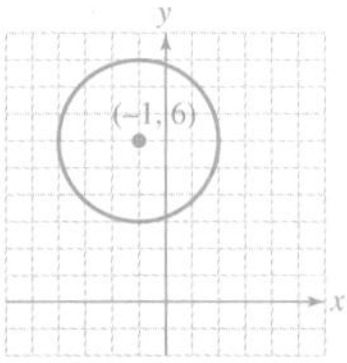

Self Check 3 Write the equation $x^2 + y^2 + 12x - 6y - 4 = 0$ in standard form and graph it. $(x + 6)^2 + (y - 3)^2 = 49$; see AIE Appendix 3.

Now Try ▶ Problem 35

Using Your Calculator ▶ Graphing Circles

Since the graphs of circles fail the vertical line test, their equations do not represent functions. It is more difficult to use a graphing calculator to graph equations that are not functions. For example, to graph the circle described by $(x - 1)^2 + (y - 2)^2 = 4$, we must split the equation into two functions and graph each one separately. We begin by solving the equation for y.

$$(x - 1)^2 + (y - 2)^2 = 4$$

$$(y - 2)^2 = 4 - (x - 1)^2$$ Subtract $(x - 1)^2$ from both sides.

$$y - 2 = \pm\sqrt{4 - (x - 1)^2}$$ Use the square root property.

$$y = 2 \pm \sqrt{4 - (x - 1)^2}$$ Add 2 to both sides.

This equation defines two functions. If we graph

$$y = 2 + \sqrt{4 - (x - 1)^2} \quad \text{and} \quad y = 2 - \sqrt{4 - (x - 1)^2}$$

we get the distorted circle shown in figure (a). To get a better circle, we can use the graphing calculator's square window feature, which gives an equal unit distance on both the x- and y-axes. (Press ZOOM, 5, ENTER.) Using this feature, we get the circle shown in figure (b). Sometimes the two arcs will not connect because of approximations made by the calculator at each endpoint.

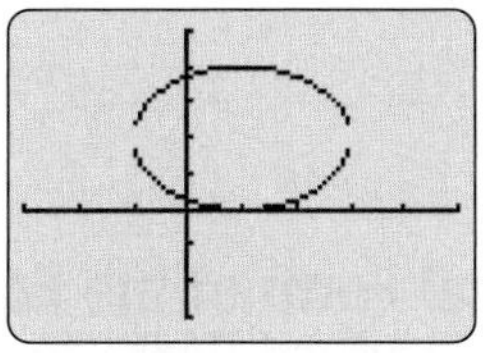

(a)

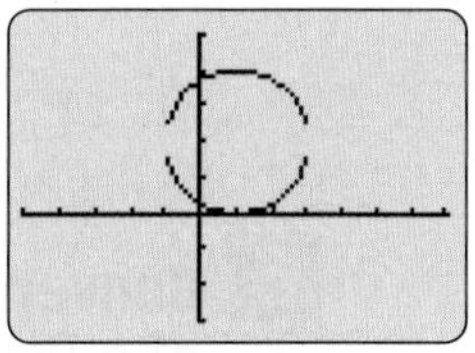

(b)

The graph of $y = 2 + \sqrt{4 - (x - 1)^2}$ is the top half of the circle.

The graph of $y = 2 - \sqrt{4 - (x - 1)^2}$ is the bottom half of the circle.

5 Solve Application Problems Involving Circles.

EXAMPLE 4 **Radio Translators.** The broadcast area of a television station is bounded by the circle $x^2 + y^2 = 3{,}600$, where x and y are measured in miles. A translator station picks up the signal and retransmits it from the center of a circular area bounded by $(x + 30)^2 + (y - 40)^2 = 1{,}600$. Find the location of the translator and the greatest distance from the main transmitter that the signal can be received.

Strategy Refer to the figure below. We will find two distances: the distance from the TV station transmitter to the translator and the distance from the translator to the outer edge of its coverage.

Why The greatest distance of reception from the main transmitter is the sum of those two distances.

Solution The coverage of the TV station is bounded by $x^2 + y^2 = 60^2$, a circle centered at the origin with a radius of 60 miles, as shown in yellow in the figure. Because the translator is at the center of the circle $(x + 30)^2 + (y - 40)^2 = 1{,}600$, it is located at $(-30, 40)$, a point 30 miles west and 40 miles north of the TV station. The radius of the translator's coverage is $\sqrt{1{,}600}$, or 40 miles.

As shown in the figure, the greatest distance of reception is the sum of d, the distance from the translator to the television station, and 40 miles, the radius of the translator's coverage.

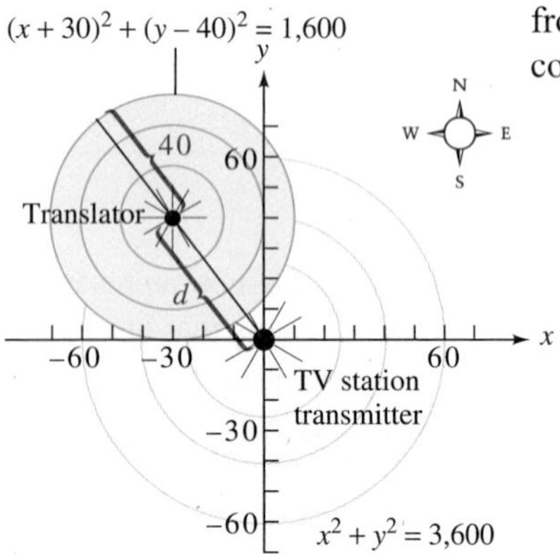

To find d, we use the distance formula to find the distance between the origin, $(x_1, y_1) = (0, 0)$, and $(x_2, y_2) = (-30, 40)$.

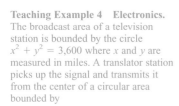

$$d = \sqrt{(x_2 - x_1)^2 + (y_2 - y_1)^2} \quad \text{The distance formula was introduced in Section 9.6.}$$

$$d = \sqrt{(-30 - 0)^2 + (40 - 0)^2} \quad \text{Substitute for } x_1, x_2, y_1, \text{ and } y_2.$$

$$d = \sqrt{(-30)^2 + 40^2} \quad \text{Simplify within the radical.}$$

$$= \sqrt{2{,}500} \quad \text{Evaluate the expression within the radical.}$$

$$= 50 \quad \text{Find the square root.}$$

The translator is located 50 miles from the television station, and it broadcasts the signal 40 miles. The greatest reception distance from the main transmitter signal is, therefore, 50 + 40, or 90 miles.

Teaching Example 4 Electronics. The broadcast area of a television station is bounded by the circle $x^2 + y^2 = 3{,}600$ where x and y are measured in miles. A translator station picks up the signal and transmits it from the center of a circular area bounded by

$$(x + 30)^2 + (y + 40)^2 = 900$$

Find the location of the translator and the greatest distance from the main transmitter that the signal can be received.
Answer: 50 miles from the station, 80 mi

Self Check 4 **Landscaping.** A landscape architect is designing a circular flower bed bounded by the circle $x^2 + y^2 = 36$ where x and y are measured in feet. Another circular flower bed in his design is bounded by $(x - 3)^2 + (y + 4)^2 = 16$. Find the length of the sidewalk from the center of the first circle to the furthest edge of the second circle. 9 ft.

Now Try Problem 87

6 Convert the General Form of the Equation of a Parabola to Standard Form to Graph It.

Another type of conic section is the parabola.

Definition of a Parabola

A **parabola** is the set of all points in a plane that are equidistant from a fixed point, called the **focus,** and a fixed line, called the **directrix.**

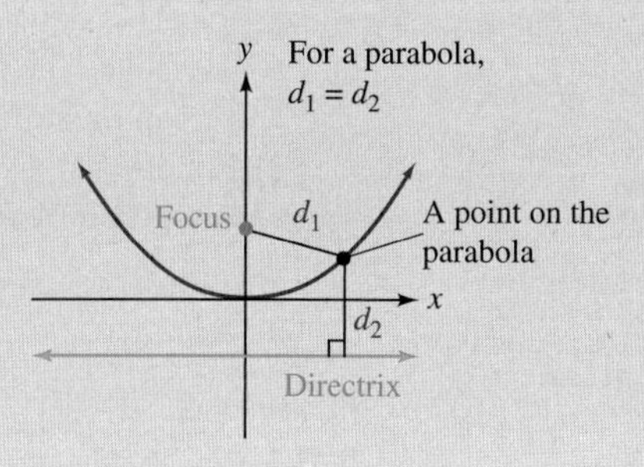

We have discussed parabolas whose graphs open upward or downward in Section 10.4: Quadratic Functions and Their Graphs.

Standard Form of the Equation of a Parabola

The graph of the quadratic function

$$y = a(x - h)^2 + k \quad \text{where } a \neq 0$$

is a parabola with vertex at (h, k). The axis of symmetry is the line $x = h$. The parabola opens upward when $a > 0$ and downward when $a < 0$.

Parabolas can also open to the right and to the left, but they do not define functions because their graphs fail the vertical line test.

The two general forms of the equation of a parabola are similar.

Equation of a Parabola

The general forms of the equation of a parabola are:

1. $y = ax^2 + bx + c$ The graph opens upward if $a > 0$ and downward if $a < 0$.
2. $x = ay^2 + by + c$ The graph opens to the right if $a > 0$ and to the left if $a < 0$.

EXAMPLE 5 Write $y = -2x^2 + 12x - 13$ in standard form and graph it.

Strategy We will complete the square on x to write the equation in standard form, $y = a(x - h)^2 + k$.

Why Standard form contains the expression $(x - h)^2$. We can obtain a perfect-square trinomial that factors into that form by completing the square on x.

Solution Because the equation is not in standard form, the coordinates of the vertex are not obvious. To write the equation in standard form, we complete the square on x.

$y = -2x^2 + 12x - 13$

$y = -2(x^2 - 6x \qquad) - 13$ Factor out -2 from $-2x^2 + 12x$.

This step adds $-2 \cdot 9$ or -18 to this side. Add 18 to counteract the addition of -18.

$y = \mathbf{-2}(x^2 - 6x \mathbf{+ 9}) - 13 \mathbf{+ 18}$ Complete the square on $x^2 - 6x$.

$y = -2(x - 3)^2 + 5$ Factor $x^2 - 6x + 9$ and add: $-13 + 18 = 5$.

$y = \mathbf{a}(x - \mathbf{h})^2 + \mathbf{k}$

Success Tip

Recall that we can find the x-coordinate of the vertex using:

$$x = -\frac{b}{2a} = -\frac{12}{2(-2)} = 3$$

To find the y-coordinate, substitute into the equation:

$$y = -2(3)^2 + 12(3) - 13 = 5$$

The vertex is at $(3, 5)$.

Success Tip

When the equation of a parabola that opens upward or downward is written in standard form, h is the number within the parentheses that *follows* the − (subtraction symbol), and k is the number outside the parentheses that follows the + (addition symbol).

This equation is written in the form $y = a(x - h)^2 + k$, where $a = -2$, $h = 3$, and $k = 5$. Thus, the graph of the equation is a parabola that opens downward with vertex at (3, 5) and an axis of symmetry $x = 3$. We can construct a table of solutions and use symmetry to plot several points on the parabola. Then we draw a smooth curve through the points to get the graph of $y = -2x^2 + 12x - 13$, as shown below.

$y = -2x^2 + 12x - 13$

x	y	
1	−3	→ (1, −3)
2	3	→ (2, 3)

Along with the vertex, a minimum of two points on each side of the axis of symmetry are needed for an accurate graph.

Because the x-coordinate of the vertex is 3, choose values for x that are close to 3 on the same side of the axis of symmetry.

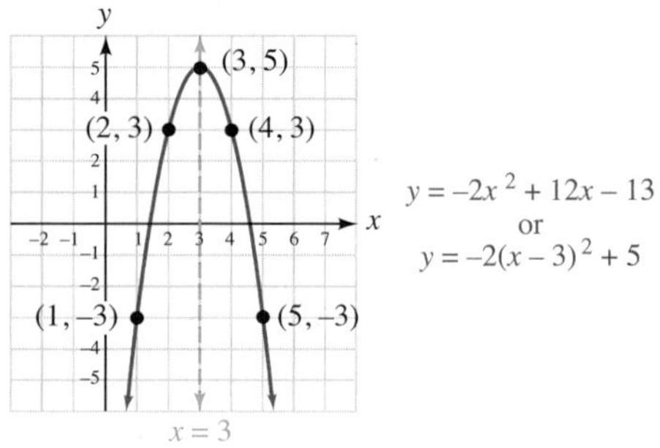

Teaching Example 5 Write $y = -3x^2 + 18x - 25$ in standard form and graph it.
Answer: $y = -3(x - 3)^2 + 2$

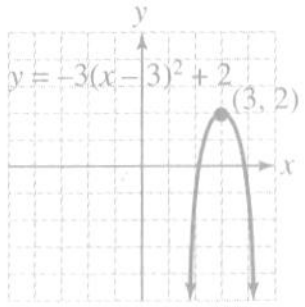

Self Check 5 Write $y = 2x^2 + 4x + 5$ in standard form and graph it.

$y = 2(x + 1)^2 + 3$; see AIE Appendix 3.

Now Try Problem 39

The **standard form for the equation of a parabola** that opens to the right or left is similar to $y = a(x - h)^2 + k$, except that the variables, x and y, exchange positions as do the constants, h and k.

Standard Form of the Equation of a Parabola

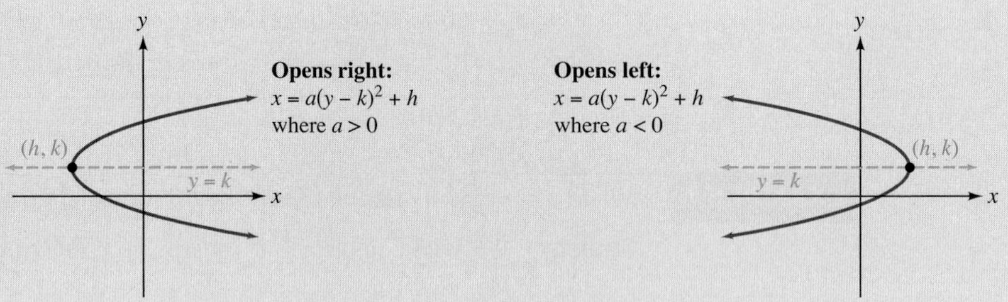

EXAMPLE 6 Graph: $x = \frac{1}{2}y^2$

Strategy We will compare the equation to the standard form of the equation of a parabola to find a, h, and k.

Why Once we know these values, we can locate the vertex of the graph. We also know whether the parabola will open to the left or to the right.

Solution This equation is written in the form $x = a(y - k)^2 + h$, where $a = \frac{1}{2}$, $k = 0$, and $h = 0$. The graph of the equation is a parabola that opens to the right with vertex at (0, 0) and an axis of symmetry $y = 0$.

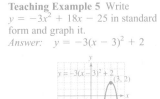

Success Tip

Unlike the equation of a circle, the equation of a parabola contains only one variable squared term.

Teaching Tip: Point out that the center is not part of the graph of a circle, but the vertex is part of the graph of a parabola.

To construct a table of solutions, we choose values of y and find their corresponding values of x. For example, if $y = 1$ and if $y = 2$, we have:

$$x = \frac{1}{2}y^2 \qquad\qquad x = \frac{1}{2}y^2$$

$$x = \frac{1}{2}(\mathbf{1})^2 \quad \text{Substitute 1 for } y. \qquad\qquad x = \frac{1}{2}(\mathbf{2})^2 \quad \text{Substitute 2 for } y.$$

$$x = \frac{1}{2} \qquad\qquad x = 2$$

The point $\left(\frac{1}{2}, 1\right)$ is on the parabola. The point (2, 2) is on the parabola.

We plot the ordered pairs from the table and use symmetry to plot three more points on the parabola. Then we draw a smooth curve through the points to get the graph of $x = \frac{1}{2}y^2$, as shown below.

Success Tip

When the equation of a parabola that opens left or right is written in standard form, k is the number within the parentheses that follows the $-$ (subtraction symbol), and h is the number outside the parentheses that follows the + (addition symbol).

Along with the vertex, three points on each side of the axis of symmetry were plotted to obtain an accurate graph.

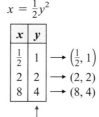

$x = \frac{1}{2}y^2$

x	y	
$\frac{1}{2}$	1	→ $\left(\frac{1}{2}, 1\right)$
2	2	→ (2, 2)
8	4	→ (8, 4)

Because the y-coordinate of the vertex is 0, choose values for y that are close to 0 on the same side of the axis of symmetry.

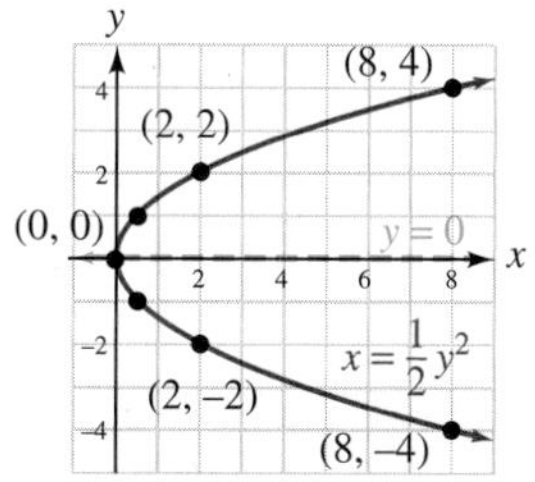

Teaching Tip: Ask if this is the graph of a function.

Teaching Example 6 Graph: $x = -\frac{1}{2}y^2$

Answer:

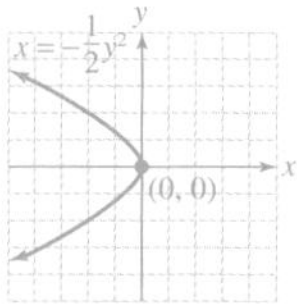

Self Check 6 Graph: $x = -\frac{2}{3}y^2$ See AIE Appendix 3.

Now Try Problem 43

EXAMPLE 7 Write $x = -3y^2 - 12y - 13$ in standard form and graph it.

Strategy We will complete the square on y to write the equation in standard form, $x = a(y - k)^2 + h$.

Why Standard form contains the expression $(y - k)^2$. We can obtain a perfect-square trinomial that factors into that form by completing the square on y.

Solution To write the equation in standard form, we complete the square.

$$x = -3y^2 - 12y - 13$$

$$x = -3(y^2 + 4y \qquad) - 13 \qquad \text{Factor out } -3 \text{ from } -3y^2 - 12y.$$

$$x = \mathbf{-3}(y^2 + 4y + \mathbf{4}) - 13 + \mathbf{12} \qquad \text{Complete the square on } y^2 + 4y. \text{ Then add 12 to the right side to counteract } -3 \cdot 4 = -12.$$

$$x = -3(y + 2)^2 - 1 \qquad \text{Factor } y^2 + 4y + 4 \text{ and add: } -13 + 12 = 1.$$

$$x = -3[y - (-2)]^2 + (-1)$$

$$x = a[y - k]^2 + h \qquad \text{Rewrite to help determine } h \text{ and } k.$$

This equation is in the standard form $x = a(y - k)^2 + h$, where $a = -3$, $k = -2$, and $h = -1$. The graph of the equation is a parabola that opens to the left with vertex at $(-1, -2)$ and an axis of symmetry $y = -2$.

Success Tip

When an equation is of the form $x = ay^2 + by + c$, we can find the y-coordinate of the vertex using

$$y = -\frac{b}{2a} = -\frac{-12}{2(-3)} = -2$$

To find the x-coordinate, substitute:

$$x = -3(-2)^2 - 12(-2) - 13 = -1$$

The vertex is at $(-1, -2)$.

We can construct a table of solutions and use symmetry to plot several points on the parabola. Then we draw a smooth curve through the points to get the graph of $x = -3y^2 - 12y - 13$, as shown below.

$$x = -3y^2 - 12y - 13$$
or
$$x = -3(y + 2)^2 - 1$$

x	y	
-4	-1	$\to (-4, -1)$
-13	0	$\to (-13, 0)$

Choose at least two values for y close to −2, and find the corresponding x-values.

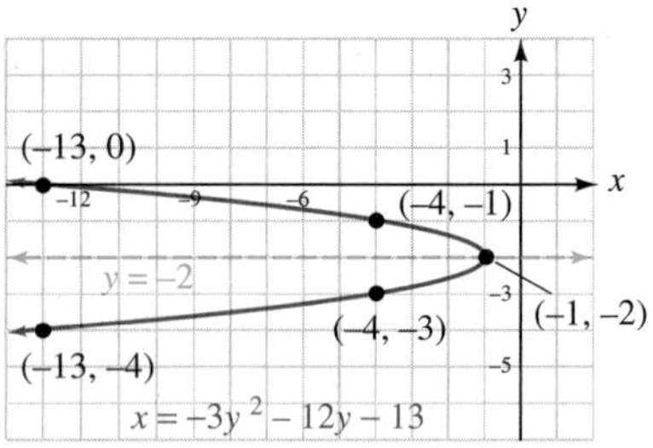

Success Tip

The equation of a circle contains an x^2 and a y^2 term. The equation of a parabola has either an x^2 term or a y^2 term, but not both.

Teaching Example 7 Write $x = -2y^2 + 12y - 17$ in standard form and graph it.
Answer:
$x = -2(y - 3)^2 + 1$

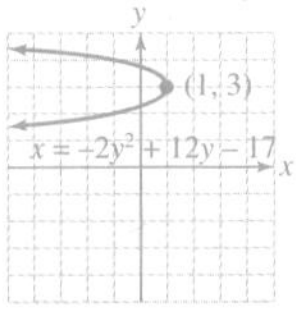

Self Check 7 Write $x = 3y^2 - 6y - 1$ in standard form and graph it. $x = 3(y - 1)^2 - 4$; see AIE Appendix 3.

Now Try Problem 49

SECTION 13.1 STUDY SET

VOCABULARY

Fill in the blanks.

1. The curves formed by the intersection of a plane with an infinite right-circular cone are called conic sections.
2. Give the name of each conic shown below.

ellipse parabola hyperbola circle

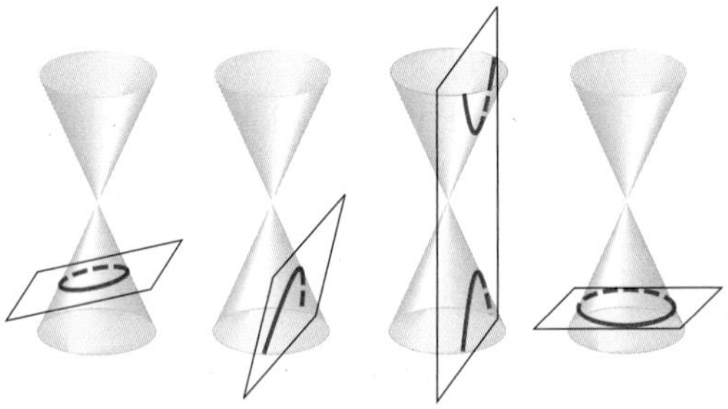

3. A circle is the set of all points in a plane that are a fixed distance from a fixed point called its center. The fixed distance is called the radius.
4. A parabola is the set of all points in a plane that are equidistant from a fixed point and a fixed line.

CONCEPTS

5. a. Write the standard form of the equation of a circle. $(x - h)^2 + (y - k)^2 = r^2$
   b. Write the standard form of the equation of a circle with the center at the origin. $x^2 + y^2 = r^2$

6. a. Find the center and the radius of the circle graphed on the right. $(0, 0)$; $r = 3$
   b. Write the equation of the circle. $x^2 + y^2 = 9$

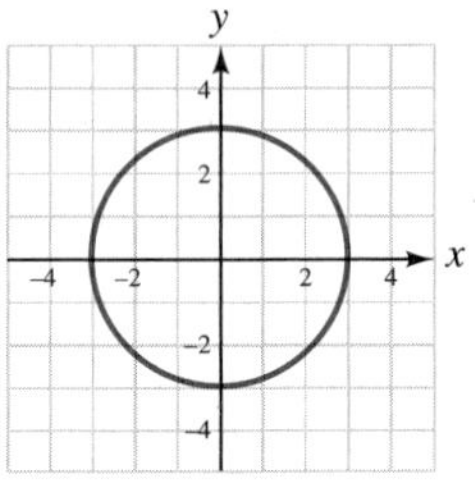

7. a. Find the center and the radius of the circle graphed on the right. $(2, -1)$; $r = 4$
   b. Write the equation of the circle. $(x - 2)^2 + (y + 1)^2 = 16$

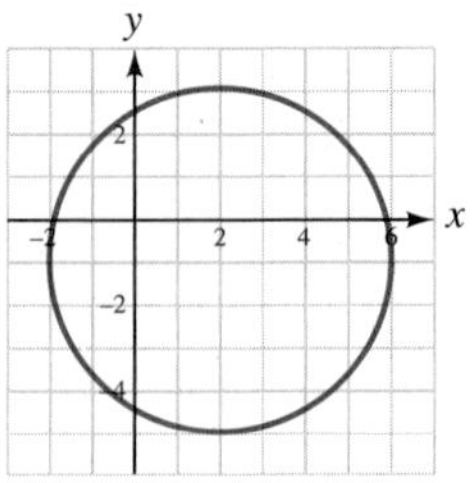

8. Fill in the blanks. To complete the square on $x^2 + 2x$ and on $y^2 - 6y$, what numbers must be added to each side of the equation?

$$x^2 + 2x + y^2 - 6y = 2$$
$$x^2 + 2x + 1 + y^2 - 6y + 9 = 2 + 1 + 9$$

9. a. What is the standard form of the equation of a parabola opening upward or downward? $y = a(x - h)^2 + k$
   b. What is the standard form of the equation of a parabola opening to the right or left? $x = a(y - k)^2 + h$
10. Fill in the blanks.
    a. To complete the square on the right side, what should be factored from the first two terms?

$$x = 4y^2 + 16y + 9$$
$$x = 4(y^2 + 4y \qquad) + 9$$

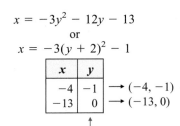

b. To complete the square on $y^2 + 4y$, what should be added within the parentheses, and what should be subtracted outside the parentheses?

$$x = 4(y^2 + 4y + 4) + 9 - 16$$

11. Determine whether the graph of each equation is a circle or a parabola.
a. $x^2 + y^2 - 6x + 8y - 10 = 0$ Circle
b. $y^2 - 2x + 3y - 9 = 0$ Parabola
c. $x^2 + 5x - y = 0$ Parabola
d. $x^2 + 12x + y^2 = 0$ Circle

12. Draw a parabola using the given facts.
- Opens right
- Vertex $(-3, 2)$
- Passes through $(-2, 1)$
- x-intercept $(1, 0)$

See AIE Appendix 3.

NOTATION

▶ 13. Find h, k, and r: $(x - 6)^2 + (y + 2)^2 = 9$ 6, −2, 3

▶ 14. a. Find a, h, and k: $y = 6(x - 5)^2 - 9$ 6, 5, −9
b. Find a, h, and k: $x = -3(y + 2)^2 + 1$ −3, 1, −2

GUIDED PRACTICE

***Find the center and radius of each circle and graph it.* See Example 1.** See AIE Appendix 3.

15. $x^2 + y^2 = 9$ $(0, 0)$, $r = 3$

16. $x^2 + y^2 = 16$ $(0, 0)$, $r = 4$

▶ 17. $x^2 + (y + 3)^2 = 1$ $(0, -3)$, $r = 1$

▶ 18. $(x + 4)^2 + y^2 = 1$ $(-4, 0)$, $r = 1$

▶ 19. $(x + 3)^2 + (y - 1)^2 = 16$ $(-3, 1)$, $r = 4$

▶ 20. $(x - 1)^2 + (y + 4)^2 = 9$ $(1, -4)$, $r = 3$

21. $x^2 + y^2 = 6$ $(0, 0)$, $r = \sqrt{6} \approx 2.4$

22. $x^2 + y^2 = 10$ $(0, 0)$, $r = \sqrt{10} \approx 3.2$

***Write the equation of a circle in standard form with the following properties.* See Example 2.**

23. Center at the origin; radius 1 $x^2 + y^2 = 1$

▶ 24. Center at the origin; radius 4 $x^2 + y^2 = 16$

▶ 25. Center at (6, 8); radius 5 $(x - 6)^2 + (y - 8)^2 = 25$

▶ 26. Center at (5, 3); radius 2 $(x - 5)^2 + (y - 3)^2 = 4$

27. Center at (−2, 6); radius 12 $(x + 2)^2 + (y - 6)^2 = 144$

▶ 28. Center at (5, −4); radius 6 $(x - 5)^2 + (y + 4)^2 = 36$

▶ 29. Center at (0, 0); radius $\frac{1}{4}$ $x^2 + y^2 = \frac{1}{16}$

▶ 30. Center at (0, 0); radius $\frac{1}{3}$ $x^2 + y^2 = \frac{1}{9}$

31. Center at $\left(\frac{2}{3}, -\frac{7}{8}\right)$; radius $\sqrt{2}$ $\left(x - \frac{2}{3}\right)^2 + \left(y + \frac{7}{8}\right)^2 = 2$

32. Center at (−0.7, −0.2); radius $\sqrt{11}$ $(x + 0.7)^2 + (y + 0.2)^2 = 11$

33. Center at the origin; diameter $4\sqrt{2}$ $x^2 + y^2 = 8$

▶ 34. Center at the origin; diameter $8\sqrt{3}$ $x^2 + y^2 = 48$

***Write each equation of a circle in standard form and graph it. Give the coordinates of its center and give the radius.* See Example 3.** See AIE Appendix 3.

▶ 35. $x^2 + y^2 - 2x + 4y = -1$
$(x - 1)^2 + (y + 2)^2 = 4$; $(1, -2)$, $r = 2$

▶ 36. $x^2 + y^2 + 6x - 4y = -12$
$(x + 3)^2 + (y - 2)^2 = 1$; $(-3, 2)$, $r = 1$

▶ 37. $x^2 + y^2 + 4x + 2y = 4$
$(x + 2)^2 + (y + 1)^2 = 9$; $(-2, -1)$, $r = 3$

▶ 38. $x^2 + y^2 + 8x + 2y = -13$
$(x + 4)^2 + (y + 1)^2 = 4$; $(-4, -1)$, $r = 2$

***Write each equation of a parabola in standard form and graph it. Give the coordinates of the vertex.* See Example 5.** See AIE Appendix 3.

39. $y = 2x^2 - 4x + 5$
$y = 2(x - 1)^2 + 3$
Vertex: (1, 3)

▶ 40. $y = x^2 + 4x + 5$
$y = (x + 2)^2 + 1$
Vertex: (−2, 1)

▶ 41. $y = -x^2 - 2x + 3$
$y = -(x + 1)^2 + 4$
Vertex: (−1, 4)

42. $y = -2x^2 - 4x$
$y = -2(x + 1)^2 + 2$
Vertex: (−1, 2)

***Graph each equation of a parabola. Give the coordinates of the vertex.* See Example 6.** See AIE Appendix 3.

43. $x = y^2$
Vertex: (0, 0)

▶ 44. $x = 2y^2$
Vertex: (0, 0)

▶ 45. $x = 2(y + 1)^2 + 3$
Vertex: (3, −1)

46. $x = 3(y - 2)^2 - 1$
Vertex: (−1, 2)

***Write each equation of a parabola in standard form and graph it. Give the coordinates of the vertex.* See Example 7.** See AIE Appendix 3.

47. $x = y^2 - 2y + 5$
$x = (y - 1)^2 + 4$;
Vertex: (4, 1)

▶ 48. $x = y^2 + 6y + 8$
$x = (y + 3)^2 - 1$;
Vertex: (−1, −3)

▶ 49. $x = -3y^2 + 18y - 25$
$x = -3(y - 3)^2 + 2$;
Vertex: (2, 3)

50. $x = -2y^2 + 4y + 1$
$x = -2(y - 1)^2 + 3$;
Vertex: (3, 1)

***Use a graphing calculator to graph each equation. (Hint: Solve for y and graph two functions.)* See Using Your Calculator: Graphing Circles.** See AIE Appendix 3.

51. $x^2 + y^2 = 7$

▶ 52. $x^2 + y^2 = 5$

53. $(x + 1)^2 + y^2 = 16$

54. $x^2 + (y - 2)^2 = 4$

Use a graphing calculator to graph each equation. (Hint: Solve for y and graph two functions when necessary.) See AIE Appendix 3.

55. $x = 2y^2$

56. $x = y^2 - 4$

57. $x^2 - 2x + y = 6$

58. $x = -2(y - 1)^2 + 2$

TRY IT YOURSELF

Write each equation in standard form, if it is not already so, and graph it. If the graph is a circle, give the coordinates of its center and its radius. If the graph is a parabola, give the coordinates of its vertex. See AIE Appendix 3.

59. $x = y^2 - 6y + 4$
$x = (y - 3)^2 - 5$;
Vertex: (−5, 3)

60. $x = y^2 - 8y + 13$
$x = (y - 4)^2 - 3$;
Vertex: (−3, 4)

61. $(x - 2)^2 + y^2 = 25$
$(2, 0)$; $r = 5$

62. $x^2 + (y - 3)^2 = 25$
$(0, 3)$; $r = 5$

63. $x^2 + y^2 - 6x + 8y + 18 = 0$
$(x - 3)^2 + (y + 4)^2 = 7$; $(3, -4)$; $r = \sqrt{7} \approx 2.6$

▶ 64. $x^2 + y^2 - 4x + 4y - 3 = 0$
$(x - 2)^2 + (y + 2)^2 = 11$; $(2, -2)$; $r = \sqrt{11} \approx 3.3$

65. $y = 4x^2 - 16x + 17$
$y = 4(x - 2)^2 + 1$; Vertex: $(2, 1)$

66. $y = 4x^2 - 32x + 63$
$y = 4(x - 4)^2 - 1$; Vertex: $(4, -1)$

67. $(x - 1)^2 + (y - 3)^2 = 15$
$(1, 3)$; $r = \sqrt{15} \approx 3.9$

68. $(x + 1)^2 + (y + 1)^2 = 8$
$(-1, -1)$; $r = 2\sqrt{2} \approx 2.8$

69. $x = -y^2 + 1$
Vertex: $(1, 0)$

70. $x = -y^2 - 5$
Vertex: $(-5, 0)$

71. $(x - 2)^2 + (y - 4)^2 = 36$
$(2, 4)$; $r = 6$

72. $(x - 3)^2 + (y - 2)^2 = 36$
$(3, 2)$; $r = 6$

▶ 73. $x = -\frac{1}{4}y^2$
Vertex: $(0, 0)$

74. $x = 4y^2$
Vertex: $(0, 0)$

75. $x = -6(y - 1)^2 + 3$
Vertex: $(3, 1)$

▶ 76. $x = -6(y + 1)^2 - 4$
Vertex: $(-4, -1)$

77. $x^2 + y^2 + 2x - 8 = 0$
$(x + 1)^2 + y^2 = 9$; $(-1, 0)$; $r = 3$

78. $x^2 + y^2 - 4y = 12$
$x^2 + (y - 2)^2 = 16$; $(0, 2)$; $r = 4$

▶ 79. $x = \frac{1}{2}y^2 + 2y$
$x = \frac{1}{2}(y + 2)^2 - 2$; Vertex: $(-2, -2)$

▶ 80. $x = -\frac{1}{3}y^2 - 2y$
$x = -\frac{1}{3}(y + 3)^2 + 3$; Vertex: $(3, -3)$

81. $y = -4(x + 5)^2 + 5$
Vertex: $(-5, 5)$

82. $y = -4(x - 4)^2 - 4$
Vertex: $(4, -4)$

Look Alikes . . .

Find the center and radius of each circle.

83. a. $(x - 4)^2 + (y + 7)^2 = 28$ $(4, -7)$; $r = 2\sqrt{7} \approx 5.3$
b. $(x + 4)^2 + (y - 7)^2 = 28$ $(-4, 7)$; $r = 2\sqrt{7} \approx 5.3$

▶ 84. a. $x^2 + y^2 + 10x - 14y - 7 = 0$ $(-5, 7)$; $r = 9$
b. $x^2 + y^2 - 10x + 14y - 7 = 0$ $(5, -7)$; $r = 9$

Find the coordinates of the vertex and the direction in which each parabola opens.

85. a. $y = 8(x - 3)^2 + 6$ $V(3, 6)$; opens upward
b. $x = 8(y - 3)^2 + 6$ $V(6, 3)$; opens to the right

▶ 86. a. $y = -x^2 + 4x + 1$ $V(2, 5)$; opens downward
b. $x = -y^2 + 4y + 1$ $V(5, 2)$; opens to the left

APPLICATIONS

▶ 87. **Broadcast Ranges.** Radio stations applying for licensing may not use the same frequency if their broadcast areas overlap. One station's coverage is bounded by $x^2 + y^2 - 8x - 20y + 16 = 0$, and the other's by $x^2 + y^2 + 2x + 4y - 11 = 0$. May they be licensed for the same frequency? No

▶ 88. **Meshing Gears.** For design purposes, the large gear is described by the circle $x^2 + y^2 = 16$. The smaller gear is a circle centered at $(7, 0)$ and tangent to the larger circle. Find the equation of the smaller gear. $(x - 7)^2 + y^2 = 9$

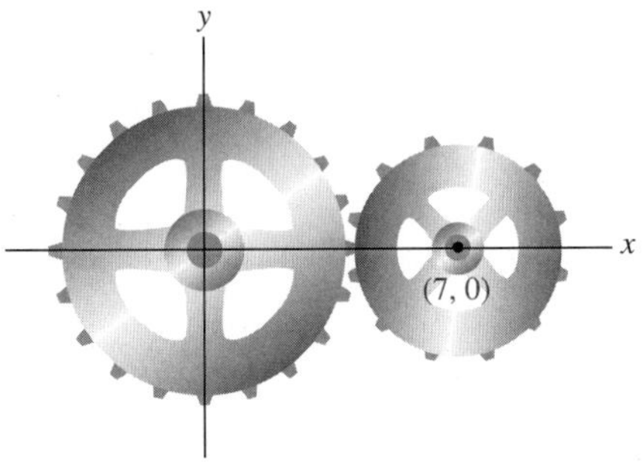

▶ 89. **from Campus to Careers**

Civil Engineer

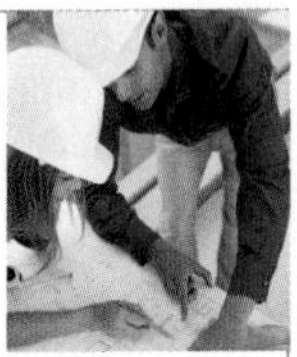

Two sections of a new freeway join with a curve that is one-quarter of a circle, as shown below. The equation of the circle is $x^2 + y^2 - 10x - 12y + 52 = 0$, where distances are measured in miles.

a. How far from City Hall will the new freeway intersect State Street? 8 mi

b. How far from City Hall will the new freeway intersect Highway 60? 9 mi

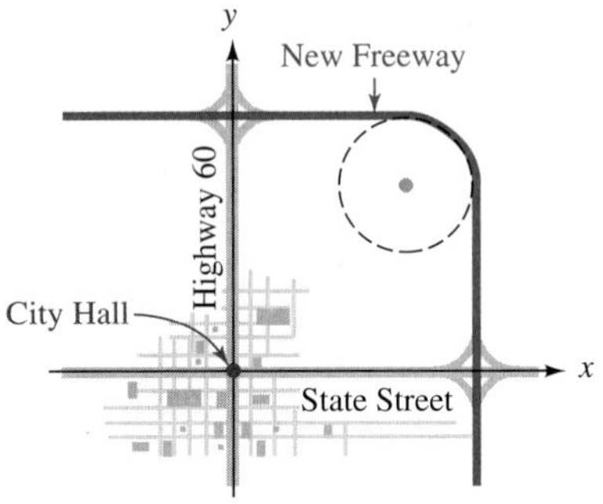

▶ 90. **Walkways.** The walkway shown is bounded by the two circles $x^2 + y^2 = 2{,}500$ and $(x - 10)^2 + y^2 = 900$, measured in feet. Find the largest and the smallest width of the walkway. 30 ft and 10 ft

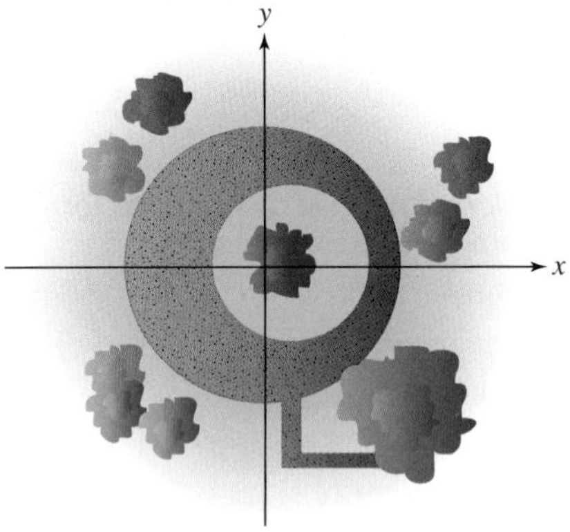

▶ 91. **Projectiles.** The cannonball in the illustration follows the parabolic path $y = 30x - x^2$. How far short of the castle does it land? 5 ft

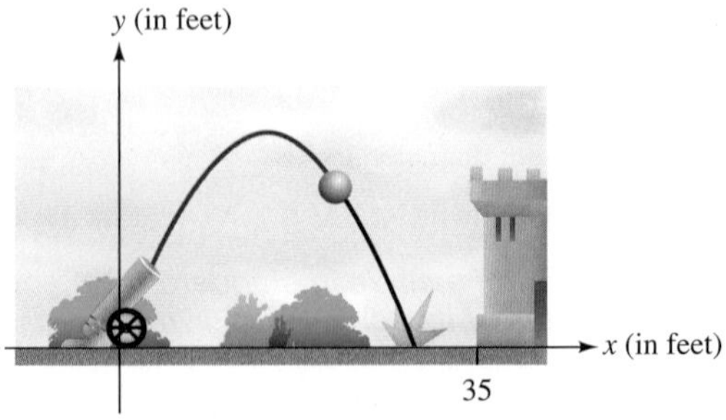

92. Projectiles. In Exercise 91, how high does the cannonball get? 225 ft

93. Comets. If the orbit of the comet is approximated by the equation $2y^2 - 9x = 18$, how far is it from the sun at the vertex V of the orbit? Distances are measured in astronomical units (AU). 2 AU

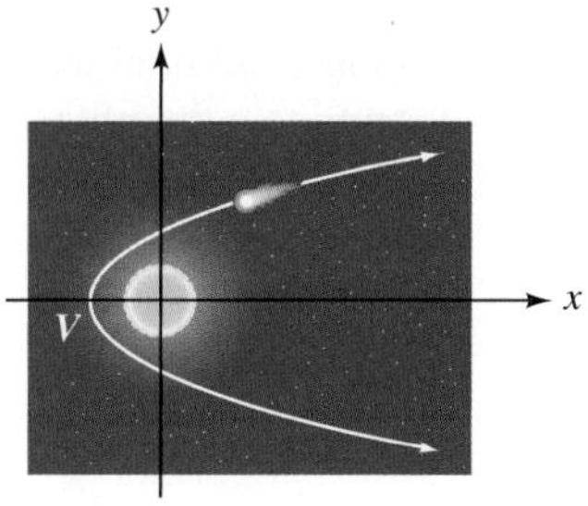

94. Satellite Antennas. The cross section of the satellite antenna in the illustration is a parabola given by the equation $y = \frac{1}{16}x^2$, with distances measured in feet. If the dish is 8 feet wide, how deep is it? 1 ft

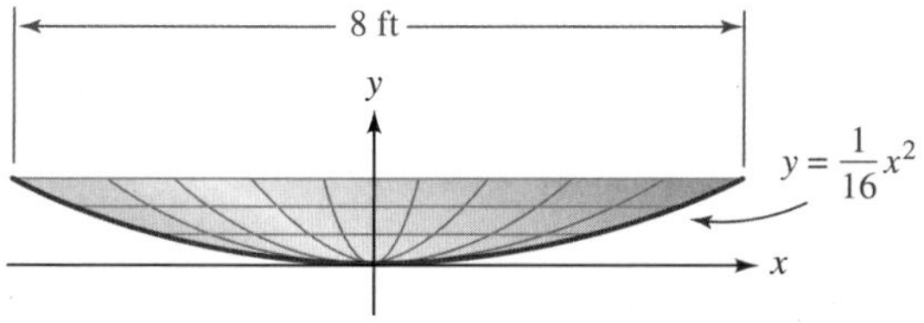

WRITING

95. Explain how to decide from its equation whether the graph of a parabola opens up, down, right, or left.

96. From the equation of a circle, explain how to determine the radius and the coordinates of the center.

97. On the day of an election, the warning "*No electioneering within a 1,000-foot radius of this polling place*" was posted in front of a school. Explain what it means.

98. What is meant by the *turning radius* of a truck?

REVIEW

Solve.

99. $|3x - 4| = 11$ $5, -\frac{7}{3}$

100. $\left|\frac{4 - 3x}{5}\right| = 12$ $\frac{64}{3}, -\frac{56}{3}$

101. $|3x + 4| = |5x - 2|$ $3, -\frac{1}{4}$

102. $|6 - 4x| = |x + 2|$ $\frac{4}{5}, \frac{8}{3}$

CHALLENGE PROBLEMS

103. Could the intersection of a plane with an infinite right-circular cone as shown on page 1077 be a single point? If so, draw a picture that illustrates this. Yes

104. Under what conditions will the graph of $x = a(y - k)^2 + h$ have no y-intercepts? $a < 0$ and $h < 0$; $a > 0$ and $h > 0$

105. Write the equation of a circle with a diameter whose endpoints are at $(-2, -6)$ and $(8, 10)$.
$(x - 3)^2 + (y - 2)^2 = 89$

106. Write the equation of a circle with a diameter whose endpoints are at $(-5, 4)$ and $(7, -3)$.
$(x - 1)^2 + \left(y - \frac{1}{2}\right)^2 = \frac{193}{4}$ $\left(\frac{193}{4} = 48.25\right)$

SECTION 13.2 The Ellipse

OBJECTIVES

1. Define an ellipse.
2. Graph ellipses centered at the origin.
3. Graph ellipses centered at (h, k).
4. Solve application problems involving ellipses.

ARE YOU READY? *Are You Ready? exercises available online at www.webassign.net/brookscole*

The following problems review some basic skills that are needed when working with ellipses.

1. Solve: $a^2 = 81$ ± 9

2. Simplify: $y - (-4)$ $y + 4$

3. Multiply: $16\left(\frac{x^2}{4} + \frac{y^2}{16}\right)$ $4x^2 + y^2$

4. Simplify: $\sqrt{8}$ $2\sqrt{2}$

A third conic section is an oval curve called an *ellipse.* Ellipses can be nearly round and look almost like a circle, or they can be long and narrow. In this section, we will learn how to construct ellipses and how to graph equations that represent ellipses.

1 Define an Ellipse.

To define a circle, we considered a fixed distance from a fixed point. The definition of an ellipse involves *two* distances from *two* fixed points.

Definition of an Ellipse

An **ellipse** is the set of all points in a plane for which the sum of the distances from two fixed points is a constant.

The figure below illustrates that any point on an ellipse is a constant distance $d_1 + d_2$ from two fixed points, each of which is called a **focus.** Midway between the **foci** is the **center** of the ellipse.

We can construct an ellipse by placing two thumbtacks fairly close together to serve as foci. We then tie each end of a piece of string to a thumbtack, catch the loop with the point of a pencil, and (keeping the string taut) draw the ellipse.

Teaching Tip: Ask if this is the graph of a function.

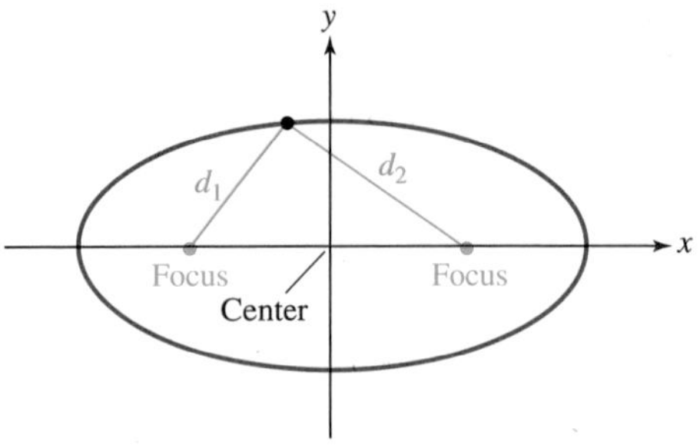

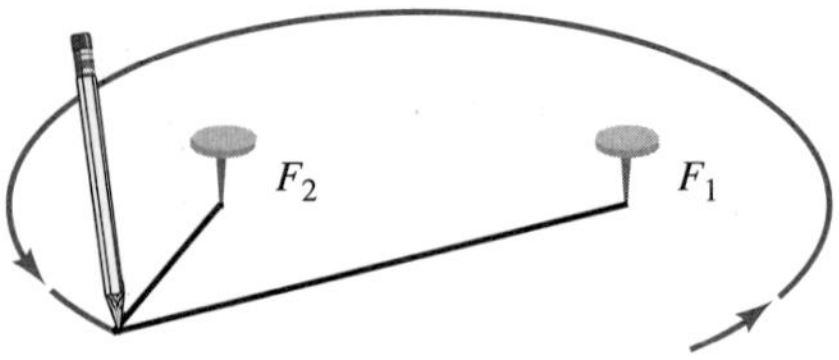

The Language of Algebra

The word **foci** (pronounced foe-sigh) is plural for the word *focus.* In the illustration on the right, the foci are labeled using subscript notation. One focus is F_1 and the other is F_2.

2 Graph Ellipses Centered at the Origin.

The definition of an ellipse can be used to develop the standard equation of an ellipse. To learn more about the derivation, see Problem 74 in the Challenge Problems section of the Study Set.

Equation of an Ellipse Centered at the Origin

The **standard form of the equation of an ellipse** that is symmetric with respect to both axes and centered at (0, 0) is

$$\frac{x^2}{a^2} + \frac{y^2}{b^2} = 1 \quad \text{where } a > 0 \text{ and } b > 0$$

To graph an ellipse centered at the origin, it is helpful to know the intercepts of the graph. To find the x-intercepts of the graph of $\frac{x^2}{a^2} + \frac{y^2}{b^2} = 1$ we let $y = 0$ and solve for x. To find the y-intercepts, we let $x = 0$ and solve for y.

***x*-intercepts:**

$\frac{x^2}{a^2} + \frac{0^2}{b^2} = 1$ Substitute 0 for y.

$\frac{x^2}{a^2} + 0 = 1$ Simplify: $\frac{0^2}{b^2} = 0$.

$x^2 = a^2$ Multiply both sides by a^2.

$x = \pm a$ Use the square root property.

The x-intercepts are $(a, 0)$ and $(-a, 0)$.

***y*-intercepts:**

$\frac{0^2}{a^2} + \frac{y^2}{b^2} = 1$ Substitute 0 for x.

$0 + \frac{y^2}{b^2} = 1$ Simplify: $\frac{0^2}{a^2} = 0$.

$y^2 = b^2$ Multiply both sides by b^2.

$y = \pm b$ Use the square root property.

The y-intercepts are $(0, b)$ and $(0, -b)$.

In general, we have the following results.

The Intercepts of an Ellipse

The graph of $\frac{x^2}{a^2} + \frac{y^2}{b^2} = 1$ is an ellipse, centered at the origin, with x-intercepts $(a, 0)$ and $(-a, 0)$ and y-intercepts $(0, b)$ and $(0, -b)$.

For $\frac{x^2}{a^2} + \frac{y^2}{b^2} = 1$, if $a > b$, the ellipse is horizontal, as shown on the next page, on the left. If $b > a$, the ellipse is vertical, as shown on the right. The points V_1 and V_2 are called the **vertices** of the ellipse. The line segment joining the vertices is called the **major axis,** and its midpoint is called the **center** of the ellipse. The line segment whose endpoints are on the ellipse and that is perpendicular to the major axis at the center is called the **minor axis** of the ellipse.

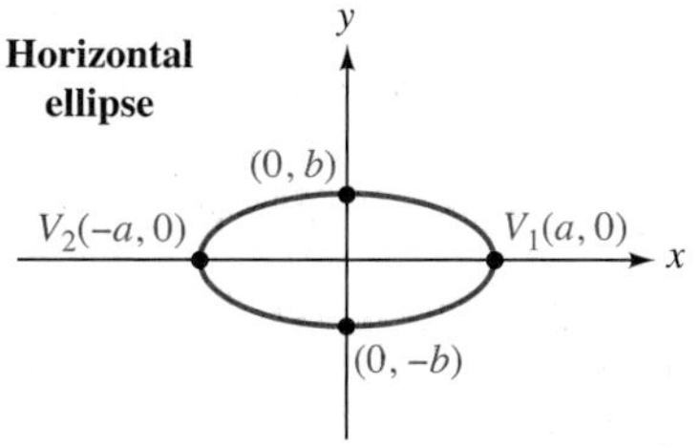

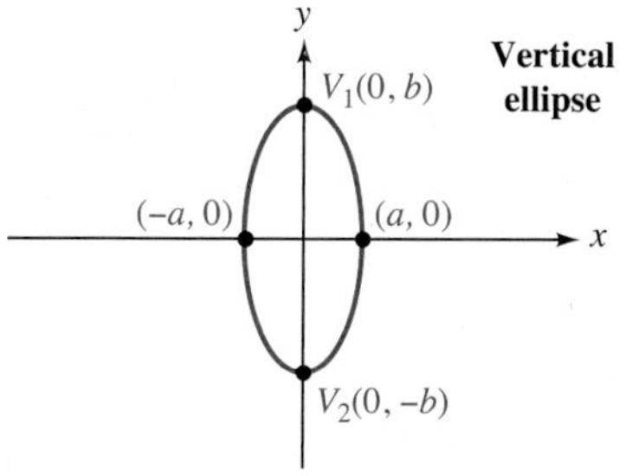

EXAMPLE 1 Graph: $\frac{x^2}{36} + \frac{y^2}{9} = 1$

Strategy This equation is in standard $\frac{x^2}{a^2} + \frac{y^2}{b^2} = 1$ form. We will identify a and b.

Why Once we know a and b, we can determine the intercepts of the graph of the ellipse.

Solution The color highlighting shows how to compare the given equation to the standard form to find a and b.

The Language of Algebra

The word **vertices** is the plural form of the word *vertex*. From the graph, we see that one vertex of this horizontal ellipse is the point (6, 0) and the other vertex is the point (−6, 0).

$$\frac{x^2}{\mathbf{36}} + \frac{y^2}{\mathbf{9}} = 1 \qquad \frac{x^2}{a^2} + \frac{y^2}{b^2} = 1$$

Since $a^2 = 36$, it follows that $a = 6$.

Since $b^2 = 9$, it follows that $b = 3$.

The center of the ellipse is (0, 0). The x-intercepts are $(a, 0)$ and $(-a, 0)$, or (6, 0) and (−6, 0). The y-intercepts are $(0, b)$ and $(0, -b)$, or (0, 3) and (0, −3). Using these four points as a guide, we draw an oval curve through them, as shown in figure (a). The result is a horizontal ellipse because $a > b$.

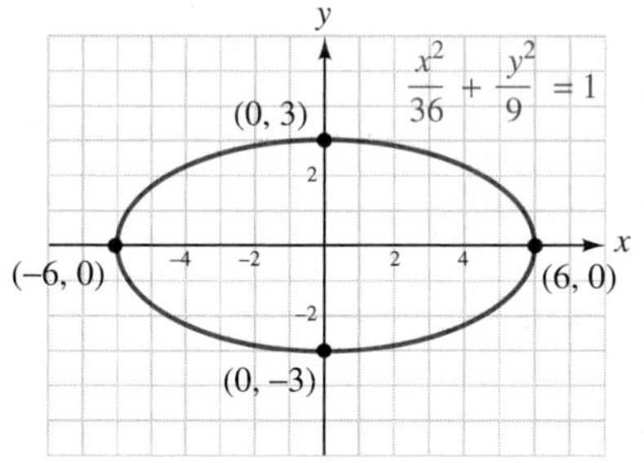

(a)

$\frac{x^2}{36} + \frac{y^2}{9} = 1$

x	y	
2	$\pm 2\sqrt{2}$	→ $(2, \pm 2.8)$
4	$\pm\sqrt{5}$	→ $(4, \pm 2.2)$

Approximate the radicals to plot these points.

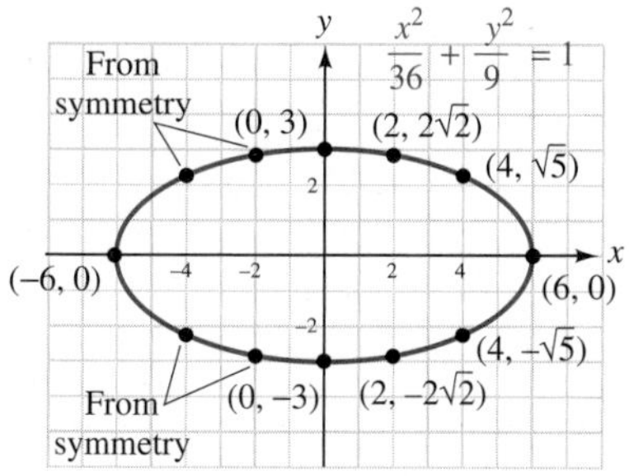

(b)

To increase the accuracy of the graph, we can find additional ordered pairs that satisfy the equation and plot them. For example, if $x = 2$, we have

Teaching Tip: Ask how we would plot the point $(2, 2\sqrt{2})$.

$\frac{2^2}{36} + \frac{y^2}{9} = 1$	Substitute 2 for x in the equation of the ellipse.
$36\left(\frac{4}{36} + \frac{y^2}{9}\right) = 36(1)$	To clear the fractions, multiply both sides by the LCD, 36.
$4 + 4y^2 = 36$	Distribute the multiplication by 36 and simplify.
$y^2 = 8$	Subtract 4 from both sides and divide both sides by 4.
$y = \pm\sqrt{8}$	Use the square root property.
$y = \pm 2\sqrt{2}$	Simplify the radical.

Since two values of y, $2\sqrt{2}$ and $-2\sqrt{2}$, correspond to the x-value 2, we have found two points on the ellipse: $(2, 2\sqrt{2})$ and $(2, -2\sqrt{2})$. To plot them, approximate the y-coordinate to the nearest tenth: (2, 2.8) and (2, −2.8).

In a similar way, we can find the corresponding values of y for the x-value 4. In figure (b) on the previous page, we record these ordered pairs in a table, plot them, use symmetry with respect to the y-axis to plot four other points, and draw the graph of the ellipse.

Teaching Example 1 Graph: $\frac{x^2}{81} + \frac{y^2}{64} = 1$

Answer:

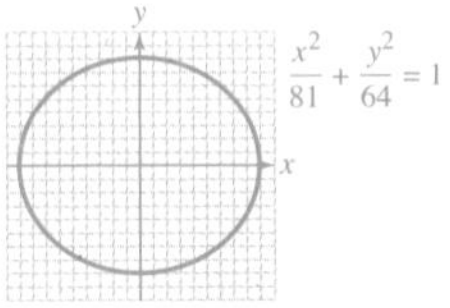

Self Check 1 Graph: $\frac{x^2}{49} + \frac{y^2}{25} = 1$ See AIE Appendix 3.

Now Try Problem 17

EXAMPLE 2 Graph: $16x^2 + y^2 = 16$

Strategy We will write the equation in standard $\frac{x^2}{a^2} + \frac{y^2}{b^2} = 1$ form.

Why When the equation is in standard form, we will be able to identify the center and the intercepts of the graph of the ellipse.

Solution The given equation is not in standard form. To write it in standard form with 1 on the right side, we divide both sides by 16.

Success Tip

Although the term $\frac{16x^2}{16}$ simplifies to x^2, we write it as the fraction $\frac{x^2}{1}$ so that it has the form $\frac{x^2}{a^2}$.

$$16x^2 + y^2 = 16$$

$$\frac{16x^2}{16} + \frac{y^2}{16} = \frac{16}{16} \quad \text{Divide both sides by 16, term-by-term.}$$

$$\frac{x^2}{1} + \frac{y^2}{16} = 1 \quad \text{Simplify the fractions: } \frac{16x^2}{16} = x^2 = \frac{x^2}{1} \text{ and } \frac{16}{16} = 1.$$

Teaching Tip: After writing the equation in standard form, and before graphing it, ask whether the ellipse will be vertical or horizontal.

The center of the ellipse is (0, 0). To determine a and b, we can write the equation in the form

$$\frac{x^2}{1^2} + \frac{y^2}{4^2} = 1 \quad \text{To find } a\text{, write 1 as } 1^2\text{. To find } b\text{, write 16 as } 4^2.$$

Since a^2 (the denominator of x^2) is 1^2, it follows that $a = 1$, and since b^2 (the denominator of y^2) is 4^2, it follows that $b = 4$. Thus, the x-intercepts of the graph are (1, 0) and (−1, 0) and the y-intercepts are (0, 4) and (0, −4). We use these four points as guides to sketch the graph of the ellipse, as shown. The result is a vertical ellipse because $b > a$.

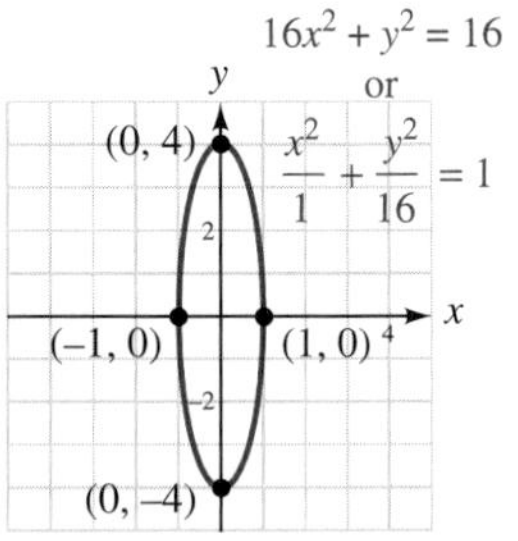

Teaching Example 2 Graph: $25x^2 + y^2 = 25$

Answer:

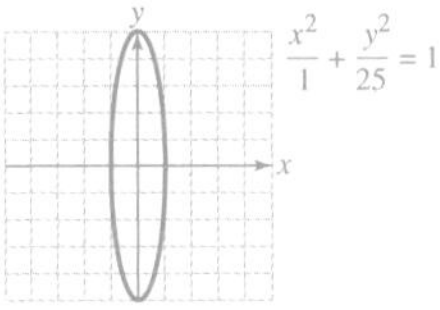

Self Check 2 Graph: $9x^2 + y^2 = 9$ $\frac{x^2}{1} + \frac{y^2}{9} = 1$; see AIE Appendix 3.

Now Try Problem 21

3 Graph Ellipses Centered at (*h*, *k*).

Not all ellipses are centered at the origin. As with the graphs of circles and parabolas, the graph of an ellipse can be translated horizontally and vertically.

Equation of an Ellipse Centered at (*h*, *k*)

The standard form of the equation of a horizontal or vertical ellipse centered at (h, k) is

$$\frac{(x-h)^2}{a^2} + \frac{(y-k)^2}{b^2} = 1 \quad \text{where } a > 0 \text{ and } b > 0$$

For a horizontal ellipse, a is the distance from the center to a vertex. For a vertical ellipse, b is the distance from the center to a vertex.

EXAMPLE 3 Graph: $\dfrac{(x-2)^2}{16} + \dfrac{(y+3)^2}{25} = 1$

Strategy The equation is in standard $\dfrac{(x-h)^2}{a^2} + \dfrac{(y-k)^2}{b^2} = 1$ form. We will identify h, k, a, and b.

Why If we know h, k, a, and b, we can graph the ellipse.

Solution To determine h, k, a, and b, we write the equation in the form

$$\frac{(x-\mathbf{2})^2}{\mathbf{4}^2} + \frac{[y-(\mathbf{-3})]^2}{\mathbf{5}^2} = 1$$

To find k, write $y + 3$ as $y - (-3)$.
To find a, write 16 as 4^2. To find b, write 25 as 5^2.

Success Tip

When the equation of an ellipse is written in standard form, h and k are the numbers within the parentheses that *follow* the − (subtraction) symbols.

We find the center of the ellipse in the same way we would find the center of a circle, by examining $(x-2)^2$ and $(y+3)^2$. Since $h = 2$ and $k = -3$, this is the equation of an ellipse centered at $(h, k) = (2, -3)$. From the denominators, 4^2 and 5^2, we find that $a = 4$ and $b = 5$. Because $b > a$, it is a vertical ellipse.

We first plot the center, as shown below. Since b is the distance from the center to a vertex for a vertical ellipse, we can locate the vertices by counting 5 units above and 5 units below the center. The vertices are the points $(2, 2)$ and $(2, -8)$.

To locate two more points on the ellipse, we use the fact that a is 4 and count 4 units to the left and to the right of the center. We see that the points $(-2, -3)$ and $(6, -3)$ are also on the graph.

Using these four points as guides, we draw the graph shown below on the left. The illustration on the right shows how the graph of $\frac{(x-2)^2}{16} + \frac{(y+3)^2}{25} = 1$ can be obtained by translating the graph of $\frac{x^2}{16} + \frac{y^2}{25} = 1$ to the right 2 units, and then down 3 units.

The center $(2, -3)$ is not part of the graph of the ellipse; it only helps us sketch the graph.

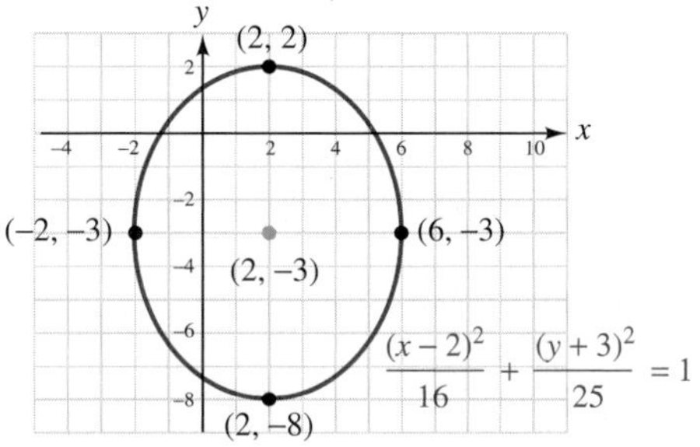

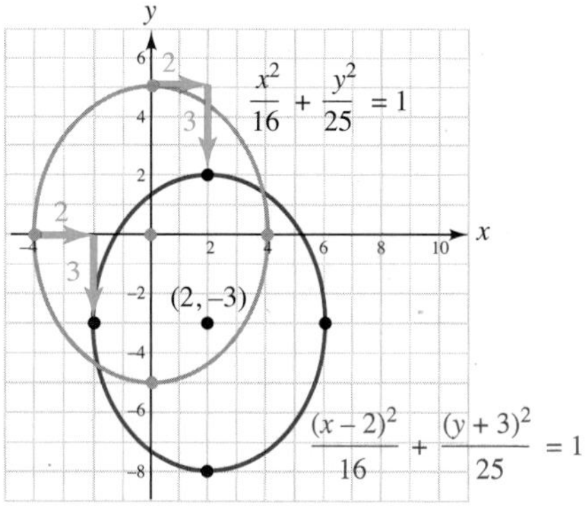

Teaching Example 3 Graph: $\dfrac{(x+3)^2}{4} + \dfrac{(y-1)^2}{16} = 1$
Answer:

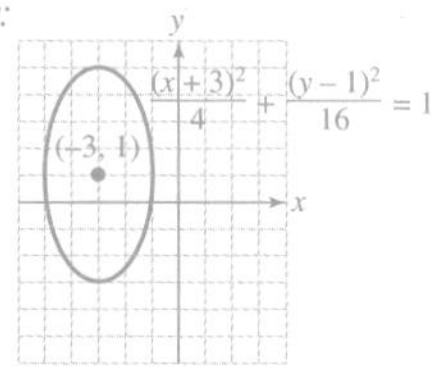

Self Check 3 Graph: $\dfrac{(x-1)^2}{9} + \dfrac{(y+2)^2}{16} = 1$ See AIE Appendix 3.

Now Try Problem 25

Using Your Calculator ▶ Graphing Ellipses

To use a graphing calculator to graph the equation from Example 3, $\dfrac{(x-2)^2}{16} + \dfrac{(y+3)^2}{25} = 1$ we clear the equation of fractions and solve for y.

$$25(x-2)^2 + 16(y+3)^2 = 400 \quad \text{Multiply both sides by 400.}$$

$$16(y+3)^2 = 400 - 25(x-2)^2 \quad \text{Subtract } 25(x-2)^2 \text{ from both sides.}$$

$$(y+3)^2 = \frac{400 - 25(x-2)^2}{16} \quad \text{Divide both sides by 16.}$$

$$y + 3 = \pm\frac{\sqrt{400 - 25(x-2)^2}}{4} \quad \text{Use the square root property.}$$

$$y = -3 \pm \frac{\sqrt{400 - 25(x-2)^2}}{4} \quad \text{Subtract 3 from both sides.}$$

The previous equation represents two functions. On a calculator, we can graph them in a square window to get the ellipse shown here.

$$y = -3 + \frac{\sqrt{400 - 25(x - 2)^2}}{4}$$

and

$$y = -3 - \frac{\sqrt{400 - 25(x - 2)^2}}{4}$$

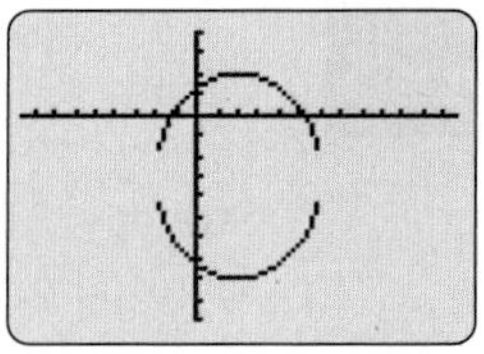

As we saw with circles, the two portions of the ellipse do not quite connect. This is because the graphs are nearly vertical there.

EXAMPLE 4 Graph: $4(x - 2)^2 + 9(y - 1)^2 = 36$

Strategy We will write the equation in standard $\frac{(x - h)^2}{a^2} + \frac{(y - k)^2}{b^2} = 1$ form. Then we will identify h, k, a, and b.

Why If we know h, k, a, and b, we can graph the ellipse.

Solution This equation is not in standard form. To write it in standard form with 1 on the right side, we divide both sides by 36.

$$4(x - 2)^2 + 9(y - 1)^2 = 36$$

$$\frac{4(x - 2)^2}{\mathbf{36}} + \frac{9(y - 1)^2}{\mathbf{36}} = \frac{36}{\mathbf{36}} \quad \text{Divide both sides by 36.}$$

$$\frac{(x - 2)^2}{9} + \frac{(y - 1)^2}{4} = 1 \quad \text{Simplify: } \frac{4}{36} = \frac{1}{9}, \frac{9}{36} = \frac{1}{4}, \text{ and } \frac{36}{36} = 1.$$

This is the standard form of the equation of a horizontal ellipse, centered at (2, 1), with $a = 3$ and $b = 2$. The graph of the ellipse is shown in the margin.

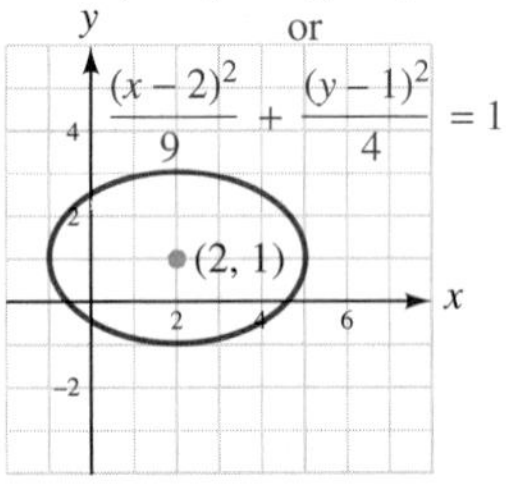

Teaching Example 4 Graph:
$4(x + 2)^2 + 25(y - 1)^2 = 100$
Answer:

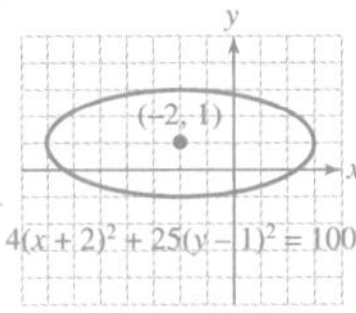

Self Check 4 Graph: $12(x - 1)^2 + 3(y + 1)^2 = 48$ $\frac{(x - 1)^2}{4} + \frac{(y + 1)^2}{16} = 1$; see AIE Appendix 3.

Now Try Problem 29

4 Solve Application Problems Involving Ellipses.

EXAMPLE 5 **Landscape Design.** A landscape architect is designing an elliptical pool that will fit in the center of a 20-by-30-foot rectangular garden, leaving 5 feet of clearance on all sides, as shown in the illustration on the next page. Find the equation of the ellipse.

Strategy We will establish a coordinate system with its origin at the center of the garden. Then we will determine the x- and y-intercepts of the edge of the pool.

Why If we know the x- and y-intercepts of the graph of the edge of the elliptical pool, we can use that information to write its equation.

Solution We place the rectangular garden in the coordinate system shown below. To maintain 5 feet of clearance at the ends of the ellipse, the x-intercepts must be the points $(10, 0)$ and $(-10, 0)$. Similarly, the y-intercepts are the points $(0, 5)$ and $(0, -5)$.

Since the ellipse is centered at the origin, its equation has the form

$$\frac{x^2}{a^2} + \frac{y^2}{b^2} = 1$$

with $a = 10$ and $b = 5$. Thus, the equation of the boundary of the pool is

$$\frac{x^2}{100} + \frac{y^2}{25} = 1$$

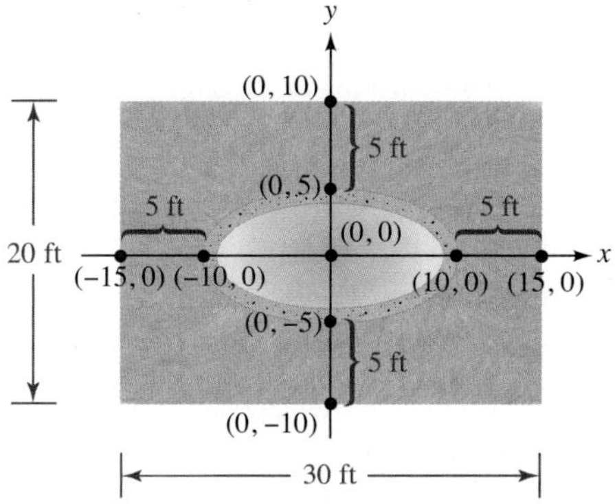

Teaching Example 5 Coffee Tables. A furniture maker is designing an elliptical pattern in the center of a 22-by-50-in. rectangular coffee table, leaving 3 in. of clearance on all sides. If he establishes a coordinate system with the length of the table being along the x-axis, and the center of the table at the origin, find the equation of the ellipse.

Answer: $\frac{x^2}{484} + \frac{y^2}{64} = 1$

Self Check 5 **Decorating.** An interior decorator is designing an elliptical mirror that will fit in the center of a 52-in. tall and 40-in. wide rectangular panel, leaving 2 in. of clearance on all sides. Find the equation of the ellipse. $\frac{x^2}{324} + \frac{y^2}{576} = 1$

Now Try Problem 59

Ellipses, like parabolas, have **reflective properties** that are used in many practical applications. For example, any light or sound originating at one focus of an ellipse is reflected by the interior of the figure to the other focus.

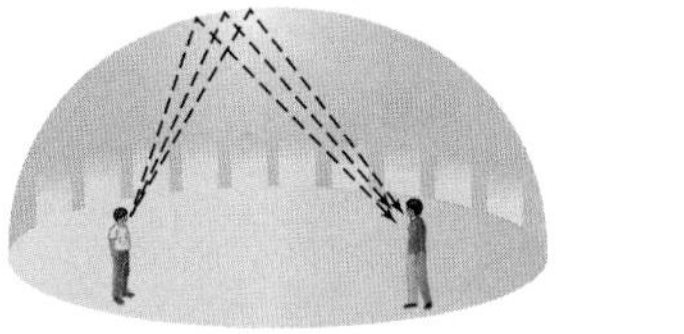

Whispering Galleries

In an elliptical dome, even the slightest whisper made by a person standing at one focus can be heard by a person standing at the other focus.

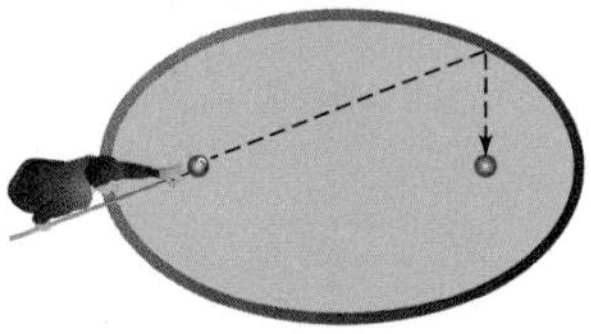

Elliptical billiards tables

When a ball is shot from one focus, it will rebound off the side of the table into a pocket located at the other focus.

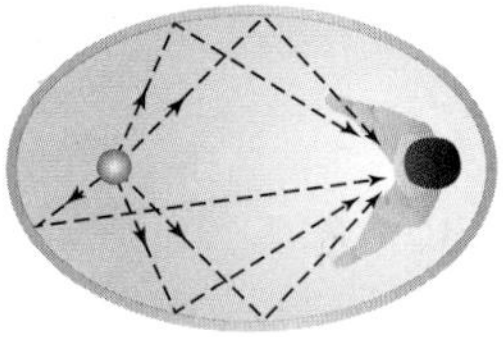

Treatment for kidney stones

The patient is positioned in an elliptical tank of water so that the kidney stone is at one focus. High-intensity sound waves generated at another focus are reflected to the stone to shatter it.

SECTION 13.2 STUDY SET

VOCABULARY

Fill in the blanks.

1. The curve graphed at the right is an ellipse.

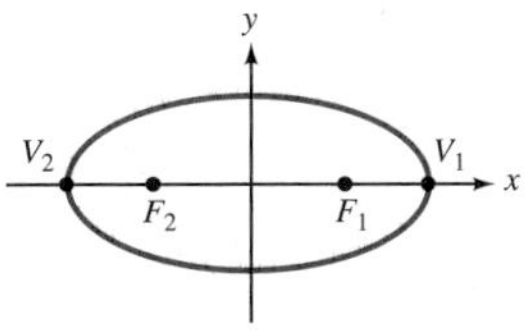

2. An ellipse is the set of all points in a plane for which the sum of the distances from two fixed points is a constant.
3. In the graph above, F_1 and F_2 are the foci of the ellipse. Each one is called a focus of the ellipse.
4. In the graph above, V_1 and V_2 are the vertices of the ellipse. Each one is called a vertex of the ellipse.
5. The line segment joining the vertices of an ellipse is called the major axis of the ellipse.
6. The midpoint of the major axis of an ellipse is the center of the ellipse.

CONCEPTS

7. Write the standard form of the equation of an ellipse centered at the origin and symmetric to both axes. $\frac{x^2}{a^2} + \frac{y^2}{b^2} = 1$
8. Write the standard form of the equation of a horizontal or vertical ellipse centered at (h, k). $\frac{(x-h)^2}{a^2} + \frac{(y-k)^2}{b^2} = 1$
9. Find the x- and the y-intercepts of the graph of $\frac{x^2}{a^2} + \frac{y^2}{b^2} = 1$.
x-intercepts: $(a, 0), (-a, 0)$; y-intercepts: $(0, b), (0, -b)$

Selected exercises available online at www.webassign.net/brookscole

▶ **10. a.** Find the center of the ellipse graphed on the right. What are a and b? $(0, 0); a = 3, b = 2$

b. Is the ellipse horizontal or vertical? Horizontal

c. Find the equation of the ellipse. $\frac{x^2}{9} + \frac{y^2}{4} = 1$

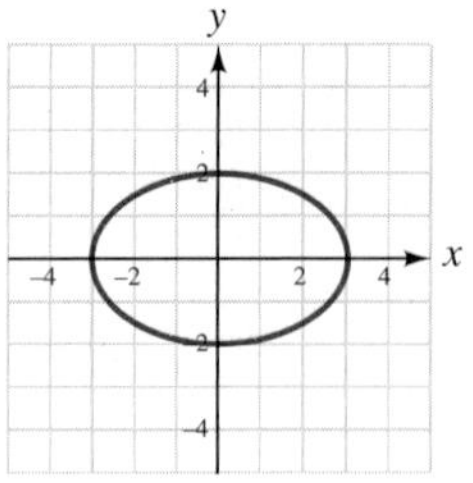

▶ **11. a.** Find the center of the ellipse graphed on the right. What are a and b? $(-2, 1); a = 2, b = 5$

b. Is the ellipse horizontal or vertical? Vertical

c. Find the equation of the ellipse. $\frac{(x + 2)^2}{4} + \frac{(y - 1)^2}{25} = 1$

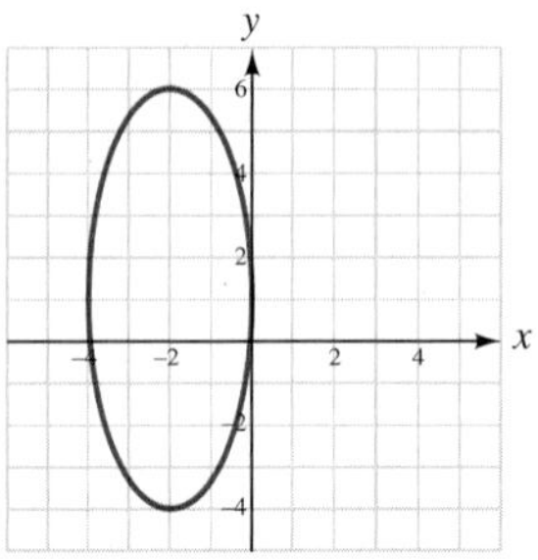

▶ **12.** Find two points on the graph of $\frac{x^2}{16} + \frac{y^2}{4} = 1$ by letting $x = 2$ and finding the corresponding values of y. $(2, \sqrt{3}), (2, -\sqrt{3})$

13. Divide both sides of the equation by 64 and write the equation in standard form:

$$4(x - 1)^2 + 64(y + 5)^2 = 64 \qquad \frac{(x - 1)^2}{16} + \frac{(y + 5)^2}{1} = 1$$

▶ **14.** Determine whether the graph of each equation is a circle, a parabola, or an ellipse.

a. $x = y^2 - 2y + 10$ Parabola

b. $\frac{x^2}{49} + \frac{y^2}{64} = 1$ Ellipse

c. $(x - 3)^2 + (y + 4)^2 = 25$ Circle

d. $2(x - 1)^2 + 8(y + 5)^2 = 32$ Ellipse

NOTATION

15. Find h, k, a, and b: $\frac{(x + 8)^2}{100} + \frac{(y - 6)^2}{144} = 1$

$h = -8, k = 6, a = 10, b = 12$

▶ **16.** Write each denominator in the equation $\frac{x^2}{81} + \frac{y^2}{49} = 1$ as the square of a number. $\frac{x^2}{9^2} + \frac{y^2}{7^2} = 1$

GUIDED PRACTICE

***Graph each equation.* See Example 1.** See AIE Appendix 3.

▶ **17.** $\frac{x^2}{25} + \frac{y^2}{4} = 1$ ▶ **18.** $\frac{x^2}{16} + \frac{y^2}{9} = 1$

▶ **19.** $\frac{x^2}{4} + \frac{y^2}{9} = 1$ ▶ **20.** $\frac{x^2}{16} + \frac{y^2}{25} = 1$

***Graph each equation.* See Example 2.** See AIE Appendix 3.

▶ **21.** $x^2 + 9y^2 = 9$ ▶ **22.** $25x^2 + 9y^2 = 225$

▶ **23.** $16x^2 + 4y^2 = 64$ ▶ **24.** $4x^2 + 9y^2 = 36$

***Graph each equation.* See Example 3.** See AIE Appendix 3.

▶ **25.** $\frac{(x - 2)^2}{9} + \frac{(y - 1)^2}{4} = 1$ ▶ **26.** $\frac{(x - 1)^2}{9} + \frac{(y - 3)^2}{4} = 1$

▶ **27.** $\frac{(x + 2)^2}{64} + \frac{(y - 2)^2}{100} = 1$ ▶ **28.** $\frac{(x - 6)^2}{36} + \frac{(y + 6)^2}{144} = 1$

***Graph each equation.* See Example 4.** See AIE Appendix 3.

▶ **29.** $(x + 1)^2 + 4(y + 2)^2 = 4$

▶ **30.** $25(x + 1)^2 + 9y^2 = 225$

31. $16(x - 2)^2 + 4(y + 4)^2 = 256$

32. $4(x - 2)^2 + 9(y - 4)^2 = 144$

***Use a graphing calculator to graph each equation.* See Using Your Calculator: Graphing Ellipses.** See AIE Appendix 3.

33. $\frac{x^2}{9} + \frac{y^2}{4} = 1$ **34.** $x^2 + 16y^2 = 16$

35. $\frac{x^2}{4} + \frac{(y - 1)^2}{9} = 1$ **36.** $\frac{(x + 1)^2}{9} + \frac{(y - 2)^2}{4} = 1$

TRY IT YOURSELF

Write each equation in standard form, if it is not already so, and graph it. The problems include equations that describe circles, parabolas, and ellipses. See AIE Appendix 3.

37. $(x + 1)^2 + (y - 2)^2 = 16$ **38.** $(x - 3)^2 + (y + 1)^2 = 25$

▶ **39.** $\frac{x^2}{16} + \frac{y^2}{1} = 1$ ▶ **40.** $\frac{x^2}{1} + \frac{y^2}{9} = 1$

41. $x = \frac{1}{2}(y - 1)^2 - 2$ **42.** $x = -\frac{1}{2}(y + 4)^2 + 5$

▶ **43.** $x^2 + y^2 - 25 = 0$ $x^2 + y^2 = 25$ **44.** $x^2 = 36 - y^2$ $x^2 + y^2 = 36$

45. $x^2 = 100 - 4y^2$ $\frac{x^2}{100} + \frac{y^2}{25} = 1$ **46.** $x^2 = 36 - 4y^2$ $\frac{x^2}{36} + \frac{y^2}{9} = 1$

47. $y = -3x^2 - 24x - 43$ $y = -3(x + 4)^2 + 5$ **48.** $y = 5x^2 - 60x + 173$ $y = 5(x - 6)^2 - 7$

49. $x^2 + y^2 - 2x + 4y - 4 = 0$ $(x - 1)^2 + (y + 2)^2 = 9$

50. $x^2 + y^2 + 4x + 6y + 9 = 0$ $(x + 2)^2 + (y + 3)^2 = 4$

51. $9(x - 1)^2 + 4(y + 2)^2 = 36$ $\frac{(x - 1)^2}{4} + \frac{(y + 2)^2}{9} = 1$

▶ **52.** $16(x - 5)^2 + 25(y - 4)^2 = 400$ $\frac{(x - 5)^2}{25} + \frac{(y - 4)^2}{16} = 1$

Look Alikes . . .

Graph the ellipses described by the equations in parts a and b on the same coordinate system. See AIE Appendix 3.

53. a. $\frac{x^2}{9} + \frac{y^2}{25} = 1$ **b.** $\frac{x^2}{25} + \frac{y^2}{9} = 1$

▶ **54. a.** $\frac{x^2}{169} + \frac{y^2}{25} = 1$ **b.** $\frac{x^2}{25} + \frac{y^2}{169} = 1$

55. a. $\frac{(x - 3)^2}{100} + \frac{(y - 2)^2}{36} = 1$ **b.** $\frac{(x + 3)^2}{100} + \frac{(y + 2)^2}{36} = 1$

56. a. $\frac{(x - 4)^2}{9} + \frac{(y + 5)^2}{4} = 1$ **b.** $\frac{(x + 4)^2}{9} + \frac{(y - 5)^2}{4} = 1$

APPLICATIONS

57. from **Campus to Careers**

Civil Engineer

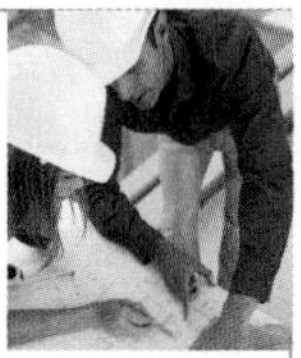

©Andresr/Shutterstock.com

The arch of an underpass shown below is part of an ellipse.

a. Find the equation of the ellipse. $\frac{x^2}{400} + \frac{y^2}{100} = 1$

b. Find the height of the arch at a point 10 feet to the right of the center line of the roadway. $5\sqrt{3}$ ft $\approx$ 8.7ft

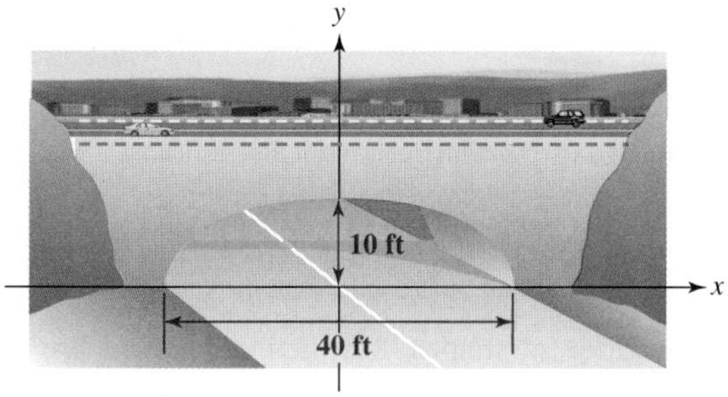

58. Fitness Equipment. With elliptical cross-training equipment, the feet move through the natural elliptical pattern that one experiences when walking, jogging, or running. Write the equation of the elliptical pattern shown below. $\frac{x^2}{144} + \frac{y^2}{25} = 1$

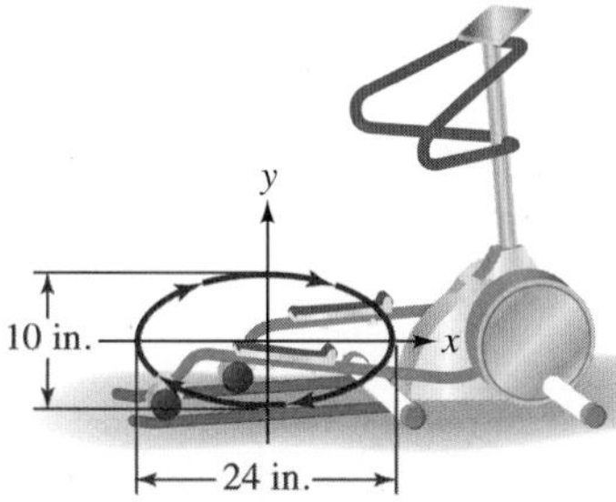

59. Koi Ponds. A landscape architect is designing an elliptical fish pond that will fit in the center of a 110-by-100-foot rectangular Japanese rock garden, leaving 15 feet of clearance on all sides. If she establishes a coordinate system with the 110-foot length along the x-axis, and the center of the pond at the origin, find the equation of the ellipse. $\frac{x^2}{1,600} + \frac{y^2}{1,225} = 1$

©Andy Lim/Shutterstock.com

60. Pool Tables. Find the equation of the outer edge of the elliptical pool table shown in the next column.
$\frac{x^2}{900} + \frac{y^2}{400} = 1$

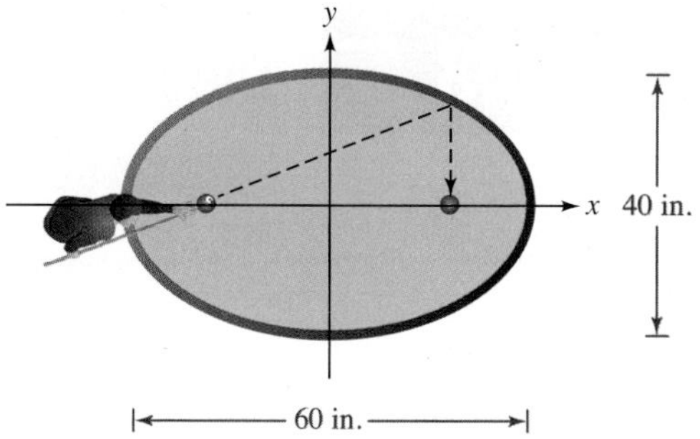

61. Area of an Ellipse. The area A bounded by an ellipse with the equation $\frac{x^2}{a^2} + \frac{y^2}{b^2} = 1$ is given by $A = \pi ab$. Find the area bounded by the ellipse described by $9x^2 + 16y^2 = 144$.
12π sq. units $\approx$ 37.7 sq. units

62. Area of a Track. The elliptical track shown in the figure is bounded by the ellipses $4x^2 + 9y^2 = 576$ and $9x^2 + 25y^2 = 900$. Find the area of the track. (See Exercise 61.) 36π sq. units $\approx$ 113.1 sq. units

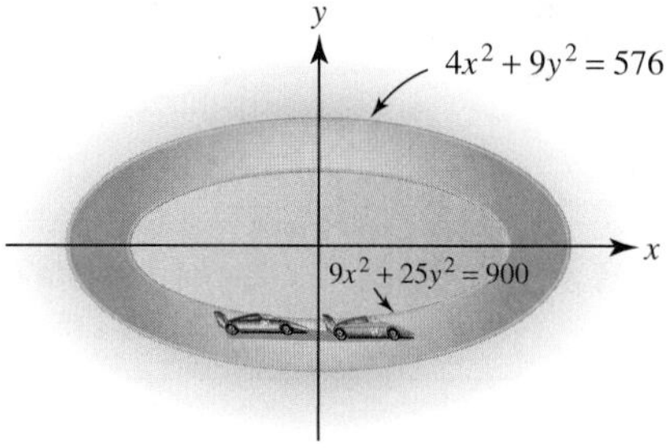

WRITING

63. What is an ellipse?

64. Explain the difference between the focus of an ellipse and the vertex of an ellipse.

65. Compare the graphs of $\frac{x^2}{81} + \frac{y^2}{64} = 1$ and $\frac{x^2}{64} + \frac{y^2}{81} = 1$. Do they have any similarities?

66. What are the reflective properties of an ellipse?

REVIEW

Find each product.

67. $3x^{-2}y^2(4x^2 + 3y^{-2})$ $\quad 12y^2 + \frac{9}{x^2}$

68. $(2a^{-2} - b^{-2})(2a^{-2} + b^{-2})$ $\quad \frac{4}{a^4} - \frac{1}{b^4}$

Simplify each expression.

69. $\dfrac{x^{-2} + y^{-2}}{x^{-2} - y^{-2}}$ $\quad \frac{y^2 + x^2}{y^2 - x^2}$

70. $\dfrac{2x^{-3} - 2y^{-3}}{4x^{-3} + 4y^{-3}}$ $\quad \frac{y^3 - x^3}{2(y^3 + x^3)}$

CHALLENGE PROBLEMS

71. What happens to the graph of the equation $\frac{x^2}{a^2} + \frac{y^2}{b^2} = 1$ when $a = b$? It is a circle.

72. Graph: $9x^2 + 4y^2 = 1$ See AIE Appendix 3.

73. Write the equation $9x^2 + 4y^2 - 18x + 16y = 11$ in the standard form of the equation of an ellipse.
$\frac{(x - 1)^2}{4} + \frac{(y + 2)^2}{9} = 1$

74. Let the foci of an ellipse be $(c, 0)$ and $(-c, 0)$. Suppose that the sum of the distances from any point (x, y) on the ellipse to the two foci is the constant $2a$. Show that the equation for the ellipse is $\frac{x^2}{a^2} + \frac{y^2}{a^2 - c^2} = 1$. Then let $b^2 = a^2 - c^2$ to obtain the standard form of the equation of an ellipse.

SECTION 13.3

The Hyperbola

OBJECTIVES

1. Define a hyperbola.
2. Graph hyperbolas centered at the origin.
3. Graph hyperbolas centered at (h, k).
4. Graph equations of the form $xy = k$.
5. Solve application problems involving hyperbolas.
6. Identify conic sections by their equations.

ARE YOU READY? *Are You Ready? exercises available online at www.webassign.net/brookscole*

The following problems review some basic skills that are needed when working with hyperbolas.

1. Solve: $b^2 = 16$ ± 4

2. Simplify: $x - (-2)$ $x + 2$

3. What is the slope of the line represented by $y = \frac{2}{3}x$? $\frac{2}{3}$

4. If $xy = 20$, find y when $x = -2$. -10

The final conic section that we will discuss, the *hyperbola,* is a curve that has two branches. In this section, we will learn how to graph equations that represent hyperbolas.

1 Define a Hyperbola.

Ellipses and hyperbolas have completely different shapes, but their definitions are similar. Instead of the *sum* of distances, the definition of a hyperbola involves a *difference* of distances.

Definition of a Hyperbola

A **hyperbola** is the set of all points in a plane for which the difference of the distances from two fixed points is a constant.

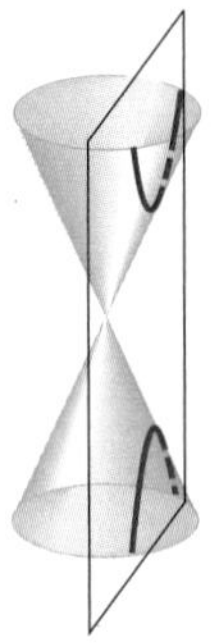

The figure below illustrates that any point P on the hyperbola is a constant distance $d_1 - d_2$ from two fixed points, each of which is called a **focus.** Midway between the **foci** is the **center** of the hyperbola.

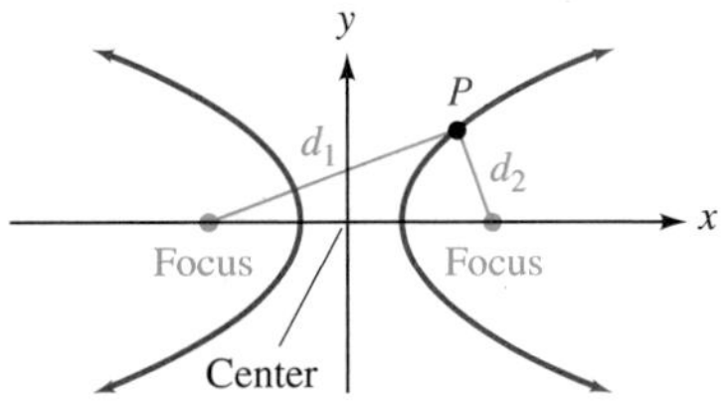

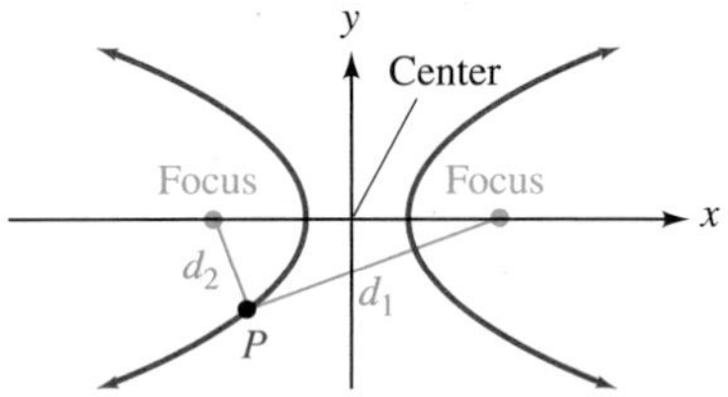

2 Graph Hyperbolas Centered at the Origin.

The graph of the equation

$$\frac{x^2}{25} - \frac{y^2}{9} = 1$$

is a hyperbola. To graph the equation, we make a table of solutions that satisfy the equation, plot each point, and join them with a smooth curve.

$\frac{x^2}{25} - \frac{y^2}{9} = 1$

x	y	
-7	± 2.9	→ $(-7, \pm 2.9)$
-6	± 2.0	→ $(-6, \pm 2.0)$
-5	0	→ $(-5, 0)$
5	0	→ $(5, 0)$
6	± 2.0	→ $(6, \pm 2.0)$
7	± 2.9	→ $(7, \pm 2.9)$

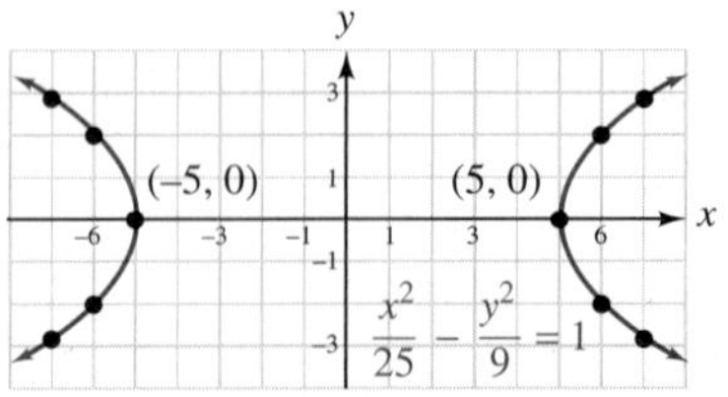

Teaching Tip: Ask if this is the graph of a function.

This graph is centered at the origin and intersects the x-axis at (5, 0) and (−5, 0). We also note that the graph does not intersect the y-axis.

It is possible to draw a hyperbola without plotting points. For example, if we want to graph the hyperbola with an equation of

$$\frac{x^2}{a^2} - \frac{y^2}{b^2} = 1$$

Teaching Tip: Point out why the slope of an asymptote is $\frac{b}{a}$ using a slope triangle.

we first find the x- and y-intercepts. To find the x-intercepts, we let $y = 0$ and solve for x:

$$\frac{x^2}{a^2} - \frac{0^2}{b^2} = 1$$

$$x^2 = a^2$$

$$x = \pm a \quad \text{Use the square root property.}$$

The hyperbola crosses the x-axis at the points $V_1(a, 0)$ and $V_2(-a, 0)$, called the **vertices** of the hyperbola.

To attempt to find the y-intercepts, we let $x = 0$ and solve for y:

$$\frac{0^2}{a^2} - \frac{y^2}{b^2} = 1$$

$$y^2 = -b^2$$

$$y = \pm\sqrt{-b^2}$$

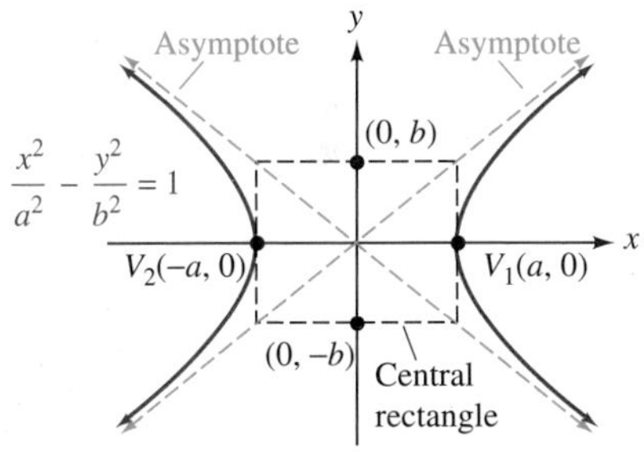

Since b^2 is always positive, $\sqrt{-b^2}$ is an imaginary number. This means that the hyperbola does not intersect the y-axis.

As shown in the illustration above, we can construct a rectangle, called the **central rectangle,** whose sides pass horizontally through $\pm b$ on the y-axis and vertically through $\pm a$ on the x-axis. The extended diagonals of the rectangle are a pair of intersecting straight lines called **asymptotes** of the hyperbola. As the hyperbola gets farther away from the origin, its branches get closer and closer to the asymptotes. Asymptotes should be drawn as dashed lines because they are not part of the hyperbola, but they do serve as a helpful guide when drawing its graph. Since the slopes of the diagonals are $\frac{b}{a}$ and $-\frac{b}{a}$, and since the diagonals pass through the origin, the equations of the asymptotes are

$$y = \frac{b}{a}x \quad \text{and} \quad y = -\frac{b}{a}x$$

The Language of Algebra

The central rectangle is also called the **fundamental rectangle.**

Standard Form of the Equation of a Horizontal Hyperbola Centered at the Origin

The equation $\frac{x^2}{a^2} - \frac{y^2}{b^2} = 1$ has a graph that is a hyperbola centered at the origin. The x-intercepts are the vertices $V_1(a, 0)$ and $V_2(-a, 0)$. There are no y-intercepts.

The asymptotes of the hyperbola are the extended diagonals of the central rectangle, and their equations are $y = \frac{b}{a}x$ and $y = -\frac{b}{a}x$.

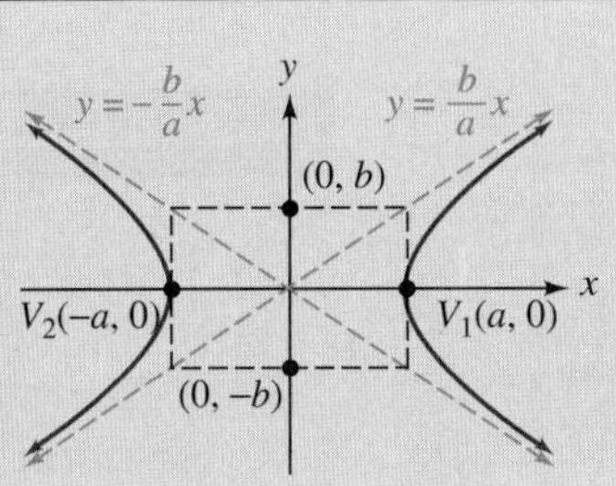

The branches of the hyperbola in previous discussions open to the left and to the right. It is possible for hyperbolas to have different orientations with respect to the x- and y-axes. For example, the branches of a hyperbola can open upward and downward. In that case, the following equation applies.

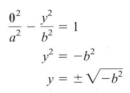

Standard Form of the Equation of a Vertical Hyperbola Centered at the Origin

The equation $\frac{y^2}{a^2} - \frac{x^2}{b^2} = 1$ has a graph that is a hyperbola centered at the origin. The y-intercepts are the vertices $V_1(0, a)$ and $V_2(0, -a)$. There are no x-intercepts.

The asymptotes of the hyperbola are the extended diagonals of the central rectangle, and their equations are $y = \frac{a}{b}x$ and $y = -\frac{a}{b}x$.

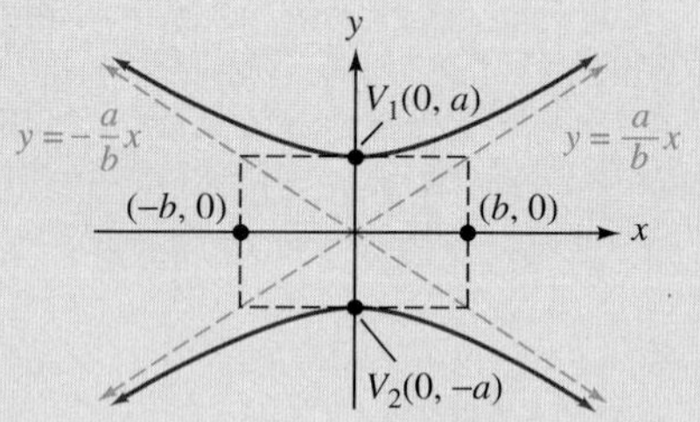

EXAMPLE 1 Graph: $\frac{x^2}{9} - \frac{y^2}{16} = 1$

Teaching Tip: Point out that when graphing an ellipse, we used a and b to determine its length and width. Here we use a and b to determine the length and width of the central rectangle.

Strategy This equation is in standard $\frac{x^2}{a^2} - \frac{y^2}{b^2} = 1$ form. We will identify a and b.

Why We can use a and b to find the vertices of the graph of the hyperbola and the location of the central rectangle.

Solution The color highlighting shows how to compare the given equation with the standard form to find a and b.

$$\frac{x^2}{\mathbf{9}} - \frac{y^2}{\mathbf{16}} = 1 \qquad \frac{x^2}{\mathbf{a^2}} - \frac{y^2}{\mathbf{b^2}} = 1$$

Since $a^2 = 9$, it follows that $a = 3$. Since $b^2 = 16$, it follows that $b = 4$.

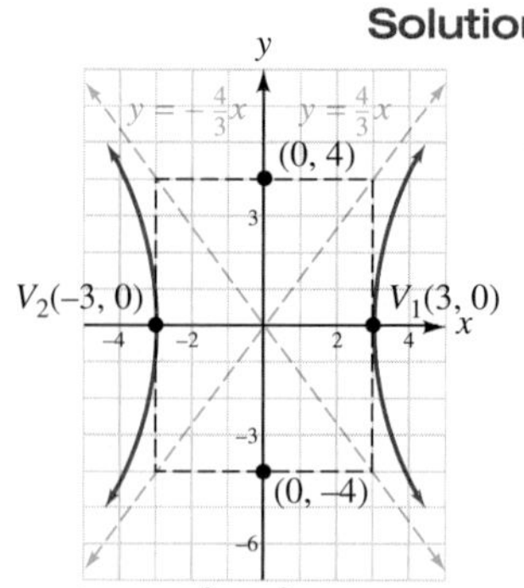

$\frac{x^2}{9} - \frac{y^2}{16} = 1$

This is the standard form of the equation of a hyperbola, centered at the origin, that opens left and right. The x-intercepts are $(a, 0)$ and $(-a, 0)$, or $(3, 0)$ and $(-3, 0)$. They are also the vertices of the hyperbola.

To construct the central rectangle, we use the values of $a = 3$ and $b = 4$. The rectangle passes through $(3, 0)$ and $(-3, 0)$ on the x-axis, and $(0, 4)$ and $(0, -4)$ on the y-axis. We draw extended diagonal dashed lines through the rectangle to obtain the asymptotes and write their equations: $y = \frac{4}{3}x$ and $y = -\frac{4}{3}x$. Then we draw a smooth curve through each vertex that gets close to the asymptotes.

Teaching Tip: Ask if this is the graph of a function.

Teaching Example 1 Graph: $\frac{x^2}{25} - \frac{y^2}{9} = 1$

Answer:

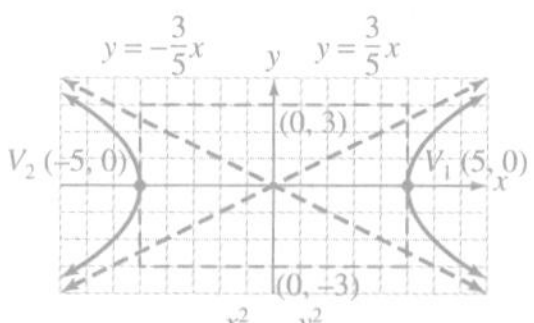

$\frac{x^2}{25} - \frac{y^2}{9} = 1$

Self Check 1 Graph: $\frac{x^2}{25} - \frac{y^2}{4} = 1$ See AIE Appendix 3.

Now Try Problem 17

EXAMPLE 2 Graph: $9y^2 - 4x^2 = 36$

Strategy We will write the equation in standard $\frac{y^2}{a^2} - \frac{x^2}{b^2} = 1$ form.

Why When the equation is in standard form, we will be able to identify the center and the vertices of the graph of the hyperbola and the location of the central rectangle.

Solution To write the equation in standard form, we divide both sides by 36.

$$9y^2 - 4x^2 = 36$$

$$\frac{9y^2}{36} - \frac{4x^2}{36} = \frac{36}{36} \qquad \text{To get a 1 on the right side, divide both sides by 36.}$$

The positive variable term in the standard form equation determines whether a hyperbola is vertical or horizontal. In this example, the positive variable term involves y, so the hyperbola is vertical.

$$\frac{y^2}{4} - \frac{x^2}{9} = 1 \qquad \text{Simplify each fraction.}$$

Caution

Although the branches of a hyperbola look like parabolas, they are not two parabolas. The curvature of each branch of a hyperbola is different from that of a parabola. A parabola is not constrained by asymptotes as are the branches of a hyperbola.

This is the standard form of the equation of a hyperbola, centered at the origin, that opens up and down. The color highlighting shows how we compare the resulting equation to the standard form to find a and b.

$$\frac{y^2}{4} - \frac{x^2}{9} = 1 \qquad \frac{y^2}{a^2} - \frac{x^2}{b^2} = 1$$

Since $a^2 = 4$, it follows that $a = 2$. Since $b^2 = 9$, it follows that $b = 3$.

The y-intercepts are $(0, a)$ and $(0, -a)$, or $(0, 2)$ and $(0, -2)$. They are also the vertices of the hyperbola.

Since $a = 2$ and $b = 3$, the central rectangle passes through $(0, 2)$ and $(0, -2)$, as well as $(3, 0)$ and $(-3, 0)$. We draw its extended diagonals and sketch the hyperbola.

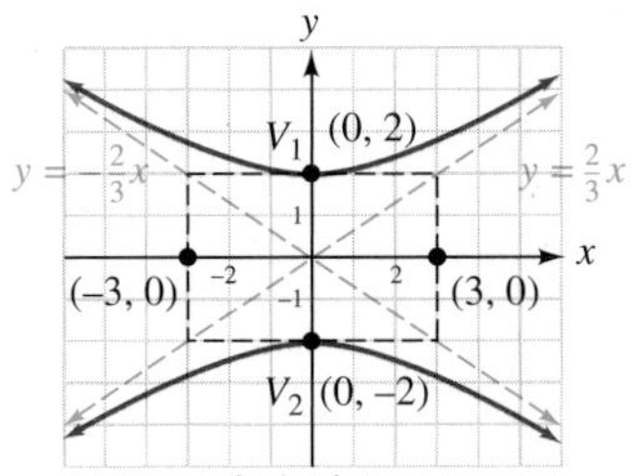

$9y^2 - 4x^2 = 36$
or
$\frac{y^2}{4} - \frac{x^2}{9} = 1$

Teaching Example 2 Graph: $16y^2 - 9x^2 = 144$
Answer:

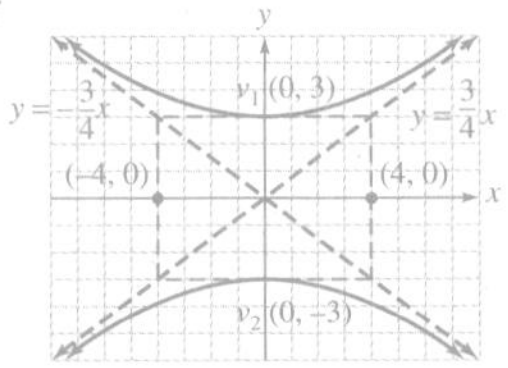

$\frac{y^2}{9} - \frac{x^2}{16} = 1$

Self Check 2 Graph: $16y^2 - x^2 = 16$ $\quad \frac{y^2}{1} - \frac{x^2}{16} = 1$; see AIE Appendix 3.

Now Try Problem 21

Using Your Calculator ▶ Graphing Hyperbolas

To graph $\frac{x^2}{9} - \frac{y^2}{16} = 1$ from Example 1 using a graphing calculator, we follow the same procedure that we used for circles and ellipses. To write the equation as two functions, we solve for y to get $y = \pm\frac{\sqrt{16x^2 - 144}}{3}$. Then we graph the following two functions in a square window setting to get the graph of the hyperbola shown at the right.

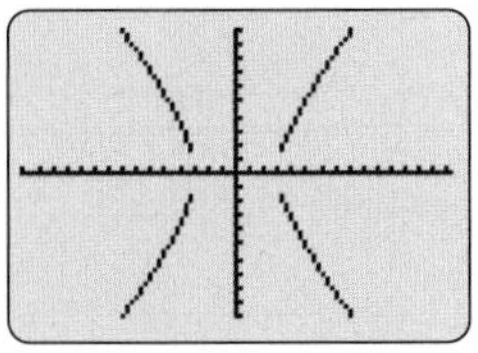

$$y = \frac{\sqrt{16x^2 - 144}}{3} \quad \text{and} \quad y = -\frac{\sqrt{16x^2 - 144}}{3}$$

3 Graph Hyperbolas Centered at (*h*, *k*).

If a hyperbola is centered at a point with coordinates (h, k), the following equations apply.

Standard Form of the Equation of a Hyperbola Centered at (*h*, *k*)

The equation $\frac{(x - h)^2}{a^2} - \frac{(y - k)^2}{b^2} = 1$ has a graph that is a hyperbola centered at (h, k) and that opens left and right.

The equation $\frac{(y - k)^2}{a^2} - \frac{(x - h)^2}{b^2} = 1$ has a graph that is a hyperbola centered at (h, k) and that opens up and down.

EXAMPLE 3 Graph: **a.** $\frac{(x-3)^2}{16} - \frac{(y+1)^2}{4} = 1$ **b.** $\frac{(y-2)^2}{9} - \frac{(x-1)^2}{9} = 1$

Strategy We will write each equation in a form that makes it easy to identify h, k, a, and b.

Why If we know h, k, a, and b, we can graph the hyperbola and the central rectangle.

Solution **a.** We can write the given equation as

$$\frac{(x-3)^2}{4^2} - \frac{[y-(-1)]^2}{2^2} = 1$$

To find k, write $y + 1$ as $y - (-1)$.
To find a, write 16 as 4^2. To find b, write 4 as 2^2.

Success Tip

When the equation of a hyperbola is written in standard form, h and k are the numbers within the parentheses that *follow* the − (subtraction) symbols.

Teaching Tip: Before graphing, ask your students to determine whether the hyperbola will be horizontal or vertical.

Because the term involving x is positive, the hyperbola opens left and right. We find the center by examining $(x-3)^2$ and $[y-(-1)]^2$. Since $h = 3$ and $k = -1$, the hyperbola is centered at $(h, k) = (3, -1)$. From the denominators, 4^2 and 2^2, we find that $a = 4$ and $b = 2$. Thus, its vertices are located 4 units to the right and left of the center, at $(7, -1)$ and $(-1, -1)$. Since $b = 2$, we can count 2 units above and below the center to locate points $(3, 1)$ and $(3, -3)$. With these four points, we can draw the central rectangle along with its extended diagonals (the asymptotes). We then can sketch the hyperbola, as shown.

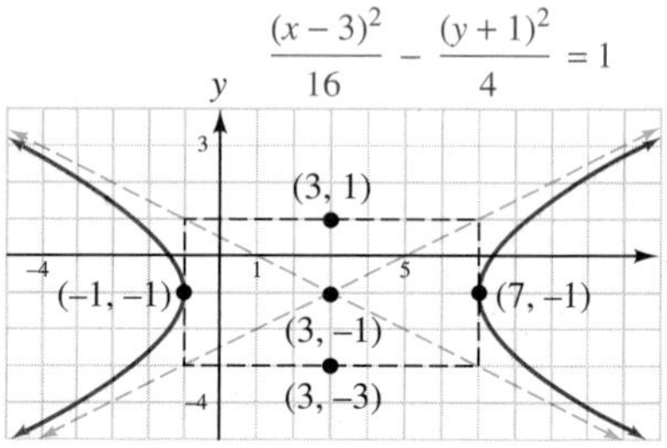

Caution

When sketching the graph of a hyperbola, the branches should not touch the asymptotes.

b. We can write the given equation as

$$\frac{(y-2)^2}{3^2} - \frac{(x-1)^2}{3^2} = 1$$

Because the term involving y is positive, the hyperbola opens up and down. We find its center by examining $(y-2)^2$ and $(x-1)^2$. Since $k = 2$ and $h = 1$, the hyperbola is centered at $(h, k) = (1, 2)$. From the denominators, 3^2 and 3^2, we find that $a = 3$ and $b = 3$, and we use that information to draw the central rectangle and its extended diagonals (the asymptotes), as shown.

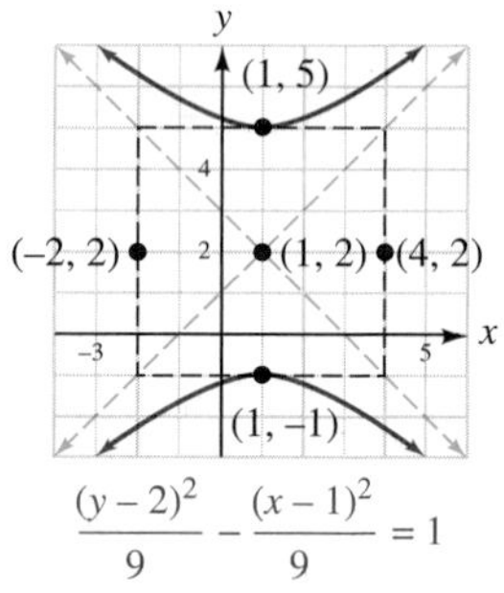

Teaching Example 3 Graph
a. $\frac{(x+2)^2}{25} - \frac{(y-1)^2}{9} = 1$
b. $\frac{(y-3)^2}{16} - \frac{(x-1)^2}{9} = 1$
Answers:
a.

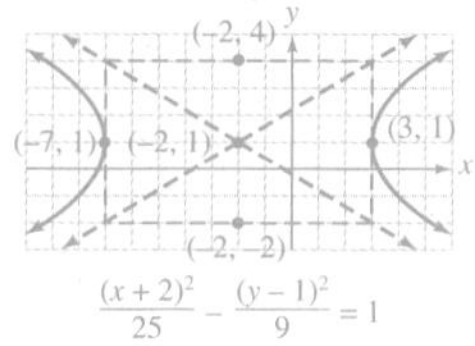

b.

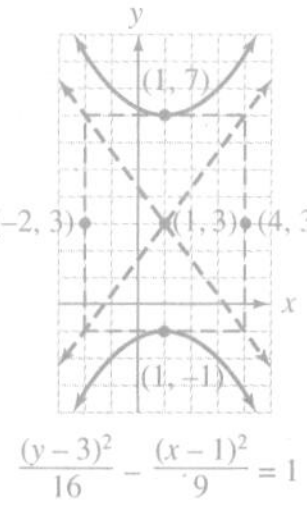

Self Check 3 Graph: **a.** $\frac{(x+2)^2}{9} - \frac{(y-1)^2}{4} = 1$ **b.** $\frac{(y+1)^2}{1} - \frac{(x+1)^2}{4} = 1$

See AIE Appendix 3.

Now Try Problems 25 and 27

4 Graph Equations of the Form $xy = k$.

There is a special type of hyperbola (also centered at the origin) that does not intersect either the x- or the y-axis. These hyperbolas have equations of the form $xy = k$, where $k \neq 0$.

EXAMPLE 4 Graph: $xy = -8$

Strategy We will make a table of solutions, plot the points, and connect the points with a smooth curve.

Why Since this equation cannot be written in standard form, we cannot use the methods used in the previous examples.

Solution To make a table of solutions, we can solve the given equation for y to get $y = \frac{-8}{x}$. Then we choose several values for x, find the corresponding values of y, and record the results in the table below. We plot the ordered pairs and join them with a smooth curve to obtain the graph of the hyperbola.

$$xy = -8 \quad \text{or} \quad y = \frac{-8}{x}$$

x	y	
1	−8	→ (1, −8)
2	−4	→ (2, −4)
3	$-\frac{8}{3}$	→ (3, −2.7)
4	−2	→ (4, −2)
8	−1	→ (8, −1)
−1	8	→ (−1, 8)
−2	4	→ (−2, 4)
−3	$\frac{8}{3}$	→ (−3, 2.7)
−4	2	→ (−4, 2)
−8	1	→ (−8, 1)

The y-values for $x = 3$ and $x = -3$ were approximated.

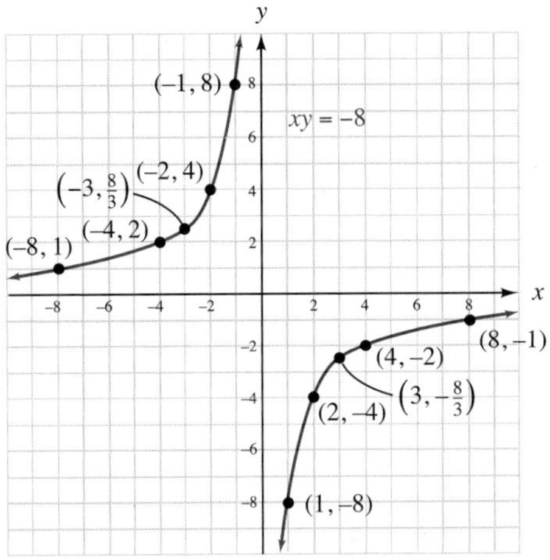

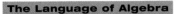

The Language of Algebra

The asymptotes of this hyperbola are the x- and y-axes. A hyperbola for which the aymptotes are perpendicular is called a **rectangular hyperbola.**

Teaching Example 4 Graph: $xy = -6$
Answer:

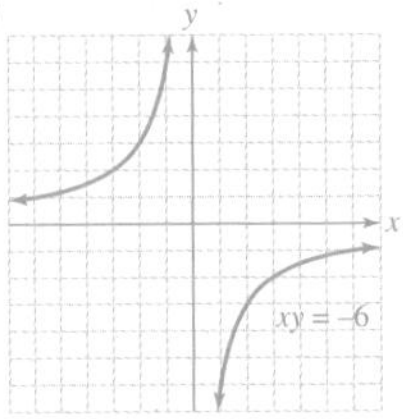

Self Check 4 Graph: $xy = 6$ See AIE Appendix 3.

Now Try Problem 33

The result in Example 4 illustrates the following general equation.

Equations of Hyperbolas of the Form $xy = k$

Any equation of the form $xy = k$, where $k \neq 0$, has a graph that is a **hyperbola,** which does not intersect either the x- or y-axis.

5 Solve Application Problems Involving Hyperbolas.

EXAMPLE 5

Atomic Structure. In an experiment that led to the discovery of the atomic structure of matter, Lord Rutherford (1871–1937) shot high-energy alpha particles toward a thin sheet of gold. Many of them were reflected, and Rutherford showed the existence of the nucleus of a gold atom. An alpha particle is repelled by the nucleus at the origin; it travels along the hyperbolic path given by $4x^2 - y^2 = 16$. How close does the particle come to the nucleus?

Strategy We will write the equation in standard form and find the coordinates of point V.

Why The distance from the origin to point V is the closest the particle comes to the nucleus.

Solution To find the distance from the nucleus at the origin, we must find the coordinates of the vertex V. To do so, we write the equation of the particle's path in standard form:

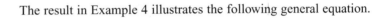

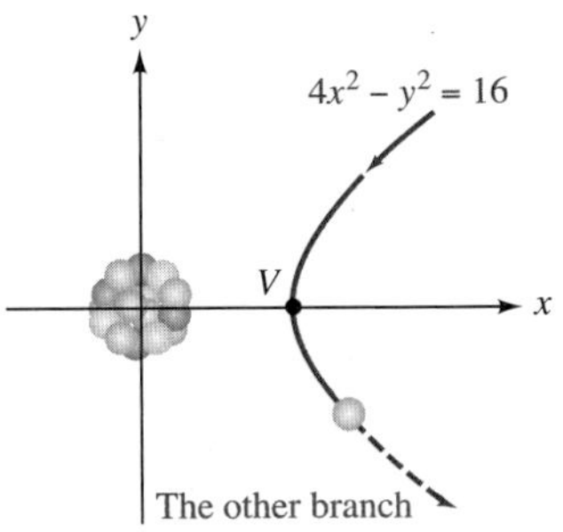

The other branch of the hyperbola is not shown.

$$4x^2 - y^2 = 16 \quad \text{This is the given equation.}$$

$$\frac{4x^2}{16} - \frac{y^2}{16} = \frac{16}{16} \quad \text{Divide both sides by 16, term-by-term.}$$

$$\frac{x^2}{4} - \frac{y^2}{16} = 1 \quad \text{Simplify the fractions.}$$

$$\frac{x^2}{2^2} - \frac{y^2}{4^2} = 1 \quad \text{To determine } a \text{ and } b\text{, write 4 as } 2^2 \text{ and 16 as } 4^2.$$

This equation is in the form $\frac{x^2}{a^2} - \frac{y^2}{b^2} = 1$ with $a = 2$. Thus, the vertex of the path is (2, 0). The particle is never closer than 2 units from the nucleus.

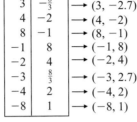

Teaching Example 5 Atomic Structure. An alpha particle is repelled by the nucleus at the origin and it travels along the hyperbolic path given by $9x^2 - y^2 = 81$. How close does the particle come to the nucleus?
Answer: 3 units

Self Check 5 **Astronomy.** Some comets have a hyperbolic orbit, with the sun as one focus and Earth at the center. For one such comet, the equation of its path is $\frac{x^2}{1 \times 10^{18}} - \frac{y^2}{2 \times 10^{18}} = 1$. The units are miles. How close does this comet come to Earth? 1×10^9 mi

Now Try Problem 61

6 Identify Conic Sections by Their Equations.

We can determine whether an equation, when graphed, will be a circle, a parabola, an ellipse, or a hyperbola by examining its variable terms.

$x^2 + y^2 = 16$ With the variable terms on the same side of the equation, we see that the coefficients of the squared terms are the same. The graph is a circle.

$4x^2 + 9y^2 = 144$ With the variable terms on the same side of the equation, we see that the coefficients of the squared terms are different, but have the same sign. The graph is an ellipse.

$4x^2 - 9y^2 = 144$ With the variable terms on the same side of the equation, we see that the coefficients of the squared terms have different signs. The graph is a hyperbola.

$x = y^2 + y - 16$ Since one variable is squared and the other is not, the graph is a parabola.

SECTION 13.3 STUDY SET

VOCABULARY

Fill in the blanks.

1. The two-branch curve graphed on the right is a hyperbola.

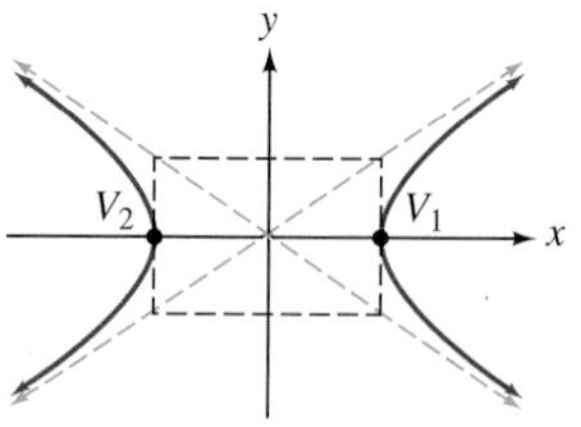

2. A hyperbola is the set of all points in a plane for which the difference of the distances from two fixed points is a constant.
3. In the graph above, V_1 and V_2 are the vertices of the hyperbola.
4. In the graph above, the figure drawn using dashed black lines is called the central rectangle.
5. The extended diagonals of the central rectangle are asymptotes of the hyperbola.
6. To write $9x^2 - 4y^2 = 36$ in standard form, we divide both sides by 36.

CONCEPTS

7. Write the standard form of the equation of a hyperbola centered at the origin that opens left and right. $\frac{x^2}{a^2} - \frac{y^2}{b^2} = 1$
8. Write the standard form of the equation of a hyperbola centered at (h, k) that opens up and down. $\frac{(y - k)^2}{a^2} - \frac{(x - h)^2}{b^2} = 1$
9. Write the standard form of the equation of a hyperbola centered at (h, k) that opens left and right. $\frac{(x - h)^2}{a^2} - \frac{(y - k)^2}{b^2} = 1$
10. a. Find the center of the hyperbola graphed on the right. What are a and b? $(0, 0)$; $a = 2$, $b = 4$
b. Find the x-intercepts of the graph. What are the y-intercepts of the graph? $(2, 0), (-2, 0)$; none
c. Find the equation of the hyperbola. $\frac{x^2}{4} - \frac{y^2}{16} = 1$
d. Find the equations of the asymptotes. $y = 2x, y = -2x$

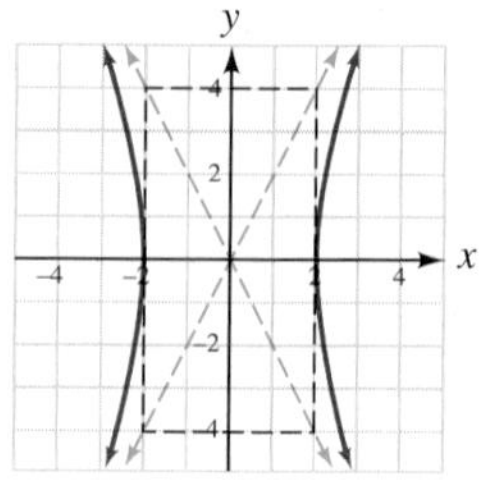

11. a. Find the center of the hyperbola graphed on the right. What are a and b? $(-1, -2)$; $a = 3$, $b = 1$
b. Find the equation of the hyperbola. $\frac{(y + 2)^2}{9} - \frac{(x + 1)^2}{1} = 1$

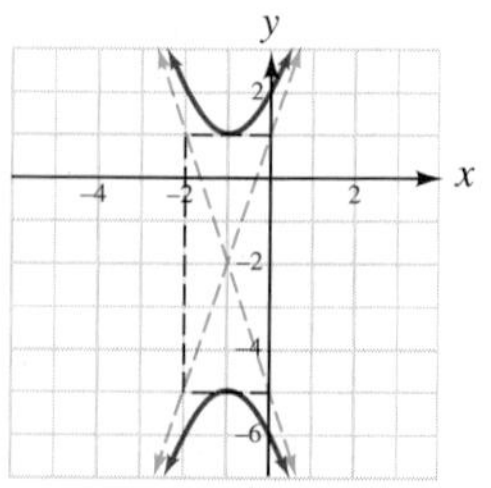

12. a. Fill in the blank: An equation of the form $xy = k$, where $k \neq 0$, has a graph that is a hyperbola that does not intersect either the x-axis or the y-axis.
b. Complete the table of solutions for $xy = 10$.

x	y
-2	-5
5	2

13. Divide both sides of the equation by 100 and write the equation in standard form:

$$100(x + 1)^2 - 25(y - 5)^2 = 100 \quad \frac{(x+1)^2}{1} - \frac{(y-5)^2}{4} = 1$$

14. Determine whether the graph of the equation will be a circle, a parabola, an ellipse, or a hyperbola.
a. $x^2 + y^2 = 10$ Circle
b. $9y^2 - 16x^2 = 144$ Hyperbola
c. $x = y^2 - 3y + 6$ Parabola
d. $4x^2 + 25y^2 = 100$ Ellipse

NOTATION

15. Find h, k, a, and b: $\frac{(x-5)^2}{25} - \frac{(y+11)^2}{36} = 1$
$h = 5, k = -11, a = 5, b = 6$

16. Write each denominator in the equation $\frac{x^2}{36} - \frac{y^2}{81} = 1$ as the square of a number. $\frac{x^2}{6^2} - \frac{y^2}{9^2} = 1$

GUIDED PRACTICE

***Graph each hyperbola.* See Example 1.** See AIE Appendix 3.

17. $\frac{x^2}{9} - \frac{y^2}{4} = 1$
18. $\frac{x^2}{4} - \frac{y^2}{4} = 1$
19. $\frac{y^2}{4} - \frac{x^2}{9} = 1$
20. $\frac{y^2}{4} - \frac{x^2}{64} = 1$

***Graph each hyperbola.* See Example 2.** See AIE Appendix 3.

21. $y^2 - 4x^2 = 16$
22. $9y^2 - 25x^2 = 225$
23. $25x^2 - y^2 = 25$
24. $9x^2 - 4y^2 = 36$

***Graph each hyperbola.* See Example 3.** See AIE Appendix 3.

25. $\frac{(x-2)^2}{9} - \frac{y^2}{16} = 1$
26. $\frac{(x+2)^2}{16} - \frac{(y-3)^2}{25} = 1$
27. $\frac{(y+1)^2}{1} - \frac{(x-2)^2}{4} = 1$
28. $\frac{(y-2)^2}{4} - \frac{(x+1)^2}{1} = 1$
29. $\frac{(x+1)^2}{9} - \frac{(y+1)^2}{9} = 1$
30. $\frac{(x-2)^2}{16} - \frac{(y-1)^2}{16} = 1$
31. $\frac{(y-3)^2}{25} - \frac{x^2}{25} = 1$
32. $\frac{(y-1)^2}{9} - \frac{x^2}{9} = 1$

***Graph each equation.* See Example 4.** See AIE Appendix 3.

33. $xy = 8$
34. $xy = 4$
35. $xy = -10$
36. $xy = -12$

***Use a graphing calculator to graph each equation.* See Using Your Calculator: Graphing Hyperbolas.** See AIE Appendix 3.

37. $\frac{x^2}{9} - \frac{y^2}{4} = 1$
38. $y^2 - 16x^2 = 16$
39. $\frac{x^2}{4} - \frac{(y-1)^2}{9} = 1$
40. $\frac{(y+1)^2}{9} - \frac{(x-2)^2}{4} = 1$

TRY IT YOURSELF

Write each equation in standard form, if it is not already so, and graph it. The problems include equations that describe circles, parabolas, ellipses, and hyperbolas. See AIE Appendix 3.

41. $(x + 1)^2 + (y - 2)^2 = 16$
42. $(x - 3)^2 + (y + 4)^2 = 1$
43. $9x^2 - 49y^2 = 441$ $\frac{x^2}{49} - \frac{y^2}{9} = 1$
44. $25y^2 - 16x^2 = 400$ $\frac{y^2}{16} - \frac{x^2}{25} = 1$
45. $4(x + 1)^2 + 9(y + 1)^2 = 36$ $\frac{(x+1)^2}{9} + \frac{(y+1)^2}{4} = 1$
46. $16x^2 + 25(y - 3)^2 = 400$ $\frac{x^2}{25} + \frac{(y-3)^2}{16} = 1$
47. $4(x + 3)^2 - (y - 1)^2 = 4$ $\frac{(x+3)^2}{1} - \frac{(y-1)^2}{4} = 1$
48. $(x + 5)^2 - 16y^2 = 16$ $\frac{(x+5)^2}{16} - \frac{y^2}{1} = 1$
49. $xy = -6$
50. $xy = 10$
51. $x = \frac{1}{2}(y - 1)^2 - 2$
52. $x = -\frac{1}{4}(y - 3)^2 + 2$
53. $\frac{y^2}{25} - \frac{(x-2)^2}{4} = 1$
54. $\frac{y^2}{36} - \frac{(x+2)^2}{4} = 1$
55. $y = -x^2 + 6x - 4$ $y = -(x - 3)^2 + 5$
56. $y = x^2 - 2x + 5$ $y = (x - 1)^2 + 4$
57. $\frac{x^2}{1} + \frac{y^2}{36} = 1$
58. $\frac{x^2}{4} + \frac{y^2}{16} = 1$
59. $x^2 + y^2 + 4x - 6y - 23 = 0$ $(x + 2)^2 + (y - 3)^2 = 36$
60. $x^2 + y^2 + 8x - 2y - 8 = 0$ $(x + 4)^2 + (y - 1)^2 = 25$

APPLICATIONS

61. **Alpha Particles.** The particle in the illustration below approaches the nucleus at the origin along the path $9y^2 - x^2 = 81$ in the coordinate system shown. How close does the particle come to the nucleus? 3 units

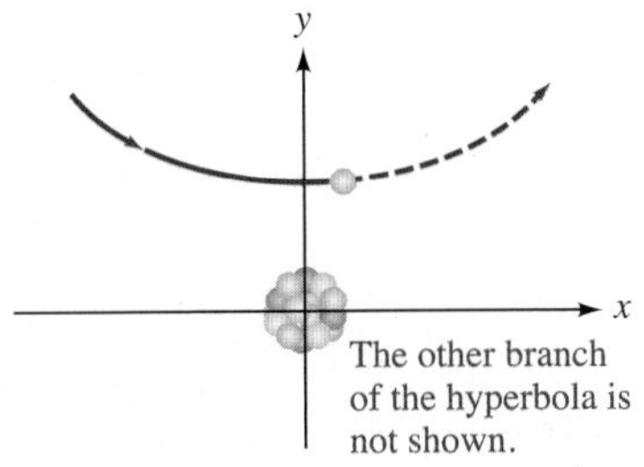
y
x
The other branch of the hyperbola is not shown.

62. **LORAN.** By determining the difference of the distances between the ship in the illustration and two radio transmitters, the LORAN navigation system places the ship on the hyperbola $x^2 - 4y^2 = 576$ in the coordinate system shown. If the ship is 5 miles out to sea, find its coordinates. (26, 5)

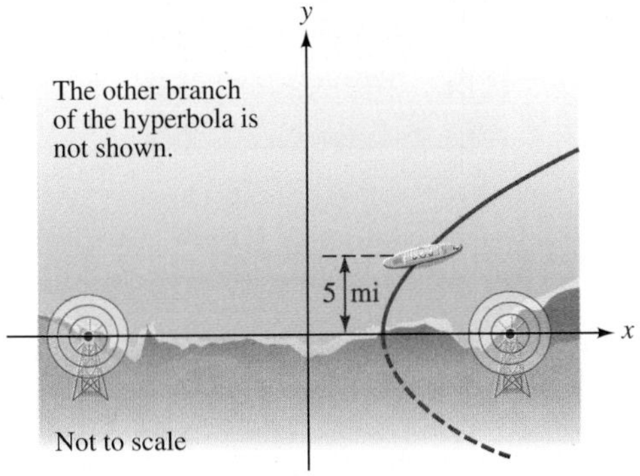
y
x
The other branch of the hyperbola is not shown.
5 mi
Not to scale

63. **from Campus to Careers**

Civil Engineer

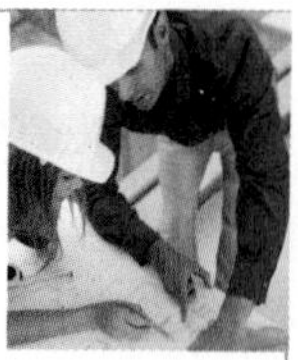

A new subdivision of homes is planned for an area that military jets often fly over. On the coordinate system shown below, a sonic boom is heard by those on the ground within one branch of the hyperbola $y^2 - x^2 = 25$. How wide is the hyperbola 5 miles from its vertex? $10\sqrt{3}$ miles $\approx$ 17.3 miles

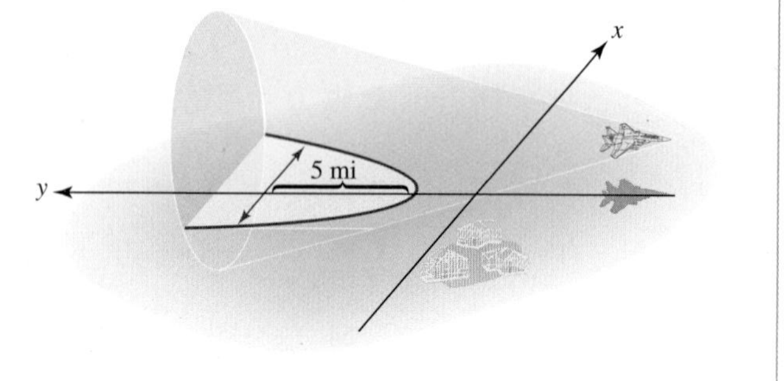

64. Lighting. Refer to the illustration. The "cones" of light emitted upward and downward by a lamp with a cylindrical shade cast an interesting pattern on the wall. What type of conic section can be seen? A hyperbola

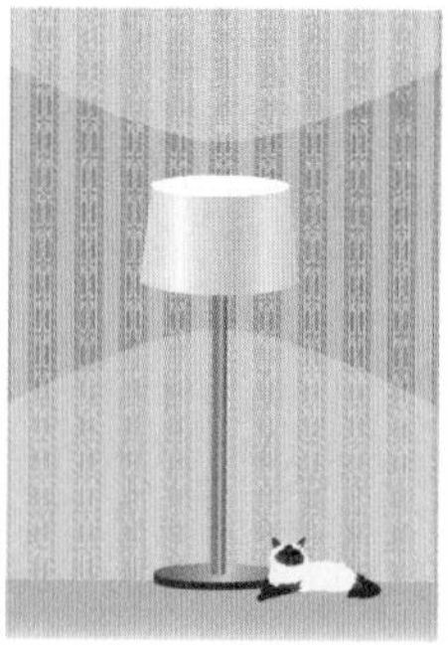

65. Nuclear Power. The photograph below is an example of a standard cooling tower of a nuclear power plant. What type of conic section appears to be used in the design? A hyperbola

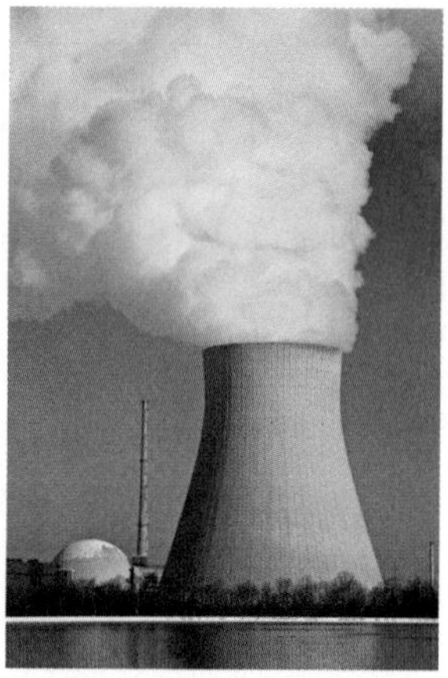

66. Fluids. See the illustration on the right. Two glass plates in contact at the left, and separated by about 5 millimeters on the right, are dipped in beet juice, which rises by capillary action to form a hyperbola. The hyperbola is modeled by an equation of the form $xy = k$. If the curve passes through the point (12, 2), what is k? 24

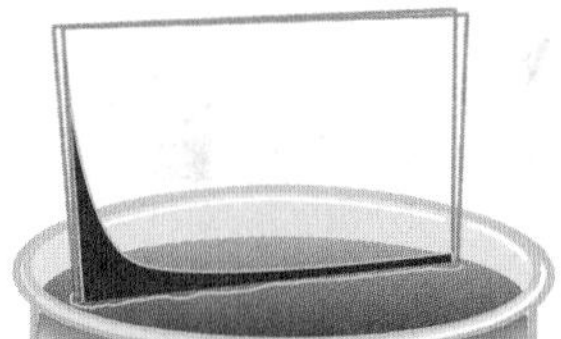

WRITING

67. What is a hyperbola?

68. Compare the graphs of $\frac{x^2}{81} - \frac{y^2}{64} = 1$ and $\frac{y^2}{81} - \frac{x^2}{64} = 1$. Do they have any similarities?

69. Explain how to determine the dimensions of the central rectangle that is associated with the graph of $\frac{x^2}{36} - \frac{y^2}{25} = 1$.

70. Explain why the graph of $\frac{x^2}{a^2} - \frac{y^2}{b^2} = 1$ has no y-intercept.

REVIEW

Find each value of x.

71. $\log_8 x = 2$ 64

72. $\log_{25} x = \frac{1}{2}$ 5

73. $\log_{1/2} \frac{1}{8} = x$ 3

74. $\log_{12} x = 0$ 1

75. $\log_x \frac{9}{4} = 2$ $\frac{3}{2}$

76. $\log_6 216 = x$ 3

77. $\log_x 1{,}000 = 3$ 10

78. $\log_2 \sqrt{2} = x$ $\frac{1}{2}$

CHALLENGE PROBLEMS

79. Write the equation $x^2 - y^2 - 2x + 4y = 12$ in standard form to show that it describes a hyperbola. $\frac{(x-1)^2}{9} - \frac{(y-2)^2}{9} = 1$

80. Write the equation $x^2 - 4y^2 + 2x - 8y = 7$ in standard form to show that it describes a hyperbola. $\frac{(x+1)^2}{4} - \frac{(y+1)^2}{1} = 1$

81. Write the equation $36x^2 - 25y^2 - 72x - 100y = 964$ in standard form to show that it describes a hyperbola. $\frac{(x-1)^2}{25} - \frac{(y+2)^2}{36} = 1$

82. Write an equation of a hyperbola whose graph has the following characteristics:

- vertices $(\pm 1, 0)$
- equations of asymptotes: $y = \pm 5x$ $\frac{x^2}{1} - \frac{y^2}{25} = 1$

83. Graph: $16x^2 - 25y^2 = 1$ See AIE Appendix 3.

84. Show that the equations of the extended diagonals of the fundamental rectangle of the hyperbola $\frac{x^2}{a^2} - \frac{y^2}{b^2} = 1$ are

$$y = \frac{b}{a}x \quad \text{and} \quad y = -\frac{b}{a}x$$

SECTION 13.4

OBJECTIVES

1. Solve systems by graphing.
2. Solve systems by substitution.
3. Solve systems by elimination (addition).

Solving Nonlinear Systems of Equations

ARE YOU READY? *Are You Ready? exercises available online at www.webassign.net/brookscole*

The following problems review some basic skills that are needed when solving nonlinear systems of equations.

1. Graph: $2x - 3y = 6$ See AIE Appendix 3.

2. Solve: $9y^2 + 4y - 5 = 0$ $-1, \frac{5}{9}$

3. Solve: $\begin{cases} 3x + 2y = 36 \\ 4x - y = 4 \end{cases}$ (4, 12)

4. Solve: **a.** $x^2 = \frac{5}{9}$ $\pm\frac{\sqrt{5}}{3}$ **b.** $y^2 = 18$ $\pm 3\sqrt{2}$

In Chapter 12, we discussed how to solve systems of linear equations by the graphing, substitution, and elimination methods. In this section, we will use these methods to solve systems in which at least one of the equations is nonlinear.

1 Solve Systems by Graphing.

A solution of a **nonlinear system of equations** is an ordered pair of real numbers that satisfies all of the equations in the system. The **solution set of a nonlinear system** is the set of all such ordered pairs. One way to solve a system of two equations in two variables is to graph the equations on the same rectangular coordinate system.

EXAMPLE 1 Solve $\begin{cases} x^2 + y^2 = 25 \\ 2x + y = 10 \end{cases}$ by graphing.

Teaching Tip: Point out that solutions will be ordered pairs.

Strategy We will graph both equations on the same coordinate system.

Why If the equations are graphed on the same coordinate system, we can see whether they have any common solutions.

Solution

Success Tip

It is helpful to sketch the possibilities before solving the system:

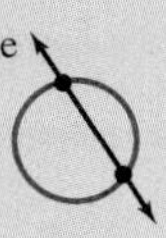

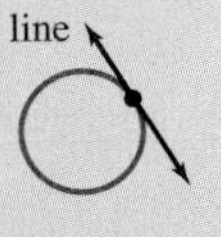

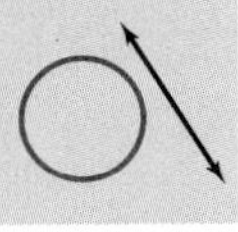

The graph of $x^2 + y^2 = 25$ is a circle with center at the origin and radius of 5. The graph of $2x + y = 10$ is a line with x-intercept (5, 0) and y-intercept (0, 10). Depending on whether the line is a **secant** (intersecting the circle at two points) or a **tangent** (intersecting the circle at one point) or does not intersect the circle at all, there are two, one, or no solutions to the system, respectively.

After graphing the circle and the line, it appears that the points of intersection are (5, 0) and (3, 4). To verify that they are solutions of the system, we need to check each one.

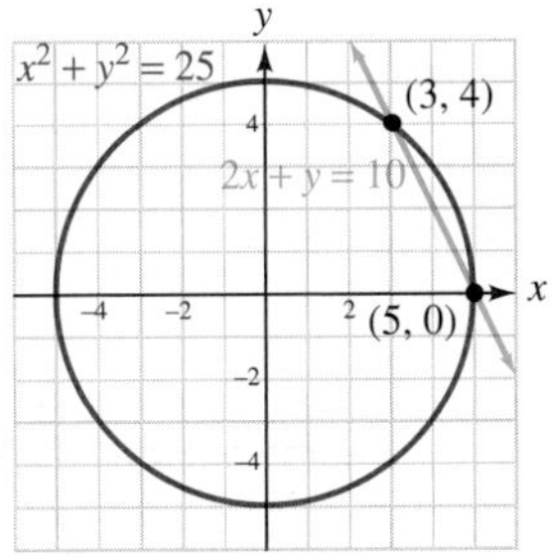

Check:

For (5, 0)		*For* (3, 4)	
$2x + y = 10$	$x^2 + y^2 = 25$	$2x + y = 10$	$x^2 + y^2 = 25$
$2(5) + 0 \stackrel{?}{=} 10$	$5^2 + 0^2 \stackrel{?}{=} 25$	$2(3) + 4 \stackrel{?}{=} 10$	$3^2 + 4^2 \stackrel{?}{=} 25$
$10 = 10$ True	$25 = 25$ True	$10 = 10$ True	$25 = 25$ True

The ordered pair (5, 0) satisfies both equations of the system, and so does (3, 4). Thus, there are two solutions, (5, 0) and (3, 4), and the solution set is {(5, 0), (3, 4)}.

Teaching Example 1 Solve $\begin{cases} x^2 + y^2 = 25 \\ x + y = 1 \end{cases}$ by graphing.

Answer:

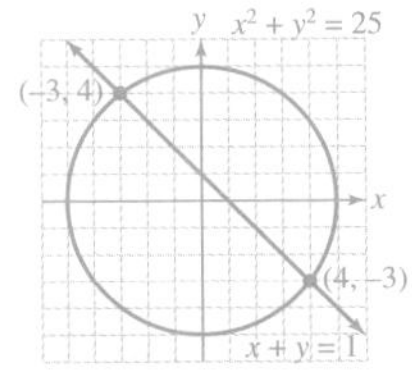

Self Check 1 Solve $\begin{cases} x^2 + y^2 = 25 \\ y = -2x - 5 \end{cases}$ by graphing. (−4, 3), (0, −5); see AIE Appendix 3.

Now Try Problem 15

Using Your Calculator ▶ Solving Systems of Equations

To solve Example 1 with a graphing calculator, we graph the circle and the line on one set of coordinate axes. See figure (a). We then trace to find the coordinates of the intersection points of the graphs. See figures (b) and (c).

We can zoom for better results.

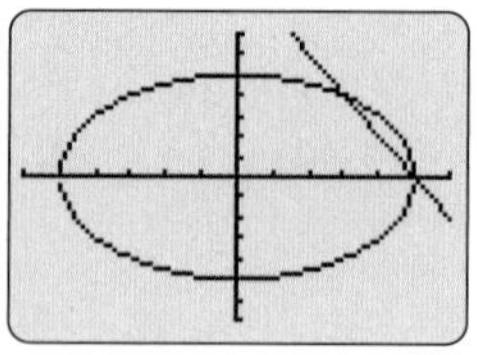

(a)

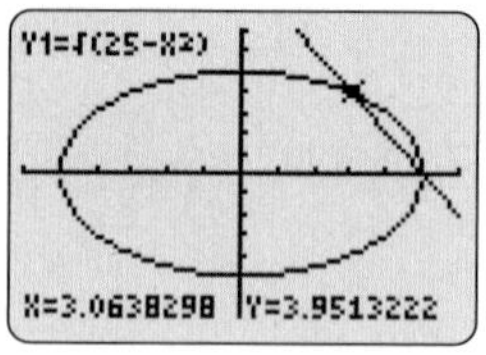

(b)

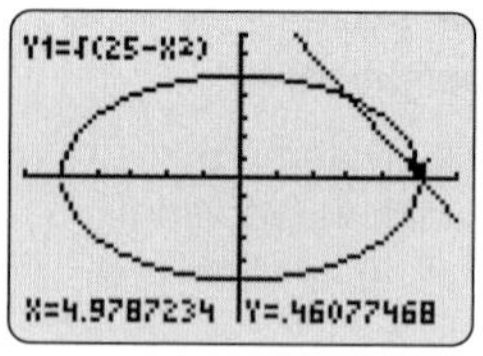

(c)

2 Solve Systems by Substitution.

When solving a system by graphing, it is often difficult to determine the coordinates of the intersection points. A more precise algebraic method called the **substitution method** can be used to solve certain systems involving nonlinear equations.

EXAMPLE 2 Solve $\begin{cases} x^2 + y^2 = 2 \\ 2x - y = 1 \end{cases}$ by substitution.

Strategy We will solve the second equation for y and substitute the result for y in the first equation.

Why We can solve the resulting equation for x and then back substitute to find y.

Solution This system has one second-degree equation and one first-degree equation. We can solve this type of system by substitution. Solving the linear equation for y gives:

Teaching Tip: Point out that substitution is most often the best method to use to solve systems consisting of a first-degree equation and a second-degree equation.

$$2x - y = 1$$
$$-y = -2x + 1 \quad \text{Subtract 2x from both sides.}$$
$$y = 2x - 1 \quad \text{Multiply both sides by −1. We call this the substitution equation.}$$

Because y and $2x - 1$ are equal, we can substitute $2x - 1$ for y in the first equation of the system.

$$y = \boxed{2x - 1} \qquad x^2 + y^2 = 2$$

Success Tip

With this method, the objective is to use an appropriate substitution to obtain *one* equation in *one* variable.

Then we solve the resulting quadratic equation for x.

$$x^2 + y^2 = 2$$
$$x^2 + (\mathbf{2x - 1})^2 = 2 \quad \text{Substitute 2x − 1 for y. This equation is in one variable.}$$
$$x^2 + 4x^2 - 4x + 1 = 2 \quad \text{Use a special-product rule to find } (2x - 1)^2.$$
$$5x^2 - 4x - 1 = 0 \quad \text{To get 0 on the right side, subtract 2 from both sides and then combine like terms.}$$
$$(5x + 1)(x - 1) = 0 \quad \text{Factor.}$$
$$5x + 1 = 0 \quad \text{or} \quad x - 1 = 0 \quad \text{Set each factor equal to 0.}$$
$$x = -\frac{1}{5} \quad \Big| \quad x = 1 \quad \text{Solve each equation.}$$

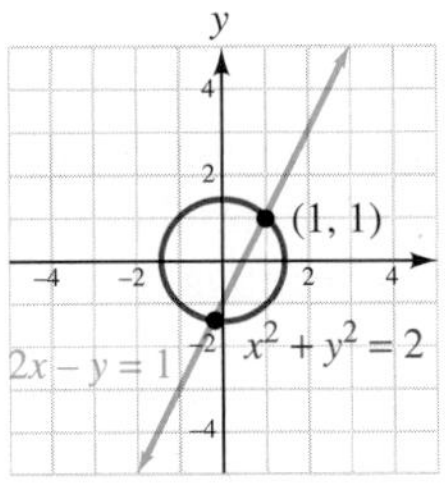

If we substitute $-\frac{1}{5}$ for x in the equation $y = 2x - 1$, we get $y = -\frac{7}{5}$. If we substitute 1 for x in $y = 2x - 1$, we get $y = 1$. Thus, the system has two solutions, $\left(-\frac{1}{5}, -\frac{7}{5}\right)$ and $(1, 1)$. Verify that each ordered pair satisfies both equations of the original system.

The graph in the margin confirms that the system has two solutions, and that one of them is $(1, 1)$. However, it would be virtually impossible to determine from the graph that the coordinates of the second point of intersection are $\left(-\frac{1}{5}, -\frac{7}{5}\right)$.

Teaching Example 2 Solve $\begin{cases} x^2 + y^2 = 25 \\ x - y = 1 \end{cases}$ by substitution.
Answer: $(-3, -4), (4, 3)$

Self Check 2 Solve $\begin{cases} x^2 + y^2 = 10 \\ y = x + 2 \end{cases}$ by substitution. $(1, 3), (-3, -1)$

Now Try Problem 23

EXAMPLE 3

Solve: $\begin{cases} 4x^2 + 9y^2 = 5 \\ y = x^2 \end{cases}$

Strategy Since $y = x^2$, we will substitute y for x^2 in the first equation.

Why This will give an equation in one variable that we can solve for y. We can then find x by back substitution.

Solution

We can solve this system by substitution.

$$4x^2 + 9y^2 = 5 \qquad \boxed{y} = x^2$$

When we substitute y for x^2 in the first equation, the result is a quadratic equation in y.

$$4x^2 + 9y^2 = 5$$

$$4y + 9y^2 = 5 \qquad \text{Substitute } y \text{ for } x^2\text{. This equation is in one variable.}$$

$$9y^2 + 4y - 5 = 0 \qquad \text{To get 0 on the right side, subtract 5 from both sides.}$$

$$(9y - 5)(y + 1) = 0 \qquad \text{Factor } 9y^2 + 4y - 5.$$

$$9y - 5 = 0 \quad \text{or} \quad y + 1 = 0 \qquad \text{Set each factor equal to 0.}$$

$$y = \frac{5}{9} \quad \Big| \quad y = -1 \qquad \text{Solve each equation.}$$

Since $y = x^2$, the values of x are found by solving the equations

$$x^2 = \frac{5}{9} \quad \text{or} \quad \cancel{x^2 = -1}$$

Because $x^2 = -1$ has no real solutions, this possibility is discarded. The solutions of $x^2 = \frac{5}{9}$ are

$$x = \sqrt{\frac{5}{9}} = \frac{\sqrt{5}}{\sqrt{9}} = \frac{\sqrt{5}}{3} \quad \text{or} \quad x = -\sqrt{\frac{5}{9}} = -\frac{\sqrt{5}}{\sqrt{9}} = -\frac{\sqrt{5}}{3}$$

Thus, the solutions of the system are $\left(\frac{\sqrt{5}}{3}, \frac{5}{9}\right)$ and $\left(-\frac{\sqrt{5}}{3}, \frac{5}{9}\right)$. Verify this by using a check.

Success Tip

$4x^2 + 9y^2 = 5$ is the equation of an ellipse centered at $(0, 0)$, and $y = x^2$ is the equation of a parabola with vertex at $(0, 0)$, opening upward. We would expect two solutions.

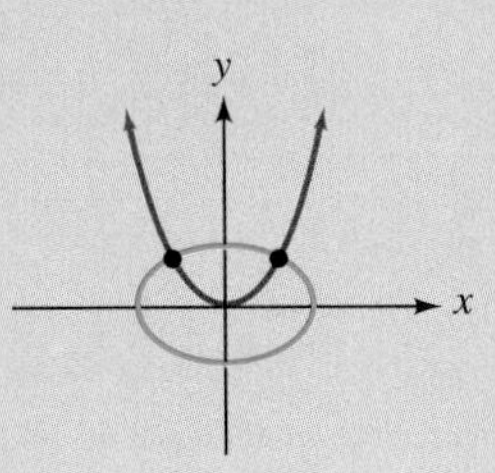

Caution

In this section, we are solving for only the real values of x and y.

Teaching Example 3 Solve: $\begin{cases} x^2 + y^2 = 25 \\ y = -x^2 + 13 \end{cases}$
Answer:
$(3, 4), (-3, 4), (4, -3), (-4, -3)$

Self Check 3 Solve: $\begin{cases} x^2 + y^2 = 20 \\ y = x^2 \end{cases}$ $(2, 4), (-2, 4)$

Now Try Problem 27

3 Solve Systems by Elimination (Addition).

Another method for solving nonlinear system of equations is the **elimination** or **addition method.** The elimination method is most often used when the equations of a nonlinear system are both second-degree equations. With this method, we combine the equations in a way that will eliminate the terms of one of the variables.

EXAMPLE 4 Solve: $\begin{cases} 3x^2 + 2y^2 = 36 \\ 4x^2 - y^2 = 4 \end{cases}$

Strategy We will multiply both sides of the second equation by 2 and add the result to the first equation.

Why This will eliminate the y^2-terms and produce an equation that we can solve for x.

Solution To solve this system of two second-degree equations, we can use either the substitution or the elimination method. We will use the elimination method because the y^2-terms can be eliminated by multiplying the second equation by 2 and adding it to the first equation.

Success Tip

The elimination method is generally better than the substitution method when both equations of the system are of the form $Ax^2 + By^2 = C$.

$$\begin{cases} 3x^2 + 2y^2 = 36 \\ 4x^2 - y^2 = 4 \end{cases} \xrightarrow[\text{Multiply by 2}]{\text{Unchanged}} \begin{cases} 3x^2 + 2y^2 = 36 \\ 8x^2 - 2y^2 = 8 \end{cases}$$

We add the two equations on the right to eliminate y^2 and solve the resulting equation for x:

$$11x^2 = 44$$

$$x^2 = 4 \quad \text{Divide both sides by 11.}$$

$$x = 2 \quad \text{or} \quad x = -2 \quad \text{Use the square root property.}$$

Success Tip

$3x^2 + 2y^2 = 36$ is the equation of an ellipse, centered at (0, 0), and $4x^2 - y^2 = 4$ is the equation of a hyperbola, centered at (0, 0), opening left and right. It is possible to have four solutions.

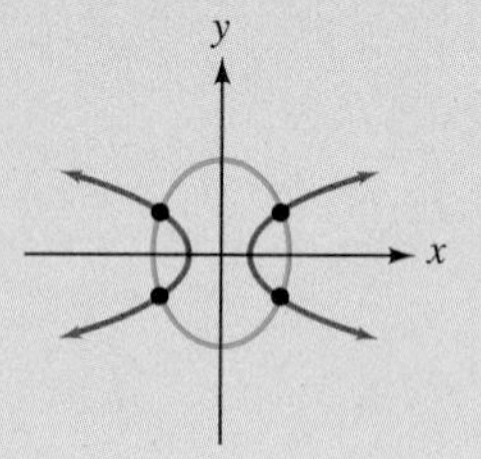

To find y, we can substitute 2 for x and then -2 for x into any equation containing both variables. It appears that the calculations will be simplest if we use $3x^2 + 2y^2 = 36$.

For x = 2:

$$\begin{aligned} 3x^2 + 2y^2 &= 36 \\ 3(2)^2 + 2y^2 &= 36 \\ 12 + 2y^2 &= 36 \\ 2y^2 &= 24 \\ y^2 &= 12 \end{aligned}$$

For x = −2:

$$\begin{aligned} 3x^2 + 2y^2 &= 36 \\ 3(-2)^2 + 2y^2 &= 36 \\ 12 + 2y^2 &= 36 \\ 2y^2 &= 24 \\ y^2 &= 12 \end{aligned}$$

Now we use the square root property to solve for y.

For x = 2:

$y = \sqrt{12}$ or $y = -\sqrt{12}$

$y = 2\sqrt{3}$ | $y = -2\sqrt{3}$

For x = −2:

$y = \sqrt{12}$ or $y = -\sqrt{12}$

$y = 2\sqrt{3}$ | $y = -2\sqrt{3}$

The four solutions of this system are $(2, 2\sqrt{3})$, $(2, -2\sqrt{3})$, $(-2, 2\sqrt{3})$, and $(-2, -2\sqrt{3})$. Verify this by using a check.

Teaching Example 4 Solve: $\begin{cases} -3x^2 + 2y^2 = 5 \\ x^2 + y^2 = 25 \end{cases}$
Answer:
(3, 4), (−3, 4), (3, −4), (−3, −4)

Self Check 4 Solve: $\begin{cases} x^2 + 4y^2 = 16 \\ x^2 - y^2 = 1 \end{cases}$ $(2, \sqrt{3}), (2, -\sqrt{3}), (-2, \sqrt{3}), (-2, -\sqrt{3})$

Now Try ▶ Problem 31

SECTION 13.4 STUDY SET

VOCABULARY

Fill in the blanks.

1. $\begin{cases} 4x^2 + 6y^2 = 24 \\ 9x^2 - y^2 = 9 \end{cases}$ is a system of two nonlinear equations.
2. The graph of $2x + y = 10$ is a line and the graph of $x^2 + y^2 = 25$ is a circle.
3. When solving a system by graphing, it is often difficult to determine the coordinates of the points of intersection of the graphs.
4. Two algebraic methods for solving systems of nonlinear equations are the substitution method and the elimination method.
5. A secant is a line that intersects a circle at two points.
6. A tangent is a line that intersects a circle at one point.

CONCEPTS

7. a. A line can intersect an ellipse in at most two points.
   b. An ellipse can intersect a parabola in at most four points.
   c. An ellipse can intersect a circle in at most four points.
   d. A hyperbola can intersect a circle in at most four points.
8. Determine whether $(1, -1)$ is a solution of the system: $\begin{cases} 2x + y - 1 = 0 \\ x^2 - y^2 = 3 \end{cases}$ No
9. Find the solutions of the system $\begin{cases} x^2 + 4y^2 = 25 \\ x^2 - 2y^2 = 1 \end{cases}$ that is graphed on the right. $(-3, 2), (3, 2), (-3, -2), (3, -2)$

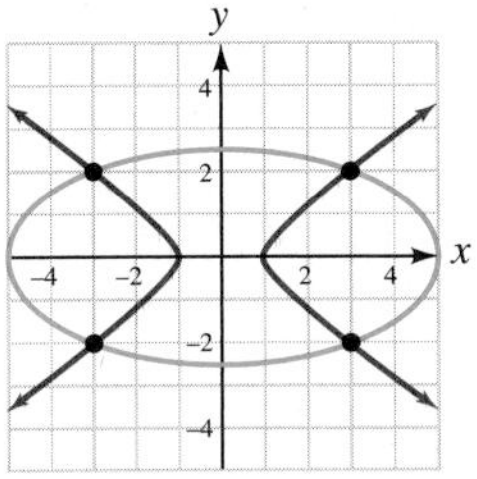

10. Find a substitution equation that can be used to solve the system: $\begin{cases} x^2 + y^2 = 9 \\ 2x - y = 3 \end{cases}$ $y = 2x - 3$
11. Consider the system: $\begin{cases} 6x^2 + y^2 = 9 \\ 3x^2 + 4y^2 = 36 \end{cases}$
    a. If the y^2-terms are to be eliminated, by what should the first equation be multiplied? -4
    b. If the x^2-terms are to be eliminated, by what should the second equation be multiplied? -2
12. Suppose you begin to solve the system $\begin{cases} x^2 + y^2 = 10 \\ 4x^2 + y^2 = 13 \end{cases}$ and find that x is ± 1. Use the first equation to find the corresponding y-values for $x = 1$ and $x = -1$. State the solutions as ordered pairs. $(1, 3), (1, -3), (-1, 3), (-1, -3)$

NOTATION

Complete each solution to solve the system.

13. Solve: $\begin{cases} x^2 + y^2 = 5 \\ y = 2x \end{cases}$

$x^2 + y^2 = 5$ This is the first equation.

$x^2 + (2x)^2 = 5$

$x^2 + 4x^2 = 5$

$5x^2 = 5$

$x^2 = 1$

$x = 1$ or $x = -1$

If $x = 1$, then $y = 2(1) = 2$. Use the second equation.

If $x = -1$, then $y = 2(-1) = -2$.

The solutions are $(1, 2)$ and $(-1, -2)$.

14. Solve: $\begin{cases} y = x^2 + 2 \\ y = -x^2 + 4 \end{cases}$

$2y = 6$ Add the equations.

$y = 3$

If $y = 3$, then

$3 = x^2 + 2$ This is the first equation.

$1 = x^2$

$\pm 1 = x$

The solutions are $(1, 3)$ and $(-1, 3)$.

GUIDED PRACTICE

Solve each system of equations by graphing. See Example 1. See AIE Appendix 3.

15. $\begin{cases} x^2 + y^2 = 9 \\ y - x = 3 \end{cases}$ $(0, 3), (-3, 0)$
16. $\begin{cases} x^2 + y^2 = 16 \\ y - x = -4 \end{cases}$ $(0, -4), (4, 0)$
17. $\begin{cases} 9x^2 + 16y^2 = 144 \\ 9x^2 - 16y^2 = 144 \end{cases}$ $(-4, 0), (4, 0)$
18. $\begin{cases} x^2 + 9y^2 = 9 \\ 9y^2 - x^2 = 9 \end{cases}$ $(0, -1), (0, 1)$
19. $\begin{cases} y = x^2 - 4x \\ x^2 + y = 0 \end{cases}$ $(0, 0), (2, -4)$
20. $\begin{cases} x^2 - y = 0 \\ y = -x^2 + 4x \end{cases}$ $(0, 0), (2, 4)$
21. $\begin{cases} x^2 + 4y^2 = 4 \\ x = 2y^2 - 2 \end{cases}$ $(-2, 0), (0, -1), (0, 1)$
22. $\begin{cases} 4x^2 + y^2 = 4 \\ y = 2x^2 - 2 \end{cases}$ $(0, -2), (1, 0), (-1, 0)$

Solve each system of equations by substitution for real values of x and y. See Examples 2 and 3.

23. $\begin{cases} x^2 + y^2 = 5 \\ x + y = 3 \end{cases}$ $(1, 2), (2, 1)$
24. $\begin{cases} x^2 - x - y = 2 \\ 4x - 3y = 0 \end{cases}$ $\left(-\frac{2}{3}, -\frac{8}{9}\right), (3, 4)$

25. $\begin{cases} y = x^2 + 6x + 7 \\ 2x + y = -5 \end{cases}$ $(-6, 7), (-2, -1)$

26. $\begin{cases} 2x + y = 1 \\ x^2 + y = 4 \end{cases}$ $(-1, 3), (3, -5)$

▶ 27. $\begin{cases} x^2 + y^2 = 13 \\ y = x^2 - 1 \end{cases}$ $(-2, 3), (2, 3)$

28. $\begin{cases} x^2 + y^2 = 10 \\ y = 3x^2 \end{cases}$ $(-1, 3), (1, 3)$

▶ 29. $\begin{cases} x^2 + y^2 = 30 \\ y = x^2 \end{cases}$ $(\sqrt{5}, 5), (-\sqrt{5}, 5)$

30. $\begin{cases} x^2 + y^2 = 20 \\ y = x^2 \end{cases}$ $(2, 4), (-2, 4)$

Solve each system of equations by elimination for real values of x and y. See Example 4.

31. $\begin{cases} x^2 + y^2 = 20 \\ x^2 - y^2 = -12 \end{cases}$ $(2, 4), (2, -4), (-2, 4), (-2, -4)$

▶ 32. $\begin{cases} x^2 + y^2 = 13 \\ x^2 - y^2 = 5 \end{cases}$ $(3, 2), (3, -2), (-3, 2), (-3, -2)$

33. $\begin{cases} 9x^2 - 7y^2 = 81 \\ x^2 + y^2 = 9 \end{cases}$ $(3, 0), (-3, 0)$

▶ 34. $\begin{cases} x^2 + y^2 = 25 \\ 2x^2 - 3y^2 = 5 \end{cases}$ $(4, 3), (-4, 3), (4, -3), (-4, -3)$

▶ 35. $\begin{cases} 2x^2 + y^2 = 6 \\ x^2 - y^2 = 3 \end{cases}$ $(\sqrt{3}, 0), (-\sqrt{3}, 0)$

▶ 36. $\begin{cases} x^2 + y^2 = 36 \\ 49x^2 + 36y^2 = 1{,}764 \end{cases}$ $(6, 0), (-6, 0)$

37. $\begin{cases} x^2 - y^2 = -5 \\ 3x^2 + 2y^2 = 30 \end{cases}$ $(-2, 3), (2, 3), (-2, -3), (2, -3)$

38. $\begin{cases} 6x^2 + 8y^2 = 182 \\ 8x^2 - 3y^2 = 24 \end{cases}$ $(3, 4), (3, -4), (-3, 4), (-3, -4)$

Solve each system. See Using Your Calculator: Solving Systems of Equations. See AIE Appendix 3.

39. $\begin{cases} x^2 - 6x - y = -5 \\ x^2 - 6x + y = -5 \end{cases}$ $(1, 0), (5, 0)$

40. $\begin{cases} x^2 - y^2 = -5 \\ 3x^2 + 2y^2 = 30 \end{cases}$ $(-2, 3), (2, 3), (-2, -3), (2, -3)$

TRY IT YOURSELF

Solve each system of equations for real values of x and y.

41. $\begin{cases} 2x^2 - 3y^2 = 5 \\ 3x^2 + 4y^2 = 16 \end{cases}$ $(2, 1), (-2, 1), (2, -1), (-2, -1)$

▶ 42. $\begin{cases} 2x^2 - y^2 + 2 = 0 \\ 3x^2 - 2y^2 + 5 = 0 \end{cases}$ $(1, 2), (-1, 2), (1, -2), (-1, -2)$

43. $\begin{cases} y = x^2 - 4 \\ x^2 - y^2 = -16 \end{cases}$ $(0, -4), (-3, 5), (3, 5)$

44. $\begin{cases} y - x = 0 \\ 4x^2 + y^2 = 10 \end{cases}$ $(-\sqrt{2}, -\sqrt{2}), (\sqrt{2}, \sqrt{2})$

▶ 45. $\begin{cases} 3y^2 = xy \\ 2x^2 + xy - 84 = 0 \end{cases}$ $(6, 2), (-6, -2), (-\sqrt{42}, 0), (\sqrt{42}, 0)$

46. $\begin{cases} x^2 + y^2 = 10 \\ 2x^2 - 3y^2 = 5 \end{cases}$ $(\sqrt{7}, \sqrt{3}), (\sqrt{7}, -\sqrt{3}), (-\sqrt{7}, \sqrt{3}), (-\sqrt{7}, -\sqrt{3})$

▶ 47. $\begin{cases} y^2 = 40 - x^2 \\ y = x^2 - 10 \end{cases}$ $(-\sqrt{15}, 5), (\sqrt{15}, 5), (-2, -6), (2, -6)$

48. $\begin{cases} 25x^2 + 9y^2 = 225 \\ 5x + 3y = 15 \end{cases}$ $(3, 0), (0, 5)$

49. $\begin{cases} 3x - y = -3 \\ 25y^2 - 9x^2 = 225 \end{cases}$ $(0, 3), \left(-\frac{25}{12}, -\frac{13}{4}\right)$

50. $\begin{cases} x - 2y = 2 \\ 9x^2 - 4y^2 = 36 \end{cases}$ $(2, 0), \left(-\frac{5}{2}, -\frac{9}{4}\right)$

51. $\begin{cases} x^2 - y = 0 \\ x^2 - 4x + y = 0 \end{cases}$ $(0, 0), (2, 4)$

52. $\begin{cases} xy = -\dfrac{9}{2} \\ 3x + 2y = 6 \end{cases}$ $\left(-1, \frac{9}{2}\right), \left(3, -\frac{3}{2}\right)$

53. $\begin{cases} x^2 - 2y^2 = 6 \\ x^2 + 2y^2 = 2 \end{cases}$ No solution, $\varnothing$

54. $\begin{cases} x^2 + 9y^2 = 1 \\ x^2 - 9y^2 = 3 \end{cases}$ No solution, $\varnothing$

55. $\begin{cases} y = x^2 - 4 \\ 6x - y = 13 \end{cases}$ $(3, 5)$

56. $\begin{cases} y = x + 1 \\ x^2 - y^2 = 1 \end{cases}$ $(-1, 0)$

▶ 57. $\begin{cases} x^2 + y^2 = 4 \\ 9x^2 + y^2 = 9 \end{cases}$ $\left(\frac{\sqrt{10}}{4}, \frac{3\sqrt{6}}{4}\right), \left(\frac{\sqrt{10}}{4}, -\frac{3\sqrt{6}}{4}\right), \left(-\frac{\sqrt{10}}{4}, \frac{3\sqrt{6}}{4}\right), \left(-\frac{\sqrt{10}}{4}, -\frac{3\sqrt{6}}{4}\right)$

58. $\begin{cases} 2x^2 - 6y^2 + 3 = 0 \\ 4x^2 + 3y^2 = 4 \end{cases}$ $\left(\frac{\sqrt{2}}{2}, \frac{\sqrt{6}}{3}\right), \left(-\frac{\sqrt{2}}{2}, \frac{\sqrt{6}}{3}\right), \left(\frac{\sqrt{2}}{2}, -\frac{\sqrt{6}}{3}\right), \left(-\frac{\sqrt{2}}{2}, -\frac{\sqrt{6}}{3}\right)$

59. $\begin{cases} xy = \dfrac{1}{6} \\ y + x = 5xy \end{cases}$ $\left(\frac{1}{2}, \frac{1}{3}\right), \left(\frac{1}{3}, \frac{1}{2}\right)$

▶ 60. $\begin{cases} xy = \dfrac{1}{12} \\ y + x = 7xy \end{cases}$ $\left(\frac{1}{4}, \frac{1}{3}\right), \left(\frac{1}{3}, \frac{1}{4}\right)$

61. $\begin{cases} x^2 = 4 - y \\ y = x^2 + 2 \end{cases}$ $(-1, 3), (1, 3)$

62. $\begin{cases} 3x + 2y = 10 \\ y = x^2 - 5 \end{cases}$ $\left(\frac{5}{2}, \frac{5}{4}\right), (-4, 11)$

63. $\begin{cases} x^2 - y^2 = 4 \\ x + y = 4 \end{cases}$ $\left(\frac{5}{2}, \frac{3}{2}\right)$

64. $\begin{cases} x - y = -1 \\ y^2 - 4x = 0 \end{cases}$ $(1, 2)$

APPLICATIONS

Use a nonlinear system of equations to solve each problem.

▶ 65. **Integer Problem.** The product of two integers is 32, and their sum is 12. Find the integers. 4, 8

▶ 66. **Number Problem.** The sum of the squares of two numbers is 221, and the sum of the numbers is 9. Find the numbers. 14, −5

67. Archery. See the illustration below. An arrow shot from the base of a hill follows the parabolic path $y = -\frac{1}{6}x^2 + 2x$, with distances measured in meters. The inclined hill has a slope of $\frac{1}{3}$ and can therefore be modeled by the equation $y = \frac{1}{3}x$. Find the coordinates of the point of impact of the arrow and then its distance from the archer. $\left(10, \frac{10}{3}\right); \frac{10}{3}\sqrt{10}$ m

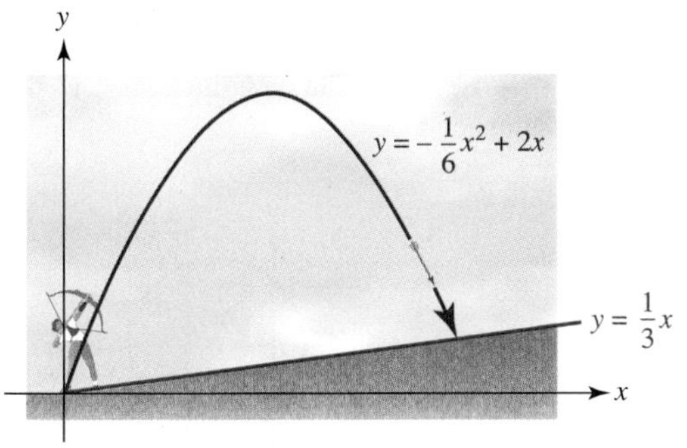

68. Geometry. The area of a rectangle is 63 square centimeters, and its perimeter is 32 centimeters. Find the dimensions of the rectangle. 7 cm by 9 cm

69. Fencing Pastures. The rectangular pasture shown here is to be fenced in along a riverbank. If 260 feet of fencing is to enclose an area of 8,000 square feet, find the dimensions of the pasture. 80 ft by 100 ft or 50 ft by 160 ft

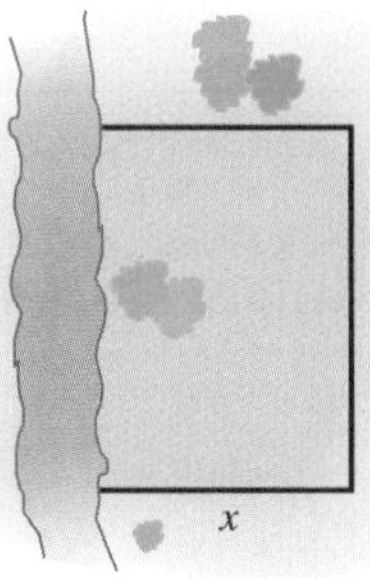

70. Driving Rates. Jim drove 306 miles. Jim's brother made the same trip at a speed 17 mph slower than Jim did and required an extra $1\frac{1}{2}$ hours. What was Jim's rate and time? 68 mph, 4.5 hr

71. Investing. Grant receives $225 annual income from one investment. Jeff invested $500 more than Grant, but at an annual rate of 1% less. Jeff's annual income is $240. What are the amount and rate of Grant's investment? $2,500 at 9%

72. Investing. Carol receives $67.50 annual income from one investment. John invested $150 more than Carol at an annual rate of $1\frac{1}{2}$% more. John's annual income is $94.50. What are the amount and rate of Carol's investment? (*Hint:* There are two answers.) Either $750 at 9% or $900 at 7.5%

WRITING

73. a. Describe the benefits of the graphical method for solving a system of nonlinear equations.

b. Describe the drawbacks of the graphical method.

74. Explain why the elimination method, not the substitution method, is the better method to solve the system:

$$\begin{cases} 4x^2 + 9y^2 = 52 \\ 9x^2 + 4y^2 = 52 \end{cases}$$

REVIEW

Solve each equation.

75. $\log 5x = 4$ 2,000

76. $\log 3x = \log 9$ 3

77. $\dfrac{\log (8x - 7)}{\log x} = 2$ 7

78. $\log x + \log (x + 9) = 1$ 1

CHALLENGE PROBLEMS

79. a. The graphs of the two independent equations of a system are parabolas. How many solutions might the system have? 0, 1, 2, 3, 4

b. The graphs of the two independent equations of a system are hyperbolas. How many solutions might the system have? 0, 1, 2, 3, 4

80. Solve the system for real solutions: $\begin{cases} \frac{1}{x} + \frac{2}{y} = 1 \\ \frac{2}{x} - \frac{1}{y} = \frac{1}{3} \end{cases}$ (3, 3)

81. Solve the system for real solutions: $\begin{cases} \frac{1}{x} + \frac{3}{y} = 4 \\ \frac{2}{x} - \frac{1}{y} = 7 \end{cases}$ $\left(\frac{7}{25}, 7\right)$

82. Solve the system $\begin{cases} x^2 - y^2 = 16 \\ x^2 + y^2 = 9 \end{cases}$ over the complex numbers.

$\left(\frac{5\sqrt{2}}{2}, \frac{i\sqrt{14}}{2}\right), \left(-\frac{5\sqrt{2}}{2}, -\frac{i\sqrt{14}}{2}\right), \left(-\frac{5\sqrt{2}}{2}, \frac{i\sqrt{14}}{2}\right), \left(\frac{5\sqrt{2}}{2}, -\frac{i\sqrt{14}}{2}\right)$

13 Summary & Review

SECTION 13.1 The Circle and the Parabola

DEFINITIONS AND CONCEPTS

A **circle** is the set of all points in a plane that are a fixed distance from a fixed point called its **center.** The fixed distance is called the **radius** of the circle.

Standard forms of the equation of a circle:

$x^2 + y^2 = r^2$ Center (0, 0), radius r

$(x - h)^2 + (y - k)^2 = r^2$ Center (h, k), radius r

EXAMPLES

The graph of the equation $x^2 + y^2 = 16$, which can be written $x^2 + y^2 = \mathbf{4}^2$, is a circle with center at (0, 0) and a radius of **4**.

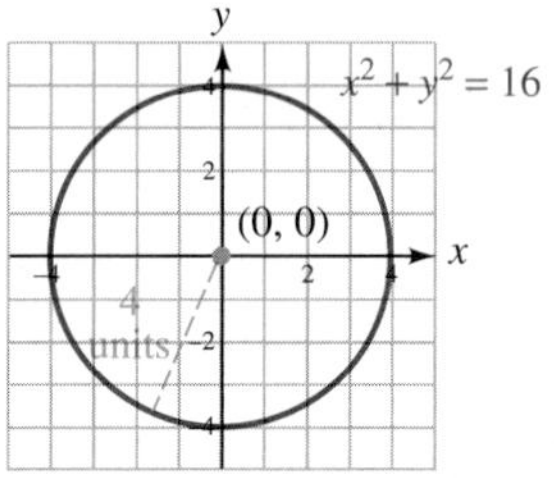

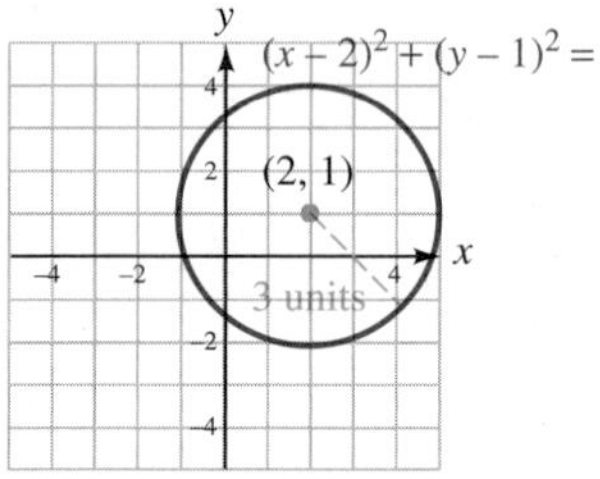

The graph of the equation $(x - 2)^2 + (y - 1)^2 = 9$, which can be written $(x - \mathbf{2})^2 + (y - \mathbf{1})^2 = 3^2$, is a circle with center at (**2**, **1**) and a radius of 3.

Because a circle is determined by its center and radius, that information is all we need to know to write its equation.

Write the equation of a circle centered at (4, −3) and with a radius of 5.

In this problem, $h = 4$, $k = -3$, and $r = 5$. We substitute these values into the standard form of the equation of a circle and simplify.

$$(x - h)^2 + (y - k)^2 = r^2$$
$$(x - 4)^2 + [y - (-3)]^2 = 5^2$$
$$(x - 4)^2 + (y + 3)^2 = 25$$

A **parabola** is the set of all points in a plane that are equidistant from a fixed point, called the **focus,** and a fixed line, called the **directrix.**

General forms of the equation of a parabola:

$y = ax^2 + bx + c$ $a > 0$: up; $a < 0$: down

$x = ay^2 + by + c$ $a > 0$: right; $a < 0$: left

Standard forms of the equation of a parabola:

$y = a(x - h)^2 + k$ $a > 0$: up; $a < 0$: down
Vertex at (h, k) Axis of symmetry is $x = h$

$x = a(y - k)^2 + h$ $a > 0$: right; $a < 0$: left
Vertex at (h, k) Axis of symmetry is $y = k$

The equation $x = 2y^2 - 4y + 5$ is the equation of a parabola that opens to the right. To find its vertex and axis of symmetry, we complete the square on y and write the equation in standard form.

$x = 2y^2 - 4y + 5$

$x = 2(y^2 - 2y \quad) + 5$ Factor out 2.

$x = 2(y^2 - 2y + \mathbf{1}) + 5 - \mathbf{2}$ Complete the square.

$x = 2(y - 1)^2 + 3$ Factor and simplify.

From the standard form, we see that $h = 3$ and $k = 1$. Thus, the vertex is at (3, 1) and the axis of symmetry is $y = 1$. To construct a table of solutions, we choose values of y and find their corresponding values of x.

$x = 2(y - 1)^2 + 3$

x	y	
5	2	→ (5, 2)
11	3	→ (11, 3)

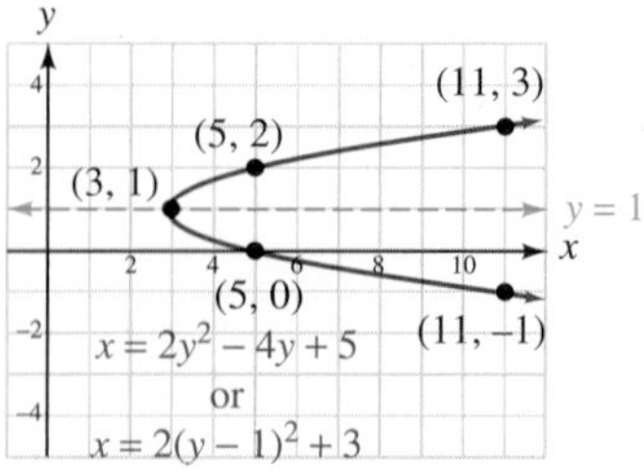

REVIEW EXERCISES

Graph each equation. See AIE Appendix 3.

1. $x^2 + y^2 = 16$

2. $(x - 4)^2 + (y + 3)^2 = 4$

3. Write the equation $x^2 + y^2 + 4x - 2y = 4$ in standard form and graph it.
$(x + 2)^2 + (y - 1)^2 = 9$; see AIE Appendix 3.

4. Art History. Leonardo da Vinci's *Vitruvian Man* (1492) is one of the most famous pen-and-ink drawings of all time. Use the coordinate system that is superimposed on the drawing to write the equation of the circle in standard form.
$(x - 9)^2 + (y - 9)^2 = 9^2$ or $(x - 9)^2 + (y - 9)^2 = 81$

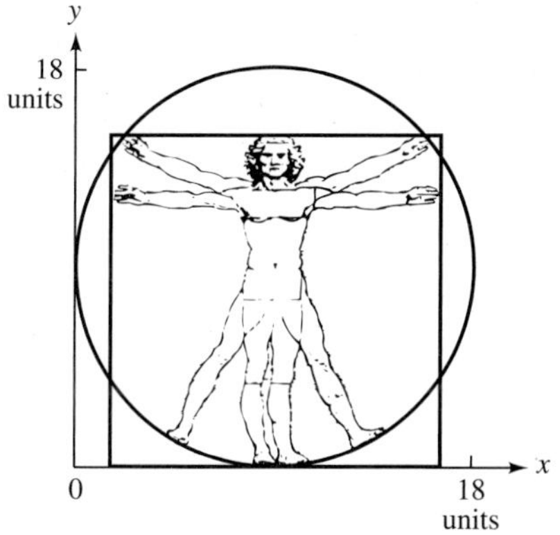

5. Find the center and the radius of the circle whose equation is $(x + 6)^2 + y^2 = 24$. $(-6, 0)$; $r = 2\sqrt{6}$

6. Fill in the blanks: A circle is the set of all points in a plane that are a fixed distance from a point called its center. The fixed distance is called the radius of the circle.

Graph each parabola and give the coordinates of the vertex. See AIE Appendix 3.

7. $x = y^2$ $(0, 0)$

8. $x = 2(y + 1)^2 - 2$ $(-2, -1)$

Write each equation in standard form and graph it. See AIE Appendix 3.

9. $x = -3y^2 + 12y - 7$ $x = -3(y - 2)^2 + 5$

10. $y = x^2 + 8x + 11$ $y = (x + 4)^2 - 5$

11. The axis of symmetry, the vertex, and two additional points on the graph of a parabola are shown. Find the coordinates of two other points on the parabola.
$(2, -2)$, $(8, -3)$

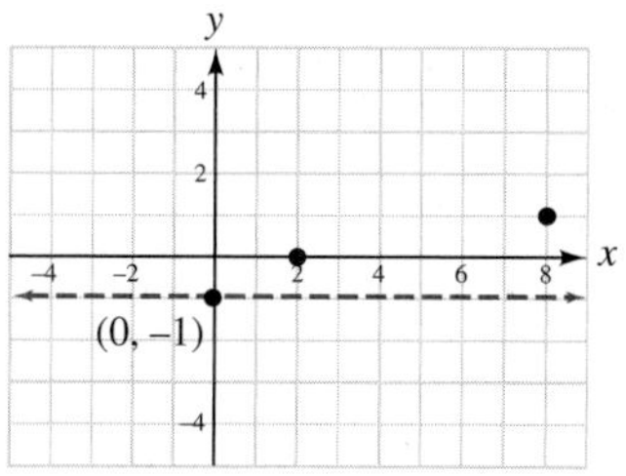

12. Long Jump. The equation describing the flight path of the long jumper is $y = -\frac{5}{121}(x - 11)^2 + 5$. Show that she will land at a point 22 feet away from the take-off board.
When $x = 22$, $y = 0$: $-\frac{5}{121}(22 - 11)^2 + 5 = 0$

SECTION 13.2 The Ellipse

DEFINITIONS AND CONCEPTS	EXAMPLES
An **ellipse** is the set of all points in a plane for which the sum of the distances from two fixed points is a constant.	The equation $\frac{x^2}{9} + \frac{y^2}{4} = 1$, which can be written $\frac{x^2}{3^2} + \frac{y^2}{2^2} = 1$, represents an ellipse that is centered at the origin. Here, $a = \mathbf{3}$ and $b = \mathbf{2}$.

Standard forms of the equation of an ellipse:

$\frac{x^2}{a^2} + \frac{y^2}{b^2} = 1$ Center (0, 0)

$\frac{(x - h)^2}{a^2} + \frac{(y - k)^2}{b^2} = 1$ Center (h, k)

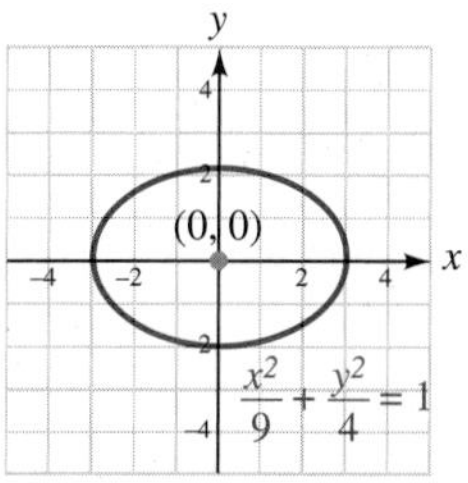

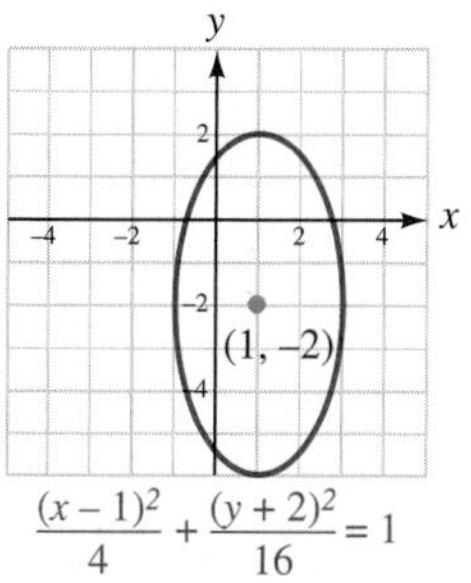

The equation $\frac{(x - 1)^2}{4} + \frac{(y + 2)^2}{16} = 1$, which can be written $\frac{(x - 1)^2}{2^2} + \frac{(y + 2)^2}{4^2} = 1$, represents an ellipse that is centered at $(1, -2)$. Here, $a = \mathbf{2}$ and $b = \mathbf{4}$.

To write the equation $25x^2 + 16y^2 = 400$ in standard form, divide both sides by 400 and simplify.

$$25x^2 + 16y^2 = 400$$

$$\frac{25x^2}{400} + \frac{16y^2}{400} = \frac{400}{400} \quad \text{To get 1 on the right side, divide both sides by 400.}$$

$$\frac{x^2}{16} + \frac{y^2}{25} = 1 \quad \text{Simplify each fraction.}$$

This result represents an ellipse that is centered at (0, 0), with $a = 4$ and $b = 5$.

REVIEW EXERCISES

Graph each ellipse. See AIE Appendix 3.

13. $\frac{x^2}{16} + \frac{y^2}{9} = 1$

14. $\frac{(x-2)^2}{4} + \frac{(y-1)^2}{25} = 1$

15. $4(x+1)^2 + 9(y-1)^2 = 36$

16. Consider the equation $\frac{x^2}{144} + \frac{y^2}{1} = 1$. Write each term on the left side with a denominator that is the square of a number. $\frac{x^2}{12^2} + \frac{y^2}{1^2} = 1$

17. Determine whether the graph of each equation is a circle, a parabola, or an ellipse.

a. $(x-1)^2 + (y+9)^2 = 100$ Circle

b. $\frac{x^2}{49} + \frac{y^2}{121} = 1$ Ellipse

c. $x = y^2 - 2y + 6$ Parabola

d. $16(x-4)^2 + 4(y+8)^2 = 16$ Ellipse

18. Salami. When a delicatessen slices a cylindrical salami at an angle, the results are elliptical pieces that are larger than circular pieces. See the illustration. Write the equation of the shape of the slice of salami shown if it was centered at the origin of a coordinate system. $\frac{x^2}{25} + \frac{y^2}{9} = 1$

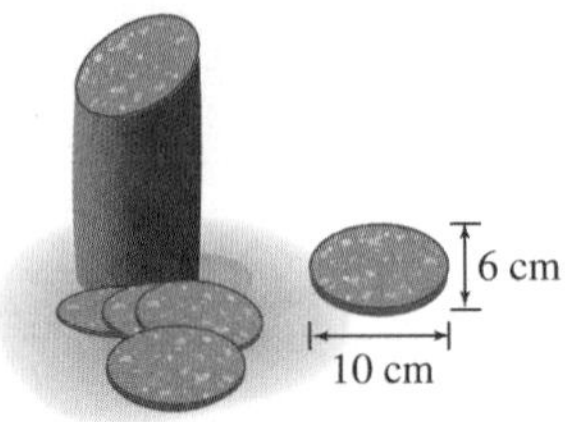

19. Fill in the blanks: An ellipse is the set of all points in a plane for which the sum of the distances from two fixed points is a constant. Each of the fixed points is called a focus.

20. Construction. Sketch the path of the sound when a person, standing at one focus, whispers something in the whispering gallery dome shown below. Answers may vary.

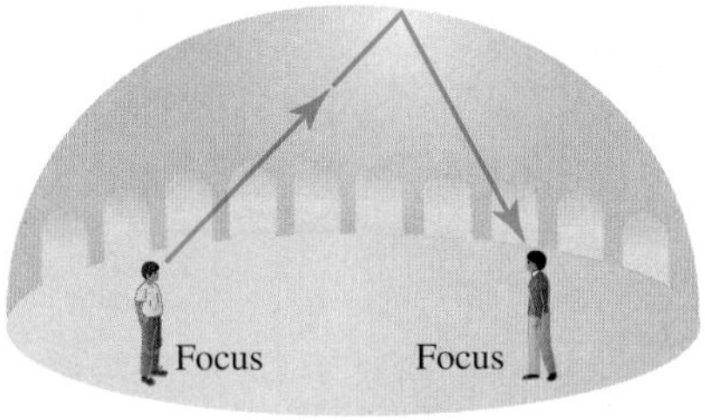

SECTION 13.3 The Hyperbola

DEFINITIONS AND CONCEPTS	EXAMPLES
A **hyperbola** is the set of all points in a plane for which the difference of the distances from two fixed points is a constant.	The equation $\frac{x^2}{4} - \frac{y^2}{9} = 1$, which can be written $\frac{x^2}{2^2} - \frac{y^2}{3^2} = 1$, represents a hyperbola, centered at (0, 0), that opens left and right. Here, $a = \mathbf{2}$ and $b = \mathbf{3}$.

Standard forms of the equation of a hyperbola:

$\frac{x^2}{a^2} - \frac{y^2}{b^2} = 1$ Center (0, 0), opens left and right

$\frac{y^2}{a^2} - \frac{x^2}{b^2} = 1$ Center (0, 0), opens up and down

$\frac{(x-h)^2}{a^2} - \frac{(y-k)^2}{b^2} = 1$ Center (h, k), opens left and right

$\frac{(y-k)^2}{a^2} - \frac{(x-h)^2}{b^2} = 1$ Center (h, k), opens up and down

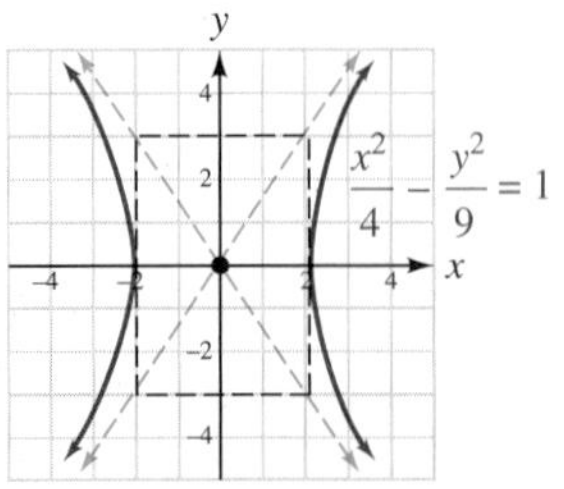

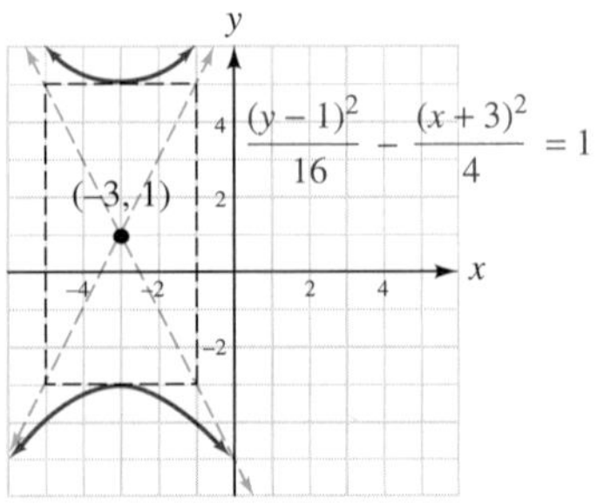

The equation $\frac{(y-1)^2}{16} - \frac{(x+3)^2}{4} = 1$, which can be written $\frac{(y-1)^2}{4^2} - \frac{(x+3)^2}{2^2} = 1$, represents a hyperbola, centered at (−3, 1), that opens up and down. Here, $a = \mathbf{4}$ and $b = \mathbf{2}$.

To write the equation $25y^2 - 9x^2 = 225$ in standard form, divide both sides by 225 and simplify.

$$25y^2 - 9x^2 = 225$$

$$\frac{25y^2}{225} - \frac{9x^2}{225} = \frac{225}{225} \quad \text{To get 1 on the right side, divide both sides by 225.}$$

$$\frac{y^2}{9} - \frac{x^2}{25} = 1 \quad \text{Simplify each fraction.}$$

This result represents a hyperbola centered at the origin that opens up and down. Here, $a = 3$ and $b = 5$.

REVIEW EXERCISES

Graph each hyperbola. See AIE Appendix 3.

21. $\frac{y^2}{9} - \frac{x^2}{1} = 1$

22. $9(x - 1)^2 - 4(y + 1)^2 = 36$

23. $\frac{(y - 2)^2}{25} - \frac{(x + 1)^2}{25} = 1$

24. $xy = 9$

25. Electrostatic Repulsion. Two similarly charged particles are shot together for an almost head-on collision, as in the illustration. They repel each other and travel the two branches of the hyperbola given by $x^2 - 4y^2 = 4$ on the given coordinate system. How close do they get? 4 units

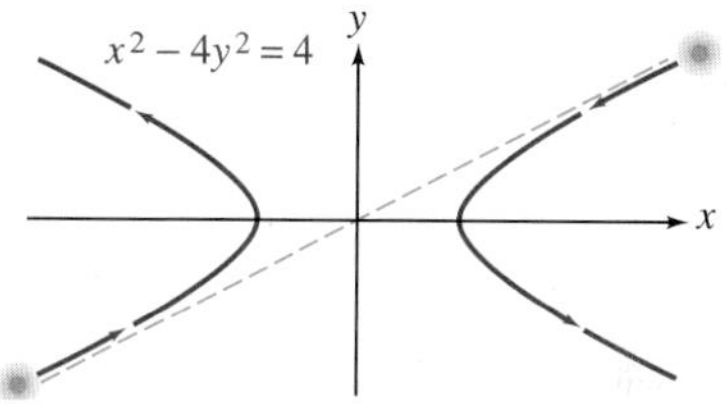

26. Determine whether the graph of each equation will be a circle, parabola, ellipse, or hyperbola.

a. $\frac{(x - 4)^2}{16} + \frac{y^2}{49} = 1$ Ellipse

b. $16(x + 3)^2 - 4(y - 1)^2 = 64$ Hyperbola

c. $x = -4y^2 - y + 1$ Parabola

d. $x^2 + 2x + y^2 - 4y = 40$ Circle

SECTION 13.4 Solving Nonlinear Systems of Equations

DEFINITIONS AND CONCEPTS	EXAMPLES
A **nonlinear system of equations** is a system that contains at least one nonlinear equation. Systems of nonlinear equations are solved by **graphing,** by **substitution,** or by **elimination (addition).**	To solve the nonlinear system $\begin{cases} x^2 + y^2 = 20 \\ y = x^2 \end{cases}$ by graphing, we graph the equations on the same rectangular coordinate system, and determine the coordinates of the points of intersection of the graphs. Since the points of intersection of the graphs are $(-2, 4)$ and $(2, 4)$, the solutions of the system are $(-2, 4)$ and $(2, 4)$ and the solution set is $\{(-2, 4), (2, 4)\}$. 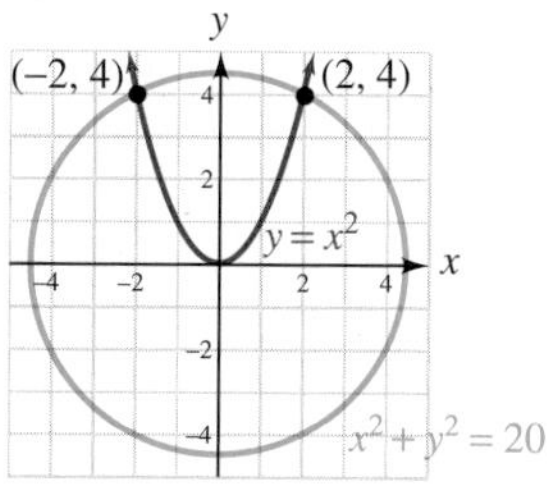

With the **substitution method,** the objective is to use an appropriate substitution to obtain *one* equation in *one* variable.

To use substitution to solve the nonlinear system $\begin{cases} y = 3x - 5 \\ x^2 + y^2 = 5 \end{cases}$, we substitute $3x - 5$ for y in the second equation and solve for x.

$$x^2 + y^2 = 5$$

$$x^2 + (3x - 5)^2 = 5 \quad \text{This is a quadratic equation in } x.$$

$$x^2 + 9x^2 - 30x + 25 = 5$$

$$10x^2 - 30x + 20 = 0 \quad \text{Combine terms and subtract 5 from both sides.}$$

$$x^2 - 3x + 2 = 0 \quad \text{Divide both sides by 10.}$$

$$(x - 2)(x - 1) = 0 \quad \text{Factor.}$$

$$x - 2 = 0 \quad \text{or} \quad x - 1 = 0 \quad \text{Set each factor equal to 0.}$$

$$x = 2 \quad | \quad x = 1$$

If $x = 2$, then $y = 3x - 5 = 3(2) - 5 = 1$.

If $x = 1$, then $y = 3x - 5 = 3(1) - 5 = -2$.

The two solutions of the system are (2, 1) and (1, −2).

With the **elimination (addition) method,** we combine the equations in a way that will eliminate the terms of one of the variables.

To use elimination to solve the nonlinear system $\begin{cases} x^2 - y = 0 \\ x + y = 0 \end{cases}$, we add the equations to get $x^2 + x = 0$. Then we factor this result to get $x = 0$ or $x = -1$. We can substitute these values into the second equation to find y.

If $x = 0$: $x + y = 0$ This is the second equation.
$0 + y = 0$ Substitute 0 for x.
$y = 0$

If $x = -1$: $x + y = 0$ This is the second equation.
$-1 + y = 0$ Substitute −1 for x.
$y = 1$

The two solutions of the system are (0, 0) and (−1, 1).

REVIEW EXERCISES

27. Determine whether $\left(-\sqrt{11}, -3\right)$ is a solution of the system: $\begin{cases} x^2 + y^2 = 20 \\ x^2 - y^2 = 2 \end{cases}$ Yes

28. The graphs of $y^2 - x^2 = 9$ and $x^2 + y^2 = 9$ are shown. Estimate the solutions of the system

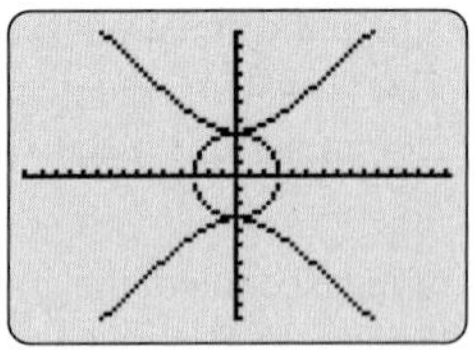

$\begin{cases} y^2 - x^2 = 9 \\ x^2 + y^2 = 9 \end{cases}$

(0, 3), (0, −3)

29. Solve the system $\begin{cases} xy = 4 \\ y = 2x - 2 \end{cases}$ by graphing.
(2, 2), (−1, −4); see AIE Appendix 3.

30. Determine the maximum number of solutions there could be for a system of equations consisting of the given curves.

a. A line and an ellipse 2 **b.** Two hyperbolas 4

c. An ellipse and a circle 4 **d.** A parabola and a circle 4

31. Suppose the x-coordinate of both points of intersection of the circle, represented by $x^2 + y^2 = 1$, and the hyperbola, defined by $4y^2 - x^2 = 4$, is 0. Without graphing, determine the y-coordinates of both points of intersection. Express the answers as ordered-pair solutions. (0, 1), (0, −1)

32. Find a substitution equation that can be used to solve $\begin{cases} x^2 + y^2 = 16 \\ 3x - y = 1 \end{cases}$. Do not solve the system. $y = 3x - 1$

Solve each system for real values of x and y.

33. $\begin{cases} y^2 - x^2 = 16 \\ y + 4 = x^2 \end{cases}$
(0, −4), (−3, 5), (3, 5)

34. $\begin{cases} y = -x^2 + 2 \\ x^2 - y - 2 = 0 \end{cases}$
$\left(\sqrt{2}, 0\right), \left(-\sqrt{2}, 0\right)$

35. $\begin{cases} x^2 + 2y^2 = 12 \\ 2x - y = 2 \end{cases}$
$(2, 2), \left(-\frac{2}{9}, -\frac{22}{9}\right)$

36. $\begin{cases} 3x^2 + y^2 = 52 \\ x^2 - y^2 = 12 \end{cases}$
(4, 2), (4, −2), (−4, 2), (−4, −2)

37. $\begin{cases} \dfrac{x^2}{16} + \dfrac{y^2}{12} = 1 \\ \dfrac{x^2}{1} - \dfrac{y^2}{3} = 1 \end{cases}$
(2, 3), (2, −3), (−2, 3), (−2, −3)

38. $\begin{cases} xy = 4 \\ \dfrac{x^2}{1} + \dfrac{y^2}{2} = 9 \end{cases}$
$\left(2\sqrt{2}, \sqrt{2}\right), \left(-2\sqrt{2}, -\sqrt{2}\right)$, (1, 4), (−1, −4)

39. $\begin{cases} y = -x^2 + 1 \\ x + y = 5 \end{cases}$
No solution; ∅

40. $\begin{cases} x = y^2 - 3 \\ x = y^2 - 3y \end{cases}$
(−2, 1)

13 Chapter Test

1. Fill in the blanks.
   a. The curves formed by the intersection of a plane with an infinite right-circular cone are called conic sections.
   b. A circle is the set of all points in a plane that are a fixed distance from a point called its center. The fixed distance is called the radius of the circle.
   c. The standard form for the equation of a hyperbola centered at the origin that opens left and right is $\frac{x^2}{a^2} - \frac{y^2}{b^2} = 1$.
   d. $\begin{cases} y = x^2 + x - 4 \\ x^2 + y^2 = 36 \end{cases}$ is a nonlinear system of equations.
   e. The standard form for the equation of an ellipse centered at the origin is $\frac{x^2}{a^2} + \frac{y^2}{b^2} = 1$.
2. Find the center and the radius of the circle represented by the equation $x^2 + y^2 = 100$ and graph it. See AIE Appendix 3. $(0, 0)$; $r = 10$
3. Find the center and the radius of the circle represented by the equation $x^2 + y^2 + 4x - 6y = 5$. $(-2, 3)$; $r = 3\sqrt{2}$
4. **TV History.** In the early days of television, stations broadcast a black-and-white test pattern like that shown here during the early morning hours. Use the given coordinate system to write an equation of the large, bold circle in the center of the pattern. $(x - 4)^2 + (y - 3)^2 = 9$

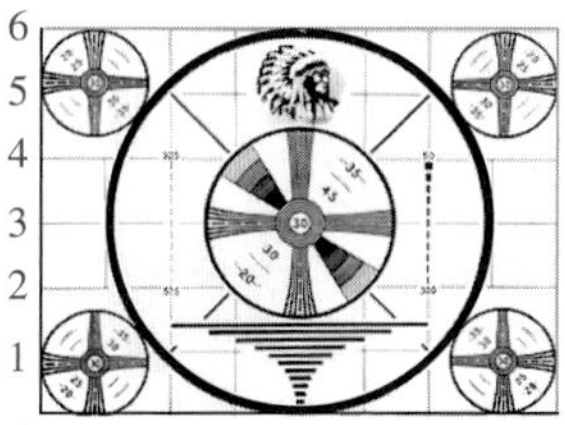

5. **Frisbee.** In the illustration, an Ultimate Frisbee is centered on a rectangular coordinate system. Determine the *diameter* of the circular disc from the given equation. $\frac{21}{2}$ in. $= 10.5$ in.

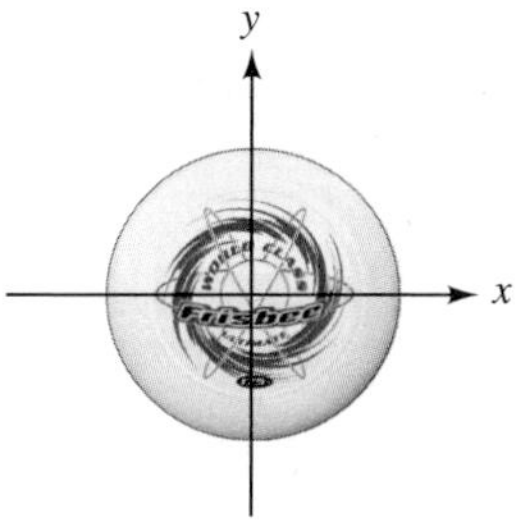

$x^2 + y^2 = \frac{441}{16}$

6. Fill in the blanks to complete the square on the right side of the following equation.

$$x = y^2 + 8y + 10$$
$$x = (y^2 + 8y + 16) + 10 - 16$$
$$x = (y + 4)^2 - 6$$

Graph each equation. See AIE Appendix 3.

7. $(x + 2)^2 + (y - 1)^2 = 9$
8. $x = y^2 - 2y + 3$
9. $y = -2x^2 - 4x + 5$
10. $xy = -4$
11. $9x^2 + 4y^2 = 36$
12. $\frac{(x - 2)^2}{9} - \frac{y^2}{1} = 1$
13. $\frac{(x - 3)^2}{49} + \frac{(y + 2)^2}{16} = 1$
14. $x^2 + y^2 = 7$
15. $x = -\frac{1}{2}y^2$
16. $\frac{y^2}{25} - \frac{x^2}{9} = 1$

17. Write the equation in standard form of the ellipse graphed here. $\frac{(x - 1)^2}{16} + \frac{(y + 2)^2}{9} = 1$

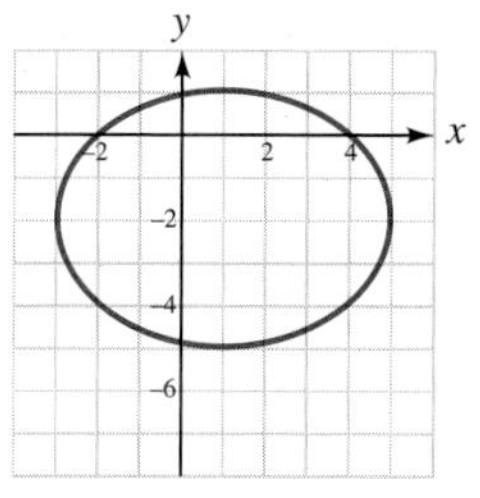

18. **Light.** The cross section of a parabolic mirror is given by the equation $x = \frac{1}{10}y^2$, with distances measured in inches. If the dish is 10 inches wide, how deep is it? 2.5 in.

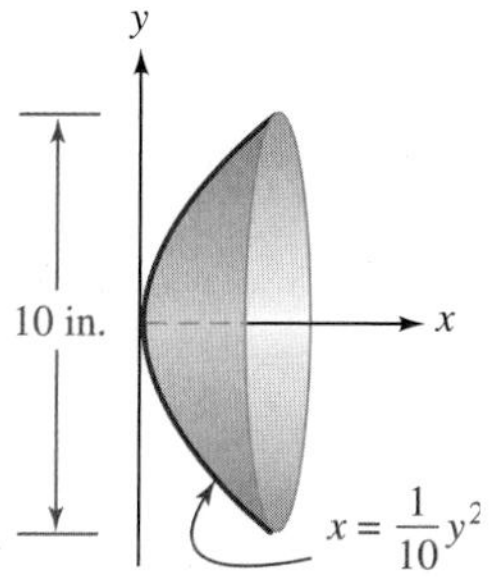

19. **Advertising.** An elliptical logo for Tom's Shoes is to be centered on a 36-by-60-inch rectangular background, leaving 2 inches of space on all sides. Find the equation of the ellipse. $\frac{x^2}{784} + \frac{y^2}{256} = 1$

20. Find the center and the length and width of the central rectangle of the graph of $(x + 1)^2 - (y - 1)^2 = 4$. $(-1, 1)$; length: 4 units, width: 4 units
21. Find the equation in standard form of the hyperbola graphed here. $\frac{y^2}{16} - \frac{x^2}{36} = 1$

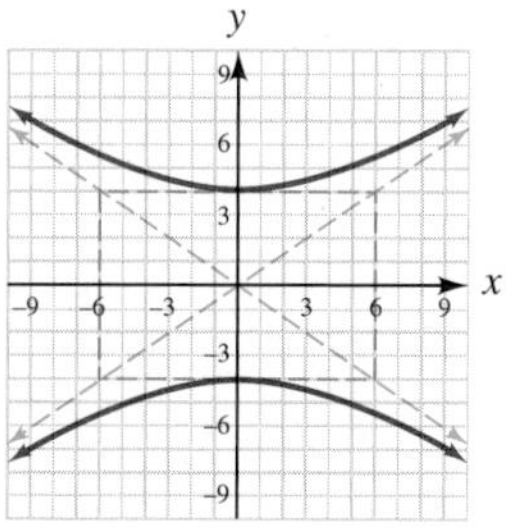

22. Determine whether the graph of each equation will be a circle, a parabola, an ellipse, or a hyperbola.
   a. $25x^2 + 100y^2 = 400$ Ellipse
   b. $9x^2 - y^2 = 9$ Hyperbola
   c. $x^2 + 8x + y^2 - 16y - 1 = 0$ Circle
   d. $x = 8y^2 - 9y + 4$ Parabola
23. Solve $\begin{cases} x^2 + y^2 = 25 \\ y - x = 1 \end{cases}$ graphically. $(-4, -3), (3, 4)$; see AIE Appendix 3.

Solve each system for real values of x and y.

24. $\begin{cases} 2x - y = -2 \\ x^2 + y^2 = 16 + 4y \end{cases}$ $(2, 6), (-2, -2)$
25. $\begin{cases} 5x^2 - y^2 - 3 = 0 \\ x^2 + 2y^2 = 5 \end{cases}$ $(1, \sqrt{2}), (1, -\sqrt{2}), (-1, \sqrt{2}), (-1, -\sqrt{2})$

26. $\begin{cases} xy = -\dfrac{9}{2} \\ 3x + 2y = 6 \end{cases}$ $\left(-1, \frac{9}{2}\right), \left(3, -\frac{3}{2}\right)$

27. $\begin{cases} y = x + 1 \\ x^2 - y^2 = 1 \end{cases}$ $(-1, 0)$

28. $\begin{cases} x^2 + 3y^2 = 6 \\ x^2 + y = 8 \end{cases}$ No solution; $\varnothing$

Group Project

Parabolas

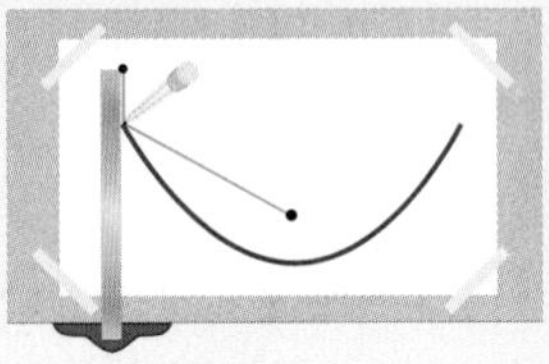

Overview: In this activity, you will construct several models of parabolas.

Instructions: Form groups of 2 or 3 students. You will need a T-square, string, paper, pencil, and a thumbtack. To construct a parabola, secure one end of a piece of string that is as long as the T-square to a large piece of paper using a brad or thumbtack, as shown at the left. Attach the other end of the string to the upper end of the T-square. Hold the string taut against the T-square with a pencil and slide the T-square along the edge of the table. As the T-square moves, the pencil will trace a parabola.

Each point on the parabola is the same distance away from a given point as it is from a given line. With this model, what is the given point, and what is the given line?

Make other models by moving the fixed point closer and farther away from the edge of the table. How is the shape of the parabola affected?

Ellipses

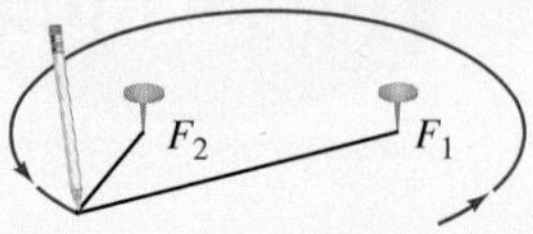

Overview: In this activity, you will construct several ellipses.

Instructions: Form groups of 2 or 3 students. You will need two thumbtacks, a pencil, and a length of string with a loop tied at one end. To construct an ellipse, place two thumbtacks (or brads) fairly close together, as shown in the illustration. Catch the loop of the string with the point of the pencil and, keeping the string taut, draw the ellipse.

Make several models by moving one of the thumbtacks farther away and then closer to the other thumbtack. How does the shape of the ellipse change?

For each point on the ellipse, the sum of the distances of the point from two given points is a constant. With this method of construction, what are the two points? What is the constant distance?

Miscellaneous Topics

14

from Campus to Careers

Real Estate Sales Agent

Buying a house is probably the biggest purchase that most people will make in their lives. The complex process of purchasing a home is much easier with the help of a real estate agent. Real estate agents use their mathematical skills in many ways. They calculate square footage, appraise property, calculate commissions, and write offer sheets. Technology is widely used in the real estate industry. Most sales agents use computers to locate and list available properties and identify sources of financing.

Problem 81 in **Study Set 14.3** involves a situation that real estate agents and their clients are always concerned with—changing property values.

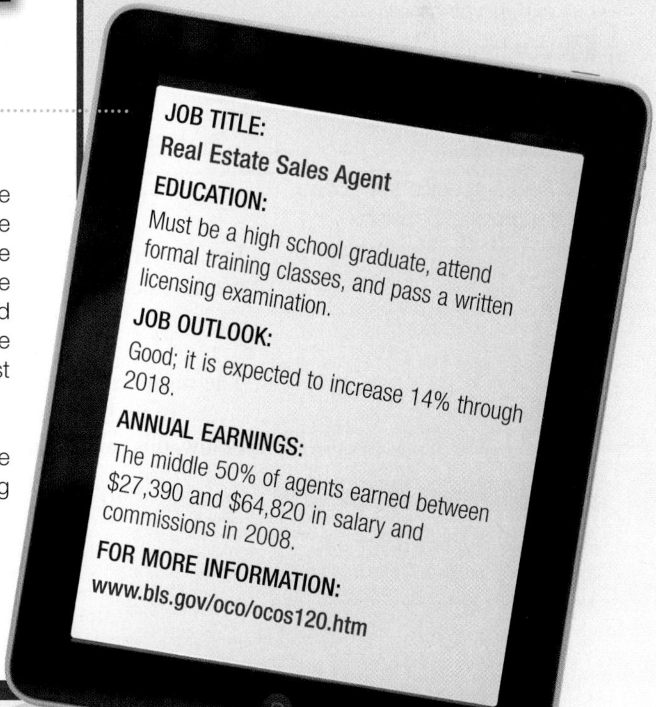

JOB TITLE:
Real Estate Sales Agent
EDUCATION:
Must be a high school graduate, attend formal training classes, and pass a written licensing examination.
JOB OUTLOOK:
Good; it is expected to increase 14% through 2018.
ANNUAL EARNINGS:
The middle 50% of agents earned between $27,390 and $64,820 in salary and commissions in 2008.
FOR MORE INFORMATION:
www.bls.gov/oco/ocos120.htm

Ultimately, your choice of career will determine the math course(s) that you need to take after Intermediate Algebra. Before the end of this term, it would be wise to have at least a general idea of your career goals.

HOW DO YOU DECIDE?: Seek the advice of a counselor, visit your school's career center, search the Internet, or read books that will help you discover your interests and possible related careers.

ONCE YOU'VE DECIDED: Talk to your counselor and consult the appropriate college catalogs to develop a long-term plan that will put you on the correct educational path.

Now Try This

1. Do you have a career goal in mind? If so, what is it?
2. Take at least two personality tests and two career-choice tests. A list of tests offered online can be found at www.cengage.com/math/tussy
3. Visit a counselor to discuss which classes you should take during your next term and beyond. Make a list of classes that your counselor suggests that you take.

SECTION 14.1

The Binomial Theorem

OBJECTIVES

1. Use Pascal's triangle to expand binomials.
2. Use factorial notation.
3. Use the binomial theorem to expand binomials.
4. Find a specific term of a binomial expansion.

ARE YOU READY? *Are You Ready? exercises available online at www.webassign.net/brookscole*

The following problems review some basic skills that are needed when working with the binomial theorem.

1. Determine the coefficient of each term: $x^4 + 4x^3 + 6x^2 + 4x + 1$ 1, 4, 6, 4, 1
2. Find the product: $(a + b)^2$ $a^2 + 2ab + b^2$
3. Evaluate: $5 \cdot 4 \cdot 3 \cdot 2 \cdot 1$ 120
4. Simplify: $\dfrac{10 \cdot 9 \cdot 8}{2 \cdot 3 \cdot 4}$ 30

In Chapter 5, we discussed how to raise binomials to positive-integer powers. For example, we have seen that

$$(a + b)^2 = a^2 + 2ab + b^2$$

The result, $a^2 + 2ab + b^2$, is called the **binomial expansion** of $(a + b)^2$.

To find the binomial expansion of $(a + b)^3$, we multiply $(a + b)^2$ by $(a + b)$. As we see below, this process involves several steps and many ways to make errors.

The Language of Algebra

Recall that two-term polynomial expressions such as $a + b$ and $3u - 2v$ are called **binomials**.

$$\begin{aligned}(a + b)^3 &= (a + b)(a + b)^2 \\ &= (a + b)(a^2 + 2ab + b^2) \\ &= a^3 + 2a^2b + ab^2 + a^2b + 2ab^2 + b^3 \\ &= a^3 + 3a^2b + 3ab^2 + b^3\end{aligned}$$

In this section, we will discuss a method for finding binomial expansions quickly, without having to perform all of the steps shown above.

1 Use Pascal's Triangle to Expand Binomials.

To **expand** a binomial of the form $(a + b)^n$, where n is a nonnegative integer, means to write it as a sum of terms. To develop a method for expanding such binomials, consider the following:

$$
\begin{array}{lcl}
(a+b)^0 = & 1 & \text{1 term} \\
(a+b)^1 = & a + b & \text{2 terms} \\
(a+b)^2 = & a^2 + 2ab + b^2 & \text{3 terms} \\
(a+b)^3 = & a^3 + 3a^2b + 3ab^2 + b^3 & \text{4 terms} \\
(a+b)^4 = & a^4 + 4a^3b + 6a^2b^2 + 4ab^3 + b^4 & \text{5 terms} \\
(a+b)^5 = & a^5 + 5a^4b + 10a^3b^2 + 10a^2b^3 + 5ab^4 + b^5 & \text{6 terms} \\
(a+b)^6 = & a^6 + 6a^5b + 15a^4b^2 + 20a^3b^3 + 15a^2b^4 + 6ab^5 + b^6 & \text{7 terms}
\end{array}
$$

The Language of Algebra

When we **expand** a power of a binomial, the result is called a **binomial expansion.** For powers greater than or equal to 2, an expansion has more terms than the original binomial.

Several patterns appear in these expansions:

1. Each expansion has one more term than the power of the binomial.
2. For each term of an expansion, the sum of the exponents on a and b is equal to the exponent of the binomial being expanded. For example, in the expansion of $(a + b)^5$, the sum of the exponents in each term is 5:

$$(a+b)^5 = a^5 + \underbrace{5a^4b}_{4+1=5} + \underbrace{10a^3b^2}_{3+2=5} + \underbrace{10a^2b^3}_{2+3=5} + \underbrace{5ab^4}_{1+4=5} + b^5$$

3. The first term in each expansion is a, raised to the power of the binomial, and the last term in each expansion is b, raised to the power of the binomial.
4. The exponents on a decrease by one in each successive term, ending with $a^0 = 1$ in the last term. The exponents on b, beginning with $b^0 = 1$ in the first term, increase by one in each successive term. For example, the expansion of $(a + b)^4$ could be written as

$$a^4b^0 + 4a^3b^1 + 6a^2b^2 + 4a^1b^3 + a^0b^4$$

Thus, the variables have the pattern

$$a^n, \quad a^{n-1}b, \quad a^{n-2}b^2, \quad \ldots, \quad ab^{n-1}, \quad b^n$$

5. The coefficients of each expansion begin with 1, increase through some values, and then decrease through those same values, back to 1.

To see another pattern, we write just the coefficients of each expansion of $a + b$ in a triangular array:

$$
\begin{array}{ccccccccccccc l}
 & & & & & & 1 & & & & & & & \text{Row for } (a+b)^0 \\
 & & & & & 1 & & 1 & & & & & & \text{Row for } (a+b)^1 \\
 & & & & 1 & & 2 & & 1 & & & & & \text{Row for } (a+b)^2 \\
 & & & 1 & & 3 & & 3 & & 1 & & & & \text{Row for } (a+b)^3 \\
 & & 1 & & 4 & & 6 & & 4 & & 1 & & & \text{Row for } (a+b)^4 \\
 & 1 & & 5 & & 10 & & 10 & & 5 & & 1 & & \text{Row for } (a+b)^5 \\
1 & & 6 & & 15 & & 20 & & 15 & & 6 & & 1 & \text{Row for } (a+b)^6
\end{array}
$$

The Language of Algebra

This array of numbers is named **Pascal's triangle** in honor of the French mathematician Blaise Pascal (1623–1662).

In this array, called **Pascal's triangle,** each entry between the 1's is the sum of the closest pair of numbers in the line immediately above it. For example, the first 15 in the bottom row is the sum of the 5 and 10 immediately above it. Pascal's triangle continues with the same pattern forever. The next two lines are

$$
\begin{array}{ccccccccccccccccc l}
 & 1 & & 7 & & 21 & & 35 & & 35 & & 21 & & 7 & & 1 & & \text{Row for } (a+b)^7 \\
1 & & 8 & & 28 & & 56 & & 70 & & 56 & & 28 & & 8 & & 1 & \text{Row for } (a+b)^8
\end{array}
$$

EXAMPLE 1 Expand: $(x + y)^5$

Strategy We will use the pattern shown on the previous page for raising binomials to powers and Pascal's triangle.

Why The pattern provides the variable expressions in the expansion and Pascal's triangle provides their coefficients.

Solution The first term in the expansion is x^5, and the exponents on x decrease by one in each successive term. A y first appears in the second term, and the exponents on y increase by one in each successive term, concluding when the term y^5 is reached. Thus, the variable expressions in the expansion are

$$x^5, \quad x^4y, \quad x^3y^2, \quad x^2y^3, \quad xy^4, \quad y^5$$

1
1 1
1 2 1
1 3 3 1
1 4 6 4 1
1 5 10 10 5 1
1 6 15 20 15 6 1
1 7 21 35 35 21 7 1
1 8 28 56 70 56 28 8 1

Since the exponent of the binomial that is being expanded is 5, the coefficients of these variables are found in row 5 of Pascal's triangle.

1 5 10 10 5 1 Remember, the 1 at the top of Pascal's triangle is labeled row 0.

Combining this information gives the following expansion:

$$(x + y)^5 = x^5 + 5x^4y + 10x^3y^2 + 10x^2y^3 + 5xy^4 + y^5$$

Teaching Example 1 Expand: $(a + b)^5$
Answer: $a^5 + 5a^4b + 10a^3b^2 + 10a^2b^3 + 5ab^4 + b^5$

Self Check 1 Expand: $(x + y)^4$ $x^4 + 4x^3y + 6x^2y^2 + 4xy^3 + y^4$

Now Try Problem 25

EXAMPLE 2 Expand: $(u - v)^4$

Strategy We will use the pattern shown on the previous page for raising binomials to powers and Pascal's triangle.

Why The pattern provides the variable expressions in the expansion and Pascal's triangle provides their coefficients.

Solution We note that $(u - v)^4$ can be written as $[u + (-v)]^4$. The variable expressions in this expansion are

$$u^4, \quad u^3(-v), \quad u^2(-v)^2, \quad u(-v)^3, \quad (-v)^4$$

and the coefficients are given in row 4 of Pascal's triangle:

1 4 6 4 1 Remember, the 1 at the top of Pascal's triangle is labeled row 0.

Thus, the required expansion is

$$(u - v)^4 = u^4 + 4u^3(-v) + 6u^2(-v)^2 + 4u(-v)^3 + (-v)^4$$

The Language of Algebra

To **alternate** means to change back and forth. In this expansion, the signs + and − alternate.

Now we simplify each term. When $-v$ is raised to an even power, the sign is positive, and when $-v$ is raised to an odd power, the sign is negative. This causes the signs of the terms in the expansion to alternate between + and −.

$$(u - v)^4 = u^4 - 4u^3v + 6u^2v^2 - 4uv^3 + v^4$$

Teaching Example 2 Expand: $(c - d)^4$
Answer: $c^4 - 4c^3d + 6c^2d^2 - 4cd^3 + d^4$

Self Check 2 Expand: $(x - y)^5$ $x^5 - 5x^4y + 10x^3y^2 - 10x^2y^3 + 5xy^4 - y^5$

Now Try Problem 27

2 Use Factorial Notation.

Although Pascal's triangle gives the coefficients of the terms in a binomial expansion, it is not the easiest way to expand a binomial. To develop a better way, we introduce **factorial notation.** The symbol $n!$ (read as "n **factorial**") is defined as follows.

Factorial Notation

$n!$ is the product of consecutively decreasing natural numbers from n to 1.
For any natural number n,

$$n! = n(n-1)(n-2)(n-3) \cdot \cdots \cdot 3 \cdot 2 \cdot 1$$

Zero factorial is defined as $0! = 1$.

EXAMPLE 3 Evaluate each expression: **a.** $4!$ **b.** $6!$ **c.** $3! \cdot 2!$ **d.** $5! \cdot 0!$

Strategy We will use the definition of $n!$.

Why The definition explains how to evaluate factorials.

Solution

a. $4! = 4 \cdot 3 \cdot 2 \cdot 1 = 24$ Read as "4 factorial."

b. $6! = 6 \cdot 5 \cdot 4 \cdot 3 \cdot 2 \cdot 1 = 720$ Read as "6 factorial."

c. $3! \cdot 2! = (3 \cdot 2 \cdot 1) \cdot (2 \cdot 1) = 6 \cdot 2 = 12$ Find each factorial and multiply the results.

d. $5! \cdot 0! = (5 \cdot 4 \cdot 3 \cdot 2 \cdot 1) \cdot 1 = 120$ Simplify: $0! = 1$.

Teaching Example 3 Evaluate each expression:
a. $5!$ **b.** $4! \cdot 2!$ **c.** $7! \cdot 0!$
Answers: **a.** 120 **b.** 48 **c.** 5,040

Self Check 3 Evaluate each expression: **a.** $7!$ 5,040 **b.** $4! \cdot 3!$ 144 **c.** $1! \cdot 0!$ 1

Now Try Problems 31 and 37

Using Your Calculator ▶ Factorials

We can find factorials using a calculator. For example, to find $12!$ with a scientific calculator, we enter

12 [x!] (You may have to use a [2nd] or [SHIFT] key first.) `479001600`

To find $12!$ on a graphing calculator, we enter

12 [MATH] [▶] to PRB 4 [ENTER] `12!` `479001600`

The following property follows from the definition of factorial.

Factorial Property

For any natural number n,

$$n(n-1)! = n!$$

We can use this property to evaluate many expressions involving factorials.

EXAMPLE 4 Evaluate each expression: **a.** $\frac{6!}{5!}$ **b.** $\frac{10!}{8!(10-8)!}$

Strategy We will use the factorial property to partially expand the factorial and then we will simplify the fraction.

Why By using this approach, we can avoid difficult multiplications and divisions.

Solution **a.** If we write 6! as $6 \cdot 5!$, we can simplify the fraction by removing the common factor 5! in the numerator and denominator.

$$\frac{6!}{5!} = \frac{6 \cdot 5!}{5!} = \frac{6 \cdot \overset{1}{\cancel{5!}}}{\underset{1}{\cancel{5!}}} = 6 \quad \text{Simplify: } \frac{5!}{5!} = 1.$$

Teaching Tip: Point out that writing some of the descending factors of a factorial expression is often helpful when simplifying a fraction such as $\frac{10!}{8!2!}$.

b. We subtract within the parentheses, write 10! as $10 \cdot 9 \cdot 8!$, and simplify.

$$\frac{10!}{8!(10-8)!} = \frac{10!}{8! \cdot 2!} = \frac{10 \cdot 9 \cdot \overset{1}{\cancel{8!}}}{\underset{1}{\cancel{8!}} \cdot 2!} = \frac{5 \cdot \overset{1}{\cancel{2}} \cdot 9}{\underset{1}{\cancel{2}} \cdot 1} = 45 \quad \text{Simplify: } \frac{8!}{8!} = 1. \text{ Factor 10 as } 5 \cdot 2 \text{ and simplify: } \frac{2}{2} = 1.$$

Teaching Example 4 Evaluate each expression:
a. $\frac{5!}{4!}$ b. $\frac{10!}{7!(10-7)!}$
Answers: a. 5 b. 120

Self Check 4 Evaluate each expression: **a.** $\frac{4!}{3!}$ 4 **b.** $\frac{7!}{5!(7-5)!}$ 21

Now Try Problems 39 and 47

3 Use the Binomial Theorem to Expand Binomials.

The following theorem summarizes our observations about binomial expansions and our work with factorials. Known as the **binomial theorem**, it is usually the best way to expand a binomial.

The Binomial Theorem

For any positive integer n,

$$(a+b)^n = a^n + \frac{n!}{1!(n-1)!}a^{n-1}b + \frac{n!}{2!(n-2)!}a^{n-2}b^2 + \frac{n!}{3!(n-3)!}a^{n-3}b^3 + \cdots + \frac{n!}{r!(n-r)!}a^{n-r}b^r + \cdots + b^n$$

In the binomial theorem, the exponents on the variables follow the familiar pattern:

- The sum of the exponents on a and b in each term is n.
- The exponents on a decrease by 1 in each successive term.
- The exponents on b increase by 1 in each successive term.

The method of finding the coefficients involves factorials. Except for the first and last terms, the numerator of each coefficient is $n!$. If the exponent on b in a particular term is r, the denominator of the coefficient of that term is $r!(n-r)!$.

EXAMPLE 5 Use the binomial theorem to expand $(a+b)^3$.

Strategy We will substitute 3 for n in the binomial theorem and simplify.

Why The binomial theorem is the fastest way to expand expressions of the form $(a+b)^n$.

Solution

$$(a + b)^3 = a^3 + \frac{3!}{1!(3-1)!}a^2b + \frac{3!}{2!(3-2)!}ab^2 + b^3$$
$$= a^3 + \frac{3!}{1! \cdot 2!}a^2b + \frac{3!}{2! \cdot 1!}ab^2 + b^3$$
$$= a^3 + \frac{\overset{1}{3 \cdot \cancel{2!}}}{\underset{1}{1! \cdot \cancel{2!}}}a^2b + \frac{\overset{1}{3 \cdot \cancel{2!}}}{\underset{1}{\cancel{2!} \cdot 1!}}ab^2 + b^3$$

Write 3! as 3 · 2! to simplify the fractions.

$$= a^3 + 3a^2b + 3ab^2 + b^3$$

Teaching Example 5 Use the binomial theorem to expand: $(x + y)^3$
Answer: $x^3 + 3x^2y + 3xy^2 + y^3$

Self Check 5 Use the binomial theorem to expand $(a + b)^4$. $a^4 + 4a^3b + 6a^2b^2 + 4ab^3 + b^4$

Now Try Problem 57

We can find expansions of binomials in variables other than a and b by making substitutions into the binomial theorem.

EXAMPLE 6 Use the binomial theorem to expand $(x - y)^4$.

Strategy First, we will write $(x - y)^4$ as $[x + (-y)]^4$. Then we will use the binomial theorem with $a = x$, $b = -y$, and $n = 4$.

Why To substitute directly into the binomial theorem, the difference within the parentheses, $x - y$, must be expressed as a sum.

Solution

$$(x - y)^4 = [x + (-y)]^4$$
$$= x^4 + \frac{4!}{1!(4-1)!}x^3(-y) + \frac{4!}{2!(4-2)!}x^2(-y)^2 + \frac{4!}{3!(4-3)!}x(-y)^3 + (-y)^4$$
$$= x^4 - \frac{4!}{1! \cdot 3!}x^3y + \frac{4!}{2! \cdot 2!}x^2y^2 - \frac{4!}{3! \cdot 1!}xy^3 + y^4$$
$$= x^4 - \frac{\overset{1}{4 \cdot \cancel{3!}}}{\underset{1}{1! \cdot \cancel{3!}}}x^3y + \frac{\overset{1}{4 \cdot 3 \cdot \cancel{2!}}}{\underset{1}{\cancel{2!} \cdot 2 \cdot 1}}x^2y^2 - \frac{\overset{1}{4 \cdot \cancel{3!}}}{\underset{1}{\cancel{3!} \cdot 1!}}xy^3 + y^4$$

Write 4! as 4 · 3! and as 4 · 3 · 2! to simplify the fractions.

$$= x^4 - 4x^3y + 6x^2y^2 - 4xy^3 + y^4$$

Note the alternating signs.

Teaching Example 6 Use the binomial theorem to expand: $(a - b)^4$
Answer: $a^4 - 4a^3b + 6a^2b^2 - 4ab^3 + b^4$

Self Check 6 Use the binomial theorem to expand $(x - y)^3$. $x^3 - 3x^2y + 3xy^2 - y^3$

Now Try Problem 59

EXAMPLE 7 Use the binomial theorem to expand $(3u - 2v)^4$.

Strategy We will write the expansion of $(a + b)^4$. Then we will substitute for a and b to find the expansion of $(3u - 2v)^4$.

Why For binomials with more complicated terms, the calculations are often easier if the general expansion is written first, followed by the appropriate substitutions.

Solution We can use the binomial theorem to expand $(a + b)^4$.

$$(a + b)^4 = a^4 + \frac{4!}{1!(4-1)!}a^3b + \frac{4!}{2!(4-2)!}a^2b^2 + \frac{4!}{3!(4-3)!}ab^3 + b^4$$
$$= a^4 + 4a^3b + 6a^2b^2 + 4ab^3 + b^4$$

If we write $(3u - 2v)^4$ as $[3u + (-2v)]^4$, we see that the expressions $3u$ and $-2v$ can be substituted for a and b respectively in the expansion of $(a + b)^4$.

$$(3u - 2v)^4 = (3u)^4 + 4(3u)^3(-2v) + 6(3u)^2(-2v)^2 + 4(3u)(-2v)^3 + (-2v)^4$$
$$= 81u^4 - 216u^3v + 216u^2v^2 - 96uv^3 + 16v^4$$

Teaching Example 7 Use the binomial theorem to expand: $(4x - 3y)^4$
Answer: $256x^4 - 768x^3y + 864x^2y^2 - 432xy^3 + 81y^4$

Self Check 7 Use the binomial theorem to expand $(4a - 5b)^3$. $64a^3 - 240a^2b + 300ab^2 - 125b^3$

Now Try Problem 69

4 Find a Specific Term of a Binomial Expansion.

To find a specific term of a binomial expansion, we don't need to write out the entire expansion. The binomial theorem and the pattern of the terms suggest the following method for finding a single term of an expansion.

Finding a Specific Term of a Binomial Expansion

The $(r + 1)$st term of the expansion of $(a + b)^n$ is

$$\frac{n!}{r!(n-r)!}a^{n-r}b^r$$

EXAMPLE 8 Find the 4th term of the expansion of $(a + b)^9$.

Strategy We will determine n and r and substitute into the formula for finding a specific term of a binomial expansion.

Why We will use the formula because it enables us to find the fourth term of the expansion without us having to write out all the terms of the expansion.

Solution To use the formula for finding a specific term of $(a + b)^9$, we must determine n and r. Since $r + 1 = 4$ in the fourth term, $r = 3$ and since this binomial is raised to the 9th power, $n = 9$.

We substitute 3 for r and 9 for n into the formula to find the fourth term.

$$\frac{n!}{r!(n-r)!}a^{n-r}b^r = \frac{9!}{3!(9-3)!}a^{9-3}b^3$$
$$= \frac{9!}{3!6!}a^6b^3 \quad \text{Evaluate: } \frac{9!}{3!6!} = \frac{9 \cdot 8 \cdot 7 \cdot \overset{1}{\cancel{6!}}}{3 \cdot 2 \cdot 1 \cdot \underset{1}{\cancel{6!}}} = 84.$$
$$= 84a^6b^3$$

Teaching Example 8 Find the 4th term of the expansion of $(x + y)^8$.
Answer: $56x^5y^3$

Self Check 8 Find the 3rd term of the expansion of $(a + b)^9$. $36a^7b^2$

Now Try Problem 77

EXAMPLE 9 Find the 6th term of the expansion of $\left(x^2 - \frac{y}{2}\right)^7$.

Strategy We will determine n, r, a, and b and substitute these values into the formula for finding a specific term of a binomial expansion.

Why We will use the formula because it enables us to find the sixth term of the expansion without us having to write out all the terms of the expansion.

Solution To use the formula for finding a specific term of $\left(x^2 - \frac{y}{2}\right)^7$, we must determine n, r, a, and b. In the sixth term, $r + 1 = 6$. So $r = 5$. By comparing $\left(x^2 - \frac{y}{2}\right)^7$ to $(a + b)^n$, we see that $a = x^2$, $b = -\frac{y}{2}$, and $n = 7$. We substitute these values into the formula as follows.

$$\begin{aligned}\frac{n!}{r!(n-r)!}a^{n-r}b^r &= \frac{7!}{5!(7-5)!}(x^2)^{7-5}\left(-\frac{y}{2}\right)^5\\ &= \frac{7!}{5!2!}(x^2)^2\left(-\frac{y^5}{32}\right)\\ &= -\frac{21}{32}x^4y^5\end{aligned}$$

Evaluate: $\frac{7!}{5!2!} = \frac{7 \cdot 6 \cdot \overset{1}{\cancel{5!}}}{\underset{1}{\cancel{5!}} \cdot 2 \cdot 1} = 21$.

Teaching Example 9 Find the 5th term of the expansion of $\left(a^2 - \frac{b}{2}\right)^6$.
Answer: $\frac{15}{16}a^4b^4$

Self Check 9 Find the 5th term of the expansion of $\left(c^2 - \frac{d}{3}\right)^7$. $\frac{35}{81}c^6d^4$

Now Try Problem 87

SECTION 14.1 STUDY SET

VOCABULARY

Fill in the blanks.

1. The two-term polynomial expression $a + b$ is called a binomial.
2. $a^4 + 4a^3b + 6a^2b^2 + 4ab^3 + b^4$ is the binomial expansion of $(a + b)^4$. To expand a binomial of the form $(a + b)^n$ means to write it as a sum of terms.
3. We can use the binomial theorem to raise binomials to positive-integer powers without doing the actual multiplication.
4. The array of numbers that gives the coefficients of the terms of a binomial expansion is called Pascal's triangle.
5. $n!$ (read as "n factorial") is the product of consecutively decreasing natural numbers from n to 1.
6. In the expansion $a^3 - 3a^2b + 3ab^2 - b^3$, the signs alternate between + and −.

CONCEPTS

Fill in the blanks.

7. The binomial expansion of $(m + n)^6$ has one more term than the power of the binomial.
8. For each term of the expansion of $(a + b)^8$, the sum of the exponents of a and b is 8.
9. The first term of the expansion of $(r + s)^{20}$ is r^{20} and the last term is s^{20}.
10. In the expansion of $(m - n)^{15}$, the exponents on m decrease and the exponents on n increase.
11. The coefficients of the terms of the expansion of $(c + d)^{20}$ begin with 1, increase through some values, and then decrease through those same values, back to 1.
12. Complete Pascal's Triangle:

1
1 1
1 2 1
1 3 3 1
1 4 6 4 1
1 5 10 10 5 1
1 6 15 20 15 6 1
1 7 21 35 35 21 7 1
1 8 28 56 70 56 28 8 1

13. $n \cdot (n - 1)! = n!$
14. $8! = 8 \cdot 7!$
15. $0! = 1$
16. According to the binomial theorem, the third term of the expansion of $(a + b)^n$ is $\frac{n!}{2!(n-2)!}a^{n-2}b^2$.
17. The coefficient of the fourth term of the expansion of $(a + b)^9$ is 9! divided by $3!(9 - 3)!$.
18. The exponent on a in the fourth term of the expansion of $(a + b)^6$ is 3 and the exponent on b is 3.
19. The exponent on a in the fifth term of the expansion of $(a + b)^6$ is 2 and the exponent on b is 4.
20. The expansion of $(a - b)^4$ is

$$a^4 - 4a^3b + 6a^2b^2 - 4ab^3 + b^4$$

21. $(x + y)^3$

$$= x^3 + \frac{3!}{1!(3-1)!}x^2y + \frac{3!}{2!(3-2)!}xy^2 + y^3$$

22. Fill in the blanks.
 a. The $(r + 1)$st term of the expansion of $(a + b)^n$ is $\frac{n!}{r!(n - r)!}a^{n-r}b^{r}$.
 b. To use the specific term formula to find the 6th term of the expansion of $\left(m + \frac{x}{2}\right)^8$, we note that $r = 5$, $n = 8$, $a = m$, and $b = \frac{x}{2}$.

NOTATION

Fill in the blanks.

23. $n! = n(n - 1)(n - 2) \cdot \cdots \cdot 3 \cdot 2 \cdot 1$

24. The symbol 5! is read as "five factorial" and it means $5 \cdot 4 \cdot 3 \cdot 2 \cdot 1$.

GUIDED PRACTICE

Use Pascal's triangle to expand each binomial. See Examples 1 and 2.

25. $(a + b)^3$ $a^3 + 3a^2b + 3ab^2 + b^3$
26. $(m + p)^4$ $m^4 + 4m^3p + 6m^2p^2 + 4mp^3 + p^4$
27. $(m - p)^5$ $m^5 - 5m^4p + 10m^3p^2 - 10m^2p^3 + 5mp^4 - p^5$
28. $(a - b)^3$ $a^3 - 3a^2b + 3ab^2 - b^3$

Evaluate each expression. See Examples 3 and 4.

29. $3!$ 6
30. $7!$ 5,040
31. $5!$ 120
32. $6!$ 720
33. $3! + 4!$ 30
34. $4! + 4!$ 48
35. $3!(4!)$ 144
36. $2!(3!)$ 12
37. $8(7!)$ 40,320
38. $4!(5)$ 120
39. $\frac{49!}{47!}$ 2,352
40. $\frac{101!}{100!}$ 101
41. $\frac{9!}{11!}$ $\frac{1}{110}$
42. $\frac{13!}{10!}$ 1,716
43. $\frac{9!}{7!0!}$ 72
44. $\frac{7!}{5!0!}$ 42
45. $\frac{5!}{1!(5 - 1)!}$ 5
46. $\frac{15!}{14!(15 - 14)!}$ 15
47. $\frac{5!}{3!(5 - 3)!}$ 10
48. $\frac{6!}{4!(6 - 4)!}$ 15
49. $\frac{5!(8 - 5)!}{4! \cdot 7!}$ $\frac{1}{168}$
50. $\frac{6! \cdot 7!}{(8 - 3)!(7 - 4)!}$ 5,040
51. $\frac{7!}{5!(7 - 5)!}$ 21
52. $\frac{8!}{6!(8 - 6)!}$ 28

Use a calculator to evaluate each expression. See Using Your Calculator: Factorials.

53. $11!$ 39,916,800
54. $13!$ 6,227,020,800
55. $20!$ $2.432902008 \times 10^{18}$
56. $55!$ $1.269640335 \times 10^{73}$

Use the binomial theorem to expand each expression. See Examples 5 and 6.

57. $(m + n)^4$ $m^4 + 4m^3n + 6m^2n^2 + 4mn^3 + n^4$
58. $(a - b)^4$ $a^4 - 4a^3b + 6a^2b^2 - 4ab^3 + b^4$
59. $(c - d)^5$ $c^5 - 5c^4d + 10c^3d^2 - 10c^2d^3 + 5cd^4 - d^5$
60. $(c + d)^5$ $c^5 + 5c^4d + 10c^3d^2 + 10c^2d^3 + 5cd^4 + d^5$
61. $(a - b)^9$ $a^9 - 9a^8b + 36a^7b^2 - 84a^6b^3 + 126a^5b^4 - 126a^4b^5 + 84a^3b^6 - 36a^2b^7 + 9ab^8 - b^9$
62. $(a + b)^7$ $a^7 + 7a^6b + 21a^5b^2 + 35a^4b^3 + 35a^3b^4 + 21a^2b^5 + 7ab^6 + b^7$
63. $(s + t)^6$ $s^6 + 6s^5t + 15s^4t^2 + 20s^3t^3 + 15s^2t^4 + 6st^5 + t^6$
64. $(s - t)^6$ $s^6 - 6s^5t + 15s^4t^2 - 20s^3t^3 + 15s^2t^4 - 6st^5 + t^6$

Use the binomial theorem to expand each expression. See Example 7.

65. $(2x + y)^3$ $8x^3 + 12x^2y + 6xy^2 + y^3$
66. $(x + 2y)^3$ $x^3 + 6x^2y + 12xy^2 + 8y^3$
67. $(2t - 3)^5$ $32t^5 - 240t^4 + 720t^3 - 1{,}080t^2 + 810t - 243$
68. $(2b + 1)^4$ $16b^4 + 32b^3 + 24b^2 + 8b + 1$
69. $(5m - 2n)^4$ $625m^4 - 1{,}000m^3n + 600m^2n^2 - 160mn^3 + 16n^4$
70. $(2m + 3n)^5$ $32m^5 + 240m^4n + 720m^3n^2 + 1{,}080m^2n^3 + 810mn^4 + 243n^5$
71. $\left(\frac{x}{3} + \frac{y}{2}\right)^3$ $\frac{x^3}{27} + \frac{x^2y}{6} + \frac{xy^2}{4} + \frac{y^3}{8}$
72. $\left(\frac{x}{2} - \frac{y}{3}\right)^3$ $\frac{x^3}{8} - \frac{x^2y}{4} + \frac{xy^2}{6} - \frac{y^3}{27}$
73. $\left(\frac{x}{3} - \frac{y}{2}\right)^4$ $\frac{x^4}{81} - \frac{2x^3y}{27} + \frac{x^2y^2}{6} - \frac{xy^3}{6} + \frac{y^4}{16}$
74. $\left(\frac{x}{2} + \frac{y}{3}\right)^4$ $\frac{x^4}{16} + \frac{x^3y}{6} + \frac{x^2y^2}{6} + \frac{2xy^3}{27} + \frac{y^4}{81}$
75. $(c^2 - d^2)^5$ $c^{10} - 5c^8d^2 + 10c^6d^4 - 10c^4d^6 + 5c^2d^8 - d^{10}$
76. $(u^2 - v^3)^5$ $u^{10} - 5u^8v^3 + 10u^6v^6 - 10u^4v^9 + 5u^2v^{12} - v^{15}$

Find the indicated term of each binomial expansion. See Examples 8 and 9.

77. $(x + y)^8$; 3rd term $28x^6y^2$
78. $(x + y)^9$; 7th term $84x^3y^6$
79. $(r + s)^6$; 5th term $15r^2s^4$
80. $(r + s)^7$; 5th term $35r^3s^4$
81. $(x - 1)^{13}$; 3rd term $78x^{11}$
82. $(x - 1)^{10}$; 5th term $210x^6$
83. $(x - 3y)^4$; 2nd term $-12x^3y$
84. $(3x - y)^5$; 3rd term $270x^3y^2$
85. $(2x - 3y)^5$; 5th term $810xy^4$
86. $(3x - 2y)^4$; 2nd term $-216x^3y$
87. $\left(\frac{c}{2} - \frac{d}{3}\right)^4$; 2nd term $-\frac{1}{6}c^3d$
88. $\left(\frac{c}{3} + \frac{d}{2}\right)^5$; 4th term $\frac{5c^2d^3}{36}$
89. $(2t - 5)^7$; 4th term $-70{,}000t^4$
90. $(2t - 3)^6$; 6th term $-2{,}916t$
91. $(a^2 + b^2)^6$; 2nd term $6a^{10}b^2$
92. $(a^2 + b^2)^7$; 6th term $21a^4b^{10}$

WRITING

93. Describe how to construct Pascal's triangle.
94. Explain why the signs alternate in the expansion of $(x - y)^9$.
95. Explain why the third term of the expansion of $(m + 3n)^9$ could not be $324m^7n^3$.
96. Using your own words, write a definition of $n!$.

REVIEW

Assume that x, y, z, and b represent positive numbers. Use the properties of logarithms to write each expression as the logarithm of a single quantity.

97. $2 \log x + \frac{1}{2} \log y$ $\log x^2y^{1/2}$ or $\log x^2\sqrt{y}$

▶ **98.** $-2\log x - 3\log y + \log z$ $\log \frac{z}{x^2y^3}$

99. $\ln(xy + y^2) - \ln(xz + yz) + \ln z$ $\ln y$

100. $\log_2 (x + 1) - \log_2 x$ $\log_2 \frac{x+1}{x}$

CHALLENGE PROBLEMS

101. Find the constant term in the expansion of $\left(x + \frac{1}{x}\right)^{10}$. 252

102. Find the coefficient of a^5 in the expansion of $\left(a - \frac{1}{a}\right)^9$. 36

103. **a.** If we applied the pattern of the coefficients to the coefficient of the first term in a binomial expansion, the coefficient would be $\frac{n!}{0!(n-0)!}$. Show that this expression is 1.

b. If we applied the pattern of the coefficients to the coefficient of the last term in a binomial expansion, the coefficient would be $\frac{n!}{n!(n-n)!}$. Show that this expression is 1.

▶ **104.** Expand $(i - 1)^7$, where $i = \sqrt{-1}$. $-8 - 8i$

SECTION 14.2 Arithmetic Sequences and Series

OBJECTIVES

1. Find terms of a sequence given the general term.
2. Find terms of an arithmetic sequence by identifying the first term and the common difference.
3. Find arithmetic means.
4. Find the sum of the first n terms of an arithmetic sequence.
5. Solve application problems involving arithmetic sequences.
6. Use summation notation.

ARE YOU READY? *Are You Ready? exercises available online at www.webassign.net/brookscole*

The following problems review some basic skills that are needed when working with arithmetic sequences and series.

1. Let $f(x) = 6x - 3$. Find $f(4)$. 21

2. Evaluate: **a.** $(-1)^4$ 1 **b.** $(-1)^5$ -1

3. Find the difference: $11 - 3$ 8

4. Evaluate: $12 + (6 - 1)8$ 52

The word *sequence* is used in everyday conversation when referring to an ordered list. For example, a history instructor might discuss the sequence of events that led up to the sinking of the *Titanic.* In mathematics, a **sequence** is a list of numbers written in a specific order.

1 Find Terms of a Sequence Given the General Term.

Each number in a sequence is called a **term** of the sequence. **Finite sequences** contain a finite number of terms and **infinite sequences** contain infinitely many terms. Two examples of sequences are:

Finite sequence: 1, 5, 9, 13, 17, 21, 25

Infinite sequence: 3, 6, 9, 12, 15, . . . The . . . indicates that the sequence goes on forever.

Sequences are defined formally using the terminology of functions.

Finite and Infinite Sequences

A **finite sequence** is a function whose domain is the set of natural numbers $\{1, 2, 3, 4, \ldots, n\}$ for some natural number n.

An **infinite sequence** is a function whose domain is the set of natural numbers: $\{1, 2, 3, 4, \ldots\}$.

Instead of using $f(x)$ notation, we use a_n (read as "a sub n") notation to write the value of a sequence at a number n. For the infinite sequence introduced earlier, we have:

1st term	2nd term	3rd term	4th term	5th term	
3,	6,	9,	12,	15, . . .	Read a_1 as "a sub 1."
↑ a_1	↑ a_2	↑ a_3	↑ a_4	↑ a_5	

To describe all the terms of a sequence specifically, we can write a formula for a_n, called the **general term** of the sequence. For the sequence 3, 6, 9, 12, 15, . . . , we note that $a_1 = 3 \cdot \mathbf{1}$, $a_2 = 3 \cdot \mathbf{2}$, $a_3 = 3 \cdot \mathbf{3}$, and so on. In general, the nth term of the sequence is found by multiplying n by 3.

$$a_n = 3n$$

We can use this formula to find any term of the sequence. For example, to find the 12th term, we substitute 12 for n.

$$a_{12} = 3(12) = 36$$

EXAMPLE 1 Given an infinite sequence with $a_n = 2n - 3$, find each of the following:
a. the first four terms **b.** a_{50}

Strategy We will substitute 1, 2, 3, 4, and 50 for n in the formula that defines the sequence.

Why To find the first term of the sequence, we let $n = 1$. To find the second term, let $n = 2$, and so on.

Solution

a. $a_1 = 2(1) - 3 = -1$ Substitute 1 for n. $a_2 = 2(2) - 3 = 1$ Substitute 2 for n.

$a_3 = 2(3) - 3 = 3$ Substitute 3 for n. $a_4 = 2(4) - 3 = 5$ Substitute 4 for n.

The first four terms of the sequence are -1, 1, 3, and 5.

b. To find a_{50}, the 50th term of the sequence, we let $n = 50$ in the formula for the nth term:

$$a_{50} = 2(50) - 3 = 97$$

Notation

The variable n represents the position of the term in the sequence, while the notation a_n represents the value of the term in that position.

Teaching Example 1 Given an infinite sequence with $a_n = 5n - 1$, find each of the following:
a. the first four terms **b.** a_{50}
Answers: **a.** 4, 9, 14, 19 **b.** 249

Self Check 1 Given an infinite sequence with $a_n = 3n + 5$, find each of the following:
a. the first three terms 8, 11, 14 **b.** a_{100} 305

Now Try Problem 17

EXAMPLE 2 Find the first four terms of the sequence whose general term is $a_n = \frac{(-1)^n}{2^n}$.

Strategy We will substitute 1, 2, 3, and 4 for n in the formula that defines the sequence.

Why To find the first term of the sequence, we let $n = 1$. To find the second term, let $n = 2$, and so on.

Solution

$$a_1 = \frac{(-1)^1}{2^1} = -\frac{1}{2} \quad (-1)^1 = -1 \qquad a_2 = \frac{(-1)^2}{2^2} = \frac{1}{4} \quad (-1)^2 = 1$$

$$a_3 = \frac{(-1)^3}{2^3} = \frac{-1}{8} = -\frac{1}{8} \quad (-1)^3 = -1 \qquad a_4 = \frac{(-1)^4}{2^4} = \frac{1}{16} \quad (-1)^4 = 1$$

The first four terms of the sequence are $-\frac{1}{2}$, $\frac{1}{4}$, $-\frac{1}{8}$, and $\frac{1}{16}$.

Success Tip

The factor $(-1)^n$ in $a_n = \frac{(-1)^n}{2^n}$ causes the signs of the terms to alternate between positive (when n is even) and negative (when n is odd).

Teaching Example 2 Find the first four terms of the sequence whose general term is $a_n = \frac{(-1)^n}{3^n}$.
Answer: $-\frac{1}{3}, \frac{1}{9}, -\frac{1}{27}, \frac{1}{81}$

Self Check 2 Find the first four terms of the sequence whose general term is $a_n = \frac{(-1)^n}{n}$. $-1, \frac{1}{2}, -\frac{1}{3}, \frac{1}{4}$

Now Try Problem 25

2 Find Terms of an Arithmetic Sequence by Identifying the First Term and the Common Difference.

A sequence in which each term is found by adding the same number to the previous term is called an **arithmetic sequence**. Two examples are

5, 12, 19, 26, 33, 40 — This is a finite arithmetic sequence in which each term is found by adding 7 to the previous term.
Add 7

3, 1, -1, -3, -5, -7, . . . — This is an infinite arithmetic sequence in which each term is found by adding -2 to the previous term.
Add -2

The Language of Algebra

We pronounce the adjective **arithmetic** in the phrase *arithmetic sequence* as:

air'-rith-met'-ic

Arithmetic Sequence

An **arithmetic sequence** is a sequence of the form

$$a_1, \quad a_1 + d, \quad a_1 + 2d, \quad a_1 + 3d, \quad \ldots, \quad a_1 + (n-1)d, \ldots$$

where a_1 is the **first term** and d is the **common difference.** The nth term is given by the **general term formula**

$$a_n = a_1 + (n-1)d \quad \text{Read } a_n \text{ as "a sub } n\text{."}$$

Teaching Tip: Point out that d is positive for a sequence whose terms increase, and negative for a sequence whose terms decrease.

We note that the second term of an arithmetic sequence has an addend of $1d$, the third term has an addend of $2d$, the fourth term has an addend of $3d$, and the nth term has an addend of $(n-1)d$. We also note that the **difference between any two consecutive terms in an arithmetic sequence is d.**

EXAMPLE 3 An arithmetic sequence has a first term 5 and a common difference 4. Write the first five terms of the sequence and find the 25th term.

Strategy To find the first five terms, we will write the first term and add 4 to each successive term until we produce five terms. To find the 25th term, we will substitute 5 for a_1, 4 for d, and 25 for n in the formula $a_n = a_1 + (n-1)d$.

Why The same number is added to each term of an arithmetic sequence to get the next term. However, successively adding 4 to find the 25th term would be time consuming. Using the formula is faster.

Solution Since the first term is 5 and the common difference is 4, the first five terms are

5, 9, 13, 17, 21

Add 4

Success Tip

The most important characteristics of an arithmetic sequence are:

a_1: the first term

d: the constant difference

n: the number of terms

Since the first term is $a_1 = 5$ and the common difference is $d = 4$, the arithmetic sequence is defined by the formula

$$a_n = 5 + (n-1)4 \quad \text{In } a_n = a_1 + (n-1)d, \text{ substitute 5 for } a_1 \text{ and 4 for } d.$$

To find the 25th term, we substitute 25 for n and simplify.

$$\begin{aligned} a_{25} &= 5 + (25 - 1)4 \\ &= 5 + (24)4 \\ &= 101 \end{aligned}$$

The 25th term is 101.

Teaching Example 3 Write the first five terms of an arithmetic sequence with a first term 4 and a common difference of 8. Then find the 40th term.
Answer: 4, 12, 20, 28, 36; 316

Self Check 3 Write the first five terms of an arithmetic sequence with a first term 10 and a common difference of 8. Then find the 30th term.

10, 18, 26, 34, 42; 242

Now Try Problem 29

EXAMPLE 4 The first three terms of an arithmetic sequence are 3, 8, and 13. Find the 100th term.

Strategy We can use the first three terms to find the common difference. Then we will know the first term of the sequence and the common difference.

Why Once we know the first term and the common difference, we can use the formula $a_n = a_1 + (n-1)d$ to find the 100th term by letting $n = 100$.

Solution The common difference d is the difference between any two successive terms. Since $a_1 = 3$ and $a_2 = 8$, we can find d using subtraction.

$$d = a_2 - a_1 = 8 - 3 = 5 \quad \text{Also note that } a_3 - a_2 = 13 - 8 = 5.$$

Success Tip

The common difference d of an arithmetic sequence is defined to be

$$d = a_{n+1} - a_n$$

To find the 100th term, we substitute 3 for a_1, 5 for d, and 100 for n in the formula for the nth term.

$$\begin{aligned} a_n &= a_1 + (n - 1)d && \text{This is the general term formula.} \\ a_{100} &= 3 + (100 - 1)5 \\ &= 3 + (99)5 && \text{Evaluate the right side.} \\ &= 498 \end{aligned}$$

Teaching Example 4 The first three terms of an arithmetic seqeuence are −6, 11, and 28. Find the 100th term.
Answer: 1,677

Self Check 4 The first three terms of an arithmetic sequence are −3, 6, and 15. Find the 99th term. 879

Now Try Problem 37

EXAMPLE 5 The first term of an arithmetic sequence is 12 and the 50th term is 3,099. Write the first six terms of the sequence.

Strategy We will find the common difference by substituting 3,099 for a_n, 12 for a_1, and 50 for n in the formula $a_n = a_1 + (n - 1)d$.

Why Once we know the first term and the common difference, we can successively add the common difference to each term to produce the first six terms.

Solution Since the 50th term of the sequence is 3,099, we substitute 3,099 for a_{50}, 12 for a_1, and 50 for n in the formula $a_n = a_1 + (n - 1)d$ and solve for d.

$$\begin{aligned} a_{50} &= a_1 + (n - 1)d \\ 3{,}099 &= 12 + (50 - 1)d && \text{Substitute 3,099 for } a_{50}\text{, 12 for } a_1\text{, and 50 for } n. \\ 3{,}099 &= 12 + 49d && \text{Simplify.} \\ 3{,}087 &= 49d && \text{Subtract 12 from both sides.} \\ 63 &= d && \text{To isolate } d\text{, divide both sides by 49.} \end{aligned}$$

Since the first term is 12 and the common difference is 63, the first six terms are

12, 75, 138, 201, 264, 327   Add 63 to a term to get the next term.

Teaching Example 5 The first term of an arithmetic sequence is 4 and the 15th term is 88. Write the first four terms of the sequence.
Answer: 4, 10, 16, 22

Self Check 5 The first term of an arithmetic sequence is 15 and the 12th term is 92. Write the first four terms of the sequence. 15, 22, 29, 36

Now Try Problem 41

3 Find Arithmetic Means.

If numbers are inserted between two numbers a and b to form an arithmetic sequence, the inserted numbers are called **arithmetic means** between a and b. If a single number is inserted, it is called **the arithmetic mean** between a and b.

EXAMPLE 6 Insert two arithmetic means between 6 and 27.

Strategy Because two arithmetic means are to be inserted between 6 and 27, we will consider a sequence of four terms, with a first term of 6 and a fourth term of 27. We will then use $a_n = a_1 + (n - 1)d$ to find the common difference d.

Why Once we know the first term and the common difference, we can add the common difference to find the two unknown terms.

Solution The first term is $a_1 = 6$ and the fourth term is $a_4 = 27$. We must find the common difference so that the terms

$$\underset{a_1}{\underset{\uparrow}{6}}, \quad \underset{a_2}{\underset{\uparrow}{6 + d}}, \quad \underset{a_3}{\underset{\uparrow}{6 + 2d}}, \quad \underset{a_4}{\underset{\uparrow}{27}}$$

form an arithmetic sequence. To find the common difference d, we substitute 6 for a_1, 4 for n, and 27 for a_4 in the formula for the 4th term:

$$a_4 = a_1 + (n - 1)d$$ This gives the 4th term of any arithmetic sequence.
$$27 = 6 + (4 - 1)d$$ Substitute.
$$27 = 6 + 3d$$ Subtract within the parentheses.
$$21 = 3d$$ Subtract 6 from both sides.
$$7 = d$$ To isolate d, divide both sides by 3.

To find the two arithmetic means between 6 and 27, we add the common difference 7, as shown:

$$6 + d = 6 + 7 = 13$$ This is a_2.

or

$$6 + 2d = 6 + 2(7) = 6 + 14 = 20$$ This is a_3.

Two arithmetic means between 6 and 27 are 13 and 20.

Teaching Example 6 Insert two arithmetic means between 10 and 73. *Answer:* 31, 52

Self Check 6 Insert two arithmetic means between 8 and 44. 20, 32

Now Try Problem 45

4 Find the Sum of the First n Terms of an Arithmetic Sequence.

To develop a formula for finding the sum of the first n terms of an arithmetic sequence, we let S_n represent the sum of the first n terms of an arithmetic sequence:

$$S_n = a_1 + [a_1 + d] + [a_1 + 2d] + \cdots + [a_1 + (n - 1)d]$$

We write the same sum again, but in reverse order:

$$S_n = [a_1 + (n - 1)d] + [a_1 + (n - 2)d] + [a_1 + (n - 3)d] + \cdots + a_1$$

Adding these equations together, term by term, we get

$$2S_n = [2a_1 + (n - 1)d] + [2a_1 + (n - 1)d] + [2a_1 + (n - 1)d] + \cdots + [2a_1 + (n - 1)d]$$

Because there are n equal terms on the right side of the preceding equation, we can write

(1) $$2S_n = n[2a_1 + (n - 1)d]$$
(2) $$2S_n = n[a_1 + a_1 + (n - 1)d]$$ Write $2a_1$ as $a_1 + a_1$.
$$2S_n = n(a_1 + a_n)$$ Substitute a_n for $a_1 + (n - 1)d$.
$$S_n = \frac{n(a_1 + a_n)}{2}$$ To isolate S_n, divide both sides by 2.

This reasoning establishes the following formula.

Teaching Tip: Remind your students that this formula is only for arithmetic sequences.

The Language of Algebra

The sum of a finite number of terms of a sequence is called a **partial sum.** The word *partial* means not complete, as in a *partial* eclipse of the moon.

Sum of the First n Terms of an Arithmetic Sequence

The sum of the first n terms of an arithmetic sequence is given by the formula

$$S_n = \frac{n(a_1 + a_n)}{2}$$

where a_1 is the first term, a_n is the nth (or last) term, and n is the number of terms in the sequence.

EXAMPLE 7 Find the sum of the first 40 terms of the arithmetic sequence: 4, 10, 16, ...

Strategy We know the first term is 4 and we can find the common difference d. We will substitute these values into the formula $a_n = a_1 + (n - 1)d$ to find the last term to be added, a_{40}.

Why To use the formula $S_n = \frac{n(a_1 + a_n)}{2}$ to find the sum of the first 40 terms, we need to know the first term, a_1, and the last term, a_{40}.

Solution We can substitute 4 for a_1, 40 for n, and $10 - 4 = 6$ for d into $a_n = a_1 + (n - 1)d$ to get $a_{40} = 4 + (40 - 1)6 = 238$. We then substitute these values into the formula for S_{40}:

$$S_n = \frac{n(a_1 + a_{40})}{2}$$ This is the formula for the sum of the terms of an artithmetic sequence.

$$S_{40} = \frac{40(4 + 238)}{2}$$ Substitute: $a_1 = 4$, $n = 40$, and $a_{40} = 238$.

$$= 20(242)$$ Add within the parentheses.

$$= 4{,}840$$ Multiply.

The sum of the first 40 terms is 4,840.

Success Tip

An alternate form of the summation formula for arithmetic sequences can be obtained from Equation (1) on the previous page above by dividing both sides by 2.

Teaching Tip: Point out that the formula in the Success Tip saves a step in the solution of Example 7 because we do not have to find a_{40}.

Teaching Example 7 Find the sum of the first 75 terms of the arithmetic sequence 9, 12, 15, ...
Answer: 9,000

Self Check 7 Find the sum of the first 50 terms of the arithmetic sequence: 3, 8, 13, . . . 6,275

Now Try Problem 53

5 Solve Application Problems Involving Arithmetic Sequences.

EXAMPLE 8 **Halftime Performances.** Each row of a formation formed by the members of a college marching band has one more person in it than the previous row. If 4 people are in the front row and 21 are in the 18th (and last) row, how many band members are there?

Strategy To find the number of band members, we will write an arithmetic sequence to model the situation and find the sum of its terms.

Why We can use an arithmetic sequence to model this situation because each row has one more person in it than the previous one. Thus, the common difference is 1.

Solution When we list the number of band members in each row of the formation, we get the arithmetic sequence 4, 5, 6, . . . , 21, where $a_1 = 4$, $d = 1$, $n = 18$, and $a_{18} = 21$. We can use the formula $S_n = \frac{n(a_1 + a_n)}{2}$ to find the sum of the terms of the sequence.

$$S_{18} = \frac{18(4 + 21)}{2} = \frac{18(25)}{2} = 225$$ 4 is the first term and 21 is the last term.

There are 225 members of the marching band.

Teaching Example 8 **Halftime Performances.** Each row of a formation formed by the members of a drill team has one more member than the previous row. If 3 drill team members are in the first row, and 27 are in the 25th (and last) row, how many drill team members are there?
Answer: 375 members

Self Check 8 **Marching Bands.** How many band members would it take to form a 10-row formation, if the first row has 5 people in it, the second row has 7 people, the third row has 9 people, and so on? 140 members

Now Try Problem 95

6 Use Summation Notation.

When the commas between the terms of a sequence are replaced with + signs, we call the sum a **series.** The sum of the terms of an arithmetic sequence is called an **arithmetic series.** Some examples are

$4 + 8 + 12 + 16 + 20 + 24$ Since this series has a limited number of terms, it is a finite arithmetic series.

$5 + 8 + 11 + 14 + 17 + \cdots$ Since this series has infinitely many terms, it is an infinite arithmetic series.

When the general term of a sequence is known, we can use a special notation to write a series. This notation, called **summation notation,** involves the Greek letter Σ (sigma). The expression

$\sum_{k=1}^{4} 3k$ Read as "the summation of $3k$ as k runs from 1 to 4."

designates the sum of all terms obtained if we successively substitute the numbers 1, 2, 3, and 4 for k, called the **index of the summation.** Thus, we have

$$\sum_{k=1}^{4} 3k = \underset{k=1}{3(1)} + \underset{k=2}{3(2)} + \underset{k=3}{3(3)} + \underset{k=4}{3(4)}$$
$$= 3 + 6 + 9 + 12$$
$$= 30$$

EXAMPLE 9 Write the series associated with each summation and find the sum:

a. $\sum_{k=1}^{3} (2k + 1)$ **b.** $\sum_{k=2}^{8} k^2$

Strategy In part (a), we will substitute 1, 2, and 3 for k and add the resulting numbers. In part (b), we will substitute 2, 3, 4, 5, 6, 7, and 8 for k and add the resulting numbers.

Why Think of k as a counter that begins with the number written at the bottom of the notation and successively increases by 1 until it reaches the number written at the top.

Solution

a. We substitute the integers 1, 2, and 3 for k and find the sum.

$$\sum_{k=1}^{3} (2k + 1) = [2(1) + 1] + [2(2) + 1] + [2(3) + 1]$$

Read as "the summation of $2k + 1$ as k runs from 1 to 3."

$$= 3 + 5 + 7$$
$$= 15$$

The Language of Algebra

In these examples, the letter k has been used as the **index of summation.** However, other letters, such as i (not the complex number), j, and n are also often used.

b. We substitute the integers from 2 to 8 for k and find the sum.

$$\sum_{k=2}^{8} k^2 = 2^2 + 3^2 + 4^2 + 5^2 + 6^2 + 7^2 + 8^2$$

Read as "the summation of k^2 as k runs from 2 to 8."

$$= 4 + 9 + 16 + 25 + 36 + 49 + 64$$
$$= 203$$

Teaching Example 9 Write the series associated with the sum $\sum_{k=1}^{3}(3k+4)$. Then find the sum.

Answer: $7 + 10 + 13 = 30$

Self Check 9 Find the sum: $\sum_{k=1}^{4}(2k^2 - 2)$ 52

Now Try Problems 57 and 61

SECTION 14.2 STUDY SET

VOCABULARY

Fill in the blanks.

1. A sequence is a function whose domain is the set of natural numbers.
2. A sequence with an unlimited number of terms is called an infinite sequence. A sequence with a specific number of terms is called a finite sequence.
3. Each term of an arithmetic sequence is found by adding the same number to the previous term.
4. 5, 15, 25, 35, 45, 55, . . . is an example of an arithmetic sequence. The first term is 5 and the common difference is 10.
5. If a single number is inserted between a and b to form an arithmetic sequence, the number is called the arithmetic mean between a and b.
6. The sum of the terms of an arithmetic sequence is called an arithmetic series.

CONCEPTS

7. Write the first three terms of an arithmetic sequence if $a_1 = 1$ and $d = 6$. 1, 7, 13
8. Given the arithmetic sequence 4, 7, 10, 13, 16, 19, . . . , find a_5 and d. 16, 3
9. a. Write the formula for a_n, the general term of an arithmetic sequence. $a_n = a_1 + (n-1)d$
   b. Write the formula for S_n, the sum of the first n terms of an arithmetic sequence. $S_n = \frac{n(a_1 + a_n)}{2}$
10. An infinite arithmetic sequence is of the form

$$a_1, \quad a_1 + d, \quad a_1 + 2d, \quad a_1 + 3d, \quad a_1 + 4d, \ldots$$

NOTATION

Fill in the blanks.

11. The notation a_n represents the nth term of a sequence.
12. To find the common difference of an arithmetic sequence, we use the formula $d = a_{n+1} - a_n$.
13. The symbol Σ is the Greek letter sigma.
14. In the notation $\sum_{k=1}^{5}(2k-5)$, k is called the index of summation.
15. We read $\sum_{k=1}^{10} 3k$ as "the summation of $3k$ as k runs from 1 to 10."
16. $\sum_{k=1}^{5} k = 1 + 2 + 3 + 4 + 5$

GUIDED PRACTICE

Write the first five terms of each sequence and then find the specified term. See Example 1.

17. $a_n = 4n - 1$, a_{40} 3, 7, 11, 15, 19; 159
18. $a_n = 5n - 3$, a_{25} 2, 7, 12, 17, 22; 122
19. $a_n = -3n + 1$, a_{30} −2, −5, −8, −11, −14; −89
20. $a_n = -6n + 2$, a_{15} −4, −10, −16, −22, −28; −88
21. $a_n = -n^2$, a_{20} −1, −4, −9, −16, −25; −400
22. $a_n = -n^3$, a_{10} −1, −8, −27, −64, −125; −1,000
23. $a_n = \frac{n-1}{n}$, a_{12} $0, \frac{1}{2}, \frac{2}{3}, \frac{3}{4}, \frac{4}{5}; \frac{11}{12}$
24. $a_n = \frac{n+1}{2n}$, a_{100} $1, \frac{3}{4}, \frac{2}{3}, \frac{5}{8}, \frac{3}{5}; \frac{101}{200}$

Write the first four terms of each sequence. See Example 2.

25. $a_n = \frac{(-1)^n}{3^n}$ $-\frac{1}{3}, \frac{1}{9}, -\frac{1}{27}, \frac{1}{81}$
26. $a_n = \frac{(-1)^n}{4^n}$ $-\frac{1}{4}, \frac{1}{16}, -\frac{1}{64}, \frac{1}{256}$
27. $a_n = (-1)^n(n + 6)$ −7, 8, −9, 10
28. $a_n = (-1)^n(7n)$ −7, 14, −21, 28

Write the first five terms of each arithmetic sequence with the given properties and find the specified term. See Example 3.

29. First term: 3, common difference: 2; find the 10th term. 3, 5, 7, 9, 11; 21
30. First term: −2, common difference: 3; find the 20th term. −2, 1, 4, 7, 10; 55
31. First term: −5, common difference: −3; find the 15th term. −5, −8, −11, −14, −17; −47
32. First term: 8, common difference: −5; find the 25th term. 8, 3, −2, −7, −12; −112
33. First term: 7, common difference: 12; find the 30th term. 7, 19, 31, 43, 55; 355
34. First term: −1, common difference: 4; find the 55th term. −1, 3, 7, 11, 15; 215
35. First term: −7, common difference: −2; find the 15th term. −7, −9, −11, −13, −15; −35
36. First term: 8, common difference: −3; find the 25th term. 8, 5, 2, −1, −4; −64

The first three terms of an arithmetic sequence are shown below. Find the specified term. See Example 4.

37. 1, 4, 7, . . . ; 30th term 88
38. 2, 6, 10, . . . ; 28th term 110
39. −5, −1, 3, . . . ; 17th term 59
40. −7, −1, 5, . . . ; 15th term 77

Write the first five terms of the arithmetic sequence with the following properties. **See Example 5.**

▶ **41.** The first term is 5 and the fifth term is 29. 5, 11, 17, 23, 29

42. The first term is 4 and the sixth term is 39. 4, 11, 18, 25, 32

▶ **43.** The first term is -4 and the sixth term is -39. $-4, -11, -18, -25, -32$

44. The first term is -5 and the fifth term is -37. $-5, -13, -21, -29, -37$

Insert the given number of arithmetic means between the numbers. **See Example 6.**

45. Two arithmetic means between 2 and 11 5, 8

46. Four arithmetic means between 5 and 25 9, 13, 17, 21

47. Four arithmetic means between 10 and 20 12, 14, 16, 18

48. Three arithmetic means between 20 and 80 35, 50, 65

▶ **49.** Three arithmetic means between 20 and 30 $\frac{45}{2}, 25, \frac{55}{2}$

50. Two arithmetic means between 10 and 19 13, 16

51. One arithmetic mean between -4.5 and 7 $\frac{5}{4}$

52. One arithmetic mean between -6.5 and 8.5 1

For each arithmetic sequence, find the sum of the specified number of terms. **See Example 7.**

▶ **53.** The first 35 terms of 5, 9, 13, . . . 2,555

▶ **54.** The first 50 terms of 7, 12, 17, . . . 6,475

▶ **55.** The first 40 terms of $-5, -1, 3, \ldots$ 2,920

56. The first 25 terms of $2, -3, -8, \ldots$ $-1{,}450$

Write the series associated with each summation. **See Example 9.**

▶ **57.** $\sum_{k=1}^{4}(3k)$ $3 + 6 + 9 + 12$

▶ **58.** $\sum_{k=1}^{4}(k-9)$ $-8 + (-7) + (-6) + (-5)$

59. $\sum_{k=2}^{4}k^2$ $4 + 9 + 16$

60. $\sum_{k=3}^{5}(-2k)$ $-6 + (-8) + (-10)$

Find each sum. **See Example 9.**

61. $\sum_{k=1}^{4}(6k)$ 60

▶ **62.** $\sum_{k=2}^{5}(3k)$ 42

63. $\sum_{k=3}^{4}k^3$ 91

64. $\sum_{k=2}^{4}(-k^2)$ -29

65. $\sum_{k=3}^{4}(k^2 + 3)$ 31

▶ **66.** $\sum_{k=2}^{6}(k^2 + 1)$ 95

67. $\sum_{k=4}^{4}(2k + 4)$ 12

68. $\sum_{k=3}^{5}(3k^2 - 7)$ 20

69. $\sum_{k=2}^{5}(5k)$ 70

70. $\sum_{k=2}^{5}(3k - 5)$ 22

71. $\sum_{k=4}^{6}(4k - 1)$ 57

72. $\sum_{k=3}^{5}(k^3)$ 216

TRY IT YOURSELF

73. Find the common difference of the arithmetic sequence with a first term of 40 if its 44th term is 556. 12

74. Find the first term of the arithmetic sequence with a common difference of -5 if its 23rd term is -625. -515

75. Find the sum of the first 12 terms of the arithmetic sequence if its second term is 7 and its third term is 12. 354

▶ **76.** Find the sum of the first 16 terms of the arithmetic sequence if its second term is 5 and its fourth term is 9. 288

77. Find the first five terms of the arithmetic sequence if the common difference is 7 and the sixth term is -83. $-118, -111, -104, -97, -90$

▶ **78.** Find the first five terms of the arithmetic sequence if the common difference is 3 and the seventh term is 12. $-6, -3, 0, 3, 6$

79. Find the first six terms of the arithmetic sequence if the common difference is -3 and the ninth term is 10. 34, 31, 28, 25, 22, 19

80. Find the first six terms of the arithmetic sequence if the common difference is -5 and the tenth term is -27. $18, 13, 8, 3, -2, -7$

81. The first three terms of an arithmetic sequence are 5, 12, and 19. Find the 200th term. 1,398

82. The first three terms of an arithmetic sequence are 10, 14, and 18. Find the 500th term. 2,006

83. Find the sum of the first 50 natural numbers. 1,275

84. Find the sum of the first 100 natural numbers. 5,050

85. Find the 37th term of the arithmetic sequence with a second term of -4 and a third term of -9. -179

86. Find the 40th term of the arithmetic sequence with a second term of 6 and a fourth term of 16. 196

▶ **87.** Find the first term of the arithmetic sequence with a common difference of 11 if its 27th term is 263. -23

▶ **88.** Find the common difference of the arithmetic sequence with a first term of -164 if its 36th term is -24. 4

89. Find the 15th term of the arithmetic sequence: $\frac{1}{2}, \frac{1}{4}, 0, \ldots$ -3

90. Find the 14th term of the arithmetic sequence $\frac{2}{3}, \frac{1}{2}, \frac{1}{3}, \ldots$ $-\frac{3}{2}$

91. Find the sum of the first 50 odd natural numbers. 2,500

▶ **92.** Find the sum of the first 50 even natural numbers. 2,550

APPLICATIONS

▶ **93. Saving Money.** Yasmeen puts \$60 in a bank safety deposit box. She decides to begin a savings plan by putting \$50 more in the box every month. Write the first six terms of an arithmetic sequence that gives the monthly amounts in her safety deposit box. Then find the amount of money that she will have in the box after 10 years of the deposits. \$60, \$110, \$160, \$210, \$260, \$310; \$6,060

▶ **94. Installment Loans.** Maria borrowed \$10,000, interest-free, from her mother. She agreed to pay back the loan in monthly installments of \$275. Write the first six terms of an arithmetic sequence that shows the balance due after each month, and find the balance due after 17 months. \$9,725, \$9,450, \$9,175, \$8,900, \$8,625, \$8,350; \$5,325

95. Designing Patios. Refer to the illustration. Each row of bricks in a triangular patio floor is to have one more brick than the previous row, ending with the longest row of 150 bricks. How many bricks will be needed? 11,325

96. Logging. Logs are stacked so that the bottom row has 30 logs, the next row has 29 logs, the next row has 28 logs, and so on.

a. If there are 20 rows in the stack, how many logs are in the top row? 11

b. How many logs are in the stack? 410

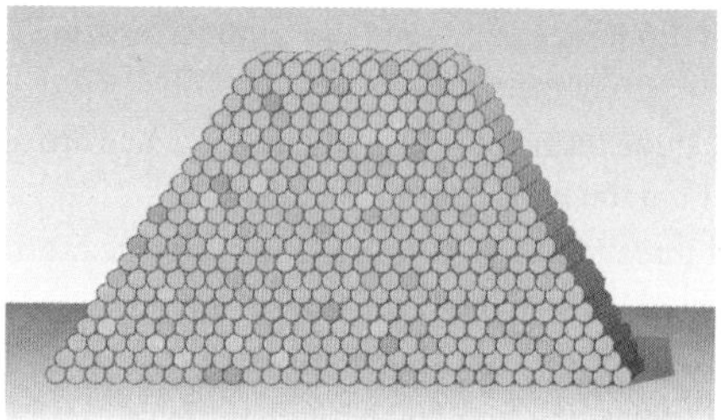

97. Holiday Songs. A popular song of European origin lists the gifts received from someone's "true love" over a 12-day span: a partridge in a pear tree, two turtle doves, three French hens, four calling birds, five gold rings, six geese a-laying, seven swans a-swimming, eight maids a-milking, nine ladies dancing, ten lords a-leaping, eleven pipers piping, and twelve drummers drumming. Use a formula from this section to determine the total number of gifts in this list. 78 gifts

98. Interior Angles. The sums of the angles of several polygons are given in the table below. Assuming that the pattern continues, complete the table.

Figure	Number of sides	Sum of angles
Triangle	3	180°
Quadrilateral	4	360°
Pentagon	5	540°
Hexagon	6	720°
Octagon	8	1,080°
Dodecagon	12	1,800°

WRITING

99. Explain why 1, 4, 8, 13, 19, 26, . . . is not an arithmetic sequence.

100. What is the difference between a sequence and a series?

101. What is the difference between a_n and S_n?

102. How is the symbol Σ used in this section?

REVIEW

Assume that x, y, z and b represent positive numbers. Use the properties of logarithms to write each expression in terms of the logarithms of x, y, and z.

103. $\log_2 \frac{2x}{y}$ $1 + \log_2 x - \log_2 y$

104. $\ln x\sqrt{z}$ $\ln x + \frac{1}{2}\ln z$

105. $\log x^3y^2$ $3 \log x + 2 \log y$

106. $\log x^3y^{1/2}$ $3 \log x + \frac{1}{2} \log y$

CHALLENGE PROBLEMS

107. Write the summation notation for

$1 + 4 + 9 + 16 + 25$ $\sum_{k=1}^{5} k^2$

108. Write the summation notation for $3 + 4 + 5 + 6$ without using $k = 1$ in your answer. $\sum_{k=3}^{6} k$

109. For what value of x will $x - 2$, $2x + 4$, and $5x - 8$, in that order, form an arithmetic sequence? 9

110. For what value of x will the arithmetic mean of $x + 4$ and $x + 8$ be 5? -1

SECTION 14.3

OBJECTIVES

1. Find terms of a geometric sequence by identifying the first term and the common ratio.
2. Find geometric means.
3. Find the sum of the first n terms of a geometric sequence.
4. Define and use infinite geometric series.
5. Solve application problems involving geometric sequences.

Teaching Tip: Contrast the sequence on the right with the arithmetic sequence 2, 8, 14, 20, ... with $d = 6$.

Geometric Sequences and Series

ARE YOU READY? *Are You Ready? exercises available online at www.webassign.net/brookscole*

The following problems review some basic skills that are needed when working with geometric sequences and series.

1. Let $f(x) = 6(2^x)$. Find $f(3)$. 48

2. Evaluate: **a.** $(5)^4$ 625 **b.** $\left(\frac{1}{3}\right)^3$ $\frac{1}{27}$

3. Simplify: $\frac{6}{24}$ $\frac{1}{4}$

4. Solve: $r^2 = 144$ ± 12

We have seen that the same number is added to each term of an arithmetic sequence to get the next term. In this section, we will consider another type of sequence in which we *multiply* each term by the same number to get the next term. This type of sequence is called a *geometric sequence.* Two examples are

2, 8, 32, 128, . . . This is an infinite geometric sequence in which each term is found by multiplying the previous term by 4.

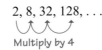

Multiply by 4

$27, 9, 3, 1, \frac{1}{3}, \frac{1}{9}$ This is a finite geometric sequence in which each term is found by multiplying the previous term by $\frac{1}{3}$.

Multiply by $\frac{1}{3}$

1 Find Terms of a Geometric Sequence by Identifying the First Term and the Common Ratio.

Each term of a geometric sequence is found by multiplying the previous term by the same number.

Geometric Sequence

A geometric sequence is a sequence of the form

$$a_1,\ a_1r,\ a_1r^2,\ a_1r^3, \ldots,\ a_1r^{n-1}, \ldots$$

where a_1 is the **first term** and r is the **common ratio.** The nth term is given by the **general form formula**

$$a_n = a_1r^{n-1}$$

We note that the second term of a geometric sequence has a factor of r^1, the third term has a factor of r^2, the fourth term has a factor of r^3, and the nth term has a factor of r^{n-1}. We also note that ***r* is the quotient obtained when any term is divided by the previous term**.

EXAMPLE 1 A geometric sequence has a first term 5 and a common ratio 3.

a. Write the first five terms of the sequence. **b.** Find the 9th term.

Strategy In part (a), we will write the first term and successively multiply each term by 3 until we produce five terms. In part (b), we will substitute 5 for a_1, 3 for r, and 9 for n in the formula for the nth term of a geometric sequence $a_n = a_1r^{n-1}$.

Why To find the terms of a geometric sequence, we multiply each term by the same number to get the next term. To answer part (b), successively multiplying by 3 to find the 9th term would be time consuming. Using the formula is faster.

Solution

a. Because the first term is $a_1 = 5$ and the common ratio is $r = 3$, the first five terms are

$$\underset{a_1}{5}, \quad \underset{a_2}{5(3)}, \quad \underset{a_3}{5(3^2)}, \quad \underset{a_4}{5(3^3)}, \quad \underset{a_5}{5(3^4)}$$

Each term is found by multiplying the previous term by 3.

or

5, 15, 45, 135, 405

b. We are given $a_1 = 5$ and $r = 3$. Because we want the ninth term, we let $n = 9$:

$a_n = a_1 r^{n-1}$ This is the general term formula.

$a_9 = 5(3)^{9-1}$ Substitute.

$= 5(3)^8$ Subtract.

$= 5(6{,}561)$ Evaluate $(3)^8$.

$= 32{,}805$ Multiply.

Caution

A common error when evaluating an expression such as $5(3)^8$ is to incorrectly multiply first. Recall that by the order of operations rule, we should evaluate $(3)^8$ first, and then multiply that result by 5.

$$5(3)^8 \neq 15^8$$

Self Check 1 A geometric sequence has a first term 3 and a common ratio 4.
a. Write the first four terms. 3, 12, 48, 192 **b.** Find the 8th term. 49,152

Now Try Problem 15

Teaching Example 1 A geometric sequence has a first term 6 and a common ratio 3.
a. Write the first four terms.
b. Find the 8th term.
Answers:
a. 6, 18, 54, 162 b. 13,122

EXAMPLE 2

The first three terms of a geometric sequence are 16, 4, and 1. Find the 7th term.

Strategy We can use the first three terms to find the common ratio. Then we will know the first term and the common ratio.

Why Once we know the first term, a_1, and the common ratio, r, we can use the formula $a_n = a_1 r^{n-1}$ to find the seventh term by letting $n = 7$.

Solution The common ratio r is the ratio between any two successive terms. Since $a_1 = 16$ and $a_2 = 4$, we can find r as follows:

$$r = \frac{a_2}{a_1} = \frac{4}{16} = \frac{1}{4}$$ Also note that $\frac{a_3}{a_2} = \frac{1}{4}$.

To find the seventh term, we substitute 16 for a_1, $\frac{1}{4}$ for r, and 7 for n in the formula for the nth term and simplify:

$a_n = a_1 r^{n-1}$ This is the general term formula.

$a_7 = 16\left(\frac{1}{4}\right)^{7-1}$ Substitute 16 for a_1, 7 for n, and $\frac{1}{4}$ for r.

$= 16\left(\frac{1}{4}\right)^6$ Do the subtraction.

$= 16\left(\frac{1}{4{,}096}\right)$ Use a calculator to evaluate $\left(\frac{1}{4}\right)^6$.

$= \frac{1}{256}$ Simplify: $\frac{16}{4{,}096} = \frac{\overset{1}{\cancel{16}}}{\underset{1}{\cancel{16}} \cdot 256}$.

Success Tip

The common ratio r of a geometric sequence is defined to be

$$r = \frac{a_{n+1}}{a_n}$$

Self Check 2 The first three terms of a geometric sequence are 25, 5, and 1. Find the 7th term. $\frac{1}{625}$

Now Try Problem 19

Teaching Example 2 The first three terms of a geometric sequence are 49, 7, and 1. Find the 6th term.
Answer: $\frac{1}{343}$

EXAMPLE 3 Find the first five terms of the geometric sequence with a first term 2, a third term 32, and a common ratio that is positive.

Strategy We will substitute $a_1 = 2, a_3 = 32$, and $n = 3$ into the formula for the nth term of a geometric sequence $a_n = a_1r^{n-1}$ and solve for r.

Why Once we know the common ratio, we can successively multiply each term by the common ratio to produce the first five terms.

Solution We will substitute 3 for n, 2 for a_1, 32 for a_3, and solve for r.

Success Tip

The most important characteristics of a geometric sequence are:

- a_1: the first term
- r: the common ratio
- n: the number of terms

$a_n = a_1r^{n-1}$ This is the formula for the nth term of a geometric sequence.

$a_3 = 2r^{3-1}$ Substitute 3 for n and 2 for a_1.

$32 = 2r^2$ Substitute 32 for a_3 and simplify.

$16 = r^2$ Divide both sides by 2.

$\pm 4 = r$ Use the square root property.

Since r is given to be positive, $r = 4$. The first five terms are produced by multiplying by the common ratio:

$$2,\ 2 \cdot 4,\ 2 \cdot 4^2,\ 2 \cdot 4^3,\ 2 \cdot 4^4$$

or

$$2,\ 8,\ 32,\ 128,\ 512$$

Teaching Example 3 Find the first five terms of the geometric sequence with a first term −12, a third term −48, and a common ratio that is positive.
Answer:
−12, −24, −48, −96, −192

Self Check 3 Find the first five terms of the geometric sequence with a first term −2 and a fourth term −54. −2, −6, −18, −54, −162

Now Try Problem 23

2 Find Geometric Means.

If numbers are inserted between two numbers a and b to form a geometric sequence, the inserted numbers are called **geometric means** between a and b. If a single number is inserted, that number is called **the geometric mean** between a and b.

EXAMPLE 4 Insert two geometric means between 7 and 1,512.

Strategy Because two geometric means are to be inserted between 7 and 1,512, we will consider a sequence of four terms, with a first term of 7 and a fourth term of 1,512. We will then use $a_n = a_1r^{n-1}$ to find the common ratio r.

Why Once we know the first term and the common ratio, we can multiply by the common ratio to find the two unknown terms.

Solution In this example, the first term is $a_1 = 7$, and the fourth term is $a_4 = 1{,}512$. To find the common ratio r so that the terms

$$\underset{a_1}{\overset{7,}{\uparrow}} \quad \underset{a_2}{\overset{7r,}{\uparrow}} \quad \underset{a_3}{\overset{7r^2,}{\uparrow}} \quad \underset{a_4}{\overset{1{,}512}{\uparrow}}$$

form a geometric sequence, we substitute 4 for n and 7 for a_1 in the formula for the nth term of a geometric sequence and solve for r.

$$a_n = a_1r^{n-1}$$ This is the general term formula.
$$a_4 = 7r^{4-1}$$ Substitute 4 for n and 7 for a_1.
$$1{,}512 = 7r^3$$ Substitute 1,512 for a_4.
$$216 = r^3$$ Divide both sides by 7.
$$6 = r$$ Take the cube root of both sides.

To find the two geometric means between 7 and 1,512, we multiply by the common ratio 6, as shown:

$$7r = 7(6) = \mathbf{42} \qquad \text{and} \qquad 7r^2 = 7(6)^2 = 7(36) = \mathbf{252}$$

The numbers 7, 42, 252, and 1,512 are the first four terms of a geometric sequence.

Teaching Example 4 Insert two geometric means between 5 and 2,560.
Answer: 5, 40, 320, 2,560

Self Check 4 Insert three positive geometric means between 1 and 16. 1, 2, 4, 8, 16

Now Try Problem 27

EXAMPLE 5 Find the geometric mean between 2 and 20.

Strategy Because one geometric mean is to be inserted between 2 and 20, we will consider a sequence of three terms, with a first term of 2 and a third term of 20. We will then use $a_n = a_1r^{n-1}$ to find the common ratio r.

Why Once we know the first term and the common ratio, we can multiply by the common ratio to find the unknown term.

Solution We want to find the middle term of the three-termed geometric sequence

$$\underset{a_1}{\underset{\uparrow}{2}}, \quad \underset{a_2}{\underset{\uparrow}{2r}}, \quad \underset{a_3}{\underset{\uparrow}{20}}$$

with $a_1 = 2$, $a_3 = 20$, and $n = 3$. To find r, we substitute these values into the formula for the nth term of a geometric sequence:

$$a_n = a_1r^{n-1}$$ This is the general term formula.
$$a_3 = 2r^{3-1}$$ Substitute 3 for n and 2 for a_1.
$$20 = 2r^2$$ Substitute 20 for a_3.
$$10 = r^2$$ Divide both sides by 2.
$$\pm\sqrt{10} = r$$ Use the square root property.

Teaching Tip: Remind your students that r can be positive or negative.

Because r can be either $\sqrt{10}$ or $-\sqrt{10}$, there are two values for the geometric mean. They are

$$2r = 2\sqrt{10} \qquad \text{and} \qquad 2r = -2\sqrt{10}$$

The sets of numbers 2, $2\sqrt{10}$, 20 and 2, $-2\sqrt{10}$, 20 both form geometric sequences. The common ratio of the first sequence is $\sqrt{10}$, and the common ratio of the second sequence is $-\sqrt{10}$.

Teaching Example 5 Find the geometric mean between 3 and 30.
Answer: $3\sqrt{10}$ and $-3\sqrt{10}$

Self Check 5 Find the positive geometric mean between 2 and 200. 20

Now Try Problem 31

3 Find the Sum of the First n Terms of a Geometric Sequence.

There is a formula that gives the sum of the first n terms of a geometric sequence. To develop this formula, we let S_n represent the sum of the first n terms of a geometric sequence.

Success Tip

If the common factor of a_1 in the numerator of $\frac{a_1 - a_1r^n}{1 - r}$ is factored out, the formula can be written in a different form:

$$S_n = \frac{a_1(1 - r^n)}{1 - r}$$

Teaching Tip: Remind your students that this formula is only for geometric sequences.

(1) $S_n = a_1 + a_1r + a_1r^2 + a_1r^3 + \cdots + a_1r^{n-1}$

We multiply both sides of Equation 1 by r to get

(2) $S_nr = \quad a_1r + a_1r^2 + a_1r^3 + \cdots + a_1r^{n-1} + a_1r^n$

We now subtract Equation 2 from Equation 1 and solve for S_n:

$$S_n - S_nr = a_1 - a_1r^n$$

$$S_n(1 - r) = a_1 - a_1r^n \quad \text{Factor out } S_n \text{ from the left side.}$$

$$S_n = \frac{a_1 - a_1r^n}{1 - r} \quad \text{Divide both sides by } 1 - r.$$

This reasoning establishes the following formula.

Sum of the First n Terms of a Geometric Sequence

The sum of the first n terms of a geometric sequence is given by the formula

$$S_n = \frac{a_1 - a_1r^n}{1 - r} \quad \text{or} \quad S_n = \frac{a_1(1 - r^n)}{1 - r} \quad \text{where } r \neq 1$$

where S_n is the sum, a_1 is the first term, r is the common ratio, and n is the number of terms.

EXAMPLE 6 Find the sum of the first six terms of the geometric sequence: 250, 50, 10, . . .

Strategy We will find the common ratio r.

Why If we know the first term and the common ratio, we can use the formula $S_n = \frac{a_1 - a_1r^n}{1 - r}$ to find the sum of the first six terms by letting $n = 6$.

Solution The common ratio r is the ratio between any two successive terms. Since $a_1 = 250$ and $a_2 = 50$, we can find r as follows:

$$r = \frac{a_2}{a_1} = \frac{50}{250} = \frac{1}{5} \quad \text{Also note that } \frac{a_3}{a_2} = \frac{10}{50} = \frac{1}{5}.$$

In this sequence, $a_1 = 250$, $r = \frac{1}{5}$, and $n = 6$. We substitute these values into the formula for the sum of the first n terms of a geometric sequence and simplify:

The Language of Algebra

In Example 6, we say we are finding a **partial sum** of a geometric series.

$$S_n = \frac{a_1 - a_1r^n}{1 - r} \quad \text{This is the formula for the sum of terms of a geometric sequence.}$$

$$S_6 = \frac{250 - 250\left(\frac{1}{5}\right)^6}{1 - \frac{1}{5}} \quad \text{Substitute.}$$

$$= \frac{250 - 250\left(\frac{1}{15{,}625}\right)}{\frac{4}{5}} \quad \text{Use the calculator to evaluate: } \left(\frac{1}{5}\right)^6.$$

$$= \left(250 - \frac{250}{15{,}625}\right) \cdot \frac{5}{4} \quad \text{Multiply the numerator by the reciprocal of the denominator.}$$

$$= 312.48 \quad \text{Use a calculator.}$$

The sum of the first six terms is 312.48.

Teaching Example 6 Find the sum of the first six terms of the geometric sequence:

$10, 2, \frac{2}{5}, \ldots$

Answer: 12.4992

Self Check 6 Find the sum of the first five terms of the geometric sequence: 100, 20, 4, . . . 124.96

Now Try Problem 35

4 Define and Use Infinite Geometric Series.

When we add the terms of a geometric sequence, we form a **geometric series.** If we form the sum of the terms of an infinite geometric sequence, we get a series called an **infinite geometric series.** For example, if the common ratio r is 3, we have

Infinite geometric sequence	***Infinite geometric series***
2, 6, 18, 54, 162, . . .	$2 + 6 + 18 + 54 + 162 + \cdots$

As the number of terms of this series gets larger, the value of the series gets larger. We can see that this is true by forming some **partial sums.**

The first partial sum of the series is $S_1 = 2$.

The second partial sum of the series is $S_2 = 2 + 6 = 8$.

The third partial sum of the series is $S_3 = 2 + 6 + 18 = 26$.

The fourth partial sum of the series is $S_4 = 2 + 6 + 18 + 54 = 80$.

The Language of Algebra

The word **partial** means only a part, not total. Have you ever seen a *partial* eclipse of the moon?

We can see that as the number of terms gets infinitely large, the value of this series gets infinitely large.

The values of some infinite geometric series get closer and closer to a specific number as the number of terms approaches infinity. One such series is

$$\frac{3}{2} + \frac{3}{4} + \frac{3}{8} + \frac{3}{16} + \frac{3}{32} + \frac{3}{64} + \cdots \qquad \text{Here, } r = \tfrac{1}{2}.$$

To see that this is true, we form some partial sums.

The first partial sum is $S_1 = \frac{3}{2} = \mathbf{1.5}$

The second partial sum is $S_2 = \frac{3}{2} + \frac{3}{4} = \frac{9}{4} = \mathbf{2.25}$

The third partial sum is $S_3 = \frac{3}{2} + \frac{3}{4} + \frac{3}{8} = \frac{21}{8} = \mathbf{2.625}$

The fourth partial sum is $S_4 = \frac{3}{2} + \frac{3}{4} + \frac{3}{8} + \frac{3}{16} = \frac{45}{16} = \mathbf{2.8125}$

The fifth partial sum is $S_5 = \frac{3}{2} + \frac{3}{4} + \frac{3}{8} + \frac{3}{16} + \frac{3}{32} = \frac{93}{32} = \mathbf{2.90625}$

The sixth partial sum is $S_6 = \frac{3}{2} + \frac{3}{4} + \frac{3}{8} + \frac{3}{16} + \frac{3}{32} + \frac{3}{64} = \frac{189}{64} = \mathbf{2.953125}$

As the number of terms in this series gets larger, the values of the partial sums approach the number 3. We say that 3 is the **limit** of S_n as n approaches infinity, and we say that 3 is the **sum of the infinite geometric series.**

To develop a formula for finding the sum of an infinite geometric series, we consider the formula that gives the sum of the first n terms.

$$S_n = \frac{a_1 - a_1 r^n}{1 - r} \quad \text{where } r \neq 1$$

If $|r| < 1$ and a_1 is constant, the term $a_1 r^n$ in the above formula approaches 0 as n becomes very large. For example,

$$a_1\left(\frac{1}{2}\right)^1 = \frac{1}{2}a_1, \qquad a_1\left(\frac{1}{2}\right)^2 = \frac{1}{4}a_1, \qquad a_1\left(\frac{1}{2}\right)^3 = \frac{1}{8}a_1$$

and so on. Thus, when n is very large, the value of $a_1 r^n$ is insignificant, and the term $a_1 r^n$ in the above formula can be ignored. This reasoning justifies the following formula.

Sum of the Terms of an Infinite Geometric Series

If a_1 is the first term and r is the common ratio of an infinite geometric sequence, and $|r| < 1$, the sum of the terms of the corresponding series is given by

$$S = \frac{a_1}{1 - r}$$

EXAMPLE 7 Find the sum of the terms of the infinite geometric series: $125 + 25 + 5 + \cdots$

Strategy We will identify a_1 and find r.

Why To use the formula $S = \frac{a_1}{1-r}$ to find the sum, we need to know the first term, a_1, and the common ratio, r.

Solution In this geometric series, $a_1 = 125$ and $r = \frac{25}{125} = \frac{1}{5}$. Since $|r| = \left|\frac{1}{5}\right| = \frac{1}{5} < 1$, we can find the sum of the series. We do this by substituting 125 for a_1 and $\frac{1}{5}$ for r in the formula $S = \frac{a_1}{1-r}$ and simplifying:

$$S = \frac{a_1}{1 - r} = \frac{125}{1 - \frac{1}{5}} = \frac{125}{\frac{4}{5}} = 125 \cdot \frac{5}{4} = \frac{625}{4}$$

The sum of the series $125 + 25 + 5 + \cdots$ is $\frac{625}{4} = 156.25$.

Notation

The sum of the terms of an infinite geometric sequence also is written as S_∞.

Teaching Example 7 Find the sum of the infinite geometric series:

$1 + \frac{1}{6} + \frac{1}{36} + \cdots$

Answer: $\frac{6}{5}$

Self Check 7 Find the sum of the infinite geometric series: $100 + 20 + 4 + \cdots$ 125

Now Try Problem 43

EXAMPLE 8 Find the sum of the infinite geometric series: $64 + (-4) + \frac{1}{4} + \cdots$

Strategy We will identify a_1 and find r.

Why To use the formula $S = \frac{a_1}{1-r}$ to find the sum, we need to know the first term, a_1, and the common ratio, r.

Solution In this geometric series, $a_1 = 64$ and $r = \frac{-4}{64} = -\frac{1}{16}$. Since $|r| = \left|-\frac{1}{16}\right| = \frac{1}{16} < 1$, we can find the sum of the series. We substitute 64 for a_1 and $-\frac{1}{16}$ for r in the formula $S = \frac{a_1}{1-r}$ and simplify:

$$S = \frac{a_1}{1 - r} = \frac{64}{1 - \left(-\frac{1}{16}\right)} = \frac{64}{\frac{17}{16}} = 64 \cdot \frac{16}{17} = \frac{1{,}024}{17}$$

The sum of the geometric series $64 + (-4) + \frac{1}{4} + \cdots$ is $\frac{1{,}024}{17}$.

Notation

This infinite geometric series also could be written:

$64 - 4 + \frac{1}{4} - \cdots$

Caution

If $|r| \geq 1$ for an infinite geometric sequence, the sum of the terms of the sequence does not exist.

Teaching Example 8 Find the sum of the infinite geometric series:

$-3 + 1 + \left(-\frac{1}{3}\right) + \cdots$

Answer: $-\frac{9}{4}$

Self Check 8 Find the sum of the infinite geometric series: $81 + (-27) + 9 + \cdots$ $\frac{243}{4}$

Now Try Problem 51

EXAMPLE 9 Write $0.\overline{8}$ in fraction form.

Strategy First, we will show that the decimal 0.888 . . . can be represented by an infinite geometric series whose terms are fractions. Then we will identify a_1, find the common ratio r, and find the sum.

Why When we use the formula $S = \frac{a_1}{1 - r}$ to find the sum, the result will be the required common fraction.

Solution The decimal $0.\overline{8}$ can be written as the infinite geometric series

$$\begin{aligned} 0.\overline{8} &= 0.888\ldots \\ &= 0.8 + 0.08 + 0.008 + \cdots \\ &= \frac{8}{10} + \frac{8}{100} + \frac{8}{1{,}000} + \cdots \end{aligned}$$

Here, $a_1 = \frac{8}{10}$ and $r = \frac{1}{10}$. Because $|r| = \left|\frac{1}{10}\right| = \frac{1}{10} < 1$, we can find the sum as follows:

$$S = \frac{a_1}{1 - r} = \frac{\frac{8}{10}}{1 - \frac{1}{10}} = \frac{\frac{8}{10}}{\frac{9}{10}} = \frac{8}{10} \cdot \frac{10}{9} = \frac{8}{9}$$

Thus, $0.\overline{8} = \frac{8}{9}$. Long division will verify that $\frac{8}{9} = 0.888\ldots$.

Teaching Example 9 Write $0.\overline{5}$ in fraction form.
Answer: $\frac{5}{9}$

Self Check 9 Write $0.\overline{6}$ in fraction form. $\frac{2}{3}$

Now Try Problem 55

5 Solve Application Problems Involving Geometric Sequences.

EXAMPLE 10 **Inheritances.** A father decides to give his son part of his inheritance early. Each year, on the son's birthday, the father will pay the son 15% of what remains in the inheritance fund. If the fund initially begins with \$100,000, how much money will be left in the fund after 20 years of payments?

Strategy We will model the facts of the problem using a geometric sequence with a first term of 100,000 and a common ratio of 0.85.

Why One of the terms of the sequence will represent the amount of money left in the inheritance fund after 20 years of payments.

Solution If 15% of the money in the inheritance fund is given to the son each year, 85% of that amount remains after each payment. To find the amount of money that remains in the fund after a payment is made, we multiply the amount that was in the fund by 0.85. Over the years, the amounts of money that are left in the fund after a payment form a geometric sequence.

Amount of money remaining in the inheritance fund

The fund begins with:	100,000	← a_1
After first payment:	$100{,}000(\mathbf{0.85}) = 100{,}000(0.85)^1$	← a_2
After second payment:	$[100{,}000(0.85)^1](\mathbf{0.85}) = 100{,}000(0.85)^2$	← a_3
After third payment:	$[100{,}000(0.85)^2](\mathbf{0.85}) = 100{,}000(0.85)^3$	← a_4
After fourth payment:	$[100{,}000(0.85)^3](\mathbf{0.85}) = 100{,}000(0.85)^4$	← a_5
⋮	⋮	⋮
After 20th payment:	???	← a_{21}

The amount of money remaining in the inheritance fund after 20 years is represented by the 21st term of a geometric sequence, where $a_1 = 100{,}000$, $r = 0.85$, and $n = 21$.

$$a_n = a_1r^{n-1} \quad \text{This is the formula for the } n\text{th term.}$$
$$a_{21} = \mathbf{a_1}r^{\mathbf{21}-1} \quad \text{Substitute 21 for } n.$$
$$= \mathbf{100{,}000(0.85)}^{21-1} \quad \text{Substitute 100,000 for } a_1 \text{ and 0.85 for } r.$$
$$= 100{,}000(0.85)^{20} \quad \text{Do the subtraction.}$$
$$\approx 3{,}876 \quad \text{Use a calculator. Round to the nearest dollar.}$$

In 20 years, approximately \$3,876 of the inheritance fund will be left.

Teaching Example 10 Inheritances. Refer to Example 10. How much money will be left in the inheritance fund after 25 years?
Answer: About \$1,720

Self Check 10 **Inheritances.** How much money will be left in the inheritance fund after 30 years of payments? About \$763

Now Try ▶ Problem 79

EXAMPLE 11 **Testing Steel.** One way to measure the hardness of a steel anvil is to drop a ball bearing onto the face of the anvil. The bearing should rebound $\frac{4}{5}$ of the distance from which it was dropped. If a bearing is dropped from a height of 10 inches onto a hard forged steel anvil, and if it could bounce forever, what total distance would the bearing travel?

Teaching Tip: You also may want to share other similar real-life applications, such as bungee jumping and a bouncing Superball or Ping Pong ball.

Strategy We will show that the facts in the problem can be modeled by an infinite geometric sequence.

Why The sum of the terms of the infinite geometric sequence will give the total distance the bearing will travel.

Solution The total distance the ball bearing travels is the sum of two motions, falling and rebounding. The bearing falls 10 inches, then rebounds $\frac{4}{5} \cdot 10 = 8$ inches, and falls 8 inches, and rebounds $\frac{4}{5} \cdot 8 = \frac{32}{5}$ inches, and falls $\frac{32}{5}$ inches, and rebounds $\frac{4}{5} \cdot \frac{32}{5} = \frac{128}{25}$ inches, and so on.

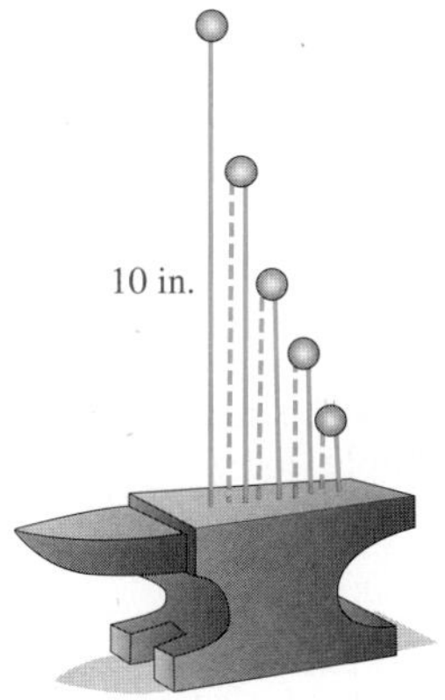

The distance the ball falls is given by the sum

$$10 + 8 + \frac{32}{5} + \frac{128}{25} + \cdots \quad \text{This is an infinite geometric series with } a_1 = 10 \text{ and } r = \tfrac{4}{5}.$$

The distance the ball rebounds is given by the sum

$$8 + \frac{32}{5} + \frac{128}{25} + \cdots \quad \text{This is an infinite geometric series with } a_1 = 8 \text{ and } r = \tfrac{4}{5}.$$

Since each of these is an infinite geometric series with $|r| < 1$, we can use the formula $S = \frac{a_1}{1 - r}$ to find each sum.

$$\text{Falling: } \frac{10}{1 - \frac{4}{5}} = \frac{10}{\frac{1}{5}} = 50 \text{ in.} \qquad \text{Rebounding: } \frac{8}{1 - \frac{4}{5}} = \frac{8}{\frac{1}{5}} = 40 \text{ in.}$$

The total distance the bearing travels is 50 inches + 40 inches = 90 inches.

Teaching Example 11 Testing Steel. Refer to Example 11. If the bearing is dropped from a height of 12 inches, and if it could bounce forever, what total distance would the bearing travel?
Answer: 108 in.

Self Check 11 **Testing Steel.** If a bearing was dropped from a height of 15 inches onto a hard forged steel anvil, and if it could bounce forever, what total distance would it travel? 135 in.

Now Try Problem 85

SECTION 14.3 STUDY SET

VOCABULARY

Fill in the blanks.

1. Each term of a geometric sequence is found by multiplying the previous term by the same number.
2. 8, 16, 32, 64, 128, . . . is an example of a geometric sequence. The first term is 8 and the common ratio is 2.
3. If a single number is inserted between a and b to form a geometric sequence, the number is called the geometric mean between a and b.
4. The sum of the terms of a geometric sequence is called a geometric series. The sum of the terms of an infinite geometric sequence is called an infinite geometric series.

CONCEPTS

5. Write the first three terms of a geometric sequence if $a_1 = 16$ and $r = \frac{1}{4}$. 16, 4, 1
6. Given the geometric sequence 1, 2, 4, 8, 16, 32, . . . , find a_5 and r. 16, 2
7. Write the formula for a_n, the general term of a geometric sequence. $a_n = a_1 r^{n-1}$
8. a. Write the formula for S_n, the sum of the first n terms of a geometric sequence. $S_n = \frac{a_1 - a_1 r^n}{1 - r}$ or $S_n = \frac{a_1(1 - r^n)}{1 - r}$
   b. Write the formula for S, the sum of the terms of an infinite geometric sequence, where $|r| < 1$. $S = \frac{a_1}{1 - r}$
9. Which of the following values of r satisfy $|r| < 1$?
   a. $r = \frac{2}{3}$ Yes
   b. $r = -3$ No
   c. $r = 6$ No
   d. $r = -\frac{1}{5}$ Yes
10. Write $0.\overline{7}$ as an infinite geometric series:

$$0.\overline{7} = 0.7 + 0.07 + 0.007 + \cdots$$
$$= \frac{7}{10} + \frac{7}{100} + \frac{7}{1{,}000} + \cdots$$

NOTATION

Fill in the blanks.

11. An infinite geometric sequence is of the form
$$a_1,\ a_1r,\ a_1r^2,\ a_1r^3,\ a_1r^4, \ldots$$
12. The first four terms of the sequence defined by $a_n = 4(3)^{n-1}$ are 4, 12, 36, 108.
13. To find the common ratio of a geometric sequence, we use the formula: $r = \frac{a_{n+1}}{a_n}$
14. S_8 represents the sum of the first eight terms of a geometric sequence.

GUIDED PRACTICE

Write the first five terms of each geometric sequence with the given properties and then find the specified term. See Example 1.

15. First term: 3, common ratio: 2; find the 9th term
3, 6, 12, 24, 48; 768
16. First term: −2, common ratio: 2; find the 8th term
−2, −4, −8, −16, −32; −256
17. First term: −5, common ratio: $\frac{1}{5}$; find the 8th term
$-5, -1, -\frac{1}{5}, -\frac{1}{25}, -\frac{1}{125}; -\frac{1}{15{,}625}$
18. First term: 8, common ratio: $\frac{1}{2}$; find the 10th term
$8, 4, 2, 1, \frac{1}{2}; \frac{1}{64}$

Find the specified term of the geometric sequence with the following properties. See Example 2.

19. The first three terms are 2, 6, 18; find the 7th term 1,458
20. The first three terms are 50, 100, 200; find the 10th term 25,600
21. The first three terms are $\frac{1}{2}, -\frac{5}{2}, \frac{25}{2}$; find the 6th term $-\frac{3{,}125}{2}$
22. The first three terms are $\frac{1}{4}, -\frac{3}{4}, \frac{9}{4}$; find the 9th term $\frac{6{,}561}{4}$

Selected exercises available online at www.webassign.net/brookscole

Write the first five terms of the geometric sequence with the following properties. **See Example 3.**

23. First term: 2, $r > 0$, third term: 18 2, 6, 18, 54, 162

24. First term: 2, $r < 0$, third term: 50 2, −10, 50, −250, 1,250

▶ **25.** First term: 3, fourth term: 24 3, 6, 12, 24, 48

26. First term: −3, fourth term: −192 −3, −12, −48, −192, −768

Find the geometric means to be inserted in each geometric sequence. **See Example 4.**

▶ **27.** Insert three positive geometric means between 2 and 162. 6, 18, 54

▶ **28.** Insert four geometric means between 3 and 96. 6, 12, 24, 48

29. Insert four geometric means between −4 and −12,500. −20, −100, −500, −2,500

▶ **30.** Insert three geometric means (two positive and one negative) between −64 and −1,024. 128, −256, 512

Find a geometric mean to be inserted in each geometric sequence. **See Example 5.**

31. Find the geometric mean between 2 and 128. −16 or 16

▶ **32.** Find the geometric mean between 3 and 243. −27 or 27

33. Find the geometric mean between 10 and 20. $-10\sqrt{2}$ or $10\sqrt{2}$

34. Find the geometric mean between 5 and 15. $-5\sqrt{3}$ or $5\sqrt{3}$

For each geometric sequence, find the sum of the specified number of terms. **See Example 6.**

▶ **35.** The first 6 terms of 2, 6, 18, . . . 728

▶ **36.** The first 6 terms of 2, −6, 18, . . . −364

37. The first 5 terms of 2, −6, 18, . . . 122

38. The first 5 terms of 2, 6, 18, . . . 242

39. The first 8 terms of 3, −6, 12, . . . −255

▶ **40.** The first 8 terms of 3, 6, 12, . . . 765

41. The first 7 terms of 3, 6, 12, . . . 381

42. The first 7 terms of 3, −6, 12, . . . 129

Find the sum of each infinite geometric series, if possible. **See Examples 7 and 8.**

▶ **43.** $8 + 4 + 2 + \cdots$ 16

44. $12 + 6 + 3 + \cdots$ 24

45. $54 + 18 + 6 + \cdots$ 81

▶ **46.** $45 + 15 + 5 + \cdots$ $\frac{135}{2}$

47. $-\frac{27}{2} + (-9) + (-6) + \cdots$ $-\frac{81}{2}$

▶ **48.** $-112 + (-28) + (-7) + \cdots$ $-\frac{448}{3}$

49. $\frac{9}{2} + 6 + 8 + \cdots$ No sum

50. $\frac{18}{25} + \frac{6}{5} + 2 + \cdots$ No sum

▶ **51.** $12 + (-6) + 3 + \cdots$ 8

52. $8 + (-4) + 2 + \cdots$ $\frac{16}{3}$

▶ **53.** $-45 + 15 + (-5) + \cdots$ $-\frac{135}{4}$

54. $-54 + 18 + (-6) + \cdots$ $-\frac{81}{2}$

Write each decimal in fraction form. Then check the answer by performing long division. **See Example 9.**

55. $0.\overline{1}$ $\frac{1}{9}$

56. $0.\overline{2}$ $\frac{2}{9}$

57. $0.\overline{3}$ $\frac{1}{3}$

▶ **58.** $0.\overline{4}$ $\frac{4}{9}$

59. $0.\overline{12}$ $\frac{4}{33}$

▶ **60.** $0.\overline{21}$ $\frac{7}{33}$

61. $0.\overline{75}$ $\frac{25}{33}$

62. $0.\overline{57}$ $\frac{19}{33}$

TRY IT YOURSELF

▶ **63.** Find the common ratio of the geometric sequence with a first term −8 and a sixth term −1,944. 3

64. Find the common ratio of the geometric sequence with a first term 12 and a sixth term $\frac{3}{8}$. $\frac{1}{2}$

65. Write the first five terms of the geometric sequence if its first term is −64, $r < 0$, and its fifth term is −4.
−64, 32, −16, 8, −4

66. Write the first five terms of the geometric sequence if its first term is −64, $r > 0$, and its fifth term is −4.
−64, −32, −16, −8, −4

67. Find a geometric mean, if possible, between −50 and 10.
No geometric mean exists.

68. Find a negative geometric mean, if possible, between −25 and −5. $-5\sqrt{5}$

▶ **69.** Find the 10th term of the geometric sequence with $a_1 = 7$ and $r = 2$. 3,584

70. Find the 12th term of the geometric sequence with $a_1 = 64$ and $r = \frac{1}{2}$. $\frac{1}{32}$

71. Write the first four terms of the geometric sequence if its first term is −64 and its sixth term is −2. −64, −32, −16, −8

72. Write the first four terms of the geometric sequence if its first term is −81 and its sixth term is $\frac{1}{3}$. −81, 27, −9, 3

73. Find the sum of the terms of the geometric sequence:
$3, \frac{3}{4}, \frac{3}{16}, \frac{3}{64}, \ldots$ 4

74. Find the sum of the terms of the geometric sequence:
$1, -\frac{1}{2}, \frac{1}{4}, -\frac{1}{8}, \ldots$ $\frac{2}{3}$

▶ **75.** Find the first term of the geometric sequence with a common ratio −3 and an eighth term −81. $\frac{1}{27}$

▶ **76.** Find the first term of the geometric sequence with a common ratio 2 and a tenth term 384. $\frac{3}{4}$

77. Find the sum of the first five terms of the geometric sequence if its first term is 3 and the common ratio is 2. 93

78. Find the sum of the first five terms of the geometric sequence if its first term is 5 and the common ratio is −6. 5,555

APPLICATIONS

Use a calculator to help solve each problem.

▶ **79. Declining Savings.** John has \$10,000 in a safety deposit box. Each year, he spends 12% of what is left in the box. How much will be in the box after 15 years? \$1,469.74

▶ **80. Savings Growth.** Sally has \$5,000 in a savings account earning 12% annual interest. How much will be in her account 10 years from now? (Assume that Sally makes no deposits or withdrawals.) \$15,529.24

81. ▶ **from Campus to Careers**

Real Estate Sales Agent

©Michael Pettigrew/Shutterstock.com

Suppose you are a real estate sales agent and you are working with a client who is considering buying a \$250,000 house in Seattle as an investment. If the property values in that area have a track record of increasing at a rate of 8% per year, what will the house be worth 10 years from now? About \$539,731

▶ **82. Boat Depreciation.** A boat that cost \$5,000 when new depreciates at a rate of 9% per year. How much will the boat be worth in 5 years? \$3,120.16

83. Inscribed Squares. Each inscribed square in the illustration joins the midpoints of the next larger square. The area of the first square, the largest, is 1 square unit. Find the area of the 12th square. $\left(\frac{1}{2}\right)^{11} \approx 0.0005$ square unit

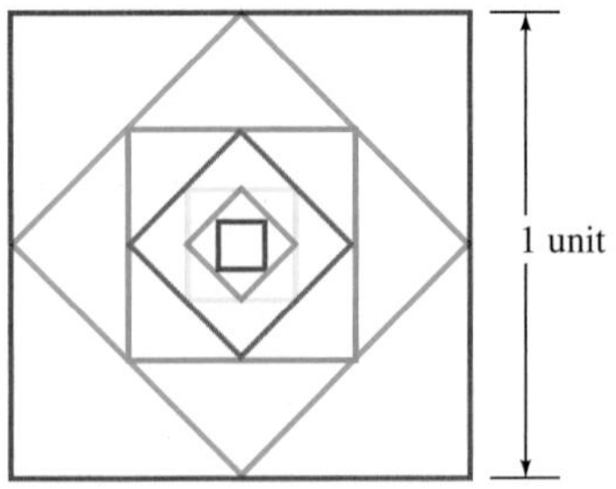

84. Genealogy. The following family tree spans 3 generations and lists 7 people. How many names would be listed in a family tree that spans 10 generations? $2^{10} - 1 = 1{,}023$

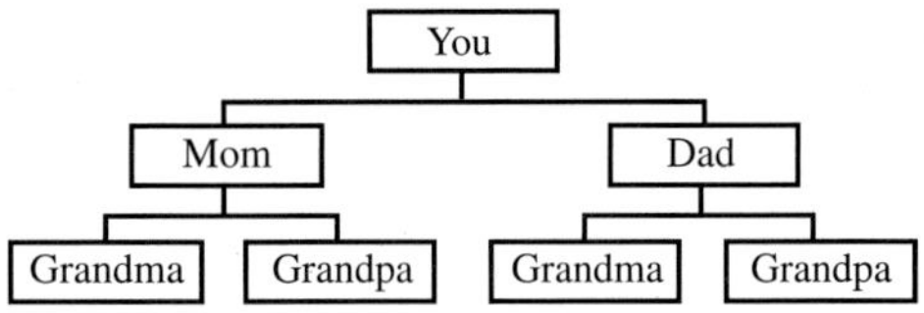

85. Bouncing Balls. On each bounce, the rubber ball in the illustration rebounds to a height one-half of that from which it fell. Find the total vertical distance the ball travels. 30 m

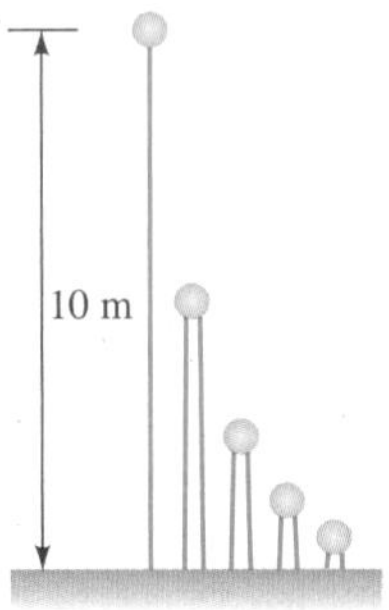

86. Bouncing Balls. A golf ball is dropped from a height of 12 feet. On each bounce, it returns to a height that is two-thirds of the distance it fell. Find the total vertical distance the ball travels. 60 ft

87. Pest Control. To reduce the population of a destructive moth, biologists release 1,000 sterilized male moths each day into the environment. If 80% of these moths alive one day survive until the next, then after a long time the population of sterile males is the sum of the infinite geometric series

$$1{,}000 + 1{,}000(0.8) + 1{,}000(0.8)^2 + 1{,}000(0.8)^3 + \cdots$$

Find the long-term population. 5,000

88. Pendulums. On its first swing to the right, a pendulum swings through an arc of 96 inches. Each successive swing, the pendulum travels $\frac{99}{100}$ as far as on the previous swing. Determine the total distance that the pendulum will travel by the time it comes to rest. 9,600 in.

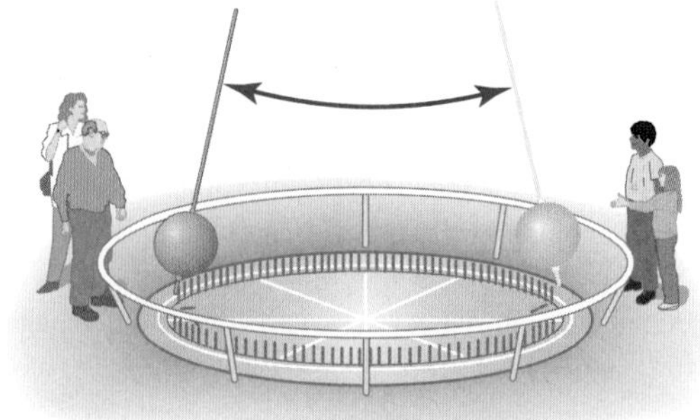

WRITING

89. Describe the real numbers that satisfy $|r| < 1$.

90. Why must the absolute value of the common ratio be less than 1 before an infinite geometric sequence can have a sum?

91. Explain the difference between an arithmetic sequence and a geometric sequence.

92. Why is $1 - \frac{1}{2} + \frac{1}{4} - \frac{1}{8} + \frac{1}{16} - \frac{1}{32} + \cdots$ called an alternating infinite geometric series?

REVIEW

Solve each inequality. Write the solution set using interval notation.

93. $x^2 - 5x - 6 \le 0$ $[-1, 6]$

94. $a^2 - 7a + 12 \ge 0$ $(-\infty, 3] \cup [4, \infty)$

95. $\frac{x - 4}{x + 3} > 0$ $(-\infty, -3) \cup (4, \infty)$

96. $\frac{t^2 + t - 20}{t + 2} < 0$ $(-\infty, -5) \cup (-2, 4)$

CHALLENGE PROBLEMS

97. If $f(x) = 1 + x + x^2 + x^3 + x^4 + \cdots$, find $f\left(\frac{1}{2}\right)$ and $f\left(-\frac{1}{2}\right)$. $f\left(\frac{1}{2}\right) = 2, f\left(-\frac{1}{2}\right) = \frac{2}{3}$

98. Find the sum:

$$\frac{1}{\sqrt{3}} + \frac{1}{3} + \frac{1}{3\sqrt{3}} + \frac{1}{9} + \cdots \qquad \frac{\sqrt{3} + 1}{2}$$

99. If $a > b > 0$, which is larger: the arithmetic mean between a and b or the geometric mean between a and b? Arithmetic mean

100. Is there a geometric mean between -5 and 5? No

14 Summary & Review

SECTION 14.1 The Binomial Theorem

DEFINITIONS AND CONCEPTS	EXAMPLES
Pascal's triangle gives the coefficients of the terms of the expansion of $(a+b)^n$.	**Pascal's triangle:** 1 — Row for $(a+b)^0$ 1 1 — Row for $(a+b)^1$ 1 2 1 — Row for $(a+b)^2$ 1 3 3 1 — Row for $(a+b)^3$ 1 4 6 4 1 — Row for $(a+b)^4$ **1 5 10 10 5 1 — Row for $(a+b)^5$** $(x+y)^5 = \mathbf{1}x^5 + \mathbf{5}x^4y + \mathbf{10}x^3y^2 + \mathbf{10}x^2y^3 + \mathbf{5}xy^4 + \mathbf{1}y^5$
The symbol ***n*!** (***n factorial***) is the product of consecutively decreasing natural numbers from n to 1. $n! = n(n-1)(n-2)\cdot \cdots \cdot 2\cdot 1$	Evaluate each expression: $4! = 4\cdot 3\cdot 2\cdot 1 = 24$ $3!\cdot 2! = 3\cdot 2\cdot 1\cdot 2\cdot 1 = 12$ $\frac{6!}{5!} = \frac{6\cdot \cancel{5!}}{\cancel{5!}} = 6$ $1! = 1$ and $0! = 1$
The **binomial theorem** is usually the best way to expand a binomial. $(a+b)^n = a^n + \frac{n!}{1!(n-1)!}a^{n-1}b + \frac{n!}{2!(n-2)!}a^{n-2}b^2 + \cdots + b^n$	Use the binomial theorem to expand $(p+q)^4$, where $n = 4$. $(p+q)^4 = p^4 + \frac{4!}{1!(4-1)!}p^3q + \frac{4!}{2!(4-2)!}p^2q^2 + \frac{4!}{3!(4-3)!}pq^3 + q^4$ $= p^4 + 4p^3q + 6p^2q^2 + 4pq^3 + q^4$
To find a specific term of an expansion: The $(r+1)$st term of the expansion of $(a+b)^n$ is $\frac{n!}{r!(n-r)!}a^{n-r}b^r$ Remember that r is always 1 less than the number of the term that you are finding.	Find the third term of the expansion of $(a+b)^5$. In the third term of $(a+b)^5$, $n = 5$ and $r = 2$. $\frac{5!}{2!(5-2)!}a^{5-2}b^2 = \frac{5\cdot 4\cdot \cancel{3!}}{2\cdot 1\cdot \cancel{3!}}a^3b^2 = 10a^3b^2$

REVIEW EXERCISES

1. Complete Pascal's triangle. List the row that gives the coefficients for the expansion of $(a+b)^5$. 1, 5, 10, 10, 5, 1

1
1 1
1 2 1
1 3 3 1
1 4 6 4 1
1 5 10 10 5 1
1 6 15 20 15 6 1

2. Consider the expansion of $(a+b)^{12}$.

a. How many terms does the expansion have? 13

b. For each term, what is the sum of the exponents on a and b? 12

c. What is the first term? What is the last term? a^{12}, b^{12}

d. How do the exponents on a and b change from term to term? a: decrease; b: increase

Evaluate each expression.

3. $4!\cdot 3!$ 144

4. $\frac{5!}{3!}$ 20

5. $\frac{6!}{2!(6-2)!}$ 15

6. $\frac{12!}{3!(12-3)!}$ 220

7. $(n-n)!$ 1

8. $\frac{8!}{7!}$ 8

Use the binomial theorem to find each expansion.

9. $(x + y)^5$ $x^5 + 5x^4y + 10x^3y^2 + 10x^2y^3 + 5xy^4 + y^5$

10. $(x - y)^9$ $x^9 - 9x^8y + 36x^7y^2 - 84x^6y^3 + 126x^5y^4 - 126x^4y^5 + 84x^3y^6 - 36x^2y^7 + 9xy^8 - y^9$

11. $(4x - y)^3$ $64x^3 - 48x^2y + 12xy^2 - y^3$

12. $\left(\frac{c}{2} + \frac{d}{3}\right)^4$ $\frac{c^4}{16} + \frac{c^3d}{6} + \frac{c^2d^2}{6} + \frac{2cd^3}{27} + \frac{d^4}{81}$

Find the specified term in each expansion.

13. $(x + y)^4$; third term $6x^2y^2$

14. $(x - y)^6$; fourth term $-20x^3y^3$

15. $(3x - 4y)^3$; second term $-108x^2y$

16. $(u^2 - v^3)^5$; fifth term $5u^2v^{12}$

SECTION 14.2 Arithmetic Sequences and Series

DEFINITIONS AND CONCEPTS	EXAMPLES
A **sequence** is a list of numbers written in a specific order.	**Finite sequence:** 1, 4, 7, 10, 13 **Infinite sequence:** 2, 6, 10, 14, 18, . . .
To describe all the terms of a sequence specifically, we can write a formula for a_n, called the **general term** of the sequence.	Find the first four terms of the sequence described by $a_n = 5n + 1$. We substitute 1, 2, 3, and 4, for n in the formula: $a_1 = 5(1) + 1 = 6$ $\quad a_2 = 5(2) + 1 = 11$ $a_3 = 5(3) + 1 = 16$ $\quad a_4 = 5(4) + 1 = 21$ The first four terms are 6, 11, 16, and 21.
A sequence in which each term is found by adding the same number to the previous term is called an **arithmetic sequence.** An arithmetic sequence has the form $a_1, a_1 + d, a_1 + 2d, \dots, a_1 + (n - 1)d, \dots$ where a_1 is the first term and d is the common difference. The ***n*th term of an arithmetic sequence** is given by $a_n = a_1 + (n - 1)d$. The **common difference *d*** of an arithmetic sequence is the difference between any two consecutive terms: $d = a_{n+1} - a_n$.	Find the 12th term of the arithmetic sequence $-4, -1, 2, 5, 8, \dots$. First, we find the common difference d. It is the difference between any two successive terms: $d = a_2 - a_1 = -1 - (-4) = 3$. Then we substitute into the formula for the nth term: $a_n = a_1 + (n - 1)d$ $a_{12} = -4 + (12 - 1)3$ In this sequence, $n = 12$, $a_1 = -4$, and $d = 3$. $= -4 + (11)3$ Do the subtraction. $= -4 + 33$ Evaluate the right side. $= 29$ The 12th term of the sequence is 29.
If numbers are inserted between two given numbers a and b to form an arithmetic sequence, the inserted numbers are **arithmetic means** between a and b.	Find three arithmetic means between -6 and 14. Since there will be five terms, $n = 5$. We also know that $a_1 = -6$ and $a_5 = 14$. We will substitute these values in the formula for the nth term and solve for d. $a_5 = a_1 + (n - 1)d$ $14 = -6 + (5 - 1)d$ Substitute. $14 = -6 + 4d$ Do the subtraction. $20 = 4d$ Add 6 to both sides. $5 = d$ Solve for d. Since the common difference is 5, we successively add 5 to the terms to get the sequence $-6, -1, 4, 9, 14$. Thus, the three arithmetic means are -1, 4, and 9.

When the commas between the terms of a sequence are replaced with + signs, we call the sum a **series.**

The **sum of the first *n* terms of an arithmetic sequence** is given by

$$S_n = \frac{n(a_1 + a_n)}{2}$$

Find the sum of the first 12 terms of the arithmetic sequence $-4, -1, 2, 5, 8, \ldots$.

Here $a_1 = -4$ and $n = 12$. On the previous page, we found that for this sequence $a_{12} = 29$. We substitute these values into the formula for S_n and simplify.

$$S_n = \frac{n(a_1 + a_n)}{2}$$

$$S_{12} = \frac{12(-4 + 29)}{2} \quad \text{Substitute.}$$

$$= \frac{300}{2} \quad \text{Evaluate the numerator.}$$

$$= 150 \quad \text{Do the division.}$$

The sum of the first 12 terms is 150.

Summation notation involves the Greek letter sigma Σ. It designates the sum of terms called a **series.**

$$\sum_{n=1}^{4}(3k - 2) = [3(1) - 2] + [3(2) - 2] + [3(3) - 2] + [3(4) - 2]$$

($k = 1$, $k = 2$, $k = 3$, $k = 4$)

$$= 1 + 4 + 7 + 10$$

$$= 22$$

REVIEW EXERCISES

17. Find the first four terms of the sequence defined by $a_n = 2n - 4$. $-2, 0, 2, 4$

18. Find the first five terms of the sequence defined by $a_n = \frac{(-1)^n}{n+1}$. $-\frac{1}{2}, \frac{1}{3}, -\frac{1}{4}, \frac{1}{5}, -\frac{1}{6}$

19. Find the 50th term of the sequence defined by $a_n = 100 - \frac{n}{2}$. 75

20. Find the eighth term of an arithmetic sequence whose first term is 7 and whose common difference is 5. 42

21. Write the first five terms of the arithmetic sequence whose ninth term is 242 and whose seventh term is 212. 122, 137, 152, 167, 182

22. The first three terms of an arithmetic sequence are 6, −6, and −18. Find the 101st term. −1,194

23. Find the common difference of an arithmetic sequence if its 1st term is −515 and the 23rd term is −625. −5

24. Find two arithmetic means between 8 and 25. $\frac{41}{3}, \frac{58}{3}$

25. Find the sum of the first ten terms of the sequence $9, 6.5, 4, \ldots$. $-\frac{45}{2}$

26. Find the sum of the first 28 terms of an arithmetic sequence if the second term is 6 and the sixth term is 22. 1,568

Find each sum.

27. $\sum_{k=4}^{6} \frac{1}{2}k$ $\frac{15}{2}$

28. $\sum_{k=2}^{5} 7k^2$ 378

29. $\sum_{k=1}^{4} (3k - 4)$ 14

30. $\sum_{k=10}^{10} 36k$ 360

31. What is the sum of the first 200 natural numbers? 20,100

32. Seating. The illustration shows the first 2 of a total of 30 rows of seats in an amphitheater. The number of seats in each row forms an arithmetic sequence. Find the total number of seats. 1,170

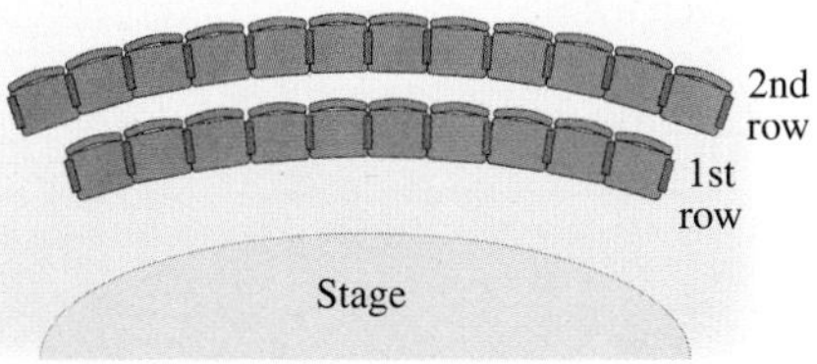

SECTION 14.3 Geometric Sequences and Series

DEFINITIONS AND CONCEPTS	EXAMPLES
Each term of a **geometric sequence** is found by multiplying the previous term by the same number. A geometric sequence has the form: $a_1, \quad a_1r, \quad a_1r^2, \quad a_1r^3, \quad \ldots, a_1r^{n-1}, \ldots$ where a_1 is the first term and r is the common ratio. The ***n*th term of a geometric sequence** is given by: $a_n = a_1r^{n-1}$ The **common ratio *r*** of a geometric sequence is the quotient obtained when any term is divided by the previous term: $r = \frac{a_{n+1}}{a_n}$	Find the 8th term of the geometric sequence 8, 4, 2, ... First, we find the common ratio r. It is the quotient obtained when any term is divided by the previous term: $r = \frac{a_2}{a_1} = \frac{4}{8} = \frac{1}{2}$ Then we substitute into the formula for the nth term: $a_n = a_1r^{n-1}$ $a_8 = 8\left(\frac{1}{2}\right)^{8-1}$ Substitute: $n = 8$, $a_1 = 8$, and $r = \frac{1}{2}$. $= 8\left(\frac{1}{2}\right)^7$ Do the subtraction. $= \frac{8}{128}$ Evaluate: $\left(\frac{1}{2}\right)^7 = \frac{1}{128}$. $= \frac{1}{16}$ Simplify the fraction: Remove a common factor of 8 in the numerator and denominator.
If numbers are inserted between a and b to form a geometric sequence, the inserted numbers are **geometric means** between a and b.	Find two geometric means between 6 and 162. Since there will be four terms, $n = 4$. We also know that $a_1 = 6$ and $a_4 = 162$. We will substitute these values in the formula for the nth term and solve for r. $a_n = a_1r^{n-1}$ $162 = 6r^{4-1}$ Substitute. $27 = r^3$ Divide both sides by 6. $3 = r$ Take the cube root of both sides. Since the common ratio is 3, we successively multiply the terms by 3 to get the sequence 6, 18, 54, 162. Thus, the two geometric means are 18 and 54.
The **sum of the first *n* terms of a geometric sequence** is given by: $S_n = \frac{a_1 - a_1r^n}{1 - r} \quad r \neq 1$	Find the sum of the first five terms of the geometric sequence 2, 6, 18, ... Here $a_1 = 2$, $r = 3$, and $n = 5$. We substitute these values into the formula for S_n and simplify. $S_n = \frac{a_1 - a_1r^n}{1 - r}$ $S_5 = \frac{2 - (2)(3)^5}{1 - 3}$ Substitute. $= \frac{2 - 2(243)}{-2}$ Evaluate: $(3)^5 = 243$. $= \frac{2 - 486}{-2}$ Do the multiplication. $= \frac{-484}{-2}$ Evaluate the numerator. $= 242$ Do the division. The sum of the first five terms is 242.

If we form the sum of the terms of an infinite geometric sequence, we get a series called an **infinite geometric series.** The sum of an infinite geometric series is given by: $S = \frac{a_1}{1 - r}$ where $\|r\| < 1$	Find the sum of the infinite geometric series: $12 + 8 + \frac{16}{3} + \cdots$ Here $a_1 = 12$ and $r = \frac{8}{12} = \frac{2}{3}$. (Note that $\|r\| < 1$.) We substitute these values into the formula for the sum of an infinite geometric series. $S = \frac{a_1}{1 - r} = \frac{12}{1 - \frac{2}{3}} = \frac{12}{\frac{1}{3}} = 12 \cdot \frac{3}{1} = 36$ The sum is 36.

REVIEW EXERCISES

33. Find the sixth term of a geometric sequence with a first term of $\frac{1}{8}$ and a common ratio of 2. 4

34. Write the first five terms of the geometric sequence whose fourth term is 3 and whose fifth term is $\frac{3}{2}$. $24, 12, 6, 3, \frac{3}{2}$

35. Find the first term of a geometric sequence if it has a common ratio of -3 and the ninth term is 243. $\frac{1}{27}$

36. Find two geometric means between -6 and 384. $24, -96$

37. Find the sum of the first seven terms of the sequence:

$162, 54, 18, \ldots$ $\frac{2{,}186}{9}$

38. Find the sum of the first eight terms of the sequence:

$\frac{1}{8}, -\frac{1}{4}, \frac{1}{2}, \ldots$ $-\frac{85}{8}$

39. Feeding Birds. Tom has 50 pounds of birdseed stored in his garage. Each month, he uses 25% of what is left in the bag to feed the birds in his yard. How much birdseed will be left in 12 months? About 1.6 lb

40. Find the sum of the infinite geometric series:

$25 + 20 + 16 + \cdots$ 125

41. Change the decimal $0.\overline{05}$ to a common fraction. $\frac{5}{99}$

42. Wham-O Toys. Tests have found that 1998 Superballs rebound $\frac{9}{10}$ of the distance from which they are dropped. If a Superball is dropped from a height of 10 feet, and if it could bounce forever, what total distance would it travel? 190 ft

Teaching Tip: Because this Chapter Test is a comprehensive collection of problems that covers all of the topics discussed in Chapter 14, it is lengthy. If your students have time to complete it, that would be optimal. If, because of time constraints, they are unable to do so, assign an appropriate subset of problems that reflects the types of problems that the students will see on your exam.

14 Chapter Test

1. Fill in the blanks.

a. The array of numbers that gives the coefficients of the terms of a binomial expansion is called Pascal's triangle.

b. In the expansion $a^3 - 3a^2b + 3ab^2 - b^3$, the signs alternate between + and −.

c. Each term of an arithmetic sequence is found by adding the same number to the previous term.

d. The sum of the terms of an arithmetic sequence is called an arithmetic series.

e. Each term of a geometric sequence is found by multiplying the previous term by the same number.

2. Find the first 4 terms of the sequence defined by: $a_n = -6n + 8$ $2, -4, -10, -16$

3. Find the first 5 terms of the sequence defined by: $a_n = \frac{(-1)^n}{n^3}$ $-1, \frac{1}{8}, -\frac{1}{27}, \frac{1}{64}, -\frac{1}{125}$

4. Evaluate: $\frac{10!}{6!(10 - 6)!}$ 210

5. Expand: $(a - b)^6$

$a^6 - 6a^5b + 15a^4b^2 - 20a^3b^3 + 15a^2b^4 - 6ab^5 + b^6$

6. Find the third term in the expansion of $(x^2 + 2y)^4$. $24x^4y^2$

7. Find the tenth term of an arithmetic sequence whose first 3 terms are 3, 10, and 17. 66

8. Find the sum of the first 12 terms of the sequence: $-2, 3, 8, \ldots$ 306

9. Find two arithmetic means between 2 and 98. 34, 66

10. Find the common difference of an arithmetic sequence if the second term is $\frac{5}{4}$ and the 17th term is 5. $\frac{1}{4}$

11. Find the sum of the first 27 terms of an arithmetic sequence if the 4th term is -11 and the 20th term is -75. $-1{,}377$

12. Plumbing. Plastic pipe is stacked so that the bottom row has 25 pipes, the next row has 24 pipes, the next row has 23 pipes, and so on until there is 1 pipe at the top of the stack. If a worker removes the top 15 rows of pipe, how many pieces of pipe will be left in the stack? 205 pipes

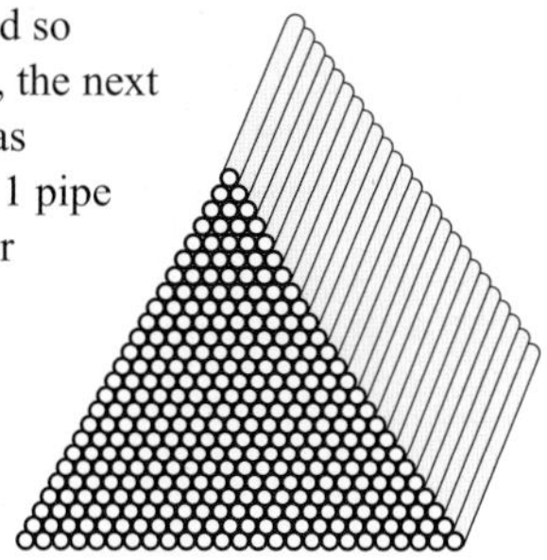

13. Falling Objects. If an object is in free fall, the sequence 16, 48, 80, . . . represents the distance in feet that object falls during the 1st second, during the 2nd second, during the 3rd second, and so on. How far will the object fall during the first 10 seconds? 1,600 ft

14. Evaluate: $\sum_{k=1}^{3}(2k - 3)$ 3

15. Find the seventh term of the geometric sequence whose first 3 terms are $-\frac{1}{9}, -\frac{1}{3}$, and -1. -81

16. Find the sum of the first 6 terms of the sequence:

$\frac{1}{27}, \frac{1}{9}, \frac{1}{3}, \ldots$ $\frac{364}{27}$

17. Find the first term of a geometric sequence if the common ratio is $-\frac{2}{3}$ and the fourth term is $-\frac{16}{9}$. 6

18. Find two geometric means between 3 and 648. 18, 108

19. Find the sum of infinite geometric series: $9 + 3 + 1 + \cdots$ $\frac{27}{2}$

20. Depreciation. A yacht that cost \$1,500,000 when new depreciates at a rate of 8% per year. How much will the yacht be worth in 10 years? About \$651,583

21. Pendulums. On its first swing to the right, a pendulum swings through an arc of 60 inches. Each successive swing, the pendulum travels $\frac{97}{100}$ as far as on the previous swing. Determine the total distance the pendulum will travel by the time it comes to rest. 2,000 in.

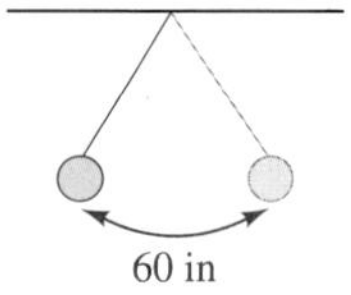

22. Change the decimal $0.\overline{7}$ to a common fraction. $\frac{7}{9}$

Group Project

The Language of Algebra

Overview: This activity will help you review for the final exam.

Instructions: Form groups of two students. Match each instruction in column I with the most appropriate item in column II. Each letter in column II is used only once.

Column I

g 1. Use the FOIL method.
o 2. Apply a rule for exponents to simplify.
i 3. Add the rational expressions.
l 4. Rationalize the denominator.
u 5. Factor completely.
y 6. Evaluate the expression for $a = -1$ and $b = -6$.
p 7. Express in lowest terms.
x 8. Solve for t.
m 9. Combine like terms.
d 10. Remove parentheses.
f 11. Solve the system by graphing.
c 12. Find $f(g(x))$.
w 13. Solve using the quadratic formula.
b 14. Identify the base and the exponent.
j 15. Write without a radical symbol.
s 16. Write the equation of the line having the given slope and y-intercept.
e 17. Solve the inequality.
z 18. Complete the square to make a perfect-square trinomial.
n 19. Find the slope of the line passing through the given points.
k 20. Use a property of logarithms to simplify.
h 21. Set each factor equal to zero and solve for x.
v 22. State the solution of the compound inequality using interval notation.
q 23. Find the inverse function, $h^{-1}(x)$.
a 24. Write using scientific notation.
r 25. Write the logarithmic statement in exponential form.
t 26. Find the sum of the first 6 terms of the sequence.

Column II

a. 2,300,000,000
b. e^3
c. $f(x) = x^2 + 1$ and $g(x) = 5 - 3x$
d. $-2x(3x^2 - 4x + 8)$
e. $4x - 7 > -3x - 7$
f. $\begin{cases} 2x = y - 5 \\ x + y = -1 \end{cases}$
g. $(x^2 - 5)(x^2 + 3)$
h. $(x + 2)(x - 10) = 0$
i. $\dfrac{x - 1}{2x^2} + \dfrac{x + 1}{8x}$
j. $\sqrt{4x^2}$
k. $\ln 6 + \ln x$
l. $\dfrac{10}{\sqrt{6} - \sqrt{2}}$
m. $2x - 8 + 6y - 14$
n. $(3, -2)$ and $(0, -5)$
o. $x^4 \cdot x^3$
p. $\dfrac{4x^2y}{16xy}$
q. $h(x) = 10^x$
r. $\log_2 8 = 3$
s. $m = \dfrac{2}{3}$ and passes through $(0, 2)$
t. $2, 6, 18, \ldots$
u. $3y^3 - 243b^6$
v. $x + 7 \geq 0$ and $-x < -1$
w. $x^2 - 3x - 4 = 0$
x. $Rt = cd + 2t$
y. $-2\pi a^2 - 3b^3$
z. $x^2 + 4x$

CUMULATIVE REVIEW Chapters 1–14

1. Give the elements of the set $\{-\frac{4}{3}, \pi, 5.6, \sqrt{2}, 0, -23, e, 7i\}$ that belong to each of the following sets. [Section 1.3]
   a. Whole numbers 0
   b. Rational numbers $-\frac{4}{3}, 5.6, 0, -23$
   c. Irrational numbers $\pi, \sqrt{2}, e$
   d. Real numbers $-\frac{4}{3}, \pi, 5.6, \sqrt{2}, 0, -23, e$
2. Solve: $6[x - (2 - x)] = -4(8x + 3)$ [Section 2.2] 0
3. Solve $A = \frac{1}{2}h(b_1 + b_2)$ for b_2. [Section 2.4]
   $b_2 = \frac{2A - b_1h}{h}$ or $b_2 = \frac{2A}{h} - b_1$
4. **Martial Arts.** Find the measure of each angle of the triangle shown in the illustration. [Section 2.5]
   85°, 80°, 15°

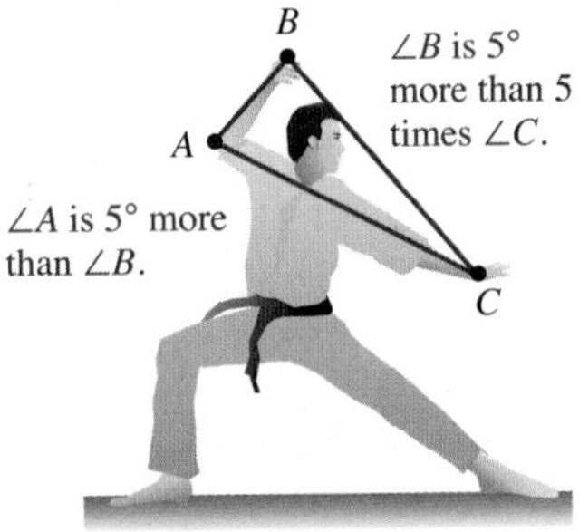

5. **Financial Planning.** Anna has some money to invest. Her financial planner tells her that if she can come up with \$3,000 more, she will qualify for an 11% annual interest rate. Otherwise, she will have to invest the money at 7.5% annual interest. The financial planner urges her to invest the larger amount, because the 11% investment would yield twice as much annual income as the 7.5% investment. How much does she originally have on hand to invest? [Section 2.6] \$8,250
6. Solve $-5x + 7 \le 12$. Write the solution set in interval notation and graph it. [Section 2.7] $[-1, \infty)$; see AIE Appendix 3.
7. **Windows.** An architect needs to design a triangular-shaped bathroom window that has an area no greater than 100 in.^{2}. If the base of the window must be 16 inches long, what window heights will meet this condition? [Section 2.7] 12.5 in. or less

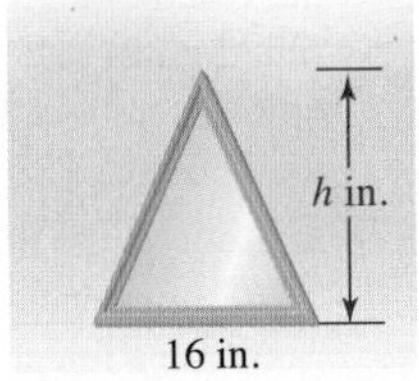

8. **Boating.** Use the following graph to determine the average rate of change in the sound level of the engine of a boat in relation to the number of revolutions per minute (rpm) of the engine. [Section 3.4] $\frac{1}{120}$ decibels/rpm

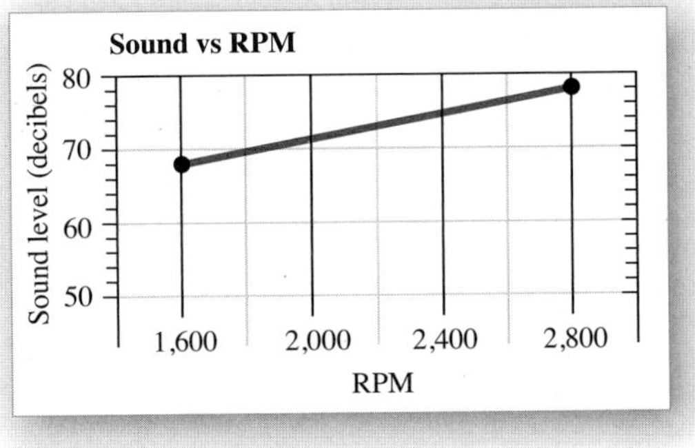

9. Decide whether the graphs of the equations are parallel or perpendicular. [Section 3.5]
   a. $3x - 4y = 12$, $y = \frac{3}{4}x - 5$ Parallel
   b. $y = 3x + 4$, $x = -3y + 4$ Perpendicular
10. **Salvage Values.** A truck was purchased for \$28,000. Its salvage value at the end of 6 years is expected to be \$7,600. Find the straight-line depreciation equation. [Section 3.6]
    $y = -3{,}400x + 28{,}000$

Find an equation of the line with the given properties. Write the equation in slope–intercept form.

11. $m = -2$, passes through $(0, 5)$ [Section 2.4] $y = -2x + 5$
12. Passes through $(8, -5)$ and $(-5, 4)$ [Section 2.4] $y = -\frac{9}{13}x + \frac{7}{13}$
13. Explain why the graph does not represent a function. [Section 3.8]
    It doesn't pass the vertical line test. The graph passes through (0, 2) and (0, −2).

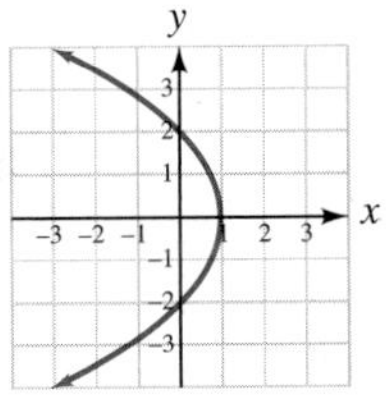

14. If $f(x) = 3x^5 - 2x^2 + 1$, find $f(-1)$ and $f(a)$. [Section 3.8]
    -4; $3a^5 - 2a^2 + 1$
15. Use substitution to solve: $\begin{cases} 2x - y = -21 \\ 4x + 5y = 7 \end{cases}$ [Section 4.3]
    $(-7, 7)$
16. Use elimination (addition) to solve: $\begin{cases} 4y + 5x - 7 = 0 \\ \frac{10}{7}x - \frac{4}{9}y = \frac{17}{21} \end{cases}$
    [Section 4.3] $\left(\frac{4}{5}, \frac{3}{4}\right)$
17. **Mixing Coffee.** How many pounds of regular coffee (selling for \$4 per pound) and how many pounds of Brazilian coffee (selling for \$11.50 per pound) must be combined to get 40 pounds of a mixture worth \$6 per pound? [Section 4.4]
    Regular: $29\frac{1}{3}$ lb; Brazilian: $10\frac{2}{3}$ lb
18. Graph the solution set: $\begin{cases} 3x - 2y \le 6 \\ y < -x + 2 \end{cases}$ [Section 4.5]
    See AIE Appendix 3.

Simplify each expression. Write answers using positive exponents.

19. $(x^2)^5y^7y^3x^{-2}y^0$
    [Section 5.2] x^8y^{10}
20. $\left(\frac{3x^5y^2}{6x^5y^{-2}}\right)^{-4}$
    [Section 5.2] $\frac{16}{y^{16}}$
21. Write 173,000,000,000,000 and 0.000000046 in scientific notation. [Section 5.3] 1.73×10^{14}; 4.6×10^{-8}
22. Write each number in scientific notation and perform the indicated operations. Give the answer in scientific notation and in standard notation. [Section 5.3]
    $$\frac{(0.00024)(96{,}000{,}000)}{(640{,}000{,}000)(0.025)}$$ 1.44×10^{-3}; 0.00144

23. Simplify the polynomial: $\frac{9}{4}rt^2 - \frac{5}{3}rt - \frac{1}{2}rt^2 + \frac{5}{6}rt$ [Section 5.5] $\frac{7}{4}rt^2 - \frac{5}{6}rt$

24. Find the sum when $(3x^2 + 4x - 7)$ is added to the difference of $(-2x^2 - 7x + 1)$ and $(-4x^2 + 8x - 1)$. [Section 5.5] $5x^2 - 11x - 5$

Perform the indicated operations and simplify, if possible.

25. $(-2x^2y^3 + 6xy + 5y^2) - (-4x^2y^3 - 7xy + 2y^2)$ [Section 5.5] $2x^2y^3 + 13xy + 3y^2$

26. $(x - 3y)(x^2 + 3xy + 9y^2)$ [Section 5.6] $x^3 - 27y^3$

27. $(2m^5 - 7)(3m^5 - 1)$ [Section 5.6] $6m^{10} - 23m^5 + 7$

28. $(9ab^2 - 4)^2$ [Section 5.7] $81a^2b^4 - 72ab^2 + 16$

Factor the expression completely.

29. $b^3 - 4b^2 - 3b + 12$ [Section 6.1] $(b - 4)(b^2 - 3)$

30. $12y^2 + 23y + 10$ [Section 6.3] $(3y + 2)(4y + 5)$

31. $144x^2 - y^2$ [Section 6.4] $(12x + y)(12x - y)$

32. $27t^3 + u^3$ [Section 6.5] $(3t + u)(9t^2 - 3tu + u^2)$

33. $a^4b^2 - 20a^2b^2 + 64b^2$ [Section 6.6] $b^2(a + 4)(a - 4)(a + 2)(a - 2)$

34. Solve: $3x^2 + 5x - 2 = 0$ [Section 6.7] $-2, \frac{1}{3}$

35. Solve: $(x + 7)^2 = -2(x + 7) - 1$ [Section 6.7] A repeated solution of -8

36. Solve: $x^3 + 8x^2 = 9x$ [Section 6.7] $0, 1, -9$

37. **Painting.** When it is spread out, a rectangular-shaped painting tarp covers an area of 84 square feet. Its length is 1 foot longer than five times its width. Find its width and length. [Section 6.8] Length: 21 ft; width: 4 ft

38. **Projectiles.** The height (in feet) of an object thrown upward from the roof of an apartment building is given by the function $h(t) = -16t^2 + 64t + 80$, where t is the number of seconds since it was thrown. How long will it take for the object to hit the ground? [Section 5.9] 5 sec

39. **Geometry.** The longer leg of a right triangle is 2 units longer than the shorter leg. If the hypotenuse is 4 units longer than the shorter leg, find the lengths of the sides of the triangle. [Section 6.8] 6, 8, 10

40. Simplify: $\frac{x^2 - 10x + 21}{6x^2 - 41x - 7}$ [Section 7.1] $\frac{x - 3}{6x + 1}$

Perform the indicated operations and simplify, if possible.

41. $\frac{p^3 - q^3}{q^2 - p^2} \cdot \frac{q^2 + pq}{p^3 + p^2q + pq^2}$ [Section 7.2] $-\frac{q}{p}$

42. $\frac{2x + 1}{x^4 - 81} + \frac{2 - x}{x^4 - 81}$ [Section 7.3] $\frac{1}{(x^2 + 9)(x - 3)}$

43. $\frac{2}{a - 2} + \frac{3}{a + 2} - \frac{a - 1}{a^2 - 4}$ [Section 7.4] $\frac{4a - 1}{(a + 2)(a - 2)}$

44. $\dfrac{\frac{y}{x} - \frac{x}{y}}{\frac{1}{x} + \frac{1}{y}}$ [Section 7.5] $y - x$

45. Solve: $\frac{3}{4n} + \frac{2}{n} = 1$ [Section 7.6] $\frac{11}{4}$

46. Solve $\frac{1}{R} = \frac{1}{R_1} + \frac{1}{R_2} + \frac{1}{R_3}$ for R. [Section 7.6] $R = \frac{R_1R_2R_3}{R_2R_3 + R_1R_3 + R_1R_2}$

47. **Printing Paychecks.** It takes a printer 6 hours to print the payroll checks for all of the employees of a large company. A faster printer can print the paychecks in 4 hours. How long will it take the two printers working together to print all of the paychecks? [Section 7.7] $2\frac{2}{5}$ hr = 2.4 hr

48. **Capture-Release Method.** To estimate the ground squirrel population on his acreage, a farmer trapped, tagged, and then released two dozen squirrels. Two weeks later, the farmer trapped 31 squirrels and noted that 8 were tagged. Use this information to estimate the number of ground squirrels on his acreage. [Section 7.8] 93 squirrels

49. Solve: $3(x - 4) + 6 = -2(x + 4) + 5x$ [Section 8.1] No solution, $\varnothing$; contradiction

50. Use a check to determine whether -4 is a solution of $\frac{3}{2}(y + 4) = \frac{20 - y}{2}$. [Section 8.1] It is not a solution.

51. **Work Schedules.** A student works two part-time jobs. She earns \$12 an hour for working at the college book store and \$22.50 an hour working graveyard shift at an airport. To save time for study, she limits her work to 25 hours a week. If she enjoys the work at the bookstore more, how many hours can she work at the bookstore and still earn at least \$450 a week? [Section 8.1] 10 hr

52. Does $x = |y|$ define a function? [Section 8.2] No; (2, 2), (2, −2)

53. Use the graph of function h to find each of the following. [Section 8.2]

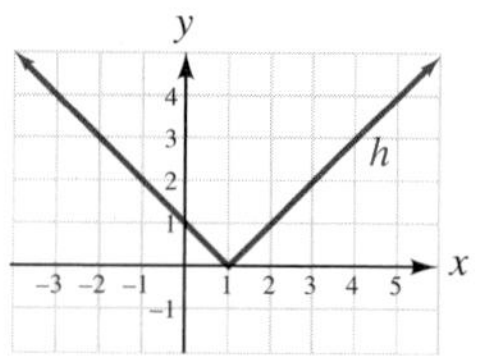

a. $h(-3)$ 4

b. $h(4)$ 3

c. The value(s) of x for which $h(x) = 1$ 0, 2

d. The value(s) of x for which $h(x) = 0$ 1

54. **Union Membership.** The percent of private-sector workers in the U.S. belonging to unions can be modeled by a linear function. In 1995, 10.3% of the private-sector workforce were union members. By the year 2000, that number had decreased to about 9%. (Source: unionstats.gsu.edu) [Section 8.2]

a. Let t be the number of years after 1990 and P be the percent of the private-sector workforce that are union members. Write a linear function $P(t)$ to model the situation. $P(t) = -0.26t + 11.6$

b. Predict the percent of private-sector workers who will belong to a union in 2030, if the trend continues. 1.2%

55. **Produce.** The polynomial function $n(c) = \frac{1}{3}c^3 + \frac{1}{2}c^2 + \frac{1}{6}c$ gives the number of cantaloupes used in a display shaped like a square pyramid, having a base formed by c cantaloupes per side. Find the number of cantaloupes needed to make the display shown in the illustration below. [Section 8.2] 91 cantaloupes

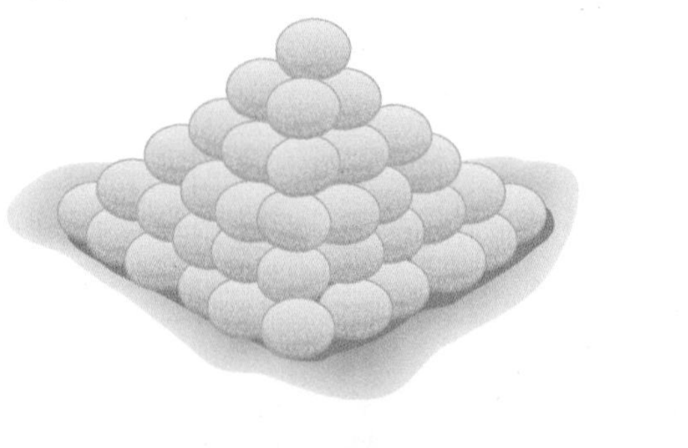

56. Consider the function $g(x) = (x - 6)^2$. First sketch the graph of its associated function. Then draw the graph of function g using a translation. Give the domain and range of g. [Section 8.3] D: the set of real numbers $(-\infty, \infty)$; R: the set of nonnegative real numbers $[0, \infty)$; see AIE Appendix 3.

Give the solution in interval notation and graph the solution set. See AIE Appendix 3.

57. $4.5x - 1 < -10$ or $6 - 2x \ge 12$ [Section 8.4] $(-\infty, -2)$

58. $|x + 5| \ge 7$ [Section 8.4] $(-\infty, -12] \cup [2, \infty)$

59. $5(x + 1) \le 4(x + 3)$ and $x + 12 < -3$ [Section 8.4] $(-\infty, -15)$

60. $|5 - 3x| - 14 \le 0$ [Section 8.5] $\left[-3, \frac{19}{3}\right]$

61. Solve: $|2x + 1| = |3(x + 1)|$ [Section 8.5] $-2, -\frac{4}{5}$

62. Use a substitution to factor: $14(x - t)^2 - 17(x - t) - 6$ [Section 8.6] $(7x - 7t + 2)(2x - 2t - 3)$

63. Solve for λ: $\dfrac{A\lambda}{2} + 1 = 2d + 3\lambda$ [Section 8.6] $\lambda = \dfrac{4d - 2}{A - 6}$

64. Factor: $x^2 + 10x + 25 - 16z^2$ [Section 8.7] $(x + 5 + 4z)(x + 5 - 4z)$

65. Factor: $64a^6 - 27b^3$ [Section 8.7] $(4a^2 - 3b)(16a^4 + 12a^2b + 9b^2)$

66. Simplify the rational function $g(x) = \dfrac{x^3 + 64}{x^3 + 4x^2 + 3x + 12}$. List any restrictions on the domain. [Section 8.8]
$g(x) = \dfrac{x^2 - 4x + 16}{x^2 + 3}$ provided $x \ne -4$

67. Solve: $\dfrac{1}{a + 5} = \dfrac{1}{3a + 6} - \dfrac{a + 2}{a^2 + 7a + 10}$ [Section 8.8] $-\frac{7}{5}$

68. Find the domain of the function $f(x) = \frac{8}{x^2 - 25}$. [Section 8.8] All real numbers except -5 and 5.

69. **Light.** The intensity of a light source is inversely proportional to the square of the distance from the source. If the intensity is 18 lumens at a distance of 4 feet, what is the intensity when the distance is 12 feet? [Section 8.9] 2 lumens

70. Find the domain of $f(x) = \sqrt{x + 3}$. [Section 9.1] $[-3, \infty)$

71. **Aquariums.** The function $L(g) = \sqrt[3]{\frac{g}{7.5}}$ gives the length (in feet) of an edge of a cube-shaped tank that holds g gallons of water. What length of an edge should a cube-shaped aquarium have if it is to hold 480 gallons of water? [Section 9.1] 4 ft

72. Graph: $f(x) = \sqrt{x} + 2$. Give the domain and range of the function. [Section 9.1] D: $[0, \infty)$; R: $[2, \infty)$; see AIE Appendix 3.

73. Evaluate: $\left(\dfrac{25}{49}\right)^{-3/2}$ [Section 9.2] $\frac{343}{125}$

74. Simplify: $\sqrt{112a^3b^5}$ [Section 9.3] $4ab^2\sqrt{7ab}$

Simplify each expression. All variables represent positive numbers.

75. $\sqrt{98} + \sqrt{8} - \sqrt{32}$ [Section 9.3] $5\sqrt{2}$

76. $12\sqrt[3]{648x^4} + 3\sqrt[3]{81x^4}$ [Section 9.3] $81x\sqrt[3]{3x}$

77. $(2\sqrt{7} + 1)(\sqrt{7} - 1)$ [Section 9.4] $13 - \sqrt{7}$

78. $3(\sqrt{5x} - \sqrt{3})^2$ [Section 9.4] $15x - 6\sqrt{15x} + 9$

Rationalize each denominator.

79. $\dfrac{\sqrt[3]{4}}{\sqrt[3]{b}}$ [Section 9.4] $\dfrac{\sqrt[3]{4b^2}}{b}$

80. $\dfrac{3t - 1}{\sqrt{3t} + 1}$ [Section 9.4] $\sqrt{3t} - 1$

Solve each equation.

81. $2x = \sqrt{16x - 12}$ [Section 9.5] 1, 3

82. $\sqrt[3]{12m + 4} = 4$ [Section 9.5] 5

83. Let $f(x) = \sqrt{x + 3} - \sqrt{3}$ and $g(x) = \sqrt{x}$. Find all values of x for which $f(x) = g(x)$. [Section 9.5] 0

84. **Accessories.** A silk scarf has sides 16 inches long. When folded over itself, as shown below, the result is a triangular scarf that can be wrapped around one's neck. Find the length of the longest side. [Section 9.6] About $22\frac{1}{2}$ in.

85. Express $\sqrt{-25}$ in terms of i. [Section 9.7] $5i$

86. Simplify: i^{42} [Section 9.7] -1

Write each expression in a + bi form.

87. $(-7 + 9i) - (-2 - 8i)$ [Section 9.7] $-5 + 17i$

88. $\dfrac{2 - 5i}{2 + 5i}$ [Section 9.7] $-\frac{21}{29} - \frac{20}{29}i$

89. Solve: $t^2 = 24$ [Section 10.1] $\pm 2\sqrt{6}$

90. Solve $m^2 + 10m - 7 = 0$ by completing the square. [Section 10.1] $-5 \pm 4\sqrt{2}$

Solve each equation.

91. $4w^2 + 6w + 1 = 0$ [Section 10.2] $\dfrac{-3 \pm \sqrt{5}}{4}$

92. $3x^2 - 4x = -2$ [Section 10.2] $\frac{2}{3} \pm \frac{\sqrt{2}}{3}i$

93. $2(2x + 1)^2 - 7(2x + 1) + 6 = 0$ [Section 10.3] $\frac{1}{4}, \frac{1}{2}$

94. $x^4 + 19x^2 + 18 = 0$ [Section 10.3] $-i, i, -3i\sqrt{2}, 3i\sqrt{2}$

95. Tire Wear. Refer to the graph. [Section 10.4]

a. What type of function does it appear would model the relationship between the inflation of a tire and the percent of service it gives? A quadratic function

b. At what percent(s) of inflation will a tire offer only 90% of its possible service?
At about 85% and 120% of the suggested inflation

Effect of Inflation on Tire Service

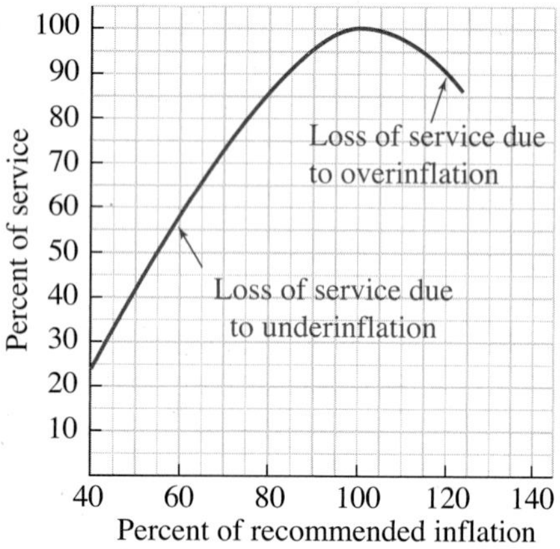

96. Graph $f(x) = -6x^2 - 12x - 8$ using the vertex formula. Then determine the x- and y-intercepts of the graph. Finally, plot several points and complete the graph. [Section 10.4]
$(-1, -2)$; $(0, -8)$; No x-intercepts; see AIE Appendix 3.

97. Solve $x^2 - 8x \le -15$. Write the solution set in interval notation and graph it. [Section 10.5] $[3, 5]$; see AIE Appendix 3.

98. If $f(x) = x^2 - 2$ and $g(x) = 2x + 1$, find: [Section 11.1]

a. $(f + g)(x)$ $x^2 + 2x - 1$

b. $(f \cdot g)(x)$ $2x^3 + x^2 - 4x - 2$

c. $(f \circ g)(x)$ $4x^2 + 4x - 1$

99. Find the inverse function of $f(x) = 2x^3 - 1$. [Section 11.2]
$f^{-1}(x) = \sqrt[3]{\frac{x+1}{2}}$

100. Graph $f(x) = \left(\frac{1}{2}\right)^x$ and give the domain and range of the function. [Section 11.3] D: $(-\infty, \infty)$; R: $(0, \infty)$; see AIE Appendix 3.

101. Graph $f(x) = e^x$ and its inverse on the same coordinate system. [Section 11.5] See AIE Appendix 3.

102. Population Growth. As of 2010, the population of Mexico was about 114 million and the annual growth rate was 1.102%. If the growth rate remains the same, estimate the population of Mexico in 25 years. (Source: CIA World Factbook) [Section 11.5] About 150 million

Find x.

103. $\log 1{,}000 = x$ [Section 11.4]
3

104. $\log_8 64 = x$ [Section 11.4]
2

105. $\log_3 x = -3$ [Section 11.4]
$\frac{1}{27}$

106. $\log_x 25 = 2$ [Section 11.4]
5

107. $\ln e = x$ [Section 11.5]
1

108. $\ln \frac{1}{e} = x$ [Section 11.5]
-1

109. Find ln 0, if possible. [Section 11.5] Undefined

110. Use the properties of logarithms to simplify $\log_6 \frac{36}{x^3}$. [Section 11.6] $2 - 3\log_6 x$

111. Write the expression $\frac{1}{2}\ln x + \ln y - \ln z$ as a single logarithm. [Section 11.6] $\ln \frac{y\sqrt{x}}{z}$

112. Bacteria Growth. The bacteria in a laboratory culture increased from an initial population of 200 to 600 in 4 hours. How long will it take the population to reach 8,000? [Section 11.7] About 13.4 hr

Solve each equation. Round to four decimal places when necessary.

113. $5^{4x} = \frac{1}{125}$ [Section 11.7] $-\frac{3}{4}$

114. $2^{x+2} = 3^x$ [Section 11.7] 3.4190

115. $\log x + \log(x + 9) = 1$ [Section 11.7] 1, -10 does not check

116. $\log_3 x = \log_3\left(\frac{1}{x}\right) + 4$ [Section 11.7] 9

117. Match each function to its graph shown below.

a. $f(x) = \frac{1}{x}$ [Section 6.1] ii. **b.** $f(x) = \log x$ [Section 9.4] iv.

c. $f(x) = x^3$ [Section 2.6] i. **d.** $f(x) = \sqrt[3]{x}$ [Section 7.1] iii.

i.

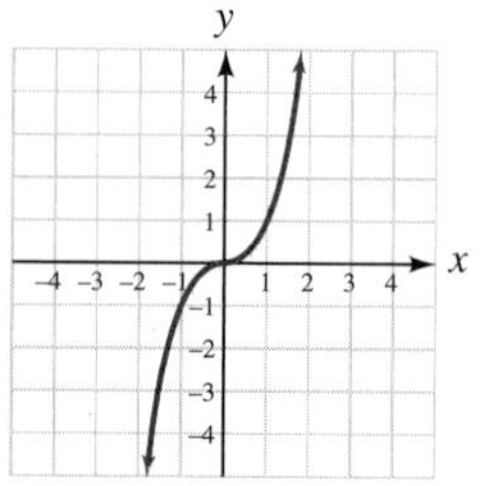

ii.

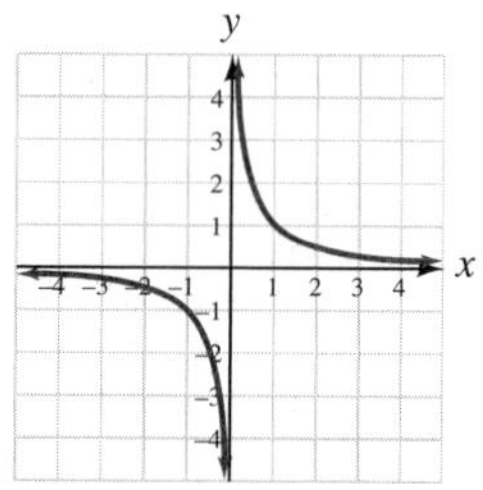

iii.

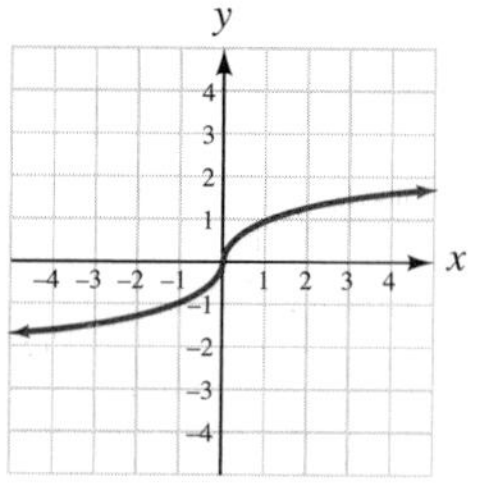

iv.

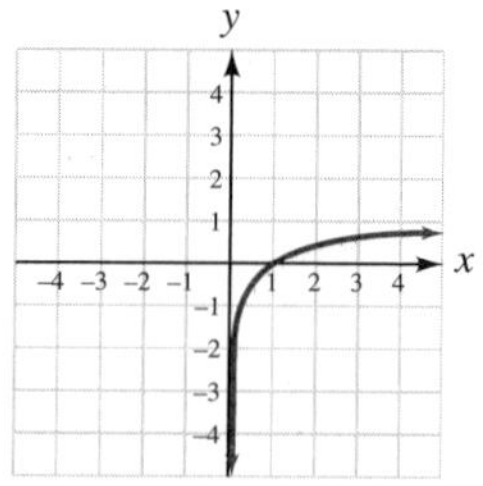

118. Use matrices to solve the system: $\begin{cases} 2x + y = 1 \\ x + 2y = -4 \end{cases}$
[Section 12.3] $(2, -3)$

119. Use Cramer's rule to solve: $\begin{cases} 2(x + y) + 1 = 0 \\ 3x + 4y = 0 \end{cases}$
[Section 12.4] $\left(-2, \frac{3}{2}\right)$

120. Solve: $\begin{cases} b + 2c = 7 - a \\ a + c = 8 - 2b \\ 2a + b + c = 9 \end{cases}$ [Section 12.2] $(3, 2, 1)$

121. **Sports Socks.** A company makes three types of high-end athletic socks: "ankle," "low cut," and "crew." The cost to make one pair of ankle socks is \$2 and they sell for \$3. The cost to make one pair of low cut socks is \$3 and they sell for \$5. The cost to make one pair of crew socks is \$4 and they sell for \$6. Each day, the cost of manufacturing 500 pairs of socks is \$1,650, and the daily revenue from their sale is \$2,550. How many pairs of each type of sock are manufactured daily? [Section 12.5] Ankle: 100 pair; low cut: 150 pair; crew: 250 pair

122. Write the equation of the circle that has its center at $(1, -3)$ and a radius of 2. Graph the equation. [Section 13.1] $(x - 1)^2 + (y + 3)^2 = 4$; See AIE Appendix 3.

123. Complete the square to write the equation $y^2 + 4x - 6y = -1$ in $x = a(y - k)^2 + h$ form. Determine the vertex and the axis of symmetry of the graph. Then plot several points and complete the graph. [Section 13.1] $x = -\frac{1}{4}(y - 3)^2 + 2$; $(2, 3)$, $y = 3$; see AIE Appendix 3.

124. Graph: $\frac{(x + 1)^2}{4} + \frac{(y - 3)^2}{16} = 1$ [Section 13.2] see AIE Appendix 3.

125. Write the equation in standard form and graph it: $(x - 2)^2 - 9y^2 = 9$ [Section 13.3] $\frac{(x - 2)^2}{9} - \frac{y^2}{1} = 1$; see AIE Appendix 3.

126. Use the binomial theorem to expand $(3a - b)^4$. [Section 14.1] $81a^4 - 108a^3b + 54a^2b^2 - 12ab^3 + b^4$

127. Find the seventh term of the expansion of $(2x - y)^8$. [Section 14.1] $112x^2y^6$

128. Evaluate: $\frac{12!}{10!(12 - 10)!}$ [Section 14.1] 66

129. Find the 20th term of an arithmetic sequence with a first term -11 and a common difference 6. [Section 14.2] 103

130. Find the sum of the first 20 terms of an arithmetic sequence with a first term 6 and a common difference 3. [Section 14.2] 690

131. Evaluate: $\sum_{k=3}^{5} (2k + 1)$ [Section 14.2] 27

132. **Boat Depreciation.** How much will a \$9,000 boat be worth after 9 years if it depreciates 12% per year? [Section 14.3] \$2,848.31

133. Find the sum of the first ten terms of the sequence: $\frac{1}{64}, \frac{1}{32}, \frac{1}{16}, \ldots$ [Section 14.3] $\frac{1,023}{64}$

134. Find the sum of the infinite series: $9 + 3 + 1 + \cdots$ [Section 14.3] $\frac{27}{2}$

APPENDIX 1 Roots and Powers

n	n^2	$\sqrt{n}$	n^3	$\sqrt[3]{n}$	n	n^2	$\sqrt{n}$	n^3	$\sqrt[3]{n}$
1	1	1.000	1	1.000	51	2,601	7.141	132,651	3.708
2	4	1.414	8	1.260	52	2,704	7.211	140,608	3.733
3	9	1.732	27	1.442	53	2,809	7.280	148,877	3.756
4	16	2.000	64	1.587	54	2,916	7.348	157,464	3.780
5	25	2.236	125	1.710	55	3,025	7.416	166,375	3.803
6	36	2.449	216	1.817	56	3,136	7.483	175,616	3.826
7	49	2.646	343	1.913	57	3,249	7.550	185,193	3.849
8	64	2.828	512	2.000	58	3,364	7.616	195,112	3.871
9	81	3.000	729	2.080	59	3,481	7.681	205,379	3.893
10	100	3.162	1,000	2.154	60	3,600	7.746	216,000	3.915
11	121	3.317	1,331	2.224	61	3,721	7.810	226,981	3.936
12	144	3.464	1,728	2.289	62	3,844	7.874	238,328	3.958
13	169	3.606	2,197	2.351	63	3,969	7.937	250,047	3.979
14	196	3.742	2,744	2.410	64	4,096	8.000	262,144	4.000
15	225	3.873	3,375	2.466	65	4,225	8.062	274,625	4.021
16	256	4.000	4,096	2.520	66	4,356	8.124	287,496	4.041
17	289	4.123	4,913	2.571	67	4,489	8.185	300,763	4.062
18	324	4.243	5,832	2.621	68	4,624	8.246	314,432	4.082
19	361	4.359	6,859	2.668	69	4,761	8.307	328,509	4.102
20	400	4.472	8,000	2.714	70	4,900	8.367	343,000	4.121
21	441	4.583	9,261	2.759	71	5,041	8.426	357,911	4.141
22	484	4.690	10,648	2.802	72	5,184	8.485	373,248	4.160
23	529	4.796	12,167	2.844	73	5,329	8.544	389,017	4.179
24	576	4.899	13,824	2.884	74	5,476	8.602	405,224	4.198
25	625	5.000	15,625	2.924	75	5,625	8.660	421,875	4.217
26	676	5.099	17,576	2.962	76	5,776	8.718	438,976	4.236
27	729	5.196	19,683	3.000	77	5,929	8.775	456,533	4.254
28	784	5.292	21,952	3.037	78	6,084	8.832	474,552	4.273
29	841	5.385	24,389	3.072	79	6,241	8.888	493,039	4.291
30	900	5.477	27,000	3.107	80	6,400	8.944	512,000	4.309
31	961	5.568	29,791	3.141	81	6,561	9.000	531,441	4.327
32	1,024	5.657	32,768	3.175	82	6,724	9.055	551,368	4.344
33	1,089	5.745	35,937	3.208	83	6,889	9.110	571,787	4.362
34	1,156	5.831	39,304	3.240	84	7,056	9.165	592,704	4.380
35	1,225	5.916	42,875	3.271	85	7,225	9.220	614,125	4.397
36	1,296	6.000	46,656	3.302	86	7,396	9.274	636,056	4.414
37	1,369	6.083	50,653	3.332	87	7,569	9.327	658,503	4.431
38	1,444	6.164	54,872	3.362	88	7,744	9.381	681,472	4.448
39	1,521	6.245	59,319	3.391	89	7,921	9.434	704,969	4.465
40	1,600	6.325	64,000	3.420	90	8,100	9.487	729,000	4.481
41	1,681	6.403	68,921	3.448	91	8,281	9.539	753,571	4.498
42	1,764	6.481	74,088	3.476	92	8,464	9.592	778,688	4.514
43	1,849	6.557	79,507	3.503	93	8,649	9.644	804,357	4.531
44	1,936	6.633	85,184	3.530	94	8,836	9.695	830,584	4.547
45	2,025	6.708	91,125	3.557	95	9,025	9.747	857,375	4.563
46	2,116	6.782	97,336	3.583	96	9,216	9.798	884,736	4.579
47	2,209	6.856	103,823	3.609	97	9,409	9.849	912,673	4.595
48	2,304	6.928	110,592	3.634	98	9,604	9.899	941,192	4.610
49	2,401	7.000	117,649	3.659	99	9,801	9.950	970,299	4.626
50	2,500	7.071	125,000	3.684	100	10,000	10.000	1,000,000	4.642

APPENDIX 2 Synthetic Division

1 Synthetic division

2 The remainder theorem

3 The factor theorem

We have discussed how to divide polynomials by polynomials using a long division process. We will now discuss a shortcut method, called **synthetic division,** that we can use to divide a polynomial by a binomial of the form $x - k$.

1 Synthetic Division.

To see how synthetic division works, we consider $(4x^3 - 5x^2 - 11x + 20) \div (x - 2)$. On the left below is the long division, and on the right is the same division with the variables and their exponents removed. The various powers of x can be remembered without actually writing them, because the exponents of the terms in the divisor, dividend, and quotient were written in descending order.

$$\begin{array}{r l}
 & 4x^2 + 3x - 5 \\
x - 2\,) & \overline{4x^3 - 5x^2 - 11x + 20} \\
 & \underline{4x^3 - 8x^2} \\
 & \quad 3x^2 - 11x \\
 & \quad \underline{3x^2 - 6x} \\
 & \qquad -5x + 20 \\
 & \qquad \underline{-5x + 10} \\
 & \qquad\qquad 10 \quad \text{(remainder)}
\end{array}
\qquad
\begin{array}{r l}
 & 4 \quad 3 - 5 \\
1 - 2\,) & \overline{4 - 5 - 11 \quad 20} \\
 & \underline{4 - 8} \\
 & \quad 3 - 11 \\
 & \quad \underline{3 - 6} \\
 & \qquad -5 \quad 20 \\
 & \qquad \underline{-5 \quad 10} \\
 & \qquad\qquad 10 \quad \text{(remainder)}
\end{array}$$

The numbers printed in color need not be written, because they are duplicates of the numbers above them. Thus, we can write the division in the form shown below on the left. We can shorten the process further by compressing the work vertically and eliminating the 1 (the coefficient of x in the divisor) as shown below on the right.

The Language of Algebra

Synthetic means devised to imitate something natural. You've probably heard of *synthetic* fuels or *synthetic* fibers. *Synthetic* division imitates the long division process.

$$\begin{array}{r l}
 & 4 \quad 3 - 5 \\
1 - 2\,) & \overline{4 - 5 - 11 \quad 20} \\
 & \underline{\;\; - 8} \\
 & \quad 3 \\
 & \quad \underline{\; - 6} \\
 & \qquad - 5 \\
 & \qquad\qquad \underline{10} \\
 & \qquad\qquad 10
\end{array}
\qquad
\begin{array}{r r r r r}
 & 4 & 3 & -5 & \\
-2\,) & 4 & -5 & -11 & 20 \\
 & & -8 & -6 & 10 \\
\hline
 & & 3 & -5 & 10
\end{array}$$

If we write the 4 in the quotient on the bottom line, that line gives the coefficients of the quotient and the remainder. If we eliminate the top line, the division appears as follows:

$$\begin{array}{r|rrrr} -2 & 4 & -5 & -11 & 20 \\ & & -8 & -6 & 10 \\ \hline & 4 & 3 & -5 & 10 \end{array}$$

The Language of Algebra

Synthetic division is used to divide a polynomial by a binomial of the form $x - k$. We call k the **synthetic divisor.** In this example, we are dividing by $x - 2$, so k is 2.

The bottom line was obtained by subtracting the middle line from the top line. If we replace the -2 in the divisor by 2, the division process will reverse the signs of every entry in the middle line, and then the bottom line can be obtained by addition. This gives the final form of the synthetic division.

$$\begin{array}{r|rrrr} 2 & 4 & -5 & -11 & 20 \\ & & 8 & 6 & -10 \\ \hline & 4 & 3 & -5 & 10 \\ & \downarrow & \downarrow & \downarrow & \downarrow \\ & 4x^2 & +\ 3x & -\ 5 & +\ \dfrac{10}{x-2} \end{array}$$

These are the coefficients of the dividend.

These are the coefficients of the quotient and the remainder.

Read the result from the bottom row.

Thus,

$$\frac{4x^3 - 5x^2 - 11x + 20}{x - 2} = 4x^2 + 3x - 5 + \frac{10}{x - 2}$$

EXAMPLE 1 Use synthetic division to find $(6x^2 + 5x - 2) \div (x - 5)$.

Solution We write the coefficients in the dividend and the 5 in the divisor in the following form:

Since we are dividing the polynomial by $x - 5$, the synthetic divisor is 5.

$$\longrightarrow \begin{array}{r|rrr} 5 & 6 & 5 & -2 \\ & & & \\ \hline \end{array} \longleftarrow \text{This represents the dividend } 6x^2 + 5x - 2.$$

Then we follow these steps:

$$\begin{array}{r|rrr} 5 & 6 & 5 & -2 \\ & \downarrow & & \\ \hline & 6 & & \end{array}$$

Begin by bringing down the 6.

Success Tip

In his process, numbers below the line are multiplied by the synthetic divisor and that product is carried above the line to the next column. Numbers above the horizontal line are added.

$$\begin{array}{r|rrr} 5 & 6 & 5 & -2 \\ & & 30 & \\ \hline & 6 & & \end{array}$$

Multiply 5 by 6 to get 30.

$$\begin{array}{r|rrr} 5 & 6 & 5 & -2 \\ & & 30 & \\ \hline & 6 & 35 & \end{array}$$

Add 5 and 30 to get 35.

$$\begin{array}{r|rrr} 5 & 6 & 5 & -2 \\ & & 30 & 175 \\ \hline & 6 & 35 & \end{array}$$

Multiply 35 by 5 to get 175.

$$\begin{array}{r|rrr} 5 & 6 & 5 & -2 \\ & & 30 & 175 \\ \hline & 6 & 35 & 173 \end{array}$$

Add −2 and 175 to get 173.

The numbers 6 and 35 represent the quotient $6x + 35$, and 173 is the remainder. Thus,

$$\frac{6x^2 + 5x - 2}{x - 5} = 6x + 35 + \frac{173}{x - 5}$$

Self Check 1 Divide $5x^2 - 4x + 2$ by $x - 3$.

EXAMPLE 2 Use synthetic division to find $\dfrac{x^3 + x^2 - 1}{x - 3}$.

Solution We begin by writing

$$\begin{array}{r|rrrr} 3 & 1 & 1 & 0 & -1 \\ \hline & & & & \end{array}$$

Write 0 for the coefficient of x, the missing term.

and complete the division as follows.

$$\begin{array}{r|rrrr} 3 & 1 & 1 & 0 & -1 \\ & & 3 & & \\ \hline & 1 & 4 & & \end{array} \qquad \begin{array}{r|rrrr} 3 & 1 & 1 & 0 & -1 \\ & & 3 & 12 & \\ \hline & 1 & 4 & 12 & \end{array} \qquad \begin{array}{r|rrrr} 3 & 1 & 1 & 0 & -1 \\ & & 3 & 12 & 36 \\ \hline & 1 & 4 & 12 & 35 \end{array}$$

Multiply, then add. *Multiply, then add.* *Multiply, then add.*

Thus,

$$\frac{x^3 + x^2 - 1}{x - 3} = x^2 + 4x + 12 + \frac{35}{x - 3}$$

Self Check 2 Use synthetic division to find $\frac{x^3 + 3x - 90}{x - 4}$.

EXAMPLE 3 Use synthetic division to divide $5x^2 + 6x^3 + 2 - 4x$ by $x + 2$.

Solution First, we write the dividend with the exponents in descending order.

$$6x^3 + 5x^2 - 4x + 2$$

Then we write the divisor in $x - k$ form: $x - (\mathbf{-2})$. Thus, $k = -2$. Using synthetic division, we begin by writing

This represents division by x + 2.

$$\begin{array}{r|rrrr} \mathbf{-2} & 6 & 5 & -4 & 2 \\ \hline & & & & \end{array}$$

and complete the division.

$$\begin{array}{r|rrrr} -2 & 6 & 5 & -4 & 2 \\ & & -12 & 14 & -20 \\ \hline & 6 & -7 & 10 & -18 \end{array}$$

The remainder is negative.

Notation

Because the remainder is negative, we can also write the result as

$$6x^2 - 7x + 10 + \frac{-18}{x + 2}$$

Thus,

$$\frac{5x^2 + 6x^3 + 2 - 4x}{x + 2} = 6x^2 - 7x + 10 - \frac{18}{x + 2}$$

Self Check 3 Divide $2x - 4x^2 + 3x^3 - 3$ by $x + 1$.

2 The Remainder Theorem.

Synthetic division is important because of the **remainder theorem.**

Remainder Theorem If a polynomial $P(x)$ is divided by $x - k$, the remainder is $P(k)$.

It follows from the remainder theorem that we can evaluate polynomials using synthetic division. We illustrate this in the following example.

Notation

Naming the function with the letter P, instead of f, stresses that we are working with a polynomial function.

Success Tip

It is often easier to find $P(k)$ by using synthetic division than by substituting k for x in $P(x)$. This is especially true if k is a decimal.

EXAMPLE 4 Let $P(x) = 2x^3 - 3x^2 - 2x + 1$. Find **a.** $P(3)$ and **b.** the remainder when $P(x)$ is divided by $x - 3$.

Solution **a.** To find $P(3)$ we evaluate the function for $x = 3$.

$$\begin{aligned} P(3) &= 2(3)^3 - 3(3)^2 - 2(3) + 1 \quad \text{Substitute 3 for } x. \\ &= 2(27) - 3(9) - 6 + 1 \\ &= 54 - 27 - 6 + 1 \\ &= 22 \end{aligned}$$

Thus, $P(3) = 22$.

b. We can use synthetic division to find the remainder when $P(x)$ is divided by $x - 3$.

$$\begin{array}{r|rrrr} 3 & 2 & -3 & -2 & 1 \\ & & 6 & 9 & 21 \\ \hline & 2 & 3 & 7 & \mathbf{22} \end{array} \qquad P(x) = 2x^3 - 3x^2 - 2x + 1$$

Thus, the remainder is 22.

The same results in parts a and b show that rather than substituting 3 for x in $P(x) = 2x^3 - 3x^2 - 2x + 1$, we can divide $2x^3 - 3x^2 - 2x + 1$ by $x - 3$ to find $P(3)$.

Self Check 4 Let $P(x) = 5x^3 - 3x^2 + x + 6$. Find **a.** $P(1)$ and **b.** use synthetic division to find the remainder when $P(x)$ is divided by $x - 1$.

3 The Factor Theorem.

If two quantities are multiplied, each is called a **factor** of the product. Thus, $x - 2$ is a factor of $6x - 12$, because $6(x - 2) = 6x - 12$. A theorem, called the **factor theorem,** tells us how to find one factor of a polynomial if the remainder of a certain division is 0.

Factor Theorem If $P(x)$ is a polynomial in x, then

$$P(k) = 0 \quad \text{if and only if} \quad x - k \text{ is a factor of } P(x)$$

If $P(x)$ is a polynomial in x and if $P(k) = 0$, k is called a **zero of the polynomial function.**

EXAMPLE 5 Let $P(x) = 3x^3 - 5x^2 + 3x - 10$. Show that **a.** $P(2) = 0$ and **b.** $x - 2$ is a factor of $P(x)$.

Solution **a.** We can use the remainder theorem to evaluate $P(2)$ by dividing $P(x)$ by $x - 2$.

2	3	−5	3	−10	$P(x) = 3x^2 - 5x^2 + 3x - 10$
		6	2	10	
	3	1	5	**0**	

The Language of Algebra

The phrase *if and only if* in the factor theorem means:

If $P(2) = 0$, then $x - 2$ is a factor of $P(x)$

and

If $x - 2$ is a factor of $P(x)$, then $P(2) = 0$.

The remainder in this division is 0. By the remainder theorem, the remainder is $P(2)$. Thus, $P(2) = 0$, and 2 is a zero of the polynomial.

b. Because the remainder is 0, the numbers 3, 1, and 5 in the synthetic division in part a represent the quotient $3x^2 + x + 5$. Thus,

$$\underbrace{(x-2)}_{\text{Divisor}} \cdot \underbrace{(3x^2 + x + 5)}_{\text{quotient}} + \underbrace{0}_{\text{remainder}} = \underbrace{3x^3 - 5x^2 + 3x - 10}_{\text{the dividend, } P(x)}$$

or

$$(x - 2)(3x^2 + x + 5) = 3x^3 - 5x^2 + 3x - 10$$

Thus, $x - 2$ is a factor of $3x^3 - 5x^2 + 3x - 10$.

Self Check 5 Let $P(x) = x^3 - 4x^2 + x + 6$. Show that $x + 1$ is a factor of $P(x)$ using synthetic division.

The result in Example 5 is true, because the remainder, $P(2)$, is 0. If the remainder had not been 0, then $x - 2$ would not have been a factor of $P(x)$.

APPENDIX 2 STUDY SET

VOCABULARY

Fill in the blanks.

1. The method of dividing $x^2 + 2x - 9$ by $x - 4$ shown below is called synthetic division.

4	1	2	−9
		4	24
	1	6	15

2. Synthetic division is used to divide a polynomial by a binomial of the form $x - k$.
3. In Exercise 1, the synthetic divisor is 4.
4. By the remainder theorem, if a polynomial $P(x)$ is divided by $x - k$, the remainder is $P(k)$.
5. The factor theorem tells us how to find one factor of a polynomial if the remainder of a certain division is 0.
6. If $P(x)$ is a polynomial and if $P(k) = 0$, then k is called a zero of the polynomial.

CONCEPTS

7. **a.** What division is represented below? $(5x^3 + x - 3) \div (x + 2)$
   b. What is the answer? $5x^2 - 10x + 21 - \frac{45}{x+2}$

−2	5	0	1	−3
		−10	20	−42
	5	−10	21	−45

Fill in the blanks.

8. In the synthetic division process, numbers below the line are multiplied by the synthetic divisor and that product is carried above the line to the next column. Numbers above the horizontal line are added.
9. Rather than substituting 8 for x in $P(x) = 6x^3 - x^2 - 17x + 9$, we can divide the polynomial $6x^3 - x^2 - 17x + 9$ by $x - 8$ to find $P(8)$.
10. For $P(x) = x^3 - 4x^2 + x + 6$, suppose we know that $P(3) = 0$. Then $x - 3$ is a factor of $x^3 - 4x^2 + x + 6$.

NOTATION

Complete each synthetic division.

11. Divide $6x^3 + x^2 - 23x + 2$ by $x - 2$.

2	6	1	−23	2
		12	26	6
	6	13	3	8

12. Divide $2x^3 - 4x^2 - 25x + 15$ by $x + 3$.

−3	2	−4	−25	15
		−6	30	−15
	2	−10	5	0

GUIDED PRACTICE

Use synthetic division to perform each division.

13. $\dfrac{x^2 + x - 2}{x - 1}$ $x + 2$

14. $\dfrac{x^2 + x - 6}{x - 2}$ $x + 3$

15. $\dfrac{x^2 - 7x + 12}{x - 4}$ $x - 3$

16. $\dfrac{x^2 - 6x + 5}{x - 5}$ $x - 1$

17. $\dfrac{x^2 + 8 + 6x}{x + 4}$ $x + 2$

18. $\dfrac{x^2 - 15 - 2x}{x + 3}$ $x - 5$

19. $\dfrac{x^2 - 5x + 14}{x + 2}$ $x - 7 + \frac{28}{x + 2}$

20. $\dfrac{x^2 + 13x + 42}{x + 6}$ $x + 7$

21. $\dfrac{3x^3 - 10x^2 + 5x - 6}{x - 3}$ $3x^2 - x + 2$

22. $\dfrac{2x^3 - 9x^2 + 10x - 3}{x - 3}$ $2x^2 - 3x + 1$

23. $\dfrac{2x^3 - 6 - 5x}{x - 2}$ $2x^2 + 4x + 3$

24. $\dfrac{4x^3 - 1 + 5x^2}{x + 2}$ $4x^2 - 3x + 6 - \frac{13}{x + 2}$

25. $\dfrac{5x^2 + 4 + 6x^3}{x + 1}$ $6x^2 - x + 1 + \frac{3}{x + 1}$

26. $\dfrac{4 - 3x^2 + x}{x - 4}$ $-3x - 11 - \frac{40}{x - 4}$

27. $\dfrac{t^3 + t^2 + t + 2}{t + 1}$ $t^2 + 1 + \frac{1}{t + 1}$

28. $\dfrac{m^3 - m^2 - m - 1}{m - 1}$ $m^2 - 1 - \frac{2}{m - 1}$

29. $\dfrac{a^5 - 1}{a - 1}$ $a^4 + a^3 + a^2 + a + 1$

30. $\dfrac{b^4 - 81}{b - 3}$ $b^3 + 3b^2 + 9b + 27$

31. $\dfrac{-5x^5 + 4x^4 + 30x^3 + 2x^2 + 20x + 3}{x - 3}$ $-5x^4 - 11x^3 - 3x^2 - 7x - 1$

32. $\dfrac{-6c^5 + 14c^4 + 38c^3 + 4c^2 + 25c - 36}{c - 4}$ $-6c^4 - 10c^3 - 2c^2 - 4c + 9$

33. $\dfrac{8t^3 - 4t^2 + 2t - 1}{t - \frac{1}{2}}$ $8t^2 + 2$

34. $\dfrac{9a^3 + 3a^2 - 21a - 7}{a + \frac{1}{3}}$ $9a^2 - 21$

35. $\dfrac{x^4 - x^3 - 56x^2 - 2x + 16}{x - 8}$ $x^3 + 7x^2 - 2$

36. $\dfrac{x^4 - 9x^3 + x^2 - 7x - 20}{x - 9}$ $x^3 + x + 2 - \frac{2}{x - 9}$

Use a calculator and synthetic division to perform each division.

37. $\dfrac{7.2x^2 - 2.1x + 0.5}{x - 0.2}$ $7.2x - 0.66 + \frac{0.368}{x - 0.2}$

38. $\dfrac{2.7x^2 + x - 5.2}{x + 1.7}$ $2.7x - 3.59 + \frac{0.903}{x + 1.7}$

39. $\dfrac{9x^3 - 25}{x + 57}$ $9x^2 - 513x + 29{,}241 - \frac{1{,}666{,}762}{x + 57}$

40. $\dfrac{0.5x^3 + x}{x - 2.3}$ $0.5x^2 + 1.15x + 3.645 + \frac{8.3835}{x - 2.3}$

Let $P(x) = 2x^3 - 4x^2 + 2x - 1$. Evaluate $P(x)$ by substituting the given value of x into the polynomial and simplifying. Then evaluate the polynomial by using the remainder theorem and synthetic division.

41. $P(1)$ −1

42. $P(2)$ 3

43. $P(-2)$ −37

44. $P(-1)$ −9

45. $P(3)$ 23

46. $P(-4)$ −201

47. $P(0)$ −1

48. $P(4)$ 71

Let $Q(x) = x^4 - 3x^3 + 2x^2 + x - 3$. Evaluate $Q(x)$ by substituting the given value of x into the polynomial and simplifying. Then evaluate the polynomial by using the remainder theorem and synthetic division.

49. $Q(-1)$ 2

50. $Q(1)$ −2

51. $Q(2)$ −1

52. $Q(-2)$ 43

53. $Q(3)$ 18

54. $Q(0)$ −3

55. $Q(-3)$ 174

56. $Q(-4)$ 473

Use the remainder theorem and synthetic division to find $P(k)$.

57. $P(x) = x^3 - 4x^2 + x - 2; k = 2$ −8

58. $P(x) = x^3 - 3x^2 + x + 1; k = 1$ 0

59. $P(x) = 2x^3 + x + 2; k = 3$ 59

60. $P(x) = x^3 + x^2 + 1; k = -2$ −3

61. $P(x) = x^4 - 2x^3 + x^2 - 3x + 2; k = -2$ 44

62. $P(x) = x^5 + 3x^4 - x^2 + 1; k = -1$ 2

63. $P(x) = 3x^5 + 1; k = -\frac{1}{2}$ $\frac{29}{32}$

64. $P(x) = 5x^7 - 7x^4 + x^2 + 1; k = 2$ 533

Use the factor theorem and determine whether the first expression is a factor of $P(x)$.

65. $x - 3$; $P(x) = x^3 - 3x^2 + 5x - 15$ Yes

66. $x + 1$; $P(x) = x^3 + 2x^2 - 2x - 3$
(*Hint:* Write $x + 1$ as $x - (-1)$.) Yes

67. $x + 2$; $P(x) = 3x^2 - 7x + 4$
(*Hint:* Write $x + 2$ as $x - (-2)$.) No

68. x; $P(x) = 7x^3 - 5x^2 - 8x$
(*Hint:* $x = x - 0$.) Yes

WRITING

69. When dividing a polynomial by a binomial of the form $x - k$, synthetic division is considered to be faster than long division. Explain why.

70. Let $P(x) = x^3 - 6x^2 - 9x + 4$. You now know two ways to find $P(6)$. What are they? Which method do you prefer?

71. Explain the factor theorem.

72. What is a *zero* of a polynomial function?

REVIEW

Evaluate each expression for $x = -3$, $y = -5$, and $z = 0$.

73. $x^2z(y^3 - z)$ 0

74. $|y^3 - z|$ 125

75. $\dfrac{x - y^2}{2y - 1 + x}$ 2

76. $\dfrac{2y + 1}{x} - x$ 6

CHALLENGE PROBLEMS

Suppose that
$P(x) = x^{100} - x^{99} + x^{98} - x^{97} + \cdots + x^2 - x + 1.$

77. Find the remainder when $P(x)$ is divided by $x - 1$. 1

78. Find the remainder when $P(x)$ is divided by $x + 1$. 101

APPENDIX 3 Answers to Graphing Exercises

STUDY SET SECTION 1.1 (page 7)

61. d.

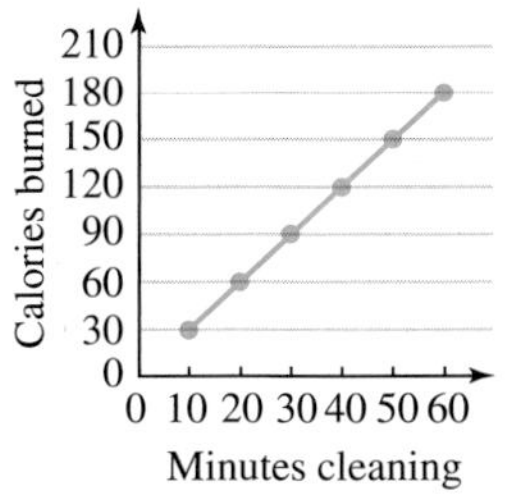

SELF CHECKS 1.3

3.

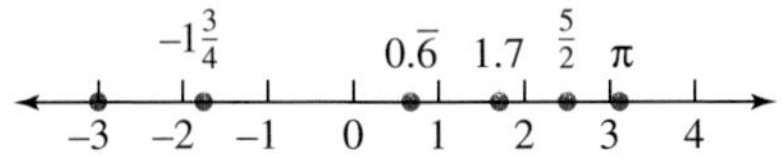

STUDY SET SECTION 1.3 (page 29)

57.

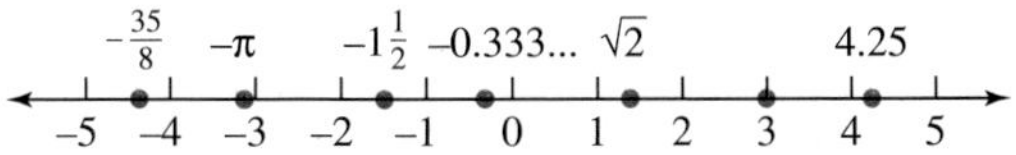

58.

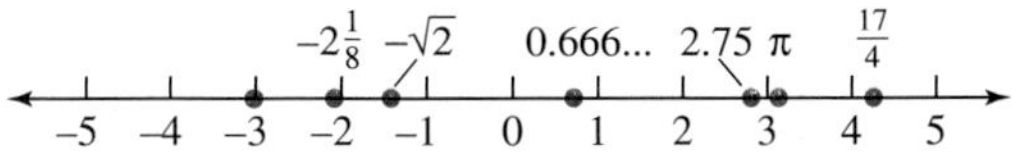

59.

60.

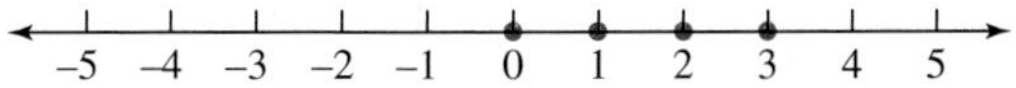

STUDY SET SECTION 1.4 (page 37)

110.

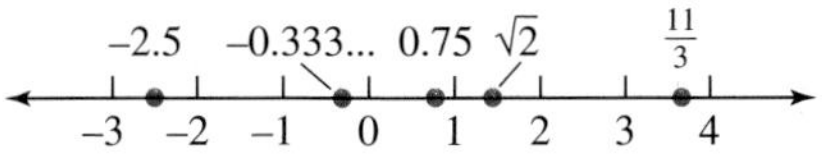

CHAPTER 1 REVIEW (page 89)

41.

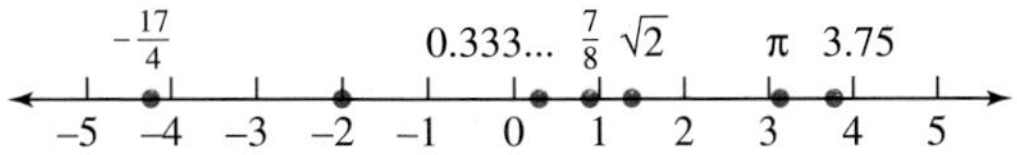

CHAPTER 1 TEST (page 98)

11.

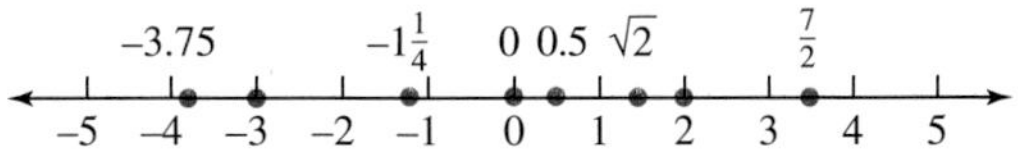

ARE YOU READY? 2.7 (page 162)

3.

SELF CHECKS 2.7

2.

3.

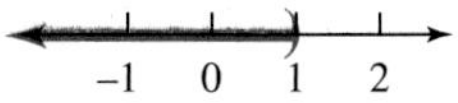

4. a. / **b.**

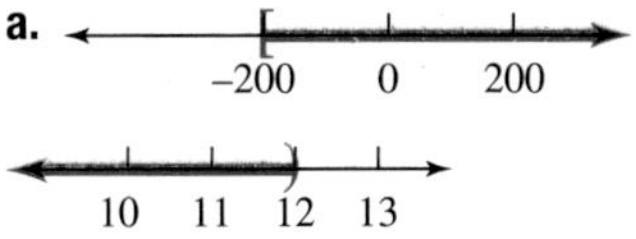

5.

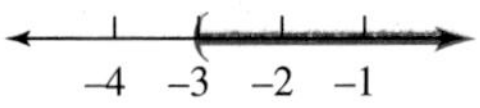

6.

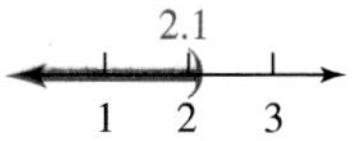

7.

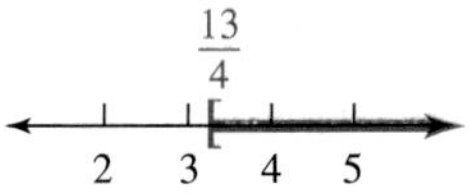

8.

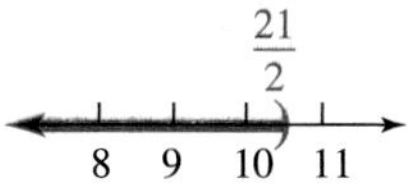

9.

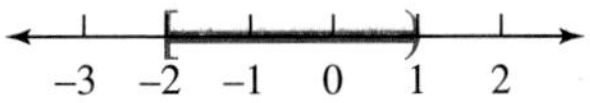

10.

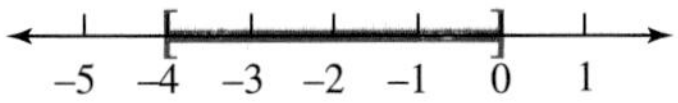

STUDY SET SECTION 2.7 (page 171)

19.

20.

21.

22.

23.

24.

25.

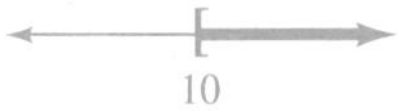

26.

27.

28.

29.

30.

31.

32.

33.

34.

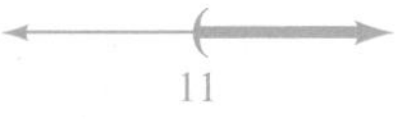

35.

36.

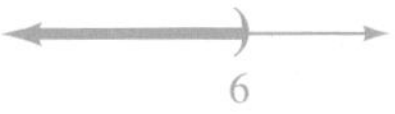

37.

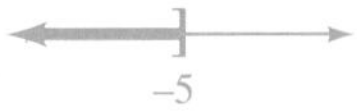

38.

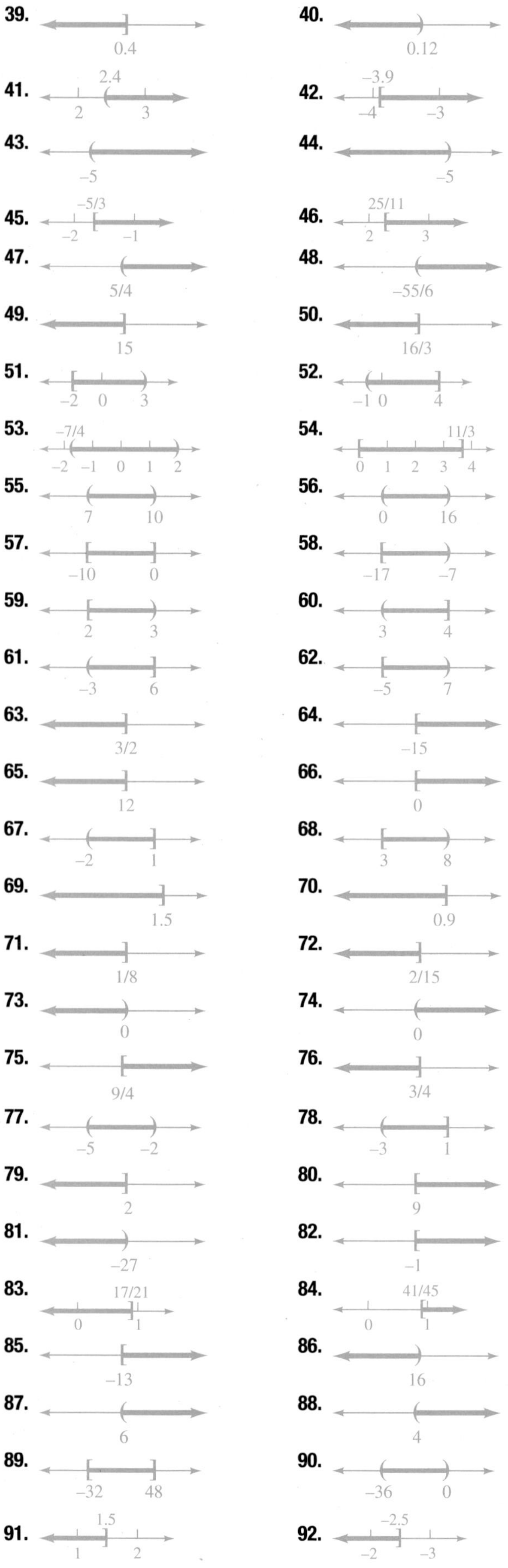

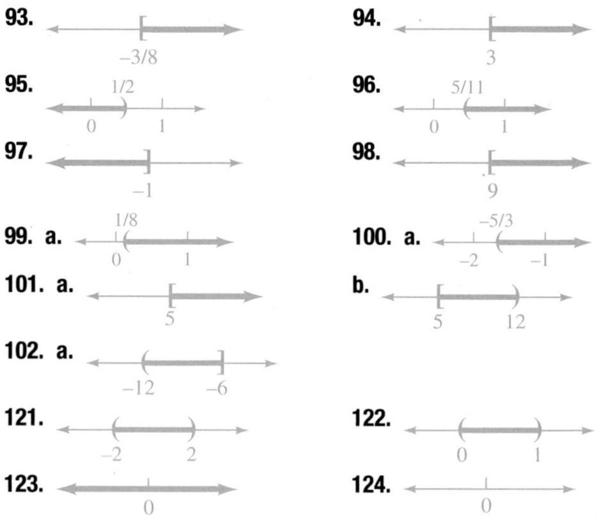

CHAPTER 2 REVIEW (page 175)

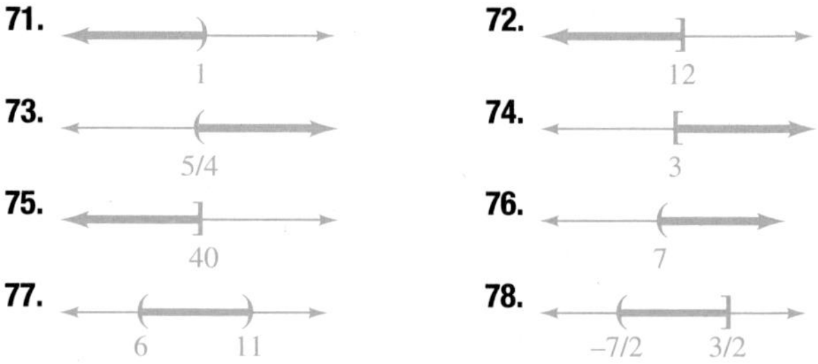

CHAPTER 2 TEST (page 182)

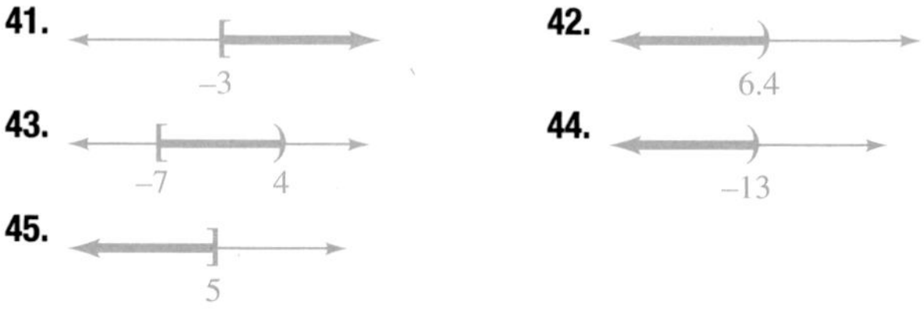

ARE YOU READY? 3.1 (page 186)

1.

SELF CHECKS 3.1

1.

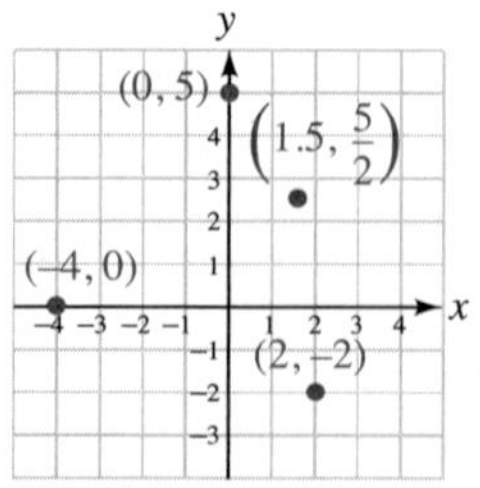

STUDY SET SECTION 3.1 (page 191)

17.

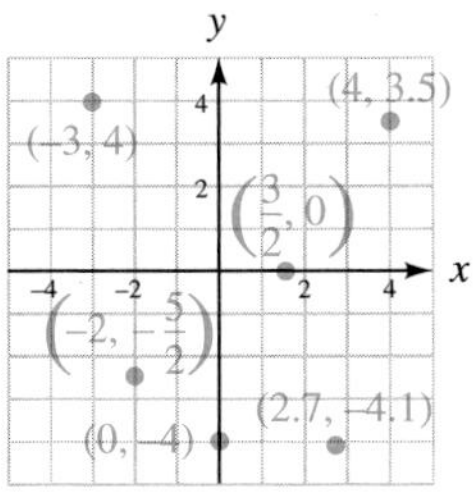

18.

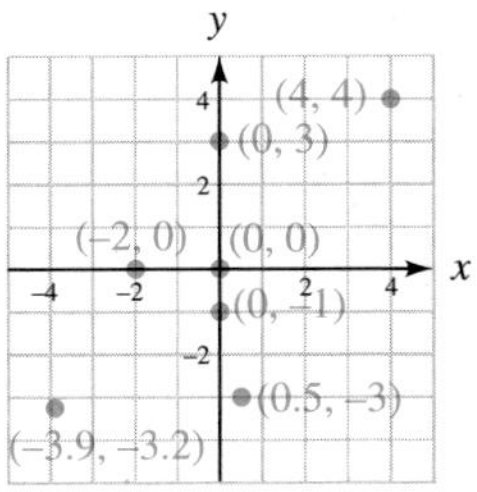

48.

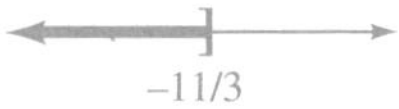

SELF CHECKS 3.2

4.

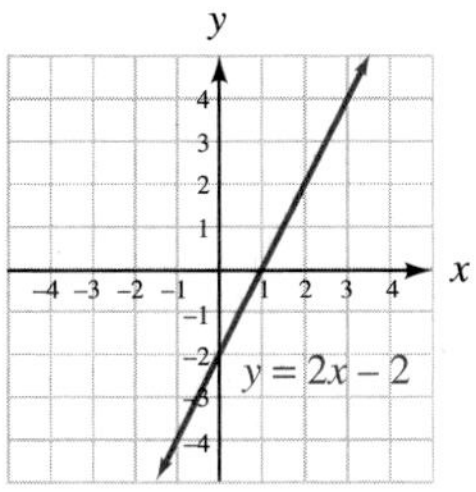

5.

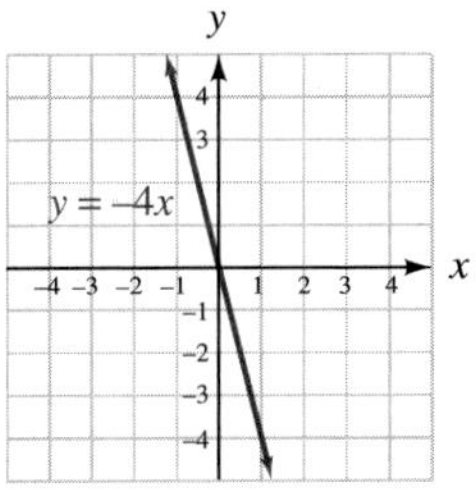

6.

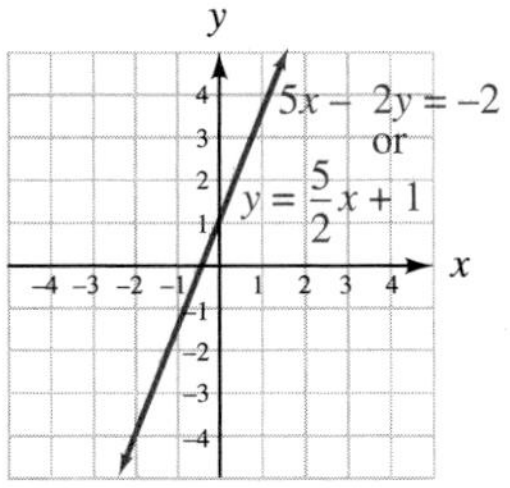

7.

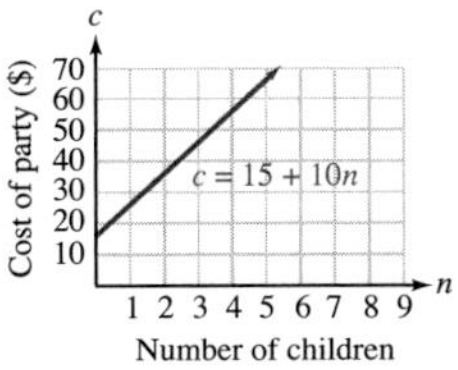

STUDY SET SECTION 3.2 (page 203)

41.

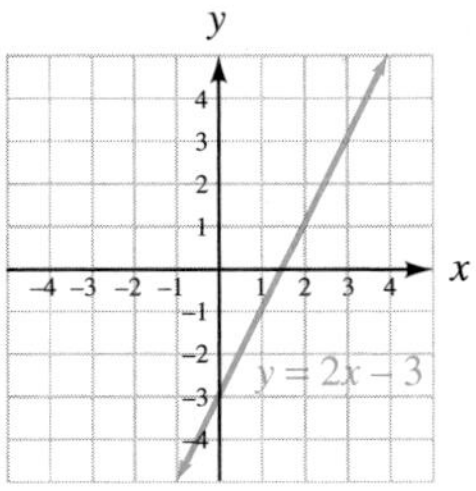

42.

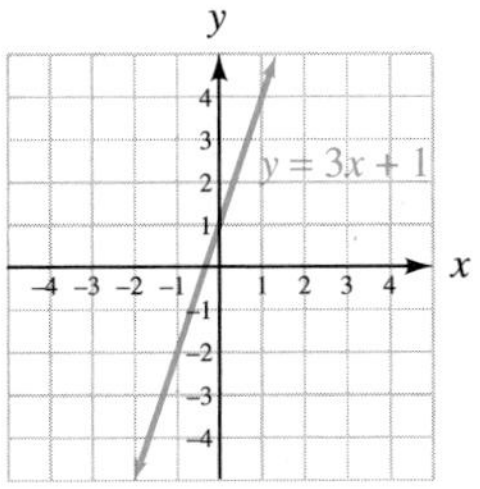

43.

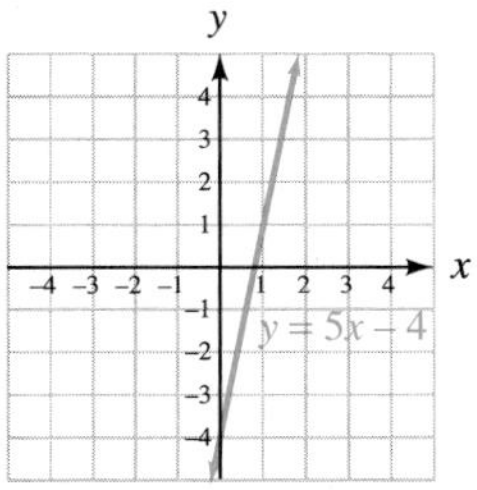

44.

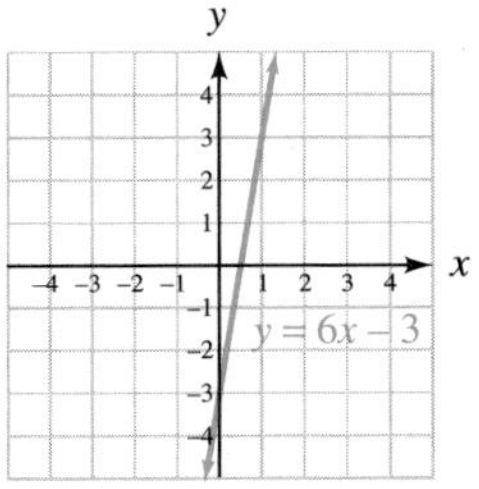

45.

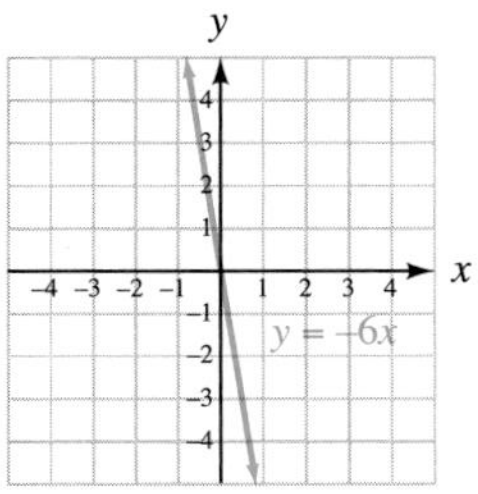

46.

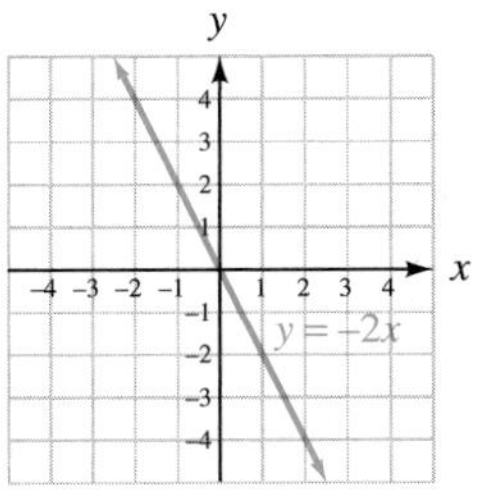

47.

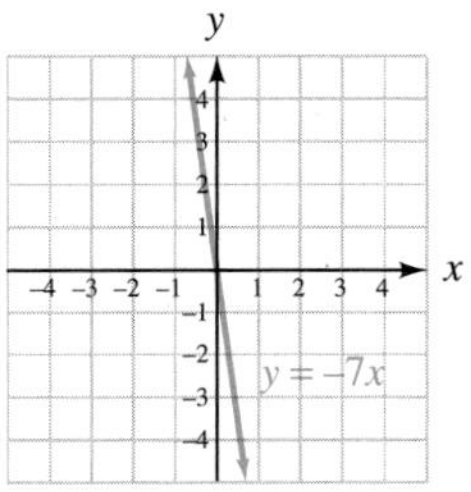

48.

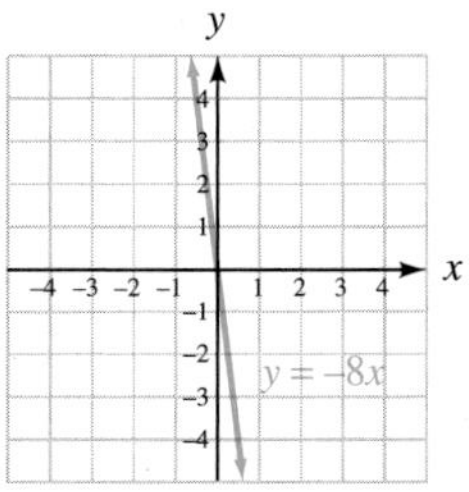

49.

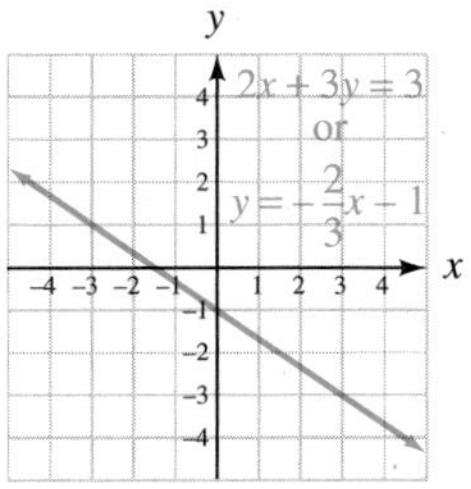

50.

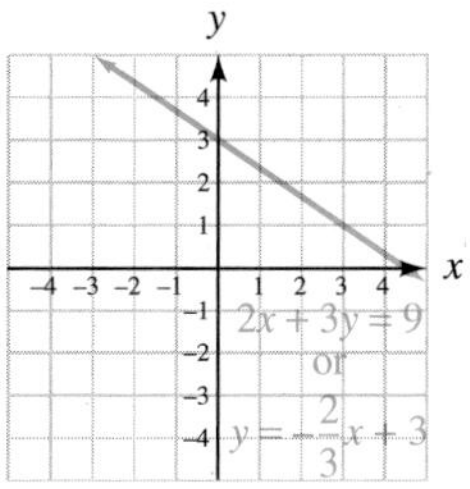

51.

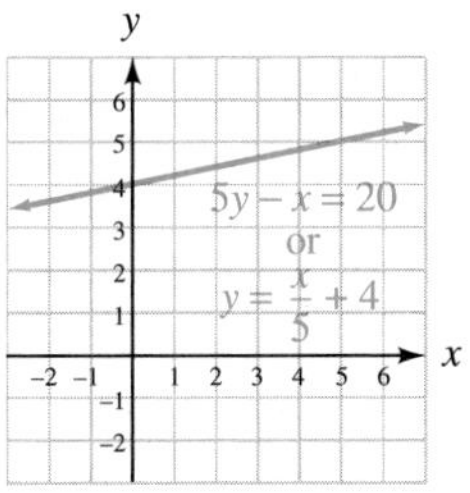

52.

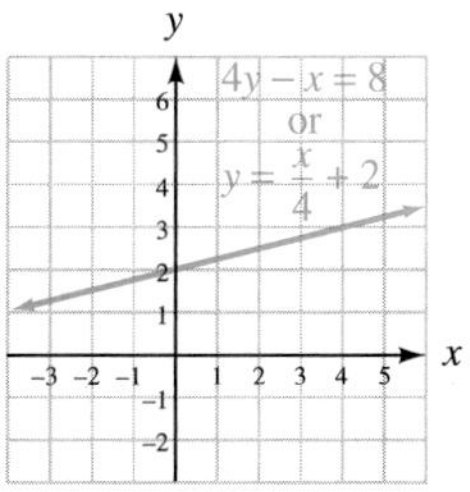

53.

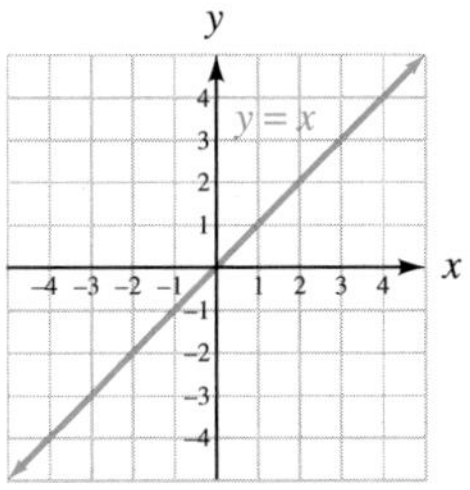

54.

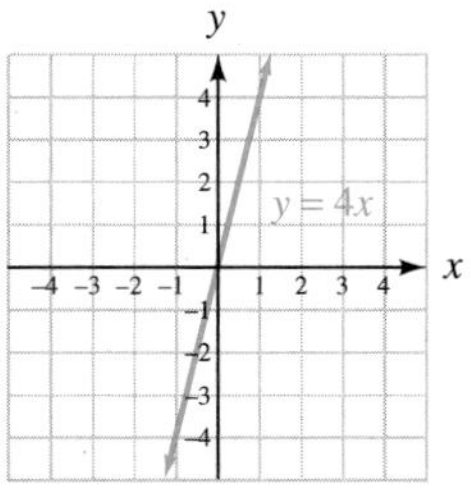

55.

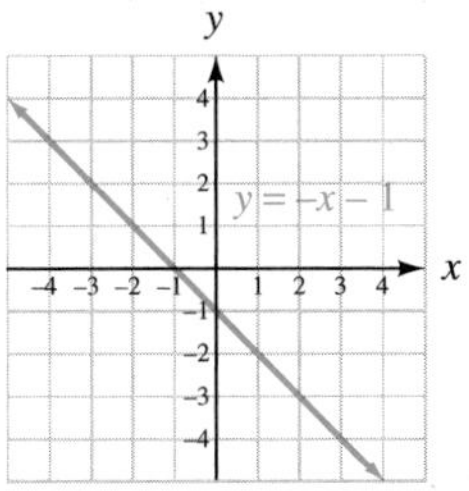

56.

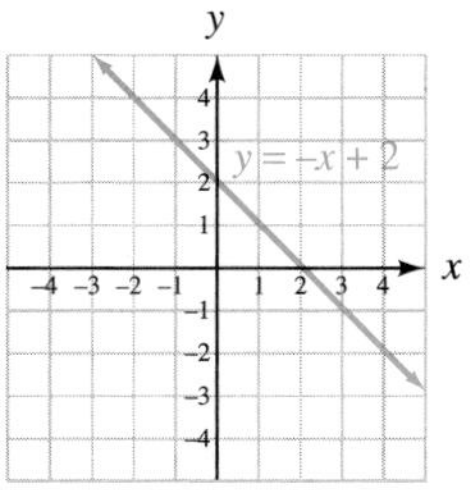

57.

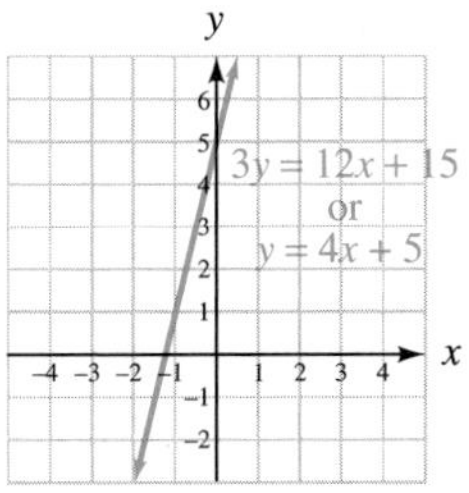

58.

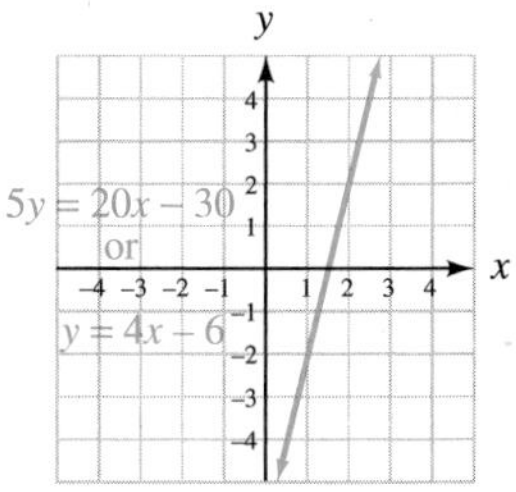

59.

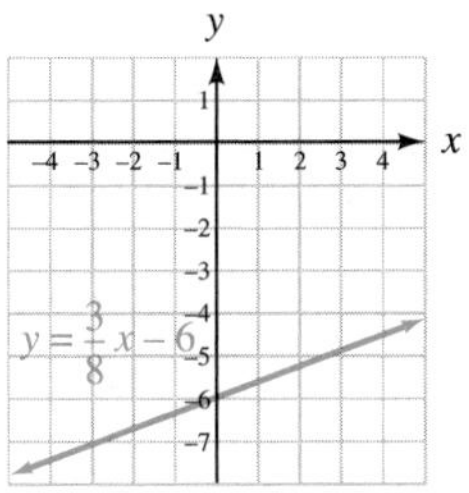

60.

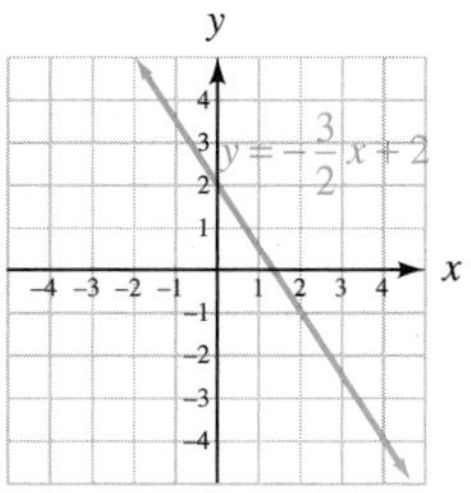

61.
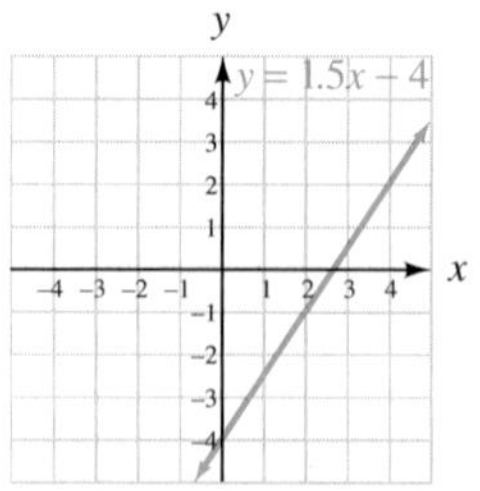

62.
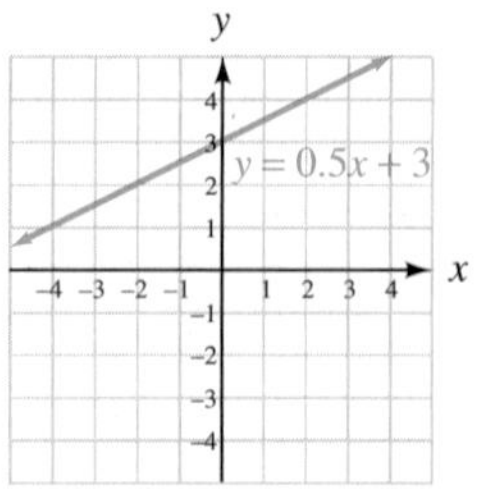

63.
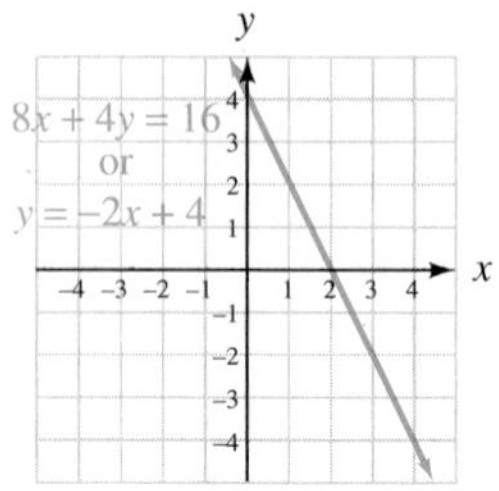

64.
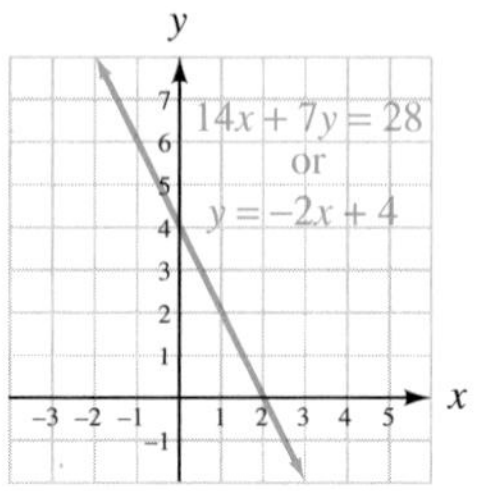

65.
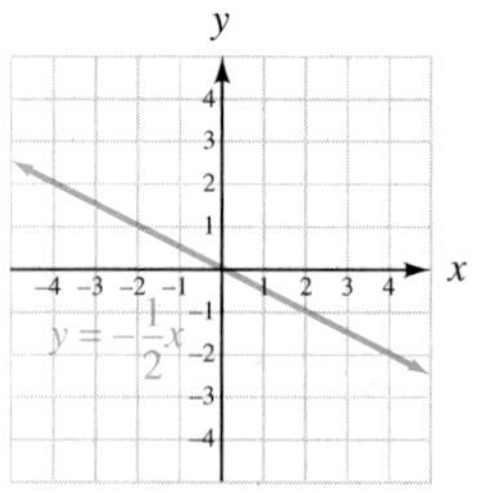

66.
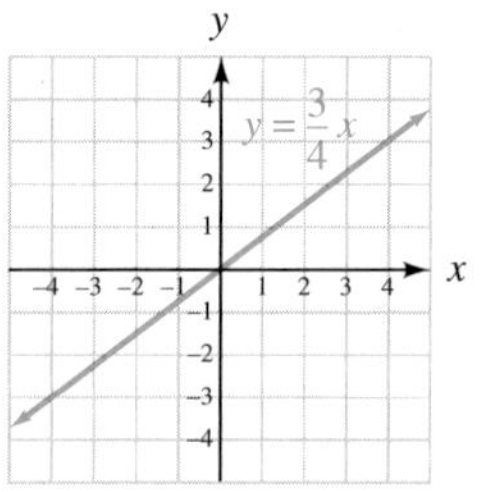

67.
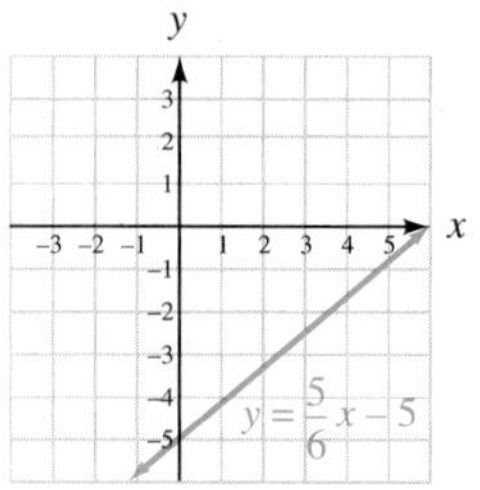

68.
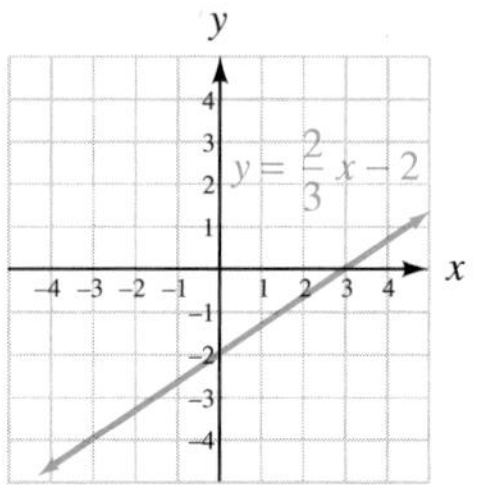

69.
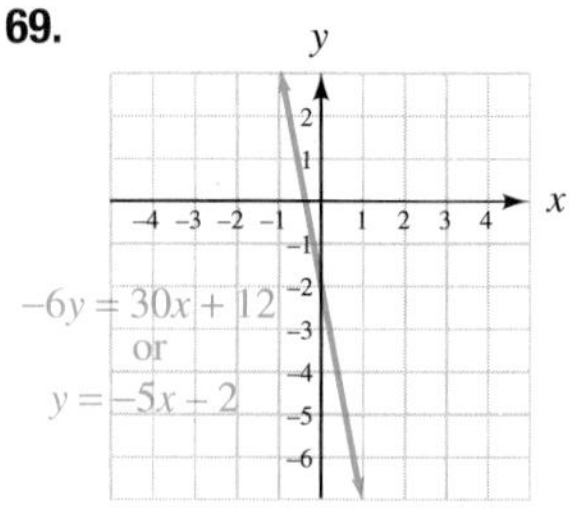

70.
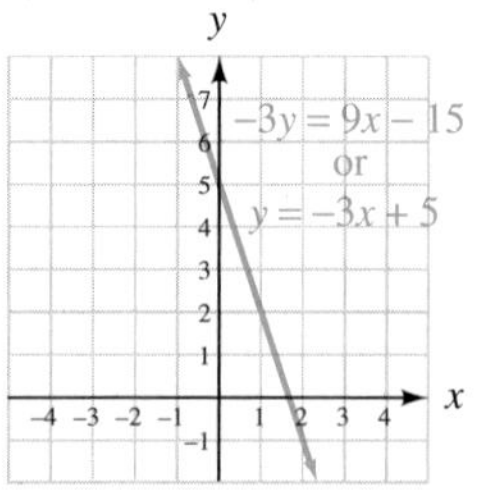

71.
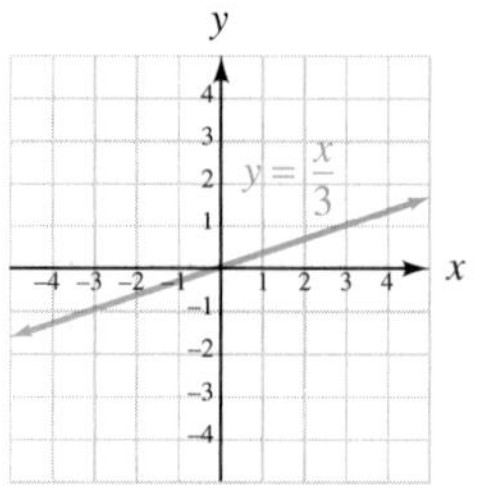

72.
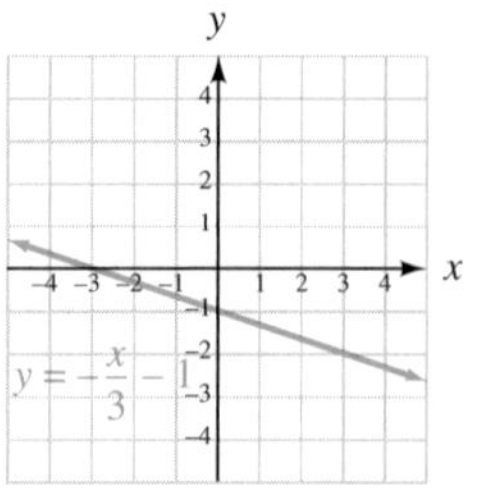

73.
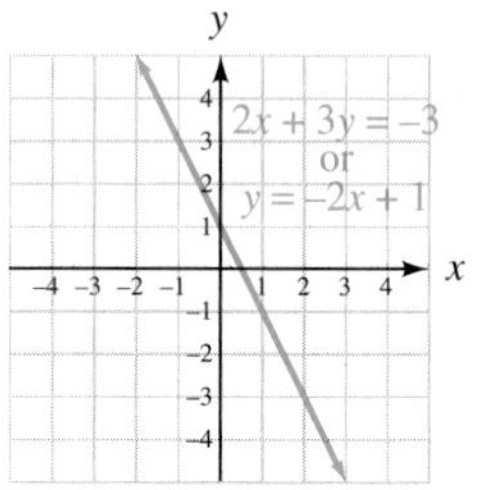

74.
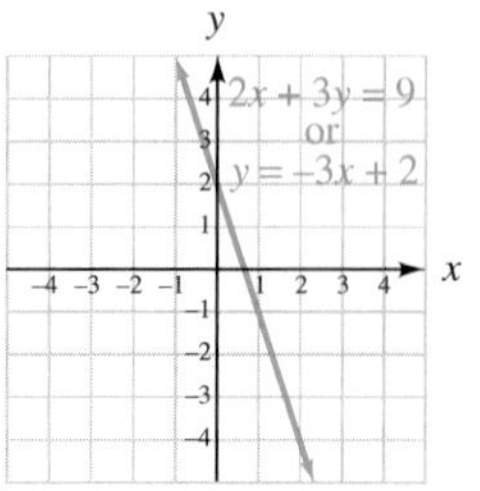

75.
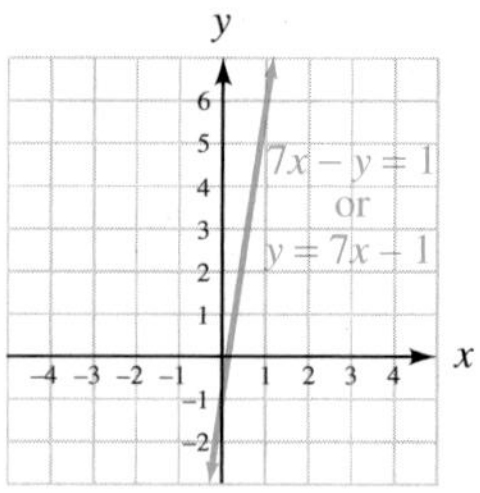

76.
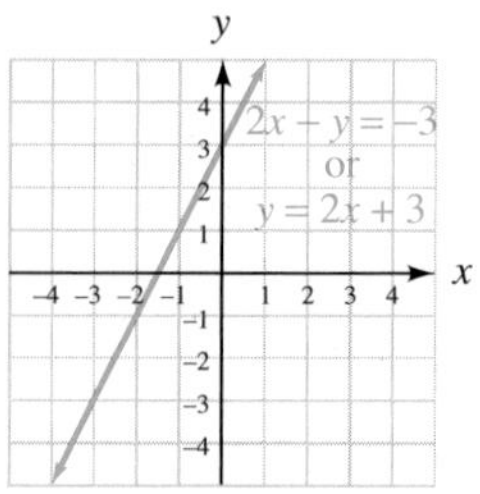

77.
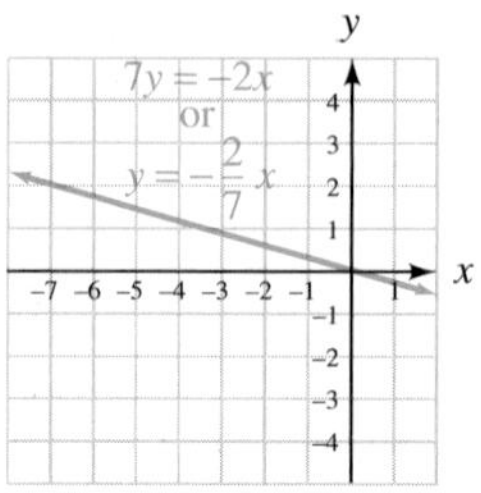

78.
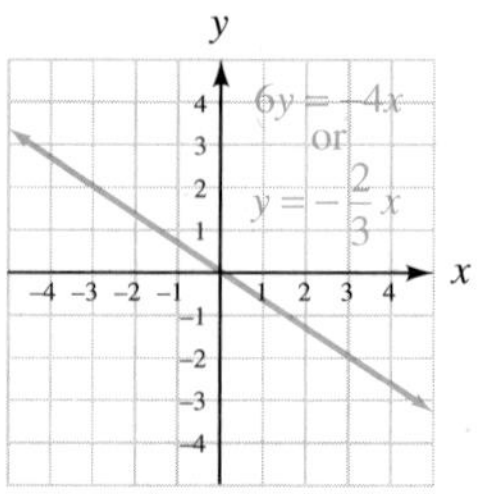

79.
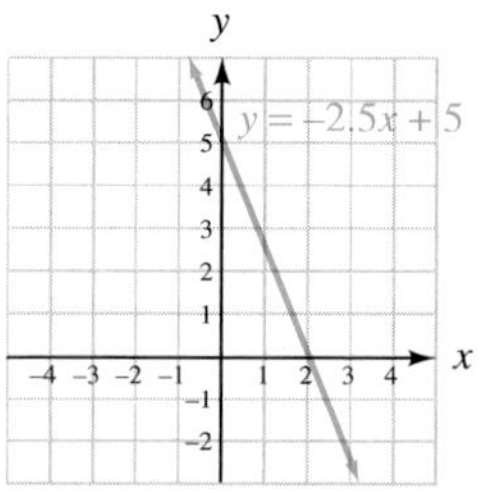

80.
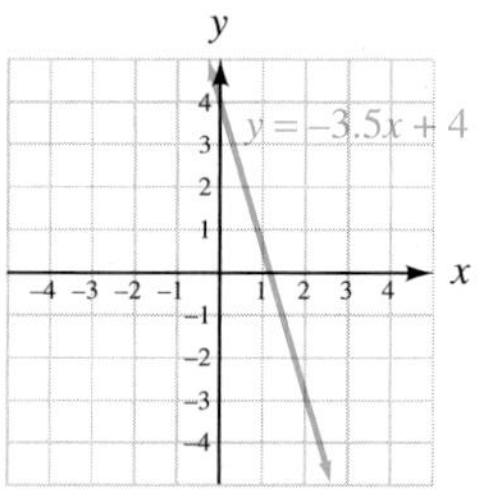

81.
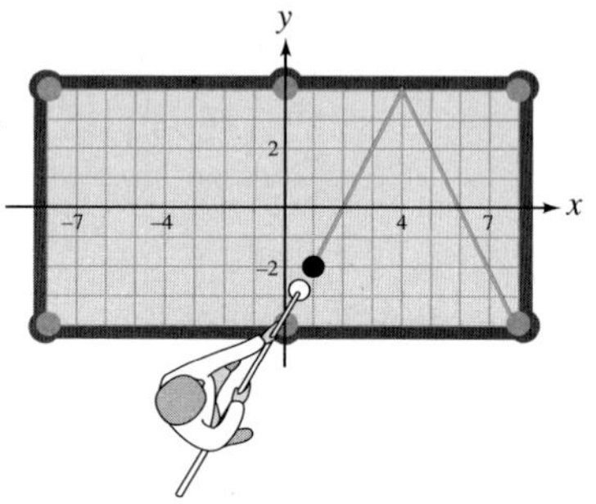

83.
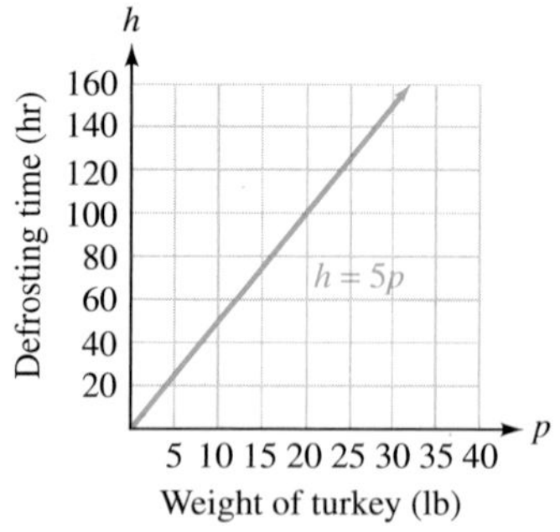

84.
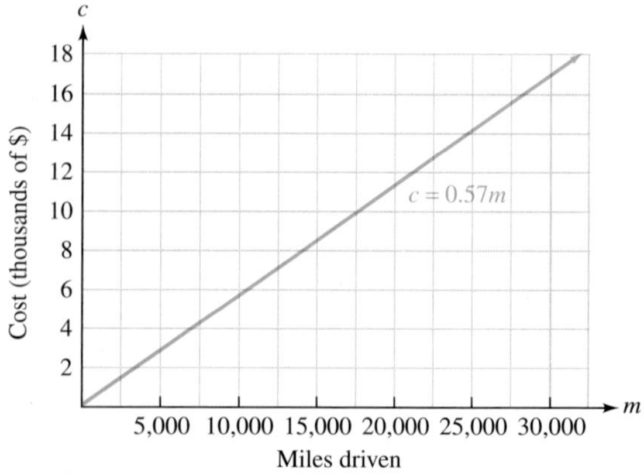

85.

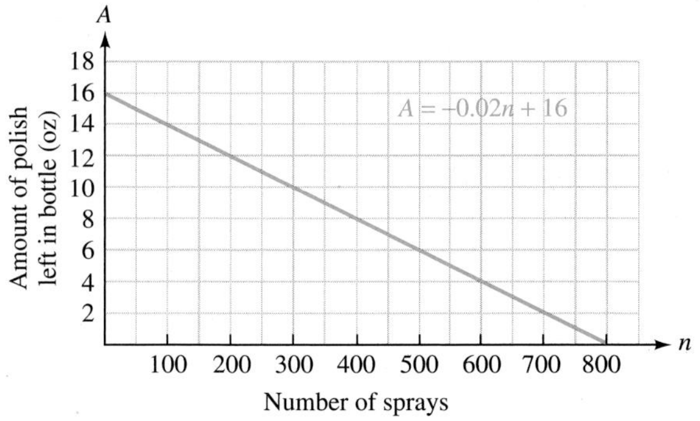

86.

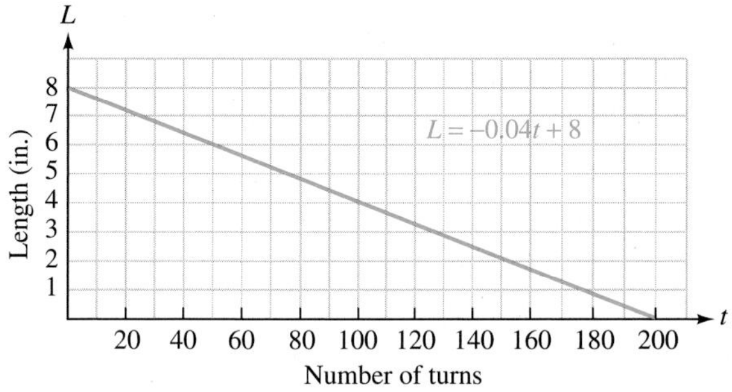

87.

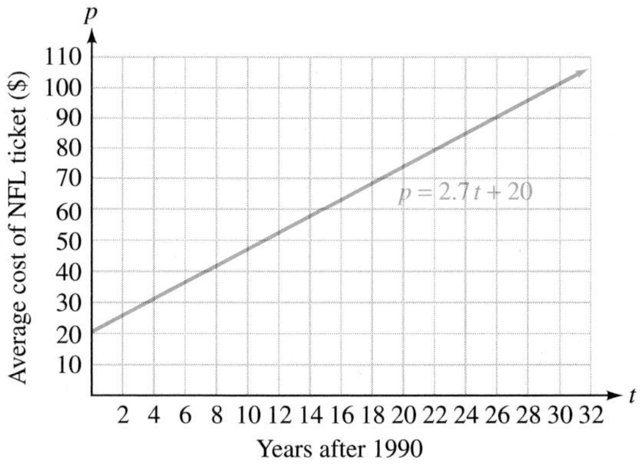

88.

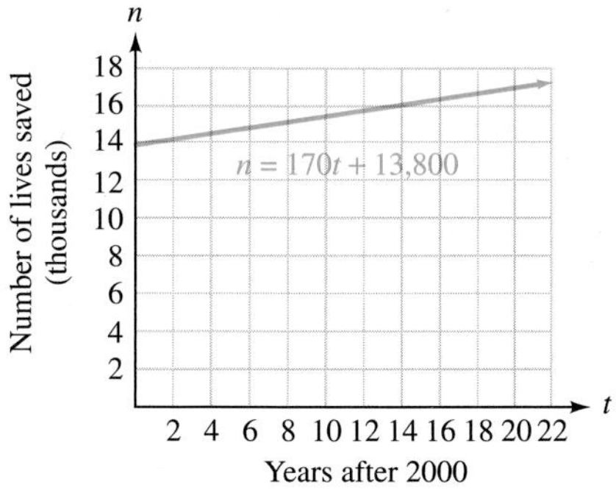

89.

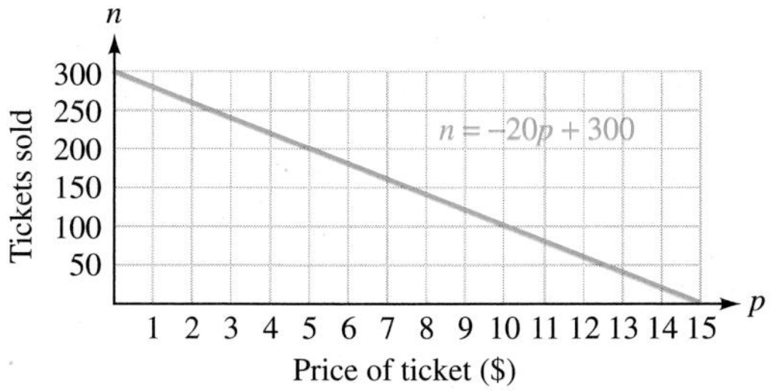

90.

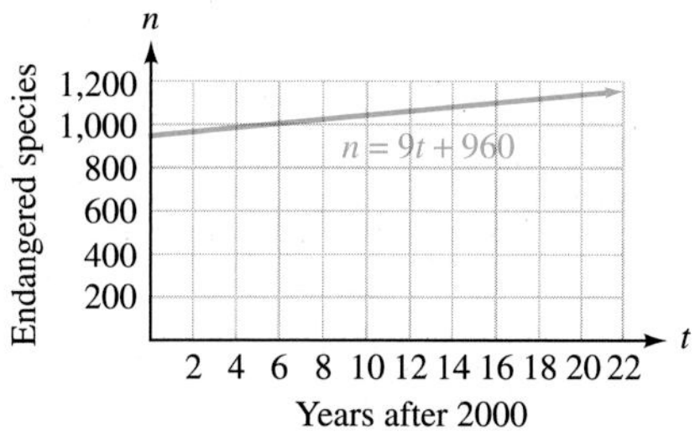

91.

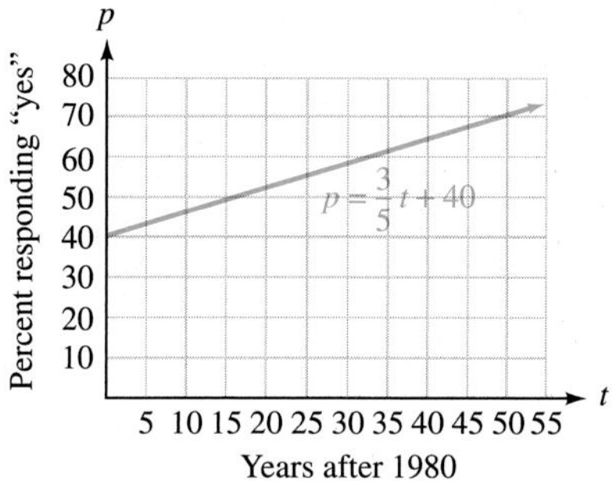

105.

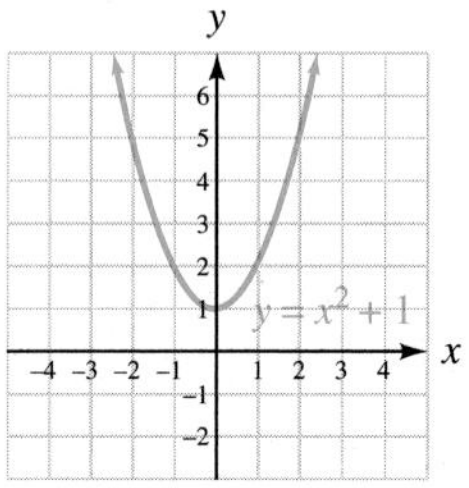

106.

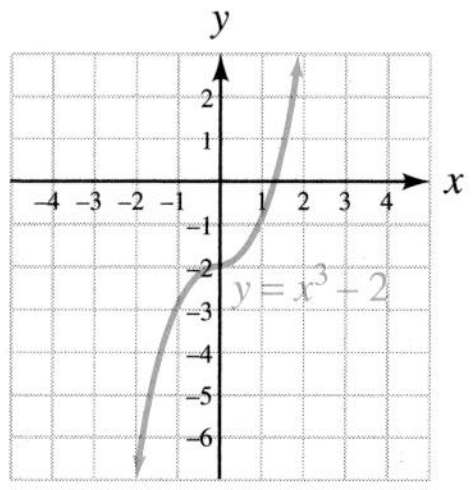

107.

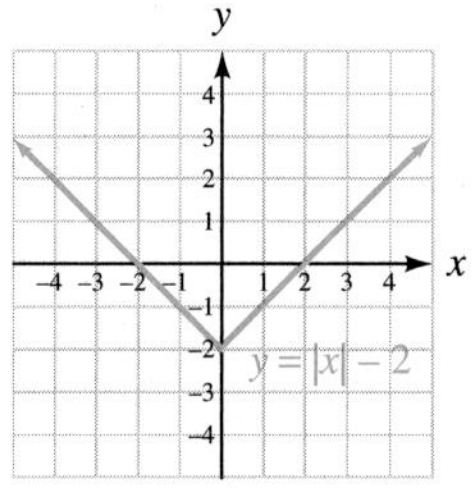

108.

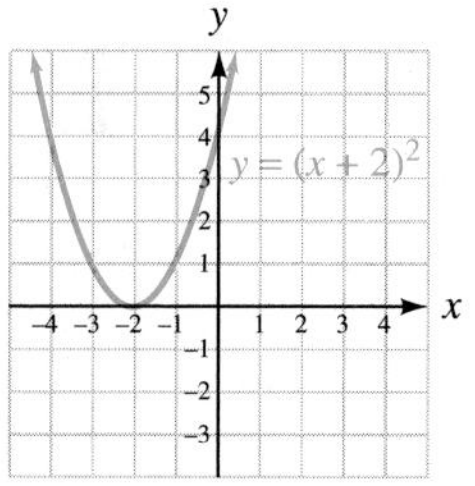

ARE YOU READY? 3.3 (page 206)

1.

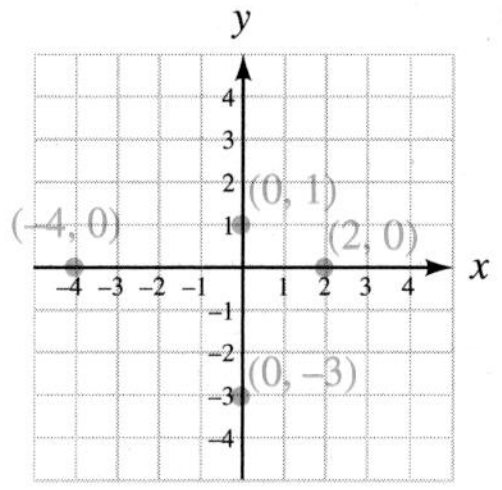

SELF CHECK 3.3

2.

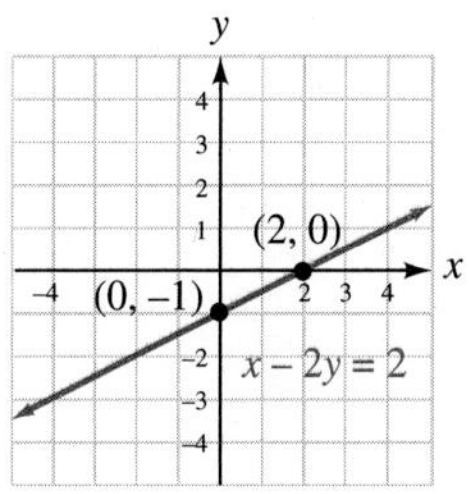

3.

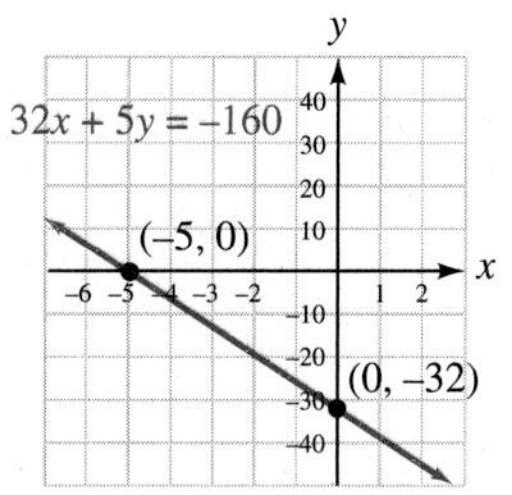

4.

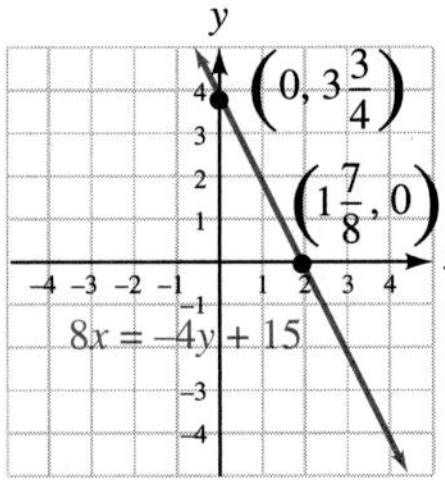

5.

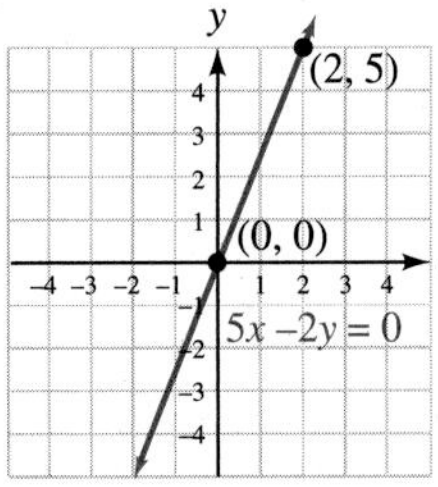

6.

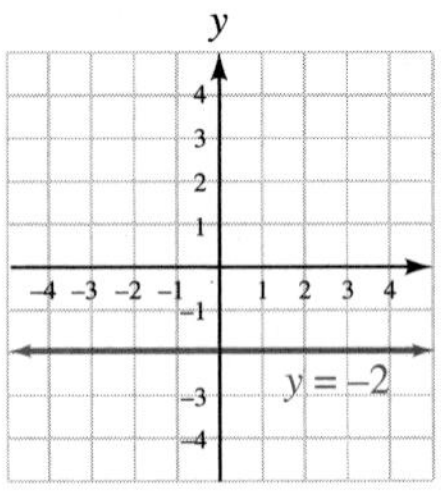

7.

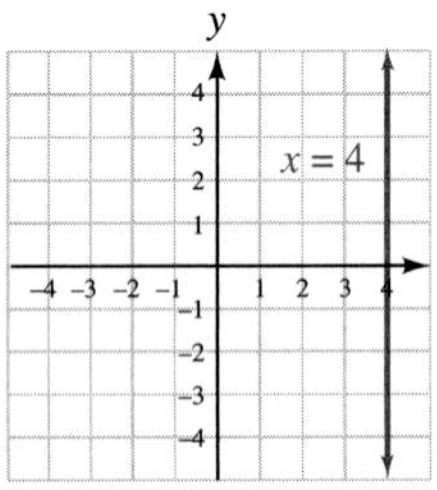

STUDY SET SECTION 3.3 (page 214)

27.

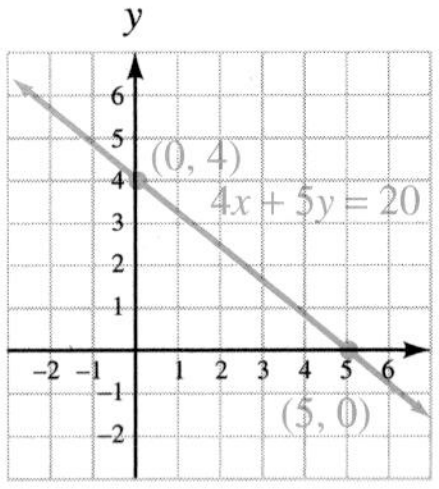

28.

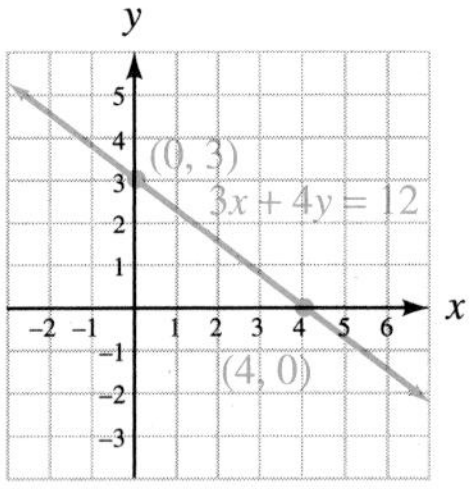

29.

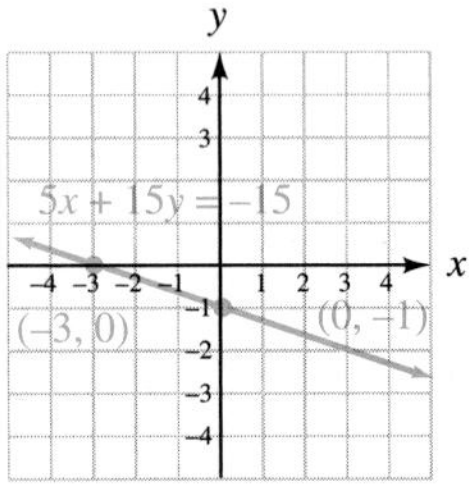

30.

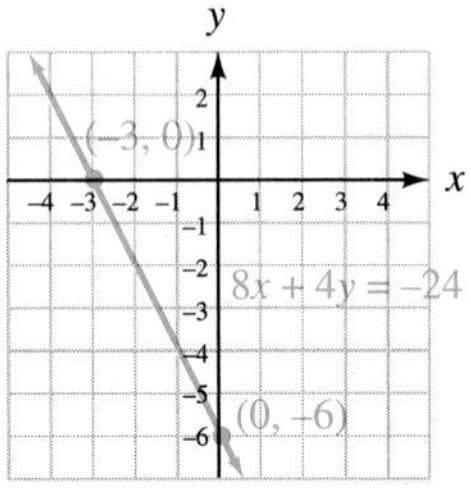

31.

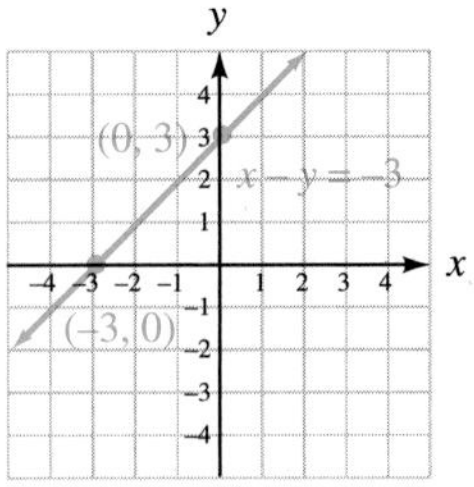

32.

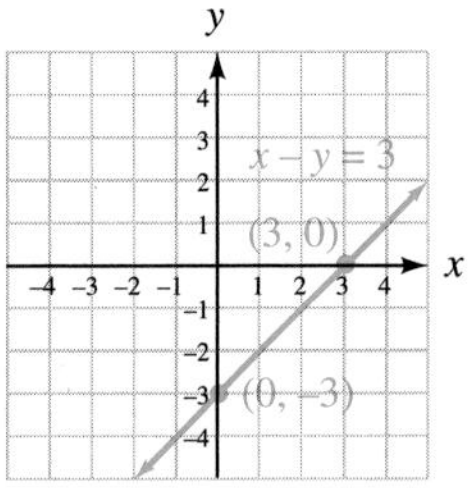

33.

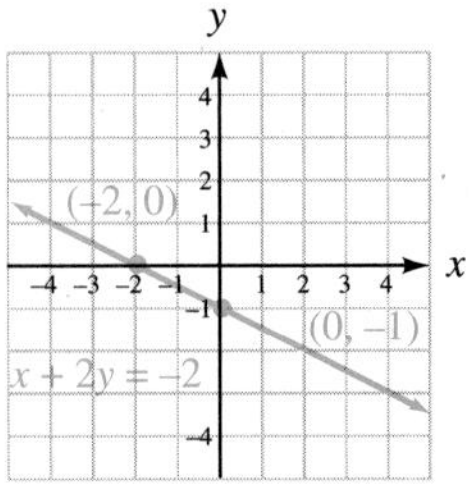

34.

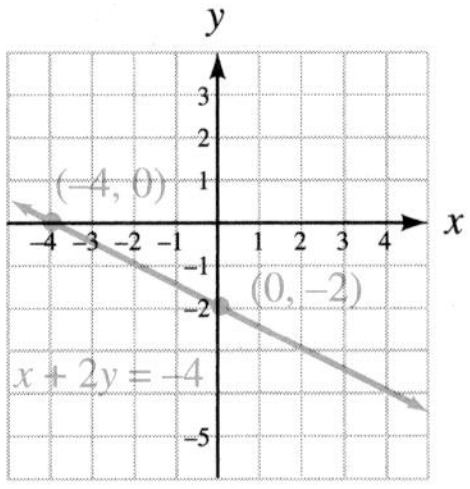

35.

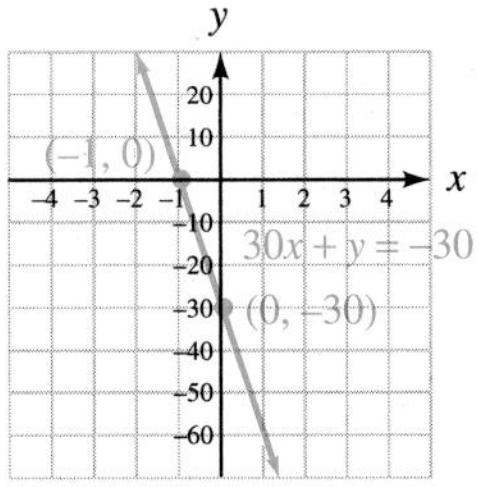

36.

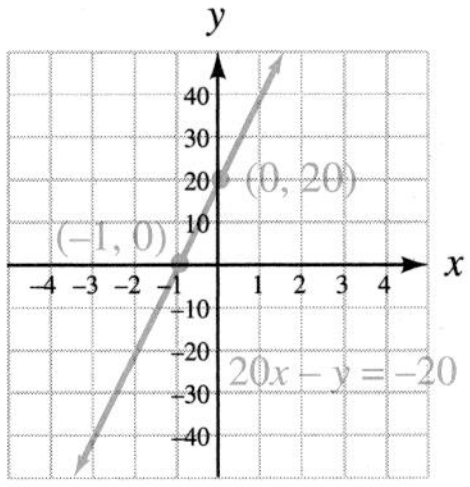

37.

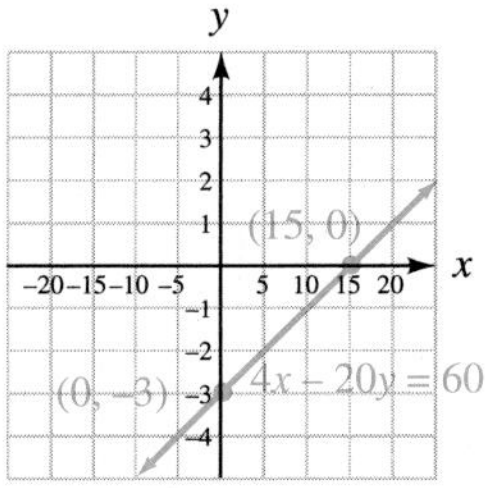

38.

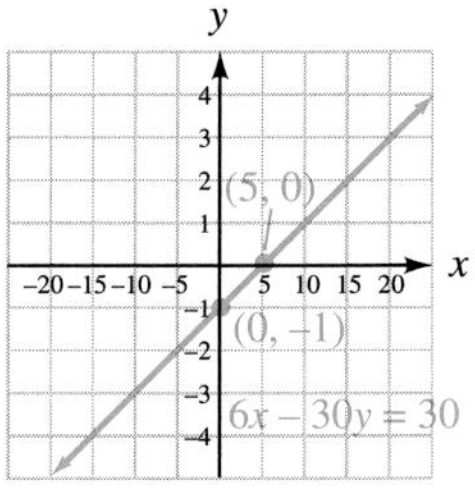

39.

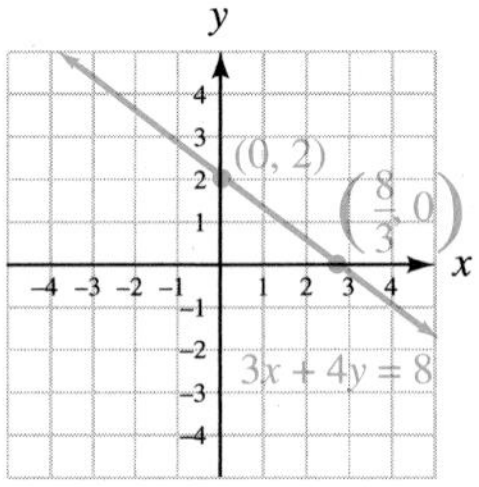

40.

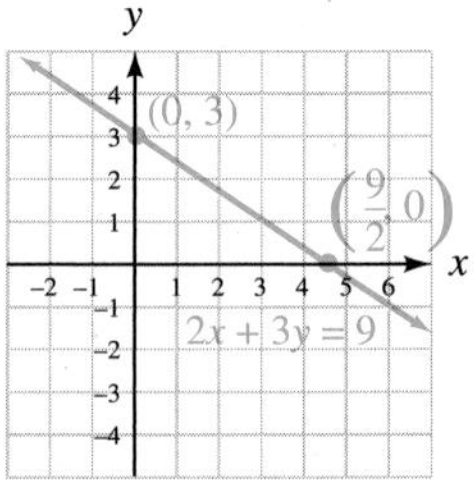

41.

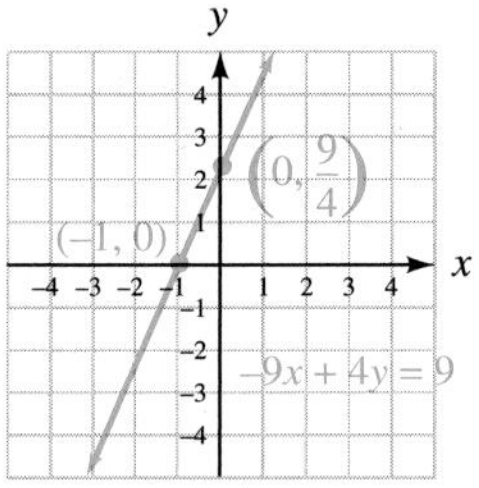

42.

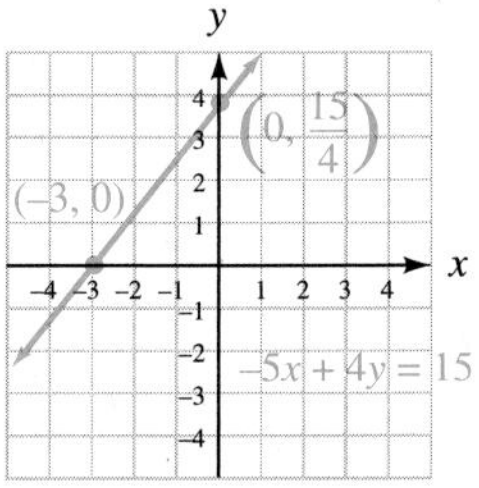

43.

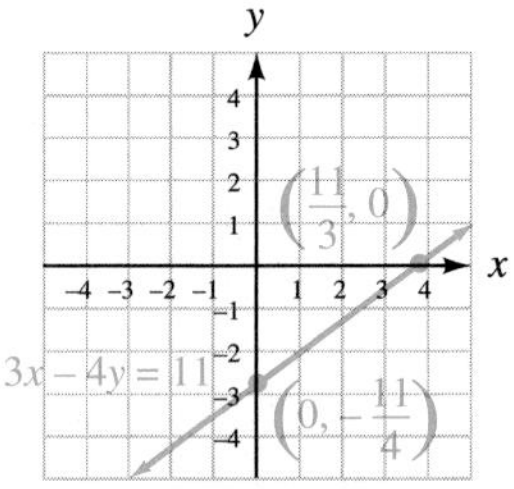

44.

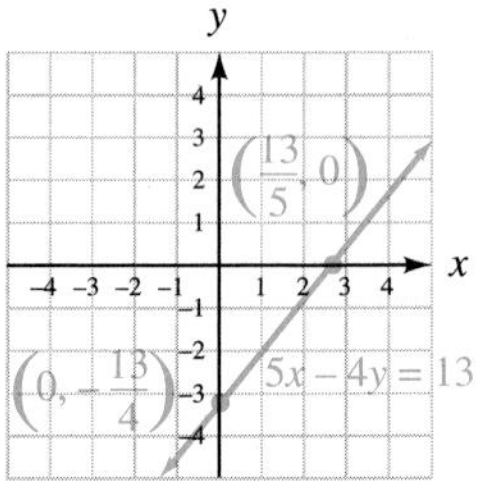

45.

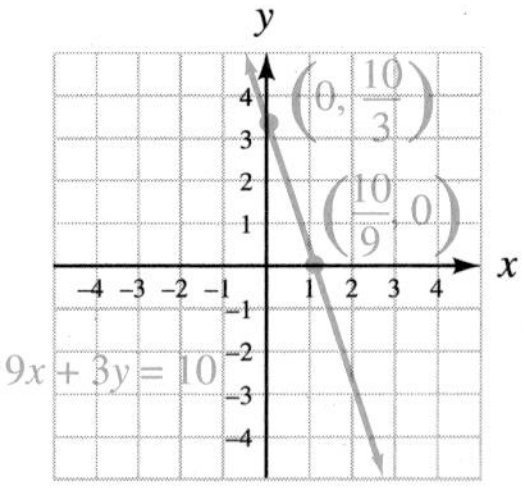

46.

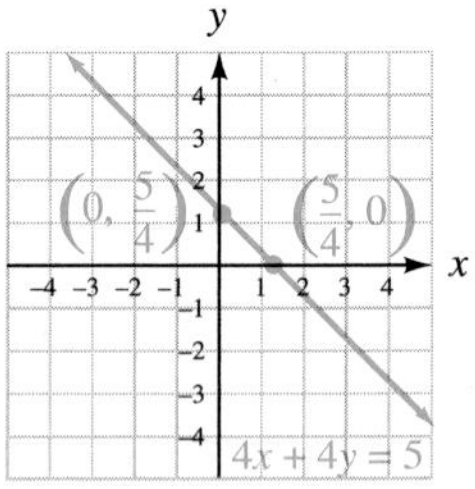

47.

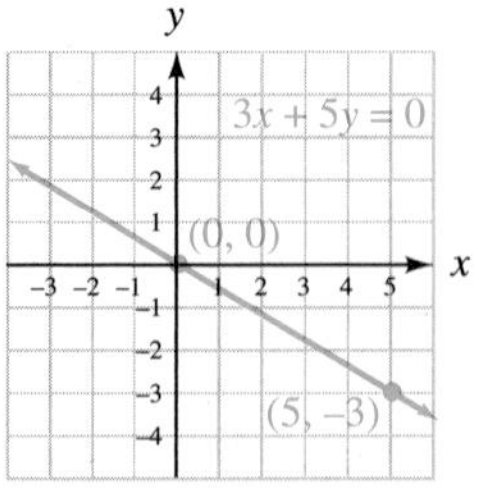

48.

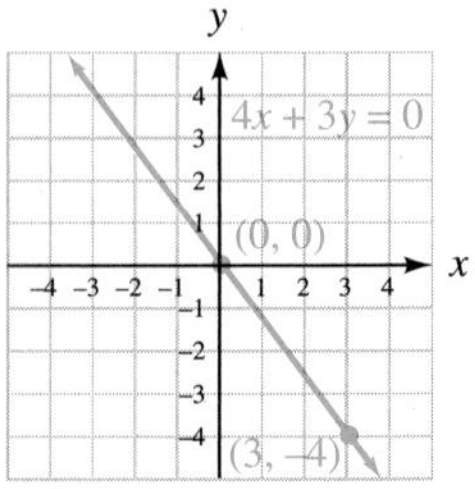

49.

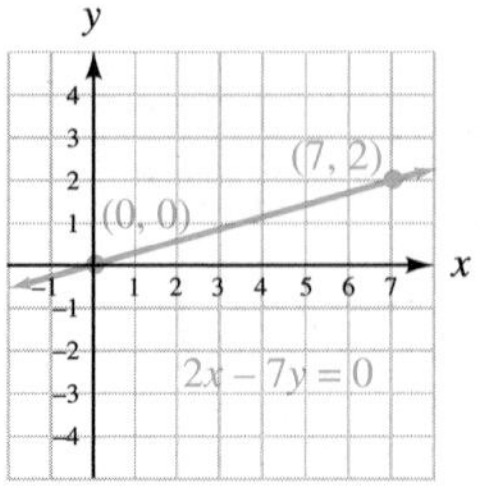

50.

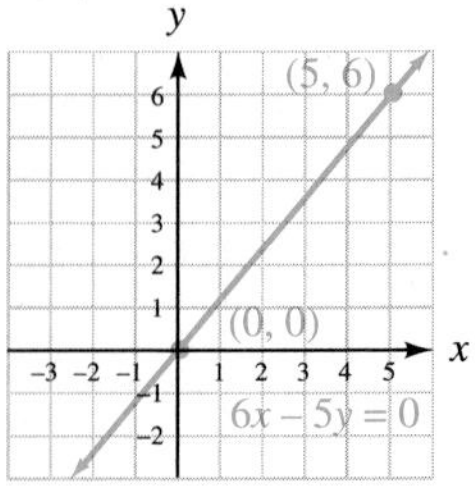

51.

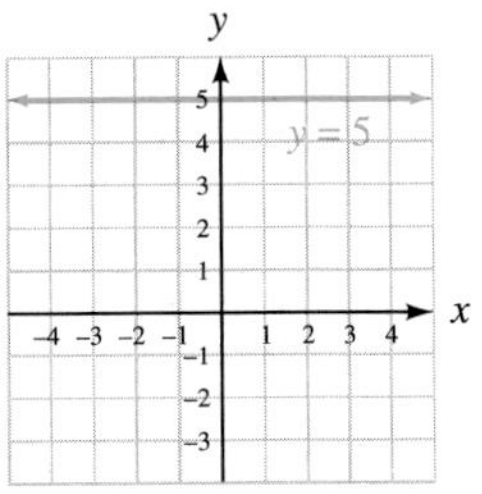

52.

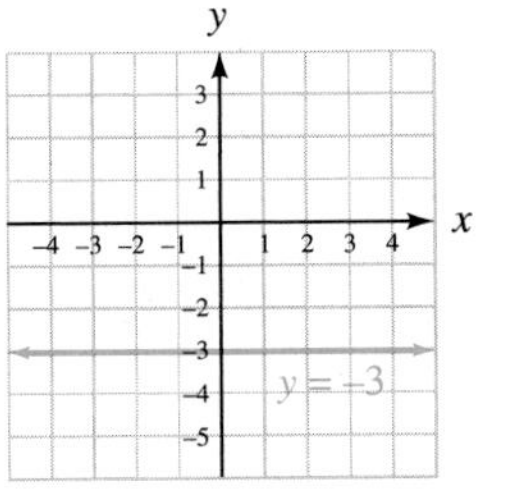

65.

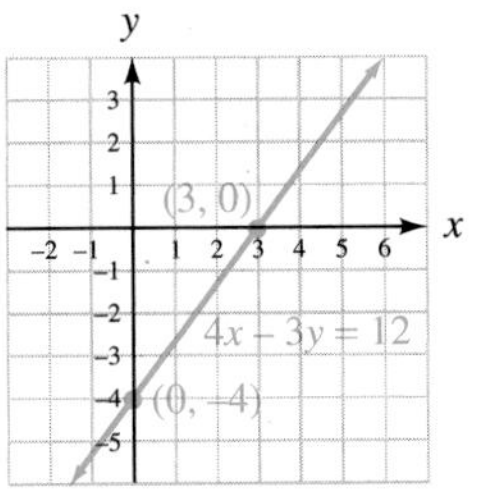

66.

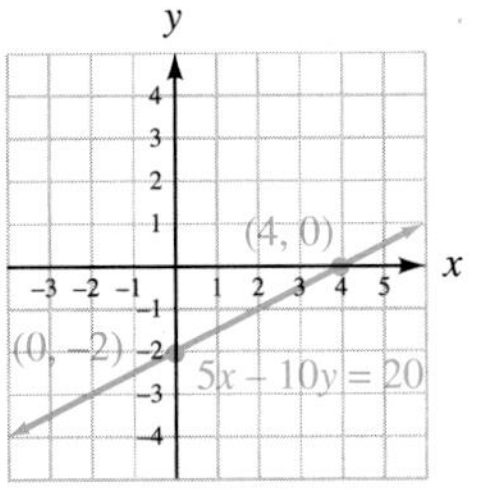

53.

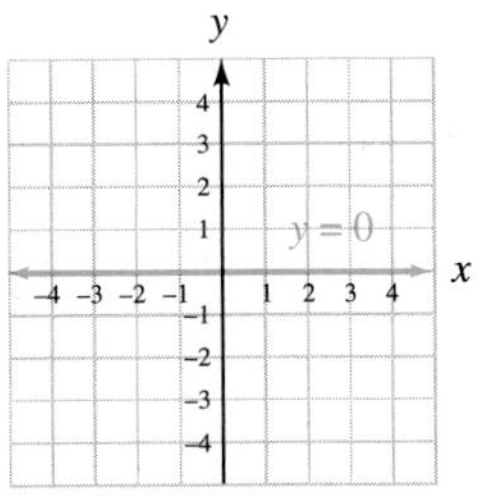

54.

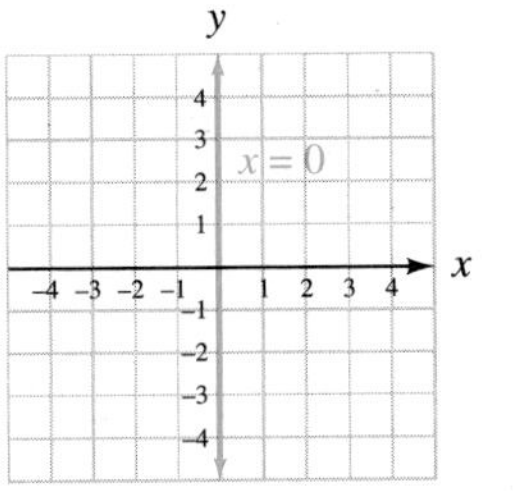

67.

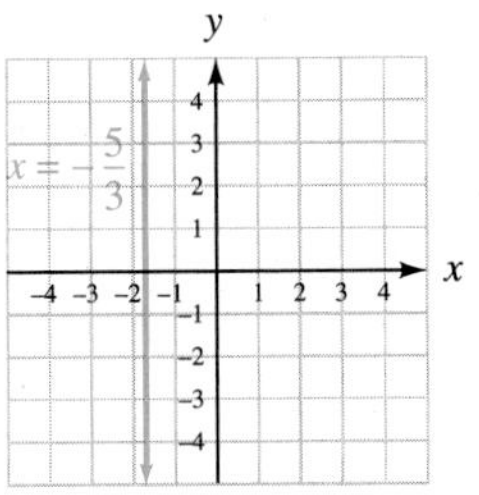

68.

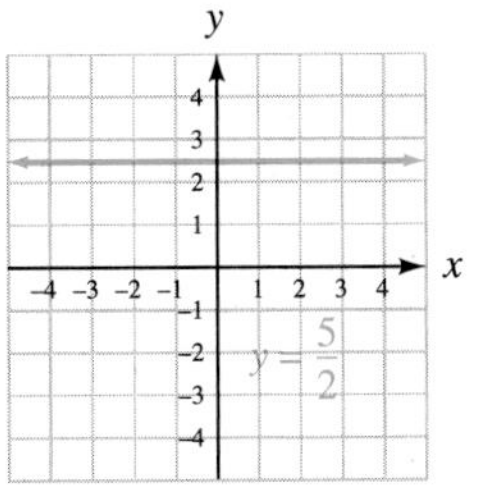

55.

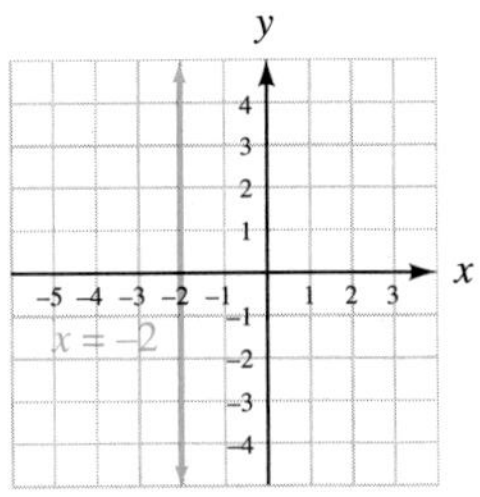

56.

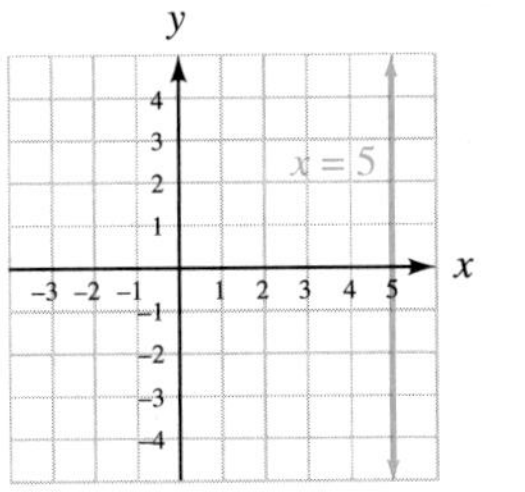

69.

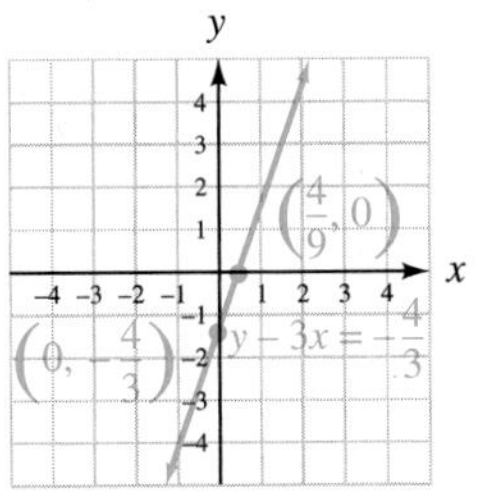

70.

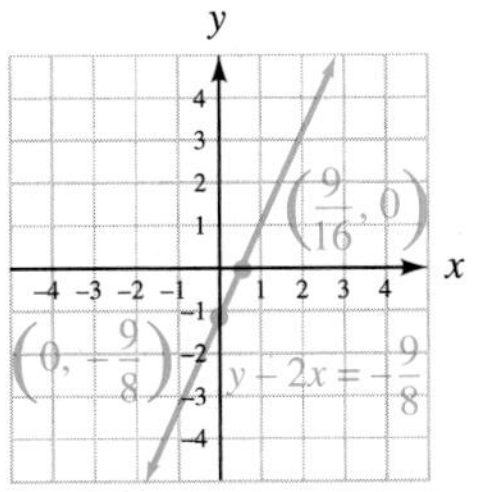

57.

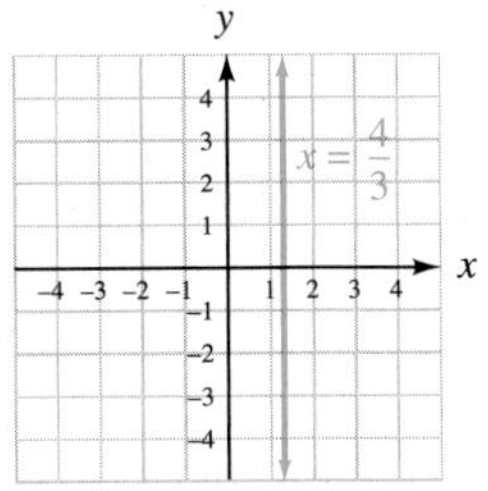

58.

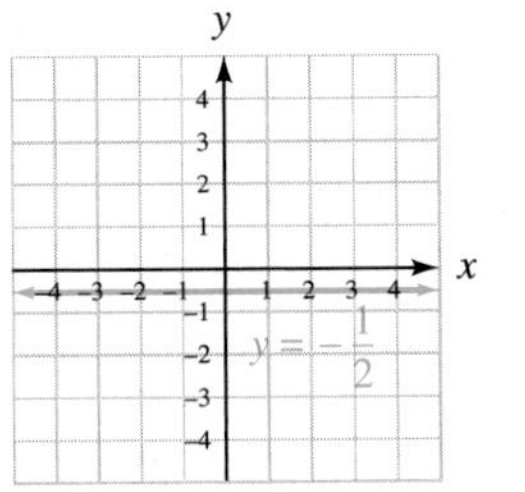

71.

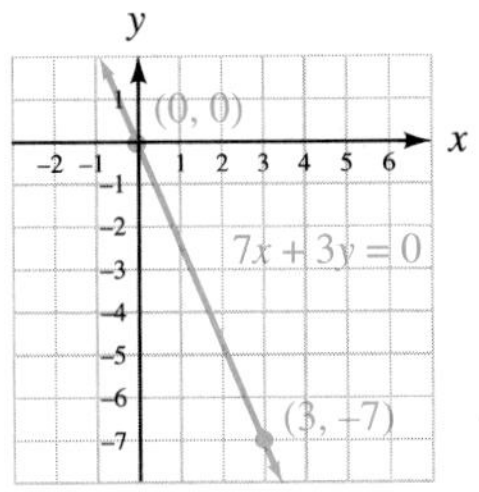

72.

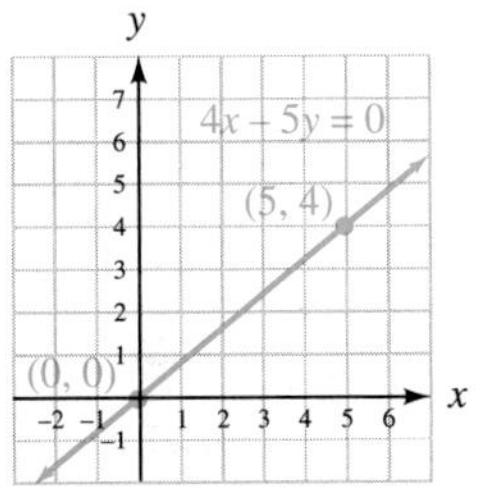

59.

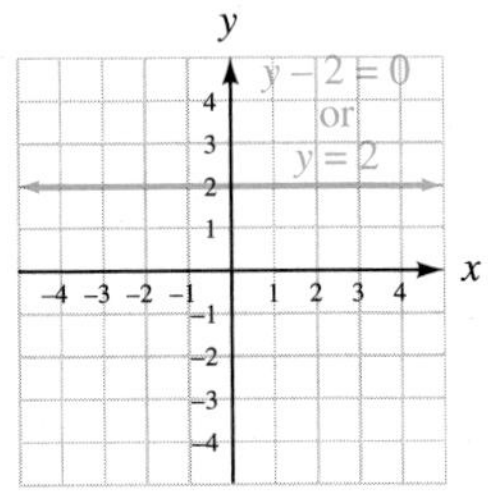

60.

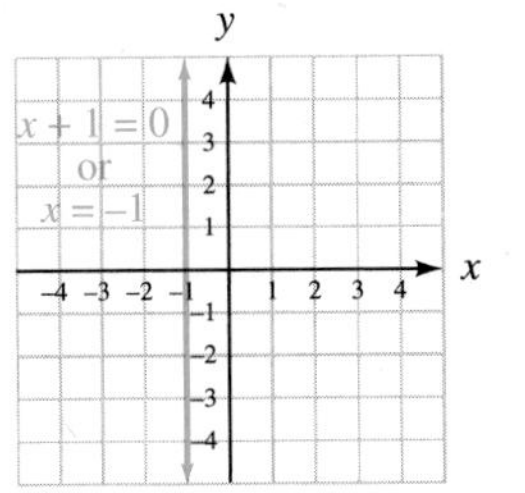

73.

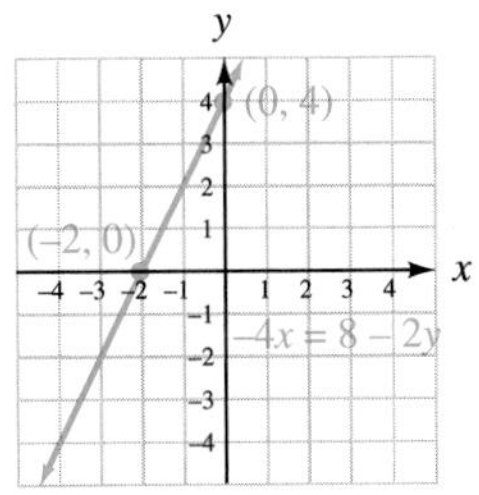

74.

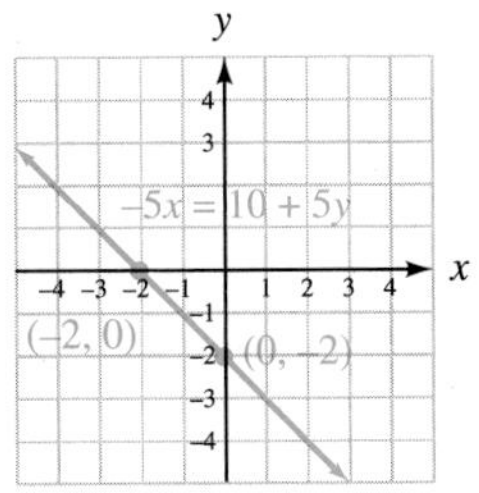

61.

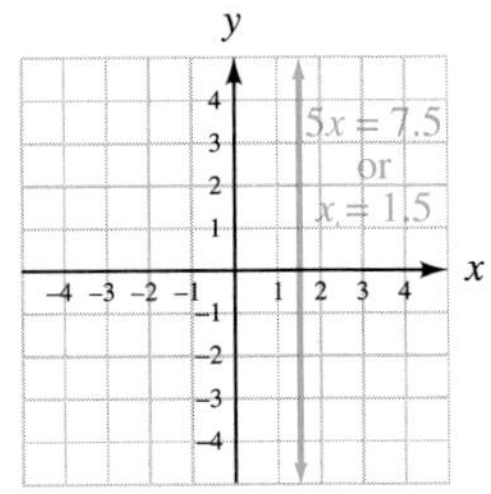

62.

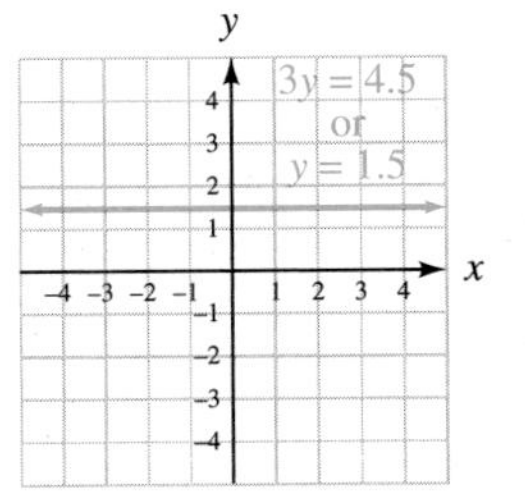

75.

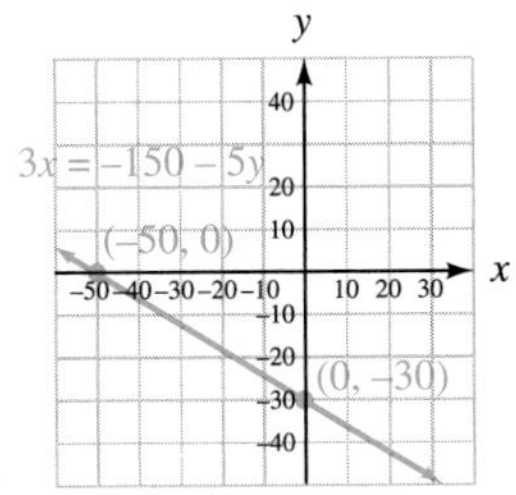

76.

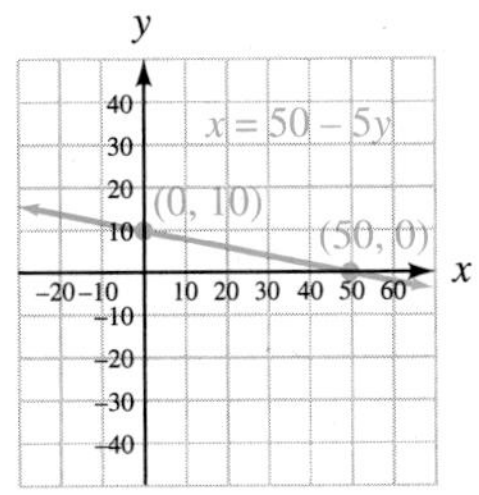

63.

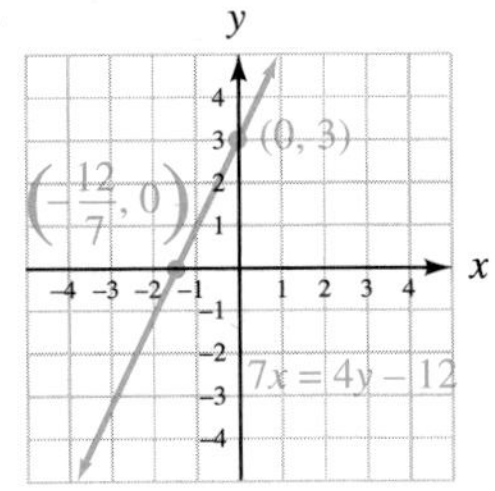

64.

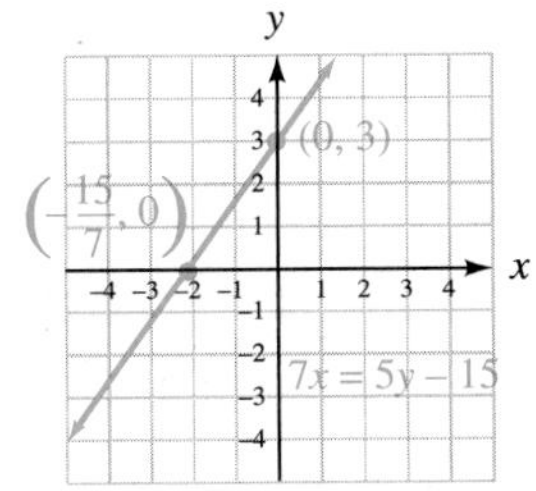

77.

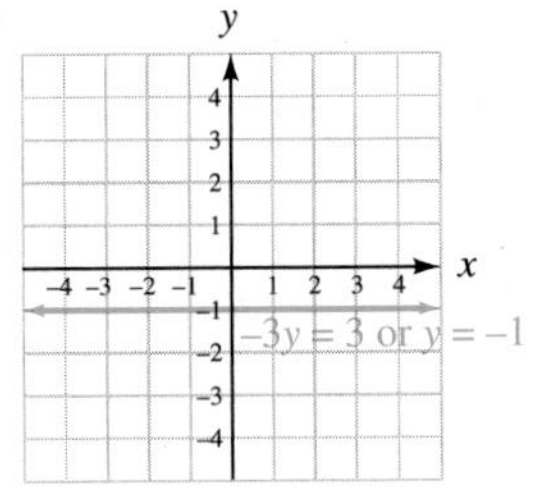

78.

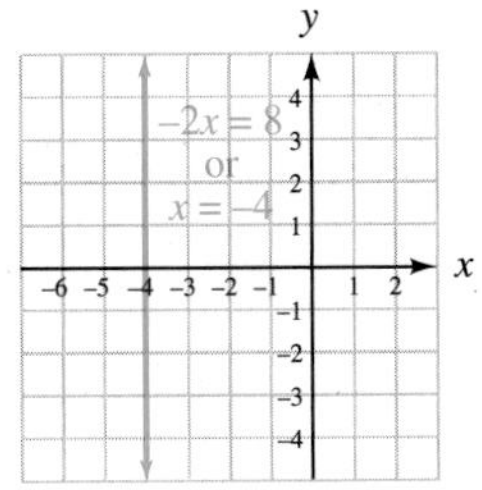

83.

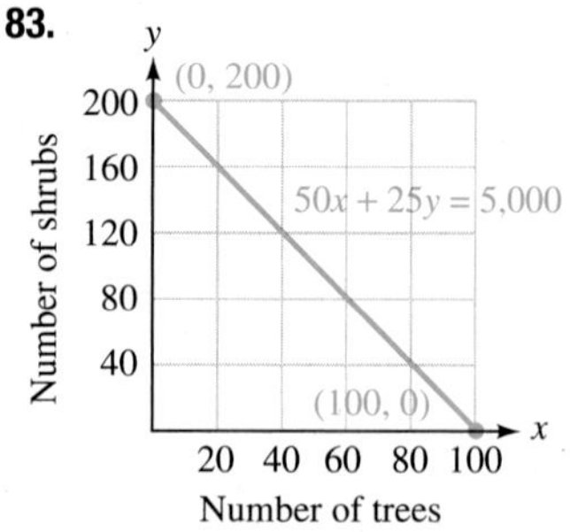

96.

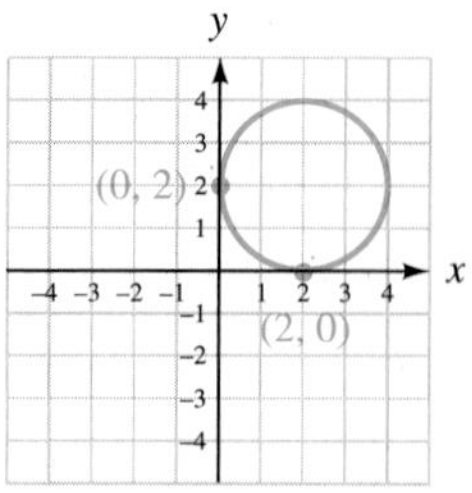

SELF CHECK 3.5

4.

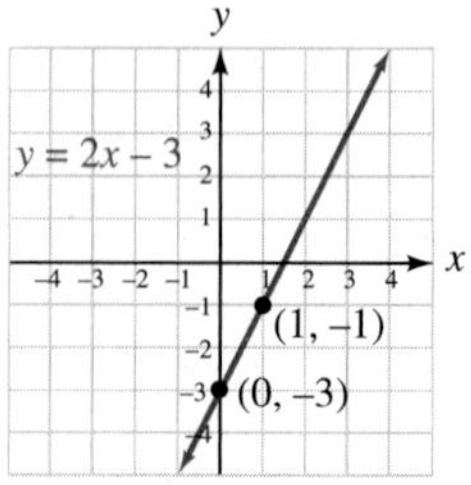

5.

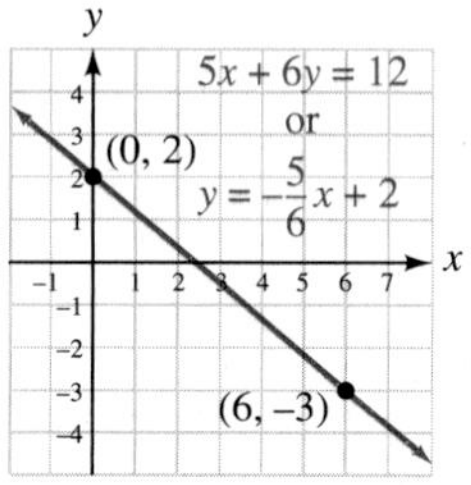

STUDY SET SECTION 3.5 (page 237)

43.

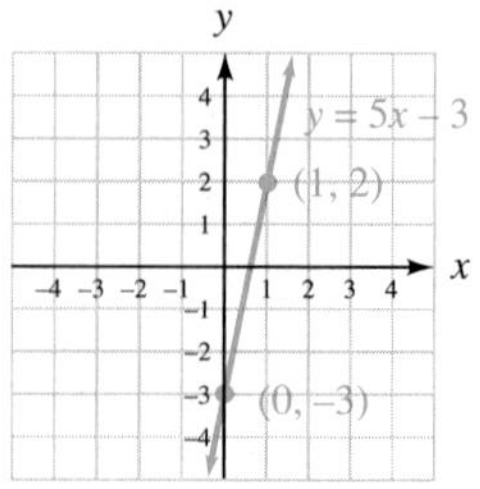

44.

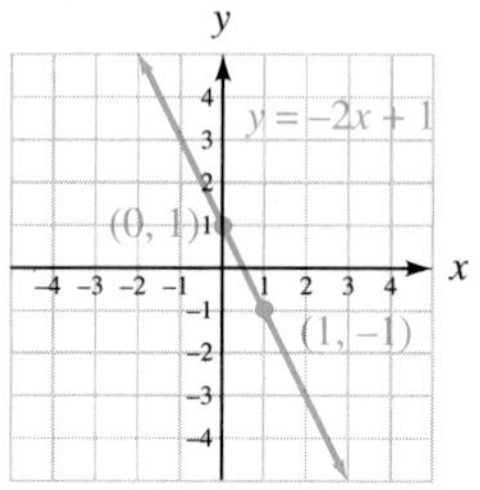

45.

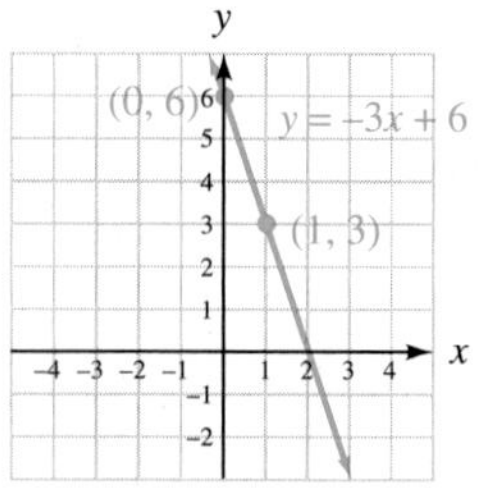

46.

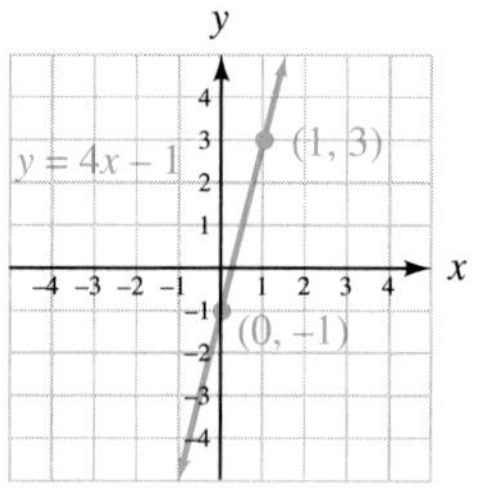

47.

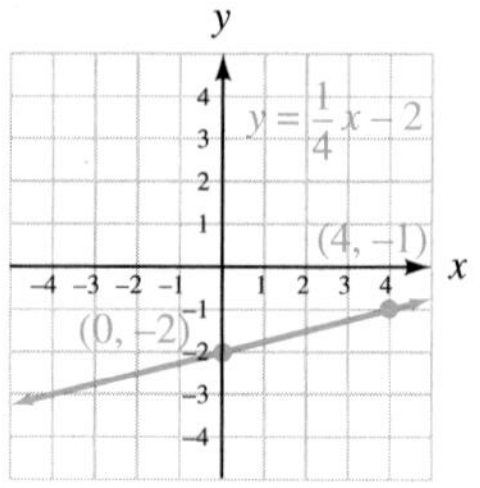

48.

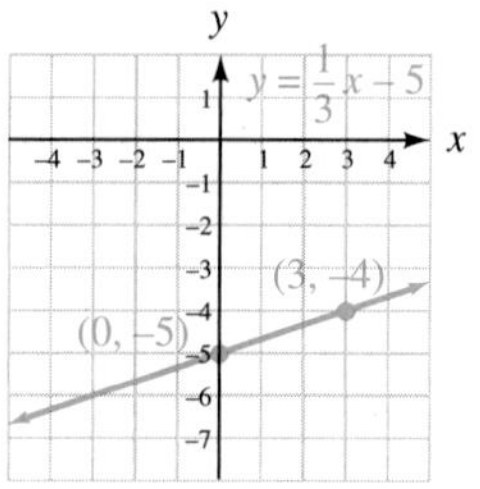

49.

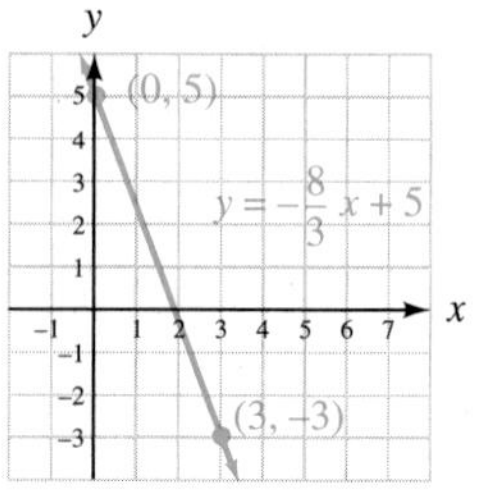

50.

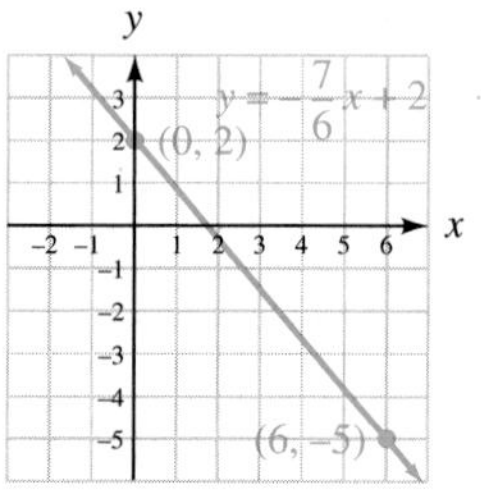

51.

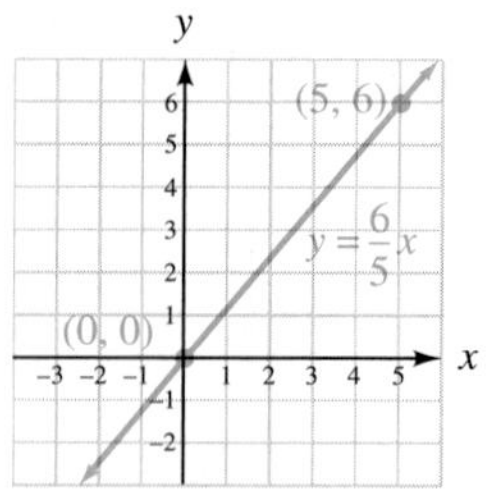

52.

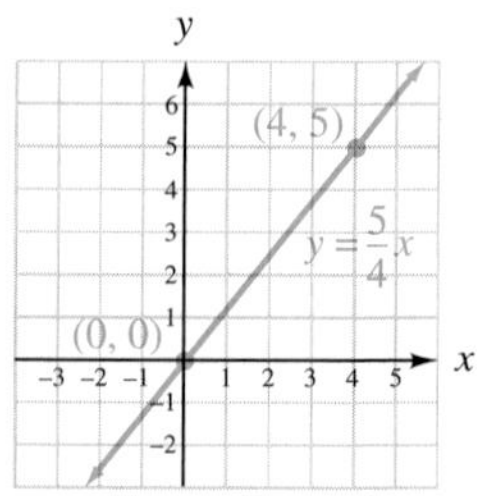

53.

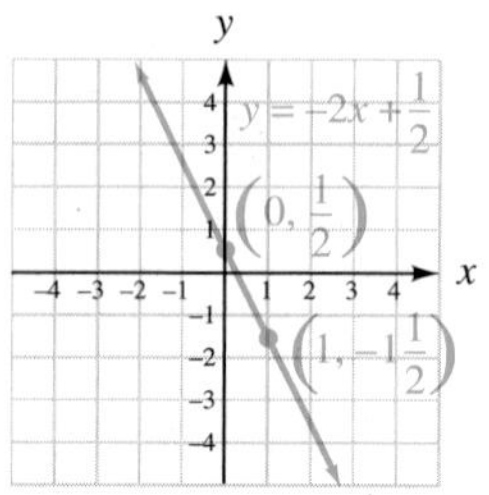

54.

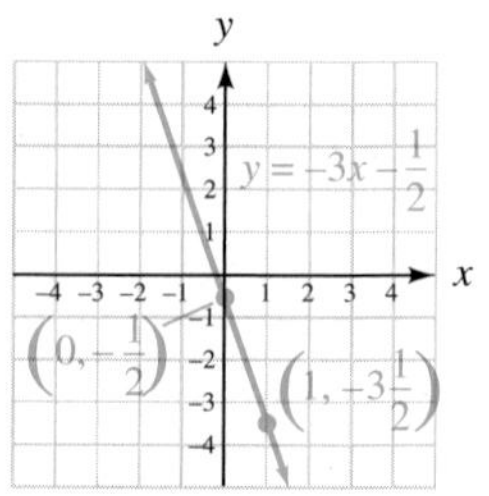

63.

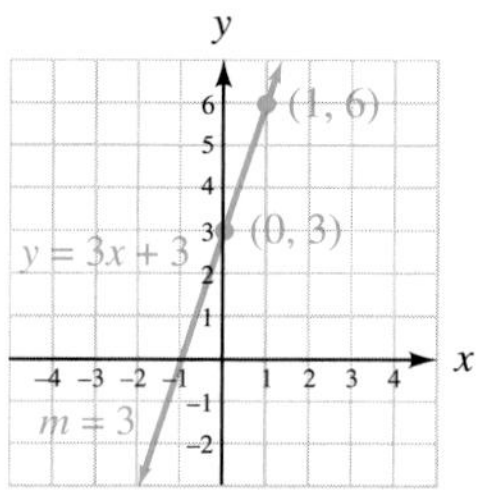

64.

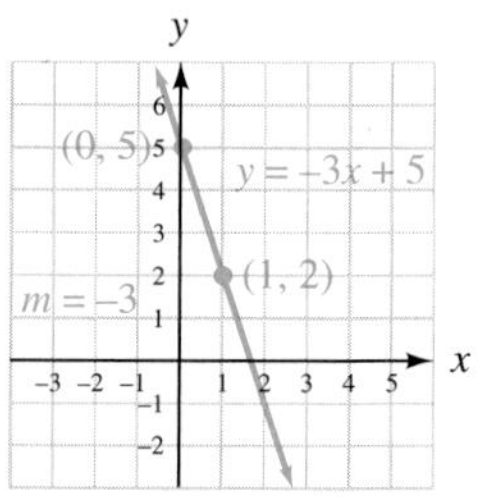

65.

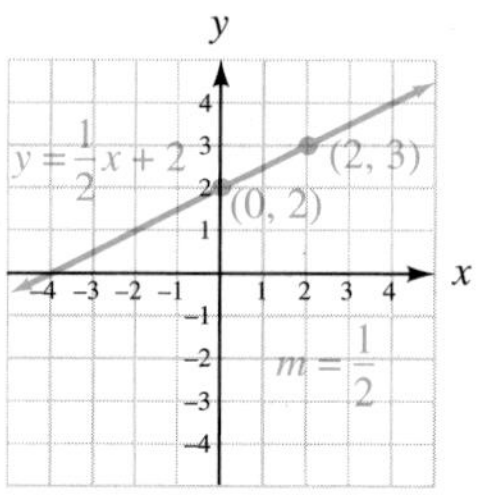

66.

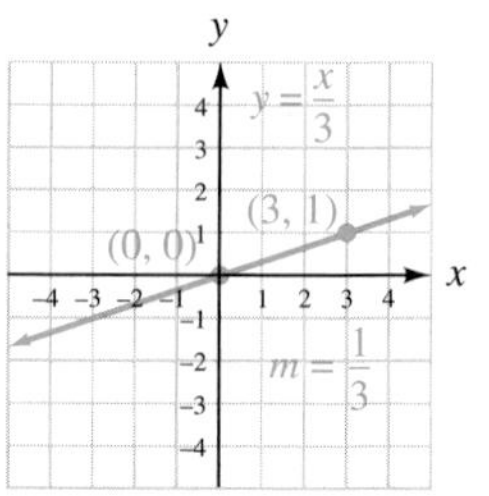

67.

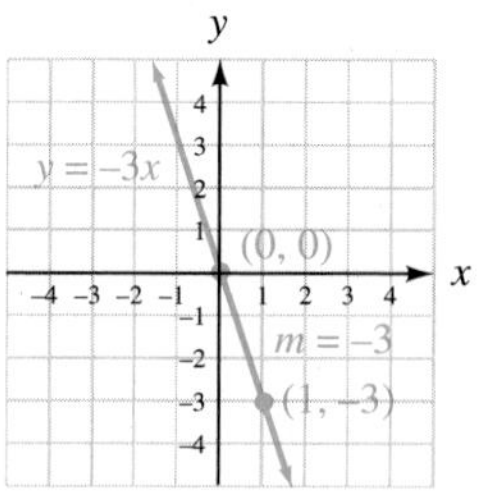

68.

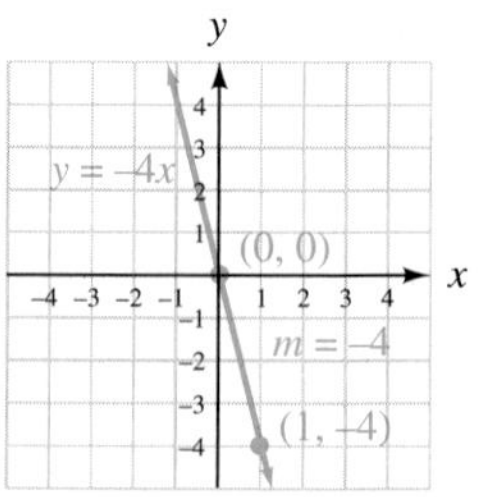

69.

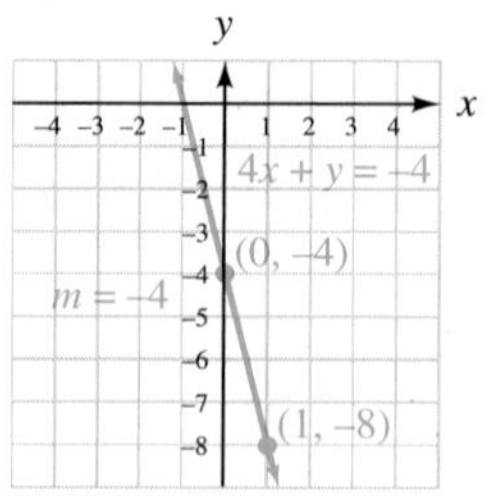

70.

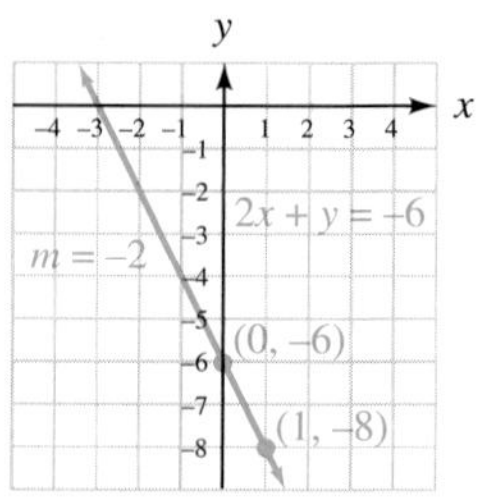

71.

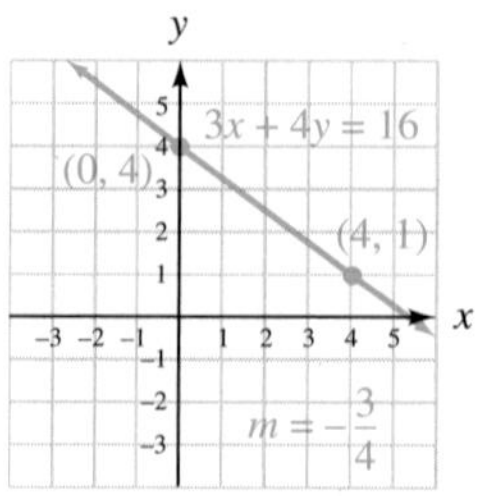

72.

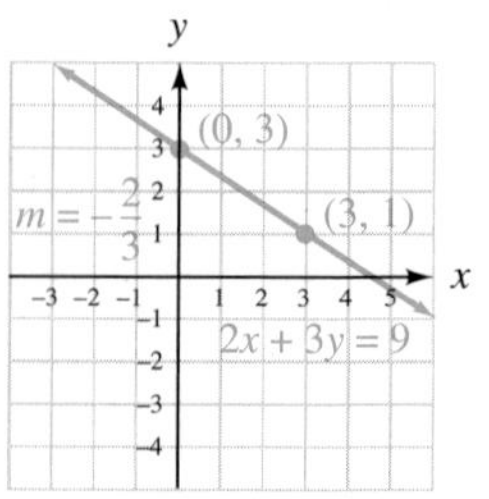

73.

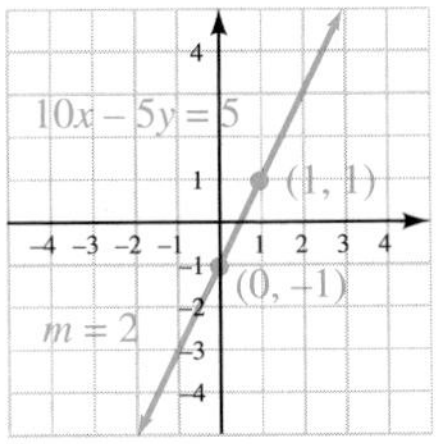

74.

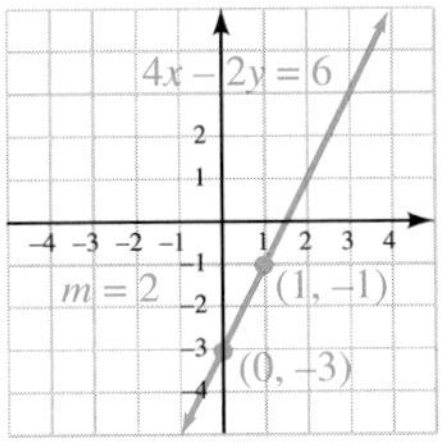

SELF CHECK 3.6

4.

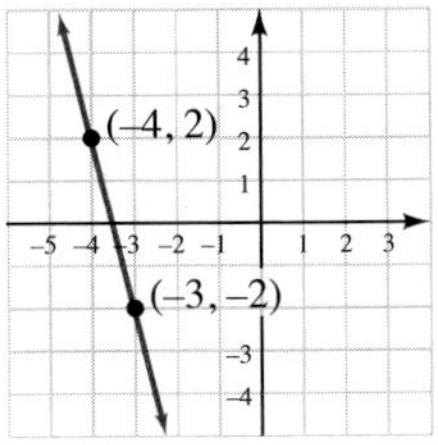

STUDY SET SECTION 3.6 (page 246)

45.

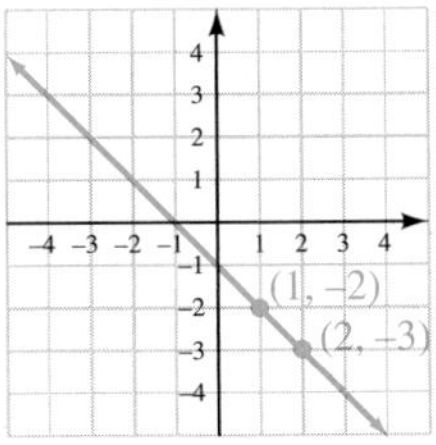

46.

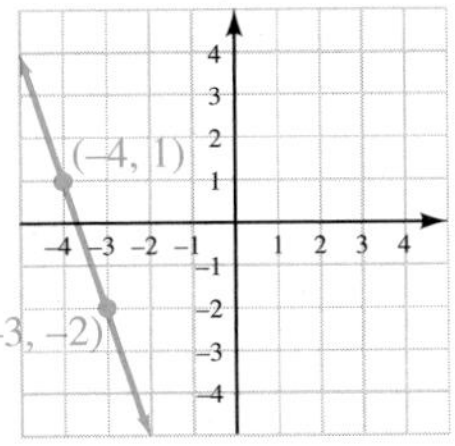

47.

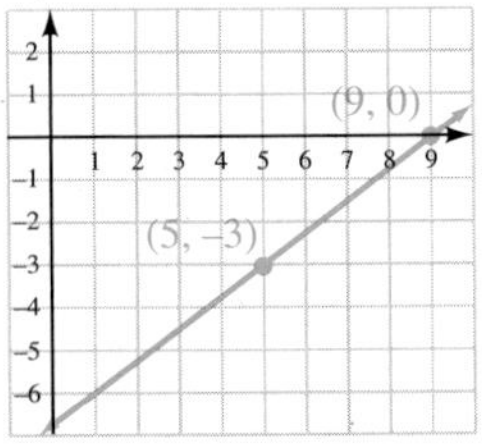

48.

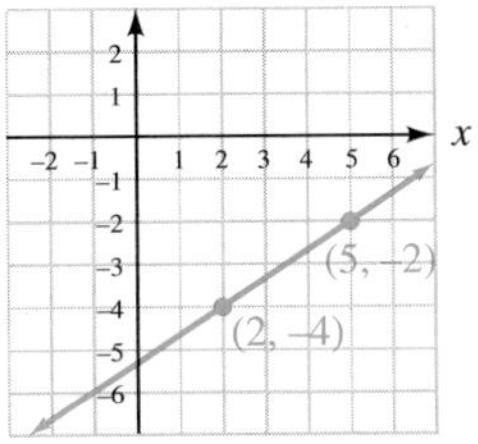

49.

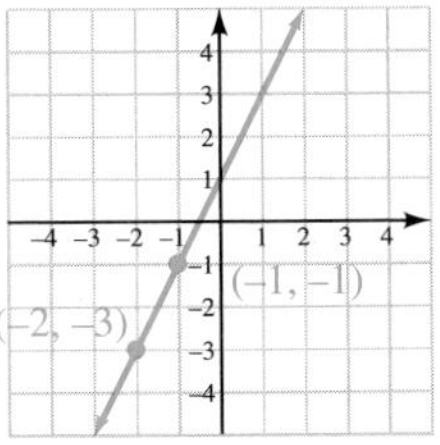

50.

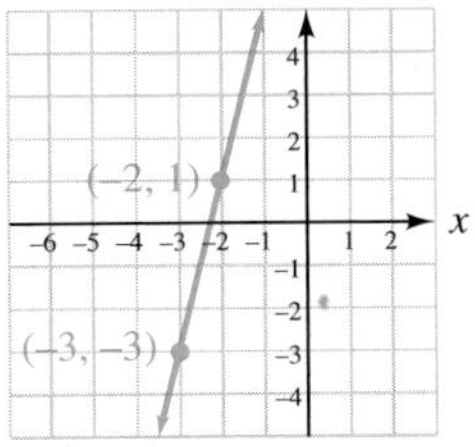

51.

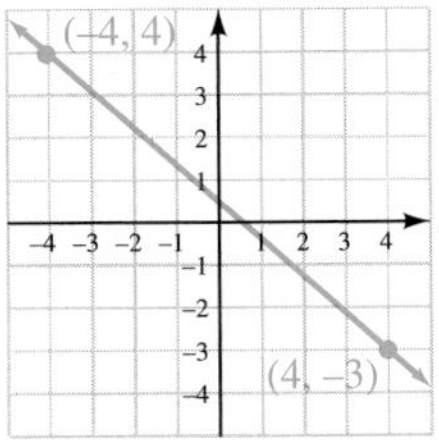

52.

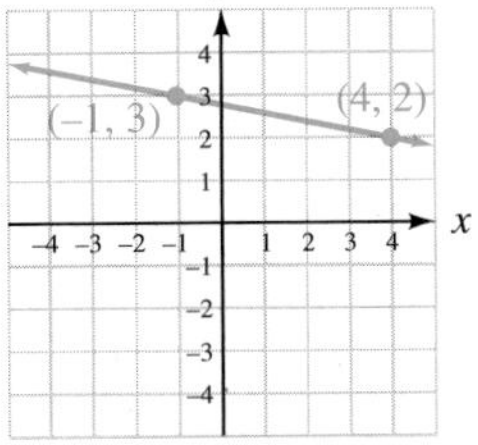

ARE YOU READY? 3.7 (page 249)

3.

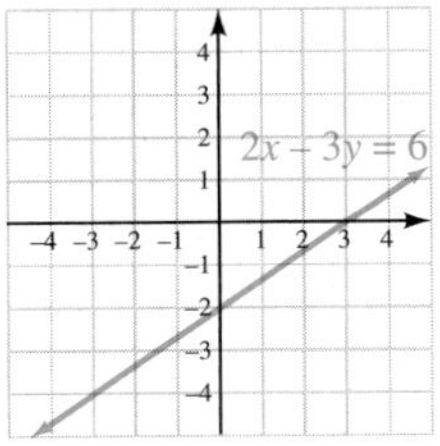

SELF CHECKS 3.7

2.

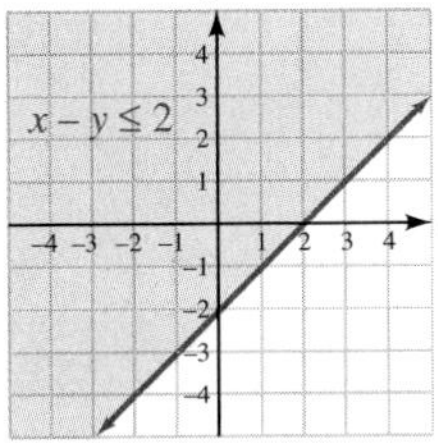

3.

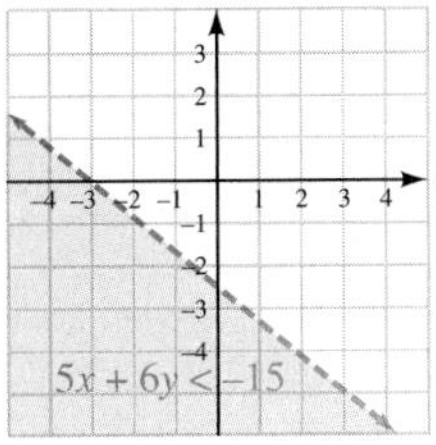

4.

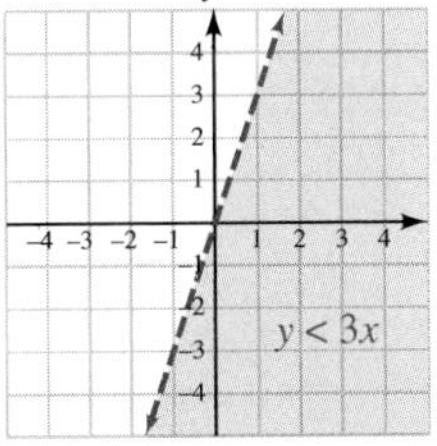

5.

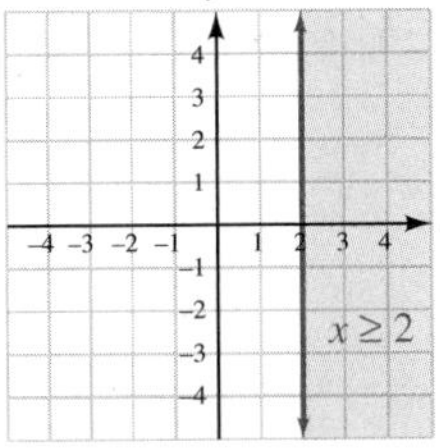

(a)

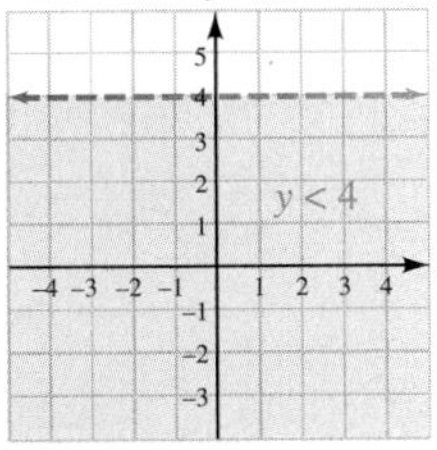

(b)

STUDY SET SECTION 3.7 (page 255)

35.

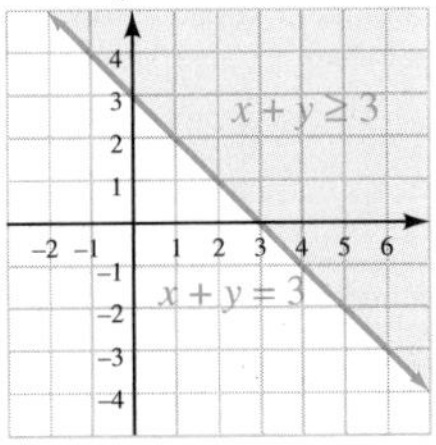

36.

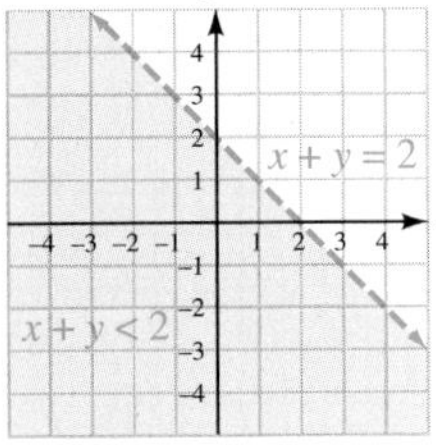

37.

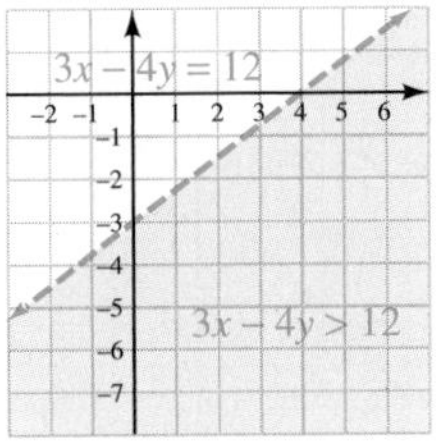

38.

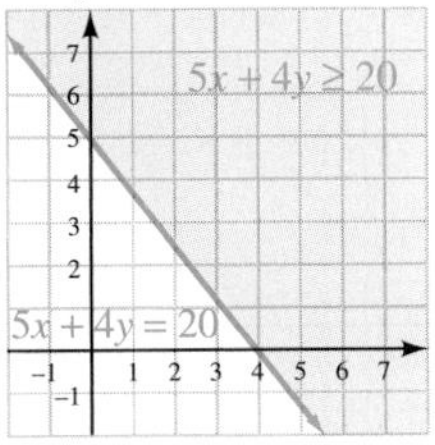

39.

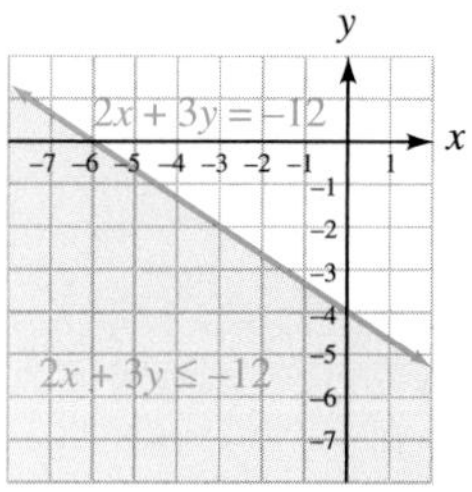

40.

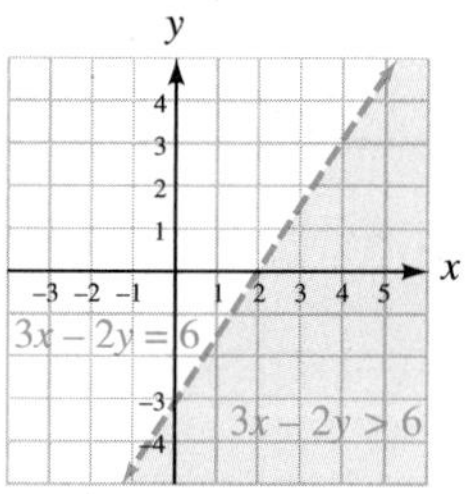

41.

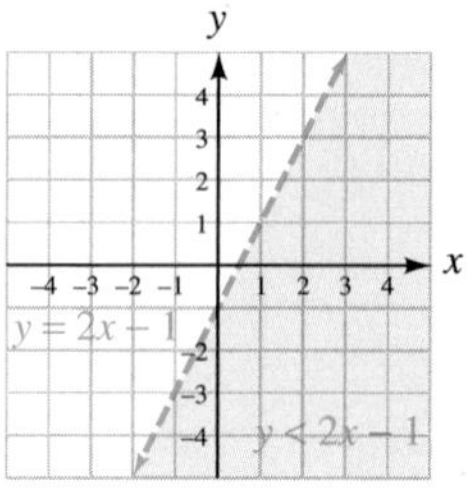

42.

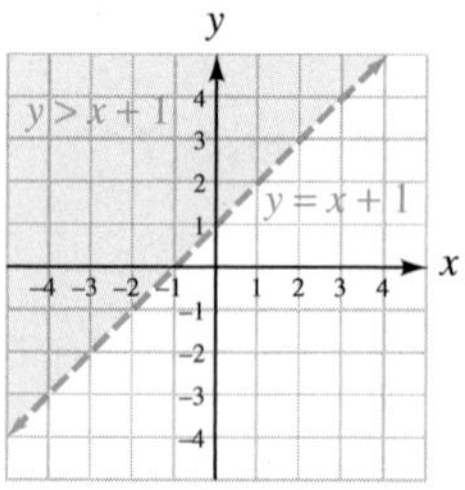

43.

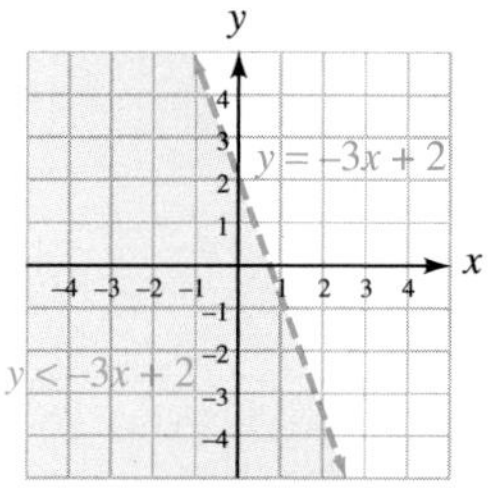

44.

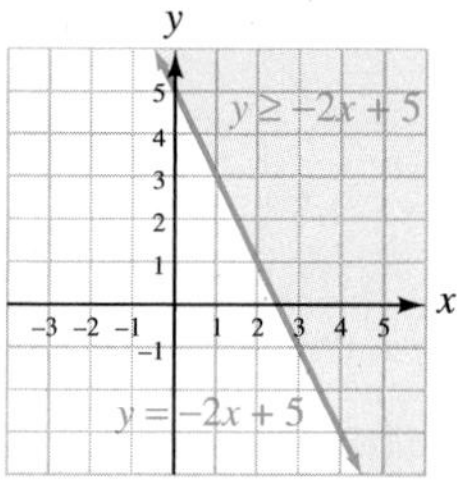

45.

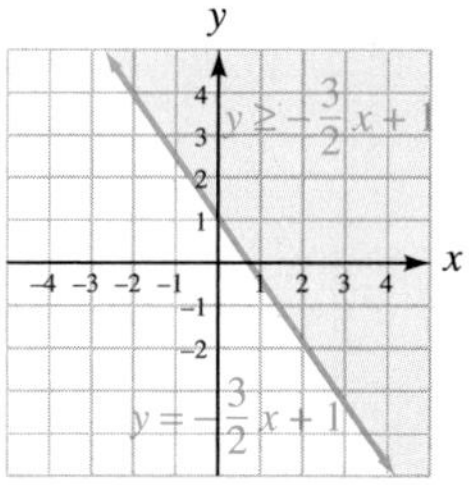

46.

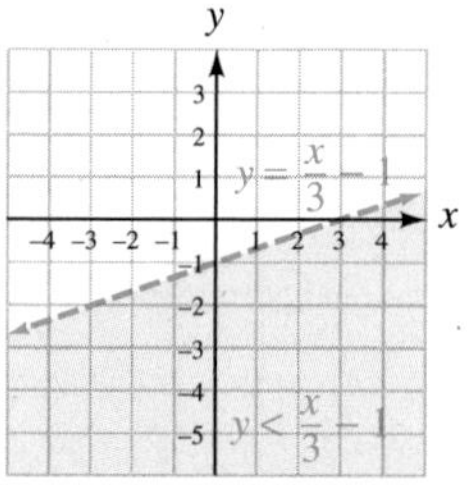

47.

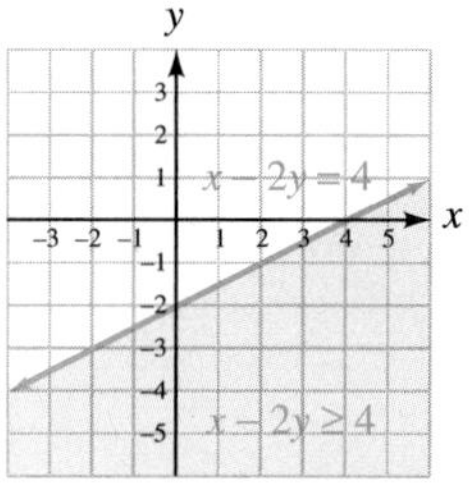

48.

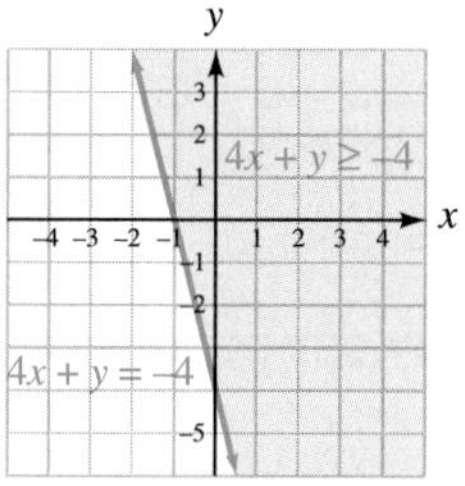

49.

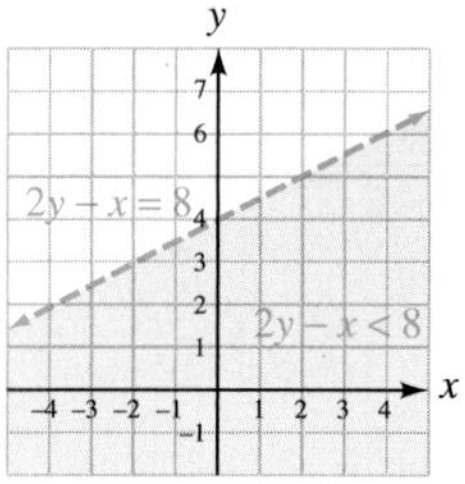

50.

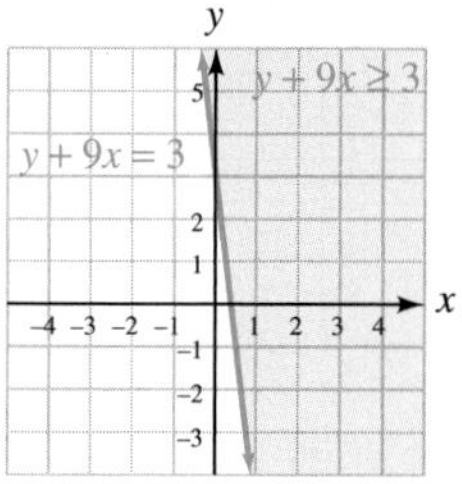

51.

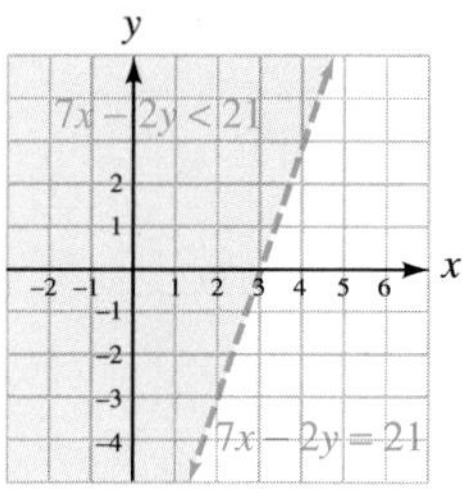

52.

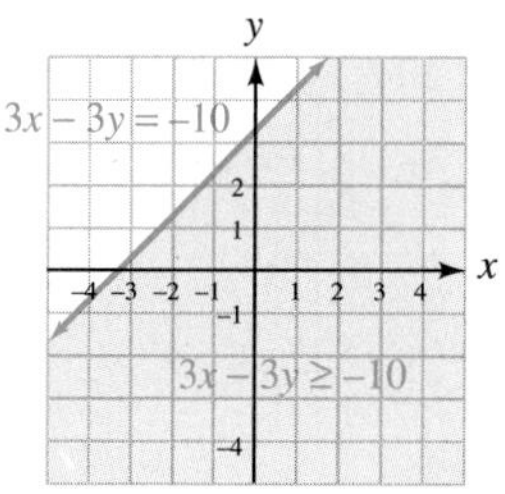

53.

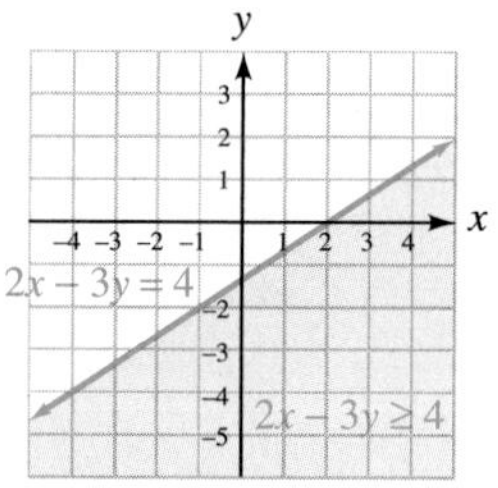

54.

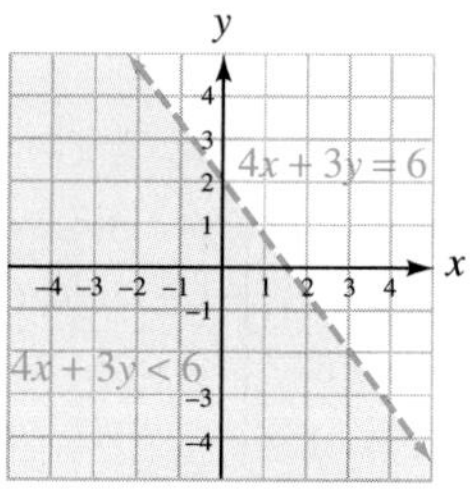

55.

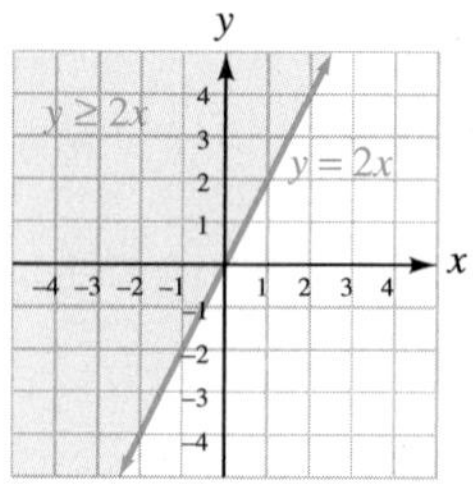

56.

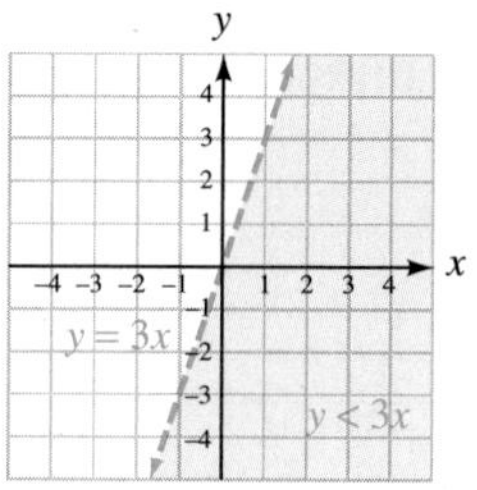

57.

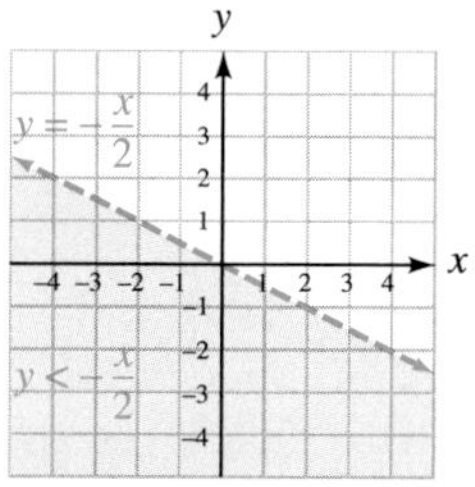

58.

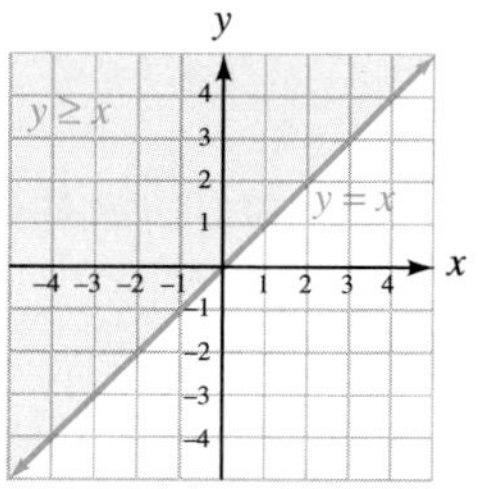

59.

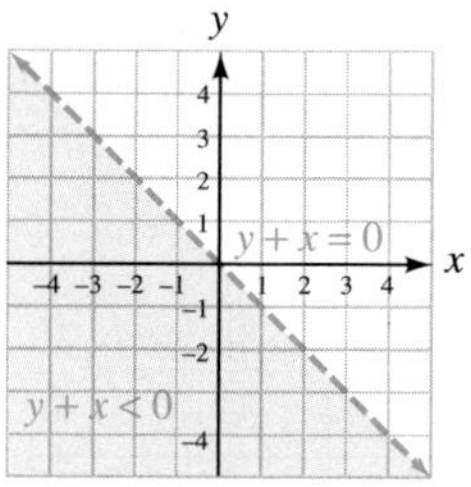

60.

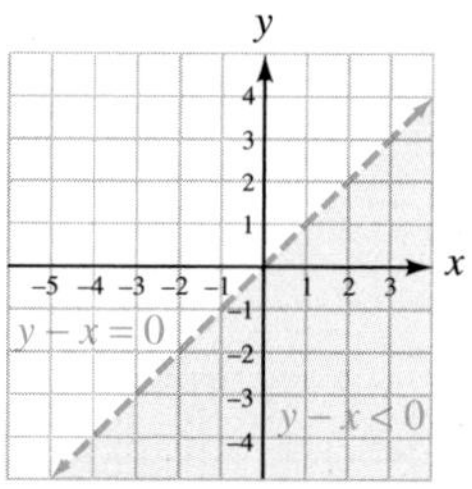

61.

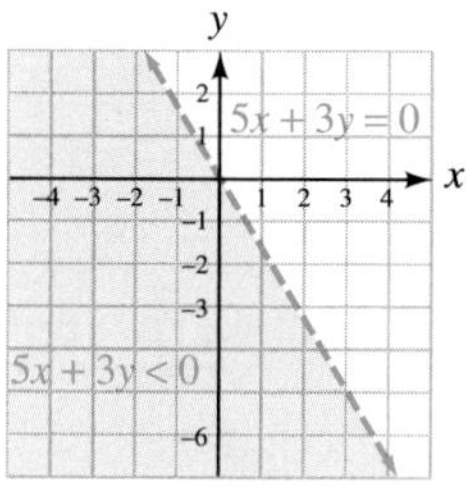

62.

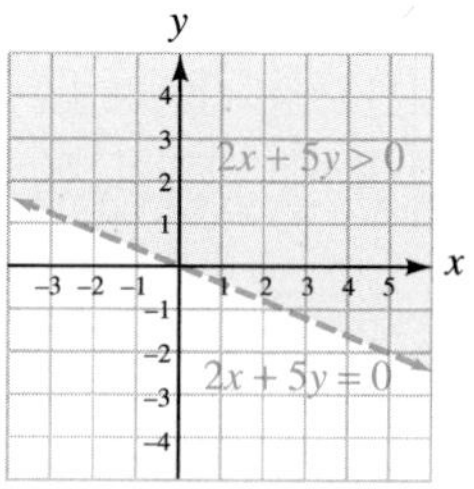

63.

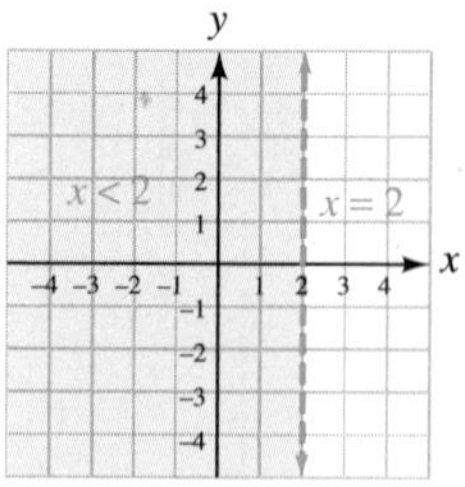

64.

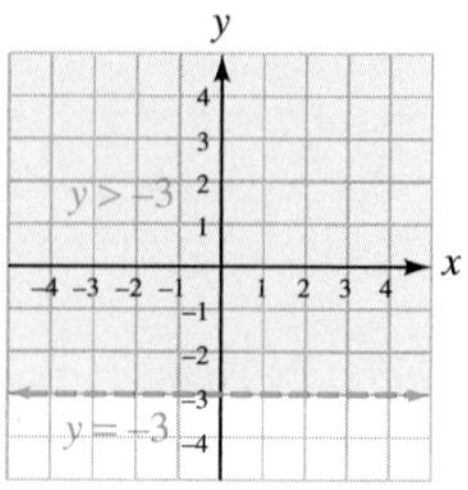

65.

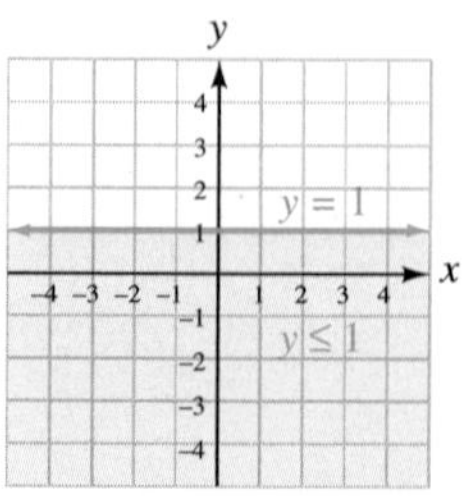

66.

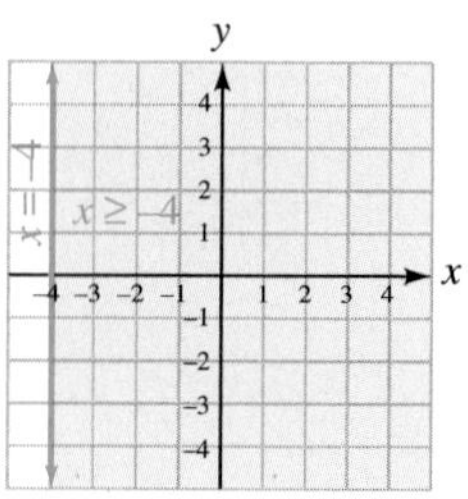

67.

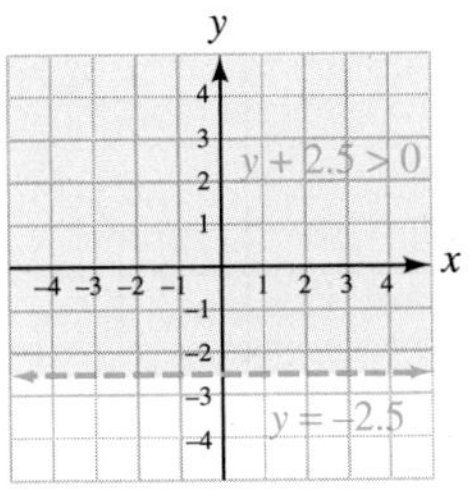

68.

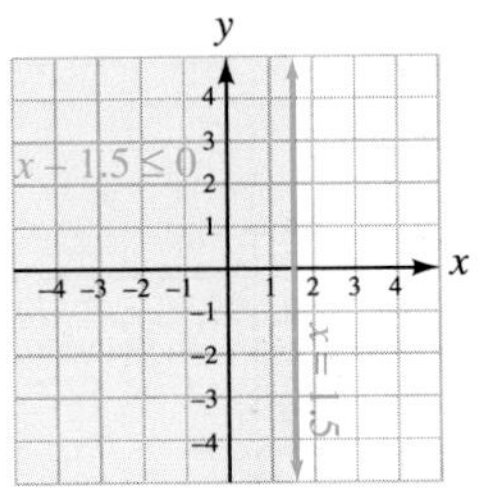

69.

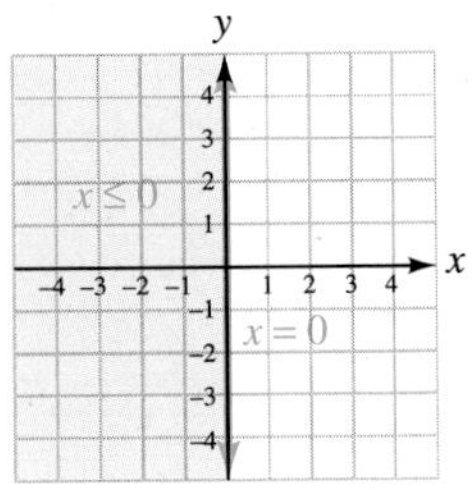

70.

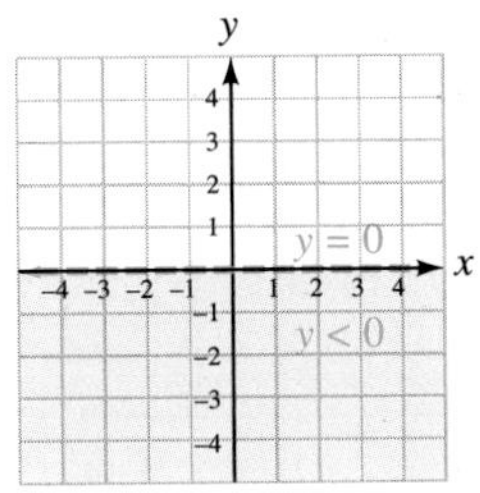

71. a.

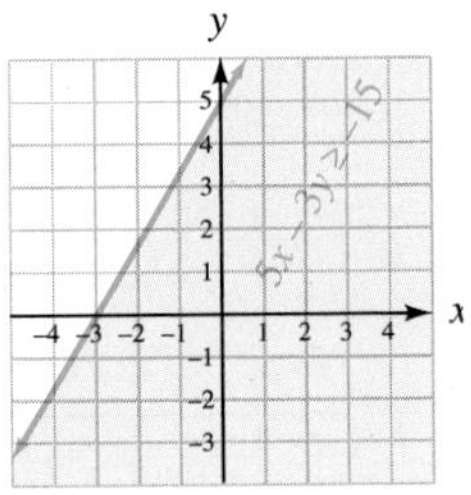

b.

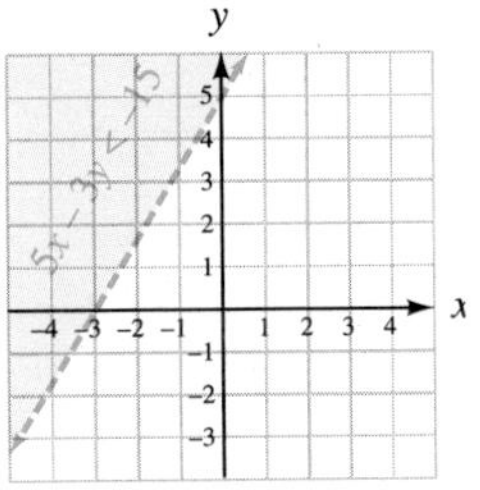

72. a.

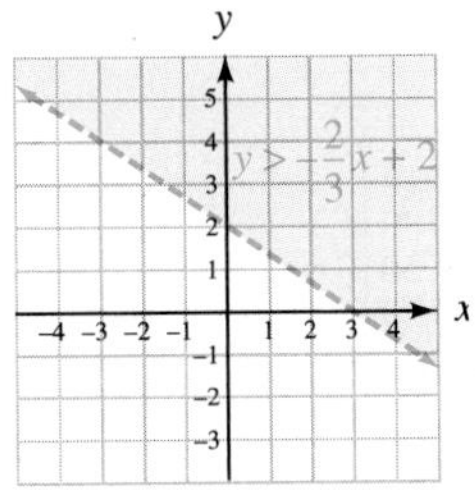

b.

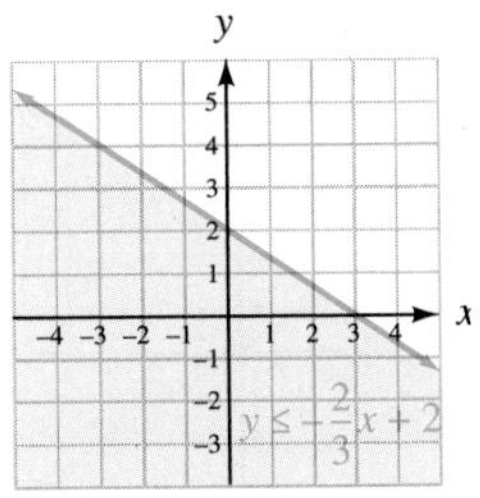

73. a.

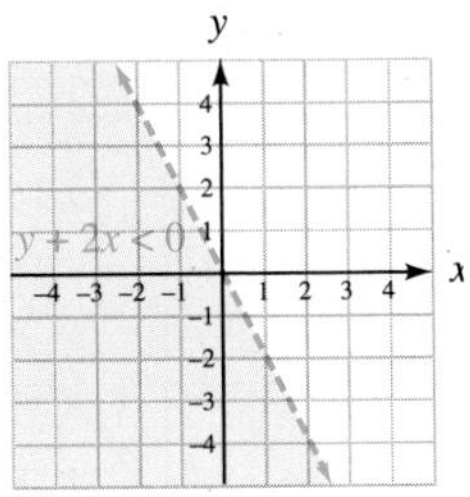

b.

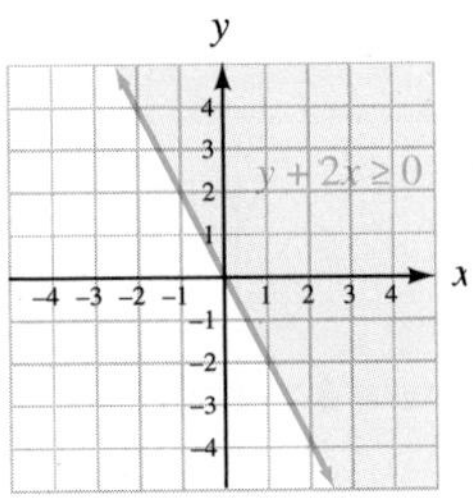

74. a.

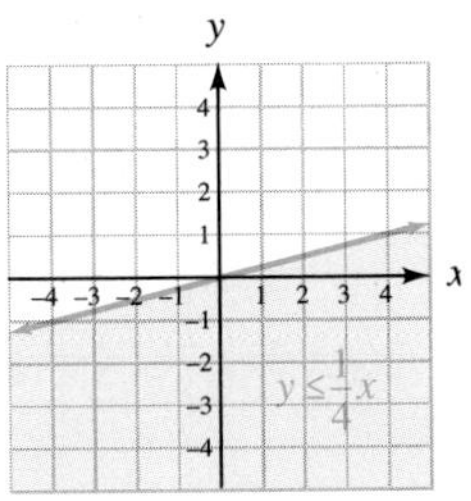

b.

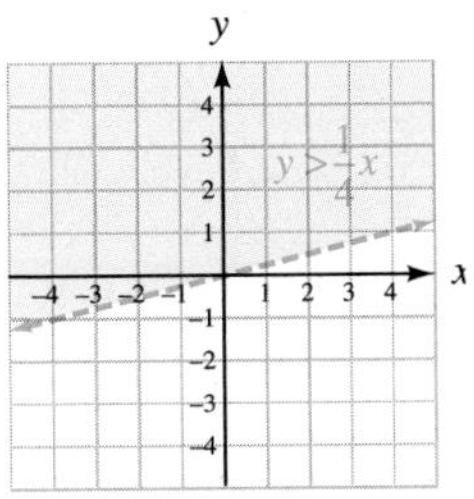

77.

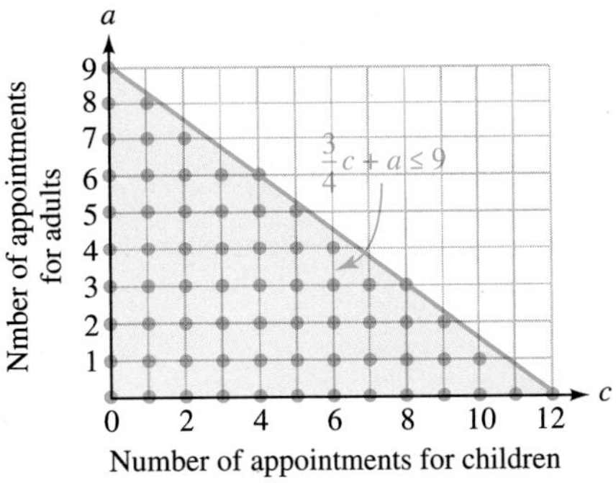

79.

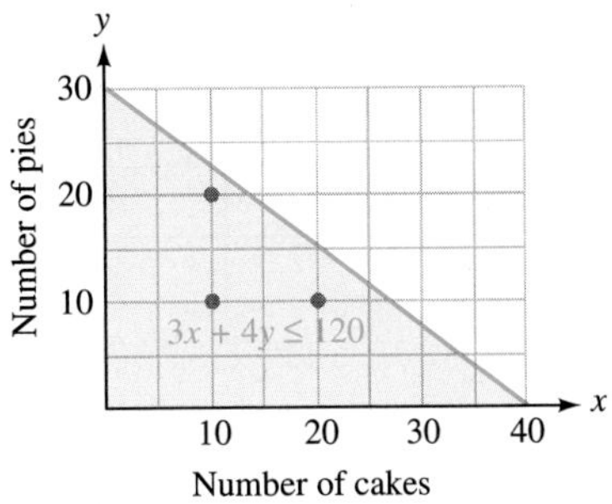

80.

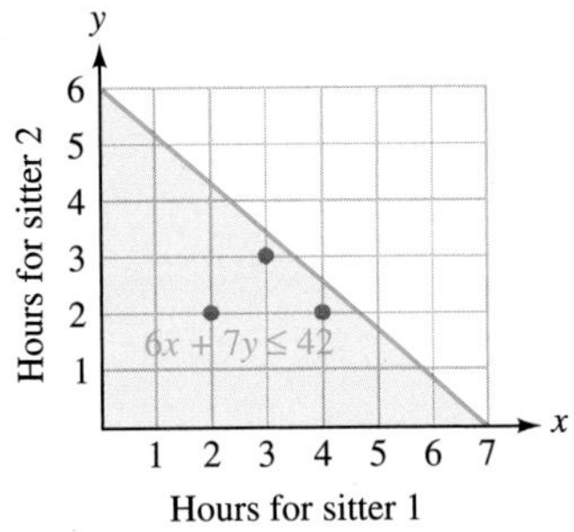

81.

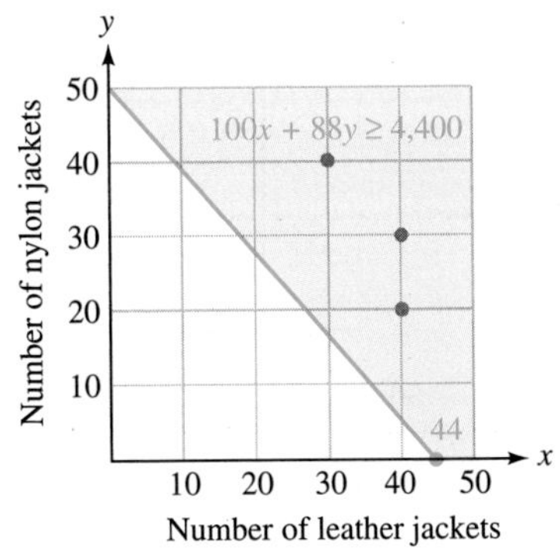

82.

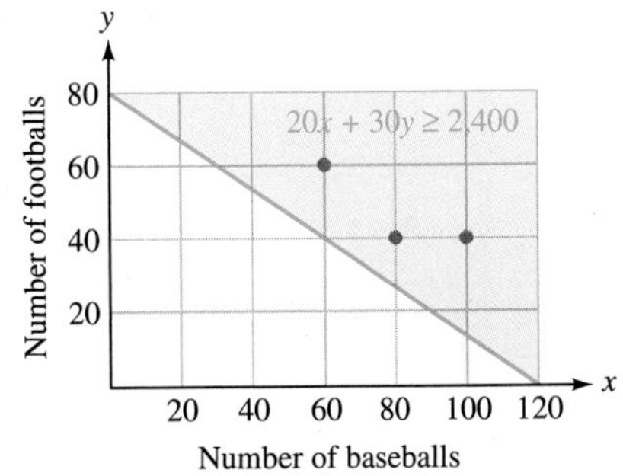

92.
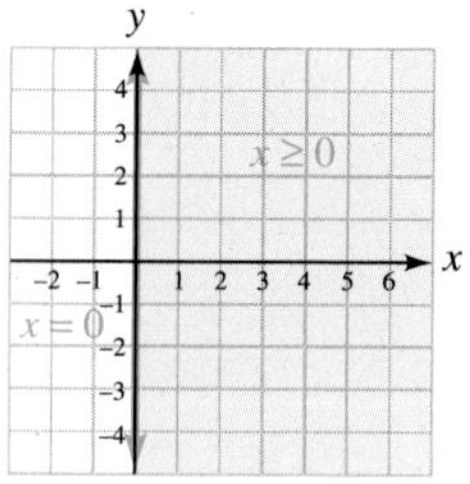

53.
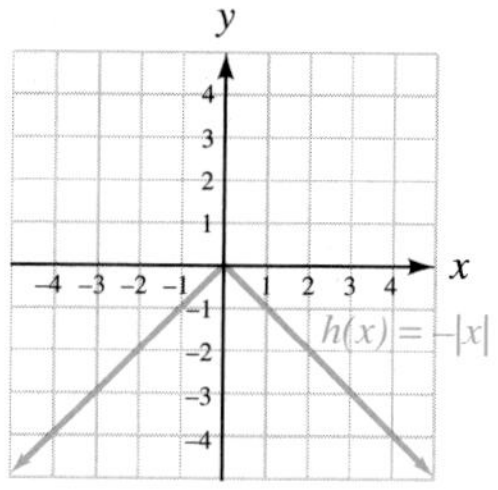

54.
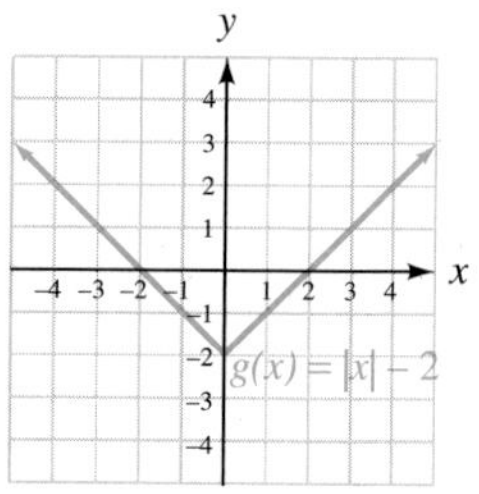

SELF CHECKS 3.8

5.
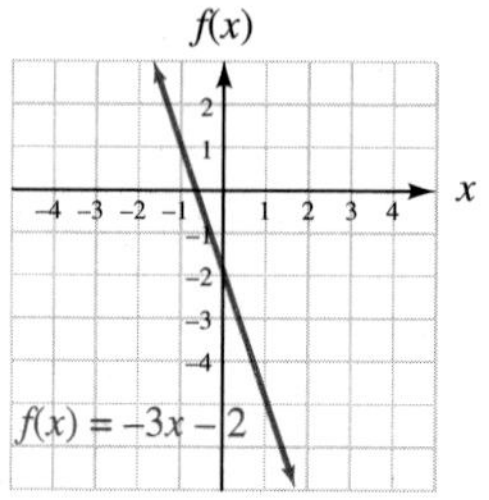

6.
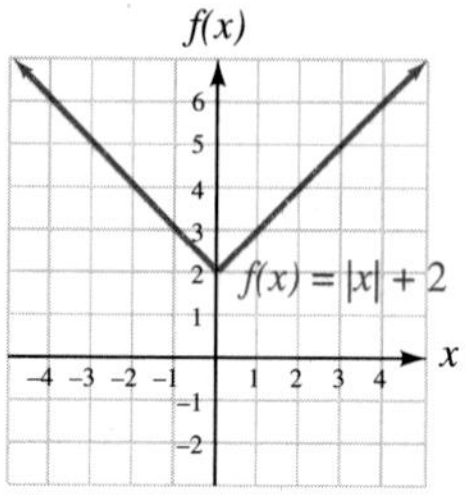

STUDY SET SECTION 3.8 (page 266)

7.
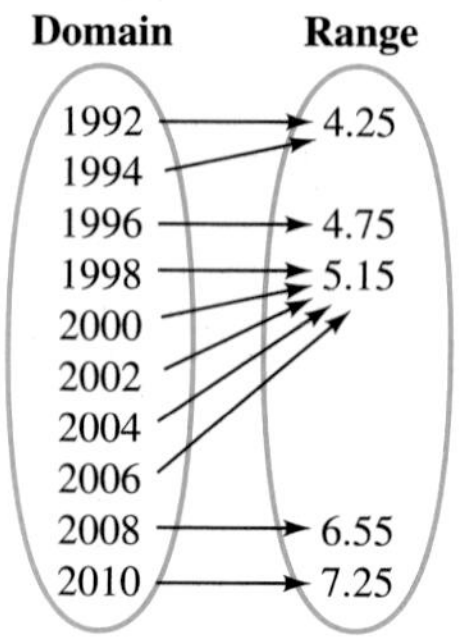

47.
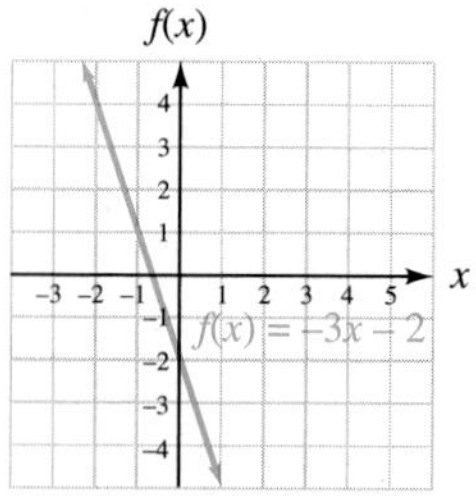

48.
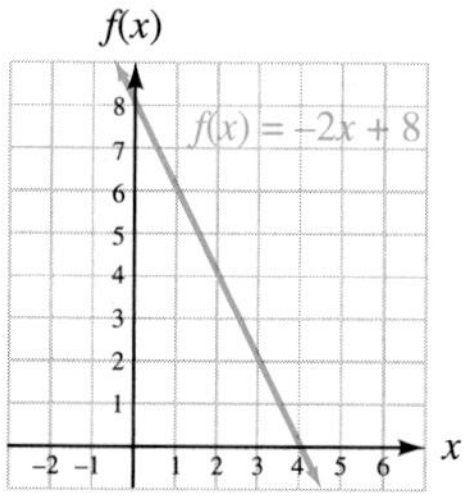

49.
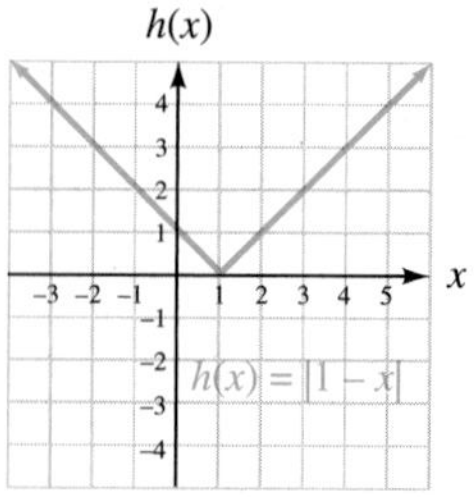

50.
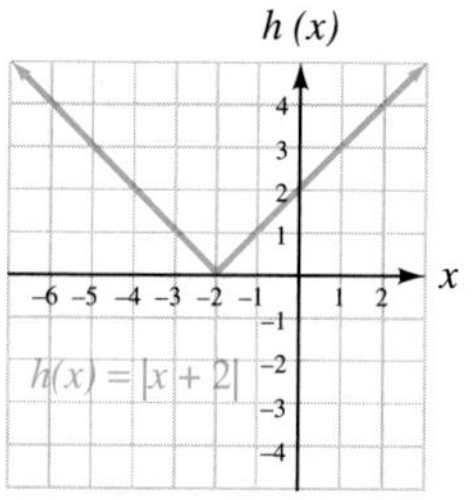

51.
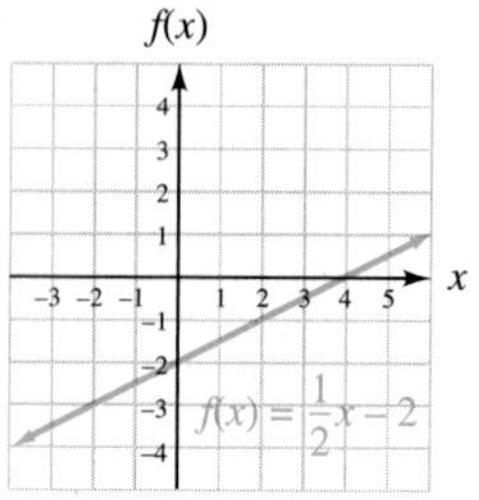

52.
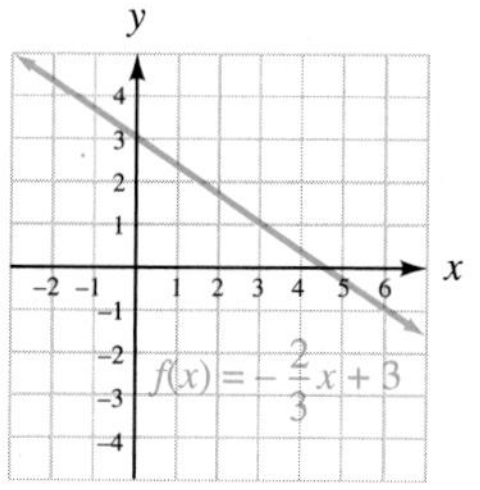

CHAPTER 3 REVIEW (page 270)

1.
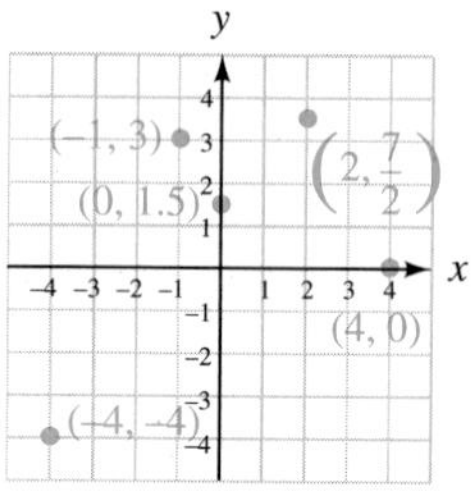

11.
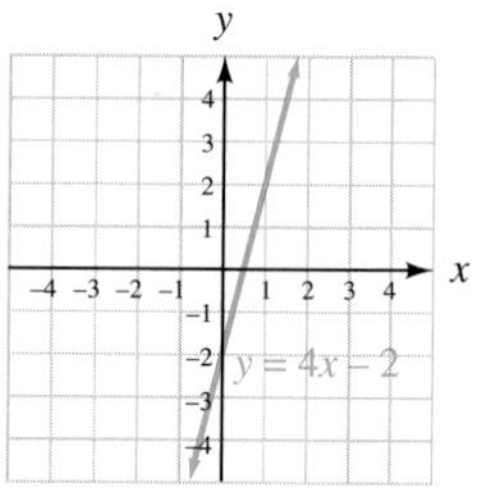

12.
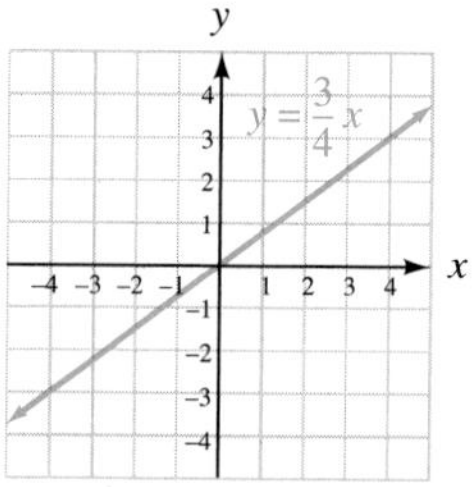

13.
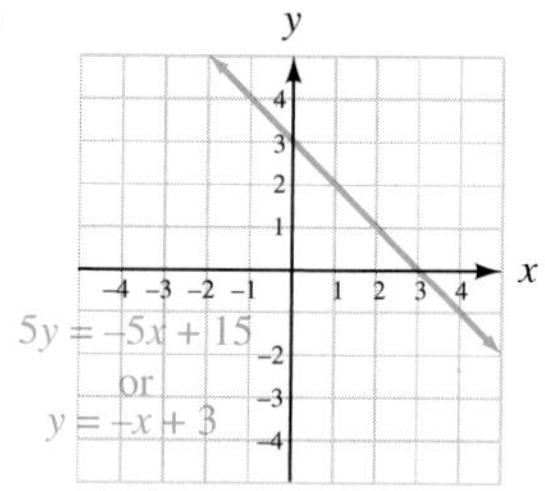

14.
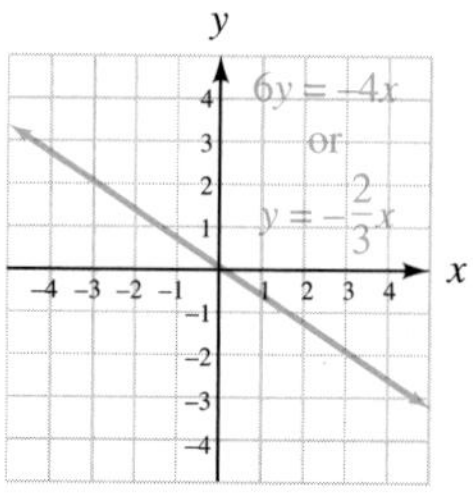

15.
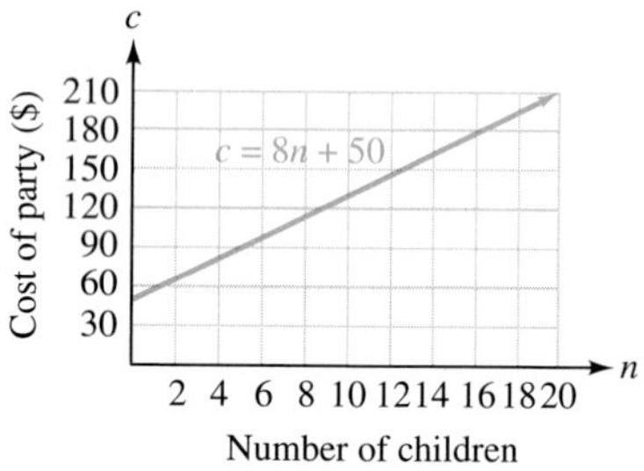

19.
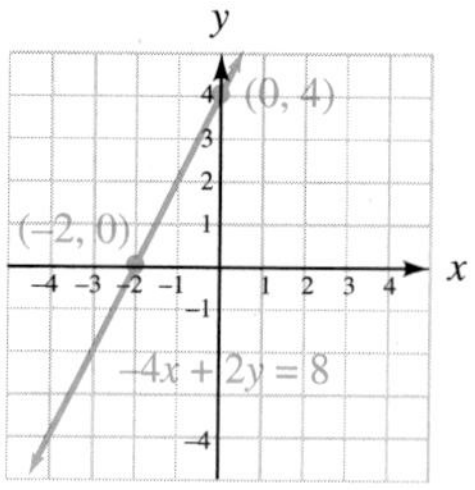

20.
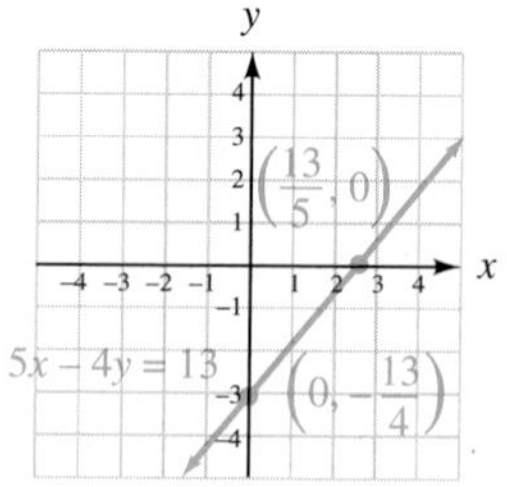

21.
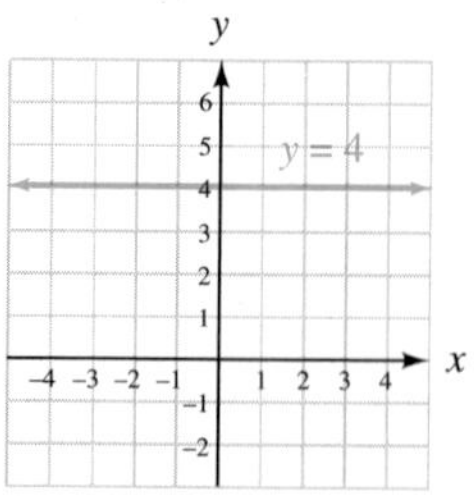

22.
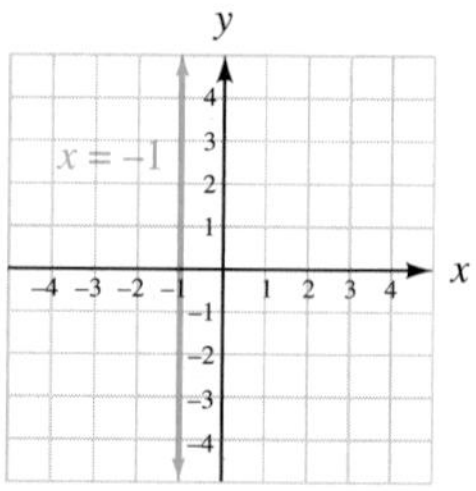

27.

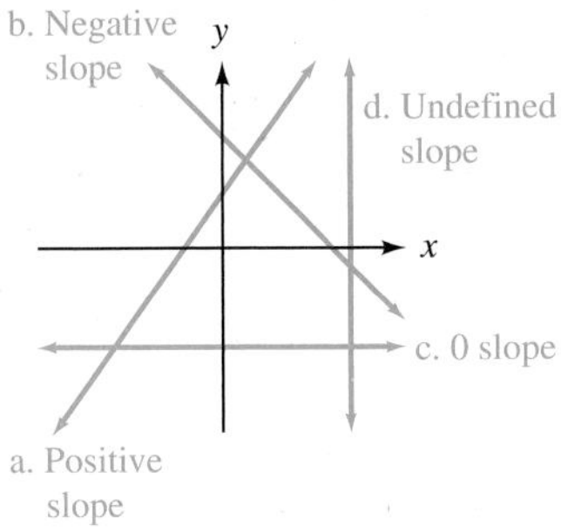

37.

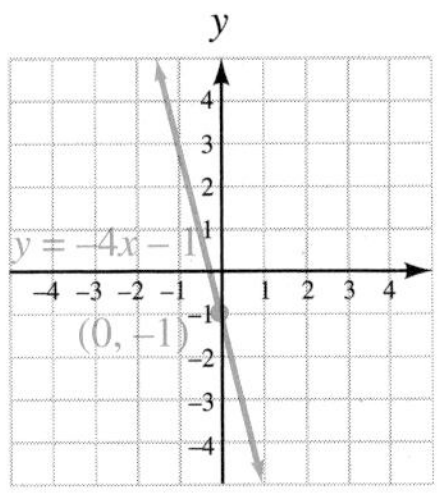

39.

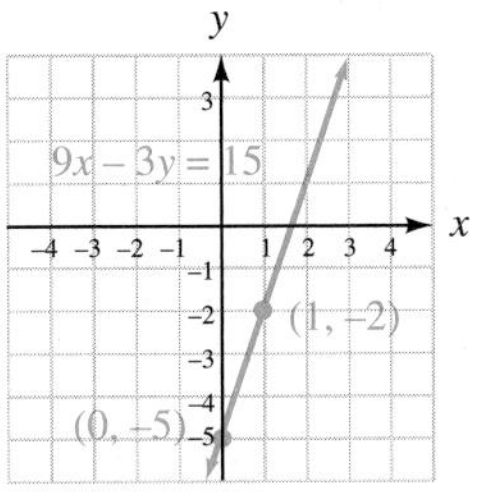

43.

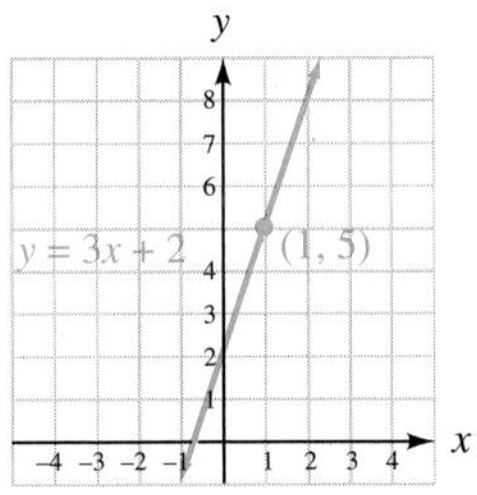

44.

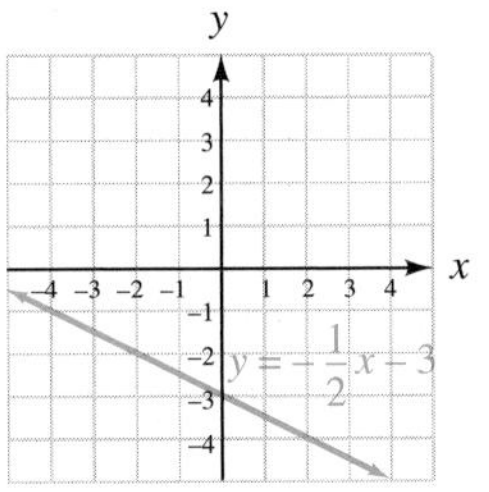

51.

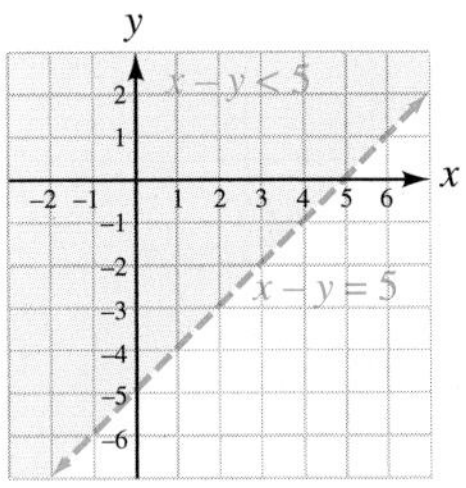

52.

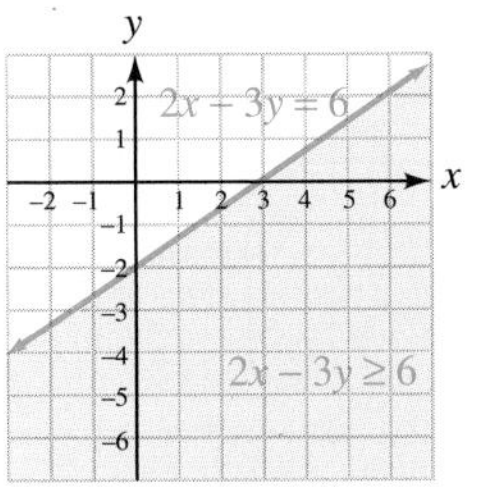

53.

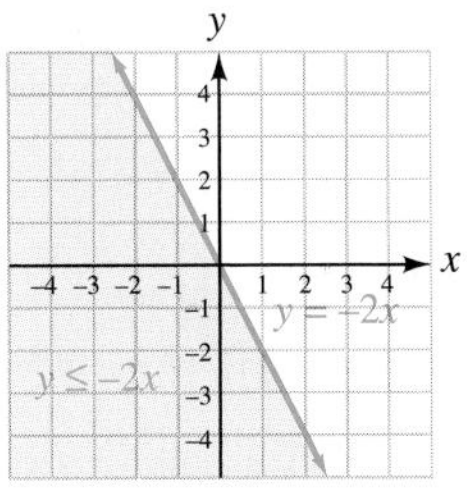

54.

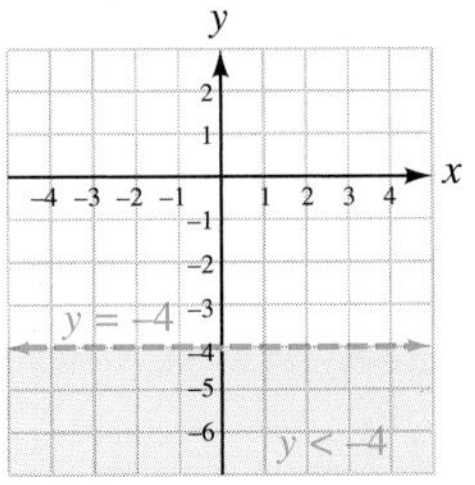

56.

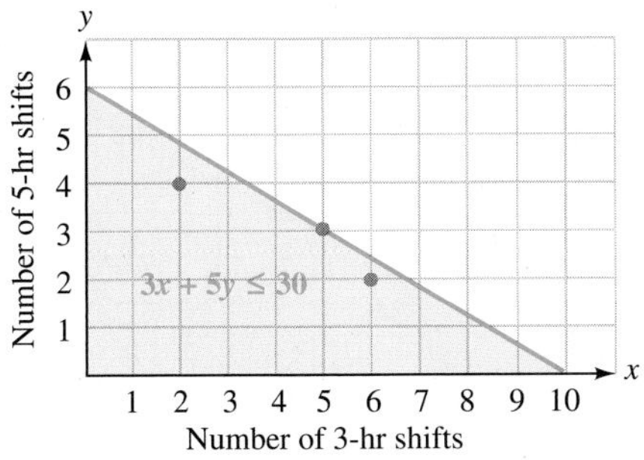

77.

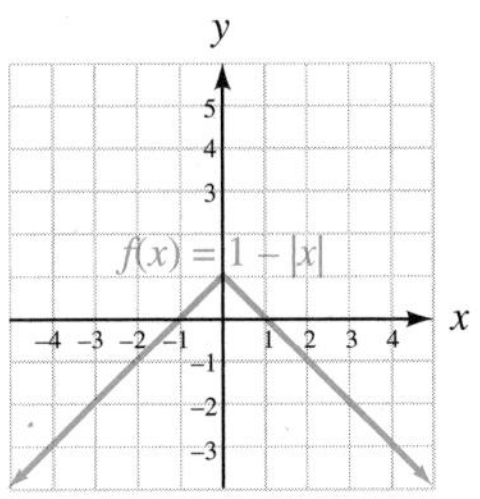

CHAPTER 3 TEST (page 279)

6.

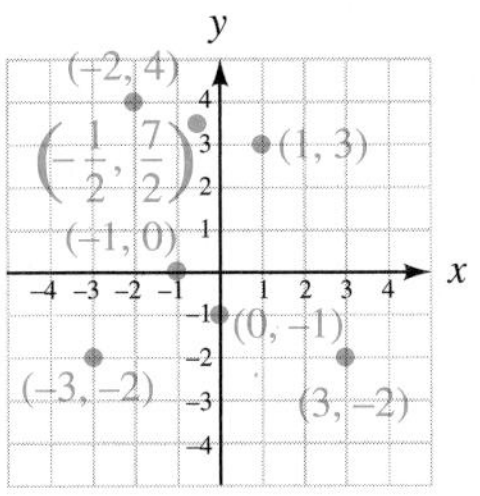

12.

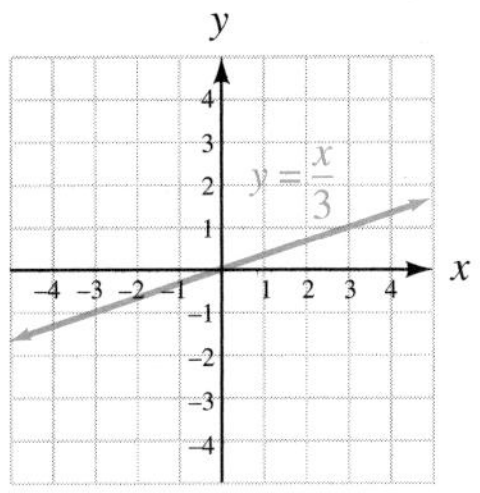

14.

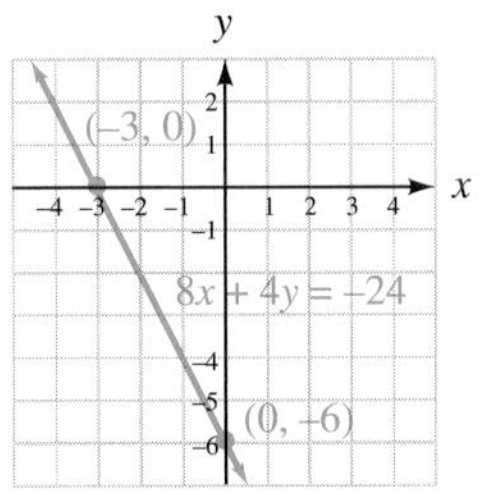

23.

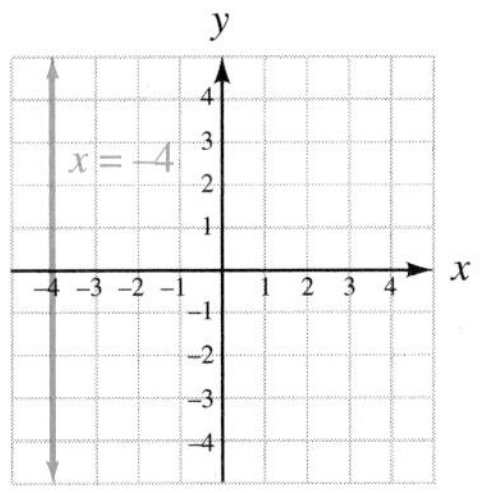

24.

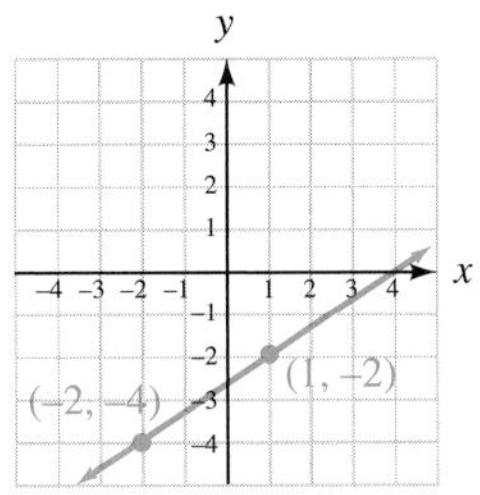

33.

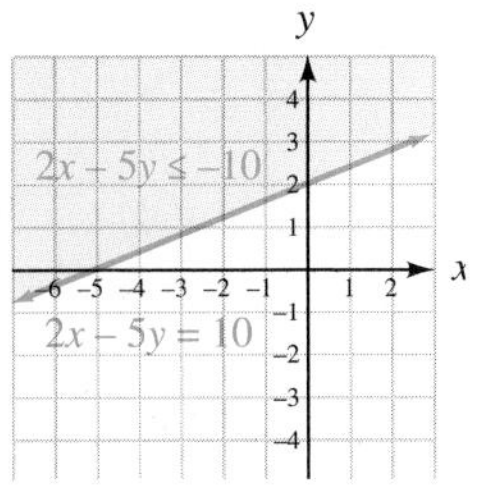

44.

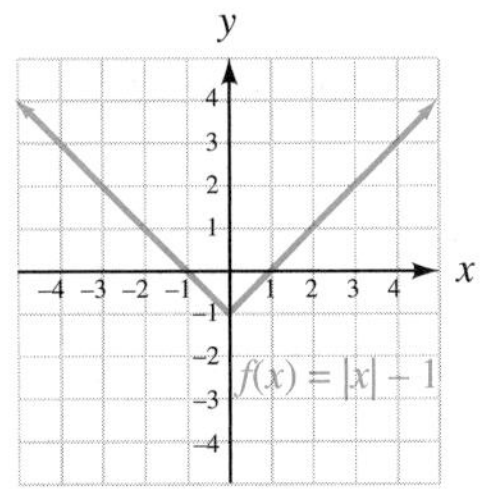

CUMULATIVE REVIEW CHAPTERS 1–3 (page 281)

38.

39.

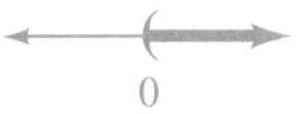

42.

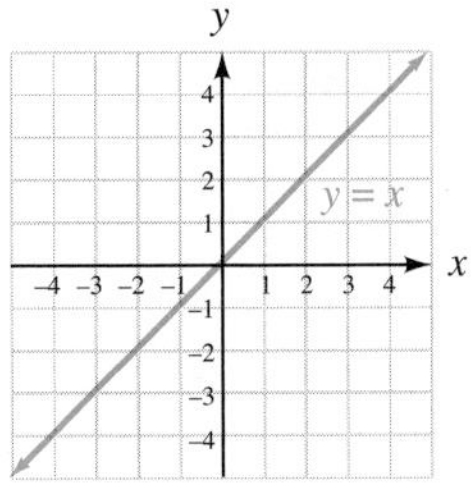

43.

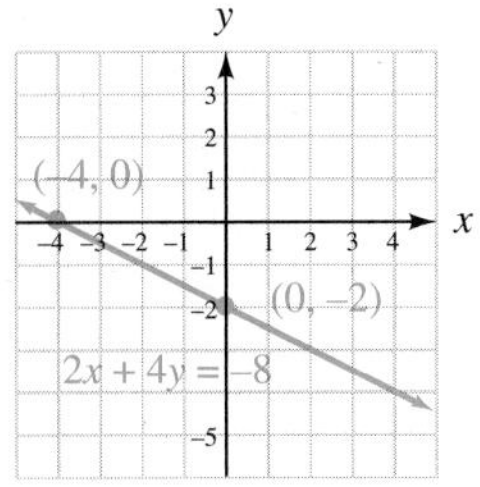

51.

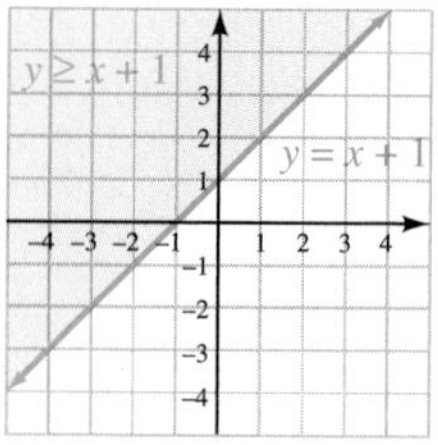

52.

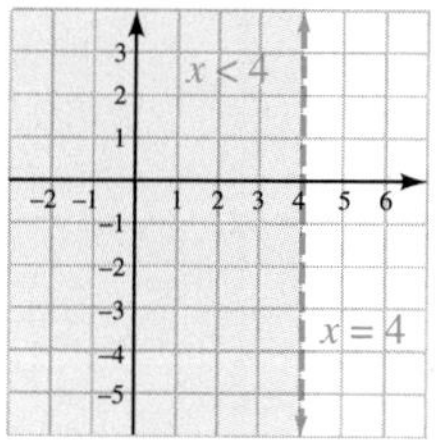

ARE YOU READY? 4.1 (page 284)

2.

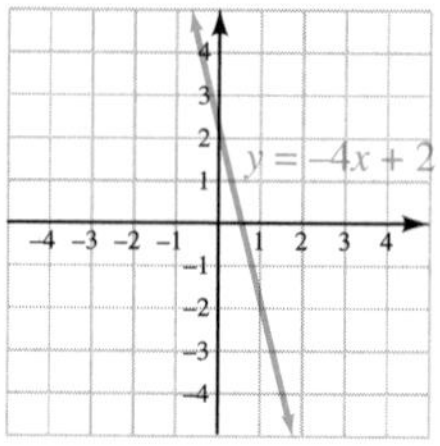

3.

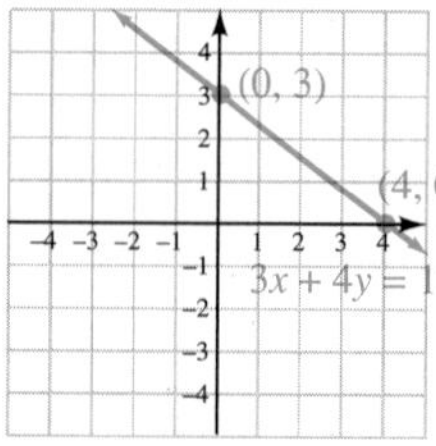

SELF CHECKS 4.1

2.

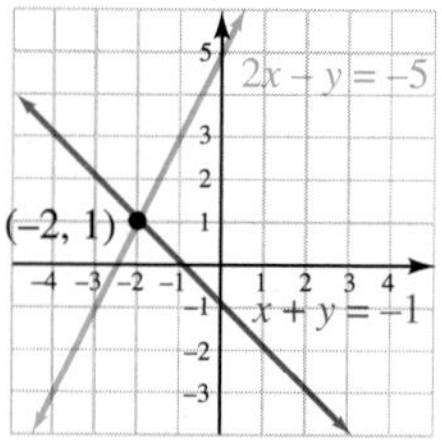

3.

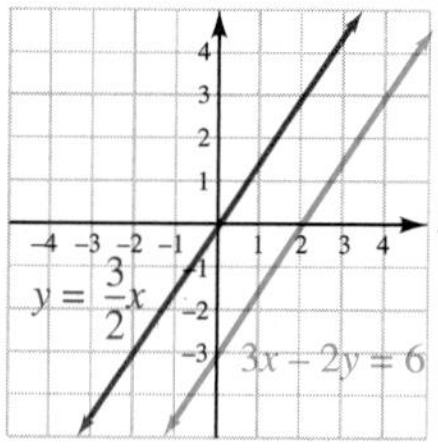

4.

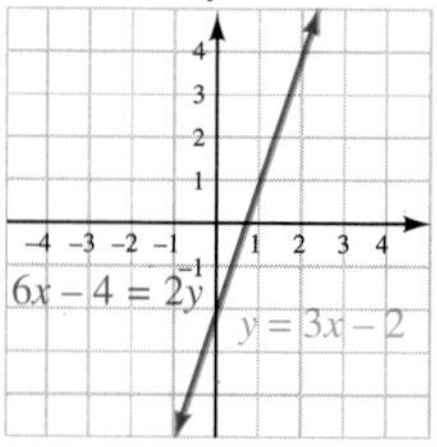

STUDY SET SECTION 4.1 (page 290)

25.

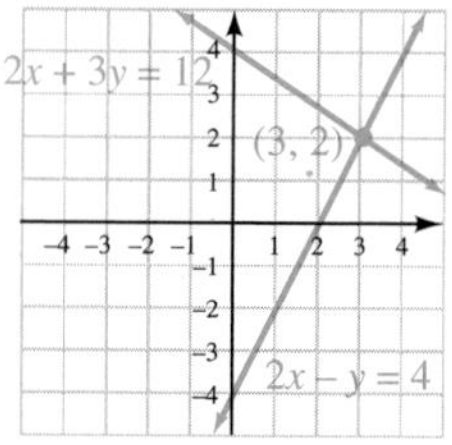

26.

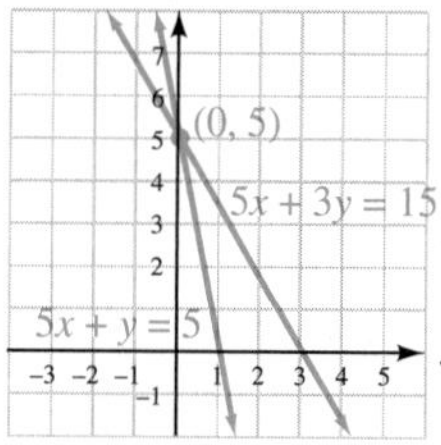

27.

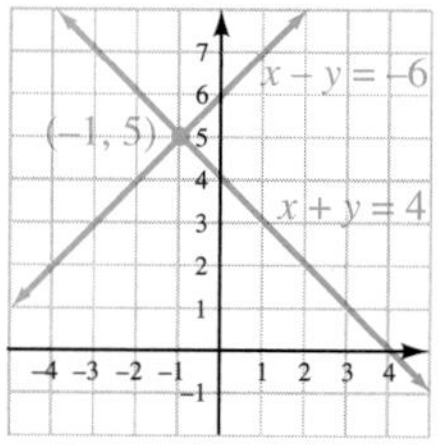

28.

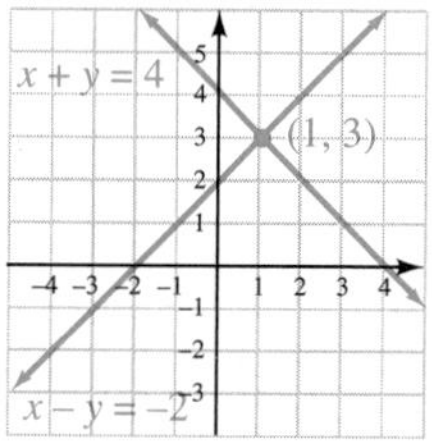

29.

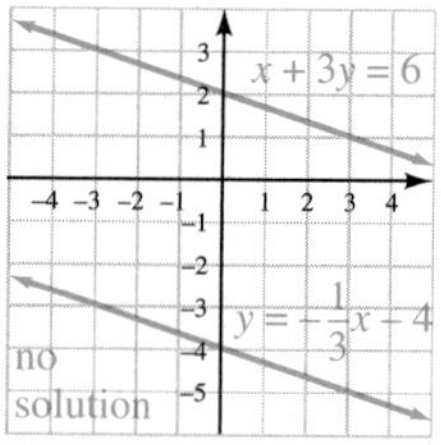

30.

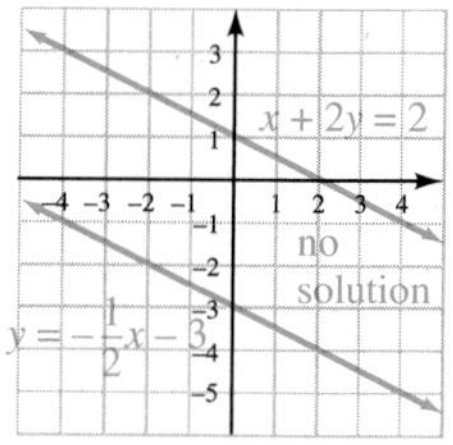

31.

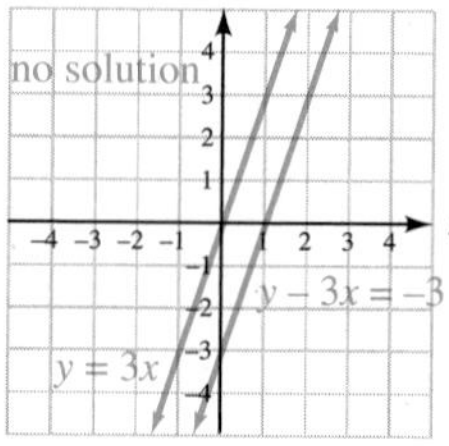

32.

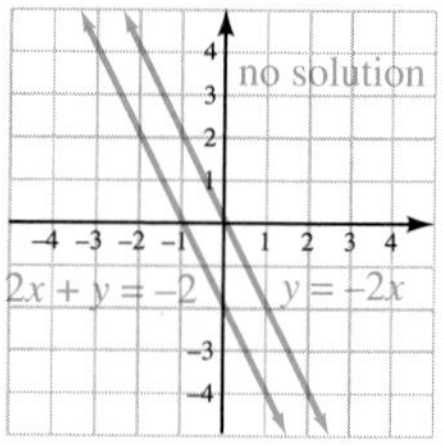

33.

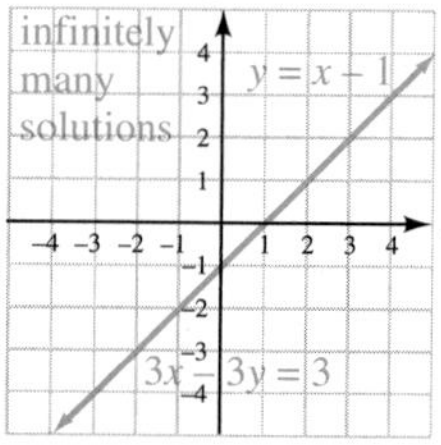

34.

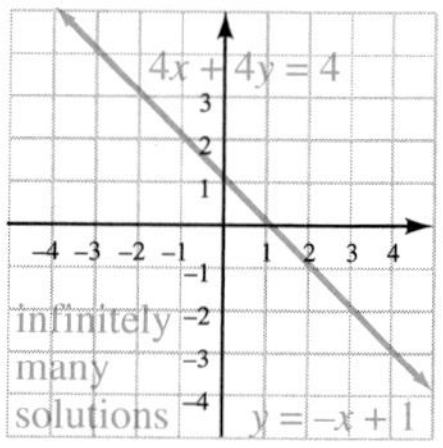

35.

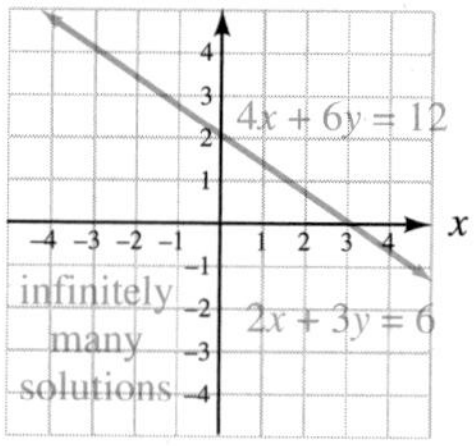

36.

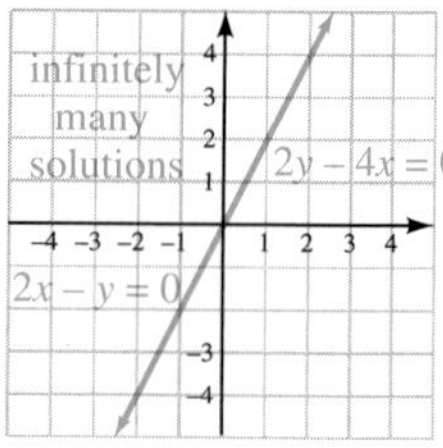

49.

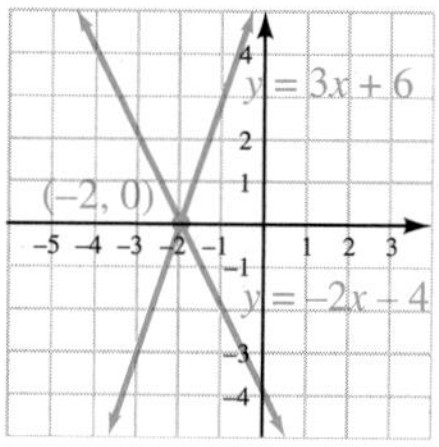

50.

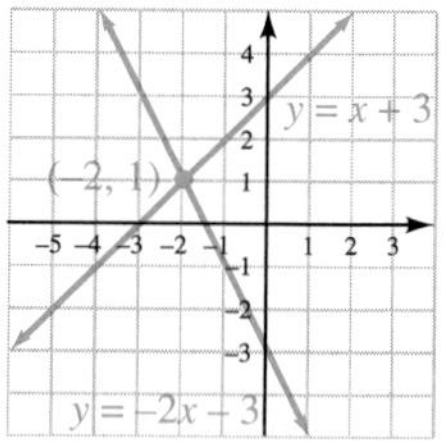

51.

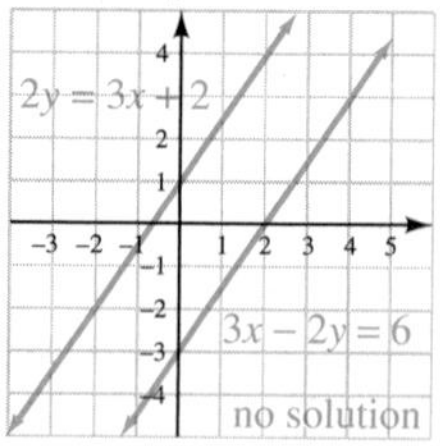

52.

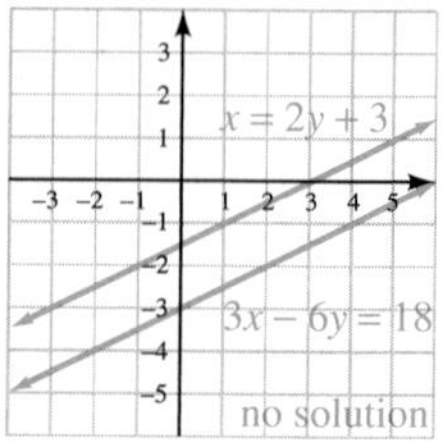

53.

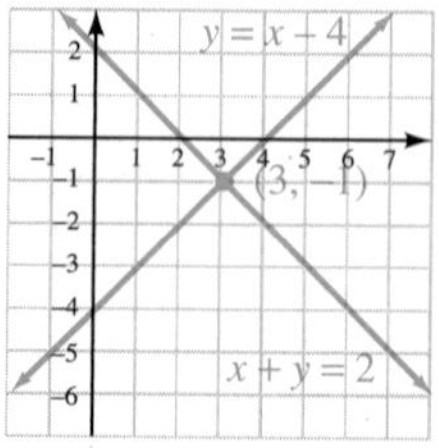

54.

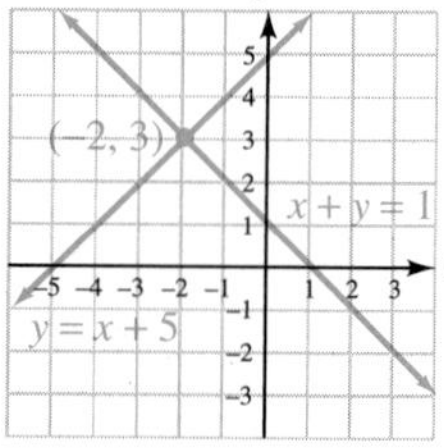

55.
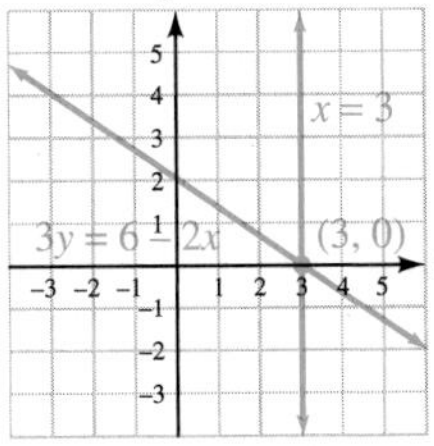

56.
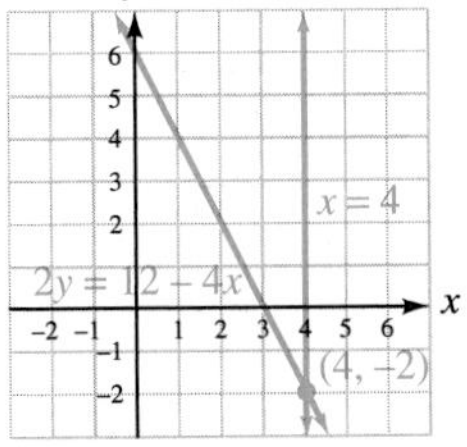

57.
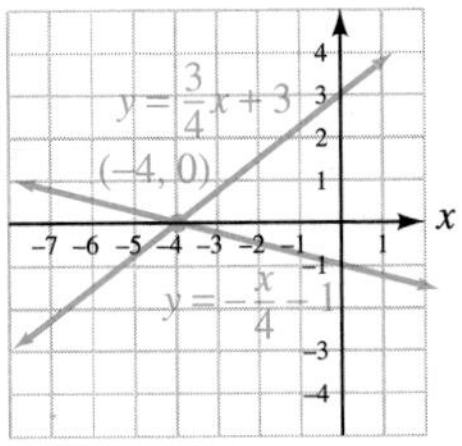

58.
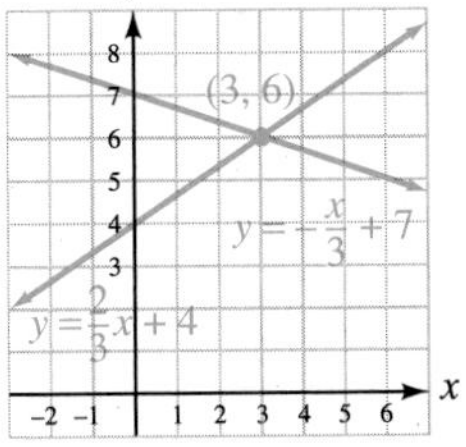

59.
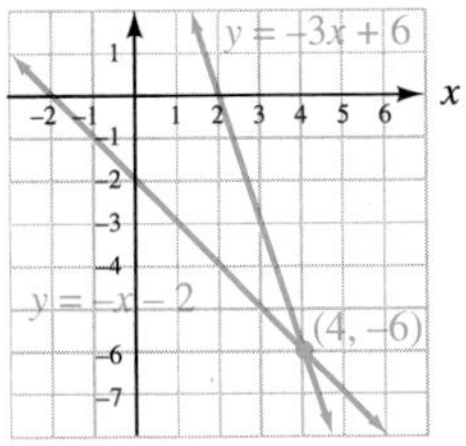

60.
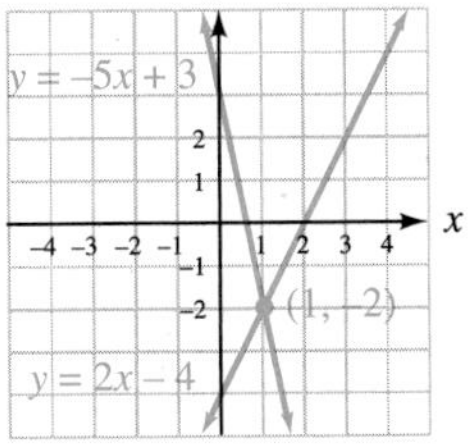

61.
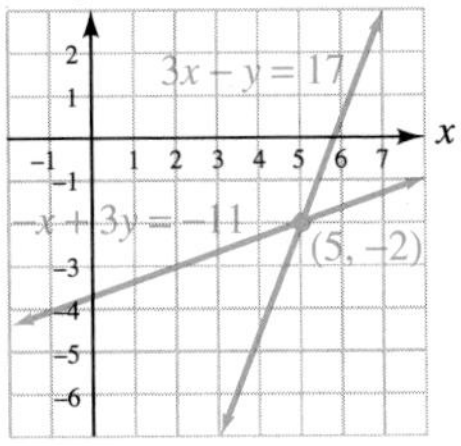

62.
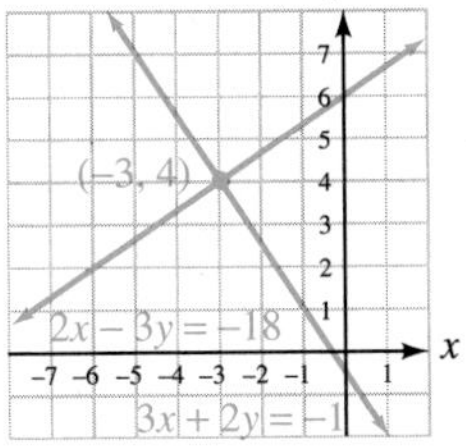

63.
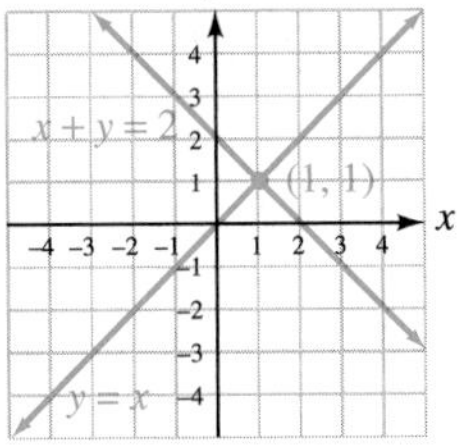

64.
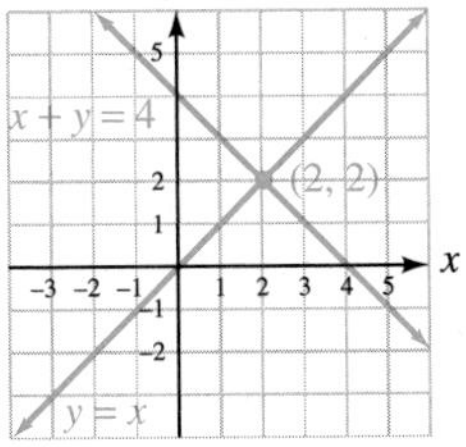

65.
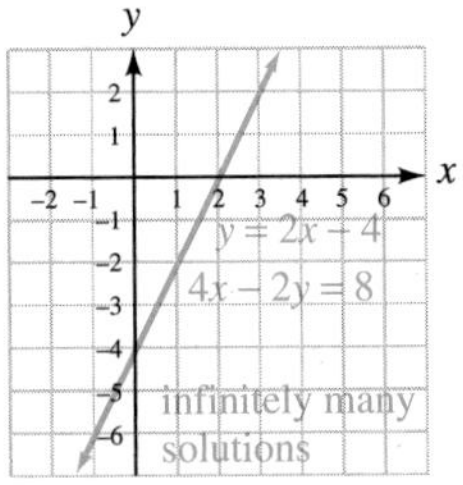

66.
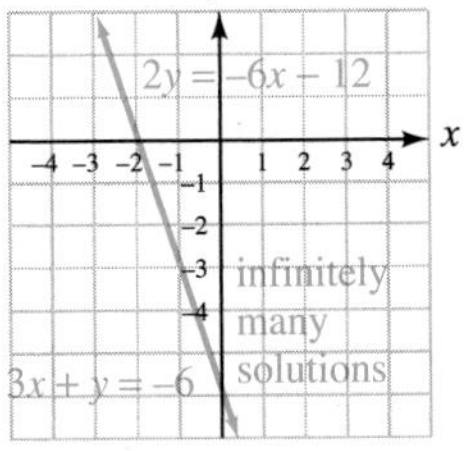

67.
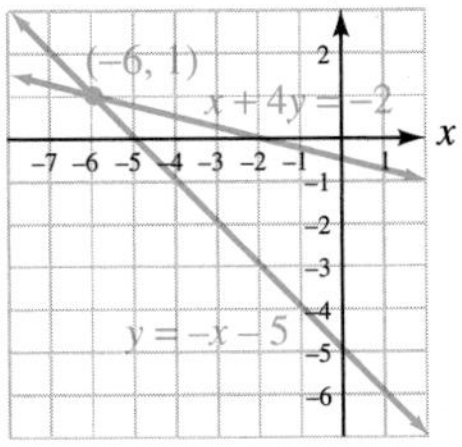

68.
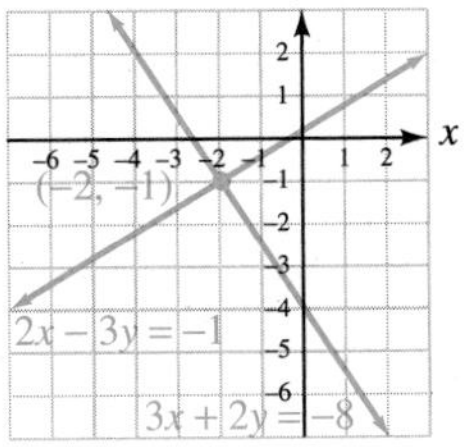

69.
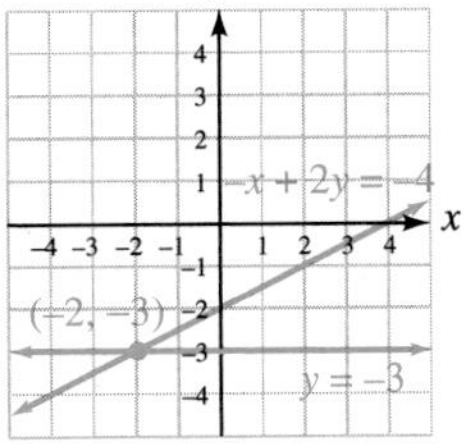

70.
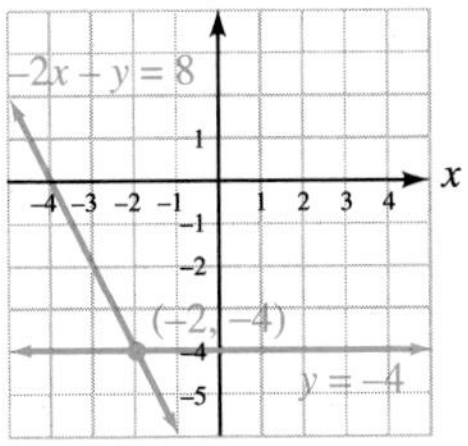

71.
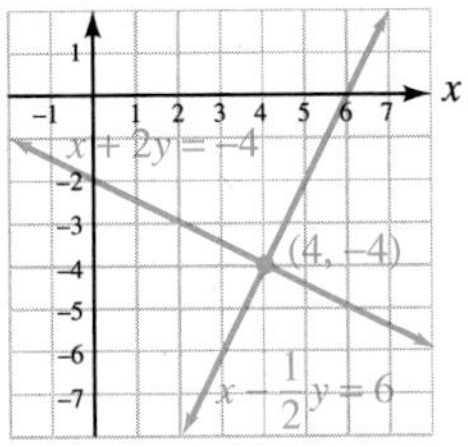

72.
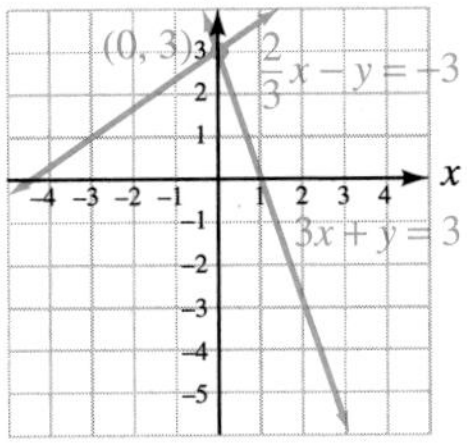

87.
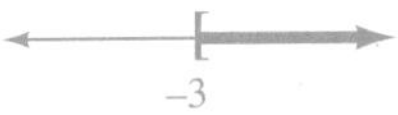

88.

89.
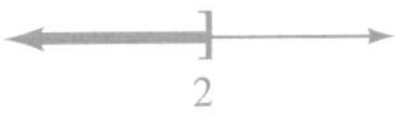

90.

94.
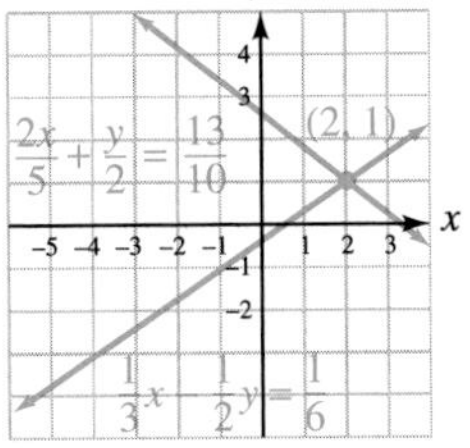

STUDY SET SECTION 4.4 (page 323)

55.

56.

57.

58.
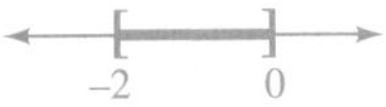

ARE YOU READY? 4.5 (page 328)

2.
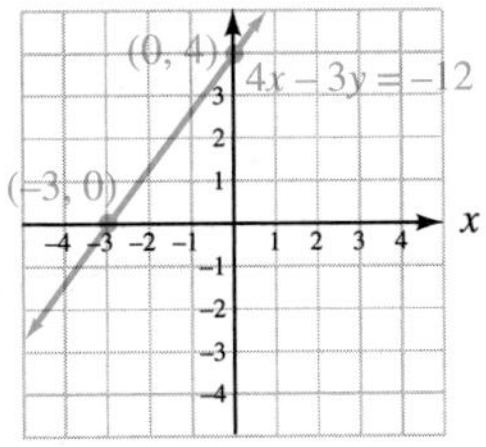

SELF CHECKS 4.5

1.
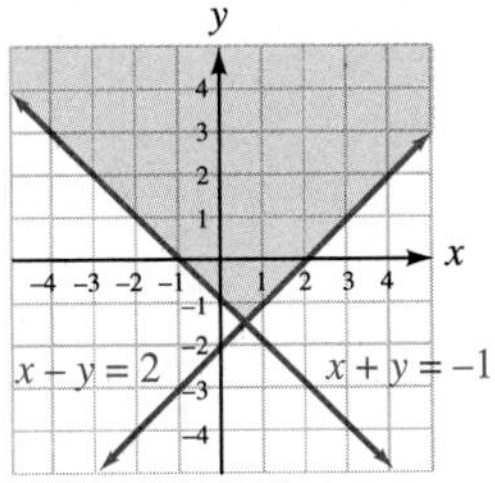

2.
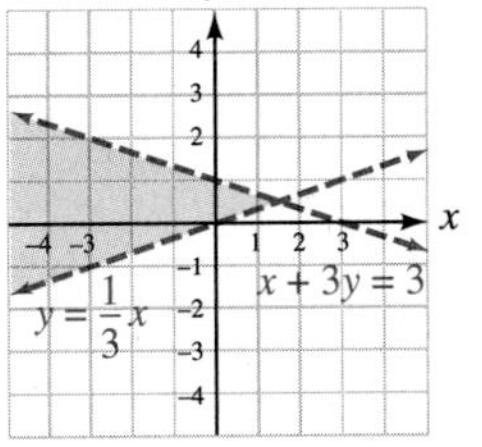

3.

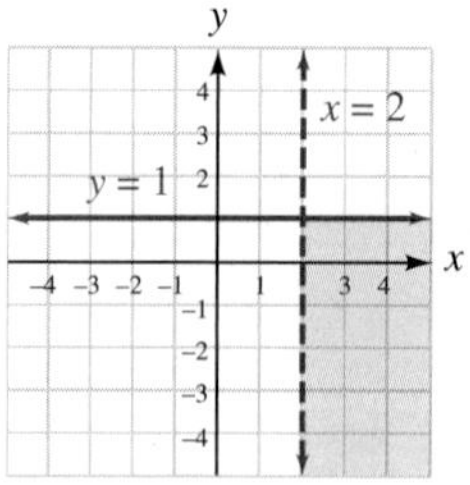

4.

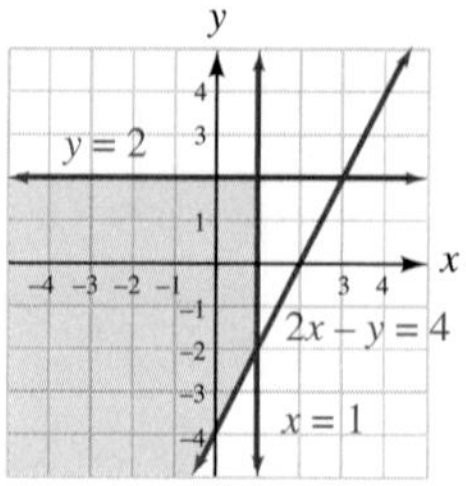

5.

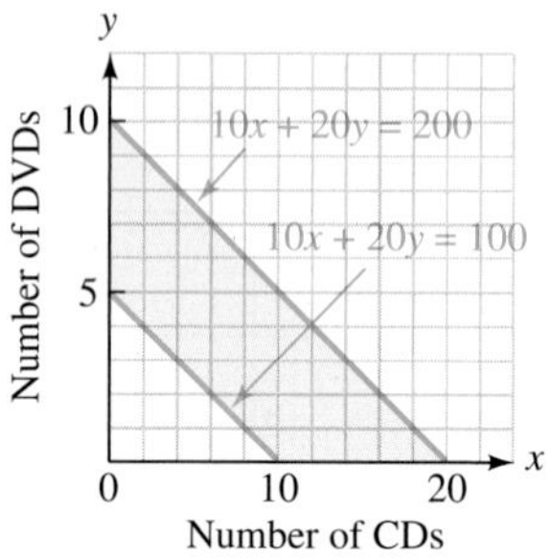

Any ordered pair in the shaded region with whole number coordinates is a possible solution

23.

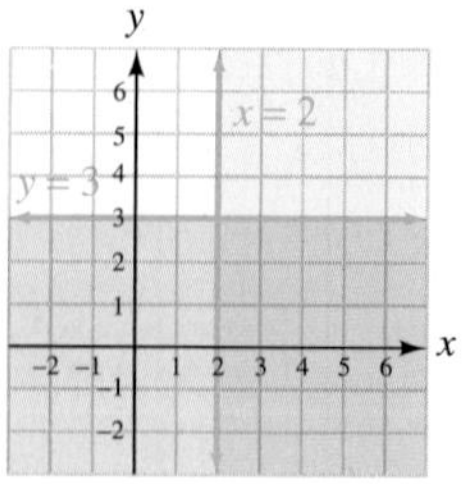

24.

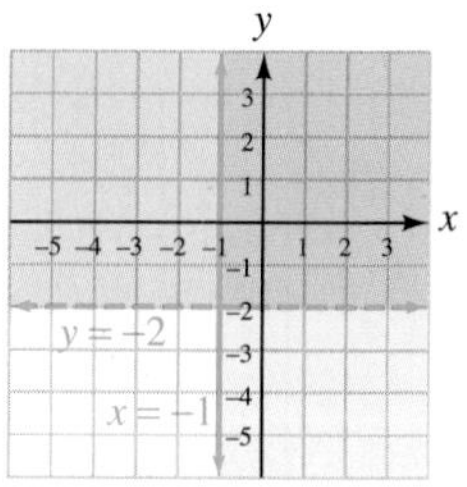

25.

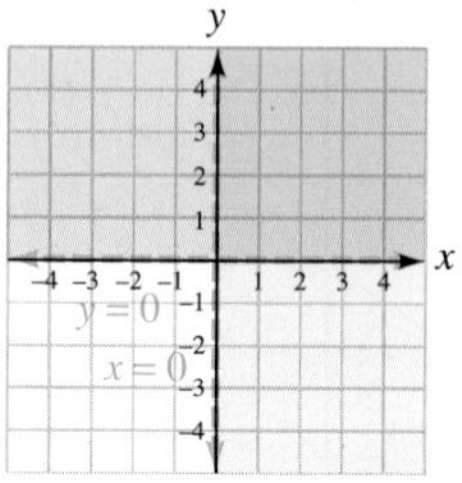

26.

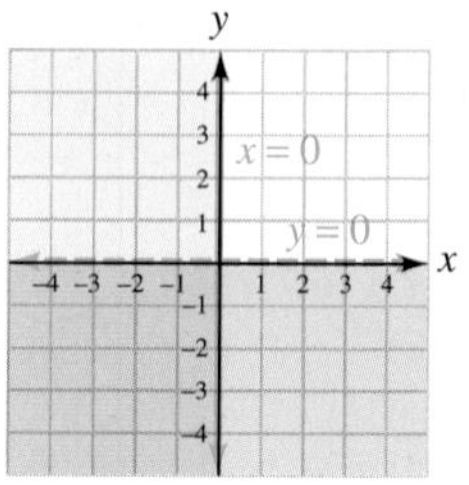

27.

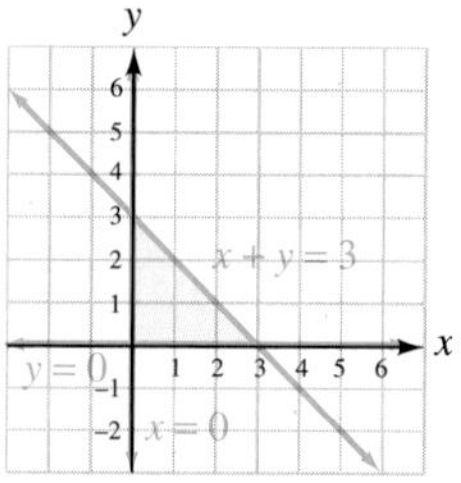

28.

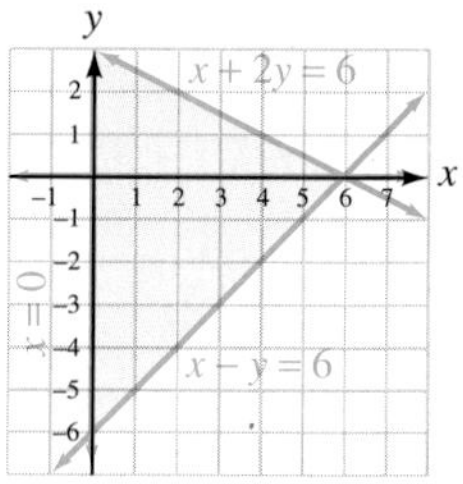

29.

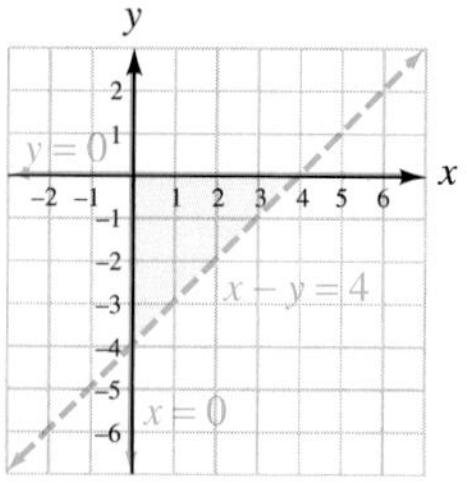

30.

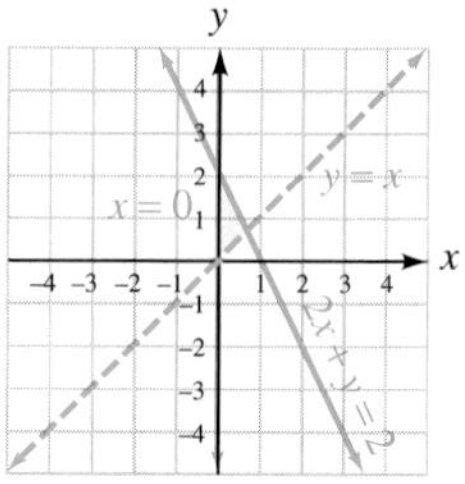

31.

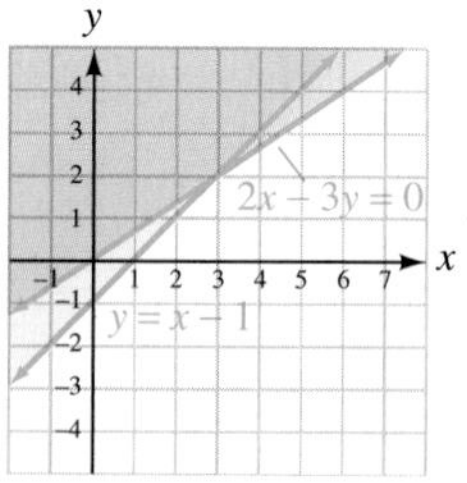

32.

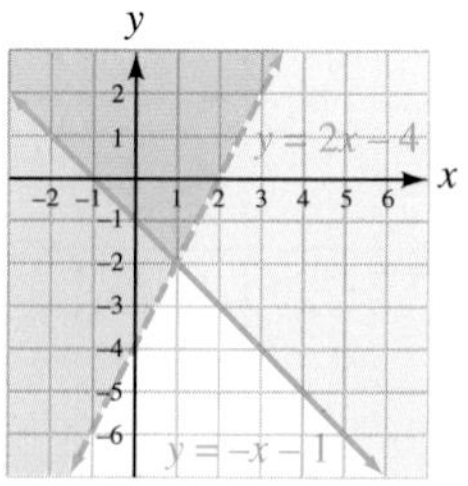

33.

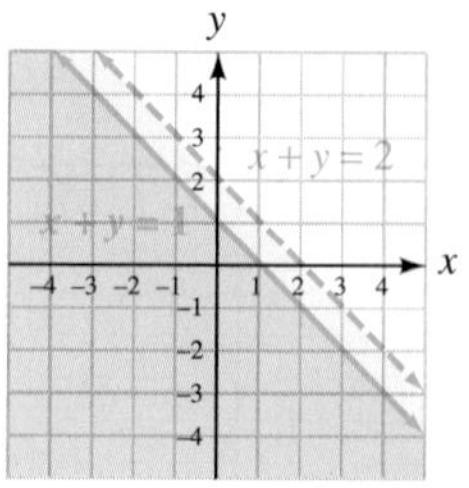

34.

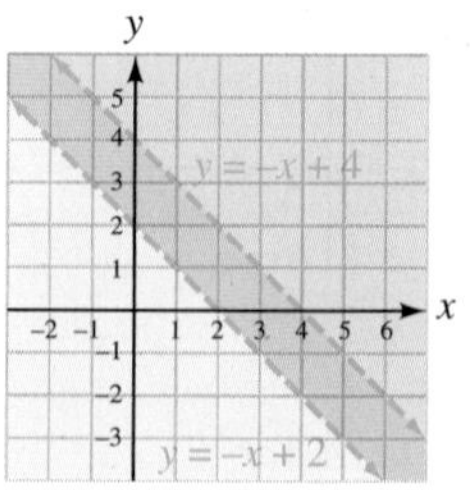

35.

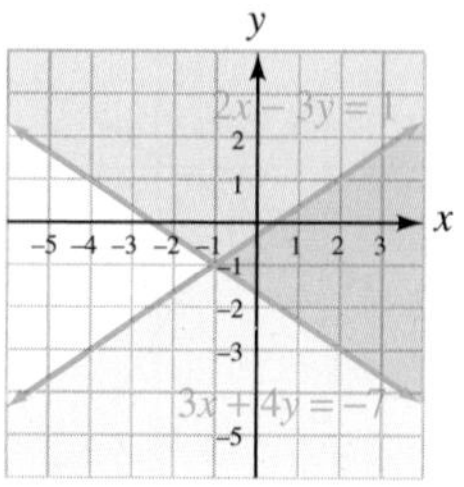

36.

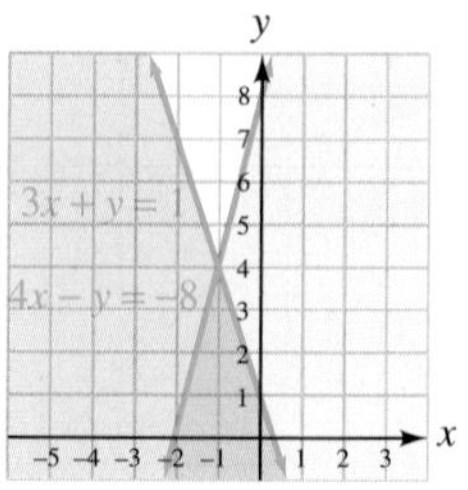

STUDY SET SECTION 4.5 (page 334)

15.

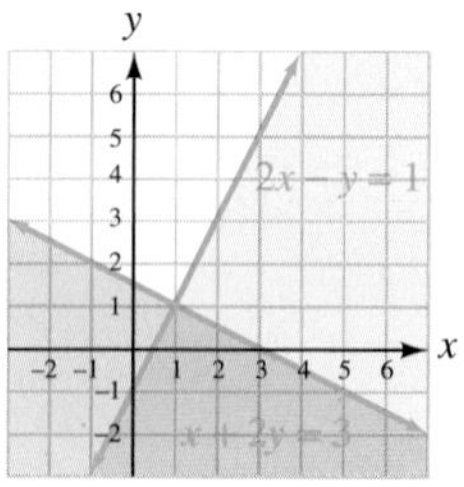

16.

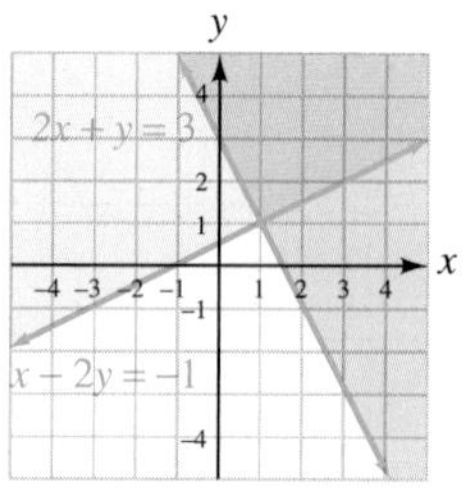

17.

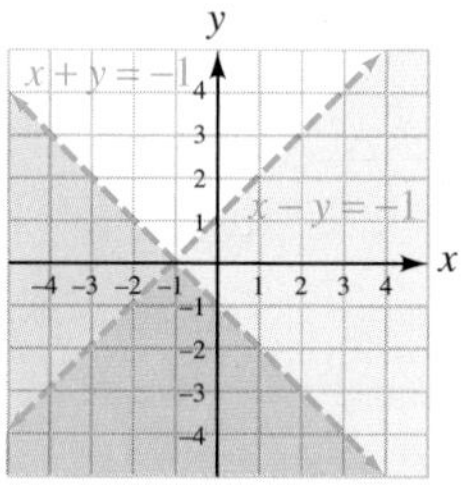

18.

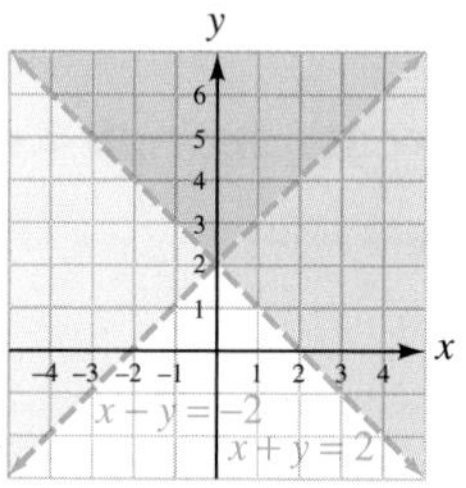

19.

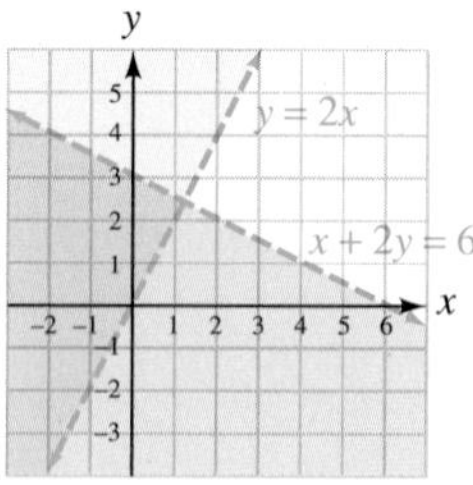

20.

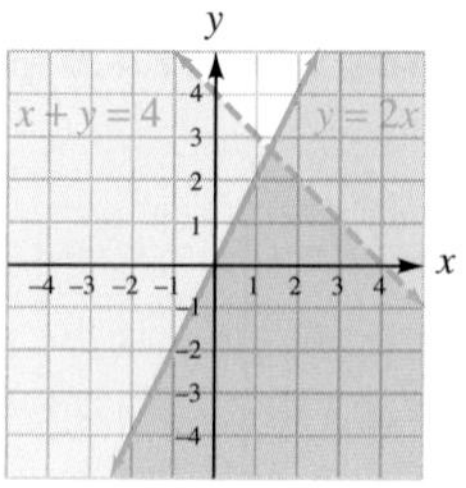

21.

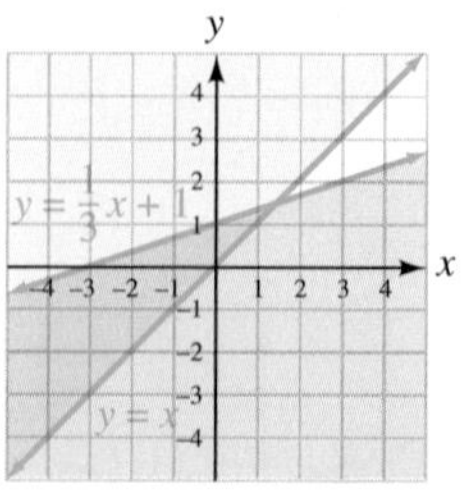

22.

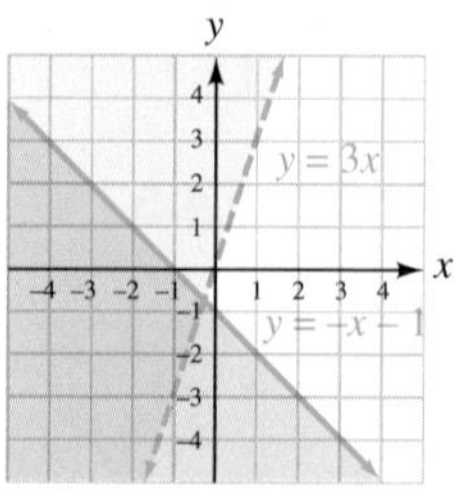

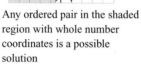

37.

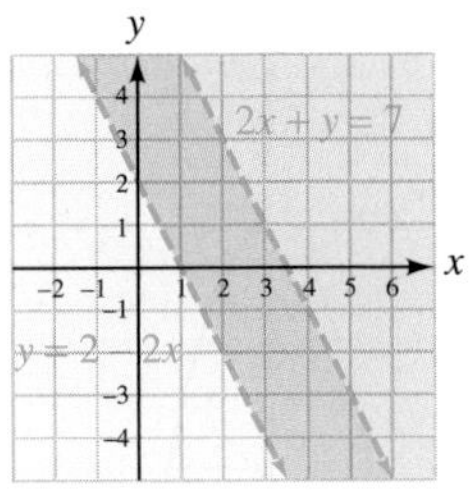

38.

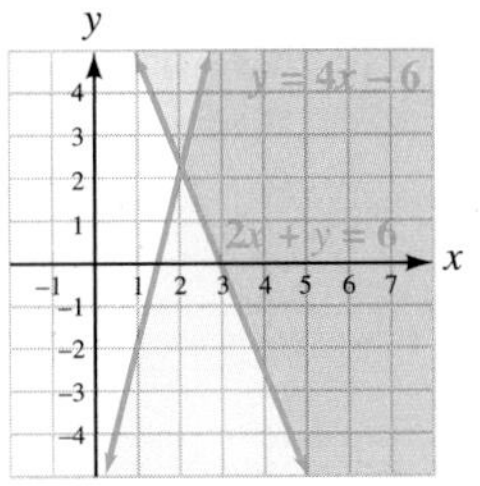

39.

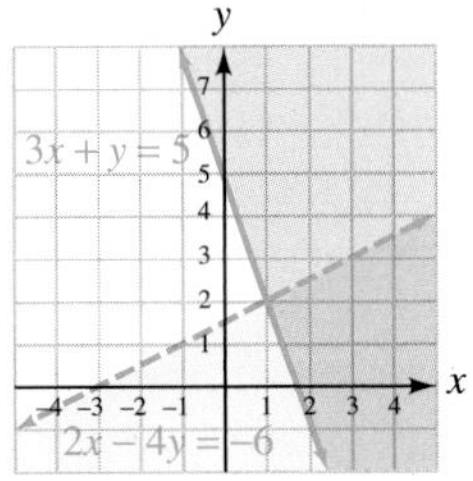

40.

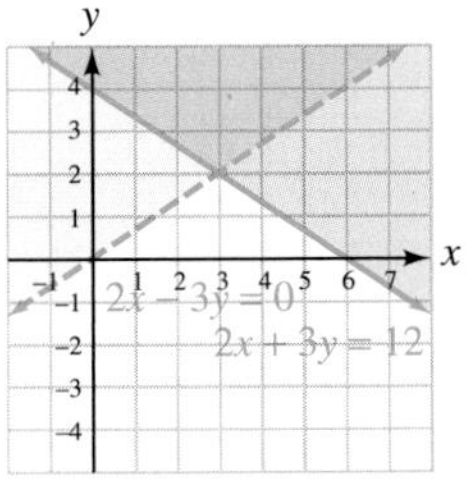

41.

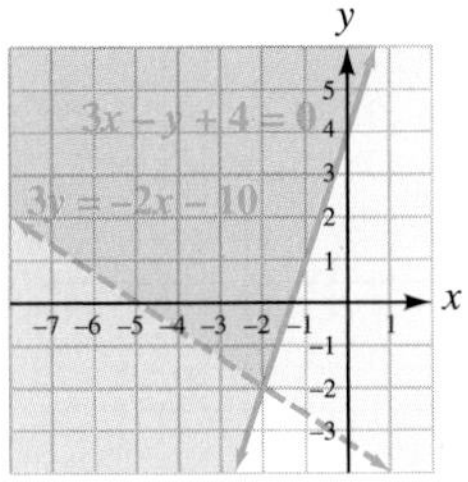

42.

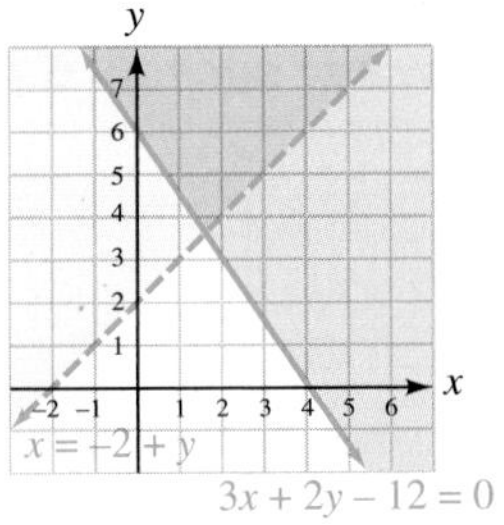

43.

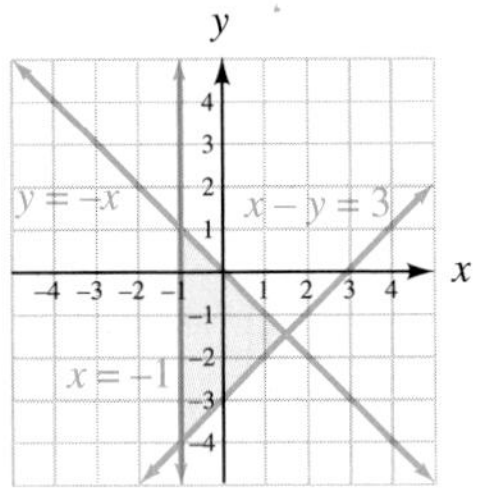

44.

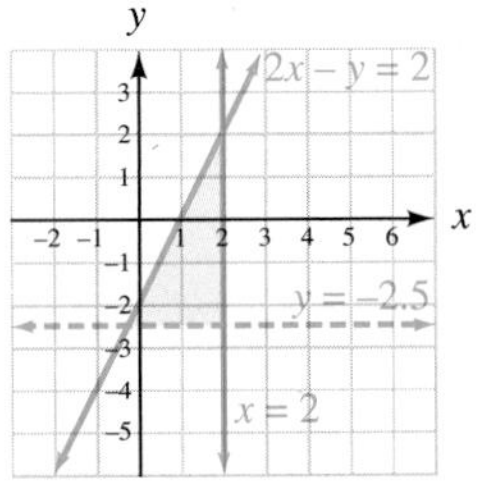

45.

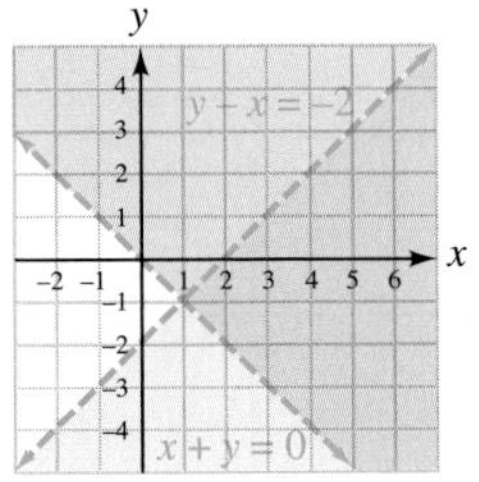

46.

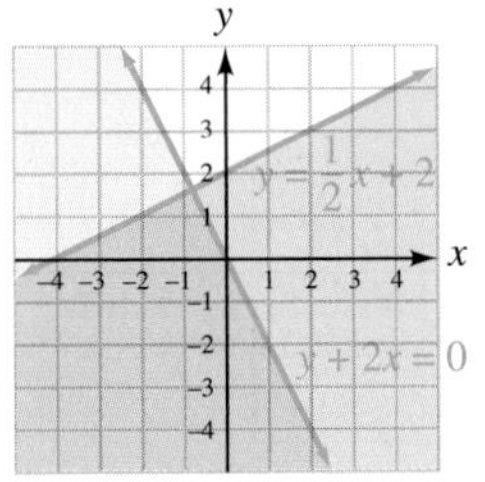

47. a.

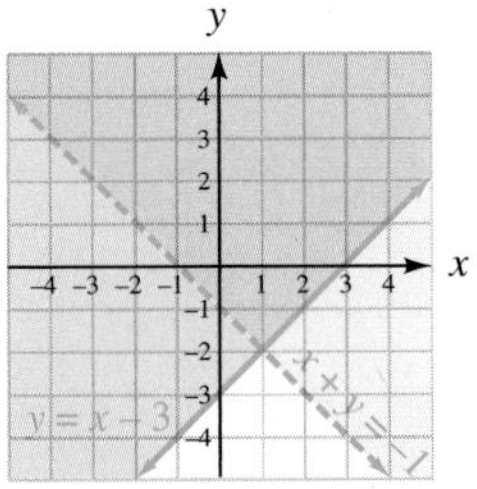

48.

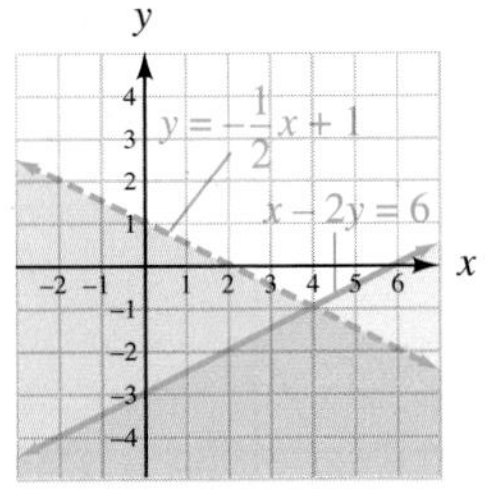

65.

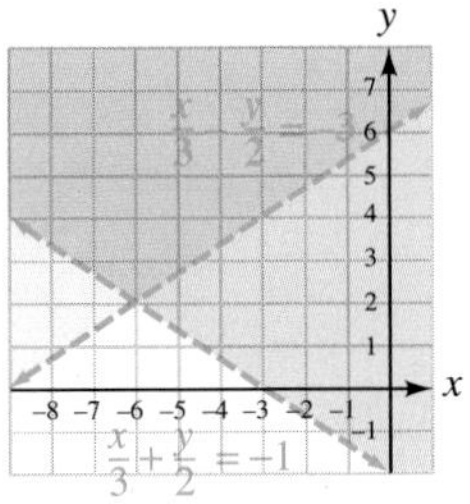

66.

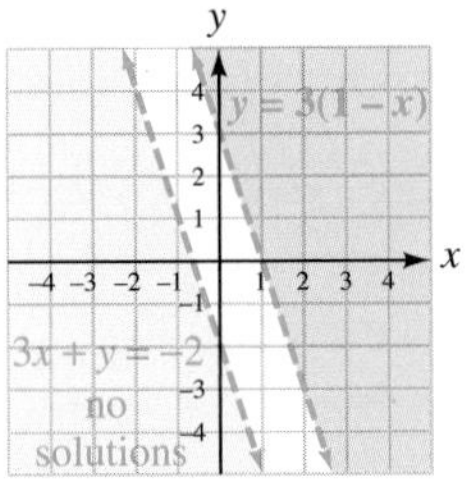

67.

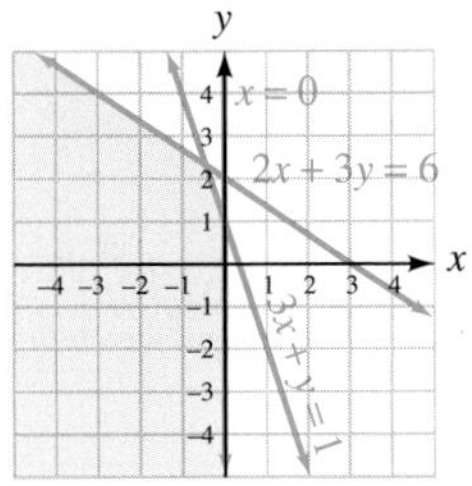

68.

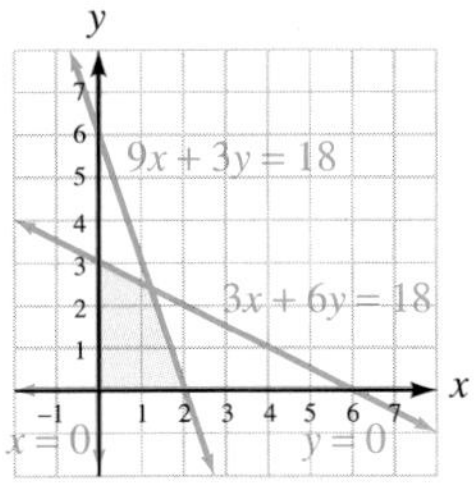

CHAPTER 4 REVIEW (page 337)

3.

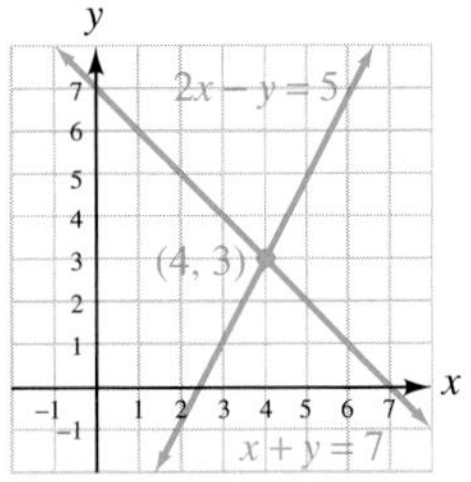

4.

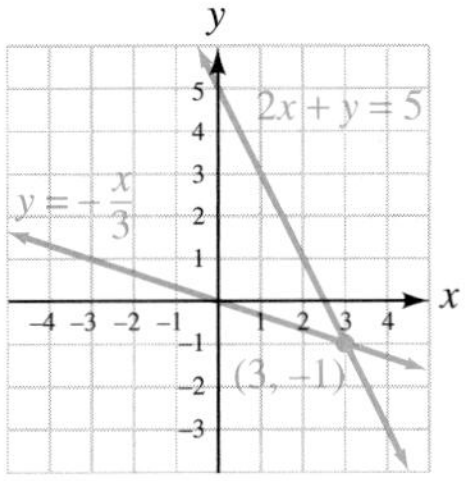

5.

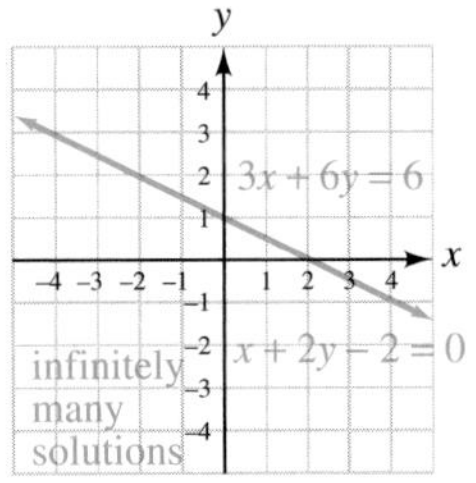

6.

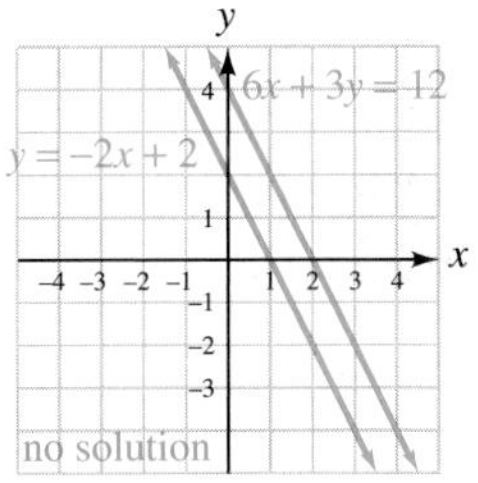

39.

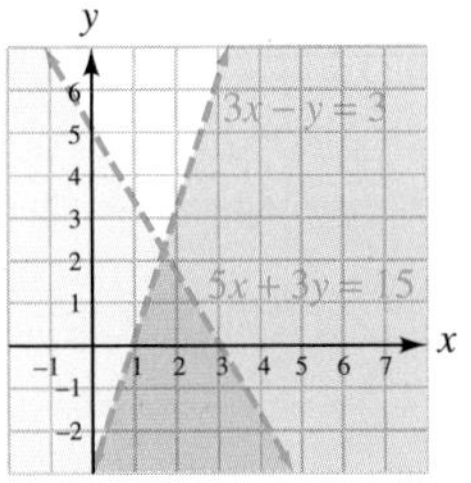

40.

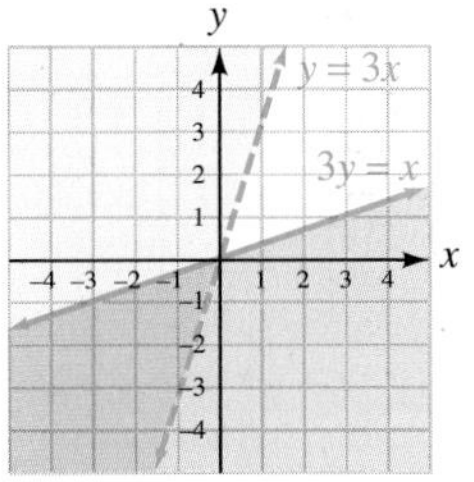

41.

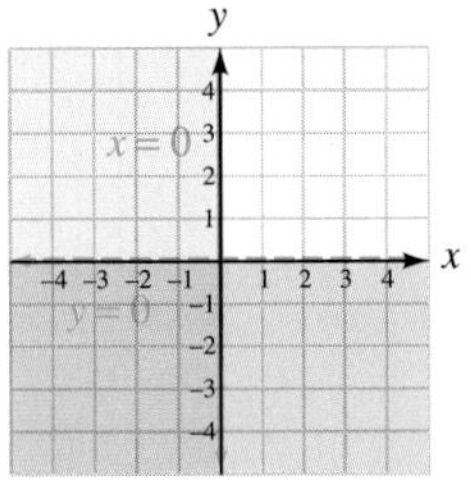

42.

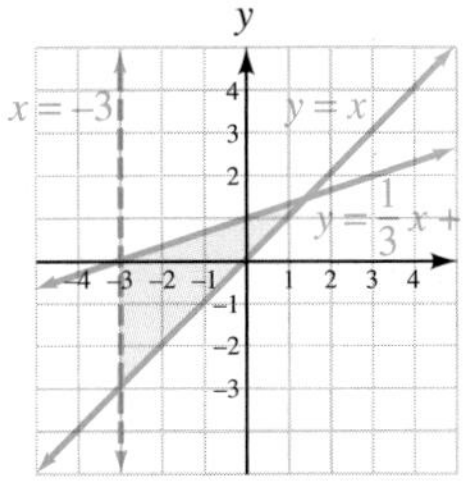

CHAPTER 4 TEST (page 342)

5.

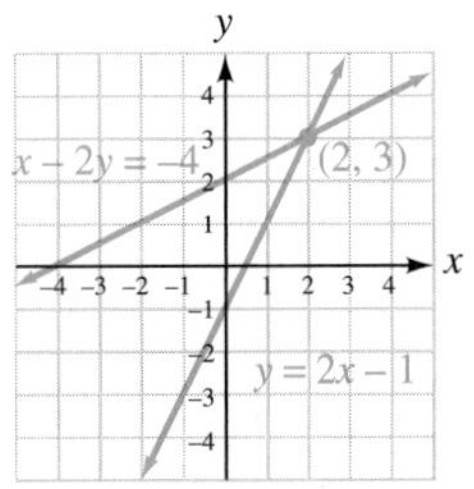

6.

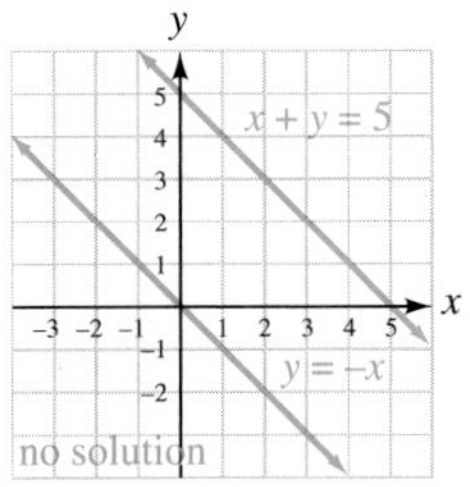

29.

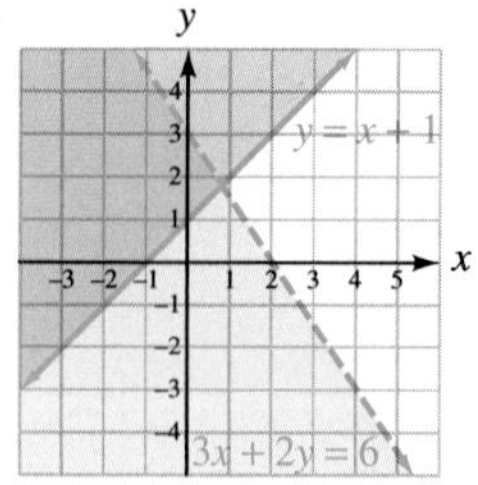

30.

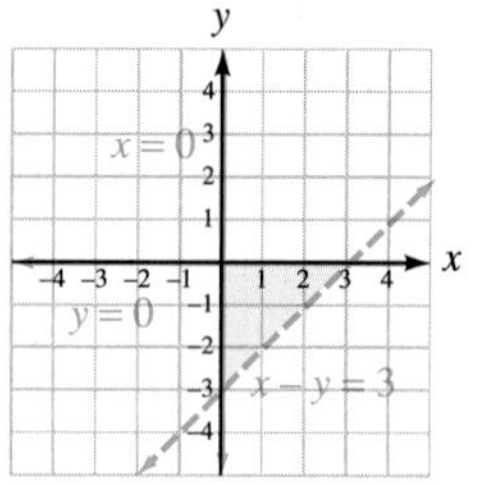

31.

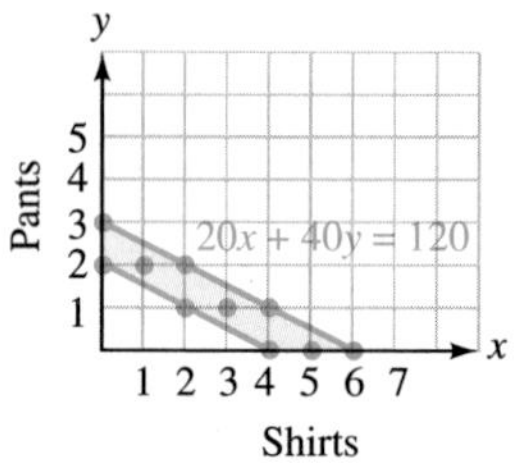

5.

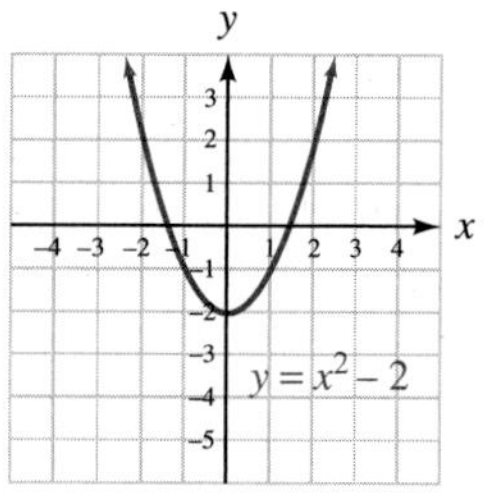

6.

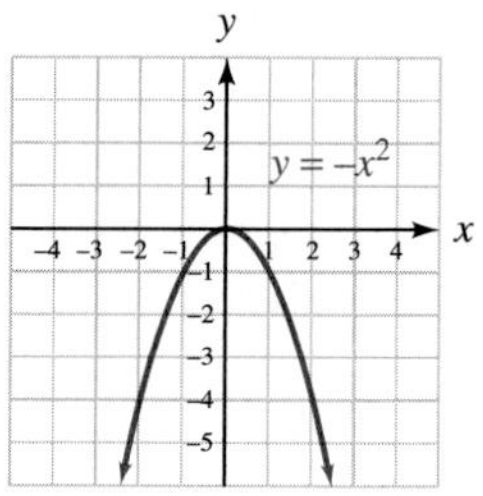

7.

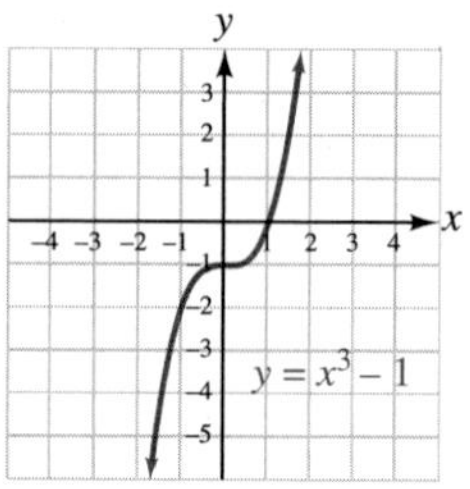

CUMULATIVE REVIEW CHAPTERS 1–4 (page 344)

6.

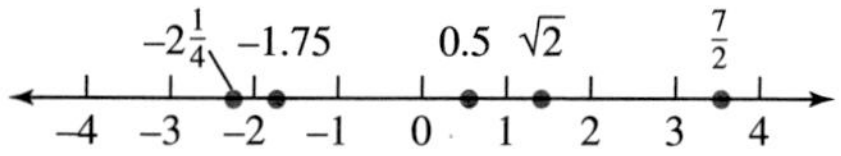

28.

33.

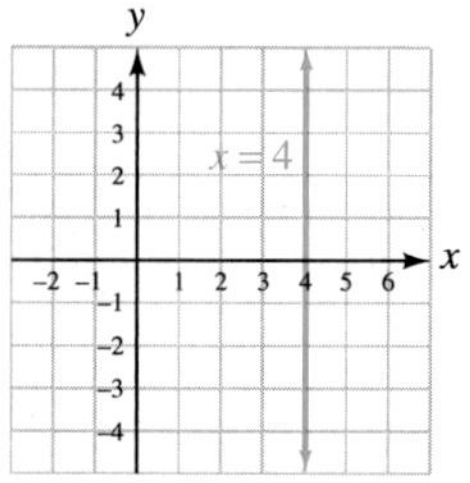

34.

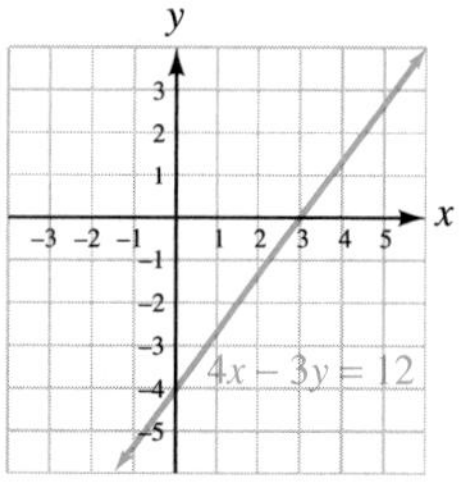

44.

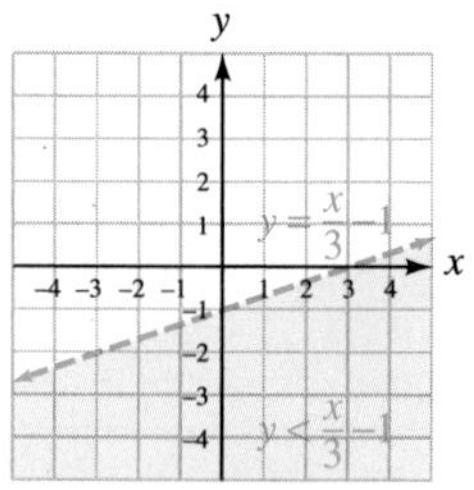

47.

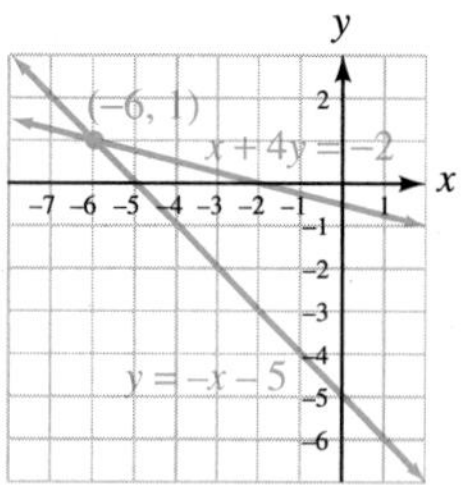

48.

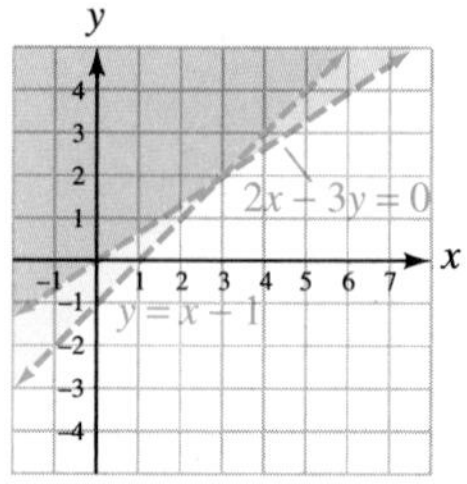

SELF CHECKS 5.4

1. a. A trinomial in one variable of degree 2 written in descending powers of x; terms: x^2, $4x$, -16; coefficients: 1, 4, -16; degree of terms: 2, 1, 0 **b.** A binomial in two variables of degree 7 written in descending powers of s and ascending powers of t; terms: $-14s^5t$, s^4t^3; coefficients: -14, 1; degree of terms: 6, 7

STUDY SET SECTION 5.4 (page 382)

69.

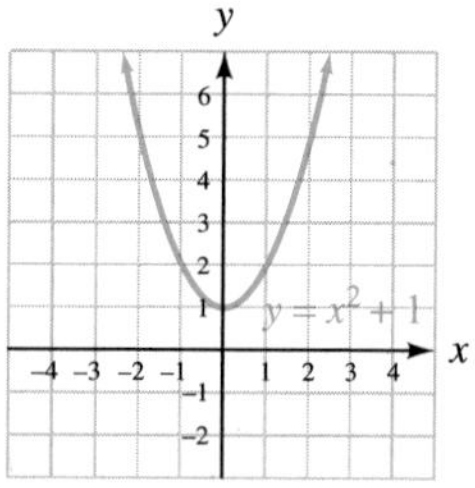

70.

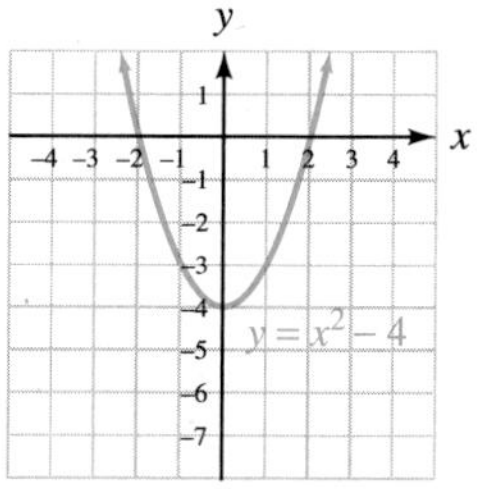

71.

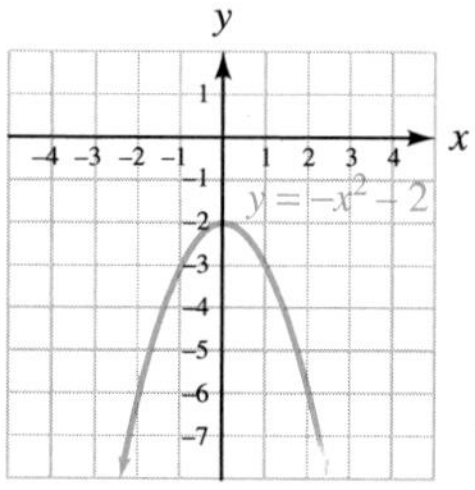

72.

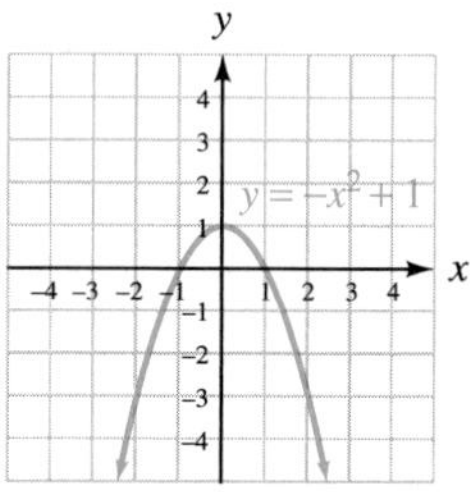

73.

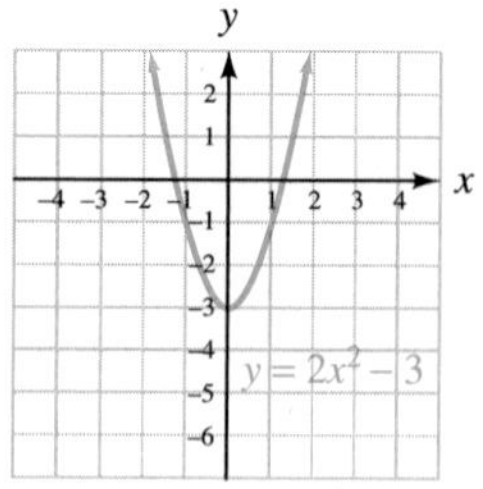

74.

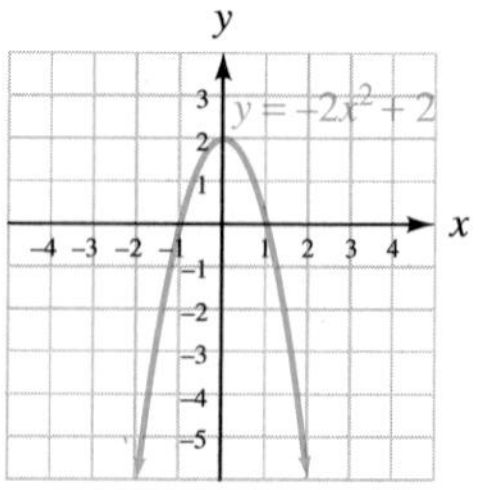

75.

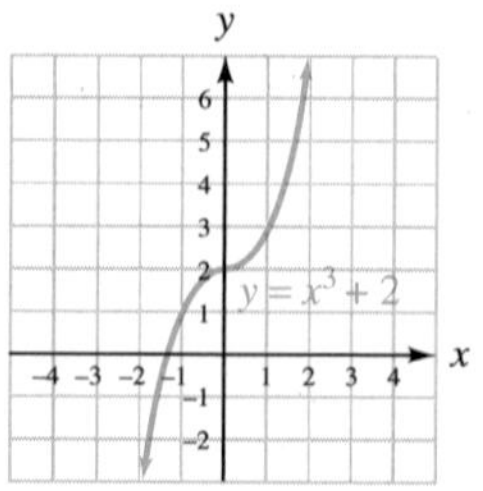

76.

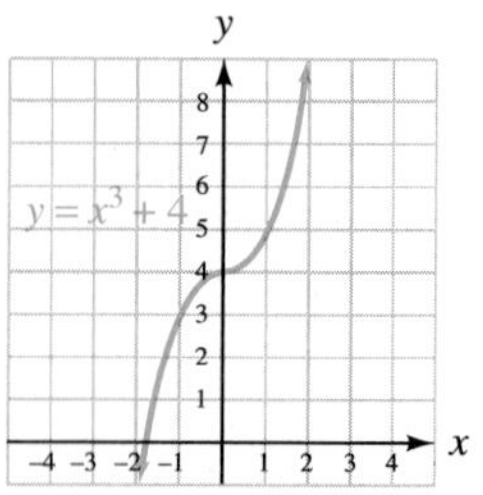

77.

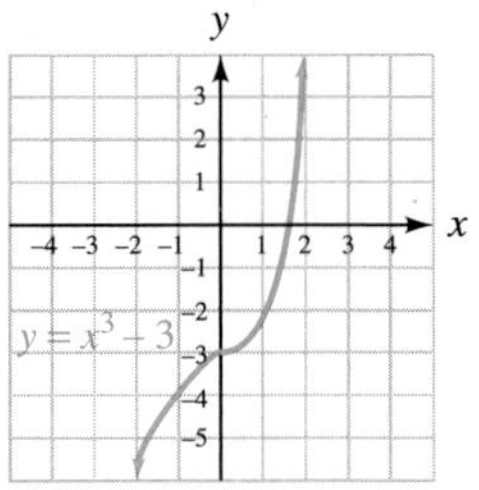

78.

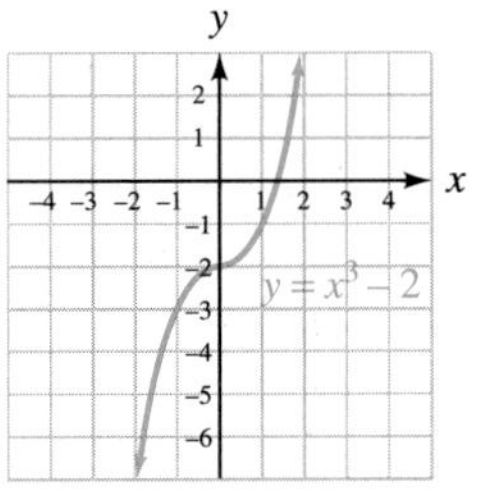

79.

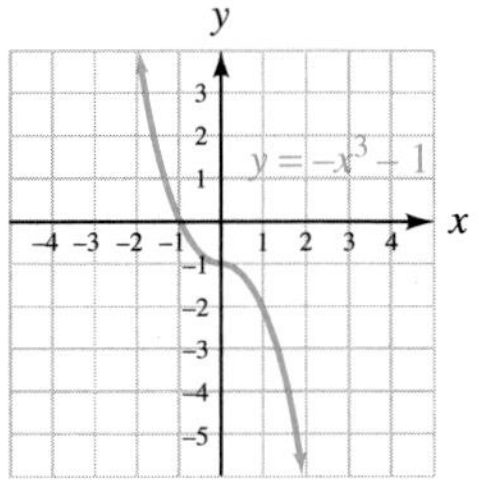

80.

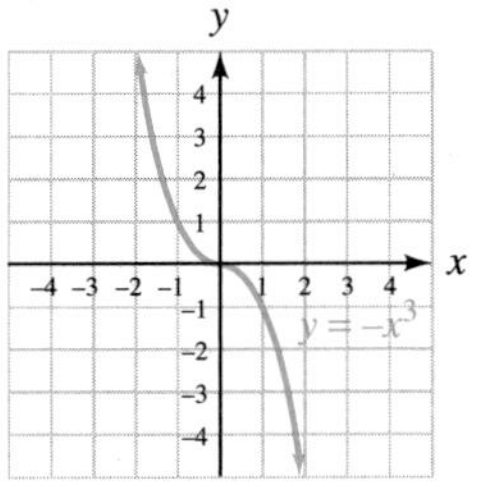

87.

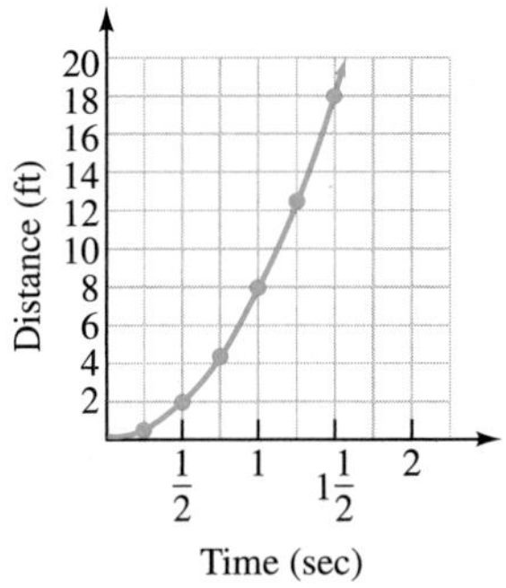

93.

94.

100.

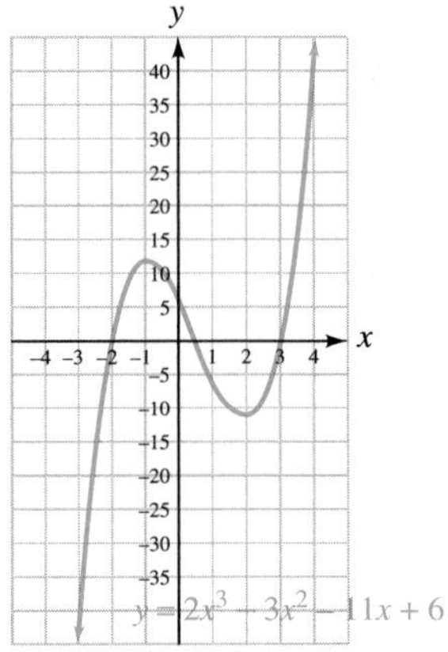

STUDY SET SECTION 5.5 (page 391)

119.

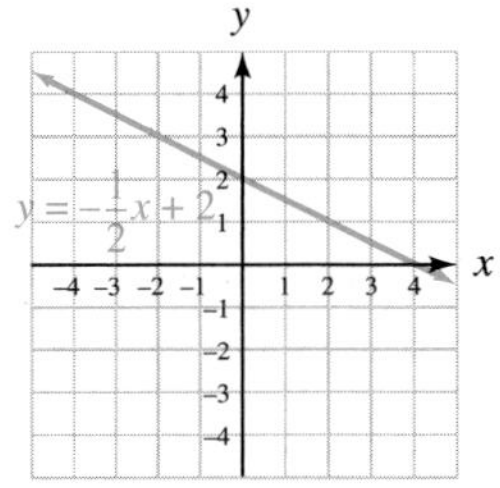

120.

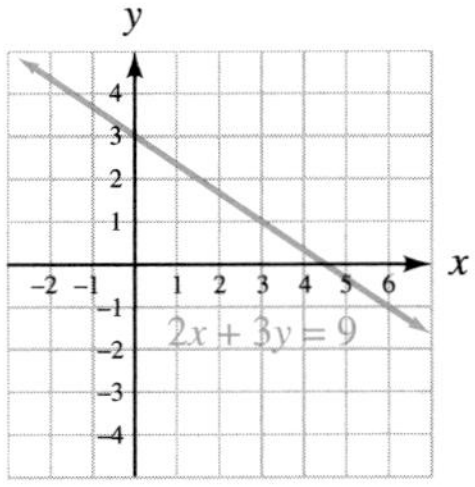

CHAPTER 5 REVIEW (page 420)

55.

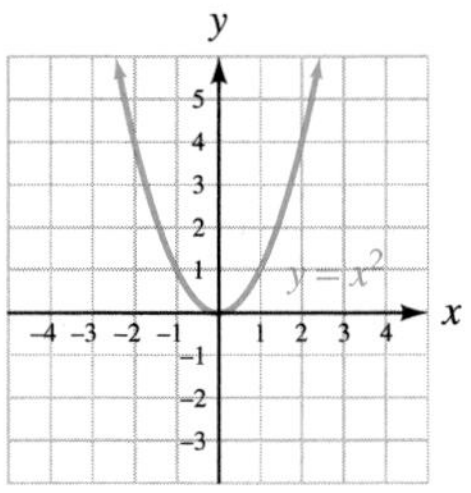

56.

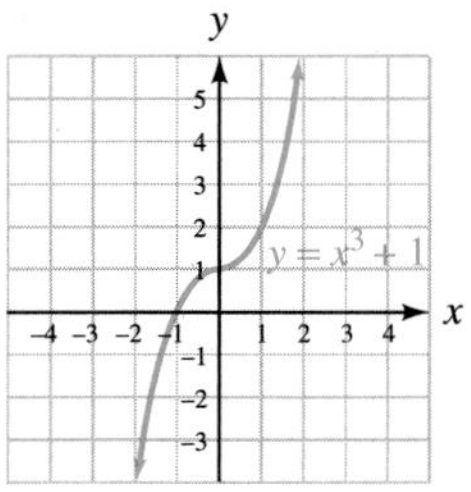

CHAPTER 5 TEST (page 427)

19.

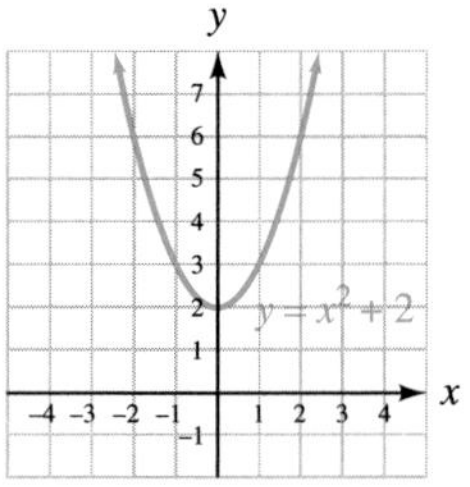

CUMULATIVE REVIEW CHAPTERS 1–5 (page 429)

14.

15.

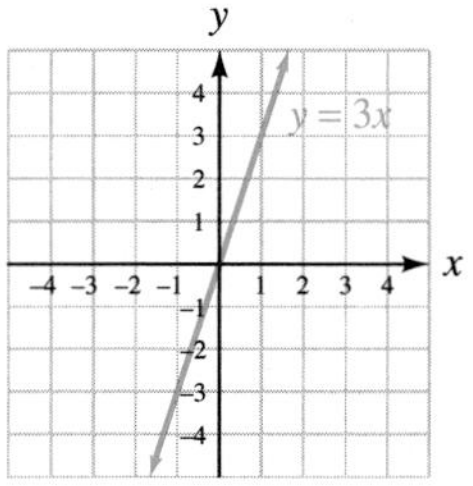

16.

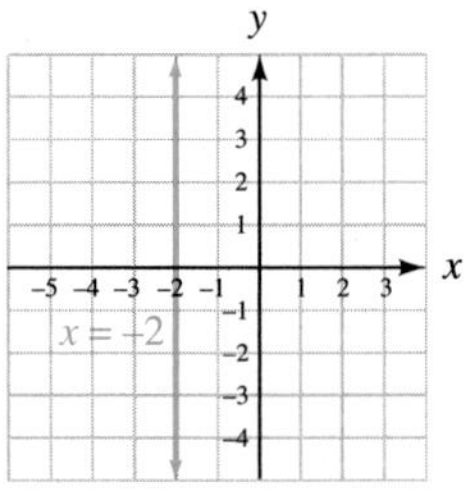

26.

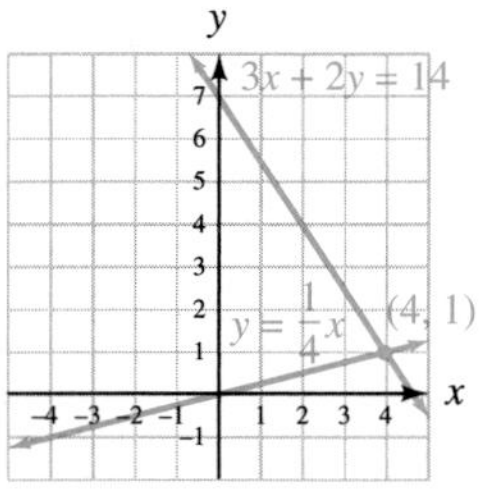

30.

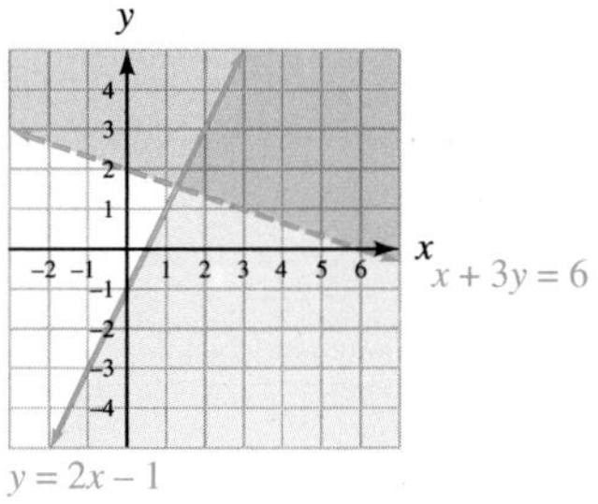

41.

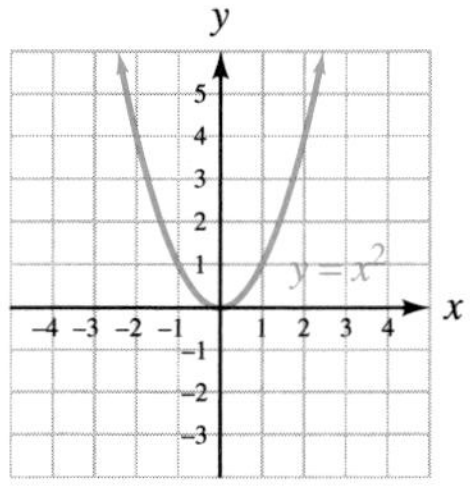

STUDY SET SECTION 6.6 (page 481)

81.

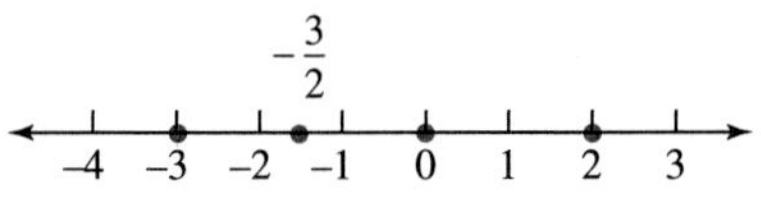

82.

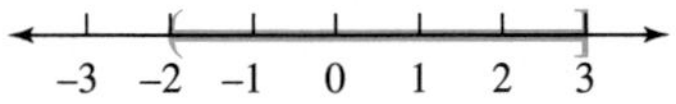

83.

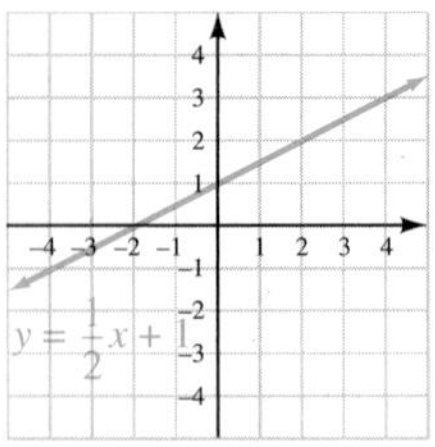

84.

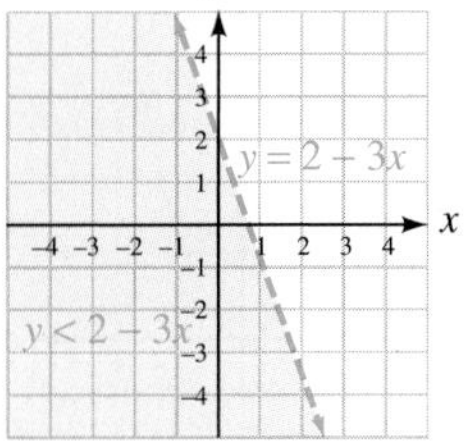

CUMULATIVE REVIEW CHAPTERS 1–6 (page 508)

28.

30.

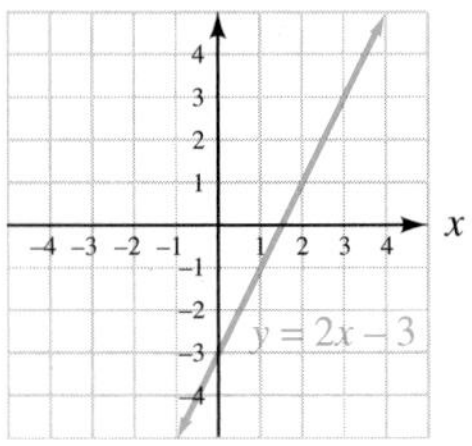

36.

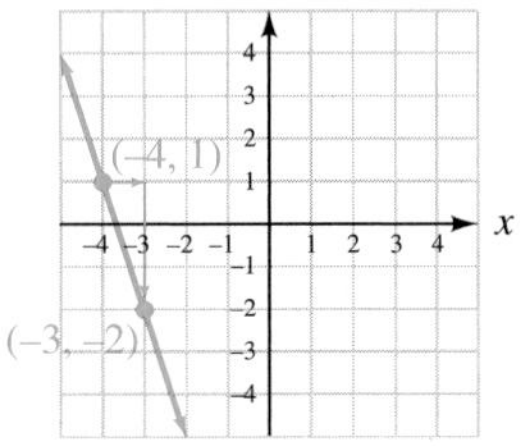

37.

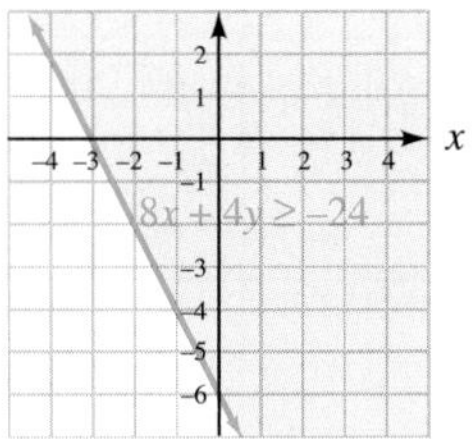

40.

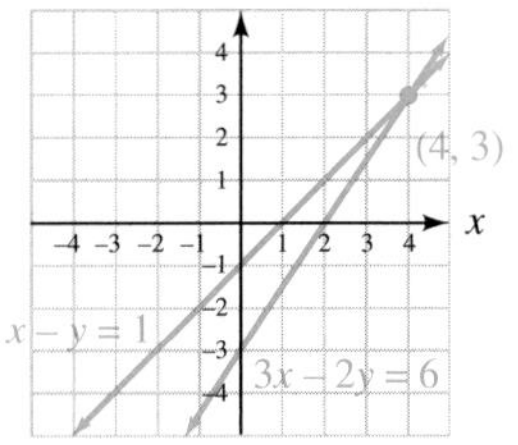

44.

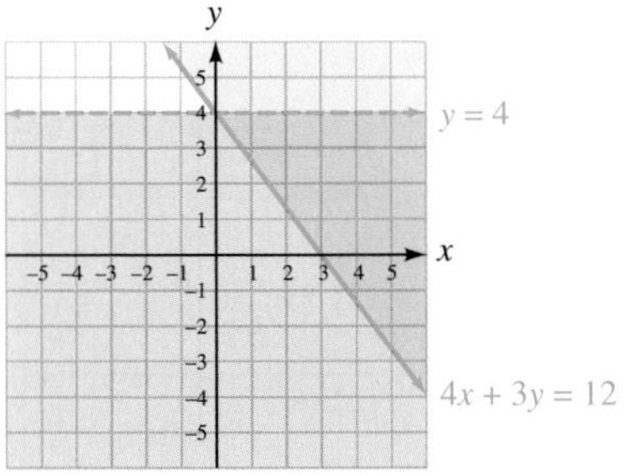

52.

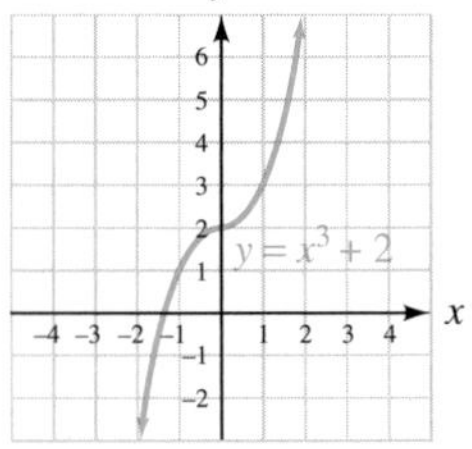

CUMULATIVE REVIEW CHAPTERS 1–7 (page 597)

14.

15.

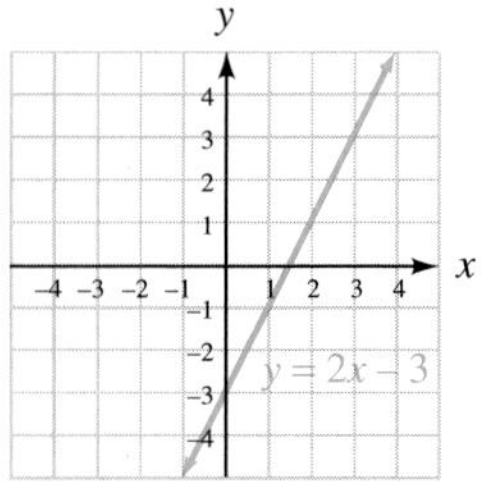

20.

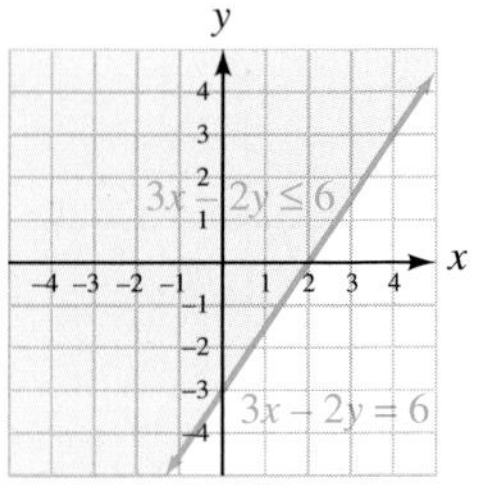

23.

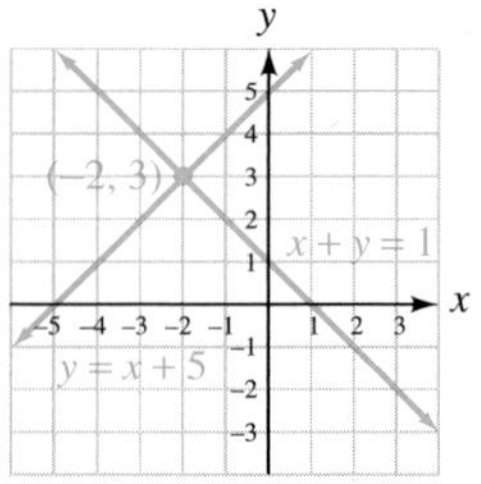

35.

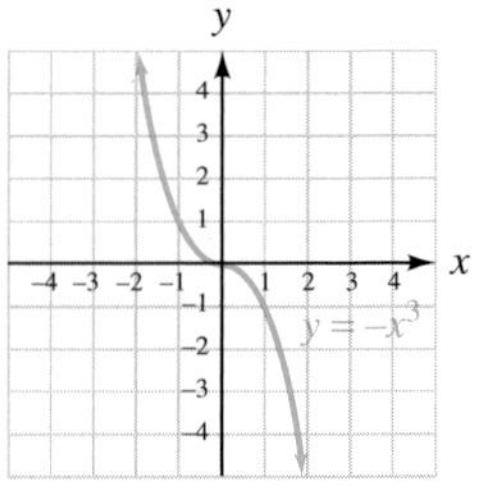

ARE YOU READY? 8.1 (page 600)

5.

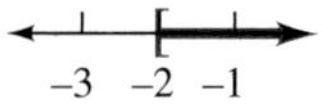

SELF CHECKS 8.1

6.

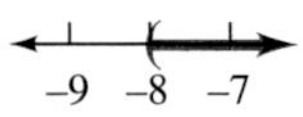

7.

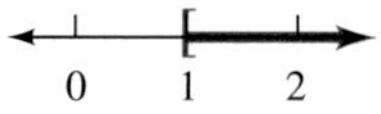

8.

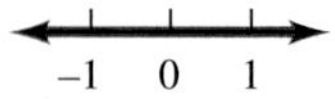

STUDY SET SECTION 8.1 (page 609)

14. b. **51.** **52.**

53. **54.** **55.**

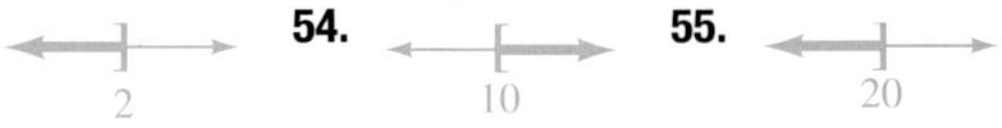

56. **57.** **58.**

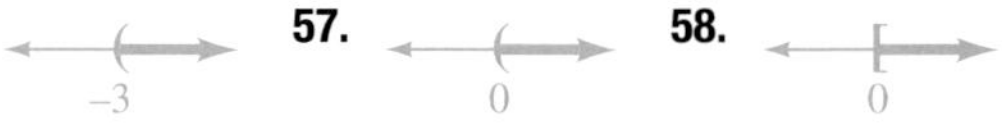

59. **60.** **61.**

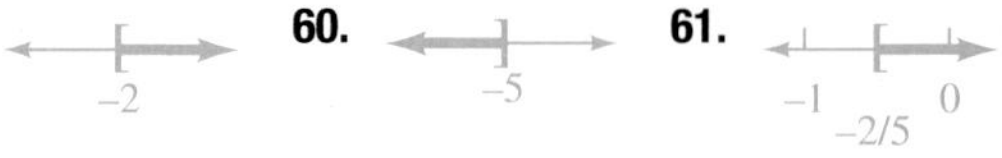

62. **63.** **64.**

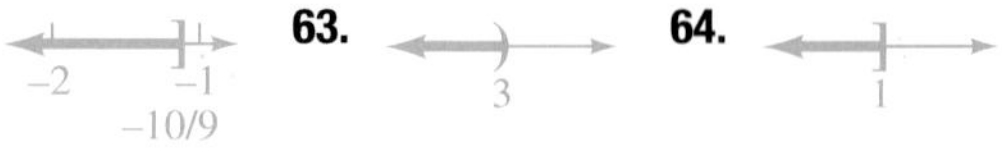

65. **66.** **67.**

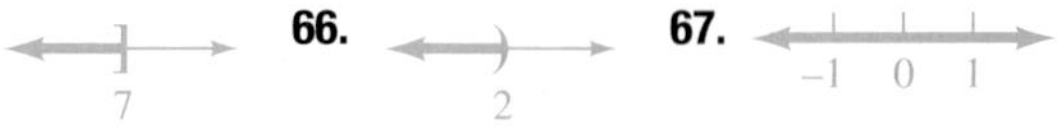

68. **91.** **92.**

93. **94.** **95.**

96. **97.** **98.**

99. **100.** **101.**

102.

STUDY SET SECTION 8.2 (page 624)

7. c.

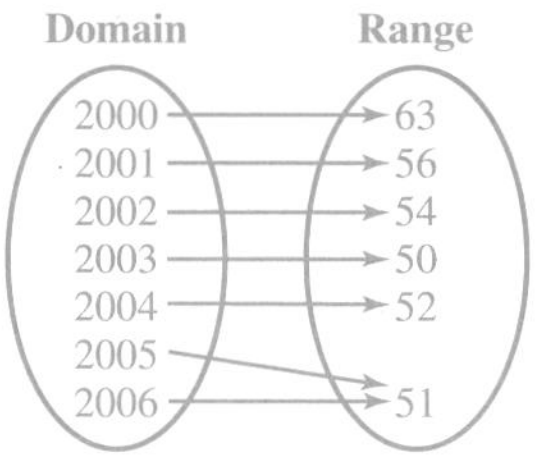

83.

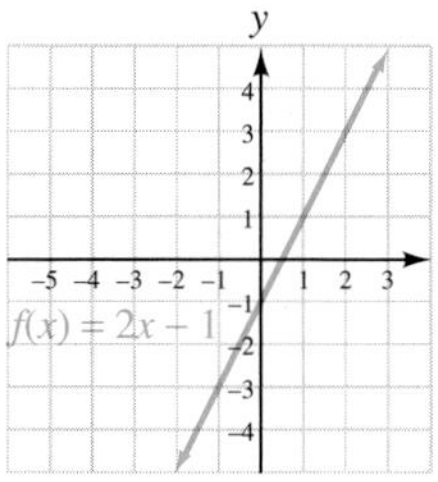

84.

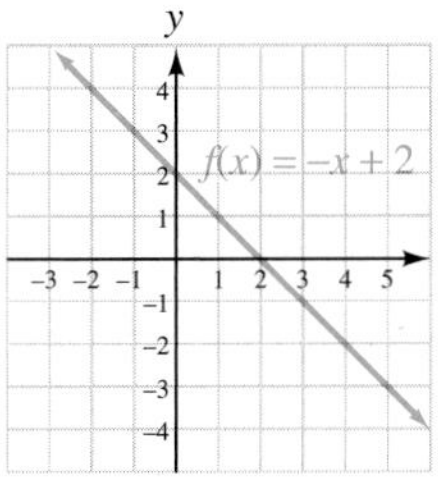

85.

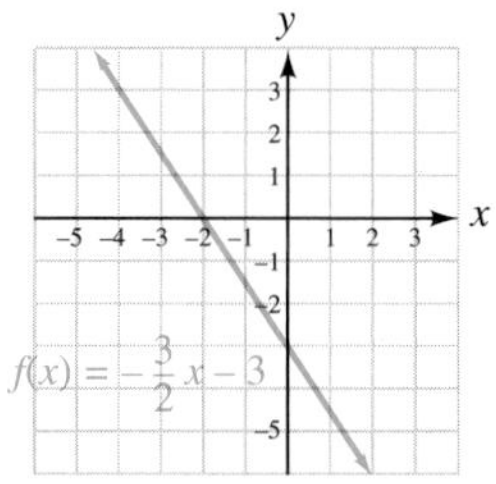

86.

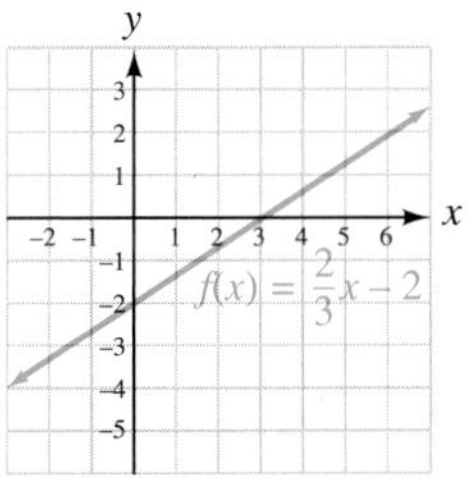

87.

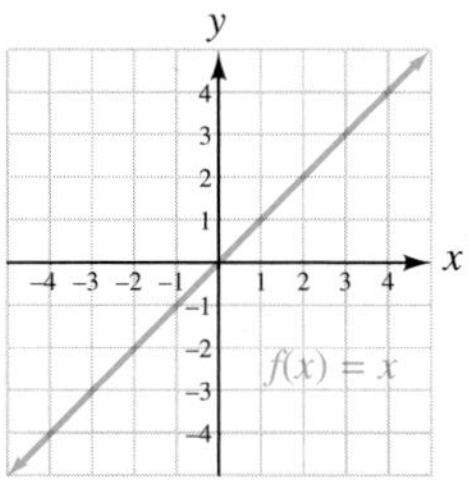

88.

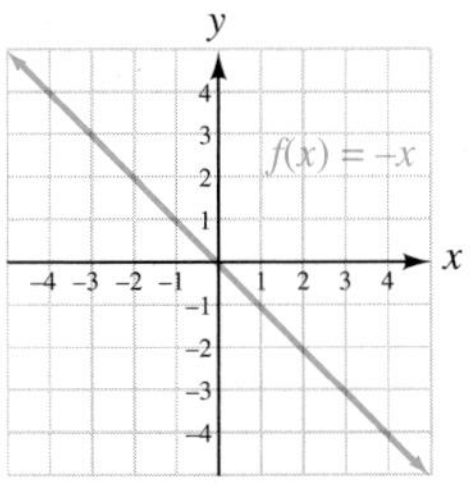

89.

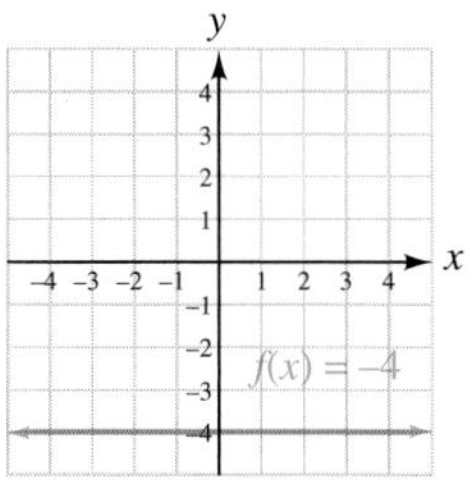

90.

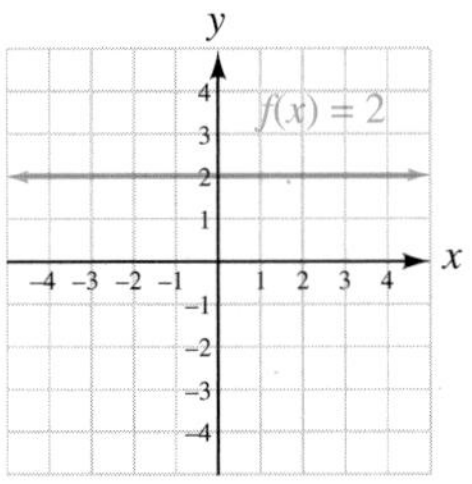

91.

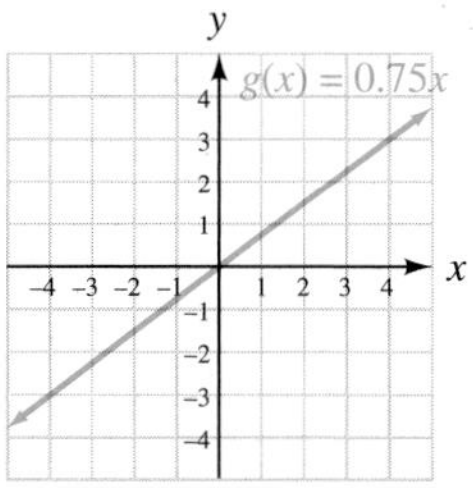

92.

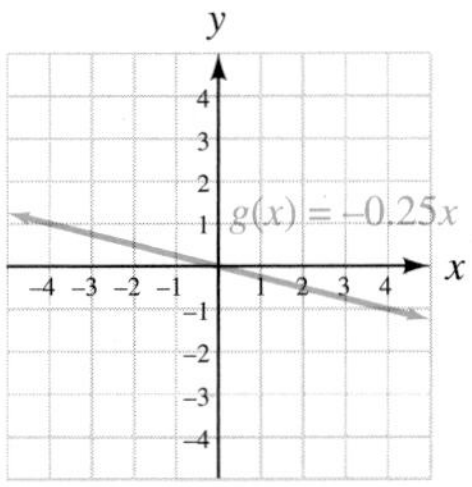

93.

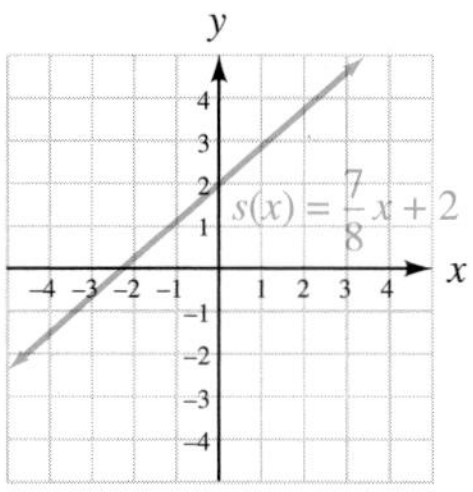

94.

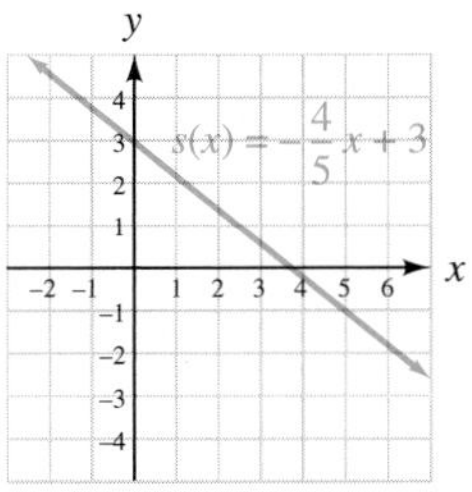

SELF CHECKS 8.3

2.

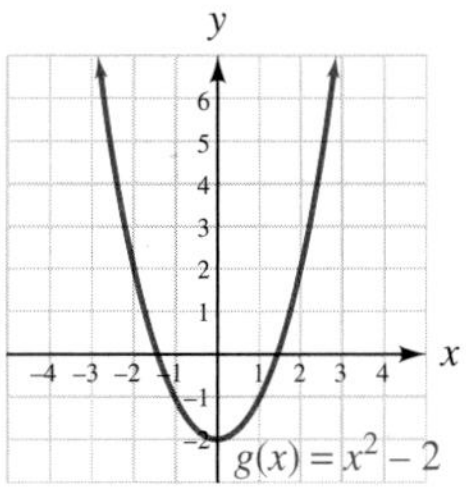

3.

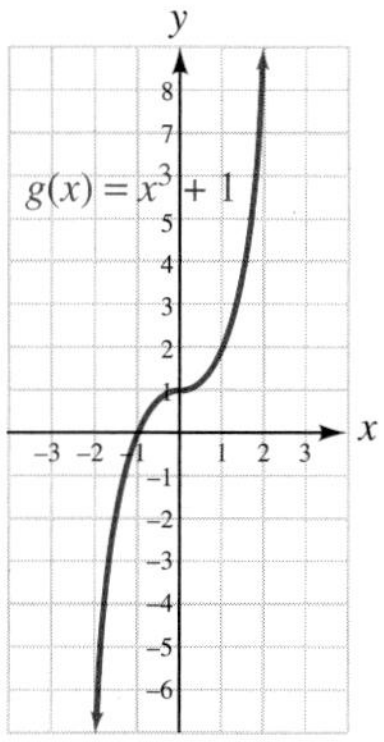

4.

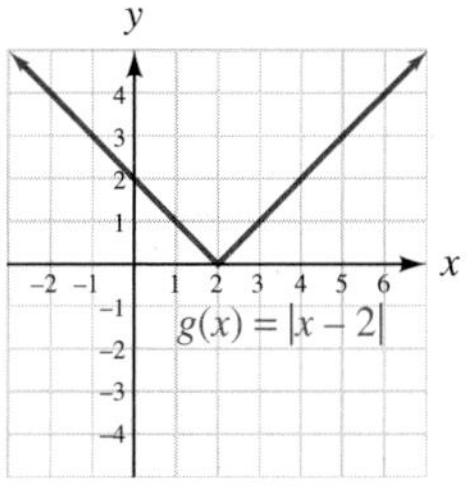

5.

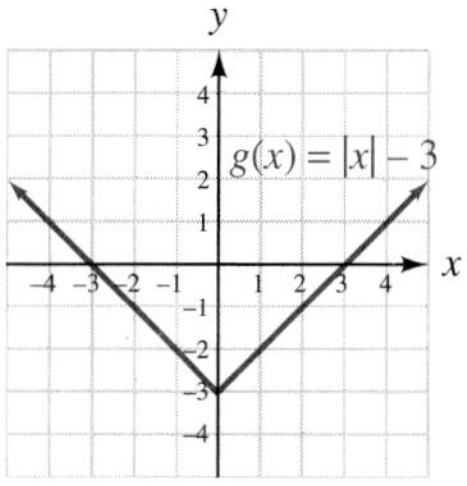

6.

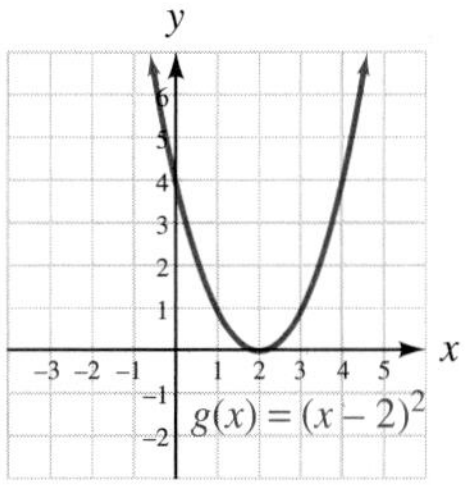

7.

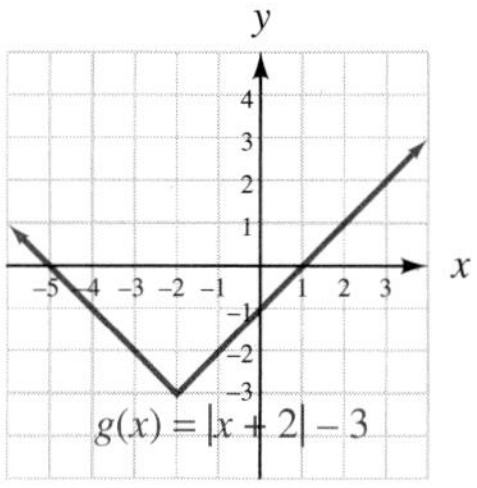

8.

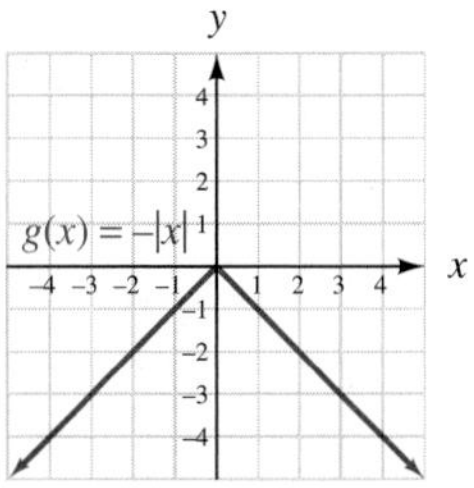

STUDY SET SECTION 8.3 (page 638)

5. a.

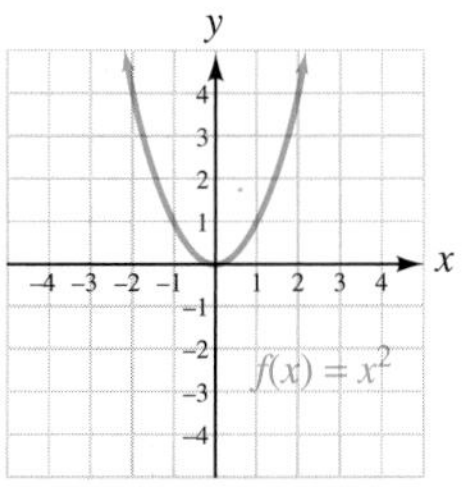

b.

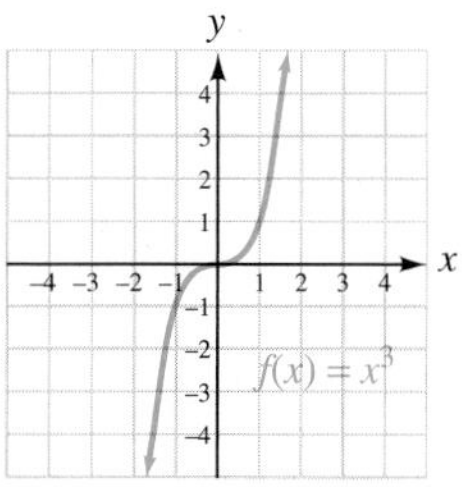

c.

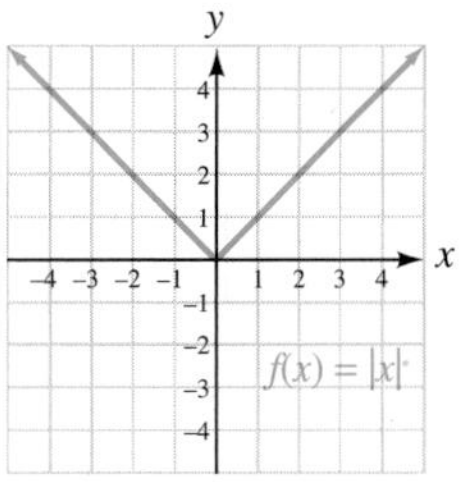

29.

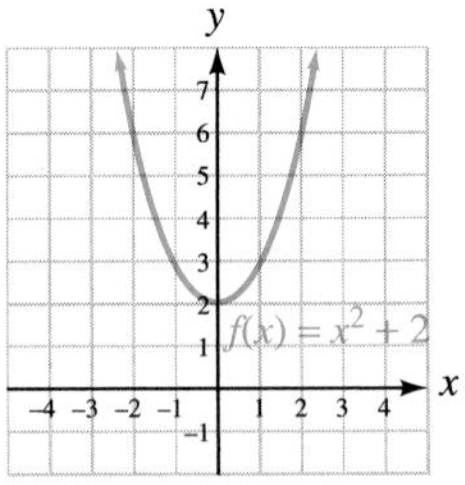

30.

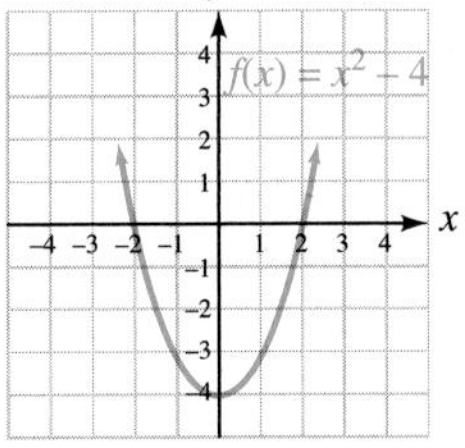

31.

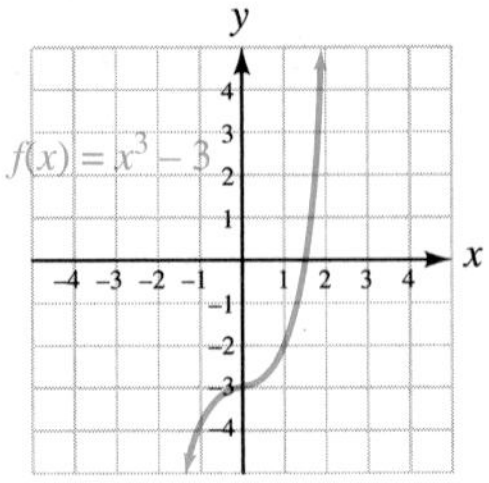

32.

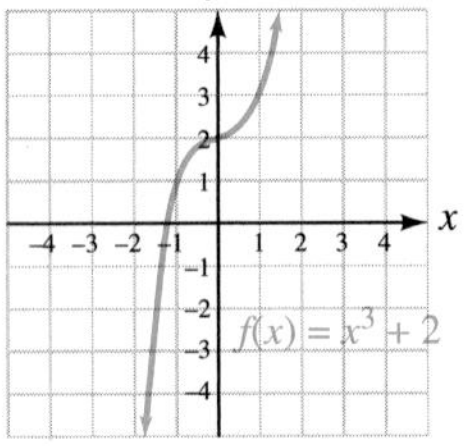

33.

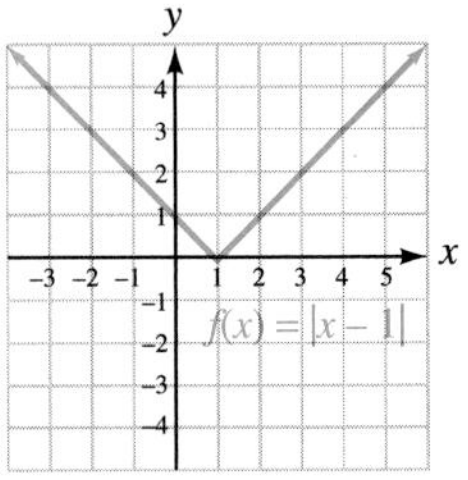

34.

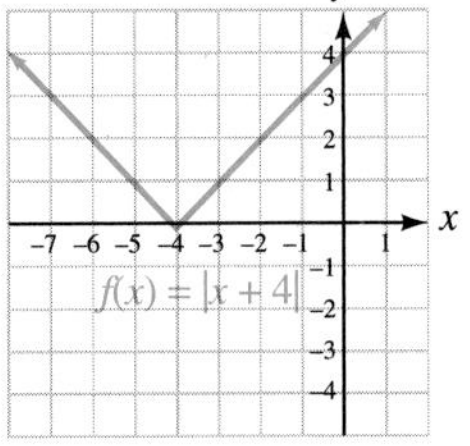

35.

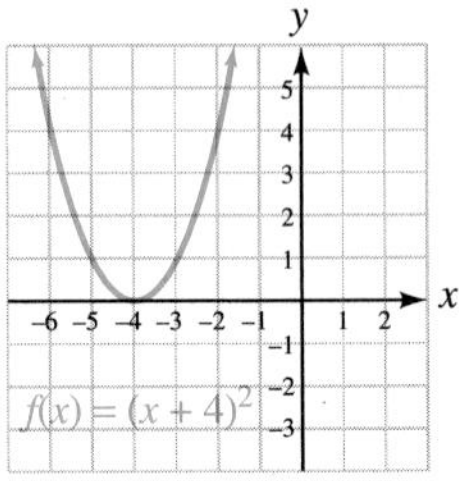

36.

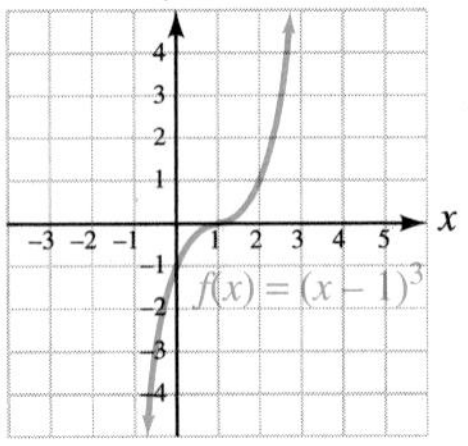

37.

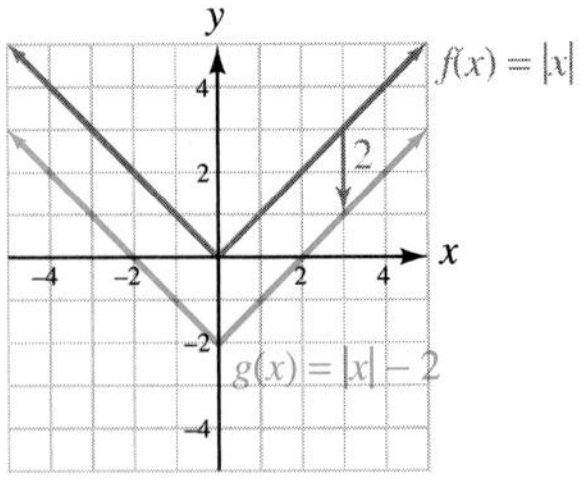

38.

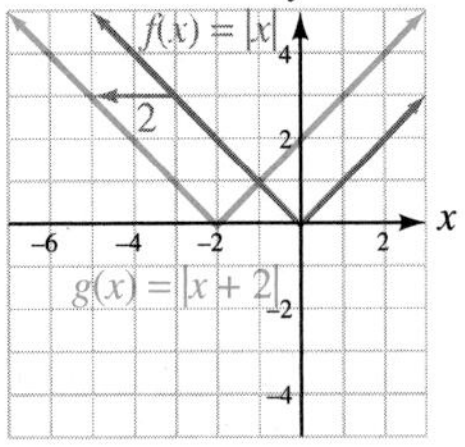

39.

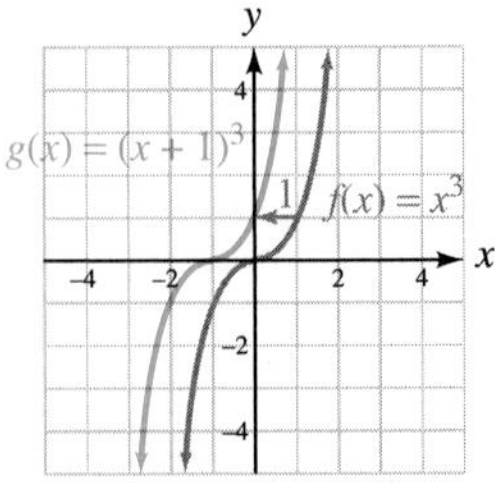

40.

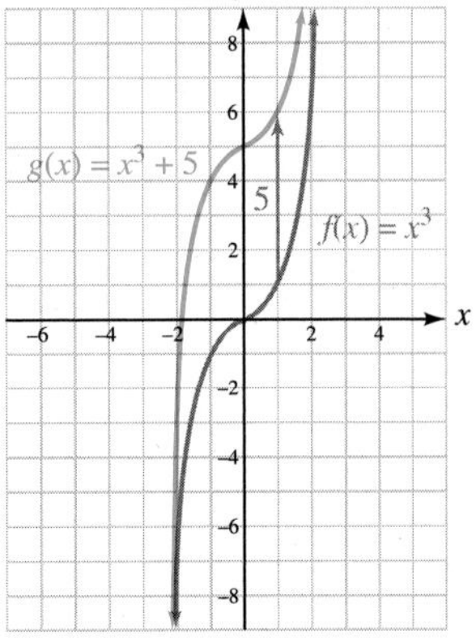

41.

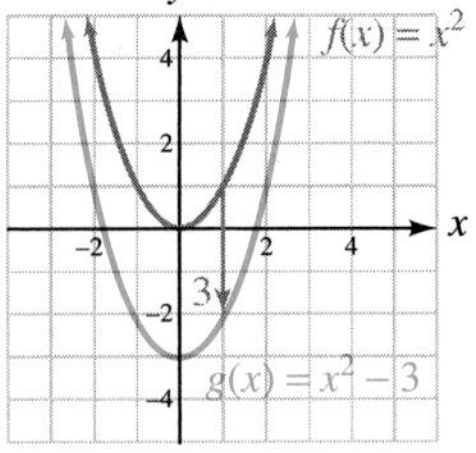

42.

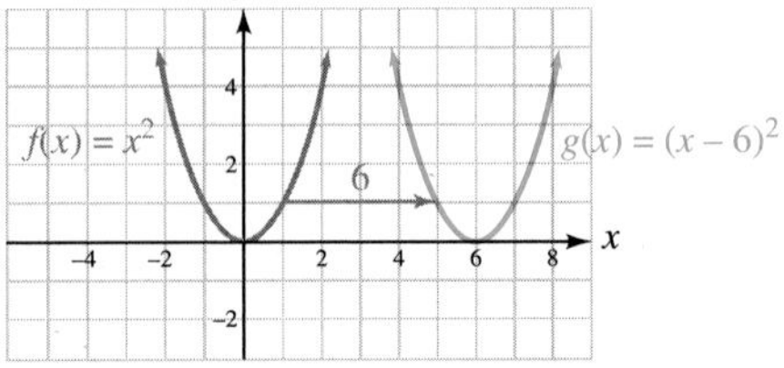

43.

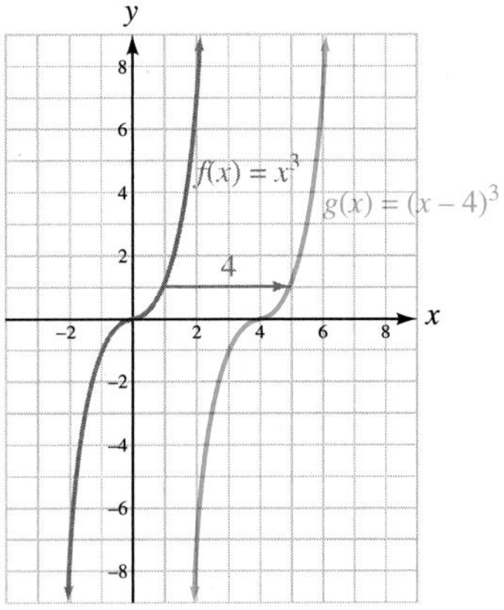

44.

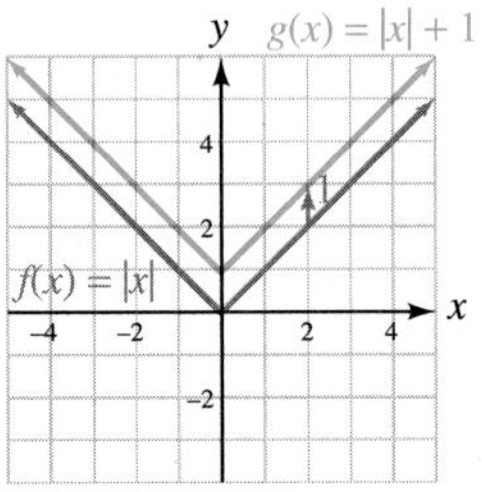

45.

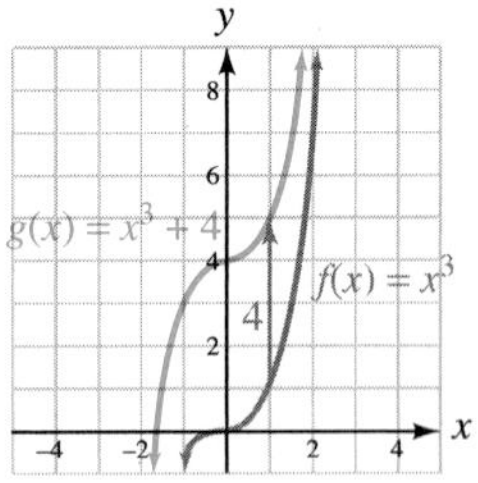

46.

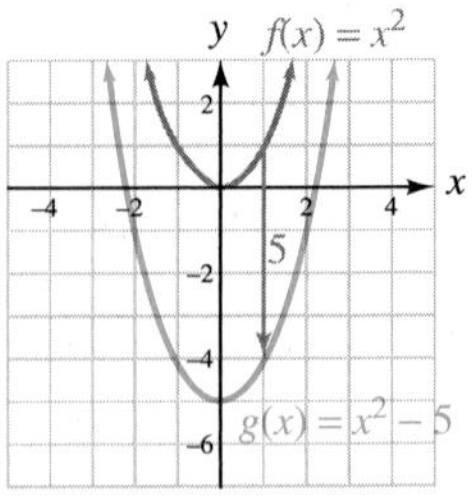

47.

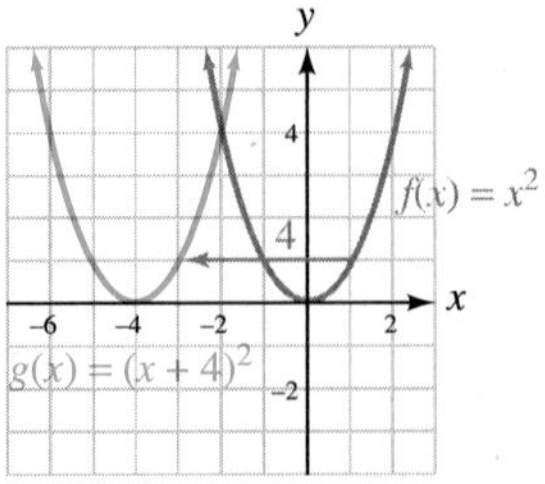

48.

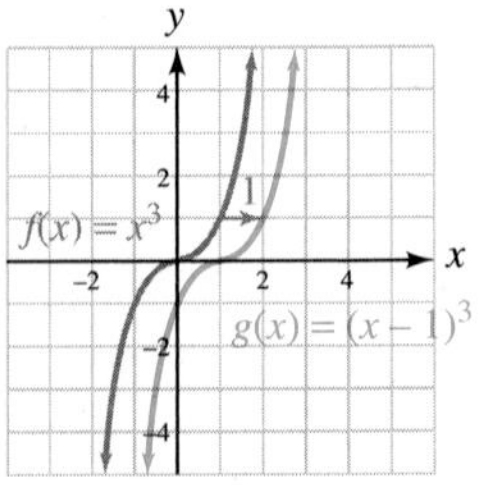

49.

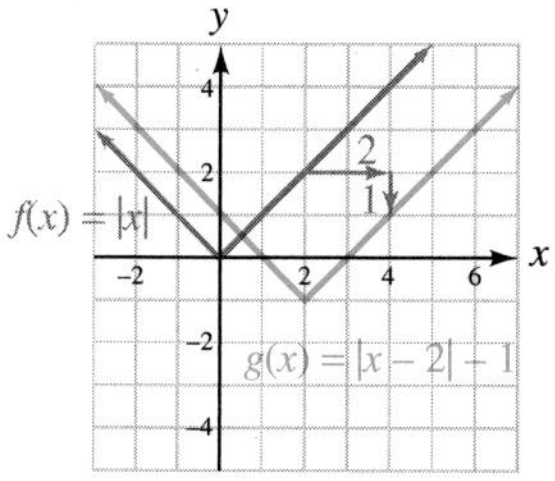

50.

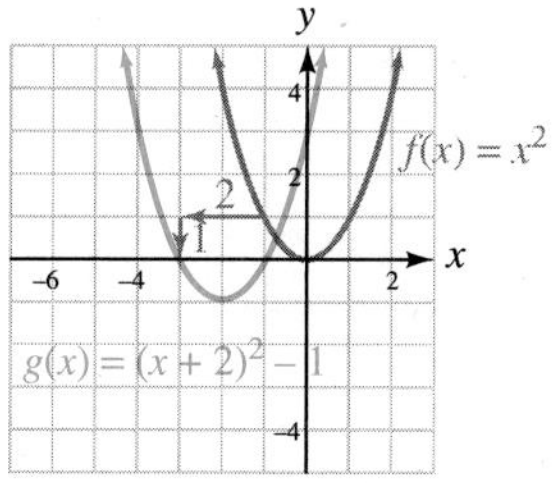

51.

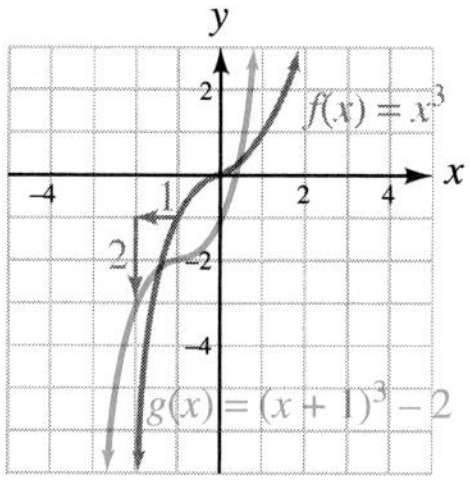

52.

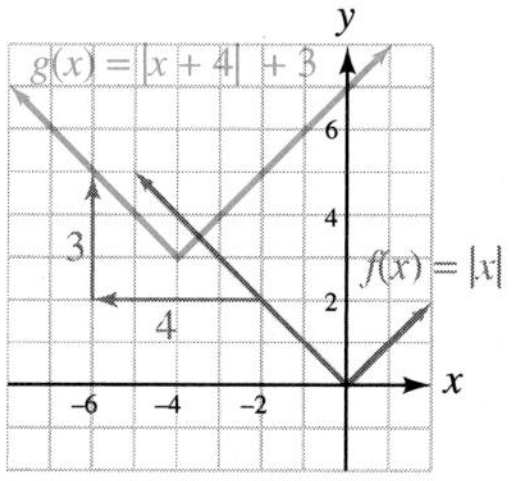

53.

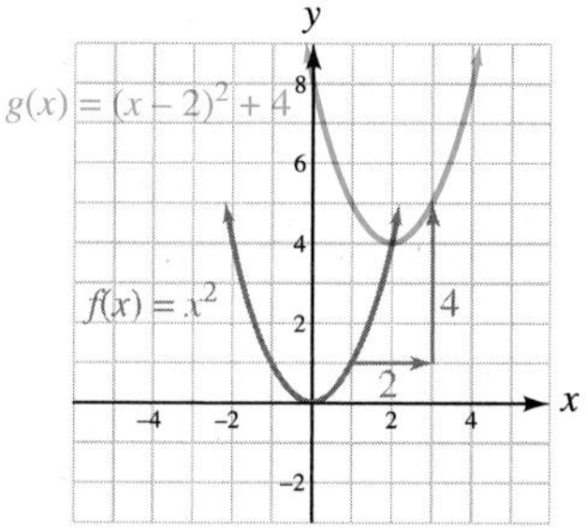

54.

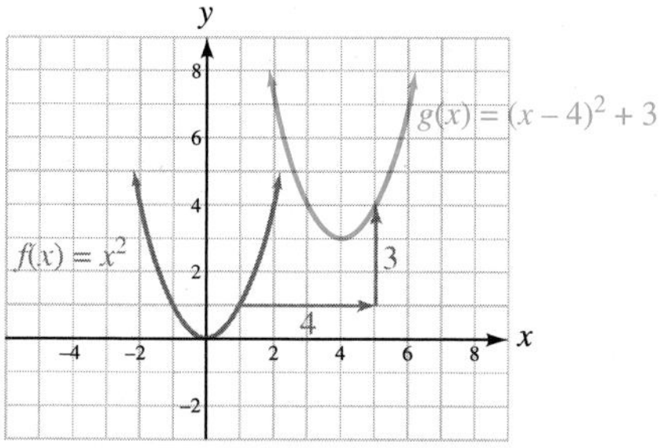

55.

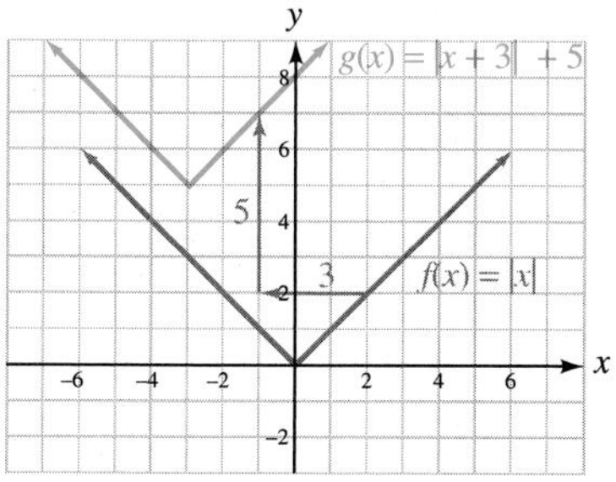

56.

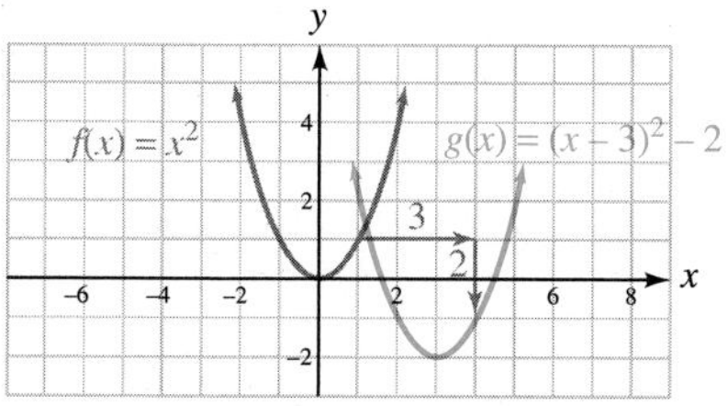

57.

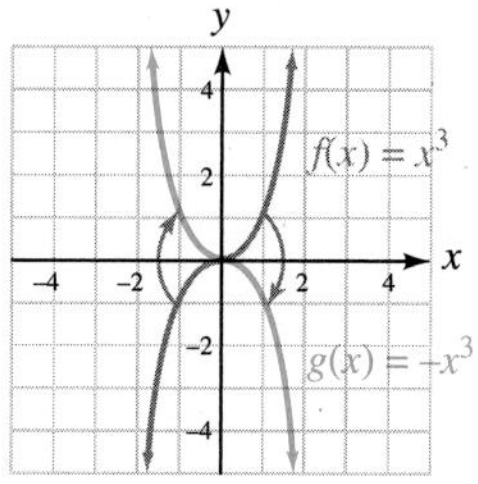

58.

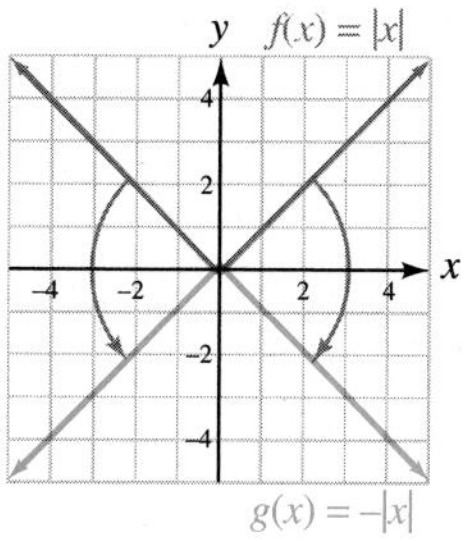

59.

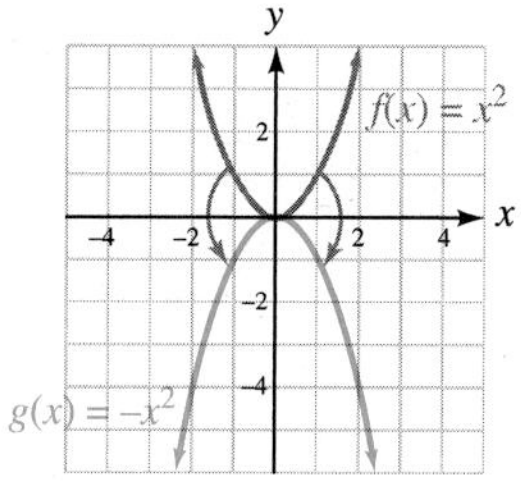

60.

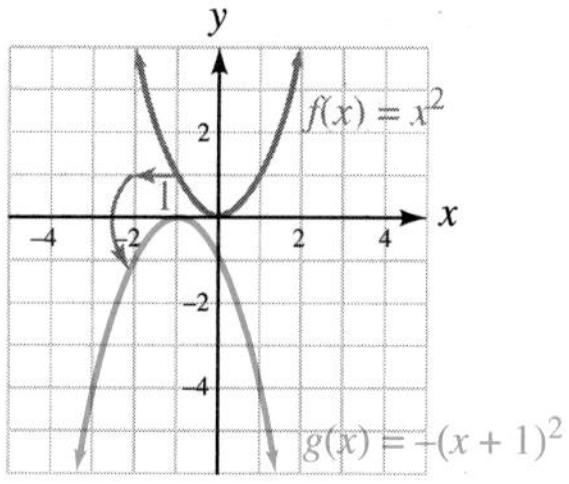

61.

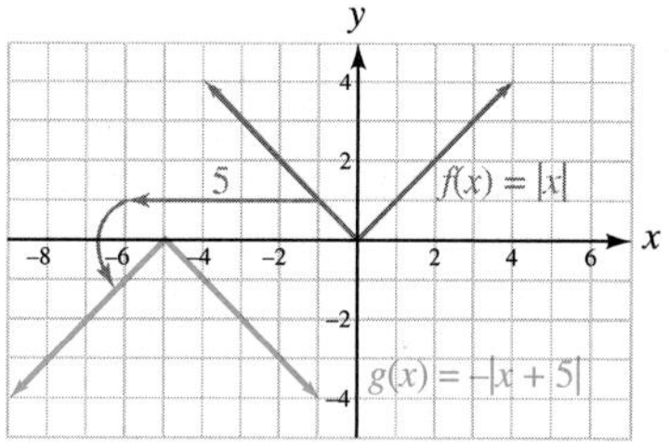

62.

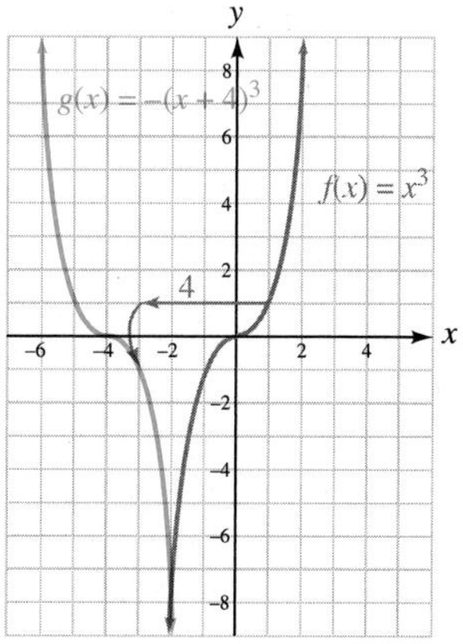

63.

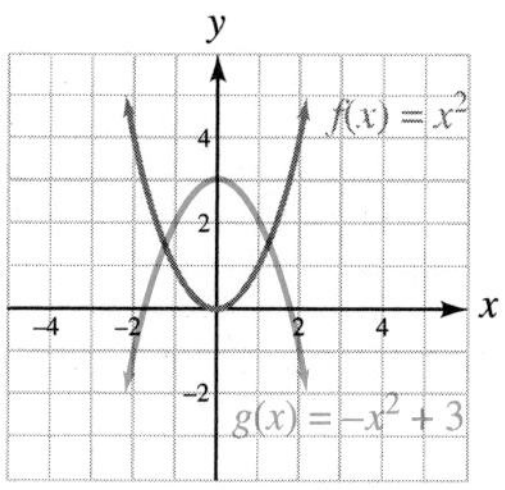

64.

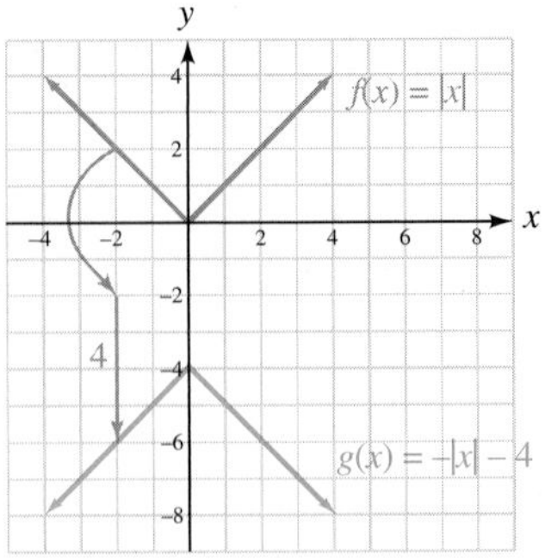

77.

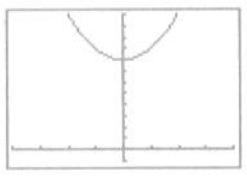

78.

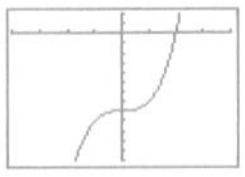

79.

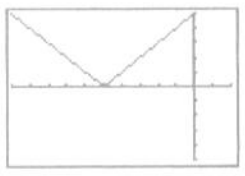

80.

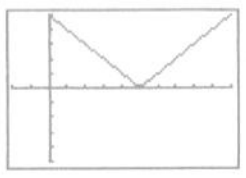

81.

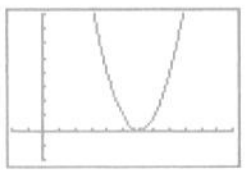

82.

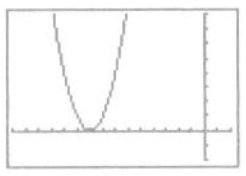

83.

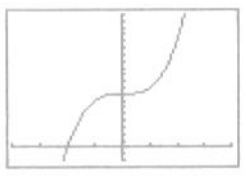

84.

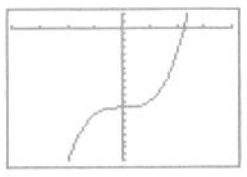

101.

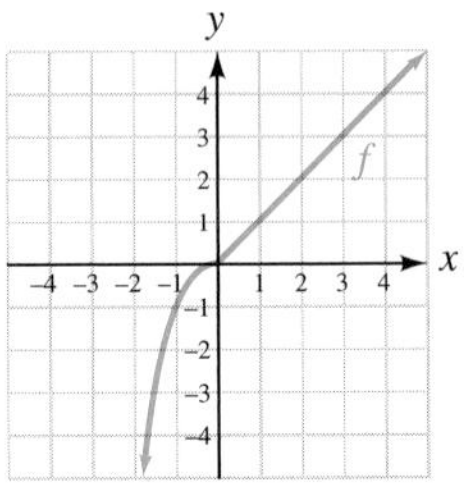

102.

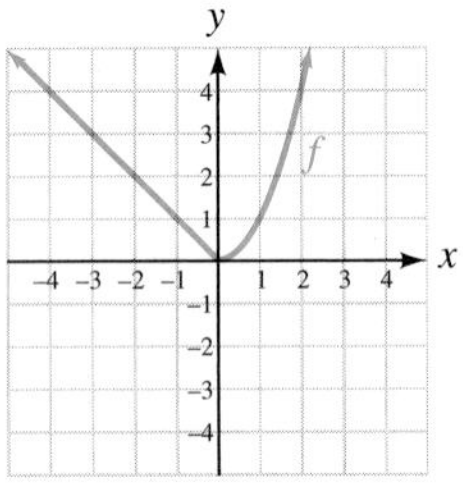

105.

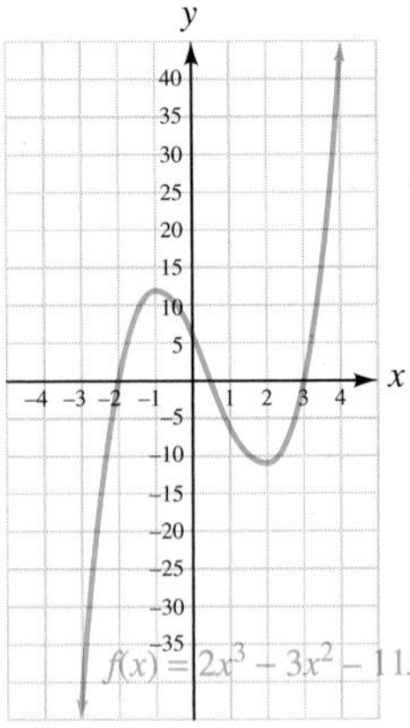

106.

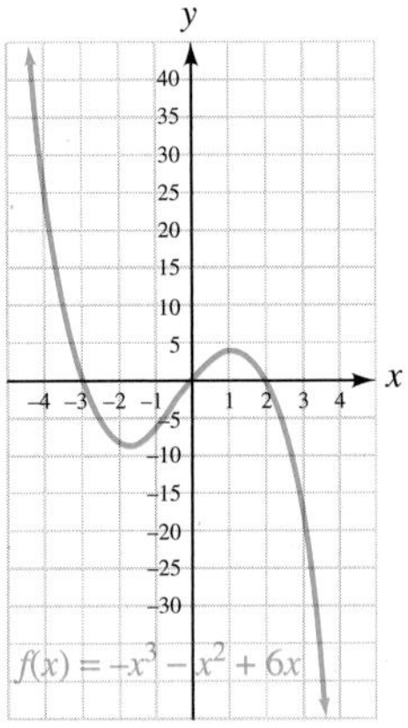

ARE YOU READY? 8.4 (page 643)

2.

4.

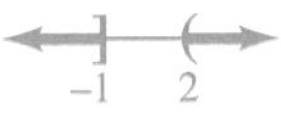

SELF CHECKS 8.4

2. $(-6, 10]$

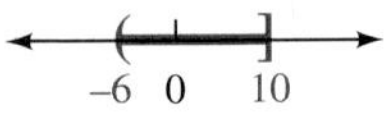

3. $\left(\frac{1}{2}, \infty\right)$

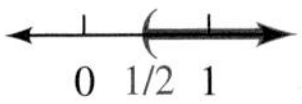

5. $[1, 5]$

6. $(-\infty, 2) \cup (4, \infty)$

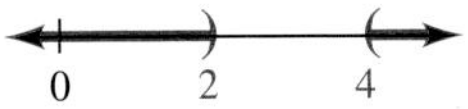

7. $(-\infty, \infty)$

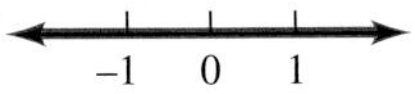

STUDY SET SECTION 8.4 (page 651)

15. **16. a.** **b.**

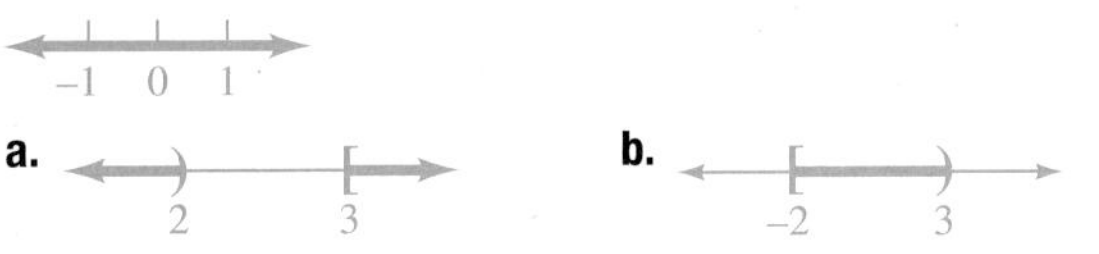

25. **26.** **27.** **28.**

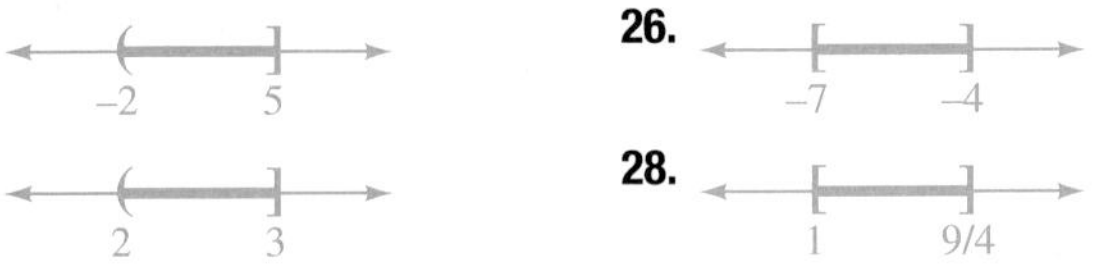

29. **30.** **33.** **34.**

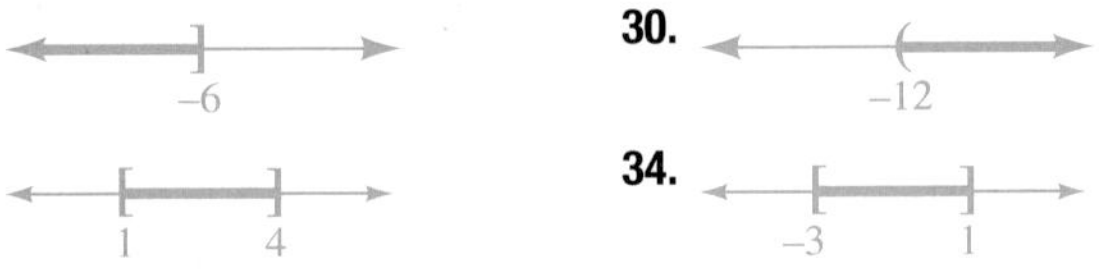

35. **36.** **37.** **38.**

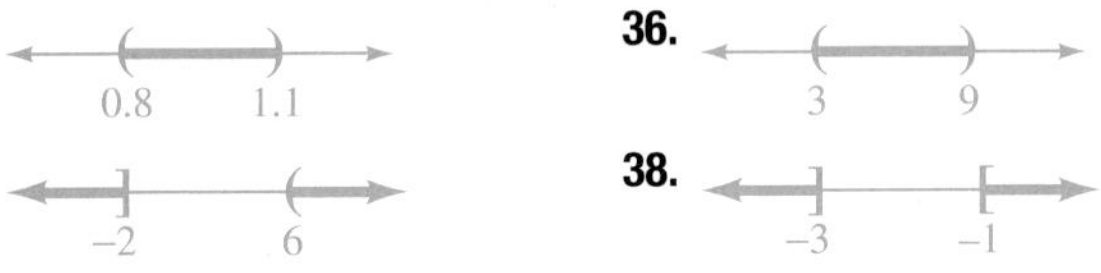

39. **40.** **41.** **42.**

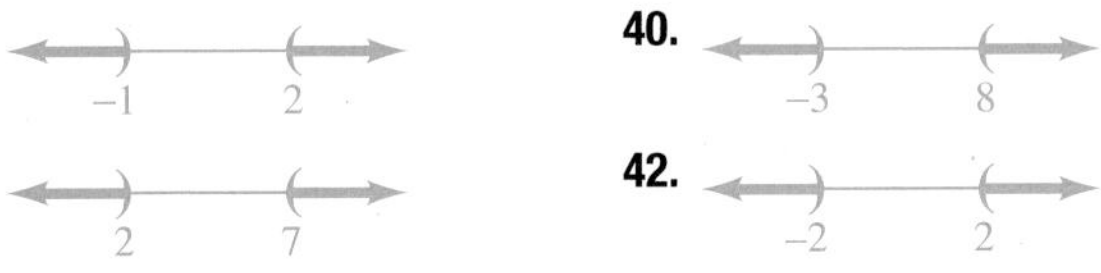

43. **44.** **45.** **46.**

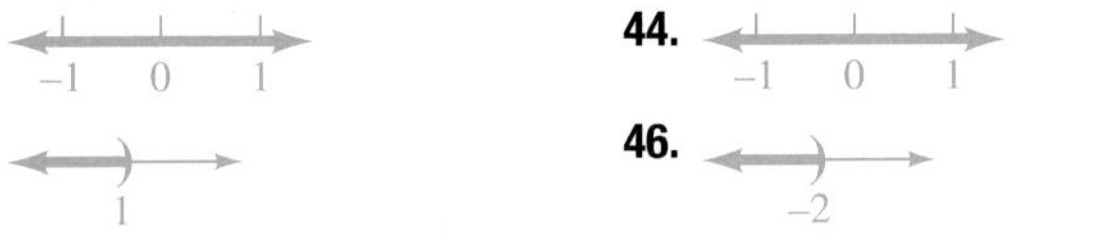

47. **48.** **49.** **50.**

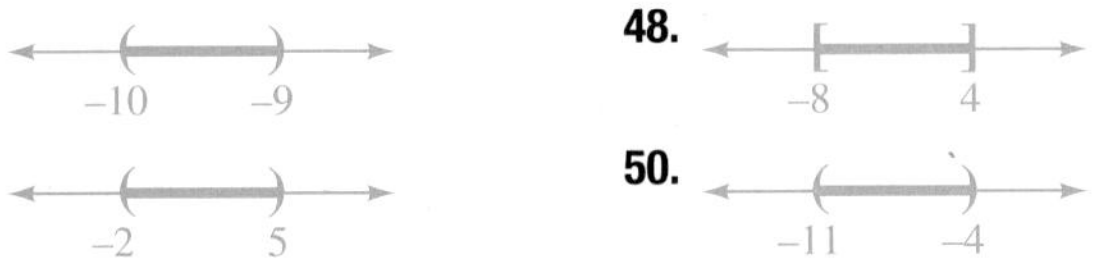

51. **52.** **53.** **54.**

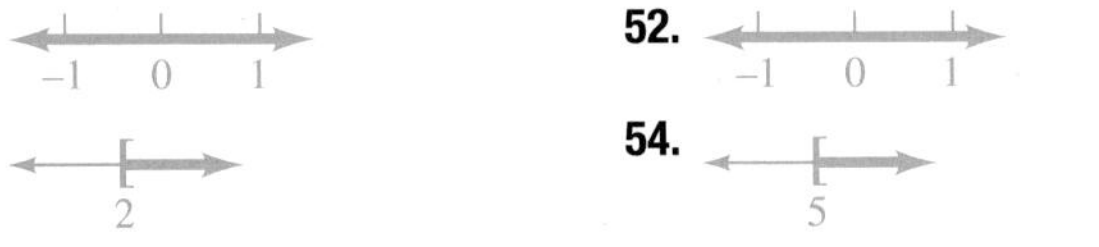

57. **58.** **59.** **60.**

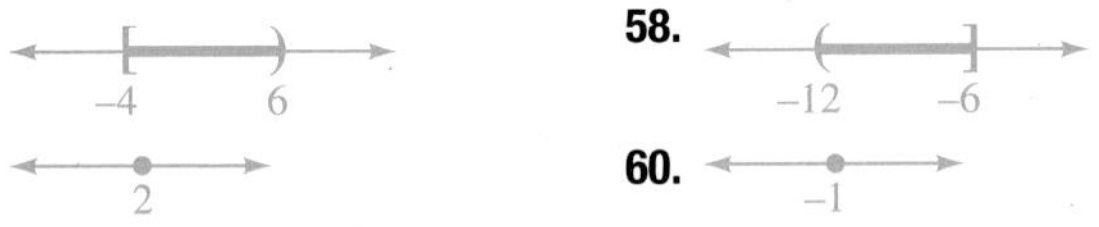

61. **62.** **63.** **64.**

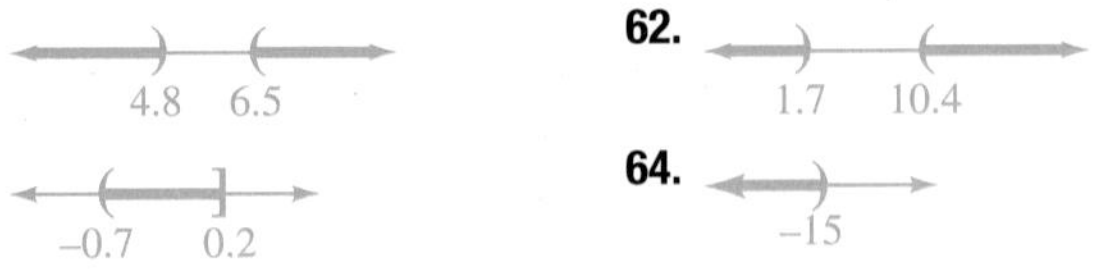

65. **66.** **67.** **68.**

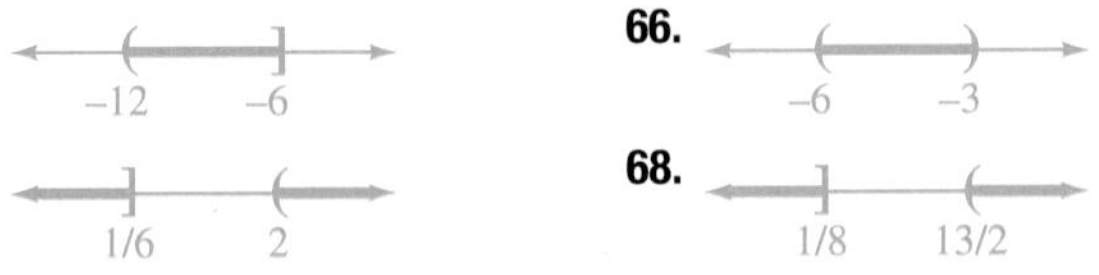

69. **70.** **71.** **72.**

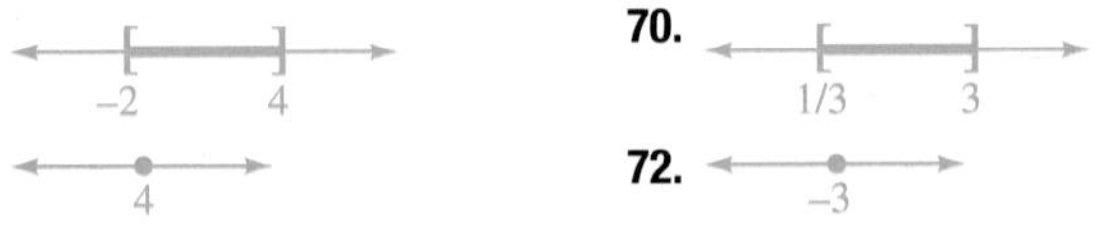

73. **74.** **75.** **76.**

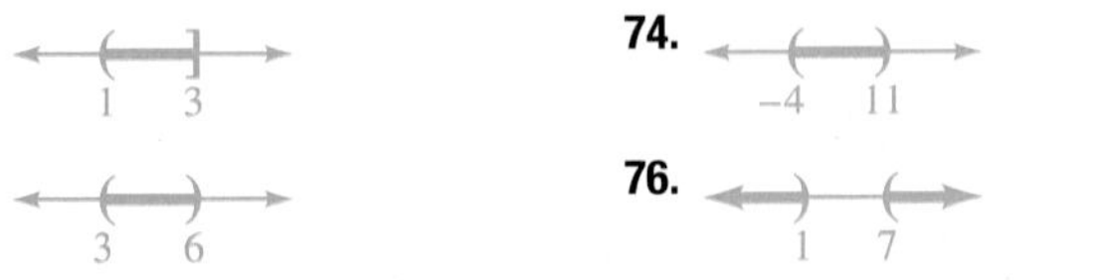

77. a. **b.**

78. a. 0 **b.** [5, 8]

79. b. 3 6

80. a. 2 3/2 11/2 **b.** 2 3/2 11/2

87. a.

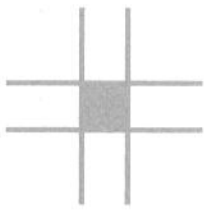

b.

97. −2 8 10 **98.** −1 4

99. −4 8 **100.** −8 6

ARE YOU READY? 8.5 (page 655)

3. −3 1 **4.** −8 4

SELF CHECKS 8.5

6. $\left(-2, \frac{2}{3}\right)$ −3 −2 0 2/3 1

9. $(-\infty, -2] \cup [6, \infty)$ −2 0 6

10. $(-\infty, -8) \cup \left(\frac{8}{3}, \infty\right)$ −8 0 8/3

11. $(-\infty, \infty)$ −1 0 1

STUDY SET SECTION 8.5 (page 664)

55. −4 4 **56.** −9 9

57. −21 3 **58.** −4 20

59. −8/3 4 **60.** −3 17/3

63. −3 3 **64.** −7 7

65. −12 36 **66.** −12 2

67. −1/5 3/5 **68.** −1/3 4/3

69. 0 **70.** 0

75. −16/3 4 **76.** −10 15

77. −10 14 **78.** −10 14

81. 0 **82.** −3 2

85. 2/3 **86.** 3

91. −2 5 **92.** −3 7/3

94. 0

95. c. 20 **d.** 0 20

96. c. 5 **d.** −5/2 5

97. c. −25 **d.** −25 31

99. a. 3 7 **b.** 3 7

100. a. 0

101. a. −8 10 **b.** −8 10

102. a. −16 4 **b.** −16 4

STUDY SET SECTION 8.8 (page 696)

139.

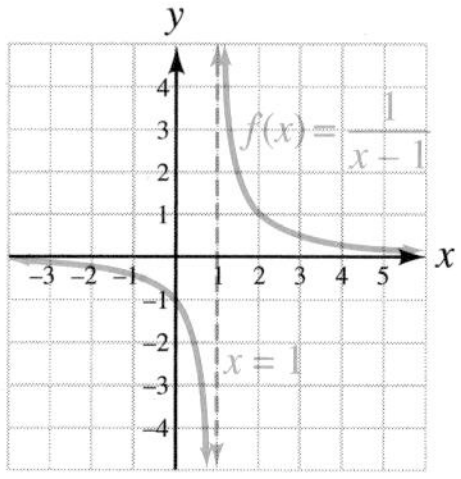

140.

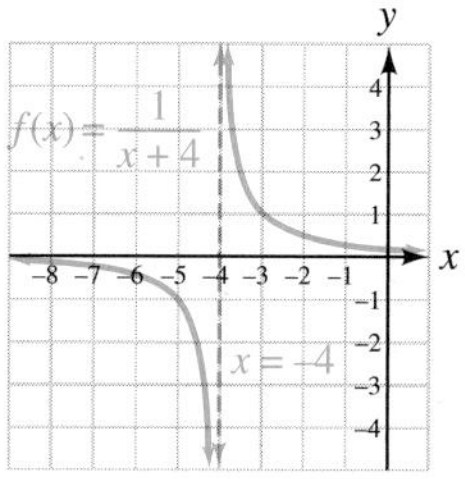

STUDY SET SECTION 8.9 (page 705)

55. 2 5 **56.** −2

57. 6 9 **58.** 6

CHAPTER 8 REVIEW (page 708)

38.

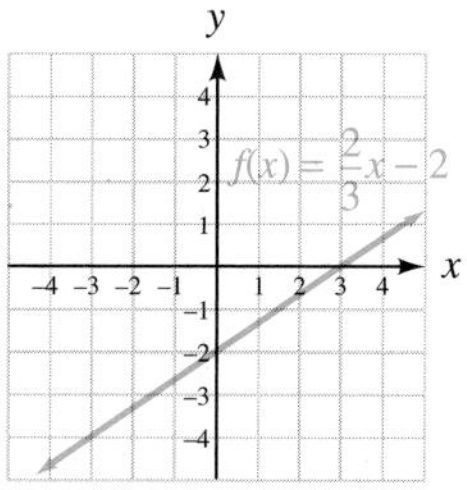

53.

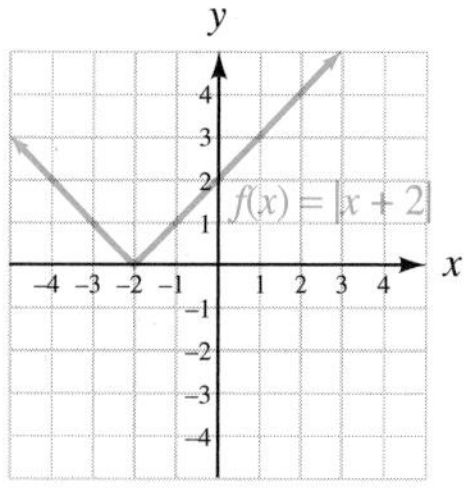

55.

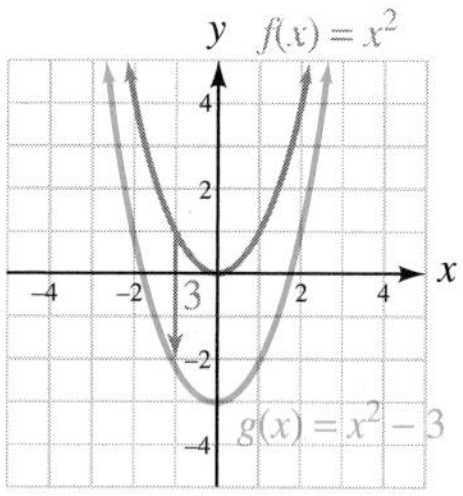

56.
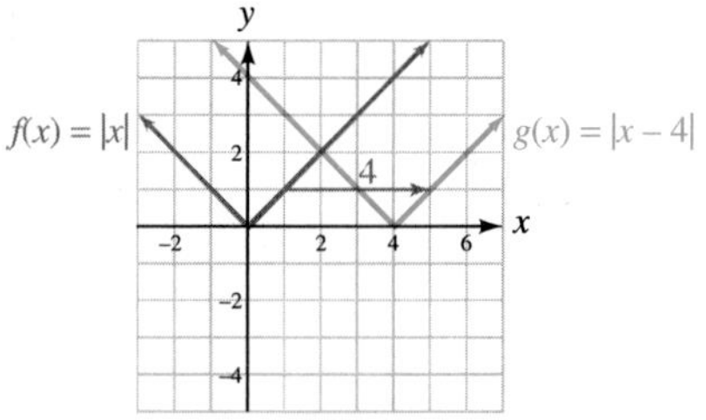

57.
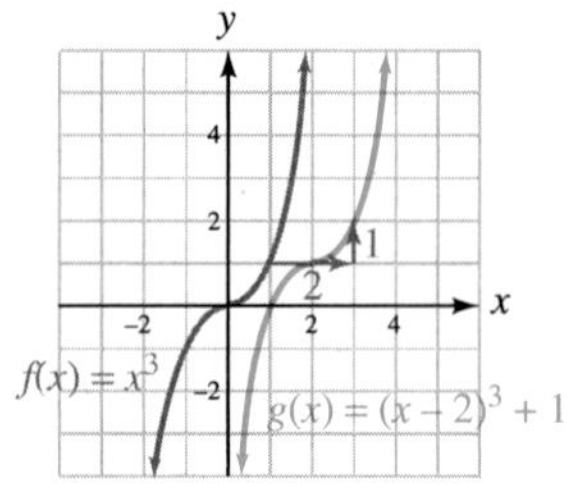

58.
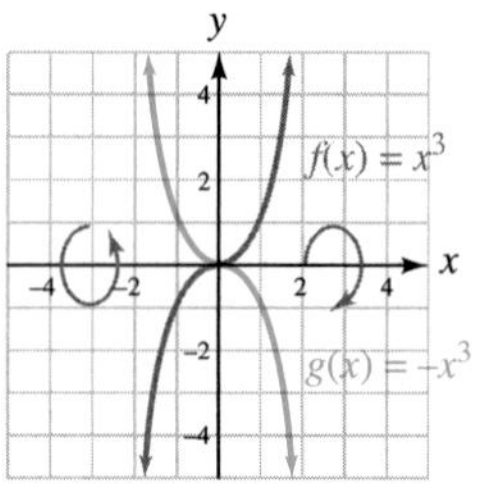

65.

66.

67.

68.

70.

71.

72.

75.

76.

79.

80.

89.

90.

91.

93.
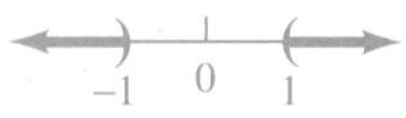

94.

95.
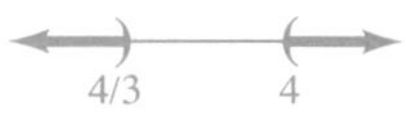

96.

100.
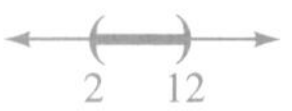

CHAPTER 8 TEST (page 723)

23.
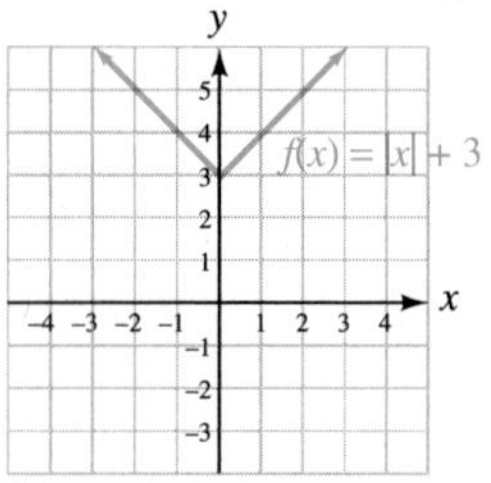

24.
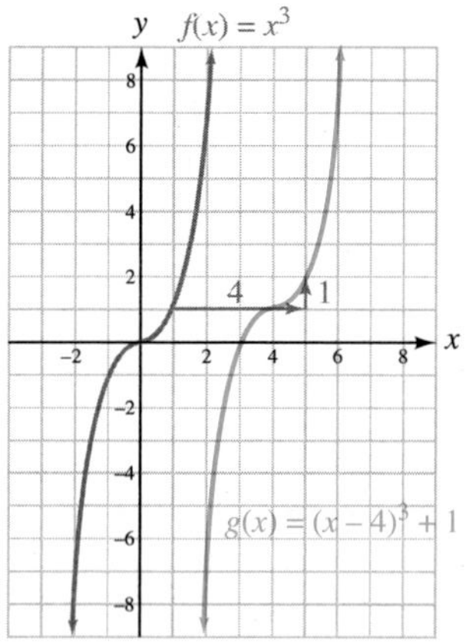

25.

26.

27.

28.

29.
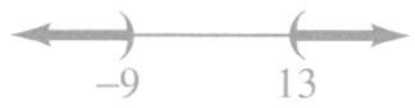

30.

SELF CHECKS 9.1

3.
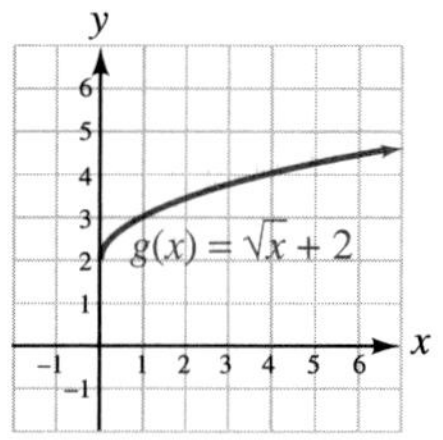

4. b.
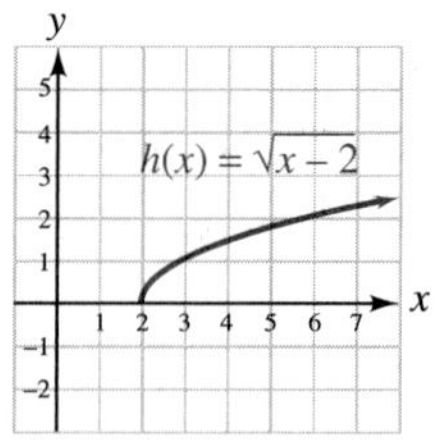

7. a.
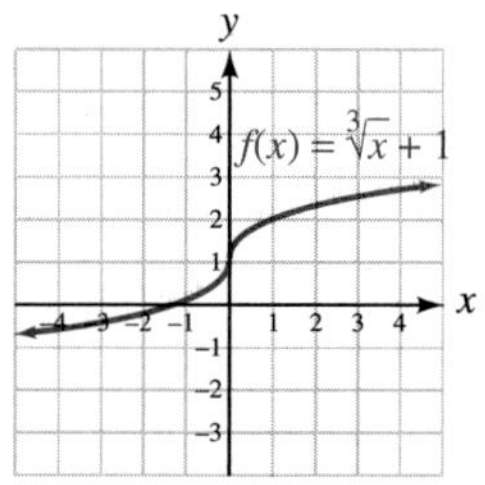

STUDY SET SECTION 9.1 (page 739)

51.
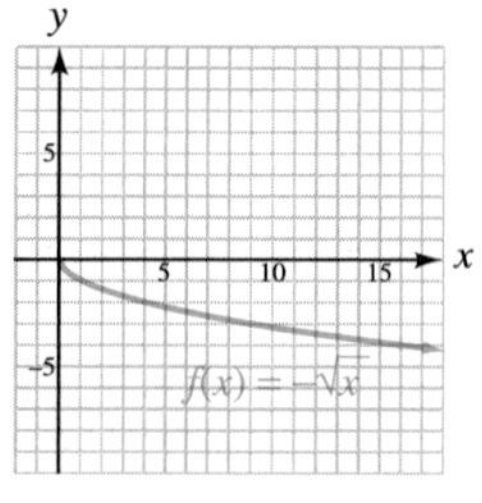

52.
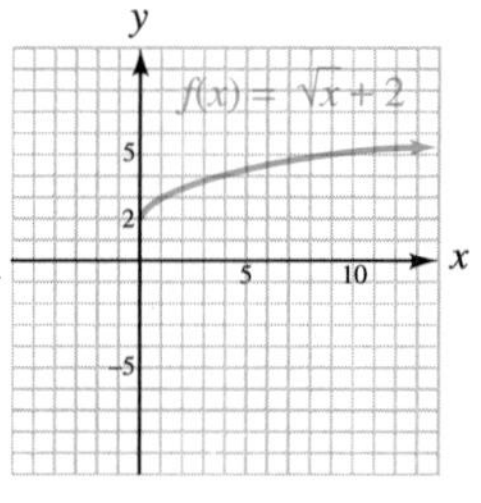

53.
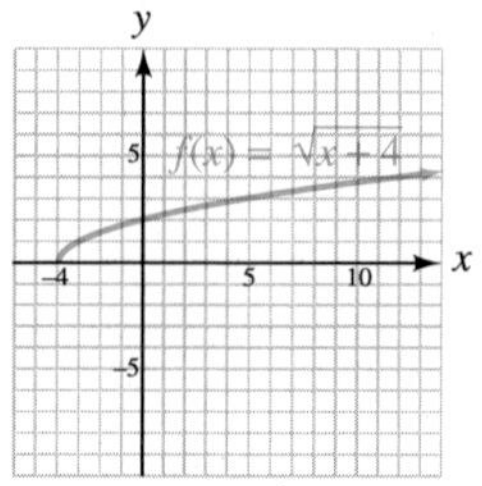

54.
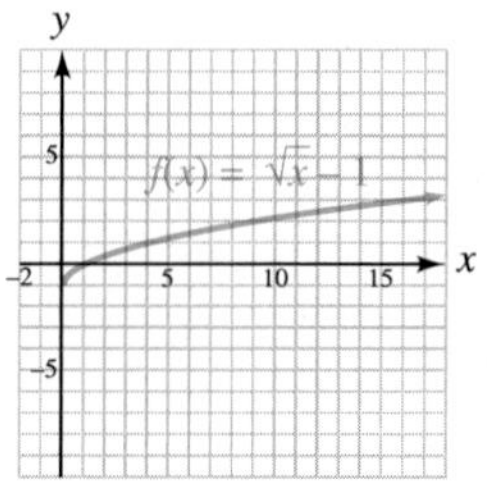

83.

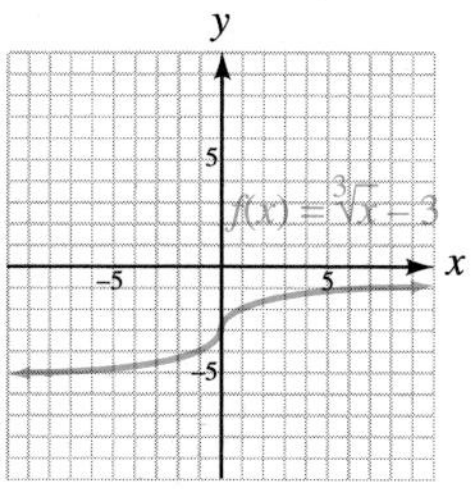

84.

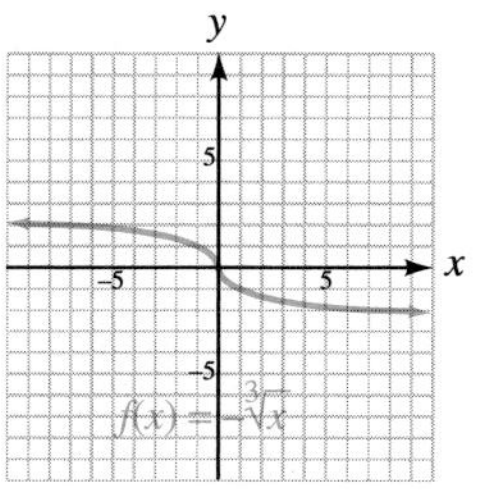

85.

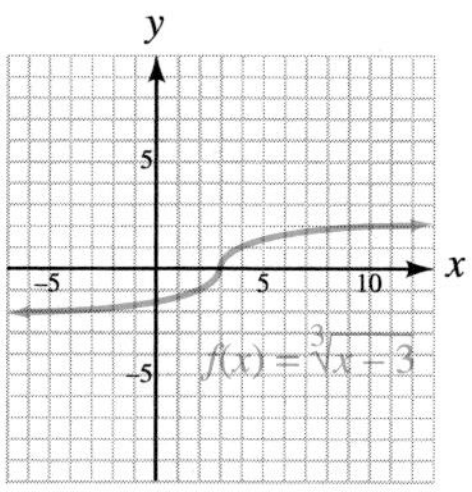

86.

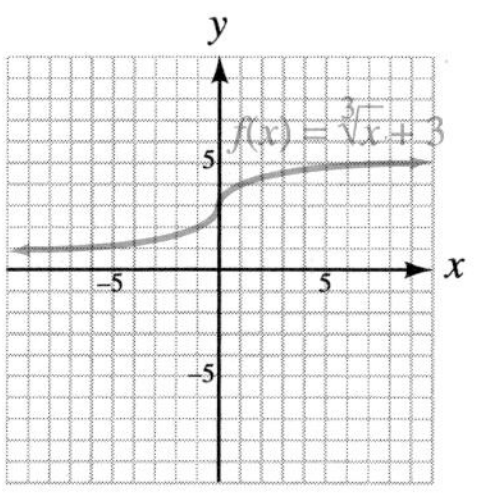

139.

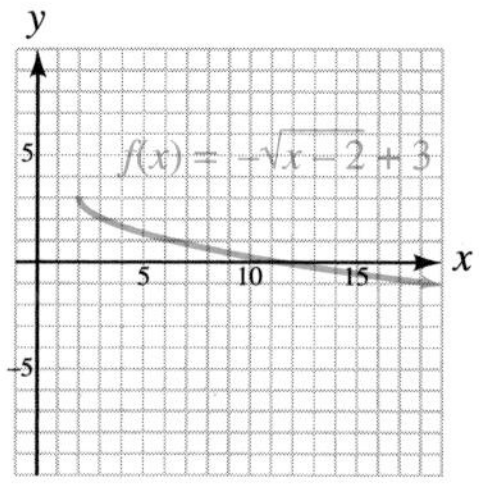

STUDY SET SECTION 9.2 (page 751)

9.

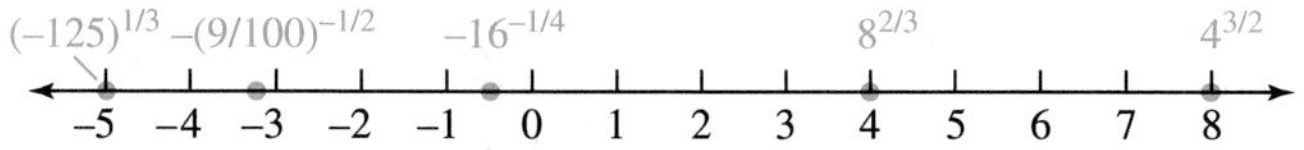

CHAPTER 9 REVIEW (page 812)

27.

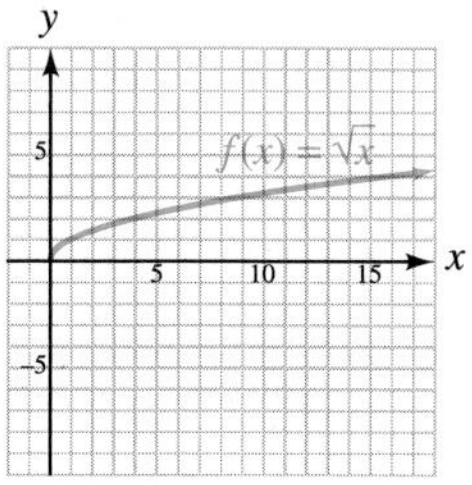

28.

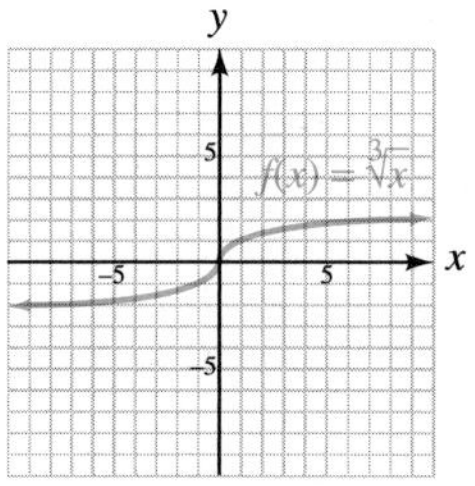

29.

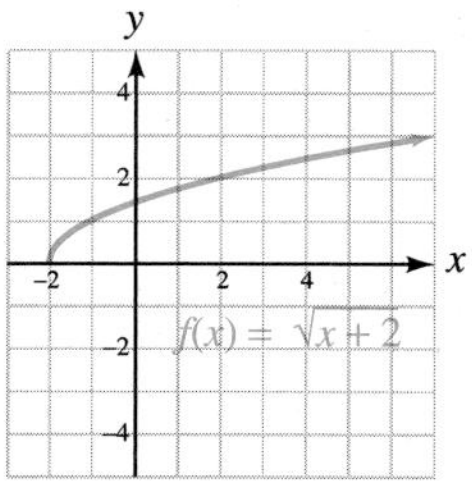

30.

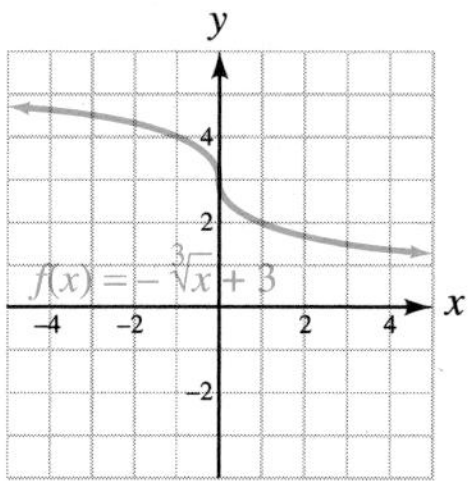

CHAPTER 9 TEST (page 821)

3.

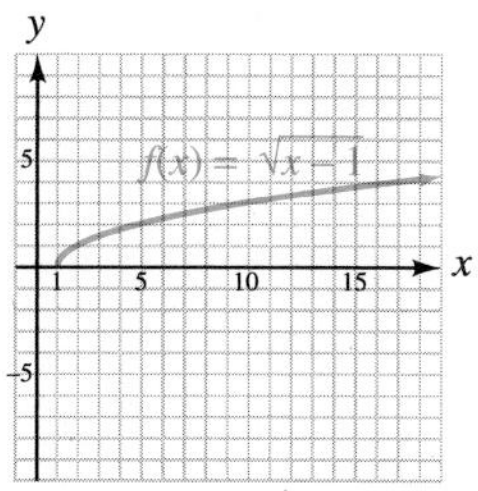

STUDY SET SECTION 10.2 (page 848)

8. b.

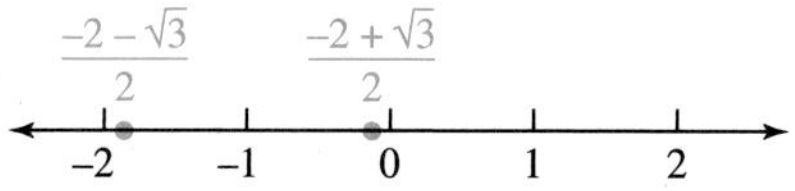

ARE YOU READY? 10.4 (page 861)

1.

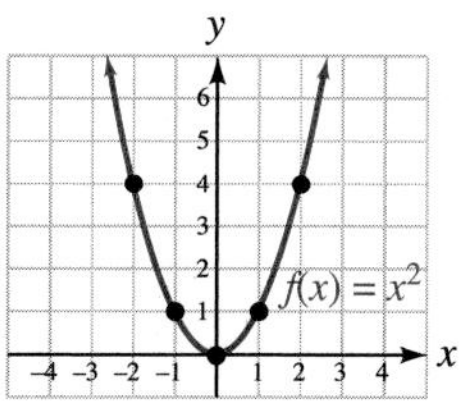

SELF CHECKS 10.4

1.

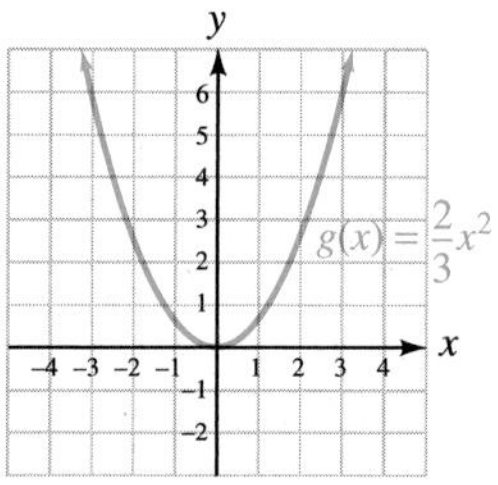

2.

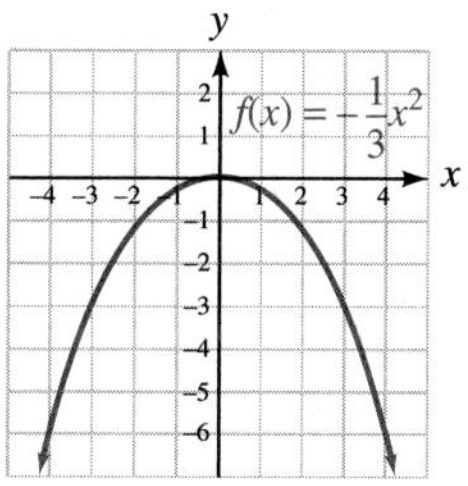

3.

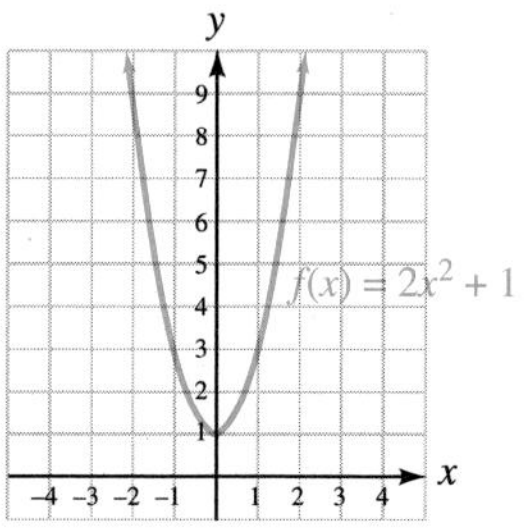

4.

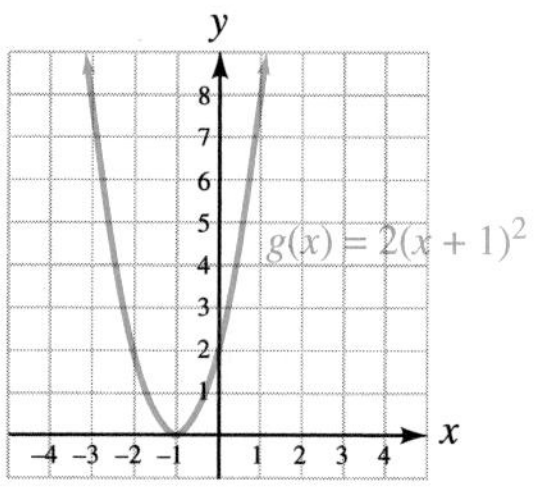

5.

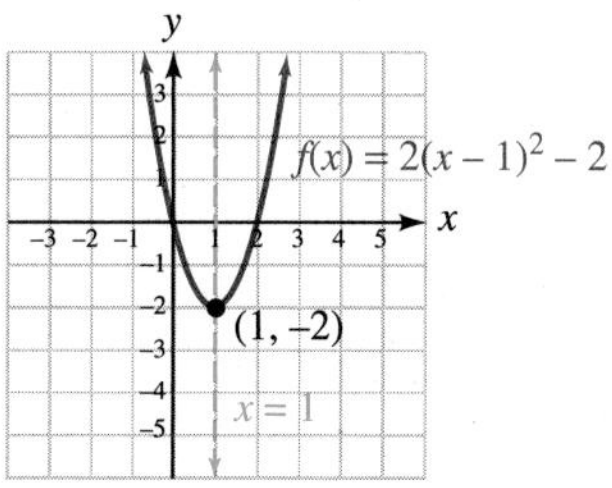

7.

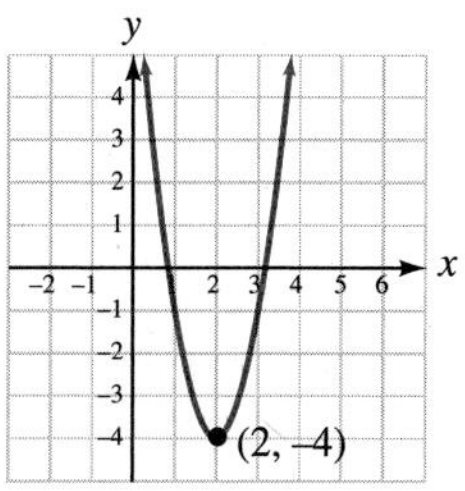

$f(x) = 3x^2 - 12x + 8$

9.

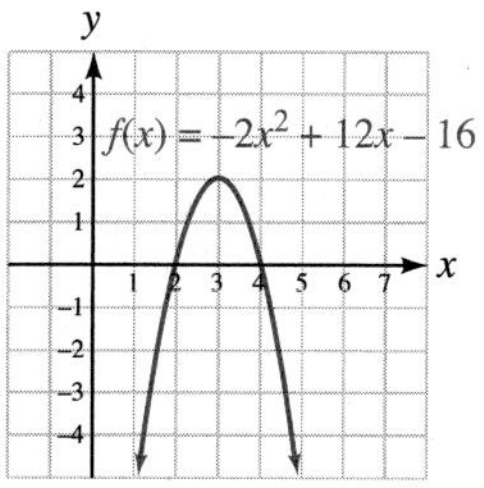

STUDY SET SECTION 10.4 (page 872)

7.

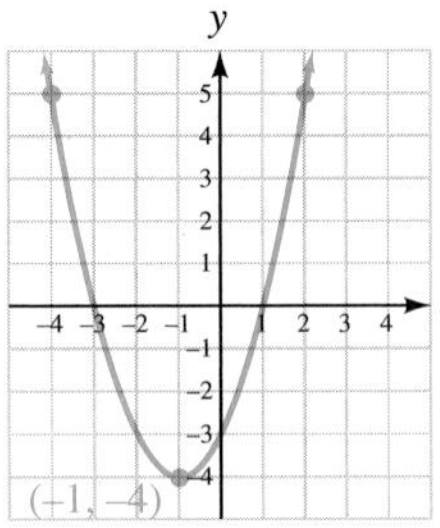

15.

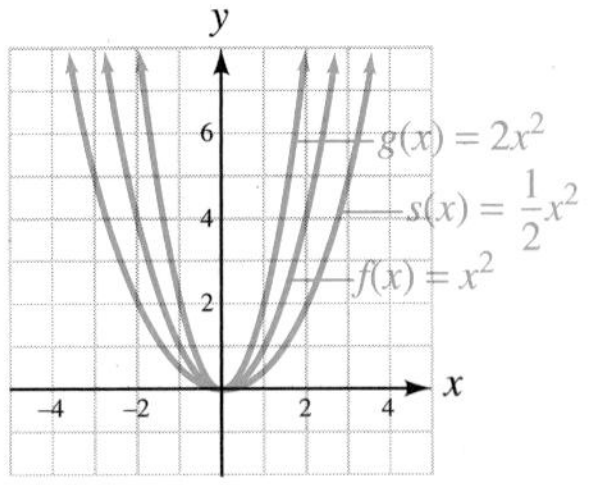

16.

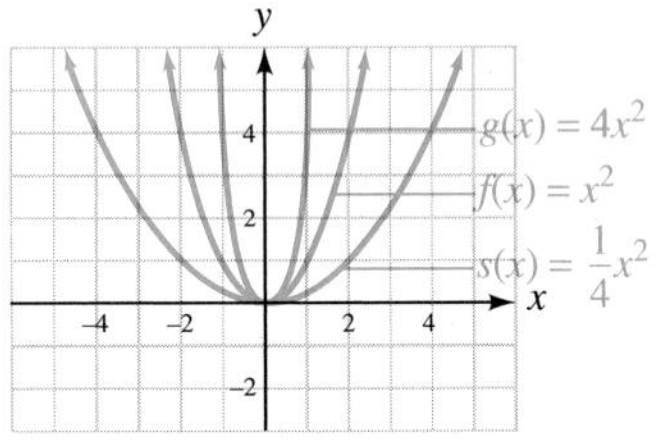

17.

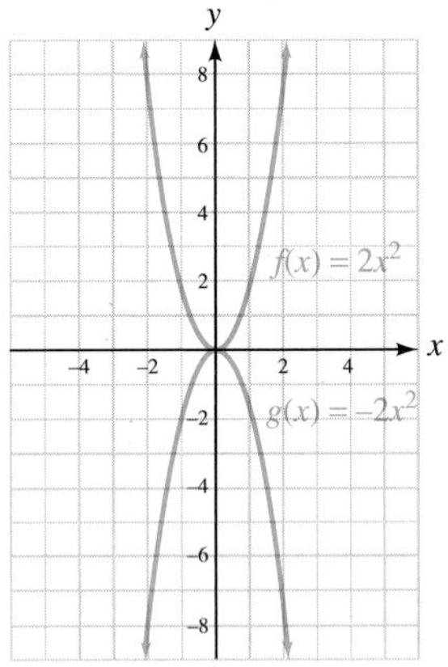

18.

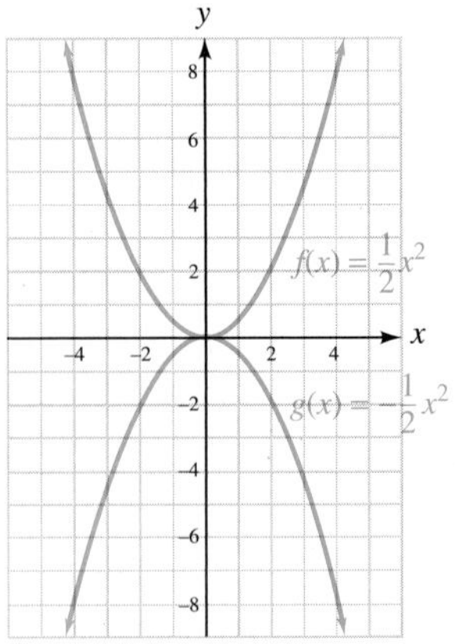

19.

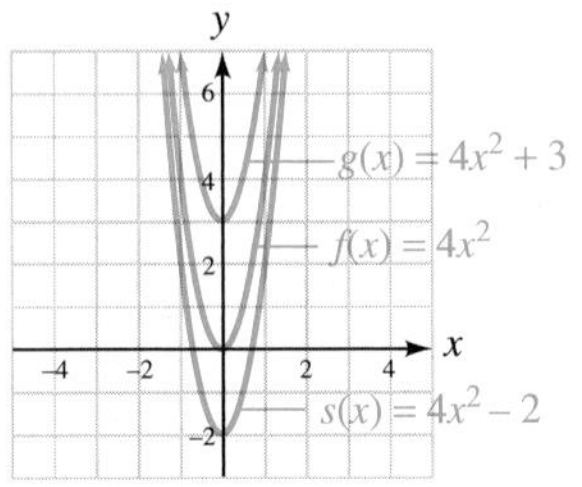

20.

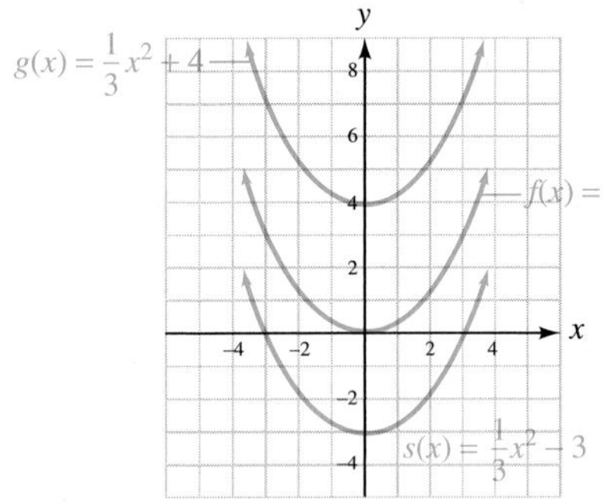

21.

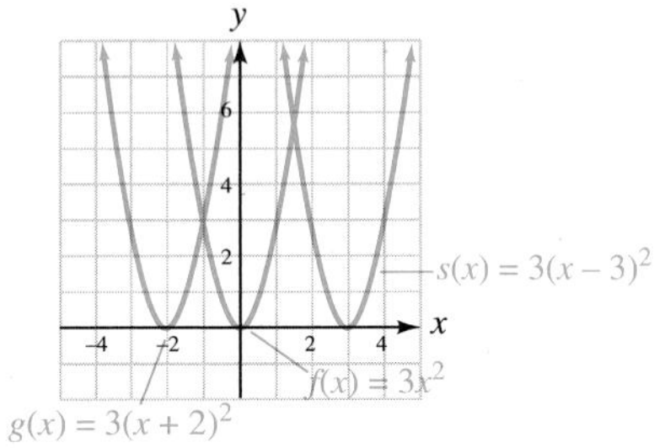

22.

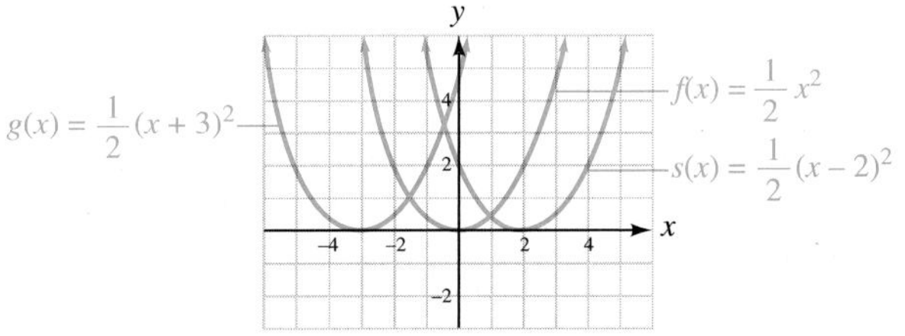

31.

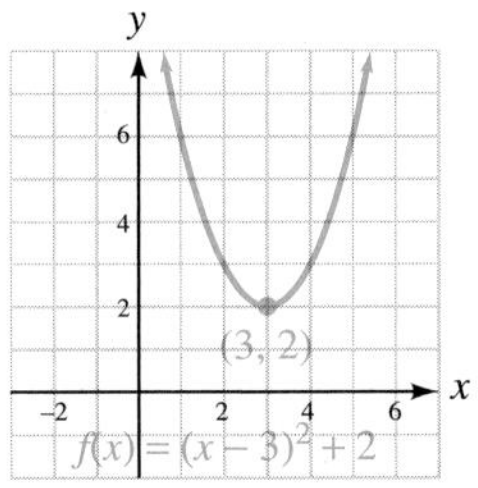

32.

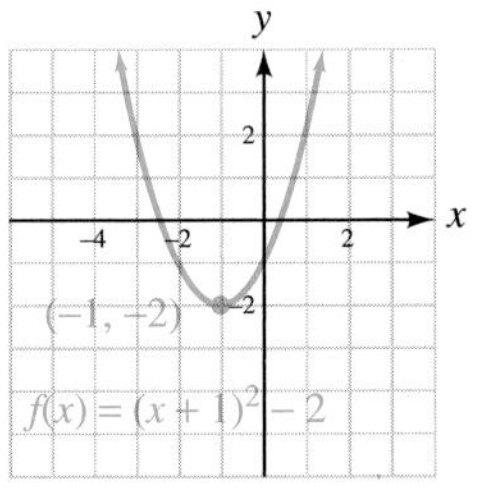

33.

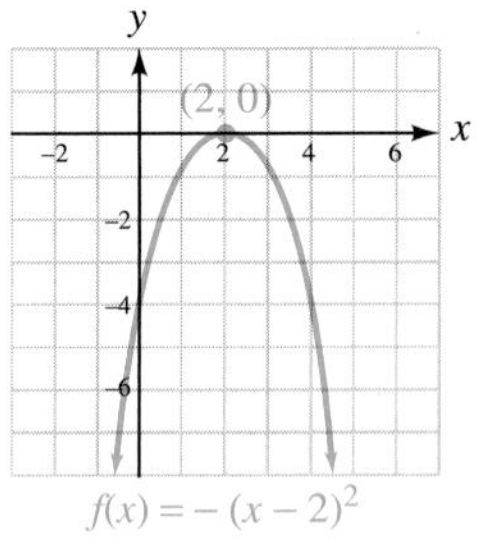

34.

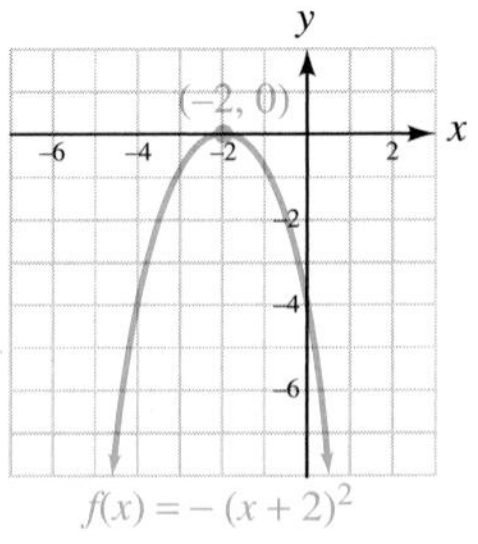

35.

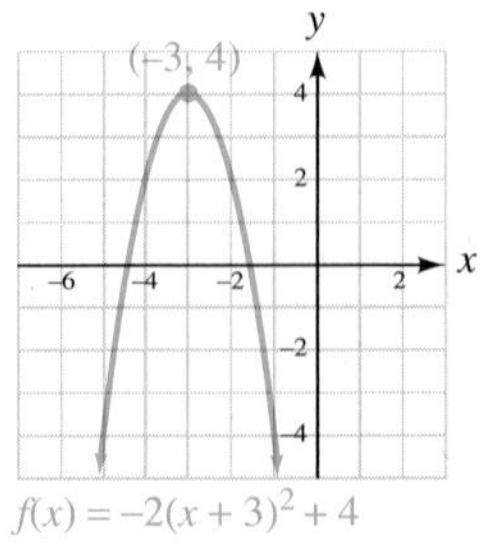

36.

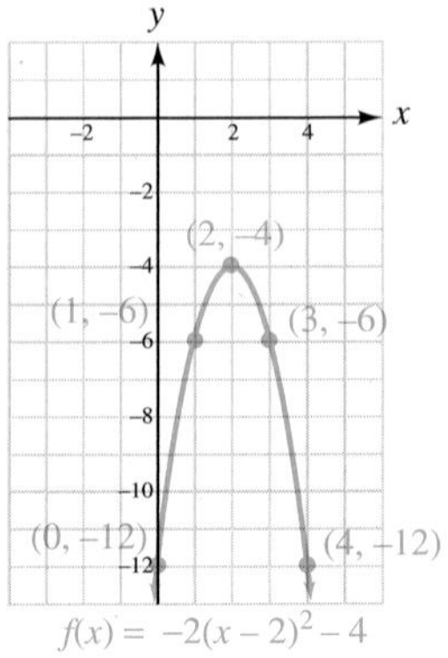

37.
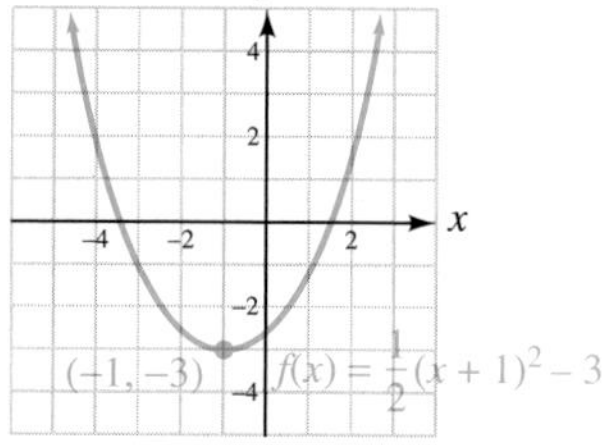

38.
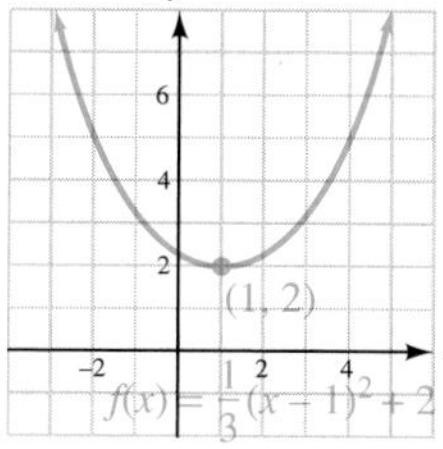

43.
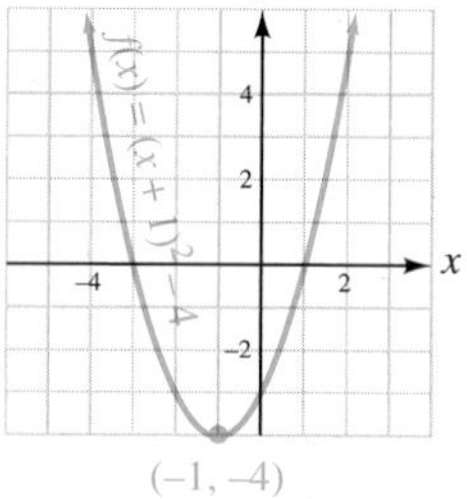

44.
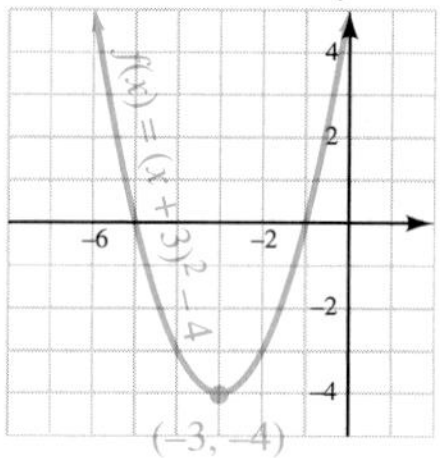

45.
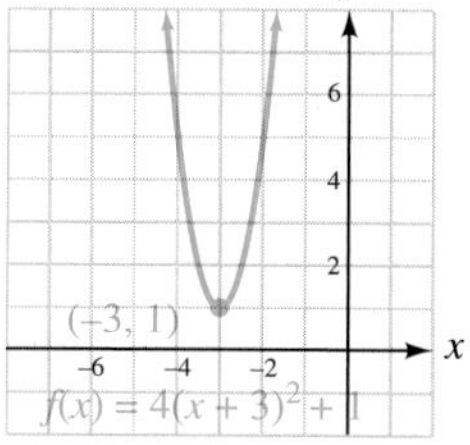

46.
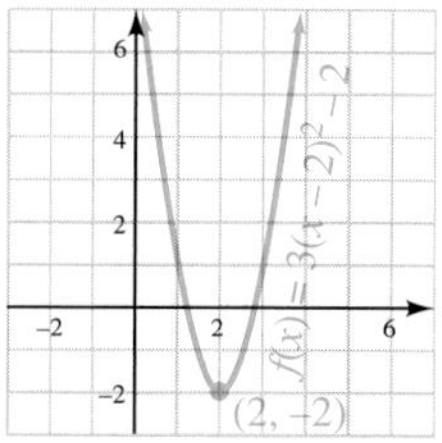

47.
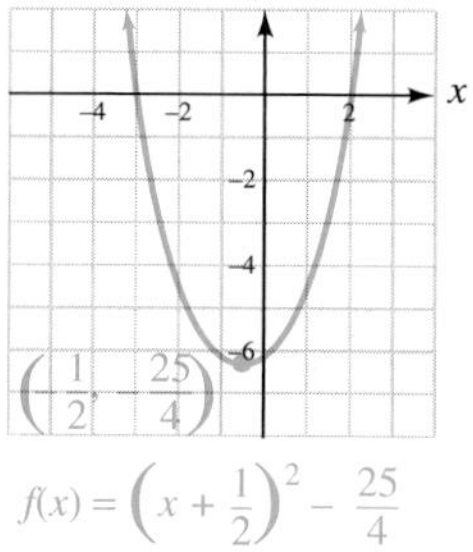

48.
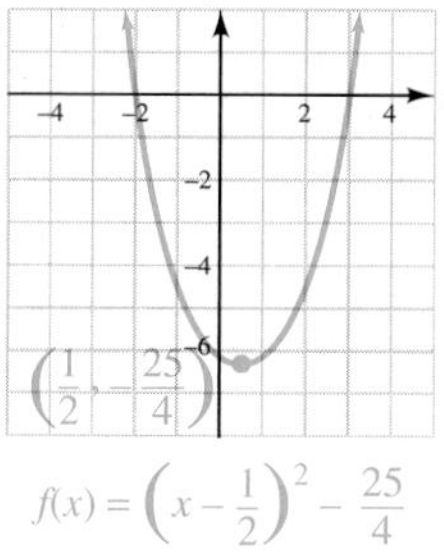

49.
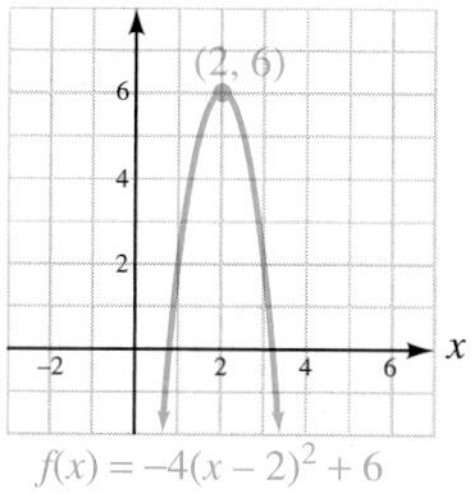

50.
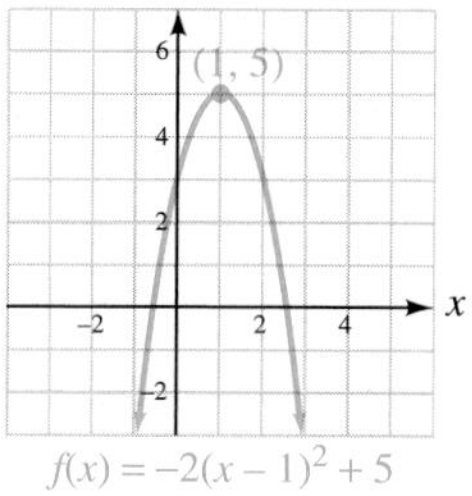

51.
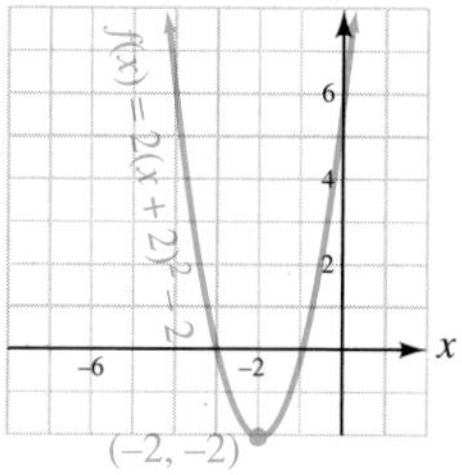

52.
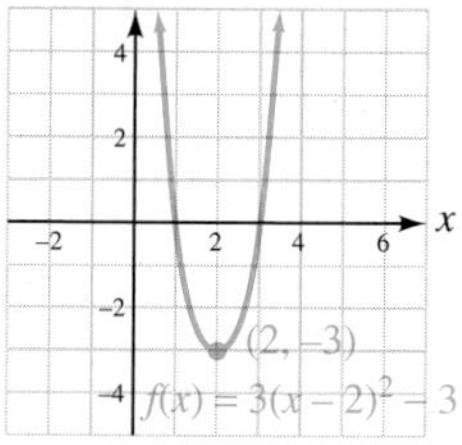

53.
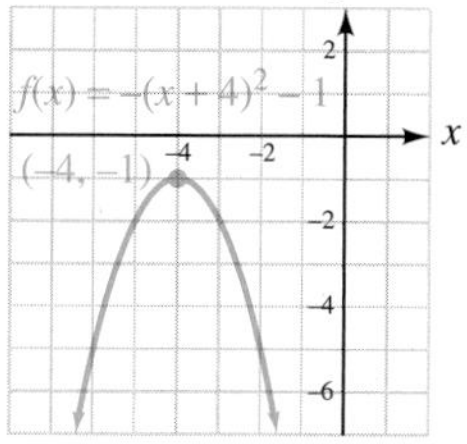

54.
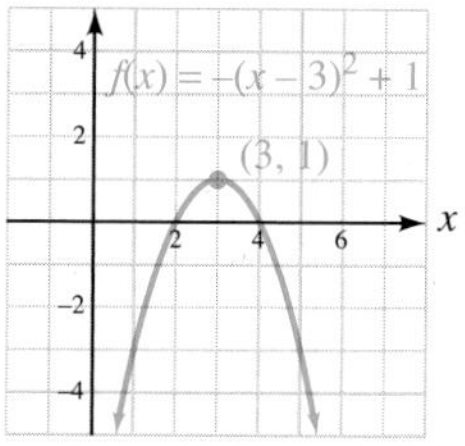

63.
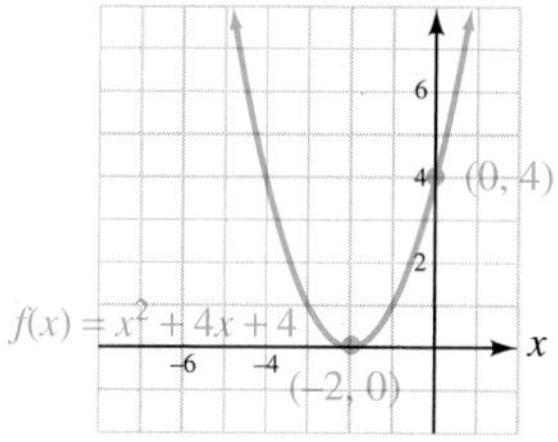

64.
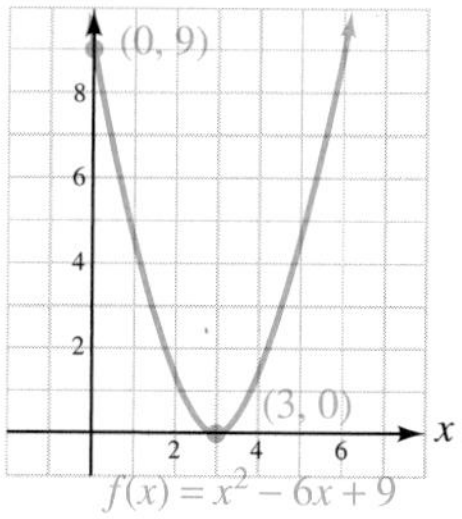

65.
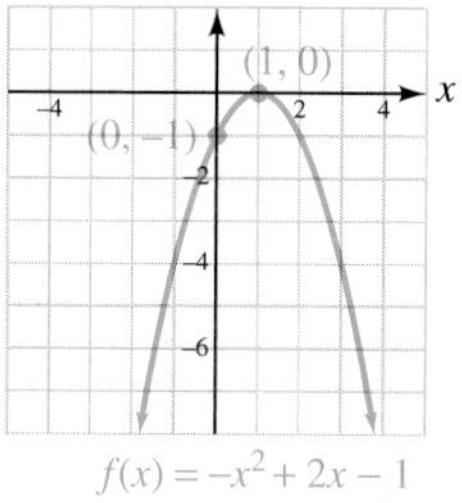

66.
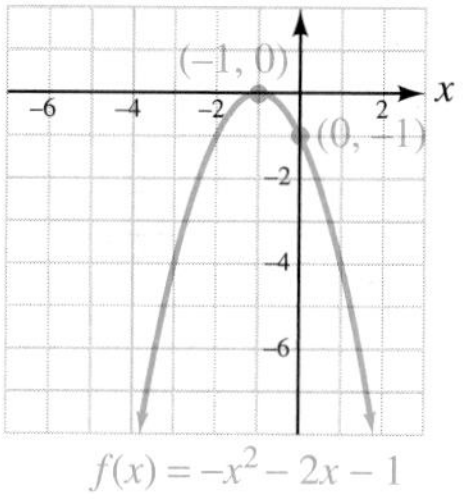

67.
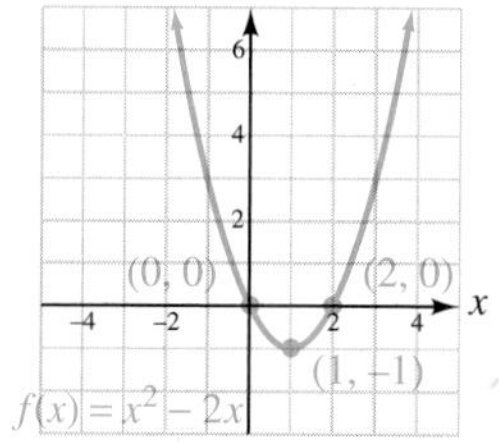

68.
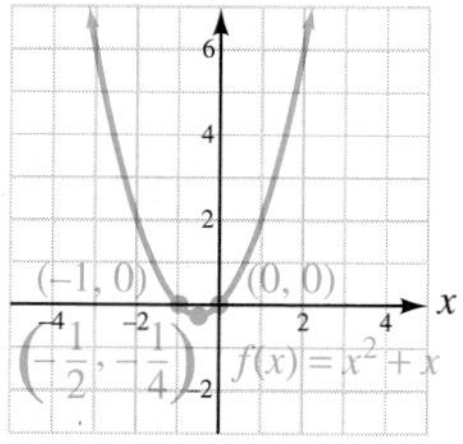

69.
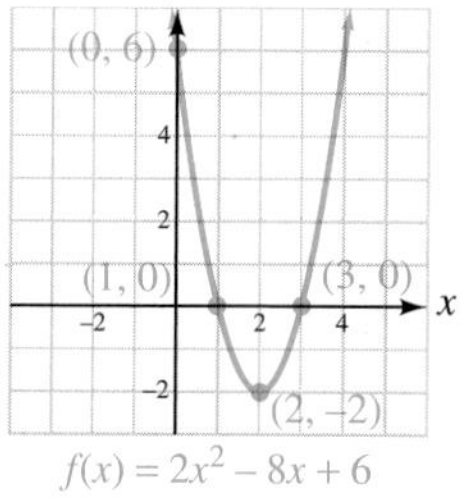

70.
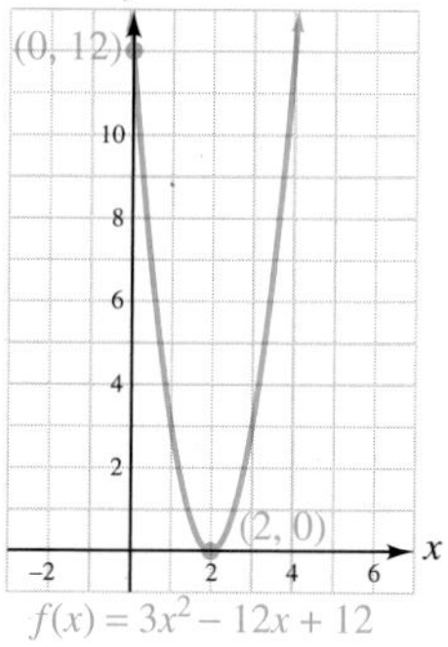

71.

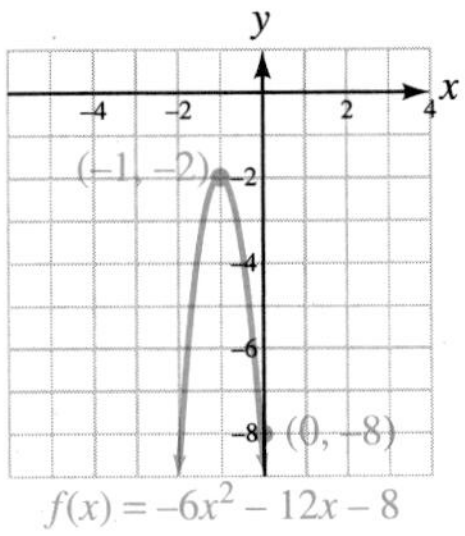

72.

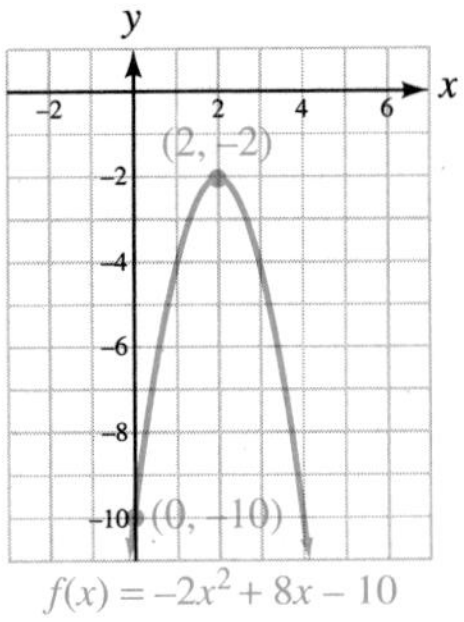

73.

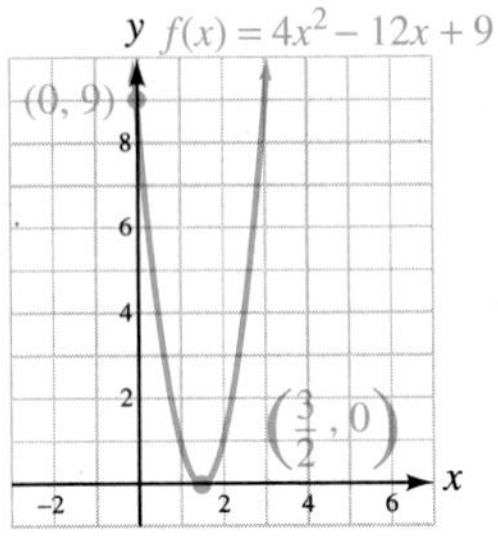

74.

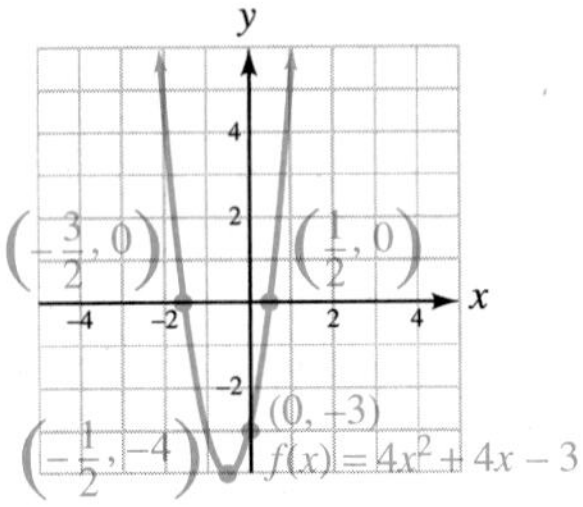

ARE YOU READY? 10.5 (page 875)

3.

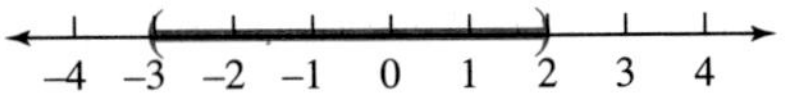

4.

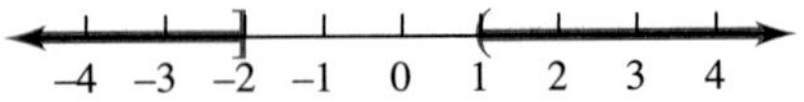

6.

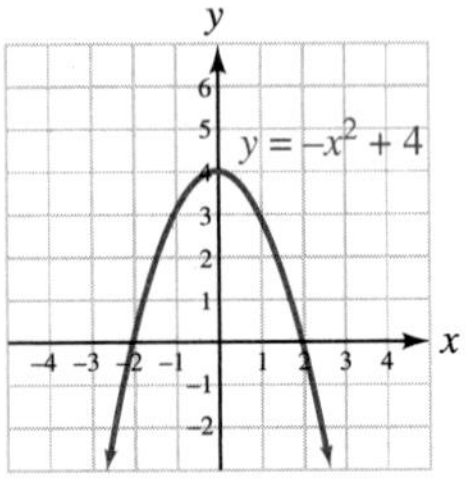

SELF CHECKS 10.5

1.

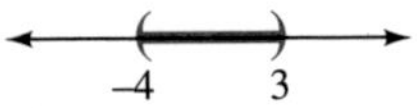

2.

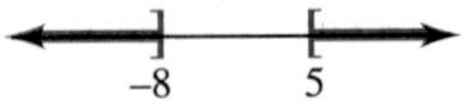

3.

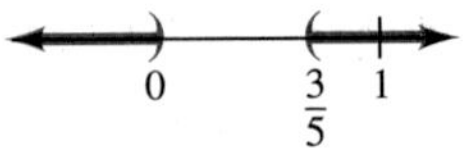

4.

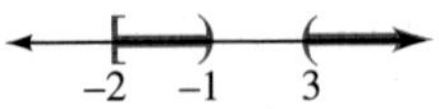

5.

6.

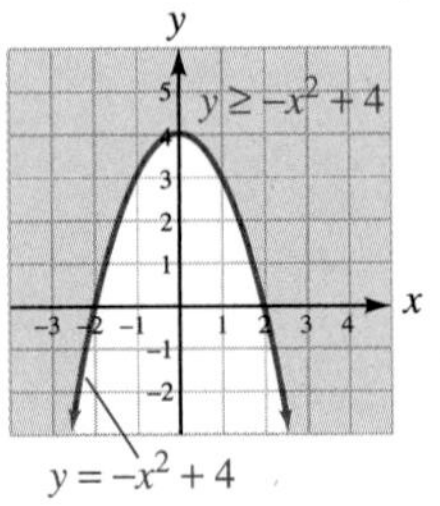

7.

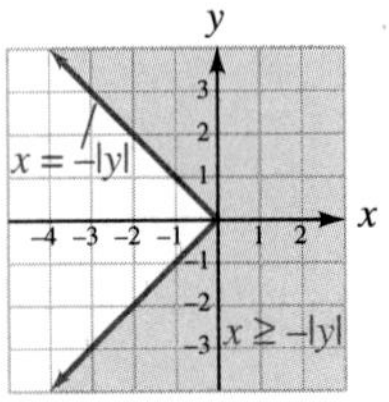

STUDY SET SECTION 10.5 (page 884)

6. a.

b.

15.

16.

17. **18.**

19. **20.**

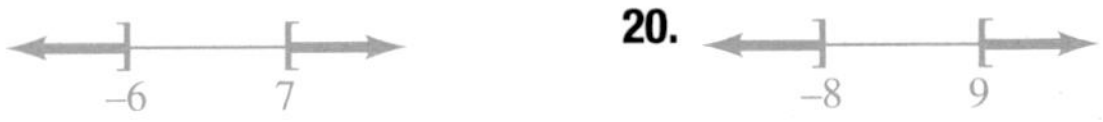

21. **22.**

23. **24.**

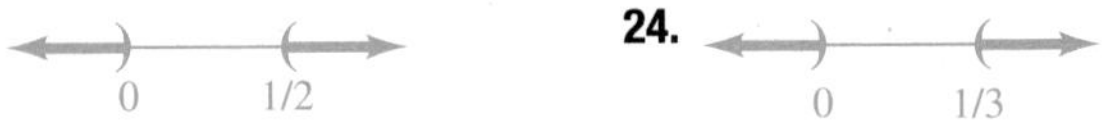

25. **26.**

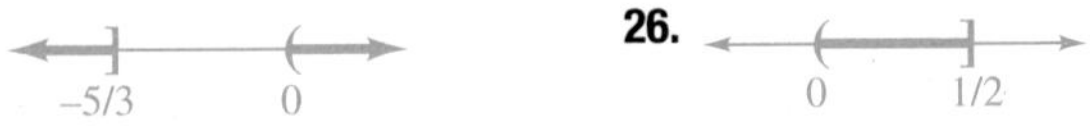

27. **28.**

29. **30.**

31. **32.**

33. **34.**

35.

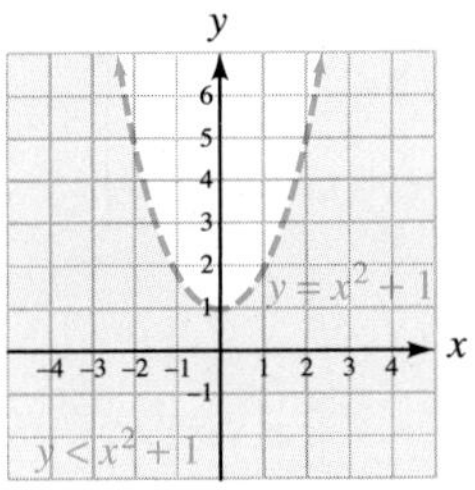

36.

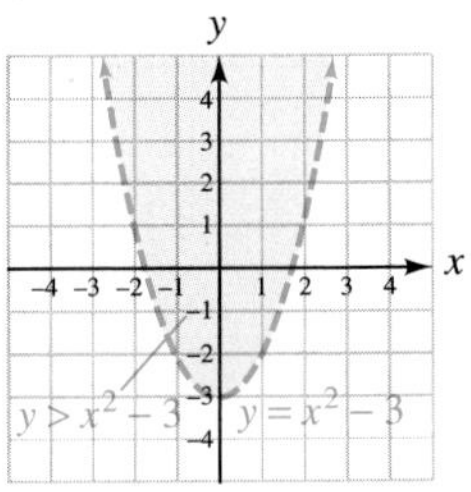

37.

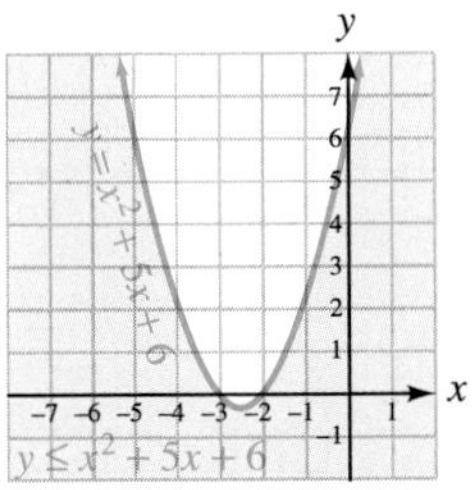

38.

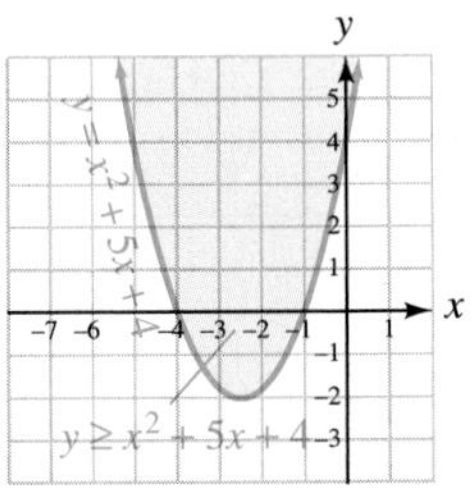

39.

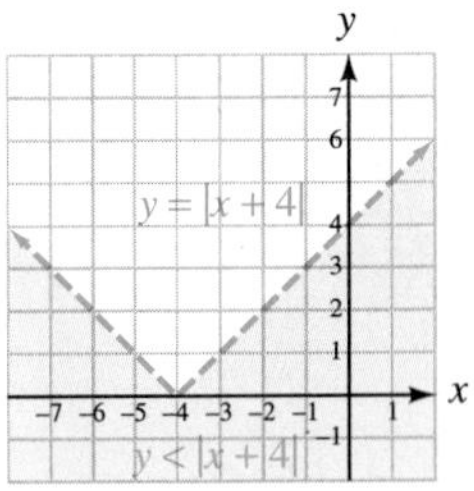

40.

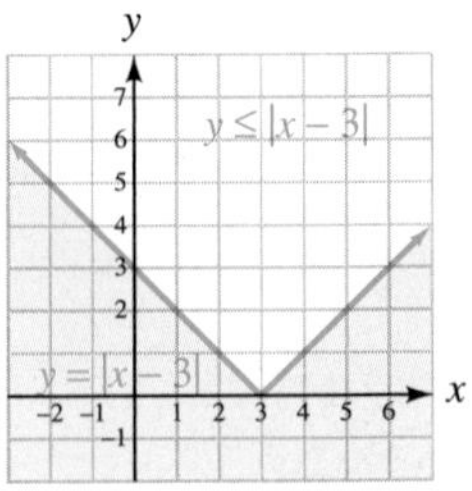

41.

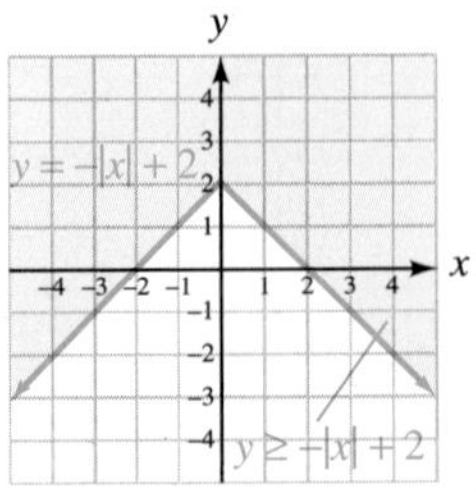

42.

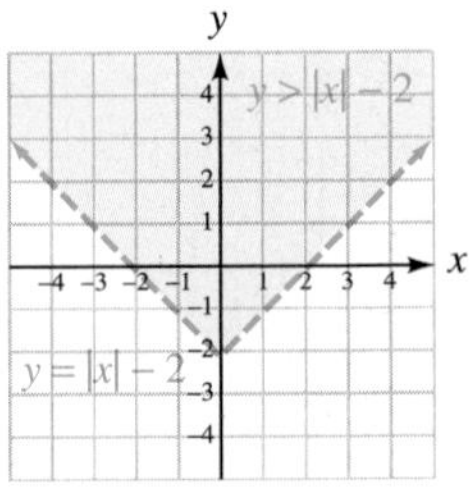

47.

48.

49.

50.

51. 

53. 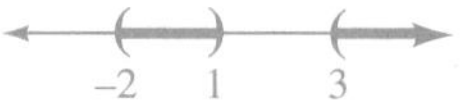

54. 

55. 

56. 

57. 

58. 

60. 

61. 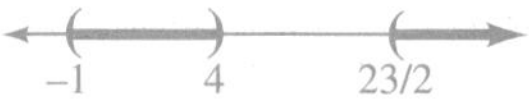

62. 

CHAPTER 10 REVIEW (page 886)

49. 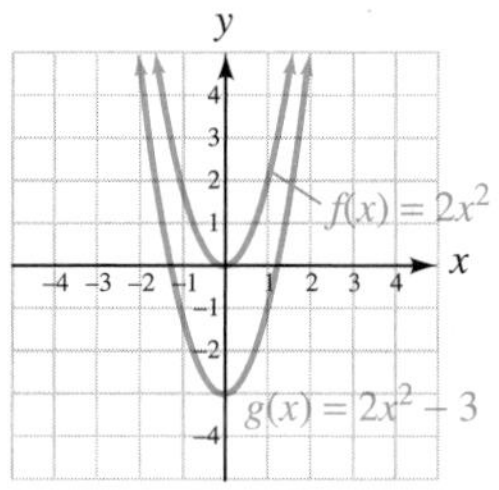

50. 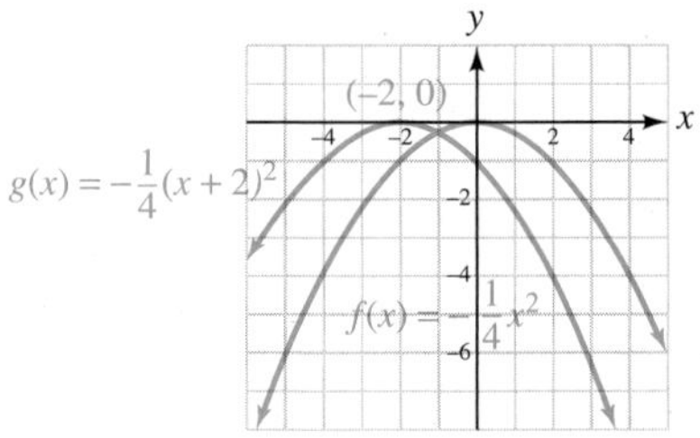

51. 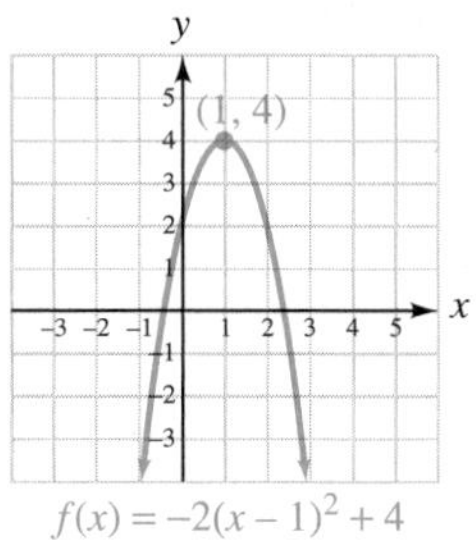

52. 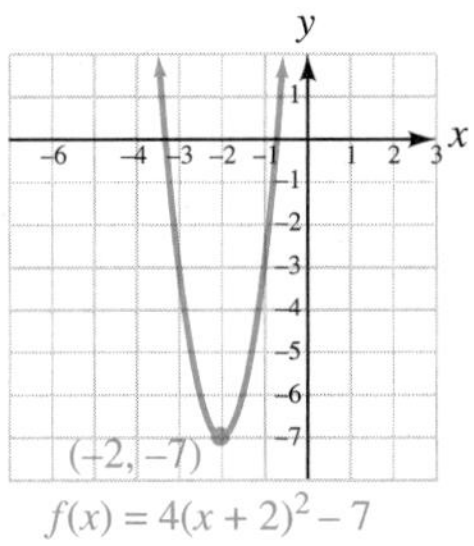

54. 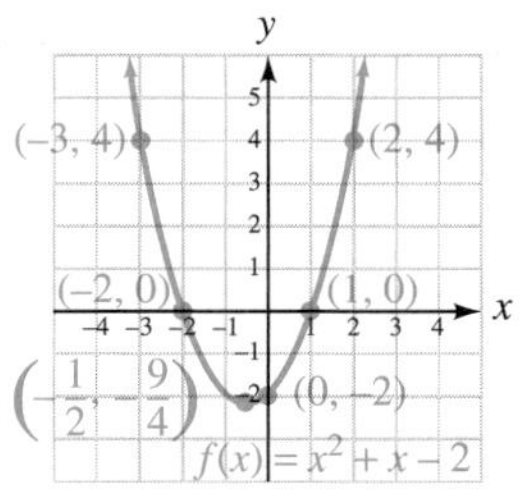

57. 

58. 

59. 

60. 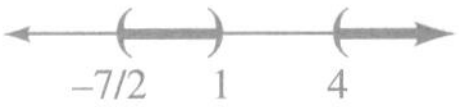

63. 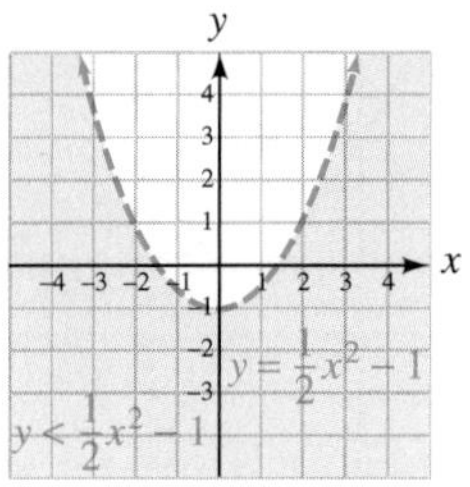

64. 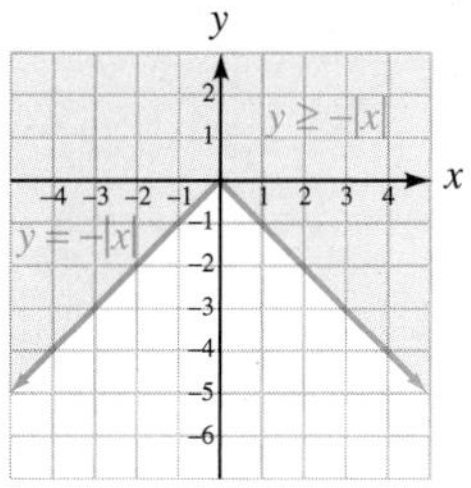

CHAPTER 10 TEST (page 892)

31. 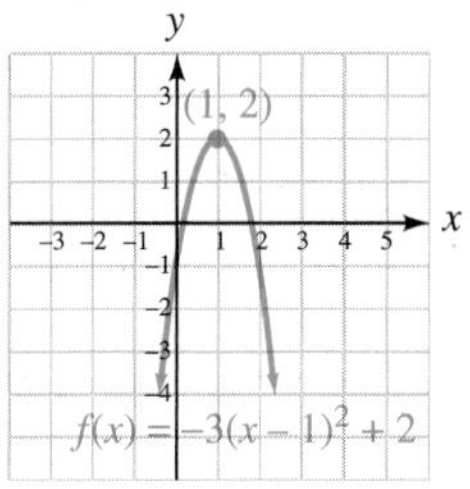

32. 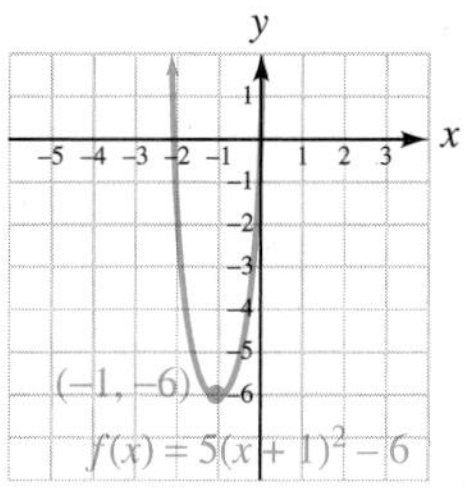

33. 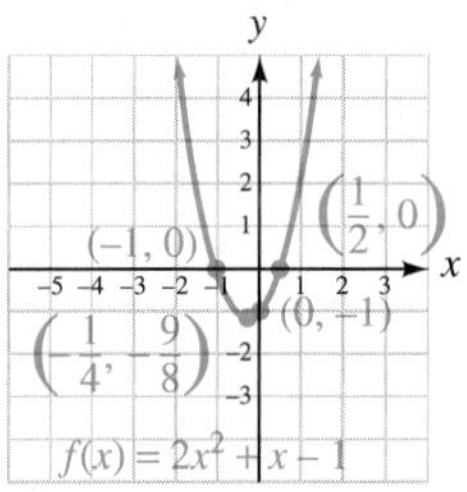

35. 

36. 

38. 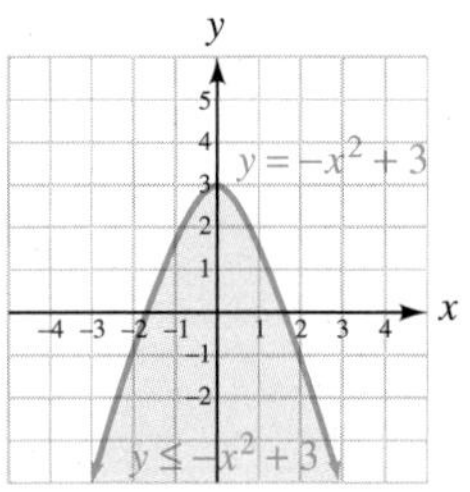

CUMULATIVE REVIEW CHAPTERS 1–10 (page 895)

13. 

15. 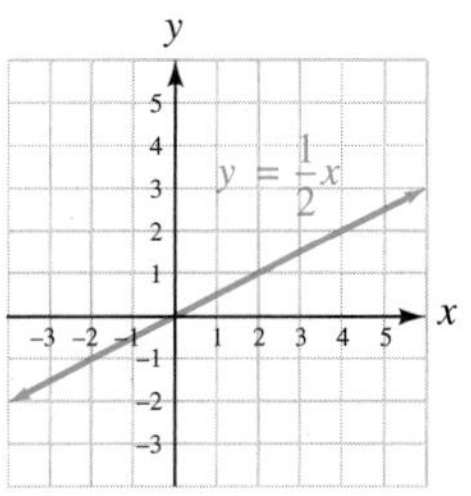

16. 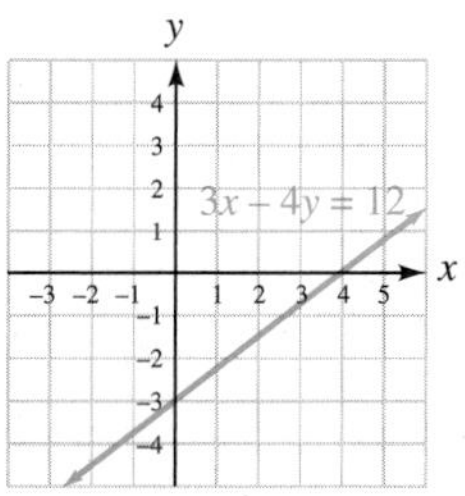

17. 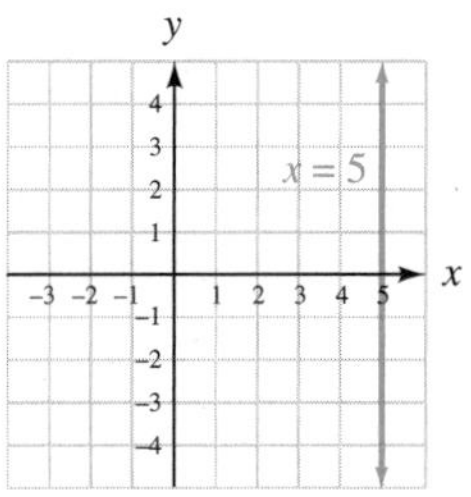

18. 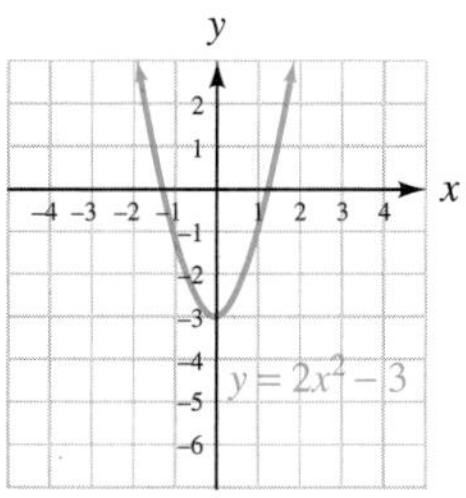

28.

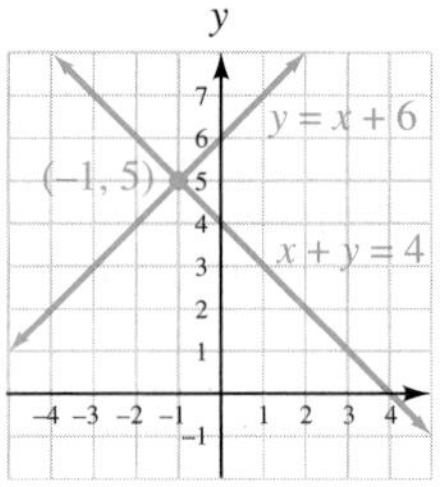

32.

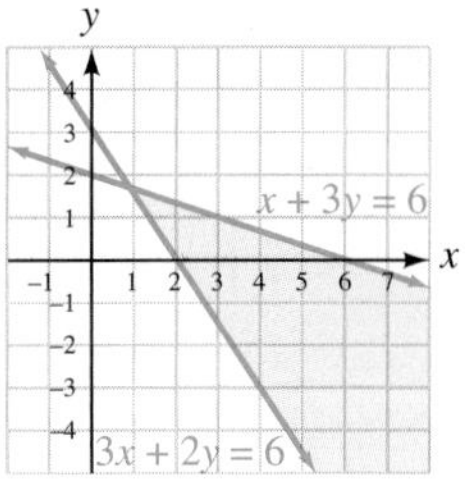

91.

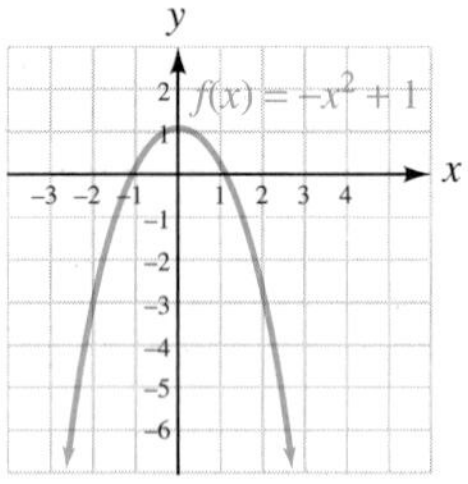

92.

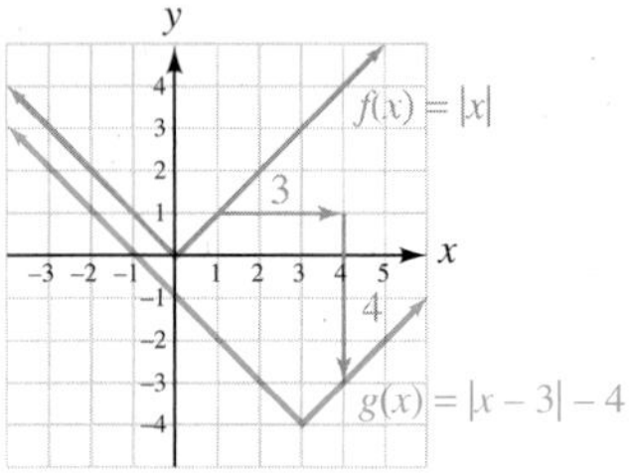

95.

96.

97.

98.

99.

107.

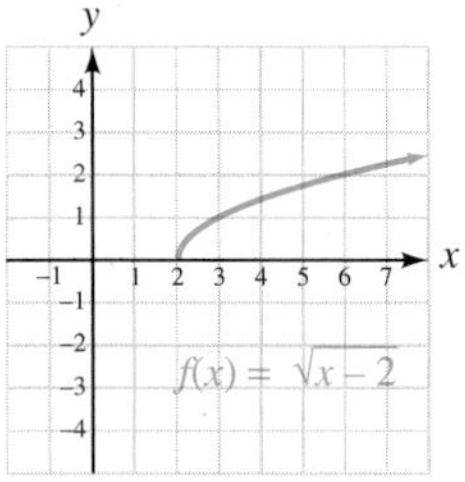

135.

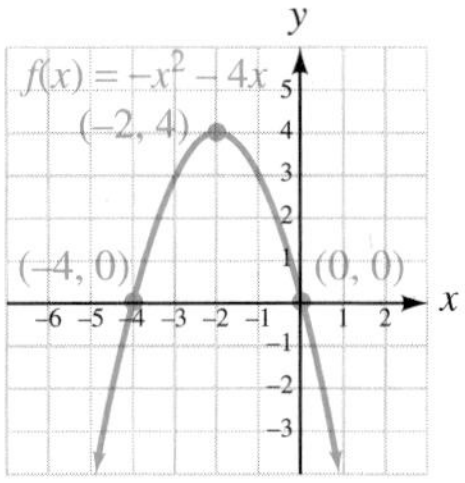

136.

137.

SELF CHECKS 11.2

5.

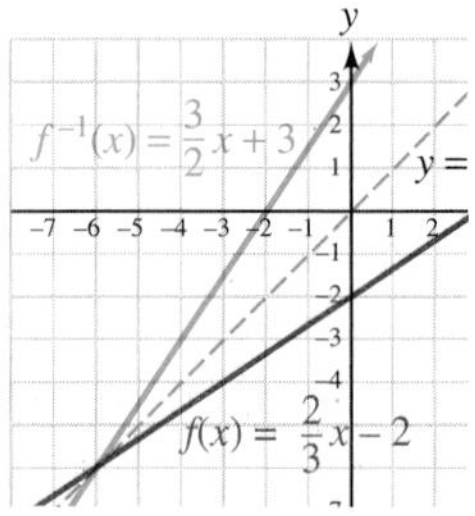

STUDY SET SECTION 11.2 (page 918)

59.

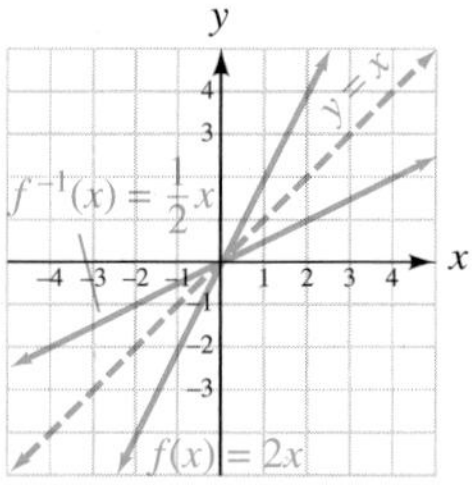

60.

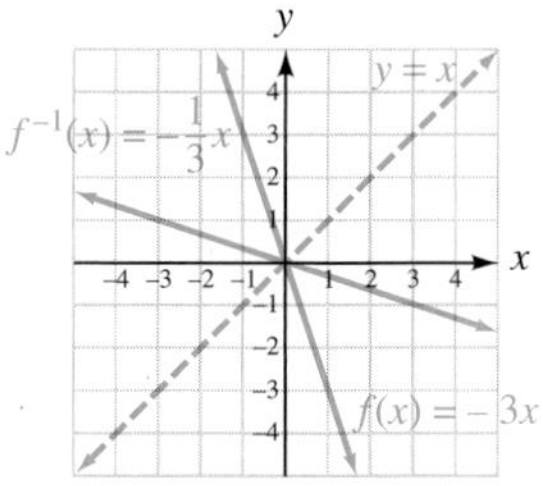

61.

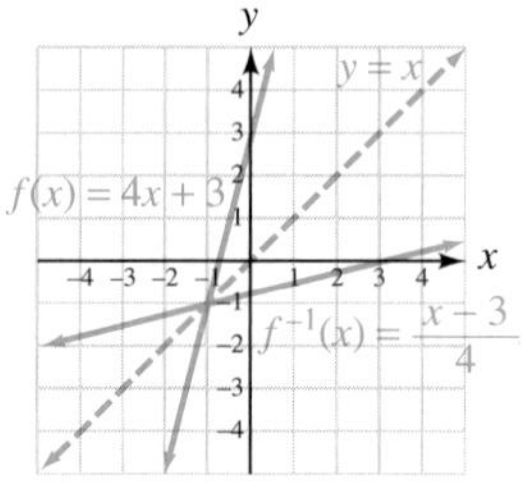

62.

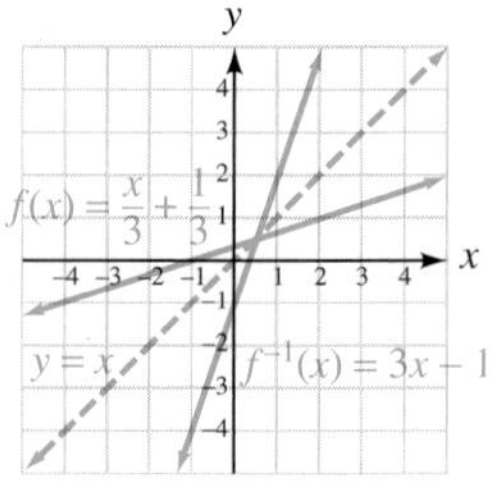

63.

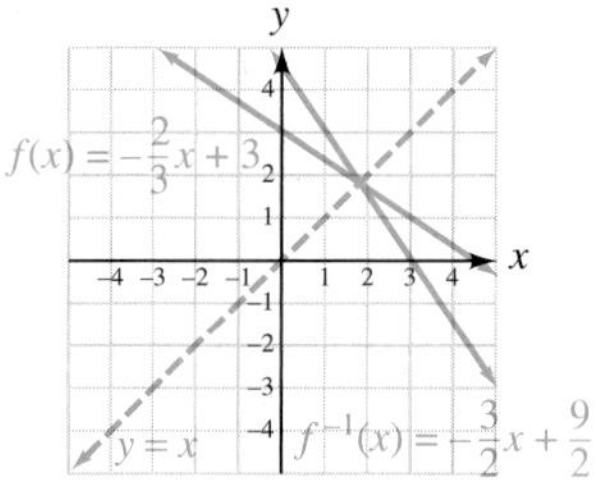

64.

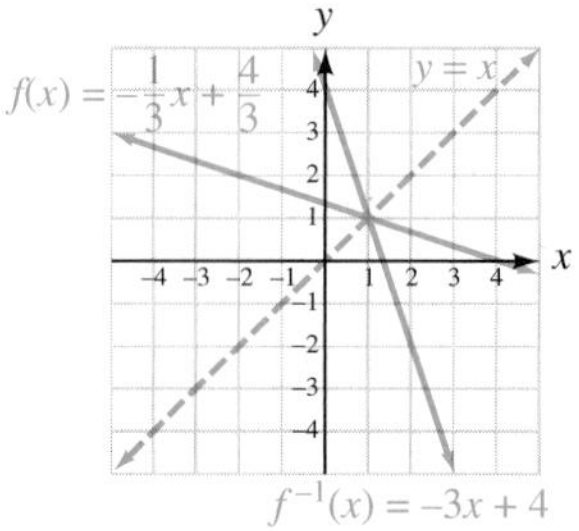

65.

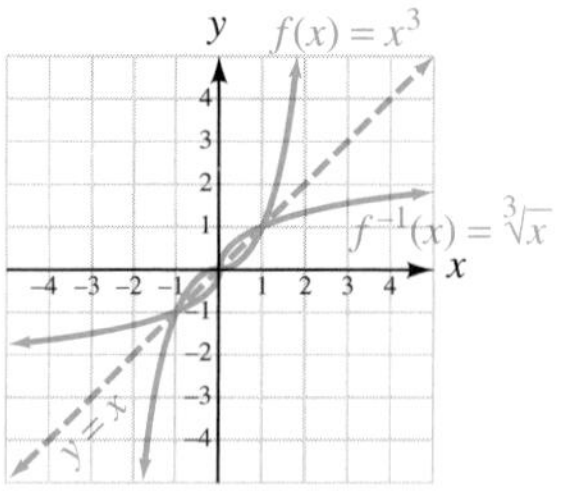

66.

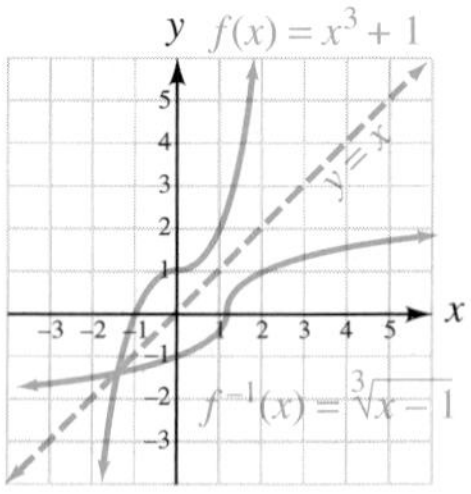

67.

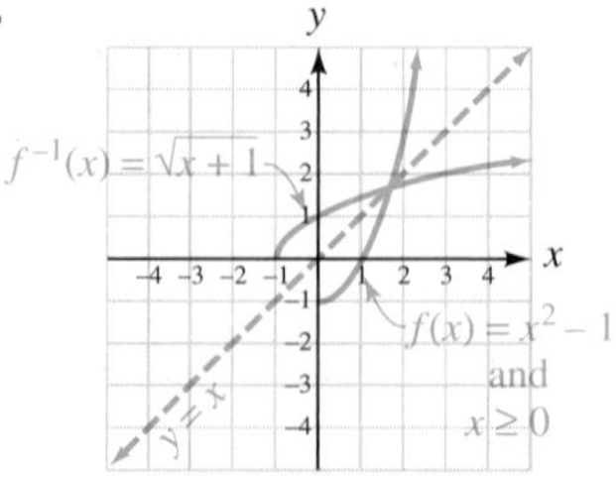

68.

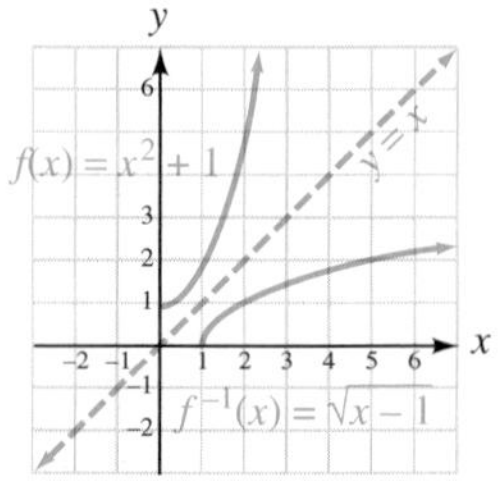

SELF CHECKS 11.3

1.

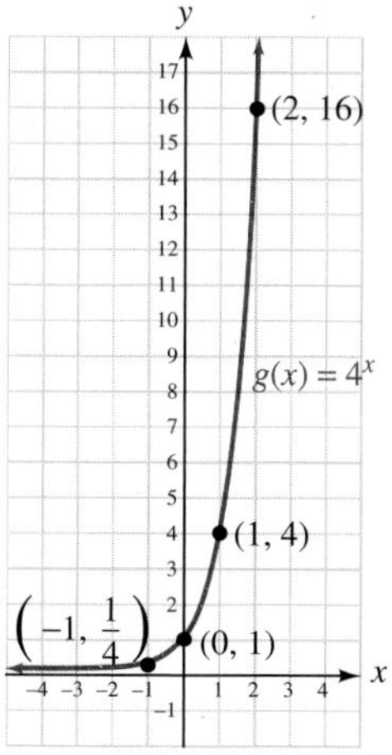

2.

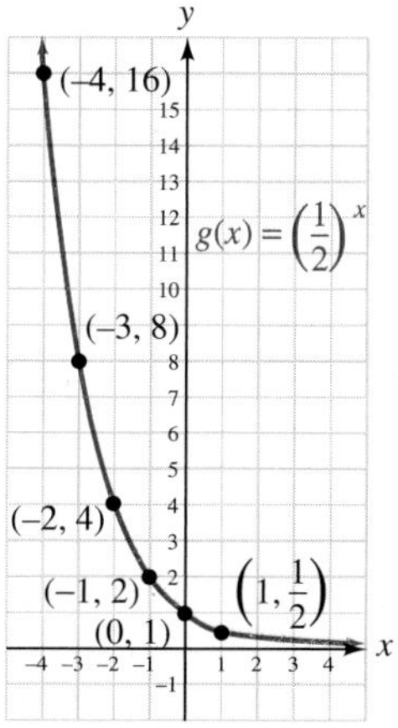

3. a.

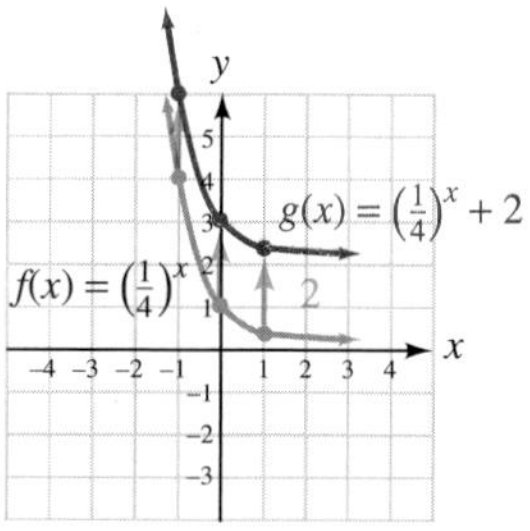

b.

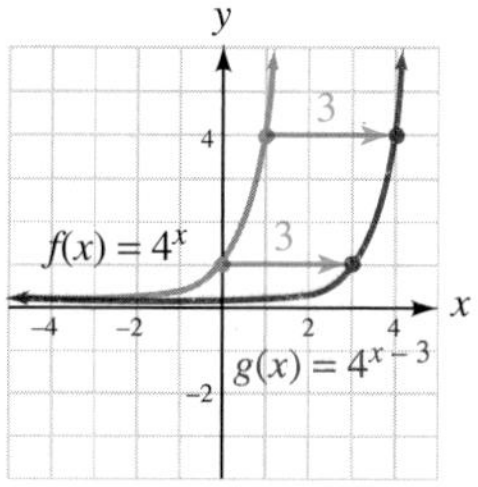

STUDY SET SECTION 11.3 (page 931)

19.

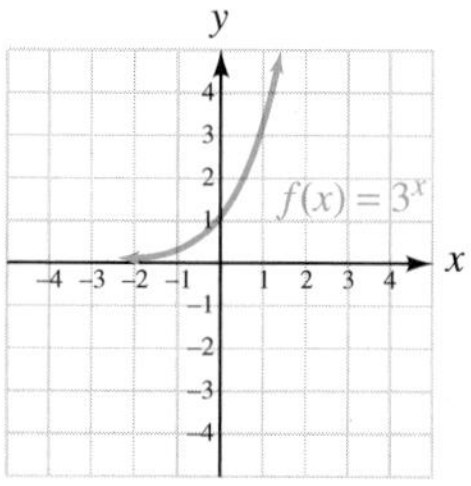

20.

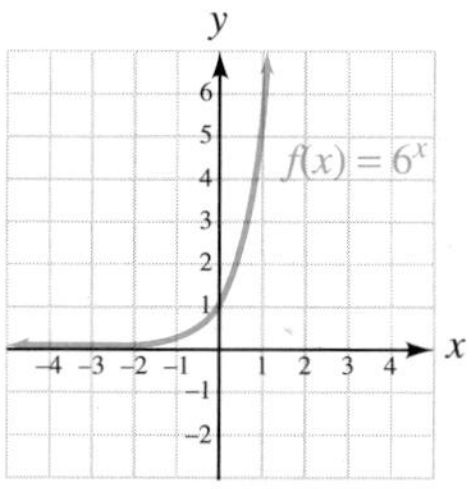

21.

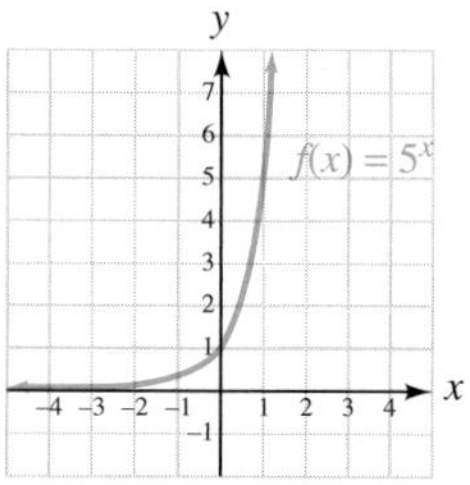

22.

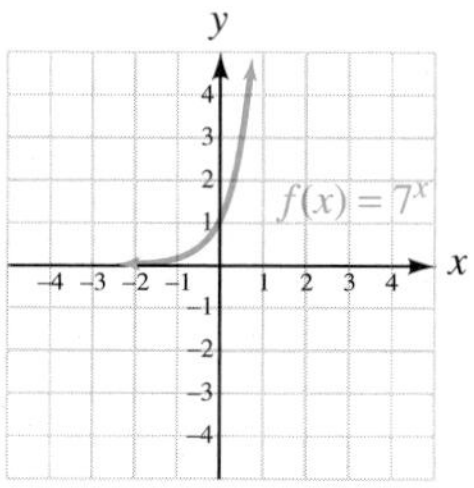

23.

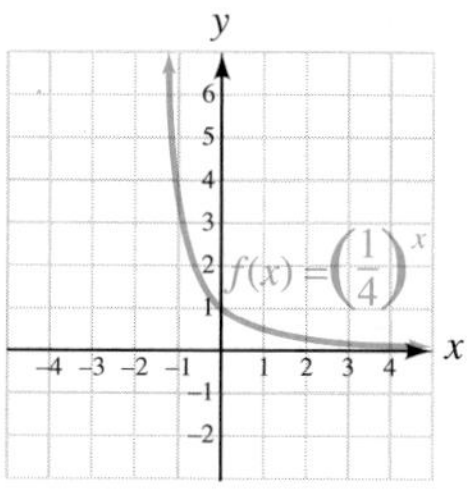

24.

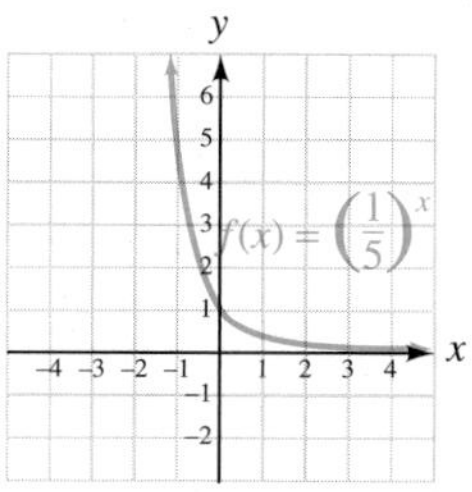

25.

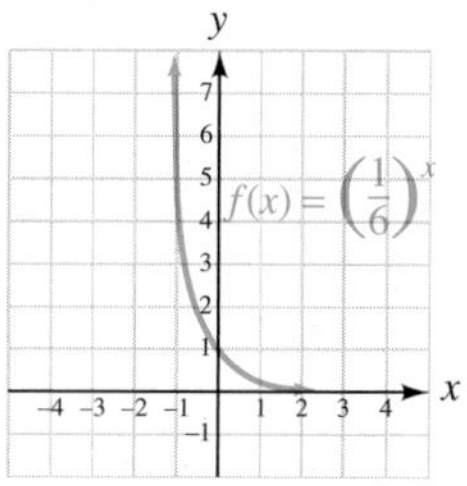

26.

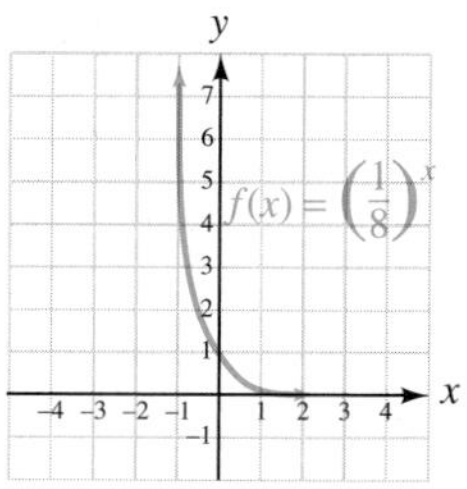

27.

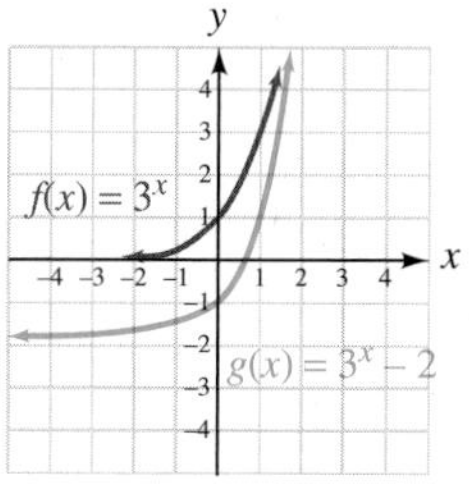

28.

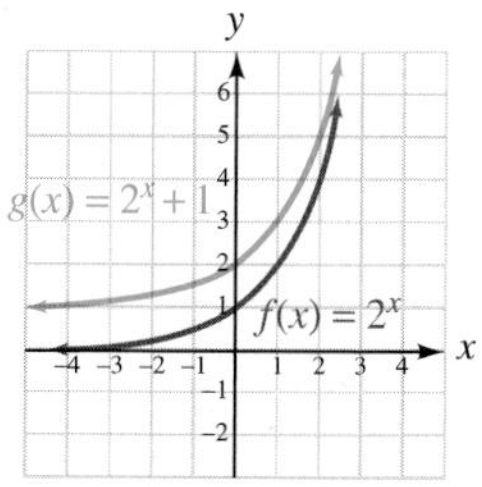

29.

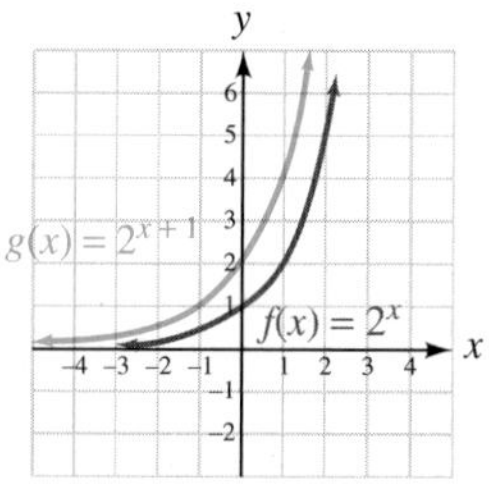

30.

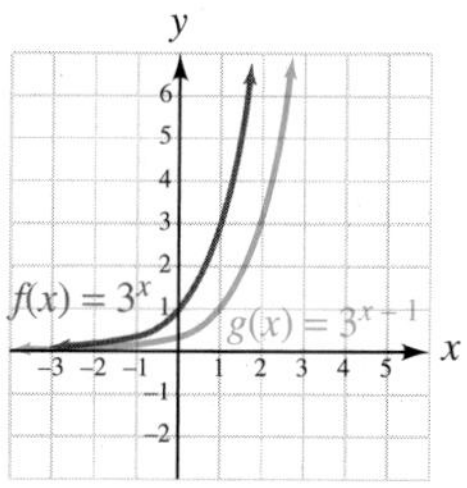

31.

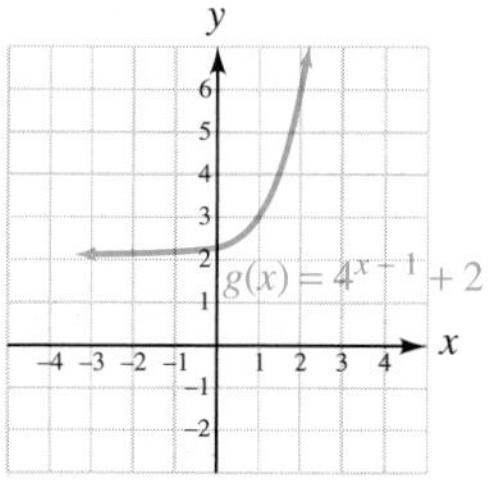

32.

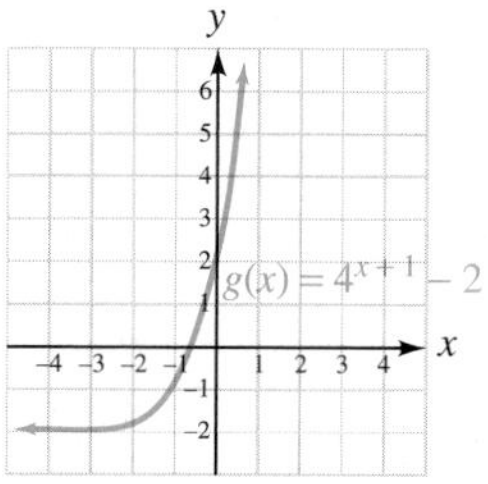

33.

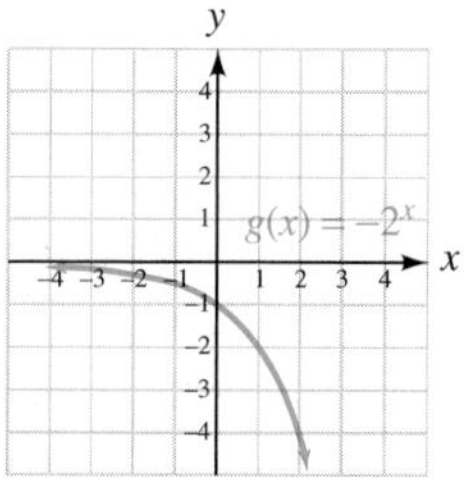

34.

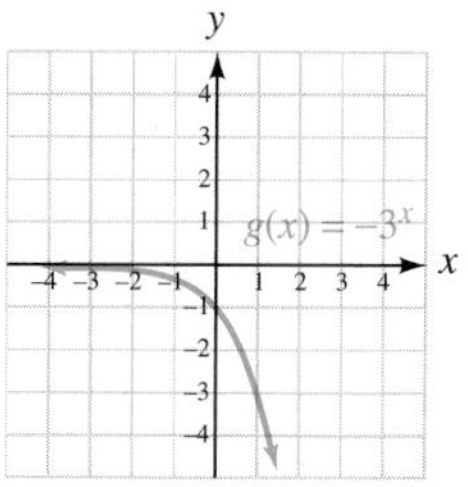

35.

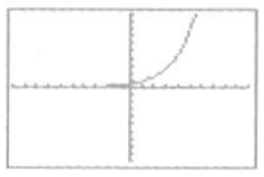

36.

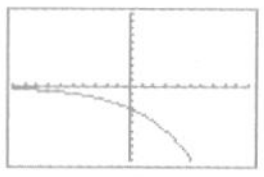

37.

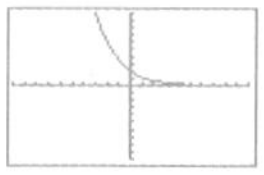

38.

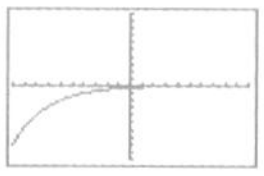

42.

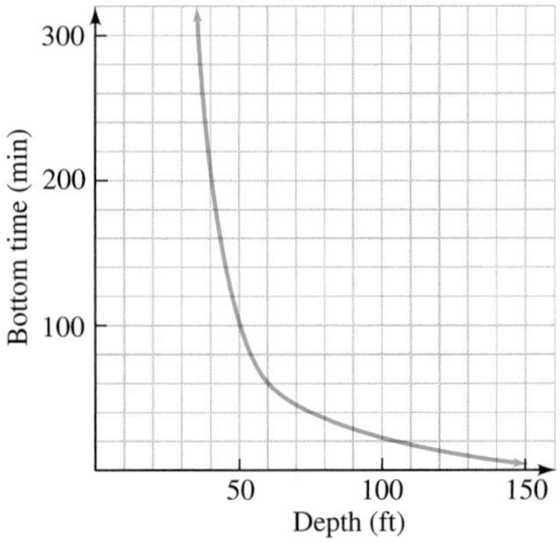

43. a.

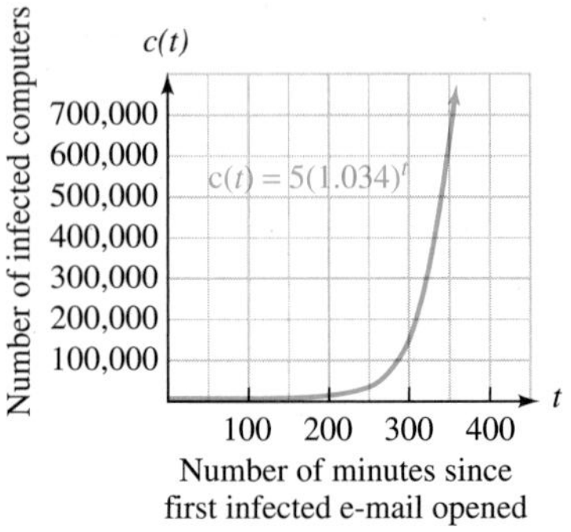

44. a.

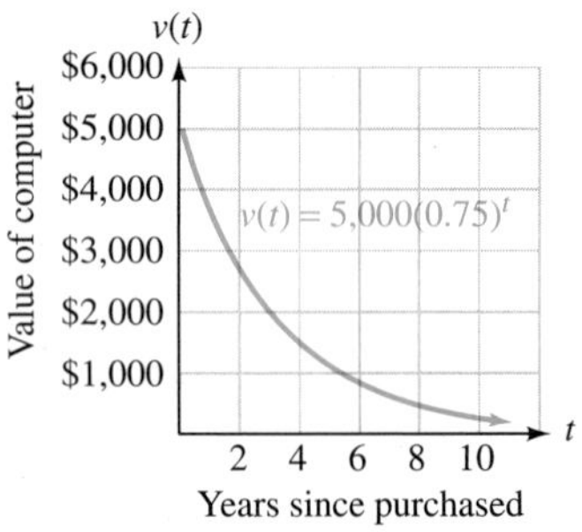

74.

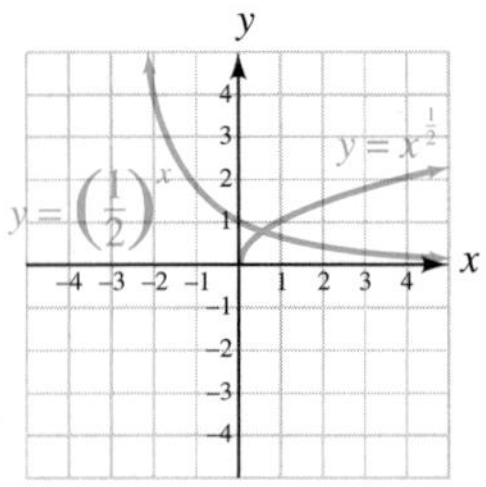

77.

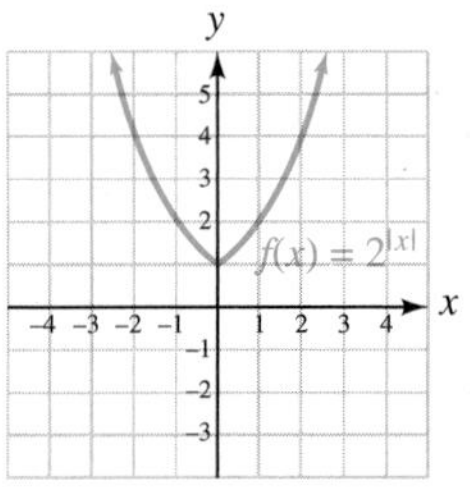

78.

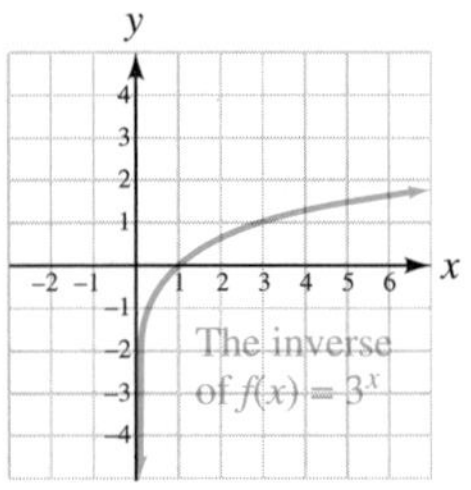

SELF CHECKS 11.4

7. a.

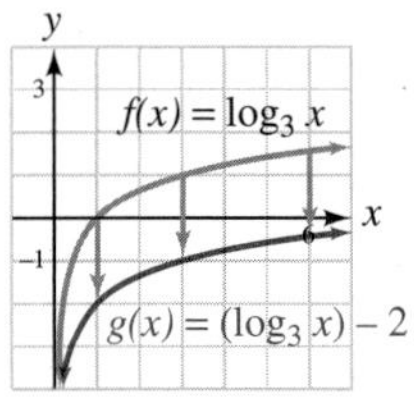

b.

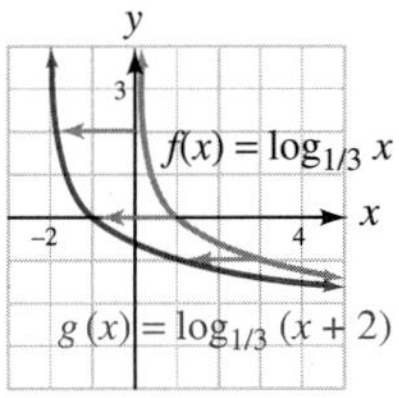

STUDY SET SECTION 11.4 (page 945)

99.

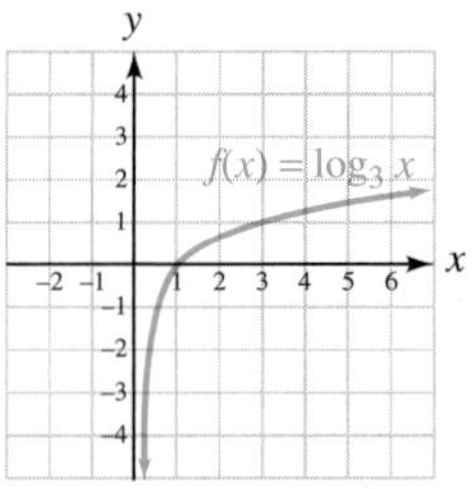

100.

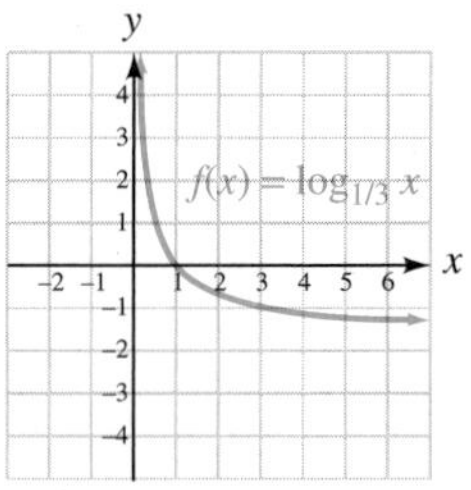

101.

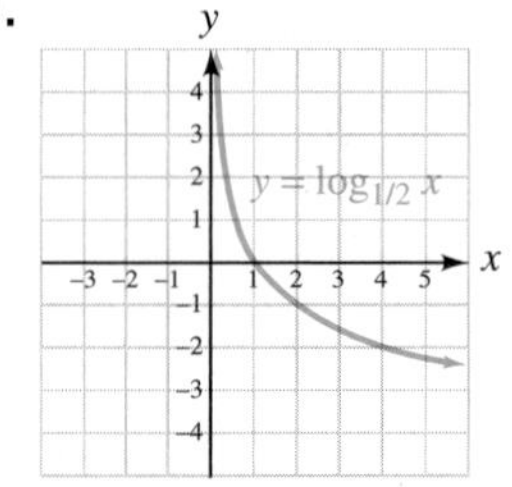

102.

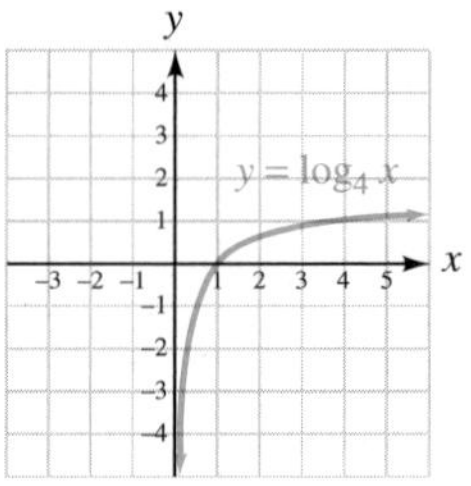

103.

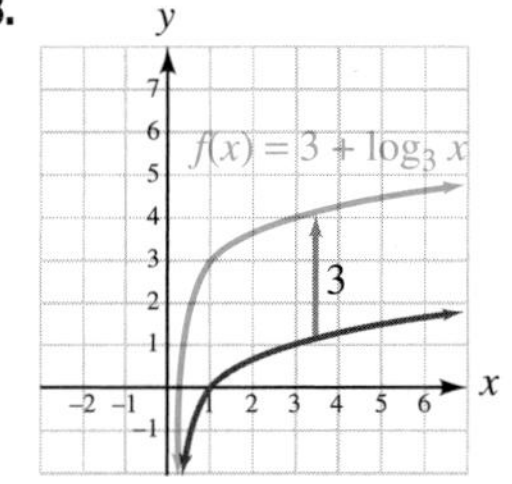

104.

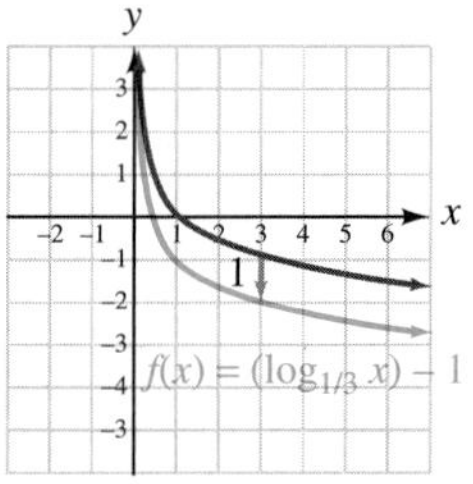

105.

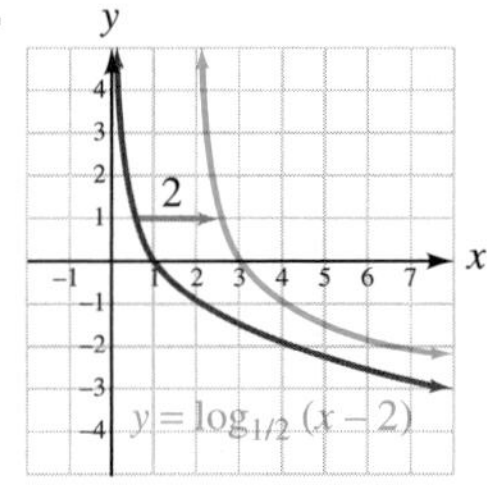

106.

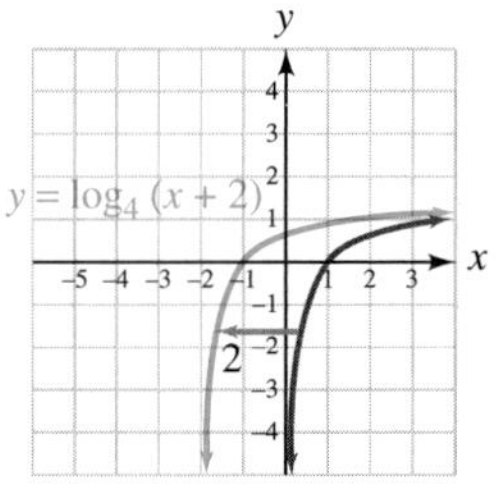

107.

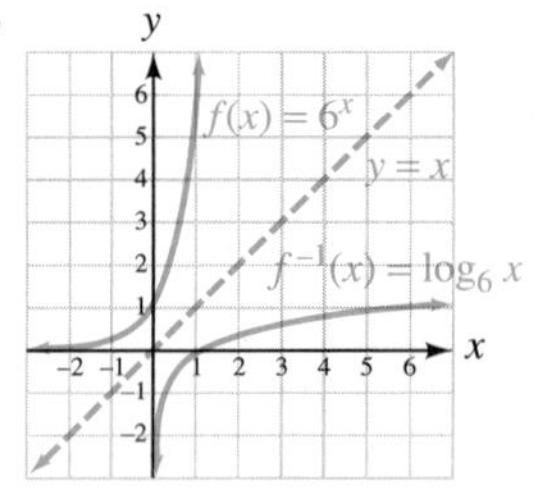

108.

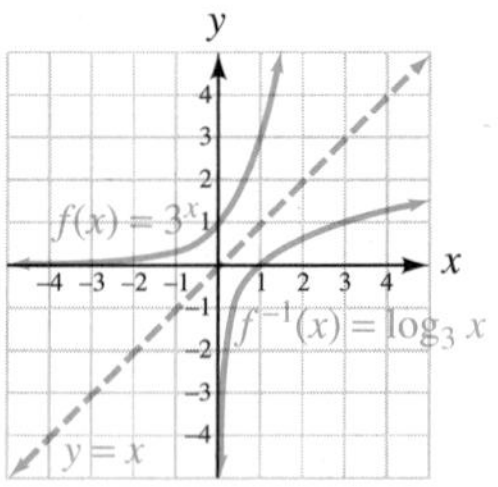

109.

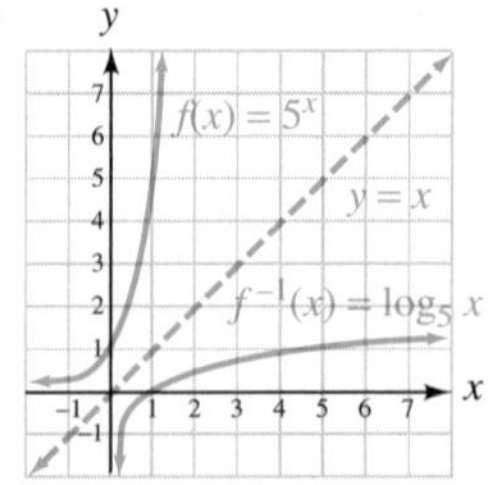

110.

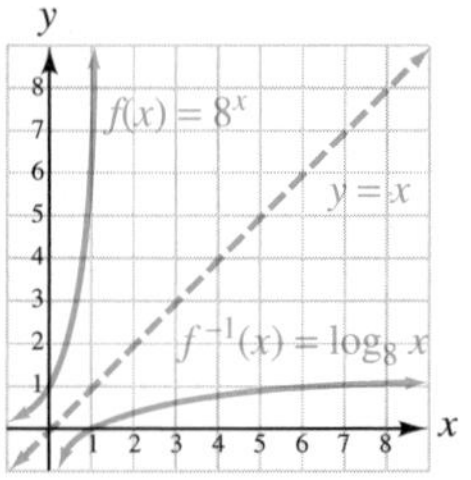

134.

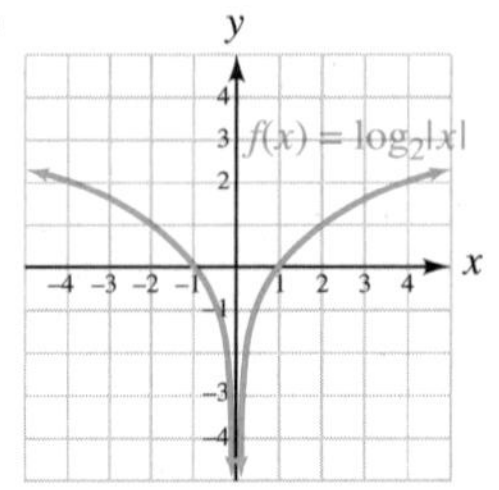

STUDY SET SECTION 11.5 (page 957)

25.

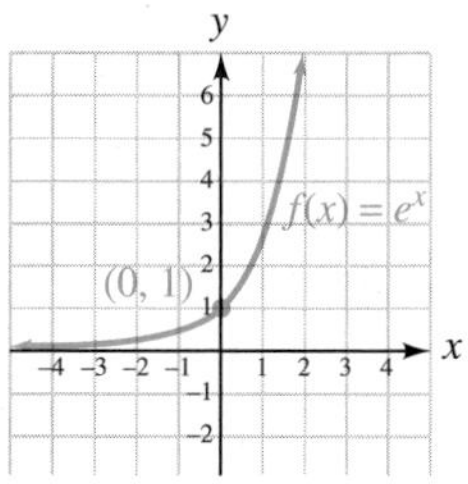

26.

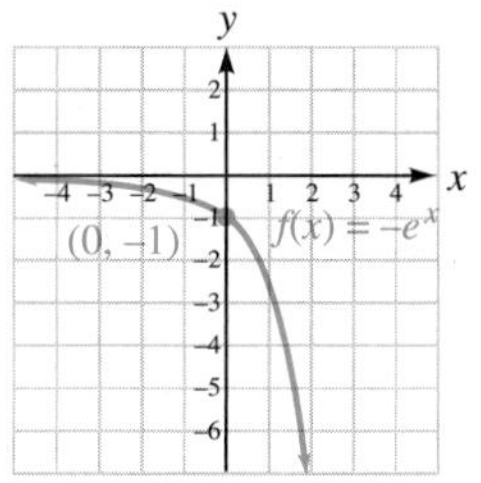

27.

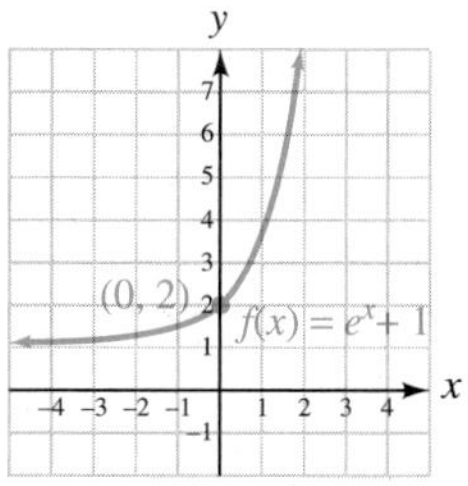

28.

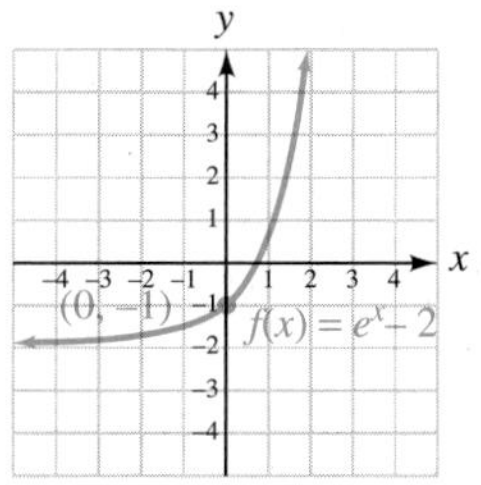

29.

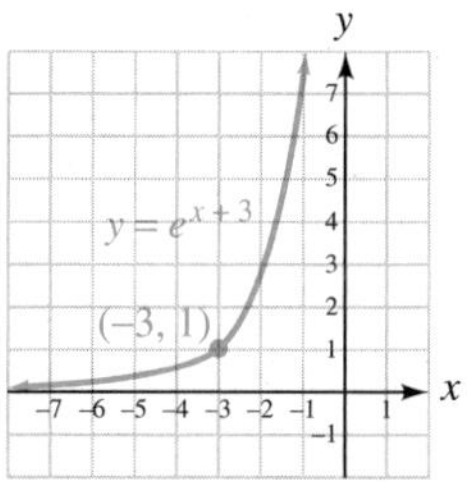

30.

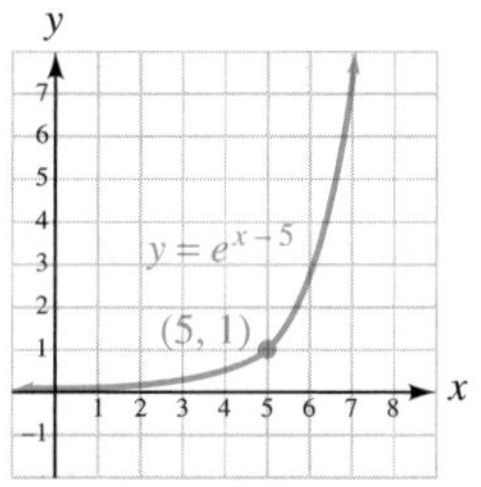

31.

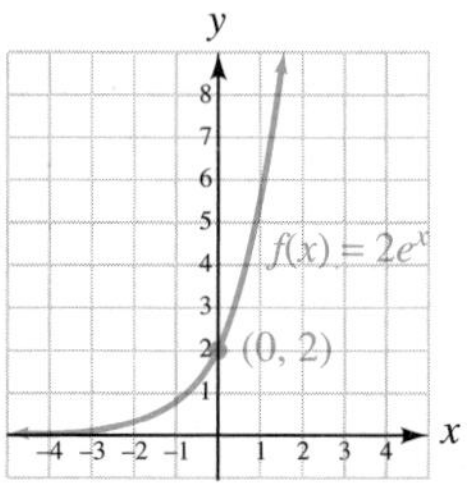

32.

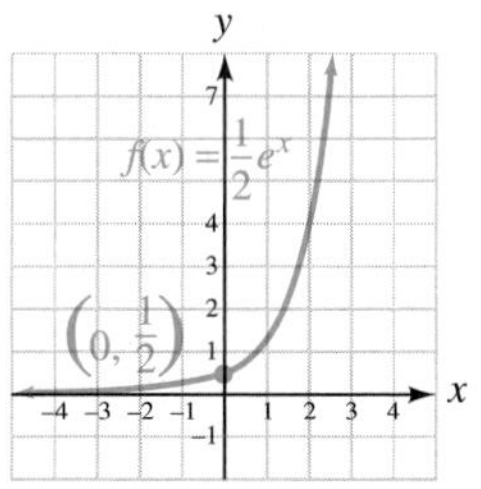

69.

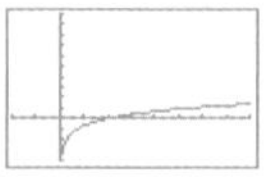

70.

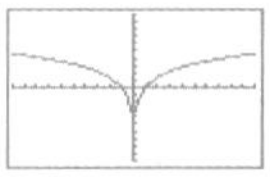

71.

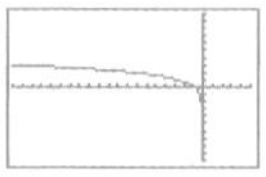

72.

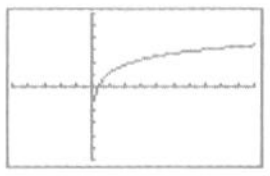

120.

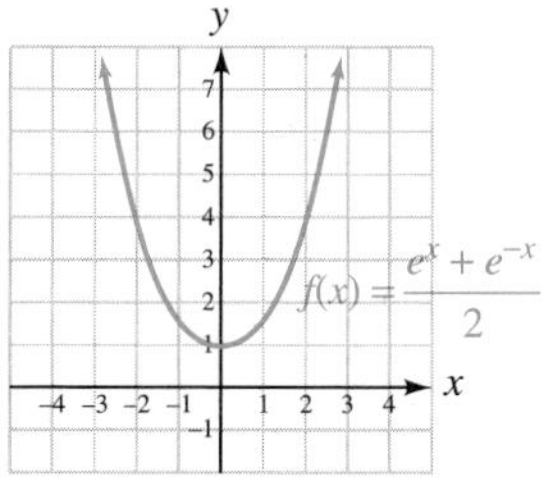

125.

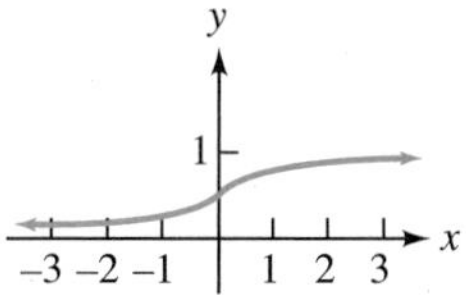

CHAPTER 11 REVIEW (page 986)

23.

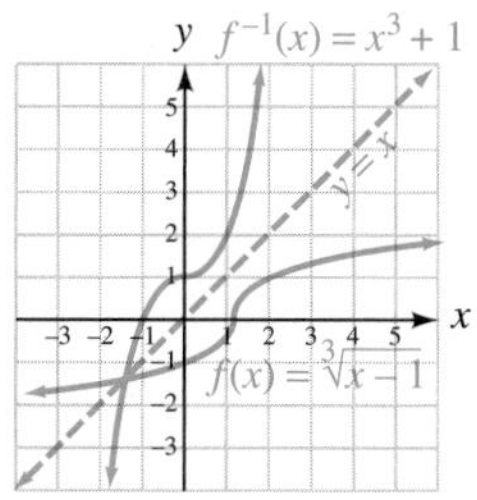

27.

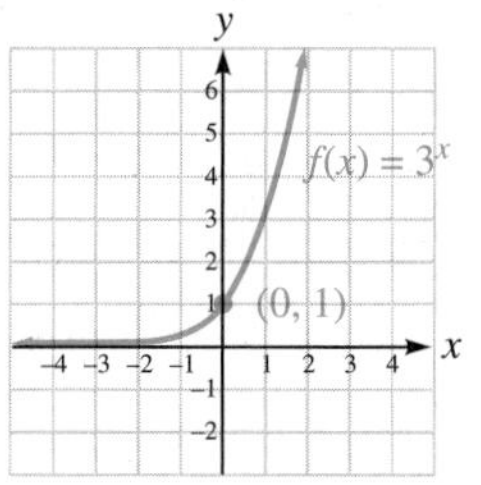

28.

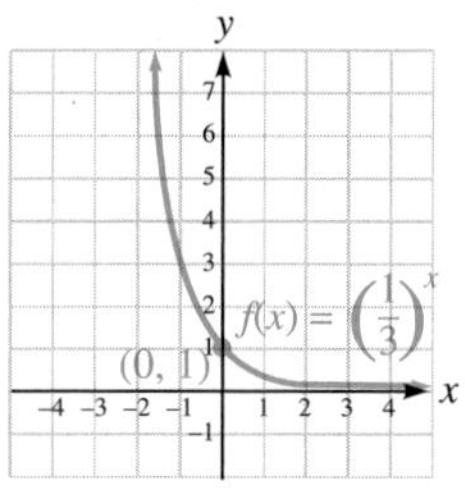

29.

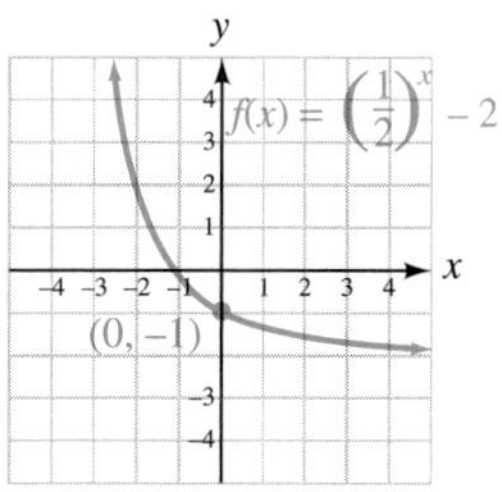

30.

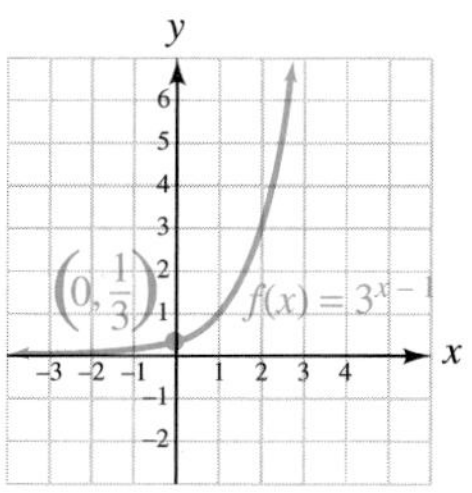

53.

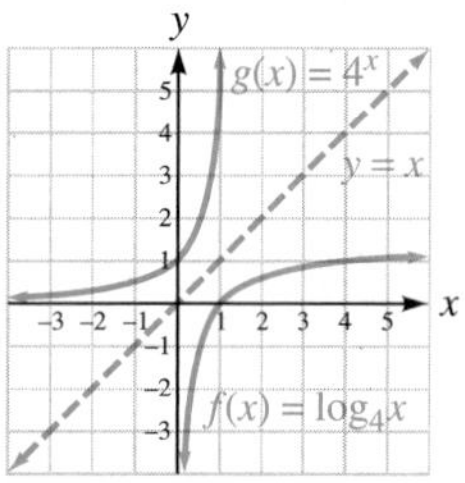

54.

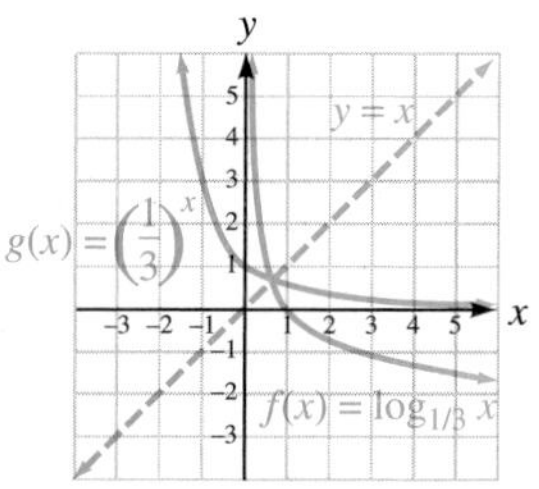

55.

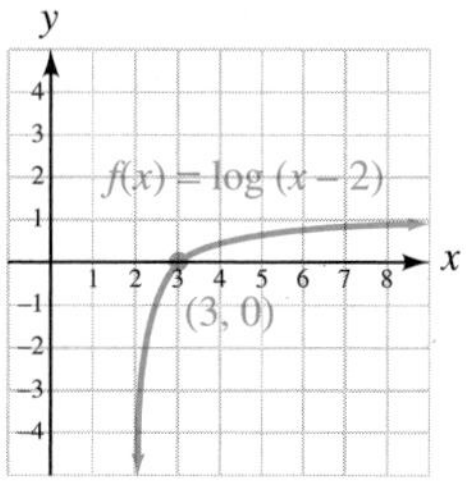

56.

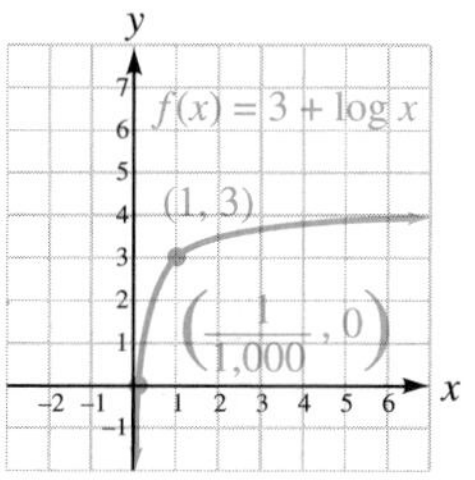

63.

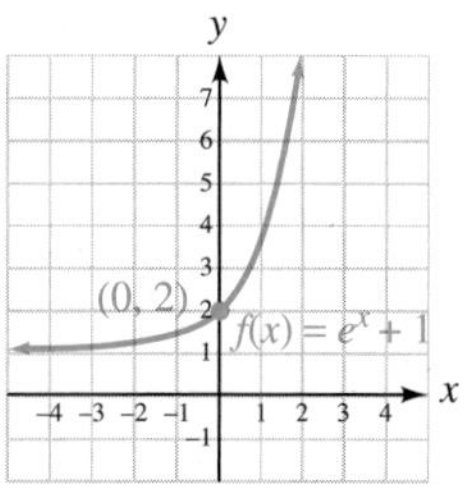

64.

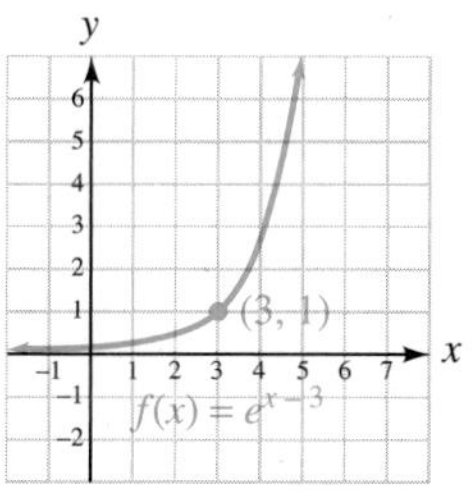

83.

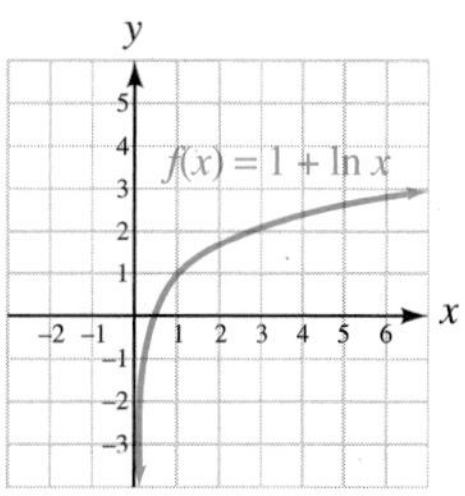

84.

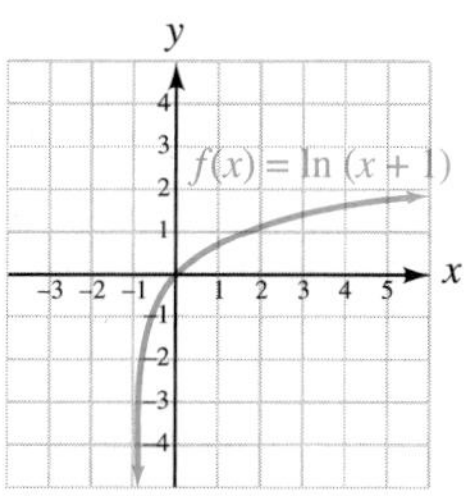

CHAPTER 11 TEST (page 997)

10.

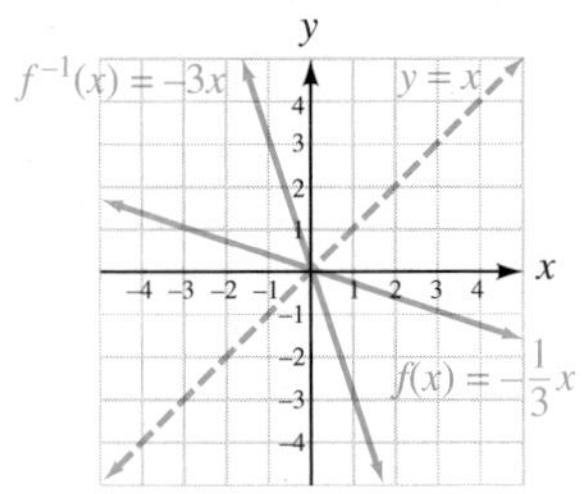

15.

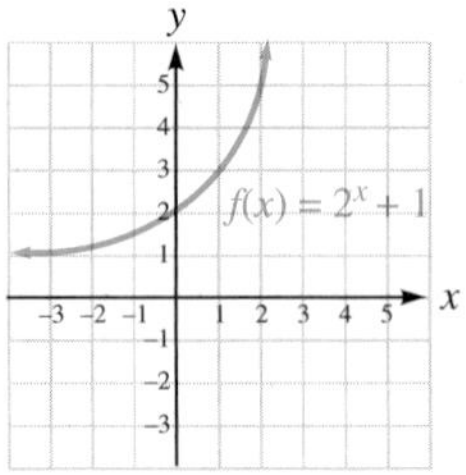

16.

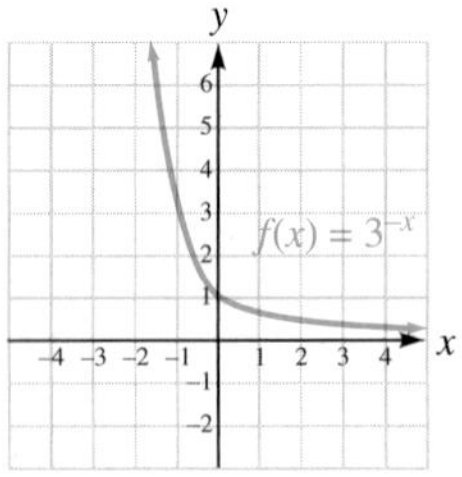

19. a.

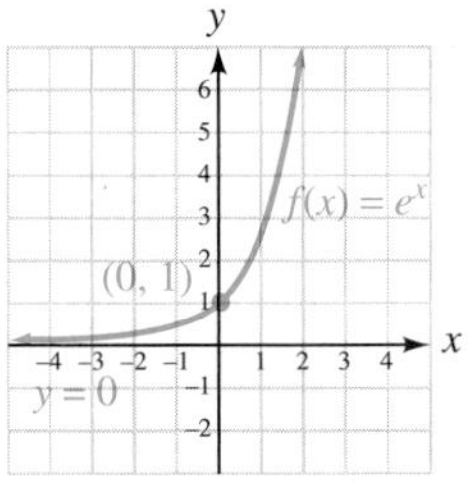

35.

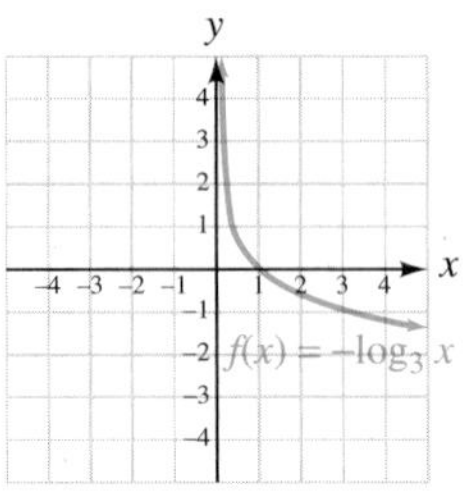

36.

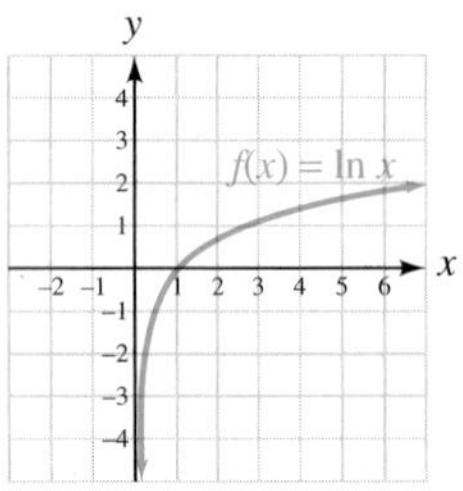

ARE YOU READY? 12.1 (page 1002)

2.

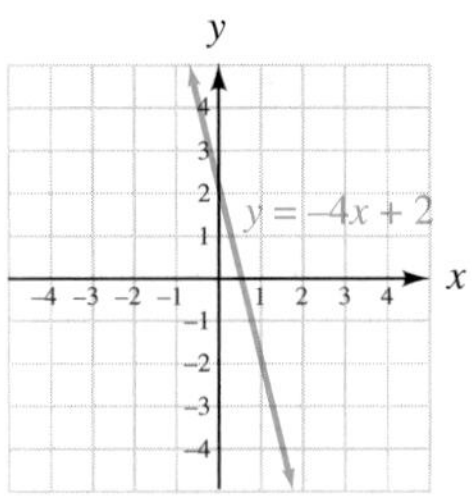

3.

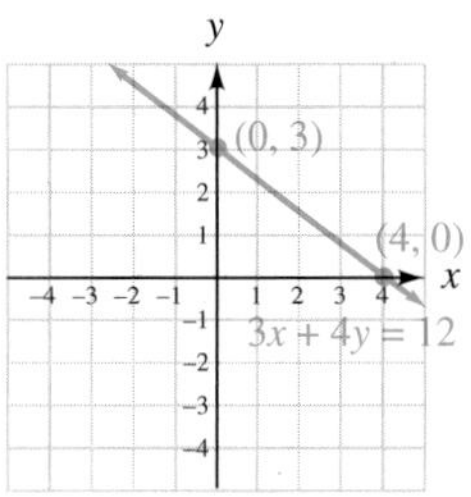

SELF CHECKS 12.1

2.

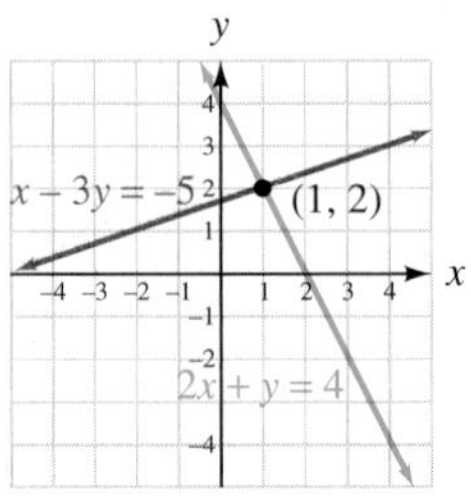

3.

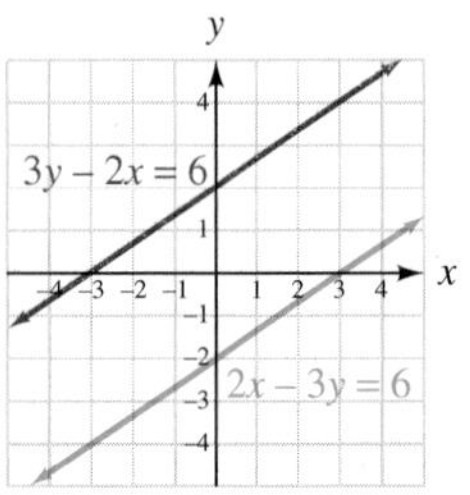

4.

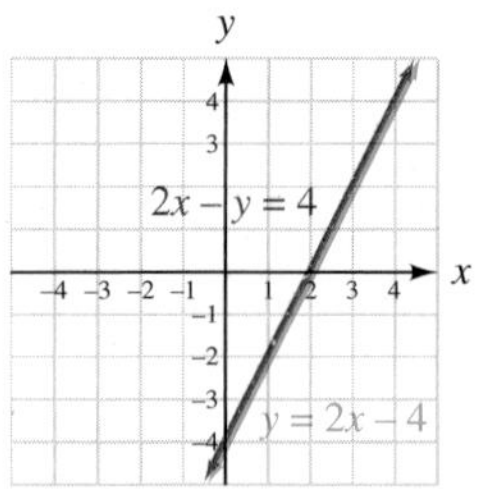

5.

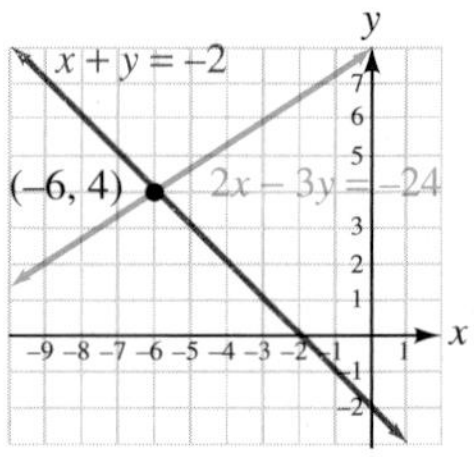

STUDY SET SECTION 12.1 (page 1014)

19.

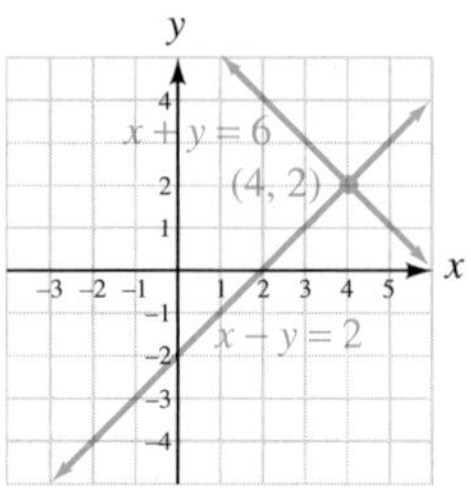

20.

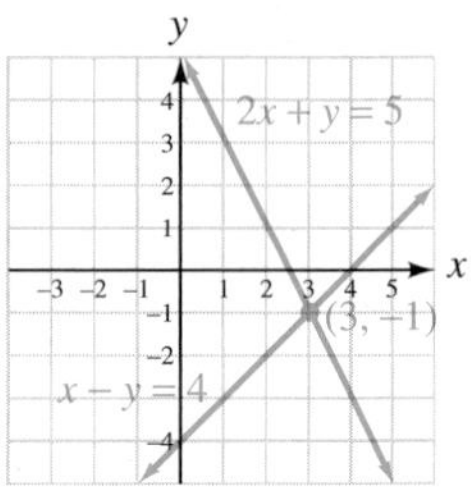

21.

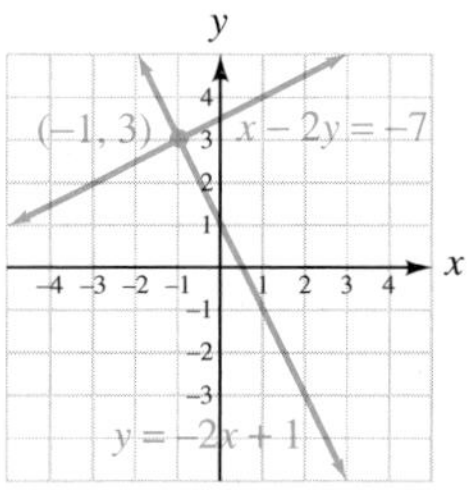

22.

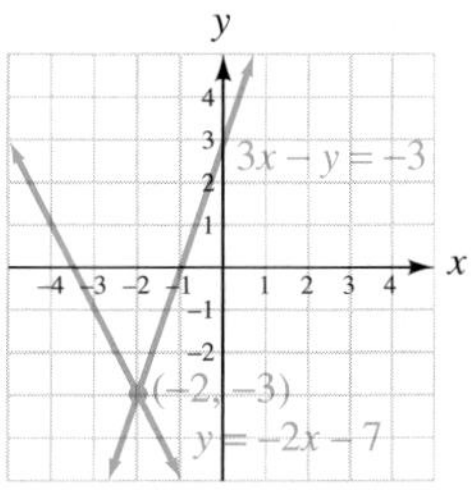

23.

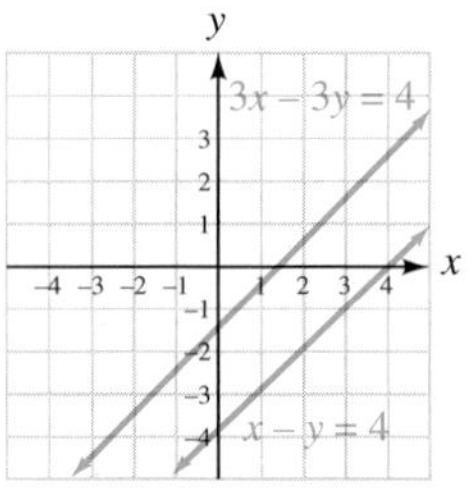

24.

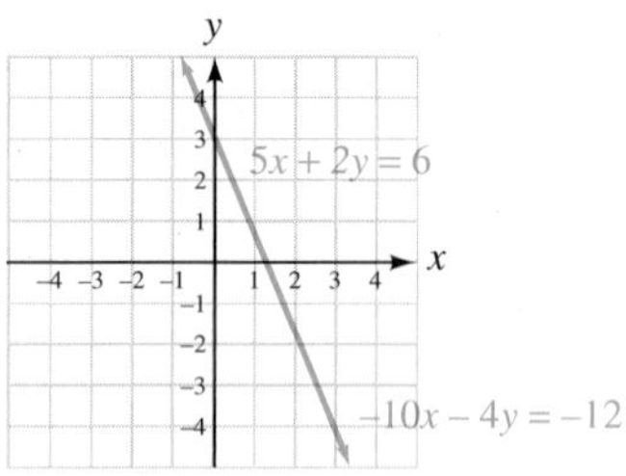

25.

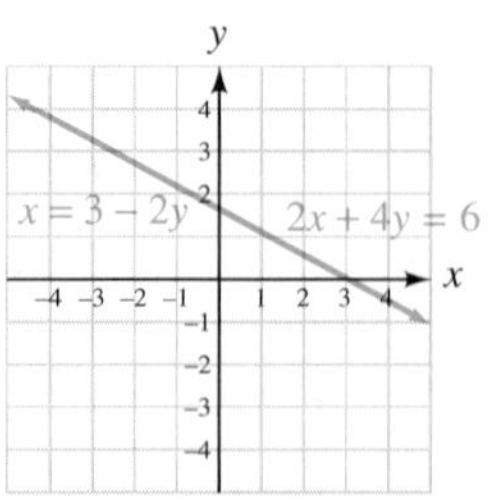

26.

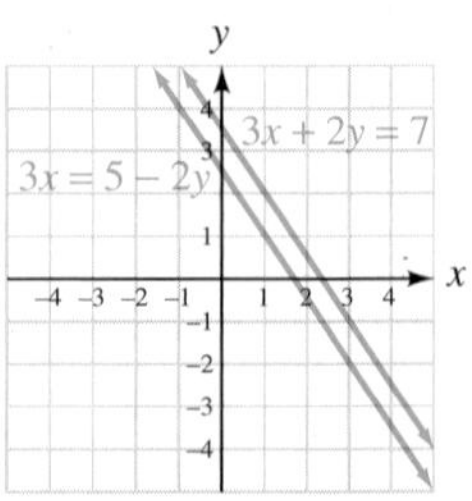

27.

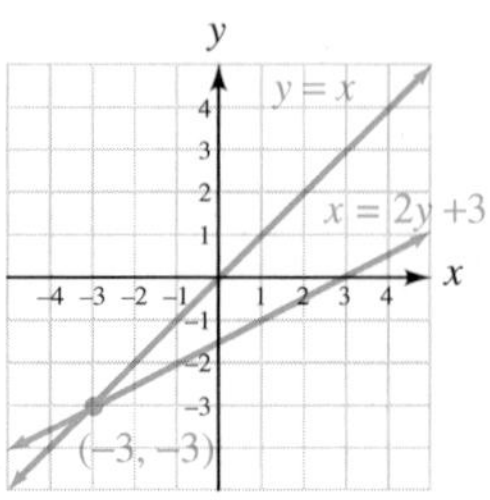

28.
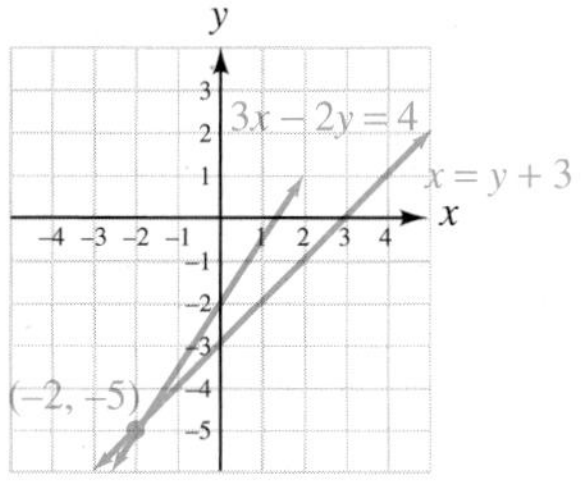

29.
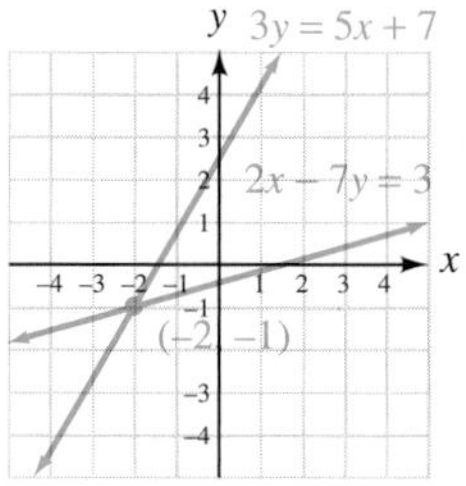

30.
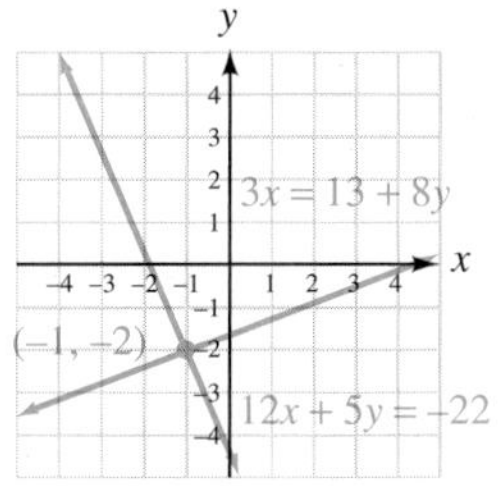

55.
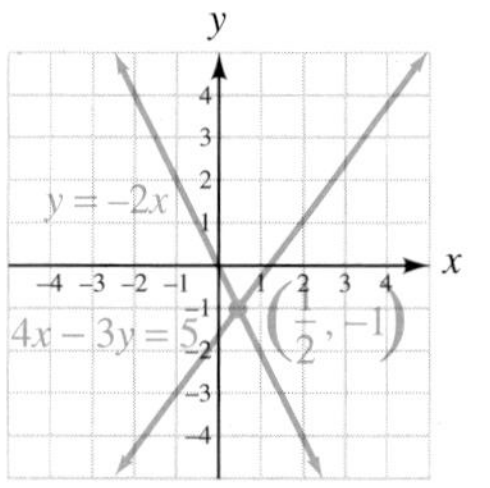

56.
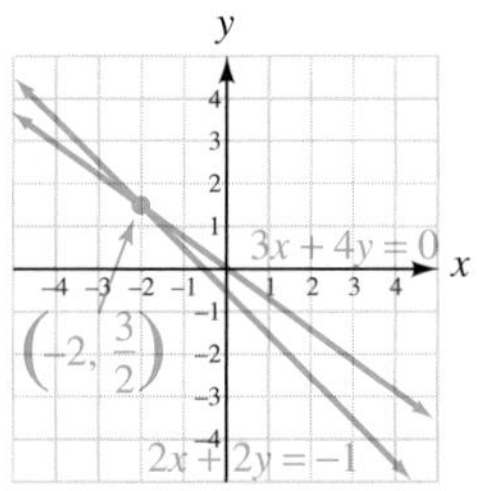

57.
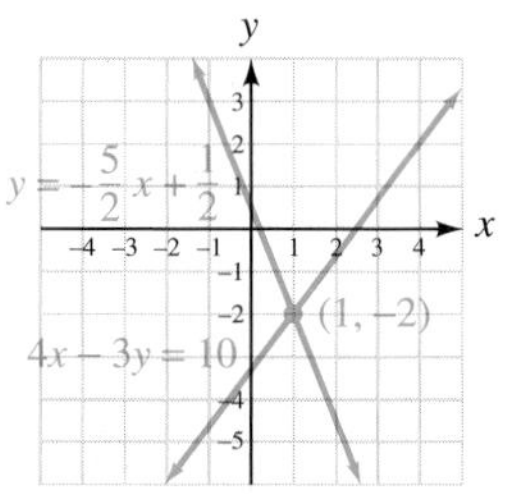

58.
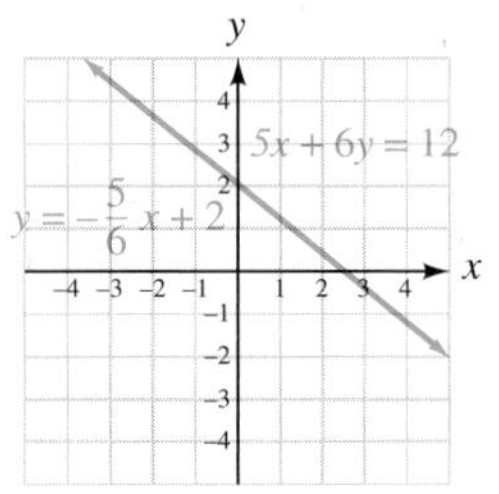

59.
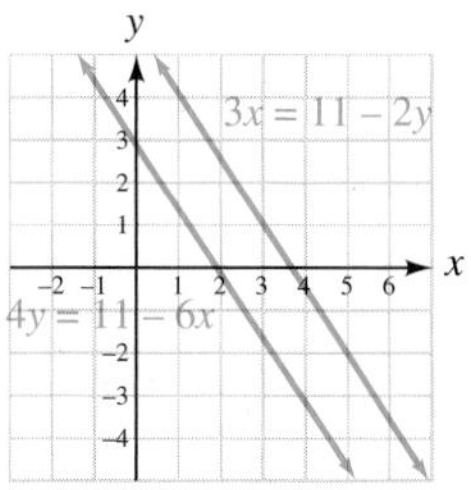

60.
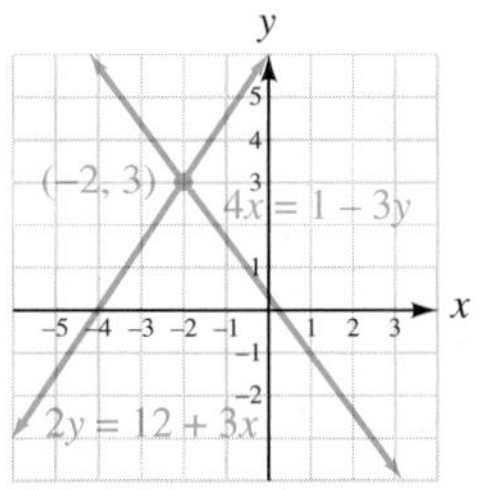

61.
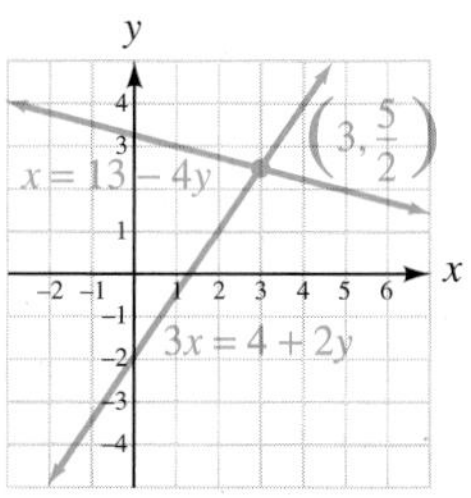

62.
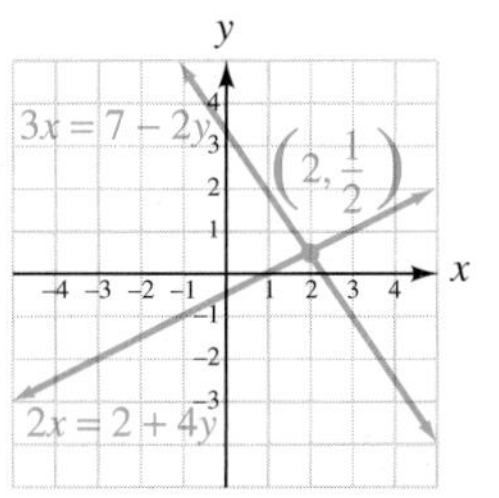

63.
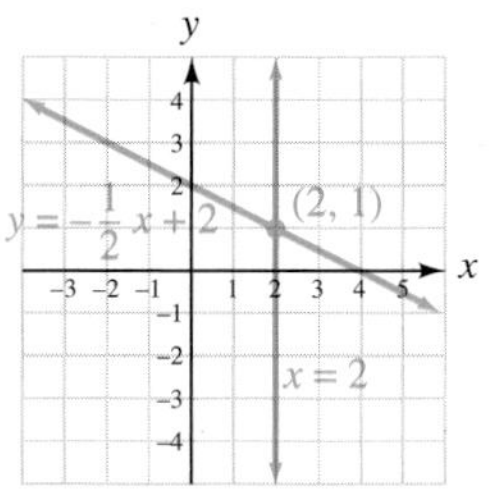

64.
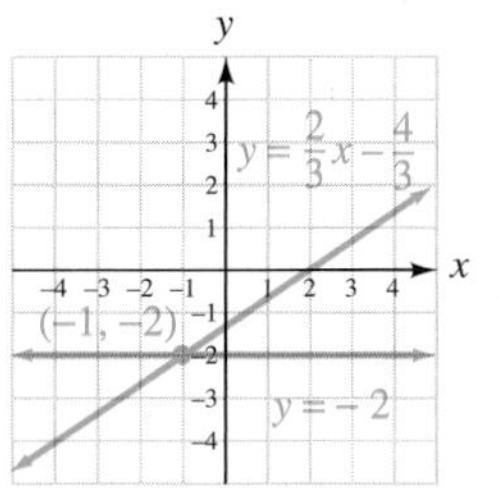

65.
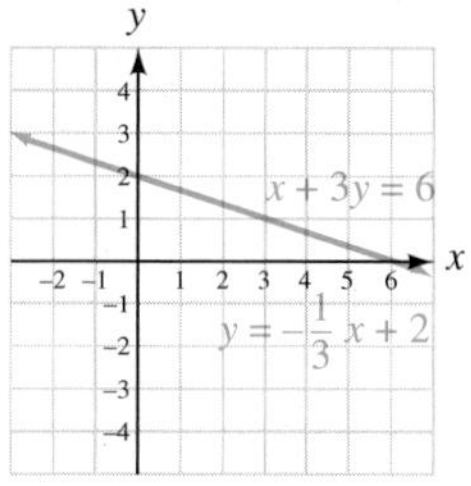

66.
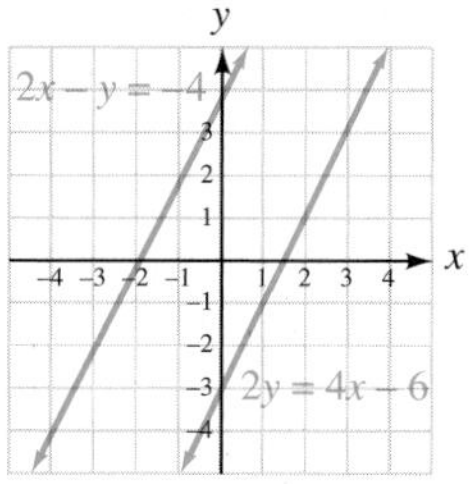

94. a.
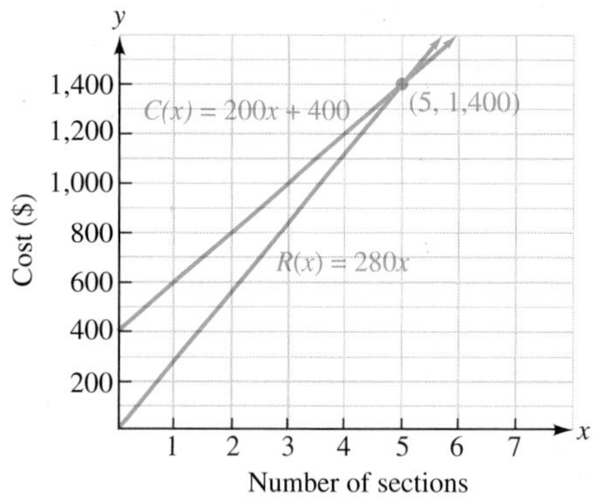

STUDY SET SECTION 12.2 (page 1027)

67.
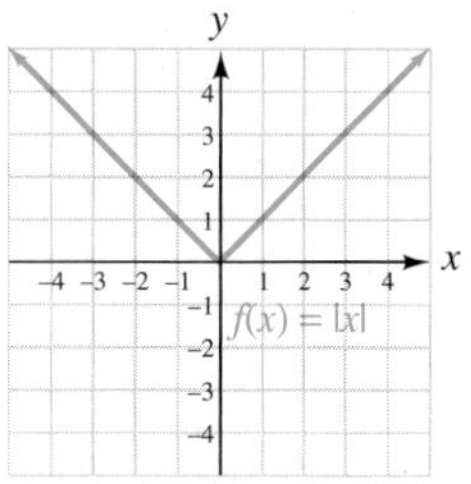

68.
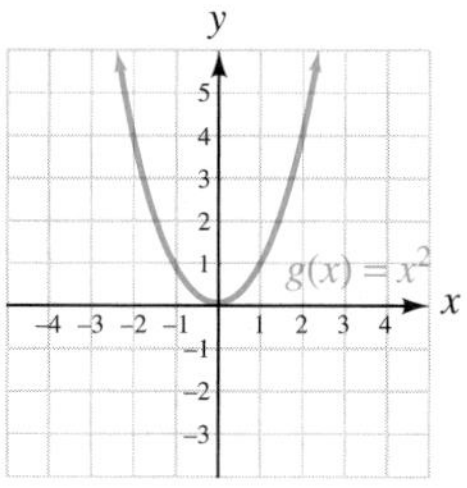

69.
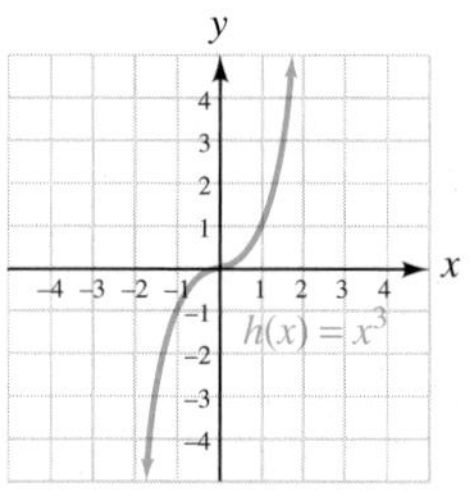

70.
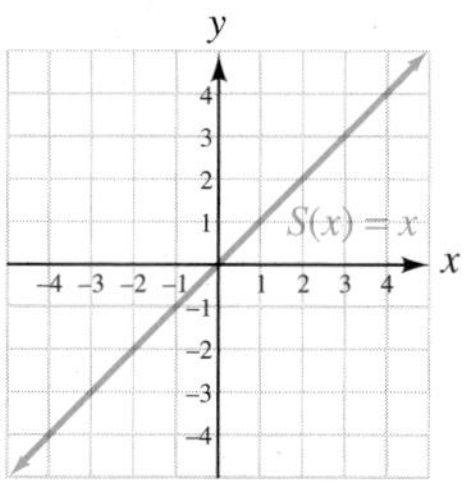

CHAPTER 12 REVIEW (page 1059)

5.
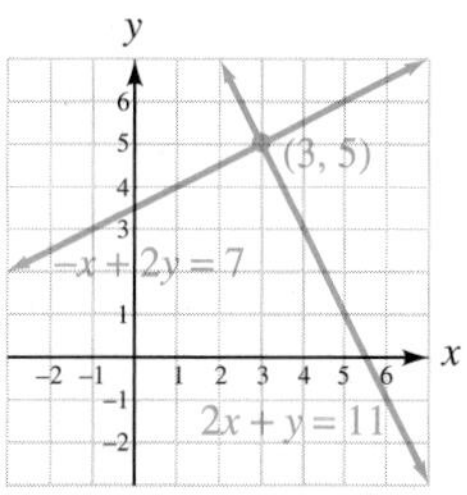

6.
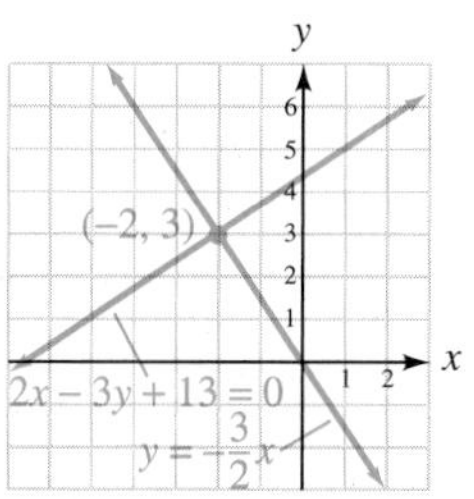

7.
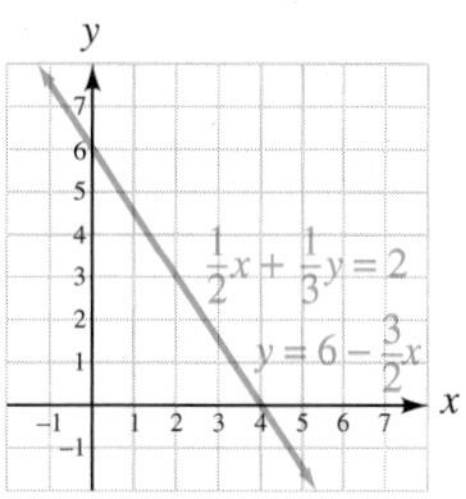

8.
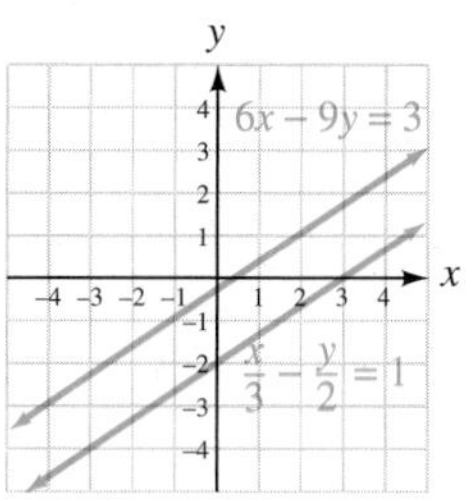

CHAPTER 12 TEST (page 1068)

2.

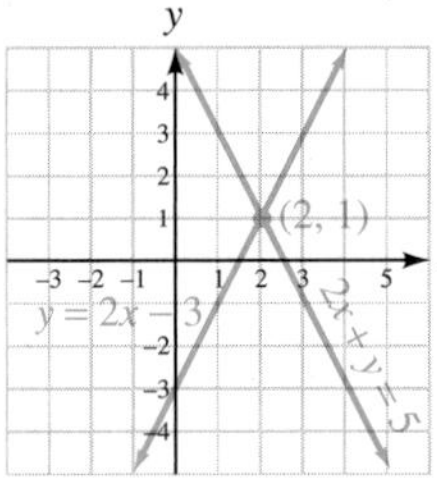

CUMULATIVE REVIEW CHAPTERS 1–12 (page 1070)

15.

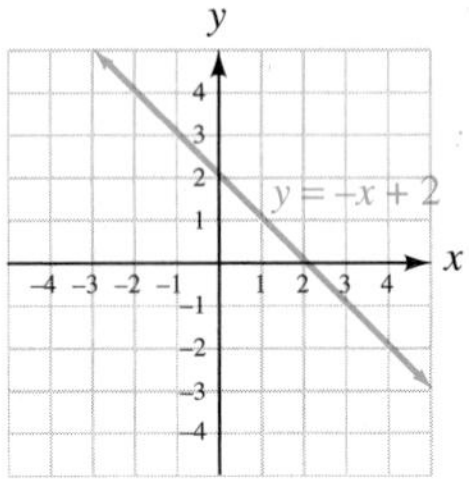

16.

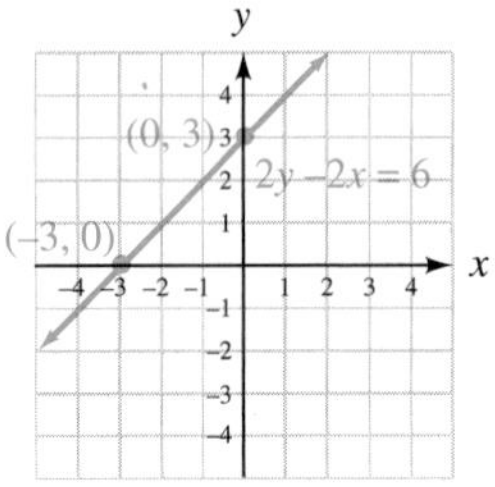

17.

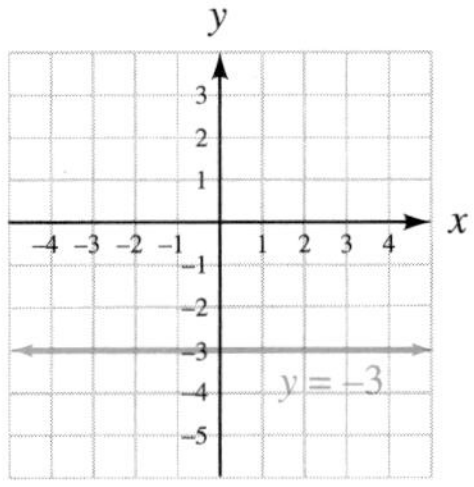

18.

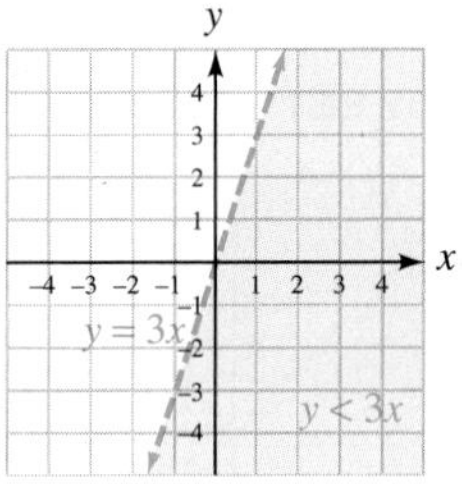

25.

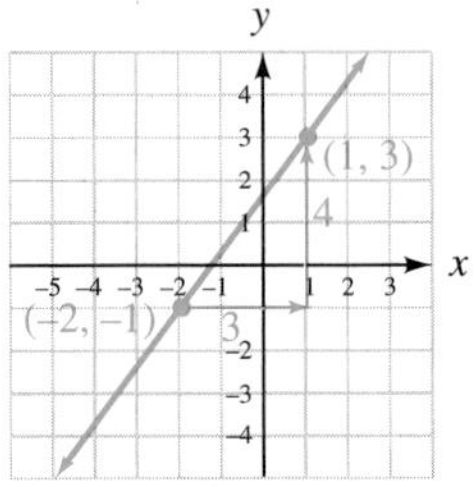

29.

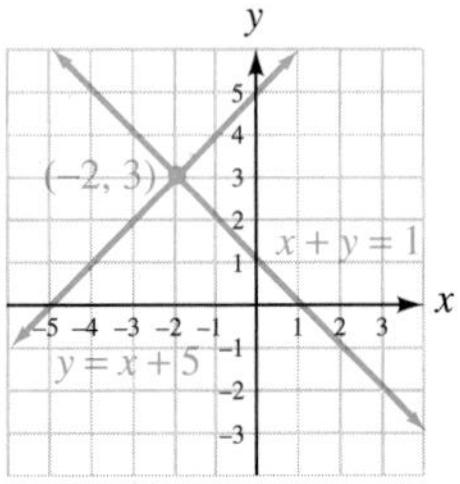

34.

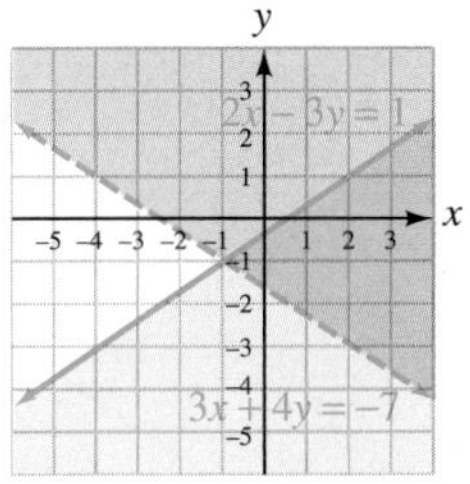

41.

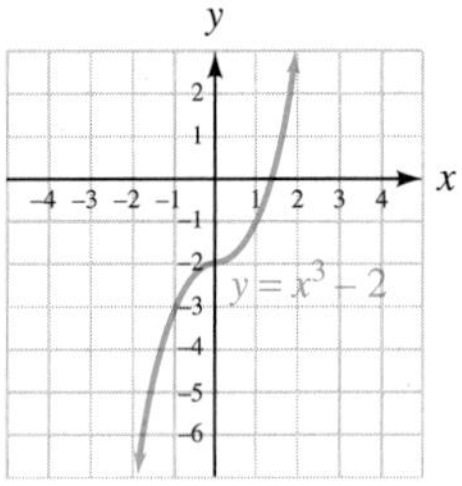

86. a.

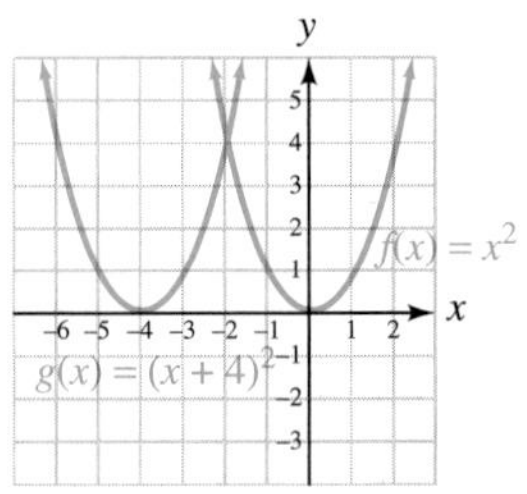

94.

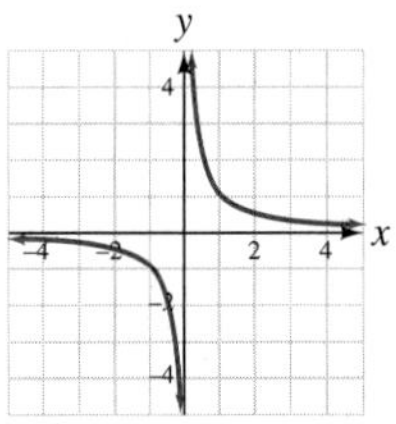

97.

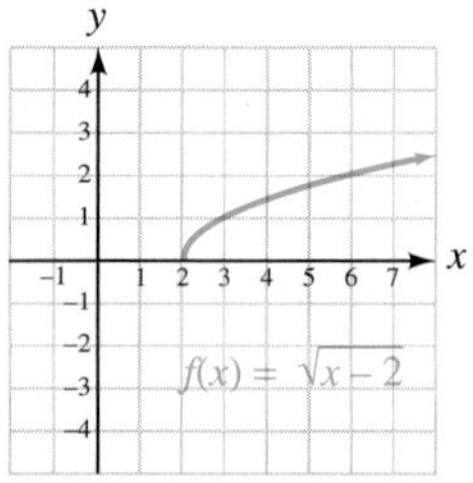

120.

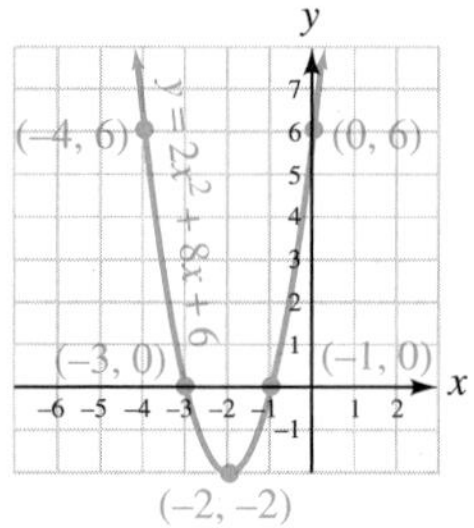

123.

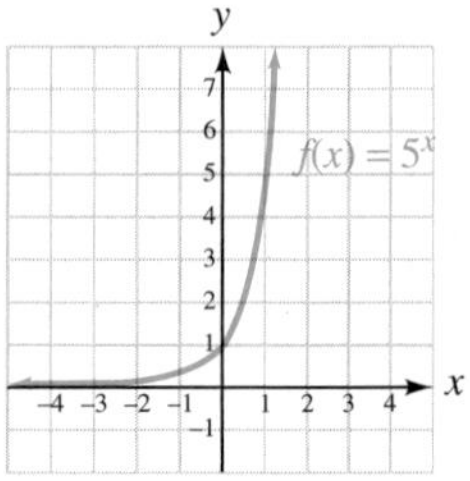

124.

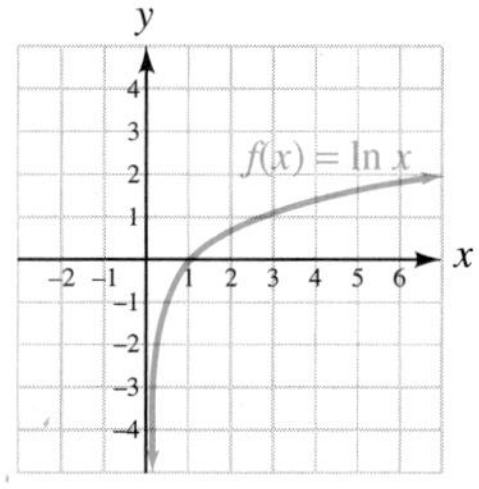

SELF CHECK 13.1

1. a.

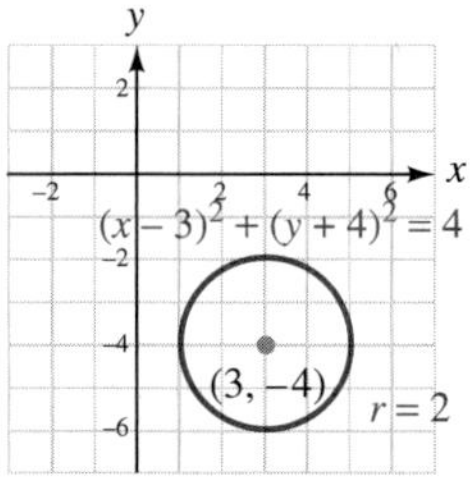

b.

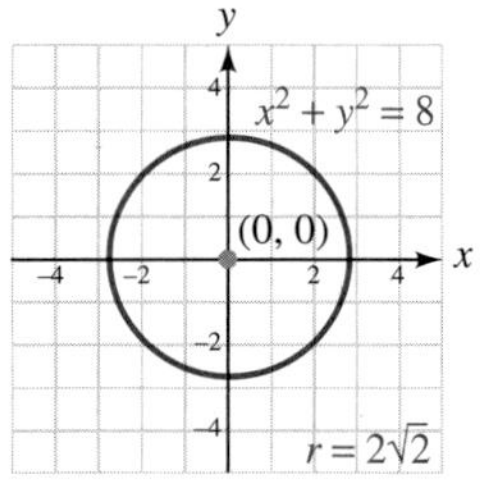

3.

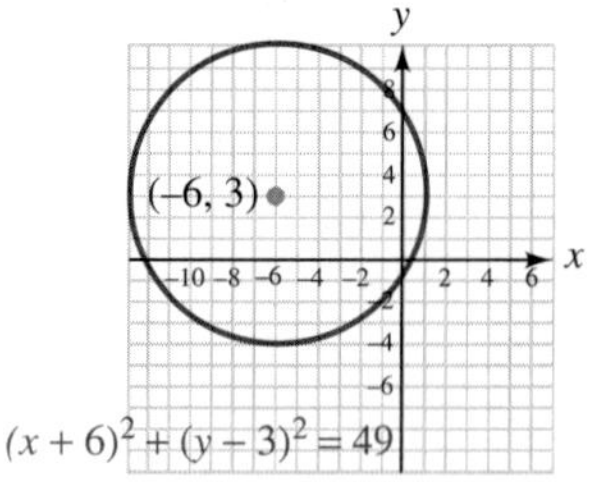

5.

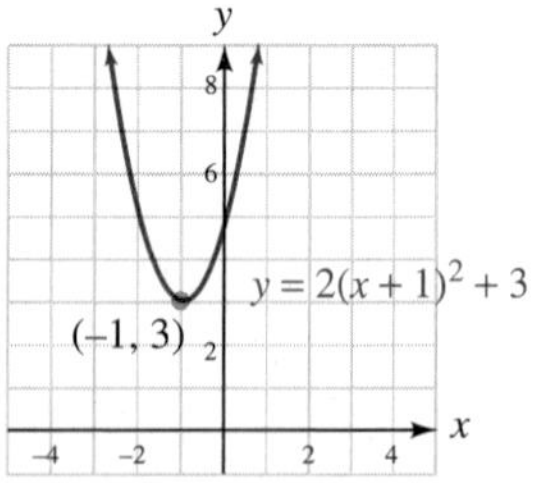

6.

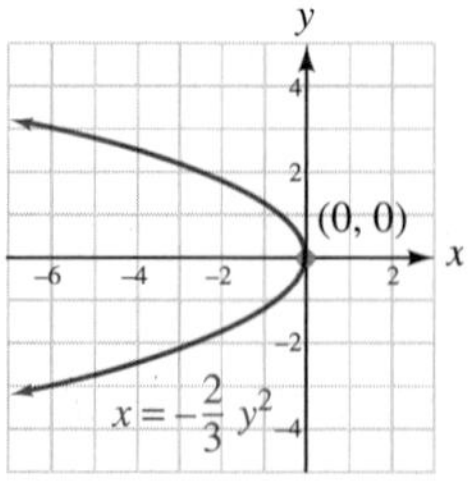

7.

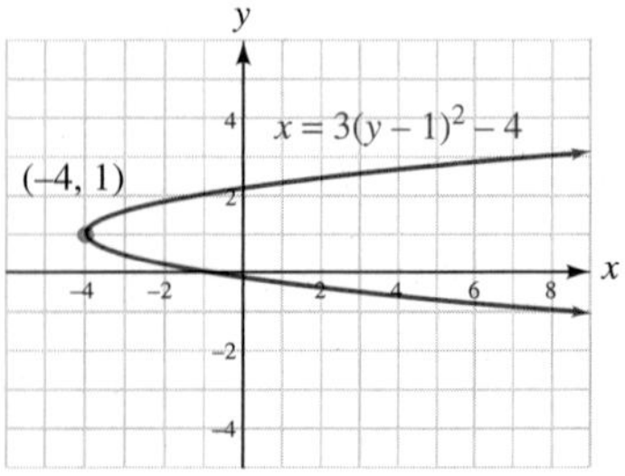

STUDY SET SECTION 13.1 (page 1086)

12.

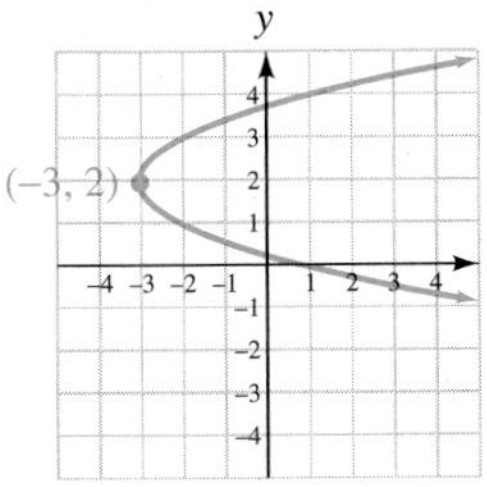

15.

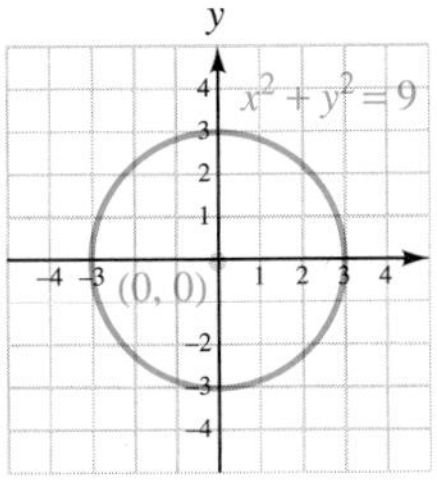

16.

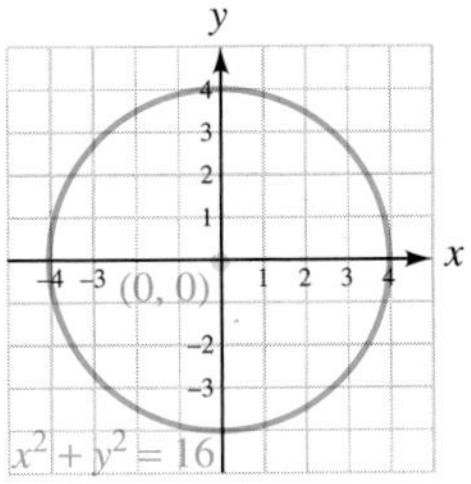

17.

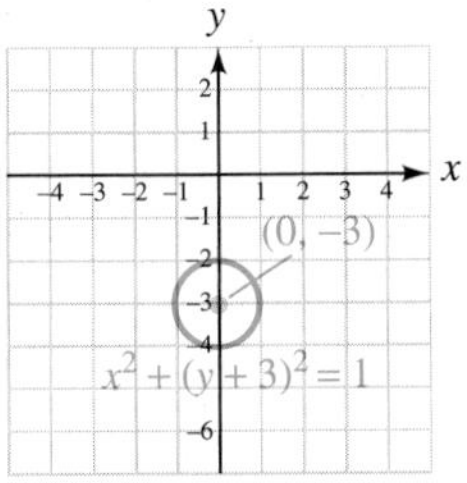

18.

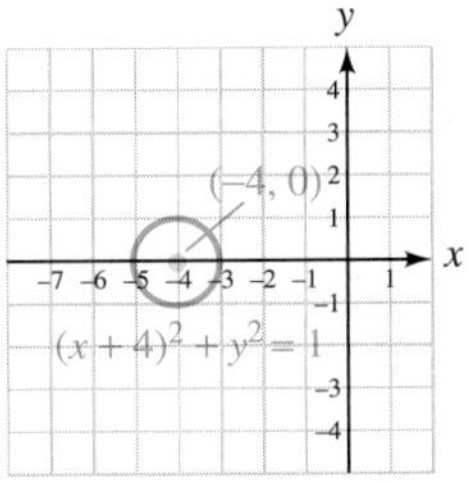

19.

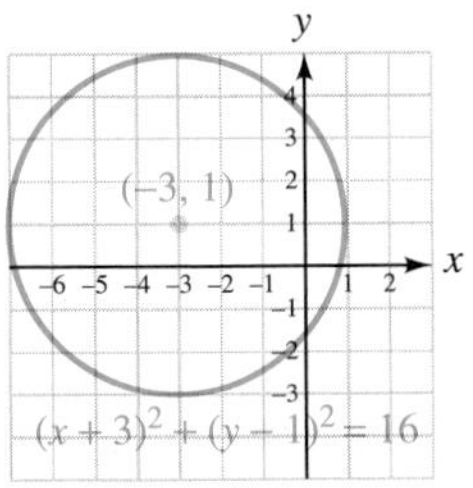

20.

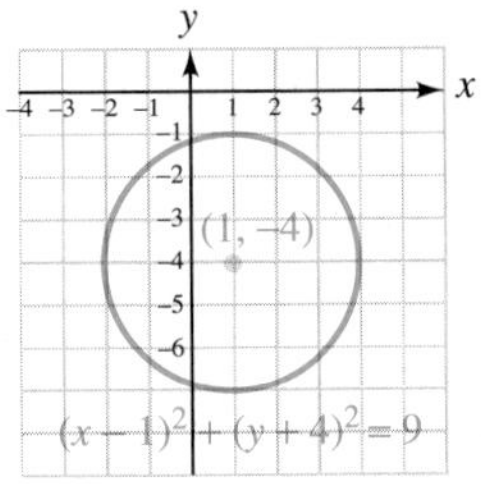

21.

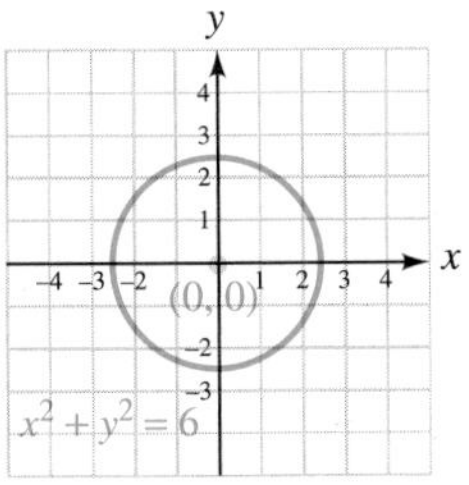

22.

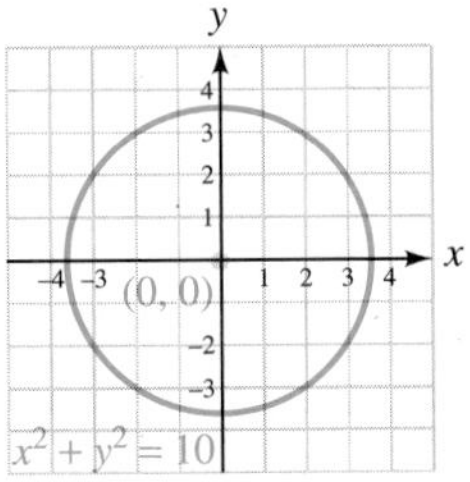

35.

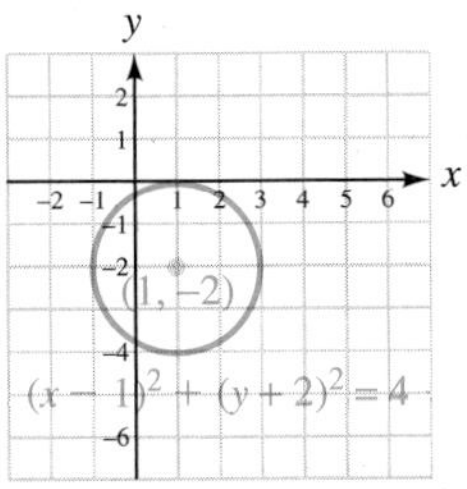

36.

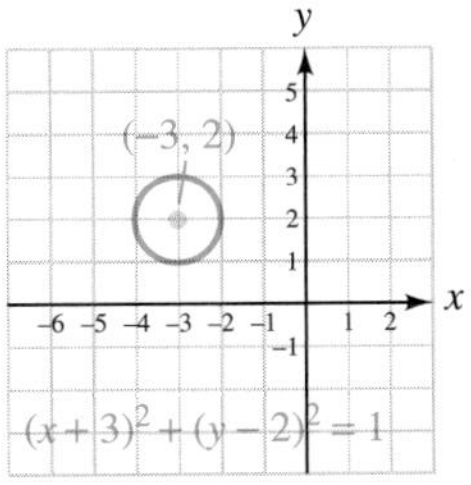

37.

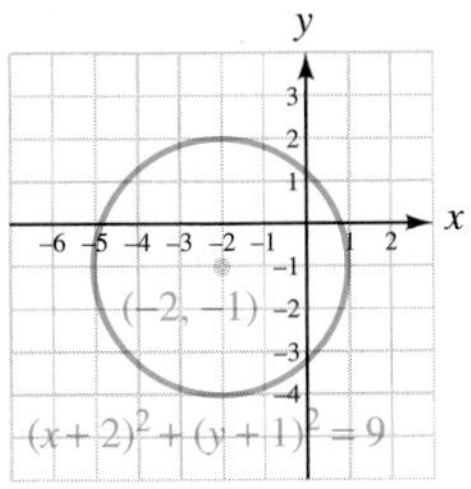

38.

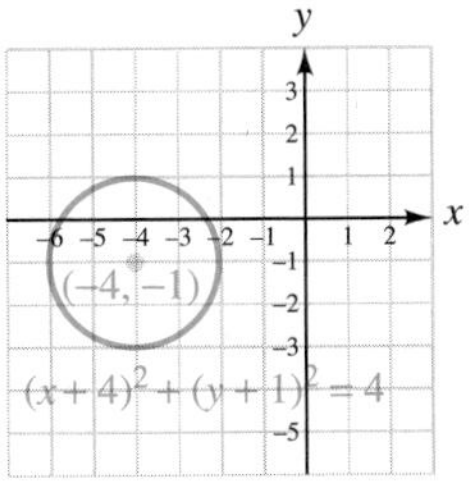

39.

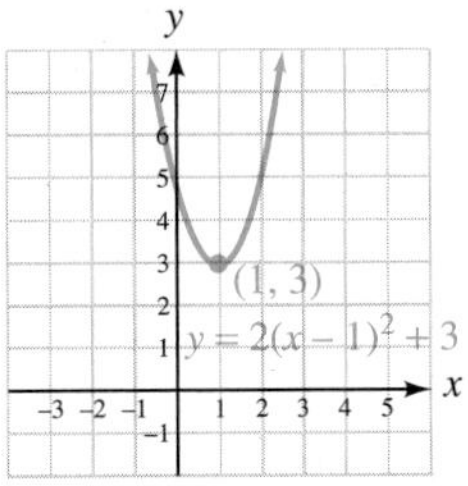

40.

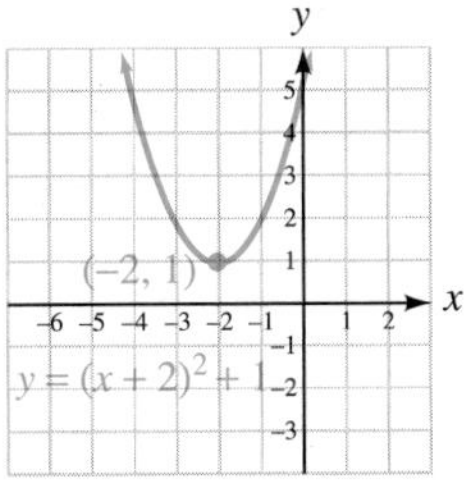

41.

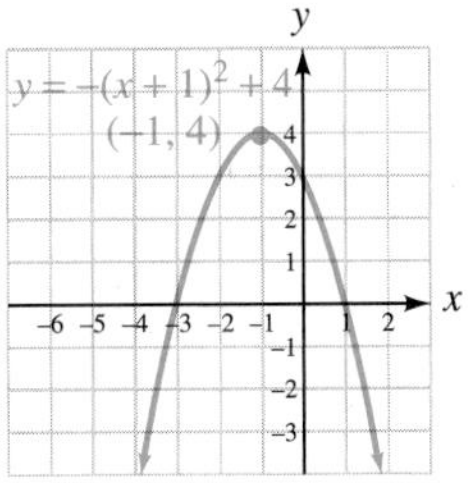

42.

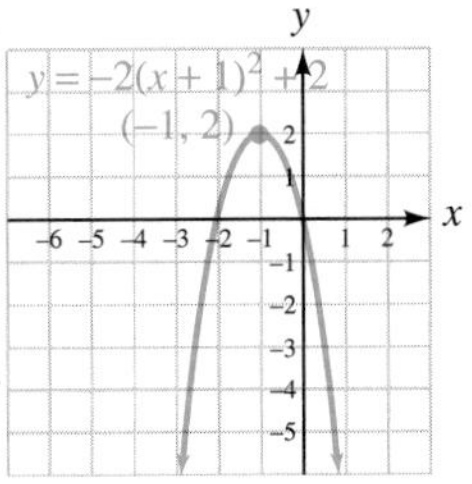

43.

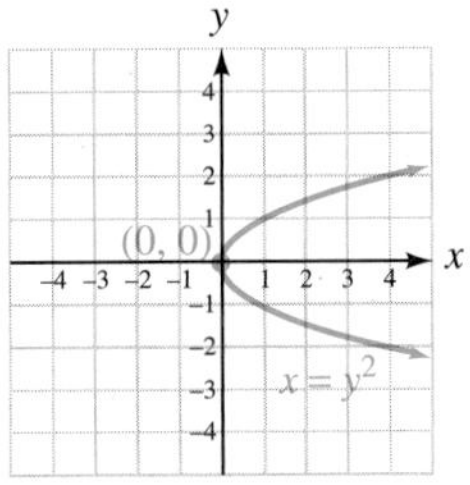

44.

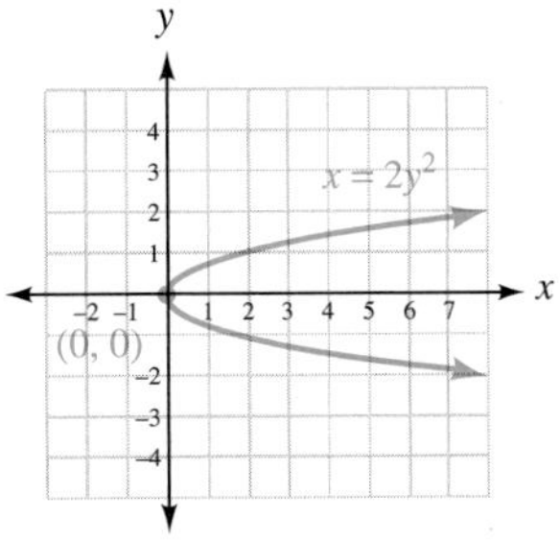

45.

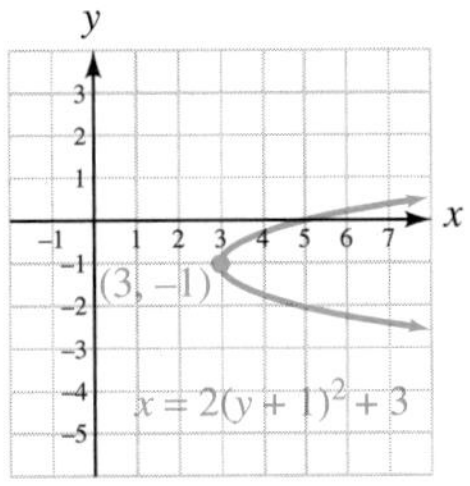

46.

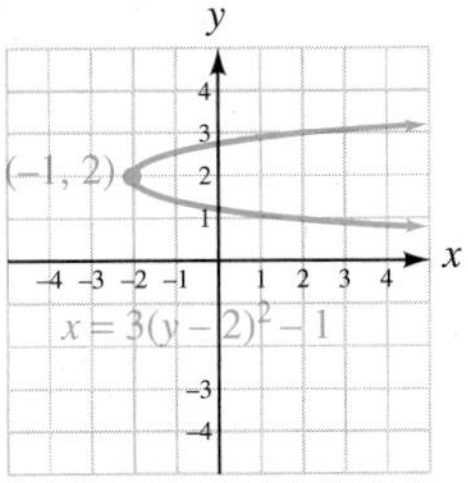

47.

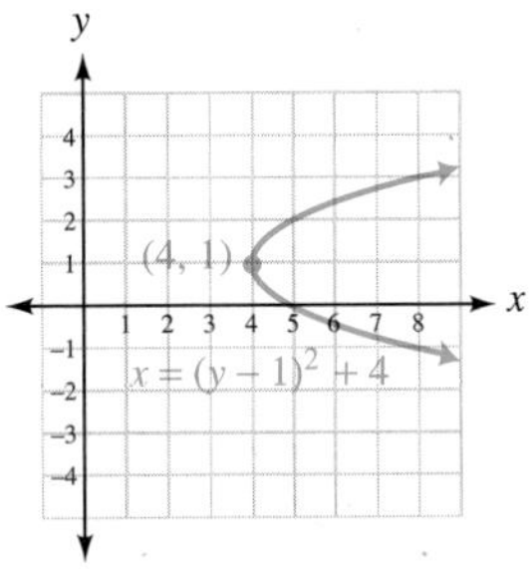

48.

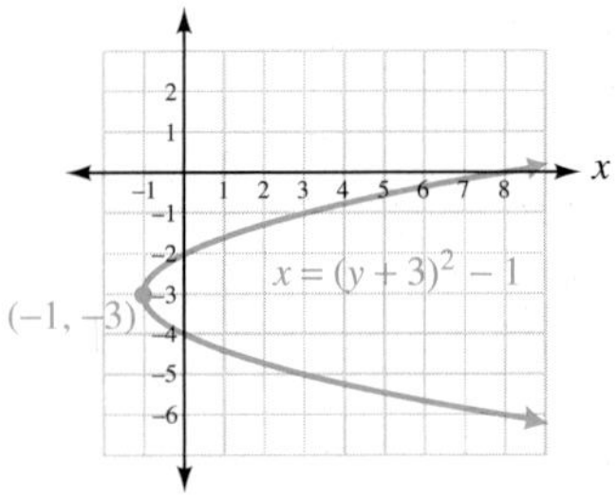

49.

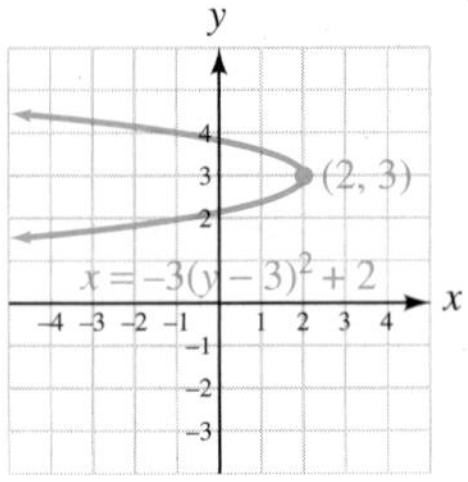

50.

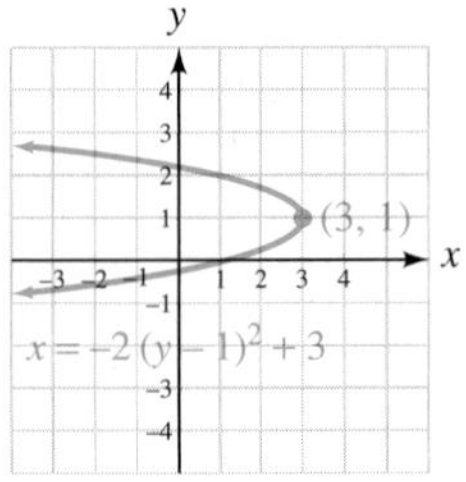

51.

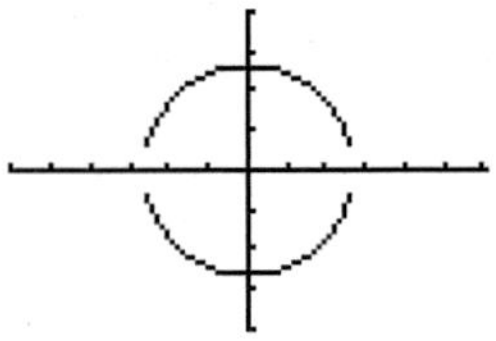

52.

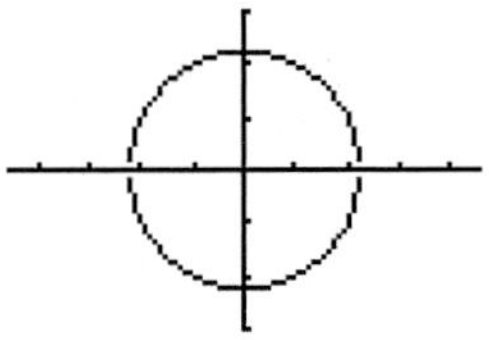

53.

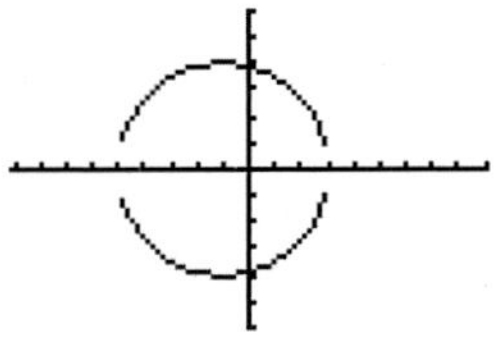

54.

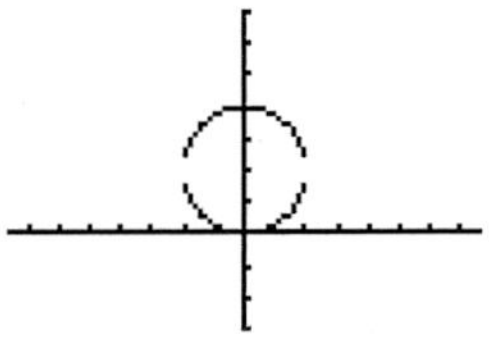

55.

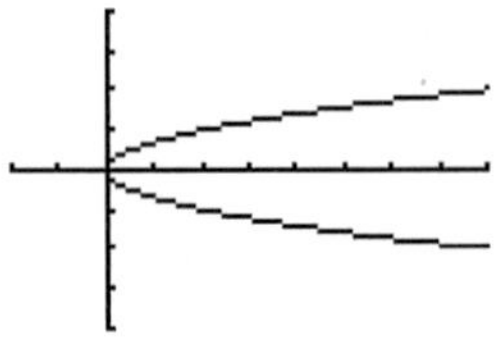

56.

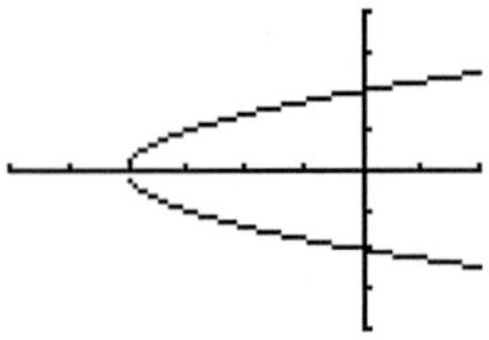

57.

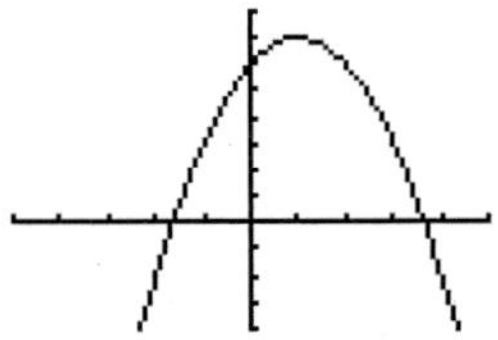

58.

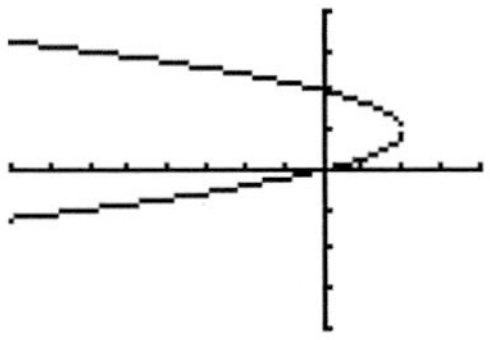

59.

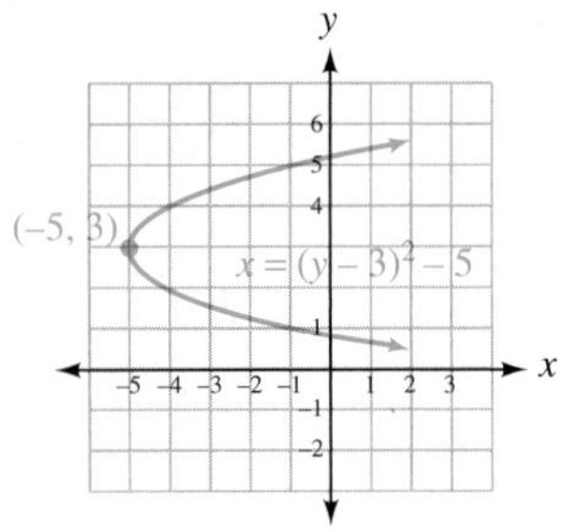

60.

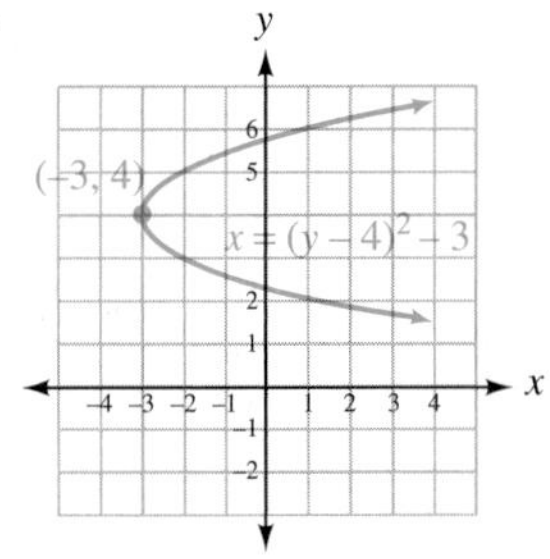

61.

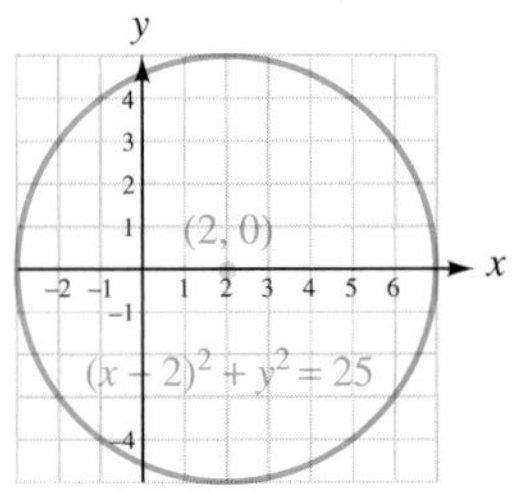

62.

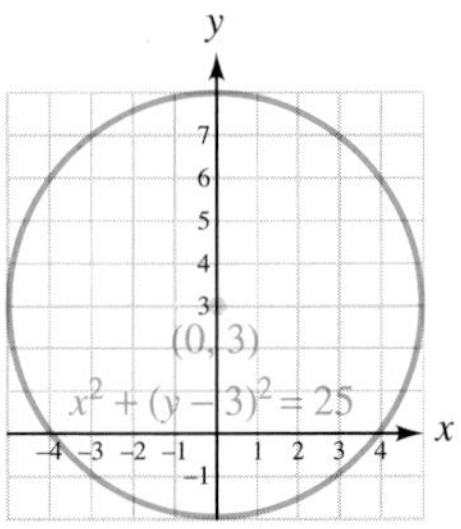

63.

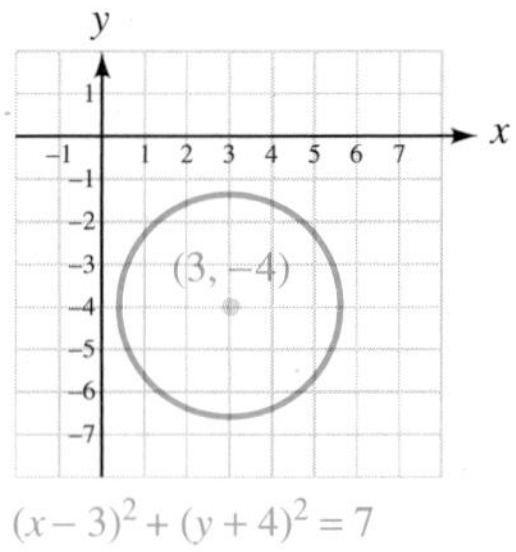

64.

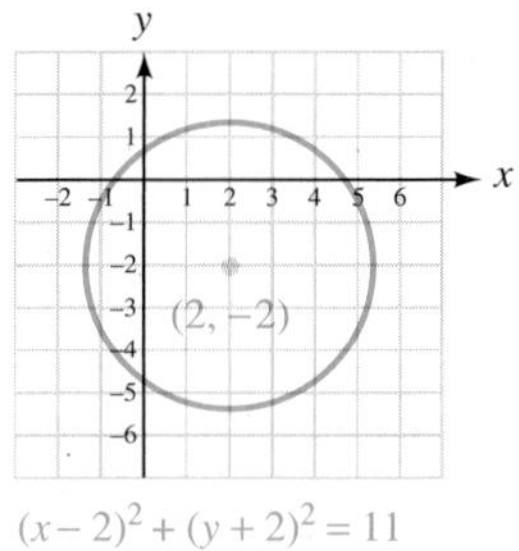

65.

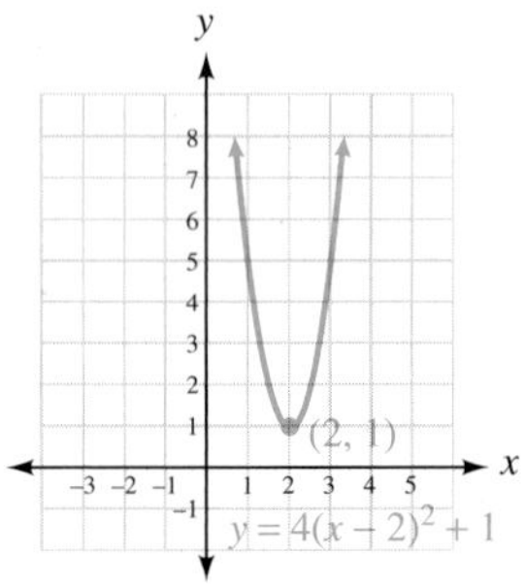

66.

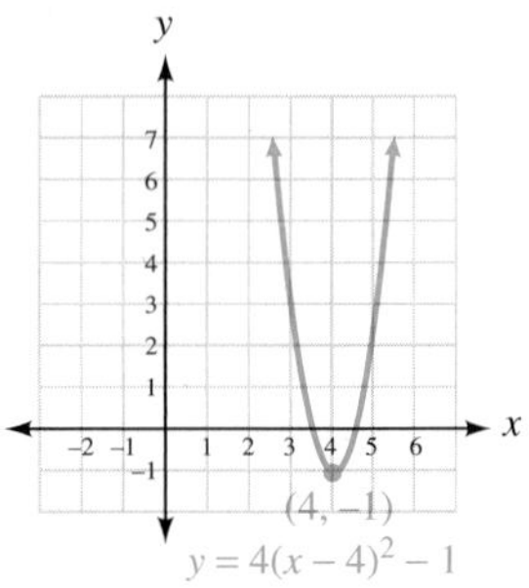

67.

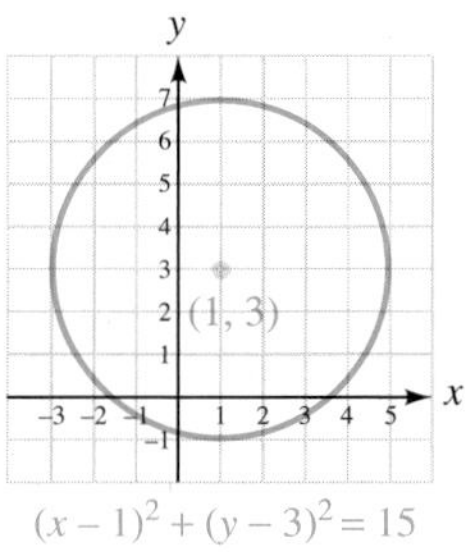

68.

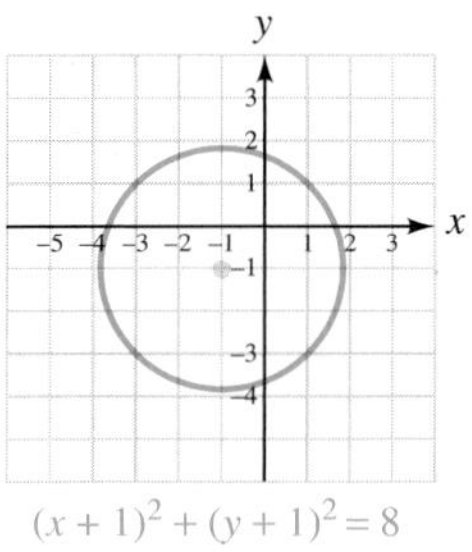

69.

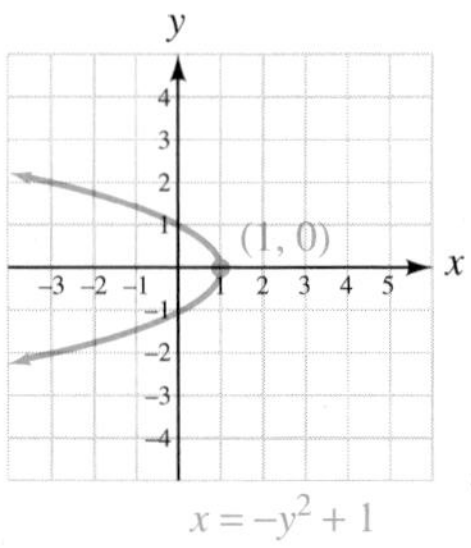

70.

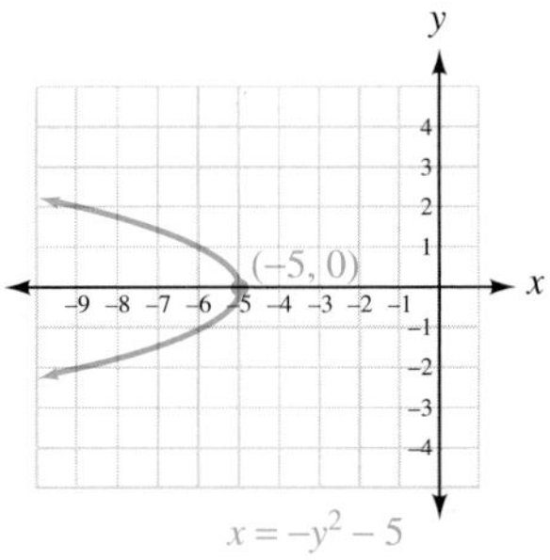

71.

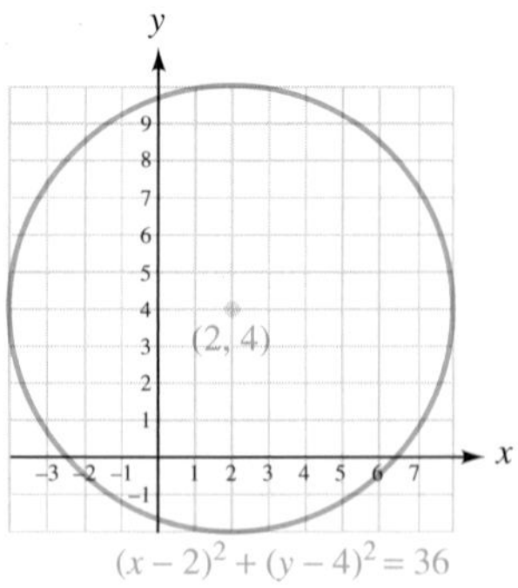

72.

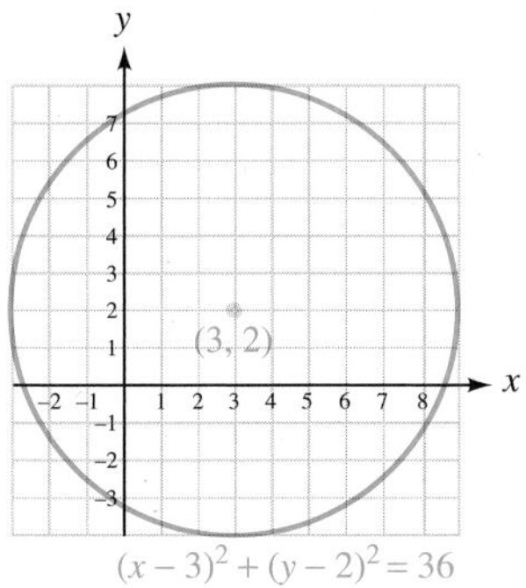

$(x-3)^2 + (y-2)^2 = 36$

73.

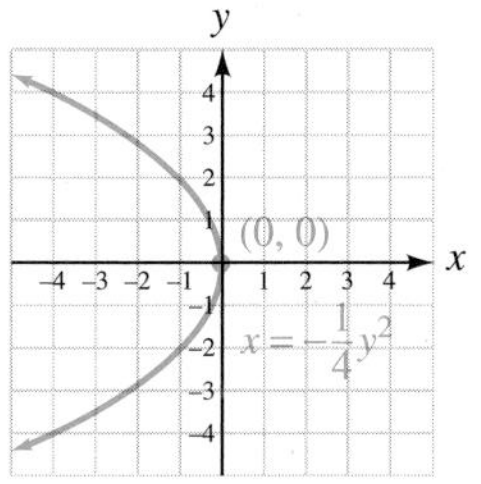

74.

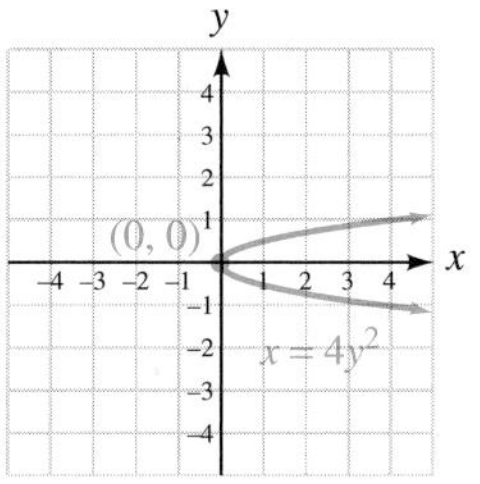

75.

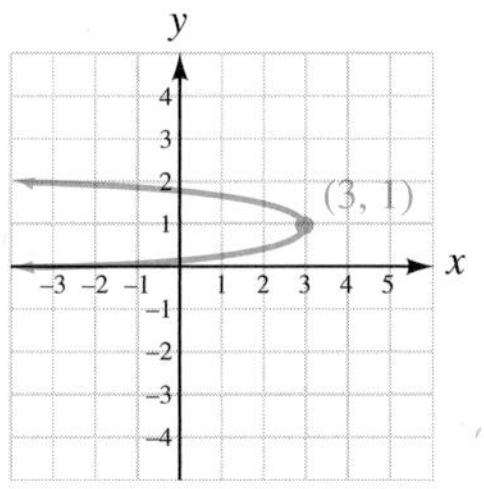

$x = -6(y-1)^2 + 3$

76.

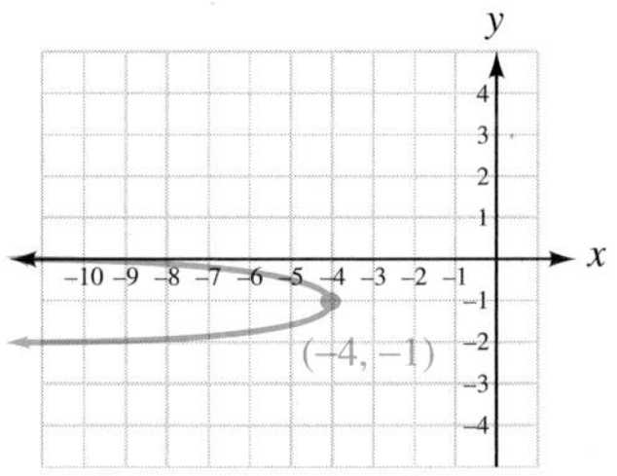

$x = -6(y+1)^2 - 4$

77.

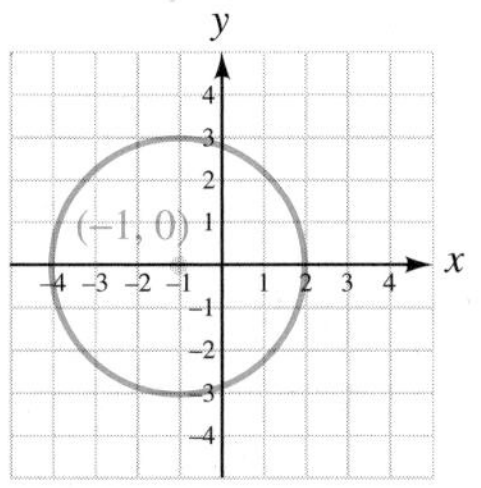

$(x+1)^2 + y^2 = 9$

78.

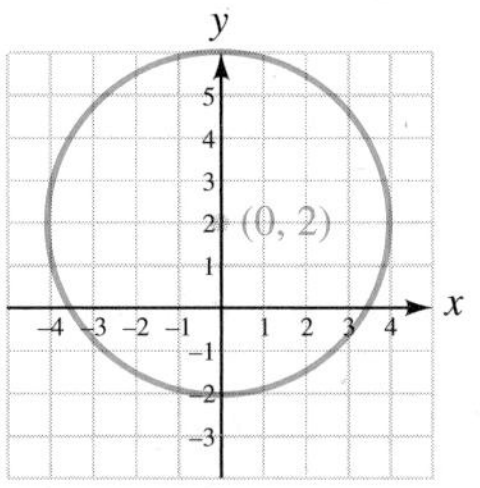

$x^2 + (y-2)^2 = 16$

79.

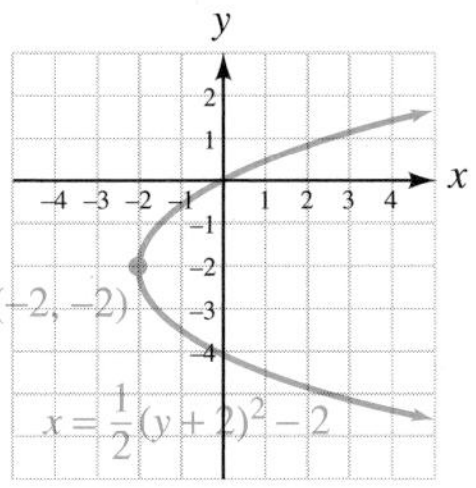

80.

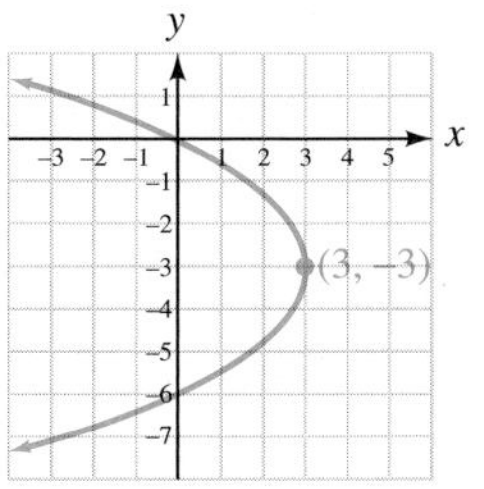

$x = -\frac{1}{3}(y+3)^2 + 3$

81.

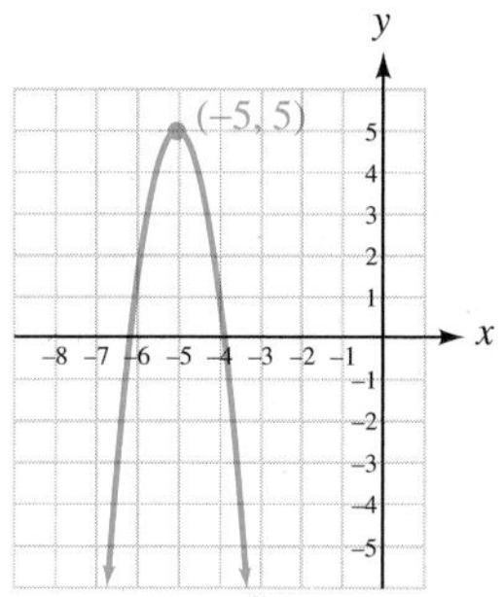

$y = -4(x+5)^2 + 5$

82.

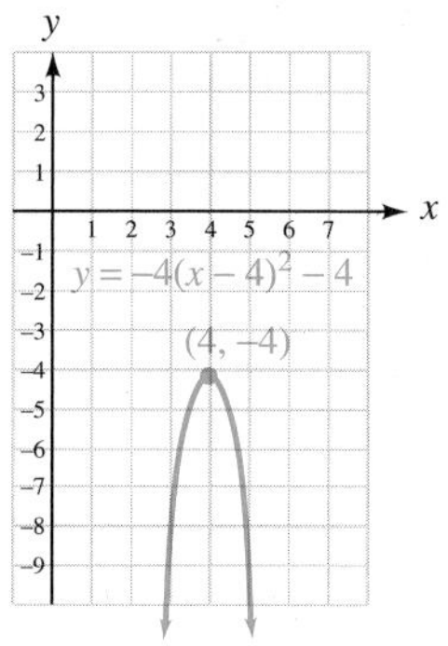

SELF CHECK 13.2

1.

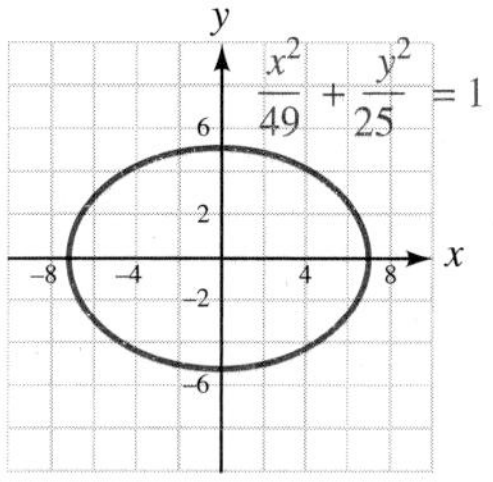

2.

$9x^2 + y^2 = 9$

or

$\frac{x^2}{1} + \frac{y^2}{9} = 1$

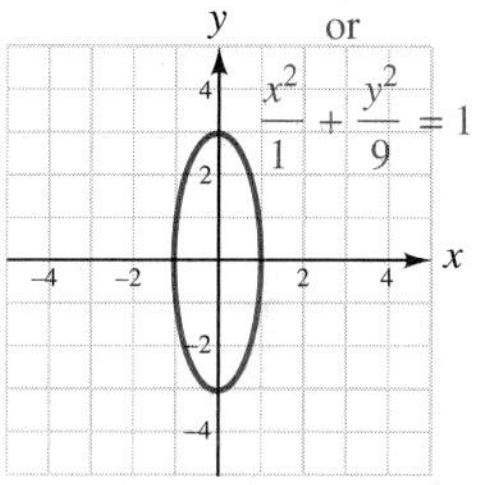

3.

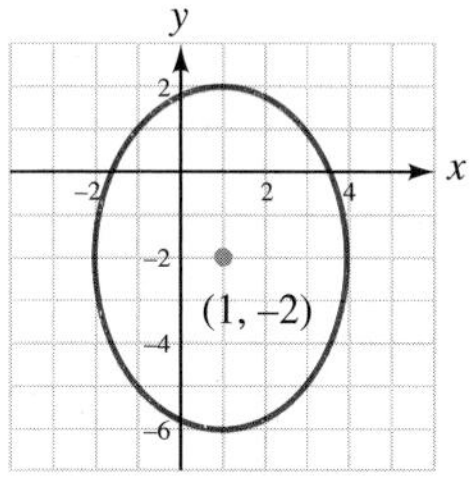

$\frac{(x-1)^2}{9} + \frac{(y+2)^2}{16} = 1$

4.

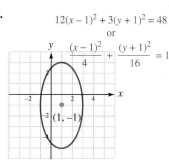

$12(x-1)^2 + 3(y+1)^2 = 48$

or

$\frac{(x-1)^2}{4} + \frac{(y+1)^2}{16} = 1$

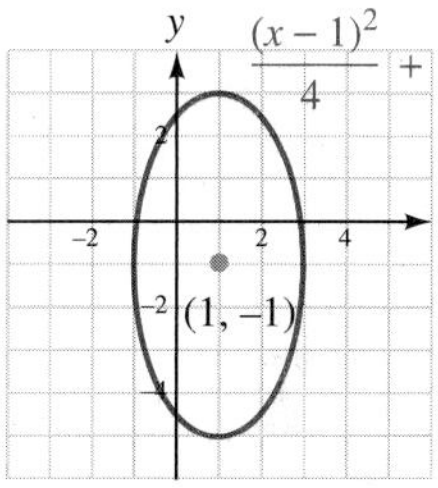

STUDY SET SECTION 13.2 (page 1095)

17.

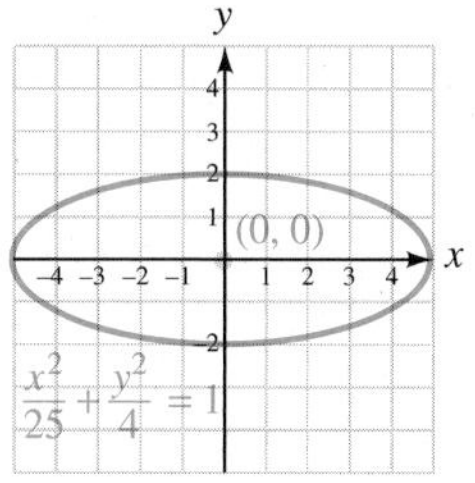

18.

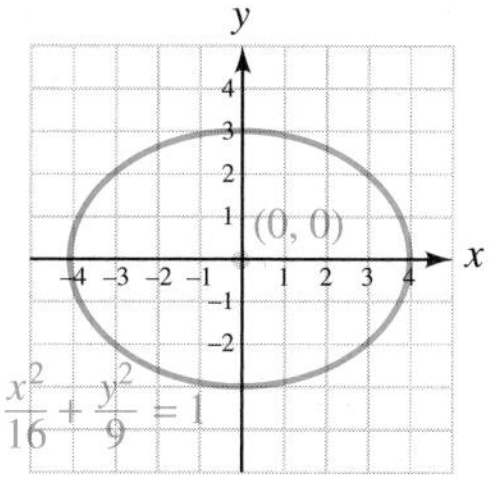

19.

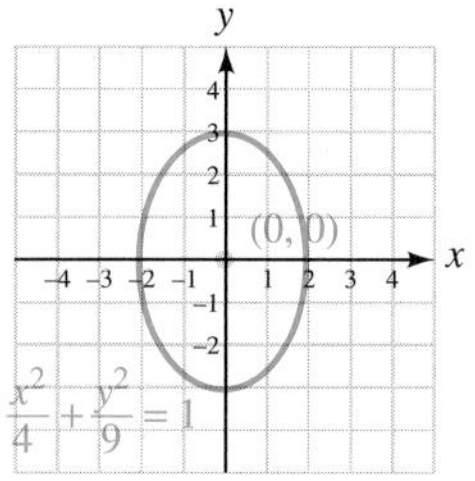

20.

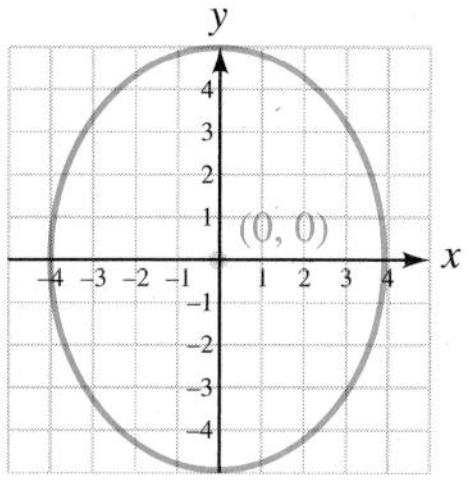

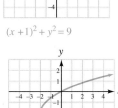

$\frac{x^2}{16} + \frac{y^2}{25} = 1$

21.

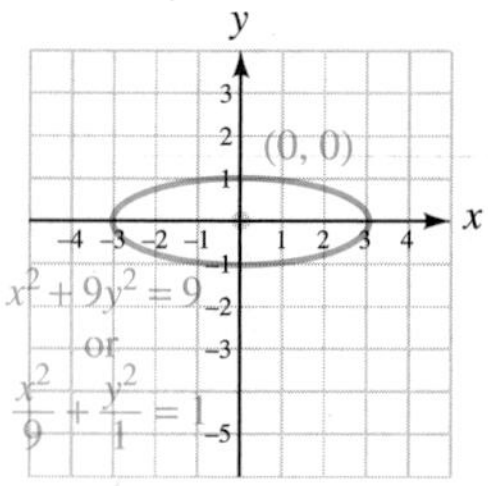

22.

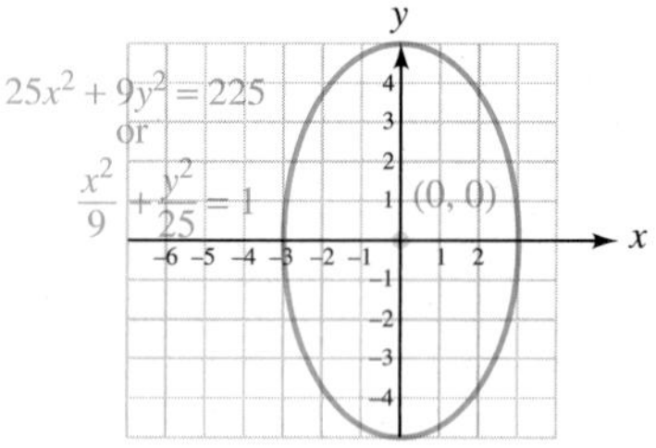

23.

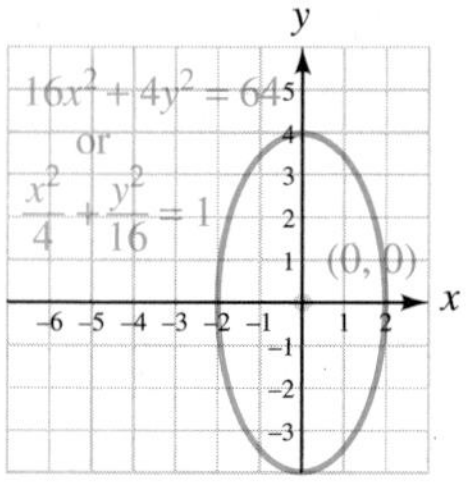

24.

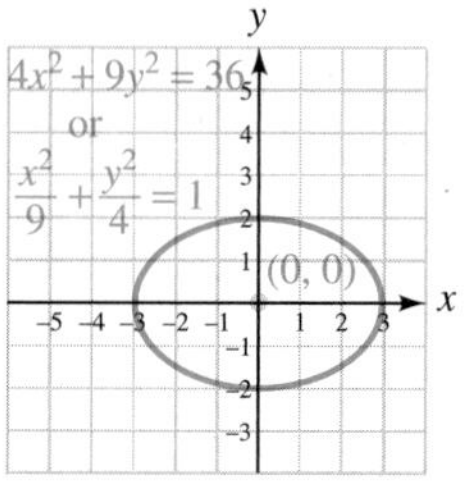

25.

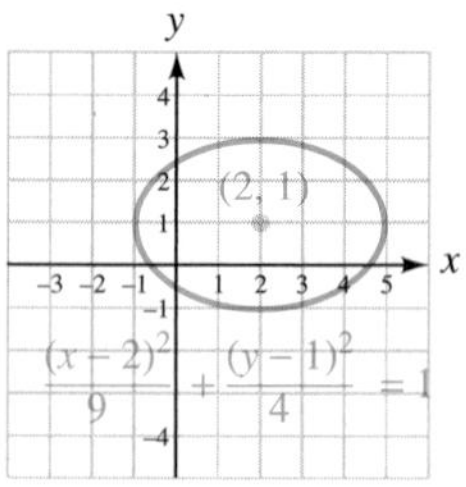

26.

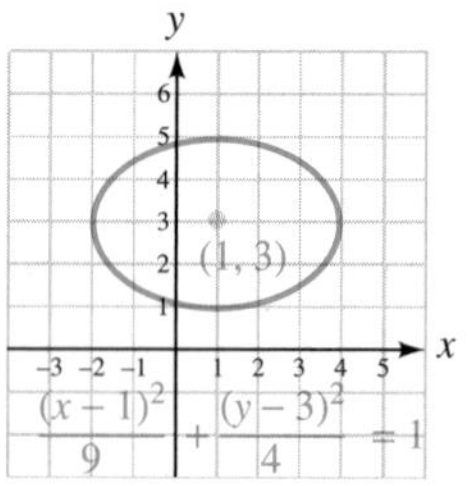

27.

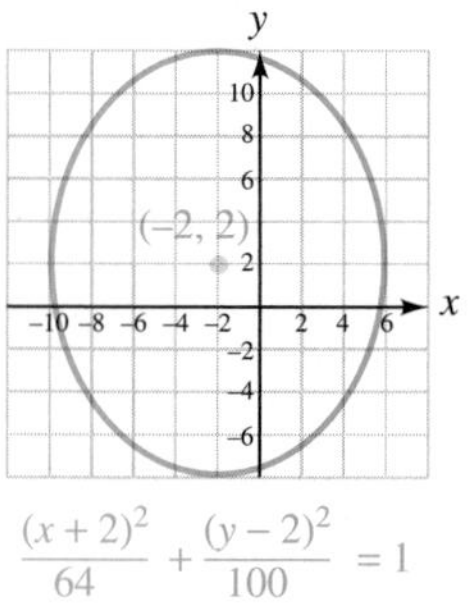

28.

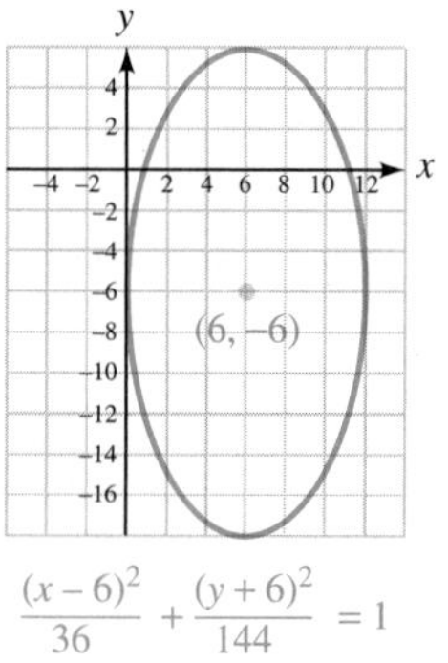

29.

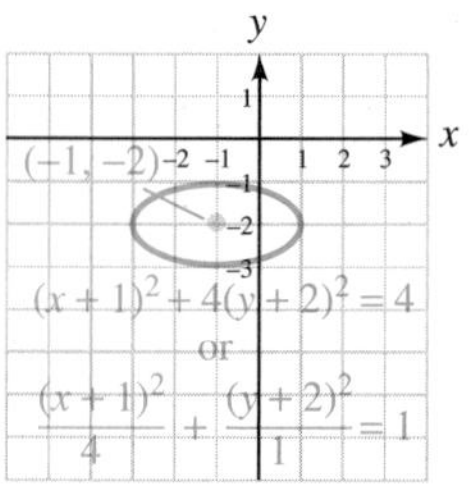

30.

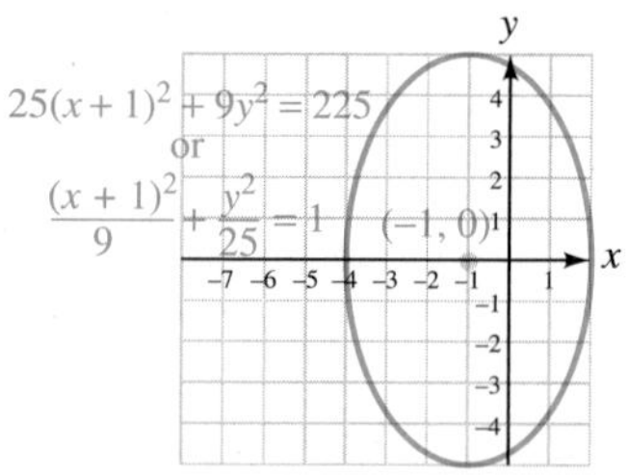

31.

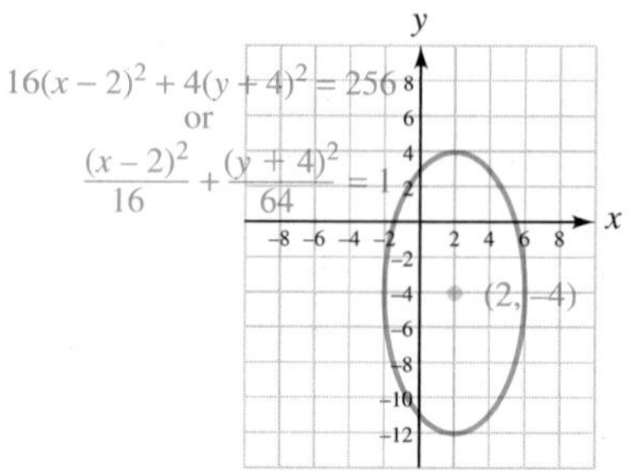

32.

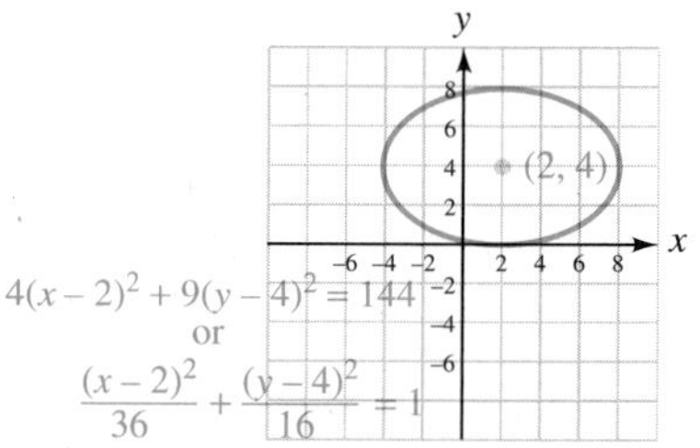

33.

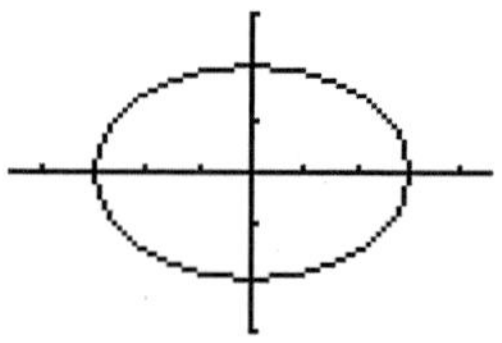

34.

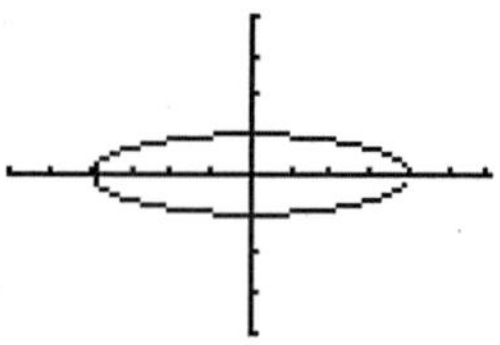

35.

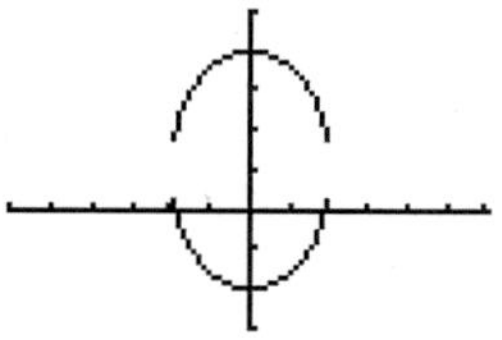

36.

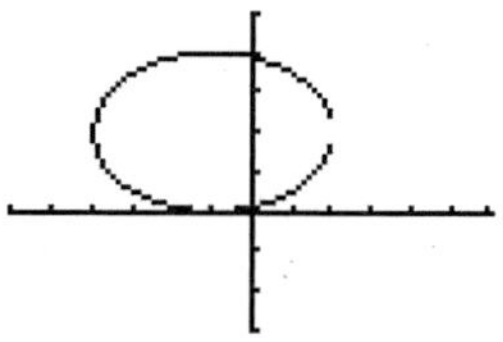

37.

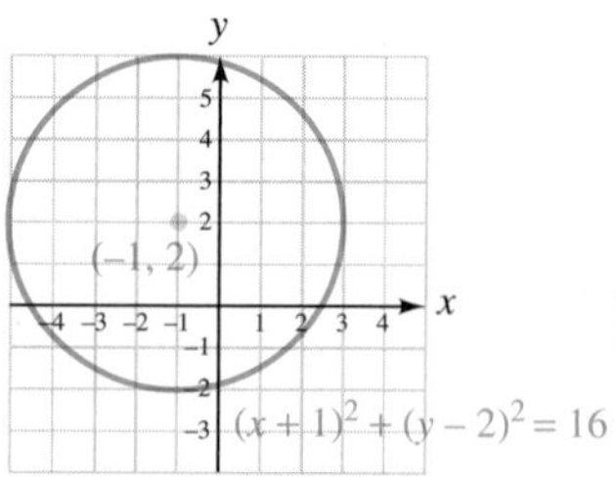

38.

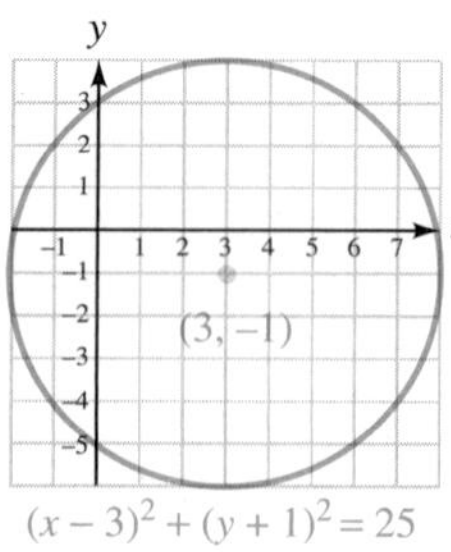

39.

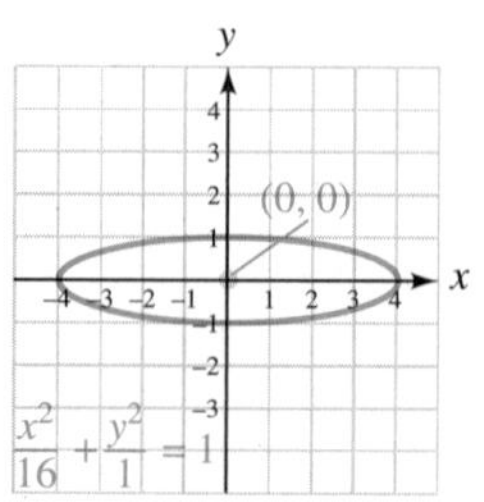

40.

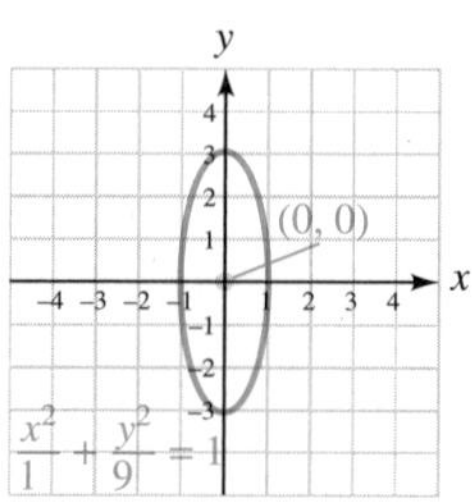

41.

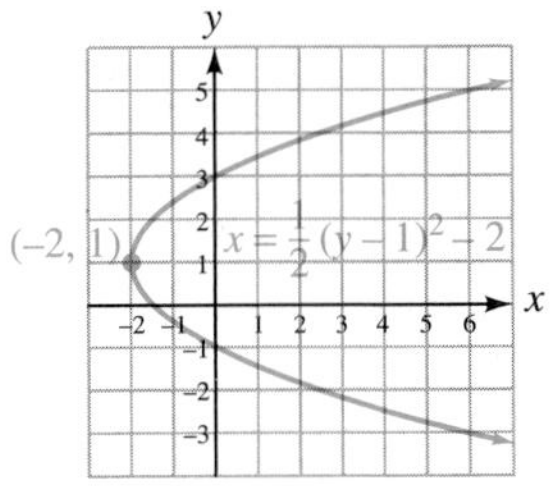

42.

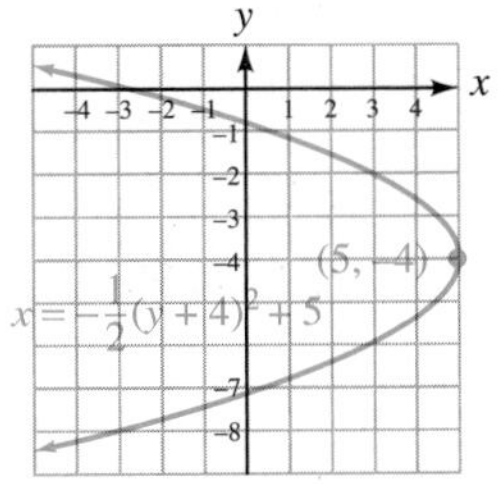

43.

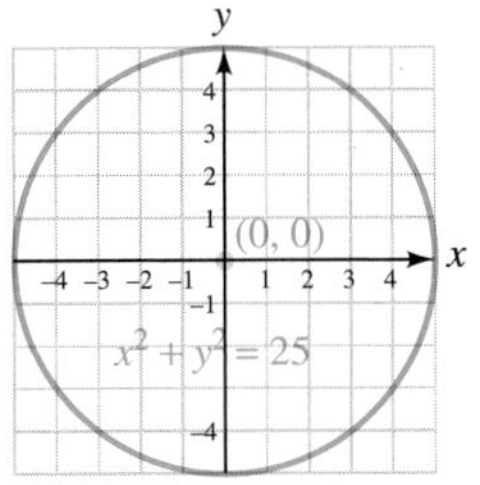

44.

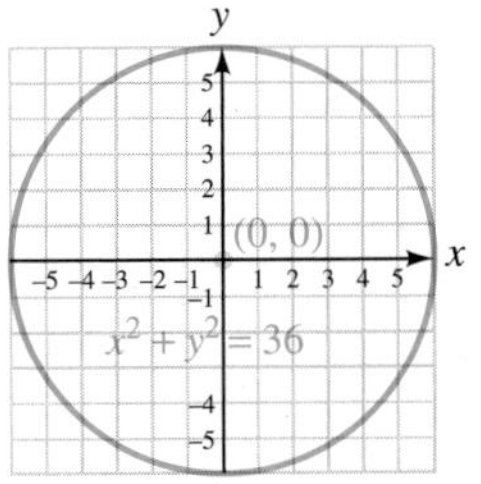

45.

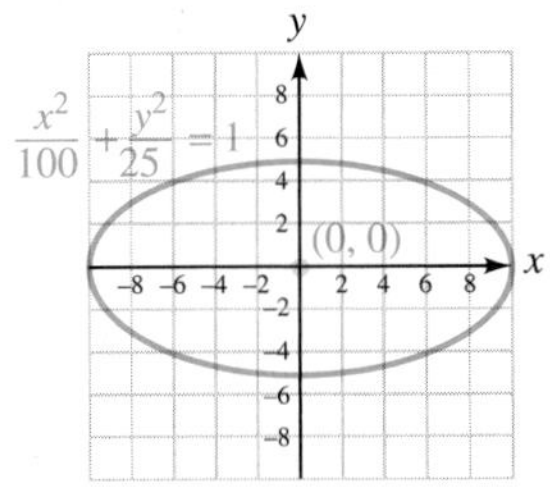

46.

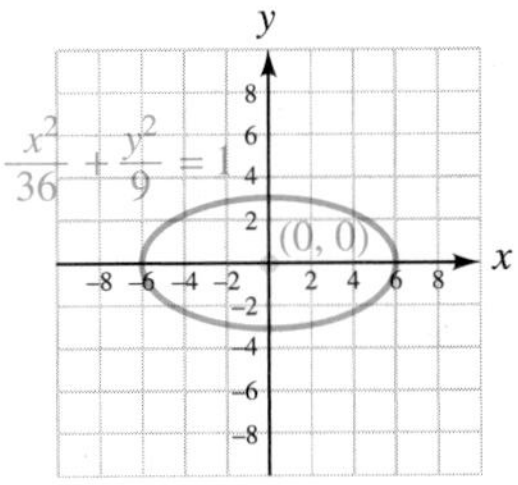

47.

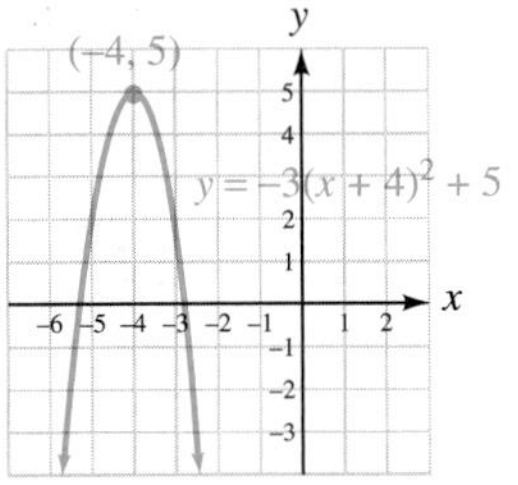

48.

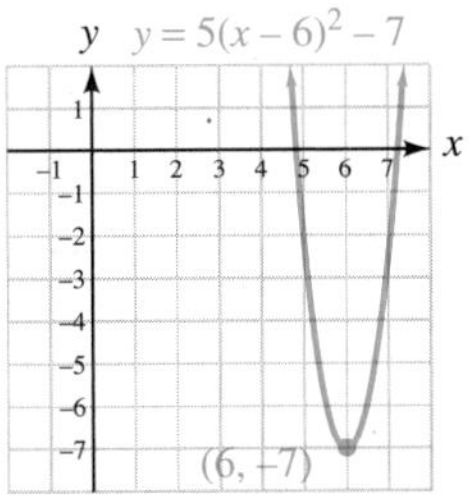

49.

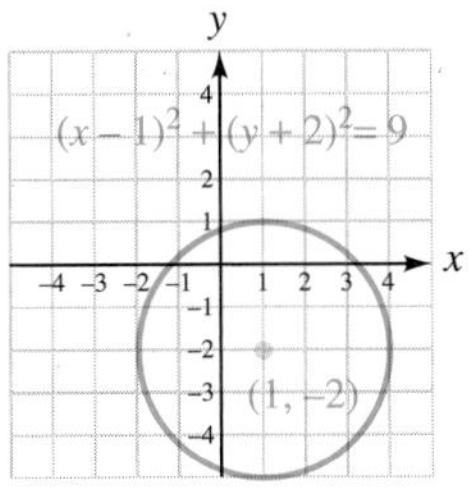

50.

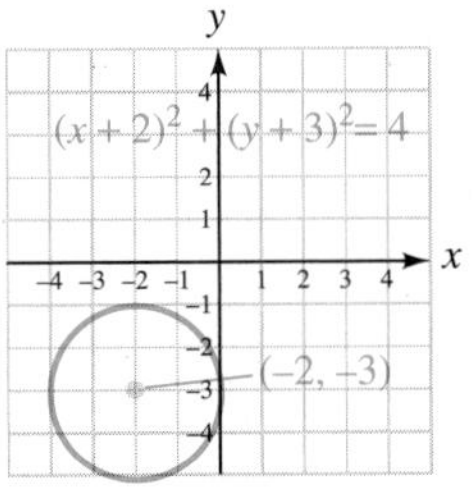

51.

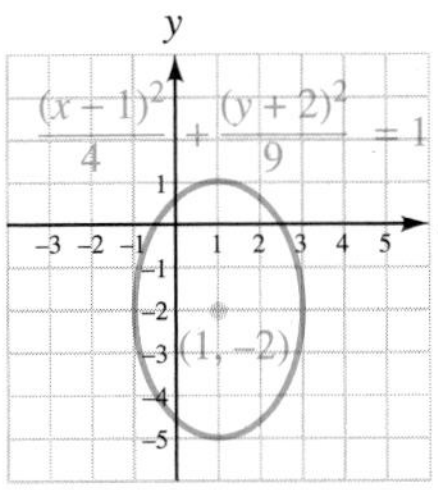

52.

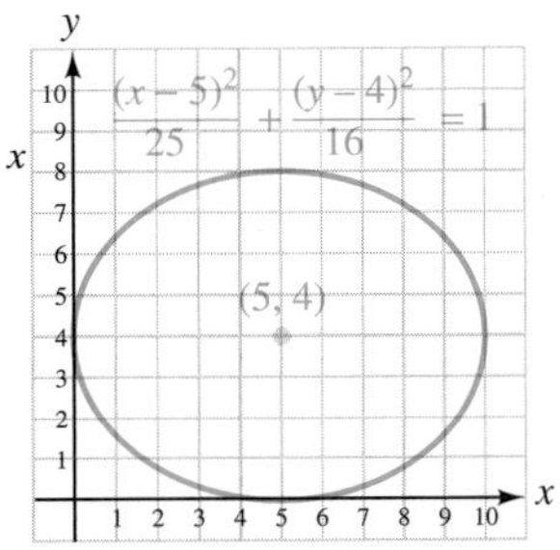

53.

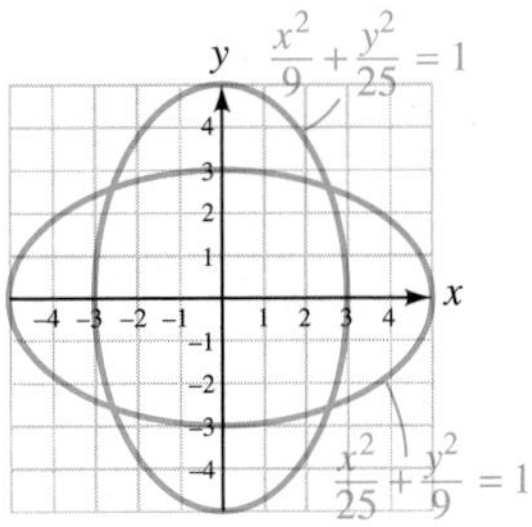

54.

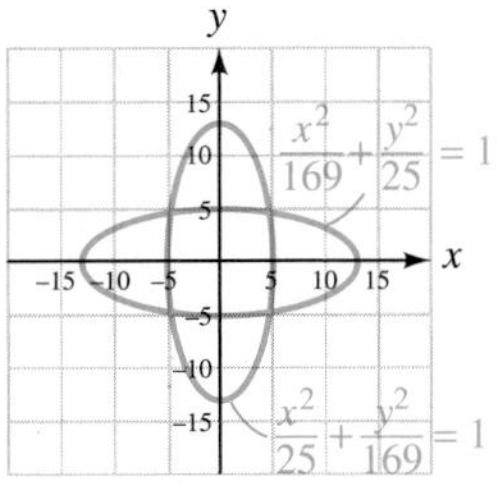

55.

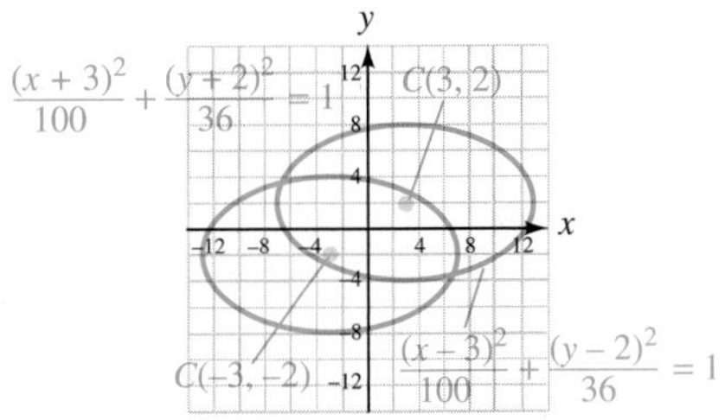

56

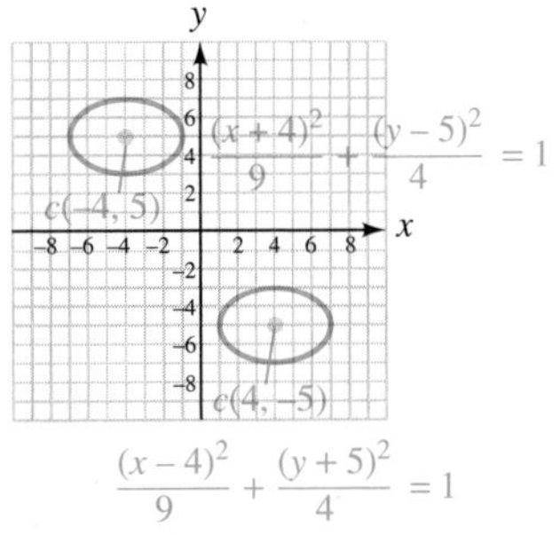

72.

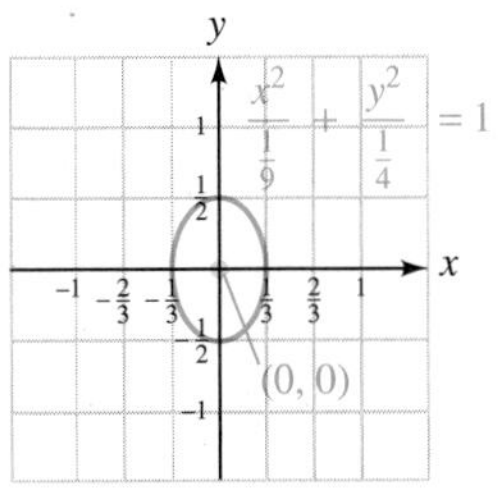

SELF CHECKS 13.3

1.

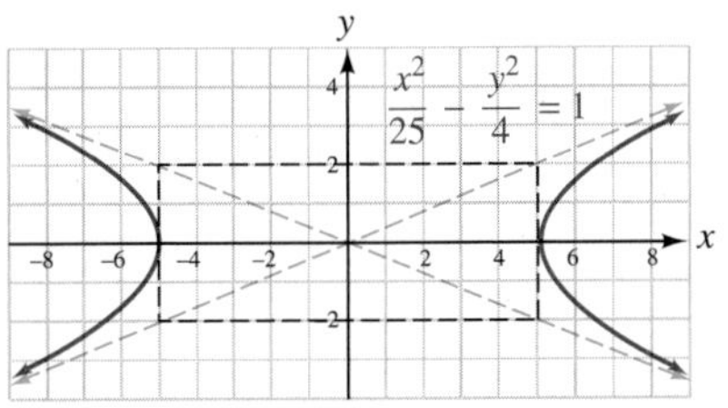

2.

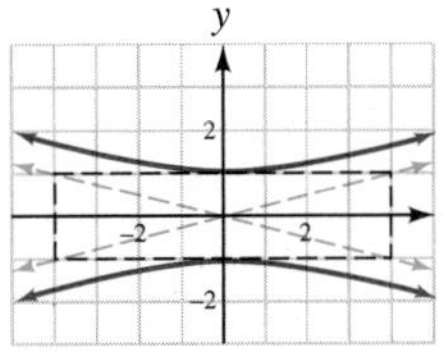

$16y^2 - x^2 = 16$

or

$\frac{y^2}{1} - \frac{x^2}{16} = 1$

3. a.

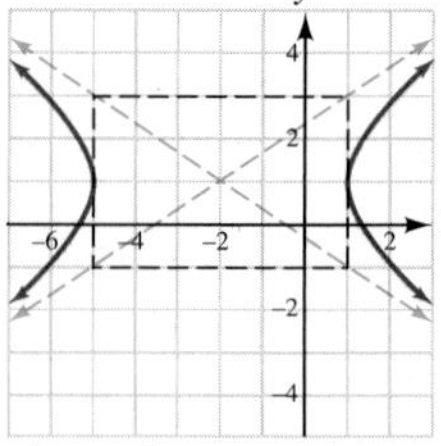

$$\frac{(x+2)^2}{9} - \frac{(y-1)^2}{4} = 1$$

b.

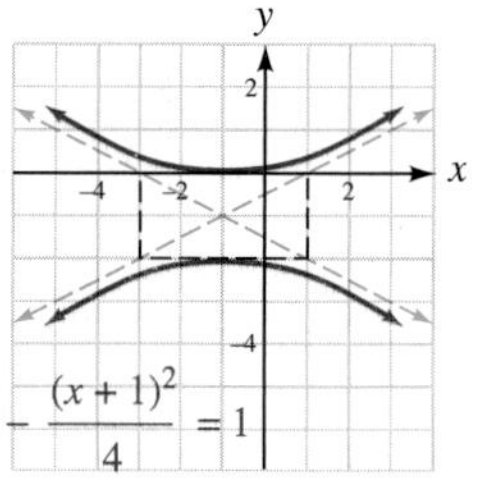

$$\frac{(y+1)^2}{1} - \frac{(x+1)^2}{4} = 1$$

4.

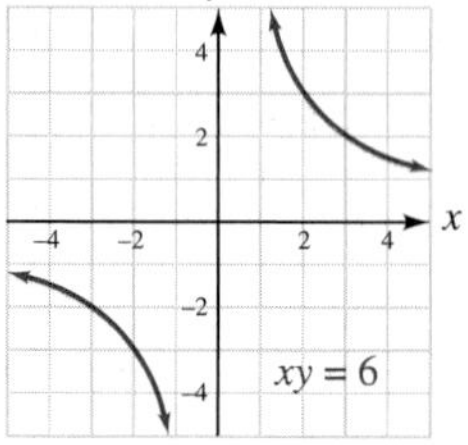

STUDY SET SECTION 13.3 (page 1104)

17.

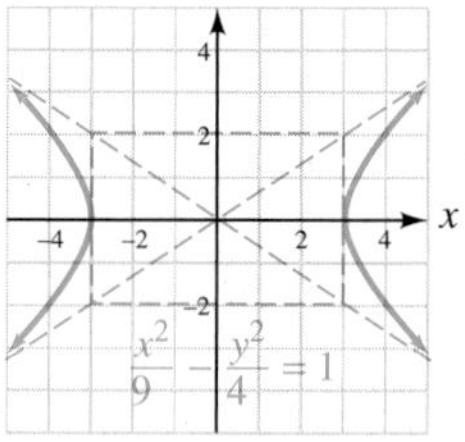

18.

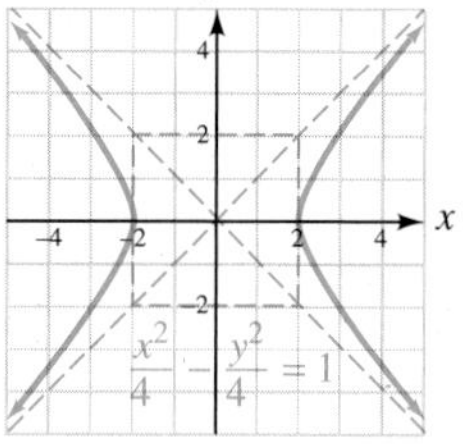

19.

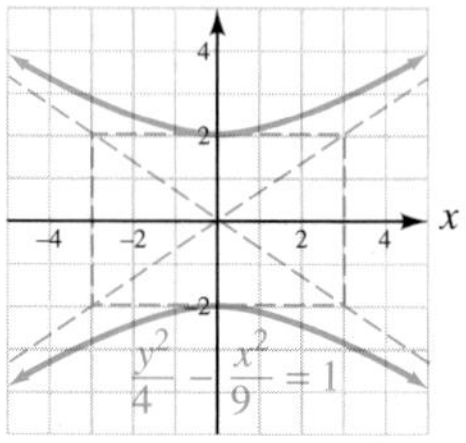

20.

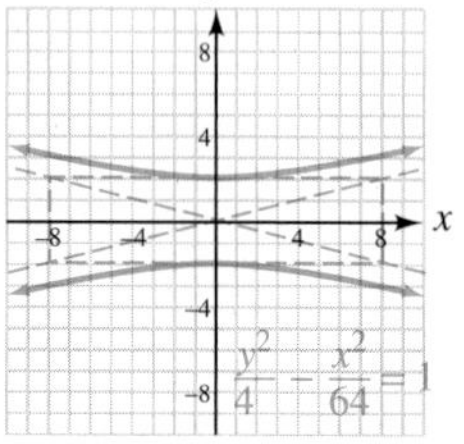

21.

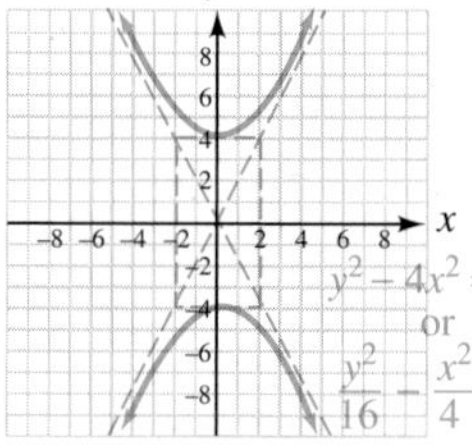

22.

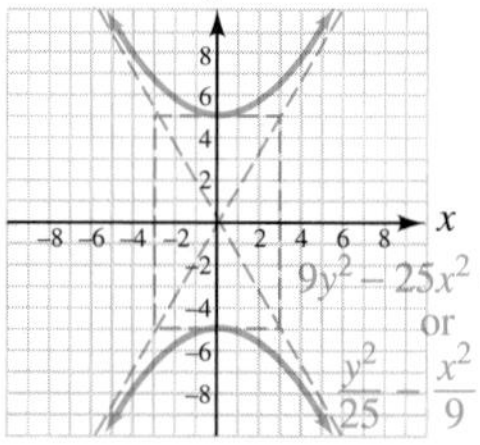

23.

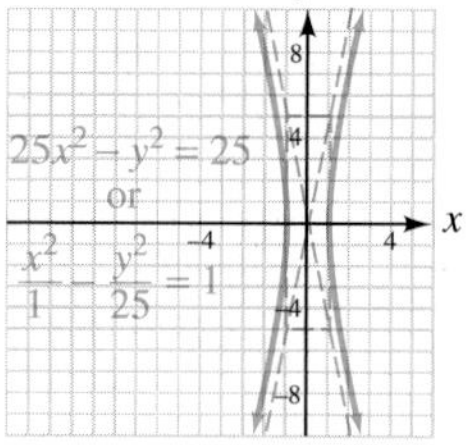

24.

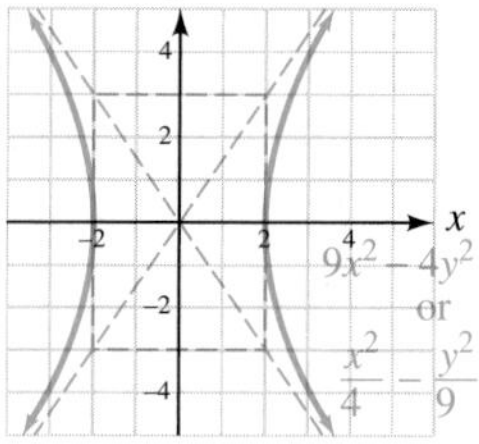

25.

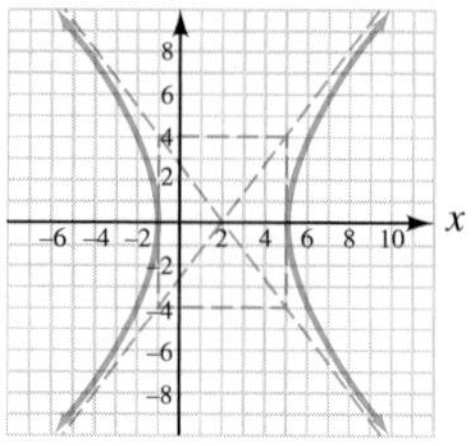

$$\frac{(x-2)^2}{9} - \frac{y^2}{16} = 1$$

26.

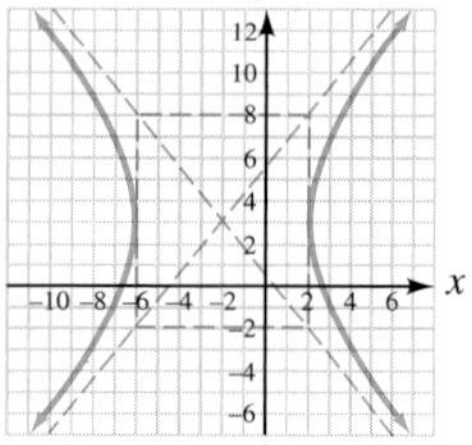

$$\frac{(x+2)^2}{16} - \frac{(y-3)^2}{25} = 1$$

27.

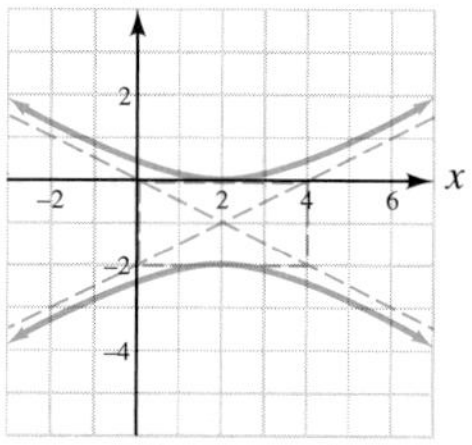

$$\frac{(y+1)^2}{1} - \frac{(x-2)^2}{4} = 1$$

28.

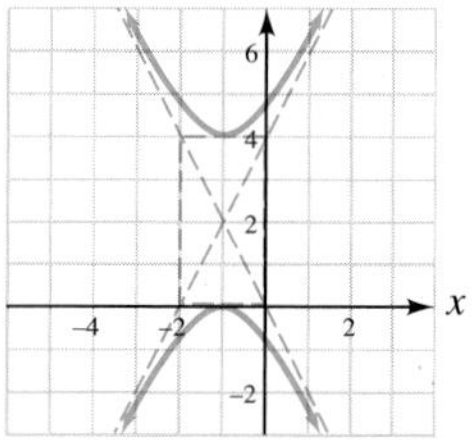

$$\frac{(y-2)^2}{4} - \frac{(x+1)^2}{1} = 1$$

29.

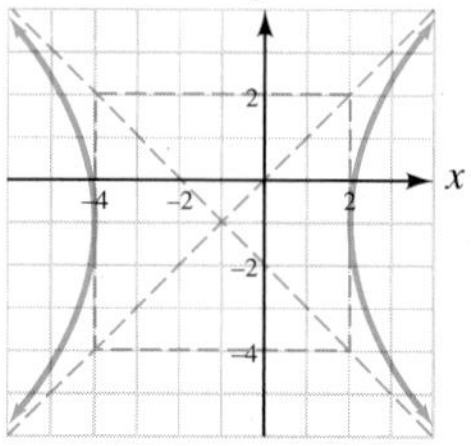

$$\frac{(x+1)^2}{9} - \frac{(y+1)^2}{9} = 1$$

30.

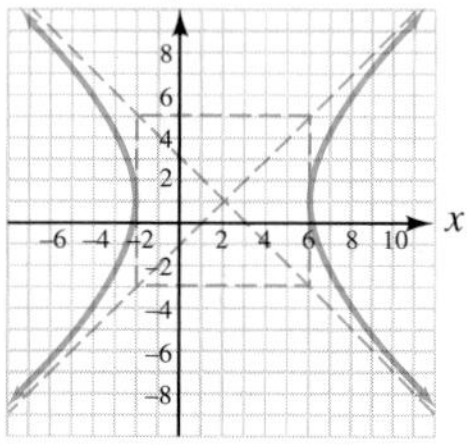

$$\frac{(x-2)^2}{16} - \frac{(y-1)^2}{16} = 1$$

31.

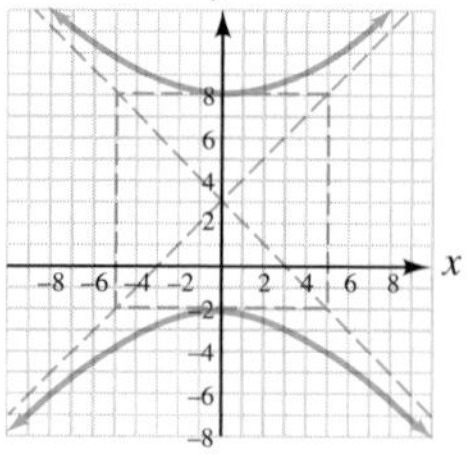

$$\frac{(y-3)^2}{25} - \frac{x^2}{25} = 1$$

32.

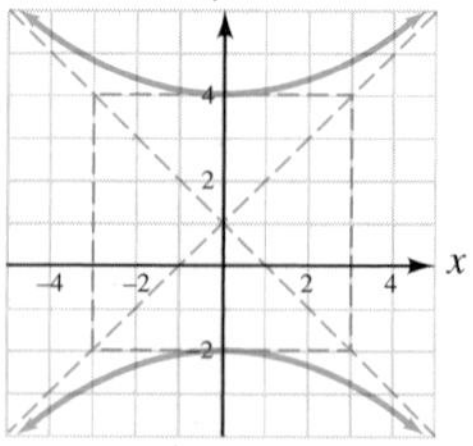

$$\frac{(y-1)^2}{9} - \frac{x^2}{9} = 1$$

33.

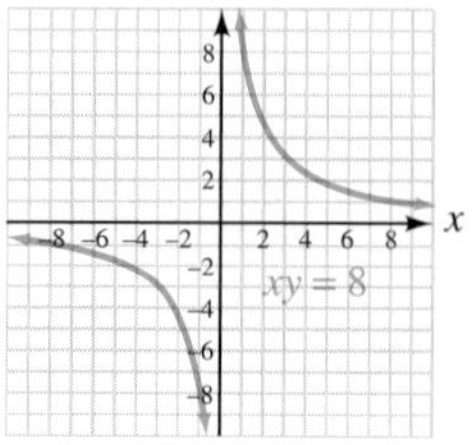

34.

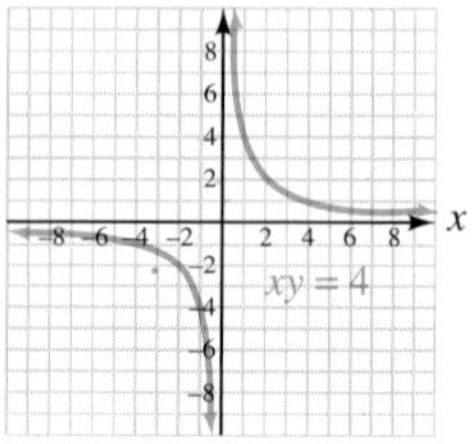

35.
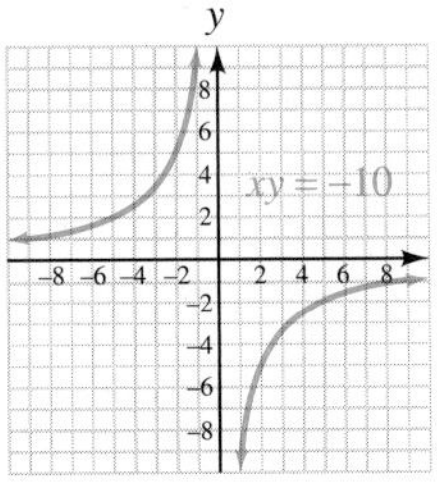

36.
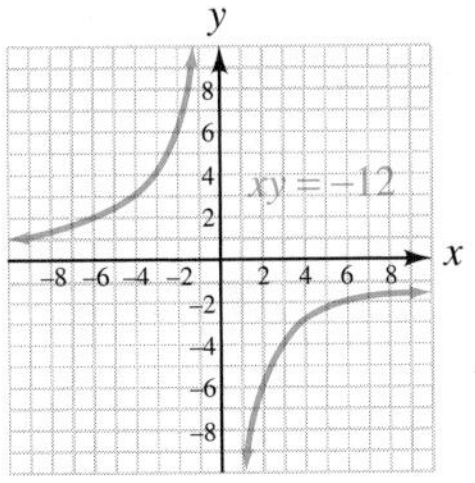

37.
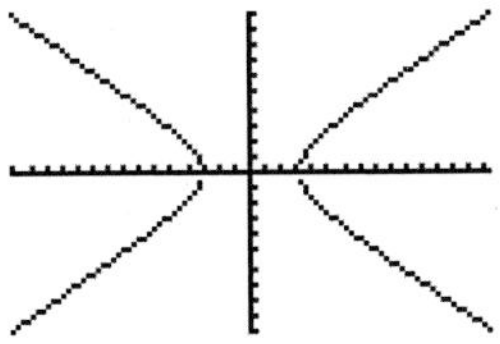

38.
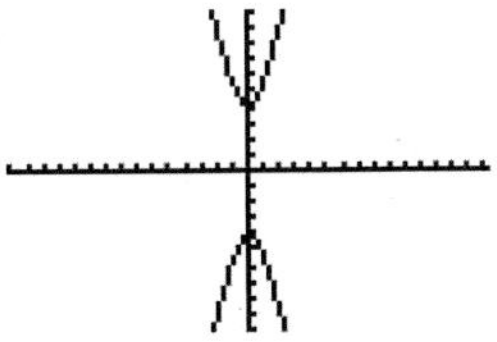

39.
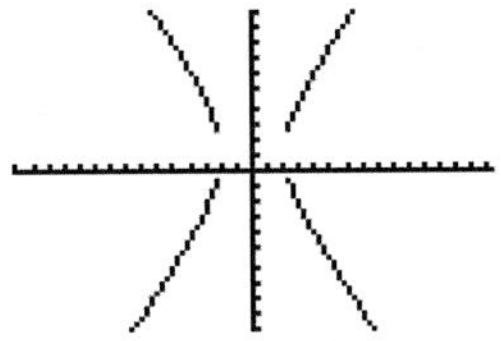

40.
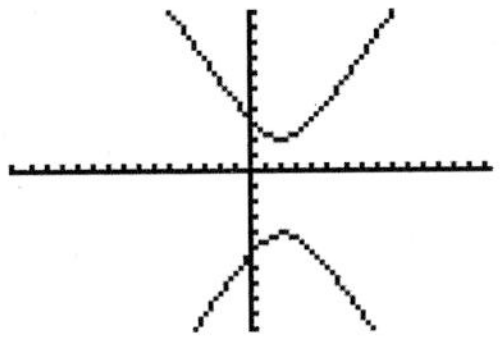

41.
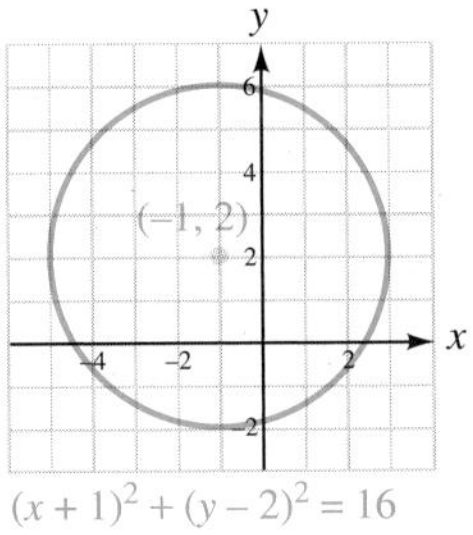

42.
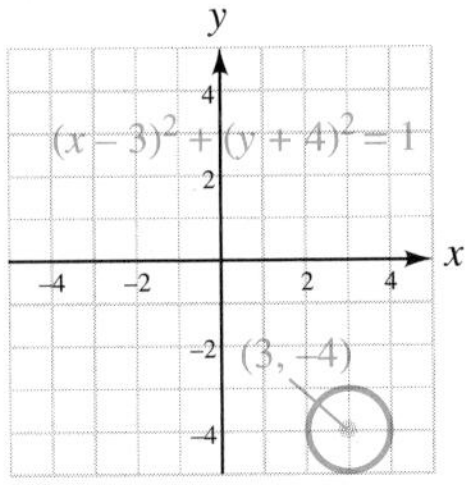

43.
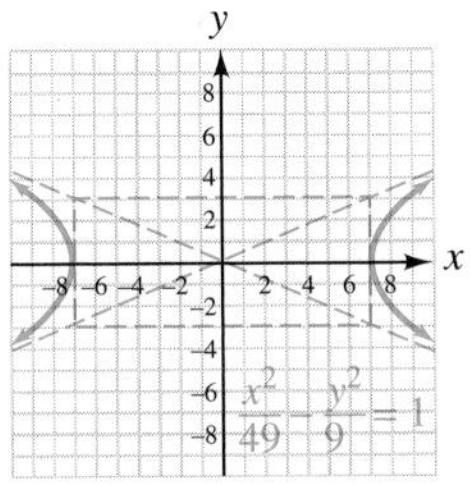

44.
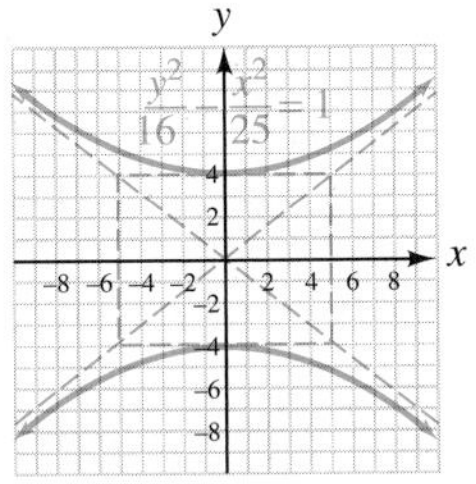

45.
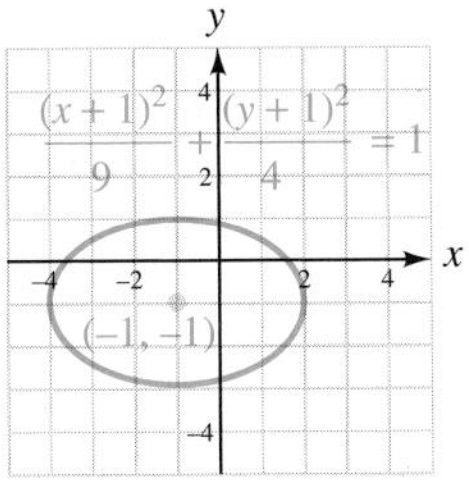

46.
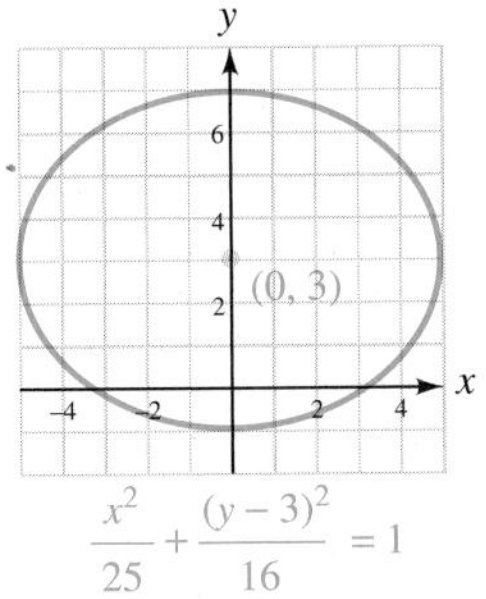

47.
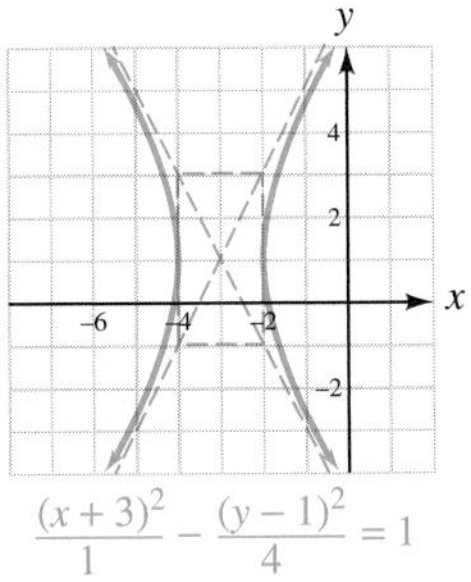

48.
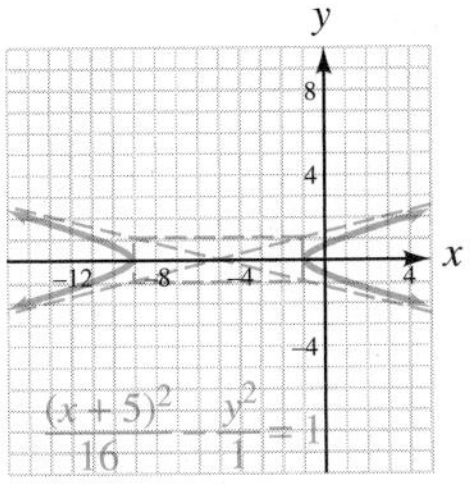

49.
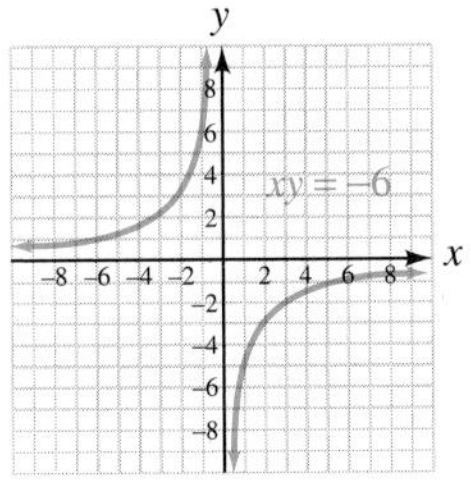

50.
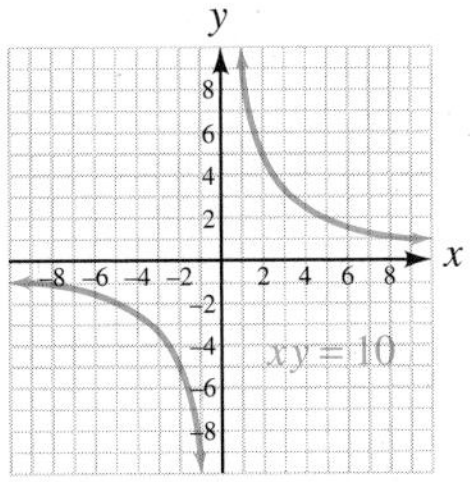

51.
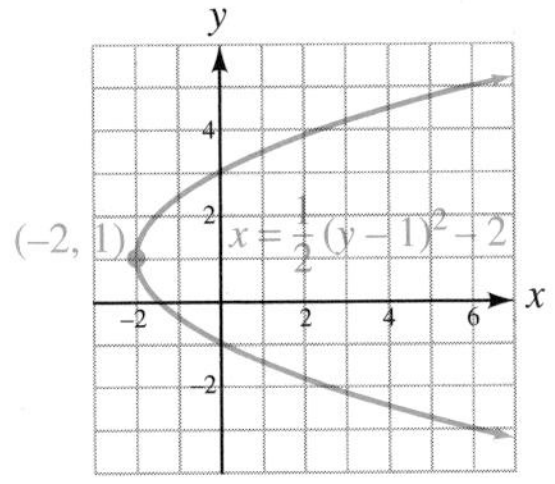

52.
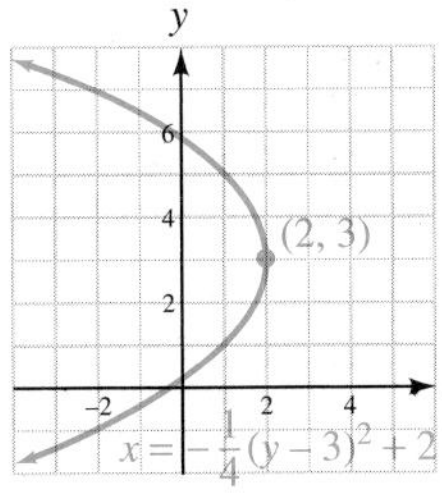

53.
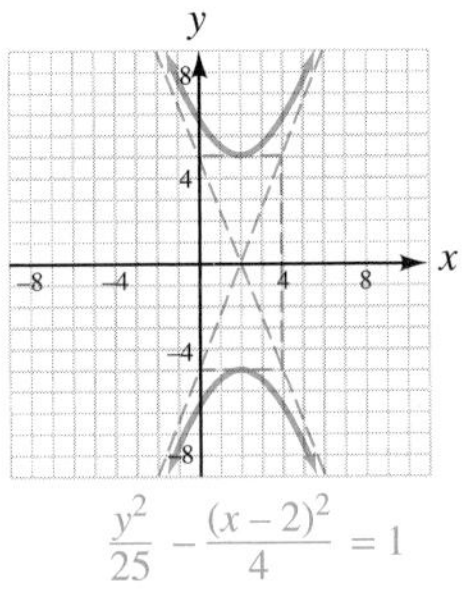

54.
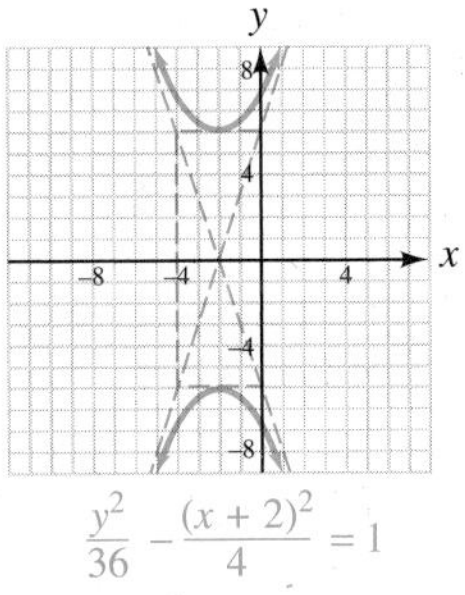

55.
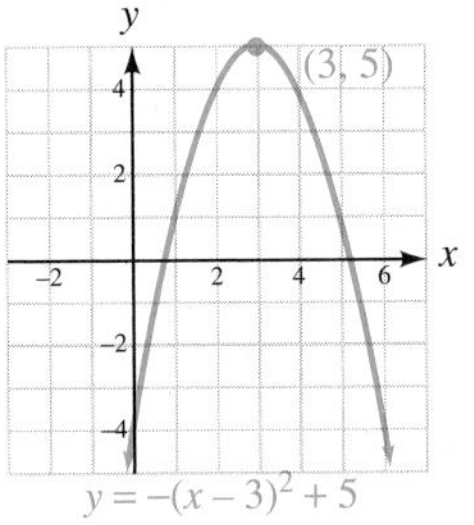

56.
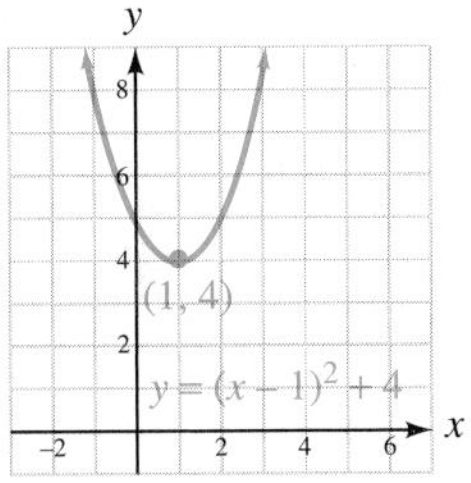

57.
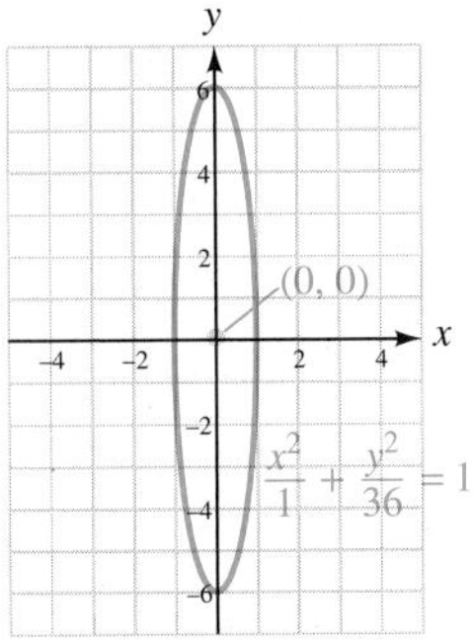

58.
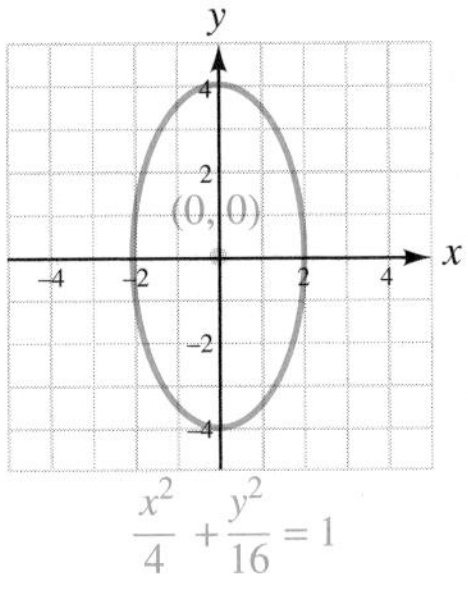

59.
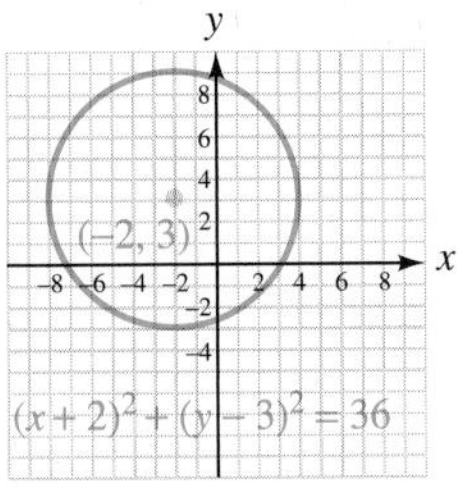

60.
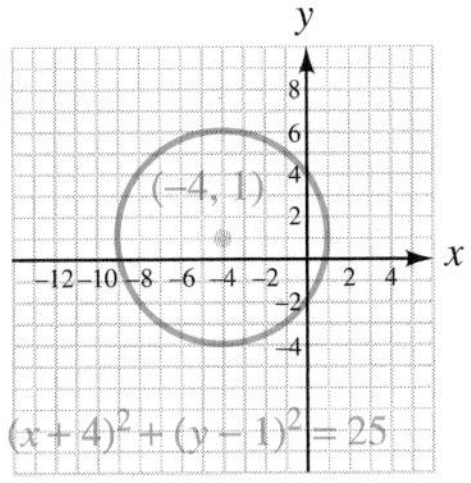

83.

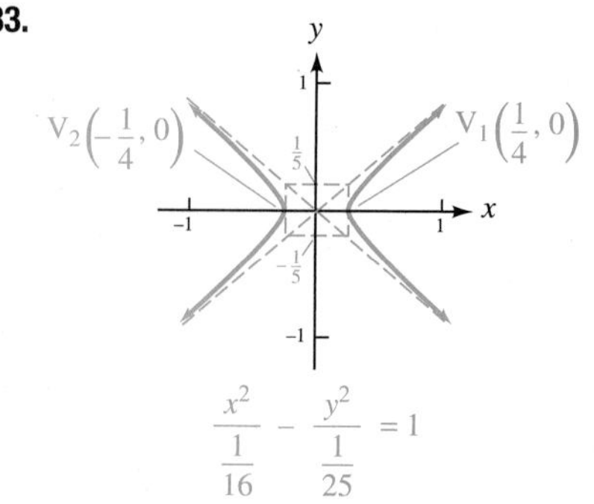

ARE YOU READY? 13.4 (page 1107)

1.

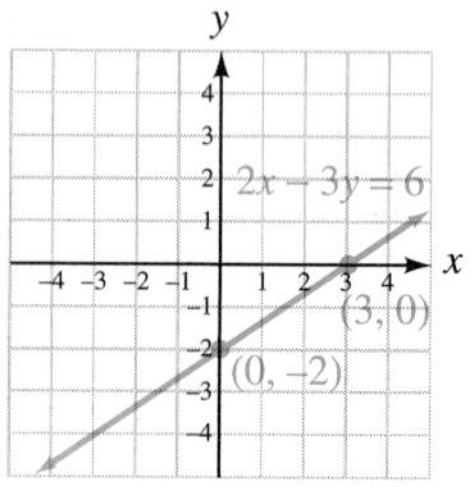

SELF CHECK 13.4

1.

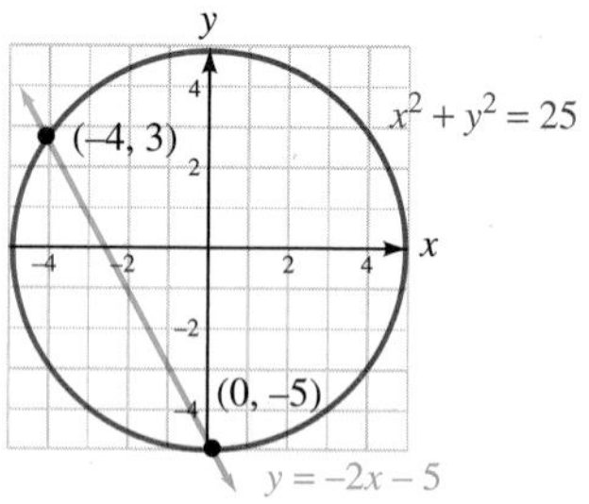

STUDY SET SECTION 13.4 (page 1111)

15.

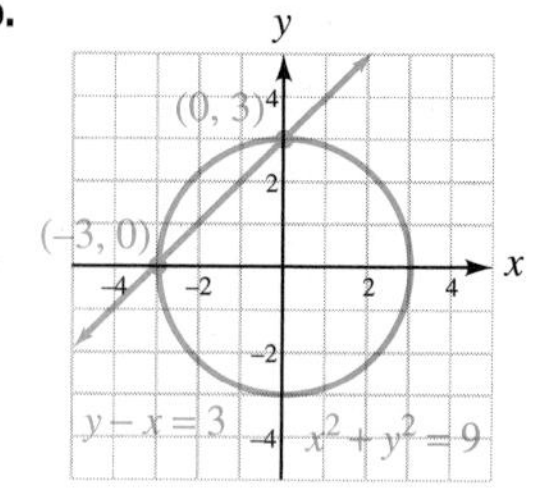

16.

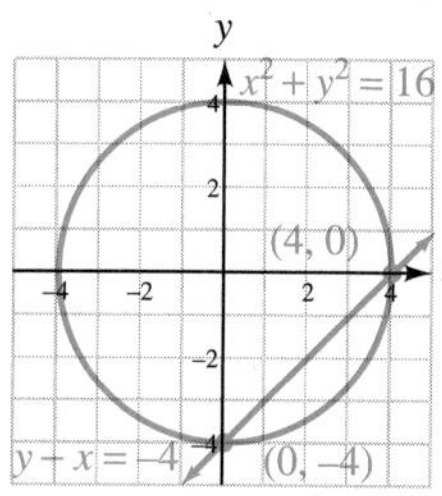

17.

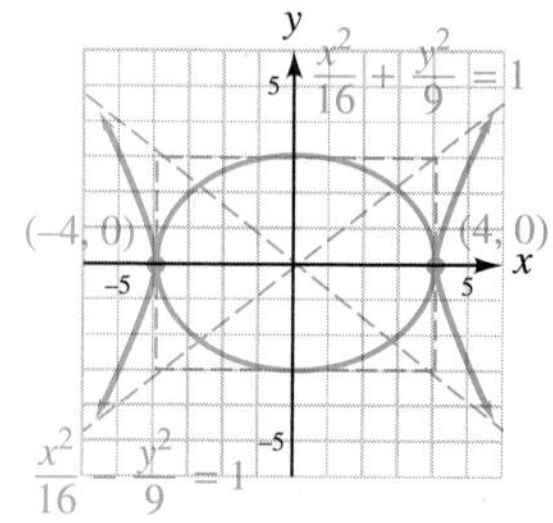

18.

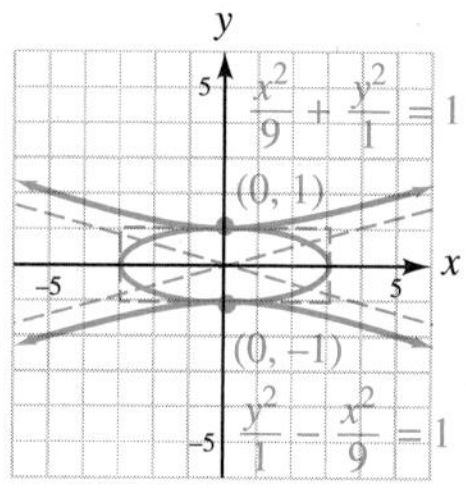

19.

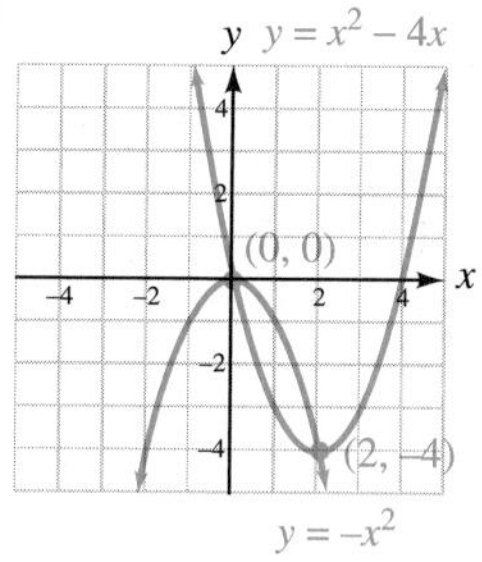

20.

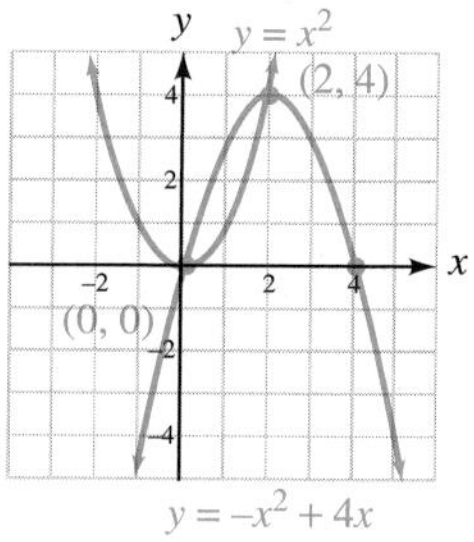

21.

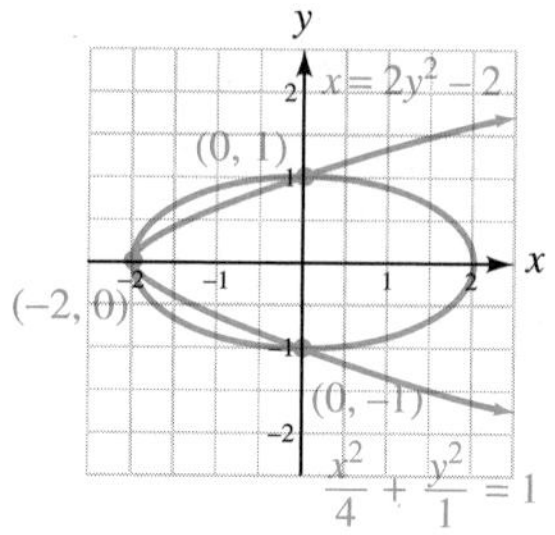

22.

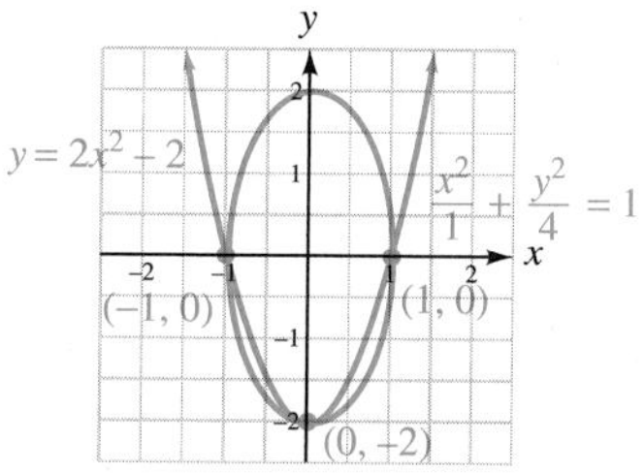

39.

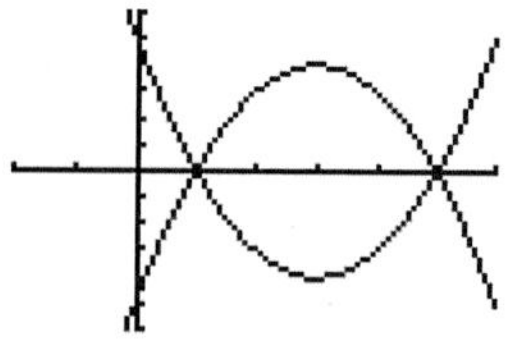

40.

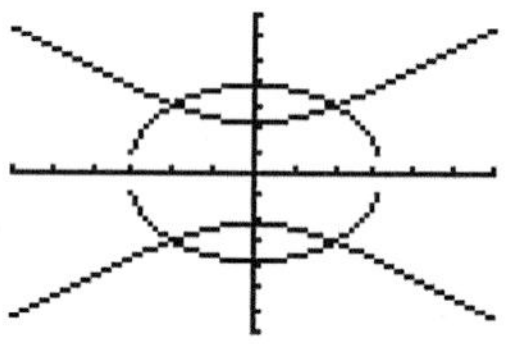

CHAPTER 13 REVIEW (page 1114)

1.

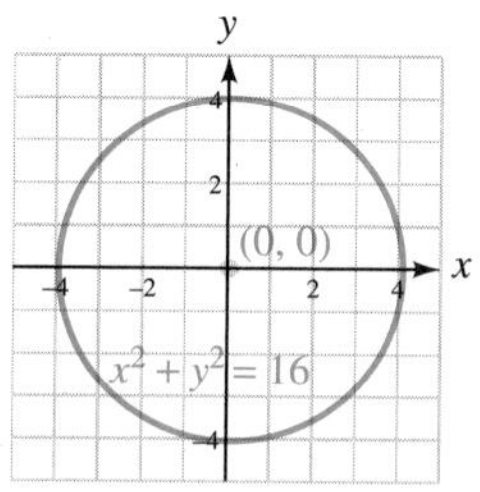

2.

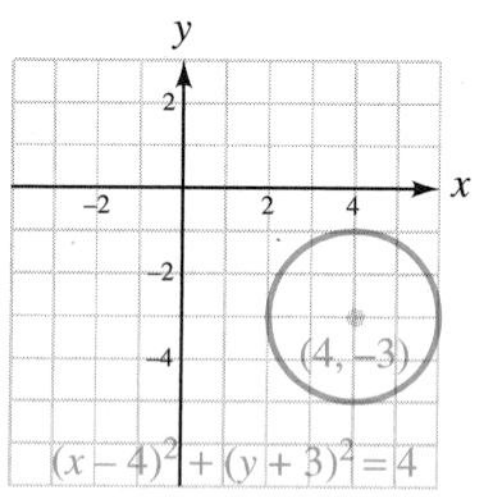

3.

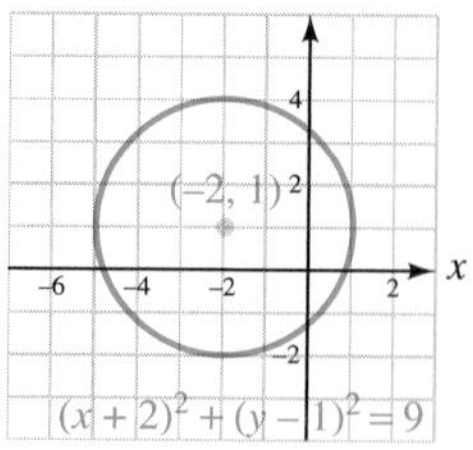

7.

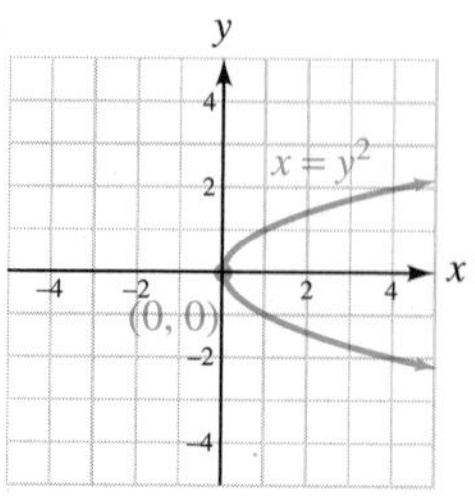

8.

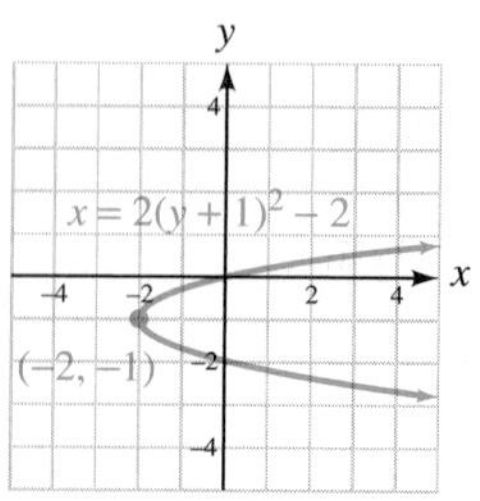

9.

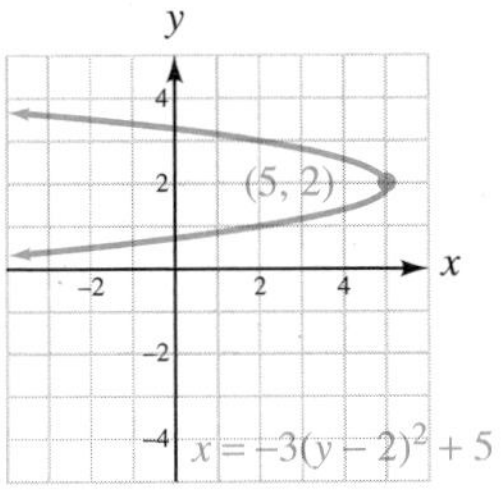

10.

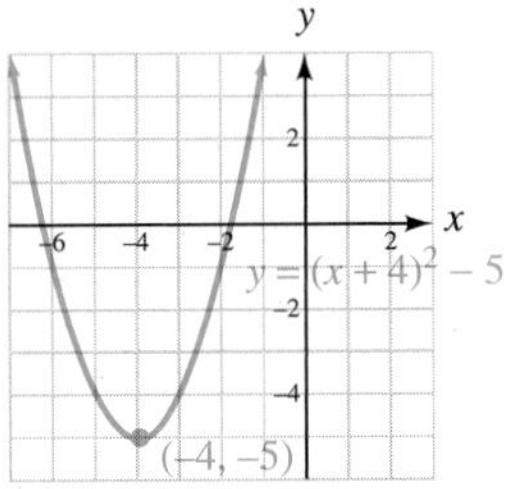

13.

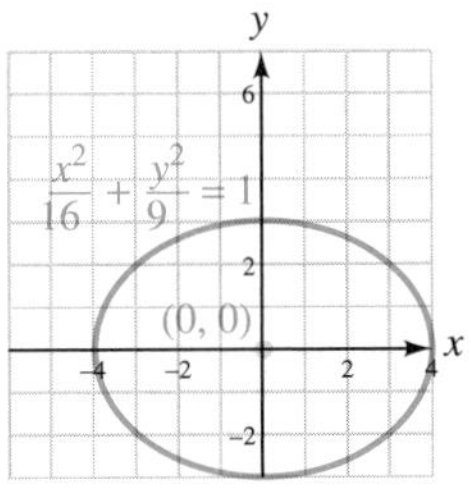

14.

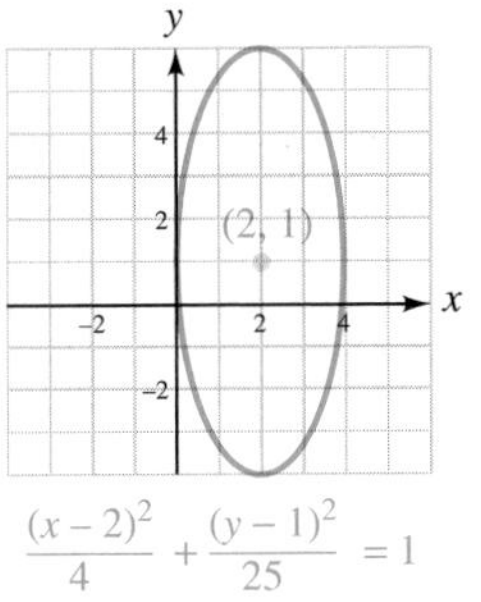

15.

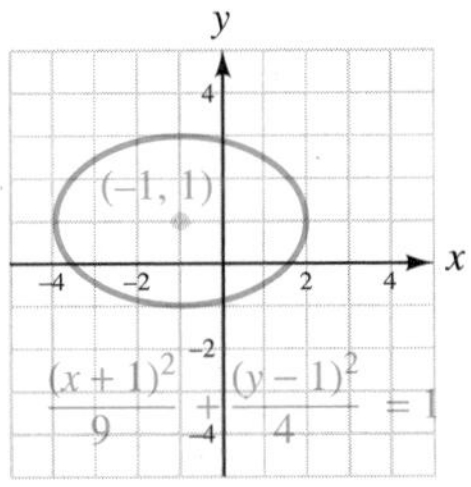

21.

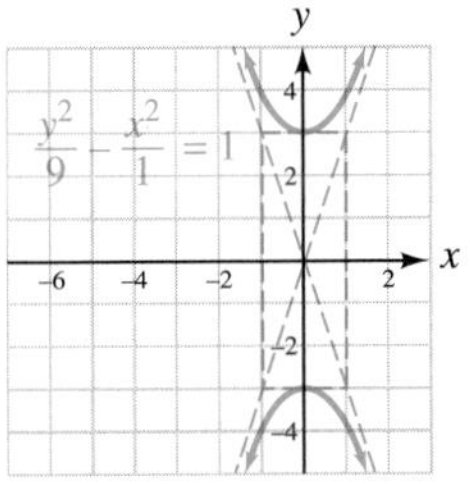

22.

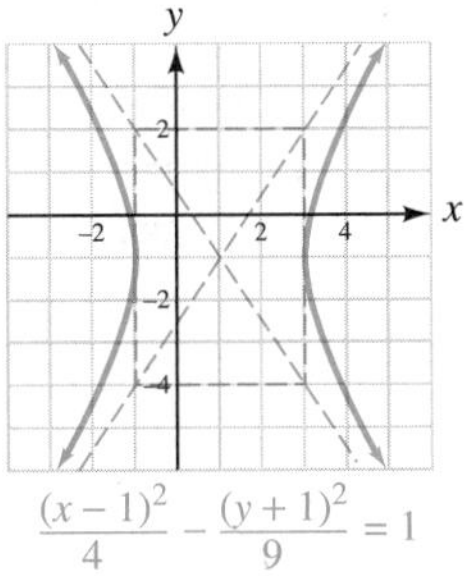

23.

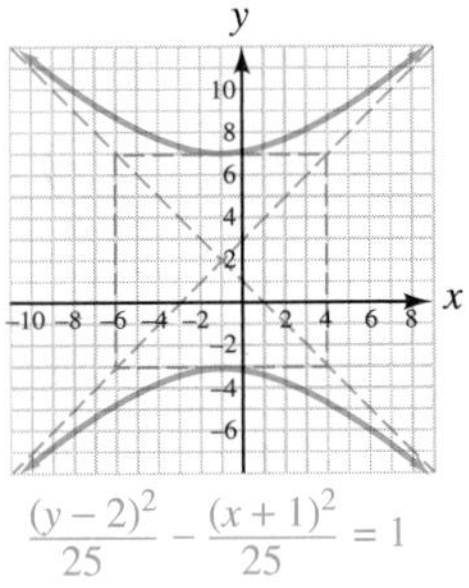

24.

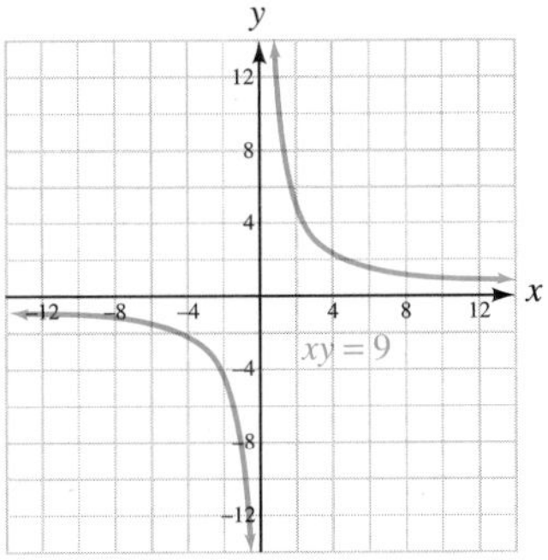

29.

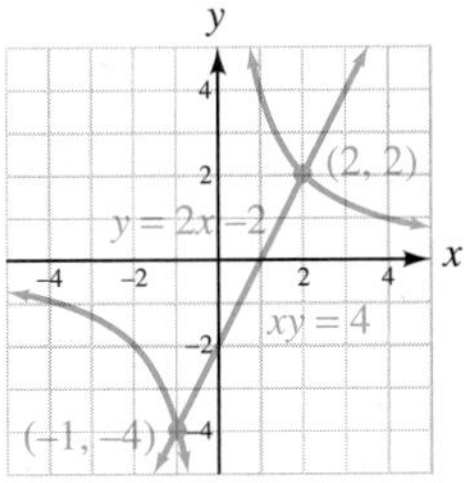

CHAPTER 13 TEST (page 1119)

2.

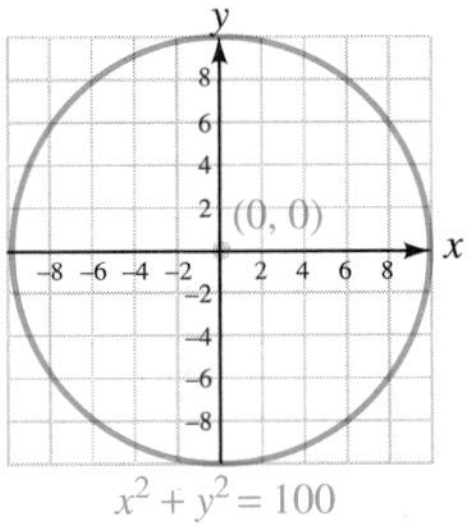

7.

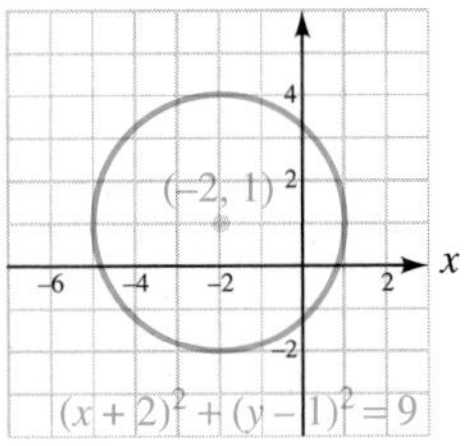

8.

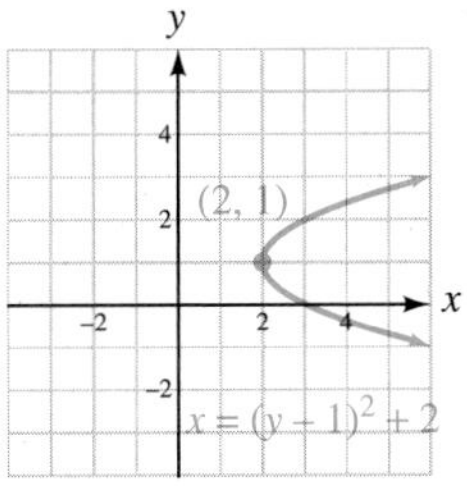

9.

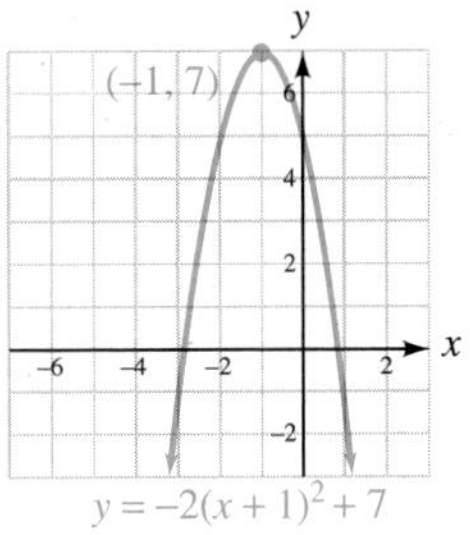

10.

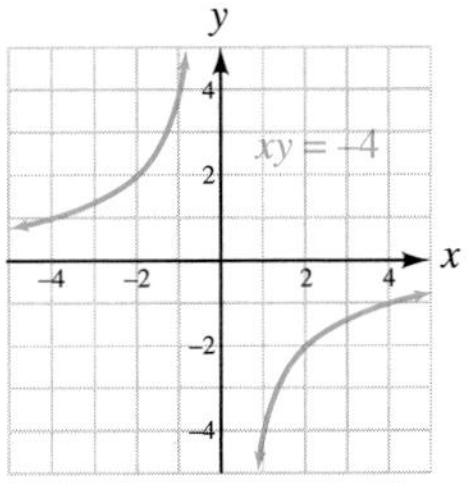

11.

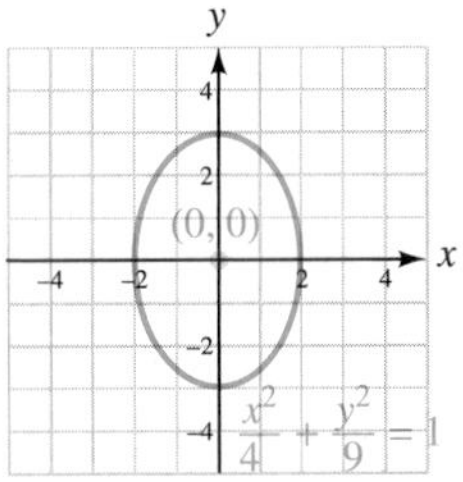

12.

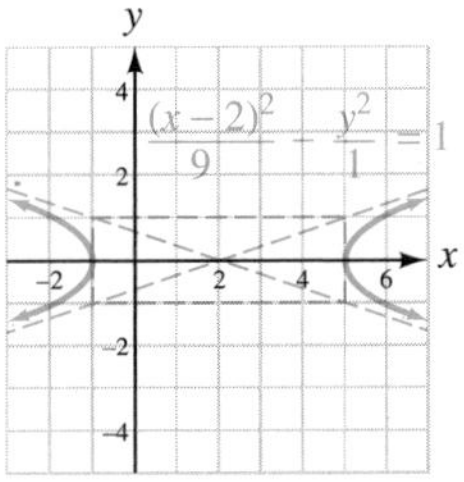

13.

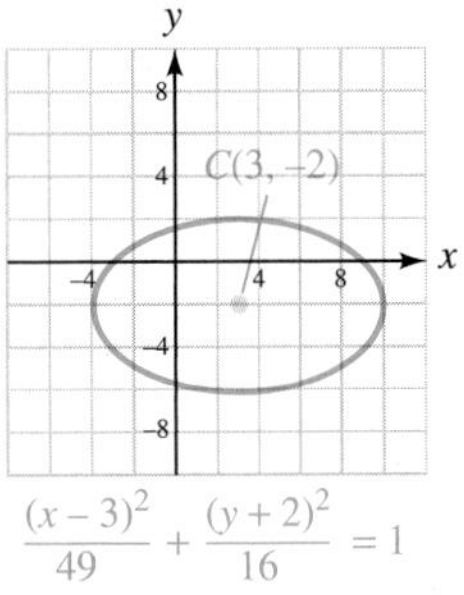

14.

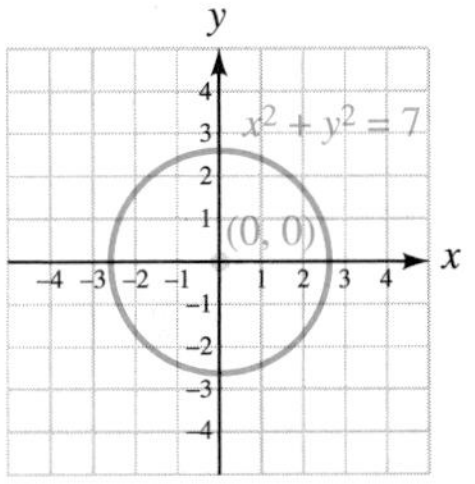

15.

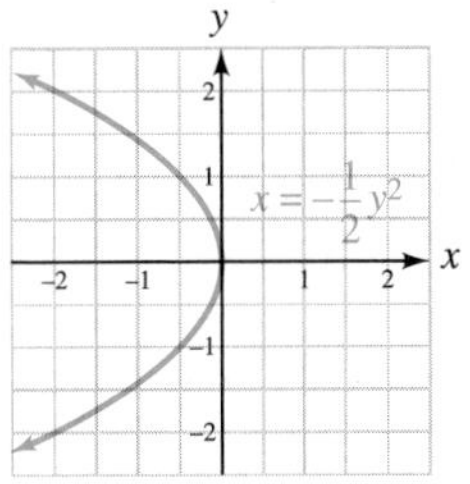

16.

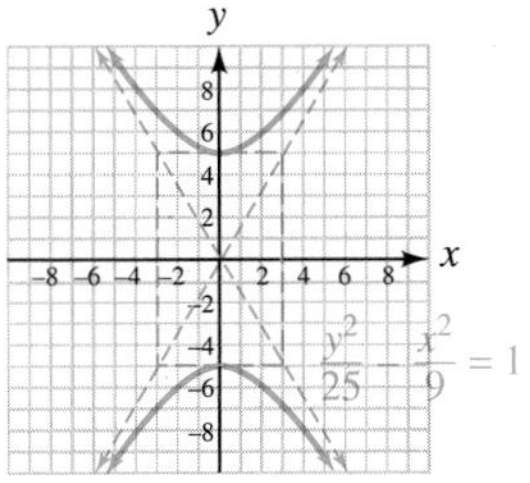

23.

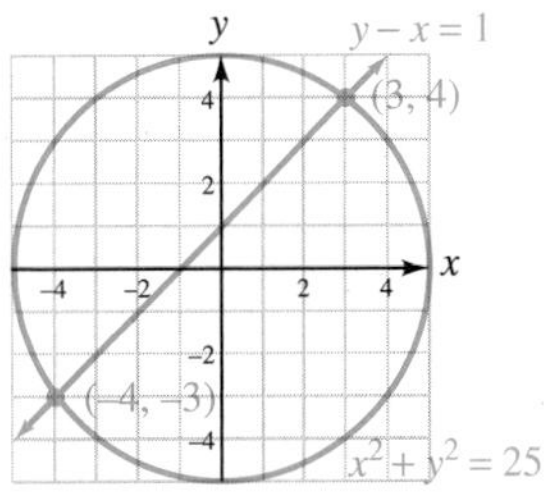

CUMULATIVE REVIEW CHAPTERS 1–14 (page 1159)

6.

18.

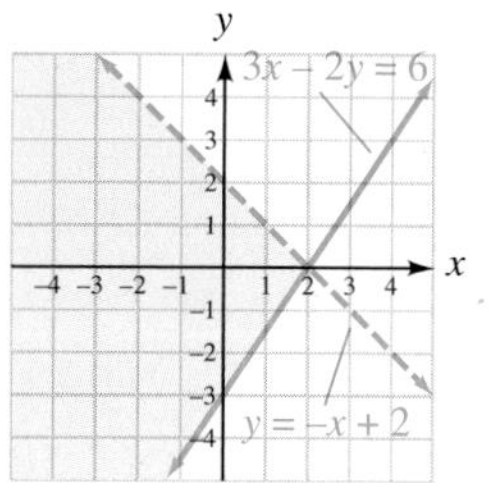

56.

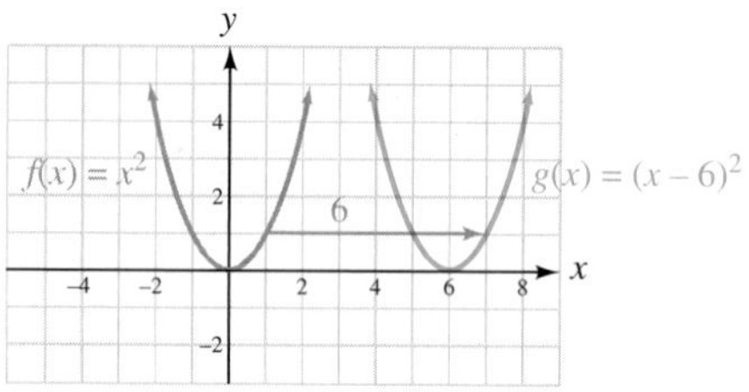

57.

58.

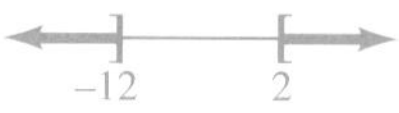

59.

60.

72.

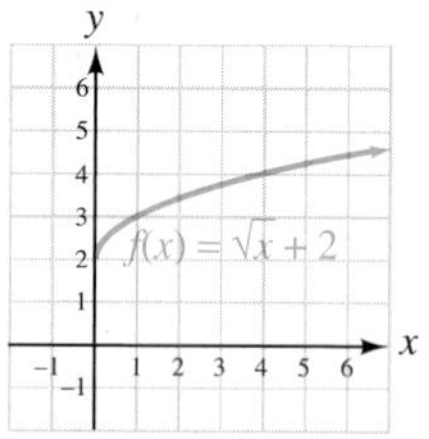

96.

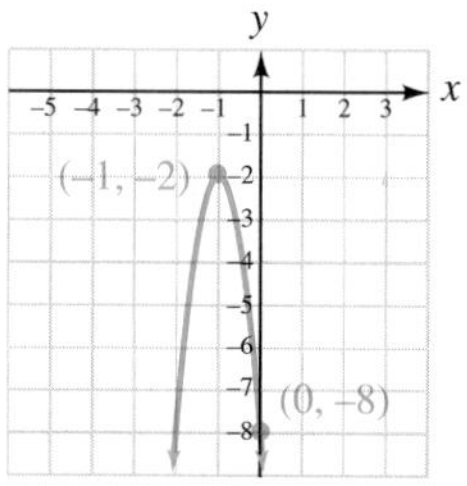

$f(x) = -6x^2 - 12x - 8$

97.

100.

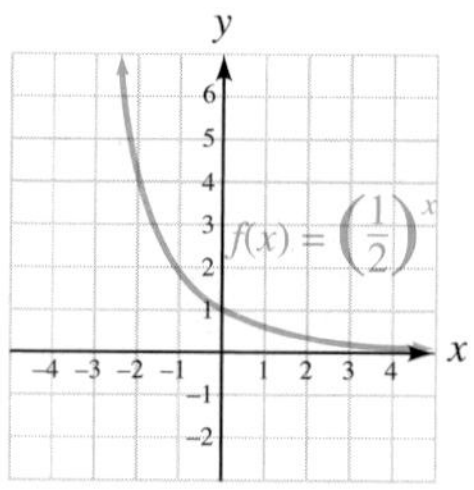

101.

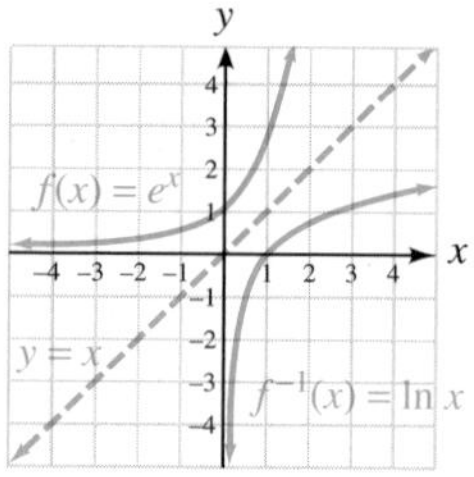

122.

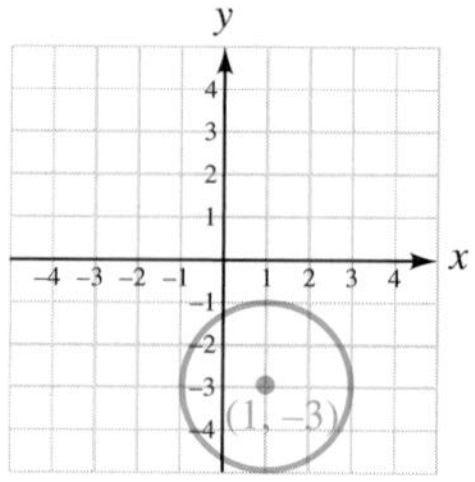

$(x - 1)^2 + (y + 3)^2 = 4$

123.

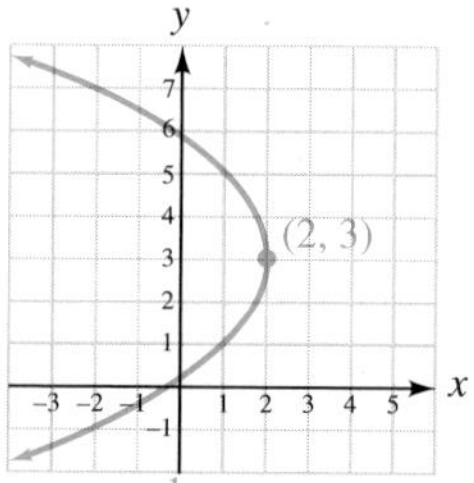

$x = -\frac{1}{4}(y - 3)^2 + 2$

124.

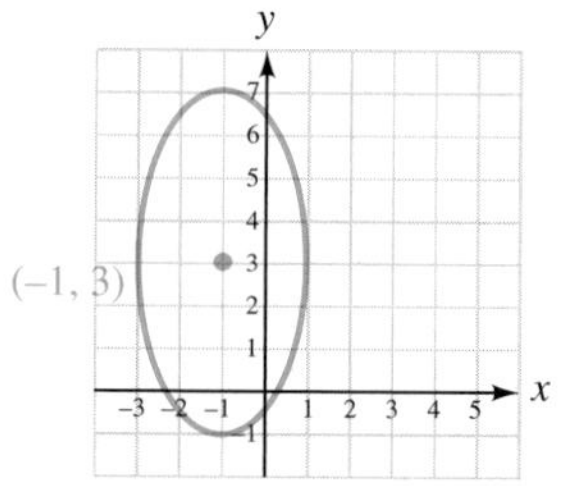

$\frac{(x + 1)^2}{4} + \frac{(y - 3)^2}{16} = 1$

125.

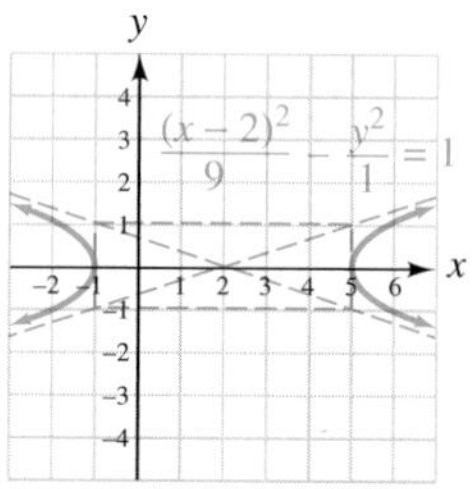

INDEX

D

E

Q

R

S

CALCULATOR INDEX